NOVA ACTA

REGIAE SOCIETATIS

SCIENTIARUM

UPSALIENSIS

SER. V:A. VOL. 1

Uppsala General Catalogue
of
Galaxies

Data for 12,921 galaxies north of $\delta = -2°30'$

by
Peter Nilson

Presented to the Royal Society of Sciences of Uppsala
September 17th, 1971

UPPSALA 1973

ISBN 91-554-0064-7
Uppsala Offset Center AB
Uppsala 1973

CONTENTS

THE UPPSALA GENERAL CATALOGUE OF GALAXIES

1. Introduction

The present Catalogue is based mainly on the Palomar Sky Survey prints. It is designed to be essentially complete to the limiting diameter 1'.0 on the blue prints; it also includes all galaxies to the limiting apparent magnitude 14.5 (even those smaller than 1'.0) in the *Catalogue of Galaxies and of Clusters of Galaxies* (CGCG) by Zwicky *et al.* It covers the sky to the north of $\delta = -02°30'$. The galaxies are ordered according to their right ascensions; the epoch used is 1950.0.

Besides descriptions of the galaxies and their surrounding areas, the Catalogue contains classifications in conventional systems. Both as concerns the use of a limiting diameter and in the classification and description technique, the present Catalogue differs from the *Morfologičeskij Katalog Galaktik* (MCG) by Voroncov-Vel'jaminov *et al.*, which was also based on the Sky Survey. Unlike most modern catalogues of galaxies, this one contains position angles for flattened galaxies. The revived interest during recent years in the orientation of galaxies in space makes it important to have such data listed for a large number of fairly bright galaxies.

For details concerning the Sky Survey, see Wilson (1952), California Institute of Technology (1954), Abell (1958) and Minkowski and Abell (1963).

Diameters have been measured on both the blue and red Sky Survey prints. The measuring procedure is described in section 2*b*. The classifications and descriptions have been designed to give as complete an account as possible of the appearance of the objects on the prints. Eighty-two plates from the Uppsala Observatory 40-in. Schmidt telescope at Kvistaberg have been used for tests of the classifications; the plates cover about 480 square degrees in areas rich in galaxies. Magnitudes are listed from CGCG or estimated from the blue prints for objects fainter than 15.7.

Special care has been taken to record peculiarities, interaction, distortion, or the occurrence of bridges, jets, plumes and other non-regular phenomena. A fixed vocabulary has been developed to describe properties not covered by the classification symbols. When possible, different classification systems have been used in a complementary way, to fully exploit the information content of the images on the prints.

The application of a limiting diameter results in the inclusion of a fairly large number of "dwarf" objects with an extremely low surface brightness. The difficulties of discovering such objects and of measuring their diameters make their inclusion in the Catalogue somewhat arbitrary. Of course, the concept of a "limiting diameter" has different significance for different types of galaxies; it is naturally the less meaningful, the lower the surface brightness of the objects.

In order to make the Catalogue as complete as possible and to avoid large inhomogeneities in the measurements, the Sky Survey prints have been searched through three separate times. The Catalogue is based mainly on the last two surveys. Each survey continued for approximately one year. It is, of course, impossible to keep a fixed limiting diameter strictly, as the outer parts of the galaxies are usually very faint and an apparent "edge" is difficult to define. Moreover, the Sky Survey prints themselves cannot be expected to be quite homogeneous. Variations in the actual limiting diameter are thus to be expected, and, consequently, the total number of galaxies down to the estimated limiting diameter is for different areas not certainly significant for the description of the distribution of galaxies: this number should, in any case, be treated with care. An analysis of data from the Catalogue will appear shortly.

The original data were recorded in a card catalogue, including approximately 15,000 cards and containing all separate measurements and classifications. Listed diameters are mean values of the last two measurements. Classifications have, in the final Catalogue, been listed in such a way as to give an account of the considerable uncertainties which are often present. Attention should be paid to the definition of the "information parameter", which may be interpreted as an estimate of this uncertainty.

All survey work, measurements, identifications and classifications were made by the author, who also prepared the final manuscript for printing by the offset method.

A catalogue of this kind may be expected to be used in three different ways: *pro primo,* as a reference work, giving elementary data for a large number of galaxies in the northern sky; *pro secundo,* as a fairly homogeneous set of data which may be used in statistical investigations; *pro tertio,* as a guide to the selection of objects in several kinds of observation programs, for example, studies of interacting galaxies.

Peter Nilson

Astronomical Observatory, Uppsala
March, 1973

2. Explanations

(a) Arrangement of the Catalogue pages

The Catalogue is divided into an Index Section, containing lists of data for 12,921 galaxies, and a Notes Section. In the Index Section, the data for each galaxy are written on two lines (hereafter denoted *a* and *b*). Each page is divided into 12 columns. The listed data are as follows:

Column	Line	Data
1	*a*	Number of the galaxy in this Catalogue
2	*a*	α, δ for the epoch 1950.0
2	*b*	l^{II}, b^{II}
3	*a*	MCG numbers (*Morfologičeskij Katalog Galaktik* by Voroncov-Vel'jaminov *et al.*)
3	*b*	NGC and IC numbers; Messier numbers; Markarian's numbers; in a few cases, other identifications
4	*a*	Number of the Palomar Sky Survey field, in which the galaxy is best visible
4	*b*	Number of cluster in CGCG (*Catalogue of Galaxies and of Clusters of Galaxies* by Zwicky *et al.*) classified by Zwicky as "near", if the galaxy is located inside the contour line of such a cluster (according to lists on pages 445–447)
5	*a*	Diameters measured on the blue Palomar Sky Survey prints
5	*b*	Diameters measured on the red prints
6	*a*	Position angle (from north eastwards)
6	*b*	Estimated inclination to the line of sight of flattened galaxies (scale 1-7; 7 = edge-on)
7	*a*	Information parameter, estimating the uncertainty in the classification (scale 0-5; 0 = most uncertain case)
8	*a*	Classification in the Hubble system
8	*b*	Luminosity class, according to van den Bergh's system (only IV, IV-V and V)
9	*a*	Classification, according to de Vaucouleurs's revised Hubble system
9	*b*	Classification, according to Holmberg's revised Hubble system
		Columns 8 and 9 are also frequently used for short descriptions (cf. sections 3 and 4)
10	*a*	Apparent photographic magnitude from CGCG (Zwicky) or estimated from the blue prints if fainter than 15.7
11	*a*	Radial velocity referred to the sun
11	*b*	Radial velocity corrected for solar motion
12		References to the Notes Section (*) or to other catalogues (cf. section 2*b*)

(b) Further discussion of the Catalogue data

Column 2, line a: Co-ordinates have been adopted from CGCG (with a few corrections accounted for in the Notes Section), to facilitate the joint use of CGCG and the present Catalogue. For galaxies fainter than 15.7 or located at low galactic latitudes not covered by CGCG, positions have been determined in relations to GC stars; the accuracy is only what is necessary for identification purposes (to tenths of minutes in right ascension and to minutes of arc in declination).

Column 2, line b: Galactic co-ordinates have been computed from listed α and δ in an HP 9100A calculator. Two decimals are given in all values, regardless the accuracy in the position determinations. The north galactic pole is situated at

$$\alpha_0 = 12^h 49^m$$
$$\delta_0 = +27° 24' \qquad (1950.0)$$

and the north celestial pole (1950.0) at galactic longitude

$$\lambda_0 = 123°00'$$

(cf. Blaauw *et al.* 1960).

Column 3, line a: Numbers in MCG. For typographical reasons, these numbers are written, for instance, 03—01—018 instead of 3—1—18. The figures in the first position give the number of the Palomar Sky Survey zone, each zone covering 6° in declination, from the equator (No. 1) to the north celestial pole (No. 15). The figures in the second position give the number of the Sky Survey field along the zone; the figures in the last position represent the number of the galaxy in this field in MCG. Galaxies not included in MCG are listed with appropriate figures in the first two positions (the field number in MCG) but have 000 in the last position.

Column 3, line b: IC numbers are — if repeated or written in connection with NGC numbers — indicated by an asterisk (*). Mark = Markarian (numbers from the lists of galaxies with ultraviolet continuum: Markarian (1967, 1969 *a, b*), Markarian and Lipovetsky (1971, 1972)). Ho = Holmberg (nine dwarf galaxies discovered by him and denoted by Roman numerals: Holmberg (1958, 1969)).

Column 4, line b: Clusters of galaxies classified as "near" in CGCG are listed in order of right ascension on pages 445—447. A number in this position gives the number of such a cluster if the galaxy is located inside its contour line; however, this location is no guarantee of membership in individual cases. Obvious foreground objects are denoted by having their cluster numbers placed within brackets.

Column 5: Diameters (major axis x minor axis) were measured directly on the Sky Survey prints by using a Leitz magnifier with a magnification of X 6 and a built-in measuring scale (graded to 0.1 mm). Listed values refer to *maximum* diameters on the prints, i.e. the maximum extent of features which may confidently be described as belonging to the galaxy. As these values are very difficult to obtain for early-type systems (E,S0 and early Sa), the diameters for such objects have been placed within brackets. Late-type objects are usually somewhat patchy in the outskirts, where the surface brightness is low, which introduces uncertainties of a different kind; uncertain values for such objects have been indicated by a colon or, for great uncertainties, a double colon.

The galaxies have always been searched for on the blue prints. The prints have been surveyed through an S.O.C. "Visolett" magnifier with a magnification of X 1.6 and a diameter of 3.5 cm. The procedure was to survey first the right-hand half of the prints from the bottom upwards, and the lens was moved in "sweeps", with so much overlap that every object was expected to enter in the field of view at least three times. The same procedure was then applied to the left-hand part of the print, with sufficient overlap between the two halves. Every galaxy expected to be close to 1'.0 or larger was measured with the Leitz magnifier. At least two separate settings were made for each diameter, with the measuring scale in opposite orientations. Approximate positions were determined by using a co-ordinate grid.

Diameter values inside square brackets denote *total* diameters for pairs, triplets or multiple systems. Such objects were included in the Catalogue if they were in some way connected with each other (contact or connection by bridge or other intergalactic features), and the total maximum diameter was found to exceed 0'.9. For such objects, usually only diameters from the blue prints are listed.

Blue and red diameters were measured at such a long interval (usually several hours) that they may be regarded as essentially independent of each other. For practical reasons, it was impossible to make separate surveys on the blue and red prints.

Column 6, line a: Position angles were measured from the north eastwards by using a protractor 25.5 cm in diameter, in combination with a triangle giving the direction perpendicular to the upper and lower edges of the prints. Corrections for the inclination of the meridian lines were computed for different zones and indicated on the co-ordinate grid. Uncertain values are denoted by a colon. Values in brackets refer to almost face-on spirals or other galaxies with small ellipticity. Square brackets denote asymmetric systems, for which the position angle gives less significant information as regards the orientation in space.

At least two separate settings were made at each measurement, with the position of the protractor reversed in order to

avoid effects from small asymmetries in the galaxy. In order to avoid selection effects in position angles (Öpik 1963, 1968), several prints were re-surveyed in different orientations. No systematic effects were noticed.

Column 6, line b: The inclination to the line of sight for spirals is estimated on a scale from 1 (face-on) to 7 (edge-on). For galaxies with a high inclination to the line of sight, the parameter was calculated from the Hubble formula $n = 10(a-b)/a$, where a = major axis and b = minor axis. The value 7 is used for objects inclined not more than a few degrees to the line of sight.

Column 7, line a: Owing to the small scale and the high contrast, classifications on the Sky Survey prints are usually regarded as difficult. However, it is the author's experience that classifications (at least approximate ones) are possible for most galaxies to the limiting diameter 1'.0, provided that the estimated uncertainties are accounted for in an appropriate way. The information parameter (0—5) is to be interpreted as follows:

0	A very small amount of information available; the object is possibly not extragalactic.
1	Low information content in the galaxy's image on the prints; classification, if tried, is only a rough estimate.
2	Classification appears quite possible but is still difficult because of over-exposure, faintness or presence of superimposed stars.
3	The galaxy appears fairly easy to classify; the classifications may in most cases be expected to be reliable.
4	The structure of the galaxy is easily visible, and the listed Hubble class is most probably correct. The classification in de Vaucouleurs's revised Hubble system is not expected to deviate more than one unit (in "stage") from classifications obtained on plates with higher information contents.
5	The galaxy is a prominent object, and has, as a rule, been classified by other observers; these classifications and those listed in the present Catalogue agree in the most important aspects.

Columns 8 and 9: The Hubble class corresponds approximately to the well-known "tuning-fork" classification, with the class S B 0 added (Hubble 1926, 1936); it agrees fairly well with Sandage's revised Hubble system (1961), although Sandage's subtypes are omitted. For spiral galaxies, the class is estimated essentially from the texture of the arms or the texture of the outer parts of the objects. It appears to be generally recognized that even early spiral systems may (in contradiction to Hubble's criterion) sometimes be very open and that late spirals may have tightly coiled arms. Colours, which are helpful in the classification work, have been estimated by comparing the blue and red prints.

It has been argued (Voroncov-Vel'jaminov 1960, 1967) that traditional classification systems are not adequate to account for the variety of structures among galaxies. In fact, it seems likely that no existing classification system *alone* is able to represent the varieties, as concerns morphology, colour and surface brightness; thus, it appeared reasonable to use in this Catalogue several systems alongside each other in a complementary way. Voroncov-Vel'jaminov (1967) has also suggested that many "peculiar" galaxies form definite morphological types. Even if this statement is true, it is clearly impossible to base a more appropriate classification system, as concerns the morphology of galaxies, on the Sky Survey alone. The observed "peculiarities" have thus been accounted for mainly by a fixed description system.

Complementary classifications have been made essentially according to the descriptions by van den Bergh (1960a), de Vaucouleurs (1959) and Holmberg (1958). Very late spirals (de Vaucouleurs's stages *dm* and *m*) are not classified in the Hubble system. The notation "dwarf" is used for objects with (1) very low surface brightness and (2) little or no central concentration of light on the red prints (cf. van den Bergh 1959).

In this context, the standard Hubble system (as concerns the spirals) may be interpreted as a division of galaxies according to colour (or texture in the outer parts). On the other hand, de Vaucouleurs's revised system accounts mainly for the morphological structure of the objects; the finer classification in "stages" makes the location of the object along the Hubble sequence (the "third dimension" in this system) more uncertain. Owing to the over-exposure on the prints, the "varieties" (central ring or spiral structure) are in many cases difficult or impossible to determine.

For the definition of the notations "singular", "peculiar" and other terms used in columns 8 and 9, see sections 3 and 4.

No classification system can be expected to be completely independent of the observer's interpretation of it. Thus, the classifications listed in the present Catalogue (UGC) may show certain individual features which distinguish them from results obtained by other observers. The aim has been, of course, to make the classifications as homogeneous as possible within the Catalogue itself.

Column 10, line a: Magnitudes adopted from CGCG. If the galaxy is a member of a pair or a system of higher order, and the total magnitude is given in CGCG, this magnitude value has been placed within brackets. For objects fainter than 15.7, the magnitudes have been estimated from the blue prints according to the following scale:

16.0:; 16.5:; 17.; 18.; 19.

For brighter objects at low galactic latitudes not covered

by CGCG, the magnitudes have been estimated to $\pm 0^m.5$.

Column 11, line a: Radial velocities listed in the *Reference Catalogue of Bright Galaxies* (BG) are adopted in the present Catalogue, taking into account the corrections listed in *Publ. astr. Soc. Pacific 78,* 262 and *84,* 461. Supplementary radial velocities have been taken from Arakelian *et al.* (1970a, b; 1971; 1972a, b), Arp (1972), Chincarini and Rood (1971), DuPuy (1970), Fairall (1971), Rudnicki and Tarraro (1969), Sargent (1970b), de Vaucouleurs and de Vaucouleurs (1967), Weedman and Khachikian (1968, 1969), and Zwicky (1971). Among other publications concerning radial velocities, attention is called particularly to Chincarini and Rood (1972), Ford *et al.* (1971), Page (1970), Sargent (1970a, 1972) and Ulrich (1971).

Column 11, line b: Radial velocity corrected for solar motion relative the Local Group according to the formula $V_0 = + 300 \cos A$, where A is the distance to the conventional solar apex at $l^1 = 55°$, $b^1 = 0°$, or $l^{II} = 87°$, $b^{II} = + 1$ (cf. de Vaucouleurs and de Vaucouleurs 1964, p. 3; Humason *et al.* 1956; Humason and Wahlquist 1955).

In some cases, only corrected velocities have been available. Attention is called to the fact that some observers (in this context Sargent and Ulrich) correct the velocities to the centre of the Galaxy, using the value 250 km s^{-1} for the solar motion (the current IAU value of galactic rotation). (Considering the uncertainties present in most redshift determinations, this difference is, in fact, usually of minor importance).

If the uncertainties in the measurements of the radial velocity were considered too large to make any correction meaningful, usually only the uncorrected value was listed.

Column 12: References according to the following system:

*	Reference to the Notes Section.
1	Reference to the notes in the *Morfologičeskij Katalog Galaktik* (Morphological Catalogue of Galaxies) by Voroncov-Vel'jaminov *et al.* (1962, 1963, 1964, 1968).
2	Galaxy listed in the *Reference Catalogue of Bright Galaxies* by de Vaucouleurs and de Vaucouleurs (1964).
2̲	Reference to note in the *Reference Catalogue of Bright Galaxies.*
3	Galaxy listed in *Die Herschel-Nebel nach Aufnahmen der Königstuhl-Sternwarte* by Reinmuth (1926). (This catalogue contains re-observations of the objects in *A General Catalogue of Nebulae and Clusters of Stars* by John Herschel (1864).)
4	Galaxy listed in *A Survey of External Galaxies brighter than the Thirteenth Magnitude* by Shapley and Ames (1932).

5	Galaxy listed in *A Photographic Photometry of Extragalactic Nebulae* by Holmberg (1958).
6	Galaxy listed in *A Preliminary Classification of the Forms of Galaxies according to their Stellar Population* by Morgan (1958).
<u>6</u>	Reference to note in the above-mentioned list.
7	Galaxy listed in *A Preliminary Classification of the Forms of Galaxies according to their Stellar Population, II,* by Morgan (1959).
<u>7</u>	Reference to note in the above-mentioned list.
8	Galaxy listed in *A Re-classification of the Shapley-Ames Galaxies* by van den Bergh (1960*b*).
9	Galaxy listed in *A Study of External Galaxies* by Holmberg (1964).
V	Galaxy reproduced and listed in the *Atlas i Katalog Vzaimodejstvujuščih Galaktik* (Atlas and Catalogue of Interacting Galaxies) by Voroncov-Vel'jaminov (1959).
A	Galaxy reproduced and listed in the *Atlas of Peculiar Galaxies* by Arp (1966).
C	Galaxy listed in the *Catalogue of Selected Compact Galaxies and of Post-Eruptive Galaxies* by Zwicky (1971).

(c) The Notes Section

The galaxies are divided into groups of ten, according to their UGC numbers. Thus, every tenth UGC number is to be found in the left-hand margin. The UGC numbers in the notes are preceded by an "U" and followed by a bracket.

A note in the form "comp 1.5, 15, 0.4 x 0.3" is to be interpreted "companion at 1'.5 from centre, in position angle 15°, with diameters 0'.4 x 0'.3". The notation "p w comp at. . ." ("pair with companion at. . .") is used only if the companion is comparable in size and/or magnitude with the listed galaxy. "Companion" galaxies are accounted for to the limiting diameter 0'.3 and to a distance of 5'.0 from the centre of the galaxy (however, "companions" are not recorded in crowded areas). Listed co-ordinates and magnitudes refer to CGCG. For the sources of radial velocities, see section 2*b*.

Two UGC galaxies were described as a "pair" whenever there were no other nearby objects of similar size or brightness and it was deemed possible that they might form a physical pair.

Data concerning supernovae have been taken from Karpowicz and Rudnicki (1968) and from Kowal and Sargent (1971).

The terminology of the Notes Section is explained under *Abbreviations* and *Terminology*.

3. Abbreviations

abs	absorption
acc	according
alm	almost
amorph	amorphous
approx	approximate
asym	asymmetry, asymmetric
B	bright
bel	belong
BG	*Reference Catalogue of Bright Galaxies*
brightn	brightness
centr	central
CGCG	*Catalogue of Galaxies and of Clusters of Galaxies*
circ	circular
cl	cluster
classif	classify, classification
comp	companion
compl	complete
cond	condensation
cont	continue
coord	co-ordinates
CSCG	*Catalogue of Selected Compact Galaxies and of Post-Eruptive Galaxies*
descr	describe, description
de V	de Vaucouleurs
DDO	David Dunlap Observatory ("DDO" precedes numbers of objects in van den Bergh's catalogue of dwarf galaxies (1959, 1966)).
diam	diameter
dif	diffuse
disr	disrupted
distr	distribution
e	extremely
eastw	eastwards
ecc	eccentric
elong	elongated
em	emission
env	envelope
evid	evidence
exp	exposure, exposed
ext	extension, extended
F	faint
f	following
Ho	Holmberg
I	IC
ident	identify, identification
incl	include, including
indiv	individual
intercon	interconnected
invis	invisible
irr	irregular
M	Messier
Mark	Markarian
MCG	*Morfologičeskij Katalog Galaktik* (Morphological Catalogue of Galaxies)
N	NGC
n	north
northw	northwards

nr	number
nucl	nucleus, nuclear
obj	object
outw	outwards
p	preceding
PA	Palomar Atlas (Palomar Sky Survey)
pec	peculiar
phys	physical
pos	position
prob	probable, probably
p w or P w	pair with
reg	region
S	small
s	south
sep	separation, separated
sev	several
SN	supernova
southw	southwards
spir	spiral
stell	stellar
str	structure
superimp	superimposed
surf	surface
sym	symmetry, symmetrical
syst	system
tend	tendency
tow	towards
transp	transparent
U	UGC (*Uppsala General Catalogue of Galaxies*)
v	very; radial velocity referred to the sun
v_0	radial velocity corrected for solar motion
var	variable
vdB	van den Bergh
vic	vicinity
vis	visible
VV	Voroncov-Vel'jaminov ("VV" precedes numbers of systems in the *Atlas and Catalogue of Interacting Galaxies*)
w	with
westw	westwards
Zw	Zwicky (quotations from "Zw" refer to *Catalogue of Selected Compact Galaxies and of Post-Eruptive Galaxies*)
*	(in the Notes Section) star
**	(in the Notes Section) stars

4. Terminology

acicular	Apparently "needle-shaped" or "cigar-shaped" galaxy (Jacobi ellipsoid). Cf. Ogorodnikov (1965, 1967).
compact	Notation adopted from CGCG, where galaxies are listed as "compact" ("very compact", "extremely compact"), if they "might be easily mistaken for stars" (Zwicky), or from the *Catalogue of Selected Compact Galaxies and of Post-Eruptive Galaxies* (Zwicky 1971).
companion	Object in the neighbourhood of a galaxy listed in the present Catalogue; the object may be either a physical or an optical companion.
disrupted	Strong tidal disturbances.
distorted	Clear evidence of tidal effects; the structure is, however, not completely dissolved or disrupted.
disturbed	Weak tidal effects or suspicion that interaction with a nearby galaxy is taking place.
dwarf	Very low surface brightness; little or no central concentration of light on the red Sky Survey prints.
eruptive	Peculiar object with evidence of violent outflow of matter. Note that Zwicky (1971, p. XXX) defines "eruptive" in a different way; galaxies listed as "eruptive" in the present Catalogue would presumably be classified as "post-eruptive" by Zwicky.
integral	Flattened object with tips bent in opposite directions.
interaction	Synonym of "disturbed", used collectively for pairs or systems of higher order.
jet	Short, thin beam of outflowing matter.
peculiar	Greatly deviating from normal types in the common classification systems; classification is in general not possible.
plume	Extensive, diffuse, sometimes irregular formation of outflowing matter.
singular	Less prominent peculiarity than in objects denoted as "peculiar"; classification is often possible.
streamer	Long, broad beam of outflowing matter.
symbiotic	Double (triple, multiple) system with main bodies intermingled.
tail	Long, thin formation of outflowing matter; long jet.
unilateral	One-armed spiral galaxy.

This work was in part financially supported by grants from the Swedish Natural Science Research Council and the Uppsala University Astronomical Observatory. The printing was made possible through the courtesy of the Royal Society of Science at Uppsala.

5. References

Abell, G.O. 1958, *Astrophys. J. Suppl., 3,* 211.
Arakelian, M. A. *et al.* 1970a, *Astrofizika, 6,* 39.
Arakelian, M.A. *et al.* 1970b, *ibid., 6,* 357.
Arakelian, M.A. *et al.* 1971, *ibid., 7,* 177.
Arakelian, M.A. *et al.* 1972a, *ibid., 8,* 33.
Arakelian, M.A. *et al.* 1972b, *ibid., 8,* 177.
Arp, H. 1966, *Atlas of Peculiar Galaxies,* California Institute of Technology, Pasadena.

Arp, H. 1972, Ejection of small compact galaxies from larger galaxies, in *External Galaxies and Quasi-Stellar Objects*, (IAU Symp. 44), Ed. D.E. Evans, D. Reidel Publishing Company, Dordrecht, p. 380.

Bergh, S. van den 1959, *Publ. David Dunlap Obs., 2*, 147.

Bergh, S. van den 1960a, *Astrophys. J., 131*, 215.

Bergh, S. van den 1960b, *Publ. David Dunlap Obs., 2*, 159.

Bergh, S. van den 1966, *Astr. J., 71*, 922.

Blauuw, A. *et al.* 1960, *Mon. Not. R. astr. Soc., 121*, 123.

California Institute of Technology, 1954, *National Geographic Society — Palomar Observatory Sky Atlas*, Pasadena.

Chincarini, G., Rood, H.J. 1971, *Astrophys. J., 168*, 321.

Chincarini, G., Rood, H.J. 1972, *Astr. J., 77*, 4.

DuPuy, D.L. 1970, *Astr. J., 75*, 1143.

Fairall, A.P. 1971, *Mon. Not. R. astr. Soc., 153*, 383.

Ford, W.K. *et al.* 1971, *Astr. J., 76*, 22.

Herschel, J.F.W. 1864, *Phil. Trans. R. Soc. London, 154*, 1.

Holmberg, E. 1958, *Medd. Lunds astr. Obs.*, Ser. II, No. 136.

Holmberg, E. 1964, *Ark. Astr., 3*, 387; *Uppsala astr. Obs. Medd.*, No. 148.

Holmberg, E. 1969, *Ark. Astr., 5*, 20; *Uppsala astr. Obs. Medd.*, No. 166.

Hubble, E. 1926, *Astrophys. J., 64*, 321.

Hubble, E. 1936, *The Realm of the Nebulae*, Yale University Press, New Haven. (Reprinted in 1958, Dover Publications Inc., New York.)

Humason, M.L. *et al.* 1956, *Astr. J., 61*, 97.

Humason, M.L., Wahlquist, H.D. 1955, *Astr. J., 60*, 254.

Karpowicz, M., Rudnicki, K. 1968, *Publ. astr. Obs. Warsaw Univ., 15*.

Kowal, C.T., Sargent, W.L.W. 1971, *Astr. J., 76*, 756.

Markarian, B.E. 1967, *Astrofizika, 3*, 55.

Markarian, B.E. 1969a, *ibid., 5*, 443.

Markarian, B.E. 1969b, *ibid., 5*, 581.

Markarian, B.E., Lipovetsky, V.A. 1971, *Astrofizika, 7*, 511.

Markarian, B.E., Lipovetsky, V.A. 1972, *ibid., 8*, 155.

Minkowski, R.L., Abell, G.O. 1963, The National Geographic Society — Palomar Observatory Sky Survey, in *Basic Astronomical Data*, Ed. K. Aa. Strand, Univ. of Chicago Press, Chicago, p. 481.

Morgan, W.W. 1958, *Publ. astr. Soc. Pacific, 70*, 364.

Morgan, W.W. 1959, *ibid., 71*, 394.

Ogorodnikov, K.F. 1965, *Dynamics of Stellar Systems*, translated from the Russian by J.B. Sykes, Pergamon Press, Oxford, p. 308.

Ogorodnikov, K.F. 1967, Acicular galaxies, in *Modern Astrophysics, A Memorial to Otto Struve*, Ed. M. Hack,

Gauthier-Villars, Paris, p. 329.

Öpik, E. 1963, *Irish astr. J., 6*, 34.

Öpik, E. 1968, *ibid., 8*, 229.

Page, T. 1970, *Astrophys, J., 159*, 791.

Reinmuth, K. 1926, *Veröff. Sternw. Heidelberg, 9*, 1.

Rudnicki, K., Tarraro, I. 1969, *Acta Astr., 19*, 171.

Sandage, A. 1961, *The Hubble Atlas of Galaxies*, Carnegie Institution of Washington, Washington, D.C.

Sargent, W.L.W. 1970a, *Astrophys. J., 159*, 765.

Sargent, W.L.W. 1970b, *ibid., 160*, 405.

Sargent, W.L.W. 1972, *ibid., 173*, 7.

Shapley, H., Ames, A. 1932, *Ann. Harv. Coll. Obs., 88*, No. 2.

Ulrich, M.-H.J. 1971, *Astrophys. J., 163*, 441.

Vaucouleurs, G. de 1959, Classification and morphology of external galaxies, in *Encyclopedia of Physics*, vol. 53, Ed. S. Flügge, Springer-Verlag, Berlin, p. 275.

Vaucouleurs, G., de, Vaucouleurs, A. de 1964, *Reference Catalogue of Bright Galaxies*, No. 1 in the series: The University of Texas Monographs in Astronomy, The University of Texas Press, Austin.

Vaucouleurs, G., de, Vaucouleurs, A. de 1966, *Publ. astr. Soc. Pacific, 78*, 262.

Vaucouleurs, A., de, Vaucouleurs, G., de 1967, *Astr. J., 72*, 730.

Vaucouleurs, G., de, Vaucouleurs, A., de 1972, *Publ. astr. Soc. Pacific, 84*, 461.

Voroncov-Vel'jaminov, B.A. 1959, *Atlas i Katalog Vzaimodejstvujuščih Galaktik* (Atlas and Catalogue of Interacting Galaxies), Sternberg Institute, Moscow State University, Moscow.

Voroncov-Vel'jaminov, B.E. 1960, *Astr. Zu., 37*, 381; *Soviet Astr., 4*, 365.

Voroncov-Vel'jaminov, B.A. *et al.* 1962, 1963, 1964, 1968, *Morfologičeskij Katalog Galaktik* (Morphological Catalogue of Galaxies), I–IV, Moscow State University, Moscow.

Voroncov-Vel'jaminov, B.A. 1967, New morphological types of galaxies, in *Modern Astrophysics, A Memorial to Otto Struve*, Ed. M. Hack, Gauthier-Villars, Paris, p. 347.

Weedman, D.W., Khachikian, E.E. 1968, *Astrofizika, 4*, 587.

Weedman, D.W., Khachikian, E.E. 1969, *ibid., 5*, 113.

Wilson, A.G. 1952, *Trans. int. astr. Un., 8*, 335.

Zwicky, F. *et al.* 1961, 1963, 1965, 1966, 1968a, 1968b, *Catalogue of Galaxies and of Clusters of Galaxies*, I–VI, California Institute of Technology, Pasadena.

Zwicky, F. 1971, *Catalogue of Selected Compact Galaxies and of Post-Eruptive Galaxies*, L. Speich, Zürich.

Catalogue

1	2	3	4	5	6	7	8	9	10	11	12
00001	00 00.0 + 16 22 106.37 - 44.65	03 - 01 - 15 + 16 IC 5378	1195	[1.5 x 1.3] ...	— ...	3	Double system Contact		14.9		*VA
00002	00 00.0 + 44 39 113.91 - 17.09	07 - 01 - 000	1243	1.1 x 0.2 1.0 x 0.2	120 7	2	... n of 2	S d - m	17.		*
00003	00 00.2 + 18 37 107.24 - 42.49	03 - 01 - 018	1195	2.0 x 1.0 1.8: x 0.8:	90: 5	3	S B a p of 2		14.8		*
00004	00 00.5 + 03 56 100.65 - 56.55	01 - 01 - 014	1465 1	1.3 x 0.8 1.1 x 0.7	155: 4	1	Sb - c		15.5		
00005	00 00.6 - 02 11 096.26 - 62.25	00 - 01 - 021	0319	1.3 x 0.8 1.4 x 0.9	45 4	3	S B b/Sc		14.3		*1
00006	00 00.6 + 21 42 108.37 - 39.52	04 - 01 - 013 Mark 334	0779	1.0 x 0.7 1.0 x 0.7	[105] ...	3	... Peculiar, loop		14.4	+ 6900	*1C
00007	00 00.7 + 15 42 106.35 - 45.34	03 - 01 - 019 IC 5381	1195	1.5 x 0.4 1.5 x 0.4	54 (6)	1	... Peculiar		14.9		*1̲2̲
00008	00 00.7 + 15 52 106.41 - 45.18	03 - 01 - 020 NGC 7814	1195	(6.5 x 2.7) (6.3 x 2.7)	135 7	2	Sa - b		12.0	+ 1047 + 1245	*1̲2̲ 3̲4̲6̲
00009	00 00.8 + 04 21 101.03 - 56.19	01 - 01 - 000	1465 1	1.3: x 1.2: 1.3: x 1.2:	— ...	1	...		16.5:		
00010	00 00.8 + 08 20b 103.21 - 52.41	01 - 01 - 015	1465	1.9 x 1.7 2. x 2.	— 1	3	Sc nf of 2, disturbed?		(15.4)		*1
00011	00 00.8 + 21 50 108.47 - 39.41	04 - 01 - 014	0779	1.1: x 0.8 0.9: x 0.6:	55: ...	1	...		16.0:		*
00012	00 00.8 + 29 31 110.64 - 31.93	05 - 01 - 028	1257	1.1 x 0.9 0.7 x 0.6	(100) 2:	2	Sc		15.7		
00013	00 00.9 + 27 05 110.02 - 34.30	04 - 01 - 015	0779	1.1 x 1.0 1.0 x 1.0	— 1	3	S B 0/a	(R´)S B(s)0/a	15.0		*
00014	00 01.0 + 22 56 108.87 - 38.35	04 - 01 - 017	0779	1.9 x 1.0 2.0: x 1.0:	32 5	4	Sc	SA(r)c	14.0		
00015	00 01.2 + 04 00 100.99 - 56.55	01 - 01 - 017	1465 1	1.5 x 0.3 1.2 x 0.2	63 6 - 7	2	Sa - b		16.0:		*
00016	00 01.2 + 07 11 102.78 - 53.53	01 - 01 - 018 NGC 7816	1465 2	2.0 x 2.0 2.1 x 2.1	— 1	3	Sb/Sc P w 21		14.0		*13
00017	00 01.3 + 14 56 106.26 - 46.11	02 - 01 - 017	0010	3. x 2. 3. x 2.	... ...	3	Dwarf V		17.		*
00018	00 01.4 + 10 35 104.50 - 50.30	02 - 01 - 000	0010	1.0 x 0.9 (1.1 x 1.0)	— ...	2	... Compact		15.7		
00019	00 01.4 + 20 28 108.22 - 40.77	03 - 01 - 021 NGC 7817	1195	4.0 x 1.1 3.8 x 1.0	45 6	2	Sb/Sc		12.7		*1̲2̲ 3
00020	00 01.5 + 80 01 120.85 + 17.61	13 - 01 - 004	1213	2.2: x 1.4: 1.2: x 1.1:	(5) ...	1	...	S dm?	16.0:		1
00021	00 01.6 + 07 05 102.89 - 53.66	01 - 01 - 019 NGC 7818	1465 2	1.1 x 1.1 1.1 x 1.1	— 1	3	Sc P w 16		15.1		*
00022	00 01.7 + 10 02 104.36 - 50.85	02 - 01 - 018	0010	1.3 x 0.8 1.3 x 0.8	25: 4	3	S0		16.0:		*
00023	00 01.7 + 10 30 104.57 - 50.40	02 - 01 - 019	0010	1.3 x 0.9 1.2 x 0.8	20: 3	3	S B b	S B(rs)b Sb -	15.2		*
00024	00 01.7 + 22 19 108.88 - 38.99	04 - 01 - 018	0779	1.2 x 0.8: 0.9 x 0.7	(50) ...	3	S B c Disturbed?		15.4		*
00025	00 01.8 + 05 53 102.33 - 54.82	01 - 01 - 020	1465 2	1.1 x 0.3 1.1 x 0.3	123 6 - 7	1	Sc		16.0:		
00026	00 01.8 + 31 12 111.31 - 30.33	05 - 01 - 029 NGC 7819	1257	2.0 x 1.8: 1.9 x 1.7:	— 1 - 2	3	S B b	S B(s)b Sb +	14.3		*3
00027	00 01.9 + 05 34 102.19 - 55.12	01 - 01 - 021	1465 1	2.2 x 1.2 2.1 x 1.1	135 4	3	Sc		15.3		*
00028	00 02.0 + 04 55 101.87 - 55.75	01 - 01 - 022 NGC 7820	1465 1	1.6 x 0.7 1.6 x 0.7	165 6	1	S0 - a Brightest in group		13.9		*3
00029	00 02.0 + 28 02 110.56 - 33.43	05 - 01 - 030	1257	(1.3 x 1.2) (1.3 x 1.2)	— ...	3	[E] Compact		15.2		
00030	00 02.0 + 33 17 111.85 - 28.30	05 - 01 - 000	1257	1.2 x 0.15 0.9 x 0.12	126 (7)	1	S ... Brightest of 3		16.0:		*
00031	00 02.3 + 16 55 107.32 - 44.26	03 - 01 - 022	1195	1.4 x 1.1 1.3 x 1.0	[55] ...	3	Irr	IA m Ir I	15.3		*
00032	00 02.4 + 11 26 105.24 - 49.56	02 - 01 - 000	0010	1.3: x 1.3: 1.2 x 1.2	— ...	1	...		17.		
00033	00 02.5 + 04 51 102.04 - 55.85	01 - 01 - 024	1465 1	1.2 x 0.9 1.3 x 1.0	— 2:	2	S B 0 - a NGC 7820 group		15.4		
00034	00 02.6 + 06 38 103.05 - 54.16	01 - 01 - 025 NGC 7824	1465 2	1.9 x 1.3 2.0 x 1.4	145 3	3	Sa/Sb P w 36	Sb -	14.5		*
00035	00 02.7 + 05 59 102.75 - 54.79	01 - 01 - 000	1465 2	1.7: x 1.5:: 1.5:: x 1.3::	— ...	3	Dwarf spiral, S m S V, sp of 2		17.		*

N G C   7814, 7816 - 20, 7824   I C   5378, 5381

1	2	3	4	5	6	7	8	9	10	11	12
00036	00 02.7 + 06 30 103.02 - 54.30	01 - 01 - 026	1465 2	1.7 x 0.8 1.7 x 0.8	18 5	3	Sa P w 34	SA:(r)a	14.7		*
00037	00 02.9 + 04 54 102.24 - 55.84	01 - 01 - 028 NGC 7825	1465 1	1.3 x 0.4 1.3 x 0.4	[27] ...	3	S B b P w 38, interaction?	S B (s)b	15.5		*3
00038	00 02.9 + 04 57 102.27 - 55.79	01 - 01 - 027 NGC 7827	1465 1	(1.4 x 1.3) (1.3 x 1.2)	--- 1	3	S B 0 P w 37, interaction?		14.6		*3
00039	00 03.0 + 53 22 116.09 - 08.62	09 - 01 - 000	0636 495	1.3 x 0.6 1.2 x 0.6	7 5	2	Sc		17.		
00040	00 03.2 + 27 10 110.65 - 34.34	04 - 01 - 021	0779	1.2 x 0.7 1.2 x 0.7	115: 4	1	S B ...		15.2		
00041	00 03.4 + 22 12 109.34 - 39.19	04 - 01 - 022	0779	1.2 x 0.3 0.8 x 0.2	30 6	2	S B c ssf of 2		16.0:		*
00042	00 03.5 + 12 51 106.21 - 48.26	02 - 01 - 000	0010	1.0: x 1.0: 0.9: x 0.9:	--- ...	2	... Disturbed?	S B d - dm	17.		*
00043	00 03.5 + 14 09 106.71 - 47.01	02 - 01 - 000	0010	1.0 x 0.3 1.2 x 0.3	175 6	2	...	S d - m	16.5:		
00044	00 03.6 + 19 28 108.57 - 41.86	03 - 01 - 000	1195	1.0 x 0.9 0.9: x 0.9:	--- 1	3	...	SAB: dm	17.		
00045	00 03.6 + 54 16 116.35 - 07.77	09 - 01 - 000	0636	1.0 x 0.4 0.9 x 0.3	150 (6)	2	Sc - Irr		15.5		
00046	00 03.8 + 17 09 107.88 - 44.12	03 - 01 - 23+24	1195	[0.8 x 0.6] ...	--- ...	2	Double system Bridge		14.5		*1
00047	00 04.0 + 17 00 107.90 - 44.28	03 - 01 - 000	1195	1.4 x 0.2 1.0: x 0.2	64 7	2	...	d - m	17.		*
00048	00 04.0 + 47 36 115.20 - 14.32	08 - 01 - 021	0839 494	2.1: x 1.9: 2. x 2.	--- 1	2	S B: c 	Sc +	16.0:		
00049	00 04.1 + 08 05 104.37 - 52.89	01 - 01 - 030 NGC 7834	1465 2	1.4: x 1.3: 1.5: x 1.5:	--- 1	3	Sc		15.4		*3
00050	00 04.1 + 25 53 110.56 - 35.63	04 - 01 - 024	0779	1.0 x 0.30 0.9 x 0.25	10 6	2	Sa - b		14.9		
00051	00 04.2 + 04 50 102.75 - 56.00	01 - 01 - 032	1465 1	1.1 x 0.7 1.1 x 0.7	30: 3	2	S B b - c		15.3		1
00052	00 04.2 + 08 21 104.54 - 52.64	01 - 01 - 034	1465 2	1.8 x 1.8 1.8 x 1.8	--- 1	3	Sc	SA c	15.0		
00053	00 04.3 + 19 02 108.65 - 42.32	03 - 01 - 025	1195	1.4: x 0.8: 1.0: x 0.7:	--- ...	2	Sb/S B c		15.5		*1
00054	00 04.3 + 41 28 114.09 - 20.36	07 - 01 - 000	1243	1.2 x 0.10 1.2 x 0.10	28 7	3	Sc		17.		
00055	00 04.4 + 46 23 115.04 - 15.53	08 - 01 - 022	0839	1.2: x 0.9: 1.1:: x 0.9::	(115) ...	3	... S B IV or IV - V	S B dm	17.		
00056	00 04.5 + 13 48 106.93 - 47.41	02 - 01 - 000	0010	1.0 x 0.8 0.9 x 0.7	(30) 2	3	Sc		15.6		
00057	00 04.6 + 27 26 111.09 - 34.14	04 - 01 - 025 NGC 1	0779	1.8 x 1.2 1.6: x 1.1:	120: 3	3	Sb P w 59		13.4		*12 3
00058	00 04.7 + 08 02 104.58 - 52.98	01 - 01 - 037 NGC 3	1465 2	1.2 x 0.7 1.1 x 0.7	113: ...	1	S0?		14.6		*3
00059	00 04.7 + 27 24 111.10 - 34.18	04 - 01 - 026 NGC 2	0779	1.1 x 0.7 0.9 x 0.5	115 3 - 4	2	Sa - b P w 57		14.8		*12 3
00060	00 04.7 + 32 20 112.27 - 29.34	05 - 01 - 032 IC 1530	1257	1.8 x 0.35 1.8 x 0.35	38 6	1	Sb		13.4		*2
00061	00 04.9 + 46 46 115.20 - 15.17	08 - 01 - 024	0839	(1.6 x 1.4) (1.6 x 1.4)	(145) 1 - 2	3	S0 Brightest of 3		14.3		*
00062	00 05.2 + 35 05 112.98 - 26.67	06 - 01 - 013 NGC 5	1247	(1.2 x 0.7) (1.1 x 0.6)	(115) ...	3	[E] Compact		14.6		*C
00063	00 05.2 + 35 41 113.11 - 26.08	06 - 01 - 014	1247	1.0: x 0.7: 0.9: x 0.5:	(55) ...	3	Irr		15.6		1
00064	00 05.2 + 40 36 114.10 - 21.24	07 - 01 - 005	1243	1.4: x 1.1: 1.3: x 0.9:	[35] ...	2	Knots in nebulosity Interacting systems?		15.5		*1
00065	00 05.4 + 32 48 112.54 - 28.91	05 - 01 - 000 Mark 338	1257	1.1 x 0.7 0.7 x 0.35	133: ...	1	...		13.8	+ 5100	*
00066	00 05.5 + 09 26 105.53 - 51.69	01 - 01 - 038	1465	1.5 x 1.3 1.5 x 1.3	85 1 - 2	3	Sb/S B b		14.9		*
00067	00 05.6 + 07 30 104.69 - 53.55	01 - 01 - 039	1465 2	1.5 x 0.5 1.5 x 0.5	24 6	1	Sa - b		15.4		
00068	00 05.6 + 26 44 111.18 - 34.87	04 - 01 - 028	0779	1.0: x 0.55 0.9 x 0.55	[145] ...	2	S B ... npp of 2, distorted		14.4		*1
00069	00 05.6 + 27 15 111.30 - 34.36	04 - 01 - 027	0779	1.3 x 0.9 1.3 x 0.9	42 4	3	Sc		14.7		*
00070	00 05.7 + 49 30 115.82 - 12.50	08 - 01 - 000	0839	1.0 x 0.8 1.0: x 1.0:	--- ...	3	... nff of 2	IAB m	16.5:		*

N G C 7825, 7827, 7834, 1 - 3, 5   I C 1530

1	2	3	4	5	6	7	8	9	10	11	12
00071	00 06.0 + 43 50 114.86 – 18.09	07 – 01 – 000	1243	1.2 x 0.7 1.0 x 0.6	(100) 4	2	Sb – c		15.7		
00072	00 06.1 – 01 00 099.92 – 61.67	00 – 01 – 028	0319	1.1 x 0.5 1.0 x 0.5	18 6	2	Sa		15.5		*
00073	00 06.1 + 37 10 113.61 – 24.65	06 – 01 – 015 NGC 11	1247	1.6 x 0.22 1.7 x 0.25	111 7	2	Sa		14.5		
00074	00 06.2 + 04 20 103.33 – 56.63	01 – 01 – 040 NGC 12	1465	2.0 x 1.7 2.1 x 1.9	(125) 1 – 2	4	Sc	SAB(rs)c Sc –	14.5		*23
00075	00 06.2 + 15 32 108.12 – 45.83	03 – 01 – 026 NGC 14	1195	3.0: x 2.5: 3. x 2.	[25] ...	3	Irr	Ir I	13.3		*13 VA
00076	00 06.2 + 24 15 110.71 – 37.33	04 – 01 – 029	0779 5	1.1 x 0.3 1.1 x 0.3	136 ...	1	...		16.0:		
00077	00 06.2 + 33 10 112.81 – 28.59	05 – 01 – 034 NGC 13	1257	2.5 x 0.7 2.4 x 0.6	53 6	2	Sb		14.2		*23
00078	00 06.3 + 23 33 110.55 – 38.02	04 – 01 – 030 NGC 9	0779 5	1.3 x 0.7: 0.9 x 0.4	[155] ...	2	Peculiar, disturbed: P w NGC 8		14.5		*12 3
00079	00 06.4 + 25 21 111.05 – 36.26	04 – 01 – 031	0779	1.7 x 1.3 1.7: x 1.3:	(85) 2:	4	Sc		15.7		
00080	00 06.4 + 27 27 111.56 – 34.20	04 – 01 – 032 NGC 16	0779	(1.8 x 1.0) (1.8 x 1.0)	16 2	2	S0		12.5	+ 3110 + 3334	*12 3468
00081	00 06.5 + 10 37 106.41 – 50.61	02 – 01 – 022	0010	1.4 x 0.4 1.5 x 0.4	47 7	3	Sb		15.7		
00082	00 06.5 + 21 21 110.01 – 40.18	03 – 01 – 027 NGC 15	1195	1.2 x 0.7 1.2 x 0.7	30 4	2	Sa		14.9		3
00083	00 06.8 + 00 20 101.19 – 60.47	00 – 01 – 000	0319	1.1 x 0.12 0.8 x 0.12	39 7	1	Sb		16.5:		*
00084	00 06.9 + 33 03 112.94 – 28.73	05 – 01 – 036 NGC 20	1257	(1.7 x 1.5) (2.0 x 1.7)	(140) ...	2	E – S0		14.5		*23
00085	00 06.9 + 47 05 115.60 – 14.91	08 – 01 – 026	0839	1.0 x 0.45 1.0 x 0.45	87 6	1	Sb – c		14.5		
00086	00 07.2 + 27 34 111.80 – 34.12	05 – 01 – 039 NGC 22	1257	1.8 x 1.4: 1.7 x 1.4	160: 2:	3	Sb or Sb/S B b		14.9		*
00087	00 07.2 + 28 05 111.92 – 33.62	05 – 01 – 038	1257	(1.2 x 1.2) (1.4 x 1.4)	–– ...	1	E?		15.1		*
00088	00 07.3 + 04 25a 103.85 – 56.63	01 – 01 – 041	1465	(1.2 x 0.9) (1.2 x 0.9)	(25) ...	1	S0 – a n of 2		(15.3)		*1
00089	00 07.3 + 25 39 111.37 – 36.01	04 – 01 – 033 NGC 23	0779	2.2 x 1.6: 2.2 x 1.5	8: ...	3	S B a P w 94		12.5	+ 4568 + 4788	*12 3468
00090	00 07.4 + 16 40 108.90 – 44.79	03 – 01 – 000	1195	1.1 x 0.3 1.0: x 0.3:	88 ...	2	Dwarf IV – V or V		18.		
00091	00 07.4 + 27 53 111.93 – 33.82	05 – 01 – 040?	1257	1.4 x 0.35 1.3 x 0.35	115 6	1	S ...		15.7		*
00092	00 07.4 + 27 56 111.94 – 33.77	05 – 01 – 041	1257	1.5: x 1.3 1.6: x 1.3:	–– 1 – 2	3	Sb Disturbed?, Sb +		14.7		*1
00093	00 07.7 + 30 35 112.61 – 31.18	05 – 01 – 042	1257	2.6: x 2.0: 2.0: x 1.7:	(60) 2:	4	... S IV	SA dm	15.6		
00094	00 07.8 + 25 33 111.48 – 36.13	04 – 01 – 034 NGC 26	0779	2.0 x 1.1 1.6 x 1.1	100: 4:	2	Sb P w 89		13.9		*12 36
00095	00 07.8 + 28 43 112.22 – 33.02	05 – 01 – 043	1257	1.9 x 0.15 1.9 x 0.15	7 7	3	Sc P w 96		15.7		*1
00096	00 07.9 + 28 44 112.25 – 33.01	05 – 01 – 044 NGC 27	1257	1.5 x 0.6 1.3 x 0.6	117 6	1	S ... P w 95		14.5		*1
00097	00 08.0 + 28 24 112.20 – 33.34	05 – 01 – 045	1257	(1.3 x 0.7) (1.1 x 0.6)	70: 4 – 5	2	S0		15.6		
00098	00 08.0 + 32 43 113.13 – 29.10	05 – 01 – 046 NGC 21	1257	1.2 x 0.6 1.2 x 0.6	42 5	3	S B b	S B(r)b	13.9		
00099	00 08.1 + 13 27 108.06 – 47.96	02 – 01 – 023	0010	3. x 3. 3. x 3.	–– 1	4	Dwarf spiral S V		16.0:		
00100	00 08.1 + 33 05 113.23 – 28.74	05 – 01 – 048 NGC 29	1257	1.6 x 0.7 1.6 x 0.7	154 6	2	Sb – c		13.5		*12 3
00101	00 08.3 + 25 17 111.55 – 36.41	04 – 01 – 035	0205	1.4 x 0.6 1.3 x 0.6	155 5 – 6	2	S B a	S B(s)a	15.4		*
00102	00 08.4 + 29 47 112.61 – 32.00	05 – 01 – 049	1257	1.6 x 1.3 1.6 x 1.3	(85) 2	3	S B a/Sa		14.8		*
00103	00 08.5 + 04 00 104.16 – 57.11	01 – 01 – 000	1465	1.1: x 1.1: 1.0: x 0.9:	–– 1	1	S ...	SAB ...	17.		
00104	00 08.6 + 07 43 105.99 – 53.55	01 – 01 – 042	1465	1.2 x 0.7 1.4 x 0.9	75: 4	3	S B a		16.0:		
00105	00 08.7 + 28 38 112.43 – 33.14	05 – 01 – 050	1257	1.2 x 1.1 1.2 x 1.1	–– 1	4	S B 0	(R)SAB(s)0 +	15.1		

N G C   9, 11 – 16, 20 – 23, 26, 27, 29

1	2	3	4	5	6	7	8	9	10	11	12
00106	00 08.8 + 06 06 105.34 - 55.12	01 - 01 - 043 NGC 36	1465	2.7 x 1.5 2.5 x 1.4	21 4 - 5	4	Sb/S B b p of 2, interaction	SAB(rs)b	14.5		*13
00107	00 09.0 + 27 40 112.30 - 34.10	05 - 01 - 000	1257	1.1 x 0.20 1.0 x 0.15	18 7	1	S ...		16.0:		
00108	00 09.2 + 28 14 112.47 - 33.55	05 - 01 - 051	1257	1.0 x 0.5 1.1 x 0.6	51 5	2	S B: b		14.8		
00109	00 09.3 - 01 22 101.26 - 62.28	00 - 01 - 037	0319	1.1: x 1.0: 1.1: x 1.0:	-- ...	2	S - Irr		17.		
00110	00 09.4 + 26 07 112.05 - 35.64	04 - 01 - 000	0205	1.3 x 0.8 1.3 x 0.8	145 4	4	Sc	SA(s)cd or SAB(s)cd	15.3		
00111	00 09.6 + 11 46 108.00 - 49.69	02 - 01 - 024	0010	1.3 x 0.4 1.3 x 0.4	76 6	2	Sc		15.4		
00112	00 09.6 + 41 28 115.13 - 20.53	07 - 01 - 000	1243	2. x 2. 1.8:: x 1.5::	-- 1	4	Dwarf spiral S V		17.		
00113	00 09.7 + 22 03 111.15 - 39.65	04 - 01 - 037 5	0205	1.2 x 0.4 1.2 x 0.5	105 6	1	Sa sff of 2		14.9		*1
00114	00 09.7 + 30 47 113.15 - 31.07	05 - 01 - 052 NGC 39	1257	1.2 x 1.1 1.3 x 1.2	-- 1	4	Sc	SA(rs)c Sc -	14.4		3
00115	00 10 + 88 06 122.64 + 25.53	15 - 01 - 001	0570	2.3 x 2.1 1.8: x 1.7:	-- ...	3	Dwarf spiral SB IV - V		16.5:		*
00116	00 10.1 + 05 14 105.47 - 56.04	01 - 01 - 045	1465	1.2 x 0.5 0.8 x 0.3	68 6	1	S ...		15.4		
00117	00 10.2 + 33 06 113.73 - 28.80	05 - 01 - 053	1257	1.7: x 1.2: 1.7: x 1.2:	110: 3	4	Sc	SAB(s)cd	15.2		
00118	00 10.3 + 21 50 111.27 - 39.89	04 - 01 - 041 NGC 42	0205 5	1.1 x 0.6 1.1 x 0.6	115 ...	2	E - S0 p of 2		15.0		*3
00119	00 10.4 + 14 08 109.10 - 47.42	02 - 01 - 025	0010	1.0 x 0.6 1.1 x 0.5	[78] ...	1	... Peculiar		14.2		
00120	00 10.4 + 30 39 113.29 - 31.22	05 - 01 - 054 NGC 43	1257	1.4 x 1.3 1.5 x 1.4	-- 1	4	S B 0		13.9		3
00121	00 10.4 + 38 58 114.86 - 23.02	06 - 01 - 017	1247	1.2 x 0.7 1.1 x 0.7	105 4	3	S B c	S B(s)c	15.4		
00122	00 10.7 + 16 45 110.01 - 44.88	03 - 01 - 028 3	1195	2.6 x 0.4 1.9: x 0.4	[109] ...	3	Irr Ir I	I m	15.2		*
00123	00 10.8 + 17 13 110.18 - 44.43	03 - 01 - 029 IC 4	1195 3	1.0 x 1.0 1.3 x 1.1	(10) ...	2	S ... Singular		14.2		*
00124	00 10.8 + 28 06 112.86 - 33.75	05 - 01 - 000	1257 4	1.1 x 0.20 1.2 x 0.20	32 7	2	Sa		15.5		
00125	00 11.1 + 47 53 116.46 - 14.24	08 - 01 - 028 IC 1534	0839	1.1 x 0.4 1.2 x 0.6	72 5 - 6	3	S0 NGC 51 group		15.2		*1C
00126	00 11.2 + 14 20 109.44 - 47.27	02 - 01 - 000	0010	1.1 x 1.0 1.0 x 1.0	-- 1	1	S ...		15.7		*
00127	00 11.2 + 26 42 112.67 - 35.14	04 - 01 - 042	0205	1.8 x 0.8: 1.7: x 0.7:	[135] ...	2	S B: c		16.5:		
00128	00 11.2 + 35 43 114.45 - 26.26	06 - 01 - 019	1247	2.3 x 2.0: 2. x 2.	(65) ...	4	... S IV - V	S dm	16.5:		
00129	00 11.3 + 03 26 105.11 - 57.85	00 - 01 - 041	0319	1.0 x 0.9 0.9 x 0.9	-- 1	2	S B b Sb -	(R´)S B(s)ab	15.5		
00130	00 11.3 + 30 37 113.51 - 31.29	05 - 01 - 000	1257	0.45 x 0.35 0.45 x 0.35	(160) ...	1	... Compact:		14.2		*
00131	00 11.3 + 47 53 116.49 - 14.24	08 - 01 - 030 IC 1535	0839	1.1 x 0.20 1.1 x 0.20	170 6 - 7	1	S ... NGC 51 group		15.2		*1
00132	00 11.4 + 12 42 108.98 - 48.88	02 - 01 - 026	0010	1.6 x 0.5 1.8: x 0.7:	14 (6)	3	... P w 134	SAB dm	15.3		*
00133	00 11.4 + 47 58 116.52 - 14.16	08 - 01 - 031 NGC 48	0839	1.7 x 1.0 1.7 x 0.8	15 4 - 5	3	Sb Disturbed?; NGC 51 group		15.0		*12
00134	00 11.5 + 12 38 108.99 - 48.95	02 - 01 - 000	0010	1.0: x 1.0: 0.9 x 0.8	-- 1:	2	... S IV - V, p w 132	S dm - m	17.		*
00135	00 11.6 + 07 09 106.97 - 54.28	01 - 01 - 000	1465	1.2 x 0.1 1.2: x 0.1	27 7	2	...	S d - dm	17.		
00136	00 11.7 + 47 58 116.58 - 14.17	08 - 01 - 033 NGC 49	0839	1.0 x 0.9 1.1 x 1.0	(165) ...	1	S0? NGC 51 group		15.3		*1
00137	00 11.8 + 48 24 116.66 - 13.74	08 - 01 - 034	0839	1.2 x 0.4 1.0 x 0.3	77 6	2	Sb - c		15.7		*
00138	00 11.9 + 47 59 116.61 - 14.16	08 - 01 - 035 NGC 51	0839	(1.4 x 1.3) (1.7 x 1.6)	-- (1)	1	(S0 - a) Brightest of 6, singular		14.6		*12
00139	00 12.0 - 01 00 102.88 - 62.15	00 - 01 - 043	0319	2.4 x 1.3 2.2 x 1.1	82 5	2	Sc		14.8		*
00140	00 12.0 + 18 17 110.86 - 43.45	03 - 01 - 030 NGC 52	1195 3	2.6 x 0.5: 2.6 x 0.5	127 6	1	S ... nf of 2, disturbed		14.6		*3

N G C   36,   39,   42,   43,   48,   49,   51,   52

1	2	3	4	5	6	7	8	9	10	11	12
00141	00 12.0 + 28 10 113.19 - 33.73	05 - 01 - 000	1257 4	1.2 x 0.6 1.2: x 0.6	145 6	2	S B 0 - a		15.7		*
00142	00 12.2 + 04 57 106.25 - 56.45	01 - 01 - 000	1465	1.3 x 0.4 1.3 x 0.4	133 6	3	S B b npp of 2		15.6		*
00143	00 12.7 + 05 37 106.77 - 55.84	01 - 01 - 046	1465	1.4 x 1.3 1.1 x 0.9	-- 1	2	S B a	S B(r)a	16.0:		
00144	00 12.8 + 15 57 110.48 - 45.77	03 - 01 - 032	1195	1.0 x 0.3 0.9 x 0.2	90 6	1	Sb - c		15.7		
00145	00 12.9 + 17 03 110.82 - 44.70	03 - 01 - 031 NGC 57	1195 3	(2.6 x 2.1) (2.8 x 2.3)	(40) ...	2	E		13.7		3
00146	00 13.1 + 27 10 113.27 - 34.76	04 - 01 - 045	0205	1.2 x 0.6 1.1 x 0.5	65 ...	2	S0		15.4		
00147	00 13.1 + 29 23 113.72 - 32.57	05 - 01 - 057	1257 4	1.7: x 0.25 1.7 x 0.30	156 7	2	S0 - a		14.6		*
00148	00 13.2 + 15 48 110.57 - 45.94	03 - 01 - 034	1195	1.5 x 0.4 1.5 x 0.4	98 6	1	S ...		14.0		*
00149	00 13.3 + 13 47 110.01 - 47.92	02 - 01 - 027	0010	1.3 x 0.3 1.1 x 0.3	60 6	1	S ...		15.6		
00150	00 13.4 - 00 34 103.86 - 61.84	00 - 01 - 048 NGC 60	0319	1.6 x 1.6 1.4 x 1.4	-- 1	3	S ... Singular, distorted?		15.4		*
00151	00 13.6 + 10 03 108.89 - 51.58	02 - 01 - 028	0010	(1.3 x 1.2) (1.5 x 1.3)	-- ...	1	S0?		14.7		
00152	00 13.9 + 29 39 113.97 - 32.34	05 - 01 - 059	1257 4	1.5 x 0.5 1.5 x 0.5	73 6	2	Sc		15.7		
00153	00 14.0 + 47 40 116.93 - 14.52	08 - 01 - 000	0839	1.2 x 0.6 0.8 x 0.3	178 ...	1	... ... B dm - m ssp of 2, disturbed?		17.		*
00154	00 14.2 + 02 11 105.80 - 59.25	00 - 01 - 000	0319	1.2: x 1.2: 0.8: x 0.8:	-- ...	3	Dwarf S: IV - V		18.		
00155	00 14.2 + 06 47 107.90 - 54.80	01 - 01 - 047	1465	1.6 x 0.4 1.5 x 0.4	3 6	1	S ...		14.6		*
00156	00 14.2 + 12 04 109.80 - 49.64	02 - 01 - 029	0010	3. x 2. 3. x 2.	(5) ...	4	Dwarf irregular Ir V		15.3		1
00157	00 14.3 + 27 35 113.68 - 34.39	05 - 01 - 060	1257	1.1 x 0.9 1.1 x 0.9	(15) 1 - 2	2	S B b: s of 2		15.1		*
00158	00 14.4 + 41 53 116.15 - 20.26	07 - 01 - 006	1243	1.9 x 0.6 1.9 x 0.5	173 6	3	Sb		15.5		
00159	00 14.5 + 17 15 111.40 - 44.57	03 - 01 - 000	1195 3	1.4:: x 1.4:: 1. x 1.	-- ...	3	Dwarf V, p w 162		18.		*
00160	00 14.5 + 34 12 114.93 - 27.86	06 - 01 - 021	1247	1.6: x 1.5: 1.4: x 1.4:	-- 1	4	... SA dm S IV - V		16.5:		
00161	00 14.6 + 06 27 107.93 - 55.14	01 - 01 - 000	1465	1.1: x 0.9: 1. x 1.	-- ...	3	Dwarf V, p w 163		18.		*
00162	00 14.6 + 17 09 111.41 - 44.68	03 - 01 - 035	1195 3	1.3 x 0.5 1.3 x 0.5	43 6	1	S ... P w 159		15.6		*
00163	00 14.7 + 06 27 107.97 - 55.15	01 - 01 - 048	1465	1.1 x 0.6 1.1 x 0.6	(103) ...	2	S B c S B(s)cd: P w 161, disturbed?		15.4		*
00164	00 14.8 + 17 47 111.64 - 44.06	03 - 01 - 036	1195 3	1.8 x 0.7 1.5 x 0.7	65 5 - 6	2	S B b/S B c Sc -		15.2		*
00165	00 15.0 + 24 23 113.22 - 37.57	04 - 01 - 047	0205 5	1.0 x 0.2 1.0 x 0.2	37 7	1	Sa - b		15.4		*
00166	00 15.0 + 29 56 114.30 - 32.09	05 - 01 - 062	1257 4	2.4 x 1.2 1.7 x 0.9	18 5	3	Sc		15.5		*
00167	00 15.1 + 11 10 109.84 - 50.57	02 - 01 - 030 NGC 63	0010	1.7 x 1.1 2.0 x 1.2	108 (4)	1	S ... Peculiar		12.6		13
00168	00 15.5 + 18 01 111.92 - 43.86	03 - 01 - 037	1195 3	1.2 x 0.3 1.3 x 0.3	162 7	2	Sa		16.0:		*
00169	00 15.7 + 19 06 112.25 - 42.81	03 - 01 - 000	1195 3	1.1 x 0.9 1.0: x 0.9	-- 1 - 2	1	S B? c		14.8		*
00170	00 15.7 + 29 48 114.46 - 32.25	05 - 01 - 065 NGC 68	1257 4	1.2 x 1.1 1.5 x 1.4	-- ...	2	[S0] In compact group		14.5	+ 5787 + 6011	*12 3VA
00171	00 15.7 + 48 27 117.33 - 13.78	08 - 01 - 036	0839	0.9 x 0.7 1.0 x 0.8	[30] ...	1	Irr		14.4		*
00172	00 15.8 + 17 34 111.91 - 44.32	03 - 01 - 000	1195 3	1.0 x 0.12 0.9 x 0.15	110 7	2	Sc		17.		
00173	00 15.8 + 29 47 114.48 - 32.27	05 - 01 - 068 NGC 71	1257 4	(1.5 x 1.2) (1.5 x 1.2)	-- ...	2	[E] In compact group		14.8	+ 6591 + 6815	*12 3VA
00174	00 15.8 + 29 49 114.48 - 32.24	05 - 01 - 067 NGC 70 = 1539*	1257 4	2.0? x 1.6: 2.0? x 1.6:	(0) ...	3	Sb SA(rs)bc In compact group		14.5		*12 3VA
00175	00 15.8 + 49 43 117.51 - 12.53	08 - 01 - 038	0839	1.3 x 0.9 1.2 x 0.8	(50) ...	3	Sc/S B c Disturbed?		15.5		*

N G C   57, 60, 63, 68, 70, 71   I C   1539

1	2	3	4	5	6	7	8	9	10	11	12
00176	00 15.9 + 29 46 114.50 - 32.29	05 - 01 - 069 NGC 72	1257 4	1.3 x 1.0 1.3 x 1.2	(15) 1 - 2	4	S B a S B(r)a In compact group		15.0	+ 6976 + 7200	*12 3
00177	00 16.0 + 79 26 121.42 + 16.93	13 - 01 - 000	1213	[1.3 x 0.5] ...	[42] ...	3	S ... + companion Bridge		15.4		*
00178	00 16.1 + 47 44 117.30 - 14.50	08 - 01 - 037	0839	1.2 x 0.6: 1.0 x 0.35	17 5 - 6	1	S ...		14.5		
00179	00 16.4 + 23 12 113.38 - 38.79	04 - 01 - 049	0205 5	1.3 x 0.7 1.3 x 0.6	3 5	2	Sc		15.3		
00180	00 16.5 + 15 28 111.60 - 46.41	02 - 01 - 032	0010	1.2 x 0.4 1.1 x 0.3	170 6	1	Sa - b		15.3		
00181	00 16.5 + 19 07 112.51 - 42.82	03 - 01 - 038	1195 3	1.3 x 0.3 1.3 x 0.4	14 6	2	Sa		15.4		
00182	00 16.8 + 06 10 108.75 - 55.54	01 - 01 - 051 NGC 75	1465	(1.9 x 1.9) (1.8 x 1.8)	-- 1	3	S0		14.8		*
00183	00 16.9 + 46 58 117.33 - 15.28	08 - 01 - 039	0839	1.6 x 0.7 1.5 x 0.6	50 6	3	Sb Sb -		13.8		
00184	00 17.0 + 12 55 111.08 - 48.94	02 - 02 - 000	0823	1. x 1. 1. x 1.	-- ...	0	Dwarf V		19.		
00185	00 17.0 + 29 40 114.76 - 32.42	05 - 01 - 072 NGC 76	1257 4, 6	(1.4 x 1.2) (1.4 x 1.2)	(80) ...	2	... Compact, p of 2		14.0		*1
00186	00 17.2 + 23 30 113.67 - 38.52	04 - 01 - 050 IC 1540	0205 5	1.1 x 0.5 1.2 x 0.6	30 5	2	S B b		14.9		*1
00187	00 17.4 + 08 52 109.98 - 52.93	01 - 01 - 000	0915	1.1 x 0.3 1.1 x 0.3	66 6 - 7	1	S ...		16.5:		
00188	00 17.4 + 19 43 112.93 - 42.27	03 - 02 - 003	0903	1.4 x 0.4 1.3: x 0.3	[13] ...	2	S B ... nnf of 2, distorted		15.7		*1
00189	00 17.5 + 14 50 111.78 - 47.08	02 - 02 - 000	0823	1.2: x 0.9: 1.0: x 0.8::	(120) ...	1	... S: d - dm		17.		
00190	00 17.5 + 42 17 116.81 - 19.94	07 - 01 - 007	1243	1.4:x 0.9: 1.0: x 0.8:	(110) ...	3	... S d - dm		16.5:		
00191	00 17.6 + 10 36 110.62 - 51.25	02 - 01 - 033	0823	2.6: x 2.1: 2.3: x 1.8:	(150) ...	4	Dwarf spiral S V		15.6		*1
00192	00 17.6 + 59 02 118.96 - 03.32	10 - 01 - 000 IC 10	1233	8. :: x 7. :: 10. :: x 10. ::	-- ...	4	Dwarf irregular, I m Ir V:, Local group?		...	- 347 - 92	*2 2
00193	00 17.8 + 00 33a 106.71 - 61.06	00 - 02 - 004 NGC 78a	0591	1.3: x 0.9 1.1: x 0.8	80: 3:	2	S B 0 - a P w 194, contact		14.5		*13
00194	00 17.8 + 00 33b 106.71 - 61.06	00 - 02 - 005 NGC 78b	0591	1.2: x 1.0 1.3: x 0.9	-- ...	1	S0: P w 193, contact				*13
00195	00 17.8 + 07 25 109.64 - 54.37	01 - 02 - 003 IC 13	0915	1.5 x 0.6 1.6 x 0.6	163 6	1	Sb - c Brightest of 3		15.0		*
00196	00 17.9 + 47 10 117.53 - 15.10	08 - 01 - 040	0839	1.5 x 1.1 1.1 x 0.8	(140) 3	3	S B c S B(s)c		14.2		
00197	00 18.2 + 19 45 113.19 - 42.26	03 - 02 - 004	0903	1.2 x 1.1 1.4 x 1.2	-- 1	3	Sa SAB a		15.1		
00198	00 18.3 + 21 36 113.61 - 40.44	04 - 02 - 002 IC 1543	1188 5	0.8 x 0.7 0.8 x 0.7	(30) ...	1	S ...		14.2		
00199	00 18.4 + 12 35 111.50 - 49.33	02 - 02 - 000	0823	1. x 1. 1. x 1.	-- ...	0	Dwarf V		19.		
00200	00 18.4 + 17 25 112.73 - 44.57	03 - 02 - 000	0903	1.1: x 1.0: 0.8: x 0.8:	-- ...	3	Dwarf irregular Ir V		18.		
00201	00 18.5 + 07 20 109.90 - 54.49	01 - 02 - 000	0915	1.2 x 0.2 0.9 x 0.15	18 7	3	Sc		16.5:		*
00202	00 18.5 + 26 57 114.68 - 35.16	04 - 02 - 000	1188	1.4 x 0.8 1.4: x 0.9:	25: 4	3	S B 0 (R)S B 0		14.5		*
00203	00 18.6 + 22 05 113.80 - 39.97	04 - 02 - 004 NGC 80	1188 5	(2.2 x 2.0) (2.3 x 2.2)	-- ...	2	S0 Brightest in group		13.7	+ 5586 + 5790	*12 3
00204	00 18.7 + 22 50 113.98 - 39.24	04 - 02 - 006 IC 1544	1188 5	1.4 x 0.9 1.2 x 0.8	150 3 - 4	4	Sc SAB(s)c		14.6		
00205	00 18.8 + 17 03 112.78 - 44.95	03 - 02 - 006	0903	[1.4 x 0.7] ...	[10] ...	2	S B ... + companion Bridge:		15.6		*1
00206	00 18.8 + 22 09 113.87 - 39.92	04 - 02 - 005 NGC 83	1188 5	(1.3 x 1.2) (1.3 x 1.2)	-- ...	3	E NGC 80 group		14.3	+ 6541 + 6745	*12 3
00207	00 18.9 + 29 23 115.20 - 32.76	05 - 02 - 004	1244 6	1.0 x 0.3 1.0 x 0.3	17 6	2	Sb		15.6		
00208	00 19.2 + 22 08 113.99 - 39.95	04 - 02 - 011 NGC 91	1188 5	2.4 x 0.9 1.8: x 0.9	[132] ...	4	S ... SAB(s) ... Distorted, NGC 80 group		14.5		*13 A
00209	00 19.4 + 22 08 114.05 - 39.95	04 - 02 - 012 NGC 93	1188 5	1.5 x 0.7 1.6 x 0.7:	48 5 - 6	2	S ... NGC 80 group		14.7		*13 A
00210	00 19.4 + 23 28 114.30 - 38.63	04 - 02 - 013	1188 5	1.3 x 0.45 1.3 x 0.45	19 6	2	Sb		15.0		*

N G C   72, 75, 76, 78, 80, 83, 91, 93   I C 10, 13, 1540, 1543, 1544

1	2	3	4	5	6	7	8	9	10	11	12
00211	00 19.6 + 20 20 113.75 – 41.74	03 – 02 – 000	0903	1.1 x 0.2 1.1 x 0.1	165 (7)	1	...	d – m	17.		
00212	00 19.8 – 01 34 106.62 – 63.24	00 – 02 – 14+15	0591	1.7 x 0.7 1.2 x 0.6	[18] ...	3	S B ... n of 2, disturbed		15.0		*1V A
00213	00 19.8 + 06 10 110.05 – 55.70	01 – 02 – 000	0915	(1.3 x 0.8) (1.2 x 0.7)	(150) ...	2	[SO] Compact		15.7		*
00214	00 19.8 + 10 13 111.35 – 51.72	02 – 02 – 003 NGC 95	0823	1.8 x 1.3 1.7 x 1.3	(75) 1 – 3	3	Sc		13.4		*12 3468
00215	00 19.8 + 29 13 115.41 – 32.95	05 – 02 – 006	1244 6	1.3 x 0.8 1.4 x 0.8	135 4	3	S B a or S B a/S B b	(R´)S B(s) ...	14.9		
00216	00 19.9 + 29 28 115.47 – 32.71	05 – 02 – 007 NGC 97	1244 6	(1.5 x 1.3) (1.5 x 1.3)	-- ...	3	[E] Compact		13.5		3
00217	00 20.0 + 27 50 115.23 – 34.33	05 – 02 – 000	1244	1.2: x 0.8: 0.9: x 0.8:	(10) ...	3	Dwarf S: IV – V or V		17.		
00218	00 20.3 + 06 42 110.45 – 55.20	01 – 02 – 005 IC 1549	0915	(1.4 x 1.4) (1.6 x 1.6)	-- 1	1	S ...		14.9		
00219	00 20.4 + 15 58 113.07 – 46.08	03 – 02 – 007	0903	1.2 x 0.5 1.2 x 0.5	148 ...	2	...	d – m	15.7		*
00220	00 20.5 + 02 10 108.80 – 59.65	00 – 02 – 000	0591	1.4: x 0.5 1.7 x 0.7	177 6	2	SO sff of 2		16.0:		*
00221	00 20.6 + 27 10 115.28 – 35.01	04 – 02 – 000	1188	0.7 x 0.5 0.7 x 0.4	155 ...	1	... Peculiar		14.5		*
00222	00 20.7 + 34 52 116.47 – 27.37	06 – 02 – 000	0398	1.2: x 0.1 0.8: x 0.1	138 ...	1	...		18.		
00223	00 20.9 + 19 58 114.08 – 42.15	03 – 02 – 000	0903	1.6: x 1.6: 1.5: x 1.5:	-- ...	3	(Dwarf) IV – V		16.5:		
00224	00 21.0 – 00 47 107.67 – 62.55	00 – 02 – 18+19	0591	[1.3 x 0.7] ...	... ...	3	Disrupted pair Bridge		15.4		*1V A
00225	00 21.0 + 20 34 114.23 – 41.56	03 – 02 – 008	0903	(1.1 x 1.0) (1.2 x 1.0)	-- ...	1	SO:		16.0:		
00226	00 21.2 + 14 24 112.98 – 47.66	02 – 02 – 005 Mark 338	0823	1.2 x 0.6 1.1: x 0.6:	4 ...	1	...		15.1	+ 5400	*
00227	00 21.2 + 26 40 115.37 – 35.52	04 – 02 – 000	1188	(1.5 x 1.1) (1.5 x 1.1)	(145) ...	2	... Compact		14.7		*
00228	00 21.3 + 24 02 114.95 – 38.13	04 – 02 – 015	1188 5	1.3 x 0.9: 1.0 x 0.8:	43 3:	1	Sb – c		14.8		
00229	00 21.3 + 28 03 115.61 – 34.15	05 – 02 – 000	1244	1.0 x 0.5 0.7 x 0.4	35: 5	2	Sb		15.3		
00230	00 21.4 + 15 30 113.30 – 46.58	02 – 02 – 006 NGC 99	0823	1.5 x 1.5 1.4 x 1.4	-- 1	3	Sc Disturbed:		14.0		*
00231	00 21.5 + 16 12 113.49 – 45.89	03 – 02 – 009 NGC 100	0903	5.8 x 0.7 5.1 x 0.7	56 6 – 7	3	Sc		14.6		12
00232	00 22.0 + 32 59 116.51 – 29.28	05 – 02 – 009	1244	1.5: x 1.5: 0.8 x 0.55	-- 1:	3	S B a	S B(r)a	14.7		
00233	00 22.1 + 14 33 113.33 – 47.55	02 – 02 – 000 Mark 339	0823	1. x 1. 1. x 1.	-- ...	1	... Peculiar		14.6	+ 5400	*
00234	00 22.2 + 29 17 116.03 – 32.95	05 – 02 – 000	1244 6	1.2 x 0.5 1.0 x 0.5	72 6	3	Sc		15.6		
00235	00 22.2 + 44 53 118.03 – 17.46	07 – 02 – 000	0851	1.1 x 0.5 1.0 x 0.5	(135) ...	1	E – S ... Disturbed?		16.0:		*
00236	00 22.3 + 43 23 117.88 – 18.95	07 – 02 – 000	0851	(1.1 x 1.1) (1.1 x 1.1)	-- ...	1	E		15.0		
00237	00 22.4 + 06 23 111.24 – 55.61	01 – 02 – 006	0915	1.0 x 0.5 1.0 x 0.5	70 5	1	S B a		15.4		
00238	00 22.4 + 31 05 116.34 – 31.17	05 – 02 – 010	1244 6	2.0 x 0.5 1.9 x 0.5	178 6	1	S ...		14.4		
00239	00 22.5 + 41 47 117.73 – 20.55	07 – 02 – 000	0851	1.1:: x 0.8:: 0.8: x 0.6:	-- ...	1	Irr		16.5:		
00240	00 22.7 + 06 13a 111.32 – 55.79	01 – 02 – 007	0915	1.0 x 0.5 1.1 x 0.6	68 5:	3	Sb/S B b spp of 2, distorted	SAB(rs)b	(14.9)		*1
00241	00 22.8 + 12 37b 113.13 – 49.48	02 – 02 – 008 NGC 105	0823	0.9 x 0.7 1.0 x 0.8	(167) 2	2	Sa s of 2, contact		(14.1)		*12
00242	00 22.8 + 19 56 114.74 – 42.24	03 – 02 – 010	0903	2.0 x 1.0 2.1 x 1.8	-- ...	4	S B c s of 2, disturbed	SAB d	14.3		*
00243	00 22.8 + 45 39 118.22 – 16.71	08 – 02 – 001	1224	1.9 x 0.4 1.9 x 0.4	2 7	2	Sb		14.9		
00244	00 22.9 + 24 33 115.50 – 37.67	04 – 02 – 017	1188 5	1.1 x 0.5 1.0 x 0.5	154 5	1	S B? b – c		14.9		*
00245	00 23.3 + 05 08 111.23 – 56.88	01 – 02 – 010	0915	1.5 x 0.5 1.3 x 0.5	3 6	2	S B a – b		15.7		

N G C   95, 97, 99, 100, 105

I C   1549

1	2	3	4	5	6	7	8	9	10	11	12
00246	00 23.3 + 28 56 116.27 − 33.33	05 − 02 − 012 NGC 108	1244 6	2.3: x 2.0: 2.3 x 2.0	-- 1	4	S B 0	(R)S B(r)0 +	13.3		3
00247	00 23.4 + 14 04 113.69 − 48.07	02 − 02 − 000	0823	1.1 x 0.2 1.0 x 0.2	19 6	3	Sc		16.5:		
00248	00 23.4 + 25 27 115.78 − 36.79	04 − 02 − 018	1188	[1.9 x 1.3] ...	[48] ...	3	Disrupted pair Contact		14.9		*1
00249	00 23.6 + 13 22b 113.60 − 48.77	02 − 02 − 009	0823	1.7 x 1.0 1.5 x 0.9	[90] ...	3	S − Irr f of 2, disrupted		(14.8)		*1
00250	00 23.6 + 16 08 114.20 − 46.03	03 − 02 − 011	0903	1.4: x 1.1: 1.3: x 1.0:	(155) 2:	2	S B: c		15.6		*
00251	00 23.6 + 21 32 115.20 − 40.68	04 − 02 − 020 NGC 109	1188	1.3 x 1.1 1.3 x 1.1	-- 1 − 2	4	S B a f of 2	S B(r)a	15.0		*13
00252	00 23.8 + 00 45 109.83 − 61.21	00 − 02 − 026	0591	1.1 x 0.8 1.0 x 0.7	(120) 3	4	Sc	SA c	16.5:		
00253	00 23.9 + 06 01 111.78 − 56.04	01 − 02 − 011	0915	1.4 x 0.8 1.1 x 0.7	30: 4	3	Sb/S B b s of 2	SAB(s)b	15.3		*
00254	00 23.9 + 39 13 117.72 − 23.13	06 − 02 − 000	0398	1.1 x 0.3 1.0 x 0.3	33 6	1	S ...		16.5:		
00255	00 24.2 + 31 25 116.83 − 30.89	05 − 02 − 013 NGC 112	1244 6	1.3 x 0.6 1.1 x 0.5	108 5	1	S ...		14.5		*
00256	00 24.2 + 49 45 118.90 − 12.65	08 − 02 − 003	1224	1.3 x 0.2 1.3 x 0.2	171 7	1	Sb − c		15.6		*
00257	00 24.3 + 01 12 110.27 − 60.79	00 − 02 − 000	0591	(1.4 x 1.2) (1.3 x 1.0)	-- ...	1	...		16.0:		*
00258	00 24.4 + 50 49 119.04 − 11.59	08 − 02 − 000	1224	1.1 x 0.5 1.1 x 0.5	167 5 − 6	1	Sa − b sff of 2		16.0:		*
00259	00 24.5 − 02 03 108.93 − 63.98	00 − 02 − 027 NGC 114	0591	(1.3 x 1.0) (1.2 x 0.9)	(165) 2	4	S B 0		15.0		*
00260	00 24.5 + 11 18 113.46 − 50.85	02 − 02 − 011	0823	3.1 x 0.7 3.0 x 0.7	21 6	3	Sc f of 2, disturbed		14.0		*1
00261	00 24.6 + 23 54 115.88 − 38.36	04 − 02 − 022	1188	1.3: x 1.0: 1.2 x 1.0	-- ...	4	... Peculiar		14.8		*1
00262	00 24.6 + 39 31 117.90 − 22.84	07 − 02 − 000	0851	1.2 x 0.6 1.0 x 0.7	[67] ...	3	... Peculiar, eruptive?		15.1	+ 10850 + 11122	*C
00263	00 24.6 + 43 46 118.36 − 18.61	07 − 02 − 004	0851	1.6 x 0.3 1.7 x 0.3	[171] 6 − 7	2	Sc − Irr Disrupted:		16.5:		1
00264	00 24.7 − 02 03 109.04 − 63.99	00 − 02 − 032 NGC 118	0591	(1.3 x 0.9) (0.9 x 0.8)	— ...	1	... Compact		14.9		*1C
00265	00 24.7 + 19 47 115.24 − 42.45	03 − 02 − 012	0903	1.3 x 1.0: 1.2 x 1.0:	(117) 2	1	S ...		15.2		
00266	00 24.8 + 10 33 113.40 − 51.60	02 − 02 − 000	0823	1.1 x 0.7: 1.2: x 0.8:	(137) ...	1	S?...		16.0:		
00267	00 24.9 − 01 47 109.28 − 63.74	00 − 02 − 033 NGC 120	0591	1.6 x 0.6 1.5 x 0.6	73 6 − 7	3	S0		14.8		*2
00268	00 25.0 + 08 36 112.99 − 53.53	01 − 02 − 013 IC 1551	0915	2.6 x 1.3: 2.7 x 1.4	[15] ...	4	S ... Singular		15.0		*
00269	00 25.1 + 03 58 111.66 − 58.11	01 − 02 − 000	0915	1.1 x 0.5 1.0 x 0.4	110 5 − 6	1	S ...		16.0:		
00270	00 25.2 + 32 30 117.22 − 29.83	05 − 02 − 000	1244	1.2:: x 1.1:: 1. x 1.	-- ...	1	...		17.		
00271	00 25.3 − 02 05 109.36 − 64.06	00 − 02 − 038 NGC 124	0591	1.6 x 0.9 1.5 x 0.9	168: 4	3	Sc		13.8		*2
00272	00 25.3 − 01 28 109.65 − 63.45	00 − 02 − 036	0591	1.8 x 0.8 1.5: x 0.8	130 5 − 6	3	Sc		15.5		
00273	00 25.3 + 25 44 116.35 − 36.56	04 − 02 − 000	1188	1.2: x 0.8: 0.8:: x 0.7::	(115) ...	2	... S? IV or IV − V	d − m	17.		
00274	00 25.3 + 30 20 116.97 − 31.99	05 − 02 − 014	1244 6	1.2 x 0.7 1.1 x 0.6	25: 4	3	S B 0 − a		15.6		*
00275	00 25.4 + 02 14 111.21 − 59.83	00 − 02 − 037 IC 17	0591 7	1.9 x 0.9 1.7 x 0.8	128 5	2	S B ... P w 281		14.9		*
00276	00 25.4 + 30 29 117.01 − 31.84	05 − 02 − 015	1244 6	1.1 x 0.2 1.2 x 0.2	33 7	2	Sa − b P w 279		15.6		*
00277	00 25.7 + 02 57 111.61 − 59.13	00 − 02 − 039	0591 7	1.7: x 1.4: 1.4: x 1.2:	(160) 1 − 2	1	Sc:		15.7		
00278	00 25.7 + 27 07 116.65 − 35.20	04 − 02 − 025	1188	1.5 x 0.4 1.4 x 0.25	5 6	2	S B b − c		15.3		
00279	00 25.7 + 30 32 117.10 − 31.80	05 − 02 − 016	1244 6	1.8 x 0.5 1.7 x 0.5	118 6	2	S B? ... P w 276, disturbed?		14.3		*
00280	00 25.8 − 00 30 110.35 − 62.53	00 − 02 − 043	0591	1.4 x 1.1 1.1 x 0.8	(55) 2 − 3	4	S B b		16.0:		*

N G C   108, 109, 112, 114, 118, 120, 124   I C   17, 1551

1	2	3	4	5	6	7	8	9	10	11	12
00281	00 25.8 + 02 15 111.41 − 59.83	00 − 02 − 042	0591 7	1.4 x 0.7 1.3 x 0.5	22 5	3	Sc P w 275		15.7		*
00282	00 25.8 + 03 07a 111.71 − 58.98	00 − 02 − 040	0591 7	2.0 x 2.0 1.9: x 1.9:	— 1	4	... P w 283, contact	SAB dm			*1
00283	00 25.8 + 03 07b 111.71 − 58.98	00 − 02 − 041	0591 7	1.4 x 1.0 1.3: x 0.8:	168: 3:	3	Sc P w 282, contact	SA cd	15.0		*1
00284	00 25.9 + 33 00 117.44 − 29.35	05 − 02 − 017	1244	1.6 x 1.1 1.5 x 1.0	23: 3	4	S B c/Sc	SAB(s)c Sc +	15.1		*
00285	00 26.2 + 28 40 116.99 − 33.67	05 − 02 − 018	1244 6	1.0 x 0.3 0.9 x 0.25	107 6 – 7	2	Sa		15.0		
00286	00 26.3 + 02 33 111.76 − 59.55	00 − 02 − 048 NGC 125	0591 7	1.6 x 1.5 1.7 x 1.6	— 1	4	S0 NGC 128 group		14.2	+ 5289 + 5425	*12 3
00287	00 26.3 + 09 52 113.83 − 52.33	02 − 02 − 012	0823	1.3 x 1.1 1.2 x 1.2	— 1	4	Sc	SA d	15.6		*1
00288	00 26.3 + 43 09 118.62 − 19.26	07 − 02 − 000	0851	1.4 x 0.9 1.3: x 0.8:	135: ...	0	... Peculiar		16.0:		*
00289	00 26.4 + 15 41 115.08 − 46.57	03 − 02 − 000	0903	1.1 x 0.2 1.0 x 0.15	20 7	1	S ... P w 290		17.		*
00290	00 26.5 + 15 37 115.10 − 46.64	03 − 02 − 000	0903	2.1: x 0.2: 2.0: x 0.3:	136 ...	1	Dwarf? P w 289		18.		*
00291	00 26.5 + 32 51 117.57 − 29.51	05 − 02 − 000	1244	1.0 x 0.9 1.0: x 0.9:	— ...	2	(Dwarf) S: IV – V		17.		
00292	00 26.7 + 02 35 111.96 − 59.54	00 − 02 − 051 NGC 128	0591 7	3.0 x 0.9 3.2 x 1.0	1 7	3	S0 in group, disturbed	Brightest	(13.2)	+ 4250 + 4386	*12 3468
00293	00 26.7 + 26 08 116.80 − 36.20	04 − 02 − 000	1188	1.1 x 0.35 1.0 x 0.3	56 ...	0	...		17.		*
00294	00 26.7 + 31 06 117.42 − 31.26	05 − 02 − 000	1244 6	1.2: x 0.10 1.2: x 0.10	121 7	3	Sc		17.		
00295	00 26.9 − 01 22 110.57 − 63.43	00 − 02 − 056	0591	1.1 x 0.6 1.1 x 0.6	154 ...	3	S B b	SAB(s)b Sb +	16.0:		*1
00296	00 27.0 + 11 46 114.52 − 50.47	02 − 02 − 000	0823	[1.2 x 0.7] ...	...		Triple system Interaction		16.5:		*
00297	00 27.0 + 21 12 116.19 − 41.11	03 − 02 − 015 IC 1552	0903	1.0 x 0.22 1.1 x 0.25	127 6 – 7	1	S ...		15.4		
00298	00 27.2 + 02 42 112.24 − 59.45	00 − 02 − 059	0591 7	1.0 x 0.3 0.8 x 0.3	111 6 – 7	2	Sb		16.0:		
00299	00 27.2 + 31 07 117.54 − 31.25	05 − 02 − 019	1244 6	1.6 x 1.0 1.4 x 0.8	74 4:	3	Sc	Sc –	15.3		
00300	00 27.5 + 03 15 112.56 − 58.91	00 − 02 − 000	0591 7	1.4:: x 1.4:: 1. x 1.	---- ...	4	Dwarf V		18.		
00301	00 27.6 + 01 49 112.15 − 60.33	00 − 02 − 063 NGC 132	0591 7	1.9 x 1.7 1.8 x 1.5	(40) 2	4	S B b/Sc	SAB(s)bc Sb +	13.8		*23
00302	00 27.7 + 24 49 116.90 − 37.53	04 − 02 − 000	1188	1.0: x 0.9: 1.0: x 0.8:	— ...	4	Dwarf spiral S IV – V		17.		
00303	00 27.8 + 41 49 118.78 − 20.61	07 − 02 − 005	0851	2.1 x 1.5 1.6: x 1.2:	175: ...	4	Irr P w 306	Ir I	16.5:		*
00304	00 28.0 + 12 24 115.03 − 49.88	02 − 02 − 015	0823	(1.1 x 0.7) (1.2 x 0.8)	(125) 3 – 4	2	S0 – a		15.7		
00305	00 28.0 + 13 05 115.16 − 49.20	02 − 02 − 014	0823	1.4 x 1.0 1.6: x 1.4:	— 1	3	Sc Disturbed?	SA c	15.3		*
00306	00 28.2 + 41 55 118.87 − 20.52	07 − 02 − 006	0851	1.0 x 0.7 1.0 x 0.7	(105) 3	3	Sc Disturbed?, p w 303		15.7		*
00307	00 28.3 + 10 23 114.74 − 51.89	02 − 02 − 016	0823	1.2 x 0.6 1.3 x 0.7	100: 5	1	S0		16.0:		
00308	00 28.4 + 04 54 113.47 − 57.32	01 − 02 − 016 NGC 138	0915 8	1.4 x 0.7 1.4 x 0.7	175 5	1	Sa: Brightest in group		14.8		*13
00309	00 28.5 + 09 55 114.72 − 52.35	02 − 02 − 017 NGC 137	0823	(1.6 x 1.6) (2.0 x 2.0)	— 1	2	S0		14.2		13
00310	00 28.6 + 28 43 117.62 − 33.67	05 − 02 − 020	1244	1.3 x 0.4 1.3 x 0.4	30 6	3	Sc		14.8		
00311	00 28.7 + 30 31 117.85 − 31.88	05 − 02 − 021 NGC 140	1244 6	1.5 x 1.3 1.4 x 1.2	(45) 1	3	Sc Singular		14.2		
00312	00 28.8 + 08 11 114.46 − 54.08	01 − 02 − 019	0915	1.7 x 0.8 1.7 x 0.8	7 ...	3	S B ... Brightest of 4, distorted		14.6		*1
00313	00 28.9 + 05 56 113.96 − 56.31	01 − 02 − 021	0915 8	1.4 x 0.8 1.4 x 0.8	10 4	1	S ...		14.3		*1
00314	00 28.9 + 08 07 114.49 − 54.15	01 − 02 − 020	0915	1.2 x 0.9 1.3 x 1.0	(165) ...	3	S ... 312 group, distorted		15.4		*1
00315	00 29.0 + 08 07 114.53 − 54.15	01 − 02 − 000	0915	1.0 x 1.0 0.7 x 0.5	— 1	1	Sb – c 312 group		16.5:		*

N G C   125, 128, 132, 137, 138, 140   I C   1552

1	2	3	4	5	6	7	8	9	10	11	12
00316	00 29.0 + 14 20 115.75 − 47.99	02 − 02 − 000	0823	1.4: x 0.2 1.2: x 0.15	40 7	2	Sc		15.7		
00317	00 29.1 + 00 39 112.51 − 61.55	00 − 02 − 000	0591	1.0 x 0.9 0.8: x 0.8:	— ...	1		d − m	17.		
00318	00 29.3 + 37 24 118.68 − 25.04	06 − 02 − 008	0398	1.1:: x 1.0:: 0.6 x 0.6	— 1	4	S B c	S B(rs)c	16.5:		*
00319	00 29.5 + 31 24 118.14 − 31.02	05 − 02 − 022	1244 6	1.6 x 1.0 1.4 x 0.9	135 4	2	Sb − c		15.1		*
00320	00 29.9 + 02 18 113.44 − 59.95	00 − 02 − 066	0591 7	1.3 x 0.2 1.3 x 0.7	172 7	3	Sc		16.5:		*
00321	00 29.9 + 23 08 117.33 − 39.26	04 − 02 − 027	1188	1.6 x 0.3 1.1 x 0.2	149 6	3	S B? c		15.4		
00322	00 30.0 + 01 13 113.16 − 61.02	00 − 02 − 067	0591	1.3 x 0.6 1.2 x 0.6	[68] 7	2	Sc Distorted		16.5:		*
00323	00 30.3 + 11 27 115.73 − 50.89	02 − 02 − 018	0823	1.0 x 0.3 0.9 x 0.3	110 6:	1	S: ... nff of 2		15.3		*
00324	00 30.4 + 02 54 113.86 − 59.37	00 − 02 − 000	0591 7	1.0 x 0.5 0.9 x 0.5	78 5	2	Sb np of 2		16.5:		*
00325	00 30.4 + 21 57 117.33 − 40.45	04 − 02 − 028	1188	1.0: x 0.8: 0.8: x 0.6:	(10) ...	1	...		16.0:		
00326	00 30.4 + 48 14 119.81 − 14.25	08 − 02 − 005 NGC 147	1224	(15. x 9.) (15. x 9.)	25: ...	5	Dwarf elliptical P w 396, Local group		12.0		*12- −9
00327	00 30.6 + 07 38 115.10 − 54.68	01 − 02 − 025	0915	1.2 x 0.5 1.2 x 0.5	121 5	2	Sb − c np of 2		15.7		*
00328	00 30.8 − 01 23 112.72 − 63.62	00 − 02 − 070	0591	1.6 x 1.6 1.3: x 1.2:	— 1:	3	...	S B m	16.0:		
00329	00 31.0 + 02 24 114.01 − 59.89	00 − 02 − 071	0591 7	2.1 x 1.5 2.1 x 1.4	(170) 3	4	Sc	SA cd	15.7		
00330	00 31.0 + 39 17 119.21 − 23.18	06 − 02 − 009	0398	1.4 x 0.5 1.4 x 0.5	140 6	3	S0 Brightest of 3		15.0		*
00331	00 31.2 + 06 58 115.21 − 55.36	01 − 02 − 027	0915	1.0 x 0.5 1.1 x 0.6	170 5	3	S B a 335 group		15.2		*
00332	00 31.2 + 30 27 118.48 − 31.99	05 − 02 − 024 NGC 149	1244 6	1.3 x 0.8 1.5 x 0.9	155 4	2	[S0] Compact		15.0		
00333	00 31.2 + 48 40 119.98 − 13.83	08 − 02 − 006	1224	1.3 x 1.1 1.1 x 0.9	— ...	1	S ...	SAB ...	15.7		*1
00334	00 31.3 + 31 10 118.57 − 31.28	05 − 02 − 025	1244	2.0: x 2.0: 2. x 2.	— ...	4	Dwarf spiral S V		15.7		
00335	00 31.4 + 06 59ab 115.30 − 55.35	01 − 02 − 28+29	0915	[2.5 x 1.2] ...	[143] ...	3	E + companion Contact, b brightest of 3		14.2		*1
00336	00 31.4 + 43 52 119.65 − 18.62	07 − 02 − 008	0851	1.7 x 1.6 1.6 x 1.6	— 1	4	Sc	SA d	15.6		
00337	00 31.6 + 24 22 117.97 − 38.06	04 − 02 − 031	1188	1.2 x 1.1 1.1: x 0.9:	(150) 1	4	Sc S IV	SAB dm	16.5:		
00338	00 31.7 + 39 20 119.36 − 23.14	06 − 02 − 010	0398	1.0 x 0.2 1.0 x 0.2	99 6	2	S B? b − c 330 group		15.5		*
00339	00 31.8 + 09 55 116.04 − 52.45	02 − 02 − 020	0823	[1.3 x 0.8] ...	[110] ...	2	Double system Contact		15.6		*1
00340	00 31.8 + 12 00 116.40 − 50.38	02 − 02 − 021 IC 31	0823	1.4 x 0.25 1.3 x 0.25	89 7	2	Sa − b		15.5		*
00341	00 31.9 + 01 24 114.18 − 60.91	00 − 02 − 076	0591 7	1.2 x 1.0 0.9 x 0.7	— ...	1	S: ... n of 2		16.0:		*1
00342	00 31.9 + 03 18 114.69 − 59.02	00 − 02 − 75	0591 7	1.1 x 0.9 1.0 x 0.8	(65) 2	4	Sc	SAB c Sc −	15.7		
00343	00 32.0 + 53 10 120.42 − 09.35	09 − 02 − 000	1237	1.4: x 1.0: 1.4:: x 0.9::	(70) ...	0	[Sc − Irr]		16.0:		*
00344	00 32.1 + 39 17 119.44 − 23.20	06 − 02 − 011	0398	1.2 x 0.7 1.2 x 0.7	110: 3 − 4	4	Sc/S B c 330 group	SAB(s)cd	15.4		*
00345	00 32.2 + 28 08 118.51 − 34.32	05 − 02 − 000	1244	1.1 x 0.25 1.1 x 0.25	51 7	3	S B? c		15.6		
00346	00 32.3 + 31 40 118.86 − 30.80	05 − 02 − 026	1244	1.1 x 0.5 1.0: x 0.45	128 6	2	Sb − c		14.8		
00347	00 32.3 + 45 15 119.92 − 17.25	08 − 02 − 000	1224	1.0 x 0.20 1.0 x 0.20	87 6	3	Sc		16.5:		
00348	00 32.8 + 02 39 114.96 − 59.70	00 − 02 − 087	0591 7	1.2 x 1.2 1.1 x 1.1	— ...	3	...	S dm − m	15.7		
00349	00 32.8 + 03 07 115.08 − 59.23	00 − 02 − 085	0591 7	1.5: x 0.6 1.3: x 0.6	136 ...	2	S B b Brightest in group	SAB(s)bc	15.6		*1
00350	00 32.8 + 04 36 115.42 − 57.76	01 − 02 − 000	0915	(1.2 x 0.6) (1.3 x 0.7)	(115) ...	1	S0 − a:		15.7		*

1	2	3	4	5	6	7	8	9	10	11	12
00351	00 33.0 + 08 51	01 - 02 - 032	0915	3.2 x 1.2	156	4	S B a	S B(r)a	13.9		*
	116.34 - 53.54	IC 34		3.4 x 1.3	6		P w 353				
00352	00 33.0 + 24 00	04 - 02 - 000	1188	1.4 :: x 0.9 ::	(140)	2	...		16.5:		*
	118.33 - 38.46			1. x 1.	...						
00353	00 33.1 + 08 43	01 - 02 - 033	0915	1.4 x 0.5	124	1	Sc		15.6		*
	116.36 - 53.68			1.5 x 0.6	6		P w 351				
00354	00 33.3 + 23 42	04 - 02 - 032	1188	1.0 x 0.3	119	2	Sb - c		16.0:		*
	118.39 - 38.76			1.0 x 0.25	6						
00355	00 33.3 + 31 36	05 - 02 - 027	1244	1.2 x 0.35	123	1	S B a - b		14.5		*
	119.10 - 30.88			1.2 x 0.35	6						
00356	00 33.4 + 23 41	04 - 02 - 033	1188	3.0 x 1.6	45	4	Sa		13.7	+ 5255	*12
	118.42 - 38.78	NGC 160		3.2: x 1.8:	5		P w 365 ab			+ 5453	3
00357	00 33.5 + 21 12	03 - 02 - 000	0903	1.3 x 0.6	177:	1	Sb - c		15.6		*
	118.19 - 41.26			1.2 x 0.6	5						
00358	00 33.6 + 01 26	00 - 02 - 088	0591	1.3 x 1.3	--	1	S ...		15.3		*
	115.06 - 60.93		7	1.2 x 1.1	1						
00359	00 33.6 + 04 22	01 - 02 - 035	0915	1.0 x 0.4	136	1	S ...		15.5		
	115.74 - 58.02			1.1 x 0.4	6						
00360	00 33.6 + 25 32	04 - 02 - 000	1188	1.7? x 0.5	...	1	...		16.0:		*
	118.65 - 36.94		9	1.1 x 0.7	...		f of 2, plume?				
00361	00 33.7 + 48 44	08 - 02 - 000	1224	(1.2 x 1.0)	(100)	1	...		15.7		
	120.40 - 13.79			(1.2 x 1.0)	...		Compact				
00362	00 33.9 + 32 29	05 - 02 - 000	1244	1.2: x 0.9:	--	3	Dwarf spiral		17.		
	119.32 - 30.01			1.0: x 0.8:	...		S B IV - V or V				
00363	00 33.9 + 43 32	07 - 02 - 000	0851	1.0 x 0.2	159	1	S ...		16.5:		
	120.11 - 18.98			1.0 x 0.2	6						
00364	00 34.1 + 21 17	03 - 02 - 017	0903	1.2 x 0.45	58	1	S ...		14.5		*
	118.39 - 41.18			1.2 x 0.35	6		Brightest of 3				
00365	00 34.2 + 23 43a	04 - 02 - 035	1188	3.5 x 1.1	88	3	Sb		(13.3)	+ 4500	*12
	118.65 - 38.76	NGC 169, Mark 341		3.3: x 1.1:	6		P w IC 1559, contact				3
00366	00 34.2 + 28 38	05 - 02 - 000	1244	(1.2 x 0.9)	--	2	...		16.0:		*
	119.09 - 33.85		10	(1.2 x 0.9)	...		Compact:				
00367	00 34.4 + 25 25	04 - 02 - 036	1188	(1.7 x ...)	--	2	...		14.8		*1
	118.87 - 37.07		9	...	...		Compact, np of 2, contact				
00368	00 34.4 + 38 50	06 - 02 - 000	0398	1.0 x 0.2	142	1	S ...		16.5:		
	119.90 - 23.68			1.0 x 0.2	7						
00369	00 34.6 + 01 40	00 - 02 - 092	0591	4.0 x 3.5	90	4	Sb/Sc	SA(rs)bc	14.5		*13
	115.63 - 60.73	NGC 173	7	3.6: x 2.8:	2 - 3			Sb +			
00370	00 34.6 + 12 47	02 - 02 - 023	0823	1.3 x 1.2	--	3		SA dm	16.0:		
	117.57 - 49.67			1.2: x 1.0:	...		S IV - V				
00371	00 34.7 + 28 53	05 - 02 - 029	1244	1.8 x 0.15	45	3	Sc		15.0		
	119.24 - 33.61		10	1.7 x 0.20	7						
00372	00 34.8 + 42 38	07 - 02 - 009	0851	2.3 :: x 2.1 ::	--	4	...	SA m	17.		
	120.22 - 19.89			1.6 :: x 1.6 ::	1		S IV - V				
00373	00 35.0 + 09 17	01 - 02 - 000	0915	1.2 x 0.5:	170	1	S? ...		16.0:		*
	117.23 - 53.16			1.2: x 0.5:	...						
00374	00 35.1 + 10 05	02 - 02 - 024	0823	1.3: x 1.2:	--	3	Sc		15.0		
	117.39 - 52.37	IC 35		1.2 x 1.2	1						
00375	00 35.1 + 25 22	04 - 02 - 038	1188	(1.2 x 1.2)	--	2	...		15.0		
	119.06 - 37.13		9	(1.1 x 1.1)	...		Compact				
00376	00 35.1 + 32 27	05 - 02 - 000	1244	1.4 x 0.1	142	3	Sc		16.0:		
	119.61 - 30.06			1.2 x 0.1	7						
00377	00 35.2 + 03 45	01 - 02 - 000	0915	1.3 x 0.7	157:	3	...	S dm - m	17.		
	116.37 - 58.67		7	1.2: x 0.6:	...						
00378	00 35.2 + 47 57	08 - 02 - 008	1224	1.0 x 0.2	[66]	1	...		16.5:		*
	120.61 - 14.59			0.9: x 0.2	...		Peculiar				
00379	00 35.3 + 04 52	01 - 02 - 037	0915	1.2 x 0.3	120	3	Sc		16.0:		
	116.63 - 57.56		8	1.3 x 0.3	7						
00380	00 35.4 + 08 22	01 - 02 - 039	0915	2.8 x 2.3	160:	4	S B c	S B(rs)bc	14.3		*3
	117.26 - 54.08	NGC 180		2.8 x 2.3	2			Sc +			
00381	00 35.5 + 30 37	05 - 02 - 030	1244	1.6: x 1.2:	(5)	1	...		15.3		
	119.58 - 31.89			1.2 x 1.1	...		Very compact				
00382	00 35.7 + 02 27	00 - 02 - 095	0591	2.3 x 1.8	(75)	4	Sa	SAB a	13.8	+ 5234	*12
	116.35 - 59.98	NGC 182	7	2.3 x 1.8	2		Brightest in group			+ 5363	3
00383	00 35.7 + 08 31	01 - 02 - 040	0915	1.3: x 1.1:	--	1	...		17.		
	117.41 - 53.94			1.0: x 0.9:	...						
00384	00 35.7 + 32 22	05 - 02 - 033	1244	1.9 x 1.7	--	4	Sc	SAB(rs)d	14.9		
	119.75 - 30.15			1.8: x 1.7:	1						
00385	00 35.8 + 13 12	02 - 02 - 000	0823	2.3 x 0.8	[64]	1	...		15.7		*
	118.07 - 49.27			2. x 0.7:	...		Peculiar				

N G C   160, 169, 173, 180, 182

1	2	3	4	5	6	7	8	9	10	11	12
00386	00 35.8 + 14 46 118.25 - 47.71	02 - 02 - 025 Mark 343	0823	1.5 x 0.9 1.7 x 1.3	(170) ...	3	S0		15.2	+ 5400	
00387	00 35.8 + 29 15 119.56 - 33.26	05 - 02 - 035 NGC 183	1244 10	(2.1 x 1.6) (1.7 x 1.5)	(130) ...	3	[E] Compact		13.8		*
00388	00 35.8 + 30 01 119.61 - 32.50	05 - 02 - 034	1244 10	1.7: x 0.8: 1.5: x 0.6:	3: 5	3	Sc		16.0:		
00389	00 35.9 + 01 42 116.29 - 60.73	00 - 02 - 099	0591 7	1.0 x 0.4 0.9 x 0.4	73 ...	1	...		15.4		
00390	00 35.9 + 02 53 116.54 - 59.55	00 - 02 - 098 NGC 186	0591 7	1.7 x 1.1 1.7 x 1.1	23: 3 - 4	1	S B 0 - a NGC 182 group		14.8		*13
00391	00 35.9 + 49 45 120.83 - 12.79	08 - 02 - 009	1224	1.1 x 0.6 1.1 x 0.6	77 5	2	S B b	SAB(s)b	16.0:		
00392	00 36 + 82 57 122.57 + 20.36	14 - 01 - 009	0568	1.8 x 0.25 1.7 x 0.22	15 (7)	1	S ... Peculiar?		14.8		*
00393	00 36.0 + 17 08 118.58 - 45.36	03 - 02 - 020	0903	2. x 0.6: 3.0: x 0.7	170 (6)	1	S - Irr		16.5:		
00394	00 36.0 + 41 43 120.41 - 20.82	07 - 02 - 013	0851	2.2 x 0.7 1.9 x 0.6	5 ...	4	... S B IV	SAB dm	15.7		
00395	00 36.2 + 31 16 119.80 - 31.25	05 - 02 - 036	1244	(1.2 x 0.7) (1.0 x 0.6)	(90) ...	2	... Compact		15.4		
00396	00 36.2 + 48 03 120.79 - 14.50	08 - 02 - 010 NGC 185	1224	(14. x 12.) (12. x 10.)	(35) ...	5	Dwarf elliptical P w 326, Local group		11.0	- 253 - 10	*12- -9-
00397	00 36.3 + 06 46c 117.40 - 55.70	01 - 02 - 041 NGC 190	0915 8	1.1 x 0.9 1.2 x 1.0	— 1 - 2	3	Sa - b + companion Contact, in group of 4		(14.4)		*1C
00398	00 36.3 + 25 22 119.40 - 37.14	04 - 02 - 040	1188 9	1.5 x 1.1 1.4 x 1.0	(25) 3	3	Sa	SA(r)a	14.6		*
00399	00 36.5 + 05 45 117.33 - 56.71	01 - 02 - 044 IC 1564	0915 8	1.2 x 0.6 1.1 x 0.6	83 5:	2	Sb/S B c:		14.8		*
00400	00 36.6 + 29 23 119.77 - 33.14	05 - 02 - 037	1244 10	1.2: x 1.1: 1.0: x 0.9:	— 1:	1	...		15.2		
00401	00 36.7 + 00 35 116.47 - 61.86	00 - 02 - 104 NGC 192	0591 7	2.4 x 1.1 2.3 x 1.1	167 5 - 6	1	S B a: Brightest of 4		13.9		*13
00402	00 36.7 + 03 41 117.07 - 58.78	01 - 02 - 046	0915 7	(1.6 x 1.3) (1.4 x 1.2)	(20) 2:	3	S0 P w 416		14.8		*
00403	00 36.7 + 12 50a 118.37 - 49.66	02 - 02 - 026	0823	1.2: x 0.6: 0.9 x 0.3	55 ...	2	S ... P w 404, distorted				*1
00404	00 36.7 + 12 50b 118.37 - 49.66	02 - 02 - 027	0823	1.2 x 0.5: 1.3: x 0.5:	[40] ...	2	Sc - Irr P w 403, distorted		15.3		*1
00405	00 36.8 + 00 38a 116.53 - 61.81	00 - 02 - 110 NGC 196	0591 7	(1.3 x 1.1) (1.3 x 1.1)	(15) 1 - 2	2	S B 0 P w 406, NGC 192 group				*13
00406	00 36.8 + 00 38b 116.53 - 61.81	00 - 02 - 107 NGC 197	0591 7	1.4: x 1.4: 1.3: x 1.3:	— 1:	1	S? ... P w 405, NGC 192 group		14.2		*13
00407	00 36.8 + 02 45 116.95 - 59.71	00 - 02 - 105 NGC 194	0591 7	(1.7 x 1.7) (2.0 x 2.0)	— ...	2	... Compact, NGC 182 group		13.9	+ 5105 + 5234	*12 3-
00408	00 36.8 + 03 03 117.01 - 59.41	00 - 02 - 103 NGC 193	0591 7	(1.7 x 1.6) (2. x 2.)	— ...	2	... P w 423 Compact, NGC 182 group		14.3		*13
00409	00 36.8 + 04 17 117.22 - 58.18	01 - 02 - 045	0915	1.2 x 0.8 1.2 x 0.8	[40] ...	1	S - Irr Interacting pair?		15.6		*1
00410	00 36.8 + 06 27 117.57 - 56.02	01 - 02 - 047 IC 1565	0915 8	(1.8 x 1.8) (2.1 x 1.9)	— ...	2	... Compact		14.9		*1
00411	00 36.8 + 25 22 119.54 - 37.15	04 - 02 - 042	1188 9	1.2 x 0.9 1.2 x 0.9	(100) ...	2	... Compact		14.5		
00412	00 36.8 + 29 29 119.83 - 33.04	05 - 02 - 000	1244 10	1.0 x 0.2 1.0 x 0.2	11 7	2	S0 - a		15.5		
00413	00 36.9 + 02 10 116.89 - 60.29	00 - 02 - 106 IC 40	0591 7	1.1 x 0.5 1.2 x 0.6	13 6	1	S ...		15.1		
00414	00 36.9 + 02 31 116.96 - 59.94	00 - 02 - 109 NGC 198	0591 7	1.3 x 1.3 1.3 x 1.3	-- 1	3	Sc NGC 182 group		14.1		*12 3
00415	00 37.0 + 02 51 117.07 - 59.61	00 - 02 - 111 NGC 199	0591 7	1.3 x 0.8 1.4 x 1.0	160 3 - 4	2	S0 NGC 182 group		15.0		*13
00416	00 37.0 + 03 40 117.21 - 58.80	01 - 02 - 049	0915 7	1.4: x 1.1: 1.3: x 1.0:	(10) ...	2	... P w 402	SAB dm:	16.5:		*
00417	00 37.0 + 03 55 117.26 - 58.55	01 - 02 - 000	0915 7	1. x 1. 1. x 1.	-- ...	3	Dwarf V		18.		
00418	00 37.0 + 08 41 117.97 - 53.80	01 - 02 - 050	0915	2.1 x 0.25 2.1 x 0.25	98 7	3	Sb P w 422 Sb +		15.2		*
00419	00 37.1 + 00 35 116.68 - 61.87	00 - 02 - 115 NGC 201	0591 7	2.3 x 2.1 2.3 x 2.0:	(155) 1	3	Sc/S B c NGC 192 group, Sc -		14.7		*13
00420	00 37.1 + 02 37 117.08 - 59.85	00 - 02 - 112 NGC 200	0591 7	2.0 x 1.2 2.0 x 1.2	161 4	3	S B c NGC 182 group, Sc -	SAB(rs)c	14.0		*12 3-

N G C   183, 185, 186, 190, 192 - 94, 196 - 201   I C   40, 1564, 1565

1	2	3	4	5	6	7	8	9	10	11	12
00421	00 37.1 + 03 15 117.19 − 59.22	00 − 02 − 113 NGC 202	0591 7	1.0 x 0.4 0.9 x 0.3	153 6	1	S ... NGC 182 group		15.5		*13
00422	00 37.1 + 08 43 118.02 − 53.77	01 − 02 − 051	0915	1.4: x 1.2: 1.2 x 1.1	— 1:	1	S B ... P w 418		15.0		*1
00423	00 37.2 + 03 01 117.20 − 59.45	00 − 02 − 116 NGC 204	0591 7	(1.4 x 1.4) (1.7 x 1.5)	— ...	2	... Compact, NGC 182 group	P w 408	14.6		*13
00424	00 37.3 + 20 16 119.29 − 42.25	03 − 02 − 000	0903	1. x 1. 1.3: x 0.8:	— ...	1			18.		*
00425	00 37.5 + 22 26 119.53 − 40.09	04 − 02 − 000	1188	0.7 x 0.5 0.7 x 0.6	— ...	1	S B ... Peculiar, interacting pair?		14.5		*
00426	00 37.6 + 41 25 120.71 − 21.13	07 − 02 − 014 NGC 205 = M 110	0851	(19. x 12.) (20. x 13.)	170 ...	5	E M 31 triplet, Local group		9.4	− 239 − 6	*12− −9
00427	00 37.6 + 50 04 121.13 − 12.49	08 − 02 − 011	1224	1.5 x 0.6 1.5: x 0.7:	64 5 − 6	3	Sc Distorted?		15.7		*
00428	00 37.7 + 29 17 120.05 − 33.25	05 − 02 − 000	1244 10	1.3: x 1.1: 1.5: x 1.3:	— ...	1	...		16.0:		*
00429	00 37.9 + 06 38 118.08 − 55.86	01 − 02 − 000	0915 8	(1.4 x 1.0) (1.6 x 1.2)	(120) ...	1	S0:		15.6		*
00430	00 38.0 + 02 58 117.58 − 59.52	00 − 02 − 119	0591 7	1.2 x 0.3 1.3 x 0.3	44 6	3	Sc NGC 182 group		16.0:		1
00431	00 38.0 + 29 53 120.17 − 32.66	05 − 02 − 000	1244 10	(1.2 x 1.0) 0.6 x 0.4	— ...	2	... Very compact		15.1		
00432	00 38.1 − 00 35 116.98 − 63.06	00 − 02 − 121 IC 1571	0591	1.3 x 1.0 1.2 x 0.9	0: 2 − 3	3	Sb		14.8		*
00433	00 38.3 + 31 27 120.34 − 31.09	05 − 02 − 038	1244	1.9 x 0.30 1.9 x 0.25	74 7	3	Sc		14.6		
00434	00 38.3 + 50 28 121.26 − 12.10	08 − 02 − 000	1224	1.2 x 0.6 1.1 x 0.5	103 5	1	Sb − c		16.5:		*
00435	00 38.5 − 01 54 116.92 − 64.38	00 − 02 − 124	0591	1.2 x 0.8 0.9 x 0.5	25: ...	1	... P w 439		15.6		*
00436	00 38.6 + 16 12 119.38 − 46.33	03 − 02 − 023 NGC 213	0903	1.8 x 1.7 1.9: x 1.7:	— 1	4	S B a 	S B(rs)a	14.8		*13
00437	00 38.8 + 03 13 118.01 − 59.28	00 − 02 − 000	0591	1. x 1. 1.1:x 0.9::	— ...	3	Dwarf spiral S IV − V or V		18.		
00438	00 38.8 + 25 13 120.10 − 37.33	04 − 02 − 044 NGC 214	1188	2.2 x 1.7 2.2 x 1.7	35 2	3	Sc 	Sc −	13.0	+ 4521 + 4720	*12 3468
00439	00 38.9 − 01 59 117.14 − 64.47	00 − 02 − 125	0591	1.2 x 1.2 1.1 x 1.1	— 1	3	Sa P w 435		14.4		*
00440	00 39.0 + 36 05 120.74 − 26.47	06 − 02 − 013 NGC 218	0398	1.3: x 1.2: 1.2: x 0.8:	— ...	2	S ... Peculiar		15.5		*3
00441	00 39.1 + 10 33 119.05 − 51.97	02 − 02 − 000	0823	1.0: x 0.8: 0.8: x 0.8:	— 1 − 2	1	S ...		16.5:		*
00442	00 39.2 + 32 43 120.62 − 29.84	05 − 02 − 000	1244	1.8 x 0.9 1.1: x 0.5:	175: 5	4	Sc 	SA(rs)d	15.4		*
00443	00 39.4 + 33 08 120.69 − 29.42	05 − 02 − 000	1244	1.1 x 0.3 1.0 x 0.2	0 6	2	Sc np of 2		15.5		*
00444	00 39.4 + 36 32 120.85 − 26.03	06 − 02 − 015	0398	1.2 x 0.8 1.0 x 0.7	(160) ...	1	S? ...		14.0		*
00445	00 39.6 + 30 02 120.59 − 32.52	05 − 02 − 000	1244	1.0 x 0.4 1.0 x 0.4	43 6	3	S B b 	(R′)S B(s)b	16.0:		
00446	00 39.6 + 32 56 120.73 − 29.63	05 − 02 − 000	1244	1.2::x 1.0:: 1. x 1.	— ...	3	Dwarf irregular Ir V		17.		
00447	00 39.7 + 10 28 119.28 − 52.07	02 − 02 − 000	0823	1.0 x 0.9 1. x 1.	— 1:	1	S ...		16.5:		
00448	00 39.7 + 29 22 120.58 − 33.19	05 − 02 − 040 IC 43	1244	2.2 x 2.1 1.9 x 1.8	— 1	3	Sc Singular, p w 449	SAB c	14.4		*
00449	00 39.7 + 29 25 120.58 − 33.14	05 − 02 − 039 IC 45?	1244	1.0 x 0.25 1.0 x 0.25	51 (7)	1	... Peculiar, p w 448		15.3		*1
00450	00 39.8 + 00 34 118.11 − 61.95	00 − 02 − 129 NGC 223 = 44*	0591	1.5 x 1.0 1.4 x 0.9	62 ...	1	... Peculiar		14.5		*3
00451	00 39.8 + 18 33 119.97 − 44.00	03 − 02 − 024	0903	1.3 x 1.3 (1.9 x 1.7)	— ...	1	...		16.5:		*
00452	00 39.9 + 40 36 121.14 − 21.57	07 − 02 − 015 NGC 221 = M 32	0851	(11. x 7.) (11. x 7.)	170: ...	5	E M 31 triplet, Local group		9.2	− 213 + 17	*12− −9
00453	00 40.0 + 29 37 120.67 − 32.94	05 − 02 − 000	1244	1.3: x 0.2 0.8 x 0.1	173 7	2	Sc		16.5:		*
00454	00 40.0 + 41 00 121.17 − 21.57	07 − 02 − 016 NGC 224 = M 31	0851	200: x 80: 200: x 80:	35 6	5	Sb Local group, Sb −	SA(s)b	4.3	− 299 − 68	*12− −9
00455	00 40.1 − 02 16 117.78 − 64.78	00 − 02 − 131	0591	(1.6 x 1.6) (1.4 x 1.3)	— ...	2	... Compact		15.0		*1

N G C   202, 204, 205, 213, 214, 218, 221, 223, 224   I C   43, 44, 45?, 1571

1	2	3	4	5	6	7	8	9	10	11	12
00456	00 40.1 - 01 48 117.86 - 64.31	00 - 02 - 135 NGC 227	0591	(1.7 x 1.4) (1.7 x 1.4)	(155) ...	3	E		13.7	+ 5315 + 5424	*123 48
00457	00 40.1 + 33 15 120.87 - 29.31	05 - 02 - 000	1244	1.2 x 0.2 1.4 x 0.25	6 6 - 7	1	Sb - c Disturbed?		15.5		*
00458	00 40.2 + 23 14 120.39 - 39.32	04 - 02 - 048 NGC 228	1188	1.2 x 1.1 1.1: x 1.1:	-- 1	3	S B a 2nd of 3	(R)S B(r)ab	14.9		*1
00459	00 40.2 + 32 18 120.85 - 30.26	05 - 02 - 000 NGC 226	1244	1.0 x 1.0 1.0 x 0.8	-- ...	1	... Peculiar		14.4		*3
00460	00 40.6 + 50 24 121.63 - 12.18	08 - 02 - 000	1224	1.2 x 0.8 1.2 x 0.8	115: 3:	1	Sc:		14.6		*
00461	00 40.9 - 00 24 118.55 - 62.93	00 - 02 - 136 NGC 237	0591	2.0 x 1.2 1.6 x 1.1	175 4	2	Sb - c		13.6		*12 48
00462	00 40.9 + 02 40 118.97 - 59.87	00 - 03 - 001 NGC 236	1196	1.3 x 1.2 1.4 x 1.4	-- 1	3	Sc SA(s)c Singular, brightest of 3		14.5		*3
00463	00 40.9 + 14 04 120.04 - 48.49	02 - 02 - 028 NGC 234	0823	1.8 x 1.8 2.0: x 2.0:	-- 1	4	Sc	SAB(rs)c	13.5		*3
00464	00 41.0 + 30 19 120.96 - 32.25	05 - 02 - 041 NGC 233	1244	(2.0 x 1.7) (2.1 x 1.7)	-- ...	1	E?		13.8		*3
00465	00 41.1 + 32 35 121.08 - 29.99	05 - 02 - 042	1244	1.8 x 0.45 1.8 x 0.45	178 6	3	S B a		14.5		
00466	00 41.3 + 00 33 118.90 - 61.99	00 - 03 - 002	1196	1.6 x 0.9 1.5 x 0.9	165 4 - 5	1	S ...		15.4		
00467	00 41.3 + 28 36 120.96 - 33.97	05 - 02 - 000	1244	1.4:: x 0.5:: 0.9: x 0.4:	(140) ...	2	... S dm S IV or IV - V		17.		
00468	00 41.4 + 01 35 119.09 - 60.96	00 - 03 - 003 IC 49	1196	1.9 x 1.7 1.9 x 1.7	-- 1	3	Sc SA c Singular, disturbed?		14.5		*
00469	00 41.4 + 25 55 120.87 - 36.65	04 - 02 - 051	1188	1.6 x 0.35 1.4: x 0.35	131 6	1	S ...		15.3		*
00470	00 41.5 + 26 35 120.93 - 35.99	04 - 02 - 052	1188	1.1 x 0.35 0.9 x 0.30	[160] ...	1	Irr or Peculiar?		14.8		*
00471	00 42.0 + 45 09 121.71 - 17.43	08 - 02 - 000	0851	2.0 x 1.5 1.7 x 1.3	(50) 2 - 3	3	Sb/Sc SA(s)bc Sc -		15.7		
00472	00 42.3 + 16 40 120.69 - 45.90	03 - 03 - 000	0930	1.5: x 0.8: 1.5:: x 0.8::	(115) ...	3	Dwarf V		17.		
00473	00 42.5 + 05 52 120.06 - 56.70	01 - 03 - 001 NGC 240	0877	(1.6 x 1.6) (1.5 x 1.5)	-- 1	2	S0 - a		14.8		*1
00474	00 42.7 + 10 15 120.46 - 52.32	02 - 03 - 000	1274	1.0 x 0.10 0.7 x 0.10	83 7	3	Sc		16.0:		
00475	00 43.4 + 50 57 122.10 - 11.64	08 - 02 - 014	1224	1.7 x 0.8 1.7 x 0.7	105 5	3	Sc		14.9		
00476	00 43.5 - 01 59 119.80 - 64.55	00 - 03 - 005 NGC 245	1196	1.3 x 1.3 1.4 x 1.4	-- ...	3	S ... Peculiar		12.9		*12 3478
00477	00 43.6 + 19 13 121.25 - 43.36	03 - 03 - 002	0930	3.5: x 0.8 3.2: x 0.8	167 (6)	3	... S dm - m P w 490		15.4		*
00478	00 43.7 + 29 58 121.64 - 32.62	05 - 02 - 044	1244	1.6 x 0.25 1.6 x 0.25	116 7	2	Sa		14.5		
00479	00 43.8 + 31 32 121.71 - 31.05	05 - 02 - 046	1244	1.1 x 0.6 1.3 x 0.7	15 4 - 5	3	S B a	(R)S B(s)ab	15.3		
00480	00 43.8 + 36 03 121.83 - 26.54	06 - 02 - 016	0398	1.6 x 1.1 1.6 x 1.1	... ...	4	S ... p of 2, disrupted		13.6		*1
00481	00 44 + 83 29 122.85 + 20.88	14 - 01 - 000	0568	1.2 x 0.20 0.9 x 0.15	112 6 - 7	1	S ...		16.0:		*
00482	00 44.0 + 08 12 120.88 - 54.38	01 - 03 - 000	0877	1.0: x 0.9: 0.9 x 0.9	-- 1	3	Sc		17.		
00483	00 44.2 + 26 13 121.66 - 36.37	04 - 03 - 000	0857	1.2: x 0.5 1.0 x 0.5	69 ...	1	... IV - V:		15.7		
00484	00 44.2 + 32 24 121.83 - 30.19	05 - 03 - 001	0601	2.8 x 0.8 2.3 x 0.8	25 6	4	S B b (R')S B(s)b spp of 2 Sb +		14.1		*
00485	00 44.4 + 30 04 121.82 - 32.52	05 - 03 - 002	0601	2.2 x 0.2 2.2 x 0.2	179 7	3	Sc		14.8		*
00486	00 44.6 + 50 37 122.29 - 11.98	08 - 02 - 015	1224	1.3 x 1.0 1.2 x 1.0	(165) 2	3	Sb/S B b	SAB(s)b	14.8		
00487	00 44.7 + 07 39 121.15 - 54.93	01 - 03 - 002 NGC 250	0877	1.4 x 0.8 1.2 x 0.7	153 4	2	S0 - a		14.9		
00488	00 44.7 + 14 26 121.44 - 48.15	02 - 03 - 001	1274	1.0 x 0.6 0.9 x 0.6	90 4	1	Sa - b		15.2		*
00489	00 44.7 + 27 33 121.84 - 35.04	05 - 03 - 004 IC 1584	0601	1.8: x 1.7: 1.3: x 1.0:	-- 1	3	S ... SAB(rs) ... Singular, NGC 252 group		15.0		*3
00490	00 45.2 + 19 18 121.77 - 43.29	03 - 03 - 003 NGC 251	0930	2.5 x 2.0 2.3: x 1.8:	(105) 2	4	Sc P w 477 Sc -		14.6		*3

N G C   226 - 28,   233,   234,   236,   237,   240, 245, 250,   251   I C   49,   1584

1	2	3	4	5	6	7	8	9	10	11	12
00491	00 45.3 + 27 21 121.99 - 35.24	04 - 03 - 004 NGC 252	0857	1.7 x 1.3 1.7 x 1.3	80: 2 - 3	3	SO Brightest in group	SA(r)0	13.4		*12 3
00492	00 45.5 - 01 49 120.98 - 64.40	00 - 03 - 006	1196	(1.4 x 0.8) (1.3 x 0.8)	128 ...	2	E		14.6		*
00493	00 45.5 + 08 03 121.51 - 54.54	01 - 03 - 003 NGC 257	0877	2.2 x 1.6 2.2 x 1.6	105: 3	3	Sc		13.7		13
00494	00 45.8 + 03 50 121.46 - 58.76	01 - 03 - 005 IC 52	0877	1.2 x 0.5 0.9 x 0.4	97 6	1	S ...		15.4		
00495	00 45.8 + 07 04 121.60 - 55.52	01 - 03 - 004	0877	1.3 x 0.6 1.0 x 0.5	175 5	1	Sa		15.7		*
00496	00 45.9 + 01 05 121.38 - 61.51	00 - 03 - 007	1196	[1.5 x 0.9] ...	... ...	2	Double system Contact		14.6		*1
00497	00 45.9 + 27 25 122.16 - 35.18	04 - 03 - 006 NGC 260	0857	0.9 x 0.9 0.8 x 0.7	— ...	1	S - Irr NGC 252 group		14.3		*12 3
00498	00 46.0 + 16 21 121.96 - 46.24	03 - 03 - 000	0930	1.0 x 0.8 0.8: x 0.5:	(170) 2:	1	Sb - c		16.5:		*
00499	00 46.1 + 31 42 122.28 - 30.50	05 - 03 - 008 NGC 262, Mark 348	0601	(1.6 x 1.5) (1.4 x 1.4)	— 1	3	SO Seyfert		15.0	+ 4600	*1
00500	00 46.2 + 11 06 121.90 - 51.49	02 - 03 - 004	1274	1.3 x 1.2 1.2 x 1.2	— 1	3	Sb/Sc	SAB bc Sc -	15.6		*
00501	00 46.3 + 27 57 122.28 - 34.65	05 - 03 - 000	0601	1.8 x 0.2 1.7 x 0.2	110 7	3	Sc		15.2		*
00502	00 46.3 + 45 55 122.51 - 16.68	08 - 02 - 000	1224	1.4:: x 1.3:: 0.9: x 0.8:	— ...	2	Dwarf: spiral S IV - V:		18.		
00503	00 46.5 + 04 15 121.81 - 58.34	01 - 03 - 000	0877	1.0 x 0.9 0.8: x 0.8:	— ...	1	Irr sf of 2		17.		*
00504	00 46.8 + 11 55 122.15 - 50.68	02 - 03 - 000	1274	1.3 x 1.0: 1.0: x 0.7:	— ...	1	S ... nf of 2		16.5:		*
00505	00 46.9 - 02 01 121.78 - 64.61	00 - 03 - 008	1196	1.5 x 1.1 1.2 x 1.0	135: 2 - 3	1	S ...		15.1		
00506	00 47.0 + 22 40 122.40 - 39.93	04 - 03 - 010	0857 12	(1.6 x 1.3) (1.7 x 1.4)	— ...	1	...		14.5		*1
00507	00 47.1 + 00 51 122.00 - 61.75	00 - 03 - 009	1196	2.2 x 0.3 2.3 x 0.3	128 7	3	Sc		15.2		
00508	00 47.1 + 32 00 122.53 - 30.60	05 - 03 - 009 NGC 266	0601	3.5 x 3.3 3.5 x 3.3	— 1	4	S B a	S B(rs)ab	12.6		*3
00509	00 47.2 + 31 20 122.55 - 31.27	05 - 03 - 000	0601	1.4 x 0.5 0.9: x 0.5:	35 6	3	Sc		16.5:		
00510	00 47.3 + 21 26 122.47 - 41.16	03 - 03 - 004	0930	1.4 x 0.4 1.3 x 0.3	20 6	3	Sc		15.3		*
00511	00 47.4 + 31 28 122.60 - 31.13	05 - 03 - 000	0601	1.8 x 0.4 1.7 x 0.3	105 6	3	Sc		15.6		
00512	00 47.5 + 07 39 122.35 - 54.95	01 - 03 - 000	0877	1.2 x 0.7 1.1 x 0.7	[135] ...	2	S - Irr interacting pair?	Peculiar,	15.1		*
00513	00 47.8 + 11 27 122.53 - 51.15	02 - 03 - 000	1274	1.1 x 0.6 1.1 x 0.6	8: 4 - 5	1	S ...		15.7		
00514	00 48.0 + 00 11 122.46 - 62.42	00 - 03 - 000	1196	1.0: x 0.9: 0.9: x 0.9:	— ...	1	...		16.0:		
00515	00 48.0 + 11 50 122.61 - 50.77	02 - 03 - 006	1274	1.1 x 0.9 1.0: x 0.9:	— ...	2	...	S dm - m	17.		
00516	00 48.1 + 10 20 122.64 - 52.27	02 - 03 - 005 IC 53	1274	(1.5 x 1.4) (1.5 x 1.4)	— ...	2	E - SO		15.5		*
00517	00 48.1 + 11 05 122.65 - 51.52	02 - 03 - 000	1274	(1.1 x 1.0) (1.0 x 1.0)	— ...	1	... Compact		15.7		*
00518	00 48.1 + 29 53 122.77 - 32.72	05 - 03 - 010	0601 20	1.1 x 1.0 1.0 x 0.9	— 1	3	S B c	S B(r)c	15.2		*
00519	00 48.2 - 02 09 122.53 - 64.75	00 - 03 - 012 NGC 271	1196	2.5 x 2.0 2.7 x 2.1	130: 2	4	S B b	S B(r)ab Sb -	13.2		3
00520	00 48.5 + 17 28 122.83 - 45.13	03 - 03 - 000	0930	1.0 x 0.6 0.8 x 0.5	(135) 4:	1	Sb - c		17.		
00521	00 48.6 + 11 45 122.84 - 50.85	02 - 03 - 008	1274	1.2 x 0.9 1.2 x 1.1	[50] ...	3	Irr	Ir I	15.4		*
00522	00 48.7 + 40 27 122.94 - 22.15	07 - 02 - 020	0851	1.1 x 0.7 1.0 x 0.7	65: 3 - 4	3	Sb Disturbed?, Sb +		14.6		*1
00523	00 48.8 + 51 50 122.97 - 10.77	09 - 02 - 000	1237	1.2 x 0.5 1.2 x 0.5	50 6	2	Sa - b		15.7		
00524	00 48.9 + 29 08 122.97 - 33.47	05 - 03 - 013	0601 20	0.9 x 0.9 0.9 x 0.9	— 1	3	S B b	(R')SB(s)b Sb -	14.5		
00525	00 48.9 + 29 27 122.97 - 33.15	05 - 03 - 012	0601 20	1.7: x 1.0: 1.9: x 1.1:	150: ...	3	S ... Singular	SAB(s) ...	15.5		

1	2	3	4	5	6	7	8	9	10	11	12
00526	00 49.1 + 02 55 123.05 − 59.68	00 − 03 − 000	1196	1.0 x 0.4 1.0 x 0.4	94 ...	2	... P w 527	dm − m	16.5:		*
00527	00 49.2 + 02 49 123.10 − 59.78	00 − 03 − 017	1196	1.5: x 1.0: 1.1 x 0.8	(15) ...	2	... P w 526	m	16.0:		*
00528	00 49.2 + 47 17 123.04 − 15.32	08 − 02 − 016 NGC 278	1224	(2.8 x 2.8) (2.6 x 2.6)	— 1	1	S ... Singular		10.5	+ 633 + 867	*123 46<u>7</u>8
00529	00 49.4 + 29 24 123.10 − 33.20	05 − 03 − 014	0601 20	1.0 x 0.22 1.0 x 0.22	93 	2	S0 − a		14.3		
00530	00 49.4 + 44 04 123.08 − 18.53	07 − 02 − 021	0851	1.1 x 0.25 1.0 x 0.35	0 6	1	Sa		15.5		*
00531	00 49.6 + 02 10 123.30 − 60.43	00 − 03 − 019	1196	1.5 x 1.0 1.4 x 0.7	... ...	3	S B ... Distorted		15.7		*
00532	00 49.7 − 02 28 123.41 − 65.07	00 − 03 − 19a NGC 279	1196	(1.6 x 1.4) (1.6 x 1.4)	— 1 − 2	3	S0		14.0		3
00533	00 49.8 + 14 15 123.29 − 48.35	02 − 03 − 009	1274	1.8 x 0.7 1.5 x 0.7	120 5 − 6	3	Sc		15.4		*1
00534	00 49.8 + 24 05 123.23 − 38.52	04 − 03 − 013 NGC 280	0857	1.8: x 1.3 1.6: x 1.1	[95] ...	3	S ... Distorted		14.6		*13
00535	00 49.8 + 25 52 123.22 − 36.73	04 − 03 − 014	0857	1.0 x 0.6 1.0: x 0.5	145: ...	2	S B b	SB(s)b	15.5		
00536	00 49.9 + 28 56 123.24 − 33.67	05 − 03 − 000	0601 20	1.0 x 0.5 1.0 x 0.5	50 ...	2	...		16.5:		
00537	00 50.0 + 05 45 123.45 − 56.85	01 − 03 − 000	0877	1. x 1. 1. x 1.	— ...	3	Dwarf irregular Ir V, 538 group		18.		*
00538	00 50.1 + 05 47 123.50 − 56.82	01 − 03 − 000	0877	1.1 x 0.5 1.2 x 0.6	75 5	2	Sc Brightest of 4		15.7		*
00539	00 50.1 + 41 42 123.22 − 20.90	07 − 03 − 002	0627	1.2 x 1.1 1.2 x 1.0	— 1 − 2	3	Sc		14.7		*
00540	00 50.2 + 28 45 123.32 − 33.85	05 − 03 − 016	0601 20	0.70 x 0.35 0.70 x 0.35	[137] ...	1	... Peculiar		14.1		*
00541	00 50.5 + 21 39 123.46 − 40.95	04 − 03 − 000	0857	1.2 x 0.2 1.2 x 0.2	102 7	3	Sc		16.0:		
00542	00 50.7 + 29 00 123.45 − 33.60	05 − 03 − 017	0601 20	2.3: x 0.3 2.0: x 0.25	160 7	1	S ...		14.8		
00543	00 50.9 + 05 30 123.87 − 57.10	01 − 03 − 000 IC 1592	0877	1.1 x 0.8 0.9 x 0.8	(165) 2 − 3	1	S ... f of 2		15.4		*
00544	00 51.0 + 02 40 124.00 − 59.93	00 − 03 − 023	1196	1.7 x 0.3 1.6 x 0.3	42 7	3	Sc		15.7		
00545	00 51.0 + 12 25 123.76 − 50.18	02 − 03 − 000	1274	0.6 x 0.5 0.5 x 0.5	... ...		... Very compact; Seyfert		14.0:	+ 18300 + 18450	*C
00546	00 51.0 + 17 48 123.67 − 44.80	03 − 03 − 005	0930	1.3 x 0.6 1.3 x 0.6	5 ...	2	...	S dm − m	15.7		
00547	00 51.2 + 24 30 123.64 − 38.10	04 − 03 − 015	0857	1.4 x 0.35 1.3 x 0.35	28 6	1	S0		15.3		
00548	00 51.7 + 31 24 123.67 − 31.20	05 − 03 − 018	0601 20	1.4 x 1.2 1.3 x 1.2	— 1	4	Sc		14.8		
00549	00 51.9 + 36 30 123.65 − 26.10	06 − 03 − 000	0406	1.1 x 0.1 1.1 x 0.1	0 7	3	Sc		15.7		
00550	00 52.0 + 21 15 123.93 − 41.34	03 − 03 − 007 IC 1596	0930	2.5: x 1.2: 1.1 x 0.5	120 ...	1	S ... Singular		15.1		*
00551	00 52.0 + 44 29 123.56 − 18.11	07 − 03 − 000	0627	1.4:: x 0.7:: 1.3:: x 0.7::	(120) ...	3	Dwarf V		18.		
00552	00 52.0 + 52 14 123.47 − 10.36	09 − 02 − 000	1237	1.5: x 0.6 1.3 x 0.5	162 6	3	Sc		15.7		
00553	00 52.1 + 05 30 124.42 − 57.09	01 − 03 − 007 IC 1598	0877	1.2 x 0.6 1.3 x 0.7	2 5	2	Sa		15.0		*
00554	00 52.1 + 28 27 123.82 − 34.15	05 − 03 − 020	0601 20	1.2 x 0.4 1.1 x 0.4	125 6	2	Sa − b		15.6		*
00555	00 52.1 + 28 35 123.82 − 34.01	05 − 03 − 021	0601 20	1.0 x 0.2 1.0 x 0.2	28 7	2	S0 − a		14.9		
00556	00 52.1 + 28 59 123.81 − 33.61	05 − 03 − 022	0601 20	1.2 x 0.7 1.1 x 0.7	100: ...	1	S ... Peculiar?		15.3		*
00557	00 52.1 + 31 05 123.78 − 31.51	05 − 03 − 023	0601 20	1.1 x 0.5 1.1 x 0.5	37 6	2	S ... Distorted		14.9		*
00558	00 52.2 + 10 16 124.29 − 52.33	02 − 03 − 011	1274	1.7 x 0.7 1.5 x 0.6	37 ...	1	S0:		15.7		1
00559	00 52.2 + 11 34 124.25 − 51.03	02 − 03 − 010 IC 57	1274 15	1.1 x 1.0 1.1 x 1.0	— ...	1	S0?		15.7		*
00560	00 52.2 + 13 25 124.19 − 49.18	02 − 03 − 000	1274	1.1: x 1.0: 1. x 1.	— ...	3	Dwarf (spiral) S: V		18.		

1	2	3	4	5	6	7	8	9	10	11	12
00561	00 52.3 + 30 32 123.84 - 32.06	05 - 03 - 000	0601 20	1.5 x 0.7 1.4 x 0.7	(165) ...	2	... Peculiar		16.0:		
00562	00 52.4 + 31 16 123.85 - 31.33	05 - 03 - 024 NGC 295	0601 20	2.2 x 0.9 2.3 x 0.9	164 6	3	S B: b		13.5		3
00563	00 52.5 + 01 07 124.83 - 61.47	00 - 03 - 024	1196	1.2: x 0.8: 1.1 x 0.7	20 6 - 7:	1	S ...		15.1		*
00564	00 52.5 + 35 10 123.80 - 27.43	06 - 03 - 000	0406	1.0 x 0.7 0.8 x 0.7	(175) 3:	1	S ...		15.7		*
00565	00 52.6 + 31 24 123.90 - 31.19	05 - 03 - 027 NGC 296	0601 20	1.2 x 0.3 1.2 x 0.3	148 6	3	Sc		15.4		3
00566	00 52.6 + 31 27 123.90 - 31.14	05 - 03 - 026	0601 20	1.5: x 1.5: 1.3:: x 1.3::	-- 1	3	... SA dm S IV:, p w 567		15.7		*
00567	00 52.6 + 31 28 123.90 - 31.13	05 - 03 - 025	0601 20	1.1 x 0.4 1.1 x 0.4	62 6	3	S0 P w 566		14.9		*
00568	00 52.7 - 01 18 125.10 - 63.88	00 - 03 - 025	1196 14	(1.3 x 0.9) (1.5 x 1.3)	-- ...	1	... Compact, np of 2		15.7		*
00569	00 52.7 + 09 11 124.53 - 53.41	01 - 03 - 000	0877	1.1 x 0.8 0.9 x 0.6	-- 3	2	Sc		16.5:		*
00570	00 53.4 - 01 10 125.49 - 63.75	00 - 03 - 029	1196 14	(1.1 x 0.7) (1.1 x 0.8)	(10) ...	1	E: + companion		15.3		*
00571	00 53.4 + 11 50 124.70 - 50.75	02 - 03 - 015	1274 15	1.2 x 0.9 1.3: x 0.9:	135: ...	4	S B b S B(s)b		15.4		
00572	00 53.4 + 13 57 124.62 - 48.64	02 - 03 - 014	1274	1.3 x 1.0 1.0 x 0.9	(140) 1 - 2	3	Sc		15.6		
00573	00 53.4 + 23 51 124.29 - 38.74	04 - 03 - 018 NGC 304	0857	1.4: x 0.9: 1.4: x 0.9:	[175] ...	1	... Eruptive?, sff of 2		14.0		*
00574	00 53.5 + 04 08 125.14 - 58.45	01 - 03 - 000	0877	1.3:: x 1.2:: 1.2:: x 1.0::	-- ...	3	Dwarf IV - V or V		17.		
00575	00 53.5 + 30 49 124.14 - 31.77	05 - 03 - 000	0601 20	1.0 x 0.10 0.9 x 0.12	161 7	1	Sb - c		16.5:		
00576	00 53.5 + 50 21 123.73 - 12.24	08 - 02 - 019	1224	1.2 x 0.7 1.1 x 0.6	125: 4	4	Sc SAB: cd		15.3		
00577	00 53.5 + 50 30 123.73 - 12.09	08 - 02 - 000	1224	(1.3 x 1.3) (1.1 x 1.1)	-- ...	1	E?		16.0:		
00578	00 53.6 + 39 33 123.96 - 23.04	07 - 03 - 007	0627	1.1: x 0.4 1.0 x 0.4	8 ...	1	...		14.8		*
00579	00 53.7 - 01 30 125.69 - 64.08	00 - 03 - 032	1196 14	(1.4 x 1.2) (1.5 x 1.3)	(40) ...	2	E 2nd brightest in cluster		15.0		*1
00580	00 53.7 + 13 37 124.74 - 48.97	02 - 03 - 016	1274 15	(1.2 x 0.9) (1.2 x 0.8)	(30) ...	1	... P w 582, contact		16.0:		*1
00581	00 53.8 + 11 36 124.87 - 50.98	02 - 03 - 000	1274 15	1.1 x 0.8 1.2 x 0.9	170 3	2	Sb		16.0:		*
00582	00 53.8 + 13 38 124.78 - 48.95	02 - 03 - 017	1274 15	1.0 x 0.7 1.0 x 0.7	(5) ...	1	S0? P w 580, contact		16.5:		*1
00583	00 53.9 - 01 30 125.80 - 64.09	00 - 03 - 034	1196 14	(1.1 x 1.1) (1.1 x 1.1)	-- ...	1	E: Brightest in cluster		14.9		*1
00584	00 54.0 - 02 02 125.91 - 64.61	00 - 03 - 035 NGC 307	1196 14	(1.7 x 0.8) (1.7 x 0.8)	85 ...	2	E - S0		14.1		*3
00585	00 54.0 + 26 44 124.38 - 35.85	04 - 03 - 022	0857 16	1.1 x 0.7 1.0 x 0.7	90 3 - 4	2	Sb - c sf of 2		16.0:		*1
00586	00 54.3 + 07 03 125.32 - 55.53	01 - 03 - 000	0877	1.0 x 0.1 0.5 x 0.12	139 7	1	S ...		16.5:		
00587	00 54.4 - 01 28 126.08 - 64.03	00 - 03 - 038	1196 14	1.2 x 1.0 1.1 x 0.9	-- ...	1	...		15.3		
00588	00 54.5 - 01 07 126.10 - 63.68	00 - 03 - 039	1196 14	1.1 x 1.0 1.2 x 1.2	-- ...	1	...		15.1		*1
00589	00 54.5 + 07 15 125.40 - 55.32	01 - 03 - 009 IC 61	0877	(1.3 x 1.2) (1.5 x 1.5)	-- ...	2	... Compact, nnf of 2		15.3		*
00590	00 54.5 + 14 38 124.99 - 47.95	02 - 03 - 000	1274	1.1 x 0.7 0.8 x 0.4	145: ...	1	S? ...		15.7		
00591	00 54.6 + 23 37 124.65 - 38.97	04 - 03 - 023 Mark 350	0857	1.0 x 0.8 0.9: x 0.7:	[165] ...	1	...		15.2	+ 5100	*
00592	00 54.8 + 30 00 124.49 - 32.58	05 - 03 - 028 NGC 311	0601 20	(1.5 x 0.8) 1.4 x 0.7	120 5	3	S0 P w 597		14.1		*3
00593	00 54.8 + 43 31a 124.11 - 19.07	07 - 03 - 009 NGC 317a	0627	1.5: x 1.4: 1.5: x 1.4:	-- ...	2	... P w 594, contact, distorted		13.8		*1C
00594	00 54.8 + 43 31b 124.11 - 19.07	07 - 03 - 010 NGC 317b	0627	1.0 x 0.5 1.1 x 0.5	105 ...	2	S B ... P w 593, contact, distorted				*1C
00595	00 55.1 - 01 38 126.50 - 64.19	00 - 03 - 041	1196 14	(1.4 x 1.3) (1.5 x 1.5)	-- ...	2	... Compact		14.8		

N G C   295, 296, 304, 307, 311, 317

I C   61

1	2	3	4	5	6	7	8	9	10	11	12
00596	00 55.1 + 07 49 125.62 - 54.75	01 - 03 - 010	0877	(1.1 x 1.1) (1.3 x 1.3)	— ...	2	... Compact		15.6		*
00597	00 55.1 + 30 05 124.56 - 32.50	05 - 03 - 031 NGC 315	0601 20	(3.0 x 2.5) (3.6 x 2.8)	40: ...	3	E P w 592		12.5		*12 3
00598	00 55.1 + 31 13 124.53 - 31.37	05 - 03 - 030	0601 20	1.0 x 0.35 1.0 x 0.35	29 6	2	S0 - a		14.4		*
00599	00 55.2 - 00 40 126.44 - 63.23	00 - 03 - 042	1196	[1.3 x 0.8] ...	... .	2	E? + companion Contact		14.7		1
00600	00 55.5 + 48 23 124.11 - 14.20	08 - 02 - 021	1224	1.6 x 1.2 1.6 x 1.2	(135) 2 - 3	4	Sb/S B b	SAB(s)b	14.5		*
00601	00 55.7 + 26 36 124.85 - 35.98	04 - 03 - 025 NGC 326	0857 16	(1.2 x 1.2) (1.6 x 1.6)	— ...	2	Pair of compacts Connected, in halo		14.9		*13 C
00602	00 55.7 + 36 28 124.50 - 26.11	06 - 03 - 003	0406	1.7 x 1.6 1.6: x 1.4:	— ...	4	Sc/S B c	SAB(s)c Sc +	14.5		
00603	00 55.8 + 11 18a 125.66 - 51.26	02 - 03 - 018	1274 15	1.0 x 0.5 1.0 x 0.5	[157] ...	2	S ... Disturbed by b, bridge		(15.5)		*
00604	00 55.8 + 44 45 124.27 - 17.83	07 - 03 - 000	0627	1.1 x 0.1 1.0 x 0.1	137 7	3	Sc		18.		
00605	00 55.9 + 22 02 125.10 - 40.54	04 - 03 - 028	0857	1.1 x 1.1 1.1 x 1.1	— (1)	3	S ... sff of 2, disturbed	SAB(s)...	16.0:		*1
00606	00 56.1 + 11 33 125.76 - 51.01	02 - 03 - 021 IC 62	1274 15	1.0 x 0.7 0.9 x 0.6	25 3:	1	S ...		15.0		*1
00607	00 56.1 + 12 29 125.70 - 50.08	02 - 03 - 020	1274 15	1.1 x 1.1 1.3 x 1.3	— 1	2	Sc		15.4		*
00608	00 56.1 + 47 45 124.23 - 14.83	08 - 03 - 001	0863	2.0 x 0.9 1.5: x 0.8:	133 5	3	...	SAB dm	15.6		
00609	00 56.2 + 06 50 126.17 - 55.72	01 - 03 - 000 NGC 332	0877	(1.3 x 1.3) (1.6 x 1.4)	— ...	2	... Compact		14.9		*
00610	00 56.2 + 12 43 125.72 - 49.85	02 - 03 - 022	1274 15	(1.8 x 1.6) (1.8 x 1.6)	— ...	3	E		14.8		*1
00611	00 56.3 + 00 20 126.92 - 62.21	00 - 03 - 047 IC 1607	1196	1.1 x 1.1 1.0 x 1.0	— ...	1	... Peculiar:		14.5		
00612	00 56.3 + 23 35 125.15 - 38.99	04 - 03 - 029	0857	1.0 x 0.45 0.8 x 0.35	90 ...	1	...		14.5		
00613	00 56.7 + 26 47 125.12 - 35.79	04 - 03 - 031 IC 64	0857 16	(1.3 x 1.1) (1.3 x 1.0)	— ...	1	...		15.6		*
00614	00 56.8 + 35 18 124.79 - 27.27	06 - 03 - 005	0406	1.6 x 1.0 1.6 x 1.0:	70 4	4	S B c	SB(s)d	14.0		*
00615	00 57.0 + 15 04 125.86 - 47.49	02 - 03 - 023	1274	1.0 x 0.5 1.1 x 0.6	23 5	3	Sa/S B b		14.2		*
00616	00 57.0 + 18 35 125.63 - 43.98	03 - 03 - 010	0930	1.5 x 1.2 1.5 x 1.2	(85) ...	4	S B b	(R')SB(r)b	15.5		*
00617	00 57.1 + 17 45 125.72 - 44.81	03 - 03 - 011	0930	1.1 x 0.9 1.0 x 0.8	— 1 - 2	3	S B a Brightest of 3		15.2		*
00618	00 57.2 + 00 40 127.35 - 61.86	00 - 03 - 049	1196	1.1 x 1.1 1.2 x 1.2	— 1	2	Sc		15.4		
00619	00 57.2 + 14 27 125.97 - 48.10	02 - 03 - 000	1247	1.3 x 0.12 0.9 x 0.1	5 7	3	Sc		16.5:		
00620	00 57.5 + 17 44 125.85 - 44.82	03 - 03 - 000	0930	1.0 x 0.4 1.0 x 0.4	10 6	2	Sb - c 617 group		16.0:		*
00621	00 57.5 + 18 01 125.84 - 44.54	03 - 03 - 012	0930	1.2 x 0.6 1.2 x 0.6	2 5	1	S ...		15.5		
00622	00 57.6 + 47 43 124.50 - 14.86	08 - 03 - 004	0863	1.1 x 0.7 1.1 x 0.7	160: 4	3	Sc		14.6		*
00623	00 57.8 + 30 32 125.24 - 32.03	05 - 03 - 033 IC 66	0601 20	1.1 x 0.5 0.55 x 0.25	125 5 - 6	1	Sa		15.0		
00624	00 57.9 + 30 24 125.27 - 32.16	05 - 03 - 034 NGC 338	0601 20	1.9 x 0.7 2.2 x 0.9	109 6	2	Sa - b		14.0		
00625	00 58.0 + 47 25 124.58 - 15.16	08 - 03 - 005 IC 65	0863	4.4 x 1.2 2.9 x 0.9	155 6	4	Sb/S B c	SAB(s)bc Sb +	13.8		
00626	00 58.1 - 01 58 128.28 - 64.47	00 - 03 - 052	1196	1.0 x 1.0 1.0 x 0.9	— 1	3	Sc np of 2, bridge	SAB(s)c	15.0		*1
00627	00 58.3 + 13 12 126.48 - 49.34	02 - 03 - 024	1274	1.3 x 0.9 1.2 x 0.8	115 3	1	S ...		15.2		
00628	00 58.3 + 19 13 126.02 - 43.33	03 - 03 - 013	0930	2.3: x 1.8: 2.0: x 1.6:	— ...	3	Dwarf spiral SB IV - V or V		17.		*
00629	00 58.3 + 29 20 125.42 - 33.22	05 - 03 - 036	0601	1.4 x 1.1 1.2 x 1.0	(110) 2:	3	Sc		14.9		
00630	00 58.3 + 29 41 125.41 - 32.87	05 - 03 - 037	0601 20	1.3: x 1.0: 1.0: x 0.9:	(100) ...	3	... S IV:	SA d - dm	17.		*

N G C   315, 326, 332, 338   I C   62, 64 - 66, 1607

1	2	3	4	5	6	7	8	9	10	11	12
00631	00 58.4 + 09 27 126.86 − 53.08	02 − 03 − 025	1274	1.7 x 0.2 1.7 x 0.2	137 7	3	Sc		16.0:		*
00632	00 58.4 + 29 52 125.42 − 32.69	05 − 03 − 038	0601 20	1.2 x 0.5 1.1 x 0.5	163 6	2	S B a − b s of 2		14.8		*1
00633	00 58.6 + 31 15 125.40 − 31.31	05 − 03 − 040	0601 20	1.7 x 0.35 1.6 x 0.35	9 6	3	Sb		14.6		1
00634	00 58.8 + 07 21 127.26 − 55.17	01 − 03 − 011	0877	1.9: x 1.1 2.0: x 1.3:	(35) ...	4	Dwarf spiral S(B) V		15.7		*1
00635	00 58.8 + 46 24 124.76 − 16.17	08 − 03 − 000	0863	1.0 x 0.6 0.9 x 0.5	15: 4	3	Sc		17.		
00636	00 59.0 + 23 47 125.93 − 38.76	04 − 03 − 034	0857	(1.5 x 1.2) (1.4 x 1.1)	(110) ...	2	E		14.7		*
00637	00 59.3 + 02 47 128.11 − 59.71	00 − 03 − 056	1196	1.3 x 0.5 1.2 x 0.5	157 6	3	S B c		16.0:		
00638	00 59.4 + 26 41 125.87 − 35.86	04 − 03 − 035	0857 16	(1.3 x 0.9) (1.7 x 1.3)	(110) ...	1	... Brightest in group		15.7		*1
00639	00 59.5 + 02 11 129.14 − 64.66	00 − 03 − 057 NGC 351	1196	1.4 x 0.7 1.5 x 0.8	142 5	3	S B 0 − a P w 641		14.3		*
00640	00 59.7 + 08 50 127.47 − 53.67	01 − 03 − 012	0877	1.1 x 1.1 1.0 x 1.0	−− 1	2	Sb		15.7		*
00641	01 00.0 − 02 12 129.43 − 64.66	00 − 03 − 058 NGC 353, Mark 353	1196	1.6 x 0.5 1.5 x 0.5	26 6	2	S B a − b P w 639		14.7	+ 4800	
00642	01 00.0 + 80 24 123.48 + 17.81	13 − 02 − 000	1214	1.1 x 0.4 1.3 x 0.5	46 6	2	Sb n of 2		16.0:		*
00643	01 00.4 + 24 43 126.28 − 37.81	04 − 03 − 036	0857	1.2 x 1.0 1.1: x 0.9:	(10) 2:	4	Sc		15.7		
00644	01 00.6 + 13 47a 127.27 − 48.72	02 − 03 − 026	1274	1.0: x 0.3 1.0 x 0.3	63 (6)	1	S − Irr Disturbed by b, bridge:		(14.5)		*1
00645	01 00.6 + 22 05 126.53 − 40.44	04 − 03 − 037 NGC 354	0857	1.0: x 0.6: 0.8 x 0.35	29 ...	1	S B ... Peculiar		14.2		
00646	01 00.7 + 31 58 125.88 − 30.57	05 − 03 − 043	0601 20	2.0? x 1.0? 0.60 x 0.25	[105] ...	1	S B ...		15.0		
00647	01 00.9 − 01 46 129.85 − 64.21	00 − 03 − 063	1196 19	1.0 x 1.0 0.9 x 0.9	−− 1	2	Sb		16.0:		*
00648	01 01.0 + 01 37 129.16 − 60.84	00 − 03 − 062	1196	1.1 x 0.5 0.7 x 0.5	(125) ...	1	S ... Unilateral?		15.6		*
00649	01 01.0 + 06 09 128.38 − 56.32	01 − 03 − 000	0877	1.2 x 0.6 1.3 x 0.7	172 5	1	Sa − b		15.7		*
00650	01 01.0 + 08 17 128.08 − 54.20	01 − 03 − 000	0877	1.1 x 0.5 1.1 x 0.5	38 5 − 6	1	S0 − a		16.0:		*
00651	01 01.1 − 00 45 129.72 − 63.19	00 − 03 − 064	1196 19	1.2 x 0.3 1.2 x 0.3	165 6	3	Sc		16.0:		*
00652	01 01.1 + 21 45 126.71 − 40.76	04 − 03 − 038	0857	1.3: x 0.6: 0.6: x 0.5:	[35] ...	3	S B [c] Singular		17.		
00653	01 01.1 + 30 56 126.05 − 31.60	05 − 03 − 000	0601 20	1.0 x 0.12 1.0 x 0.12	174 7	3	Sc		16.5:		
00654	01 01.2 + 20 58 126.81 − 41.54	03 − 03 − 014	0930	(2. x 2.) (2. x 2.)	−− ...	2	... Compact		15.0		
00655	01 01.2 + 41 35 125.44 − 20.96	07 − 03 − 015	0627	3. x 3. 3. x 3.	−− 1	4	... S IV − V	S m	15.2		
00656	01 01.3 + 02 01 129.24 − 60.43	00 − 03 − 065	1196	1.1: x 0.6 0.9 x 0.5	[115] ...	1	S B ... Singular, distorted?		15.6		*
00657	01 01.3 + 32 42 125.98 − 29.83	05 − 03 − 045	0601 20	1.2: x 1.1: 1.0: x 0.9:	−− ...	3	Dwarf spiral S V		17.		
00658	01 01.3 + 35 41 125.80 − 26.85	06 − 03 − 008	0406 18	1.2 x 1.1 1.0: x 1.1:	−− ...	1	...		16.0:		
00659	01 01.6 + 18 26 127.16 − 44.07	03 − 03 − 015	0930	1.1 x 0.35 1.1 x 0.3	[159] ...	1	... Double system?		15.5		*
00660	01 01.7 + 06 21 128.67 − 56.11	01 − 03 − 000	0877	1. x 1. 1. x 1.	−− ...	3	Dwarf V		18.		*
00661	01 01.8 − 01 19 130.35 − 63.74	00 − 03 − 067	1196 19	1.0 x 0.6 1.2 x 0.7	108 4	2	S0 − a n of 2		14.9		
00662	01 01.8 − 01 01 130.17 − 63.44	00 − 03 − 066 NGC 359	1196 19	(2.0 x 1.3) (2.1 x 1.4)	(135) ...	2	E − S0 P w 666		14.8		*3
00663	01 02.0 − 00 26 130.14 − 62.85	00 − 03 − 000	1196 19	(1.1 x 0.5) (1.2 x 0.6)	(7) ...	3	E		16.0:		
00664	01 02.0 + 09 38 128.31 − 52.83	02 − 03 − 000	1274	(1.3 x 1.3) (1.2 x 1.2)	−− ...	1	...		16.0:		*
00665	01 02.0 + 41 57 125.58 − 20.58	07 − 03 − 017	0627	1.1 x 0.9 0.9 x 0.8	(45) ...	1	S B ...		15.4		

N G C   351, 353, 354, 359

1	2	3	4	5	6	7	8	9	10	11	12
00666	01 02.2 – 01 30 130.40 – 63.46	00 – 03 – 069 NGC 364	1196 19	(1.5 x 1.2) (1.6 x 1.4)	(30) ...	3	S0 P w 662		14.6		*3
00667	01 02.2 + 05 21 129.06 – 57.10	01 – 03 – 000	0877	1.2: x 0.8: 1. x 0.5:	(75) ...	1	...	d – dm	17.		*
00668	01 02.4 + 01 53 129.82 – 60.54	00 – 03 – 070 IC 1613	1196	22. x 20. 20. x 18.	— ...	5	Dwarf irregular, I B m Ir V, Local group		10.7	– 238 – 129	*12 59
00669	01 02.5 + 31 25 126.36 – 31.10	05 – 03 – 046	0601 20	1.6 x 0.4 1.4 x 0.4	125 6	3	Sc		15.7		*
00670	01 03.0 + 75 20 123.91 + 12.76	12 – 02 – 002	1214	1.8 x 1.7 1.7 x 1.7	— 1	2	S B? b		15.2		1
00671	01 03.2 + 32 09 126.48 – 30.35	05 – 03 – 000 IC 1618	0601 20	1.2 x 0.5 0.9 x 0.4	159 6	3	S0 nf of 2		15.6		*
00672	01 03.3 + 44 43 125.67 – 17.81	07 – 03 – 000	0627	1.5:: x 1.4:: 1.3:: x 1.3::	— ...	4	Dwarf irregular Ir V		18.		
00673	01 03.4 + 31 08 126.61 – 31.37	05 – 03 – 047	0601 20	2.0 x 0.6 1.0 x 0.3	33 6	3	Sc		15.5	+ 4500:	*1
00674	01 03.5 + 21 07 127.51 – 41.36	03 – 03 – 000	0930	1.3: x 1.3:: 1. x 1.	— ...	3	Dwarf V		18.		
00675	01 03.6 + 00 31 130.76 – 61.86	00 – 03 – 071	1196 19	1.6:: x 1.6:: 1.2: x 1.2:	— ...	2	(Dwarf) IV – V:		17.		
00676	01 03.7 + 00 58 130.70 – 61.41	00 – 03 – 000	1196 19	1.4 x 0.5 1.3 x 0.5	22: 6	1	S ...		15.6		
00677	01 03.7 + 14 19 128.34 – 48.13	02 – 03 – 032	1274	1.2 x 1.0 0.8 x 0.7	— 1 – 2	1	S ...		15.4		*1
00678	01 03.8 + 03 19 130.20 – 59.08	00 – 03 – 072	1196	2.0 x 1.9 1.5: x 1.4:	— 1	3	Sc	SA(rs)cd	15.2		*
00679	01 04.3 + 32 07 126.75 – 30.37	05 – 03 – 049	0601 20	1.2 x 0.4 1.0 x 0.4	95 6	2	... IV or IV – V		16.5:		*1
00680	01 04.3 + 32 32 126.72 – 29.96	05 – 03 – 048 NGC 374	0601 20	1.3 x 0.5 1.0 x 0.4	175 6	3	S0 – a		14.3		*3
00681	01 04.5 + 13 42b 128.71 – 48.73	02 – 03 – 034 IC 1620	1274	1.2 x 0.8 1.2 x 0.8	90 3 – 4	3	Sb/S B c f of 2, disturbed:		(14.2)		*1
00682	01 04.5 + 32 12 126.80 – 30.27	05 – 03 – 051 NGC 380	0601 20	(1.3 x 1.1) (1.5 x 1.3)	— ...	3	E NGC 383 group		13.9	+ 4341 + 4539	*12 3
00683	01 04.5 + 32 15 126.79 – 30.24	05 – 03 – 050 NGC 379	0601 20	(1.5 x 0.8) (1.6 x 0.8)	0 5:	3	S0 NGC 383 group		14.0	+ 5374 + 5573	*12 3
00684	01 04.6 + 10 35 129.21 – 51.83	02 – 03 – 035 IC 75	1274	1.6 x 1.2 1.5 x 0.8	30: 2 – 3:	2	Sb		15.0		
00685	01 04.7 + 16 25 128.43 – 46.02	03 – 03 – 017	0930	1.9: x 1.3: 1.7: x 1.3:	[120] ...	4	Irr 	SA m Ir I	14.5		*1
00686	01 04.7 + 32 01 126.86 – 30.47	05 – 03 – 055 NGC 384	0601 20	(1.1 x 0.8) 1.1 x 0.8	135 3	3	S0 NGC 383 group		14.3	+ 4401 + 4599	*12 3
00687	01 04.7 + 32 03 126.86 – 30.43	05 – 03 – 056 NGC 385	0601 20	(1.3 x 1.0) (1.3 x 1.0)	— ...	2	E NGC 383 group		14.3	+ 4845 + 5043	*12 3
00688	01 04.7 + 32 08 126.85 – 30.35	05 – 03 – 052 NGC 382	0601 20	0.25 x 0.25 0.25 x 0.25	— ...	1	E? NGC 383 group, p w 689		14.2	+ 5156 + 5354	*12 3VĀC
00689	01 04.7 + 32 09 126.85 – 30.33	05 – 03 – 053 NGC 383	0601 20	(2.0 x 1.7) (2.4 x 2.1)	— ...	3	S0 Brightest in group, p w 688		13.6	+ 4888 + 5086	*12 3VĀC
00690	01 04.7 + 39 08 126.32 – 23.37	06 – 03 – 011	0406 17	2.5 x 1.9 2.3 x 1.8	105 2 – 3	4	Sc		13.8		
00691	01 04.8 + 31 49 126.90 – 30.66	05 – 03 – 000	0601 20	1.4 x 0.15 0.9 x 0.15	136 6 – 7	1	S ...		16.0:		
00692	01 04.8 + 32 40 126.83 – 29.82	05 – 03 – 058	0601 20	1.1 x 1.0 1.1: x 0.9:	— 1 – 2	4	S B c	SB(s)cd	15.1		*
00693	01 04.9 + 00 40 131.40 – 61.67	00 – 03 – 075 NGC 391	1196 19	(1.2 x 0.8) (1.1 x 0.8)	(45) ...	2	E		14.6		*3
00694	01 05.0 + 01 55a 131.12 – 60.43	00 – 03 – 076	1259 19	1.1 x 0.5 1.0 x 0.5	[145] ...	4	S B b ssp of 2, disturbed	SB(s)b	(15.5)		*
00695	01 05.3 + 00 48 131.57 – 61.53	00 – 04 – 003	1259 19	1.3 x 1.1 1.0 x 1.0	— ...	1	... Peculiar		15.4		*
00696	01 05.3 + 20 50 128.10 – 41.61	02 – 04 – 003	1251	(1.1 x 0.9) (1.1 x 0.9)	(150) ...	1	E – S0		15.0		*
00697	01 05.3 + 33 11 126.91 – 29.29	05 – 03 – 061	0601 20	1.0 x 0.6 1.1 x 0.7	(85) 4:	1	S B ...		14.7		*
00698	01 05.4 + 02 17 131.23 – 60.06	00 – 04 – 004	1259	1.5 x 0.5 1.1 x 0.3	86 6	1	Sa – b		15.5		*
00699	01 05.5 + 20 49 128.16 – 41.62	02 – 04 – 000	1251	1. x 1. ...	— ...	3	Dwarf spiral S V		19.		*
00700	01 05.6 + 32 52 127.01 – 29.60	05 – 03 – 062 NGC 392	0601 20	(1.2 x 0.9) (1.2 x 0.9)	(50) ...	2	E – S0 Brightest of 3		13.9		*13

N G C   364, 374, 379, 380, 382 – 85, 391, 392   I C   75, 1613, 1618, 1620

1	2	3	4	5	6	7	8	9	10	11	12
00701	01 05.7 + 01 56 131.47 - 60.39	00 - 04 - 005	1259 19	[1.6 x 1.3] ...	--- ...	1	Double system Contact		15.5		*1
00702	01 05.7 + 08 08 130.08 - 54.24	01 - 04 - 000	1201	[1.1: x 0.9:] ...	--- ...	1	Double system Symbiotic, disrupted		18.		*
00703	01 05.7 + 39 26 126.50 - 23.05	06 - 03 - 014 NGC 389	0406 18	1.4 x 0.4 0.9 x 0.35	54 6	2	S0 P w 707		15.0		*
00704	01 05.8 + 05 29 130.66 - 56.87	01 - 04 - 000	1201	1.1 x 0.4 1.0 x 0.5:	22 5 - 6	1	S ... Brightest of 3		16.0:		*
00705	01 05.8 + 06 12 130.51 - 56.16	01 - 04 - 001	1201	1.3 x 1.1 1.1 x 1.1	--- 1	3	S B c	S B(r)c	16.0:		*1
00706	01 05.8 + 08 04 130.14 - 54.30	01 - 04 - 000	1201	1.1 x 1.1 1.1 x 1.1	--- 1	3	S B b	S B(r)ab Sb -	15.4		*
00707	01 05.8 + 39 23 126.53 - 23.10	06 - 03 - 015 NGC 393	0406 18	(1.7 x 1.4) (1.7 x 1.4)	(20) ...	3	E or S0 P w 703		13.3		*3C
00708	01 05.9 + 16 37 128.81 - 45.79	03 - 04 - 005	1251	1.0 x 0.6 0.9 x 0.6	48 4	2	S B: c		14.9		
00709	01 06.0 + 01 05 131.85 - 61.22	00 - 04 - 009	1259 19	1.4 x 0.6 1.2: x 0.5:	142 ...	3	... p of 2	S dm	16.0:		*
00710	01 06.0 + 33 12 127.08 - 29.27	05 - 03 - 000	0601 20	1.5 x 0.5 1.4 x 0.4	56 6	2	Sb - c		15.6		
00711	01 06.1 + 01 23 131.82 - 60.92	00 - 04 - 008	1259 19	4.0 x 0.5 3.4 x 0.5	118 7	3	Sc		14.8		*1
00712	01 06.2 + 32 22 127.20 - 30.09	05 - 03 - 067 NGC 399	0601 20	1.2 x 1.0 1.1 x 0.8	(40) 2 - 3	2	S B a:	S B(r) ...	14.5		*3
00713	01 06.3 + 37 35 126.78 - 24.89	06 - 03 - 000	0406 18	1.1: x 1.0: 1.0 x 0.7	--- ...	2	...	S dm:	16.5:		
00714	01 06.5 + 31 53 127.31 - 30.57	05 - 03 - 069	0601 20	1.3 x 0.9 1.4 x 0.9	10 3	3	Sc	SA c	14.5		*
00715	01 06.5 + 32 29 127.26 - 29.97	05 - 03 - 068 NGC 403	0601 20	2.0 x 0.7 1.8: x 0.7	86 6	2	S0 - a		13.3		*23
00716	01 06.7 + 12 55 129.64 - 49.46	02 - 04 - 001	0635	(1.1 x 0.9) (1.4 x 1.1)	(15) ...	1	...		15.7		*
00717	01 06.7 + 14 05 129.46 - 48.30	02 - 04 - 003	0635	1.6 x 1.2 1.6 x 1.1	(160) ...	3	S B b P w 719, disturbed		14.8		*1V A
00718	01 06.7 + 35 27 127.05 - 27.01	06 - 03 - 018 NGC 404	0406 (18)	(6. x 6.) ...	--- ...	2	E - S0		11.3	- 36 + 169	*12- -568
00719	01 06.8 + 14 06 129.49 - 48.28	02 - 04 - 05	0635	1.0 x 0.9 1.0 x 0.9	--- 1	3	S B b P w 717, disturbed	S B(r)b	15.2		*1V A
00720	01 06.8 + 01 57 132.02 - 60.34	00 - 04 - 013	1259 19	[1.4 x 0.9] ...	--- ...	1	Double system Contact		15.4		*1
00721	01 07.0 + 01 34 132.23 - 60.71	00 - 04 - 014	1259 19	1.0 x 0.25 0.9 x 0.3	35 6	1	S ...		15.6		
00722	01 07.1 + 13 02 129.77 - 49.33	02 - 04 - 006	0635	1.9 x 1.2 2. x 1.2:	(150) ...	4	...	SB(s)dm	16.0:		
00723	01 07.2 + 20 30 128.73 - 41.90	03 - 04 - 006	1251	1.8 x 0.15 1.8 x 0.15	165 7	3	Sc		15.7		
00724	01 07.2 + 32 05 127.47 - 30.36	05 - 03 - 075	0601 20	2.3: x 1.9: 1.5: x 1.2:	(15) 2:	1	S ...		14.0		*1
00725	01 07.3 + 42 50 126.56 - 19.64	07 - 03 - 019	0627	2.2 x 0.6 2.1 x 0.6	43 6	3	S B? c P w 728		14.5		*
00726	01 07.4 - 02 01 133.63 - 64.24	00 - 04 - 018	1259	2.0 x 1.1 2.1: x 1.2:	144 4 - 5	3	... Disturbed?	SAB? dm	15.1		*
00727	01 07.4 + 13 43 129.77 - 48.64	02 - 04 - 007	0635	[1.6 x 1.6] ...	--- ...		Multiple system		16.5:		*1
00728	01 07.6 + 43 01 126.60 - 19.45	07 - 03 - 020	0627	1.5 x 1.5 1.5 x 1.5	--- ...	4	Sc/S B c P w 725	SAB(s)c Sc +	14.2		*1
00729	01 07.8 + 03 15 132.14 - 59.02	00 - 04 - 020	1259	(1.1 x 0.7) (1.0 x 0.8)	--- ...	1	... Compact		15.4		
00730	01 07.8 + 32 51 127.54 - 29.58	05 - 03 - 077 NGC 407	0601 20	2.0 x 0.35 2.0 x 0.35	0 7	3	S0 - a P w 735		14.3		*12 3
00731	01 07.8 + 49 21 126.14 - 13.14	08 - 03 - 000	0863 22	2.4: x 2.2: 2. x 2.	--- ...	0	Dwarf irregular? Ir V?		17.		*
00732	01 07.9 + 33 18 127.52 - 29.13	05 - 03 - 078	0601 20	1.7: x 0.9: 1.6: x 0.9:	80: 4 - 5	4	Sc	SA(r)d	14.7		*
00733	01 08.0 + 16 20 129.57 - 46.02	03 - 04 - 007	1251 21	1.0: x 0.8: 0.8 x 0.8	--- ...	2	S B b	S B(s)b Sb -	15.7		
00734	01 08.1 - 00 32 133.47 - 62.75	00 - 04 - 022	1259 19	1.2 x 0.3 1.2 x 0.3	175 6	1	S ...		15.6		*
00735	01 08.2 + 32 53 127.63 - 29.54	05 - 03 - 080 NGC 410	0601 20	(2.3 x 1.8) (2.4 x 2.0)	(30) ...	3	E P w 730		12.6		*12 3

N G C   389, 393, 399, 403, 404, 407, 410

1	2	3	4	5	6	7	8	9	10	11	12
00736	01 08.3 – 02 01 134.14 – 64.20	00 – 04 – 025	1259	1.4 x 1.0 1.2 x 1.0	(125) 2 – 3	4	Sc		15.5		
00737	01 08.3 – 00 04 133.42 – 62.28	00 – 04 – 024	1259 19	1.3: x 1.0: 0.9 x 0.5	— ...	1	S ...		15.7		
00738	01 08.3 + 33 35 127.59 – 28.85	06 – 03 – 019	0406 20	1.3: x 1.0: 1.2 x 0.8:	— 2 – 3	4	... S dm		15.5		
00739	01 08.4 + 17 22 129.55 – 44.99	03 – 04 – 009 IC 1635	1251	[1. x 1.] ...	— ...	1	Multiple system Contact w 740		15.7		*1
00740	01 08.4 + 17 23 129.55 – 44.97	03 – 04 – 008 IC 1634	1251	[1. x 1.] ...	— ...	1	Multiple system Contact w 739		15.7		1
00741	01 08.5 + 08 30 131.18 – 53.79	01 – 04 – 000	1201	(1.0 x 0.9) (1.1 x 1.0)	— ...	2	[E] Compact		15.5		
00742	01 08.5 + 31 28 127.85 – 30.95	05 – 03 – 081	0601 20	1.3 x 1.3 1.3 x 1.3	— 1	3	... S dm		15.4		
00743	01 08.5 + 31 37 127.83 – 30.80	05 – 03 – 000	0601 20	1.5 x 0.7 1.4 x 0.7	8 5	3	Sa		14.8		*
00744	01 08.5 + 32 50 127.71 – 29.59	05 – 03 – 000 NGC 414	0601 20	[0.8 x 0.4] ...	[35] ...	1	Double system		14.5		*3C
00745	01 08.7 + 36 25 127.41 – 26.02	06 – 03 – 000	0406 18	1.0 x 0.4 0.8 x 0.4	170 6	3	Sc		16.5:		
00746	01 08.7 + 48 51 126.33 – 13.62	08 – 03 – 008	0863	(1.6 x 1.2) (1.6 x 1.2)	(100) ...	2	E		14.8		
00747	01 08.8 + 01 01 133.31 – 61.19	00 – 04 – 028	1259 19	1.0 x 0.2 0.9 x 0.2	... ...	1	... P w 749, distorted		15.7		*
00748	01 08.9 + 35 00 127.59 – 27.42	06 – 03 – 020	0406	2.0 x 0.7 1.9: x 0.6:	79 6	1	S ...		15.3		
00749	01 09.0 + 01 04 133.40 – 61.13	00 – 04 – 030	1259 19	1.3 x 0.5 1.3 x 0.5	[136] ...	2	S – Irr P w 747, distorted		14.2		*
00750	01 09.2 – 00 55 134.21 – 63.08	00 – 04 – 031 IC 1639	1259 19	(0.9 x 0.8) (0.7 x 0.7)	— ...	1	...		14.2		*
00751	01 09.3 + 17 02 129.91 – 45.29	03 – 04 – 011	1251	(1.4 x 1.4) (1.5 x 1.4)	— ...	1	...		16.0:		*
00752	01 09.3 + 31 52 128.00 – 30.54	05 – 03 – 083 NGC 420	0601 20	(2.0 x 2.0) (2.0 x 2.0)	— (1)	2	S0:		13.4		3
00753	01 09.5 – 00 31 134.22 – 62.67	00 – 04 – 000	1259 19	1.2 x 0.7 1.2 x 0.8	(60) ...	1	... nnf of 2		16.5:		*
00754	01 09.6 + 50 21 126.36 – 12.12	08 – 03 – 009	0863 22	1.1 x 0.8 1.3 x 0.9	160: 3	3	S B 0 (R)SAB 0 +		16.0:		
00755	01 09.8 + 38 14 127.48 – 24.19	06 – 03 – 021	0406 18	1.5 x 1.5 1.4 x 1.3	— 1	4	Sc/S B c SAB(rs)c Sc +		14.6		*
00756	01 09.9 + 32 43 128.06 – 29.68	05 – 03 – 084	0601 20	1.0 x 0.8 1.0 x 0.8	(120) 2	3	Sc		14.7		
00757	01 10.0 + 00 01 134.29 – 62.12	00 – 04 – 034	1259 19	1.1 x 0.9 1.0 x 0.9	— 1 – 2	1	S B ...		16.0:		*
00758	01 10.2 + 38 30 127.54 – 23.91	06 – 03 – 023 NGC 425	0406 18	1.0 x 0.9 0.9 x 0.9	— 1	1	S ... Singular		13.5		*
00759	01 10.2 + 49 24 126.54 – 13.05	08 – 03 – 000	0863 22	1.0 x 0.6 1.0 x 0.6	(170) 4	2	S B 0 – a 1st of 3		16.5:		*
00760	01 10.3 – 00 33 134.66 – 62.67	00 – 04 – 035 NGC 426	1259 19	(1.7 x 1.5) (1.8 x 1.6)	(140) ...	2	E NGC 430 group		14.4		*3
00761	01 10.3 + 50 25 126.47 – 12.04	08 – 03 – 010	0863 22	1.1 x 1.1 1.2 x 1.2	— ...	1	E?		16.0:		*
00762	01 10.4 – 00 36 134.73 – 62.71	00 – 04 – 037 NGC 429	1259 19	1.6 x 0.25 1.5 x 0.25	19 7	2	S0 NGC 430 group		14.4		*3
00763	01 10.4 + 00 43 134.24 – 61.42	00 – 04 – 036 NGC 428	1259 19	5.0 x 3.7 5.1 x 3.6	120 1 – 3	5	... SAB(s)m		11.9	+ 1078 + 1173	*12– –9
00764	01 10.4 + 34 43 127.96 – 27.68	06 – 03 – 000	0406	1.1 x 0.1 1.1 x 0.1	12 7	2	Sc		17.		
00765	01 10.5 – 00 31 134.75 – 62.63	00 – 04 – 039 NGC 430	1259 19	(1.8 x 1.5) (1.9 x 1.5)	— ...	1	E: Brightest of 3		13.6		*3
00766	01 10.5 + 49 21 126.59 – 13.10	08 – 03 – 000	0863 22	1.2 x 0.6 1.2 x 0.6	170: ...	2	Sc – Irr 2nd of 3		16.5:		*
00767	01 10.6 – 01 59 135.42 – 64.06	00 – 04 – 041	1259	1.1 x 0.3 0.9 x 0.3	171 6	1	S ...		15.7		
00768	01 10.6 + 02 01 133.88 – 60.13	00 – 04 – 040	1259 19	(1.2 x 1.1) (1.7 x 1.3)	— ...	1	... Compact, disturbed?		15.4		*
00769	01 10.6 + 12 30 131.18 – 49.76	02 – 04 – 012	0635	1.1 x 1.0 1.1 x 0.9	— ...	1	Sc/Irr:		16.0:		
00770	01 10.9 + 02 23 133.91 – 59.75	00 – 04 – 000	1259	1.0: x 1.0: 0.7 x 0.7	— 1:	1	S ... 3rd of 3		16.5:		*

N G C   414, 420, 425, 426, 428 – 30

I C   1634, 1635, 1639

1	2	3	4	5	6	7	8	9	10	11	12
00771	01 11.1 – 00 21 135.01 – 62.43	00 – 04 – 044	1259 19	1.3 x 0.7 1.4 x 0.7	105 5	1	Sa – b		15.0		*1
00772	01 11.1 + 00 37 134.63 – 61.48	00 – 04 – 043	1259 19	1.6: x 1.4: 1.4:: x 1.2::	— ...	3	Dwarf irregular Pair of dwarfs?		17.		*1
00773	01 11.2 + 02 07 134.15 – 60.00	00 – 04 – 000	1259 19	1.4:: x 1.0:: 1.5: x 1.0:	— ...	1	...		16.0:		
00774	01 11.2 + 13 00 131.30 – 49.24	02 – 04 – 014	0635	1.0 x 0.8 1.1 x 0.8	(95) ...	2	S ... + companion Connected		15.0		*1
00775	01 11.2 + 18 55 130.23 – 43.37	03 – 04 – 000	1251	1.0 x 0.12 1.1 x 0.15	140 7	3	Sb		16.5:		
00776	01 11.3 + 33 27 128.31 – 28.92	05 – 04 – 002 NGC 431	0030 20	1.4 x 0.8 1.3 x 0.7	20: 4 – 5	4	S B 0		14.0		3
00777	01 11.3 + 41 59 127.42 – 20.43	07 – 03 – 023	0627	1.0 x 0.8 0.9 x 0.7	(70) 2	3	Sc		15.4		
00778	01 11.3 + 49 58 126.67 – 12.47	08 – 03 – 000	0863 22	1.1 x 0.10 0.9 x 0.10	28 7	2	...	S d – m	17.		*
00779	01 11.4 + 01 48 134.36 – 60.31	00 – 04 – 046 NGC 435	1259 19	1.4 x 0.5 1.3 x 0.5	20 6	1	S – Irr		15.0		3
00780	01 11.4 + 37 41 127.88 – 24.70	06 – 03 – 000	0406 18	1.1 x 0.18 1.1 x 0.18	110 7	2	Sb		15.5		
00781	01 11.4 + 37 52 127.86 – 24.52	06 – 03 – 000	0406 18	1.1 x 0.20 1.1 x 0.20	48 7	2	Sb – c		16.5:		*
00782	01 11.4 + 50 54 126.60 – 11.54	08 – 03 – 000	0863 22	1.4 x 0.4 1.1 x 0.3	140 6	1	Sc?		16.5:		
00783	01 11.5 + 42 17 127.43 – 20.12	07 – 03 – 024	0627	1.6 x 0.9 1.7 x 0.9	153 4 – 5	4	Sc	SAB(rs)c Sc –	14.6		
00784	01 11.6 – 02 00 135.98 – 64.03	00 – 04 – 047	1259	1.5 x 0.9 1.4 x 0.8	50 4	3	Sb	Sb –	15.0		*
00785	01 11.6 + 15 37 130.94 – 46.64	03 – 04 – 012	1251 21	1.1 x 0.8 1.1 x 0.8	(155) 3:	2	S0	(R:)SAB 0	15.2		*
00786	01 11.6 + 50 12 126.70 – 12.24	08 – 03 – 000	0863 22	1.0 x 0.2 0.9 x 0.2	51 7	2	Sc		17.		*
00787	01 11.7 + 12 07 131.67 – 50.10	02 – 04 – 015	0635	1.0 x 0.5 1.1 x 0.5	... ...	2	Sb	SAB(s)b Sb –	16.0:		*
00788	01 11.8 + 05 40 133.31 – 56.48	01 – 04 – 005 NGC 437	1201	(1.7 x 1.5) (1.9 x 1.6)	(130) 1 – 2	2	S0/Sa		14.0		
00789	01 12.1 – 01 17 135.94 – 63.30	00 – 04 – 054 NGC 442	1259	(1.0 x 0.7) (1.7 x 1.2)	[157] ...	1	Peculiar Eruptive?, Ir II		14.5		*1<u>2</u>
00790	01 12.1 + 00 55 135.03 – 61.14	00 – 04 – 053	1259 19	1.0 x 0.9 1.0 x 0.9	— 1	2	Sc np of 2		15.0		*
00791	01 12.1 + 01 34 134.79 – 60.50	00 – 04 – 052	1259 19	1.1 x 0.8 0.9 x 0.8	130: 1 – 3	2	S B a – b		15.0		*
00792	01 12.2 + 31 42 128.74 – 30.64	05 – 04 – 003 IC 1652	0030 20	1.4: x 0.22 1.3 x 0.22	169 7	1	S0 – a		14.3		
00793	01 12.3 – 00 45 135.81 – 62.77	00 – 04 – 055	1259	[0.70 x 0.45] ...	[87] ...	2	Double system Contact, bridge:		14.5		*1
00794	01 12.3 + 03 55 134.07 – 58.18	01 – 04 – 006 NGC 446	1201	1.3 x 0.22 1.3 x 0.22	6 7	1	S ...		15.3		3
00795	01 12.3 + 12 07 131.90 – 50.08	02 – 04 – 016	0635	1.0 x 1.0 0.9 x 0.9	— 1	2	Sb	SA(r)b	16.5:		
00796	01 12.3 + 33 07 128.59 – 29.23	05 – 04 – 005 IC 1653	0030 20	0.9 x 0.8 0.9 x 0.8	— ...	1	...		13.9		*C
00797	01 12.4 + 00 10 135.48 – 61.87	00 – 04 – 057	1259 19	(1.2 x 1.1) (1.3 x 1.1)	— ...	1	E:		15.5		*
00798	01 12.4 + 29 56 129.00 – 32.39	05 – 04 – 004 IC 1654	0030	1.6: x 1.2 1.6 x 1.2	(60) 2 – 3	3	S B a	(R)S B(r)a or (R')S B(r)a	14.3		*
00799	01 12.5 + 06 35 133.35 – 55.54	01 – 04 – 000	1201	1.4: x 0.4: 1.3: x 0.5:	103 6	2	Sc Distorted?		16.5:		*
00800	01 12.6 + 28 13 129.28 – 34.10	05 – 04 – 05a	0030	1.3: x 1.0: 1.2: x 1.0:	(110) 2:	3	Sc s of 2		15.2		*
00801	01 12.8 – 01 53 136.60 – 63.85	00 – 04 – 060 NGC 448	1259	(1.9 x 1.0) (1.9 x 1.1)	116 5 – 6	3	S0		13.2		*
00802	01 12.8 + 07 06 133.34 – 55.02	01 – 04 – 000	1201	1.0 x 0.7 0.8 x 0.6	55: 3	3	Sc		16.0:		
00803	01 12.8 + 07 50 133.14 – 54.30	01 – 04 – 008	1201	1.8: x 1.5: 2.1: x 1.8:	— 1 – 2	3	...	SA dm	16.5:		
00804	01 12.8 + 32 48 128.75 – 29.53	05 – 04 – 006 NGC 449 = 1656*	0030 20	(2.6 x 2.4) (2.6 x 2.6)	— 1	4	S B 0/S B a Mark 1	(R:)S B(r)0 +	14.0 + 4800		*
00805	01 12.9 + 06 38 133.51 – 55.48	01 – 04 – 000	1201	1.4: x 0.3 1.4: x 0.3:	142 6	1	Sc		16.5:		*

N G C   431, 435, 437, 442, 443, 446, 448
449   I C   1652 – 54, 1656

1	2	3	4	5	6	7	8	9	10	11	12
00806	01 13.0 – 01 07a 136.35 – 63.09	00 – 04 – 062 NGC 450	1259	3.5 x 2.8 3.5 x 2.9	72: 2	4	Sc 807 superimposed	SAB(s)d			*123 478
00807	01 13.0 – 01 07b 136.35 – 63.09	00 – 04 – 064	1259	1.1 x 0.5 1.1 x 0.5	40 5 – 6	1	S ... Superimposed 806		13.0		*1
00808	01 13.0 + 09 02 132.90 – 53.11	01 – 04 – 009	1201	2.8 x 2.1 2.5: x 1.8:	60 2 – 3	4	Sc	SA(r)c Sc –	15.0		1
00809	01 13.0 + 33 33 128.70 – 28.70	06 – 03 – 027	0406 20	1.4 x 0.20 1.2 x 0.18	23 7	3	Sc		14.8		
00810	01 13.1 + 30 49 129.07 – 31.50	05 – 04 – 007 NGC 444 = 1658*	0030	2.1 x 0.4 2.1 x 0.4	157 6	3	Sc P w 820	Sc +	14.7		*3
00811	01 13.2 + 37 23 128.30 – 24.97	06 – 03 – 028	0406 18	1.1: x 0.9: 0.8 x 0.6	— 2:	2	S0 – a		15.7		*
00812	01 13.3 + 30 05 129.21 – 32.22	05 – 04 – 008 IC 1659	0030	(1.6 x 1.1) (1.6 x 1.1)	(20) ...	3	E		14.6		*
00813	01 13.3 + 46 29 127.35 – 15.91	08 – 03 – 012	0863	1.4 x 0.6 1.3 x 0.6	[110] ...	1	... P w 816, distorted		15.0		*1
00814	01 13.3 + 40 54 127.93 – 21.47	07 – 03 – 000	0627	1. x 1. 1: x 1.	— ...	1	...		17.		
00815	01 13.4 + 04 55 134.26 – 57.15	01 – 04 – 011 NGC 455	1201	(2.7 x 1.6) (2.7 x 1.7)	[165] ...	2	... Peculiar, eruptive?		13.9		*13 A
00816	01 13.4 + 46 29 127.36 – 15.91	08 – 03 – 013	0863	1.8 x 0.9 1.4 x 0.8	... ...	3	S ... P w 813, distorted	SAB(s) ...	14.6		*1
00817	01 13.5 + 01 18 135.59 – 60.70	00 – 04 – 065	1259	1.2 x 0.9 1.1 x 0.8	— ...	2	Sb/S B b:		15.0		*
00818	01 13.5 + 04 02 134.59 – 58.02	01 – 04 – 012 IC 89	1201	(2.0 x 1.6) (2.1 x 1.8)	— 1 – 2	4	S B 0	(R)SAB 0	13.8		
00819	01 13.5 + 06 23 133.85 – 55.70	01 – 04 – 000	1201	1.0: x 0.2 1.0: x 0.3:	76 7	2	Sc – Irr		17.		
00820	01 13.5 + 30 46 129.18 – 31.54	05 – 04 – 010 NGC 452	0030	2.7 x 1.1 2.6: x 1.0	43 6	3	S B a – b P w 810		14.0		*13
00821	01 13.6 – 00 17 136.30 – 62.25	00 – 04 – 066	1259	1.0 x 0.3 1.1 x 0.3	70 6	1	S ...		16.0:		*
00822	01 13.6 + 38 47 128.23 – 23.57	06 – 03 – 029	0406 18	1.2 x 0.6 1.3 x 0.7	1 5	4	S B a		15.1		
00823	01 14.0 + 00 57 135.98 – 61.02	00 – 04 – 000	1259	1.0: x 0.9: 0.8: x 0.7:	— ...	2	Dwarf: spiral S IV – V or V		18.		*
00824	01 14.3 + 12 45 132.50 – 49.38	02 – 04 – 017	0635	1.2 x 0.25 1.2 x 0.25	73 7	2	Sa		14.9		
00825	01 14.6 + 48 45 127.34 – 13.63	08 – 03 – 000	0863	1.4 x 0.12 1.4 x 0.12	145 7	2	...	S d – m	17.		
00826	01 14.7 + 43 22 127.94 – 18.99	07 – 03 – 026	0627	1.3 x 0.8 1.2 x 0.8	165: ...	2	S ... disrupted pair?, f of 2	Singular,	14.8		*1C
00827	01 14.8 + 14 26 132.30 – 47.70	02 – 04 – 018	0635	[1.1 x 0.5] ...	... ...	1	Double system Contact		15.4		*1V A
00828	01 15.0 + 21 09 131.05 – 41.04	03 – 04 – 016	1251	1.2 x 0.4 1.1 x 0.4	150 6	2	Sa – b		15.7		
00829	01 15.1 + 09 56 133.51 – 52.13	02 – 04 – 000	0635	1.4 x 1.0 1.6 x 1.2	— ...	2	... Peculiar		15.7		*
00830	01 15.3 – 02 13 138.15 – 64.04	00 – 04 – 072	1259 25	2.8 x 2.0 2.7: x 2.0:	(120) 2 – 3	3	Sc		15.6		*
00831	01 15.3 + 38 11 128.66 – 24.13	06 – 04 – 001	1189 18	1.4 x 1.0 0.5 x 1.0	— ...	2	S ... Disturbed?	SAB ...	15.1		*1
00832	01 15.4 + 17 17 131.91 – 44.86	03 – 04 – 017 NGC 459	1251	1.1 x 1.0 1.0 x 1.0	— 1	2	Sb – c		15.7		3
00833	01 15.5 + 11 07 133.35 – 50.95	02 – 04 – 019	0635	2.5 x 1.7 2.6 x 1.7	50 3:	4	S B c	SB(r)cd	14.3		
00834	01 15.8 + 04 29 135.49 – 57.47	01 – 04 – 017	1201	1.0 x 0.9 1.0 x 0.9	— 1:	3	Sb/S B c	SAB(s)bc Sc –	15.7		*
00835	01 15.9 + 30 47 129.77 – 31.46	05 – 04 – 013	0030	1.1 x 0.8 1.2: x 1.0:	75: 2 – 3	3	Sc		14.8		
00836	01 16.0 + 44 02 128.11 – 18.30	07 – 03 – 028	0627	(1.4 x 1.4) (1.1 x 1.1)	— ...	1	E: Brightest of 3		16.0:		*1
00837	01 16.1 – 01 28 138.20 – 63.27	00 – 04 – 073	1259 25	1.1 x 0.5 1.0 x 0.5	167 ...	1	E – S ...		16.5:		
00838	01 16.1 + 14 45 132.69 – 47.34	02 – 04 – 020	0635	0.65 x 0.45 0.65 x 0.5	— ...	1	...		14.2		*
00839	01 16.3 + 12 41 133.26 – 49.37	02 – 04 – 000	0635	1.0 x 0.8 0.8 x 0.8	— ...	2	S ... Disturbed?		16.5:		*
00840	01 16.3 + 16 03 132.47 – 46.05	03 – 04 – 019 NGC 463	1251	1.2 x 0.4 1.3 x 0.4	[4] ...	1	Double system?		15.2		*3

N G C   444, 450, 452, 455, 459, 463   I C   89, 1658, 1659

1	2	3	4	5	6	7	8	9	10	11	12
00841	01 16.3 + 32 47 129.59 – 29.47	05 – 04 – 015	0030 20	1.6: x 0.4 1.5 x 0.4	54 6	2	Sb – c		15.0		
00842	01 16.4 – 01 16 138.26 – 63.05	00 – 04 – 077	1259 25	(1.5 x 1.2) (1.9 x 1.4)	(85) ...	1	...		15.4	+ 4855 + 4938	*
00843	01 16.4 + 05 18 135.47 – 56.64	01 – 04 – 018	1201	1.2 x 1.1 1.2 x 1.1	— 1:	2	... sf of 2	S dm	16.5:		*
00844	01 16.4 + 09 46 134.07 – 52.24	02 – 04 – 000	0635	1.0 x 0.8 0.9 x 0.8	(35) 1 – 2	1	Sc		16.5:		*
00845	01 16.5 + 21 29 131.44 – 40.67	03 – 04 – 000	1251	1.2 x 1.0 1.1 x 0.9	(40) 1 – 2	3	Sb		15.0		
00846	01 16.5 + 48 56 127.64 – 13.42	08 – 03 – 000	0863	1.1 x 0.9 1.2: x 1.0:	— 1 – 2	2	Sc/S B c	Sc +	16.5:		
00847	01 16.6 – 00 25 137.93 – 62.22	00 – 04 – 078	1259 25	1.6 x 0.12 1.6 x 0.1	67 7	3	...	S d – m	16.5:		*
00848	01 16.6 + 03 02 136.41 – 58.85	00 – 04 – 079 NGC 467	1259	(2.5 x 2.5) (2.4 x 2.4)	— 1	3	S0 NGC 470 group		13.3	+ 5568 + 5667	*12 36
00849	01 16.7 + 12 11 133.53 – 49.85	02 – 04 – 021	0635	1.3 x 0.5 1.2 x 0.5	[120] ...	2	Sc – Irr s of 2, distorted		15.1		*1V A
00850	01 16.7 + 20 53 131.62 – 41.25	03 – 04 – 000	1251	1.0 x 0.8 1.1 x 0.8	(70) 2	3	Sb	SA(r)bc Sb +	15.6		*
00851	01 16.8 + 19 28 131.92 – 42.65	03 – 04 – 020	1251	1.1 x 0.9 0.9 x 0.6	— 1 – 2	3	S B b	(R')S B(s)bc Sb +	16.5:		*
00852	01 16.9 + 17 18 132.40 – 44.79	03 – 04 – 000	1251	1.2 x 0.4: 0.9 x 0.3	74 ...	1	...		16.0:		
00853	01 16.9 + 31 24 129.93 – 30.83	05 – 04 – 000	0030 20	1.1 x 0.6 1.2 x 0.3	70: ...	3	S B b Disturbed	S B(r)b	15.6		*
00854	01 17.0 + 03 39 136.35 – 58.22	01 – 04 – 020	1201	1.0 x 0.5 0.9 x 0.3	146 5	2	Sc		16.5:		
00855	01 17.0 + 07 55 134.86 – 54.04	01 – 04 – 019	1201	1.2 x 1.1 1.2 x 1.1	— 1	4	S B b	S B(rs)b Sb +	14.9		
00856	01 17.1 – 01 58 139.01 – 63.69	00 – 04 – 000	1259	1.3 x 1.0: 1.2: x 1.0:	(25) 2:	4	...	S B(s)m	16.5:		
00857	01 17.1 + 32 12 129.87 – 30.03	05 – 04 – 019 IC 1666	0030 20	1.1 x 1.1 1.2 x 1.1	— 1	3	Sc		14.4		
00858	01 17.2 + 03 08 136.65 – 58.72	00 – 04 – 084 NGC 470	1259	3.3 x 1.9 3.1 x 1.9	155 4	4	Sb/Sc Brightest of 3		12.4		*123 4678
00859	01 17.2 + 16 17 132.73 – 45.78	03 – 04 – 022 NGC 473	1251	(2.6 x 1.6) (2.4 x 1.5)	153 4	3	S0		13.2		*123 468
00860	01 17.3 + 12 40 133.63 – 49.35	02 – 04 – 026	0635	1.1 x 1.0 1.0 x 0.9	— 1 – 2	3	Sc		15.6		*
00861	01 17.3 + 14 31 133.17 – 47.53	02 – 04 – 024 NGC 471	0635	1.2 x 0.7 1.3 x 0.8	(85) 3 – 4	1	S0 Brightest in multiple system		(14.0)		*13
00862	01 17.4 + 33 15 129.79 – 28.98	05 – 04 – 021	0030 20	(0.9 x 0.9) (1.0 x 1.0)	— ...	1	...		14.5		
00863	01 17.4 + 78 22 124.48 + 15.85	13 – 02 – 000	1214	2. x 2. 1.7: x 1.7:	— (1)	1	Sc?		17.		
00864	01 17.5 + 03 09 136.79 – 58.69	00 – 04 – 085 NGC 474	1259	10. x 9. 8. x 8.	— (1)	4	S0 (R:)SA 0 – NGC 470 group, disturbed?		12.9	+ 2306 + 2405	*123 478A
00865	01 17.5 + 14 18 133.29 – 47.73	02 – 04 – 027	0635	(1.3 x 1.1) (1.5 x 1.4)	— ...	1	...		16.0:		*
00866	01 17.6 – 00 28 138.48 – 62.21	00 – 04 – 087	1259 25	1.5: x 0.5 1.7 x 0.5	57 6	2	...	S d – m	15.6		*
00867	01 17.7 – 00 36 138.60 – 62.33	00 – 04 – 088	1259 25	1.0 x 1.0 0.9 x 0.8	— 1	1	Sb – c		15.5		*
00868	01 17.7 + 09 03 134.80 – 52.89	01 – 04 – 000	1201	1.2 x 0.2 1.0 x 0.15	103 7	1	S ...		16.0:		*
00869	01 17.7 + 30 30 130.27 – 31.70	05 – 04 – 000	0030	1.1 x 0.2 1.1 x 0.20	13 7	1	S ...		16.0:		
00870	01 17.7 + 32 27 129.97 – 29.76	05 – 04 – 022 NGC 472	0030 20	(1.3 x 1.1) (1.3 x 1.1)	— ...	2	... Very compact		14.2		3
00871	01 17.8 + 05 35 135.99 – 56.29	01 – 04 – 022	1201	2. x 2. 2. x 2.	— ...	4	Dwarf irregular Ir V		17.		
00872	01 17.9 + 29 26 130.48 – 32.75	05 – 04 – 024 IC 1672	0030	1.8 x 1.3 1.7: x 1.3:	140: 3	1	S ... n of 2		14.0		*
00873	01 17.9 + 30 08 130.37 – 32.05	05 – 04 – 000	0030	1.1: x 1.0: 1.0: x 0.8:	— 1 – 2	1	Sb – c		15.5		
00874	01 18.2 + 01 02 138.05 – 60.71	00 – 04 – 089	1259 23	1.3: x 0.18 1.1 x 0.2	34 7	1	S0?		15.5		*
00875	01 18.2 + 01 10 137.99 – 60.58	00 – 04 – 090	1259 23	1.0 x 0.9 1.2 x 1.1	— 1	3	Sc	SAB(s)c	15.2		*

N G C   467,   470 – 74   I C   1666, 1672

1	2	3	4	5	6	7	8	9	10	11	12
00876	01 18.2 + 03 56 136.79 – 57.88	01 – 04 – 023	1201	1.0 x 0.15 1.1 x 0.2	117 7	1	...	d – m	17.		
00877	01 18.2 + 28 59 130.63 – 33.18	05 – 04 – 000	0030	1.0 x 0.15 1.0 x 0.15	71 7	1	Sb – c		15.7		
00878	01 18.2 + 33 38 129.92 – 28.58	06 – 04 – 003	1189 20	(1.2 x 0.9) 1.1 x 0.8	(95) ...	1	S0?		14.7		
00879	01 18.2 + 34 00 129.87 – 28.21	06 – 04 – 004 IC 1675	1189 20	0.8 x 0.25 0.8 x 0.22	70 ...	1	...		14.3		
00880	01 18.2 + 42 35 128.70 – 19.69	07 – 03 – 030	0627	1.2 x 0.9 1.0 x 0.8	(60) 2 – 3	2	S ... Disturbed?		15.0		*1
00881	01 18.3 + 04 32 136.60 – 57.29	01 – 04 – 024	1201	(1.3 x 1.1) (1.2 x 1.0)	(55) ...	1	E		15.4		*
00882	01 18.3 + 06 18 135.95 – 55.56	01 – 04 – 025	1201	1.6 x 1.2 2.0: x 1.6:	(85) ...	4	Irr I m Ir IV – V Ir I		15.7		
00883	01 18.3 + 16 49 132.98 – 45.22	03 – 04 – 024	1251	1.2 x 1.0 1.5: x 1.2:	(50) 1 – 2	4	... S B dm S B IV – V		16.5:		*1
00884	01 18.3 + 25 18 131.29 – 36.83	04 – 04 – 000	1229	1.5 x 0.5 0.9 x 0.6	[175] ...	2	... f of 2, disturbed		16.5:		*
00885	01 18.4 + 02 43 137.39 – 59.06	00 – 04 – 000	1259	1.1 x 0.4 1.1 x 0.5:	138 6	2	Sc		16.5:		
00886	01 18.4 + 40 13 129.05 – 22.04	07 – 03 – 032 NGC 477	0627	2.5: x 1.5: 2.1: x 1.3:	135: 4	4	Sc/S B c SAB(s)c Sc –		14.0		*13
00887	01 18.5 + 04 53 136.56 – 56.94	01 – 04 – 030	1201	1.2 x 1.0 0.7 x 0.8	— 1 – 2	2	Sb – c		15.7		
00888	01 18.5 + 05 10 136.45 – 56.66	01 – 04 – 028	1201	1.5:: x 1.4:: 1.4:: x 1.4::	— ...	4	Dwarf irregular Ir V		17.		*
00889	01 18.5 + 17 36 132.87 – 44.44	03 – 04 – 000	1251	1.0 x 0.5 1.1 x 0.6	178: 5	2	S B c		16.0:		
00890	01 18.6 + 01 06 138.22 – 60.62	00 – 04 – 093	1259 23	2.3 x 0.7 2.4 x 0.7	177 6	3	Sb Sb +		14.3		*
00891	01 18.6 + 12 09 134.25 – 49.80	02 – 04 – 030	0635	3.2: x 1.5: 3.0: x 1.5:	[47] ...	3	Dwarf IV – V		15.7		*1
00892	01 18.7 – 00 48 139.23 – 62.46	00 – 04 – 095	1259 25	2.1 x 2.1 2.1 x 2.0	— 1	3	S B a Disturbed?		14.5		*1A
00893	01 18.7 + 03 36 137.16 – 58.18	01 – 04 – 031 NGC 479	1201	1.2 x 1.0 1.1 x 0.9	— 2	2	Sb		15.1		3
00894	01 18.8 – 00 10 138.95 – 61.84	00 – 04 – 097 IC 1681	1259 25	1.1 x 0.5 1.2 x 0.7:	99 ...	1	S: ... Peculiar, disturbed?		14.8		*
00895	01 18.8 + 06 45 136.01 – 55.10	01 – 04 – 032 NGC 485	1201	2.0 x 0.7 2.0 x 0.7	3 6	1	S ...		14.2		*3
00896	01 18.8 + 08 55 135.28 – 52.97	01 – 04 – 000	1201 24	1.2 x 0.2 1.1 x 0.2	101 6 – 7	3	... S d – m		16.5		*
00897	01 18.8 + 15 41 133.41 – 46.31	03 – 04 – 025	1251	1.0 x 1.0 1.0 x 1.0	— 1	1	Sb – c		16.0:		
00898	01 18.8 + 19 36 132.53 – 42.45	03 – 04 – 000	1251	1.0 x 0.5 0.9 x 0.5	140 5	1	Sb		15.7		
00899	01 18.8 + 20 39 132.31 – 41.41	03 – 04 – 000	1251	1.0 x 0.5 0.9 x 0.4	63 5	1	Sb – c		15.2		
00900	01 18.8 + 23 30 131.75 – 38.59	04 – 04 – 006	1229	1.6? x 1.6? (1.4 x 1.4)	— ...	2	... P w 905		15.7		*
00901	01 18.8 + 32 21 130.25 – 29.83	05 – 04 – 000	0030 20	0.7 x 0.20 0.8 x 0.22	145 (7)	1	...		14.5		*
00902	01 18.9 + 49 47 127.94 – 12.53	08 – 03 – 016	0863 22	(1.4 x 1.3) (1.6 x 1.4)	— ...	2	[E] Compact		15.5		*
00903	01 19.1 + 17 20 133.13 – 44.68	03 – 04 – 026	1251	2.0 x 0.4 1.7 x 0.4	52 7	1	S ...		14.7		
00904	01 19.1 + 18 00 132.97 – 44.02	03 – 04 – 027	1251	1.2 x 0.7 1.3 x 0.7	48 4 – 5	3	S B a (R´)S B(s)a		15.3		*
00905	01 19.1 + 23 31 131.83 – 38.57	04 – 04 – 007	1229	1.5: x 1.2: 1.5: x 1.2:	115: ...	2	S: ... P w.900		16.0:		*
00906	01 19.1 + 33 16 130.19 – 28.91	05 – 04 – 029 NGC 483	0030 20	0.8 x 0.8 0.9 x 0.8	— ...	1	...		14.0		*3
00907	01 19.2 + 05 00 136.83 – 56.79	01 – 04 – 033 NGC 488	1201	6.0 x 4.3 5.6 x 4.2	15 3	5	Sb SA(r)b Sb –		11.6	+ 2180 + 2284	*12– –8$^-$
00908	01 19.2 + 08 56 135.44 – 52.94	01 – 04 – 034 NGC 489	1201 24	1.8 x 0.4 1.7 x 0.35	120 6 – 7	1	S ...		13.4		*3
00909	01 19.2 + 37 09 129.64 – 25.06	06 – 04 – 006	1189	1.7 x 1.1 1.7: x 1.1	45: 3 – 4	3	Sc Sc +		14.4		*1
00910	01 19.3 + 15 31 133.63 – 46.46	03 – 04 – 028	1251	1.4 x 1.0: 1.3 x 1.0:	0: 2 – 3	4	Sc SA(s)c		15.0		*

N G C   477, 479, 483, 485, 488, 489

I C   1675, 1681

1	2	3	4	5	6	7	8	9	10	11	12
00911	01 19.3 + 31 58 130.43 - 30.20	05 - 04 - 030	0030 20	1.1 x 0.9 1.0 x 0.8	5 1 - 2	3	S B b		15.0		
00912	01 19.4 + 33 00 130.30 - 29.17	05 - 04 - 032 IC 1682	0030 20	0.9 x 0.4 0.9 x 0.35	122 ...	1	...		14.3		*
00913	01 19.4 + 34 25 130.08 - 27.77	06 - 04 - 007	1189	0.55 x 0.30 0.5 x 0.25	53 ...	1	... Peculiar		14.4		
00914	01 19.6 + 00 41 138.92 - 60.97	00 - 04 - 099 NGC 493	1259 23	4.3 x 1.6 4.3 x 1.5	58 6	4	Sc		13.0		*3
00915	01 19.8 - 01 08 139.99 - 62.72	00 - 04 - 100 NGC 497	1259 25	2.5 x 1.1 2.3 x 1.0	132 5 - 6	4	S B b/S B c	SAB(rs)bc Sb +	14.1	+ 8030 + 8107	*1A
00916	01 19.8 + 34 11 130.21 - 27.99	06 - 04 - 008 IC 1683	1189 20	1.7 x 0.7 1.4 x 0.6	177: 6	1	S ...		14.2		
00917	01 20.0 + 18 47 133.09 - 43.21	03 - 04 - 000	1251 26	1.1 x 1.0 1.1 x 0.9	— 1 - 2	1	SO:		16.0:		*
00918	01 20.1 + 19 02 133.06 - 42.96	03 - 04 - 029	1251	1.2 x 0.4 1.1 x 0.4	70 6	2	Sc		15.6		1
00919	01 20.1 + 32 55 130.48 - 29.23	05 - 04 - 034 NGC 494	0030 20	2.1 x 0.8 2.1 x 0.8	100 6	2	Sa - b		13.8		3
00920	01 20.1 + 33 13 130.43 - 28.94	05 - 04 - 035 NGC 495	0030 20	1.3 x 0.8 1.4 x 0.9	170: 4	4	S B O/S B a		14.0	+ 4114 + 4304	*12 3
00921	01 20.2 - 01 39 140.50 - 63.19	00 - 04 - 105	1259 25	2.3 x 0.8 2.3 x 0.8	108 6	1	S ...		14.6	+ 5651 + 5728	
00922	01 20.2 + 08 47 135.89 - 53.04	01 - 04 - 043 NGC 502	1201 24	(1.5 x 1.5) (1.9 x 1.9)	— 1	3	SO	SA(r)0	13.8		*3
00923	01 20.2 + 12 52 134.64 - 49.03	02 - 04 - 000	0635	1.0 x 0.3 1.0 x 0.3	97 6	1	Sb - c		16.5:		
00924	01 20.3 + 09 12 135.79 - 52.62	01 - 04 - 041 NGC 505	1201 24	(1.1 x 0.7) (1.2 x 0.9)	— 3 - 4	2	SO		15.1		*3
00925	01 20.3 + 42 43 129.09 - 19.51	07 - 04 - 000	0824	1.0 x 0.20 0.9 x 0.15	141 6	3	Sc		16.5:		
00926	01 20.4 + 33 13 130.50 - 28.93	05 - 04 - 038 NGC 499 = 1686*	0030 20	(1.7 x 1.3) (2.0 x 1.5)	82: 2 - 3	3	SO		13.0	+ 4375 + 4565	*12 3
00927	01 20.4 + 33 17 130.49 - 28.86	05 - 04 - 036 NGC 496	0030 20	1.7 x 0.9 1.7 x 0.9	28 5	2	Sb - c		14.3		*13
00928	01 20.6 - 00 54 140.28 - 62.44	00 - 04 - 108	1259 25	1.2 x 0.7 1.3 x 0.7	45 4 - 5	1	SO P w 931		15.2		*
00929	01 20.6 - 00 39 140.14 - 62.20	00 - 04 - 107	1259 25	1.5 x 1.5 1.6: x 1.6:	— 1	2	Sc		14.9		*
00930	01 20.6 + 49 54 128.20 - 12.38	08 - 03 - 021	0863 22	1.4 x 0.6 1.4 x 0.6	125 (6)	1	...		16.0:		
00931	01 20.7 - 00 57 140.36 - 62.48	00 - 04 - 109	1259 25	1.3 x 0.3 1.3: x 0.2	6 7	1	S ... P w 928		15.7		*
00932	01 20.7 + 09 10 135.96 - 52.64	01 - 04 - 045 NGC 509	1201 24	(1.6 x 0.7) (1.6 x 0.8)	82 5 - 6	1	SO?		14.7		*23
00933	01 20.7 + 14 57 134.25 - 46.96	02 - 04 - 032	0635	1.1 x 0.9 1.0 x 0.8	(45) 2:	3	S B a		15.5		*
00934	01 20.7 + 30 32b 131.01 - 31.57	05 - 04 - 42	0030	2.2: x 0.7 ...	[150] ...	3	S ... nff of 2, distorted		(14.5)		*1V A
00935	01 20.7 + 32 57 130.62 - 29.18	05 - 04 - 041 NGC 504	0030 20	1.6 x 0.4 1.6 x 0.4	47 7	3	SO		14.0		*3
00936	01 20.8 + 11 02 135.40 - 50.80	02 - 04 - 033 NGC 511	0635 24	(1.2 x 1.2) (1.1 x 1.1)	— ...	1	E:		15.4		3
00937	01 20.8 + 32 23 130.73 - 29.74	05 - 04 - 043	0030 20	(1.2 x 1.0) 1.0 x 0.9	— ...	1	... Peculiar?	B ...	15.0		*
00938	01 20.8 + 33 00 130.63 - 29.13	05 - 04 - 044 NGC 507	0030 20	(4. x 4.) (4. x 4.)	— ...	3	E P w 939		13.0	+ 4929 + 5118	*12 3 VA
00939	01 20.8 + 33 02 130.63 - 29.10	05 - 04 - 045 NGC 508	0030 20	(1.3 x 1.3) (1.2 x 1.2)	— ...	2	E P w 938		14.5		*13 VA
00940	01 20.8 + 34 19 130.42 - 27.83	06 - 04 - 011	1189 20	1.1 x 0.8 1.0 x 0.7	85: 3	4	Sc f of 2	SA(s)c	14.9		*
00941	01 21.0 + 06 42 136.96 - 55.03	01 - 04 - 046	1201	1.3 x 0.9: 1.2 x 0.9:	— ...	4	Irr	I m Ir I	15.7		*1
00942	01 21.1 + 06 27 137.09 - 55.27	01 - 04 - 000	1201	1.2 x 1.3 1.2: x 1.1	— (1)	3	... S IV - V	S m	16.0:		
00943	01 21.1 + 11 15 135.45 - 50.57	02 - 04 - 034	0635 24	1.2 x 1.1 1.3: x 1.3:	— 1	2	Sb	SAB(r)b Sb -	16.0:		
00944	01 21.2 + 33 39 130.62 - 28.47	06 - 04 - 013 NGC 512	1189 20	1.7 x 0.35 1.5 x 0.4	116 6	2	Sa - b n of 2		14.0		*3
00945	01 21.2 + 42 58 129.23 - 19.24	07 - 04 - 000	0824	(1.3 x 1.0) (0.8 x 0.6)	(175) ...	2	[SO] Compact:		16.0:		

N G C   493 - 97, 499, 502, 504, 505, 507 - 09, 511, 512   I C   1682, 1683, 1686

1	2	3	4	5	6	7	8	9	10	11	12
00946	01 21.4 + 09 18 136.19 - 52.47	01 - 04 - 048 NGC 516	1201 24	1.5 x 0.5 1.7 x 0.6	44 7	2	S0		14.3		23
00947	01 21.4 + 12 39 135.14 - 49.19	02 - 04 - 035 NGC 514	0635	4.3 x 3.1 4.3 x 3.0	(110) 3	5	Sc	SAB(rs)c Sc -	12.8	+ 2541 + 2670	*12 3468
00948	01 21.4 + 26 48 131.86 - 35.24	04 - 04 - 000	1229	1.1 x 0.4 1.0: x 0.3:	[171] ...	2	Dwarf p of 2		17.		*
00949	01 21.5 + 09 40 136.11 - 52.11	02 - 04 - 036 IC 101	0635 24	1.6 x 0.8 1.6 x 0.8	127 5	1	S ...		15.1		
00950	01 21.5 + 31 58 130.97 - 30.13	05 - 04 - 051	0030 20	1.1 x 0.4 1.0 x 0.25	68 6	1	Sa - b		15.2		*
00951	01 21.6 + 00 47 139.86 - 60.75	00 - 04 - 115	1259 23	1.3 x 0.4 1.2 x 0.4	28 ...	1	... Disturbed		15.7		
00952	01 21.6 + 09 04 136.35 - 52.69	01 - 04 - 049 NGC 518	1201 24	1.7 x 0.6 1.8 x 0.6	98 6	2	Sa		14.4		*23
00953	01 21.7 + 33 33 130.75 - 28.56	06 - 04 - 016 NGC 513	1189 20	0.75 x 0.35 0.65 x 0.35	75 ...	1	...		13.4		*3
00954	01 21.8 + 09 38 136.24 - 52.12	02 - 04 - 000 IC 102	0635 24	1.0 x 0.3 1.0 x 0.4	112 6	1	S0 - a		15.6		*
00955	01 21.8 + 15 30 134.49 - 46.37	03 - 04 - 032	1251	1.1 x 0.15 1.1 x 0.15	168 7	2	S B? c		15.7		
00956	01 21.8 + 33 13 130.83 - 28.89	05 - 04 - 052 NGC 515	0030 20	(1.3 x 1.0) (1.6 x 1.4)	— 2:	3	S0 P w 960		14.3		*13
00957	01 21.9 + 03 38 138.62 - 57.97	01 - 04 - 000	1201	1.3: x 1.0: 1. x 1.	— ...	3	Dwarf V		18.		*
00958	01 21.9 + 16 17 134.32 - 45.59	03 - 04 - 033	1251	1.8 x 0.2 1.7 x 0.2	26 7	3	S B: c		15.4		
00959	01 21.9 + 31 55 131.07 - 30.17	05 - 04 - 053	0030 20	1.2: x 1.0: 1.0: x 0.9:	(60) 1 - 2	1	Sa: Disturbed?		14.2		*1
00960	01 21.9 + 33 11 130.86 - 28.92	05 - 04 - 054 NGC 517	0030 20	2.1 x 1.0 1.9 x 0.9	20 5	3	S0 P w 956		13.6		*13 V
00961	01 22.0 - 00 18 140.66 - 61.77	00 - 04 - 000	1259 23,25	1.2 x 0.3 1.0 x 0.25	54 7	1	S ...		16.0:		
00962	01 22.0 + 01 28 139.70 - 60.06	00 - 04 - 118 NGC 521	1259 23	3.8 x 3.6 3.6: x 3.6:	— 1	4	S B b P w 992	SAB(rs)bc Sb +	12.9		*123 4678
00963	01 22.0 + 01 46 139.55 - 59.77	00 - 04 - 117 IC 103	1259 23	(1.1 x 0.8) (1.4 x 1.1)	(130) ...	1	E - S0		15.3		*
00964	01 22.0 + 07 28 137.09 - 54.23	01 - 04 - 050	1201	1.5 x 0.5 1.4 x 0.4	91 6	1	S ...		15.5		*
00965	01 22.1 - 01 50 141.63 - 63.23	00 - 04 - 119 NGC 530 = 106*	1259 25	1.7 x 0.5 1.8 x 0.45	134 7	2	Sa P w 973		14.0	+ 5016 + 5093	*
00966	01 22.1 + 03 28 138.75 - 58.04	01 - 04 - 052 NGC 520	1201	5. x 2. ...	... ...	4	Strongly peculiar Ir II		12.4	+ 2223 + 2320	*12- -678VA
00967	01 22.1 + 08 32 136.74 - 53.18	01 - 04 - 051	1201 24	1.1 x 1.1 0.6 x 0.6	— 1	3	Sc		16.0:		
00968	01 22.1 + 09 17 126.48 - 52.45	01 - 04 - 053 NGC 524	1201 24	(3.5 x 3.5) (3.5 x 3.5)	— 1	4	S0		11.5	+ 2470 + 2587	*12 3478
00969	01 22.1 + 09 30 136.40 - 52.24	02 - 04 - 037	0635 24	1.3: x 0.9 0.9 x 0.6	135: 3:	2	SBb P w 972		15.7		*
00970	01 22.1 + 09 44 136.32 - 52.01	02 - 04 - 038 NGC 522	0635 24	2.7 x 0.5 2.6 x 0.35	33 7	2	Sb - c		14.2		*23
00971	01 22.2 + 00 46 140.17 - 60.72	00 - 04 - 000	1259 23	1.3 x 0.2 0.9: x 0.2:	148 ...	2	Dwarf		18.		
00972	01 22.2 + 09 26 136.47 - 52.30	01 - 04 - 054 NGC 525	1201 24	1.6 x 0.7 1.5 x 0.7	5 6	2	S0 P w 969		14.5		*3
00973	01 22.3 - 01 52 141.75 - 63.25	00 - 04 - 122 IC 1696	1259 25	(1.1 x 1.0) (1.5 x 1.3)	— ...	3	E P w 965		14.7	+ 5768 + 5845	*
00974	01 22.3 - 01 45 141.68 - 63.14	00 - 04 - 121	1259 25	1.1 x 0.22 1.0 x 0.2	160 7	1	S0 - a		15.0	+ 4788 + 4865	
00975	01 22.4 + 34 06 130.83 - 27.99	06 - 04 - 017	1189 20	1.2 x 0.8 0.7 x 0.5	115: 3	1	S ...		15.0		
00976	01 22.5 + 00 10 140.65 - 61.28	00 - 04 - 125 IC 1697	1259 23	1.2 x 0.8 1.2: x 0.7	110 ...	1	...		14.9		*
00977	01 22.5 + 08 26 136.94 - 53.26	01 - 04 - 055 IC 1695	1201 24	(1.8 x 1.6) (1.8 x 1.7)	— ...	2	... Compact		15.4		*1
00978	01 22.5 + 14 37 134.97 - 47.20	02 - 04 - 039 IC 107	0635	1.2 x 1.0 1.2 x 0.9	(100) 2 - 3	4	Sc IC 1700 group		15.4		*1
00979	01 22.5 + 33 46 130.91 - 28.32	06 - 04 - 018 NGC 523	1189 20	3.2 x 0.8 2.6 x 0.8	[108] ...	1	... Peculiar		13.5	+ 4750 + 4798	*13 AC
00980	01 22.6 + 01 48 139.82 - 59.70	00 - 04 - 128 IC 109	1259 23	1.2 x 1.0 1.3 x 1.1	— 2	1	S B 0		14.9		*

N G C   513 - 18, 520 - 25, 530   I C   101 - 103, 106, 107, 109, 1695 - 97

1	2	3	4	5	6	7	8	9	10	11	12
00981	01 22.6 + 02 02 139.70 – 59.48	00 – 04 – 127 	1259 23	1.1: x 0.8: 1.0 x 0.7	(25) 3	1	S ...		15.5		
00982	01 22.6 + 09 00 136.77 – 52.70	01 – 04 – 056 NGC 532	1201 24	2.8: x 0.9: 3.0 x 1.1	28 6	2	Sa? Singular		13.5		*23
00983	01 22.6 + 14 35 135.02 – 47.23	02 – 04 – 040 IC 1698	0635 	1.7 x 0.8 2.0 x 0.8	120 6	2	S0 IC 1700 group		14.9		*1
00984	01 22.7 – 01 46 141.90 – 63.13	00 – 04 – 129 	1259 25	1.3 x 0.4 1.4 x 0.5	130 6	3	S0		14.8	+ 5118 + 5191	*
00985	01 22.7 + 14 16 135.14 – 47.54	02 – 04 – 042 	0635 	1.8 x 0.5 1.8 x 0.5	19 6	3	Sb		15.2		
00986	01 22.7 + 14 36 135.05 – 47.21	02 – 04 – 041 IC 1700	0635 	(2. x 2.) (2.2 x 2.1)	— ...	2	E Brightest of 3, disturbed		14.3		*1
00987	01 22.7 + 31 53 131.27 – 30.18	05 – 04 – 058 	0030 20	2.6 x 0.8 2.4 x 0.7	32 6	2	Sa		14.0		
00988	01 22.7 + 33 25 131.01 – 28.66	05 – 04 – 057 NGC 528	0030 20	1.8 x 1.1 1.8 x 1.1	55 4	3	S0		13.7		*3
00989	01 22.8 + 07 17 137.49 – 54.37	01 – 04 – 057 	1201 	1.6: x 1.3: 1.6: x 1.3:	(15) 2:	3	...	SAB(s)m	16.0:		
00990	01 22.8 + 10 33 136.32 – 51.18	02 – 04 – 044 	0635 24	1.1 x 1.1 1.5: x 1.4:	— ...	3	Dwarf: S B IV – V		17.		
00991	01 22.9 – 01 48 142.03 – 63.14	00 – 04 – 130 NGC 538	1259 25	1.5: x 0.9: 1.8: x 1.2:	(40) 3 – 4	2	Sa		14.7	+ 5398 + 5473	*
00992	01 22.9 + 01 29 140.13 – 59.99	00 – 04 – 131 NGC 533	1259 23	(4. x 3.) (5. x 4.)	(50) ...	4	E P w 962		13.1	+ 5003 + 5091	*123 46$\overline{7}$8
00993	01 22.9 + 07 44 137.36 – 53.92	01 – 04 – 058 	1201 	[1.2 x 0.9] ...	... ...	3	Double system Disrupted		15.6		*1
00994	01 22.9 + 17 48 134.26 – 44.06	03 – 04 – 000 	1259 26	1.0 x 0.8 0.9 x 0.6	(20) 2:	3	Sc		16.5:		
00995	01 22.9 + 34 28 130.88 – 27.62	06 – 04 – 019 NGC 529	1189 20	(2.4 x 2.1) (2.4 x 2.1)	(160) ...	3	E – S0 P w 1013		13.1		*13
00996	01 23.0 – 01 45 142.05 – 63.09	00 – 04 – 000 	1259 25	1.0 x 0.22 1.1 x 0.25	89 7	1	S0 – a		14.8	+ 6591 + 6666	*
00997	01 23.0 – 01 39 141.99 – 62.99	00 – 04 – 133 NGC 535	1259 25	0.9 x 0.25 0.9 x 0.25	58 ...	1	S0		14.9	+ 4939 + 5014	*3
00998	01 23.0 – 00 08 141.07 – 61.54	00 – 04 – 134 	1259 23,25	1.4 x 0.3 1.0: x 0.3:	[136] ...	1	Irr		16.5:		
00999	01 23.0 + 14 43 135.12 – 47.08	02 – 04 – 000 	0635 	1.1 x 0.3 1.0 x 0.3	62 6	1	Sc		16.5:		*
01000	01 23.0 + 50 22 128.53 – 11.87	08 – 03 – 000 	0863 22	1.1 x 0.4 1.0 x 0.3	81 6	2	Sc		16.5:		
01001	01 23.1 + 00 12 140.93 – 61.21	00 – 04 – 136 	1259 23	1.2 x 0.9 1.3 x 1.0	(150) 3	1	S ...		16.5:		
01002	01 23.1 + 17 55 134.30 – 43.94	03 – 04 – 035 IC 1701	1251 26	1.5 x 1.0 (1.6 x 1.1)	(90) ...	1	... Very compact		15.3		*
01003	01 23.2 – 01 42 142.12 – 63.03	00 – 04 – 000 	1259 25	1.0 x 0.3 0.8 x 0.4	124 6	2	S0		15.0	+ 5321 + 5396	*
01004	01 23.2 – 01 37 142.07 – 62.95	00 – 04 – 137 NGC 541	1259 25	(2.4 x 2.4) (2.6 x 2.6)	— ...	3	E P w 1007 + 1009		14.0	+ 5392 + 5465	*13 A
01005	01 23.2 + 16 20 134.74 – 45.49	03 – 04 – 036 IC 1702	1251 	1.5 x 1.2 1.4 x 1.1	(170) 2	2	Sc		14.5		*
01006	01 23.3 + 01 14 140.46 – 60.20	00 – 04 – 000 	1259 23	1.2 x 1.1 1.2 x 1.1	— 1 – 2	2	S0 – a		15.3		*
01007	01 23.4 – 01 35 142.16 – 62.90	00 – 04 – 142 NGC 545	1259 25	(3. x 2.) ...	... ...	3	S0 P w 1009		13.7	+ 5316 + 5389	*1$\underline{2}$ 3A
01008	01 23.4 + 11 11 136.34 – 50.53	02 – 04 – 047 IC 112	0635 24	0.9 x 0.45 0.8 x 0.40	[128] ...	1	S – Irr		14.2		
01009	01 23.5 – 01 36 142.22 – 62.91	00 – 04 – 143 NGC 547	1259 25	(1.5 x 1.5) ...	— ...	2	E P w 1007		13.4	+ 5472 + 5545	*1$\underline{2}$ 3A
01010	01 23.5 – 01 29 142.14 – 62.80	00 – 04 – 141 NGC 548	1259 25	(1.2 x 1.1) (1.3 x 1.2)	— ...	2	E – S0		15.1	+ 5332 + 5405	*1
01011	01 23.5 + 00 04 141.20 – 61.31	00 – 04 – 000 	1259 23	1.1 x 0.3 1.0 x 0.2	175 ...	1	S – Irr		16.5:		
01012	01 23.5 + 34 30 131.01 – 27.56	06 – 04 – 020 	1189 20	1.9 x 0.5 1.8 x 0.5	34 6	3	S B: 0 – a		14.9		*
01013	01 23.6 + 34 27 131.05 – 27.61	06 – 04 – 021 NGC 536	1189 20	3.4 x 1.7 3.5: x 1.7	62 5	4	S B b	S B(r)b Sb –	13.2		*13
01014	01 23.7 + 06 01 138.38 – 55.55	01 – 04 – 059 	1201 	1.6 x 1.6 1.6: x 1.6:	— (1)	4	Irr	S m	15.2		1
01015	01 23.7 + 09 40 136.97 – 51.99	02 – 04 – 048 IC 114	0635 24	(1.5 x 0.8) (1.6 x 0.8)	150 5 – 6	2	S0		15.7		

N G C   528, 529, 532, 533, 535, 536, 538   I C   112, 114, 1698, 1700 – 02
      541, 545, 547, 548

1	2	3	4	5	6	7	8	9	10	11	12
01016	01 23.8 - 01 54 142.57 - 63.17	00 - 04 - 144 IC 1703	1259 25	2.0: x 1.2: 2.0: x 1.3:	45: 4:	2	S0 - a		14.9	+ 5688 + 5764	*
01017	01 23.8 + 46 31 129.21 - 15.66	08 - 03 - 000	0863	1.1 x .0.8 1.0: x 0.5:	(30) ...	3	...	S B(s)dm:	17.		
01018	01 24.0 + 00 18 141.32 - 61.05	00 - 04 - 000	1259 23	1.1 x 1.0 1. x 1.	— ...	3	Dwarf V		18.		
01019	01 24.0 + 10 01 136.97 - 51.63	02 - 04 - 050	0635 24	1.2: x 0.7: 1.3: x 1.1:	— ...	2	... S B IV - V	S B m	16.5:		
01020	01 24.0 + 17 00 134.83 - 44.80	03 - 04 - 038	1251	1.4: x 0.6 1.4 x 0.6	52 6	1	S ...		15.0		*
01021	01 24.1 + 01 45 140.56 - 59.65	00 - 04 - 146 NGC 550	1259 23	(2.0 x 1.0) (2.2 x 1.1)	120 5	1	S ...		13.6		*23
01022	01 24.1 + 38 45 130.45 - 23.34	06 - 04 - 025	1189	1.2 x 1.0 1.0 x 0.8	(110) 2:	1	S ...		14.8		*
01023	01 24.2 + 11 46 136.45 - 49.91	02 - 04 - 049	0635	(1.5 x 0.9) (1.6 x 1.1)	(60) 3	2	S0 1st of 3		15.7		*1
01024	01 24.2 + 12 47 136.12 - 48.92	02 - 04 - 051	0635	(1.1 x 0.9) (1.2 x 1.0)	— ...	1	...		15.6		
01025	01 24.4 + 18 20 134.61 - 43.47	03 - 04 - 040	1251 26	1.1 x 1.1 1.2 x 1.1	— 1	2	S0 - a		15.0		
01026	01 24.5 + 13 20 136.06 - 48.37	02 - 04 - 000	0635	2.0:x 1.5:: 1.7:x 1.3::	— ...	4	Dwarf spiral S V		17.		
01027	01 24.5 + 14 31 135.70 - 47.21	02 - 04 - 052 IC 1706	0635	1.2 x 0.8 1.2 x 0.8	(165) 3 - 4	2	S ... Peculiar?		14.2		1
01028	01 24.6 + 02 00 140.67 - 59.38	00 - 04 - 000	1259 23	1.0 x 0.5 1.2 x 0.5	12 5	1	S ...		15.0		
01029	01 24.6 + 11 44 136.61 - 49.93	02 - 04 - 000	0635	1.1 x 0.25 1.1 x 0.20	24 7	1	Sa 3rd of 3		15.7		*
01030	01 24.7 - 01 31 142.79 - 62.74	00 - 04 - 150	1259 25	(1.1 x 0.6) (1.1 x 0.6)	177: ...	2	E		14.9	+ 4794 + 4869	*
01031	01 24.7 + 40 02 130.36 - 22.06	07 - 04 - 000	0824	1.3 x 0.9 1.3: x 0.9:	(135) ...	1	S B ...		15.6		*
01032	01 24.8 + 18 55 134.59 - 42.88	03 - 04 - 041 Mark 359	1251 26	0.7 x 0.5 0.7 x 0.45	(10) ...	1	... Peculiar		13.8	+ 5100	*
01033	01 24.8 + 31 18 131.89 - 30.68	05 - 04 - 060	0030 20	3.4 x 0.7 2.7 x 0.4	133 6	2	Sc		14.4		*
01034	01 24.8 + 36 55 130.90 - 25.14	06 - 04 - 027 NGC 551	1189	2.0 x 0.8 1.7 x 0.7	140 6	2	S B b - c		13.5		3
01035	01 24.8 + 48.33 129.08 - 13.63	08 - 03 - 023	0863 22	1.5 x 0.8 1.5 x 0.8	20 5	3	Sb	Sb -	15.4		
01036	01 24.9 - 02 10 143.33 - 63.34	00 - 04 - 151 NGC 560	1259 25	2.0 x 0.6 2.2 x 0.7	178 6 - 7	4	S0		14.0	+ 5503 + 5576	*12 3
01037	01 24.9 + 42 54 129.95 - 19.22	07 - 04 - 001	0824	1.2 x 0.6 1.2 x 0.6	92 5	3	Sb P w 1038		15.4		*
01038	01 24.9 + 43 00 129.94 - 19.12	07 - 04 - 000	0824	1.4: x 0.6: 1.0: x 0.5:	(30) ...	2	Dwarf: IV: or IV - V:, p w 1037		17.		*
01039	01 25 + 84 46 123.88 + 22.23	14 - 02 - 001	1277	1.7 x 0.9 1.7: x 0.7	95 5	2	Sa - b		15.4		
01040	01 25.0 - 01 21 142.84 - 62.56	00 - 04 - 152	1259 25	1.4 x 0.20 1.0 x 0.20	34 7	1	S0 - a		14.8	+ 4661 + 4736	*
01041	01 25.0 + 11 58 136.68 - 49.68	02 - 04 -(055)	0635	[1.7 x ...] ...	... ...		Chain of 6 galaxies		(15.6)		*1
01042	01 25.0 + 48 59 129.05 - 13.19	08 - 03 - 024	0863 22	1.5 x 0.7 1.4 x 0.7	95 5	2	S0		14.9		
01043	01 25.1 - 01 23 142.91 - 62.58	00 - 04 - 153	1259 25	(1.0 x 0.9) (1.0 x 0.9)	— ...	1	E:		15.1	+ 5174 + 5249	*
01044	01 25.2 - 02 08 143.47 - 63.29	00 - 04 - 154 NGC 564	1259 25	(1.9 x 1.5) (2.0 x 1.6)	(145) ...	4	E		13.8	+ 5851 + 5923	*12 3
01045	01 25.2 + 31 47 131.90 - 30.19	05 - 04 - 061	0030 20	1.3 x 0.35 1.3 x 0.4	143 6	1	S ...		15.1		
01046	01 25.2 + 37 58 130.81 - 24.09	06 - 04 - 028	1189	1.6: x 1.6: 1.3: x 1.3:	— 1	4	... S IV - V	SA m	15.7		
01047	01 25.3 - 02 17 143.63 - 63.42	00 - 04 - 157 IC 119	1259 25	1.6 x 0.8 1.6 x 0.8	77 5	2	S0 - a		15.0	+ 6203 + 6276	*
01048	01 25.4 + 34 03 131.53 - 27.95	06 - 04 - 029 NGC 561	1189 20	1.7 x 1.5 1.7 x 1.5	— 1	4	S B a	(R)S B(s)a	14.1		3
01049	01 25.4 + 48 07 129.25 - 14.04	08 - 03 - 025 NGC 562	0863 22	1.4 x 1.2 1.4 x 1.2	(20) 1 - 2	4	Sc	SA(rs)c	14.5		*
01050	01 25.5 + 10 10 137.49 - 51.41	02 - 04 - 000	0635	1.2: x 0.6: 1.3: x 0.7:	— ...	2	Sc - Irr		16.5:		

N G C   550, 551, 560 - 62, 564   I C   119, 1703, 1706

1	2	3	4	5	6	7	8	9	10	11	12
01051	01 25.5 + 28 45 132.56 – 33.17	05 – 04 – 000	0030	1.2:: x 0.9:: 1.0:: x 0.9::	— ...	3	Dwarf IV – V or V		17.		
01052	01 25.6 – 01 33 143.28 – 62.70	00 – 04 – 158 NGC 565	1259 25	1.5 x 0.5 1.4 x 0.5	36 6	2	Sa		14.5	+ 4464 + 4539	*
01053	01 25.8 + 02 15 141.10 – 59.05	00 – 04 – 159 IC 121	1259	1.1 x 0.6 1.2 x 0.7	108 4 – 5	2	Sb – c		14.3		
01054	01 25.9 + 34 05 131.64 – 27.90	06 – 04 – 000	1189 20	1.4: x 0.12 0.7 x 0.10	124 7	2	Sc		17.		
01055	01 26.0 – 01 59 143.79 – 63.08	00 – 04 – 160	1259 25	1.6 x 0.6 1.5 x 0.6	157 6	2	S B a		15.1	+ 6271 + 6344	
01056	01 26.1 + 16 25 135.69 – 45.27	03 – 04 – 044	1251 26	[1.0 x 0.9] ...	— ...	1	Irr + companion		15.0		*
01057	01 26.2 + 13 31 136.61 – 48.10	02 – 04 – 056	0635	1.6 x 0.5 1.6 x 0.4	153 6	2	Sb – c		14.9		
01058	01 26.2 + 32 05 132.08 – 29.86	05 – 04 – 062 NGC 566	0030 20	1.5 x 0.35 1.6 x 0.35	178 6	3	S0		14.6		3
01059	01 26.3 + 39 10 130.84 – 22.87	06 – 04 – 031	1189	1.6 x 1.0 1.1: x 0.7	[97] ...	3	S B? ... nf of 2, disturbed		14.7		*
01060	01 26.4 – 02 04 144.05 – 63.13	00 – 04 – 000	1259 25	1.0 x 0.25 1.0 x 0.25	81 6	1	S ...		15.7		
01061	01 26.4 – 01 12 143.46 – 62.31	00 – 04 – 162 NGC 570	1259 25	2.5: x 1.7: (2.3 x 1.7)	175 3:	3	S B 0/S B a	S B(rs)0/a	14.2	+ 5502 + 5578	*
01062	01 26.4 – 00 49 143.21 – 61.94	00 – 04 – 163	1259 25	1.6: x 0.9: 1.8: x 1.0:	60 4 – 5	1	S ...		14.0		
01063	01 26.5 + 10 53 137.61 – 50.65	02 – 04 – 053 NGC 569	0635 24	1.1 x 0.5 1.1 x 0.5	[163] ...	1	... P w 1065, disrupted		14.7		*13
01064	01 26.5 + 50 08 129.13 – 12.02	08 – 03 – 000	0863 22	1.5:: x 1.5:: 1.5:: x 1.5::	— ...	3	Dwarf spiral S IV – V or V		18.		
01065	01 26.6 + 10 53 137.65 – 50.65	02 – 04 – 054	0635 24	1.3 x 0.7 1.3 x 0.8	15 ...	3	Sc P w 1063, distorted	SAB(s)bc	15.7		*1
01066	01 26.6 + 31 49 132.23 – 30.11	05 – 04 – 000	0030 20	1.4 x 0.3 1.1: x 0.3:	44 6	2	Sc		16.5:		*
01067	01 26.7 + 51 41 128.93 – 10.48	09 – 03 – 000	0945 22	1.2 x 0.25 1.2 x 0.30	87 6	2	S B: a – b		16.5:		
01068	01 26.8 + 45 21 129.92 – 16.74	08 – 03 – 027	0863	1.9 x 1.2 1.7 x 1.2	30 3 – 4	3	Sc p of 2	Sc –	13.7		*
01069	01 27.1 + 32 15 132.27 – 29.67	05 – 04 – 063 NGC 571	0030 20	1.8: x 1.8: 1.3: x 1.3:	— 1	2	S ... Singular	SA ...	15.0		3
01070	01 27.1 + 40 43 130.74 – 21.31	07 – 04 – 002	0824	2.1 x 1.6 2.2: x 1.4:	55 3 – 4	3	Sc Disturbed		14.3		*1
01071	01 27.2 – 02 14 144.59 – 63.22	00 – 04 – 000 IC 126	1259 25	(1.5 x 1.3) (1.2 x 1.2)	— ...	1	...		15.7		
01072	01 27.2 – 01 30 144.07 – 62.53	00 – 04 – 164	1259 25	(1.7 x 1.1) (1.6 x 1.0)	(68) 3 – 4	4	S0		14.7		*
01073	01 27.3 + 25 36 133.71 – 36.20	04 – 04 – 011	1229	1.9 x 1.1 1.8: x 1.1	(163) (4)	4	Dwarf spiral S B V:		15.3		*
01074	01 27.3 + 33 04 132.16 – 28.85	05 – 04 – 000	0030 20	1.2:: x 1.2:: 1. x 1.	— ...	4	Dwarf irregular Ir V		17.		
01075	01 27.5 + 02 36 141.70 – 58.60	00 – 04 – 000	1259	1.0 x 0.4 1.1: x 0.4	94 (6)	2	...	d – m	17.		
01076	01 27.6 + 16 46 136.09 – 44.85	03 – 04 – 049	1251 26	1.1 x 0.3 1.1 x 0.3	133 7	1	S ...		15.5		
01077	01 27.8 + 19 21 135.41 – 42.32	03 – 04 – 050	1251 26	(1.2 x 0.8) (1.3 x 0.9)	(50) ...	1	E – S0		15.5		*
01078	01 27.9 + 41 00 130.85 – 21.01	07 – 04 – 000 NGC 573	0824	0.45 x 0.40 0.45 x 0.35	— ...	1	... Peculiar		13.5		*
01079	01 27.9 + 48 48 129.57 – 13.30	08 – 03 – 000	0863 22	1.1 x 0.5 0.9 x 0.4	100 5 – 6	1	S ...		16.0:		
01080	01 28.1 – 02 15 145.08 – 63.17	00 – 04 – 165 NGC 577 = 580	1259 25	2.2 x 1.8 2.0 x 1.8	(140) 2	3	S B a	S B(rs)a	14.2		
01081	01 28.1 + 21 11 135.02 – 40.50	03 – 04 – 051 IC 1710	1251	1.9 x 1.8 1.8 x 1.8	— 1	4	S B c	S B(rs)c Sc –	13.8		*
01082	01 28.2 + 16 56 136.23 – 44.66	03 – 04 – 052 IC 1711	1251 26	2.7 x 0.5 2.6 x 0.5	43 7	3	Sb		14.8		
01083	01 28.3 + 13 50 137.25 – 47.68	02 – 04 – 057	0635	1.6 x 0.8 1.6 x 0.8	125 5	3	Sb		14.7		
01084	01 28.6 + 23 42 134.53 – 38.01	04 – 04 – 000	1229	1.4:: x 1.2:: 0.6:: x 0.6::	— ...	4	Dwarf (spiral) V		17.– 18.		
01085	01 28.7 + 07 32 139.80 – 53.77	01 – 04 – 000	1201	1.4: x 0.9: 1. x 0.8:	[0] ...	3	Irr	I m Ir I	16.5:		

N G C   565, 566, 569 – 71, 573, 577, 580   I C   121, 126, 1710, 1711

1	2	3	4	5	6	7	8	9	10	11	12
01086	01 28.7 + 34 32 132.20 - 27.36	06 - 04 - 034	1189	1.0 x 0.35 1.0 x 0.4	38 6	1	S ... Distorted?		15.0		*
01087	01 28.8 + 14 01 137.36 - 47.47	02 - 04 - 058	0635	1.6 x 1.6 1.6 x 1.6	— 1	4	Sc	SA(rs)c Sc -	15.1		
01088	01 28.8 + 36 35 131.83 - 25.33	06 - 04 - 033	1189	1.7: x 1.6: 1.7: x 1.5:	— 1	3	Sc		15.7		
01089	01 28.9 + 33 22 132.47 - 28.50	05 - 04 - 064 NGC 579	0030 20	1.7: x 1.6: 1.6: x 1.6:	— 1	2	Sc P w 1094, disturbed?		13.6		*3
.01090	01 28.9 + 38 27 131.51 - 23.49	06 - 04 - 035	1189	1.2 x 1.1 1.1: x 1.1:	— 1	3	Sb		14.8		*
01091	01 29.0 + 08 11 139.64 - 53.12	01 - 05 - 000	0021	1. x 1. 1. x 1.	— ...	3	Dwarf V		18.		
01092	01 29.1 - 01 11 144.82 - 62.08	00 - 05 - 001 NGC 585	0362 25	2.7 x 0.8 2.7 x 0.8	86 7	3	Sa		14.2		*3
01093	01 29.1 + 17 19b 136.41 - 44.24	03 - 05 - 002	1275 26	1.8: x 1.4: 1.9: x 1.4:	(10) ...	2	... Brightest of 3, contact with a		(14.8)		*1
01094	01 29.1 + 33 14 132.55 - 28.62	05 - 04 - 065 NGC 582	0030 20	2.2 x 0.6 2.2 x 0.6	58 6	3	S B b P w 1089, Sb -		13.7		*3
01095	01 29.3 + 31 51b 132.88 - 29.98	05 - 04 - 067	0030 20	[1.0 x 0.4] ...	... ...	2	S ... f of 2, distorted		(15.0)	+ 12532 + 12711	*1V AC
01096	01 29.5 + 12 13 138.24 - 49.18	02 - 05 - 000	0015	1.1 x 0.3 1.0 x 0.3	150 6	1	S ...		16.0:		
01097	01 29.5 + 17 45 136.41 - 43.80	03 - 05 - 003	1275 26	(1.2 x 1.0) (1.1 x 0.9)	(35) ...	1	S: ...		16.0:		*
01098	01 29.5 + 21 10 135.44 - 40.46	03 - 05 - 004	1275	1.1 x 0.6 1.0 x 0.5	90 5	2	S B b - c	S B(r) ...	14.8		*
01099	01 29.7 + 19 23 136.00 - 42.19	03 - 05 - 005	1275 26	1.1 x 1.1 1.1 x 1.1	— ...	2	... Compact		15.3		*
01100	01 29.7 + 35 06 132.32 - 26.76	06 - 04 - 037 NGC 587 = 1713*	1189	2.2 x 0.8 2.3 x 0.9	67 6	3	Sb/S B b	SAB(s)b Sb +	13.7		13
01101	01 29.7 + 41 44 131.08 - 20.23	07 - 04 - 000	0824	(1.2 x 1.2) (1.0 x 1.0)	— 1	3	S0		15.5		*C
01102	01 30.0 + 04 20 141.89 - 56.76	01 - 05 - 002	0021	[1.6 x 1.4] ...	— ...	3	Double system Symbiotic, p w 1105		14.6		*1V A
01103	01 30.0 + 11 35 138.66 - 49.77	02 - 05 - 000	0015	1.1 x 0.15 1.0 x 0.12	156 7	2	Sc		17.		
01104	01 30.0 + 18 04 136.48 - 43.46	03 - 05 - 006	1275 26	1.2 x 0.8 1.1 x 0.8	(5) ...	1	Irr	Ir I	14.3		*
01105	01 30.2 + 04 23 141.95 - 56.70	01 - 05 - 003	0021	1.3:: x 1.0:: 1. x 1.	— ...	4	Dwarf irregular Ir V, p w 1102		17.		*1V A
01106	01 30.4 - 00 57 145.30 - 61.75	00 - 05 - 003 IC 138	0362 25	1.4 x 1.1 1.4 x 1.1	(30) 2	3	Sc	SAB(s)c Sc -	14.9		*
01107	01 30.5 - 00 04 144.74 - 60.91	00 - 05 - 000	0362	1.1: x 1.1: 1. x 1.	— 1	3	Dwarf S IV - V	S m	17.		
01108	01 30.6 + 17 21 136.89 - 44.13	03 - 05 - 000	1275 26	1.1 x 0.2 1.0 x 0.2	155 6	1	S ...		16.0:		*
01109	01 30.6 + 44 40 130.73 - 17.31	07 - 04 - 003 NGC 590	0824	3.6: x 1.8:: 3.2: x 1.8:	150: 4 - 5	4	S B 0/S B a sff of 2		14.2		*3
01110	01 30.7 + 13 05 138.36 - 48.27	02 - 05 - 001	0015	2.0 x 0.6 1.9 x 0.6	174 6	4	Sc		15.3		1
01111	01 30.7 + 35 25 132.48 - 26.42	06 - 04 - 038 NGC 591	1189	1.4 x 1.1 1.4 x 1.1	(5) 1 - 2	4	S B 0/S B a	(R')S B 0/a	14.5		*
01112	01 30.8 + 02 50 143.07 - 58.13	00 - 05 - 004	0362	1.5 x 0.7 1.5 x 0.7	108 5	1	Sc		15.0		*
01113	01 30.8 + 17 10 137.01 - 44.30	03 - 05 - 007	1275 26	1.3 x 1.3 1.1: x 1.1:	— 1	1	Sc		16.0:		
01114	01 30.8 + 18 46 136.52 - 42.74	03 - 05 - 008	1275 26	1.3 x 0.7 1.1 x 0.6	60 4 - 5	2	Sc		15.6		*
01115	01 30.9 + 12 19 138.71 - 49.00	02 - 05 - 002 IC 1715	0015	0.7 x 0.5 0.8 x 0.6	[100] ...	3	Irr	Ir I	14.5		*
01116	01 31.0 - 01 21 145.89 - 62.07	00 - 05 - 007	0362 25	1.0 x 0.22 1.0 x 0.22	118 6	1	Sc - Irr 1123 group		15.2		*
01117	01 31.0 + 30 24 133.60 - 31.34	05 - 04 - 069 NGC 598 = M 33	0030	73. x 45. 71. x 40.	23: 4	5	Sc Local group	SA(s)cd	6.5	- 186 - 11	*12 -9
01118	01 31.2 + 03 18 142.98 - 57.66	00 - 05 - 008	0362	2.5 x 0.7 2.3 x 0.7	66 6	4	Sc	SA dm	14.8		
01119	01 31.2 + 16 58 137.21 - 44.47	03 - 05 - 000	1275 26	1.8 x 1.3 1.2 x 1.1	(125) ...	3	...	S dm	16.5:		*
01120	01 31.5 - 01 20 146.13 - 62.01	00 - 05 - 009	0362 25	2.3 x 0.6 1.9 x 0.6	139 6 - 7	2	S B: a 1123 group		14.9		*1

N G C   579, 582, 585, 587, 590, 591, 598   I C   138, 1713, 1715

1	2	3	4	5	6	7	8	9	10	11	12
01121	01 31.5 + 13 15 138.58 – 48.06	02 – 05 – 003	0015	1.1 x 0.7 1.1 x 0.7	(115) 3 – 4	2	Sb 2nd of 3		16.0:		*1
01122	01 31.5 + 29 04 134.03 – 32.62	05 – 04 – 000	0030	1.1: x 0.9:: 1.0: x 1.0:	— 1 – 2	2	Sc		17.		*
01123	01 31.6 – 01 17 146.14 – 61.96	00 – 05 – 010	0362 25	1.5 x 0.5 1.4 x 0.4	71 6	1	Sa – b Brightest of 3		14.4		*1
01124	01 31.7 + 55 10 129.12 – 06.93	09 – 03 – 000	0945	1.3 x 0.3 1.0 x 0.3	40 6	3	Sc sff of 2		17.		*
01125	01 32.0 + 33 47 133.11 – 27.97	06 – 04 – 039	1189	0.65 x 0.55 0.65 x 0.55	— ...	1	S ... Peculiar		14.3		*1
01126	01 32.1 + 21 10 136.22 – 40.33	03 – 05 – 010 NGC 606	1275	1.7 x 1.5 1.8 x 1.6	— 1	4	S B c	S B(r)c Sc –	14.5		*
01127	01 32.1 + 39 40 131.94 – 22.19	07 – 04 – 000	0824	1.2: x 1.2: 0.8 x 0.7	— 1	2	...	SA dm	16.5:		*
01128	01 32.1 + 41 00 131.69 – 20.87	07 – 04 – 004 NGC 605	0824	(2.2 x 1.1) (2.2 x 1.1)	145 5	3	S0		14.3		
01129	01 32.2 + 11 50 139.37 – 49.39	02 – 05 – 005	0015	1.2 x 0.7 1.2 x 0.7	95 4	2	[S0] Compact		15.3		1
01130	01 32.4 + 01 05 144.88 – 59.67	00 – 05 – 011	0362	1.1 x 0.7 1.4: x 0.15	— 6 – 7	2	S ...		15.6		
01131	01 32.4 + 34 14 133.11 – 27.52	06 – 04 – 041	1189	1.4: x 1.2 1.1: x 1.1:	— 1 – 2	3	...	SAB: dm – m	15.4		
01132	01 32.4 + 47 18 130.58 – 14.66	08 – 04 – 001	0861	1.9: x 1.0 1.6 x 0.7	13 5 – 6	4	...	S dm	15.2		*
01133	01 32.5 + 04 06 143.10 – 56.80	01 – 05 – 005	0021	4. x 2. 4. x 2.	... ...	4	Dwarf (irregular) V		16.5:		*
01134	01 32.5 + 36 04 132.75 – 25.71	06 – 04 – 000	1189	1.1: x 0.5: 0.8: x 0.3:	[17] ...	1	Dwarf? spiral S V?		17.		
01135	01 32.6 + 33 25 133.33 – 28.31	05 – 04 – 073 NGC 608	0030	0.9 x 0.6 0.9 x 0.6	32 ...	1	... P w 1140		14.0		*3
01136	01 32.6 + 41 17 131.73 – 20.58	07 – 04 – 000	0824	1.5:: x 1.4:: 1.3:: x 1.3::	— ...	3	Dwarf (irregular) Ir: IV – V or V		17.		*
01137	01 32.7 + 05 14 142.57 – 55.70	01 – 05 – 000	0021	1.3: x 1.3: 1. x 1.	— ...	3	Dwarf V		18.		*
01138	01 32.8 + 04 37 142.94 – 56.28	01 – 05 – 006	0021	1.3 x 1.1 1.5: x 1.1	— ...	4	Irr	IAB m Ir I	16.0:		*1
01139	01 32.9 + 11 41 139.68 – 49.49	02 – 05 – 006	0015	1.1 x 0.9 1.1 x 0.9	— 2	1	S B ...		15.0		1
01140	01 33.0 + 33 26 133.42 – 28.28	05 – 04 – 075 NGC 614	0030	(1.6 x 1.5) (1.6 x 1.5)	— ...	1	S0? P w 1135		13.9		*3
01141	01 33.0 + 80 24 124.92 + 17.96	13 – 02 – 000	1214	1.5: x 0.6: 0.9 x 0.5	120 6:	1	Sb		16.0:		
01142	01 33.3 + 47 08 130.77 – 14.80	08 – 04 – 002	0861	(1.3 x 1.3) (1.5 x 1.5)	— ...	2	... Compact		15.7		
01143	01 33.4 + 00 25 145.80 – 60.21	00 – 05 – 014 NGC 622	0362	2.1 x 1.7 2.2 x 1.7	(45) 2	4	S B b	S B(r)b Sb –	14.1		*3
01144	01 33.5 + 11 26 140.00 – 49.70	02 – 05 – 007	0015	1.7 x 0.7 1.8: x 0.7:	3 6	4	Sc		16.0:		
01145	01 33.6 + 39 40 132.25 – 22.13	07 – 04 – 005	0824	1.6 x 0.7 1.6: x 0.7	157: 6	1	S B b		15.2		*
01146	01 33.7 + 10 17 140.56 – 50.79	02 – 05 – 000	0015	1.0 x 0.10 1.0 x 0.10	75 7	1	...	S d – m	17.		
01147	01 33.9 + 35 16 133.22 – 26.44	06 – 04 – 045 NGC 621	1189	(1.8 x 1.6) (1.6 x 1.4)	— 1	4	S B 0		14.2		*C
01148	01 34 + 83 43 124.31 – 21.23	14 – 02 – 005	1277	1.2 x 1.2 1.1 x 1.1	— ...	1	...		15.5		
01149	01 34.0 + 15 32 138.61 – 45.70	03 – 05 – 011 NGC 628 = M 74	1275	12. x 12. 11. x 10.	— 1	5	Sc	SA(s)c Sc –	10.5	+ 653 + 782	*12– –689
01150	01 34.0 + 42 04 131.85 – 19.76	07 – 04 – 006 NGC 620	0824	1.0 x 1.0 1.0 x 0.9	— (1)	1	... Peculiar		13.9		*3C
01151	01 34.1 – 01 38 147.65 – 62.06	00 – 05 – 015	0362	(1.1 x 0.9) (1.1 x 0.9)	(15) ...	2	E – S0 nff of 2		15.1		*1
01152	01 34.1 + 31 44 134.06 – 29.90	05 – 04 – 076	0030	1.2 x 0.6 1.1 x 0.6	177 5	2	Sb		14.6		*
01153	01 34.2 + 05 35 143.01 – 55.26	01 – 05 – 007 NGC 631	0021	(1.7 x 1.5) (1.6 x 1.4)	— ...	3	E P w 1157		15.0		*3
01154	01 34.4 + 28 38 134.87 – 32.92	05 – 04 – 077	0030	0.9 x 0.8 0.8 x 0.8	— ...	1	S? ...		14.4		*1
01155	01 34.6 + 04 38 143.70 – 56.13	01 – 05 – 009	0021	(0.9 x 0.7) (1.0 x 0.8)	[166] ...	1	...		14.5		*

N G C   605, 606, 608, 614, 620 – 22, 628, 631

1	2	3	4	5	6	7	8	9	10	11	12
01156	01 34.6 + 14 13 139.30 − 46.94	02 − 05 − 000	0015	1.2 x 1.0 0.9 x 0.8	— 1	1	Sc	SAB(rs)cd: Sc +	17.		*
01157	01 34.7 + 05 37 143.20 − 55.19	01 − 05 − 010 NGC 632	0021	1.7 x 1.3 1.5 x 1.2	170: 2 − 3	2	S0 P w 1153		13.5		*3
01158	01 34.8 + 32 25 134.07 − 29.20	05 − 04 − 000	0030	1.0: x 0.7: 0.7:x 0.6::	(155) 3:	1	S ... Compact core		16.5:		*C
01159	01 35.1 − 00 12 147.04 − 60.64	00 − 05 − 016	0362	1.1 x 0.35 1.0 x 0.3	37 (6)	1	S: ...		15.5		
01160	01 35.2 + 32 14 134.21 − 29.36	05 − 04 − 000	0030	1.7 x 0.2 1.9 x 0.2	102 7	3	Sc		16.0:		
01161	01 35.3 + 36 21 133.30 − 25.32	06 − 04 − 047	1189	(1.0 x 0.9) (0.8 x 0.7)	— ...	2	... Compact		15.0		
01162	01 35.3 + 41 24 132.24 − 20.37	07 − 04 − 007	0824	1.6 x 1.6 1.6 x 1.6	— 1	4	S B b	S B(r)b Sb +	15.1		
01163	01 35.4 + 00 45 146.50 − 59.72	00 − 05 − 017	0362	(1.0 x 0.9) (1.2 x 1.1)	— ...	2	... Compact		15.5		*
01164	01 35.4 + 35 07 133.59 − 26.53	06 − 04 − 048 NGC 634	1189	2.2 x 0.7 1.7 x 0.25	167 6	3	Sa		14.0		3
01165	01 35.7 + 28 29 135.25 − 33.01	05 − 04 − 000	0030	1.4 x 0.2 1.4 x 0.15	51 7	3	Sc		15.7		
01166	01 35.7 + 34 45 133.74 − 26.88	06 − 04 − 049	1189	1.5 x 0.45 1.4 x 0.35	69 6	2	S0 − a		14.0		
01167	01 35.8 + 07 16 142.78 − 53.53	01 − 05 − 013	0021	2.8: x 2.2: 2.6: x 2.1:	(120) 2:	4	Sc	SA(rs)cd	14.8		1
01168	01 36.2 + 48 31 131.00 − 13.35	08 − 04 − 003	0861 30	1.6 x 1.2 1.6: x 1.2:	(85) 2 − 3	1	Sb?		15.7		
01169	01 36.3 + 00 50 146.86 − 59.56	00 − 05 − 021	0362	1.4 x 0.6 1.5 x 0.7	74 5 − 6	2	S0		14.2		
01170	01 37.0 + 06 58 143.41 − 53.72	01 − 05 − 014 NGC 638	0021	0.9 x 0.6 1.0 x 0.7	20 ...	1	...		14.4		*
01171	01 37.0 + 15 39 139.57 − 45.40	03 − 05 − 000	1275	1.4: x 1.3: 1.3: x 1.2:	— ...	3	Dwarf: IV − V:, p w 1176		17.		*
01172	01 37.1 + 05 32 144.24 − 55.08	01 − 05 − 015	0021	1.0 x 0.22 0.9 x 0.3	27 (7)	1	... P w 1177		15.0		*
01173	01 37.1 + 46 18 131.59 − 15.50	08 − 04 − 000	0861	1.1 x 0.3 0.9: x 0.2:	96 6	2	...	S dm − m	17.		
01174	01 37.3 − 02 17 149.77 − 62.35	00 − 05 − 025	0362	1.3: x 0.6 1.1 x 0.6	103: 5	2	Sb − c		15.1		*
01175	01 37.3 + 10 51 141.64 − 50.00	02 − 05 − 000	0015	1.2 x 1.1 1. x 1.	— (1)	3	... S IV − V		17.		
01176	01 37.4 + 15 39 139.70 − 45.38	03 − 05 − 000	1275	5. x 4. 4. x 3.	— ...	4	Dwarf irregular Ir V, p w 1171		17.		*
01177	01 37.6 + 05 28 144.48 − 55.10	01 − 05 − 016 NGC 645	0021	3.0 x 1.3 3.1 x 1.4	125 5 − 6	2	S B: b P w 1172		13.8		*3
01178	01 37.6 + 34 23 134.26 − 27.15	06 − 04 − 053	1189	2.0 x 0.35 1.9 x 0.3	55 7	2	Sc		14.8		*
01179	01 37.7 + 43 37 132.24 − 18.11	07 − 04 − 000	0824	1.7: x 0.9: (1.7 x 1.0)	— ...	1	... Compact		15.6		*C
01180	01 37.8 − 00 33 148.59 − 60.71	00 − 05 − 027	0362	1.1 x 1.0 (1.1 x 1.1)	— ...	1	... Compact:		16.0:		*
01181	01 37.8 + 14 17 140.36 − 46.67	02 − 05 − 000	0015	1.1 x 0.2 1.0 x 0.2	133 6 − 7	2	Sc np of 2		17.		*
01182	01 37.8 + 49 00 131.17 − 12.83	08 − 04 − 000	0861 30	1.1 x 1.1 1.0 x 1.0	— 1	2	Sc		15.6		*
01183	01 37.9 + 02 26 146.48 − 57.93	00 − 05 − 028	0362	1.5 x 0.8 1.6: x 0.9:	75: 4 − 5	1	S ...		16.0:		
01184	01 38.1 + 07 43 143.46 − 52.93	01 − 05 − 017 NGC 652	0021	1.2 x 0.7 1.1 x 0.6	55 4 − 5	1	S ...		14.7		*1
01185	01 38.2 + 44 45 132.10 − 16.98	07 − 04 − 008	0824	(1.5 x 0.9) (1.5 x 0.9)	(135) ...	2	... Compact		15.0		*
01186	01 38.7 + 48 47 131.37 − 13.01	08 − 04 − 004	0861 30	1.6 x 1.3 1.6 x 1.3	— ...	1	... Peculiar		15.7		*
01187	01 38.8 + 08 15 143.45 − 52.37	01 − 05 − 019 IC 1721	0021	1.2 x 0.5 1.0 x 0.4	100 6	1	S: ...		14.3		*
01188	01 39.2 + 22 26 137.92 − 38.71	04 − 05 − 000	0896	1.1: x 0.5: 1.0 x 0.3	[22] ...	1	S? ... ssp of 2		15.7		*
01189	01 39.3 + 06 52 144.38 − 53.64	01 − 05 − 021	0021	1.3 x 0.5 1.3 x 0.4	19 6	1	S ...		15.5		*
01190	01 39.5 + 02 33 147.10 − 57.67	00 − 05 − 031	0362	1.0 x 0.2 0.9 x 0.15	100 6	1	Sc		16.5:		

N G C 632, 634, 638, 645, 652

1	2	3	4	5	6	7	8	9	10	11	12
01191	01 39.5 + 07 26 / 144.15 – 53.09	01 – 05 – 023	0021	[1.0 x 1.0] / ...	— / ...	2	Double system		16.0:		*1V
01192	01 39.5 + 12 20 / 141.77 – 48.42	02 – 05 – 009 / NGC 658	0015 / 29	3.5 x 1.7 / 3.3: x 1.7:	20: / 5	3	Sb	Sb +	13.6		*1
01193	01 39.5 + 35 23 / 134.44 – 26.09	06 – 04 – 058 / NGC 653	1189	1.6 x 0.22 / 1.6 x 0.2	39 / 7	3	Sa – b		14.1		
01194	01 39.6 + 25 53 / 136.98 – 35.34	04 – 05 – 002 / NGC 656	0896	1.5 x 1.3 / 1.5 x 1.3	(35) / 1 – 2	3	S B 0		13.5		3
01195	01 39.7 + 13 43 / 141.24 – 47.08	02 – 05 – 011	0015 / 29	3.5 x 1.1 / 3.5 x 1.1:	[50] / ...	3	Irr / Distorted?		13.9		*1
01196	01 39.8 + 06 28 / 144.80 – 53.97	01 – 05 – 000	0021	1.1 x 1.0 / 1.0: x 1.0:	— / 1	3	Sb	SA(s)b	16.5:		*
01197	01 39.8 + 18 03 / 139.58 – 42.91	03 – 05 – 000	1275	2.0 x 0.6 / 2.0 x 0.6	[56] / ...	2	Irr	Ir I	15.3		*
01198	01 40 + 85 01 / 124.19 + 22.53	14 – 02 – 006	1277	1.0 x 0.8 / 0.9 x 0.8	(70) / ...	1	... / Eruptive		15.1		*C
01199	01 40.1 + 07 55 / 144.13 – 52.58	01 – 05 – 024	0021	1.1 x 0.4 / 1.0 x 0.4	78 / 6	1	Sa – b		15.6		
01200	01 40.1 + 12 54 / 141.73 – 47.84	02 – 05 – 012	0015 / 29	2.1 x 1.5 / 1.9 x 1.5:	[170] / ...	4	Irr	I B: m / Ir I	14.3		*1
01201	01 40.3 + 13 23 / 141.59 – 47.36	02 – 05 – 013 / NGC 660	0015 / 29	10. x 4.5 / 10. x 4.	[170] / ...	4	S B [a] / Singular, distorted?		12.8		*12 / 35
01202	01 40.4 + 03 57 / 146.56 – 56.29	01 – 05 – 026 / IC 150	0021	1.0 x 0.55 / 1.0 x 0.5	143 / 4 – 5	1	Sb / NGC 664 group		14.9		*
01203	01 40.4 + 05 35 / 145.56 – 54.76	01 – 05 – 000	0021	1.6: x 0.8: / 1.6: x 0.8:	(35) / ...	1	...		16.5:		
01204	01 40.6 + 04 02 / 146.59 – 56.19	01 – 05 – 027	0021	1.4 x 0.4 / 1.6 x 0.4	41 / 6 – 7	2	Sb – c / NGC 664 group		15.5		*
01205	01 40.6 + 08 38 / 143.94 – 51.86	01 – 05 – 028 / IC 1723	0021	3.6 x 0.9 / 3.5 x 0.9	29 / 6	2	Sb		14.2		*
01206	01 40.7 + 11 39 / 142.50 – 48.99	02 – 05 – 014	0015 / 29	(1.8 x 1.0) / (1.5 x 1.0)	(30) / ...	1	...		15.3		
01207	01 41 + 81 58 / 124.91 + 19.56	14 – 02 – 000	1277	1.8:: x 1.5:: / 1.7:: x 1.3::	— / ...	2	Dwarf		16.5:		
01208	01 41.0 + 01 50 / 148.26 – 58.20	00 – 05 – 032	0362	1.1 x 0.5 / 1.2 x 0.6	12 / 5	1	Sa – b		15.7		
01209	01 41.1 + 11 55 / 142.52 – 48.70	02 – 05 – 015	0015 / 29	1.5 x 0.7 / 1.4 x 0.7	65 / (5)	1	S – Irr / Disturbed?		15.0		*
01210	01 41.2 + 03 59 / 146.88 – 56.18	01 – 05 – 029 / NGC 664	0021	1.8 x 1.5 / 2.0: x 1.6:	(65) / 1 – 2	1	Sb: / Brightest of 3		13.9		*3
01211	01 41.2 + 13 33 / 141.83 – 47.14	02 – 05 – 016	0015 / 29	3. x 2. / 2. x 2.	— / ...	3	Dwarf / IV – V		17.		
01212	01 41.2 + 34 09 / 135.13 – 27.22	06 – 04 – 059	1289	1.7 x 1.4: / 1.5: x 1.2:	(95) / 2:	2	Sb		14.5		
01213	01 41.3 + 31 05 / 135.95 – 30.20	05 – 05 – 004	0413 / 32	1.0: x 0.9: / (1.3 x 1.1)	(120) / (1 – 2)	2	S0		14.9		
01214	01 41.4 + 02 06 / 148.24 – 57.92	00 – 05 – 033	0362	(1.7 x 1.6) / (1.9 x 1.8)	— / 1	4	S0	SAB 0	14.0		*
01215	01 41.4 + 28 28 / 136.70 – 32.74	05 – 05 – 005 / NGC 661	0413 / 32	(2.0 x 1.6) / (2.0 x 1.6)	(60) / ...	3	E		13.0		*23
01216	01 41.4 + 40 41 / 133.60 – 20.84	07 – 04 – 000	0824	1.0:: x 0.8:: / 0.8:: x 0.8::	— / ...	2	Dwarf: spiral / S IV – V:		18.		
01217	01 41.4 + 42 47 / 133.12 – 18.79	07 – 04 – 010	0824	1.0 x 0.8 / 1.1 x 0.9	— / 2:	1	S ...	SAB ...	15.6		*
01218	01 41.6 + 12 00 / 142.66 – 48.59	02 – 05 – 017	0015 / 29	1.0 x 0.5 / 0.9 x 0.4	140 / 5	2	S B: b – c		15.5		
01219	01 41.6 + 17 13 / 140.46 – 43.60	03 – 05 – 014	1275	1.4 x 0.6 / 1.4 x 0.6	102 / 5 – 6	2	S B ... / Peculiar	(R')S B(s) ...	13.6		*1
01220	01 41.6 + 37 26 / 134.41 – 24.00	06 – 04 – 060 / NGC 662	1189	0.8 x 0.45 / 0.9 x 0.50	(20) / ...	1	S ... / Peculiar		13.6		*C
01221	01 41.7 + 37 57 / 134.30 – 23.49	06 – 04 – 061	1189	1.3 x 0.9 / 1.0 x 0.7	145: / 3	3	Sb – c		15.0		*
01222	01 41.9 + 04 25 / 146.89 – 55.72	01 – 05 – 000	0021	1.2 x 0.5 / 0.7 x 0.5	85 / ...	1	...		16.5:		
01223	01 42.3 + 10 10 / 143.79 – 50.27	02 – 05 – 019 / NGC 665	0015	(2.4 x 1.6) / (2.6 x 1.7)	(125) / 3 – 4:	2	S0 / Brightest in group	(R?) ...	13.5		*13
01224	01 42.3 + 31 54 / 135.97 – 29.36	05 – 05 – 000	0413 / 32	1.1 x 0.10 / 1.1 x 0.1	125 / 7	3	Sc		17.		*
01225	01 42.4 – 00 32 / 150.72 – 60.24	00 – 05 – 034	0362	[1.0 x 0.7] / ...	[20] / ...	3	Double system / Disrupted		16.5:		

N G C 653, 656, 658, 660 – 62, 664, 665   I C 150, 1723

1	2	3	4	5	6	7	8	9	10	11	12
01226	01 42.4 + 10 07 143.85 - 50.31	02 - 05 - 021	0015	1.1 x 0.8 1.1 x 0.8	(80) 3:	3	... S IV - V	S m	16.5:		*1
01227	01 42.4 + 31 49 136.02 - 29.43	05 - 05 - 006	0413 32	1.0: x 1.0: 0.9: x 0.8:	— 1	3	Sc		15.6		*
01228	01 42.5 + 28 29 136.98 - 32.66	05 - 05 - 007	0413 32	1.8 x 0.8 1.4 x 0.7	[25] ...	2	S - Irr Interacting systems?		15.1		*
01229	01 42.6 + 10 24 143.78 - 50.03	02 - 05 - 023 IC 154	0015	1.3: x 0.22 1.3: x 0.25	66 7	2	Sb NGC 665 group		14.8		*1
01230	01 42.7 + 25 16 138.00 - 35.77	04 - 05 - 005	0896	2.3:: x 2.3:: 1.5:: x 1.3::	— (1)	4	Dwarf spiral S V		17.		
01231	01 42.8 + 10 18 143.91 - 50.11	02 - 05 - 025 IC 156	0015	1.8 x 1.6 1.8 x 1.6	— 1	1	S ... NGC 665 group		15.0		*1
01232	01 42.8 + 40 58 133.81 - 20.50	07 - 04 - 000	0824	1.1: x 0.3: 1.0: x 0.3:	10 6	3	Sc		17.		
01233	01 42.9 + 28 34 137.05 - 32.56	05 - 05 - 008	0413 32	1.0 x 0.7 0.9 x 0.6	110 3	3	Sc		14.7		*
01234	01 42.9 + 34 52 135.33 - 26.44	06 - 05 - 001	1189 31	1.2 x 0.7 1.0: x 0.7:	(170) ...	3	Sc/S B c Distorted?	SAB(s)c	14.8		*
01235	01 43.0 + 03 07 148.21 - 56.82	00 - 05 - 035	0362	1.1 x 0.5 1.0 x 0.5	143 5	2	Sb - c		14.9		
01236	01 43.2 + 34 08 135.59 - 27.14	06 - 05 - 002 NGC 666	1189	0.8 x 0.5 0.6 x 0.35	80 ...	1	...		13.6		*C
01237	01 43.3 + 09 34 144.46 - 50.76	02 - 05 - 026	0015	1.8 x 0.8 2.6: x 0.9:	45 6	1	Sa - b Brightest of 4		15.5		*1
01238	01 43.4 + 36 12 135.10 - 25.12	06 - 05 - 003 NGC 668	1225	2.2 x 1.6 2.2: x 1.5:	30: 3	2	Sb		13.5		1
01239	01 43.5 + 06 10 146.45 - 53.94	01 - 05 - 031	0021	1.3 x 1.1 1.2 x 0.9:	(155) ...	1	Sc - Irr		15.1		
01240	01 43.7 + 04 01 147.89 - 55.92	01 - 05 - 032	0021	1.3 x 0.5 1.0 x 0.3	91 (6)	1	S - Irr Distorted:		14.9		*
01241	01 43.8 + 48 10 132.35 - 13.44	08 - 04 - 000	0861 30	1.3: x 0.6: 1.0: x 0.5:	17: (5)	2	...	S d - dm	17.		
01242	01 43.9 + 14 26 142.35 - 46.10	02 - 05 - 027	0015 29	1.3 x 1.3 1.3: x 1.2:	— (1)	3	...	S: m	15.7		*
01243	01 44.0 + 18 19 140.79 - 42.38	03 - 05 - 015	1275	1.5 x 0.8 1.0 x 0.5	135 ...	1	S0?		15.0		*
01244	01 44.0 + 24 13 138.69 - 36.71	04 - 05 - 000	0896 28	1.4 x 0.6 1.3 x 0.6:	80 ...	1	S: ...		15.5		
01245	01 44.2 + 11 10 143.98 - 49.18	02 - 05 - 028	0015 29	1.8 x 1.1 1.7 x 1.1	165: 3 - 4	1	S ... Compact nucleus		15.0		*
01246	01 44.3 + 12 10 143.52 - 48.22	02 - 05 - 030	0015 29	1.6 x 1.6 1.7: x 1.7:	— (1)	4	Irr 	IA m Ir I	15.1		*1V
01247	01 44.3 + 12 52 143.19 - 47.56	02 - 05 - 029 NGC 671	0015 29	1.7: x 0.6 1.1 x 0.5	55 6	1	S ...		14.3		
01248	01 44.3 + 35 18 135.53 - 25.95	06 - 05 - 004 NGC 669	1225 31	3.3 x 0.7 3.0 x 0.5	36 6 - 7	2	Sa - b		12.9		*
01249	01 44.6 + 27 05 137.93 - 33.90	04 - 05 - 009 IC 1727	0896	8.0 x 3.0 7.0: x 2.8:	150 ...	5	... NGC 672 group, Ir I	S B(s)m	12.2	+ 362 + 518	*12 59$\overline{V}$
01250	01 44.6 + 27 38 137.76 - 33.37	05 - 05 - 012 NGC 670	0413 32	(2.2 x 0.8) (2.2 x 0.8)	172 (6)	3	S0		13.1		*12 3478
01251	01 44.6 + 35 47 135.47 - 25.47	06 - 05 - 000	1225 31	1.0 x 0.5 0.9 x 0.5	[44] ...	1	... Peculiar, drop-shaped		15.0		*
01252	01 44.7 + 16 03 141.92 - 44.50	03 - 05 - 016	1275	1.2 x 0.9 1.1 x 0.9	(5) 2 - 3	3	... S(B) IV - V:	SAB(rs)dm	16.5:		*
01253	01 44.8 + 11 52 143.84 - 48.47	02 - 05 - 031	0015 29	1.6 x 0.7 1.3 x 0.5	18 6	1	S ...		15.5		*1
01254	01 44.9 + 48 23 132.48 - 13.19	08 - 04 - 006	0861 30	1.0 x 0.8 1.0 x 0.8	(5) 2	2	Sc		16.5:		
01255	01 45.0 + 11 06 144.29 - 49.18	02 - 05 - 032	0015 29	1.2 x 0.8 1.3 x 0.8	85 4	4	...	S dm	15.7		
01256	01 45.0 + 27 11 138.01 - 33.78	04 - 05 - 011 NGC 672	0896	7.2 x 2.8 7.3 x 2.8	65 6	4	S B c Brightest in group	S B(s)d	11.4	+ 340 + 496	*12- -9
01257	01 45.2 + 36 12 135.49 - 25.04	06 - 05 - 006	1225 31	1.1 x 0.55 1.2 x 0.6	107 5	3	Sa - b		15.0		
01258	01 45.3 + 35 10 135.78 - 26.04	06 - 05 - 000	1225 31	[1.2 x 0.2] ...	[2] ...	1	Double system Bridge		16.0:		*
01259	01 45.7 + 11 17 144.45 - 48.95	02 - 05 - 033 NGC 673	0015 29	2.3 x 1.7 2.3 x 1.7	(0) 2 - 3	4	Sc	SAB(s)c Sc -	13.3		*3
01260	01 45.8 + 12 21 143.96 - 47.93	02 - 05 - 000	0015 29	0.9 x 0.7 1.0 x 1.0	— 1	4	S B a	(R')S B(s)a	14.0		*

N G C   666, 668 - 73

1	2	3	4	5	6	7	8	9	10	11	12
01261	01 45.8 + 13 10 143.57 - 47.16	02 - 05 - 034	0015 29	1.4 x 0.3 1.4 x 0.3	92 6	2	Sc		15.5		*
01262	01 45.8 + 13 27 143.47 - 46.89	02 - 05 - 000	0015 29	1.4: x 0.8: 1.2: x 0.7:	--- ...	2	Irr P w 1264		16.0:		*
01263	01 45.9 + 10 19 145.02 - 49.84	02 - 05 - 035	0015 29	1.1 x 0.9 1.2 x 0.9	(140) 2 - 3	1	S ... SAB(s) ... IC 162 group		16.0:		*1
01264	01 45.9 + 13 31 143.44 - 46.82	02 - 05 - 000	0015 29	1.2 x 0.3 1.0 x 0.3	77 6	2	Sc P w 1262		16.0:		*
01265	01 46.0 + 20 00 140.75 - 40.63	03 - 05 - 017	1275	[1.1 x 0.6] ...	[170] ...	3	S B ... + companion Connected		14.5		*1
01266	01 46.1 + 10 15 145.12 - 49.89	02 - 05 - 036 IC 162	0015 29	1.0 x 0.7 1.1 x 0.7	[65] ---	1	... Brightest in group		14.2		*1 V
01267	01 46.2 + 10 16 145.15 - 49.86	02 - 05 - 038	0015 29	(1.7 x 1.7) (1.9 x 1.9)	--- 1	3	SO		14.8		*1 V A
01268	01 46.2 + 10 20 145.12 - 49.80	02 - 05 - 037	0015 29	1.0 x 0.25 1.2 x 0.3	46 6	1	S ... IC 162 group		15.3		*1
01269	01 46.2 + 34 44 136.10 - 26.41	06 - 05 - 007	1225 31	(1.1 x 0.7) (1.2 x 0.7)	105: ...	1	E - SO 1272 group		15.4		*1
01270	01 46.3 + 05 40 147.86 - 54.15	01 - 05 - 034 NGC 676	0021	5.0 x 1.8 4.8 x 1.8	172 6	1	SO - a Brightest of 3		10.5		*12 3
01271	01 46.3 + 12 57 143.84 - 47.32	02 - 05 - 040	0015 29	1.7 x 1.1 1.8 x 1.2	95: 3 - 4	3	S B O		14.6		
01272	01 46.3 + 34 50 136.09 - 26.31	06 - 05 - 009	1225 31	(1.5 x 0.9) (1.1 x 0.8)	23: (4)	1	SO: Brightest of 4		14.3		*
01273	01 46.4 + 12 48 143.94 - 47.46	02 - 05 - 041 NGC 675	0015 29	1.2 x 0.5 1.1 x 0.5	99 6	1	S ... P w 1275		15.5		*1
01274	01 46.5 + 12 36 144.07 - 47.64	02 - 05 - 043	0015 29	1.7 x 0.7 1.6 x 0.6	108 6	1	Sa:		15.2		*
01275	01 46.5 + 12 48 143.98 - 47.45	02 - 05 - 042 NGC 677	0015 29	(2.0 x 2.0) (2.1 x 2.1)	--- ...	3	E P w 1273		14.3		*1
01276	01 46.5 + 20 27 140.72 - 40.17	03 - 05 - 018 IC 163	1275	2.0 x 0.9 1.9 x 0.9	95 (5 - 6)	3	... S B dm		13.8		*
01277	01 46.5 + 35 12 136.04 - 25.94	06 - 05 - 011	1225 31	2.0: x 1.2: 1.7: x 1.0:	75 4:	1	SO - a		14.5		1
01278	01 46.6 + 12 28 144.17 - 47.76	02 - 05 - 044	0015 29	1.0 x 0.5 1.1: x 0.5	105 5	3	Sc		15.7		*
01279	01 46.7 + 13 07 143.89 - 47.13	02 - 05 - 045	0015 29	1.0 x 0.25 1.0 x 0.25	148 6	1	S ...		15.3		
01280	01 46.7 + 21 45 140.30 - 38.91	04 - 05 - 014 NGC 678	0896	5.0 x 1.1 4.9 x 1.3	78 7	3	Sb NGC 697 group, p w 1286		13.3		*12 3
01281	01 46.7 + 32 20 136.89 - 28.71	05 - 05 - 014	0413 32	4.7 x 1.0 4.7: x 1.0:	38 7	3	... S dm S IV:		13.0		*1
01282	01 46.8 + 12 15 144.35 - 47.95	02 - 05 - 046	0015 29	1.6 x 0.7 1.6 x 0.7	55 6	2	SO - a		14.2		*
01283	01 46.8 + 35 32 136.01 - 25.61	06 - 05 - 012 NGC 679	1225 31	(1.8 x 1.8) (2.1 x 2.1)	--- ...	2	E - SO		13.1		*3 C
01284	01 46.9 + 28 14 138.16 - 32.66	05 - 05 - 000	0413 32	1.0 x 0.1 0.6 x 0.1	42 7	3	Sc Sc +		17.		
01285	01 47 + 86 27 123.97 + 23.96	14 - 02 - 007	1277	1.5 x 1.2 1.5 x 1.2	170: 2	3	Sb		14.5		
01286	01 47.0 + 21 43 140.40 - 38.92	04 - 05 - 015 NGC 680	0896	(2.7 x 2.4) (2.4 x 2.4)	--- ...	2	E NGC 697 group, p w 1280		13.0		*12 3
01287	01 47.0 + 22 08 140.24 - 38.52	04 - 05 - 000	0896	1.2:: x 1.2:: 1. x 1.	--- ...	3	Dwarf V, NGC 697 group		18.		
01288	01 47.1 + 11 27 144.85 - 48.68	02 - 05 - 047 NGC 683	0015 29	1.0 x 1.0 1.0 x 1.0	--- 1	1	S ... spp of 2		14.8		*13
01289	01 47.2 + 16 10 142.67 - 44.20	03 - 05 - 019	1275	1.4 x 0.9 1.3 x 0.9	(120) 3 - 4	3	Sc		16.0:		
01290	01 47.3 + 11 34 144.87 - 48.55	02 - 05 - 049	0015 29	1.0 x 0.7 1.0 x 0.7	(140) 3	1	Sb - c		15.1		
01291	01 47.3 + 26 57 138.67 - 33.87	04 - 05 - 018 IC 1731	0896	1.7 x 1.1 1.7 x 1.1	140: 3 - 4	4	Sc NGC 672 group		14.2		2
01292	01 47.3 + 27 24 138.53 - 33.44	04 - 05 - 017 NGC 684 = 165*	0896	3.4 x 0.9 3.5 x 0.9	90 6	3	Sb NGC 672 group		13.2		*23
01293	01 47.4 + 01 50 151.02 - 57.56	00 - 05 - 036	0362	1.1 x 0.7 1.2 x 0.7	143: 4	1	Sa - b		15.0		
01294	01 47.4 + 22 54 140.08 - 37.76	04 - 05 - 000	0896 28	1.2:: x 1.2:: 1. x 1.	--- ...	3	Dwarf V		18.		
01295	01 47.4 + 33 15 136.78 - 27.79	05 - 05 - 015	0413 32	1.0: x 1.0: 1.4: x 1.3:	--- 1	2	SO		15.6		*

N G C   675 - 80, 683, 684   I C   162, 163, 165, 1731

1	2	3	4	5	6	7	8	9	10	11	12
01296	01 47.5 – 00 59 153.42 – 60.09	00 – 05 – 037	0362	1.0 x 0.22 1.0 x 0.3	105 7	1	S ...		15.5		
01297	01 47.6 + 02 03 150.93 – 57.35	00 – 05 – 038	0362	1.2 x 0.9 1.3 x 1.1:	— ...	4	Irr	IB(s)m Ir I	15.2		
01298	01 47.6 + 36 10 136.03 – 25.00	06 – 05 – 014 NGC 687	1225 31	(1.4 x 1.4) (1.6 x 1.6)	— 1	2	S0		13.3		3
01299	01 47.7 + 35 07 136.33 – 25.96	06 – 05 – 013	1225 31	1.0 x 0.6 1.0: x 0.6:	[85] ...	3	Dwarf IV – V or V		15.7		*
01300	01 47.8 + 08 09 146.91 – 51.71	01 – 05 – 000	0021	[1.5 x 0.7] ...	... ...	3	Double system Bridge		16.5:		*
01301	01 47.8 + 32 50 137.00 – 28.17	05 – 05 – 016 IC 1733	0413 32	(1.5 x 1.4) (1.7 x 1.6)	— ...	3	[E] Compact		14.7		*
01302	01 47.8 + 35 02 136.37 – 26.04	06 – 05 – 015 NGC 688	1225 31	2.6 x 1.7 2.4 x 1.3	145 3 – 4	4	S B b	S B(s)b Sb +	13.3		*13
01303	01 47.8 + 48 06 133.03 – 13.35	08 – 04 – 007	0861 30	1.5: x 1.3 1.6: x 1.3	— 1 – 2	2	S B a		15.6		
01304	01 47.9 + 05 54 148.34 – 53.78	01 – 05 – 035 NGC 693	0021	(3.2 x 1.6) (3.2 x 1.7)	106 5 – 6	1	S ... NGC 676 group		13.5		*12 3
01305	01 47.9 + 21 30 140.73 – 39.07	04 – 05 – 019 NGC 691	0896	3.8 x 2.7 3.7 x 2.6	95 3	4	Sb/Sc NGC 697 group, Sb +	SA(rs)bc	13.5		*12 3
01306	01 47.9 + 32 18 137.18 – 28.68	05 – 05 – 017	0413 32	1.3 x 1.1 1.2 x 1.0	(70) 2	2	S0	(R)SAB 0	15.0		*
01307	01 47.9 + 35 40 136.22 – 25.42	06 – 05 – 016 IC 1732	1225 31	1.7 x 0.5 1.6 x 0.4	62 6 – 7	1	S ...		15.1		
01308	01 47.9 + 36 01 136.12 – 25.08	06 – 05 – 017	1225 31	(2.3 x 2.3) (2.3 x 2.3)	— ...	3	E		14.5		
01309	01 48.2 + 18 03 142.18 – 42.34	03 – 05 – 020 IC 1736	1275	1.4 x 0.35 1.2 x 0.30	33 6	1	Sb – c		15.0		
01310	01 48.2 + 21 45 140.73 – 38.81	04 – 04 – 020 NGC 694, Mark 363	0896	0.55 x 0.35 0.55 x 0.35	[160] ...	1	... Peculiar		13.9	+ 3000	*12 3C
01311	01 48.2 + 29 34 138.07 – 31.30	05 – 05 – 019	0413 32	1.7: x 1.0: 1.4: x 0.9:	(20) ...	3	Dwarf IV – V or V		17.		
01312	01 48.3 + 13 03 144.47 – 47.07	02 – 05 – 051	0015 29	1.5 x 0.4 1.5 x 0.4	96 6	1	S ...		15.1		
01313	01 48.3 + 21 40 140.79 – 38.88	04 – 05 – 021 IC 167	0896	3.0 x 1.7 3.0 x 1.6	[95] ...	4	S B [c] Singular, NGC 697 group	SAB(s) ...	14.0		*2A
01314	01 48.4 + 12 21 144.85 – 47.72	02 – 05 – 052	0015 29	1.5 x 1.0: 1.6 x 0.9	160 (3 – 4)	4	... S B IV, nf of 2	S B(s)dm	16.0:		*
01315	01 48.4 + 22 20 140.57 – 38.24	04 – 05 – 000 NGC 695	0896	0.50 x 0.45 0.50 x 0.45	[40] ...	1	... Peculiar, NGC 697 group		13.7		*23 C
01316	01 48.4 + 34 36 136.63 – 26.43	06 – 05 – 019	1225 31	1.4: x 0.8: 0.9: x 0.4:	178: 4 – 5	1	S ... s of 2		16.5:		*1
01317	01 48.5 + 22 06 140.68 – 38.45	04 – 05 – 022 NGC 697	0896	5.1 x 1.7 5.1 x 1.7	105 6	4	Sc Brightest in group		12.7		*12 3
01318	01 48.5 + 32 47 137.17 – 28.18	05 – 05 – 020	0413 32	(1.2 x 0.9) (1.5 x 1.2)	(135) ...	3	[E] Compact		15.4		
01319	01 48.5 + 35 49 136.31 – 25.25	06 – 05 – 020	1225 31	0.9 x 0.6 0.9 x 0.6	(155) ...	1	S? ... Peculiar		14.5		*1
01320	01 48.5 + 41 35 134.77 – 19.66	07 – 04 – 000	0824	1.2 x 0.12 1.4 x 0.1	153 7	3	Sc n of 2		16.5:		*
01321	01 48.6 – 01 17 154.18 – 60.23	00 – 05 – 000	0362	(1.1 x 1.1) (1.1 x 1.1)	— 1	2	S0 nnp of 2		16.0:		*
01322	01 48.7 + 12 53 144.68 – 47.19	02 – 05 – 053	0015 29	1.5 x 1.3 1.7: x 1.7:	— 1	4	Sc	SA(s)c Sc –	15.5		*1
01323	01 48.8 + 11 38 145.35 – 48.36	02 – 05 – 000	0015	1.6 x 0.5 1.4: x 0.5:	[147] ...	3	Dwarf IV – V or V		17.		
01324	01 48.8 + 18 51 142.04 – 41.53	03 – 05 – 021	1275	1.6 x 1.1 1.5 x 1.0	90: 3	2	S B b	Sb +	14.6		*
01325	01 48.9 + 08 00 147.41 – 51.75	01 – 05 – 038	0021	(1.8 x 1.8) (2.0 x 2.0)	— ...	2	E P w 1326		14.2		*
01326	01 49.0 + 08 03 147.42 – 51.69	01 – 05 – 037	0021	(1.1 x 1.1) (1.2 x 1.2)	— ...	1	E: P w 1325		15.0		*
01327	01 49.0 + 30 18 138.04 – 30.55	05 – 05 – 023	0413 32	1.4: x 0.7: 1.4: x 0.7:	100 ...	2	E – S0 p of 2		14.9		*
01328	01 49.1 + 16 48 142.99 – 43.46	03 – 05 – 023	1275	1.6 x 1.1 1.6 x 1.1	125 3	3	Sc	SAB:(rs)c	14.7		
01329	01 49.1 + 17 55 142.51 – 42.40	03 – 05 – 022	1275	2.4 x 0.8 2.1 x 0.9:	128 ...	3	S B: c Distorted?		15.7		*
01330	01 49.1 + 34 48 136.72 – 26.20	06 – 05 – 000	1225 31	1.3 x 1.3 1.3: x 1.3:	— 1	4	...	SA(s)dm	16.0:		

N G C　687, 688, 691, 693 – 95, 697　　　　　　　　　　　　　　　I C　167, 1732, 1733, 1736

1	2	3	4	5	6	7	8	9	10	11	12
01331	01 49.1 + 42 40 143.60 – 18.58	07 – 04 – 011	0824	1.0 x 0.6 1.0 x 0.6	20 4	4	Sc		14.7		
01332	01 49.1 + 47 50 133.31 – 13.56	08 – 04 – 008	0861 30	(1.8 x 1.6) (2.5 x 2.3)	(95) ...	3	E P w 1340		15.7		*
01333	01 49.2 + 00 01 153.28 – 59.01	00 – 05 – 041	0362	1.5 x 1.0 1.3 x 0.8	(105) 3 – 4	3	S ... Peculiar		15.7		*
01334	01 49.2 + 06 03 148.74 – 53.52	01 – 05 – 040 NGC 706	0021	2.1 x 1.7 2.2 x 1.8	— 2	1	Sc: Singular, NGC 676 group		13.2		*23
01335	01 49.2 + 17 18 142.81 – 42.98	03 – 05 – 000	1275	1.1 x 0.8 1.1 x 0.8	— 2 – 3	2	... P w 1342	S B dm	17.		*
01336	01 49.2 + 35 50 136.45 – 25.19	06 – 05 – 000 NGC 700	1225 31	1.2 x 0.3 1.2 x 0.3	10 7:	3	SO		15.6		*3C
01337	01 49.3 + 08 35 147.21 – 51.17	01 – 05 – 039	0021	1.1 x 0.3 1.1 x 0.3	32 6	3	Sc	Sc +	16.5:		1
01338	01 49.4 + 35 33 136.58 – 25.46	06 – 05 – 025	1225 31	1.0 x 0.8 0.9 x 0.7	(75) 2	3	Sb P w 1339		15.2		*1
01339	01 49.5 + 35 36 136.58 – 25.40	06 – 05 – 026	1225 31	1.2: x 1.1: 1.3: x 1.1:	— 1	3	S B 0 P w 1338	SB(r)0 +	15.0		*1
01340	01 49.5 + 47 51 133.38 – 13.53	08 – 04 – 009	0861 30	(1.0 x 0.9) (1.0 x 0.9)	— ...	2	E P w 1332		16.5:		*
01341	01 49.6 + 31 45 137.74 – 29.11	05 – 05 – 025	0413 32	1.3 x 1.2 1.3 x 1.2	— 1	4	Sc/S B c	SAB(s)d	15.6		*
01342	01 49.7 + 17 16 142.98 – 42.97	03 – 05 – 024 NGC 711	1275	1.7 x 0.8 1.7 x 0.8	15 5	3	SO P w 1335		14.5		*
01343	01 49.7 + 35 52 136.55 – 25.13	06 – 05 – 028 NGC 704	1225 31	[0.6 x 0.5] ...	— ...	1	Double system Symbiotic, NGC 708 group		14.1		*13 C
01344	01 49.7 + 36 15 136.44 – 24.76	06 – 05 – 027	1225 31	1.7 x 0.8 1.9 x 0.9	22 5	4	S B a	(R)S B a	14.0		1
01345	01 49.8 + 35 55 136.56 – 25.10	06 – 05 – 030 NGC 705	1225 31	1.2 x 0.25 1.2 x 0.25	117 7	2	SO – a NGC 708 group		14.5		*13 C
01346	01 49.8 + 35 56 136.55 – 25.07	06 – 05 – 029 NGC 703	1225 31	(1.2 x 0.9) (1.2 x 0.9)	(50) ...	2	E – SO NGC 708 group		14.5		*13
01347	01 49.8 + 36 22 136.43 – 24.65	06 – 05 – 032	1225 31	1.4 x 1.2 1.3 x 1.2	— 1	4	Sc/S B c	SAB(rs)c	13.9		*C
01348	01 49.9 + 35 55 136.58 – 25.08	06 – 05 – 031 NGC 708	1225 31	(3.0 x 2.5) (3.5 x 3.0)	(35) ...	2	E Brightest in group		14.8		*13
01349	01 50.0 + 35 48 136.64 – 25.18	06 – 05 – 033 NGC 710	1225 31	1.7: x 1.7: 1.4: x 1.3:	— 1	4	Sc		14.3		*3
01350	01 50.0 + 36 15 136.51 – 24.75	06 – 05 – 034	1225 31	2.1: x 1.5: 1.9 x 1.5:	(55) 3	4	S B b	S B(r)b	14.5		*
01351	01 50.2 + 12 28 145.40 – 47.46	02 – 05 – 054 IC 1743	0015	2.1 x 0.9 2.1 x 0.9	57 5 – 6	2	S B: a		14.0		
01352	01 50.2 + 36 34 136.46 – 24.43	06 – 05 – 035 NGC 712	1225 31	(1.3 x 1.0) (1.2 x 0.9)	85 3	2	SO		13.9		*3
01353	01 50.4 + 36 43 136.46 – 24.28	06 – 05 – 036	1225 31	(1.2 x 1.0) (1.2 x 1.0)	(110) ...	2	E – SO		14.4		*C
01354	01 50.5 – 01 20 155.08 – 60.06	00 – 05 – 043	0362	1.0 x 0.4 1.0 x 0.4	67 6	2	Sa – b		15.7		*
01355	01 50.5 + 43 43 134.59 – 17.50	07 – 05 – 001	0931	1.7 x 1.6 1.6 x 1.3	— 1	3	Sb/S B b	SAB(s)b	14.2		
01356	01 50.6 + 03 57 150.74 – 55.31	01 – 05 – 041 NGC 718	0021	(2.6 x 2.2) (3.0 x 2.7)	45: 1 – 2	4	S B a	(R')SAB(s)a	12.5	+ 1802 + 1876	*12 3468
01357	01 50.6 + 19 42 142.22 – 40.60	03 – 05 – 025	1275	1.3 x 0.3 1.2 x 0.3	46 7	2	Sa		15.4		
01358	01 50.6 + 35 58 136.72 – 24.99	06 – 05 – 037 NGC 714	1225 31	1.6 x 0.35 2.1 x 0.35	112 6 – 7	2	SO – a		13.9		3
01359	01 50.8 + 29 41 138.67 – 31.03	05 – 05 – 026	0413 32	0.9 x 0.8 0.9 x 0.8	— 1	2	S B ...	S B(s) ...	14.2		
01360	01 50.9 + 19 35 142.36 – 40.68	03 – 05 – 026 NGC 719 = 1744*	1275	(1.4 x 1.1) (1.4 x 1.1)	(150) ...	1	SO?		14.7		3
01361	01 50.9 + 36 20 136.68 – 24.62	06 – 05 – 000	1225 31	1.2 x 0.4 1.2 x 0.4	130 6	3	Sc nnf of 2		15.7		*
01362	01 51.0 + 14 32 144.64 – 45.45	02 – 05 – 000	0015	1. x 1. 1. x 1.	— ...	3	Dwarf (spiral) S: V		18.		
01363	01 51.0 + 35 59 136.80 – 24.95	06 – 05 – 041 NGC 717	1225 31	1.5 x 0.22 1.4 x 0.2	117 7	2	SO – a		14.7		3
01364	01 51.3 + 14 41 144.66 – 45.29	02 – 05 – 000	0015	1.5: x 0.9: 1.3: x 0.8:	(30) 4:	4	S B c		16.5:		
01365	01 51.4 – 01 00 155.16 – 59.66	00 – 05 – 044	0362	1.2 x 0.45 1.1 x 0.5	118 6	1	S ... P w 1367		14.6		*

N G C   700, 703 – 06, 708, 710 – 12, 714, 717 – 19   I C   1743, 1744

1	2	3	4	5	6	7	8	9	10	11	12
01366	01 51.4 + 36 22 136.77 – 24.56	06 – 05 – 042	1225 31	1.7 x 0.45 1.7 x 0.5	140 6	2	S B c		14.7		*
01367	01 51.5 – 01 00 155.20 – 59.64	00 – 05 – 000	0362	1.1 x 0.4 1.0: x 0.3:	10 ...	3	Dwarf: IV – V or V		17.		*
01368	01 51.5 + 07 37 148.61 – 51.86	01 – 05 – 042	0021	1.6 x 0.6 1.5 x 0.5	53 6	1	Sa – b		15.0		*1
01369	01 51.5 + 17 51 143.28 – 42.28	03 – 05 – 027	1275	1.6 x .1.2 1.7 x 1.3	(50) 2 – 3	2	Sc Brightest of 3		14.9		*
01370	01 51.6 + 13 17 145.45 – 46.58	02 – 05 – 055	0015	1.0 x 0.4 1.0 x 0.4	153 6	1	Sa – b		15.6		
01371	01 51.7 + 04 33 150.74 + 54.65	01 – 05 – 043 IC 1746	0021	1.5 x 0.5 1.5 x 0.5	93 6	2	S0		15.1		
01372	01 51.7 + 17 25 143.53 – 42.68	03 – 05 – 029	1275	1.2 x 0.5 1.1 x 0.4	14 6	1	S ...		15.1		
01373	01 51.8 + 05 12 150.32 – 54.04	01 – 05 – 044	0021	1.2 x 0.4 1.0 x 0.3	80 6	1	S: ... 1st of 3		16.0:		*
01374	01 51.8 + 16 37 143.92 – 43.42	03 – 05 – 030	1275	1.2 x 0.8 1.0: x 0.8	0 2 – 3	3	S B c		15.7		
01375	01 51.8 + 19 49 142.53 – 40.40	03 – 05 – 028	1275	1.2 x 1.1 1.4: x 1.2	— 1	3	S B c		15.0		*1
01376	01 51.8 + 39 09 136.07 – 21.86	06 – 05 – 043 NGC 721	1225	1.8 x 1.0 1.8 x 1.1	135: 4 – 5	4	S B b	S B(rs)bc Sb +	13.8		*12 3
01377	01 51.9 + 16 48 143.87 – 43.24	03 – 05 – 000	1275	1.3 x 0.15 1.1 x 0.1	78 6 – 7	1	S B: c		16.5:		
01378	01 51.9 + 73 02 127.62 + 11.00	12 – 03 – 001	1230	5.0: x 3.4: 3.5 x 1.4	5: ...	2	S B a	(R:)S B(rs)a	14.8		1
01379	01 52.0 + 20 27 142.33 – 39.78	03 – 05 – 032 NGC 722	1275	2.1 x 0.7 1.8 x 0.6	138 6	1	S ...		14.6		*1
01380	01 52.0 + 37 05 136.69 – 23.84	06 – 05 – 000	1225 31	1.0: x 0.6: 0.7 x 0.7	(165) ...	1	S ...		15.6		*
01381	01 52.0 + 48 47 133.56 – 12.52	08 – 04 – 000	0861	1.2: x 0.7: 0.9 x 0.6	₁07 4:	2	Sc npp of 2 Sc +		16.5:		*
01382	01 52.1 – 00 24 154.91 – 59.04	00 – 05 – 048	0362	(1.4 x 1.4) (1.6 x 1.5)	— ...	1	E?		14.7		*
01383	01 52.1 + 05 45 150.03 – 53.48	01 – 05 – 046	0021	1.4 x 0.2 1.7 x 0.22	35 7	2	Sc		16.0:		*
01384	01 52.1 + 17 52 143.45 – 42.22	03 – 05 – 000	1275	1.7 x 0.7 1.4: x 0.6:	8 6	2	...	S B: d – dm	17.		
01385	01 52.1 + 36 41 136.83 – 24.22	06 – 05 – 044 Mark 2	1225 31	0.8 x 0.6 0.8 x 0.6	(170) 2 – 3	3	S B a Brightest of 3	(R)S B 0/a	14.2	+ 5400	*1
01386	01 52.2 + 13 14b 145.68 – 46.57	02 – 05 – 056	0015	1.0 x 0.15 1.0 x 0.18	166 7	1	Sb – c n of 2		(15.7)		*1
01387	01 52.2 + 36 01 137.05 – 24.86	06 – 05 – 049	1225 31	1.1 x 0.9 1.0 x 0.7	(175) ...	2	S – Irr Distorted?		15.4		*
01388	01 52.3 + 35 02 137.36 – 25.80	06 – 05 – 050 IC 171	1225 31	(2.5 x 2.2) (2.2 x 1.7)	(105) ...	1	...		13.8		*1
01389	01 52.3 + 47 43 133.88 – 13.55	08 – 04 – 010	0861 30	(1.8 x 1.6) (2.0 x 1.8)	— ...	2	E		15.6		
01390	01 52.4 + 36 03 137.08 – 24.81	06 – 05 – 051	1225 31	1.0 x 0.22 1.0 x 0.20	45 7	1	S ...		15.5		*
01391	01 52.5 + 09 46 147.67 – 49.77	02 – 05 – 059	0015	1.5 x 0.15 1.4 x 0.15	177 7	2	Sc		15.5		
01392	01 52.5 + 36 19 137.02 – 24.55	06 – 05 – 052	1225 31	1.2 x 0.6 (1.1 x 0.6)	43: ...	1	S0?		16.0:		*
01393	01 52.6 + 21 02 142.26 – 39.18	03 – 05 – 033	1275	(1.0 x 0.5) (0.9 x 0.5)	65: ...	1	E – S0 P w 1396		15.6		*
01394	01 52.6 + 46 34 134.23 – 14.65	08 – 04 – 011	0861 30	2.3 x 1.3 2.3 x 1.3	70 4 – 5	3	Sa		15.0		*
01395	01 52.7 + 06 21 149.88 – 52.90	01 – 05 – 048	0021	1.7 x 1.5 1.6 x 1.5	— 1	3	Sb	SA(rs)b	14.5		*
01396	01 52.8 + 21 05 142.30 – 39.12	03 – 05 – 034	1275	1.1 x 0.5 1.1 x 0.5	94 5 – 6	2	S0 P w 1393		15.1		*
01397	01 53.0 + 32 45 138.22 – 27.96	05 – 05 – 027	0413 32	1.1 x 0.5 1.1 x 0.4	36 6	2	S0		15.0		*
01398	01 53.0 + 36 53 136.96 – 23.98	06 – 05 – 054	1225 31	1.4 x 1.2 1.2 x 1.1	— 1	1	Sc:		14.9		
01399	01 53.1 + 17 48 143.78 – 42.21	03 – 06 – 001	1202	1.0 x 0.7 0.8 x 0.4	2 3	2	S0 nf of 2		15.1		*
01400	01 53.2 + 35 53 137.30 – 24.93	06 – 05 – 055	1225 31	2.5 x 0.35 2.6 x 0.35	156 7	3	Sb		13.8		*

1	2	3	4	5	6	7	8	9	10	11	12
01401	01 53.4 - 00 34 155.62 - 59.04	00 - 05 - 000	0852	1.0 x 0.12 1.0 x 0.12	145 7	2	Sc		16.5:		*
01402	01 53.4 + 01 02 154.19 - 57.62	00 - 06 - 001 IC 173	0852	1.1 x 0.9 1.1 x 0.9	(110) 2:	3	S B b	S B(s)b	14.9		*
01403	01 53.4 + 17 24 144.05 - 42.56	03 - 06 - 002 IC 1748	1202	1.2 x 0.8 0.9 x 0.7	130 3	2	Sb/S B c	Sb +	14.7		
01404	01 53.4 + 36 59 137.02 - 23.86	06 - 05 - 056	1225 31	1.4 x 0.8 0.7 x 0.6	100: 6:	3	S B b	S B(s)b	15.6		*
01405	01 53.4 + 37 12 136.95 - 23.65	06 - 05 - 000	1225 31	1.2 x 0.12 1.1 x 0.1	125 7	3	Sc		15.7		*
01406	01 53.5 + 36 34 137.16 - 24.26	06 - 05 - 057 NGC 732	1225 31	1.6 x 1.1 1.7: x 1.2:	(10) 3	3	S0		14.9		
01407	01 53.6 + 06 30 150.12 - 52.67	01 - 06 - 001 IC 1749	1282	1.2 x 0.9 1.1 x 0.8	155: ...	1	S0:		14.8		
01408	01 53.6 + 12 55 146.31 - 46.75	02 - 06 - 002	0209	1.2 x 0.6 1.0 x 0.6	90 5	3	Sc	Sc +	15.7		
01409	01 53.7 + 03 31 152.29 - 55.37	01 - 06 - 008 IC 174	1282	(1.7 x 1.2) (1.5 x 1.0)	(95) ...	1	S0?		14.6		*
01410	01 53.7 + 04 24 151.63 - 54.57	01 - 06 - 007	1282	1.8 x 0.22 1.5 x 0.2	93 6 - 7	3	Sc	Sc +	15.5		
01411	01 53.7 + 33 56 138.00 - 26.78	06 - 05 - 058 NGC 735	1225	1.8 x 0.8 1.8 x 0.8	138 6	2	Sb		13.9		*13 C
01412	01 53.8 + 03 50 152.09 - 55.07	01 - 06 - 000 IC 1750	1282	1.0 x 0.22 1.1 x 0.22	64 7	2	S0 - a		15.4		*
01413	01 53.8 + 05 23 150.96 - 53.67	01 - 06 - 003 NGC 741	1282	(3. x 3.) (3. x 3.)	— ...	3	E Brightest in group		13.2	+ 5559 + 5636	*123 46̲78 VC
01414	01 53.8 + 32 48 138.38 - 27.86	05 - 05 - 028 NGC 736	0413 32	(1.8 x 1.6) (2.0 x 1.7)	— ...	3	E P w 1421		13.6	+ 4366 + 4531	*12̲ 3 C̲
01415	01 53.8 + 36 08 137.36 - 24.66	06 - 05 - 060	1225 31	1.1 x 0.25 1.4 x 0.25	1 6	3	S0 - a		14.5		
01416	01 53.8 + 36 39 137.20 - 24.16	06 - 05 - 059	1225 31	1.3 x 0.7 1.2 x 0.7	65 4 - 5	1	S ...		14.9		*
01417	01 53.9 + 17 28 144.18 - 42.46	03 - 06 - 000	1202	1.3 x 0.08 1.1 x 0.05	153 7	2	...	S d - m	18.		
01418	01 53.9 + 40 06 136.22 - 20.83	07 - 05 - 002	0931	(1.7 x 1.3) (1.9 x 1.6)	50 2:	3	S0		14.5		
01419	01 54.0 + 10 03 148.03 - 49.37	02 - 06 - 000	0209	1.2: x 0.8: 1.1: x 0.8:	(5) 4 - 5:	2	...	SAB(s)dm	17.		
01420	01 54.0 + 14 46 145.49 - 44.98	02 - 06 - 003	0209	1.0 x 0.6 1.0 x 0.6	3 4:	1	S ... Disturbed?		14.9		*1
01421	01 54.0 + 32 46 138.44 - 27.88	05 - 05 - 031 NGC 740	0413 32	1.7 x 0.3 1.1 x 0.25	137 7:	1	S ... P w 1414		14.9		*2̲3
01422	01 54.2 + 32 33 138.55 - 28.08	05 - 05 - 000	0413 32	1.3 x 0.6 1.2 x 0.7	[90] ...	1	S: ... Peculiar, double system?		14.3		*
01423	01 54.2 + 74 52 127.31 + 12.81	12 - 03 - 000	1230	1.3: x 0.5: 1.3 x 0.5	72 6	1	S: ...		17.		
01424	01 54.3 + 03 47 152.33 - 55.06	01 - 06 - 000	1282	(1.1 x 1.1) (1.0 x 1.0)	— ...	2	... Compact		15.4		
01425	01 54.3 + 05 32 151.04 - 53.48	01 - 06 - 010	1282	[1.7? x 0.5] ...	... ...	1	... Long jet, disturbed	Compact	14.6		*1
01426	01 54.4 - 02 16 157.71 - 60.39	00 - 06 - 004 IC 176	0852	2.4 x 0.6 2.1 x 0.6	94 6	3	Sc		15.0		*
01427	01 54.4 + 05 48 150.90 - 53.23	01 - 06 - 012	1282	1.1 x 0.7 1.0 x 0.6	65 4	2	S0		14.7		
01428	01 54.5 + 14 19 145.88 - 45.36	02 - 06 - 005 IC 1755	0209	1.6 x 0.4 1.6 x 0.35	154 6 - 7	2	Sa		14.8		*
01429	01 54.6 - 00 43 156.27 - 59.02	00 - 06 - 005 IC 1756	0852	1.5 x 0.25 1.5 x 0.25	155 7	3	Sc p of 2		15.5		*
01430	01 54.6 + 32 58a 138.51 - 27.65	05 - 05 - 034 NGC 750	0413 32	[2.5 x 1.7]	...	2	E + E		12.9	+ 5130 + 5295	*123VA 46̲78 C
01431	01 54.6 + 32 58b 138.51 - 27.65	05 - 05 - 035 NGC 751	0413 32	...	...	2	Compact cores, bridge			+ 5126 + 5291	*12̲VA 36̲C
01432	01 54.7 + 16 58 144.65 - 42.86	03 - 06 - 004	1202	1.0 x 0.7 1.0 x 0.7	[65] ...	2	Sb - c nf of 2, disturbed, bridge		14.8		*1V
01433	01 54.7 + 19 41 143.43 - 40.30	03 - 06 - 005	1202	1.1: x 1.1: 0.6 x 0.6	— 1	2	Sc Brightest of 3, disturbed:		15.7		*
01434	01 54.7 + 36 00 137.59 - 24.74	06 - 05 - 064	1225 31	1.5: x 0.5: 1.5 x 0.5	164 6	2	S0 - a		15.4		*
01435	01 54.8 + 05 22 151.35 - 53.58	01 - 06 - 014	1282	1.4 x 0.7 0.9 x 0.5	40: 5	1	S ...		15.0		

N G C   732, 735, 736, 740, 741, 750, 751

I C   173, 174, 176, 1748 - 1750
        1755, 1756

1	2	3	4	5	6	7	8	9	10	11	12
01436	01 54.8 + 19 39 143.48 – 40.33	03 – 06 – 007	1202	1.1: x 0.9: 0.7 x 0.4	— 2:	2	Sc 1433 group, disturbed		16.0:		*1
01437	01 54.8 + 35 40 137.71 – 25.05	06 – 05 – 066 NGC 753	1225 31	3.3 x 2.1 3.2: x 2.1:	125 3 – 4	4	Sc P w 1440	SAB:(rs)bc Sc –	12.6	+ 4766 + 4938	*123 4568
01438	01 54.8 + 44 41 135.11 – 16.37	07 – 05 – 003 NGC 746	0931	2.2 x 1.6 2.1 x 1.6	[90] ...	4	Irr	I B m Ir I	13.8		
01439	01 54.9 + 33 08 138.53 – 27.48	05 – 05 – 036 NGC 761	0413 32	1.7 x 0.5 1.9 x 0.7	143 6	1	S B 0 – a		14.5		*23
01440	01 54.9 + 36 05 137.61 – 24.65	06 – 05 – 067 NGC 759	1225 31	(1.8 x 1.8) (1.7 x 1.7)	— ...	3	E P w 1437		13.7		*23
01441	01 54.9 + 37 07 137.29 – 23.65	06 – 05 – 065	1225 31	1.1 x 0.18 1.1 x 0.15	134 6 – 7	2	Sb		15.5		
01442	01 55.0 – 02 20 158.04 – 60.38	00 – 06 – 006	0852	1.6: x 1.2: 1.5: x 1.2:	— 2:	4	...	SAB(s)dm	15.7		*
01443	01 55.0 + 03 29 152.83 – 55.26	00 – 06 – 000	0852	1.2 x 0.8 0.6 x 0.4	167: 3 – 4	1	S ...		16.5:		
01444	01 55.4 + 04 09 152.48 – 54.61	01 – 06 – 017	1282	2.1 x 0.7 1.8 x 0.6	140 (7)	1	S? ...		15.6		*
01445	01 55.4 + 18 52 144.00 – 41.02	03 – 06 – 009	1202	1.2 x 0.8 1.1 x 0.7	90 3 – 4	3	S0		15.1		*
01446	01 55.5 + 03 08 153.30 – 55.52	00 – 06 – 008	0852	[1.5 x 0.5] ...	[51] ...	3	Irr + companion Contact		14.9		*
01447	01 55.5 + 44 20 135.34 – 16.67	07 – 05 – 004	0931	1.2: x 1.2: 1.1: x 1.0:	— 1	3	Sc	SAB cd	15.1		*
01448	01 55.6 + 01 50 154.40 – 56.66	00 – 06 – 011	0852	1.3 x 0.35 1.3 x 0.3	110 6	3	Sc		15.1		
01449	01 55.6 + 02 50b 153.58 – 55.77	00 – 06 – 009	0852	2.2: x 0.9: 2. x 0.8:	... ...	4	Irr Contact, disrupted	f of 2	(14.0)		*1V A
01450	01 55.6 + 14 03 146.37 – 45.51	02 – 06 – 006	0209	(1.3 x 1.3) (1.3 x 1.3)	— ...	1	...		15.6		*
01451	01 55.6 + 25 07 141.48 – 35.09	04 – 05 – 024	0896	1.3 x 0.7 1.4 x 0.7	(85) ...	1	S B b? P w 1462		14.3		*
01452	01 55.7 + 21 54 142.78 – 38.13	04 – 05 – 000	0896	1.4: x 0.3: 1.2 x 0.3	128 (6)	1	...		17.		*
01453	01 55.9 + 24 25 141.83 – 35.73	04 – 05 – 000	0896	1.7:: x 1.4:: 1.4:: x 1.4::	— ...	3	Dwarf V		17.		
01454	01 56.0 + 03 01 153.59 – 55.56	00 – 06 – 014	0852	1.9: x 1.6: 1.7: x 1.6:	— 1 – 2	3	... S B IV – V	S B m	16.5:		
01455	01 56.0 + 24 39 141.77 – 35.50	04 – 05 – 025 NGC 765	0896	3. x 3. 3. x 3.	— 1	3	S B b/Sc	SAB(rs)bc Sb +	14.2		*3
01456	01 56.0 + 36 26 137.73 – 24.25	06 – 05 – 070 IC 178	1225 31	1.4 x 0.9 1.4 x 0.9	170: 3 – 4	2	Sa – b		14.0		*
01457	01 56.1 + 00 17 155.97 – 57.97	00 – 06 – 016 NGC 768	0852	2.1 x 1.2 2.1 x 1.1	30 5	3	Sb sp of 2, disturbed?		14.3		*
01458	01 56.1 + 08 06 149.98 – 50.96	01 – 06 – 019 NGC 766	1282	(2. x 2.) (2. x 2.)	— ...	3	E Brightest of 4		14.4		*13
01459	01 56.1 + 35 49 137.94 – 24.83	06 – 05 – 072	1225 31	1.6 x 0.20 1.5 x 0.18	109 7	3	Sc		15.4		
01460	01 56.1 + 36 01 137.88 – 24.64	06 – 05 – 071	1225 31	1.5 x 0.9 1.5 x 0.9	145 4	2	Sa		15.0		
01461	01 56.2 + 05 21 151.89 – 53.45	01 – 06 – 022	1282	1.3 x 0.2 1.2 x 0.2	172 7	2	Sc		16.0:		
01462	01 56.3 + 25 09 141.66 – 35.01	04 – 05 – 026	0896	1.7 x 1.4 1.7 x 1.4	65: 2:	3	S B c P w 1451		15.6		*
01463	01 56.4 + 18 43 144.36 – 41.08	03 – 06 – 010 NGC 770	1202	1.1 x 0.8 1.0 x 0.6	15 ...	1	... P w 1466		14.2		*12 3
01464	01 56.5 + 01 40 154.91 – 56.70	00 – 06 – 000	0852	1.8:: x 1.2:: 1.6:: x 1.1::	— ...	4	Dwarf irregular Ir V		17.		
01465	01 56.5 + 17 47 144.82 – 41.95	03 – 06 – 000	1202	1.0 x 0.10 0.9 x 0.1	8 (6 – 7)	1	S B? ...		17.		
01466	01 56.5 + 18 46 144.37 – 41.02	03 – 06 – 011 NGC 772	1202	8. x 5. 7. x 4.	130 3 – 4	5	Sb Singular, p w 1463, Sb +		11.3	+ 2431 + 2552	*12– –8A̅
01467	01 56.7 + 30 40 139.76 – 29.73	05 – 05 – 037 NGC 769	0413 32	0.9 x 0.45 1.0 x 0.45	73 5	2	S ... n of 2		13.4		*1
01468	01 56.8 + 13 42 146.94 – 45.73	02 – 06 – 007	0209	1.7 x 1.2 1.6 x 1.1	145: 3:	3	Sc P w 1469	Sc +	16.0:		*
01469	01 56.9 + 13 46 146.94 – 45.66	02 – 06 – 008 NGC 774	0209	(2.0 x 1.5) (2.1 x 1.6)	165 2 – 3	2	S0 P w 1468		14.4		*3
01470	01 56.9 + 31 50 139.41 – 28.60	05 – 05 – 000	0413 32	1.7 x 0.2 1.7: x 0.2	138 7	3	...	S d – m	16.5:		*

N G C   746, 753, 759, 761, 765, 766, 768 – 70
772, 774

I C   178

1	2	3	4	5	6	7	8	9	10	11	12
01471	01 57.1 + 23 24 142.56 − 36.61	04 − 05 − 028 NGC 776	0896	1.9 x 1.9 1.9 x 1.9	— 1	4	S B b/Sb	SAB(rs)b Sb +	13.4		*13
01472	01 57.1 + 34 06 138.71 − 26.42	06 − 05 − 000	1225	1.3: x 1.2: 1. x 1.	— ...	3	Dwarf V		18.		
01473	01 57.2 + 07 10 150.99 − 51.70	01 − 06 − 026 IC 182	1282	2.0:: x 1.7:: 1.5: x 1.3:	— 1 − 2	2	S B b	Sb +	14.6		*
01474	01 57.2 + 37 21 137.70 − 23.30	06 − 05 − 076	1225 31	1.6 x 1.2 1.2: x 1.2:	15: 1 − 3	4	... S B III − IV	S B(s)dm	15.0		
01475	01 57.2 + 37 47 137.57 − 22.88	06 − 05 − 075 IC 179	1225 31	(1.8 x 1.5) (1.8 x 1.5)	(110) ...	3	E		13.4		
01476	01 57.3 + 31 10 139.73 − 29.21	05 − 05 − 038 NGC 777	0413 32	(2.8 x 2.2) (2.9 x 2.2)	(155) ...	3	E	P w 1480	12.7		*12 3478
01477	01 57.4 + 06 43 151.37 − 52.08	01 − 06 − 000	1282	1.2 x 1.0 1.1 x 1.0	— 1 − 2	3	...	SAB dm	15.7		
01478	01 57.4 + 24 00 142.40 − 36.02	04 − 05 − 032	0896	1.0 x 0.9 1.0 x 1.0	— 1	2	S B c sf of 2	S B(rs)c	14.6		*
01479	01 57.4 + 24 14 142.30 − 35.80	04 − 05 − 034	0896	1.1 x 0.25 1.4 x 0.25	176 7	1	S ...		14.8		*
01480	01 57.4 + 31 03 139.79 − 29.32	05 − 05 − 039 NGC 778	0413 32	1.1 x 0.5 1.0 x 0.5	150 ...	1	S0: P w 1476		14.2		*3
01481	01 57.5 − 00 31 157.29 − 58.49	00 − 06 − 000	0852	1.2 x 0.3 1.0 x 0.3	147 6	1	Sa − b		15.6		*
01482	01 57.5 + 12 24 147.88 − 46.87	02 − 06 − 010 NGC 781	0209	1.7 x 0.5 1.6 x 0.4	13 7	1	S ...		14.0		3
01483	01 57.5 + 22 39 142.97 − 37.29	04 − 05 − 000	0896	1.3:: x 1.3:: ...	— ...	3	Dwarf V		18.		
01484	01 57.6 + 15 44 146.14 − 43.77	03 − 06 − 000	1202 35	1.0 x 0.15 0.8 x 0.1	107 7	1	...	d − m	17.		
01485	01 57.6 + 20 52 143.75 − 38.96	03 − 06 − 012	1202	1.1 x 0.8 1.4 x 0.9	[25] ...	3	... Peculiar, disrupted pair?		14.6		*1
01486	01 57.6 + 24 20 142.32 − 35.69	04 − 05 − 033 IC 1764	0896	1.7 x 1.7 1.6: x 1.6:	— 1	3	S B b	S B(r)b Sb +	14.5		*
01487	01 57.6 + 24 50 142.12 − 35.21	04 − 05 − 035	0896	1.2 x 0.9 1.1 x 0.9	(60) ...	2	[S0] Compact		15.7		
01488	01 57.7 + 27 59 140.95 − 32.22	05 − 05 − 041 NGC 780	0413 32	(1.8 x 1.0) (1.9 x 1.1)	170: ...	1	...		14.6		*3C
01489	01 57.7 + 42 14 136.34 − 18.59	05 − 05 − 000	0413 32	1.2: x 0.9: 1.1: x 0.8:	(120) 2 − 3	3	Sc	,	17.		
01490	01 57.8 + 21 03 143.73 − 38.77	03 − 06 − 000	1202	0.50 x 0.22 0.45 x 0.20	85 ...	1	... Peculiar; ssf of 2		14.1		*
01491	01 57.8 + 29 29 140.44 − 30.78	05 − 05 − 000	0413 32	1. x 1. 1. x 1.	— ...	3	Dwarf (spiral) S: V		18.		
01492	01 57.8 + 45 55 135.30 − 15.04	08 − 04 − 012	0861	1.5 x 0.4 1.5 x 0.4	3 6	2	S0 − a		14.9		
01493	01 57.9 + 37 58 137.66 − 22.67	06 − 05 − 077	1225 31	2.3 x 0.8 2.2 x 0.8	87 6	2	S B? a − b		14.0		*
01494	01 58.1 + 17 04 145.64 − 42.48	03 − 06 − 014	1202 35	1.3 x 0.7 1.1 x 0.5	83 5	1	S ...		16.0:		
01495	01 58.1 + 17 29 145.44 − 42.09	03 − 06 − 015	1202	1.0 x 0.8 0.9 x 0.6	25: 2:	2	Sc		15.6		
01496	01 58.2 + 14 56 146.74 − 44.46	02 − 06 − 011	0209 35	(2.0 x 1.2) 2.2 x 1.2	165 4 − 5	2	S0		14.6		
01497	01 58.2 + 31 38 139.78 − 28.71	05 − 05 − 042 NGC 783 = 1765*	0413 32	1.7 x 1.4 1.6 x 1.3	(35) 2 − 3	4	Sc	Sc −	12.8		*3
01498	01 58.3 + 08 04 150.79 − 50.77	01 − 06 − 027	1282	1.0 x 0.5 0.9 x 0.4	26 5	1	S ...		14.4		*
01499	01 58.3 + 31 43 139.77 − 28.62	05 − 05 − 000	0413 32	1.1 x 0.3 1.0 x 0.25	84 6	2	Sa − b		16.5:		*
01500	01 58.4 + 19 26 144.62 − 40.24	03 − 06 − 016	1202 33	1.1 x 0.2 1.2 x 0.2	168 7	2	Sc		16.5:		*
01501	01 58.4 + 28 35 140.91 − 31.60	05 − 05 − 045 NGC 784	0413 32	6.8 x 1.8 6.5: x 1.7:	0 6 − 7	4	...	S dm	12.1		*12 3459
01502	01 58.4 + 30 07 140.35 − 30.14	05 − 05 − 000	0413 32	2. x 1. 2. x 1.	— ...	3	Dwarf V		18.		
01503	01 58.4 + 33 05 139.33 − 27.31	05 − 05 − 044	0413 32	(1.0 x 0.9) (0.9 x 0.8)	— ...	2	E		14.4		
01504	01 58.4 + 44 46 135.73 − 16.12	07 − 05 − 005	0931 47	(1.3 x 0.9) (1.4 x 1.1)	155: 3	2	S0 − a		14.8		*
01505	01 58.5 + 06 17 152.08 − 52.36	01 − 06 − 028	1282	1.1 x 0.4 1.2 x 0.4	100 6	2	S0		14.7		

N G C   776 − 78, 780, 781, 783, 784   I C   179, 182, 1764, 1765

1	2	3	4	5	6	7	8	9	10	11	12
01506	01 58.6 + 15 24 146.62 – 43.99	02 – 06 – 012 NGC 786	0209 35	0.8 x 0.7 0.9 x 0.9	— 1	1	S ...		14.3		3
01507	01 58.7 + 26 15 141.85 – 33.79	04 – 05 – 037	0896	2.1 x 0.8 2.1 x 0.8	70 6	3	S B a P w 1510		13.9		*
01508	01 58.7 + 44 40 135.81 – 16.20	07 – 05 – 007	0931 47	2.2: x 0.7: 2.1: x 0.7:	7 (6)	1	...		15.6		*
01509	01 58.8 + 31 35 139.93 – 28.72	05 – 05 – 046 NGC 785 = 1766*	0413 32	(1.5 x 0.9) (1.8 x 1.1)	80: ...	2	E – S0		13.9		3
01510	01 58.9 + 26 18 141.89 – 33.73	04 – 05 – 038	0896	0.6 x 0.30 0.6 x 0.30	[45] ...	1	... P w 1507		14.4		*
01511	01 59.1 + 08 15 150.95 – 50.52	01 – 06 – 031 NGC 791	1282	(1.6 x 1.6) (1.5 x 1.5)	— ...	2	E Brightest of 3		14.8		*13
01512	01 59.1 + 16 01 146.46 – 43.37	03 – 06 – 019	1202 35	1.5 x 0.5 1.4 x 0.4	53 6	2	S0 – a		14.6		
01513	01 59.2 + 08 14 150.99 – 50.52	01 – 06 – 032	1282	1.0 x 0.8 1.0 x 0.8	(110) 2:	1	S B 0 – a NGC 791 group		15.7		*1
01514	01 59.2 + 20 52 144.20 – 38.83	03 – 06 – 020	1202	1.5: x 1.4: 1:2: x 1.2:	— ...	1	...		14.7		*
01515	01 59.3 + 11 29 149.00 – 47.55	02 – 06 – 000	0209	1.3: x 1.1: 1. x 1.	— ...	3	Dwarf V		18.		
01516	01 59.4 + 44 44 135.92 – 16.10	07 – 05 – 000	0931 47	1.1: x 0.15 0.9 x 0.15	36 7	2	Sc		16.5:		*
01517	01 59.5 + 15 28 146.87 – 43.85	02 – 06 – 015 NGC 792	0209 35	(1.8 x 1.0) (1.7 x 0.9)	130 4 – 5	2	S0		14.6		*3
01518	01 59.5 + 18 50 145.21 – 40.71	03 – 06 – 023	1202 33	1.7 x 0.9 2.0 x 1.1	100: 5	1	...		16.0:		*1
01519	01 59.5 + 18 57 145.16 – 40.60	03 – 06 – 022	1202 33	1.6: x 1.0: 1.5: x 0.9:	(40) 4:	2	...	S dm – m	17.		*
01520	01 59.5 + 31 50 140.01 – 28.43	05 – 05 – 047 NGC 789	0413 32	0.7 x 0.35 0.8 x 0.6	[3] ...	1	... Peculiar		14.0		3
01521	01 59.5 + 39 53 137.39 – 20.74	07 – 05 – 000	0931	1.2 x 0.4 1.0 x 0.3	140 6	1	S ...		16.5:		
01522	01 59.6 + 09 44 150.17 – 49.12	02 – 06 – 013 IC 1770	0209	(1.5 x 1.4) (1.4 x 1.3)	— (1)	2	S0 nnp of 2		15.4		*1
01523	01 59.6 + 19 25 144.97 – 40.16	03 – 06 – 021	1202 33	1.2 x 0.9 1.0 x 0.8	150: 2 – 3	3	Sb		15.7		*
01524	01 59.6 + 23 24 143.23 – 36.42	04 – 05 – 041	0896	1.3: x 0.5: 1.3: x 0.6:	[0] ...	1	...		16.5:		*1
01525	01 59.7 – 01 21 159.02 – 58.93	00 – 06 – 025	0852	(1.5 x 1.1) (2.0 x 1.6)	— ...	2	E sp of 2		15.7		*
01526	01 59.7 – 00 22 158.05 – 58.08	00 – 06 – 024 NGC 800	0852	1.3 x 1.0 1.4 x 1.0	(10) 2 – 3	3	Sc P w 1527	Sc –	14.7		*1
01527	01 59.7 – 00 20 158.02 – 58.06	00 – 06 – 023 NGC 799	0852	2.5 x 2.0 2.8: x 2.1	(100) 1 – 3	4	S B a P w 1526		14.2		*1
01528	01 59.7 + 18 08 145.60 – 41.35	03 – 06 – 024 NGC 794 = 191*	1202 33	(1.3 x 1.1) (1.4 x 1.2)	(45) ...	2	E – S0		14.0		*3
01529	01 59.8 + 10 51 149.55 – 48.08	02 – 06 – 016 IC 193	0209	1.9: x 1.9: 2.0: x 2.0:	— 1	2	Sc	SA(rs)c Sc –	14.7		*
01530	01 59.8 + 15 47 146.80 – 43.53	03 – 06 – 025 IC 192	1202 35	1.0 x 0.8 1.1 x 0.9	— 2:	2	S0		14.8		
01531	01 59.8 + 16 55 146.23 – 42.48	03 – 06 – 026	1202 35	1.4 x 0.5 1.1 x 0.4	114 6	2	Sb		15.1		
01532	01 59.8 + 44 47 135.97 – 16.03	07 – 05 – 000	0931 47	1.0 x 0.8 0.9 x 0.7	(90) ...	1	E – S		15.3		*
01533	01 59.9 + 26 20 142.13 – 33.63	04 – 05 – 042	0896	1.1 x 0.9 0.9 x 0.9	— 1 – 2	2	S B c/Sc		15.4		
01534	01 59.9 + 46 14 135.57 – 14.63	08 – 04 – 000	0861	1.0: x 0.9: 0.9: x 0.8:	— 1 – 2	1	S B 0		16.5:		
01535	02 00.2 + 36 04 138.74 – 24.35	06 – 05 – 000	1225	1.1: x 0.9: 1. x 1.	— ...	4	Dwarf irregular Ir V		17.		
01536	02 00.3 + 07 17 152.03 – 51.26	01 – 06 – 034	1282	1.2 x 0.6 1.0 x 0.5	35 5	2	Sb – c		15.4		
01537	02 00.3 + 48 13 135.07 – 12.71	08 – 04 – 013	0861	1.4 x 0.25 1.4 x 0.25	97 6 – 7	2	Sc		16.5:		*1
01538	02 00.4 + 23 31 143.40 – 36.25	04 – 05 – 000	0896	1.4: x 0.9: 0.8: x 0.7:	[55] ...	3	Dwarf (irregular) Ir: V		17.		
01539	02 00.4 + 31 50 140.22 – 28.37	05 – 05 – 048 NGC 798	0413 32	(1.2 x 0.5) (1.3 x 0.5)	137 ...	2	E		14.7		3
01540	02 00.4 + 33 24 139.67 – 26.88	05 – 05 – 000	0413	1.0: x 0.6: 1.3: x 0.7:	(175) (4 – 5)	3	Dwarf spiral S IV – V or V		17.		

N G C   785, 786, 789, 791, 792, 794, 798 – 800   I C 191 – 93, 1766, 1770

1	2	3	4	5	6	7	8	9	10	11	12
01541	02 00.4 + 37 53 138.19 – 22.61	06 – 05 – 078 NGC 797	1225 31	1.9: x 1.4 1.7 x 1.3	65: 3:	3	Sa/S B a nff of 2, connected	SAB(s)a	(13.1)		*13 C
01542	02 00.5 + 02 22 155.89 – 55.61	00 – 06 – 026 IC 194	0852	1.7 x 0.22 1.6 x 0.2	13 7	3	Sb		15.4		
01543	02 00.5 + 19 29 145.20 – 40.02	03 – 06 – 027 33	1202	(1.0 x 0.6) (1.1 x 0.7)	(155) ...	1	E		15.7		*
01544	02 00.5 + 47 52 135.20 – 13.04	08 – 04 – 015	0861	1.5: x 1.5: 1.4: x 1.4:	— 1	4	S B c 1552 group	SAB(rs)d	16.5:		*
01545	02 00.6 + 05 25 153.48 – 52.90	01 – 06 – 035	1282	1.1 x 0.3 0.9 x 0.3	124 6	1	S0 – a		15.0		
01546	02 00.6 + 18 24 145.74 – 41.02	03 – 06 – 000 33	1202	1.2 x 1.2 1.2 x 1.1	— 1	4	Sc	SAB(s)c	14.8		
01547	02 00.6 + 21 48 144.19 – 37.84	04 – 05 – 043	0896	2.2 x 2.0 1.4: x 1.4:	— ...	4	Irr Ir IV – V:	I B m Ir I	15.0		*1
01548	02 00.6 + 42 46 136.72 – 17.92	07 – 05 – 010	0931	1.1: x 0.7: 1.1 x 0.7	80: 3 – 4	2	Sc		16.0:		
01549	02 00.7 + 26 04 142.44 – 33.82	04 – 05 – 044	0896	1.0: x 0.22 0.9 x 0.2	36 7	2	S0 – a		14.6		*
01550	02 00.7 + 38 01 138.21 – 22.46	06 – 05 – 079 NGC 801	1225 31	3.3 x 0.7 3.4 x 1.0	150 6	3	Sc P w 1541	Sc –	13.5		*
01551	02 00.8 + 23 50 143.37 – 35.92	04 – 05 – 045	0896	3.0: x 1.5: 3. x 2.	(135) 1 – 2	4	... S B IV – V	S B(rs)dm	14.1		*1
01552	02 00.9 + 47 44 135.31 – 13.15	08 – 04 – 016	0861	1.1 x 0.9 (1.4 x 1.2)	— 1 – 2	4	S B 0 Brightest in group	S B(r)0	14.7		*1C
01553	02 01.0 + 04 33 154.30 – 53.63	01 – 06 – 036	1282	1.4 x 0.9 1.3: x 0.9:	(145) 3 – 4	3	Sc	Sc +	15.5		
01554	02 01.0 + 15 48 147.16 – 43.41	03 – 06 – 028 NGC 803	1202 35	3.5 x 1.6 2.8: x 1.2:	8 6	4	Sc	Sc –	13.5		*12 359
01555	02 01.0 + 14 28 147.87 – 44.64	02 – 06 – 017 IC 195	0209 35	1.5 x 0.8 1.4 x 0.7	135 5	3	S0 P w 1556, disturbed	SAB 0	14.3		*1V A
01556	02 01.1 + 14 30 147.88 – 44.60	02 – 06 – 018 IC 196	0209 35	3.0 x 1.5 3.0 x 1.3	5: ...	4	S B [b] P w 1555, distorted		14.2		*1V A
01557	02 01.1 + 30 35 140.83 – 29.52	05 – 05 – 049 NGC 804 = 1773*	0413 32	1.4 x 0.25 1.1 x 0.35	7 7	2	S0		14.7		*
01558	02 01.2 + 11 43 149.49 – 47.15	02 – 06 – 019	0209	1.0: x 1.0: 1.2 x 1.0:	— ...	2	Dwarf S IV – V or V	SAB: m	16.0:		*
01559	02 01.2 + 15 04 147.61 – 44.07	02 – 06 – 020 IC 1774	0209 35	1.9 x 1.7 2.1 x 1.7	(140) 1 – 2	4	Sc/S B c	SAB(s)d	15.2		*1
01560	02 01.2 + 19 25 145.43 – 40.03	03 – 06 – 030	1202 33	1.4: x 0.7 0.9 x 0.7	65: 4 – 5	4	S B b nf of 2, disturbed?, Sb –	(R´:)S B(s)b	14.9		*
01561	02 01.2 + 23 58 143.52 – 35.76	04 – 05 – 048	0896	1.5: x 1.0: 1.2: x 0.8	[100] ...	4	Irr	I m Ir I	14.7		*1C
01562	02 01.2 + 45 32 136.00 – 15.24	08 – 04 – 18+19	0861	[1.4 x 0.7] ...	... ...	3	E: + S ... Contact		15.0		*1V C
01563	02 01.4 + 47 42 135.40 – 13.15	08 – 04 – 021	0861	(1.3 x 1.3) (1.3 x 1.3)	— ...	1	E: 1552 group		15.5		*
01564	02 01.5 + 02 33 156.12 – 55.33	00 – 06 – 027 IC 197	0852	1.0 x 0.5 1.0 x 0.5	55 5	2	S B: b – c s of 2		14.3		*1
01565	02 01.6 + 27 41 142.03 – 32.23	05 – 05 – 051	0413	1.6 x 0.6 1.4: x 0.5:	4 ...	2	...	S B dm	15.4		
01566	02 01.6 + 28 34 141.69 – 31.39	05 – 05 – 050 NGC 805	0413 32	1.1 x 0.7 1.0 x 0.7	115: 3 – 4	1	S B 0		14.7		3
01567	02 01.7 + 42 55 136.88 – 17.72	07 – 05 – 000	0931	1.3 x 0.20 1.3 x 0.2	119 7	3	Sc	Sc +	16.0:		*
01568	02 01.8 + 35 26 139.29 – 24.86	06 – 05 – 000	1225	1.1 x 1.1 0.7: x 0.6:	— (1)	3	...	S dm – m	17.		
01569	02 01.9 + 34 55 139.48 – 25.34	06 – 05 – 081	1225 37	1.2 x 0.12 1.1 x 0.15	146 7	2	Sb n of 2		16.0:		*1
01570	02 02.0 + 21 12 144.84 – 38.29	03 – 06 – 000	1202	1.1 x 0.15 1.1 x 0.18	25 7	2	Sa – b		15.5		
01571	02 02.0 + 28 45 141.72 – 31.19	05 – 06 – 001 NGC 807	1248 32	(2.2 x 1.6) (2.5 x 1.7)	145: ...	2	E		13.8		3
01572	02 02.1 + 08 18 151.97 – 50.15	01 – 06 – 037	1282	1.5 x 0.4 1.4 x 0.4	62 6	1	S: ...		14.3		
01573	02 02.2 + 01 40 157.16 – 56.01	00 – 06 – 028 34	0852	1.0 x 0.5 1.0 x 0.15	148 6 – 7	2	Sc f of 2		16.5:		*
01574	02 02.2 + 37 24 138.72 – 22.96	06 – 05 – 000	1225 37	1.2 x 0.35 0.8 x 0.2	145 6	2	Sc		15.7		*
01575	02 02.3 + 24 26 143.52 – 35.24	04 – 06 – 000	0858	1.2: x 0.1 1.1 x 0.1	6 (7)	1	Dwarf?		18.		

N G C   797, 801, 803 – 05, 807   I C   194 – 97, 1773, 1774

1	2	3	4	5	6	7	8	9	10	11	12
01576	02 02.4 + 29 45 / 141.44 – 30.22	05 – 06 – 000	1248 / 32	1.2 x 0.10 / 1.0: x 0.1	153 / 7	3	Sc	Sc +	17.		
01577	02 02.5 + 30 56 / 141.02 – 29.09	05 – 06 – 002	1248	2.4 x 1.7 / 2.3: x 1.6	(85) / 3:	4	S B b	S B(r)bc / Sb +	14.0		*
01578	02 02.5 + 50 30 / 134.76 – 10.42	08 – 04 – 000	0861	1.2 x 0.12 / 1.2 x 0.12	152 / 7	2	Sc		17.		
01579	02 02.6 + 05 52 / 153.88 – 52.28	01 – 06 – 038 / IC 1776	1282	2.3 x 2.2 / 2.0 x 2.0	— / 1	4	S B c	S B(s)d	14.4		*
01580	02 02.6 + 09 41 / 151.22 – 48.85	02 – 06 – 024	0209	1.5 x 0.3 / 1.4 x 0.3	137 / 6	1	S ...		15.2		
01581	02 02.6 + 34 40 / 139.72 – 25.57	06 – 05 – 084	1225 / 37	1.8 x 0.8 / 1.6 x 0.7	[160] / ...	1	S – Irr / Peculiar?		15.2		*
01582	02 02.6 + 39 36 / 138.08 – 20.84	07 – 05 – 011	0931	1.4 x 0.5 / 1.2 x 0.5	37 / 6	3	Sc	Sc +	14.3		
01583	02 02.7 + 13 00 / 149.23 – 45.82	02 – 06 – 026 / NGC 810	0209 / 35	(1.7 x 1.3) / (1.8 x 1.5)	[25] / ...	1	E? / Companion superimposed		15.4		*3
01584	02 02.7 + 13 08 / 149.15 – 45.70	02 – 06 – 025	0209 / 35	1.5: x 0.6 / 1.5: x 0.7	82 / 5 – 6	2	Sc / NGC 810 group		16.0:		*
01585	02 02.9 + 44 58 / 136.47 – 15.70	07 – 05 – 012	0931 / 47	1.2 x 0.9 / 1.0 x 0.8	(3) / 2 – 3	3	Sa/S B b / f of 2	SAB(s)ab / Sb –	14.6		*
01586	02 02.9 + 49 40 / 135.07 – 11.20	08 – 04 – 000	0861	1.2: x 1.0: / 1.2: x 1.0:	— / 1 – 2	3	Sb/S B b	Sb +	15.6		*
01587	02 03.0 + 06 32 / 153.53 – 51.64	01 – 06 – 039	1282	1.1 x 0.4 / 1.0 x 0.4	24 / (6)	1	S ...		14.6		*
01588	02 03.1 – 00 56 / 159.98 – 58.12	00 – 06 – 029	0852	1.4 x 0.2 / 1.4 x 0.2	150 / 7	2	Sc		15.7		
01589	02 03.2 + 14 41 / 148.44 – 44.23	02 – 06 – 027	0209 / 35	1.0 x 0.6 / 0.9 x 0.6	95: / 4:	2	Sa – b		15.6		*
01590	02 03.2 + 29 33 / 141.70 – 30.35	05 – 06 – 004	1248 / 32	(1.9 x 1.4) / (2.0 x 1.4)	(130) / ...	2	S0 – E		14.0		*
01591	02 03.3 + 29 44 / 141.66 – 30.17	05 – 06 – 005	1248 / 32	1.5 x 0.22 / 1.7 x 0.2	153 / 7	1	S ... / P w 1596		14.5		*
01592	02 03.4 + 09 03 / 151.91 – 49.34	01 – 06 – 040 / IC 198	1282	1.2 x 0.8 / 1.2 x 0.8	55 / 3	2	S ... / P w 1594		14.8		*
01593	02 03.4 + 13 02 / 149.43 – 45.72	02 – 06 – 028	0209 / 35	1.3 x 1.1 / 1.0: x 1.0:	— / 1 – 2	4	Sc / NGC 810 group	SAB(rs)c	15.6		*
01594	02 03.6 + 08 59 / 152.02 – 49.38	01 – 06 – 041 / IC 199 = 1778	1282	1.6 x 0.9 / 1.5 x 0.8	25 / 4 – 5	2	Sa – b / P w 1592		15.4		*
01595	02 03.6 + 26 48 / 142.88 – 32.91	04 – 06 – 003	0858	1.4 x 0.8 / 1.2 x 0.7	75: / 4	3	Sc		15.5		
01596	02 03.6 + 29 45 / 141.72 – 30.13	05 – 06 – 006	1248 / 32	1.3 x 0.9 / 1.3 x 0.8	(125) / 3	3	S0 / P w 1591	SAB 0	14.8		*
01597	02 03.7 – 00 32 / 159.82 – 57.71	00 – 06 – 030	0852	(1.2 x 1.0) / (1.3 x 1.1)	(0) / ...	1	S0		14.6		
01598	02 03.7 + 44 20 / 136.80 – 16.26	07 – 05 – 014 / NGC 812	0931 / 47	3.2 x 1.5 / 2.8 x 1.0	[160] / ...	4	S ... / Peculiar		12.8		*13
01599	02 03.8 + 37 11 / 139.12 – 23.07	06 – 05 – 000	1225	1.1 x 0.12 / 0.5 x 0.10	26 / 7	2	Sc – Irr		16.5:		
01600	02 04.0 + 01 17 / 158.20 – 56.12	00 – 06 – 031	0852 / 34	1.3 x 0.2 / 1.2 x 0.2	111 / 7	3	Sc	Sc +	17.		
01601	02 04.0 + 42 44 / 137.36 – 17.77	07 – 05 – 015	0931	1.3 x 0.20 / 1.1 x 0.20	169 / 7	3	S B b – c		15.5		
01602	02 04.1 + 43 32 / 137.12 – 17.00	07 – 05 – 000	0931	1.5: x 0.3 / 1.3: x 0.3:	105 / 6 – 7	3	...	S d – m	16.5:		
01603	02 04.2 – 01 06 / 160.59 – 58.12	00 – 06 – 032	0852	1.6 x 1.3 / 1.5: x 1.2:	— / (1 – 2)	1	...	SA dm:	15.6		*
01604	02 04.3 + 36 52 / 139.33 – 23.34	06 – 05 – 085	1225	(1.5 x 0.9) / (1.4 x 0.8)	48 / ...	1	S0?		15.0		*C
01605	02 04.4 + 07 54 / 153.04 – 50.26	01 – 06 – 042	1282	1.6 x 0.6 / 1.3 x 0.6	60 / ...	1	...		15.4		*
01606	02 04.5 + 08 45 / 152.49 – 49.49	01 – 06 – 000	1282	1.2 x 1.1 / 1.0 x 1.0	— / 1	2	Sc	Sc +	16.5		*
01607	02 04.5 + 45 23 / 136.62 – 15.22	08 – 05 – 001	0861	2.1 x 0.8 / 1.8: x 0.6	72 / 6	4	S B b	S B(s)ab / Sb –	14.2		
01608	02 04.6 + 30 13 / 141.78 – 29.62	05 – 06 – 000	1248	1.1: x 1.0: / 0.8: x 0.8:	— / (1)	3	...	SA dm	16.5:		
01609	02 04.7 + 44 37 / 136.89 – 15.94	07 – 05 – 016	0931 / 47	1.3 x 0.25 / 1.3 x 0.3	81 / 6	3	Sb		15.3		
01610	02 04.8 + 08 56 / 152.46 – 49.29	01 – 06 – 043 / IC 202	1282	1.6 x 0.25 / 1.6 x 0.25	132 / 6	2	Sb		15.3		*

N G C 810, 812

I C 198, 199, 202, 1776, 1778

1	2	3	4	5	6	7	8	9	10	11	12
01611	02 04.8 + 16 58 147.70 – 41.98	03 – 06 – 033 NGC 817	1202 35	0.8 x 0.35 0.8 x 0.30	[27] ...	1	... Peculiar		13.9		*
01612	02 04.8 + 43 21 137.31 – 17.14	07 – 05 – 017	0931	1.4 x 1.1 1.0: x 0.9:	(55) 1 – 2	1	Sc	SA c Sc –	15.5		
01613	02 04.9 – 02 20 162.18 – 59.05	00 – 06 – 034 IC 205	0852	1.1 x 1.1 1.1 x 1.1	--- 1	2	S B a	S B(s)a	14.8		
01614	02 04.9 + 15 07 148.72 – 43.67	02 – 06 – 033	0209 35	1.5 x 0.4 1.5 x 0.4	91 6 – 7	2	S0 – a		15.7		
01615	02 04.9 + 41 05 138.05 – 19.29	07 – 05 – 000	0931	1.1 x 0.9 0.9: x 0.5:	--- (1 – 2)	2	...	S B(s)dm:	17.		
01616	02 05 + 85 05 124.73 + 22.74	14 – 02 – 000	1277	1.5: x 0.4 0.8 x 0.2	10 ...	1	S B ... ssf of 2		16.5:		*
01617	02 05.0 + 01 52 158.05 – 55.49	00 – 06 – 035	0852 34	(1.1 x 1.1) (1.2 x 1.2)	--- ...	2	E – S0 2nd of 6		15.1		*1
01618	02 05.0 + 01 56a 157.99 – 55.43	00 – 06 – 037	0852 34	(1.2 x 0.9) (1.1 x 0.9)	--- ...	2	E – S0 3rd of 6		(15.3)		*1
01619	02 05.0 + 34 58 140.13 – 25.10	06 – 05 – 000	1225 37	[1.1 x 0.8] ...	... ...	3	Double system Plumes, bridge		16.0:		*C:
01620	02 05.1 + 01 54 158.06 – 55.45	00 – 06 – 038	0852 34	(1.1 x 1.1) (1.1 x 1.1)	--- ...	2	E – S0 5th of 6		15.1		*1
01621	02 05.2 + 02 27 157.61 – 54.96	00 – 06 – 039	0852 34	1.2: x 0.7: 1.1: x 0.6:	(20) 4 – 5	3	Sc	SAB cd	16.5:		
01622	02 05.2 + 15 59 148.34 – 42.85	03 – 06 – 035	1202 35	1.2 x 0.4 1.2 x 0.4	165 6	3	Sc		15.0		*
01623	02 05.2 + 18 56 146.81 – 40.13	03 – 06 – 034	1202 33	1.1 x 0.7 1.1 x 0.7:	5: 3 – 4	1	S B ...		15.7		*
01624	02 05.3 + 01 55 158.12 – 55.41	00 – 06 – 040	0852 34	(1.4 x 0.8) (1.5 x 0.8)	(130) ...	1	S0: 6th of 6		15.2		*1
01625	02 05.3 + 28 06 142.78 – 31.56	05 – 06 – 009	1248	1.1 x 0.20 1.1 x 0.20	169 7	2	Sb – c		16.5:		*
01626	02 05.3 + 41 15 138.07 – 19.11	07 – 05 – 018	0931	1.7 x 1.7 1.8 x 1.8	--- 1	4	Sc/S B c Disturbed	SAB(rs)c Sc –	14.5		*1
01627	02 05.5 + 01 39 158.44 – 55.61	00 – 06 – 041	0852 34	1.1 x 0.35 1.0 x 0.4	70 6	1	S ...		14.9		*
01628	02 05.5 + 42 46 137.62 – 17.66	07 – 05 – 019	0931	1.0 x 0.4 1.0 x 0.4	137 6	1	S ...		15.4		
01629	02 05.6 + 14 06 149.51 – 44.53	02 – 06 – 036 NGC 820	0209 35	1.5 x 0.9 1.4 x 0.9	72 4	3	Sb		13.7		3
01630	02 05.6 + 14 44 149.15 – 43.95	02 – 06 – 035	0209 35	1.2 x 0.6 1.1 x 0.5	[43] ...	1	S ... Peculiar		14.3		*
01631	02 05.7 + 10 45 151.57 – 47.56	02 – 06 – 034 NGC 821	0209	(3.3 x 2.2) (3.5 x 2.5)	25 ...	4	E		12.6	+ 1778 + 1864	*12 34<u>78</u>
01632	02 05.7 + 29 00 142.51 – 30.68	05 – 06 – 000 NGC 819	1248	0.65 x 0.45 0.65 x 0.45	(10) ...	1	... Peculiar		14.1		*3
01633	02 05.7 + 38 32 139.05 – 21.67	06 – 05 – 086 NGC 818	1225	3.5 x 1.4 3.3: x 1.4:	113 6	2	S B: b – c		12.7		*23
01634	02 05.7 + 46 59 136.33 – 13.63	08 – 05 – 002	0907	3.2: x 3.0: 2.8: x 2.8:	--- 1	4	Sc	SAB(r)cd	15.0		*
01635	02 05.8 + 06 10 154.79 – 51.64	01 – 06 – 044 IC 208	1282	2.1 x 2.0 2.1: x 1.9:	--- 1	3	Sb P w 1636	SA bc Sb +	14.8		*1
01636	02 05.9 + 06 05 154.89 – 51.70	01 – 06 – 045 NGC 825	1282	2.5 x 1.1 2.4 x 1.0	53 7	2	Sa P w 1635		14.5		*13
01637	02 06.0 + 44 03 137.30 – 16.41	07 – 05 – 020	0931 47	1.0 x 0.8 1.0 x 0.7	(105) 2 – 3	4	Sc		15.6		
01638	02 06.1 + 25 48 143.92 – 33.66	04 – 06 – 004	0858	1.0 x 0.5 1.0 x 0.5	15 5	2	Sc		15.4		
01639	02 06.1 + 34 52 140.41 – 25.12	06 – 05 – 087	1225 37	1.0 x 0.25 0.8 x 0.25	120 6	1	S ...		15.7		
01640	02 06.2 + 07 44 153.78 – 50.21	01 – 06 – 046 NGC 827	1282	2.7: x 1.4 2.4 x 1.1	85 5	1	S ...		14.0		3
01641	02 06.2 + 31 46 141.57 – 28.04	05 – 06 – 010	1248	1.5 x 1.5 1.7 x 1.6	--- 1	4	...	SA(s)dm	14.7		1
01642	02 06.2 + 36 58 139.69 – 23.13	06 – 05 – 000	1225 37	1.4 x 0.1 1.1 x 0.1	178 7	3	... 2nd of 3	S d – m	17.		*
01643	02 06.3 + 02 28 158.01 – 54.81	00 – 06 – 044	0852	1.2 x 0.5 1.3 x 0.5	13 6	3	S B b	Sb +	15.4		*
01644	02 06.3 + 30 06 142.23 – 29.61	05 – 06 – 000	1248	1. x 1. 1. x 1.	--- ...	3	Dwarf V		18.		
01645	02 06.3 + 34 02 140.75 – 25.90	06 – 05 – 000	1225 37	1. x 1. 0.8: x 0.5:	--- ...	1	S ...		17.		*

N G C 817 – 21, 825, 827

I C 205, 208

1	2	3	4	5	6	7	8	9	10	11	12
01646	02 06.4 + 04 53 156.01 − 52.70	01 − 06 − 048	1282	2.2 x 0.7 2.0 x 0.8	173 6	2	S B b − c		15.0		
01647	02 06.4 + 05 54 155.21 − 51.81	01 − 06 − 000	1282	1.3 x 0.4 0.9: x 0.3	30 6	2	S B: c	Sc +	16.0:		
01648	02 06.4 + 25 20 144.19 − 34.07	04 − 06 − 006	0858 38	1.2: x 0.8: 1.1: x 0.7:	75 ...	1	...		14.5		
01649	02 06.5 + 06 23 154.88 − 51.37	01 − 06 − 047	1282	1.8: x 1.7: 1.5: x 1.5:	— 1	3	...	SA(s)dm	16.5:		
01650	02 06.5 + 37 01 139.73 − 23.06	06 − 05 − 000	1225 37	2.2 x 0.10 1.5: x 0.10	28 7	2	S[c] 3rd of 3		17.		*
01651	02 06.7 + 35 34 140.28 − 24.42	06 − 05 − 088	1225 37	(2.3 x 2.0) (2.2 x 2.0)	— ...	1	E? Distorted		14.9		*C
01652	02 06.8 + 21 00 146.26 − 38.07	03 − 06 − 038	1202	1.6 x 0.7 1.3 x 0.6	30 5 − 6	3	Sc Disturbed?, Sc +		15.0		*
01653	02 07.1 + 07 05 154.56 − 50.68	01 − 06 − 000	1282	1.0 x 0.4 0.9 x 0.4	[138] ...	3	... Peculiar, disrupted?		15.6		*
01654	02 07.1 + 36 28 140.05 − 23.54	06 − 05 − 091	1225 37	1.1 x 0.7 1.1 x 0.7	48 3 − 4	3	Sc	Sc −	15.1		
01655	02 07.1 + 38 57 139.19 − 21.19	06 − 05 − 092 NGC 828	1225	3.5 x 2.7 3.3: x 2.2:	— ...	4	... Peculiar		13.0		*13 C
01656	02 07.1 + 41 20 138.38 − 18.93	07 − 05 − 000	0931	1.4 x 0.10 1.1 x 0.1	63 7	3	... 1661 group	S d − m	18.		*
01657	02 07.1 + 45 22 137.08 − 15.09	08 − 05 − 000	0907	1.5: x 0.5: 1.5: x 0.5:	160 ...	1	...		16.0:		*
01658	02 07.2 + 10 44 152.07 − 47.42	02 − 06 − 038	0209	1.0 x 0.4 1.0 x 0.5	92 5 − 6	1	S ...		15.4		
01659	02 07.2 + 15 48 149.04 − 42.82	03 − 06 − 039	1202 35	1.7 x 0.7 1.5 x 0.7	38 6	3	S B c		14.9		*
01660	02 07.2 + 38 21 139.41 − 21.75	06 − 05 − 093	1225	1.0 x 0.7 0.9 x 0.5	25 3 − 4	4	Sc	Sc +	15.4		
01661	02 07.2 + 41 17 138.42 − 18.97	07 − 05 − 022	0931	(1.4 x 1.2) (1.1 x 1.0)	— ...	3	[E] Compact, brightest of 3		15.5		*
01662	02 07.3 + 10 32 152.23 − 47.59	02 − 06 − 039	0209	1.0 x 0.35 1.0 x 0.4	153 6	2	Sa − b		15.3		
01663	02 07.4 + 07 25 154.42 − 50.35	01 − 06 − 050	1282	2.1 x 1.0 1.8 x 0.8	13 5 − 6	4	Sc		15.1		
01664	02 07.6 + 07 36 154.36 − 50.17	01 − 06 − 049 NGC 840	1282	2.6 x 1.4 2.3 x 1.3	73 4 − 5	4	S B b	S B(r)b	14.7		*3
01665	02 07.8 + 00 58 159.94 − 55.89	00 − 06 − 000	0852 34	1. x 1. ...	— ...	3	Dwarf V		18.		
01666	02 07.8 + 34 45 140.81 − 25.12	06 − 05 − 096	1225 37	(1.1 x 1.1) (1.1 x 1.1)	— ...	1	... Disturbed		16.0:		*1
01667	02 07.8 + 34 54 140.76 − 24.98	06 − 05 − 095	1225 37	1.0 x 0.5 0.8: x 0.6:	[55] ...	3	... S IV:	S dm	16.5:		*
01668	02 07.8 + 34 59 140.73 − 24.90	06 − 05 − 094	1225 37	1.1: x 1.0: 1.0: x 0.9:	— ...	1	... Peculiar		15.7		*1C
01669	02 08.0 + 05 38 155.98 − 51.85	01 − 06 − 051	1282	1.1 x 0.8 1.0 x 0.7	173 3	1	S ...		14.5		
01670	02 08.0 + 06 31 155.30 − 51.07	01 − 06 − 052	1282	2.5: x 2.5: 2. x 2.	— (1)	4	Dwarf (spiral) S: V		16.0:		*
01671	02 08.0 + 32 28 141.71 − 27.26	05 − 06 − 011	1248	1.1 x 0.5 1.1 x 0.5	98 6	1	S B ... s of 2		15.5		*1
01672	02 08.0 + 37 26 139.89 − 22.57	06 − 05 − 099 NGC 834	1225 37	1.2 x 0.5 1.1 x 0.5	20 6	1	S ...		13.2		*3
01673	02 08.0 + 37 35 139.84 − 22.43	06 − 05 − 098	1225 37	1.1 x 0.25 1.0 x 0.22	73 7	1	S ...		15.2		*
01674	02 08.1 + 38 31 139.53 − 21.54	06 − 05 − 100	1225	1.1 x 0.7 1.1 x 0.7	[60] ...	3	Sc	Sc +	15.1		
01675	02 08.3 + 25 35 144.57 − 33.68	04 − 06 − 007	0858 38	1.2 x 0.6: 1.2 x 0.6:	[125] ...	1	S B: ... Peculiar?		15.5		*
01676	02 08.3 + 37 16 140.01 − 22.71	06 − 05 − 101 NGC 841	1225 37	2.0 x 1.0 2.1 x 1.1	135 5	3	S B a/Sb	(R')SAB ab Sb −	12.8		*1C
01677	02 08.4 + 06 27 155.49 − 51.09	01 − 06 − 000	1282	1.2 x 0.1 1. x 0.1	135 (7)	1	...	d − m	18.		
01678	02 08.5 + 03 38 157.80 − 53.52	01 − 03 − 053 IC 211	1282	2.6 x 2.1 2.5 x 2.0	45: 2	4	Sc P w 1680	SAB(s)c Sc −	14.5		*1
01679	02 08.6 − 01 44 162.99 − 58.03	00 − 06 − 049 NGC 850	0852	(1.5 x 1.4) (1.7 x 1.7)	— 1	3	S0 − a		14.1		
01680	02 08.6 + 03 33 157.91 − 53.58	01 − 06 − 054 NGC 851	1282	1.3 x 0.8 1.2 x 0.8	135 4	2	S0 P w 1678		14.7		*

N G C   828,   834,   840,   841,   850,   851

1	2	3	4	5	6	7	8	9	10	11	12
01681	02 08.6 + 14 16 150.34 - 44.08	02 - 06 - 000	0209 35	1.0 x 0.2 1.0 x 0.2	78 7	2	Sc p of 2		16.5:		*
01682	02 08.6 + 31 17 142.30 - 28.32	05 - 06 - 013	1248	1.2 x 0.5 1.2 x 0.5	107 6	3	Sc		14.7		*
01683	02 08.7 + 15 41 149.55 - 42.78	03 - 06 - 000	1202 35	1.0 x 0.1 0.9 x 0.1	3 7	2	Sb - c 1684 group		16.5:		*
01684	02 08.7 + 15 44 149.53 - 42.74	03 - 06 - 043	1202 35	1.6 x 1.6 1.7 x 1.6	— 1	4	S B 0 Brightest of 4	(R)SAB 0 +	14.8		*1
01685	02 08.7 + 33 49 141.35 - 25.93	06 - 05 - 102	1225	1.1 x 0.25 1.0 x 0.25	72 6	2	Sa - b		14.9		
01686	02 08.9 + 46 08 137.14 - 14.27	08 - 05 - 003	0907	1.2 x 1.2 1.0 x 1.0	— 1	2	...	SAB dm - m	17.		
01687	02 09.0 + 14 04 150.58 - 44.22	02 - 06 - 042	0209 35	(1.0 x 0.8) 1.1 x 0.8	(0) ...	1	S0: f of 2		15.3		*
01688	02 09.0 + 44 20 137.74 - 15.97	07 - 05 - 024 NGC 846 = 847	0931 47	2.3 x 2.1 2.3 x 2.0	(140) 1	4	S B b Sb -	S B(rs)ab	13.2		*3
01689	02 09.1 + 13 46 150.78 - 44.48	02 - 06 - 043	0209 35	1.4 x 1.1: 1.3 x 0.9:	(175) ...	4	Dwarf: S IV - V, p w 1693		16.5:		*
01690	02 09.1 + 29 05 143.29 - 30.35	05 - 06 - 000	1248	1.6 x 0.15 1.6 x 0.15	106 7	1	S ...		15.7		
01691	02 09.1 + 39 00 139.56 - 21.02	06 - 05 - 103	1225	(1.8 x 1.6) (1.4 x 1.1)	— 1 - 2	3	S0		14.6		*C
01692	02 09.1 + 42 09 138.49 - 18.03	07 - 05 - 025	0931 47	1.8::x 1.7:: 1.6:x 1.6:	— 1	1	Sb - c		16.5:		*
01693	02 09.2 + 13 52 150.76 - 44.38	02 - 06 - 044	0209 35	1.6:x 1.4: 1.4::x 1.4::	— 1:	4	Dwarf spiral S V, p w 1689		17.		*
01694	02 09.3 + 09 19 153.71 - 48.45	01 - 06 - 000	1282	1.6:x 1.6: 1.5:x 0.9::	— ...	4	... nnf of 2	SAB(s)dm	17.		*
01695	02 09.3 + 37 15 140.22 - 22.66	06 - 05 - 104 NGC 845	1225 37	1.8 x 0.5 1.7 x 0.30	149 7	3	Sb		14.5		*3
01696	02 09.6 + 29 37 143.19 - 29.81	05 - 06 - 015	1248	1.3 x 1.3 1.1 x 1.1	— (1)	2	...	S dm	15.6		
01697	02 09.7 - 02 23 164.12 - 58.40	00 - 06 - 052	0852	1.4 x 0.6 1.3 x 0.5	135 6	2	Sb		15.0		*
01698	02 09.7 - 01 03 162.68 - 57.31	00 - 06 - 051	0852	1.3 x 1.2 1.0:x 0.8:	— 1	3	S B a	(R:)S B(r)a	15.3		*
01699	02 09.7 + 53 12 135.04 - 07.54	09 - 04 - 004	1245	1.5 x 0.8 1.5 x 0.8:	76 5:	2	S0 - a		15.1		*1
01700	02 09.8 + 16 37a 149.34 - 41.79	03 - 06 - 000	1202 35	1.0 x 0.2 1. x 0.2	130 7	2	Sc np of 2, disturbed:		(15.7)		*
01701	02 09.8 + 36 05 140.75 - 23.72	06 - 05 - 105	1225 37	1.2 x 1.1 1.3 x 1.2	— 1 - 2	2	Sb: Peculiar	(R')SAB ...	15.0		*C
01702	02 10.0 + 13 44 151.08 - 44.42	02 - 06 - 047	0209 35	1.0 x 0.9 1.0 x 0.9	— (1)	1	...	SA d?	16.0:		*
01703	02 10.0 + 32 37 142.09 - 26.97	05 - 06 - 000	1248	1.8 ::x 1.0:: 1. x 1.	— ...	3	Dwarf V		18.		
01704	02 10.2 + 41 01 139.08 - 19.04	07 - 05 - 026	0931 47	1.6 x 0.7 1.3 x 0.6	35 5 - 6	3	Sc 	Sc +	15.5		
01705	02 10.4 + 09 16 154.11 - 48.37	01 - 06 - 000	1282	1.0 x 0.5 0.6 x 0.35	... ...	3	S B b nnf of 2, disturbed	SAB(s)b	15.6		*
01706	02 10.7 + 25 37 145.16 - 33.45	04 - 06 - 012	0858 38	1.2 x 0.4 1.1 x 0.4	154 6	3	Sc		14.7		
01707	02 10.8 + 04 31 157.87 - 52.47	01 - 06 - 055	1282	1.1 x 0.8 1.3 x 0.8	(0) 3 - 4	3	S B b	S B(s)b	15.5		
01708	02 10.8 + 09 23 154.16 - 48.22	01 - 06 - 000	1282	1.3 x 0.8 1.2 x 0.8	153 4 - 5	1	S ...		16.0:		
01709	02 10.9 + 03 45 158.56 - 53.12	01 - 06 - 000	1282	1.1 x 0.6 1.0:x 0.5:	[45] ...	2	... np of 2		16.0:		*
01710	02 10.9 + 13 29 151.51 - 44.55	02 - 06 - 048	0209 35	1.0:x 0.9: 0.9:x 0.9:	— 1	3	S B a	S B(r)a	16.0:		
01711	02 10.9 + 24 40 145.64 - 34.32	04 - 06 - 000	0858 38	1.0 x 0.4 1.0 x 0.4	[127] ...	2	Dwarf:		16.5:		
01712	02 10.9 + 33 23 142.00 - 26.19	05 - 06 - 000	1248	1.0:x 1.0: 1.1:x 1.0:	— 1:	3	...	SA dm	17.		
01713	02 11.0 - 00 58 163.09 - 57.06	00 - 06 - 054 NGC 856	0852	1.2 x 0.9 1.4 x 1.1	20 2 - 3	2	Sa P w 1727		14.4		*
01714	02 11.0 + 10 06 153.73 - 47.56	02 - 06 - 000	0209	1.3 x 0.2 1. x 0.2	77 7	3	...	d - m	17.		
01715	02 11.0 + 49 47 136.30 - 10.69	08 - 05 - 000	0907 36	1.2 x 0.9 1.2:x 0.9:	148: 2 - 3	3	Sc/S B c		16.0:		

N G C   845 - 47

1	2	3	4	5	6	7	8	9	10	11	12
01716	02 11.1 + 03 53 158.52 – 52.98	01 – 06 – 056	1282	(0.9 x 0.7) (0.8 x 0.5)	— ...	2	... Extremely compact		14.3	+ 3430 + 3485	*C
01717	02 11.1 + 16 34 149.77 – 41.74	03 – 06 – 045	1202 35	1.1 x 0.6 1.1 x 0.6:	73 5	3	S B c 4th of 4		15.7		*
01718	02 11.2 + 27 39 144.39 – 31.52	05 – 06 – 016 NGC 855	1248	(3.3 x 1.4) (3.1 x 1.3)	67 ...	3	E		13.0		*3
01719	02 11.3 + 16 14 150.01 – 42.02	03 – 06 – 046 IC 213	1202 35	2.3 x 1.7 2.2: x 1.6:	150 2 – 3	4	Sb	SAB(rs)b	15.4		
01720	02 11.4 + 04 57 157.72 – 52.02	01 – 06 – 057 IC 214	1282	0.9 x 0.6 0.9 x 0.6	... ...	3	Peculiar Disrupted pair?, f of 2		14.4		*1
01721	02 11.5 + 37 10 140.70 – 22.59	06 – 05 – 107	1225 37	2.3: x 2.3: 2. x 2.	— 1	4	S B b/S B c Sc –	S B(rs)bc	14.5		
01722	02 11.5 + 43 09 138.58 – 16.95	07 – 05 – 000	0931 47	1.0 x 0.8 1.1 x 0.9	(145) 2:	3	...	SA d – dm	16.5:		
01723	02 11.6 + 07 20 155.91 – 49.93	01 – 06 – 058	1282	1.3 x 1.0 1.4 x 1.0	(50) 2 – 3	3	Sc	SAB cd	16.0:		
01724	02 11.7 + 07 37 155.73 – 49.67	01 – 06 – 059	1282	1.2 x 1.2 1.2 x 1.1	— 1	4	Sc	SAB(s)c Sc –	15.7		
01725	02 11.9 + 01 15 161.19 – 55.10	00 – 06 – 055	0852	1.1 x 0.15 1.0 x 0.15	[84] ...	3	Sc – Irr Distorted,	sp of 2 bridge:	16.5:		*
01726	02 11.9 + 31 15 143.06 – 28.11	05 – 06 – 017	1248	1.9 x 0.4: 1.7 x 0.4	63 6	1	Sb – c		14.6		*
01727	02 12.0 – 01 00 163.50 – 56.94	00 – 06 – 056 NGC 863	0852	1.3 x 1.3 1.4 x 1.4	— 1	3	Sa P w 1713	SA a	14.0		*3
01728	02 12.0 + 49 38 136.51 – 10.78	08 – 05 – 000	0907 36	1.8: x 0.12 1.5 x 0.12	37 7	2	Sc Sc +		17.		
01729	02 12.1 + 32 30 142.60 – 26.93	05 – 06 – 018	1248	1.7 x 1.0 1.7: x 1.0:	35: 4	3	S B c	SAB(s)cd	15.6		*
01730	02 12.3 + 32 28 142.66 – 26.94	05 – 06 – 000	1248	1.1 x 0.7 0.9:: x 0.6:	160: ...	1	S B ... Singular		16.0:		*
01731	02 12.5 + 18 05 149.33 – 40.22	03 – 06 – 048	1202 35	(1.3 x 1.1) (1.4 x 1.2)	— ...	1	E?		15.0		
01732	02 12.5 + 18 27 149.14 – 39.89	03 – 06 – 000	1202	1.1 x 0.1 1.1 x 0.1	23 7	2	Sc		16.5:		
01733	02 12.5 + 21 47 147.42 – 36.84	04 – 06 – 013	0858	1.7 x 0.20 1.8 x 0.22	128 7	3	Sc		15.6		
01734	02 12.5 + 31 40 143.03 – 27.67	05 – 06 – 000	1248	1.0 x 0.15 0.9 x 0.15	39 7	1	Sa – b		15.7		
01735	02 12.6 + 35 17 141.63 – 24.28	06 – 06 – 002	0407 37	(1.4 x 1.1) (1.1 x 1.0)	— ...	1	E – S0		14.0		
01736	02 12.8 + 05 46 157.54 – 51.14	01 – 06 – 061 NGC 864	1282	4.8 x 3.5 5.2: x 3.8:	20 3	4	Sc Sc –	SAB(rs)c	12.0	+ 1583 + 1644	*12 3478
01737	02 12.8 + 35 41 141.52 – 23.89	06 – 06 – 003 NGC 861	0407 37	1.5 x 0.6 1.5 x 0.4	38 6	2	Sb		14.8		*3
01738	02 12.8 + 42 35 139.02 – 17.40	07 – 05 – 030	0931 47	1.5 x 0.25 0.9 x 0.25	105 6 – 7	3	Sc P w 1743		16.5:		*
01739	02 12.9 + 24 59 146.00 – 33.85	04 – 06 – 014	0858 38	1.3 x 0.25 1.0 x 0.25	37 6 – 7	1	S ...		14.9		*
01740	02 12.9 + 43 16 138.80 – 16.75	07 – 05 – 000	0931 47	1.0: x 0.9: 1.0: x 0.9:	— ...	2	Dwarf: IV – V:		17.		
01741	02 13.0 + 01 25 161.43 – 54.81	00 – 06 – 057	0852	1.0 x 0.3 1.0 x 0.25	178 6	2	S B b – c P w 1746		15.1		*
01742	02 13.0 + 15 10 151.13 – 42.81	02 – 06 – 049	0209 35	1.0 x 0.9 0.9 x 0.9	— 1	4	Sc Brightest of 3	SAB cd	15.7		*
01743	02 13.1 + 42 35 139.07 – 17.38	07 – 05 – 031	0931 47	1.3 x 1.2 1.3 x 1.2	— 1	3	S B b P w 1738	(R:)S B b	15.7		*
01744	02 13.2 + 32 25 142.88 – 26.92	05 – 06 – 019 IC 1784	1248	2.3:: x 1.2: 1.0 x 0.7	92 4:	2	Sb – c P w IC 1785		14.5		*12 5
01745	02 13.2 + 50 27 136.43 – 09.95	08 – 05 – 000	0907 36	1.4: x 0.7: 1.1 x 0.6	(92) ...	1	...		16.5:		
01746	02 13.3 + 01 32 161.43 – 54.67	00 – 06 – 058	0852	1.5 x 1.2 1.3: x 1.2:	— 1 – 2	3	Sc P w 1741		15.2		*
01747	02 13.3 + 28 22 144.59 – 30.68	05 – 06 – 020 NGC 865	1248	2.0 x 0.35 1.8 x 0.35	158 6	1	S ...		14.0		*3
01748	02 13.4 – 00 58 163.98 – 56.71	00 – 06 – 000 NGC 868	0852	1.4: x 1.3: 1.3:: x 1.3::	— ...	1	...		15.6		
01749	02 13.4 + 18 05 149.59 – 40.13	03 – 06 – 051	1202 35	1.1 x 0.3 1.2 x 0.35	135 6	2	Sa – b		15.7		
01750	02 13.4 + 31 47 143.18 – 27.50	05 – 06 – 022	1248	1.4 x 0.4 1.3 x 0.4	71 6	2	Sa – b		14.6		*

N G C 855, 861, 863 – 65, 868

I C 213, 214, 1784

1	2	3	4	5	6	7	8	9	10	11	12
01751	02 13.5 + 16 21 150.59 - 41.69	03 - 06 - 000	1202 35	1.0 x 0.4 0.9 x 0.3	23 6	2	S B b:		16.0:		
01752	02 13.5 + 24 40 146.30 - 34.09	04 - 06 - 000	0858 38	1.7: x 1.7: 1.5:: x 1.3::	-- 1	3	Sc	SA(s)cd	16.5:		
01753	02 13.6 + 27 59 144.83 - 31.02	05 - 06 - 023	1248 38	1.1 x 0.3 1.2 x 0.4	[16] ...	3	Irr or disrupted spiral		15.1		
01754	02 14.0 + 30 43 143.75 - 28.44	05 - 06 - 000	1248 38	1.3 x 0.15 1.5 x 0.2	42 ...	2	S B? b - c Disturbed		15.6		*
01755	02 14.2 + 11 34 153.75 - 45.90	02 - 06 - 051	0209	1.2 x 0.9: 1.1 x 0.9	162: 2 - 3	2	Sb - c		16.0:		
01756	02 14.3 + 01 57 161.39 - 54.18	00 - 06 - 059	0852	1.1 x 0.4 1.0 x 0.4	52 6	1	S ...		15.2		
01757	02 14.3 + 38 11 140.88 - 21.44	06 - 06 - 006	0407	0.9 x 0.22 0.9 x 0.25	87 ...	1	S: ...		13.6		
01758	02 14.3 + 46 49 137.82 - 13.32	08 - 05 - 000	0907	1.0: x 0.8: 1.0 x 0.8	-- 2:	3	Sc/S B c		17.		
01759	02 14.4 + 14 19 152.06 - 43.43	02 - 06 - 053 NGC 871	0209 35	1.1 x 0.35 1.1 x 0.35	[4] ...	1	... NGC 877 group		13.6	+ 3731 + 3821	*12 3
01760	02 14.5 + 01 00 162.38 - 54.94	00 - 06 - 060 NGC 875	0852	1.5 x 1.5 (1.4 x 1.4)	-- 1	3	S0 - a P w IC 218		14.2		*3
01761	02 14.7 + 14 21 152.13 - 43.36	02 - 06 - 054	0209 35	1.2 x 1.2 1. x 1.	-- ...	4	Dwarf irregular Ir V	NGC 877 group	16.5:		*
01762	02 14.9 + 12 17 153.50 - 45.18	02 - 06 - 055 IC 1790	0209	1.1 x 0.3 1.0 x 0.3	65 7	1	S ... P w 1764		15.7		*
01763	02 14.9 + 32 10 143.36 - 27.36	05 - 06 - 024 IC 1789	1248 38	2.7: x 0.4 2.5: x 0.4	29 7	2	Sa:		14.8		
01764	02 15.0 + 12 15 153.55 - 45.20	02 - 06 - 056 IC 1791	0209	(1.0 x 1.0) (1.2 x 1.2)	-- 1	3	S0 P w 1762		14.6		*
01765	02 15.0 + 35 32 142.04 - 23.88	06 - 06 - 008	0407 37	1.2 x 0.7 1.0 x 0.7	46 ...	1	S B? c - Irr sf of 2		15.4		*
01766	02 15.1 + 14 18 152.28 - 43.36	02 - 06 - 057 NGC 876	0209 35	1.7: x 0.6: 1.7: x 0.5	20 7	1	S ... NGC 877 group, p w 1768		16.5:		*23
01767	02 15.1 + 37 51 141.17 - 21.70	06 - 06 - 009	0407 37	1.2 x 1.2 1.2 x 1.1	-- ...	3	Irr P w 1772	Ir I	14.0		*1
01768	02 15.2 + 14 19 152.30 - 43.34	02 - 06 - 058 NGC 877	0209 35	2.3 x 1.8 2.2 x 1.8	140 2	4	Sc Brightest in group, Sc -	SAB(rs)c	12.5	+ 4016 + 4106	*123 4678
01769	02 15.2 + 36 51 141.57 - 22.63	06 - 06 - 010	0407 37	1.0 x 0.6 1.0 x 0.6	123 4	2	Sb/Sc		13.9		*
01770	02 15.3 + 36 17 141.81 - 23.15	06 - 06 - 000	0407 37	1.0 x 0.22 1.1 x 0.25	49 6	1	S0 - a		16.5:		
01771	02 15.3 + 37 41 141.27 - 21.84	06 - 06 - 011	0407 37	2.2:: x 1.8:: 1.8:: x 1.4::	-- ...	4	Dwarf spiral S V		17.		
01772	02 15.4 + 37 48 141.25 - 21.73	06 - 06 - 000	0407 37	1.1 x 0.5 0.7 x 0.5	143: ...	1	... P w 1767		13.7		*
01773	02 15.5 + 12 58 153.24 - 44.51	02 - 06 - 059	0209	1.6 x 0.3 1.5: x 0.3	48 6 - 7	1	S B ...		15.3		*
01774	02 15.7 + 05 19 158.90 - 51.16	01 - 06 - 000	1282	1.0 x 0.3 0.9 x 0.2	63 6	2	Sb - c		16.0:		*
01775	02 15.7 + 05 25 158.81 - 51.07	01 - 06 - 062	1282	1.9 x 1.8 2.0: x 1.9:	-- ...	4	S ... Peculiar		14.2		*1A
01776	02 15.7 + 35 14 142.30 - 24.10	06 - 06 - 013	0407 37	1.6 x 1.3 1.5 x 1.3	(20) 1 - 3	4	S B b	S B(s)b Sb -	15.3		
01777	02 15.8 + 30 17 144.35 - 28.70	05 - 06 - 000	1248 38	1.0 x 0.6 1.0 x 0.6	100: ...	4	S B c	SAB(s)cd	15.7		
01778	02 15.8 + 33 30 143.01 - 25.71	06 - 06 - 014	0407	1.4 x 1.1 1.2: x 1.1:	(160) ...	2	...	SA dm:	15.6		
01779	02 15.8 + 38 28 141.07 - 21.08	06 - 06 - 000	0407	1.1 x 0.9 0.8 x 0.5	(40) ...	3	... n of 2	S dm	17.		*
01780	02 15.9 + 40 20 140.40 - 19.32	07 - 05 - 035	0931	2.0 x 0.4 1.8: x 0.4:	[158] ...	4	Irr Ir IV - V	IB: m Ir I	15.6		
01781	02 16.0 + 34 15 142.75 - 25.00	06 - 06 - 015 IC 1792	0407 37	(1.6 x 1.5) (1.7 x 1.6)	-- ...	2	... Compact		15.0		*
01782	02 16.0 + 42 31 139.62 - 17.27	07 - 05 - 037	0931 47	1.8: x 0.8: 1.7: x 0.8::	63 6	3	... S IV or IV - V	S B: dm	17.		
01783	02 16.0 + 43 54 139.12 - 15.97	07 - 05 - 000	0931 47	1.5:: x 1.1:: 0.7 x 0.6	-- ...	2	Dwarf: IV - V:		17.		
01784	02 16.2 + 36 27 141.93 - 22.93	06 - 06 - 000	0407 37	1.2: x 0.7: 0.6:: x 0.7::	-- ...	1	Irr or Peculiar		17.		*
01785	02 16.4 + 04 35 159.76 - 51.69	01 - 06 - 000	1282	1.4 x 0.7: 1.2: x 0.5:	150 5 - 6	3	Sc	Sc +	16.0:		

N G C   871, 875 - 77   I C   1789 - 92

1	2	3	4	5	6	7	8	9	10	11	12
01786	02 16.5 + 36 53 141.82 − 22.51	06 − 06 − 016	0407 37	(1.0 x 0.5) (0.9 x 0.5)	13: ...	1	[E] Compact, np of 2		15.6		*
01787	02 16.6 + 37 43 141.52 − 21.72	06 − 06 − 017	0407 37	1.4 x 0.5 1.3 x 0.4	117 ...	2	S − Irr Distorted		14.7		*1
01788	02 16.7 + 36 24 142.05 − 22.94	06 − 06 − 000	0407 37	(1.1 x 0.9) (1.2 x 0.9)	(25) ...	1	E:		16.0:		
01789	02 16.9 + 15 35 152.03 − 42.02	03 − 06 − 052 NGC 882	1202 35	(1.1 x 0.6) 1.2 x 0.6	82 5:	2	S0		14.9		
01790	02 16.9 + 41 29 140.16 − 18.18	07 − 05 − 000	0931 47	1.2: x 0.9: 1.3 x 1.1	— ...	1	...		15.4		*
01791	02 17.0 + 28 02 145.61 − 30.68	05 − 06 − 000	1248 38	1. x 1. 1.1: x 1.1:	— ...	2	Dwarf: IV: or IV − V:		17.		
01792	02 17.0 + 28 49 145.26 − 29.96	05 − 06 − 026	1248 38	2.6 x 1.5 2.2 x 1.4:	0: 4	4	S B c	SAB(r)c Sc −	14.3		*
01793	02 17.0 + 37 41 141.61 − 21.72	06 − 06 − 000	0407 38	1.1 x 0.22 1.0 x 0.2	163 7	3	Sc		15.7		*
01794	02 17.2 − 00 29 164.86 − 55.76	00 − 07 − 001	1283	1.7 x 1.7 1.3 x 0.7	— 1	3	S B b	SAB(s)b Sb +	14.6		
01795	02 17.3 + 35 41 142.45 − 23.57	06 − 06 − 000	0407 37	1.0 x 0.9: 0.8 x 0.5	— ...	2	S B c	Sc +	16.5:		
01796	02 17.3 + 40 34 140.58 − 19.01	07 − 05 − 038	0931 47	1.6: x 1.6: 1.5:: x 1.5::	0: 1	4	... S IV	SAB(s)dm	15.5		*
01797	02 17.4 + 01 43 162.71 − 53.94	00 − 07 − 003	1283	1.3 x 0.8 1.4 x 0.9	135 3 − 5	2	S0 − a sf of 2		15.2		*1
01798	02 17.4 + 38 18 141.45 − 21.12	06 − 06 − 000	0407 37	1.0 x 0.18 1.0 x 0.20	21 6	2	Sa − b p of 2		15.6		*
01799	02 17.4 + 48 13 137.84 − 11.83	08 − 05 − 000	0907	1.3 x 0.8 1.2: x 0.8:	23 ...	3	S B b/Sc sff of 2,	SAB(s)bc disturbed	16.0:		*
01800	02 17.5 + 34 59 142.77 − 24.20	06 − 06 − 000	0407 37	1.6 x 0.8 1.9 x 0.9	160 5	1	S ...		16.0:		
01801	02 17.7 + 07 48 157.57 − 48.77	01 − 07 − 001	1306	1.2 x 0.8 1.1 x 0.8	(80) ...	3	S B b	S B(s)b	15.7		*
01802	02 17.8 + 48 44 137.72 − 11.32	08 − 05 − 004	0907	(1.6 x 1.5) (1.7 x 1.6)	— ...	3	E		15.6		
01803	02 17.9 + 06 35 158.59 − 49.79	01 − 07 − 002	1306	2.6 x 0.6 2.0: x 0.5:	[57] ...	3	Irr		15.1		
01804	02 17.9 + 38 26 141.50 − 20.96	06 − 06 − 022	0407 37	1.7 x 0.8 1.4: x 0.8:	[95] ...	1	...	[SA m]?	15.5		*
01805	02 18.0 + 32 37 143.85 − 26.36	05 − 06 − 000	1248 38	1.0 x 0.22 0.9 x 0.22	164 (7)	1	...		14.5		
01806	02 18.0 + 33 10 143.62 − 25.85	05 − 06 − 000	1248 38	1.2:: x 0.5: 1. x 0.5:	[15] ...	2	Dwarf np of 2		18.		*
01807	02 18.0 + 42 33 139.97 − 17.11	07 − 05 − 040	0931 47	2.5:: x 2.5:: 2. x 2.	— ...	4	Dwarf IV − V or V		16.5:		
01808	02 18.2 + 23 22 148.12 − 34.86	04 − 06 − 017	0858 38	1.2 x 1.1 1.1 x 1.1	— 1	2	Sb		14.7		1
01809	02 18.3 + 00 20a 164.40 − 54.94	00 − 07 − 005	1283	1.1 x 0.8 1.2 x 0.9	60 2 − 3	2	S0 − a Brightest of 2		(15.1)		*
01810	02 18.4 + 39 09 141.32 − 20.26	06 − 06 − 023	0407	2.1 x 1.4 1.9 x 1.3	[50] ...	4	S[b] P w 1813, distorted		13.7		*1V AC
01811	02 18.5 + 13 30 153.80 − 43.70	02 − 07 − 002	1300	1.3: x 0.8: 1.0: x 0.7:	145: 3 − 4	2	Sc		16.0:		
01812	02 18.5 + 25 11 147.30 − 33.17	04 − 06 − 018	0858 38	1.1 x 0.5 1.3 x 0.5	20 6	3	Sc		14.6		
01813	02 18.5 + 39 08 141.35 − 20.27	06 − 06 − 024	0407	1.4 x 0.4 1.4 x 0.4	[93] ...	1	S B ... P w 1810, distorted		15.3		*1
01814	02 18.6 + 16 20 152.06 − 41.17	03 − 07 − 002	443 35	2.7 x 1.3 2.7 x 1.4	157 ...	4	Sb/S B c Disturbed:	SAB(s)bc	14.0		*
01815	02 18.6 + 30 49 144.75 − 27.98	05 − 06 − 028	1248 38	1.4 x 1.3 1.3 x 1.3	— 1:	3	... S B IV		15.7		
01816	02 18.6 + 32 20 144.10 − 26.58	05 − 06 − 027 IC 1793	1248 38	1.8 x 0.8 1.7 x 0.8	33 6	2	Sa − b		14.8		
01817	02 18.8 + 13 59 153.57 − 43.24	02 − 07 − 003	1300	2.6 x 0.4 2.7 x 0.3	163 7	3	Sc	Sc +	14.8		*
01818	02 18.8 + 15 07 152.86 − 42.23	02 − 07 − 000	1300 35	[1.0 x 0.5] 	... ...	3	S B: ... + companion Contact, distorted		16.5:		*
01819	02 18.8 + 16 09 152.23 − 41.31	03 − 07 − 000	0443 35	1. x 1. 1. x 1.	— ...	3	Dwarf V		18.		*
01820	02 18.8 + 32 48 143.95 − 26.13	05 − 06 − 029	1248 38	1.8 x 0.35 1.8 x 0.35	124 7	3	Sc		15.1		

N G C   882   I C   1793

1	2	3	4	5	6	7	8	9	10	11	12
01821	02 19.0 + 06 33 158.98 - 49.67	01 - 07 - 000	1306	1.1 x 1.0 1. x 1.	-- ...	2	Dwarf: S: IV - V:		17.		
01822	02 19.0 + 16 39 151.99 - 40.84	03 - 07 - 004	0443 35	1.3 x 0.2 1.3 x 0.20	78 7	3	Sc		15.7		
01823	02 19.0 + 33 03 143.89 - 25.88	05 - 06 - 030 NGC 890	1248 38	(2.9 x 2.3) (2.8 x 2.1)	55: 2:	3	S0		12.5	+ 4043 + 4190	*12 34 6 8
01824	02 19.1 + 00 02 164.99 - 55.07	00 - 07 - 006	1283	1.0 x 0.6 1.1 x 0.6	30 4	2	S0 - a		15.6		
01825	02 19.1 + 32 00 144.35 - 26.85	05 - 06 - 031	1248 38	1.3 x 0.9 1.1 x 0.7	(130) 3	3	Sc	SA(s)c	15.0		
01826	02 19.1 + 33 43 143.63 - 25.26	06 - 06 - 025	0407	1.2 x 0.35 1.0 x 0.35	45 ...	1	S B: ... Distorted?		14.6		*1
01827	02 19.1 + 43 19 139.88 - 16.32	07 - 05 - 044	0931 47	1.0: x 0.7 0.7 x 0.6	[0] ...	1	S - Irr		15.7		
01828	02 19.3 + 28 31 145.93 - 30.04	05 - 06 - 032	1248 38	1.4 x 0.15 1.1 x 0.15	122 7	1	S ... P w 1833, disturbed		15.5		*1
01829	02 19.3 + 33 31 143.76 - 25.43	06 - 06 - 026	0407	[1.1: x 0.8:] ...	-- ...	3	Irr + companion Contact		15.7		1
01830	02 19.3 + 47 37 138.36 - 12.28	08 - 05 - 006	0907	3.2 x 2.6 1.0 x 1.5	(105) 1:	4	S B 0/S B a		14.8		*1C
01831	02 19.4 + 42 07 140.38 - 17.42	07 - 05 - 046 NGC 891	0931 47	14. x 3. 14. x 3.	22 7	4	Sb NGC 1023 group, Sb +		10.8	+ 72 + 243	*12- -9
01832	02 19.4 + 42 50 140.12 - 16.75	07 - 05 - 045	0931 47	1.6 x 0.9: 1.7 x 0.9	163: 4 - 5	2	Sa		15.4		
01833	02 19.5 + 28 30 145.99 - 30.04	05 - 06 - 033	1248 38	1.6: x 1.3: 1.4: x 1.2:	-- ...	4	S B c P w 1828, disturbed	S B(s)d	15.6		*1
01834	02 19.6 + 11 49a 155.25 - 45.05	02 - 07 - 000	1300	1.5 x 0.8 1.2 x 0.6	124 5	3	S B [c] Peculiar, sp of 2, disturbed		(15.3)		*
01835	02 19.8 + 28 02 146.27 - 30.44	05 - 06 - 034 IC 221	1248 38	2.2 x 1.2 2.2: x 1.2	30: 4 - 5	4	Sc	Sc -	13.9		*
01836	02 19.8 + 37 52 142.09 - 21.35	06 - 06 - 000	0407 37	1.1:: x 0.6:: 0.6: x 0.4:	(170) ...	1	Dwarf:		18.		*
01837	02 19.8 + 42 47 140.21 - 16.77	07 - 06 - 001	0449 47	(1.2 x 0.9) (1.3 x 0.9)	(25) ...	2	S0 P w 1841, disturbed?		15.2		*
01838	02 19.9 + 24 08 148.17 - 34.00	04 - 06 - 000	0858 38	1.3 x 0.3 1.2 x 0.2	169 6 - 7	3	Sc		16.5:		
01839	02 20.0 - 00 51 166.24 - 55.64	00 - 07 - 08a	1283	3.0: x 0.4 2.3: x 0.4:	45 7	3	...	S d - m	15.6		
01840	02 20.0 + 41 09 140.86 - 18.28	07 - 06 - 002	0449 47	[1.8 x 1.7] ...	... ...	3	... Peculiar, loop		14.1	+ 5240 + 5388	*1A C
01841	02 20.0 + 42 46a 140.25 - 16.77	07 - 06 - 003	0449 47	(3. x 3.) (3. x 3.)	-- ...	3	E P w 1837		(15.0)		*1C
01842	02 20.2 + 41 44 140.67 - 17.72	07 - 06 - 004 NGC 898	0449 47	2.0 x 0.5 2.0 x 0.5	170 7	3	Sa		13.8		*12 3
01843	02 20.7 + 26 17 147.31 - 31.96	04 - 06 - 020 NGC 900	0858 38	1.1 x 0.7 0.9 x 0.6	(30) 3 - 4	2	S0		15.0		*3
01844	02 20.8 + 26 56 147.02 - 31.36	04 - 06 - 021	0858 38	1.6: x 1.3: 1.0 x 0.8	-- 2:	3	...	S B dm	15.7		
01845	02 20.8 + 47 45 138.55 - 12.07	08 - 05 - 007	0907	1.5 x 0.8 1.6 x 0.9	145 4 - 5	1	Sa - b		15.7		1
01846	02 20.9 + 28 38 146.25 - 29.79	05 - 06 - 000	1248 38	(1.1 x 0.8) (1.3 x 1.1)	(145) ...	1	... Compact		15.5		
01847	02 21.0 + 02 18 163.39 - 52.95	00 - 07 - 000	1283	[1.1 x 0.2] ...	[117] ...	1	Double system Bridge		15.7		*
01848	02 21.0 + 27 16 146.91 - 31.04	04 - 06 - 023	0858 38	1.2 x 0.5 1.2 x 0.5	22 6	2	Sb		15.5		
01849	02 21.0 + 43 04 140.31 - 16.43	07 - 06 - 005	0449 47	1.4 x 0.3 1.6 x 0.3	152 6	1	S ... P w 1854		16.0:		*
01850	02 21.1 + 26 53 147.12 - 31.38	04 - 06 - 000	0858 38	1.1: x 0.9: 0.8 x 0.8	(35) ...	3	Dwarf spiral S B IV - V:		16.5:		
01851	02 21.1 + 49 17 138.05 - 10.61	08 - 05 - 000	0907	1.2: x 1.0: 1.1: x 0.9:	(5) ...	2	...	S dm	17.		
01852	02 21.2 + 27 07 147.03 - 31.16	04 - 06 - 024 NGC 904	0858 38	(1.5 x 1.0) (1.7 x 1.1)	(130) ...	3	E sf of 2		15.0		*
01853	02 21.3 + 33 25 144.23 - 25.36	05 - 06 - 000	1248	1.2 x 0.10 1.0 x 0.10	160 7	3	Sc	Sc +	17.		
01854	02 21.3 + 43 02 140.38 - 16.44	07 - 06 - 007	0449 47	1.9: x 0.5 1.7: x 0.5	45 7	1	S ... P w 1849		16.5:		*
01855	02 21.4 + 40 39 141.31 - 18.65	07 - 06 - 008	0449 47	1.5 x 0.7 1.5 x 0.7	97 5	3	S B a	S B(s)a	15.1		

N G C   890, 891, 898, 900, 904

I C   221

1	2	3	4	5	6	7	8	9	10	11	12
01856	02 21.6 + 31 23 145.17 – 27.21	05 – 06 – 038	1248 38	2.2 x 0.35 2.2 x 0.3	123 7	3	Sc	Sc +	14.9		
01857	02 21.6 + 32 57 144.49 – 25.77	05 – 06 – 037	1248	2.4 x 0.7 2.1: x 0.7	44 6	4	...	S d – dm	15.3		
01858	02 21.6 + 41 28 141.03 – 17.88	07 – 06 – 010	0449 47	1.7: x 1.3: 1.8: x 1.5:	— ...	2	S B ...		15.7		
01859	02 21.6 + 42 24 140.67 – 17.01	07 – 06 – 009	0449 47	(1.7 x 1.1) (1.6 x 1.0)	47 ...	3	... Compact		14.3		*1
01860	02 21.8 + 25 20 148.04 – 32.73	04 – 06 – 025	0858 38	1.5 x 0.9 1.5 x 0.9	30: 4	4	S B b n of 2	S B(rs)b	14.7		*1
01861	02 21.8 + 30 37 145.56 – 27.90	05 – 06 – 000	1248 38	1.1: x 0.7: 1.1: x 0.5:	[90] ...	3	Dwarf: irregular Ir IV – V:		17.		
01862	02 21.9 – 02 23 168.62 – 56.54	00 – 07 – 009	1283	2.3 x 1.8 2.2: x 1.8:	(10) 1 – 2	3	Sc	SAB(s)d: Sc +	14.0		*
01863	02 21.9 + 01 37 164.36 – 53.38	00 – 07 – 010	1283	1.1 x 1.1 0.9 x 0.8	— 1	3	Sc	SAB(s)c	15.3		
01864	02 21.9 + 37 15 142.76 – 21.77	06 – 06 – 000	0407 37	1.4 x 0.7 1.2 x 0.6	88 5	1	Sb		15.7		
01865	02 22.0 + 35 49 143.36 – 23.09	06 – 06 – 028	0407	3.5: x 2.8: 2.5:: x 2.0::	(80) ...	4	Dwarf spiral S V		16.5:		*
01866	02 22.0 + 41 38 141.04 – 17.69	07 – 06 – 011	0449 47	1.1 x 0.6 1.1 x 0.6	30 5	2	S B a		14.9		*
01867	02 22.0 + 45 15 139.67 – 14.33	08 – 05 – 000	0907	2.1 x 0.3 2.2 x 0.3	57 7	3	Sc	Sc +	15.5		*
01868	02 22.1 + 41 52 140.97 – 17.47	07 – 06 – 012 NGC 906	0449 47	2.1 x 2.0 2.3 x 2.2	— 1	4	S B a	SAB(s)ab	14.4		*1
01869	02 22.1 + 50 36 137.72 – 09.33	08 – 05 – 008	0907	1.0 x 0.15 0.9 x 0.18	151 7	1	Sa – b		15.7		
01870	02 22.2 + 19 27 151.25 – 38.00	03 – 07 – 000	0443	1.5: x 0.2 1.1 x 0.2	24 7	1	S ... s of 2		18.		*
01871	02 22.2 + 22 00 149.85 – 35.71	04 – 06 – 027	0858 38	1.0 x 0.7 1.2 x 0.9	37 3:	1	S ...		14.9		*1
01872	02 22.2 + 41 49 141.01 – 17.51	07 – 06 – 013 NGC 909	0449 47	(0.9 x 0.9) (1.3 x 1.3)	— ...	3	E		14.5		*
01873	02 22.3 + 16 52 152.79 – 40.29	03 – 07 – 009	0443	(1.0 x 1.0) (0.8 x 0.8)	— ...	1	...		16.0:		*
01874	02 22.3 + 40 29 141.54 – 18.74	07 – 06 – 000	0449 47	1.4:: x 1.0:: 1. x 1.	— ...	3	...	SAB d – dm	17.		
01875	02 22.3 + 41 36 141.11 – 17.70	07 – 06 – 014 NGC 910	0449 47	(2. x 2.) (4. x 4.)	— ...	3	E		14.5		*3
01876	02 22.5 – 00 49 167.07 – 55.23	00 – 07 – 000	1283	1.1: x 0.9: 0.8 x 0.6	— 1 – 2	1	S ...		16.5:		
01877	02 22.6 + 36 45 143.10 – 22.18	06 – 06 – 030	0407 37	(1.6 x 1.2) (1.7 x 1.3)	— ...	2	E		15.0		*
01878	02 22.6 + 41 45 141.10 – 17.54	07 – 06 – 016 NGC 911	0449 47	(1.7 x 0.9) (1.8 x 1.0)	115: ...	3	E		14.0		*
01879	02 22.7 + 10 52 156.85 – 45.50	02 – 07 – 006	1300	1.3: x 1.2 1.4: x 1.3:	— 1	3	Sc s of 2	SAB c Sc –	15.6		*1
01880	02 22.7 + 20 10 150.98 – 37.30	03 – 07 – 010 IC 1797	0443	1.1: x 0.4: 1.2: x 0.5:	138 ...	1	S B ...		15.3		*
01881	02 22.7 + 26 31 147.68 – 31.57	04 – 06 – 030	0858 38	[1.0 x 0.9] ...	— 1:	2	Sb + companion Connected		14.8		*
01882	02 22.7 + 37 00 143.02 – 21.94	06 – 06 – 031	0407 37	1.5 x 1.3 1.6 x 1.4	— 1	4	Sc	Sc –	14.9		
01883	02 22.8 + 11 15 156.61 – 45.16	02 – 07 – 007	1300	2.5 x 0.7 1.8 x 0.6	157 6	3	...	S d – dm	15.7		*1
01884	02 22.8 + 37 40 142.76 – 21.32	06 – 06 – 032	0407 37	1.5: x 1.4: 1.4: x 1.4:	— 1:	3	S B a	S B(r) ...	15.7		*
01885	02 22.9 + 27 11 147.40 – 30.94	04 – 06 – 035	0858 38	1.0: x 0.9: 1.2: x 1.1:	— 1:	2	S B b	S B(r)b	15.6		1
01886	02 22.9 + 39 15 142.14 – 19.84	06 – 06 – 033	0407	4.5 x 2.3 3.8 x 1.8	35 5	4	Sb	SAB(rs)bc Sb +	13.1		
01887	02 22.9 + 41 55 141.09 – 17.37	07 – 06 – 017 NGC 914	0449 47	2.1 x 1.4 2.1 x 1.4	117 3 – 4	4	Sc	SA(s)c Sc +	13.9		*1
01888	02 23.1 + 18 16 152.17 – 38.96	03 – 07 – 011 NGC 918	0443	3.6 x 2.1 3.8 x 2.2	157 4	4	Sc		14.3		*12 3
01889	02 23.1 + 29 57 146.15 – 28.40	05 – 06 – 040	1248 38	(1.2 x 1.0) 1.3 x 1.2	— 1:	3	S B 0		15.0		
01890	02 23.2 + 31 41 145.39 – 26.80	05 – 06 – 039	1248 38	2.5: x 1.3 2.4: x 1.2	55 5	2	Sa – b		14.5		*

1	2	3	4	5	6	7	8	9	10	11	12
01891	02 23.3 + 25 50 148.16 – 32.13	04 – 06 – 038	0858 38	(1.1 x 1.1) (1.7 x 1.7)	— ...	3	E ssp of 2		15.5		*
01892	02 23.3 + 27 23 147.40 – 30.72	04 – 06 – 037	0858 38	1.3: x 1.3: 1.3: x 1.3:	— 1	2	S B b	S B(r) ...	15.3		
01893	02 23.3 + 49 50 138.18 – 09.97	08 – 05 – 009	0907	1.5: x 1.1: 1.7 x 1.3	45 2 – 3	3	S B 0/S B a		15.5		
01894	02 23.4 + 26 59 147.62 – 31.08	04 – 06 – 039 NGC 919	0858 38	1.3 x 0.25 1.4 x 0.25	138 7	2	Sa – b		15.5		3
01895	02 23.4 + 28 17 146.99 – 29.89	05 – 06 – 000	1248 38	1.5 x 1.1 1.4: x 0.8:	125: ...	3	S B c	S B(s)c	15.7		*
01896	02 23.4 + 30 13 146.10 – 28.13	05 – 06 – 041	1248 38	1.5: x 1.0: 1.4: x 1.0:	5 3 – 4:	2	S0	SA(r)0	14.6		
01897	02 23.5 + 12 14 156.13 – 44.22	02 – 07 – 008	1300	1.1 x 0.9 1.0: x 0.9:	(40) 2:	3	Sc	Sc +	16.0:		
01898	02 23.5 + 22 47 149.77 – 34.87	04 – 06 – 000	0858 38	1.5: x 0.3 1.1 x 0.15	50 7	3	Sc	Sc +	16.5:		*
01899	02 23.5 + 27 26 147.42 – 30.66	04 – 06 – 040	0858 38	1.3: x 1.2: 1.3: x 1.3:	— ...	1	... Disturbed?		15.0		*1
01900	02 23.5 + 37 11 143.10 – 21.71	06 – 06 – 035	0407 37	1.5 x 0.5 1.3 x 0.4	55 6	2	Sa – b		16.0:		*
01901	02 23.6 – 00 33 167.16 – 54.85	00 – 07 – 011 NGC 926	1283	2.5 x 1.3 2.2 x 1.1	36 5	3	S B b	S B(s) ... Sb +	13.9		23
01902	02 23.6 + 27 35 147.37 – 30.51	05 – 06 – 043	1248 38	1.2 x 0.6 0.9 x 0.4	132: 5	3	Sc		16.5:		*
01903	02 23.6 + 29 36 146.42 – 28.67	05 – 06 – 042	1248 38	1.0 x 0.20 1.0 x 0.20	75 7	2	Sb		15.7		*
01904	02 23.6 + 36 56 143.22 – 21.93	06 – 06 – 000	0407 37	1.1 x 0.20 1.4 x 0.2	117 6	2	Sb – c		16.0:		
01905	02 23.8 – 01 46 168.57 – 55.77	00 – 07 – 012	1283	1.0 x 0.9 0.9 x 0.8	— 1	3	S0		15.6		
01906	02 23.8 + 22 32 149.97 – 35.07	04 – 06 – 000	0858 38	1.4 x 0.2 1.2 x 0.15	152 7	2	Sb		16.5:		
01907	02 23.9 + 00 56 165.72 – 53.63	00 – 07 – 013 IC 225	1283	(1.5 x 1.3) (1.8 x 1.4)	— ...	3	E		14.7		
01908	02 23.9 + 11 56 156.45 – 44.43	02 – 07 – 009 NGC 927	1300	1.3 x 1.3 1.4 x 1.3	— 1	4	S B c f of 2	S B(r)c Sc –	14.5		*1
01909	02 23.9 + 31 52 145.46 – 26.57	05 – 06 – 044	1248 38	1.2 x 0.7 1.2: x 0.7	132 4	3	Sc	Sc +	15.4		*1
01910	02 23.9 + 34 58 144.11 – 23.72	06 – 06 – 036	0407	2.3 x 0.6 2.0 x 0.5	175 6	1	S ...		14.8		*
01911	02 24.0 + 00 25 166.28 – 54.02	00 – 07 – 014	1283	1.0 x 0.1 1.0 x 0.12	131 7	2	Sc		16.5:		
01912	02 24.0 + 20 16 151.27 – 37.08	03 – 07 – 012 NGC 924	0443	(2.3 x 1.3) (2. x 1.4)	53: 4 – 5:	2	S0 sp of 2, NGC 932 group		13.8		*13
01913	02 24.3 + 33 21 144.89 – 25.18	05 – 06 – 045 NGC 925	1248 38	13. x 8. 12. x 7.	115: 4	5	Sc/S B c NGC 1023 group	SAB(s)d	10.5	+ 568 + 712	*12– –9
01914	02 24.3 + 40 50 141.78 – 18.28	07 – 06 – 000	0449 47	(1.2 x 1.2) (1.4 x 1.4)	— ...	3	[E] Very compact		15.2		
01915	02 24.4 + 41 45 141.43 – 17.42	07 – 06 – 022 NGC 923	0449 47	0.9 x 0.6 0.9 x 0.7	95: ...	2	S[b] 2nd of 3		14.4		*
01916	02 24.4 + 50 16 138.19 – 09.51	08 – 05 – 010	0907	(1.1 x 0.6) (1.3 x 0.7)	(5) ...	2	E		16.0:		
01917	02 24.6 + 24 02 149.39 – 33.63	04 – 06 – 045	0858 38	1.0 x 0.7 1.1 x 0.8	105: 3	3	S B b	(R′)S B(s)ab Sb –	15.5		
01918	02 24.6 + 25 26 148.67 – 32.37	04 – 06 – 046	0858 38	1.3 x 0.7 1.7 x 0.7	118 6	2	S B a – b		14.7		1
01919	02 24.6 + 35 56 143.84 – 22.78	06 – 06 – 038	0407	1.6 x 1.1 1.4 x 1.1	(45) ...	4	S B b	S B(s)b	15.0		
01920	02 24.6 + 45 44 139.92 – 13.71	08 – 05 – 011 NGC 920	0907	1.9 x 1.3 1.7 x 1.1	10: 3 – 4	3	S B b IC 1799 group, Sb –	(R′)S B(s)ab	15.6		*
01921	02 24.8 + 26 22 148.25 – 31.51	04 – 06 – 051	0858 38	1.1 x 1.1 1.1 x 1.1	— 1:	3	S B b Disturbed?	S B(s)b	14.6		*1C
01922	02 24.8 + 27 58 147.47 – 30.05	05 – 06 – 046 IC 226	1248 38	2.3: x 1.8: 1.8: x 1.5:	— ...	1	S ... Peculiar		16.0:		*
01923	02 24.9 + 01 54 165.08 – 52.70	00 – 07 – 015	1283	1.1 x 0.8 0.9 x 0.7	— ...	1	S – Irr		17.		
01924	02 24.9 + 31 30 145.84 – 26.82	05 – 06 – 047	1248 38	1.9 x 0.25 1.9: x 0.25	3 7	3	Sc		15.4		
01925	02 24.9 + 37 14 143.36 – 21.56	06 – 06 – 000	0407 37	[1.1: x 0.4] ...	[48] ...	1	Double system Plumes		16.5:		*C

1	2	3	4	5	6	7	8	9	10	11	12
01926	02 25.0 − 00 28 167.55 − 54.57	00 − 07 − 016 NGC 934	1283	(1.8 x 1.6) (1.8 x 1.6)	— 1 − 2	3	S0		14.4		3
01927	02 25.0 + 43 23 140.90 − 15.86	07 − 06 − 000	0449 47	1.1 x 0.1 1.2 x 0.10	78 7	2	...	S d − m	17.		
01928	02 25.0 + 43 38 140.80 − 15.63	07 − 06 − 000	0449 47	1.5: x 0.3 1.8 x 0.4	113 7	1	S ...		16.5:		
01929	02 25.1 − 01 23 168.59 − 55.26	00 − 07 − 017 NGC 936	1283	(5.6 x 4.5) (5.8 x 4.8)	135: 2	4	S B 0 or S B 0/S B a P w 1954		11.3	+ 1348 + 1371	*12− −9
01930	02 25.0 + 51 14 137.92 − 08.57	09 − 05 − 001	0907	2.3: x 2.3: 2.3: x 2.3:	— 1	3	S B 0	SB(r)0 +	15.2		*
01931	02 25.1 + 20 06 151.65 − 37.11	03 − 07 − 014 NGC 930 = 932	0443	1.8 x 1.7 1.9 x 1.9	— 1	4	Sa Brightest of 5	SA a	13.7		*3
01932	02 25.1 + 27 57 147.55 − 30.04	·05 − 06 − 048 IC 227	1248 38	(2.0 x 1.8) (2.3 x 2.0)	— ...	2	E:		15.5		*
01933	02 25.2 + 38 13 143.01 − 20.63	06 − 06 − 039	0407 37	1.4: x 1.0: 1.2: x 1.0:	(85) ...	1	S ... Disturbed?	SAB(s)...	16.0:		*
01934	02 25.3 + 00 16 166.87 − 53.94	00 − 07 − 018	1283	1.9 x 0.35 1.7 x 0.35	112 6 − 7	2	Sb − c		15.5		*
01935	02 25.3 + 31 05 146.12 − 27.17	05 − 06 − 049 NGC 931	1248 38	4.5 x 1.3: 3.7: x 1.2:	73 6	3	Sb	SA bc Sb +	13.9		*13
01936	02 25.4 + 19 21 152.16 − 37.75	03 − 07 − 016 IC 1801	0443	1.4 x 0.6 1.4 x 0.6	30 6	3	S B b: P w 1937, contact		14.8		*1V A
01937	02 25.4 + 19 22 152.15 − 37.73	03 − 07 − 015 NGC 935	0443	1.9 x 1.2 1.7 x 1.1	155 3 − 4	4	Sc P w 1936, contact		13.9		*13 459V
01938	02 25.4 + 23 00 150.13 − 34.48	04 − 06 − 054	0858 38	1.5 x 0.35 1.3 x 0.3	154 6	1	Sb − c		15.0		
01939	02 25.4 + 26 05 148.54 − 31.71	04 − 06 − 053	0858 38	1.2: x 0.8 1.1: x 0.9:	65: ...	2	S B a		14.9		*
01940	02 25.5 + 27 55 147.66 − 30.03	05 − 06 − 000	1248 38	1.1 x 0.35 1.2 x 0.5	101 6 − 7	2	S0		16.5:		
01941	02 25.5 + 37 45 143.26 − 21.03	06 − 06 − 040	0407 37	1.6: x 1.6: 1.6:: x 1.6::	— ...	2	...		16.0:		*
01942	02 25.5 + 38 24 142.99 − 20.44	06 − 06 − 000	0407 37	1.0 x 0.12 1.1 x 0.15	17 7	2	Sb		16.5:		
01943	02 25.5 + 45 45 140.07 − 13.64	08 − 05 − 012 IC 1799	0907 47	1.3 x 0.4 1.3 x 0.4	34 6	1	S ... Brightest of 3		15.0		*
01944	02 25.6 + 29 47 146.79 − 28.33	05 − 06 − 000	1248 38	1.5:: x 1.0:: 1.5:: x 1.0::	(100) ...	3	Dwarf spiral S IV − V or V		17.		
01945	02 25.7 − 01 34 168.99 − 55.31	00 − 07 − 020	1283	2.4: x 0.9: 2.3: x 0.9:	[12] ...	3	Sc − Irr		14.4		*
01946	02 25.7 + 15 25 154.64 − 41.18	02 − 07 − 000	1300	1.9 x 0.2 1.4: x 0.1	137 7	3	Sc	Sc +	16.0:		
01947	02 25.7 + 20 03 151.83 − 37.09	03 − 07 − 017 NGC 938	0443 38	(1.6 x 1.2) (1.7 x 1.3)	100 ...	3	E NGC 932 group		13.8		*3
01948	02 25.7 + 37 57 143.21 − 20.83	06 − 06 − 041	0407 37	1.0 x 0.4 0.9 x 0.35	114 6	2	Sa − b Disturbed:		15.2		*1
01949	02 25.8 − 00 48 168.18 − 54.70	00 − 07 − 000	1283	1. x 1. 1. x 1.	— ...	3	Dwarf V		18.		
01950	02 25.8 + 23 35 149.92 − 33.92	04 − 06 − 056	0858 38	1.1 x 0.7 1.1 x 0.7	145: ...	1	...		15.7		
01951	02 25.8 + 37 30 143.42 − 21.24	06 − 06 − 000	0407 37	(1.2 x 1.1) (1.2 x 1.1)	— ...	2	E: + companion		16.0:		*
01952	02 25.9 + 40 15 142.31 − 18.70	07 − 06 − 000	0449 47	1.2: x 1.0: 1.2: x 1.0:	(0) 2:	3	Sc np of 2		16.5:		*
01953	02 25.9 + 44 51 140.48 − 14.44	07 − 06 − 000	0449 47	1.2: x 1.0: 1.1: x 1.0:	— 1 − 2	2	S ... Peculiar, f of 2		16.5:		*
01954	02 26.0 − 01 22 168.87 − 55.11	00 − 07 − 022 NGC 941	1283	3.5 x 2.6: 3.3: x 2.6:	170 2 − 3	4	Sc P w 1929	SAB cd	13.4		*123 4589
01955	02 26.0 + 25 07 149.17 − 32.52	04 − 06 − 055	0858 38	1.6 x 0.7 1.5 x 0.7	165: ...	1	S ...		15.4		*
01956	02 26.0 + 45 42 140.17 − 13.65	08 − 05 − 013 NGC 933	0907 47	(2.2 x 1.1) (2.1 x 1.1)	35 ...	1	... IC 1799 group		15.5		*
01957	02 26.0 + 47 17 139.56 − 12.18	08 − 05 − 000	0907	1.7 x 0.20 1.7 x 0.20	175 7	2	Sc		15.7		
01958	02 26.1 + 27 55 147.80 − 29.98	05 − 06 − 000	1248 38	1.5 x 0.3 1.3 x 0.3	18 7	3	Sc sff of 2, disturbed		15.7		*
01959	02 26.1 + 34 19 144.85 − 24.14	06 − 06 − 000	0407	1.2: x 0.8: 1.0: x 0.7:	[45] ...	3	Dwarf V; sp of 2		18.		*
01960	02 26.1 + 38 10 143.20 − 20.60	06 − 06 − 044	0407 37	1.1 x 0.5 1.1 x 0.5	5 ...	1	... np of 2		15.5		*

N G C   930 − 36,   938,   941   I C   227,   1799,   1801

1	2	3	4	5	6	7	8	9	10	11	12
01961	02 26.3 + 42 02 141.66 - 17.02	07 - 06 - 024 NGC 937	0449 47	1.1 x 0.6 1.0 x 0.5	117 5	2	S B? c		15.0		
01962	02 26.4 + 00 09 167.36 - 53.87	00 - 07 - 024	1283	1.4 x 1.0 1.6 x 1.1	105 3	3	Sc		14.6		
01963	02 26.4 + 31 15 146.29 - 26.92	05 - 06 - 051	1248 38	1.3 x 0.9 1.4 x 0.9	(65) 3	3	Sa/S B b	SAB(rs)ab Sb -	14.5		*
01964	02 26.4 + 31 25 146.21 - 26.77	05 - 06 - 050 NGC 940	1248 38	(1.7 x 1.2) (1.9 x 1.3)	20: ...	1	SO:		13.4		3
01965	02 26.6 + 20 00 152.10 - 37.04	03 - 07 - 018	0443 38	1.7 x 0.4 1.7 x 0.35	92 7	1	S ... NGC 932 group		15.7		*
01966	02 26.7 + 09 57 158.73 - 45.78	02 - 07 - 011	1300	1.6 x 1.1 1.5: x 1.0:	40 3	3	Sb Brightest in group, Sb -	(R´)SAB ab	14.9		*
01967	02 26.8 + 44 57 140.59 - 14.29	07 - 06 - 000	0449 47	1.0 x 0.2 1.3 x 0.4	122 6 - 7	3	...	S d - dm	16.5:		
01968	02 26.9 + 29 32 147.20 - 28.44	05 - 06 - 000	1248 38	1.5:: x 1.0:: 1.4:: x 0.8::	(80) ...	3	Dwarf V		17.		
01969	02 27.0 + 15 50 154.73 - 40.67	03 - 07 - 000	0443	1. x 1. 1. x 1.	— ...	3	Dwarf V		18.		
01970	02 27.0 + 25 02 149.45 - 32.50	04 - 06 - 059	0858 38	2.2 x 0.3 2.1 x 0.25	22 7	3	Sc		15.3		
01971	02 27.0 + 28 25 147.76 - 29.44	05 - 06 - 052	1248 38	1.4 x 1.1 0.9: x 0.9:	— ...	1	...		14.9		*
01972	02 27.0 + 37 53 143.49 - 20.80	06 - 06 - 000	0407 37	1.6:: x 1.6:: 1.4:: x 1.4::	— ...	2	[E] Very compact		15.7		*
01973	02 27.1 + 06 40 161.47 - 48.49	01 - 07 - 000	1306	1.1 x 1.0 1.1 x 1.0	— ...	1	...		16.0:		
01974	02 27.2 + 09 38 159.12 - 45.98	02 - 07 - 013	1300	1.2 x 0.6 1.1 x 0.6	42 5	3	Sa		15.7		
01975	02 27.2 + 32 55 145.70 - 25.33	05 - 06 - 000	1248 38	1.3 x 0.20 1.0 x 0.20	143 6 - 7	1	S ...		15.3		
01976	02 27.3 + 35 07 144.74 - 23.31	06 - 06 - 046	0407	1.7 x 0.8 1.7 x 0.8	113 5	3	Sb		15.0		
01977	02 27.3 + 49 55 138.76 - 09.66	08 - 05 - 000	0907	1.2 x 0.9 1.1 x 0.8	(165) 3	3	Sb/S B c		16.0:		*
01978	02 27.4 + 00 57 166.86 - 53.08	00 - 07 - 025 IC 231	1283	1.1 x 0.8 1.1 x 0.8	(160) ...	1	[SO] Compact		15.0		
01979	02 27.5 + 42 01 141.88 - 16.95	07 - 06 - 026 NGC 946	0449 47	1.5 x 1.0 1.7 x 1.4	65: ...	2	[SO] Compact		14.5		*
01980	02 27.6 + 31 57 146.22 - 26.18	05 - 06 - 054	1248 38	1.1 x 0.4 1.1 x 0.4	147 6	1	S ...		14.5		
01981	02 27.7 + 00 44 167.18 - 53.21	00 - 07 - 026	1283	1.7: x 1.5: 1.5:: x 1.5::	— ...	3	Dwarf V		17.		1
01982	02 27.8 + 08 29 160.20 - 46.88	01 - 07 - 006	1306	1.2 x 0.9 1.0: x 0.8:	— ...	3	Irr	Ir I	16.5:		
01983	02 27.8 + 36 55 144.06 - 21.62	06 - 06 - 048 NGC 949	0407 37	(3.6 x 2.3) (3.5 x 2.2)	145 3 - 4	2	S ...		12.0		*12 3478
01984	02 27.8 + 43 08 141.49 - 15.90	07 - 06 - 000	0449 47	1.2 x 0.08 1.2 x 0.12	129 7	3	Sc 1997 group		16.5:		*
01985	02 28 + 81 59 126.56 + 20.08	14 - 02 - 010	1277	1.1 x 0.6 1.0 x 0.5	165 ...	2	S B ... In group		16.5:		*
01986	02 28.0 - 01 20 170.50 - 54.76	00 - 07 - 27a NGC 955	1283	3.4 x 1.0 3.2 x 1.0	19 6 - 7	2	Sa - b		13.0		*12 348
01987	02 28.0 + 43 07 141.53 - 15.90	07 - 06 - 028	0449 47	(1.0 x 0.9) (1.3 x 1.2)	— ...	1	E? 1997 group		15.7		*1
01988	02 28.1 + 40 10 142.75 - 18.61	07 - 06 - 027	0449 47	1.1 x 0.4 1.2 x 0.5	123 6	2	Sa - b		14.7		
01989	02 28.1 + 42 40 141.73 - 16.31	07 - 06 - 000	0449 47	1.6: x 1.6: 1.6: x 1.6:	— 1	1	S: ... 1997 group		16.0:		*
01990	02 28.2 + 27 29 148.49 - 30.17	04 - 07 - 000	0222 38	1.3: x 0.15 0.9: x 0.15	[37] ...	1	... Peculiar, streamer		15.7		*
01991	02 28.2 + 29 22 147.57 - 28.47	05 - 07 - 001 NGC 953	1301 38	(1.5 x 1.5) (1.8 x 1.8)	— ...	2	E		14.5		3
01992	02 28.2 + 42 42 141.73 - 16.27	07 - 06 - 000	0449 47	1.1:: x 1.0:: ...	— ...	0	Dwarf V, 1997 group		18.-19.		*
01993	02 28.5 + 39 10 143.20 - 19.50	06 - 06 - 049	0407	2.4 x 0.6 2.2 x 0.5	140 6	3	Sb		14.3		
01994	02 28.6 + 01 02 167.17 - 52.83	00 - 07 - 028 IC 232	1283 39	1.7 x 1.2 1.7 x 1.2	155 3	3	SO nnp of 2		14.2		*
01995	02 28.8 + 01 07 167.15 - 52.73	00 - 07 - 030	1283 39	1.7 x 0.9 1.7 x 0.9	62 5	3	Sc		14.6		*

N G C   937, 940, 946, 949, 953, 955   I C 231, 232

1	2	3	4	5	6	7	8	9	10	11	12
01996	02 28.8 + 22 42 151.14 – 34.40	04 – 07 – 004 IC 1809	0222 38	1.1 x 0.9 1.2 x 1.1	(50) 1 – 2	3	S B b	(R´)S B(s)ab Sb –	15.0		*C
01997	02 29.0 + 43 14 141.66 – 15.72	07 – 06 – 029	0449 47	1.3 x 0.25 1.4 x 0.30	73 6 – 7	3	Sb Brightest in group		15.4		
01998	02 29.1 + 00 05 168.32 – 53.49	00 – 07 – 032	1283 39	1.3 x 0.8 1.3 x 0.8	143 4	3	Sc		16.0:		*
01999	02 29.1 + 18 55 153.39 – 37.72	03 – 07 – 021	0443	3.1 x 0.5 2.6: x 0.5	96 6	3	Sc		15.1		
02000	02 29.2 + 05 24 163.21 – 49.24	01 – 07 – 000	1306	1.4 x 1.2 1.0: x 0.8:	— ...	3	... p of 2	SAB d – dm	16.5:		*
02001	02 29.2 + 41 59 142.20 – 16.86	07 – 06 – 030	0449 47	1.6 x 1.1 1.7 x 1.2	35 3	2	Sa – b		14.6		
02002	02 29.3 + 35 17 145.07 – 22.99	06 – 06 – 051 NGC 959	0407	2.6 x 1.5 2.8: x 1.5:	65 ...	3	Sc/Irr	Sc + or Ir I	12.5		*3
02003	02 29.3 + 39 12 143.38 – 19.41	06 – 06 – 050	0407	1.1 x 0.5 1.1 x 0.4	140 6	1	S ...		15.0		
02004	02 29.4 + 00 42 167.77 – 52.97	00 – 07 – 036	1283 39	1.4 x 0.9 1.3 x 0.9	95: 3 – 4	3	S B c	SAB(s)c Sc –	14.8		*
02005	02 29.4 + 01 01 167.45 – 52.72	00 – 07 – 035	1283 39	(2.3 x 2.3) (3. x 3.)	— ...	1	[E]		14.8		*
02006	02 29.4 + 42 12 142.15 – 16.65	07 – 06 – 031	0449 47	(1.1 x 1.1) (1.5 x 1.5)	— ...	1	E		15.7		
02007	02 29.5 + 09 30 159.91 – 45.79	02 – 07 – 000	1300	(1.1 x 1.1) (1.3 x 1.3)	— 1	3	S B 0		15.3		
02008	02 29.5 + 31 23 146.90 – 26.52	05 – 07 – 002	1301 38	1.0 x 0.9 0.8 x 0.6	— 1 – 2	3	... P w 2011	S d – dm	16.5:		*
02009	02 29.6 – 01 36 170.33 – 54.70	00 – 07 – 038	1283	1.0 x 0.5 1.1 x 0.6	120 5	1	S ... P w 2010		16.0:		*
02010	02 29.6 – 01 35 170.31 – 54.69	00 – 07 – 037	1283	1.7: x 1.1 1.6: x 1.1	85 3 – 4	4	S B b P w 2009, disturbed, Sb +	S B(s)b	14.3		*
02011	02 29.6 + 31 20 146.94 – 26.56	05 – 07 – 003	1301 38	1.7 x 0.8 1.5 x 0.8	110: 5	3	Sb P w 2008	SAB(r)ab	15.6		*
02012	02 29.8 + 20 48 152.46 – 35.98	03 – 07 – 022	0443 38	1.2 x 0.9 1.2 x 0.9	140: 2 – 3	2	Sb – c		15.1		*
02013	02 29.8 + 27 51 148.68 – 29.69	05 – 07 – 004 NGC 962	1301 38	(1.7 x 1.2) (1.7 x 1.2)	(170) ...	2	E		14.2		*3
02014	02 29.8 + 38 28 143.78 – 20.04	06 – 06 – 054	0407	2.4: x 1.3: 2. x 1.	[176] ...	4	Dwarf V		17.		*
02015	02 29.9 + 34 39 145.48 – 23.52	06 – 06 – 052	0407 40	(1.3 x 1.3) (1.3 x 1.3)	— ...	1	S0?		16.0:		
02016	02 30.0 + 20 25 152.74 – 36.30	03 – 07 – 000 Mark 368	0443	0.45 x 0.35 0.45 x 0.35	— ...	1	... Peculiar		14.5	+ 9000	
02017	02 30.0 + 28 36 148.35 – 28.99	05 – 07 – 000	1301 38	2.5:: x 2.0:: 2. x 1.	— ...	4	Dwarf irregular Ir V		17.		
02018	02 30.1 + 00 03 168.68 – 53.36	00 – 07 – 040	1283 39	1.6 x 0.8 1.6 x 0.8	75 5	1	S ...		14.3		
02019	02 30.1 + 00 24 168.31 – 53.09	00 – 07 – 039	1283 39	0.55 x 0.45 0.55 x 0.40	[65] ...	1	... Peculiar		14.4		
02020	02 30.1 + 23 07 151.23 – 33.90	04 – 07 – 007	0222 38	1.6 x 0.25 1.4 x 0.25	109 7	2	Sc		15.6		
02021	02 30.2 + 05 29 163.45 – 49.03	01 – 07 – 008	1306	1.2 x 1.1 1.3 x 1.1	— 1 – 2	2	S0		14.9		
02022	02 30.3 + 32 32 146.53 – 25.41	05 – 07 – 005	1301 40	(1.0 x 0.8) (1.1 x 0.9)	(170) ...	3	[E] Very compact		15.2		
02023	02 30.3 + 33 17 146.18 – 24.73	05 – 07 – 007	1301 40	3.0: x 2.8: 3. x 3.	— ...	3	Dwarf V		14.9		*
02024	02 30.4 + 00 12 168.62 – 53.20	00 – 07 – 041	1283 39	1.2 x 0.8 1.2 x 0.8	153 3 – 4	2	Sa – b		14.5		
02025	02 30.4 + 25 17 150.14 – 31.93	04 – 07 – 010	0222 38	1.3 x 0.15 1.4 x 0.15	32 7	2	Sc		15.7		
02026	02 30.4 + 27 05 149.21 – 30.32	04 – 07 – 000	0222 38	1.5: x 1.1: 1.3: x 1.	— ...	2	Dwarf IV – V or V		17.		
02027	02 30.5 + 08 42 160.84 – 46.33	01 – 07 – 000	1306	1.5 x 1.2 ...	— ...	0	...		16.0:		
02028	02 30.5 + 22 10 151.86 – 34.70	04 – 07 – 011	0222 38	1.5 x 0.7 1.5 x 0.7	110 5	3	Sc		14.6		*
02029	02 30.5 + 22 32 151.66 – 34.37	04 – 07 – 000	0222 38	1.1 x 0.6 1.0 x 0.5	(20) ...	2	... IV? or IV – V?		17.		
02030	02 30.6 + 09 23 160.33 – 45.74	01 – 07 – 009 NGC 975	1306	1.3 x 0.9 1.2 x 0.7	0 3	2	S0 – a Brightest in group		14.2		*

N G C   959, 962, 975   I C   1809

1	2	3	4	5	6	7	8	9	10	11	12
02031	02 30.6 + 20 03 153.11 – 36.56	03 – 07 – 025	0443 38	1.2 x 0.8 1.5 x 1.0	10: 3 – 4	2	S0		15.5		*
02032	02 30.6 + 20 22 152.92 – 36.28	03 – 07 – 000	0443 38	1.3 x 0.25 0.9 x 0.2	54 6:	2	Sc		16.5:		*
02033	02 30.6 + 37 27 144.38 – 20.91	06 – 06 – 055	0407 37	1.6 x 0.8 1.6 x 0.7	160 5	4	S B b	S B(s)b Sb +	14.7		*
02034	02 30.6 + 40 19 143.15 – 18.29	07 – 06 – 032	0449 47	3.5:: x 3.0:: 3.5:: x 3.0::	(170) ...	4	Dwarf irregular Ir IV – V		15.0		*1
02035	02 30.7 + 44 08 141.58 – 14.78	07 – 06 – 033	0449 47	2.1 x 1.4 1.9 x 1.2	125 3 – 4	2	Sb		14.3		
02036	02 30.8 + 41 15 142.79 – 17.42	07 – 06 – 000	0449 47	1.0: x 0.7: 1. x 0.6::	(85) ...	2	...	S B: d – m	17.		*
02037	02 31.0 + 06 15 163.03 – 48.29	01 – 07 – 010	1306 41	1.3 x 0.3 1.2 x 0.3	103 6	1	S ... sf of 2		15.6		*
02038	02 31.0 + 40 18 143.23 – 18.27	07 – 06 – 000	0449 47	1.1: x 0.5: 1.0: x 0.5:	115 5 – 6	2	...	S B d – m	17.		*
02039	02 31.1 + 32 44 146.61 – 25.16	05 – 07 – 008 NGC 969	1301 40	2.0 x 1.5 2.0 x 1.5	5: 2 – 3	4	S0 P w 2049	SAB 0	13.5		*13
02040	02 31.1 + 34 16 145.90 – 23.77	06 – 06 – 056 NGC 968	0407 40	(3.6 x 1.8) (4.3 x 2.0)	60: ...	3	E npp of 2		13.8		*
02041	02 31.2 + 09 25 160.48 – 45.64	01 – 07 – 011	1306	1.4 x 0.8 1.2: x 0.8:	170 4:	1	Sa – b NGC 975 group		15.3		*
02042	02 31.2 + 20 45 152.85 – 35.87	03 – 07 – 027 NGC 976	0443 38	1.7 x 1.6 1.8 x 1.7	— 1	3	Sb	Sb +	12.9		*123 4678
02043	02 31.2 + 44 48 141.40 – 14.13	07 – 06 – 000	0449 47	1.0 x 0.9 1.1 x 1.1	— 1	4	S B c P w 2050	S B(s)c	15.0		*
02044	02 31.3 + 00 42 168.38 – 52.67	00 – 07 – 043	1283 39	1.0 x 0.6 1.0 x 0.6	30 4	2	S B c	S B(s) ...	15.7		
02045	02 31.3 + 29 05 148.41 – 28.43	05 – 07 – 010 NGC 972	1301 38	3.7 x 1.7 4.0 x 2.0	152 5	2	Sa – b Singular		12.1	+ 1555 + 1681	*12 3468
02046	02 31.3 + 29 47 148.06 – 27.80	05 – 07 – 011	1301 38	1.4 x 1.1 1.2 x 1.0	65: ...	3	...	S dm – m	16.5:		
02047	02 31.3 + 32 13 146.89 – 25.61	05 – 07 – 014 IC 1815	1301 40	1.7 x 1.6 2.0 x 1.8	— 1	3	S B 0 P w 2048		14.3		*1
02048	02 31.3 + 32 17 146.86 – 25.55	05 – 07 – 013 NGC 973	1301 40	4.0 x 0.6 3.6 x 0.5	48 7	4	Sb P w 2047	Sb +	13.7		*
02049	02 31.4 + 32 45 146.66 – 25.11	05 – 07 – 012 NGC 974	1301 40	4.0: x 3.5: 3.5:: x 3.5::	— 1	3	S ... P w 2039	SAB(rs)[b]	13.9		*13
02050	02 31.4 + 44 42 141.47 – 14.21	07 – 06 – 000	0449 47	1.1 x 0.7 1.2 x 0.8	2: 3 – 4	2	Sc/S B c P w 2043	SAB(s)c	15.2		*
02051	02 31.5 + 01 08 168.00 – 52.30	00 – 07 – 044	1283 39	(1.6 x 1.5) (1.5 x 1.5)	— (1)	1	S0		14.4		
02052	02 31.5 + 25 01 150.54 – 32.06	04 – 07 – 000	0222 38	1.5: x 0.2 1.0: x 0.1	[174] ...	1	Dwarf?		18.		
02053	02 31.5 + 29 32 148.23 – 28.01	05 – 07 – 015	1301 38	2.3 x 1.3 2. x 1.	[43] ...	4	Dwarf irregular Ir V		15.7		*
02054	02 31.5 + 33 44 146.23 – 24.21	06 – 06 – 058	0407 40	1.7 x 1.0 1.6: x 1.0:	80: (4)	1	S – Irr		15.3		
02055	02 31.5 + 36 56 144.78 – 21.31	06 – 06 – 000	0407 37	1.4:: x 1.0:: 1.2:: x 0.8:	(15) ...	3	Dwarf (spiral) S: IV – V or V		18.		
02056	02 31.6 + 02 24 166.76 – 51.29	00 – 07 – 045	1283 39	1.1 x 0.9 0.9 x 0.7	— 2	2	S B b/Sc	Sb +	15.0		*
02057	02 31.8 + 32 38a 146.80 – 25.18	05 – 07 – 016 NGC 978	1301 40	(2.0 x 1.7) (2.0 x 1.7)	(80) ...	2	E – S0 np of 2, contact		(13.3)		*13
02058	02 31.8 + 40 55 143.11 – 17.64	07 – 06 – 000	0449 47	1.3 x 1.2 1.7 x 1.5	— 1	3	Sb/S B c sff of 2	Sc –	15.6		*
02059	02 31.9 + 23 12 151.63 – 33.63	04 – 07 – 012 NGC 984	0222 38	(3. x 2.) (2.6 x 1.6)	120: 2 – 4	2	S0	SA 0 +	14.5		*3C
02060	02 31.9 + 41 09 143.03 – 17.42	07 – 06 – 037	0449 47	1.5 x 0.6 1.6 x 0.7	177 6	1	S B a – b		14.7		*
02061	02 32.0 – 01 11 170.64 – 54.00	00 – 07 – 047	1283	1.7: x 0.4 1.6 x 0.4	56 6 – 7	1	S ...		16.0:		
02062	02 32.0 + 01 08 168.16 – 52.22	00 – 07 – 046	1283 39	1.0 x 0.5 1.0 x 0.5	98 5	1 5	Sb – c		15.0		
02063	02 32.2 + 40 43 143.27 – 17.80	07 – 06 – 038 NGC 982	0449 47	1.7 x 0.9 1.7 x 0.9	110 5	2	S0 P w 2066		14.3		*13
02064	02 32.3 + 20 38 153.20 – 35.85	03 – 07 – 028	0443 38	2.3 x 1.7 2.1 x 1.5	(165) 3	4	S B b/Sc	SAB(s)bc Sb +	15.0		*
02065	02 32.3 + 37 16 144.79 – 20.94	06 – 06 – 060	0407 37	1.7: x 1.7: 1.4: x 1.4:	— 1	3	... P w 2067	S m	14.6		*1

N G C   968, 969, 972 – 74, 976, 978, 982, 984

I C   1815

1	2	3	4	5	6	7	8	9	10	11	12
02066	02 32.3 + 40 40 143.31 - 17.83	07 - 06 - 039 NGC 980	0449 47	1.8 x 0.7 2.1 x 0.9	132 6	2	Sa P w 2063		13.2		*13
02067	02 32.4 + 37 18 144.79 - 20.90	06 - 06 - 061	0407 37	2.5 x 0.5 2.2 x 0.4	158 6 - 7	1	Sa - b P w 2065		14.8		*1
02068	02 32.4 + 40 42 143.31 - 17.80	07 - 06 - 000	0449 47	1.2 x 0.2 0.8 x 0.2	97 6 - 7	2	...	S dm - m	17.		*
02069	02 32.5 + 37 25 144.76 - 20.78	06 - 06 - 062	0407 37	2.5 x 1.7 2.7: x 1.9:	65 3	4	S B c	SAB(s)d	13.2		*1
02070	02 32.6 + 12 36 158.50 - 42.77	02 - 07 - 016 IC 238	1300	(1.8 x 1.2) (1.7 x 1.1)	30: 3 - 4	3	S0		14.1		*
02071	02 32.6 + 19 43 153.82 - 36.62	03 - 07 - 029	0443 38	1.3 x 0.7 1.2 x 0.5	130 5 - 6	2	Sb - c		15.6		
02072	02 32.8 + 45 28 141.39 - 13.40	08 - 06 - 006	1249	1.2 x 1.2 1.2 x 1.2	-- 1	1	S ...		16.5:		
02073	02 32.9 + 42 12 142.77 - 16.39	07 - 06 - 041	0449 47	(1.8 x 1.3) (2.1 x 1.7)	(105) ...	1	S0?		14.0		*
02074	02 32.9 + 42 24 142.68 - 16.20	07 - 06 - 000	0449 47	(1.0 x 0.7) (1.2 x 0.9)	(130) ...	2	... Very compact		15.0		*
02075	02 33.0 + 06 03 163.81 - 48.16	01 - 07 - 012	1306 41	(1.4 x 1.3) (1.5 x 1.3)	-- 1 - 2	2	S0 - a		15.0		*
02076	02 33.0 + 11 15 159.60 - 43.86	02 - 07 - 000	1300	1.2: x 0.7 1.1: x 0.7:	25: 4:	1	Sb - c		16.0:		*
02077	02 33.1 + 34 23 146.25 - 23.49	06 - 06 - 063	0407 40	1.1 x 0.9 1.0 x 0.8	-- 2:	3	Sc P w 2090		15.5		*
02078	02 33.3 + 10 13 160.47 - 44.68	02 - 07 - 017	1300	(1.3 x 1.1) (1.5 x 1.3)	-- ...	1	...		16.0:		*1
02079	02 33.3 + 23 41 151.70 - 33.05	04 - 07 - 014	0222 38	1.8 x 0.8 2.2 x 0.8	157 ...	3	Sc/S B c	SAB(s)c	14.8		*1
02080	02 33.3 + 38 45 144.33 - 19.50	06 - 06 - 065 IC 239	0407	6.0: x 5.8: 5.0: x 5.0:	-- 1	4	Sc NGC 1023 group, Sc -	SAB(s)c	12.1		*12 59
02081	02 33.4 + 00 12 169.58 - 52.72	00 - 07 - 048	1283 39	3.0 x 2.0 2.5: x 1.8:	80: 3:	4	Sc	SAB(s)cd	15.1		1
02082	02 33.4 + 25 12 150.90 - 31.70	04 - 07 - 016	0222 38	6.3 x 1.1 6.0: x 1.0	133 6 - 7	3	Sc		14.0		
02083	02 33.4 + 32 30 147.20 - 25.16	05 - 07 - 019	1301 40	1.7 x 0.20 1.5 x 0.20	49 7	2	Sb - c		15.0		
02084	02 33.4 + 35 55 145.61 - 22.07	06 - 06 - 000	0407	1.1 x 0.7 1.0 x 0.6	85 4:	1	S ... P w 2094		15.7		*
02085	02 33.5 + 01 20 168.42 - 51.83	00 - 07 - 049	1283 39	1.3 x 0.5 1.2: x 0.4	176 6	2	Sc		15.7		*
02086	02 33.5 + 13 52 157.85 - 41.57	02 - 07 - 000	1300	1.1 x 0.8 1.0 x 0.7	(155) ...	1	S B: a - b		16.0:		*
02087	02 33.5 + 31 23 147.76 - 26.16	05 - 07 - 020	1301 38	1.1 x 0.35 1.0 x 0.35	48 6	3	Sc		15.1		
02088	02 33.6 + 00 55 170.85 - 53.53	00 - 07 - 000	1283	1.1 x 0.12 1.1 x 0.1	78 7	2	Sc spp of 2		16.5:		*
02089	02 33.6 + 11 26 159.64 - 43.62	02 - 07 - 018 NGC 990	1300	(1.8 x 1.5) (2.0 x 1.7)	-- ...	3	E		13.9		3
02090	02 33.6 + 34 24 146.35 - 23.43	06 - 06 - 066	0407 40	1.1 x 1.0 0.9 x 0.9	-- 1	3	Sb	SAB(s)b	15.4		*
02091	02 33.8 + 00 30 169.38 - 52.42	00 - 07 - 050	1283 39	1.8 x 0.5 1.5 x 0.3	165 6	1	Sc		15.1		*
02092	02 33.8 + 07 05 163.16 - 47.20	01 - 07 - 014	1306 41	3.3 x 0.3 2.5 x 0.3	32 7	3	Sc NGC 997 group		15.6		*
02093	02 33.8 + 33 06 147.00 - 24.58	05 - 07 - 021 NGC 987	1301 40	1.4 x 1.0 1.8 x 1.2	30: 3 - 4	3	S B 0/S B a		13.4		3
02094	02 33.8 + 35 54 145.70 - 22.05	06 - 06 - 067	0407	2.3 x 2.2 2.1 x 1.9	-- 1	4	S B c P w 2084	S B(rs)c	13.8		*
02095	02 34.1 + 01 50 168.10 - 51.35	00 - 07 - 052 NGC 993 = 994	1283 39	(1.1 x 1.0) (1. x 1.)	-- ...	1	E?		14.9		3
02096	02 34.2 + 10 51 160.25 - 44.03	02 - 07 - 019	1300	1.2 x 0.3 1.1 x 0.3	164 6	2	Sb - c		16.0:		
02097	02 34.3 + 05 14 164.93 - 48.63	01 - 07 - 000	1306	1.1 x 0.1 0.4 x 0.1	6 7	2	Sc		17.		
02098	02 34.4 + 19 44 154.27 - 36.40	03 - 07 - 034	0443 38	1.3 x 0.4 1.0: x 0.4	127 6	1	S ...		16.5:		
02099	02 34.4 + 21 21 153.30 - 34.99	03 - 07 - 033	0443 38	(1.1 x 0.5) (1.4 x 0.7)	140 5:	1	S0:		15.5		
02100	02 34.4 + 33 25 146.97 - 24.24	05 - 07 - 000	1301 40	1.1 x 0.8 1.1 x 0.8	70: 2 - 3	4	...	S dm	15.6		

N G C   980, 987, 990, 993, 994

1	2	3	4	5	6	7	8	9	10	11	12
02101	02 34.4 + 42 25 142.94 – 16.08	07 – 06 – 042	0449 47	2.1 x 0.22 2.1 x 0.25	162 7	2	Sb Disturbed?, Sb +		15.1		*
02102	02 34.6 + 07 05 163.40 – 47.09	01 – 07 – 016 NGC 997	1306 41	(1.2 x 1.2) (1.3 x 1.2)	— ...	2	E or S0 + companion Brightest in group		14.6		*3
02103	02 34.6 + 20 53 153.63 – 35.38	03 – 07 – 035 NGC 992	0443 38	0.9 x 0.7 0.9 x 0.6	10 ...	1	S? ... Singular		13.5		*
02104	02 34.6 + 23 05 152.35 – 33.44	04 – 07 – 018	0222 38	1.0 x 0.8 1.0 x 0.8	(55) 2:	3	Sc		15.5		
02105	02 34.6 + 34 14 146.63 – 23.49	06 – 06 – 068	0407 40	1.6 x 1.3 1.7 x 1.7	— 1 – 2	4	S B a	(R´)S B(s)a	13.9	+ 4800 + 4939	*2
02106	02 34.8 – 02 02 172.49 – 54.17	00 – 07 – 000	1283	1.0 x 0.3 0.9 x 0.3	30 6	1	S ...		15.4		*
02107	02 34.8 + 12 58 158.85 – 42.17	02 – 07 – 000	1300	0.9 x 0.3 0.9 x 0.3	162 6	2	S0 – a		16.0:		
02108	02 34.8 + 42 21 143.04 – 16.11	07 – 06 – 000	0449 47	1.4: x 1.2: 1.1: x 1.1:	— ...	2	...	SA dm	17.		*
02109	02 34.9 + 34 02 146.78 – 23.64	06 – 06 – 069	0407 40	1.9 x 1.7 2.2: x 1.9:	— 1	4	Sc	SAB(s)d	14.5		
02110	02 34.9 + 37 56 144.99 – 20.11	06 – 06 – 000	0407 37	1.1 x 0.9 1.1: x 1.0:	— 1 – 2	2	S B b – c		16.5:		*
02111	02 34.9 + 41 35 143.38 – 16.80	07 – 06 – 043	0449 47	2.3: x 0.4 2.8 x 0.8	119 7	3	Sa – b		14.8		
02112	02 35.0 + 01 45 168.47 – 51.27	00 – 07 – 057 NGC 1004	1283 39	(1.8 x 1.8) (1.9 x 1.8)	— ...	3	E		14.3		*1
02113	02 35.3 – 01 54 172.50 – 53.98	00 – 07 – 000	1283	1.3 x 0.7 1.3: x 0.7	(160) ...	2	...	S B dm	16.5:		
02114	02 35.3 + 01 52 168.44 – 51.13	00 – 07 – 060 NGC 1008	1283 39	(1.0 x 0.7) (0.9 x 0.5)	85 ...	2	E		14.9		*13
02115	02 35.3 + 02 07 168.19 – 50.94	00 – 07 – 058 IC 241	1283 39	(1.2 x 0.8) (1.1 x 0.7)	(150) ...	1	...		14.5		*1
02116	02 35.3 + 30 37 148.52 – 26.68	05 – 07 – 022	1301 38	2.0: x 1.7: 1.7: x 1.5:	— ...	4	... S IV	S dm	16.0:		
02117	02 35.3 + 32 59 147.37 – 24.55	05 – 07 – 000	1301 40	1.1 x 0.4 1.0 x 0.4	58 ...	2	Sc – Irr		16.5:		*
02118	02 35.3 + 41 20 143.56 – 16.99	07 – 06 – 044 NGC 995	0449 47	1.7 x 1.2 1.9 x 1.3	35: 3	3	S0		14.9		*3
02119	02 35.4 – 02 04 172.72 – 54.09	00 – 07 – 061	1283	2.1: x 0.8 2.2 x 1.0	2 5 – 6	4	S B b	S B(s)b	14.5		*
02120	02 35.5 + 01 10 169.22 – 51.64	00 – 07 – 064	1283 39	1.6 x 0.4 1.5 x 0.4	149 6	1	S B: b – c		15.1		
02121	02 35.4 + 01 28 168.88 – 51.43	00 – 07 – 062	1283 39	1.7 x 1.6 1.6: x 1.4:	— 1	4	Sc	SAB(s)c Sc –	15.0		*
02122	02 35.5 + 29 32 149.11 – 27.63	05 – 07 – 023	1301 38	1.1 x 1.1 1.2 x 1.2	— 1	4	Sc/S B c	SAB(s)c Sc –	14.7		
02123	02 35.5 + 41 28 143.54 – 16.86	07 – 06 – 045 NGC 996	0449 47	(1.4 x 1.4) (1.7 x 1.7)	— ...	3	E P w 2127		14.5		*13
02124	02 35.6 – 01 32 172.17 – 53.66	00 – 07 – 066 NGC 1015	1283	3.3 x 3.3 3.0 x 3.0	— 1	4	S B a	S B(r:)a	13.5		3
02125	02 35.6 + 31 51 147.98 – 25.54	05 – 07 – 024 IC 1823	1301	2.3 x 2.0 2.3 x 2.0	— 1 – 2	4	S B c	S B(r)c Sc –	14.8		*
02126	02 35.6 + 40 30 143.98 – 17.73	07 – 06 – 046	0449 47	1.4 x 1.4 1.7 x 1.7	— 1	1	...	SAB: dm?	15.4		*
02127	02 35.6 + 41 28 143.56 – 16.85	07 – 06 – 047 NGC 999 = 240*	0449 47	1.1 x 1.0 1.1 x 0.9	— 1 – 2	4	Sa/S B a P w 2123	(R´)SAB(s)a	14.5		*13
02128	02 35.7 + 01 55 168.52 – 51.03	00 – 07 – 067 NGC 1016	1283 39	(2.5 x 2.5) (2.8 x 2.8)	— ...	4	E		13.3		*13
02129	02 35.7 + 02 06 168.33 – 50.89	00 – 07 – 065 NGC 1009	1283 39	1.6 x 0.25 1.7 x 0.3	124 7	3	Sb		15.4		*1
02130	02 35.7 + 07 47 163.13 – 46.36	01 – 07 – 017	1306 41	1.2 x 0.3 1.1 x 0.3	124 6	1	S ...		15.4		*
02131	02 35.7 + 33 14 147.33 – 24.29	05 – 07 – 025	1301 40	1.7 x 0.7 1.5 x 0.7	93 ...	4	S B c	S B(r)d	15.6		
02132	02 35.8 + 01 42 168.76 – 51.18	00 – 07 – 068 NGC 1019	1283 39	1.1 x 1.0 1.3 x 1.1	— 1	4	S B b	S B(s)b Sb –	14.6		*1
02133	02 35.9 + 34 25 146.81 – 23.21	06 – 06 – 070 NGC 1002	0407	1.6 x 1.0: 1.5 x 0.9	140 4	1	S B b:	S B(r) ...	14.0		*
02134	02 36.0 + 27 37 150.21 – 29.28	05 – 07 – 026	1301 38	1.8 x 0.8 1.8 x 0.8	105 6	3	Sb		14.4		
02135	02 36.0 + 41 02 143.82 – 17.21	07 – 06 – 049	0449 47	1.5: x 1.0: 1.5 x 1.0	5: 3 – 4	3	Sc	Sc +	15.7		*

I C   240, 241, 1823

N G C   992, 995 – 97, 999, 1002, 1004, 1008
 1009, 1015, 1016, 1019

1	2	3	4	5	6	7	8	9	10	11	12
02136	02 36.1 + 06 42 164.17 − 47.18	01 − 07 − 000	1306	1.1 x 0.15 0.7 x 0.15	25 7	2	Sb		16.0:		
02137	02 36.1 + 40 40 144.00 − 17.54	07 − 06 − 051 NGC 1003	0449 47	7. x 3. 6.2: x 2.5:	97 6	4	Sc NGC 1023 group		12.1	+ 585 + 741	*12 359
02138	02 36.2 + 08 53 162.37 − 45.39	01 − 07 − 000 IC 1825	1306 41	1.4 x 1.0 1.4 x 0.9	15 3 − 4	3	Sc		14.9		*
02139	02 36.2 + 36 12 146.03 − 21.57	06 − 06 − 000	0407 42	1. x 1. 1. x 1.	— ...	3	Dwarf V		18.		
02140	02 36.3 + 18 10a 155.73 − 37.54	03 − 07 − 037	0443	1.9 x 1.0 2. x 1.	[155] ...	4	Irr np of 2 Strongly disrupted		(14.6)		*12 V($\overline{\text{A}}$)
02141	02 36.3 + 29 56 149.08 − 27.19	05 − 07 − 027 NGC 1012	1301 38	2.9 x 1.5 2.8 x 1.5	[24] ...	2	S − Irr Peculiar, Ir II?		13.1		*3
02142	02 36.5 + 10 38 161.07 − 43.90	02 − 07 − 020 NGC 1024	1300	4.8 x 1.7 5.0 x 1.7	155 6	4	Sb (R')SA(r)ab Brightest in group, Sb −		13.8		*13 A
02143	02 36.5 + 35 52 146.25 − 21.85	06 − 06 − 000	0407 42	0.55 x 0.55 0.55 x 0.55	— ...	1	... Peculiar		14.0		*C
02144	02 36.6 + 29 01 149.62 − 27.98	05 − 07 − 000	1301 38	1.2 x 1.1 1.2: x 1.1	— ...	3	Dwarf spiral S V		17.		
02145	02 36.7 + 06 20 164.67 − 47.39	01 − 07 − 018 NGC 1026	1306 41	(2.0 x 1.8) (1.8 x 1.7)	— (1)	3	S0		14.1		*13
02146	02 36.7 + 42 53 143.14 − 15.48	07 − 06 − 000	0449 47	1.7: x 1.7: 1.6:: x 1.6::	— ...	3	Dwarf: IV − V:		17.		
02147	02 36.8 + 00 53 169.91 − 51.65	00 − 07 − 073 NGC 1032	1283	3.8 x 1.3 4.1 x 1.4	68 7	3	Sa		13.2		*3
02148	02 36.8 + 12 28 159.77 − 42.33	02 − 07 − 022	1300	1.1 x 0.3 1.1 x 0.3	17 6	2	S: ... Disturbed?		16.0:		*1
02149	02 36.9 + 10 35 161.22 − 43.89	02 − 07 − 024 NGC 1029	1300	1.6 x 0.5 1.6 x 0.5	70 6 − 7	1	S0 − a NGC 1024 group		14.9		*13
02150	02 37.0 + 09 41 161.96 − 44.62	02 − 07 − 000	1300	1.2: x 1.0: 1. x 1.	— ...	3	Dwarf (spiral) S: V		18.		
02151	02 37.0 + 28 06 150.18 − 28.75	05 − 07 − 028	1301 38	2.0 x 0.9 1.7 x 0.7	117 5 − 6	4	Sc SA c Sc +		14.8		
02152	02 37.1 + 01 20 169.54 − 51.26	00 − 07 − 075 IC 1827	1283 39	1.2 x 0.22 1.2 x 0.25	154 7	3	Sa P w 2158		14.5		*
02153	02 37.1 + 17 49 156.17 − 37.74	03 − 07 − 039 NGC 1030	0443	1.7 x 0.7 1.7 x 0.7	[8] ...	1	S? ... Peculiar		14.5		*3
02154	02 37.2 + 38 51 145.01 − 19.09	06 − 06 − 073 NGC 1023	0407	8.5 x 4.0 8.0 x 3.5	87 6	4	S B 0 Brightest in group		10.5	+ 578 + 729	*123 45789A
02155	02 37.3 + 14 10 158.68 − 40.83	02 − 07 − 000	1300	1.1 x 0.7 0.7 x 0.6	(20) ...	3	S B b S B(s)b		16.0:		*
02156	02 37.3 + 32 02 148.25 − 25.22	05 − 07 − 029	1301 40	2.2: x 2.0: 1.7: x 1.7:	— 1	4	Sc SA(r)c		14.5		
02157	02 37.3 + 38 21 145.26 − 19.54	06 − 06 − 071	0407	2.0: x 0.6: 1.9 x 0.5	39 (6)	2	... S d − m		14.9		
02158	02 37.5 + 01 18 169.69 − 51.22	00 − 07 − 076 NGC 1038	1283 39	1.3 x 0.45 1.4 x 0.5	61 6	3	Sa		14.4		*
02159	02 37.5 + 29 52 149.38 − 27.13	05 − 07 − 000	1301 38	1.1 x 0.10 1.0 x 0.1	129 7	3	Sc Sc +		16.5:		
02160	02 37.6 + 19 05 155.49 − 36.59	03 − 07 − 041 NGC 1036=1828−9*	0443	(1.6 x 1.1) (1.5 x 1.2)	(5) ...	1	... Peculiar?; Mark 370		13.5	+ 900 + 985	*3
02161	02 37.6 + 43 36 142.98 − 14.76	07 − 06 − 054	0449 47	(1.9 x 1.7) (2.2 x 2.2)	— 1	3	S0 SAB: 0		15.4		
02162	02 37.8 + 01 01 170.08 − 51.38	00 − 07 − 077	1283 39	2. x 2. ...	— ...	3	Dwarf irregular Ir V		18.		*
02163	02 37.9 + 15 30 157.91 − 39.63	02 − 07 − 025	1300	1.0 x 0.6 1.0 x 0.7	15 3 − 4	2	Sb? (R')SAB(s)...		16.0:		
02164	02 37.9 + 43 28 143.09 − 14.86	07 − 06 − 000	0449 47	1.3 x 0.4 1.7 x 0.5	90 6	1	Sb − c		15.6		
02165	02 38.0 + 38 32 145.30 − 19.31	06 − 06 − 074	0407	(1.9 x 1.4) (1.9 x 1.4)	(160) ...	1	...		16.0:		
02166	02 38.1 + 35 39 146.66 − 21.90	06 − 06 − 000	0407 42	1.4 x 0.7 1.3 x 0.6	132 5	2	Sc Sc +		15.7		
02167	02 38.2 + 08 23 163.35 − 45.51	01 − 07 − 021	1306 41	1.3 x 0.4 1.1 x 0.4	140 6	2	S B: c		15.1		1
02168	02 38.3 + 17 13 156.87 − 38.11	03 − 07 − 043	0443	1.4: x 0.6: 1.4:: x 0.5::	(23) ...	3	Dwarf IV − V or V		16.5:		
02169	02 38.5 + 37 01 146.10 − 20.64	06 − 06 − 075	0407 42	(1.5 x 1.1) 2. x 1.	50: ...	3	S0 SA 0		15.5		*1C
02170	02 38.6 + 17 35 156.70 − 37.76	03 − 07 − 044 IC 248	0443	1.1 x 0.8 1.3 x 1.0	(145) 2 − 3	1	Sa:		14.4		*

N G C   1003, 1012, 1023 − 26, 1029, 1030
        1032, 1036, 1038

I C   248, 1825, 1827 − 29

1	2	3	4	5	6	7	8	9	10	11	12
02171	02 38.9 + 31 52 148.66 - 25.21	05 - 07 - 000	1301	1.3: x 0.1 1.5: x 0.2	154 7	3	Sc	Sc +	17.		
02172	02 38.9 + 43 09 143.40 - 15.07	07 - 06 - 055	0449 47	1.8: x 1.1: 1.8: x 1.1:	30 ...	1	Irr: Peculiar?, Ir II?		14.6		*
02173	02 39.1 + 00 13 171.32 - 51.77	00 - 07 - 081 NGC 1055	1283	8. x 3.5: 8. x 3.5:	105 6 - 7	4	Sb NGC 1068 group, Sb +		12.5		*123 45789
02174	02 39.1 + 32 09 148.56 - 24.94	05 - 07 - 031	1301 40	2.3 x 2.2 2.3 x 2.2	-- 1	4	Sc/S B c	SAB(s)c Sc +	15.2		
02175	02 39.1 + 42 11 143.86 - 15.93	07 - 06 - 056	0449 47	1.4 x 1.3 1.2 x 1.1	-- 1	3	Sb - c	SAB(s)...	15.2		*
02176	02 39.2 + 05 45 165.92 - 47.48	01 - 07 - 000	1306	1.4 x 0.5 1.3 x 0.5	14 6	2	Sb		14.6		
02177	02 39.5 - 01 06 172.88 - 52.68	00 - 07 - 000	1283	1.0 x 0.15 1.1 x 0.15	50 7	2	Sc		16.0:		
02178	02 39.5 + 34 34 147.46 - 22.75	06 - 06 - 078 NGC 1050	0407	1.8 x 1.2 2.0 x 1.5	(110) 2 - 4	3	S B a	(R')S B(s)a	13.5		3
02179	02 39.5 + 35 00 147.25 - 22.36	06 - 06 - 076	0407 42	1.7 x 1.7 1.7 x 1.7	-- 1	4	S B c	S B(s)cd	15.1		*
02180	02 39.5 + 39 20 145.22 - 18.47	06 - 06 - 079	0407	1.0 x 0.4 0.9 x 0.3	37 6	1	S ...		14.6		*
02181	02 39.6 + 02 13 169.39 - 50.18	00 - 07 - 082	1283	1.1 x 1.1 0.9 x 0.9	-- 1	1	S ...		15.2		
02182	02 39.8 + 32 28 148.55 - 24.59	05 - 07 - 000	1301 40	1.4: x 0.4: 1.2: x 0.4:	[161] ...	4	Dwarf irregular Ir V		17.		
02183	02 39.9 + 28 21 150.69 - 28.24	05 - 07 - 032 NGC 1056	1301 38	2.5: x 1.7: 1.9: x 1.5:	160: 6:	1	Sa:		13.5		*13
02184	02 40.0 + 32 17 148.68 - 24.74	05 - 07 - 033 NGC 1057	1301 40	1.2 x 0.7 1.1 x 0.7	115 3 - 4	2	S0		15.7		*3
02185	02 40.0 + 40 14 144.90 - 17.61	07 - 06 - 059	0449 47	3.4 x 1.0 3.5 x 1.1	144 6	4	Sc sf of 2		13.8		*
02186	02 40.0 + 40 51 144.61 - 17.05	07 - 06 - 058	0449 47	1.5 x 0.8 1.7 x 0.8	142 5	2	S0 - a		14.5		*
02187	02 40.0 + 41 18 144.41 - 16.65	07 - 06 - 060 NGC 1053	0449 47	1.7 x 0.8 (1.7 x 0.8)	40 5	2	S0 Brightest in group		14.0		*1
02188	02 40.1 - 00 13 172.09 - 51.93	00 - 07 - 083 NGC 1068	1283	9. x 8. 8. x 7.	(70) 1 - 2	5	Sb (R)SA(rs)b, Sb - Seyfert, brightest in group		9.7	+ 1080 + 1094	*12- -9A
02189	02 40.1 + 02 52 168.90 - 49.60	00 - 07 - 085 IC 1834	1283	1.1 x 1.1 1.1 x 1.0	-- 1	2	Sb		15.2		*
02190	02 40.1 + 37 36 146.13 - 19.97	06 - 07 - 000	1307 47	1.0 x 0.4 0.9 x 0.4	172 6	2	S0 f of 2		16.0:		*
02191	02 40.2 + 32 12 148.77 - 24.79	05 - 07 - 035 NGC 1060	1301 40	(2.3 x 1.7) (2.5 x 2.0)	(75) ...	2	E - S0		13.4		*3
02192	02 40.2 + 47 37 141.67 - 10.92	08 - 06 - 000	0845 47	1.1 x 0.20 0.9 x 0.22	74 6 - 7	1	S0 - a		15.2		
02193	02 40.3 + 37 08 146.39 - 20.37	06 - 07 - 001 NGC 1058	1307 47	3.8 x 3.7 3.3 x 3.0	-- 1	4	Sc SA(rs:)c NGC 1023 group, Sc -		11.8	+ 521 + 665	*12- -789
02194	02 40.3 + 41 13 144.50 - 16.70	07 - 06 - 061	0449 47	1.7 x 0.8 1.7 x 0.8	78 5	2	S B: b NGC 1053 group		14.6		*
02195	02 40.3 + 41 44 144.27 - 16.24	07 - 06 - 000	0449 47	1.1 x 0.6 1.0 x 0.6	95 4 - 5	4	Sc	Sc +	15.7		*
02196	02 40.3 + 44 10 143.19 - 14.04	07 - 06 - 062	0449 47	1.1: x 1.1: 1.1 x 1.1	-- 1:	1	S ...		15.6		*
02197	02 40.4 + 31 15 149.29 - 25.62	05 - 07 - 037	1301 40	2.0: x 1.4: 2.0: x 1.4:	(160) 3:	4	Sc		15.6		
02198	02 40.5 + 31 35 149.14 - 25.31	05 - 07 - 038	1301 40	1.1 x 0.25 1.1 x 0.25	25 7	3	Sc	Sc +	15.7		
02199	02 40.7 + 01 33 170.40 - 50.51	00 - 07 - 086	1283	(1.3 x 0.9) (1.2 x 0.8)	153 3 - 4	1	S0 - a		14.8		
02200	02 40.7 + 04 46 167.26 - 48.03	01 - 07 - 026 NGC 1070	1306	2.8: x 2.5: 2.7: x 2.3:	(175) 1	3	Sb	Sb -	13.0		*3
02201	02 40.7 + 32 17 148.83 - 24.67	05 - 07 - 041	1301 40	1.7 x 0.3 1.6 x 0.3	101 7	3	Sc	Sc +	15.7		*1
02202	02 40.8 + 32 10 148.91 - 24.76	05 - 07 - 040	1301 40	1.0 x 0.9: 1.0 x 0.9:	-- ...	3	Dwarf: IV - V:		16.5:		
02203	02 40.8 + 32 15 148.87 - 24.69	05 - 07 - 042 NGC 1066	1301 40	(1.7 x 1.6) (2.0 x 1.7)	-- ...	2	[E] Compact		14.9		*1
02204	02 40.8 + 32 18 148.84 - 24.64	05 - 07 - 043 NGC 1067	1301 40	1.1 x 1.0 1.2 x 1.1	-- 1	4	Sc SAB(s)c Sc -		14.6		*13
02205	02 40.8 + 32 57 148.52 - 24.07	05 - 07 - 045	1301 40	1.2 x 0.8 1.2 x 0.8	30: 3 - 4	3	Sc	SAB: c	15.3		

N G C   1050, 1053, 1055 - 58, 1060
1066 - 68, 1070

I C   1834

1	2	3	4	5	6	7	8	9	10	11	12
02206	02 40.8 + 33 08 148.42 – 23.90	05 – 07 – 044	1301 40	1.1 x 0.8 1.0 x 0.7	105 ...	1	...		15.0		
02207	02 40.8 + 38 36 145.80 – 19.01	06 – 07 – 000	1307	1.1 x 0.4 1.1 x 0.3	105 6	1	S B ... s of 2		16.5:		*
02208	02 41.0 + 00 05 172.03 – 51.56	00 – 07 – 088 NGC 1072	1283	1.5 x 0.5 1.3 x 0.5	11 6	2	Sa – b		14.3		2
02209	02 41.0 + 49 22 141.05 – 09.28	08 – 06 – 000	0845 47	1.0 x 0.6 1.0 x 0.6	142: 4	1	S ...		16.5:		
02210	02 41.1 + 01 10 170.92 – 50.73	00 – 08 – 001 NGC 1073	1453	5.5: x 5.5: 5.5: x 5.5:	-- 1	5	S B c NGC 1068 group	S B(rs)cd	12.5	+ 1874 + 1893	*12- -9-
02211	02 41.2 + 06 26 165.89 – 46.64	01 – 08 – 000	1320 41	1.3 x 0.2 1.3 x 0.2	174 7	3	Sc 1st of 3		16.5:		*
02212	02 41.4 + 33 31 148.36 – 23.51	06 – 07 – 002	1307 47	1.1 x 0.5 1.0 x 0.5	91 (5 – 6)	1	...		15.4		*
02213	02 41.4 + 37 47 146.29 – 19.69	06 – 07 – 000	1307 47	1.1 x 1.0 1.0 x 1.0	-- 1	3	Sb		15.7		
02214	02 41.7 + 09 31 163.41 – 44.09	02 – 08 – 000	16,15	1.3 x 0.7 1.3 x 0.6:	10: ...	2	...	S dm	16.5:		
02215	02 41.7 + 41 47 144.49 – 16.08	07 – 06 – 063	0449 47	1.0 x 0.20 1.0 x 0.2	3 ...	1	...		14.8		
02216	02 41.9 + 00 28 171.89 – 51.12	00 – 08 – 003	1453	1.8: x 0.5 1.8: x 0.6:	[15] ...	2	Irr		16.0:		*
02217	02 41.9 + 16 31 158.27 – 38.25	03 – 08 – 003	0425	1.4 x 0.3 1.3 x 0.3	152 7:	2	Sb		15.6		*
02218	02 41.9 + 32 30 148.97 – 24.36	05 – 07 – 047	1301 40	(1.1 x 1.1) (1.1 x 1.1)	-- ...	2	... Very compact		15.4		
02219	02 41.9 + 38 22 146.11 – 19.13	06 – 07 – 000	1307 47	1.3 x 1.0 1.1: x 0.8:	(0) 2 – 3	2	Sc		16.5:		
02220	02 42.0 + 15 02 159.32 – 39.49	02 – 08 – 001 IC 1839	16,15	1.1 x 0.5 1.0 x 0.5	97 5	2	Sb – c		15.3		1
02221	02 42.0 + 30 10 150.19 – 26.42	05 – 07 – 000	1301	1.7 x 0.1 0.8 x 0.2:	80 7	1	...	S dm?	17.		
02222	02 42.1 + 32 47 148.86 – 24.09	05 – 07 – 048	1301 40	1.2: x 0.7: 1.5: x 0.7:	[105] ...	1	... P w 2225, distorted		14.9		*1
02223	02 42.1 + 34 59 147.77 – 22.13	06 – 07 – 000	1307 42	1.2 x 0.35 1.2 x 0.35	47 6	2	Sc		14.8		
02224	02 42.1 + 45 47 142.77 – 12.44	08 – 06 – 000	0845 47	1.2: x 0.8: 0.9 x 0.5	(0) ...	1	S ...		16.5:		
02225	02 42.2 + 32 46 148.89 – 24.09	05 – 07 – 049	1301 40	1.1 x 0.3 1.1 x 0.3	6 6	2	S ... P w 2222, distorted		15.2		*1
02226	02 42.4 + 41 57 144.54 – 15.87	07 – 06 – 066	0449 47	(1.3 x 0.7) (1.5 x 0.9)	(50) ...	2	E f of 2		15.7		*
02227	02 42.7 + 42 36 144.30 – 15.26	07 – 06 – 067	0449 47	2.3 x 2.3 2.3 x 2.3	-- 1	4	S B b 	S B(r)b Sb +	14.7		*
02228	02 42.8 + 02 40 169.89 – 49.32	00 – 08 – 004 IC 1843	1453	1.7 x 0.9 1.6 x 0.8	70: 5	1	S B a:		14.2		
02229	02 42.9 + 00 43b 171.92 – 50.77	00 – 08 – 006	1453	(1.5 x 0.7) (1.7 x 0.8)	175 ...	1	E – S ... nf of 2		(14.8)		*
02230	02 42.9 + 39 53a 145.58 – 17.68	07 – 06 – 069 NGC 1077a	0449 47	1.3 x 0.9 1.3 x 0.9	165: 3	3	Sb spp of 2,	Sb +	(14.6)		*1
02231	02 42.9 + 41 54 144.65 – 15.87	07 – 06 – 000	0449 47	1.2: x 0.8: 1.1: x 1.0:	(115) ...	3	...	S B(s)m Ir I	17.		
02232	02 43.0 + 36 42 147.11 – 20.52	06 – 07 – 004	1307 42	(1.4 x 1.4) (1.4 x 1.4)	-- ...	2	...		16.0:		*
02233	02 43.0 + 44 45 143.38 – 13.30	07 – 06 – 000	0449 47	1.7 x 0.3 1.7 x 0.30	108 7	3	Sb		15.5		*
02234	02 43.1 + 44 56 143.31 – 13.13	07 – 06 – 000	0449 47	1.4 x 0.4 1.4 x 0.4	126 6	3	Sc P w 2235		15.4		*
02235	02 43.2 + 45 00 143.30 – 13.06	08 – 06 – 000	0449 47	1.1: x 0.8: 1.2: x 0.9:	(125) ...	3	... S: IV – V or V, p w 2234		17.		*
02236	02 43.3 + 26 50 152.37 – 29.21	04 – 07 – 020	0222 38	1.3 x 0.7 1.3 x 0.7	26 4 – 5	1	S ...		15.5		
02237	02 43.3 + 35 01 147.99 – 21.99	06 – 07 – 005	1307 42	1.2 x 0.5 1.3 x 0.5	170: 6:	1	S ... Disturbed?		15.5		*1
02238	02 43.6 + 12 53 161.31 – 41.06	02 – 08 – 003	16,15 44	1.8 x 1.1: 1.7 x 1.0:	(165) ...	1	Irr?		15.2		1
02239	02 43.6 + 32 14 149.45 – 24.43	05 – 07 – 051	1301 40	1.8: x 0.3 1.9 x 0.3	13 6	2	S B? b – c		15.6		*
02240	02 43.6 + 48 00 142.03 – 10.33	08 – 06 – 000	0845 47	(1.7 x 1.2) (1.7 x 1.2)	(5) 3:	3	S B 0		15.4		

N G C   1072, 1073, 1077a   I C   1839, 1843

1	2	3	4	5	6	7	8	9	10	11	12
02241	02 43.8 + 03 24 169.46 - 48.60	00 - 08 - 010 NGC 1085	1453	2.5: x 1.8: 2.3: x 1.7:	15: 2 - 3	3	Sb	SA b Sb -	13.6		*3
02242	02 43.8 + 20 10 156.35 - 34.91	03 - 08 - 000	0425	1.2 x 1.2 1. x 1.	— ...	3	Dwarf: IV - V or V		17.		
02243	02 43.8 + 38 49 146.25 - 18.56	06 - 07 - 008	1307 47	1.4: x 0.45 1.0 x 0.45	100 6	2	S B b - c		14.9		*
02244	02 43.8 + 39 25b 145.96 - 18.02	06 - 07 - 006	1307 47	(1.0 x 0.9) (1.0 x 1.0)	— ...	1	... s of 2		(15.4)		*1
02245	02 43.9 - 00 42 173.75 - 51.64	00 - 08 - 009 NGC 1087	1453 (43)	4. x 3. 3.8: x 2.8:	0 2 - 3	5	Sc NGC 1068 group, Sc -	SAB(rs)c	11.4	+ 1824 + 1833	*12- -9
02246	02 43.9 + 43 00 144.32 - 14.81	07 - 06 - 000	0449 47	1.0 x 0.3 1.2 x 0.3	103 6	2	Sc		17.		
02247	02 44.0 - 00 27 173.50 - 51.44	00 - 08 - 011 NGC 1090	1453 (43)	4.3: x 2.1: 4.5 x 2.0:	102 5 - 6	5	S B b NGC 1068 group, Sb +	S B(rs)bc	12.8		*12- -9
02248	02 44.0 + 23 24 154.41 - 32.11	04 - 07 - 021	0222	1.1 x 0.8 1.2 x 0.8	(145) 3:	1	SO:		15.7		
02249	02 44.0 + 45 20 143.28 - 12.70	08 - 06 - 000	0845 47	2.0: x 1.5: 2.0 x 1.5:	(40) 2 - 3	3	SO	SAB 0 +	16.0:		
02250	02 44.0 + 46 08 142.92 - 11.98	08 - 06 - 000	0845 47	1.3:: x 1.3:: 1. x 1.	— ...	3	Dwarf IV - V or V		18.		
02251	02 44.1 + 24 39 153.70 - 31.02	04 - 07 - 022	0222	1.0 x 0.4 1.1 x 0.7	[150] ...	1	Dwarf? nnf of 2		17.		*
02252	02 44.2 + 15 13 159.77 - 39.05	02 - 08 - 004	16,15	(1.1 x 1.1) (1.1 x 1.1)	— ...	2	... Compact		14.8		*
02253	02 44.2 + 15 59 159.23 - 38.40	03 - 08 - 009 NGC 1088	0425	1.2 x 0.7 1.3 x 0.8	105: 4	2	SO - a Companion superimposed		14.8		*3
02254	02 44.2 + 37 19 147.04 - 19.86	06 - 07 - 000	1307 47	1.1: x 0.9: 0.8: x 1.0:	— ...	2	Dwarf irregular Ir V		17.		
02255	02 44.4 + 08 26 165.06 - 44.57	01 - 08 - 000	1320 41	0.9 x 0.4 (1.3 x 0.8)	85 4 - 5	2	SO - a		14.3		
02256	02 44.5 + 40 18 145.67 - 17.17	07 - 06 - 070	0449 47	1.3 x 0.7 1.4 x 0.8	85 4 - 5	2	SO		15.1		
02257	02 44.6 + 48 34 141.94 - 09.75	08 - 06 - 005	0845	(1.2 x 1.2) (1.3 x 1.3)	— ...	1	SO?		16.5:		
02258	02 44.7 + 41 03 145.36 - 16.49	07 - 06 - 071 NGC 1086	0449 47	1.7 x 1.0 1.7 x 1.0	35: 4:	3	Sc Singular		13.6		*1
02259	02 44.8 + 37 20 147.14 - 19.79	06 - 07 - 009	1307 47	2.8 x 2.2 3.0: x 2.0:	155: ...	4	... s of 2	S B(s)dm	14.3		*1
02260	02 44.8 + 42 49 144.56 - 14.90	07 - 06 - 000	0449 47	(1.0 x 0.9) (1.0 x 1.0)	— ...	1	... Compact		16.0:		
02261	02 44.8 + 50 36 141.07 - 07.90	08 - 06 - 006	0845	(1.7 x 1.2) (2.0 x 1.5)	(70) ...	2	E P w 2270		15.2		*
02262	03 44.9 - 00 30 173.82 - 51.32	00 - 08 - 015 NGC 1094	1453 43	1.3 x 1.0 1.6 x 1.3	85 2 - 3	3	Sa/Sb NGC 1068 group, s of 2		13.5		*23
02263	02 44.9 + 16 38 158.96 - 37.77	03 - 08 - 010	0425	1.0 x 0.7 1.0 x 0.7	85: ...	1	SO?		16.0:		
02264	02 45.0 + 04 26 168.81 - 47.61	01 - 08 - 001 NGC 1095	1320	1.4 x 0.9 1.4 x 0.9	45 3 - 4	4	S B c	S B(s)c	14.2		13
02265	02 45.0 + 13 03 161.56 - 40.73	02 - 08 - 006 IC 1846	16,15 44	(1.1 x 0.9) (0.9 x 0.9)	— ...	2	... Compact		15.4		*
02266	02 45.0 + 14 11 160.72 - 39.80	02 - 08 - 000	16,15 44	1.2: x 1.2: 1.0: x 1.0:	— 1	3	...	SA: dm	17.		
02267	02 45.0 + 23 12 154.76 - 32.17	04 - 07 - 023	0222	1.8:: x 1.8:: 0.9 x 0.8	— 1	3	S B b	(R)S B(r)b	15.6		
02268	02 45.0 + 25 53 153.19 - 29.85	04 - 07 - 000	0222	1.1: x 1.1: 1.0: x 1.0:	— ...	2	Dwarf:		17.		
02269	02 45.0 + 44 27 143.84 - 13.42	07 - 06 - 000	0449 47	1.7:: x 1.5:: 1.3: x 0.9:	(100) ...	3	Dwarf V		18.		
02270	02 45.0 + 50 33 141.12 - 07.93	08 - 06 - 008	0845	2.2 x 2.1 2.2 x 2.2	— 1	1	Sb: P w 2261		14.8		*1
02271	02 45.1 + 00 13 173.09 - 50.77	00 - 08 - 016	1453 43	1.6: x 0.9: 1.6: x 0.9:	10: 4 - 5	2	S B b		15.6		
02272	02 45.1 + 26 54 152.64 - 28.95	04 - 07 - 024	0222 38	2.0 x 0.35 2.0 x 0.35	65 6	2	Sb - c		14.7		
02273	02 45.1 + 47 47 142.36 - 10.42	08 - 06 - 007	0845 47	1.5 x 0.8 1.5 x 0.8	120 ...	3	S B c nnf of 2, disturbed?		15.7		*
02274	02 45.2 + 34 13 148.76 - 22.52	06 - 07 - 011 NGC 1093	1307	2.2 x 1.4 1.9 x 1.3	100: 3 - 4	1	S ...	SAB ab?	14.3		
02275	02 45.4 + 03 41 169.64 - 48.12	01 - 08 - 000	1320	6. x 6. 4. x 3.	— ...	4	Dwarf (spiral) S: V		15.7		

1	2	3	4	5	6	7	8	9	10	11	12
02276	02 45.5 + 17 32 158.49 - 36.93	03 - 08 - 012	0425	1.4 x 0.5: 1.1 x 0.3	52 6	1	S ...		16.0:		
02277	02 45.6 + 40 29 145.78 - 16.92	07 - 06 - 072	0449 47	(1.1 x 1.0) (1.1 x 1.0)	— ...	2	... Compact		15.5		
02278	02 45.7 + 04 22 169.07 - 47.55	01 - 08 - 003 NGC 1101	1320	(1.4 x 1.2) (1.6 x 1.3)	— 1 - 2	3	S0		14.8		3
02279	02 45.7 + 18 45 157.73 - 35.88	03 - 08 - 000	0425	1.0 x 0.2 0.9 x 0.2	[15] ...	2	S ... distorted,	npp of 2, integral	16.5:		*
02280	02 45.7 + 41 35 145.28 - 15.93	07 - 06 - 073	0449 47	2.4 x 0.8 2.1 x 0.8	44 6	3	Sb	SA bc Sb +	15.0		
02281	02 45.7 + 46 57 142.82 - 11.12	08 - 06 - 009	0845 47	1.3 x 1.1 1.3 x 1.1	(100) ...	1	S ...		16.0:		*
02282	02 45.9 + 14 06 161.01 - 39.74	02 - 08 - 010	16,15 44	1.1 x 0.9 1.2 x 1.0	(150) 1 - 2	4	S B c nnf of 2,	S B(rs)bc Sc -	14.8		*
02283	02 45.9 + 17 00 158.96 - 37.33	03 - 08 - 014	0425	1.0 x 0.5 1.0 x 0.5	110: 5	2	S B: c: Singular		15.3		*
02284	02 46.0 - 02 18 176.18 - 52.42	00 - 08 - 000	1453	1.4:x 1.2: 1.2:x 1.0	— ...	3	...	S dm	17.		
02285	02 46.0 + 06 19 167.35 - 46.00	01 - 08 - 004	1320	1.4 x 0.5 1.3:x 0.5	[85] ...	2	Sc - Irr		15.3		*
02286	02 46.0 + 20 39 156.58 - 34.23	03 - 08 - 015	0425	(1.1 x 0.9) (1.1 x 1.1)	— ...	2	... Compact		15.4		
02287	02 46.1 - 00 28 174.13 - 51.09	00 - 08 - 019 NGC 1104	1453 43	1.4 x 1.4 1.5 x 1.4	— 1	3	S B a		14.8		23
02288	02 46.1 + 35 08 148.48 - 21.62	06 - 07 - 012	1307 42	1.3 x 0.7 0.9 x 0.5	175 ...	1	S0 - a		15.5		*
02289	02 46.2 + 15 31 160.07 - 38.53	03 - 08 - 016	0425	1.0 x 0.3 0.9 x 0.22	130 6	1	S ...		15.4		*
02290	02 46.2 + 22 48 155.29 - 32.37	04 - 07 - 000	0222	1.7::x 1.7:: 1.7::x 1.7::	— ...	3	Dwarf (spiral) V		18.		
02291	02 46.3 - 00 58 174.74 - 51.42	00 - 08 - 021 IC 1856	1453 43	1.1 x 0.6 1.2 x 0.6	60 5	2	S B: b		14.5		
02292	02 46.3 + 00 47 172.84 - 50.15	00 - 08 - 020	1453 43	1.1 x 1.0 1.2 x 1.2	— 1	2	Sa	SA a	15.4		
02293	02 46.3 + 13 01 161.92 - 40.58	02 - 08 - 011 IC 1852	16,15 44	1.3 x 1.0 1.2 x 0.8	(3) 2 - 3	3	S B c	SAB(s)c Sc -	14.9		1
02294	02 46.3 + 46 57 142.92 - 11.08	08 - 06 - 000	0845 47	1.2 x 0.4 0.9 x 0.4	57 ...	1	S ... n of 2, disturbed		16.0:		*
02295	02 46.4 + 02 58 170.63 - 48.50	00 - 08 - 022	1453	1.6 x 0.8 1.6 x 0.7:	85 5 - 6	1	S ...		14.0		*
02296	02 46.4 + 18 07 158.33 - 36.33	03 - 08 - 017	0425	(0.9 x 0.9) (1.0 x 0.9)	— ...	2	... Extremely compact		13.1		*
02297	02 46.4 + 30 27 150.97 - 25.72	05 - 07 - 000	1301	1.1 x 0.6 1.0 x 0.6	135 4 - 5	2	S0 - a P w 2300		16.0:		*
02298	02 46.4 + 46 47 143.01 - 11.22	08 - 06 - 011 IC 257	0845 47	(2.2 x 1.6) (2.5 x 1.7)	(155) ...	2	E - S0		14.5		*
02299	02 46.5 + 10 54 163.62 - 42.27	02 - 08 - 000	16,15	1.0 x 0.9 0.9: x 0.8:	— 1	3	...	SA(s)dm or SAB(s)dm	17.		
02300	02 46.5 + 30 29 150.97 - 25.68	05 - 07 - 000	1301	1.1 x 0.7 1.0 x 0.6	100 4	2	Sa - b P w 2297		16.0:		*
02301	02 46.5 + 38 04 147.10 - 18.99	06 - 07 - 000	1307 47	1.0: x 0.4: 1.0: x 0.4:	[128] ...	3	Dwarf: IV - V or V		18.		
02302	02 46.6 + 01 56 171.72 - 49.25	00 - 08 - 023	1453	6. x 6. 5. : x 5. :	— 1	4	Dwarf spiral S V		15.3		*
02303	02 46.6 + 17 27 158.83 - 36.86	03 - 08 - 018	0425	1.3 x 1.3 1.4 x 1.4	— 1	3	Sb	SAB b Sb -	14.5		*
02304	02 46.6 + 21 55 155.92 - 33.08	04 - 07 - 025	0222	1.4 x 0.25 1.0 x 0.25	61 7	1	S ...		16.0:		
02305	02 46.6 + 37 51 147.23 - 19.17	06 - 07 - 000	1307 47	1.5 x 1.0 1.5: x 0.9	20 3 - 4	4	Sc	SAB cd	15.5		
02306	02 46.6 + 40 51 145.79 - 16.50	07 - 06 - 000 IC 259	0449 47	(1.4 x 1.2) (1.7 x 1.4)	— 1 - 2	4	S B 0		15.6		*
02307	02 46.7 + 07 54 166.15 - 44.64	01 - 08 - 006 NGC 1107	1320 41	(1.8 x 1.5) (2.3 x 2.0)	(140) 2:	3	S0		14.1		*13
02308	02 46.7 + 46 52 143.02 - 11.12	08 - 06 - 000	0845 47	[1.5: x 1.3:] ...	— ...	1	Multiple system		16.0:		*
02309	02 46.8 + 21 00 156.55 - 33.84	03 - 08 - 019	0425	1.1 x 0.8 (1.7 x 1.4)	160: ...	1	S0		15.6		
02310	02 46.8 + 35 40 148.35 - 21.09	06 - 07 - 000	1307 42	1.0: x 0.4 1.1: x 0.5	[95] ...	1	...		16.5:		

N G C   1101, 1104, 1107   I C   257, 259, 1852, 1856

1	2	3	4	5	6	7	8	9	10	11	12
02311	02 46.9 - 01 05 175.04 - 51.40	00 - 08 - 024	1453 43	1.5 x 1.5 1.7 x 1.6	— 1	3	S B b	Sb -	13.9		
02312	02 46.9 + 14 25 161.04 - 39.35	02 - 08 - 013 IC 1857	16,15 44	1.0 x 0.5 1.0 x 0.5	150 5	1	S ...		15.1		
02313	02 46.9 + 41 11 145.68 - 16.18	07 - 06 - 000	0449 47	1.5 x 0.25 1.4 x 0.2	141 6 - 7	3	Sc sf of 2		15.7		*
02314	02 46.9 + 47 20 142.84 - 10.69	08 - 06 - 000	0845 47	1.6:: x 1.6:: 1.4:: x 1.4::	... ...	3	Dwarf: IV - V:		18.		*
02315	02 47.0 + 22 31 155.65 - 32.52	04 - 07 - 000	0222	1.1 x 0.25 1.1 x 0.25	140 7	2	Sb - c		17.		
02316	02 47.0 + 39 00 146.74 - 18.11	06 - 07 - 000	1307 47	(1.1 x 0.8) (0.8 x 0.6)	(10) ...	2	SO		16.0:		*
02317	02 47.0 + 46 30 143.23 - 11.42	08 - 06 - 012	0845 47	1.7 x 0.8 1.8 x 0.9	20 5	2	Sa		15.6		*
02318	02 47.2 + 43 27 144.67 - 14.13	07 - 06 - 075	0449 47	1.1 x 1.1 1.1 x 1.0	-- 1	3	Sb - c		15.7		
02319	02 47.4 - 01 12 175.32 - 51.39	00 - 08 - 026	1453 43	1.4 x 0.35 1.2 x 0.4	47 6	3	Sc		15.6		
02320	02 47.4 + 12 40 162.48 - 40.71	02 - 08 - 015	16,15 44	[1.2 x 0.6] ...	[30] ...	3	S ... + S ... Contact, plume		15.2		*1 V A
02321	02 47.4 + 18 09 158.55 - 36.17	03 - 08 - 020	0425	1.3 x 0.3 1.2 x 0.3	80 6	1	Sa - b		15.7		
02322	02 47.4 + 41 29 145.63 - 15.87	07 - 06 - 076 NGC 1106	0449 47	1.7 x 1.7 1.9 x 1.9	-- 1	4	SO	SA 0 +	13.7		*3
02323	02 47.6 + 07 58 166.34 - 44.45	01 - 08 - 007	1320 41	1.1 x 1.0 1.0 x 1.0	-- 1	3	S B? c nff of 2	S (r)c	15.7		*
02324	02 47.7 - 00 06b 174.18 - 50.55	00 - 08 - 029	1453 43	1.1 x 0.9: 1.0 x 0.9	(5) 1 - 2	2	S B b f of 2	S B(r)b	(14.9)		*1
02325	02 47.7 + 46 45 143.23 - 11.15	08 - 06 - 014 IC 260	0845 47	(1.4 x 0.9) (1.8 x 1.3)	(175) ...	2	[E] Compact		15.2		*
02326	02 47.8 + 13 07 162.24 - 40.29	02 - 08 - 017 NGC 1116	16,15 44	1.5 x 0.3 1.5 x 0.4	27 6 - 7	1	Sa - b		15.4		*3
02327	02 47.8 + 20 29 157.11 - 34.15	03 - 08 - 022	0425	1.2 x 0.9 1.1 x 1.0:	(65) 1 - 3	2	Sb: Singular		15.0		*
02328	02 47.9 + 37 15 147.77 - 19.58	06 - 07 - 014	1307 47	(1.7 x 1.4) (1.5 x 1.1)	(70) ...	3	E		13.8		
02329	02 48.0 + 15 48 160.33 - 38.05	03 - 08 - 023	0425	1.3: x 0.9 1.0 x 0.7	177: 3:	1	Sa - b Brightest in group		15.0		*
02330	02 48.0 + 41 22 145.78 - 15.92	07 - 06 - 077	0449 47	(1.3 x 0.8) (1.4 x 0.9)	(80) ...	3	E		16.0:		
02331	02 48.0 + 47 00 143.16 - 10.90	08 - 06 - 000	0845 47	1.1 x 0.22 1.1 x 0.25	42 6	1	Sa		16.5:		*
02332	02 48.2 - 01 56 176.38 - 51.77	00 - 08 - 030 NGC 1121	1453 43	1.1 x 0.4 1.0 x 0.4	10 6 - 7	1	S ...		13.7		
02333	02 48.3 + 25 12 154.33 - 30.06	04 - 07 - 027	0222	1.0 x 0.5 0.9 x 0.4	32: 5 - 6	1	Sb - c		15.7		
02334	02 48.3 + 40 32 146.24 - 16.64	07 - 06 - 078	0449 47	1.1 x 1.1 (1.4 x 1.3)	-- ...	1	... Brightest of 3, disturbed		16.0:		*
02335	02 48.4 + 42 38 145.25 - 14.76	07 - 06 - 080 IC 262	0449 47	(2.1 x 1.7) (2.3 x 1.9)	(50) 2:	4	S B O		14.8		
02236	02 48.5 + 07 52 166.67 - 44.40	01 - 08 - 009	1320 41	1.9: x 1.2: 1.7 x 0.9	165 4 - 5	4	S B c	SB(s)cd	15.7		*
02337	02 48.5 + 12 59 162.52 - 40.30	02 - 08 - 19+20	16,15 44	[1.0 x 0.6] ...	[174] ...	2	Triple system		14.9		*1
02338	02 48.6 + 02 23 171.83 - 48.58	00 - 08 - 033	1453	1.7 x 0.9 1.7: x 0.9	67 5	4	Sc		14.6		1
02339	02 48.6 + 15 50 160.45 - 37.94	03 - 08 - 024	0425	1.6: x 1.3: 1.0: x 0.8	(45) ...	3	... 2334 group	S dm	16.5:		*
02340	02 48.8 + 13 22 162.31 - 39.94	02 - 08 - 021	16,15 44	1.2 x 0.6 1.1 x 0.6	80 5	2	S B c		16.0:		
02341	02 48.9 + 17 28 159.39 - 36.55	03 - 08 - 025	0425	1.1 x 0.8 1.1 x 0.8:	(155) 3:	2	Sb/S B b	SAB(s)b	16.5:		*
02342	02 49.0 + 43 52 144.77 - 13.62	07 - 06 - 000	0449 47	(1.1 x 1.0) (1.3 x 1.3)	-- ...	2	... Very compact		15.2		
02343	02 49.1 - 00 57 175.52 - 50.91	00 - 08 - 034	1453 43	1.1 x 0.8 1.0 x 0.8	— 2 - 3	1	Sa:		15.2		*
02344	02 49.2 + 13 43 162.15 - 39.60	02 - 08 - 000	16,15 44	1.0 x 0.2 0.9 x 0.18	125 7	2	Sb - c		16.5:		*
02345	02 49.3 - 01 22 176.04 - 51.17	00 - 08 - 035	1453 43	4.5:: x 3.5:: 3. x 3.	— ...	4	Dwarf spiral SB V		16.0:		*

1	2	3	4	5	6	7	8	9	10	11	12
02346	02 49.3 + 00 23 174.11 – 49.92	00 – 08 – 036	1453	1.3 x 0.6 1.0 x 0.5	107 5	1	Sb – c		15.7		
02347	02 49.4 + 11 56 163.57 – 41.02	02 – 08 – 022	16,15 44	1.3 x 0.6 1.1 x 0.5	3 5	1	S ...		16.0:		
02348	02 49.4 + 13 47 162.15 – 39.52	02 – 08 – 023	16,15 44	(1.3 x 1.3) (1.6 x 1.6)	— (1)	1	SO?		15.2		*
02349	02 49.4 + 41 07 146.15 – 16.02	07 – 06 – 082	0449 47	1.3 x 0.6 1.8 x 0.9	48 5	3	SO P w 2350		16.0:		*
02350	02 49.4 + 41 12 146.11 – 15.95	07 – 06 – 081	0449 47	1.2 x 1.1 1.2 x 1.1	— 1	2	Sb	Sb +	14.9		*
02351	02 49.4 + 46 45 143.49 – 11.02	08 – 06 – 018	0845 47	2.3: x 2.3: 2.3: x 2.3:	— 1	4	S B b	S B(s)b	14.9		*
02352	02 49.5 + 04 11 170.31 – 47.08	01 – 08 – 000	1320	1.1 x 0.2 1.1 x 0.2	25 ...	1	Dwarf		17.		
02353	02 49.6 + 42 00 145.76 – 15.22	07 – 06 – 083 NGC 1122=1123	0449 47	2.2 x 1.7 2.2 x 1.7	40: 2 – 3	3	Sb/S B b		13.0		*3
02354	02 49.7 + 42 02 145.76 – 15.19	07 – 06 – 000	0449 47	1.0 x 0.10 1.0 x 0.10	41 7	3	Sc		16.5:		*
02355	02 49.8 + 01 47 172.78 – 48.82	00 – 08 – 000	1453	1.5:: x 0.9:: 1. x 1.	— ...	3	Dwarf V		18.		
02356	02 50.1 + 13 03 162.89 – 40.02	02 – 08 – 024 NGC 1127	16,15 44	1.0 x 1.0 0.9 x 0.8	— 1	2	S B a – b	(R:)S B(r) ...	15.7		3
02357	02 50.1 + 25 17 154.68 – 29.78	04 – 07 – 028 IC 1861	0222	1.8 x 1.3 1.7 x 1.1	150: 3 – 4	4	SO	SA O	14.7		
02358	02 50.2 + 74 57 130.77 + 14.20	12 – 03 – 003	1230	2.2 x 1.3 1.7 x 0.8	25 4 – 5	4	S B b	S B(r)b	15.7		*
02359	02 50.3 – 01 28 176.43 – 51.07	00 – 08 – 040 NGC 1132	1453 43	(2.0 x 1.3) (2.9 x 2.0)	(140) ...	3	E		13.9		*3
02360	02 50.3 + 31 28 151.23 – 24.42	05 – 07 – 000	1301	1.3:: x 1.1:: 1.0: x 0.9:	— 1 – 2	4	...	SA d – dm	16.0:		
02361	02 50.3 + 41 42 146.02 – 15.43	07 – 06 – 084	0449 47	1.6 x 1.4 1.5 x 1.2	(60) 2	3	S B b	SAB(s)b	14.4		
02362	02 50.4 + 12 50 163.13 – 40.15	02 – 08 – 025	16,15 44	1.7: x 1.2: 1.7: x 1.1:	— ...	4	Irr NGC 1134 group,	I m Ir I	15.5		*1
02363	02 50.6 + 39 23 147.20 – 17.45	06 – 07 – 017	1307 47	(1.6 x 1.4) (1.6 x 1.4)	— ...	2	... Compact		14.6		
02364	02 50.7 + 06 20 168.63 – 45.24	01 – 08 – 011	1320 45	1.6 x 0.5 1.4 x 0.4	[103] ...	2	... np of 2		16.5:		*
02365	02 50.9 + 12 48 163.29 – 40.11	02 – 08 – 027 NGC 1134	16,15 44	2.5 x 0.9 2.6 x 0.9	148 6	2	S ... Singular Brightest in group		13.2		*13 A
02366	02 50.9 + 36 13 148.85 – 20.21	06 – 07 – 018	1307 47	1.4 x 0.8 1.4 x 0.8	63: 4 – 5	2	SO		15.7		*
02367	02 51.0 + 06 04 168.96 – 45.40	01 – 08 – 012	1320 45	2.0: x 0.8: 2.3 x 0.8	145 6	2	SO P w 2375		15.3		*
02368	02 51.1 + 12 38 163.47 – 40.21	02 – 08 – 028 IC 267	16,15 44	2.1 x 1.4 2.1 x 1.5	(15) 3 – 4	4	S B b NGC 1134 group,	(R′)S B(s)b Sb +	14.1		*1
02369	02 51.2 + 14 46 161.88 – 38.46	02 – 08 – 29+30	16,15 44	[1.0 x 0.8] ...	... ...	2	Double system Contact, distorted		14.6		*1
02370	02 51.2 + 42 27 145.81 – 14.69	07 – 07 – 000	1618 47	3.0 x 0.10 2.5 x 0.10	127 (7)	2	... ssf of 2 Flat, IV – V? or V?		15.7		*
02371	02 51.3 + 02 05 172.89 – 48.34	00 – 08 – 042	1453	1.5 x 0.5 1.4 x 0.4	17 6	1	Sa – b		15.7		
02372	02 51.3 + 05 47 169.30 – 45.57	01 – 08 – 013	1320 45	1.0 x 0.9 1.1 x 0.9	— 1 – 2	3	Sc	SAB(rs:)cd	15.5		*
02373	02 51.3 + 41 23 146.35 – 15.62	07 – 07 – 004 NGC 1129	1618 47	(4. x 3.) (5. x 4.)	(90) ...	3	E Singular		14.6		*13
02374	02 51.4 + 02 45 172.24 – 47.84	00 – 08 – 043 NGC 1137	1453	2.5 x 0.6 2.7 x 1.6	20 4	3	Sb	Sb –	13.5		
02375	02 51.4 + 06 03 169.08 – 45.35	01 – 08 – 014	1320 45	1.1 x 0.3 1.0 x 0.3	114 6	1	S ... P w 2367		15.5		*
02376	02 51.4 + 31 05 151.66 – 24.64	05 – 07 – 052	1301	2.0: x 1.7: 1.8: x 1.6:	— 1 – 2	4	Sc		16.0:		
02377	02 51.5 + 39 11 147.46 – 17.54	06 – 07 – 000	1307 47	1.0 x 0.20 1.2 x 0.2	93 7	2	Sa – b		15.6		
02378	02 51.7 + 11 32 164.48 – 41.01	02 – 08 – 032	16,15 44	1.6 x 0.3 1.4 x 0.3	174 7	3	Sc		16.0:		
02379	02 51.7 + 13 57 162.62 – 39.06	02 – 08 – 000	16,15 44	[1.4 x 0.6] ...	... ...	3	S ... + companion Connected		15.5		*
02380	02 51.7 + 51 43 141.55 – 06.42	09 – 05 – 000	1617	1.6 x 0.7 1.6 x 0.6	9 6	2	Sb		16.0:		

N G C   1122,   1123,   1127,   1129,   1132,   1134,   1137   I C   267,   1861

1	2	3	4	5	6	7	8	9	10	11	12
02381	02 51.9 + 11 49 164.31 – 40.75	02 – 08 – 033	16,15 44	1.0 x 0.4: 1.0 x 0.3	[135] ...	1	S ...		16.0:		
02382	02 52.0 + 09 10 166.50 – 42.84	01 – 08 – 000	1320 41	1.1 x 0.15 1.1 x 0.15	35 7	3	Sc		17.		
02383	02 52.0 + 46 09 144.18 – 11.34	08 – 06 – 000	0845 47	1.2: x 0.8: 1.1: x 0.9:	(0) ...	1	...	S B dm:	18.		
02384	02 52.2 + 15 49 161.37 – 37.47	03 – 08 – 028 NGC 1145	0425 46	1.0 x 0.4 0.9 x 0.3	132 6	1	S ...		15.1		*
02385	02 52.5 – 01 23 176.95 – 50.61	00 – 08 – 046	1453	1.0 x 0.9 1.1 x 1.0	— 1	2	Sc	Sc –	15.5		
02386	02 52.5 + 09 09 166.65 – 42.78	01 – 08 – 016	1320 41	1.2 x 1.1 1.2 x 1.1	— 1	1	S ...		16.0:		1
02387	02 52.5 + 12 01 164.31 – 40.50	02 – 08 – 035	16,15 44	1.7 x 0.5 1.5 x 0.5	12 6	3	Sc		15.7		
02388	02 52.6 – 00 23a 175.86 – 49.89	00 – 08 – 047 NGC 1143	1453	[1.4 x 1.2]	—	3	E + disrupted S ...		13.2		*12 3VA
02389	02 52.6 – 00 23b 175.86 – 49.89	00 – 08 – 048 NGC 1144	1453	...	...						
02390	02 52.6 + 36 29 149.03 – 19.81	06 – 07 – 022	1307 47	1.1 x 0.9 1.1 x 1.1	— 1 – 2	1	S B b		16.0:		
02391	02 52.7 + 08 37 167.15 – 43.16	01 – 08 – 017 IC 1865	1320 41	1.6: x 1.2: 1.5: x 1.2:	— 2 – 3	2	Sb – c		15.0		*
02392	02 52.7 + 33 35 150.57 – 22.33	06 – 07 – 021	1307 47	2.0 x 0.5 1.9 x 0.5	12 6	1	Sc?		15.0		1
02393	02 52.8 + 32 08 151.37 – 23.58	05 – 07 – 000	1301	1.1 x 0.5 1.0 x 0.4	87 6	4	Sc	Sc +	15.5		
02394	02 52.8 + 37 56 148.33 – 18.52	06 – 07 – 000	1307 47	1.4 x 0.2 1.3 x 0.15	31 7	2	Sc In group		18.		*
02395	02 52.9 + 41 26 146.60 – 15.44	07 – 07 – 000	1618 47	1.1 x 0.20 1.1 x 0.22	153 7	2	S0 – a		15.2	•	*
02396	02 53.0 + 36 52 148.91 – 19.43	06 – 07 – 000	1307 47	1.2 x 0.10 1.4 x 0.10	138 7	3	... 1st of 3	S d – dm	16.5:		*
02397	02 53.0 + 36 57 148.87 – 19.36	06 – 07 – 000	1307 47	1.1 x 0.5 1.1: x 0.6:	[0] ...	1	Peculiar, interacting systems?, 2nd of 3		17.		*
02398	02 53.1 + 48 30 143.24 – 09.18	08 – 06 – 023	0845 47	(1.2 x 1.0) (1.2 x 1.0)	... ...	2	E f of 2		15.7		*
02399	02 53.2 + 06 00 169.61 – 45.10	01 – 08 – 018	1320 45	1.6 x 1.6 1.6: x 1.5:	— 1	4	Sc	SAB(s)cd	15.3		
02400	02 53.2 + 09 06 166.87 – 42.71	01 – 08 – 019 IC 1867	1320 41	(1.5 x 1.2) (2.2 x 1.6)	(15) ...	1	...		15.0		*
02401	02 53.2 + 36 49 148.97 – 19.46	06 – 07 – 000	1307 47	(1.1 x 1.0) (1.2 x 1.2)	— ...	2	S0? 3rd of 3		16.0:		*
02402	02 53.2 + 45 55 144.48 – 11.45	08 – 06 – 000	0845 47	1.1 x 0.7: 1.2 x 0.8	(145) ...	1	S B: ...		16.5:		
02403	02 53.3 + 00 29 175.11 – 49.16	00 – 08 – 050	1453	1.5: x 0.9: 1.8: x 0.9:	155: ...	1	S B: ... nnp of 2, disturbed:		14.9		*1
02404	02 53.3 + 00 54 174.67 – 48.86	00 – 08 – 000	1453	1.3 x 0.10 1.2 x 0.10	9 7	2	Sc		16.5:		
02405	02 53.3 + 06 17 169.38 – 44.87	01 – 08 – 020	1320 45	1.8 x 0.9 1.8 x 0.8	167 5	3	Sc		15.1		
02406	02 53.3 + 15 35 161.81 – 37.50	03 – 08 – 030	0425 46	1.2 x 0.7 1.1 x 0.7	65: 3 – 4	3	Sa	(R′)SAB(s)a	15.7		
02407	02 53.3 + 31 42 151.71 – 23.90	05 – 07 – 053	1301	1.0 x 0.20 1.0 x 0.20	158 6	1	S B? a – b				
02408	02 53.3 + 42 51 145.97 – 14.15	07 – 07 – 012 NGC 1138	1618 47	(1.4 x 1.4) (2.1 x 2.1)	— 1	4	S B 0		14.1		3
02409	02 53.3 + 50 26 142.36 – 07.45	08 – 06 – 000	0845	1.1 x 0.3 1.0 x 0.25	67 (6)	1	S: ... np of 2		17.		*
02410	02 53.4 + 41 08 146.83 – 15.66	07 – 07 – 000	1618 47	1.2 x 0.9 1.2 x 0.9	(70) (2 – 3)	1	S: ...		15.6		*
02411	02 53.4 + 75 33 130.66 + 14.83	13 – 03 – 000	1226	4.5: x 0.3 4. x 0.3	12 ...	1	...		16.5:		
02412	02 53.5 – 01 55 177.83 – 50.80	00 – 08 – 000	1453	1.1 x 1.1 1.0: x 1.0:	— 1	3	Sc	S B(s)d	16.5:		
02413	02 53.6 + 15 43 161.79 – 37.35	03 – 08 – 031	0425 46	(1.0 x 0.8) (1.2 x 0.9)	(70) ...	2	... Compact, sff of 2		15.2		*1
02414	02 53.7 + 04 20 171.31 – 46.27	01 – 08 – 022	1320	1.2 x 0.25 1.2 x 0.25	159 6 – 7	2	Sc		15.5		*
02415	02 53.7 + 05 57 169.79 – 45.06	01 – 08 – 000	1320 45	1.1 x 0.4 1.0: x 0.3	1 6	2	S B b – c	S B(s) ...	15.5		*

N G C   1138, 1143, 1144   I C   1865, 1867

1	2	3	4	5	6	7	8	9	10	11	12
02416	02 53.8 + 18 30 159.87 – 35.04	03 – 08 – 000	0425	1.0: x 0.9: 1. x 1.	— 1:	2	Sb:		17.		
02417	02 53.8 + 36 08 149.44 – 20.00	06 – 07 – 000	1307 47	1.2: x 1.1: 1.0: x 0.7:	— ...	4	Dwarf irregular Ir V		18.		
02418	02 53.9 + 00 40 175.08 – 48.92	00 – 08 – 051	1453	1.2 x 0.8 1.3 x 0.9	(97) 3 – 4	3	S B b/Sc	Sc –	15.2		
02419	02 54.0 + 07 08 168.80 – 44.11	01 – 08 – 023	1320 45	1.3 x 1.1 1.3 x 1.1	(5) 1 – 2	4	S B a	(R´)S B (s)a	14.8		*
02420	02 54.0 + 15 46 161.85 – 37.26	03 – 08 – 000	0425 46	1.2 x 0.2 1.4 x 0.18	150 7	3	Sb		16.0:		*
02421	02 54.0 + 44 33 145.26 – 12.59	07 – 07 – 000	1618 47	1.3: x 1.1: 1.3: x 1.1:	— 1 – 2	3	Sc 1st of 3		17.		*
02422	02 54.0 + 44 57 145.07 – 12.24	07 – 07 – 000	1618 47	1.4 x 0.4 1.0 x 0.4	120 6	2	...	S d – dm	17.		
02423	02 54.1 + 04 47 170.99 – 45.87	01 – 08 – 024	1320 45	1.4 x 0.4 1.5: x 0.4	141 6	2	Sc		15.7		
02424	02 54.3 + 17 19 160.81 – 35.95	03 – 08 – 033	0425	2.5 x 0.6 2.2 x 0.5	158 6	2	Sb – c		14.6		
02425	02 54.5 + 02 35 173.25 – 47.43	00 – 08 – 052 IC 273	1453	1.8 x 0.7 1.8 x 0.6	31 6	1	S B: 0 – a		14.4		*
02426	02 54.5 + 05 07 170.78 – 45.56	01 – 08 – 025	1320 45	1.0 x 0.8 1.1 x 0.9	65 2	3	Sb		15.1		
02427	02 54.5 + 46 42 144.30 – 10.66	08 – 06 – 000	0845 47	1.0 x 0.5 0.9 x 0.5	166 5	1	Sb – c		16.5:		
02428	02 54.6 – 00 25 176.44 – 49.56	00 – 08 – 056	1453	1.5: x 1.3: 1.4: x 1.1:	— ...	2	...	SAB: dm	16.5:		*
02429	02 54.6 + 01 09 174.76 – 48.45	00 – 08 – 055	1453	1.4: x 0.8: 1.3 x 0.8:	[60] ...	3	Irr		16.0:		
02430	02 54.7 + 10 17 166.27 – 41.55	02 – 08 – 036	16,15	1.0 x 0.8 1.1 x 0.9	75: 2	2	Sa P w 2433		15.3		*
02431	02 54.7 + 47 58 143.73 – 09.52	08 – 06 – 000	0845 47	1.2: x 1.0: 1.2: x 1.0:	— 2:	3	...	SA m	17.		
02432	02 54.8 + 09 55 166.60 – 41.82	02 – 08 – 000	16,15	1.2: x 1.0: 1.1: x 1.0:	— ...	3	Dwarf spiral S V		17.		
02433	02 54.8 + 10 16 166.31 – 41.55	02 – 08 – 038	16,15	1.4: x 0.8 1.4 x 0.8	140: 4:	2	S B a – b P w 2430		16.0:		*1
02434	02 54.9 + 07 55 168.34 – 43.36	01 – 08 – 026	1320	1.0 x 0.9 1.0 x 0.9	— 1	4	Sc	SA(s)cd or SAB(s)cd	16.0:		
02435	02 54.9 + 35 04 150.21 – 20.81	06 – 07 – 023	1307 47	2.5: x 2.0: 2.3: x 1.9:	145: 2	4	Sc	SA(s)cd	14.6		
02436	02 55.1 + 36 18 149.59 – 19.72	06 – 07 – 000	1307 47	1.4: x 0.4: 1.3 x 0.3	87 7	3	...	S d – m	17.		
02437	02 55.2 + 10 59 165.83 – 40.92	02 – 08 – 040	16,15	1.6: x 1.5: 1.4: x 1.4:	— 1	3	Sc	Sc +	15.4		*
02438	02 55.2 + 12 50 164.36 – 39.45	02 – 08 – 039	16,15	1.1: x 0.9: 1.2: x 1.1:	— ...	1	...		16.5:		*1
02439	02 55.5 + 03 10 172.94 – 46.84	00 – 08 – 059 NGC 1153	1453	(1.4 x 1.4) (1.8 x 1.7)	— ...	1	E – S ...		13.5		*3
02440	02 55.7 – 02 28 179.06 – 50.78	00 – 08 – 000	1453	1.0 x 0.25 0.9 x 0.22	178 6	1	Sb – c		15.7		*
02441	02 55.7 + 03 40 172.50 – 46.43	01 – 08 – 029	1320	2.2 x 0.4 2.3 x 0.4	13 7	3	Sc sff of 2, Sc +		15.0		*
02442	02 55.7 + 25 06 156.03 – 29.27	04 – 08 – 005	1276	1.3 x 1.1 1.3 x 1.1	(150) ...	2	...		16.0:		1
02443	02 55.8 – 02 14 178.81 – 50.60	00 – 08 – 060	1453	1.5 x 0.8 1.4 x 0.8	163 4 – 5	2	Sc		14.6		
02444	02 55.8 + 06 06 170.21 – 44.60	01 – 08 – 031	1320 45	1.1 x 0.9 (0.9 x 0.8)	(45) ...	1	S ...		15.2		*
02445	02 55.8 + 25 34 155.77 – 28.87	04 – 08 – 000	1276	1.2 x 0.3 1.1 x 0.3	45 ...	1	Dwarf?		17.		
02446	02 56.0 + 03 14 173.00 – 46.70	00 – 08 – 061	1453	0.5 x 0.35 0.55 x 0.45	(105) ...	1	... Peculiar		14.4		
02447	02 56.0 + 32 26 151.84 – 22.98	05 – 08 – 000	0847	1.2 x 0.1 0.8 x 0.12	14 7	2	Sb – c		17.		*
02448	02 56.0 + 33 43b 151.14 – 21.87	06 – 07 – 025	1307 47	1.3 x 0.4 1.1: x 0.4:	[97] ...	2	... f of 2, distorted		(15.6)		*1
02449	02 56.0 + 46 53 144.45 – 10.37	08 – 06 – 000	0845 47	1.8: x 1.8: 1.6: x 1.5:	— ...	4	... Peculiar		15.5		*
02450	02 56.2 + 13 24 164.17 – 38.86	02 – 08 – 044	16,15	(1.1 x 0.9) (1.8 x 1.1)	(30) ...	1	...		16.5:		*1

1	2	3	4	5	6	7	8	9	10	11	12
02451	02 56.3 + 36 49 149.54 – 19.15	06 – 07 – 000	1307 47	1.3 x 0.3 0.9 x 0.12	177 6	3	Sc		16.5:		
02452	02 56.3 + 44 20 145.74 – 12.59	07 – 07 – 000	1618 47	1.3:: x 1.1:: 1. x 1.	— ...	4	Dwarf V		19.		
02453	02 56.4 + 15 50 162.39 – 36.87	03 – 08 – 038	0425 46	1.6: x 1.3: 1.5: x 1.1	— 2 – 3	3	Sb	SAB(s)b Sb –	15.6		
02454	02 56.7 + 07 06 169.54 – 43.70	01 – 08 – 032	1320 45	1.3 x 0.4 1.3 x 0.35	118 6	3	Sc	Sc +	16.0:		*
02455	02 56.8 + 25 03 156.31 – 29.18	04 – 08 – 006 NGC 1156	1276	3.7 x 3.0 3.7: x 3.0:	[25] ...	5	Irr	I B m Ir I	12.0	+ 405 + 497	*12– –689
02456	02 56.8 + 36 37 149.74 – 19.28	06 – 07 – 027	1307 47	(2.0 x 1.5) (1.8 x 1.5)	(90) 2?	4	S B 0	(R)SB(s)0 +	13.8		
02457	02 57.0 + 24 02 156.98 – 30.01	04 – 08 – 007	1276	1.3 x 1.2 1.2 x 1.2	— 1	4	Sc	SA cd	15.5		
02458	02 57.1 + 44 27 145.81 – 12.42	07 – 07 – 000	1618 47	1.2: x 1.1: 1.1: x 0.9:	— 1 – 2	3	Sc	Sc +	17.		
02459	02 57.1 + 48 52 143.66 – 08.54	08 – 06 – 000	0845	2.7: x 0.5: 3.0 x 0.4	62 6	1	...	S d – m	18.		
02460	02 57.2 + 02 34 173.99 – 46.98	00 – 08 – 064 IC 277	1453	1.7: x 1.7: 1.6: x 1.6:	— 1	2	S B b – c Disturbed?	S B(rs) ...	13.8		*
02461	02 57.2 + 33 53 151.28 – 21.60	06 – 07 – 028	1307 47	1.2 x 0.5 0.9 x 0.4	27 6	4	Sc P w 2462	SA d	15.7		*
02462	02 57.3 + 33 51 151.32 – 21.62	06 – 07 – 029	1307 47	1.0 x 0.9 0.7 x 0.7	— 1	4	S B c	SAB cd	16.0:		*
02463	02 57.4 + 40 03 148.07 – 16.24	07 – 07 – 013	1618 47	3. x 2. 3. x 2.	(95) ...	4	... S IV – V	SAB m	15.0		
02464	02 57.5 + 32 28 152.12 – 22.79	05 – 08 – 001	0847	1.1 x 0.5 1.0 x 0.4	30 6	1	S0 – a		16.0:		
02465	02 57.5 + 34 59 150.74 – 20.62	06 – 07 – 031	1307 47	1.3 x 0.35 1.3 x 0.35	144 7	1	Sa:		14.7		* C
02466	02 57.5 + 35 27 150.49 – 20.22	06 – 07 – 030	1307 47	1.7:: x 0.9:: 1.4:: x 0.6::	[172] ...	4	Irr	IA m Ir I	16.5:		*
02467	02 57.5 + 42 58 146.61 – 13.68	07 – 07 – 000 NGC 1159	1618 47	0.5 x 0.45 0.5 x 0.45	— ...	1	...		14.2		
02468	02 57.5 + 44 15 145.97 – 12.56	07 – 07 – 000	1618 47	(1.5 x 1.0) (2.0 x 1.4)	(170) ...	1	E – S ...		16.5:		*
02469	02 57.7 + 05 31 171.25 – 44.73	01 – 08 – 037	1320 45	1.1 x 1.0 1.5: x 1.1	— ...	1	... Peculiar?		15.2		*
02470	02 57.7 + 42 50 146.71 – 13.78	07 – 07 – 000	1618 47	1.4 x 0.20 1.2 x 0.20	169 7	2	Sa nf of 2		14.7		*
02471	02 57.8 + 11 39 165.95 – 40.01	02 – 08 – 046 NGC 1166	16,15	1.2 x 1.1 1.3 x 1.3	— 1	2	Sa – b P w 2476		15.4		*3
02472	02 57.8 + 22 59 157.83 – 30.79	04 – 08 – 000	1276	1.0 x 0.20 1.1 x 0.20	67 7	2	Sa		16.0:		
02473	02 57.9 + 41 12 147.57 – 15.19	07 – 07 – 000	1618 47	1.7: x 0.4 1.4: x 0.4	36 ...	2	S[B:c] Distorted		15.4		*
02474	02 57.9 + 44 43 145.80 – 12.12	07 – 07 – 015 NGC 1161	1618 47	(2.8 x 2.0) (3.0 x 2.5)	23: 3	3	S0 P w 2475		12.6		*13
02475	02 57.9 + 44 46 145.78 – 12.07	07 – 07 – 014 NGC 1160	1618 47	1.5 x 0.7 1.7 x 0.8	50 5	4	Sc P w 2474		13.0		*13
02476	02 58.0 + 11 35 166.05 – 40.03	02 – 08 – 047 NGC 1168	16,15	1.3 x 0.8 1.5 x 0.8	18 4 – 5	2	Sb P w 2471	SAB(rs:)b	15.4		*3
02477	02 58.0 + 41 31 147.42 – 14.90	07 – 07 – 000	1618 47	1.0 x 0.5 1.7 x 0.9	150: ...	1	S ...		16.0:		*
02478	02 58.0 + 74 28 131.47 + 14.02	12 – 04 – 000	0865	1.4 x 0.7 1.2 x 0.6	23: 5:	2	S0 – a		15.6		*
02479	02 58.2 – 00 10 177.13 – 48.74	00 – 08 – 065	1453 49	1.8 x 0.7 1.7 x 0.6	25 ...	2	...	S d – dm	16.0:		
02480	02 58.2 + 23 05 157.85 – 30.66	04 – 08 – 000	1276	1.1 x 0.6 1.1 x 0.8:	97 ...	2	... contact, disturbed	sff of 2,	16.5:		*
02481	02 58.3 + 37 35 149.51 – 18.29	06 – 07 – 032 IC 278	1307 47	(2.3 x 2.3) (2.3 x 2.3)	— ...	3	[E] Compact		14.8		
02482	02 58.5 – 00 56 178.05 – 49.22	00 – 08 – 067	1453 49	1.1 x 0.9 1.2: x 1.1:	— ...	2	...	dm – m	16.5:		
02483	02 58.5 + 31 39 152.77 – 23.38	05 – 08 – 000	0847	1.6: x 0.4 1.4 x 0.2	137 ...	1	...		16.0:		
02484	02 58.5 + 41 55 147.30 – 14.51	07 – 07 – 000	1618 47	1.0 x 0.5 0.6 x 0.3	40: 5:	3	Sc		17.		*
02485	02 58.5 + 74 14 131.62 + 13.84	12 – 04 – 000	0865	1.6: x 1.4: 1.3: x 1.2:	— 1	3	Sc		17.		

N G C   1156, 1159 – 61, 1166, 1168

1	2	3	4	5			6	7	8	9	10	11	12
02486	02 58.6 + 17 39 161.61 – 35.08	03 – 08 – 042	0425	1.0 x 0.9 1.1 x 1.0			––– 1	4	Sb/Sc	SA(r)bc Sb +	14.8		*
02487	02 58.6 + 35 01 150.93 – 20.93	06 – 07 – 033 NGC 1167	1307 47	(3.3 x 2.3) (3.5 x 2.5)			(70) ...	4	S0	SA 0 –	14.0		*3
02488	02 58.7 + 28 32 154.61 – 26.01	05 – 08 – 000	0847	1.2:: x 1.2:: 1. x 1.			––– ...	3	Dwarf V		18.		
02489	02 58.7 + 35 39 150.61 – 19.92	06 – 07 – 034	1307	[1.1 x 0.6] ...			... ...	2	Compact group of 7 galaxies		16.0:		*1C
02490	02 58.7 + 42 24 147.09 – 14.07	07 – 07 – 016 NGC 1164	1618 47	1.6 x 1.3 1.6 x 1.3			(145) 2	4	S B a	(R´)SAB(s)ab	14.4		*3
02491	02 58.8 + 35 33 150.68 – 20.00	06 – 07 – 035	1307 47	2.2 x 1.8 2.2 x 1.8			(0) ...	4	S B a P w 2493	SB(r)0/a	14.8		*1
02492	02 58.8 + 47 47 144.43 – 09.36	08 – 06 – 000	0845 47	1.0: x 0.8: 1.0: x 0.8:			––– ...	1	...	S m:	17.–18.		
02493	02 58.9 + 35 35 150.68 – 19.96	06 – 07 – 036	1307 47	[1.4 x 0.7] ...			... ...	2	Double system, strongly disrupted, p w 2491		15.5		*1
02494	02 58.9 + 35 55 150.50 – 19.67	06 – 07 – 037	1307 47	1.1 x 0.5 1.0 x 0.3			29 6	1	S ...		14.7		*
02495	02 58.9 + 41 25 147.62 – 14.91	07 – 07 – 017	1618 47	(1.3 x 1.1) (1.2 x 1.0)			(105) 2	2	S0		14.9		
02496	02 59.1 + 46 15 145.23 – 10.67	08 – 06 – 000	0845 47	1.1 x 0.1 1.1 x 0.1			113 7	1	... 2nd of 5	d – m	18.		*
02497	02 59.2 + 28 55 154.49 – 25.63	05 – 08 – 003	0847	3.4 x 0.8 3.0: x 0.8:			70 (6)	4	...	S dm	14.8		
02498	02 59.3 + 17 09 162.13 – 35.39	03 – 08 – 043	0425	1.7 x 0.8 1.5: x 0.8			65 5	3	S B c	SAB cd	15.7		
02499	02 59.3 + 39 35 148.64 – 16.46	07 – 07 – 000	1618 47	1.4: x 1.3: 1.5: x 1.4:			––– ...	2	Dwarf irregular or pair in contact		18.		*
02500	02 59.5 + 41 26 147.72 – 14.84	07 – 07 – 000	1618 47	1.6: x 1.0: 1.7: x 1.0:			(165) 4:	3	S B a	S B(s)a	15.4		
02501	02 59.9 + 02 00 175.28 – 46.92	00 – 08 – 072	1453	1.4 x 0.6 1.4 x 0.7			[133] ...	2	Irr Disrupted?		14.0		*
02502	03 00.0 + 00 47 176.58 – 47.76	00 – 08 – 071	1453 49	1.0 x 0.3 0.9 x 0.3			[34] ...	3	S ... Integral		16.0:		
02503	03 00.1 + 46 12 145.41 – 10.63	08 – 06 – 025 NGC 1169	0845 47	5.2 x 3.0 5.0: x 2.8:			28: 2	4	Sb/ S B b	Sb –	13.2		*12 3468
02504	03 00.1 + 46 45 145.14 – 10.15	08 – 06 – 026	0845 47	1.7 x 0.7 1.6 x 0.7			[125] ...	2	... Peculiar		15.5		*
02505	03 00.4 – 01 17 178.94 – 49.11	00 – 08 – 073	1453 49	1.1: x 0.8: 1.1: x 0.8:			––– ...	3	Dwarf irregular Ir IV – V or V		18.		
02506	03 00.5 + 27 30 155.60 – 26.67	05 – 08 – 000	0847	(1.3 x 1.1) (1.4 x 1.2)			––– ...	2	... Compact		15.0		
02507	03 00.6 + 30 26 153.89 – 24.18	05 – 08 – 000	0847	1.1 x 0.4 1.0 x 0.4			10 ...	3	S B? c		16.5:		
02508	03 00.7 – 02 10 180.02 – 49.65	00 – 08 – 074	1453 49	1.2 x 0.2 1.0 x 0.15			172 6 – 7	2	Sc		16.5:		
02509	03 00.7 + 04 18 173.19 – 45.12	01 – 08 – 038	1320	1.0 x 0.4 1.0 x 0.3			54 6	1	S ...		15.3		*
02510	03 00.7 + 43 12 147.01 – 13.19	07 – 07 – 018 NGC 1171	1618 47	3.2 x 1.4 3.5 x 1.5			147 6	3	Sc		13.6		*
02511	03 00.8 + 48 05 144.58 – 08.93	08 – 06 – 027	0845	1.5 x 0.4 1.4 x 0.4			150 6	2	S B: a – b		16.0:		
02512	03 00.9 + 30 04 154.16 – 24.46	05 – 08 – 000	0847	1.1 x 0.8 0.7 x 0.5			0: 3	2	Sc	SA c Sc +	16.5:		
02513	03 01.0 + 01 45 175.82 – 46.90	00 – 08 – 000	1453	1.0 x 0.7 1.6: x 1.0:			(165) ...	1	Sa:		15.0		
02514	03 01.2 – 01 17 179.15 – 48.96	00 – 08 – 078 NGC 1194	1453 49	(1.7 x 1.3) (2.3 x 1.4)			140: ...	2	S0 – a Disturbed?, p w 2517		14.7		*
02515	03 01.3 + 42 09 147.65 – 14.05	07 – 07 – 019 NGC 1175	1618 47	2.3 x 0.8 2.3 x 0.8			153 6	1	S ... sp of 2		13.8		*12 3468
02516	03 01.3 + 47 07 145.14 – 09.73	08 – 06 – 000	0845 47	1.5: x 1.1: 1.5: x 1.1:			(0) ...	3	...	S B dm	17.		
02517	03 01.6 – 01 23 179.36 – 48.96	00 – 08 – 080	1453 49	1.3 x 0.7 1.2 x 0.7			138 ...	1	S B: ... P w 2514		14.9		*
02518	03 01.7 + 01 34 176.19 – 46.91	00 – 08 – 081	1453 49	1.2 x 0.5: 1.0 x 0.4			79 6	3	Sc		15.6		*
02519	03 02.0 + 79 57 128.80 + 18.88	13 – 03 – 001	1226	1.5 x 0.9 1.7: x 1.1:			75: 4	1	Sc:		14.3		
02520	03 02.1 + 21 12 159.96 – 31.71	03 – 08 – 000	0425	1.4: x 1.0: 1.5 x 1.0:			––– ...	3	S B: c	Sc +	16.5:		

N G C   1164, 1167, 1169, 1171, 1175, 1194

1	2	3	4	5	6	7	8	9	10	11	12
02521	03 02.2 + 42 39 147.54 − 13.53	07 − 07 − 021 NGC 1186	1618 47	4.0 x 1.6 4.5 x 1.7	122 6	4	S B c	S B(s)c Sc +	12.5		123
02522	03 02.3 + 00 01 177.99 − 47.88	00 − 08 − 083	1453 49	1.6: x 0.6 1.7: x 0.6	179 6	2	Sb		16.0:		
02523	03 02.5 + 00 54 177.10 − 47.23	00 − 08 − 085	1453	1.9 x 0.4 1.8 x 0.4	175 6	3	Sc	Sc +	15.3		
02524	03 02.6 + 05 03 172.95 − 44.25	01 − 08 − 000	1320	1. x 1. ...	— ...	0	Dwarf V		18.		
02525	03 02.6 + 33 12 152.70 − 21.61	05 − 08 − 004	0847	1.0: x 0.3 0.8 x 0.3	105 6	1	S ...		15.6		
02526	03 02.6 + 36 35 150.82 − 18.72	06 − 07 − 038	1307 47	4.5: x 0.8 4.5 x 0.8	136 6	3	Sb		13.5		
02527	03 02.7 + 02 11 175.81 − 46.30	00 − 08 − 087	1453	1.2: x 0.8 1.1: x 0.8:	65 ...	1	S B ... Distorted		14.9		*1
02528	03 02.7 + 41 25 148.26 − 14.55	07 − 07 − 022	1618 47	1.8 x 0.8 2.0 x 0.9	18 5 − 6	2	S0 − a		14.8		*
02529	03 02.9 − 00 34 178.78 − 48.16	00 − 08 − 090	1453 49	1.0 x 0.3 1.0 x 0.3:	[143] ...	3	S ... Distorted, integral	n of 2	16.0:		*1
02530	03 02.9 + 22 00 159.60 − 30.95	04 − 08 − 000	1276	1.8: x 0.9: 1.7: x 0.9:	37 5	3	...	S dm:	16.5:		
02531	03 02.9 + 42 11 147.90 − 13.87	07 − 07 − 023 IC 284	1618 47	5.0 x 2.5 5.2 x 2.6	13 5	4	...	SA dm	13.8		*C
02532	03 03.0 + 15 55 163.91 − 35.86	03 − 08 − 000	0425	1.2 x 0.7 1.5 x 0.9	28: ...	1	S0?		16.5:		
02533	03 03.0 + 41 40 148.18 − 14.31	07 − 07 − 024 NGC 1198	1618 47	(2.5 x 1.3) (2.8 x 1.5)	120 ...	2	E − S0		14.0		
02534	03 03.3 + 41 17 148.43 − 14.61	07 − 07 − 025	1618 47	(1.3 x 1.1) (1.4 x 1.2)	— ...	1	... Peculiar?		15.7		*
02535	03 03.5 + 01 25 176.81 − 46.69	00 − 08 − 000	1453	1. x 1. 0.8: x 0.8::	— ...	3	Dwarf (irregular) Ir: V		18.		
02536	03 03.6 + 41 20 148.46 − 14.54	07 − 07 − 000	1618 47	1.3 x 0.3 1.6 x 0.5	124 6 − 7	2	S0		15.7		*
02537	03 03.6 + 46 26 145.83 − 10.13	08 − 06 − 028	0845 47	1.4 x 1.4 1.3 x 1.2	— 1	4	Sc	SA(s)cd	15.1		
02538	03 03.8 + 41 34 148.37 − 14.32	07 − 07 − 026	1618 47	1.7: x 0.8: 1.7: x 0.8:	137: 5:	1	S B a		15.6		
02539	03 03.9 + 07 39 170.88 − 42.11	00 − 08 − 000	1320	1.0 x 0.2 1.0 x 0.15	20 6	1	Sc		17.		
02540	03 03.9 + 35 59 151.39 − 19.09	06 − 07 − 040	1307 47	1.9 x 0.20 1.8 x 0.22	69 7	2	Sa − b		15.7		*
02541	03 04.0 + 34 10 152.42 − 20.63	06 − 07 − 000	1307 47	1.2: x 0.8: 1.1: x 0.8:	28: ...	3	...	S dm	17.		
02542	03 04.1 + 70 22 133.96 + 10.69	12 − 04 − 000	0865	1.6 x 0.2 1.6 x 0.2	145 7	3	Sb		17.		
02543	03 04.2 + 37 39 150.53 − 17.64	06 − 07 − 041	1307 47	2.1 x 0.7 2.0 x 0.7	105 6	4	...	SA dm	14.9		
02544	03 04.2 + 42 12 148.10 − 13.73	07 − 07 − 027 IC 288	1618 47	1.2 x 0.3 1.3 x 0.4	42 6	1	S ...		15.0		*
02545	03 04.3 − 00 59 179.60 − 48.19	00 − 08 − 093 NGC 1211	1453 49	2.9 x 2.4 2.8 x 2.5	(30) 1 − 2	4	S B 0/S B a	(R:)S B(r)0/a	13.5		*
02546	03 04.3 + 39 05 149.77 − 16.40	06 − 07 − 042	1307 47	1.0 x 1.0 1.0 x 0.9	— 1	3	S B b	(R?)S B(r)b	15.6		
02547	03 04.8 + 02 20 176.19 − 45.82	00 − 08 − 094	1453	1.0 x 0.5 1.1 x 0.5	148 5 − 6	2	S B a − b		15.4		
02548	03 05.0 + 38 11 150.38 − 17.10	06 − 07 − 043 NGC 1207	1307 47	2.8 x 1.8 2.0 x 1.6	123: 3:	4	Sb Singular	SA(rs)b or SAB(rs)b	13.7		*3
02549	03 05.1 + 23 27 159.13 − 29.47	04 − 08 − 009	1276	1.2 x 0.5 1.1 x 0.4	135 6	1	Sb		16.0:		
02550	03 05.1 + 36 15 151.46 − 18.74	06 − 07 − 044	1307 47	1.7 x 0.9 2.1: x 1.3:	95 5	4	Sc		15.1		*
02551	03 05.2 + 02 58 175.65 − 45.30	00 − 09 − 002 IC 1182	0363	1.0 x 0.5 0.9 x 0.4	20 5 − 6	1	S ...		15.1		
02552	03 05.4 + 42 41 148.04 − 13.21	07 − 07 − 000	1618 47	1.1: x 1.0: 1.1: x 1.0:	— ...	1	S ... Disturbed?		16.5:		*
02553	03 05.6 + 20 35 161.16 − 31.74	03 − 09 − 002	1460 48	1.3 x 1.1 1.2 x 0.9	— ...	3	S B b Brightest of 3, disturbed	S B(rs)b	15.1		*
02554	03 05.6 + 41 17 148.82 − 14.39	07 − 07 − 028	1618 47	(1.3 x 0.9) (1.6 x 1.2)	(165) ...	1	S0?		15.7		*
02555	03 05.8 + 03 55 174.86 − 44.52	01 − 09 − 001 NGC 1218	1471	1.2 x 1.0 (1.4 x 1.2)	(155) 2	3	S0 − a		14.0		

N G C   1186, 1198, 1207, 1211, 1218   I C   284, 288, 1182

1	2	3	4	5	6	7	8	9	10	11	12
02556	03 05.9 + 01 57 176.86 − 45.89	00 − 09 − 006 NGC 1219	0363	1.1 x 1.1 1.3 x 1.2	— 1	3	Sc	SA(s)bc Sc −	13.5		13
02557	03 06.0 + 38 27 150.41 − 16.77	06 − 07 − 045 NGC 1213=1881*	1307 47	3.0: x 2.3: 2.2:: x 1.8::	(60) ...	4	...	SA(s)dm	15.7		*
02558	03 06.0 + 39 30 149.84 − 15.88	07 − 07 − 000	1618 47	1.1: x 1.0: 1.1:: x 0.9::	— ...	3	Dwarf V		18.		
02559	03 06.0 + 42 47 148.09 − 13.06	07 − 07 − 029	1618 47	1.6 x 0.7 2.2 x 0.8	50 6 − 7	2	SO		14.5		
02560	03 06.4 + 40 43 149.25 − 14.80	07 − 07 − 000 IC 290?	1618 47	1.0: x 0.5: 1.1 x 0.5	22 5	1	SO − a		16.0:		*
02561	03 06.4 + 40 48 149.21 − 14.72	07 − 07 − 000 IC 1884	1618 47	1.2 x 0.18 1.2 x 0.22	131 7	1	Sb		15.5		*
02562	03 06.7 + 41 55 148.66 − 13.74	07 − 07 − 000	1618 47	(1.0 x 0.8) (1.2 x 1.0)	(135) 2:	2	SO		16.0:		*
02563	03 06.8 + 18 19 163.03 − 33.39	03 − 09 − 000	1460 48	1.6 x 0.2 1.6 x 0.2	18 7	3	Sc		15.5		*
02564	03 06.9 + 41 59 148.66 − 13.66	07 − 07 − 000	1618 47	1.0 x 0.9 1.0 x 0.9	— 1	2	SO		16.0:		*
02565	03 07.0 + 31 44 154.39 − 22.35	05 − 08 − 006	0847	1.5: x 1.0: 1.4: x 0.9:	20: ...	3	...	S B dm p of 2	16.5:		*1
02566	03 07.0 + 33 40 153.27 − 20.72	06 − 07 − 000	1307	1.1 x 0.7 1.0 x 0.7	(65) ...	3	...	S B(s)dm	17.		
02567	03 07.0 + 40 35 149.43 − 14.85	07 − 07 − 030 IC 1887 = 292*	1618 47	1.3 x 0.7 1.3 x 0.7	75 ...	1	S − Irr		14.3	+ 3140	
02568	03 07.0 + 42 02 148.65 − 13.61	07 − 07 − 000	1618 47	(1.5 x 0.7) (1.5 x 0.7)	157 5	3	SO		15.5		*
02569	03 07.3 + 41 20 149.07 − 14.18	07 − 07 − 000	1618 47	1.0: x 0.9: 1.1 x 0.9	— ...	1	SO?		16.5:		
02570	03 07.5 + 20 13 161.84 − 31.77	03 − 09 − 000	1460 48	1.6 x 0.3 1.4 x 0.3	117 6	1	S ...		16.5:		
02571	03 07.5 + 22 42 160.14 − 29.76	04 − 08 − 000	1276	1.1 x 0.15 0.9 x 0.15	25 7	2	Sb		16.0:		*
02572	03 07.5 + 44 00 147.69 − 11.88	07 − 07 − 000	1618 47	1.1: x 0.9: 1.0: x 0.9:	— 1 − 2	1	S B: b − c		16.5:		*
02573	03 07.6 + 44 50 147.27 − 11.16	07 − 07 − 032	1618 47	1.7 x 0.7 1.7 x 0.7	82 6	3	S B b	Sb +	15.7		
02574	03 07.8 + 40 26 149.64 − 14.90	07 − 07 − 033 IC 294 = 296*	1618 47	(2.3 x 1.7) (2.8 x 2.5)	— ...	4	S B 0/S B a	(R:)S B(rs)0/a	15.6		*
02575	03 07.9 + 35 12 152.56 − 19.33	06 − 08 − 001 NGC 1226	1333	(2.3 x 1.8) (2.6 x 2.1)	(95) ...	3	E		14.5		*
02576	03 08.0 − 01 28 181.06 − 47.82	00 − 09 − 000	0363 49	1.0 x 0.2 0.9 x 0.2	141 6 − 7	1	Sb − c		16.5:		
02577	03 08.0 + 35 08 152.61 − 19.37	06 − 08 − 000 NGC 1227	1333 47	1.1 x 1.0 1.3 x 1.0	— 1 − 2	3	S B 0/S B a	(R)S B(s)0 +	15.7		*
02578	03 08.0 + 41 11 149.27 − 14.24	07 − 07 − 034 NGC 1224	1618 47	(1.4 x 1.2) (1.8 x 1.6)	— ...	2	E − SO		15.5	+ 5174	*
02579	03 08.3 + 35 17 152.58 − 19.21	06 − 08 − 000	1333 47	(1.2 x 0.8) (1.2 x 0.8)	(50) ...	1	E?		15.7		*
02580	03 08.9 + 06 31 173.15 − 42.11	01 − 09 − 000	1471	1. x 1. 1.2: x 0.7:	— ...	1	... Peculiar:, nf of 2		18.		*
02581	03 08.9 + 40 04 150.02 − 15.10	07 − 07 − 000	1618 47	1.3 x 0.2 1.2 x 0.2	90 6 − 7	3	Sc		17.		
02582	03 09.0 + 27 56 157.08 − 25.27	05 − 08 − 000	0847	1.0: x 0.9: 0.9: x 0.9:	— 1	2	Sc:		16.5:		
02583	03 09.0 + 80 37 128.70 + 19.60	13 − 03 − 002 NGC 1184	1226	2.8 x 0.7 3.6 x 0.9	168 7	3	SO − a		13.4		*3
02584	03 09.1 + 00 53 178.77 − 46.04	00 − 09 − 000	0363	1.1 x 0.1 1.1 x 0.1	134 7	1	Dwarf?		18.		
02585	03 09.2 − 00 35 180.37 − 47.01	00 − 09 − 018	0363 49	1.8 x 1.4 1.7: x 1.4	165: 2:	4	S B b	S B(r)b Sb +	14.4		
02586	03 09.3 + 39 08 150.60 − 15.85	06 − 08 − 003 NGC 1233	1333 47	2.3: x 0.9: 2.2: x 0.8:	27 6	3	Sb		13.9		*3
02587	03 09.4 − 01 21 181.27 − 47.48	00 − 09 − 021	0363 49	1.5 x 0.3 1.5 x 0.3	28 7	1	Sb:		16.0:		
02588	03 09.7 + 14 14 166.77 − 36.18	02 − 09 − 000	12,11	1. x 1. 1. x 1.	— ...	1	Irr		17.		
02589	03 09.8 + 12 13 168.41 − 37.71	02 − 09 − 001	12,11	1.4: x 1.0: 1.2 x 1.0	(40) 2 − 3	1	Sb − c		16.0:		
02590	03 09.8 + 42 16 148.98 − 13.14	07 − 07 − 000	1618 47	(1.1 x 0.9) (1.2 x 1.0)	— ...	1	...		15.6		*

N G C   1184, 1213, 1219, 1224, 1226, 1227, 1233   I C   290?, 292, 294, 296
                                                                       1881, 1884, 1887

1	2	3	4	5	6	7	8	9	10	11	12
02591	03 09.9 + 42 35 148.82 – 12.86	07 – 07 – 000	1618 47	1.2 x 0.5 1.3 x 0.5	22 6	2	Sb – c		16.5:		
02592	03 10.0 + 18 36 163.55 – 32.71	03 – 09 – 006	1460 48	1.0 x 0.6 1.1 x 0.6	93 4 – 5	2	Sb		15.7		*
02593	03 10.0 + 28 27 156.96 – 24.72	05 – 08 – 007	0847	1.2 x 0.5 1.2 x 0.6	170 ...	1	S ...		16.0:		
02594	03 10.2 – 00 30 180.52 – 46.77	00 – 09 – 024	0363 49	1.0 x 0.9 1.1 x 0.9	— 1 – 2	2	S B: 0 – a		14.5		
02595	03 10.2 + 04 31 175.36 – 43.32	01 – 09 – 002 IC 302	1471	2.2: x 2.0: 2.5: x 2.3:	— 1	4	S B b/S B c	S B(rs)bc Sc –	14.0		1
02596	03 10.2 + 43 57 148.14 – 11.67	07 – 07 – 035	1618 47	2.2 x 1.2 2.2: x 1.2:	17 ...	4	... S IV – V	SA: m	14.5		*
02597	03 10.4 + 17 48 164.22 – 33.28	03 – 09 – 007	1460 48	1.5: x 0.7 1.2 x 0.6	170: 5	1	Sb		16.0:		
02598	03 10.8 + 41 06 149.78 – 14.03	07 – 07 – 000	1618 47	(1.5 x 0.9) (1.5 x 0.9)	(25) ...	3	[S0] Compact		15.1		*C
02599	03 11.1 + 02 35 177.49 – 44.51	00 – 09 – 025	0363	1.2 x 0.2 1.1 x 0.25	149 7	1	Sb		15.6		*
02600	03 11.2 – 00 24 180.65 – 46.51	00 – 09 – 027 IC 307	0363 49	1.8 x 0.8 1.9 x 0.9	73 5 – 6	1	Sa:		14.6		*
02601	03 11.3 + 39 36 150.69 – 15.25	07 – 07 – 000	1618 47	1.3: x 0.7: 1.2: x 0.5:	102 ...	2	Dwarf: IV – V:	S m:	17.		
02602	03 11.4 + 16 18 165.57 – 34.31	03 – 09 – 008	1460 50	1.6 x 1.5 1.5 x 1.5	— 1	4	S B c sp of 2	SAB(s)c Sc –	15.0		*
02603	03 11.4 + 81 10 128.46 + 20.11	14 – 02 – 000	1277	1.9: x 1.6 1.9: x 1.6:	[55] ...	3	Dwarf irregular Ir V		17.		*
02604	03 11.5 + 39 27 150.80 – 15.36	06 – 08 – 004	1333 47	1.6 x 1.2 1.4 x 0.9	(140) 3:	3	S B c	SAB(s)c	14.8		
02605	03 11.5 + 40 06 150.44 – 14.81	07 – 07 – 000	1618 47	1.5:: x 0.6:: 1.4:: x 0.5:	63 ...	3	Dwarf IV – V or V		18.		
02606	03 11.5 + 42 02 149.38 – 13.17	07 – 07 – 036 IC 301	1618 47	(1.7 x 1.6) (1.7 x 1.7)	— ...	2	E		15.2		
02607	03 11.6 – 01 49 182.32 – 47.36	00 – 09 – 029	0363 49	1.7 x 1.5 1.5 x 1.5	— 1	4	S B b	S B(s)b	14.8		
02608	03 11.7 + 41 51 149.51 – 13.31	07 – 07 – 037	1618 47	1.0 x 0.9 1.1 x 0.9	— 1	3	S B b P w 2612, disturbed?	(R')S B(s)b	14.0		*
02609	03 11.8 + 37 43 151.83 – 16.79	06 – 08 – 005 IC 304	1333 47	1.3 x 0.8 1.3 x 0.8	25 4	2	Sb nnp of 2, disturbed		14.8		*1
02610	03 11.8 + 39 11 151.00 – 15.55	06 – 08 – 000	1333 47	1.2: x 0.5: 1.2: x 0.5:	34 6	2	Sb		15.7		
02611	03 11.9 – 02 05 182.69 – 47.48	00 – 09 – 032	0363 49	1.1 x 1.0 (1.2 x 1.1)	— (1)	2	S0		14.8		
02612	03 11.9 + 41 48 149.57 – 13.33	07 – 07 – 038	1618 47	1.0 x 0.7 1.1 x 0.8	(110) 3:	3	Sc P w 2608		15.4		*
02613	03 12.0 + 41 10 149.94 – 13.86	07 – 07 – 040 NGC 1250	1618 47	2.4 x 0.9 2.5 x 1.0	159 6:	3	S0		14.2	+ 6317	*
02614	03 12.0 + 42 30 149.21 – 12.73	07 – 07 – 039	1618 47	1.7 x 0.6 2.0: x 0.7	90 6	1	S0 – a		14.8		
02615	03 12.1 + 30 45 155.96 – 22.57	05 – 08 – 000	0847	1.2 x 0.7 1.2 x 0.7	(40) 4	2	S B 0 – a		16.0:		
02616	03 12.5 + 41 40 149.74 – 13.38	07 – 07 – 000	1618 47	(1.1 x 0.5) (1.3 x 0.7)	(45) ...	1	E – S0		16.5:		
02617	03 12.7 + 40 43 150.30 – 14.17	07 – 07 – 041	1618 47	3.0 x 1.0 2.9 x 1.0	176 6	4	Sc n of 2	SAB(s)d	14.3		*
02618	03 12.7 + 41 53 149.66 – 13.18	07 – 07 – 042	1618 47	1.4 x 0.6 1.5 x 0.7	167 6	2	Sa – b		14.9		*
02619	03 13.0 + 41 00 150.19 – 13.90	07 – 07 – 000	1618 47	1.2: x 1.2: 1.6: x 1.5:	— (1)	1	S0		16.5:	+ 6411	*
02620	03 13.0 + 80 04 129.16 + 19.23	13 – 03 – 004	1226	2.3:: x 1.8:: 2.3:: x 1.8::	(160) ...	1	... S(B) IV?		15.5		
02621	03 13.2 + 41 21 150.03 – 13.58	07 – 07 – 044	1618 47	1.6 x 0.22 1.7 x 0.30	68 7	2	Sa		14.7	+ 4867	
02622	03 13.3 + 08 59 171.98 – 39.56	01 – 09 – 003	1471	1.6: x 1.0: 1.5: x 0.9	(75) ...	4	S B b	S B(s)b	16.0:		1
02623	03 13.3 + 34 53 153.72 – 18.99	06 – 08 – 009	1333 47	2.7: x 2.5: 2.2: x 2.0:	— ...	4	S B c	S B(s)d	15.7		
02624	03 13.4 + 41 08 150.18 – 13.74	07 – 07 – 045 IC 310	1618 47	1.3 x 1.3 1.8 x 1.8	— 1	3	S0		14.3	+ 5328	*
02625	03 13.5 + 39 50 150.92 – 14.83	07 – 07 – 000	1618 47	1.2: x 1.2: 1.9: x 1.9:	[113] ...	1	S? ... Peculiar?		15.7		*

N G C   1250   I C   301, 302, 304, 307, 310

1	2	3	4	5	6	7	8	9	10	11	12
02626	03 13.7 + 41 10	07 – 07 – 000	1618	1.3: x 0.20	108	1	Sa?		15.7		*
	150.21 – 13.68		47	1.4: x 0.22	7					+ 6477	
02627	03 13.9 + 31 23	05 – 08 – 008	0847	2.0 x 1.7	—	4	Sc	SA(s)c	14.9		*1
	155.92 – 21.82			1.9 x 1.7	1		P w 2629,	disturbed			
02628	03 14.0 – 00 39	00 – 09 – 040	0363	1.7 x 0.6	127	2	Sb – c		14.9		
	181.59 – 46.15		47	1.7 x 0.6	6						
02629	03 14.0 + 31 24	05 – 08 – 009	0847	1.0 x 0.8	—	3	S B c		15.6		*1
	155.93 – 21.80			0.9 x 0.8	...		P w 2627, disturbed				
02630	03 14.0 + 36 57	06 – 08 – 010	1333	1.0: x 0.25	40	1	Sb – c		15.5		
	152.65 – 17.19		47	1.0 x 0.30	6						
02631	03 14.1 + 05 50	01 – 09 – 000	1471	1.3 x 0.2	150	1	S ...		17.		
	175.03 – 41.70			1.0 x 0.4:	6						
02632	03 14.2 – 02 13	00 – 09 – 000	0363	1.0 x 0.2	147	1	...		16.5:		
	183.38 – 47.12		49	1.0 x 0.18	(7)						
02633	03 14.2 + 36 23	06 – 08 – 000	1333	1.6 x 0.2	81	2	Sc		17.		
	153.01 – 17.64		47	1.5 x 0.20	7						
02634	03 14.2 + 41 13	07 – 07 – 047	1618	1.4: x 0.6	86	1	S0 – a		14.2		*
	150.27 – 13.59	NGC 1260	47	1.4: x 0.6	6		Disturbed?			+ 5640	
02635	03 14.3 – 01 43	00 – 09 – 041	0363	1.0 x 0.2	173	1	Sb – c		16.5:		
	182.84 – 46.78		49	1.0 x 0.20	6						
02636	03 14.3 + 36 52	06 – 08 – 000	1333	1.8: x 0.1	10	2	Sc		17.		
	152.75 – 17.23		47	1.5 x 0.15	7						
02637	03 14.3 + 37 52	06 – 08 – 000	1333	1.5 x 0.2	47	2	Sc		17.		*
	152.17 – 16.39		47	1.5 x 0.22	7						
02638	03 14.5 + 01 05	00 – 09 – 042	0363	1.0 x 0.15	8	2	Sa – b		16.0:		
	179.85 – 44.91			1.0 x 0.20	7						
02639	03 14.5 + 41 47	07 – 07 – 000	1618	1.2 x 0.15	92	2	Sa – b		15.6		*
	150.00 – 13.08		47	1.3 x 0.18	7						
02640	03 14.5 + 43 07	07 – 07 – 049	1618	1.3 x 0.8	70:	3	S B b		14.8		*
	149.27 – 11.27		47	1.3 x 0.8	4		P w 2655				
02641	03 14.6 + 03 25	00 – 09 – 000	0363	1.2: x 0.5:	[115]	2	...	S d – m	17.		*
	177.49 – 43.31			1.2 x 0.4:	...		4th of 5				
02642	03 14.7 + 40 45	07 – 07 – 000	1618	1.0: x 0.7	—	1	S B b;		16.5:		*
	150.61 – 13.93		47	1.0 x 1.0	...						
02643	03 14.7 + 41 20	07 – 07 – 050	1618	1.2 x 1.1	—	2	S B a – b		16.0:		*
	150.28 – 13.44	NGC 1264	47	1.2 x 1.2	1					+ 3365	
02644	03 14.8 + 41 34	07 – 07 – 051	1618	(1.2 x 0.6)	(125)	2	E		14.9		*
	150.17 – 13.24	IC 312	47	(1.6 x 0.9)	...		np of 2			+ 4923	
02645	03 14.9 – 00 15	00 – 09 – 045	0363	1.2 x 0.25	27	1	Sb:		15.7		*1
	181.37 – 45.72		49	1.3 x 0.3	7		Disturbed				
02646	03 14.9 + 40 11	07 – 07 – 000	1618	1.1: x 0.7:	73:	1	S0:		16.5:		*
	150.96 – 14.39		47	1.2 x 0.8	...						
02647	03 15.0 + 38 55	06 – 08 – 000	1333	1.0: x 0.9:	—	3	S B c		16.5:		
	151.69 – 15.44		47	1.1: x 1.1:	1						
02648	03 15.0 + 47 15	08 – 07 – 001	1249	1.5: x 1.4:	—	3	S B c		16.5:		
	147.09 – 08.42			1.5: x 1.4:	1						
02649	03 15.1 – 02 00	00 – 09 – 048	0363	1.1 x 1.1	—	2	[S0]		14.5		
	183.35 – 46.81		49	(1.4 x 1.4)	...		Compact				
02650	03 15.1 – 01 52	00 – 09 – 046	0363	1.1 x 0.20	64	2	Sa – b		15.6		*
	183.20 – 46.72		49	1.2 x 0.22	7		1st of 3				
02651	03 15.1 + 41 40	07 – 07 – 052	1618	(1.8 x 1.6)	(165)	3	E		14.7		*
	150.16 – 13.12	NGC 1265	47	(2.3 x 2.0)	...					+ 7656	
02652	03 15.4 – 00 20	00 – 09 – 050	0363	1.1 x 1.0	—	2	S B ...		13.9		*
	181.58 – 45.68	NGC 1280	49	0.9 x 0.9	1		Singular				
02653	03 15.4 + 37 26	06 – 08 – 012	1333	2.3:: x 0.8::	155	3	Sc		16.5:		
	152.61 – 16.63		47	2.2:: x 0.8::	6			Sc +			
02654	03 15.4 + 42 07	07 – 07 – 053	1618	1.6 x 0.6	[177]	2	...		14.6		*1
	149.96 – 12.71		47	1.6 x 0.6	..		Peculiar, disrupted?				
02655	03 15.4 + 43 03	07 – 07 – 054	1618	2.1 x 0.9	175	4	S B c	SAB(s)d	14 1		*
	149.44 – 11.93		47	2.1 x 0.9	6		P w 2640				
02656	03 15.5 + 40 54	07 – 07 – 000	1618	(1.0 x 0.7)	(30)	2	E		16.0:		*
	150.66 – 13.72		47	(1.1 x 0.7)	...						
02657	03 15.5 + 41 17	07 – 07 – 055	1618	(1.1 x 0.9)	—	1	E?		15.4		*3
	150.44 – 13.40	NGC 1267	47	(1.1 x 1.1)	...					+ 5230	
02658	03 15.5 + 41 18	07 – 07 – 056	1618	1.1 x 0.9	(120)	2	Sb	SAB b	14.5		*3
	150.43 – 13.39	NGC 1268	47	1.1 x 0.9	2					+ 3243	
02659	03 15.6 + 40 25	07 – 07 – 000	1618	1.3 x 0.5	65	2	Sb – c		14.9		
	150.95 – 14.12		47	1.4 x 0.5	6						
02660	03 15.7 + 41 17	07 – 07 – 057	1618	(0.9 x 0.7)	(15)	2	E		14.4	+ 4905	*12
	150.47 – 13.38	NGC 1270	47	(1.3 x 1.1)	...					+ 5036	348

N G C   1260, 1264, 1265, 1267, 1268, 1270, 1280   I C   312

1	2	3	4	5	6	7	8	9	10	11	12
02661	03 16.0 + 39 17 151.65 – 15.02	06 – 08 – 000	1333 47	1.3 x 0.25 1.4 x 0.35	147 7	3	S0		14.7		
02662	03 16.1 + 41 18 150.53 – 13.33	07 – 07 – 058 NGC 1272	1618 47	(2.2 x 2.0) (3. x 3.)	— ...	2	E		14.5	+ 4172 + 4303	*12 3
02663	03 16.2 + 08 37 172.98 – 39.34	00 – 09 – 000	1471	1.1 x 0.22 1.1 x 0.20	10 7	1	Sc:		17.		*
02664	03 16.2 + 40 34 150.96 – 13.93	07 – 07 – 000	1618 47	1.4:: x 1.0:: 1.1:: x 0.8::	— ...	3	Dwarf V		18.		
02665	03 16.2 + 41 27 150.46 – 13.19	07 – 07 – 060	1618 47	1.1 x 0.5 1.1 x 0.5	122 5	1	Sc?		15.5	+ 7980	*
02666	03 16.3 – 02 09 183.80 – 46.67	00 – 09 – 054 NGC 1289 = 314*	0363 49	(2.0 x 1.5) (2.1 x 1.6)	100 ...	3	E – S0		13.8		*1
02667	03 16.4 + 38 15 152.31 – 15.84	06 – 08 – 000	1333 47	1.1: x 1.1: 1.0: x 0.9:	— 1	3	Sc Disturbed	SAB(s)c	17.		*
02668	03 16.5 + 40 17 151.17 – 14.14	07 – 07 – 000	1618 47	1.3: x 1.3: 1.4: x 1.4:	— ...	1	... IV?		16.5:		*
02669	03 16.5 + 41 20 150.58 – 13.26	07 – 07 – 063 NGC 1275	1618 47	(3.5 x 2.5) (3.5 x 2.5)	(110) ...	3	Peculiar Seyfert Ir II		13.0	+ 5160 + 5291	*12 37
02670	03 16.7 + 41 22 150.59 – 13.21	07 – 07 – 065 NGC 1278	1618 47	(1.4 x 1.1) (1.8 x 1.7)	— ...	2	E sf of 2, disturbed		14.4	+ 6115 + 6246	*12 3
02671	03 16.8 + 07 58 173.70 – 39.71	00 – 09 – 000	1471	1.2: x 1.2: 0.9: x 0.9:	— ...	3	...	SAB(s)d – dm	17.		
02672	03 16.8 + 40 44 150.96 – 13.73	07 – 07 – 000	1618 47	1.0 x 0.3 1.1 x 0.3	178 6:	1	Sa?		15.7		
02673	·03 16.8 + 41 04 150.77 – 13.45	07 – 07 – 066	1618 47	(1.6 x 1.4) (1.7 x 1.5)	(135) ...	3	[S0] Compact		15.6	+ 4337	*
02674	03 17.0 + 09 34 172.33 – 38.51	02 – 09 – 000	12,11	(1.0 x 0.9) (1.1 x 1.1)	— ...	1	S0?		15.3		
02675	03 17.0 + 41 11 150.74 – 13.33	07 – 07 – 068 NGC 1282	1618 47	(1.4 x 1.1) (1.6 x 1.3)	(25) ...	3	E P w 2676		14.3	+ 2321	*12 3
02676	03 17.0 + 41 13 150.72 – 13.30	07 – 07 – 069 NGC 1283	1618 47	(1.0 x 0.6) 1.2 x 0.6	(70) ...	2	... Compact, p w 2675		15.6	+ 6845	*1
02677	03 17.3 + 03 24 178.14 – 42.83	00 – 09 – 059	0363	1.6 x 0.4 1.7 x 0.5	122 6	2	Sb		15.2		
02678	03 17.3 + 37 19 153.01 – 16.52	06 – 08 – 013	1333 47	1.7: x 1.5: 2.2: x 2.0:	— 1	1	S B b:		16.0:		
02679	03 17.5 + 01 11 180.45 – 44.28	00 – 09 – 060	0363	2.3 x 0.7 2.1: x 0.7	120 6	3	Sb – c		15.4		
02680	03 17.6 – 02 03 183.98 – 46.36	00 – 09 – 061	0363 49	(1.4 x 0.9) (1.4 x 1.1)	(20) ...	1	... Peculiar?		14.9		
02681	03 17.6 + 38 55 152.13 – 15.16	06 – 08 – 000	1333 47	1.1: x 0.3: 0.8: x 0.3:	123 6	2	...	S d – m	17.		
02682	03 17.6 + 41 42 150.55 – 12.84	07 – 07 – 073 IC 313	1618 47	(1.3 x 0.9) (1.4 x 1.1)	— ...	2	E P w 2688		15.1	+ 4548	*
02683	03 17.7 – 02 17 184.27 – 46.48	00 – 08 – 062 NGC 1298	0363 49	(1.2 x 1.1) (1.3 x 1.1)	— ...	2	E:		14.2		*3
02684	03 17.7 + 17 08 166.34 – 32.71	03 – 09 – 000	1460 50	2. x 1. ...	... ...	1	Dwarf?		18.–19.		
02685	03 17.7 + 38 05 152.63 – 15.84	06 – 08 – 014	1333 47	2.2 x 1.7 2.2 x 1.7	(0) 1 – 2	3	S B b	SAB(s)b Sb –	14.8		
02686	03 17.7 + 40 37 151.18 – 13.73	07 – 07 – 000	1618 47	1.6 x 0.9 1.3 x 0.6	43: 5	2	S B a	SAB a	16.0:		
02687	03 17.8 – 02 15 184.25 – 46.44	00 – 09 – 063	0363 49	1.1: x 0.9 (1.3 x 1.0)	— ...	1	S B a	SAB(rs)a	15.2		*
02688	03 18.0 + 41 45a 150.58 – 12.76	07 – 07 – 000 IC 316	1618 47	1.5 x 0.8 1.4: x 0.8	[64] ...	1	... Disturbed, b superimposed, p w 2682		(15.0)		*
02689	03 18.1 + 40 38 151.23 – 13.68	07 – 07 – 000	1618 47	1.5: x 1.0: 1.7: x 0.9:	(127) ...	1	S0?		16.0:		*
02690	03 18.2 – 01 17 183.26 – 45.76	00 – 09 – 066	0363 49	1.5 x 0.6 1.5 x 0.7	143 6	2	Sb – c Brightest in group, disturbed?		14.8		*
02691	03 18.2 – 01 13 183.19 – 45.71	00 – 09 – 067	0363 49	[1.4: x 0.8:] ...	... ...	2	Double system, bridge 2690 group		14.9		*1
02692	03 18.4 – 00 32 182.49 – 45.24	00 – 09 – 068	0363 49	1.4 x 1.3 1.3 x 1.3	— 1	3	Sc		14.1		*
02693	03 18.5 + 17 41 166.10 – 32.17	03 – 09 – 000	1460	1.1 x 1.0 1.1 x 1.1	— ...	3	...	(R) ...	16.5:		*
02694	03 18.5 + 41 11 150.98 – 13.18	07 – 07 – 076 NGC 1294	1618 47	(1.5 x 1.3) (1.7 x 1.4)	— ...	1	S0? ssf of 2		15.1	+ 6667	*23
02695	03 18.7 + 07 15 174.79 – 39.89	01 – 09 – 000	1471 51	1.6 x 0.2 1.4 x 0.7	22 7	1	Sc?		16.5:		

N G C 1272, 1275, 1278, 1282, 1283, 1289, 1294, 1298   I C   313, 314, 316

1	2	3	4	5	6	7	8	9	10	11	12
02696	03 18.7 + 42 00 150.55 – 12.48	07 – 07 – 000	1618 47	1.0 x 0.12 1.0 x 0.12	175 7	1	S? ... Acicular?		15.7		*
02697	03 18.8 – 02 30 184.76 – 46.40	00 – 09 – 069 NGC 1305	0363 49	(1.6 x 1.1) (1.7 x 1.2)	(130) ...	1	S0?		14.8		23
02698	03 18.8 + 40 42 151.31 – 13.55	07 – 07 – 077	1618 47	(1.0 x 0.9) (1.5 x 1.3)	— ...	3	E		14.4		*
02699	03 19.5 – 01 14 183.50 – 45.47	00 – 09 – 000	0363 49	1.1 x 0.2 1.1 x 0.15	135 7	2	Sc		16.5:		
02700	03 19.6 + 42 22 150.49 – 12.08	07 – 07 – 079	1618 47	1.8 x 0.5 2.0 x 0.5	132 6	2	S B? b		15.5		*
02701	03 19.7 + 09 16 173.21 – 38.27	01 – 09 – 000	1471	1.0 x 0.2 1.4 x 0.15	143 7	1	Sb – c		16.0:		
02702	03 19.7 + 36 51 153.69 – 16.64	06 – 08 – 016	1333 47	(3.0 x 2.8) (3.0 x 2.8)	— ...	3	...		15.5		*
02703	03 19.8 + 14 44 168.66 – 34.22	02 – 09 – 005	12,11	1.5 x 0.5 1.5 x 0.4	23 6	2	Sb s of 2 Brightest in group		15.7		*
02704	03 20.2 – 02 05 184.60 – 45.87	00 – 09 – 000	0363 49	1.0 x 1.0 0.9 x 0.9	— ...	1	...		14.3		
02705	03 20.2 – 00 01 182.34 – 44.56	00 – 09 – 072	0363 49	1.1 x 1.1 1.1 x 1.0	— 1	4	Sc	SAB cd	15.0		
02706	03 20.5 + 36 45 153.89 – 16.63	06 – 08 – 017	1333 47	1.2: x 1.2: 1.4 x 1.2	— ...	2	S – Irr Peculiar?		15.7		*
02707	03 20.5 + 36 52 153.82 – 16.53	06 – 08 – 000	1333 47	1.4:: x 0.7: 1.3: x 0.6	(150) ...	3	...	S B(s)dm:	18.		*
02708	03 20.6 + 40 23 151.78 – 13.62	07 – 07 – 080	1618 47	1.4 x 1.4 1.4 x 1.4	— 1	3	S0		15.7		*
02709	03 20.7 + 38 30 152.89 – 15.16	06 – 08 – 018	1333 47	3.5:: x 1.3: 3.5:: x 1.3:	3 6	2	Sb/S B b		14.8		
02710	03 20.8 + 37 35 153.45 – 15.91	06 – 08 – 019	1333 47	(1.4 x 1.2) (1.7 x 1.5)	— 1 – 2	3	S0		15.0		*
02711	03 20.9 + 01 09 181.26 – 43.67	00 – 09 – 075 NGC 1312	0363	1.5 x 0.25 1.5 x 0.3	167 6	2	S B: b		15.7		3
02712	03 21.0 + 06 24 176.10 – 40.09	01 – 09 – 004	1471 51	1.7:: x 1.7:: 1.7: x 1.7:	— 1	3	...	S B dm:	17.		1
02713	03 21.1 – 00 17 182.83 – 44.56	00 – 09 – 77+78	0363 49	[1.2 x 0.7] ...	... ...	2	Triple system Contact		15.3		1
02714	03 21.1 + 37 51 153.34 – 15.66	06 – 08 – 000	1333 47	1.0: x 0.4: 1.0: x 0.4:	47 ...	2	...	S dm:	17.		
02715	03 21.1 + 40 38 151.72 – 13.36	07 – 08 – 000	0643 47	1.0 x 0.20 0.9 x 0.20	37 7	2	Sb – c		16.0:		*
02716	03 21.2 + 17 34 166.77 – 31.85	03 – 09 – 014	1460	2. x 1. 1.5: x 1.0:	88 ...	2	S – Irr f of 2, disturbed		15.4		*1
02717	03 21.4 + 40 31 151.83 – 13.42	07 – 08 – 01a	0643 47	(1.0 x 0.9) (1.0 x 0.9)	— ...	3	E		15.0		*
02718	03 21.5 + 41 48 151.11 – 12.35	07 – 08 – 000	0643 47	1.2: x 0.4 1.2 x 0.4	78 6	2	Sc		16.0:		
02719	03 21.6 + 38 46 152.89 – 14.85	06 – 08 – 000	1333 47	1.1 x 0.1 0.8 x 0.1	3 7	2	Sc		16.5:		*
02720	03 21.7 + 38 57 152.79 – 14.68	06 – 08 – 020	1333 47	1.1 x 0.20 1.0 x 0.22	90 6 – 7	1	Sa – b		15.5		
02721	03 21.8 – 00 04 182.75 – 44.29	00 – 09 – 080	0363 49	1.1: x 1.1: 1.0: x 0.9:	— ...	2	...	S dm – m	17.		
02722	03 21.9 + 42 14 150.92 – 11.95	07 – 08 – 002	0643 47	1.3 x 1.3 1.3 x 1.3	— 1	1	Sc		16.5:		
02723	03 21.9 + 42 30 150.77 – 11.73	07 – 08 – 003	0643 47	1.6: x 1.4: 1.6: x 1.3:	(165) 1 – 2	3	S B c	S B(r)c	16.5:		*
02724	03 22.1 + 40 28 151.90 – 13.28	07 – 08 – 000	0643 47	1.1: x 0.9: 0.9: x 0.8:	— ...	1	Sc – Irr		16.0:		*
02725	03 22.2 + 41 04 151.64 – 12.88	07 – 08 – 005	0643 47	(1.4 x 0.9) (1.4 x 0.9)	(162) 4	3	S0 P w 2734		15.6		*
02726	03 22.3 + 07 15 175.61 – 39.26	00 – 09 – 000	1471 51	1.5 x 0.7 1.3: x 0.7	[165] ...	2	S ... + companion Distorted, bridge		16.5:		*
02727	03 22.4 + 05 02 177.71 – 40.78	00 – 09 – 000	1471 51	1.3 x 0.2 1.1 x 0.15	172 7	2	Sc	Sc +	17.		
02728	03 22.4 + 41 54 151.19 – 12.17	07 – 08 – 000	0643 47	1.2 x 0.2 1.1 x 0.2	51 7	3	Sc P w 2731	Sc +	17.		*
02729	03 22.5 + 68 23 136.42 + 09.87	11 – 05 – 000	0973	... 3.5:: x 2.5::	(130) ...	1	...		18.		*
02730	03 22.6 + 40 35 151.99 – 13.24	07 – 08 – 006	0643 47	1.7 x 0.35 1.9 x 0.35	127 7	3	Sb		15.3		*

N G C   1305, 1312

1	2	3	4	5		6	7	8	9	10	11	12
02731	03 22.7 + 41 52 151.26 – 12.17	07 – 08 – 000	0643 47	1.2 x 0.2 1.1 x 0.15		142 7	3	Sc P w 2728	Sc +	17.		*
02732	03 22.8 + 40 37 152.00 – 13.19	07 – 08 – 007 IC 320	0643 47	1.6 x 1.5 1.6 x 1.5		— 1	3	S B b	S B(rs)ab Sb –	15.4		*
02733	03 22.8 + 41 05 151.73 – 12.81	07 – 08 – 008	0643 47	(1.2 x 1.0) (1.0 x 0.9)		(120) ...	1	E? P w 2725		15.3		*
02734	03 22.8 + 46 27 148.66 – 08.37	08 – 07 – 000	1249	1.3: x 1.0: 0.8: x 0.7:		— ...	2	...	S d – dm	17.		
02735	03 23.1 – 01 12 184.27 – 44.75	00 – 09 – 084	0363 49	1.1 x 0.5 1.0 x 0.45		[87] ...	3	S B ... Distorted, bridge	p of 2	15.6		*
02736	03 23.2 + 40 20 152.23 – 13.38	07 – 08 – 009	0643 47	1.8 x 0.3 1.8 x 0.3		69 6 – 7	2	Sa – b		14.7		*
02737	03 23.6 + 46 49 148.57 – 07.99	08 – 07 – 000	1249	1.1 x 0.3 1.1 x 0.3		143 6	2	S0 – a 2nd of 4		15.7		*
02738	03 23.7 + 39 23 152.87 – 14.11	06 – 08 – 021	1333 47	1.4: x 1.1: 0.9: x 0.8:		(115) 2:	3	Sc	SAB c	16.5:		*
02739	03 23.8 + 49 49 146.91 – 05.48	08 – 07 – 000	1249	1.1 x 0.3 1.1 x 0.3		36 6	1	Sc:		17.–18.		
02740	03 24.0 + 07 32 175.73 – 38.76	01 – 09 – 005	1471 51	1.5 x 0.9 1.5 x 0.9		55 4	1	Sb:		15.2		
02741	03 24.2 + 06 56 177.04 – 39.15	00 – 09 – 000	1471 51	1.3:: x 1.3:: 1.2: x 1.2:		— ...	1	S ...		16.5:		
02742	03 24.4 + 40 44 152.19 – 12.92	07 – 08 – 010	0643 47	1.1 x 0.8 0.9 x 0.7		(110) ...	3	S B c	S B(s)bc Sc –	15.5		1
02743	03 24.5 + 39 52 152.71 – 13.62	07 – 08 – 000	0643 47	1.2 x 0.15 1.0 x 0.12		84 7	3	Sc		16.0:		. *
02744	03 24.7 + 02 26 180.79 – 42.11	00 – 09 – 085	0363	1.1: x 0.8: 1.1: x 0.7:		37 3 – 4:	2	Sc P w 2748, in group		15.5		*
02745	03 24.7 + 36 13 154.93 – 16.58	06 – 08 – 022	1333 47	(1.2 x 0.8) (1.5 x 1.0)		(65) 3	1	S ...		15.5		
02746	03 25.2 + 36 37 154.77 – 16.20	06 – 08 – 025	1333 47	(1.6 x 1.1) (1.8 x 1.3)		(175) ...	2	E – S0		16.0:		*
02747	03 25.2 + 37 51 154.03 – 15.19	06 – 08 – 000	1333 47	(1.1 x 1.1) (1.2 x 1.2)		— ...	1	E?		16.0:		*
02748	03 25.3 + 02 23 180.98 – 42.03	00 – 09 – 086	0363	(1.2 x 0.8) (1.7 x 1.0)		(155) ...	1	E: P w 2744, in group		15.5		*
02749	03 25.3 + 39 54 152.82 – 13.51	07 – 08 – 000	0643 47	1.3 x 0.4 1.1: x 0.4		165 ...	2	S – Irr		16.0:		*
02750	03 25.5 + 36 22 154.98 – 16.36	06 – 08 – 000	1333 47	1.6 x 0.15 1.5 x 0.15		156 7	2	Sb – c		16.5:		
02751	03 25.6 + 39 52 152.89 – 13.50	07 – 08 – 000	0643 47	1.1 x 1.1 1.0 x 1.0		— (1)	0	S0? Planetary nebula?	[(R)SA 0]?	16.0:		*
02752	03 25.8 + 40 39 152.46 – 12.84	07 – 08 – 012	0643 47	2.0: x 0.8: 2.0: x 0.8:		80 6	1	S0: P w 2756		15.6		*
02753	03 26 + 82 15 128.25 + 21.30	14 – 02 – 012	1277	1.0 x 0.15 0.9 x 0.15		90 6 – 7	1	Sb – c		16.0:		
02754	03 26.2 + 37 56 154.15 – 15.01	06 – 08 – 000	1333 47	1.1 x 0.4 1.1 x 0.4		78 6	1	Sb – c Brightest in group		16.5:		
02755	03 26.2 + 39 37a 153.14 – 13.64	07 – 08 – 014	0643 47	(1.5 x 1.5) (1.5 x 1.5)		— ...	2	... ssp of 2		(14.9)		*
02756	03 26.3 + 40 42 152.51 – 12.74	07 – 08 – 15+16	0643 47	[1.5 x 0.7]		... ...	2	Double system Contact, p w 2752		15.0		*1C
02757	03 26.3 + 43 08 151.08 – 10.75	07 – 08 – 000	0643 47	1.2 x 0.2 1.2 x 0.2		143 7	2	Sc		17.		
02758	03 26.7 + 38 54 153.65 – 14.17	06 – 08 – 000	1333 47	(1.3 x 1.3) (1.5 x 1.5)		— 1	3	S B 0		15.7		
02759	03 26.7 + 41 40 152.00 – 11.91	07 – 08 – 018 NGC 1334	0643 47	1.8 x 1.1 1.7 x 0.9		[115] ...	1	Peculiar Disrupted spiral?		14.8		*3
02760	03 26.8 + 42 47 151.36 – 10.99	07 – 08 – 000	0643 47	1.1 x 1.1 1.1 x 1.0		— 1	3	Sc	SAB ...	16.5:		*
02761	03 26.9 + 38 17 154.05 – 14.64	06 – 08 – 026	1333 47	1.4 x 0.20 1.3 x 0.20		45 6 – 7	2	S0 – a		15.4		
02762	03 27.0 + 41.25 152.20 – 12.08	07 – 08 – 019	0643 47	(1.3 x 1.0) (1.2 x 0.9)		(165) (2)	1	E – S0		15.7		
02763	03 27.2 + 36 47 155.01 – 15.83	06 – 08 – 000	1333 47	1.5 x 0.20 1.6 x 0.2		165 7	2	S0 – a		15.6		
02764	03 27.4 + 41 51 152.00 – 11.68	07 – 08 – 000	0643 47	1.4 x 0.4 1.4 x 0.4		31 6	1	S0 – a		16.5:		
02765	03 27.5 + 68.12 136.92 + 09.98	11 – 05 – 000	0973	... 4.0: x 1.6:		165: ...	1	...		18.		*

1	2	3	4	5	6	7	8	9	10	11	12
02766	03 27.7 + 40 39 152.76 - 12.63	07 - 08 - 021	0643 47	(1.2 x 1.1) (1.2 x 1.1)	-- 1	3	S0	SAB 0	16.0:		
02767	03 27.8 + 79 56 129.80 + 19.49	13 - 03 - 000	1226	1.7:: x 1.5:: 1.5:: x 1.5::	-- ...	3	Dwarf (spiral?) S? V		18.		
02768	03 28.0 + 39 15 153.65 - 13.73	06 - 08 - 000	1333 47	1.1 x 0.15 1.1 x 0.2	46 7	2	Sb - c		17.		
02769	03 28.0 + 42 38 151.63 - 10.98	07 - 08 - 000	0643 47	(1.2 x 0.5) (1.3 x 0.8)	137 6	2	S0 npp of 2		16.0:		*
02770	03 28.1 + 35 18 156.08 - 16.92	06 - 08 - 028	1333 47	1.7: x 1.5: 1.7: x 1.5:	(95) ...	3	S B b	S B(rs)b Sb +	14.9		
02771	03 28.1 + 39 35 153.46 - 13.45	07 - 08 - 023	0643 47	(2.2 x 1.1) (2.1 x 1.0)	12 5	3	S0		15.4		*
02772	03 28.3 + 72 01 134.71 + 13.13	12 - 04 - 000	0865	1.2 x 0.1 1.2 x 0.3	36 ...	1	Sc		17.		
02773	03 28.6 + 47 38 148.81 - 06.83	08 - 07 - 003	1249	[1.7 x 1.3]	... ...	2	Double system Contact, distorted		15.0		*
02774	03 28.8 + 04 12 179.94 - 40.17	00 - 09 - 006 NGC 1349	1471	(1. x 1.) 0.9 x 0.9	--	2	S0		15.0		
02775	03 28.8 + 41 26 152.47 - 11.87	07 - 08 - 024	0643 47	1.8 x 1.6 1.8 x 1.6	-- 1	2	Sc	SAB(s) ...	16.0:		*
02776	03 29.0 + 10 21 174.32 - 35.90	02 - 10 - 000	0940	1.0 x 0.9 1.0 x 0.9	-- ...	2	Sc		17.		
02777	03 29.4 + 40 10 153.32 - 12.83	07 - 08 - 000	0643 47	1.4 x 0.4 1.3 x 0.3	34 6	2	...	S d - m	16.5:		
02778	03 29.5 + 36 05 155.83 - 16.12	06 - 08 - 029	1333 47	1.1: x 0.9: 1.1: x 1.0:	(55) ...	1	S B b		16.5:		*
02779	03 29.6 + 00 11 184.19 - 42.62	00 - 10 - 002 IC 330	0932	1.0 x 0.3 0.9 x 0.25	78 7	1	Sa - b IC 331 group		15.0		*
02780	03 29.9 + 36 00 155.95 - 16.13	06 - 08 - 031	1333 47	1.8: x 1.7: 1.7: x 1.7:	-- ...	3	...		16.5:		*
02781	03 30.3 + 15 42 170.14 - 31.81	03 - 10 - 000	0444	1.5 x 0.1 1.2 x 0.1	158 7	3	Sc	Sc +	17.		
02782	03 30.3 + 40 12 153.44 - 12.71	07 - 08 - 000	0643 47	1.4 x 0.2 1.4 x 0.2	93 7	2	... nff of 2	S d - m	17.		*
02783	03 31.0 + 39 12 154.16 - 13.43	06 - 08 - 034	1333 47	(1.6 x 1.4) (1.7 x 1.5)	-- ...	1	E?		14.2		*
02784	03 31.0 + 39 23 154.05 - 13.29	06 - 08 - 033	1333 47	(1.6 x 1.2) (1.7 x 1.3)	(160) ...	2	[S0] Compact		14.3		*
02785	03 31.1 + 40 45 153.24 - 12.17	07 - 08 - 000	0643 47	1.2 x 0.20 1.4 x 0.20	39 7	3	Sc nf of 2		16.5:		*
02786	03 31.5 + 02 56 181.77 - 40.50	00 - 10 - 000	0932	1.4: x 1.0: 1. x 1.	-- ...	3	...	S B(s)d - dm	17.		*
02787	03 31.5 + 41 10 153.05 - 11.79	07 - 08 - 000	0643 47	1.1 x 0.4 1.1 x 0.4	17 ...	1	... Brightest of 3		16.0:		*
02788	03 31.6 + 36 03 156.20 - 15.89	06 - 08 - 035	1333 47	(1.4 x 0.8) (2.1 x 1.7)	(90) ...	1	E? Disturbed?		15.7		*1
02789	03 31.7 + 67 24 137.71 + 09.56	11 - 05 - 001	0973	2.0:: x 1.4:: 2.3 x 1.6	13: 3:	1	S B ...		17.		
02790	03 32.0 + 41 10 153.12 - 11.74	07 - 08 - 000	0643 47	1.0 x 0.5 1.2 x 0.7	165: 5	3	Sc 2787 group		16.5:		*
02791	03 32.2 + 41 15 153.11 - 11.65	07 - 08 - 000	0643 47	1.5:: x 1.4:: 1.2: x 1.1:	-- ...	3	... 2787 group	S dm - m	17.		*
02792	03 32.4 + 72 24 134.74 + 13.63	12 - 04 - 001 NGC 1343	0865	2.7 x 1.8 2.7 x 1.9	80 ...	2	... Peculiar		14.1	+ 300: + 500:	*13 C
02793	03 32.8 + 36 30 156.12 - 15.38	06 - 08 - 000	1333 47	1.1 x 0.7 1.2 x 0.8	155 ...	1	...		16.5:		
02794	03 32.9 + 48 20 148.99 - 05.84	08 - 07 - 000	1249	1.6:: x 1.6:: 1.5: x 1.5:	-- ...	1	S ...		18.		
02795	03 33.0 + 04 55 180.15 - 38.91	01 - 10 - 001	1499	1.1 x 0.5 1.4 x 0.5	34 6	1	S B: b - c		15.6		1
02796	03 34.0 + 13 15 172.91 - 32.97	02 - 10 - 000	0940	1.5 x 0.4: 1.6 x 0.4:	149 6	1	Sb - c		16.5:		
02797	03 34.6 + 23 07 165.40 - 25.58	04 - 09 - 002	0441	1.2 x 0.9 1.1 x 1.0	(140) ...	1	...		16.0:		1
02798	03 34.8 + 40 50 153.76 - 11.69	07 - 08 - 026	0643 47	2.3 x 0.8 2.3 x 0.8	70 6	3	Sb/S B c Brightest of 3, Sb +		14.8		*
02799	03 34.9 + 39 45 154.44 - 12.54	07 - 08 - 000	0643 47	1.0 x 0.2 0.7 x 0.2	66 6 - 7	1	S ...		16.5:		
02800	03 34.9 + 71 14 135.62 + 12.81	12 - 04 - 002	0865	3.0: x 1.5: 3.0: x 1.5:	100 ...	2	Dwarf: V:		16.5:		1

N G C   1343, 1349   I C   330

1	2	3	4	5	6	7	8	9	10	11	12
02801	03 35.0 + 41 07 153.62 – 11.44	07 – 08 – 000	0643 47	1.5 x 0.18 1.2 x 0.20	107 6 – 7	3	Sc		16.0:		*
02802	03 35.3 + 04 47 180.76 – 38.57	01 – 10 – 004	1499	1.3: x 1. 1.4 x 0.9	90: ...	1	S ... f of 2		15.6		*
02803	03 35.3 + 12 04 174.17 – 33.59	02 – 10 – 000	0940	1.0: x 0.7 1.0 x 0.8	(90) ...	1	S B b:		17.		
02804	03 35.5 + 01 46 183.79 – 40.48	00 – 10 – 000	0932	1.1 x 0.15 1.0 x 0.15	173 7	2	Sc		16.5:		
02805	03 35.5 + 40 50 153.87 – 11.61	07 – 08 – 027	0643 47	1.2 x 0.8 1.1 x 0.8	(40) 3:	2	Sc/S B c 2798 group	SAB c	15.7		*
02806	03 35.7 + 07 28 178.32 – 36.71	01 – 10 – 000	1499	1.0 x 0.3 1.1 x 0.3	102 7	1	S ... spp of 2		16.0:		*
02807	03 35.9 + 30 06 160.80 – 20.03	05 – 09 – 000	1457	1.6: x 0.9: 1.4 x 0.9	(140) ...	2	Dwarf: (spiral) (SB) IV – V:		17.		
02808	03 36.3 + 38 32 155.42 – 13.35	06 – 09 – 001	1302 47	1.1 x 0.7 0.9 x 0.7	(160) 4	4	Sc	SA cd	15.2		*
02809	03 36.6 + 19 34 168.42 – 27.94	03 – 10 – 004	0444	1.8: x 1.7: 1.4: x 1.4:	--- ...	3	Dwarf		16.5:		
02810	03 36.8 + 39 27 154.93 – 12.56	06 – 09 – 000	1302 47	1.2 x 1.1 1.2 x 1.1	--- 1 – 2	4	Sc	SA(rs)c	15.0		
02811	03 36.9 + 43 27 152.48 – 09.36	07 – 08 – 000	0643	1.2 x 0.9: 1.2: x 0.9:	(60) ...	1	... S IV?	S d – m	17.		
02812	03 37.4 – 02 16 188.44 – 42.56	00 – 10 – 009	0932	[1.0 x 0.7] ...	... ...	3	S B ... + companion Distorted, p w 2814		14.9		*1
02813	03 37.5 + 71 09 135.84 + 12.87	12 – 04 – 000	0865	2.3: x 1.0: 2.5: x 1.2:	35 ...	3	Dwarf IV – V or V		16.5:		
02814	03 37.6 – 02 14 188.45 – 42.50	00 – 10 – 010	0932	1.0: x 1.0: 0.9 x 0.8	--- ...	1	... P w 2812		15.0		*
02815	03 37.9 + 17 35 170.21 – 29.19	03 – 10 – 005 IC 1977	0444	1.7 x 0.8 1.7 x 0.9	177 5	3	S B b		14.6		
02816	03 38.2 + 23 51 165.56 – 24.49	04 – 09 – 003	0441	1.4 x 0.4 1.1 x 0.3	[142] ...	1	...		15.7		
02817	03 38.4 + 37 03 156.69 – 14.26	06 – 09 – 000	1302	2.3: x 0.8: 2.5: x 0.8:	132 ...	1	...		17.		*
02818	03 38.4 + 39 11 155.35 – 12.58	06 – 09 – 002	1302 47	1.1 x 0.8 1.3 x 0.8	(143) ...	4	S ... Peculiar		14.5		*
02819	03 38.5 + 39 48 154.97 – 12.08	07 – 08 – 029	0643 47	(1.3 x 1.1) (1.4 x 1.2)	--- 	2	E		16.0:		*
02820	03 38.6 + 05 28 180.79 – 37.50	01 – 10 – 006	1499	1.4: x 1.3: 1.4: x 1.4:	--- 1	1	S ...		16.5:		
02821	03 38.7 – 01 27 187.82 – 41.82	00 – 10 – 11+12 NGC 1409 + 1410	0932	[1.2 x 1.1] ...	--- ...	2	Double system Distorted		14.7	+ 7378	*13 C
02822	03 38.7 + 40 59 154.27 – 11.12	07 – 08 – 000	0643 47	1.1 x 0.4 0.9 x 0.3	121 6	1	Sb – c		16.5:		
02823	03 38.8 + 15 51 171.76 – 30.30	03 – 10 – 000	0444 52	1.6 x 0.3 1.6 x 0.25	8 7	1	Sb:		16.0:		
02824	03 38.8 + 76 29 132.50 + 17.12	13 – 03 – 007 IC 334	1226	6. x 6. 5. x 5.	--- ...	3	... Peculiar, Ir II		13.2		*
02825	03 38.9 – 01 46 188.20 – 41.97	00 – 10 – 013	0932	1.3 x 0.4 1.3 x 0.5	69 6	1	S ...		15.5		*1
02826	03 39.0 + 68 08 137.83 + 10.56	11 – 05 – 002	0973	1.4 x 0.8 1.6 x 1.0	120: ...	1	...		15.0		
02827	03 39.0 + 69 10 137.19 + 11.38	12 – 04 – 003	0865	1.2: x 1.0: 1.5: x 1.0:	(60) ...	1	S B ...		17.		
02828	03 39.1 + 39 05 155.52 – 12.58	06 – 09 – 003	1302 47	1.5 x 1.4 1.4 x 1.3	--- ...	4	S B b	S B(rs)bc Sb +	15.3		*
02829	03 39.2 + 08 00 178.56 – 35.71	01 – 10 – 007	1499	1.0 x 0.3 (1.4 x 0.5)	162 6 – 7	2	S0		15.0		
02830	03 39.2 + 15 50 171.86 – 30.24	03 – 10 – 000	0444 52	1.1 x 0.7 1.2 x 0.7	85 ...	2	[S0] Compact		15.7		
02831	03 39.5 + 03 02 183.34 – 38.90	00 – 10 – 014	0932	1.6 x 0.9 1.4: x 0.8:	60 ...	3	... sf of 2	S dm	17.		*
02832	03 39.8 – 00 30 187.02 – 41.03	00 – 10 – 015	0932	1.0 x 0.12 0.8 x 0.2	105 7	1	Sc – Irr		16.5:		
02833	03 40.0 + 15 50 172.02 – 30.11	03 – 10 – 000	0444 52	1.0 x 0.3 0.9 x 0.3	10 6	1	Sc:		16.5:		
02834	03 40.1 + 42 37 153.46 – 09.67	07 – 08 – 000	0643	1. x 1. 0.8:: x 0.8::	--- ...	2	Dwarf (irregular) Ir: IV – V or V		18.		
02835	03 40.2 + 40 55 154.54 – 11.00	07 – 08 – 030	0643 47	1.4 x 0.20 1.4 x 0.20	123 7	3	Sb np of 2		16.0:		*

1	2	3	4	5	6	7	8	9	10	11	12
02836	03 40.6 + 39 09 155.71 – 12.35	06 – 09 – 006	1302 47	(1.2 x 1.0) (1.4 x 1.1)	(0) ...	2	E – S0 Brightest of 3		13.8		*
02837	03 40.6 + 39 52 155.26 – 11.78	07 – 08 – 031	0643 47	1.8 x 0.9 1.8 x 0.9	155 ...	3	S B b	S B(rs)b Sb +	15.1		*
02838	03 40.8 + 23 53 166.03 – 24.07	04 – 09 – 005	0441	1.4 x 0.15 1.3 x 0.15	116 7	1	Sc		16.0:		
02839	03 41.0 + 14 10 173.58 – 31.13	02 – 10 – 000	0940	1.0 x 0.3 1.0 x 0.3	78 ...	2	Sc – Irr s of 2		18.		*
02840	03 41.0 + 22 29 167.09 – 25.09	04 – 09 – 006	0441	1.8: x 1.5: 1.8: x 1.5:	— ...	2	... IV		16.0:		1
02841	03 41.0 + 39 11 155.75 – 12.27	06 – 09 – 000	1302 47	1.1 x 0.3 1.1 x 0.35	12 6	1	S0: 2836 group		16.0:		*
02842	03 41.3 + 00 27 186.32 – 40.16	00 – 10 – 016	0932	1.1 x 1.0 0.9 x 0.9	— 1 – 2	3	Sc		15.3		
02843	03 41.3 + 46 04 151.50 – 06.81	08 – 07 – 009	1249	1.1 x 0.3 1.0 x 0.35	66 6	1	Sb – c		16.5:		
02844	03 41.5 + 45 50 151.67 – 06.97	08 – 07 – 012	1249	(1.2 x 1.0) (1.4 x 1.3)	— ...	2	[E] Compact		15.3		
02845	03 42.0 + 02 41 184.19 – 38.64	00 – 10 – 017 NGC 1431	0932	(1.0 x 0.9) (1.0 x 0.9)	— ...	1	S0?		15.5		*3
02846	03 42.0 + 38 36 156.28 – 12.61	06 – 09 – 000	1302 47	1.0 x 0.12 0.9 x 0.12	7 6 – 7	1	Sb – c		17.		
02847	03 42.0 + 67 57 138.17 + 10.59	11 – 05 – 003 IC 342	0973	24. x 24. 28. x 25.	— 1	4	Sc	SA(rs)cd	10.5	+ 52 + 239	*12
02848	03 42.0 + 72 30 135.27 + 14.15	12 – 04 – 000	0865	1.2 x 0.5 1.1 x 0.5	140 ...	2	Sc – Irr		16.5:		
02849	03 42.1 + 44 43 152.45 – 07.79	07 – 08 – 000	0643	2.0 x 1.2 1.8: x 0.8:	73 4	3	Sc		17.		
02850	03 42.2 + 20 34 168.75 – 26.31	03 – 10 – 006	0444	1.4: x 0.5 1.3: x 0.5:	[120] ...	3	Irr		16.5:		
02851	03 42.2 + 46 32 151.34 – 06.34	08 – 07 – 014	1249	2.1 x 0.9 1.9 x 0.9	125 ...	1	Sb: s of 2		15.4		*
02852	03 42.3 + 05 45 181.27 – 36.62	01 – 10 – 000	1499	1.7 x 0.2 1.7 x 0.2	120 7	3	...	S d – m	16.0:		
02853	03 43.0 + 41 44 154.44 – 10.03	07 – 08 – 000	0643 47	1.1:: x 1.1:: 1. x 1.	— ...	3	Dwarf V		18.		
02854	03 43.3 + 36 35 157.78 – 14.02	06 – 09 – 000	1302	1.2 x 0.4 1.2 x 0.4	176 6	2	Sc		17.		
02855	03 43.3 + 70 00 136.96 + 12.27	12 – 04 – 004	0865	4.6 x 2.2 4.6 x 2.2	112 5	4	Sc/S B c P w 2866	SAB c	14.6		*1
02856	03 43.6 + 15 17 173.18 – 29.89	02 – 09 – 000	0940	1.0 x 0.3 1.0 x 0.3	23 6	1	Sb – c		17.		
02857	03 43.7 + 38 29 156.62 – 12.49	06 – 09 – 007	1302 47	1.5 x 1.1 1.5 x 1.0	(35) 3	2	S B b or Sb/S B b		15.0		
02858	03 43.9 + 45 48 152.03 – 06.74	08 – 07 – 016	1249	2.0: x 1.4: 1.7: x 1.1:	(160) ...	2	...	S dm	16.5:		
02859	03 44.0 + 40 43 155.24 – 10.71	07 – 08 – 032	0643 47	1.4 x 1.4 1.2 x 1.2	— 1	3	Sc		16.0:		*
02860	03 44.0 + 78 10 131.60 + 18.60	13 – 03 – 000	1226 53	1.3: x 0.4 1.2 x 0.3	88 6	2	Sc		16.0:		
02861	03 44.4 + 39 12 156.27 – 11.85	06 – 09 – 008	1302 47	2.3: x 1.3: 2.3: x 1.3:	50: 4:	3	Sb	SA(r)b	16.0:		*
02862	03 44.8 + 13 05 175.24 – 31.23	02 – 10 – 001	0940	1.3 x 1.0 1.6 x 0.9	35 ...	2	Sa: P w 2871	(R´)SAB(s) ...	15.6		*
02863	03 44.9 + 41 46 154.70 – 09.79	07 – 08 – 034	0643 47	1.3 x 1.3 1.7 x 1.6	— 1	2	S B b P w 2867		16.0:		*
02864	03 44.9 + 73 58 134.47 + 15.42	12 – 04 – 000	0865	1.0: x 0.4: 1.2 x 0.5	92 6	1	Sa?		15.2		
02865	03 45.0 + 72 52 135.21 + 14.57	12 – 04 – 005	0865	1.3 x 0.7 1.7 x 0.8	20 ...	1	S B ...		15.1		
02866	03 45.2 + 69 57 137.12 + 12.33	12 – 04 – 006	0865	1.1: x 0.9: 1.5: x 1.3:	(75) ...	1	... P w 2855		15.5		*
02867	03 45.3 + 41 43 154.80 – 09.78	07 – 08 – 000	0643 47	1.4 x 0.35 1.3 x 0.30	97 6	1	S0 – a P w 2863		16.0:		*
02868	03 45.4 + 34 59 159.17 – 14.98	06 – 09 – 000	1302	1.8 x 0.18 1.6 x 0.20	140 7	2	Sb – c		17.		*
02869	03 45.5 + 42 12 154.52 – 09.38	07 – 08 – 000	0643 47	1.3: x 0.2 1.4 x 0.3	140 6	1	Sa – b 1st of 3		17.		*
02870	03 45.6 + 42 11 154.54 – 09.38	07 – 08 – 000	0643 47	1.3 x 0.6 1.4 x 0.7	136 5	2	S0 2nd of 3		16.5:		*

1	2	3	4	5	6	7	8	9	10	11	12
02871	03 46.0 + 12 58 175.57 - 31.10	02 - 10 - 002	0940	1.3 x 1.0 1.1 x 1.1	— 1 - 2	3	Sb P w 2862		15.5		*
02872	03 46.5 + 01 02 186.75 - 38.77	00 - 10 - 019	0932	1.1: x 0.8 1. x 0.8:	155: ...	3	... p of 2	SAB(s)dm	17.		*
02873	03 47.1 + 17 41 171.93 - 27.60	03 - 10 - 007	0444	1.4 x 0.5: 1.3 x 0.4	40 6	2	Sb P w 2874		16.0:		*
02874	03 47.3 + 17 32 172.08 - 27.67	03 - 10 - 000	0444	1.1 x 0.20 1.1 x 0.18	171 7	3	S0 - a P w 2873		15.7		*
02875	03 47.3 + 36 23 158.55 - 13.66	06 - 09 - 000	1302	1.1 x 0.3 1.0 x 0.3	129 6	1	S ...		16.5:		*
02876	03 47.7 + 37 28 157.90 - 12.78	06 - 09 - 000	1302	1.4 :: x 1.3 :: 1.3 :: x 1.2 ::	— ...	1	S ...		17.		
02877	03 47.9 + 36 45 158.41 - 13.30	06 - 09 - 000	1302	1.6: x 0.9: 1.6: x 0.9:	(25) ...	3	...	S B dm - m	17.		
02878	03 48.5 + 42 40 154.65 - 08.67	07 - 08 - 000	0643 47	1.2 x 0.3 1.1 x 0.3	150 6	1	Sc		17.		
02879	03 48.6 + 32 52 161.12 - 16.17	05 - 10 - 000	1468	1.5 :: x 1.2 :: 1.5 :: x 1.2 ::	— ...	2	Dwarf		18.		
02880	03 48.9 + 24 46 166.87 - 22.18	04 - 10 - 001	0031	1.4 x 0.2 1.0 x 0.12	146 7	2	Sc		16.0:		1
02881	03 49.0 + 36 05 159.02 - 13.67	06 - 09 - 011	1302	1.5 x 1.0 1.5 x 1.0	148 3	4	S0	(R)SAB 0 +	15.0		*
02882	03 49.5 + 34 32 160.14 - 14.78	06 - 09 - 000	1302	1.8 x 0.7 1.7: x 0.7:	94 6	2	... S IV or IV - V	S dm - m	17.		
02883	03 49.7 - 01 40 190.17 - 39.72	00 - 10 - 025	0932	1.2 x 1.1 1.1 x 1.0	— ...	1	...		15.0		*
02884	03 49.7 + 02 14 186.15 - 37.41	00 - 10 - 000	0932	1.1 x 0.1 1.2 x 0.1	17 7	2	Sc	Sc +	16.5:		
02885	03 49.8 + 35 27 159.57 - 14.05	06 - 09 - 012	1302	5.5: x 2.5: 4.7: x 2.3:	40 6	3	Sc	SA(rs)c	14.4		*
02886	03 49.9 + 34 49 160.01 - 14.51	06 - 09 - 000	1302	1.7 :: x 1.3 :: 1.6 :: x 1.3 ::	— ...	2	... S IV - V:, p w 2887		18.		*
02887	03 50.0 + 34 48 160.04 - 14.51	06 - 09 - 000	1302	1.1 :: x 1.0 :: 1. x 1.	— ...	2	Dwarf: V:, p w 2886		18.		*
02888	03 50.2 + 32 10 161.86 - 16.48	05 - 10 - 002	1468	1.1 x 0.8 1.0 x 0.7	20: ...	1	S ... nnf of 2, NGC 1465 group		15.2		*1
02889	03 50.3 + 37 06 158.55 - 12.72	06 - 09 - 013	1302	1.6 x 0.8 1.6 x 0.8	(100) ...	3	Sb/S B b or Sb/S B c Brightest of 3		15.7		*
02890	03 50.3 + 72 46 135.58 + 14.75	12 - 04 - 007	0865	2.6 x 0.7 2.6 x 0.8	33 7	1	S - Irr		14.7		
02891	03 50.4 + 32 20 161.78 - 16.32	05 - 10 - 003 NGC 1465	1468	1.9 x 0.5 2.2 x 0.4	165 7	3	S0 - a Brightest of 3		14.9		*
02892	03 50.7 + 18 57 171.61 - 26.10	03 - 10 - 009	0444	1.6 x 1.5 1.6 x 1.5:	— 1	4	S B b	S B(rs)bc Sb +	16.0:		*
02893	03 51.0 + 36 22 159.15 - 13.19	06 - 09 - 000	1302	1.2 x 1.0 1.2: x 1.0:	— ...	3	S B c Disturbed?, Sc +		17.		*
02894	03 51.3 + 15 50 174.21 - 28.18	03 - 10 - 011	0444	1.2 x 1.0 1.1 x 1.0	— 1 - 2	2	Sc nnp of 2		15.6		*
02895	03 51.4 + 17 26 172.94 - 27.05	03 - 10 - 000	0444	1.3 x 0.7 1.3 x 0.8	97: ...	2	S ... + companion Distorted, nnp of 2		16.0:		*
02896	03 51.4 + 79 25 131.02 + 19.77	13 - 04 - 001	1322 53	1.7 x 0.35 1.1 x 0.25	54 6	2	Sc		15.4		
02897	03 51.5 + 42 05 155.46 - 08.76	07 - 08 - 000	0651 47	1.0: x 0.7: 1.0: x 0.7:	(70) ...	1	...	S d - m	17.		
02898	03 51.7 + 10 34 178.75 - 31.70	02 - 10 - 003 IC 2002	0940	1.4 x 1.1 1.3 x 1.1	0: 2:	3	S B a		15.1		
02899	03 51.8 + 06 27 182.49 - 34.36	01 - 10 - 013	1499	1.5 x 0.2 1.3 x 0.2	103 7	2	...	S d - m	16.0:		
02900	03 52.0 + 18 14 172.42 - 26.39	03 - 10 - 000	0444	1.1: x 0.3 1.0: x 0.3	117 6	1	S ...		16.0:		
02901	03 53.5 + 34 43 160.66 - 14.10	06 - 09 - 014	1302	1.1 x 1.1 1.0 x 1.0	— (1)	1	S: ...		16.0:		*
02902	03 53.9 + 34 01 161.20 - 14.58	06 - 09 - 015	1302	(1.4 x 0.9) (1.7 x 1.1)	(15) 4	2	S0		15.7		
02903	03 54.1 + 07 57 181.54 - 32.96	01 - 11 - 000	1308	1.1 x 0.5 1.1 x 0.5	120 6 - 7	2	S0 - a		15.5		*
02904	03 54.2 + 16 20 174.35 - 27.34	03 - 11 - 001	0223	1.6: x 1.3: 0.9 x 0.8	— 1 - 2	1	Sb: P w 2905		15.7		*1
02905	03 54.2 + 16 22 174.33 - 27.31	03 - 11 - 002	0223	1.1 x 0.7 1.0 x 0.5	[153] ...	3	Irr P w 2904		15.7		*1

N G C   1465   I C   2002

1	2	3	4	5	6	7	8	9	10	11	12
02906	03 55.1 + 73 56 135.05 + 15.85	12 – 04 – 009	0865	3.3 x 2.0 3.3 x 2.0	5 4	2	Sb:		14.3		
02907	03 55.4 + 78 08 132.09 + 18.96	13 – 04 – 003	1322 53	1.8 x 1.8 1.8 x 1.8	— ...	3	... Peculiar		15.2		*1C
02908	03 55.5 + 43 13 155.29 – 07.42	07 – 09 – 002	0651 47	2.0 x 1.0 1.8: x 0.9:	115 5	1	S ... 2nd of 3		16.0:		*
02909	03 55.5 + 68 26 138.84 + 11.76	11 – 05 – 004 NGC 1469	0973	2.1 x 0.7 2.5 x 0.8	153 ...	2	SO:		14.5		2
02910	03 55.7 + 06 35 183.11 – 33.53	01 – 11 – 000	1308	1.0 x 0.2 1.0 x 0.2	134 6	1	Sb – c		16.5:		
02911	03 56.0 + 43 11 155.38 – 07.39	07 – 09 – 000	0651 47	1.1 x 0.3 0.9 x 0.25	86 6	1	S B: ... 3rd of 3		16.5:		*
02912	03 56.2 + 42 29 155.87 – 07.90	07 – 09 – 000	0651 47	1.2 x 1.1 1.4 x 1.3	— 1	2	SO – a		16.0:		
02913	03 56.4 + 01 14 188.41 – 36.67	00 – 11 – 003	0232	1.0 x 1.0 1.0 x 1.0	— 1	2	S B a:	S B(s) ...	16.0:		*
02914	03 56.7 + 06 33 183.32 – 33.36	01 – 11 – 002	1308	1.3 x 0.9 1.3 x 1.0	15: ...	4	S B O Brightest in group		14.8		*1
02915	03 56.8 + 32 29 162.73 – 15.31	05 – 10 – 000	1468	1.2 x 0.2 1.4 x 0.2	23 7	1	Sc		17.		
02916	03 57.1 + 71 34 136.81 + 14.19	12 – 04 – 000	0865	1.9: x 1.8: 1.4: x 1.3:	— 1	1	Sa – b p of 2		14.9		*
02917	03 57.2 + 79 43 131.00 + 20.17	13 – 04 – 000	1322	1.1: x 0.9: 1.2:: x 0.9:	— ...	3	Dwarf (spiral) S: V		18.		*
02918	03 57.3 + 22 45 169.88 – 22.31	04 – 10 – 002	0031	1.5 x 1.4 1.5: x 1.4:	— 1:	3	... S IV	S dm	15.7		
02919	03 57.4 + 05 34 184.38 – 33.84	01 – 11 – 005	1308	1.2 x 0.9 1.2: x 0.8:	(160) 2 – 3	2	Sc		16.0:		
02920	03 57.5 + 34 53 161.18 – 13.43	06 – 09 – 016	1302	2.3 x 0.4 2.2 x 0.25	63 7	3	Sc		15.3		
02921	03 57.6 + 00 37 189.25 – 36.78	00 – 11 – 005	0232	1.5:: x 1.5:: 1.2: x 1.2:	— 1:	2	...	S d – m	17.		
02922	03 57.9 + 17 27 174.13 – 25.92	03 – 11 – 005	0223	1.1 x 0.7 1.0 x 0.6	62 4	2	Sb		16.5:		
02923	03 58.0 + 78 32 131.90 + 19.34	13 – 04 – 000	1322 53	1.1 x 0.5 0.9 x 0.5	64 ...	2	Sb/S B b spp of 2	SAB(s)b Sb +	16.0:		*
02924	03 58.3 + 26 42 167.11 – 19.32	04 – 10 – 003	0031	1.0 x 0.5 0.9 x 0.5	(150) ...	1	S ...		16.5:		
02925	03 58.5 – 00 51 190.92 – 37.44	00 – 11 – 006	0232	1.3 x 1.1 1.7 x 1.4	145: 1 – 2	3	...	S B(r)d – m	17.		
02926	03 58.5 + 05 25 184.72 – 33.72	01 – 11 – 006	1308	1.0 x 0.8 1.2 x 0.9	125 2 – 3	3	S B b		15.5		1
02927	03 58.7 + 22 58 169.96 – 21.93	04 – 10 – 006	0031	2.4 x 1.5 2.4: x 1.5:	80: ...	3	S B a NGC 1497 group	(R)S B(s)a	15.2		*
02928	03 58.7 + 23 04 169.89 – 21.86	04 – 10 – 005	0031	1.0 x 0.7 0.9 x 0.7	140 ...	2	Sa/S B a NGC 1497 group	(R´)SAB(s)a	15.5		*1
02929	03 59.1 + 23 00 170.01 – 21.84	04 – 10 – 008 NGC 1497	0031	1.8 x 1.2 (2.0 x 1.3)	60 ...	3	SO Brightest of 4		14.5		*3
02930	03 59.3 + 05 30 184.79 – 33.51	01 – 11 – 007 IC 2019	1308	[1.1 x 0.9] ...	... ...	2	Double system		15.3		*1
02931	03 59.3 + 25 40 168.04 – 19.91	04 – 10 – 010	0031	1.4: x 1.0 1.2 x 1.0	(15) 1 – 2	3	Sc	SA(s)c Sc –	14.8		1
02932	03 59.7 + 33 44 162.32 – 13.98	06 – 09 – 000	1302	1.1: x 1.0: 1.0: x 1.0:	— ...	2	...	S dm:	18.		
02933	03 59.7 + 70 52 137.45 + 13.82	12 – 04 – 010 NGC 1485	0865	2.4 x 0.8 2.7 x 0.9	22 6	1	S ...		13.6		2
02934	03 59.9 + 30 55 164.33 – 16.02	05 – 10 – 000	1468 55	1.5: x 0.3 1.7 x 0.4	38 6	1	S ...		17.		
02935	04 00 + 83 20 128.35 + 22.82	14 – 03 – 000	1328	1.1 x 1.1 1.0: x 1.0:	— 1	3	Sc	Sc +	17.		*
02936	04 00.2 + 01 49 188.52 – 35.55	00 – 11 – 007	0232	2.6 x 0.9 2.8 x 0.9	30 6	3	Sc	Sc +	15.7		
02937	04 00.2 + 46 35 153.70 – 04.34	08 – 08 – 001	1253	1.4 x 0.3 1.4 x 0.30	175 7	2	SO – a		16.0:		1
02938	04 00.4 + 04 25 186.02 – 33.96	01 – 11 – 008	1308	1.3 x 0.7 1.2 x 0.6	175 5	3	S B a – b		15.7		*
02939	04 00.5 + 46 05 154.07 – 04.68	08 – 08 – 000	1253	1.4: x 1.0: 1.4: x 1.0	(90) ...	1	Sc – Irr		17.		
02940	04 00.8 + 19 45 172.82 – 23.84	03 – 11 – 006 IC 358	0223	1.0 x 0.4 1.4 x 0.30	64 6 – 7	2	SO Brightest in group		15.3		*

1	2	3	4	5	6	7	8	9	10	11	12
02941	04 00.8 + 22 01 171.06 – 22.26	04 – 10 – 016 IC 357	0031	1.3 x 1.0 1.3 x 1.1	(175) 2:	3	S B b	S B(s)ab Sb –	14.3		
02942	04 01.2 + 22 00a 171.14 – 22.20	04 – 10 – 018	0441	1.4: x 0.5 1.2 x 0.25	77 ...	1	S ... P w b, contact		15.7		*1
02943	04 01.2 + 22 00b 171.14 – 22.20			1.1: x 0.1 1.7: x 0.4	12 ...	1	... P w a, contact				
02944	04 01.4 + 33 10 162.98 – 14.16	05 – 10 – 000	1468	1.7: x 1.2: 1.3 x 0.7:	(105) ...	1	S ...		17.		*
02945	04 01.4 + 33 40 162.63 – 13.79	06 – 09 – 017	1302	1.5: x 1.1: (1.6 x 1.2)	(60) 3	3	S0 – a		15.2		
02946	04 01.5 + 25 50 168.29 – 19.44	04 – 10 – 000	0031	1.1 x 0.4 1.1 x 0.4	3 6	3	...	S B d – m	16.5:		
02947	04 01.9 – 02 19 193.05 – 37.56	00 – 11 – 009 NGC 1507	0232	3.6 x 1.1 3.5 x 1.1	11 6 – 7	3	...	S d – m	13.1	+ 898 + 832	*123 478
02948	04 02.0 + 20 41 172.30 – 22.99	03 – 11 – 000	0223	1.4 x 0.4 1.3 x 0.3	167 6	2	Sa – b		15.7		
02949	04 02.0 + 25 07 168.91 – 19.87	04 – 10 – 019	0031	1.2 x 0.6 1.1 x 0.6	119 5	2	S B: a – b		14.6		
02950	04 02.0 + 30 54 164.69 – 15.72	05 – 10 – 000	1468 55	1.8: x 0.3 1.9 x 0.3	167 7	1	S ...		17.		
02951	04 02.2 + 02 18 188.40 – 34.87	00 – 11 – 010	0232	1.0 x 0.2 1.0 x 0.25	155 6 – 7	2	Sc		17.		
02952	04 02.6 + 36 56 160.55 – 11.22	06 – 09 – 018	1302	1.2 x 0.9 1.3 x 0.9	30: 3	1	S: ...		16.0:		
02953	04 02.6 + 69 41 138.46 + 13.12	12 – 04 – 011 IC 356	0865	5. x 4.5:: 5. x 4.5::	(90) 1:	3	Sb – c		13.3		*2A
02954	04 02.9 + 04 19 186.57 – 33.52	01 – 11 – 009	1308	1.0 x 0.4 1.0 x 0.4	15 6	3	Sc n of 2		15.4		*1
02955	04 03.2 + 69 32 138.61 + 13.04	12 – 04 – 000	0865	1.3: x 0.4: 1.3 x 0.4	98 6	1	S ...		17.		
02956	04 03.3 + 31 10 164.70 – 15.34	05 – 10 – 006	1468 55	2.2: x 0.5: 1.7 x 0.5	124 6	2	Sc		17.		
02957	04 03.4 + 36 59 160.63 – 11.08	06 – 09 – 019 IC 2027	1302	(1.0 x 1.0) (1.2 x 1.2)	— ...	2	E		16.0:		
02958	04 03.9 + 22 44 171.05 – 21.24	04 – 10 – 022	0031	1.4 x 0.3 1.5 x 0.35	20 7	2	Sb		15.0		*
02959	04 04.0 + 34 19 162.58 – 12.95	06 – 09 – 020	1302	1.4 x 1.2 1.2: x 1.2:	— ...	1	S: ... Peculiar?, n of 2		16.0:		*
02960	04 04.3 + 24 28 169.80 – 19.96	04 – 10 – 023	0031	1.4: x 0.9: 1.4 x 0.9	[5] ...	2	... IV – V:		16.5:		
02961	04 04.3 + 26 35 168.21 – 18.47	04 – 10 – 000	0031	1.5: x 1.4: 1.2:: x 1.2::	— ...	3	Dwarf V		17.		
02962	04 04.7 + 25 39 168.97 – 19.07	04 – 10 – 000	0031	1.5: x 0.1 1.7 x 0.12	19 7	1	Dwarf?		18.		
02963	04 05.1 + 03 51 187.41 – 33.36	01 – 11 – 012	1308	2.3: x 0.8: 2.4: x 0.8:	145 6	3	Sc Disturbed?, Sc +		15.3		*
02964	04 05.9 + 27 04 168.12 – 17.87	04 – 10 – 025	0031	1.3 x 1.2 1.1 x 1.0	— 1	4	Sc	SA cd	15.5		
02965	04 06.0 + 03 00 188.39 – 33.69	00 – 11 – 000	0232	1. x 1. 1. x 1.	— ...	2	Dwarf (irregular) Ir: IV – V:		18.		
02966	04 06.0 + 08 25 183.32 – 30.39	01 – 11 – 000	1308	1.1 x 0.5: 1.0: x 0.4:	8 5 – 6	3	Sc		16.5:		
02967	04 06.2 + 36 52 161.13 – 10.78	06 – 10 – 000	1458	1.1: x 0.9: 1.1: x 0.9:	— ...	1	S B ... 2971 group		17.		*
02968	04 06.3 + 16 58 176.03 – 24.78	03 – 11 – 007	0223	2.0 x 0.9: 2.1 x 0.9	22 5 – 6	1	Sb:		16.0:		
02969	04 06.5 – 01 17 192.77 – 36.03	00 – 11 – 018	0232	1.4 x 0.45 1.5 x 0.6	127 6	2	Sb – c Brightest in group		14.8		
02970	04 06.5 + 08 31 183.32 – 30.23	01 – 11 – 000 NGC 1517	1308	1.2 x 1.1 1.0 x 0.8	— 1	3	Sc		14.3		
02971	04 06.5 + 36 53 161.16 – 10.73	06 – 10 – 001	1458	1.8 x 1.2 1.7 x 1.1	143: ...	2	... Brightest of 3		16.5:		*
02972	04 06.6 + 36 23 161.53 – 11.08	06 – 10 – 000	1458	1.6 x 1.0 1.6 x 1.0	170: ...	1	...		17.		
02973	04 07.4 – 01 30 193.14 – 35.96	00 – 11 – 022	0232	1.1 x 0.35 1.1 x 0.4	74 6	1	Sa – b 2969 group		15.3		
02974	04 08.0 + 26 02 169.24 – 18.27	04 – 10 – 000	0031	1.1:: x 1.1:: 1.0:: x 0.9::	— ...	2	... IV:		17.		
02975	04 08.0 + 26 30 168.89 – 17.94	04 – 10 – 000	0031	1.4:: x 1.3:: 1.1 x 1.0	— ...	2	... IV – V		17.		

1	2	3	4	5			6	7	8	9	10	11	12
02976	04 08.3 + 26 45 168.75 – 17.72	04 – 10 – 026	0031	1.8 1.8	x x	0.7 0.7	149 6	2	Sc		14.6		
02977	04 09.0 + 27 49 168.07 – 16.86	05 – 10 – 008	1468	1.2 1.2	x x	0.7 0.7	85: 4	2	Sa – b		16.0:		
02978	04 09.0 + 34 47 163.01 – 11.90	06 – 10 – 000	1458	1.6 1.5	x x	0.4 0.4	117 6	1	S B? ...		17.		*
02979	04 09.2 + 06 20 185.80 – 31.05	01 – 11 – 000	1308	1.4: 1.3	x x	1.0: 0.9	(30) ...	1	S ...		16.0:		
02980	04 09.4 + 27 35 168.31 – 16.96	05 – 10 – 009 IC 359	1468	(1.3 (2.3	x x	1.3) 2.3)	— ...	2	E – S0		15.4		
02981	04 09.5 + 36 46 161.69 – 10.40	06 – 10 – 000	1458	1.3:: 1.3::	x x	1.2:: 1.3::	— ...	1	S: ...		18.		
02982	04 09.7 + 05 26 186.72 – 31.50	01 – 11 – 013	1308	1.0 0.9	x x	0.5 0.5	[111] ...	1	S ...		15.5		
02983	04 10.2 + 02 14 189.87 – 33.28	00 – 11 – 026	0232	2.3 2.8	x x	0.8 1.0	135 6	3	Sb		15.3		*
02984	04 10.4 + 13 19 179.80 – 26.44	02 – 11 – 000	0854	1.8: 1.8	x x	1.5: 1.6:	— ...	3	... n of 2	S B dm – m	17.		*
02985	04 10.4 + 27 26 168.58 – 16.90	05 – 10 – 011	1468	1.4 1.1	x x	0.4 0.45	48 6	2	Sc P w 2986		16.0:		*
02986	04 10.5 + 27 28 168.57 – 16.87	05 – 10 – 010?	1468	1.0 1.0	x x	0.4 0.5	[70] ...	2	Irr P w 2985		16.5:		*
02987	04 10.5 + 79 41 131.47 + 20.57	13 – 04 – 000	1322 53	1.6 1.3:	x x	0.6 0.6	113 ...	2	...	S B dm:	16.5:		*
02988	04 10.6 + 25 21 170.18 – 18.32	04 – 10 – 027	0031	3.? 3.0	x x	1.? 0.9	5 6	1	Sb:		15.5		
02989	04 10.7 + 29 03 167.43 – 15.72	05 – 10 – 012	1468	1.7 1.7	x x	0.8 0.8	30 5	1	S B? ...		16.0: + 5290 + 5347		*C
02990	04 10.7 + 31 20 165.75 – 14.11	05 – 10 – 000	1468 55	1.3: 1.4	x x	0.5 0.6	0 6	2	Sa – b		16.5:		*
02991	04 10.7 + 36 44 161.89 – 10.26	06 – 10 – 002	1458	2.2 2.0	x x	1.7 1.6	113: 2	4	S B b	S B(r)b	16.5:		
02992	04 11.1 + 01 38 190.62 – 33.44	00 – 11 – 000	0232 56	[1.5: ...	x	0.5] ...	[55] ...	2	Double system Bridge, plumes		15.4		*
02993	04 11.5 + 35 03 163.20 – 11.35	06 – 10 – 000	1458	1.2:: 1.4::	x x	1.1:: 1.3::	— ...	1	Sc:		17.		
02994	04 11.8 + 02 35 189.81 – 32.75	00 – 11 – 028	0232 56	1.7 1.7	x x	1.0 1.1	35 4	3	Sc		15.7		
02995	04 13.0 + 00 22 192.19 – 33.76	00 – 11 – 000	0232	1.1 0.9	x x	0.2 0.2	6 7	2	Sc		17.		
02996	04 13.0 + 01 02 191.53 – 33.39	00 – 11 – 034	0232 56	1.0 1.2	x x	0.25 0.35	124 7	2	S0 – a f of 2		15.3		*
02997	04 13.3 + 08 04 184.92 – 29.19	01 – 11 – 015	1308	1.2 1.2	x x	0.6 0.6	178 5	2	S0 – a		15.0		
02998	04 13.9 + 02 37 190.13 – 32.30	00 – 11 – 037	0232 56	1.6 1.8	x x	1.5 1.7	— 1	3	S B b	S B(rs)b	14.9		
02999	04 13.9 + 23 23 172.24 – 19.13	04 – 11 – 000	0228	1.1 1.1	x x	0.8 0.8	0: ...	1	S? ...		16.0:		
03000	04 14.3 + 36 10 162.81 – 10.15	06 – 10 – 003	1458 57	1.4 1.5	x x	0.4 0.6	17 6	1	S0 – a		16.0:		
03001	04 14.4 + 00 42 192.10 – 33.29	00 – 11 – 040 NGC 1541	0232 56	1.3 1.5	x x	0.6 0.7	77 5 – 6	2	S0 – a		14.9		3
03002	04 14.4 + 02 58 189.88 – 32.00	00 – 11 – 039	0232 56	1.2 1.2:	x x	0.9 0.9:	(25) ...	2	S B c		16.0:		
03003	04 14.6 + 04 40 188.29 – 30.97	01 – 11 – 016 NGC 1542	1308	1.4 1.5	x x	0.6 0.7	128 5 – 6	1	Sa – b		15.1		3
03004	04 14.7 + 02 17a 190.59 – 32.33	00 – 11 – 042	0232 56	1.5: 1.4	x x	1.0: 1.0	— ...	2	S B ... 1st of 3		(14.9)		*1
03005	04 14.7 + 02 17b 190.59 – 32.33	00 – 11 – 041	0232 56	1.1: 1.2:	x x	0.15 0.2	3 7	2	... 2nd of 3	d – m	17.		*1
03006	04 14.8 + 02 14 190.66 – 32.34	00 – 11 – 044	0232 56	1.8 1.9	x x	0.8 0.8	157 6	2	S0 3rd of 3		15.0		*
03007	04 15.2 + 33 39 164.76 – 11.80	06 – 10 – 000	1458	1.3 1.4	x x	1.1 1.2	— ...	4	Sc	SAB(s)d	17.		
03008	04 16.0 + 02 25 190.68 – 31.99	00 – 11 – 049	0232 56	(1.3 (2.0	x x	0.7) 1.0)	75: ...	2	S B? 0 – a		14.8		
03009	04 16.0 + 26 05 170.50 – 16.94	04 – 11 – 001	0228	1.8 1.8	x x	0.4 0.4	107 6	3	Sc		16.0:		
03010	04 16.3 + 05 20 187.96 – 30.96	01 – 11 – 000	1308	1.1 1.2	x x	0.7 0.7	132 3 – 4	4	S B c		15.4		*

N G C   1541,   1542   I C   359

1	2	3	4	5	6	7	8	9	10	11	12
03011	04 16.7 + 02 29 190.73 - 31.81	00 - 11 - 053	0232 56	1.1 x 0.6 1.1 x 0.6	124 5 - 6	1	S ...		15.5		
03012	04 17.0 + 02 18 190.96 - 31.85	00 - 11 - 055 NGC 1550	1524 56	(1.7 x 1.7) (1.9 x 1.9)	-- ...	3	E nnp of 2		14.0		*3
03013	04 17.0 + 75 12 135.20 + 17.78	13 - 04 - 004 NGC 1530	1322	5.2 x 2.8 5.2 x 3.0	[17] ...	4	S B b Compact core	SB(r)b (~)	13.4		*12 3C
03014	04 17.3 + 01 59 191.32 - 31.97	00 - 12 - 002	1524 56	1.4 x 0.8 1.0 x 0.7	45: ...	3	S ... Singular		14.7		*
03015	04 17.7 - 00 48 194.14 - 33.42	00 - 12 - 007 NGC 1552	1524	(2.0 x 1.5) (2.1 x 1.5)	110: ...	2	S0:		14.4		*3
03016	04 18.0 + 36 39 163.00 - 09.29	06 - 10 - 000	1458 57	1.6:: x 0.3 1.5 x 0.25	91 7	1	Sc P w 3019		17.		*
03017	04 18.2 + 05 33 188.07 - 29.73	01 - 12 - 000	0887	1.2: x 0.2 1.0: x 0.2	42 6	1	...	S d - m	16.5:		
03018	04 18.4 - 00 17 193.74 - 33.00	00 - 12 - 012	1524	1.3 x 0.9 1.0 x 0.9	(170) ...	1	S ...		15.3		
03019	04 18.4 + 36 37 163.08 - 09.25	06 - 10 - 004	1458 57	1.2: x 0.7: 1.1 x 0.8	(115) ...	1	S ... P w 3016		17.		*
03020	04 18.5 - 02 25 195.91 - 34.12	00 - 12 - 013	1524	1.2 x 0.6 1.2 x 0.5	158 5 - 6	3	...	S d - dm	16.5:		
03021	04 18.6 + 36 00 163.56 - 09.66	06 - 10 - 005	1458 57	(1.5 x 1.1) (2.3 x 1.4)	(35) ...	2	E Disturbed?		17.		*1
03022	04 19.4 - 00 56 194.55 - 33.14	00 - 12 - 018	1524	1.1 x 0.3 0.9 x 0.25	164 6	1	S ...		16.0:		
03023	04 19.4 + 01 43 191.93 - 31.68	00 - 12 - 16a	1524	(1.4 x 1.0) (2.2 x 1.5)	(20) ...	3	S B 0		14.9		*1
03024	04 19.4 + 27 11 170.21 - 15.63	05 - 10 - 000	1454	1.2: x 0.9: 1.5 x 1.0	(150) ...	1	E - S0:		16.5:		
03025	04 20.5 + 05 28 188.54 - 29.31	01 - 12 - 000	0887	1.2 x 0.4 1.0 x 0.3	73 ...	1	...	d - m	16.0:		
03026	04 20.7 + 70 04 139.32 + 14.49	12 - 05 - 001	0866	1.2 x 0.7 1.1 x 0.5	0 ...	1	S ...		15.6		
03027	04 21.1 + 30 48 167.75 - 12.89	05 - 11 - 000	1454	1.2: x 0.1 1.2 x 0.12	60 7	2	Sc		18.		
03028	04 21.3 + 33 45 165.59 - 10.82	06 - 10 - 007	1458	1.7 x 0.8 1.6 x 0.7	143 ...	1	S ...		16.0:		
03029	04 21.4 - 00 57 194.89 - 32.73	00 - 12 - 024	1524 59	1.3 x 0.9 1.3 x 0.9	160 3	3	Sb		15.5		
03030	04 21.7 + 66 13 142.26 + 11.92	11 - 06 - 000	1303	1.0: x 0.5: 1.2: x 0.8:	(50) ...	1	Dwarf?		18.		
03031	04 21.8 - 00 51a 194.85 - 32.59	00 - 12 - 026 NGC 1568a	1524 59	[1.6: x 0.5] ...	[135] ...	3	... P w b, bridge + plume		(14.9)		*1C
03032	04 21.8 - 00 51b 194.85 - 32.59	00 - 12 - 027 NGC 1568b	1524 59	(1.7 x 1.3) (1.6 x 1.2)	(135) ...	1	E - S ... P w a, bridge				*1C
03033	04 21.8 + 10 47 183.95 - 25.87	02 - 12 - 000	1504	1.1: x 1.0: 1.2: x 1.2:	-- 1	1	S B ...		17.		
03034	04 21.9 + 09 34 185.04 - 26.59	02 - 12 - 000	1504	1.1:: x 0.8:: 1. x 1.	-- ...	1	Dwarf?		17.		
03035	04 22.0 + 07 06 187.28 - 28.05	01 - 12 - 002	0887	(1.7 x 1.1) (1.7 x 1.1)	(20) ...	3	S B 0 P w 3038		14.8		*
03036	04 22.2 + 70 47 138.86 + 15.07	12 - 05 - 002	0866	1.3: x 0.2 1.1: x 0.2	102 6	1	S ... P w 3043		16.5:		*
03037	04 22.3 - 01 09 195.23 - 32.64	00 - 12 - 000	1524 59	1.4 x 0.4 1.1 x 0.3	99 6	2	Sb - c		15.7		
03038	04 22.4 + 07 04 187.38 - 27.99	01 - 12 - 003	0887	1.3 x 0.3 1.1 x 0.3	35 6	1	Sb: P w 3035		15.5		*
03039	04 22.5 + 02 55 191.28 - 30.37	00 - 12 - 000	1524	1.1 x 0.18 1.1 x 0.20	33 7	2	S0		15.7		
03040	04 22.8 + 35 05 164.83 - 09.68	06 - 10 - 000	1458	1.4:: x 1.4:: 0.8:: x 0.7::	-- ...	1	...		18.		
03041	04 23.2 - 00 43 194.94 - 32.22	00 - 12 - 000	1524 59	1.0 x 0.4 0.8 x 0.3	67 6	2	Sc		16.5:		
03042	04 23.2 + 70 15 139.33 + 14.77	12 - 05 - 003	0866	2.1 x 0.9 2.1 x 1.2	70 ...	4	Sb/ S B c P w 3048	SAB(s)bc Sc -	15.6		*1
03043	04 23.4 + 70 49 138.90 + 15.16	12 - 05 - 000	0866	1.3: x 0.1 1.2 x 0.15	75 6 - 7	2	Sc P w 3036		17.		*
03044	04 23.7 + 29 51 168.85 - 13.13	05 - 11 - 000	1454	1.2:: x 0.8:: 0.9: x 0.7:	-- ...	1	Dwarf?		18.		
03045	04 23.8 + 20 18 176.29 - 19.47	03 - 12 - 000	1501	1.4 x 0.3 1.4 x 0.4	47 6	1	Sb? spp of 2		16.5:		*

N G C   1530, 1550, 1552, 1568

1	2	3	4	5	6	7	8	9	10	11	12
03046	04 23.9 + 69 26 139.99 + 14.26	12 – 05 – 004	0866	1.7 x 1.4 1.7 x 1.4	— 1:	3	Sb sf of 2		15.3		*
03047	04 24.0 + 32 32 166.89 – 11.25	05 – 11 – 000	1454	1.6: x 1.4 1.5: x 1.4:	(30) 1 – 2	3	S B c		17.		
03048	04 24.0 + 70 20 139.31 + 14.87	12 – 05 – 03a	0866	1.3 x 0.8 1.1 x 0.8	125: ...	3	S B b – c P w 3042	SAB(rs)bc:	15.3		*
03049	04 24.3 + 01 35 192.85 – 30.74	00 – 12 – 000	1524 59	1.1 x 0.5 1.0 x 0.5	50 5 – 6	1	Sb – c		15.4		
03050	04 24.3 + 30 51 168.19 – 12.35	05 – 11 – 000	1454	1.0: x 0.8: 0.8: x 0.7:	— ...	1	S ...		18.		
03051	04 24.7 + 02 15 192.27 – 30.29	00 – 12 – 029	1524	1.1 x 1.0 1.1 x 1.0	— 1	4	Sc nff of 2		16.0:		*
03052	04 25.0 + 39 30 161.91 – 06.31	07 – 10 – 000	0644	1.6 x 1.1 1.6 x 1.1	10: 3:	1	S B b?		16.0:		
03053	04 25.2 + 21 33 175.52 – 18.41	04 – 11 – 002	0228	1.0 x 0.8 1.0 x 0.8	(170) 1 – 2	2	Sc		14.9		
03054	04 25.6 + 00 56 193.69 – 30.82	00 – 12 – 000	1524 59	1.6 x 0.15 1.5: x 0.2	30 7	2	... nnp of 2	d – m	17.		*
03055	04 25.8 + 10 57 184.47 – 24.99	02 – 12 – 000	1504	1.4 x 0.7 (1.4 x 0.7)	60: ...	1	S0?		16.0:		
03056	04 26.1 + 64 45 143.68 + 11.25	11 – 06 – 001 NGC 1569	1303	3.3: x 2.0: 3.0: x 1.7:	[120] ...	4	Irr	I m Ir I	11.8	– 42 + 121	*123A 456̄8C
03057	04 26.3 + 76 28 134.60 + 19.04	13 – 04 – 005	1322	1.2 x 1.1 1.2 x 0.8	— ...	2	Sb/S B b Sb +	SAB(s)b	15.1		*
03058	04 27.0 + 00 39 194.19 – 30.68	00 – 12 – 032	1524 59	1.3 x 0.6 1.2 x 0.5	[82] ...	2	S B ... + superimposed companion, bridge		15.6		*
03059	04 27.1 + 03 35 191.39 – 29.04	01 – 12 – 005	0887	2.4: x 1.0: 2.5: x 1.0:	40 6	4	...	SA dm	15.4		
03060	04 27.1 + 71 47 138.36 + 16.03	12 – 05 – 005 NGC 1560=2062*	0866	10. x 1.8 10.0 x 1.8	23 7	4	...	S dm	12.1		*12 59̄
03061	04 27.5 + 06 50 188.43 – 27.10	01 – 12 – 006	0887	2.2: x 2.0: 2. x 2.	— ...	3	Multiple system Interaction		15.5		*1
03062	04 28.1 – 00 25 195.41 – 31.03	00 – 12 – 036 NGC 1586	1524 59	2.1 x 1.0 1.7 x 1.0	155 5	3	Sb ssp of 2	Sb –	14.3		*3
03063	04 28.1 + 00 33 194.46 – 30 51	00 – 12 – 035 NGC 1587	1524 59	(1.8 x 1.7) (2.0 x 1.8)	— ...	4	E P w 3064		13.3	+ 3890 + 3812	*12 3C̄
03064	04 28.2 + 00 33 194.48 – 30.49	00 – 12 – 037 NGC 1588	1524 59	1.5 x 0.8 1.4 x 0.8	[175] ...	2	... P w 3063 Compact, disturbed		14.1	+ 3275	*13 C
03065	04 28.2 + 00 45 194.28 – 30.38	00 – 12 – 038 NGC 1589	1524 59	3.4 x 1.1 3.5: x 1.1	160 6	3	Sb		13.8		*3
03066	04 28.3 + 05 26 189.85 – 27.75	01 – 12 – 007	0887	2.0: x 1.5: 2.0: x 1.5:	(110) 1 – 2	3	Sc/S B c Sc +	SAB(r)d:	15.0		
03067	04 28.3 + 07 59 187.51 – 26.27	01 – 12 – 000	0887	1.1 x 1.1 0.9 x 0.9	— ...	1	E? P w 3075		15.3		*
03068	04 28.4 + 05 48 189.52 – 27.52	01 – 12 – 000	0887	[1.7 x 1.1] ...	... ...	2	Double or triple system Distorted		17.		*
03069	04 28.4 + 73 05 137.40 + 16.95	12 – 05 – 007	0866 64	(1.3 x 0.7) (1.3 x 0.7)	55: ...	2	E – S0 NGC 1573 group		15.1		*
03070	04 28.5 – 02 07 197.17 – 31.83	00 – 12 – 039	1524 59	1.6 x 1.0 1.6 x 1.0	160 4	1	Sb? Disturbed?		14.9		*
03071	04 28.5 + 07 31 187.97 – 26.50	01 – 12 – 008 NGC 1590	0887	1.2 x 1.1 1.2 x 1.1	— ...	1	... Peculiar		14.6	+ 3738	*3C
03072	04 28.6 + 00 44 194.36 – 30.30	00 – 12 – 000	1524 59	1.4: x 0.8: 1. x 0.7:	— ...	3	Dwarf V		18.		*
03073	04 28.6 + 02 35 192.57 – 29.29	00 – 12 – 000	1524	1.0 x 0.3 1.0 x 0.3	38 6	1	S ...		16.5:		
03074	04 28.7 + 06 32 188.90 – 27.03	01 – 12 – 000	0887	1.3 x 1.1 1.2: x 1.0:	— ...	2	...	S: d – dm	15.6		
03075	04 28.7 + 08 03 187.52 – 26.15	01 – 12 – 000	0887	1.1 x 0.3 0.8 x 0.3	108 6 – 7	1	S ... P w 3067		16.5:		*
03076	04 29.0 + 08 20 187.31 – 25.92	01 – 12 – 000	0887	1.1 x 0.2 0.6 x 0.2	109 7	2	Sb		16.0:		
03077	04 29.0 + 73 09 137.38 + 17.03	12 – 05 – 008 NGC 1573	0866 64	1.9 x 1.3 (2.3 x 1.7)	35 ...	2	E Brightest of 4		13.3		*C
03078	04 29.1 + 33 08 167.18 – 10.06	05 – 10 – 000	1468	1.6 x 0.4 1.8 x 0.4	9 6	2	Sb – c		17.		
03079	04 29.2 + 01 40 193.55 – 29.67	00 – 12 – 040	1524 59	1.4 x 0.3 1.4 x 0.3	120 6	1	S ... Brightest in group		15.7		
03080	04 29.3 + 01 05 194.13 – 29.97	00 – 12 – 042	1524 59	2.2 x 2.1 2.2: x 2.2:	— 1	4	Sc Sc +	SAB(s)c	14.9		

N G C   1560, 1569, 1573, 1586 – 90

1	2	3	4	5	6	7	8	9	10	11	12
03081	04 29.4 + 02 16 193.00 – 29.30	00 – 12 – 041	1524	1.6: x 0.9: 1.5: x 0.8:	10 4 – 5	3	S B a	S B(r)a	15.7		
03082	04 29.5 + 00 27 194.78 – 30.27	00 – 12 – 044 IC 2077	1524 59	1.7 x 0.6 1.6 x 0.6	130 6 – 7	3	S0		14.8		
03083	04 29.7 + 01 37 193.68 – 29.59	00 – 12 – 000	1524 59	1.4 x 0.2 1.1 x 0.2	168 6	2	Sc 3079 group		16.5:		
03084	04 30.0 + 10 17 185.74 – 24.57	02 – 12 – 000	1504	1.9 x 0.4 2.3 x 0.5	158 6	3	Sc		16.0:		
03085	04 30.0 + 72 50 137.68 + 16.88	12 – 05 – 000	0866 64	1.4 x 0.3 0.9 x 0.3	62 6	2	Sc		16.5:		*
03086	04 30.3 + 00 26 194.92 – 30.11	00 – 12 – 000	1524 59	1. x 1. 1. x 1.	— ...	3	Dwarf (spiral)		18.		
03087	04 30.5 + 05 15 190.37 – 27.40	01 – 12 – 009	887	0.8 x 0.6 0.9 x 0.7	(130) ...	1	[S0] Compact; Seyfert		14.2	+ 9927	*C
03088	04 30.6 + 07 36 188.23 – 26.03	01 – 12 – 000	0887	1.3: x 0.9: 1.3:: x 0.9::	(160) ...	2	Dwarf: IV – V:		18.		*
03089	04 31.0 + 16 48 180.32 – 20.39	03 – 12 – 002	1501	1.1 x 1.0 1.3 x 1.1	— 1 – 2	2	Sb – c		15.6		1
03090	04 31.0 + 71 27 138.83 + 16.04	12 – 05 – 000	0866	1.6: x 1.2: 2.0: x 1.2:	(30) ...	3	Dwarf IV – V or V		17.		
03091	04 31.3 + 01 00 194.52 – 29.59	00 – 12 – 045	1524 59	1.5 x 1.5 1.3 x 1.3	— 1	3	Sc		15.4		*
03092	04 32.2 + 72 13 138.28 + 16.61	12 – 05 – 000	0866 64	1.6 x 0.10 1.4 x 0.12	162 7	3	Sc	Sc +	17.		
03093	04 32.3 + 07 53 188.25 – 25.52	01 – 12 – 012	0887	1.1 x 0.2 1.0: x 0.2	118 6	2	Sc		15.6		
03094	04 32.6 + 19 03 178.71 – 18.68	03 – 12 – 000	1501	1.2 x 0.4 1.1 x 0.4	0 6	1	S ...		16.5:		
03095	04 32.7 + 18 14 179.40 – 19.18	03 – 12 – 004	1501	(1.1 x 0.9) (1.3 x 1.1)	(135) ...	1	E – S0		16.0:		
03096	04 33.0 + 19 50 178.14 – 18.11	03 – 12 – 005 NGC 1615	1501	(1.5 x 0.8) (1.7 x 0.9)	115: ...	3	E – S0		15.0		*2
03097	04 33.1 + 02 09 193.69 – 28.59	00 – 12 – 000	1524	1.2 x 0.6 1.1 x 0.6	115 5	2	S0		15.6		*
03098	04 33.4 + 43 51 159.80 – 02.19	07 – 10 – 000	0644	1.0 x 0.4 1.4: x 0.4	2 6	1	Sb – c np of 2		17.		*
03099	04 33.5 + 20 30 177.67 – 17.60	03 – 12 – 006	1501	1.6 x 0.3 1.6 x 0.3	8 6	1	S ...		16.0:		
03100	04 33.5 + 65 13 143.85 + 12.14	11 – 06 – 002	1303	1.3 x 1.2 1.0 x 1.0	— 1	2	Sc		16.0:		
03101	04 33.5 + 79 53 132.02 + 21.48	13 – 04 – 006	1322	1.6 x 1.3: 1.5 x 1.2	115: 2	4	... P w 3112	SA: dm	16.0:		*
03102	04 33.7 + 14 14 182.91 – 21.46	02 – 12 – 000	1504	1.4: x 0.7 1.1 x 0.6	170: 5:	1	S ...		16.0:		
03103	04 34.0 – 00 15 196.15 – 29.69	00 – 12 – 052 NGC 1620	1524	3.2 x 1.2 2.8 x 1.2	25 6	4	Sc	Sc –	13.6		23
03104	04 34.1 – 02 24 198.29 – 30.78	00 – 12 – 055	1524 59a	1.2 x 0.6 1.5 x 0.6	157 5 – 6	2	Sa nf of 2		15.4		*
03105	04 34.5 – 02 25 198.37 – 30.70	00 – 12 – 056	1524 59a	(1.4 x 0.9) (2.0 x 1.5)	75: ...	3	S0		14.5		*
03106	04 34.6 – 01 58 197.94 – 30.45	00 – 12 – 057	1524 59a	1.6: x 0.6: 1.4 x 0.4	55 6	1	Sa – b		15.6		
03107	04 34.6 + 09 27 187.21 – 24.15	02 – 12 – 000	1504	1.1 x 0.3 1.1: x 0.25	72 6	1	Sb – c		16.5:		
03108	04 34.8 + 43 57 159.89 – 01.94	07 – 10 – 000	0644	2. x 2. 1.8 x 1.8	— ...	1	S ...		18.		*
03109	04 35.0 – 00 24 196.45 – 29.55	00 – 12 – 059	1524	1.6: x 1.0:: 1.2: x 0.6:	10: ...	2	... sf of 2	S dm:	16.5:		*
03110	04 35.0 + 73 34 137.33 + 17.62	12 – 05 – 015	0866 64	2.5 x 1.0 2.2 x 1.0	115 6	3	S B c		15.0		
03111	04 35.3 + 08 47 187.92 – 24.40	01 – 12 – 000	0887	1.3 x 0.4 1.0 x 0.3	148 6	2	Sa		16.5:		*
03112	04 35.4 + 79 55 132.04 + 21.56	13 – 04 – 000	1322	1.0: x 0.9: 0.8 x 0.7	— ...	3	Dwarf (spiral) S: IV – V, p w 3101		16.5:		*
03113	04 35.5 – 01 03 197.16 – 29.79	00 – 12 – 000	1524	1.1: x 0.4 0.9: x 0.3:	41 6	2	Sc		16.5:		
03114	04 35.5 + 66 32 142.97 + 13.16	11 – 06 – 003	1303	2.8: x 1.0: 2.5 x 0.6	72 6	4	Sc		15.5		
03115	04 35.9 + 18 44 179.49 – 18.27	03 – 12 – 007	1501	1.4 x 0.7 1.4 x 0.8	15 4 – 5	3	S0		15.7		

N G C 1615, 1620   I C 2077

1	2	3	4	5	6	7	8	9	10	11	12
03116	04 36.2 + 00 02 196.21 – 29.07	00 – 12 – 062	1524	1.0 x 0.3 1.1 x 0.22	145 6	1	S ... Brightest of 3		15.7		*
03117	04 36.2 + 02 44 193.62 – 27.62	00 – 12 – 061	1524	1.6: x 1.6: 1.7: x 1.5:	— 1 – 2	1	S ...		15.1		
03118	04 36.3 + 05 31 191.04 – 26.06	01 – 12 – 000	0887	1.3: x 1.3: 1.3:: x 1.2::	— ...	3	Dwarf IV – V or V		18.		
03119	04 36.3 + 11 27 185.72 – 22.63	02 – 12 – 000	1504	1.2 x 0.4 1.1 x 0.4	101 6	1	Sb – c		16.5:		
03120	04 36.8 + 40 10 162.96 – 04.20	07 – 09 – 000	0651	1.1 x 0.7 1.1 x 0.7	115: 4	3	Sc	SAB(s)c	17.		
03121	04 37.1 + 02 55 193.58 – 27.33	00 – 12 – 000	1524	1.4 x 0.4 1.1 x 0.3	104 6	1	S ...		15.6		
03122	04 37.2 + 06 57 189.87 – 25.07	01 – 12 – 013	0887	2.4 x 1.3 2.8 x 1.7	25: 5	4	S B c	SAB(rs)c Sc –	14.8		*1
03123	04 37.3 + 12 25 185.04 – 21.86	02 – 12 – 000	1504	1.0 x 0.8 0.8 x 0.7	— ...	1	Double system Contact		17.		
03124	04 37.3 + 69 09 141.01 + 14.97	12 – 05 – 016	0866	1.0 x 0.6 1.1 x 0.6	20 4 – 5	2	S B a – b		15.4		
03125	04 37.4 + 07 15 189.62 – 24.85	01 – 12 – 014 NGC 1633	0887	1.2 x 1.1 1.4 x 1.2	— 1	4	Sa/S B b P w NGC 1634	SAB(s)ab	14.6		*13
03126	04 37.5 – 00 40 197.09 – 29.16	00 – 12 – 063 NGC 1635	1524	1.6 x 1.4 1.6 x 1.4	(5) 1 – 2	4	S B 0/S B a	(R)S B(r)0/a	13.5		23
03127	04 37.8 – 02 08 198.57 – 29.85	00 – 12 – 064	1524 59a	2.1 x 0.5 2.0 x 0.5	24 6	2	Sc Distorted?		15.1		*
03128	04 37.8 + 04 06 192.58 – 26.54	01 – 12 – 016	0887	(1.3 x 1.0) (1.3 x 1.0)	— ...	3	S0	SAB 0 +	15.3	+ 4600 + 4528	*12
03129	04 37.8 + 17 00 181.22 – 18.99	03 – 12 – 000	1501	2.5:: x 1.3:: 2. x 1.	(45) ...	3	Dwarf IV – V or V		16.5:		
03130	04 37.8 + 75 34 135.80 + 19.01	13 – 04 – 007 IC 381	1322	3.0 x 1.7 3.0 x 1.7	170 ...	4	Sb/S B b	SAB(s)b	14.5		*2
03131	04 38.9 + 72 43 138.21 + 17.31	12 – 05 – 017	0866 64	1.5 x 0.8 1.5 x 0.8	77 5	1	S ...		16.0:		
03132	04 39 + 82 34 129.82 + 23.24	14 – 03 – 005	1328	1.6 x 0.6 1.5 x 0.6	162 6	2	S0 – a		15.4		
03133	04 39.0 – 01 55 198.54 – 29.48	00 – 12 – 069 NGC 1638	1524	(2.3 x 1.7) (2.3 x 1.8)	70: ...	3	E – S0		13.6		*23 48
03134	04 39.2 – 01 24 198.06 – 29.17	00 – 12 – 070	1524	1.3 x 1.1 1.2 x 1.1	— 1 – 2	4	S B b/S B c nf of 2	S B(s)bc Sb +	14.4		*
03135	04 39.2 + 19 57 179.00 – 16.90	03 – 12 – 000	1501	1.6:: x 1.2:: 1.6: x 1.3:	(65) ...	2	S B c		17.		
03136	04 39.3 + 00 58 195.77 – 27.92	00 – 12 – 000	1524	1.1: x 0.6 1.2: x 0.6	[30] ...	3	Sb n of 2, disturbed		15.3		*
03137	04 39.4 + 76 20 135.22 + 19.56	13 – 04 – 008	1322	3.9 x 0.8 4.0 x 0.9	75 7	1	Sb?		15.5		
03138	04 39.9 + 18 28 180.33 – 17.69	03 – 12 – 008	1501	1.1 x 0.4 1.2 x 0.5	23: 6	1	Sb: sf of 2		16.0:		*
03139	04 40.2 + 40 02 163.49 – 03.80	07 – 10 – 000	0644	1.5: x 1.1: 1.4: x 0.9:	(75) ...	1	S ...		17.		
03140	04 40.3 + 00 30 196.37 – 27.95	00 – 12 – 072 NGC 1642	1524	2.1 x 2.0 2.0 x 2.0	— 1	4	Sc	SA c Sc –	13.6		*12 3
03141	04 40.5 + 00 38 196.27 – 27.84	00 – 12 – 073	1524	1.0 x 1.0 0.9 x 0.9	— 1	3	S ... + superimposed companion, bridge:		15.3		*1C
03142	04 40.8 + 28 54 172.10 – 10.96	05 – 12 – 000	1314	1.1 x 0.9 1.2 x 0.9	(165) ...	2	S B 0 – a	(R)S B ...	16.5:		
03143	04 41.3 + 70 02 140.53 + 15.80	12 – 05 – 000	0866	1.9:: x 0.2 1.8: x 0.2	135 7	3	Sc		17.		
03144	04 41.4 + 74 50 136.57 + 18.75	12 – 05 – 018	0866	2.1 x 1.1 2.1: x 2.1:	117: (1)	3	Dwarf irregular Ir IV – V		15.7		*
03145	04 41.7 + 00 15 196.82 – 27.79	00 – 13 – 001	0969	1.5 x 1.3 1.5: x 1.1:	(80) 1 – 3	1	S B b – c		15.4		
03146	04 41.7 + 63 46 145.57 + 11.88	11 – 06 – 004	1303	1.4 x 0.8 1.4: x 0.8:	65: ...	1	S ...		17.		
03147	04 41.8 + 72 46 138.31 + 17.51	12 – 05 – 019	0866 64	1.4 x 0.7 1.6 x 0.9	42 5	2	S ... Disturbed?		14.8		
03148	04 42.0 + 18 30 180.62 – 17.28	03 – 13 – 000	1481	1.0 x 0.5 1.2 x 0.8	30: ...	1	S ...		16.5:		
03149	04 42.3 + 65 57 143.88 + 13.31	11 – 06 – 000	1303	1.8 x 1.0 (1.7 x 1.0)	110 5	2	S ...	S (r) ...	15.1		
03150	04 42.3 + 73 22 137.83 + 17.91	12 – 05 – 020	0866 64	1.8 x 1.3 1.8 x 1.3	165: 3	3	Sb	SAB bc Sb +	14.6		

N G C 1633, 1635, 1638, 1642

1	2	3	4	5	6	7	8	9	10	11	12
03151	04 42.5 + 10 59 187.09 – 21.67	02 – 13 – 000	1534	1.1 x 0.25 1.1 x 0.3	94 6 – 7	1	S ...		16.5:		
03152	04 43.0 + 79 19 132.78 + 21.48	13 – 04 – 000	1322	1.0 x 0.15 1.1 x 0.15	150 7	2	Sc		16.0:		*
03153	04 43.3 – 02 30 199.73 – 28.85	00 – 13 – 003 NGC 1653	0969 62a	(1.7 x 1.7) (2.3 x 2.3)	— ...	4	E		12.9		3
03154	04 43.3 – 02 11 199.42 – 28.69	00 – 13 – 000 NGC 1654	0969 62a	0.9 x 0.9 0.9 x 0.9	— 1	1	S ...		14.4		*
03155	04 43.3 + 71 27 139.47 + 16.80	12 – 05 – 021	0866	1.1 x 0.5 1.5:: x 0.5	171 ...	1	... Peculiar?		15.3		1
03156	04 43.6 – 02 10 199.45 – 28.62	00 – 13 – 004 NGC 1657	0969 62a	1.5 x 1.1 1.3 x 1.0	150 2 – 3	3	Sb/S B b		15.0		*
03157	04 43.6 + 18 23 180.96 – 17.05	03 – 13 – 001	1481	1.4 x 1.3 1.7: x 1.4:	— 1 – 2	2	S B ... + companions		15.7		*
03158	04 43.8 + 03 25 194.13 – 25.66	01 – 13 – 001	0974	1.5 x 1.1 1.0 x 0.7	170: 3:	2	S0		14.8		1
03159	04 43.8 + 72 53 138.31 + 17.70	12 – 05 – 000	0866 64	1.0 x 0.5 1.1 x 0.5	5 5	2	Sb – c		15.6		
03160	04 44 + 86 09 126.64 + 25.36	14 – 03 – 006 NGC 1544	1328	1.7 x 1.1 1.7 x 1.1	130 ...	2	...		14.2		3
03161	04 44.0 + 00 31 196.87 – 27.14	00 – 13 – 000	0969	1.3 x 0.15 1.1 x 0.15	68 7	2	Sc		16.5:		
03162	04 44.2 + 08 13 189.80 – 22.92	01 – 13 – 002	0974 62	1.2 x 0.8 1.7: x 0.9:	140: ...	1	Sc:		16.0:		*
03163	04 44.2 + 69 24 141.21 + 15.60	12 – 05 – 023	0866	1.6 x 0.35 1.5 x 0.35	86 6 – 7	2	Sa		15.3		
03164	04 44.5 – 00 01 197.49 – 27.33	00 – 13 – 007	0969	1.5: x 1.5: 1.4: x 1.4:	— 1	3	... S IV – V	SA m	15.6		
03165	04 44.5 + 23 53 176.61 – 13.49	04 – 12 – 000	1461	2.5: x 1.5: 2.7: x 1.7:	(135) ...	3	Irr Ir IV – V:	IA m	16.0:		
03166	04 44.6 – 02 09 199.58 – 28.39	00 – 13 – 008 NGC 1661	0969 62a	1.8: x 1.3: 1.6 x 1.1	35: ...	2	Sb – c		14.3		
03167	04 44.8 + 63 51 145.72 + 12.20	11 – 06 – 000	1303	1.6: x 0.25 1.8 x 0.25	167 7	2	Sb – c		17.		*
03168	04 45.0 – 02 24 199.88 – 28.43	00 – 13 – 010	0969 62a	1.1 x 1.0 1.1 x 1.1	— 1	3	Sc	Sc +	15.3		
03169	04 45.0 + 09 30 188.78 – 22.03	02 – 13 – 000	1534 62	1.2 x 0.2 0.8 x 0.1	73 6 – 7	1	Sc		17.		
03170	04 45.1 – 01 15 198.77 – 27.83	00 – 13 – 011	0969	1.0 x 0.3 1.1 x 0.45	117 6 – 7	1	Sa: nnf of 2		15.5		*
03171	04 45.1 + 01 44 195.90 – 26.28	00 – 13 – 012	0969	1.5 x 1.1 1.3: x 1.2:	— 1 – 2:	2	S B: c Singular		15.2		*
03172	04 45.3 + 08 41 189.55 – 22.43	01 – 13 – 000	0974 62	1.7 x 0.2 1.5 x 0.15	138 7	3	Sc	Sc +	16.5:		
03173	04 45.8 + 07 58 190.27 – 22.73	01 – 13 – 000	0974 62	1.0: x 0.8: 0.9: x 0.7:	— ...	2	...	S d – dm	16.5:		
03174	04 46.0 + 00 10 197.53 – 26.91	00 – 13 – 014	0969	2.0:: x 1.6:: 1.5:: x 1.5::	— ...	3	Dwarf (irregular) Ir: V		17.		*
03175	04 46.1 + 74 24 137.13 + 18.74	12 – 05 – 000	0866	1.6 x 0.25 1.8 x 0.25	173 7	2	Sb – c		15.2		
03176	04 46.5 + 67 07 143.21 + 14.37	11 – 06 – 000	1303	(1.4 x 1.2) (1.4 x 1.2)	— ...	2	E – S0		14.8		
03177	04 46.8 + 21 35 178.81 – 14.50	04 – 12 – 000	1461	1.4:: x 1.4:: 1.4:: x 1.4::	— ...	3	Dwarf V		17.		
03178	04 47.0 + 00 10 197.67 – 26.70	00 – 13 – 015 IC 395	0969	1.1 x 0.9 (1.6 x 1.2)	(130) ...	2	S0:		13.9		1
03179	04 47.2 + 03 15 194.79 – 25.03	01 – 13 – 000	0974	1.1 x 0.9 0.8 x 0.4	— ...	1	... Very compact, nff of 2		14.8	+ 8320	*C
03180	04 47.7 + 08 37 189.97 – 21.98	01 – 13 – 003	0974 62	1.0 x 0.7 1.1 x 0.8	(170) ...	1	S B ...		15.0		
03181	04 48.0 + 05 55 192.45 – 23.42	01 – 13 – 004	0974	2.4 x 1.0 2.2 x 0.8	97 6	3	S B b		14.5		1
03182	04 48.2 + 72 14 139.06 + 17.58	12 – 05 – 026	0866 64	(1.4 x 1.4) (1.5 x 1.4)	— ...	2	S0		14.5		
03183	04 48.3 + 23 06 177.79 – 13.29	04 – 12 – 000	1461	1.0 x 0.8 1.2 x 0.9	120: 2	1	Sb:		16.0:		
03184	04 49.0 + 05 23 193.08 – 23.50	01 – 13 – 000	0974	1.5: x 0.5 1.4 x 0.5	13 (6)	1	S: ...		16.0:		*
03185	04 49.0 + 64 24 145.57 + 12.90	11 – 06 – 000	1303	1.3: x 1.0: 1.2 x 0.8	(120) ...	2	Sc		16.5:		

N G C   1544, 1653, 1654, 1657, 1661   I C   395

1	2	3	4	5	6	7	8	9	10	11	12
03186	04 49.1 + 03 35 194.76 – 24.45	01 – 13 – 000	0974	1.7 x 0.25 1.7 x 0.2	48 7	3	Sc		16.0:		
03187	04 49.1 + 06 45 191.86 – 22.73	01 – 13 – 000	0974	1.5 x 0.9 1.3 x 0.8	0 4	3	Sc	Sc +	15.7		
03188	04 49.1 + 08 45 190.06 – 21.62	01 – 13 – 005	0974	1.1 x 0.6 1.0 x 0.7	85: ...	1	S ... Singular		15.0		*1
03189	04 49.2 + 69 39 141.28 + 16.10	12 – 05 – 027	0866	2.2 x 0.6 1.7 x 0.6	40 6	3	Sc	Sc +	15.4		
03190	04 49.5 + 78 07 134.03 + 21.06	13 – 04 – 011 IC 391	1322	1.8: x 1.7: 1.7: x 1.7:	— 1	2	S ... Peculiar		12.8	+ 1607 + 1800	*12
03191	04 50.1 + 04 19 194.23 – 23.85	01 – 13 – 006	0974	1.5 x 0.9 1.5 x 0.9	75 ...	2	Sb: P w 3195		15.3		*
03192	04 50.2 + 01 10 197.18 – 25.49	00 – 13 – 020 63	0969	(1.0 x 0.8) (1.6 x 1.4)	(55) ...	2	E – S0 Brightest in group		14.6		*
03193	04 50.3 + 02 58 195.51 – 24.53	00 – 13 – 021 63	0969	1.6 x 0.7 1.8 x 0.7	170 6	3	S B b	Sb –	14.7		*
03194	04 50.4 + 01 15 197.13 – 25.41	00 – 13 – 022 63	0969	1.8 x 0.7 1.8 x 0.7	0 6:	4	S B b 3192 group	S B(s)b	15.4		*
03195	04 50.5 + 04 18 194.30 – 23.77	01 – 13 – 007	0974	1.5: x 1.0: 1.2 x 0.8:	(110) ...	2	Sc P w 3191		15.6		*
03196	04 50.6 + 67 40 143.00 + 15.01	11 – 07 – 001	1246	1.2 x 0.3 0.9: x 0.3	163 6	3	Sc nf of 2		16.5:		*
03197	04 51.0 + 80 05 132.32 + 22.21	13 – 04 – 012	1322	(1.4 x 1.3) (1.4 x 1.4)	— ...	1	E?		15.5		*
03198	04 51.7 + 01 33 197.04 – 24.97	00 – 13 – 027 NGC 1690 63	0969	1.1 x 1.1 (1.2 x 1.2)	— ...	3	E P w 3199 Brightest in group		15.0		*3
03199	04 51.7 + 01 35 197.00 – 24.96	00 – 13 – 026 63	0969	1.1 x 0.30 1.0 x 0.30	116 6	3	S B b S B(s)b NGC 1690 group, p w 3198		15.5		*
03200	04 51.7 + 02 03 196.57 – 24.71	00 – 13 – 028 63	0969	1.2 x 0.9: 1.3: x 0.9:	(110) 1 – 3	3	S B b	S B(rs)b	15.2		
03201	04 52.0 + 03 11 195.55 – 24.05	01 – 13 – 009 NGC 1691	0974	1.8 x 1.7 (1.7 x 1.7)	— 1	4	S B 0/S B a (R)S B(s)0/a		13.2		3
03202	04 52.7 – 01 16 199.86 – 26.20	00 – 13 – 034 65	0969	1.3 x 1.2 1.2 x 1.2	— 1	3	S B a – b		15.1		
03203	04 52.8 + 68 14 142.66 + 15.52	11 – 07 – 002 IC 396	1246	2.4: x 1.7: 2.5: x 1.8:	85 3	3	S ... Peculiar		13.2		
03204	04 53.1 + 64 55 145.43 + 13.56	11 – 07 – 003	1246	1.2 x 0.3 1.4 x 0.3	99 6	2	Sc		15.7		
03205	04 53.2 + 29 59 172.97 – 08.18	05 – 12 – 002	1314	2.3: x 0.9: 2.0: x 0.8:	53 6	2	Sa – b		16.0:		
03206	04 53.3 + 02 51 196.05 – 23.95	00 – 13 – 036 63	0969	1.3 x 0.8 1.0 x 0.6	120 4	3	Sb P w 3209		15.3		*
03207	04 53.5 + 02 05 196.79 – 24.31	00 – 13 – 037 63	0969	2.8 x 0.8 2.7 x 0.7	124 6	3	Sb sf of 2		14.3		*1
03208	04 53.7 + 01 31 197.35 – 24.56	00 – 13 – 038 63	0969	1.3 x 1.2 1.2 x 1.1	— 1	3	Sc	Sc –	15.6		
03209	04 53.8 + 02 57 196.03 – 23.79	00 – 13 – 039 63	0969	1.7 x 0.7 1.4: x 0.7	32 6	2	Sb – c P w 3206		15.3		*
03210	04 54 + 82 01 130.62 + 23.36	14 – 03 – 000	1328	1.2 x 0.6 1.0 x 0.6	68 ...	1	S: ...		16.5:		*
03211	04 54.7 + 01 49 197.21 – 24.20	00 – 13 – 044	0969	1.4 x 0.35 1.4: x 0.35	170 6	2	Sc		15.3		
03212	04 55.0 + 71 08 140.32 + 17.37	12 – 05 – 000	0866	1.2 x 0.4 0.5 x 0.2	13 ...	1	Dwarf: spiral S IV – V: or V:		18.		
03213	04 55.3 + 02 51 196.34 – 23.53	00 – 13 – 049	0969	1.1 x 0.4 1.1 x 0.4	48 6	1	Sa:		15.7		
03214	04 55.4 – 00 12 199.22 – 25.08	00 – 13 – 048 65	0969	3.3 x 0.6 3.3 x 0.7	57 6	4	Sb	Sb +	14.4		*1
03215	04 55.7 + 00 40 198.44 – 24.57	00 – 13 – 050 65	0969	1.3 x 1.2 1.7 x 1.4	(90) 2:	3	S0 sf of 2		14.8		*
03216	04 56.0 – 01 32 200.58 – 25.62	00 – 13 – 000 65	0969	1.2: x 0.5: 0.9: x 0.5:	15 ...	2	Dwarf: IV or IV – V		17.		
03217	04 56.0 + 53 45 154.66 + 07.04	09 – 09 – 000	1309	1.4: x 0.6: 1.4: x 0.6:	160: ...	1	...		17.		
03218	04 56.0 + 62 10 147.88 + 12.17	10 – 08 – 001	0993	1.8 x 1.0 1.8 x 1.0	145 4	2	Sb	SA b	15.0:		
03219	04 56.1 + 06 54 192.75 – 21.20	01 – 13 – 011 68	0974	1.4 x 0.6 1.2: x 0.5:	100 5 – 6	3	Sc P w 3220	Sc +	15.7		*
03220	04 56.1 + 07 02 192.63 – 21.12	01 – 13 – 010 68	0974	1.1 x 0.7 1.0 x 0.5	110: ...	3	Sc P w 3219	Sc +	15.7		*

N G C   1690, 1691   I C   391, 396

1	2	3	4	5	6	7	8	9	10	11	12
03221	04 56.2 - 00 58 200.06 - 25.29	00 - 13 - 055	0969 65	1.1 x 0.25 1.1 x 0.25	162 6	1	S ...		15.5		
03222	04 56.3 - 00 34 199.69 - 25.07	00 - 13 - 056 NGC 1713	0969 65	(2.2 x 1.8) (2.6 x 2.2)	(45) ...	3	E f of 2		13.9		*3
03223	04 56.5 + 04 54 194.62 - 22.19	01 - 13 - 012	0974	1.6 x 0.8 1.5 x 0.8	80 5	4	S B a		14.0		
03224	04 56.6 + 05 33 194.04 - 21.82	01 - 13 - 013	0974	1.7 x 1.3 1.6 x 1.2	15: 2 - 3	3	Sb		14.4		1
03225	04 56.6 + 12 47 187.64 - 17.82	02 - 13 - 000	1534	[1.1 x 0.5] ...	... ...	2	S ... + S ... Connected		16.5:		
03226	04 57.0 - 00 20 199.57 - 24.80	00 - 13 - 060 NGC 1719	0969 65	1.2 x 0.4 1.3 x 0.45	102 6	1	Sa:		14.5		*13
03227	04 57.4 + 76 48 135.46 + 20.68	13 - 04 - 013	1322	1.4 x 0.5 1.3 x 0.5	27 6	2	Sb - c		16.0:		
03228	04 58.5 + 75 58 136.24 + 20.28	13 - 04 - 014	1322	1.2 x 0.2 1.2 x 0.2	3 7	1	Sb - c		16.5:		
03229	04 59.6 + 07 33 192.67 - 20.11	01 - 13 - 000	0974 68	1.3 x 1.3 1.2 x 1.2	— 1	3	Sc	SAB c	15.7		
03230	04 59.8 + 75 32 136.67 + 20.11	13 - 04 - 016	1322	1.4 x 1.0 1.7 x 1.3	(10) 2 - 3	2	S0 - a		14.1		*
03231	05 00.0 + 00 10 199.51 - 23.90	00 - 13 - 063	0969 65	1.9 x 0.8 1.7 x 0.7	138 5	3	Sc P w 3233		14.9		*
03232	05 00.4 + 18 23 183.42 - 13.85	03 - 13 - 000	1481	1.1: x 0.5: 1.2 x 0.5	175: ...	1	S ...		17.		
03233	05 00.5 + 00 11 199.56 - 23.79	00 - 13 - 064	0969 65	1.3 x 1.1 1.2 x 1.1	(25) 1 - 2	2	Sc P w 3231		15.3		*
03234	05 00.5 + 16 20 185.16 - 15.01	03 - 13 - 000	1481	2. x 2. 2. x 2.	— ...	2	Dwarf (irregular) Ir: IV - V or V		16.5:		*
03235	05 00.5 + 66 57 144.19 + 15.37	11 - 07 - 000	1246	2.5: x 1.0: 2.0: x 0.8:	150 6	3	Sc	SAB(s)cd: Sc +	16.0:		*
03236	05 00.8 + 06 35 193.71 - 20.38	01 - 13 - 014	0974 68	1.1 x 0.4 1.1 x 0.4	18 6	2	Sb - c		15.6		
03237	05 00.9 + 00 35 199.24 - 23.50	00 - 13 - 065	0969	1.6 x 0.35 1.7 x 0.35	112 6	2	Sc		15.2		
03238	05 01.0 + 01 30 198.39 - 23.01	00 - 13 - 067 NGC 1762	0969	1.8 x 1.1 1.6 x 1.1	175 3 - 4	2	Sc		13.5		3
03239	05 01.7 - 00 29 200.36 - 23.86	00 - 13 - 000	0969	1.2 x 0.4 1.0 x 0.3	154 6	1	Sa - b		15.7		
03240	05 01.7 + 04 35 195.65 - 21.26	01 - 13 - 000	0974 68	1.1 x 0.4 1.0 x 0.25	0 6	2	Sb - c		15.7		
03241	05 02.1 + 75 22 136.91 + 20.15	13 - 04 - 017	1322	1.8 x 1.0 1.6 x 0.9	172: 4	3	S0		14.6		
03242	05 02.2 + 04 25 195.87 - 21.24	01 - 13 - 000	0974 68	1.4 x 0.3 1.3 x 0.25	123 7	3	Sb		15.7		
03243	05 02.2 + 20 56 181.55 - 12.01	03 - 13 - 000	1481	1.0 x 0.9 1.0 x 1.0	— 1	3	Sc/S B c	SAB(s) ...	16.5:		
03244	05 02.3 + 33 57 171.01 - 04.22	06 - 12 - 000	1315	1.4:: x 1.2:: 1.3:: x 1.1::	— ...	1	...		18.		
03245	05 03.1 + 70 26 141.32 + 17.52	12 - 05 - 028	0866	1.6 x 1.6 1.3 x 1.3	— 1	4	Sc	SAB(s)c Sc -	15.0		
03246	05 03.4 + 00 29 199.68 - 23.01	00 - 13 - 070	0969	1.0 x 0.9 1.1 x 1.0	— 1 - 2	1	S ...		15.0		
03247	05 03.9 + 08 35 192.36 - 18.66	01 - 13 - 000	0974	1.0 x 1.0 0.9 x 0.9	— 1	1	S ...		15.3		
03248	05 04.5 + 03 55 196.65 - 21.01	01 - 13 - 015	0974 68	1.8 x 0.9 1.7 x 0.7	135 5 - 6	2	Sb		15.0		1
03249	05 05.4 + 75 52 136.57 + 20.58	13 - 04 - 000	1322	1.4 x 1.0 1.5 x 1.2	75 ...	4	... nff of 2	S dm	15.6		*
03250	05 05.8 + 63 07 147.77 + 13.64	11 - 07 - 005	1246	2.3 x 1.5 1.8 x 1.4	100: 3	3	S B b		14.5		
03251	05 06.0 + 13 13 188.61 - 15.68	02 - 14 - 000	1267	1.1 x 0.6: 1.1 x 0.8	100: ...	1	...		16.0:		
03252	05 06.6 + 67 25 144.13 + 16.12	11 - 07 - 006	1246	2.3 x 1.8 1.8 x 1.3	(30) ...	4	Sc/S B c sf of 2	SAB(s)c	14.1		*
03253	05 07 + 84 00 128.96 + 24.69	14 - 03 - 007	1328	1.9 x 1.0 1.7 x 0.9	93 5	4	S B b	S B(r)b	13.1		
03254	05 07.0 + 66 42 144.77 + 15.75	11 - 07 - 000	1246	1.5 x 0.7 1.1 x 0.5	45 ...	1	S B ...		16.0:		
03255	05 07.2 + 07 25 193.86 - 18.60	01 - 14 - 001	1270 68	1.3 x 0.3 1.5 x 0.3	17 6	1	S B? b:		15.6		

N G C   1713, 1719, 1762

1	2	3	4	5	6	7	8	9	10	11	12
03256	05 07.6 + 00 50 199.92 – 21.92	00 – 14 – 007	0909	1.2 x 1.1 1.1 x 1.1	— 1	3	Sb/S B b f of 2		14.6		*
03257	05 08 + 84 26 128.55 + 24.92	14 – 03 – 008	1328	1.7 x 0.8 1.6 x 0.7	157 5	3	S B a		14.9		
03258	05 08.1 + 00 21 200.44 ⟳ 22.06	00 – 14 – 000	0909	0.7 x 0.6 0.8 x 0.7	— ...	3	S B ... Singular		13.9		*
03259	05 08.2 + 72 16 139.93 + 18.86	12 – 05 – 029	0866 64	2.1 x 1.5 2.1: x 1.5:	120: 3	4	Sc	SAB d	15.6		
03260	05 08.5 + 51 15 157.82 + 07.06	09 – 09 – 000	1309 69	1.3: x 0.3 1.5: x 0.3	116 6 – 7	1	S0?		15.2		
03261	05 08.6 + 16 59 185.75 – 13.05	03 – 14 – 000	0639 66	1.3 x 0.8 1.4 x 1.0	55: ...	2	Sc – Irr		15.5		
03262	05 08.9 – 00 38 201.47 – 22.37	00 – 14 – 012	0909 67	1.0 x 0.2 0.9 x 0.2	147 6 – 7	1	S ... P w 3264		15.6		*
03263	05 09.0 – 00 44 201.58 – 22.40	00 – 14 – 014	0909 67	1.1 x 1.0 1.1: x 1.0:	— 1	3	S B c		16.0:		1
03264	05 09.0 – 00 38 201.48 – 22.35	00 – 14 – 013	0909 67	1.2 x 0.6 1.0 x 0.6	110 4 – 5	2	S0 – a P w 3262		14.7		*1
03265	05 09.1 + 05 08 196.18 – 19.40	01 – 14 – 002 NGC 1819	1270 68	1.7 x 1.2 1.5 x 1.1	120: ...	3	S B 0		13.7		1
03266	05 09.3 + 13 50 188.54 – 14.67	02 – 14 – 000	1267	1.1 x 0.7 1.0 x 0.7	20: ...	1	S ...		16.0:		
03267	05 10.0 + 71 25 140.77 + 18.53	12 – 05 – 030	0866	1.7 x 1.1 1.7 x 1.1	15: 4	3	Sc		15.2		*
03268	05 10.8 + 68 12 143.67 + 16.88	11 – 07 – 008	1246	1.5: x 0.9: 0.8 x 0.6	(160) ...	1	...		16.5:		
03269	05 11.5 + 06 27 195.32 – 18.20	01 – 14 – 000	1270 68	1.0 x 0.6 0.9 x 0.5	175 4:	2	Sb – c		14.9		
03270	05 12.7 + 06 25 195.52 – 17.97	01 – 14 – 008	1270 68	1.0 x 0.4 0.8 x 0.3	177 6	1	S ...		15.5		
03271	05 13.6 – 00 12 201.69 – 21.14	00 – 14 – 018	0909	1.4 x 1.0 (1.2 x 1.0)	(175) ...	1	...		14.6		
03272	05 13.6 + 06 04 195.95 – 17.96	01 – 14 – 000	1270 68	1.1 x 0.3 1.0 x 0.3	121 6	1	S ...		16.0:		
03273	05 13.7 + 53 30 156.43 + 09.01	09 – 09 – 001	1309	3.3: x 1.0: 3. x 1.	45 ...	3	... S IV – V	S m	15.2		
03274	05 14.0 + 06 23 195.72 – 17.71	01 – 14 – (16)	1270 68	[1.3 x 0.5] ...	... ...	1	Chain of 5 – 6 galaxies		14.5:		*1V
03275	05 14.1 + 06 34 195.57 – 17.59	01 – 14 – 017	1270 68	1.9 x 0.4 1.8 x 0.3	37 6	1	S ...		16.0:		*
03276	05 14.1 + 76 18 136.45 + 21.25	13 – 04 – 020	1322	1.7 x 0.35 1.7 x 0.35	80 7	3	Sb		15.7		*
03277	05 14.2 + 65 25 146.31 + 15.67	11 – 07 – 009	1246 70	2.4 x 0.8 1.9 x 0.7	156 6	2	Sc		15.4		
03278	05 14.4 + 06 08 196.00 – 17.75	01 – 14 – 000	1270 68	1.0 x 0.3 0.9 x 0.2	... ...	2	S: ... Brightest of 3, distorted		16.0:		*
03279	05 14.5 + 06 52 195.36 – 17.35	01 – 14 – 020	1270 68	1.4 x 0.3 1.4 x 0.25	56 7	2	Sb		15.3		
03280	05 14.6 + 30 25 175.41 – 04.22	05 – 13 – 000	1455	1.0 x 0.2 1.2 x 0.2	[160] ...	0	...		17.		*
03281	05 14.6 + 72 29 140.00 + 19.38	12 – 06 – 001	0975 64	1.6 x 1.0 1.5 x 0.9	150: 4	2	S0		14.9		
03282	05 15.0 + 06 45 195.53 – 17.31	01 – 14 – 023	1270 68	1.2 x 0.9 1.1: x 0.8:	(25) ...	2	S B c:		15.7		
03283	05 15.2 – 01 13 202.85 – 21.28	00 – 14 – 000	0909	1.6 x 0.2 1.5 x 0.15	105 7	3	Sc p of 2		16.5:		*
03284	05 15.2 + 72 40 139.86 + 19.51	12 – 06 – 000	0975 64	1.7 x 1.1 0.8 x 0.7	125 4	4	Sc	SAB(s)c	16.0:		
03285	05 16.0 + 19 09 184.93 – 10.38	03 – 14 – 000	0639	1.2: x 0.9: 1.7: x 1.5:	— ...	0	...		16.5:		
03286	05 16.4 + 16 50 186.95 – 11.58	03 – 14 – 000	0639	1.2 x 0.4 1.2 x 0.4	57 6	2	S B b – c		16.0:		
03287	05 16.7 + 01 16b 200.74 – 19.74	00 – 14 – 000	0909 68	1.0 x 0.8 0.9 x 0.8	— ...	3	S ... P w a, distorted		(14.7)		*
03288	05 16.9 + 04 04 198.20 – 18.28	01 – 14 – 000	1270 68	1.1 x 0.7 0.7: x 0.6:	— ...	1	...	S d – m	17.		
03289	05 17.4 + 06 31 196.07 – 16.92	01 – 14 – 025	1270 68	1.1 x 0.35 1.0 x 0.4	50 6	2	S B b 2nd of 3		15.6		*
03290	05 17.4 + 17 40 186.37 – 10.92	03 – 14 – 000	0639	1.2 x 0.5 1.1 x 0.4	23: ...	1	S ...		17.		

N G C   1819

1	2	3	4	5	6	7	8	9	10	11	12
03291	05 17.7 + 06 31 196.11 − 16.85	01 − 14 − 026	1270 68	1.1 x 0.5 1.0 x 0.5	94 5 − 6	2	S B: c 3rd of 3		15.4		*
03292	05 17.8 + 52 48 157.37 + 09.12	09 − 09 − 000	1309	1.1 x 0.2 1.1 x 0.2	130 ...	1	S ...		17.		
03293	05 18.0 + 08 45 194.16 − 15.62	01 − 14 − 000	1270	1.8 x 0.9 1.5 x 0.9	140 4 − 5	2	Sc		16.0:		*
03294	05 18.4 + 03 58 198.49 − 18.01	01 − 14 − 029	1270 68	3.4 x 1.7 3.4 x 1.7:	133: 5	4	Sb	SA(rs)b	14.5		
03295	05 18.4 + 15 12 188.61 − 12.08	03 − 14 − 000	0639	1.1 x 0.5 1.2 x 0.5	107 ...	1	S: ...		17.		
03296	05 18.7 + 04 50 197.75 − 17.51	01 − 14 − 030	1270 68	1.9 x 1.4 1.5 x 1.2	140: 2 − 3	3	Sa/Sb		14.3		
03297	05 18.9 + 73 15 139.46 + 20.04	12 − 06 − 000	0975 64	1.1 x 0.8 0.8: x 0.8	[162] ...	1	... Multiple system?		16.5:		
03298	05 19.3 + 03 26 199.10 − 18.09	01 − 14 − 034 IC 412 = 2123*	1270 68	1.4 x 0.7: 1.6 x 0.8	... ...	2	... P w 3299, distorted				
03299	05 19.4 + 03 26 199.11 − 18.07	01 − 14 − 035 IC 413 = 2124*	1270 68	1.1 x 1.0: 1.2 x 1.0	— ...	2	... P w 3298, distorted		14.5		*1V
03300	05 19.9 + 43 30 165.30 + 04.17	07 − 11 − 000	0668	1.1: x 0.9: 1.0: x 0.8:	— ...	1	Sc:		18.		
03301	05 20.4 − 00 11 202.56 − 19.65	00 − 14 − 021	0909	1.4: x 1.0: 1.3 x 0.9	20: 3	2	S B O − a		15.5		
03302	05 21.0 + 76 37 136.36 + 21.76	13 − 04 − 021	1322	1.8 x 1.6: 1.8: x 1.6:	(160) 1	3	Sc	SAB:(r)bc Sc −	14.9		1
03303	05 22.3 + 04 27 198.57 − 16.93	01 − 14 − 000	1270	4. x 3. 4. x 3.	— ...	0	Dwarf?		16.0:		
03304	05 22.4 + 21 49 183.53 − 07.65	04 − 13 − 000	0454	1.3: x 1.3: 1.2 x 0.6	— ...	1	S ...		17.		
03305	05 22.6 + 54 23 156.43 + 10.59	09 − 09 − 000	1309	1.0 x 0.3 1.0 x 0.2	[126] ...	2	Dwarf:		17.		
03306	05 22.7 + 00 22 202.35 − 18.88	00 − 14 − 022	0909	(1.2 x 0.6) (1.3 x 0.7)	52 ...	2	E − S0		15.0		
03307	05 23.1 + 63 50 148.22 + 15.65	11 − 07 − 010	1246 70	(1.0 x 0.8) (0.9 x 0.7)	(85) ...	2	E		14.8		
03308	05 23.3 + 08 55 194.73 − 14.42	01 − 14 − 000	1270	1.0 x 1.0 1.0: x 1.0:	— 1	2	Sc		16.5:		
03309	05 24.0 + 67 19 145.14 + 17.49	11 − 07 − 011	1246	1.5 x 0.8 1.1 x 0.7	10 ...	1	...		15.3		
03310	05 24.4 + 49 56 160.36 + 08.40	08 − 10 − 006	0224	1.0 x 0.2 1.1 x 0.4	84 ...	1	S ...		17.		
03311	05 24.5 + 79 21 133.81 + 23.15	13 − 05 − 000	1256	1.3: x 0.3: 1.0: x 0.2:	64 ...	1	...	S d − m	16.5:		
03312	05 25.6 + 22 04 183.73 − 06.89	04 − 13 − 000	0454	1.5 x 0.6 1.4 x 0.5	2 6	1	S ...		16.5:		
03313	05 25.6 + 79 11 133.99 + 23.12	13 − 05 − 000	1256	1.2:: x 0.8:: 1.0:: x 0.8::	(160) ...	2	Dwarf (spiral?) S? V		18.		
03314	05 25.9 + 55 53 155.39 + 11.79	09 − 09 − 000	1309	1.8: x 0.2 1.6 x 0.2	120 ...	1	...		18.		
03315	05 26.3 + 53 21 157.61 + 10.49	09 − 09 − 003	1309	1.1: x 0.7: 1.2: x 0.7:	160: ...	1	...		17.		
03316	05 26.8 + 55 49 155.52 + 11.86	09 − 09 − 000	1309	1.5: x 1.0: 1.5: x 1.0:	(100) ...	1	Dwarf?		18.		
03317	05 27.5 + 73 42 139.35 + 20.79	12 − 06 − 006	0975 64	2.3: x 2.3: 2. x 2.	— ...	4	Dwarf irregular Ir V		17.		*
03318	05 27.8 + 76 46 136.41 + 22.18	13 − 05 − 000	1256	1.7: x 1.4: 1.5:: x 1.2::	(45) ...	3	Dwarf spiral S(B) V		17.		
03319	05 28.7 + 70 09 142.74 + 19.23	12 − 06 − 000	0975	1.8 x 0.3 1.7 x 0.2	127 7	2	Sc		16.5:		*
03320	05 30.0 + 79 33 133.72 + 23.46	13 − 05 − 001	1256 76	1.3 x 0.7 1.1 x 0.7	90: ...	3	S B b	S B(s)b	15.3		
03321	05 30.9 + 06 14 198.10 − 14.18	01 − 15 − 001	1493 71	1.3 x 0.7 1.5 x 0.8	147 4 − 5	2	Sa − b		16.0:		
03322	05 31.6 + 06 47 197.70 − 13.75	01 − 15 − 000	1493 71	1.1 x 0.6 1.1: x 0.6	25: ...	1	S: ... nnf of 2		17.		*
03323	05 31.7 + 49 43 161.19 + 09.27	08 − 11 − 001	0637 74	1.1: x 0.7: 1.4: x 1.0:	(172) ...	3	...	S d − dm	17.		
03324	05 31.7 + 63 34 148.96 + 16.34	11 − 07 − 000	1246 70	1.2 x 0.10 1.1 x 0.10	40 7	3	Sc		17.		*
03325	05 31.8 + 40 53 168.71 − 04.55	07 − 12 − 001	0645	1.8 x 1.1 1.4 x 1.0	(30) 3 − 4	1	Sb − c		16.5:		1

I C   412, 413, 2123, 2124

1	2	3	4	5	6	7	8	9	10	11	12
03326	05 32.1 + 77 17 136.01 + 22.62	13 – 05 – 002	1256	3.4 x 0.35 3.3 x 0.35	69 7	3	Sc		15.6		
03327	05 33.5 + 31 54 176.45 – 00.04	05 – 14 – 001	1459	(2.5 x 2.5) (2.5 x 2.5)	— ...	0	E?		13.:		
03328	05 33.6 + 07 18 197.50 – 13.06	01 – 15 – 002	1493 71	1.1 x 0.35 1.2 x 0.35	169 6	1	Sb		15.6		
03329	05 33.7 + 16 33 189.46 – 08.25	03 – 15 – 000	0427	1.4 x 0.6 1.6 x 0.9	10 ...	2	Sb – c		16.0:		
03330	05 33.9 + 14 24 191.34 – 09.34	02 – 15 – 000	0415	1.2: x 1.1: 1.2 x 1.1	— 1 – 2	1	S ...		17.		
03331	05 35.3 + 00 05 204.21 – 16.26	00 – 15 – 001	0435 72	1.2 x 1.1 1.3: x 1.2:	— 1 – 2	2	Sb		16.0:		1
03332	05 36.2 + 15 33 190.64 – 08.27	03 – 15 – 000	0427	1.2: x 0.3: 1.5 x 0.4	30 6	1	S ...		16.5:		
03333	05 36.2 + 49 18 161.93 + 09.68	08 – 11 – 000	0637 74	1.1 x 0.5 0.9 x 0.3	59 ...	1	...		17.		
03334	05 36.6 + 69 21 143.83 + 19.47	12 – 06 – 007 NGC 1961 = 2133*	0975	4.6 x 4.4 4.6 x 3.3:	(85) ...	3	Sb	SAB bc Sb +	12.2	+ 3870 + 4027	*123 4568A
03335	05 36.7 + 79 13 134.19 + 23.61	13 – 05 – 000	1256 76	1.2 x 0.5 1.1 x 0.4	122 6	2	SO		15.4		
03336	05 37 + 85 54 127.34 + 26.07	14 – 03 – 011	1328	1.7 x 1.1 1.1: x 1.0:	45: ...	3	S B b	S B(s)b Sb –	15.6		
03337	05 37.2 + 61 01 151.60 + 15.66	10 – 09 – 000	0959	1.3 x 0.3 1.1 x 0.2	117 7	1	S ...		16.5:		*
03338	05 37.8 + 16 27 190.07 – 07.47	03 – 15 – 000	0427	1.1:: x 0.9:: 1.0: x 0.7:	— ...	1	S ...		16.5:		
03339	05 38.2 + 76 53 136.56 + 22.76	13 – 05 – 000	1256	1.1 x 0.25 1.0 x 0.25	177 6	3	Sc	Sc +	16.5:		
03340	05 38.7 + 79 37 133.83 + 23.85	13 – 05 – 004	1256 76	1.2 x 0.45 1.4 x 0.45	152 6	1	Sa – b		14.7		
03341	05 39.0 + 18 29 188.47 – 06.16	03 – 15 – 000	0427	2. x 2. 1.8: x 1.8:	— 1:	2	S B a – b		15.5:		
03342	05 39.0 + 69 17 143.99 + 19.63	12 – 06 – 008	0975	1.8 x 0.4 1.8 x 0.4	42 6	4	Sc		15.4		*1
03343	05 39.2 + 72 19 141.08 + 20.95	12 – 06 – 012	0975	2.5 x 0.6 2.1 x 0.6	80 6	1	S ...		13.9		
03344	05 39.5 + 60 09 144.14 + 19.62	12 – 06 – 010	0975	2.7: x 1.7: 1.7 x 0.9	25: 4	2	Sb	SAB bc Sb +	14.7		*
03345	05 40 + 85 49 127.45 + 26.09	14 – 03 – 000	1328	1.0 x 0.4: 0.8 x 0.3	84 6?	1	S ...		16.5:		
03246	05 40.1 + 51 11 160.61 + 11.17	09 – 10 – 001	0664 74	1.4 x 0.5: 1.5 x 0.6:	135 ...	1	S – Irr Disturbed?		15.5		*1
03347	05 40.2 + 79 32 133.94 + 23.88	13 – 05 – 000	1256 76	1.1 x 1.0 1.0 x 0.9	— 1	1	S B b:		16.0:		
03348	05 40.3 + 16 29 190.35 – 06.94	03 – 15 – 000	0427	1.4 x 0.5 1.1 x 0.4	9 6	1	S ...		16.0:		
03349	05 41.1 + 69 02 144.32 + 19.69	12 – 06 – 014	0975	1.2 x 0.8 1.3 x 0.8	85: 3	1	Sa – b		14.4		*
03350	05 41.6 + 58 34 154.11 + 14.98	10 – 09 – 001	0959	1.0 x 0.4 0.7 x 0.3	8 6	1	Sa – b P w 3351		16.0:		*
03351	05 41.7 + 58 40 154.02 + 15.04	10 – 09 – 002	0959	1.7 x 0.3 1.7 x 0.3	163 6	2	Sa – b P w 3350		16.0:		*
03352	05 41.9 + 16 45 190.32 – 06.47	03 – 15 – 000	0427	1.3 x 0.45 1.5 x 0.6	177 ...	1	S B: ... 2nd of 3		16.0:		*
03353	05 42.7 + 79 41 133.83 + 24.04	13 – 05 – 005	1256 76	(1.4 x 1.2) (2.1 x 1.7)	(20) 1 – 2	4	S B O		14.8		
03354	05 43.1 + 56 05 156.46 + 13.96	09 – 10 – 002	0664	2.1 x 0.7 2.2 x 0.7	164 7	2	Sa – b		15.3		
03355	05 43.3 + 50 50 161.18 + 11.43	08 – 11 – 004	0637 74	1.1: x 1.1: 1.2: x 1.1:	— 	2	[SO] Compact		15.6		*
03356	05 44.2 + 17 33 189.92 – 05.59	03 – 15 – 000	0427	1.0 x 0.7 0.9 x 0.7	— 2 – 3	2	Sb – c spp of 2		16.0:		*
03357	05 44.9 + 74 15 139.36 + 22.11	12 – 06 – 015	0974	1.4 x 0.6 1.7 x 0.8	58 6	2	SO – a		15.0		*
03358	05 46.0 + 50 22 161.81 + 11.58	08 – 11 – 007	0637 74	1.1: x 0.8: 1.1 x 0.8	(0) ...	1	S ...		16.0:		
03359	05 46.0 + 51 05 161.17 + 11.93	09 – 10 – 000	0664 74	1.5 x 0.2 1.5 x 0.2	162 7	3	S B: c		15.7		
03360	05 46.4 + 17 50 189.95 – 04.99	03 – 15 – 000	0427	1.3 x 0.5 1.3 x 0.5	147 6	1	S ... 3363 group		16.0:		*

1	2	3	4	5	6	7	8	9	10	11	12
03361	05 46.5 + 49 42 162.44 + 11.32	08 – 11 – 008	0637	1.3: x 1.0: 1.1: x 1.0:	(110) ...	2	S B: c – Irr		16.5:		
03362	05 46.6 + 17 39 190.13 – 05.05	03 – 15 – 000	0427	1.3 x 0.4 1.5 x 0.6	136 6	1	S ... 3363 group, p w 3363		16.0:		*
03363	05 46.6 + 17 41 190.10 – 05.03	03 – 15 – 000	0427	(1.6 x 0.7) (1.7 x 1.1)	35: ...	1	... of 3, p w 3362	Brightest	15.0:		*
03364	05 46.8 + 76 41 136.96 + 23.14	13 – 05 – 006	1256	1.9: x 1.9: 1.9: x 1.9:	— 1	4	Sc	SA d	16.0:		
03365	05 47.4 + 66 47 146.74 + 19.27	11 – 08 – 000	0677	2.4 x 0.3 2.0 x 0.35	154 7	1	Sa:		15.0		
03366	05 47.6 + 48 50 163.30 + 11.05	08 – 11 – 000	0637	1.3: x 1.0: 1.3 x 1.0	(50) ...	2	Sc		16.5:		
03367	05 48.0 + 37 52 172.94 – 05.63	06 – 13 – 000	1290	1.1 x 0.3 1.7 x 0.5	90 6	1	S ...		17.		
03368	05 48.0 + 39 49 171.25 – 06.62	07 – 12 – 003	0645	1.5 x 0.4 1.5 x 0.4	141 6	3	Sc		17.		
03369	05 48.7 + 41 46 169.62 – 07.72	07 – 12 – 000	0645	1.7: x 0.7: 1.4: x 0.5:	(165) ...	2	...	S dm – m	17.		
03370	05 48.8 + 78 31 135.14 + 23.90	13 – 05 – 000	1256 76	1.4:: x 0.2 0.50 x 0.15	134 ...	1	Sb – c		17.		*
03371	05 49.8 + 75 19 138.42 + 22.81	13 – 05 – 07a	1256	5. x 4. 4. x 3.	— ...	3	Dwarf V		17.		*
03372	05 49.9 + 76 31 137.20 + 23.25	13 – 05 – 008	1256	1.1 x 0.35 1.0 x 0.35	40 6	2	Sb		16.0:		
03373	05 51.0 + 78 30 135.20 + 23.99	13 – 05 – 009	1256 76	1.6 x 1.2 1.5 x 1.1	100 3	4	Sc	SAB c	15.0		
03374	05 51.1 + 46 26 165.72 + 10.40	08 – 11 – 011	0637	2.8: x 2.5: 2.8: x 2.5:	(90) 1	3	S B ...		13.9		*1
03375	05 51.4 + 51 55 160.83 + 13.06	09 – 10 – 003	0664 74	2.3 x 1.1 2.3 x 1.1	45 5	4	Sc	SA(rs)c Sc –	14.6		
03376	05 51.8 + 15 10 192.91 – 05.24	03 – 15 – 000	0427	1.1 x 0.8: 1.1 x 0.8	170: ...	1	S B ...		15.5:		
03377	05 52.9 + 64 44 148.96 + 18.93	11 – 08 – 003	0677	1.1 x 0.7 1.1 x 0.7	110: ...	2	Sc		16.0:		
03378	05 53 + 83 50 129.65 + 25.75	14 – 03 – 017	1328	(1.7 x 1.7) (1.5 x 1.5)	— ...	2	...		14.6		*1
03379	05 53.0 + 68 26 145.37 + 20.43	11 – 08 – 002	0677	1.7 x 1.2 1.7 x 1.1	115: ...	4	S B b/Sb	(R')SAB b Sb –	13.8		
03380	05 54.6 + 11 58 196.04 – 06.25	02 – 16 – 000	1502	(1.2 x 1.0) (1.7 x 1.3)	145: ...	2	E		15.0:		
03381	05 55 + 83 43 129.79 + 25.77	14 – 03 – 016	1328	1.5:: x 1.1:: 1.2:: x 1.0::	(15) ...	1	...		16.5:		*
03382	05 55.1 + 62 08 151.55 + 18.07	10 – 09 – 004	0959	1.2 x 1.2 1.2 x 1.2	— 1	3	S B a	S B(rs)a	14.8		
03383	05 55.2 + 54 27 158.78 + 14.74	09 – 10 – 000	0664	1.1 x 0.7 1.1 x 0.7	155: ...	1	...		16.0:		
03384	05 55.5 + 73 07 140.79 + 22.38	12 – 06 – 000	0975	1.9: x 1.9: 1.8: x 1.6:	— ...	3	Dwarf spiral S V		16.0:		
03385	05 56.0 + 80 08 133.58 + 24.74	13 – 05 – 000	1256 76	1.4 x 0.5 1.5 x 0.7	168 6	2	S0		15.1		
03386	05 57.5 + 65 23 148.54 + 19.63	11 – 08 – 008	0677	1.2 x 0.7 1.1 x 0.6	43 6	1	Sa:		14.4		
03387	05 58.0 + 64 50 149.10 + 19.46	11 – 08 – 009	0677	1.5 x 0.5 1.4 x 0.4	155 6	3	Sc		16.0:		
03388	05 58.2 + 75 23 138.55 + 23.33	13 – 05 – 010	1256	1.2 x 0.9 0.9: x 0.7:	(0) ...	1	S ...		16.0:		1
03389	05 58.5 + 66 18 147.68 + 20.08	11 – 08 – 000	0677	1.0 x 0.6 0.9:: x 0.5::	153: ...	3	Sc sff of 2		17.		*
03390	05 58.8 + 36 07 175.54 + 06.63	06 – 14 – 001	0942	2.3: x 1.3: 2.3: x 1.3:	35: 4	3	...	SAB dm	16.5:		
03391	05 59.4 + 57 31 156.19 + 16.61	10 – 09 – 009	0959 75	1.2 x 1.0 1.1 x 0.9	— 2:	3	Sc	SA c – d	16.5:		
03392	06 00.3 + 57 39 156.11 + 16.77	10 – 09 – 010 NGC 2128	0959 75	(1.5 x 1.1) (1.5 x 1.1)	60: ...	2	S0 or E		13.7		
03393	06 00.4 + 07 50 200.37 – 07.03	01 – 16 – 000	0229	0.35 x 0.35 0.6 x 0.6	— ...	1	... Compact		14.5	+ 5201	*C
03394	06 00.6 + 56 09 157.55 + 16.17	09 – 10 – 000	0664	2. x 2. 1.7:: x 1.7::	— 1:	2	S B ...	S B(r) ...	15.4		
03395	06 01.3 + 08 40 199.75 – 06.43	01 – 16 – 000	0229	1.2: x 1.2: 1.3: x 1.3:	— 1	2	Sb	SAB(s)b	16.5:		

N G C   2128

1	2	3	4	5	6	7	8	9	10	11	12
03396	06 01.3 + 79 56 133.87 + 24.90	13 – 05 – 011	1256 76	(1.2 x 1.0) (1.2 x 1.0)	(50) ...	3	E P w 3397		15.0		*
03397	06 01.9 + 79 52 133.95 + 24.91	13 – 05 – 012	1256 76	1.2 x 0.7 1.3 x 0.8	168 4	2	Sa – b P w 3396		15.6		*
03398	06 02.0 + 60 35 153.41 + 18.18	10 – 09 – 011	0959 75	1.5 x 0.7 1.2 x 0.6	112 5	3	Sc		16.0:		
03399	06 02.4 + 54 44 158.99 + 15.80	09 – 10 – 000	0664	1.0 x 0.6 0.8: x 0.6	170: ...	3	S B c		16.5:		
03400	06 02.5 + 80 38 133.14 + 25.15	13 – 05 – 000	1256 76	1.2 x 0.2 1.1 x 0.2	48 6	2	...	S d – m	16.5:		
03401	06 04 + 81 05 132.68 + 25.33	14 – 03 – 018	1328	1.9 x 0.22 1.9 x 0.22	20 7	2	Sb P w 3413		15.7		*
03402	06 04.5 + 67 59 146.24 + 21.25	11 – 08 – 013	0677	1.1 x 0.9 1.0 x 0.8	(135) 2:	3	Sc	Sc +	16.0:		1
03403	06 04.7 + 71 23 142.81 + 22.44	12 – 06 – 016	0975	2.7 x 0.8 2.7 x 0.8	27 6	2	S B? c		14.6		1
03404	06 05.0 + 80 01 133.83 + 25.08	13 – 05 – 016	1256 76	(1.3 x 1.0) (1.2 x 0.9)	... ...	1	SO		15.2		*
03405	06 05.0 + 80 29 133.33 + 25.20	13 – 05 – 014	1256 76	1.6 x 0.35 1.5 x 0.35	127 6 – 7	2	Sb – c P w 3410		15.5		*
03406	06 05.4 + 67 16 147.00 + 21.07	11 – 08 – 000	0677	1.3 x 0.1 1.3 x 0.1	27 (7)	2	...		17.		
03407	06 05.5 + 42 05 170.86 + 10.61	07 – 13 – 007	0669	1.3 x 0.9 1.3 x 0.9	18: 3	3	Sa		14.6		
03408	06 05.5 + 79 43 134.15 + 25.02	13 – 05 – 017	1256 76	1.6 x 0.6 1.5 x 0.5	[12] ...	1	Triple system? Arrow – shaped, p w 3412		16.0:		*
03409	06 05.8 + 64 36 149.67 + 20.14	11 – 08 – 015	0677	1.6 x 0.4 0.8:: x 0.2::	10 ...	2	Dwarf spiral S IV – V or V		17.		
03410	06 06.0 + 80 28 133.36 + 25.24	13 – 05 – 015	1256 76	2.3 x 0.8 2.3 x 0.8	118 6	1	Sb P w 3405		15.1		*
03411	06 06.2 + 62 01 152.23 + 19.21	10 – 09 – 013	0959	1.7: x 1.1 1.5: x 1.1	(35) ...	1	Sc:		15.7		
03412	06 06.8 + 79 44 134.15 + 25.08	13 – 05 – 018	1256 76	1.6 x 0.8 1.5 x 0.7	153 5	4	Sc P w 3408		16.0:		*
03413	06 07 + 81 10 132.63 + 25.46	14 – 03 – 019	1328	2.6 x 1.8 2.3: x 1.7:	85: ...	3	S B c P w 3401	S B(r)bc Sc –	13.8		*1
03414	06 07.4 + 64 18 150.03 + 20.19	11 – 08 – 018	0677	0.7 x 0.7 0.7 x 0.7	— ...	2	... Peculiar		13.9		*1C
03415	06 07.5 + 70 15 144.05 + 22.28	12 – 06 – 000	0975	1.7 x 0.25 1.2: x 0.2:	154 ...	2	Dwarf:		16.5:		
03416	06 08.0 + 69 45 144.58 + 22.15	12 – 06 – 017	0975	2.4 x 1.2 1.7 x 1.0	90 5	3	Sc		15.2		*1
03417	06 08.4 + 51 32 162.40 + 15.25	09 – 11 – 000	0984 80	1.3: x 0.2 1.1 x 0.20	165 7	2	Sb		15.7		
03418	06 08.8 + 44 27 168.99 + 12.22	07 – 13 – 009	0669	1.1 x 1.0 1.0 x 1.0	— 1	3	S B b	S B(rs)b	15.0		*
03419	06 09.0 + 66 21 148.05 + 21.08	11 – 08 – 022	0677	1.0: x 0.15 0.9 x 0.15	79 7	1	Sa – b		16.0:		
03420	06 09.0 + 75 58 138.17 + 24.15	13 – 05 – 019	1256	3.4 x 0.9 2.9 x 0.9	152 6	4	Sb		14.4		*
03421	06 09.2 + 53 05 160.99 + 16.01	09 – 11 – 000	0984	1.0 x 0.1 1.1 x 0.1	12 (7)	1	...		18.		
03422	06 09.3 + 71 09 143.18 + 22.71	12 – 06 – 018	0975	2.4 x 1.8 2.1 x 1.8	43 ...	3	Sb/S B b P w 3426	SAB(rs)b	14.5		*
03423	06 09.4 + 78 52 135.11 + 24.97	13 – 05 – 020	1256 76	1.6 x 0.2 1.5 x 0.12	156 7	2	...	S d – m	16.5:		
03424	06 09.5 + 53 05 161.01 + 16.05	09 – 11 – 000	0984	1.5: x 1.0: 1.5: x 1.0:	(145) ...	1	Dwarf?		17.		
03425	06 09.8 + 66 35 147.84 + 21.24	11 – 08 – 023	0677	2.5 x 0.35 2.3 x 0.35	90 7	3	Sb		15.3		
03426	06 09.8 + 71 03 143.30 + 22.72	12 – 06 – 019 Mark 3	0975	(1.8 x 1.6) (1.8 x 1.6)	— ...	1	SO: P w 3422		13.8	+ 4050 + 4200	*
03427	06 10.0 + 80 05 133.82 + 25.30	13 – 05 – 021 IC 440	1256 76	2.1 x 1.2 2.1 x 1.2	33 5	3	Sb	SA(r)ab Sb –	14.3		
03428	06 10.1 + 67 45 146.68 + 21.66	11 – 08 – 024	0677	1.6 x 0.4 1.2 x 0.3	77 6	2	...	S dm – m	16.5:		*
03429	06 10.5 + 78 22 135.66 + 24.89	13 – 05 – 022 NGC 2146	1256 76	5.9 x 4.9 5.9 x 4.5:	[123] ...	4	S [B a] Peculiar, twisted plane	S [AB(s)ab]	11.1	+ 812 + 994	*1234 5789
03430	06 10.9 + 64 27 150.03 + 20.60	11 – 08 – 026	0677	1.0 x 0.4 0.9 x 0.35	115 (6)	1	...		15.1		

1	2	3	4	5	6	7	8	9	10	11	12
03431	06 11.9 + 76 50 137.31 + 24.55	13 – 05 – 023	1256	(1.8 x 1.3) (1.8 x 1.3)	173 ...	3	E		13.6		
03432	06 12.0 + 57 04 157.34 + 17.98	10 – 09 – 000	0959 75	1.8 x 0.4 1.2 x 0.2	136 6	2	Sc		15.5		
03433	06 12.6 + 00 13 208.59 – 07.97	00 – 16 – 001	1487	2.7 x 0.8 2.5 x 0.6	92 6	1	S ...		16.0:		
03434	06 13.3 + 32 14 180.42 + 07.41	05 – 15 – 000	0411	1. x 1. 1.1: x 0.9:	— ...	1	S B ...		17.		
03435	06 14 + 82 21 131.42 + 25.98	14 – 03 – 020	1328	1.3 x 0.5 1.2 x 0.5	122 ...	1	... Peculiar?		14.6		
03436	06 14.0 + 63 28 151.14 + 20.57	11 – 08 – 031	0677	1.6: x 1.5: 1.0: x 0.9:	— ...	3	Sc/S B c	SAB(s)c	16.0:		
03437	06 14.7 + 64 21 150.28 + 20.95	11 – 08 – 034	0677	1.6 x 1.1 1.2: x 0.7:	35 3	3	Sc		16.5:		
03438	06 15.1 + 66 36 148.02 + 21.74	11 – 08 – 035	0677	1.0 x 0.6 1.0 x 0.6	101 4:	2	Sb/S B b	SAB(s) ...	14.9		
03439	06 15.6 + 78 33 135.53 + 25.18	13 – 05 – 025	1256 76	3.5 x 1.3 3.3 x 1.2	30 6	3	Sc P w 3429		14.2		*2
03440	06 15.9 + 71 15 143.26 + 32.25	12 – 06 – 020	0975	1.1 x 0.8 1.1 x 0.8	130: 3	3	Sc Disturbed?	S cd	15.7		*
03441	06 16 + 83 20 130.36 + 26.25	14 – 03 – 021	1328	1.2: x 1.0: 0.9 x 0.7	55: 2	3	Sa	(R?)SAB a	15.3		
03442	06 16 + 84 57 128.59 + 26.56	14 – 03 – 022	1328 102	1.2: x 0.35 1.2: x 0.35	110 6 – 7	3	S0 – a		14.8		
03443	06 16.0 + 51 24 163.03 + 16.27	09 – 11 – 000	0984 80	1.1: x 0.2 0.5 x 0.2	166 ...	1	Dwarf?		17.		
03444	06 16.5 + 73 44 140.67 + 23.99	12 – 06 – 000	0975	1.5: x 0.8: 1.0: x 0.5:	60 ...	3	Dwarf IV – V or V		17.		
03445	06 17.1 + 59 08 155.59 + 19.38	10 – 09 – 016	0959	1.5: x 0.35 1.5 x 0.35	101 7	1	S0 – a P w 3446, contact		13.9		*1C
03446	06 17.2 + 59 08 155.59 + 19.39	10 – 09 – 017	0959	1.4: x 1.0: 1.3 x 0.9	(150) ...	1	S0: P w 3445, contact		14.1		*1C
03447	06 17.4 + 28 00 184.61 + 06.22	05 – 15 – 000	0411	(1.2 x 0.6) (1.7 x 1.1)	(90) ...	1	E:		15.6		
03448	06 17.4 + 66 36 148.10 + 21.96	11 – 08 – 038	0677	1.3 x 0.3 1.2 x 0.3	45 7	2	S0 – a		14.5		*
03449	06 17.5 + 51 08 163.38 + 16.38	09 – 11 – 009	0984 80	1.3 x 0.2 1.1 x 0.15	146 7	2	Sc		15.7		
03450	06 18.0 + 27 53 184.78 + 06.28	05 – 15 – 000	0411	1.0 x 0.5 1.6 x 0.9	162 ...	3	S B b		15.6		
03451	06 18.2 + 77 54 136.27 + 25.15	13 – 05 – 026	1256 76	1.3: x 1.2: 1.2: x 0.9:	— 1	3	S B b	S B(r)b	15.7		*
03452	06 18.5 + 51 57 162.66 + 16.85	09 – 11 – 010 NGC 2208	0984 80	(1.6 x 0.9) (2.0 x 1.3)	110: ...	2	S0:		14.0		1
03453	06 18.6 + 73 08 141.35 + 23.97	12 – 07 – 001	0665	1.3 x 0.8 1.2 x 0.7	50: ...	1	S B ... Peculiar		14.6		
03454	06 18.8 + 72 07 142.42 + 23.71	12 – 07 – 002	0665	1.2 x 1.0 0.8: x 0.7:	— ...	3	Dwarf IV – V or V		16.5:		*1
03455	06 19 + 85 23 128.12 + 26.70	14 – 04 – 000	0692	1.1 x 0.10 0.9 x 0.08	23 7	3	Sc 2nd of 3		17.		*
03456	06 19.0 + 77 26 136.78 + 25.08	13 – 05 – 027	1256 76	1.3 x 1.0: 1.2 x 0.8	(160) 2	2	S0 – a		15.4		1
03457	06 19.2 + 00 24 209.19 – 06.45	00 – 17 – 001	1483	1.3 x 0.9 1.5 x 1.2	65 ...	1	S0 – a		15.0:		
03458	06 21.0 + 64 46 150.10 + 21.73	11 – 08 – 047	0677	2.9 x 0.7 2.7 x 0.7	122 6	3	Sb		15.1		*1
03459	06 21.4 + 04 43 205.62 – 03.93	01 – 17 – 000	0923	1.2: x 0.9: 1.1 x 0.7	135: ...	1	Sc:		17.		
03460	06 21.5 + 74 18 140.17 + 24.47	12 – 07 – 003 Mark 4	0665	1.7 x 0.8 1.7 x 0.8	12 5	3	S B b	S B(s)b Sb –	15.0	+ 3480 + 3630	*1
03461	06 22 + 82 55 130.85 + 26.35	14 – 04 – 000	0692	1.2: x 1.1: 1.2: x 1.1:	— ...	3	Dwarf spiral S V		16.5:		
03462	06 22.0 + 28 34 184.58 + 07.38	05 – 15 – 000	0411	1.2 x 0.9 1.0 x 0.9	(0) ...	1	Sc		16.5:		
03463	06 22.5 + 59 07 155.86 + 20.02	10 – 10 – 001 IC 2166	1264	2.6 x 1.8 2.7: x 2.0:	115: 3:	4	Sc	SAB(s)bc Sc –	13.5		
03464	06 22.6 + 74 30 139.98 + 24.59	12 – 07 – 004	0665	1.4 x 0.4 1.4 x 0.35	46 6 – 7	1	Sa?		15.5		
03465	06 22.8 + 52 29 162.42 + 17.66	09 – 11 – 014	0984 80	1.0 x 0.7 1.0 x 0.7	5: 3	2	S B a – b	(R')S B(s) ...	15.4		

N G C   2208   I C   2166

1	2	3	4	5	6	7	8	9	10	11	12
03466	06 23 + 83 34 130.15 + 26.49	14 – 04 – 001	0692	1.2 x 0.5 1.2: x 0.5	133 6	2	Sb – c		16.0:		
03467	06 23.5 + 49 38 165.21 + 16.67	08 – 12 – 000	0225	1.8 x 1.2 1.7 x 1.1	20: ...	2	Sa – b		14.7		
03468	06 24.3 + 18 51 193.47 + 03.34	03 – 17 – 000	1470	1.4 x 0.4 1.3 x 0.4	96 (6)	2	Sc – Irr		17.		
03469	06 24.5 + 56 04 159.00 + 19.20	09 – 11 – 000	0984	1.3 x 0.1 1.0 x 0.1	10 7	3	Sc		16.5:		
03470	06 25 + 83 00 130.78 + 26.45	14 – 04 – 003 IC 442	0692	1.3 x 1.3 1.4 x 1.4	-- ...	1	... Peculiar		13.7		*
03471	06 25.0 + 74 27 140.07 + 24.74	12 – 07 – 005	0665	1.7 x 1.7 1.7 x 1.7	-- 1	4	Sc	SA(s)c or SAB(s)c	14.9		*
03472	06 26.1 + 75 55 138.51 + 25.14	13 – 05 – 028	1256	2.1 x 1.1 2.1 x 1.1	60 5	3	S0		15.5		
03473	06 26.4 + 59 37 155.54 + 20.66	10 – 10 – 002	1264	1.2 x 0.15 1.3 x 0.2	43 7	2	Sc	Sc –	16.0:		*
03474	06 26.7 + 71 36 143.14 + 24.17	12 – 07 – 006	0665	2.6 x 0.25 2.1 x 0.25	158 7	3	Sc		15.6		
03475	06 27.0 + 39 31 175.05 + 13.13	07 – 14 – 002	0696 86	2.8:: x 1.5:: 1.8: x 1.2:	85: ...	4	Dwarf spiral S IV – V		15.5		•
03476	06 27.2 + 33 20 180.77 + 10.51	06 – 15 – 000	1268	1.1 x 0.3 0.9 x 0.3	[63] ...	3	Irr		15.7		
03477	06 27.5 + 50 07 165.00 + 17.45	08 – 12 – 000	0225	2.1 x 0.4 2.1 x 0.4	0 7	3	Sc	Sc +	15.7		
03478	06 28.0 + 63 43 151.43 + 22.13	11 – 08 – 054	0677 83	1.9 x 0.50 1.9 x 0.50	42 6	2	Sb		13.3		*
03479	06 29.0 + 35 14 179.19 + 11.67	06 – 15 – 000	1268	1.2 x 0.3 1.0 x 0.3	179 6	1	Sa – b		15.4		
03480	06 29.0 + 53 35 161.70 + 18.93	09 – 11 – 018	0984 80	1.2 x 1.2 1.0 x 1.0	-- 1	4	S B c	SAB(rs)c	15.3		
03481	06 29.1 + 40 14 174.55 + 13.80	07 – 14 – 000	0696 86	1.7? x 0.3 0.55 x 0.20	96 ...	1	...		15.7		
03482	06 29.3 + 19 15 193.67 + 04.58	03 – 17 – 000	1470	1.2: x 0.3: 1.2 x 0.25	30 6	1	S ...		17.		
03483	06 29.3 + 51 23 163.88 + 18.18	09 – 11 – 000	0984 80	1.2 x 0.2 1.2 x 0.2	97 7	2	Sb – c		16.0:		
03484	06 29.7 + 58 54 156.41 + 20.82	10 – 10 – 005	1264	1.8 x 0.22 1.9 x 0.20	72 7	2	Sb – c f of 2		15.6		*
03485	06 29.7 + 65 55 149.20 + 22.93	11 – 08 – 000	0677	1.0: x 0.1 0.7: x 0.1	87 ...	1	...		18.		
03486	06 29.8 + 74 21 140.26 + 25.03	12 – 07 – 008	0665 85	1.6: x 0.6 1.7 x 0.6	130 6	1	Sa – b		15.5		*
03487	06 30.1 + 40 43 174.18 + 14.17	07 – 14 – 004	0696 86	1.8: x 1.1: 1.3 x 0.7	110: ...	3	Sc n of 2	S d	15.3		*1
03488	06 30.2 + 52 11 163.15 + 18.60	09 – 11 – 019	0984 80	1.1 x 0.7 1.0 x 0.6	50: 4	2	S0 – a		15.1		*
03489	06 30.5 + 21 04 192.18 + 05.66	04 – 16 – 000	0023	2.2: x 0.3 2.1 x 0.3	123 7	2	Sb – c		17.		
03490	06 30.6 + 12 05 200.17 + 01.53	02 – 17 – 000	0445	1.0 x 1.0 1.4 x 1.2	-- ...	0	S? ...		15.0:		
03491	06 31.5 + 75 27 139.09 + 25.37	13 – 05 – 000	1256	1. x 1. 0.9:: x 0.8::	-- ...	2	Dwarf V		17.		
03492	06 31.5 + 75 28 139.08 + 25.37	13 – 05 – 000	1256	1.0 x 1.0 1.0: x 1.0:	-- 1	3	S B b	S B(r)b	16.5:		*
03493	06 31.6 + 48 53 166.47 + 17.60	08 – 12 – 030	0225	2.3: x 1.1: 1.7 x 0.8	100: ...	1	S ...		14.8		
03494	06 31.6 + 56 22 159.06 + 20.23	09 – 11 – 020	0984	0.9 x 0.9 0.9 x 0.9	-- 1	4	S B b	(R')S B(s)b	16.0:		
03495	06 31.8 + 76 52 137.56 + 25.66	13 – 05 – 000	1256	1.1 x 0.22 1.1 x 0.20	127 7	2	Sa – b		16.0:		
03496	06 32 + 85 53 127.61 + 27.02	14 – 04 – 000	0692 102	2.3:: x 1.7:: 0.8: x 0.8:	-- ...	3	Dwarf		17.		
03497	06 32.0 + 67 54 147.19 + 23.68	11 – 09 – 001 IC 445	0686	1.2 x 0.9: 0.8 x 0.7	25: ...	1	S0?		14.3		*1
03498	06 32.5 + 15 00 197.80 + 03.29	03 – 17 – 000	1470	1.2 x 0.3 1.0 x 0.2	114 6	1	S ...		17.		
03499	06 33.4 + 37 37 177.36 + 13.48	06 – 15 – 003	1268	1.2 x 0.8 0.9 x 0.8	(30) ...	1	S ...		14.8		*
03500	06 34 + 84 13 129.47 + 26.87	14 – 04 – 006	0692 102	1.9 x 0.4 1.8 x 0.4	[3] ...	2	... Peculiar, disrupted?		14.7		*

1	2	3	4	5	6	7	8	9	10	11	12
03501	06 34.6 + 49 17 166.26 + 18.21	08 – 12 – 031	0225	1.2 x 0.4 1.0: x 0.4:	101 ...	2	Dwarf		18.		
03502	06 34.8 + 65 30 149.79 + 23.33	11 – 09 – 003	0686	1.3 x 0.45 1.2 x 0.45	119 6	2	S B: b		14.8		
03503	06 35.0 + 22 41 191.20 + 07.32	04 – 16 – 000	0023 81	1.8: x 0.4 1.3: x 0.3	115 (6)	1	Sc – Irr		15.6		
03504	06 35.6 + 60 07 155.42 + 21.91	10 – 10 – 009	1264	2.9 x 2.5 2.8: x 2.5:	(135) 1 – 2	4	Sc	SAB(s)cd	13.3		1
03505	06 36.6 + 20 49 193.05 + 06.82	03 – 17 – 000	1470	1.5 x 0.3 1.4 x 0.2	55 6	2	Sc		17.		
03506	06 36.7 + 50 09 165.54 + 18.83	08 – 12 – 032	0225	0.8 x 0.7 (0.9 x 0.7)	— ...	2	E – S0		14.4		
03507	06 37.3 + 53 47 161.94 + 20.15	09 – 11 – 024	0984 80	1.1 x 0.9 1.1 x 0.9	120: 2	2	Sb	(R?)SA b	15.4		*
03508	06 37.5 + 33 40 181.40 + 12.59	06 – 15 – 000	1268	1.0: x 0.2 0.9 x 0.2	13 7	1	S ...		16.0:		
03509	06 38 + 85 42 127.83 + 27.11	14 – 04 – 000	0692 102	1.1 x 0.45 1.4 x 0.5	142 ...	1	...		15.5		
03510	06 38.0 + 40 13b 175.27 + 15.34	07 – 14 – 010	0696 86	1.4: x 0.9 1.1 x 0.8	[100] ...	3	S ... P w a, distorted		(15.0)		*1
03511	06 38.7 + 65 15 150.17 + 23.65	11 – 09 – 008	0686	1.6 x 1.2 1.5 x 1.1	135 3:	2	Sc Singular		13.0		*
03512	06 39.2 + 42 28 173.21 + 16.42	07 – 14 – 000	0696 86	1.6: x 0.7: 1.3: x 0.6:	120 ...	3	Dwarf IV – V or V		18.		
03513	06 39.3 + 41 28 174.17 + 16.06	07 – 14 – 000	0696 86	1.1 x 0.4 1.0 x 0.3	82 6	2	Sc		15.6		
03514	06 39.4 + 57 41 158.10 + 21.65	10 – 10 – 010	1264	1.2 x 0.3 1.3 x 0.3	12 6	1	Sa – b		15.4		*
03515	06 39.8 + 71 24 143.61 + 25.14	12 – 07 – 014 IC 449	0665	(1.7 x 1.3) (1.7 x 1.3)	— ...	3	E		13.7		
03516	06 40.1 + 22 55 191.52 + 08.48	04 – 16 – 000	0023	1.4 x 0.5 1.5: x 0.4	6 6	2	...	dm – m	17.		
03517	06 40.2 + 71 11 143.85 + 25.13	12 – 07 – 000	0665	1.5:: x 1.4:: 1.4: x 1.1:	— ...	1	S ...		16.0:		
03518	06 40.5 + 28 30 186.46 + 10.99	05 – 16 – 000	0920	1.1 x 0.3 1.1 x 0.4	164 6	1	S0 – a		15.5		
03519	06 40.7 + 74 17 140.49 + 25.74	12 – 07 – 015 NGC 2256	0665 85	(2.3 x 2.0) (2.3 x 2.3)	— ...	1	E?		14.0		*2
03520	06 40.8 + 62 36 153.02 + 23.20	10 – 10 – 012	1264	1.4 x 0.8 1.0 x 0.6	58: ...	3	... S B IV	S B dm:	17.		1
03521	06 41 + 84 06 129.62 + 27.04	14 – 04 – 008	0692 102	1.7: x 0.9: 1.1 x 0.7	75 ...	2	S ... P w 3528		15.5		*
03522	06 41 + 84 58 128.65 + 27.11	14 – 04 – 009	0692 102	2.3: x 1.2: 1.8: x 1.1:	[132] ...	3	... Peculiar; compact core		15.7		*C
03523	06 41.1 + 74 30 140.26 + 25.80	12 – 07 – 016 NGC 2258	0665 85	2.3: x 1.5: 2.5: x 1.7:	150: ...	2	S0		13.2		*12
03524	06 41.2 + 12 28 201.03 + 04.01	02 – 18 – 001	0428	1.5 x 0.7 1.7 x 0.7	10: 5 – 6	1	S ...		16.0:		
03525	06 41.5 + 40 28 175.30 + 16.06	07 – 14 – 012	0696 86	1.5: x 1.0: 1.4 x 0.8	[70] ...	1	... Peculiar		14.9		
03526	06 41.6 + 53 31 162.43 + 20.67	09 – 11 – 029	0984 80	1.4 x 0.3 1.3 x 0.2	54 7	2	Sb		15.7		*
03527	06 41.9 + 72 48 142.13 + 25.56	12 – 07 – 017	0665 85	1.2 x 0.6 0.9 x 0.4	45 6	1	S ... ssf of 2		15.7		*
03528	06 42 + 84 08 129.59 + 27.07	14 – 04 – 013	0692 102	1.5 x 0.8 1.7 x 0.9	40 5	2	S B a – b P w 3521		14.5		*
03529	06 42.0 + 34 32 180.98 + 13.80	06 – 15 – 004	1268	1.0 x 0.4 1.0 x 0.4	165 6	1	S ...		14.9		
03530	06 42.0 + 60 24 155.39 + 22.75	10 – 10 – 013	1264	2.5 x 1.5 2.5 x 1.3	55 ...	4	S B c	SAB(s)d	14.1		*2
03531	06 42.5 + 25 52 189.07 + 10.26	04 – 16 – 000	0023 84	1.4 x 0.9 1.5 x 1.0	150 3 – 4	3	S B b/Sc	SAB bc Sb +	15.4		*
03532	06 42.5 + 43 50 172.11 + 17.49	07 – 14 – 013	0696 86	1.8: x 0.7 1.5: x 0.6	100 6	2	S B b 3535 group, Sb –	S B(s)b	15.3		*1
03533	06 42.5 + 53 08 162.87 + 20.68	09 – 11 – 030	0984 80	1.1: x 1.1: 0.8 x 0.7	— 1	2	S ...	SA(r) ...	15.4		*
03534	06 42.6 + 22 28 192.19 + 08.80	04 – 16 – 000	0023	1.6:: x 0.6:: 1.7:: x 0.8::	... ...	1	S ...		18.		
03535	06 42.7 + 43 52 172.10 + 17.54	07 – 14 – 014	0696 86	(1.2 x 1.2) (1.0 x 0.9)	— ...	2	Compact Brightest of 3, plume		14.7		*1

1	2	3	4	5	6	7	8	9	10	11	12
03536	06 42.8 + 29 24 185.85 + 11.83	05 – 16 – 003	0920	(1.2 x 0.6) (1.2 x 0.7)	142 4 – 5	2	S0		14.4		*
03537	06 43.4 + 33 41 181.90 + 13.72	06 – 15 – 006	1268 87	1.0 x 0.9 1.0 x 0.9	--- 1 – 2	3	S B c NGC 2274 group		15.2		*
03538	06 43.8 + 47 43 168.38 + 19.08	08 – 13 – 011	0691 89	1.6 x 0.7 1.5 x 0.7	0 5 – 6	3	S B a	S B(s)a	14.9		
03539	06 43.9 + 66 18 149.19 + 24.42	11 – 09 – 013	0686	2.1 x 0.20 2.1 x 0.22	116 7	2	S B? b – c		15.4		
03540	06 44 + 82 33 131.36 + 26.99	14 – 04 – 012	0692	1.8 x 0.4 1.6 x 0.3	128 6	2	Sc		15.7		
03541	06 44.0 + 33 38 182.00 + 13.81	06 – 15 – 008 NGC 2274	1268 87	(1.8 x 1.8) (1.9 x 1.9)	--- ...	2	E Brightest of 4, p w 3542, disturbed:		13.6		*13
03542	06 44.0 + 33 40 181.97 + 13.83	06 – 15 – 007 NGC 2275	1268 87	1.4 x 1.0 (1.5 x 1.1)	20: ...	3	S ... P w 3541, distorted		14.5		*13
03543	06 44.8 + 05 46 207.40 + 01.72	01 – 18 – 002	0929	(1.4 x 1.4) (1. x 1.)	--- ...	0	...		15.0:		*
03544	06 44.8 + 33 39 182.06 + 13.97	06 – 15 – 000	1268 87	1.2 x 0.2 1.1 x 0.15	155 7	3	Sc NGC 2274 group		16.5:		*
03545	06 45.3 + 61 40 154.15 + 23.47	10 – 10 – 014	1264	1.4: x 1.4: 1.1: x 1.1:	--- ...	3	...		17.		1
03546	06 45.6 + 60 54 154.99 + 23.31	10 – 10 – 015 NGC 2273	1264	3.6 x 2.4 3.4 x 1.8	50 3	4	S B a	S B(r)a	12.5		*12 3
03547	06 45.6 + 74 28 140.35 + 26.10	12 – 07 – 018 IC 450, Mark 6	0665 85	1.0 x 0.6 1.0 x 0.6	130: ...	2	S0 – a P w 3550		14.8	+ 5400	*1
03548	06 45.9 + 77 28 137.04 + 26.53	13 – 05 – 034	1256	1.1 x 1.0 1.0 x 0.8	--- 1 – 2	2	Sa		14.8		*1
03549	06 46 + 81 02 133.07 + 26.92	14 – 04 – 016	0692	0.6 x 0.6 (1.0 x 0.9)	--- ...	1	E nf of 2		14.4		*1
03550	06 46.3 + 74 31 140.30 + 26.15	12 – 07 – 019 IC 451	0665 85	1.6 x 1.4 1.6: x 1.4:	(155) ...	2	Sb P w 3547		14.9		*12
03551	06 46.5 + 29 35 186.02 + 12.64	05 – 16 – 005	0920 88	1.1 x 0.10 1.1 x 0.10	38 7	2	... S d – dm p of 2		15.7		*1
03552	06 46.6 + 28 26 187.10 + 12.18	05 – 16 – 006	0920 88	1.1 x 0.6 1.0 x 0.5	75 5	2	S B c		15.3		
03553	06 46.8 + 20 08 194.74 + 08.65	03 – 18 – 000	0417	1.1 x 0.5 1.1 x 0.5	158 ...	1	Sc 3558 group		15.6		*
03554	06 46.8 + 43 06 173.13 + 17.95	07 – 14 – 017	0696 86	1.2 x 0.7 1.2 x 0.7	148 ...	2	S B? b		14.7		*
03555	06 46.9 + 25 41b 189.67 + 11.08	04 – 16 – 004	0023	1.2 x 1.1 1.2 x 1.1	--- 1	4	Sb/S B c SAB(rs)bc ssf a, contact; Sc –		(14.3)		*1
03556	06 47.0 + 53 05c 163.14 + 21.30	09 – 12 – 002	0971 80	1.3 x 0.4 1.0 x 0.3	75 6	2	Sb – c		(15.7)		*
03557	06 47.1 + 71 34 143.55 + 25.74	12 – 07 – 020	0665	1.3 x 0.9 1.3 x 0.9	5: 3	4	... S d – dm		16.5:		
03558	06 47.3 + 20 12 194.73 + 08.78	03 – 18 – 000	0417	(1.0 x 0.9) (1.5 x 1.2)	(130) ...	1	E Brightest in group		15.4		*
03559	06 47.4 + 34 07 181.84 + 14.65	06 – 15 – 000	1268 87	1.1 x 0.3 0.9 x 0.2	65	3	Sc		16.0:		
03560	06 47.5 + 33 32 182.42 + 14.43	06 – 15 – 010 NGC 2289	1268 87	(1.1 x 0.8) (1.1 x 0.8)	125: ...	3	S0 NGC 2289 – 90 group		14.6		*13
03561	06 47.5 + 49 56 166.37 + 20.39	08 – 13 – 017	0691 89	0.8 x 0.40 0.7 x 0.30	68 ...	1	S? ...		14.5		
03562	06 47.6 + 33 30 182.44 + 14.45	06 – 15 – 012 NGC 2290	1268 87	1.3 x 0.7 1.3 x 0.6	50 5:	2	Sa NGC 2289 – 90 group		14.6		*12 3
03563	06 47.6 + 63 09 152.65 + 24.09	11 – 09 – 016	0686	1.2: x 0.7 1.0: x 0.6	[15] ...	1	... Disturbed; multiple?		15.6		*1C
03564	06 47.7 + 16 25 198.20 + 07.19	03 – 18 – 000	0417	1.1 x 1.1 1.0: x 1.0:	--- 1:	2	... S dm – m		17.		
03565	06 47.8 + 09 43 204.22 + 04.19	02 – 18 – 000	0428	1.0 x 0.2 1.1 x 0.15	164 6	2	Sc		17.		
03566	06 48.0 + 41 49 174.47 + 17.70	07 – 14 – 000	0696 86	1. x 1. 0.7:: x 0.5::	--- ...	3	Dwarf spiral S IV – V		18.		
03567	06 48.0 + 48 34 167.78 + 20.03	08 – 13 – 018	0691 89	1.0 x 0.8 1.1 x 0.9	(100) 2:	2	S0		14.2		*
03568	06 48.1 + 50 28 165.86 + 20.65	08 – 13 – 019	0691 89	1.5 x 0.2 1.0 x 0.2	16 7	2	Sb – c		15.7		
03569	06 48.2 + 57 14 158.91 + 22.66	10 – 10 – 016	1264	1.6 x 1.5 1.2 x 1.1	--- 1:	3	... S dm S IV – V		15.7		
03570	06 48.3 + 12 58 201.37 + 05.77	02 – 18 – 002 IC 454	0428	1.9 x 1.0 2.1 x 1.0	140 5	2	S B a – b		14.5:		1

N G C 2273 – 2275, 2289, 2290   I C 450, 451, 454

1	2	3	4	5	6	7	8	9	10	11	12
03571	06 48.5 + 29 08 186.63 + 12.85	05 - 16 - 008	0920 88	1.4 x 0.8 1.2 x 0.8	105: 4	3	Sc		15.1		
03572	06 48.5 + 49 47 166.58 + 20.50	08 - 13 - 000	0691 89	1.1 x 0.7 1.1 x 0.7	150: 3:	3	Sc		16.5:		
03573	06 48.7 + 27 33 188.12 + 12.23	05 - 16 - 009	0920 88	2.0 x 0.3 1.5: x 0.20	140 7	2	Sb npp of 2		15.2		*
03574	06 48.9 + 57 15 157.54 + 18.19	10 - 10 - 017	1264	4.5: x 4.5: 4.0:: x 4.0::	-- 1	4	Sc	SA(s)cd	13.9		*1
03575	06 49.1 + 70 49 144.40 + 25.77	12 - 07 - 021	0665	1.8 x 0.15 1.7 x 0.15	166 7	2	Sc		16.5:		
03576	06 49.3 + 50 05 166.32 + 20.71	08 - 13 - 023	0691 89	1.7 x 0.8 1.6 x 0.8	128 ...	3	S B b	S B(s)b	14.6		
03577	06 49.3 + 65 16 150.43 + 24.74	11 - 09 - 018	0686	1.6 x 1.2: 1.1 x 0.6	(90) ...	2	S B b	S B(r)b or SAB(r)b; Sb -	15.3		
03578	06 49.5 + 15 19 199.39 + 07.09	03 - 18 - 001	0417	1.7 x 0.8 1.8 x 1.0	20 4 - 5	3	S B a - b nff of 2	S B(s) ...	14.1		*1
03579	06 49.5 + 47 15 169.19 + 19.83	08 - 13 - 024	0691 89	1.2 x 0.2 1.2 x 0.2	60 7	2	Sa - b		15.7		
03580	06 50.0 + 69 39 145.69 + 25.65	12 - 07 - 022	0665	4.4 x 2.0 3.6 x 1.6	3 6	3	Sa	SA a	12.9		
03581	06 50.0 + 80 04 134.16 + 27.00	13 - 05 - 035	1256	1.8 x 1.4 1.7 x 1.3	100 2	4	S B c	SAB(rs:)c Sc -	14.4		
03582	06 50.2 + 12 15 202.22 + 05.87	02 - 18 - 000	0428	1.1 x 1.0 1.0 x 0.7	-- ...	2	S B b Disturbed:	S B(s)b	16.5:		*
03583	06 50.2 + 16 59 197.96 + 07.98	03 - 18 - 002	0417	4. x 4. 4. x 4.	-- ...	0	Resolved dwarf system? (V)		14.3		*1
03584	06 50.2 + 27 09 188.63 + 12.36	05 - 17 - 000	1337 88	1.7 x 0.25 1.8 x 0.25	14 7	2	Sb - c		15.4		
03585	06 50.5 + 27 23 188.45 + 12.52	05 - 17 - 003	1337 88	1.1 x 0.9 0.9 x 0.6	(0) ...	2	Sc		15.5		*
03586	06 50.6 + 14 49 199.96 + 07.10	02 - 18 - 000	0428	1.2: x 0.5 1.2 x 0.6	40 ...	2	... 3rd of 3	S d - dm	17.		*
03587	06 50.9 + 19 22 195.87 + 09.18	03 - 18 - 003	0417	3.0: x 1.1: 2.6: x 0.9:	107 6	1	S ...		14.3		
03588	06 50.9 + 45 47 170.74 + 19.57	08 - 13 - 026	0691 89	1.7 x 0.5 1.6 x 0.5	30 6	3	S0		14.9		
03589	06 51.3 + 26 45 189.11 + 12.42	04 - 17 - 000	1590 88	1.1: x 0.8: 0.9: x 0.7:	-- ...	1	S ... np of 2, contact		16.5:		*
03590	06 51.4 + 30 08 185.95 + 13.83	05 - 17 - 000	1337	1.0 x 0.4 1.0 x 0.5	45 5 - 6	2	S B a - b		15.3		
03591	06 51.4 + 34 56 181.39 + 15.73	06 - 15 - 000	1268	1.1 x 0.3 0.8: x 0.3	157 6	1	S ...		16.0:		
03592	06 51.7 + 40 24 176.12 + 17.83	07 - 14 - 020	0696 86	1.0 x 1.0 0.8 x 0.4	-- 1	3	S B a	(R)S B(s)a	15.1		
03593	06 51.7 + 41 01 175.52 + 18.05	07 - 14 - 021	0696 86	1.8 x 1.0 1.2 x 0.7	40 4	4	Sc ssf of 2	SAB: c	14.7		*
03594	06 51.8 + 23 34 192.11 + 11.18	04 - 17 - 002	1590	1.3 x 0.6 1.1 x 0.6	137 5	3	Sb/S B b	SAB(s)b Sb +	15.2		
03595	06 51.9 + 55 28 160.90 + 22.67	09 - 12 - 010	0971 80	1.5 x 1.4 1.1 x 1.3	-- 1	4	S B b	(R:)S B(r)b	15.2		
03596	06 52.0 + 39 50 176.70 + 17.68	07 - 15 - 001	0988 86	1.1 x 1.0 0.8 x 0.8	-- (1)	1	S0?		13.5		
03597	06 52.0 + 48 00 168.58 + 20.47	08 - 13 - 028	0691 89	2.1 x 0.3 1.8 x 0.3	102 7	2	S ... npp of 2, bridge:		15.2		*
03598	06 52.0 + 60 43 155.90 + 24.02	10 - 10 - 019	1264	2.2 x 1.2 1.7: x 1.2	[25] ...	4	Irr Ir IV	I B m Ir I	15.5		1
03599	06 52.1 + 24 17 191.48 + 11.55	04 - 17 - 003	1590	(1.1 x 0.9) (1.0 x 1.0)	-- ...	2	... Compact, ssp of 2		14.5		*
03600	06 52.2 + 39 08 177.40 + 17.46	07 - 15 - 000	0988 86	1.2: x 0.5: 1.2:: x 0.3::	35 ...	3	Dwarf V		18.		
03601	06 52.3 + 40 04 176.49 + 17.82	07 - 15 - 000	0988 86	0.6 x 0.45 0.5 x 0.4	(5) ...	1	...		14.5		
03602	06 52.6 + 16 00 199.11 + 08.06	03 - 18 - 004	0417	1.7: x 1.7: 1.6: x 1.6:	-- 1	2	...	SA dm:	14.8		
03603	06 52.6 + 45 34 171.06 + 19.78	08 - 13 - 031	0691 89	(1.5 x 1.5) (1.6 x 1.5)	-- ...	3	E Disturbed?		13.9		*
03604	06 53 + 81 01 133.11 + 27.19	14 - 04 - 018	0692	0.8 x 0.8 0.8 x 0.8	-- 1	2	S ...		14.2		*
03605	06 53.0 + 14 00 200.96 + 07.26	02 - 18 - 000	0428	1.2 x 0.1 1.1 x 0.15	142 7	3	Sc		16.5:		

1	2	3	4	5	6	7	8	9	10	11	12
03606	06 53.1 + 63 45 152.16 + 24.82	11 – 09 – 022	0686	1.6 x 0.9 1.6 x 0.9	12 4	3	... S IV	S dm	15.0		
03607	06 53.7 + 06 20 207.91 + 03.95	01 – 18 – 000	0929	1.1 x 1.0 1.0: x 1.0:	— 1	3	Sc/SBc	SAB(s)c	16.5:		
03608	06 53.9 + 46 29 170.22 + 20.29	08 – 13 – 033	0691 89	1.6: x 1.1: 1.6: x 1.0:	20 ...	1	S ... Peculiar		13.7		*
03609	06 53.9 + 69 21 146.08 + 25.93	12 – 07 – 000	0665	1.1 x 0.7 1.1 x 0.7	130 4	3	Sb		15.6		
03610	06 54.0 + 58 03 158.27 + 23.62	10 – 10 – 000	1264	1.1: x 0.6 1.0: x 0.7:	(90) ...	2	S B b:	S B(s) ...	16.0:		
03611	06 54.2 + 20 30 195.17 + 10.37	03 – 18 – 000	0417	1.0 x 0.5 1.1 x 0.5	65 5 – 6	2	SO – a		15.1		
03612	06 54.2 + 39 09 177.53 + 17.83	07 – 15 – 003	0988 86	0.8 x 0.6 0.8 x 0.6	25 ...	3	S B b	S B(s)b	14.2		
03613	06 54.4 + 13 37 201.45 + 07.39	02 – 18 – 000	0428	1.6: x 0.5 0.6: x 0.3:	85 ...	3	... P w 3617	S B dm	16.5:		*
03614	06 54.6 + 45 08 171.62 + 19.97	08 – 13 – 036	0691 89	1.4 x 1.1 1.4 x 1.1	60: 2:	3	S B a – b		14.7		
03615	06 54.7 + 35 48 180.83 + 16.68	06 – 16 – 003	0889	1.5 x 0.3 1.4 x 0.25	31 7	2	Sa – b		15.4		*
03616	06 54.8 + 22 57 192.98 + 11.54	04 – 17 – 004	1590	1.4: x 1.0: 1.1: x 0.8:	(170) ...	3	...	S d – dm	17.		*
03617	06 54.9 + 13 40 201.46 + 07.52	02 – 18 – 000	0428	1.0 x 1.0 0.9: x 0.8:	— 1	1	S B a – b P w 3613		16.0:		*
03618	06 55.0 + 45 17 171.49 + 20.08	08 – 13 – 037 NGC 2308	0691 89	2.1 x 1.2 1.7 x 1.3	170: 3	3	Sa – b		14.4		3
03619	06 55.2 + 34 17 182.33 + 16.20	06 – 16 – 005	0889	1.0 x 1.0 0.8 x 0.6	— 1	3	S B c		15.3		
03620	06 55.8 + 76 40 137.99 + 26.99	13 – 05 – 000	1256	1.1 x 0.12 1.1: x 0.15	40 7	1	Sa – b		16.5:		*
03621	06 56.3 + 14 22 200.98 + 08.14	02 – 18 – 005	0428	1.6:: x 1.5:: 1.5:: x 1.3::	— ...	3	Dwarf (irregular) Ir: IV – V or V		17.		
03622	06 56.3 + 33 05 183.58 + 15.96	06 – 16 – 006	0889	1.5 x 0.3 1.4 x 0.3	67 6	1	S ...		15.7		
03623	06 56.5 + 35 31 181.25 + 16.92	06 – 16 – 007 IC 2175	0889	2.0 x 0.9 1.0 x 0.6	66 5	3	S B: c	Sc –	15.0		
03624	06 56.7 + 27 38 188.79 + 13.88	05 – 17 – 000	1337 88	1.2 x 0.2 1.3 x 0.2	146 7	2	Sb		16.0:		
03625	06 57.0 + 51 25 165.34 + 22.28	09 – 12 – 000	0971	[1.1 x 0.2] ...	[157] ...	2	Double system Distorted, bridge		16.5:		*
03626	06 57.0 + 71 02 144.27 + 26.44	12 – 07 – 023	0665	1.3 x 0.4 1.3 x 0.4	155 6	1	S B? b – c		15.3		
03627	06 57.1 + 51 21 165.41 + 22.27	09 – 12 – 018	0971	1.3 x 0.9 1.2 x 0.8	80: 3	3	Sc	Sc +	14.7		
03628	06 57.3 + 38 42 178.20 + 18.23	06 – 15 – 000	1268	1.0 x 0.2 0.5 x 0.2	153 ...	2	Sc – Irr		18.		
03629	06 58.0 + 42 43 174.26 + 19.74	07 – 15 – 000	0988 86	1.0 x 0.2 1.0 x 0.2	134 6	2	Sc		16.5:		
03630	06 58.4 + 01 59 212.32 + 03.00	00 – 18 – 001	1346	1.6 x 0.8 1.7 x 0.9	0 5	1	S ...		15.5:		*1
03631	06 58.5 + 29 58 186.75 + 15.17	05 – 17 – 006	1337	1.4 x 0.7 1.3 x 0.7	[107] ...	1	S B ... Singular, sp of 2		15.7		*
03632	06 58.6 + 77 58 136.55 + 27.25	13 – 05 – 000	1256	1.1: x 0.8: 1.1: x 0.8:	(170) ...	3	Sc		16.5:		*
03633	06 58.7 + 50 40 166.20 + 22.32	08 – 13 – 045 NGC 2315	0691 89	1.4 x 0.3 1.7 x 0.4	116 7	2	SO – a		14.5		*3
03634	06 58.8 + 14 12 201.41 + 08.61	02 – 18 – 006	0428	1.1 x 0.8 (1.4 x 0.9)	(135) 3 – 4:	3	S B a	S B(r)a	15.5:		
03635	06 58.8 + 17 15 198.63 + 09.95	03 – 18 – 000	0417	1.0 x 0.6 1.1 x 0.7:	[115] ...	1	S – Irr Peculiar?		15.4		
03636	06 59.1 + 75 32 139.28 + 27.09	13 – 05 – 037	1256	1.2 x 0.35 1.0 x 0.25	0 6	1	Sa – b		14.9		
03637	06 59.2 + 05 00 209.72 + 04.56	01 – 18 – 000	0929	1.2: x 0.4 1.2: x 0.4	125 (6)	2	Sc – Irr		16.5:		
03638	06 59.2 + 49 30 167.43 + 22.06	08 – 13 – 047	0691 89	1.1 x 0.7 1.1 x 0.7	48 4	2	S B: a – b		14.4		
03639	06 59.5 + 20 02 196.15 + 11.29	03 – 18 – 000	0417 90	1.0 x 0.25 1.0 x 0.25	93 6	1	S ...		15.4		
03640	06 59.5 + 37 12 179.84 + 18.10	06 – 16 – 009	0889	1.1 x 0.3 1.1 x 0.2	102 6 – 7	1	SO – a		15.1		*

1	2	3	4	5	6	7	8	9	10	11	12
03641	06 59.6 + 11 18 204.12 + 07.49	02 - 18 - 007	0428	1.1 x 1.1 1.1 x 1.0	--- 1	2	Sb		16.0:		1
03642	06 59.6 + 64 06 151.93 + 25.59	11 - 09 - 030	0686	1.4 x 1.0 1.5 x 1.3	(30) ...	3	S0	SA 0	13.5		*
03643	06 59.8 + 53 52 162.91 + 23.35	09 - 12 - 000	0971 96	1.5: x 0.3 1.3: x 0.3	35 ...	1	...		16.0:		
03644	06 59.9 + 71 08 144.19 + 26.69	12 - 07 - 024	0665	2.2 x 1.3 2.2: x 1.3	140 4	4	S B c	SAB(s)d	14.9		
03645	07 00.1 + 50 45 166.18 + 22.56	08 - 13 - 048	0691 89	1.6: x 0.2 1.8 x 0.2	33 7	1	S ...		15.5		
03646	07 00.2 + 54 18 162.47 + 23.52	09 - 12 - 026	0971 96	(1.2 x 1.0) (1.0 x 0.8)	(55) ...	3	E		15.0		
03647	07 00.6 + 56 36 160.05 + 24.14	09 - 12 - 027	0971 96	1.4: x 1.2: 1.3: x 1.2:	(40) ...	4	Irr	I B m Ir I	16.0:		*
03648	07 00.7 + 64 11 151.87 + 25.72	11 - 09 - 032	0686	1.4 x 0.25 1.2 x 0.25	104 6	2	Sb - c		15.3		*
03649	07 00.8 + 29 20 187.55 + 15.38	05 - 17 - 008	1337	1.0 x 0.5 1.1 x 0.6	63 5	2	Sa p of 2		14.6		*
03650	07 00.8 + 37 46 179.37 + 18.54	06 - 16 - 015	0889	1.1 x 0.5 0.9 x 0.5	17 ...	1	S ...		15.5		
03651	07 00.8 + 45 25 171.71 + 21.09	08 - 13 - 049	0691 89	1.0 x 0.2 1.0 x 0.2	23 7	2	Sa		15.6		
03652	07 00.9 + 22 27 194.05 + 12.61	04 - 17 - 005	1590	2.1 x 0.7 1.9 x 0.7	115 6	3	Sc		14.7		1
03653	07 01 + 84 28 129.24 + 27.55	14 - 04 - 022 NGC 2268	0692 102	3.7 x 2.5 3.7 x 2.5	63 3	3	Sc	Sc -	12.1	+ 2337 + 2536	*123 4568
03654	07 01 + 85 48 127.73 + 27.54	14 - 04 - 021	0692 102	1.3 x 0.7 1.3 x 0.7	17: ...	2	... Compact		15.2		
03655	07 01.2 + 46 13 170.91 + 21.40	08 - 13 - 050	0691 89	0.55 x 0.4 0.6 x 0.5	[9] ...	2	... Peculiar		14.5		
03656	07 01.4 + 18 37 197.65 + 11.09	03 - 18 - 008	0417 90	1.1 x 0.4 1.1 x 0.5	174 6	2	S0 - a 1st of 3		15.6		*
03657	07 01.7 + 71 16 144.06 + 26.85	12 - 07 - 025	0665	1.2 x 0.12 1.1 x 0.12	148 7	2	Sc		16.5:		
03658	07 01.8 + 17 41 198.55 + 10.78	03 - 18 - 000	0417	2.0:: x 0.8:: 2. x 0.7::	[65] ...	3	Dwarf (irregular) Ir: V		17.		
03659	07 01.8 + 50 40 166.35 + 22.79	08 - 13 - 051 NGC 2320	0691 89	(1.4 x 0.8) (1.7 x 1.1)	140 ...	3	E		13.9		*3
03660	07 01.8 + 63 57 152.14 + 25.80	11 - 09 - 033	0686	1.8 x 1.0 1.8 x 1.0	110 5	3	Sa/S B a or Sa/S B b		13.6		
03661	07 02 + 85 51 127.67 + 27.56	14 - 04 - 024	0692 102	1.0 x 0.2 1.8: x 0.2	14 7:	1	S ...		15.6		
03662	07 02.1 + 50 35 166.45 + 22.81	08 - 13 - 054 NGC 2322	0691 89	1.3 x 0.4 1.3 x 0.5	136 6	2	S B a:		14.6		*3
03663	07 02.1 + 50 50 166.19 + 22.88	08 - 13 - 053 NGC 2321	0691 89	(1.5 x 1.1) 1.3 x 1.0	(135) 3	2	S B a		14.8		
03664	07 02.2 + 33 54 183.27 + 17.40	06 - 16 - 017	0889 92	1.0 x 0.7 1.0 x 0.7	80 3	3	Sc		15.4		
03665	07 02.2 + 71 38 143.66 + 26.93	12 - 07 - 026	0665	1.1 x 0.3 1.2 x 0.3	105 7	2	Sa		15.3		
03666	07 02.2 + 75 27 139.39 + 27.28	13 - 06 - 002 IC 2174	1295	1.1 x 1.1 0.8 x 0.8	--- 1	3	Sa P w 3679		15.0		*2
03667	07 02.8 + 39 15 178.05 + 19.42	07 - 15 - 000	0988	1.4 x 0.5 1.0: x 0.3	125 ...	1	S ...		16. ?		
03668	07 03 + 81 04 133.07 + 27.58	14 - 04 - 023	0692	1.3 x 0.9 1.3 x 0.9	(160) 3	3	Sc		16.0:		
03669	07 03 + 85 33 128.01 + 27.58	14 - 04 - 000	0692 102	1.3:: x 1.2:: 1.4:: x 1.3::	--- ...	2	...		17.		
03670	07 03 + 85 41 127.86 + 27.58	14 - 04 - 026	0692 102	1.3 x 0.7 1.2 x 0.6	40 ...	1	...		15.4		
03671	07 03.0 + 78 30 135.96 + 27.50	13 - 06 - 000	1295	1.0: x 1.0: 0.9: x 0.8:	--- ...	3	Dwarf (irregular) Ir: V		17.		
03672	07 03.3 + 30 26 186.72 + 16.31	05 - 17 - 000	1337	1.3: x 0.8: 1.1: x 0.6::	[150] ...	4	Irr	I m	17.		
03673	07 03.3 + 44 55 172.36 + 21.35	07 - 15 - 005	0988 89	1.1 x 0.9 0.8 x 0.5	(35) ...	3	Sc/S B c	SAB cd	15.7		*
03674	07 03.6 + 25 32 191.42 + 14.44	04 - 17 - 000	1590	1.1 x 0.5 1.1 x 0.5	110 5 - 6	3	Sc		15.5		
03675	07 03.6 + 74 59 139.92 + 27.33	12 - 07 - 027	0665	1.6 x 0.2 1.6 x 0.2	18 7	2	Sc		15.5		

N G C   2268, 2320 · - 2322   I C   2174

1	2	3	4	5	6	7	8	9	10	11	12
03676	07 03.7 + 23 58 192.90 + 13.82	04 – 17 – 007	1590	1.1 x 0.7 1.0 x 0.7	120: 3 – 4	2	Sb npp of 2		15.4		*
03677	07 03.7 + 75 25 139.44 + 27.37	13 – 06 – 003 NGC 2314	1295	(1.8 x 1.6) (2.0 x 1.8)	(25) ...	2	E – S0 P w 3666		13.1	+ 3860 + 4024	*12 468$\overline{9}$
03678	07 03.8 + 19 27 197.12 + 11.96	03 – 18 – 000	0417 90	1.0 x 0.8 0.9 x 0.7	(0) ...	1	S ...		15.4		
03679	07 03.8 + 44 52 172.44 + 21.42	07 – 15 – 007	0988 89	1.7: x 1.1: 0.8 x 0.4	145: ...	2	S ...		15.5		*
03680	07 04.0 + 53 00 164.00 + 23.73	09 – 12 – 000	0971 96	1.4: x 0.3 1.5: x 0.3:	51 ...	2	Dwarf IV or IV – V		17.		
03681	07 04.3 + 50 45 166.38 + 23.19	08 – 13 – 062 NGC 2326	0691 89	1.8 x 1.8 2.3 x 2.1	— 1	4	S B b	S B(rs)b Sb –	14.3		*3
03682	07 04.4 + 14 15 201.96 + 09.85	02 – 18 – 000	0428	1.0 x 0.1 0.9 x 0.1	71 7	1	...	d – m	17.		
03683	07 04.5 + 46 12 171.12 + 21.94	08 – 13 – 065	0691 89	(2.0 x 1.3) (2.2 x 1.6)	(50) ...	3	S0		14.1		
03684	07 04.5 + 51 20 165.77 + 23.38	09 – 12 – 036	0971 89	1.2: x 0.9 1.2: x 0.9	(170) ...	4	S B b	(R')S B(s)b Sb –	14.8		
03685	07 04.5 + 61 41 154.68 + 25.70	10 – 11 – 002	1291 100	4.0 x 4.0 2.0 x 1.9	— 1	4	S B b	S B(rs)b Sb +	13.1		
03686	07 04.7 + 46 44 170.58 + 22.13	08 – 13 – 068	0691 89	1.0 x 0.5 0.9 x 0.5	155 5	3	Sc	SA cd	16.0:		
03687	07 04.7 + 50 42 166.45 + 23.24	08 – 13 – 067	0691 89	1.0 x 0.5 1.1 x 0.5	15 ...	2	... Peculiar		15.5		*
03688	07 05.0 + 13 10 203.02 + 09.50	02 – 19 – 000	1354	1.1 x 0.4 1.1 x 0.3	43 6	2	Sb Brightest in group		15.6		*
03689	07 05.0 + 35 15 182.17 + 18.43	06 – 16 – 020 NGC 2333	0889	1.3: x 0.8: 1.1 x 0.8	35: 2 – 3	2	Sa		14.1		3
03690	07 05.0 + 53 33 163.45 + 24.01	09 – 12 – 040	0971 96	1.1: x 1.1: 1.5:: x 1.5::	— ...	4	Dwarf irregular Ir V		17.		*1
03691	07 05.1 + 15 15 201.13 + 10.44	03 – 19 – 001	0641	2.4 x 1.1 2.2 x 1.1	65 5	3	Sc	SA cd	13.4		*1
03692	07 05.3 + 29 58 187.34 + 16.53	05 – 17 – 000	1337	1.5: x 0.6: 1.6 x 0.6	13 ...	1	S0?		15.7		
03693	07 05.4 + 18 51 197.84 + 12.05	03 – 19 – 002 NGC 2339	0641	2.6 x 1.9 2.7: x 2.1:	175: 2 – 3	4	Sb/S B c	SAB(rs)bc Sb +	12.3	+ 2361 + 2259	*123 467$\overline{8}$
03694	07 05.4 + 32 47 184.62 + 17.61	05 – 17 – 013	1337 92	1.6 x 0.8 1.5 x 0.7	95 5	3	Sc	Sc +	15.3		*
03695	07 05.4 + 48 42 168.57 + 22.81	08 – 13 – 073 NGC 2329	0691 89	1.3 x 1.1 (1.7 x 1.4)	(175) ...	2	E – S0 P w 3696		13.7		*3
03696	07 05.6 + 48 43 168.57 + 22.84	08 – 13 – 077	0691 89	(1.0 x 0.7) (1.1 x 0.7)	77: ...	2	...		13.8		*
03697	07 05.6 + 71 55 143.37 + 27.22	12 – 07 – 028	0665	3.3 x 0.2 3.3 x 0.2	[76] ...	3	... Integral, disturbed		13.1		*1
03698	07 05.7 + 44 28 172.96 + 21.62	07 – 15 – 000	0988 94	1.1: x 0.7:: 1.1: x 0.7:	[0] ...	3	Dwarf V		17.		
03699	07 05.7 + 50 15 166.97 + 23.28	08 – 13 – 079 NGC 2332	0691 89	(1.5 x 1.0) (1.7 x 1.1)	(60) ...	2	S0:		14.0		*13
03700	07 05.8 + 55 03 161.89 + 24.48	09 – 12 – 042	0971 96	1.0 x 0.2 1.0 x 0.2	156 7	1	S ...		15.4		*
03701	07 05.8 + 72 15 143.00 + 27.27	12 – 07 – 029	0665	2.0 x 2.0 2.0 x 2.0	— 1	4	Sc	SA(rs:)cd	15.2		*
03702	07 06.0 + 28 47 188.54 + 16.22	05 – 17 – 000	1337 88	1.1 x 1.1 1.1 x 1.0	— 1	3	Sc P w 3710	Sc +	15.7		*
03703	07 06.0 + 36 22 181.14 + 19.02	06 – 16 – 022	0889 91	1.2 x 1.2 1.2 x 1.2	— 1	4	Sc	SA(s)c	15.4		*
03704	07 06.0 + 61 52 154.52 + 25.90	10 – 11 – 004	1291 100	1.4 x 0.3 1.4 x 0.3	50 7	2	Sa		14.5		
03705	07 06.0 + 73 32 141.56 + 27.39	12 – 07 – 031	0665	1.5 x 1.4 1.3 x 1.3	— 1	1	S ...		14.4		
03706	07 06.1 + 47 59 169.35 + 22.72	08 – 13 – 83+84	0691 89	[1.1 x 0.2] ...	... ...	2	Double system Contact, disrupted		14.7		*1
03707	07 06.3 + 07 01 208.72 + 07.05	01 – 19 – 000	0999	1.0 x 0.8 1.0 x 0.8	45: 2 – 3	1	Sa – b		15.6		
03708	07 06.3 + 20 40 196.25 + 13.01	03 – 19 – 003 NGC 2341	0641	0.9 x 0.9 0.8 x 0.8	— ...	1	... Peculiar, p w 3709		13.7		*13
03709	07 06.4 + 20 43 196.21 + 13.05	03 – 19 – 004 NGC 2342	0641	1.4 x 1.3: 1.3 x 1.3	— ...	3	S ... Peculiar P w 3708, distorted:		12.6		*13
03710	07 06.4 + 28 45 188.61 + 16.28	05 – 17 – 000	1337 88	(1.3 x 1.0) (1.4 x 1.0)	... ...	3	... P w 3702		15.6		*

N G C   2314,   2326,   2329,   2332,   2333,   2339
2341,   2342

1	2	3	4	5	6	7	8	9	10	11	12
03711	07 06.6 + 44 32 172.95 + 21.80	07 – 15 – 010 NGC 2337	0988 94	2.7:: x 2.2:: 2.5: x 1.8:	[120] ...	3	Irr	I B m	13.1		3
03712	07 06.6 + 63 17 152.98 + 26.21	11 – 09 – 035	0686	1.1: x 0.9: 0.7 x 0.7	(80) ...	2	S ...		16.0:		
03713	07 06.7 + 50 11 167.08 + 23.41	08 – 13 – 085 IC 458	0691 89	(0.9 x 0.4) (1.0 x 0.5)	170 ...	2	E – S0		14.4		*
03714	07 06.8 + 71 50 143.48 + 27.31	12 – 07 – 032	0665	2.0: x 1.7: 2.0: x 1.7:	35: ...	1	S? ... Singular, disturbed:		12.7		*
03715	07 07 + 86 19 127.15 + 27.63	14 – 04 – 027	0692 102	2.3: x 1.1: 2.2: x 1.1:	90: ...	2	...		15.5		*
03716	07 07.0 + 39 49 177.77 + 20.38	07 – 15 – 011	0988 94	1.1 x 0.6 0.8 x 0.5	55: 4 – 5	4	Sc P w 3718		16.0:		*1
03717	07 07.0 + 73 55 141.14 + 27.48	12 – 07 – 033	0665	2.3 x 1.0 2.3 x 1.0	103 6	2	Sb – c		13.9		1
03718	07 07.1 + 39 48 177.79 + 20.39	07 – 15 – 012	0988 94	1.0 x 0.4 0.8: x 0.4:	118 ...	2	S0		15.2		*1
03719	07 07.2 + 48 35 168.78 + 23.06	08 – 13 – 094	0691 89	1.2 x 0.2 1.3 x 0.25	160 7	2	Sa – b		15.1		*
03720	07 07.2 + 50 15 167.04 + 23.51	08 – 13 – 096 NGC 2340	0691 89	(1.8 x 1.2) (2.3 x 1.4)	80: ...	3	E		13.9		*3
03721	07 07.2 + 59 48 156.80 + 25.67	10 – 11 – 005	1291	1.0 x 0.10 0.9 x 0.12	20 7	2	Sc		16.5:		
03722	07 07.4 + 54 28 162.57 + 24.57	09 – 12 – 000	0971 96	1.0: x 0.3: 0.7: x 0.2	158 ...	2	...	S B: dm:	17.		
03723	07 07.5 + 34 30 183.10 + 18.64	06 – 16 – 024	0889	1.6 x 0.8 1.6 x 0.8	3 5	3	S B 0		14.4		
03724	07 07.7 + 48 19 169.09 + 23.07	08 – 13 – 100	0691 89	1.7: x 0.9 1.7: x 0.9	55: ...	3	S B b		14.5		
03725	07 07.9 + 49 56 167.40 + 23.53	08 – 13 – 101	0691 89	(1.4 x 0.8) (1.7 x 1.4)	(140) ...	2	E – S0		14.2		
03726	07 08.0 + 26 00 191.38 + 15.53	04 – 17 – 008	1590	1.1 x 0.4 1.1 x 0.5	110 ...	1	... Peculiar?		14.8		
03727	07 08.2 + 26 28 190.96 + 15.76	04 – 17 – 000	1590	1.2 x 0.9 1.0 x 0.9	(0) 1 – 2	1	Sa – b		15.0		
03728	07 08.2 + 30 15 187.32 + 17.22	05 – 17 – 000	1337 92	1.1 x 0.6 1.1 x 0.6	140 4 – 5	2	S B b: sp of 2		15.3		*
03729	07 08.2 + 39 39 178.02 + 20.54	07 – 15 – 013	0988 94	1.0 x 0.3 1.0 x 0.3	78 6	3	Sc		15.5		
03730	07 08.2 + 73 33 141.56 + 27.54	12 – 07 – 035	0665	2.7: x 1.0 2.7 x 1.0	[165] ...	3	... peculiar, jets + plume	Strongly	13.7		*1V A
03731	07 08.5 + 29 16 188.29 + 16.91	05 – 17 – 016	1337	1.4 x 0.5 1.3 x 0.4	158 (6)	1	S – Irr		14.8		
03732	07 08.6 + 41 51 175.81 + 21.32	07 – 15 – 000	0988 94	1.3 x 0.2 1.1 x 0.2	175 7	1	Sa – b		15.2		
03733	07 08.6 + 51 55 165.33 + 24.14	09 – 12 – 000	0971 96	1.1 x 0.2 1.0 x 0.2	157 7	2	Sc		16.0:		
03734	07 08.7 + 47 15 170.25 + 22.93	08 – 13 – 103 NGC 2344	0691 89	2.2: x 2.0: 2.1: x 2.0:	— 1	4	Sb	SA bc Sb +	13.2	+ 914 + 949	*2
03735	07 08.8 + 33 10 184.52 + 18.41	06 – 16 – 025	0889	1.0 x 0.7 1.0: x 0.7	55: 3:	3	Sc	Sc +	16.0:		
03736	07 09.3 + 23 28 193.91 + 14.79	04 – 17 – 000	1590	1.3 x 0.3 1.3 x 0.3	3 6 – 7	1	S ...		15.7		
03737	07 09.4 + 23 49 193.59 + 14.95	04 – 17 – 000	1590	[1.1 x 0.8] ...	... ...	2	Triple system Contact, disrupted		15.3		
03738	07 09.5 + 07 20 208.80 + 07.90	01 – 19 – 000	0999	1.1 x 0.6 0.9 x 0.5	30 4 – 5	1	S ...		15.2		
03739	07 09.8 + 75 51 138.97 + 27.77	13 – 06 – 004	1295	1.2:: x 1.2:: 1.2:: x 1.1::	— ...	4	Dwarf V		16.5:		
03740	07 10 + 85 51 127.67 + 27.70	14 – 04 – 028 NGC 2276	0692 102	2.8 x 2.5 3.1 x 2.5	(20) 1 – 2	4	Sc P w 3798, disturbed, Sc +	SAB(rs)c	12.3	+ 2391 + 2595	*123A 4568C
03741	07 10.0 + 50 20 167.07 + 23.96	08 – 13 – 108	0691 89	1.1 x 0.5 1.0 x 0.5	60 5 – 6	3	Sc		15.5		
03742	07 10.1 + 35 12 182.61 + 19.39	06 – 16 – 026	0889 91	1.5 x 0.7 1.4 x 0.7	73 ...	3	...	S dm	14.3		1
03743	07 10.2 + 35 44 182.09 + 19.59	06 – 16 – 027	0889 91	1.3 x 1.3 1.2: x 1.2:	— 1	3	Sc	SAB(s)cd	16.0:		*
03744	07 10.3 + 19 05 198.13 + 13.20	03 – 19 – 005 IC 2181	0641	1.0 x 0.5 0.9 x 0.5	140 5	1	Sa – b		14.8		
03745	07 10.3 + 27 37 190.04 + 16.64	05 – 17 – 018	1337	1.4 x 1.1 1.2 x 1.0	— 1:	2	S B b:		14.9		

N G C 2276, 2337, 2340, 2344   I C 458, 2181

1	2	3	4	5	6	7	8	9	10	11	12
03746	07 10.3 + 55 14 161.85 + 25.15	09 – 12 – 045	0971 96	1.2 x 0.9 0.6 x 0.3	(165) 2 – 3	2	Sb	Sb –	15.5		
03747	07 10.4 + 12 21 204.35 + 10.33	02 – 19 – 001 NGC 2350	1354	1.5 x 0.8 1.2 x 0.7	110 4 – 5	2	S0 – a		14.1		3
03748	07 10.5 + 65 31 150.57 + 26.95	11 – 09 – 038	0686	1.0 x 0.4 0.9: x 0.4:	83 ...	2	Dwarf: IV or IV – V		16.0:		
03749	07 10.6 + 68 04 147.73 + 27.26	11 – 09 – 037	0686	1.7 x 0.2 1.6 x 0.2	21 7	3	Sc		15.2		
03750	07 10.7 + 65 02 151.11 + 26.91	11 – 09 – 38a IC 2179	0686	0.8 x 0.8 1.0 x 0.9	-- ...	2	... P w 3759		13.4		*2
03751	07 10.8 + 23 10 194.34 + 14.99	04 – 17 – 011	1590	1.4 x 0.4 1.5: x 0.4	30 6 – 7	1	S ...		14.8		
03752	07 10.8 + 35 22 182.50 + 19.58	06 – 16 – 028	0889 91	0.45 x 0.40 0.45 x 0.4	-- ...	1	...		14.5		
03753	07 10.9 + 23 20 194.19 + 15.07	04 – 17 – 012	1590	1.6 x 0.7 1.7 x 0.8	35: 5 – 6	1	Sa:		15.1		
03754	07 11.0 + 12 25 204.35 + 10.49	02 – 19 – 000	1354	1.1 x 0.8 1.0 x 0.7	(135) ...	2	Sc/S B c	SAB cd: Sc +	16.0:		*
03755	07 11.1 + 10 35 206.04 + 09.70	02 – 19 – 002	1354	1.8 x 1.1 1.7: x 1.0:	[160] ...	4	Irr	I m Ir I	14.9		
03756	07 11.1 + 35 29 182.40 + 19.68	06 – 16 – 000	0889 91	1.2 x 0.2 1.2 x 0.2	149 7	2	Sb:		15.5		
03757	07 11.2 + 48 49 168.73 + 23.76	08 – 13 – 000	0691 89	1.4 x 0.1 1.4 x 0.1	166 7	3	Sc		17.		
03758	07 11.2 + 50 37 166.83 + 24.22	08 – 13 – 111	0691 89	(0.5 x 0.5) (0.9 x 0.8)	-- ...	2	... Compact		14.2		*1
03759	07 11.3 + 64 49 151.36 + 26.95	11 – 09 – 039 NGC 2347	0686	2.0 x 1.5 1.9 x 1.4	175: 3	3	Sb P w 3750	SA(r)b	13.2	+ 4521 + 4640	*123 4$\overline{7}$8
03760	07 11.7 + 05 09 211.01 + 07.40	01 – 19 – 000	0999	(1.3 x 1.3) (1.2 x 1.2)	-- ...	1	Double system		15.3		*
03761	07 11.7 + 38 13 179.70 + 20.71	06 – 16 – 000	0889	1.3 x 0.2 1.1 x 0.2	161 7	3	Sc		16.0:		
03762	07 11.7 + 73 57 141.12 + 27.81	12 – 07 – 000	0665	1.0 x 0.2 0.7 x 0.2	175 6	1	S ...		15.7		
03763	07 11.8 + 34 53 183.05 + 19.60	06 – 16 – 000	0889 91	1.0 x 0.2 0.9 x 0.2	38 7	2	Sc		17.		
03764	07 11.8 + 67 12 148.71 + 27.28	11 – 09 – 040	0686	1.3 x 0.9 1.2 x 0.8	120: 3:	2	S B ... Singular	SB(r) ...	14.1		*
03765	07 11.9 + 56 54 160.09 + 25.71	09 – 12 – 046	0971	0.8 x 0.8 0.8 x 0.8	-- ...	2	S0		14.3		*
03766	07 12.0 + 17 03 200.19 + 12.71	03 – 19 – 000	0641	1.2 x 0.5 1.2: x 0.6	145 5 – 6	3	Sc		15.5		
03767	07 12.1 + 06 53 209.50 + 08.27	01 – 19 – 000	0999	1.4: x 0.9: 1.3: x 0.9:	140: ...	3	...	S dm	16.0:		
03768	07 12.2 + 06 40 209.71 + 08.19	01 – 19 – 000	0999	1.0 x 0.2 0.9 x 0.2	152 6	1	S ...		15.3		
03769	07 12.4 + 00 51 214.94 + 05.58	00 – 19 – 000	1491	1.0: x 0.4 0.9 x 0.4	115 6	1	S ...		15.7		
03770	07 12.4 + 23 31 194.16 + 15.46	04 – 17 – 013	1590	1.3 x 0.8: 1.1: x 0.8:	27: ...	2	Irr		15.4		
03771	07 12.4 + 70 37 144.88 + 27.66	12 – 07 – 036	0665	1.2 x 0.9 1.2 x 0.9	35: ...	2	S B b	SB(s)b	15.0		*
03772	07 12.7 + 15 13 201.96 + 12.08	03 – 19 – 000	0641	(1.4 x 0.9) (1.4 x 0.9)	(40) ...	1	...		15.2		*
03773	07 12.9 + 39 51 178.13 + 21.46	07 – 15 – 000	0988	1.3 x 0.4 0.9: x 0.3	58 ...	2	Dwarf IV or IV – V		17.		
03774	07 13.0 + 33 01 185.00 + 19.18	06 – 16 – 033	0889	1.5 x 0.3 1.4 x 0.3	60 6 – 7	2	Sa – b		15.3		*
03775	07 13.1 + 12 13 204.77 + 10.86	02 – 19 – 000	1354	1.7:: x 1.5:: 1.5:: x 1.3::	-- ...	4	Dwarf spiral S V		17.		
03776	07 13.4 + 34 05 183.97 + 19.64	06 – 16 – 034	0889 95	1.8 x 0.4 1.7 x 0.3	66 6	2	S ... Brightest of 3		14.3		*
03777	07 13.5 + 29 57 188.06 + 18.17	05 – 17 – 020	1337	2.0 x 0.35 1.9 x 0.3	152 7	2	Sc		14.5		
03778	07 13.8 + 28 38 189.37 + 17.74	05 – 17 – 000	1337	1.1: x 1.1: 1. x 1.	-- ...	3	Dwarf spiral S IV – V		16.5:		
03779	07 14.2 + 34 04 184.05 + 19.78	06 – 16 – 036	0889 95	1.4 x 0.2 1.2 x 0.2	88 7	1	S ... 3776 group		15.5		*
03780	07 14.2 + 34 10 183.95 + 19.82	06 – 16 – 035	0889 95	1.1 x 0.4 1.1 x 0.3	59 7	1	S ... 3776 group		14.9		*

1	2	3	4	5	6	7	8	9	10	11	12
03781	07 14.5 + 41 05 176.97 + 22.13	07 - 15 - 000	0988	[1.7 x 0.9] ...	... ...	3	Double system Bridge + plume		14.9		*
03782	07 14.6 + 23 27 194.43 + 15.90	04 - 17 - 014 NGC 2357	1590	3.7 x 0.5 3.3 x 0.5	122 6 - 7	3	Sc		14.6		*
03783	07 14.9 + 24 50 193.14 + 16.51	04 - 18 - 000	1310	1.4 x 0.12 1.2 x 0.12	24 7	2	Sc		15.6		
03784	07 14.9 + 26 44 191.31 + 17.24	04 - 17 - 000	1590	1.1 x 0.12 0.5 x 0.1	158 7	2	Sc		17.		
03785	07 15.0 + 08 02a 208.79 + 09.43	01 - 19 - 004	0999	1.4 x 0.6 1.4 x 0.7	12 5 - 6	1	Sa: p of 2, contact		(14.7)		*1
03786	07 15.0 + 30 48 187.35 + 18.78	05 - 17 - 000	1337 95	1.3 x 0.25 1.2 x 0.3:	27 (7)	2	...	d - m	15.6		
03787	07 15.2 + 09 46a 207.23 + 10.25	02 - 19 - 000	1354	(1.2 x 0.6) (1.2 x 0.6)	(0) ...	1	E - S0: p of 2		(15.6)		*
03788	07 15.2 + 31 39 186.53 + 19.13	05 - 18 - 002	0678 95	1.4 x 0.4 1.3 x 0.4	155 6	1	S ...		15.0		
03789	07 15.2 + 59 28 157.37 + 26.60	10 - 11 - 021	1291 100	1.6 x 1.4 1.6 x 1.4	(170) 1	4	Sa Compact core	(R)SA(r)ab	13.3		*C
03790	07 15.3 + 31 28 186.72 + 19.08	05 - 18 - 3 +4	0678 95	[1.3 x 0.5] ...	... ...	3	Double system Contact		15.2		*1
03791	07 15.4 + 27 15 190.85 + 17.54	05 - 18 - 001	0678	1.3 x 0.22 1.2 x 0.2	15 (7)	2	Sc - Irr		15.3		
03792	07 15.4 + 51 24 166.16 + 25.04	09 - 12 - 000	0971	(1.8 x 1.3) (1.9 x 1.4)	65: ...	3	S0	SA 0/a	14.0		
03793	07 15.4 + 57 15 159.81 + 26.25	10 - 11 - 000	1291	1.0 x 0.12 0.9 x 0.12	71 7	3	Sc P w 3795		16.5:		*
03794	07 15.4 + 77 55 136.64 + 28.12	13 - 06 - 005	1295	1.8: x 1.2: 0.8: x 0.5:	120: ...	3	Dwarf spiral SB IV - V		16.5:		*
03795	07 15.7 + 57 13 159.86 + 26.28	10 - 11 - 000	1291	1.0 x 0.1 0.7 x 0.1	121 7	3	Sc P w 3793		17.		*
03796	07 15.8 + 60 21 156.40 + 26.82	10 - 11 - 026	1291 100	1.1 x 0.2 1.0 x 0.2	50 7	2	Sb - c nf of 2		16.0:		*
03797	07 15.9 + 59 28 157.38 + 26.69	10 - 11 - 028	1291 100	1.1 x 0.7 1.0 x 0.7	[55] ...	3	S B c	S B(s)cd	16.0:		*
03798	07 16 + 85 49 127.70 + 27.81	14 - 04 - 031 NGC 2300	0692 102	(3.2 x 2.7) (3.5 x 3.0)	-- ...	4	E P w 3740		12.2	+ 1986 + 2189	*123 4568A
03799	07 16.1 + 53 06 164.35 + 25.52	09 - 12 - 056	0971 96	1.1 x 1.0 1.1 x 1.0	-- 1	4	Sc/S B c	SAB(s)d	14.9		*
03800	07 16.3 + 56 30 160.66 + 26.23	09 - 12 - 058	0971 96	1.2: x 0.9:: 1.0: x 0.8:	(45) ...	1	... Disturbed		16.5:		*
03801	07 16.4 + 63 44 152.65 + 27.36	11 - 09 - 046	0686	1.3 x 0.2 1.2 x 0.2	79 7	2	Sc ssf of 2, disturbed:		16.5:		*
03802	07 16.6 + 31 01 187.27 + 19.18	05 - 18 - 006	0678 95	1.2 x 0.20 1.1 x 0.20	168 7	1	Sa:		15.1		
03803	07 17.0 + 22 11 195.87 + 15.90	04 - 18 - 002	1310	1.0 x 0.9 0.9 x 0.9	-- 1	1	Sa:		15.1		
03804	07 17.0 + 71 42 143.69 + 28.09	12 - 07 - 037	0665	1.9 x 1.3 1.9 x 1.3	13 3	3	Sc		13.0		1
03805	07 17.1 + 18 02 199.80 + 14.22	03 - 19 - 006	0641	1.3 x 0.4 1.3 x 0.35	147 6	1	S ...		14.9		*
03806	07 17.4 + 23 00 195.13 + 16.31	04 - 18 - 003	1310 97	1.0 x 0.9 1.1 x 1.0	-- 1	2	S B: c		15.3		1
03807	07 17.4 + 54 47 162.57 + 26.06	09 - 12 - 000	0971 96	1.0 x 0.5: 0.8 x 0.4	65 ...	1	S ...		16.5:		*
03808	07 18.0 + 25 16 193.01 + 17.32	04 - 18 - 006	1310	2.0: x 1.0 1.5: x 0.8:	50 ...	3	S B c	S B(s)d: Sc +	15.5		*1
03809	07 18.0 + 80 16 133.97 + 28.20	13 - 06 - 006 NGC 2336	1295	7.0 x 3.6 7.2 x 3.7	178: 5	4	S B c P w 3834	S B(r)bc Sc -	11.3	+ 2252 + 2435	*12- -8
03810	07 18.1 + 46 56 171.05 + 24.39	08 - 14 - 007	0670 89	1.0 x 0.9 0.7 x 0.8	-- 1	2	Sc Distorted?		15.3		*1
03811	07 18.4 + 35 50 182.59 + 21.19	06 - 16 - 037	0889	1.4: x 1.0 1.0: x 0.6:	(50) ...	2	Sc or Sc/S B c	Sc +			
03812	07 18.5 + 49 23 168.45 + 25.06	08 - 14 - 008	0670 89	(0.7 x 0.7) (0.8 x 0.8)	-- ...	3	E		14.4		*
03813	07 18.6 + 13 21 204.32 + 12.56	02 - 19 - 000	1354	1.4:: x 1.0:: 1.3:: x 1.1::	(90) ..	3	Dwarf V		15.7		
03814	07 18.6 + 49 22 168.48 + 25.07	08 - 14 - 000	0670 89	1.0 x 0.1 0.9 x 0.10	178 (7)	1	Sc - Irr		17.		*
03815	07 19 + 85 39 127.88 + 27.88	14 - 04 - 033 IC 455	0692 102	1.1 x 0.7 1.2 x 0.8	82 4	2	S0		14.3		

1	2	3	4	5	6	7	8	9	10	11	12
03816	07 19.0 + 58 11 158.87 + 26.88	10 – 11 – 034	1291 100	1.0 x 0.7 1.0 x 0.7	112 ...	3	SO		13.4		*1
03817	07 19.1 + 45 12 172.94 + 24.11	08 – 14 – 000	0670	2. x 1. 1. x 1.	-- ...	3	Dwarf V		17.		
03818	07 19.2 + 41 45 176.56 + 23.17	07 – 15 – 016	0988	1.1 x 0.3 1.1 x 0.25	126 6	1	S ...		15.0		
03819	07 19.4 + 05 14 211.82 + 09.14	01 – 19 – 000	0999	1.1 x 0.3 1.0: x 0.3:	[130] ...	2	Irr		15.7		
03820	07 19.5 + 17 22 200.67 + 14.47	03 – 19 – 008	0641	1.6 x 0.3 1.6 x 0.3	100 7	2	Sc		15.1		*
03821	07 19.5 + 22 10 196.13 + 16.42	04 – 18 – 008 NGC 2365	1310 (97)	3.0 x 1.6 3.3 x 1.7	170 5	4	Sa/S B a nnf of 2		13.8		*3
03822	07 19.8 + 37 33 180.93 + 22.01	06 – 17 – 001	1336	1.5 x 0.8 1.1 x 0.7	15: ...	3	S B c	S B(rs)bc Sc –	15.0		
03823	07 19.9 + 19 00 199.18 + 15.23	03 – 19 – 009	0641	1.2 x 0.7 1.3: x 0.7	103: 4 – 5	2	Sb		15.1		
03824	07 19.9 + 22 40 195.69 + 16.71	04 – 18 – 012	1310 97	(0.9 x 0.8) (0.9 x 0.9)	-- 1	2	SO		14.4		
03825	07 20.0 + 41 31 176.85 + 23.25	07 – 15 – 017	0988	1.4 x 1.3 1.1 x 1.1	-- 1	3	Sb/S B b	SAB(s)bc Sb –	15.3		
03826	07 20.0 + 61 47 154.89 + 27.52	10 – 11 – 038	1291	4.0 x 3.5 2.5: x 2.5:	(85) 1	4	Sc	SAB(s)d	15.0		
03827	07 20.2 + 22 18 196.07 + 16.62	04 – 18 – 013	1310 97	1.1 x 1.1 1.1 x 1.0	-- 1	1	S B? ... Singular		14.8		
03828	07 20.4 + 58 05 159.02 + 27.05	10 – 11 – 039	1291 100	1.8 x 0.9 1.8 x 0.9	0 ...	3	Sb/S B b	SAB(rs)b Sb +	12.7		
03829	07 20.5 + 33 31 185.08 + 20.83	06 – 17 – 002	1336	[1.3 x 1.3] ...	-- ...	3	Double system Contact, distorted		13.7		*1
03830	07 20.9 + 02 43 214.26 + 08.33	00 – 19 – 001	1491	1.4 x 0.5 1.5 x 0.5	116 6	1	S ...		15.1		
03831	07 21.2 + 49 35 168.35 + 25.53	08 – 14 – 014	0670 89	1.2 x 1.0 1.1 x 0.9	(85) 1	4	S B b Singular	S B(s)bc Sb +	13.6		*1
03832	07 21.3 + 53 32 164.06 + 26.37	09 – 12 – 074	0971 96	1.3 x 0.7 1.1 x 0.6	107 5	2	S B a – b		15.2		*
03833	07 21.5 + 32 55 185.76 + 20.82	05 – 18 – 010	0678 95	1.0: x 0.9: 0.4 x 0.3	-- 1:	1	S ...		15.3		
03834	07 21.7 + 79 58 134.30 + 28.37	13 – 06 – 007 IC 467	1295	3.5 x 1.4 3.5 x 1.4	80 6	4	Sc P w 3809	SAB: cd	12.7		*12
03835	07 22.0 + 23 53 194.71 + 17.63	04 – 18 – 015 NGC 2370	1310 97	1.1 x 0.6 1.1 x 0.7	43 ...	1	S ...		14.3		3
03836	07 22.0 + 64 55 151.40 + 28.09	11 – 09 – 047	0686 100	1.3 x 1.1: 1.2 x 1.1:	-- 1:	3	S B c	S B(rs)cd	15.5		
03837	07 22.2 + 36 46 181.90 + 22.22	06 – 17 – 003	1336	1.0 x 0.5 1.0 x 0.5	105 5	1	S ...		15.1		
03838	07 22.3 + 72 40 142.60 + 28.53	12 – 07 – 038 Mark 7	0665	0.9 x 0.35 0.9 x 0.35	[17] ...	1	... Peculiar, horseshoe		13.9		*C
03839	07 22.4 + 09 36 208.19 + 11.76	02 – 19 – 003	1354	1.1 x 1.0 1.1 x 1.0	-- 1 – 2	2	S B b		15.2		
03840	07 22.4 + 19 16 199.17 + 15.88	03 – 19 – 010	0641 98	(1.1 x 1.1) (1.7 x 1.7)	-- ...	3	E P w 3842		14.4		*
03841	07 22.6 + 30 04 188.90 + 20.05	05 – 18 – 012	0678	[1.0 x 0.4] ...	... ...	2	Double system Bridge, 2nd and 3rd of 3		15.2		*1
03842	07 22.7 + 19 13 199.25 + 15.92	03 – 19 – 011	0641 98	1.1 x 0.8 1.1 x 0.8	120: 3:	3	SO P w 3840		14.9		*
03843	07 22.8 + 20 11 198.34 + 16.33	03 – 19 – 000	0641	1.0 x 0.2 0.9 x 0.2	147 6	1	Sc		15.5		
03844	07 23.0 + 43 23 175.06 + 24.30	07 – 16 – 002	0701	(1.7 x 1.1) (1.8 x 1.2)	(5) ...	1	E?		14.3		
03845	07 23.0 + 47 12 170.99 + 25.27	08 – 14 – 016	0670 89	1.6 x 1.1 1.5 x 1.0	176 3	4	S B b	S B(s)bc Sb +	13.6		*
03846	07 23.0 + 75 37 139.25 + 28.58	13 – 06 – 000	1295	1.4: x 0.6: 1.4:: x 0.6::	57 6	3	...	SA: dm	16.5:		
03847	07 23.1 + 69 17 146.46 + 28.49	12 – 07 – 039 NGC 2363	0665	1.7: x 1.1 1.7: x 1.1:	[20] ...	1	Irr P w 3851		15.5:		*13
03848	07 23.3 + 33 54 184.90 + 21.50	06 – 17 – 004 NGC 2373	1336 95	0.9 x 0.6 0.5 x 0.6	(0) ...	1	S ...		14.5		13
03849	07 23.3 + 37 28 181.25 + 22.64	06 – 17 – 000	1336	[1.0 x 0.9] ...	-- ...	3	Double system Strongly distorted		14.6		*
03850	07 23.6 + 63 22 153.16 + 28.11	11 – 09 – 049	0686 100	1.7 x 1.7 1.7 x 1.7	-- 1	4	Sa/S B a	(R')SAB(s)a	13.9		*1

N G C   2363, 2365, 2370, 2373   I C   467

1	2	3	4	5	6	7	8	9	10	11	12
03851	07 23.6 + 69 18 146.44 + 28.53	12 - 07 - 040 NGC 2366	0665	9: x 3.5:: 5: x 2.5::	[25] ...	5	Irr I m P w 3847 Ir I		11.6	+ 145 + 281	*123 45689
03852	07 23.6 + 72 14 143.10 + 28.62	12 - 07 - 041 IC 2184, Mark 8	0665	[1.2 x 1.0] ...	— ...	1	Double or triple system Contact		13.8	+ 3420 + 3570	*1C
03853	07 23.8 + 48 33 169.57 + 25.71	08 - 14 - 018	0670 89	1.0 x 0.1 0.9 x 0.1	16 7	3	Sc		17.		
03854	07 23.9 + 33 55 184.93 + 21.63	06 - 17 - 005 NGC 2375	1336 95	1.4 x 0.9 1.4 x 0.9	(170) ...	4	S B b S B(s)b P w 3857, NGC 2389 group		14.7		*13
03855	07 24.0 + 58 38 158.49 + 27.60	10 - 11 - 052	1291 100	2.0 x 0.45 2.2 x 0.45	52 6	2	Sa - b		14.2		
03856	07 24.1 + 20 28 198.20 + 16.73	03 - 19 - 012	0641 98	1.1 x 0.9 0.8 x 0.8	(90) 1 - 2	3	Sc		16.5:		
03857	07 24.2 + 33 54 184.95 + 21.69	06 - 17 - 006 NGC 2378 = 2379	1336 95	(1.0 x 1.0) (1.1 x 1.0)	— ...	1	[S0] Very compact P w 3854, NGC 2389 group		14.9	+ 4030 + 3994	*12 3
03858	07 24.7 + 73 44 141.39 + 28.71	12 - 07 - 043	0665	1.2 x 1.0 1.4 x 1.1	(120) 2:	2	Sa P w 3859		13.8		*
03859	07 24.7 + 73 49 141.30 + 28.71	12 - 07 - 045	0665	1.6 x 1.1 1.7 x 1.2	45: 3	3	Sa P w 3858		13.3		*
03860	07 24.8 + 40 51 177.83 + 23.92	07 - 16 - 003	0701	1.8: x 1.2: 1.7: x 1.0:	[0] ...	3	Dwarf irregular Ir IV - V or V		15.5		*
03861	07 24.8 + 58 19 158.86 + 27.66	10 - 12 - 140	0960 100	1.0 x 0.4 0.9 x 0.3	90 6	3	... S d - dm		16.0:		*
03862	07 24.9 + 20 29 198.26 + 16.90	03 - 19 - 016	0641 98	1.6: x 0.4 1.6 x 0.5:	13 6	1	S ...		15.6		
03863	07 25.1 + 49 14 168.89 + 26.07	08 - 14 - 020	0670	1.3 x 0.6 1.4 x 0.7	76 5	3	S B a (R´)S B a		13.9		
03864	07 25.1 + 72 37 142.66 + 28.74	12 - 08 - 001	0680	1.7 x 1.1 1.5 x 0.8	... ...	2	...		15.0		*V
03865	07 25.3 + 51 08 166.82 + 26.50	09 - 13 - 004	0985	1.0: x 0.3 0.7 x 0.25	40 6	2	Sb - c		16.0:		
03866	07 25.4 + 31 02 187.96 + 20.95	05 - 18 - 000	0678	1.1 x 0.25 1.0 x 0.2	178 6	1	S ... Disturbed?		15.4		*
03867	07 25.4 + 57 15 160.06 + 27.58	10 - 11 - 056	1291 100	1.0 x 0.8 1.4: x 1.0:	(135) ...	1	... Pair of compacts		15.3		*C
03868	07 25.5 + 40 18 178.45 + 23.89	07 - 16 - 004	0701	1.5:: x 1.1:: 1.2:: x 0.9::	[155] ...	3	Dwarf irregular Ir V		17.		
03869	07 25.6 + 24 35 194.36 + 18.65	04 - 18 - 019	1310 97	1.0 x 0.4 1.0 x 0.4	25 (6)	1	S0?		15.1		
03870	07 25.6 + 33 55 185.05 + 21.96	06 - 17 - 010 NGC 2388	1336 95	1.0 x 0.6 0.9 x 0.6	65: ...	1	... NGC 2389 group		14.7		*13
03871	07 25.7 + 42 22 176.28 + 24.50	07 - 16 - 005	0701	1.1 x 0.9 1.0 x 0.9	45 2 - 3	3	Sc SAB(s)c		14.9		
03872	07 25.8 + 33 57 185.03 + 22.01	06 - 17 - 011 NGC 2389	1336 95	2.0 x 1.5 1.8 x 1.4	83 3	3	Sc Brightest in group		13.5	+ 3816 + 3780	*12 3
03873	07 25.9 + 20 40 198.18 + 17.19	03 - 19 - 017	0641	1.9 x 0.4 1.8 x 0.4	144 6	2	Sc		14.6		
03874	07 25.9 + 27 01 192.00 + 19.62	05 - 18 - 014	0678	1.2 x 0.8 1.2 x 0.8	... ...	3	... Peculiar, plumes		15.2		*1
03875	07 25.9 + 52 54 164.90 + 26.93	09 - 13 - 009	0985 96	1.4: x 1.2: 1.2: x 1.1:	(50) 1 - 2	4	Sc/SBc SAB(rs)d or SAB(s)d		15.3		1
03876	07 26.1 + 28 01 191.03 + 20.03	05 - 18 - 015	0678	2.5: x 1.5: 2.3 x 1.3	2 4	4	Sc SA(s)d		14.4		
03877	07 26.4 + 14 31 204.07 + 14.77	02 - 19 - 000	1354	1.0 x 0.7 (0.9 x 0.6)	48 3:	2	S0 sf of 2		15.2		*
03878	07 26.4 + 73 02 142.19 + 28.83	12 - 08 - 000	0680	1.4 x 0.8 1.2: x 0.8	102 ...	3	... S dm		16.0:		
03879	07 26.5 + 33 47 185.26 + 22.09	06 - 17 - 012	1336 95	2.3 x 0.3 2.3: x 0.2	103 (7)	2	Sc - Irr NGC 2389 group		15.3		*
03880	07 26.5 + 37 33 181.38 + 23.27	06 - 17 - 013 IC 2190	1336	1.0 x 0.5 0.9 x 0.5	21 5	2	S B: b - c		14.8		
03881	07 26.6 + 49 37 168.53 + 26.39	08 - 14 - 022	0670 103	1.5: x 0.7: 1.0 x 0.6	(115) ...	3	... SA dm		15.6		
03882	07 26.6 + 55 40 161.85 + 27.50	09 - 13 - 011	0985 96	1.7 x 0.9 1.2 x 0.5	170: 6	3	Sc SA(r:)cd		16.0:		
03883	07 26.8 + 07 12 210.88 + 11.67	01 - 19 - 000	0999	1.4:: x 0.5: 1. x 0.4::	5 ...	2	Dwarf V		18.		
03884	07 26.8 + 34 07 184.94 + 22.26	06 - 17 - 014 NGC 2393	1336 95	1.1 x 0.7 1.1 x 0.7	103 4	3	Sc NGC 2398 group, Sc -		14.9		*
03885	07 26.8 + 59 36 157.45 + 28.08	10 - 11 - 066	1291 100	1.0 x 0.9 0.9 x 0.9	— 1:	1	S ...		13.9		*

N G C   2366, 2375, 2378 = 2379, 2388, 2389, 2393   I C   2184, 2190

1	2	3	4	5	6	7	8	9	10	11	12
03886	07 26.9 + 62 35 154.09 + 28.41	10 – 11 – 067	1291	1.2 x 1.1 1.1 x 1.1	-- 1	4	Sc	SA(s)c	16.0:		
03887	07 27.1 + 36 12 182.82 + 22.98	06 – 17 – 015	1336	1.3: x 1.1: 1.1: x 1.0:	(140) 1	3	S B c	S B(s)cd	15.3		
03888	07 27.6 + 39 07 179.81 + 23.94	07 – 16 – 006	0701	1.4 x 0.7 1.0 x 0.3	115 5 – 6	2	S B b		15.5		
03889	07 27.8 + 73 45 141.37 + 28.93	12 – 08 – 003	0680	1.3 x 0.2 1.2 x 0.2	112 7	2	Sa – b		15.1		
03890	07 28 + 83 54 129.83 + 28.27	14 – 04 – 034	0692	1.3 x 0.25 1.3 x 0.3	135 (7)	1	... Spindle with jets		14.9		*
03891	07 28.0 + 09 45 208.67 + 13.07	02 – 19 – 004 NGC 2402	1354	[1.2 x 1.2] ...	-- ...	2	Double system Contact		15.4		*13
03892	07 28.3 + 13 10 205.53 + 14.62	02 – 19 – 005	1354	1.4 x 0.4 1.4 x 0.4	80 6:	1	Sc: Disturbed Brightest in group		15.1		*
03893	07 28.3 + 65 07 151.23 + 28.77	11 – 10 – 002	1330 100	1.1 x 1.0 0.9 x 0.9	-- 1	3	Sb:	SA(r) ...	14.7		*
03894	07 28.4 + 65 12 151.13 + 28.78	11 – 10 – 003	1330 100	(1.2 x 1.2) (1.7 x 1.7)	-- ...	2	E		14.5		*
03895	07 28.7 + 00 09 217.47 + 08.87	00 – 19 – 000	1491	1. x 1. 0.7: x 0.7:	-- ...	3	Dwarf V		17.		
03896	07 29.0 + 18 27 200.61 + 16.97	03 – 20 – 001 NGC 2407	1507 98	(1.0 x 0.8) (1.3 x 1.0)	(75) ...	2	E – S0 Brightest in group		14.9		*
03897	07 29.0 + 59 45 157.32 + 28.37	10 – 11 – 074	1291 100	1.1 x 0.9 1.5 x 1.1	(45) ...	3	S B 0	S B(r)0 +	14.2		*
03898	07 29.0 + 65 34 150.72 + 28.87	11 – 10 – 004	1330	1.2: x 0.7: 1.2: x 0.7	18: ...	2	Sb	SAB(rs)b	15.3		
03899	07 29.3 + 35 43 183.46 + 23.25	06 – 17 – 016	1336	1.1 x 0.3 1.1 x 0.25	46 6 – 7	2	Sa – b		15.4		
03900	07 29.7 + 11 00 207.70 + 13.99	02 – 20 – 001	1003	1.0 x 0.8 1.0 x 0.8	150: 2:	2	S B b		15.0		*
03901	07 30.1 + 49 24 168.89 + 26.90	08 – 14 – 025	0670 103	1.0 x 1.0 0.9 x 0.9	-- 1	2	S B ...	S B(s) ...	15.3		
03902	07 30.2 + 31 36 187.76 + 22.11	05 – 18 – 018 IC 2193	0678 99	1.6 x 0.9 1.4 x 0.9	90 4 – 5	2	Sb IC 2199 group		14.7		
03903	07 30.3 + 19 18 199.93 + 17.59	03 – 20 – 004	1507 98	1.7 x 0.6 1.7 x 0.6	174 6	4	S B a	(R´)SAB(rs)b	14.7		
03904	07 30.3 + 30 40 188.71 + 21.81	05 – 18 – 019	0678 99	1.4 x 1.4 1.2 x 1.2	-- 1	4	S B 0/S B a (R)S B(r)0/a nnf of 2		14.6		*
03905	07 30.3 + 62 39 154.05 + 28.81	10 – 11 – 085	1291 100	0.9 x 0.7 1.0 x 0.7	80 2 – 3	1	Sa – b		14.4		1
03906	07 30.4 + 74 34 140.43 + 29.08	12 – 08 – 005	0680	[1.7 x 1.0] ...	... ...	3	Double system Contact, disrupted		14.8		*1
03907	07 30.5 + 37 07 182.09 + 23.90	06 – 17 – 018	1336	1.4 x 0.7 0.9 x 0.7	80: ...	3	Sc		15.7		*1
03908	07 30.5 + 66 31 149.64 + 29.06	11 – 10 – 006	1330	1.1 x 0.2 1.0 x 0.2	50 7	3	Sb		16.5:		
03909	07 30.8 + 73 49 141.29 + 29.13	12 – 08 – 007	0680	2.8 x 0.6 2.5 x 0.6	82 6	3	S B: c		14.9		*
03910	07 30.9 + 31 31 187.89 + 22.22	05 – 18 – 021 IC 2196	0678 99	(1.4 x 1.1) (1.5 x 1.2)	(150) ...	3	E IC 2199 group		14.0		*
03911	07 31.1 + 23 13 196.20 + 19.30	04 – 18 – 000	1310	1.1 x 0.8 (1.3 x 1.0)	(160) 2 – 3:	3	S0		15.2		
03912	07 31.6 + 04 39 213.75 + 11.59	01 – 20 – 001	1527	2.1: x 1.2 2.1 x 1.2	(120) ...	4	Irr	I B: m	14.5		
03913	07 31.6 + 33 38 185.78 + 23.05	06 – 17 – 019	1336	1.6: x 0.5: 1.2: x 0.2	160 6	1	S ...		15.7		
03914	07 31.7 + 18 23a 200.95 + 17.53	03 – 20 – 005 NGC 2411	1507 98	(1.0 x 0.6) (1.1 x 0.7)	50 ...	1	E: sp of 2		(14.6)		*
03915	07 31.7 + 31 23 188.09 + 22.34	05 – 18 – 022 IC 2199	0678 99	1.2 x 0.6 1.1 x 0.6	25 ...	1	S ... Brightest in group		13.6		*
03916	07 31.8 + 22 41 196.79 + 19.24	04 – 18 – 027	1310	1.2 x 0.3 1.1 x 0.3	85 6	3	Sc	Sc +	15.7		
03917	07 31.8 + 32 56 186.51 + 22.86	05 – 18 – 023 NGC 2410	0678	2.5: x 0.8 2.3: x 0.6	31 6	3	S B? b	Sb +	13.9		3
03918	07 32.0 + 65 43 150.57 + 29.18	11 – 10 – 007 NGC 2403	1330	28. x 14. 22. x 13.	127 5	5	Sc M 81 group	SAB(s)cd	9.3	+ 136 + 255	*12- -9
03919	07 32.2 + 64 40 151.77 + 29.15	11 – 10 – 008	1330 100	1.6 x 0.12 1.6: x 0.15	81 7	3	Sc		16.5:		
03920	07 32.4 + 19 10 200.26 + 18.00	03 – 20 – 006	1507 98	1.4: x 1.3 1.3 x 1.2	-- 1	2	Sb	SAB b	14.5		

N G C   2402,   2403,   2407,   2410,   2411   I C   2193,   2196,   2199

1	2	3	4	5	6	7	8	9	10	11	12
03921	07 32.5 + 70 59 144.53 + 29.32	12 - 08 - 000	0680	1.2 x 0.12 1.1 x 0.12	53 7	2	Sc		16.5:		
03922	07 32.6 + 55 10 167.56 + 28.27	09 - 13 - 020	0985 96	1.7 x 0.2 1.4: x 0.2	164 7	2	Sc Disturbed?		16.0:		*
03923	07 32.7 + 55 25 162.28 + 28 32	09 - 13 - 021	0985 96	1.0 x 0.6 0.7 x 0.5	153 4	2	Sb		14.9		1
03924	07 32.8 + 11 39 207.44 + 14.96	02 - 20 - 000	1003	1.1: x 1.0: 1.0:: x 1.0::	--- ...	3	Dwarf spiral S V		15.7		*
03925	07 32.9 + 11 43 207.39 + 15.01	02 - 20 - 002 NGC 2416	1003	1.1 x 0.7 1.0 x 0.9	110 ...	2	Sc Brightest in group		14.3		*3
03926	07 33.0 + 33 14 186.29 + 23.20	06 - 17 - 020 IC 2201	1336	1.3 x 0.3 1.2 x 0.3	67 7	2	Sa		14.9		*
03927	07 33.1 + 59 49 157.31 + 28.89	10 - 11 - 099	1291 100	(1.3 x 0.8) (1.2 x 0.7)	15 ...	3	S0		15.5		
03928	07 33.4 + 59 27 157.73 + 28.89	10 - 11 - 100	1291 100	(1.8 x 0.9) (1.8 x 0.9)	60: ...	2	... Peculiar, jets		14.9		*
03929	07 33.4 + 75 33 139.30 + 29.23	13 - 06 - 008	1295	1.6? x 1.3? ...	... ...	1	... Compact		15.3		*C
03930	07 33.6 + 35 21 184.14 + 23.97	06 - 17 - 021 NGC 2415	1336	1.0 x 1.0 0.9 x 0.9	--- ...	1	... Peculiar		12.5	+ 3822 + 3791	*12<u>3</u>
03931	07 33.7 + 18 00 201.52 + 17.81	03 - 20 - 008 NGC 2418	1507 98	(1.8 x 1.8) (1.8 x 1.8)	--- ...	2	E Disturbed?		13.7		*3
03932	07 34.0 + 22 28 197.21 + 19.63	04 - 18 - 028	1310	1.6 x 0.3 1.4 x 0.25	150 7	2	Sa - b		14.9		
03933	07 34.0 + 42 03 177.06 + 25.90	07 - 16 - 008	0701	1.3 x 0.9 1.3 x 0.9	0 ...	4	Sb/SBc	SAB(s)bc Sb +	14.5		
03934	07 34.2 + 35 46 183.74 + 24.21	06 - 17 - 000	1336	1.1: x 0.9: 0.7: x 0.7:	--- ...	2	Dwarf (spiral) S: V		17.		*
03935	07 34.2 + 46 31 172.22 + 26.98	08 - 14 - 028	0670	1.2 x 0.9: 1.1 x 0.8	15: 3:	3	Sc	SAB c	15.7		
03936	07 34.3 + 13 43 205.66 + 16.17	02 - 20 - 004	1003	1.2 x 1.0 1.3 x 1.1	(100) 1 - 2	2	S B b - c		14.1		
03937	07 34.3 + 35 43 183.80 + 24.22	06 - 17 - 023	1336	2.1 x 0.5 1.8 x 0.5	151 6	1	S B? ...		14.2		*
03938	07 34.4 + 10 01 209.13 + 14.60	02 - 20 - 005	1003	1.1 x 0.9 1.0 x 0.9	--- 2:	3	Sc Brightest in group, Sc -		15.1		*
03939	07 34.4 + 20 05 199.57 + 18.79	03 - 20 - 009	1507 98	(1.1 x 0.9) (1.2 x 1.0)	(25) ...	3	S0		14.9		*
03940	07 34.4 + 71 17 144.18 + 29.47	12 - 08 - 000	0680	1.5: x 1.5: 1.3: x 1.0:	--- ...	3	Dwarf irregular Ir IV - V		16.5:		
03941	07 34.5 + 14 12 205.22 + 16.42	02 - 20 - 000	1003	1.6 x 0.4 1.1 x 0.3	74 6	2	S B: c		15.4		
03942	07 34.6 + 27 09 192.61 + 21.48	05 - 18 - 000	0678	1.2 x 0.2 1.1 x 0.2	151 7	3	Sb		15.3		
03943	07 34.9 + 59 17 157.94 + 29.07	10 - 11 - 114	1291 100	1.7 x 0.6 1.6 x 0.6	108 6	3	S B c	S B(s)c	14.9		*
03944	07 35.2 + 37 45 181.72 + 24.98	06 - 17 - 024	1336	2.0 x 0.9 1.6 x 0.8	130 5	3	Sc		15.0		
03945	07 35.2 + 69 11 146.60 + 29.56	12 - 08 - 000	0680	1.3 x 0.6 0.9 x 0.2	2 5 - 6	1	Sb		15.7		
03946	07 35.4 + 03 25 215.31 + 11.87	01 - 20 - 003	1527	1.4 x 0.9 1.3 x 0.8	[15] ...	4	Irr I m Ir IV - V Ir I		14.4		*
03947	07 35.8 + 34 02 185.66 + 24.00	06 - 17 - 000	1336	1.0: x 0.7: 0.8 x 0.4	(10) ...	2	Dwarf: spiral S IV - V:		17.		
03948	07 35.8 + 54 26 163.46 + 28.62	09 - 13 - 025	0985 96	1.1: x 0.8: 1.1 x 0.8:	(155) ...	3	S B c	S B(r)cd or SAB(rs)cd	15.6		
03949	07 36.0 + 48 51 169.71 + 27.74	08 - 14 - 031	0670 103	1.2 x 1.2 1.2 x 1.2	--- 1	4	S B c	S B(s)d	14.9		
03950	07 36.1 + 02 20 216.38 + 11.52	00 - 20 - 002	0933	1.1 x 0.25 1.1 x 0.25	176 6 - 7	1	S0 - a		14.8		
03951	07 36.1 + 67 00 149.11 + 29.63	11 - 10 - 000	1330	1.1 x 0.12 0.9 x 0.15	20 7	2	Sb		16.0:		
03952	07 36.2 + 19 37 200.20 + 19.00	03 - 20 - 000	1507 98	1.1 x 0.20 0.8 x 0.2	58 6 - 7	1	Sb - c		15.4		
03953	07 36.2 + 58 01 159.40 + 29.11	10 - 11 - 119	1291 100	1.1 x 0.7 1.0 x 0.5	2 4	3	Sb/SBb	SAB(s)b	15.6		*
03954	07 36.5 + 39 08 180.32 + 25.60	07 - 16 - 000	0701	1.1: x 0.8: ...	--- 	3	Dwarf irregular Ir V		18.		
03955	07 36.7 + 09 01 210.32 + 14.67	02 - 20 - 000	1003	1.6 x 0.2 1.6 x 0.15	71 7	3	...	S d - dm	15.3		

N G C   2415, 2416, 2418   I C   2201

1	2	3	4	5	6	7	8	9	10	11	12
03956	07 36.8 + 24 35 195.37 + 21.02	04 - 18 - 000	1310 101	1.1 x 0.3 0.9 x 0.3	22 6	3	...	S d - dm	16.5:		*
03957	07 36.9 + 55 32 162.24 + 28.92	09 - 13 - 030	0985 96	(1.2 x 1.2) (1.5 x 1.5)	-- ...	3	E Brightest in multiple system		(14.2)		*
03958	07 37.3 + 34 20 185.45 + 24.39	06 - 17 - 025 IC 2203	1336	1.2 x 1.0 1.2 x 0.9	(160) 2:	3	S B c P w 3965	SB cd	14.5		*
03959	07 37.3 + 39 20 180.16 + 25.81	07 - 16 - 009 NGC 2424	0701 106	3.9 x 0.5 3.4 x 0.5	81 7	3	Sb		13.9		*12
03960	07 37.4 + 23 23 196.62 + 20.70	04 - 18 - 030	1310 101	(1.3 x 1.1) (1.5 x 1.2)	(50) ...	3	E		15.0		
03961	07 37.5 + 47 47 170.95 + 27.78	08 - 14 - 000	0670	1.1 x 0.9 0.9: x 0.8:	(140) 1 - 2	3	S B c	SB d	15.6		
03962	07 37.7 + 13 59 205.77 + 17.03	02 - 20 - 006	1003	2.3 x 0.7 2.2 x 0.7	133 6	3	Sa/S B b	Sb -	14.7		
03963	07 37.7 + 51 49 166.45 + 28.52	09 - 13 - 032	0985	1.6: x 0.3 1.6: x 0.3	54 7	2	...	S dm:	16.5:		
03964	07 37.8 - 01 28 220.00 + 10.12	00 - 20 - 004	0933	2.0: x 1.5: 1.6: x 1.2:	(125) ...	3	...	SAB: dm	15.5		
03965	07 38.0 + 34 20 185.50 + 24.53	06 - 17 - 026 IC 2204	1336	1.1 x 1.1 1.1 x 1.1	-- 1	4	S B b	(R)S B(r)ab Sb -	14.6		*
03966	07 38.0 + 40 13 179.25 + 26.17	07 - 16 - 011	0701 106	2.2: x 2.0: 1.7:x 1.7:	-- ...	3	Dwarf irregular Ir V		16.0:		*
03967	07 38.0 + 63 03 153.66 + 29.71	11 - 10 - 021	1330 100	1.5:: x 1.1:: 0.8: x 0.8:	-- ...	3	S B? c Singular Sc +		16.5:		*
03968	07 38.0 + 66 22 149.84 + 29.81	11 - 10 - 019	1330	1.5: x 1.5: 0.7 x 0.6	-- 1	4	S B c	SB(r)c	15.2		
03969	07 38.1 + 27 44 192.32 + 22.41	05 - 18 - 028	0678	1.1 x 0.15 1.0 x 0.15	135 6 - 7	2	Sc		15.4		*
03970	07 38.3 + 61 40 155.25 + 29.67	10 - 11 - 124	1291 100	1.0 x 0.1 1.0 x 0.1	154 7	2	Sc		16.0:		
03971	07 38.6 + 60 17 156.84 + 29.62	10 - 11 - 127	1291 100	1.2 x 0.6 1.2 x 0.6	103 ...	2	SO - a		15.6		*
03972	07 38.6 + 73 56 141.12 + 29.67	12 - 08 - 08a	0680	1.3 x 0.7 1.1 x 0.6	160 ...	2	S B ... Disturbed		14.7		*1V
03973	07 38.8 + 49 55 168.62 + 28.38	08 - 14 - 033 Mark 79	0670 103	1.4 x 1.4 1.5 x 1.3:	-- ...	4	S B b Seyfert		13.3	+ 6600	
03974	07 39.0 + 16 55 203.10 + 18.53	03 - 20 - 010	1507	5.: x 4.: 3.: x 2.:	-- ...	4	Dwarf irregular Ir V		15.4		*
03975	07 39.0 + 72 55 142.29 + 29.76	12 - 08 - 009	0680	2.3: x 1.8: 1.1: x 1.0	(160) ...	4	...	S B dm	15.1		
03976	07 39.2 + 41 02 178.43 + 26.59	07 - 16 - 012	0701 106	1.1: x 1.0: 0.5 x 0.5	-- 1:	2	S ...		15.4		
03977	07 39.4 + 52 27 165.78 + 28.87	09 - 13 - 038 NGC 2426	0985	1.1 x 1.1 1.2 x 1.2	-- ...	3	E		14.4		*3
03978	07 39.5 + 62 11 154.67 + 29.84	10 - 11 - 131	1291 100	1.2 x 1.0 1.2 x 1.0	(85) ...	4	S B b	(R´)S B(s)b	15.0		*C
03979	07 39.7 + 67 23 148.67 + 29.98	11 - 10 - 024	1330	1.9 x 0.3 1.8 x 0.3	153 6	2	Sb - c		14.8		
03980	07 39.8 + 18 27 201.69 + 19.32	03 - 20 - 000	1507	1.4 x 0.25 1.2 x 0.2	18 6 - 7	1	S ...		15.2		
03981	07 39.8 + 57 06 160.50 + 29.50	10 - 11 - 130 Mark 81	1291	1.5 x 0.9 1.5 x 0.9	73 ...	3	SO		14.3		
03982	07 39.9 + 49 46 168.82 + 28.53	08 - 14 - 035 IC 471	0670 103	0.6 x 0.6 0.8 x 0.8	-- ...	3	E P w 3985		14.2		*
03983	07 39.9 + 52 29 165.76 + 28.95	09 - 13 - 039 NGC 2429	0985	1.8 x 0.4 1.8 x 0.4	145 7	2	S ... np of 2, disturbed		14.7		*13 V
03984	07 39.9 + 70 10 145.46 + 29.95	12 - 08 - 010	0680	2.1 x 1.0 2.2 x 1.0	55 5	4	S B b	S B(s)b or SAB(s)b	14.2		
03985	07 40.1 + 49 43 168.88 + 28.55	08 - 14 - 036 IC 472	0670 103	1.9 x 1.0 1.8 x 1.0	167 5	4	Sb/S B b P w 3982	SAB(rs)b	14.5		*1
03986	07 40.3 + 31 39 188.46 + 24.16	05 - 19 - 000	1344	1.1 x 0.1 0.9 x 0.1	-- 7	2	Sc ssf of 2		15.7		*
03987	07 40.5 + 23 03 197.23 + 21.24	04 - 18 - 032	1310	1.0 x 0.18 0.9 x 0.18	49 7	1	S ...		15.3		
03988	07 40.7 + 47 52 170.97 + 28.33	08 - 14 - 000	0670	1.1 x 0.1 1.0 x 0.1	36 7	3	Sc		16.5:		*
03989	07 40.7 + 53 58 164.09 + 29.27	09 - 13 - 000	0985	1.1: x 1.1: 0.3: x 0.2:	-- ...	2	Dwarf spiral S IV - V:		18.		*
03990	07 40.9 + 40 28 179.13 + 26.76	07 - 16 - 013	0701 106	1.1 x 0.1 1.0 x 0.1	172 7	1	S ... P w 3997		15.7		*

N G C   2424, 2426, 2429   I C   471, 472, 2203, 2204

1	2	3	4	5	6	7	8	9	10	11	12
03991	07 40.9 + 65 53 150.40 + 30.10	11 – 10 – 025	1330	1.2 x 0.6 1.2 x 0.6	162 5	2	Sc		14.9		
03992	07 41 + 84 45 128.82 + 28.46	14 – 04 – 036	0692 102	1.7: x 1.7: 1.7: x 1.7:	--- ...	0	...		16.0:		*
03993	07 41 + 85 04 128.47 + 28.40	14 – 04 – 037	0692 102	1.6? x 1.2? 1.2 x 1.0	(35) ...	1	SO? Compact nucleus		14.0		*
03994	07 41 + 85 18 128.21 + 28.36	14 – 04 – 038 IC 469	0692 102	2.3 x 1.2 2.7 x 1.4	90 5	3	Sa		13.6		
03995	07 41.0 + 29 21 190.90 + 23.56	05 – 19 – 001	1344	2.5 x 1.1 2.5: x 1.1	85 ...	3	S ... Peculiar, disturbed		13.6		*1
03996	07 41.0 + 31 46 188.39 + 24.34	05 – 19 – 002 NGC 2435	1344	2.6 x 0.6 2.5 x 0.7	36 6	3	Sa		13.5		3
03997	07 41.2 + 40 29 179.13 + 26.82	07 – 16 – 014	0701 106	1.5: x 1.5: 1.2: x 1.0:	--- ...	2	Dwarf: IV – V:		16.0:		*
03998	07 41.2 + 56 18 161.44 + 29.61	09 – 13 – 043	0985	1.3: x 0.7: 0.8: x 0.6:	(75) ...	3	...	S B m	16.5:		1
03999	07 41.4 + 53 12 164.98 + 29.28	09 – 13 – 042 NGC 2431	0985	0.9 x 0.8 1.0 x 0.8	(35) 2:	3	S B a	(R´:)S B(s)a	14.3		3
04000	07 41.5 + 46 11 172.88 + 28.14	08 – 14 – 037	0670	1.6 x 0.5 1.5 x 0.5	40 6	2	S B? b		14.8		
04001	07 41.9 + 62 35 154.22 + 30.13	10 – 11 – 134	1291 100	(1.3 x 0.6) (1.5 x 0.8)	135 ...	2	E		14.6		
04002	07 42.1 + 51 07 167.36 + 29.08	09 – 13 – 048	0985 103	1.3 x 0.5 1.0 x 0.4	50 ...	3	... S IV – V	S m	17.		
04003	07 42.2 + 59 05 158.26 + 29.98	10 – 11 – 136	1291 100	1.4 x 0.7 1.4 x 0.7	87 5	3	Sa or SO/Sa, 4020 group		14.8		*
04004	07 42.4 – 00 15 219.46 + 11.71	00 – 20 – 000	0933	1.3 x 0.2 1.2 x 0.15	73 6 – 7	1	S ...		15.6		
04005	07 42.5 + 08 03 211.87 + 15.53	01 – 20 – 004	1527	2.0 x 0.5 1.9 x 0.4	15 6:	1	S ... Singular, disturbed?	f of 2	15.1		*
04006	07 42.5 + 11 11 208.94 + 16.90	02 – 20 – 007	1003	1.0 x 0.7 1.0 x 0.7	(165) ...	1	S ...		14.8		
04007	07 42.7 + 48 28 170.40 + 28.76	08 – 14 – 042	0670 103	1.1 x 0.4 1.0 x 0.4	129 6	2	S B a		14.6		*
04008	07 43.0 + 44 54 174.36 + 28.14	08 – 14 – 043	0670	... ...	... ...	1	SO – a + companion		14.3		*
04009	07 43.0 + 54 11 163.89 + 29.63	09 – 13 – 051	0985	1.1 x 0.3 0.9: x 0.2	... ...	1	... Peculiar, distorted?		16.5:		*
04010	07 43.1 + 05 05 214.69 + 14.34	01 – 20 – 000	1527	1.5: x 0.4 1.1 x 0.3	90 6 – 7	1	Sa?		15.1		
04011	07 43.1 + 52 23 165.95 + 29.42	09 – 13 – 053	0985	1.0 x 0.12 1.0 x 0.12	143 7	3	Sc		16.0:		*
04012	07 43.1 + 59 08 158.21 + 30.10	10 – 11 – 137	1291 100	1.3 x 0.3 1.1 x 0.3	86 7	1	S ... 4020 group		15.4		*
04013	07 43.2 + 61 03 156.00 + 30.22	10 – 11 – 138 Mark 10	1291 100	1.9 x 0.7 1.7 x 0.7	130 6	3	Sb Seyfert		14.0	+ 8700	*
04014	07 43.2 + 74 27 140.50 + 29.94	12 – 08 – 011 Mark 11	0680	1.0 x 0.5 0.9 x 0.5	122 ...	2	E – SO		14.4		*
04015	07 43.4 + 62 26 154.40 + 30.30	10 – 11 – 140	1291	1.2 x 1.1 1.2 x 1.1	--- 1	4	S B c	S B(s)cd	15.4		
04016	07 43.5 + 39 08a 180.72 + 26.92	07 – 16 – 016 NGC 2444	0701 105							+ 3965 + 3951	
04017	07 43.5 + 39 08b 180.72 + 26.92	07 – 16 – 017 NGC 2445	0701 105	[3.0 x 2.3] ...	... ...	4	Double system, contact Extreme distortion		13.1		*12 3VA
04018	07 43.5 + 41 40b 177.95 + 27.53	07 – 16 – 018	0701 106	(1.0 x 0.8) 0.8 x 0.7	--- ...	1	E: nnf of 2, disturbed		(14.8)		*1
04019	07 43.5 + 67 03 149.05 + 30.35	11 – 10 – 026	1330	1.1 x 0.2 0.9 x 0.2	123 6	1	S ...		15.3		
04020	07 43.8 + 59 08 158.22 + 30.19	10 – 11 – 142	1291 100	2.2 x 1.0 1.8 x 0.9	18 5	3	Sb/S B b Brightest of 3		14.7		*
04021	07 43.8 + 65 28 150.88 + 30.40	11 – 10 – 027	1330	1.2 x 0.9 1.1 x 0.9	(50) ...	2	S B a – b		15.1		
04022	07 44.0 + 48 20 170.56 + 28.95	08 – 14 – 046	0670 103	1.1 x 0.4 1.0 x 0.4	133 6	3	Sc		15.5		
04023	07 44.0 + 66 19 149.90 + 30.41	11 – 10 – 028	1330	1.6 x 0.7 1.6 x 0.7	136 6	3	SO		14.2		*
04024	07 44.1 + 62 19 154.54 + 30.38	10 – 11 – 143	1291 100	1.7 x 0.4 1.7 x 0.4	115 6	3	Sc		14.8		
04025	07 44.2 + 07 25 212.66 + 15.63	01 – 20 – 006	1527	1.2 x 0.4 1.2 x 0.5	37 6	2	S B: b		14.7		

N G C   2431, 2435, 2444, 2445   I C   469

1	2	3	4	5	6	7	8	9	10	11	12
04026	07 44.3 + 27 03 193.53 + 23.47	05 – 19 – 007 NGC 2449	1344 108	1.6 x 0.7 1.5 x 0.7	137 5 – 6	2	Sa – b sf of 2		14.3		*13
04027	07 44.7 + 54 44 163.30 + 29.94	09 – 13 – 058 NGC 2446	0985 112	2.1 x 1.0 1.8 x 0.9	130: 5	4	Sb		13.9		*3
04028	07 44.8 + 74 28 140.47 + 30.05	12 – 08 – 013 Mark 12	0680	1.1 x 0.9 1.0 x 0.9	(10) ...	2	S ...	SAB ...	12.7		*
04029	07 45.0 + 34 28 185.83 + 25.95	06 – 17 – 028	1336	2.7 x 0.5 2.3: x 0.4	63 6	2	S B: b – c		14.6		
04030	07 45.1 + 28 20 192.27 + 24.07	05 – 19 – 009	1344 108	... ...	... ...	2	Double system		14.4		*1
04031	07 45.4 + 23 21 197.38 + 22.39	04 – 19 – 002	0226 107	1.0 x 0.7 0.9 x 0.7	165 3:	3	Sc		15.3		
04032	07 45.4 + 30 16 190.29 + 24.76	05 – 19 – 000	1344 108	1.1 x 0.15 1.2 x 0.15	122 7	2	Sc		15.6		
04033	07 45.4 + 61 28 155.53 + 30.51	10 – 11 – 145	1291 100	1.1 x 1.0 1.1 x 1.0	-- 1	3	Sb	SAB(r)ab Sb –	14.9		*
04034	07 46.1 + 31 01 189.55 + 25.13	05 – 19 – 000	1344	1.2 x 0.3 0.8: x 0.3:	34 6	3	Sc	Sc +	15.7		
04035	07 46.1 + 55 30 162.44 + 30.22	09 – 13 – 068	0985 112	(0.8 x 0.8) (1.1 x 1.1)	-- ...	3	E sp of 2		14.1		*
04036	07 46.1 + 73 08 142.00 + 30.26	12 – 08 – 015 NGC 2441	0680	2.3 x 2.2 2.2 x 2.2	-- 1	4	Sc/SBc	SAB(r)c Sc –	12.7	+ 3623 + 3774	*12 4689
04037	07 46.4 + 73 43 141.32 + 30.23	12 – 08 – 000	0680	1.2 x 0.2 1.0 x 0.2	157 7	2	Sc		16.5:		
04038	07 46.5 + 27 02 193.72 + 23.92	05 – 19 – 013	1344 108	1.1 x 0.4 1.1 x 0.4	83 6	1	S ...		15.1		
04039	07 46.5 + 30 04 190.58 + 24.92	05 – 19 – 014	1344 108	1.4 x 0.25 1.2 x 0.2	140 7	1	Sb: P w 4042		15.2		*
04040	07 46.6 + 34 05 186.34 + 26.15	06 – 17 – 029 IC 2207	1336	2.0 x 0.3 1.9 x 0.3	124 7	3	Sc		15.4		
04041	07 46.7 + 73 38 141.42 + 30.26	12 – 08 – 016	0680	0.9 x 0.35 0.9 x 0.35	132 6	1	S ...		13.6		*
04042	07 46.8 + 30 09 190.51 + 25.01	05 – 19 – 015	1344 108	1.0 x 0.8 1.1 x 0.8	(25) 2:	4	S B b P w 4039	(R:)SB(r)b Sb –	15.0		*
04043	07 46.8 + 54 29 163.62 + 30.22	09 – 13 – 072	0985	2.2 x 0.2 2.1 x 0.2	5 7	3	Sc	Sc +	14.8		
04044	07 46.9 + 18 57 201.91 + 21.06	03 – 20 – 014	1507 105	1.5 x 0.7 1.2: x 0.7:	147 4 – 5	4	Sc		15.1		*
04045	07 46.9 + 34 33 185.86 + 26.34	06 – 17 – 000	1336	(1. x 1.) (1.0 x 0.9)	-- ...	1	... Very compact, jet		15.5		*
04046	07 47.0 + 30 48 189.85 + 25.25	05 – 19 – 000	1344 108	1.0 x 0.6 0.9: x 0.5:	[175] ...	2	S B ...		17.		
04047	07 47.0 + 30 51 189.79 + 25.27	05 – 19 – 017	1344 108	1.8 x 1.0 2.2: x 1.1	[120] ...	4	S B b sff of 2, distorted		14.5		*1
04048	07 47.1 + 19 42 201.19 + 21.39	03 – 20 – 000	1507 105	1.0 x 0.5 0.9 x 0.5	160: ...	1	S ...		15.7		
04049	07 47.3 + 57 03 160.66 + 30.51	10 – 11 – 153	1291	1.4 x 0.7 1.4: x 0.7:	10: ...	1	S? ...		14.6		1
04050	07 47.3 + 72 10 143.11 + 30.43	12 – 08 – 017	0680	0.7 x 0.6 0.7 x 0.5	(70) ...	1	S? ... f of 2		14.3		*
04051	07 47.6 + 50 17 168.45 + 29.83	08 – 15 – 006	1317 103	0.45 x 0.30 0.7 x 0.55	... ...	1	...		14.4		*
04052	07 47.6 + 50 21 168.38 + 29.84	08 – 15 – 007	1317 103	[0.8 x 0.4] ...	... ...	2	Double system Contact		14.1		*1
04053	07 47.7 + 16 30 204.39 + 20.26	03 – 20 – 015 NGC 2454	1507	(1.1 x 0.6) 1.1 x 0.6	100 ...	1	...		14.7		*3
04054	07 47.9 + 24 01 197.09 + 23.55	04 – 19 – 005	0226 107	2.2 x 0.5 2.1 x 0.5	175 6	2	Sb		15.1		
04055	07 47.9 + 34 11 186.32 + 26.44	06 – 17 – 030	1336	1.2 x 0.9 1.1: x 1.0:	(135) 1 – 3	3	Sc	Sc +	15.6		
04056	07 48.0 + 43 00 176.69 + 28.63	07 – 16 – 020	0701	1.1 x 0.7 1.1 x 0.6	30 4	3	Sc	SAB(s)c	15.0		*
04057	07 48.0 + 74 32 140.37 + 30.25	12 – 08 – 018	0680	2.6 x 0.7 2.7 x 0.8	55 6	3	Sa		13.4		1
04058	07 48.1 + 18 06 202.87 + 20.99	03 – 20 – 000	1507	1.1: x 1.1: 1. x 1.	-- 1:	2	... S IV	S dm	16.5:		
04059	07 48.2 + 62 40 154.14 + 30.86	10 – 12 – 003	0960	1.5: x 0.9: 1.3: x 0.7:	50: 4:	3	Sc		15.4		1
04060	07 48.7 + 14 09 206.78 + 19.53	02 – 20 – 009	1003	1.2 x 0.35 1.1 x 0.4	104 6	1	S ...		14.6		

N G C   2441,   2446,   2449,   2454   I C   2207

1	2	3	4	5	6	7	8	9	10	11	12
04061	07 48.8 + 27 26 193.50 + 24.54	05 – 19 – 000	1344 108	1.3: x 0.12 1.2 x 0.15	161 7	1	Sb		15.7		
04062	07 48.9 + 50 32 168.20 + 30.07	08 – 15 – 000	1317 103	1.3 x 0.2 1.0 x 0.2	135 6 – 7	1	Sb – c		16.0:		
04063	07 49 + 85 57 127.45 + 28.38	14 – 04 – 000	0692 102	1.2: x 1.0: 1.0: x 1.0:	— ...	2	... S B IV – V	S B m	17.		
04064	07 49.0 + 30 30 190.31 + 25.57	05 – 19 – 000	1344 108	1.5: x 0.6: 1. x 0.5::	[154] ...	3	Dwarf V		17.		
04065	07 49.0 + 55 21 162.65 + 30.61	09 – 13 – 077	0985 112	1.6 x 0.2 1.6 x 0.2	159 7	3	Sc	S d	15.5		
04066	07 49.0 + 78 08 136.24 + 29.84	13 – 06 – 012	1295	2.1 x 2.0 1.9 x 1.9	— 1	3	Sc		14.2		*
04067	07 49.2 + 72 45 142.42 + 30.52	12 – 08 – 019	0680	1.5 x 0.9 1.5: x 0.9:	165 4	3	S B c		15.4		
04068	07 49.5 + 40 17 179.77 + 28.31	07 – 16 – 000	0701 106	1.6 x 0.2 1.3 x 0.2	14 7	2	Sc		17.		
04069	07 49.7 + 39 53 180.22 + 28.26	07 – 16 – 021	0701 106	1.0 x 0.2 0.8 x 0.2	9 6 – 7	1	S ...		15.7		
04070	07 49.7 + 50 21 168.43 + 30.17	08 – 15 – 016	1317 103	1.0 x 0.8 1.0 x 0.7	(135) 2 – 3	2	Sa/S B a	SAB(r:)a	14.9		*
04071	07 49.8 + 51 56 166.61 + 30.39	09 – 13 – 079	0985	1.0: x 0.8: 0.8 x 0.6	(50) ...	1	...		15.0		
04072	07 49.8 + 53 01 165.36 + 30.51	09 – 13 – 080	0985	1.4 x 0.5 1.4: x 0.4	17 6	3	S B b	S B(s)b	14.6		*
04073	07 50.1 + 55 37 162.36 + 30.79	09 – 13 – 082 NGC 2456	0985 112	(1.1 x 0.8) (1.3 x 0.9)	30: ...	3	E		14.3		3
04074	07 50.3 + 54 23 163.79 + 30.71	09 – 13 – 084	0985	1.1 x 1.1 1.1 x 1.0	— 1	4	Sc	SA(rs)cd	13.8		1
04075	07 50.3 + 60 19 156.89 + 31.06	10 – 12 – 009	0960 100	1.1 x 0.10 1.2 x 0.10	160 7	2	Sc		16.0:		
04076	07 50.4 + 63 08 153.60 + 31.11	11 – 10 – 043	1330 100	1.0 x 0.35 1.0 x 0.35	126 6	2	S B? b – c		15.6		
04077	07 50.6 + 14 45 206.40 + 20.19	02 – 20 – 010	1003	1.2: x 1.0: 1.1: x 1.0:	— 1 – 2	3	S B c	S B(rs:)d	15.1		*
04078	07 51 + 84 46 128.75 + 28.68	14 – 04 – 042	0692 102	2.3 x 0.5 2.3 x 0.5	82 6	2	Sb – c		15.4		
04079	07 51.0 + 55 50 162.12 + 30.93	09 – 13 – 090	0985 112	1.0 x 0.5 1.0 x 0.5	3 ...	1	S ... Peculiar		13.6		*
04080	07 51.0 + 73 55 141.05 + 30.53	12 – 08 – 021	0680	1.1 x 0.8 1.0 x 0.8	60 3:	2	Sa		13.9		
04081	07 51.1 + 38 47 181.51 + 28.26	07 – 17 – 000	1329 106	1.1 x 0.4 0.8:: x 0.3::	[58] ...	3	Dwarf irregular Ir IV – V		17.		
04082	07 51.1 + 50 10 168.67 + 30.37	08 – 15 – 018	1317 102	(1.1 x 1.0) (0.9 x 0.9)	— ...	2	E:		14.5		*1
04083	07 51.3 + 56 21 161.52 + 31.01	09 – 13 – 000	0985 112	1.1: x 0.1 0.6 x 0.1	35 7	2	Sc P w 4088		17.		*
04084	07 51.4 + 29 50 191.19 + 25.85	05 – 19 – 000	1344 108	1.2 x 0.3 0.9 x 0.3	[150] ...	3	Irr		15.6		
04085	07 51.4 + 53 28 164.87 + 30.79	09 – 13 – 091	0985	1.0 x 0.7 0.8 x 0.6	(65) ...	1	S ...		14.4		*
04086	07 51.5 + 16 20 204.95 + 21.03	03 – 20 – 000	1507	1. x 1. 0.8: x 0.7:	— ...	3	Dwarf IV – V		18.		*
04087	07 51.5 + 43 37 176.14 + 29.38	07 – 16 – 024	0701	1.0: x 0.7: 1.1: x 0.7:	100: 3:	3	Sc	Sc +	15.5		
04088	07 51.5 + 56 20 161.54 + 31.04	09 – 13 – 000	0985 112	1.1 x 0.1 1.0 x 0.1	169 7	2	Sc P w 4083		17.		*
04089	07 51.6 + 13 52 207.36 + 20.05	02 – 20 – 000	1003	1.2 x 0.25 1.1 x 0.25	35 7	2	Sa		15.3		
04090	07 51.6 + 16 56 204.37 + 21.29	03 – 20 – 016	1507	(1.0 x 0.8) 0.9 x 0.6	(30) ...	1	S? ... Companion superimposed		14.8		1
04091	07 51.7 + 04 35 216.16 + 16.01	01 – 20 – 009 NGC 2470	1527	2.2 x 0.6 2.1 x 0.6	128 6	1	Sa – b		14.2		
04092	07 51.9 + 58 34 158.94 + 31.21	10 – 12 – 015	0960	1.2 x 0.45 1.2 x 0.5	56 6	1	Sa – b		14.3		
04093	07 52.0 + 60 26 156.76 + 31.28	10 – 12 – 017 IC 2209, Mark 13	0960 100	1.1 x 0.8 0.7 x 0.7	(145) ...	1	S ... P w 4097		14.5 + 15600		*12
04094	07 52.1 + 63 44 152.90 + 31.30	11 – 10 – 045	1330 100	1.4: x 1.1: 1.1: x 0.8	[110] ...	2	Sa/SBb		14.9		
04095	07 52.1 + 66 44 149.39 + 31.21	11 – 10 – 044	1330	0.9 x 0.35 0.9 x 0.35	[51] ...	1	... Peculiar		14.4		*

N G C   2456, 2470   I C   2209

1	2	3	4	5	6	7	8	9	10	11	12
04096	07 52.3 + 26 52 194.37 + 25.08	04 – 19 – 000 IC 480	0226	1.8 x 0.3 1.8 x 0.3	168 7	2	Sb – c		15.2		
04097	07 52.6 + 60 30 156.68 + 31.35	10 – 12 – 021 NGC 2460	0960 100	4.0: x 2.8: 1.9 x 1.5	40: 3:	3	Sb P w 4093	Sb –	12.5	+ 1442 + 1534	*12 468
04098	07 52.7 + 66 34 149.58 + 31.27	11 – 10 – 046	1330	0.9 x 0.55 0.8 x 0.55	105 ...	1	S ... Peculiar		14.5		*
04099	07 52.8 + 24 50 196.52 + 24.50	04 – 19 – 008	0226	1.9 x 1.6 1.8 x 1.4	(100) 2:	4	S B b	S B(r)b	14.7		
04100	07 53 + 84 55 128.57 + 28.69	14 – 04 – 043	0692 102	1.8: x 1.5: 1.5: x 1.2:	(135) ...	4	Irr	IB m Ir I	16.0:		1
04101	07 53.0 + 45 41 173.86 + 30.02	08 – 15 – 022	1317 110	1.3 x 0.2 1.2 x 0.2	30 6	1	S ...		16.0:		*
04102	07 53.0 + 74 02 140.90 + 30.65	12 – 08 – 000	0680	1.0 x 0.3 1.0 x 0.3	23 6	2	Sc		16.5:		
04103	07 53.0 + 79 30 134.65 + 29.81	13 – 06 – 013	1295	1.5 x 0.6 1.1 x 0.5	105 6	3	Sc	Sc +	16.0:		
04104	07 53.1 + 58 40 158.83 + 31.37	10 – 12 – 030	0960	1.4 x 0.2 1.4 x 0.2	25 7	2	Sa		15.5		
04105	07 53.3 + 27 09 194.16 + 25.39	05 – 19 – 022	1344 108	1.1 x 0.7 1.1 x 0.7	85 4	1	S ...		15.0		
04106	07 53.3 + 40 04 180.20 + 28.97	07 – 17 – 003 NGC 2476	1329 106	1.4 x 0.8 1.2 x 0.7	135: ...	1	E?		13.4		
04107	07 53.3 + 49 42 169.26 + 30.66	08 – 15 – 027	1317 103	1.7 x 1.6 1.6 x 1.6	--- 1	4	Sc	SA(rs)c	13.9		
04108	07 53.3 + 59 14 158.17 + 31.41	10 – 12 – 032	0960	1.2 x 0.3 1.1 x 0.3	135 6	1	S ...		15.0		
04109	07 53.5 + 11 48 209.57 + 19.60	02 – 21 – 000	0247	1.2: x 1.1: 1.1 x 1.1	--- 1	3	S B b	S B(r)b	15.2		*
04110	07 54.0 + 56 30 161.37 + 31.39	09 – 13 – 095 NGC 2468	0985 112	[1.0 x 0.6] ...	... ...	2	Double system Contact		14.9		*23
04111	07 54.0 + 56 49 161.00 + 31.41	10 – 12 – 035 NGC 2469	0960 112	1.0 x 0.7 1.0 x 0.7	160 3:	1	S ...		13.2		23
04112	07 54.1 + 07 37 213.61 + 17.92	01 – 21 – 001 NGC 2485	0027	1.6 x 1.6 1.8 x 1.7	--- 1	4	Sa f of 2		13.3		*13
04113	07 54.1 + 31 37 189.49 + 26.94	05 – 19 – 000	1344	1.0 x 0.1 0.7 x 0.1	66 7	3	Sc		16.5:		
04114	07 54.1 + 53 00 165.46 + 31.15	09 – 13 – 96+97 NGC 2474 + 2475	0985	[0.9 x 0.7] ...	... ...	2	Double system Contact		13.9	+ 5019 + 5076	*12 3
04115	07 54.2 + 14 32 206.99 + 20.90	02 – 21 – 000	0247	2.0: x 1.1: 2. x 1.	[145] ...	4	Irr	IA m Ir I	16.0:		
04116	07 54.2 + 23 55 197.59 + 24.48	04 – 19 – 009 NGC 2481	0226	1.6: x 1.1: 1.4 x 0.9	(160) ...	2	SB ... P w 4118, distorted		15.2		*13
04117	07 54.2 + 36 04 184.67 + 28.19	06 – 18 – 003	0989	1.5 x 0.7 1.2 x 0.7	[136] ...	3	Irr	IB m	15.5		
04118	07 54.3 + 23 54 197.62 + 24.49	04 – 19 – 010 NGC 2482	0226	1.6 x 0.5 1.5 x 0.5	18 ...	1	... P w 4116, distorted		13.4		*1
04119	07 54.5 + 32 42 188.35 + 27.34	05 – 19 – 023 IC 2211	1344	0.8 x 0.4 0.8 x 0.4	140 5:	1	Sa:		14.5		*1
04120	07 54.8 + 39 58 180.38 + 29.23	07 – 17 – 004	1329 106	1.0 x 0.1 0.8 x 0.1	12 7	3	Sc		15.6		
04121	07 54.8 + 58 12 159.38 + 31.57	10 – 12 – 046	0960	2.5: x 0.8: 2. x 0.8::	10 ...	3	Dwarf spiral SB IV – V		16.0:		*
04122	07 54.8 + 59 15 158.15 + 31.60	10 – 12 – 040	0960	1.6 x 1.1 1.6 x 1.1	3 ...	3	E P w 4124		14.7		*
04123	07 54.9 + 25 17 196.24 + 25.10	04 – 19 – 011 NGC 2486	0226	1.9 x 1.1 1.7 x 1.0	100 4 – 5	3	Sa P w 4126		14.3		*13
04124	07 54.9 + 59 17 158.11 + 31.62	10 – 12 – 041	0960	1.1 x 0.9: 1.0 x 0.9:	150: ...	1	SO P w 4122		15.4		*
04125	07 55.1 + 37 55 182.68 + 28.82	06 – 18 – 004 NGC 2484	0989	(1.3 x 0.8) (1.7 x 1.0)	(145) ...	2	SO:		14.9		
04126	07 55.3 + 25 16 196.29 + 25.18	04 – 19 – 012 NGC 2487	0226	2.5 x 2.2 2.5: x 2.3:	(115) 1 – 2	3	S B b P w 4123	S B(r)b Sb +	14.0		*13
04127	07 55.9 + 08 10 213.30 + 18.56	01 – 21 – 002 NGC 2496	0027 111	(1.1 x 0.9) (1.4 x 1.2)	(2) ...	3	E		14.8		*
04128	07 55.9 + 60 25 156.78 + 31.76	10 – 12 – 049	0960	1.8 x 1.1 1.7 x 0.9	45 4	2	Sc		13.9		
04129	07 56 + 86 18 127.03 + 28.41	14 – 04 – 44+45	0692	[1.5: x 1.3::] ...	... ...	3	Double system Disrupted		16.5:		*1
04130	07 56.0 + 24 17 197.37 + 24.99	04 – 19 – 013 IC 481	0226	1.0 x 0.2 1.1 x 0.2	3 6 – 7	1	S ...		15.1		

N G C 2460, 2468, 2469, 2474 – 2476, 2481,
2482, 2484 – 2487, 2496

I C 480, 481, 2211

1	2	3	4	5	6	7	8	9	10	11	12
04131	07 56.0 + 31 56 189.28 + 27.42	05 - 19 - 026	1344	1.4 x 0.9 1.0 x 0.7	(140) ...	3	S B c	SAB c Sc(-)	15.7		
04132	07 56.0 + 33 03 188.07 + 27.74	06 - 18 - 005	0989	1.9: x 0.4 2.0 x 0.4	28 6	1	Sb - c		13.5		*
04133	07 56.0 + 56 31 161.37 + 31.67	09 - 13 - 102	0985 112	1.8 x 0.1 1.9 x 0.15	164 7	3	Sc P w 4134		16.0:		*1
04134	07 56.2 + 56 30 161.39 + 31.69	09 - 13 - 101	0985 112	1.0 x 0.4: 0.9 x 0.4	[25] ...	1	S B ... P w 4133		15.5		*1
04135	07 56.3 + 24 35 197.08 + 25.16	04 - 19 - 014	0226	1.0 x 0.25 1.0 x 0.2	105 6 - 7	1	S ...		15.5		
04136	07 56.3 + 47 33 171.82 + 30.86	08 - 15 - 034	1317 110	1.5 x 0.3 1.6 x 0.3	142 7	3	Sa		15.0		
04137	07 56.3 + 73 10 141.87 + 30.99	12 - 08 - 023	0680	1.9: x 1.4: 1.8: x 1.4:	145: ...	3	... S IV - V	S dm	15.7		
04138	07 56.4 + 27 10 194.39 + 26.04	05 - 19 - 028 NGC 2492	1344 108	(1.0 x 0.9) (1.4 x 1.1)	(95) ...	2	E - S0		14.4		*3
04139	07 56.5 + 16 34 205.23 + 22.22	03 - 21 - 004	0236	1.4 x 1.3 1.2 x 1.2	-- 1	4	Sc NGC 2507 group	SA(s)c	15.0		
04140	07 56.5 + 18 15 203.56 + 22.88	03 - 21 - 003	0236	1.5 x 0.25 1.4 x 0.25	144 6 - 7	2	Sb - c		14.9		
04141	07 56.6 - 00 30 221.41 + 14.72	00 - 21 - 001 IC 487	1298	1.1 x 1.1 1.1 x 1.0	-- 1 - 2	3	S B a		14.2		
04142	07 56.6 + 25 06 196.57 + 25.40	04 - 19 - 015 NGC 2498	0226	1.2 x 0.8 1.3 x 0.8	113 3	2	S B a: Singular, loop		14.4		*
04143	07 56.7 + 33 26 187.70 + 27.99	06 - 18 - 007 IC 2214	0989	0.8 x 0.7 0.8 x 0.7	-- 1 - 2	2	S B a - b		14.4		
04144	07 56.8 + 07 35 213.96 + 18.50	01 - 21 - 000	0027 111	1.0 x 0.12 1.2 x 0.12	5 7	2	Sc		15.6		
04145	07 56.8 + 15 31 206.30 + 21.87	03 - 21 - 005	0236	1.6 x 0.7 (1.5 x 0.7)	140 5 - 6	3	Sa NGC 2507 group		14.0		*
04146	07 56.8 + 59 16 158.14 + 31.86	10 - 12 - 054	0960	1.3 x 0.10 1.1 x 0.10	56 7	3	Sc		16.0:		
04147	07 56.9 + 18 33 203.30 + 23.09	03 - 21 - 006	0236	1.2 x 0.25 1.3 x 0.25	66 6	1	Sa - b		15.0		*1
04148	07 56.9 + 42 19 177.83 + 30.10	07 - 17 - 006	1329	2.5 x 0.3 1.8: x 0.2	10 7	3	...	S d - dm	15.5		
04149	07 57.0 - 01 32 222.40 + 14.31	00 - 21 - 000	1298	1.4 x 0.7 1.3: x 0.8:	(105) ...	3	... IV - V	m	16.5:		
04150	07 57.0 + 39 58 180.49 + 29.64	07 - 17 - 007 NGC 2493	1329 114	(2.2 x 2.2) (2.2 x 2.2)	-- 1	3	S B 0		13.1		*13
04151	07 57.0 + 77 58 136.35 + 30.28	13 - 06 - 016	1295	1.7 x 1.6 1.6: x 1.5:	-- (1)	2	Sc - Irr		13.1		
04152	07 57.2 + 05 45 215.73 + 17.76	01 - 21 - 004 NGC 2504	0027	0.45 x 0.35 0.45 x 0.45	-- ...	1	... Peculiar		14.1		3
04153	07 57.2 + 61 33 155.45 + 31.91	10 - 12 - 055	0960 100	1.0: x 0.7: 0.7: x 0.7:	(45) ...	3	...	S B dm	16.5:		*
04154	07 57.3 + 13 17 208.55 + 21.07	02 - 21 - 001	0247	1.4 x 0.6 1.4 x 0.5	147 6	3	Sc	Sc +	15.1		*
04155	07 57.3 + 26 45 194.90 + 26.10	04 - 19 - 018 IC 486	0226 108	1.0 x 0.7 1.0 x 0.7	145: 3	3	S B a Brightest in group		14.7		*
04156	07 57.3 + 26 50 194.82 + 26.13	05 - 19 - 000 IC 485	1344 108	1.3: x 0.3 1.4 x 0.25	153 6	1	Sa: f of 2, IC 486 group		15.5		*
04157	07 57.4 + 37 20 183.45 + 29.12	06 - 18 - 000	0989	1.0 x 0.8 1.0: x 0.8:	[120] ...	3	Irr		15.5		
04158	07 57.6 + 22 32 199.32 + 24.72	04 - 19 - 019 NGC 2503	0226	1.1 x 1.1 1.1 x 1.0	-- 1	3	S B b/Sc	SAB(rs)bc Sb +	15.0		13
04159	07 57.6 + 61 32 155.47 + 31.96	10 - 12 - 056	0960 100	0.9 x 0.50 0.9 x 0.4	90 ...	1	S ... P w 4169		13.9		*
04160	07 57.7 + 27 38 194.00 + 26.47	05 - 19 - 031 IC 2217	1344 108	0.6 x 0.4 0.5 x 0.4	(80) ...	1	S? ... Peculiar		14.2		
04161	07 57.8 + 56 42 161.16 + 31.92	09 - 13 - 109 NGC 2488	0985 112	(1.4 x 0.8) (1.6 x 1.1)	100 ...	2	E - S0		14.2		*3
04162	07 57.9 + 16 40 205.28 + 22.57	03 - 21 - 008	0236	1.0 x 0.2 1.0 x 0.2	28 7	1	S ...		14.9		
04163	07 57.9 + 44 19 175.59 + 30.64	07 - 17 - 000	1329	1.1: x 0.3 1.0 x 0.3	1 (6 - 7)	1	...		15.5		
04164	07 58.0 + 56 47 161.07 + 31.95	10 - 12 - 058	0960 112	1.4 x 0.6 1.4 x 0.6	140 6	2	S0 - a		15.2		*
04165	07 58.2 + 50 54 167.97 + 31.58	09 - 13 - 110 NGC 2500	0985	2.8 x 2.8 2.7: x 2.7:	-- 1	4	... S IV	SAB dm	12.3	+ 470 + 513	*12- -689

N G C   2488, 2492, 2493, 2498, 2500, 2503, 2504   I C   485 - 487, 2214, 2217

1	2	3	4	5	6	7	8	9	10	11	12
04166	07 58.2 + 74 11 140.67 + 30.98	12 – 08 – 024	0680	1.2 x 0.8 1.1 x 0.8	95 ...	2	S B c S B(s)c Disturbed?		14.5		*2
04167	07 58.3 + 25 25 196.39 + 25.87	04 – 19 – 020	0226	1.1 x 0.7 1.0 x 0.7	0: ...	3	Sc ssf of 2, disturbed, Sc +		(15.5)		*1
04168	07 58.3 + 57 05 160.72 + 32.00	10 – 12 – 061 NGC 2497	0960 112	(1.4 x 1.2) (1.1 x 1.1)	-- ...	3	... Compact		14.5		*
04169	07 58.3 + 61 31 155.49 + 32.04	10 – 12 – 062	0960 100	1.7 x 0.8 1.8 x 0.8	140 5	4	Sc SA cd P w 4159		13.5		*
04170	07 58.6 + 15 30 206.51 + 22.26	03 – 21 – 009	0236	(1.1 x 0.9) (1.3 x 1.1)	(140) ...	1	E: NGC 2507 group		14.5		
04171	07 58.8 + 09 51 212.03 + 19.94	02 – 21 – 003	0247 113	2.4 x 0.4 2.0 x 0.3	113 7	3	Sb NGC 2513 group		14.6		1
04172	07 58.8 + 15 51 206.18 + 22.45	03 – 21 – 010 NGC 2507	0236	(1.8 x 1.7) (2.5 x 2.2)	-- (1)	3	S ... Peculiar Brightest in group		14.0		*12 3
04173	07 59.0 + 80 15 133.74 + 29.92	13 – 06 – 017	1295	3.3: x 0.7: 2. x 0.8	[136] ...	3	Irr Disrupted		15.7		*1
04174	07 59.2 + 08 42 213.18 + 19.53	02 – 21 – 004 NGC 2508	0247 113	(2.0 x 1.6) (2.0 x 1.7)	(130) ...	1	E? 		14.2		3
04175	07 59.2 + 15 11 206.88 + 22.27	03 – 21 – 000	0236	(1.0 x 0.9) 0.8 x 0.7	-- 	1	S B? 0		15.2		
04176	07 59.3 + 40 49 179.63 + 30.24	07 – 17 – 009	1329 114	1.8 x 0.5 1.2 x 0.4	112 6	3	S B c Sc +		15.2		
04177	07 59.4 + 07 49 214.04 + 19.18	01 – 21 – 000	0027	1.2 x 0.4 1.0 x 0.3	90. 6	1	Sa – b		15.3		
04178	07 59.4 + 09 37 212.32 + 19.97	02 – 21 – 007 NGC 2510	0247	(1.0 x 0.7) (1.0 x 0.7)	120: (3)	2	S0: NGC 2513		14.7		*1
04179	07 59.5 + 00 57 220.45 + 16.05	00 – 21 – 002	1298	1.3 x 1.1 1.1 x 1.0	-- 1 – 2	2	S B b Brightest in multiple system		15.5		*
04180	07 59.5 + 27 35 194.20 + 26.83	05 – 19 – 035 IC 2219	1344 108	1.5 x 0.6 1.4 x 0.6	175 6	3	Sc Sc –		14.5		
04181	07 59.5 + 65 02 151.33 + 32.05	11 – 10 – 054	1330	1.0 x 0.35 1.0 x 0.35	56 6	2	S B? b – c		16.0:		
04182	07 59.6 + 61 29 155.52 + 32.20	10 – 12 – 068	0960 100	0.9 x 0.5 0.8 x 0.4	[170] ...	1	... Disturbed, n of 2		14.5		*1
04183	07 59.7 + 07 01 214.83 + 18.89	01 – 21 – 007	0027	2.0 x 1.2 1.9 x 1.2	73 4	4	S B b S B(r)ab Sb –		14.8		*
04184	07 59.7 + 09 33 212.42 + 20.01	02 – 21 – 009 NGC 2513	0247 113	(2.5 x 2.0) (2.8 x 2.3)	(170) ...	4	E Brightest in group		13.7		*13
04185	07 59.7 + 63 07a 153.59 + 32.16	11 – 10 – 056	1330 100	1.4 x 0.15 1.4 x 0.2	144 7	1	S ... 1st of 3		(15.5)		*1
04186	07 59.7 + 63 07c 153.59 + 32.16	11 – 10 – 058	1330 100	1.2 x 0.15 1.4 x 0.2	129 7	1	S ... 3rd of 3				*1
04187	07 59.8 + 66 55 149.12 + 31.95	11 – 10 – 057	1330	1.2 x 0.3 0.9 x 0.2	75 6	3	Sc P w 4195		16.0:		*
04188	07 59.9 + 42 03 178.25 + 30.59	07 – 17 – 010	1329	(1.5 x 0.7) (1.2 x 0.6)	115 5:	3	S0		14.7		
04189	08 00.0 + 15 57 206.21 + 22.75	03 – 21 – 011 NGC 2514	0236	1.2 x 1.2 1.2 x 1.2	-- 1	4	S B c S B(s)bc NGC 2507 group, Sc –		14.4		2
04190	08 00.1 + 16 27 205.72 + 22.97	03 – 21 – 012	0236	(1.2 x 0.4) 1.2 x 0.4	32 7:	1	S0: ssp of 2, NGC 2507 group		14.7		*1
04191	08 00.1 + 23 32 198.51 + 25.61	04 – 19 – 021 Mark 384	0226	1.4 x 1.0 1.5 x 1.0	113 3 – 4	3	S B b S B(r)b Sb –		14.2		
04192	08 00.2 + 43 29 176.62 + 30.90	07 – 17 – 000	1329	1.2 x 0.15 1.1 x 0.1	174 7	3	Sc Sc +		16.0:		
04193	08 00.2 + 53 42 164.72 + 32.11	09 – 13 – 115 NGC 2505	0985	1.3 x 0.6 1.4 x 0.7	0 5	3	S B a		14.0		
04194	08 00.3 + 72 41 142.38 + 31.35	12 – 08 – 025	0680	1.1 x 0.5 1.0 x 0.4	15 (6)	1	Sc – Irr		16.5:		
04195	08 00.4 + 66 55 149.11 + 32.01	11 – 10 – 060	1330	2.1: x 1.1 1.9: x 0.9	20 5	4	S B b S B(r)b P w 4187 Sb +		14.6		*
04196	08 00.5 + 61 41 155.28 + 32.30	10 – 12 – 071	0960 100	1.1 x 0.8 1.0 x 0.7	(165) 2 – 3	3	Sc Sc +		15.1		*
04197	08 00.7 + 10 11 211.93 + 20.51	02 – 21 – 011	0247 113	1.9 x 0.3 1.9 x 0.3	133 7	1	Sb: P w 4198		14.5		*
04198	08 01.0 + 10 06 212.04 + 20.54	02 – 21 – 012	0247 113	(1.5 x 1.1) (1.5 x 1.1)	(155) ...	1	S0? P w 4197		15.1		*
04199	08 01.0 + 73 29 141.44 + 31.28	12 – 08 – 029	0680	1.3 x 0.4 1.2 x 0.4	99 6	1	S ... nff of 2		14.6		*
04200	08 01.1 + 40 21 180.24 + 30.49	07 – 17 – 011	1329 114	1.0 x 0.7 0.8 x 0.7	(5) 1 – 3	3	S B b (R')S B(s)bc Sb +		14.6		*

N G C   2497,   2505,   2507,   2508,   2510,   2513,   2514   I C   2219

1	2	3	4	5	6	7	8	9	10	11	12
04201	08 01.4 + 35 34 185.64 + 29.47	06 - 18 - 000	0989	1.2: x 0.15 0.9 x 0.2	164 7	2	Sc		16.5:		
04202	08 01.4 + 74 44 140.00 + 31.10	12 - 08 - 027	0680	1.0 x 0.1 0.9 x 0.1	175 7	2	Sc		16.0:		
04203	08 01.5 + 05 15 216.71 + 18.49	01 - 21 - 009	0027	(0.9 x 0.9) (0.9 x 0.9)	-- ...	1	...		14.4		
04204	08 01.5 + 56 07 161.87 + 32.47	09 - 14 - 000	0679 112	1.2: x 0.9: 1.0: x 0.7:	(130) ...	3	Dwarf V		17.		
04205	08 01.6 + 47 12 172.37 + 31.71	08 - 15 - 036	1317	1.5 x 0.3 1.5 x 0.3	122 6	2	Sb		15.3		1
04206	08 01.7 + 67 12 148.76 + 32.11	11 - 10 - 061	1330	1.6 x 0.2 1.5 x 0.2	132 7	2	Sb - c		16.0:		
04207	08 01.8 + 20 50 201.45 + 25.01	04 - 19 - 022	0226	1.7 x 1.0 1.7 x 1.0	125: 4	1	Sb/Sc?		14.8		
04208	08 01.8 + 25 00 197.12 + 26.47	04 - 19 - 000	0226	1.0: x 0.6: 0.9: x 0.5:	160: ...	2	Sc	Sc +	16.0:		
04209	08 01.9 + 46 51 172.78 + 31.71	08 - 15 - 037	1317	(1.2 x 0.8) (1.1 x 0.7)	80: ...	2	E - S0		14.7		*
04210	08 02.0 + 25 12 196.93 + 26.58	04 - 19 - 023	0226	1.1 x 0.12 1.1 x 0.15	35 7	1	Sc		15.3		
04211	08 02.1 + 10 55 211.38 + 21.13	02 - 21 - 013	0247 113	1.4 x 0.6 1.2 x 0.5	33 (6)	1	S - Irr		14.9		*
04212	08 02.6 + 26 18 195.81 + 27.07	04 - 19 - 024 IC 492	0226	1.1 x 1.0 1.1 x 1.0	-- 1	3	S B b	S B(s)b	14.6		*2
04213	08 02.6 + 51 07 167.80 + 32.29	09 - 14 - 003	0679	1.1 x 0.5 0.8 x 0.4	118 5 - 6	3	Sc	SA(s)c or SAB(s)c	15.5		1
04214	08 02.7 + 55 17 162.87 + 32.55	09 - 14 - 005	0679 112	1.4 x 0.3 1.0 x 0.3	157 6	2	Sb		15.4		
04215	08 02.8 + 10 32 211.83 + 21.12	02 - 21 - 015	0247 113	1.2 x 1.0 1.0 x 0.9	(30) 1 - 2	1	S ...		15.1		
04216	08 03.1 + 12 37 209.83 + 22.07	02 - 21 - 000	0247	1.0 x 0.6 1.2 x 0.6	150 4 - 5	1	Sa - b P w 4220		15.3		*
04217	08 03.2 + 06 13 216.00 + 19.31	01 - 21 - 000	0027	1.3 x 0.2 1.2 x 0.3	125 6 - 7	1	S ...		16.0:		
04218	08 03.3 + 17 51 204.64 + 24.22	03 - 21 - 014 NGC 2522	0236	1.1 x 0.35 1.2 x 0.2	32 7	2	S0 - a Brightest in group		14.4		*3
04219	08 03.3 + 39 14 181.61 + 30.67	07 - 17 - 013	1329 114	2.7 x 2.0: 1.8: x 1.6	(150) ...	3	Sb	SA(rs)b	15.5		
04220	08 03.4 + 12 41 209.80 + 22.17	02 - 21 - 016 IC 2226	0247	1.2 x 1.0 1.3 x 1.1	-- 1 - 2	3	Sb P w 4216	SAB ab Sb -	14.9		*
04221	08 03.5 + 51 17 167.61 + 32.44	09 - 14 - 006 NGC 2518 = 2519	0679	(1.2 x 1.0) (1.3 x 1.1)	(35) ...	2	E - S0		14.2		*1
04222	08 03.7 + 05 46 216.49 + 19.21	01 - 21 - 000	0027	1.0 x 0.4 1.0 x 0.45	40 6	2	S B b:		15.4		
04223	08 03.7 + 62 55 153.81 + 32.63	11 - 10 - 000	1330 100	1.2 x 0.12 1.1 x 0.15	129 7	2	Sc 3rd of 3		16.0:		*
04224	08 03.8 + 01 11 220.77 + 17.10	00 - 21 - 004 IC 494	1298	1.3 x 0.6 1.4 x 0.6	50 5 - 6	3	S0		14.3		
04225	08 03.9 + 22 59 199.42 + 26.24	04 - 19 - 025	0226	1.2 x 0.4 1.1 x 0.4	172 6	1	Sa - b		15.3		
04226	08 03.9 + 40 33 180.13 + 31.05	07 - 17 - 014	1329	2.1: x 1.5: 1.6: x 1.2:	35: 2 - 3	3	Sc	SA(r)cd	15.2		
04227	08 04.0 + 39 20 181.53 + 30.83	07 - 17 - 015	1329 114	1.5 x 1.5 1.5 x 1.5	-- 1	3	Sb	SAB(rs)b	13.5		
04228	08 04.1 + 05 27 216.83 + 19.16	01 - 21 - 011	0027	(2.0 x 1.2) (2.8 x 1.7)	145: ...	3	S0		13.9		
04229	08 04.2 + 39 09 181.75 + 30.83	07 - 17 - 000	1329 114	0.6 x 0.5 0.5 x 0.5	-- ...	1	...		14.4		
04230	08 04.2 + 55 32 162.58 + 32.77	09 - 14 - 008	0679 112	1.2 x 0.2 1.3: x 0.2	167 7	1	S ...		15.5		
04231	08 04.3 + 08 09 214.29 + 20.42	01 - 21 - 012 NGC 2526	0027	1.0 x 0.5 0.9 x 0.4	[140] ...	1	... sp of 2		14.6		*3
04232	08 04.5 + 17 54 204.71 + 24.50	03 - 21 - 019	0236	1.0 x 0.3 1.0 x 0.35	57 6	2	Sb - c NGC 2522 group		15.1		*
04233	08 04.6 + 09 00 213.51 + 20.86	02 - 21 - 000	0247	1.2 x 0.20 1.1 x 0.20	114 6	1	S ...		15.1		
04234	08 04.7 + 39 18 181.60 + 30.95	07 - 17 - 016 NGC 2524	1329 114	(1.5 x 1.1) (1.5 x 1.1)	125 3:	2	S0 - a		13.7		3
04235	08 04.7 + 57 55b 159.75 + 32.88	10 - 12 - 077 NGC 2521	0960 112	(1.2 x 0.7) 0.8 x 0.8	(45) ...	1	E? ssf of 2, disturbed		(14.2)		*12 3C

N G C   2518, 2519, 2521, 2522, 2524, 2526   I C   492, 494, 2226

1	2	3	4	5	6	7	8	9	10	11	12
04236	08 04.9 + 26 10 196.14 + 27.52	04 – 19 – 027	0226	1.1: x 0.5 1.0: x 0.4	145 (6)	2	... S IV:	SA dm	15.7		
04237	08 05.0 + 17 58 204.69 + 24.64	03 – 21 – 020 NGC 2529 = 2531	0236	1.4 x 0.9 1.6 x 1.1	170: ...	4	S B c NGC 2522 group	SB(s)d	14.7		*1
04238	08 05.0 + 76 35 137.83 + 30.98	13 – 06 – 018	1295	2.8 x 1.6 2.6 x 1.4	83 4:	3	S B c	SB d	13.5		*1
04239	08 05.3 + 05 21 217.07 + 19.38	01 – 21 – 000	0027 115	1.1 x 1.1 1.1 x 1.1	-- 1	1	S ...		15.2		*
04240	08 05.3 + 14 59 207.73 + 23.53	03 – 21 – 021	0236	1.5 x 0.3 1.4: x 0.25	177 6	1	S ... In group		15.4		
04241	08 05.3 + 57 54 159.77 + 32.96	10 – 12 – 080	0960 112	1.1 x 0.4 1.3 x 0.5	146 6	2	Sa – b		14.8		*
04242	08 05.4 + 72 57 142.00 + 31.68	12 – 08 – 000 Mark 14	0680	(1.0 x 0.8) (1.2 x 0.9)	(165) ...	1	...		14.4	+ 3150 + 3300	
04243	08 05.6 + 67 23 148.50 + 32.47	11 – 10 – 066	1330	1.5 x 0.8 1.3 x 0.7	158 5	2	Sa – b		14.5		*
04244	08 05.7 + 11 19 211.39 + 22.10	02 – 21 – 000	0247	1.0 x 0.7 0.9 x 0.7	(55) 2 – 3	2	Sa – b		15.1		
04245	08 05.9 + 18 20 204.41 + 24.98	03 – 21 – 023	0236	1.6 x 0.4 1.6 x 0.4	110 6	3	S B b NGC 2522 group		14.9		*
04246	08 05.9 + 44 05 176.12 + 32.02	07 – 17 – 018	1329	1.3 x 0.3 1.0 x 0.3	75 6	2	Sa – b		15.5		
04247	08 06.2 + 16 49 205.98 + 24.46	03 – 21 – 000	0236	1.2 x 0.10 1.1 x 0.10	18 7	2	S B? c	Sc +	15.5		*
04248	08 06.3 + 00 27 221.75 + 17.30	00 – 21 – 005	1298	1.3 x 1.2 (1.2 x 1.1)	-- 1	3	Sa/S B a P w 4251	SAB(rs)a	14.9		*
04249	08 06.4 + 17 08 205.68 + 24.63	03 – 21 – 024	0236	1.0 x 0.2 1.0 x 0.2	52 6 – 7	1	S ...		14.9		
04250	08 06.5 + 46 20 173.51 + 32.43	08 – 15 – 044	1317	1.1 x 1.1 0.6 x 0.6	-- 1	3	Sc		15.4		1
04251	08 06.6 + 00 25 221.82 + 17.35	00 – 21 – 006	1298	1.1 x 0.7 1.1 x 0.7	145 ...	2	E – S0 P w 4248		14.7		*
04252	08 06.6 + 40 15 180.59 + 31.49	07 – 17 – 020	1329	2.1 x 0.7 1.8: x 0.5	17 6	3	Sc		15.4		
04253	08 06.7 – 00 13 222.42 + 17.07	00 – 21 – 007	1298	1.4 x 0.35 1.4 x 0.35	98 6	2	Sb – c		14.5		
04254	08 06.8 + 00 45 221.54 + 17.56	00 – 21 – 008	1298	1.7 x 0.8 1.5 x 0.8	[92] ...	2	S – Irr		14.4		
04255	08 06.8 + 05 25 217.19 + 19.74	01 – 21 – 015 IC 498	0027 115	1.4 x 1.3 1.3 x 1.2	-- 1	3	S ... Peculiar		14.6		*1
04256	08 07.0 + 34 06 187.61 + 30.23	06 – 18 – 013 NGC 2532	0989	2.2 x 1.7 2.2 x 1.8	10: 1 – 2	4	Sc	SAB(rs)c Sc +	12.9	+ 5153 + 5108	*123
04257	08 07.1 + 25 01 197.55 + 27.61	04 – 20 – 002	1364	2.3 x 0.2 1.7 x 0.2	33 7	3	Sc 3rd of 3		15.4		*1
04258	08 07.2 + 47 04 172.66 + 32.64	08 – 15 – 046	1317	1.7 x 0.5 1.6: x 0.4	122 6	3	Sc	Sc +	15.7		
04259	08 07.2 + 73 43 141.09 + 31.67	12 – 08 – 030	0680	2.1 x 0.25 2.1 x 0.25	92 7	3	Sb		14.8		*2
04260	08 07.6 + 46 37 173.20 + 32.65	08 – 15 – 047	1317	1.8 x 1.7 1.7:x 1.3::	-- ...	4	Irr		14.3		*1
04261	08 07.7 + 36 58 184.41 + 31.04	06 – 18 – 000	0989	1.0 x 0.45 1.0 x 0.4	... ...	2	Peculiar Plume + jet		14.7		*
04262	08 08 + 83 25 130.12 + 29.44	14 – 04 – 046	0692	2.4 x 1.9 2.2 x 1.6	155: 2	4	Sb Companion superimposed		14.2		*1
04263	08 08.0 + 79 00 135.04 + 30.60	13 – 06 – 000	1295	(0.9 x 0.9)? 0.50 x 0.35	(45) ...	1	... Peculiar		14.0		*C
04264	08 08.2 + 25 20 197.30 + 27.95	04 – 20 – 004 NGC 2535	1364	3.6 x 2.5 ...	... ...	4	S[c] P w NGC 2536, bridge	SA(r)[c]	13.5	+ 4135 + 4046	*12 3VA
04265	08 08.4 + 05 14a 217.56 + 20.01	01 – 21 – 018 IC 2231	0027 115	(1.6 x 1.6) (2. x 2.)	-- ...	2	E: Companion superimposed		15.0		*1
04266	08 08.8 + 03 46 218.99 + 19.42	01 – 21 – 019 NGC 2538	0027	1.7 x 1.3 1.7 x 1.3	25: 2 – 3	4	S B a	(R´)S B a	13.8		13
04267	08 09.0 + 55 07 163.10 + 33.45	09 – 14 – 015	0679 121	1.0 x 0.5 0.8 x 0.4	35 6	1	...		14.2		*
04268	08 09.0 + 55 49 162.26 + 33.46	09 – 14 – 014 NGC 2534, Mark 85	0679 121	1.4 x 1.2 (0.9 x 0.9)	-- ...	2	... Peculiar		13.8		*23
04269	08 09.1 + 19 31 203.51 + 26.12	03 – 21 – 026	0236	1.2 x 0.6 1.1 x 0.5	90 5 – 6	1	S ...		14.6		
04270	08 09.3 + 58 00 159.64 + 33.49	10 – 12 – 095	0960 116	1.7 x 1.3 1.4 x 1.0	130: 2	4	Sb/S B c	SAB(rs)bc Sb +	14.9		*

N G C   2529 = 2531, 2532, 2534, 2535, 2538

I C   498, 2231

1	2	3	4	5	6	7	8	9	10	11	12
04271	08 09.3 + 73 44 141.04 + 31.81	12 – 08 – 031 NGC 2523	0680	3.0 x 2.2 3.0 x 2.2	57 3	5	S B b	S B(r)bc Sb +	12.4	+ 3448 + 3600	*12 468A
04272	08 09.4 + 44 31 175.72 + 32.70	07 – 17 – 022	1329	1.0 x 0.5 1.0 x 0.6	170: 4 – 5:	1	S ...		16.0:		
04273	08 09.7 + 36 24 185.15 + 31.30	06 – 18 – 014 NGC 2543 = 2232*	0989	2.7 x 1.2 2.7 x 1.2	45 ...	4	S B b	S B(s)b Sb +	12.7		*12 3
04274	08 09.7 + 46 09 173.80 + 32.96	08 – 15 – 050 NGC 2537, Mark 86	1317	1.7: x 1.5: 1.5: x 1.5:	--- 1	1	S ... Peculiar, p w 4278		11.7	+ 404 + 422	*12V 3468A
04275	08 09.8 + 26 31 196.16 + 28.67	05 – 20 – 004 NGC 2540	1351	1.4 x 1.0 1.4 x 0.9	125: 3 – 4	3	S B c	S B(rs) ...	14.5		1
04276	08 10.1 + 09 32 213.63 + 22.31	02 – 21 – 017	0247	1.1 x 0.8 1.1 x 0.8	147: ...	2	S B a	(R´:)S B a	15.1		
04277	08 10.1 + 52 48 165.88 + 33.54	09 – 14 – 019	0679	3.8 x 0.4 3.5 x 0.4	110 7	3	Sc		15.4		
04278	08 10.4 + 45 54 174.12 + 33.05	08 – 15 – 052 IC 2233	1317	4.7 x 0.5 4.5 x 0.5	172 7	3	Sc P w 4274		13.6		*12 5
04279	08 10.4 + 73 50 140.90 + 31.87	12 – 08 – 000	0680	1.1 x 0.8 0.9 x 0.4	100 ...	3	... S B IV	S B dm	16.5:		
04280	08 10.6 + 54 56 163.32 + 33.67	09 – 14 – 021	0679 121	1.5 x 0.3 1.4 x 0.3	3 6	2	Sa		14.4		*
04281	08 10.6 + 58 23b 159.18 + 33.65	10 – 12 – 101	0960 116	1.1 x 0.25 1.1 x 0.25	81 6 – 7	1	Sa – b nff of 2, contact, bridge:		(15.2)		*1
04282	08 10.8 + 77 40 136.51 + 31.05	13 – 06 – 020	1295	1.3 x 0.8 1.2 x 0.6	147 4 – 5	2	Sb – c n of 2		15.1		*
04283	08 11.0 + 39 25 181.74 + 32.17	07 – 17 – 025	1329	1.3 x 0.9 0.9: x 0.8	130: ...	3	S B b	S B(s)b	15.2		
04284	08 11.0 + 49 13 170.18 + 33.47	08 – 15 – 054 NGC 2541	1317	8.1 x 3.5 5.2: x 2.8:	165 5 – 6	5	Sc	SA(s)cd	13.0	+ 601 + 634	*123 45689
04285	08 11.2 + 00 48 222.05 + 18.54	00 – 21 – 010	1289	1.5 x 0.9 1.4 x 0.9	(175) ...	3	Sc		15.2		
04286	08 11.4 + 18 36 204.68 + 26.29	03 – 21 – 027	0236	1.3: x 0.4 1.3: x 0.4	45 6	1	S ... p of 2, disturbed?		14.6		*1
04287	08 11.4 + 21 30 201.66 + 27.34	04 – 20 – 007 NGC 2545	1364 118	2.5 x 1.2 2.3 x 1.1	170 5	1	S ...	(R´)S (r) ...	13.2		*123 4678
04288	08 11.7 + 19 30 203.78 + 26.69	03 – 21 – 000	0236	1.0:: x 0.8:: 1. x 1.	--- ...	2	...	S dm:	17.		
04289	08 11.7 + 58 29 159.05 + 33.79	10 – 12 – 111	0960 116	(1.0 x 0.8) (1.1 x 0.9)	(105) ...	2	E		14.6		*2
04290	08 12.1 + 73 28 141.30 + 32.06	12 – 08 – 032	0680	1.5 x 0.8 1.4 x 0.8	95: ...	1	E?		14.1		*2
04291	08 12.2 + 01 42 221.34 + 19.19	00 – 21 – 000	1298	1.1 x 0.15 1.0 x 0.15	115 7	2	Sb – c		15.6		
04292	08 12.2 + 79 23 134.55 + 30.69	13 – 06 – 021	1295	1.2 x 0.4 1.0 x 0.4	90 6	1	S0 – a		14.6		
04293	08 12.5 + 52 35 166.16 + 33.90	09 – 14 – 022	0679	1.1: x 0.8 0.5 x 0.5	55: ...	1	S ...		15.4		
04294	08 12.6 + 60 39 156.44 + 33.80	10 – 12 – 116	0960	1.0 x 0.15 1.1 x 0.15	8 7	1	S ...		16.0:		
04295	08 12.6 + 64 40 151.63 + 33.47	11 – 10 – 071	1330	1.3 x 0.5 0.7 x 0.4	53 6	2	S B: c P w 4302	Sc +	16.0:		*
04296	08 12.8 + 08 30 214.95 + 22.46	01 – 21 – 000	0027	1.1 x 0.1 1.1 x 0.1	112 7	2	Sc		15.4		
04297	08 13 + 85 47 127.49 + 28.84	14 – 04 – 048	0692 102	1.8 x 0.7 2.1 x 0.7	83 6	3	Sa		14.5		
04298	08 13.0 + 05 25 217.94 + 21.11	01 – 21 – 000	0027	1. x 1. 1. x 1.	--- ...	3	Dwarf spiral S B IV – V or V		17.		
04299	08 13.0 + 23 20 199.85 + 28.32	04 – 20 – 008	1364 118	2.0 x 0.25 1.8 x 0.25	150 7	2	Sb – c nnp of 2		15.2		*
04300	08 13.0 + 27 14 195.63 + 29.57	05 – 20 – 005	1351	1.6: x 1.3: 1.3: x 1.2:	--- 1 – 2	2	Sc		14.7		
04301	08 13.0 + 28 47 193.93 + 30.03	05 – 20 – 006	1351	1.3 x 0.55 1.2 x 0.5	146 6	2	Sb – c		14.6		
04302	08 13.0 + 64 43 151.57 + 33.50	11 – 10 – 072	1330	1.2 x 1.1: 1.2 x 1.1:	(40) 1	3	S B: b P w 4295		15.5		*
04303	08 13.4 + 26 09 196.85 + 29.32	04 – 20 – 000	1364	1.0 x 0.7 1.0 x 0.7	(120) ...	1	S ... s of 2		15.7		*
04304	08 13.5 + 23 58 199.22 + 28.64	04 – 20 – 010	1364	1.5 x 0.3 1.3: x 0.3	168 6	2	S0		15.6		
04305	08 14.1 + 70 52 144.28 + 32.71	12 – 08 – 033 Ho II	0680	9.3:: x 6.6:: 9. x 6.	[15] ...	5	Dwarf irregular, I m Ir IV – V	Ir I	11.3	+ 164 + 303	*12 5AC

N G C   2523, 2537, 2540, 2541, 2543, 2545   I C   2232, 2233

1	2	3	4	5	6	7	8	9	10	11	12
04306	08 14.4 + 35 36 186.32 + 32.05	06 − 18 − 015	0989	1.3: x 0.8: 1.2: x 0.7:	135 4:	1	S ...		15.1		
04307	08 14.4 + 48 00 171.69 + 33.94	08 − 15 − 060	1317	1.5 x 0.3 1.4 x 0.3	65 7:	1	S ...		14.4		
04308	08 14.5 + 21 50 201.59 + 28.13	04 − 20 − 013	1364 118	2.2 x 1.8 2.2 x 1.7	110: 2	4	S B c	SB(rs)c	14.5		*1
04309	08 14.6 + 49 23 170.03 + 34.07	08 − 15 − 061	1317	1.0 x 1.0 0.9 x 0.9	— 1	1	S ...	(R')SAB: ...	15.4		
04310	08 14.7 + 01 22 221.97 + 19.58	00 − 21 − 011	1298	1.3 x 1.2 1.2: x 1.1:	— 1:	3	... S IV	SA dm	15.3		
04311	08 14.8 + 37 33 184.08 + 32.54	06 − 19 − 000	0648	1.1 x 0.2 0.9 x 0.2	135 7	1	Sb:		16.0:		*
04312	08 14.9 + 23 37 199.72 + 28.83	04 − 20 − 015 NGC 2554	1364 118	3.5 x 2.6: (3.5 x 3.)	(0) ...	3	S0/Sa Singular		13.5		*13
04313	08 14.9 + 57 58 159.66 + 34.23	10 − 12 − 124 NGC 2549	0960 116	3.6 x 1.1 3.7 x 1.1	177 7	3	S0		12.1	+ 1082 + 1159	*12 3468
04314	08 14.9 + 58 26 159.09 + 34.21	10 − 12 − 125	0960	[1. x 0.6] ...	... ...	1	... Double system?	Peculiar?	14.9		*1C
04315	08 15.0 + 24 53 198.36 + 29.26	04 − 20 − 016 IC 2267	1364	2.3 x 0.35 1.8 x 0.3	153 6 − 7	2	S B? c + companion Contact		15.1		*
04316	08 15.1 + 04 46 218.82 + 21.28	01 − 21 − 020	0027	1.6 x 1.5 1.3: x 1.3:	— 1	4	... S IV − V	SAB dm	15.4		*
04317	08 15.1 + 13 03 210.73 + 24.91	02 − 21 − 000	0247	1.0 x 0.4 1.0 x 0.4	23 6	3	S B b	SB(s)b	15.1		
04318	08 15.3 + 03 32 220.01 + 20.74	01 − 21 − 000	0027	1.0 x 0.4 0.9 x 0.4	... ...	2	S B: ... sp of 2, distorted		15.3		*
04319	08 15.4 + 00 54 222.49 + 19.51	00 − 21 − 012 NGC 2555	1298	(2.1 x 1.6) 2.2 x 1.7	115: 2 − 3	4	S B a	SB(s)a	13.5		*13
04320	08 15.5 + 59 52 157.36 + 34.21	10 − 12 − 127	0960	1.1 x 0.4 1.2 x 0.4	90 (6)	1	S: ...		15.7		*
04321	08 15.6 + 11 47 212.05 + 24.50	02 − 21 − 000	0247	1.1 x 0.2 1.1 x 0.2	171 7	1	Sa − b		15.7		*
04322	08 15.6 + 62 59 153.60 + 33.97	11 − 10 − 076	1330	(1.4 x 1.0) (1.4 x 1.0)	45: ...	2	E		15.0		
04323	08 15.6 + 67 07 148.67 + 33.46	11 − 10 − 074	1330	(1.8 x 1.3) (2.0 x 1.6)	50: ...	1	E? Eruptive?		14.4		*
04324	08 15.7 + 20 55 202.67 + 28.07	04 − 20 − 017	1364 118	1.6: x 0.5 1.3 x 0.4	27 6	1	S ... ssf of 2		15.3		*
04325	08 15.7 + 50 10 169.10 + 34.29	08 − 15 − 062 NGC 2552	1317	3.9: x 2.6: 3.4: x 2.7:	(45) ...	4	Irr		13.5		*123 468
04326	08 15.7 + 68 47 146.70 + 33.21	11 − 10 − 075	1330	1.5 x 0.6 1.5 x 0.6	152 6	2	Sb − c		13.8		
04327	08 15.9 + 74 08 140.46 + 32.18	12 − 08 − 034 NGC 2544, Mark 87	0680	1.0 x 0.8 1.0 x 0.8	70 2	3	S B a spp of 2	SB(s)a	13.4		*2
04328	08 16.0 + 79 20 134.55 + 30.87	13 − 06 − 022	1295	1.7 x 0.8: 1.5: x 0.8:	10: 5	2	... S B IV	S B dm:	15.5		
04329	08 16.2 + 21 20 202.28 + 28.33	04 − 20 − 018	1364 118	2.5 x 1.5 2.5: x 1.5:	125: 4:	4	Sc	SA(r)c − d Sc +	15.0		*
04330	08 16.3 + 21 36 202.00 + 28.44	04 − 20 − 021 NGC 2557	1364 118	1.1 x 1.0 1.2 x 1.0	(55) 1 − 2	3	S B 0		14.6		*3
04331	08 16.4 + 20 40 203.00 + 28.14	04 − 20 − 022 NGC 2558	1364 118	2.2 x 1.7 1.8: x 1.3:	(160) 2 − 3	3	Sa − b	SAB(rs) ...	14.6		3
04332	08 16.8 + 21 16 202.40 + 28.44	04 − 20 − 025	1364 118	1.4 x 0.8 1.4 x 0.8	[53] ...	3	... Peculiar		15.5		
04333	08 16.8 + 52 42 166.04 + 34.55	09 − 14 − 000	0679	1.0 x 0.3 0.9 x 0.3	[161] ...	3	Dwarf IV − V or V		17.		*
04334	08 16.9 + 22 11 201.43 + 28.78	04 − 20 − 026 NGC 2565, Mark 386	1364 118	1.8 x 0.7 1.7 x 0.7	167 6	3	S B b sff of 2	(R')SB b	13.8		*12
04335	08 16.9 + 54 16 164.13 + 34.58	09 − 14 − 030	0679 121	1.4: x 1.2: (1.1 x 1.0)	(125) 1 − 2	3	S B a sff of 2	(R:)SB(r)a	14.8		*
04336	08 17.0 + 04 49 219.01 + 21.72	01 − 22 − 001 NGC 2561	0642 120	1.2 x 0.6 1.2 x 0.6	138 5:	1	S B ... in cluster, singular?	Brightest	14.0		*1
04337	08 17.0 + 21 08 202.56 + 28.43	04 − 20 − 027 NGC 2560	1364 118	1.5 x 0.4 1.7 x 0.5	93 7	2	S0 − a		14.9		*3
04338	08 17.0 + 55 25 162.74 + 34.59	09 − 14 − 033	0679 121	1.0 x 0.8 1.0 x 0.8	120: 2	2	S B a		15.6		*
04339	08 17.0 + 62 48 153.80 + 34.15	10 − 12 − 129	0960	1.4 x 0.9 1.2 x 0.7	50: ...	2	S ...	SAB ...	15.5		*
04340	08 17.1 + 26 10 197.12 + 30.12	04 − 20 − 028	1364	1.4: x 1.1: 1.1: x 0.9:	120: ...	3	Sc	SAB: d	15.5		*

N G C 2544, 2549, 2552, 2554, 2555
2557, 2558, 2560, 2561, 2565

I C 2267

1	2	3	4	5	6	7	8	9	10	11	12
04341	08 17.1 + 27 15 195.93 + 30.44	05 – 20 – 009	1351	1.1 x 0.6 1.1 x 0.6	43 5	1	SO – a		15.0		
04342	08 17.2 + 40 27 180.78 + 33.52	07 – 17 – 000	1329	1.3 x 0.4 0.9 x 0.3	43 6 – 7	2	SO – a		15.3		*
04343	08 17.4 + 17 30 206.43 + 27.19	03 – 22 – 000	1311	1.1 x 0.8: 1.1: x 0.6:	[145] ...	4	Dwarf irregular Ir V		17.		
04344	08 17.4 + 21 02 202.71 + 28.49	04 – 20 – 030	1364 118	1.7: x 1.7: 1.5:: x 1.5::	-- 1	4	...	SA dm	15.5		*1
04345	08 17.5 + 21 17 202.45 + 28.59	04 – 20 – 031 NGC 2562	1364 118	(1.3 x 1.0) (1.1 x 0.9)	(3) 2:	2	SO – a		14.0	+ 4963 + 4853	*12 3
04346	08 17.6 + 26 04 197.27 + 30.19	04 – 20 – 032	1364	1.0 x 0.9 0.9 x 0.8	-- 1 – 2	3	S B a – b	S B(s) ...	15.0		*
04347	08 17.7 + 21 14 202.52 + 28.62	04 – 20 – 033 NGC 2563	1364 118	(2.0 x 1.7) (2.2 x 1.8)	(80) 1 – 2	2	SO		13.7	+ 4775 + 4664	*12 3
04348	08 18 + 86 07 127.10 + 28.81	14 – 04 – 050	0692 102	1.8 x 1.7 1.8 x 1.7	-- (1)	4	... Peculiar, s of 2		15.7		*
04349	08 18.0 – 01 13 224.79 + 19.05	00 – 22 – 001	1305 119	1.3 x 0.3 1.4 x 0.3	97 6	2	S B? b – c 4352 group, disturbed?		15.1		*1
04350	08 18.0 + 16 48a 207.21 + 27.05	03 – 22 – 002	1311	1.0 x 0.45 1.0 x 0.5	115 5	4	S B b spp of 2	(R´)S B(s)b Sb –	(15.0)		*1
04351	08 18.0 + 35 58 186.08 + 32.84	06 – 19 – 000	0648	1.0: x 0.7:: 0.9 x 0.6	60: ...	1	...		16.0:		*
04352	08 18.2 – 01 15 224.84 + 19.07	00 – 22 – 2+3	1305 119	[1.3 x 1.1] ...	-- ...	2	Triple system Contact, in group of 4		15.0		*1
04353	08 18.2 + 67 06 148.65 + 33.71	11 – 10 – 078	1330	1.2 x 0.5 1.1 x 0.5	28 6	2	Sc		15.3		1
04354	08 18.5 + 21 05 202.76 + 28.74	04 – 20 – 036 NGC 2570	1364 118	1.3 x 0.7 0.9: x 0.4:	75 5	1	Sa – b P w NGC 2569		15.4		*1
04355	08 18.6 + 19 18 204.66 + 28.13	03 – 22 – 004 NGC 2572	1311	1.7 x 0.5 1.5: x 0.5	133 6	1	Sa?		14.8		3
04356	08 18.8 + 03 20 220.64 + 21.42	01 – 22 – 002 IC 2327	0642	1.5 x 0.45 1.2 x 0.4	168 6	1	Sa?		13.9		3
04357	08 18.8 + 56 28 161.45 + 34.81	09 – 14 – 035	0679	(1.2 x 1.0) 1.4 x 1.4	-- 1:	3	SO	(R)SA(r:) ...	14.9		*
04358	08 18.9 – 00 15 224.01 + 19.72	00 – 22 – 000	1305 119	1.0 x 0.4 1.0: x 0.3:	... ...	1	... Brightest of 3, disturbed		16.0:		*
04359	08 18.9 + 74 10 140.37 + 32.37	12 – 08 – 037 NGC 2550	0680	1.1 x 0.4 1.1 x 0.4	103 6	1	S ... Peculiar?		13.1		2
04360	08 18.9 + 78 24 135.56 + 31.27	13 – 06 – 000	1295	1.1 x 0.3 0.8: x 0.3:	80 6:	2	...	S d – m	16.5:		
04361	08 19.0 + 22 48 200.96 + 29.44	04 – 20 – 037	1364 118	1.2 x 0.45 1.0 x 0.4	60 6	1	Sc		15.6		*
04362	08 19.1 + 73 35 141.03 + 32.52	12 – 08 – 038 NGC 2551	0680	1.8 x 1.2 1.9 x 1.4	55 3	1	S ...		12.7	+ 2296 + 2447	*12 48
04363	08 19.3 + 74 36 139.86 + 32.29	12 – 08 – 039	0680	1.5 x 1.5 1.4 x 1.3	-- 1	3	S B c s of 2, disturbed?	(R´)S B(s)d:	14.9		*1
04364	08 19.4 + 25 39 197.88 + 30.45	04 – 20 – 039	1364	1.5: x 1.5: 1.4: x 1.4:	-- 1	1	Sa/S B b:		15.4		
04365	08 19.5 + 19 34 204.47 + 28.42	03 – 22 – 005 IC 2329	1311	2.2 x 0.4 2.0: x 0.4	117 (7)	3	Sc – Irr		15.0		*
04366	08 19.6 + 03 25 220.67 + 21.64	01 – 22 – 004 IC 503	0642	1.1 x 1.0 0.9 x 0.8	-- 1 – 2	3	S B a		14.0		
04367	08 19.8 + 22 43 201.12 + 29.59	04 – 20 – 042 NGC 2577	1364 118	(1.8 x 1.1) (1.7 x 1.1)	105 ...	2	E – SO P w 4375		13.8		*3
04368	08 19.8 + 24 27 199.23 + 30.16	04 – 20 – 040 NGC 2575	1364 118	2.6 x 2.0 2.5: x 2.0:	(145) 2 – 3	4	Sc	SA(rs:)cd	14.3		*
04369	08 19.8 + 65 47 150.18 + 34.07	11 – 10 – 080	1330	1.0 x 0.4 0.7 x 0.6	-- 1	3	S B c		16.0:		1
04370	08 19.9 – 00 53 224.73 + 19.62	00 – 22 – 004	1305 119	1.3 x 0.7 1.2 x 0.7	135 ...	1	Sc:		14.9		
04371	08 19.9 + 25 54 197.64 + 30.63	04 – 20 – 041 NGC 2576	1364	1.9 x 0.3 1.7 x 0.3	41 6 – 7	2	Sb		15.4		3
04372	08 20.0 + 04 25 219.76 + 22.19	01 – 22 – 005 IC 504	0642 120	(1.1 x 0.8) (1.1 x 0.8)	140: ...	3	SO Second brightest in cluster		14.3		1
04373	08 20.0 + 27 52 195.46 + 31.24	05 – 20 – 010	1351	1.3 x 0.7 1.0 x 0.6	55: 5:	2	S B: c f of 2		15.5		*1
04374	08 20.2 + 03 44 220.44 + 21.92	01 – 22 – 006	0642	1.7 x 1.3 1.5 x 1.2	55: 2 – 3	4	S B c	S B(s)cd	14.7		
04375	08 20.2 + 22 49 201.05 + 29.71	04 – 20 – 043	1364 118	2.7 x 1.8 2.5: x 1.7:	0 3 – 4	3	Sc P w 4367	SAB: c Sc –	14.6		*

N G C   2550, 2551, 2562, 2563, 2570   I C   503, 504, 2327, 2329
2572, 2575 – 77

1	2	3	4	5	6	7	8	9	10	11	12
04376	08 20.2 + 67 00 148.73 + 33.92	11 – 11 – 001	1286	(1.1 x 0.8) (1.0 x 0.7)	150: ...	1	...		15.1		
04377	08 20.2 + 67 29 148.15 + 33.84	11 – 11 – 002	1286	1.4: x 0.7: 1.0 x 0.4	5 5 – 6	3	Sc		16.5:		
04378	08 20.3 + 70 04 145.10 + 33.37	12 – 08 – 000	0680	1.0 x 0.2 1.0 x 0.2	145 ...	1	Dwarf:		17.		
04379	08 20.5 + 61 42 155.08 + 34.67	10 – 12 – 136	0960	1.3: x 0.5 1.1 x 0.4	118 6	2	S B b	SAB(s) ...	15.4		
04380	08 20.6 + 55 00 163.23 + 35.11	09 – 14 – 037	0679 121	1.3 x 1.3 1.1: x 1.1:	-- 1	3	Sc		15.1		*
04381	08 20.7 – 00 42 224.66 + 19.89	00 – 22 – 006	1305 119	1.5 x 0.35 1.5 x 0.4	33 6	2	Sb – c		14.8		
04382	08 20.7 + 04 32 219.74 + 22.40	01 – 22 – 008 IC 505	0642 120	1.3 x 1.0 1.4 x 1.0	143: ...	1	S ...		14.8		
04383	08 20.7 + 21 30 202.51 + 29.37	04 – 20 – 44 + 45 IC 2338 + 2339	1364 118	[1.5 x 0.9] ...	... ...	2	Double system Bridge		14.7		*1A
04384	08 20.8 + 21 36 202.42 + 29.43	04 – 20 – 046 IC 2341	1364 118	(1.3 x 0.6) (1.3 x 0.7)	1 ...	2	E – S0		14.9		*
04385	08 21.1 + 14 55 209.49 + 27.00	03 – 22 – 009	1311	0.9 x 0.8 0.8 x 0.8	-- ...	1	Peculiar Eruptive?		14.4		*
04386	08 21.1 + 21 12 202.87 + 29.35	04 – 20 – 048	1364 118	1.9 x 0.5 1.9 x 0.5	21 6	3	Sb nnf of 2		14.8		*1
04387	08 21.3 + 47 04 172.93 + 35.03	08 – 16 – 001	0671	1.5 x 0.2 1.2 x 0.2	97 7	3	Sc		16.0:		
04388	08 21.7 + 18 46 205.53 + 28.61	03 – 22 – 010 IC 2351	1311	1.1 x 0.9 1.0 x 0.8	(10) 2	1	S ...		14.4		*
04389	08 22.1 + 73 49 140.71 + 32.67	12 – 08 – 041	0680	1.4 x 0.7 1.4 x 0.7	170: ...	1	S ... Disturbed?		15.5		*1V
04390	08 22.2 + 73 41 140.86 + 32.71	12 – 08 – 042	0680	2.2: x 2.0: 2.2:: x 2.0::	-- 1:	4	S B c	S B d	15.6		
04391	08 22.4 + 20 30 203.75 + 29.39	04 – 20 – 050 NGC 2582 = 2359*	1364 118	1.1 x 1.1 1.1 x 1.1	-- 1	4	S B b	(R')SAB(s)ab Sb –	14.3		*3
04392	08 22.5 – 00 26 224.65 + 20.41	00 – 22 – 010 NGC 2590	1305 119	2.5 x 0.8 2.3 x 0.7	77 6	4	Sb		14.0		
04393	08 22.6 + 46 08 174.09 + 35.18	08 – 16 – 003	0671	2.5 x 1.7 1.5 x 1.0	(45) 2 – 3	1	S B ...		13.6		
04394	08 22.7 + 28 03 195.46 + 31.86	05 – 20 – 012 IC 2361	1351	1.3 x 0.5 (1.7 x 0.6)	78 6	1	... IC 2365 group		14.9		*
04395	08 22.7 + 28 17 195.19 + 31.93	05 – 20 – 013	1351	1.8 x 0.2 1.6 x 0.2	154 7	3	Sc IC 2365 group		15.5		*
04396	08 23 + 84 29 128.81 + 29.48	14 – 04 – 051	0692 102	2.2 x 0.4 2.1 x 0.4	32 6	2	SBb – c	SAB(s) ...	15.3		*
04397	08 23.0 + 73 55 140.57 + 32.70	12 – 08 – 043	0680	1.7 x 1.4 1.7 x 1.4	(0) 1 – 2	4	Sc	SA(rs)c Sc +	13.5		*
04398	08 23.1 + 64 23 151.80 + 34.62	11 – 11 – 004	1286	1.3: x 1.1: 0.9 x 0.8	(170) 1	3	Sb/Sc	SA(rs) ... Sb +	15.1		
04399	08 23.2 + 21 37 202.62 + 29.95	04 – 20 – 051	1364 118	1.1 x 0.35 0.9 x 0.35	30 (6)	1	S – Irr		15.5		*1
04400	08 23.2 + 21 50 202.38 + 30.03	04 – 20 – 000	1364 118	1.7 x 0.1 1.3: x 0.1	5 7	2	Sc		15.7		
04401	08 23.2 + 48 58 170.63 + 35.45	08 – 16 – 004	0671	1.4: x 1.2: 1.0: x 0.9:	-- ...	2	Dwarf		17.		
04402	08 23.3 + 28 00 195.56 + 31.98	05 – 20 – 014 IC 2365	1351	(1.4 x 1.1) (1.3 x 1.0)	(45) ...	3	S0 Brightest in group		14.7		*
04403	08 23.5 + 11 40 213.07 + 26.20	02 – 22 – 000	0456	1.0 x 0.35 1.0 x 0.3	79 6	1	S ...		15.4		
04404	08 23.6 + 22 26 201.77 + 30.32	04 – 20 – 000	1364 118	1.1 x 0.15 1.0 x 0.15	151 (7)	2	Sc – Irr		15.5		
04405	08 23.6 + 23 21 200.76 + 30.62	04 – 20 – 052	1364 118	1.5 x 0.4 1.6 x 0.6	28 6	2	S0 – a		15.2		*1
04406	08 23.7 + 23 07 201.03 + 30.57	04 – 20 – 053	1364 118	1.2 x 1.0 1.1 x 0.9	(125) 2:	2	S B 0	(R)S B 0	15.5		*
04407	08 23.8 + 18 00 206.55 + 28.79	03 – 22 – 000	1311 122	1.0 x 0.2 0.9 x 0.20	9 7	1	S ...		15.6		
04408	08 24.0 + 17 32 207.07 + 28.66	03 – 22 – 012 NGC 2593	1311 122	1.1 x 0.5 1.0 x 0.5	172 5 – 6	2	S0 – a		14.9		*3
04409	08 24.0 + 20 32 203.86 + 29.75	04 – 20 – 054 IC 2373	1364 118	1.4: x 1.4: 1.1: x 0.9:	-- 1:	1	Sc?		15.5		
04410	08 24.0 + 55 18 162.85 + 35.58	09 – 14 – 039	0679 121	1.2 x 0.5 1.2 x 0.5	0 6	1	S ... Disturbed?		15.1		*1

N G C   2582, 2590, 2593

I C   505, 2338, 2339, 2341, 2351
2359, 2361, 2365, 2373

1	2	3	4	5	6	7	8	9	10	11	12
04411	08 24.1 + 26 08 197.72 + 31.60	04 - 20 - 055 NGC 2592	1364	(1.7 x 1.4) (1.8 x 1.5)	(45) ...	3	E Brightest of 3		13.6		*13
04412	08 24.1 + 37 56 184.04 + 34.42	06 - 19 - 004	0648	1.3 x 0.4 1.4 x 0.4	53 7	1	S ...		15.2		1
04413	08 24.1 + 74 01 140.43 + 32.75	12 - 08 - 000	0680	1.7 x 0.35 1.6 x 0.3	105 7	3	Sc		15.6		*
04414	08 24.2 + 21 49 202.49 + 30.24	04 - 20 - 058	1364 118	1.1 x 1.0 1.1 x 1.0	-- 1 - 2	4	S B 0/S B a	(R′)S B(s)0/a	15.0		*
04415	08 24.2 + 54 36 163.70 + 35.64	09 - 14 - 040	0679 121	1.3: x 1.1: 1.3 x 0.8:	(155) ...	4	S B c	S B(s)d	15.3		
04416	08 24.3 + 23 02 201.17 + 30.67	04 - 20 - 060	1364 118	2.5 x 0.9 2.3 x 0.9	165 6	4	S B b	S B(s)b	14.5		*1
04417	08 24.3 + 55 52 162.15 + 35.60	09 - 14 - 000 Mark 88	0679 121	0.2 x 0.2 0.2 x 0.2	-- ...	1	... Extremely compact		14.3	+ 9222	*C
04418	08 24.5 + 25 53 198.03 + 31.61	04 - 20 - 000	1364	1.1 x 0.5 1.1: x 0.5:	107 6	2	Sc NGC 2592 group, Sc +		15.7		*
04419	08 24.6 + 17 27 207.21 + 28.76	03 - 22 - 013 NGC 2596	1311 122	1.5 x 0.6 1.5 x 0.6	65 6	3	Sb		14.2		3
04420	08 24.7 + 63 30 152.83 + 34.92	11 - 11 - 009	1286	(1.6 x 0.8) (1.3 x 0.7)	120: ...	2	S0		14.6		*
04421	08 24.8 + 02 01 222.66 + 22.11	00 - 22 - 000	1305	1.1 x 0.8 0.9: x 0.7:	(0) ...	3	Sb/S B b	SAB(s)...	15.3		
04422	08 24.8 + 21 39 202.73 + 30.32	04 - 20 - 062 NGC 2595	1364 118	3.2 x 2.8 3.2: x 2.8:	(45) ...	4	S B b/Sc	SAB(rs)bc Sb +	13.9	+ 4028	*12 3C
04423	08 24.9 + 74 51 139.46 + 32.58	12 - 08 - 045	0680	1.3 x 0.9: 1.4: x 1.0:	100: 3:	3	Sc	Sc +	15.4		*
04424	08 25.2 + 20 25 204.10 + 29.98	03 - 22 - 000	1311 118	1.3 x 0.3 1.2 x 0.2	0 6 - 7	1	S ...		16.0:		*
04425	08 25.2 + 28 14 195.43 + 32.44	05 - 20 - 015	1351	1.5 x 0.6 1.5: x 0.6	130 (6)	1	SB ... IC 2365 group		15.2		*
04426	08 25.2 + 42 02 179.14 + 35.23	07 - 18 - 004	0707 127	2. x 1. ...	... ...	1	Dwarf V		18.		*1
04427	08 25.3 + 55 40 162.38 + 35.75	09 - 14 - 043	0679 121	1.2 x 0.5 1.2 x 0.5	7 6	2	Sb - c npp of 2		15.2		*1
04428	08 25.4 + 63 08 153.26 + 35.05	11 - 11 - 011	1286	1.4 x 0.2 0.9 x 0.2	152 7	3	Sc		16.5:		
04429	08 25.5 + 40 50 180.60 + 35.13	07 - 18 - 007	0707 127	1.1 x 0.5 1.0 x 0.5	18 5 - 6	3	Sc		15.5		*
04430	08 25.6 - 01 22 225.93 + 20.62	00 - 22 - 011	1305	1.1 x 0.6 1.1 x 0.6	142 5	2	S0 - a Brightest in group		15.3		*
04431	08 25.6 + 00 11 224.49 + 21.39	00 - 22 - 013	1305	1.1 x 0.9 1.0 x 0.8	(145) 2:	3	Sb - c		15.1		
04432	08 25.6 + 01 10 223.57 + 21.87	00 - 22 - 012	1305	1.7 x 1.2 1.6: x 1.0:	(135) ...	2	Sb - c		15.2		
04433	08 25.7 + 17 38 207.13 + 29.07	03 - 22 - 014	1311 122	1.1 x 0.8 1.1 x 0.8	(15) ...	3	S ... Peculiar		14.3		*
04434	08 25.7 + 34 50 187.80 + 34.15	06 - 19 - 005	0648	1.0 x 0.4 1.1 x 0.4	[105] ...	2	... Peculiar		14.6		1
04435	08 25.9 + 04 53 220.06 + 23.71	01 - 22 - 000	0642	1.1 x 0.3 0.6 x 0.2	143 6	1	S ...		16.0:		
04436	08 26.2 + 48 56 170.69 + 35.94	08 - 16 - 007	0671	1.3 x 0.3 1.3 x 0.3	42 6	2	Sb - c		15.5		
04437	08 26.2 + 52 08 166.74 + 35.98	09 - 14 - 048	0679 124	1.3 x 0.3 1.2 x 0.3	33 6	3	Sc		16.0:		
04438	08 26.2 + 52 52 165.83 + 35.98	09 - 14 - 047 Mark 90	0679 124	0.8 x 0.7 0.8 x 0.7	(15) ...	1	S ... Peculiar:		13.9	+ 4200	*
04439	08 26.7 + 02 48 222.16 + 22.90	01 - 22 - 012	0642	1.5: x 0.8: 1.1 x 0.6	7: 5:	3	Sc		15.7		
04440	08 26.8 + 15 14 209.78 + 28.39	03 - 22 - 000	1311	1.1 x 0.35 0.9 x 0.3	5 6	1	S B: ... sp of 2		(15.2)		*
04441	08 26.9 + 40 03 181.60 + 35.29	07 - 18 - 010	0707 127	1.1 x 0.2 1.0 x 0.2	38 7	2	Sb sp of 2		15.6		*1
04442	08 27.0 + 52 28 166.32 + 36.10	09 - 14 - 049	0679 124	1.1 x 0.9 1.1 x 0.9	110: ...	3	Sb/S B c	SAB(s)bc	14.5		
04443	08 27.1 + 21 40 202.92 + 30.82	04 - 20 - 065 NGC 2597 = 2598	1364 118	1.2 x 0.5 1.4 x 0.5	3 6	1	S ...		15.1		23
04444	08 27.2 + 17 26 207.50 + 29.33	03 - 22 - 015	1311 122	1.6 x 0.9 1.5: x 0.9:	125 4 - 5	3	S B c	S B(s) ...	14.4		
04445	08 27.3 + 61 10 155.61 + 35.54	10 - 12 - 142	0960	1.7 x 1.7 1.4: x 1.3:	-- 1	4	Sc	SA(s)c	15.3		

N G C   2592, 2595 - 98

1	2	3	4	5	6	7	8	9	10	11	12
04446	08 27.8 + 20 47 203.95 + 30.67	04 – 20 – 000	1364 118	1.2 x 0.10 1.0 x 0.1	135 7	3	Sc		15.7		
04447	08 28.0 + 20 00 204.82 + 30.44	03 – 22 – 000	1311	1.1: x 0.4: 1.0: x 0.3:	[30] ...	1	... nf of 2	d – m	16.5:		*
04448	08 28.0 + 74 35 139.70 + 32.85	12 – 08 – 046	0680	(0.8 x 0.5) 0.9 x 0.5	120 ...	2	E		14.3		
04449	08 28.1 + 41 07 180.34 + 35.66	07 – 18 – 015	0707 127	1.3: x 1.1: 1.3 x 1.1	-- ...	2	... Peculiar		15.7		1
04450	08 28.4 + 27 45 196.21 + 32.99	05 – 20 – 000	1351	1.2 x 0.15 1.1 x 0.12	132 7	3	Sc		15.3		
04451	08 28.4 + 75 20a 138.84 + 32.66	13 – 06 – 024	1295	0.9 x 0.2 1.0 x 0.2	135 6	2	Sa – b np of 2; comp = Mark 15		(14.5)		*
04452	08 28.5 + 09 47 215.56 + 26.50	02 – 22 – 000	0456	(1.2 x 1.1) (0.9 x 0.8)	-- ...	1	S0? n of 2		15.2		*
04453	08 28.6 + 22 12 202.47 + 31.33	04 – 20 – 000	1364	1.1: x 0.3 0.9: x 0.2	156 6 – 7	2	S0		15.4		
04454	08 29 + 80 44 132.80 + 30.97	13 – 07 – 000	0693	1.0 x 0.25 1.0 x 0.25	[28] ...	1	... Peculiar:		15.3		*
04455	08 29.0 – 01 02 226.08 + 21.52	00 – 22 – 014	1305 126	1.0 x 0.8 0.9 x 0.7	(20) 2:	3	S B a	S B(r)a	15.0		
04456	08 29.0 + 24 10 200.32 + 32.05	04 – 20 – 066 IC 509	1364	1.9 x 1.7 1.8: x 1.6:	-- 1 – 2	4	Sc	SA(rs)c Sc +	14.6		
04457	08 29.1 + 19 23 205.60 + 30.46	03 – 22 – 017	1311	1.7: x 0.9 1.7: x 0.9	[125] ...	3	S ... Singular, disturbed?		14.9		*
04458	08 29.2 + 22 44 201.94 + 31.64	04 – 20 – 067 NGC 2599, Mark 389	1364 118	(2.6 x 2.6) (2. x 2.)	-- 1	4	Sa	SA a	13.4		*13
04459	08 29.5 + 66 20 149.32 + 34.95	11 – 11 – 013	1286 129	1.8 x 1.5 1.6: x 1.5:	[120] ...	4	Dwarf irregular Ir V		15.4		*1C
04460	08 29.6 – 02 00 227.05 + 21.16	00 – 22 – 015 IC 510	1305 126	1.1 x 0.9 1.1 x 0.7	(150) ...	3	SB ... + companion Contact, distorted		15.2		*1
04461	08 29.6 + 52 42 166.03 + 36.49	09 – 14 – 057	0679 124	1.9 x 0.5 1.7 x 0.5	43 6	2	Sb – c Disturbed?		14.4		*
04462	08 29.9 + 09 57 215.56 + 26.88	02 – 22 – 000	0456	[1.2 x 0.5] ...	... ...	2	Double system U – shaped, bridge		15.7		
04463	08 30 + 85 55 127.23 + 29.08	14 – 04 – 054 IC 499	0692 102	2.3 x 1.4 2.3 x 1.3	80 4	4	Sa		14.5		*
04464	08 30.0 + 26 11 198.13 + 32.88	04 – 20 – 068	1364	1.7 x 0.3 1.3 x 0.3	110 6 – 7	2	S B? c n of 2, disturbed		15.5		*
04465	08 30.0 + 41 26 180.00 + 36.05	07 – 18 – 019	0707 127	1.2 x 0.8 0.9 x 0.6	160: 3	2	Sa		15.0		*1
04466	08 30.0 + 78 00 135.80 + 31.93	13 – 07 – 000	0693	1.5: x 1.2: 1.3:: x 1.2::	(30) ...	3	Dwarf (spiral) S? V		17.		
04467	08 30.2 + 00 24 224.90 + 22.50	00 – 22 – 000	1305 126	1.5 x 0.6 1.3 x 0.6	170 ...	1	S: ...		15.5		
04468	08 30.2 + 41 42 179.68 + 36.12	07 – 18 – 020	0707 127	(1.5 x 0.9) (1.4 x 0.8)	165 ...	2	S0		14.7		
04469	08 30.3 + 29 43 194.08 + 33.91	05 – 20 – 022 NGC 2604	1351 123	2.1: x 2.0: 2.2: x 2.0:	-- 1	4	S B c np of 2	S B(rs)cd	13.5		*13
04470	08 30.3 + 55 47 162.19 + 36.45	09 – 14 – 065	0679 121	1.3: x 0.4 0.8 x 0.3	128 6	1	S ...		15.7		*
04471	08 30.4 + 41 36 179.81 + 36.14	07 – 18 – 022	0707 127	1.1 x 0.2 1.0 x 0.3	63 7:	1	S ...		15.6		*
04472	08 30.6 + 78 12 135.57 + 31.89	13 – 07 – 001 NGC 2591	0693	3.2 x 0.7 3.0 x 0.7	32 6	3	Sc		12.8		*23
04473	08 30.9 + 27 09 197.09 + 33.35	05 – 20 – 025 NGC 2607	1351	1.6: x 1.6: 0.9: x 0.9:	-- 1	1	S ...	SA ...	14.9		*13
04474	08 31 + 84 49 128.38 + 29.53	14 – 04 – 053	0692 102	1.5 x 0.22 1.7 x 0.25	70 7	2	Sb		15.7		
04475	08 31.0 + 52 54 165.77 + 36.70	09 – 14 – 068 NGC 2600	0679 124	1.3 x 0.4 1.2 x 0.4	78 6	2	Sb		15.1		
04476	08 31.0 + 67 30 147.89 + 34.85	11 – 11 – 000	1286	1.0:: x 0.8:: 1.0:: x 0.8::	... ...	1	Dwarf?		18.		
04477	08 31.3 + 63 47 152.35 + 35.60	11 – 11 – 15a	1286 129	1.0 x 0.25 0.9 x 0.2	17 6	2	S B: b nff of 2		16.5:		*
04478	08 31.4 + 55 59 161.93 + 36.59	09 – 14 – 070	0679 121	1.3: x 0.8: 1.4 x 0.8	(50) ...	3	S B a	(R)S B(s)a	15.2		*
04479	08 31.8 + 39 51 182.00 + 36.19	07 – 18 – 027	0707 127	1.0 x 0.15 1.0 x 0.2	12 7	2	S B? c		15.6		
04480	08 31.9 + 01 50 223.77 + 23.57	00 – 22 – 17+18	1305	[1.4 x 1.0] ...	... ...	3	S B b + companion Contact		14.7		*1

N G C 2591, 2599, 2600, 2604, 2607

I C 499, 509, 510

1	2	3	4	5	6	7	8	9	10	11	12
04481	08 32.0 – 02 22 227.72 + 21.49	00 – 22 – 019 NGC 2615	1305 126	2.3 x 1.4: 2.2: x 1.6:	40 3 – 4	4	S B b	S B(rs)b Sb –	13.5		
04482	08 32.1 + 28 56 195.11 + 34.09	05 – 20 – 000	1351	1.1 x 0.2 0.9 x 0.2	125 7	1	Sc – Irr		15.6		
04483	08 32.1 + 69 58 144.95 + 34.37	12 – 08 – 048	0680	1.4 x 0.7 1.1: x 0.6:	[162] ...	3	Dwarf Companion superimposed		15.3		*1
04484	08 32.2 + 28 39 195.45 + 34.04	05 – 20 – 027 NGC 2608	1351	2.5: x 1.8: 2.0 x 1.3	60 3 – 4:	2	S B b/S B c: P w 4503		13.2	+ 2119 + 2043	*12 3468A
04485	08 32.2 + 61 54 154.62 + 36.01	10 – 13 – 001	1296	1.0 x 0.8: 0.8 x 0.4	[10] ...	1	S ...		16.5:		
04486	08 32.5 + 17 36 207.87 + 30.56	03 – 22 – 000	1311	1.0 x 0.25 0.9: x 0.2	165 6	1	Sa – b sf of 2		16.0:		*
04487	08 32.5 + 38 44a 183.39 + 36.17	07 – 18 – 000	0707 125	1.0 x 0.1 1.2 x 0.1	124 7	2	S B? c n of 2, disturbed		(15.5)		*
04488	08 33.0 – 01 44 227.27 + 22.03	00 – 22 – 000 IC 515	1305 126	1.2 x 0.6 1.1 x 0.6	30: ...	1	... NGC 2616 group		15.6		*
04489	08 33.0 – 01 40 227.21 + 22.06	00 – 22 – 021 NGC 2616	1305 126	(1.2 x 1.2) (1.5 x 1.4)	-- ...	1	... Compact, brightest of 3		15.2		*
04490	08 33.1 + 66 20 149.23 + 35.30	11 – 11 – 000 Mark 93	1286 129	1.1 x 0.3 1.0 x 0.2	77 6 – 7	1	S ...		16.0:	+ 4800	*
04491	08 33.2 + 01 54 223.88 + 23.89	00 – 22 – 022	1305	2.7 x 0.9 2.6: x 0.9	60 6	3	Sa – b		13.9		*1
04492	08 33.3 + 00 52 224.87 + 23.40	00 – 22 – 023 NGC 2618	1305	3.0: x 2.5: (2.8 x 2.5)	(140) 1 – 2	3	Sa	SA ab	13.9		*13
04493	08 33.5 + 00 47 224.98 + 23.40	00 – 22 – 000	1305	1.0 x 0.3 0.9: x 0.3	88 6	1	S ...		15.7		*
04494	08 33.5 + 02 04 223.76 + 24.03	00 – 22 – 000	1305	1.4 x 0.7 1.3 x 0.6	70: 5	1	Sa – b		15.2		*
04495	08 33.5 + 69 13 145.79 + 34.68	12 – 08 – 049	0680	1.0 x 0.7 1.0 x 0.7	70 3	3	Sc		15.2		
04496	08 33.7 + 25 19 199.41 + 33.42	04 – 21 – 000	1355 128	1.0 x 0.1 0.9 x 0.1	23 7	3	Sc		15.7		*
04497	08 33.9 + 67 33 147.76 + 35.11	11 – 11 – 018	1286	1.3 x 1.0: 0.9:: x 0.9::	-- ...	3	Dwarf V		16.5:		*
04498	08 34.0 + 40 13 181.61 + 36.65	07 – 18 – 030	0707 127	1.1 x 0.7 1.0 x 0.7	10: ...	2	S B a		14.8		
04499	08 34.0 + 51 50 167.08 + 37.19	09 – 14 – 078	0679	3.0 x 2.4 2.3: x 2.3:	(140) ...	4	... S IV; Mark 94 superimposed	SAB dm	14.0		*
04500	08 34.1 + 71 53 142.66 + 34.03	12 – 09 – 001	1325	1.6 x 0.9 1.5: x 0.9	45: 4	3	Sc/S B c	SAB(s)cd	15.4		*
04501	08 34.5 + 25 07 199.70 + 33.54	04 – 21 – 001 NGC 2620	1355 128	2.2 x 0.6 2.1 x 0.5	93 6	1	S ...		14.8		*3
04502	08 34.5 + 73 55 140.31 + 33.47	12 – 09 – 002	1325	1.7 x 0.4 1.5: x 0.4	75 6	2	S B c		15.6		
04503	08 34.6 + 28 53 195.34 + 34.61	05 – 21 – 002 NGC 2619	0924	2.8 x 1.7 2.2 x 1.6	35 4	3	Sb/Sc P w 4484		13.6		*13
04504	08 34.8 + 20 41 204.73 + 32.17	04 – 21 – 002	1355	1.5 x 1.0 1.3 x 0.8	30 3 – 4	4	Sc	S d	15.3		
04505	08 35.2 + 50 36 168.63 + 37.41	09 – 14 – 000	0671	1.3 x 0.2 1.3 x 0.2	59 6	1	S ...		15.0		
04506	08 35.3 + 19 54 205.64 + 32.01	03 – 22 – 019 NGC 2624	1311	0.8 x 0.7 0.7 x 0.5	-- ...	1	... P w NGC 2624		14.5		*3
04507	08 35.3 + 43 43 177.28 + 37.24	07 – 18 – 032	0707 127	1.9 x 0.4 1.8 x 0.4	43 6	2	Sb		15.0		*1
04508	08 35.4 – 02 17 228.11 + 22.26	00 – 22 – 025	1305 126	0.35 x 0.35 0.35 x 0.35	-- ...	1	... Extremely compact		14.5		
04509	08 35.4 + 25 56 198.83 + 33.97	04 – 21 – 009 NGC 2623	1355 128	[2.2 x 0.7] ...	... ...	3	Triple system, contact Strongly distorted, tails		14.4	+ 5435 + 5345	*12 VA
04510	08 35.4 + 73 40 140.57 + 33.60	12 – 09 – 003 IC 511	1325	1.4 x 0.5 1.6 x 0.6	143 6	3	Sa		14.5		2
04511	08 35.5 + 30 58 192.94 + 35.31	05 – 21 – 003 IC 2387	0924	1.2 x 0.5 1.2 x 0.5	18 6	3	Sc		14.8		*
04512	08 35.5 + 61 08 155.48 + 36.52	10 – 13 – 005	1296	1.5 x 0.7 1.0 x 0.6	110 5	3	Sc		15.3		
04513	08 35.6 + 40 32 181.27 + 36.99	07 – 18 – 033	0707 127	(1.0 x 1.0) (1.0 x 1.0)	-- ...	1	... Compact		15.0		
04514	08 35.9 + 53 38 164.81 + 37.40	09 – 14 – 081	0679	2.3: x 1.1: 1.8 x 0.8:	70: ...	2	S B c:		14.1		
04515	08 36.4 + 52 38 166.06 + 37.53	09 – 14 – 082	0679	1.7 x 0.7 1.6: x 0.8	175 5 – 6	2	S B: b	S B:(r)b	14.5		

N G C 2608, 2615, 2616, 2618 – 20, 2623, 2624   I C 499, 509, 511, 515, 2387

1	2	3	4	5	6	7	8	9	10	11	12
04516	08 36.6 + 67 02 148.30 + 35.48	11 – 11 – 019	1286	1.7 x 0.3 1.4: x 0.3	123 (6)	2	Sc – Irr		15.2		*
04517	08 36.9 + 73 52 140.30 + 33.64	12 – 09 – 004	1325	1.1 x 1.1 (0.9 x 0.8)	-- ...	1	...		15.5		
04518	08 37.4 + 42 07 179.33 + 37.49	07 – 18 – 000	0707 127	1.0 x 0.3 0.9 x 0.2	[140] ...	1	...		16.0:		*
04519	08 37.5 + 23 43 201.55 + 33.76	04 – 21 – 012 NGC 2628	1355	1.2 x 1.1 1.1 x 1.1	-- 1	1	Sc?	SAB:(r)c?	14.1		3
04520	08 37.5 + 37 05 185.61 + 36.90	06 – 19 – 014	0648	1.2: x 0.5 0.9 x 0.5	[165] ...	1	S ...		15.7		*
04521	08 37.5 + 43 01 178.20 + 37.58	07 – 18 – 034	0707 127	1.2 x 0.3 1.1 x 0.3	112 6	2	Sb – c		15.5		
04522	08 37.5 + 71 09 143.42 + 34.49	12 – 09 – 006	1325	1.0 x 1.0 0.9: x 0.9:	-- 1	2	Sc		15.0		1
04523	08 37.5 + 73 09 141.11 + 33.90	12 – 09 – 005 NGC 2614	1325	2.8 x 2.4 2.8 x 2.4	150: 1 – 2	4	Sc	SA c	14.0		2<u>3</u>
04524	08 37.6 + 05 49 220.65 + 26.72	01 – 22 – 000	0642	1.4 x 0.2 1.4: x 0.2	49 7	3	Sc	Sc +	15.2		
04525	08 37.9 + 51 26 167.56 + 37.81	09 – 14 – 085	0679	1.3 x 0.8 1.2 x 0.8	88: ...	4	S B b	S B(s)b	15.1		
04526	08 38.0 + 19 32 206.31 + 32.48	03 – 22 – 021	1311	1.4 x 0.35 1.3 x 0.35	59 6 – 7	1	Sa – b		14.8		
04527	08 38.0 + 77 06 136.63 + 32.64	13 – 07 – 000	0693	1.5:: x 1.0:: 0.8: x 0.5:	[40] ...	3	Dwarf (irregular) Ir: V		17.		
04528	08 38.2 + 16 22 209.80 + 31.36	03 – 22 – 022	1311	1.0 x 0.9 1.0: x 0.9:	-- 1 – 2	3	...	S dm	16.0:		
04529	08 38.2 + 46 58 173.22 + 37.90	08 – 16 – 000	0671	1.0 x 0.2 0.7 x 0.1	123 6	2	Sb – c		16.5:		
04530	08 38.2 + 57 46 159.59 + 37.33	10 – 13 – 012	1296	1.1 x 0.7 1.1 x 0.7	37 4	3	Sb/Sc	SAB: bc Sc –	16.0:		
04531	08 38.7 + 33 03 190.61 + 36.42	06 – 19 – 015	0648	1.2 x 0.4 1.2 x 0.4	40 6	2	S B b		14.5		
04532	08 38.8 + 19 02 206.94 + 32.48	03 – 22 – 023	1311	1.6 x 0.18 1.4: x 0.2	34 7	2	Sb – c		15.4		
04533	08 38.9 + 05 09 221.48 + 26.70	01 – 22 – 016 NGC 2644	0642	1.8 x 0.7 1.9 x 0.7	[14] ...	3	... Peculiar		13.4		*
04534	08 39.2 + 37 23 185.30 + 37.28	06 – 19 – 016 NGC 2638	0648	1.9 x 0.7 1.9 x 0.5	72 6	2	S0 – a		13.7		
04535	08 39.2 + 64 26 151.35 + 36.32	11 – 11 – 025	1286 129	1.4 x 0.6 0.9 x 0.4	130 6	3	Sb		15.7		
04536	08 39.3 + 65 22 150.22 + 36.13	11 – 11 – 021	1286 129	2.3: x 2.2: 2.0: x 2.0:	-- 1	4	S B c	S B(r)c Sc –	15.5		1
04537	08 39.5 + 35 58 187.07 + 37.12	06 – 19 – 000	0648	1.0 x 0.1 0.9 x 0.1	149 7	2	Sc		16.5:		
04538	08 39.5 + 65 14 150.38 + 36.18	11 – 11 – 023	1286 129	1.5: x 1.3: 1.3: x 1.1:	(140) 1 – 2	4	Sc	SAB(s)c	16.0:		*1
04539	08 39.5 + 67 08 148.10 + 35.73	11 – 11 – 024	1286	1.0 x 0.7 1.0 x 0.7	18 3	2	Sc Singular		14.5		*
04540	08 39.9 + 10 46 215.93 + 29.45	02 – 22 – 004	0456	1.7 x 1.6 1.3 x 1.2	-- ...	3	...	S B dm	14.6		*
04541	08 39.9 + 14 28 212.03 + 30.99	02 – 22 – 005 NGC 2648	0456 128	3.6 x 1.1 (3.6 x 1.2)	148 6	4	Sa np of 2, disturbed		13.0		*13 A
04542	08 39.9 + 25 15 199.98 + 34.74	04 – 21 – 014	1355 128	1.5:: x 1.5:: 1.3:: x 1.3::	-- 1	4	...	SA m	15.5		
04543	08 39.9 + 45 55 174.56 + 38.16	08 – 16 – 000	0671	3.6 x 2.0: 2.0:: x 1.7::	160: ...	4	...	SA dm	15.1		
04544	08 40.0 + 50 22 168.90 + 38.18	08 – 16 – 024 NGC 2639	0671	2.0: x 1.5: 1.7: x 1.2:	140 ...	1	Sa?		12.4	+ 3314 + 3351	*12 347<u>8</u>
04545	08 40.2 + 13 50 212.74 + 30.80	02 – 22 – 007	0456	1.3 x 1.1 1.1 x 0.7	-- ...	3	S ... Peculiar, double system?		14.5		*
04546	08 40.2 + 52 11 166.59 + 38.13	09 – 15 – 003	0982	1.4 x 0.2 1.2 x 0.25	25 7	3	Sa		14.9		
04547	08 40.2 + 73 11 141.00 + 34.08	12 – 09 – 009	1325	1.4 x 0.2 1.3 x 0.2	26 7	2	Sb – c		15.2		*
04548	08 40.3 + 18 24 207.80 + 32.58	03 – 22 – 024	1311	1.2 x 0.2 1.0 x 0.2	50 6 – 7	1	S ...		15.1		*
04549	08 40.4 + 59 00 158.00 + 37.46	10 – 13 – 014	1296	1.7 x 1.4 1.3 x 1.1	10: ...	1	Sc? or Irr?		14.2		
04550	08 40.5 + 13 16 213.38 + 30.64	02 – 22 – 008	0456	2.8 x 0.4 2.2 x 0.35	4 7	3	Sb		14.8		

N G C   2614, 2628, 2638, 2639, 2644, 2648

1	2	3	4	5	6	7	8	9	10	11	12
04551	08 40.5 + 49 58 169.41 + 38.27	08 - 16 - 025	0671	1.7 x 0.6 1.8 x 0.6	113 ...	1	SO?		13.1		1
04552	08 40.6 + 10 55 215.86 + 29.67	02 - 22 - 009	0456	1.1 x 0.8 1.0 x 0.6	40 3:	2	Sb		15.0		
04553	08 40.8 + 03 48 223.06 + 26.47	01 - 22 - 017	0642	1.3 x 0.6 1.0 x 0.4	30 5 - 6	1	S ...		14.8		*
04554	08 40.8 + 22 17 203.48 + 34.02	04 - 21 - 015	1355	1.4 x 0.4 1.3 x 0.3	90 6	4	Sc		15.4		
04555	08 40.9 + 34 54 188.45 + 37.22	06 - 19 - 018 NGC 2649	0648	1.7 x 1.6 1.7 x 1.6	-- 1	4	S B b/Sc	SAB(rs)bc Sc -	13.1		*23
04556	08 40.9 + 41 55 179.65 + 38.12	07 - 18 - 041	0707 127	0.8 x 0.25 0.8 x 0.25	[21] ...	1	... Peculiar, jet		14.1		*
04557	08 41 + 84 29 128.62 + 29.88	14 - 04 - 000	0692 102	1.1 x 0.5 1.1 x 0.4	[15] ...	3	Irr	Ir I	16.5:		*
04558	08 41.0 + 33 42 189.93 + 37.02	06 - 19 - 019	0648	1.8 x 0.3 1.6 x 0.3	17 7	2	Sb - c		15.2		
04559	08 41.1 + 30 18 194.08 + 36.32	05 - 21 - 004	0924	3. x 0.6 2.8: x 0.6	50 7	2	Sa - b		14.7		
04560	08 41.2 + 52 40 165.96 + 38.25	09 - 15 - 014	0982	1.2 x 0.3 1.1 x 0.3	122 7	2	Sa - b		15.7		
04561	08 41.3 + 01 01 225.82 + 25.21	00 - 23 - 001	0469	1.1 x 0.25 1.1 x 0.25	2 6	1	S ...		15.0		
04562	08 41.4 + 41 55 179.66 + 38.21	08 - 16 - 000	0671	1.3 x 0.9 1.1 x 0.7	127 3:	3	S B c	SB(s)c	15.4		
04563	08 41.5 + 78 45 134.73 + 32.21	13 - 07 - 000	0693	1.1 x 0.8 0.8: x 0.4:	[130] ...	3	Dwarf V, p w 4566, distorted?		16.5:		*
04564	08 41.6 + 55 17 162.64 + 38.08	08 - 15 - 017	0982	1.0 x 0.2 1.0 x 0.2	37 6 - 7	1	S ... Disturbed		15.0		*1
04565	08 41.7 + 09 43 217.23 + 29.40	02 - 23 - 000	0438	1.0 x 0.10 0.9 x 0.10	114 7	2	Sc		16.0:		
04566	08 41.8 + 78 44 134.74 + 32.23	13 - 07 - 000	0693	1.1: x 0.9: 0.9: x 0.8:	-- ...	3	Dwarf V, p w 4563		16.5:		*
04567	08 42.0 + 09 59 217.00 + 29.58	02 - 23 - 000	0438	1.0 x 0.2 0.8 x 0.15	167 6 - 7	1	S ... npp of 2		15.3		*
04568	08 42.0 + 10 40 216.29 + 29.88	02 - 23 - 001	0438	1.7 x 0.4 1.5 x 0.4	80 (7)	2	Irr: Peculiar?		14.9		*
04569	08 42.0 + 73 10 140.97 + 34.21	12 - 09 - 010 NGC 2629	1325	(2.3 x 1.8) (2.3 x 1.8)	105 ...	2	E		12.8		*23
04570	08 42.2 + 28 00 196.92 + 35.99	05 - 21 - 000	0924 130	1.1 x 0.9: 1.0: x 0.8:	-- ...	3	... Brightest of 3, disturbed:	SAB dm	16.5:		*
04571	08 42.3 + 01 20 225.66 + 25.58	00 - 23 - 000	0469	1.0 x 0.4 1. x 0.4:	160 6:	1	Sc		15.7		
04572	08 42.4 + 37 07 185.75 + 37.87	06 - 19 - 021	0648	0.7 x 0.7 0.6 x 0.6	-- ...	2	... Compact		13.8		
04573	08 42.6 + 09 50 217.23 + 29.65	02 - 23 - 002 NGC 2657	0438	1.6: x 1.4: 1.5: x 1.5:	-- 1 - 2	4	Sc	SA:(rs)d	14.0		
04574	08 42.6 + 74 17 139.68 + 33.87	12 - 09 - 013 NGC 2633	1325	2.8 x 1.7 2.7 x 1.6	175: 4	3	S B b	S B(s)b Sb +	12.4	+ 2228 + 2382	*12 4678A
04575	08 42.7 + 24 03 201.61 + 34.99	04 - 21 - 018	1355	1.2 x 0.5 1.1 x 0.5	118 6	2	Sb		15.6		
04576	08 42.7 + 73 43 140.32 + 34.07	12 - 09 - 011 IC 2389	1325	1.6 x 0.3 1.5 x 0.3	126 6 - 7	1	S ... Disturbed		13.2	+ 2632 + 2783	*12
04577	08 42.8 + 73 05 141.04 + 34.29	12 - 09 - 012 NGC 2641	1325	1.3 x 1.0 1.4 x 1.2	(5) 2 - 3	3	[SO] Compact		15.0		*3
04578	08 42.9 + 41 46 179.88 + 38.48	07 - 18 - 046	0707 127	1.3 x 0.8 1.1 x 0.8	65: 2 - 3	3	Sb		15.1		1
04579	08 43.0 + 35 53 187.31 + 37.80	06 - 19 - 000	0648	1.0 x 0.1 0.7: x 0.1	81 7	3	Sc		16.5:		*
04580	08 43.0 + 48 37 171.12 + 38.71	08 - 16 - 000	0671	0.9 x 0.45 0.9 x 0.45	7 5	2	Sb - c		14.5		*
04581	08 43.0 + 74 09 139.82 + 33.95	12 - 09 - 015 NGC 2634	1325	(2.0 x 2.0) (2.0 x 2.0)	-- ...	3	E P w 4585, disturbed		12.6		*2
04582	08 43.1 + 12 58 214.00 + 31.09	02 - 23 - 003	0438	1.3 x 0.4 1.2 x 0.4	150 6	1	S ... p of 2		15.0		*1
04583	08 43.1 + 73 50 140.17 + 34.06	12 - 09 - 000 NGC 2636	1325	0.35 x 0.35 0.35 x 0.35	-- ...	2	... (Compact)		14.4		*2
04584	08 43.2 + 12 49 214.17 + 31.05	02 - 23 - 004 NGC 2661	0438	1.6 x 1.5 1.5 x 1.5	-- 1	2	Sc Singular		13.9		*13
04585	08 43.3 + 74 07 139.85 + 33.98	12 - 09 - 016	1325	1.7: x 0.35 1.5: x 0.35	73 ...	1	S ... P w 4581, disturbed		14.3		*2

N G C   2629, 2633, 2634, 2636, 2641, 2649   I C   2389
        2657, 2661

1	2	3	4	5	6	7	8	9	10	11	12
04586	08 43.5 + 75 29 138.30 + 33.51	13 - 07 - 000	0693	1.1 x 0.15 1.3 x 0.15	166 7	3	Sc		16.5:		
04587	08 43.7 + 49 44 169.68 + 38.79	08 - 16 - 029	0671	(1.7 x 1.0)? (1.7 x 1.2)	8: 3 - 4	1	SO?		13.8	+ 3060	
04588	08 43.8 + 19 12 207.26 + 33.64	03 - 23 - 002	0062 131	1.7 x 0.3 1.5: x 0.2	159 6 - 7	2	Sc		15.0		*
04589	08 43.8 + 28 22 196.59 + 36.43	05 - 21 - 007 IC 2393	0924 130	(1.3 x 0.9) (1.4 x 1.0)	20 ...	3	E Brightest of 3		14.6		*
04590	08 43.9 + 13 24 213.63 + 31.45	02 - 23 - 006	0438	1.4 x 0.4 1.1 x 0.4	20 6	1	S ...		15.0		*
04591	08 43.9 + 28 27 196.49 + 36.47	05 - 21 - 008	0924 130	1.4 x 0.15 1.4 x 0.15	18 7	2	Sc IC 2393 group		15.6		*1
04592	08 44.0 + 21 55 204.19 + 34.61	04 - 21 - 020	1355	1.6 x 0.5 1.5: x 0.5	[143] ...	2	S B ... Singular	SAB(s) ...	15.5		*
04593	08 44.0 + 70 18 144.22 + 35.25	12 - 09 - 000	1325	0.7 x 0.45 0.6 x 0.4	(155) ...	1	... Peculiar, double nucleus		13.4		*
04594	08 44.1 + 07 09 220.16 + 28.77	01 - 23 - 003	1358	1.0 x 0.5 1.1 x 0.6	70 5	2	Sa - b		14.8		
04595	08 44.1 + 28 27 196.51 + 36.51	05 - 21 - 009 IC 2394	0924 130	1.7 x 0.7 1.6 x 0.7	90 5 - 6	3	S B b IC 2393 group, Sb -	SB(s)b	15.2		*1
04596	08 44.3 + 19 49 206.61 + 33.97	03 - 23 - 004	0062 131	1.2 x 1.0 1.1 x 1.0	(155) 1 - 2	4	SO	(R)SA(r)0 +	15.0		1
04597	08 44.7 + 26 05 199.38 + 36.02	04 - 21 - 021	1355	1.1 x 0.5 1.1 x 0.5	70 5 - 6	1	S ... Brightest in group, disturbed?		14.9		*1
04598	08 44.8 + 42 03 179.55 + 38.85	07 - 18 - 048	0707 127	1.0: x 1.0: 0.7: x 0.7:	-- 1	3	Sc	Sc +	16.5:		
04599	08 44.9 + 13 36 213.53 + 31.75	02 - 23 - 007	0438	2.1 x 2.1 2.1 x 2.0	-- 1	4	SO	(R)SA 0	14.9		
04600	08 44.9 + 37 56 184.80 + 38.47	06 - 20 - 005 IC 2401	1342	(1.2 x 0.7) (1.3 x 0.8)	(110) ...	2	[SO] Compact		15.0		
04601	08 45 + 85 00 128.04 + 29.73	14 - 05 - 001	0727 102	1.6 x 0.7 1.6 x 0.7	140 ...	1	...		14.9		
04602	08 45.0 + 26 00 199.50 + 36.06	04 - 21 - 022	1355	1.3 x 0.5 1.4 x 0.5	130 6	2	Sa - b 4597 group		15.1		*
04603	08 45.0 + 70 30 143.95 + 35.27	12 - 09 - 020 NGC 2650	1325	1.8 x 1.3 1.8 x 1.3	82 3	4	S B b Sb -	S B(rs)ab	14.3		*12 3
04604	08 45.1 + 73 39 140.33 + 34.25	12 - 09 - 019 NGC 2646	1325	1.7 x 1.7 1.7 x 1.7	-- 1	3	S B 0		13.0	+ 3546 + 3697	*12 48
04605	08 45.2 + 60 25 156.09 + 37.82	10 - 13 - 017 NGC 2654	1296	4.2 x 0.7 4.3 x 0.7	63 6	2	Sa - b		12.8	+ 1360 + 1448	*124 78
04606	08 45.3 + 17 54 208.87 + 33.50	03 - 23 - 005 IC 2406	0062 131	1.6 x 0.8 1.2 x 0.6	173 5	2	SO - a P w 4607		14.4		*
04607	08 45.4 + 17 48 208.99 + 33.49	03 - 23 - 006 IC 2407	0062 131	1.3 x 0.25 1.2 x 0.2	86 6 - 7	2	Sb - c P w 4608		15.1		*
04608	08 45.6 + 18 31 208.21 + 33.79	03 - 23 - 008 IC 2409	0062 131	1.0 x 0.7 0.9 x 0.7	165: 2 - 3	4	S B a	SAB(s)a	14.4		3
04609	08 45.7 + 41 23 180.42 + 38.97	07 - 18 - 049	0707 127	2 ? x 1.2:: ...	... ...	1	... Peculiar, compact		15.4		*1
04610	08 45.8 + 01 15 226.22 + 26.30	00 - 23 - 000	0469	1.1 x 0.3 1.1 x 0.2	145 6	2	Sb - c P w 4613		15.7		*
04611	08 45.8 + 30 04 194.66 + 37.26	05 - 21 - 011	0924	1.8 x 0.2 1.4: x 0.2	107 7	2	Sc	Sc +	15.6		
04612	08 46 + 85 44 127.27 + 29.42	14 - 05 - 000	0727 102	1.5:: x 1.0:: 1.0:: x 0.5::	(175) ...	3	Dwarf (spiral) S: V		17.		*
04613	08 46.1 + 01 13 226.30 + 26.35	00 - 23 - 003	0469	1.7 x 1.5 1.7 x 1.5	-- 1 - 2	4	S B b P w 4610	(R)SB(s)ab	14.9		*
04614	08 46.1 + 36 17 186.93 + 38.48	06 - 20 - 000	1342	1.2 x 0.8 0.7 x 0.7	... ...	1	... Peculiar		14.3		
04615	08 46.1 + 41 58 179.68 + 39.09	07 - 18 - 051	0707 127	1.0 x 0.3 1.1 x 0.3	14 6	2	S B c		15.6		*1
04616	08 46.2 + 36 53 186.18 + 38.59	06 - 20 - 007 NGC 2668	1342	1.6: x 0.9: 1.0 x 0.7	155: ...	1	Sa - b		14.9		
04617	08 46.4 + 29 43 195.12 + 37.30	05 - 21 - 012	0924	1.6: x 0.6: 1.6: x 0.6:	57 6	4	Sc	SA d	15.5		
04618	08 46.4 + 66 29 148.66 + 36.56	11 - 11 - 000	1286	1.0 x 0.6 0.8: x 0.6:	100 ..	2	... S: IV - V		16.5:		
04619	08 46.5 + 19 16a 207.46 + 34.26	03 - 23 - 010 NGC 2672	0062 131	[2.3 x 2.3] [2.8 x 2.8]	-- ...	3	E + companion Contact		13.4	+ 4223 + 4100	*12 3478A
04620	08 46.5 + 19 16b 207.46 + 34.26	03 - 23 - 011 NGC 2673	0062 131						14.4	+ 3792 + 3669	*12 3A

N G C   2646, 2650, 2654, 2668, 2672, 2673

I C   2393, 2394, 2401, 2406, 2407, 2409

1	2	3	4	5	6	7	8	9	10	11	12
04621	08 47.0 + 35 15 188.27 + 38.51	06 - 20 - 008	1342	1.3 x 0.7 1.0 x 0.6	[140] ...	1	... Peculiar		13.6		
04622	08 47.0 + 41 29 180.32 + 39.22	07 - 18 - 056	0707 127	1.4: x 1.1: 0.9 x 0.8	-- 1 - 2	4	Sc	SA d	15.3		
04623	08 47.1 + 76 41 136.87 + 33.27	13 - 07 - 007	0693	3.5 x 1.0 3.3 x 0.9	60 6	2	Sc		13.5		
04624	08 47.4 + 26 08 199.53 + 36.61	04 - 21 - 023	1355	1.1 x 0.5 1.0 x 0.4	30 6	2	Sa - b		15.4		1
04625	08 47.6 + 03 40 224.12 + 27.89	01 - 23 - 000	1358	1.7 x 0.15 1.7 x 0.15	124 7	3	Sc		15.4		
04626	08 48.0 + 24 30 201.53 + 36.28	04 - 21 - 000	1355	1. x 1. ...	-- ...	3	Dwarf V		18.		
04627	08 48.0 + 47 44 172.24 + 39.56	08 - 16 - 032 NGC 2676	0671	1.2 x 1.1 0.8 x 0.8	-- 1	1	S0:		14.3	+ 6010	*
04628	08 48.0 + 51 19 167.60 + 39.39	09 - 15 - 035	0982 135	1.8: x 0.3 1.4: x 0.2	74 6 - 7	3	Sc		15.5		
04629	08 48.3 + 53 48 164.40 + 39.21	09 - 15 - 037 NGC 2675	0982 135	(1.5 x 1.1) (1.6 x 1.2)	(80) 1	2	E		14.4		3
04630	08 48.4 + 73 41 140.19 + 34.46	12 - 09 - 026 IC 520	1325	(2.6 x 2.3) (2.5 x 2.2)	(0) 1	3	Sa P w 4634		11.9		*12
04631	08 48.5 + 19 32 207.35 + 34.80	03 - 23 - 014	0062 131	(1.1 x 1.0) (1.2 x 1.1)	-- (1)	2	S0		15.3		
04632	08 48.5 + 31 04 193.58 + 38.04	05 - 21 - 014 NGC 2679 + 2680	0924	(1.8 x 1.8) (2. x 2.)	-- ...	2	S B : 0 Companion superimposed		14.3		*13
04633	08 48.5 + 53 04 165.33 + 39.31	09 - 15 - 036	0982 135	1.1 x 0.5 1.1 x 0.5	87 6	3	S B b		15.5		*1
04634	08 48.5 + 73 44 140.13 + 34.45	12 - 09 - 000	1325	1.3: x 1.3: 1.1: x 1.1:	-- ...	3	Dwarf (irregular) Ir: IV - V or V, p w 4630		16.5:		*
04635	08 48.6 + 41 02 180.93 + 39.49	07 - 18 - 058	0707 127	[1.8 x 0.8] ...	... ...	3	... Peculiar, plume		15.2		*1
04636	08 48.8 + 29 29 195.56 + 37.76	05 - 21 - 000	0924	1.0 x 0.15 0.9 x 0.15	173 6 - 7	1	S ...		15.6		
04637	08 49.0 + 78 25 134.92 + 32.68	13 - 07 - 010 NGC 2655	0693	(6.5 x 5.8) (6.8 x 6.8)	-- 1	4	S0/Sa		10.8	+ 1299 + 1471	*12- -689A
04638	08 49.1 - 02 10 229.95 + 25.25	00 - 23 - 5 + 6	0469 134	[2. x 1.] ...	... ...	3	Double system Disrupted		14.5		*1V A
04639	08 49.1 + 17 08 210.13 + 34.06	03 - 23 - 015	0062 131	(1.4 x 1.3) (1.3 x 1.2)	-- ...	1	S0?		14.5		
04640	08 49.2 - 01 58 229.78 + 25.37	00 - 23 - 007	0469 134	3.4 x 1.2 3.0: x 1.2:	3 6	4	Sc		14.2		
04641	08 49.6 + 33 36 190.47 + 38.76	06 - 20 - 011 NGC 2683	1342	9.2 x 2.6 9.0 x 2.6	44 6	4	Sb	Sb -	9.7	+ 338 + 287	*12- -689
04642	08 49.6 + 42 36 178.91 + 39.77	07 - 18 - 060	0707	0.7 x 0.7 0.7 x 0.7	-- 1	1	S ...		14.1		1
04643	08 49.8 + 21 37 205.07 + 35.78	04 - 21 - 024	1355	1.7: x 1.3: 1.7: x 1.2:	(90) 1 - 2:	3	Sb/Sc	SA(s)bc Sc -	14.6		*1
04644	08 49.8 + 79 53 133.32 + 32.11	13 - 07 - 012	0693	1.2 x 1.1 1.1 x 1.0	-- 1	3	S B b	SAB(s)b	15.4		1
04645	08 49.9 + 51 31 167.31 + 39.67	09 - 15 - 041 NGC 2681	0982 135	(3.8 x 3.8) (4.0 x 4.0)	-- 1	4	Sa	SAB 0/a	10.4	+ 709 + 751	*12- -689
04646	08 50 + 85 42 127.27 + 29.50	14 - 05 - 002 IC 512	0727 102	3.6 x 2.7 2.8 x 2.2	(175) 3	3	Sc	SAB(s)cd	13.2		*
04647	08 50.1 - 02 25 230.33 + 25.33	00 - 23 - 008 NGC 2690	0469 134	2.0 x 0.4 2.0 x 0.4	19 6 - 7	2	Sa - b		14.1		
04648	08 50.2 + 45 31 175.11 + 39.96	08 - 16 - 033	0671	1.2: x 1.0: 0.7: x 0.6:	(135) ...	3	Dwarf IV - V		17.		
04649	08 50.3 + 40 55 181.11 + 39.80	07 - 18 - 062	0707 127	(1.3 x 0.8) 0.8 x 0.8	(25) ...	1	...		14.9		
04650	08 50.4 + 39 20 183.16 + 39.69	07 - 18 - 061	0707	1.9 x 0.8 1.5 x 0.5	90 6	2	Sa - b		14.6		*
04651	08 50.5 - 00 51 228.91 + 26.24	00 - 23 - 000	0469	1. x 1. 0.9: x 0.9:	-- 1:	3	...	S dm - m	17.		
04652	08 50.5 + 09 20 218.74 + 31.17	02 - 23 - 009 IC 523	0438	1.4: x 1.0 1.0 x 1.0	-- ...	1	Sa - b		15.0		
04653	08 50.7 + 35 20 188.32 + 39.27	06 - 20 - 012	1342	[1.8 x 0.8] ...	... ...	3	Triple system, SB(s)b + companions, contact, plume		15.0		*1V A
04654	08 50.7 + 57 21 159.80 + 39.05	10 - 13 - 031 IC 522	1296	1.0 x 0.8 1.0 x 0.8	(165) 2:	2	S0		13.9		
04655	08 50.9 + 04 58 223.28 + 29.24	01 - 23 - 000	1358	1.2 x 0.1 1.0 x 0.1	156 7	2	Sc		15.6		

N G C   2655, 2675, 2676, 2679 - 81, 2683, 2690   I C   512, 520, 522, 523

1	2	3	4	5	6	7	8	9	10	11	12
04656	08 51.0 + 18 22 208.94 + 34.94	03 - 23 - 000	0062	1.5: x 0.3 1.1 x 0.25	162 6	2	Sb nf of 2		15.3		*
04657	08 51.1 + 18 52 208.38 + 35.14	03 - 23 - 000	0062	1.5 x 0.15 1.2 x 0.15	18 7	1	Sc:		15.6		
04658	08 51.2 + 32 52 191.47 + 38.96	06 - 20 - 013 IC 2421	1342	2.4 x 2.2 2.2 x 2.0	-- 1	4	Sc	SA(rs)c Sc -	14.9		*12
04659	08 51.2 + 47 17 172.80 + 40.11	08 - 16 - 034	0671	2.4: x 0.9: 1.7: x 0.5:	115 ...	4	...	SA: dm	15.5		
04660	08 51.3 + 34 44 189.30 + 39.30	06 - 20 - 014	1342	1.7: x 1.3: 1.1:: x 1.1::	-- ...	3	Dwarf (spiral) S: IV - V		17.		*
04661	08 51.3 + 36 38 186.67 + 39.57	06 - 20 - 015	1342	1.0 x 0.3 0.9 x 0.3	11 6 - 7	1	S ...		15.1		*1
04662	08 51.3 + 49 20 170.13 + 40.04	08 - 16 - 035 NGC 2684	0671 132	0.9 x 0.8 1.0 x 0.8	-- 1 - 2	1	S ... Peculiar		13.4		*13
04663	08 51.5 + 20 34 206.46 + 35.81	04 - 21 - 000	1355	1.8:: x 0.4: ...	[0] ...	0	Dwarf		18.		
04664	08 51.5 + 39 44 182.67 + 39.94	07 - 18 - 064 NGC 2691, Mark 391	0707 127	1.6 x 1.0 0.9 x 0.7	165: ...	1	Sa?		13.9		*3
04665	08 51.6 + 68 28 146.12 + 36.46	11 - 11 - 033	1286 138	1.1 x 0.25 1.2: x 0.30	4 6 - 7	1	S ...		15.5		*
04666	08 51.7 + 58 55 157.79 + 38.90	10 - 13 - 039 NGC 2685	1296	4.7: x 2.5: 3.0 x 2.5	[38] ...	2	... Peculiar, acicular?		12.1	+ 877 + 957	*124 56789A
04667	08 51.9 + 20 25 206.67 + 35.85	03 - 23 - 017 IC 2423	0062	1.1 x 0.9 0.9 x 0.7	(100) 2	3	Sb/S B b	SAB(s)b	14.7		*3
04668	08 52.1 + 57 52 159.11 + 39.15	10 - 13 - 044	1296	1.1 x 0.15 1.1 x 0.15	84 7	2	Sb - c		15.6		
04669	08 52.3 + 19 07 208.21 + 35.49	03 - 23 - 018	0062	1.6: x 1.4: 1.3: x 1.2:	-- (1)	2	Sc/Irr		15.5		
04670	08 53.1 + 13 25 214.70 + 33.50	02 - 23 - 010	0438	(1.6 x 1.2) (1.7 x 1.4)	-- ...	1	S0: 1st of 3		14.2		*
04671	08 53.1 + 52 18 166.23 + 40.09	09 - 15 - 053	0982 135	1.6 x 1.4 1.6: x 1.4:	-- ...	3	S ... P w 4675, disturbed		13.6		*
04672	08 53.2 - 02 06 230.49 + 26.15	02 - 23 - 013	0469 134	1. x 1. 0.9 x 0.9	-- 1:	1	S ...		14.8		*
04673	08 53.3 + 02 43 225.85 + 28.66	01 - 23 - 005	1358	1.5: x 1.2 1.4: x 1.2:	(20) 1 - 2	4	S B c	S B(s)cd	14.9		
04674	08 53.3 + 51 33 167.20 + 40.19	09 - 15 - 055 NGC 2693	0982 135	(1.8 x 1.4) (2.2 x 1.6)	(160) ...	3	E		13.1	+ 4956 + 4998	*12 3468
04675	08 53.3 + 52 16 166.27 + 40.12	09 - 15 - 057 NGC 2692	0982 135	1.3 x 0.4 1.4 x 0.4	165 6	1	S B: a - b P w 4671		14.1		*23
04676	08 53.5 + 52 00 166.61 + 40.18	09 - 15 - 058	0982 135	1.3 x 0.5 1.3 x 0.5	55 6	3	Sc	Sc +	16.0:		
04677	08 53.6 + 13 22 214.81 + 33.59	02 - 23 - 012	0438	1.5 x 1.1 1.4: x 1.1:	(120) 2 - 3	2	Sc 2nd of 3		15.1		*
04678	08 53.6 + 39 34 182.93 + 40.33	07 - 19 - 005 IC 2424	0721 127	1.1 x 1.1 1.1 x 1.0	-- 1	2	S B b	S B(r) ... Sb -	14.4		*1
04679	08 53.6 + 51 40 167.05 + 40.23	09 - 15 - 059	0982 135	1.4 x 0.2 1.1 x 0.15	91 7	2	Sc		16.5:		*
04680	08 53.7 - 02 23 230.83 + 26.11	00 - 23 - 017 NGC 2706	0469 134	1.8 x 0.6 1.7 x 0.5	167 6 - 7	1	Sa?		13.8		
04681	08 53.7 + 11 45 216.58 + 32.93	02 - 23 - 013	0438	1.0 x 0.2 0.9 x 0.25	162 ...	2	S ... Brightest in multiple system, disturbed		15.6		*1
04682	08 54 + 86 20 126.59 + 29.26	14 - 05 - 003	0727	1.3 x 0.7 1.1 x 0.5	67: 5	3	Sc	Sc +	16.5:		
04683	08 54.0 + 59 16 157.27 + 39.12	10 - 13 - 046	1296	2.0: x 1.1: 1.6: x 1.1:	[110] ...	3	Dwarf irregular Ir V		16.5:		*
04684	08 54.1 + 00 33 228.09 + 27.73	00 - 23 - 018	0469 133	1.4 x 1.1 1.3 x 1.0	(175) 2 - 3	4	Sc	SA d	14.7		
04685	08 54.3 + 13 24 214.86 + 33.76	02 - 23 - 014	0438	1.4 x 0.6 1.4 x 0.5	150 6	1	S ... 3rd of 3		14.2		*
04686	08 54.4 + 43 18 178.03 + 40.68	07 - 19 - 007	0721	0.8 x 0.45 0.8 x 0.5	90 ...	1	S ...		14.5		*
04687	08 54.4 + 66 40 148.15 + 37.27	11 - 11 - 000	1286 138	0.7 x 0.6 0.7 x 0.6	[45] ...	1	... Peculiar		14.2		*
04688	08 54.5 + 17 29 210.31 + 35.39	03 - 23 - 020 NGC 2711	0062 140	1.0 x 0.7 0.9 x 0.6	(170) ...	1	S B ...		14.6		3
04689	08 54.5 + 39 43 182.75 + 40.51	07 - 19 - 006	0721 127	1.0 x 0.25 0.9 x 0.2	24 6	1	S ...		16.0:		
04690	08 54.5 + 52 23 166.09 + 40.29	09 - 15 - 061	0982 135	1.2 x 0.6 1.1 x 0.5	120 5	2	Sa - b		14.9		

N G C 2684, 2685, 2691 - 93, 2706, 2711   I C 2421, 2423, 2424

1	2	3	4	5	6	7	8	9	10	11	12
04691	08 54.7 + 03 06 225.67 + 29.15	01 – 23 – 006 NGC 2713	1358	4.0 x 1.4 4.1 x 1.4	107 6	4	S B b P w 4692	S B(rs)ab Sb –	12.9		*12 346$\overline{8}$
04692	08 54.9 + 03 16 225.53 + 29.28	01 – 23 – 007 NGC 2716	1358	1.6 x 1.2 (1.6 x 1.2)	(30) 2 – 3:	3	S0 P w 4691	S 0 +	13.7	+ 3537 + 3344	*12 36$\overline{}$
04693	08 55.0 + 66 25 148.43 + 37.40	11 – 11 – 034	1286 138	1.3 x 0.7 0.9: x 0.5:	110: ...	3	... S IV	S dm	16.5:		
04694	08 55.1 + 12 41 215.74 + 33.64	02 – 23 – 015	0438	1.0 x 0.6 0.9 x 0.6	55: 4:	3	S B c f of 2		15.1		*
04695	08 55.3 + 53 57 164.04 + 40.21	09 – 15 – 063 NGC 2701	0982 135	2.0 x 1.3 1.8 x 1.3	23 3 – 4	4	Sc		12.3		*12 346$\overline{8}$
04696	08 55.4 + 53 49 164.21 + 40.25	09 – 15 – 064	0982 135	1.1 x 0.1 1.1 x 0.1	136 7	2	Sc		15.7		*
04697	08 55.4 + 70 13 143.93 + 36.19	12 – 09 – 028	1325	1.7 x 0.3 1.5 x 0.3	57 6 – 7	2	S B c		16.0:		
04698	08 55.5 + 28 29 197.23 + 38.95	05 – 21 – 015	0924	1.7 x 0.8 1.6 x 0.8	55 5	2	S B a – b		14.6		
04699	08 55.6 + 39 42 182.79 + 40.72	07 – 19 – 008	0721 127	1.3 x 1.2 1.2 x 1.2	--- ...	1	... Peculiar, jet		14.7		*1
04700	08 55.6 + 41 46 180.07 + 40.84	07 – 19 – 010	0721	1.2 x 1.1 1.1 x 1.0	--- 1	3	Sb		14.6		1
04701	08 55.6 + 78 29 134.69 + 32.95	13 – 07 – 013	0693	2.2 x 1.0 2.1 x 0.9	110 5	3	...	SA d – dm	15.7		
04702	08 55.7 + 39 00 183.72 + 40.69	07 – 19 – 009	0721	(1.5 x 1.4) (1.0 x 0.9)	--- ...	1	S0?		15.0		
04703	08 55.8 + 06 31 222.37 + 31.05	01 – 23 – 013	1358	[1.9 x 0.4] ...	... ...	2	Double system Long bridge		15.5	(+ 3556) (+ 3375)	*1C
04704	08 55.8 + 39 24 183.19 + 40.74	07 – 19 – 011	0721	4.1 x 0.4 4.0: x 0.3	115 (7)	2	...	d – m	15.5		*1
04705	08 56.0 + 55 53 161.52 + 40.02	09 – 15 – 066 NGC 2710	0982	2.2 x 1.1 2.2 x 1.1	125 5	4	S B b	S B(rs)b Sb –	13.8		3
04706	08 56.0 + 65 06 149.98 + 37.89	11 – 11 – 035	1286 138	1.4 x 1.1 1.3 x 1.1	(165) 1 – 2	4	S B b	S B(s)b	15.5		*
04707	08 56.2 + 06 30 222.44 + 31.13	01 – 23 – 015 NGC 2718	1358	2.3: x 2.3: 2.2: x 2.1:	--- 1	4	Sa/S B b	(R´)SAB(s)ab Sb –	13.3		3
04708	08 56.2 + 45 06 175.66 + 41.01	08 – 17 – 003 NGC 2712	1004 136	3.5 x 1.7 3.0: x 1.5:	178 5	3	S B b		12.3	+ 1840 + 1849	*12 346$\overline{8}$
04709	08 56.2 + 46 08 174.29 + 40.99	08 – 17 – 005	1004 136	0.55 x 0.45 0.55 x 0.45	(10) ...	1	S ...		14.5		*
04710	08 56.4 + 11 22 217.33 + 33.37	02 – 23 – 016 NGC 2720	0438	(1.2 x 1.1) (1.4 x 1.2)	--- ...	2	E – S0 P w 4712		14.2		*3
04711	08 56.6 + 78 57 134.16 + 32.79	13 – 07 – 000	0693	1.1 x 1.0 0.8 x 0.8	--- ...	1	...		15.4		
04712	08 56.7 + 11 20 217.41 + 33.42	02 – 23 – 000	0438	1.0 x 0.25 1.1: x 0.25	131 6	1	S ... P w 4710		15.2		*
04713	08 56.7 + 52 42 165.63 + 40.59	09 – 15 – 070	0982 135	1.9 x 1.3 1.7 x 1.3	177: 2 – 3	3	Sb		13.7		
04714	08 56.7 + 78 46 134.35 + 32.88	13 – 07 – 014	0693	1.4 x 1.1 1.2 x 1.1	105: 2	2	Sb	SAB b	13.7		
04715	08 56.8 + 53 16 164.89 + 40.53	09 – 15 – 071	0982 135	[1.0: x 0.5] ...	... ...	1	Strongly peculiar, plume Disrupted pair?		16.0:		*1
04716	08 57.0 + 40 30 181.76 + 41.04	07 – 19 – 012	0721	1.4 x 0.9 1.4 x 0.9	--- ...	4	S B b	S B(s)b Sb +	15.5		
04717	08 57.0 + 51 25 167.30 + 40.78	09 – 15 – 073	0982 135	1.1 x 0.5 0.6 x 0.2	40 5:	2	Sa – b		15.1		
04718	08 57.1 + 35 55a 187.80 + 40.64	06 – 20 – 017 NGC 2719	1342	1.4 x 0.3 1.3 x 0.3	133 (6 – 7)	1	Peculiar n of 2, disrupted:		(13.7)	+ 3113 + 3073	*12 3A$\overline{}$
04719	08 57.1 + 50 53 168.00 + 40.85	09 – 15 – 072	0982	2.3 x 0.3 2.2 x 0.3	95 7	3	Sc Disturbed:, Sc –		15.0		*
04720	08 57.2 + 34 51 189.19 + 40.51	06 – 20 – 016	1342	1.1 x 0.4 1.0 x 0.3	178 6	1	S ...		15.5		
04721	08 57.3 + 17 07 211.04 + 35.88	03 – 23 – 000	0062 140	1.1 x 1.0 1.0 x 0.9	--- 1 – 2	3	S B c		15.5		
04722	08 57.5 + 25 47 200.73 + 38.70	04 – 21 – 026	1355	1.6 x 0.25 1.6 x 0.3	32 7	2	S – Irr Peculiar, plume		15.2		*
04723	08 57.6 + 03 22 225.82 + 29.91	01 – 23 – 017 NGC 2723	1358	(1.1 x 1.1) (1.1 x 1.1)	--- (1)	3	S0		14.5	+ 3725 + 3532	*12 3
04724	08 57.7 + 17 34 210.56 + 36.13	03 – 23 – 026	0062 140	1.1 x 0.2 1.2 x 0.3	[52] ...	2	Irr 4th of 4		15.3		*
04725	08 57.8 + 32 12 192.65 + 40.21	05 – 21 – 000	0924	1.5 x 0.15 1.0: x 0.1	67 7	3	Sc		15.7		

N G C   2701, 2710, 2712, 2713, 2716, 2718 – 20
2723

1	2	3	4	5	6	7	8	9	10	11	12
04726	08 57.8 + 35 57 187.78 + 40.78	06 − 20 − 019	1342	1.8: x 1.1 1.6 x 1.1	(2) ...	2	S B c/Sc	SAB(s)c	14.8		*
04727	08 57.8 + 60 21 155.77 + 39.34	10 − 13 − 051	1296	1.3 x 0.12 1.1 x 0.1	81 7	2	Sc P w 4730		16.0:		*
04728	08 57.9 + 17 23 210.80 + 36.11	03 − 23 − 000	0062 140	1.1 x 0.3 1.1 x 0.3	55 6	2	Sa − b f of 2		16.0:		*
04729	08 57.9 + 17 49 210.30 + 36.27	03 − 23 − 027	0062 140	1.0 x 0.9 1.1 x 1.0	— 1	3	S B c	S B(s)c − d	14.7		1
04730	08 58.0 + 60 21 155.76 + 39.37	10 − 13 − 052	1296	0.9 x 0.3 0.9 x 0.3	93 ...	1	... P w 4727		14.3		*
04731	08 58.3 + 10 49 218.17 + 33.55	02 − 23 − 017	0438	1.4 x 1.4 1.5 x 1.4	— 1	2	S0 − a		14.7		
04732	08 58.3 + 11 17 217.66 + 33.75	02 − 23 − 018 NGC 2725	0438	0.7 x 0.6 0.8 x 0.7	— ...	1	... Peculiar		14.1		3
04733	08 58.6 + 04 18 225.02 + 30.59	01 − 23 − 000	1358	1.3 x 0.1 1.0 x 0.1	178 7	3	Sc		15.7		
04734	08 58.6 + 19 10 208.80 + 36.90	03 − 23 − 000	0062 140	1.0 x 0.3 1.0 x 0.3	158 6	1	S ...		15.4		
04735	08 58.8 − 01 40 230.91 + 27.57	00 − 23 − 019 IC 525	0469	1.0 x 0.25 0.9 x 0.2	11 6	1	S ...		14.9		
04736	08 58.8 + 75 07 138.26 + 34.55	13 − 07 − 000	0693	1.6 x 1.3 1.0: x 0.8:	(160) ...	3	...	SA dm or SAB dm	16.5:		
04737	08 58.9 + 03 55 225.45 + 30.47	01 − 23 − 018 NGC 2729	1358	0.8 x 0.5 0.8 x 0.5	0: ...	1	S0?		14.0		*13
04738	08 59.0 + 11 16 217.77 + 33.90	02 − 23 − 020 NGC 2728	0438	1.2 x 1.0 1.2 x 0.9	(60) 2 − 3	2	Sb		14.9		3
04739	08 59.1 + 69 42 144.39 + 36.67	12 − 09 − 000	1325	1.4 x 0.6 1.0 x 0.5	(105) ...	1	... Peculiar		15.2		
04740	08 59.2 + 23 35 203.57 + 38.45	04 − 22 − 001	0237	1.2 x 0.5 1.3 x 0.5	166 6	1	S ...		14.9		
04741	08 59.5 + 08 30 220.80 + 32.78	02 − 23 − 021 NGC 2731	0438	0.8 x 0.5 0.8 x 0.5	70 ...	1	S ...		14.2		3
04742	08 59.5 + 14 44 214.00 + 35.45	03 − 23 − 000	0062	1.3 x 0.3 1.3 x 0.4	128 6	2	Sa − b Brightest of 3		15.5		*
04743	08 59.5 + 17 02 211.38 + 36.33	03 − 23 − 028 NGC 2730	0062 140	1.7 x 1.3 1.7 x 1.3	(80) 2 − 3:	4	S B c − Irr Singular, disturbed?		13.7		*13
04744	08 59.7 + 26 08a 200.47 + 39.27	04 − 22 − 002 NGC 2735	0237	1.1 x 0.4 1.1 x 0.45	[94] ...	1	... p of 2, bridge		(14.2)	+ 2670 + 2578	*1V AC
04745	08 59.8 + 25 17 201.53 + 39.07	04 − 22 − 000	0237	1.0 x 0.15 1.0 x 0.15	170 7	2	Sc		15.7		
04746	08 59.8 + 25 36 201.14 + 39.15	04 − 22 − 004	0237	1.4 x 0.7 1.4 x 0.7	145 5	2	Sb − c		15.3		
04747	09 00.2 + 30 47 194.60 + 40.44	05 − 22 − 001 IC 2428	1365	1.9 x 0.45 1.9 x 0.4	75 6	2	Sc		14.7		
04748	09 00.6 + 10 13 219.12 + 33.79	02 − 23 − 000	0438	[1.3 x 0.35] ...	[10] ...	2	Sc + companion Contact, in group		15.4		*
04749	09 01.0 + 51 50 166.66 + 41.35	09 − 15 − 083 Mark 101	0982 135	0.7 x 0.7 0.7 x 0.7	— 1	1	S ...		13.6		
04750	09 01.0 + 60 08 155.92 + 39.78	10 − 13 − 054 NGC 2726, Mark 18	1296	1.8 x 0.5 1.7 x 0.5	87 6	1	Sa?		13.1		23
04751	09 01.1 + 22 06 205.55 + 38.42	04 − 22 − 005 NGC 2737	0237	1.1 x 0.4 1.2 x 0.45	61 6	1	Sa − b NGC 2738 group		14.8		*3
04752	09 01.1 + 22 10 205.47 + 38.44	04 − 22 − 006 NGC 2738	0237	1.5 x 0.7 1.5 x 0.7	55 5	1	S ... Brightest in group		13.8		*13
04753	09 01.2 + 45 29 175.12 + 41.88	08 − 17 − 018	1004	1.5 x 0.2 1.5 x 0.2	33 7	3	Sc	S d	15.7		
04754	09 01.3 − 00 18 229.98 + 28.82	00 − 23 − 000	0469	1.1 x 0.2 1.1 x 0.22	135 7	1	Sb − c		16.0:		
04755	09 01.5 + 28 09 198.05 + 40.16	05 − 22 − 005 IC 2430	1365	1.1 x 0.6 1.0 x 0.5	43 5	2	S0 − a		14.4		1
04756	09 01.7 + 14 48 214.18 + 35.96	03 − 23 − 030 IC 2431	0062	[0.6 x 0.45] ...	... ...	2	Quadrupel system		14.3		1
04757	09 01.8 + 18 39 209.74 + 37.43	03 − 23 − 031 NGC 2744	0062 140	[1.6 x 1.0] ...	... ...	3	Double system Disrupted, NGC 2749 group		13.7	+ 3450 + 3324	*12 3
04758	09 01.9 + 22 13 205.48 + 38.63	04 − 22 − 008	0237	1.6: x 0.7 1.2 x 0.6	15: 5 − 6	2	S B b NGC 2738 group		15.4		*
04759	09 01.9 + 78 17 134.73 + 33.33	13 − 07 − 015 NGC 2715	0693	5.0 x 1.7 5.0 x 1.7	22 6	5	Sc	SAB(rs)c Sc −	11.9	+ 1158 + 1329	*12− −689
04760	09 02.0 + 25 12 201.81 + 39.52	04 − 22 − 009 NGC 2743	0237	1.2 x 0.8 1.3 x 0.9	105: ...	2	Sc − Irr		14.3		3

N G C   2715, 2725, 2726, 2728 − 31, 2735   I C   525, 2428, 2430, 2431
          2737, 2738, 2743, 2744

1	2	3	4	5	6	7	8	9	10	11	12
04761	09 02.1 + 17 39 210.95 + 37.14	03 - 23 - 034 	0062 140	(1.0 x 0.9) (1.1 x 1.0)	— ...	3	E		14.8		*1
04762	09 02.3 + 45 31 175.07 + 42.07	08 - 17 - 000 	1004	1.1: x 0.5: 0.7:: x 0.3::	[5] ...	4	Dwarf irregular Ir V		18.		
04763	09 02.5 + 18 30 209.99 + 37.53	03 - 23 - 036 NGC 2749	0062 140	(1.8 x 1.7) (2.2 x 2.0)	— ...	3	E Brightest in group		13.3	+ 4203 + 4076	*12 3468
04764	09 02.5 + 25 45 201.16 + 39.78	04 - 22 - 011 	0237	1.0 x 0.7 0.7: x 0.6:	25: ...	1	... Peculiar?		15.4		*
04765	09 02.5 + 47 23b 172.56 + 42.02	08 - 17 - 023 	1004	1.1: x 1.0: 1.0: x 0.9:	— ...	2	Strongly peculiar nnf of 2, disrupted:		(14.5)		*1
04766	09 02.5 + 72 03 141.56 + 36.04	12 - 09 - 030 	1325	1.0 x 0.8 1.0: x 0.8:	(145) 2:	3	Sc	SA cd	16.0:		
04767	09 02.6 + 36 32 187.15 + 41.81	06 - 20 - 022 	1342 139	(1.3 x 1.1) (1.3 x 1.1)	25: ...	1	S0		14.0		
04768	09 02.7 - 00 18 230.19 + 29.12	00 - 23 - 000 	0469	1.0 x 0.1 1.0 x 0.12	38 7	2	Sb - c		16.5:		
04769	09 02.8 + 25 38 201.33 + 39.81	04 - 22 - 012 NGC 2750	0237	2.2 x 2.0 2.4 x 2.2	— 1	4	Sc	SAB c Sc -	12.7		*13
04770	09 02.8 + 35 34 188.45 + 41.74	06 - 20 - 023 NGC 2746	1342	1.8 x 1.7 1.6 x 1.6	— 1	3	S B a	S B(rs)a	14.4		*3
04771	09 02.8 + 50 17 168.67 + 41.81	08 - 17 - 025 	1004	1.4 x 0.7 1.4 x 0.7	30 5	3	S B b	(R')S B(rs)b Sb -	15.6		
04772	09 02.9 + 18 32 209.99 + 37.63	03 - 23 - 038 NGC 2752	0062 140	1.8 x 0.4 2.1 x 0.4	58 6 - 7	3	Sb NGC 2749 group		14.8		*12 36
04773	09 03.1 + 18 58 209.50 + 37.83	03 - 23 - 039 	0062 140	1.5 x 0.5 1.5 x 0.4	67 6	1	S ... NGC 2749 group:		14.8		*1
04774	09 03.3 + 25 47 201.18 + 39.96	04 - 22 - 013 	0237	1.0: x 0.6: 0.9: x 0.7:	— 1:	4	...	S B dm	15.7		
04775	09 03.3 + 66 46 147.67 + 38.07	11 - 11 - 037 	1286 138	(1.4 x 0.9) (1.5 x 1.0)	0: ...	2	E - S0		14.3		
04776	09 03.3 + 79 34 133.34 + 32.79	13 - 07 - 000 	0693	1.4:: x 1.0:: 1.3:: x 1.0::	— ...	3	Dwarf (spiral) S: V		16.5:		
04777	09 03.6 + 34 49 189.47 + 41.81	06 - 20 - 024 	1342	2.4: x 0.4 1.8: x 0.4	140 (6 - 7)	3	Irr	dm - m Ir I	15.2		
04778	09 03.6 + 50 55 167.81 + 41.86	09 - 15 - 000 	0982	0.9 x 0.9 0.9 x 0.9	— 1	3	Sb		14.2		*
04779	09 03.6 + 60 40 155.14 + 39.96	10 - 13 - 057 NGC 2742	1296	3.3 x 1.7 3.2 x 1.7	87 5	4	Sc	SA(s)c Sc -	12.0	+ 1291 + 1380	*123 4678
04780	09 03.8 + 19 32 208.90 + 38.18	03 - 23 - 040 	0062 140	1.7 x 1.3 1.6: x 1.2:	(60) ...	4	... NGC 2749 group:	SAB dm	15.0		*1
04781	09 03.9 + 06 30 223.51 + 32.81	01 - 23 - 022 	1358	1.9 x 0.7 1.7: x 0.6:	127 6	3	Sc		14.9		
04782	09 03.9 + 26 28 200.36 + 40.27	04 - 22 - 014 IC 2435	0237	(1.0 x 0.4) 0.9 x 0.35	120 ...	2	E		15.2		
04783	09 04.0 + 39 09 183.67 + 42.30	07 - 19 - 032 	0721	1.0 x 0.15 0.8 x 0.2	162 7	1	S ...		15.2		
04784	09 04.0 + 46 26 173.82 + 42.33	08 - 17 - 031 	1004	1.6: x 0.8 1.5: x 0.7:	... ...	3	S ... 3rd in chain of 3 Distorted		16.0:		*1
04785	09 04.1 + 37 24 186.03 + 42.19	06 - 20 - 025 IC 2434	1342 139	1.9 x 0.6 1.5: x 0.6	13 6	2	S B ... NGC 2759 group Peculiar, disturbed:		14.5		*1
04786	09 04.2 + 28 30 197.78 + 40.82	05 - 22 - 000 	1365	1.0 x 0.10 0.8 x 0.1	81 7	2	Sc		16.5:		
04787	09 04.5 + 33 28 191.30 + 41.80	06 - 20 - 028 	1342	2.3 x 0.5 2.3: x 0.5	6 6	3	...	S dm	14.3		
04788	09 04.5 + 59 16a 156.87 + 40.43	10 - 13 - 059 	1296	1.7 x 0.9 1.3 x 0.7	5: ...	1	S B ... nnp of 2		(15.5)		*1
04789	09 04.7 + 41 55 179.93 + 42.54	07 - 19 - 034 NGC 2755	0721	1.2 x 0.8 1.2 x 0.8	130: 3 - 4	1	S ...		14.2		*3
04790	09 04.7 + 51 51 166.53 + 41.91	09 - 15 - 096 	0982 135	1.1 x 0.6 1.1 x 0.6	0 5	2	S B:a - b		15.3		*
04791	09 05.0 + 03 35 226.68 + 31.62	01 - 24 - 001 NGC 2765	0028	(1.7 x 1.0) (2.0 x 1.0)	107 6	3	S0		13.3		3
04792	09 05.2 + 20 42 207.64 + 38.88	04 - 22 - 000 	0237 140	1.1 x 0.1 1.0 x 0.12	44 7	2	Sb - c		15.6		
04793	09 05.3 + 19 02 209.65 + 38.34	03 - 24 - 000 	0057	1.0 x 0.5 0.8 x 0.3	100 5 - 6	1	S ...		15.7		
04794	09 05.4 + 21 39 206.50 + 39.23	04 - 22 - 017 NGC 2764	0237	(1.4 x 0.9) (1.7 x 0.9)	15 ...	1	S? ...		13.9		*12 348
04795	09 05.4 + 37 49 185.50 + 42.48	06 - 20 - 033 NGC 2759	1342 139	(1.0 x 0.7) (1.3 x 0.9)	50 ...	2	E - S0 Brightest in group		14.2		3

N G C   2742, 2746, 2749, 2750, 2752, 2755   I C   2434, 2435
      2759, 2764, 2765

1	2	3	4	5	6	7	8	9	10	11	12
04796	09 05.4 + 54 03 163.60 + 41.66	09 – 15 – 098 NGC 2756	0982 135	1.7 x 1.1 1.5 x 1.0	0 3 – 4	3	Sb		13.2		*3
04797	09 05.5 + 06 08 224.12 + 32.98	01 – 24 – 003	0028	2.5:: x 2.5:: 2.5:: x 2.5::	-- (1)	4	Dwarf spiral S V		15.7		*
04798	09 05.5 + 45 00 175.74 + 42.66	08 – 17 – 038	1004	1.1 x 1.1 1.0: x 0.9:	-- 1	3	Sc	SA cd	15.2		
04799	09 05.6 + 37 42 185.66 + 42.51	06 – 20 – 034	1342 139	1.2 x 0.3 1.5 x 0.3	39 (6)	1	S B ... Peculiar, disturbed	NGC 2759 group	15.2		*1
04800	09 05.7 + 55 06 162.21 + 41.51	09 – 15 – 099	0982	1.7 x 0.5 1.6 x 0.5	120 6	2	S B c:	S B(s) ...	14.8		
04801	09 05.8 + 30 04 195.85 + 41.49	05 – 22 – 009 NGC 2766	1365	1.5 x 0.5 1.4 x 0.5	132 6	2	Sa – b		14.6		*
04802	09 05.9 – 01 25 231.76 + 29.20	00 – 24 – 001	0430	1.1 x 0.9 1.0 x 0.8	-- 2:	3	Sc 4804 group	SAB(s)c	15.3		*
04803	09 06.0 + 62 27 152.81 + 39.73	10 – 13 – 060	1296	1.8 x 0.8 1.7: x 0.8:	90 5	2	S ... Singular, disturbed:		14.0		*2
04804	09 06.1 – 01 33 231.92 + 29.17	00 – 24 – 002	0430	1.0 x 0.9 0.8 x 0.8	-- 1 – 2	2	S B b – c Brightest of 3		15.2		*
04805	09 06.2 + 46 10 174.15 + 42.72	08 – 17 – 044	1004	1.0 x 0.8 1.0 x 0.8	115 2	1	S ...		14.6		
04806	09 06.4 + 33 19 191.58 + 42.17	06 – 20 – 038 NGC 2770	1342	3.6 x 1.0 3.4 x 0.9	148 6	4	Sc	Sc +	12.1		*12 3
04807	09 06.4 + 54 46 162.62 + 41.67	09 – 15 – 101	0982	1.1 x 1.0 1.0 x 1.0	-- 1	4	Sc	SA:(rs)cd	14.5		
04808	09 06.4 + 60 10 155.65 + 40.42	10 – 13 – 061	1296	1.4 x 0.7 1.3 x 0.5	165 5	3	Sc		15.7		
04809	09 06.5 + 20 55 207.50 + 39.24	04 – 22 – 019	0237 140	1.5 x 0.6: 1.5 x 0.5	103 6	2	S B c:		15.1		
04810	09 06.5 + 37 47 185.57 + 42.70	06 – 20 – 039 IC 527	1342 139	1.9: x 1.8: 1.6: x 1.5:	-- 1	1	S ... NGC 2759 group	SAB ...	14.6		*
04811	09 06.6 + 16 00 213.38 + 37.52	03 – 24 – 001 IC 528	0057 140	1.7 x 0.8 1.6 x 0.7	163 5 – 6	2	Sb	Sb –	14.6		*1
04812	09 06.7 + 50 15 168.62 + 42.44	08 – 17 – 000	1004 143	[1.3 x 0.6] ...	... ...	3	Sc + companion		14.4		*
04813	09 06.7 + 50 37 168.12 + 42.39	08 – 17 – 046 NGC 2767	1004 143	0.6 x 0.5 0.7 x 0.6	-- ...	1	... NGC 2769 group		14.4		*3
04814	09 06.8 + 44 49 175.98 + 42.89	08 – 17 – 000?	1004	1.1 x 0.12 0.9 x 0.12	4 7	2	Sb		16.5:		*
04815	09 07.0 + 07 23 223.02 + 33.91	01 – 24 – 004 NGC 2773	0028	0.8 x 0.35 0.7 x 0.30	[83] ...	1	...		14.5		*3
04816	09 07.0 + 50 39 168.07 + 42.43	08 – 17 – 050 NGC 2769	1004 143	2.0 x 0.5 1.9 x 0.4	146 7	2	Sa Brightest in group	P w 4817	13.8		*3
04817	09 07.1 + 50 36 168.13 + 42.45	08 – 17 – 051 NGC 2771	1004 143	2.0 x 1.8 2.0 x 1.8:	-- 1	4	S B a P w 4816, 4816 group	(R′)S B(r)ab	14.0		*3
04818	09 07.1 + 79 24 133.41 + 33.02	13 – 07 – 016 NGC 2732	0693	1.8 x 0.8 1.9 x 0.8	67 6 – 7	1	S0 P w 4832		12.6	+ 2121 + 2297	*12 3468
04819	09 07.2 + 71 58 141.47 + 36.40	12 – 09 – 000	1325	1.4 x 0.2 1.4 x 0.2	175 7	2	Sc		16.5:		
04820	09 07.7 + 07 15 223.26 + 34.00	01 – 24 – 005 NGC 2775	0028	5. x 4. (5.5 x 5.0)	155 2:	4	Sa	SA(r:)ab	11.4	+ 1135 + 958	*12 3468
04821	09 07.7 + 60 15 155.49 + 40.56	10 – 13 – 065 NGC 2768	1296	(6.5 x 3.0) (7.0 x 3.5)	95 ...	3	E – S0		11.1	+ 1408 + 1495	*123 4678
04822	09 07.8 + 19 41 209.13 + 39.12	03 – 24 – 000	0057 140	1.1: x 0.5: 0.9: x 0.4	165: ...	3	...	S B dm	16.5:		*
04823	09 08.0 + 07 25 223.13 + 34.14	01 – 24 – 006 NGC 2777	0028	0.7 x 0.5 0.8 x 0.7	... ...	1	...		13.9		*23
04824	09 08.0 + 51 28 166.94 + 42.47	09 – 15 – 105	0982	2.5 x 0.6 2.5 x 0.6	98 6	4	Sc	SA d	14.5		1
04825	09 08.0 + 76 41 136.25 + 34.36	13 – 07 – 019 NGC 2748	0693	3.1 x 1.2 2.8 x 1.3	38 6	1	Sb – c		11.7	+ 1489 + 1653	*12 3468
04826	09 08.2 + 03 10 227.58 + 32.10	01 – 24 – 000	0028	1.0 x 0.15 0.9 x 0.20	35 7	2	Sa – b		15.3		
04827	09 08.3 + 13 38 216.31 + 36.96	02 – 24 – 001	1508	1.4 x 0.5 1.3 x 0.3	80 (6)	1	S: ...		15.0		
04828	09 08.3 + 19 52 208.96 + 39.29	03 – 24 – 000	0057 140	1.6 x 0.15 1.6 x 0.15	5 7	3	Sc		15.7		
04829	09 08.4 + 46 51 173.18 + 43.06	08 – 17 – 053 Mark 102	1004	0.8 x 0.7 0.6 x 0.5	-- ...	1	... Compact		14.3		
04830	09 08.5 + 19 30 209.43 + 39.21	03 – 24 – 006	0057 140	1.2 x 0.35 1.2 x 0.25	104 6	1	Sa – b s of 2		15.3		*

N G C 2732, 2748, 2756, 2766 – 71
2773, 2775, 2777

         I C 527, 528

1	2	3	4	5	6	7	8	9	10	11	12
04831	09 08.5 + 33 02 192.05 + 42.57	06 – 20 – 040	1342	1.1 x 1.1 1.1: x 1.0:	— 1	4	Sc	SAB d	15.4		
04832	09 08.5 + 79 25 133.36 + 33.07	13 – 07 – 017	0693	1.5? x 0.4 1.4: x 0.4:	[118] ...	1	Peculiar P w 4818, disturbed?		15.0		*1
04833	09 08.6 + 30 16 195.74 + 42.12	05 – 22 – 000	1365 141	1.0 x 0.15 0.9 x 0.18	141 7	1	Sa: sff of 2		15.4		*
04834	09 08.7 + 35 10 189.18 + 42.89	06 – 20 – 041	1342	1.3: x 0.8 1.2: x 0.7:	90: ...	3	... NGC 2778 group	S B dm	16.5:		
04835	09 08.7 + 75 49 137.16 + 34.80	13 – 07 – 018	0693	1.0: x 0.1 0.9 x 0.1	144 7	1	S ...		17.		
04836	09 08.9 + 70 48 142.72 + 37.01	12 – 09 – 031	1325	1.4 x 0.7 1.5 x 0.7	57 5	3	S B b	S B(s)ab Sb –	15.3		*C
04837	09 09.0 + 35 44 188.41 + 43.01	06 – 20 – 042	1342	2.1: x 1.1: 2. x 1.	160 ...	3	Dwarf (spiral) S? V		15.5		*
04838	09 09.0 + 45 10 175.48 + 43.27	08 – 17 – 056 NGC 2776	1004	3.3 x 3.3 3.2: x 3.0:	— 1	5	Sc	SAB(rs)c Sc –	12.1	+ 2673 + 2682	*12 3468
04839	09 09.1 + 16 29 213.10 + 38.26	03 – 24 – 010	0057 140	1.4 x 0.2 1.2 x 0.2	60 6	2	Sb – c		15.0		
04840	09 09.2 + 35 13 189.12 + 42.99	06 – 20 – 043 NGC 2778	1342	1.4 x 1.0 1.5 x 1.1	40: ...	3	E Brightest in group		13.1		*13
04841	09 09.4 + 74 26 138.64 + 35.47	12 – 09 – 032 Ho III	1325 147	4.3: x 3.2: 4. x 3.	(150) ...	5	S B c	SAB(s)d	14.4		*12 5
04842	09 09.6 + 12 43 217.52 + 36.86	02 – 24 – 000	1508	1.2 x 0.3 1.2 x 0.3	144 6	2	Sa – b		15.2		
04843	09 09.6 + 35 07 189.27 + 43.06	06 – 20 – 047 NGC 2780	1342	1.0 x 0.7 1.0 x 0.7	150: 3:	3	S B ... NGC 2778 group	S B(rs) ...	14.2		*3
04844	09 09.6 + 49 50 169.10 + 42.95	08 – 17 – 058	1004 143	1.7 x 1.2 1.2 x 1.0	170: 3	2	Sc	SAB: bc Sc –	14.1		
04845	09 09.7 + 10 10 220.38 + 35.78	02 – 24 – 000	1508	1.6 x 0.7 1.5 x 0.4	98 6	3	S B: c	Sc +	15.1		
04846	09 09.7 + 62 32 152.53 + 40.11	10 – 13 – 068	1296	1.0 x 0.5 1.0 x 0.5	77 5	3	S B a		15.1		
04847	09 09.7 + 79 29 133.26 + 33.08	13 – 07 – 020	0693	1.1 x 0.12 1.2 x 0.12	106 7	1	Sc		16.0:		*
04848	09 09.8 + 34 15 190.46 + 43.00	06 – 20 – 000	1342	1.0 x 0.15 0.9 x 0.15	6 7	3	Sc		16.5:		
04849	09 09.8 + 43 14 178.13 + 43.48	07 – 19 – 000	0721	1.5:: x 0.6:: 0.8: x 0.4:	40 (6)	2	...	dm – m	17.		
04850	09 09.9 + 33 37 191.32 + 42.94	06 – 20 – 000	1342	1.0: x 0.7 0.8: x 0.5:	(100) ...	3	...	S m	17.		
04851	09 09.9 + 53 11 164.58 + 42.48	09 – 15 – 108	0982	0.9 x 0.6 (1.1 x 0.7)	145: ...	2	E – S0		14.2		*
04852	09 10.0 + 80 15 132.46 + 32.71	13 – 07 – 021	0693	1.5: x 0.9: 1.5 x 0.9:	[95] ...	3	Irr np of 2	I m Ir I	15.3		*
04853	09 10.2 + 20 35 208.27 + 39.95	04 – 22 – 023	0237	1.2 x 1.0 1.1 x 0.9	25: 2:	3	S B c		15.3		
04854	09 10.2 + 32 00 193.51 + 42.76	05 – 22 – 013 IC 2445	1365 141	1.0 x 0.8 0.7 x 0.3	(10) ...	1	S B ...		15.2		
04855	09 10.5 + 29 10 197.31 + 42.31	05 – 22 – 015 IC 2446	1365 141	1.6 x 0.7 1.7 x 0.7	148 6	3	Sa		15.0		
04856	09 10.5 + 30 12 195.93 + 42.51	05 – 22 – 017	1365 141	1.7 x 0.22 1.7 x 0.22	77 7	3	Sb P w 4859, NGC 2789 group		15.2		*1
04857	09 10.6 + 03 26 227.67 + 32.75	01 – 24 – 007	0028	1.5 x 0.8 1.5 x 0.8	63 5	3	S B c nf of 2, contact		14.7		*1
04858	09 10.6 + 19 35 209.54 + 39.70	03 – 24 – 012	0057 140	1.6 x 1.5 1.5: x 1.3:	— ...	4	Dwarf irregular Ir V		15.5		*
04859	09 10.6 + 30 11 195.96 + 42.53	05 – 22 – 019 NGC 2783	1365 141	(2.1 x 1.5) (2.0 x 1.5)	168 ...	3	E P w 4856, NGC 2789 group		13.9		*13
04860	09 10.8 + 29 23 197.04 + 42.41	05 – 22 – 021	1365 141	1.0 x 0.35 0.9 x 0.4	[17] ...	1	Double system Disrupted, p of 2		15.0		*
04861	09 10.9 + 12 39 217.76 + 37.12	02 – 24 – 002	1508	1.0 x 0.7 0.9 x 0.6	77	1	Sa?		13.8		
04862	09 10.9 + 40 20 182.13 + 43.68	07 – 19 – 036 NGC 2782	0721	4.2 x 3.2: 3.9 x 3.0:	... ...	4	S ... Peculiar		12.3	+ 2530 + 2514	*12 3468A
04863	09 11.2 + 41 00 181.21 + 43.75	07 – 19 – 038	0721	1.1 x 0.2 1.1 x 0.2	68 6 – 7	2	Sb – c		16.5:		
04864	09 11.4 + 16 57 212.81 + 38.94	03 – 24 – 014	0057	1.9 x 1.9 1.8 x 1.7	— 1	4	Sa	SA(r)ab	14.7		*1
04865	09 11.4 + 35 55 188.23 + 43.51	06 – 20 – 051	1342	1.1 x 0.15 1.0 x 0.15	99 7	1	S ... s of 2		15.6		*

N G C   2776, 2778, 2780, 2782, 2783   I C   2445, 2446

1	2	3	4	5	6	7	8	9	10	11	12
04866	09 11.5 + 36 17 187.73 + 43.57	06 – 20 – 052	1342	1.2 x 0.25 1.0 x 0.25	51 6	1	Sb: f of 2		15.5		*
04867	09 11.5 + 41 05 181.10 + 43.81	07 – 19 – 039	0721	1.9 x 1.4 2.0 x 1.4	85: 2 – 3	4	S B c/Sc P w 4876	SAB(s)d	15.2		*
04868	09 11.5 + 48 48 170.45 + 43.38	08 – 17 – 062	1004 143	1.0: x 0.8: 1. x 0.8:	-- ...	3	... P w 4874	S B m	16.5:		*
04869	09 11.6 + 30 20 195.82 + 42.77	05 – 22 – 022	1365 141	(2. x 0.7) (2. x 1.)	35 ...	1	SO? sp of 2, NGC 2789 group		14.5		*1
04870	09 11.6 + 47 07 172.76 + 43.58	08 – 17 – 063	1004	1.1 x 0.5 1.1 x 0.4	142 5 – 6	1	S ...		14.0		*
04871	09 11.7 + 39 28 183.34 + 43.81	07 – 19 – 040	0721	1.8 x 0.6 1.6: x 0.6	20 (6)	4	Dwarf spiral S(B) IV – V	SB(s)m	16.0:		*
04872	09 11.8 + 40 16 182.23 + 43.85	07 – 19 – 041	0721	1.9 x 0.2 1.8 x 0.2	12 7	2	S B b		16.0:		
04873	09 11.9 + 15 40 214.39 + 38.57	03 – 24 – 015	0057 140	1.2 x 0.9 1.5 x 1.1	45 ...	1	S? ... Peculiar		14.2		
04874	09 11.9 + 48 53 170.33 + 43.44	08 – 17 – 064	1004 143	1.1 x 0.9 1.0:: x 0.7::	-- ...	3	... P w 4868	S B m	16.5:		*1
04875	09 12.0 + 29 56 196.37 + 42.78	05 – 22 – 026 NGC 2789	1365 141	(1.7 x 1.7) (1.7 x 1.6)	-- 1	2	SO – a Brightest in group		13.8		
04876	09 12.0 + 41 07 181.05 + 43.90	07 – 19 – 042 NGC 2785	0721	1.5 x 0.6 1.7 x 0.6	[120] ...	2	Irr or Peculiar P w 4867		14.9		*1
04877	09 12.0 + 43 15 178.09 + 43.88	07 – 19 – 043	0721	1.2 x 0.3 1.1 x 0.3	5 6	1	S ...		16.0:		
04878	09 12.4 + 31 36 194.15 + 43.16	05 – 22 – 000	1365 141	1.9 x 0.7 1.5: x 0.5:	25 6	3	... 3rd of 3	S B m	16.0:		*
04879	09 12.5 + 53 03 164.66 + 42.88	09 – 15 – 113	0982	2.5: x 1.5: 2.3:: x 1.4::	[85] ...	4	Irr	IA m Ir I	14.0		*1V
04880	09 12.6 + 12 06 218.61 + 37.26	02 – 24 – 003 IC 530	1508	2.1: x 0.7 2. x 0.6	90 6	2	Sa – b		14.3		*1
04881	09 12.7 + 44 33 176.28 + 43.95	08 – 17 – 065	1004	[1.0 x 0.9] ...	-- ...	3	Double system Contact, disrupted		14.9		*1
04882	09 13.1 + 43 12a 178.15 + 44.08	07 – 19 – 047	0721	1.1 x 0.7 1.0 x 0.6	... ...	3	... s of 2, disrupted		(15.3)		*1
04883	09 13.1 + 74 33 138.37 + 35.64	12 – 09 – 034	1325 147	1.4 x 0.8 1.3 x 0.5	27 4 – 5	1	S ...		12.8		
04884	09 13.2 + 10 20 220.67 + 36.62	02 – 24 – 004	1508	1.0 x 0.6 1.0 x 0.6	35 4	2	Sa – b		14.7		*
04885	09 13.2 + 17 48 211.99 + 39.65	03 – 24 – 018 NGC 2794	0057 140	1.2: x 1.1: 1.6: x 1.5:	-- ...	2	S ... P w 4887, disturbed, in group		14.0		*13
04886	09 13.2 + 18 02 211.71 + 39.74	03 – 24 – 019 IC 2454	0057 140	1.2 x 1.1 1.1 x 1.0	-- ...	1	Sa? Singular, NGC 2794 group		14.4		*1
04887	09 13.3 + 17 50 211.96 + 39.69	03 – 24 – 020 NGC 2795	0057 140	(1.4 x 1.0) (1.7 x 1.2)	170: ...	1	E P w 4885, NGC 2794 group		14.1		*13
04888	09 13.3 + 73 57 139.02 + 35.93	12 – 09 – 035 IC 529	1325 147	3.9 x 1.8 3.8: x 1.8	145 5	4	Sc		12.0		*12 5
04889	09 13.4 + 40 05 182.50 + 44.15	07 – 19 – 049	0721	1.6 x 1.5 1.4: x 1.4:	-- 1	3	S ... Peculiar		15.0		*
04890	09 13.5 + 06 33 224.84 + 34.92	01 – 24 – 009	0028	1.2 x 0.7 1.1 x 0.6	55 4 – 5	2	S B c sp of 2		14.9		*1
04891	09 13.6 + 17 56 211.88 + 39.79	03 – 24 – 023 NGC 2797	0057 140	0.55 x 0.45 0.55 x 0.50	-- ...	1	... Peculiar, NGC 2794 group		14.3		*13
04892	09 13.6 + 45 55 174.37 + 44.02	08 – 17 – 068	1004	1.2 x 0.7 1.1 x 0.6	... ...	2	... Peculiar, eruptive?		14.8		*1
04893	09 13.7 + 31 07b 194.87 + 43.35	05 – 22 – 029 NGC 2796	1365 141	(1.4 x 0.9) (1.3 x 0.9)	80: 3 – 4:	1	Sa? Brightest of 3		(14.6)		*13
04894	09 13.7 + 34 39 190.04 + 43.85	06 – 21 – 002 NGC 2793	0925 142	1.3 x 1.0 1.2 x 1.0	-- (1 – 2)	4	SB ... Peculiar		13.9	+ 1669 + 1623	*123 4678
04895	09 13.8 + 27 42 199.46 + 42.70	05 – 22 – 031	1365	1.3 x 0.4 1.1 x 0.4	163 6	2	Sb		14.9		
04896	09 13.8 + 70 01 143.40 + 37.71	12 – 09 – 036	1325	2.1 x 0.5 1.9 x 0.4	145 7	2	Sa – b Brightest in group		14.8		*
04897	09 13.9 + 19 10a 210.40 + 40.29	03 – 24 – 026 NGC 2802	0057 140	[1.7 x 0.8] ...	... ...	3	Double system Bridge		14.3		*13
04898	09 13.9 + 19 10b 210.40 + 40.29	03 – 24 – 027 NGC 2803	0057 140								
04899	09 13.9 + 20 08 209.20 + 40.62	03 – 24 – 025 NGC 2801	0057 140	1.2 x 1.0 1.2: x 1.0:	35: 2:	3	Sc NGC 2809 group, Sc –	SA(s)c	15.4		*3
04900	09 14.0 + 07 29 223.91 + 35.48	01 – 24 – 010	0028	1.2 x 0.8 1.4 x 0.8	122 3 – 4	2	Sb – c		14.8		1

N G C   2785, 2789, 2793 – 97, 2801 – 03   I C   529, 530, 2454

1	2	3	4	5	6	7	8	9	10	11	12
04901	09 14.0 + 20 24	03 - 24 - 028	0057	(1.5 x 1.3)	(60)	2	S0		14.0		3
	208.88 + 40.73	NGC 2804	140	(1.7 x 1.5)	2:		NGC 2809 group				
04902	09 14.1 + 25 38	04 - 22 - 029	0237	(1.4 x 0.8)	157	1	S0?		14.0		
	202.20 + 42.27		146	(1.5 x 0.7)	4 - 5:						
04903	09 14.1 + 37 10	06 - 21 - 003	0925	1.5: x 0.4	38	3	...	S m	16.5:		
	186.57 + 44.16			1.2: x 0.3	6		IV - V:				
04904	09 14.1 + 42 07	07 - 19 - 054	0721	1.1 x 0.6	137	1	S B ...		15.0		*
	179.66 + 44.29			0.9 x 0.6	...		NGC 2798 group				
04905	09 14.1 + 42 12	07 - 19 - 055	0721	2.8 x 0.9	160	3	S B a, p w 4909, disturbed		12.9	+ 1708	*12V
	179.54 + 44.29	NGC 2798		2.7 x 0.9	6		Brightest in group			+ 1702	348A
04906	09 14.1 + 53 12	09 - 15 - 114	0982	2.2 x 0.5	48	3	Sa		13.4		
	164.40 + 43.09			2.2 x 0.5	7						
04907	09 14.2 + 16 31	03 - 24 - 032	0057	1.1: x 0.8	---	1	Sc		15.5		
	213.65 + 39.40		140	0.8 x 0.6	...						
04908	09 14.2 + 32 13	05 - 22 - 033	1365	1.0: x 0.6:	25:	2	Sb		15.7		
	193.40 + 43.63			0.8 x 0.4	4 - 5:						
04909	09 14.2 + 42 12	07 - 19 - 056	0721	2.1 x 0.5	125	1	S ...	P w 4905,	14.4		*12
	179.54 + 44.31	NGC 2799		2.0 x 0.5	6 - 7		disturbed, in group				36V̄A
04910	09 14.3 + 20 16	03 - 24 - 033	0057	(1.6 x 1.4)	---	3	S0		13.9		*3
	209.08 + 40.75	NGC 2809	140	(1.7 x 1.5)	(1)		Brightest in group				
04911	09 14.4 + 54 26	09 - 15 - 000	0982	1.0 x 0.1	29	3	Sc		16.5:		
	162.74 + 42.88			1.0 x 0.1	7						
04912	09 14.5 + 26 10	04 - 22 - 030	0237	0.6 x 0.4	...	1	...		14.4		
	201.54 + 42.49			0.8 x 0.5	...						
04913	09 14.6 + 55 52	09 - 15 - 116	0982	1.2 x 0.4	68	2	Sc		15.7		*
	160.83 + 42.57			1.0 x 0.4	6						
04914	09 14.8 + 69 25	12 - 09 - 039	1325	(3.5 x 2.3)	117	4	S B 0		11.7	+ 621	*123
	144.04 + 38.04	NGC 2787		(3.7 x 2.4)	3 - 4					+ 753	4678
04915	09 14.9 - 00 25	00 - 24 - 005	0430	1.1 x 0.25	119	1	Sb - c		15.0		
	232.20 + 31.63			1.0 x 0.22	6 - 7						
04916	09 14.9 + 20 06	03 - 24 - 037	0057	(1.0 x 0.8)	(145)	3	S0		15.4		*3
	209.35 + 40.83	NGC 2813	140	(1.1 x 0.9)	2:		sf of 2, NGC 2809 group				
04917	09 14.9 + 48 10	08 - 17 - 073	1004	1.0 x 0.8	---	3	Sc		15.5		*
	171.23 + 44.02		143	0.8 x 0.7	1 - 2						
04918	09 14.9 + 70 00	12 - 09 - 038	1325	1.0 x 0.6	85	1	S: ...		16.0:		*
	143.37 + 37.80			1.0 x 0.6	(4)		4896 group				
04919	09 15.0 + 45 52	08 - 17 - 072	1004	1.5 x 1.2	5:	3	Sc	SAB(rs)c	14.4		
	174.42 + 44.27			1.2 x 0.9	2			Sc -				
04920	09 15.0 + 52 43	09 - 15 - 117	0982	(1.4 x 0.9)	(15)	3	E		14.0		3
	165.01 + 43.32	NGC 2800		(1.5 x 1.0)	...						
04921	09 15.1 + 49 46	08 - 17 - 075	1004	1.1 x 0.2	26	3	Sc		16.5:		
	169.02 + 43.84		143	1.1: x 0.2	7						
04922	09 15.2 + 48 05	08 - 17 - 074	1004	4. x 2.	55	4	...	SA m	15.1		
	171.34 + 44.08		143	3.5:: x 2.	5:		S IV - V				
04923	09 15.3 - 00 04	00 - 24 - 006	0430	1.8 x 0.5	60	1	S B: a - b		14.9		
	231.92 + 31.90	IC 531		1.8: x 0.5	6						
04924	09 15.4 + 16 25	03 - 24 - 040	0057	(1.4 x 1.3)	---	2	E		14.3		3
	213.91 + 39.63	NGC 2819	140	(1.5 x 1.3)	...						
04925	09 15.5 + 17 58	03 - 24 - 042	0057	1.7 x 0.2	78	3	Sc		15.4		
	212.05 + 40.22		140	1.8: x 0.2	7			Sc +			
04926	09 15.5 + 34 46	06 - 21 - 007	0925	1.7 x 0.3	154	1	Sb?		15.4		*1
	189.94 + 44.23		142	1.3 x 0.25	6						
04927	09 15.5 + 50 13	08 - 17 - 076	1004	(1.0 x 0.8)	(135)	2	[E - S0]		14.7		*
	168.39 + 43.83		143	(1.3 x 1.1)	...		Compact				
04928	09 15.5 + 51 32	09 - 15 - 119	0982	1.0 x 0.9	---	3	S B b	S B(s)bc	15.3		*
	166.59 + 43.61			0.8: x 0.7:	1			Sb +			
04929	09 15.6 + 67 22	11 - 12 - 000	0708	1.3 x 0.1	138	2	Sb - c		17.		
	146.39 + 38.95		138	0.9 x 0.1	7						
04930	09 15.8 + 49 10	08 - 17 - 078	1004	1.0 x 0.8	(115)	2	Sc		15.4		
	169.82 + 44.04		143	0.9: x 0.7:	2:						
04931	09 16.0 + 14 02	02 - 24 - 006	1508	(1.1 x 0.8)	165:	1	S0 - a		15.0		*
	216.82 + 38.82			(1.0 x 0.8)	2 - 3						
04932	09 16.0 + 51 19	09 - 15 - 122	0982	1.8 x 0.6	140	3	...	dm - m	15.4		
	166.87 + 43.73			1.6: x 0.6:	(6)						
04933	09 16.1 + 26 29	04 - 22 - 031	0237	(1.0 x 0.6)	160:	3	S0		14.3		3
	201.24 + 42.92	NGC 2824, Mark 394		(1.2 x 0.6)	...						
04934	09 16.1 + 51 05	09 - 15 - 121	0982	1.0 x 0.15	4	2	Sb		15.7		*
	167.18 + 43.78		143	1.0 x 0.15	7		nnp of 2				
04935	09 16.2 + 34 13	06 - 21 - 008	0925	1.1 x 0.6	30:	2	S B a		15.7		3
	190.72 + 44.31	NGC 2823	142	1.0 x 0.5	5						

N G C   2787, 2798 - 2800, 2804, 2809, 2813
      2819, 2823, 2824

I C   531

1	2	3	4	5	6	7	8	9	10	11	12
04936	09 16.3 + 64 19 150.01 + 40.19	11 – 12 – 003 NGC 2805	0708	7.5 x 5.1 7.3 x 4.7	(125) 1 – 2	5	Sc Brightest of 4, disturbed	SAB(rs)d	11.9	+ 1916 + 2023	*12 35⁻
04937	09 16.3 + 75 21 137.39 + 35.44	13 – 07 – 024	0693	1.4 x 0.7 1.2 x 0.7	165: 5:	3	...	S d – dm	15.2		
04938	09 16.4 + 20 27 209.06 + 41.28	03 – 24 – 045	0057 140	1.3 x 0.3 1.2 x 0.25	147 6	2	Sb		15.4		
04939	09 16.4 + 33 50 191.26 + 44.31	06 – 21 – 011 NGC 2826	0925 142	1.5 x 0.25 1.3 x 0.25	143 7	2	SO – a		14.6		23
04940	09 16.6 + 27 40 199.69 + 43.30	05 – 22 – 036	1365	1.3 x 0.7 1.2 x 0.6	58 5	3	Sa/S B b		14.8		
04941	09 16.7 + 33 57 191.11 + 44.38	06 – 21 – 014 NGC 2830	0925 142	1.2 x 0.25 1.2 x 0.25	112 7	1	S ...		15.4		*12 3
04942	09 16.8 + 33 58 191.09 + 44.41	06 – 21 – 015 NGC 2832	0925 142	(3.0 x 2.0) (3.5 x 2.3)	160: ...	3	E, NGC 2831 superimposed Brightest in cluster		13.6	+ 6946 + 6897	*12 3478A
04943	09 16.8 + 37 23 186.31 + 44.71	06 – 21 – 019 IC 2461	0925	2.7 x 0.4 2.1 x 0.4	143 7	3	Sb		15.1		
04944	09 16.8 + 69 37 143.72 + 38.12	12 – 09 – 040	1325	1.4 x 0.2 1.4 x 0.2	93 7	3	Sc		15.7		
04945	09 16.8 + 76 00 136.68 + 35.14	13 – 07 – 000	0693	1.5: x 1.4: 1.4: x 1.4:	-- ...	4	Dwarf irregular Ir V		17.		
04946	09 16.9 + 06 06 225.83 + 35.44	01 – 24 – 012	0028	1.5: x 0.9: 1.3: x 0.9:	30: ...	3	S B b	S B(r) ...	15.7		
04947	09 16.9 + 33 08 192.25 + 44.32	06 – 21 – 016	0925 142	1.1 x 0.5 0.9 x 0.5	[5] ...	1	S B ...		15.3		
04948	09 17 + 85 32 127.12 + 30.03	14 – 05 – 004	0727	1.2 x 1.2: 1.0: x 1.0:	-- ...	2	Dwarf (spiral) S? IV – V:		16.5:		
04949	09 17.0 + 33 18 192.02 + 44.37	06 – 21 – 020	0925 142	1.1 x 0.15 1.1 x 0.15	144 7	2	Sb – c f of 2		16.0:		*1
04950	09 17.0 + 39 22 183.52 + 44.83	07 – 19 – 060	0721	1.0 x 0.5 1.0 x 0.5	62 5	2	Sa		15.1		
04951	09 17.0 + 71 45 141.30 + 37.18	12 – 09 – 041 Mark 20 = Mark 107	1325	1.2 x 0.9 1.0 x 0.8	145: ...	1	...		14.7	+ 3600	
04952	09 17.1 + 64 28 149.79 + 40.22	11 – 12 – 004 NGC 2814	0708	1.3 x 0.25 1.2 x 0.25	179 ...	1	... NGC 2805 group	I 0?	14.0	+ 1672 + 1780	*12 3
04953	09 17.3 + 36 27 187.63 + 44.74	06 – 21 – 022	0925	1.6 x 0.4 1.1: x 0.4	122 6	3	...	S d – dm	16.0:		
04954	09 17.3 + 72 03 140.95 + 37.07	12 – 09 – 042 NGC 2810	1325	(1.7 x 1.7) (1.8 x 1.8)	-- ...	3	E		13.4		13
04955	09 17.4 + 25 30 202.64 + 42.96	04 – 22 – 000	0237 146	1.0: x 0.5: 0.9: x 0.5:	100: 5:	3	...	S dm	16.5:		
04956	09 17.5 + 01 15 230.96 + 33.08	00 – 24 – 007	0430	(1.8 x 1.3) (2.0 x 1.5)	15: ...	4	E		14.1		*
04957	09 17.5 + 09 00 222.76 + 36.96	02 – 24 – 000	1508	1.9: x 0.2 1.4 x 0.2	152 7	3	...	S d – dm	15.6		
04958	09 17.6 + 49 48 168.89 + 44.23	08 – 17 – 082	1004 143	1.5 x 0.45 1.4 x 0.4	122 6	1	S B ...		15.0		
04959	09 17.8 + 07 17 224.69 + 36.21	01 – 24 – 013	0028	1.0 x 0.9 0.7 x 0.6	— 1 – 2:	1	Sb – c		15.0		
04960	09 17.8 + 35 35 188.86 + 44.78	06 – 21 – 025 NGC 2840	0925	1.0 x 0.9 1.0 x 0.9	— 1 – 2	3	S B b	S B(rs)bc Sb +	14.8		3
04961	09 17.8 + 64 29 149.73 + 40.28	11 – 12 – 006 NGC 2820	0708	4.4 x 0.35 4.3 x 0.4	59 7	2	Sc, NGC 2805 group P w IC 2458, contact		13.1	+ 1691 + 1799	*12 35⁻
04962	09 17.9 + 15 19 215.53 + 39.76	03 – 24 – 048	0057	1.4 x 1.3 1.1 x 1.1	-- 1	3	SBa sp of 2		15.2		*1
04963	09 17.9 + 18 29 211.68 + 40.94	03 – 24 – 000	0057 140	(1.1 x 0.5) 1.2 x 0.3	130 ...	2	SO In group		16.0:		*
04964	09 17.9 + 29 05 197.86 + 43.87	05 – 22 – 000	1365	1.1: x 0.2 0.8 x 0.15	173 (6 – 7)	1	S – Irr sp of 2		15.7		*
04965	09 18.1 + 24 31 203.99 + 42.86	04 – 22 – 000	0237 146	1.1 x 0.25 1.1: x 0.25	60 6	1	S ...		15.2		
04966	09 18.5 + 51 12 166.93 + 44.13	09 – 16 – 005 NGC 2841	0681	7.4 x 3.5 7.6 x 3.6	147 6	5	Sb	SA(r)b Sb –	9.9	+ 631 + 671	*12– –689
04967	09 18.5 + 72 16 140.65 + 37.05	12 – 09 – 044	1325	1.2 x 1.0 1.1 x 0.9	(0) 2	3	Sa		14.9		
04968	09 18.6 + 03 22 228.97 + 34.43	01 – 24 – 000 IC 534	0028	1.9 x 0.25 1.7 x 0.3	148 7	1	Sb		15.1		
04969	09 18.6 + 19 46 210.15 + 41.54	03 – 24 – 000	0057 140	1.4 x 0.15 1.2 x 0.20	93 7	3	Sb		15.7		
04970	09 18.6 + 39 45 182.99 + 45.14	07 – 19 – 063	0721	1.6 x 0.20 1.5 x 0.20	105 7	3	Sc		16.0:		

N G C   2805, 2810, 2814, 2820, 2826, 2830
2832, 2840, 2841

I C   534, 2461

1	2	3	4	5	6	7	8	9	10	11	12
04971	09 18.6 + 40 22 182.11 + 45.15	07 - 19 - 064 NGC 2844	0721 145	1.9 x 0.9 1.9 x 0.9	13 5	2	S0 - a Brightest of 3		13.6		*12 3468
04972	09 18.8 + 33 37 191.64 + 44.78	06 - 21 - 026	0925	(1.5 x 1.0) (1.7 x 1.3)	0: ...	2	[S0] Very compact		14.6		
04973	09 18.8 + 61 05 153.88 + 41.60	10 - 14 - 004	0703	1.0 x 0.9 1.0 x 0.9	— 1	2	Sb - c		14.9		
04974	09 19.1 + 34 03 191.05 + 44.89	06 - 21 - 027	0925	(1.5 x 1.2) 1.4 x 1.2	(0) ...	2	[S0] Compact		14.5		*
04975	09 19.2 + 50 27 167.93 + 44.38	08 - 17 - 089	1004 143	1.2 x 0.15 1.2 x 0.15	31 7	2	Sc		16.0:		
04976	09 19.3 + 46 52 172.92 + 44.91	08 - 17 - 088	1004	1.0 x 0.6 0.8 x 0.6	120 ...	2	...	SAB: d - dm	15.5		
04977	09 19.4 + 03 55 228.87 + 34.71	01 - 24 - 000	0028	1.1 x 0.35 0.9 x 0.3	145: ...	1	S ...		15.3		
04978	09 19.5 + 04 06 228.34 + 35.00	01 - 24 - 014	0028	1.6: x 1.6: 1.3: x 1.3:	— 1	4	Sc	SA d	15.6		
04979	09 19.5 + 36 53 187.06 + 45.21	06 - 21 - 000	0925	1.1: x 0.8: 0.5: x 0.4:	(135) ...	2	Dwarf IV - V or V		17.		
04980	09 19.7 + 04 55 227.50 + 35.46	01 - 24 - 016	0028	1.0 x 0.4 0.9 x 0.35	172 6	1	S B: b		14.8		
04981	09 19.7 + 68 47 144.54 + 38.71	12 - 09 - 045	1325	1.2 x 0.4 1.2 x 0.4	146 6	2	Sb		14.6		*
04982	09 19.8 + 44 46 175.86 + 45.20	08 - 17 - 091	1004	1.3 x 0.35 1.2 x 0.3	4 6	1	S - Irr		14.0		
04983	09 19.9 + 29 32 197.36 + 44.39	05 - 22 - 042	1365	1.0 x 0.7 0.8: x 0.7	... ...	4	Sc/S B c	SAB(s)c	15.7		
04984	09 19.9 + 54 43 162.11 + 43.59	09 - 16 - 012	0681	1.1 x 0.8 1.0: x 0.7:	140: ...	3	Dwarf (spiral) S? IV - V		16.5:		
04985	09 20.0 + 22 12 207.18 + 42.62	04 - 22 - 033	0237 146	1.7 x 0.6 1.7 x 0.6	177 6	4	Sb/S B b Disturbed?, Sb +	SAB(s)b	15.0		*1
04986	09 20.0 + 40 23 182.09 + 45.41	07 - 19 - 065 NGC 2852	0721 145	(1.1 x 1.1) (1.1 x 1.0)	— ...	1	... P w 4987, NGC 2844 group	Compact	14.0		*12 3
04987	09 20.1 + 40 25 182.04 + 45.43	07 - 19 - 066 NGC 2853	0721 145	1.9 x 0.9 1.2 x 0.8	25: 5:	3	Sa P w 4986, NGC 2844 group		14.6		*12 3
04988	09 20.2 + 34 56 189.84 + 45.20	06 - 21 - 028	0925	1.1 x 0.9 1.2 x 0.9:	70: ...	3	... S(B) IV - V	SAB m	15.7		
04989	09 20.3 + 03 21 229.25 + 34.78	01 - 24 - 017 NGC 2858	0028	1.8 x 0.9 1.9 x 1.0	117 5	2	S0 - a		13.8		3
04990	09 20.3 + 62 14 152.36 + 41.37	10 - 14 - 006	0703	1.1 x 0.6 1.0 x 0.6	175 4 - 5	2	S B c		15.2		
04991	09 20.5 + 22 32 206.80 + 42.83	04 - 22 - 37+38	0237 146	[1.2 x 0.8] ...	... ...	3	Double system, contact In chain of 3		15.2		*1V
04992	09 20.5 + 42 24 179.21 + 45.46	07 - 19 - 068	0721	1.2 x 0.1 1.1 x 0.1	0 7	2	Sc		15.6		
04993	09 20.6 + 02 49 229.85 + 34.56	01 - 24 - 000	0028	1.1 x 0.4 0.9 x 0.4	120 6	1	S ... n of 2		15.1		*
04994	09 20.6 + 24 58 203.60 + 43.52	04 - 22 - 041	0237 146	1.1 x 0.4 1.1 x 0.45	33 6	2	Sa - b		14.9		*
04995	09 20.6 + 49 25 169.31 + 44.77	08 - 17 - 092 NGC 2854	1004 143	1.9: x 0.55 1.7 x 0.50	50 6	4	S B b P w 4997, brightest of 3	S B(s)b	13.8		*13 A
04996	09 20.7 - 00 31 234.23 + 32.23	00 - 24 - 009	0430	1.5 x 0.5 1.3 x 0.5	130 6	2	Sb - c		14.2		*
04997	09 20.8 + 49 27 169.26 + 44.80	08 - 17 - 093 NGC 2856	1004 143	1.1 x 0.5 1.1 x 0.5	134 5 - 6	1	S ... P w 4995, NGC 2854 group		13.9		*13 A
04998	09 20.8 + 68 37 144.67 + 38.87	12 - 09 - 047	1325	2. x 1.3:: 1.8:: x 1.	80: ...	3	Dwarf V		15.4		1
04999	09 21.0 + 02 20 230.41 + 34.39	00 - 24 - 010 NGC 2861	0430	1.7 x 1.5 1.6 x 1.5	— 1	3	S B b nnp of 2	S B(r)b	14.0		*13
05000	09 21.1 + 49 34 169.08 + 44.84	08 - 17 - 095 NGC 2857	1004 143	2.4 x 2.2 2.5 x 2.3	— 1	4	Sc NGC 2854 group, Sc -	SA(rs)c	14.3		*13 A
05001	09 21.3 + 34 44 190.15 + 45.41	06 - 21 - 030 NGC 2859	0925	4.5 x 4.0 4.7 x 4.2	85: 1 - 2	5	S B 0 	(R)SB(r)0 +	11.8	+ 1694 + 1649	*12 3468
05002	09 21.4 + 28 30 198.87 + 44.51	05 - 22 - 044	1365	1.0 x 0.5 0.9 x 0.5	170: ...	1	S B ...		14.8		1
05003	09 21.5 + 11 20 220.69 + 38.89	02 - 24 - 000	1508	1.1 x 0.5 0.9 x 0.4	30 6	2	Sc		15.1		
05004	09 21.5 + 34 52 189.97 + 45.46	06 - 21 - 031	0925	1.4: x 1.1: 1.2: x 0.9:	(130) ...	3	... IV - V	... AB m	16.5:		1
05005	09 21.6 + 22 29 206.96 + 43.06	04 - 22 - 000	0237 146	1.6:: x 1.3:: 1.2:: x 1.2::	— ...	4	Dwarf irregular Ir V		18.		

N G C   2844, 2852 - 54, 2856 - 59, 2861

1	2	3	4	5	6	7	8	9	10	11	12
05006	09 21.7 + 25 20 203.19 + 43.86	04 – 22 – 045 IC 536	0237 146	1.2 x 0.2 1.2 x 0.25	23 6 – 7	1	Sa?		15.3		
05007	09 21.7 + 41 16 180.82 + 45.72	07 – 20 – 003 NGC 2860	0661	1.4 x 0.6 1.4 x 0.6	108 6	3	S B a		14.8		
05008	09 21.7 + 47 12 172.38 + 45.27	08 – 17 – 000	1004	1.0:: x 0.5:: 0.8:: x 0.3::	[55] ...	2	Dwarf (irregular) Ir: V		18.		
05009	09 21.9 + 20 14 209.90 + 42.42	03 – 24 – 056	0057 140	1.1 x 0.9 1.1 x 1.0	(150) 1 – 2	3	Sc		15.5		
05010	09 21.9 + 26 59 200.98 + 44.29	05 – 22 – 045 NGC 2862	1365	2.4 x 0.6 2.5 x 0.5	114 7	1	S ...		13.8		3
05011	09 22.2 + 34 20 190.74 + 45.56	06 – 21 – 034	0925	1.2 x 0.9 0.5 x 0.4	165: 2 – 3:	2	S B c		16.0:		
05012	09 22.4 + 17 49 213.02 + 41.70	03 – 24 – 057	0057 140	1.1 x 0.25 1.1 x 0.3	145 7	2	Sa – b		14.6		
05013	09 22.4 + 61 35 153.06 + 41.83	10 – 14 – 007	0703	1.8 x 1.0 1.7 x 1.0	120 4	3	Sb		15.5		
05014	09 22.6 + 35 04 189.71 + 45.71	06 – 21 – 000	0925	1.1 x 0.1 1.0 x 0.1	38 7	2	... d – m		17.		
05015	09 22.7 + 34 30 190.52 + 45.67	06 – 21 – 035	0925	2.0:x 1.8: 1.7:x 1.6:	(20) 1:	4	... SAB dm S IV or IV – V		15.7		
05016	09 22.7 + 49 30 169.12 + 45.10	08 – 17 – 097	1004 143	1.1 x 0.5 1.1 x 0.6	30 5 – 6	3	S B a: Brightest in group		15.4		*
05017	09 22.8 + 11 17 220.93 + 39.16	02 – 24 – 000	1508	1.1 x 0.15 1.1 x 0.15	140 7	3	Sc		15.7		
05018	09 23.0 + 11 39 220.53 + 39.36	02 – 24 – 008 NGC 2872	1508	(1.8 x 1.7) (2.0 x 1.8)	-- ...	3	E P w 5021		13.0	+ 2960 + 2803	*12 3A
05019	09 23.0 + 24 21 204.62 + 43.89	04 – 22 – 000	0237 146	1.0: x 0.2 0.9 x 0.2	168 6	1	S ...		15.7		
05020	09 23.0 + 34 52 190.01 + 45.77	06 – 21 – 037	0925 149	2.3 x 0.5 2.3 x 0.5	79 6	3	Sc nnf of 2		15.3		*1
05021	09 23.1 + 11 39 220.54 + 39.38	02 – 24 – 010 NGC 2874 = 2875	1508	2.4 x 0.8 2.2 x 0.7	43 6	2	Sc P w 5018, disturbed		13.5	+ 3600 + 3443	*12 3A
05022	09 23.2 + 02 18 230.80 + 34.84	00 – 24 – 014 NGC 2878	0430	1.0 x 0.3 1.1 x 0.3	174 6	1	Sa – b		14.9		
05023	09 23.2 + 19 36 210.85 + 42.50	03 – 24 – 058 Mark 400	0057	0.8 x 0.7 0.7 x 0.6	-- ...	1	...		14.4		
05024	09 23.2 + 70 27 142.46 + 38.23	12 – 09 – 000	1325	1.1 x 0.7 1.0 x 0.7	55 4:	3	Sc n of 2		16.5:		*
05025	09 23.3 + 12 57 219.05 + 39.99	02 – 24 – 011	1508	0.6 x 0.5 0.7 x 0.6	(70) ...	1	S0? (R) ... Singular		13.9		*
05026	09 23.3 + 46 05 173.91 + 45.68	08 – 17 – 100	1004	1.1 x 0.8 1.0 x 0.7	15 3:	3	S0		14.2		
05027	09 23.6 + 03 20 229.79 + 35.47	01 – 24 – 000	0028	1.1 x 0.2 1.1 x 0.15	173 7	2	Sb – c		15.4		
05028	09 23.6 + 68 39 144.49 + 39.09	12 – 09 – 049 Mark 111	1325	0.60 x 0.35 0.6 x 0.35	145 ...	1	... Peculiar, p w 5029		13.9	+ 3600	*1C
05029	09 23.8 + 68 40 144.46 + 39.10	12 – 09 – 050	1325	1.8 x 1.1 1.7: x 1.1	13: 4:	3	Sc SAB(s)c P w 5028, distorted		14.3		*1
05030	09 23.9 + 08 10 224.62 + 37.96	01 – 24 – 021 NGC 2882	0028	1.7 x 0.9 1.7 x 0.9	80 5	1	S ...		13.5		3
05031	09 24.0 + 77 25 134.94 + 34.76	13 – 07 – 029	0693 172	1.1 x 0.4 0.9 x 0.25	62 6	1	S ...		15.4		
05032	09 24.1 + 01 22 231.91 + 34.53	00 – 24 – 000	0430	1.0 x 0.2 1.1 x 0.2	122 7	2	Sa – b		15.7		
05033	09 24.1 + 65 09 148.58 + 40.64	11 – 12 – 011	0708	1.6 x 0.1 1.5: x 0.1	167 7	3	Sc		16.5:		
05034	09 24.2 + 57 36 158.06 + 43.37	10 – 14 – 013 NGC 2870	0703	2.7 x 0.7 2.7 x 0.7	123 6	2	Sb – c		13.9		3
05035	09 24.3 + 21 49 208.09 + 43.45	04 – 22 – 056	0237 146	1.1 x 1.0 1.1 x 1.0	-- 1	3	S B a (R)S B(s)a		15.6		
05036	09 24.3 + 41 03 181.12 + 46.22	07 – 20 – 012	0661	1.2 x 0.3 0.7 x 0.3	5 (6)	1	Sc – Irr		16.5:		
05037	09 24.4 + 23 14 206.23 + 43.89	04 – 22 – 057 IC 2474	0237	1.1 x 0.8 1.6 x 0.9	(80) 4 – 5	3	S0 Brightest in group		14.8		*1
05038	09 24.4 + 30 39 196.05 + 45.54	05 – 22 – 047 IC 2473	1365 148	1.6 x 1.6 0.9 x 0.8	-- 1	4	S B b S B(r)bc NGC 2893 group, Sb +		14.6		
05039	09 24.5 + 22 28 207.26 + 43.69	04 – 22 – 000	0237 146	1.1 x 0.4 0.9 x 0.2	147 ...	1	S ...		16.0:		
05040	09 24.7 + 29 01 198.35 + 45.32	05 – 22 – 048	1365	2.2: x 2.1: 1.8:: x 1.7::	-- (1)	4	... IV – V		15.7		*

N G C   2860,   2862,   2870, 2872,   2874 = 2875   I C   536,   2473,   2474
2878,   2882

1	2	3	4	5	6	7	8	9	10	11	12
05041	09 24.8 + 32 52 192.92 + 45.93	06 – 21 – 039	0925	1.3 x 0.3 1.2 x 0.3	118 6	3	Sc		15.4		
05042	09 24.8 + 66 42 146.69 + 40.05	11 – 12 – 012	0708	1.2 x 1.2 1.2 x 1.2	--- 1	3	... S B dm Disturbed?		16.0:		*
05043	09 24.9 + 30 13 196.68 + 45.57	05 – 23 – 001 IC 2476	0466 148	(1.3 x 1.1) (1.3 x 1.1)	--- ...	2	E – SO NGC 2893 group Brightest in subgroup		14.5		*1
05044	09 25.0 + 12 30 219.81 + 40.17	02 – 24 – 13+14	1508	[1.1 x 1.0] ...	--- ...	3	Double system Symbiotic, in group of 4		14.7		*1A
05045	09 25.0 + 44 53 175.58 + 46.11	08 – 17 – 104	1004	1.2 x 1.0 1.3 x 1.0	--- 2	4	Sc/S B c SAB(r)c		14.3		
05046	09 25.3 + 17 25 213.85 + 42.20	03 – 24 – 060	0057	1.0 x 0.25 1.1 x 0.25	16 6 – 7	1	S? ...		14.5		*
05047	09 25.3 + 51 47 165.83 + 45.05	09 – 16 – 023	0681	1.6 x 0.2 1.5: x 0.2	160 7	2	... d – m		16.5:		
05048	09 25.5 + 38 45 184.45 + 46.47	07 – 20 – 013	0661	1.3 x 0.3 1.1 x 0.2	11 (7)	2	Irr ssp of 2		15.2		*1
05049	09 25.5 + 49 26 169.10 + 45.55	08 – 17 – 106	1004 143	1.2 x 0.2 1.2 x 0.2	158 7	3	Sc		15.7		
05050	09 25.6 + 76 41 135.63 + 35.24	13 – 07 – 030	0693 172	1.2 x 0.25 1.2 x 0.25	51 ...	1	S? ... or Peculiar, ssp of 2		15.4		*1V A
05051	09 25.7 + 62 43 151.45 + 41.77	10 – 14 – 015 NGC 2880	0703	(2.4 x 1.5) (2.4 x 1.5)	140 4	4	S B O s of 2		12.6	+ 1514 + 1614	*12 3468
05052	09 26.2 + 74 01 138.41 + 36.67	12 – 09 – 053	1325 147	1.7 x 1.3 1.7 x 1.4	120: 2	4	S B a SB(s)a		14.2		
05053	09 26.3 + 60 22 154.36 + 42.70	10 – 14 – 000	0703	1.2 x 0.5 1.0: x 0.4:	67 (6)	3	... S: dm – m		17.		
05054	09 26.6 – 02 20 236.02 + 32.98	00 – 24 – 017 IC 539	0430	1.2 x 1.1 1.0: x 0.9:	--- 1 – 2	2	Sc		14.3		
05055	09 26.6 + 56 04 159.95 + 44.14	09 – 16 – 024 Mark 114	0681	1.5 x 1.3 1.5 x 1.3	155: 1 – 2	4	S B b (R´)SB(s)b		14.5	+ 6000	*1
05056	09 26.8 + 07 57 225.30 + 38.49	01 – 24 – 024 NGC 2894	0028	2.3 x 1.1 (2.2 x 1.1)	27 5	2	Sa		13.4		*13
05057	09 26.9 – 01 07 234.87 + 33.73	00 – 24 – 000	0430	[1.1 x ...] ...	... ...	1	Chain of 4 galaxies		16.5:		*
05058	09 27.1 + 68 00 145.05 + 39.67	11 – 12 – 013	0708	1.2 x 0.5 1.2 x 0.4	124 6	2	Sc NGC 2892 group, Sc –		15.4		*
05059	09 27.3 + 20 17 210.41 + 43.64	03 – 24 – 061 IC 2487	0057	1.9 x 0.45 1.9 x 0.4	164 6	2	Sb		14.2		1<u>2</u>
05060	09 27.3 + 29 46 197.44 + 46.01	05 – 23 – 005 NGC 2893, Mark 401	0466 148	1.2 x 1.2 1.1 x 1.1	--- 1	3	S B a (R)SB 0/a Brightest in group		13.6		13
05061	09 27.3 + 67 50 145.23 + 39.76	11 – 12 – 014	·0708	1.2 x 0.6 1.1 x 0.5	110 5	2	S B a NGC 2892 group		15.3		*
05062	09 27.4 + 26 52 201.52 + 45.46	05 – 23 – 006 IC 2486	0466 154	1.0 x 0.9 1.1 x 1.0:	--- 1 – 2	2	S B b – c SAB(rs) ...		15.1		
05063	09 27.4 + 49 28 168.97 + 45.85	08 – 17 – 000 Mark 115	1004	1.3 x 0.8 1.3 x 0.6	[50] ...	2	... Peculiar, very compact		15.4	+ 7800	
05064	09 27.5 + 08 07 225.22 + 38.72	01 – 24 – 025 IC 540	0028	1.1 x 0.25 1.4: x 0.25	170 7	1	S ...		14.8		
05065	09 27.7 + 04 22 229.36 + 36.88	01 – 24 – 026 NGC 2900	0028	1.6 x 1.3 1.5 x 1.4	--- 1 – 2	3	S B c Singular		14.6		*
05066	09 27.8 + 46 37 173.01 + 46.39	08 – 17 – 111	1004	1.2 x 0.7 0.9 x 0.6	70 4:	2	S B b – c		15.5		1
05067	09 27.9 + 68 40 144.24 + 39.43	12 – 09 – 055	1325	1.4: x 1.2: 1.4:: x 1.2::	--- 1 – 2	4	Dwarf spiral S IV – V		16.5:		*
05068	09 28 + 81 22 130.88 + 32.70	14 – 05 – 005	0727	0.9 x 0.8 0.9 x 0.8	--- 1 – 2	1	S ...		14.3		
05069	09 28.0 + 06 20 227.28 + 37.95	01 – 24 – 000	0028	1.1 x 0.9 1.2 x 1.0	(15) 1 – 2	2	SO		15.0:		*
05070	09 28.3 + 30 15 196.81 + 46.30	05 – 23 – 007	0466 148	1.1 x 0.45 1.1 x 0.45	43 6	1	S ... NGC 2893 group		14.5		
05071	09 28.3 + 42 34 178.86 + 46.89	07 – 20 – 000	0661	... ...	... ...	0	Dwarf? Non – existing?		18. ?		*
05072	09 28.4 + 41 32 180.37 + 46.97	07 – 20 – 015	0661	1.0 x 0.5 0.9 x 0.5	45 5	2	S B b:		15.7		
05073	09 28.6 + 67 50 145.16 + 39.87	11 – 12 – 015 NGC 2892	0708	(1.4 x 1.4) (1.5 x 1.5)	--- ...	3	E Brightest in group		14.4		*2
05074	09 28.9 + 30 01 197.17 + 46 39	05 – 23 – 008	0466 148	1.0 x 0.8 1.0 x 0.8	45: 2:	1	Sa? NGC 2893 group		15.0		1
05075	09 29.0 + 04 43 229.20 + 37.34	01 – 25 – 001	0233	(1.1 x 1.1) 1.0 x 0.9	--- ...	1	SO – a?		14.6		

N G C   2880, 2892 – 94, 2900   I C   539, 540, 2476, 2486, 2487

1	2	3	4	5	6	7	8	9	10	11	12
05076	09 29.1 + 52 06 165.20 + 45.55	09 – 16 – 026	0681	1.5: x 1.4: 1.1: x 1.1:	-- ...	3	Dwarf IV – V or V		15.5		1
05077	09 29.1 + 59 58 154.70 + 43.17	10 – 14 – 020	0703	1.7 x 0.8 1.6 x 0.7	77 5 – 6	3	S B b		15.2		*1
05078	09 29.2 + 03 57 230.05 + 36.99	01 – 25 – 000	0233	1.0 x 0.2 1.0 x 0.2	170 6 – 7	3	Sc		16.0:		*
05079	09 29.4 + 21 44 208.70 + 44.56	04 – 23 – 009 NGC 2903 = 2905	0025	13.3 x 6.0 12.5 x 6.	17 5 – 6	5	Sb/Sc	SAB bc Sc –	9.8	+ 644 + 507	*12– –689
05080	09 29.4 + 42 00 179.67 + 47.13	07 – 20 – 000	0661	1.0 x 1.0 0.9 x 0.8	-- ...	1	...		16.5:		
05081	09 29.5 + 08 40 224.91 + 39.42	02 – 25 – 001 NGC 2906	0990	1.5 x 0.9 1.5 x 0.9	75 4:	3	Sc		13.1		3
05082	09 29.7 + 11 02 222.20 + 40.56	02 – 25 – 002	0990	1.2 x 0.3 1.1 x 0.3	127 6	1	S ... np of 2		14.8		*
05083	09 29.8 + 00 33 233.69 + 35.27	00 – 25 – 000	1318	1.0 x 0.2 0.9 x 0.15	74 6 – 7	1	S ...		15.3		
05084	09 30.0 + 27 44 200.48 + 46.21	05 – 23 – 009 154	0466	1.1 x 0.7 1.0 x 0.7	15 3 – 4	1	S ... ssp of 2		15.0		*1
05085	09 30.0 + 42 35 178.81 + 47.20	07 – 20 – 000	0661	1.0:: x 0.8:: 0.6:: x 0.6::	-- ...	3	Dwarf IV – V or V		17.		
05086	09 30.1 + 21 43 208.79 + 44.71	04 – 23 – 000	0025	1. x 1. 1.0: x 1.0:	-- ...	4	Dwarf V		18.		*
05087	09 30.1 + 30 10 197.02 + 46.67	05 – 23 – 010 IC 2490	0466 148	1.5 x 0.9 1.4 x 1.0	175 4	4	Sb NGC 2893 group, Sb +		14.7		
05088	09 30.4 + 34 16 191.06 + 47.24	06 – 21 – 047	0925 149	1.2: x 1.0: 1.3: x 1.0:	-- ...	1	...		15.2		
05089	09 30.6 + 40 13 182.28 + 47.44	07 – 20 – 021	0661	1.1 x 0.7 1.0 x 0.6	150 4	2	S B b – c		15.5		
05090	09 30.6 + 47 05 172.23 + 46.79	08 – 18 – 006	0672	1.0: x 1.0: 0.9 x 0.8:	-- 1	3	Sc	SAB c	15.6		
05091	09 30.7 + 44 29 176.01 + 47.16	08 – 18 – 018	0661	1.1 x 0.4 0.8 x 0.3	77 (6)	2	Sc – Irr		15.1		*
05092	09 31.1 + 10 22 223.19 + 40.56	02 – 25 – 003 NGC 2911	0990	(4. x 3.) (4. x 3.)	(140) ...	3	... Brightest in group		13.6	+ 3140 + 2978	*12 3468A
05093	09 31.3 + 10 15 223.36 + 40.55	02 – 25 – 004	0990	1.1 x 0.4 1.0 x 0.35	153 6	1	S ... NGC 2911 group		14.9		*1
05094	09 31.3 + 24 26 205.20 + 45.73	04 – 23 – 000 154	0025	1.0 x 0.6 (0.9 x 0.5)	150: ...	3	S0 p of 2		15.5		*
05095	09 31.4 + 09 42 224.01 + 40.32	02 – 25 – 005 NGC 2913	0990	1.2 x 0.8 1.1 x 0.7	140 3 – 4	1	S ...		14.1		3
05096	09 31.4 + 10 20 223.27 + 40.61	02 – 25 – 006 NGC 2914	0990	1.1 x 0.6 1.0 x 0.6	15 4 – 5	1	Sa? NGC 2911 group		13.7	+ 3370 + 3208	*12 3A
05097	09 31.6 + 00 29a 234.07 + 35.61	00 – 25 – 001	1318	0.6 x 0.4 0.6 x 0.35	55 ...	1	... Peculiar, s of 2		(13.9)		*1
05098	09 31.9 – 02 16 236.87 + 34.10	00 – 25 – 002 NGC 2917	1318	1.5 x 0.4 1.4 x 0.35	169 6	1	S0 – a		14.5		3
05099	09 32.0 + 00 19 234.31 + 35.60	00 – 25 – 003	1318	1.2 x 0.7 1.0 x 0.5	90: ...	1	S ... Double system?		14.7		*
05100	09 32.0 + 06 04 228.22 + 38.68	01 – 25 – 002	0233	1.6: x 0.9 1.1 x 0.8	30: ...	4	S B b	S B(s)b Sb +	14.9		
05101	09 32.0 + 61 34 152.49 + 42.89	10 – 14 – 025	0703	1.2 x 0.6 1.2 x 0.8	[87] ...	3	... Peculiar, eruptive?		15.5		*1
05102	09 32.1 + 10 30 223.18 + 40.84	02 – 25 – 007 NGC 2919	0990	1.7 x 0.55 1.7 x 0.6	159 6	2	Sb – c NGC 2911 group		13.6		*23
05103	09 32.1 + 21 56 208.70 + 45.21	04 – 23 – 011 NGC 2916	0025	2.5 x 1.7 2.3 x 1.7:	20 ...	2	S ... Singular		12.3		*23
05104	09 32.2 + 34 57 190.10 + 47.66	06 – 21 – 053 IC 2491	0925	1.1 x 0.8 0.7 x 0.5	75: ...	1	S0:		15.0		
05105	09 32.3 + 36 08 188.34 + 47.74	06 – 21 – 054	0925	1.2 x 0.8 1.1 x 0.8	(140) ...	3	Dwarf V		16.5:		
05106	09 32.4 + 54 49 161.28 + 45.29	09 – 16 – 033	0681	(1.4 x 0.9) 1.4 x 0.9	(95) ...	1	...		14.8		1
05107	09 32.5 + 05 20 229.10 + 38.41	01 – 25 – 004	0233	2.2 x 0.6 1.8: x 0.5	47 6	3	S B c	Sc +	15.2		*
05108	09 32.5 + 30 03 197.31 + 47.17	05 – 23 – 018 148	0466	1.4 x 0.7 1.7 x 0.6	143 5 – 6	2	S B a – b NGC 2893 group		15.0		*1
05109	09 32.5 + 48 23 170.28 + 46.88	08 – 18 – 011	0672	1.3 x 0.2 1.4 x 0.2	35 (7)	1	...		16.5:		*
05110	09 32.5 + 73 36 138.56 + 37.27	12 – 09 – 057 147	1325	1.8: x 1.4: 1. x 1.	[135] ...	4	Irr	IAB m Ir I	15.6		

N G C   2903, 2905, 2906, 2911, 2913, 2914   I C   2490, 2491
2916, 2917, 2919

1	2	3	4	5	6	7	8	9	10	11	12
05111	09 32.7 + 67 01	11 – 12 – 016	0708	1.6 x 0.25	120	2	Sb – c		15.5		
	145.86 + 40.60			1.6 x 0.3	6 – 7						
05112	09 32.8 + 31 56	05 – 23 – 019	0466	(1.4 x 1.0)	(65)	2	E		13.6		*3
	194.57 + 47.50	NGC 2918		(1.6 x 1.2)	...						
05113	09 32.8 + 66 47	11 – 12 – 017	0708	1.2 x 0.25	43	2	Sa – b		15.5		
	146.13 + 40.72			1.2 x 0.25	7						
05114	09 33 + 82 21	14 – 05 – 006	0727	1.9: x 0.7	[140]	3	Irr	IB: m	16.0:		
	129.82 + 32.25			1.3 x 0.6	...			Ir I			
05115	09 33.0 + 76 33	13 – 07 – 032	0693	2.4 x 1.3:	105:	4	S B c	SB(rs)cd	14.4		3
	135.48 + 35.67	NGC 2938	172	2.2 x 1.2:	5						
05116	09 33.5 + 03 14	01 – 25 – 000	0233	1.1 x 0.2	178:	1	S ...		15.5		*
	231.53 + 37.51			1.2 x 0.3	6:		nnp of 2, disturbed				
05117	09 33.7 + 04 49	01 – 25 – 000	0233	[1.0 x 0.4]	...	2	Double system, bridge		16.5:		*
	229.86 + 38.40			...	...		2nd and 3rd of 3				
05118	09 33.8 + 37 55	06 – 21 – 057	0925	1.1 x 0.45	103	2	Irr		14.6		
	185.70 + 48.09	NGC 2922		1.1 x 0.4	(6)		or Peculiar				
05119	09 34.0 + 38 18	06 – 21 – 059	0925	0.7 x 0.7	—	1	...		14.5		
	185.12 + 48.13			0.8 x 0.7	...		Peculiar, jets?				
05120	09 34.2 + 36 50	06 – 21 – 000	0925	1.0: x 0.15	53	1	Sa – b		16.5:		
	187.32 + 48.15			0.8 x 0.15	7						
05121	09 34.3 + 14 07	02 – 25 – 000	0990	1.5 x 0.15	131	3	Sc		15.7		
	219.13 + 42.91			1.3: x 0.15	7			Sc +			
05122	09 34.3 + 23 49	04 – 23 – 016	0025	1.8 x 1.1	155:	4	Sb/S B b	SAB(rs)b	14.1		*3
	206.32 + 46.23	NGC 2927	154	1.7 x 1.1	4			Sb +			
05123	09 34.4 + 20 03	03 – 25 – 006	1509	0.7 x 0.5	(155)	1	...		14.3		1
	211.48 + 45.13			0.7 x 0.5	...						
05124	09 34.4 + 21 54	04 – 23 – 015	0025	1.1: x 0.5:	175:	1	S ...		15.7		
	208.97 + 45.71			1.0 x 0.5:	...						
05125	09 34.5 + 33 04	06 – 21 – 060	0925	0.9 x 0.8	(120)	1	S ...	SAB(s) ...	14.4		
	192.95 + 47.98	NGC 2926		0.8 x 0.6	...		Peculiar				
05126	09 34.6 + 23 23	04 – 23 – 017	0025	1.3 x 0.35	144	1	S ...		14.4		*13
	206.95 + 46.18	NGC 2929	154	1.2 x 0.35	6		Brightest in group				
05127	09 34.6 + 37 18	06 – 21 – 062	0925	1.5 x 0.5	147	3	Sc	SAB: d	16.0:		*1
	186.62 + 48.24			1.0 x 0.4	6		Brightest of 3				
05128	09 35 + 84 02	14 – 05 – 007	0727	1.2 x 1.0	—	3	Sc		15.4		
	128.23 + 31.26			0.9 x 0.7	1 – 2						
05129	09 35.0 + 25 43	04 – 23 – 020	0025	1.8 x 0.8	103	2	Sa		14.3		
	203.70 + 46.86		154	1.5 x 0.8	5 – 6						
05130	09 35.1 + 02 58a	01 – 25 – 005	0233			3	E + strongly distorted		14.4		*13
	232.09 + 37.71	NGC 2936		[1.6 x 1.5]	...		multiple system				VA
05131	09 35.1 + 02 58b	01 – 25 – 006	0233	...	...						
	232.09 + 37.71	NGC 2937									
05132	09 35.2 + 17 15	03 – 25 – 008	1509	1.0 x 0.35	30	1	S ...	s of 2,	14.9		*1
	215.26 + 44.33	NGC 2933+2934	150	0.9 x 0.3	6		contact, NGC 2943 group				
05133	09 35.2 + 43 44	07 – 20 – 028	0661	1.1 x 0.25	149	3	SO		14.7		
	176.98 + 48.04			1.3 x 0.25	7						
05134	09 35.4 + 09 45	02 – 25 – 011	0990	2.6 x 0.9	154	1	Sb – c		13.5		*3
	224.56 + 41.21	NGC 2939		2.5 x 0.8	6		Brightest of 3				
05135	09 35.5 + 43 24	07 – 20 – 000	0661	1.2: x 1.0:	—	1	S ...		15.3		
	177.47 + 48.13			0.8 x 0.7	1 – 2						
05136	09 35.8 + 17 16	03 – 25 – 011	1509	(2.2 x 1.2)	130	3	E		14.0		*13
	215.31 + 44.47	NGC 2943	150	(2.3 x 1.5)	...		Brightest in group				
05137	09 35.8 + 20 53	04 – 23 – 023	0025	1.0 x 0.5	153	2	S B ...		15.7		
	210.50 + 45.71			1.0 x 0.5	5						
05138	09 35.9 + 02 47	01 – 25 – 000	0233	1.2 x 0.2	170	2	Sb		15.4		
	232.42 + 37.77			1.2 x 0.2	7						
05139	09 36.0 + 71 25	12 – 09 – 059	1325	4.0: x 4.0:	—	5	Dwarf irregular, I m		15.5		*12
	140.72 + 38.65	Ho I		3.5:: x 3.5::	...		Ir V, M 81 group, Ir I				59
05140	09 36.1 + 34 14	06 – 21 – 065	0925	2.1 x 1.6	165:	4	Sc	SA(s)c	14.1		*12
	191.25 + 48.41	NGC 2942		1.9 x 1.5	2						3468
05141	09 36.3 + 07 11	01 – 25 – 007	0233	1.7 x 0.9	7	3	S B b/S B c		13.8		3
	227.66 + 40.16	NGC 2948		1.7 x 1.0	4 – 5			Sc –			
05142	09 36.3 + 08 00	01 – 25 – 000	0233	1.0: x 1.0:	—	2	...	SA d:	16.0:		
	226.72 + 40.57			0.8: x 0.8:	1						
05143	09 36.3 + 17 15	03 – 25 – 013	1509	1.2 x 0.35	13	1	S B: ...		14.8		*3
	215.40 + 44.58	NGC 2946	150	1.2 x 0.35	6						
05144	09 36.3 + 32 33	06 – 21 – 067	0925	[1.1 x 0.4]	...	3	Double system, contact,		14.7		*12
	193.78 + 48.30	NGC 2944		...	...		disrupted, p w 5146				VAC
05145	09 36.3 + 48 39	08 – 18 – 019	0672	1.8 x 0.4	46	1	S ...		15.1		*
	169.71 + 47.44			1.7 x 0.4	7		Peculiar?				

N G C   2918, 2922, 2926, 2927, 2929, 2933,
        2934, 2936 – 39, 2942 – 44, 2946, 2948

1	2	3	4	5	6	7	8	9	10	11	12
05146	09 36.4 + 32 36 193.71 + 48.33	06 – 21 – 71+72	0925	[1.1 x 0.9] ...	... ...	3	Double system, contact, disrupted, p w 5144		14.3	+ 6800 + 6747	*1V AC
05147	09 36.4 + 38 40 184.56 + 48.59	07 – 20 – 000	0661	1.2 x 0.1 1.0 x 0.1	147 7	2	Sc Disturbed?		16.5:		*
05148	09 36.8 + 11 44 222.41 + 42.43	02 – 25 – 013	0990 151	1.1 x 0.2 1.1 x 0.2	54 7	3	Sc		15.0		1
05149	09 37.0 + 21 14 210.15 + 46.09	04 – 23 – 000	0025	1.0 x 0.10 1.1 x 0.10	141 7	3	Sc		16.0:		
05150	09 37.2 + 47 44 171.00 + 47.77	08 – 18 – 020	0672	1.3: x 0.6 0.8 x 0.5	160 ...	2	...	S d – dm	16.0:		
05151	09 37.2 + 48 34 169.79 + 47.61	08 – 18 – 021	0672	0.8 x 0.7 0.8 x 0.7	— ...	1	...		13.5		
05152	09 37.4 + 79 55 132.00 + 33.87	13 – 07 – 034 NGC 2908	0693	1.0 x 1.0 0.9 x 0.8	— 1	1	S ...		14.2		3
05153	09 37.5 + 61 17 152.48 + 43.61	10 – 14 – 030	0703	1.1 x 1.0 0.9 x 0.9	— 1	2	Sb – c		15.6		*1
05154	09 37.6 + 29 13 198.79 + 48.13	05 – 23 – 000	0466	1.0 x 0.25 1.0 x 0.3	2 6	2	Sa – b		15.6		
05155	09 37.7 + 15 09 218.29 + 44.08	03 – 25 – 019 NGC 2954	1509	(1.7 x 1.1) (1.7 x 1.2)	160: ...	3	E		13.5		3
05156	09 37.7 + 25 43 203.91 + 47.45	04 – 23 – 000	0025 154	1.2 x 0.35 1.1 x 0.25	130 6	1	S B ... p of 2		15.4		*
05157	09 37.7 + 47 50 170.83 + 47.84	08 – 18 – 022	0672	1.9 x 0.4 1.6 x 0.35	18 6	1	S ...		14.5		1
05158	09 37.9 + 53 27 162.81 + 46.47	09 – 16 – 036	0681	1.0 x 0.1 1.0 x 0.15	55 7	1	S ...		16.0:		
05159	09 38.0 + 03 48 231.69 + 38.76	01 – 25 – 009 NGC 2960	0233	1.7 x 1.3 (2.7 x 1.4)	40 ...	1	Sa?		13.6		3
05160	09 38.0 + 12 07 222.12 + 42.86	02 – 25 – 015 NGC 2958	0990 151	1.1 x 0.8 1.1 x 0.7	10 ...	1	S B : b – c S (r) ...		13.9		
05161	09 38.0 + 28 00a 200.61 + 48.00	05 – 23 – 025	0466 154	1.0: x 0.5: 0.9 x 0.6:	[150] ...	2	S ... P w 5162, distorted				
05162	09 38.0 + 28 00b 200.61 + 48.00	05 – 23 – 025	0466 154	1.1: x 0.5 0.8 x 0.3	[160] ...	2	S ... P w 5161, distorted		15.5		*1
05163	09 38.1 + 68 38 143.68 + 40.25	12 – 09 – 060	1325	1.0 x 0.4 0.9 x 0.4	169 (6)	2	Sc – Irr nnf of 2, in group?		15.6		*
05164	09 38.2 + 11 47a 222.56 + 42.76	02 – 25 – 016	0990 151	1.1 x 0.15 1.0 x 0.12	22 7	2	S – Irr n of 2, contact, distorted		(14.9)		*1
05165	09 38.2 + 21 25 210.02 + 46.41	04 – 23 – 024	0025 146	1.1 x 0.3 1.2 x 0.35	107 6 – 7	1	S ... In group		15.7		*1
05166	09 38.2 + 36 07 188.44 + 48.93	06 – 21 – 073 NGC 2955	0925	1.5 x 0.7 1.5 x 0.7	162 5	4	Sb		13.9		*12 346$\overline{8}$
05167	09 38.3 + 05 24 229.99 + 39.68	01 – 25 – 011 NGC 2962	0233	3.0 x 2.2 3.0 x 2.2	3: 2 – 3	4	S0/Sa		13.1		*123 46$\overline{8}$
05168	09 38.3 + 07 10 228.01 + 40.58	01 – 25 – 012 IC 551	0233	0.9 x 0.8 0.9 x 0.8	— ...	1	...		14.5		
05169	09 38.5 + 47 12 171.72 + 48.09	08 – 18 – 024	0672	1.1 x 1.1 1.1: x 1.0:	— 1	3	...	SAB: dm	15.5		
05170	09 38.6 + 03 02 232.62 + 38.47	01 – 25 – 000	0233	1.0 x 0.5 1.0 x 0.6:	177: ...	1	S ...		16.0:		
05171	09 38.6 + 10 52 223.73 + 42.43	02 – 25 – 017 IC 552	0990 151	1.0 x 0.6 1.1 x 0.6	175 4 – 5	3	S0 p of 2		14.5		*
05172	09 38.6 + 48 54 169.23 + 47.76	08 – 18 – 027	0672	2. x 2. 1.5:: x 1.5::	— 1:	4	... Brightest of 3	SA(s)m	15.6		*
05173	09 38.8 + 11 38 222.83 + 42.82	02 – 25 – 018	0990 151	2.2 x 0.3 2.2 x 0.25	131 7	3	Sb		15.1		*
05174	09 38.8 + 33 44 192.07 + 48.93	06 – 21 – 000	0925	1.0 x 0.1 0.9 x 0.1	109 7	2	Sc		16.5:		
05175	09 38.8 + 75 05 136.72 + 36.79	13 – 07 – 035 NGC 2977	0693	1.6 x 0.6 1.5 x 0.6	145 6	1	S ...		12.7		$\underline{235}$
05176	09 39.0 + 59 05 155.18 + 44.67	10 – 14 – 032 NGC 2950	0703	(3.3 x 2.3) (3.5 x 2.5)	145: 3:	4	S B 0		11.8	+ 1393 + 1475	*123 46$\overline{8}$
05177	09 39.2 + 11 52 222.60 + 43.01	02 – 25 – 000	0990 151	1.1 x 0.4 0.9: x 0.4:	165 (6)	1	S? ...		15.2		
05178	09 39.2 + 12 31 221.80 + 43.30	02 – 25 – 020 IC 555	0990 151	(1.3 x 0.6) 1.1 x 0.5	18 7	3	S0		14.4		
05179	09 39.4 + 59 12 155.00 + 44.67	10 – 14 – 033	0703	1.3: x 0.7 1.3 x 0.7	150: ...	1	S ...		15.1		
05180	09 39.5 + 00 34 235.37 + 37.28	00 – 25 – 007 NGC 2967	1318	2.8: x 2.8: 2.6: x 2.6:	— 1	5	Sc Sc +	SA(s)c	12.2	+ 2245 + 2044	*123 47$\overline{8}$

N G C 2908, 2950, 2954, 2955, 2958, 2960
2962, 2967, 2977

I C 551, 552, 555

1	2	3	4	5	6	7	8	9	10	11	12
05181	09 39.6 + 04 54 230.77 + 39.69	01 – 25 – 013 NGC 2966	0233	2.3 x 0.9 2.3 x 0.9	72 6	1	S B? ...		14.0		*
05182	09 39.8 + 04 30 231.24 + 39.52	01 – 25 – 014	0233	1.2 x 0.9 (1.2 x 1.0)	135: ...	3	S0 Companion superimposed		14.1		*1
05183	09 39.9 + 32 05 194.60 + 49.01	05 – 23 – 027 NGC 2964 (+ Mark 404)	0466	3.5 x 1.9 3.6 x 2.0	97 4 – 5	2	Sb/Sc: Singular, p w 5190, brightest of 3		12.0	+ 1340 + 1284	*123 45789
05184	09 39.9 + 38 03 185.49 + 49.29	06 – 22 – 001	1345	1.2 x 0.7 1.0 x 0.7	70: ...	2	S B b		14.9		
05185	09 40.0 + 29 13b 198.92 + 48.64	05 – 23 – 028	0466	1.9 x 0.6 1.6 x 0.6	168 6	2	Sc nff of 2 Sc –		(14.5)		*1
05186	09 40.0 + 33 28 192.50 + 49.16	06 – 22 – 000	1345	1.4: x 0.3 0.7: x 0.2	[43] ...	1	Irr Ir I		16.5:		*
05187	09 40.0 + 41 20 180.46 + 49.15	07 – 20 – 034	0661	1.1 x 0.8 1.0 x 0.8	135: 2:	1	S B ... Peculiar		14.1		1
05188	09 40.0 + 66 13 146.34 + 41.62	11 – 12 – 022 NGC 2909, Mark 119	0708	0.45 x 0.35 0.45 x 0.35	... ...	2	... Peculiar, loop		14.1	+ 3000	3
05189	09 40.2 + 09 43 225.35 + 42.24	02 – 25 – 22+23	0990	[3.3 x 1.1] ...	... ...	4	Irregular multiple system Connected		13.8		*1
05190	09 40.2 + 32 10 194.49 + 49.08	05 – 23 – 029 NGC 2968	0466	(2.4 x 1.7) (3.3 x 2.5)	[45:] ...	3	... I 0 P w 5183 Ir II		13.1	+ 1608 + 1552	*123 45789
05191	09 40.2 + 36 28 187.91 + 49.34	06 – 22 – 003 NGC 2965	1345	(1.2 x 0.9) (1.2 x 0.9)	(85) ...	3	S0 Brightest in group		14.7		*3
05192	09 40.3 + 21 25 210.24 + 46.88	04 – 23 – 027	0025	2.1: x 0.9 1.9: x 1.0:	90 5:	2	S B: b		14.8		
05193	09 40.4 + 39 39 183.03 + 49.34	07 – 20 – 036	0661 152	(1. x 1.) (... x 1.)	-- ...	1	E: n of 2, in common envelope?		15.5		*1
05194	09 40.5 + 34 16 191.29 + 49.32	06 – 22 – 000	1345	1.1 x 0.15 0.9 x 0.15	49 7	3	Sc		16.0:		
05195	09 40.6 + 00 39 235.48 + 37.56	00 – 25 – 000	1318	1.1 x 0.22 1.0 x 0.2	68 6 – 7	1	Sb – c		15.4		
05196	09 40.7 + 17 10 216.06 + 45.52	03 – 25 – 020	1509	1.0: x 1.0: ---	-- ...	0	? Non – existing?		15.7		*1
05197	09 40.7 + 36 24 188.02 + 49.44	06 – 22 – 005 NGC 2971	1345	1.2 x 0.8 1.0 x 0.7	135 3:	3	S B b S B(r)b NGC 2965 group		15.0		*
05198	09 40.8 + 41 55 179.55 + 49.25	07 – 20 – 042	0661	1.1 x 0.7 1.0 x 0.7	150 ...	2	S B a		15.0		*
05199	09 40.8 + 42 54 178.06 + 49.15	07 – 20 – 039	0661	1.1 x 0.45 1.1 x 0.5	65 6	2	Sa – b nf of 2		(15.0)		*1
05200	09 41.0 + 11 18 223.57 + 43.15	02 – 25 – 025 IC 556	0990	0.7 x 0.7 0.9 x 0.9	-- ...	1	...		14.3		*
05201	09 41.0 + 56 00 159.11 + 46.06	09 – 16 – 038	0681 155	1.5 x 0.9 1.4: x 0.9:	45 3	4	Sc SAB(s)c Sc +		15.3		
05202	09 41.0 + 68 49 143.30 + 40.38	12 – 09 – 062 NGC 2959	1325	1.4 x 1.4 1.5 x 1.5	-- 1	3	Sa SAB: a P w NGC 2961		13.7		*23
05203	09 41.0 + 80 02 131.77 + 33.92	13 – 07 – 037	0693	2.3 x 0.18 2.6 x 0.20	80 7	3	Sc Sc +		15.5:		*
05204	09 41.2 + 09 53 225.31 + 42.54	02 – 25 – 026	0990	1.1 x 0.9 (1.0 x 1.0)	-- ...	1	...		14.8		
05205	09 41.6 – 00 26 236.78 + 37.13	00 – 25 – 011	1318	1.2 x 0.8 1.0 x 0.7	[152] ...	3	Irr sf of 2		15.4		*1
05206	09 41.7 + 14 21 219.86 + 44.63	02 – 28 – 002 NGC 3357	0976	(1.1 x 0.9) (1.3 x 1.0)	(90) ...	1	E:		14.3		23
05207	09 41.8 + 53 55 161.91 + 46.87	09 – 16 – 000	0681 155	1.1 x 0.6 0.7 x 0.4	55: 5 – 6	2	S B a – b		15.0		
05208	09 42.0 + 31 20 195.82 + 49.37	05 – 23 – 032 NGC 2981	0466	1.2 x 1.0 1.2: x 1.0:	(95) 2:	3	Sb/Sc SAB:(rs)bc		15.0		*1
05209	09 42.0 + 32 20 194.06 + 49.49	05 – 23 – 000	0466	1. x 1. 1. x 1.	-- ...	4	Dwarf irregular Ir V, NGC 2964 group?		19.		
05210	09 42.0 + 69 12 142.81 + 40.26	12 – 09 – 064	1325	2.1 x 0.2 1.8 x 0.2	153 7	3	Sc		16.0:		*
05211	09 42.1 + 00 01 236.41 + 37.50	00 – 25 – 012	1318	(1.0 x 1.0) (1.0 x 0.9)	-- ...	1	...		15.5		*
05212	09 42.1 + 39 41 182.95 + 49.66	07 – 20 – 046	0661 152	1.0 x 0.7 1.0 x 0.7	70: 3:	2	Sc		16.0:		
05213	09 42.2 + 16 56 216.56 + 45.77	03 – 25 – 024	1509	1.2 x 1.1 1.1 x 1.0	-- 1 – 2	3	S B a S B(rs) ...		14.3		*
05214	09 42.5 + 23 19 207.76 + 47.91	04 – 23 – 000	0025 154	1.1: x 0.6: 0.8: x 0.5:	[65] ...	3	Dwarf IV – V		17.		
05215	09 42.6 + 09 20 226.19 + 42.58	02 – 25 – 030	0990	1.7 x 0.9 1.7 x 0.9:	23 5	2	Sb – c		13.8		

N G C   2909, 2959, 2964 – 66, 2968, 2971, 2981   I C   556

1	2	3	4	5	6	7	8	9	10	11	12
05216	09 42.7 + 09 59 225.43 + 42.91	02 – 25 – 031	0990	1.8: x 0.8 1.7: x 0.8:	165 5 – 6	3	...	SAB: d – dm	14.8		
05217	09 42.8 + 34 05 191.62 + 49.78	06 – 22 – 008	1345	1.0 x 0.5 1.0 x 0.5	5 5	3	S B c	S B (s)cd	15.5		
05218	09 42.9 + 06 37 229.41 + 41.29	01 – 25 – 016	0233	2.0 x 1.0 1.4 x 0.7	125 5	4	Sc	SAB d	15.0		
05219	09 42.9 + 28 43 199.83 + 49.19	05 – 23 – 036	0466	1.2: x 0.9: 0.8: x 0.8:	(170) 1 – 2	3	Sc/S B c	SAB(s)d	16.5:		
05220	09 43.0 + 05 10 231.06 + 40.55	01 – 25 – 017 NGC 2987	0233	1.5 x 0.6 1.5 x 0.8	160 5 – 6	2	Sa – b		13.9		
05221	09 43.1 + 68 09 143.92 + 40.90	11 – 12 – 025 NGC 2976	0708	5.5: x 3.0: 6. x 3.	143 6	3	Sc Singular		10.9	+ 42 + 169	*12– –689
05222	09 43.2 + 73 12 138.44 + 38.12	12 – 10 – 003 NGC 2963, Mark 122	0685	1.2 x 0.6 1.2 x 0.6	165 5	2	S B a – b P w NGC 2957		14.3		*3
05223	09 43.3 – 00 02 236.68 + 37.71	00 – 25 – 013 IC 560	1318	1.5 x 0.7 1.5 x 0.7	15 5	2	S0 – a		14.6		
05224	09 43.3 + 03 12 233.27 + 39.55	01 – 25 – 018	0233	1.7: x 0.8: 1.7: x 0.9:	70 ...	4	...	S B dm	15.3		1
05225	09 43.3 + 46 00 173.27 + 49.13	08 – 18 – 030	0672	1.1 x 1.1 1.1 x 1.1	--- ...	2	... Compact		14.9	+ 4955	*C
05226	09 43.4 + 04 38 231.72 + 40.35	01 – 25 – 020	0233	(1.0 x 0.9) (1.2 x 1.0)	--- ...	1	S0:		14.0		
05227	09 43.4 + 42 26 178.69 + 49.68	07 – 20 – 047	0661	1.2 x 0.3 1.2 x 0.6	68 6	2	Sc		16.0:		
05228	09 43.5 + 01 54 234.70 + 38.86	00 – 25 – 014	1318	2.6 x 0.8 2.5 x 0.7	127 6	2	S B ...		14.1		
05229	09 43.6 + 05 56 230.30 + 41.08	01 – 25 – 021 NGC 2990	0233	1.1 x 0.55 1.1 x 0.55	85 ...	1	S? ...		12.5		*12 3468
05230	09 43.8 + 03 18 233.25 + 39.71	01 – 25 – 023 IC 564	0233	1.7 x 0.45 1.8 x 0.5	68 (6)	3	Irr P w IC 563		14.1	+ 6110 + 5921	*12
05231	09 43.9 + 42 45 178.18 + 49.73	07 – 20 – 049	0661	1.0 x 0.5 0.9 x 0.4	40 5	2	S B: a – b		15.4		*1
05232	09 44.0 + 14 00 220.64 + 44.99	02 – 25 – 033	0990	1.0 x 0.5 1.1 x 0.5	0 5	1	S ... s of 2, contact		14.5		*1
05233	09 44.0 + 22 15b 209.44 + 47.94	04 – 23 – 033 NGC 2991	0025 154	(1.4 x 1.1) (1.7 x 1.5)	--- ...	2	S0 P w NGC 2988, contact		(14.3)		*13
05234	09 44.1 + 16 17 217.67 + 45.94	03 – 25 – 026	1509 153	1.7 x 1.1 1.7: x 1.1:	115: 3 – 4	3	Sc	S (r)c Sc –	14.9		*1
05235	09 44.1 + 23 16 207.99 + 48.25	04 – 23 – 000	0025 154	1.4 x 0.15 1.4 x 0.15	151 7	3	Sc		16.0:		*
05236	09 44.2 + 21 58 209.86 + 47.90	04 – 23 – 034	0025 154	1.3: x 1.2: 1.2: x 1.1:	--- ...	4	Dwarf spiral S V		16.5:		
05237	09 44.2 + 46 51 171.95 + 49.11	08 – 18 – 031	0672	1.6 x 0.8 1.4 x 0.7	155 5	4	Sc		15.0		
05238	09 44.3 + 00 44 236.07 + 38.36	00 – 25 – 015	1318	2.5: x 0.9 2.0: x 0.7	55 6	3	Sc	SAB: d	15.3		1
05239	09 44.5 + 22 20 209.37 + 48.08	04 – 23 – 035 NGC 2994	0025 154	(1.2 x 0.9) (1.3 x 0.9)	125 2 – 3	1	S0 P w 5233ab		14.4		*13
05240	09 44.5 + 25 59 204.03 + 49.00	04 – 23 – 000	0025 154	1.2 x 0.2 1.2 x 0.15	109 7	3	Sc In group of 3		15.6		*
05241	09 44.5 + 54 15 161.25 + 47.14	09 – 16 – 051	0681 155	0.8 x 0.8 0.8 x 0.7	--- 1	2	S ... p of 2		14.0		*
05242	09 44.6 + 01 12 235.64 + 38.69	00 – 25 – 016	1318	1.5 x 0.9 1.2: x 0.6:	10 ...	3	Dwarf IV – V		16.0:		
05243	09 44.9 + 58 13 155.88 + 45.72	10 – 14 – 041 Mark 21	0703	1.1? x 1.0 1.1 x 1.1	--- 1	1	S B ...		15.1		*
05244	09 44.9 + 64 25 148.12 + 42.94	11 – 12 – 026	0708	1.7 x 0.25 1.8 x 0.25	31 6 – 7	2	Sc sp of 2		15.2		*
05245	09 45.0 – 01 48 238.79 + 37.01	00 – 25 – 017	1318	2.8 x 0.5 2.4 x 0.4	160 (6 – 7)	3	...	d – m	15.5		1
05246	09 45.0 + 23 57 207.07 + 48.63	04 – 23 – 036	0025 154	1.1 x 0.8 0.9 x 0.8	140: 2 – 3:	1	S ...		15.1		
05247	09 45.0 + 69 39 142.13 + 40.24	12 – 10 – 005	0685	1.8 x 0.9 1.7 x 0.8	155 5	4	...	SA dm	15.6		
05248	09 45.1 + 16 05 218.07 + 46.08	03 – 25 – 028 IC 565	1509 153	1.7 x 0.2 1.6 x 0.2	50 7	1	Sc: Companion superimposed		15.1		*
05249	09 45.2 + 02 51 233.99 + 39.75	01 – 25 – 025	0233	2.4 x 0.8 2.3 x 0.7	15 6	3	S B: c	Sc +	14.4		
05250	09 45.5 + 44 18 175.74 + 49.80	07 – 20 – 051 NGC 2998	0661	2.6 x 1.3 2.6 x 1.3	53 5	5	Sc Brightest in group, Sc +	SAB(rs)c	13.3		*12 3468

N G C   2963, 2976, 2987, 2990, 2991, 2994, 2998   I C   560, 564, 565

1	2	3	4	5	6	7	8	9	10	11	12
05251	09 45.6 + 33 39 192.35 + 50.33	06 - 22 - 013 NGC 3003	1345	5.7 x 1.7 5.6 x 1.6	79 6	3	S B? c		12.3	+ 1476 + 1429	*12- -68$\overline{9}$
05252	09 45.8 + 41 42 179.74 + 50.20	07 - 20 - 053	0661	1.1: x 0.5: 0.8 x 0.3	[15] ...	1	... Double system?		16.0:		
05253	09 45.8 + 72 31 139.01 + 38.67	12 - 10 - 006 NGC 2985	0685	4.3 x 3.3 4.5: x 3.5:	0 2	4	Sb P w 5316	SA bc Sb +	11.1	+ 1277 + 1424	*12- -9
05254	09 46.1 + 43 35 176.82 + 50.02	07 - 20 - 056	0661	1.0 x 0.2 0.4 x 0.2	96 7	1	S ...		16.5:		
05255	09 46.2 + 35 16 189.82 + 50.53	06 - 22 - 014	1345	1.0 x 0.2 0.8 x 0.15	157 7	1	S ...		15.7		
05256	09 46.3 + 35 24 189.61 + 50.56	06 - 22 - 015	1345	1.0 x 0.4 0.7 x 0.3	130 6	2	S B ...		16.0:		
05257	09 46.6 + 21 55 210.18 + 48.42	04 - 23 - 039	0025 154	1.0 x 0.5 1.0 x 0.5	73 5	1	S ...		15.5		
05258	09 46.7 + 14 54 219.86 + 45.96	03 - 25 - 029	1509	1.4 x 0.15 1.2 x 0.15	129 7	3	Sc		15.4		*
05259	09 46.7 + 32 27 194.25 + 50.47	05 - 23 - 038 NGC 3011, Mark 409	0466	0.9 x 0.8 0.9 x 0.8	--- 2:	3	S0		14.2		
05260	09 46.7 + 64 44 147.61 + 42.96	11 - 12 - 029	0708	1.2 x 0.15 1.0 x 0.18	65 7	2	Sb - c		16.0:		
05261	09 46.8 + 01 22 235.87 + 39.24	00 - 25 - 020 NGC 3015	1318	0.5 x 0.4 0.45 x 0.35	(95) ...	1	... Brightest of 3		14.2		*13
05262	09 46.8 + 34 56 190.35 + 50.64	06 - 22 - 017 NGC 3012	1345	(1.1 x 1.0) (1.1 x 1.1)	--- ...	2	[E] Compact		14.9		3
05263	09 46.8 + 46 35 172.21 + 49.60	08 - 18 - 037	0672	1.0 x 0.2 0.8 x 0.2	[28] ...	1	... Peculiar		15.2		*
05264	09 47.0 + 44 32 175.32 + 50.03	07 - 20 - 062 NGC 3009	0661	0.8 x 0.7 0.8 x 0.7	--- ...	2	S ...		14.5		*13
05265	09 47.1 + 00 51 236.47 + 39.00	00 - 25 - 021 NGC 3018	1318	1.3 x 1.0 1.1 x 0.6	[27] ...	1	S: ... P w 5269, distorted, bridge?		14.2		*1
05266	09 47.1 + 12 56 222.46 + 45.22	02 - 25 - 040 NGC 3016	0990	1.2 x 1.0 1.3 x 1.1	70: 2:	3	Sb NGC 3020 group		13.7		*13
05267	09 47.2 + 09 19 226.96 + 43.56	02 - 25 - 041	0990	1.6 x 0.25 1.6 x 0.25	44 7	2	Sb 5270 group		14.9		*
05268	09 47.2 + 62 25 150.37 + 44.13	10 - 14 - 043	0703	1.5 x 0.5 1.4 x 0.4	134 6	2	Sa - b		15.2		
05269	09 47.3 + 00 51 236.51 + 39.04	00 - 25 - 022 NGC 3023	1318	3.2 x 1.6 3.0: x 1.5	[70] ...	3	S ... P w 5265, distorted, bridge?		13.5		*12
05270	09 47.3 + 09 23b 226.90 + 43.61	02 - 25 - 043	0990	(1.0 x 0.6) (1.1 x 0.6)	3: ...	1	S0 nnf of 2 Brightest in group		(14.6)		*
05271	09 47.4 + 13 03 222.36 + 45.33	02 - 25 - 045 NGC 3020	0990	3.2 x 1.8 3.0: x 1.7:	105 4 - 5	4	S B c S B(r)cd Brightest in group		13.2		*12 3
05272	09 47.4 + 31 43 195.42 + 50.55	05 - 23 - 041	0466	2.2 x 0.8 2.0: x 0.7	[115] ...	4	Irr I m Ir IV Ir I		14.7		*1
05273	09 47.4 + 44 34 175.25 + 50.09	07 - 20 -65+66 NGC 3010 +67	661	[1.9 x 0.6] ...	... ...	2	Triple system		14.3		*13
05274	09 47.5 + 16 32 217.80 + 46.79	03 - 25 - 030	1509	1.1 x 1.1 0.9 x 0.8	--- 1	4	Sc		14.9		
05275	09 47.8 + 13 00 222.48 + 45.40	02 - 25 - 046 NGC 3024	0990	2.0 x 0.5 1.7 x 0.5	125 6	1	S ... NGC 3020 group		13.7		*12 3
05276	09 47.8 + 30 44 196.98 + 50.53	05 - 23 - 042	0466	1.7 :: x 1.1 :: 1.7: x 1.2:	(125) ...	1	... Brightest of 3		15.1		*
05277	09 47.8 + 65 44 146.36 + 42.55	11 - 12 - 030	0708	1.8 x 1.7 1.6 x 1.6	--- 1	4	S B b S B(rs)bc Sb +		15.3		
05278	09 47.9 + 22 59 208.76 + 49.01	04 - 23 - 040	0025 154	1.0 x 0.45 1.1 x 0.5	58 6	3	S B b S B(s) ...		15.0		
05279	09 48.0 + 28 47 200.02 + 50.30	05 - 23 - 043 NGC 3026	0466	2.6: x 0.7 2.5 x 0.7	82 (6)	4	Irr I m Ir I		13.8		
05280	09 48.0 + 33 47 192.19 + 50.84	06 - 22 - 019 NGC 3021	1345	1.5 x 0.9 1.4 x 0.8	110 4:	1	S ...		12.6	+ 1529 + 1483	*12 346$\overline{8}$
05281	09 48.1 + 30 49 196.86 + 50.60	05 - 23 -44+45	466	[2.1 x 0.3] ...	[164] ...	3	Double system Bridge, 5276 group		15.1		*1
05282	09 48.2 + 33 22 192.85 + 50.86	06 - 22 - 020	1345	1.0 x 0.5 1.0: x 0.5:	40 (5)	3	... S: m		16.0:		
05283	09 48.2 + 45 10 174.29 + 50.12	08 - 18 - 041	0672	1.5 x 0.9 1.5 x 0.9	145 4	4	Sc SA d np of 2		15.5		*
05284	09 48.3 + 04 31 232.72 + 41.31	01 - 25 - 000	0233	1.1: x 0.9: 1. x 1.	--- ...	2	... dm - m		16.5:		
05285	09 48.4 + 15 58 218.68 + 46.77	03 - 25 - 031 IC 568	1509	1.4 :: x 0.8 :: 1.5: x 1.1:	... ...	3	S B b S B(rs)b Sb -		14.8		*

N G C   2985,   3003,   3009 - 12,   3015,   3016,   3018,   I C   568
        3020,   3021,   3023,   3024,   3026

1	2	3	4	5	6	7	8	9	10	11	12
05286	09 48.5 + 09 14 227.28 + 43.80	02 − 25 − 047	0990	2.1 x 1.5 1.7: x 1.4:	35: 3	4	Sc	SA d	14.4		
05287	09 48.5 + 33 10 193.17 + 50.91	06 − 22 − 021	1345	1.5 x 1.0 1.4 x 0.9	15: 3:	4	S B c	S B(s)cd	14.7		*
05288	09 48.7 + 08 04 228.71 + 43.26	01 − 25 − 028	0233	1.5 x 1.0 1.2 x 0.9:	155	2	S B − Irr	... B ...	14.4		
05289	09 49.0 − 01 32 239.28 + 37.97	00 − 25 − 000	1318	[1.4 x 0.6] ...	... ...	3	Double system Contact		16.0:		*
05290	09 49.0 + 41 06 180.58 + 50.86	07 − 20 − 070	0661	1.1 x 0.7 1.0 x 0.6	0: ...	1	S0 − a?		14.9		
05291	09 49.2 + 13 11 222.46 + 45.78	02 − 25 − 050	0990	(1.0 x 0.8) (1.1 x 0.8)	(0) ...	2	E − S0		14.8		*1
05292	09 49.2 + 29 28 199.02 + 50.66	05 − 23 − 046 NGC 3032	0466	(2.0 x 1.7) (2.3 x 1.9)	(95) 1 − 2	4	S0	SA(r)0	13.0	+ 1568 + 1500	*123 4678
05293	09 49.2 + 41 30 179.94 + 50.85	07 − 20 − 000	0661	1.0 x 0.5 0.9: x 0.5:	130 5:	2	...	S d − dm	16.5:		
05294	09 49.3 + 33 17 193.00 + 51.08	06 − 22 − 023	1345	1.1 x 0.3 1.5: x 0.4	[137] ...	1	... Jet?		15.2		*
05295	09 49.7 + 43 05 177.44 + 50.74	07 − 20 − 073	0661	2.3 x 1.0 2.1 x 1.0	150 5 − 6	4	S B b/Sb Brightest in group, Sb +	SAB(s)b	14.5		*1
05296	09 49.7 + 58 43 154.85 + 46.08	10 − 14 − 046	0703	1.3 x 1.2 1.0 x 1.0	-- (1)	3	Dwarf spiral S IV − V, 5306 group		16.5:		*
05297	09 49.9 + 02 22 235.38 + 40.44	00 − 25 − 027 NGC 3039	1318	1.1 x 0.6 1.1 x 0.6	12 5	2	Sa − b		14.4		*12 3
05298	09 50 + 82 32 129.23 + 32.56	14 − 05 − 000	0727	1.0 x 0.2 1.0 x 0.2	147 (6)	1	Sc − Irr		16.5:		
05299	09 50.1 + 00 02 237.89 + 39.13	00 − 25 − 000	1318	1.0: x 0.2 0.6: x 0.2	148 ...	1	...	d − m	17.		
05300	09 50.2 + 19 40 213.80 + 48.51	03 − 25 − 037 NGC 3040	1509	[1.6 x 0.6] ...	... ...	3	Double system Contact, in chain of 4		14.2		*1
05301	09 50.2 + 43 05 177.42 + 50.83	07 − 20 − 074	0661	1.4 x 0.15 1.4 x 0.15	159 7	3	Sc 5295 group		16.0:		*
05302	09 50.3 + 68 34 142.98 + 41.23	12 − 10 − 008	0685	2.2 x 1.7 1.8 x 1.6	-- 1 − 2	4	... M 81 group	SA(s)dm	15.1		*
05303	09 50.4 + 16 55 217.67 + 47.58	03 − 25 − 039 NGC 3041	1509	3.7 x 2.4 3.8 x 2.4	95 3 − 4	5	Sc	SAB(rs)c	13.1		*123 468
05304	09 50.5 + 08 07 228.96 + 43.67	01 − 25 − 030	0233	[1.2 x 0.8] ...	... ...	3	S ... + companion Distorted, bridge, in group		14.8		*1V A
05305	09 50.5 + 10 40 225.86 + 44.92	02 − 25 − 000	0990	1.1 x 0.2 0.5 x 0.18	177 7	1	S ...		16.0:		
05306	09 50.6 + 58 35 154.95 + 46.24	10 − 14 − 048	0703	1.2: x 0.7 1.0 x 0.7	30 4:	3	S B b s of 2, brightest in group	S B(s)b	15.5		*
05307	09 50.8 + 00 56 237.08 + 39.80	00 − 25 − 030 NGC 3042	1318	1.1 x 0.7 1.1 x 0.6	111 6 − 7	2	S0		13.8		*13
05308	09 50.8 + 08 07 229.01 + 43.74	01 − 25 − 031	0233	1.2 x 1.2 1.1 x 1.1	-- 1	2	S ... 5304 group, disturbed:		15.0		*1
05309	09 51 + 80 59 130.58 + 33.62	14 − 05 − 008	0727	1.4 x 1.3 1.4 x 1.3	-- 1	3	...	SA m	16.0:		
05310	09 51.0 + 79 32 131.90 + 34.58	13 − 07 − 000	0693 172	1.2 x 0.9 1.0 x 0.9	-- ...	3	S B b	S B(s)b	15.7		
05311	09 51.1 + 01 48 236.21 + 40.36	00 − 25 − 031 NGC 3044	1318	4.7 x 0.8 4.6 x 0.7	13 6 − 7	3	Sc		12.4	+ 1326 + 1132	*123 4678
05312	09 51.1 + 02 37 235.33 + 40.83	01 − 25 − 032	0233	1.0: x 0.15 0.8 x 0.15	85 7	3	Sc		15.5		
05313	09 51.1 + 23 37 208.12 + 49.89	04 − 24 − 001 154	1352	0.8 x 0.6 0.7 x 0.6	(130) ...	1	... P w 5320		14.5		*
05314	09 51.2 + 09 07 227.88 + 44.32	02 − 25 − 000	0990	1.1 x 0.1 1.0 x 0.1	168 7	2	Sc		15.7		
05315	09 51.2 + 37 32 186.21 + 51.53	06 − 22 − 025	1345 150	1.2 x 0.15 1.2 x 0.15	59 7	3	Sc		15.6		*
05316	09 51.2 + 72 27 138.78 + 39.04	12 − 10 − 009 NGC 3027	0685	5. x 2.5: 4.5: x 2.3:	130 5	4	S B c P w 5253	S B(rs)d	12.3	+ 1079 + 1226	*12 35
05317	09 51.3 + 75 15 135.95 + 37.33	13 − 07 − 000	0693	1.4 x 0.25 1.1 x 0.2	123 6	1	S ...		15.5		
05318	09 51.4 + 69 18 142.10 + 40.90	12 − 10 − 010 NGC 3031 = M 81	0685	26. x 14. 24. x 13.5	157 4 − 5	5	Sb Brightest in group, Sb −	SA(s)ab	8.1	− 44 + 88	*12− −9−
05319	09 51.4 + 76 06 135.12 + 36.81	13 − 07 − 040 NGC 3061	0693 172	2.0 x 1.8 1.8 x 1.7	-- 1	4	S B c	S B(rs)c	13.9		235
05320	09 51.5 + 23 31 208.31 + 49.95	04 − 24 − 002 154	1352	2.0 x 0.8 2.1: x 0.8	102 6	4	S B c P w 5313	S B(r)cd	14.6		*

N G C   3027,   3031,   3032,   3039 − 42,   3044,   3061

1	2	3	4	5	6	7	8	9	10	11	12
05321	09 51.6 + 37 38 186.05 + 51.61	06 – 22 – 027 IC 2515	1345	1.1 x 0.2 1.1 x 0.2	173 7	2	Sb		15.1		
05322	09 51.7 + 69 55 141.41 + 40.57	12 – 10 – 011 NGC 3034 = M 82	0685	13. x 6. 12. x 8.	65 7	5	Irr I 0 M 81 group, Ir II, eruptive		9.2 + 186 + 322		*12– –789A
05323	09 52.0 – 01 04 239.39 + 38.85	00 – 25 – 32 + 33 NGC 3047	1318	[0.9 x 0.45] ...	... ...	1	Double system		14.2		*1
05324	09 52.1 + 13 37 222.34 + 46.61	02 – 25 – 000	0990	1.2: x 0.5 0.7: x 0.3:	160 6:	1	S ...		15.3		
05325	09 52.2 + 09 30 227.58 + 44.73	02 – 25 – 055 NGC 3049	0990	2.5 x 1.6 2.4 x 1.5	25 3 – 4	4	S B b	S B(rs)ab Sb –	13.5		
05326	09 52.4 + 33 30 192.72 + 51.74	06 – 22 – 029	1345	1.1 x 0.9 1.1 x 0.9	— ...	4	Irr I m Ir I		14.5		*
05327	09 52.6 + 59 32 153.56 + 46.05	10 – 14 – 052 NGC 3043	0703	1.9 x 0.5 1.8 x 0.5	84 6	1	S ...		13.3		234 8
05328	09 52.7 + 04 31 233.53 + 42.23	01 – 25 – 034 NGC 3055	0233	2.2 x 1.2 2.2 x 1.3	63 4 – 5	3	Sc	Sc –	12.3 + 1913 + 1730		*123 468
05329	09 52.8 + 16 40 218.34 + 48.01	03 – 25 – 040 NGC 3053	1509	1.8 x 0.8 1.7 x 0.8	140 5 – 6	2	S B? a P w 5332	S (r) ...	13.7		*3
05330	09 52.8 + 19 33 214.28 + 49.05	03 – 25 – 041	1509	2. x 2. 1.5:: x 1.4::	— ...	2	... Strongly peculiar		14.8		*
05331	09 52.8 + 32 40 194.07 + 51.77	06 – 22 – 033	1345	1.0 x 0.5 1.0 x 0.3	167 6	2	Sc		15.6		
05332	09 53.0 + 16 39 218.39 + 48.05	03 – 25 – 000	1509	1.2: x 0.3: 0.8:: x 0.2::	[22] ...	3	Dwarf P w 5329		17.		*
05333	09 53.2 + 46 25 172.08 + 50.72	08 – 18 – 043	0672	1.0 x 0.15 0.9 x 0.15	[13] ...	1	... p of 2, disturbed		15.7		*1
05334	09 53.4 + 10 44 226.25 + 45.58	02 – 26 – 001 IC 577	0074	0.55 x 0.50 0.7 x 0.5	— ...	1	S ... p of 2		14.4		*
05335	09 53.5 + 27 28 202.39 + 51.27	05 – 24 – 003 IC 2520	1396	0.7 x 0.6 0.6 x 0.6	— ...	1	... Peculiar		14.3		*
05336	09 53.5 + 69 17 141.97 + 41.06	12 – 10 – 012 Ho IX	0685	3.0:: x 2.7:: 2. x 2.	— ...	4	Dwarf irregular Ir V, M 81 group		16.5:		*1
05337	09 53.6 + 10 43 226.31 + 45.62	02 – 26 – 002 IC 578	0074	1.1 x 0.45 1.3 x 0.6	72 5 – 6	1	S B: a f of 2		14.7		*
05338	09 53.6 + 17 05 217.87 + 48.35	03 – 26 – 002 NGC 3060	1356	2.4 x 0.45 2.1 x 0.45	78 6	3	Sb P w 5343		13.8		*3
05339	09 53.7 + 20 43 212.69 + 49.63	04 – 24 – 4 + 5	1352 154	[0.9 x 0.5] ...	... ...	2	Double system Connected, in group		14.4		*1
05340	09 53.8 + 29 04 199.88 + 51.60	05 – 24 – 004	1396	2.6 x 1.1: 2. x 1.	[0] ...	4	Dwarf peculiar Plumes		15.2		*1
05341	09 53.9 + 20 53 212.47 + 49.73	04 – 24 – 006	1352 154	2.9 x 0.2 2.7: x 0.2	57 7	3	Sc		15.5		
05342	09 54.0 + 15 53 219.58 + 47.97	03 – 26 – 003	1356	1.1 x 0.5 1.0 x 0.4	22 5 – 6	1	S ... P w 5344		14.3		*
05343	09 54.0 + 17 03 217.97 + 48.42	03 – 26 – 004	1356	1.2: x 0.5 1.1 x 0.4:	115 6	3	... P w 5338	S dm	15.5		*
05344	09 54.4 + 15 48 219.76 + 48.02	03 – 26 – 005	1356	1.1 x 1.0 0.9: x 0.9:	— 1	4	Sc P w 5342	SAB d	15.1		*
05345	09 54.4 + 45 29 173.45 + 51.13	08 – 18 – 047	0672	1.5 x 0.25 1.3 x 0.25	65 6 – 7	1	S ... Brightest of 4		14.4		*
05346	09 54.5 + 45 30 173.42 + 51.14	08 – 18 – 048	0672	1.0 x 0.3 1.0 x 0.3	28 6	1	S ... 5345 group		15.2		*
05347	09 54.7 + 04 46 233.62 + 42.78	01 – 26 – 003	1359	1.5 x 0.15 1.2 x 0.15	18 7	3	Sc	Sc +	15.0		
05348	09 54.8 + 47 35 170.19 + 50.69	08 – 18 – 050	0672	1.1 x 0.3: 0.9 x 0.2	120 6	1	...	d – m	16.0:		*
05349	09 55.1 + 37 31 186.19 + 52.30	06 – 22 – 043	1345 156	2.6 x 0.9 2.5: x 0.9:	38 6	3	...	S dm – m	14.5		1
05350	09 55.4 + 10 36 226.76 + 45.95	02 – 26 – 006 NGC 3070	0074 157	(1.6 x 1.6) (2.1 x 2.0)	— ...	4	E		13.2		*3
05351	09 55.4 + 32 37 194.21 + 52.37	06 – 22 – 046 NGC 3067	1345	2.2 x 0.8 2.2 x 0.8	105 6	2	Sa – b		12.7 + 1506 + 1455		*123 468
05352	09 55.5 + 16 10b 219.41 + 48.41	03 – 26 – 008 IC 581	1356	1.0 x 0.4: 1.1 x 0.8:	130 ...	1	S B: a – b Brightest in group		(15.0)		*1
05353	09 55.7 + 29 07b 199.90 + 52.02	05 – 24 – 006 NGC 3068	1396	(1.1 x 0.9) 0.9 x 0.7	— ...	1	E – S0 nf of 2, bridge, very long plume?	(15.1) + 6409 + 6340		*12 3
05354	09 55.7 + 47 58 169.55 + 50.74	08 – 18 – 051	0672	2.3 x 1.3 1.7 x 1.1	75 4:	3	S ... Companions superimposed, disturbed?		14.4		*1
05355	09 55.9 + 01 18 237.67 + 41.05	00 – 26 – 000	0470 158	1.2 x 0.3 1.3: x 0.25	108 6	1	Sc: Brightest in group		16.0:		*

N G C   3034,   3043,   3047,   3049,   3053   I C   577,   578,   581,   2515,   2520
        3055,   3060,   3067,   3068,   3070

1	2	3	4	5	6	7	8	9	10	11	12
05356	09 55.9 + 51 15 164.64 + 49.77	09 – 17 – 001	1331 163	1.2 x 0.6 1.1 x 0.6	58 5:	2	S B b – c S B(rs) ...		15.2		*
05357	09 56.0 + 05 30 233.02 + 43.46	01 – 26 – 000	1359	1.1 x 0.3 1.0 x 0.2	50 6 – 7	3	Sc		16.0:		
05358	09 56.1 + 11 38 225.56 + 46.59	02 – 26 – 009	0074 157	1.7 x 1.1 1.5 x 0.7	77 4 – 5:	2	S B b S B(s)b or SAB(s)b		14.9		*
05359	09 56.1 + 19 27 214.82 + 49.75	03 – 26 – 010	1356	1.3 x 0.45 1.3 x 0.4	97 6	2	Sb – c		15.0		
05360	09 56.2 + 14 39 221.59 + 47.95	03 – 26 – 009 NGC 3075	1356	1.2 x 0.8 1.2 x 0.8	135 3	3	Sc Sc –		14.5		*3
05361	09 56.2 + 25 28 205.73 + 51.46	04 – 24 – 008	1352	1.0: x 1.0: (1.1 x 1.0)	-- ...	1	...		16.0:		*
05362	09 56.3 + 18 03 216.87 + 49.30	03 – 26 – 011 IC 582	1356	1.1 x 1.0 1.0 x 0.9	-- 1	1	S ... Brightest of 3, disturbed:		14.7		*1
05363	09 56.4 + 18 04 216.86 + 49.33	03 – 26 – 012 IC 583	1356	1.0 x 0.2 1.0 x 0.2	114 7	2	S ... IC 582 group		15.2		*1
05364	09 56.4 + 30 59 196.90 + 52.39	05 – 24 – 008 Leo A	1396	5.5: x 3.5: 5. x 3.	... ...	5	Dwarf irregular Ir V, Local group?		14.6		*12 5
05365	09 56.6 + 03 32 235.37 + 42.48	01 – 26 – 000	1359	1.0: x 0.9: 0.8: x 0.8:	-- ...	3	Dwarf IV – V or V, 5376 group		17.		*
05366	09 56.7 + 35 38 189.27 + 52.67	06 – 22 – 047 NGC 3074	1345	2.4 x 2.2 2.6 x 2.2	(145) 1 – 2	4	Sc SAB(rs)c Sc –		14.8		*3
05367	09 56.7 + 45 31 173.26 + 51.51	08 – 18 – 52 +53	0672	[1.1 x 0.9] ...	... ...	2	Double system Distorted, 2 bridges		15.6		*1V
05368	09 56.8 + 46 29 171.75 + 51.30	08 – 18 – 000	0672	1.1 x 1.0 1.0 x 0.8	-- ...	1	...		14.8		
05369	09 56.9 + 54 47 159.52 + 48.61	09 – 17 – 004	1331 163	1.5 x 0.25 1.5 x 0.25	157 7	1	S ...		14.9		
05370	09 57.0 + 00 48 238.42 + 40.97	00 – 26 – 000	0470	1.0 x 0.6 0.8 x 0.4	160: ...	1	Sb: s of 2		16.0:		*
05371	09 57.1 + 13 13 223.63 + 47.50	02 – 26 – 014 IC 585	0990	(1.0 x 1.0) (1.4 x 1.1)	-- ...	2	E – S0 P w 5372		14.8		*
05372	09 57.3 + 13 17 223.59 + 47.60	02 – 26 – 015 NGC 3080	0074	1.0 x 0.9 0.9 x 0.9	-- 1	2	Sa P w 5371		14.5		*3
05373	09 57.4 + 05 34 233.21 + 43.78	01 – 26 – 005 Sextans B	1359	6. x 4. 6. x 4.	[110] ...	5	Dwarf irregular Ir IV – V		12.2		*12 5
05374	09 57.4 + 55 51 158.01 + 48.24	09 – 17 – 007 NGC 3073, Mark 131	1331	1.2 x 1.1 1.2 x 1.1	-- (1)	3	S0 NGC 3079 group		13.8		*12 3
05375	09 57.6 + 72 25 138.43 + 39.44	12 – 10 – 014 NGC 3065	0685	(1.8 x 1.7) (1.9 x 1.8)	-- 1	4	S0 P w 5379		12.9	+ 2051 + 2198	*123 4678C
05376	09 57.8 + 03 37 235.51 + 42.78	01 – 26 – 006	1359	2.1 x 0.7 1.8 x 0.7	151 7	1	S – Irr Brightest in group		14.5		*
05377	09 57.9 + 03 27 235.71 + 42.70	01 – 26 – 007	1359	1.6 x 1.2 1.5: x 1.0:	75: ...	4	... SAB: dm 5376 group		14.8		*
05378	09 57.9 + 04 39 234.36 + 43.38	01 – 26 – 008	1359	1.8 x 0.8 1.3 x 0.7	103 5 – 6	3	Sb		14.2		
05379	09 57.9 + 72 22 138.46 + 39.49	12 – 10 – 015 NGC 3066, Mark 133	0685	1.2 x 1.2 1.4 x 1.3	-- 1	2	S ... P w 5375, disturbed:		12.8	+ 2132 + 2279	*12 3C
05380	09 58.0 – 01 55 241.44 + 39.51	00 – 26 – 006	0470	1.1 x 0.8 1.1 x 0.8	155: 2 – 3	3	S B a – b S B(r) ...		14.3		
05381	09 58.0 + 22 33 210.41 + 51.13	04 – 24 – 009	1352	1.1 x 0.6 1.0 x 0.4	103 5 – 6	1	S ... P w 5384ab		15.4		*
05382	09 58.0 + 36 51 187.25 + 52.91	06 – 22 – 050	1345 156	1.0 x 0.6 0.9 x 0.6	65 4	2	Sa – b		14.7		*
05383	09 58.1 + 04 58 234.03 + 43.60	01 – 26 – 009	1359	1.2 x 0.9 (1.3 x 1.0)	155: 2 – 3:	2	S0		14.6		*
05384	09 58.3 + 22 38 210.31 + 51.22	04 – 24 – 10 +11 NGC 3088	1352	[1.4 x 0.7] ...	... ...	3	Double system Contact, p w 5381		14.7		*13
05385	09 58.4 + 17 10 218.41 + 49.44	03 – 26 – 014	1356	1.4: x 1.0: 1.1 x 0.9:	... ...	2	Sc		15.3		
05386	09 58.4 + 70 59 139.85 + 40.38	12 – 10 – 016	0685	1.2 x 0.8 1.2 x 0.8	178 3	2	Sb		15.2		1
05387	09 58.5 + 55 55 157.82 + 48.35	09 – 17 – 010 NGC 3079	1331 163	8.7 x 1.6 8.1 x 1.5	165 7	4	... S m Brightest of 3		11.2	+ 1171 + 1240	*12 -689
05388	09 58.6 + 00 28 239.09 + 41.09	00 – 26 – 010	0470 158	1.1 x 0.9 (1.1 x 1.0)	-- ...	1	S? ...		15.0		*
05389	09 58.6 + 39 52 182.25 + 52.81	07 – 21 – 005	0711	2.0 x 0.15 1.9 x 0.15	122 7	3	Sc		15.7		
05390	09 58.7 + 16 01 220.08 + 49.06	03 – 26 – 015 NGC 3094	1356	(1.6 x 1.2) 1.4 x 0.8	75: ...	4	S B a S B(s)a		13.5		

N G C   3065,   3066,   3073 – 75,   3079   I C   582,   583,   585
        3080,   3088,   3094

1	2	3	4	5	6	7	8	9	10	11	12
05391	09 58.7 + 37 29 186.19 + 53.02	06 – 22 – 054	1345 156	2.3 x 0.9 2.2 x 0.9:	170 6	4	... P w IC 2530	S m	14.9		*1
05392	09 58.8 + 21 51 211.56 + 51.11	04 – 24 – 000	1352	1.2 x 0.12 1.2 x 0.12	11 7	3	Sc		16.0:		
05393	09 58.8 + 33 22 193.04 + 53.07	06 – 22 – 055	1345	2.3: x 1.3 2.1: x 1.1:	... ...	3	...	S B(r:)dm:	14.9		
05394	09 58.8 + 36 45 187.40 + 53.07	06 – 22 – 000	1345 156	1.6 x 0.2 1.2 x 0.2	56 7	2	Sc		16.5:		*
05395	09 58.9 + 13 00 224.23 + 47.82	02 – 26 – 018	0074	(1.0 x 0.8) (1.2 x 0.9)	40: ...	2	E – S0		14.6		*
05396	09 59.0 + 11 00 226.86 + 46.92	02 – 26 – 019	0074 157	1.7 x 0.45 1.6 x 0.5	162 6	3	Sc		14.6		*
05397	09 59.4 + 24 57 206.80 + 52.05	04 – 24 – 012 NGC 3098	1352	2.3 x 0.6 2.3 x 0.5	90 7	1	S0 – a		13.0		*12 3468
05398	09 59.4 + 68 58 141.90 + 41.67	12 – 10 – 017 NGC 3077	0685	(6.0 x 4.5) (6.5 x 5.0)	45 ...	5	Irr I 0 M 81 group Ir II		10.7	– 158 – 26	*12– –789
05399	09 59.6 + 03 17 236.23 + 42.95	01 – 26 – 010 IC 588	1359	1.1 x 1.0 0.9 x 0.9	— 1	4	S B a 5376 group	(R)S B(r)a	14.9		*
05400	09 59.6 + 13 56 223.09 + 48.39	02 – 26 – 020	0074	(1.0 x 0.8) (1.2 x 1.0)	(160) 2:	2	S0		14.3		
05401	09 59.8 + 19 15 215.58 + 50.50	03 – 26 – 017	1356	1.4: x 0.6 1.1: x 0.6:	[65] ...	3	Dwarf		17.		
05402	09 59.8 + 79 30 131.60 + 34.90	13 – 07 – 043	0693 172	1.6 x 1.1 1.4 x 0.9	175: 3	3	Sb		14.8		
05403	09 59.9 + 19 25 215.35 + 50.58	03 – 26 – 018	1356	1.3 x 0.6 1.3 x 0.6	80 6	2	S0 – a		14.6		*
05404	10 00 + 80 32 130.68 + 34.20	14 – 05 – 010 NGC 3057	0727	2.8: x 2.3: 2.4: x 1.6:	(5) ...	4	... S IV	S B(s)dm	14.2		*3
05405	10 00.0 + 54 10 160.11 + 49.27	09 – 17 – 015	1331 163	1.2 x 0.5 0.8 x 0.4	60 6	1	S0 – a		15.6		*
05406	10 00.1 + 63 50 147.63 + 44.66	11 – 12 – 033	0708	1.3 x 0.3 1.0 x 0.2	62 6 – 7	2	Sc		16.5:		
05407	10 00.2 + 33 17 193.20 + 53.36	06 – 22 – 060	1345	1.1 x 0.4 1.1 x 0.4	176 6	1	S ...		16.0:		
05408	10 00.3 + 59 40 152.72 + 46.84	10 – 15 – 004 Mark 25	0962	0.55 x 0.45 0.5 x 0.4	— ...	2	... Very compact		14.2	+ 27300	*C
05409	10 00.4 + 10 59 227.13 + 47.21	02 – 26 – 021	0074 157	1.3 x 0.3 1.1 x 0.3	165 6	1	S ...		15.0		
05410	10 00.5 + 50 38 165.16 + 50.66	08 – 18 – 061	0672 163	1.2 x 0.5 1.0 x 0.5	25 5	3	Sc		16.0:		*1
05411	10 00.6 – 02 08 242.19 + 39.88	00 – 26 – 012	0470	1.5 x 0.7 1.4 x 0.6	107 5:	1	S ...		14.3		
05412	10 00.6 + 50 36 165.19 + 50.71	08 – 18 – 061	0672 163	1.2 x 0.5 1.0 x 0.5	25 5	3	Sc		16.0:		1
05413	10 00.8 + 13 20 224.09 + 48.38	02 – 26 – 000	0074	1.1 x 0.1 0.8 x 0.10	15 7	2	Sc		17.		*
05414	10 00.9 + 41 00 180.29 + 53.10	07 – 21 – 007 NGC 3104	0711	3.6 x 2.3 3.0: x 2.0:	[35:] ...	4	Irr IA m Ir I		14.2	+ 725 + 718	*13 VA
05415	10 00.9 + 72 26 138.21 + 39.63	12 – 10 – 018	0685	1.1 x 0.20 1.2 x 0.22	97 6	1	S ...		14.5		
05416	10 01.0 + 39 23 182.97 + 53.32	07 – 21 – 008	0711	1.0: x 0.4 0.8 x 0.3	1 6?	1	S B? ...		15.7		
05417	10 01.0 + 53 38 160.77 + 49.62	09 – 17 – 017	1331 163	1.5 x 1.1 0.7 x 0.5	(135) ...	4	S B b	S B(s)b	15.7		*
05418	10 01.0 + 60 21 151.79 + 46.58	10 – 15 – 007 NGC 3102	0962	0.9 x 0.9 0.9 x 0.9	— ...	2	E – S0		14.3		3
05419	10 01.1 + 31 25 196.34 + 53.43	05 – 24 – 009 NGC 3106	1396	1.8 x 1.8 1.7 x 1.7	— 1	3	S0		14.0		*13
05420	10 01.2 + 22 31b 210.79 + 51.83	04 – 24 – 013	1352	1.1 x 0.8 1.1 x 0.8	(75) ...	1	S0: n of 2		(14.5)		*1
05421	10 01.3 + 55 34 158.05 + 48.86	09 – 17 – 018	1331 163	2.0 x 0.3 2.0: x 0.2	172 (7)	2	Sc – Irr		16.5:		
05422	10 01.4 + 57 50 154.99 + 47.85	10 – 15 – 009	0962	1.1 x 0.8 1.0 x 0.8	(170) 1 – 2	4	S B b	S B(s)b	15.6		
05423	10 01.4 + 70 37 140.03 + 40.80	12 – 10 – 021	0685	1.3: x 1.0: 1.3: x 1.0:	(140) ...	4	Irr I m Ir IV: Ir I, nf of 2		15.3		*
05424	10 01.7 + 03 55 235.93 + 43.75	01 – 26 – 000	1359	1.1 x 0.5 1.0 x 0.3	79 6	3	Sc		15.3		
05425	10.01.7 + 13 52 223.51 + 48.82	02 – 26 – 022	0074	0.8 x 0.7 0.9 x 0.7	(140) ...	1	...		13.6		

N G C   3057, 3077, 3098, 3102, 3104, 3106   I C   588

1	2	3	4	5		6	7	8	9	10	11	12
05426	10 01.7 + 15 00 221.95 + 49.30	03 - 26 - 000	1356 159	1.5 x 0.10 1.4 x 0.12		177 7	3	Sc		15.4		*
05427	10 01.7 + 29 37 199.37 + 53.38	05 - 24 - 010	1396	1.3 x 0.9 1.4 x 1.0		120: 3:	2	...	S dm:	15.2		1
05428	10 01.7 + 66 47 144.12 + 43.14	11 - 12 - 000	0708	1. x 1. 1. x 1.		-- ...	4	Dwarf V		18.		*
05429	10 01.8 + 44 57 173.82 + 52.52	08 - 19 - 000	1348	1.1 x 0.9 1.1: x 0.9:		(0) ...	2	Sc - Irr		17.		
05430	10 02.1 + 44 45 174.12 + 52.61	08 - 18 - 066	0672	1.4 x 0.1 0.9 x 0.1		40 7	3	Sc	Sc +	17.		*
05431	10 02.2 + 21 45 212.09 + 51.83	04 - 24 - 014	1352	1.3 x 0.2 1.3 x 0.2		48 7	3	Sc P w 5434		16.0:		*
05432	10 02.3 + 05 18 234.46 + 44.65	01 - 26 - 013	1359	(1.1 x 1.1) (1.2 x 1.1)		-- ...	1	E:		14.7		
05433	10 02.3 + 22 22 211.14 + 52.03	04 - 24 - 000	1352	1.4: x 0.4 0.8 x 0.2		[37] ...	3	Dwarf f of 2		17.		*
05434	10 02.5 + 21 42 212.20 + 51.89	04 - 24 - 015	1352	1.4 x 1.0 1.4 x 1.0:		25: 3	3	Sb/S B b P w 5431	SAB(s)b	14.8		*
05435	10 02.5 + 59 03 153.31 + 47.39	10 - 15 - 013	0962	0.8 x 0.4 0.7 x 0.35		40 ...	1	...		13.6		
05436	10 02.6 + 19 31 215.54 + 51.22	03 - 26 - 023	1356	1.5: x 0.6 0.9 x 0.25		[130] ...	1	S ... nnp of 2, distorted		14.3		*1
05437	10 02.7 + 27 45 202.51 + 53.33	05 - 24 - 000	1396	1.3 x 0.4 1.0: x 0.4:		112 (6)	1	...	S dm:	17.		
05438	10 03 + 85 04 126.81 + 31.04	14 - 05 - 011	0727	1.3 x 0.4 0.8 x 0.3		100 6	1	S b - c		16.5:		
05439	10 03.0 + 00 19 240.15 + 41.87	00 - 26 - 017 Sextans C	0470	3.: x 3.: 4.: x 4.:		-- ...	5	Dwarf V		15.5		1
05440	10 03.0 + 04 31 235.51 + 44.36	01 - 26 - 000	1359	1.3 x 0.5 1.1 x 0.3		97 6	3	Sc	Sc +	16.5:		
05441	10 03.0 + 47 30 169.71 + 52.05	08 - 19 - 002 NGC 3111	1348	(0.9 x 0.8) (1.0 x 1.0)		-- ...	2	E - S0		14.0		*3
05442	10 03.0 + 68 05 142.59 + 42.46	11 - 12 - 000	0708	2. x 1. 2. x 1.		... ...	4	Dwarf V		18.		
05443	10 03.3 + 00 53 239.60 + 42.28	00 - 26 - 018 IC 590	0470 158	[1.1 x 1.0] ...		... ...	2	Double system Contact, common envelope		14.2		*1
05444	10 03.5 + 14 53 222.40 + 49.65	03 - 26 - 000	1356 159	1.0 x 0.1 0.9 x 0.12		28 7	1	Sb		16.0:		
05445	10 03.6 + 03 08 237.20 + 43.68	01 - 26 - 014 NGC 3117	1359	(1.0 x 1.0) (1.0 x 1.0)		-- ...	1	E?		14.6		
05446	10 03.6 + 33 11 193.41 + 54.06	06 - 22 - 071	1345	1.5 x 0.3 1.2 x 0.3		51 6 - 7	3	Sc		15.3		
05447	10 03.7 - 01 18 241.98 + 41.00	00 - 26 - 019	0470	[1.0 x 0.7] ...		... ...	2	Double system Contact		15.2		*1
05448	10 03.9 + 14 41b 222.74 + 49.65	03 - 26 - 026	1356	1.1 x 0.8 1.0 x 0.7		175: 3:	1	Sb: nf of 2, contact, p w 5450		(15.1)		*1
05449	10 04.0 + 68 36 141.96 + 42.22	12 - 10 - 000	0685	1.0 x 0.7 0.9 x 0.6		70 3	2	S B? c		15.3		*
05450	10 04.1 + 14 37 222.86 + 49.66	03 - 26 - 027 NGC 3121	1356 159	(1.1 x 0.8) (1.1 x 0.9)		(20) ...	3	E Companion superimposed, p w 5436		14.2		*13
05451	10 04.2 + 47 16 169.98 + 52.31	08 - 19 - 004	1348	1.6 x 0.8 1.6 x 0.8		[103] ...	2	Irr	Ir I	14.1		
05452	10 04.3 + 33 16 193.27 + 54.21	06 - 22 - 074 NGC 3118	1345	2.5 x 0.35 2.2 x 0.3		41 7	1	Sb - c		14.4		
05453	10 04.4 + 16 14 220.62 + 50.40	03 - 26 - 028	1356	1.1: x 0.9: 0.9 x 0.8		(80) ...	3	Irr	Ir I	15.1		
05454	10 04.5 + 12 54 225.29 + 48.99	02 - 26 - 000	0074	1.2 x 0.8 1.1: x 0.8:		40: ...	3	...	SAB dm - m	15.3		
05455	10 04.6 + 70 53 139.53 + 40.84	12 - 10 - 000	0685	2. x 2. 2. x 2.		-- ...	4	Dwarf irregular Ir V		17.		
05456	10 04.7 + 10 36 228.38 + 47.95	02 - 26 - 024	0074	1.7 x 0.8 1.5 x 0.7		148 ...	2	... Strongly peculiar, Ir II?		13.5		*
05457	10 04.7 + 67 22 143.24 + 43.03	11 - 12 - 036	0708	1.0 x 0.2 0.8 x 0.2		122 7:	1	S ...		16.0:		
05458	10 04.8 + 12 31 225.86 + 48.88	02 - 26 - 025	0074	1.3 x 0.8 1.3 x 0.8		170: 4	1	S ...		14.0		*
05459	10 04.8 + 53 20 160.84 + 50.26	09 - 17 - 027	1331 163	4.3 x 0.5 4.4 x 0.5		132 7	3	Sc		13.8		
05460	10 04.9 + 52 06 162.61 + 50.77	09 - 17 - 028	1331 163	2.6 x 2.4 2.3: x 2.3:		-- 1	4	S B c	S B(rs)d	13.9		*1

1	2	3	4	5	6	7	8	9	10	11	12
05461	10 05.0 + 29 43 199.34 + 54.10	05 – 24 – 015	1396	1.1 x 0.2 0.9 x 0.2	1 7	1	Sb – c P w 5464		15.4		*1
05462	10 05.1 + 42 40 177.32 + 53.59	07 – 21 – 011	0711	1.0 x 0.25 0.9 x 0.25	107 6	2	Sa		14.9		
05463	10 05.2 + 78 25 132.34 + 35.84	13 – 08 – 006	1326 172	1.2 x 0.4? 1.1 x 0.4	32 (6)	1	...		15.6		
05464	10 05.3 + 29 48 199.21 + 54.17	05 – 24 – 017	1396	1.5 x 0.9 1.5: x 1.1:	(135) ...	3	Dwarf spiral S IV – V or V, p w 5461		16.5:		*
05465	10 05.4 – 02 15 243.31 + 40.73	00 – 26 – 020 IC 592	0470	0.9 x 0.7 1.0 x 0.8	-- ...	1	Sb – c P w 5469, disturbed:		14.0		*
05466	10 05.4 + 32 06 195.29 + 54.39	05 – 24 – 019 NGC 3126	1396	2.9 x 0.5 2.8 x 0.6	123 6	3	Sb		13.5		13
05467	10 05.5 + 18 57 216.77 + 51.66	03 – 26 – 031	1356	0.8 x 0.8 0.6 x 0.5	-- ...	1	S0?		14.3		1
05468	10 05.6 + 10 13 229.04 + 47.95	02 – 26 – 026 NGC 3130	0074	1.1 x 0.6 1.1 x 0.6	30 4 – 5	2	S0 – a		14.3		*3
05469	10 05.8 – 02 17 243.43 + 40.78	00 – 26 – 021 IC 593	0470	0.8 x 0.6 0.8 x 0.7	-- ...	1	S ... P w 5465, disturbed:		14.2		*
05470	10 05.8 + 12 33 225.99 + 49.11	02 – 26 – 027 Leo I	0074	(12. x 9.) (15. x 11.)	(80) ...	5	Dwarf elliptical Local group		11.3		*12
05471	10 05.9 + 18 29 217.53 + 51.59	03 – 26 – 033 NGC 3131	1356	2.3 x 0.7 2.5 x 0.8	54 6	3	S B: b		14.0		*13
05472	10 06.0 – 00 25 241.55 + 42.01	00 – 26 – 023 IC 594	0470	1.1 x 0.5 1.1 x 0.55	127 5	3	S B b – c S B(r) ...		14.7		
05473	10 06.1 + 09 42 229.80 + 47.80	02 – 26 – 028	0074	1.1 x 0.5 1.0 x 0.30	65 6	1	S ... spp of 2		14.9		*
05474	10 06.1 + 32 45 194.18 + 54.57	06 – 22 – 077	1345	1.5 x 1.4 1.2 x 1.2	-- 1	4	Sc/S B c SAB(rs)cd sf of 2		15.0		*
05475	10 06.1 + 58 42 153.41 + 47.97	10 – 15 – 019	0962	1.7 x 0.8 0.9 x 0.7	30 5	3	S B c S B(s)c Brightest of 3		15.5		*
05476	10 06.2 + 40 37 180.68 + 54.15	07 – 21 – 012	0711	1.4 x 1.3 1.3 x 1.3	-- 1	3	Sb/S B b Disturbed?		14.6		*1
05477	10 06.5 + 15 15 222.35 + 50.46	03 – 26 – 034	1356 159	(0.8 x 0.8) (1.0 x 1.0)	-- ...	3	E npp of 2		14.5		*1
05478	10 06.6 + 30 24 198.24 + 54.52	05 – 24 – 020	1396	1.8: x 1.6: 1.7: x 1.6:	-- ...	4	Dwarf irregular Ir V, p w 5481		15.5		*
05479	10 06.8 + 54 44 158.67 + 49.93	09 – 17 – 034	1331 163	1.2 x 0.7 1.2 x 0.7	[100] ...	3	S ... 1st in compact group		14.8		*1
05480	10 06.8 + 58 44 153.30 + 48.03	10 – 15 – 022	0962	1.0 x 0.7 1.0 x 0.4	170 ...	2	... S d – m 5475 group, np of 2		15.6		*
05481	10 06.9 + 30 34 197.96 + 54.60	05 – 24 – 021	1396	1.6 x 0.5 1.7 x 0.7	91 6	1	Sa: P w 5478		15.1		*
05482	10 07.3 + 32 32 194.58 + 54.81	06 – 22 – 000	1345	1.3 x 0.8 1.1 x 0.7	175: 4:	3	Sc		16.5:		
05483	10 07.5 – 02 13 243.72 + 41.15	00 – 26 – 024	0470 170?	1.0 x 1.0 1.0 x 1.0	-- 1	1	S ...		15.5		*
05484	10 07.6 + 28 12 202.05 + 54.47	05 – 24 – 023 IC 2550	1396	1.1 x 0.9 1.1 x 0.9	135: 2:	2	Sb: SAB(rs) ...		14.6		
05485	10 07.7 + 65 31 145.04 + 44.37	11 – 13 – 003	0662	1.2 x 0.4 1.2 x 0.4	127 ...	1	... Peculiar, eruptive?		15.1		*
05486	10 07.8 + 46 12 171.37 + 53.21	08 – 19 – 007 NGC 3135	1348	1.0 x 0.6 0.8 x 0.5	90 4:	1	S ...		14.3		3
05487	10 07.9 + 02 29 238.81 + 44.16	00 – 26 – 025	0470	1.1 x 0.5 1.0 x 0.5	70 5 – 6	2	S B b – c Brightest of 3, disturbed:		15.2		*1
05488	10 07.9 + 24 40 208.00 + 53.86	04 – 24 – 016 IC 2551	1352	1.3 x 1.1 1.1 x 1.0	-- ...	1	...		14.6		
05489	10 08.0 + 20 18 215.02 + 52.67	03 – 26 – 037	1356	2.2 x 1.1 2.2 x 1.1	0 5	3	Sa		14.0		
05490	10 08.6 + 31 02 197.21 + 55.00	05 – 24 – 024	1396	1.4 x 1.1 1.3 x 1.1	10: 2	4	Sc		15.0		
05491	10 08.6 + 59 07 152.63 + 48.03	10 – 15 – 028 Mark 26	0962	1.0 x 0.5 0.9 x 0.5	72 5	2	Sb sf of 2		15.5		*
05492	10 08.6 + 59 58 151.55 + 47.59	10 – 15 – 027	0962	1.4 x 0.5 1.4 x 0.5	25 6	2	Sa – b np of 2		16.0:		*
05493	10 08.8 + 00 41 240.97 + 43.24	00 – 26 – 026	0470	1.7 x 1.4 1.6 x 1.3	(15) ...	4	Sb/S B c SAB(rs)bc f of 2 Sb +		14.0		*1
05494	10 09.0 + 67 40 142.57 + 43.17	11 – 13 – 000	0662	1.1: x 0.2 0.8 x 0.2	[172] ...	1	Interacting pair? or Peculiar?		16.0:		
05495	10 09.1 + 16 41 220.68 + 51.61	03 – 26 – 039	1356	3.0 x 0.35 2.6 x 0.35	101 6 – 7	2	Sc		14.7		

N G C 3126, 3130, 3131, 3135   I C 592 – 594, 2550, 2551

1	2	3	4	5	6	7	8	9	10	11	12
05496	10 09.1 + 46 33 170.70 + 53.32	08 – 19 – 009	1348	1.4 x 0.3 0.9 x 0.25	152 (6)	2	Sc – Irr		15.5		
05497	10 09.1 + 64 22 146.22 + 45.17	11 – 13 – 007	0662	1.1: x 1.1: 1.1: x 1.1:	-- (1)	1	...		15.7		
05498	10 09.3 + 23 20 210.32 + 53.85	04 – 24 – 017	1352 164	1.6 x 0.25 1.6 x 0.25	63 (7)	2	Sa: + companion Connected		15.3		*1
05499	10 09.4 + 28 07 202.29 + 54.85	05 – 24 – 025	1396	3.0 x 0.8 2.7 x 0.8	42 6	2	S B b:		14.3		
05500	10 09.5 + 78 03 132.47 + 36.25	13 – 08 – 009 NGC 3197	1326 172	1.4 x 1.1 1.4 x 1.1	155 2	3	Sb	Sb +	14.5		*3
05501	10 09.6 + 05 10 236.08 + 46.08	01 – 26 – 018	1359	0.6 x 0.6 0.55 x 0.50	-- ...	1	...		14.5		1
05502	10 09.9 + 43 24 175.79 + 54.29	07 – 21 – 016 IC 598	0711	1.6 x 0.45 1.5 x 0.45	8 6	2	SO – a		13.8		
05503	10 10.1 + 03 22 238.27 + 45.13	01 – 26 – 019 NGC 3156	1359	1.8 x 1.1 2.0 x 1.1	47 4 – 5	3	SO NGC 3166 group		12.8		*12 34$\underline{8}$
05504	10 10.1 + 07 21 233.54 + 47.40	01 – 26 – 000	1359	1.5: x 0.2 1.0 x 0.2	81 7	3	...	S d – dm	15.6		
05505	10 10.2 + 12 55 226.26 + 50.23	02 – 26 – 032 NGC 3153	0074	2.3 x 1.0 2.2 x 0.9	170 5 – 6	3	Sc p of 2		13.6		*3
05506	10 10.3 + 05 02 236.38 + 46.14	01 – 26 – 021	1359	1.4 x 0.7 1.3 x 0.7	17 5	2	S ... Companion superimposed, sf of 2		14.9		*1
05507	10 10.3 + 17 17b 219.96 + 52.11	03 – 26 – 040 NGC 3154	1356	1.0 x 0.45 0.9 x 0.35	124 6	1	S ... Brightest in group		(14.3)		*1
05508	10 10.3 + 69 22 140.67 + 42.18	12 – 10 – 022	0685	1.3 x 1.3 1.2 x 1.1	-- 1	3	Sc		15.4		*
05509	10 10.8 + 20 25 215.19 + 53.33	03 – 26 – 041	1356	1.6 x 0.25 1.5 x 0.20	66 7	2	Sb		14.9		
05510	10 10.8 + 22 59 211.05 + 54.09	04 – 24 – 019 NGC 3162	1352 164	3.4 x 2.8 3.2: x 2.8	-- 2:	5	Sc	SAB(rs)bc Sc –	12.2	+ 1456 + 1361	*123 46$\underline{78}$
05511	10 10.8 + 39 01 183.21 + 55.25	07 – 21 – 020 NGC 3158	0711 161	(2.3 x 2.2) (2.5 x 2.3)	-- ...	4	E Brightest in group		13.4	+ 7024 + 7009	*12 34$\underline{78}$
05512	10 10.9 + 03 38 238.14 + 45.45	01 – 26 – 023 NGC 3165	1359	1.6 x 0.8 1.1 x 0.6	[177] (5)	3	Irr: dm – m NGC 3166 group		14.5		*12 3
05513	10 10.9 + 39 06 183.06 + 55.25	07 – 21 – 023 NGC 3160	0711 161	1.4 x 0.25 1.4 x 0.25	140 7	1	S ... NGC 3158 group, disturbed		15.2		*3
05514	10 11.0 + 18 22 218.42 + 52.67	03 – 26 – 042	1356	1.2 x 0.15 1.2 x 0.1	139 7	2	Sc p of 2		15.5		*1
05515	10 11.1 – 00 41 242.92 + 42.83	00 – 26 – 030	0470 162	(1.4 x 1.1) (1.6 x 1.3)	(90) ...	3	E or D Brightest in cluster		14.4		*
05516	10 11.1 + 03 40 238.14 + 45.51	01 – 26 – 024 NGC 3166	1359	(5.0 x 2.8) (5.0 x 2.8)	87 5	4	SO/Sa Brightest in group		11.1	+ 1381 + 1200	*12– –6$\underline{89}$
05517	10 11.1 + 38 55 183.37 + 55.32	07 – 21 – 026 NGC 3163	0711 161	(1.1 x 1.1) (1.3 x 1.2)	-- ...	2	E – SO NGC 3158 group		14.4	+ 6245 + 6229	*12 3
05518	10 11.2 + 39 43 181.98 + 55.23	07 – 21 – 028	0711 161	2.1: x 1.2: 1.5:: x 1.0::	[60] ...	4	Dwarf irregular Ir V, sp of 2		17.		*
05519	10 11.2 + 74 29 135.60 + 38.29	12 – 10 – 023 NGC 3144 = 3174	0685 172	1.4 x 0.8 1.3 x 0.7	0 4 – 5	2	S B a – b		14.3		23
05520	10 11.3 + 65 24 144.85 + 44.74	11 – 13 – 013	0662	2.2 x 1.4 2.0 x 1.2	100: 4	3	Sc		14.4		
05521	10 11.4 + 00 47 241.42 + 43.81	00 – 26 – 032	0470 162	1.0 x 0.9 1.1 x 1.0	-- 1	3	Sc	SAB(s)c	15.0		
05522	10 11.4 + 07 16 233.90 + 47.62	01 – 26 – 025	1359	3.0 x 1.8 1.7 x 0.7	145: 4:	3	Sc	Sc +	14.4		*
05523	10 11.4 + 30 26 198.35 + 55.55	05 – 24 – 000	1396	1.0 x 0.7 0.9 x 0.7	50: 2 – 3:	1	S: ...		15.6		
05524	10 11.6 + 22 22 212.15 + 54.10	04 – 24 – 020	1352 164	1.7 x 0.3 1.8: x 0.3	43 7	2	Sc		15.7		
05525	10 11.7 + 03 43 238.21 + 45.66	01 – 26 – 026 NGC 3169	1359	5.5: x 3.0: 5. x 3.	45: ...	3	Sa NGC 3166 group P w 5516, distorted		11.9	+ 1298 + 1116	*12– –6$\underline{89}$
05526	10 11.7 + 16 09 221.87 + 51.97	03 – 26 – 047	1356	1.0 x 0.45 0.8 x 0.35	98 6	1	S ...		14.8		
05527	10 11.8 + 56 55 155.18 + 49.52	10 – 15 – 036 NGC 3164	0962	0.9 x 0.7 0.9 x 0.6	0 2	1	S ...		14.5		*3
05528	10 12.1 – 00 35 243.03 + 43.08	00 – 26 – 033	0470 162	1.1 x 0.9 1.0 x 1.0	-- ...	1	... 2nd brightest in cluster		14.8		*1
05529	10 12.3 + 21 25 213.77 + 53.98	04 – 24 – 021	1352 164	1.4 x 0.3 1.3 x 0.3	21 (6)	2	Sc – Irr		15.4		
05530	10 12.5 + 66 10 143.90 + 44.37	11 – 13 – 014	0662	1.4 x 0.5 1.3 x 0.4	95 6	1	S ...		15.1		

N G C   3144, 3153, 3154, 3156, 3158, 3160, 3162 – 66
3169, 3174, 3197

I C   598

1	2	3	4	5	6	7	8	9	10	11	12
05531	10 12.7 + 44 14 174.18 + 54.57	07 – 21 – 033	0711	1.1 x 0.2 1.2 x 0.2	27 7	1	S B: ...		15.4		
05532	10 12.7 + 73 38 136.30 + 39.48	12 – 10 – 025 NGC 3147	0685	4.7 x 4.0 4.7 x 4.0	(155) 2	5	Sb/Sc	SA(rs)bc Sb +	11.3	+ 2721 + 2875	*123 4678
05533	10 12.8 + 14 17 224.78 + 51.42	02 – 26 – 000	0074	1.2 x 0.25 0.9 x 0.25	98 6	2	Sb		15.4		
05534	10 12.8 + 58 40 152.78 + 48.73	10 – 15 – 048	0962	1.6 x 0.7 1.4 x 0.7	98 5 – 6	2	Sb/S B b P w 5541	SAB ...	14.9		*1
05535	10 13.0 + 19 11 217.43 + 53.40	03 – 26 – 049	1356	1.2 x 0.2 1.1 x 0.25	127 7	1	S0 – a		15.2		
05536	10 13.0 + 60 29 150.47 + 47.77	10 – 15 – 052 NGC 3168	0962	1.0 x 0.9 1.2 x 1.1	–– ...	1	E Brightest in group	Compact	14.6		*3
05537	10 13.1 + 07 34 233.87 + 48.14	01 – 26 – 027	1359	2.5 x 0.2 2.3 x 0.2	147 7	3	Sc	Sc +	15.1		
05538	10 13.2 + 74 36 135.37 + 38.83	12 – 10 – 026 NGC 3155 = 3194	1326 172	1.6 x 1.1 1.5 x 0.8	35 3:	1	S ...		13.9		3
05539	10 13.3 + 02 56 239.45 + 45.51	01 – 26 – 029	1359	2.2: x 1.1: 1.1:: x 0.9::	... ...	3	Dwarf I:B m IV – V or V		15.3		*1
05540	10 13.4 + 38 01 184.86 + 55.87	06 – 23 – 004	1032 167	1.7 x 0.3 1.5 x 0.2	111 7	1	Sc:		14.6		
05541	10 13.5 + 58 39 152.73 + 48.82	10 – 15 – 055	0962	1.5 x 0.7 1.3 x 0.6	[20] ...	3	Irr P w 5534	Ir I	15.6		*1
05542	10 13.5 + 60 32 150.36 + 47.79	10 – 15 – 054	0962	1.0 x 1.0 1.0 x 1.0	–– ...	2	[E] Compact, NGC 3168 group		14.9		*1
05543	10 13.7 + 05 04 237.05 + 46.85	01 – 26 – 031	1359	1.4 x 0.8 1.2 x 0.7	167 4	3	Sc	SAB c	14.6		1
05544	10 13.8 + 21 22 214.03 + 54.29	04 – 24 – 023 NGC 3177	1352 164	1.6 x 1.3 1.4 x 1.1	(135) 1 – 2	3	Sb NGC 3189 – 90 group		12.8	+ 1220 + 1118	*123 4678
05545	10 13.8 + 49 53 165.04 + 52.92	08 – 19 – 014	1348	1.0 x 0.8 0.9 x 0.7	130: 2:	1	S ...		14.4		*1
05546	10 13.8 + 55 04 157.48 + 50.67	09 – 17 – 047	1331 163	1.2 x 0.8 1.2 x 0.8	150 4	3	S B b	Sb –	15.2		
05547	10 13.9 + 17 13 220.61 + 52.88	03 – 26 – 050	1356	(1.1 x 0.7) 0.9 x 0.5	125: ...	1	S0:		14.9		1
05548	10 14.0 + 12 50 227.05 + 51.02	02 – 26 – 000	0074	1.1 x 0.2 1.2 x 0.35	67 6	1	S ... s of 2		15.1		*
05549	10 14.0 + 53 43 159.35 + 51.33	09 – 17 – 048	1331 163	1.2 x 0.9 1.1 x 0.8	55: 3	3	Sb	Sb +	15.3		
05550	10 14.1 + 64 40 145.42 + 45.42	11 – 13 – 000	0662	1.0 x 0.1 1.0 x 0.1	101 7	2	Sc		17.		
05551	10 14.6 + 04 35 237.82 + 46.75	01 – 26 – 000	1359	1.0: x 0.8: 0.9: x 0.7:	–– ...	2	Dwarf sf of 2		17.		*
05552	10 14.7 + 17 20 220.55 + 53.10	03 – 26 – 051	1356	1.6 x 0.35 1.4 x 0.3	32 6	2	Sb – c		15.0		
05553	10 14.7 + 60 36 150.15 + 47.87	10 – 15 – 057	0962	1.1 x 0.2 1.2 x 0.2	78 7	2	Sa – b		16.0:		
05554	10 14.9 + 21 56 213.23 + 54.71	04 – 24 – 024 NGC 3185	1352 164	2.0 x 1.2 2.8 x 1.8	130 4 – 5	5	S B a NGC 3189 – 90 group	(R)S B(r)a	12.9	+ 1241 + 1142	*12– -789
05555	10 15.0 + 41 22 178.89 + 55 65	07 – 21 – 036 NGC 3179	0711	2.0 x 0.5 1.9 x 0.5	48 6	3	S0		14.2		
05556	10 15.1 + 22 07 212.95 + 54.80	04 – 24 – 025 NGC 3187	1352 164	3.5 x 1.5 3.5 x 1.5	... ...	3	S ... NGC 3189 – 90 group Distorted, P w 5559		13.8	+ 1589 + 1491	*12V 3569A
05557	10 15.3 + 41 40 178.35 + 55.64	07 – 21 – 037 NGC 3184	0711	8.5 x 7.8 8.0 x 7.5	(135) 1	4	Sc	SA(rs)cd	10.4	+ 418 + 418	*12– -9
05558	10 15.3 + 46 13 170.68 + 54.44	08 – 19 – 015	1348	1.2 x 0.25 1.1 x 0.2	12 7	2	Sa		15.3		
05559	10 15.4 + 22 05 213.04 + 54.86	04 – 24 – 026 NGC 3189 = 3190	1352 164	4.5 x 1.7 4.5 x 1.7	125 6	4	Sa Brightest in group Disturbed, p w 5556		11.9	+ 1354 + 1255	*12–V -789A
05560	10 15.5 + 34 52 190.45 + 56.53	06 – 23 – 005	1032	1.0 x 0.5 1.0: x 0.5	86 5	1	S ...		15.6		
05561	10 15.7 + 07 18 234.74 + 48.53	01 – 26 – 034 IC 602	1359	0.9 x 0.55 0.9 x 0.5	177: ...	1	S: ... P w IC 601		13.4		*1
05562	10 15.7 + 22 09 212.97 + 54.95	04 – 24 – 027 NGC 3193	1352 164	(2.5 x 2.5) (2.8 x 2.6)	–– ...	5	E NGC 3189 – 90 group		12.4	+ 1371 + 1273	*12– -9
05563	10 15.9 + 38 44 183.47 + 56.27	07 – 21 – 038	0711 167	1.2 x 0.45 0.7 x 0.3	3 6	1	S ...		14.6		
05564	10 15.9 + 59 23 151.54 + 48.68	10 – 15 – 061	0962	1.4 x 0.5 1.3 x 0.5	75 6	3	Sb		16.0:		*
05565	10 16.0 + 46 43 169.80 + 54.39	08 – 19 – 018 NGC 3191	1348	0.7 x 0.45 0.7 x 0.45	5 3 – 4	1	S ... f of 2		13.9		*3

N G C   3147,   3155,   3168,   3177,   3179,   3184,   I C   602
        3185,   3187,   3189 – 91,   3193,   3194

1	2	3	4	5	6	7	8	9	10	11	12
05566	10 16.0 + 58 45 152.33 + 49.04	10 - 15 - 060	0962	1.3 x 0.5 0.8 x 0.4	80: ...	1	...		15.5		*1
05567	10 16.2 + 34 55 190.35 + 56.67	06 - 23 - 007 IC 2561	1032 167	1.1 x 0.6 0.8 x 0.5	17 5	1	S ... nnf of 2		14.9		*1
05568	10 16.2 + 58 28 152.68 + 49.21	10 - 15 - 062 NGC 3182	0962	(2.3 x 1.8) (2.1 x 1.9)	(155) 1 - 2	2	S0 - a		13.0		*23
05569	10 16.4 + 57 40 153.69 + 49.66	10 - 15 - 065 NGC 3188, Mark 31	0962	1.2 x 1.2 1.2 x 1.2	-- 1	4	S B b (R)S B(r)ab nff of 2; comp = Mark 30		14.7		*3
05570	10 16.4 + 73 32 136.16 + 39.73	12 - 10 - 027	0685	1.1 x 0.1 1.1 x 0.1	160 7	2	Sb - c p of 2		16.5:		*
05571	10 16.6 + 52 20 161.06 + 52.30	09 - 17 - 000	1331 163	1.3: x 0.4 1.1 x 0.4	[172] ...	3	Dwarf spiral S V		17.		
05572	10 16.9 + 45 49 171.19 + 54.83	08 - 19 - 020 NGC 3198	1348	10.0 x 3.8 8.0 x 3.3	35 6	4	Sc SAB(rs)c Sc -	+ 649 + 670	10.7	*12- -689
05573	10 17.0 + 06 35 235.91 + 48.40	01 - 27 - 001	1399	1.5 x 0.8 1.6 x 0.8	128 5	3	Sc		14.6		
05574	10 17.0 + 22 43 212.16 + 55.39	04 - 25 - 001	1380 164	1.4 x 0.3 1.2 x 0.2	93 6	1	Sc: P w 5575		16.0:		*
05575	10 17.0 + 22 51 211.93 + 55.43	04 - 25 - 000	1380	1.2 x 0.5 1.1 x 0.5	55 ...	2	Dwarf IV - V or V, p w 5574		16.5:		*
05576	10 17.0 + 65 26 144.29 + 45.19	11 - 13 - 022	0662	1.5 x 0.9 1.1 x 0.7	0: ...	1	S? ... Peculiar?		14.1		
05577	10 17.1 + 38 53 183.14 + 56.48	07 - 21 - 039	0711 167	1.0 x 0.8 1.1 x 0.9	(100) ...	3	... Peculiar		14.3		*1
05578	10 17.3 + 43 05 175.75 + 55.67	07 - 21 - 040	0711	1.4 x 0.2 1.1 x 0.2	136 6 - 7	1	Sc NGC 3202 group		15.6		
05579	10 17.3 + 57 28 153.86 + 49.87	10 - 15 - 066	0962 180	1.0 x 0.5 0.9 x 0.4	122 5	2	Sa - b		15.5		
05580	10 17.4 + 28 05 202.80 + 56.59	05 - 25 - 001 NGC 3204	1387	1.5 x 1.0 1.3: x 0.9:	110: 3	2	S B b SAB(r)b		14.8		*3
05581	10 17.5 + 43 16 175.42 + 55.66	07 - 21 - 041 NGC 3202	0711	1.3 x 0.9 1.3 x 0.9	20: 3	4	S B a S B(r)a Brightest in group		14.2		*13
05582	10 17.5 + 74 26 135.27 + 39.15	12 - 10 - 028 NGC 3183 = 3218	0685 172	2.4 x 1.3 2.5 x 1.4	170 4 - 5	3	S B b S B(s)b		12.5		*12 35-
05583	10 17.6 + 25 38 207.17 + 56.21	04 - 25 - 000	1380 166	1.0 x 0.2 1.0 x 0.2	67 7	2	Sb NGC 3209 group		16.0:		*
05584	10 17.8 + 25 45 206.98 + 56.28	04 - 25 - 002 NGC 3209	1380 166	(1.3 x 1.1) (1.8 x 1.5)	(80) ...	3	E Brightest in group		13.9		*3
05585	10 17.8 + 43 13 175.48 + 55.73	07 - 21 - 042 NGC 3205	0711	1.5 x 1.1 1.6 x 1.6	-- ...	3	... P w 5587 Peculiar, NGC 3202 group		14.4		*13
05586	10 18.0 - 02 15 246.08 + 43.11	00 - 27 - 000	0467 170	[1.7 x 0.9] ...	... ...	2	Double system Envelopes in contact		16.0:		*
05587	10 18.0 + 43 14 175.44 + 55.76	07 - 21 - 043 NGC 3207	0711	(1.4 x 1.0) (1.5 x 1.1)	(73) ...	3	... P w 5585 Peculiar, NGC 3202 group		14.3		*13
05588	10 18.1 + 25 37 207.24 + 56.32	04 - 25 - 000	1380 166	0.55 x 0.35 0.55 x 0.35	40: ...	1	... Peculiar, NGC 3209 group		14.0		*
05589	10 18.5 + 57 12 154.07 + 50.15	10 - 15 - 069 NGC 3206	0962 180	3.0 x 2.0 3.0: x 1.9	0 3:	4	S B c S B(s)cd Disturbed	+ 1192 + 1270	12.7	*12 3-
05590	10 18.6 + 19 53 217.08 + 54.88	03 - 27 - 004 NGC 3213	0058 169	1.1 x 0.8 1.2 x 0.9	133 2 - 3	1	Sb - c		14.3		2
05591	10 18.6 + 24 36 209.08 + 56.22	04 - 25 - 000	1380 166	1.0 x 0.25 1.0 x 0.3	15 6 - 7	1	S0 - a		15.3		
05592	10 18.8 + 22 48 212.22 + 55.81	04 - 25 - 006	1380 164	1.1 x 0.8 1.1: x 0.8	70 3	3	Sc Sc +		15.7		
05593	10 18.9 + 24 10 209.86 + 56.18	04 - 25 - 007 NGC 3216	1380 166	(1.4 x 1.2) (1.6 x 1.3)	(0) ...	1	E: Brightest in group		15.1		*3
05594	10 18.9 + 48 17 166.99 + 54.30	08 - 19 - 022	1348	1.2 x 0.3 1.2 x 0.3	122 6	1	Sb		15.4		
05595	10 19.0 + 12 49 228.00 + 52.08	02 - 27 - 002	0238 168	1.1 x 1.0 0.7 x 0.5	-- 1:	2	Sb - c		15.3		1
05596	10 19.0 + 79 07 131.15 + 35.80	13 - 08 - 015	1326 172	(1.3 x 1.3) (1.2 x 0.7)	-- ...	1	E?		14.7		*
05597	10 19.1 + 24 07 209.97 + 56.21	04 - 25 - 008	1380 166	1.3 x 0.2 1.2 x 0.2	168 6 - 7	1	Sa - b NGC 3216 group		15.5		*
05598	10 19.5 + 20 51 215.60 + 55.40	04 - 25 - 000	1380 169	1.6 x 0.35 1.4: x 0.3	35 6	1	S ...		15.5		
05599	10 19.5 + 44 07 173.80 + 55.78	07 - 21 - 050	0711	1.5: x 0.7 1.4 x 0.6	175 5 - 6	3	Sc Sc +		16.0:		
05600	10 19.5 + 78 52 131.33 + 36.00	13 - 08 - 16a	1326 172	1.4 x 1.1 (1.1 x 0.9)	(170) ...	1	S0? P w 5609, disturbed		14.4		*1

N G C   3182,   3183,   3188,   3198,   3202,   3204 - 07   I C   2561
        3209,   3213,   3216,   3218

1	2	3	4	5	6	7	8	9	10	11	12
05601	10 19.6 + 21 50 213.96 + 55.72	04 - 25 - 013 NGC 3221	1380 169	3.3 x 0.8 3.2 x 0.7	167 6	2	S B c - Irr		14.3		*12 3
05602	10 19.6 + 22 41 212.50 + 55.96	04 - 25 - 000	1380 164	1.1 x 0.5 1.1 x 0.5	128 5 - 6	1	S B? ...		16.0:		
05603	10 19.6 + 36 51a 186.72 + 57.22	06 - 23 - 009 IC 2568	1032 167	1.2 x 0.6 1.2 x 0.6	98 5	3	S B a P w IC 2566		(15.0)		*1
05664	10 19.7 + 46 30 169.79 + 55.06	08 - 19 - 023	1348	2.6 x 1.3 2.6: x 1.3:	47 5	4	Sc	SAB(rs)c	14.8		
05605	10 19.7 + 48 53 165.96 + 54.19	08 - 19 - 024	1348	1.1 x 0.8 1.1 x 0.7	60: 3 - 4:	1	S ...		15.5		
05606	10 19.8 + 01 27 242.55 + 45.86	00 - 27 - 003 IC 605	0467 166a	0.7 x 0.6 0.6 x 0.6	-- ...	1	S ...		14.5		
05607	10 19.8 + 04 15 239.35 + 47.59	01 - 27 - 004	1399	1.9 x 0.6 1.8 x 0.6	62 6	3	Sb		14.2		*
05608	10 19.8 + 27 37 203.78 + 57.05	05 - 25 - 002	1387	1.1 x 0.5 1.1 x 0.45	70 6	2	Sb		16.0:		*
05609	10 19.8 + 78 51 131.33 + 36.02	13 - 08 - 16b	1326 172	1.1 x 0.7 1.2 x 0.6	15 ...	3	... P w 5600, distorted		14.5		*1
05610	10 19.9 + 20 07 216.87 + 55.25	03 - 27 - 011 NGC 3222	0058 169	0.9 x 0.8 (1.2 x 1.1)	-- ...	2	E - S0		14.5	+ 5577 + 5471	*12 3
05611	10 20.1 + 42 06 177.23 + 56.41	07 - 21 - 053	0711 171	1.1 x 0.6 1.0 x 0.5	45 5	3	S B c		15.7		*
05612	10 20.1 + 71 08 138.17 + 41.61	12 - 10 - 032	0685 184	3.5 x 2.3 3.5: x 2.0:	165 3 - 4	4	... SB IV - V	S B(s)dm	14.8		*1
05613	10 20.4 + 52 36 160.24 + 52.69	09 - 17 - 060	1331 163	1.0 x 0.2 0.9 x 0.2	4 (7)	1	S - Irr		15.0		
05614	10 20.4 + 57 17 153.75 + 50.32	10 - 15 - 073 NGC 3220	0962 180	1.3 x 0.6 1.1 x 0.4	97 ...	1	...		13.7		*12 3
05615	10 20.6 + 53 21 159.13 + 52.36	09 - 17 - 61 + 62	1331 163	[1.2 x 0.7] ...	... ...	2	Double system Contact		14.2		*1V
05616	10 20.7 + 10 12 231.99 + 51.14	02 - 27 - 003	0238	1.3 x 0.10 1.2 x 0.12	140 7	2	Sc		15.7		1
05617	10 20.7 + 20 08 216.96 + 55.43	03 - 27 - 015 NGC 3226	0058 169	(2.5 x 2.2) (3.0 x 2.5)	(15) ...	4	E NGC 3189 - 90 group P w 5620, contact		13.3	+ 1338 + 1232	*12V 347̲8̲A
05618	10 20.7 + 21 22 214.88 + 55.82	04 - 25 - 000	1380 169	1.0: x 0.15 0.8 x 0.15	128 7	1	S ...		16.5:		
05619	10 20.8 + 18 22 219.86 + 54.83	03 - 27 - 017	0058 173	1.0 x 0.45 1.0 x 0.5	85 5 - 6	1	S0 - a		15.1		*
05620	10 20.8 + 20 06 217.02 + 55.44	03 - 27 - 016 NGC 3227	0058 169	[6.5 x 4.5] 5. ? x 4.5	155 5:	4	Sb, Seyfert; p w 5617 NGC 3189 - 90 group, Sb -		12.2	+ 1111 + 1005	*12 347̲8̲
05621	10 20.8 + 28 35 202.06 + 57.39	05 - 25 - 003	1387	1.1 x 0.9 1.1 x 0.8:	90: 2:	3	Sb/Sc	SAB(s)bc or SA(s)bc, Sc -	15.7		1
05622	10 20.8 + 34 01 191.95 + 57.65	06 - 23 - 011	1032	1.3 x 0.9 1.0 x 0.8	... ...		Sb/S B c P w 5623	SAB(s)bc Sb +	15.4		*1
05623	10 20.9 + 34 04 191.85 + 57.67	06 - 23 - 013	1032	1.0 x 0.7 0.9: x 0.5:	125: 3:	2	Sc P w 5622		16.5:		*1
05624	10 21.1 + 12 49 228.40 + 52.53	02 - 27 - 007 NGC 3230	0238 168	2.3 x 1.1 2.0 x 0.9	115 5:	2	S0		14.9		*3
05625	10 21.1 + 12 53 228.30 + 52.56	02 - 27 - 000	0238 168	1.0 x 0.4 0.9 x 0.25	20 6	1	S ...		15.2		*
05626	10 21.2 + 57 39 153.17 + 50.21	10 - 15 - 075	0962 180	1.7 x 0.7 1.0 x 0.4	53 (6)	4	Irr	Ir I	15.2		*
05627	10 21.5 + 13 49 227.02 + 53.09	02 - 27 - 008	0238 168	1.0 x 0.5 0.9 x 0.4	62 5	1	S ... s of 2		14.8		*
05628	10 21.5 + 17 00 222.15 + 54.46	03 - 27 - 018 IC 607	0058 173	1.5 x 1.3 1.5 x 1.4	105 1 - 2	4	S B b Disturbed? Sb +	S B(rs)bc	14.9		*1
05629	10 21.5 + 21 19 215.07 + 55.99	04 - 25 - 000	1380	1.4: x 1.1: 1.3: x 1.0:	... ...	2	Dwarf (spiral?) V		17.		*
05630	10 21.7 + 70 20 138.82 + 42.26	12 - 10 - 033	0685 184	[1.4 x 0.8] ...	... ...	2	Double system Connected, disrupted		14.7		*
05631	10 21.8 + 58 25 152.12 + 49.85	10 - 15 - 077 NGC 3225	0962	2.3 x 1.2 2.2 x 1.1	155 5	3	Sc		13.3		3
05632	10 21.9 + 20 22 216.73 + 55.78	03 - 27 - 022	0058 169	1.0 x 0.5 0.8 x 0.5	163: 5:	2	S B b	(R´)SAB(s) ...	15.5		
05633	10 22.0 + 15 00 225.34 + 53.73	03 - 27 - 020	0058	2.6 x 1.7: 2.5:: x 1.6::	175: ...	4	... SB IV - V	S B dm	15.4		*
05634	10 22.0 + 71 27 137.72 + 41.50	12 - 10 - 034	0685 184	1.3 x 0.7 1.3 x 0.7	152 5	2	Sb/S B b		15.6		
05635	10 22.1 + 28 17 202.68 + 57.64	05 - 25 - 007 NGC 3235	1387	(1.2 x 0.9) (1.3 x 1.0)	85: ...	2	E - S0 Brightest in group		14.7		*3

N G C 3220 - 22, 3225 - 27, 3230, 3235   I C 605, 607, 2568

1	2	3	4	5	6	7	8	9	10	11	12
05636	10 22.2 + 28 21	05 - 25 - 008	1387	1.0 x 0.5	27	2	Sa		15.6		*
	202.56 + 57.67	IC 2572		0.9 x 0.4	5		NGC 3235 group				
05637	10 22.4 + 17 25	03 - 27 - 025	0058	6. x 4.	...	4	Strongly peculiar		13.5	+ 880	*12
	221.64 + 54.82	NGC 3239	173	5.0: x 4.0:	...		Ir I pec			+ 761	3V$\overline{A}$
05638	10 22.6 + 26 43	05 - 25 - 000	1387	1.3 x 0.7	117	1	S0		15.5		
	205.60 + 57.52			1.4 x 0.7	5						
05639	10 22.8 + 17 30	03 - 27 - 029	0058	1.2 x 0.8	125:	3	Sc	SAB c	15.2		*1
	221.57 + 54.94		173	1.1 x 0.7	3 - 4		In group				
05640	10 22.8 + 39 55	07 - 22 - 003	0690	1.1 x 1.1	—	3	S0	(R)SAB 0	14.2		*3
	180.93 + 57.39	NGC 3237	177	1.1 x 1.1	1						
05641	10 23.0 - 01 58	00 - 27 - 009	0467	1.6 x 0.7	10:	3	S B b		14.4		*VA
	246.95 + 44.22	IC 609	170	1.6 x 0.8	5:		Singular, disturbed?				
05642	10 23.0 + 11 59	02 - 27 - 011	0238	1.7: x 0.22	98	1	Sb - c		14.8		
	229.97 + 52.53			1.5 x 0.25	7						
05643	10 23.0 + 80 03	13 - 08 - 021	1326	1.7: x 0.9	...	3	S B ...		14.3		*13
	130.23 + 35.21	NGC 3212	172	1.1 x 0.8	...		P w 5659, distorted				VA
05644	10 23.1 + 13 58a	02 - 27 - 012	0238	1.0 x 0.6	(15)	3	S ...	SA:(r) ...	(14.6)		*1
	227.09 + 53.50			1.1 x 0.7:	...		sp of 2, brightest of 4				
05645	10 23.1 + 71 41	12 - 10 - 035	0685	1.2 x 0.35	142	3	S B b	S B(s)b	16.0:		
	137.42 + 41.39		184	1.1 x 0.35	6						
05646	10 23.2 + 14 37	03 - 27 - 031	0058	2.5 x 0.7	165	1	S ...		14.2		1
	226.14 + 53.82			2.0 x 0.6	6						
05647	10 23.4 + 28 09	05 - 25 - 011	1387	1.1 x 0.2	170	1	S ...		15.7		
	203.00 + 57.91			1.2 x 0.25	6 - 7						
05648	10 23.5 + 04 37	01 - 27 - 008	1399	1.4 x 0.4	16	1	S ...		15.0		
	239.74 + 48.55		175	0.7 x 0.25	6						
05649	10 23.5 + 57 29	10 - 15 - 080	0962	1.3 x 1.3	—	2	E - S0		14.1		*23
	153.12 + 50.56	NGC 3238	180	1.3 x 1.3	(1)						
05650	10 23.6 + 45 30	08 - 19 - 031	1348	1.0 x 0.15	52	2	Sb		16.5:		
	171.05 + 56.04			1.1 x 0.12	7						
05651	10 23.7 + 17 46	03 - 27 - 033	0058	1.4 x 1.1	20:	3	Sc	SAB cd:	14.4		
	221.29 + 55.25		173	1.5 x 1.2	2			Sc +			
05652	10 23.8 - 02 22	00 - 27 - 012	0467	(1.6 x 1.3)	—	2	E - S0		14.0		
	247.55 + 44.10	NGC 3243	170	(1.7 x 1.4)	...						
05653	10 23.8 + 20 29	03 - 27 - 034	0058	1.9 x 0.25	29	3	Sb		14.8		2
	216.79 + 56.23	IC 610 = 611	169	1.8 x 0.25	7						
05654	10 23.9 + 15 36	03 - 27 - 035	0058	1.2 x 0.6	128	2	Sb - c		15.5		*
	224.76 + 54.40			1.1 x 0.5	5						
05655	10 23.9 + 16 24	03 - 27 - 036	0058	1.0 x 0.25	49	1	S ...		15.1		*
	223.50 + 54.74		173	1.0 x 0.25	6		1st in group				
05656	10 23.9 + 35 11	06 - 23 - 017	1032	1.5 x 0.8	135	3	Sb/S B c	SAB(s)bc	15.1		1
	189.69 + 58.23			1.1 x 0.6	5			Sb +			
05657	10 23.9 + 42 09	07 - 22 - 007	0690	1.7 x 0.2	87	1	S ...		15.4		*
	176.83 + 57.08		171	1.4 x 0.2	6 - 7						
05658	10 23.9 + 71 29	12 - 10 - 000	0685	1.0: x 0.5	95	3	Dwarf		17.		
	137.55 + 41.58		184	1.0: x 0.5	...		V:				
05659	10 24.0 + 80 02	13 - 08 - 022	1326	1.1 x 1.0	—	3	S ...	SAB(s) ...	14.0		*13
	130.20 + 35.25	NGC 3215	172	1.1 x 1.0	...		P w 5643, distorted				VA
05660	10 24.1 + 01 15	00 - 27 - 000	0467	1.1 x 0.25	72	1	S ...		15.3		*
	243.77 + 46.56			1.1 x 0.25	6						
05661	10 24.1 + 04 07	01 - 27 - 009	1399	2.3 x 1.4	100:	4	...	SAB dm	13.8		3
	240.48 + 48.36	NGC 3246	175	2.5: x 1.4:	...						
05662	10 24.2 + 28 54	05 - 25 - 012	1387	3.6 x 0.3	150	3	S B b		15.4		*12
	201.63 + 58.17			2.8 x 0.3	7		P w 5663				34$^-$
05663	10 24.5 + 28 46	05 - 25 - 013	1387	(2.9 x 1.9)	177	4	S0		11.6	+ 1261	*12
	201.90 + 58.22	NGC 3245		(2.9 x 1.8)	3 - 4					+ 1198	347$\overline{8}$
05664	10 24.6 + 01 13	00 - 27 - 000	0467	1.0 x 0.9	—	1	S ...		15.4		
	243.93 + 46.63			0.8 x 0.8	1						
05665	10 24.7 + 11 20	02 - 27 - 020	0238	1.1 x 0.30	143	1	S ...		15.1		*1
	231.24 + 52.56	IC 615	174	1.4 x 0.3	7						
05666	10 24.8 + 68 40	12 - 10 - 038	0685	15. x 9. :	[50]	5	Dwarf spiral, SAB(s)m		11.2	+ 46	*12
	140.20 + 43.61	IC 2574		14. : x 6. :	...		S IV - V	Ir I		+ 179	59$\overline{C}$
05667	10 24.9 + 01 31	00 - 27 - 18+19	0467	[1.5 x 0.7]	...	2	Double system, distorted,		14.9		*1
	243.66 + 46.88			...			bridge, in group of 3				
05668	10 24.9 + 50 15	08 - 19 - 000	1348	1.3 x 0.8	5	3	Sc		16.5:		
	163.23 + 54.39			1.1 x 0.5	4:						
05669	10 25.0 + 23 06	04 - 25 - 020	1380	(2.5? x 1.1)	135	3	S0		13.9		3
	212.38 + 57.27	NGC 3248	176	(1.7 x 0.9)	...						
05670	10 25.0 + 27 23	05 - 25 - 000	1387	1.4 x 0.15	115	1	Sb - c		17.		
	204.53 + 58.15			0.9 x 0.15	7						

N G C   3212,   3215,   3237 - 39,   3243,   3245   I C   609 - 611, 615, 2572, 2574
      3246,   3248

1	2	3	4	5	6	7	8	9	10	11	12
05671	10 25.2 + 67 05 141.77 + 44.73	11 - 13 - 000	0662	1.5: x 0.6: 1.2:: x 0.6::	[155] ...	3	Dwarf V		17.		
05672	10 25.6 + 22 50 212.92 + 57.33	04 - 25 - 021	1380 176	2.3 x 0.6 2.0 x 0.5	158 6	1	S ...		14.9		*1
05673	10 25.6 + 59 14 150.64 + 49.78	10 - 15 - 085	0962	1.0 x 0.8 0.9 x 0.7	(145) ...	1	S B ...		14.8		
05674	10 25.8 + 12 57 229.12 + 53.60	02 - 27 - 021 NGC 3253	0238	1.2 x 1.1 1.2 x 1.2	--- 1	4	Sb/S B c	SAB(rs)bc Sb +	14.4		*
05675	10 25.8 + 19 50 218.18 + 56.46	03 - 27 - 047	0058	2. x 1.5:: 1.5:: x 1.2::	... ...	3	Dwarf (spiral) S: V, 5690 group		18.		
05676	10 25.9 + 54 58 156.21 + 52.21	09 - 17 - 066	1331 163	1.6: x 1.0 1.5: x 1.0	(15) ...	3	S B c/Irr		14.8		*1
05677	10 26.0 + 03 49 241.28 + 48.55	01 - 27 - 000	1399 175	1.6 x 0.15 1.3 x 0.2	6 7	2	... P w 5678	d - m	15.2		*
05678	10 26.1 + 03 56 241.16 + 48.64	01 - 27 - 012	1399	1.1 x 0.8 0.9 x 0.8	50: 3:	1	Sb - c P w 5677		15.0		
05679	10 26.1 + 26 36 206.07 + 58.27	05 - 25 - 015	1387	1.7 x 0.6 1.4 x 0.5	[122] (6)	1	S ... Disturbed?		15.4		*
05680	10 26.1 + 60 33 148.97 + 49.03	10 - 15 - 087	0962	1.0 x 0.2 1.2 x 0.25	165 6 - 7	1	S ...		14.9		
05681	10 26.2 + 20 00 217.95 + 56.61	03 - 27 - 048	0058 181	1.3 x 0.2 1.3 x 0.2	165 6	3	S B b p of 2, 5690 group		15.6		*
05682	10 26.3 + 79 07 130.82 + 36.02	13 - 08 - 025	1326 172	1.2 x 1.2 1.2: x 1.2:	--- 1	4	S B c s of 2	SAB(rs)c	15.5		*
05683	10 26.5 + 23 59 210.95 + 57.82	04 - 25 - 024	1380	[1.6 x 0.35] ...	... ...	3	Double system Bridge, in group		15.5		*1
05684	10 26.5 + 26 21 206.57 + 58.31	04 - 25 - 023 IC 2579	1380	2.0 x 0.45 2.0 x 0.45	55 6	2	S B ... Singular, disturbed?		14.2		*1
05685	10 26.5 + 29 45 200.10 + 58.75	05 - 25 - 018 NGC 3254	1387	5.3 x 1.5 5.3 x 1.5	46 6	5	Sb	SA(s)bc Sb +	12.4	+ 1228 + 1170	*123 4678
05686	10 26.5 + 74 29 134.66 + 39.53	12 - 10 - 047	0685	1.8: x 1.3: 1.3:: x 1.0::	(140) ...	3	Dwarf V		17.		
05687	10 26.6 + 06 23 238.28 + 50.23	01 - 27 - 013	1399	2.1 x 0.22 1.9 x 0.25	111 7	2	Sc		14.9		
05688	10 26.6 + 70 19 138.44 + 42.57	12 - 10 - 044	0685 184	3.6: x 2.1: 3.5:: x 2.	[145] ...	3	... + companions, in chain	S B m: +	(14.6)		*1V C
05689	10 26.6 + 74 08 134.95 + 39.79	12 - 10 - 042 NGC 3252	0685 184	1.7 x 0.2 1.6 x 0.2	154 7	3	Sc	Sc +	15.4		*23
05690	10 26.8 + 19 52 218.27 + 56.69	03 - 27 - 050	0058 181	0.8 x 0.7 0.9 x 0.8	--- ...	2	S B ... Brightest in group	S B(s) ...	14.5		1
05691	10 26.8 + 22 15 214.10 + 57.44	04 - 25 - 025	1380 176	1.1 x 0.8 0.9 x 0.8	(140) ...	1	S ...		15.6		
05692	10 26.8 + 70 53 137.89 + 42.17	12 - 10 - 045	0685 184	4. x 2.5:: 3. x 2.	(0) ...	3	Dwarf spiral S V:		15.1		*
05693	10 26.9 + 39 16 181.85 + 58.29	07 - 22 - 009	0690 177	1.4 x 0.25 1.5 x 0.25	75 6	1	Sb - c		15.4		
05694	10 27.1 + 02 10 243.45 + 47.72	00 - 27 - 000	0467	[1.1 x 0.5] ...	... ...	2	Double system Bridge		15.3		*
05695	10 27.1 + 13 16 228.91 + 54.03	02 - 27 - 022	0238	1.4 x 0.5 1.5 x 0.6	96 6	1	S ...		14.9		
05696	10 27.1 + 20 05 217.94 + 56.83	03 - 27 - 052	0058 181	1.3 x 0.7 1.3 x 0.7	37 5	2	Sb		14.8		*
05697	10 27.3 + 38 43 182.85 + 58.46	07 - 22 - 011	0690	1.2: x 0.2 1.1 x 0.2	136 6	1	S ...		16.5:		
05698	10 27.3 + 44 23 172.57 + 57.03	07 - 22 - 013	0690 179	1.9 x 0.2 1.7 x 0.2	127 7	1	Sb - c		16.0:		
05699	10 27.3 + 60 16 149.17 + 49.33	10 - 15 - 000	0962	1.0 x 0.12 0.8 x 0.12	95 7	2	Sb - c		16.5:		
05700	10 27.3 + 72 22 136.48 + 41.13	12 - 10 - 046	0685 184	1.4 x 1.3 1.7 x 1.6	--- 1	2	Sb/S B b?		15.2		
05701	10 27.4 + 78 04 131.61 + 36.86	13 - 08 - 028	1326 172	1.7: x 0.9: 1.3:: x 0.7::	[0] ...	2	Dwarf: IV - V?		16.0:		
05702	10 27.5 + 05 18 239.83 + 49.76	01 - 27 - 000	1399	1.0 x 0.5 1.0 x 0.5	106 5:	1	S ...		15.2		
05703	10 27.5 + 53 46 157.69 + 53.04	09 - 17 - 067	1331 163	1.1 x 0.9 1.1 x 0.9	--- 1 - 2	3	Sb		15.6		*1
05704	10 27.6 + 23 00 212.85 + 57.82	04 - 25 - 000	1380	1.5 x 0.1 1.2 x 0.15	71 7	2	Sc		16.5:		*
05705	10 28.2 + 29 04 201.49 + 59.06	05 - 25 - 019 NGC 3265	1387	0.9 x 0.7 0.9 x 0.7	73: ...	1	E:		14.1		*23

N G C   3252 - 54, 3265   I C   2579

1	2	3	4	5	6	7	8	9	10	11	12
05706	10 28.3 + 34 46 190.37 + 59.16	06 – 23 – 000	1032	1.5? x 1.2? ...	-- ...	3	Dwarf irregular Ir V		18.		
05707	10 28.3 + 43 23 174.22 + 57.51	07 – 22 – 016	0690 179	2.6: x 1.9: 2.3: x 1.8:	155: 3:	4	S B c	SAB(s)cd	14.7		1
05708	10 28.6 + 04 44 240.79 + 49.63	01 – 27 – 014	1399 175	3.4 x 0.6 3.2 x 0.5:	168 6	3	...	S B d – dm	14.6		
05709	10 28.6 + 19 38 218.94 + 57.01	03 – 27 – 054	0058 181	1.3 x 0.8 1.2: x 0.6:	115: 4 – 5	2	...	S d – dm	15.7		
05710	10 28.7 + 24 24 210.40 + 58.41	04 – 25 – 000	1380	1.8: x 0.5 1.6 x 0.5	70 6	1	S ...		16.0:		
05711	10 28.7 + 25 07 209.06 + 58.56	04 – 25 – 029 NGC 3270	1380	3.1 x 0.8 3.0 x 0.8	10 6	3	Sb	SAB:(r)b	14.1		*3
05712	10 28.7 + 32 38 194.54 + 59.32	06 – 23 – 000	1032	1.1 x 0.12 1.1 x 0.12	43 7	3	Sc		16.0:		
05713	10 28.8 + 26 14 206.96 + 58.80	04 – 25 – 030	1380	1.8 x 0.5 1.6 x 0.4	5 6	2	Sb – c		15.0		
05714	10 28.8 + 46 55 168.09 + 56.36	08 – 19 – 036	1348	1.1 x 0.8 1.0 x 0.7	85: 3:	3	S B c	S B(s)c	14.7		*1
05715	10 29.0 + 00 43 245.53 + 47.13	00 – 27 – 026	0467	1.1 x 0.7 1.1 x 0.6	155: 4:	2	Sb – c		14.3		
05716	10 29.0 + 25 35 208.21 + 58.72	04 – 25 – 000	1380	1.4 x 0.9 0.9 x 0.8	... ...	3	Dwarf spiral S IV – V		17.		
05717	10 29.0 + 65 19 143.25 + 46.21	11 – 13 – 027 NGC 3259	0662	2.3 x 1.2 2.0 x 1.0	20 5	3	Sb	Sb +	12.9	+ 1866 + 1984	*12 3468
05718	10 29.2 + 21 10 216.34 + 57.65	04 – 25 – 000	1380 181	1.0 x 0.15 1.0 x 0.15	101 7	1	Sa – b		16.5:		
05719	10 29.2 + 56 19 153.96 + 51.86	09 – 17 – 069 NGC 3264	1331	3.5 x 1.5: 2.9 x 1.3:	177: ...	4	S B c/Irr Disturbed?		14.3		*13
05720	10 29.4 + 54 38 156.23 + 52.82	09 – 17 – 070 Mark 33	1331 163	1.0 x 0.9 (0.9 x 0.8)	-- (1)	3	Peculiar		13.2	+ 1620 + 1680	*1
05721	10 29.5 + 27 56 203.74 + 59.21	05 – 25 – 020 NGC 3274	1387	2.1 x 1.1 2.0 x 0.8	100 ...	2	Irr? or Peculiar		13.3		*12 3478
05722	10 29.5 + 59 07 150.32 + 50.25	10 – 15 – 097	0962	1.4 x 1.0 1.2 x 0.9	40: 3	1	S ...		15.1		
05723	10 29.6 – 01 55b 247.79 + 45.92	00 – 27 – 000	0467 170	1.0: x 0.7 1.8: x 0.7	... ...	3	S ... nf of 2, distorted, long streamer		(15.1)		*
05724	10 29.6 + 67 53 140.55 + 44.48	11 – 13 – 28+29	0662	[2.5:: x 0.6:] ...	... ...	2	Double system Common envelope, in group		16.0:		*1
05725	10 29.7 + 65 01 143.50 + 46.47	11 – 13 – 030 NGC 3266	0662	1.4 x 1.2 1.4 x 1.2	(105) 1 – 2	4	S B 0	SAB 0	13.5		*23
05726	10 29.8 + 02 49 243.35 + 48.66	01 – 27 – 000	1399	1.1 x 0.4 1.0 x 0.3	157 ...	1	Dwarf?		17.		
05727	10 29.8 + 64 46 143.76 + 46.64	11 – 13 – 032	0662	1.5 x 0.45 1.5 x 0.4	29 6	2	Sb – c		15.0		
05728	10 29.8 + 79 25 130.43 + 35.89	13 – 08 – 000	1326 172	1.5: x 1.0: 1.2: x 0.9:	(0) ...	3	Dwarf spiral S B V		17.		
05729	10 30.0 + 20 11 218.19 + 57.51	03 – 27 – 059	0058 181	1.1 x 0.5 1.1 x 0.55	5 5	2	S B a – b		15.2		*
05730	10 30.1 + 16 07 225.06 + 55.97	03 – 27 – 060 IC 616	0058	1.0 x 0.9 1.1 x 0.9	-- 1	3	Sc		14.6		1
05731	10 30.1 + 28 47 202.12 + 59.44	05 – 25 – 022 NGC 3277	1387	(2.4 x 2.2) (2.6 x 2.3)	-- 1	4	Sa/Sb		12.3	+ 1460 + 1399	*123 4678
05732	10 30.4 + 74 01 134.80 + 40.05	12 – 10 – 049	0685 184	2.0 x 0.7 2.3 x 0.7	35 (6)	3	Sc – Irr		14.2		
05733	10 30.6 + 52 37 158.95 + 54.03	09 – 17 – 073	1331	1.0 x 0.8 1.0 x 0.8	(140) ...	1	S B ...		14.9		
05734	10 31.1 + 53 07 158.15 + 53.84	09 – 17 – 074	1331	1.7 x 0.25 1.5 x 0.25	156 6 – 7	2	S0 – a Disturbed?		14.0		*
05735	10 31.2 + 13 08 229.94 + 54.84	02 – 27 – 025	0238	1.0: x 0.8: 0.9 x 0.7	0: 2:	1	Sb – c		14.8		
05736	10 31.3 – 00 18 247.20 + 46.88	00 – 27 – 029	0467	1.1 x 0.9 0.9 x 0.7	(80) 2:	1	Sb – c		15.6		
05737	10 31.3 + 11 28 232.44 + 54.02	02 – 27 – 000	0238	1.0 x 0.9 1.1 x 0.9	(110) 1 – 2	2	Sc P w 5741		15.2		*
05738	10 31.6 + 35 30 188.82 + 59.77	06 – 23 – 019	1032	1.0 x 0.6 0.9 x 0.6	30 4	1	S ...		14.1		
05739	10 31.7 + 14 00 228.72 + 55.36	02 – 27 – 026	0238	1.1 x 0.3 1.2 x 0.35	152 ...	1	Irr? or Peculiar?		13.9		
05740	10 31.7 + 51 01 161.19 + 54.98	09 – 17 – 075	1331	2.2 x 1.3 2.0: x 1.1:	140: ...	4	Dwarf spiral SAB m S IV – V		15.7		*

N G C   3259,   3264,   3266,   3270,   3274,   3277   I C   616

1	2	3	4	5	6	7	8	9	10	11	12
05741	10 32.0 + 11 27 232.62 + 54.16	02 – 27 – 027 IC 622	0238	2.8 x 0.35 2.9 x 0.35	152 6 – 7	3	Sc P w 5737 Sc +		14.1		*1
05742	10 32.0 + 21 55 215.36 + 58.49	04 – 25 – 032 NGC 3287	1380 185	2.1 x 1.0 2.0 x 1.0	20 ...	3	Irr P w 5767 Ir I		12.9		*123 4678
05743	10 32.0 + 25 47 208.08 + 59.43	04 – 25 – 000	1380	1.1 x 0.35 1.1 x 0.3	147 6	2	Sc		16.0:		
05744	10 32.1 + 46 49 167.85 + 56.92	08 – 19 – 000 Mark 146	1348	0.45 x 0.40 0.55 x 0.45	(110) ...	2	... Compact		14.1	+ 3600	*
05745	10 32.2 – 01 43 248.92 + 46.07	00 – 27 – 030	0467	1.1 x 0.9 1.1 x 0.9	(135) 2:	3	S B 0 – a		13.8		
05746	10 32.2 + 45 21 170.33 + 57.50	08 – 19 – 040	1348 179	1.7 x 1.1 1.5 x 1.0	40 3 – 4	3	S B a – b		14.2		
05747	10 32.7 + 44 35 171.61 + 57.86	07 – 22 – 000	0690 179	1.3 x 0.22 1.3 x 0.20	133 7?	1	... Peculiar?		15.4		*
05748	10 32.8 + 03 49 242.91 + 49.87	01 – 27 – 017 IC 623	1399	1.0 x 0.4 1.1 x 0.35	152 6	1	S ...		15.0		
05749	10 32.8 + 28 50 202.14 + 60.04	05 – 25 – 22a	1387 183	0.45 x 0.40 0.40 x 0.35	--- ...	1	... Compact		14.2		*
05750	10 33.0 + 21 16 216.68 + 58.52	04 – 25 – 000	1380 185	1.2 x 0.6 1.0: x 0.5:	163: 5:	3	S B dm – m P w 5751, in group		17.		*
05751	10 33.0 + 21 19 216.59 + 58.54	04 – 25 – 033	1380 185	2.1 x 0.4 1.8 x 0.40	172 6	3	S ... P w 5750, disturbed, in group		14.9		*1
05752	10 33.2 + 58 49 150.23 + 50.81	10 – 15 – 114 NGC 3288	0962	1.1: x 0.7 1.0 x 0.7	(175) 3 – 4	1	Sb: P w NGC 3286		15.0		*23
05753	10 33.3 + 37 35 184.63 + 59.82	06 – 23 – 021 NGC 3294	1032	3.8 x 1.8 3.6 x 1.8	122 5	5	Sc SA(s)c Sc –		11.5	+ 1469 + 1453	*123 4678
05754	10 33.3 + 42 56 174.49 + 58.52	07 – 22 – 030	0690	1.1 x 0.8 0.8 x 0.7	[150] ...	3	Sc SAB(s)cd: Distorted, Sc +		16.5:		*1
05755	10 33.5 + 22 22 214.72 + 58.96	04 – 25 – 000	1380 185	1.0 x 0.1 0.9 x 0.12	173 7	1	Sa – b		16.5:		
05756	10 33.5 + 27 14 205.35 + 60.00	05 – 25 – 024 IC 2590	1387	(1.1 x 1.1) (0.9 x 0.9)	--- (1)	3	S0 nff of 2		14.7		*1
05757	10 33.5 + 79 38 130.11 + 35.83	13 – 08 – 029	1326 172	2.3 x 0.9 1.7 x 0.7	115 6	2	Sa? Eruptive?		14.9		*
05758	10 33.6 + 13 42 229.57 + 55.62	02 – 27 – 000	0238	1.0: x 0.4: 0.5: x 0.4:	... ...	3	Dwarf V, p of 2		19.		*
05759	10 33.6 + 38 42 182.43 + 59.67	07 – 22 – 031	0690	(1.0 x 0.8) (1.0 x 0.8)	(25) ...	2	E – S0		14.5		
05760	10 33.7 + 13 58 229.18 + 55.77	02 – 27 – 028	0238	1.4 x 0.30 1.5 x 0.30	1 6	1	S ...		14.2		*1
05761	10 33.8 + 12 57 230.77 + 55.30	02 – 27 – 029 NGC 3299	0238	2.0: x 1.7: 2.2: x 1.9:	(3) ...	4	... SAB(s)dm NGC 3306 group		14.1		*12 3
05762	10 33.8 + 23 28 212.69 + 59.32	04 – 25 – 000	1380	1.0 x 0.3 1.0 x 0.2	87 (6)	1	S? ... nff of 2		16.5:		*
05763	10 33.8 + 35 18 189.13 + 60.23	06 – 23 – 022 IC 2591	1032	1.5 x 0.8 1.1 x 0.7	137 4 – 5	1	S ...		14.5		
05764	10 33.9 + 31 49 196.16 + 60.42	05 – 25 – 025	1387	2.0 x 1.2 1.8: x 0.9:	[60] ...	3	Dwarf irregular Ir IV – V		15.6		*1A
05765	10 33.9 + 69 05 138.99 + 43.89	12 – 10 – 059	0685	1.8 x 0.25 1.6 x 0.25	129 6	2	Sb – c		15.1		
05766	10 34.0 + 14 26 228.51 + 56.06	02 – 27 – 030 NGC 3300	0238	(1.8 x 0.9) (1.7 x 0.9)	173 5	4	S B 0		13.4		*12 348
05767	10 34.2 + 22 09 215.21 + 59.05	04 – 25 – 035 NGC 3301	1380 185	3.4 x 1.1 3.2 x 1.1	52 6	4	S B 0 P w 5742		12.2	+ 1333 + 1241	*123 4678
05768	10 34.2 + 60 15 148.36 + 50.00	10 – 15 – 115	0962	1.3 x 0.3 1.2 x 0.25	39 7	1	S ... f of 2		15.7		*
05769	10 34.3 + 20 43a 217.87 + 58.63	04 – 25 – 000	1380 181	1.1 x 1.0 0.9: x 0.9:	--- (1)	1	... P w 5770				*
05770	10 34.3 + 20 43b 217.87 + 58.63	04 – 25 – 000	1380 181	1.0 x 1.0 1.0 x 1.0	--- (1)	1	... P w 5769		15.4		*
05771	10 34.3 + 43 51 172.72 + 58.39	07 – 22 – 033	0690	1.7 x 1.2 (1.7 x 1.2)	60: 3	2	S0 – a ssp of 2		14.6		*
05772	10 34.4 + 00 29 247.13 + 47.98	00 – 27 – 031	0467	1.0 x 0.6 0.9 x 0.5	88 4 – 5	1	S ...		14.7		
05773	10 34.4 + 18 24 222.00 + 57.84	03 – 27 – 066 NGC 3303	0058	2.5: x 2.0: 2.7: x 2.7:	--- ...	3	Strongly peculiar double system, extensive plumes		14.5		*13 VA
05774	10 34.5 + 12 54 230.99 + 55.42	02 – 27 – 032 NGC 3306	0238	1.4 x 0.5 1.4 x 0.4	141 6	1	Sc? Brightest in group		13.7		*2
05775	10 34.5 + 37 20 185.04 + 60.10	06 – 23 – 025	1032	1.5 x 0.2 1.3: x 0.2	112 7	2	Sc		16.0:		

N G C   3287, 3288, 3294, 3299, 3300, 3301   I C   622, 623, 2590, 2591
          3303, 3306

1	2	3	4	5	6	7	8	9	10	11	12
05776	10 34.5 + 64 32 143.50 + 47.17	11 – 13 – 000 Mark 149	0662	0.5 x 0.5 0.4 x 0.4	-- ...	2	... Very compact		14.4	+ 600	*C
05777	10 34.7 + 37 43 184.27 + 60.07	06 – 23 – 026 NGC 3304	1032 182	1.5 x 0.5 1.5 x 0.5	158 6	3	Sa		14.4		*1<u>23</u>
05778	10 34.7 + 46 20 168.33 + 57.52	08 – 19 – 045	1348	1.0 x 0.10 1.1 x 0.12	111 7	3	Sc f of 2		17.		*
05779	10 34.9 + 05 53 240.88 + 51.57	01 – 27 – 021	1399	1.0: x 0.2 1.1 x 0.18	123 6	1	S ... P w 5780, disturbed:		15.1		*1
05780	10 35.0 + 05 51 240.94 + 51.57	01 – 27 – 022 IC 628	1399	1.0 x 0.8 0.9 x 0.7	115: ...	2	Sa – b SAB(s) ... P w 5779		14.8		*1
05781	10 35.3 + 13 59 229.48 + 56.12	02 – 27 – 000	0238	1.1 x 0.5 0.6: x 0.4	[110] ...	3	Dwarf V		18.		
05782	10 35.4 + 78 53 130.58 + 36.47	13 – 08 – 030	1326 172	1.1 x 1.0 0.8 x 0.8	-- 1	1	Sb – c		16.0:		
05783	10 35.5 + 02 00 245.71 + 49.20	00 – 27 – 033	0467	1.2 x 0.45 1.2 x 0.4	54 6	1	S B b:		15.0		
05784	10 35.5 + 05 09 241.95 + 51.23	01 – 27 – 023	1399	1.4 x 0.3 1.0 x 0.25	125 6	1	S ...		15.0		
05785	10 35.6 + 30 25 199.03 + 60.75	05 – 25 – 029	1387	1.0: x 0.5 0.6 x 0.4	5 5:	1	S ...		15.6		
05786	10 35.7 + 53 45 156.62 + 54.07	09 – 18 – 008 NGC 3310	0673 163	3.8 x 3.5 3.4 x 3.4	-- 1:	4	S[b] Peculiar		11.0	+ 1026 + 1090	*123 4<u>678</u>A
05787	10 35.9 – 02 19 250.48 + 46.31	00 – 27 – 034	0467	2.0 x 0.7 1.7 x 0.7	136 6	3	S B b nf of 2		14.7		*
05788	10 36.2 + 05 57 241.11 + 51.87	01 – 27 – 024	1399	1.3 x 0.8 1.1 x 0.6	10: 4	2	Sb np of 2		14.6		*
05789	10 36.2 + 41 57 175.98 + 59.33	07 – 22 – 036 NGC 3319	0690	7.5 x 4.3 7.3: x 4.3:	37 4	5	S B c SB(rs)cd		12.0	+ 826 + 832	*12– –689
05790	10 36.5 + 04 54 242.51 + 51.27	01 – 27 – 000	1399	1.2:: x 1.1:: 1.5: x 1.3:	-- ...	1	S? ...		15.6		
05791	10 36.5 + 48 12 164.97 + 57.00	08 – 20 – 004	0709	1.6 x 0.5: 1.7: x 0.8:	[43] ...	2	... P w 5798, distorted		14.4		*1
05792	10 36.6 – 00 09 248.38 + 47.95	00 – 27 – 035 IC 632	0467	1.0 x 0.6 0.9 x 0.5	30 4:	1	Sa? P w 5796, NGC 3325 group		14.8		*1
05793	10 36.6 + 15 06 227.96 + 56.92	03 – 27 – 000	0058	1.0 x 0.15 1.0 x 0.15	171 7	1	Sa – b		15.2		
05794	10 36.6 + 47 39 165.86 + 57.26	08 – 20 – 010 NGC 3320	0709	2.2 x 1.0 2.2 x 1.0	20 5	4	Sc Sc +		13.1		*12 3468
05795	10 36.7 + 00 03 248.19 + 48.10	00 – 27 – 036 NGC 3325	0467	(1.0 x 0.9) (1.1 x 1.0)	-- ...	1	E: Brightest of 3		14.0		*
05796	10 36.8 – 00 08 248.42 + 48.00	00 – 27 – 037 IC 633	0467	0.55 x 0.25 0.55 x 0.3	102 ...	1	... P w 5792, NGC 3325 group		14.5		*1
05797	10 36.8 + 01 58 246.08 + 49.42	00 – 27 – 038	0467	1.0: x 0.9: 0.8 x 0.8	-- ...	3	Irr		14.7		*
05798	10 36.8 + 48 11 164.96 + 57.05	08 – 20 – 011	0709	1.0 x 0.22 1.0 x 0.20	45 ...	1	... P w 5791, interaction		14.5		*1
05799	10 36.9 + 05 22 242.03 + 51.64	01 – 27 – 025 NGC 3326	1399	0.7 x 0.6 1.0 x 0.7	-- ...	2	Sa		14.2		3
05800	10 36.9 + 25 35 208.89 + 60.47	04 – 25 – 036 NGC 3323	1380	1.4 x 0.7 1.0 x 0.4	-- ...	2	S B ... SB(s) ... Peculiar		14.3		
05801	10 37.1 + 22 07 215.66 + 59.68	04 – 25 – 037	1380 185	1.1 x 1.1 1.0: x 1.0:	-- ...	1	...		15.7		
05802	10 37.2 + 15 52 226.82 + 57.40	03 – 27 – 000	0058	1.0 x 0.1 0.9 x 0.12	2 7	1	Sc		16.0:		
05803	10 37.2 + 24 21 211.35 + 60.28	04 – 25 – 038 NGC 3327	1380 194	1.4 x 0.9 1.3 x 0.8	85 3 – 4	3	Sb SA:(r)b Sb –		14.2		13
05804	10 37.4 + 38 45 182.02 + 60.39	07 – 22 – 041	0690	1.6 x 1.0 1.5 x 0.9	110: 3 – 4	3	Sb SAB b Sb –		15.2		
05805	10 37.6 + 21 53 216.17 + 59.73	04 – 25 – 039	1380 185	1.7 x 1.0: 1.6: x 1.0:	90 ...	4	... S B dm		15.4		
05806	10 37.7 + 37 36 184.28 + 60.68	06 – 24 – 002	0731	1.1 x 1.0 1.0 x 0.9	-- 1	3	Sc		16.0:		*
05807	10 37.8 + 09 27 236.81 + 54.26	02 – 27 – 038 NGC 3332	0238	(1.4 x 1.4) (3. x 3.)	-- 1	4	S0 (R)SA 0 –		13.7		3
05808	10 37.8 + 12 33 232.25 + 55.94	02 – 27 – 039	0238	1.2 x 1.0 0.9 x 0.9	-- 1	3	Sb SA(s)b Sb +		14.6		
05809	10 37.8 + 69 58 137.82 + 43.47	12 – 10 – 000	0685	1.1 x 0.10 0.8 x 0.10	93 7	2	Sc		17.		
05810	10 37.9 + 40 04 179.39 + 60.17	07 – 22 – 000	0690 188	1.1 x 0.15 1.2 x 0.20	39 7	3	Sb		16.5:		

N G C   3304, 3310, 3319, 3320, 3323, 3325 – 27   I C   628, 632, 633
3332

1	2	3	4	5	6	7	8	9	10	11	12
05811	10 38.3 + 06 15 241.25 + 52.46	01 – 27 – 000 IC 634	1399	1.3: x 0.45 1.0 x 0.4	116 ...	1	... Peculiar?		15.1		
05812	10 38.3 + 12 44 232.08 + 56.14	02 – 27 – 040	0238	1.0 x 0.5 0.9 x 0.3	[27] ...	3	Irr I m Ir I		15.6		
05813	10 38.3 + 36 38 186.19 + 60.96	06 – 24 – 003	0731	1.8 x 0.8 1.7 x 0.8	35 5	4	Sc		15.3		
05814	10 38.3 + 77 46 131.29 + 37.45	13 – 08 – 031	1326 172	1.6 x 0.9: 1.6 x 0.9	[130] ...	3	... Peculiar		15.0		*1
05815	10 38.5 + 51 05 160.13 + 55.87	09 – 18 – 011	0673	1.0 x 0.10 1.1 x 0.10	75 7	3	Sc		17.		
05816	10 38.6 + 06 32 240.96 + 52.70	01 – 27 – 000 IC 634	1399	1.0 x 0.2 1.0 x 0.20	138 6	1	S B: ... In group		15.6		*
05817	10 38.6 + 37 34 184.28 + 60.86	06 – 24 – 004 NGC 3334	0731	1.1 x 1.0 (1.4 x 1.4)	--- ...	1	S0?		14.1		*3
05818	10 38.7 + 06 37 240.88 + 52.77	01 – 27 – 027	1399	1.5: x 0.9 1.3: x 0.7:	137 4	4	Sc sf of 2, contact, in group		15.4		*1
05819	10 38.9 + 38 59b 181.42 + 60.62	07 – 22 – 045	0690 188	1.6 x 0.35 1.6 x 0.35	120 6 – 7	2	Sb – c Brightest of 3		(14.5)		*1
05820	10 39 + 81 10 128.76 + 34.73	14 – 05 – 013	0727	1.8 x 0.7 1.5 x 0.5	75 6	2	Sb – c		15.2		
05821	10 39.0 + 15 55 227.10 + 57.81	03 – 27 – 069 IC 635	0058 187	1.3: x 0.3 1.2 x 0.3	5 6	1	S ...		15.2		1
05822	10 39.1 + 21 31 217.07 + 59.95	04 – 25 – 041	1380 185	1.4 x 1.3 1.2: x 1.1:	--- 1	3	S B a (R)S B(r)a 3rd of 3		15.5		*
05823	10 39.3 + 01 03 247.77 + 49.26	00 – 27 – 041	0467	1.0 x 0.8 1.0 x 0.8	(165) ...	3	Irr or Peculiar		14.9		*1
05824	10 39.3 + 04 36 243.59 + 51.62	01 – 27 – 028 IC 636	1399	1.1 x 0.4 1.0 x 0.45	48 6	1	S B: ...		14.9		
05825	10 39.4 + 24 00 212.28 + 60.69	04 – 25 – 042	1380 194	1.2 x 0.5 1.2 x 0.5	92 6	1	Sa?		15.1		
05826	10 39.5 + 14 00 230.35 + 57.02	02 – 27 – 041 NGC 3338	0238	5.2 x 3.5 5.2 x 3.6:	100 3	5	Sc SA(s)c Leo group Sc –		12.1	+ 1330 + 1202	*12– –689
05827	10 39.7 – 00 07 249.15 + 48.53	00 – 27 – 042 NGC 3339 = 3340	0467	0.9 x 0.8 1.0 x 0.8	--- ...	1	S ...		13.6		13
05828	10 39.8 + 16 01b 227.09 + 58.03	03 – 27 – 073	0058 187	1.2 x 0.2 1.0 x 0.2	28 7	1	S ... s of 2, in group of 3		(15.7)		*1
05829	10 39.8 + 34 43 190.06 + 61.51	06 – 24 – 006	0731	5.3 x 4.5 4.7: x 4.0:	... ...	4	Dwarf irregular, IB(s)m Ir V Ir I		15.1		*
05830	10 39.9 + 24 13 211.90 + 60.85	04 – 25 – 043	1380 194	1.3 x 0.7 1.1 x 0.7	30: 4 – 5	3	Sb SAB:(r)b		15.2		
05831	10 40.0 + 05 18 242.90 + 52.20	01 – 27 – 031 NGC 3341	1399	1.4 x 0.4 1.0 x 0.5	[24] ...	2	... Peculiar Ir II?		14.9		3
05832	10 40.2 + 13 43 230.96 + 57.03	02 – 27 – 042	0238	1.1 x 1.0 1.2 x 0.9	(95) ...	2	S B? ... Peculiar		13.8		*1V A
05833	10 40.3 + 20 41 218.83 + 59.95	04 – 25 – 044 Mark 416	1380	1.6 x 0.5 1.3 x 0.5	148 6	3	S0		15.1		
05834	10 40.3 + 69 58 137.59 + 43.61	12 – 10 – 065	0685	1.4 x 0.25 1.1 x 0.25	49 6	2	S B b P w 5835		16.0:		*
05835	10 40.3 + 70 01 137.55 + 43.58	12 – 10 – 066	0685	1.1 x 0.8 1.1 x 0.8	140 2 – 3	2	Sa? P w 5834		14.6		*
05836	10 40.4 + 21 55 216.49 + 60.36	04 – 25 – 045	1380 185	1.0 x 0.8 0.9 x 0.7	(170) 2:	3	Sb SAB(r)bc Sb +		15.6		
05837	10 40.4 + 77 05 131.69 + 38.07	13 – 08 – 033 NGC 3329 = 3397	1326 172	2.2 x 1.3 (2.3 x 1.4)	140 4	1	Sa? Brightest in group		12.9		*12 3468
05838	10 40.5 + 41 03 177.21 + 60.37	07 – 22 – 048	0690	1.3 x 0.5 1.2 x 0.5	130 6	3	S B b		14.0		*1
05839	10 40.6 + 39 57 179.34 + 60.70	07 – 22 – 049	0690 188	1.2 x 0.3 (1.3 x 0.5)	147 6	2	S0 – a		14.8		1
05840	10 40.7 + 25 10 210.07 + 61.23	04 – 25 – 046 NGC 3344	1380	7.5 x 7.0 7. x 7.	--- 1	4	Sc (R)SAB(r)c Sc –		11.1	+ 579 + 504	*12– –689
05841	10 40.8 + 76 58 131.76 + 38.17	13 – 08 – 034	1326 172	1.7 x 0.9 1.7 x 0.9	130 5	4	S B c/Sc SAB(s)c NGC 3329 – 97 group, Sc –		15.1		*
05842	10 41.0 + 15 08 228.82 + 57.88	03 – 28 – 001 NGC 3346	0463	2.6 x 2.5 3.0 x 2.6	--- 1	5	S B c S B(rs)cd		12.8		*12 3468
05843	10 41.0 + 69 20 138.10 + 44.13	12 – 10 – 72a	0685	1.1 x 0.25 1.0 x 0.2	7 6 – 7	1	S ...		15.5		
05844	10 41.1 + 28 25 203.35 + 61.81	05 – 25 – 036	1387 192	1.6 x 0.3 1.5: x 0.3	121 6 – 7	3	Sc Sc +		16.0:		
05845	10 41.2 + 44 10 171.28 + 59.43	07 – 22 – 051	0690	1.1 x 0.8 0.7 x 0.6	(160) ...	1	...		14.8		

N G C   3329, 3334, 3338 – 41, 3344, 3346, 3397   I C   634 – 36

1	2	3	4	5	6	7	8	9	10	11	12
05846	10 41.2 + 60 38 147.02 + 50.40	10 – 16 – 002	0712	2.0:: x 2.0:: 1.5:: x 1.5::	-- ...	4	Dwarf irregular Ir V		15.4		*
05847	10 41.3 – 02 15 251.82 + 47.30	00 – 28 – 001	1397	1.3 x 0.6 1.2 x 0.6	5 5	2	S B b: p of 2	(R´)SB(s) ...	15.5		*1
05848	10 41.3 + 56 41 151.84 + 52.98	09 – 18 – 015	0962 189	2.1: x 1.4: 1.4: x 0.9:	115 ...	4	Dwarf spiral S V, np of 2		15.3		*
05849	10 41.4 – 01 01 250.57 + 48.20	00 – 28 – 003	1397	1.6: x 0.7: 1.0 x 0.4	... ...	1	...		14.1		*1
05850	10 41.4 + 11 58 233.97 + 56.38	02 – 28 – 001 NGC 3351 = M 95	0976	8.5 x 5.0 9.0 x 5.5	13 4	5	S B b Leo group	SB(r)b Sb +	11.2	+ 780 + 643	*12– 6789
05851	10 41.5 + 22 38 215.24 + 60.81	04 – 25 – 048 NGC 3352	1380 185	(1.5 x 1.1) (1.6 x 1.2)	(0) ...	3	S0		14.1		
05852	10 41.6 + 07 01 241.09 + 53.58	01 – 28 – 004 NGC 3356	0722	1.8 x 0.8 2.1 x 0.8	102: 5:	3	Sb Disturbed, p w 5857, Sb +		13.3		*13
05853	10 41.6 + 58 43 149.25 + 51.71	10 – 16 – 003	0712 189	1.5 x 0.2 1.5 x 0.2	38 7	3	Sc P w 5862		15.5		*
05854	10 41.6 + 77 22 131.41 + 37.88	13 – 08 – 035	1326 172	1.1 x 1.0 0.9 x 0.9	-- 1	2	Sc – Irr NGC 3329 – 97 group		15.2		*
05855	10 41.9 + 26 27 207.53 + 61.73	04 – 26 – 001	1366	1.4 x 0.35 1.5 x 0.35	97 6	1	S ...		15.0		
05856	10 42.0 + 38 26 182.24 + 61.34	06 – 24 – 008	0731 188	1.0 x 0.2 1.0 x 0.2	97 7	1	S ...		15.1		*1
05857	10 42.2 + 06 52 241.44 + 53.61	01 – 28 – 005 NGC 3362	0722	1.4 x 1.1 1.3 x 1.1	(90) 1 – 2	3	Sc P w 5852	SAB c Sc –	13.6		*3
05858	10 42.2 + 16 13 227.24 + 58.63	03 – 28 – 003	0463	1.6: x 1.2: 0.8 x 0.7	(160) ...	3	...	S dm	15.6		
05859	10 42.2 + 43 58 171.51 + 59.67	07 – 22 – 055	0690	1.5 x 0.8 1.6 x 0.8	25 5	3	S B a		14.9		
05860	10 42.2 + 56 13 152.32 + 53.37	09 – 18 – 022 NGC 3353, Mark 35	0673 189	1.5 x 1.1 1.4 x 1.0	... ...	3	... Peculiar		12.9	+ 1020 + 1098	*12 3478
05861	10 42.3 + 39 15 180.56 + 61.20	07 – 22 – 056	0690 188	1.5 x 0.15 1.5 x 0.15	151 7	2	Sc		16.5:		
05862	10 42.3 + 58 50 149.01 + 51.70	10 – 16 – 004	0712 189	1.3 x 0.6 1.4 x 0.6	120 5 – 6	2	S ... P w 5853, disturbed?		15.3		*
05863	10 42.3 + 73 37 134.30 + 40.89	12 – 10 – 073 NGC 3343	0685 184	(1.3 x 0.9) (1.6 x 1.2)	(55) ...	4	E		14.7		3
05864	10 42.4 + 10 27 236.49 + 55.75	02 – 28 – 000	0976 186	1.0: x 0.3 0.7 x 0.3	125 (6)	1	S: ...		15.5		
05865	10 42.5 + 05 12 243.68 + 52.61	01 – 28 – 006	0722	1.0 x 1.0 0.9: x 0.6:	-- 1:	1	Sc?		15.7		
05866	10 42.5 + 22 20 215.96 + 60.94	04 – 26 – 002 NGC 3363	1366 185	1.6 x 0.9 1.6 x 0.9	0 4	1	S ...		14.7		*
05867	10 42.6 + 00 23 249.38 + 49.40	00 – 28 – 004	1397	1.0 x 1.0 1.0 x 0.9	-- 1	2	S B b np of 2, brightest in chain of 4		15.0		*
05868	10 42.9 + 37 28 184.14 + 61.72	06 – 24 – 009	0731 188	1.4 x 1.2 1.4 x 1.2	-- 1 – 2	3	Sb P w 5871		15.0		*
05869	10 43.1 + 11 37 234.91 + 56.55	02 – 28 – 003	0976 186	1.4 x 0.7 1.1 x 0.7	95 5:	3	Sb	SAB:(rs)b:	14.9		
05870	10 43.1 + 35 14 188.81 + 62.13	06 – 24 – 010	0731	1.1 x 1.1 1.1 x 1.0	-- (1)	1	S0? Companion superimposed		14.3		*1
05871	10 43.1 + 37 26 184.19 + 61.76	06 – 24 – 000	0731 188	1.2 x 0.3 1.3 x 0.3	89 6	1	S ... P w 5868		16.0:		*
05872	10 43.1 + 49 48 161.43 + 57.16	08 – 20 – 020	0709	1.6: x 0.6: 0.9 x 0.4	74 6	2	Sb – c		15.1		
05873	10 43.3 + 63 30 143.59 + 48.58	11 – 13 – 037 NGC 3359	0662	8.0 x 4.8 6.8 x 4.0	170 4	5	S B c	SB(rs)c Sc –	11.0	+ 1008 + 1120	*12– –689
05874	10 43.4 + 26 10 208.24 + 62.02	04 – 26 – 005	1366	1.1 x 0.6 0.9 x 0.6	155 4 – 5	1	S ... 5881 group		15.5		*
05875	10 43.4 + 73 06 134.64 + 41.35	12 – 10 – 077 NGC 3348	0685 184	(2.0 x 2.0) (2.5 x 2.5)	-- ...	5	E		12.0	+ 2855 + 3010	*123 4678
05876	10 43.6 + 52 25 157.38 + 55.81	09 – 18 – 025	0673	1.4 x 0.8 1.3: x 0.9:	103 3 – 4:	3	Sc		15.7		
05877	10 43.6 + 77 07 131.48 + 38.15	13 – 08 – 038	1326 172	1.0 x 0.35 0.9 x 0.20	115 6	1	S ... np of 2, NGC 3329 – 97 group		15.3		*
05878	10 43.7 + 02 05 247.76 + 50.77	00 – 28 – 006 NGC 3365	1397	4.6 x 0.8 4.5 x 0.8	159 6	3	Sc		13.6		3
05879	10 43.8 + 60 11 147.20 + 50.94	10 – 16 – 007	0712	1.5 x 0.15 1.4 x 0.12	28 7	3	Sc	Sc +	15.7		*
05880	10 43.9 + 14 02 231.27 + 57.97	02 – 28 – 005 NGC 3367	0976	2.3 x 2.2 2.2 x 2.2	-- 1	5	S B c	SB(rs)c	12.0	+ 2879 + 2753	*123 4678

N G C   3343, 3348, 3351 – 53, 3356, 3359, 3362
3363, 3365, 3367

1	2	3	4	5	6	7	8	9	10	11	12
05881	10 44.0 + 26 12 208.21 + 62.15	04 - 26 - 006	1366	1.2 x 0.5 1.1 x 0.5	55 6	2	Sa Brightest of 3		15.0		*1
05882	10 44.1 + 12 05 234.43 + 57.01	02 - 28 - 006 NGC 3368 = M 96	0976	7.5 x 5.0 8.0 x 5.0	5: 3 - 4	5	Sa/Sb Leo group	SAB(rs)ab	10.0	+ 935 + 800	*12- -789
05883	10 44.2 + 54 18 154.60 + 54.77	09 - 18 - 026	0673	1.1 x 1.0 0.8: x 0.9:	-- ...	4	Dwarf irregular, I m Ir IV - V		15.6		
05884	10 44.3 + 26 48 206.97 + 62.31	05 - 26 - 006	1357 192	1.2 x 0.8 1.2 x 0.8	97: 3:	2	Sb: Brightest in group	SA(s)b:	14.4		*1
05885	10 44.3 + 30 19 199.38 + 62.63	05 - 26 - 005	1357	1.5 x 0.5 1.5 x 0.5	138 6	2	Sb ssp of 2		15.5		*
05886	10 44.4 - 01 07 251.48 + 48.65	00 - 28 - 009	1397	1.2 x 0.7 0.7 x 0.6	115: ...	1	S ...		14.4		
05887	10 44.4 + 17 32 225.36 + 59.67	03 - 28 - 008 NGC 3370	0463	2.9 x 1.7 2.8 x 1.7	148 4	5	Sc	SA(s)c	12.4	+ 1400 + 1290	*12 3468
05888	10 44.6 + 56 20 151.82 + 53.56	09 - 18 - 028	0673 189	1.6: x 1.5: 1.3: x 1.1:	-- ...	4	Irr	Ir I	15.2		
05889	10 44.7 + 14 20 230.96 + 58.28	02 - 28 - 007	0976	1.7: x 1.6: 1.6:: x 1.5::	-- ...	4	Dwarf spiral S V, Leo group?		15.0		*2
05890	10 44.7 + 72 40 134.89 + 41.75	12 - 10 - 082 NGC 3364	0685 184	1.8 x 1.7 1.8 x 1.7	-- 1	4	Sc	SAB(s)c Sc -	13.8		*12 35
05891	10 44.8 + 06 19 242.85 + 53.77	01 - 28 - 007 NGC 3376	0722	0.8 x 0.25 1.0 x 0.35	167 (6)	1	S: ...		14.4		3
05892	10 44.8 + 07 31 241.25 + 54.52	01 - 28 - 008	0722	1.0 x 0.9 1.1 x 0.9	-- 1 - 2	3	S B b Brightest of 3		14.5		*1
05893	10 44.8 + 39 12b 180.40 + 61.68	07 - 22 - 063	0690 188	(1.1 x 1.0) (1.1 x 0.9)	-- ...	1	E? nf of 2, contact, in group		(14.4)		*1
05894	10 44.9 + 26 33 207.54 + 62.41	05 - 26 - 008	1357 192	1.6 x 0.7 1.6 x 0.7	155 6	3	S B a 5884 group	SAB(s)ab	14.7		*1
05895	10 44.9 + 30 02 200.01 + 62.75	05 - 26 - 009	1357	1.2 x 0.25 1.2 x 0.25	73 6	1	S B? a - b		15.7		
05896	10 45.0 - 01 13 251.75 + 48.68	00 - 28 - 010	1397	1.5 x 0.5 1.4 x 0.5	132 6	2	Sa - b np of 2		15.3		*
05897	10 45.0 + 11 22 235.75 + 56.80	02 - 28 - 008	0976	2.8 x 0.9 2.8 x 0.9	75 6	4	Sc	SA c	14.0		
05898	10 45.0 + 33 59 191.39 + 62.65	06 - 24 - 014	0731	1.6 x 0.4 1.5 x 0.3	158 (6)	3	... p of 2	d - m	16.0:		*
05899	10 45.1 + 14 15 231.19 + 58.33	02 - 28 - 009 NGC 3377	0976	(4.0 x 2.5) (4.2 x 2.7)	35 ...	5	E Leo group		10.7	+ 718 + 593	*123 4678
05900	10 45.1 + 41 02 176.70 + 61.20	07 - 22 - 067	0690	1.4 x 0.25 1.3 x 0.25	38 6	1	S ...		15.6		
05901	10 45.1 + 43 26 172.09 + 60.36	07 - 22 - 066 NGC 3374	0690	1.4 x 1.0 1.2 x 0.9	160: 3	3	S B c nnf of 2	S B(rs)c	14.6		*13
05902	10 45.2 + 12 51 233.49 + 57.64	02 - 28 - 011 NGC 3379	0976	(3.8 x 3.8) (4.5 x 4.5)	-- ...	5	E Leo group		9.6	+ 877 + 746	*123 4678
05903	10 45.2 + 28 31 203.32 + 62.71	05 - 26 - 011	1357 192	1.1 x 0.7 1.2 x 0.7	15 4:	2	S B a - b	SAB(s) ...	15.1		
05904	10 45.2 + 66 37b 140.23 + 46.43	11 - 13 - 38a	0662	2.2: x 0.4 2.1: x 0.35	152 7	3	Sb ssf of 2, contact		(14.5)		*1C
05905	10 45.4 + 18 27 223.88 + 60.26	03 - 28 - 010 IC 642	0463	(1.4 x 1.3) (1.6 x 1.6)	-- ...	1	...		14.0		
05906	10 45.5 + 28 53 202.54 + 62.81	05 - 26 - 012 NGC 3380	1357 192	1.6 x 1.3 1.7 x 1.7	-- 1	1	S ...		13.6		*12 3
05907	10 45.5 + 66 24 140.40 + 46.62	11 - 13 - 000	0662	1.8: x 1.4: 1.4: x 1.0:	... ...	3	Dwarf		17.		
05908	10 45.6 + 05 11 244.53 + 53.19	01 - 28 - 009 NGC 3385	0722 190	1.4 x 0.8 1.7 x 1.1	97: ...	3	S0 s of 2		13.7		*3
05909	10 45.6 + 34 58 189.23 + 62.67	06 - 24 - 015 NGC 3381	0731	2.3 x 2.1 2.3 x 2.1	-- 1	3	S B ... Peculiar		12.8		*13
05910	10 45.6 + 38 40 181.41 + 61.96	07 - 22 - 068	0690	1.5: x 1.5: 1.4 x 1.0	-- 1 - 3	4	S B b	S B(r)b Sb +	14.7		
05911	10 45.7 + 12 54 233.53 + 57.77	02 - 28 - 012 NGC 3384	0976	(5.4 x 2.8) 5.6 x 2.5	53 5 - 6	4	S0 Leo group	S 0 -	10.0	+ 767 + 636	*123 46-9
05912	10 45.7 + 26 51 206.96 + 62.63	05 - 26 - 014	1357 192	1.3 x 1.2 1.4 x 1.3	-- 1	4	Sc Disturbed, 5884 group	SA(s)c	14.5		*
05913	10 45.8 + 01 07 249.43 + 50.48	00 - 28 - 11a	1397	0.8 x 0.30 0.8 x 0.30	110 ...	1	...		14.3		
05914	10 45.8 + 12 48 233.71 + 57.74	02 - 28 - 013 NGC 3389	0976	2.9 x 1.2 2.8 x 1.3	112 6	4	Sc Leo group	Sc +	12.0	+ 1334 + 1203	*123 4678
05915	10 45.8 + 48 18 163.42 + 58.31	08 - 20 - 027	0709	1.1 x 0.8 1.0 x 0.8	(25) ...	1	Sa - b		16.0:		

N G C   3364, 3368, 3370, 3374, 3376, 3377
3379 - 81, 3384, 3385, 3389

I C   642

1	2	3	4	5	6	7	8	9	10	11	12
05916	10 46.0 + 21 59 217.15 + 61.62	04 – 26 – 008	1366 194	1.1 x 1.0 0.9: x 0.9:	-- 1	3	Sc	SAB: cd	16.0:		
05917	10 46.0 + 46 59 165.60 + 58.98	08 – 20 – 029	0709	1.2 x 0.6 1.1 x 0.6	[163] ...	4	Irr	Ir I	15.0		
05918	10 46.1 + 65 50 140.89 + 47.08	11 – 13 – 039	0662	3. x 3. 2. x 2.	-- --	3	Dwarf		17.		*
05919	10 46.2 + 58 37 148.72 + 52.22	10 – 16 – 012	0712	1.0 x 0.20 1.0 x 0.22	80 7	2	Sa		15.5		
05920	10 46.3 + 14 29 231.07 + 58.70	02 – 28 – 014 NGC 3391	0976	1.0 x 0.7 0.9 x 0.5	[35:] ...	1	...		13.1		3
05921	10 46.4 + 28 11 204.10 + 62.95	05 – 26 – 016	1357 192	1.7 x 0.4 1.5 x 0.4	161 6	3	Sc – Irr		15.3		
05922	10 46.5 – 00 22 251.26 + 49.55	00 – 28 – 012	1397	1.1 x 0.7 1.1 x 0.7	10: ...	1	S B: ... npp of 2		14.7		*1
05923	10 46.5 + 07 11 242.15 + 54.64	01 – 28 – 011	0722	0.9 x 0.4 1.0 x 0.4	173 (6)	1	S0 – a?		14.4		
05924	10 46.5 + 22 17 216.62 + 61.82	04 – 26 – 011	1366 194	1.6 x 0.5 1.7 x 0.6	52 6	2	Sa		14.9		
05925	10 46.5 + 43 34 171.64 + 60.54	07 – 22 – 070	0690	1.4 x 0.15 1.2 x 0.15	157 7	3	Sc	Sc +	15.6		*
05926	10 46.5 + 77 13 131.24 + 38.16	13 – 08 – 042	1326 172	1.2 x 0.9 1.0 x 0.7	75 ...	1	S: ... P w 5939 NGC 3329 – 97 group		14.4		*
05927	10 46.6 + 33 02 193.41 + 63.06	06 – 24 – 016 IC 2604	0731	1.4 x 1.0 1.3: x 1.0:	40: ...	3	Irr NGC 3395 group		15.0		*12
05928	10 46.8 + 52 10 157.24 + 56.35	09 – 18 – 034	0673	1.0 x 1.0 1.0 x 1.0	-- --	2	E – S0		15.0		*
05929	10 46.9 + 05 04 245.03 + 53.36	01 – 28 – 000	0722 190	1.2 x 0.5 0.9 x 0.5	68 6	2	Sb – c		15.1		
05930	10 46.9 + 22 15 216.74 + 61.89	04 – 26 – 000	1366 194	[1.1 x 0.45] ...	[167] ...	3	S ... + companion distorted, bridge		16.5:		*
05931	10 47.0 + 33 14 192.95 + 63.13	06 – 24 – 017 NGC 3395	0731	1.8 x 1.0 1.8 x 1.0	[50] ...	3	S[c], p w 5935, contact, disrupted, brightest of 7		12.1	+ 1656 + 1622	*123V 4678A
05932	10 47.0 + 65 05 141.53 + 47.70	11 – 13 – 040	0662	1.7 x 0.7 1.5 x 0.6	52 6	2	Sb – c		15.2		
05933	10 47.1 + 16 30 227.76 + 59.81	03 – 28 – 14+15 NGC 3405	0463 193	[0.9 x 0.9] ...	... ...	2	Double system, contact, brightest in cluster		14.4		*13
05934	10 47.1 + 32 10 195.31 + 63.22	05 – 26 – 018	1357	1.8: x 1.1 1.4: x 1.2:	... ...	4	...	S B(s)dm	15.7		1
05935	10 47.1 + 33 15 192.91 + 63.15	06 – 24 – 018 NGC 3396	0731	3.7: x 1.4 3.7: x 1.4:	100 (6)	4	Irr, p w 5931, contact, disturbed, NGC 3395 group		12.6	+ 1645 + 1611	*123V 4678A
05936	10 47.2 + 36 36 185.61 + 62.73	06 – 24 – 019	0731	1.2 x 0.9 1.2 x 0.9	83: 2 – 3	4	S0	(R)SA: 0 +	14.4		1
05937	10 47.2 + 65 59 140.62 + 47.04	11 – 13 – 041 NGC 3394	0662	2.0 x 1.6 2.0 x 1.6	35: 2	4	Sc SA(rs)c sp of 2 Sc –		13.1		*3
05938	10 47.3 + 77 52 130.73 + 37.65	13 – 08 – 000	1326 172	1.0 x 0.15 1.0 x 0.12	[42] ...	3	Peculiar, tail P w 5942, disrupted?		16.0:		*
05939	10 47.4 + 77 12 131.20 + 38.20	13 – 08 – 046	1326 172	1.2: x 0.5: 1.0 x 0.4	[7] ...	1	Peculiar P w 5926 NGC 3329 – 97 group		14.9		*
05940	10 47.5 + 05 35 244.53 + 53.81	01 – 28 – 000	0722	1.2 x 0.3 1.2 x 0.3	113 6	3	S B b		15.2		
05941	10 47.5 + 41 44 175.01 + 61.39	07 – 22 – 071	0690	[0.8 x 0.7] ...	... ...	2	Double system Connected		14.5		*1
05942	10 47.6 + 77 50 130.73 + 37.69	13 – 08 – 000	1326 172	1.2 x 0.25 0.8 x 0.12	[167] ...	3	Peculiar, tail P w 5938, disrupted?		16.5:		*
05943	10 47.7 – 01 00 252.26 + 49.31	00 – 28 – 018	1397	1.1 x 0.9 1.1 x 1.0	-- --	3	Sb/S B c SAB bc Disturbed Sc –		14.8		*
05944	10 47.7 + 13 32 232.98 + 58.51	02 – 28 – 000	0976	1.1: x 1.1: 0.9: x 0.8:	-- ...	4	Dwarf V, Leo group?		15.7		*
05945	10 47.7 + 17 50 225.47 + 60.51	03 – 28 – 016	0463	1.9 x 0.8: 1.9 x 0.8	[95] ...	3	Irr IB m Ir I		15.1		
05946	10 47.7 + 79 25ab 129.61 + 36.38	13 – 08 – 45+46	1326 172	[1.2 x 0.6] ...	... ...	2	Double system Envelopes in contact		(14.3)		*
05947	10 47.8 + 19 55 221.53 + 61.34	03 – 28 – 017	0463	1.4 x 0.8: 1.4 x 0.6	[25] ...	2	Dwarf peculiar		15.0		*1
05948	10 48.0 + 16 00 228.84 + 59.78	03 – 28 – 000	0463	1.4 x 0.4 1.1: x 0.3	[35] ...	1	Dwarf?		17. – 18.		
05949	10 48.0 + 28 44 202.96 + 63.35	05 – 26 – 020 NGC 3400	1357 192	1.2 x 0.7 1.4 x 0.7	100 4 – 5	2	S B 0 – a		14.3		*12 3
05950	10 48.1 + 15 35 229.60 + 59.60	03 – 28 – 000	0463	2. x 1. ...	... ...	0	Pair of dwarfs?		18.		

N G C   3391, 3394 – 96, 3400, 3405

1	2	3	4	5	6	7	8	9	10	11	12
05951	10 48.2 + 36 28 185.81 + 62.95	06 - 24 - 000	0731	1.2 x 0.12 1.2 x 0.12	161 7	2	Sc	Sc +	16.0:		
05952	10 48.3 + 13 41 232.87 + 58.71	02 - 28 - 016 NGC 3412	0976	3.3 x 1.9 3.3 x 1.9	155 5	4	S0 Leo group		10.8	+ 861 + 735	*123 4678
05953	10 48.4 + 44 50 169.02 + 60.33	08 - 20 - 033 Mark 155	0709	0.45 x 0.25 0.45 x 0.25	85 ...	2	... Peculiar		13.2	+ 1800	
05954	10 48.4 + 55 39 152.13 + 54.40	09 - 18 - 038 IC 644	0673	1.0: x 0.35 0.9 x 0.35	78 6	1	S ...		14.4		*1
05955	10 48.4 + 72 02 135.11 + 42.42	12 - 10 - 087	0685 184	1.2 x 1.2 1.2 x 1.2	-- ...	2	E		14.4		*
05956	10 48.5 - 01 52 253.39 + 48.81	00 - 28 - 020 IC 651	1397	0.7 x 0.7 0.7 x 0.7	-- ...	1	... Peculiar		12.9		
05957	10 48.5 + 20 26 220.63 + 61.67	03 - 28 - 018	0463	1.2 x 0.25 1.1 x 0.2	122 6 - 7	2	S ...		14.7		1
05958	10 48.5 + 28 07 204.35 + 63.40	05 - 26 - 022	1357 192	1.6 x 0.3 1.6 x 0.3	179 7	2	Sb - c NGC 3414		15.6		*
05959	10 48.5 + 28 15 204.05 + 63.41	05 - 26 - 021 NGC 3414	1357 192	(3.0 x 2.6) (3.2 x 2.8)	... ...	2	[SB0]? or Peculiar? Brightest of 3		12.1	+ 1449 + 1391	*123 4678A
05960	10 48.5 + 33 01 193.38 + 63.46	06 - 24 - 024 NGC 3413	0731	1.9 x 0.8 1.9 x 0.9	178 7?	1	Sa? NGC 3395 group		13.1		*12 36
05961	10 48.5 + 48 13 163.12 + 58.74	08 - 20 - 000	0709	1.5 x 0.3 1.3: x 0.35	58 6 - 7	2	Sa - b		16.0:		*
05962	10 48.6 + 06 06 244.16 + 54.36	01 - 28 - 012 NGC 3423	0722	4.3 x 3.6 4.3 x 3.6	(10) 2:	5	Sc	SA(s)cd	12.1		*12- -689
05963	10 48.6 + 28 23 203.76 + 63.45	05 - 26 - 023 NGC 3418	1357 192	1.2 x 1.0 1.3 x 1.0	75: ...	3	S0/Sa		14.5		*12 3
05964	10 48.7 + 14 14 232.05 + 59.08	02 - 28 - 018 NGC 3419	0976	0.7 x 0.6 0.7 x 0.6	(115) ...	3	S0 P w 5965		13.4	+ 2982 + 2859	*12 3
05965	10 48.7 + 14 18 231.94 + 59.11	02 - 28 - 019	0976	1.7 x 0.15 1.6 x 0.15	137 7	2	S B b - c P w 5964		14.9		*
05966	10 48.8 + 08 34 240.85 + 55.94	02 - 28 - 020 NGC 3427	0976 195	(1.3 x 0.9) 1.3 x 0.8	77: ...	2	S0 - a		14.0		
05967	10 48.8 + 08 50 240.47 + 56.10	02 - 28 - 021 NGC 3425	0976 195	1.1 x 1.1 1.1 x 1.0	-- 1	3	S0 nnf of 2		14.5		*13
05968	10 48.8 + 09 33 239.44 + 56.53	02 - 28 - 022 NGC 3428	0976 195	1.6 x 0.7 1.5 x 0.6	170 6	4	Sb/S B b	SAB(s)b	14.1		3
05969	10 48.8 + 43 59 170.52 + 60.76	07 - 22 - 072 NGC 3415	0690	2.1 x 1.2 2.0 x 1.0	10 4 - 5	2	S0 - a		13.2		*12 348
05970	10 48.8 + 51 18 158.20 + 57.10	09 - 18 - 040 NGC 3406	0673	[1.1 x 1.0] ...	-- ...	2	Double system Contact, in group		13.7		*13
05971	10 48.8 + 67 02 139.44 + 46.35	11 - 13 - 000	0662	1.1 x 0.35 1.0 x 0.35	7 6	1	Sa - b		15.4		
05972	10 48.9 + 33 10 193.03 + 63.03	06 - 24 - 025 NGC 3424	0731	2.7 x 0.8 2.9 x 0.9	112 6	1	S ... NGC 3395 group		13.2		*12 3
05973	10 49.0 + 00 32 250.97 + 50.63	00 - 28 - 000	1397 191	1.1 x 0.4 1.0 x 0.4	112 6	2	S B? a - b		16.0:		*
05974	10 49.0 + 04 51 245.88 + 53.61	01 - 28 - 013	0722 190	2.1 x 0.8 1.9 x 0.7	130 6	2	Sc		14.7		
05975	10 49.0 + 18 44 224.03 + 61.16	03 - 28 - 020 NGC 3426	0463	(1.8 x 1.5) (1.8 x 1.6)	-- ...	1	...		13.9		*
05976	10 49.0 + 55 52 151.76 + 54.33	09 - 18 - 043 NGC 3398	0673	1.5 x 1.2 1.3 x 1.1	(30) 2	4	Sc	SAB cd:	15.3		*23
05977	10 49.1 + 58 43 148.18 + 52.43	10 - 16 - 016 NGC 3408	0712	0.9 x 0.7 0.9 x 0.7	175: ...	1	S ...		14.1		23
05978	10 49.1 + 61 39 144.84 + 50.37	10 - 16 - 017 NGC 3407	0712	1.4 x 0.7 1.3 x 0.7	15 ...	2	E - S0		14.8		3
05979	10 49.2 + 68 14 138.30 + 45.45	11 - 13 - 043	0662	1.8: x 1.4: 2.0: x 1.6:	-- ...	4	Dwarf irregular, IA m Ir IV - V		15.6		
05980	10 49.4 + 04 04 246.97 + 53.16	01 - 28 - 015 NGC 3434	0722	2.3 x 1.9 2.5 x 2.0	(5) 1 - 2	4	Sb Unilateral? Brightest of 3	SA(r)b	13.4		*13
05981	10 49.4 + 10 25 238.31 + 57.16	02 - 28 - 023 NGC 3433	0976	4.0: x 3.6: 3.5: x 3.0:	50: 1 - 2	4	Sc Leo group?	SAB(s)c	13.6		*123 4678
05982	10 49.4 + 33 13 192.90 + 63.64	06 - 24 - 026 NGC 3430 = 2613*	0731	4.5 x 2.4 4.5 x 2.4	30 5	5	Sc NGC 3395 group, Sc -	SAB(rs)c	12.2	+ 1742 + 1709	*123 4678
05983	10 49.4 + 36 52 184.84 + 63.11	06 - 24 - 027	0731	1.2: x 1.0: 1.2: x 1.0:	-- ...	4	Dwarf V, Pw 5986		17.		*1
05984	10 49.5 + 30 20 199.41 + 63.75	05 - 26 - 24+25	1357	[2.4 x 1.8] ...	... ...	3	Double system, strongly distorted, a) unilateral		14.6		*1V A
05985	10 49.6 - 00 17 252.03 + 50.15	00 - 28 - 022 IC 653	1397	2.2 x 1.1 2.3 x 1.1	55 5	2	Sa		14.2		1

N G C   3398, 3406 - 08, 3412 - 15, 3418, 3419   I C 644, 651, 653, 2613
         3423 - 28, 3430, 3433, 3434

1	2	3	4	5	6	7	8	9	10	11	12
05986	10 49.7 + 36 53 184.78 + 63.16	06 – 24 – 028 NGC 3432	0731	7.5 x 2.0 7.3 x 2.0	38 6	4	... P w 5983	S B m	11.7	+ 609 + 594	*12–V −789A
05987	10 49.7 + 39 32 179.16 + 62.49	07 – 22 – 074	0690	1.0 x 0.3 0.7 x 0.25	45 6	2	S B b – c		15.7		
05988	10 49.8 + 10 50 237.78 + 57.48	02 – 28 – 025 NGC 3438	0976	0.9 x 0.8 0.7 x 0.7	–– (1)	1	S? ... n of 2		14.3		*13
05989	10 49.8 + 20 03 221.62 + 61.82	03 – 28 – 024	0463	1.5 x 0.5 1.5 x 0.5	[127] ...	1	Irr or Peculiar, tails		14.2		
05990	10 49.8 + 34 45 189.46 + 63.55	06 – 24 – 029	0731	1.5 x 0.4 1.5 x 0.4	14 6	2	Sa – b		15.1		1
05991	10 49.8 + 49 52 160.24 + 58.04	08 – 20 – 036	0709	(1.1 x 0.7) (1.3 x 1.1)	155: ...	2	E – S0 6013 group		14.7		*
05992	10 49.8 + 66 00 140.31 + 47.20	11 – 13 – 000	0662	1.0 x 0.3 0.9 x 0.2	127 (6)	2	...	d – m	16.5:		
05993	10 49.9 + 07 30 242.64 + 55.50	01 – 28 – 017 NGC 3441	0722	0.8 x 0.4 0.8 x 0.4	5 5	1	S ...		13.9		
05994	10 49.9 + 10 17 238.64 + 57.18	02 – 28 – 000	0976	1.7 x 0.1 1.4 x 0.15	142 7	2	Sc		15.6		*
05995	10 49.9 + 23 12 215.18 + 62.82	04 – 26 – 016 NGC 3437	1366	2.8 x 0.9 2.4 x 0.8	122 6	2	Sc		12.6		*12 3468
05996	10 50.0 + 40 39 176.83 + 62.20	07 – 22 – 075	0690	1.1: x 0.8: 0.9: x 0.7:	–– 1 – 2:	2	...	S dm:	16.5:		
05997	10 50.0 + 73 57 133.47 + 40.95	12 – 10 – 089 NGC 3403	0685 184	3.5 x 1.5 3.2: x 1.5:	73 6	3	Sc	Sc –	13.3	+ 1244 + 1403	*12 3478
05998	10 50.2 + 50 34 159.08 + 57.70	08 – 20 – 037 Mark 156	0709	0.9 x 0.35 0.8 x 0.25	[110] ...	3	... Peculiar, disrupted pair?		14.5	+ 1500	
05999	10 50.3 + 07 54 242.20 + 55.82	01 – 28 – 000	0722	1.6:: x 1.4:: 1. x 1.	–– ...	4	Dwarf irregular Ir V		18. – 19.		
06000	10 50.3 + 17 50 225.99 + 61.07	03 – 28 – 025 NGC 3443	0463	2.6 x 1.2 2.3 x 1.2	145 5 – 6	4	Sc	SA d	14.3		1
06001	10 50.3 + 34 10 190.73 + 63.73	06 – 24 – 000 NGC 3442, Mark 418	0731	0.65 x 0.50 0.55 x 0.50	(30) ...	1	... Peculiar		13.2		
06002	10 50.3 + 37 41 182.99 + 63.10	06 – 24 – 031	0731 201	1.3 x 1.1 1.3 x 0.9	(110) ...	3	S B b	S B(s)b Sb –	15.5		
06003	10 50.4 + 04 54 246.21 + 53.91	01 – 28 – 018	0722	0.5 x 0.5 0.7 x 0.6	–– ...	1	...		14.1		
06004	10 50.4 + 10 29 238.47 + 57.40	02 – 28 – 000 NGC 3444	0976	1.1 x 0.12 0.9 x 0.15	19 7	2	Sb – c		15.4		*
06005	10 50.5 + 46 16 166.12 + 60.00	08 – 20 – 000	0709	1.0: x 0.7: 1.0: x 0.7:	(140) ...	2	...	d – m	17.		*
06006	10 50.7 + 17 02a 227.57 + 60.82	03 – 28 – 027 NGC 3447a	0463	4.2 x 2.3 3.6: x 2.2:	0 ...	4	... P w 6007, contact	S B(s)m	14.3	+ 965 + 855	*12 3V̄
06007	10 50.7 + 17 02b 227.57 + 60.82	03 – 28 – 028 NGC 3447b	0463	1.7 x 0.8 1.5 x 0.8	[110] ...	4	Irr P w 6006, contact, Ir I	I		+ 1014 + 904	*12 3V̄
06008	10 50.7 + 51 03 158.26 + 57.49	09 – 18 – 047	0673	1.9 x 0.25 1.8 x 0.25	154 6 – 7	3	S B b		15.5		
06009	10 50.7 + 57 24 149.54 + 53.49	10 – 16 – 019 NGC 3440	0712 229	2.3 x 0.5 2.3 x 0.5	48 6	1	S ... NGC 3445 group		14.0		23
06010	10 50.7 + 62 47 143.42 + 49.68	10 – 16 – 000	0712	1.1 x 0.9: 1.1:: x 0.9::	(175) ...	3	Dwarf (spiral) S: V		17.		*
06011	10 50.9 – 00 20 252.46 + 50.34	00 – 28 – 000	1397	1.5: x 0.4 0.7 x 0.2	143 (6)	1	... In group	d – m	16.5:		*
06012	10 51.0 + 27 10 206.62 + 63.84	05 – 26 – 026	1357 192	1.1 x 0.12 1.2 x 0.15	157 7	2	Sb – c In group		16.0:		*
06013	10 51.0 + 49 55 159.96 + 58.17	08 – 20 – 041	0709	0.9 x 0.7 0.8 x 0.7	(45) ...	2	S0 Brightest in group	n of 2	13.9		*
06014	10 51.1 + 10 00 239.38 + 57.25	02 – 28 – 000 NGC 3444	0976 200	1.6: x 0.9: 1.3: x 0.7:	70: ...	2	...	SAB: dm:	15.2		
06015	10 51.1 + 46 17 165.99 + 60.09	08 – 20 – 039	0709 203	1.0 x 0.8 1.0 x 0.8	0: 2:	4	Sc		14.9		
06016	10 51.2 + 54 34 153.13 + 55.40	09 – 18 – 052	0673	2.2: x 1.6: 1.8:: x 1.2::	[30] ...	4	Dwarf irregular Ir V, P w 6024, bridge:		17.		*1
06017	10 51.3 + 44 53 168.47 + 60.77	08 – 20 – 042	0709	1.4 x 0.15 1.2 x 0.15	80 7	3	Sc		16.0:		
06018	10 51.4 + 20 55 220.15 + 62.47	04 – 26 – 000	1366	1.2 x 1.2 1.1: x 1.1:	–– ...	3	Dwarf irregular Ir IV – V or V		17.		
06019	10 51.4 + 59 18 147.16 + 52.24	10 – 16 – 021	0712	1.1 x 0.5 1.0 x 0.5	55 5 – 6	3	Sa		15.6		
06020	10 51.5 + 21 24 219.18 + 62.65	04 – 26 – 019	1366	1.4 x 0.2 1.3 x 0.2	55 7	3	Sb spp of 2		16.0:		*

N G C   3403, 3432, 3438, 3440 – 44, 3447

1	2	3	4	5	6	7	8	9	10	11	12
06021	10 51.5 + 57 15 149.61 + 53.67	10 – 16 – 023 NGC 3445	0712	1.7 x 1.6 1.4 x 1.4	-- (1)	5	Irr SAB(s)m Brightest in group, Ir I		12.8	+ 1984 + 2069	*123V 46̲78A
06022	10 51.6 + 18 04 225.82 + 61.45	03 – 28 – 000	0463	1.3: x 0.7: 1.2: x 0.7:	[10] ...	3	Dwarf IV – V or V		17.		
06023	10 51.6 + 27 30 205.90 + 64.02	05 – 26 – 028 NGC 3451	1357 192	1.9 x 0.9 1.9: x 0.8	50 5 – 6	3	Sc 	Sc +	13.5		3
06024	10 51.6 + 54 35 153.05 + 55.44	09 – 18 – 055 NGC 3448	0673	5.3: x 1.5 3.0 x 1.5	[65] ...	4	... I 0 P w 6016, bridge, Ir II		12.2	+ 1404 + 1476	*123 46̲78A
06025	10 51.6 + 61 34 144.59 + 50.64	10 – 16 – 022 NGC 3435	0712	1.9 x 1.3 2.0 x 1.4	35 3	3	Sb Singular		14.2		*3
06026	10 51.8 + 17 36 226.74 + 61.30	03 – 28 – 030 NGC 3454	0463	2.4 x 0.5: 2.4 x 0.5	116 ...	1	... P w 6028		14.1	+ 1167 + 1060	*12 36̲
06027	10 51.8 + 64 17 141.75 + 48.64	11 – 13 – 000	0662	1.4:: x 0.8:: 1. x 1.	-- 	3	Dwarf irregular Ir V		17.		
06028	10 51.9 + 17 33 226.86 + 61.30	03 – 28 – 031 NGC 3455	0463	2.8 x 1.8 2.6: x 1.7:	80 3 – 4	2	Sb P w 6026		13.1	+ 1113 + 1006	*12 346̲8
06029	10 52.0 + 49 59 159.68 + 58.27	08 – 20 – 046 Mark 157	0709	1.0 x 0.8 0.9 x 0.8	(35) ...	2	... Peculiar, 6013 group		14.0	+ 1500	
06030	10 52.1 + 17 53 226.27 + 61.48	03 – 28 – 032 NGC 3457	0463	1.0 x 1.0 1.0 x 0.9	-- 	1	...		13.0		*3
06031	10 52.2 + 29 48 200.66 + 64.32	05 – 26 – 030	1357	1.1: x 1.1: 1.1: x 0.7:	-- 1:	1	S ...		15.7		*
06032	10 52.4 + 73 10 133.89 + 41.68	12 – 11 – 002	0714 184	1.7 x 0.7 1.5: x 0.6:	78 6	3	Sc 	Sc +	16.5:		
06033	10 52.6 + 42 34 172.64 + 61.96	07 – 23 – 002	1349	1.3 x 0.7 1.3 x 0.7	135 5	3	Sb p of 2		15.5		*
06034	10 52.7 + 07 58 242.77 + 56.33	01 – 28 – 019 NGC 3462	0722	(1.7 x 1.2) (2.6 x 1.7)	(60) ...	3	S0: Singular, plume		13.4		*3
06035	10 52.8 + 17 24 227.33 + 61.43	03 – 28 – 035	0463	1.3 x 1.2 1.1: x 1.1:	-- ...	2	Dwarf irregular, IB m Ir IV – V		15.2		*1
06036	10 53.0 + 37 07 183.94 + 63.75	06 – 24 – 036	0731 201	1.7 x 0.35 1.7 x 0.35	101 6	2	Sa – b		14.6		
06037	10 53.0 + 57 24 149.19 + 53.71	10 – 16 – 026 NGC 3458	0712 229	1.4 x 0.8 1.4 x 0.8	5: 4 – 5	3	S0 NGC 3445 group		13.2		*12 347̲8
06038	10 53.2 + 47 40 163.25 + 59.70	08 – 20 – 051	0709	1.2 x 1.1 0.8 x 0.7	-- 1	3	Sb		15.5		
06039	10 53.3 + 57 02 149.59 + 53.99	10 – 16 – 029	0712	1.7 x 0.6 1.6 x 0.5	25 6	4	... S d – dm		16.5:		
06040	10 53.4 + 02 46 249.68 + 52.99	01 – 28 – 021	0722	1.0 x 0.55 1.0 x 0.6	155: 4 – 5	1	S ...		15.0		
06041	10 53.4 + 41 53 173.87 + 62.36	07 – 23 – 000	1349	1.1 x 0.25 1.1 x 0.2	64 ...	2	... dm – m		17.		
06042	10 53.6 + 10 01 240.04 + 57.76	02 – 28 – 028 NGC 3466	0976 200	1.1 x 0.6 1.2 x 0.7	55 4 – 5	1	Sa – b NGC 3467 group		14.6		*12 3
06043	10 53.6 + 15 29 231.04 + 60.72	03 – 28 – 037	0463	1.3 x 0.6 1.0 x 0.6	73 5:	3	Sc		15.5		
06044	10 53.9 + 20 40 221.10 + 62.94	04 – 26 – 000	1366	1.0 x 0.15 0.8 x 0.15	47 7	2	Sc		17.		
06045	10 54.1 + 10 02 240.15 + 57.87	02 – 28 – 030 NGC 3467	0976 200	(0.9 x 0.8) (0.9 x 0.8)	-- (1)	2	S0 Brightest in group		14.2		*3
06046	10 54.2 + 07 10 244.30 + 56.11	01 – 28 – 023	0722 198	1.0 x 0.7 0.9 x 0.6	60: ...	1	S B c Distorted:		14.9		*1V
06047	10 54.3 + 37 30 182.96 + 63.91	06 – 24 – 037	0731 201	1.2 x 0.6 1.1 x 0.6	[25] ...	1	S B ...		15.7		
06048	10 54.6 + 41 13 175.03 + 62.82	07 – 23 – 006 NGC 3468	1349	(1.6 x 1.0) (1.8 x 1.2)	8 4	4	S0 nff of 2		14.2		*3
06049	10 54.9 + 05 58 246.12 + 55.45	01 – 28 – 000	0722 198	1.0 x 0.6 1.3 x 0.7	67 4 – 5	2	Sa – b		15.2		
06050	10 55.2 + 40 28 176.48 + 63.19	07 – 23 – 007	1349	1.1 x 0.15 1.1 x 0.15	46 7	2	Sc		16.0:		*
06051	10 55.2 + 76 43 131.08 + 38.85	13 – 08 – 049	1326 172	1.1 x 0.6 0.9 x 0.4	140: 5:	3	Sc n of 2, contact:, Sc +		16.0:		*
06052	10 55.4 + 17 23 227.94 + 61.98	03 – 28 – 041 NGC 3473	0463 196	1.1 x 0.9 1.2 x 0.9	(40) 2 – 3	2	S B b: P w NGC 3474		14.8		*13
06053	10 55.5 + 06 18 245.85 + 55.78	01 – 28 – 024	0722 198	1.1 x 1.0 1.1 x 1.1	-- 1	3	Sc		15.1		
06054	10 55.5 + 20 21 222.04 + 63.18	03 – 28 – 000	0463	1.2 x 0.1 1.0 x 0.1	12 7	2	Sc		15.6		
06055	10 55.5 + 49 03 160.55 + 59.27	08 – 20 – 057	0709	1.2: x 0.2: 0.9: x 0.2:	153 ...	2	Irr		17.		

N G C   3435, 3445, 3448, 3451, 3454, 3455, 3457
3458, 3462, 3466 – 68, 3473

1	2	3	4	5	6	7	8	9	10	11	12
06056	10 55.6 + 75 28 131.94 + 39.91	13 - 08 - 048 NGC 3465 = 3500	1326 172	1.5 x 1.2 1.3 x 1.2	-- 1	3	Sb NGC 3523 group, Sb -		14.6		*3
06057	10 55.7 + 01 54 251.37 + 52.78	00 - 28 - 024	1397 197	[1.1 x 0.7] ...	... ...	1	Triple system		14.9		*1
06058	10 55.7 + 24 30 213.06 + 64.42	04 - 26 - 022 NGC 3475	1366	1.7 x 1.2 1.6 x 1.1	65 3	4	Sa Brightest in group		14.6		*13
06059	10 55.7 + 55 52 150.68 + 55.03	09 - 18 - 062	0673	1.7 x 0.7 1.7 x 0.7	[15] ...	3	Sb/Sc In group, distorted	SAB(s)bc	15.7		*1
06060	10 55.7 + 59 47 145.97 + 59.28	10 - 16 - 038 NGC 3470	0712 229	1.7 x 1.3 1.3 x 1.2	(170) 2	4	Sa n of 2	SA a	14.3		*23
06061	10 55.8 + 55 32 151.09 + 55.27	09 - 18 - 061	0673	1.1 x 0.3 1.2 x 0.4	12 6	2	S B c		16.0:		
06062	10 56.0 + 09 19 241.76 + 57.81	02 - 28 - 034	0976 200	(1.4 x 0.7) 1.5 x 0.8	25 5	3	S B 0		13.7		
06063	10 56.0 + 25 25 211.00 + 64.67	04 - 26 - 024	1366	1.5 x 0.20 1.4 x 0.20	65 7	1	S ...		15.4		
06064	10 56.0 + 61 48 143.73 + 50.82	10 - 16 - 039 NGC 3471, Mark 158	0712	1.9 x 0.9 2.0 x 1.1	14 5	3	Sa sp of 2		13.0	+ 3500	*13
06065	10 56.3 + 77 12 130.68 + 38.47	13 - 08 - 051	1326 172	1.3 x 0.35 1.1 x 0.3	105 6	2	Sb		15.6		*
06066	10 56.4 + 06 47 245.47 + 56.27	01 - 28 - 026	0722 198	1.6: x 0.3 1.5 x 0.35	38 6	1	Sa?		15.0		
06067	10 56.6 + 02 10 251.33 + 53.13	00 - 28 - 025	1397 197	[1.4 x 0.7] ...	... ...	1	Triple system		15.2		1
06068	10 56.6 + 05 33 247.17 + 55.49	01 - 28 - 000	0722	1.3 x 0.6 (1.5 x 0.7)	5: ...	1	S0?		15.4		
06069	10 56.6 + 46 23 164.85 + 60.86	08 - 20 - 059 NGC 3478	0709 203	2.6 x 1.3 2.3 x 1.2	132 5	4	S B b	S B(rs)b Sb +	13.7		*12 348
06070	10 56.9 + 33 39 191.54 + 65.15	06 - 24 - 000	0731	0.60 x 0.50 0.55 x 0.45	-- ...	1	... Peculiar		13.3		
06071	10 57.0 + 50 19 158.27 + 58.73	08 - 20 - 063	0709	(1.4 x 0.8) (1.4 x 0.8)	135: ...	2	E Brightest in group		14.4		*
06072	10 57.1 + 10 20 240.53 + 58.64	02 - 28 - 035	0976 200	1.0 x 0.9 1.0 x 1.0	-- 1	2	Sa - b 6081 group		14.9		*
06073	10 57.1 + 17 55 227.28 + 62.58	03 - 28 - 043	0463	[1.3 x 0.9] ...	... ...	2	Double system, contact, distorted, NGC 3487 group		15.3		*1V A
06074	10 57.1 + 51 12 156.91 + 58.21	09 - 18 - 065	0673	0.9 x 0.8 0.50 x 0.45	-- ...	1	... Peculiar		14.3		*
06075	10 57.3 + 46 00 165.41 + 61.16	08 - 20 - 065	0709 203	1.8 x 0.8 1.8 x 0.8	107 5 - 6	2	Sb		14.4		
06076	10 57.3 + 46 11 165.08 + 61.07	08 - 20 - 064	0709 203	0.8 x 0.6 0.9 x 0.8	(150) ...	1	E?		14.5		
06077	10 57.4 + 15 06 232.66 + 61.33	03 - 28 - 044 NGC 3485	0463	2.6 x 2.4 2.3 x 2.3	-- 1	4	S B b Leo group?, Sb +	S B(r)bc	12.8		*12 3468
06078	10 57.5 + 12 30 237.19 + 59.97	02 - 28 - 037	0976 199	1.2 x 0.6 1.3 x 0.6	126 5	2	Sb - c		15.1		*
06079	10 57.6 + 29 15 202.06 + 65.47	05 - 26 - 032 NGC 3486	1357	7.2 x 5.1 7.2 x 5.1	80 3	5	Sc	SAB(r)c Sc -	11.2	+ 1116 + 1067	*12- -9
06080	10 57.6 + 61 36 143.72 + 51.09	10 - 16 - 042	0712	2.1 x 0.2 1.8: x 0.2	129 7	3	Sc	Sc +	15.6		
06081	10 57.7 + 10 19 240.73 + 58.75	02 - 28 - 038	0976 200	1.1 x 0.9: 0.9 x 0.7:	... ...	1	Double system Disrupted		14.8		*1
06082	10 57.7 + 14 10 234.40 + 60.92	02 - 28 - 039 NGC 3489	0976	3.2 x 1.9 3.5 x 2.1	70 4	4	S0 Leo group	S 0 +	10.9	+ 690 + 570	*123 4678
06083	10 57.7 + 16 57 229.29 + 62.28	03 - 28 - 046	0463	1.4 x 0.15 1.4 x 0.15	143 7	2	Sb - c		15.2		1
06084	10 57.7 + 29 58 200.33 + 65.51	05 - 26 - 033	1357 206	1.0 x 0.5 0.7 x 0.3	175 5:	1	Sa - b		15.2		
06085	10 57.7 + 35 55 186.17 + 64.94	06 - 24 - 000	0731 210	1.0 x 0.12 0.9 x 0.15	53 7	1	Sb - c		16.5:		
06086	10 57.9 + 10 13 240.94 + 58.73	02 - 28 - 000	0976 200	1.0 x 0.15 1.0 x 0.15	14 7	3	Sc 6081 group		15.5		*
06087	10 58.0 + 02 24 251.48 + 53.54	00 - 28 - 027	1397	1.2: x 1.1: 1.1 x 1.1	-- 1	3	S B b	(R')S B(rs)b	15.0		*
06088	10 58.0 + 12 26 237.43 + 60.04	02 - 28 - 041 NGC 3491	0976 199	0.8 x 0.8 1.0 x 1.0	-- ...	2	E - S0		14.1		*3
06089	10 58.0 + 38 21 180.63 + 64.38	06 - 24 - 043	0731 201	1.1 x 0.9 0.9 x 0.8	-- 1 - 2	3	Sc		15.4		
06090	10 58.0 + 75 29 131.76 + 39.98	13 - 08 - 052	1326 172	1.5 x 0.7 1.0 x 0.5	54 5	2	Sa - b NGC 3523 group		14.8		*

N G C   3465, 3470, 3471, 3475, 3478, 3485
3486, 3489, 3491, 3500

1	2	3	4	5	6	7	8	9	10	11	12
06091	10 58.1 + 10 09 241.10 + 58.73	02 – 28 – 043	0976 200	1.0 x 0.35 1.1 x 0.35	2 6	3	S B b 6081 group		15.1		*
06092	10 58.1 + 17 51a 227.63 + 62.76	03 – 28 – 047 NGC 3487	0463	1.0 x 0.4 1.0 x 0.4	153 6	1	S ... spp of 2 Brightest in group		(14.6)		*1
06093	10 58.2 + 11 00 239.81 + 59.25	02 – 28 – 044	0976 202	1.2 x 1.0 1.4 x 1.1	-- 1 – 2	3	Sb/S B c SAB(rs)bc NGC 3492 group, Sb +:		14.8		*
06094	10 58.3 + 10 46 240.20 + 59.13	02 – 28 – 045 NGC 3492	0976 202	1.1 x 0.7 1.1 x 0.7	[100] ...	1	Double system Brightest in group		14.0		*
06095	10 58.3 + 19 21 224.65 + 63.42	03 – 28 – 000	0463	1.5: x 0.5 1.0: x 0.4:	5 (6)	1	Dwarf:		17.		
06096	10 58.4 + 57 57 147.67 + 53.84	10 – 16 – 045 NGC 3488	0712 229	2.1 x 1.4 1.9 x 1.3	175: 3	4	Sc		13.7		1̲2̲3
06097	10 58.6 + 30 04 200.10 + 65.71	05 – 26 – 035	1357 206	1.1 x 0.3 0.9 x 0.3	157 6	1	S0 – a		15.6		*
06098	10 58.7 + 03 54 249.89 + 54.73	01 – 28 – 027 NGC 3495	0722	4.8 x 1.0 4.8 x 1.1	20 6	4	Sc 	Sc +	13.1	+ 994 + 831	*12̲ 346̲8
06099	10 58.7 + 28 00 205.11 + 65.63	05 – 26 – 036 NGC 3493	1357 206	1.1 x 0.35 1.0 x 0.35	84 6	1	S ...		15.3		*3
06100	10 58.7 + 45 55 165.30 + 61.41	08 – 20 – 066	0709 203	0.9 x 0.6 0.9 x 0.6	10 3 – 4:	1	Sa?		14.1		
06101	10 58.9 + 47 23 162.69 + 60.66	08 – 20 – 068	0709 209	1.6: x 0.15 1.4 x 0.15	64 7	2	Sb – c		16.0:		
06102	10 59.0 + 28 58 202.78 + 65.76	05 – 26 – 000	1357 206	1.1 x 0.9: 1.2: x 0.9:	[140] ...	3	Dwarf irregular Ir IV – V		15.6		
06103	10 59.1 + 45 30 165.98 + 61.68	08 – 20 – 069 Mark 161	0709 203	0.7 x 0.45 0.8 x 0.7	... ...	2	... Peculiar		13.4	+ 6000	*
06104	10 59.2 + 16 52 229.80 + 62.56	03 – 28 – 049	0463	1.5 x 0.4 1.5 x 0.4	50 6	2	Sb – c P w 6112		14.4		*
06105	10 59.2 + 75 24 131.74 + 40.09	13 – 08 – 053 NGC 3523	1326 172	1.6 x 1.5 1.5 x 1.4	-- 1	3	Sb Brightest of 3, Sb +		13.8		*3
06106	10 59.3 + 46 09 164.77 + 61.38	08 – 20 – 070	0709 203	1.7 x 0.6 1.6 x 0.5	160 6	2	S B? b		14.5		*
06107	10 59.5 + 38 46 179.50 + 64.54	07 – 23 – 013 IC 2620	1349 201	1.1 x 1.0 1.1 x 0.9	-- ...	2	... Compact		14.9		*
06108	10 59.6 + 37 18 182.78 + 64.98	06 – 24 – 000	0731 201	1.0 x 0.12 0.8 x 0.15	89 7	1	S ...		16.5:		
06109	10 59.6 + 50 52 156.95 + 58.72	09 – 18 – 073	0673	1.4 x 1.4 1.2 x 1.2	-- 1	4	Sc SAB(r)c P w 6114 Sc –		15.3		*
06110	10 59.6 + 59 24 145.81 + 52.89	10 – 16 – 047	0712 229	(1.7 x 0.8) (1.7 x 0.8)	(65) ...	1	...		15.0		
06111	10 59.9 + 03 22 250.90 + 54.56	01 – 28 – 000	0722	1.0 x 0.22 1.0 x 0.2	2 6 – 7	1	S ...		15.2		
06112	10 59.9 + 17 00 229.71 + 62.77	03 – 28 – 050	0463	2.3 x 0.8 2.2: x 0.8	123 6	2	Sc – Irr S d? P w 6104		14.5		*
06113	10 59.9 + 52 26 154.59 + 57.77	09 – 18 – 000	0673	1. x 1. ...	-- ...	2	Dwarf V		19.		
06114	11 00.0 + 50 57 156.75 + 58.72	09 – 18 – 74+75	673	[1.5 x 0.9] ...	... ...	2	Double system Contact, p w 6109		14.5		*1
06115	11 00.0 + 56 29 149.18 + 55.04	09 – 18 – 080 NGC 3499	0673	0.8 x 0.7 0.8 x 0.7	-- ...	1	...		14.3		23
06116	11 00.2 + 18 15 227.30 + 63.38	03 – 28 – 051 NGC 3501	0463	3.8 x 0.5 3.8 x 0.5	27 7	3	Sc P w 6123		13.8		*1
06117	11 00.2 + 50 29 157.42 + 59.03	08 – 20 – 073	0709	1.0 x 0.7 0.9 x 0.7	143 3:	3	S B a		14.5		
06118	11 00.4 + 28 15 204.58 + 66.02	05 – 26 – 039 NGC 3504	1357 206	2.5 x 2.5 2.6 x 2.6	-- 1	5	Sa/S B b (R)SAB(s)ab P w 6128		11.5	+ 1526 + 1473	*12̲3 46̲7̲8
06119	11 00.6 + 03 36 250.84 + 54.85	01 – 28 – 031	0722	0.5 x 0.20 0.6 x 0.25	17 ...	1	...		14.4		
06120	11 00.6 + 11 21 239.94 + 59.93	02 – 28 – 047 NGC 3506	0976	1.3 x 1.3 1.2 x 1.1	-- 1	1	Sc? Singular, p w 6122		12.9		*12̲ 346̲8
06121	11 00.6 + 39 31 177.69 + 64.49	07 – 23 – 016	1349 201	1.7: x 0.4 0.8 x 0.2	97 (6 – 7)	1	...		15.5		*1
06122	11 00.8 + 11 24 239.91 + 60.00	02 – 28 – 000	0976	1.3: x 1.1: 1.0: x 1.0:	-- ...	2	Dwarf: spiral, SAB m: P w 6120		16.0:		*
06123	11 00.8 + 18 24 227.13 + 63.58	03 – 28 – 053 NGC 3507	0463	3.4 x 2.8 3.4 x 2.8	(110) 2	4	S B b S B(s)b P w 6116		11.4		*13
06124	11 00.8 + 32 08 195.00 + 66.11	05 – 26 – 000	1357 204	1.5 x 0.2 1.3 x 0.25	92 7	1	S ...		16.5:		
06125	11 00.8 + 45 27 165.75 + 61.96	08 – 20 – 074	0709 203	1.4 x 0.55 1.4 x 0.6	33 6	3	S B a – b (R')S B(s) ... nf of 2		14.8		*

N G C   3487, 3488, 3492, 3493, 3495, 3499
       3501, 3504, 3506, 3507, 3523

I C   2620

1	2	3	4	5	6	7	8	9	10	11	12
06126	11 00.9 + 29 10 202.34 + 66.19	05 – 26 – 040 NGC 3510	1357 204	4.2 x 0.9 4.3 x 0.9	163 6 – 7	4	...	S B m	13.6	+ 719 + 670	*123 467̄8
06127	11 01.1 + 50 05 157.87 + 59.39	08 – 20 – 076	0709	0.9 x 0.7 0.9 x 0.7	(140) 1 – 2	3	Sc		14.4		*
06128	11 01.3 + 28 18 204.49 + 66.23	05 – 26 – 041 NGC 3512	1357 206	1.7 x 1.6 1.6 x 1.6	-- 1	5	Sc P w 6118	SAB(rs)c	12.9	+ 1502 + 1449	*123 467̄8
06129	11 01.3 + 50 17 157.52 + 59.29	08 – 20 – 079	0709	1.0: x 0.9: 0.6 x 0.6	-- 1	2	Sb – c		15.2		
06130	11 01.5 + 08 38 244.38 + 58.43	02 – 28 – 048	0976	1.1 x 1.1 1.1 x 1.0	-- 1	4	Sc	SAB(s)cd	15.0		
06131	11 01.6 + 44 18 167.74 + 62.66	07 – 23 – 020	1349	(1.3 x 0.8) (1.3 x 0.9)	95 3 – 4	1	S0		14.7		*1
06132	11 01.7 + 38 28 179.86 + 65.04	06 – 24 – 000 Mark 421	0731	0.8 x 0.6 0.7 x 0.6	(90) ...	2	... P w 6140 Extremely compact		13.1		
06133	11 01.7 + 64 16 140.50 + 49.34	11 – 14 – 02a	0977	1.8:: x 1.8:: 1. x 1.	-- ...	3	Dwarf V		17.		
06134	11 01.8 + 05 06 249.33 + 56.12	01 – 28 – 033 NGC 3509	0722	2.1 x 0.9 2.0 x 0.9	[40] ...	4	S ... Strongly peculiar, loop		14.0	+ 7600 + 7443	*12 3VĀ
06135	11 01.8 + 45 24 165.65 + 62.14	08 – 20 – 081	0709 203	1.0 x 1.0 1.0 x 1.0	-- 1	1	S ...		13.0		
06136	11 01.8 + 46 15 164.11 + 61.69	08 – 20 – 082	0709 203	1.9 x 0.5 1.9 x 0.5	7 ...	1	S? ...		14.7		1
06137	11 01.9 + 16 20 231.49 + 62.88	03 – 28 – 055	0463	1.2 x 0.7 (1.3 x 0.9)	105: ...	1	S0?		14.2		
06138	11 01.9 + 28 00 205.26 + 66.33	05 – 26 – 045	1357 204	2. x 1. 1.8: x 1.1:	... ...	2	...	S m:	15.7		*
06139	11 01.9 + 28 30 204.02 + 66.37	05 – 26 – 044 NGC 3515	1357 206	1.0 x 0.7 1.0 x 0.7	(55) 3:	2	Sb – c		14.8		2̲
06140	11 01.9 + 38 29 179.79 + 65.07	06 – 24 – 000	0731	1.4: x 0.3 1.0 x 0.2	16 6 – 7	1	S ... P w 6132		16.0:		*
06141	11 02.0 + 05 28 248.91 + 56.41	01 – 28 – 000	0722 208	1.2 x 0.8 0.5 x 0.4	13 3 – 4	3	Sb/S B b	SAB(rs:)b	15.2		
06142	11 02.1 + 04 33 250.12 + 55.79	01 – 28 – 034	0722 208	0.8 x 0.4 0.9 x 0.45	150 ...	1	S ... 1st of 3		14.5		*1
06143	11 02.2 + 35 38 186.37 + 65.89	06 – 24 – 046	0731 210	1.4 x 0.3 1.3 x 0.3	12 6	1	Sb:		15.3		*
06144	11 02.6 + 56 48a 148.35 + 55.06	10 – 16 – 057 NGC 3517	0712 229	1.0 x 0.9 1.0 x 0.9	(120) 1 – 2	2	Sa – b s of 2, contact		(13.8)		*12 3̲
06145	11 03.0 + 04 26 250.55 + 55.87	00 – 28 – 000	1397	1.5:: x 1.3:: 1.5:: x 1.	... ...	4	Dwarf irregular Ir V		17. – 18.		
06146	11 03.0 + 04 29 250.49 + 55.90	01 – 28 – 000	0722 208	1.3 x 0.3 1.2 x 0.3	[140] ...	2	Irr		17.		
06147	11 03.1 + 29 05 202.59 + 66.66	05 – 26 – 053	1357 206	1.4 x 0.2 1.2 x 0.25	176 7	1	S ...		16.0:		
06148	11 03.1 + 36 40 183.81 + 65.82	06 – 25 – 002	0695	1.5 x 0.22 1.5 x 0.22	176 7	3	Sb		15.3		
06149	11 03.2 + 43 36 168.78 + 63.25	07 – 23 – 023	1349	1.0 x 0.3 1.1 x 0.3	110 6	1	S ...		14.6		
06150	11 03.3 + 00 15 255.53 + 52.85	00 – 28 – 030 NGC 3521	1397	13.5 x 7.0 13. x 7.	163 5	4	Sb	S bc Sb +	10.1	+ 790 + 614	*12 –689
06151	11 03.3 + 20 06 224.09 + 64.79	03 – 28 – 057	0463	1.8:: x 1.8:: 1.1:: x 0.9::	-- ...	4	Dwarf spiral S V		17.		*
06152	11 03.3 + 30 12 199.77 + 66.73	05 – 26 – 054	1357 206	1.3 x 0.5 1.3 x 0.5	154 6	2	S B b		15.3		
06153	11 03.3 + 72 49 133.26 + 42.42	12 – 11 – 009 NGC 3516	0714 212	(2.1 x 1.8) (2.3 x 2.0)	(55) 2	3	S B 0 Seyfert		12.3	+ 2621 + 2777	*12 347̄8
06154	11 03.4 + 76 58 130.40 + 38.88	13 – 08 – 055	1326 172	1.3 x 0.5 0.9 x 0.5	65 5	2	S B a		14.8		*C
06155	11 03.5 + 04 42 250.37 + 56.14	01 – 28 – 037	0722 208	1.3 x 1.1 1.4 x 1.2	(145) 1 – 2	3	Sb/S B c nnp of 2	SAB(rs)bc Sb +	14.4		*
06156	11 03.6 + 48 55 159.22 + 60.42	08 – 20 – 086	0709	1.3 x 0.3 1.3 x 0.3	3 6	1	S ... nf of 2, comp = Mark 163		15.1		*
06157	11 03.7 + 17 46 229.10 + 63.92	03 – 28 – 058	0463	1.8 x 1.6 1.7: x 1.5:	-- 1:	4	...	SA(s)dm	15.0		1
06158	11 03.9 + 11 40 240.39 + 60.77	02 – 28 – 050 NGC 3524	0976	1.7 x 0.5 1.7 x 0.45	14 6	2	S0 – a sf of 2		13.4		*3
06159	11 04.0 + 20 20 223.72 + 65.03	03 – 28 – 060 NGC 3522	0463	1.2 x 0.7 1.2 x 0.7	117 ...	3	E		14.2		*
06160	11 04.0 + 28 59b 202.87 + 66.86	05 – 26 – 056	1357 206	1.0 x 0.15 0.8 x 0.15	153 7	1	Sb – c sf of 2		(15.1)		*1

N G C   3509, 3510, 3512, 3515 – 17, 3521
3522, 3524

1	2	3	4	5	6	7	8	9	10	11	12
06161	11 04.0 + 44 00 167.85 + 63.18	07 - 23 - 024	1349	3.0 x 1.6: 2.8: x 1.4:	40 5:	4	...	S B dm	14.4		*1
06162	11 04.0 + 51 30 155.16 + 58.85	09 - 18 - 086	0673	2.5 x 1.2 2.2 x 1.4	88 5:	4	Sc	Sc +	14.2		
06163	11 04.1 + 23 18 216.95 + 65.99	04 - 26 - 028	1366 207	1.1 x 0.8 1.1 x 0.8	90 3	2	Sa P w 6164, in group		14.9		*1
06164	11 04.2 + 23 17 217.00 + 66.01	04 - 26 - 029	1366 207	1.0 x 0.5 0.9 x 0.5	175: 5:	1	S ... P w 6163, in group	SAB: ...	15.7		*1
06165	11 04.2 + 46 06 163.90 + 62.12	08 - 20 - 088	0709 203	1.5 x 0.35 1.5 x 0.3	59 7	2	Sa - b		15.3		*
06166	11 04.3 + 28 52 203.17 + 66.92	05 - 26 - 057	1357 206	1.1 x 0.9 1.0 x 0.8	115: ...	1	Sb - c P w 6170		15.5		*
06167	11 04.4 + 07 26 247.00 + 58.19	01 - 28 - 039 NGC 3526	0722	2.0 x 0.5 2.5 x 0.5	55 6 - 7	2	Sc		13.7		3
06168	11 04.5 + 08 04 246.13 + 58.63	01 - 28 - 000	0722	1.4: x 0.4 1.2 x 0.3	128 6	1	Sb		15.1		
06169	11 04.5 + 12 20 239.45 + 61.28	02 - 28 - 052	0976	1.9 x 0.3 1.8 x 0.3	0 6 - 7	1	Sb:		14.6		
06170	11 04.5 + 28 48 203.35 + 66.96	05 - 26 - 059 NGC 3527	1357 206	1.2 x 1.1 1.2 x 1.2	-- 1	4	S B a P w 6166	(R:)S B(r)ab	15.1		*3
06171	11 04.6 + 18 50 227.09 + 64.58	03 - 28 - 061	0463	2.5 x 0.6 2.2 x 0.6	68 (6)	3	Irr	IB: m	15.1		1
06172	11 04.6 + 23 13 217.22 + 66.08	04 - 26 - 000	1366 207	[1.1 x 0.7] ...	... ...	2	Double system		15.4		*
06173	11 04.6 + 23 45 215.94 + 66.22	04 - 26 - 030	1366 207	1.0 x 0.7 1.0 x 0.7	160: ...	3	S ... loop, disturbed?	Peculiar,	14.4		*
06174	11 04.7 + 06 34 248.30 + 57.66	01 - 28 - 040 IC 669	0722 208	1.6 x 0.8 (1.4 x 0.7)	165: 5:	2	SO - a		14.3		
06175	11 04.7 + 18 42 227.40 + 64.54	03 - 28 - 62+63	463	[1.9 x 1.1] ...	... ...	2	Double system Disrupted, plumes		14.7	+8211, 7982 *1 +8115, 7866 2VA	
06176	11 04.7 + 21 56 220.24 + 65.73	04 - 26 - 031	1366 207	1.2 x 0.5 1.1 x 0.5	20 6	2	S B 0		14.8		
06177	11 04.7 + 72 20 133.49 + 42.88	12 - 11 - 010	0714 212	1.4 x 0.4 1.2 x 0.4	29 7	1	S ...		16.0:		*
06178	11 04.8 + 06 59 247.75 + 57.96	01 - 28 - 041 IC 670	0722	1.1 x 0.8 1.3 x 1.1	... ...	2	SO		14.7		
06179	11 04.8 + 64 12 140.15 + 49.59	11 - 14 - 004	0977	1.8 x 0.8 1.6: x 0.7:	155 5 - 6	1	S ...		15.7		
06180	11 05.0 + 01 03 255.19 + 53.73	00 - 28 - 031 IC 671	1397	1.4 x 1.4 1.6 x 1.4	-- 1:	3	Sa	SAB(r)ab	14.8		
06181	11 05.1 + 19 49 225.09 + 65.08	03 - 29 - 002	0051	1.0: x 0.9: 0.9: x 0.8:	-- ...	3	Dwarf IV - V or V		15.5		
06182	11 05.1 + 53 54 151.59 + 57.36	09 - 18 - 090	0673	1.0 x 0.9 0.9 x 0.8	-- ...	2	... Peculiar, jet, p w 6186		14.1		*
06183	11 05.2 + 35 44 185.79 + 66.46	06 - 25 - 007	0695 210	1.5 x 0.8 1.0 x 0.7	5: 5:	2	Sb	SAB b	14.9		*
06184	11 05.2 + 46 15 163.43 + 62.19	08 - 20 - 091	0709	1.1 x 0.35 1.1 x 0.35	2 6	3	Sc		15.2		
06185	11 05.3 + 08 16 246.09 + 58.91	01 - 29 - 002	1392	1.7 x 0.9 1.7: x 0.8:	160 ...	3	...	SAB(s)dm	14.6		
06186	11 05.4 + 54 06 151.26 + 57.25	09 - 18 - 091	0673	1.3 x 0.3 1.3 x 0.3	5 6	1	S ... P w 6182		14.7		*
06187	11 05.6 + 45 24 164.89 + 62.71	08 - 20 - 000	0709 203	0.7 x 0.6 0.7 x 0.6	-- ...	1	S? ...		14.0		
06188	11 05.7 + 57 30 146.98 + 54.84	10 - 16 - 064 NGC 3530	0712 229	0.7 x 0.25 0.7 x 0.25	99 ...	1	...		14.4		3
06189	11 05.9 + 05 06 250.63 + 56.85	01 - 29 - 004 NGC 3535	1392 208	(1.2 x 0.5) (1.2 x 0.5)	178 6	2	Sa		14.3		3
06190	11 06.1 + 26 54 208.28 + 67.14	05 - 26 - 062 NGC 3534	1357 206	1.4 x 0.4 1.5 x 0.4	92 7	1	S ... P w 6193, in group of 3		15.4		*13
06191	11 06.1 + 28 45 203.52 + 67.31	05 - 26 - 061 NGC 3536	1357 206	1.8:: x 1.4:: 1.0: x 0.9:	(155) ...	3	S B ...	S B(r) ...	15.2		*3
06192	11 06.1 + 61 41 142.39 + 51.68	10 - 16 - 069	0712	1.1: x 0.9: 0.8 x 0.5	(155) ...	1	...		15.7		
06193	11 06.2 + 26 53 208.33 + 67.16	05 - 26 - 063	1357 206	1.0 x 0.35 1.1 x 0.4	168 6	2	Sb P w 6190, in group of 3		15.6		*1
06194	11 06.3 + 23 12 217.52 + 66.45	04 - 26 - 034	1366 207	1.6 x 1.6 1.5 x 1.5	-- 1	2	S ... Peculiar?		14.7		*1
06195	11 06.4 + 34 13 189.43 + 67.01	06 - 25 - 000	0695	1.0 x 0.3 0.9 x 0.2	141 6	2	Sc		16.0:		

N G C   3526, 3527, 3530, 3534 - 36   I C   669 - 71

1	2	3	4	5	6	7	8 9	10	11	12
06196	11 06.5 + 36 17 184.29 + 66.58	06 − 25 − 011 NGC 3540	0695 210	(1.3 x 1.3) (1.5 x 1.5)	— 1:	4	S B 0	14.6		3
06197	11 06.6 + 08 28 246.21 + 59.28	01 − 29 − 005	1392	[1.4 x 0.45] ...	... ...	3	S ... + companion Distorted, bridge:	15.2		*1
06198	11 06.6 + 29 50 200.71 + 67.44	05 − 26 − 067	1357 206	1.0 x 0.7 1.0 x 0.7	130: 3:	1	S0?	15.0		*
06199	11 06.7 + 62 34 141.43 + 51.02	11 − 14 − 007	0977	1.4 x 0.35 1.4 x 0.30	157 6	1	S ... Brightest in group	14.9		*
06200	11 06.8 + 00 10 256.73 + 53.35	00 − 29 − 003	1400	2.0: x 0.8 1.4: x 0.5	165 6	2	Sa	14.5		*
06201	11 06.8 + 47 05 161.63 + 61.94	08 − 20 − 093	0709	0.7 x 0.5 0.7 x 0.5	110 ...	1	...	14.3		
06202	11 06.8 + 51 13 155.02 + 59.37	09 − 18 − 094	0673	1.0 x 0.15 0.9 x 0.15	70 7	1	S ...	15.4		
06203	11 07.0 + 22 02 220.42 + 66.27	04 − 26 − 035	1366 207	(1.4 x 1.2) (1.4 x 1.2)	(30) ...	1	E?	14.8		*1
06204	11 07.1 + 24 32 214.36 + 66.96	04 − 26 − 036	1366 207	1.0 x 0.5 1.0 x 0.45	177: ...	1	S ... P w 6207, contact, distorted	14.5	+ 6314 + 6246	*12
06205	11 07.1 + 46 22 162.83 + 62.39	08 − 20 − 095	0709	1.8 x 1.4 1.7: x 1.3:	(50) ...	3	Dwarf spiral, S S IV − V:	15.6		
06206	11 07.2 + 13 02 239.05 + 62.23	02 − 29 − 006	0066 205	0.9 x 0.5 0.7 x 0.35	... ...	1	Double system?	14.2		*1
06207	11 07.2 + 24 31 214.42 + 66.98	04 − 26 − 037	1366 207	1.5 x 0.20 1.5 x 0.25	64 7	1	S B? ... P w 6204, contact	14.6	+ 5996 + 5928	*12
06208	11 07.3 + 07 30 247.83 + 58.76	01 − 29 − 000	1392	1.0 x 0.2 1.1 x 0.2	107 7	1	Sb Brightest of 3	15.1		*
06209	11 07.3 + 11 00 242.52 + 61.03	02 − 29 − 007 NGC 3547	0066 209	2.1 x 0.9 1.8 x 0.8	7 5 − 6	1	S ... Peculiar, Leo group:	12.8		*123 4678
06210	11 07.4 + 05 35 250.48 + 57.45	01 − 29 − 006	1392 208	1.7 x 0.8 1.8 x 0.8	15 5	3	Sb	14.7		*
06211	11 07.6 + 55 29 149.07 + 56.49	09 − 18 − 096	0673 214	1.2 x 0.8 1.1 x 0.8	(125) ...	3	Sb/S B c SAB bc Sb +	16.0:		*
06212	11 07.8 + 05 06b 251.24 + 57.18	01 − 29 − 008	1392 208	1.0 x 0.6 1.0 x 0.7	75 3 − 4	2	Sb nnf of 2	(14.0)		*1
06213	11 07.8 + 61 38 142.19 + 51.84	10 − 16 − 075 NGC 3543	0712	1.4 x 0.25 1.3 x 0.25	8 6	1	S ... Peculiar:	14.8		*3
06214	11 07.9 + 29 02 202.82 + 67.71	05 − 27 − 002 NGC 3550	0099 206	1.1 x 1.0 (1.2 x 1.1)	— ...	3	... Peculiar, double nucleus	14.2		*13
06215	11 08.0 + 53 40 151.34 + 57.83	09 − 18 − 097 NGC 3549	0673	3.3 x 1.0 2.8 x 1.0	38 6	4	Sc Sc −	12.8		*123 4678
06216	11 08.1 + 05 07 251.32 + 57.25	01 − 29 − 010	1392 208	1.3 x 0.3 1.3 x 0.22	173 7	1	S0 − a?	14.9		*1
06217	11 08.1 + 12 17 240.62 + 61.97	02 − 29 − 008 NGC 3559	0066	1.3 x 0.9 1.4 x 0.9	55 ...	4	S ... Strongly peculiar	13.7		*3
06218	11 08.1 + 28 33 204.10 + 67.73	05 − 27 − 000	0099 206	1.1 x 0.15 0.8 x 0.1	26 7	1	S ...	16.0:		*
06219	11 08.3 + 19 27 226.61 + 65.63	03 − 29 − 005	0051	1.6 x 0.25 1.6 x 0.25	83 6 − 7	1	S ... spp of 2, disturbed	14.9		*1
06220	11 08.3 + 29 37 201.29 + 67.81	05 − 27 − 000	0099 206	[1.7 x 0.8:] ...	... ...	2	Double system Distorted, bridge:	16.5:		*
06221	11 08.3 + 43 54 167.17 + 63.90	07 − 23 − 027 IC 674	1349	1.8 x 0.8 1.7 x 0.7	120 6	2	Sa − b	14.5		
06222	11 08.4 + 34 50 187.66 + 67.30	06 − 25 − 000	0695	1.5:: x 1.5:: 1. x 1.	— ...	3	Dwarf V, spp of 2	17.		*
06223	11 08.4 + 62 28 141.28 + 51.22	10 − 16 − 078	0712	1.4: x 0.8: 1.3: x 0.6:	150 ...	1	Sc? 6199 group	15.6		*
06224	11 08.5 + 28 58 203.01 + 67.84	05 − 27 − 10+11 NGC 3561	0099 206	[5.3 x 1.0] ...	[0] ...	3	Double system, bridge, very long plume	14.7	See Notes	*13 VAC
06225	11 08.5 + 55 56 148.35 + 56.25	09 − 18 − 098 NGC 3556	0673	8.8 x 2.2 8.2 x 2.1	80 6	4	Sc Sc +	10.7	+ 680 + 763	*12− −789
06226	11 08.6 + 06 54 249.10 + 58.59	00 − 29 − 000	1400	1.0 x 0.1 0.9 x 0.12	157 7	1	Sc	17.		
06227	11 08.6 + 47 19 160.85 + 62.05	08 − 20 − 096	0709 216	0.8 x 0.7 0.9 x 0.8	— ...	1	E − S0	14.3		
06228	11 08.6 + 56 49 147.27 + 55.61	10 − 16 − 079	0712 229	1.3 x 1.1 1.1 x 1.0	— ...	3	S B [c] nf of 2, disrupted	15.4		*1
06229	11 08.6 + 62 37 141.11 + 51.12	11 − 14 − 010	0977	1.4:: x 1.0:: 1.0:: x 0.8::	— ...	3	Dwarf V	17.		
06230	11 08.7 + 06 06 250.22 + 58.05	01 − 29 − 011 NGC 3567	1392	(1.2 x 0.9) (1.5 x 1.0)	... ...	2	... np of 2, contact	14.4		*13

N G C 3540, 3543, 3547, 3549, 3550, 3556
3559, 3561, 3567

I C 674

1	2	3	4	5	6	7	8	9	10	11	12
06231	11 08.7 + 11 51 241.55 + 61.82	02 – 29 – 000	0066 209	1.0: x 0.7 1.0 x 0.5	55? ...	1	...		15.3		
06232	11 08.7 + 44 10 166.57 + 63.83	07 – 23 – 028	1349	1.5 x 0.4 1.8 x 0.4	146 6 – 7	3	Sa		15.2		
06233	11 08.8 + 07 11 248.77 + 58.82	01 – 29 – 013	1392	1.1 x 0.35 1.0 x 0.3	3 6	2	Sa		14.6		*
06234	11 08.8 + 27 14b 207.60 + 67.78	05 – 27 – 014 NGC 3563	0099 206	1.1 x 0.8 1.1 x 0.7	15: ...	2	S B: 0 f of 2, contact		(14.6)		*13
06235	11 08.9 + 79 05 128.77 + 37.17	13 – 08 – 000	1326	1.2 x 0.22 1.0 x 0.22	82 6 – 7	1	Sb – c		16.0:		
06236	11 09.0 + 37 00 182.20 + 66.86	06 – 25 – 019	0695 210	1.0 x 0.20 1.3 x 0.22	49 7	3	Sa		15.5		
06237	11 09.1 + 65 30 138.42 + 48.81	11 – 14 – 10a	0977	1.0: x 1.0: 1.0: x 1.0:	--- ...	2	... Disturbed?	S dm – m	16.5:		*1
06238	11 09.3 + 35 43 185.33 + 67.27	06 – 25 – 020 NGC 3569	0695 210	(1.1 x 1.0) (1.2 x 1.2)	--- 1	3	S0	SA 0	14.5		*3
06239	11 09.4 + 03 25 253.88 + 56.23	01 – 29 – 014	1392	(1.1 x 0.9) (1.1 x 0.9)	(70) ...	3	E p of 2		14.6		*
06240	11 09.4 + 27 51 206.00 + 67.97	05 – 27 – 019 NGC 3570	0099 206	(1.0 x 1.0) (1.1 x 1.1)	--- (1)	2	S0		15.0		*
06241	11 09.4 + 28 45 203.61 + 68.03	05 – 27 – 018	0099 206	1.0 x 0.20 0.9 x 0.20	42 6 – 7	1	S ...		14.9		
06242	11 09.5 + 73 09 132.50 + 42.37	12 – 11 – 011 NGC 3562	0714 212	(1.7 x 1.3) (1.8 x 1.4)	165: ...	3	E		13.2		*3
06243	11 09.7 + 31 41 195.76 + 68.03	05 – 27 – 023	0099 215	1.0 x 0.7 0.9 x 0.7	163: 3	3	Sc	SA c Sc –:	16.0:		*1
06244	11 09.8 + 28 18 204.82 + 68.09	05 – 27 – 23a	0099 206	[1.4 x 0.4] ...	... ...	2	Double system Bridge		15.7		*1
06245	11 10.1 + 09 20 246.05 + 60.50	02 – 29 – 009 IC 676	0066	(2.2 x 1.6) (2.2 x 1.6)	10: 3:	4	S B 0	SB 0 +	13.4		
06246	11 10.2 + 23 32 217.30 + 67.41	04 – 27 – 003	1353 207	1.5 x 0.2 1.5: x 0.2	45 7	3	Sc In group		15.7		*1
06247	11 10.2 + 27 43 206.40 + 68.14	05 – 27 – 026	0099 206	1.5 x 0.7 1.3 x 0.7	125 5	2	Sa – b		15.1		
06248	11 10.3 + 10 28 244.33 + 61.27	02 – 29 – 000	0066 209	1.6:: x 1.3:: 1. x 1.	--- ...	4	Dwarf irregular Ir V		18.		
06249	11 10.3 + 60 11 143.26 + 53.17	10 – 16 – 083	0712 229	1.8 x 1.8 1.8: x 1.7:	--- 1	4	Sc 5 companions, disturbed:	SA cd	14.6		*1
06250	11 10.5 + 28 05 205.43 + 68.23	05 – 27 – 027	0099 206	(1.2 x 1.0) (1.3 x 1.1)	--- ...	1	...		14.7		
06251	11 10.5 + 53 52 150.58 + 57.95	09 – 19 – 006	0059	1.9: x 1.4: 1.7: x 1.2:	60: ...	4	Dwarf spiral, SAB: m S IV – V		15.7		*
06252	11 10.6 + 26 08 210.64 + 68.03	04 – 27 – 004	1353 213	1.0 x 0.8 0.9 x 0.7	175: ...	3	Sc		15.4		
06253	11 10.8 + 22 26 220.14 + 67.23	04 – 27 – 005 Leo II	1353	15. : x 12. : 15. : x 12. :	... ...	5	Dwarf elliptical Local group		[12.9]		*12 5
06254	11 10.8 + 23 27 217.61 + 67.52	04 – 27 – 007	1353 207	1.1 x 0.5 1.0 x 0.5	90: 5 – 6	1	S ... In group		16.0:		
06255	11 10.8 + 47 51 159.46 + 62.02	08 – 21 – 005	0700 216	0.8 x 0.45 0.8 x 0.45	35 (4 – 5)	1	S: ...		13.6		*
06256	11 10.8 + 65 27 138.24 + 48.95	11 – 14 – 011	0977	1.4 x 0.18 1.2 x 0.18	89 7	3	Sc		16.0:		
06257	11 10.9 + 48 33 158.29 + 61.60	08 – 21 – 006 NGC 3577	0700	1.6 x 1.6 1.5 x 1.5	--- 1	4	S B b P w 6263	S B(r)ab Sb –	14.7		*12 36
06258	11 11.1 + 21 48 221.75 + 67.10	04 – 27 – 008	1353	2.1 x 0.5 1.8 x 0.5	[175] ...	3	Irr	Ir I	15.6		*1
06259	11 11.2 + 09 52 245.58 + 61.06	02 – 29 – 011 IC 2637	0066 209	0.9 x 0.8 0.9 x 0.7	(70) ...	1	...		13.9		
06260	11 11.3 + 04 33 253.11 + 57.39	01 – 29 – 020	1392	1.3 x 0.4 1.3 x 0.35	43 6 – 7	1	Sb: NGC 3640 group?		14.9		
06261	11 11.3 + 10 50 244.07 + 61.70	02 – 29 – 012 IC 2638	0066 209	1.1 x 0.6 1.0 x 0.7	170: ...	1	S B 0		15.1		
06262	11 11.3 + 12 35 241.10 + 62.77	02 – 29 – 013 IC 677	0066	1.6 x 0.6 1.6 x 0.7	45 6	2	Sb Disturbed? Sb +		13.6		*
06263	11 11.3 + 48 36 158.12 + 61.62	08 – 21 – 008 NGC 3583	0700	2.7 x 2.0: 2.7 x 1.9:	125 3:	2	S B b P w 6257	Sb +	11.6		*123 4678
06264	11 11.4 + 20 40 224.52 + 66.77	04 – 27 – 009 NGC 3588	1353	[1.1 x 0.9] ...	... ...	2	Double system Contact, crossing plumes		15.3		*1
06265	11 11.6 + 63 38 139.73 + 50.49	11 – 14 – 012	0977 219	1.1 x 0.3 1.1 x 0.3	136 6	1	S ...		15.0		*

N G C   3562, 3563, 3569, 3570, 3577, 3583   I C   676, 677, 2637, 2638
3588

1	2	3	4	5	6	7	8	9	10	11	12
06266	11 11.7 + 43 30	07 – 23 – 035	1349	1.6: x 0.9:	55:	2	Dwarf (spiral), S: m		17.		*
	167.23 + 64.64			1.0: x 0.8:	...		S: IV – V:				
06267	11 11.8 + 17 32	03 – 29 – 011	0051	2.0 x 0.5	120	1	Sc:		14.8		*3
	231.66 + 65.53	NGC 3592		2.2 x 0.5	7		P w 6278				
06268	11 11.8 + 22 45	04 – 27 – 000	1353	1.2 x 0.12	125	3	Sc		16.0:		
	219.53 + 67.54			1.2 x 0.12	7						
06269	11 11.8 + 70 56	12 – 11 – 013	0714	1.3 x 0.5	110	1	S ...		15.7		
	133.84 + 44.37			1.0 x 0.4	6						
06270	11 11.9 + 29 49	05 – 27 – 000	0099	1.1: x 0.4	165	1	S B 0 – a		16.0:		
	200.75 + 68.59		206	0.8 x 0.4	6						
06271	11 11.9 + 30 35	05 – 27 – 031	0099	1.2 x 0.5	52	2	Sa		14.6		
	198.65 + 68.57		215	1.1 x 0.4	6						
06272	11 12.0 + 13 05	02 – 29 – 014	0066	5.2 x 2.1	92	3	S0 S 0 +		11.8	+ 547	*123
	240.44 + 63.21	NGC 3593		4.7 x 2.1	6		Leo group			+ 429	4678
06273	11 12.1 + 35 46a	06 – 25 – 026	0695	1.7 x 0.30	29	1	Sa – b		(14.4)		*1
	184.82 + 67.81		210	1.7 x 0.35	6		1st of 3				
06274	11 12.2 + 34 06	06 – 25 – 028	0695	1.0 x 0.8	(145)	2	Sc		14.7		*
	189.14 + 68.21			0.9 x 0.7	...		Disturbed?				
06275	11 12.2 + 60 58	10 – 16 – 096	0712	1.7 x 0.9	48	4	... S d – dm		14.5		
	142.16 + 52.69	NGC 3589		1.6 x 0.8	5						
06276	11 12.4 + 31 18	05 – 27 – 032	0099	(1.5 x 1.3)	––	1	S0		14.4		
	196.67 + 68.64		215	1.6 x 1.4	1 – 2						
06277	11 12.5 + 15 03	03 – 29 – 013	0051	4.4 x 4.4	––	5	Sc SAB(rs)c		11.7		*123
	236.92 + 64.42	NGC 3596		3.6: x 2.8:	1		Leo group:				4678
06278	11 12.6 + 17 32	03 – 29 – 014	0051	(1.6 x 1.2)	35	2	E – S0		13.5		*13
	231.87 + 65.70	NGC 3598		(1.6 x 1.2)	...		P w 6267				
06279	11 12.6 + 35 46	06 – 25 – 030	0695	1.4 x 0.6	7	2	Sa		14.8		*
	184.75 + 67.91		210	1.4 x 0.6	6		P w 6273ab				
06280	11 12.6 + 47 43	08 – 21 – 009	0700	1.6 x 0.7	176	1	...		13.0		3
	159.28 + 62.34	NGC 3595	216	1.5 x 0.7	...						
06281	11 12.8 + 18 23	03 – 29 – 015	0051	(2.5 x 2.5):	––	2	E/S0:		13.0		*12
	230.08 + 66.14	NGC 3599		(3.0 x 2.7):	...		NGC 3607 group				3
06282	11 12.9 + 05 23	01 – 29 – 024	1392	0.6 x 0.5	––	1	S ...		14.1		*3
	252.59 + 58.27	NGC 3601		0.55 x 0.50	...						
06283	11 13.0 + 41 52	07 – 23 – 038	1349	4.2 x 1.0	3	1	Sa?		12.6		*13
	170.31 + 65.66	NGC 3600		3.5 x 1.0	7						
06284	11 13.2 + 01 08	00 – 29 – 000	1400	[1.5 x 0.7]	...	2	Double system		15.4		
	257.79 + 55.12			...	...		Contact, plume				
06285	11 13.3 + 41 40	07 – 23 – 000	1349	(1.3 x 1.0)	––	1	...		16.0:		
	170.67 + 65.81			(0.7 x 0.7)	...						
06286	11 13.3 + 55 58	09 – 19 – 022	0059	(1.3 x 1.1)	(10)	1	S B 0:		15.2		*3
	147.40 + 56.67	NGC 3594	214	1.4 x 1.0	2 – 3:						
06287	11 13.4 + 24 12	04 – 27 – 000	1353	1.2: x 0.9:	...	4	Dwarf irregular		17. – 18.		
	216.07 + 68.28			0.5: x 0.5:	...		Ir V				
06288	11 13.5 + 10 26	02 – 29 – 017	0066	1.0 x 0.8	35:	3	Sc		15.3		1
	245.45 + 61.86	IC 2672		0.9 x 0.8	2:						
06289	11 13.6 + 03 09	01 – 29 – 025	1392	0.8 x 0.7	––	1	S ...		14.5		
	255.62 + 56.73		218	0.8 x 0.7	1						
06290	11 13.6 + 11 19	02 – 29 – 000	0066	1.4: x 0.7:	20:	3	Sc		15.3		
	244.03 + 62.44	IC 2674		1.3: x 0.7:	5:						
06291	11 13.6 + 69 35	12 – 11 – 014	0714	1.0: x 0.5:	137	1	S ...		15.7		*
	134.63 + 45.61			1.1: x 0.5	5						
06292	11 13.7 + 29 36	05 – 27 – 038	0099	1.8 x 0.8	...	4	S B b S B(s)b		15.2		*1
	201.34 + 68.98		215	1.6 x 0.8	...		Disturbed, in group, Sb +				
06293	11 13.7 + 41 21	07 – 23 – 039	1349	[1.4 x 0.7]	...	2	Sc + companion		15.3		*1
	171.26 + 66.02			...	...		Connected				
06294	11 14.0 + 26 38	05 – 27 – 000	0099	1.2 x 0.15	104	2	Sb – c		16.0:		
	209.53 + 68.86		206	1.2 x 0.18	7						
06295	11 14.1 + 18 17	03 – 29 – 019	0051	1.2 x 0.6	17	2	E – S0		12.7	+ 693	*12
	230.64 + 66.37	NGC 3605		1.4 x 0.7	...		NGC 3607 group			+ 599	348
06296	11 14.2 + 18 04	03 – 29 – 021	0051	1.3 x 0.4	166	1	S ...		14.3		
	231.15 + 66.29			1.7 x 0.4	6						
06297	11 14.3 + 18 19	03 – 29 – 020	0051	(4.5 x 4.0)	(120)	3	E		10.2	+ 934	*123
	230.62 + 66.42	NGC 3607		(5.0 x 4.5)	...		Brightest in group			+ 840	4678
06298	11 14.3 + 36 25	06 – 25 – 035	0695	1.1 x 0.5	75:	1	S ...		15.0		*
	182.82 + 68.05		210	0.9 x 0.4	6		P w 6303				
06299	11 14.4 + 18 25	03 – 29 – 022	0051	(3.0 x 2.5)	(75)	3	E		11.7	+ 1210	*123
	230.43 + 66.49	NGC 3608		(3.2 x 2.8)	...		P w 6297, NGC 3607 group			+ 1117	4678
06300	11 14.6 + 16 36	03 – 29 – 023	0051	1.2 x 0.7	80:	1	...		15.7		
	234.42 + 65.66			1.2 x 0.7	...						

N G C   3589,   3592 – 96,   3598,   3599,   3600,   3601   I C   2672,   2674
       3605,   3607,   3608

1	2	3	4	5	6	7	8	9	10	11	12
06301	11 14.7 + 22 37 220.40 + 68.14	04 - 27 - 011	1353	1.6 x 0.3: 1.7 x 0.3	139 6 - 7	1	Sc		15.7		*
06302	11 14.7 + 27 57 205.98 + 69.15	05 - 27 - 000	0099 206	1.1 x 0.15 1.0 x 0.18	166 7	1	S ...		15.7		
06303	11 14.8 + 36 20 182.96 + 68.17	06 - 25 - 038	0695 210	1.3 x 1.1 1.3 x 1.0	-- 2:	3	Sc P w 6298		15.1		*
06304	11 14.8 + 58 38 144.12 + 54.74	10 - 16 - 103	0712 229	1.4: x 1.2: 1.0: x 0.9:	-- ...	3	Dwarf ... B m V		16.5:		
06305	11 14.9 + 04 50 253.99 + 58.20	01 - 29 - 026 NGC 3611	1392	(2.7 x 2.7) (2.3 x 1.8)	-- ...	4	Sa, singular, p w 6306, disturbed?, NGC 3640 group		12.4	+ 1754 + 1603	*123 4678
06306	11 14.9 + 04 53 253.93 + 58.24	01 - 29 - 000	1392	1.3: x 0.5: 0.6 x 0.3	[45] ...	3	Dwarf P w 6305 IV - V or V, NGC 3640 group?		17.		*
06307	11 14.9 + 38 19 177.99 + 67.52	06 - 25 - 039	0695	1.3 x 0.35 1.3 x 0.35	160 6	1	S - Irr Peculiar?, p w 6326		14.8		*
06308	11 15.0 + 27 22 207.63 + 69.16	05 - 27 - 000	0099 206	1.1 x 0.2 0.9 x 0.2	94 6 - 7	2	Sb - c		16.0:		
06309	11 15.0 + 51 46 152.50 + 59.92	09 - 19 - 029	0059	1.6 x 0.8 1.6 x 0.8	[125] ...	3	S B ... S B(rs) ... Peculiar		13.7		*
06310	11 15.2 + 26 54 208.94 + 69.16	05 - 27 - 043 NGC 3609	0099 206	1.2 x 1.0 1.1 x 0.9	(50) 2	2	Sa - b P w 6321		14.1		*13
06311	11 15.4 - 01 49 261.60 + 53.11	00 - 29 - 011	1400 215a	1.5 x 1.4 1.3 x 1.2	-- ...	4	Sc SAB(s)c Sc -		14.3		1
06312	11 15.4 + 08 07 249.67 + 60.65	01 - 29 - 027	1392	1.2 x 0.6 1.4 x 0.6	45 5:	1	S B? 0 - a		14.9		
06313	11 15.4 + 23 40 217.78 + 68.59	04 - 27 - 012 NGC 3615	1353 220	(1.4 x 0.9) (1.7 x 1.1)	40: ...	3	E P w 6327		14.0		*3
06314	11 15.4 + 30 41 198.27 + 69.32	05 - 27 - 050	0099 214	1.0 x 0.7 1.0 x 0.7	45 ...	2	S ... Companion superimposed, nnp of 3		15.1		*1
06315	11 15.4 + 54 02 149.37 + 58.32	09 - 19 - 34 +35 Mark 38 + 39	59 214	[1.1 x 0.7] ...	... ...	2	Double system Connected		15.2		*1
06316	11 15.4 + 65 19 137.74 + 49.32	11 - 14 - 14 +15	977	[1.7 x 0.6] ...	... ...	2	Double system		14.4		*
06317	11 15.5 + 04 54 254.12 + 58.35	01 - 29 - 000	1392	1.1 x 0.6 1.0: x 0.5:	[40] ...	2	Dwarf irregular NGC 3640 group?		16.5:		*
06318	11 15.5 + 46 02 161.54 + 63.77	08 - 21 - 015 NGC 3614	0700 216	4.6 x 2.8 4.3: x 2.5:	80 4	4	Sc SAB(r:)c nf of 2 Sc +		12.7		*123 4678
06319	11 15.5 + 59 05 143.52 + 54.44	10 - 16 - 107 NGC 3610	0712 229	(3.2 x 3.2) (3.2 x 3.2)	-- (1)	2	S0		11.4	+ 1765 + 1864	*123 4678
06320	11 15.6 + 19 07 229.15 + 67.05	03 - 29 - 028	0051	1.1 x 1.0 1.1 x 0.9	-- ...	1	... Peculiar, p w 6324		13.6		*
06321	11 15.6 + 26 54 208.97 + 69.25	05 - 27 - 051 NGC 3612	0099 206	1.0 x 0.8 1.0 x 0.9	(160) ...	2	S - Irr P w 6310		15.0		*13
06322	11 15.6 + 28 33 204.32 + 69.38	05 - 27 - 053	0099 215	1.3 x 1.3 1.4 x 1.3	-- 1	3	S0 SA 0 + or 0/a n of 2		14.7		*1
06323	11 15.6 + 58 17 144.35 + 55.08	10 - 16 - 109 NGC 3613	0712 229	(3.4 x 1.8) (3.7 x 2.0)	102 ...	2	E - S0 P w 6330		11.6	+ 2054 + 2150	*12 3478
06324	11 15.7 + 19 00 229.45 + 67.02	03 - 29 - 029	0051	1.3 x 0.5 1.3 x 0.5	167 6	3	S0 P w 6320		14.9		*
06325	11 15.8 + 25 35 212.65 + 69.10	04 - 27 - 013	1353 213	1.6 x 0.5 1.4 x 0.4	10 6	1	Sc:		15.5		1
06326	11 15.8 + 38 13 178.06 + 67.72	06 - 25 - 041	0695	1.1: x 0.9: 1.1: x 0.9:	(100) ...	2	... S d - dm P w 6307		16.0:		*
06327	11 15.9 + 23 45 217.64 + 68.72	04 - 27 - 014 NGC 3618	1353 220	1.0 x 0.8 0.8 x 0.7	(175) 2:	2	Sb SAB: ... P w 6313		14.4		*3
06328	11 16.3 + 13 22 241.32 + 64.22	02 - 29 - 018 NGC 3623 = M 65	0066	9.5 x 2.3 9.5 x 2.3	174 6	4	Sa Leo group		9.6	+ 755 + 640	*12- -9
06329	11 16.4 + 00 27 259.63 + 55.08	00 - 29 - 014	1400	1.2 x 1.1 1.1 x 1.0	-- 1	3	Sc		15.4		
06330	11 16.5 + 58 02 144.45 + 55.35	10 - 16 - 115 NGC 3619	0712 229	(4. x 3.) (4.0 x 3.6)	-- ...	3	S0/Sa		12.6	+ 1649 + 1744	*12 3478
06331	11 16.6 + 03 30 256.24 + 57.48	01 - 29 - 000	1392 218	1.3 x 0.3 1.3 x 0.25	57 (7)	1	Sc - Irr NGC 3640 group		15.1		*
06332	11 16.6 + 21 05 224.67 + 68.05	04 - 27 - 019	1353	1.3 x 1.1 1.3 x 1.1	(130) 2:	3	S B a (R)S B a		14.5		
06333	11 16.6 + 23 10 219.31 + 68.72	04 - 27 - 018	1353	1.2 x 0.6 1.2 x 0.6	120 5	3	Sc - Irr 2nd of 3		15.4		*
06334	11 16.8 + 28 56 203.24 + 69.65	05 - 27 - 055	0099 215	2.7 x 1.4 2.9 x 1.4	82 5	2	S0 - a (R:) ... Singular		14.4		*
06335	11 16.8 + 59 34 142.80 + 54.15	10 - 16 - 117	0712 229	1.9 x 1.7 2.0: x 1.8:	-- 1	4	Sc SA(s)cd		15.1		

N G C   3609 - 15, 3618, 3619, 3623

1	2	3	4	5	6	7	8	9	10	11	12
06336	11 17.0 + 25 13 213.81 + 69.30	04 – 27 – 021	1352 220	1.4 x 0.15 1.4 x 0.18	158 7	2	Sa – b		15.6		*
06337	11 17.1 + 33 21 190.64 + 69.36	06 – 25 – 000	0695 217	(1.0 x 0.9) (1.0 x 0.9)	— ...	2	... Very compact		15.2		*
06338	11 17.2 + 36 22 182.47 + 68.62	06 – 25 – 044	0695 210	1.1 x 0.5 1.2 x 0.5	150 6	3	Sc		15.0		*
06339	11 17.2 + 67 31 135.76 + 47.55	11 – 14 – 000 NGC 3622	0977	1.3 x 0.5 1.3 x 0.5	7 (6)	1	S? ...		13.7		3
06340	11 17.4 – 00 36 261.07 + 54.39	00 – 29 – 016	1400	1.9 x 0.8 1.8 x 0.8	155: 5	3	Sb	Sb(+)	14.1		*1
06341	11 17.4 + 18 32 230.98 + 67.18	03 – 29 – 031	0051	1.1 x 0.15 0.9 x 0.15	152 7	2	...	d – m	16.0:		*
06342	11 17.4 + 30 46 197.95 + 69.74	05 – 27 – 057	0099 215	(1.0 x 0.8) (1.1 x 0.9)	(155) ...	2	E – S0		15.2		1
06343	11 17.5 + 18 38 230.78 + 67.24	03 – 29 – 032 NGC 3626	0051	2.8 x 2.0 3.2 x 2.5	157: 3	4	S0/Sa	S (r)0/a	11.2	+ 1452 + 1361	*123 46$\overline{7}$8
06344	11 17.5 + 58 01 144.29 + 55.45	10 – 16 – 000	0712 229	1.2:: x 1.0:: 1.2:: x 0.8::	(75) ...	4	Dwarf spiral S B V		17.		*
06345	11 17.6 + 02 48 257.43 + 57.10	01 – 29 – 030	1392 218	2.5 x 1.5 2.5 x 1.5	[75] ...	4	Irr IB m NGC 3640 group, Ir I		14.4		*1
06346	11 17.6 + 13 16 241.94 + 64.41	02 – 29 – 019 NGC 3627 = M 66	0066	9.0 x 4.2 9.0 x 4.0	173 6	4	Sb Brightest in Leo group, Sb +		8.9	+ 706 + 591	*12– –9$\overline{V}$A
06347	11 17.6 + 34 22 187.76 + 69.25	06 – 25 – 045	0695 217	1.1 x 0.6 1.0 x 0.6	85 4 – 5	1	Sa – b		15.3		
06348	11 17.6 + 58 04 144.21 + 55.42	10 – 16 – 120 NGC 3625	0712 229	2.1 x 0.5 2.0 x 0.5	148 6	2	S B b: SAB ...		13.9		*12 36$\overline{}$
06349	11 17.7 + 03 14 256.95 + 57.45	01 – 29 – 031 NGC 3630	1392 218	1.9 x 0.8 1.9 x 0.8	37 6:	2	S0 – a NGC 3640 group		12.8		*12 347$\overline{8}$
06350	11 17.7 + 13 52 240.85 + 64.79	02 – 29 – 020 NGC 3628	0066	15.5 x 4.3: 15.5: x 4.5:	104 7	3	Sb Leo group Disturbed, Sb +		11.5	+ 842 + 730	*12– –7$\overline{8}$9
06351	11 17.9 + 03 51 256.28 + 57.96	01 – 29 – 032 NGC 3633	1392 218	1.2 x 0.4 1.2 x 0.4	72 6	1	Sa NGC 3640 group		14.3		*$\underline{2}$
06352	11 17.9 + 27 15 208.14 + 69.79	05 – 27 – 058 NGC 3629	0099 206	2.1 x 1.6 2.1 x 1.6	30: 2 – 3	4	Sc SA cd		12.9		*123 468
06353	11 17.9 + 56 59 145.34 + 56.30	10 – 16 – 121	0712 229	1.2 x 0.6 1.1 x 0.6	5 5	2	Sb: Companion superimposed		15.4		*
06354	11 17.9 + 63 41 138.77 + 50.84	11 – 14 – 019	0977 219	1.1 x 0.4 1.1 x 0.4	113 6	1	S ...		15.3		
06355	11 18.0 + 31 30 195.81 + 69.81	05 – 27 – 059	0099 215	2.1 x 0.3 2.1 x 0.3	102 6 – 7	3	Sc P w 6367, Sc +		14.8		*1
06356	11 18.0 + 32 34 192.75 + 69.67	06 – 25 – 047	0695	1.1: x 0.2 0.8 x 0.2	120 6 – 7	1	S ...		15.3		
06357	11 18.0 + 33 23 190.44 + 69.63	06 – 25 – 000	0695 217	1.0 x 0.10 1.0 x 0.10	18 7	1	Sc		16.0:		
06358	11 18.0 + 73 05 131.81 + 42.73	12 – 11 – 017	0714 212	[1.3 x 0.8] ...	... ...	3	Double system Connected, in group		15.5		*
06359	11 18.1 – 01 13a 261.93 + 53.99	00 – 29 – 020	1400 215a	1.1 x 0.7 1.1 x 0.8	75 ...	3	S0 sp of 2		(14.7)		*1
06360	11 18.2 + 53 27 149.53 + 59.03	09 – 19 – 047 NGC 3631	0059 229	6.0: x 5.0: 5.0 x 5.0	— 1:	5	Sc SA(s)c Sc –		11.0	+ 1087 + 1162	*12– –7$\overline{8}$9A
06361	11 18.3 + 00 44 259.99 + 55.59	00 – 29 – 022	1400	[1.3 x 0.5] ...	... ...	2	Triple system		14.3		*1
06362	11 18.3 + 19 54 227.99 + 67.96	03 – 29 – 035	0051	1.1 x 0.25 1.0 x 0.25	63 6	1	S ...		14.6		*1
06363	11 18.3 + 21 37 223.70 + 68.61	04 – 27 – 022	1353	1.1 x 0.25 1.1 x 0.25	57 6	1	Sc: P w 6366		15.2		*1
06364	11 18.3 + 34 37 186.98 + 69.33	06 – 25 – 048 IC 2735	0695 217	1.2 x 0.22 1.2 x 0.22	100 7	1	Sa – b		15.4		*1
06365	11 18.3 + 63 18 139.04 + 51.19	11 – 14 – 020	0977 219	1.0 x 0.3 0.9 x 0.3	117 (6)	2	... d – m		16.5:		
06366	11 18.4 + 21 38 223.68 + 68.64	04 – 27 – 023	1353	1.2 x 0.2 1.1 x 0.2	18 6	2	Sb – c P w 6363		15.7		*1
06367	11 18.4 + 31 31 195.73 + 69.89	05 – 27 – 060	0099 215	1.2 x 0.8 1.0 x 0.7	170 3 – 4	2	S B a – b SB(s) ... P w 6355		15.0		*1
06368	11 18.5 + 03 30 256.92 + 57.79	01 – 29 – 033 NGC 3640	1392 218	(4.5 x 4.0) (4.5 x 4.0)	(100) ...	5	E P w 6370 Brightest in group		11.8	+ 1354 + 1199	*123 46$\overline{7}$8
06369	11 18.5 + 58 03 144.07 + 55.55	10 – 16 – 125	0712 229	1.0 x 0.9 1.0: x 0.9:	— ...	4	Dwarf spiral S V		17.		*
06370	11 18.6 + 03 28 257.00 + 57.78	01 – 29 – 034 NGC 3641	1392 218	(0.8 x 0.8) (0.9 x 0.9)	— ...	4	E P w 6368 NGC 3640 group		14.0:		*12 3$\overline{}$

N G C   3622, 3625 – 31, 3633, 3641   I C   2735

1	2	3	4	5	6	7	8	9	10	11	12
06371	11 18.6 + 70 55 133.17 + 44.66	12 - 11 - 019	0714	1.5 x 0.5 1.1 x 0.5	5 6	1	S ...		15.4		*
06372	11 18.7 + 39 33 174.24 + 67.69	07 - 23 - 000	1349	1.1 x 0.3 1.0 x 0.3	97 6	2	Sc	Sc +	16.5:		
06373	11 18.9 + 03 05 257.56 + 57.53	01 - 29 - 037 NGC 3644 = 684*	1392 218	1.6 x 0.7 1.6 x 0.6	63 6	1	Sa: NGC 3640 group?		15.2		*3
06374	11 18.9 + 18 44 230.94 + 67.58	03 - 29 - 036	0051	0.9: x 0.8: 1.0: x 0.9:	-- ...	1	S? ... Peculiar		14.0		
06375	11 18.9 + 46 30 159.89 + 63.94	08 - 21 - 025	0700 216	1.2 x 0.25 1.2 x 0.25	97 7	1	Sa - b		15.1		
06376	11 19.0 + 20 26 226.86 + 68.32	03 - 29 - 037 NGC 3646	0051	3.8 x 2.2 3.8 x 2.2	50 4:	4	Sc Singular	P w 6386 Sc -	11.5 + 4279 + 4198		*12- -78
06377	11 19.0 + 41 30 169.73 + 66.82	07 - 23 - 000	1349	1.0 x 0.8 0.8: x 0.8:	-- ...	3	Dwarf spiral S IV - V or V		17.		
06378	11 19.0 + 69 55 133.82 + 45.55	12 - 11 - 020	0714	2.4 x 0.35 2.1 x 0.3	142 7	3	Sc	Sc +	14.7		
06379	11 19.1 + 02 42 258.08 + 57.26	01 - 29 - 000	1392 218	1. x 1. 1. x 1.	-- ...	3	Dwarf V		18. - 19.		
06380	11 19.2 + 50 53 152.81 + 61.01	09 - 19 - 050	0059	1.6 x 0.8 1.6 x 0.8	58 5	2	SO - a		14.7		
06381	11 19.2 + 69 24 134.15 + 46.01	12 - 11 - 000	0714	1.7:: x 1.3:: 1.4:: x 0.6::	(5) ...	3	Dwarf spiral S V		17.		
06382	11 19.3 + 35 59 183.11 + 69.14	06 - 25 - 000	0695 221	1.1 x 0.10 0.9 x 0.10	132 7	3	Sc		17.		
06383	11 19.3 + 43 05 166.29 + 66.03	07 - 23 - 042	1349	1.2 x 0.20 1.2 x 0.20	59 6 - 7	3	Sc	Sc +	15.3		*1
06384	11 19.4 + 35 13 185.17 + 69.39	06 - 25 - 053	0695 221	1.3 x 0.10 1.0 x 0.10	143 7	3	Sc		16.5:		
06385	11 19.4 + 59 22 142.54 + 54.51	10 - 16 - 128 NGC 3642	0712 229	6.2 x 5.0 6.2 x 5.0:	105: 2	4	Sb/Sc	SA bc Sc -	11.9 + 1623 + 1725		*12- -9
06386	11 19.6 + 20 28 226.92 + 68.47	03 - 29 - 038 NGC 3649	0051	1.5 x 0.5 1.5 x 0.5	140 6	3	SO P w 6376		14.7		*12 3
06387	11 19.7 + 13 21a 242.51 + 64.87	02 - 29 - 021 IC 2763	0066	1.5 x 0.2 1.0 x 0.2	99 7	1	Sc? P w IC 2767		(14.9)		*1
06388	11 19.8 + 24 34 216.01 + 69.78	04 - 27 - 028 NGC 3651	1353 220	(1.1 x 1.1) (1.3 x 1.3)	-- ...	3	E + companion Brightest in group		14.6		*1
06389	11 19.8 + 40 09 172.58 + 67.61	07 - 23 - 043 NGC 3648	1349	(1.3 x 0.8) 1.3 x 0.8	75 4	3	SO		13.5		3
06390	11 19.8 + 64 21 137.93 + 50.39	11 - 14 - 021	0977 219	2.2 x 0.35 1.9: x 0.3	130 6 - 7	3	Sc	Sc +	14.7		
06391	11 19.9 + 20 59 225.69 + 68.73	04 - 27 - 031 NGC 3650	1353	1.7 x 0.25 1.7 x 0.25	54 7	3	Sb		14.6		*2
06392	11 19.9 + 38 02 177.68 + 68.54	06 - 25 - 055 NGC 3652	0695	2.9 x 0.8: 2.8 x 0.8	150 6	2	S [B c] Singular		12.6		*13
06393	11 20.1 + 34 37 186.71 + 69.68	06 - 25 - 056	0695 217	1.0 x 0.7 1.0 x 0.7	40: 3:	3	S B a		14.9		*
06394	11 20.2 + 34 23 187.35 + 69.76	06 - 25 - 057	0695 217	(1.1 x 1.0) (1.4 x 1.3)	-- ...	1	E:		14.6		*
06395	11 20.3 + 13 43 242.03 + 65.21	02 - 29 - 023 IC 2782	0066	1.1 x 0.9 1.1 x 1.0	-- ...	1	...	d - m	15.2		
06396	11 20.3 + 16 52 235.58 + 66.97	03 - 29 - 039 NGC 3655	0051	1.6 x 1.0 1.6 x 1.0	30: 4:	1	S ...		11.6		*123 4678
06397	11 20.3 + 34 47 186.22 + 69.68	06 - 25 - 058	0695 217	1.8 x 0.25 1.8 x 0.25	0 7	2	Sa - b p of 2		15.0		*1
06398	11 20.5 + 29 53 200.45 + 70.45	05 - 27 - 062	0099 215	1.0 x 0.2 1.1 x 0.25	140 7	2	Sb - c		15.7		
06399	11 20.6 + 51 12 152.03 + 60.94	09 - 19 - 060	0059	3.3 x 0.8 3.3: x 0.8:	142 6	3	...	S: m	14.9		
06400	11 20.6 + 53 58 148.35 + 58.88	09 - 19 - 062	0059 229	1.1 x 0.4 0.9 x 0.4	98 6	2	Sb - c		16.0:		*
06401	11 20.7 + 13 55 241.78 + 65.40	02 - 29 - 000 IC 2787	0066	1.0: x 1.0: 0.9 x 0.8	-- ...	1	Sc?		15.5		
06402	11 20.8 - 00 39 262.31 + 54.85	00 - 29 - 024	1400 215a	0.9 x 0.25 0.9 x 0.25	93 ...	1	...		14.4		
06403	11 20.8 + 54 07 148.12 + 58.78	09 - 19 - 063 NGC 3656	0059 229	1.7 x 1.7 1.8 x 1.8	-- ...	4	... Peculiar, loop, Ir II?	(R?) ...	13.4		*12 3VA
06404	11 20.8 + 77 44 128.86 + 38.65	13 - 08 - 000	1326	1.2 x 0.18 0.8 x 0.20	115 6	1	S ...		16.0:		
06405	11 21.1 + 18 05 233.09 + 67.74	03 - 29 - 040 NGC 3659	0051	2.0 x 1.0 1.9: x 0.9:	60 (5)	2	Sc - Irr	S m:	12.7		*123 4678

N G C   3642,   3644,   3646,   3648 - 52,   3655
      3656,   3659

I C   684,   2763,   2782,   2787

1	2	3	4	5	6	7	8	9	10	11	12
06406	11 21.1 + 53 12 149.21 + 59.51	09 – 19 – 065 NGC 3657	0059	(2.9 x 2.9) (1.5 x 1.5)	-- ...	3	... Very compact		13.1		*12 3
06407	11 21.1 + 69 42 133.74 + 45.83	12 – 11 – 022 NGC 3654	0714	1.8: x 1.1: 1.7 x 0.8	27 4 – 5	1	S ...		13.4		3
06408	11 21.1 – 00 50 262.63 + 54.75	00 – 29 – 025 NGC 3662	1400 215a	1.5 x 0.8 1.4 x 0.8	(25) ...	1	S ... + companions Triple system		13.8		123
06409	11 21.3 + 38 50 175.38 + 68.46	07 – 24 – 002 NGC 3658	0719	(1.8 x 1.7) (1.7 x 1.6)	-- ...	2	E – S0 P w 6426		13.3		*23
06410	11 21.3 + 46 06 159.99 + 64.51	08 – 21 – 030	0700 216	1.2 x 0.9 1.2 x 0.9	5: 2 – 3	3	Sc	SAB c	14.8		
06411	11 21.3 + 48 58 155.19 + 62.60	08 – 21 – 031	0700	1.1 x 0.2 1.0 x 0.2	79 6	2	Sb – c		15.7		
06412	11 21.3 + 58 53 142.68 + 55.05	10 – 16 – 130	0712 229	1.1: x 0.6: 0.8: x 0.5:	110 5:	2	...	S dm:	17.		
06413	11 21.4 + 02 58 258.61 + 57.83	01 – 29 – 040	1392 218	1.1 x 1.1 0.9 x 0.9	-- 1	1	Sb: f of 2		14.8		*
06414	11 21.5 + 24 54 215.30 + 70.23	04 – 27 – 000	1353	1.8 x 0.15 1.4 x 0.15	51 7	2	Sb – c		16.0:		
06415	11 21.5 + 27 18 208.22 + 70.60	05 – 27 – 064	0099 222	2.2 x 0.5 1.8 x 0.4	20 6	2	Sa – b		15.3		
06416	11 21.7 + 39 31 173.63 + 68.22	07 – 24 – 000	0719	1.3: x 0.9: 1.1:: x 0.6::	... ...	2	Dwarf IV – V or V		17.		
06417	11 21.8 + 03 25 258.22 + 58.24	01 – 29 – 000	1392 218	1.0 x 0.1 1.0 x 0.1	144 7	2	Sc		15.6		
06418	11 21.8 + 03 30 258.12 + 58.31	01 – 29 – 000	1392 218	1.0 x 1.0 0.6: x 0.6:	-- ...	3	Irr NGC 3640 group, Ir I		15.4		*
06419	11 21.8 + 03 36 258.00 + 58.39	01 – 29 – 041 NGC 3664	1392 218	1.9 x 1.7 1.8 x 1.7	-- 1	4	S B ... Peculiar, NGC 3640 group		13.6	+ 1406 + 1253	*12V 46̲7̲8A
06420	11 21.8 + 11 37 246.39 + 64.17	02 – 29 – 025 NGC 3666	0066	4.5: x 1.2 4.5: x 1.3:	100 6	3	Sc Leo group:		12.5		*123 4678
06421	11 21.8 + 27 44 206.93 + 70.70	05 – 27 – 065	0099 222	1.3: x 1.1: 1.3: x 0.9:	100: ...	2	...	d – m	16.0:		1
06422	11 21.8 + 54 01 148.03 + 58.95	09 – 19 – 070	0059 229	1.3: x 0.4 1.1 x 0.4	38 (6)	1	...		16.5:		
06423	11 21.9 + 60 54 140.62 + 53.43	10 – 16 – 132	0712 229	1.0 x 0.5 1.0 x 0.5	37: 5	2	S B: c		16.0:		
06424	11 22.0 + 15 13 239.65 + 66.42	03 – 29 – 041	0051	1.3 x 0.2 1.2 x 0.2	121 7	1	S – Irr		15.5		1
06425	11 22.0 + 23 53 218.29 + 70.11	04 – 27 – 032	1353	1.3 x 1.2 1.3 x 1.1	-- 1	3	Sb	SA bc Sb +	15.4		
06426	11 22.0 + 39 02 174.73 + 68.49	07 – 24 – 003 NGC 3665	0719	(3.5 x 3.0) (4.0 x 3.5)	30: ...	2	E – S0 P w 6409		11.6	+ 2002 + 2010	*12 34̲6̲78
06427	11 22.1 + 24 13 217.36 + 70.21	04 – 27 – 033 NGC 3670	1353 220	1.3 x 0.7 1.1 x 0.7	35 4 – 5	2	S B 0 – a		14.6		*13
06428	11 22.2 + 38 11 176.80 + 68.89	06 – 25 – 000	0695	1.5: x 0.3: 1.3: x 0.2	62 7	2	...	dm – m	17.		
06429	11 22.4 + 64 01 137.83 + 50.82	11 – 14 – 022	0977 219	2.9 x 2.9 2.6: x 2.6:	-- 1	4	Sc P w 6430	SA(rs)c	14.1		*1
06430	11 22.5 + 63 43 138.06 + 51.09	11 – 14 – 023 NGC 3668	0977 219	2.1: x 1.3 2.1: x 1.3	137 4:	3	Sb/Sc P w 6429		13.1		*13
06431	11 22.6 + 58 00 143.34 + 55.86	10 – 16 – 135 NGC 3669	0712 229	2.2 x 0.5 2.0 x 0.5	153 (6)	1	Irr:		12.9		*23
06432	11 22.7 + 00 37 261.69 + 56.15	00 – 29 – 030	1400	1.0: x 0.7 0.7 x 0.6	(105) ...	1	S ...		15.2		
06433	11 22.8 + 38 19 176.32 + 68.94	06 – 25 – 066	0695	1.1 x 0.8: 1.1 x 0.8:	[79] ...	3	... Peculiar		14.7		1
06434	11 22.9 + 32 36 192.16 + 70.69	06 – 25 – 067	0695	1.0 x 0.18 0.9 x 0.18	25 6 – 7	1	S ... Brightest in group		15.7		*1
06435	11 23.0 – 00 30 262.94 + 55.28	00 – 29 – 031	1400 215a	1.1 x 1.1 1.1 x 1.0	-- (1)	3	S0 Jet?		14.2		*
06436	11 23.1 + 14 57 240.58 + 66.48	03 – 29 – 043 IC 2810	0051	1.2 x 0.5: 1.1 x 0.4	30: 5 – 6	1	S B ... npp of 2		15.4		*1
06437	11 23.1 + 19 12 231.02 + 68.68	03 – 29 – 045	0051	1.3 x 0.5 1.2 x 0.6	35 5 – 6	2	Sb – c		15.3		
06438	11 23.3 + 10 16 249.22 + 63.54	02 – 29 – 027 IC 692	0066	0.8 x 0.6 0.8 x 0.5	(125) ...	1	...		14.1		
06439	11 23.4 + 43 52 163.65 + 66.19	07 – 24 – 004 NGC 3675	0719	6.8: x 3.5: 6.5: x 3.3:	178 5	3	Sb	Sb –	10.4	+ 696 + 727	*12 346̲8̲
06440	11 23.5 + 02 15 260.21 + 57.58	00 – 29 – 032	1400	1.1 x 0.8 1.1 x 0.7	43 3:	3	Sb		14.7		

N G C   3654, 3657, 3658, 3662, 3664 – 66   I C   692, 2810
3668 – 70, 3675

1	2	3	4	5	6	7	8	9	10	11	12
06441	11 23.5 + 47 15 157.42 + 64.04	08 - 21 - 035 NGC 3677	0700 216	(1.6 x 1.4) (1.7 x 1.5)	(130) 1 - 2	2	Sa: sp of 2, disturbed?		13.5		*23
06442	11 23.6 + 08 07 252.68 + 62.08	01 - 29 - 000	1392	1.5 x 0.1 1.4 x 0.10	176 7	3	Sc	Sc +	15.5		*
06443	11 23.6 + 28 09 205.73 + 71.12	05 - 27 - 071 NGC 3678	0099 222	0.8 x 0.7 0.7 x 0.7	--- 1	2	Sb - c		14.2		3
06444	11 23.6 + 57 20 143.85 + 56.48	10 - 16 - 138 NGC 3674	0712 229	(1.5 x 0.5) 1.5 x 0.5	33 6 - 7	3	SO		13.1		*23
06445	11 23.8 + 17 08 236.12 + 67.82	03 - 29 - 048 NGC 3681	0051	2.7:: x 2.7:: 2.3: x 2.3:	--- 1	4	Sb/Sc SAB(r)bc NGC 3686 group, Sb +		12.2	+ 1314 + 1220	*123 4678
06446	11 23.8 + 54 01 147.59 + 59.14	09 - 19 - 079	0059 229	4.1 x 2.6 3.5: x 2.5:	10: 3 - 4	4	Sc SA d		14.5		
06447	11 23.9 + 59 26 141.66 + 54.78	10 - 16 - 139 IC 691, Mark 169	0712 229	0.8 x 0.5 0.5 x 0.4	(150) ...	1	... Compact, jet		14.2	+ 1200	*
06448	11 23.9 + 64 25 137.28 + 50.56	11 - 14 - 025 Mark 170	0977	2. x 1. 1.9: x 0.9:	[160:] ...	3	... Peculiar		14.9	+ 1050	*A
06449	11 24.0 + 11 43 247.03 + 64.64	02 - 29 - 000 IC 2822	0066	1.7 x 0.8 1.5 x 0.6	130: 5:	1	S B b - c		15.2		
06450	11 24.0 + 79 54 127.55 + 36.74	13 - 08 - 060	1326	1.1: x 0.7 0.6:: x 0.5::	[90] ...	2	... P w 6451, contact, 6481 group		16.5:		*1
06451	11 24.2 + 79 54 127.54 + 36.74	13 - 08 - 061	1326	1.4:: x 0.8: 1.1:: x 0.7::	(170) ...	2	... P w 6450, contact, 6481 group		16.5:		*1
06452	11 24.5 + 59 55 141.09 + 54.42	10 - 16 - 142	0712 229	1.1 x 0.5 1.1 x 0.5	30 5 - 6	1	S ...		14.0		*
06453	11 24.6 + 17 18 236.00 + 68.07	03 - 29 - 050 NGC 3684	0051	3.5 x 2.4 2.5: x 1.7:	130: 3	4	Sc SA(rs)c NGC 3686 group, Sc -		12.1	+ 1422 + 1329	*123 4678
06454	11 24.6 + 38 56 174.35 + 68.99	07 - 24 - 007	0719	1.7 x 0.12 1.5 x 0.12	34 7	3	Sc		16.0:		
06455	11 24.6 + 40 16 171.13 + 68.35	07 - 24 - 008	0719	1.9 x 0.8 1.8 x 0.7	83 6	1	S ...		15.6		*
06456	11 24.6 + 79 16 127.84 + 37.33	13 - 08 - 058	1326	1.6: x 0.9: 1.3:: x 0.8::	... ...	3	Strongly peculiar Eruptive? or multiple?		14.7		*C
06457	11 24.7 - 00 43 263.78 + 55.35	00 - 29 - 034	1400	1.3: x 1.2: 0.9: x 0.9:	--- ...	2	Irr		15.0		
06458	11 24.7 + 57 10 143.80 + 56.71	10 - 16 - 143 NGC 3683	0712 229	2.1: x 0.8 1.8 x 0.8	128 (6)	2	...		12.7		*123 68
06459	11 24.7 + 66 52 135.30 + 48.47	11 - 14 - 027 NGC 3682	0977	(2.3 x 1.6) (2.2 x 1.5)	95 3	2	SO - a		13.4		*123
06460	11 25.1 + 17 30 235.71 + 68.28	03 - 29 - 051 NGC 3686	0051	3.1 x 2.5 3.0 x 2.5	15: 2 - 3	5	S B b/S B c S B(s)bc Brightest in group, Sb +		11.6	+ 1022 + 930	*12 3478
06461	11 25.1 + 21 39 225.19 + 70.10	04 - 27 - 000	1353 220	1.2 x 0.4 1.1 x 0.3	2 6	2	Sa - b		16.0:		
06462	11 25.2 + 08 16 253.07 + 62.46	01 - 29 - 000	1392	1.7: x 0.25 1.4: x 0.2	3 6	1	S ... Brightest of 3		15.1		*
06463	11 25.3 + 29 47 200.64 + 71.50	05 - 27 - 073 NGC 3687	0099 222	1.8 x 1.8 1.8 x 1.8	--- 1	3	Sb SAB:(r)b Sb +		13.0		*12 3478
06464	11 25.5 + 17 12 236.53 + 68.21	03 - 29 - 053 NGC 3691	0051	1.1 x 0.9 1.2 x 0.9	15: ...	2	... Peculiar, NGC 3686 group		13.1		*12 348
06465	11 25.5 + 22 16 223.57 + 70.40	04 - 27 - 038	1353 220	1.2 x 0.9 1.3 x 0.9	120: ...	3	S B ... Peculiar, distorted?		15.0		*1
06466	11 25.6 + 04 36 258.22 + 59.76	01 - 29 - 043	1392	1.3 x 0.4 1.2 x 0.45	5: 6:	3	... SAB: d - dm sp of 2		15.5		*1
06467	11 25.6 + 25 56 212.73 + 71.33	04 - 27 - 037 NGC 3689	1353	1.6: x 1.0 1.6 x 1.0	97 4:	3	Sc SAB c		12.9		*12 3468
06468	11 25.6 + 71 07 132.33 + 44.74	12 - 11 - 000	0714	1.2 x 0.3 1.0 x 0.2	153 6	3	Sc	Sc +	16.5:		
06469	11 25.7 + 02 56 260.25 + 58.46	01 - 29 - 044	1392	0.9 x 0.35 0.9 x 0.35	123 6	1	S ...		14.1		
06470	11 25.7 + 09 25 251.51 + 63.37	01 - 29 - 031 IC 2853	0066 223	1.0 x 0.45 1.0 x 0.5	15 5	2	S B a - b IC 689 group		14.6		*
06471	11 25.7 + 58 50a 141.91 + 55.41	10 - 17 - 02a IC 694	0723 229	[2.9: x 2.1]	...	3	Double system Mark 171		11.8	+ 3111 + 3212	*12 VA
06472	11 25.7 + 58 50b 141.91 + 55.41	10 - 17 - 003 NGC 3690	0723 229	...	...		Contact, disrupted			+ 2996 + 3097	*123V 4678A
06473	11 25.7 + 73 18 131.00 + 42.78	12 - 11 - 026	0714 212	1.6 x 0.6 1.6 x 0.6	20 6	2	Sa - b		15.2		*
06474	11 25.8 + 09 41 251.13 + 63.58	02 - 29 - 032 NGC 3692	0066 223	3.1 x 0.7 2.9 x 0.7	95 6	3	Sb Leo group?		12.9		*3
06475	11 25.9 + 09 23 251.64 + 63.38	02 - 29 - 033 IC 2857	0066 223	1.7 x 0.15 1.6 x 0.15	161 7	3	Sc P w 6477, IC 698 group		15.3		*

N G C 3674, 3677, 3678, 3681 - 84, 3686   I C 691, 694, 2822, 2853, 2857
3687, 3689 - 92

1	2	3	4	5	6	7	8	9	10	11	12
06476	11 25.9 + 23 40 219.62 + 70.91	04 – 27 – 040	1353 220	1.5 x 1.1 1.2 x 0.9	115 2 – 3	3	Sb	Sb +	14.4		
06477	11 26.0 + 09 22 251.71 + 63.34	02 – 29 – 034 IC 696	0066 223	1.0 x 0.9 0.9 x 0.7	— 1	2	S B c – Irr P w 6475, IC 698 group		14.5		*1
06478	11 26.0 + 20 00 229.84 + 69.65	03 – 29 – 056	0051	1.0 x 0.3 1.0 x 0.3	24 6	1	S ... spp of 3		15.5		*
06479	11 26.1 + 21 04 227.03 + 70.10	04 – 27 – 042 NGC 3697	1353 220	2.5 x 0.7 2.5 x 0.7	93 6	3	Sb/S B b		14.1		*13
06480	11 26.2 + 35 41 182.64 + 70.56	06 – 25 – 076 NGC 3694	0695 221	(1.1 x 0.9) 1.1 x 0.9	(120) ...	1	...		13.5		*3
06481	11 26.2 + 79 52 127.45 + 36.81	13 – 08 – 063	1326	1.5 x 0.8: 1.3 x 0.6	120 ...	3	S ... Peculiar? Brightest of 3		16.0:		*
06482	11 26.4 + 09 23 251.84 + 63.47	02 – 29 – 035 IC 698	0066 223	1.0 x 0.5 0.9 x 0.5	147 ...	1	S0? Brightest in group		14.4		*
06483	11 26.4 + 17 31 236.10 + 68.55	03 – 29 – 057	0051	2.1 x 0.2 1.8 x 0.2	68 7	3	Sc		15.3		*1
06484	11 26.4 + 57 25 143.20 + 56.63	10 – 17 – 006	0723	2.4 x 1.8 2.2 x 1.7	75: 2 – 3	4	S B c	S B(rs)c Sc +	12.6		*<u>2</u>
06485	11 26.5 + 09 16 252.06 + 63.40	02 – 29 – 036 IC 699	0066 223	1.3 x 0.35 1.3 x 0.35	12 6	1	Sb: IC 698 group		14.6		
06486	11 26.5 + 12 09 247.22 + 65.39	02 – 29 – 000	0066	1.4:x 1.2:: 1. x 1.	— ...	4	Dwarf irregular Ir V		18.		
06487	11 26.5 + 20 51 227.72 + 70.10	04 – 27 – 047 IC 700	1353 220	[1.4: x 0.7:] ...	... ...	2	Quadrupel system Connected		15.0		1
06488	11 26.5 + 42 03 166.59 + 67.70	07 – 24 – 012	0719	(1.1 x 0.5) (1.1 x 0.5)	20 ...	1	E – S p of 2		16.0:		*1
06489	11 26.5 + 42 47 165.02 + 67.28	07 – 24 – 013	0719	1.6 x 0.8 1.6 x 0.8	55 (5)	1	S: ...		15.7		
06490	11 26.6 + 35 51 182.08 + 70.58	06 – 25 – 078 NGC 3695	0695 221	1.1 x 0.7 1.1 x 0.7	(10) 4:	1	S ... P w 6494		14.9		*3
06491	11 26.7 + 35 08 184.11 + 70.84	06 – 25 – 077	0695 221	2.0: x 1.0 1.6: x 0.9:	0: 5:	4	... S dm		15.6		
06492	11 26.7 + 52 31 148.80 + 60.57	09 – 19 – 091	0059	1.0 x 0.12 1.0 x 0.15	74 7	1	S ... In group		16.0:		*1
06493	11 26.8 + 24 21 217.74 + 71.28	04 – 27 – 048 NGC 3701	1353	2.0 x 0.9 2.1 x 1.0	145 5 – 6	2	Sb – c		14.1		3
06494	11 26.9 + 35 47 182.21 + 70.66	06 – 25 – 079 NGC 3700	0695 221	1.1 x 0.7 1.1 x 0.8	130: 3 – 4	2	S B a – b (R´)S B(r) ... P w 6490		15.1		*3
06495	11 27.0 + 22 24 223.53 + 70.77	04 – 27 – 049	1353 220	1.0 x 0.9 1.0 x 0.8	— 1 – 2	2	Sb/S B b: SAB(s) ...		14.9		
06496	11 27.2 + 25 13 215.15 + 71.56	04 – 27 – 000	1353	(0.6 x 0.4) (0.7 x 0.5)	... ...	1	... Very compact		14.5		*
06497	11 27.4 + 38 54 173.73 + 69.49	07 – 24 – 015	0719	2.3 x 0.3: 1.8 x 0.15	86 7	3	... S d – dm		16.0:		
06498	11 27.5 + 09 31 252.06 + 63.75	02 – 29 – 039 NGC 3705	0066 223	5.0 x 2.2 4.5 x 1.6	122 6	4	Sb SAB(r)b Leo group? Sb –		11.5		*12 34<u>7</u>8
06499	11 27.6 + 36 08 181.07 + 70.67	06 – 25 – 000	0695 221	1.7: x 0.6: 1.3:: x 0.6::	5 ...	3	Dwarf V, sp of 2		17.		*
06500	11 28.0 + 44 26 161.27 + 66.48	07 – 24 – 019	0719 227	1.0 x 0.35 0.9 x 0.30	178 6	2	Sa – b Brightest in group		15.3		*
06501	11 28.0 + 60 47 139.67 + 53.92	10 – 17 – 008	0723	1.2: x 1.0: 1.0: x 0.9:	— ...	2	... S d:		16.5:		
06502	11 28.0 + 77 33 128.51 + 38.97	13 – 08 – 000	1326	1.4: x 0.10 1.0 x 0.10	43 7	2	Sc		17.		*
06503	11 28.2 + 20 45 228.46 + 70.43	04 – 27 – 051 IC 701	1353 234	1.2 x 0.6 1.1 x 0.6	[47] ...	2	S B ... n of 2 Peculiar, streamer		14.7		*1V A
06504	11 28.4 + 23 02 222.00 + 71.28	04 – 27 – 052 NGC 3710	1353 220	(1.0 x 0.8) (1.0 x 0.8)	(105) ...	2	E		14.5		3
06505	11 28.4 + 51 59 149.08 + 61.15	09 – 19 – 101	0059	1.1 x 1.0 1.0 x 0.9	— 1	3	Sc SAB: ... Sc +		16.0:		
06506	11 28.5 + 28 50 203.65 + 72.21	05 – 27 – 082	0099 222	2.3: x 1.2: 1.5: x 0.5:	[160] ...	2	... B ... Peculiar, plumes		15.7		*
06507	11 28.5 + 34 29 185.69 + 71.39	06 – 25 – 080	0695 225	1.5 x 1.0 1.1 x 0.9	175: 3	4	S B c S B(s)c Sc +		15.6		
06508	11 28.6 + 26 35 210.99 + 72.09	05 – 27 – 083	0099 222	1.0 x 0.9 0.8 x 0.7	— 1 – 2	3	Sb 3rd of 3		15.5		*1
06509	11 28.7 + 23 23 221.02 + 71.45	04 – 27 – 054	1353 220	1.9 x 0.15 1.9 x 0.15	79 7	3	Sc S d		16.0:		*1
06510	11 29.0 – 02 01 266.59 + 54.84	00 – 30 – 001	0471 231	2.1 x 1.8 2.0 x 1.8	(10) 1 – 2	4	Sc SAB(rs)cd		14.0		

N G C   3694, 3695, 3697, 3700, 3701, 3705,
3710

1	2	3	4	5	6	7	8	9	10	11	12
06511	11 29.0 + 28 25 205.02 + 72.32	05 – 27 – 084 NGC 3713	0099 222	1.2 x 0.8 1.1 x 0.7	125 4 – 5	2	E – S0		14.4		*3
06512	11 29.0 + 34 36 185.24 + 71.45	06 – 25 – 081	0695 225	1.2 x 0.9 1.2: x 0.8:	150: ...	1	S ...		15.3		
06513	11 29.1 + 03 46 260.60 + 59.64	01 – 30 – 001 NGC 3716	0495	0.7 x 0.6 0.7 x 0.6	(150) ...	1	S0?		14.5		3
06514	11 29.1 + 71 05 131.99 + 44.90	12 – 11 – 028	0714	[1.1 x 0.3] ...	... ...	2	Chain of 5 galaxies		15.2	(+ 15580) (+ 15733)	*12 VA
06515	11 29.1 + 74 54 129.82 + 41.43	13 – 08 – 064 NGC 3752	1326	1.8 x 0.8 1.8 x 0.7	155 6	2	Sa – b		13.7		3
06516	11 29.2 + 28 38 204.31 + 72.37	05 – 27 – 085 NGC 3714	0099 222	(2.2 x 1.0) (1.2 x 0.9)	[68] ...	1	... Peculiar		14.3		*13
06517	11 29.3 + 36 58 178.35 + 70.67	06 – 25 – 082	0695	2.0: x 1.2: 1.6: x 0.9:	40 4:	3	Sb	Sb +	14.0		*
06518	11 29.5 + 54 11 146.09 + 59.52	09 – 19 – 109	0059 229	1.1 x 0.6 1.1 x 0.5	18 5	1	S ...		14.6		
06519	11 29.6 + 01 30 263.34 + 57.87	00 – 30 – 000	0471 231	1.3 x 0.15 1.1 x 0.15	93 7	3	Sc	Sc +	15.6		
06520	11 29.6 + 62 48 137.70 + 52.28	11 – 14 – 028 Mark 175	0977 219	1.5 x 0.8 1.2 x 0.8	165: ...	1	S ... nnp of 2		14.1		*
06521	11 29.7 + 01 06 263.80 + 57.55	00 – 30 – 005 NGC 3719	0471 231	2.1 x 1.5 1.8 x 1.3	15 3	3	Sb P w 6523	SAB:(rs)b Sb +	13.8		*12 3
06522	11 29.7 + 28 19 205.37 + 72.47	05 – 27 – 087	0099 222	(1.4 x 1.4) 1.5 x 1.3	— ...	1	S0? Disturbed?		15.0		*1
06523	11 29.8 + 01 05 263.86 + 57.55	00 – 30 – 006 NGC 3720	0471 231	1.0 x 0.9 1.0 x 1.0	— ...	1	... P w 6521		13.7		*12 3478
06524	11 29.8 + 53 21 147.02 + 60.21	09 – 19 – 114 NGC 3718	0059 229	11. x 5. 11. x 4.	15 6	3	S0/S B a P w 6547,	Singular disturbed	11.8	+ 1050 + 1128	*12– –9 VA
06525	11 29.9 + 20 43 229.04 + 70.78	04 – 27 – 058	1353 234	1.4 x 1.3 1.4 x 1.3	— 1	4	S B b	(R:)S B(r)b Sb +	14.9		*
06526	11 29.9 + 35 36 182.09 + 71.30	06 – 25 – 083	0695 221	1.9: x 0.4 1.9: x 0.4	90 ...	1	...		14.1		
06527	11 29.9 + 53 14 147.13 + 60.31	09 – 19 – (111) Mark 176	0059 229	[1.1 x 0.6] ...	... ...	2	Triple system Connected, in group of 5		(14.7)	+ 7950 + 8025	*1V AC
06528	11 29.9 + 62 07 138.20 + 52.89	10 – 17 – 013	0723	1.2 x 1.1 1.2 x 1.1	— 1	4	Sc P w 6542	SA(rs)c Sc +	14.1		*
06529	11 30.0 + 41 07 167.67 + 68.77	07 – 24 – 020	0719	1.1 x 0.9 0.9 x 0.7	(70) 2:	2	Sb – c		15.4		
06530	11 30.0 + 72 23 131.14 + 43.75	12 – 11 – 030	0714	1.2 x 0.7 1.2 x 0.7	135 4	2	Sb		15.1		*
06531	11 30.1 + 39 21 171.87 + 69.73	07 – 24 – 022	0719	2.0 x 2.0 1.8: x 1.8:	— 1	4	...	S B(r)dm	15.7		
06532	11 30.2 + 70 40 132.13 + 45.31	12 – 11 – 032	0714	1.2 x 1.0 1.1 x 1.1	— 1:	2	S B c	S B(r) ...	15.3		*
06533	11 30.3 + 06 00 258.28 + 61.58	01 – 30 – 000	0495	1.0 x 0.2 0.9 x 0.20	178 7	1	Sb		15.2		
06534	11 30.4 + 63 34 136.96 + 51.65	11 – 14 – 030	0977 219	3.0 x 0.8 2.1 x 0.4	60 6	3	Sc	Sc +	13.3		
06535	11 30.5 + 50 35 150.44 + 62.44	08 – 21 – 050	0700	1.3 x 0.15 1.2 x 0.15	67 7	3	Sc P w IC 705		15.1		*
06536	11 30.6 + 24 43 217.24 + 72.21	04 – 27 – 061 NGC 3728	1353	2.5 x 1.6 2.0: x 1.5:	25: 3 – 4:	3	Sb		14.7		3
06537	11 30.6 + 47 18 155.40 + 64.88	08 – 21 – 051 NGC 3726	0700	6.1 x 4.1 5.8 x 4.0	10 3	4	Sc	SAB(r)c	11.2	+ 948 + 998	*12– –9
06538	11 30.6 + 49 34 151.87 + 63.22	08 – 21 – 052	0700 226	1.0 x 1.0 0.6: x 0.6:	— 1	1	S ... P w 6541		16.0:		*1
06539	11 30.7 + 32 52 190.32 + 72.24	06 – 25 – 085	0695	1.5 x 0.3 1.4 x 0.25	25 6 – 7	2	Sb – c P w 6545		15.4		*1
06540	11 30.8 + 34 35 184.94 + 71.81	06 – 25 – 086 IC 2928	0695 225	1.2 x 0.9 0.9 x 0.8	(155) 2 – 3	2	Sb – c P w 6550		14.7		*1
06541	11 30.8 + 49 31 151.89 + 63.28	08 – 21 – 053 Mark 178	0700 226	1.3 x 0.7 1.2 x 0.7	[133] ...	3	Irr (+ companion:) P w 6538 Ir I		13.9		*1
06542	11 30.8 + 62 10 138.01 + 52.90	10 – 17 – 015 NGC 3725, Mark 179	0723	1.6: x 1.2 1.5 x 1.1	145: ...	3	S B c P w 6528	Sc –	13.6	+ 4500	*3
06543	11 31.0 + 21 39 226.70 + 71.38	04 – 27 – 064 IC 707	1353 235	0.6 x 0.5 0.6 x 0.5	— 1 – 2	1	S ...		14.4		
06544	11 31.0 + 23 41 220.57 + 72.04	04 – 27 – 065	1353	1.2 x 0.6 1.0 x 0.5	95 5	2	Sb – c		15.1		
06545	11 31.0 + 32 54 190.17 + 72.30	06 – 26 – 001	0109	1.1 x 0.35 1.1 x 0.35	133 ...	1	... P w 6539		14.5		*1

N G C   3713, 3714, 3716, 3718 – 20, 3725,
      3726, 3728, 3752

I C   707, 2928

1	2	3	4	5	6	7	8	9	10	11	12
06546	11 31.1 + 17 40 237.38 + 69.59	03 - 30 - 006	1406	1.2 x 0.15 1.2 x 0.15	7 7	2	Sb - c		15.5		*
06547	11 31.1 + 53 25 146.62 + 60.27	09 - 19 - 117 NGC 3729	0059 229	3.4 x 2.4 3.4 x 2.4	15 3:	3	S B ... SB(r) ... P w 6524, disturbed		12.2		*123 45789
06548	11 31.2 + 23 55b 219.88 + 72.15	04 - 27 - 066	1353	1.0 x 0.3 0.9 x 0.22	40 6 - 7	1	S0 n of 2		(15.0)		*1
06549	11 31.2 + 49 20 152.05 + 63.46	08 - 21 - 056 IC 708	0700 226	(1.4 x 0.9) (1.7 x 1.3)	95: ...	2	E		14.2		*1
06550	11 31.5 + 34 35 184.79 + 71.94	06 - 26 - 004 IC 2933	0109 225	1.4 x 0.4 1.3 x 0.4	8 6	2	Sa - b P w 6540		15.2		*
06551	11 31.5 + 36 57 177.84 + 71.08	06 - 26 - 002	0109	1.4 x 0.2 1.2 x 0.15	113 7	3	Sc Sc +		16.0:		1
06552	11 31.5 + 71 49 131.32 + 44.31	12 - 11 - 034	0714	1.2 x 0.45 1.2 x 0.45	46 6	1	S ...		14.3		
06553	11 31.6 + 12 48a 248.06 + 66.75	02 - 30 - 001 NGC 3731	0468	(1.0 x 0.9) (1.0 x 0.9)	(50) ...	2	E n of 2		(14.3)		*13
06554	11 32.3 + 55 08 144.37 + 58.98	09 - 19 - 123 NGC 3733	0059 229	4.8 x 2.2 4.3: x 2.0:	170 6	4	Sc Sc +		13.2		*23
06555	11 32.3 + 51 31 148.71 + 61.89	09 - 19 - 122	0059	(1.3 x 1.0) (1.4 x 1.1)	(90) ...	2	E - S0		14.4		*
06556	11 32.4 + 16 23 240.90 + 69.14	03 - 30 - 010	1406	1.5 x 0.2 1.5 x 0.15	91 6 - 7	3	Sc P w 6559 Sc +		15.6		*
06557	11 32.4 + 30 10 199.08 + 73.01	05 - 27 - 000	0099	1.0 x 0.9 0.9 x 0.9	— ...	2	... Peculiar, ring		16.0:		*
06558	11 32.5 + 02 49 263.05 + 59.36	01 - 30 - 000	0495	1.0 x 0.3 1.0 x 0.35	123 6	3	S B c Sc +		15.4		
06559	11 32.5 + 16 14 241.29 + 69.07	03 - 30 - 013	1406	1.7 x 0.10 0.9: x 0.10	176 7	3	Sc P w 6556		15.7		*
06560	11 32.6 + 73 44 130.15 + 42.59	12 - 11 - 035	0714 212	1.4: x 0.8: 1.6: x 0.8:	155: ...	1	...		15.6		*
06561	11 32.8 + 33 34 187.72 + 72.50	06 - 26 - 006	0109	1.1 x 0.20 1.0 x 0.15	139 (6 - 7)	2	Sc - Irr		15.6		1
06562	11 32.8 + 43 45 161.19 + 67.58	07 - 24 - 000	0719 227	1.1 x 0.12 0.9 x 0.10	150 7	2	Sc		16.5:		
06563	11 32.8 + 55 13 144.16 + 58.95	09 - 19 - 128 NGC 3737	0059 229	(1.1 x 1.0) (1.6 x 1.4)	— ...	1	...		13.9		*12 3
06564	11 33.0 + 25 20 215.61 + 72.87	04 - 27 - 071 NGC 3739	1353	1.3 x 0.3 1.3 x 0.3	17 6	2	Sb - c		15.3		3
06565	11 33.0 + 54 47 144.59 + 59.32	09 - 19 - 130 NGC 3738	0059 229	3.5 x 3.0 3.5 x 3.0	[155:] ...	4	Irr Ir I		11.5		*12- -9A
06566	11 33.0 + 58 28 140.84 + 56.23	10 - 17 - 020	0723 229	1.3: x 1.1: 1.1: x 0.8:	(110) ...	3	Dwarf spiral, S B m SB IV - V or V		16.5:		
06567	11 33.0 + 70 48 131.75 + 45.29	12 - 11 - 036 NGC 3735	0714	4.3 x 0.9 4.5 x 0.9	131 6	3	Sc		12.4		*123 4678
06568	11 33.1 + 00 25 265.82 + 57.45	00 - 30 - 014	0471 231	1.0 x 0.45 0.7 x 0.5	2 ...	1	S: ...		14.4		
06569	11 33.2 + 17 43 238.03 + 70.04	03 - 30 - 000	1406	1.0 x 0.4 0.9 x 0.4	7 6	1	S B ...		15.7		
06570	11 33.2 + 35 36 181.33 + 71.93	06 - 26 - 007	0109 235	1.2 x 0.6 1.1 x 0.5	123 5	2	S0 - a		14.5		
06571	11 33.4 + 08 35 255.94 + 64.06	02 - 30 - 000	0468	1.0 x 0.20 1.0 x 0.22	132 7	2	Sa		15.3		
06572	11 33.4 + 45 34 157.56 + 66.44	08 - 21 - 068 NGC 3741	0700	1.7 x 1.1: 1.5: x 1.0:	[5] ...	3	Irr Ir I		14.2		*3
06573	11 33.4 + 60 15 139.16 + 54.72	10 - 17 - 023 NGC 3740	0723	1.0 x 0.4 0.9 x 0.35	110 ...	1	S: ...		14.9		3
06574	11 33.6 + 30 03 199.41 + 73.28	05 - 27 - 000	0099	1.1 x 0.2 0.9 x 0.2	132 7	1	Sb - c		16.5:		
06575	11 33.6 + 58 29 140.70 + 56.26	10 - 17 - 025	0723 229	2.2 x 0.4 2.1: x 0.4	9 6	3	Sc		14.3		*
06576	11 33.8 + 48 06 153.19 + 64.66	08 - 21 - 071	0700	1.2 x 1.1 1.2 x 1.1	— 1	1	S B 0		15.6		*
06577	11 33.9 + 36 40 178.01 + 71.63	06 - 26 - 008 NGC 3755	0109	3.6 x 1.4 3.6 x 1.4	133 6	4	Sc SAB(rs) ... Singular, Sc +		13.9		*12 3
06578	11 34.0 + 01 06 265.49 + 58.14	00 - 30 - 000	0471	1.0: x 0.3 0.6 x 0.3	[160] ...	1	Double system		15.1		*
06579	11 34.0 + 54 34 144.60 + 59.58	09 - 19 - 134 NGC 3756	0059 229	5.0 x 2.5 4.7 x 2.3	177 5	4	Sc SAB(rs)bc Sc -		12.1		*12- -9
06580	11 34.0 + 70 22 131.90 + 45.72	12 - 11 - 000	0714	1.0 x 0.15 0.9 x 0.15	143 7	1	Sb - c		15.7		*

N G C   3729, 3731, 3733, 3735, 3737 - 41   I C   708, 2933
      3755, 3756

1	2	3	4	5	6	7	8 9	10	11	12
06581	11 34.1 + 55 06 143.99 + 59.15	09 – 19 – 136 NGC 3759	0059 229	(1.2 x 1.2) (1.3 x 1.3)	––– (1)	2	S0 P w IC 2943	14.3		*23
06582	11 34.2 + 55 26 143.61 + 58.88	09 – 19 – 135	0059 229	1.4 x 1.2 1.4 x 1.2	––– ...	2	S B b – c S B(r) ... n of 2, disturbed?	14.5		*2
06583	11 34.3 + 20 14 231.76 + 71.50	03 – 30 – 019 Mark 181	1406 234	0.7 x 0.35 0.7 x 0.35	[5] ...	2	... Peculiar Brightest of 3	13.9		*1
06584	11 34.3 + 58 42 140.36 + 56.12	10 – 17 – 026 NGC 3757	0723 229	1.1 x 1.1 1.1 x 0.9	––– ...	1	S0?	13.5		3
06585	11 34.5 + 13 12 248.51 + 67.54	02 – 30 – 004 IC 2945	0468	(1.1 x 0.9) 1.2 x 0.9	... ...	1	Double system Halo	15.0		*1
06586	11 34.5 + 15 51 242.95 + 69.23	03 – 30 – 022	1406	2.2 x 1.0 1.4 x 0.8	175 5	3	Sb/S B c P w 6588	14.5		*
06587	11 34.6 + 03 06 263.58 + 59.89	01 – 30 – 004	0495	1.0 x 0.5 0.8 x 0.45	[150] ...	3	S B b SAB(s)b sf of 2, distorted	15.0		*1
06588	11 34.6 + 15 42 243.33 + 69.16	03 – 30 – 025	1406	1.4 x 0.5 1.4 x 0.5	144 6	2	Sb – c P w 6586	14.9		*
06589	11 34.6 + 18 07 237.55 + 70.53	03 – 30 – 024 NGC 3768	1406	(1.5 x 0.9) (1.6 x 0.9)	155: 4	3	S0	13.7		*3
06590	11 34.7 + 17 09 239.99 + 70.03	03 – 30 – 023 NGC 3767	1406	0.9 x 0.8 (1.2? x 1.1)	––– 1	3	S B 0 (R?)S B(r)0	14.5		*3
06591	11 34.7 + 62 02 137.45 + 53.24	10 – 17 – 027 NGC 3762	0723	2.3 x 0.5 2.2 x 0.45	167 7	2	Sa	13.3		3
06592	11 34.7 + 73 02 130.34 + 43.29	12 – 11 – 000	0714	1.0 x 0.5 1.1 x 0.5	163 5	1	S ...	15.3		
06593	11 34.8 + 26 02 213.47 + 73.39	04 – 28 – 002	0103	(1.1 x 0.9) 1.1 x 0.9	––– ...	1	... Peculiar, streamer?	14.5		*
06594	11 35.0 + 16 50 240.87 + 69.91	03 – 30 – 027	1406	2.7 x 0.35 2.3 x 0.3	134 7	3	Sc Sc +	14.8		*
06595	11 35.0 + 48 10 152.74 + 64.75	08 – 21 – 076 NGC 3769	0700	3.3 x 1.0 3.0 x 1.0	152 6	1	Sb: np of 2, bridge?	11.7	+ 748 + 804	*123 4678A
06596	11 35.0 + 56 25 142.41 + 58.11	09 – 19 – 140	0059 229	1.2 x 0.7 1.1 x 0.7	[140] ...	4	Irr IB m Ir I	15.5		*1
06597	11 35.1 + 22 17 225.91 + 72.49	04 – 28 – 005	0103 234	1.2 x 0.6 1.3 x 0.6	127 5	2	S B a – b 6602 group	15.3		*1
06598	11 35.2 + 22 58 223.79 + 72.74	04 – 28 – 006 NGC 3772	0103	1.2 x 0.6 1.2 x 0.6	16 5	2	S B a	14.4		3
06599	11 35.2 + 24 25 219.04 + 73.16	04 – 28 – 000	0103	1.3: x 0.8: 1.3: x 0.8:	... ...	4	Dwarf irregular, IB m Ir V	17.		*
06600	11 35.2 + 59 53 139.13 + 55.15	10 – 17 – 028 NGC 3770	0723 229	1.1 x 0.7 1.0 x 0.7	107 3:	2	S B a	13.5		3
06601	11 35.4 + 22 05 226.62 + 72.49	04 – 28 – 000	0103 234	(1.1 x 0.5) 0.8 x 0.4	5 ...	1	S0? 6602 group	15.3		*
06602	11 35.4 + 22 16a 226.05 + 72.55	04 – 28 – 010	0103 234	2.0 x 0.5: 1.8 x 0.4	[120] ...	1	... Brightest in group, distorted	(14.6)		*1
06603	11 35.4 + 35 28 181.19 + 72.39	06 – 26 – 009	0109 235	2.3 x 0.5 2.2 x 0.5	78 6	3	Sc	15.0		
06604	11 35.4 + 59 02 139.84 + 55.90	10 – 17 – 029	0723 229	1.0 x 1.0 0.8 x 0.8	––– ...	1	...	14.0		1
06605	11 35.6 + 12 24 250.50 + 67.20	02 – 30 – 005 NGC 3773	0468	1.4 x 1.1 (1.4 x 1.1)	(165) 2:	3	S0	13.1		*123 478
06606	11 35.7 + 47 43 153.24 + 65.16	08 – 21 – 083	0700	[1.7 x 0.9] ...	... ...	3	Double system Connected, disrupted	14.9		*1
06607	11 35.9 + 21 01 230.00 + 72.18	04 – 28 – 012	0103 234	1.6 x 0.9 1.5 x 0.7	75 4 – 5	4	Sc Sc +	15.2		*
06608	11 36.0 – 00 55 268.24 + 56.69	00 – 30 – 017	0471 231	1.1 x 0.7 1.1 x 0.7	36 4	1	S ...	14.5		
06609	11 36.0 + 20 49 230.63 + 72.11	04 – 28 – 013	0103 234	(1.4 x 0.8) (1.4 x 0.9)	(100) ...	2	E Companion superimposed	14.9		*1
06610	11 36.1 + 34 04 185.44 + 73.01	06 – 26 – 011	0109 235	2.3 x 0.4 2.2 x 0.4	13 6	3	Sc ssf of 2	15.3		*
06611	11 36.2 + 43 26 160.75 + 68.26	07 – 24 – 026	0719	1.7 x 0.4 1.3 x 0.3	9 6	1	S ... ssp of 2	15.4		*
06612	11 36.5 + 00 04 267.51 + 57.60	00 – 30 – 018 IC 716	0471 231	1.6 x 0.20 1.7 x 0.22	132 7	2	Sb	14.9		
06613	11 36.5 + 39 36 169.32 + 70.64	07 – 24 – 028	0719	1.2 x 0.5 1.1 x 0.4	57 6	1	S B a	15.2		*1
06614	11 36.6 + 17 25 240.09 + 70.55	03 – 30 – 029	1406	3.2: x 2.8: 1.2 x 1.1	––– ...	3	Sa? (R?)SA(r)a?	14.8		
06615	11 36.6 + 56 32 141.93 + 58.13	09 – 19 – 150 NGC 3780	0059 229	3.1 x 2.5 3.1 x 2.5	90: 2	4	Sc SA c P w 6640 Sc +	12.2		*123 4678

· N G C   3757, 3759, 3762, 3767 – 70, 3772
            3773, 3780

I C   716, 2945

1	2	3	4	5	6	7	8	9	10	11	12
06616	11 36.6 + 58 33 140.03 + 56.40	10 – 17 – 035	0723 229	3.0: x 2.4: 2.7: x 2.2:	(80) ...	4	Sc	SA(s:)d	14.2		1
06617	11 36.7 + 10 14a 254.78 + 65.83	02 – 30 – 006	0468 228	0.9 x 0.3 0.9 x 0.3	161 ...	1	S0? n of 2		(14.5)		*
06618	11 36.7 + 46 47 154.45 + 65.97	08 – 21 – 087 NGC 3782	0700	1.4 x 0.7 1.4 x 0.7	[0] ...	1	Irr?		13.1		*123 46\underline{7}8
06619	11 36.7 + 60 49 138.07 + 54.42	10 – 17 – 037	0723 229	1.4 x 0.3 1.3 x 0.3	8 7	1	S ...		15.4		
06620	11 36.9 + 26 35 211.77 + 73.93	05 – 28 – 007 NGC 3785	1379 230	1.0 x 0.4 1.0 x 0.4	25 (6?)	1	S0 sf of 2, disturbed		15.3		*1
06621	11 37.0 + 32 11 191.61 + 73.69	05 – 28 – 008 NGC 3786	1379	2.2 x 1.1 2.0 x 1.0	77: 5	5	Sa P w 6623, disturbed	SAB(rs)a	13.0	+ 2755 + 2737	*12 3 V\overline{A}
06622	11 37.1 + 04 45 262.71 + 61.61	01 – 30 – 000	0495	1.1 x 0.4 0.8 x 0.35	141 6	1	S ...		15.3		
06623	11 37.1 + 32 13 191.48 + 73.70	05 – 28 – 009 NGC 3788	1379	1.8 x 0.6 1.7 x 0.5	178 6	2	S ... P w 6621, disturbed		13.2	+ 2345 + 2327	*12 3 V\overline{A}
06624	11 37.2 + 17 59 238.90 + 70.98	03 – 30 – 032 NGC 3790	1406	1.1 x 0.25 1.1 x 0.25	154 6 – 7	1	S0 – a NGC 3801 group		14.5		3
06625	11 37.2 + 20 12 232.83 + 72.09	03 – 30 – 031	1406 234	0.8 x 0.7 0.7 x 0.5	(55) ...	1	S ...		14.2		*
06626	11 37.3 + 09 09 256.77 + 65.11	02 – 30 – 007 IC 718	0468	1.2 x 0.45 1.1 x 0.5	[178] ...	2	Irr m: 6730 group		14.6		*
06627	11 37.3 + 13 45 248.63 + 68.42	02 – 30 – 000	0468	1.1 x 0.3 1.1 x 0.25	41 6	2	...	S d – m	15.7		
06628	11 37.4 + 46 13 155.18 + 66.47	08 – 21 – 089	0700	3.5: x 3.5: 3.5:: x 3.5::	-- 1	4	...	SA m	14.6		
06629	11 37.4 + 58 54 139.56 + 56.14	10 – 17 – 038 NGC 3795	0723 229	2.3 x 0.6 2.5 x 0.5	53 6	1	S ...		14.1		3
06630	11 37.6 + 15 36 244.81 + 69.67	03 – 30 – 037 NGC 3799	1406	0.7 x 0.5 0.7 x 0.4	... ...	1	S ... P w 6634, contact, distorted		14.4	+ 3518 + 3434	*12 3 V\overline{A}
06631	11 37.6 + 17 35 240.08 + 70.84	03 – 30 – 036	1406	0.7 x 0.4 0.8 x 0.5	117 ...	1	...		14.3		
06632	11 37.6 + 24 59 217.54 + 73.82	04 – 28 – 018 NGC 3798	0103	(2.6 x 1.4) 2.3 x 1.3	60 4 – 5	3	S B 0 1 of 2 brightest in group		13.9		*3
06633	11 37.7 + 09 17 256.75 + 65.28	02 – 30 – 008 IC 719	0468	1.3 x 0.35 1.4 x 0.35	52 6 – 7	1	S0? P w 6626, 6730 group		13.6		
06634	11 37.7 + 15 37 244.82 + 69.70	03 – 30 – 039 NGC 3800	1406	1.9: x 0.45 1.9: x 0.45	52 6	1	S ... P w 6630, contact, distorted		13.1	+ 3562 + 3469	*12 3 V\overline{A}
06635	11 37.7 + 18 00 239.05 + 71.09	03 – 30 – 040 NGC 3801	1406	(2.3 x 1.3) (2.2 x 1.1)	120 ...	1	S0? Brightest in group, singular		13.3		*13
06636	11 37.7 + 18 03 238.92 + 71.12	03 – 30 – 041 NGC 3802	1406	1.2 x 0.25 1.1 x 0.25	85 7	1	S ... NGC 3801 group		14.7		*13
06637	11 37.8 + 28 40 204.22 + 74.25	05 – 28 – 010	1379 230	0.9 x 0.4 0.7 x 0.4	72 ...	1	...		14.5		*1
06638	11 37.8 + 60 34 138.07 + 54.71	10 – 17 – 039 NGC 3796	0723 229	1.4 x 0.9 1.4 x 0.9	127 3 – 4	1	S ...		13.4		3
06639	11 37.9 + 36 24 177.68 + 72.47	06 – 26 – 014	0109 235	1.0 x 0.3 0.9 x 0.3	172 (6)	1	...		15.4		
06640	11 38.1 + 56 28 141.66 + 58.30	09 – 19 – 153 NGC 3804	0059 229	2.5 x 1.7 2.3 x 1.7	120: 3	4	Sc P w 6615 Sc +		13.8		*\underline{2}3
06641	11 38.2 + 18 04 239.08 + 71.23	03 – 30 – 042 NGC 3806 = 3807	1406	1.5 x 1.4 1.1 x 1.0	-- 1	3	Sb/S B b NGC 3801 group		14.6		*13
06642	11 38.2 + 20 38 231.90 + 72.50	04 – 28 – 019 NGC 3805	0103 234	(1.4 x 1.1) (1.6 x 1.3)	(60) ...	3	E – S0		13.8		3
06643	11 38.2 + 22 43 225.38 + 73.31	04 – 28 – 20 +21 103 NGC 3808	103 234	[2.5 x 0.8] ...	... ...	3	Double system Distorted, twisted bridge		14.1	See Notes	*13 VA
06644	11 38.3 + 11 45 252.89 + 67.20	02 – 30 – 010 NGC 3810	0468	4.1 x 2.8 4.1 x 2.8	15 3	5	Sc Leo group Sc –	SA(rs)c	11.4	+ 988 + 878	*12– –6\underline{8}9
06645	11 38.3 + 26 04 213.79 + 74.17	04 – 28 – 022	0103	1.7 x 1.3 0.9 x 0.6	-- 1 – 2	3	Sb	SAB(r)b Sb +	14.8		1
06646	11 38.4 + 05 20 262.55 + 62.27	01 – 30 – 009	0495	1.0 x 0.35 0.9 x 0.35	140 6	1	S – Irr		15.2		
06647	11 38.5 + 10 30 255.15 + 66.32	02 – 30 – 011	0468 228	1.4: x 1.2: 0.9 x 0.9:	-- 1	3	Sc S d: NGC 3869 group, Sc +		15.5		1
06648	11 38.5 + 25 07 217.23 + 74.04	04 – 28 – 023 NGC 3812	0103	(1.7 x 1.6) (2.0 x 1.8)	-- ...	4	E 1 of 2 brightest in group		13.9		*13
06649	11 38.5 + 60 10 138.26 + 55.10	10 – 17 – 040 NGC 3809	0723 229	(1.3 x 1.2) (1.4 x 1.3)	-- (1)	1	S0		13.6		3
06650	11 38.6 + 47 58 151.97 + 65.29	08 – 21 – 091 NGC 3811, Mark 185	0700	2.5: x 1.8 2.2: x 1.7	[160] ...	3	S B c Singular	SB(r) ...	13.0	+ 3000	*3

N G C   3782, 3785, 3786, 3788, 3790, 3795,
3796, 3798 – 3802, 3804 – 12

I C   718, 719

1	2	3	4	5	6	7	8	9	10	11	12
06651	11 38.7 + 36 49 176.20 + 72.43	06 - 26 - 019 NGC 3813	0109 235	2.1 x 1.2 2.1 x 1.1	87 ...	1	S ...		12.6		123 4678
06652	11 38.8 + 33 58 185.15 + 73.57	06 - 26 - 000	0109 235	1.5: x 0.5: 1.5: x 0.5:	97 6:	1	S B ... f of 2 Peculiar, distorted?		16.0:		*
06653	11 39.0 + 16 15 243.92 + 70.33	03 - 30 - 044	1406	1.5 x 0.35 1.5 x 0.3	11 7	2	S B ? 0 - a 6666 group		14.6		*1
06654	11 39.1 + 25 05 217.45 + 74.17	04 - 28 - 025 NGC 3815	0103	1.9 x 1.0 1.8 x 0.8	72 5	2	Sa - b NGC 3798 - 3812 group		14.2		*13
06655	11 39.2 + 16 15 244.01 + 70.37	03 - 30 - 047	1406	0.5 x 0.3 0.7 x 0.5	(20) ...	1	... 6666 group		14.5		*1
06656	11 39.2 + 20 23 233.00 + 72.59	03 - 30 - 046 NGC 3816	1406 234	1.9 x 1.1 1.9 x 1.1	70 4 - 5	3	S0 Disturbed?		13.6		*3
06657	11 39.3 + 10 35 255.37 + 66.52	02 - 30 - 012 NGC 3817	0468 228	1.1 x 0.8 1.1 x 0.8	140: ...	2	S B 0 - a + companion NGC 3869 group		14.4		*13
06658	11 39.4 + 32 17 190.85 + 74.16	05 - 28 - 016	1379	1.0 x 0.4 1.0 x 0.4	129 6	1	S ...		15.0		1
06659	11 39.5 + 32 49 188.94 + 74.05	06 - 26 - 020	0109 235	1.0 x 0.9 0.8 x 0.8	-- 1:	3	Sc		16.0:		
06660	11 39.5 + 62 27 136.31 + 53.12	10 - 17 - 042	0723	1.3 x 0.2 1.0 x 0.15	176 7	2	... d - m		17.		*1
06661	11 39.6 + 10 33 255.57 + 66.54	02 - 30 - 015 NGC 3822	0468 228	1.3 x 0.7 1.3 x 0.7	178 ...	1	S0? NGC 3869 group		13.7		*13
06662	11 39.6 + 13 13 250.73 + 68.46	02 - 30 - 000	0468	1.0 x 0.4 0.9: x 0.4	47 (6)	1	Sc - Irr		15.4		
06663	11 39.6 + 20 36 232.49 + 72.78	04 - 28 - 030 NGC 3821	0103 234	1.5 x 1.3 1.5 x 1.3	-- 1 - 2	3	S B a (R)SAB(s)ab		13.8		*3
06664	11 39.6 + 30 31 197.26 + 74.52	05 - 28 - 017	1379 233	1.6 x 0.25 1.4 x 0.25	102 6 - 7	2	Sc		15.5		
06665	11 39.7 + 00 37 268.28 + 58.48	00 - 30 - 019	0471 231	0.40 x 0.40 0.5 x 0.40	-- ...	1	... Peculiar, jet, plume?		13.7		*
06666	11 39.7 + 16 18 244.11 + 70.50	03 - 30 - 050	1406	1.1 x 0.4 1.0 x 0.4	14 6	1	S ... Brightest in group		14.3		*
06667	11 39.7 + 51 53 146.28 + 62.28	09 - 19 - 157	0059	3.5 x 0.5 3.3 x 0.5	89 7	3	Sc		14.8		
06668	11 39.8 + 10 33 255.66 + 66.58	02 - 30 - 018 NGC 3825	0468 228	1.4 x 1.1 1.4 x 1.1	(160) 1 - 2	4	S B a NGC 3869 group		13.8		*13
06669	11 39.8 + 15 15 246.57 + 69.86	03 - 30 - 000	1406	1.6: x 0.8: 1.5:: x 0.7::	... ...	4	Dwarf irregular Ir V		17.		
06670	11 39.8 + 18 36 238.30 + 71.83	03 - 30 - 053	1406	3.1 x 0.9 2.1 x 0.8:	153 (6)	4	Irr IB m Ir I		14.3		*1
06671	11 39.8 + 26 46 211.37 + 74.60	05 - 28 - 018 NGC 3826	1379 230	0.9 x 0.7 1.0 x 0.8	(65) ...	2	E		14.4		*3
06672	11 39.8 + 33 55 185.08 + 73.78	06 - 26 - 000	0109 235	1.0 x 0.2 0.9 x 0.15	106 6	2	Sc		16.0:		
06673	11 40.0 + 19 07 236.96 + 72.15	03 - 30 - 054 NGC 3827	1406 234	0.9 x 0.7 0.8 x 0.6	(65) ...	1	...		13.6		*3
06674	11 40.0 + 25 07 217.49 + 74.38	04 - 28 - 034	0103	1.2 x 0.8 1.1 x 0.7	153: 3 - 4	3	Sc NGC 3798 - 3812 group		15.2		*
06675	11 40.0 + 77 40 127.68 + 39.10	13 - 09 - 001 NGC 3901	1339	1.9 x 0.8 1.9 x 0.8	165 6	3	Sc		14.6		3
06676	11 40.1 + 53 04 144.77 + 61.33	09 - 19 - 161 NGC 3824	0059 229	1.4 x 0.8 1.5 x 0.8	118 ...	1	... P w 6690		14.6		*23
06677	11 40.2 + 26 50 211.16 + 74.69	05 - 28 - 019	1379 230	1.3 x 0.2 1.3 x 0.2	142 7	2	Sa - b		15.6		*1
06678	11 40.3 + 26 33 212.24 + 74.68	05 - 28 - 022	1379 230	1.0 x 0.5 0.7 x 0.3:	85: 5:	1	Sb: s of 3		15.5		*1
06679	11 40.3 + 41 07 164.37 + 70.33	07 - 24 - 030	0719	1.3 x 0.15 1.3 x 0.15	109 7	3	Sc Sc +		16.0:		
06680	11 40.4 + 19 55 234.82 + 72.62	03 - 30 - 058	1406 234	1.2 x 0.6 1.2 x 0.6	110 5	1	Sb		15.5		
06681	11 40.4 + 24 13 220.81 + 74.25	04 - 28 - 035	0103	1.2 x 0.3 1.1 x 0.3	156 7	1	S ... Brightest of 3		15.4		*
06682	11 40.5 + 59 23 138.52 + 55.91	10 - 17 - 049	0723 229	1.7 x 1.4: 1.7: x 1.3:	(100) ...	3	Dwarf spiral S V		16.5:		*
06683	11 40.6 + 20 01 234.60 + 72.71	03 - 30 - 059	1406 234	1.0 x 0.25 1.0 x 0.25	71 7	2	S0 - a P w 6688		15.2		*
06684	11 40.7 + 31 45 192.55 + 74.55	05 - 28 - 000	1379	1.0 x 0.7: 0.9 x 0.6:	[160] ...	3	Dwarf p of 2		15.5		*
06685	11 40.7 + 55 47 141.74 + 59.07	09 - 19 - 163	0059 229	1.4 x 0.2 1.4 x 0.25	6 6 - 7	2	Sc		16.0:		

N G C   3813,   3815 - 17,   3821,   3822,   3824 - 27
3901

1	2	3	4	5	6	7	8	9	10	11	12
06686	11 40.8 + 16 46 243.47 + 70.99	03 – 30 – 062	1406	2.9 x 0.3 2.8: x 0.3	51 7	3	Sb P w 6712		15.1		*
06687	11 40.8 + 18 28 239.08 + 71.96	03 – 30 – 060	1406	1.6 x 0.3 1.5: x 0.2	25 6 – 7	2	...	S d – m	16.0:		
06688	11 40.8 + 20 01 234.68 + 72.75	03 – 30 – 061 IC 2951	1406 234	1.4 x 0.7 1.3 x 0.6	80 5	2	Sa P w 6683		15.0		*
06689	11 40.8 + 21 56 228.71 + 73.59	04 – 28 – 038	0103 234	1.0 x 0.2 1.1 x 0.2	12 6	1	S ...		15.5		*
06690	11 40.8 + 53 00 144.66 + 61.44	09 – 19 – 164 NGC 3829	0059 229	1.0 x 0.55 1.1 x 0.6	95: ...	1	S B ... P w 6676		15.0		*23
06691	11 40.8 + 60 58 137.19 + 54.52	10 – 17 – 050	0723 229	1.3: x 0.9: 1.1: x 0.9:	120: ...	3	...	S d – m	16.5:		
06692	11 40.9 + 10 26 256.37 + 66.67	02 – 30 – 020 NGC 3833	0468 228	1.5 x 0.7 1.5 x 0.7	27 5	3	Sc NGC 3869 group, Sc –		14.7		13
06693	11 40.9 + 23 00 225.18 + 74.00	04 – 28 – 040 NGC 3832	0103 234	2.2 x 1.8 2.2 x 2.1	— 1	4	S B c	S B(rs)bc Sc –	14.0		*13
06694	11 41 + 83 45 125.18 + 33.36	14 – 06 – 002	1340	1.4 x 0.7 1.4 x 0.7	150 ...	3	...	SA m	16.5:		
06695	11 41.0 + 09 13 258.36 + 65.75	02 – 30 – 022 IC 724	0468	2.5 x 1.0 (2.5 x 1.0)	60 6	4	Sa 6730 group		13.8		*
06696	11 41.0 + 33 41 185.59 + 74.09	06 – 26 – 000	0109 235	1.0 x 0.5 1.1 x 0.5	102 5	1	S B a np of 2		15.7		*
06697	11 41.2 + 20 15 234.13 + 72.95	03 – 30 – 066	1406 234	1.7 x 0.30 1.7 x 0.20	137 (7)	2	Irr		14.3		*
06698	11 41.2 + 71 30 130.50 + 44.90	12 – 11 – 000	0714	1.1: x 0.9: 1.0 x 0.7	(5) ...	1	S ...		15.1		
06699	11 41.3 + 08 12 260.01 + 65.00	01 – 30 – 011 NGC 3843	0495	1.0 x 0.4 0.9 x 0.35	42 6	2	S0 – a		14.1		
06700	11 41.3 + 11 04 255.48 + 67.21	02 – 30 – 024 NGC 3839	0468	1.0 x 0.5 1.0 x 0.5	87 (5)	1	S – Irr NGC 3869 group		13.6		
06701	11 41.3 + 20 10 234.42 + 72.93	03 – 30 – 068 NGC 3837	1406 234	(0.8 x 0.7) (0.9 x 0.8)	— ...	1	E		14.2		*3
06702	11 41.3 + 20 21 233.87 + 73.02	03 – 30 – 070 NGC 3840	1406 234	1.1 x 0.7 1.1 x 0.7	67 3 – 4	2	Sa		14.7		*3
06703	11 41.3 + 60 23 137.55 + 55.07	10 – 17 – 055 NGC 3835	0723 229	2.3 x 0.7 2.1 x 0.7	60 6	2	Sa – b		13.0		23
06704	11 41.4 + 20 13 234.31 + 72.97	03 – 30 – 072 NGC 3842	1406 234	(1.4 x 1.0) (1.5 x 1.2)	(5) ...	3	E Brightest in cluster		13.3		*13
06705	11 41.4 + 20 18 234.06 + 73.01	03 – 30 – 069 NGC 3844	1406 234	1.4 x 0.25 1.3 x 0.25	28 7	2	S0 – a		14.9		*3
06706	11 41.5 + 55 18 142.03 + 59.54	09 – 19 – 169	0059 229	2.1 x 1.8 2.3: x 1.9:	[40] ...	1	S – Irr Peculiar, f of 2		14.2		*12 V –
06707	11 41.5 + 58 14 139.29 + 56.99	10 – 17 – 056 NGC 3838	0723 229	1.4 x 0.5 1.5 x 0.7	141 6:	1	S0?		12.7		*23
06708	11 41.6 + 33 47 185.09 + 74.18	06 – 26 – 023 NGC 3847	0109 235	(1.1 x 1.1) (1.3 x 1.3)	— ...	2	E		14.6		*3
06709	11 41.7 + 33 37 185.65 + 74.25	06 – 26 – 025 IC 2953	0109 235	1.2 x 1.0 1.2 x 0.9	(65) 2 – 3	4	S B b	(R´)S B(r)b	15.1		*1
06710	11 41.7 + 55 55 141.37 + 59.02	09 – 19 – 171 NGC 3846	0059 229	1.0 x 0.8 1.1 x 0.8	135 2:	2	Sb – c NGC 3898 group		14.7		*23
06711	11 41.7 + 70 01 131.23 + 46.30	12 – 11 – 039	0714	0.8 x 0.45 0.8 x 0.40	[132] ...	1	...		13.8		1
06712	11 41.8 + 16 50 243.75 + 71.22	03 – 30 – 081 NGC 3853	1406	(1.5 x 0.9) (1.6 x 1.0)	140 ...	3	E nff of 2, p w 6686		13.5		*3
06713	11 41.8 + 49 07 149.29 + 64.72	08 – 21 – 094	0700	2.0: x 1.7: 1.8:x 1.5::	— ...	4	Dwarf spiral, S m S V		15.5		
06714	11 41.8 + 68 13 132.22 + 47.97	11 – 14 – 037	0977	1.0 x 0.2 0.8 x 0.3	148 (7)	1	...		14.9		
06715	11 41.9 + 11 03 255.80 + 67.29	02 – 30 – 025 IC 727	0468	1.6 x 0.25 1.6 x 0.25	161 7	2	Sb p of 2, disturbed:, NGC 3869 group		15.0		*1
06716	11 42.0 + 36 14 176.93 + 73.29	06 – 26 – 027	0109 235	1.8 x 1.7 1.6: x 1.5:	— 1	2	Sb	Sb +	15.1		
06717	11 42.2 + 09 28 258.53 + 66.13	02 – 30 – 027	0468	1.6:: x 1.6:: 1.4::x 1.4::	— ...	4	Dwarf (spiral), S:B m S: V, 6730 group		17.		*
06718	11 42.2 + 20 05 235.01 + 73.08	03 – 30 – 088 NGC 3860	1406 234	1.2 x 0.6 1.4 x 0.7	38 5	2	Sa – b		14.5		*13
06719	11 42.2 + 20 24 234.05 + 73.23	03 – 30 – 089	1406 234	1.2 x 0.8 1.1 x 0.7	30: 3:	1	S ...		14.6		*
06720	11 42.3 – 01 19 271.11 + 57.11	00 – 30 – 021 IC 728	0471 231	1.3 x 0.6 1.2 x 0.6	65 5	3	S B b		14.7		

N G C   3829, 3832, 3833, 3835, 3837 – 40   I C   724, 727, 728, 2951, 2953
3842 – 44, 3846, 3847, 3853, 3860

1	2	3	4	5	6	7	8	9	10	11	12
06721	11 42.3 + 19 44 236.10 + 72.92	03 - 30 - 091 NGC 3859	1406 234	1.0 x 0.25 1.0 x 0.2	58 ...	1	... Peculiar		14.9		*
06722	11 42.5 + 08 45 259.77 + 65.62	02 - 30 - 028 NGC 3863	0468	2.8 x 0.6 2.8 x 0.6	75 6	2	Sb - c 6730 group		14.0		13
06723	11 42.5 + 19 53 253.73 + 73.04	03 - 30 - 095 NGC 3862	1406 234	(1.0 x 1.0) (1.3 x 1.3)	-- ...	3	E		14.0		*13
06724	11 42.5 + 20 15 234.63 + 73.22	03 - 30 - 093 NGC 3861	1406 234	2.4 x 1.4 2.4 x 1.4	77 4	4	Sb npp of 2, contact	(R´)SAB(r)b	14.0		*13
06725	11 42.5 + 20 44 233.13 + 73.44	04 - 28 - 046	0103 234	1.5 x 1.2 1.7 x 1.3	(40) 2:	4	S0		14.5		
06726	11 42.5 + 50 00 147.87 + 64.07	08 - 21 - 095	0700	(1.0 x 0.5) (1.0 x 0.5)	65 ...	1	E - S0		14.8		
06727	11 42.5 + 61 59 136.12 + 53.69	10 - 17 - 060	0723	0.9 x 0.25 0.9 x 0.25	125 ...	1	S ...		14.5		
06728	11 42.5 + 79 58 126.58 + 36.97	13 - 09 - 002	1339	(1.0 x 0.7) (1.3 x 0.9)	(160) 3:	2	S B 0 - a		14.8		
06729	11 42.6 + 27 03 210.53 + 75.25	05 - 28 - 027	1379 230	1.3 x 0.8 0.7 x 0.7	65 4	2	Sb/S B c	SAB(s)bc	14.9		*
06730	11 42.8 + 09 26 258.87 + 66.20	02 - 30 - 029	0468	1.1 x 0.4 1.2 x 0.7	105 3 - 4	2	S[a - b] P w 6734 Brightest in group		13.4		*
06731	11 42.8 + 19 40 236.50 + 72.99	03 - 30 - 103 NGC 3867	1406 234	1.2 x 0.4 1.2 x 0.7	173 ...	1	S ...		14.6		*1
06732	11 42.8 + 59 15 138.16 + 56.16	10 - 17 - 061	0723 229	(0.9 x 0.8) (1.1 x 1.1)	-- ...	1	...		13.5		*C
06733	11 42.9 + 56 09 140.86 + 58.90	09 - 19 - 174 NGC 3850	0059 229	2.2 x 1.0 2.2 x 1.1	130 5	4	S B c NGC 3898 group		14.4		*2̲3
06734	11 43.0 + 09 24 259.02 + 66.20	02 - 30 - 000	0468	1.6 x 0.3 1.5 x 0.3	33 6	2	Sb P w 6730, 6730 group		15.4		*
06735	11 43.1 + 20 03 235.47 + 73.24	03 - 30 - 106 NGC 3873	1406 234	(0.9 x 0.8) (1.0 x 1.0)	-- ...	2	E P w 6739, contact		14.2		*13
06736	11 43.2 + 03 18 267.03 + 61.23	01 - 30 - 014	0495	1.5 x 0.35 1.3 x 0.25	178 6	2	Sc		14.9		
06737	11 43.2 + 11 06 256.33 + 67.54	02 - 30 - 032 NGC 3869	0468	1.9 x 0.45 1.7 x 0.45	135 7	3	Sa Brightest in group		13.5		3
06738	11 43.2 + 14 03 250.74 + 69.67	02 - 30 - 033 NGC 3872	0468	(2.3 x 1.7) (2.7 x 2.2)	... ...	4	E P w 6758		12.9	+ 3109 + 3012	*12 3478̲
06739	11 43.2 + 20 02 235.56 + 73.26	03 - 30 - 105 NGC 3875	1406 234	1.0 x 0.25 1.2 x 0.25	87 7	2	S0 - a P w 6735, contact		14.8		*13
06740	11 43.3 + 10 45 256.98 + 67.29	02 - 30 - 034	0468	1.1 x 0.55 0.9 x 0.5	80 5	1	S ... NGC 3869 group		14.8		
06741	11 43.3 + 17 28 242.82 + 71.89	03 - 30 - 000	1406	[1.6 x 0.3] ...	[172] ...	2	Double system Long bridge		16.0:		*
06742	11 43.3 + 50 30 146.98 + 63.74	08 - 22 - 001 NGC 3870, Mark 186	1338	0.9 x 0.7 1.0 x 0.7	25: ...	1	S0?		13.2	+ 600 + 660	*13
06743	11 43.4 + 21 19 231.60 + 73.89	04 - 28 - 049	0103 234	1.6 x 1.5 1.6 x 1.6	-- 1	3	Sb/Sc	SAB bc Sc -	14.8		
06744	11 43.5 + 33 23 186.02 + 74.68	06 - 26 - 031 NGC 3871	0109 235	1.0 x 0.2 1.1 x 0.22	105 6 - 7	1	S ...		15.4		3
06745	11 43.5 + 47 46 150.72 + 65.96	08 - 22 - 002 NGC 3877	1338	5.6 x 1.2 5.6 x 1.2	35 6	4	Sc		11.8		*123̲ 4678
06746	11 43.6 + 20 41 233.71 + 73.65	04 - 28 - 051 NGC 3884	0103 234	(1.9 x 1.4) (2.1 x 1.7)	10: 2:	3	Sa	SA(r)0/a	14.0		*3
06747	11 43.8 + 14 06 250.93 + 69.81	02 - 30 - 000	0468	1.3 x 0.2 0.9 x 0.1	70 (7)	2	Irr		15.7		*
06748	11 43.9 + 36 00 177.04 + 73.73	06 - 26 - 000	0109 235	1.0 x 0.8 0.9: x 0.7:	... ...	2	S ... In group, disrupted		16.0:		*
06749	11 44.0 + 65 39 133.43 + 50.42	11 - 15 - 002	0674 255	1.0 x 0.35 0.9 x 0.35	20 6	2	Sb - c		16.0:		
06750	11 44.1 - 01 43 272.19 + 56.96	00 - 30 - 022	0471 231	1.2 x 0.35 1.2 x 0.3	13 6	1	S ...		15.0		
06751	11 44.1 + 24 15 221.54 + 75.07	04 - 28 - 052	0103	1.4 x 0.5 1.3 x 0.4	97 6	2	Sb - c nff of 2		15.4		*
06752	11 44.1 + 69 40 131.14 + 46.70	12 - 11 - 040 NGC 3879	0714	2.5 x 0.45 2.5 x 0.4	130 (6 - 7)	2	Sc - Irr		13.5		3
06753	11 44.2 + 14 49 249.58 + 70.37	03 - 30 - 110	1406	1.1 x 0.6 1.1 x 0.6	90 ...	1	S B: ...		15.3		*
06754	11 44.2 + 20 58 233.03 + 73.90	04 - 28 - 053 NGC 3883	0103 234	3.1 x 2.8 3.3: x 3.0:	-- 1	4	Sb	SA(rs)b Sb +	14.2		*13
06755	11 44.3 + 50 59 146.08 + 63.43	09 - 19 - 177	0059	1.1 x 0.6 1.0 x 0.5	67 5	1	S ...		14.7		

N G C   3850,   3859,   3861 - 63,   3867,   3869 - 73
3875,   3877,   3879,   3883,   3884

1	2	3	4	5	6	7	8	9	10	11	12
06756	11 44.4 + 57 23 139.40 + 57.92	10 – 17 – 065	0723 229	1.0 x 0.4 1.0 x 0.4	67 6	1	S ...		15.2		
06757	11 44.4 + 61 38 136.03 + 54.11	10 – 17 – 067	0723 229	1.1 x 0.6 1.0: x 0.5:	155: ...	1	Dwarf:	m	16.5:		
06758	11 44.5 + 13 59 251.51 + 69.85	02 – 30 – 036	0468	2.0: x 1.8: 1.7: x 1.5:	-- 1	3	S ... P w 6738	SAB(s) ...	14.4		*1
06759	11 44.5 + 16 32 245.74 + 71.55	03 – 30 – 112	1406	1.2 x 0.3 0.9 x 0.3	71 6	1	S ...		16.0:		
06760	11 44.5 + 20 07 235.83 + 73.56	03 – 30 – 111 NGC 3886	1406 234	1.2 x 0.8 1.3 x 0.8	132 ...	2	E – S0		14.3		3
06761	11 44.5 + 29 52 199.29 + 75.65	05 – 28 – 029	1379 233	1.2 x 0.5 1.0 x 0.4	97 6	2	S0 – a		15.2		*1
06762	11 44.6 + 60 34 136.77 + 55.08	10 – 17 – 068	0723 229	1.1 x 1.1 1.0 x 0.9	-- (1)	1	S: ...		13.9		
06763	11 44.8 + 50 43 146.26 + 63.69	09 – 19 – 000	0059	1.0: x 0.2 0.7 x 0.2	8 6	1	Sb – c		15.6		
06764	11 44.8 + 69 24 131.20 + 46.97	12 – 11 – 041	0714	1.8: x 0.7: 1.3: x 0.4:	70: (6)	2	...	d – m	16.5:		
06765	11 44.9 + 56 14 140.31 + 58.96	09 – 19 – 189 NGC 3888, Mark 188	0059 229	1.7 x 1.4 1.8 x 1.4	120: 2	2	Sc NGC 3898 group		12.6	+ 2400	*123 4678
06766	11 45.0 + 54 47a 141.68 + 60.24	09 – 19 – 185	0059 229	1.1 x 0.20 1.0 x 0.20	135 7	2	Sb – c nnp of 2		(15.6)		*1
06767	11 45.0 + 57 55 138.80 + 57.48	10 – 17 – 070	0723 229	[0.8 x 0.6] ...	... ...	2	Double system Symbiotic		13.9		*1
06768	11 45.1 + 44 01 156.52 + 69.00	07 – 24 – 031	0719 248	1.0 x 0.45 0.9 x 0.4	42 6	2	S B: a – b Brightest of 3		15.2		*
06769	11 45.2 + 02 06 269.16 + 60.45	00 – 30 – 023	0471	1.1 x 0.5 1.1 x 0.5	73 5 – 6	2	S B: a – b		15.0		
06770	11 45.2 + 18 50 239.94 + 73.04	03 – 30 – 000	1406 234	[1.0 x 0.9] ...	... ...	3	Double system Contact		15.6		*
06771	11 45.4 + 04 46 266.38 + 62.77	01 – 30 – 015	0495	1.9: x 1.7: 2.1: x 1.8:	-- 1 – 2	4	S B a	S B(r)a	14.4		
06772	11 45.4 + 30 39 196.05 + 75.74	05 – 28 – 031 NGC 3891	1379	2.3 x 1.8 1.8 x 1.6	(70) 1 – 2	3	Sb/Sc	Sb +	13.7		13
06773	11 45.4 + 50 05 146.89 + 64.27	08 – 22 – 004	1338	1.8 x 0.7 1.8 x 0.7	168 (6)	3	Irr	Ir I	15.2		
06774	11 45.4 + 55 19 141.06 + 59.80	09 – 19 – 195	0059 229	2.1 x 0.2 1.8 x 0.2	155 7	1	Sc? ssf of 2		16.0:		*
06775	11 45.7 + 13 29 253.13 + 69.71	02 – 30 – 038	0468	1.4 x 0.8 1.4 x 0.8	103: ...	1	S ...		15.0		
06776	11 45.9 + 44 00 156.26 + 69.11	07 – 24 – 033	0719 248	1.5 x 0.2 1.5 x 0.2	33 7	3	Sc 6768 group		16.5:		
06777	11 46.0 + 32 55 187.10 + 75.33	06 – 26 – 040	0109	(0.9 x 0.8) 0.8 x 0.6	-- ...	1	... Compact		15.5		
06778	11 46.0 + 48 59 148.16 + 65.23	08 – 22 – 007 NGC 3893	1338	4.6 x 2.5 4.5 x 2.3	165 5:	4	Sc P w 6781, disturbed, Sc –	SAB c	10.6	+ 1001 + 1065	*12 3478
06779	11 46.1 + 59 42 137.13 + 55.94	10 – 17 – 078 NGC 3894	0723 229	(2.0 x 1.4) (2.5 x 1.7)	20 3:	3	S0 P w 6785		12.9		*12 3
06780	11 46.3 – 01 45 273.12 + 57.18	00 – 30 – 024	0471 231	3.5 x 1.0 3.5: x 1.1	20 6	3	Sc Singular?	SA d	14.7		*
06781	11 46.3 + 48 56 148.14 + 65.30	08 – 22 – 008 NGC 3896	1338	1.8 x 1.0 1.7 x 1.2	125: ...	1	... P w 6778, disturbed:		14.0		*12 3
06782	11 46.4 + 24 07 222.61 + 75.54	04 – 28 – 057	0103 234	2. x 2. 1.8:: x 1.8::	-- ...	4	Dwarf irregular Ir V		17.		*1
06783	11 46.4 + 31 35 192.16 + 75.77	05 – 28 – 000	1379	1.1 x 0.8 0.9: x 0.9:	-- ...	4	Dwarf irregular Ir V		17.		
06784	11 46.4 + 35 18 178.49 + 74.50	06 – 26 – 041 NGC 3897	0109 235	2.1 x 2.1 1.9 x 1.9	-- 1	3	Sb/Sc	Sb +	14.2		*3
06785	11 46.4 + 59 43 137.06 + 55.95	10 – 17 – 080 NGC 3895	0723 229	1.3 x 0.9 1.4 x 0.9	125 3	1	S B a P w 6779		14.0		*12 3
06786	11 46.5 + 27 19 209.73 + 76.13	05 – 28 – 034 NGC 3900	1379	3.2 x 1.6 (2.8 x 1.6)	2 5	4	Sa	SA(r)0/a	12.5	+ 1702 + 1666	*123 4678
06787	11 46.5 + 56 20 139.84 + 58.98	09 – 19 – 204 NGC 3898	0059 229	3.6 x 2.1 (3.5 x 2.2)	107 4	4	Sa Brightest in group		11.7	+ 1038 + 1135	*123 4678
06788	11 46.6 + 74 36 128.53 + 42.13	13 – 09 – 003 NGC 3890	1339	1.0 x 1.0 1.0 x 0.9	-- 1	1	S ...		14.1		3
06789	11 46.6 + 76 17 127.81 + 40.53	13 – 09 – 004	1339	1.1 x 0.7 1.1 x 0.7	128 3 – 4	2	S B b		14.9		
06790	11 46.7 + 26 24 213.55 + 76.08	04 – 28 – 055 NGC 3902	0103	1.7 x 1.3 1.4 x 1.1	85 2 – 3	3	S B b/Sc Sb +	SAB(s)bc	14.0		23

N G C 3886, 3888, 3890, 3891, 3893 – 98
3900, 3902

1	2	3	4	5	6	7	8	9	10	11	12
06791	11 46.7 + 27 01 211.00 + 76.15	05 – 28 – 035	1379	1.9 x 0.3 1.7 x 0.25	1 7	3	Sc	Sc +	15.2		
06792	11 46.7 + 40 03 164.55 + 71.95	07 – 24 – 034	0719	2.6 x 0.4 2.5 x 0.4	172 6 – 7	1	Sc		14.8		
06793	11 46.8 – 00 48 272.54 + 58.08	00 – 30 – 026	0471 231	1.2 x 0.30 1.1 x 0.22	76 6 – 7	3	Sb P w 6796		14.8		*1
06794	11 46.8 + 16 55 245.89 + 72.22	03 – 30 – 118	1406	1.4 x 0.5 1.2 x 0.4	72 6	2	S B c – Irr		14.7		
06795	11 46.8 + 25 13 218.38 + 75.90	04 – 28 – 056 NGC 3911	0103 242	1.2 x 0.9 1.0 x 0.8	(110) ...	2	S ... Peculiar, p w 6803		15.4		*13
06796	11 46.9 – 00 48 272.58 + 58.09	00 – 30 – 028 NGC 3907	0471 231	1.2 x 0.8 1.2 x 0.8	(40) 3:	3	S0 P w 6793		14.4		*13
06797	11 47.0 + 48 42 148.23 + 65.56	08 – 22 – 012 NGC 3906	1338	1.7 x 1.7 1.7 x 1.7	– 1	3	S B ... Peculiar		14.1		*12 3
06798	11 47.0 + 78 07 127.05 + 38.79	13 – 09 – 005	1339	1.2 x 0.25 1.3 x 0.25	13 6	2	Sa – b		15.2		
06799	11 47,3 + 29 17 201.47 + 76.30	05 – 28 – 000	1379	1.0 x 0.10 0.9 x 0.12	23 7	1	Sb – c		16.0:		
06800	11 47.4 + 21 38 232.02 + 74.86	04 – 28 – 058 NGC 3910	0103 234	(1.6 x 1.2) (1.9 x 1.5)	(150) ...	2	E – S0		14.4		*3
06801	11 47.4 + 26 45 212.18 + 76.28	05 – 28 – 037 NGC 3912	1379	1.8 x 1.0 1.3 x 0.7	5 ...	1	S – Irr I 0? Ir II?		13.2		123 $4\overline{7}8$
06802	11 47.4 + 52 09 143.81 + 62.70	09 – 19 – 209	0059 229	2.3 x 0.15 2.1 x 0.2	160 7	3	Sc P w 6815		16.0:		*
06803	11 47.5 + 25 13 218.52 + 76.06	04 – 28 – 059 NGC 3920	0103 242	1.3 x 1.2 1.3 x 1.2	-- ...	1	... P w 6795		14.1		*13
06804	11 47.6 + 07 16 264.31 + 65.16	01 – 30 – 000	0495	1.8 x 0.12 1.5 x 0.15	104 7	3	Sc	Sc +	15.3		
06805	11 47.6 + 42 20 158.95 + 70.53	07 – 24 – 000	0719	0.35 x 0.25 0.35 x 0.25	... ...	1	...		14.2		
06806	11 47.7 + 26 15 214.30 + 76.28	04 – 28 – 060	0103 242	2.0: x 0.6 2.0 x 0.5	45 6	2	S ... Disturbed, p w 6807		14.4		*1
06807	11 47.8 + 26 17 214.18 + 76.31	04 – 28 – 061	0103 242	1.0: x 0.9: 0.9: x 0.5:	-- ...	3	Dwarf IV – V, p w 6806		16.5:		*1
06808	11 47.8 + 35 32 177.22 + 74.65	06 – 26 – 000	0109	1.5 x 0.1 1.5 x 0.12	140 7	1	Sc		16.5:		
06809	11 48.0 + 06 51 265.06 + 61.87	01 – 30 – 017 NGC 3914	0495 239	1.1 x 0.6 1.1 x 0.6	40 5	1	S B: ...		13.8		13
06810	11 48.0 + 20 17 236.78 + 74.36	03 – 30 – 119 NGC 3919	1406 234	(0.9 x 0.9) (0.9 x 0.9)	-- ...	2	E		14.5		*3
06811	11 48.0 + 50 50 145.16 + 63.87	09 – 20 – 003	1389	1.2 x 0.4 1.1 x 0.4	9 6 – 7	2	S0 – a ssp of 2		14.8		*1
06812	11 48.0 + 50 53 145.10 + 63.83	09 – 20 – 000	1389	1.6: x 0.4: 0.3 x 0.2	54 (6)	1	...	d – m	17.		*
06813	11 48.0 + 55 37 140.13 + 59.71	09 – 20 – 001 NGC 3913 = 740*	1389 229	3.0 x 2.8 2.5: x 2.0:	-- 1	4	Sc	SA d	14.2		*12 $3\overline{\ }$
06814	11 48.0 + 64 37 133.44 + 51.53	11 – 15 – 000	0674 255	1.2: x 0.5: 1. x 0.4	15: (6)	2	... npp of 2	d – m	17.		*
06815	11 48.1 + 52 07 143.64 + 62.78	09 – 20 – 008 NGC 3917	1389 229	5.1 x 1.1 5.1 x 1.1	77 6	4	Sc P w 6802		12.5		*123 $46\overline{7}8$
06816	11 48.1 + 56 44 139.11 + 58.73	10 – 17 – 086	0723 229	1.6 x 1.3 1.6 x 1.3	-- ...	4	Irr IB m Ir I		15.1		*1V
06817	11 48.2 + 39 09 166.25 + 72.74	07 – 24 – 035	0719	4.7: x 2.0: 3. x 2.	[65] ...	4	Dwarf irregular Ir V		15.1		*1
06818	11 48.2 + 46 05 151.75 + 67.78	08 – 22 – 015	1338	2.1 x 0.9 2.0 x 0.8	[78] ...	1	... ssf of 2, contact, disturbed?		14.6		*1
06819	11 48.2 + 55 25 140.26 + 55 25	09 – 20 – 005 NGC 3916	1389 229	1.6 x 0.35 1.6 x 0.35	45 6	2	Sb P w 6823		14.8		*23
06820	11 48.3 + 20 47 235.27 + 74.67	04 – 28 – 064	0103 234	1.2 x 0.35 1.3 x 0.4	138 6	2	S0 – a P w 6821		15.4		*
06821	11 48.4 + 20 41 235.64 + 74.64	04 – 28 – 069	0103 234	1.4 x 1.2 1.3 x 1.3	-- 1	2	Sb/S B c: SAB(rs) ... P w 6820		14.6		*
06822	11 48.4 + 21 05 234.31 + 74.83	04 – 28 – 068 IC 742	0103 234	1.3 x 1.2 1.2 x 1.1	-- 1	2	S B a – b		15.1		
06823	11 48.5 + 55 21 140.25 + 59.98	09 – 20 – 009 NGC 3921, Mark 430	1389 229	2.2 x 1.3 2.2 x 1.3	[20] ...	4	... Peculiar, loop, p w 6819		13.4	+ 5930 + 6023	*12 $3V\!A\overline{C}$
06824	11 48.6 + 50 27 145.44 + 64.24	08 – 22 – 017 NGC 3922	1338	1.9 x 0.9 2.1 x 0.9	38 6	2	S0 – a		13.8		*3
06825	11 48.6 + 52 17 143.31 + 62.68	09 – 20 – 011 NGC 3931	1389 229	1.1 x 0.9 1.1 x 0.9	160: ...	2	E – S0		14.6		*23

N G C   3906, 3907, 3910 – 14, 3916, 3917
         3919 – 22,   3931

I C   740, 742

1	2	3	4	5	6	7	8	9	10	11	12
06826	11 48.8 + 32 50 186.67 + 75.91	06 - 26 - 000	0109	1.3 x 0.1 1.2 x 0.15	152 7	2	Sc		16.5:		
06827	11 48.8 + 35 43 176.24 + 74.74	06 - 26 - 042	0109	1.2 x 0.5 1.3 x 0.6	127 6	2	SO - a		14.3		
06828	11 48.8 + 53 43 141.75 + 61.44	09 - 20 - 012 Mark 431	1389 229	1.4 x 1.2 1.2 x 1.1	-- 1	2	Sb	Sb +	15.5		
06829	11 48.9 + 22 19 230.16 + 75.46	04 - 28 - 73 +74 103 NGC 3926	234	[1.0 x 0.8] ...	... ...	2	Double system Common envelope		14.7		*13
06830	11 48.9 + 24 09 223.15 + 76.10	04 - 28 - 072 IC 739	0103 234	1.2 x 0.8 1.2 x 0.8	150: 3	2	S B: a - b NGC 3951 group		15.1		
06831	11 49.0 + 15 45 249.87 + 71.86	03 - 30 - 121	1406	1.0 x 0.9 0.9 x 0.8	-- 1 - 2	3	SO		14.9		
06832	11 49.1 + 21 17 233.91 + 75.06	04 - 28 - 076 NGC 3929	0103 234	0.5 x 0.35 0.7 x 0.5	(80) ...	1	...		14.5		3
06833	11 49.1 + 38 16 168.30 + 73.41	06 - 26 - 045 NGC 3930	0109	4.5: x 3.5: 4. x 3.	30: 2 - 3	4	Sc/S B c SAB(s)c Sc +		13.5		*12 3
06834	11 49.1 + 48 57 147.20 + 65.55	08 - 22 - 019 NGC 3928, Mark 190	1338	(1.5 x 1.5) (1.6 x 1.6)	-- ...	1	...		13.1	+ 1500	*12 3
06835	11 49.2 + 17 08 246.53 + 72.80	03 - 30 - 000	1406	1.1 x 0.15 0.8 x 0.1	60 7	1	S ... NGC 3933 group		16.0:		*
06836	11 49.2 + 36 52 172.39 + 74.39	06 - 26 - 046	0109	1.0 x 0.6 0.9 x 0.5	[90] ...	1	S ... f of 2, disturbed:		15.6		*1
06837	11 49.3 + 18 49 241.90 + 73.83	03 - 30 - 000	1406	1.0 x 0.10 0.9 x 0.10	154 7	2	Sc 1st of 3		15.7		*
06838	11 49.4 - 02 22 274.89 + 56.95	00 - 30 - 000	0471 231	1.1 x 0.6 0.8 x 0.25	113 ...	1	...		15.4		
06839	11 49.4 + 17 05 246.76 + 72.80	03 - 30 - 122 NGC 3933	1406	1.4 x 0.7 0.9 x 0.6	83 ...	1	S ... Brightest in group		14.2		3
06840	11 49.5 + 52 23 142.95 + 62.66	09 - 20 - 019	1389 229	3. x 2.5:: 1.4 x 0.4	-- ...	4	Dwarf spiral, S B m SB IV - V		15.6		*
06841	11 49.6 + 17 08 246.73 + 72.87	03 - 30 - 123 NGC 3934	1406	1.3 x 1.1 1.0 x 0.7	... ...	1	... Acicular? NGC 3933 group		15.0		*3
06842	11 49.7 + 35 04 178.13 + 75.21	06 - 26 - 048	0109 235	1.2: x 0.9: 1.1 x 0.9	(20) ...	1	S ...		16.5:		
06843	11 49.8 + 32 41 186.99 + 76.16	06 - 26 - 049 NGC 3935	0109	1.2 x 0.7 1.1 x 0.6	114 4 - 5	1	S ...		14.0		3
06844	11 49.8 + 70 10 130.22 + 46.40	12 - 11 - 000	0714	1.1 x 0.15 0.9 x 0.18	35 7	1	Sa - b		15.5		
06845	11 50.0 + 13 45 254.82 + 70.62	02 - 30 - 000	0468	1.1 x 0.2 1.1 x 0.2	9 6 - 7	1	Sb?		15.2		
06846	11 50.0 + 23 53 224.53 + 76.26	04 - 28 - 079	0103 234	(1.0 x 0.9) (1.2 x 1.0)	-- ...	1	E? NGC 3951 group		14.8		*1
06847	11 50.0 + 24 35 221.68 + 76.46	04 - 28 - 080	0103	1.7 x 0.35 1.1 x 0.3	151 6	2	Sb - c NGC 3951 group		15.4		
06848	11 50.0 + 35 39 176.02 + 74.99	06 - 26 - 000	0109	1.0 x 0.2 0.9 x 0.2	61 6	2	... d - m f of 2		17.		*
06849	11 50.0 + 50 18 145.19 + 64.49	08 - 22 - 026 NGC 3924	1338	1.8: x 1.7: 1.6: x 1.6:	-- 1:	4	Dwarf spiral S IV - V		15.7		
06850	11 50.1 - 02 11 275.04 + 57.19	00 - 30 - 032	0471 231	0.7 x 0.6 0.7 x 0.5	-- ...	1	... Peculiar		14.1		
06851	11 50.1 + 20 55 235.59 + 75.10	04 - 28 - 081 NGC 3937	0103 234	(1.8 x 1.6) (2.7 x 2.5)	(15) ...	2	E - SO		14.0		*13
06852	11 50.1 + 21 17 234.33 + 75.27	04 - 28 - 082 NGC 3940	0103 234	(1.3 x 1.2) (1.7 x 1.5)	-- ...	2	E		14.3		3
06853	11 50.1 + 29 37 199.80 + 76.88	05 - 28 - 042	1379	1.2 x 0.6 1.1 x 0.6	35 5	2	SO		14.9		*1
06854	11 50.2 + 02 01 249.64 + 51.90	00 - 30 - 033	0471	1.1 x 1.0 1.0 x 0.9	-- 1:	3	S B ... Disturbed		14.4		*1V
06855	11 50.2 + 23 45 225.12 + 76.26	04 - 28 - 000	0103 234	(1.0 x 0.6) 1.2 x 0.7	(130) ...	1	SO NGC 3951 group		16.0:		
06856	11 50.2 + 44 23 153.91 + 69.33	07 - 25 - 001 NGC 3938	1367 248	5.4 x 5.1 5.2 x 4.7	-- 1	5	Sc SA(s)c Sc -		11.0	+ 874 + 919	*12- -9
06857	11 50.3 + 37 15 170.77 + 74.20	06 - 26 - 051 NGC 3941	0109	(3.6 x 2.5) (3.6 x 2.5)	10: 3	4	SO		11.3	+ 957 + 969	*123 4678
06858	11 50.3 + 61 30 135.02 + 54.51	10 - 17 - 095	0723 229	1.6 x 0.9 1.6 x 0.9	15 4 - 5	4	Sc		15.1		
06859	11 50.5 + 26 30 213.60 + 76.94	04 - 28 - 085 NGC 3944	0103 242	(1.4 x 1.1) (1.2 x 0.9)	(25) ...	2	E - SO		14.3		3
06860	11 50.5 + 60 56 135.36 + 55.04	10 - 17 - 096 NGC 3945	0723 229	(5.8 x 3.6) 5.6 x 3.5	15 4	4	S B 0		11.6	+ 1220 + 1337	*123 468

N G C   3924, 3926, 3928 - 30, 3933 - 35   I C 739
         3937, 3938, 3940, 3941, 3944, 3945

1	2	3	4	5	6	7	8	9	10	11	12
06861	11 50.6 + 25 43 217.03 + 76.84	04 – 28 – 087	0103 242	1.0 x 0.7 0.8 x 0.7	115: ...	1	S ...		15.1		
06862	11 50.7 + 11 56 258.69 + 69.36	02 – 30 – 000	0468	1.5 x 0.15 1.1 x 0.1	104 7	2	Sc	Sc +	16.0:		
06863	11 50.8 + 21 02 235.50 + 75.30	04 – 28 – 088 NGC 3947	0103 234	1.4 x 1.4 1.4 x 1.4	— 1	4	S B b	(R)S B(rs)b Sb –	14.2		*13
06864	11 51.0 + 11 09 260.22 + 68.80	02 – 30 – 000	0468	1.2 x 0.15 1.2 x 0.15	110 7	2	Sb – c sf of 2		15.3		*
06865	11 51.0 + 43 44 154.80 + 69.92	07 – 25 – 002	1367 248	[1.3 x 0.4] ...	[35] ...	3	Double system Connected		14.7		*1 V AC
06866	11 51.1 + 21 10 235.16 + 75.42	04 – 28 – 091 NGC 3954	0103 234	(0.9 x 0.8) (1.4 x 1.3)	— ...	1	... Compact		14.4		3
06867	11 51.1 + 23 40 225.73 + 76.43	04 – 28 – 090 NGC 3951	0103 234	1.2 x 0.6 1.1 x 0.5	172 (6)	1	S? ... Brightest in group		14.5		3
06868	11 51.1 + 29 33 199.99 + 77.10	05 – 28 – 000	1379	1.0 x 0.3 0.7 x 0.2	106 6	1	Sb – c		16.0:		
06869	11 51.1 + 48 08 147.64 + 66.41	08 – 22 – 029 NGC 3949	1338	2.8 x 1.7 2.5 x 1.6	120 4:	1	S ...		10.9	+ 681 + 743	*123 467̲8
06870	11 51.1 + 52 36 142.25 + 62.59	09 – 20 – 026 NGC 3953	1389 229	6.5 x 3.4 6.5 x 3.5	13 4 – 5	5	S B b	S B(r)bc Sb +	10.8	+ 958 + 1041	*12– –9
06871	11 51.2 + 10 41 261.10 + 68.46	02 – 30 – 043	0468	1.0 x 0.8 1.0 x 0.8	143: 2:	3	S B 0		14.6		1
06872	11 51.2 + 33 38 182.85 + 76.10	06 – 26 – 052 IC 2973	0109	1.5 x 0.8 1.5 x 0.8	125 5	3	S B c	S B(s)d: Sc +	14.5		
06873	11 51.3 + 21 02 235.72 + 75.40	04 – 28 – 000	0103 234	1.0 x 0.12 0.9 x 0.15	166 7	1	S ...		16.0:		*
06874	11 51.3 + 50 29 144.55 + 64.44	08 – 22 – 031	1338	1.1: x 0.5: 1.0 x 0.5:	105 5:	1	S ...		15.6		
06875	11 51.4 + 06 37b 267.04 + 65.12	01 – 30 – 000	0495	1.2: x 0.5: 0.7 x 0.3	105 6	1	S ... nff of 2		(15.3)		*
06876	11 51.4 + 20 52 236.34 + 75.34	04 – 28 – 093	0103 234	1.0 x 0.8 1.1 x 0.9	— 2	2	S B: a – b	SAB ...	14.9		*1
06877	11 51.7 + 00 25 273.62 + 59.72	00 – 30 – 034 IC 745	0471 231	0.8 x 0.7 0.7 x 0.6	— ...	1	...		13.7		*1
06878	11 51.8 + 33 08 184.60 + 76.40	06 – 26 – 053	0109	1.0 x 0.15 0.9 x 0.15	161 6 – 7	1	Sb – c		16.0:		
06879	11 51.9 – 02 02 275.69 + 57.52	00 – 30 – 035	0471 231	1.9 x 0.6 1.6: x 0.4	168 6	2	Sc – Irr		14.4		
06880	11 51.9 + 58 40 136.69 + 57.19	10 – 17 – 098 NGC 3958	0723 229	1.4 x 0.55 1.4 x 0.55	28 6	2	Sa P w 6884		13.1		*12 36̲
06881	11 52.1 + 20 22 238.37 + 75.23	03 – 30 – 000	1406 234	1.6: x 0.7: 1.0:: x 0.5::	125 ...	3	Dwarf V		17.		*
06882	11 52.2 + 33 48 181.86 + 76.22	06 – 26 – 000	0109	1.1 x 0.15 1.1 x 0.18	124 7	1	S ... sff of 2, contact		16.5:		*
06883	11 52.3 + 26 30 213.83 + 77.34	04 – 28 – 094	0103	1.4 x 1.4 0.7: x 0.7:	— 1	3	S B c	SAB(s)cd: Sc +	15.3:		*
06884	11 52.4 + 58 46 136.51 + 57.19	10 – 17 – 100 NGC 3963	0723 229	2.9 x 2.7 2.8 x 2.5	— 1	4	S B b P w 6880	SAB bc Sb +	12.2		*12 3468̲
06885	11 52.4 + 69 36 130.18 + 47.01	12 – 11 – 000 NGC 3961	0714	1.4 x 1.4 1.6 x 1.5	— 1	3	S B a	(R:)S B(r)a	14.7		3
06886	11 52.6 + 06 27 267.85 + 65.14	01 – 30 – 020	0495	1.3 x 0.9 1.3 x 0.9	80: 3	4	S B b	S B(s)b Sb +	14.5		
06887	11 52.6 + 22 59 228.98 + 76.51	04 – 28 – 095	0103 245	1.2 x 0.5: 1.2 x 0.3:	[157] ...	1	S ...		15.5		*
06888	11 52.6 + 50 26 144.20 + 64.59	08 – 22 – 039	1338	1.0 x 0.2 0.9 x 0.2	154 6 – 7	1	S ...		15.5		
06889	11 52.6 + 67 41 131.08 + 48.82	11 – 15 – 000	0674	1.0 x 0.1 0.9: x 0.1	132 7	2	Sc		16.5:		
06890	11 52.7 + 00 46 273.76 + 60.14	00 – 31 – 000	1401	1.3: x 0.4: 1.0: x 0.4:	[5] ...	4	Dwarf irregular Ir V		17. – 18.		
06891	11 52.7 + 17 46 246.65 + 73.85	03 – 30 – 126	1406	1.7 x 0.35 1.5: x 0.3	113 6	2	Sa – b sf of 2		15.4		*
06892	11 52.8 + 33 24 183.67 + 76.49	06 – 26 – 055	0109	1.1 x 0.3 1.1 x 0.3	82 ...	1	... sf of 2		15.3		*1
06893	11 52.8 + 39 30 163.38 + 73.19	07 – 25 – 003	1367	1.2 x 1.1 1.0 x 1.0	— 1	4	Sc p of 2	Sc +	15.7		*
06894	11 52.8 + 54 56 139.52 + 60.63	09 – 20 – 029	1389 229	1.4 x 0.20 1.3 x 0.2	89 7	3	Sc		15.6		
06895	11 52.9 + 12 15 259.31 + 69.95	02 – 30 – 045 NGC 3968	0468	2.9 x 1.9 2.5 x 1.8	10: 3 – 4	4	Sb/S B c	SAB(rs)bc Sb +	13.3		*13

N G C   3947,   3949,   3951,   3953,   3954,   3958
3961,   3963,   3968

I C   745,   2973

1	2	3	4	5	6	7	8	9	10	11	12
06896	11 53 + 80 30 125.85 + 36.59	14 – 05 – 000	0727	[1.1 x ...] ...	... ...	3	Triple system Bridges		15.2		*C
06897	11 53.0 + 10 04 263.06 + 68.22	02 – 31 – 000	1385	1.2: x 0.5: 0.6: x 0.4:	8 5 – 6	2	S B c		16.0:		
06898	11 53.0 + 26 10 215.44 + 77.45	04 – 28 – 096 IC 746	0103 242	1.3 x 0.35 1.0 x 0.3	169 6	1	S ...		14.5		
06899	11 53.0 + 30 16 196.52 + 77.41	05 – 28 – 047 NGC 3971	1379 244	(1.4 x 1.2) (1.3 x 1.1)	-- ...	3	S0 P w 6905		13.9		*3
06900	11 53.0 + 31 48 189.75 + 77.07	05 – 28 – 049	1379	2.0: x 1.2: 1.8: x 1.1:	(115) ...	4	Dwarf V		15.7		*
06901	11 53.0 + 43 19 154.77 + 70.47	07 – 25 – 004	1367 248	1.4 x 0.7 1.4 x 0.7	32: 5:	1	S ...		14.4		1
06902	11 53.0 + 46 30 149.32 + 67.94	08 – 22 – 000	1338	[1.0 x 0.35] ...	... ...	3	Double system Connected, distorted		15.6		*
06903	11 53.1 + 01 32 273.27 + 60.87	00 – 31 – 002	1401	2.5 x 2.2 2.7: x 2.4:	(150) 1	4	S B c	SAB(s)cd	14.1		
06904	11 53.1 + 55 35 138.87 + 60.06	09 – 20 – 032 NGC 3972	1389 229	4.1 x 1.0 3.5 x 0.9	120 6	3	Sc	Sc –	12.9		*12 36̲
06905	11 53.2 + 30 13 196.71 + 77.47	05 – 28 – 050	1379 244	1.1 x 0.2 1.1 x 0.15	41 7	2	Sb P w 6899		16.0:		*
06906	11 53.3 + 07 02 267.48 + 65.73	01 – 31 – 001 NGC 3976	1611	3.6 x 1.1 3.6 x 1.1	53 6	4	Sb 	SAB b Sb +	12.8		*12 346̲8̲
06907	11 53.5 – 02 26 276.67 + 57.31	00 – 31 – 003 NGC 3979	1401 231	1.1 x 1.0 1.1 x 0.9	... ...	2	S0? 2 companions superimposed		14.2		*1
06908	11 53.5 + 37 26 168.85 + 74.61	06 – 26 – 000	0109	[1.1 x 0.3] ...	... ...	2	S B ... ssp of 2, distorted		16.0:		*
06909	11 53.5 + 55 40 138.70 + 60.01	09 – 20 – 034 NGC 3977 = 3980	1389 229	1.8 x 1.7 1.7 x 1.6	-- 1	4	Sa	(R)SA ab	14.7		*12 3
06910	11 53.5 + 60 47 134.87 + 55.32	10 – 17 – 105 NGC 3978	0723 229	1.7 x 1.6 1.7 x 1.7	-- 1	2	Sb P w NGC 3975, Sb +		13.2		*12 3
06911	11 53.6 + 15 24 253.20 + 72.20	03 – 31 – 002	0089	1.3 x 0.35 1.4 x 0.3	18 6	2	Sb – c		15.2		
06912	11 53.6 + 58 29 136.45 + 57.45	10 – 17 – 104	0723 229	2.4: x 1.1: 2.0: x 1.0:	... ...	2	Strongly peculiar Disrupted multiple system?		15.1		*1C V
06913	11 53.7 + 17 18 248.48 + 73.73	03 – 31 – 003	0089	1.6 x 0.8 1.6 x 0.7	0 5	2	Sb		15.3		
06914	11 53.8 + 24 09 224.57 + 77.16	04 – 28 – 098 NGC 3983	0103 242	1.2 x 0.3 1.0 x 0.30	114 7	2	S0 – a		14.8		*3
06915	11 53.8 + 32 18 187.39 + 77.08	05 – 28 – 051 IC 2978	1379 244	1.1 x 0.5 1.1 x 0.4	126 6	2	Sc P w 6920 NGC 3995 group, Sc +		15.4		*
06916	11 53.8 + 40 01 161.63 + 72.98	07 – 25 – 006	1367	1.9 x 0.45 1.5 x 0.5	34 6	2	S B: b – c		14.7		*
06917	11 53.9 + 50 43 143.46 + 64.44	09 – 20 – 038	1389	4.8: x 3.0: 4. x 2.5::	130: ...	4	...	S B m	14.0		
06918	11 53.9 + 55 23 138.84 + 60.29	09 – 20 – 036 NGC 3982	1389 229	2.4 x 2.2 1.8 x 1.7	-- 1	1	S ...		11.6		123̲ 467̲8̲
06919	11 54.0 + 55 54 138.38 + 59.83	09 – 20 – 037	1389 229	1.4 x 0.4 1.5: x 0.5:	87 (6)	1	S – Irr		15.4		*
06920	11 54.1 + 32 17 187.37 + 77.15	05 – 28 – 053 NGC 3986	1379 244	2.5 x 0.5 2.8 x 0.5	110 7	3	Sa P w 6915, NGC 3995 group		14.0		*12 3
06921	11 54.1 + 48 36 145.96 + 66.29	08 – 22 – 045 NGC 3985	1338	1.1 x 0.7 1.1 x 0.7	73 3 – 4	1	S ...		13.0		*123̲ 467̲8̲
06922	11 54.2 + 51 07 142.91 + 64.12	09 – 20 – 042	1389 229	2. x 2. 1.3: x 1.2:	-- 1	1	S ...		15.4		
06923	11 54.2 + 53 27 140.51 + 62.05	09 – 20 – 040	1389 229	2.2 x 0.8 2.0 x 0.7	175 ...	3	Irr		14.1		*
06924	11 54.3 + 13 03 258.59 + 70.78	02 – 31 – 000	1385	1.1: x 0.3 1.0 x 0.2	104 6	2	Sc		15.3		
06925	11 54.3 + 32 25 186.74 + 77.14	05 – 28 – 054	1379 244	0.8 x 0.7 0.7 x 0.5	(0) ...	3	S B 0		14.5		*
06926	11 54.3 + 57 47 136.81 + 58.13	10 – 17 – 111	0723 229	1.4 x 0.6 1.3 x 0.5	98 6	2	...	S d – m	16.0:		*
06927	11 54.5 + 30 40 194.42 + 77.66	05 – 28 – 055	1379 244	0.9 x 0.7 1.1 x 0.8	-- ...	2	S0		14.5		
06928	11 54.7 + 25 29 218.89 + 77.70	04 – 28 – 099 NGC 3987	0103 242	2.3 x 0.4 2.2 x 0.35	58 7	3	Sb NGC 4005 group		14.4		*13
06929	11 54.7 + 36 41 170.64 + 75.24	06 – 26 – 058	0109	1.1 x 0.9 1.2: x 1.0:	(60) 2:	2	Sc 1st in group		15.3		
06930	11 54.7 + 49 33 144.56 + 65.52	08 – 22 – 046	1338	4.5: x 4.3: 4. x 4.	-- 1	4	S B c	SAB(s)d	14.2		1

N G C   3971, 3972, 3976 – 80, 3982, 3983
3985 – 87

I C   746, 2978

1	2	3	4	5	6	7	8	9	10	11	12
06931	11 54.7 + 58 13 136.40 + 57.75	10 – 17 – 115	0723 229	1.7: x 1.1 1.5: x 1.0:	(80) ...	3	... S:B m		14.5		*1
06932	11 54.9 + 31 21 191.22 + 77.58	05 – 28 – 056	1379 244	1.1 x 0.4 0.9 x 0.3	[170] ...	3	Irr or Peculiar		15.5		
06933	11 54.9 + 32 36 185.77 + 77.20	06 – 26 – 060 NGC 3991	0109 244	1.4 x 0.35 1.2 x 0.25	[33] ...	2	Strongly peculiar Multiple?, NGC 3995 group		13.8	+ 3308 + 3301	*12 36V̄A
06934	11 55.0 – 00 58 276.22 – 58.80	00 – 31 – 004	1401 231	1.8 x 0.2 1.4: x 0.15	141 7	3	Sc		15.2		*
06935	11 55.0 + 25 31 218.80 + 77.77	04 – 28 – 101 NGC 3993	0103 242	1.6 x 0.4 1.7 x 0.35	141 6	1	S ... NGC 4005 group		14.8	+ 4831 + 4792	*12 3
06936	11 55.0 + 32 33 185.95 + 77.24	06 – 26 – 059 NGC 3994	0109 244	1.1 x 0.6 1.0 x 0.5	10 (5)	1	S? ... P w 6944 NGC 3995 group		13.7	+ 3133 + 3126	*12 36V̄A
06937	11 55.0 + 53 39 140.10 + 61.93	09 – 20 – 044 NGC 3992 = M 109	1389 229	8.3 x 4.6 8.0 x 4.5	68 5	5	S B b SB(rs)bc Sb +		10.7	+ 1059 + 1147	12– –689
06938	11 55.0 + 55 43 138.27 + 60.06	09 – 20 – 043 NGC 3990	1389 229	1.4 x 0.8 1.3 x 0.8	40 4:	1	S0? P w 6946		13.6	+ 720 + 817	*12 36
06939	11 55.0 + 57 50 136.61 + 58.12	10 – 17 – 117	0723 229	(1.1 x 0.9) 1.1 x 0.9	— ...	1	...		14.6		
06940	11 55.1 + 53 32 140.18 + 62.04	09 – 20 – 045	1389 229	1.0 x 0.15 0.8 x 0.1	136 7	1	S ...		16.0:		
06941	11 55.2 + 14 35 255.96 + 72.09	03 – 31 – 004 NGC 3996	0089	0.9 x 0.7 0.9 x 0.7	(50) ...	1	S ...		14.4		3
06942	11 55.2 + 25 33 218.69 + 77.83	04 – 28 – 102 NGC 3997	0103 242	1.6 x 1.3: 1.7 x 1.4:	-- ...	2	S B ... Singular, distorted? NGC 4005 group		14.3	+ 4769 + 4730	*12 3
06943	11 55.2 + 29 19 200.64 + 78.02	05 – 28 – 058 NGC 3984	1379	1.2 x 1.1 1.1 x 1.0	-- 1	3	S B b SB(rs)b		14.8		13
06944	11 55.2 + 32 34 185.81 + 77.27	06 – 26 – 061 NGC 3995	0109 244	2.8 x 1.0 2.8 x 1.0	33 6	3	S ... Brightest in group, disrupted		12.9	+ 3362 + 3356	*123V 4678A
06945	11 55.3 + 36 40 170.42 + 75.35	06 – 26 – 062	0109	[1.2 x 0.9] ...	... ...	4	Double system, bridge Disrupted, in group		15.1		*1V A
06946	11 55.3 + 55 43 138.20 + 60.07	09 – 20 – 046 NGC 3998	1398 229	(3.0 x 2.5) (3.0 x 2.5)	140: 2:	2	S0 P w 6938		11.2	+ 1080 + 1177	*123 4678
06947	11 55.4 + 22 28 232.16 + 76.90	04 – 28 – 000	0103 245	1.0 x 0.1 1.0 x 0.12	97 7	1	Sc		16.5:		*
06948	11 55.4 + 23 24 228.32 + 77.26	04 – 28 – 105 NGC 4003 = 4007	0103	(1.5 x 1.1) 1.7 x 1.1	(10?) ...	3	S B 0 s of 2		14.8		*3
06949	11 55.4 + 25 25 219.35 + 77.84	04 – 28 – 103 NGC 4000	0103 242	1.1 x 0.20 1.0 x 0.22	3 6 – 7	1	S ... NGC 4005 group		15.2		*
06950	11 55.4 + 28 09 206.26 + 78.13	05 – 28 – 060 NGC 4004, Mark 43	1379 2 243	2.0 x 0.5 1.8 x 0.7	[8] ...	3	Strongly peculiar f of 2, disrupted?		14.0		*13 V
06951	11 55.6 – 01 49 277.12 + 58.08	00 – 31 – 006 NGC 4006 = 2983*	1401 231	(1.3 x 0.9) (1.4 x 1.0)	(20) ...	2	E P w 6958		14.2		*3
06952	11 55.6 + 25 24 219.47 + 77.88	04 – 28 – 107 NGC 4005	0103 242	1.1 x 0.6 0.9 x 0.7	92: ...	1	S ... Brightest in group		14.1		*3
06953	11 55.6 + 28 28 204.72 + 78.16	05 – 28 – 061 NGC 4008	1379 243	(2.4 x 1.3) (2.1 x 1.2)	167 ...	3	E – S0		13.1		*123 4678
06954	11 55.8 + 27 48 207.97 + 78.21	05 – 28 – 063 NGC 4016?	1379 243	1.5 x 0.8 1.1 x 0.5	175: ...	3	S B c – Irr P w 6967		14.6		*13 A
06955	11 55.8 + 38 20 165.15 + 74.40	06 – 26 – 063	0109	6. x 3. 4. x 2.	70 ...	4	Dwarf irregular Ir V		15.2		*
06956	11 55.8 + 51 13 142.30 + 64.15	09 – 20 – 050	1389 229	2.7:x 2.5:: 2. x 2.	-- ...	4	Dwarf spiral, S B m S B V		17.		*
06957	11 55.8 + 57 52 136.40 + 58.13	10 – 17 – 119	0723 229	2.0: x 1.0: 1.9 x 1.1	60 ...	2	S B ...		14.8		*
06958	11 55.9 – 01 59 277.37 + 57.96	00 – 31 – 007	1401 231	1.3 x 1.3 1.3 x 1.3	-- 1	3	Sb P w 6951		14.5		*
06959	11 55.9 + 01 00 275.03 + 60.69	00 – 31 – 8+9	1401	[1.2 x 0.3] ...	... ...	2	Chain of 3 galaxies Bridges?		15.4		*1
06960	11 55.9 + 10 18 264.29 + 68.82	02 – 31 – 006 NGC 4012	1385	2.1 x 0.5 2.0 x 0.5	153 6	2	Sb		14.6		3
06961	11 56.0 + 16 27 252.03 + 73.56	03 – 31 – 005 NGC 4014	0089	(1.9 x 1.1) (2.0 x 1.3)	120 3 – 4	2	S0 – a		13.5		*3
06962	11 56.0 + 43 00 154.10 + 71.06	07 – 25 – 008 IC 749	1367 248	2.5 x 2.1 2.6 x 2.2	150: 2	4	S B c SAB(rs)cd P w 6973		13.4		*12 4
06963	11 56.0 + 44 13 151.86 + 70.10	07 – 25 – 009 NGC 4013	1367 248	5.1 x 1.1 5.3 x 1.2	66 7	4	Sb		12.4		*123 4678
06964	11 56.0 + 47 32 146.70 + 67.36	08 – 22 – 049 NGC 4010	1338	4.0 x 0.9 4.1 x 0.9	66 7	2	Sc – Irr		13.1		*12 3
06965	11 56.1 + 25 19 219.98 + 77.97	04 – 28 – 109+ NGC 4015 110	0103 242	[1.4 x 1.4] ...	-- ...	2	Double system Contact, NGC 4005 group		14.2		*1V A

N G C 3984, 3990 – 3998, 4000, 4003 – 08   I C 749, 2983
      4010, 4012 – 16

1	2	3	4	5	6	7	8	9	10	11	12
06966	11 56.1 + 25 36 218.65 + 78.03	04 – 28 – 108 NGC 4018	0103 242	1.8 x 0.3 1.6 x 0.25	163 7	1	Sa – b NGC 4005 group		14.7		*
06967	11 56.1 + 27 43 208.39 + 78.28	05 – 28 – 065 NGC 4017?	1379 243	1.8 x 1.5 1.5 x 1.4	-- ...	3	Sb/S B c SAB(s)bc P w 6954, distorted		13.5		*13 A
06968	11 56.1 + 28 34 204.21 + 78.27	05 – 28 – 064	1379 243	3.0 x 1.0 2.5: x 1.0	85 6	4	S ...		15.2		*1
06969	11 56.1 + 53 42 139.74 + 61.95	09 – 20 – 048	1389 229	1.7 x 0.5 1.5 x 0.4	152 ...	3	Irr Ir I		15.5		*
06970	11 56.2 – 01 10a 276.90 + 58.74	00 – 31 – 011	1401 231	1.6 x 0.8 1.4 x 0.7	[75] ...	3	Irr dm – m nnp of 2, contact, Ir I		(14.6)		*1
06971	11 56.3 + 30 41 193.95 + 78.03	05 – 28 – 066 NGC 4020	1379	2.0 x 1.0 1.9 x 0.8	15 5	2	... S d – m		13.2		123
06972	11 56.3 + 42 50 154.29 + 71.23	07 – 25 – 011 IC 751	1367 248	1.4 x 0.4 1.1 x 0.4	30 6	1	S ... p of 2		15.1		*2
06973	11 56.3 + 42 59 154.00 + 71.11	07 – 25 – 010 IC 750	1367 248	3.1 x 1.5 3.1 x 1.2	43 5 – 6	1	S ... P w 6962		12.7		*12 47
06974	11 56.4 + 21 11 237.58 + 76.51	04 – 28 – 000	0103 245	1.0 x 0.25 0.8 x 0.20	137 6	1	S ...		16.0:		
06975	11 56.4 + 25 30 219.19 + 78.08	04 – 28 – 111 NGC 4022	0103 242	(1.3 x 1.2) (1.3 x 1.3)	-- 1:	3	SO SAB: 0 NGC 4005 group		14.4		*
06976	11 56.5 + 22 06 234.11 + 76.97	04 – 28 – 114	0103 245	(1.4 x 1.4) (1.6 x 1.5)	-- 1	1	SO:		15.0		
06977	11 56.5 + 25 16 220.31 + 78.05	04 – 28 – 113 NGC 4023	0103 242	1.1 x 0.7 0.7 x 0.7	25: ...	1	... NGC 4005 group		14.6		*
06978	11 56.6 – 02 18 277.90 + 57.93	00 – 31 – 000	1401 231	1.3: x 0.8: 1.2: x 0.8:	(130) ...	3	Dwarf V		17.		
06979	11 56.6 – 00 13 276.35 + 59.65	00 – 31 – 012 IC 753	1401 231	0.35 x 0.25 0.6 x 0.4	(30) ...	1	... Peculiar		14.3		
06980	11 56.6 + 24 45 222.75 + 77.95	04 – 28 – 000	0103 242	1.4: x 0.6: 1.0: x 0.6:	[153] ...	3	Dwarf irregular Ir IV – V or V		17.		
06981	11 56.6 + 31 00 192.39 + 78.02	05 – 28 – 069 IC 2985	1379 244	1.1 x 0.6 1.1: x 0.6:	142 ...	1	S B ... nnf of 2		15.2		*
06982	11 56.6 + 38 03 165.58 + 74.70	06 – 26 – 064 NGC 4025	0109	2.8 x 1.7 1.6 x 1.4	40: 4:	4	S B c S B(s)cd		14.9		*12 3
06983	11 56.6 + 52 59 140.27 + 62.63	09 – 20 – 051	1389 229	4.4 x 3.2 4.0: x 3.0:	85 3	4	S B c S B(rs)cd		14.5		*
06984	11 56.8 – 01 21 277.30 + 58.63	00 – 31 – 013 IC 754	1401 231	0.9 x 0.7 0.9 x 0.7	(40) ...	3	E		14.5		*
06985	11 56.8 + 51 15 141.94 + 64.19	09 – 20 – 052 NGC 4026	1389 229	4.5 x 1.1 4.6 x 1.1	178 7	3	SO		11.5	+ 878 + 956	12– –8
06986	11 56.9 + 18 02 248.26 + 74.78	03 – 31 – 008	0089	1.1 x 0.35 1.1 x 0.25	129 6	1	S ...		15.3		
06987	11 57.0 + 30 26 194.98 + 78.23	05 – 28 – 070	1379 244	1.3 x 0.7 0.9 x 0.5	10 5:	3	S B b S B(s)b		15.3		
06988	11 57.3 + 55 59 137.47 + 59.94	09 – 20 – 054	1389 229	1.3 x 0.4 1.1 x 0.4	134 ...	2	Dwarf		16.5:		
06989	11 57.4 + 21 55 235.26 + 77.07	04 – 28 – 118	0103 245	1.0 x 0.5 0.9 x 0.5	15 ...	1	Sc:		16.0:		*
06990	11 57.5 + 08 28 267.85 + 67.49	01 – 31 – 008 NGC 4029	1611	1.2 x 0.7 1.1 x 0.7	150 4:	3	Sb/S B b: SAB ... Singular, disturbed?		14.5		*13
06991	11 57.6 + 19 30 244.12 + 75.82	03 – 31 – 009	0089	1.0 x 0.15 1.0 x 0.15	64 6 – 7	2	Sb – c		15.7		
06992	11 57.7 + 50 57 141.97 + 64.52	09 – 20 – 059	1389 229	1.5 x 0.7 1.4 x 0.6	63 ...	1	S – Irr		14.9		
06993	11 57.8 – 00 48 277.34 + 59.23	00 – 31 – 016 NGC 4030	1401 231	4.2 x 3.2 4.0 x 3.0	27 2 – 3	5	Sb SA(s)bc Sb +		12.4	+ 1509 + 1361	*123 468
06994	11 58.0 + 09 08 267.21 + 68.12	02 – 31 – 000	1385	1.2 x 0.35 1.1 x 0.3	74 6	3	Sc Sc +		15.6		
06995	11 58.0 + 20 21 241.45 + 76.39	03 – 31 – 010 NGC 4032	0089 251	1.9 x 1.8 1.9 x 1.8	-- ...	3	Irr		12.7		*12 3478
06996	11 58.0 + 79 08 126.03 + 37.97	13 – 09 – 000	1339	1.5:: x 1.2:: 1.2:: x 1.0::	... ...	3	Dwarf V		17.		
06997	11 58.2 + 32 09 186.61 + 78.01	05 – 28 – 076	1379	1.1: x 0.9: 0.9: x 0.8:	-- ...	4	Dwarf spiral S V		15.5		
06998	11 58.3 + 00 16 276.75 + 60.26	00 – 31 – 000	1401	1.8 x 1.4 1. x 1.	-- ...	4	Dwarf spiral S V		17.		
06999	11 58.3 + 50 11 142.61 + 65.25	08 – 22 – 053	1338	1.1 x 0.4 1.0 x 0.4	[28] ...	3	Dwarf		18.		
07000	11 58.6 – 01 00 277.85 + 59.12	00 – 31 – 017	1401 231	1.1 x 0.9 1.1 x 0.9	[50] ...	4	Irr IB: m Ir I		14.4		1

N G C 4017, 4018, 4020, 4022, 4023, 4025   I C 750, 751, 753, 754, 2985
4026, 4029, 4030, 4032

1	2	3	4	5	6	7	8	9	10	11	12
07001	11 58.6 + 14 23 258.43 + 72.48	02 - 31 - 014 IC 755	1385	2.7 x 0.25 2.2 x 0.25	145 7	1	S ...		13.9		*
07002	11 58.8 + 13 40 260.04 + 71.95	02 - 31 - 015 NGC 4037	1385	2.8: x 2.5: 2.8:: x 2.8::	-- 1:	3	S B ... Singular	S B(rs) ...	13.8		*12 3478
07003	11 58.8 + 14 43 257.83 + 72.76	03 - 31 - 000	0089	1.8:: x 1.4:: 1. x 1.	-- ...	4	Dwarf V		18.		*
07004	11 58.9 - 00 25 277.55 + 59.69	00 - 31 - 018	1401	1.3 x 1.1 1.3 x 1.1	(60) 1 - 2	3	S B c	S B(r)c Sc +	15.0		
07005	11 58.9 + 62 10 132.98 + 54.25	10 - 17 - 125 NGC 4036	0723 229	(3.7 x 1.8) (4.1 x 1.8)	85 6:	4	S0 Brightest in group	P w 7014	11.5	+ 1382 + 1507	*12- -689
07006	11 58.9 + 69 37 129.38 + 47.16	12 - 11 - 044 NGC 4034	0714	1.8 x 1.3 1.6 x 1.0	5: 3 - 4	2	Sc		14.5		*3
07007	11 59.0 + 33 37 179.93 + 77.58	06 - 26 - 068	0109	2. x 2. 1. x 1.	-- ...	4	Dwarf (spiral) S: V		17.		
07008	11 59.1 + 05 02 272.80 + 64.67	01 - 31 - 000	1611	1.3 x 0.35 1.2 x 0.35	117 6 - 7	1	S ...		15.3		
07009	11 59.2 + 62 37 132.67 + 53.84	11 - 15 - 010	0674	1.7 x 0.5 1.7 x 0.4	175 ...	4	Irr NGC 4036 group, Ir I		14.5		*
07010	11 59.3 + 22 49 232.43 + 77.86	04 - 28 - 000	0103 245	1.2 x 0.20 1.3 x 0.22	37 7	2	Sa		15.5		
07011	11 59.4 - 00 46 278.04 + 59.41	00 - 31 - 019	1401 231	1.1 x 0.20 1.1 x 0.20	40 7	2	Sb		15.3		*1
07012	11 59.5 + 30 07 196.01 + 78.82	05 - 28 - 077	1379	2.1 x 1.1 1.1 x 0.7	8 5:	2	Sc P w 7017		14.3		*1
07013	11 59.6 + 18 06 249.69 + 75.31	03 - 31 - 018 NGC 4040	0089	(1.0 x 0.7) (1.2 x 0.8)	(145) ...	3	E		15.0		*1
07014	11 59.6 + 62 25 132.71 + 54.04	10 - 17 - 129 NGC 4041	0723 229	2.8 x 2.8 2.8 x 2.8	-- 1	4	Sc	Sc -	11.6	+ 1186 + 1312	*12- -689
07015	11 59.8 + 04 37 273.59 + 64.38	01 - 31 - 012 NGC 4043	1611	0.6 x 0.5 0.7 x 0.6	(135) ...	1	S ...		14.1		3
07016	11 59.8 + 15 07 257.56 + 73.22	03 - 31 - 019	0089	1.6 x 0.35 1.6 x 0.4	137 6 - 7	1	S B: a - b		14.7		1
07017	11 59.8 + 30 08 195.86 + 78.88	05 - 28 - 078	1379	1.9 x 0.4 1.8 x 0.4	70 6	2	Sb P w 7012		14.4		*1
07018	11 59.9 + 00 05 277.63 + 60.25	00 - 31 - 020 NGC 4044	1401	1.1 x 1.0 1.2 x 1.1	-- ...	2	E - S0		14.6		3
07019	11 59.9 + 62 42 132.50 + 53.79	11 - 15 - 011	0674	1.8 x 0.9 1.7: x 0.8	[85] ...	4	Irr NGC 4036 group, Ir I	I m	15.6		*1
07020	12 00.0 + 41 20 155.65 + 72.82	07 - 25 - 014	1367	2.1 x 1.7 2.2 x 1.7	-- 2:	3	Sb	SA(rs)bc Sb +	14.4		*
07021	12 00.2 + 02 16 276.00 + 62.29	00 - 31 - 022 NGC 4045	1401 249	3.0 x 2.0 (3.0 x 2.2)	95 3:	3	Sa n of 2	SAB:(r)a	13.5		*12 3468
07022	12 00.2 + 45 27 148.05 + 69.51	08 - 22 - 057	1338	1.0 x 0.2 0.9 x 0.2	55 7	1	S ...		16.0:		
07023	12 00.3 + 18 17 249.58 + 75.56	03 - 31 - 020 NGC 4048	0089	0.55 x 0.45 0.55 x 0.5	92 ...	1	... in quadrupel system	Brightest	14.4		*13
07024	12 00.3 + 22 23 233.70 + 77.99	04 - 29 - 000	0135	1.2 x 0.12 1.1 x 0.15	101 7	2	Sc		16.0:		
07025	12 00.3 + 48 55 143.35 + 66.52	08 - 22 - 058 NGC 4047	1338	1.2 x 1.0 1.4 x 1.1	(105) 2:	1	S ...		12.8		*123 4678
07026	12 00.4 + 05 07 273.39 + 64.89	01 - 31 - 015 IC 756	1611	2.1 x 0.8 2.0 x 0.8	102 6	4	Sc	Sc +	14.8		
07027	12 00.4 + 19 02 247.31 + 76.06	03 - 31 - 021 NGC 4049	0089	0.9 x 0.7 0.9 x 0.7	52 ...	1	...		14.2		3
07028	12 00.6 + 16 45 254.09 + 74.56	03 - 31 - 023	0089	1.0 x 0.2 1.2 x 0.3	169 7	2	Sa - b 7032 group		16.0:		*1
07029	12 00.6 + 20 00 244.18 + 76.69	03 - 31 - 024 NGC 4053	0089 251	1.1 x 0.4 1.1 x 0.5	109 6:	1	S: ...		14.6		3
07030	12 00.6 + 44 48 148.90 + 70.09	08 - 22 - 059 NGC 4051	1338	6.0 x 5.0 6.5 x 5.2	135: 2	5	Sb/S B c Seyfert	SAB(rs)bc Sb +	11.5	+ 658 + 709	*12- -9
07031	12 00.8 + 29 42 197.90 + 79.17	05 - 29 - 001	1398	1.7 x 0.25 1.7 x 0.25	57 7	2	Sb - c		15.2		1
07032	12 00.9 + 16 46 254.24 + 74.62	03 - 31 - 027	0089	0.6 x 0.5 0.6 x 0.6	-- ...	1	... Brightest in group		14.0		*1
07033	12 00.9 + 47 38 144.71 + 67.69	08 - 22 - 060	1338	1.4 x 0.5 1.4 x 0.4	168 6	2	S B ...		15.6		
07034	12 01.1 + 02 20 276.38 + 62.44	00 - 31 - 024	1401 249	1.1 x 0.6 1.1 x 0.7:	90: ...	1	... Peculiar?		14.8		*
07035	12 01.1 + 02 55 275.87 + 62.97	01 - 31 - 016	1611	1.2 x 0.3 1.3 x 0.35	149 6	1	S ...		14.6		

N G C 4034, 4036, 4037, 4040, 4041, 4043 - 45   I C 755, 756
4047 - 49, 4051, 4053

1	2	3	4	5	6	7	8	9	10	11	12
07036	12 01.2 + 03 50 275.08 + 63.82	01 – 31 – 017 NGC 4058	1611	1.3 x 0.7 1.5 x 0.8	165 5	2	Sa		14.0		
07037	12 01.3 + 04 30 274.49 + 64.43	01 – 31 – 018	1611	1.0 x 0.3 1.0 x 0.3	138 6	1	S ...		14.8		
07038	12 01.3 + 14 49 259.19 + 73.23	03 – 31 – 000	0089	1.5:: x 0.8:: 1.3:: x 0.9::	... ...	4	Dwarf V		17.		
07039	12 01.3 + 16 50 254.32 + 74.73	03 – 31 – 031	0089	1.2: x 0.8: 1.2 x 0.8	25: ...	1	Sc: 7032 group		15.6		*
07040	12 01.3 + 25 43 219.26 + 79.21	04 – 29 – 005	0135 256	1.8 x 1.1 1.8 x 1.1	143: ...	3	...	S B dm	14.9		*1
07041	12 01.3 + 29 59 196.31 + 79.23	05 – 29 – 002	1398	1.4 x 0.15 1.5 x 0.15	156 7	2	Sc		16.0:		*1
07042	12 01.4 + 02 11 276.66 + 62.33	00 – 31 – 026	1401 249	1.2 x 0.3 1.2 x 0.3	132 6	1	S ...		15.4		*
07043	12 01.5 + 13 38 261.80 + 72.33	02 – 31 – 000	1385	1.6: x 0.2 1.2: x 0.3	49 ...	2	... Peculiar, plume		16.0:		
07044	12 01.5 + 20 30 242.92 + 77.15	04 – 29 – 006 NGC 4061	0135 251	(1.2 x 0.9) (1.3 x 1.0)	— ...	3	[E] Compact, p w 7050		14.4		*13 V
07045	12 01.5 + 32 10 185.30 + 78.65	05 – 29 – 004 NGC 4062	1398 263	4.8 x 2.0 4.7 x 1.9	100 6	4	Sc	SA(s)c Sc –	11.9		*12 3478
07046	12 01.5 + 49 23 142.38 + 66.19	08 – 22 – 062	1338	1.1 x 0.2 1.1 x 0.2	5 6 – 7	1	Sb – c		15.6		
07047	12 01.5 + 52 52 138.91 + 63.05	09 – 20 – 079 NGC 4068	1389 229	3.0 x 1.8 3.0: x 2.0:	[30] ...	4	Irr	IA m Ir I	13.3		123
07048	12 01.6 + 11 08 266.32 + 70.29	02 – 31 – 019 NGC 4067	1385 252	1.2 x 0.9 1.2 x 0.9	35: 2 – 3	2	Sb		13.2		*23
07049	12 01.6 + 20 27 243.16 + 77.14	03 – 31 – 000	0089 251	1.0 x 0.10 1.1 x 0.15	41 7	2	Sc		15.7		*
07050	12 01.6 + 20 30 242.98 + 77.17	04 – 29 – 007 NGC 4065	0135 251	(1.2 x 1.1) (1.4 x 1.2)	— ...	3	E Very compact Brightest in cluster		14.0		*13
07051	12 01.6 + 20 38 242.48 + 77.25	04 – 29 – 008 NGC 4066	0135 251	(1.1 x 1.0) (1.2 x 1.2)	— ...	3	E		14.4		*3
07052	12 01.6 + 20 42 242.23 + 77.28	04 – 29 – 009 NGC 4070	0135 251	(1.2 x 1.2) (1.4 x 1.4)	— ...	3	E		14.3		*3
07053	12 01.7 – 01 15 279.43 + 59.17	00 – 31 – 000	1401	1.8:: x 1.5:: 1. x 1.	— ...	4	Dwarf irregular Ir V		17.		*
07054	12 01.7 + 18 43 249.14 + 76.09	03 – 31 – 033 NGC 4064	0089	(4.2 x 1.8) 4.3 x 1.8	150 6	3	S B a		12.5	+ 1033 + 967	*123 4678
07055	12 01.7 + 24 25 226.02 + 78.96	04 – 29 – 010	0135 256	[1.0 x 0.8] ...	... ...	1	Double system		15.7		*1V
07056	12 01.7 + 62 47 132.12 + 53.77	11 – 15 – 014 IC 758	0674	2.0 x 1.9 2.0 x 1.8	... 1	4	S B c	S B(rs)cd: Sc +	14.4		*
07057	12 01.8 + 01 51 277.13 + 62.06	00 – 31 – 027	1401 249	2.0 x 0.4 1.8 x 0.4	100 6	1	S ...		14.6		*
07058	12 01.8 + 79 42a 125.64 + 37.46	13 – 09 – 008	1339	1.0 x 0.4 1.1 x 0.6	142 5:	2	Sb P w 7059				
07059	12 01.8 + 79 42b 125.64 + 37.46	13 – 09 – 009	1339	1.1 x 0.8 1.3 x 0.8	30: 3:	3	Sb/S B b SAB b P w 7058		15.4		*1
07060	12 01.9 + 02 11 276.91 + 62.38	00 – 31 – 029 NGC 4073	1401 249	(2.1 x 1.6) (2.3 x 1.8)	105: ...	3	E Companion superimposed		13.8		*12 348
07061	12 02.0 + 20 28 243.34 + 77.23	03 – 31 – 034 NGC 4076	0089 251	0.9 x 0.9 0.9 x 0.9	— 1	1	S ...		14.3		*13
07062	12 02.0 + 64 43 131.11 + 51.94	11 – 15 – 015 NGC 4081	0674 255	1.8 x 0.8 1.5 x 0.6	135 6	1	S ...		13.6		*2
07063	12 02.1 + 02 04 277.10 + 62.29	00 – 31 – 031 NGC 4077	1401 249	(1.0 x 0.7) (1.1 x 0.7)	15 ...	2	E – S0 P w IC 2989		14.5		*3
07064	12 02.1 + 31 27 188.55 + 79.02	05 – 29 – 005	1389 263	0.8 x 0.8 0.8 x 0.8	— ...	1	S? ... Brightest of 4		14.0		*1
07065	12 02.2 – 02 26 280.45 + 58.11	00 – 31 – 033	1401	1.7 x 1.5: 1.1 x 0.55	— 1	3	S B b	(R´)S B(r)ab Sb –	14.9		
07066	12 02.2 + 10 52 267.11 + 70.14	02 – 31 – 023 NGC 4078	1385 252	1.3 x 0.4 1.2 x 0.4	18 6	1	S0?		13.9		*3
07067	12 02.3 – 02 05 280.27 + 58.45	00 – 31 – 034 NGC 4079	1401	2.9 x 1.7 2.8: x 1.6	125 4:	2	Sb	SAB:(rs)b	14.0		123
07068	12 02.4 + 27 16 211.05 + 79.65	05 – 29 – 006 NGC 4080	1398	1.3 x 0.55 1.3 x 0.6	125 ...	1	Irr?		14.0		3
07069	12 02.4 + 43 25 150.41 + 71.42	07 – 25 – 017	1367 248	1.7 x 0.3 1.5 x 0.3	169 6	2	S B c		15.7		*
07070	12 02.4 + 58 23 134.51 + 57.97	10 – 17 – 137	0723 257	[2.0: x 1.5:] ...	... ...	3	S ... + companion Disrupted, bridge		15.4		*1V

N G C   4058,   4061,   4062,   4064 – 68,   4070
4073,   4076 – 81

I C   758

1	2	3	4	5	6	7	8	9	10	11	12
07071	12 02.6 + 38 31 161.17 + 75.25	07 – 25 – 019	1367 250	1.0 x 0.7 1.0 x 0.7	155: 3	2	Sb		15.3		
07072	12 02.7 + 29 03 201.10 + 79.67	05 – 29 – 007	1398	1.6 x 1.5: 1.4: x 1.2:	-- 1	3	...	SA dm	15.1		
07073	12 02.8 + 18 09 251.60 + 75.91	03 – 31 – 038	0089	1.1 x 0.8 1.2 x 0.8	115 3:	2	Sc P w 7074		14.1		*1
07074	12 02.8 + 18 12 251.45 + 75.94	03 – 31 – 037	0089	1.0 x 0.25 1.2 x 0.3	98 6	1	S ... P w 7073		14.6		*1
07075	12 02.8 + 50 39 140.58 + 65.15	09 – 20 – 086 NGC 4085	1389 229	2.7 x 0.8 2.8 x 0.7	78 6	1	Sc P w 7081		12.8		*12– –789
07076	12 02.9 + 20 31 243.71 + 77.43	04 – 29 – 016 NGC 4086	0135 251	(1.4 x 1.0) (1.0 x 0.9)	(85) ...	2	S0		15.1		*3
07077	12 02.9 + 20 35 243.46 + 77.47	04 – 29 – 015 NGC 4090 = 2997*	0135 251	1.2 x 0.5 1.1 x 0.5	38 6	1	Sa – b		15.0		*3
07078	12 02.9 + 47 03 144.67 + 68.37	08 – 22 – 065	1338	1.3 x 0.25 1.1 x 0.25	165 6	1	S ...		15.5		1
07079	12 02.9 + 63 26 131.58 + 53.19	11 – 15 – 019	0674 255	1.1 x 1.1 0.7 x 0.6	-- 1	1	S ... Brightest in group		14.9		*
07080	12 03.0 + 25 23 221.48 + 79.51	04 – 29 – 018	0135 256	1.2 x 0.20 1.2 x 0.25	27 6 – 7	1	Sb – c		15.4		
07081	12 03.0 + 50 50 140.33 + 64.99	09 – 20 – 089 NGC 4088	1389 229	5.9 x 2.2 6.5: x 2.5	43 6:	4	Sc Singular, p w 7075, Sc –	SAB bc	11.2	+ 733 + 812	*12– –9Ā
07082	12 03.0 + 51 48 139.40 + 64.11	09 – 20 – 088	1389 229	1.7 x 0.6 1.7 x 0.6	53 6	2	Sb		15.3		*
07083	12 03.1 + 20 50 242.63 + 77.65	04 – 29 – 019 NGC 4091	0135 251	1.1 x 0.25 1.1 x 0.25	43 6	1	S ... f of 2		15.2		*13
07084	12 03.1 + 33 23 179.03 + 78.45	06 – 27 – 001	1599 254	1.6 x 1.0 1.4 x 0.9	70 ...	1	S ...		14.9		
07085	12 03.2 + 09 16 270.04 + 68.89	02 – 31 – 028	1385	1.9: x 0.6 1.5: x 0.5	[70] ...	1	S ... f of 2, distorted, bridge		14.8		*1
07086	12 03.2 + 77 47 126.11 + 39.34	13 – 09 – 011	1339	2.6 x 0.6 2.3 x 0.6	72 7	2	Sb		15.4		
07087	12 03.3 + 20 46 243.01 + 77.65	04 – 29 – 020 NGC 4092	0135 251	1.1 x 1.1 1.2 x 1.1	-- 1	1	S ...		14.4		*3
07088	12 03.3 + 67 31 129.67 + 49.29	11 – 15 – 021	0674 255	1.5 x 0.6 1.5 x 0.6	7 6	3	S B b NGC 4108 group		14.7		*2
07089	12 03.4 + 43 25 149.94 + 71.52	07 – 25 – 020	1367 248	3.5 x 0.9 3.0 x 0.8	36 6	3	...	dm – m	14.8		*
07090	12 03.4 + 47 45 143.58 + 67.79	08 – 22 – 067 NGC 4096	1338	7.2 x 1.7 6.7 x 1.7	20 6	4	Sc	Sc +	11.6		*12– –9
07091	12 03.5 + 20 53 242.69 + 77.75	04 – 29 – 23 +24 NGC 4098	0135 251	1.1 x 1.1 1.2 x 1.1	-- ...	1	S ... Companion superimposed		14.5		*13 V
07092	12 03.5 + 37 08 164.73 + 76.34	06 – 27 – 004 NGC 4097	1599	1.2 x 0.8 1.2 x 0.8	98: 3	3	S0		14.6		3
07093	12 03.6 + 25 50 219.20 + 79.74	04 – 29 – 025 NGC 4101	0135 256	1.1 x 0.9 (1.4 x 1.0)	(60) ...	2	S0 – a		14.7		3
07094	12 03.6 + 43 14 150.17 + 71.70	07 – 25 – 022	1367 248	1.3 x 0.5 1.2 x 0.4	40 (6)	2	...	S dm:	15.6		1
07095	12 03.6 + 49 52 141.11 + 65.91	08 – 22 – 068 NGC 4100	1338	5.6 x 1.9 5.4 x 1.8	167 6	4	Sb/Sc	SA bc Sb +	11.7		*123 4678
07096	12 03.8 + 53 00 138.09 + 63.06	09 – 20 – 094 NGC 4102	1389 229	3.2 x 1.9 3.3 x 1.9	38 4	2	Sb		11.8	+ 897 + 985	*123 4678
07097	12 04 + 81 12 125.11 + 36.01	14 – 06 – 000	1340	1.7 x 0.4 1.6 x 0.3	82 7	2	Sb – c		16.0:		
07098	12 04.0 + 39 30 157.82 + 74.71	07 – 25 – 023	1367 250	1.6 x 1.0 1.0 x 0.7	... ...	3	... Peculiar, plumes		14.5		*
07099	12 04.1 + 28 27 204.34 + 80.03	05 – 29 – 016 NGC 4104	1398 263	(2.6 x 1.5) (2.8 x 1.7)	35 4:	3	S0		13.7		*13
07100	12 04.2 + 17 59 253.05 + 76.03	03 – 31 – 039	0089	1.9 x 0.35 1.8 x 0.35	8 6	1	S ...		14.7		
07101	12 04.2 + 67 26 129.58 + 49.39	11 – 15 – 023 NGC 4108	0674 255	1.8 x 1.6 1.7 x 1.5	(105) 1	2	Sc Brightest of 3		13.0		*23
07102	12 04.5 + 18 48 250.73 + 76.65	03 – 31 – 040 NGC 4110	0089	1.2 x 0.7 1.3 x 0.7	128 5	2	S B b		14.7		3
07103	12 04.5 + 43 20 149.56 + 71.70	07 – 25 – 026 NGC 4111	1367 248	4.3 x 0.8 4.5 x 0.9	150 7	3	S0		11.4	+ 794 + 841	*12– –9
07104	12 04.6 + 17 11 255.60 + 75.53	03 – 31 – 043	0089	1.2 x 0.2 1. x 0.2	92 6	1	S ... 2nd in group		16.0:		*
07105	12 04.7 + 36 56 164.68 + 76.65	06 – 27 – 012	1599	1.3 x 0.8 1.1 x 0.7	160: 4	1	S B a		15.1		

N G C 4085, 4086, 4088, 4090 – 92, 4096 – 98   I C 2997
       4100 – 02, 4104, 4108, 4110, 4111

1	2	3	4	5		6	7	8	9	10	11	12
07106	12 04.7 + 67 30 129.48 + 49.34	11 - 15 - 025	0674 255	1.6 x 1.4 1.3 x 1.1		(125) 1:	3	S[B c] Singular, NGC 4108 group		14.5		*2C
07107	12 04.8 + 17 32 254.77 + 75.82	03 - 31 - 044	0089	1.4 x 0.25 1.5 x 0.3		109 7	2	Sb		15.5		
07108	12 04.8 + 20 52 243.57 + 77.99	04 - 29 - 000	0135 251	1.0 x 0.1 0.8 x 0.15		11 7	1	S - Irr		16.5:		
07109	12 04.9 + 17 10 255.86 + 75.57	03 - 31 - 045	0089	1.2 x 0.2 1.2 x 0.2		62 6 - 7	1	S ... In group		15.5		*
07110	12 04.9 + 65 41 130.20 + 51.10	11 - 15 - 000	0674 255	1.2: x 1.0: 1.2: x 1.0:		-- ...	3	Dwarf spiral S V		17.		
07111	12 05.1 + 02 58 277.86 + 63.41	01 - 31 - 022 NGC 4116	1611	3.8 x 2.5: 3.5: x 2.3:		155 3 - 4	4	... SB(rs)dm: P w 7116		13.0	+ 1304 + 1175	*12- -789
07112	12 05.2 + 43 24 149.10 + 71.72	07 - 25 - 027 NGC 4117	1367 248	(2.5 x 0.9) (2.1 x 0.9)		18 6	3	S0		14.3		*12 3
07113	12 05.3 + 20 02 247.06 + 77.59	03 - 31 - 000	0089 251	1.0 x 0.35 0.7 x 0.3		62 6	3	Sc		16.0:		
07114	12 05.5 + 10 02ab 270.39 + 69.84	02 - 31 - 33 + 34	1385	[1.0 x 0.5] ...		... ...	2	Double system		(15.6)		*1
07115	12 05.5 + 25 31 221.49 + 80.09	04 - 29 - 031	0135 256	(1.0 x 0.9) (1.1 x 1.1)		-- ...	2	E		14.4		*
07116	12 05.6 + 03 10 277.96 + 63.64	01 - 31 - 023 NGC 4123	1611	5. x 4. 4.0: x 3.2:		135: 2:	5	S B c SB(r)c P w 7111 Sc +		13.1		*12- -789
07117	12 05.6 + 10 39 269.56 + 70.39	02 - 31 - 036 NGC 4124 = 3011*	1385	(4.1 x 1.8) (3.9 x 1.7)		114 6	4	S0		12.7		*123 4678
07118	12 05.6 + 65 27 130.19 + 51.34	11 - 15 - 027 NGC 4125	0674 255	(6.0 x 5.0) (6.0 x 4.5)		95 ...	4	E		10.9	+ 1365 + 1504	*12 3468
07119	12 06.0 + 39 06 157.70 + 75.27	07 - 25 - 28a IC 3014	1367	1.1 x 0.9 0.9 x .0.7		... ...	1	S B ...		14.4		
07120	12 06.0 + 69 03 128.71 + 47.87	12 - 12 - 02a NGC 4128	1411	2.5 x 0.9 2.5 x 0.9		58 6	3	S0		12.7	+ 2395 + 2548	*12 3468
07121	12 06.0 + 69 50 128.42 + 47.11	12 - 12 - 001 NGC 4120	1411	1.8 x 0.4 1.8 x 0.4		166 (6)	1	S - Irr		14.1		23
07122	12 06.0 + 77 05 126.12 + 40.06	13 - 09 - 012 NGC 4127	1339	3.2 x 1.6 2.7: x 1.5:		140 5	3	Sc or Sb/Sc		13.5		*3
07123	12 06.1 + 16 25 258.68 + 75.20	03 - 31 - 047 NGC 4126	0089	1.0 x 0.7 1.0 x 0.7		0 3:	2	S0		14.6		13
07124	12 06.1 + 25 14 223.28 + 80.15	04 - 29 - 036	0135 256	1.0 x 0.35 1.0 x 0.35		158 6	1	S0		15.2		*
07125	12 06.1 + 37 05 163.37 + 76.75	06 - 27 - 014	1599	4.7 x 0.8 2.5: x 0.8:		85 6	3	... S m		14.7		
07126	12 06.2 + 29 35 197.41 + 80.35	05 - 29 - 019 NGC 4131	1398 263	1.5 x 0.8 1.4 x 0.7		73 5	1	S ... NGC 4134 group		14.1		*13
07127	12 06.2 + 75 12 126.65 + 41.90	13 - 09 - 013 NGC 4133	1339 269	2.0 x 1.6 2.0 x 1.6		(125) 2	3	Sb/ S B b:		13.1		3
07128	12 06.3 + 62 39 131.34 + 54.06	11 - 15 - 000	0674	1.2: x 0.7: 0.7: x 0.5:		... ...	3	Dwarf V		17.		
07129	12 06.4 + 42 01 151.00 + 72.99	07 - 25 - 029	1367 248	1.3 x 0.8 1.2 x 0.9		75 4	1	Sa - b		14.3		*1
07130	12 06.6 + 29 27 198.10 + 80.46	05 - 29 - 023 NGC 4134	1398 263	2.3 x 0.8 2.1 x 0.8		150 6	3	Sb Brightest in group		13.8		*13
07131	12 06.6 + 31 11 188.20 + 80.02	05 - 29 - 021	1398 263	1.8 x 0.5 1.8 x 0.5		25 6	2	... d - m		15.7		
07132	12 06.6 + 31 51 184.66 + 79.77	05 - 29 - 022	1398 263	(1.2 x 1.2) (1.3 x 1.3)		-- ...	3	E		14.4		
07133	12 06.8 + 19 16 250.81 + 77.36	03 - 31 - 048	0089	1.6 x 1.0 1.6 x 1.0		170: 4	4	Sc SAB d NGC 4155 group		15.5		
07134	12 06.8 + 30 12 193.65 + 80.34	05 - 29 - 025 NGC 4136	1398 263	4.2 x 4.2 4.0: x 4.0:		-- 1	4	S B c SAB(r)c		12.1	+ 445 + 434	*12 3478
07135	12 06.8 + 44 21 146.78 + 71.05	07 - 25 - 033 NGC 4137	1367	[1.1 x 0.8] ...		[100] ...	2	S ... + companion Bridge		15.0		*12
07136	12 06.9 + 14 16 264.14 + 73.61	02 - 31 - 038 IC 3019	1385	(1.8 x 1.4) (1.8 x 1.6)		-- ...	2	Dwarf elliptical		15.0		
07137	12 06.9 + 22 23 238.42 + 79.22	04 - 29 - 39 + 40	0135 253	[1.8 x 0.55] ...		... ...	2	Double system Bridge		15.7		*1
07138	12 06.9 + 26 30 216.07 + 80.58	05 - 29 - 026	1398 256	1.8 x 1.2 1.8 x 1.2		85 3:	3	Sc Sc +		15.3		*1
07139	12 07.0 + 43 57 147.31 + 71.42	07 - 25 - 035 NGC 4138	1367	(2.9 x 1.7) (3.0 x 1.8)		150 4	3	S0		12.1	+ 1039 + 1090	*123 4678
07140	12 07.0 + 53 24 136.78 + 62.86	09 - 20 - 102 NGC 4142	1389 229	2.2 x 1.1 2.2 x 1.1		175 5	3	S B c Sc +		14.3		23

N G C   4116,   4117,   4120,   4123 - 28,   4131   I C   3011, 3014, 3019
        4133,   4134,   4136 - 38,   4142

1	2	3	4	5	6	7	8	9	10	11	12
07141	12 07.1 + 23 34 232.75 + 79.80	04 – 29 – 041	0135 253	1.7 x 0.35 1.7 x 0.3	146 6	1	S ...		15.7		
07142	12 07.1 + 42 48 149.19 + 72.41	07 – 25 – 036 NGC 4143	1367 248	(2.9 x 1.8) (2.7 x 1.8)	144 4	4	S0		12.0	+ 784 + 830	123 4678
07143	12 07.2 + 25 19 223.18 + 80.41	04 – 29 – 042	0135 256	1.0 x 0.45 1.0 x 0.4	107 ...	1	...		14.6		*
07144	12 07.2 + 56 48 134.34 + 59.67	10 – 17 – 153	0723 257	1.0 x 0.8 1.0 x 0.8	(60) 2	3	Sb		14.3		
07145	12 07.3 + 38 30 158.51 + 75.89	07 – 25 – 037	1367	1.5 x 0.5 1.4 x 0.5	152 6	3	Sc		15.4		
07146	12 07.3 + 43 15 148.31 + 72.05	07 – 25 – 000	1367	1.5: x 1.0: 1.0: x 0.6:	... ...	3	Dwarf V		16.5:		
07147	12 07.3 + 59 07 132.95 + 57.47	10 – 17 – 152 NGC 4141	0723	1.4 x 0.9 1.2: x 0.9	(75) ...	3	S B c		14.5		*13
07148	12 07.4 + 01 13 280.78 + 61.78	00 – 31 – 000	1401	1.3 x 0.25 0.8 x 0.2	57 6	1	S B ...		15.4		
07149	12 07.4 + 13 18 266.37 + 72.88	02 – 31 – 000 IC 3021	1385	1.8: x 1.0: 1.3: x 0.9:	0: ...	3	...	S m	15.4		
07150	12 07.4 + 14 39 263.70 + 73.99	03 – 31 – 051 IC 3023	0089	1.2 x 0.8: 1.0: x 0.6:	140: ...	3	Dwarf		16.0:		
07151	12 07.4 + 46 43 143.22 + 69.02	08 – 22 – 077 NGC 4144	1338	6.9 x 1.7 6.7 x 1.7	104 6	4	Sc	Sc +	12.3		*123 4678
07152	12 07.4 + 59 43 132.60 + 56.90	10 – 17 – 154	0723 259	1.2 x 0.3 1.0 x 0.3	146 7:	1	S ...		15.4		*
07153	12 07.4 + 62 33 131.18 + 54.19	11 – 15 – 031	0674	2.1 x 0.3 1.9 x 0.3	166 7	3	Sc		15.1		
07154	12 07.5 + 40 10 154.26 + 74.62	07 – 25 – 040 NGC 4145	1367	6.5 x 4.3 6.0: x 4.0:	100 3 – 4	4	Sc P w 7166	SAB(rs)d	12.2		*12 3478
07155	12 07.5 + 49 33 140.01 + 66.46	08 – 22 – 079	1338	1.1 x 0.5 0.9 x 0.5	40: 5 – 6	2	Sc nf of 2		16.0:		*
07156	12 07.5 + 78 02 125.76 + 39.14	13 – 09 – 000	1339	1.3 x 0.12 1.2 x 0.12	165 7	3	Sc		16.5:		
07157	12 07.6 + 25 35 221.75 + 80.56	04 – 29 – 043	0135 256	1.1 x 0.5 1.2 x 0.6	79 5	1	S0		14.5		
07158	12 07.6 + 36 09 165.53 + 77.61	06 – 27 – 018 NGC 4148	1599 258	(1.6 x 0.9) (1.8 x 1.1)	165: 4:	3	S0	S 0 +	14.6	+ 4750	*3
07159	12 07.6 + 39 20 156.17 + 75.29	07 – 25 – 042	1367	1.2: x 1.1: 0.9: x 0.8:	–– 1	3	Sc		15.6		
07160	12 07.6 + 70 42 127.93 + 46.30	12 – 12 – 000	1411	1.0:: x 1.0:: 1. x 1.	–– ...	3	Dwarf		17.		*
07161	12 07.7 + 12 35 267.85 + 72.31	02 – 31 – 039 IC 3024	1385	1.2 x 0.35 0.9 x 0.3	167 (6)	1	S? ...		15.1		
07162	12 07.7 + 24 10 229.87 + 80.16	04 – 29 – 000	0135	1.4 x 0.2 1.3 x 0.2	124 7	1	S ...		15.7		*
07163	12 07.8 + 26 42 215.00 + 80.80	05 – 29 – 028 NGC 4146	1398	1.7 x 1.6 1.6 x 1.6	–– 1	4	S B a	SAB(s)a	13.8		*12 3
07164	12 07.9 + 70 48 127.86 + 46.21	12 – 12 – 000	1411	1.8: x 1.5: 1.5:: x 1.5::	(135) ...	4	Dwarf spiral S(B)V		17.		
07165	12 08.0 + 30 40 190.53 + 80.47	05 – 29 – 029 NGC 4150	1398 263	(2.1 x 1.5) (2.1 x 1.5)	147 3	3	S0		12.6	+ 244 + 236	*123 478
07166	12 08.0 + 39 41 155.09 + 75.06	07 – 25 – 044 NGC 4151	1367	7. x 6. 4.0 x 3.0	–– 1	4	Sa/S B b Seyfert, p w 7154	SAB(rs)ab	11.2	+ 957 + 989	*123 4678
07167	12 08.0 + 58 35 133.08 + 58.01	10 – 17 – 155 NGC 4149	0723 257	1.4 x 0.25 1.8 x 0.3	87 7	1	S ...		13.9		*3
07168	12 08.0 + 70 40 127.90 + 46.34	12 – 12 – 000	1411	1.3 x 0.25 1.1 x 0.2	0 ...	1	... Peculiar?		15.4		*
07169	12 08.1 + 16 18 260.43 + 75.41	03 – 31 – 052 NGC 4152	0089	2.2 x 1.9 2.2 x 1.9	115: 1 – 2	4	Sc		12.5		*123 4678
07170	12 08.1 + 19 06 252.33 + 77.48	03 – 31 – 055	0089	3.0 x 0.2 2.8 x 0.20	12 7	3	Sc NGC 4155 group		15.2		*
07171	12 08.2 + 13 36 266.37 + 73.24	02 – 31 – 041 IC 3029	1385	1.5 x 0.35 1.5 x 0.35	28 6	1	S B: ... n of 2, disturbed?		15.0		*1
07172	12 08.2 + 19 19 251.67 + 77.64	03 – 31 – 058 NGC 4155	0089	(1.1 x 1.0) (1.3 x 1.1)	–– ...	3	E Brightest in group		14.7		*1
07173	12 08.3 + 39 45 154.75 + 75.04	07 – 25 – 045 NGC 4156	1367	1.4 x 1.3 1.4 x 1.2	–– 1:	4	S B b	S B(rs)b Sb +	14.3		*12 3
07174	12 08.3 + 76 25 126.14 + 40.73	13 – 09 – 015 NGC 4159	1339 269	1.5 x 0.6 1.5 x 0.6	35 6	1	S ...		14.3		3
07175	12 08.4 + 40 02b 154.04 + 74.83	07 – 25 – 046	1367	2.1 x 0.5 1.7: x 0.5:	75 ...	2	S – Irr nnf of 2, bridge:		(15.3)		*1

N G C   4141, 4143 – 46, 4148 – 52, 4155   I C   3021, 3023, 3024, 3029
4156, 4159

1	2	3	4	5	6	7	8	9	10	11	12
07176	12 08.4 + 50 36 138.68 + 65.55	09 - 20 - 000	1389 229	1.6 x 0.5 1.4: x 0.5:	[80] ...	3	Dwarf V		17.		*
07177	12 08.5 + 01 15 280.93 + 62.11	00 - 31 - 000	1401	(1.0 x 1.0) (1.0 x 1.0)	-- ...	2	... Very compact		15.1		
07178	12 08.5 + 02 18 280.16 + 63.09	00 - 31 - 036	1401	1.5: x 1.0: 1. x 1.	-- ...	4	Dwarf irregular Ir V		17.		*
07179	12 08.5 + 64 12 130.24 + 52.63	11 - 15 - 033	0674 255	1.9 x 0.3 1.9 x 0.3	135 7	1	S ...		14.0		*
07180	12 08.5 + 67 12 129.05 + 49.72	11 - 15 - 034	0674 255	1.1 x 0.25 1.1 x 0.25	103 6	1	S ...		15.0		
07181	12 08.6 + 13 52 266.14 + 73.54	02 - 31 - 043	1385	1.2 x 0.8 1.2 x 0.7:	2: ...	1	...	S dm?	15.1		
07182	12 08.6 + 20 27 247.83 + 78.45	03 - 31 - 060 NGC 4158	0089	1.7 x 1.6 (1.7 x 1.5)	-- 1	1	S ...		13.1		*123 468
07183	12 08.6 + 50 47 138.44 + 65.39	09 - 20 - 106 NGC 4157	1389 229	7.7 x 1.3 6.7 x 1.2	66 6	3	Sb	Sb +	11.9		*123 4678
07184	12 08.8 + 01 46 280.71 + 62.62	00 - 31 - 037	1401	1.7 x 0.6 1.5 x 0.5	148 6	3	S B c	Sc +	14.9		
07185	12 08.8 + 03 12 279.61 + 63.96	01 - 31 - 024	1611	1.3: x 1.3: 1. x 1.	-- ...	4	Dwarf V		15.5		*
07186	12 08.8 + 18 18 255.43 + 77.03	03 - 31 - 062	0089	1.7 x 0.3 1.9: x 0.3	83 6	3	...	dm - m	15.6		*
07187	12 08.8 + 36 07 164.87 + 77.81	06 - 27 - 023	1599 258	1.0 x 0.5 1.0 x 0.5	108 5	1	S B a:		15.0		*1
07188	12 08.8 + 39 41 154.61 + 75.16	07 - 25 - 048	1367	1.1 x 1.0 0.9: x 0.8:	-- 1:	3	...	S B m	16.5:		
07189	12 08.8 + 75 06 126.46 + 42.03	13 - 09 - 016	1339 269	2.0 x 1.0 2.0: x 1.0	162 5	2	...	S B dm:	14.8		
07190	12 09.0 + 29 22 198.05 + 80.99	05 - 29 - 030	1398 263	1.1 x 0.9 (1.2 x 1.0)	-- ...	1	SO?		15.1		
07191	12 09.0 + 58 00 133.17 + 58.60	10 - 18 - 002 NGC 4161	1427 257	1.2 x 0.8 1.2 x 0.8	50: 3	1	S ...		13.7		3
07192	12 09.3 + 12 25 269.24 + 72.37	02 - 31 - 044 IC 768	1385	1.6 x 0.8 1.4 x 0.7	115 5	2	Sc P w 7209		14.8		*
07193	12 09.3 + 24 24 229.36 + 80.59	04 - 29 - 046 NGC 4162	0135 260	2.5 x 1.4 2.3 x 1.4	174 4 - 5	3	Sc	Sc -	12.6	+ 2546 + 2510	*123 4678
07194	12 09.4 + 16 30 260.91 + 75.76	03 - 31 - 065	0089	1.1 x 1.1 1.1 x 1.1	-- 1	3	Sc		15.2		
07195	12 09.4 + 68 12 128.56 + 48.77	11 - 15 - 035	0674	1.0 x 0.5 1.0 x 0.5	105 5:	1	S B ...		15.4		
07196	12 09.5 + 15 41 262.96 + 75.13	03 - 31 - 066	0089	1.4 x 0.3 1.5 x 0.35	57 6 - 7	1	S ...		15.2		1
07197	12 09.5 + 61 37 131.20 + 55.15	10 - 18 - 005	1427 261	1.0 x 0.25 0.8 x 0.25	105 6	1	S ...		16.0:		*
07198	12 09.6 + 18 02 256.86 + 76.96	03 - 31 - 068 NGC 4166	0089	1.1 x 0.9 1.1 x 1.0	20: 2:	3	S B 0		14.3		
07199	12 09.6 + 36 27 163.21 + 77.69	06 - 27 - 026 NGC 4163	1599 258	2.1: x 2.0: 2.2: x 2.0:	-- ...	4	Irr Ir I	IA m	13.7		123
07200	12 09.7 + 12 45 268.96 + 72.71	02 - 31 - 000	1385	1.8:: x 1.5:: 1.4:: x 1.2::	-- ...	3	...	S m	15.3		
07201	12 09.7 + 13 31a 267.59 + 73.36	02 - 31 - 045 NGC 4165 = 3035*	1385	1.2 x 0.8 1.3 x 0.8	160 ...	2	Sb s of 2, p w 7203		(14.8)		*12 359
07202	12 09.7 + 29 27 197.34 + 81.12	05 - 29 - 032 NGC 4169	1398 263	1.8 x 0.9 2.0 x 1.1	153 5:	3	SO Brightest in group		12.9		*13
07203	12 09.8 + 13 29 267.72 + 73.35	02 - 31 - 046 NGC 4168	1385	(2.8 x 2.5) (3.0 x 2.7)	-- ...	3	E P w 7201		12.7		*12- -5789
07204	12 09.8 + 29 29 197.10 + 81.14	05 - 29 - 033 NGC 4173 = 70 = 71	1398 263	5.1 x 0.8 4.1 x 0.6	134 6	2	... S B: d - dm NGC 4169 group		13.7		*13
07205	12 09.8 + 56 27 133.88 + 60.12	09 - 20 - 109 NGC 4172	1389	1.4 x 1.3 1.3 x 1.3	-- 1	1	S ...		14.4		*3
07206	12 09.9 + 29 25 197.50 + 81.17	05 - 29 - 034 NGC 4174	1398 263	0.8 x 0.3 0.7 x 0.25	50 ...	1	... NGC 4169 group		14.3		*13
07207	12 09.9 + 37 17 160.36 + 77.14	06 - 27 - 027	1599 258	2.5: x 1.5: 1.5:: x 1.5::	-- ...	4	Dwarf irregular Ir V		17.		
07208	12 09.9 + 39 23 154.66 + 75.52	07 - 25 - 049	1367	1.8: x 1.5: 0.8 x 0.8	-- ...	1	... Compact nucleus		14.6		*1C
07209	12 10.0 + 12 24 269.76 + 72.44	02 - 31 - 047 IC 769	1385	2.5 x 1.7 2.5 x 1.7	43 3	3	Sb/Sc P w 7192	SA(rs)bc	14.1		*
07210	12 10.0 + 15 33 263.65 + 75.09	03 - 31 - 072	0089	1.0 x 0.3 0.8 x 0.25	144 6 - 7	1	S ...		15.2		

N G C   4157, 4158, 4161 - 63, 4165, 4166   I C   768, 769, 3035
       4168 - 74

1	2	3	4	5	6	7	8	9	10	11	12
07211	12 10.0 + 29 26 197.37 + 81.19	05 - 29 - 036 NGC 4175	1398 263	2.1 x 0.4 1.9 x 0.35	130 7	1	S ... NGC 4169 group		14.2		*13
07212	12 10.1 + 34 59 168.21 + 78.75	06 - 27 - 029	1599 238	1.3 x 0.1 1.3 x 0.1	43 7	2	Sc s of 2		(15.4)		*1V
07213	12 10.1 + 40 50 151.30 + 74.36	07 - 25 - 050	1367	1.4 x 1.2 1.1 x 1.1	--- 1	4	S B c	Sc +	15.5		
07214	12 10.3 + 01 35 281.61 + 62.57	00 - 31 - 038 NGC 4179	1401	3.8 x 1.0 4.0 x 1.1	143 7	4	S0		12.8	+ 1279 + 1148	123 468
07215	12 10.3 + 11 08 271.93 + 71.37	02 - 31 - 050 NGC 4178	1385	5.5 x 1.7 5.7 x 1.8	30 6	4	S B c	Sc +	12.9	+ 233 + 140	*12- -9
07216	12 10.3 + 14 15 266.63 + 74.06	02 - 31 - 051 IC 3044	1385	2.2 x 0.8 2.2 x 0.8	68 6	1	S ...		14.7		*2
07217	12 10.4 + 25 34 222.80 + 81.17	04 - 29 - 000	0135	1.7 x 0.2 1.7 x 0.2	1 7	1	S ...		15.6		
07218	12 10.4 + 52 33 136.37 + 63.84	09 - 20 - 113	1389 229	1.4 x 0.7 1.4 x 0.7	[170] ...	3	Irr or Peculiar		15.0		*1
07219	12 10.5 + 07 19 276.80 + 67.94	01 - 31 - 025 NGC 4180	1611	1.7 x 0.5 1.5 x 0.5	22 6	1	S ...		13.2		23
07220	12 10.6 + 13 12 268.81 + 73.20	02 - 31 - 052 IC 3046	1385	1.2 x 0.25 1.2 x 0.2	130 6 - 7	2	S ... ssf of 2, bridge		15.1		*1
07221	12 10.7 + 29 07 199.25 + 81.40	05 - 29 - 037	1398 263	1.5 x 0.5 1.1 x 0.4	65 6	1	S ...		15.0		
07222	12 10.7 + 43 58 145.45 + 71.73	07 - 25 - 051 NGC 4183	1367	5.5 x 0.6 5. x 0.6	166 6 - 7	3	Sc	Sc +	13.5		*12 3
07223	12 10.8 + 15 03 265.36 + 74.79	03 - 31 - 076 NGC 4186	0089	1.4: x 1.1: 1.0: x 0.9:	--- ...	3	...	SA dm	16.0:		*12
07224	12 10.8 + 21 55 243.22 + 79.74	04 - 29 - 047	0135	(1.0 x 0.8) (1.2 x 1.0)	--- ...	2	E		14.6		*1
07225	12 10.8 + 28 47 201.44 + 81.46	05 - 29 - 038 NGC 4185	1398 263	2.9 x 2.2 2.6 x 1.7	165 2 - 3	3	Sb	Sb +	13.5		3
07226	12 10.9 + 75 20 126.22 + 41.83	13 - 09 - 018	1339 269	1.9 x 1.6 1.9: x 1.6:	(15) ...	4	... S m S IV - V		15.4		*
07227	12 11.0 + 14 46 266.11 + 74.58	03 - 31 - 000 IC 3049	0089	[1.1: x 0.8:] ...	... ...	1	Double system		15.7		
07228	12 11.0 + 46 45 141.62 + 69.26	08 - 22 - 085	1338	1.3 x 0.3 1.3 x 0.3	178 ...	1	...		15.1		
07229	12 11.0 + 51 02 137.38 + 65.30	09 - 20 - 117 NGC 4187	1389 229	(1.2 x 0.8) (1.5 x 0.9)	145: ...	3	E		14.5		*13
07230	12 11.1 + 16 24 262.48 + 75.93	03 - 31 - 077	0089	[1.6 x 0.6] ...	... ...	3	Multiple system Strongly disrupted		14.8		*1V A
07231	12 11.2 + 15 10 265.41 + 74.94	03 - 31 - 079 NGC 4192 = M 98	0089	9.9 x 2.2 10.2 x 2.2	155 6	4	Sb	SAB b	11.0	- 124 - 199	*12- -9
07232	12 11.2 + 36 55 160.62 + 77.58	06 - 27 - 030 NGC 4190	1599 258	1.8 x 1.7 1.7 x 1.7	--- ...	4	Irr I m Ir I		13.5		*12 348V
07233	12 11.3 + 07 28 277.13 + 68.16	01 - 31 - 026 NGC 4191	1611	1.2 x 0.9 (1.3 x 1.1)	(5) 2 - 3	2	S0		13.9		3
07234	12 11.3 + 13 27 268.88 + 73.51	02 - 31 - 053 NGC 4193 = 3051*	1385	2.3 x 1.1 2.3 x 1.1	93 5	2	Sb		13.4		*12 359
07235	12 11.3 + 13 42 268.42 + 73.72	02 - 31 - 054 NGC 4189 = 3050*	1385	2.7 x 2.2 2.6 x 2.0	85: 2:	4	Sc	Sc -	12.7		*12- -5789
07236	12 11.3 + 24 33 229.48 + 81.06	04 - 29 - 000	0135 260	1.2 x 1.0 1. x 1.	--- ...	3	Dwarf V		16.5:		
07237	12 11.5 + 18 11 257.92 + 77.38	03 - 31 - 080	0089	1.0 x 0.7 1.0 x 0.7	50: ...	3	... S m		16.0:		
07238	12 11.5 + 74 47 126.32 + 42.37	13 - 09 - 019	1339 269	1.4 x 0.15 1.4 x 0.20	47 7	3	Sc		16.0:		
07239	12 11.6 + 08 03 276.69 + 68.72	01 - 31 - 028	1611	2.5: x 2.3: 2. x 2.	--- ...	4	Dwarf V		15.0		
07240	12 11.6 + 15 00 266.08 + 74.86	03 - 31 - 081	0089	1.0 x 0.8 1.0 x 0.8	60 2:	2	Sa		14.9		*1
07241	12 11.6 + 54 48 134.42 + 61.77	09 - 20 - 119 NGC 4194, Mark 201	1389	2.3 x 1.6 2.3 x 1.5	... ...	3	... Peculiar, large plume		13.0	+ 2585 + 2684	*12 3CVA
07242	12 11.7 + 66 22 128.87 + 50.61	11 - 15 - 037	0674 255	1.7? x 0.8 ... x 0.8	172 5:	3	Sc		15.0		*
07243	12 11.8 + 32 42 177.37 + 80.37	06 - 27 - 036	1599	1.0 x 0.8 1.0 x 0.8	(25) 2:	2	S0 Compact Brightest of 3		14.6		*1
07244	12 11.8 + 59 53 131.54 + 56.90	10 - 18 - 010 NGC 4195	1427 259	1.7 x 1.5 1.7 x 1.5	--- 1	4	S B c	S B(s)cd	15.5		*3
07245	12 11.9 + 28 42 201.81 + 81.71	05 - 29 - 040 NGC 4196	1398 263	(1.5 x 1.2) (2.0 x 1.5)	60: ...	1	S0? Peculiar?		13.7		3

N G C 4175, 4178 - 80, 4183, 4185 - 87   I C 3044, 3046, 3049 - 51
4189 - 96

1	2	3	4	5	6	7	8	9	10	11	12
07246	12 11.9 + 56 17 133.42 + 60.36	09 – 20 – 123 NGC 4198	1398	1.1 x 0.6 1.0 x 0.6	130 4 – 5	2	S0 – a		14.6		3
07247	12 12.0 + 06 05 278.91 + 66.94	01 – 31 – 029 NGC 4197	1611	3.8 x 0.6 3.5 x 0.7	36 6	3	Sc	Sc +	13.8	+ 2039 + 1927	*12 3$^-$
07248	12 12.0 + 24 35 229.64 + 81.22	04 – 29 – 049	0135 260	1.0 x 0.5 1.0 x 0.5	12 5:	2	S B b		15.6		
07249	12 12.1 + 13 05 270.11 + 73.29	02 – 31 – 056	1385	1.4 x 0.5 1.3 x 0.5	[60] ...	4	Irr	I m Ir I	15.3		*1
07250	12 12.2 + 01 01 282.97 + 62.18	00 – 31 – 000	1401	1.0 x 0.2 0.9 x 0.2	51 6 – 7	1	S ... sf of 2		15.3		*
07251	12 12.2 + 12 27 271.24 + 72.75	02 – 31 – 057 NGC 4200	1385	(1.5 x 0.8) (1.6 x 0.9)	98 5	4	S0		14.1		3
07252	12 12.2 + 35 48 163.68 + 78.52	06 – 27 – 000	1599 258	1.2 x 0.1 1.0: x 0.1	144 7	1	Sc		16.5:		
07253	12 12.4 + 60 10 131.27 + 56.64	10 – 18 – 011 NGC 4199	1427	[1.1 x 0.6] ...	... ...	2	Double system Common envelope		15.5		*1V
07254	12 12.5 + 13 44 269.26 + 73.90	02 – 31 – 062 IC 3059	1385	2.0: x 1.6: 2.0: x 1.5:	(0) ...	4	Dwarf V		15.3		
07255	12 12.5 + 14 19 268.16 + 74.40	02 – 31 – 063 IC 3061	1385	2.4 x 0.4 2.0 x 0.35	122 6	1	Sc P w 7275		14.9		*12 5
07256	12 12.5 + 33 28 173.12 + 80.07	06 – 27 – 040 NGC 4203	1599	(3.5 x 3.3) (3.7 x 3.4)	(10) 1	4	S0		11.8	+ 1001 + 1008	*123 4678
07257	12 12.5 + 36 14 161.95 + 78.26	06 – 27 – 039	1599	1.6 x 1.0 1.5 x 0.9	160 ...	2	S – Irr ssf of 2		14.2	+ 3980	*1
07258	12 12.5 + 64 04 129.60 + 52.86	11 – 15 – 038 NGC 4205	0674 255	2.1 x 0.8 1.9 x 0.7	28 (6)	1	S: ...		13.8		*3
07259	12 12.6 + 12 17 271.80 + 72.65	02 – 31 – 064 IC 3063	1385	1.0 x 0.5 1.0 x 0.5	20 5	3	Sa		14.9		
07260	12 12.7 + 13 18 270.18 + 73.55	02 – 31 – 066 NGC 4206 = 3064*	1385	5.8 x 0.9 5.1 x 0.9	0 6	3	Sc	Sc –	13.8		*12 359
07261	12 12.7 + 20 56 249.05 + 79.50	04 – 29 – 051 NGC 4204	0135	4.7: x 4.7: 4.5: x 4.5:	-- 1	4	...	S B(s)dm	14.3		3
07262	12 12.8 + 13 45 269.45 + 73.95	02 – 31 – 068 IC 3066	1385	1.0 x 0.2 0.9 x 0.25	138 6	1	S ...		15.2		*
07263	12 12.8 + 19 34 254.45 + 78.60	03 – 31 – 083	0089	1.1 x 0.9 1.0 x 0.8	130: 2	1	S B? b – c		14.9		
07264	12 12.8 + 66 15 128.75 + 50.75	11 – 15 – 039 NGC 4210	0674 255	2.1 x 1.5 2.1 x 1.5	105: ...	4	S B b	S B(r)b Sb +	13.4		*12 36$^-$
07265	12 12.8 + 76 32 125.77 + 40.67	13 – 09 – 020	1339 269	1.2 x 0.7 1.2 x 0.7	100: ...	2	...	S dm	16.5:		
07266	12 12.9 + 24 22 231.47 + 81.34	04 – 29 – 052	0135 260	(1.2 x 0.6) 1.3 x 0.6	40 5	1	S0		15.6		*
07267	12 12.9 + 51 39 136.22 + 64.82	09 – 20 – 128	1389 229	2.0 x 0.8 1.9 x 0.8	53 ...	3	...	S: dm – m	14.6		1
07268	12 13.0 + 09 52 275.47 + 70.52	02 – 31 – 069 NGC 4207	1385	1.6 x 0.8 1.6 x 0.8	[124] ...	1	...		13.7		*3
07269	12 13.0 + 15 17 266.57 + 75.28	03 – 31 – 000	0089	1.0: x 0.8: 1.0: x 1.0:	-- ...	3	Dwarf V		17.		
07270	12 13.0 + 22 06 244.01 + 80.26	04 – 29 – 053	0135	1.7 x 0.4 1.2 x 0.4	162 6	1	S ... s of 2		15.3		*1
07271	12 13.0 + 43 42 144.66 + 72.16	07 – 25 – 052	1367	2.3 x 0.8 2.4 x 0.8	160 6	2	...	S B d – dm	15.4		
07272	12 13.0 + 52 13 135.76 + 64.28	09 – 20 – 129	1389 229	1.0 x 0.12 1.0 x 0.1	63 7	3	Sc		16.0:		
07273	12 13.1 + 08 25 277.23 + 69.21	01 – 31 – 000	1611	1.0 x 0.3 0.9 x 0.2	87 6	1	S ...		15.3		
07274	12 13.1 + 13 53 269.44 + 74.10	02 – 31 – (65) IC 3073	1385	1.0: x 0.8: 0.9: x 0.6:	... ...	3	Dwarf IV – V or V		15.6		*
07275	12 13.1 + 14 11 268.87 + 74.36	02 – 31 – 070 NGC 4212	1385	2.9 x 1.9 3.0 x 2.0	75 3 – 4	3	Sc P w 7255	Sc –	11.9	+ 2125 + 2047	*12– –57$\overline{8}$9
07276	12 13.1 + 24 15 232.30 + 81.33	04 – 29 – 054 NGC 4213	0135 260	(1.7 x 1.7) (1.7 x 1.7)	-- ...	3	E Brightest in cluster		14.3		*13
07277	12 13.1 + 28 27 203.37 + 82.00	05 – 29 – 42 + 43 NGC 4211	1398 263	[2.0 x 1.0] ...	... ...	3	Double system Bridge, plume		14.4	+6571, 6618 +6556, 6603	*1 2VA
07278	12 13.1 + 36 37 160.25 + 78.06	06 – 27 – 042 NGC 4214	1599	11. x 9. 10. x 8.	... ...	5	Irr	IAB m Ir I	10.3	+ 290 + 311	*12– –9$^-$
07279	12 13.2 + 10 58 274.16 + 71.54	02 – 31 – 071 IC 3074	1385	2.2 x 0.2 2.2 x 0.2	160 7	3	...	dm – m	15.1		
07280	12 13.3 + 00 40 283.75 + 61.93	00 – 31 – 043	1401	1.3 x 0.7 1.1 x 0.6	160: 5:	1	S ...		14.8		

N G C   4197 – 4200, 4203 – 07, 4210 – 14

I C   3059,   3061,   3063,   3064,   3066
3073,   3074

1	2	3	4	5	6	7	8	9	10	11	12
07281	12 13.3 + 06 40 279.15 + 67.60	01 - 31 - 031 NGC 4215	1611	1.6 x 0.6 1.7 x 0.6	174 7	2	S0 - a		13.0		*123 4$\underline{7}$8
07282	12 13.3 + 47 21 139.95 + 68.86	08 - 22 - 087 NGC 4217	1338	5.5 x 1.6 5.5 x 1.7	50 7	3	Sb		12.4		*123 4678
07283	12 13.3 + 48 23 138.90 + 67.90	08 - 22 - 088 NGC 4218	1338	1.0 x 0.55 0.9 x 0.6	142 ...	1	...		13.2		*$\underline{2}$3
07284	12 13.4 + 13 26 270.47 + 73.75	02 - 31 - 072 NGC 4216	1385	8.5 x 1.7 8.3 x 1.7	19 6	4	Sb		11.2	+ 38 - 43	*12- -689
07285	12 13.4 + 14 42 268.10 + 74.84	03 - 31 - 085	0089	1.4: x 0.7 1.0 x 0.7	0: ...	2	Sc	Sc +	15.4		1
07286	12 13.4 + 27 43 208.64 + 82.11	05 - 29 - 044	1398 263	1.5 x 0.5 1.5 x 0.4	108 6	2	Sa - b		15.3		1
07287	12 13.6 + 28 24 203.65 + 82.12	05 - 29 - 046	1398 263	1.1 x 0.7 1.1 x 0.7	75: 3 - 4	3	S B c		15.5		
07288	12 13.6 + 66 30 128.54 + 50.52	11 - 15 - 040 NGC 4221	0674 255	1.9 x 1.3 1.8 x 1.3	25 3:	3	S B 0		13.6		*$\underline{2}$3
07289	12 13.7 + 29 02 199.07 + 82.06	05 - 29 - 047	1398 263	1.5 x 0.25 1.5 x 0.2	9 7	1	S ...		15.5		
07290	12 13.7 + 48 09 138.96 + 68.14	08 - 22 - 089 NGC 4220	1338	3.8 x 1.4 3.9 x 1.4	141 6	3	Sa		12.4	+ 979 + 1051	*123 46$\underline{7}$8
07291	12 13.9 + 13 35 270.59 + 73.94	02 - 31 - 075 NGC 4222 = 3087*	1385	3.1 x 0.4 2.9 x 0.4	56 7	3	Sc		14.6		*1$\underline{2}$ 3
07292	12 14.0 + 07 44 278.53 + 68.66	01 - 31 - 034 NGC 4224	1611	2.0 x 0.9 2.3 x 1.1	57 6	4	Sa		13.3		*12- 5$\underline{7}$89
07293	12 14.0 + 17 45 261.29 + 77.42	03 - 31 - 089	0089	1.0 x 0.35 1.1 x 0.3	92 6	1	S ...		15.4		*1
07294	12 14.0 + 30 07 191.43 + 81.87	05 - 29 - 000	1398 263	1.1 x 0.1 1.0 x 0.1	53 7	2	Sc		16.0:		
07295	12 14.0 + 33 42 170.94 + 80.19	06 - 27 - 000	1599	1.0 x 0.25 1.0 x 0.25	1 6	2	...	S d - dm	16.5:		*
07296	12 14.0 + 33 48 170.48 + 80.12	06 - 27 - 043 NGC 4227	1599	1.6 x 1.0 1.6 x 1.0	70: 4:	2	S0 - a P w 7299	SAB: ...	13.8	+ 4820	*13
07297	12 14.0 + 47 17 139.71 + 68.97	08 - 22 - 090 NGC 4226	1338	1.1 x 0.5 1.2 x 0.5	127 (6)	1	S: ... Disturbed?		14.4		*12 3
07298	12 14.0 + 52 32 135.20 + 64.03	09 - 20 - 137	1389 229	1.2: x 0.7: 1.0: x 0.6:	[140] ...	3	Dwarf V		17.		
07299	12 14.1 + 33 50 170.25 + 80.12	06 - 27 - 044 NGC 4229	1599	1.3 x 0.9 1.4 x 0.8	3: 3 - 4	1	S ...		14.3		*13
07300	12 14.2 + 29 00 199.18 + 82.17	05 - 29 - 048	1398 263	1.7: x 1.7: 1.6: x 1.6:	-- ...	4	Dwarf irregular Ir V		17.		*
07301	12 14.2 + 46 21 140.64 + 69.84	08 - 22 - 092	1338	2.1 x 0.3 2.1 x 0.4	82 7	3	Sc	Sc +	15.7		
07302	12 14.3 + 30 33 188.39 + 81.79	05 - 29 - 000	1398 263	1.0: x 1.0: 1. x 1.	-- ...	3	Dwarf V		17.		
07303	12 14.3 + 47 42 139.15 + 68.60	08 - 22 - 093 NGC 4232	1338	1.5 x 0.8 1.3 x 0.6	155 5	2	S B a - b P w 7304, disturbed		14.6		*1$\underline{2}$ 3
07304	12 14.3 + 47 43 139.14 + 68.58	08 - 22 - 094 NGC 4231	1338	1.3 x 1.1 0.9 x 0.8	-- ...	1	... P w 7303		14.5		*1$\underline{2}$ 3
07305	12 14.4 + 13 54 270.41 + 74.27	02 - 31 - 078 IC 3094	1385	0.6 x 0.5 0.6 x 0.5	-- ...	1	...		14.1		
07306	12 14.4 + 69 45 127.41 + 47.35	12 - 12 - 004 NGC 4236	1411	23. x 8. 22. x 7.	162 6	5	...	S B(s)dm	10.7	+ 27 + 185	*12- -9$\underline{}$
07307	12 14.5 + 10 17 275.97 + 71.05	02 - 31 - 000	1385	2. x 2. 1. x 1.	-- ...	4	Dwarf V		16.5:		
07308	12 14.5 + 63 42 129.38 + 53.27	11 - 15 - 041 NGC 4238	0674	1.8 x 0.45 1.7 x 0.45	36 6	3	Sc	Sc +	14.2		*3
07309	12 14.6 + 03 58 282.19 + 65.17	01 - 31 - 035 NGC 4234	1611	1.2: x 1.1: 1.2 x 1.1	-- ...	4	Irr	Ir I	13.4	+ 2143 + 2024	*123 4$\underline{7}$8
07310	12 14.6 + 07 28 279.18 + 68.46	01 - 31 - 036 NGC 4235 = 3098*	1611	(3.9 x 0.8) (3.8 x 0.8)	48 7	3	Sa		13.2		*12- -$\underline{7}$89
07311	12 14.6 + 07 54 278.74 + 68.87	01 - 31 - 037 NGC 4233	1611	(2.0 x 0.9) (2.0 x 0.9)	174 5:	3	S0		13.2		*12- -5$\underline{7}$89
07312	12 14.6 + 12 34 272.82 + 73.12	02 - 31 - 000 IC 3100	1385	1.7: x 0.5 1.6: x 0.6:	58 ...	1	... IC 3105 group		15.3		*
07313	12 14.6 + 12 44 272.55 + 73.27	02 - 31 - 079 IC 3099	1385	2.1 x 0.25 1.9 x 0.25	172 7	1	Sc: IC 3105 group		15.1		*1
07314	12 14.7 + 09 14 277.37 + 70.11	02 - 31 - 000	1385	1.1: x 0.7: 1.0: x 0.6:	150: ...	2	...	dm - m	15.5		
07315	12 14.7 + 15 36 267.26 + 75.77	03 - 31 - 091 NGC 4237	0089	2.2 x 1.4 2.2 x 1.4	108 3 - 4	1	Sb		12.3		*123 478

N G C   4215 - 18, 4220 - 22, 4224, 4226, 4227   I C   3087, 3094, 3098 - 3100
4229, 4231 - 37

1	2	3	4	5	6	7	8	9	10	11	12
07316	12 14.7 + 16 48 264.43 + 76.76	03 − 31 − 092 NGC 4239	0089	1.6 x 1.0 1.5 x 0.9	120 ...	2	E Disturbed?		13.5		*1
07317	12 14.7 + 41 35 147.14 + 74.17	07 − 25 − 053	1367	1.1: x 0.7: 1.2: x 0.7	... ...	1	...		15.7		*1
07318	12 14.7 + 49 46 137.08 + 66.69	08 − 22 − 000	1338	1.0 x 0.1 0.9 x 0.1	140 7	3	Sc		17.		*
07319	12 14.9 + 06 58 279.85 + 68.02	01 − 31 − 038 NGC 4241 = 3102*	1611	(2.5 x 1.5) (2.5 x 1.5)	128 4:	3	S0 P w 7333		13.6		*12 359
07320	12 14.9 + 45 05 141.81 + 71.05	08 − 22 − 095	1338	1.1: x 0.6: 1.0: x 0.5:	75 ...	2	...	d − m	16.5:		
07321	12 15.0 + 22 50 241.82 + 81.05	04 − 29 − 060	0135	5.5 x 0.3 5.2 x 0.25	82 7	3	Sc	Sc +	14.0		1
07322	12 15.0 + 38 05 154.57 + 77.16	06 − 27 − 045 NGC 4244	1599	18.5 x 2.3 17. x 2.2	48 7	4	Sc	S d	10.8	+ 240 + 269	*12 −9⁻
07323	12 15.0 + 45 53 140.80 + 70.33	08 − 22 − 098 NGC 4242	1338	5.7 x 3.8: 5.5: x 4.0:	25 3:	4	...	SAB(s)dm	11.9	+ 661 + 724	*12 −9⁻
07324	12 15.0 + 46 50 139.74 + 69.45	08 − 22 − 096	1338	1.1 x 0.4 1.0 x 0.3	[120] ...	1	... Disrupted?		15.4		*
07325	12 15.0 + 47 05 139.48 + 69.22	08 − 22 − 097	1338	1.6 x 0.6 1.5 x 0.5	145 6	2	Sa − b		15.3		
07326	12 15.1 + 12 40 273.04 + 73.27	02 − 31 − 080 IC 3105	1385	1.8 x 0.5 1.2 x 0.3:	27 ...	3	Irr Brightest of 3		15.0		*1
07327	12 15.1 + 17 00 264.27 + 76.98	03 − 31 − 095	0089	1.9 x 0.2 1.7: x 0.18	91 7	3	Sc	Sc +	15.7		
07328	12 15.1 + 29 53 192.54 + 82.16	05 − 29 − 049 NGC 4245	1398 263	(3.5 x 3.3) (3.5 x 3.2)	−− 1 − 2	4	S B 0/S B a	S B(r)0/a	12.4	+ 890 + 882	*123 4678
07329	12 15.1 + 71 05 126.95 + 46.06	12 − 12 − 005 NGC 4250	1411	(2.5 x 1.9) 2.3 x 1.8	168 ...	4	S0/S B a Compact core	SAB(r)0/a	13.0		*12 3C
07330	12 15.3 + 11 07 275.44 + 71.89	02 − 31 − 082 IC 3107	1385	1.5 x 0.7 1.4 x 0.8	133 4 − 5	1	S ...		14.5		
07331	12 15.3 + 17 42 262.56 + 77.57	03 − 31 − 096	0089	1.6 x 0.8 1.6 x 0.8	0 5:	4	S B b	(R′)S B(s)b	14.7		*1
07332	12 15.4 + 00 43 284.79 + 62.13	00 − 31 − 045	1401	2.3:x 1.5:: 2. x 1.	... ...	4	Dwarf irregular Ir V		15.4		
07333	12 15.4 + 06 56 280.19 + 68.04	01 − 31 − 040 IC 3115	1611	1.7 x 1.3 1.7 x 1.3	−− 2:	3	S B c P w 7319	Sc +	14.4		*12 59
07334	12 15.4 + 07 28 279.68 + 68.54	01 − 31 − 041 NGC 4246 = 3113*	1611	2.4 x 1.2 2.7 x 1.4	83 5	4	Sc	SA(s)c	14.0		*12 359
07335	12 15.4 + 47 41 138.70 + 68.68	08 − 22 − 099 NGC 4248	1338	3.2 x 1.3 3.2 x 1.2	108 6	1	S ...		13.9		123 59
07336	12 15.5 + 16 14 266.49 + 76.40	03 − 31 − 097	0089	1.0 x 0.20 0.9 x 0.20	169 6	1	S ...		15.4		
07337	12 15.6 − 00 47 285.74 + 60.70	00 − 31 − 046	1401	1.1 x 0.6 1.0 x 0.5	127 5	2	Sb − c		14.7		
07338	12 15.6 + 28 28 202.83 + 82.55	05 − 29 − 050 NGC 4251	1398 263	(3.6 x 2.0) (3.3 x 1.8)	100 ...	3	S0		11.5	+ 1014 + 1000	*12 3478
07339	12 15.7 + 09 47 277.41 + 70.71	02 − 31 − 083 IC 3118	1385	1.6: x 0.8: 1.6: x 0.8:	170: 5:	3	Dwarf IV − V or V		15.2		
07340	12 15.7 + 44 27 142.21 + 71.69	07 − 25 − 055 Mark 203	1367	[1.5 x 0.7] ...	... ...	3	Double system Bridge		14.2	+ 7500	*1
07341	12 15.8 + 25 30 225.47 + 82.33	04 − 29 − 062 IC 3122	0135	1.5 x 0.8 1.5 x 0.8	150 5	3	Sb		14.7		
07342	12 15.8 + 29 33 194.63 + 82.40	05 − 29 − 000	1398 263	1.1 x 0.7 1.0 x 0.6	140: ...	1	... f of 2		16.0:		*
07343	12 15.9 + 05 50 281.46 + 67.04	01 − 31 − 045 NGC 4252	1611	1.2 x 0.3 1.2 x 0.3	48 6	1	S ...		15.0		23
07344	12 15.9 + 30 05 190.75 + 82.27	05 − 29 − 051 NGC 4253	1398 263	0.9 x 0.9 0.9 x 0.8	−− 1 − 2	3	S B a		13.7		2̲3̲
07345	12 16.2 + 14 42 270.35 + 75.18	03 − 31 − 099 NGC 4254 = M 99	0089	5.0 x 4.7 4.6 x 4.5	−− 1	5	Sc	SA(s)c Sc −	10.2	+ 2471 + 2397	*12 −9⁻
07346	12 16.2 + 17 59 262.55 + 77.93	03 − 31 − 098	0089	(1.2 x 1.2) (1.2 x 1.1)	−− ...	1	...		15.5		
07347	12 16.3 + 12 45 273.82 + 73.47	02 − 31 − 000	1385	1.2: x 0.1 1.1: x 0.1	44 7	2	Sc		16.5:		*
07348	12 16.4 + 05 04 282.38 + 66.36	01 − 31 − 047 NGC 4255	1611	1.2 x 0.6 1.1 x 0.5	115 5	3	S0		13.5		
07349	12 16.4 + 06 28 281.23 + 67.68	01 − 31 − 048 IC 3136	1611	1.0 x 0.25 1.0 x 0.25	32 6	1	S ...		14.7		*2
07350	12 16.4 + 13 11 273.21 + 73.87	02 − 31 − 087 IC 775	1385	(1.1 x 0.8) (1.4 x 0.9)	(10) ...	2	E − S0		(14.8)		*12 3

N G C   4239, 4241, 4242, 4244 − 46, 4248   I C   775, 3102, 3105, 3107, 3113, 3115
4250 − 55   3118, 3122, 3136

1	2	3	4	5	6	7	8	9	10	11	12
07351	12 16.4 + 66 10 128.21 + 50.90	11 - 15 - 045 NGC 4256	0674 255	4.5 x 0.8 4.3 x 0.8	42 6	3	Sb		12.7	+ 2583 + 2728	*12 346$\overline{8}$
07352	12 16.5 + 09 08 278.69 + 70.19	02 - 31 - 088 IC 776	1385	2.1 x 1.1 1.8 x 1.0	98 4 - 5	3	... S dm P w IC 3134		14.9		*1
07353	12 16.5 + 47 35 138.31 + 68.84	08 - 22 - 104 NGC 4258 = M 106	1338	22. x 9. 20. x 9.	150 6	5	Sb	SAB(rs)bc Sb +	9.6	+ 472 + 543	*12- -9
07354	12 16.6 + 04 08 283.20 + 65.48	01 - 31 - 050 Mark 49	1611	0.5 x 0.4 0.7 x 0.6	... ...	1	...		14.5	+ 1716 + 1599	*12 C
07355	12 16.6 + 14 15 271.52 + 74.84	02 - 31 - 89+90 IC 3142	1385	[1.6 x 1.0] ...	... ...	3	Double system Contact		15.4		*1
07356	12 16.6 + 47 22 138.48 + 69.05	08 - 22 - 105	1338	1.1: x 0.8: 1.0 x 0.9	--- ...	1	Dwarf?		17.		
07357	12 16.7 + 22 43 243.86 + 81.32	04 - 29 - 063	0135	1.7 x 1.6 1.4 x 1.4	--- 1	3	Sc	SAB(s)c	15.5		
07358	12 16.7 + 49 38 136.42 + 66.92	08 - 22 - 106	1338	1.7 x 0.5 1.7 x 0.5	37 6	4	Sc		14.3		
07359	12 16.8 + 05 39 282.16 + 66.94	01 - 31 - 051 NGC 4259	1611	0.9 x 0.3 1. x 0.4	143 6 - 7	3	S0		14.5		*12 359
07360	12 16.8 + 06 06 281.79 + 67.37	01 - 31 - 052 NGC 4261	1611	(3.5 x 3.0) (3.5 x 3.0)	(160) ...	4	E		12.0	+ 2202 + 2093	*123 467$\overline{8}$
07361	12 16.8 + 06 22 281.56 + 67.62	01 - 31 - 054 NGC 4260	1611	2.2 x 1.2 2.6 x 1.4	58 4 - 5	4	S B a	S B(s)a	13.1	+ 1935 + 1827	*123 467$\overline{8}$
07362	12 16.8 + 26 13 220.52 + 82.72	04 - 29 - 000	0135	1.1 x 0.1 1.1 x 0.1	150 7	1	Sb - c		16.5:		
07363	12 16.9 + 28 35 201.65 + 82.82	05 - 29 - 052 IC 777	1398 263	1.3 x 0.7 1.1 x 0.6	140 5	1	S B ...		14.5		
07364	12 17.0 + 06 07 281.89 + 67.40	01 - 32 - 001 NGC 4264	1560	0.8 x 0.6 0.8 x 0.7	--- ...	1	S B 0		13.9		*12 3
07365	12 17.0 + 15 09 270.13 + 75.67	03 - 31 - 101 NGC 4262	0089	(1.9 x 1.8) (1.9 x 1.8)	--- 1	4	S B 0		12.3	+ 1351 + 1280	*123 47$\overline{8}$
07366	12 17.0 + 17 30 264.63 + 77.65	03 - 32 - 001	1576	1.1: x 0.6: 0.8: x 0.5:	140: ...	1	...		15.5		1
07367	12 17.0 + 50 05 135.95 + 66.51	08 - 23 - 001	1408	1.6 x 0.9 1.6 x 0.9	93 4 - 5	3	S0 - a		13.7		
07368	12 17.1 + 05 49 282.20 + 67.12	01 - 32 - 002 NGC 4266	1560	1.8 x 0.4 2.1 x 0.4	76 7	1	S ...		15.0		12$\overline{3}$
07369	12 17.1 + 30 10 189.52 + 82.49	05 - 29 - 053 IC 779	1398 263	1.1 x 1.1 1.0 x 1.0	--- ...	1	...		15.4		*
07370	12 17.2 + 02 22 284.74 + 63.84	00 - 32 - 001	1405	1.2 x 0.3 1.2 x 0.3	151 6	2	Sb - c		14.6		*
07371	12 17.2 + 05 34 282.47 + 66.89	01 - 32 - 004 NGC 4268	1560	1.3 x 0.5 1.2 x 0.5	48 6	2	S0 - a		13.9		*12 359
07372	12 17.2 + 06 18 281.86 + 67.59	01 - 32 - 005 NGC 4269	1560	1.3 x 0.8 1.4 x 0.9	137 4	2	S0 - a P w IC 3135		13.9		*12 3
07373	12 17.2 + 13 05 273.99 + 73.87	02 - 32 - 004 NGC 4267	1563	(3.6 x 3.6) (3.5 x 3.5)	--- 1	4	S0		12.4	+ 1260 + 1180	*123 47$\overline{8}$
07374	12 17.2 + 29 09 197.10 + 82.78	05 - 29 - 057	1398 263	1.1 x 0.7 1.0 x 0.6	157 3 - 4	1	S ...		15.3		*
07375	12 17.2 + 57 01 131.62 + 59.85	10 - 18 - 025 NGC 4271	1427	(1.5 x 1.3) (1.4 x 1.4)	(55) ...	2	E - S0		13.7		*13
07376	12 17.3 + 05 45 282.38 + 67.08	01 - 32 - 007 NGC 4270	1560	1.7 x 0.7 1.6 x 0.7	110 6	4	S0		13.3	+ 2347 + 2237	*1234 5789
07377	12 17.3 + 29 54 191.36 + 82.61	05 - 29 - 060 NGC 4274	1398 263	7.3 x 2.7 7.3 x 2.7	102 6	4	S B a	S B(r)ab	11.1	+ 767 + 761	*12- -789
07378	12 17.3 + 30 27 186.22 + 82.36	05 - 29 - 059 NGC 4272	1398 263	(1.0 x 0.9) (1.1 x 0.9)	--- ...	3	E		14.2		*3
07379	12 17.3 + 53 23 133.57 + 63.36	09 - 20 - 145	1389	1.1 x 0.8 0.8 x 0.7	5: 3:	3	Sc	Sc +	16.0:		
07380	12 17.4 + 05 37 282.55 + 66.96	01 - 32 - 008 NGC 4273	1560	2.5 x 1.2 2.3 x 1.2	10: 4 - 5	4	Sc Disturbed, Sc -		12.3	+ 2302 + 2192	*12- -9
07381	12 17.4 + 26 03 222.04 + 82.82	04 - 29 - 064 IC 780	0135	(1.1 x 0.8) (1.3 x 0.9)	7 ...	2	E - S0		14.5		
07382	12 17.4 + 27 54 207.08 + 82.99	05 - 29 - 058 NGC 4275	1398 263	0.8 x 0.8 0.8 x 0.8	--- (1)	1	S? ...		13.4		3
07383	12 17.5 + 08 53 279.63 + 70.04	02 - 32 - 005	1563	1.2 x 0.8 0.7 x 0.3	130 3:	2	Sa - b Brightest of 3		14.8		*
07384	12 17.5 + 28 15 204.21 + 82.99	05 - 29 - 061 IC 3165	1398 263	1.9 x 1.1 2.0 x 1.1	5 4 - 5	2	S B b:		14.9		*
07385	12 17.6 + 07 58 280.62 + 69.19	01 - 32 - 010 NGC 4276	1560	1.7 x 1.7 1.5: x 1.4:	--- 1:	1	S? ...		14.1		

N G C   4256, 4258 - 62, 4264, 4266 - 76   I C   776, 777, 779, 780, 3142, 3165

1	2	3	4	5	6	7	8	9	10	11	12
07386	12 17.6 + 29 33 193.85 + 82.77	05 – 29 – 062 NGC 4278	1398 263	(3.5 x 3.5) (3.6 x 3.6)	-- ...	4	E		11.2	+ 630 + 622	*12‾ –9‾
07387	12 17.7 + 04 29 283.59 + 65.90	01 – 32 – 011	1560	2.1 x 0.15 2.0 x 0.15	16 7	3	Sc		15.2		
07388	12 17.7 + 33 55 166.99 + 80.64	06 – 27 – 047	1599	1.0 x 0.8 0.8 x 0.7	[110] ...	1	S B ...		16.0:		
07389	12 17.8 + 05 40 282.75 + 67.03	01 – 32 – 012 NGC 4281	1560	(2.5 x 1.2) (2.7 x 1.5)	88 5	4	S0		12.5	+ 2602 + 2492	*12‾ 5789
07390	12 17.8 + 29 35 193.50 + 82.81	05 – 29 – 063 NGC 4283	1398 263	1.1 x 1.1 (1.3 x 1.3)	-- ...	4	E		13.1	+ 1085 + 1078	*12‾ –9‾
07391	12 17.8 + 46 12 139.11 + 70.21	08 – 23 – 004	1408 265	[1.1 x 0.15] ...	[171] ...	2	Double system Connected		16.5:		
07392	12 17.8 + 48 25 136.99 + 68.13	08 – 23 ? – 003	1408	1.3 x 0.2 1.3: x 0.2	5 7	2	S B ? c		16.5:		
07393	12 17.8 + 58 22 130.84 + 58.56	10 – 18 – 026 NGC 4284	1427	3.0 x 1.3 2.2 x 1.1	102 5	2	Sb – c P w 7402		14.7		*13
07394	12 17.9 + 01 46 285.48 + 63.31	00 – 32 – 000	1405	1.8 x 0.15 1.8 x 0.15	146 7	3	Sc	Sc +	15.2		
07395	12 17.9 + 31 27 180.25 + 82.11	05 – 29 – 64a	1398 263	0.50 x 0.45 0.55 x 0.45	-- ...	1	...		14.5		
07396	12 18.0 + 01 05 285.93 + 62.65	00 – 32 – 002	1405	1.7 x 0.40 1.7 x 0.4	101 6	2	Sc – Irr		15.0		
07397	12 18.0 + 75 40 125.56 + 41.58	13 – 09 – 024 NGC 4291	1339 269	(2.0 x 1.7) (2.0 x 1.7)	(110) ...	3	E P w 7429		12.3	+ 1815 + 1994	*123‾ 4678‾
07398	12 18.2 + 29 38 192.92 + 82.88	05 – 29 – 065 NGC 4286 = 3181*	1398 263	1.7 x 0.9 1.6 x 0.9	150 4 – 5	1	S0?		14.7		*23‾
07399	12 18.2 + 46 35 138.53 + 69.88	08 – 23 – 006 NGC 4288	1408 265	3.0: x 1.8 2.5 x 1.7	130: 4	4	S B c	S B (s)d	13.6		*12‾ 3
07400	12 18.3 + 63 22 128.80 + 53.69	11 – 15 – 047	0674	1.0: x 0.8: 0.7: x 0.5:	-- ...	2	...	S d – m	17.		
07401	12 18.4 + 48 07 137.00 + 68.45	08 – 23 – 008	1408	1.1 x 0.7 1.1: x 0.7:	[25] ...	3	Dwarf irregular Ir IV – V or V		16.5:		
07402	12 18.4 + 58 22 130.70 + 58.57	10 – 18 – 029 NGC 4290	1427	2.5 x 1.9 2.3 x 1.8	90: 2 – 3	4	S B b Sb –	S B (rs)ab	12.8		*12‾ 3468‾
07403	12 18.5 + 04 00 284.39 + 65.49	01 – 32 – 015 NGC 4289	1560	3.9 x 0.35 3.6: x 0.35	1 7	3	Sc		15.0		1
07404	12 18.7 + 04 52 283.90 + 66.34	01 – 32 – 016 NGC 4292	1560	(1.9 x 1.3) (1.9 x 1.3)	7: 3:	3	S0		14.1		*12‾ 3
07405	12 18.7 + 18 40 262.85 + 78.83	03 – 32 – 006 NGC 4293	1576	5.8 x 3.3 5.8 x 3.3	72 4	4	S B 0/SB a		11.6	+ 750 + 695	*123‾ 4678‾
07406	12 18.7 + 61 22 129.43 + 55.65	10 – 18 – 031	1427 261	1.1 x 0.9 1.1 x 0.8	(30) 2:	2	S B c		14.5		*
07407	12 18.8 + 11 47 277.13 + 72.85	02 – 32 – 009 NGC 4294	1563	3.0 x 1.1 2.8 x 1.0	155 6	4	Sc P w 7414		12.6	+ 390 + 306	*12‾ –5789
07408	12 18.8 + 46 06 138.73 + 70.37	08 – 23 – 013	1408 265	2.7: x 1.6: 2.0: x 1.5::	[100] ...	4	Dwarf irregular, IA m Ir IV – V Ir I		14.8		*1
07409	12 18.9 + 06 56b 282.39 + 68.32	01 – 32 – 017 NGC 4296	1560	(1.5 x 0.9) 1.7 x 0.9	(15) 4 – 5	3	S0 P w NGC 4297		(14.0)		*13
07410	12 19 + 82 58 124.11 + 34.37	14 – 06 – 005	1340	1.4 x 0.7 1.6 x 0.8	3: (5)	1	S0?		16.0:		
07411	12 19.0 + 05 03 283.94 + 66.54	01 – 32 – 019	1560	1.3 x 0.4 0.9 x 0.3	132 6	1	S0 – a		14.9		1
07412	12 19.0 + 14 53 272.35 + 75.67	03 – 32 – 007 NGC 4298	1576	3.0 x 1.7 3.2 x 1.7	140 4	4	Sc P w 7418, disturbed		12.2		*12‾ –9‾
07413	12 19.1 + 05 39 283.55 + 67.12	01 – 32 – 021 NGC 4300	1560	1.5 x 0.6 1.7 x 0.7	42 6	3	Sa		13.9		3
07414	12 19.1 + 11 47 277.36 + 72.88	02 – 32 – 010 NGC 4299	1563	1.7 x 1.7 1.5 x 1.5	-- ...	4	Irr P w 7407 Ir I		12.8	+ 187 + 103	*12‾ 3458‾
07415	12 19.1 + 16 00 270.18 + 76.66	03 – 32 – 008 IC 783	1576	1.2 x 1.1 1.1 x 1.0	-- 1	1	...	SA d?	15.0		*12‾
07416	12 19.1 + 41 08 145.18 + 74.96	07 – 26 – 001	0115	1.5 x 1.3 1.4 x 1.2	-- 1	4	S B b Sb +	S B (r)b	14.2		
07417	12 19.2 + 10 53 278.58 + 72.06	02 – 32 – 000 IC 3199	1563	1.0: x 0.5: 1.1: x 0.7:	0: ...	1	S ...		15.4		
07418	12 19.2 + 14 53 272.53 + 75.69	03 – 32 – 009 NGC 4302	1576	5.1 x 0.9 5.5 x 0.7	178 7	3	Sc P w 7412, disturbed		13.4		*12‾ –689‾
07419	12 19.2 + 26 10 221.81 + 83.24	04 – 29 – 067 IC 3203	0135	1.5 x 0.20 1.5 x 0.22	145 7	2	Sb		15.7		
07420	12 19.3 + 04 45 284.34 + 66.27	01 – 32 – 022 NGC 4303 = M 61	1560	6.6 x 6.4 5.5 x 5.5	-- 1	5	Sb/SB c Sc –	SAB(rs)bc	10.9	+ 1671 + 1559	*12‾ –9‾

N G C 4278, 4281, 4283, 4284, 4286, 4288 – 94,
4296, 4298 – 4300, 4302, 4303

I C 783, 3181, 3199, 3203

1	2	3	4	5	6	7	8	9	10	11	12
07421	12 19.3 + 12 14 276.91 + 73.31	02 – 32 – 000	1563	1.1: x 0.4: 1. x 0.4::	... ...	3	Dwarf V		16.5:		
07422	12 19.4 + 05 23 283.94 + 66.88	01 – 32 – 023	1560	1.0 x 0.3 1.0 x 0.3	168 (6)	1	S: ...		14.9		
07423	12 19.4 + 06 43 282.89 + 68.16	01 – 32 – 024	1560	1.3: x 0.7: 1. x 0.7:	5: ...	2	...	S m:	15.3		
07424	12 19.4 + 08 57 280.86 + 70.27	02 – 32 – 011	1563	1.2: x 1.1: 1.1: x 1.0:	— ...	4	Dwarf		15.6		
07425	12 19.4 + 15 55 270.63 + 76.62	03 – 32 – 010	1576	1.2: x 0.8: 0.9: x 0.8:	... ...	2	...	S d – dm	16.5:		1
07426	12 19.4 + 30 20 186.89 + 82.89	05 – 29 – 069 NGC 4308	1398 263	0.8 x 0.7 0.7 x 0.6	— ...	1	E:		14.3		*
07427	12 19.4 + 35 18 159.79 + 79.89	06 – 27 – 049	1599	1.3 x 0.7 1.2: x 0.6:	[25] ...	3	Dwarf irregular, I m Ir IV – V:		16.5		
07428	12 19.5 + 32 22 173.37 + 81.91	05 – 29 – 072	1398	1.3 x 1.2 1.4 x 1.3	— ...	3	Irr		14.3		
07429	12 19.5 + 75 37 125.45 + 41.64	13 – 09 – 025 NGC 4319	1339 269	3.1 x 2.3 3.4 x 2.3	160 ...	4	S B b P w 7397	S B(r)ab Sb –	13.0	+ 1700 + 1878	*12 36
07430	12 19.6 + 09 16 280.68 + 70.58	02 – 32 – 012 IC 3211	1563	1.1 x 0.9 0.9 x 0.9	— 1:	3	Sc P w 7431	Sc +	15.5		*1
07431	12 19.6 + 09 19 280.63 + 70.63	02 – 32 – 12a NGC 4307	1536	3.5 x 0.7 3.4 x 0.7	24 6	3	Sb P w 7430		13.4		*12 378
07432	12 19.6 + 13 01 276.01 + 74.06	02 – 32 – 013 NGC 4305	1563	1.9 x 1.0 2.1 x 1.0	32 5	3	Sa P w 7433		13.8		*12 3
07433	12 19.6 + 13 04 275.93 + 74.11	02 – 32 – 014 NGC 4306	1563	(1.3 x 0.9) (1.4 x 1.0)	140: ...	2	S0 P w 7432		14.4		*12 3
07434	12 19.6 + 26 20 220.56 + 83.36	04 – 29 – 068 IC 3215	0135	2.0 x 0.5 1.8 x 0.5	92 6	3	...	S dm	15.7		
07435	12 19.7 + 07 25 282.49 + 68.84	01 – 32 – 025 NGC 4309	1560	1.8 x 1.0 1.7 x 0.9	85: 4 – 5	2	S0 – a		14.3		*12 3
07436	12 19.8 + 15 02 272.77 + 75.89	03 – 32 – 012	1576	1.3 x 0.8 1.1 x 0.7	120: ...	1	S: ... np of 2		15.2		*1
07437	12 19.8 + 29 07 196.34 + 83.35	05 – 29 – 073 IC 3222	1398 263	1.1 x 0.55 0.9 x 0.5	17 5:	2	S B c		16.0:		*
07438	12 19.8 + 30 21 186.49 + 82.96	05 – 29 – 000	1398 263	1.5 x 0.4 1.3 x 0.4	5 ...	1	...	d – m	16.0:		*
07439	12 19.9 + 04 50 284.63 + 66.39	01 – 32 – 027 NGC 4301	1560	1.6 x 1.4 1.4 x 1.3	— 1	1	S B ...		14.9		*12 359
07440	12 19.9 + 29 29 193.25 + 83.27	05 – 29 – 074 NGC 4310=4311	1398 263	2.3 x 1.2 (2.2 x 1.0)	128 5:	1	S ...		13.5		23
07441	12 20.0 + 06 57 283.08 + 68.42	01 – 32 – 028 IC 3225	1560	1.8 x 0.7 1.8 x 0.5	40 ...	2	... P w 7448	dm – m	14.9		*1
07442	12 20.0 + 15 49 271.39 + 76.61	03 – 32 – 014 NGC 4312	1576	4.2 x 0.9 4.2 x 1.0	170 6	2	Sa		12.9		*12 35
07443	12 20.0 + 30 10 187.75 + 83.07	05 – 29 – 075 NGC 4314	1398 263	4.6 x 4.5 4.6 x 4.6	— 1	4	S B a		11.5	+ 884 + 879	*123 4678
07444	12 20.0 + 45 20 138.95 + 71.16	08 – 23 – 015	1408 265	1.4 x 0.6 1.2 x 0.5	44 6	2	Sa		14.7		
07445	12 20.1 + 12 05 277.74 + 73.25	02 – 32 – 016 NGC 4313	1563	3.5 x 0.8 3.5 x 0.8	143 7	2	Sa – b		13.2		*12 3
07446	12 20.2 + 08 29 281.86 + 69.90	02 – 32 – 015 NGC 4318	1563	0.8 x 0.6 0.6 x 0.4	(65) ...	1	E?		14.1		3
07447	12 20.2 + 09 37 280.74 + 70.96	02 – 32 – 017 NGC 4316	1563	2.7 x 0.5 2.7 x 0.5	113 7	1	Sc?		14.0		
07448	12 20.3 + 06 57 283.28 + 68.45	01 – 32 – 030 IC 3229	1560	1.1 x 0.35 0.9 x 0.3	47 6	2	Sb – c P w 7441		15.0		*1
07449	12 20.3 + 76 28 125.21 + 40.81	13 – 09 – 026 NGC 4331	1338 269	2.2 x 0.6 2.4 x 0.6	[2] ...	3	Irr or Peculiar		14.8		*3C
07450	12 20.4 + 16 06 271.15 + 76.90	03 – 32 – 015 NGC 4321=M 100	1576	6.8 x 5.8 6.8 x 6.0	(30) 1 – 2	5	Sc	SAB(s)bc Sc –	10.6	+ 1617 + 1552	*12– 6789
07451	12 20.5 + 05 32 284.50 + 67.11	01 – 32 – 032 NGC 4324	1560	2.1 x 0.9 2.8 x 1.1	53 6	3	S0		12.5	+ 1714 + 1605	*123 4678
07452	12 20.5 + 10 49 279.63 + 72.11	02 – 32 – 018 NGC 4320	1563	1.1 x 0.9 1.2: x 0.7:	[170] ...	3	Strongly peculiar ssp of 2		15.3		*13
07453	12 20.5 + 66 07 127.58 + 51.03	11 – 15 – 048 NGC 4332	0674 255	2.4 x 1.7 2.2 x 1.6	130: ...	4	S B a	(R′)S B(s)a	13.2		*23
07454	12 20.6 + 06 21 283.95 + 67.90	01 – 32 – 033 NGC 4326	1560	1.7 x 1.1 1.5 x 1.2	(145) ...	1	S ... NGC 4339 group		13.1		*12 3
07455	12 20.6 + 58 44 130.01 + 58.28	10 – 18 – 035 NGC 4335	1427	(1.9 x 1.5) (1.8 x 1.6)	(145) ...	3	E		13.7		*13

N G C   4301, 4305 – 14, 4316, 4318 – 21   I C   3211, 3215, 3222, 3225, 3229
          4324, 4326, 4331, 4332, 4335

1	2	3	4	5	6	7	8	9	10	11	12
07456	12 20.7 + 11 39 278.77 + 72.91	02 - 32 - 020 NGC 4330	1563	4.5 x 0.9 4.0 x 0.7	59 7	1	Sc?		14.0		3
07457	12 20.7 + 29 38 191.57 + 83.40	05 - 29 - 076	1398 263	1.2 x 0.5 1.1 x 0.5	137 5 - 6	1	S ...		16.0:		
07458	12 20.8 + 07 45 282.93 + 69.25	01 - 32 - 035 NGC 4334	1560	2.3 x 1.0 2.0 x 0.9	135 5 - 6	4	S B 0		14.9		*12 3
07459	12 20.8 + 29 10 195.46 + 83.55	05 - 29 - 077 IC 3247	1398 263	2.4 x 0.2 2.0 x 0.2	175 7	3	Sc	Sc +	15.6		
07460	12 20.9 + 57 45 130.33 + 59.25	10 - 18 - 034	1427	1.2 x 0.5 1.2 x 0.5	135 ...	1	S ...		15.7		
07461	12 21.0 + 06 21 284.20 + 67.92	01 - 32 - 036 NGC 4339	1560	(2.0 x 2.0) (2.3 x 2.2)	-- ...	4	E Brightest of 3		13.1	+ 1278 + 1173	*123 4678
07462	12 21.0 + 19 43 261.67 + 79.99	03 - 32 - 020 NGC 4336 = 3254*	1576	1.6 x 0.9 (1.6 x 0.9)	162 4	2	S B 0/S B a		13.6		3
07463	12 21.0 + 47 16 136.59 + 69.39	08 - 23 - 016 NGC 4346	1408	3.2 x 1.3 3.2 x 1.3	99 6	3	S0		12.3		*123 4678
07464	12 21.1 + 03 14 286.37 + 64.93	01 - 32 - 037	1560	1.6 x 0.3 1.5 x 0.25	147 6	2	Sb		15.2		
07465	12 21.1 + 07 14 283.56 + 68.78	01 - 32 - 038 NGC 4343	1560	2.6 x 0.7 2.5 x 0.7	133 6 - 7	1	Sb:		13.5		*123 569
07466	12 21.1 + 07 20 283.48 + 68.87	01 - 32 - 039 NGC 4342 = 3256*	1560	1.0 x 0.4 1.1 x 0.5	168 ...	1	S? ...		13.0	+ 714 + 613	*12- -789
07467	12 21.1 + 17 00 269.77 + 77.77	03 - 32 - 021 NGC 4340	1576	(3.6 x 3.0) (3.6 x 2.8)	102 2:	4	S B 0 SB(r)0 + P w 7473		12.4		*123 4678
07468	12 21.1 + 17 49 267.69 + 78.46	03 - 32 - 022 NGC 4344	1576	(1.4 x 1.3) (1.3 x 1.3)	-- ...	1	S0?		13.7		123
07469	12 21.2 + 07 28 283.43 + 69.01	01 - 32 - 040 IC 3259	1560	1.8 x 0.9 1.7 x 0.8	15 (5)	1	...		14.7		*12 59
07470	12 21.2 + 12 45 277.69 + 73.97	02 - 32 - 021 IC 3258	1563	1.9: x 1.6: 1.3: x 1.3:	-- ...	4	Irr Ir I		14.3		*2
07471	12 21.2 + 32 02 173.81 + 82.39	05 - 29 - 000	1398 267	1.2 x 0.12 0.8 x 0.15	31 7	1	S ...		16.5:		
07472	12 21.4 + 07 23 283.64 + 68.94	01 - 32 - 042 NGC 4341 = 3260*	1560	1.6 x 0.5 1.5 x 0.5	96 6	2	S0		14.5		*23 569
07473	12 21.4 + 16 58 270.14 + 77.77	03 - 32 - 023 NGC 4350	1576	(2.5 x 0.9) (2.7 x 1.1)	28 6	3	S0 P w 7467		11.5	+ 1184 + 1123	*12 3468
07474	12 21.5 + 07 19 283.76 + 68.89	01 - 32 - 044 IC 3267	1560	1.0 x 1.0 1.1 x 1.1	-- 1	1	S ...		14.6		*12 59
07475	12 21.5 + 11 30 279.57 + 72.84	02 - 32 - 023 NGC 4352	1563	(1.6 x 0.8) (1.6 x 0.8)	102 5:	4	S0		14.0		23
07476	12 21.5 + 12 29 278.31 + 73.75	02 - 32 - 024 NGC 4351	1563	1.9 x 1.4 1.7 x 1.4	80: 2 - 3:	1	S ...		13.5		*12 3
07477	12 21.6 + 06 53 284.17 + 68.48	01 - 32 - 045 IC 3268	1560	0.7 x 0.7 0.6 x 0.6	-- ...	1	... Peculiar		14.2		1
07478	12 21.6 + 49 03 134.90 + 67.71	08 - 23 - 017 NGC 4357	1408	3.9 x 1.5 3.6 x 1.4	77 6	4	Sb SA bc Sb +		13.5		1
07479	12 21.6 + 58 40b 129.79 + 58.37	10 - 18 - 038 NGC 4364	1427	1.3 x 1.2 1.1 x 1.0	-- 1	2	S0 P w NGC 4358		(14.3)		*13
07480	12 21.6 + 71 34 126.09 + 45.67	12 - 12 - 000	1411	1.3 x 0.3 1.0 x 0.2	64 6 - 7	3	Sc		17.		*
07481	12 21.7 + 08 14 283.11 + 69.77	01 - 32 - 047 IC 3271	1560	1.1 x 1.1 0.9: x 0.7:	-- 1	3	Sc SAB c		15.0		1
07482	12 21.7 + 08 49 282.58 + 70.33	02 - 32 - 026 IC 3273	1563	2.6 x 0.5 2.5 x 0.4	40 7	3	Sc		14.3		
07483	12 21.7 + 31 47 174.90 + 82.62	05 - 29 - 079 NGC 4359	1398 263	3.6 x 1.0: 3.5: x 0.9:	108 6	1	S ...		13.9		23
07484	12 21.8 + 09 34 281.93 + 71.05	02 - 32 - 028 NGC 4360	1563	(1.4 x 1.1) (1.4 x 1.1)	(145) ...	3	E		13.9		*1
07485	12 21.8 + 21 26 255.48 + 81.40	04 - 29 - 000	0135	(1.0 x 0.6) (1.0 x 0.6)	(120) ...	1	...		15.6		
07486	12 21.8 + 45 13 138.14 + 71.38	08 - 23 - 019	1408 265	1.0 x 0.8 0.8 x 0.7	(130) 1 - 2	3	Sc		16.0:		
07487	12 21.9 + 00 27 288.31 + 62.28	00 - 32 - 006	1405	1.0: x 0.7: 0.9 x 0.7	(175) 2 - 3	1	S B ...		14.9		
07488	12 21.9 + 07 35 283.80 + 69.17	01 - 32 - 048 NGC 4365	1560	(5.5 x 4.5) (5.5 x 4.5)	(40) ...	4	E		11.5	+ 1183 + 1083	*123 4678
07489	12 22.1 + 39 40 145.76 + 76.52	07 - 26 - 004 NGC 4369, Mark 439	0115	2.4 x 2.4 (2.5 x 2.5)	-- 1	4	S0/Sa SA 0/a		12.3		*123 4678
07490	12 22.1 + 70 37 126.24 + 46.61	12 - 12 - 008	1411	4. x 4. 3.5: x 3.5:	-- ...	4	Dwarf spiral, SA m S V		14.6		*1

N G C 4330, 4334, 4336, 4339 - 44, 4346, 4350 - 52   I C   3247, 3254, 3256, 3258 - 60, 3267
4357, 4359, 4360, 4364, 4365, 4369   3268, 3271, 3273

1	2	3	4	5	6	7	8	9	10	11	12
07491	12 22.1 + 75 49 125.19 + 41.47	13 - 09 - 027 NGC 4386	1339 269	2.8 x 1.7 2.8 x 1.7	135 4	2	S0		12.6	+ 1811 + 1990	*12 3478
07492	12 22.4 + 07 43 284.03 + 69.33	01 - 32 - 051 NGC 4370	1560	1.5 x 0.8 1.5 x 0.9	83 7	1	Sa:		14.1		*12 3
07493	12 22.4 + 11 59 279.68 + 73.37	02 - 32 - 033 NGC 4371	1563	(4.5 x 2.1) (4.5 x 2.1)	95 6:	4	S B 0		12.1	+ 977 + 896	*12- -5789
07494	12 22.5 + 13 10 278.18 + 74.48	02 - 32 - 034 NGC 4374 = M 84	1563	(5. x 4.) (6. x 5.)	(135) ...	4	S0 Markarian's chain		10.8	+ 954 + 878	*12- -789
07495	12 22.5 + 26 14 222.65 + 83.97	04 - 29 - 070 IC 3300	0135	1.1 x 0.25 1.0 x 0.25	79 6	1	Sc:		15.3		
07496	12 22.5 + 28 50 197.67 + 83.98	05 - 29 - 080 NGC 4375	1398 263	1.4 x 1.0 1.4 x 1.0	5: 3	1	Sa - b		13.9		*12 3
07497	12 22.7 + 05 12 286.08 + 66.93	01 - 32 - 052 NGC 4378	1560	3.3 x 3.1 (3.3 x 3.0)	(167) 1	4	Sa	(R)SA a	13.2		*123 4678
07498	12 22.7 + 06 01 285.52 + 67.72	01 - 32 - 053 NGC 4376	1560	1.6 x 0.9 1.4 x 0.8	[157] ...	3	Irr	Ir I	13.9		13
07499	12 22.7 + 12 07 279.75 + 73.52	02 - 32 - 036 IC 3305	1563	1.1: x 0.4: 1.1 x 0.4	44 (6)	1	S: ...		15.4		
07500	12 22.7 + 13 00 278.58 + 74.34	02 - 32 - 035 IC 3303	1563	1.1 x 0.7 (0.9 x 0.5)	73 ...	1	S0?		15.1		12
07501	12 22.7 + 15 02a 275.37 + 76.21	03 - 32 - 025 NGC 4377	1576	1.5 x 1.2 1.5 x 1.1	177: 2	3	S0		(12.5)	+ 1270	*123 4678C
07502	12 22.7 + 15 54 273.73 + 76.98	03 - 32 - 026 NGC 4379	1576	1.5 x 1.3 1.6 x 1.4	105: 1 - 2	4	S0		12.6		*12 3478
07503	12 22.8 + 10 17 281.93 + 71.81	02 - 32 - 037 NGC 4380	1563	3.5 x 2.2 3.5 x 2.2	153 4	2	Sa	SA a:	13.4		*12 3478
07504	12 22.8 + 16 43 272.09 + 77.72	03 - 32 - 027	1576	1.0: x 0.3: 1.0 x 0.4	168 ...	1	... P w 7507		16.0:		
07505	12 22.8 + 26 59 215.59 + 84.16	05 - 29 - 081 IC 3308	1398 263	1.5 x 0.25 1.3 x 0.3	65 ...	1	S - Irr nnf of 2, disturbed?		15.4		*1
07506	12 22.8 + 54 47 131.11 + 62.20	09 - 20 - 168 NGC 4384, Mark 207	1389	1.4 x 1.1 1.3 x 1.0	90: 3	3	Sa Singular		13.5	+ 2400	3
07507	12 22.9 + 16 45 272.12 + 77.76	03 - 32 - 030 NGC 4383	1576	1.8 x 0.9 (1.9 x 1.1)	[28] ...	1	... Peculiar		12.3		*12 3478
07508	12 22.9 + 18 28 267.74 + 79.24	03 - 32 - 029 NGC 4382 = M 85	1576	(7.4 x 5.5) (8. x 6.)	... 2 - 3:	4	S0 P w 7523		10.2	+ 765 + 712	*12- -9
07509	12 22.9 + 28 40 199.08 + 84.10	05 - 29 - 082 IC 3309	1398 263	1.4 x 1.1 1.3 x 1.0	85: ...	1	S ...	SAB(s) ...	15.7		1
07510	12 23.0 + 12 32 279.46 + 73.94	02 - 32 - 038 IC 3311	1563	2.1: x 0.35 1.7 x 0.3	135 7	3	Sc - Irr ssp of 2		15.0		*2
07511	12 23.0 + 65 13 127.42 + 51.96	11 - 15 - 053 NGC 4391	0674 255	1.2 x 1.2 0.9 x 0.9	-- 1	4	[S0] Compact		13.8		*23 C
07512	12 23.1 + 02 27 287.95 + 64.29	00 - 32 - 008	1405	1.1 x 0.4 1.2: x 0.5:	[130] ...	2	Dwarf (irregular) Ir: V		15.7		
07513	12 23.1 + 07 30 284.67 + 69.18	01 - 32 - 054	1560	3.5 x 0.4 3.5 x 0.4	157 7	3	Sc		14.4		*2
07514	12 23.1 + 45 58 136.75 + 70.74	08 - 23 - 028 NGC 4389	1408 265	2.6: x 1.8: 2.7: x 1.8:	105 ...	4	S B ... Peculiar		12.8		*123 4678
07515	12 23.2 + 00 52 288.80 + 62.75	00 - 32 - 009 NGC 4385, Mark 52	1405	1.9 x 1.1 2.0: x 1.1	82 5:	3	S B 0		12.4	+ 1350 + 1225	*123 4678
07516	12 23.2 + 04 47 286.65 + 66.56	01 - 32 - 055	1560	1.3 x 0.7 1.2 x 0.6	163 5	3	Sc		14.4		
07517	12 23.2 + 13 05 278.89 + 74.47	02 - 32 - 039 NGC 4387	1563	1.5 x 0.9 1.5 x 0.9	140 ...	3	E		13.2	+ 511 + 435	*12 3
07518	12 23.3 + 07 50 284.54 + 69.51	01 - 32 - 057 IC 3322	1560	2.4 x 0.5 2.2 x 0.5	156 6 - 7	2	Sc - Irr		14.7		*2
07519	12 23.3 + 10 44 281.83 + 72.27	02 - 32 - 040 NGC 4390	1563	1.8 x 1.4 1.8: x 1.2	95: ...	3	Sc		13.7		23
07520	12 23.3 + 12 56 279.18 + 74.34	02 - 32 - 041 NGC 4388	1563	6.2 x 1.7 5.7 x 1.5	92 6	1	Sb:		12.2		*123 4578
07521	12 23.3 + 27 50 207.16 + 84.29	05 - 29 - 083 NGC 4393 = 3323*	1398 263	3.5 x 3.0 3.3: x 2.8:	(0) 1 - 2	4	Sc	SAB d	13.8		13
07522	12 23.4 + 03 42 287.43 + 65.52	01 - 32 - 058	1560	2.8 x 0.25 2.8 x 0.25	130 7	3	Sc		15.0		
07523	12 23.4 + 18 30 268.18 + 79.33	03 - 32 - 035 NGC 4394	1576	3.6 x 3.4 3.9 x 3.5	-- 1	4	S B b P w 7508	(R)S B(r)ab Sb -	11.9	+ 772 + 719	*12- -9
07524	12 23.4 + 33 49 162.05 + 81.55	06 - 27 - 053 NGC 4395	1599	16. x 13. 13. x 13.	-- 1:	5	... S IV - V	SA(s)m	11.7		*12- -6789
07525	12 23.4 + 45 33 136.98 + 71.15	08 - 23 - 033	1408 265	1.2 x 1.0 1.1 x 0.9	(25) 1 - 2	4	S B c sff of 2, disturbed:		15.5		*

N G C   4370, 4371, 4374 - 80, 4382 - 91, 4393 - 95   I C   3300, 3303, 3305, 3308, 3309, 3311
                                                                                           3322, 3323

1	2	3	4	5	6	7	8	9	10	11	12
07526	12 23.5 + 15 57 274.40 + 77.12	03 − 32 − 034 NGC 4396	1576	3.5 x 1.2 3.5 x 0.9	125 6	2	...	S dm:	13.7		12
07527	12 23.5 + 31 06 177.84 + 83.32	05 − 29 − 084 IC 3330	1398 267	1.2 x 0.6 1.1 x 0.6	103 5	1	S B ...		15.1		12
07528	12 23.6 + 13 23 278.80 + 74.78	02 − 32 − 044 NGC 4402	1563	3.8 x 1.1 3.6 x 0.9	90 6	1	S ...		13.6		123 5
07529	12 23.6 + 16 28 273.43 + 77.59	03 − 32 − 036 NGC 4405 = 788*	1576	1.7 x 1.1 1.7 x 1.2	20 3	2	S0 − a		12.9		123
07530	12 23.6 + 46 11 136.30 + 70.56	08 − 23 − 034	1408 265	1.1 x 0.3 1.1 x 0.3	10 6	2	Sb ssf of 2		16.0:		*1
07531	12 23.7 − 01 01 289.89 + 60.94	00 − 32 − 011	1405	[1.1 x 0.8] ...	... ...	2	Multiple system		14.9		*1
07532	12 23.7 + 13 14 279.10 + 74.65	02 − 32 − 046 NGC 4406 = M 86	1563	(12. x 9.) (12. x 9.)	130: ...	4	E Markarian's chain		10.9 − −	202 367	*12− −9
07533	12 23.8 + 07 44 284.56 + 69.45	01 − 32 − 061 IC 789	1560	(1.0 x 0.5) 1.1 x 0.6	140: 5:	1	S0?		15.2		
07534	12 23.8 + 58 35 129.29 + 58.51	10 − 18 − 045	1427	3.5: x 2.0: 2.0:x 1.5::	... ...	4	Irr IB m Ir IV − V Ir I		15.3		*
07535	12 23.9 + 09 17 283.71 + 70.94	02 − 32 − 047 NGC 4410	1563 266	1.0 x 0.6 1.0 x 0.7	[110] ...	2	Double system Contact		13.6		*12 35
07536	12 24.0 + 04 14 287.47 + 66.08	01 − 32 − 062 NGC 4412	1560	1.6 x 1.4 1.7 x 1.4	−− 1 − 2	2	S B c? Singular		13.2		*23 4678
07537	12 24.0 + 09 09 283.90 + 70.82	02 − 32 − 048 NGC 4411a	1563 266	2.3 x 2.2 2.2:x 2.0:	−− 1	4	S B c SAB d P w 7546		14.4		*12 59
07538	12 24.0 + 12 53 279.83 + 74.35	02 − 32 − 049 NGC 4413	1563	2.4 x 1.7 2.3 x 1.7	60 4	3	S B a		13.6		*12 3
07539	12 24.0 + 31 30 174.50 + 83.19	05 − 29 − 085 NGC 4414	1398 267	4.8:x 3.2: 4.5:x 3.0:	155 3:	3	Sc		10.9 + +	715 720	*123 4678
07540	12 24.1 + 08 42 284.37 + 70.40	02 − 32 − 052 NGC 4415	1563	1.4 x 1.2 1.3 x 1.1	(0) 2:	2	S0 − a		14.2		3
07541	12 24.2 + 08 12 284.86 + 69.92	01 − 32 − 063 NGC 4416	1560	1.8 x 1.6 1.7 x 1.7	−− 1	4	S B c SB(rs)cd: Sc +		13.5		3
07542	12 24.2 + 09 51 283.40 + 71.50	02 − 32 − 053 NGC 4417	1563	3.2 x 1.2 (2.9 x 1.2)	49 7	4	S0		12.2		*12− −5789
07543	12 24.2 + 53 18 131.37 + 63.70	09 − 20 − 177	1389	1.0 x 0.45 0.9 x 0.4	98 6	2	Sb		16.0:		
07544	12 24.2 + 62 39 127.90 + 54.51	11 − 15 − 055	0674	1.1: x 1.1: 0.8: x 0.8:	−− ...	3	Dwarf V, sf of 2		17.		*
07545	12 24.3 − 00 35 290.02 + 61.39	00 − 32 − 012 NGC 4418	1405	1.4 x 0.7 1.4 x 0.7	59 5	2	Sa np of 2		14.2		*3
07546	12 24.3 + 09 09 284.12 + 70.84	02 − 32 − 055 NGC 4411b	1563 266	3.2 x 3.2 2.5: x 2.5:	−− 1	4	Sc SA(s)d P w 7537		14.4		*12
07547	12 24.3 + 11 51 281.37 + 73.41	02 − 32 − 000 IC 3356	1563	1.6::x 1.1:: 1. x 1.	... ...	4	Dwarf irregular Ir V		15.5		*
07548	12 24.3 + 13 27 279.31 + 74.91	02 − 32 − 056 IC 3355	1563	1.3 x 0.5 1.2 x 0.5	[172] ...	4	Irr I m Ir I		15.2		1
07549	12 24.4 + 02 46 288.51 + 64.67	01 − 32 − 064 NGC 4420	1560	2.2 x 1.0 2.0 x 0.9	8 6	3	Sc		12.7		*123 4678
07550	12 24.4 + 11 56 281.35 + 73.49	02 − 32 − 057 IC 3358	1563	(1.2 x 1.0) 1.1 x 0.9	−− ...	1	...		15.0		*1
07551	12 24.4 + 15 19 276.45 + 76.63	03 − 32 − 038 NGC 4419	1576	3.1 x 1.0 3.2 x 1.1	133 6	2	Sa		11.6		*123 4678
07552	12 24.4 + 53 29 131.22 + 63.52	09 − 20 − 178	1389	1.0 x 0.2 1.0 x 0.2	179 ...	1	...		16.0:		
07553	12 24.5 − 01 15 290.39 + 60.75	00 − 32 − 000	1405	1.1 x 0.1 1.0: x 0.1	60 7	2	... d − m		17.		
07554	12 24.5 + 15 45 275.75 + 77.04	03 − 32 − 039 NGC 4421	1576	(2.5 x 1.8) (2.8 x 2.2)	20: 2 − 3	2	S B 0/S B a		12.9 + +	1692 1628	*12 3
07555	12 24.5 + 22 55 250.62 + 82.87	04 − 29 − 071 IC 791	0135	1.1 x 1.1 1.2 x 1.1	−− 1	4	S B a (R)SB(r)a		14.2		
07556	12 24.6 + 06 09 286.64 + 67.97	01 − 32 − 065 NGC 4423	1560	2.2 x 0.35 2.2 x 0.3	18 7	3	Sc − Irr Sc + or Ir I		14.4		3
07557	12 24.6 + 07 32 285.65 + 69.31	01 − 32 − 066	1560	3.2: x 2.6: 3. x 2.5::	−− ...	4	Dwarf spiral, S m S V		15.3		
07558	12 24.6 + 16 37 274.12 + 77.83	03 − 32 − 040 IC 792	1576	1.8 x 0.6 1.6 x 0.5	59 6	2	Sb		15.1		*1
07559	12 24.6 + 37 25 148.62 + 78.74	06 − 27 − 055	1599	4.5::x 3.0:: 4. x 2.	... ...	4	Dwarf irregular, IB m Ir V Ir I		15.6		
07560	12 24.6 + 48 33 133.98 + 68.33	08 − 23 − 037 Mark 210	1408	1.0 x 0.9 0.9 x 0.8	−− ...	2	S B ...		14.7 +	7800	*1

N G C   4396, 4402, 4405, 4406, 4410 − 21, 4429   I C   788, 789, 791, 792, 3330 3355, 3356, 3358

1	2	3	4	5	6	7	8	9	10	11	12
07561	12 24.7 + 09 42 283.91 + 71.40	02 – 32 – 058 NGC 4424	1563	(3.6 x 1.9) (3.5 x 1.7)	95: 5:	2	S ... Singular		13.1		*12– –578
07562	12 24.7 + 13 01 280.25 + 74.54	02 – 32 – 059 NGC 4425	1563	2.9 x 0.9 2.7 x 0.8	27 6	1	S0 – a		13.3	+ 1883 + 1808	*12– –589
07563	12 24.7 + 16 12 275.08 + 77.47	03 – 32 – 041 IC 3365	1576	2.1: x 1.2: 2.0: x 0.9:	[72] ...	4	Irr I m Ir I		15.0		1
07564	12 24.8 + 03 34 288.32 + 65.48	01 – 32 – 000	1560	1.1 x 0.3 0.9 x 0.3	64 6	2	S B b nnp of 2		15.4		*
07565	12 24.8 + 11 09 282.55 + 72.78	02 – 32 – 060 IC 3371	1563	1.9 x 0.15 1.8: x 0.15	55 7	3	Sc		15.0		
07566	12 24.9 + 06 32 286.57 + 68.36	01 – 32 – 067 NGC 4430	1560	2.9 x 2.3 (3.0 x 2.7)	(80) 1 – 2	2	S B b: SB(rs) ... P w 7570, disturbed		13.4		*13
07567	12 24.9 + 07 55 285.56 + 69.70	01 – 32 – 000	1560	1.1 x 0.7: 1.1: x 0.7:	55 ...	3	Dwarf		15.3		*
07568	12 24.9 + 11 23 282.38 + 73.01	02 – 32 – 061 NGC 4429	1563	(5.5 x 2.5) (6. x 3.)	99 5	4	S0		11.4	+ 1114 + 1032	*12– –9
07569	12 24.9 + 12 34 281.00 + 74.13	02 – 32 – 062 NGC 4431	1563	1.7 x 1.0 1.5 x 0.9	177 4:	3	S0 NGC 4440 group		14.5		*12 3
07570	12 25.0 + 06 30 286.66 + 68.34	01 – 32 – 068 NGC 4432	1560	1.0 x 0.7 0.9 x 0.5	... ...	3	Sb P w 7566 Sb +		15.0		*13
07571	12 25.0 + 08 26 285.22 + 70.20	01 – 32 – 069 NGC 4434	1560	(1.3 x 1.3) (1.4 x 1.4)	— ...	2	E		13.2		3
07572	12 25.0 + 65 05 127.11 + 52.13	11 – 15 – 056 NGC 4441	0674 255	(4.5 x 3.5) (4.0 x 3.0)	... ...	4	... Peculiar		13.5		*23
07573	12 25.2 + 12 35 281.23 + 74.17	02 – 32 – 066 NGC 4436	1563	(1.6 x 0.8) (1.5 x 0.7)	118 5 – 6	3	S0 NGC 4440 group		14.8		*12 3
07574	12 25.2 + 13 17 280.33 + 74.83	02 – 32 – 065 NGC 4438	1563	9.7 x 3.9 9.7 x 4.0:	[27] ...	4	S ... P w 7575, disrupted, Markarian's chain	12.0	– 32 – 105	*12– –9VA
07575	12 25.2 + 13 21 280.24 + 74.89	02 – 32 – 064 NGC 4435	1563	(3.2 x 1.9) (2.6 x 1.8)	13 ...	4	S B 0 P w 7574, disturbed, Markarian's chain	11.9	+ 869 + 796	*12–V –6789A
07576	12 25.2 + 28 58 194.98 + 84.53	05 – 29 – 086	1398 263	1.4 x 0.5 1.2: x 0.5	[54] ...	1	... Double system?		16.0:		1
07577	12 25.2 + 43 46 137.79 + 72.95	07 – 26 – 006	0115	4.5: x irregular 4.0:x 3.0:	[130] ...	4	Dwarf irregular Ir IV – V, p w 7608		14.7		*1
07578	12 25.3 + 27 16 213.09 + 84.74	05 – 29 – 087 IC 3376	1398 263	1.8 x 1.4 1.8 x 1.4	60: 2	3	S B a (R)SB a		14.4		1
07579	12 25.4 + 06 00 287.25 + 67.88	01 – 32 – 070	1560	1.3 x 0.2 1.2 x 0.2	130 7	1	S – Irr		15.0		1
07580	12 25.4 + 08 22 285.56 + 70.17	01 – 32 – 071	1560	1.3 x 1.2 0.8 x 0.4	— ...	1	...		15.2		
07581	12 25.4 + 12 34 281.42 + 74.17	02 – 32 – 067 NGC 4440	1563	1.7 x 1.6 1.8 x 1.7	— 1	4	S B a SB(rs)a Brightest of 3		13.0		*12 3
07582	12 25.4 + 50 41 132.32 + 66.29	09 – 20 – 187	1389	1.2 x 0.3 1.2 x 0.3	177 6	3	S B b		15.6		*1
07583	12 25.5 + 10 05 284.16 + 71.82	02 – 32 – 068 NGC 4442	1563	(4.5 x 1.8) (4.5 x 1.8)	87 6	4	S B 0		11.2	+ 580 + 493	*12– –9
07584	12 25.5 + 22 52 252.16 + 83.01	04 – 29 – 000	0135	1.0 x 0.5 0.9 x 0.4	120 ...	2	Sc – Irr		16.0:		
07585	12 25.6 + 12 22 281.83 + 74.00	02 – 32 – 070 IC 794	1563	(1.4 x 1.0) (1.5 x 1.0)	(110) ...	2	E		15.1		*2
07586	12 25.6 + 14 11 279.41 + 75.70	02 – 32 – 069 NGC 4446	1563	1.1 x 1.1 1.1 x 1.0	— 1 – 2	3	Sc P w NGC 4447, disturbed, bridge:		15.0		*1
07587	12 25.7 + 09 43 284.64 + 71.48	02 – 32 – 072 NGC 4445	1563	2.6 x 0.45 2.5 x 0.45	106 7	1	S ...		13.7		*12 35
07588	12 25.7 + 13 50 280.01 + 75.39	02 – 32 – 000	1563	1.0 x 0.1 1.0 x 0.1	149 7	3	Sc		16.0:		*
07589	12 25.7 + 12 04 282.26 + 73.72	02 – 32 – 074 IC 3381	1563	(1.3 x 1.0) (1.3 x 1.0)	(110) ...	1	E P w 7601		15.1		*2
07590	12 25.8 + 09 00 285.33 + 70.80	02 – 32 – 076	1563 268	1.5 x 0.4 1.1 x 0.4	168 6	1	S ... P w 7596		15.0		*
07591	12 25.8 + 28 54 195.30 + 84.67	05 – 29 – 089 NGC 4448	1398 263	4.0 x 1.5 4.0 x 1.5	94 6	3	Sa		11.9	+ 693 + 687	*123 4678
07592	12 25.8 + 44 22 136.83 + 72.41	07 – 26 – 009 NGC 4449	0115	6.0 x 4.5 6. x 4.	[45] ...	5	Irr IB m Ir I		10.0	+ 207 + 269	*12– –8
07593	12 25.8 + 44 43 136.49 + 72.07	08 – 23 – 38+39 Mark 212	1408	[1.2 x 0.9] ...	... ...	3	Double system Connected		14.8	+ 6900	*1C
07594	12 25.9 + 17 21 273.86 + 78.62	03 – 32 – 048 NGC 4450	1576	5.5: x 3.7: 6.5: x 4.5:	175 3	4	Sb SA ab Sb –		11.2	+ 2048 + 1994	*12– –9
07595	12 25.9 + 18 41 270.45 + 79.80	03 – 32 – 047 IC 3391	1576	1.1 x 0.9 1.1 x 0.9	63 2	3	Sc		13.9		

N G C   4424, 4425, 4429 – 32, 4434 – 36, 4438   I C   794, 3365, 3371, 3376, 3381, 3391
        4440 – 42, 4445, 4446, 4448 – 50

1	2	3	4	5	6	7	8	9	10	11	12
07596	12 26.0 + 08 54 285.56 + 70.72	02 – 32 – 078	1563	1.9 x 0.7 1.4 x 0.6	140 ...	3	Dwarf P w 7590		15.3		*
07597	12 26.0 + 28 57 194.65 + 84.70	05 – 30 – 001	0064 263	1.3 x 0.5 1.2 x 0.4	23 (6)	1	S: ...		16.0:		*1
07598	12 26.0 + 32 50 164.06 + 82.64	06 – 27 – 056	1599	1.5: x 1.5: 1.0 x 0.9	-- 1	3	S B c		15.6		
07599	12 26.0 + 37 30 147.16 + 78.80	06 – 27 – 057	1599	2.0 x 1.0 2.0: x 1.2:	135 (5)	3	Dwarf spiral, S m S V		15.6		*
07600	12 26.1 + 09 32 285.10 + 71.33	02 – 32 – 079 NGC 4451	1563	(1.5 x 1.0) 1.4 x 0.8	162 ...	1	...		13.4		123
07601	12 26.2 + 12 02 282.71 + 73.73	02 – 32 – 080 NGC 4452	1563	2.7 x 0.6 2.3 x 0.5	32 7	1	S0? P w 7588		13.1		*12 348
07602	12 26.2 + 15 16 278.27 + 76.76	03 – 32 – 049 IC 3392	1576	2.1 x 0.9 2.0 x 0.9	40 6	2	Sa – b		13.3		12
07603	12 26.2 + 23 06 251.58 + 83.29	04 – 30 – 001 NGC 4455	1435	2.7 x 1.0 2.7 x 1.0	16 (6)	3	Sc – Irr		13.0		123 478
07604	12 26.2 + 31 46 170.24 + 83.40	05 – 30 – 000	0064 267	1.2 x 0.1 1.1 x 0.1	12 7	2	Sc		16.5:		
07605	12 26.2 + 36 00 150.96 + 80.13	06 – 27 – 058	1599	1.6:: x 1.1: 1.4: x 1.0:	[30] ...	3	Dwarf irregular, I m Ir IV – V		15.3		
07606	12 26.3 – 01 39 291.44 + 60.44	00 – 32 – 014 NGC 4454	1405	(2.5 x 2.2) (2.7 x 2.5)	(100) 1	4	S B 0/S B a (R)SB(r)0/a		13.5		*12 34$\overline{7}\underline{8}$
07607	12 26.3 + 04 34 288.67 + 66.53	01 – 32 – 074	1560	2.1 x 0.10 1.8 x 0.12	54 7	3	Sc Sc +		15.5		
07608	12 26.3 + 43 31 137.40 + 73.24	07 – 26 – 010	0115	4.0: x 3.5: 4. x 3.	-- ...	4	Dwarf irregular Ir V, p w 7577		16.0:		*1
07609	12 26.4 + 03 51 289.11 + 65.84	01 – 32 – 075 NGC 4457	1560	(3.4 x 2.8) 3.2 x 2.9	-- 1 – 2	4	S0/S B a (R)SAB 0/a		11.9	+ 738 + 627	*123 46$\overline{7}\underline{8}$
07610	12 26.4 + 13 31 281.07 + 75.15	02 – 32 – 082 NGC 4458	1563	(1.5 x 1.5) (1.5 x 1.5)	-- ...	4	E P w 7613 Markarian´s chain		13.3	+ 383 + 311	*12 3$\underline{}$
07611	12 26.4 + 45 09 135.77 + 71.69	08 – 23 – 041 NGC 4460	1408 265	4.2 x 1.2 4.0 x 1.2	40 6	3	S0		12.5		*123 46$\overline{7}\underline{8}$
07612	12 26.5 + 03 00 289.60 + 65.01	01 – 32 – 077	1560	2.2: x 1.1 2.0: x 1.0:	[145] ...	4	Dwarf spiral S B: V		15.2		*
07613	12 26.5 + 13 28 281.23 + 75.11	02 – 32 – 084 NGC 4461	1563	(3.6 x 1.3) (3.6 x 1.3)	9 6	4	S0 P w 7610 Markarian´s chain		12.2	+ 1887 + 1815	*123 46$\overline{7}\underline{8}$
07614	12 26.5 + 14 15 280.14 + 75.84	02 – 32 – 083 NGC 4459	1563	(3.5 x 2.7) (3.5 x 2.7)	110 2:	4	S0		11.6	+ 1111 + 1042	*123 46$\overline{7}\underline{8}$
07615	12 26.5 + 28 04 204.07 + 84.98	05 – 30 – 005 IC 3407	0064 263	1.1 x 0.5 1.2 x 0.7	145: ...	1	S ... Disturbed?		14.7		
07616	12 26.5 + 29 08 192.40 + 84.75	05 – 29 – 000 IC 3402	1398 263	1.0 x 0.15 1.0 x 0.15	8 7	2	Sb – c		15.7		
07617	12 26.7 + 44 55 135.81 + 71.93	08 – 23 – 042	1408 265	2.5 x 0.3 2.5 x 0.3	68 7	3	Sc Sc +		15.6		*
07618	12 26.7 + 58 12 128.70 + 58.95	10 – 18 – 053	1427	1.4 x 0.15 1.1 x 0.12	45 7	3	Sc		16.0:		*
07619	12 26.8 + 08 26 286.50 + 70.32	01 – 32 – 078 NGC 4464	1560	1.0 x 0.7 0.9 x 0.7	0: ...	1	S? ...		13.5	+ 1199 + 1106	*12 3
07620	12 26.8 + 11 43 283.55 + 73.47	02 – 32 – 088 IC 3413	1563	(1.1 x 0.7) (1.2 x 0.7)	160: ...	1	E: P w 7630		15.2		*2:
07621	12 26.9 + 07 03 287.55 + 68.99	01 – 32 – 079 IC 3414	1560	1.7 x 1.0 1.9: x 1.2:	[35:] ...	2	S B c/Irr SAB dm?		14.2		
07622	12 26.9 + 09 01 286.11 + 70.89	02 – 32 – 089 NGC 4469	1563	3.2 x 1.1 3.1 x 1.1	89 6 – 7	2	S0 – a		12.6		*123 46$\overline{7}\underline{8}$
07623	12 26.9 + 16 41 276.37 + 78.13	03 – 32 – 051 IC 796	1576	1.4 x 0.5 1.4 x 0.5	145 6	2	S0 – a		13.9		1
07624	12 27 + 82 10 123.92 + 35.19	14 – 06 – 000	1340	1.2: x 0.9: 1.2: x 0.9:	... ...	2	Sc		17.		
07625	12 27.0 + 01 07 290.73 + 63.19	00 – 32 – 000	1405	1.0 x 0.1 0.9 x 0.1	73 7	2	... m		17.		*
07626	12 27.0 + 07 58 286.98 + 69.88	01 – 32 – 081 NGC 4466	1560	1.2 x 0.35 1.1 x 0.3	101 6	1	S ...		14.7		1$\underline{2}$3
07627	12 27.0 + 08 06 286.88 + 70.01	01 – 32 – 082 NGC 4470	1560	1.4 x 0.9 1.5 x 1.1	0: ...	1	S ...		12.9		23
07628	12 27.0 + 14 20 280.49 + 75.96	02 – 32 – 090 NGC 4468	1563	1.2 x 0.9 (1.5 x 0.9)	73: ...	1	E?		14.2		*12 3$\underline{}$
07629	12 27.2 + 08 16 286.90 + 70.18	01 – 32 – 083 NGC 4472 = M 49	1563	(8. x 7.) (8. x 7.)	(155) ...	5	E		10.2	+ 948 + 855	*12– –9$\overline{\text{A}}$
07630	12 27.2 + 11 40 283.93 + 73.46	02 – 32 – 000	1563	1.5: x 1.0: 1.3:: x 1.	... ...	4	Dwarf irregular Ir V		16.5:		*

N G C   4451, 4452, 4454, 4455, 4457 – 61
4464, 4466, 4468 – 70, 4472

I C   796, 3392, 3402, 3407, 3413, 3414

1	2	3	4	5	6	7	8	9	10	11	12
07631	12 27.3 + 13 42 281.64 + 75.39	02 - 32 - 093 NGC 4473	1563	(3.6 x 2.3) (3.5 x 2.3)	100 ...	4	E Markarian's chain		11.2	+ 2241 + 2171	*123 4678
07632	12 27.3 + 27 32 210.16 + 85.19	05 - 30 - 008 NGC 4475	0064 263	2.0 x 1.0 2.0 x 1.1	5 5	3	Sb	SA bc Sb +	14.6		3
07633	12 27.4 + 10 53 284.87 + 72.72	02 - 32 - 095 IC 3425	1563	1.9 x 0.8 (0.9 x 0.3)	35 6:	1	S0?		15.3		
07634	12 27.4 + 14 21 280.84 + 76.01	02 - 32 - 094 NGC 4474	1563	1.9 x 1.1 (2.1 x 1.2)	80 ...	3	S0		12.6	+ 1526 + 1459	*12 3478
07635	12 27.4 + 57 06 128.88 + 60.05	10 - 18 - 055	1427	(1.2 x 0.7) (1.2 x 0.7)	100: 4:	3	S0		14.6		*
07636	12 27.5 + 08 12 287.16 + 70.14	01 - 32 - 084	1560	1.2:: x 0.8:: 1. x 1.	... ...	4	Dwarf irregular Ir V		15.4		*1
07637	12 27.5 + 12 37 283.15 + 74.38	02 - 32 - 096 NGC 4476	1563	1.7 x 1.1 (1.7 x 1.1)	25 3 - 4	3	S0 P w 7645		13.3		*12 348
07638	12 27.5 + 13 55 281.53 + 75.62	02 - 32 - 097 NGC 4477	1563	3.8 x 3.5 3.8 x 3.5	(15) 1	4	S B 0 Markarian's chain		11.9	+ 1263 + 1194	*12 3478
07639	12 27.5 + 47 48 133.19 + 69.17	08 - 23 - 043	1408 271	3.6: x 1.7: 2.7: x 1.6:	153 (5)	4	Irr	I m Ir I	14.9		
07640	12 27.6 + 11 03 284.87 + 72.89	02 - 32 - 098 IC 3427	1563	(1.6 x 0.9) (1.6 x 0.9)	145: ...	3	E		14.2		*
07641	12 27.6 + 37 48 145.03 + 78.67	06 - 27 - 000	1599	1.4 x 0.7 1.2: x 0.5:	5: 5:	2	...	S B dm:	17.		
07642	12 27.7 + 02 54 290.34 + 64.98	01 - 32 - 085	1560	1.1 x 0.9 1. x 0.5:	130: ...	3	Dwarf V		15.3		
07643	12 27.7 + 09 21 286.44 + 71.26	02 - 32 - 000 IC 3430	1563	1.2: x 0.3 1.1 x 0.3	115 ...	1	...		15.4		*
07644	12 27.8 + 04 01 289.87 + 66.08	01 - 32 - 086	1560	2.2 x 0.5 2.1 x 0.4	50 6	4	Sc	Sc +	15.1		
07645	12 27.8 + 12 36 283.43 + 74.39	02 - 32 - 099 NGC 4478	1563	(1.7 x 1.4) (1.7 x 1.4)	(140) ...	3	E P w 7637		12.2	+ 1482 + 1407	*12 3478
07646	12 27.8 + 13 51 281.90 + 75.58	02 - 32 - 100 NGC 4479	1563	(1.5 x 1.4) (1.4 x 1.2)	— ...	3	S B 0		13.9	+ 822 + 753	*12 3
07647	12 27.9 + 04 31 289.67 + 66.57	01 - 32 - 087 NGC 4480	1560	2.5 x 1.2 2.6 x 1.2	175 5 - 6	4	Sc	Sc -	13.4		*12 3569
07648	12 28.0 + 41 59 138.04 + 74.80	07 - 26 - 013 NGC 4485	0115	3.0 x 2.5 2.7 x 2.3	[15] ...	3	Irr P w 7651, disrupted		12.4	+ 795 + 848	*12-V -5789A
07649	12 28.1 + 09 17 286.79 + 71.22	02 - 32 - 103 NGC 4483	1563	1.5 x 0.9 1.7 x 0.9	65 ...	2	S B 0		13.4		*12 348
07650	12 28.1 + 15 24 279.92 + 77.06	03 - 32 - 053 IC 3455	1576	1.2 x 0.2 1.0 x 0.2	141 7	1	S ...		15.7		
07651	12 28.1 + 41 55 138.05 + 74.87	07 - 26 - 014 NGC 4490	0115	7.0 x 3.5 7.0 x 3.5	[125] 5:	3	... S dm: P w 7648, disrupted		10.1	+ 570 + 622	*12- -9VA
07652	12 28.2 + 12 39 283.72 + 74.47	02 - 32 - 000	1563	[1.1 x 0.2] ...	... ...	2	Double system Bridge		16.0:		*
07653	12 28.3 + 08 38 287.42 + 70.61	02 - 32 - 104 NGC 4488	1563	4.6: x 1.4 (4.3 x 1.3)	5: 6:	1	S B ...		13.8		23
07654	12 28.3 + 12 40 283.78 + 74.49	02 - 32 - 105 NGC 4486 = M 87	1563	(7. x 7.) (7.0 x 6.5)	— ...	5	E Brightest in Virgo cluster, eruptive		10.4	+ 1261 + 1187	*12- -9A
07655	12 28.3 + 17 02 277.17 + 78.58	03 - 32 - 054 NGC 4489	1576	(1.8 x 1.7) (1.9 x 1.8)	— ...	3	E P w 7669		13.2		*3
07656	12 28.4 + 08 21 287.70 + 70.34	01 - 32 - 089 NGC 4492 = 3428*	1560	(2.1 x 2.1) (2.5 x 2.3)	— 1	1	S ...		14.1	+ 1735 + 1643	123
07657	12 28.4 + 11 46 284.82 + 73.64	02 - 32 - 107 NGC 4491	1563	1.8 x 1.0 1.7 x 1.0	148 ...	1	...		13.7		*12 3
07658	12 28.4 + 12 33 284.00 + 74.39	02 - 32 - 110	1563	1.0 x 0.9 1.0 x 0.9	— ...	1	[E] Very compact		11.2		
07659	12 28.4 + 57 34 128.46 + 59.61	10 - 18 - 061	1427	1.1 x 0.3 1.0 x 0.3	140 6	1	S ... sf of 2, disturbed:		14.9		*
07660	12 28.7 + 36 53 146.06 + 79.59	06 - 28 - 003	0105	1.1 x 0.4 1.1 x 0.4	121 6	2	Sa - b		15.3		*
07661	12 28.7 + 52 42 130.13 + 64.42	09 - 21 - 010	0729	1.3 x 0.6 1.3 x 0.6	50 5	4	S B b (R')S B(s)b p of 2 Sb +		15.7		*
07662	12 28.9 + 26 03 228.60 + 85.31	04 - 30 - 002 NGC 4494	1435	(4.5 x 4.3) (4.5 x 4.3)	— ...	5	E		10.7	+ 1321 + 1305	*123 4678
07663	12 28.9 + 29 25 187.29 + 85.14	05 - 30 - 012 NGC 4495	0064 263	1.6 x 0.9 1.6 x 0.9	130 4:	2	Sa - b		14.1		*13
07664	12 29 + 85 04 123.51 + 32.31	14 - 06 - 006	1340	1.1: x 0.4 0.9 x 0.5	82 ...	1	S ...		16.5:		
07665	12 29.0 + 11 54 285.19 + 73.81	02 - 32 - 113 NGC 4497	1563	(2.0 x 1.1) (2.0 x 1.1)	65 5	3	S0		13.8		23

N G C 4473 - 80, 4483, 4485, 4486, 4488 - 92   I C 3425, 3427, 3428, 3430, 3455
4494, 4495, 4497

1	2	3	4	5	6	7	8	9	10	11	12
07666	12 29.0 + 15 08	03 - 32 - 057	1576	1.3 x 0.15	166	3	Irr		15.2		
	281.24 + 76.88	IC 3453		0.9 x 0.15	7			Ir I			
07667	12 29.0 + 58 15	10 - 18 - 062	1427	1.7 x 1.1	130	3	S B a		13.2	+ 3000	*12
	128.10 + 58.95	NGC 4500, Mark 213		1.7 x 1.1	...						3
07668	12 29.1 + 04 13	01 - 32 - 090	1560	4.0 x 3.0	[70]	4	S B c + companion		13.3	+ 1883	*12-
	290.56 + 66.34	NGC 4496		3.5: x 3.0:	...		Contact			+ 1775	-6789V
07669	12 29.1 + 17 08	03 - 32 - 056	1576	3.5: x 1.7	133	4	S B c	S B d	12.8		*12
	277.86 + 78.75	NGC 4498		3.0 x 1.7	4 - 5		P w 7655				3
07670	12 29.1 + 27 47	05 - 30 - 013	0064	1.3 x 0.2	26	1	Sb:		15.7		
	206.88 + 85.57	IC 3454	263	1.3 x 0.2	7						
07671	12 29.2 + 33 52	06 - 28 - 004	0105	1.1 x 0.35	18	1	S ...		15.6		
	155.14 + 82.26			0.9 x 0.3	6						
07672	12 29.3 + 12 56	02 - 32 - 114	1563	1.1: x 0.8:	--	3	Dwarf		15.4		*
	284.37 + 74.82	IC 3457		0.9: x 0.8:	...		V				
07673	12 29.3 + 30 00	05 - 30 - 014	0064	2.0: x 1.6:	--	4	Dwarf irregular		17.		*1
	180.79 + 84.96		263	1.8:: x 1.8::	...		Ir V				
07674	12 29.4 + 12 27	02 - 32 - 115	1563	1.2: x 1.1:	--	3	Dwarf	m	15.5		
	284.98 + 74.36	IC 3459		1.2: x 1.0:	...						
07675	12 29.4 + 14 42	03 - 32 - 059	1576	6.7 x 3.0	140	4	Sb/Sc	SA bc	10.6	+ 2120	*12-
	282.26 + 76.51	NGC 4501		6.7 x 3.3	5 - 6			Sb +		+ 2056	-9
07676	12 29.4 + 15 24	03 - 32 - 058	1576	1.2 x 0.8	108	1	S B c:		13.9		
	281.24 + 77.17	IC 797		1.2 x 0.8	3						
07677	12 29.5 + 16 58	03 - 32 - 060	1576	1.3 x 0.6	40	2	Sc		14.8		3
	278.64 + 78.64	NGC 4502		1.1 x 0.5	5						
07678	12 29.5 + 40 07	07 - 26 - 019	0115	1.6 x 1.5	--	1	S B? ...		13.9		
	139.36 + 76.66			1.4 x 1.2	...						
07679	12 29.5 + 64 31	11 - 15 - 058	0674	(1.5 x 0.9)	153	2	E		14.2		*3
	126.46 + 52.75	NGC 4510	255	(1.4 x 0.8)	...						
07680	12 29.6 + 11 27	02 - 32 - 118	1563	(3.5 x 1.7)	12	4	S0		12.4		*123
	286.13 + 73.41	NGC 4503		(3.6 x 1.7)	5						4678
07681	12 29.7 + 10 31	02 - 32 - 119	1563	(1.4 x 1.3)	--	3	E		14.6		
	287.02 + 72.52	IC 3468		(1.1 x 1.0)	...						
07682	12 29.7 + 13 42	02 - 32 - 120	1563	(1.7 x 1.5)	(110)	1	S B? ...		14.2		23
	283.84 + 75.58	NGC 4506		(1.7 x 1.4)	...						
07683	12 29.7 + 66 36	11 - 15 - 059	0674	(2.4 x 1.7)	15:	4	S0	(R)SA 0	14.1		*3
	126.02 + 50.69	NGC 4513	255	(1.8 x 1.2)	...						
07684	12 29.8 + 18 31	03 - 32 - 062	1576	1.1: x 0.7:	35:	2	...	S d - dm	15.4		
	275.57 + 80.08	IC 3473		1.1 x 0.7	...						
07685	12 29.9 + 00 40	00 - 32 - 019	1405	4.7 x 3.0	30	4	...	S B dm	14.1		*12
	292.48 + 62.88	Reinmuth 80		3.8: x 3.0:	...		P w 7694				5789
07686	12 29.9 + 12 04	02 - 32 - 121	1563	1.3 x 0.25	72	1	Sc		15.4		
	285.79 + 74.03			1.0 x 0.2	7						
07687	12 30.0 + 02 56	01 - 32 - 091	1560	2.2 x 0.2	36	3	Sc		15.0		*2
	291.67 + 65.12	IC 3474		2.1 x 0.2	7			Sc +			
07688	12 30.0 + 08 19	01 - 32 - 092	1560	1. x 1.	--	2	Dwarf		15.3		*
	288.87 + 70.39			1. x 1.	...		V:				
07689	12 30.0 + 39 52	07 - 26 - 020	0115	1.8 x 0.2	64	3	Sc		16.0:		*1
	139.32 + 76.93			1.8 x 0.2	7						
07690	12 30.0 + 42 59	07 - 26 - 021	0115	2.3: x 1.8	[20]	3	Irr		13.7		
	135.66 + 73.95			1.9: x 1.7:	...						
07691	12 30.0 + 56 56	10 - 18 - 064	1427	0.8 x 0.35	...	1	...		14.3		*
	128.23 + 60.27			0.8 x 0.4	...						
07692	12 30.1 + 13 03	02 - 32 - 123	1563	2.5: x 2.3:	--	3	Dwarf (elliptical)		15.4		*12
	284.96 + 74.99	IC 3475		2. x 2.	...						
07693	12 30.1 + 30 00	05 - 30 - 015	0064	1.2 x 1.1	--	2	Sb - c		14.2		13
	179.78 + 85.11	NGC 4514	263	1.2 x 1.1	1						
07694	12 30.2 + 00 24	00 - 32 - 020	1405	10.8 x 1.5:	83	4	Sc		12.4	+ 1218	*12-
	292.74 + 62.63	NGC 4517		10. x 1.5:	7		P w 7685	Sc +		+ 1095	-789
07695	12 30.2 + 14 20	02 - 32 - 125	1563	2.7 x 2.1	[30]	4	Irr	S m	13.5		*12
	283.53 + 76.22	IC 3476		2.3 x 1.8	...			Ir I			
07696	12 30.2 + 14 28	02 - 32 - 126	1563	(1.0 x 0.9)	--	1	S0?		15.0		2
	283.36 + 76.35	IC 3478		(1.0 x 0.9)	...						
07697	12 30.4 + 20 27	03 - 32 - 064	1576	2.1: x 0.4	99	3	Sc		15.3		
	270.58 + 81.86			1.6: x 0.3	6						
07698	12 30.4 + 31 49	05 - 30 - 016	0064	7. :: x 5. ::	...	5	Dwarf irregular, I m		15.6		*1
	164.34 + 84.01			7. :: x 5. ::	...		Ir V				
07699	12 30.4 + 37 54	06 - 28 - 008	0105	4.0 x 1.1	32	3	S B c		13.4		
	142.23 + 78.80			4.0 x 1.0	6						
07700	12 30.4 + 64 10	11 - 15 - 060	0674	2.2 x 1.6	80:	4	...	S B(s)dm	14.7		*3
	126.37 + 53.11	NGC 4512	255	1.8 x 1.1	...		P w 7706				

N G C   4496, 4498, 4500 - 03, 4506, 4510   I C   797, 3453, 3454, 3457, 3459
           4512 - 14, 4517   3468, 3473 - 76, 3478

1	2	3	4	5	6	7	8	9	10	11	12
07701	12 30.5 + 16 32 280.55 + 78.32	03 - 32 - 065 NGC 4515	1576	1.3 x 1.1 1.4 x 1.3	— ...	2	E - S0		13.3		13
07702	12 30.5 + 65 33 126.09 + 51.74	11 - 15 - 000	0674 255	1.0 x 0.12 1.0 x 0.1	134 7	1	Sb - c		16.5:		
07703	12 30.6 + 14 51 283.25 + 76.74	03 - 32 - 067 NGC 4516	1576	1.7 x 0.7 (2.0 x 0.8)	0 6	1	S ...		13.9		123
07704	12 30.6 + 32 22 160.64 + 83.63	05 - 30 - 018 NGC 4509	0064	1.0 x 0.6 0.9 x 0.6	(155) ...	1	... sf of 2		14.1		*23
07705	12 30.6 + 57 09 128.00 + 60.07	10 - 18 - 065	1427	1.0 x 0.8 1.0 x 0.8	(140) 1 - 2	3	Sc	Sc +	16.0:		*
07706	12 30.6 + 64 13 126.33 + 53.06	11 - 15 - 061 NGC 4521	0674 255	2.7 x 0.6 2.6 x 0.6	167 7	2	S0 - a P w 7700		13.0		*3
07707	12 30.8 + 17 45 278.56 + 79.48	03 - 32 - 000	1576	1.0: x 0.6: 0.9: x 0.5:	... ...	2	Dwarf P w IC 3484		17.		*
07708	12 30.8 + 48 21 131.37 + 68.75	08 - 23 - 053	1408 271	1.0 x 0.3 1.0 x 0.2	176 7	2	Sb		16.0:		*
07709	12 31.0 + 08 55 289.20 + 71.03	02 - 32 - 135 NGC 4519	1563	3.8 x 2.5 3.0 x 2.5	(145) ...	4	Sc/S B c	Sc +	12.8	+ 1213 + 1125	*12- -9
07710	12 31.1 - 02 23 294.06 + 59.91	00 - 32 - 021	1405	1.4 x 1.3 1.3: x 1.1:	— ...	4	Dwarf irregular Ir V		16.5:		*1
07711	12 31.1 + 09 26 288.93 + 71.54	02 - 32 - 137 NGC 4522	1563	4.2 x 1.1 3.7: x 1.0	33 6	3	Sc		13.6		123 478
07712	12 31.2 + 11 16 287.61 + 73.33	02 - 32 - 138 IC 3499	1563	1.6 x 0.6 1.5 x 0.5	125 6 - 7	2	S0 - a		14.5		
07713	12 31.3 + 15 26 283.15 + 77.35	03 - 32 - 068 NGC 4523	1576	2.5: x 2.2: 2.5:: x 2.3::	— ...	4	Irr	SAB m Ir I	15.1		*12 3
07714	12 31.3 + 30 34 172.66 + 85.00	05 - 30 - 020 NGC 4525	0064	3.0 x 1.6 3.2 x 1.7	53 5	3	Sc		13.0		3
07715	12 31.4 + 03 49 292.13 + 66.05	01 - 32 - 099	1560	1.0 x 1.0 1.0: x 1.0:	— ...	3	Dwarf		15.3		
07716	12 31.4 + 15 38 282.97 + 77.55	03 - 32 - 069 IC 800	1576	1.5 x 1.2 1.8: x 1.2	148: 2:	2	S B c:	Sc +?	14.3		*12
07717	12 31.4 + 52 32 129.26 + 64.65	09 - 21 - 017 IC 801	0729	1.2 x 1.0 (1.2 x 1.0)	(55) 2:	1	Sa?		14.8		
07718	12 31.5 + 07 58 290.16 + 70.13	01 - 32 - 100 NGC 4526	1560	(7.0 x 2.7) (7.5 x 3.0)	113 6 - 7	5	S0		10.6	+ 487 + 396	*123 4678
07719	12 31.5 + 39 18 138.90 + 77.56	07 - 26 - 025	0115	1.9 x 0.6 1.6 x 0.5	163 ...	2	...	S dm	15.4		
07720	12 31.6 - 00 05 293.64 + 62.20	00 - 32 - 022	1405	1.1 x 1.0 1.1 x 0.9	(30) 1 - 2	3	S B a - b	(R':)S B(rs) ..	14.8		
07721	12 31.6 + 02 56 292.60 + 65.18	01 - 32 - 101 NGC 4527	1560	6.5 x 2.2 6.7 x 2.2	67 6	3	Sb		12.4	+ 1727 + 1616	*12- -9
07722	12 31.6 + 11 36 287.67 + 73.68	02 - 32 - 140 NGC 4528	1563	1.5 x 1.0 1.6 x 1.0	5 3	3	S0		12.9		*12 3
07723	12 31.6 + 35 48 145.69 + 80.82	06 - 28 - 010 NGC 4534	0105	4.5: x 3.3: 2.8: x 2.5:	(125) ...	4	...	S dm Ir I	13.2		*12 3
07724	12 31.7 + 27 43 207.28 + 86.15	05 - 30 - 022 IC 3516	0064 263	1.3 x 0.15 1.0 x 0.15	72 7	2	Sb - c		16.0:		
07725	12 31.8 + 02 36 292.84 + 64.86	01 - 32 - 102 NGC 4533	1560	2.5:: x 0.3 1.6 x 0.3	161 7	2	S - Irr		14.7		2
07726	12 31.8 + 06 45 291.04 + 68.95	01 - 32 - 103 NGC 4532	1560	3.5: x 1.2 3.3: x 1.2	[160] ...	4	Irr	Ir I	12.3	+ 2154 + 2058	*12- -9
07727	12 31.8 + 08 28 290.08 + 70.63	01 - 32 - 104 NGC 4535	1560	7.8 x 7.0 7.5: x 6.0:	(0) 2:	5	Sc/S B c	SAB(s)c Sc -	11.1	+ 1943 + 1854	*12- -9
07728	12 31.8 + 11 21 288.04 + 73.45	02 - 32 - 142 IC 3510	1563	1.1 x 0.8 0.9 x 0.8	0 ...	1	S ...		15.2		
07729	12 31.8 + 13 21 286.19 + 75.39	02 - 32 - 141 NGC 4531	1563	(3.5 x 2.3) (3.6 x 2.5)	155 3	3	S B: 0		13.3		23
07730	12 31.8 + 64 49 126.00 + 52.48	11 - 15 - 063	0674	1.6: x 1.1:: 1.2:: x 0.8::	(30) ...	3	Dwarf (spiral) S: V		16.0:		
07731	12 32 + 82 52 123.64 + 34.51	14 - 06 - 000	1340	0.45 x 0.45 0.45 x 0.45	— ...	0	... Compact? or *?		14. ?		*
07732	12 32.0 + 02 28 293.01 + 64.74	00 - 32 - 023 NGC 4536	1405	7.0 x 2.8 7.7 x 3.1	130 6	5	Sb/Sc	SAB(rs)bc Sc -	12.3	+ 1927 + 1814	*12- -9
07733	12 32.0 + 09 26 289.61 + 71.59	02 - 32 - 143 IC 3517	1563	1.5: x 0.9: 1.4: x 0.9:	15: ...	3	...	S d - m	15.3		
07734	12 32.0 + 09 54 289.30 + 72.05	02 - 32 - 000 IC 3518	1563	1.3 x 0.8 1.1 x 0.5	[45] ...	2	...	m	15.3		
07735	12 32.0 + 18 28 278.48 + 80.25	03 - 32 - 071 NGC 4539	1576	3.0 x 1.1 3.2 x 1.2	95 6	2	S B? 0 - a		13.5		*12 359

N G C 4509, 4515, 4516, 4519, 4521 - 23
4525 - 28, 4531 - 36, 4539

I C 800, 801, 3499, 3510, 3516 - 18

1	2	3	4	5	6	7	8	9	10	11	12
07736	12 32.1 + 07 26 290.88 + 69.64	01 - 32 - 106 IC 3521	1560	1.1 x 0.8 1.3: x 1.0:	— ...	4	Irr	I B m	14.2		1
07737	12 32.1 + 15 29 283.92 + 77.45	03 - 32 - 072 IC 3522	1576	1.2: x 0.6:: 1.1: x 0.6:	... ...	3	Dwarf irregular Ir V		17.		*2
07738	12 32.1 + 42 44 134.53 + 74.29	07 - 26 - 000	0115	1.3 x 0.12 1.2 x 0.1	10 7	3	Sc		16.0:		
07739	12 32.2 + 06 34 291.40 + 68.79	01 - 32 - 107 Ho VII	1560	1.4: x 1.3: 1.2: x 1.0:	— ...	4	Dwarf irregular, I m Ir V		15.3		*12
07740	12 32.2 + 49 38 130.13 + 67.53	08 - 23 - 060	1408	1.0 x 0.15 0.9 x 0.15	19 7	1	S ...		16.0:		*
07741	12 32.2 + 70 08 125.10 + 47.20	12 - 12 - 009	1411	1.1 x 0.15 1.1 x 0.15	19 7	2	Sb		16.0:		*
07742	12 32.3 + 15 50 283.64 + 77.80	03 - 32 - 074 NGC 4540	1576	2.1 x 1.7 2.3: x 2.0:	(40) ...	2	... S dm: P w IC 3528		12.5		*123 4678
07743	12 32.3 + 27 55 204.08 + 86.27	05 - 30 - 000	0064 263	1.0 x 0.1 1.0 x 0.1	110 7	2	Sc		17.		
07744	12 32.3 + 47 15 131.26 + 69.88	08 - 23 - 059	1408	1.0 x 0.10 1.0 x 0.12	27 7	2	Sc		16.5:		
07745	12 32.3 + 73 58 124.59 + 43.38	12 - 12 - 000	1411 269	1.1:: x 0.7:: 0.7: x 0.5:	(15) ...	3	Dwarf V		17.		*
07746	12 32.4 + 51 06 129.44 + 66.09	09 - 21 - 021 NGC 4542	0729 270	1.1 x 0.5 1.2 x 0.6	28 5:	1	S ...		15.5		*3
07747	12 32.4 + 63 48 126.08 + 53.50	11 - 15 - 064 NGC 4545	0674	2.9 x 1.6 2.8 x 1.5	8 5	4	S B c S B(s)cd		13.1		*12 3
07748	12 32.5 + 68 38 125.28 + 48.69	12 - 12 - 010	1411	1.4 x 0.8 1.0: x 0.6:	35 ...	2	Sc - Irr		15.6		
07749	12 32.6 + 00 03 294.13 + 62.37	00 - 32 - 024 NGC 4541	1405	1.6 x 0.6 1.6 x 0.6	91 6	2	Sb		14.0		23
07750	12 32.6 + 30 02 175.81 + 85.54	05 - 30 - 023	0064 263	1.4 x 0.5 1.4 x 0.5	145 6	1	S ...		15.1		
07751	12 32.7 + 41 20 135.45 + 75.67	07 - 26 - 029	0115	1.1 x 0.25 0.8 x 0.2	[3] (7)	2	Irr		16.5:		
07752	12 32.9 - 02 03 294.86 + 60.30	00 - 32 - 000	1405	1.7 x 0.2 1.4 x 0.3	6 7	1	S ...		15.4		
07753	12 32.9 + 14 46 285.68 + 76.82	03 - 32 - 075 NGC 4548 = M 91	1576	5.5 x 4.5 6.5 x 5.0	150 2	5	S B b S B(rs)b Sb +	11.5	+ 433 + 371	*12- -9
07754	12 32.9 + 29 47 178.05 + 85.74	05 - 30 - 025	0064 263	1.1 x 0.35 1.1 x 0.3	106 6	1	S ...		14.9		1
07755	12 33.0 + 00 04 294.34 + 62.40	00 - 32 - 000	1405	1.0 x 0.10 0.9 x 0.12	42 7	1	S ... 1st of 3		16.5:		*
07756	12 33.0 + 03 19 293.29 + 65.62	01 - 32 - 110 NGC 4544	1560	2.4 x 0.7 1.9 x 0.5	161 6 - 7	1	S ...		14.4		23
07757	12 33.0 + 12 30 288.10 + 74.64	02 - 32 - 147 NGC 4550	1563	3.3 x 0.9 3.0 x 0.9	178 6 - 7	3	S0 P w 7759		12.5	+ 350 + 279	*12 3478
07758	12 33.0 + 26 08 231.62 + 86.21	04 - 30 - 004 NGC 4562	1435	2.4 x 0.8 2.3 x 0.8	48 6	3	Sc Sc +		14.6		12
07759	12 33.1 + 12 32 288.17 + 74.67	02 - 32 - 148 NGC 4551	1563	1.7 x 1.5 1.8 x 1.5	(70) ...	2	E P w 7757		13.1	+ 978 + 907	*12 3
07760	12 33.1 + 12 50 287.90 + 74.97	02 - 32 - 149 NGC 4552	1563	(3.4 x 3.4) (3.6 x 3.6)	— ...	4	E		11.1	+ 265 + 195	*123 4678
07761	12 33.1 + 72 30 124.68 + 44.85	12 - 12 - 000	1411	0.7 x 0.5 0.7 x 0.5	(100) ...	2	S B ... Singular		14.3		*
07762	12 33.2 + 26 48 221.78 + 86.43	05 - 30 - 026 NGC 4555	0064 263	(1.4 x 1.1) (1.6 x 1.4)	125: ...	3	E		13.5		*23
07763	12 33.3 - 01 35 294.94 + 60.77	00 - 32 - 000	1405	1.0 x 0.7 0.8 x 0.5	(80) ...	1	S ...		15.6		
07764	12 33.3 + 26 33 225.76 + 86.40	05 - 30 - 000 IC 3543	0064 263	1.0 x 0.1 0.9: x 0.1	142 7	3	Sc P w IC 3546		16.5:		*
07765	12 33.3 + 27 12 215.38 + 86.51	05 - 30 - 027 NGC 4556	0064 263	(1.2 x 1.0) (1.4 x 1.3)	(80) ...	2	E - S0		14.4		*13
07766	12 33.4 + 28 14 198.51 + 86.45	05 - 30 - 030 NGC 4559	0064 (263)	13. x 5.2 10. x 5.0	150 5 - 6	5	Sc SAB(rs)cd Sc +	10.7	+ 856 + 852	*12- -9
07767	12 33.4 + 73 57 124.48 + 43.41	12 - 12 - 011	1411 269	0.9 x 0.9 0.9 x 0.9	— ...	3	E		13.5		*
07768	12 33.6 + 19 36 277.85 + 81.44	03 - 32 - 076 NGC 4561 = 3569*	1576	1.4 x 1.1 1.4 x 1.0	30: 2:	3	S B c Sc +		12.7		*12 3468
07769	12 33.6 + 54 30 127.89 + 62.75	09 - 21 - 024 NGC 4566	0729	1.2 x 0.8 1.0 x 0.8	80: ...	1	S ...		13.9		13
07770	12 33.8 + 21 15 272.81 + 82.94	04 - 30 - 000	1435	1.1 x 0.12 1.1 x 0.12	152 7	3	Sc		16.5:		

N G C   4540 - 42, 4544, 4545, 4548, 4550 - 52
          4555, 4556, 4559, 4561, 4562, 4566   I C   3521, 3522, 3543, 3569

1	2	3	4	5	6	7	8	9	10	11	12
07771	12 33.8 + 23 27 261.18 + 84.77	04 – 30 – 000	1435	1.0 x 0.3 1.0 x 0.3	75 6	1	S ...		16.0:		
07772	12 33.8 + 26 15 230.87 + 86.42	04 – 30 – 006 NGC 4565	1435	15.5 x 1.9 15.2 x 2.0	136 7	4	Sb	Sb +	10.3	+ 1183 + 1171	*12– –9
07773	12 33.9 + 11 43 289.54 + 73.92	02 – 32 – 150 NGC 4564	1563	2.6 x 1.6 3.0 x 1.7	47 ...	3	E		12.2	+ 1015 + 941	*1234 5789
07774	12 33.9 + 40 17 135.65 + 76.74	07 – 26 – 031	0115	3.7: x 0.5 3.3: x 0.4	102 6 – 7	3	Sc	S d	15.2		
07775	12 33.9 + 74 31 124.37 + 42.84	12 – 12 – 012 NGC 4572	1411 269	1.8: x 1.0:: 1.8: x 1.	(155) ...	3	S ... Singular, p w 7797	SAB(s) ...	14.9		*3
07776	12 34.0 + 11 31 289.78 + 73.73	02 – 32 – 152 NGC 4568	1563	5.1 x 2.4 4.8 x 2.0	23 5	4	Sc P w 7777, contact, Sc –		12.5	+ 2278 + 2204	*12– –689V
07777	12 34.0 + 11 32 289.76 + 73.75	02 – 32 – 151 NGC 4567	1563	3.0: x 2.5: 3.0: x 2.3::	85: 2:	4	Sc P w 7776, contact, Sc –		12.5	+ 2253 + 2179	*12– –6789V
07778	12 34.0 + 26 28 227.74 + 86.53	04 – 30 – 008 IC 3582	1435	0.40 x 0.22 0.40 x 0.20	45 ...	1	... Very compact		14.3		
07779	12 34.0 + 32 21 155.38 + 84.08	05 – 30 – 034	0064	1.1 x 0.3 1.0 x 0.35	35 6	1	S ... sp of 2		15.4		*
07780	12 34.1 + 03 23 293.92 + 65.72	01 – 32 – 000	1560	1.7 x 0.3 0.93 x 0.3	48 ...	2	Dwarf:	d – m	16.5:		
07781	12 34.1 + 06 54 292.54 + 69.20	01 – 32 – 112 IC 3576	1560	3. x 3. 2.5:: x 2.5::	— 1:	4	Dwarf spiral, S m S V		15.2		*
07782	12 34.1 + 11 23 289.96 + 73.61	02 – 32 – 153 IC 3578	1563	1.0 x 0.5 0.9 x 0.4	135 5:	2	Sc		15.1		1
07783	12 34.1 + 27 07 217.04 + 86.68	05 – 30 – 035 IC 3585	0064 263	(1.1 x 1.0) (1.1 x 1.0)	— (1)	1	SO:		15.0		
07784	12 34.2 + 13 32 288.28 + 75.71	02 – 32 – 154 IC 3583	1563	2.5 x 1.5 2.1 x 1.5	[0] ...	4	Irr P w 7786	I B m Ir I	15.0		*2
07785	12 34.3 + 07 31 292.39 + 69.81	01 – 32 – 114 NGC 4570	1560	(4.2 x 1.2) (4.2 x 1.2)	159 7	3	SO		11.8	+ 1730 + 1639	*123 4678
07786	12 34.3 + 13 26 288.46 + 75.62	02 – 32 – 155 NGC 4569 = M 90	1563	11.4 x 4.7 11. x 4.5	23 6	4	Sb P w 7784	SAB ... Sb +	11.8	– 300 – 367	*12– –6789A
07787	12 34.3 + 27 49 204.86 + 86.72	05 – 30 – 036 IC 3587	0064 263	1.4 x 0.15 1.5 x 0.15	122 7	3	Sc		16.0:		*1
07788	12 34.4 + 14 29 287.52 + 76.64	02 – 32 – 156 NGC 4571	1563	4.5 x 3.7 4.0 x 3.5	(55) 2	4	Sc	SA(r)cd	13.6		*12– –5789
07789	12 34.4 + 28 09 199.08 + 86.68	05 – 30 – 038 IC 3592	0064 263	1.3 x 0.4 0.9: x 0.3	126 6	1	Sa?		15.3		1
07790	12 34.5 + 07 12 292.68 + 69.51	01 – 32 – 115 IC 3589 = 3591*	1560	1.0 x 0.7 1.0 x 0.7	... ...	3	Irr	I B m Ir I	14.6		1
07791	12 34.8 + 28 30 192.84 + 86.68	05 – 30 – 040 IC 3598	0064 263	1.5 x 0.4 1.6 x 0.4	140 6	2	Sa – b		15.0		*2
07792	12 35.0 + 04 38 294.04 + 66.99	01 – 32 – 116 NGC 4576	1560	1.2 x 0.8 1.3 x 0.9	159 3	3	Sb		14.7		125 9
07793	12 35.0 + 09 50 291.70 + 72.13	02 – 32 – 159 NGC 4578	1563	(3.2 x 2.5) (3.2 x 2.5)	35: 2	4	SO		12.9	+ 2282 + 2201	*12 378
07794	12 35.2 + 05 38 293.81 + 67.99	01 – 32 – 117 NGC 4580	1560	2.6 x 1.8 2.6 x 1.8	165: 3	3	Sa – b	S (rs) ...	13.1		*12 3478
07795	12 35.2 + 07 23 293.09 + 69.72	01 – 32 – 000	1560	1.3 x 1.0 1.3: x 1.0:	... ...	4	Dwarf irregular, I m Ir V		15.4		*
07796	12 35.2 + 12 05 290.40 + 74.35	02 – 32 – 160 NGC 4579 = M 58	1563	6.0: x 5.0: 6.3: x 5.3:	95: 2	4	Sb	SAB(rs)b Sb –	11.5	+ 1752 + 1680	*12– –9
07797	12 35.4 + 74 28 124.24 + 42.90	12 – 12 – 013 NGC 4589	1411 269	(3.0 x 2.6) (3.3 x 2.8)	(90) ...	4	E		12.0	+ 1825 + 2003	*123 4678
07798	12 35.5 – 02 00 296.15 + 60.42	00 – 32 – 026	1405	0.9 x 0.5 0.9 x 0.45	[57] ...	1	Irr		14.1		
07799	12 35.5 + 34 17 144.98 + 82.53	06 – 28 – 015	0105	1.6 x 0.3 1.6 x 0.3	123 7	1	S ...		15.5		*1
07800	12 35.6 + 00 14 295.69 + 62.64	00 – 32 – 027	1405	1.1 x 0.35 1.1 x 0.3	133 6	1	S ...		14.9		
07801	12 35.6 + 01 44 295.31 + 64.13	00 – 32 – 028 NGC 4581	1405	(1.8 x 1.1) (1.8 x 1.1)	173 ...	2	E		13.4		
07802	12 35.8 + 08 10 293.16 + 70.51	01 – 32 – 121?	1560	1.8 x 0.3 1.8: x 0.2	56 7	3	Sc		15.3		*
07803	12 35.8 + 13 23 289.94 + 75.65	02 – 32 – 162 NGC 4584	1563	1.5 x 1.1 1.6 x 1.2	5 3	1	S B ...		14.2		*12 3
07804	12 35.9 + 04 35 294.63 + 66.97	01 – 32 – 122 NGC 4586	1560	3.9 x 1.2 3.7 x 1.2	115 6	3	Sa		13.5		*12– –9
07805	12 36.0 + 02 56 295.20 + 65.34	01 – 32 – 123 NGC 4587	1560	(1.2 x 0.7) 1.2 x 0.7	48 ...	1	SO		14.4		

N G C 4564, 4565, 4567 – 72, 4576, 4578 – 81, 4584, 4586, 4587, 4589

I C 3576, 3578, 3582, 3583, 3585, 3587, 3589, 3591, 3592, 3598

1	2	3	4	5	6	7	8	9	10	11	12
07806	12 36.1 + 01 39 295.61 + 64.07	00 – 32 – 030	1405	1.9 x 0.2 1.7 x 0.2	148 7	3	Sc		15.2		*
07807	12 36.1 + 01 44 295.59 + 64.15	00 – 32 – 000	1405	1.0 x 0.3 0.7 x 0.25	164 6	1	S ...		15.3		
07808	12 36.1 + 10 45 292.06 + 73.08	02 – 32 – 000 IC 3608	1563	3.5 x 0.4 2.5: x 0.3	95 7	1	Sb:		15.4		
07809	12 36.1 + 71 42 124.45 + 45.66	12 – 12 – 014	1411	1.1: x 1.1: 0.8 x 0.7	— 1	4	S B c	S B(s)c	15.5		
07810	12 36.2 + 07 02 293.94 + 69.41	01 – 32 – 124 NGC 4588	1560	1.4 x 0.5 1.2 x 0.5	57 6	3	Sc		15.1		3
07811	12 36.4 + 32 16 151.54 + 84.42	05 – 30 – 043	0064	(1.4 x 0.7) (1.6 x 0.7)	... ...	1	S? ... Peculiar?		14.6		*
07812	12 36.4 + 32 23 150.95 + 84.32	05 – 30 – 044	0064	1.0 x 0.8 1.0 x 0.8	(15) 2:	1	S ...		14.0		
07813	12 36.5 + 00 38 296.09 + 63.07	00 – 32 – 031	1405	(0.8 x 0.8) (0.9 x 0.9)	— ...	2	... Compact		14.4		*
07814	12 36.5 + 15 00 289.18 + 77.26	03 – 32 – 000 IC 3621	1576	1.1 x 0.7 1.0 x 0.6	50: ...	1	S: ...		15.4		
07815	12 36.5 + 18 27 284.54 + 80.60	03 – 32 – 079 IC 3615	1576	1.1 x 0.3 1.3 x 0.4	11 6	3	Sc		15.4		
07816	12 36.5 + 38 22 135.63 + 78.73	06 – 28 – 000	0105	[1.8 x 1.2] ...	... ...	3	Strongly peculiar Streamers		15.4		
07817	12 36.6 + 13 38 290.51 + 75.93	02 – 32 – 164 IC 3611	1563	1.9 x 1.1 (1.3 x 0.9)	137 ...	1	S ...		14.7		
07818	12 36.6 + 30 41 161.80 + 85.74	05 – 30 – 045	0064	1.1 x 0.4 0.7 x 0.2	162 6	1	Sa?		15.2		
07819	12 36.7 – 00 15 296.40 + 62.19	00 – 32 – 032 NGC 4592	1405	5.1: x 1.5: 4.5: x 1.3	97 6	3	Sc – Irr		12.6		*123 4$\overline{7}$8
07820	12 36.7 + 00 59 296.12 + 63.42	00 – 32 – 033	1405	1.8 x 1.7 1.6 x 1.5	— 1	4	Sc	SAB: d Sc +	14.6		
07821	12 36.7 + 06 17 294.57 + 68.68	01 – 32 – 125 NGC 4591	1560	1.8 x 0.9 1.7 x 0.8	37 5	1	S ...		14.1		3
07822	12 36.9 + 08 14 293.94 + 70.62	01 – 32 – 127 IC 3617	1560	1.2 x 0.7 1.2 x 0.8:	[65] ...	4	Irr	I m Ir I	14.8		*
07823	12 37.2 + 47 54 128.62 + 69.37	08 – 23 – 070	1408	1.1 x 0.5 1.1 x 0.5	75 6	3	Sc		14.9		
07824	12 37.3 + 01 56 296.23 + 64.38	00 – 32 – 000	1405	1.6: x 0.8: 1.5: x 0.7:	77: 5:	3	...	S: m	15.5		
07825	12 37.3 + 13 15 291.49 + 75.59	02 – 32 – 169 IC 3631	1563	1.0 x 0.7 0.8 x 0.5	90 ...	1	...		14.5		
07826	12 37.3 + 15 34 289.48 + 77.86	03 – 32 – 081 NGC 4595	1576	1.7 x 1.1 1.6 x 1.0	110 3 – 4	1	S ...		12.8		123 48
07827	12 37.3 + 45 07 129.74 + 72.13	08 – 23 – 071	1408	1.2 x 0.7 1.0: x 0.5:	[122] ...	3	Irr		17.		1
07828	12 37.4 + 10 27 293.30 + 72.83	02 – 32 – 170 NGC 4596	1563	(4.5 x 4.0) (4.0 x 3.5)	(135) 1	4	S B O P w 7842	S B(r)0 +	12.4		*12- $\overline{9}$
07829	12 37.7 + 08 39 294.37 + 71.06	02 – 32 – 171 NGC 4598	1563	(1.7 x 1.5) (1.5 x 1.3)	— 1:	2	S B O		14.1		3
07830	12 37.7 + 13 09 291.94 + 75.51	02 – 32 – 000 IC 3635	1563	1.2: x 1.1: 0.6 x 0.5	— 1	1	Sc?		15.5		
07831	12 37.7 + 61 53 125.35 + 55.47	10 – 18 – 074 NGC 4605	1427	7.0 x 2.5 (6.5 x 2.5)	125 ...	2	S ... Peculiar		10.8	+ 140 + 276	*12 347$\overline{8}$
07832	12 37.8 + 03 24 296.15 + 65.85	01 – 32 – 128 NGC 4600	1560	1.3 x 0.9 (1.4 x 1.0)	60: 3:	3	SO		13.7		13
07833	12 37.9 + 01 27 296.68 + 63.91	00 – 32 – 034 NGC 4599	1405	1.8 x 0.8 2.0 x 0.8	144 6	1	Sa:		13.7		13
07834	12 38.3 + 10 46 293.88 + 73.18	02 – 32 – 173 IC 3647	1563	1.8: x 1.1: 1.3: x 0.9:	140: ...	3	Dwarf V		15.3		
07835	12 38.3 + 27 00 221.93 + 87.59	05 – 30 – 054 IC 3651	0064	1.0 x 1.0 1.0 x 1.0	— (1)	1	SO p of 2		14.4		*
07836	12 38.4 + 29 45 167.09 + 86.69	05 – 30 – 056	0064	1.4 x 0.35 1.4 x 0.35	35 6	3	Sc		14.9		*
07837	12 38.4 + 63 37 124.99 + 53.74	11 – 16 – 001	0717	1.1 x 0.9 1.2 x 1.0	— 1 – 2	3	Sc	SAB(s)cd: Sc +	15.4		
07838	12 38.5 + 11 27 293.71 + 73.86	02 – 32 – 175 IC 3652	1563	(1.0 x 1.0) (0.8 x 0.8)	— ...	2	E		15.1		
07839	12 38.5 + 12 11 293.30 + 74.59	02 – 32 – 174 NGC 4606	1563	(2.8 x 1.6) (2.6 x 1.5)	33 4:	1	S ... P w 7843		12.7		*12 35$\overline{9}$
07840	12 38.6 – 01 20 297.61 + 61.16	00 – 32 – 000	1405	1.0 x 0.6 1.0 x 0.5	75 4 – 5	3	Sc		15.7		

N G C 4588, 4591, 4592, 4595, 4596, 4598 – 4600   I C 3608, 3611, 3615, 3617, 3621, 3631
4605, 4606   3635, 3647, 3651, 3652

1	2	3	4	5	6	7	8	9	10	11	12
07841	12 38.7 + 01 40 297.09 + 64.15	00 – 32 – 035	1405	1.0 x 0.4 0.7 x 0.1	150 ...	1	...		14.3		
07842	12 38.7 + 10 25 294.38 + 72.84	02 – 32 – 177 NGC 4608	1563	(3. x 3.) (3.5 x 3.3)	— 1	4	S B O P w 7828	S B(r)0	12.6		*12– –9
07843	12 38.7 + 12 09 293.50 + 74.56	02 – 32 – 176 NGC 4607	1536	3.2: x 0.7 2.5 x 0.7	2 6	1	S ... P w 7839		14.7		*12 5
07844	12 38.7 + 74 00 123.98 + 43.38	12 – 12 – 000	1411 269	1.2 x 0.12 1.2 x 0.12	165 7	3	Sc		16.5:		
07845	12 38.8 + 28 08 194.40 + 87.63	05 – 30 – 058	0064	1.3 x 0.25 1.2 x 0.25	75 6	2	Sb – c		15.6		
07846	12 38.8 + 48 38 127.64 + 68.67	08 – 23 – 073	1408	1.4: x 0.6: 1.0: x 0.5:	150 5 – 6	2	S ...		15.6		
07847	12 38.8 + 50 42 127.07 + 66.62	09 – 21 – 028 NGC 4617	0729	3.1 x 0.5 2.9 x 0.5	179 7	3	Sb		14.2		3
07848	12 38.8 + 63 47 124.90 + 53.58	11 – 16 – 002	0717	2.6 x 1.2 2.6: x 1.2:	60 5	4	Sc	SAB cd	14.8		
07849	12 38.9 + 14 00 292.53 + 76.39	02 – 32 – 179 NGC 4611	1563	1.3 x 0.3 1.0 x 0.25	126 7	1	S ... P w 7857		15.1		*
07850	12 39.0 + 07 35 295.72 + 70.04	01 – 32 – 134 NGC 4612	1560	(1.6 x 1.3) (1.9 x 1.6)	145: 2:	3	S0		12.9		*123 4678
07851	12 39.0 + 26 18 238.68 + 87.51	04 – 30 – 012 NGC 4614	1435	1.1 x 0.9 1.1 x 1.0	175: 2	2	S B O – a NGC 4615 group		14.2		*13
07852	12 39.1 + 26 20 238.22 + 87.55	04 – 30 – 013 NGC 4615	1435	1.6 x 1.0: 1.5 x 0.9	125 ...	2	Sc of 3, disturbed	Brightest	13.8		*13
07853	12 39.1 + 41 26 130.59 + 75.82	07 – 26 – 037 NGC 4618	0115	4.5 x 3.5 4.5 x 3.5	(25) 1 – 2	4	... P w 7861	S B m	11.5	+ 562 + 618	*12– –9VA
07854	12 39.3 + 09 41 295.19 + 72.14	02 – 32 – 000	1563	(1.0 x 0.6) (0.7 x 0.4)	38 ...	1	...		15.3		
07855	12 39.3 + 11 46 294.25 + 74.20	02 – 32 – 180 IC 3665	1563	1.0: x 0.7: 0.8: x 0.6:	90: ...	2	Dwarf IV – V or V		15.3		
07856	12 39.3 + 35 20 137.01 + 81.80	06 – 28 – 018 NGC 4619	0105	1.5 x 1.5 1.6 x 1.6	— 1	2	S B b: 	S B(r)...	13.5		23
07857	12 39.4 + 14 03 293.01 + 76.46	02 – 32 – 181	1563	1.3: x 1.1: 1.3: x 0.9:	(60) ...	3	Sc P w 7849	Sc +	15.3		*
07858	12 39.5 + 11 55 294.35 + 74.36	02 – 32 – 183 NGC 4621 = M 59	1563	(4.5 x 3.5) (4.5 x 3.5)	165 ...	4	E		11.0	+ 414 + 345	*123 4678
07859	12 39.5 + 13 13 293.64 + 75.64	02 – 32 – 182 NGC 4620	1563	(1.8 x 1.7) (1.6 x 1.6)	— 1	3	S0		14.0		3
07860	12 39.5 + 32 51 143.07 + 84.18	06 – 28 – 019 NGC 4627	0105	2.1 x 1.6 2.0 x 1.6	10: ...	1	E? P w 7865		13.3		*12 35
07861	12 39.5 + 41 34 130.21 + 75.70	07 – 26 – 038 NGC 4625 = 3675*	0115	1.5 x 1.4 1.5 x 1.4	— 1	4	... P w 7853	SAB m	13.0		*12 359
07862	12 39.6 + 07 57 296.04 + 70.42	01 – 32 – 135 NGC 4623	1560	2.2 x 0.7 2.2 x 0.6	176 6 – 7	1	S0 – a		13.6		123 48
07863	12 39.6 + 12 01 294.39 + 74.46	02 – 32 – 184 IC 3672	1563	(1.1 x 1.1) (1.0 x 1.0)	— ...	3	E		15.1		
07864	12 39.6 + 12 52 293.94 + 75.30	02 – 32 – 185 IC 810	1563	1.7 x 0.6 1.5 x 0.4	166 ...	1	...		14.7		
07865	12 39.7 + 32 49 142.80 + 84.22	06 – 28 – 020 NGC 4631	0105	17. x 3.5 15.4 x 3.2	86 7	3	Sc P w 7860	Sc +	9.8	+ 626 + 646	*12– –689A
07866	12 39.7 + 38 46 132.10 + 78.47	07 – 26 – 039 IC 3687	0115	4. x 2.5:: 2.5:: x 2.0::	... ...	4	Dwarf irregular, I m Ir V		15.5		*1
07867	12 39.9 + 42 48 129.25 + 74.49	07 – 26 – 040	0115 273	1.0 x 0.25 0.9 x 0.2	68 6 – 7	1	S ...		15.6		
07868	12 39.9 + 74 43 123.82 + 42.67	13 – 09 – 029 NGC 4648	1339 269	(1.7 x 1.3) (1.7 x 1.3)	70: ...	3	E		12.6		*12 3
07869	12 40.0 – 01 05 298.29 + 61.43	00 – 32 – 037	1405	1.5 x 1.5 1.3 x 1.3	— ...	4	Irr Ir I	I m	14.3		1
07870	12 40.0 + 00 10 298.09 + 62.68	00 – 32 – 038 NGC 4632	1405	3.2 x 1.2 3.0 x 1.1	63 6	4	Sc		12.6	+ 1688 + 1572	*123 4678
07871	12 40.0 + 04 14 297.31 + 66.73	01 – 32 – 136 NGC 4630	1560	1.8 x 1.3 1.8 x 1.5:	10: ...	2	Irr:	[I B m] ?	13.4		123 48
07872	12 40.0 + 75 35 123.75 + 41.80	13 – 09 – 030	1339 269	2.0: x 1.5:: 1.8:: x 1.3::	... ...	4	Dwarf irregular, I m Ir V		16.0:		
07873	12 40.1 + 01 38 297.89 + 64.15	00 – 33 – 001	1405	1.8 x 1.0 1.4 x 0.9	150 4 – 5	4	Sc		15.3		*
07874	12 40.1 + 14 38 293.34 + 77.07	03 – 32 – 085 NGC 4633 = 3688*	1576	2.0 x 0.8 1.8 x 0.8	30 6	4	... P w 7875	S dm	14.7		*12 59
07875	12 40.2 + 14 34 293.49 + 77.00	03 – 32 – 086 NGC 4634	1576	2.6 x 0.5 2.8 x 0.7	156 6 – 7	2	Sc P w 7874		13.6		*12 35

N G C   4607, 4608, 4611, 4612, 4614, 4615
4617 – 21, 4623, 4625, 4627, 4630 – 34, 4648

I C   810, 3665, 3672, 3675, 3687, 3688

1	2	3	4	5	6	7	8	9	10	11	12
07876	12 40.2 + 20 12 286.92 + 82.52	03 - 32 - 087 NGC 4635	1576	1.8 x 1.4 2.0: x 1.6:	170 2	4	Sc	SA d	13.7		123 478
07877	12 40.2 + 27 32 208.59 + 88.04	05 - 30 - 000	0064	1.2 x 0.2 1.0 x 0.2	64 7	1	S ...		16.5:		
07878	12 40.3 + 02 58 297.76 + 65.48	01 - 32 - 137 NGC 4636	1560	(7. x 5.) (7. x 5.)	150: ...	5	E		11.8	+ 883 + 778	*123 46̄78̄
07879	12 40.3 + 10 38 295.62 + 73.11	02 - 33 - 000 IC 3690	0041	1.3 x 0.35 0.9: x 0.3	6 6	1	S ...		15.3		
07880	12 40.3 + 11 43a 295.16 + 74.18	02 - 32 - 187 NGC 4638	1563	(2.9 x 1.9) (2.8 x 1.8)	125 ...	3	S0 P w 7881		(12.2)	+ 1080 + 1010	*123 46̄78̄
07881	12 40.3 + 11 43b 295.16 + 74.18	02 - 32 - 188 NGC 4637	1563	1.2: x 0.7: 1.4: x 0.8:	97 ...	1	... P w 7880		16.0:		*12̄ 3
07882	12 40.3 + 33 33 139.41 + 83.57	06 - 28 - 021	0105	1.5 x 0.2 1.1 x 0.2	57 7	1	Sc - Irr		15.7		
07883	12 40.4 - 00 57 298.48 + 61.57	00 - 33 - 002	1405	3.2 x 0.9 3.0: x 0.8	71 6	4	Sc	Sc +	14.3		*
07884	12 40.4 + 13 32 294.34 + 75.99	02 - 32 - 189 NGC 4639	1563	3.2 x 2.2 2.8 x 2.0	123 3	4	Sb	SAB(rs)bc	12.4		*12̄- -578̄9̄
07885	12 40.4 + 21 16 284.83 + 83.56	04 - 30 - 016 IC 3692	1435	1.0 x 0.7 1.0 x 0.8	100: 2 - 3	2	S B a:		14.8		
07886	12 40.4 + 32 10 143.85 + 84.88	05 - 30 - 000	0064	1.0 x 0.1 1.0 x 0.1	29 7	2	Sc		16.5:		
07887	12 40.4 + 55 25a 125.59 + 61.94	09 - 21 - 030 NGC 4644	0729	1.7 x 0.7: 1.7 x 0.7:	53 6	2	S B: b p of 2, disturbed, NGC 4686 group		(14.8)		*13
97888	12 40.5 + 12 34a 294.94 + 75.03	02 - 32 - 190 NGC 4640	1563	1.6: x 1.1: 1.5 x 1.0	45: 3	1	S ... p of 2		(15.2)		*1
07889	12 40.6 + 12 19 295.16 + 74.79	02 - 32 - 191 NGC 4641	1563	(1.3 x 0.9) (1.2 x 0.9)	... 3:	3	S0		14.9		
07890	12 40.6 + 28 00 194.63 + 88.05	05 - 30 - 062	0064	0.6 x 0.4 0.6 x 0.4	... ...	1	...		14.5		
07891	12 40.6 + 30 40 151.84 + 86.25	05 - 30 - 063	0064	[1.1 x 0.5] ...	... ...	3	Double system Contact, disrupted		15.5		*1V
07892	12 40.6 + 55 07 125.58 + 62.24	09 - 21 - 031 NGC 4646	0729	0.6 x 0.3 0.7 x 0.3	18 ...	1	... NGC 4686 group		13.8		*3
07893	12 40.7 - 00 22 298.55 + 62.16	00 - 33 - 004 NGC 4642	1578	2.0 x 0.5 2.0 x 0.5	37 6	1	S ... P w 7900		13.8		*23
07894	12 40.7 + 72 11 123.90 + 45.20	12 - 12 - 000	1411	1.1 x 0.2 0.9 x 0.2	62 7	2	S B? c		16.5:		
07895	12 40.8 + 02 16 298.19 + 64.79	00 - 33 - 005 NGC 4643	1578	(3. x 3.) (4.0 x 3.5)	-- 1	4	S B 0		11.9	+ 1432 + 1325	*123 46̄78̄
07896	12 41.0 + 11 52 295.72 + 74.35	02 - 33 - 001 NGC 4647	0041	2.9 x 2.4 3.0 x 2.4	125: 2:	3	Sc P w 7898, contact, disturbed		12.5	+ 1396 + 1328	*12̄- -9̄
07897	12 41.0 + 29 45 159.30 + 87.07	05 - 30 - 065	0064	1.0 x 0.35 0.9 x 0.3	48 6	1	S ...		15.7		
07898	12 41.1 + 11 50 295.83 + 74.32	02 - 33 - 002 NGC 4649 = M 60	0041	(7. x 6.) (9. x 7.)	(105) ...	5	E P w 7896, contact		10.3	+ 1269 + 1200	*12̄- -9̄V̄Ā
07899	12 41.2 + 11 03 296.23 + 73.55	02 - 33 - 003 IC 3704	0041	1.3 x 0.35 1.3 x 0.35	43 6	1	Sb - c Disturbed?		14.7		*
07900	12 41.3 - 00 16 298.86 + 62.27	00 - 33 - 006 NGC 4653	1578	2.5: x 2.3: 2.5: x 2.2:	(30) 1 - 2	4	Sc SAB:(rs)cd P w 7893		13.7		*123 458̄9̄
07901	12 41.3 + 16 40 293.18 + 79.12	03 - 33 - 001 NGC 4651	1572	3.9 x 2.5 4.0 x 2.7	80 3 - 4	3	Sc		11.3	+ 733 + 685	*12̄-V -578̄9̄A
07902	12 41.4 + 13 25 295.39 + 75.90	02 - 33 - 004 NGC 4654	0041	5.3 x 2.9 5.5: x 3.2	128 4 - 5	4	Sc Disturbed?, Sc +		11.8	+ 1022 + 960	*12̄- -9̄
07903	12 41.4 + 54 16 125.45 + 63.10	09 - 21 - 000	0729	1. x 1. 1. x 1.	-- ...	3	Dwarf V		17.		
07904	12 41.5 + 41 01 128.98 + 76.30	07 - 26 - 045	0115 273	1.3 x 0.12 1.0 x 0.1	144 7	2	S B? c		15.7		
07905	12 41.5 + 55 10 125.30 + 62.20	09 - 21 - 33+34 Mark 220 + 221	0729	[1.9 x 1.0] ...	... ...	3	Double system Disrupted, NGC 4686 group		14.1	+ 4950 + 5062	*1C
07906	12 41.6 + 12 25 125.82 + 50.14	02 - 33 - 000	0041	1.3:: x 1.0:: 1.4:: x 1.0::	... ...	4	Dwarf irregular Ir V		16.5:		
07907	12 41.6 + 32 27 140.16 + 84.70	05 - 30 - 066 NGC 4656	0064	22. x 3. 13. x 3.	33 6	5	... S B(s)m NGC 4657 superimposed		(10.6)	+ 755 + 775	*12̄- -9̄
07908	12 41.7 + 73 54 123.70 + 43.49	12 - 12 - 017	1411 269	1.6 x 0.4 1.5 x 0.4	50 7	3	Sc		15.3		
07909	12 41.8 + 04 42 298.36 + 67.24	01 - 33 - 2+3	0104	[1.1 x 0.4] ...	... ...	3	Double system Bridge		15.3		*1V
07910	12 41.8 + 45 17 127.11 + 72.06	08 - 23 - 085	1408	[1.1 x 0.8] ...	... ...	3	Multiple system Connected, disrupted		15.4		*1

N G C   4635 - 44, 4646, 4647, 4649, 4651
4653, 4654, 4656, 4657

I C   3690, 3692, 3704

1	2	3	4	5	6	7	8	9	10	11	12
07911	12 41.9 + 00 45 299.05 + 63.30	00 - 33 - 007	1578	2.9: x 2.1: 2.6:: x 2.0::	20: ...	4	Dwarf spiral, S B m S B V		14.9		*
07912	12 41.9 + 03 48 298.58 + 66.34	01 - 33 - 004	0104	[1.5 x 0.7] ...	... ...	2	Double system Contact		15.3		*1
07913	12 42.0 - 02 02 299.44 + 60.52	00 - 33 - 000	1578	1.2: x 0.7: 0.8:: x 0.5::	... ...	2	Dwarf		15.7		
07914	12 42.0 + 11 28 296.77 + 73.98	02 - 33 - 006 NGC 4660	0041	2.4 x 2.1 2.4 x 1.9	100 ...	1	E		12.1	+ 1017 + 948	*12 3478
07915	12 42.0 + 13 47 295.81 + 76.29	02 - 33 - 007 NGC 4659	0041	1.8 x 1.3 (2.3 x 1.4)	173 3:	2	S0 - a		13.3		13
07916	12 42.0 + 34 40 134.22 + 82.58	06 - 28 - 024	0105	2.6:: x 2.0:: 2.5:: x 2.0::	... ...	3	Dwarf irregular Ir V		17.		*1 V
07917	12 42.1 + 37 23 130.85 + 79.91	06 - 28 - 025 NGC 4662	0105	2.5 x 2.0 2.3 x 2.2	(55) 1 - 2	4	S B b	S B(rs)bc	14.1		3
07918	12 42.1 + 64 04 124.26 + 53.32	11 - 16 - 004	0717	1.5 x 1.0: 1.2: x 0.8:	[60] ...	3	Dwarf irregular Ir V		16.5:		*1
07919	12 42.3 + 12 21 296.72 + 74.87	02 - 33 - 000 IC 3720	0041	(2.5 x 1.5) (2.3 x 1.5)	125: ...	1	Dwarf elliptical?		16.0:		*2
07920	12 42.3 + 12 37 296.62 + 75.13	02 - 33 - 008 IC 3718	0041	2.7 x 1.0 2.7 x 0.9	72 6	1	S ...		14.7		
07921	12 42.3 + 40 58 128.37 + 76.36	07 - 26 - 049 IC 3726	0115 273	1.7 x 0.3 1.3 x 0.3	117 7	3	Sc		15.6		*
07922	12 42.3 + 56 25 124.91 + 60.96	09 - 21 - 037	0729	2.1: x 0.25 1.8: x 0.3	136 6	2	S B?b - c		15.3		1
07923	12 42.4 + 19 01 292.41 + 81.48	03 - 33 - 002 IC 3725	1572	1.0 x 0.35 1.0 x 0.4	136 6	1	S ...		14.4		
07924	12 42.5 + 03 20 299.03 + 65.88	01 - 33 - 005 NGC 4665	0104	(4.5 x 4.5) (4.8 x 4.8)	--- 1	4	S B 0	S B(s)0/a	12.4	+ 785 + 684	*123 468
07925	12 42.5 + 55 09 124.99 + 62.22	09 - 21 - 038 NGC 4669	0729	1.7 x 0.4 2.0 x 0.4	177 7	1	S ... NGC 4686 group		15.1		*13
07926	12 42.6 - 00 10 299.55 + 62.39	00 - 33 - 008 NGC 4666	1578	4.5 x 1.4 4.7 x 1.5	42 6	4	Sc P w 7931	Sc -	12.0	+ 1645 + 1530	*12- -9
07927	12 42.6 + 11 10 297.40 + 73.70	02 - 33 - 009 IC 3727	0041	1.5: x 1.5: 1.6: x 1.6:	--- 1	2	Sc		15.6		
07928	12 42.7 + 23 19 283.44 + 85.68	04 - 30 - 019 IC 813 = 3734*	1435	1.1 x 1.0 1.0 x 1.0	--- (1)	1	S: ... f of 2, contact, disturbed:		14.4		*1
07929	12 42.8 + 21 43 288.74 + 84.14	04 - 30 - 000	1435	1.0:: x 0.7:: 1. x ...	--- ...	3	Dwarf V		17. - 18.		
07930	12 42.8 + 27 24 212.64 + 88.62	05 - 30 - 072 NGC 4670	0064	1.6 x 1.4 (1.7 x 1.4)	(90) ...	3	... Peculiar, p w 7933		12.6	+ 1210 + 1209	*123 4678A
07931	12 43.0 - 00 15 299.77 + 62.31	00 - 33 - 009 NGC 4668	1578	1.2 x 0.6 1.1 x 0.6	[5] ...	3	Irr P w 7926 Ir I		13.5		*12- -5789
07932	12 43.0 + 13 36 296.91 + 76.13	02 - 33 - 011 IC 3742	0041	1.6 x 0.8 1.7 x 0.8	45 5:	1	S B ... p of 2		14.6		*1
07933	12 43.1 + 27 20 215.57 + 88.69	05 - 30 - 073 NGC 4673	0064	1.0 x 0.8 1.0 x 0.8	(170) ...	1	S0: P w 7930		13.7	+ 6991 + 6990	*12 3
07934	12 43.2 + 35 23 131.44 + 81.92	06 - 28 - 027	0105	1.2 x 0.6 1.2 x 0.5	85 5	1	S ...		16.0:		*1
07935	12 43.3 + 55 00 124.76 + 62.38	09 - 21 - 039 NGC 4675	0729	1.6 x 0.5 1.6 x 0.5	97 6	2	S B b: NGC 4686 group		15.4		*13
07936	12 43.6 + 45 28 126.05 + 71.90	08 - 23 - 089	1408	[1.3 x 0.8] ...	... ...	3	Multiple system Connected, disrupted		15.2		*1
07937	12 43.7 + 08 37 298.94 + 71.17	02 - 33 - 013 IC 3754	0041	1.4 x 0.7 1.4 x 0.7	128 5	2	Sa - b		14.9		1
07938	12 43.7 + 31 00a 140.49 + 86.22	05 - 30 - 076 NGC 4676a = 819*	0064	2.2 x 0.35 ...	...					+ 6500 + 6515	
07939	12 43.7 + 31 00b 140.49 + 86.22	05 - 30 - 077 NGC 4676b = 820*	0064	1.9 x 0.8: ...	...	4	Double system Bridge, long plumes		14.1	+ 6605 + 6620	*12 3VA
07940	12 43.8 + 46 59 125.64 + 70.39	08 - 23 - 091	1408	1.3 x 0.5 1.1 x 0.4	23 ...	1	S - Irr		15.5		
07941	12 44.0 + 64 50 123.87 + 52.56	11 - 16 - 006	0717	4.7? x 0.8 2.2 x 0.5	8 6	2	... S d - dm		15.0		
07942	12 44.1 + 09 35 299.06 + 72.15	02 - 33 - 016	0041	1.1 x 1.0 1.1: x 1.0:	--- 1	3	... SAB: d - dm		15.6		
07943	12 44.2 + 06 14 299.70 + 68.80	01 - 33 - 007	0104	2.5 x 2.0 2.1 x 1.6	30: 2	3	Sc		14.7		
07944	12 44.2 + 10 07 299.03 + 72.68	02 - 33 - 019 IC 816	0041	1.0 x 0.7 1.0 x 0.7	35 3:	3	S B ... S B(r) ... p of 2		15.1		*1
07945	12 44.4 - 01 17 300.61 + 61.30	00 - 33 - 010	1578	1.4: x 0.9: 1.5:: x 0.9::	... ...	3	Dwarf (spiral), S: m S: IV - V or V		15.6		

N G C   4659, 4660, 4662, 4665, 4666, 4668 - 70   I C   813, 816, 819, 820, 3718, 3720
       4673, 4675, 4676   3725 - 27, 3734, 3742, 3754

1	2	3	4	5	6	7	8	9	10	11	12
07946	12 44.4 + 54 49 124.44 + 62.57	09 – 21 – 044 NGC 4686	0729	2.1 x 0.7 2.2 x 0.7	3 6	3	Sa Brightest in group		13.7		3
07947	12 44.5 + 36 33 128.66 + 80.80	06 – 28 – 029 IC 3774	0105	1.1 x 0.5 1.0 x 0.4	110 6	1	S ...		16.0:		*
07948	12 44.6 + 09 17 299.51 + 71.85	02 – 33 – 000	0041	1.0: x 0.1 1.0 x 0.15	162 7	2	Sc		16.0:		
07949	12 44.6 + 36 45 128.53 + 80.60	06 – 28 – 030	0105	2.1: x 1.2: 2. x 1.	... ...	3	Dwarf		17.		*
07950	12 44.6 + 51 55 124.63 + 65.47	09 – 21 – 047	0729	1.8: x 1.3: 1.3: x 1.2:	[5] ...	3	Irr	Ir I	14.1	+ 322 + 424	*1
07951	12 44.7 – 02 27 300.84 + 60.13	00 – 33 – 011 NGC 4684	1578	(2.6 x 0.9) 2.8 x 1.1	23 6	3	S0		12.4		123 47$\overline{8}$
07952	12 44.7 + 10 29 299.37 + 73.05	02 – 33 – 021 IC 3773	0041	2.0 x 0.7 2.1 x 0.8	20 ...	3	E		14.3		*
07953	12 44.7 + 12 03 299.03 + 74.62	02 – 33 – 000	0041	1.0: x 0.4: 0.9 x 0.4	54 6	2	Sc p of 2	Sc +	16.0:		*
07954	12 44.7 + 19 44 295.46 + 82.27	03 – 33 – 004 NGC 4685	1572	(1.5 x 0.9) (1.6 x 1.0)	155 ...	1	E – S0		13.8		3
07955	12 44.7 + 26 58 237.13 + 88.95	05 – 30 – 080	0064 276	1.4 x 0.2 1.4: x 0.2	28 7	3	Sc		16.0:		*1
07956	12 45 + 83 42 123.13 + 33.70	14 – 06 – 008	1340	1.6: x 1.1: 1.9: x 1.3:	... ...	2	S ... Disturbed?		15.5		*
07957	12 45.0 + 30 04 140.96 + 87.19	05 – 30 – 083 IC 821	0064	1.2 x 1.2 1.2 x 1.1	— 1	3	S B b/Sc	SAB(s)bc	14.5		
07958	12 45.0 + 35 37 128.67 + 81.74	06 – 28 – 031 NGC 4687, Mark 442	0105	0.9 x 0.7 1.0 x 0.9	— ...	1	...		14.3		23
07959	12 45.1 + 27 16 221.53 + 89.12	05 – 30 – 084	0064 276	1.2 x 0.35 1.2 x 0.35	63 7	3	S0		14.7	+ 7118 + 7118	*2
07960	12 45.2 + 04 05 300.61 + 66.67	01 – 33 – 012	0104	1.1 x 0.5 1.1 x 0.5	50 5 – 6	2	Sa – b		15.2		
07961	12 45.2 + 04 36 300.56 + 67.18	01 – 33 – 013 NGC 4688	0104	4.4: x 4.4: 4.5: x 4.0:	— 1:	4	S B c	S B(s)cd	14.5		*12 347$\overline{8}$
07962	12 45.2 + 71 28 123.43 + 45.93	12 – 12 – 018 NGC 4693	1411	2.4 x 0.6 2.4 x 0.6	34 6	4	Sc	Sc +	14.0		*13
07963	12 45.3 – 00 55 301.05 + 61.67	00 – 33 – 000	1578 274	1.5 x 0.1 0.9 x 0.1	52 7	3	Sc	Sc +	15.7		
07964	12 45.3 – 01 22 301.08 + 61.22	00 – 33 – 012 NGC 4690	1578	1.1 x 0.8 (1.1 x 0.9)	150: ...	2	E – S0		14.0		3
07965	12 45.3 + 14 02 299.12 + 76.61	02 – 33 – 022 NGC 4689	0041	4.0: x 3.5: 4. x 4.	— 1 – 2	4	Sc	S bc Sc –	12.8		*12– –$\overline{9}$
07966	12 45.3 + 54 39 124.17 + 62.74	09 – 21 – 048 NGC 4695	0729	1.2 x 0.7 1.2 x 0.7	80: ...	1	S ... NGC 4686 group		14.5		*3
07967	12 45.4 + 27 30 205.66 + 89.20	05 – 30 – 086 NGC 4692 =823*	0064 276	1.1 x 1.0 1.2 x 1.1	0 (1)	2	E – S0		14.0	+ 7911 + 7912	*23
07968	12 45.7 + 60 34 123.74 + 56.83	10 – 18 – 083	1427	1.2 x 0.8 1.0: x 0.7:	0 3:	3	...	S dm – m	16.5:		*1
07969	12 45.8 + 11 15 300.18 + 73.83	02 – 33 – 023 NGC 4694	0041	3.5 x 1.6 3.4: x 1.7:	140 5	2	S B? 0		12.4		*123 467$\overline{8}$
07970	12 45.9 + 08 45 300.61 + 71.34	02 – 33 – 024 NGC 4698	0041	3.5 x 1.8: (4.3 x 2.3)	170 5:	3	Sa		12.1	+ 1032 + 955	*12– –$\overline{9}$
07971	12 46.1 + 51 27 124.11 + 65.94	09 – 21 – 050 NGC 4707	0729	2.5 x 2.2 2.5: x 2.2:	(25) ...	3	Dwarf spiral S IV – V or V		15.2		*13 C
07972	12 46.3 + 42 12 124.96 + 75.19	07 – 26 – 054 NGC 4704	0115 273	1.0 x 0.9 1.0 x 1.0	— 1	2	S B b		14.8		123
07973	12 46.4 + 35 36 126.70 + 81.78	06 – 28 – 033 IC 3804	0105	1.4 x 0.8 1.3 x 0.7	40 4 – 5	1	S ...		14.4		1$\overline{2}$
07974	12 46.5 + 15 10 300.16 + 77.75	03 – 33 – 006 IC 3806	1572	1.4 x 0.5 1.4 x 0.4	177 6	1	S ...		14.6		2
07975	12 46.7 + 03 40 301.57 + 66.26	01 – 33 – 015 NGC 4701	0104	3.6 x 3.0 3.5: x 2.5:	45: 2 – 3	4	Sc	Sc +	13.1		*123 467$\overline{8}$
07976	12 46.7 + 04 56 301.50 + 67.53	01 – 33 – 016	0104	1.0 x 1.0 1.0 x 0.9	— 1	2	...	S dm:	15.5		
07977	12 47.1 + 25 44 288.59 + 88.28	04 – 30 – 021 NGC 4712	1435	2.4 x 0.9 2.6: x 1.0	160 6	3	Sc		13.5		*123 467$\overline{8}$
07978	12 47.1 + 31 07 129.25 + 86.26	05 – 30 – 093	0064	1.3 x 0.7 1.2 x 0.6	8 5	3	Sc		14.8		
07979	12 47.2 + 12 54 301.25 + 75.49	02 – 33 – 027	0041	1.6 x 0.3 1.5 x 0.3	38 ...	3	...	d – m	15.3		1
07980	12 47.2 + 15 26 300.91 + 78.03	03 – 33 – 009 NGC 4710	1572	4.3 x 1.3 4.5 x 1.3	27 7	2	S0 – a		11.6	+ 1125 + 1076	*123 467$\overline{8}$

N G C   4684 – 90,   4692 – 95,   4698,   4701,   4704   I C   821,   823,   3773,   3774,   3804,   3806
        4707,   4710,   4712

1	2	3	4	5	6	7	8	9	10	11	12
07981	12 47.2 + 31 01 129.09 + 86.36	05 – 30 – 000	0064	1.1 x 0.15 0.9 x 0.15	145 7	2	Sc		16.0:		
07982	12 47.3 + 03 08 301.97 + 65.73	01 – 33 – 017	0104	3.3 x 0.8 3.4 x 0.8	0 7	1	S ...		14.6		
07983	12 47.3 + 04 07 301.93 + 66.71	01 – 33 – 000	0104	1.3: x 0.9: 0.8 x 0.6	... ...	2	Dwarf (irregular)		16.5:		
07984	12 47.3 + 69 54 123.22 + 47.50	12 – 12 – 000	1411	1.1 x 0.12 1.0 x 0.12	70 7	1			16.5:		
07985	12 47.4 + 05 35 301.93 + 68.18	01 – 33 – 018 NGC 4713	0104	3.2 x 2.0 2.7 x 2.0:	100 4	3	Sc	Sc +	12.3	+ 664 + 575	*123 4678
07986	12 47.4 + 28 05 150.29 + 89.23	05 – 30 – 096 NGC 4715	0064 276	(1.6 x 1.3) (1.8 x 1.5)	(20) 2:	4	S0	SA 0 +	15.4		13
07987	12 47.7 + 33 25 125.59 + 83.98	06 – 28 – 035 NGC 4719, Mark 446	0105	1.7 x 1.5 1.5 x 1.3	— 1:	3	S B b	S B(s)b	14.2		3
07988	12 47.9 + 53 08 123.38 + 64.27	09 – 21 – 053 NGC 4732	0729	(1.2 x 0.6) 1.2 x 0.6	8 ...	2	E		15.2		3
07989	12 48.0 + 25 46 295.15 + 88.35	04 – 30 – 022 NGC 4725	1435	12. x 9. 11. x 7. ?	35 ...	4	Sb/S B b	SAB(r)b Sb +	10.2	+ 1114 + 1109	*12– –9
07990	12 48.0 + 28 37 133.22 + 88.76	05 – 30 – 000	0064 276	1.1: x 0.7: 1.0: x 0.6:	[90] ...	4	Dwarf irregular Ir V		17.		
07991	12 48.1 + 01 44 302.48 + 64.33	00 – 33 – 014	1578	1.8 x 0.2 1.9 x 0.25	170 7	3	Sc	Sc +	14.8		1
07992	12 48.1 + 27 43 155.16 + 89.63	05 – 30 – 100	0064 276	1.0 x 0.4 1.0 x 0.4	37 6	1	S ... sff of 2		16.0:		*1
07993	12 48.2 + 52 24 123.29 + 65.00	09 – 21 – 054	0729	2.3 x 0.25 2.3: x 0.25	144 7	2	Sc		15.3		
07994	12 48.2 + 73 09 123.08 + 44.25	12 – 12 – 019 NGC 4750	1411	2.3 x 2.2 2.4 x 2.3	— 1	3	Sb	S ab Sb –	11.8	+ 1647 + 1823	*123 4678
07995	12 48.4 + 78 40 123.04 + 38.73	13 – 09 – 000	1339 275	2.0: x 1.7: 1.7: x 1.5:	— ...	3	Dwarf (irregular) Ir: V		16.5:		
07996	12 48.5 + 41 23 123.39 + 76.02	07 – 26 – 058 NGC 4736 = M 94	0115	14. x 12. 15. x 13.	105: 1 – 2	5	Sb	(R)SA(r)ab Sb –	8.7	+ 300 + 362	*12– –789
07997	12 48.6 + 11 11 302.65 + 73.78	02 – 33 – 028 NGC 4733	0041	(2.0 x 1.8) (2.2 x 2.0)	— ...	3	E		13.2		*12 3
07998	12 48.7 + 05 08 302.80 + 67.73	01 – 33 – 019 NGC 4734	0104	1.1 x 0.8 1.0 x 0.8	(145) 3:	1	Sc:		14.3		*12 3
07999	12 48.7 + 29 04 125.25 + 88.33	05 – 30 – 103 NGC 4738	0064 276	2.2 x 0.25 2.1 x 0.25	34 7	1	Sc?		14.9		3
08000	12 48.7 + 47 56 123.14 + 69.47	08 – 23 – 098 NGC 4741	1408	1.5 x 0.8 1.1 x 0.8	165 5:	3	Sc		14.5		3
08001	12 48.9 + 10 52 302.91 + 73.47	02 – 33 – 000	0041	1.0: x 0.7: 1.0: x 0.7:	70: ...	2	... f of 2	S: m	16.5:		*
08002	12 49 + 80 53 123.00 + 36.52	14 – 06 – 009	1340	1.4 x 0.5 1.5 x 0.6	175 6	2	Sb		15.6		*
08003	12 49.0 + 53 58 123.00 + 63.43	09 – 21 – 055 IC 830	0729	0.8 x 0.4 1.1 x 0.5	165 ...	1	... NGC 4686 group		14.3		*
08004	12 49.2 + 31 37 122.42 + 85.78	05 – 30 – 107	0064	1.9 x 0.6 1.8 x 0.6	5 6	4	Sc	S d	15.1		
08005	12 49.3 + 26 02 305.82 + 88.63	04 – 30 – 023 NGC 4747	1435	3.5 x 1.5 3.5: x 1.5:	[30] ...	4	Irr	I m Ir I	13.2		*12– –5789A
08006	12 49.3 + 71 55 122.97 + 45.48	12 – 12 – 020 NGC 4749	1411	1.7 x 0.35 1.9 x 0.35	158 7	1	S ...		14.2		23
08007	12 49.4 + 12 21 303.38 + 74.95	02 – 33 – 029 NGC 4746	0041	2.2 x 0.5 2.2 x 0.5	120 6 – 7	1	S ...		13.3		23
08008	12 49.5 + 16 33 303.64 + 79.15	03 – 33 – 014 IC 827	1572	1.0 x 0.45 1.0 x 0.4	104 6	1	S ...		15.0		
08009	12 49.8 – 00 56 303.42 + 61.67	00 – 33 – 016 NGC 4753	1578	(4.5 x 2.5) (6. x 3.)	[80:] ...	3	...	I 0 Ir II	11.7	+ 847 + 735	*123 4678
08010	12 49.8 + 11 35 303.72 + 74.18	02 – 33 – 030 NGC 4754	0041	4.5 x 2.5 (5.0 x 2.8)	23 ...	4	S B 0 P w 8016		11.6	+ 1461 + 1398	*12 3478
08011	12 49.8 + 21 55 304.94 + 84.51	04 – 30 – 000	1435	1.5:: x 1.2:: ...	— ...	4	Dwarf irregular Ir V		18.		
08012	12 49.9 + 51 58 122.67 + 65.43	09 – 21 – 057	0729	0.8 x 0.45 0.7 x 0.50	115: ...	1	S ...		14.3		*
08013	12 50.1 + 27 01 335.60 + 89.55	05 – 30 – 112	0064 276	1.4 x 0.4 1.2 x 0.4	97 6	1	S ...		15.7		
08014	12 50.2 + 16 07 304.47 + 78.71	03 – 33 – 015 NGC 4758	1572	3.0 x 0.8 3.0: x 0.8	[160] ...	3	Irr		14.1		23
08015	12 50.4 + 10 16 304.17 + 72.86	02 – 33 – 032	0041	1.8 x 1.1 1.7: x 1.0	60 4	2	Sa – b		15.1		

N G C 4713, 4715, 4719, 4725, 4732 – 34, 4736
4738, 4741, 4746, 4747, 4749, 4750, 4753, 4754, 4758

I C 827, 830

1	2	3	4	5	6	7	8	9	10	11	12
08016	12 50.4 + 11 30 304.25 + 74.10	02 - 33 - 033 NGC 4762	0041	9.0 x 2.0 9.5 x 2.0	32 7	4	S B? 0 P w 8010		11.1	+ 939 + 876	*123 4678
08017	12 50.4 + 28 39 109.20 + 88.71	05 - 30 - 114	0064 276	0.9 x 0.30 0.9 x 0.30	33 ...	1	...		14.5		
08018	12 50.7 + 04 44 304.10 + 67.33	01 - 33 - 020 NGC 4765	0104	1.5 x 1.1 1.6: x 1.1:	80: ...	1	...		13.0		123 478
08019	12 50.7 + 36 03 120.72 + 81.34	06 - 28 - 038 IC 3852	0105	1.4 x 0.7 1.3 x 0.7	... ...	2	S B c	Sc +	16.0:		
08020	12 50.8 + 01 32 304.03 + 64.13	00 - 33 - 017 NGC 4771	1578	4.0 x 0.9 3.8: x 0.9	133 6	3	Sc	Sc +	13.3		*123 4678
08021	12 50.9 + 02 26 304.12 + 65.03	00 - 33 - 018 NGC 4772	1578	2.9 x 1.4 (4.5 x 2.4)	147 5	4	Sa		12.9		*123 478
08022	12 51.3 + 10 00 304.89 + 72.59	02 - 33 - 034 NGC 4779	0041	2.1 x 1.8 2.1 x 1.7	70: 1 - 2	3	S B c	S (rs)bc Sc -	13.5		13
08023	12 51.4 + 36 22 119.90 + 81.02	06 - 28 - 039	0105	[1.0 x 0.5] ...	[157] ...	3	Double system Contact, disrupted		16.0:		*1V A
08024	12 51.6 + 27 25 034.80 + 89.42	05 - 30 - 120	0064	2.5: x 2.2: 2.0:: x 1.2::	[35] ...	4	Dwarf irregular, I m Ir IV - V		14.9		*2
08025	12 51.6 + 29 52 110.13 + 87.47	05 - 30 - 122	0064 276	2.0 x 0.25 2.1 x 0.25	74 7	3	Sb f of 2		14.8		*1
08026	12 51.7 + 27 20 026.81 + 89.40	05 - 30 - 121 NGC 4787	0064 276	1.0 x 0.25 1.0 x 0.2	2 6	2	S0 - a		15.5		*3
08027	12 51.8 + 42 58 121.09 + 74.42	07 - 27 - 000	0133	1.0 x 0.2 0.9 x 0.2	34 6	2	Sb - c		16.5:		
08028	12 51.9 + 27 20 027.26 + 89.35	05 - 30 - 124 NGC 4789	0064 276	(1.4 x 1.1) (1.4 x 1.1)	— ...	2	E - S0		13.3	+ 8372 + 8377	*12 3 -
08029	12 52 + 82 40 122.88 + 34.73	14 - 06 - 011	1340	1.3 x 0.35 1.3 x 0.3	2 6	2	Sb		16.0:		
08030	12 52.0 + 26 34 341.90 + 88.93	05 - 31 - 000	1393 276	1.0: x 0.5: ...	[85] ...	3	Dwarf (irregular) Ir: V		18.		
08031	12 52.0 + 35 34 118.71 + 81.81	06 - 28 - 040	0105	1.1 x 0.6 0.9 x 0.6	140: 5:	1	S ...		16.0:		
08032	12 52.2 + 13 30 306.24 + 76.08	02 - 33 - 035	0041	2.7 x 0.6 2.9 x 0.5	167 6	1	S ...		14.8		
08033	12 52.2 + 29 12 101.82 + 88.07	05 - 31 - 003 NGC 4793	1393 276	3.4 x 1.8 2.7 x 1.6	50 5	3	Sc		12.3		*12 3478
08034	12 52.3 + 02 56 304.99 + 65.52	01 - 33 - 023 NGC 4810	0104	1.8 x 0.6 1.8 x 0.7	[67] ...	4	Irr I m P w NGC 4809, interaction		14.9	+ 890 + 794	*12 3VA
08035	12 52.3 + 46 48 121.30 + 70.59	08 - 24 - 004 NGC 4800	1350	1.7 x 1.2 1.7 x 1.2	25: ...	1	S ...		12.0	+ 746 + 832	*123 478
08036	12 52.4 + 19 27 308.78 + 82.01	03 - 33 - 016 IC 3881	1572	4.0 x 0.9 2.5: x 0.8	30 6	3	S B c		14.1		
08037	12 52.5 + 08 20 305.65 + 70.92	01 - 33 - 024 NGC 4795	0104	2.1 x 1.8 2.4 x 1.9	(115) ...	3	S B 0/S B a NGC 4796 superimposed		13.5		*12 348
08038	12 52.5 + 27 41 053.26 + 89.17	05 - 31 - 004 NGC 4798	1393 276	(1.0 x 0.7) (1.0 x 0.7)	(30) ...	1	...		14.3	+ 7673 + 7680	*12 3
08039	12 52.5 + 44 25 120.87 + 72.97	07 - 27 - 004	0133	1.1 x 0.4 1.0 x 0.4	107 6	1	S ...		16.0:		
08040	12 52.5 + 59 02 122.14 + 58.36	10 - 18 - 090	1427	1.7 x 0.20 1.4 x 0.2	57 6 - 7	1	S ... P w 8046 Brightest in group		14.4		*
08041	12 52.6 + 00 23 304.98 + 62.97	00 - 33 - 021 NGC 4771	1578	3.7 x 2.2: 3.5: x 2.3:	165: 3 - 4:	4	Sc	SAB d	13.6		*12
08042	12 52.6 + 08 11 305.70 + 70.76	01 - 33 - 000	0104	1.4 x 0.9 1.4: x 0.9:	(170) 3 - 4:	2	S ... P w 8045		15.5		*
08043	12 52.7 + 03 10 305.25 + 65.75	01 - 33 - 025 NGC 4799	0104	1.6 x 0.6 1.6 x 0.7	91 6	1	S ...		14.4		3
08044	12 52.8 + 39 30 119.51 + 77.87	07 - 27 - 005 IC 3892	0133	(1.6 x 0.8) 1.5 x 0.7	175 (5)	1	S? ... P w IC 3895		15.7		*1
08045	12 52.9 + 08 11 305.93 + 70.76	01 - 33 - 000	0104	1.0: x 0.8: 0.8: x 0.8:	— ...	2	Irr I B: m P w 8042 Ir I		15.3		*
08046	12 52.9 + 59 03 122.04 + 58.34	10 - 19 - 002	1427	1.1 x 1.0 1.0 x 0.8	— ...	2	... S B dm: P w 8040, 8040 group		15.5		*1
08047	12 53.0 + 55 25 121.79 + 61.97	09 - 21 - 063	0729	1.4: x 1.4: 1.2: x 1.2:	— 1	2	Sb - c	SAB(r) ...	15.2		*
08048	12 53.1 00 00 305.23 + 62.58	00 - 33 - 000	1578	1.5:: x 0.2 2. x 0.3:	60 ...	3	Dwarf V		17. - 18.		*
08049	12 53.1 + 27 47 056.12 + 89.01	05 - 31 - 006 NGC 4807	1393 276	0.8 x 0.6 0.8 x 0.6	— ...	1	... Compact		14.4		*13
08050	12 53.2 + 52 32 121.50 + 64.85	09 - 21 - 062	0729	1.4 x 1.4 1.3 x 1.3	— 1	4	S B b	S B(rs)bc Sb +	15.1		

N G C   4762, 4765, 4771, 4772, 4779, 4787, 4789   I C   3852, 3881, 3892
         4793, 4795, 4798 - 4800, 4807, 4810

1	2	3	4	5	6	7	8	9	10	11	12
08051	12 53.2 + 58 36 121.94 + 58.79	10 - 19 - 003 NGC 4814	0704	3.5 x 2.5 2.8 x 2.2	(135) 3	3	Sb		12.4	+ 2531 + 2661	*12 3468
08052	12 53.2 + 73 28 122.59 + 43.93	12 - 12 - 021	1411 269	2.0 x 0.5 1.8 x 0.4	155 6	1	S ...		14.4		
08053	12 53.3 + 04 17 305.73 + 66.86	01 - 33 - 027	0104	1.7 x 1.3 1.8 x 1.2	15: 2 - 3	3	...	SAB dm	15.0		
08054	12 53.3 + 04 34 305.76 + 67.14	01 - 33 - 028 NGC 4808	0104	2.6 x 1.0 2.6 x 1.0	127 6	3	Sc	Sc +	12.5	+ 738 + 650	*123 4678
08055	12 53.5 + 04 05 305.83 + 66.66	01 - 33 - 000	0104	1.0: x 0.8: 1. x 1.	-- ...	4	Dwarf irregular Ir V		17.		
08056	12 53.8 + 10 28 307.05 + 73.03	02 - 33 - 000	0041	1.3 x 0.7 1.3 x 0.6:	162 ...	3	S B c	S B(s)d:	15.3		*
08057	12 53.8 + 28 01 063.41 + 88.77	05 - 31 - 010 NGC 4816	1393 276	(1.3 x 1.1) (1.3 x 1.1)	-- ...	2	E - S0		14.8		*3
08058	12 54.0 + 57 09 121.63 + 60.24	10 - 19 - 004 Mark 230	0704	1.7 x 1.1 0.7 x 0.4	(10) ...	2	S ... Very compact nucleus		14.1	+ 21300	*1C
08059	12 54.0 + 63 53 122.07 + 53.51	11 - 16 - 007	0717 279	1.4 x 0.25 1.4 x 0.25	73 7	1	S ...		14.6		*
08060	12 54.1 + 27 15 025.75 + 88.86	05 - 31 - 014 NGC 4819	1393 276	1.1 x 0.9 1.0 x 0.8	160: 2	2	S B a n of 2		14.1	+ 6696 + 6702	*12 3
08061	12 54.2 + 12 12 307.83 + 74.72	02 - 33 - 000	0041	1. x 1. 0.8:: x 0.8::	-- ...	3	Dwarf V		16.5:		
08062	12 54.2 + 21 57 315.51 + 84.42	04 - 31 - 001 NGC 4826 = M 64	1581	10. x 5. 11. x 6.	115 5:	3	Sb	S ab Sb -	8.9	+ 368 + 352	*12- -689
08063	12 54.2 + 38 53 117.93 + 78.47	07 - 27 - 008 IC 3916	0133 277	1.0 x 0.6 1.1 x 0.7	135: ...	1	S ...		15.6		*
08064	12 54.3 + 11 21 307.69 + 73.90	02 - 33 - 000	0041	1.0 x 0.12 0.9 x 0.15	151 7	1	Sb: s of 2		15.7		*
08065	12 54.3 + 27 26 034.93 + 88.82	05 - 31 - 016 NGC 4827	1393 276	1.4 x 1.2 (1.1 x 1.0)	-- 1 - 2	2	S0		14.1	+ 7650 + 7657	*23
08066	12 54.5 + 01 18 306.12 + 63.87	00 - 33 - 000	1578	1.8: x 1.4: 1.4: x 1.1:	(170) ...	4	Dwarf spiral, S B m S B V		17.		
08067	12 54.6 - 01 26 305.90 + 61.19	00 - 33 - 022	1578	2.1 x 0.4 2.0 x 0.4	138 6	1	S ...		14.3		
08068	12 54.6 + 48 34 120.44 + 68.81	08 - 24 - 11+(12)	1350	1.3 x 0.5 1.2 x 0.6	[70] ...	3	Strongly peculiar Disrupted pair?		14.4	+ 8450 + 8545	*1C
08069	12 54.8 + 29 18 089.46 + 87.71	05 - 31 - 024	1393 276	1.5:: x 0.6: 0.7 x 0.25	... 6:	1	S B? ...		14.8		*
08070	12 55.0 + 27 46 048.76 + 88.62	05 - 31 - 025 NGC 4839	1393 276	(4.0 x 1.8) (3.5 x 1.7)	65: ...	3	E		13.6	+ 7446 + 7455	*12 3
08071	12 55.1 + 28 28 071.72 + 88.28	05 - 31 - 000	1393 276	1.1 x 0.3 0.9 x 0.3	158 6	1	S ... nnf of 2		15.4		*
08072	12 55.1 + 28 45a 078.45 + 88.09	05 - 31 - 026 NGC 4841a	1393 276	[1.6 x 1.4] ...	-- ...	3	E + E Common envelope		13.5		*12 3
08073	12 55.1 + 28 45b 078.45 + 88.09	05 - 31 - 027 NGC 4841b	1393 276								
08074	12 55.2 + 02 58 306.74 + 65.52	01 - 33 - 031	0104	1.6: x 1.5: 1.3: x 1.3:	-- ...	4	Dwarf spiral, S m S V, p w 8084		15.4		*1
08075	12 55.2 + 41 44 118.34 + 75.61	07 - 27 - 000	0133	1.0 x 0.12 1.1 x 0.12	134 7	2	Sc sf of 2		16.5:		*
08076	12 55.4 + 29 55 094.22 + 87.12	05 - 31 - 034	1393 276	1.3 x 0.8 0.9 x 0.7	95: 4:	3	Sc	SAB: ... Sc +	15.2		*
08077	12 55.4 + 70 28 122.22 + 46.92	12 - 12 - 022 NGC 4857	1411	1.3 x 0.7 1.2 x 0.7	110 4 - 5	2	Sb/S B b		14.7		*3
08078	12 55.5 + 01 50 306.10 + 64.39	00 - 33 - 025 NGC 4845	1578	5.2 x 1.3 5.3 x 1.3	89 6	3	Sb	Sb -	12.9		*123 4678
08079	12 55.5 + 36 38 114.94 + 80.66	06 - 29 - 002 NGC 4846	0110	1.5 x 0.7 1.3 x 0.5	62 5 - 6	1	S ...		14.6		3
08080	12 55.6 + 27 08 023.07 + 88.51	05 - 31 - 038	1393 276	1.1 x 0.55 1.1 x 0.55	56 5	3	S0 - a		15.2	+ 5700	*C
08081	12 55.7 + 15 07 310.56 + 77.62	03 - 33 - 017	1572	1.0:: x 0.8:: 1. x 1.	-- ...	3	Dwarf V		17.		*
08082	12 55.7 + 28 31 070.43 + 88.15	05 - 31 - 039 NGC 4848	1393 276	1.7 x 0.4 1.6 x 0.4	158 6	2	S B: a - b		14.2	+ 7209 + 7221	123
08083	12 55.7 + 75 10 122.42 + 42.23	13 - 09 - 000	1339	1.1 x 0.15 0.8 x 0.18	43 7	1	S ...		16.0:		
08084	12 55.8 + 03 04 307.11 + 65.61	01 - 33 - 032	0104	1.7 x 1.5 1.7 x 1.5	-- 1:	4	... P w 8074	S B(s)dm	15.0		*1
08085	12 55.8 + 14 50 310.52 + 77.33	03 - 33 - 018	1572	2.5 x 0.6 2.5 x 0.5	115 6	2	S B? c		14.9		

N G C   4808, 4814, 4816, 4819, 4826, 4827
4839, 4841, 4845, 4846, 4848, 4857

I C   3916

1	2	3	4	5	6	7	8	9	10	11	12
08086	12 55.8 + 26 40 007.55 + 88.32	05 – 31 – 044 NGC 4849 = 3935*	1393 276	(1.7 x 1.1) (1.5 x 1.0)	(175) 3:	2	S0 P w IC 838		14.5	+ 5823 + 5828	*12 3
08087	12 55.8 + 78 53 122.58 + 38.51	13 – 09 – 040	1339 275	1.4 x 0.35 1.4 x 0.3	137 6 – 7	1	S ...		15.6		
08088	12 55.9 + 14 09 310.26 + 76.65	02 – 33 – 039	0041	1.0 x 0.5 1.0 x 0.4	90 5 – 6	1	S ...		15.2		
08089	12 56.0 + 09 48 308.69 + 72.32	02 – 33 – 000	0041	1.0: x 1.0: 0.9: x 0.9:	-- ...	2	... 8093 group	d – m	16.0:		*
08090	12 56.2 + 10 53 309.20 + 73.40	02 – 33 – 040 IC 840	0041	1.0 x 1.0 1.0 x 0.9	-- 1	1	S B: a:		15.1		
08091	12 56.2 + 14 29 310.76 + 76.98	02 – 33 – 041	0041	1.1 x 0.9: 1.1: x 1.0:	-- ...	4	Irr	I m Ir I	15.3		*1
08092	12 56.2 + 27 51 049.17 + 88.34	05 – 31 – 048 NGC 4853	1393 276	0.8 x 0.7 0.6 x 0.5	-- ...	1	... (R?) ... Very compact		14.2	+ 7550 + 7560	*123 C
08093	12 56.3 + 09 55 308.97 + 72.43	02 – 33 – 042	0041	1.0 x 0.5 1.0 x 0.5	160 5:	1	S ... Brightest of 3		15.0		*
08094	12 56.4 + 50 15 119.96 + 67.11	08 – 24 – 017	1350 278	1.0 x 0.4 1.0 x 0.4	78 6	1	S ...		16.0:		
08095	12 56.4 + 71 42 122.17 + 45.69	12 – 12 – 023	1411	1.3 x 0.5 1.0 x 0.4	118 6	1	Sa – b		15.2		
08096	12 56.5 + 28 06 057.00 + 88.23	05 – 31 – 052 IC 3949	1393 276	1.0 x 0.15 0.9 x 0.15	73 7	1	S ...		14.9	+ 7526 + 7537	*12
08097	12 56.6 + 27 05 022.82 + 88.28	05 – 31 – 053 NGC 4859	1393 276	1.6 x 0.7 1.5 x 0.7	95 5 – 6	2	S0 – a		14.8		3
08098	12 56.7 + 35 08 111.48 + 82.09	06 – 29 – 003 NGC 4861 + 3961*, Ma 59	0110	4.5 x 1.7 4.0 x 1.6:	[15] ...	4	Double system		12.8	(+ 790) (+ 831)	*123A 4678C
08099	12 56.8 + 37 34 114.32 + 79.70	06 – 29 – 004 NGC 4868	0110	1.6 x 1.5 1.6 x 1.5	-- 1	2	Sa: P w 8125		12.9		*12 348
08100	12 56.9 + 28 21 062.01 + 88.01	05 – 31 – 064 NGC 4865	1393 276	(1.3 x 0.7) 1.0 x 0.4	... ...	1	E?		14.6	+ 4643 + 4655	*12 3
08101	12 57 + 84 23 122.77 + 33.01	14 – 06 – 013	1340	1.3 x 0.6 1.3 x 0.6	130 5:	1	S? ...		15.4		
08102	12 57.0 + 14 27 311.59 + 76.92	02 – 33 – 045 NGC 4866	0041	6.0 x 1.3 6.3 x 1.3	87 6	3	Sa SA(r)a		11.9	+ 1910 + 1864	*12 3478
08103	12 57.2 + 28 14 058.15 + 88.00	05 – 31 – 070 NGC 4874	1393 276	(2.4 x 2.4) (2.4 x 2.4)	-- 1	4	S0 Second brightest in Coma cluster		13.7	+ 7171 + 7183	*12 3C
08104	12 57.3 + 43 02 117.40 + 74.28	07 – 27 – 000	0133	1.1 x 0.1 1.0 x 0.1	47 7	3	Sc		17.		
08105	12 57.4 + 02 18 307.94 + 64.82	00 – 33 – 000	1578	1.0: x 0.6: 1.3: x 0.8:	... ...	1	Double system		15.5		*
08106	12 57.5 + 28 31 064.24 + 87.82	05 – 31 – 075 NGC 4881	1393 276	(1.1 x 1.1) (1.0 x 1.0)	-- ...	3	E		14.7	+ 6691 + 6705	*12 3
08107	12 57.5 + 53 37 120.15 + 63.74	09 – 21 – 077	0729 285	2.6 x 1.1 2.7 x 1.1	[53] ...	4	Irr I B: m		15.0		1
08108	12 57.6 + 27 10 026.53 + 88.08	05 – 31 – 078 NGC 4892	1393 276	1.6 x 0.25 1.5 x 0.25	13 6 – 7	1	S ...		14.7		3
08109	12 57.7 + 12 45 311.34 + 75.21	02 – 33 – 047 NGC 4880	0041	3.4 x 2.7 (3.5 x 2.8)	165: 2	3	S0 SA 0		13.3		*12 348
08110	12 57.7 + 28 15 057.35 + 87.90	05 – 31 – 077 NGC 4889	1393 276	(2.8 x 2.0) (2.6 x 1.8)	80: 2	4	E Brightest in Coma cluster		13.0	+ 6443 + 6456	*12 378
08111	12 57.7 + 37 27 113.22 + 79.78	06 – 29 – 8 + 9 NGC 4893	0110	[1.0 x 0.6] ...	... ...	2	Double system Contact		15.6		*13 V
08112	12 57.7 + 47 28 118.73 + 69.86	08 – 24 – 019 NGC 4901	1350	(1.2 x 1.2) (1.1 x 1.1)	-- ...	1	...		15.5		13
08113	12 57.8 + 28 21 059.55 + 87.84	05 – 31 – 081 NGC 4895	1393 276	1.8 x 0.7 1.8 x 0.7	153 6	3	S0		14.3	+ 8406 + 8420	*12 3
08114	12 57.9 + 13 57 312.21 + 76.39	02 – 33 – 048	0041	1.6 x 0.6 1.1: x 0.5	[80] ...	3	... Strongly peculiar, loop		15.4		*
08115	12 57.9 + 36 31 111.88 + 80.69	06 – 29 – 012 IC 4028	0110	1.2 x 1.1 1.0 x 0.9	-- 1	3	S B c SAB: c		15.5		
08116	12 58.1 + 02 46 308.44 + 65.27	01 – 33 – 035 NGC 4900	0104	2.5 x 2.5 2.7 x 2.6	-- 1	4	Sc		12.8	+ 1054 + 962	*123 4678
08117	12 58.1 + 28 37 064.74 + 87.65	05 – 31 – 084 NGC 4896	1393 276	1.1 x 0.6 0.7 x 0.5	5: ...	1	E – S0		15.1	+ 5820 + 5834	*12
08118	12 58.2 + 29 17 076.47 + 87.24	05 – 31 – 087 IC 842	1393 276	1.3 x 0.7 1.2 x 0.7	57 5	1	S ...		14.6		
08119	12 58.2 + 48 34 118.80 + 68.76	08 – 24 – 020	1350	1.7 x 0.4 1.7 x 0.4	46 7	1	S ...		15.7		*
08120	12 58.3 + 73 58 122.12 + 43.42	12 – 12 – 024	1411	2.1 x 1.2 1.7 x 0.9	25 4	4	Sc Sc +		14.7		

N G C   4849, 4853, 4859, 4861, 4865, 4866, 4868, 4874   I C   840, 842, 3935, 3949, 3961, 4028
4880, 4881, 4889, 4892, 4893, 4895, 4896, 4900, 4901

1	2	3	4	5	6	7	8	9	10	11	12
08121	12 58.4 + 00 14 308.14 + 62.74	00 – 33 – 026 NGC 4904	1578	2.4 x 1.5 2.5 x 1.7	25 3:	4	S B c	S B d	13.2		*123 4678
08122	12 58.4 + 27 40 040.83 + 87.90	05 – 31 – 091	1393 276	1.3 x 0.7 1.0 x 0.6	160: 4 – 5:	1	S0 – a		15.5		
08123	12 58.4 + 37 22 112.36 + 79.84	06 – 29 – 000	0110	1.0 x 0.3 1.0: x 0.3	160 ...	1	Sc – Irr		16.5:		*
08124	12 58.4 + 36 37 111.41 + 80.57	06 – 29 – 013 IC 4049	0110	(1.2 x 1.2) (1.0 x 1.0)	— (1)	1	S0?		15.3		
08125	12 58.4 + 37 35 112.60 + 79.63	06 – 29 – 014 NGC 4914	0110	(3.5 x 2.0) (3.8 x 2.0)	155 ...	4	E P w 8099		12.7		*123 4678
08126	12 58.4 + 40 01 114.83 + 77.23	07 – 27 – 012 IC 4056	0133 277	1.1 x 0.9 0.9 x 0.8	— 2:	2	S B b – c		16.0:		*1
08127	12 58.5 – 01 42b 307.87 + 60.81	00 – 33 – 027	1578	1.7 x 0.4 1.7: x 0.4:	17 ...	3	Dwarf m nnf of 2		(15.3)		*1
08128	12 58.5 + 28 04 051.15 + 87.79	05 – 31 – 093 NGC 4911	1393 276	1.3 x 1.1 1.2 x 1.1	— ...	1	S ... Disturbed:		13.7	+ 8006 + 8018	*12 3
08129	12 58.5 + 28 17 056.36 + 87.72	05 – 31 – 090 IC 4051	1393 276	(1.1 x 0.9) (1.0 x 0.8)	— ...	3	E		14.8	+ 4932 + 4945	*12
08130	12 58.7 + 47 30 118.25 + 69.81	08 – 24 – 023 NGC 4917	1350 282	1.7 x 1.0 1.5 x 0.9	160 4	4	S B b	(R´)S B(s)b Sb +	15.0		3
08131	12 58.8 + 40 07 114.57 + 77.12	07 – 27 – 015 IC 4064	0133 277	(1.6 x 1.4) (1.7 x 1.5)	— 1	3	S0 Brightest in cluster		15.0		*1
08132	12 58.8 + 55 07 119.99 + 62.23	09 – 21 – 082	0729 285	1.0 x 0.2 1.0 x 0.2	19 7	1	S ...		16.0:		*
08133	12 58.9 + 28 04 050.50 + 87.71	05 – 31 – 097 NGC 4919	1393 276	1.1 x 0.6 1.0 x 0.6	140 4 – 5	3	S0		14.9		3
08134	12 59.0 + 28 08 051.92 + 87.67	05 – 31 – 098 NGC 4921	1393 276	2.5 x 2.0 2.5 x 2.2	165: 1 – 2	4	Sa	SAB a	13.7	+ 5459 + 5472	*12 3
08135	12 59.0 + 29 35 078.41 + 86.90	05 – 31 – 099 NGC 4922	1393 276	[2.3 x 1.0] ...	— ...	4	Double system Symbiotic, streamer		14.2	+ 7357 + 7376	*12 3
08136	12 59.0 + 77 51 122.32 + 39.54	13 – 09 – 000	1339 275	1.2 x 0.9 1.1 x 0.6	150 3:	2	Sa		16.0:		*
08137	12 59.2 + 29 24 075.32 + 86.99	05 – 31 – 100 IC 843	1393 276	1.0 x 0.5 1.0 x 0.5	137 5:	2	S0		14.8		
08138	12 59.3 + 05 16 309.78 + 67.73	01 – 33 – 000	0104	1.0 x 0.35 0.9 x 0.3	22 6	1	Sa – b		15.4		*
08139	12 59.3 + 36 38 110.36 + 80.51	06 – 29 – 016	0110	1.1 x 1.0 1.1 x 0.8	— 1:	1	S ...		16.5:		*
08140	12 59.4 + 29 19 073.56 + 87.02	05 – 31 – 102 IC 4088	1393 276	1.7 x 0.5 1.6 x 0.4	89 6	2	Sa – b		14.8		*1
08141	12 59.4 + 36 54 110.63 + 80.25	06 – 29 – 017 IC 4086	0110	1.0 x 0.9 0.9 x 0.8	— 1	3	Sc		15.7		
08142	12 59.5 + 27 53 045.35 + 87.62	05 – 31 – 103 NGC 4926	1393 276	1.1 x 1.0 1.1 x 1.0	— (1)	2	E – S0		14.1	+ 7661 + 7673	*23
08143	12 59.5 + 54 39 119.69 + 62.68	09 – 21 – 086	0729 285	1.0 x 0.5 1.0 x 0.5	150 5	2	Sb – c		15.4		
08144	12 59.7 + 40 41 114.26 + 76.53	07 – 29 – 019 IC 4100	0133	1.5 x 1.1 1.4 x 1.0	110 3	3	Sc		15.1		
08145	13 00.0 + 33 09 101.24 + 83.78	06 – 29 – 018	0110	1.3 x 0.3 1.6 x 0.3	104 6	2	Sa – b		15.5		*1
08146	13 00.0 + 58 58 120.30 + 58.38	10 – 19 – 015	0704	3.8 x 0.5 3.5 x 0.4	30 6	3	Sc	Sc +	14.6		
08147	13 00.0 + 80 19 122.42 + 37.07	13 – 09 – 042	1339 275	1.1 x 0.5 1.1 x 0.5	100 5	1	S ...		15.4		*
08148	13 00.2 + 80 23 122.41 + 37.00	13 – 09 – 041	1339 275	1.2 x 0.7 1.1 x 0.7	70: 3	2	Sb		15. :		*
08149	13 00.3 + 07 03 311.02 + 69.47	01 – 33 – 000	0104	1.1 x 0.10 1.0 x 0.10	134 7	3	Sc	Sc +	15.6		
08150	13 00.4 + 50 43 118.46 + 66.58	09 – 21 – 089 NGC 4932	0729	1.8 x 1.4 1.5: x 1.0:	0: 2 – 3	4	Sc	SA(r)c	14.7		*3
08151	13 00.5 + 52 03b 118.77 + 65.26	09 – 21 – 090	0729 278	1.1 x 1.1 1.0 x 0.9	— 1	3	Sc	SAB ...	(15.6)		*
08152	13 00.5 + 67 22 121.28 + 49.99	11 – 16 – 009	0717	1.2 x 0.5 0.9 x 0.5	146 6	2	Sc		16.0:		1
08153	13 00.6 + 04 16 310.33 + 66.70	01 – 33 – 039	0104	1.8 x 1.8 1.9 x 1.8	— 1:	4	Sc	Sc +	15.2		
08154	13 00.6 + 28 17 052.68 + 87.29	05 – 31 – 114 NGC 4931	1393 276	1.6 x 0.6 1.5 x 0.7	78 6 – 7	3	S0		14.4		*12 3
08155	13 00.7 + 08 04 311.69 + 70.47	01 – 33 – 040	0104	3.2 x 2.8 2.5:: x 2.5::	— ...	3	S ... Singular		14.9		*1

N G C 4904, 4911, 4914, 4917, 4919, 4921   I C 843, 4049, 4051, 4056, 4064
4922, 4926, 4931, 4932   4086, 4088, 4100

1	2	3	4	5	6	7	8	9	10	11	12
08156	13 00.8 + 04 52b 310.63 + 67.29	01 – 33 – 042	0104	1.2: x 1.1: 0.9 x 0.9	-- 1	2	Sb – c f of 2, in group		(15.4)		*
08157	13 00.8 + 75 40 122.02 + 41.71	13 – 09 – 044 NGC 4954 = 4972	1339	0.8 x 0.6 1.1 x 0.7	62 ...	1	S0? Brightest in quadrupel system		(14.2)		*13
08158	13 00.8 + 80 13 122.37 + 37.17	13 – 09 – 000	1339 275	1.1 x 0.6 1.4 x 0.9	... ...	1	...		16.0:		
08159	13 00.9 + 14 39 315.85 + 76.95	03 – 33 – 023 NGC 4935	1572	1.2 x 1.0 1.2 x 1.0	-- 2:	3	S B b	(R´)SAB(s)b	13.9		*1
08160	13 00.9 + 28 17 052.25 + 87.23	05 – 31 – 115 NGC 4934	1393 276	1.0 x 0.20 0.9 x 0.15	104 (6)	1	S: ...		15.0		13
08161	13 01.0 + 26 49 021.36 + 87.27	05 – 31 – 116	1393 276	1.0 x 0.4 1.1 x 0.4	138 6	1	S ...		15.5		
08162	13 01.1 + 50 54 118.23 + 66.39	09 – 21 – 000	0729 278	1.2: x 1.0: 0.6 x 0.6	-- 1:	2	Sc		15.6		
08163	13 01.3 + 40 31 112.83 + 76.64	07 – 27 – 000	0133 280	1.0 x 0.8 0.6 x 0.3	(40) ...	1	S ...		16.5:		
08164	13 01.3 + 78 48 122.24 + 38.58	13 – 09 – 045	1339 275	1.9 x 1.0 2.0: x 1.0	5 5	3	Sb	SA bc Sb +	15.2		*
08165	13 01.4 + 05 58 311.39 + 68.36	01 – 33 – 000	0104	1.1 x 0.3 1.0 x 0.25	78 6	1	S ...		15.5		
08166	13 01.4 + 11 15 313.82 + 73.59	02 – 33 – 000	0041	1.5 x 0.1 1.3 x 0.1	157 7	3	Sc		15.7		
08167	13 01.5 + 28 28a 054.85 + 87.04	05 – 31 – 118 NGC 4944	1393 276	1.6: x 0.6 1.5 x 0.5	89 6	1	Sa – b p of 2, bridge:		(13.3) + 6993 + 7009		*12 3
08168	13 01.5 + 51 48 118.34 + 65.49	09 – 21 – 000	0729	1.1: x 0.4 1.0 x 0.3	[8] ...	1	... Peculiar		16.0:		
08169	13 01.6 + 37 13 108.65 + 79.83	06 – 29 – 023 IC 4144	0110	1.1 x 0.15 1.0 x 0.2	6 7	1	Sb:		16.0:		
08170	13 01.8 + 09 29 313.17 + 71.83	02 – 33 – 052	0041	1.0 x 0.9 1.2 x 1.1	-- 1 – 2	3	Sc		14.8		
08171	13 02.2 + 18 42 322.91 + 80.79	03 – 33 – 028	1572	1.0 x 0.9 0.9 x 0.9	-- 1	4	Sc	S d	15.7		1
08172	13 02.2 + 61 28 120.19 + 55.86	10 – 19 – 000	0704	1.0 x 0.8 0.9 x 0.7	5: 2	3	S B b	S B(s)b	16.0:		
08173	13 02.4 + 04 05 311.39 + 66.47	01 – 33 – 049	0104	1.0 x 0.5 1.0 x 0.4	[175] ...	1	S ... Peculiar?		15.3		*1
08174	13 02.4 + 55 52 119.07 + 61.43	09 – 21 – 099	0729 285	(1.2 x 0.7) (1.2 x 0.7)	(10) ...	2	E – S0		15.6		*1
08175	13 02.6 + 29 23 067.35 + 86.41	05 – 31 – 121 NGC 4952	1393 276	1.3 x 0.8 (1.4 x 0.9)	23 ...	2	E		13.6 + 5865 + 5886		*12 3
08176	13 02.7 – 00 05 310.39 + 62.32	00 – 33 – 000	1578	1.0 x 0.1 0.9 x 0.15	99 7	2	...	d – m	18.		
08177	13 02.7 + 35 27 103.92 + 81.44	06 – 29 – 025 NGC 4956	0110 281	(1.5 x 1.5) (1.7 x 1.6)	-- 1	2	S0		13.5		3
08178	13 02.8 + 27 50 041.87 + 86.91	05 – 31 – 124 NGC 4957	1393 276	1.1 x 0.9 (1.3 x 1.0)	(100) ...	3	E		14.2		*12 36
08179	13 02.9 + 32 16 092.08 + 84.28	05 – 31 – 123	1393	1.5 x 0.45 1.5 x 0.4	15 6	2	Sb n of 2, disturbed		15.1		*1
08180	13 03.0 + 31 43 088.65 + 84.72	05 – 31 – 122 IC 4166	1393	1.0 x 0.6 1.0 x 0.5	0 ...	1	S B? a – b		15.2		
08181	13 03.1 + 33 10 096.04 + 83.48	06 – 29 – 027	0110	1.6 x 0.4 1.5 x 0.3	80 6	2	Sc – Irr		16.0:		1
08182	13 03.1 + 36 23 105.49 + 80.53	06 – 29 – 026 IC 4171	0110 281	1.0 x 0.6 0.8: x 0.4:	85: 4:	3	Sc	Sc +	16.0:		
08183	13 03.1 + 78 38 122.11 + 38.74	13 – 10 – 001	1374 275	1.6 x 0.4 1.6 x 0.35	133 6	2	Sb		15.5		
08184	13 03.2 + 56 36 119.01 + 60.69	09 – 22 – 007 NGC 4964	0675 285	1.3 x 0.6 1.2 x 0.5	134 5	1	S ...		14.0		3
08185	13 03.3 + 28 00 044.56 + 86.78	05 – 31 – 126 NGC 4961	1393 276	1.6 x 1.1 1.4 x 1.0	100: 3	2	Sc		13.5 + 2559 + 2574		*12 3468
08186	13 03.4 + 04 13 312.06 + 66.56	01 – 33 – 051	0104	1.8: x 0.3 1.8 x 0.3	107 6 – 7	2	Sb – c		15.4		
08187	13 03.4 + 36 16 104.90 + 80.62	06 – 29 – 030 IC 4178	0110 281	1.2 x 0.4 1.2 x 0.3	[123] ...	3	Irr	Ir I	16.0:		*1
08188	13 03.5 + 37 52 107.70 + 79.10	06 – 29 – 031 IC 4182	0110	7. x 6. 7. x 6.	-- ...	5	Dwarf spiral, S m S IV – V		14.0		*12
08189	13 03.5 + 46 44 115.56 + 70.46	08 – 24 – 037	1350 282	1.7 x 0.2 1.5 x 0.2	133 7	2	Sb – c		15.7		
08190	13 03.6 + 41 59 112.38 + 75.12	07 – 27 – 030 NGC 4963	0133	0.9 x 0.9 0.9 x 0.9	-- ...	1	...		14.2		*3

N G C   4934, 4935, 4944, 4952, 4954, 4956   I C   4144, 4166, 4171, 4178, 4182
4957, 4961, 4963, 4964, 4972

1	2	3	4	5	6	7	8	9	10	11	12
08191	13 03.7 + 36 14 104.48 + 80.63	06 – 29 – 034 IC 4189	0110 281	1.6 x 1.1 1.6 x 1.1	0 3	4	Sc		14.5		
08192	13 03.8 + 10 38 315.44 + 72.88	02 – 33 – 054	0041	1.2 x 1.0 1.6 x 1.5	— 1 – 2	2	Sa – b ssp of 2		15.0		*1
08193	13 03.8 + 30 29 077.48 + 85.53	05 – 31 – 129	1393	1.0 x 0.15 0.9 x 0.15	175 6 – 7	2	Sc		16.0:		1
08194	13 03.9 + 29 20b 064.41 + 86.19	05 – 31 – 131 NGC 4966	1393 276	1.0 x 0.45 0.9 x 0.45	143 5	1	S ... f of 2		(13.9)		*13
08195	13 04.0 + 29 57 071.67 + 85.84	05 – 31 – 130	1393 276	1.4 x 0.12 1.2 x 0.1	88 7	3	...	S d – m	16.0:		
08196	13 04.0 + 55 55 118.61 + 61.36	09 – 22 – 010 NGC 4977	0675 285	2. x 2. 1.1 x 1.0	— 1	3	Sb	SA(r)b	14.5		3
08197	13 04.0 + 59 04 119.34 + 58.23	10 – 19 – 024	0704	1.1 x 0.2 0.9 x 0.2	68 7	1	S ... sp of 2, disturbed		16.5		*
08198	13 04.1 + 49 14 116.42 + 67.97	08 – 24 – 039	1350	1 0: x 1.0: 0.9: x 0.9:	— 1	2	Sc – Irr		17.		*
08199	13 04.4 + 35 22 101.56 + 81.38	06 – 29 – 035	0110 281	1.4 x 0.4 0.9 x 0.4	[142] ...	1	... P w 8200		14.7		*1V
08200	13 04.6 + 35 24 101.40 + 81.34	06 – 29 – 036	0110 281	1.5 x 1.1 1.0 x 0.8	(175) 2 – 3:	3	... P w 8199	S dm	16.0:		*1
08201	13 04.6 + 67 58 120.75 + 49.37	11 – 16 – 010	0717	3.7:: x 2.5:: 3.5:: x 2.5::	... ...	4	Dwarf irregular Ir IV – V		14.1		*C
08202	13 05.0 – 00 40 311.45 + 61.67	00 – 34 – 002 IC 849	1595	1.2 x 1.1 1.1 x 1.0	— 1	3	Sc sp of 2		14.0		*
08203	13 05.0 + 33 07 092.84 + 83.32	06 – 29 – 038	0110	1.7 x 0.12 1.8 x 0.12	163 7	3	Sc P w 8210		16.0:		*
08204	13 05.1 + 06 36 314.14 + 68.85	01 – 34 – 001	1561	2.0 x 1.1 0.8 x 0.6	70 5	3	S0	(R)SAB:(r)0+	15.3		1
08205	13 05.1 + 58 24 118.92 + 58.87	10 – 19 – 034	0704	1.1 x 0.7 0.8 x 0.7	105: 4	2	Sb		14.8		*
08206	13 05.2 + 27 45 039.50 + 86.39	05 – 31 – 000	1393 276	1.0 x 0.3 1.0 x 0.22	110 6	2	S0 – a spp of 2		15.7		*
08207	13 05.2 + 36 40 103.72 + 80.12	06 – 29 – 040	0110	1.1: x 1.0: 0.6 x 0.4	— 1:	2	S B c		16.5:		
08208	13 05.3 + 03 28 312.95 + 65.75	01 – 34 – 000	1561	1.0: x 0.4 0.9: x 0.3	... ...	1	S B ...		15.6		
08209	13 05.3 + 25 05 001.55 + 85.67	04 – 31 – 007 NGC 4979 = 4198*	1581 283	1.1 x 0.7 1.0 x 0.8	100: ...	1	S B ... P w 8220	S B(r) ...	15.3		*3
08210	13 05.3 + 33 02 092.00 + 83.36	06 – 29 – 041	0110	1.3 x 1.0 1.2 x 1.0	95 2:	3	Sc P w 8203	Sc +	16.0:		*
08211	13 05.3 + 52 40 117.24 + 64.55	09 – 22 – 014	0675 285	1.3 x 1.1 1.0: x 0.9:	(75) 1 – 2	3	Sc		15.7		
08212	13 05.4 + 18 41 327.24 + 80.50	03 – 34 – 002 NGC 4978	0080	1.6 x 0.8 1.8 x 0.8	142 5	2	S0 – a		14.4		*3
08213	13 05.6 + 60 25 119.25 + 56.86	10 – 19 – 035 IC 852	0704	(1.1 x 0.9) (1.1 x 0.9)	(20) ...	1	E – S0		14.8		
08214	13 05.7 + 62 29 119.65 + 54.81	10 – 19 – 036	0704 279	0.8 x 0.7 0.8 x 0.7	(150) 1	2	S B ...	S B(s) ...	14.3		*
08215	13 05.8 + 47 06 114.61 + 70.03	08 – 24 – 048	1350 282	1.0: x 0.8: 1.0: x 0.8:	... ...	3	Dwarf irregular Ir IV – V or V		17.		
08216	13 05.8 + 52 12 116.90 + 65.00	09 – 22 – 015 NGC 4987	0675 285	(1.2 x 0.7) (1.2 x 0.7)	35: ...	2	E		14.6		3
08217	13 05.9 + 04 38 313.78 + 66.88	01 – 34 – 000	1561	[1.1 x 0.5] ...	... ...	2	Double system Bridge?		15.5		
08218	13 05.9 + 41 57 110.74 + 75.05	07 – 27 – 032 NGC 4985	0133	(1.3 x 1.0) (1.3 x 1.0)	(135) 2 – 3	2	S0		15.1		3
08219	13 06.0 + 21 19 336.38 + 82.79	04 – 31 – 009 IC 851	1581 283	1.0 x 0.4 1.1 x 0.5	150 6	1	S ...		14.9		1
08220	13 06.0 + 24 58 001.41 + 85.48	04 – 31 – 008 IC 4202	1581 283	1.8 x 0.20 1.7 x 0.22	143 7	2	Sb – c P w 8209		15.2		*
08221	13 06.0 + 35 28 099.88 + 81.16	06 – 29 – 044 NGC 4986	0110 281	1.7 x 1.0 1.7 x 1.0	70: 4:	3	S B b	S B(r)b	14.2		*3
08222	13 06.0 + 52 17 116.86 + 64.91	09 – 22 – 018	0675 285	1.0: x 0.7: 0.9: x 0.6:	15: ...	3	...	S m	17.		*
08223	13 06.1 – 01 52 311.69 + 60.45	00 – 34 – 004	1595	1.0 x 0.9 0.9 x 0.9	— 1	3	S B c		14.7		
08224	13 06.1 + 39 43 108.02 + 77.18	07 – 27 – 000	0133 280	1.5 x 0.15 1.2 x 0.15	43 7	2	S B? c		16.0:		
08225	13 06.1 + 45 38 113.57 + 71.45	08 – 24 – 049	1350 282	1.2 x 1.1 (1.2 x 1.1)	— 1	2	S0		14.6		

N G C   4966, 4977 – 79, 4985 – 87   I C   849, 851, 852, 4189, 4198, 4202

1	2	3	4	5	6	7	8	9	10	11	12
08226	13 06.2 + 61 50 119.42 + 55.45	10 – 19 – 038	0704 279	1.2 x 0.9 1.2 x 0.9	(5) ...	2	S B 0 – a		15.4		
08227	13 06.3 + 34 15 095.62 + 82.21	06 – 29 – 045	0110 281	1.2 x 0.25 1.3 x 0.25	78 7	1	S ...		15.7		
08228	13 06.4 + 44 00 112.22 + 73.03	07 – 27 – 033	0133 282	1.1 x 0.8 1.4 x 0.9	145 3	2	Sa		15.3		*
08229	13 06.5 + 28 27 049.21 + 85.99	05 – 31 – 142	1393 276	1.6 x 1.2 1.4 x 1.0	60 3:	4	S B b	S B(r)b	14.3		
08230	13 06.5 + 53 02 116.96 + 64.16	09 – 22 – 019 IC 853	0675 285	1.1 x 1.0 1.1 x 1.0	— 1	3	Sb/S B b	(R')SAB(s)ab Sb –	15.0		
08231	13 06.5 + 54 20 117.40 + 62.88	09 – 22 – 021	0675 285	1.6 x 0.5 1.7 x 0.5	78 6	1	S B ... Disturbed?		14.2		*
08232	13 06.6 + 11 54 318.77 + 73.96	02 – 34 – 001 NGC 4992	1420	1.2 x 0.7 1.1 x 0.6	10 4 – 5	2	Sa		14.6		3
08233	13 06.8 + 07 16 315.65 + 69.43	01 – 34 – 003	1561	1.7 x 0.7 1.7 x 0.7	138 6	2	Sc		15.1		
08234	13 06.8 + 62 33 119.45 + 54.73	10 – 19 – 040	0704 279	1.2 x 0.5 1.5 x 0.9	137 4:	1	S0 – a P w 8237, 8237 group		13.8		*
08235	13 07.0 + 01 08 313.08 + 63.38	00 – 34 – 009 NGC 4996	1595	2.0 x 1.6 (2.0 x 1.7)	40 2:	4	S B 0/S B a		14.4		3
08236	13 07.0 + 01 57 313.37 + 64.19	00 – 34 – 010 NGC 4999	1578	2.3 x 2.0 2.4 x 2.2	— 1	4	S B b	S B(r)b	13.5		*12 3468
08237	13 07.0 + 62 35 119.42 + 54.69	10 – 19 – 041	0704 279	1.0 x 0.8 1.0 x 0.8	(130) ...	2	S B b: P w 8234, brightest in group	(R')S B ...	13.7		*1
08238	13 07.1 – 00 46 312.52 + 61.50	00 – 34 – 011	1595	1.1 x 0.5 0.9 x 0.5	50 5:	3	Sc	Sc +	14.9		
08239	13 07.1 + 46 30 113.60 + 70.57	08 – 24 – 054	1350 282	1.1 x 0.12 1.2 x 0.12	159 7	2	Sc		16.5:		*
08240	13 07.3 + 62 26 119.33 + 54.84	10 – 19 – 042 NGC 5007	0704	(0.9 x 0.6) 0.8 x 0.8	(135) ...	1	E – S0 npp of 2, 8237 group		14.2		*3
08241	13 07.4 + 29 10 057.64 + 85.58	05 – 31 – 144 NGC 5000	1393 276	1.7 x 1.4 1.7 x 1.4	... ...	4	S B b npp of 2, disturbed, Sb +	S B(rs)bc	14.0		*13
08242	13 07.4 + 35 15 097.58 + 81.22	06 – 29 – 047	0110 281	1.0 x 0.3 1.1 x 0.25	137 6	1	S ...		15.6		*
08243	13 07.4 + 53 45 116.91 + 63.43	09 – 22 – 022 NGC 5001	0675 285	1.3 x 0.4 1.1 x 0.4	160 (6)	1	S B: ...		14.6		*3
08244	13 07.6 + 28 39 051.02 + 85.71	05 – 31 – 146	1393 276	1.3 x 0.6 1.2 x 0.5	75 5 – 6	4	Sc	Sc +	16.0:		
08245	13 07.6 + 79 12 121.89 + 38.16	13 – 10 – 002	1374 275	1.7 x 0.7 1.8 x 0.8	[75] ...	4	Irr	I m Ir I	14.7		
08246	13 07.7 + 34 27 094.54 + 81.89	06 – 29 – 048	0110 281	3.3 x 0.8 3.2: x 0.7	83 6	4	S B c	S B(s)cd	15.1		
08247	13 07.7 + 55 45 117.49 + 61.45	09 – 22 – 000	0675 285	(1.1 x 0.4) (1.1 x 0.4)	135 6 – 7	2	S0		15.7		
08248	13 08.0 + 18 43 330.69 + 80.28	03 – 34 – 004	0080	1.1 x 0.35 1.0 x 0.4	101 6	1	S ...		14.8		*
08249	13 08.0 + 25 12 006.75 + 85.21	04 – 31 – 000	1581 283	1.0 x 0.3: 0.8 x 0.2	[148] ...	3	Dwarf irregular Ir IV – V:		17.		
08250	13 08.0 + 32 45 086.63 + 83.26	06 – 29 – 049	0110	1.6 x 0.25 1.5 x 0.20	13 6	1	Sc		15.6		
08251	13 08.2 + 34 54 095.50 + 81.45	06 – 29 – 050	0110 281	1.2 x 0.7 1.2 x 0.6	147 5:	2	S0		15.7		*
08252	13 08.2 + 44 15 111.36 + 72.71	07 – 27 – 036	0133 282	1.0 x 0.2 1.0 x 0.2	174 6	2	Sc		16.5:		*
08253	13 08.3 + 11 58 320.28 + 73.92	02 – 34 – 000	1420	1.6 x 1.5 1. x 1.	— ...	4	Dwarf spiral, S m S V		17.		*
08254	13 08.3 + 36 53 100.95 + 79.68	06 – 29 – 051 NGC 5002	0110	1.7 x 1.1 1.7 x 1.1	173	4	...	S:B m	14.7		*13
08255	13 08.4 + 11 44 320.14 + 73.69	02 – 34 – 002	1420	1.7 x 1.2 1.6: x 1.1	95 3	3	Sc		14.4		
08256	13 08.6 + 37 19 101.62 + 79.26	06 – 29 – 052 NGC 5005	0110	6.3 x 3.0 6.0 x 2.8	65 5	3	Sb P w 8307	Sb +	10.6	+ 1021 + 1078	*12– –789
08257	13 08.6 + 50 10 114.97 + 66.93	08 – 24 – 063	1350 278	1.2 x 0.1 1.2 x 0.12	142 7	3	Sc		16.0:		*
08258	13 08.6 + 50 21 115.06 + 66.75	08 – 24 – 061 NGC 5009	1350 278	1.3 x 0.8 1.0 x 0.7	75: 4:	2	S B b		15.6		3
08259	13 08.7 + 29 15 063.54 + 85.04	05 – 31 – 150	1393	1.4 x 0.6: 1.2 x 0.5	172 5 – 6:	1	S B a – b NGC 5004 group		15.3		*1
08260	13 08.7 + 29 54 064.21 + 85.01	05 – 31 – 149 NGC 5004	1393	1.6 x 1.3 (1.3 x 1.0)	170 2	3	S0 Brightest of 3		14.3		*13

N G C 4992, 4996, 4999 – 5002, 5004, 5005
5007, 5009

I C 853

1	2	3	4	5	6	7	8	9	10	11	12
08261	13 08.7 + 35 46 097.63 + 80.64	06 - 29 - 053	0110 281	1.0 x 0.5 1.0 x 0.6	[100] ...	4	Irr n of 2	I m Ir I	16.0:		*
08262	13 08.8 + 00 01 313.66 + 62.21	00 - 34 - 014	1595	(1.3 x 0.7) (1.2 x 0.8)	160: 4 - 5:	1	S ...		14.3		
08263	13 08.8 + 03 40 315.13 + 65.80	01 - 34 - 005	1561	1.2 x 0.15 1.1: x 0.12	113 6 - 7	2	S B c		15.4	+ 2990 + 2910	*12
08264	13 09 + 84 53 122.47 + 32.50	14 - 06 - 016	1340	1.7 x 1.5 2.0: x 1.8:	... ...	3	... Peculiar, large plume		14.5		*C
08265	13 09.0 + 00 56 314.11 + 63.10	00 - 34 - 015	1595	0.8 x 0.35 0.8 x 0.4	25 ...	1	S B ...	S B(r) ...	14.4		
08266	13 09.0 + 34 41 093.80 + 81.55	06 - 29 - 054	0110 281	1.1 x 0.9 1.4 x 1.0	... ...	2	E - S0		15.5		
08267	13 09 + 44 00 110.68 + 72.92	07 - 27 - 040	0133 282	1.3 x 0.25 1.3 x 0.25	40 6 - 7	1	S ...		15.7		*
08268	13 09.0 + 45 56 112.24 + 71.05	08 - 24 - 067	1350 282	1.1 x 0.3 1.1 x 0.3	27 6	3	S B b		15.4		
08269	13 09.0 + 46 58 112.96 + 70.04	08 - 24 - 065	1350 282	1.0 x 0.15 1.0 x 0.15	122 6 - 7	1	S B: ... 3rd of 3		16.0:		*
08270	13 09.1 + 23 12 351.35 + 83.81	04 - 31 - 012 NGC 5012	1581 283	2.9 x 1.7 3.0 x 1.7	10 4	4	S B c P w 8290	SAB(rs)c Sc +	13.6		*12 3468
08271	13 09.2 + 36 32 099.17 + 79.92	06 - 29 - 055 NGC 5014, Mark 449	0110 281	1.7 x 0.6 1.7 x 0.6	102 6	1	S ...		13.5		23
08272	13 09.2 + 47 49 113.42 + 69.21	08 - 24 - 069	1350	1.7 x 1.1 0.9 x 0.9	5	3	S B c	(R')S B(rs)c Sc -	16.0:		*
08273	13 09.5 + 21 04 340.51 + 82.13	04 - 31 - 000	1581	1.0 x 0.4 0.9 x 0.4	164 6	1	S0		15.5		
08274	13 09.5 + 50 32 114.80 + 66.54	09 - 22 - 000	0675	1.3 x 0.5 1.2: x 0.5:	150 6	2	...	S B dm	17.		
08275	13 09.6 + 03 15 315.41 + 65.35	01 - 34 - 006	1561	1.2 x 0.4 1.1 x 0.4:	32 6	2	...	d - m	15.6		
08276	13 09.6 + 05 44 316.67 + 67.79	01 - 34 - 000	1561	1. x 0.5 ...	... ...	0	Dwarf V		18. - 19.		
08277	13 09.6 + 41 09 107.21 + 75.61	07 - 27 - 000	0133 284	1.1 x 0.12 1.0 x 0.1	78 7	3	Sc		16.5:		
08278	13 09.7 + 21 40 343.50 + 82.58	04 - 31 - 000	1581 283	1.3: x 0.6: 1.0: x 0.5:	... ...	2	Dwarf IV or IV - V		17.		*
08279	13 09.7 + 24 21 000.92 + 84.43	04 - 31 - 013 NGC 5016	1581	1.9 x 1.4 2.0 x 1.5	50 2 - 3	2	Sb - c		14.3		*12 3468
08280	13 09.8 + 35 56 096.92 + 80.39	06 - 29 - 057 IC 4213	0110 281	2.7 x 0.5 2.8: x 0.5	174 6 - 7	2	Sc		14.1		
08281	13 09.8 + 45 05 111.15 + 71.84	08 - 24 - 080	1350 282	1.2 x 0.2 1.1 x 0.15	62 7	2	Sc contact, 8283 group	nff of 2,	16.0:		*
08282	13 09.8 + 60 30 118.34 + 56.71	10 - 19 - 050	0704 294	2.0 x 0.5 1.7: x 0.5	92 6	1	Sc		15.0		
08283	13 09.9 + 44 57 110.99 + 71.96	08 - 24 - 081	1350 282	1.0 x 0.5 1.0 x 0.5	90 5	3	Sb Brightest in group		15.5		*
08284	13 09.9 + 46 28 112.17 + 70.49	08 - 24 - 084 NGC 5021	1350 282	1.6 x 0.7 1.5 x 0.7	78 6	2	S B b		14.3		13
08285	13 10.0 + 07 27 317.97 + 69.44	01 - 34 - 008	1561	2.0 x 0.5 1.5 x 0.5	55 (6)	1	S - Irr		15.2		
08286	13 10.0 + 44 18 110.36 + 72.58	07 - 27 - 043 NGC 5023	0133 (282)	7.5 x 0.8 7.3 x 0.8	28 7	3	Sc		13.2		123
08287	13 10.0 + 78 40 121.68 + 38.68	13 - 10 - 003	1374 275	2.1? x 0.9 1.2 x 0.8	155: ...	3	S B a	(R')S B(s)a	15.0		
08288	13 10.2 + 05 00 316.64 + 67.04	01 - 34 - 009 NGC 5019	1561	0.9 x 0.7 0.9 x 0.7	(105) 2	3	S ... Peculiar		14.5		*3
08289	13 10.2 + 12 52 322.88 + 74.64	02 - 34 - 003 NGC 5020	1420	3.3 x 2.8 3.3 x 2.8	(85) 1 - 2	4	Sb/S B c	SAB(rs)bc Sb +	13.4		3
08290	13 10.2 + 23 06 352.27 + 83.56	04 - 31 - 014	1581 283	1.5 x 1.1 1.6 x 1.1	[60] ...	3	... Peculiar, p w 8270	[S m]	14.8		*1
08291	13 10.4 + 22 04 346.49 + 82.79	04 - 31 - 000	1581 283	1. x 1. 0.6: x 0.6:	-- ...	3	Dwarf V		17. - 18.		
08292	13 10.4 + 32 04 079.43 + 83.42	05 - 31 - 155 NGC 5025	1393	2.2 x 0.6 2.2 x 0.6	57 6	3	Sb		14.6		3
08293	13 10.4 + 47 20 112.54 + 69.63	08 - 24 - 087 NGC 5029	1350	(1.7 x 1.1) (1.7 x 1.1)	(150) ...	3	E		14.5		3
08294	13 10.6 + 31 31 075.51 + 83.75	05 - 31 - 156	1393	1.0 x 0.8 1.0 x 0.8	110: 2:	1	Sc Singular?		15.2		*
08295	13 10.6 + 70 55 120.44 + 46.38	12 - 13 - 001 NGC 5034	1341	1.0 x 0.8 0.8 x 0.7	15: 2:	1	S ...		14.1		3

N G C 5012, 5014, 5016, 5019 - 21, 5023, 5025
 5029, 5034

I C 4213

1	2	3	4	5	6	7	8	9	10	11	12
08296	13 10.7 + 09 57 320.36 + 71.82	02 - 34 - 004	1420	1.1 x 0.25 1.1 x 0.25	178 6 - 7	2	S0 - a nnp of 2		15.0		*1
08297	13 10.8 + 06 20 317.80 + 68.31	01 - 34 - 010 NGC 5027	1561	1.5 x 1.3 1.3 x 1.1	--- 1 - 2	3	S B:b	S B:(r)b	14.8		3
08298	13 10.8 + 10 27 320.88 + 72.29	02 - 34 - 000	1420	1.1: x 1.0: 1.0 x 0.5	[100] ...	3	Irr	I m Ir I	15.2		
08299	13 10.8 + 34 15 090.02 + 81.71	06 - 29 - 060	0110 281	1.0 x 0.7 1.0 x 0.7	30: 3:	2	S B b		15.0		*
08300	13 11.0 + 28 04 042.07 + 85.09	05 - 31 - 160 NGC 5032	1392	2.2 x 1.1 2.1 x 1.2	22 5	4	S B b n of 2	S B(r)b	13.6		*13
08301	13 11.0 + 30 37 068.08 + 84.21	05 - 31 - 157	1393	1.0 x 0.5 0.8 x 0.3	175 ...	2	...	S: d - m	16.5:		*
08302	13 11.0 + 33 33 086.74 + 82.24	06 - 29 - 000	0110	1.1 x 0.12 1.0 x 0.1	173 7	2	...	d - m	16.5:		
08303	13 11.0 + 36 28 097.18 + 79.81	06 - 29 - 061 Ho VIII	0110 (281)	2.8: x 2.8: 2.7:: x 2.5::	--- ...	5	Dwarf irregular, I m Ir IV - V Ir I		14.7		*12
08304	13 11.0 + 50 52 114.39 + 66.16	09 - 22 - 028	0675	2.3 x 0.5 1.3: x 0.5	17 6	3	Sb Disturbed?		15.0		
08305	13 11.0 + 71 28 120.49 + 45.83	12 - 13 - 002	1341	1.2 x 0.15 1.2 x 0.15	28 7	2	Sb - c		15.4		*
08306	13 11.1 + 16 15 328.79 + 77.73	03 - 34 - 005	0080	1.6 x 0.25 1.5 x 0.25	37 6	2	Sb		14.9		1
08307	13 11.2 + 36 51 097.99 + 79.45	06 - 29 - 062 NGC 5033	0110	11.5 x 5.5: 11.5: x 5.5:	170 5	5	Sc P w 8256	SA(s)c Sc -	10.9	+ 899 + 956	*12- -9
08308	13 11.2 + 46 35 111.61 + 70.32	08 - 24 - 090	1350	1.3: x 0.8: 1.1: x 0.8:	[155] ...	3	Dwarf irregular Ir IV - V or V		17.		*
08309	13 11.2 + 51 45 114.76 + 65.29	09 - 22 - 030	0675 285	1.1 x 0.6 0.8 x 0.5	138 5	2	Sa - b		16.5:		*1
08310	13 11.4 + 17 20 331.34 + 78.68	03 - 34 - 006 IC 857	0080	1.0 x 0.8 1.3 x 1.1	125: 2:	4	S B b	S B(s)b	14.7		
08311	13 11.4 + 23 31 356.71 + 83.63	04 - 31 - 000	1581 283	1.0 x 0.8 1.0 x 0.7	... ...	1	S? ...		15.7		
08312	13 11.5 + 72 52 120.68 + 44.43	12 - 13 - 000	1341 284	1.3 x 0.3 1.3 x 0.3	15 6	1	...		16.0:		
08313	13 11.6 + 42 28 107.48 + 74.25	07 - 27 - 048	0133	1.8 x 0.5 2.0 x 0.7	35 6	1	S ...		14.9		
08314	13 11.8 + 36 35 096.71 + 79.63	06 - 29 - 000	0110	1.1: x 1.0: 1.0: x 0.9:	--- ...	2	Dwarf (irregular) Ir: IV - V:		17.		
08315	13 11.9 + 39 24 102.90 + 77.10	07 - 27 - 052	0133 280	[1.3 x 0.3] ...	[42] ...	2	Double system Bridge		15.6		*1
08316	13 11.9 + 48 25 112.58 + 68.52	08 - 24 - 000	1350	1.0 x 0.25 1.0 x 0.25	167 6	1	S ...		15.7		*
08317	13 12.0 + 30 45 068.06 + 83.96	05 - 31 - 161	1393	1.1 x 0.4 1.0 x 0.4	140 6	1	S ...		15.0		
08318	13 12.1 + 35 38 093.59 + 80.41	06 - 29 - 064	0110 281	2.2 x 1.1: 2.0: x 1.1:	(70) ...	2	S B c:	S B(s) ...	15.4		
08319	13 12.2 + 30 58 069.55 + 83.81	05 - 31 - 162 NGC 5041	1393 287	1.7 x 1.5 1.5 x 1.4	(150) 1 - 2	2	Sc		14.2		*3
08320	13 12.2 + 46 11 110.80 + 70.66	08 - 24 - 093	1350 282	4.3: x 1.7: 3.6: x 1.5:	[150] ...	4	Irr	I B m Ir I	14.0		
08321	13 12.4 + 17 29 332.77 + 78.71	03 - 34 - 007 IC 858	0080	(1.5 x 1.1) (1.7 x 1.3)	(100) ...	1	S0? P w IC 859, disturbed		14.7		*1
08322	13 12.5 + 12 58 325.00 + 74.56	02 - 34 - 005	1420	1.3 x 0.25 1.2 x 0.25	37 6	1	Sa?		14.7		
08323	13 12.5 + 35 08 091.49 + 80.78	06 - 29 - 065 Mark 450	0110 281	1.1 x 0.8 1.1 x 0.8	... ...	2	[Irr] Disrupted multiple system?		14.9		1
08324	13 12.6 + 03 18 317.19 + 65.25	01 - 34 - 011	1561	1.3 x 0.15 1.3 x 0.15	52 7	3	Sc		15.3		*
08325	13 12.6 + 27 17 033.08 + 84.76	05 - 31 - 163	1393	1.4 x 0.25 0.9 x 0.2	144 7	1	Sa - b		16.0:		
08326	13 12.8 + 34 35 089.12 + 81.19	06 - 29 - 068 IC 861	0110	1.0 x 0.6 1.1 x 0.7	80: ...	1	S0 - a		15.6		
08327	13 13.0 + 44 40 109.02 + 72.08	08 - 24 - 95 + 96 Mark 248	1350 282	[1.6 x 1.1] ...	... ...	3	Double system Two bridges		14.9	(+ 10800)	*1
08328	13 13.1 + 27 34 036.18 + 84.65	05 - 31 - 000	1393	1.1 x 0.25 0.7 x 0.3	57 6	1	Sc		16.5:		
08329	13 13.2 + 03 08 317.45 + 65.05	01 - 34 - 012 NGC 5050	1561	(1.4 x 0.5) 1.3 x 0.5	35 6	2	S0 - a		14.7		3
08330	13 13.2 + 29 55 059.79 + 84.13	05 - 31 - 165 NGC 5052	1393 287	1.4 x 0.9 1.3 x 0.9	160 3 - 4	2	S0 - a		14.6		*13

N G C   5027,   5032,   5033,   5041,   5050,   5052   I C   857,   858,   861

1	2	3	4	5	6	7	8	9	10	11	12
08331	13 13.3 + 47 45 111.49 + 69.10	08 - 24 - 097	1350	3.0 x 1.0 3.0: x 1.0:	[140] ...	4	Irr	IA m Ir I	15.6		*
08332	13 13.3 + 52 16 114.28 + 64.72	09 - 22 - 000	0675 285	1.0 x 0.1 0.8 x 0.1	176 7	2	Sb - c		17.		
08333	13 13.5 + 25 42 017.15 + 84.26	04 - 31 - 016	1581	1.3: x 0.4: 1.2: x 0.4:	[27] ...	3	Dwarf V, p w 8336		17.		*
08334	13 13.5 + 42 17 106.04 + 74.30	07 - 27 - 054 NGC 5055 = M 63	0133 (284)	15. x 9. 16. x 10.	105 4	4	Sb	S bc Sb +	9.7	+ 520 + 600	*12- -689
08335	13 13.6 + 62 23 118.06 + 54.78	10 - 19 - 56 + 57	0704 279	[1.7 x 0.7] ...	... ...	4	Double system Disrupted, bridge + streamers		14.4		*1V AC
08336	13 13.8 + 25 40 017.05 + 84.19	04 - 31 - 017 IC 4215	1581	1.6 x 0.25 1.8 x 0.25	45 7	2	Sa - b P w 8333		15.0		*
08337	13 13.8 + 31 12 069.59 + 83.39	05 - 31 - 166 NGC 5056	1393 287	1.9 x 0.9 1.9 x 1.0	0 5	4	Sc Brightest in cluster		13.6		*3
08338	13 13.8 + 35 18 090.74 + 80.50	06 - 29 - 069	0110 281	1.2 x 0.45 1.1 x 0.45	83 6	2	Sb - c		14.6		
08339	13 13.8 + 57 05 116.26 + 59.99	10 - 19 - 058	0704	1.7 x 0.3 1.6 x 0.3	95 6	2	Sc		15.4		
08340	13 14.0 - 01 50 315.63 + 60.15	00 - 34 - 021	1595	1.2 x 1.1 1.1 x 1.1	--- 1	1	Sc? P w 8348		14.5		*
08341	13 14.0 + 36 20 093.89 + 79.62	06 - 29 - 071	0110 281	1.1 x 0.4 0.9 x 0.4	5 6	3	Sc		16.0:		
08342	13 14.1 + 31 17 069.88 + 83.29	05 - 31 - 169 NGC 5057	1393 287	1.2 x 1.1 1.3 x 1.1	--- 1 - 2	3	S0		14.6		*3
08343	13 14.2 + 22 15 352.35 + 82.31	04 - 31 - 018	1581 283	1.3 x 0.7 1.3 x 0.8	30: 4:	1	S ...		15.5		
08344	13 14.4 + 08 06 321.48 + 69.79	01 - 34 - 000 NGC 5059	1561	1.0 x 0.2 0.9 x 0.15	8 7	1	S ...		15.5		3
08345	13 14.4 + 12 49 326.42 + 74.25	02 - 34 - 006 NGC 5058	1420	1.0 x 0.7 0.7 x 0.6	... ...	1	S? ... Companion superimposed		14.6		*1
08346	13 14.4 + 31 39 072.10 + 83.03	05 - 31 - 000	1393 287	1.1 x 0.5 0.8 x 0.3	90: 6	1	S B ...		16.0:		
08347	13 14.4 + 53 12 114.38 + 63.77	09 - 22 - 039	0675 285	(1.1 x 0.8) (1.2 x 0.9)	(150) ...	3	[S0] Compact		15.5		*
08348	13 14.5 - 02 00 315.81 + 59.97	00 - 34 - 022 IC 4218	1595	1.4 x 0.35 1.3 x 0.3	158 6	1	S ... P w 8340		14.4		*
08349	13 14.5 + 06 37 320.40 + 68.35	01 - 34 - 014	1561	1.4 x 1.1 1.2 x 1.1	--- 1 - 2	3	S B b	SAB(rs)b	15.0		*
08350	13 14.6 + 08 40 322.09 + 70.31	02 - 34 - 000	1420	1.0 x 0.2 0.7 x 0.2	138 6	3	Sc		16.0:		
08351	13 14.7 + 06 18 320.30 + 68.03	01 - 34 - 015 NGC 5060	1561	1.4 x 0.8 1.3 x 0.8	55 4:	4	S B b	S B(s)b	14.2		3
08352	13 14.7 + 34 22 086.25 + 81.12	06 - 29 - 074	0110 286	1.2: x 0.5: 1.2 x 0.4	75 6:	1	S ...		15.4		
08353	13 14.8 + 20 55 346.60 + 81.25	04 - 31 - 020 IC 867	1581	1.5 x 1.0 1.4: x 1.0:	30: 3	3	Sc/S B c IC 868 - 70 group	SAB(s)c	15.5		*
08354	13 14.8 + 20 58 346.82 + 81.29	04 - 31 - 019 IC 866	1581	1.1 x 0.35 1.0 x 0.4	32 6	1	S B: ... IC 868 - 70 group		15.6		*
08355	13 15.1 + 57 48 116.20 + 59.25	10 - 19 - 059 IC 875, Mark 249	0704	(1.6 x 1.2) (1.6 x 1.0)	150 ...	2	S0		13.9		
08356	13 15.2 + 31 20 069.14 + 83.07	05 - 31 - 170 NGC 5065	1393 287	1.5 x 0.8 1.2 x 0.7	90 4 - 5	4	Sc nf of 2	Sc +	14.3		*3
08357	13 15.4 - 00 03 317.09 + 61.83	00 - 34 - 025	1595	[1.7 x 0.5] ...	... ...	3	Triple system Disrupted, bridges		14.5		*1
08358	13 15.4 + 04 40 319.64 + 66.41	01 - 34 - 016 IC 871	1561	2.1 x 0.9 1.8 x 0.9	70 5	3	Sb		14.8		
08359	13 15.4 + 27 49 038.60 + 84.14	05 - 31 - 171	1393	1.6 x 0.6 0.9 x 0.5	112 6	1	Sa - b		15.0		
08360	13 15.6 - 00 58 316.78 + 60.92	00 - 34 - 026	1595	1.5 x 1.3 1.4 x 1.3	--- 1	3	Sc	SAB:(r)c	14.3		
08361	13 15.8 + 06 35 321.21 + 68.23	01 - 34 - 000	1561	1.0 x 0.3 1.0 x 0.3	150 6	1	Sa - b		15.1		
08362	13 16.1 + 33 07 079.00 + 81.82	06 - 29 - 000	0110	1.2 x 0.8 1.2 x 0.8	5 3:	3	S B b	(R')S B(r)b	16.0:		
08363	13 16.4 + 28 00 040.24 + 83.91	05 - 31 - 173	1393	1.3 x 0.9 0.9: x 0.6:	... ...	4	Irr	I B m Ir I	15.6		
08364	13 16.4 + 47 24 109.80 + 69.29	08 - 24 - 106	1350	1.4 x 0.1 1.0 x 0.1	53 7	2	Sc		17.		1
08365	13 16.5 + 42 12 104.02 + 74.17	07 - 27 - 058	0133	2.5 x 1.5 2.2: x 1.5:	115 4	4	S B c	S B(s)d	15.5		*

N G C   5055 - 60, 5065   I C   866, 867, 871, 875, 4215, 4218

1	2	3	4	5	6	7	8	9	10	11	12
08366	13 16.8 + 28 46 047.19 + 83.72	05 – 31 – 174 NGC 5081	1393	2.2 x 0.7 2.2 x 0.8	103 6	2	S B b Companion superimposed		14.3		*3
08367	13 16.8 + 39 52 099.97 + 76.27	07 – 27 – 059 NGC 5083	0133	1.4 x 1.2 1.2 x 1.1	(130) 1	4	S B c	S B(r)cd	15.4		
08368	13 16.8 + 43 29 105.56 + 72.97	07 – 27 – 000	0133	1.0 x 0.25 1.0 x 0.25	26 6	2	S ... Disturbed		16.0:		*
08369	13 17 + 81 39 121.75 + 35.68	14 – 06 – 000	1340	1.1 x 0.2 1.1 x 0.2	135 7	2	Sc		16.5:		*
08370	13 17.1 + 17 15 337.04 + 77.96	03 – 34 – 015	0080	1.0 x 1.0 1.0 x 0.9	— 1	2	Sb – c		15.2		
08371	13 17.3 + 30 31 061.41 + 83.07	05 – 31 – 175 NGC 5089	1393	2.1 x 1.0 1.6:: x 0.7	[120] ...	2	S ... Peculiar		14.4		*3
08372	13 17.4 + 06 04 321.84 + 67.62	01 – 34 – 000	1561	[1.0 x 0.4] ...	... ...	3	... Peculiar, streamer		15.3		
08373	13 17.4 + 40 39 100.98 + 75.52	07 – 27 – 060 NGC 5093	0133	1.6 x 0.7 1.2 x 0.7	143 5?	2	Sa		14.8		3
08374	13 17.4 + 73 02 120.12 + 44.21	12 – 13 – 004	1341	1.4 x 0.7 1.4 x 0.7	3 5	3	S B b/Sc	SAB(s)bc Sb +	14.8		
08375	13 17.5 + 16 07 334.79 + 76.93	03 – 34 – 016 IC 881	0080	1.7 x 0.35 1.7 x 0.35	11 7	2	Sa P w IC 882		14.8		*
08376	13 17.5 + 23 16 001.87 + 82.35	04 – 31 – 023 NGC 5092	1581 283	(1.0 x 1.0) (1.1 x 1.0)	— ...	3	E		14.7		3
08377	13 17.6 + 30 23 060.18 + 83.07	05 – 31 – 176	1393 287	(1.0 x 0.9) (0.8 x 0.6)	— ...	1	...		15.2		*
08378	13 17.7 + 32 14 072.57 + 82.12	05 – 31 – 177 IC 4225	1393 287	1.1 x 0.25 1.0 x 0.25	133 7	2	S0 – a		15.1		
08379	13 17.8 + 06 36 322.50 + 68.11	01 – 34 – 000	1561	1.3 x 0.3 1.2: x 0.3	79 6	1	Sb – c		15.3		
08380	13 18 + 84 57 122.25 + 32.41	14 – 06 – 000	1340	1.2 x 0.15 1.0 x 0.2	47 7	1	Sc sf of 2		16.5:		*
08381	13 18.0 – 02 01 317.51 + 59.77	00 – 34 – 029 NGC 5095	1595	1.2: x 0.4: 1.4 x 0.4	126 6	1	S ...		14.8		3
08382	13 18.0 + 05 40 321.91 + 67.20	01 – 34 – 018	1561	1.1 x 0.6 1.1: x 0.5:	[140] ...	4	Irr	I B m Ir I	15.3		
08383	13 18.0 + 14 48 332.62 + 75.71	03 – 34 – 018	0080	1.1 x 0.9 1.0 x 0.8	0: 2	3	Sb		15.4		
08384	13 18.1 + 41 44 102.32 + 74.48	07 – 27 – 061	0133	1.0 x 0.5 1.0 x 0.5	60 5	1	S ...		15.7		
08385	13 18.2 + 10 03 326.02 + 71.34	02 – 34 – 008	1420	2.6:: x 1.1 2.3 x 1.1	102 6	4	... S(B) IV or IV – V	SAB m	14.6		*
08386	13 18.2 + 51 00 111.77 + 65.76	09 – 22 – 052	0675	1.3 x 0.25 1.3 x 0.25	72 6 – 7	2	Sc		16.0:		
08387	13 18.3 + 34 25 082.98 + 80.59	06 – 29 – 000 IC 883	0110	[1.7 x 1.2] ...	... ...	3	Strongly peculiar Eruptive?, two jets		14.8	+ 6952	*AC
08388	13 18.3 + 43 20 104.51 + 73.00	07 – 27 – 062 NGC 5103	0133 282	1.5 x 1.0 1.5 x 1.0	143 ...	2	... Peculiar, acicular?		13.6		*3
08389	13 18.5 + 09 14 325.36 + 70.55	02 – 34 – 009 NGC 5100	1420	[1.3 x 0.9] ...	... ...	3	Double system Contact:, distorted		15.1		*13
08390	13 18.7 + 36 02 088.71 + 79.31	06 – 29 – 000	0110	1.1 x 0.3 1.0 x 0.3	155 (6)	2	Sc – Irr		16.5:		
08391	13 18.8 + 00 36 319.18 + 62.26	00 – 34 – 031 NGC 5104	1595	1.3 x 0.4 1.1 x 0.35	170 6	2	Sa		14.5		3
08392	13 18.9 + 31 29 066.90 + 82.32	05 – 32 – 003	0131 287	1.6 x 1.0 1.6 x 1.1	105: 4	3	Sc		15.3		1
08393	13 18.9 + 57 55 115.28 + 59.03	10 – 19 – 061 NGC 5109	0704 294	1.8 x 0.45 1.7 x 0.45	153 6	1	S ...		13.6		*3
08394	13 19 + 84 46 122.19 + 32.59	14 – 06 – 20 + 21	1340	[1.0 x 0.35] ...	... ...	2	Double system Connected		15.7		*C
08395	13 19.1 + 12 27 329.71 + 73.47	02 – 34 – 000	1420	1.0 x 0.7 0.8 x 0.5	80 ...	1	Sa – b		15.7		*
08396	13 19.1 + 38 48 096.05 + 76.99	07 – 28 – 001 NGC 5107	0154	1.8 x 0.5 1.6 x 0.4	128 (6)	1	...	d – m	13.7		236
08397	13 19.3 + 31 37 067.43 + 82.18	05 – 32 – 005	0131 287	1.4 x 0.6 1.4 x 0.6	13 6	1	Sb – c		14.8		*
08398	13 19.3 + 42 59 103.45 + 73.25	07 – 28 – 000	0154	1.0:: x 0.7:: 0.8:: x 0.4::	... ...	2	Dwarf V		17. – 18.		
08399	13 19.4 + 31 30 066.61 + 82.22	05 – 32 – 006	0131 287	1.1 x 0.8 0.9 x 0.55	(85) 3:	2	S B b	S B(r)b	14.9		
08400	13 19.4 + 42 32 102.74 + 73.65	07 – 28 – 002	0154	0.8 x 0.7 0.8 x 0.7	— ...	1	[S B ...]?		14.5		

N G C   5081,   5083,   5089,   5092,   5093,   5095
5100,   5103,   5104,   5107,   5109

I C   881,   883,   4225

1	2	3	4	5	6	7	8	9	10	11	12
08401	13 19.5 + 28 57 047.77 + 83.10	05 – 32 – 000	0131	1.0: x 0.8: 0.9: x 0.8:	— ...	2	Dwarf peculiar?		17.		*
08402	13 19.7 + 27 00 031.41 + 83.16	05 – 32 – 008 IC 4230	0131	1.1 x 0.25 1.1 x 0.25	1 7	1	Sa?		15.2		
08403	13 19.7 + 39 00 096.07 + 76.75	07 – 28 – 003 NGC 5112	0154	4.0 x 2.8 3.6 x 2.5	130: 3	5	S B c	S B(rs)cd	12.5		*123 4678
08404	13 19.8 – 02 09 318.31 + 59.53	00 – 34 – 032 IC 4239	1595	1.0 x 0.7 0.9 x 0.7	(115) ...	1	S B ...		14.2		
08405	13 20.0 + 20 34 350.66 + 80.17	04 – 32 – 000	0125	1.0 x 0.15 0.7 x 0.2	57 ...	1	...		17.		
08406	13 20.3 + 17 56 341.92 + 78.10	03 – 34 – 020	0080	1.1 x 0.25 0.9 x 0.2	97 7	1	S ...		15.6		*
08407	13 20.4 + 19 57 348.63 + 79.65	03 – 34 – 021	0080	1.2 x 0.15 1.1 x 0.15	49 7	3	Sc		15.7		
08408	13 20.5 + 14 13 333.68 + 74.91	02 – 34 – 010 NGC 5115	1420 288	1.6 x 0.7 1.3 x 0.6	97 5 – 6	4	S B c		14.8		
08409	13 20.6 + 23 34 006.49 + 81.91	04 – 32 – 009	0125	2.6 x 1.1: 1.7: x 1.0:	80 5 – 6:	4	...	SA dm	15.4		*1
08410	13 20.6 + 27 15 033.59 + 82.98	05 – 32 – 009 NGC 5116	0131	2.3 x 0.7 1.9 x 0.7	40 6	1	Sc:		13.7		*12 3468
08411	13 20.6 + 28 35 044.47 + 82.93	05 – 32 – 010 NGC 5117	0131	2.2 x 1.0 2.2 x 1.0	154 5 – 6	1	S B c?		14.5		13
08412	13 20.6 + 35 24 084.87 + 79.55	06 – 29 – 085	0110	1.1 x 0.25 1.0 x 0.3	82 6 – 7	3	Sc		16.0:		
08413	13 20.9 + 06 39 324.50 + 67.91	01 – 34 – 019 NGC 5118	1561	0.9 x 0.8 0.9 x 0.8	(100) 1 – 2	2	Sc		14.4		*3
08414	13 21 + 84 42 122.13 + 32.65	14 – 06 – 022	1340	1.5 x 0.4 1.5 x 0.4	140	3	...	S m	17.		
08415	13 21.0 + 43 20 102.98 + 72.80	07 – 28 – 005 NGC 5123	0154	1.4 x 1.2 1.3 x 1.1	— 1 – 2	3	Sc		13.5		3
08416	13 21.2 + 52 55 111.98 + 63.79	09 – 22 – 55 + 56	0675 285	[1.5 x 0.55] ...	... ...	3	Double system Contact, twisted streamer		15.6		*1V C
08417	13 21.4 + 13 58 333.97 + 74.59	02 – 34 – 000	1420 288	1.1 x 0.2 1.1 x 0.2	174 6 – 7	2	Sb – c		15.6		
08418	13 21.4 + 30 49 060.70 + 82.14	05 – 32 – 000	0131 287	1.1 x 0.10 1.0 x 0.10	147 7	3	Sc	Sc +	16.5:		
08419	13 21.4 + 31 50 067.15 + 81.68	05 – 32 – 013 NGC 5127	0131 287	(2.3 x 1.7) (2.5 x 1.8)	75 ...	4	E brightest in cluster	Second	13.9		*12 3
08420	13 21.4 + 70 47 119.15 + 46.37	12 – 13 – 005 NGC 5144, Mark 256	1341	1.3 x 0.8 1.3 x 0.8	150 4:	3	S ... + superimposed companion, disturbed		13.2	+ 3000	*12 3C
08421	13 21.5 + 09 58 328.26 + 70.96	02 – 34 – 011 NGC 5125	1420	1.9 x 1.4 2.0 x 1.4	170 3	3	Sb np of 2		13.5		*3
08422	13 21.6 + 31 15 063.40 + 81.92	05 – 32 – 014 NGC 5131	0131 287	2.2 x 0.35 2.1 x 0.35	81 7	3	Sa		14.4		*3
08423	13 21.7 + 14 15 334.72 + 74.80	02 – 34 – 012 NGC 5129	1420 288	(1.7 x 1.4) (1.7 x 1.4)	10 ...	4	E Brightest in cluster		13.3		*13
08424	13 21.7 + 22 58 003.86 + 81.38	04 – 32 – 010	0125	1.0 x 0.5 0.8 x 0.5	140: 5:	1	S: ...		15.7		*
08425	13 21.8 + 15 52 338.07 + 76.19	03 – 34 – 023	0080	1.4 x 0.2 1.2 x 0.2	151 7	1	S ... n of 2		15.6		*1
08426	13 21.9 + 31 36 065.38 + 81.70	05 – 32 – 000	0131 287	1.5 x 0.25 1.5 x 0.25	6 7	2	S B? c		15.3		
08427	13 22.0 + 06 47 325.30 + 67.95	01 – 34 – 020	1561	1.5 x 0.8 1.3 x 0.8	15 4 – 5	1	S ...		14.7		
08428	13 22.0 + 14 21 335.15 + 74.86	02 – 34 – 014 NGC 5132	1420 288	1.2 x 0.8 (1.5 x 1.1)	(75) 3:	4	S B 0	(R:)S B(r)0 +	14.3		3
08429	13 22.0 + 17 21 341.86 + 77.39	03 – 34 – 000	0080	1.0 x 0.10 1.0 x 0.10	178 7	2	Sc		15.6		
08430	13 22.0 + 42 04 100.47 + 73.86	07 – 28 – 007	0154	1.3 x 0.9 1.2 x 0.7	35 3	3	Sc	Sc +	16.0:		
08431	13 22.1 + 33 06 073.68 + 80.86	06 – 30 – 001	0116 287	1.0 x 0.7 1.0 x 0.7	100: 3:	2	S0 – a		15.6		
08432	13 22.1 + 36 10 086.37 + 78.77	06 – 30 – 002	0116	1.1 x 0.4 0.9 x 0.35	104 6	1	S ...		15.3		
08433	13 22.6 + 36 38 087.51 + 78.35	06 – 30 – 004 NGC 5141	0116	(1.5 x 1.1) (1.4 x 1.0)	80 ...	1	S0 P w 8435		13.9		*13
08434	13 22.7 + 70 48 119.00 + 46.34	12 – 13 – 006	1341	1.0 x 0.12 1.0 x 0.10	155 7	2	Sc		16.5:		
08435	13 22.8 + 36 40 087.47 + 78.30	06 – 30 – 006 NGC 5142, Mark 452	0116	1.0 x 0.7 1.0 x 0.6	5 ...	1	S0 P w 8433		14.0		*13

N G C 5112, 5115 – 18, 5123, 5125, 5127, 5129   I C 4230, 4239
     5131, 5132, 5141, 5142, 5144

1	2	3	4	5	6	7	8	9	10	11	12
08436	13 22.8 + 63 21 116.62 + 53.63	11 – 16 – 018	0717	1.1 x 1.0 1.1 x 1.0	— 1	3	Sc 	Sc +	15.1		
08437	13 22.9 + 18 43 346.67 + 78.34	03 – 34 – 029	0080	1.0 x 0.25 1.0 x 0.25	65 7	2	Sa		15.0		*
08438	13 23.0 + 16 27 340.45 + 76.52	03 – 34 – 030	0080	1.0 x 0.4 0.9 x 0.4	81 6	1	S ...		15.7		
08439	13 23.0 + 43 31 102.14 + 72.48	07 – 28 – 009 NGC 5145	0154	2.3: x 1.8: 1.2 x 0.9	90: ...	1	S ...		13.6		*3
08440	13 23.4 + 36 39 086.96 + 78.23	06 – 30 – 008	0116	1.0 x 0.6 0.9 x 0.5	... ...	2	S B c Disrupted?		16.5:		
08441	13 23.5 + 58 05 114.22 + 58.73	10 – 19 – 070	0704 294	3.5: x 2.5: 2.5: x 2.5:	... ...	4	Dwarf irregular, I m Ir V		15.6		*
08442	13 23.7 + 22 11 001.36 + 80.56	04 – 32 – 000	0125	1.0 x 0.10 1.0 x 0.1	125 7	3	Sc		16.5:		
08443	13 23.8 + 02 21 322.87 + 63.60	00 – 34 – 033 NGC 5147	1595	1.8 x 1.6 1.7 x 1.5	(120) 1	3	...	S B dm	12.7	+ 1115 + 1042	*123 4678
08444	13 23.9 + 36 12 085.11 + 78.50	06 – 30 – 010 NGC 5149	0116	1.6 x 0.9 1.5 x 0.9	155 ...	2	S B b P w 8447	Sb +	13.8		*13
08445	13 24.0 + 04 43 324.73 + 65.83	01 – 34 – 000	1561	1.0 x 0.2 0.8 x 0.15	74 6	1	S ...		15.3		*
08446	13 24.0 + 51 35 110.11 + 64.93	09 – 22 – 060	0675	1.0 x 0.6 0.8 x 0.6	163 4	2	Sc		15.7		
08447	13 24.2 + 36 16 085.11 + 78.41	06 – 30 – 011 NGC 5154	0116	1.4 x 1.4 1.1 x 0.9	— 1	3	Sc P w 8444		14.9		*13
08448	13 24.3 + 20 13 353.20 + 79.20	03 – 34 – 033	0080	1.1 x 0.9 1.0 x 0.9	(45) 2:	3	Sb/Sc sp of 2		14.9		*
08449	13 24.4 + 43 01 100.62 + 72.82	07 – 28 – 010	0154	1.4 x 0.5 1.3 x 0.4	165 (6)	2	...	dm – m	16.0:		*
08450	13 24.5 + 10 14 330.66 + 70.91	02 – 34 – 000	1420	1.1: x 0.9: 0.9:: x 0.7::	— ...	4	Dwarf irregular, I m Ir V		17.		
08451	13 24.6 + 32 28 068.47 + 80.78	05 – 32 – 019	0131 287	1.8 x 1.0 1.7 x 0.8	65 5	4	Sc	SA c	14.6		*
08452	13 24.7 + 15 21 339.34 + 75.38	03 – 34 – 035	0080	0.7 x 0.6 0.8 x 0.7	— 1	2	S0 P w 8456		14.1		*
08453	13 24.9 + 53 01 110.85 + 63.53	09 – 22 – 062 NGC 5163	0675 285	(1.1 x 0.7) (1.1 x 0.7)	10: ...	3	E		14.9		13
08454	13 25 + 84 47 122.03 + 32.55	14 – 06 – 24 + 25	1340	[2.5 x 0.5] ...	... ...	3	Double system Long bridge		15.6		*1V AC
08455	13 25.0 + 32 17 067.21 + 80.80	05 – 32 – 021 NGC 5157	0131 287	(1.6 x 1.1) 1.6 x 1.3	... 2:	3	S B a	SAB(r)a	14.4		*3
08456	13 25.1 + 15 26 339.84 + 75.39	03 – 34 – 000	0080	1.5 x 0.15 1.3 x 0.15	101 7	2	Sc P w 8452		15.7		*
08457	13 25.1 + 21 09 357.71 + 79.67	04 – 32 – 000	0125	1.2 x 0.20 1.3 x 0.20	38 7	2	Sb – c		15.6		
08458	13 25.2 + 55 45 112.51 + 60.91	09 – 22 – 063 NGC 5164, Mark 257	0675 285	1.0 x 1.0 1.0 x 0.9	— 1	4	S B b 	S B(s)b Sb +	14.6	+ 4800	*3
08459	13 25.3 + 18 02 346.64 + 77.45	03 – 34 – 038 NGC 5158	0080	1.4 x 1.3 1.3 x 1.3	1 1	2	S B? a – b		13.8		*3
08460	13 25.7 + 03 15 324.54 + 64.31	01 – 34 – 022 NGC 5159	1561	1.5 x 0.4 1.3 x 0.3	162 6	3	Sc 	Sc +	15.2		3
08461	13 25.8 + 34 35 077.62 + 79.34	06 – 30 – 012	0116	1.8:: x 1.1:: 2.1: x 1.3:	... ...	3	... Peculiar, n of 2		16.0:		*1
08462	13 25.8 + 55 04 111.92 + 61.54	09 – 22 – 064	0675 285	1.2 x 0.35 1.2 x 0.35	87 6	2	Sa – b		15.7		
08463	13 25.9 + 32 17 066.63 + 80.63	05 – 32 – 026 NGC 5166	0131 287	2.4 x 0.35 2.3: x 0.35	67 7	3	Sb		14.3		*13
08464	13 26.0 + 30 17 054.78 + 81.40	05 – 32 – 000	0131 287	1.2 x 0.12 0.9 x 0.12	70 7	3	Sc		17.		
08465	13 26.0 + 46 56 105.03 + 69.16	08 – 25 – 004 NGC 5169	1593	2.1 x 0.7 2.1 x 0.7	103 6	4	Sc P w 8468		14.7		*23
08466	13 26.1 + 31 05 059.69 + 81.11	05 – 32 – 028	0131 287	1.3 x 0.6 1.1 x 0.6	125: ...	1	Sb – c		15.5		*1
08467	13 26.1 + 63 01 115.89 + 53.87	11 – 16 – 019	0717	1.0 x 0.8 1.0 x 0.8	15: 2:	3	S B b		14.9		
08468	13 26.3 + 46 51 104.80 + 69.21	08 – 25 – 005 NGC 5173	1593	(1.0 x 1.0) (1.2 x 1.2)	— ...	1	E? P w 8465		13.5	+ 2404 + 2508	*12 3
08469	13 26.5 + 10 18 332.10 + 70.76	02 – 34 – 000	1420	1.0 x 0.25 0.9 x 0.2	36 6	2	Sc – Irr		15.6		
08470	13 26.5 + 47 12 105.11 + 68.88	08 – 25 – 007 IC 4563	1593	1.9 x 0.4 1.8 x 0.3	105 6 – 7	3	S B c		15.4		2

N G C   5145, 5147, 5149, 5154, 5157 – 59
         5163, 5164, 5166, 5169, 5173,

I C   4563

1	2	3	4	5	6	7	8	9	10	11	12
08471	13 26.6 + 38 50a 090.93 + 76.14	07 - 28 - 016	0154	1.3 x 0.6 1.1 x 0.4	13 6	1	S ... npp of 2		(15.0)		*1
08472	13 26.7 + 11 32 333.97 + 71.83	02 - 34 - 000	1420 290	1.0 x 0.3 1.0 x 0.4	45 6:	2	S0		15.3		
08473	13 26.9 - 00 08 322.85 + 61.00	00 - 34 - 000	1595	1.1 x 0.4 0.9: x 0.3:	[2] ...	2	Dwarf		17.		
08474	13 26.9 + 01 10 323.69 + 62.24	00 - 34 - 000	1595	1. x 1. ...	-- ...	2	Dwarf V		18. - 19.		
08475	13 26.9 + 11 16 333.72 + 71.58	02 - 34 - 018 NGC 5174 = 5175	1420 290	3.7: x 2.0: 3.0: x 1.7:	160 4 - 5	4	Sc		13.7		*3
08476	13 26.9 + 12 00 334.83 + 72.22	02 - 34 - 020 NGC 5171	1420 290	(1.1 x 0.8) (1.1 x 0.9)	(10) ...	2	E - S0		14.7		*1
08477	13 26.9 + 17 19 345.91 + 76.65	03 - 34 - 041 NGC 5172	0080	3.3 x 1.9 3.2: x 1.8	103 4	4	Sc	Sc -	12.7		*12 346$\overline{8}$
08478	13 27.0 + 11 53 334.72 + 72.11	02 - 34 - 022 NGC 5178	1420 290	1.1 x 0.7 1.3 x 0.7	95 4 - 5	3	S0 - a	S (r)...	15.0		1
08479	13 27.0 + 17 05 345.35 + 76.46	03 - 34 - 042 NGC 5180	0080	1.5 x 1.0 1.7 x 1.3	25: ...	1	S0? Peculiar?		14.3		*3
08480	13 27.2 + 53 20 110.34 + 63.12	09 - 22 - 069 NGC 5201	0675 285	1.8 x 1.1 1.8 x 1.0	145 4:	3	S ... Peculiar		14.3		*3
08481	13 27.3 - 00 02 323.11 + 61.07	00 - 34 - 000	1595	1.0: x 0.4 1.0 x 0.3	123 6	2	Sc	Sc +	16.5:		
08482	13 27.3 + 26 40a 030.28 + 81.44	05 - 32 - 000	0131	1.0 x 0.2 1.0 x 0.18	70 6 - 7	1	S: ... sp of 2		(15.7)		*
08483	13 27.3 + 30 02 052.68 + 81.20	05 - 32 - 000	0131	1.0: x 1.0: 0.9: x 0.9:	-- 1	1	S ...		15.5		
08484	13 27.3 + 32 40 067.76 + 80.19	06 - 30 - 016	0116	1.1 x 0.45 1.1 x 0.45	145 6	3	S B b		15.1		
08485	13 27.5 - 01 28 322.34 + 59.68	00 - 34 - 039 NGC 5183	1595	2.0 x 1.0 2.2 x 0.8	122 5 - 6	1	S ... P w 8487, distorted		13.6		*13
08486	13 27.5 + 11 20 334.23 + 71.57	02 - 34 - 000	1420 290	1.2 x 0.5 0.9 x 0.4	155 5	3	Sc	Sc +	15.6		
08487	13 27.6 - 01 25 322.41 + 59.72	00 - 34 - 041 NGC 5184	1595	2.2 x 1.2 2.2 x 1.2	135 4 - 5	3	Sb P w 8485	Sb +	13.7		*13
08488	13 27.6 + 13 41 338.21 + 73.59	02 - 34 - 025 NGC 5185	1420	1.9 x 0.7 1.9 x 0.6	58 6	3	Sb P w NGC 5181		14.7		*3
08489	13 27.6 + 45 39 102.76 + 70.21	08 - 25 - 011	1593	2.3 x 0.7 1.8: x 0.7	106 6	4	...	SAB dm	15.2		*
08490	13 27.7 + 58 40 113.50 + 58.02	10 - 19 - 078 NGC 5204	0704 294	5.3 x 3.3 5.3 x 3.0:	5 4	5	... M 101 group	SA(s)m	11.7		*12- -6$\overline{8}$9
08491	13 27.7 + 63 10 115.64 + 53.68	11 - 17 - 002	1428	1.0 x 0.2 1.1: x 0.1	165 7	1	Irr:		16.5:		*
08492	13 27.8 + 31 39 062.06 + 80.56	05 - 32 - 030	0131	(1.0 x 0.9) (0.8 x 0.8)	-- ...	2	... Compact		15.0		*
08493	13 27.8 + 47 27 104.84 + 68.56	08 - 25 - 012 NGC 5194 = M 51a	1593	9. ? x 7.5 9. ? x 8.	... 1?	5	Sc P w 8494, bridge; Sc -	SA(s)bc	8.8	+ 445 + 552	*12- -9$\overline{V}$A
08494	13 27.9 + 47 31 104.87 + 68.49	08 - 25 - 014 NGC 5195 = M 51b	1593	(7. x 5.) (8. x 7.)	-- ...	5	... P w 8493, bridge; Ir II	I 0	10.6	+ 527 + 634	*12- -9$\overline{V}$A
08495	13 27.9 + 51 10 108.42 + 65.12	09 - 22 - 073	0675	1.0: x 0.5 0.6 x 0.4	76 ...	1	...		15.7		
08496	13 28.0 + 31 35 061.59 + 80.54	05 - 32 - 031	0131	[1.3 x 0.7] ...	... ...	3	... Disrupted multiple system		14.9		*1V
08497	13 28.1 + 30 17 053.93 + 80.97	05 - 32 - 032	0131	1.1 x 0.7 1.1 x 0.6	65 ...	1	S ... Companion superimposed		15.1		*1
08498	13 28.1 + 31 53 063.18 + 80.40	05 - 32 - 033	0131	2.8 x 0.9 2.4 x 0.9	3 6	3	Sb companion superimposed	Compact	14.2		*A
08499	13 28.1 + 46 56 104.12 + 69.01	08 - 25 - 015 NGC 5198	1593	(2.0 x 1.7) (2.2 x 1.8)	-- ...	3	E		13.2	+ 2499 + 2605	*123 46$\overline{7}$8C
08500	13 28.2 + 18 24 350.10 + 77.26	03. - 34 - 043 NGC 5190	0080	1.0 x 0.8 1.1 x 0.9	-- 1 - 2	1	S B: b		13.7		3
08501	13 28.2 + 62 46 115.38 + 54.05	11 - 17 - 003 NGC 5205	1428	3.5: x 2.0 2.3 x 1.6	10: 3 - 4	1	S ... NGC 5218 group		13.5		*
08502	13 28.3 + 31 32 061.15 + 80.51	05 - 32 - 34 + 35 Mark 455	131	[1.0 x 0.5] ...	... ...	2	Double system Contact		14.6		*1V
08503	13 28.4 + 33 00 068.75 + 79.82	06 - 30 - 000	0116	1.1: x 0.7: 0.6: x 0.4:	... ...	3	Irr	Ir I	17.		
08504	13 28.4 + 35 05 077.85 + 78.61	06 - 30 - 024 NGC 5199	0116	(1.0 x 1.0) (1.2 x 1.2)	-- ...	2	... Compact		15.1		3
08505	13 28.4 + 34 09 074.02 + 79.18	06 - 30 - 023	0116	1.2 x 0.2 1.3 x 0.2	64 7	1	S ...		16.5:		

N G C   5171, 5172, 5174, 5175, 5178, 5180, 5183 - 85
      5190, 5194, 5195, 5198, 5199, 5201, 5204, 5205

1	2	3	4	5			6	7	8	9	10	11	12
08506	13 28.5 + 49 24 106.59 + 66.73	08 - 25 - 016	1593	1.1 0.9	x x	0.10 0.1	159 7	3	Sc		17.		
08507	13 28.6 + 19 42 354.77 + 78.11	03 - 34 - 044	0080	1.6 1.5	x x	0.9 0.9	[12] ...	3	[Irr] Peculiar		14.0		*1V
08508	13 28.8 + 55 10 111.14 + 61.31	09 - 22 - 075	0675 285	1.7 1.7	x x	1.0 1.0	[120] ...	4	Irr	IA m Ir I	14.8		*C
08509	13 28.8 + 67 55 117.31 + 49.04	11 - 17 - 000	1428	1.1: 1.1:	x x	0.6: 0.6:	[170] ...	3	Dwarf (irregular) Ir: V		17.		
08510	13 28.9 + 29 38 049.65 + 80.96	05 - 32 - 036	0131	1.2 1.0	x x	1.1 1.0	-- 1	2	Sb - c	SAB ...	15.6		
08511	13 29.1 + 62 29a 115.08 + 54.30	10 - 19 - 081	0704	1.0 0.8	x x	0.2 0.12	191 7	1	S ... s of 2		(15.6)		*
08512	13 29.2 - 02 28 322.55 + 58.60	00 - 35 - 001 IC 892	0465	(1.2 (1.3	x x	0.9) 0.9)	15: ...	2	S0 P w 8513		14.5		*
08513	13 29.2 - 02 21 322.62 + 58.71	00 - 35 - 002 IC 893	0465	1.2 1.1	x x	0.25 0.22	52 6 - 7	1	Sa - b P w 8512		14.9		*
08514	13 29.3 + 62 17 114.95 + 54.48	10 - 19 - 082	0704	1.3 1.3	x x	0.2 0.2	21 7	2	Sa - b		14.9		
08515	13 29.4 + 11 32 335.82 + 71.52	02 - 35 - 000	1079 290	1.1 1.0	x x	0.15 0.15	3 7	2	Sa - b		15.1		*
08516	13 29.5 + 20 15 357.48 + 78.31	03 - 35 - 001	1019	1.1 1.2	x x	0.8 0.8	30: 3	2	Sc		13.8		
08517	13 29.7 + 31 17 059.08 + 80.32	05 - 32 - 037	0131	1.0 0.9	x x	0.4 0.3	158 6	1	S ...		16.0:		
08518	13 29.8 + 14 10 340.74 + 73.71	02 - 35 - 001 NGC 5207	1079	1.9 2.0	x x	1.1 1.1	140: 4 - 5	3	Sb	SAB(r)b	14.7		3
08519	13 29.9 + 07 35 330.94 + 67.94	01 - 35 - 001 NGC 5208	0090	1.6 1.7	x x	0.6 0.7	162 6	2	S0 NGC 5208 - 10 group		14.4		*13
08520	13 30 + 80 45 120.96 + 36.49	14 - 06 - 026	1340	1.3 1.3	x x	0.7 0.7	50 5	4	Sc	SAB(s)cd	15.5		
08521	13 30.0 + 02 06 325.95 + 62.86	00 - 35 - 006	0465	1.1 1.0	x x	1.0 0.8	-- 1 - 2	1	S B a - b		14.5		
08522	13 30.1 + 07 36 331.08 + 67.94	01 - 35 - 002 NGC 5209	0090	(1.2 (1.3	x x	1.1) 1.1)	-- ...	3	E NGC 5208 - 10 group		14.7		*13
08523	13 30.3 + 07 26 331.01 + 67.76	01 - 35 - 003 NGC 5210	0090	(1.8 (1.7	x x	1.7) 1.7)	-- 1	2	Sa NGC 5208 - 10 group		14.4		*13
08524	13 30.3 + 37 06 083.66 + 76.97	06 - 30 - 028	0116	1.2 1.2	x x	0.35 0.4	160 6	2	S B: b		15.7		
08525	13 30.3 + 71 44 118.42 + 45.30	12 - 13 - 007	1341	(1.1 0.9	x x	0.8) 0.6	[150] ...	1	E - S0? Jet?		14.4		*
08526	13 30.4 - 00 54 324.06 + 60.00	00 - 35 - 008	0465	0.7 0.9	x x	0.6 0.7	... ...	1	...		14.5		*
08527	13 30.4 + 09 47 333.95 + 69.86	02 - 35 - 002	1079	1.1 1.0	x x	0.22 0.20	17 (7)	2	Sc - Irr		15.0		
08528	13 30.4 + 62 57 115.05 + 53.81	11 - 17 - 004 NGC 5216	1428	3. (2.5	x x	2. 2.0)	... ...	3	... Peculiar P w 8529, long bridge		14.0		*12 3VÄ
08529	13 30.5 + 63 01 115.06 + 53.74	11 - 17 - 005 NGC 5218	1428	2.0 1.7	x x	1.6 1.3	[100] ...	3	S ... P w 8528 Long bridge, plume		13.1		*12 3VÄ
08530	13 30.6 - 00 46 324.24 + 60.11	00 - 35 - 009 NGC 5211	0465	2.4 2.3	x x	1.7 1.7	(30) ...	4	Sb/S B b	(R')SAB(s)b	13.9		*3
08531	13 30.6 + 42 08b 095.75 + 73.00	07 - 28 - 030 NGC 5214	0154	1.1 1.0	x x	0.9 0.7	(140) 2	3	Sc nf of 2		(14.4)		*13
08532	13 30.7 + 06 01 329.75 + 66.43	01 - 35 - 004	0090	1.1 1.0	x x	0.9 0.6	-- ...	1	S ... Disturbed?		15.3		*1
08533	13 30.9 + 05 44 329.58 + 66.15	01 - 35 - 000	0090	1.1 1.0	x x	0.4 0.35	126 6	1	S ... p of 2		15.4		*
08534	13 31.0 + 01 34 326.05 + 62.28	00 - 35 - 000	0465	1.3 1.0	x x	0.5 0.4	120 6	2	S B c		15.2		
08535	13 31.0 + 17 44 350.23 + 76.32	03 - 35 - 007	1019	1.4 1.1	x x	0.6 0.5	97 5 - 6	2	S0 - a NGC 5217 group		15.1		
08536	13 31.0 + 34 48 075.08 + 78.37	06 - 30 - 036	0116	1.1 1.0	x x	0.3 0.3	98 ...	1	... NGC 5223 group		16.0:		*1
08537	13 31.1 - 01 38 323.92 + 59.25	00 - 35 - 000	0465	1.1 1.0	x x	0.15 0.18	78 6 - 7	1	S ...		15.3		*
08538	13 31.1 + 46 07 101.79 + 69.51	08 - 25 - 000	1593	1.2 0.8	x x	0.1 0.1	3 7	3	Sc	Sc +	17.		
08539	13 31.2 + 33 18 068.53 + 79.16	06 - 30 - 037	0116	1.4 1.2	x x	0.5 0.5	2 6	1	S ... Brightest of 3		14.4		*1
08540	13 31.3 + 51 46 107.78 + 64.37	09 - 22 - 078 NGC 5225	0675	0.7 0.7	x x	0.7 0.7	-- ...	1	...		14.4		3

N G C   5207 - 11,   5214,   5216,   5218,   5225   I C   892, 893

1	2	3	4	5	6	7	8	9	10	11	12
08541	13 31.3 + 77 06 119.93 + 40.05	13 - 10 - 000	1374	1.0 x 0.45 0.9 x 0.4	71 6	1	S ...		15.5		
08542	13 31.4 + 38 52 139.64 + 77.96	07 - 28 - 031	0154	1.2: x 0.8: 0.9: x 0.8:	... ...	1	S - Irr		15.7		
08543	13 31.5 + 05 00 329.20 + 65.42	01 - 35 - 006	0090	2.4 x 0.3 2.1 x 0.3	141 6 - 7	3	Sb P w 8545		15.3		*
08544	13 31.5 + 17 58 351.28 + 76.41	03 - 35 - 008 IC 897	1019	1.0 x 0.2 0.8 x 0.15	77 (7)	1	... NGC 5217 group		15.4		
08545	13 31.6 + 05 07 329.37 + 65.52	01 - 35 - 005 IC 896	0090	(1.0 x 0.7) (0.9 x 0.6)	(35) ...	3	E P w 8543		15.0		*1
08546	13 31.7 + 18 07 351.87 + 76.48	03 - 35 - 009 NGC 5217	1019	(1.1 x 0.9) (1.1 x 1.0)	--- ...	2	E Brightest in group		14.0		*13
08547	13 31.9 + 35 00 075.33 + 78.10	06 - 30 - 000	0116	1.0 x 0.15 0.9 x 0.2	32 7	1	Sb		16.0:		
08548	13 32.0 + 31 41 060.20 + 79.72	05 - 32 - 038	0131	1.3 x 0.6 1.3 x 0.6	[12] ...	1	S B? ... or disrupted pair?		15.5		*1V A
08549	13 32.0 + 32 28 064.15 + 79.40	05 - 32 - 000	0131	1.2 x 0.12 1.1 x 0.18	175 7	1	Sb		16.5:		
08550	13 32.0 + 48 10 103.93 + 67.61	08 - 25 - 019 NGC 5229	1593	3.7: x 0.6 3. x 0.6	167 7	3	S B c		14.6		2
08551	13 32.0 + 52 59 108.58 + 63.20	09 - 22 - 080	0675	1.6 x 0.15 1.6 x 0.15	18 7	3	Sb		16.0:		
08552	13 32.1 + 04 23 328.96 + 64.79	01 - 35 - 008 NGC 5213	0090	1.0 x 0.9 1.0 x 0.9	--- ...	2	S B b Disturbed?		14.9		*1V
08553	13 32.1 + 34 57 075.01 + 78.10	06 - 30 - 040 NGC 5223	0116	(1.5 x 1.3) (1.7 x 1.3)	--- ...	3	E Brightest in group		14.4		*13
08554	13 32.1 + 37 27 083.64 + 76.47	06 - 30 - 039	0116	1.6 x 0.3 0.9 x 0.2	139 6 - 7	1	Sb - c		16.5:		
08555	13 32.2 + 09 36 334.83 + 69.49	02 - 35 - 004 IC 900	1079	1.4 x 0.9 1.4 x 1.0	27 3 - 4	4	Sc	SAB(s)bc Sc -	14.3		
08556	13 32.3 + 35 02 075.22 + 78.02	06 - 30 - 043 NGC 5228	0116	(1.0 x 0.9) (1.0 x 0.9)	--- ...	2	E - S0 NGC 5223 group		14.5		*3
08557	13 32.3 + 78 32 120.26 + 38.64	13 - 10 - 008	1374 297	1.1 x 0.4 1.0 x 0.35	95 6	2	Sa - b n of 2		16.0:		*
08558	13 32.5 + 14 00 342.29 + 73.20	02 - 35 - 005 NGC 5222	1079 291	[1.6 x 1.2] ...	... ...	3	E + disrupted spiral Contact, NGC 5230 group		14.1		*13 VA
08559	13 32.5 + 14 05 342.46 + 73.27	02 - 35 - 006 NGC 5221	1079 291	3. x 0.7 2. x 0.7	... ...	3	S [b] Long streamers NGC 5230 group		14.5		*13 VA
08560	13 32.6 + 31 39 059.76 + 79.61	05 - 32 - 039	0131	1.3: x 1.0 1.3: x 1.0	... ...	3	Sb P w 8548		15.1		*1A
08561	13 32.6 + 34 18 072.11 + 78.39	06 - 30 - 046	0116	1.1 x 0.8 1.1 x 0.9	... 2 - 3:	3	Sc sff of 2, disturbed		13.8		*1
08562	13 32.7 + 10 56 337.06 + 70.59	02 - 35 - 007	1079						15.2		*1V
08563	13 32.7 + 10 57 337.08 + 70.61	02 - 35 - 008	1079	[1.0 x 0.8] ...	... ...	2	Double system Contact		14.9		*1V
08564	13 32.7 + 38 43 086.89 + 75.47	07 - 28 - 034	0154	(1.7 x 0.7) (1.7 x 0.7)	160 6	2	S0		14.7		*
08565	13 32.7 + 51 53 107.42 + 64.18	09 - 22 - 082 NGC 5238	0675	2.0: x 1.6: 1.8: x 1.5:	[160] ...	3	...	SAB(s)dm	14.2		*3C
08566	13 32.8 + 01 40 327.03 + 62.21	00 - 35 - 010 NGC 5227	0465	1.6 x 1.5 1.5 x 1.3	--- 1 - 2	3	S B b		14.6		3
08567	13 32.9 + 27 54 038.49 + 80.27	05 - 32 - 000	0131	1.1 x 0.6 0.8: x 0.5:	160: ...	2	...	S dm	16.5:		
08568	13 32.9 + 34 56 074.48 + 77.97	06 - 30 - 047 NGC 5233	0116	1.2 x 0.6 1.4 x 0.6	80 6	1	Sa - b NGC 5223 group		14.8		3
08569	13 33.0 + 06 27 331.50 + 66.59	01 - 35 - 010	0090	1.0 x 0.8 0.9 x 0.8	(155) 2	2	Sa - b		15.4		
08570	13 33.0 + 26 41 031.34 + 80.18	05 - 32 - 040 IC 4297	0131	1.1 x 0.6 1.0 x 0.3	65 ...	1	...		15.5		
08571	13 33.0 + 62 15 114.22 + 54.40	10 - 19 - 095	0704 294	1.2 x 0.3 1.2 x 0.3	67 6	1	S ...		14.1		
08572	13 33.1 + 07 48 333.09 + 67.79	01 - 35 - 000	0090	1.7 x 0.5 1.7 x 0.4	57 6	3	Sb		15.3		
08573	13 33.1 + 13 56 342.57 + 73.06	02 - 35 - 009 NGC 5230	1079 291	2.1 x 1.8 1.8 x 1.6	--- 1	5	Sc	SA(s)c Brightest in group, Sc -	13.4		*123 4678
08574	13 33.2 + 03 15 328.53 + 63.64	01 - 35 - 011 NGC 5231	0090	1.2 x 1.1 1.3 x 1.0	--- 1 - 2	1	S B a		14.7		*13
08575	13 33.2 + 09 14 334.95 + 69.05	02 - 35 - 010	1079	2.6 x 0.5 1.7: x 0.3	30 ...	3	Dwarf: irregular, I m Ir IV or IV - V , Ir I		15.5		

N G C 5213, 5217, 5221 - 23, 5227 - 31   I C 896, 897, 900, 4297
      5233, 5238

1	2	3	4	5	6	7	8	9	10	11	12
08576	13 33.2 + 13 35 341.94 + 72.76	02 – 35 – 000	1079	1.3 x 0.3 0.8: x 0.15	130 ...	2	S ... IC 909 group, disturbed		16.0:		*
08577	13 33.2 + 45 00 099.28 + 70.31	08 – 25 – 022	1593	1.3 x 0.3 1.3: x 0.2	124 7	2	Sc – Irr		16.5:		
08578	13 33.3 + 29 29 047.68 + 80.05	05 – 32 – 041	0131	1.1 x 0.25 0.9 x 0.25	23 ...	1	...		15.1	+ 855 + 897	*1
08579	13 33.3 + 33 33 068.52 + 78.66	06 – 30 – 050 IC 4301	0116	1.0 x 0.2 1.0 x 0.2	130 7	1	S ... IC 4304 group		16.0:		*
08580	13 33.3 + 33 44 069.32 + 78.56	06 – 30 – 051 IC 4302	0116	1.3 x 0.12 1.3 x 0.15	128 7	3	Sc IC 4304 group		16.5:		*
08581	13 33.4 + 02 00 327.60 + 62.47	00 – 35 – 000	0465	[1.1 x 0.5] ...	[35] ...	1	Double system		16.0:		
08582	13 33.5 + 06 50 332.21 + 66.88	01 – 35 – 012 NGC 5235	0090	1.2 x 0.5 1.3 x 0.7	120 5	1	S B: ...		14.9		*13
08583	13 33.5 + 35 15 075.35 + 77.69	06 – 30 – 052	0116	1.2 x 0.8 1.1 x 0.8	155: 3	1	S B a – b NGC 5223 group		14.6		*
08584	13 33.7 – 00 47 325.69 + 59.89	00 – 35 – 11 + 12	465	[1.2 x 0.3] ...	... ...	3	Triple system, disrupted Bridges, streamer		15.4		*1
08585	13 33.7 + 10 44 337.39 + 70.30	02 – 35 – 013	1079	1.3 x 0.9 1.0 x 0.8	120: 2 – 3	2	S B c:	S B(r) ...	15.3		*1
08586	13 33.7 + 33 41 068.89 + 78.52	06 – 30 – 055 IC 4304	0116	1.3 x 0.4 1.2 x 0.35	42 6	1	Sa – b Brightest in group		15.0		*
08587	13 33.7 + 35 50 077.38 + 77.30	06 – 30 – 056 NGC 5240	0116	2.1 x 1.6 2.1 x 1.7	60: 2	2	S B c:	S B(s) ...	14.1		*3
08588	13 33.7 + 46 11 100.75 + 69.24	08 – 25 – 023	1593	1.4 x 1.4 1.3: x 1.3:	— 1	4	Dwarf spiral S V		15.3		*
08589	13 33.9 + 07 38 333.35 + 67.55	01 – 35 – 015 NGC 5239	0090	2.1 x 2.0: 2.1 x 2.0:	— 1:	3	S B b Singular	S B(rs)bc Sb +	14.7		*13
08590	13 33.9 + 37 18 082.13 + 76.31	06 – 30 – 000	0116	1.5 x 0.10 1.4 x 0.12	44 7	3	Sc	Sc +	17.		
08591	13 34.0 + 16 20 348.58 + 74.80	03 – 35 – 000	1019	1.3 x 0.2 1.1 x .0.2	154 6 – 7	1	S ...		15.7		*
08592	13 34.0 + 38 36 085.84 + 75.38	07 – 28 – 036 NGC 5243	0154	1.5 x 0.4 1.5 x 0.4	126 6	1	S ...		14.0		13
08593	13 34.0 + 50 13 105.40 + 65.62	08 – 25 – 024 IC 902	1593	2.2 x 0.4 2.2 x 0.4	162 7	3	Sb		14.7		
08594	13 34.0 + 51 30 106.64 + 64.45	09 – 22 – 085 NGC 5250	0675	1.0 x 0.9 (0.9 x 0.9)	— 1	2	S0		14.0		3
08595	13 34.1 + 75 17 119.18 + 41.78	13 – 10 – 000	1374	1.0 x 0.35 1.1 x 0.35	118 6	2	Sa – b P w 8606		15.7		*
08596	13 34.2 + 06 45 332.51 + 66.73	01 – 35 – 000	0090	1.1 x 0.7 1.0 x 0.6	155 4	1	S ...		15.1		
08597	13 34.2 + 46 28 100.93 + 68.95	08 – 25 – 025	1593	1.8 x 1.5 1.8: x 1.5:	40: 2:	4	S B c	S B(s)d	15.2		
08598	13 34.3 + 20 27 001.53 + 77.54	03 – 35 – 010	1019	1.7 x 0.20 1.5 x 0.25	169 7	2	Sb – c		14.8		
08599	13 34.3 + 75 03 119.09 + 42.01	13 – 10 – 000	1374	1.0 x 0.15 0.8 x 0.15	44 6 – 7	2	Sc		16.0:		
08600	13 34.4 + 35 00 073.90 + 77.69	06 – 30 – 059	0116	1.7 x 0.20 1.6 x 0.22	56 7	1	Sb NGC 5223 group		15.4		*
08601	13 34.4 + 48 00 102.80 + 67.58	08 – 25 – 027	1593	1.1: x 0.5: 1.1: x 0.5:	[165] ...	3	Dwarf (spiral?) V		17.		*
08602	13 34.5 + 32 21 062.41 + 78.97	05 – 32 – 000	0131	1.2: x 1.2: 0.7: x 0.7:	— ...	3	Dwarf V, p w 8605		18.		*
08603	13 34.5 + 44 51 098.46 + 70.32	08 – 25 – 026	1593	(1.1? x 0.8?) 0.5 x 0.2	(170) ...	1	... p of 2		15.3		*
08604	13 34.5 + 66 33 115.95 + 50.22	11 – 17 – 006	1428	1.6 x 1.3 1.2: x 1.2:	(90) 1 – 2	4	Sc/S B c	SAB(s)c	15.6		
08605	13 34.6 + 32 21 062.37 + 78.95	05 – 32 – 000	0131	1.8: x 0.8: 1. x ...	... ...	3	Dwarf (spiral) S: V, p w 8602		18.		*
08606	13 34.6 + 75 18 119.14 + 41.76	13 – 10 – 000 NGC 5262	1374	(1.2 x 0.7) (1.4 x 0.9)	170 ...	2	E – S0 P w 8595		15.0		*3
08607	13 34.7 – 01 55 325.39 + 58.69	00 – 35 – 000	0465	1.0: x 0.6: 0.8 x 0.6	(80) ...	1	S0?		15.3		
08608	13 34.7 + 32 01 060.71 + 79.06	05 – 32 – 000	0131 293	1.4 x 0.15 1.5 x 0.15	102 7	2	Sc		15.7		
08609	13 34.7 + 33 48 068.88 + 78.28	06 – 30 – 060	0116	1.1 x 0.9 0.9 x 0.9	— 1 – 2	2	Sc/S B c		16.0		
08610	13 34.8 + 06 44 332.83 + 66.65	01 – 35 – 016	0090	1.0 x 0.25 0.9 x 0.25	58 6	1	Sa 8613 group		15.4		*1V A

N G C   5235, 5239, 5240, 5243, 5250, 5262   I C   902, 4301, 4302, 4304

1	2	3	4	5	6	7	8	9	10	11	12
08611	13 34.8 + 45 09 098.79 + 70.04	08 – 25 – 029	1593	1.6 x 1.5 1.4: x 1.3	— 1	4	Sc	SAB(s)d	15.5		
08612	13 34.9 + 04 21 330.42 + 64.48	01 – 35 – 017 NGC 5246	0090	1.0 x 0.8 1.0 x 0.8	— 2	3	S B b	(R′)S B(s)b	14.8		13
08613	13 34.9 + 06 41 332.83 + 66.59	01 – 35 – 018	0090	1.5 x 0.4 1.3 x 0.4	[110] 6	1	S B ... Brightest in group, disturbed		15.3		*1V A
08614	13 34.9 + 07 54 334.25 + 67.68	01 – 35 – 020	0090	4.0: x 2.0: 4. x 2.	[0] ...	4	Dwarf irregular Ir IV – V Ir I		15.4		*
08615	13 35 + 86 16 122.13 + 31.06	14 – 06 – 027	1340	1.6 x 0.35 1.6 x 0.35	28 6	3	Sc	Sc +	16.0:		
08616	13 35.0 + 09 08 335.90 + 68.75	02 – 35 – 015 NGC 5248	1079	6.8 x 5.0 7.0 x 5.0	(110) 2 – 3	5	Sc	SAB(rs)bc Sc –	11.4	+ 1181 + 1144	*12– –689
08617	13 35.2 + 05 30 331.72 + 65.50	01 – 35 – 021	0090	1.0 x 0.6 0.9 x 0.6	30 4:	1	S B ... p of 2, disturbed?		15.0		*
08618	13 35.2 + 16 14 349.14 + 74.54	03 – 35 – 015 NGC 5249	1019	1.2 x 0.8 (1.6 x 1.0)	170: ...	1	SO?		14.5		*13
08619	13 35.3 + 38 53b 085.88 + 75.00	07 – 28 – 040	0154	1.4 x 0.3 1.0 x 0.25	17 ...	1	S ... nff of 2		(15.1)		*1
08620	13 35.3 + 39 14 086.80 + 74.74	07 – 28 – 038	0154	1.1 x 0.4 1.1 x 0.5	170 6	2	Sa – b		15.3		
08621	13 35.5 + 39 25 087.16 + 74.58	07 – 28 – 041	0154	0.8 x 0.8 0.8 x 0.7	— (1)	1	S ...		14.2		
08622	13 35.7 + 04 47 331.26 + 64.80	01 – 35 – 022 NGC 5252	0090	(1.5 x 0.9) (1.8 x 1.2)	10 ...	2	SO		14.5		*3
08623	13 35.7 + 06 43 333.31 + 66.54	01 – 35 – 000	0090	1.4 x 0.10 1.3 x 0.12	54 7	3	... S d – m sp of 2		16.0:		*
08624	13 35.8 + 18 39 356.33 + 76.14	03 – 35 – 000	1019	1.0 x 0.25 0.9 x 0.25	7 6	2	Sb – c		15.2		
08625	13 35.9 + 00 01 327.29 + 60.39	00 – 35 – 013 IC 903	0465	2.1 x 0.7 1.6 x 0.7	178 6	3	Sb		14.7		
08626	13 35.9 + 07 08 333.90 + 66.88	01 – 35 – 023	0090	1.5 x 0.7 1.5 x 0.7	165 5	4	S B b	Sb +	15.0		
08627	13 35.9 + 33 05 065.16 + 78.40	06 – 30 – 062	0116	1.1 x 0.9 1.0 x 0.8	(120) 2	2	Sb – c		15.0		*
08628	13 36.0 + 00 47 327.91 + 61.09	00 – 35 – 014 IC 904	0465	1.1 x 0.7 1.1 x 0.6	135 4 – 5	1	Sa? Brightest of 3		15.1		*
08629	13 36.0 + 08 42 335.91 + 68.25	02 – 35 – 000	1079	1.4: x 0.4 1.0 x 0.4	42 (6)	2	Dwarf		15.5		
08630	13 36.1 + 33 23 066.39 + 78.22	06 – 30 – 063	0116	1.7 x 0.5 1.8 x 0.4	[95] ...	2	... Peculiar		14.3		*
08631	13 36.2 + 00 48 328.02 + 61.09	00 – 35 – 000	0465	1.2 x 0.8 0.7 x 0.3	90: ...	2	... dm – m IC 904 group		15.6		*
08632	13 36.2 + 48 32 102.75 + 66.97	08 – 25 – 031 NGC 5256, Mark 266	1593	1.2 x 1.1 0.9 x 0.8	— ...	2	Peculiar Eruptive? or disrupted pair?		14.1	See Notes	*13 C
08633	13 36.4 + 15 00 347.02 + 73.41	03 – 35 – 000	1019	1.2 x 0.12 1.3 x 0.12	139 7	2	Sb – c		15.4		
08634	13 36.5 + 02 25 329.49 + 62.55	00 – 35 – 000	0465	1.6: x 0.2 1.4:: x 0.2	61 (7)	1	S? ...		15.7		
08635	13 36.7 + 04 52 331.88 + 64.76	01 – 35 – 024	0090	1.0 x 0.9 1.0 x 0.9	— ...	2	Sb – c + companion Interaction		14.9		*1V
08636	13 36.9 + 29 13 045.59 + 79.31	05 – 32 – 051	0131 293	1.0 x 0.9 0.9 x 0.8	— 1	2	Sb – c	SAB ...	15.4		*
08637	13 37.0 + 06 25b 333.69 + 66.12	01 – 35 – 025	0090	1.1 x 1.1 1.1 x 1.1	— 1	4	Sc ssf of 2	SA(s)d	(15.2)		*1
08638	13 37.0 + 25 01 023.22 + 78.98	04 – 32 – 018	0125 292	1.3: x 0.9: 1.2: x 0.9:	[70] ...	4	Irr	Ir I	14.8		*1
08639	13 37.0 + 51 42 105.86 + 64.08	09 – 22 – 089	0675	1.6 x 1.2 1.6: x 1.2:	[170] ...	3	Dwarf V		15.5		
08640	13 37.1 + 31 28 057.11 + 78.77	05 – 32 – 053	0131 293	1.2 x 0.5 1.1 x 0.4	107 6	2	S B: a – b		15.3		*
08641	13 37.4 + 01 05 328.82 + 61.24	00 – 35 – 015 NGC 5257	0465	1.8: x 0.8 1.7: x 0.9	... ...	3	S ... P w 8645 Disrupted, bridge		13.7	+ 6810 + 6744	*12 3VA
08642	13 37.4 + 46 16 099.32 + 68.85	08 – 25 – 034	1593	1.5 x 0.2 1.4 x 0.2	113 7	3	Sc	Sc +	16.5:		
08643	13 37.4 + 51 19 105.35 + 64.40	09 – 22 – 090 IC 907	0675	1.3 x 0.2 1.1 x 0.2	20 7	2	S B? b		15.3		*
08644	13 37.5 + 07 39 335.42 + 67.15	01 – 35 – 000	0090	1.1 x 1.0 1.0: x 0.8:	— 1:	3	... S dm		16.0:		
08645	13 37.5 + 01 05 328.87 + 61.23	00 – 35 – 016 NGC 5258	0465	1.7: x 1.4 1.7: x 1.3	... ...	3	S ... P w 8641 Disrupted, bridge		13.8	+ 6635 + 6569	*12 3VA

N G C   5246, 5248, 5249, 5252, 5256 – 58   I C   903, 904, 907

1	2	3	4	5	6	7	8	9	10	11	12
08646	13 37.5 + 30 23 051.58 + 78.98	05 – 32 – 057	0131 293	1.0 x 0.5 0.8 x 0.5	40: 5	2	S B a – b Brightest of 3		15.7		*
08647	13 37.5 + 31 33 057.38 + 78.67	05 – 32 – 059	0131 293	1.1 x 0.4 1.1 x 0.4	[48] ...	4	Irr In group		16.5:		*
08648	13 37.6 + 28 40 042.57 + 79.21	05 – 32 – 058 NGC 5263	0131	1.6 x 0.35 1.5 x 0.35	26 6	1	S: ... Peculiar?		14.0		*3
08649	13 37.6 + 61 45 113.07 + 54.71	10 – 20 – 002	0705 294	(1.3 x 1.0) (1.2 x 0.9)	(55) ...	2	E		15.2		
08650	13 37.7 + 02 43 330.35 + 62.71	01 – 35 – 027	0090	1.9 x 0.35 1.8 x 0.4	151 6	3	Sb nnp of 2		14.9		*
08651	13 37.8 + 41 00 089.72 + 73.10	07 – 28 – 046	0154	3.0: x 1.5: 2.0:: x 1.2::	[80] ...	4	Dwarf irregular Ir IV – V		15.3		*
08652	13 38.0 + 26 37 031.70 + 79.06	05 – 32 – 000	0131 295	1.1 x 0.25 1.0 x 0.2	15 6 – 7	1	S ...		15.6		*
08653	13 38.0 + 31 39 057.69 + 78.53	05 – 32 – 061	0131 293	1.2 x 0.25 1.1 x 0.22	73 6 – 7	1	Sb – c In group		15.7		*
08654	13 38.0 + 76 15 119.18 + 40.79	13 – 10 – 000	1374	1.2 x 0.45 1.1 x 0.4	13 6	3	Sc	Sc +	16.0:		
08655	13 38.4 + 39 02 084.64 + 74.47	07 – 28 – 049 NGC 5267	0154	1.4 x 0.5 1.6 x 0.5	56 6	2	S B b		14.3		3
08656	13 38.4 + 43 15a 093.99 + 71.27	07 – 28 – 050	0154	1.0 x 0.2 0.9 x 0.2	37 7	1	S ... spp of 2		(15.4)		*1
08657	13 38.5 + 05 21 333.33 + 65.00	01 – 35 – 029	0090	1.6 x 1.2 0.9 x 0.9	(30) ...	3	Sb	Sb +	15.3		*1
08658	13 38.7 + 54 35 107.93 + 61.34	09 – 22 – 091 Ho V	0675	2.7 x 1.6 2.2 x 1.2	110 4	4	S B b/Sc	SAB bc Sc –	14.4		*12
08659	13 38.7 + 55 41 108.82 + 60.33	09 – 23 – 000	1409	1.1: x 1.0: 0.9:: x 0.9::	-- ...	3	Dwarf V		17. – 18.		
08660	13 39.0 + 14 06 346.73 + 72.32	02 – 35 – 000	1079	1.1 x 0.10 0.8 x 0.1	8 7	3	Sc nnf of 2		15.7		*
08661	13 39.0 + 24 45 022.53 + 78.47	04 – 32 – 000	0125	1.4: x 1.1: 0.6 x 0.5	100: 2:	1	S B a?	(R?)S B(r)a?	16.0:		*
08662	13 39.0 + 34 00 067.64 + 77.40	06 – 30 – 069	0116	1.4 x 0.15 1.1: x 0.15	74 7	3	Sc		16.0:		1
08663	13 39.1 + 05 17 333.57 + 64.87	01 – 35 – 030	0090	(1.0 x 0.9) (1.1 x 1.0)	-- ...	2	E – S0		14.6		*
08664	13 39.1 + 23 25 016.38 + 78.01	04 – 32 – 029 IC 913	0125 292	1.0 x 0.8 0.8 x 0.7	... ...	2	S B b In group		15.6		*
08665	13 39.1 + 23 30 016.75 + 78.04	04 – 32 – 27 +28 IC 911 + 912	125 292	[1.4 x 0.6] ...	... ...	2	Double system Bridge, in group		15.6		*1
08666	13 39.4 + 02 19 330.83 + 62.17	00 – 35 – 017	0465	[1.5 x 0.3] ...	... ...	2	Chain of 3 galaxies		15.5		*1V
08667	13 39.4 + 38 44 083.33 + 74.54	07 – 28 – 052	0154	1.2: x 0.6: 1.2: x 0.4:	170 ...	2	...	S dm	16.5:		
08668	13 39.5 + 18 24 357.88 + 75.31	03 – 35 – 016	1019	1.1 x 0.5 1.1 x 0.5	34 5 – 6	3	Sc	Sc +	15.4		
08669	13 39.5 + 26 38 031.99 + 78.73	05 – 32 – 63 +64 	131 295	[1.0 x 0.7] ...	... ...	3	Double system Common envelope		15.3		*1
08670	13 39.5 + 41 08 089.72 + 72.79	07 – 28 – 053	0154	1.1 x 1.1 1.1 x 1.1	-- 1	4	Sc	SAB d	15.5		
08671	13 39.5 + 55 54 108.78 + 60.09	09 – 22 – 096	0675	0.55 x 0.35 0.4 x 0.25	80 ...	1	...		14.1		*1
08672	13 39.6 + 67 55 115.82 + 48.77	11 – 17 – 007 NGC 5283, Mark 270	1428	1.1 x 1.0 1.1 x 1.1	-- (1)	1	S0? Seyfert		14.3	+ 2700	3
08673	13 39.7 + 04 30 333.06 + 64.11	01 – 35 – 031 NGC 5270	0090	1.2 x 0.9 1.2 x 0.9	(20) 2 – 3	3	S B b	(R')S B(s)b	14.7		*3
08674	13 39.8 + 32 17 059.99 + 77.96	05 – 32 – 000	0131	1.1 x 0.25 1.1 x 0.2	99 6	1	S B: ...		15.7		*
08675	13 39.8 + 35 54 074.39 + 76.27	06 – 30 – 072 NGC 5273	0116	(2.8 x 2.3) (2.7 x 2.3)	(10) 2	4	S0 P w 8679		12.7	+ 1022 + 1094	*12 346̄8
08676	13 39.8 + 48 09 100.93 + 67.02	08 – 25 – 038	1593	1.1 x 0.3 1.0: x 0.3:	152 ...	1	...	d – m	17.		
08677	13 39.8 + 55 56a 108.72 + 60.04	09 – 22 – 101 NGC 5278	0675	 [1.0 x 0.7]	 ...	2	Double system Mark 271 Connecting spiral arm		13.6	+ 7523 + 7665	*12 3VAC̄
08678	13 39.8 + 55 56b 108.72 + 60.04	09 – 22 – 102 NGC 5279	0675	...	...					+ 7566 + 7708	*12 3VAC̄
08679	13 40.0 + 38 47 083.16 + 74.42	07 – 28 – 054	0154	1.4:: x 1.4:: 1.4:: x 1.4::	-- 1:	3	Dwarf spiral S IV – V		17.		
08680	13 40.1 + 35 53 074.18 + 76.23	06 – 30 – 074 NGC 5276	0116	1.1 x 0.7 0.8 x 0.4	153 4	2	S B a P w 8675		14.6		*12 3̄

N G C 5263, 5267, 5270, 5273, 5276, 5278   I C 911 – 13
5279, 5283

1	2	3	4	5	6	7	8	9	10	11	12
08681	13 40.3 + 35 16 071.90 + 76.53	06 – 30 – 075	0116	1.0 x 0.5 1.0 x 0.5	84 5	2	Sa – b		15.2		
08682	13 40.4 + 30 06 049.52 + 78.42	05 – 32 – 070	0131 293	1.2 x 0.2 1.1 x 0.2	57 7	1	S ...		16.0:		*1
08683	13 40.4 + 39 55 085.90 + 73.57	07 – 28 – 055	0154	2.3:: x 2.3:: 2. x 2.	-- ...	3	Dwarf irregular Ir IV – V		15.7		*
08684	13 40.6 + 61 02 112.06 + 55.26	10 – 20 – 008	0705 294	1.7 x 0.4 1.7 x 0.4	30 6	3	Sc		14.5		*
08685	13 40.9 + 30 36 051.84 + 78.22	05 – 32 – 074	0131 293	1.3 x 1.0 1.3 x 1.1	(20) 2	3	S B b	S B(r)bc Sb +	14.5		
08686	13 41.1 + 04 08 333.40 + 63.62	01 – 35 – 032	0090	1.4 x 0.35 1.7 x 0.35	61 6	1	S ... npp of 2 Brightest of 3		15.0		*1
08687	13 41.1 + 30 20 050.52 + 78.23	05 – 32 – 075 NGC 5282	0131 293	(1.6 x 1.3) 1.4 x 1.1	... ...	2	... Very compact Asymmetric envelope		15.0		*
08688	13 41.4 + 43 43 093.50 + 70.57	07 – 28 – 056	0154	1.1 x 1.0 1.0 x 1.0	-- ...	2	S ... Peculiar		14.8		
08689	13 41.8 + 06 02 335.81 + 65.22	01 – 35 – 034	0090	1.6 x 1.1 1.6 x 1.1	155: 3:	3	S0		14.7		
08690	13 42.0 + 05 02 334.79 + 64.32	01 – 35 – 035	0090	1.1 x 1.0 1.0 x 0.8	-- 1 – 2	2	Sc		15.3		
08691	13 42.0 + 20 40 006.74 + 76.17	04 – 32 – 034	0125	1.1 x 0.4 1.1 x 0.5	[40] ...	2	S ... + two companions Connected, n of 4		15.0		*1V
08692	13 42.1 + 30 10 049.52 + 78.05	05 – 32 – 077	0131 293	1.2 x 0.4 1.1 x 0.4	137 6	2	Sa – b In chain of 4		15.6		*1
08693	13 42.2 + 35 26 071.63 + 76.12	06 – 30 – 076	0116	1.3 x 0.4 1.3 x 0.4	167 6	1	S ... P w 8698		14.5		*1
08694	13 42.4 + 31 28 055.49 + 77.69	05 – 32 – 000	0131 293	1.0 x 0.15 0.9 x 0.15	14 7	1	Sa – b		15.7		
08695	13 42.6 + 37 25 077.97 + 74.92	06 – 30 – 077	0116	1.1 x 0.45 1.0 x 0.4	15 6	1	S ... f of 2		15.2		*
08696	13 42.8 + 56 08 108.12 + 59.69	09 – 23 – 004 Mark 273	1409	1.2 x 0.25 1.2 x 0.3	... ...	3	Strongly peculiar Long streamer; Seyfert		15.0	+ 11400	*1C
08697	13 42.9 + 23 28 018.07 + 77.22	04 – 32 – 036 IC 933	0125 298	1.2 x 0.8 1.3 x 1.0	(155) 3:	2	S0 Brightest in cluster		14.7		*
08698	13 42.9 + 35 27 071.38 + 75.99	06 – 30 – 078	0116	1.1 x 0.25 1.1 x 0.25	126 6	1	S ... P w 8693		15.2		*
08699	13 43.0 + 41 45 088.91 + 71.90	07 – 28 – 058 NGC 5289	0154	1.9 x 0.5 1.8 x 0.5	100 6	1	Sa – b		13.5		*23
08700	13 43.1 + 41 58 089.33 + 71.72	07 – 28 – 061 NGC 5290	0154	3.6 x 0.9 3.8 x 0.8	95 6	2	Sb – c		13.0		*12 3
08701	13 43.4 + 22 20 013.65 + 76.67	04 – 32 – 037	0068 298	1.8 x 0.35 1.8 x 0.35	26 6 – 7	2	S ... Disturbed, in group		15.2		*1
08702	13 43.5 + 48 11 099.63 + 66.67	08 – 25 – 040	1593	1.7 x 0.8 1.3: x 0.8:	70: 5	4	Sc	Sc +	15.1		
08703	13 43.8 + 22 09 013.11 + 76.51	04 – 32 – 038	0068 298	1.1 x 0.6 1.1 x 0.6	30 5	1	S ...		15.1		
08704	13 43.8 + 56 52 108.46 + 58.96	10 – 20 – 026	0705 294	1.0 x 0.2 1.2 x 0.2	70 7	1	Sa		14.8		*
08705	13 44.2 + 21 06 009.38 + 75.94	04 – 33 – 001	0068	1.1 x 0.6 1.2 x 0.6	75 5	2	Sc		14.8		
08706	13 44.2 + 23 21 018.06 + 76.90	04 – 33 – 000	0068 298	1.0 x 0.12 0.9 x 0.12	174 6 – 7	1	Sb – c n of 2		16.0:		*
08707	13 44.2 + 55 58 107.63 + 59.76	09 – 23 – 007	1409	1.4 x 0.6 1.0 x 0.5	143 6	3	Sb	Sb +	15.4		*1
08708	13 44.3 + 07 38 339.08 + 66.29	01 – 35 – 000	0090	1.6: x 0.9: 1.6: x 0.8:	... ...	2	Irr or Peculiar		15.2		*
08709	13 44.3 + 44 07 093.01 + 69.93	07 – 28 – 063 NGC 5297	0154	5.8 x 1.0 5.6 x 0.9	148 6	3	Sc P w NGC 5296, Sc –		12.3		*12 3478
08710	13 44.4 + 16 31 355.52 + 73.19	03 – 35 – 024 NGC 5293	1019	1.7 x 1.5 1.7 x 1.5	120: 1 – 2	4	Sc	SA(r)c	14.3		*12 3
08711	13 44.4 + 46 22 096.70 + 68.11	08 – 25 – 041 NGC 5301	1593	4.2 x 0.8 4.2 x 0.9	151 6	3	Sc		13.0	+ 1702 + 1816	*12 3478
08712	13 44.7 + 50 43 102.35 + 64.40	09 – 23 – 008	1409	1.2: x 0.12 0.7: x 0.1	68 7	2	Sc		17.		
08713	13 44.8.+ 34 09 065.84 + 76.26	06 – 30 – 081	0116	1.9 x 0.3 2.0 x 0.3	87 6 – 7	3	Sc P w 8715 Sc +		15.5	+ 4912 + 4981	*12 V
08714	13 44.8 + 60 39 110.98 + 55.43	10 – 20 – 027	0705 294	1.7: x 1.7: 1.5: x 1.5:	-- ...	4	Dwarf irregular Ir IV – V		16.5:		*
08715	13 44.9 + 34 08 065.74 + 76.25	06 – 30 – 083	0116	1.4 x 1.1 1.4 x 1.1	-- 1 – 2	4	S B c P w 8713	S B(s)cd	14.8	+ 4448 + 4517	*12 V

N G C   5282, 5289, 5290, 5293, 5297, 5301   I C   933

1	2	3	4	5	6	7	8	9	10	11	12
08716	13 44.9 + 60 35	10 – 20 – 000	0705	1.2: x 0.4:	[158]	2	...		17.		*
	110.92 + 55.49		294	1.2 x 0.4:	...		nf of 2, distorted				
08717	13 45.0 + 30 35	05 – 33 – 001	0086	1.2 x 0.25	128	2	Sa		14.8		
	050.91 + 77.36			1.3 x 0.25	7						
08718	13 45.1 + 34 24	06 – 30 – 085	0116	0.8 x 0.6	125	1	S ...		14.5		*1
	066.67 + 76.10	Mark 461		0.7 x 0.5	2 – 3:		nnf of 2				
08719	13 45.1 + 77 05	13 – 10 – 012	1374	1.5 x 0.45	163	2	Sa – b		14.3		*13
	118.95 + 39.87	NGC 5323		1.5 x 0.45	6						
08720	13 45.3 + 17 58	03 – 35 – 025	1019	1.0 x 0.5	35	1	S ...		15.2		*1
	359.91 + 73.97			1.0 x 0.4	5:		Disturbed?				
08721	13 45.4 + 40 36	07 – 28 – 065	0154	1.2 x 0.8	177	2	...	S B dm	16.0:		
	085.25 + 72.42		306	1.2: x 0.8:	3:						
08722	13 45.4 + 61 13	10 – 20 – 029	0705	2.8 x 0.45	60	2	S0 – a		12.5	+ 2039	*123
	111.24 + 54.88	NGC 5308	294	3.0 x 0.5	7					+ 2199	4678
08723	13 45.6 + 25 10	04 – 33 – 000	0068	1.1 x 0.3	63	1	S ...		16.0:		*
	026.21 + 77.13			1.1 x 0.2	6						
08724	13 45.6 + 37 59	06 – 30 – 086	0116	1.2 x 0.4	25	1	S B ...		15.4		*
	078.29 + 74.10		306	0.6 x 0.3	...		P w 8729				
08725	13 45.6 + 38 33	07 – 28 – 067	0154	1.0 x 0.45	92	2	...		12.9		*13
	079.90 + 73.75	NGC 5303	306	0.8 x 0.4	(6)		Peculiar, jets?, n of 2				
08726	13 45.6 + 40 44	07 – 28 – 068	0154	2.3 x 0.5	127	3	Sc		15.4		1
	085.48 + 72.29		306	2.2 x 0.5	7			Sc +			
08727	13 45.7 + 04 12	01 – 35 – 038	0090	3.8 x 2.3	150	4	Sc	SA c	13.7		12
	335.73 + 63.15	NGC 5300		3.8 x 2.5	4			Sc +			3478
08728	13 45.7 + 07 38	01 – 35 – 039	0090	1.5 x 0.8	85:	1	S B ...		14.9		*1 V
	339.81 + 66.10			1.4: x 0.8:	...		3rd of 3, disturbed				
08729	13 45.7 + 38 04	06 – 30 – 087	0116	1.6 x 1.1	30	4	S B b	S B(r)b	14.7		*3
	078.49 + 74.04	NGC 5305	306	1.5 x 1.0	3		P w 8724	Sb –			
08730	13 45.9 + 25 02	04 – 33 – 000	0068	1.2 x 1.1	—	4	S B 0	(R)S B 0	16.0:		
	025.70 + 77.03			1.1 x 1.1	1						
08731	13 45.9 + 36 22	06 – 30 – 088	0116	1.5 x 0.5	19	1	S: ...		15.7		
	073.20 + 74.99			1.6 x 0.5	6						
08732	13 46.1 + 72 18	12 – 13 – 010	1341	1.1 x 0.7	126	2	Sb		15.1		*
	116.97 + 44.43			1.1 x 0.7	4		P w IC 945, disturbed:				
08733	13 46.5 + 43 39	07 – 28 – 070	0154	2.6 x 1.5	12	3	S B c		14.6		*1
	091.25 + 70.04			2.7 x 1.5	4		Singular				
08734	13 46.5 + 62 03	10 – 20 – 033	0705	1.1 x 0.5	30	2	S0 – a		15.2		
	111.57 + 54.06			1.1 x 0.5	6						
08735	13 46.8 + 40 14	07 – 28 – 072	0154	(2.2 x 1.7)	110:	1	S0 – a		13.7		13
	083.76 + 72.47	NGC 5311	306	(2.7 x 2.2)	2:						
08736	13 46.9 + 39 45	07 – 28 – 073	0154	1.4 x 0.6	175	1	S ...		14.3		
	082.51 + 72.78		306	1.3 x 0.5	6						
08737	13 46.9 + 68 20	11 – 17 – 008	1428	2.4 x 0.5	152	2	Sb – c		14.8		*
	115.05 + 48.16			2.3 x 0.4	7						
08738	13 47.0 + 28 22	05 – 33 – 006	0086	1.0 x 0.4	162	1	E – S		15.6		*1
	040.70 + 77.15		299	1.1 x 0.5	...						
08739	13 47.0 + 35 30	06 – 30 – 090	0116	2.0 x 0.4	122	1	S ...		14.7		*1
	069.86 + 75.24			1.9 x 0.4	7		Disturbed?				
08740	13 47.1 + 04 29	01 – 35 – 040	0090	1.7 x 1.6	—	3	Sa/S B b		14.4		*1
	336.71 + 63.23			1.9 x 1.7	1						
08741	13 47.3 + 60 05	10 – 20 – 034	0705	1.2: x 1.0:	(45)	2	S B ...	S B(r) ...	15.3		
	110.09 + 55.83		294	0.9 x 0.7	...						
08742	13 47.4 + 39 10	07 – 28 – 000	0154	1.6: x 1.4::	—	3	Dwarf irregular		17.		
	080.79 + 73.09		306	1.2:: x 1.2::	...		Ir V				
08743	13 47.4 + 58 55	10 – 20 – 037	0705	1.3 x 0.5	35	2	S B c		16.0:		
	109.21 + 56.90		294	1.3: x 0.5	6			Sc +			
08744	13 47.6 + 40 14	07 – 28 – 074	0154	1.6 x 0.9	40	1	S ...		12.4		12
	083.42 + 72.36	NGC 5313	306	1.9 x 1.0	5						3478
08745	13 47.6 + 60 26	10 – 20 – 035	0705	(6. x 4.)	95:	4	E		11.3	+ 1902	*123
	110.27 + 55.50	NGC 5322	294	(5.5 x 3.5)	...					+ 2061	4678
08746	13 48.0 + 18 24	03 – 35 – 028	1019	1.0 x 0.4	88	1	S ...		15.4		*
	002.55 + 73.72		300	1.0 x 0.3	6		nnp of 2				
08747	13 48.0 + 74 30	12 – 13 – 016	1341	1.7 x 1.7	—	4	Sc	SA(rs)c	15.1		*
	117.72 + 42.29			1.4: x 1.3:	1			or SAB(s)c			
08748	13 48.2 + 28 24	05 – 33 – 009	0086	1.0 x 0.4	139	1	S ...		14.9		*
	040.83 + 76.89		299	0.9 x 0.35	6						
08749	13 48.2 + 41 37	07 – 28 – 076	0154	3.6 x 1.8	18	3	Sc		13.1		23
	086.38 + 71.33	NGC 5320		3.6 x 1.8	5						
08750	13 48.3 + 02 34	01 – 35 – 043	0090	[1.3 x 0.4]	...	3	S ... + companion		15.3		*1
	335.28 + 61.40		301	...	...		Interaction, in group				

N G C   5300, 5303, 5305, 5308, 5311, 5313
5320, 5322, 5323

1	2	3	4	5	6	7	8 9	10	11	12
08751	13 48.3 + 33 57 063.85 + 75.69	06 – 30 – 096 NGC 5318	0116	(1.5 x 0.9) (1.6 x 1.1)	(165) ...	1	S0? Brightest in group	13.5		*13
08752	13 48.3 + 35 23 068.97 + 75.06	06 – 30 – 098	0116	1.0 x 0.35 1.0 x 0.35	68 6	1	Sb – c P w 8754	16.0:		*
08753	13 48.4 + 25 25 027.88 + 76.56	04 – 33 – 011	0068 307	1.0 x 0.25 1.1 x 0.20	139 6	2	Sb – c	15.4		*1
08754	13 48.4 + 35 17 068.59 + 75.09	06 – 30 – 100	0116	(1.0 x 1.0) (1.0 x 1.0)	— ...	2	E – S0	14.8		*
08755	13 48.5 + 17 18 359.76 + 72.96	03 – 35 – 000	1019 300	1.1 x 0.20 1.0 x 0.20	105 7	3	Sb	16.0:		
08756	13 48.5 + 42 47 088.74 + 70.45	07 – 28 – 079	0157	1.9 x 0.4 1.9 x 0.35	87 6	2	S0 – a	14.5		
08757	13 48.5 + 56 17 106.83 + 59.21	09 – 23 – 011	1409	1.0 x 0.5 0.6: x 0.4:	170 ...	1	...	16.0:		
08758	13 48.6 + 21 47 013.72 + 75.35	04 – 33 – 000	0068 307	1.2 x 0.22 1.2 x 0.22	104 7	2	Sa p of 2	15.4		*
08759	13 48.6 + 22 14 015.35 + 75.54	04 – 33 – 000	0068 303	1.0 x 0.9 1.2 x 1.1	— 1 – 2	2	S0	15.5		*
08760	13 48.6 + 38 15 077.78 + 73.47	06 – 30 – 105	0116 306	2.3 x 0.6 2.1 x 0.5	[33] ...	4	Dwarf irregular, I m Ir IV – V Ir I	15.4		*1
08761	13 48.6 + 39 57 082.30 + 72.41	07 – 28 – 077 IC 4336	0154 306	1.5 x 0.4 1.6 x 0.4	158 6	3	S B b	14.6		
08762	13 48.7 + 24 19 023.41 + 76.23	04 – 33 – 012	0068 307	1.7: x 1.1: 1.4: x 1.1:	(165) ...	4	... S B(s)d nnp of 2	17.		*
08763	13 48.7 + 25 20 027.59 + 76.48	04 – 33 – 13 + 14	68 307	[1.2 x 0.8] ...	... ...	1	Double system Bridge?	14.6		*1
08764	13 48.7 + 39 49 081.92 + 72.48	07 – 28 – 082 NGC 5326	0154 306	2.2 x 1.0 2.2 x 1.1	137 5	3	Sa	12.9		12 348
08765	13 48.9 + 71 25 116.29 + 45.20	12 – 13 – 018 IC 954	1341	1.2: x 0.6: 1.0: x 0.6:	[70] ...	1	... Peculiar, double nucleus	14.5		*1 C
08766	13 49.1 + 14 20 352.99 + 70.86	02 – 35 – 019 IC 944	1079 312	1.7 x 0.6 1.7 x 0.5	108 6	1	Sa ssp of 2, IC 948 group	14.7		*1
08767	13 49.1 + 43 47 090.47 + 69.63	07 – 29 – 001	1386	1.0 x 0.2 1.0 x 0.2	106 6 – 7	2	Sc	15.6		
08768	13 49.5 – 01 58 331.79 + 57.22	00 – 35 – 021 NGC 5327	0465	2.1 x 1.7 2.0 x 1.7	(90) 1 – 2	4	S B b SAB(rs)b Brightest in group	14.2		*3
08769	13 49.5 + 09 46 344.84 + 67.32	02 – 35 – 000	1079	(1.3 x 0.8) (1.3 x 0.8)	(50) ...	2	E – S0	15.3		
08770	13 49.5 + 62 05 111.04 + 53.89	10 – 20 – 040	0705	1.0 x 0.4: 1.0 x 0.3	140 6	2	S0	15.5		
08771	13 49.6 + 02 34 335.87 + 61.25	01 – 35 – 044 NGC 5329	0090 301	(1.5 x 1.5) (1.6 x 1.6)	— ...	3	E	14.4		*13
08772	13 49.7 + 14 21 353.34 + 70.77	02 – 35 – 021 IC 946	1079 312	0.9 x 0.7 1.0 x 0.8	110: 2	3	Sa IC 948 group	14.5		
08773	13 49.7 + 17 13 000.15 + 72.68	03 – 35 – 030 NGC 5332	1019 300	(0.9 x 0.9) (1.0 x 0.9)	— ...	2	E – S0	14.1		*
08774	13 49.8 + 02 20 335.73 + 61.02	00 – 35 – 022 NGC 5331	0465 301	[1.1 x 0.8] ...	... ...	3	Double system Contact, streamers	14.3		*13 V
08775	13 49.8 + 51 15 101.40 + 63.53	09 – 23 – 012 IC 951	1409	1.4 x 1.4 1.2 x 1.2	— 1	3	Sc Sc +	14.4		*
08776	13 49.8 + 60 07 109.60 + 55.68	10 – 20 – 041 NGC 5342	0705 294	0.9 x 0.3 1.0 x 0.3	152 6	3	S0	14.4		3
08777	13 49.9 + 22 46 017.78 + 75.47	04 – 33 – 015 IC 949	0068	1.2 x 0.4 1.1 x 0.35	140 (6)	1	S – Irr	15.4		
08778	13 49.9 + 38 19 077.44 + 73.23	06 – 31 – 001	0106 306	1.3 x 0.22 0.9 x 0.22	120 7	1	S ...	14.8		
08779	13 50.0 + 14 20 353.46 + 70.70	02 – 35 – 023 IC 948	1079 312	(1.2 x 0.7) (1.1 x 0.8)	152 ...	3	E Brightest in group	14.4		*1
08780	13 50.0 + 14 44b 354.33 + 70.98	03 – 35 – 032 IC 950	1019 312	1.4 x 0.5 0.9 x 0.5	... ...	1	... nf of 2, contact, disturbed	(15.2)		*1
08781	13 50.0 + 21 47 014.27 + 75.05	04 – 33 – 016	0068 303	1.7 x 0.8 1.7 x 0.8	160 5	4	S B b	14.5		*
08782	13 50.0 + 31 42 054.64 + 76.07	05 – 33 – 012	0086 305	1.0 x 0.6 1.1 x 0.5	68 ...	1	...	15.6		*1
08783	13 50.1 – 01 46 332.20 + 57.34	00 – 35 – 000	0465	1.0 x 0.15 0.9 x 0.18	38 7	1	Sb NGC 5327 group	15.4		
08784	13 50.1 + 01 03 334.63 + 59.85	00 – 35 – 023 IC 947	0465	(1.5? x 1.1) (1.7? x 1.3)	60 ...	2	E – S0	14.2		*1
08785	13 50.1 + 43 29 089.50 + 69.74	07 – 29 – 003 NGC 5336	1386	1.6 x 1.3 1.6: x 1.3:	115: 1 – 2	2	Sc	13.6		13

N G C   5318, 5326, 5327, 5329, 5331, 5332   I C   944, 946 – 51, 954, 4336
        5336, 5342

1	2	3	4	5	6	7	8	9	10	11	12
08786	13 50.3 − 01 39 332.38 + 57.42	00 − 35 − 000	0465	1.8 x 0.3 1.9 x 0.3	83 7	2	Sa − b NGC 5327 group		15.1		*
08787	13 50.3 + 02 30 336.12 + 61.10	01 − 35 − 045	0090 301	1.9: x 0.4 1.5 x 0.35	146 6	2	Sb − c p of 2		15.2		*1
08788	13 50.3 + 25 00 026.56 + 76.05	04 − 33 − 017	0068 307	1.2 x 0.8 1.3 x 0.9	35: 3	2	S0 − a		14.9		
08789	13 50.3 + 39 56 081.56 + 72.17	07 − 29 − 004 NGC 5337	1386 306	1.8 x 0.8 1.8 x 0.8	20 6	1	S ...		13.4		3
08790	13 50.4 − 00 52 333.07 + 58.11	00 − 35 − 024 NGC 5334	0465	4.5: x 3.5: 4.5: x 3.5:	15: 2:	4	S B c	S B(rs)cd	13.7		*12 3478
08791	13 50.4 + 03 04 336.75 + 61.59	01 − 35 − 046 NGC 5335	0090 301	2.3 x 1.8 2.6: x 2.0:	90: 2:	4	S B b	S B(r)b Sb −	14.5		3
08792	13 50.4 + 38 04 076.54 + 73.29	06 − 31 − 002 NGC 5341	0106 306	1.4 x 0.6 1.2 x 0.5	164 6	1	S ...		14.1		3
08793	13 50.5 + 38 56 078.89 + 72.76	07 − 29 − 005	1386 306	1.9 x 0.6 1.4 x 0.4	46 6	3	Sc	Sc +	16.0:		*
08794	13 50.6 + 21 10 012.38 + 74.66	04 − 33 − 018	0068 303	1.8: x 0.6: 2.1 x 0.7	68 6	2	Sb		15.4		
08795	13 50.6 + 37 45 075.55 + 73.44	06 − 31 − 003	0106 306	1.4 x 0.3 1.3 x 0.3	2 6	1	Sc		15.5		*
08796	13 50.7 + 05 13 339.26 + 63.40	01 − 35 − 000	0090 308	1.0 x 0.10 0.9 x 0.10	68 7	2	Sc	Sc +	15.7		
08797	13 50.7 + 24 48 025.84 + 75.92	04 − 33 − 000	0068 307	1.1 x 0.4 1.1 x 0.4	118 6	1	S ...		15.6		*
08798	13 50.7 + 44 04 090.39 + 69.23	07 − 29 − 006	1386	1.6: x 0.8 1.8: x 0.5	147 ...	2	...	S dm	16.0:		
08799	13 50.8 + 06 00 340.26 + 64.05	01 − 35 − 000	0090 308	1.3:: x 0.9:: 0.8: x 0.8:	— ...	1	Dwarf: IV − V:		16.5:		
08800	13 50.9 + 05 27 339.64 + 63.57	01 − 35 − 048 NGC 5338	0090 308	(2.3 x 1.3) (2.8 x 1.4)	97 5:	4	S B: 0		14.3		*
08801	13 51.0 − 00 58 333.24 + 57.96	00 − 35 − 025	0465	1.3 x 0.35 1.3 x 0.35	101 6	1	S ...		15.0		
08802	13 51.0 + 35 58 069.92 + 74.30	06 − 31 − 004	0106	1.3 x 0.6 1.1 x 0.5	10 5	3	Sc	Sc +	17.		
08803	13 51.0 + 38 08 076.49 + 73.16	06 − 31 − 005 NGC 5349	0106 306	1.8 x 0.5 1.8: x 0.5	82 6	2	S B b P w 8809		15.1		*13
08804	13 51.0 + 39 49 080.99 + 72.14	07 − 29 − 007 NGC 5346	1386 306	2.4 x 0.9 2.4 x 0.9	158 6	2	Sc		14.9		1
08805	13 51.1 + 33 45 062.24 + 75.22	06 − 31 − 007 NGC 5347	0106 304	1.7 x 1.4 1.7 x 1.4	130: ...	2	S B a − b	S B(rs)...	13.3		12 348
08806	13 51.1 + 38 28 077.38 + 72.95	06 − 31 − 006	0106 306	2.0 x 0.5 1.1 x 0.4	80 6	1	Sb		15.0		
08807	13 51.1 + 41 02 083.89 + 71.33	07 − 29 − 008	1386 306	[1.0 x 0.4] ...	[132] ...	1	Double system		15.4		*1
08808	13 51.2 + 03 37 337.69 + 61.96	01 − 35 − 049 IC 952	0090 301	1.4 x 0.40 1.4 x 0.45	93 6	2	S B b − c		14.9		
08809	13 51.3 + 38 10 076.47 + 73.09	06 − 31 − 008 NGC 5351	0106 306	2.9 x 1.6 3.0 x 1.7	100 4 − 5	4	Sb P w 8803	SA:(r)bc Sb +	13.1		*12 346$\overline{8}$
08810	13 51.3 + 40 36 082.79 + 71.59	07 − 29 − 009 NGC 5350	1386 306	3.4 x 2.7 2.8 x 2.2	40: 2	2	S B b − c NGC 5353 group	S B(r) ...	12.4		*123 4678
08811	13 51.3 + 72 58 116.76 + 43.67	12 − 13 − 021 Mark 278	1341	1.7 x 1.3 1.6 x 1.2	(15) ...	3	S B b	S B(s)b Sb +	14.8	+ 10800	*
08812	13 51.4 + 36 23 071.11 + 74.03	06 − 31 − 011 NGC 5352	0106 306	(1.2 x 1.0) (1.2 x 1.0)	— ...	1	E − S0		14.4		3
08813	13 51.4 + 40 31 082.55 + 71.63	07 − 29 − 010 NGC 5353	1386 306	(2.8 x 1.8) (2.6 x 1.6)	145 3 − 4:	2	S0 P w 8814, contact Brightest in group		11.8	+ 2188 + 2285	*12 347$\overline{8}$
08814	13 51.4 + 40 32 082.59 + 71.62	07 − 29 − 011 NGC 5354	1386 306	(2.2 x 2.0) (2.5 x 2.0)	— 1 − 2	2	S0 P w 8813, contact NGC 5353 group		12.3		*12 3$\overline{}$
08815	13 51.4 + 68 42 114.66 + 47.68	12 − 13 − 000	1341	1.1 x 0.1 0.7 x 0.1	147 7	2	Sb − c		16.5:		
08816	13 51.5 + 04 12 338.47 + 62.43	01 − 35 − 000	0090	1.2 x 0.5 1.3 x 0.5	25 6	3	Sc		15.4		
08817	13 51.5 + 33 28 061.08 + 75.25	06 − 31 − 012	0106 304	[1.6: x 1.5:] ...	... ...	2	Triple system Contact		15.3		*1V
08818	13 51.6 + 05 36 340.15 + 63.60	01 − 35 − 050	0090 308	1.0 x 0.9 1.1 x 1.1	— (1)	2	Sb − c + companion Contact		15.2		*
08819	13 51.6 + 40 35 082.64 + 71.56	07 − 29 − 012 NGC 5355	1386 306	1.2 x 0.7 1.1 x 0.6	35: ...	1	S0? NGC 5353 group		14.0		3
08820	13 51.7 − 01 11 333.35 + 57.68	00 − 35 − 026	0465	2. x 2. (1.8 x 1.8)	— 1	2	Sa		13.8		*

N G C 5334, 5335, 5337, 5338, 5341, 5346
5347, 5349 − 55

I C 952

1	2	3	4	5	6	7	8	9	10	11	12
08821	13 51.7 + 05 29 340.05 + 63.49	01 - 35 - 051 NGC 5348	0090 308	3.5 x 0.6 3.5 x 0.6	177 7	3	Sb		14.5		2
08822	13 51.8 + 61 54 110.49 + 53.95	10 - 20 - 043	0705	1.3 x 0.2 1.2 x 0.2	179 6 - 7	1	S ...		15.6		
08823	13 51.8 + 69 34 115.06 + 46.86	12 - 13 - 022 Mark 279	1341	0.9 x 0.5 0.9 x 0.5	33 4	2	SO Brightest of 3; Seyfert		14.5	+ 9600	*
08824	13 51.9 + 31 20 052.79 + 75.76	05 - 33 - 016	0086 305	1.1 x 0.9 1.0 x 0.5	[40] ...	2	... Peculiar, jets		15.7		1
08825	13 51.9 + 33 50 062.30 + 75.04	06 - 31 - 000	0106 304	[0.7 x 0.4] ...	... ...	2	... Compact, plumes		14.4	+ 13640 + 13709	*C
08826	13 51.9 + 40 31 082.36 + 71.56	07 - 29 - 013 NGC 5358	1386 306	1.3 x 0.35 1.1 x 0.3	138 6 - 7	2	SO - a NGC 5353 group		14.6		
08827	13 52.1 + 15 17 356.62 + 70.99	03 - 35 - 037	1019	1.1 x 1.0 1.0 x 0.9	— 1 - 2	4	S B O	(R)S B 0	14.1		
08828	13 52.1 + 22 05 016.09 + 74.73	04 - 33 - 022	0068 303	1.2 x 0.4 1.1 x 0.45	175 7?	1	S? ...		15.4		*
08829	13 52.1 + 33 11 059.87 + 75.22	06 - 31 - 021	0106 304	1.0 x 0.9 1.1 x 0.9	(80) 1 - 2	3	S B a		14.8		*
08830	13 52.2 + 36 28 071.08 + 73.85	06 - 31 - 022	0106 306	1.1 x 0.2 1.1 x 0.2	115 6 - 7	1	S ...		15.6		
08831	13 52.4 + 05 35 340.50 + 63.48	01 - 35 - 052 NGC 5356	0090 308	3.0 x 0.8 3.0 x 0.8	15 6	2	Sb		14.1		23
08832	13 52.5 + 60 55 109.66 + 54.81	10 - 20 - 044 NGC 5370	0705	(1.4 x 1.4) (1.4 x 1.4)	— 1	3	S B O	S B(r)0	14.3		3
08833	13 52.6 + 36 05 069.73 + 73.96	06 - 31 - 023	0106	1.1: x 0.9: 1.1: x 0.8:	... ...	3	Irr		16.5:		
08834	13 52.6 + 54 35 104.23 + 60.44	09 - 23 - 014 NGC 5368	1409	0.9 x 0.7 0.9 x 0.7	10: 2	3	Sa		13.8		23
08835	13 52.8 + 41 33 084.41 + 70.75	07 - 29 - 016 NGC 5362	1386 306	2.4 x 1.0 2.1 x 0.9	88 6	1	S ... Peculiar?		13.2		*12 348
08836	13 52.8 + 58 39 107.85 + 56.84	10 - 20 - 045	0705 294	1.2 x 1.0: 0.8 x 0.6	— ...	2	S B ...	S B(rs) ...	15.7		1
08837	13 52.9 + 54 09 103.71 + 60.80	09 - 23 - 017 Ho IV	1409	5. x 1.3 4.0: x 1.2:	[18] ...	4	Irr M 101 group	I m	14.2		*12
08838	13 53.0 + 05 14 340.36 + 63.11	01 - 36 - 001 NGC 5360	0096 308	1.8 x 1.0 1.7 x 0.8	70 ...	1	... Peculiar	I B 0?	14.9		*12 3
08839	13 53.0 + 18 01 003.87 + 72.54	03 - 36 - 001	0081	3.3:: x 2.5: 1. x 1.	— ...	3	Dwarf irregular Ir V		15.7		*
08840	13 53.0 + 40 10 081.08 + 71.62	07 - 29 - 017	1386 306	1.1 x 0.4 0.7 x 0.2	6 ...	1	...		15.3		
08841	13 53.0 + 40 24 081.65 + 71.47	07 - 29 - 018	1386 306	1.7 x 1.0 1.7 x 1.0:	115 ...	4	S B b	Sb +	15.1		
08842	13 53.1 + 25 18ab 028.31 + 75.50	04 - 33 - 27 +28	68 307	[1.4 x 0.45] ...	... ...	2	... 1st and 2nd of 3, contact		(15.3)		*1V
08843	13 53.1 + 58 55 108.00 + 56.58	10 - 20 - 046 NGC 5372	0705 294	0.7 x 0.4 0.7 x 0.4	140 ...	1	... Peculiar		13.7		3
08844	13 53.4 - 01 01 334.20 + 57.64	00 - 36 - 001	1424	1.8 x 0.6 1.7 x 0.6	30 6	3	S B: c		15.1		
08845	13 53.4 + 79 03 119.17 + 37.87	13 - 10 - 013	1374 297	1.0 x 0.8 0.8 x 0.7:	— 2:	3	S B b	S B(s)b	15.7		1
08846	13 53.5 + 40 42 082.18 + 71.21	07 - 29 - 020 NGC 5371	1386 306	4.5 x 3.5 4.5 x 3.5	0 2	4	S B b	S B(rs)bc Sb +	11.5	+ 2583 + 2682	*123 4678
08847	13 53.6 + 05 30 340.96 + 63.25	01 - 36 - 002 NGC 5363	0096 (308)	(5.5 x 3.5) (5.5 x 3.5)	135 ...	4	... P w 8853	I 0 Ir II	11.4	+ 1138 + 1103	*12— —689
08848	13 53.6 + 13 45 354.06 + 69.67	02 - 36 - 001 IC 959	1051 312	1.8 x 1.0 1.2 x 1.0	0 4 - 5	3	Sa		14.4		*
08849	13 53.6 + 17 45 003.43 + 72.27	03 - 36 - 3 +4 IC 960	0081 312	[1.5 x 0.8] ...	... ...	3	Double system Bridge, disrupted		14.8		*1V
08850	13 53.6 + 18 37 005.80 + 72.76	03 - 36 - 005 Mark 463	0081	1.0 x 0.45 0.5 x 0.4	... ...	2	... Peculiar, streamer		14.8		*
08851	13 53.6 + 39 57 080.32 + 71.67	07 - 29 - 019	1386 306	1.6:: x 0.5 1. x 0.4:	45: 6	2	...	S dm - m	16.5:		
08852	13 53.6 + 59 45 108.56 + 55.81	10 - 20 - 047 NGC 5376	0705 294	1.7 x 1.1 1.7 x 1.1	70 3 - 4	2	Sa - b		12.9		*123 4678
08853	13 53.7 + 05 16 340.72 + 63.04	01 - 36 - 003 NGC 5364	0096 (308)	7.2 x 5.5: 7.0 x 4.8:	30 2 - 3:	5	Sb/Sc P w 8847	SA(rs)bc Sb +	13.2	+ 1393 + 1357	*12— —689
08854	13 53.7 + 37 26 073.46 + 73.10	06 - 31 - 026	0106 306	(1.0 x 0.8) 1.1 x 0.9	(115) 2:	2	SO		14.9		*
08855	13 53.8 + 24 44 026.26 + 75.22	04 - 33 - 000	0068	1.0 x 0.2 1.0 x 0.2	8 6	1	S ...		15.5		

N G C 5348, 5356, 5358, 5360, 5362 - 64
5370 - 72, 5376

I C   959, 960

1	2	3	4	5	6	7	8	9	10	11	12
08856	13 53.8 + 30 20 048.54 + 75.53	05 - 33 - 21 + 22	86 305	[4.5: x 0.6] ...	... ...	3	Double system Bridge, streamer		15.6		*1
08857	13 53.9 + 04 38 340.06 + 62.48	01 - 36 - 000	0096	1.0 x 0.25 1.0 x 0.2	128	3	Sa		15.3		
08858	13 54.0 + 38 32 076.47 + 72.45	07 - 29 - 021	1386 306	1.6 x 0.6 1.6: x 0.5	53 (6)	1	S - Irr		15.3		1
08859	13 54.0 + 59 53 108.59 + 55.66	10 - 20 - 050	0705 294	1.6: x 1.0: 1.0: x 0.7:	160: ...	3	...	S m	16.5:		
08860	13 54.0 + 59 59 108.66 + 55.57	10 - 20 - 049 NGC 5379	0705 294	2.0: x 0.8 2.0: x 0.8	60 6	2	S0 P w 8866		14.1		*12 3
08861	13 54.1 + 10 27 348.23 + 67.16	02 - 36 - 000	1051	1.3 x 0.5 1.0: x 0.6:	30: ...	3	...	S dm	15.6		
08862	13 54.3 + 20 24 011.37 + 73.54	03 - 36 - 006	0081	1.0 x 0.12 0.9 x 0.12	136 7	2	Sb - c np of 2		15.6		*
08863	13 54.3 + 47 29 094.91 + 66.21	08 - 25 - 052 NGC 5377	1593	4.5 x 2.5 4.5: x 2.5:	20 4 - 5	4	S B a	(R)S B(s)a	12.5	+ 1830 + 1953	*123 4678
08864	13 54.5 + 23 28 021.74 + 74.71	04 - 33 - 000	0068 309	1.3 x 0.7 1.3 x 0.7	125: 5	1	Sa		16.0:		
08865	13 54.5 + 29 25 044.83 + 75.47	05 - 33 - 027 NGC 5375	0086	3.5 x 2.7 3.7 x 3.0	0 2	4	S B b	S B(r)ab Sb -	13.2		13
08866	13 54.5 + 59 59 108.56 + 55.55	10 - 20 - 051 NGC 5389	0705 294	4.5: x 1.1 4.5: x 1.1	3 6	2	S0 P w 8860		13.2		*12 3
08867	13 54.5 + 78 28 118.88 + 38.40	13 - 10 - 014 NGC 5452	1374 297	2.3 x 1.8 2.3: x 1.8:	(120) 2	4	Sc/S B c	SAB(s)d	14.2		13
08868	13 54.7 + 12 16 351.70 + 68.42	02 - 36 - 003 IC 962	1051	0.8 x 0.8 0.5 x 0.5	— ...	2	... Brightest of 3	Very compact	14.0		*1
08869	13 54.7 + 38 02 074.83 + 72.61	06 - 31 - 027 NGC 5378	0106 306	2.2 x 1.7 2.4 x 1.8	90: 2	4	S B a	S B(r)a	13.8		23
08870	13 54.8 + 37 51 074.27 + 72.69	06 - 31 - 028 NGC 5380	0106 306	(1.8 x 1.8) (2.0 x 2.0)	— 1	4	S0		13.5		*12 348
08871	13 54.9 + 06 12 342.43 + 63.65	01 - 36 - 000	0096 308	1.1: x 0.5: 0.9 x 0.4	120 ...	1	...		15.5		
08872	13 54.9 + 15 41 358.90 + 70.75	03 - 36 - 008	0081 312	(1.4 x 0.9) ...	... ...	1	S0?		14.5		*1
08873	13 54.9 + 24 29 025.56 + 74.91	04 - 33 - 033	0068 307	1.5 x 0.30 1.5 x 0.30	30 6	1	S ...		15.2		*1
08874	13 55.0 + 06 21 342.67 + 63.76	01 - 36 - 004 NGC 5374	0096 308	1.7 x 1.6 1.8: x 1.6	— 1	1	Sb:		13.7		23
08875	13 55.0 + 42 05 084.77 + 70.09	07 - 29 - 023 NGC 5383, Mark 281	1386	3.5: x 2.4 3.0: x 2.4	(85) 2 - 3	4	S B b	S B(rs)b	12.5	+ 2264 + 2369	*123 4678
08876	13 55.0 + 46 13 092.65 + 67.11	08 - 25 - 053	1593	0.9 x 0.25 1.0 x 0.25	27 7	2	S0 - a		14.3		
08877	13 55.1 + 42 02 084.62 + 70.11	07 - 29 - 022	1386	1.3 x 1.1 1.3: x 1.1:	— ...	3	...	S B dm	16.5:		*1
08878	13 55.3 + 10 08 348.29 + 66.72	02 - 36 - 005	1051	1.0 x 0.8 1.1 x 0.8	160 2:	2	S B b		14.5		
08879	13 55.4 + 26 00 031.40 + 75.12	04 - 33 - 000	0068 307	1.6 x 0.25 1.5 x 0.20	163 7	2	Sc		15.7		
08880	13 55.4 + 50 40 099.02 + 63.54	09 - 23 - 019	1409	1.2 x 0.2 0.4 x 0.2	55 7	1	Sa		16.0:		*
08881	13 55.5 + 07 25 344.32 + 64.55	01 - 36 - 05a	0096	1.2 x 0.20 1.1 x 0.15	163 7	2	Sc In group		15.2		
08882	13 55.5 + 54 21 103.25 + 60.43	09 - 23 - 020	1409	1.1 x 0.8 0.9 x 0.7	(80) ...	1	...		16.0:		1
08883	13 55.6 + 15 32 358.88 + 70.52	03 - 36 - 012	0081 312	1.0 x 0.9 1.0 x 0.9	— 1	1	S ... (+ companions?)		14.3		*1
08884	13 55.7 + 05 39 342.11 + 63.08	01 - 36 - 006 IC 966	0096 308	1.6 x 1.3 (1.5 x 1.3)	— 2:	3	S0		15.0		
08885	13 55.7 + 06 30 343.19 + 63.78	01 - 36 - 007 NGC 5382	0096 308	(1.5 x 1.1) (1.8 x 1.3)	25 ...	3	S0 P w 8890		14.0		*3
08886	13 55.7 + 06 46 343.54 + 64.00	01 - 36 - 008 NGC 5384	0096 308	1.7 x 0.8 (1.8 x 0.8)	56 5	2	S0		14.0		3
08887	13 55.7 + 20 39 012.68 + 73.36	04 - 33 - 034	0068	1.2 x 0.5 1.0 x 0.4	122 6	2	Sa - b		15.2		
08888	13 55.7 + 20 52 013.36 + 73.46	04 - 33 - 000	0068	[1.8 x 0.9] ...	... ...	1	Quadrupel system Bridges?		16.0:		*
08889	13 55.7 + 22 02 017.14 + 73.95	04 - 33 - 035	0068 309	1.7 x 0.8 1.1 x 0.5	30 5	1	S ...		15.4		*
08890	13 55.8 + 06 35 343.34 + 63.83	01 - 36 - 010 NGC 5386	0096 308	1.1 x 0.5 1.1 x 0.45	51 6	2	S0 - a P w 8885		13.7		*3

N G C   5352,   5374,   5375,   5377 - 80,   5382 - 84
      5386,   5389

I C   962,   966

1	2	3	4	5	6	7	8	9	10	11	12
08891	13 55.9 + 06 19	01 – 36 – 011	0096	1.6 x 0.25	22	2	Sb – c		14.8		23
	343.04 + 63.60	NGC 5387	308	1.5 x 0.25	6						
08892	13 56.0 + 57 15	10 – 20 – 052	0705	2.4 x 1.8:	...	3	Irr I m		15.6		*1
	105.95 + 57.88			1.5: x 1.5:	...		Disturbed				
08893	13 56.1 + 36 54	06 – 31 – 032	0106	1.3 x 1.0:	(130)	2	Sc/S B c		15.7		
	071.06 + 72.95		306	1.4: x 0.9:	...						
08894	13 56.1 + 63 42	11 – 17 – 011	1428	2.2: x 1.4:	20:	4	Dwarf spiral, S m		16.5:		
	110.98 + 52.11			1.6: x 1.2:	...		S IV – V				
08895	13 56.2 + 02 42	01 – 36 – 000	0096	1.3 x 0.12	79	2	Sc		15.6		
	338.94 + 60.54			1.2 x 0.12	7						
08896	13 56.2 + 07 28	01 – 36 – 012	0096	1.5 x 0.20	71	1	S ...		15.0		*
	344.72 + 64.49			1.4 x 0.20	7		In group				
08897	13 56.2 + 12 49	02 – 36 – 000	1051	1.0 x 0.10	29	2	Sc		15.5		
	353.46 + 68.57		312	0.7 x 0.10	7						
08898	13 56.4 + 37 42	06 – 31 – 033	0106	1.9 x 0.8	...	3	S ... P w 8900		13.7	+ 3558	*12 V
	073.29 + 72.50	NGC 5394	306	1.9 x 1.6:	...		Contact, distorted			+ 3648	36$\overline{A}$C
08899	13 56.5 + 22 14	04 – 33 – 000	0068	1.3 x 0.5	94	1	Sb – c		15.6		
	018.06 + 73.85		309	1.2 x 0.4	6						
08900	13 56.5 + 37 40	06 – 31 – 034	0106	3.0 x 1.3	167	4	Sb P w 8898		12.6		*12 V
	073.16 + 72.50	NGC 5395	306	3.0 x 1.3	5 – 6		Contact, distorted				346$\overline{8}$AC
08901	13 56.5 + 65 10	11 – 17 – 012	1428	(1.2 x 1.0)	(45)	2	E		14.4		3
	111.89 + 50.76	NGC 5413		(1.2 x 1.0)	...						
08902	13 56.6 + 15 48	03 – 36 – 019	0081	1.3 x 0.45	155	1	Sb:		14.5		*1
	359.96 + 70.51		312	1.2 x 0.45	6		ssf of 2, disturbed				
08903	13 56.6 + 60 03	10 – 20 – 054	0705	1.2 x 0.3	167	1	...		14.6		3
	108.20 + 55.37	NGC 5402	294	1.2 x 0.2	...						
08904	13 56.7 + 26 21	04 – 33 – 000	0068	1.8? x 1.3?	...	1	...		16.0:		*
	032.90 + 74.88		307	(0.7 x 0.6)	...						
08905	13 56.7 + 73 50	12 – 13 – 000	1341	(1.2 x 1.0)	(20)	2	E – S0		14.7		*
	116.67 + 42.73	NGC 5412		(1.2 x 1.0)	...						
08906	13 56.9 + 05 47	01 – 36 – 013	0096	1.5 x 0.4	111	4	Sc		15.2		1
	342.81 + 63.02		308	1.5 x 0.4	6			Sc +			
08907	13 57.0 + 13 02	02 – 36 – 000	1051	1.6 x 0.9	...	1	...		15.2		*
	354.26 + 68.58		312	0.8 x 0.6	...		np of 2				
08908	13 57.0 + 43 21	07 – 29 – 025	1386	1.1 x 0.2	39	2	Sb – c		16.0:		
	086.70 + 68.95			1.0 x 0.2	7						
08909	13 57.0 + 61 01	10 – 20 – 056	0705	1.8 x 1.6	(60)	3	Sc/S B c		14.7		*
	108.88 + 54.48			1.7: x 1.5:	1			Sc +			
08910	13 57.1 + 17 43	03 – 36 – 024	0081	1.0 x 1.0	—	3	...	S dm	16.0:		
	004.89 + 71.57		312	0.9 x 0.8	1:						
08911	13 57.3 + 28 19	05 – 33 – 036	0086	1.2 x 0.8	[165]	2	Sc		15.3		*1
	040.46 + 74.89		310	0.9 x 0.7	...		Disturbed				
08912	13 57.3 + 35 01	06 – 31 – 039	0106	1.3 x 0.35	88	1	S ...		14.7		3
	064.81 + 73.56	NGC 5399		1.3 x 0.35	6						
08913	13 57.3 + 39 02	07 – 29 – 026	1386	1.5 x 0.4	0	2	Sc		16.0:		
	076.63 + 71.63		306	1.5: x 0.3	6 – 7						
08914	13 57.4 + 52 36	09 – 23 – 000	1409	1.0: x 1.0:	—	3	Dwarf		17.		
	100.82 + 61.76			0.8 :: x 0.8 ::	...		V				
08915	13 57.5 + 26 28	04 – 33 – 000	0068	1.1: x 0.7:	...	3	Dwarf		17.		
	033.43 + 74.72			1.0: x 0.6:	...						
08916	13 57.5 + 36 29	06 – 31 – 040	0106	1.6 x 0.3	81	1	Sa		14.6		3
	069.36 + 72.90	NGC 5401	306	1.5 x 0.3	6 – 7						
08917	13 57.6 + 40 37	07 – 29 – 028	1386	1.8: x 0.2	123	2	Sc		15.1		*1
	080.49 + 70.66		306	1.8: x 0.2	7		sff of 2, contact				
08918	13 57.7 + 09 13	02 – 36 – 006	1051	1.7 x 0.3	33	1	S ...		14.8		
	347.97 + 65.64		313	1.4 x 0.3	7						
08919	13 57.8 + 38 25	06 – 31 – 041	0106	3.2 x 0.6	145	3	Sb		14.9		*12
	074.81 + 71.89	NGC 5403	306	3.2 x 0.7	7		sp of 2, disturbed?				3$\overline{V}$
08920	13 57.9 + 13 12	02 – 36 – 008	1051	1.2 x 0.35	[138]	2	S[c] Double nucleus?		15.5		*1 V
	355.00 + 68.54		312	0.7 x 0.5	...		f of 2, disturbed				
08921	13 57.9 + 61 32	10 – 20 – 000	0705	1.0 x 0.1	28	1	...		16.5:		*
	109.11 + 53.97			1.0 x 0.1	(7)		Peculiar?				
08922	13 58.0 + 38 55	07 – 29 – 029	1386	1.1 x 0.7	20	3	S B a		15.3		
	076.08 + 71.59		306	1.0 x 0.7	3						
08923	13 58.1 + 38 44	07 – 29 – 030	1386	1.2 x 0.3	141	2	S0 – a		15.0		*1
	075.56 + 71.67		306	1.2 x 0.3	6 – 7		nff of 2				
08924	13 58.2 + 02 16	00 – 36 – 009	1424	1.7 x 0.3	122	3	Sc		14.7		
	339.34 + 59.90			1.6 x 0.25	6						
08925	13 58.2 + 39 09	07 – 29 – 031	1386	1.9 x 1.4	120:	3	S B b	S B(rs)b	13.1		12
	076.62 + 71.42	NGC 5406	306	2.1 x 1.6	3						3478

N G C   5387,   5394,   5395,   5399,   5401 – 03
5406,   5412,   5413

1	2	3	4	5	6	7	8	9	10	11	12
08926	13 58.5 + 32 08 054.63 + 74.21	05 – 33 – 040 IC 4357	0086 311	1.2 x 0.6 1.2 x 0.6	70 5	1	S ...		14.9		*1
08927	13 58.7 + 07 44 346.23 + 64.33	01 – 36 – 000	0096	1.1 x 0.4 1.0 x 0.4	171 6	2	Sb – c		15.3		
08928	13 58.7 + 07 57 346.54 + 64.50	01 – 36 – 014 NGC 5405	0096	0.9 x 0.9 0.9 x 0.7	--- 1:	1	S ...		14.5		
08929	13 58.7 + 21 28 016.28 + 73.08	04 – 33 – 038	0068	[1.0 x 0.4] ...	... ...	3	Double system Bridge		15.1		*1V
08930	13 58.7 + 39 23 077.06 + 71.21	07 – 29 – 033 NGC 5407	1386 306	(1.4 x 1.0) (1.3 x 0.9)	(100) ...	1	...		14.5		3
08931	13 58.7 + 41 14 081.55 + 70.11	07 – 29 – 034 NGC 5410	1386 306	1.6 x 0.8 1.5 x 0.7	75 ...	3	S B ... P w 8932, disrupted		14.1		*13 V
08932	13 58.8 + 41 15 081.55 + 70.09	07 – 29 – 035	1386 306	1.1 x 0.4 0.7 x 0.2	173 ...	1	Irr: P w 8931, disrupted:		15.4		*1V
08933	13 58.8 + 48 41 095.29 + 64.80	08 – 26 – 001 NGC 5425	0120	1.9 x 0.45 1.7 x 0.45	127 6	3	Sc	Sc +	14.3		
08934	13 58.9 + 10 43b 350.94 + 66.58	02 – 36 – 000	1051 313	1.0 x 0.55 1.0 x 0.5	60 4 – 5	3	Sc ssf of 2		(15.2)		*
08935	13 58.9 + 55 24 103.50 + 59.28	09 – 23 – 024 NGC 5422	1409	3.3 x 0.5 3.5 x 0.5	152 7	2	S0 – a		13.1		*123 4678
08936	13 59.0 + 49 38 096.59 + 64.03	08 – 26 – 000	0120	1.4: x 1.1: 1.4:: x 0.9::	... ...	3	Dwarf V		17.		*
08937	13 59.1 + 59 33 107.30 + 55.67	10 – 19 – 062 NGC 5430	0705 294	2.3 x 1.6 2.3 x 1.5	0 ...	3	S B b	S B(s)b	12.7		*123 4678
08938	13 59.3 + 09 44 349.51 + 65.78	02 – 36 – 009 NGC 5409	1051 313	1.7 x 1.1 1.7 x 1.1	50 3 – 4	3	Sb/S B b Sb –	(R')SAB(s)b	14.4		*
08939	13 59.5 – 01 07 336.59 + 56.83	00 – 36 – 012	1424	1.4 x 1.3 1.3 x 1.1	--- 1	3	Sb Brightest of 3	SAB(r) ...	14.8		*1
08940	13 59.5 + 09 11b 348.74 + 65.33	02 – 36 – 011 NGC 5411	1051 313	1.4 x 0.8 1.4 x 0.8	140: ...	2	E – S0 sf of 2		(14.6)		*1
08941	13 59.5 + 34 04 061.11 + 73.48	06 – 31 – 045 NGC 5421	0106	[1.1 x 1.1] ...	... ...	3	Double system Contact, distorted		14.3		*1V AC
08942	13 59.6 + 10 11c 350.37 + 66.07	02 – 36 – 013 NGC 5414	1051 313	1.0 x 0.7 0.8 x 0.15	172 ...	2	... Peculiar, brightest of 3		(13.8)		*1
08943	13 59.7 + 08 17 347.48 + 64.61	01 – 36 – 015 NGC 5417	0096	1.5 x 0.6 1.7 x 0.9	120 6	3	Sa		13.8		3
08944	13 59.7 + 09 41 349.61 + 65.68	02 – 36 – 014 NGC 5416	1051 313	1.5 x 0.8 1.5 x 0.8	110 5	2	Sc		13.6		*3
08945	13 59.7 + 37 15 070.93 + 72.16	06 – 31 – 048 Mark 466	0106 306	1.1 x 0.8 1.0 x 0.7	(175) 2 – 3	2	S0 – a		14.6		
08946	13 59.8 + 07 56 347.01 + 64.32	01 – 36 – 016 NGC 5418	0096	1.1 x 0.55 1.1 x 0.55	44 ...	1	S B ...		14.4		*3
08947	14 00.0 + 46 33 091.57 + 66.29	08 – 26 – 002 NGC 5439	0120	1.1 x 0.3 1.3 x 0.3	9 6	1	S ...		14.6		
08948	14 00.1 + 09 19 349.22 + 65.34	02 – 36 – 015	1051 313	1.5 x 0.7 1.5 x 0.7	45 5	2	S B b		15.0		*
08949	14 00.1 + 14 47 359.26 + 69.21	03 – 36 – 028 IC 970	0081 312	1.1 x 0.25 1.2 x 0.25	53 6 – 7	1	S0 – a n of 2		14.7		1
08950	14 00.2 + 09 25 349.42 + 65.40	02 – 36 – 000	1051 313	1.2 x 0.12 0.9 x 0.1	73 7	3	Sc	Sc +	15.7		*
08951	14 00.3 + 09 01 348.84 + 65.08	02 – 36 – 000	1051 313	1.1 x 0.2 0.9 x 0.2	123 6	3	Sc		15.6		
08952	14 00.3 + 09 35 349.73 + 65.51	02 – 36 – 017 NGC 5423	1051 313	(1.5 x 0.9) (1.5 x 0.9)	75: ...	2	E – S0		13.9		*1
08953	14 00.3 + 18 45 008.93 + 71.49	03 – 36 – 030	0081	1.3 x 0.20 1.2 x 0.20	36 7	1	S ...		15.2		*
08954	14 00.3 + 32 45 056.48 + 73.70	06 – 31 – 050 NGC 5433	0106 311	1.7 x 0.35 1.5 x 0.35	3 (6)	1	S – Irr 3rd of 3		14.0		*13
08955	14 00.4 + 35 06 064.24 + 72.94	06 – 31 – 051	0106	1.2 x 0.5 1.2 x 0.4	178 6	1	S ... NGC 5444 group		15.5		*
08956	14 00.5 + 09 40 349.95 + 65.54	02 – 36 – 019 NGC 5424	1051 313	(1.6 x 1.3) (1.4 x 1.2)	(110) 2	4	S0		14.3		
08957	14 00.5 + 24 04 025.31 + 73.57	04 – 33 – 000	0068	1.1: x 0.8: 1.0: x 0.6:	(70) ...	2	Dwarf		16.5:		
08958	14 00.5 + 56 03 103.79 + 58.61	09 – 23 – 026 NGC 5443	1409	3.3 x 1.1 3.1 x 1.1	34 6	1	S ...		13.2		<u>123</u>
08959	14 00.5 + 69 08 113.78 + 46.97	12 – 13 – 027	1341	1.8: x 1.4: 1.2: x 0.8:	(170) ...	4	Dwarf spiral S V		16.5:		
08960	14 00.6 + 39 24 076.47 + 70.90	07 – 29 – 039	1386 306	1.7 x 0.2 1.7 x 0.2	178 7	1	S ... P w 8962		15.2		*1

N G C   5405,   5407,   5409 – 11,   5414,   5416 – 18
5421 – 25,   5430,   5433,   5439,   5443

I C   970, 4357

1	2	3	4	5	6	7	8	9	10	11	12
08961	14 00.7 + 28 16 040.32 + 74.14	05 − 33 − 042	0086 310	1.5 x 0.25 1.5 x 0.3	76 6	1	S ...		14.7		
08962	14 00.7 + 39 22 076.35 + 70.90	07 − 29 − 040	1386 306	1.3 x 1.2 1.0: x 1.0:	--- 1	1	S B ... P w 8960		15.7		*1
08963	14 00.8 + 35 00 063.82 + 72.90	06 − 31 − 052 NGC 5440	0106	3.2 x 1.6 3.0 x 1.5	50 5	4	Sa NGC 5444 group		13.4		*13
08964	14 00.8 + 79 05 118.77 + 37.72	13 − 10 − 015	1374 297	1.5 x 0.5 1.4 x 0.4	46 6	3	...	S dm	15.7		*
08965	14 00.9 + 09 41 350.15 + 65.49	02 − 36 − 022 NGC 5434	1051 313	1.8 x 1.8 1.5 x 1.5	--- 1	4	Sc P w 8967	SA c	14.3		*1
08966	14 01.0 + 07 01 346.36 + 63.41	01 − 36 − 000	0096	1.1 x 0.55 1.1 x 0.6	25 5	4	Sc		15.3		
08967	14 01.0 + 09 43 350.25 + 65.49	02 − 36 − 024	1051 313	1.8 x 0.3 1.7 x 0.3	72 6	2	Sb − c P w 8965		14.7		*1
08968	14 01.0 + 35 36 065.64 + 72.64	06 − 31 − 000	0106	1.0 x 0.15 1.1 x 0.2	... ...	2	S? ... disrupted, bridge	nnp of 2,	16.5:		*
08969	14 01.0 + 49 25 095.70 + 64.00	08 − 26 − 003 NGC 5448	0120	4.3 x 2.0 4.3 x 2.0	115 5	4	S B b	(R′)SAB(rs)b	12.7	+ 1970 + 2103	*12 346$\overline{8}$
08970	14 01.0 + 58 13 105.75 + 56.71	10 − 20 − 066	0705 294	1.6 x 1.0 1.6 x 1.0	170 4	4	...	SA dm	16.0:		
08971	14 01.2 + 09 49 350.50 + 65.54	02 − 36 − 025 NGC 5436	1051 313	1.1 x 0.5 1.2 x 0.6	126 5 − 6	1	S0 − a		14.9		*
08972	14 01.2 + 11 37 353.54 + 66.85	02 − 36 − 026	1051	1.4 x 0.7 1.1 x 0.4	157 5	3	Sb P w 8973, contact		15.1		*1
08973	14 01.2 + 11 38 353.57 + 66.86	02 − 36 − 027	1051	1.2 x 0.6 1.0 x 0.5	133 5	3	Sb P w 8972, contact		15.1		*1
08974	14 01.2 + 35 23 064.92 + 72.69	06 − 31 − 054 NGC 5444	0106	(2.5 x 2.0) (2.6 x 2.1)	90 ...	3	E Brightest in group		12.8		*12 348$\overline{}$
08975	14 01.2 + 38 46 074.63 + 71.14	07 − 29 − 043	1386 306	1.1 x 0.5 1.1 x 0.5	22 5	1	S ...		14.2		
08976	14 01.3 + 35 16 064.53 + 72.71	06 − 31 − 055 NGC 5445	0106	1.7 x 0.7 1.7 x 0.5	27 6	1	S0? NGC 5444 group		14.1		*3
08977	14 01.4 + 15 38 001.69 + 69.50	03 − 36 − 035	0081 312	1.2 x 0.7 1.2 x 0.6	102 4 − 5	1	S ...		15.6		
08978	14 01.4 + 15 58 002.44 + 69.70	03 − 36 − 036	0081 312	1.1 x 0.25 0.7 x 0.2	178 6	1	S ... 3rd of 3		15.5		*
08979	14 01.5 + 06 43 346.54 + 63.10	01 − 36 − 000	0096	1.3 x 0.1 0.9 x 0.15	15 7	1	Sb − c		16.0:		*
08980	14 01.5 + 39 17 075.88 + 70.82	07 − 29 − 044	1386 306	1.1 x 0.9 1.1 x 1.0	155: 1 − 2	3	S B b	(R′)S B(s)b	14.4		*
08981	14 01.5 + 54 35 102.02 + 59.77	09 − 23 − 028 NGC 5457 = M 101	1409	28. x 28. 23. x 20.	--- 1:	5	Sc P w 9013, disturbed?	SAB(rs)cd	8.7	+ 266 + 415	*12− −9$\overline{V}$A
08982	14 01.5 + 61 59 108.82 + 53.38	10 − 20 − 067	0705	1.4∷x 1.0∷ 0.8 x 0.5	... ...	2	... Peculiar		15.5		
08983	14 01.6 + 12 14 354.84 + 67.22	02 − 36 − 000	1051	1.1 x 0.1 1.1 x 0.1	64 7	1	Sc 1st of 3		16.0:		*
08984	14 01.6 + 35 59 066.65 + 72.37	06 − 31 − 056	0106	1.3 x 0.3 1.3 x 0.3	35 6 − 7	1	S: ...		14.2		*
08985	14 01.6 + 59 42 106.94 + 55.38	10 − 20 − 068	0705 294	1.5 x 0.6 1.2 x 0.5	127 6	3	...	SAB: dm	16.0:		
08986	14 01.7 + 04 20 343.13 + 61.15	01 − 36 − 017	0096	1.3 x 1.3 1.2 x 1.2	--- (1)	1	S0?		15.2		
08987	14 01.8 + 16 34 004.01 + 69.99	03 − 36 − 037	0081 312	3.7: x 0.5 3.5: x 0.4	101 6	4	... np of 2, bridge	S B m	15.7		*1
08988	14 01.8 + 61 13 108.16 + 54.04	10 − 20 − 069	0705	1.6 x 0.2 1.5 x 0.2	174 7	2	Sc		15.6		
08989	14 01.9 + 11 59 354.51 + 66.99	02 − 36 − 000	1051	1.1 x 0.2 1.1 x 0.2	73 7	2	S0 − a		15.3		
08990	14 01.9 + 33 35a 058.98 + 73.15	06 − 31 − 059 IC 4370	0106	1.5 x 0.25 1.4 x 0.25	142 7	1	Sa P w IC 4371, interaction, in group		(15.0)		*1
08991	14 02.1 + 15 43 002.17 + 69.42	03 − 36 − 039	0081 312	1.0 x 0.35 1.2 x 0.35	40 ...	1	... spp of 2		(15.2)		*1
08992	14 02.2 + 73 19 115.91 + 43.06	12 − 13 − 029	1341	1.0 x 0.6 1.0 x 0.6	85 4	3	S B b	S B(s)b	15.6		
08993	14 02.3 − 00 24 338.36 + 57.09	00 − 36 − 016	1424	1.1 x 0.9 1.1 x 0.7	0 2:	3	S B a P w 8994	S B(r)a	14.8		*1
08994	14 02.3 − 00 22 338.39 + 57.12	00 − 36 − 017	1424	1.1 x 0.5 1.1 x 0.6	132 5 − 6	3	Sb/S B c P w 8993	SAB(s)bc Sb +	14.7		*1
08995	14 02.3 + 09 04 349.81 + 64.79	02 − 36 − 031	1051 313	2.3 x 1.2 2.0 x 1.1	3 5	3	...	SAB dm	14.9		

N G C   5434, 5436, 5440, 5444, 5445, 5448, 5457   I C   4370

1	2	3	4	5	6	7	8	9	10	11	12
08996	14 02.3 + 14 30 359.61 + 68.62	03 – 36 – 041	0081 312	1.7 x 0.2 1.3: x 0.2	107 7	3	Sc P w 8997		15.5		*
08997	14 02.3 + 14 37 359.85 + 68.69	03 – 36 – 042 NGC 5454	0081 312	(1.6 x 1.0) (1.6 x 1.0)	110: 4:	3	S0 P w 8996		14.4		*3
08998	14 02.3 + 26 02 032.42 + 73.59	04 – 32 – 000	0068	1.0 x 0.2 0.9 x 0.2	146 6 – 7	1	Sc		16.0:		
08999	14 02.3 + 29 26 044.52 + 73.77	05 – 33 – 045	0086 310	1.0 x 0.8 1.0 x 0.8	(5) 2	3	Sc n of 2		15.5		*1
09000	14 02.4 + 11 02a 353.05 + 66.23	02 – 36 – 032a	1051	1.1 x 0.1 1.0 x 0.1	118 7	1	S ... P w 9001, contact		15.1		*1
09001	14 02.4 + 11 02b 353.05 + 66.23	02 – 36 – 032b	1051	1.1 x 0.4 1.0 x 0.4	... ...	1	S B ... P w 9000, contact, distorted				*1
09002	14 02.4 + 12 58 356.57 + 67.58	02 – 36 – 034	1051	1.3 x 1.1 1.1 x 0.8	... ...	2	S B ... n of 2, disturbed		15.3		*1V
09003	14 02.4 + 35 47 065.83 + 72.31	06 – 31 – 067	0106	1.4 x 0.7 1.0 x 0.5	25 5	1	S0		15.4		*
09004	14 02.5 + 12 07 355.02 + 66.98	02 – 36 – 036 NGC 5456	1051	1.2 x 1.0 (1.5 x 1.3)	(175) 1 – 2	4	S0		14.2		3
09005	14 02.5 + 13 22 357.39 + 67.84	02 – 36 – 037 NGC 5459	1051	1.1 x 1.0 1.3 x 1.1	(10) 1 – 2	3	S0		14.5		
09006	14 02.6 + 00 10 339.02 + 57.54	00 – 36 – 000	1424	1.3 x 0.7 1.1 x 0.6	90: 4 – 5	1	S ...		15.3		
09007	14 02.6 + 09 35 350.74 + 65.13	02 – 36 – 000	1051 313	1.0 x 0.8 0.9 x 0.8	(70) 2:	3	... S dm		15.6		
09008	14 02.6 + 11 15 353.51 + 66.35	02 – 36 – 000	1051	1.1 x 1.1 1.1 x 1.1	-- 1	4	Sc SA d		15.3		
09009	14 02.7 + 16 00 003.06 + 69.48	03 – 36 – 045	0081 312	1.0 x 0.3 0.9 x 0.2	171 6	1	S ...		15.7		
09010	14 03.0 + 31 02 050.12 + 73.47	05 – 33 – 049	0086	1.0 x 0.8 0.9 x 0.8	115: 2:	4	Sc 9012 group Sc +		15.7		*
09011	14 03.0 + 55 08 102.26 + 59.19	09 – 23 – 031 NGC 5473	1409 316	2.2 x 1.7 2.2 x 1.7	160 2	3	S0		12.5	+ 2044 + 2195	*123 4678
09012	14 03.2 + 31 00 049.98 + 73.43	05 – 33 – 050	0086	1.1 x 0.6 1.1 x 0.7	90 ...	2	S0 Brightest in group		15.5		*
09013	14 03.2 + 53 54 100.84 + 60.20	09 – 23 – 032 NGC 5474	1409	6.5 x 5.3 3.6 x 3.6	... ...	4	S[c] Singular P w 8981, disturbed?		11.9	+ 247 + 395	*12– -789V
09014	14 03.3 + 35 10 063.72 + 72.37	06 – 31 – 068	0106	1.2 x 0.2 1.2 x 0.2	63 7	1	S ...		15.7		
09015	14 03.5 + 09 16 350.64 + 64.75	02 – 36 – 000	1051 313	1.2 x 0.9 1.1 x 0.9	-- 2 – 3:	1	S ...		15.2		
09016	14 03.5 + 55 59 103.04 + 58.44	09 – 23 – 033 NGC 5475	1409 316	2.1 x 0.45 1.9 x 0.45	166 7	1	Sa?		13.4		*23
09017	14 03.7 + 09 36a 351.25 + 64.96	02 – 36 – 040 NGC 5463	1051 313	1.2 x 0.5 1.1 x 0.4	49 6	1	S ... sp of 2		(14.1)		*13
09018	14 03.8 + 54 42 101.60 + 59.49	09 – 23 – 034 NGC 5477	1409 316	1.9 x 1.4 1.7: x 1.3:	95: 3	2	Sc – Irr		14.5		*12 3
09019	14 03.9 + 20 22 014.69 + 71.52	03 – 36 – 046	0081	1.0 x 0.45 0.9 x 0.4	23 6	2	Sb – c		15.4		
09020	14 04.0 + 06 16 346.54 + 62.36	01 – 36 – 019 NGC 5470	0096	2.6 x 0.4 2.6 x 0.4	63 7	4	Sb		14.5		3
09021	14 04.1 + 12 57 357.27 + 67.27	02 – 36 – 000	1051 314	1.2 x 0.2 1.2 x 0.2	84 6	3	Sc Sc +		15.7		
09022	14 04.1 + 36 01 066.09 + 71.90	06 – 31 – 069	0106	1.8: x 1.8: 1.5: x 1.5:	-- 1:	3	... S dm		15.1		
09023	14 04.4 + 09 34 351.50 + 64.82	02 – 36 – 000	1051 313	1.2 x 0.5 1.0 x 0.5	157 6	2	Sc		15.4		*
09024	14 04.4 + 22 16 020.44 + 72.16	04 – 33 – 000	0068	2.3 x 2.2 2. x 2.	-- 2	4	S ... Peculiar		16.0:		*
09025	14 04.5 + 12 48 357.15 + 67.10	02 – 36 – 000	1051 314	1.1 x 0.7 1.0 x 0.6	75: 4	3	Sb/S B c SAB(s)bc sp of 2 Sb +		15.3		*
09026	14 04.5 + 50 39 096.45 + 62.69	09 – 23 – 035 NGC 5480	1409	1.7 x 1.0 1.6 x 0.9	[0] ...	1	Sc? Singular P w 9029, disturbed?		12.6	+ 1720 + 1859	*123 4678
09027	14 04.6 + 10 53 353.75 + 65.75	02 – 36 – 000	1051	1.0 x 0.2 0.8 x 0.2	121 6	1	S ...		15.4		
09028	14 04.7 + 67 49 112.45 + 48.01	11 – 17 – 020	1428	1.0 x 0.9 0.9: x 0.9:	-- 1:	3	... S dm		16.5:		
09029	14 04.9 + 50 59 096.79 + 62.39	09 – 23 – 036 NGC 5481	1409	(1.8 x 1.3) (1.7 x 1.2)	115: ...	3	E P w 9026		13.5	+ 2025 + 2165	*12 3
09030	14 05.2 + 09 56 352.42 + 64.96	02 – 36 – 000	1051 313	1.1 x 0.15 1.0 x 0.15	179 7	1	Sb – c		15.6		

N G C   5454,   5456,   5459,   5463,   5470,   5473 – 75
5477,   5480,   5481

1	2	3	4	5	6	7	8	9	10	11	12
09031	14 05.2 + 15 06 002.09 + 68.45	03 – 36 – 053	0081 312	1.1 x 0.2 1.1 x 0.20	40 7	1	S ...		15.5		*
09032	14 05.4 + 56 47 103.44 + 57.63	10 – 20 – 073	0705	1.0 x 0.5 0.9 x 0.5	2 5	2	Sb – c		15.0		*
09033	14 05.5 + 55 15 101.81 + 58.90	09 – 23 – 037 NGC 5485	1409 316	(2.7 x 2.0) (2.7 x 2.0)	(170) 2 – 3	3	SO		12.4	+ 1985 + 2137	*123 46̲7̲8
09034	14 05.6 – 01 28 338.64 + 55.75	00 – 36 – 019 NGC 5478	1424	1.2 x 1.1 0.9 x 0.9	--- 1	2	Sb		14.7		
09035	14 05.7 + 30 07 046.74 + 72.99	05 – 33 – 052	0086	1.5 x 1.1 1.2 x 0.9	135: 3	3	S B b	S B(rs)b	15.2		
09036	14 05.7 + 55 20 101.85 + 58.81	09 – 23 – 038 NGC 5486	1409 316	1.5 x 0.9 1.5 x 0.9	80 (4)	2	Sc – Irr		14.0		*12 3̲
09037	14 06.0 + 07 18 348.79 + 62.86	01 – 36 – 020	0096	1.8 x 0.9 1.5 x 0.8	67 5	2	Sc		14.5		1
09038	14 06.0 + 09 10 351.55 + 64.26	02 – 36 – 043 NGC 5482	1051	1.2 x 0.9 1.2 x 0.9	88 ...	2	SO		14.2		3
09039	14 06.0 + 72 23 115.05 + 43.81	12 – 13 – 030	1341	1.6 x 0.12 1.6 x 0.12	176 7	3	Sc s of 2		16.0:		*
09040	14 06.1 – 00 55 339.35 + 56.16	00 – 36 – 020 IC 976	1424	(1.3 x 0.6) 1.5 x 0.7	172 5	2	SO – a		14.1		
09041	14 06.1 + 12 03 356.43 + 66.30	02 – 36 – 044	1051	1.2 x 0.6 1.1 x 0.6	50 5	3	S B a		14.9		
09042	14 06.1 + 35 58 065.44 + 71.54	06 – 31 – 071	0106	1.0 x 0.8 0.9 x 0.8	165 ...	2	... sf of 2	S d – dm	16.5:		*
09043	14 06.3 + 16 20 005.25 + 68.97	03 – 36 – 056	0081 312	1.1 x 0.8 1.1 x 0.8	93 ...	1	SO:		14.9		
09044	14 06.6 + 14 33 001.49 + 67.84	03 – 36 – 058	0081 312	1.1 x 0.6 1.1 x 0.6	165 5	1	S ...		14.6		
09045	14 06.6 + 35 51 064.98 + 71.49	06 – 31 – 073	0106	1.2 x 1.1 0.9 x 0.8	--- ...	1	...		15.5		
09046	14 06.6 + 48 13 092.30 + 64.31	08 – 26 – 000	0120	1.2:: x 0.8:: 1.0:: x 0.5::	... ...	3	Dwarf IV – V or V		17.		
09047	14 06.6 + 55 32 101.87 + 58.58	09 – 23 – 040	1409 316	1.2 x 0.7 1.0 x 0.4	35 4	3	S B b	S B(r)b	15.5		1
09048	14 06.7 + 33 46 058.63 + 72.14	06 – 31 – 072	0106	1.5 x 0.3 1.7 x 0.25	63 6	2	Sb – c		15.6		
09049	14 06.7 + 60 28 106.64 + 54.40	10 – 20 – 074	0705 294	1.1 x 0.45 0.9 x 0.4	32 6	2	Sb		15.6		*
09050	14 06.8 + 51 22 096.81 + 61.90	09 – 23 – 000	1409	1.3: x 1.3: 1.0: x 0.6:	--- ...	3	Dwarf V		17.		
09051	14 06.8 + 55 09 101.40 + 58.87	09 – 23 – 045	1409 316	1.3: x 0.7 1.2: x 0.5:	40: ...	2	...	dm – m	16.5:		
09052	14 07.0 + 76 10 116.99 + 40.31	13 – 10 – 016	1374	1.1 x 1.1 1.1 x 1.1	--- 1	3	Sc	SA d	15.4		
09053	14 07.1 + 15 04 002.77 + 68.07	03 – 36 – 061 IC 979	0081 312	1.1 x 0.7 1.1 x 0.7	172 4	2	S B a – b P w 9055		14.5		*
09054	14 07.3 + 10 06 353.57 + 64.72	02 – 36 – 000	1051	1.0 x 0.4 0.8 x 0.4	50 ...	1	S B ...		15.3		
09055	14 07.5 + 15 06 003.00 + 68.01	03 – 36 – 062	0081 312	1.0 x 0.7 0.9 x 0.6	20 3	3	S B b P w 9053		15.0		*
09056	14 07.5 + 49 16 093.66 + 63.43	08 – 26 – 005	0120	1.1 x 0.2 0.9 x 0.2	143 (7)	1	... Peculiar		14.4		*
09057	14 07.6 – 02 20 338.61 + 54.75	00 – 36 – 023	1424	3.0 x 1.0 3.0 x 1.0	173 (6)	4	...	S B(r?)dm	14.4		1
09058	14 07.6 + 17 47 009.15 + 69.51	03 – 36 – 065 NGC 5490	0081 312	(2.3 x 1.8) (2.5 x 2.0)	(5) ...	4	E Brightest in group		13.4		*12 3̲
09059	14 07.6 + 17 56 009.52 + 69.59	03 – 36 – 066 IC 982	0081 312	1.1 x 1.1 1.3 x 1.2	--- 1	3	SO NGC 5490 group	P w 9061	14.6		*12 A̲
09060	14 07.6 + 55 31 101.63 + 58.51	09 – 23 – 044	1409 316	1.0 x 0.8 0.9 x 0.7	(45) 2	2	Sc	Sc –	16.0		*
09061	14 07.7 + 17 58 009.64 + 69.58	03 – 36 – 068 IC 983	0081 312	6.0: x 5.5: 4.5: x 3.5:	(120) 1 – 2	4	S B a/S B b P w 9059, NGC 5490 group	S B(r)ab	14.3		*12 A̲
09062	14 07.7 + 18 36 011.22 + 69.90	03 – 36 – 070 IC 984	0081 312	1.9 x 0.4 1.9 x 0.4	35 6	2	Sb		14.5		1
09063	14 07.8 + 05 49 347.52 + 61.42	01 – 36 – 000	0096	1.3: x 0.5: 1.3: x 0.5:	22 6:	3	...	m	16.0:		
09064	14 08.0 + 16 35 006.46 + 68.78	03 – 36 – 071	0081 312	1.2 x 0.15 1.1 x 0.2	10 7	2	Sb		15.6		*
09065	14 08.2 + 19 51 014.63 + 70.38	03 – 36 – 074 NGC 5492	0081 312	1.8 x 0.4 1.8 x 0.3	150 (6)	1	S: ... Peculiar?		13.7		*13

N G C 5478, 5482, 5485, 5486, 5490, 5492   I C 976, 979, 982 – 84

1	2	3	4	5	6	7	8	9	10	11	12
09066	14 08.2 + 46 41 089.33 + 65.22	08 – 26 – 006	0120	1.2 x 0.1 1.0 x 0.1	134 7	3	Sc		16.5:		
09067	14 08.3 + 15 27 004.06 + 68.07	03 – 36 – 076	0081 312	1.6 x 0.7 1.4 x 0.5	12 6	2	Sa – b		14.7		
09068	14 08.3 + 48 33 092.35 + 63.87	08 – 26 – 007	0120	1.7 x 1.0 1.7: x 1.0:	160 4	4	...	SA dm	16.0:		
09069	14 08.4 + 39 07 073.41 + 69.76	07 – 29 – 048 NGC 5497	1386	1.4 x 0.9 1.3: x 0.8:	75 ...	3	S B b f of 2	S B(s)b	15.1		*1
09070	14 08.4 + 48 47 092.68 + 63.69	08 – 26 – 008 NGC 5500	0120	(1.0 x 0.9) (1.0 x 0.9)	— ...	3	E		14.5		
09071	14 08.4 + 54 27 100.24 + 59.31	09 – 23 – 046	1409 316	2.2 x 0.45 2.0 x 0.4	37 6	2	Sc		15.1		*1
09072	14 08.5 + 06 36 348.85 + 61.92	01 – 36 – 022 NGC 5491	0096	1.6 x 1.0 1.5 x 0.9	78 4	1	S ... s of 2, contact		13.9		*13
09073	14 08.7 + 25 44 032.20 + 72.12	04 – 32 – 042 IC 4381	0068 317	1.7 x 1.3 1.6 x 1.1	(135) 1 – 2	2	Sc in group, disturbed	Brightest	14.8		*1
09074	14 08.7 + 36 08 065.31 + 70.99	06 – 31 – 076 NGC 5499	0106 320	1.1 x 0.8 1.2 x 0.8	150 (3)	1	...		14.5		
09075	14 08.8 + 25 56 032.86 + 72.13	04 – 33 – 043 NGC 5498	0068 317	1.0 x 0.8 1.3 x 0.9	120: ...	2	E – S0		15.0		
09076	14 08.8 + 38 59 072.96 + 69.75	07 – 29 – 049	1386	1.3 x 0.7 1.0: x 0.6:	115: 5:	1	S ...		16.0:		*
09077	14 08.8 + 40 00 075.48 + 69.24	07 – 29 – 050	1386	1.0 x 0.25 1.0 x 0.25	86 6 – 7	1	S ...		14.9		
09078	14 08.9 + 17 44 009.49 + 69.22	03 – 36 – 077	0081 312	1.5 x 1.1 1.5 x 1.1	20 3	4	S B a NGC 5490 group	(R´)S B(s)a	14.6		
09079	14 09.0 – 00 55 340.44 + 55.77	00 – 36 – 026 NGC 5496	1424	4.5 x 0.8 4.0: x 0.7	172 6	4	Sc	Sc +	13.4		*12 478
09080	14 09.0 + 59 35 105.42 + 55.01	10 – 20 – 082	0705 294	1.3 x 0.2 1.2 x 0.2	42 7	2	Sc ssf of 2		15.7		*
09081	14 09.5 + 39 53 075.00 + 69.19	07 – 29 – 051	1386	1.7 x 1.1 1.3 x 1.0	60: 3	1	Sa:		13.9		
09082	14 09.6 + 27 21 037.51 + 72.13	05 – 33 – 054 IC 4384	0086	(1.0 x 0.8) 1.1 x 0.9	170: 2:	2	S0 – a		15.0		
09083	14 09.6 + 50 27 094.82 + 62.32	08 – 26 – 009	0120	1.0 x 0.8 1.0 x 0.8	75 ...	3	...	S dm	15.1		*1 V
09084	14 09.7 + 08 54 352.66 + 63.45	02 – 36 – 046	1051	1.1 x 0.9 1.1 x 1.0	(5) 2:	3	Sb		14.9		*
09085	14 09.9 + 16 04 006.00 + 68.11	03 – 36 – 081 NGC 5504	0081 312	1.4 x 1.1 1.3 x 1.1	130: 2	4	Sb/S B c Brightest in group	SAB(s)bc	13.9		*1
09086	14 09.9 + 16 07 006.11 + 68.14	03 – 36 – 082 IC 4383	0081 312	1.1 x 0.4 1.0 x 0.4	80 6	1	Sc – Irr nf of 2, NGC 5504 group		15.5		*1
09087	14 09.9 + 18 32 011.79 + 69.41	03 – 36 – 083	0081 312	1.0 x 0.7 1.1 x 0.7	(30) ...	1	S0? P w 9090		14.9		*1
09088	14 09.9 + 38 25 071.22 + 69.83	06 – 31 – 078	0106	2.5 x 0.7 2.3 x 0.7	65 6	2	S0 – a		14.8		
09089	14 10.0 + 11 29 357.01 + 65.21	02 – 36 – 047	1051	1.1 x 0.4 1.0 x 0.4	52 6	1	S ...		14.9		
09090	14 10.0 + 18 39 012.11 + 69.45	03 – 86 – 085	0081	(1.0 x 0.8) (1.0 x 0.8)	(70) ...	2	E – S0 P w 9087		14.9		*
09091	14 10.0 + 30 09 046.63 + 72.06	05 – 34 – 000	0070	1.0 x 0.10 0.9 x 0.10	30 7	3	Sc		16.5:		
09092	14 10.1 + 13 32 000.83 + 66.55	02 – 36 – 048 NGC 5505	1051	1.0 x 0.7 0.9 x 0.7	130 3:	2	S B a		14.1		
09093	14 10.2 + 12 16 358.48 + 65.70	02 – 36 – 000	1051	1.0 x 0.3 0.7 x 0.2	162 6	1	Sc		15.7		
09094	14 10.2 + 24 53 029.73 + 71.62	04 – 34 – 002 NGC 5508	0061 317	(2.1 x 1.3) (2.0 x 1.4)	... ...	2	S0		15.2		*1
09095	14 10.2 + 78 50 118.12 + 37.79	13 – 10 – 000 NGC 5547+4404*	1374	[1.0 x 0.4] ...	... ...	1	Double system Bridge		14.5		*3
09096	14 10.5 + 39 32 073.87 + 69.20	07 – 29 – 052 NGC 5515	1386	1.5 x 0.8 1.5 x 0.8	108 5	2	Sa – b		13.7		*3
09097	14 10.5 + 50 36 094.80 + 62.11	08 – 26 – 013 NGC 5520	0120	1.8 x 1.1 1.7 x 0.9	66 4	2	Sb		13.3		3
09098	14 10.6 + 45 55 087.31 + 65.44	08 – 26 – 11 + 12	120	[1.2 x 0.5] ...	... ...	2	Double system Contact		14.1		*1
09099	14 10.8 + 20 40 017.59 + 70.18	04 – 34 – 005 NGC 5513	0061	1.9 x 1.1 (2.3 x 1.3)	115 4	3	S0		14.1		*13
09100	14 10.8 + 35 56 064.28 + 70.67	06 – 31 – 079 NGC 5517	0106 320	1.1 x 0.8 1.2 x 0.9	125 3	2	S0 – a		15.0		

N G C 5491, 5496 – 5500, 5504, 5505, 5508,
5513, 5515, 5517, 5520, 5547

I C 4381, 4383, 4384, 4404

1	2	3	4	5	6	7	8	9	10	11	12
09101	14 11.0 + 27 15 037.28 + 71.81	05 - 34 - 001	0070	2.1 x 1.1 1.7 x 1.0	145 5	3	Sb		14.7		
09102	14 11.2 + 07 54 351.76 + 62.46	01 - 36 - 023 NGC 5514	0096	[2.3 x 1.0] ...	...	3	Double system, contact Strongly distorted, plume		14.5		*13 V
09103	14 11.2 + 08 27 352.57 + 62.86	01 - 36 - 000	0096	1.1: x 0.2 1.2 x 0.3	162 6	1	S ... 1st of 3, disturbed		16.5:		*V
09104	14 11.2 + 12 44 359.73 + 65.83	02 - 36 - 056	1051	1.0 x 0.8 0.7 x 0.6	... 2:	1	S ...		14.4		
09105	14 11.6 + 44 06 083.63 + 66.47	07 - 29 - 054	1386	1.2 x 0.8 0.9 x 0.8	145: 	1	S ... nff of 2		15.1		*
09106	14 11.7 + 26 09 033.87 + 71.52	04 - 34 - 000	0061	1.1 x 0.15 1.0 x 0.15	165 6 - 7		Sb - c		15.7		
09107	14 11.7 + 31 48 051.76 + 71.52	05 - 34 - 004	0070	1.2 x 0.7 1.2 x 0.7	... ...	3	S B b		15.5		*
09108	14 11.8 + 03 12 345.86 + 58.75	01 - 36 - 000	0096	[1.2 x 0.5] ...	... ...	2	Triple system Contact, bridge. In group		15.6		*
09109	14 11.8 + 12 34 359.66 + 65.61	02 - 36 - 000	1051	[1.0 x 0.3] ...	[165] 6	2	Double system Contact		15.5		*
09110	14 11.8 + 15 51 006.22 + 67.61	03 - 36 - 088	0081 312	2.0 x 0.7 2.3 x 0.7	17 6	3	S B b		14.0		*1
09111	14 11.9 + 07 45 351.82 + 62.23	01 - 36 - 025 NGC 5519	0096	2.1 x 1.2 2.1 x 1.3	75 4	2	Sa P w 9102		14.6		*3
09112	14 11.9 + 47 53 090.32 + 63.93	08 - 26 - 016	0120	1.2 x 0.3 1.0 x 0.25	142 6	1	S ...		15.0		
09113	14 12.1 + 35 39 063.21 + 70.52	06 - 31 - 080	0106 320	2.3 x 0.6 1.8 x 0.5	58 7	1	S ...		15.7		*
09114	14 12.3 + 03 21 346.23 + 58.79	00 - 36 - (27) IC 989	0096	(1.3 x 1.1) (1.3 x 1.1)	-- ...	3	E Brightest in group		14.4		*
09115	14 12.3 + 58 00 103.26 + 56.10	10 - 20 - 085 NGC 5526	0705 294	2.0 x 0.2 1.7 x 0.2	136 7	2	Sb - c f of 2		(14.2)		*13
09116	14 12.4 + 15 23 005.43 + 67.22	03 - 36 - 089 NGC 5522	0081	1.8 x 0.35 1.9 x 0.4	50 6	3	Sb		14.1		3
09117	14 12.5 + 14 21 003.35 + 66.59	02 - 36 - 059	1051	2.0 x 1.0 1.3 x 0.8	7 5	3	Sa		14.3		
09118	14 12.6 + 03 20 346.33 + 58.73	01 - 36 - 000	0096	1.2 x 0.35 1.2 x 0.3	88 7:	2	Sa - b IC 989 group		16.0:		*
09119	14 12.6 + 25 33 032.12 + 71.22	04 - 34 - 008 NGC 5523	0061	4.7 x 1.4 4.5 x 1.1	99 6	4	Sc	Sc +	13.4		*12 3468
09120	14 12.7 + 05 03 348.47 + 60.06	01 - 36 - 028	0096	1.1 x 0.8 1.2 x 0.9	0: 3	4	Sc Brightest in group		14.6		*1
09121	14 12.8 + 15 58 006.82 + 67.48	03 - 36 - 091	0081 312	1.6 x 0.7 1.7 x 0.8	140 6	2	Sb - c		14.8		1
09122	14 12.9 + 04 38 348.02 + 59.71	01 - 36 - 030 NGC 5521	0096	0.7 x 0.6 0.7 x 0.7	-- ...	1	...		14.3		13
09123	14 12.9 + 36 42 065.95 + 70.00	06 - 31 - 083	0106	1.2 x 0.8 1.0: x 1.0:	... ...	4	...	S m	16.5:		
09124	14 13.2 + 14 31 003.94 + 66.56	03 - 36 - 096 NGC 5525	0081	1.2 x 0.7 1.3 x 0.8	23 7	3	S0		14.1		*1
09125	14 13.2 + 45 50 086.45 + 65.15	08 - 26 - 017	0120	1.2 x 0.6 1.0 x 0.5	120 5	1	S ...		14.8		
09126	14 13.3 + 16 46 008.76 + 67.81	03 - 36 - 098	0081 312	1.8 x 1.6 1.8: x 1.5:	-- ...	4	Dwarf m V		16.0:		*
09127	14 13.4 + 36 27 065.17 + 70.00	06 - 31 - 085 NGC 5529	0106 320	6.2 x 0.7 6.0 x 0.6	115 7	3	Sc		12.9	+ 3781 + 3879	*12 3
09128	14 13.6 + 23 17 025.55 + 70.47	04 - 34 - 009	0061	1.8 x 1.6 1.8: x 1.6:	... ...	4	Dwarf irregular Ir IV - V. Brightest of 3		15.3		*
09129	14 13.7 + 40 20 074.95 + 68.27	07 - 29 - 055	1386	1.1 x 0.2 1.0 x 0.2	107 7	2	Sc - c		16.0:		
09130	14 13.8 + 23 15 025.50 + 70.42	04 - 34 - 010	0061	[1.0 x 0.9] ...	... ...	1	Double system 9128 group		16.0:		*1
09131	14 13.8 + 63 57 108.39 + 50.98	11 - 17 - 000	1428	1.7: x 0.3 1.3: x 0.2	160 7	3	Sc	Sc +	16.5:		*
09132	14 13.9 + 33 57 057.99 + 70.65	06 - 31 - 088	0106	1.0 x 0.2 0.9 x 0.2	77 6	2	Sc		16.5:		
09133	14 14.0 + 35 34 062.60 + 70.18	06 - 39 - 089 NGC 5533	0106 320	3.7 x 2.0 (3.2 x 1.8)	30 5	3	Sb	SA ab Sb -	13.0		*12 678
09134	14 14.1 + 10 13 356.47 + 63.61	02 - 36 - 000	1051	1.5 x 0.6 1.2: x 0.5:	20 6	3	Sc		15.5		
09135	14 14.3 + 24 44 029.90 + 70.68	04 - 34 - 011	0061	1.2 x 0.5 1.1 x 0.5	40 6	1	S ...		15.5		

N G C   5514, 5519, 5521 - 23, 5525, 5526
5529, 5533

I C   989

1	2	3	4	5	6	7	8	9	10	11	12
09136	14 14.3 + 39 44 073.38 + 68.46	07 − 29 − 057 NGC 5536	1386	1.0 x 1.0 1.0 x 0.9	-- 1	2	S B a NGC 5541 group		14.5		*3
09137	14 14.4 + 11 02a 357.94 + 64.12	02 − 36 − 062 NGC 5532	1051	(1.6 x 1.6) (2.0 x 2.0)	-- 1	3	S0 Companion superimposed		(13.3)		*13
09138	14 14.4 + 23 14 025.57 + 70.28	04 − 34 − 012	0061	1.9 x 0.2 1.9 x 0.2	169 7	3	Sc 9128 group		15.5		
09139	14 14.4 + 39 49 073.55 + 68.41	07 − 29 − 059 NGC 5541	1386	0.9 x 0.7 0.8 x 0.6	... ...	2	S? ... Peculiar Brightest in group		13.4		*13
09140	14 15.0 + 01 21 344.98 + 56.79	00 − 36 − 000	1424	1.1 x 0.2 0.9: x 0.2	107 6	2	Sc 3rd of 3		16.5:		*
09141	14 15.0 + 27 05 037.03 + 70.91	05 − 34 − 007 IC 4395	0070 324	1.5 x 1.5 1.5 x 1.4	-- 1	3	S ... f of 2, contact, disturbed		15.1		*1
09142	14 15.0 + 36 48a 065.78 + 69.57	06 − 31 − 090 NGC 5544	0106 320	1.0 x 1.0 1.1 x 1.0	-- 1	2	Sa P w 9143, contact		13.2	+ 3165 + 3265	*12 3$\overline{VA}$
09143	14 15.0 + 36 48b 065.78 + 69.57	06 − 31 − 091 NGC 5545	0106 320	1.1 x 0.35 1.1 x 0.35	58 6	2	Sb − c P w 9142, contact			+ 3175 + 3275	*12 3$\overline{VA}$
09144	14 15.0 + 42 55 080.31 + 66.68	07 − 29 − 061	1386	1.5 x 1.5 1.0 x 1.0	-- 1	3	Sc		16.0:		
09145	14 15.0 + 58 02 102.77 + 55.87	10 − 20 − 091 IC 995	0705 294	1.2 x 0.3 1.4 x 0.3	147 ...	2	Sc − Irr		14.5		*
09146	14 15.3 + 56 07 100.64 + 57.39	09 − 23 − 000	1409	1.0: x 0.9: 0.3: x 0.3:	-- ...	2	Dwarf V		17. − 18.		
09147	14 15.7 + 01 07 344.99 + 56.50	00 − 36 − 033	1424 318	1.6: x 1.0 1.1 x 0.8	(130) ...	2	Sb − c		14.9		*1
09148	14 15.7 + 07 48 353.35 + 61.63	01 − 36 − 035 NGC 5546	0096 319	(1.3 x 1.1) (1.8 x 1.6)	-- ...	4	E		14.1		*3
09149	14 15.7 + 25 22 031.95 + 70.50	04 − 34 − 013 NGC 5548	0061	1.7 x 1.5 1.7 x 1.5	(110) 1	3	Sa Seyfert		13.1	+ 4930 + 4990	*12 34$\overline{68}$
09150	14 15.7 + 26 39 035.77 + 70.70	05 − 34 − 012 IC 4397	0070 324	1.1 x 0.8 1.1 x 0.8	165 3	2	S ... Peculiar		14.2		*
09151	14 15.7 + 58 56 103.55 + 55.08	10 − 20 − 093	0705 294	1.4 x 0.5 1.0 x 0.4	133 (6)	3	... S m		16.0:		*
09152	14 15.8 + 57 51 102.42 + 55.95	10 − 20 − 092 IC 996	0705 294	1.4 x 0.2 1.4 x 0.2	155 7	2	Sb − c		14.6		
09153	14 15.9 + 11 25 359.15 + 64.10	02 − 36 − 064 IC 994	1051	1.4 x 0.5 1.4 x 0.5	10 6	2	Sa sf of 2		14.8		*1
09154	14 16.0 + 13 06 002.21 + 65.16	02 − 36 − 065 NGC 5550	1051	1.3 x 0.9 1.1 x 0.8	100 3	1	S ...		14.2		3
09155	14 16.0 + 29 12 043.51 + 70.79	05 − 34 − 015	0070 326	1.3 x 1.2 1.0 x 0.8	-- 1 − 2	3	Sc SAB(rs)c Sc −		15.6		
09156	14 16.1 + 07 36 353.22 + 61.41	01 − 36 − 036 NGC 5549	0096 330	1.5 x 0.7 1.8 x 0.8	120 5	4	S0		14.2		*3
09157	14 16.1 + 26 37 035.71 + 70.61	05 − 34 − 014 IC 4399	0070 324	1.2 x 0.8 1.1 x 0.7	140 3	1	Sb		15.5		*
09158	14 16.1 + 31 53 051.61 + 70.59	05 − 34 − 016 IC 4403	0070 326	1.4 x 0.5 1.4 x 0.6	144 6	1	S ...		14.9		
09159	14 16.2 + 14 29 004.94 + 65.95	03 − 36 − 099	0081	1.1 x 0.35 1.0 x 0.3	108 6	2	Sb − c		15.4		
09160	14 16.2 + 26 31 035.42 + 70.57	05 − 34 − 017 NGC 5553	0070 324	1.3 x 0.25 1.2 x 0.3	88 7	1	Sa:		14.8		*3
09161	14 16.3 + 36 43 065.29 + 69.36	06 − 31 − 093 NGC 5557	0106	2.2 x 2.0 (3.0 x 2.5)	(105) ...	4	E		12.2	+ 3195 + 3296	*12 347$\overline{8}$
09162	14 16.4 + 11 05 358.77 + 63.78	02 − 37 − 000	0065	1.2 x 0.45 1.1 x 0.45	[135] ...	3	S B b Disturbed?, Sb +		15.4		*
09163	14 16.4 + 11 31 359.51 + 64.07	02 − 37 − 000	0065	1.3 x 0.3 1.2 x 0.3	75 6	3	Sc		15.6		
09164	14 16.4 + 22 03 022.71 + 69.48	04 − 34 − 014	0061	1.1 x 0.8 0.8 x 0.6	... ...	1	... Strongly peculiar, f of 2		15.4		*1
09165	14 16.5 + 25 10 031.47 + 70.29	04 − 34 − 015	0061 324	1.5 x 0.35 1.2 x 0.3	54 7	1	S ...		15.3		*1
09166	14 16.9 + 25 02 031.14 + 70.17	04 − 34 − 017 NGC 5559	0061 324	1.6 x 0.45 1.5 x 0.45	67 6	2	S B b		15.0		3
09167	14 17.2 + 17 52 012.50 + 67.58	03 − 37 − 002	1417 323	1.5 x 0.4 1.3 x 0.3	48 6	1	Sa		15.3		
09168	14 17.2 + 18 06 013.04 + 67.69	03 − 37 − 001 IC 999	1417 323	0.8 x 0.4 0.8 x 0.5	142 ...	1	S0? IC 1000 group		14.5		*
09169	14 17.3 + 09 36 356.68 + 62.62	02 − 37 − 001	0065	4.5 x 0.8 3.5: x 0.7	53 (6)	4	Irr I m Ir I		14.7		
09170	14 17.3 + 18 05 013.03 + 67.66	03 − 37 − 003 IC 1000	1417 323	0.7 x 0.4 0.8 x 0.4	23 ...	2	S0 Brightest of 3		14.4		*

N G C 5532, 5536, 5541, 5544 − 46, 5548 − 50
5553, 5557, 5559

I C   994 − 96, 999, 1000, 4395
4397, 4399, 4403

1	2	3	4	5	6	7	8	9	10	11	12
09171	14 17.4 + 18 04 013.02 + 67.63	03 - 37 - 004	1417 323	1.0 x 0.15 0.9 x 0.18	132 7	3	Sb IC 1000 group		16.0:		*
09172	14 17.5 + 04 13 349.23 + 58.66	01 - 37 - 001 NGC 5560	1418	4.0 x 0.9 4.0 x 0.9	115 6	3	S B b NGC 5566 group, disturbed		13.7		*12 3A
09173	14 17.5 + 36 22 064.13 + 69.25	06 - 31 - 099 NGC 5572	0106 320	1.2 x 0.9 1.1 x 0.9	--- 2 - 3	2	Sb		15.2		
09174	14 17.7 + 10 29 358.25 + 63.15	02 - 37 - 002	0065	0.7 x 0.6 0.7 x 0.7	--- ...	1	...		14.5		
09175	14 17.8 + 04 09 349.25 + 58.56	01 - 37 - 002 NGC 5566	1418	6.2 x 2.3 6.2 x 2.3	35 6	4	S B a S B(r)ab Brightest of 3		12.0	+ 1601 + 1581	*12 3468A
09176	14 18.0 + 04 12 349.39 + 58.56	01 - 37 - 003 NGC 5569	1418	2.0: x 1.6: 1.6: x 1.2:	--- 1 - 3	2	Sc NGC 5566 group		14.9		*12 3A
09177	14 18.0 + 10 40 358.66 + 63.21	02 - 37 - 003	0065	1.4 x 1.2 1.4 x 1.2	--- 1 - 2	4	Sc		14.9		
09178	14 18.1 + 52 08 095.10 + 60.18	09 - 23 - 64	1409	[1.8 x 0.7] ...	... ...	3	Double system Bridge		15.5		*1V A
09179	14 18.2 + 56 57 100.99 + 56.48	10 - 20 - 094 NGC 5585	0705 294	6.1 x 4.0 6.0: x 3.5:	30: 3	4	Sc SAB(rs)d M 101 group		11.7	+ 304 + 467	*12- -689
09180	14 18.3 + 35 25 061.43 + 69.39	06 - 32 - 002 NGC 5579	0127 320	1.9 x 1.3 1.9 x 1.3	165: 6	4	Sc SAB cd Singular, disturbed?		14.7		*13 VA
09181	14 18.4 + 03 28 348.64 + 57.93	01 - 37 - 006 NGC 5574	1418	1.1 x 0.8 1.1 x 0.8	63 ...	1	S0?		13.4	+ 1716 + 1694	*12 3468
09182	14 18.4 + 22 10 023.46 + 69.08	04 - 34 - 019	0061	2.6 x 0.5 2.6 x 0.45	123 6	4	Sc Sc +		14.7		1
09183	14 18.5 + 03 30 348.72 + 57.94	01 - 37 - 007 NGC 5576	1418	(3.0 x 2.3) (3.3 x 2.5)	95 ...	4	E Brightest of 3		12.3	+ 1528 + 1507	*12 3468
09184	14 18.5 + 06 26 352.47 + 60.15	01 - 37 - 008 NGC 5575	1418	1.0 x 1.0 (1.2 x 1.2)	--- 1	3	S0		14.5		13
09185	14 18.5 + 15 13 007.20 + 65.92	03 - 37 - 006	1417	1.1 x 0.35 0.9 x 0.30	172 6	1	S ...		15.0		
09186	14 18.6 + 23 50 028.01 + 69.52	04 - 34 - 020	0061	1.1 x 0.3 1.0 x 0.25	140 6	1	Sa		15.6		
09187	14 18.7 + 03 40 348.99 + 58.04	01 - 37 - 009 NGC 5577	1418	3.2 x 0.9 3.2 x 0.9	56 6	3	Sb NGC 5576 group, Sb +		13.6		*23
09188	14 18.7 + 39 56 072.80 + 67.62	07 - 29 - 063 NGC 5582	1386 322	(2.8 x 1.7) (2.3 x 1.6)	25 ...	3	E		13.0		3
09189	14 18.7 + 71 50 113.50 + 43.88	12 - 14 - 001 NGC 5607, Mark 286	1442	0.9 x 0.9 0.9 x 0.9	--- ...	3	... Peculiar		13.9	+ 7800	*3C
09190	14 18.9 + 05 17 351.09 + 59.24	01 - 37 - 000	1418	1.3 x 0.5 1.1 x 0.4	40 6	2	Sb - c		15.3		*
09191	14 19.0 + 30 13 046.50 + 70.12	05 - 34 - 024 IC 4408	0070 326	1.0 x 0.4 1.0 x 0.35	131 6	1	S ... n of 2		15.0		*1
09192	14 19.0 + 51 23 093.86 + 60.63	09 - 24 - 001	0715	1.1 x 0.3 1.0: x 0.2	145 ...	2	Irr		16.5:		
09193	14 19.1 + 36 58 065.40 + 68.75	06 - 32 - 000	0127 320	1. x 1. 1. x 1.	--- ...	3	Dwarf V		17.		
09194	14 19.1 + 39 44 072.24 + 67.64	07 - 30 - 002	0145	1.3 x 0.6 1.0 x 0.6	115: ...	1	S ...		15.6		1
09195	14 19.2 + 24 10 029.03 + 69.47	04 - 34 - 022	0061	1.2 x 0.45 1.1 x 0.4	90 6	2	Sb		15.5		*
09196	14 19.3 + 13 27 004.02 + 64.74	02 - 37 - 004 NGC 5583	0065	0.8 x 0.6 0.8 x 0.7	80 ...	1	...		14.2		*
09197	14 19.3 + 35 29 061.44 + 69.18	06 - 32 - 005 NGC 5589	0127 320	1.3 x 1.3 1.1 x 0.9	--- 1	3	S B a P w 9200		14.3		*13
09198	14 19.3 + 36 40 064.58 + 68.81	06 - 32 - 000	0127 320	1.0 x 0.15 0.8 x 0.15	54 7	1	Sb		16.5:		
09199	14 19.5 + 11 20 000.31 + 63.37	02 - 37 - 000	0065	1.4 x 0.20 1.4 x 0.20	128 7	1	S ...		15.3		
09200	14 19.5 + 35 25 061.23 + 69.15	06 - 32 - 006 NGC 5590	0127 320	1.8 x 1.8 (2.0 x 2.0)	--- 1	3	S0 P w 9197		13.6		*13
09201	14 19.8 - 00 09 345.11 + 54.87	00 - 37 - 001 NGC 5584	1440	3.5: x 2.8: 3.5: x 2.8:	140 2	4	Sc SAB(s)cd		12.8		*12 478
09202	14 19.8 + 14 09 005.52 + 65.05	02 - 37 - 005 NGC 5587	0065	2.7 x 0.8 2.2 x 0.7	162 6	2	S0 - a		14.0		3
09203	14 19.9 + 40 13 073.18 + 67.29	07 - 30 - 003	0145 322	1.0 x 0.9 1.0 x 0.9	--- 1	3	S B b (R')S B(s)b		15.0		
09204	14 20 + 82 52 119.68 + 33.94	14 - 07 - 000	1363	1.3 x 0.1 0.9 x 0.1	55 7	3	Sc f of 2		16.5:		*
09205	14 20 + 86 03 121.22 + 31.03	14 - 07 - 000	1363	1.2 x 1.1 1.2: x 0.9:	--- ...	1	...		17.		

N G C   5560, 5566, 5569, 5572, 5574 - 77, 5579
5582 - 5585, 5587, 5589, 5590, 5607

I C   4408

1	2	3	4	5	6	7	8	9	10	11	12
09206	14 20.0 + 15 18 007.86 + 65.67	03 – 37 – 010	1417	0.8 x 0.4 0.8 x 0.4	165 ...	1	... Peculiar		14.2		*
09207	14 20.2 + 13 57 005.27 + 64.86	02 – 37 – 006 NGC 5591	0065	[1.4 x 0.8] ...	... ...	4	Double system Strongly disrupted		14.5		*
09208	14 20.4 + 37 20 066.09 + 68.37	06 – 32 – 010 NGC 5596, Mark 470	0127	1.2 x 0.9 1.2 x 0.9	100 2 – 3	2	S0		14.5		*3
09209	14 20.5 + 40 33 073.81 + 67.03	07 – 30 – 004 NGC 5598	0145 322	(1.5 x 1.0) (1.5 x 1.0)	50 3:	2	S0 NGC 5603 group		14.3		*3
09210	14 20.5 + 50 45 092.60 + 60.92	09 – 24 – 002 NGC 5602	0715	1.4 x 0.7 1.4 x 0.7	166 5	3	Sa		13.5		3
09211	14 20.6 + 45 37 084.16 + 64.28	08 – 26 – 029	0120	2.5:: x 2.0:: 2. x 2.	... ...	3	Dwarf irregular Ir V		15.7		*1
09212	14 20.7 + 18 30 014.97 + 67.14	03 – 37 – 011	1417	1.1 x 0.6 (1.3 x 0.8)	120: ...	3	S0		14.7		
09213	14 20.7 + 38 13 068.24 + 67.99	06 – 32 – 013	0127	2.0 x 0.8 2.0 x 0.8	160 6	3	Sb ssf of 2, contact		(15.0)		*1
09214	14 20.8 + 33 04 054.58 + 69.43	06 – 32 – 014 Mark 471	0127	0.9 x 0.5 0.9 x 0.6	35: 4:	2	S B a		14.5		
09215	14 20.9 + 01 57 347.77 + 56.36	00 – 37 – 002	1440	2.5: x 1.3 2.3 x 1.3	165 5	4	S B c	Sc +	13.6		*
09216	14 21.0 + 40 39 073.92 + 66.90	07 – 30 – 007	0145 322	2.0: x 2.0: 2.0: x 2.0:	-- 1	4	Sc SA cd P w 9217, NGC 5603 group		15.1		*1
09217	14 21.1 + 40 36 073.78 + 66.91	07 – 30 – 008 NGC 5603	0145 322	(1.4 x 1.4) (1.4 x 1.4)	-- 1	4	S0 P w 9216 Brightest in group, compact		14.0		*13 C
09218	14 21.3 + 06 48 353.99 + 59.94	01 – 37 – 010 NGC 5599	1418	1.5 x 0.6 1.5 x 0.6	160 6	2	Sb		14.7		13
09219	14 21.3 + 42 00 076.81 + 66.19	07 – 30 – 009 NGC 5608	0145	3.0: x 1.7: 2.9: x 1.7:	[95] ...	4	Irr		14.3		*3
09220	14 21.5 + 14 52 007.47 + 65.12	03 – 37 – 013 NGC 5600	1417	1.4 x 1.4 1.4 x 1.3	-- 1	3	S ... Peculiar		11.9		*12 3468
09221	14 21.6 + 34 14 057.72 + 69.03	06 – 32 – 018	0129	0.9 x 0.8 0.9 x 0.9	-- 1	1	S ... P w 9222		14.4		*1
09222	14 21.7 + 34 15 057.75 + 69.01	06 – 32 – 019	0127	1.0 x 0.25 1.0 x 0.25	40 6	1	S ... P w 9221		15.1		*1
09223	14 21.7 + 40 15 072.85 + 66.97	07 – 30 – 010	0145 322	1.0 x 0.4 1.1 x 0.35	171 7	2	S0		14.7		*
09224	14 21.8 + 34 57 059.63 + 68.83	06 – 32 – 000	0127	1.0 x 0.10 1.0 x 0.10	37 7	2	Sc		17.		
09225	14 21.9 + 08 30 356.65 + 61.04	02 – 37 – 000	0065	1.1 x 0.5 0.9 x 0.4	0 6	3	... S: m		15.4		
09226	14 22.0 + 35 05 059.96 + 68.75	06 – 32 – 022 NGC 5614	0127	2.8 x 2.2 2.8 x 2.2	... ...	3	Sa NGC 5615 super- imposed, plume; p w 9228		12.6	+ 3872 + 3971	*12V 3468A
09227	14 22.0 + 33 15 054.97 + 69.15	06 – 32 – 020 NGC 5611	0106	1.3 x 0.6 1.5 x 0.7	63 5	2	S0		13.5		*3
09228	14 22.0 + 35 07 060.05 + 68.74	06 – 32 – 021 NGC 5613	0127	1.0 x 0.9 1.0 x 0.8	-- 1 – 2	1	S0 – a P w 9226		16.0:		*V A
09229	14 22.1 + 01 24a 347.58 + 55.75	00 – 37 – 004	1440	1.0 x 0.9 1.0 x 0.9	-- 1:	2	Sb npp of 2, disturbed		(14.9)		*1
09230	14 22.1 + 24 50 031.28 + 68.98	04 – 34 – 025 NGC 5610	0061	2.2 x 0.7 2.2 x 0.7	108 6	4	S B a S B(s)ab		14.5		*13
09231	14 22.2 + 36 40 064.07 + 68.26	06 – 32 – 026 NGC 5616	0127 320	2.4 x 0.4 2.5 x 0.5	157 6	2	Sb – c Disturbed?		14.8		*3
09232	14 22.3 + 33 10 054.71 + 69.10	06 – 32 – 027	0127	1.1 x 0.8 1.0 x 0.7	... ...	3	... S dm		16.5:		
09233	14 22.5 + 35 30 060.99 + 68.54	06 – 32 – 028	0127 320	1.3 x 0.5 1.3 x 0.5	132 6	1	S ... 1st of 3		15.4		*
09234	14 22.6 + 26 22 035.55 + 69.13	04 – 34 – 000	0061 324	2.1 x 1.0 2.0: x 1.0:	162 ...	2	S ... s of 2, disturbed?		15.7		*
09235	14 22.6 + 35 29 060.93 + 68.53	06 – 32 – 029	0127 320	1.1 x 0.4 1.0 x 0.4	75 6	1	S B a 2nd of 3		15.4		*
09236	14 22.7 + 25 15 032.49 + 68.93	04 – 34 – 026	0061 324	1.4 x 0.35 1.1 x 0.35	132 6	2	S B?b – c ssf of 2		15.4		*
09237	14 22.7 + 28 03 040.28 + 69.29	05 – 34 – 035	0070 326	1.0 x 0.25 0.9 x 0.2	67 6	1	S ... 3rd of 5		15.5		*
09238	14 22.7 + 35 30 060.96 + 68.51	06 – 32 – 030	0127 320	1.3 x 0.5 1.2 x 0.5	144 6	3	Sc 3rd of 3		16.0:		*
09239	14 22.8 + 13 58 006.16 + 64.35	02 – 37 – 000	0065	1.3 x 0.20 1.4 x 0.22	20 7	1	Sb – c		15.3		
09240	14 22.8 + 44 45 082.01 + 64.48	08 – 26 – 030	0120	2. x 2. 2. x 2.	-- ...	4	Irr IA m		13.9		*C

N G C   5591,   5596,   5598 – 5600,   5602,   5603
        5608,   5610,   5611,   5613,   5614,   5616

1	2	3	4	5	6	7	8	9	10	11	12
09241	14 23.2 + 32 42	06 – 32 – 033	0127	0.55 x 0.50	—	1	...		14.2		*
	053.34 + 68.99			0.55 x 0.45	...		Peculiar				
09242	14 23.3 + 39 45	07 – 30 – 011	0145	5.4 x 0.3	71	3	Sc		14.8		*C
	071.36 + 66.91			4.5 x 0.3	7			Sc +			
09243	14 23.4 + 34 04	06 – 32 – 034	0127	1.5 x 0.8	147	2	Sc		15.6		*
	057.05 + 68.71			1.4 x 0.6	5						
09244	14 23.6 + 05 27	01 – 37 – 011	1418	1.3 x 0.7	3	3	S B b		14.8		*1
	352.98 + 58.57			1.3 x 0.7	5:		ssp of 2	Sb +			
09245	14 23.8 + 56 33	09 – 24 – 005	0715	2.1 x 1.6	5:	4	S B c	S B(s)d	15.4		*
	099.46 + 56.31			2.0: x 1.5:	2 – 3						
09246	14 24.0 + 06 12	01 – 37 – 000	1418	1.3 x 0.4	152	1	S ...		15.1		
	354.12 + 59.05			1.0 x 0.3	6						
09247	14 24.1 + 26 28	04 – 34 – 028	0061	1.1 x 0.6	73	1	S ...		15.5		*1
	035.96 + 68.82	IC 4423	324	1.1 x 0.6	5		sf of 2				
09248	14 24.4 + 48 47	08 – 26 – 032	0120	1.8 x 1.0	90	3	Sb		14.2		3
	088.70 + 61.79	NGC 5622		1.8 x 0.9	5						
09249	14 24.5 + 08 54	02 – 37 – 000	0065	2.3 x 0.3	86	3	...	S d – dm	15.1		
	358.15 + 60.84			2.2: x 0.3	7						
09250	14 24.6 – 02 03	00 – 37 – 005	1440	1.8 x 1.2	10	4	S B c	S B(rs)c	14.8		3
	344.84 + 52.63	NGC 5618		1.7 x 1.1	3			Sc –			
09251	14 24.6 + 45 05	08 – 26 – 033	0120	1.0 x 0.2	96	3	Sc		15.7		*1V
	082.22 + 64.03			1.0 x 0.2	6						
09252	14 24.7 + 05 21	01 – 37 – 000	1418	1.1 x 0.2	175	3	Irr		15.7		
	353.23 + 58.31			1.0 x 0.2	...			Ir I			
09253	14 24.7 + 31 45	05 – 34 – 042	0070	1.8 x 0.4	48	2	Sb – c		14.7		1
	050.61 + 68.79		326	1.9 x 0.35	6						
09254	14 24.8 + 01 15	00 – 37 – 006	1440	2.3 x 2.2	—	4	S B b	S B(r)ab	14.8		*
	348.34 + 55.21	IC 1010		(1.5 x 1.5)	1			Sb –			
09255	14 24.8 + 05 01	01 – 37 – 012	1418	2.4 x 1.1	8	4	Sb	SAB(rs)b	14.0		*13
	352.83 + 58.05	NGC 5619		2.4 x 1.1	6		Brightest of 3				
09256	14 24.8 + 51 50	09 – 24 – 006	0715	1.1 x 0.7	3	1	S ...		14.1		1
	093.21 + 59.68	NGC 5624	328	1.1 x 0.7	4		Peculiar?				
09257	14 24.9 + 31 10	05 – 34 – 043	0070	1.3 x 0.9	105	1	S ...		14.8		
	049.00 + 68.80	IC 1012, 4431		1.3 x 0.9	3						
09258	14 25.0 + 05 00	01 – 37 – 013	1418	1.0 x 0.9	—	2	Sb – c		15.3		1
	352.88 + 58.00			0.9 x 0.9	1		NGC 5619 group				
09259	14 25.0 + 11 16	02 – 37 – 000	0065	1.0 x 0.5	165	1	S ...		15.2		
	002.09 + 62.29		325	1.1 x 0.6	5						
09260	14 25.0 + 33 28	06 – 32 – 035	0127	(1.6 x 1.1)	17	3	E		13.7		3
	055.26 + 68.49	NGC 5623		(1.8 x 1.2)	...						
09261	14 25.0 + 56 49	10 – 21 – 002	1425	(2.0 x 2.0)	—	3	SO/Sa		12.4	+ 1979	*12
	099.54 + 56.00	NGC 5631		(2.0 x 2.0)	1					+ 2146	3468
09262	14 25.1 + 36 08	06 – 32 – 036	0127	1.2 x 0.15	51	1	S ...		16.0:		*1
	062.23 + 67.86			1.3 x 0.18	7		nf of 2				
09263	14 25.2 – 00 23	00 – 37 – 007	1440	1.1 x 0.9	—	2	...	S d – dm	15.6		
	346.72 + 53.87			1.0: x 0.8:	2:						
09264	14 25.2 + 06 16	01 – 37 – 000	1418	1.0 x 0.2	135	2	Sc		15.5		
	354.62 + 58.89			1.0 x 0.2	6						
09265	14 25.2 + 25 44	04 – 34 – 029	0061	1.0 x 0.3	167	1	S ...		15.4		
	034.08 + 68.46		324	1.0 x 0.3	6						
09266	14 25.2 + 30 10	05 – 34 – 044	0070	1.0 x 0.6	165	2	S B c – Irr		15.4		*1
	046.23 + 68.78		326	1.0 x 0.6	...						
09267	14 25.4 + 11 47	02 – 37 – 010	0065	1.1 x 1.0:	—	3	S B b		14.9		
	003.09 + 62.54		325	1.0 x 0.8	1:						
09268	14 25.4 + 32 22	05 – 34 – 000	0070	1.3 x 0.6	20	3	Dwarf (spiral)		17.		
	052.25 + 68.57			1.3 x 0.6	...		S: IV – V or V				
09269	14 25.4 + 50 48	09 – 26 – 000	0715	1.3 x 0.12	155	3	Sc		16.5:		
	091.58 + 60.32			1.3 x 0.12	7						
09270	14 25.6 + 41 29	07 – 30 – 014	0145	2.5 x 0.7	98	2	Sc – Irr		13.6		13
	074.74 + 65.76	NGC 5630		2.2 x 0.7	(6)						
09271	14 25.6 + 46 22	08 – 26 – 034	0120	2.5 x 1.1	10	3	Sb		12.9	+ 2346	*12
	084.33 + 63.14	NGC 5633		1.4 x 0.8	6		Compact core			+ 2484	3468C
09272	14 25.6 + 61 26	10 – 21 – 000	1425	1.0 x 0.3	29	2	Sc		16.0:		
	104.33 + 52.33			1.0 x 0.3	6						
09273	14 25.7 + 13 46	02 – 37 – 000	0065	1.1 x 0.5	10	3	Irr		15.1		
	006.71 + 63.66			0.9 x 0.4	...						
09274	14 25.7 + 21 32	04 – 34 – 031	0061	1.4 x 0.8	[35]	4	S B ...		14.7		*1
	023.37 + 67.27			1.4 x 0.9	...		Peculiar, p w 9282				
09275	14 25.9 + 14 00	02 – 37 – 012	0065	2.8 x 1.8	90	4	...	SAB dm	14.1		
	007.20 + 63.76	IC 1014		2.8: x 1.8:	3 – 4						

N G C   5618, 5619, 5622 – 24, 5630, 5631, 5633   I C   1010, 1012, 1014, 4423, 4431

1	2	3	4	5	6	7	8	9	10	11	12
09276	14 25.9 + 26 05 035.09 + 68.36	04 – 34 – 032 IC 1017	0061 324	1.0 x 0.6 1.1 x 0.5	128 5:	1	SO?		14.9		*
09277	14 26.0 + 03 29 351.33 + 56.71	01 – 37 – 015	1418 327	1.5 x 0.25 1.5 x 0.25	45 6	2	Sb		15.0		
09278	14 26.0 + 18 09 015.62 + 65.86	03 – 37 – 019 NGC 5628	1417 323	(1.1 x 0.7) 1.0 x 0.6	175 ...	3	E		14.5		
09279	14 26.0 + 34 02 056.67 + 68.18	06 – 32 – 000	0127	1.0 x 0.3 0.7 x 0.2	110 6	2	Sc		16.0:		
09280	14 26.1 + 11 36 003.01 + 62.29	02 – 37 – 013 NGC 5627	0065 325	1.6 x 0.9 (1.8 x 1.1)	120 4	3	SO		14.7		*13
09281	14 26.1 + 26 04 035.06 + 68.31	04 – 34 – 034 NGC 5629	0061 324	(1.8 x 1.8) (2.3 x 2.3)	--- (1)	3	SO Brightest in cluster		14.2		*13
09282	14 26.3 + 21 34 023.57 + 67.15	04 – 34 – 000	0061	1.6: x 1.0: 1. x 1.	(110) ...	3	Dwarf V, p w 9274		15.7		*
09283	14 26.3 + 27 38 039.30 + 68.46	05 – 34 – 049 NGC 5635	0070	2.4 x 1.2 (2.5 x 1.3)	65 6	1	S ... Singular		13.9	+ 3819 + 3894	*3
09284	14 26.3 + 33 28 055.13 + 68.23	06 – 32 – 038	0127	1.6 x 0.3 1.5 x 0.30	137 7	1	Sa?		14.9		
09285	14 26.5 + 03 22 351.36 + 56.54	01 – 37 – 016	1418 327	1.3 x 0.3 1.3 x 0.4	87 6 – 7	1	Sc?		15.3		
09286	14 26.5 + 11 25 002.83 + 62.10	02 – 37 – 015	0065 325	1.1 x 0.5 1.1 x 0.6	146 5	3	S B b		15.2		
09287	14 26.5 + 29 12 043.57 + 68.50	05 – 34 – 050 IC 4442	0070 326	1.1 x 0.6 1.1 x 0.6	25 4 – 5	3	S B a		15.2		*1V
09288	14 26.6 + 14 05 007.58 + 63.66	02 – 37 – 15a	0065	(1.1 x 1.1) 1.2 x 1.2	--- 1	3	SO		14.4		
09289	14 26.6 + 26 15 035.60 + 68.23	04 – 34 – 035 IC 1020	0061 324	1.0 x 0.25 1.2 x 0.3	176 7	3	SO		15.2		*1
09290	14 26.6 + 30 38 047.48 + 68.46	05 – 34 – 051 NGC 5639	0070 326	1.4 x 0.9 1.4 x 0.9	98 4	4	Sc		14.6		13
09291	14 26.6 + 39 13 069.48 + 66.54	07 – 30 – 015	0145	2.8 x 1.5 2.8: x 1.5:	115 5	4	Sc	Sc +	14.0		
09292	14 26.7 + 02 30 350.40 + 55.86	00 – 37 – 011	1440	1.6: x 0.25 1.5 x 0.25	16 7	1	S ...		15.2		
09293	14 26.7 + 23 25 028.23 + 67.62	04 – 34 – 037 NGC 5637	0061 335	1.0 x 0.5 0.9 x 0.45	10 5	1	S ... Peculiar		14.6		*3
09294	14 26.7 + 25 46 034.32 + 68.13	04 – 34 – 036	0061 324	1.3 x 0.8 1.3 x 0.8	5 4	3	S B b	SAB(s)bc Sb +	15.1		
09295	14 26.7 + 70 10 111.56 + 45.00	12 – 14 – 005	1442	0.35 x 0.22 0.35 x 0.2	90 ...	1	... Peculiar, nf of 2		14.5		*1C
09296	14 26.9 + 20 53 022.05 + 66.78	04 – 34 – 038 IC 1021	0061	1.1 x 0.8 1.2 x 0.9	130: 3	3	S B a		15.2		1
09297	14 26.9 + 69 56 111.36 + 45.20	12 – 14 – 006 NGC 5671	1442	1.7 x 1.2 1.5 x 1.0	45: 3:	4	S B b Singular	S B(r)b	14.4		*3
09298	14 27 + 84 42 120.41 + 32.20	14 – 07 – 006	1363	1.2: x 0.9: 0.9: x 0.8:	(125) ...	2	Dwarf:		16.5:		
09299	14 27.0 + 00 11 347.92 + 54.03	00 – 37 – 012	1440	1.7 x 0.9 1.7 x 0.8	85 5	3	Sc	Sc +	14.7		*
09300	14 27.0 + 29 03 043.17 + 68.39	05 – 34 – 055 NGC 5641	0070 326	2.6 x 1.4 2.8 x 1.5	158 5	4	S B b	S B(r)ab Sb –	13.6		*12 48
09301	14 27.0 + 30 15 046.43 + 68.39	05 – 84 – 052 NGC 5642	0070 326	(1.8 x 1.3) (3. x 2.)	(130) ...	3	E f of 2		14.3		*3
09302	14 27.0 + 32 00 051.16 + 68.28	05 – 34 – 054	0070	1.0 x 0.2 1.1 x 0.15	7 7	1	S – Irr		15.6		
09303	14 27.0 + 33 07 054.14 + 68.14	06 – 32 – 043	0127	1.0 x 0.2 1.0 x 0.2	69 6	1	S ...		16.0:		
09304	14 27.1 + 03 29 351.70 + 56.53	01 – 37 – 017 NGC 5636	1418 327	1.5: x 1.1 1.5 x 1.1	40 3:	3	S B a NGC 5638 group	S B(r)0/a	14.6		*12 36
09305	14 27.1 + 04 16 352.67 + 57.11	01 – 37 – 000	1418	1.1 x 0.2 0.8 x 0.2	164 ...	2	Sc – Irr		16.0:		
09306	14 27.1 + 31 03 048.59 + 68.33	05 – 34 – 056 IC 4447	0070 326	1.1 x 0.7 1.3 x 0.9	175 4	2	SO – a		14.7		
09307	14 27.1 + 42 03 075.63 + 65.23	07 – 30 – 016	0145	1.0 x 0.4 1.0 x 0.4	20 6	1	S ...		15.3		
09308	14 27.2 + 03 27 351.70 + 56.49	01 – 37 – 018 NGC 5638	1418 327	(2.3 x 2.1) (2.5 x 2.2)	(150) ...	4	E Brightest in cluster		12.5	+ 1677 + 1663	*12 3468
09309	14 27.2 + 10 48 002.04 + 61.58	02 – 37 – 000	0065 325	1.0 x 0.4 1.0 x 0.4	50 6	2	Sb – c		15.3		
09310	14 27.5 + 03 26 351.78 + 56.43	01 – 37 – 000	1418 327	2.2: x 0.6 2.2: x 0.6:	163 6	3	... NGC 5638 group	S B dm	15.3		*

N G C   5627 – 29,   5635 – 39,   5641,   5642,   5671   I C   1017,   1020,   1021,   4442,   4447

1	2	3	4	5	6	7	8	9	10	11	12
09311	14 27.5 + 03 59 352.45 + 56.83	01 – 37 – 000 IC 1022	1418	1.2 x 0.4 1.0 x 0.3	161 6	1	S ...		15.3		
09312	14 27.5 + 35 40 060.71 + 67.52	06 – 32 – 045 NGC 5646	0127	1.6 x 0.45 1.6 x 0.45	81 6	1	S B:b		15.2		
09313	14 27.5 + 38 53 068.54 + 66.51	07 – 30 – 017	0145	1.1 x 0.4 1.1 x 0.4	120 6	2	Sb – c		15.5		
09314	14 27.5 + 45 18 081.97 + 63.48	08 – 26 – 037	0120	1.3 x 0.45 1.3 x 0.5	148 6	1	S ...		15.1		*
09315	14 27.7 + 35 47a 060.98 + 67.45	06 – 32 – 046	0127	1.0 x 0.2 1.0 x 0.2	70 6 – 7	1	S ... npp of 2		(15.4)		*1
09316	14 27.9 + 21 59 024.89 + 66.94	04 – 34 – 000	0061 335	1.8 x 0.5 0.9 x 0.5	28 ...	1	SO?		15.6		
09317	14 27.9 + 27 45 039.69 + 68.12	05 – 34 – 057	0070 324	1.3 x 1.1 1.1 x 1.1	— 1	4	S B c	S B(rs)c	15.1		
09318	14 27.9 + 31 26 049.59 + 68.14	05 – 34 – 058 NGC 5653	0070	1.8 x 1.5 1.9 x 1.6	125 2	1	S ... Peculiar		12.7	+ 3557 + 3648	*12 3468
09319	14 27.9 + 36 35 062.94 + 67.19	06 – 32 – 050 NGC 5654	0127	1.6 x 1.0 1.6 x 1.0	145 ...	1	... Peculiar, double nucleus		14.1		3
09320	14 27.9 + 37 05 064.17 + 67.05	06 – 32 – 049	0127	1.2 x 1.2 1.2 x 1.1	— 1	4	...	S B dm	15.5		
09321	14 28.0 + 12 09 004.56 + 62.26	02 – 37 – 016 NGC 5644	0065 325	(1.2 x 1.2) (1.8 x 1.8)	— 1	3	SO Brightest in cluster	P w 9329	14.1		*
09322	14 28.0 + 23 17 028.10 + 67.30	04 – 34 – 041	0061	1.0 x 0.8 1.0 x 0.8	65 2	3	Sb/S B c	SAB(s)bc Sb +	15.1		*1
09323	14 28.0 + 40 09 071.33 + 65.92	07 – 30 – 019	0145	1.0 x 0.15 1.0 x 0.15	49 7	2	Sb – c		16.0:		
09324	14 28.0 + 44 40 080.67 + 63.76	08 – 26 – 000	0120	2.6: x 1.5: 2.0: x 1.5:	... ...	4	...	S B m	15.7		*
09325	14 28.0 + 49 50 089.56 + 60.66	08 – 26 – 039 NGC 5660	0120	3.2: x 2.8: 2.8: x 2.5:	(90) 1	4	Sc sff of 2, p w 9366	SAB(rs)cd	12.2		*12 3468
09326	14 28.2 + 00 28a 348.62 + 54.06	00 – 37 – 000	1440	1.0 x 0.2 0.8 x 0.2	140 ...	3	... strongly disrupted	P w 9327,	15.5		*
09327	14 28.2 + 00 28b 348.62 + 54.06	00 – 37 – 000	1440	1.3 x 0.6 1.1 x 0.5	172 5:	3	S ... P w 9326, disrupted				*
09328	14 28.2 + 07 30 357.35 + 59.22	01 – 37 – 019 NGC 5645	1418	3.2: x 1.7 2.8: x 1.6:	80 5:	4	...	S B dm	12.8		*12 3468
09329	14 28.2 + 12 06a 004.53 + 62.19	02 – 37 – 017 NGC 5647	0065 325	1.3 x 0.25 1.0 x 0.25	0 6	1	Sa p of 2, p w 9321		(15.3)		*1
09330	14 28.2 + 14 15 008.37 + 63.43	02 – 37 – 019 NGC 5648	0065	1.0 x 0.7 1.0 x 0.7	172: ...	1	S ... P w 9333		14.1		*1
09331	14 28.2 + 54 12 095.72 + 57.64	09 – 24 – 009 IC 1027	0715 328	1.0 x 0.8 (1.0 x 0.8)	(170) ...	1	...		15.4		*1
09332	14 28.3 + 35 32 060.27 + 67.40	06 – 32 – 053 NGC 5656	0127	1.8 x 1.4 1.8 x 1.5	50 2	2	Sa – b		12.7		*3
09333	14 28.4 + 14 11 008.31 + 63.36	02 – 37 – 020 NGC 5649	0065	0.9 x 0.8 0.9 x 0.8	— 1	1	Sc? P w 9330		14.0		*13
09334	14 28.5 + 06 12 355.65 + 58.26	01 – 37 – 020 NGC 5652	1418	2.1 x 1.5 2.1 x 1.5	117 3	4	Sb	SAB bc Sb +	13.8		*13
09335	14 28.5 + 29 24 044.13 + 68.07	05 – 34 – 060 NGC 5657	0070 326	1.9 x 0.7 2.0 x 0.7	163 6	3	S B: b		14.4		1
09336	14 28.5 + 40 15 071.46 + 65.80	07 – 30 – 020	0145	1.2 x 0.25 1.1 x 0.25	1 7	2	Sa – b		15.5		*1
09337	14 28.5 + 41 02 073.18 + 65.46	07 – 30 – 021	0145	1.0 x 0.5 1.0 x 0.5	123 5	1	S ...		15.6		
09338	14 28.7 + 05 31 354.80 + 57.74	01 – 37 – 000	1418	1.1 x 0.4 1.2 x 0.5	49 6	2	Sb sf of 2		15.2		*
09339	14 28.8 + 08 10 358.51 + 59.56	01 – 37 – 021	1418	1.1 x 0.5 0.9 x 0.5	88 5:	1	S ...		14.9		
09340	14 28.8 + 25 43 034.40 + 67.66	04 – 34 – 043	0061 324	1.0 x 0.9 1.0 x 0.9	— 1	3	S B c	Sc +	15.3		*1
09341	14 28.9 + 03 13 351.98 + 56.03	01 – 37 – 022 IC 1024	1418 327	1.6 x 0.6 1.5 x 0.5	30 6	1	SO?		14.0		3
09342	14 28.9 + 25 35 034.07 + 67.61	04 – 34 – 044 NGC 5659	0061 324	1.7 x 0.45 1.7 x 0.45	43 6	3	Sb		15.0		*3
09343	14 29.0 + 40 56 072.86 + 65.42	07 – 30 – 022	0145	1.2 x 0.25 1.0 x 0.25	87 6 – 7	1	S ...		15.7		
09344	14 29.0 + 59 42 102.02 + 53.45	10 – 21 – 004 NGC 5667	1425 329	1.8 x 0.9 1.8 x 0.9	[168] ...	3	... Peculiar		13.1		13
09345	14 29.2 + 06 22 356.10 + 58.25	01 – 37 – 000	1418	1.2 x 0.2 1.0 x 0.15	142 7	3	Sc	Sc +	15.4		

N G C   5644 – 49, 5652 – 54, 5656, 5657, 5659   I C   1022, 1024, 1027
5660, 5667

1	2	3	4	5	6	7	8	9	10	11	12
09346	14 29.5 + 06 28 356.34 + 58.27	01 − 37 − 023 NGC 5661	1418	1.6 x 0.6 1.7 x 0.6	23 ...	2	S B b: nf of 2, contact, disturbed		14.2		*13
09347	14 29.8 + 50 10 089.69 + 60.22	08 − 26 − 041 NGC 5673	0120	2.6 x 0.5 2.5 x 0.4	136 6	4	Sc P w 9361		14.0		*23
09348	14 29.9 + 00 30 349.21 + 53.82	00 − 37 − 016	1440	1.8 x 0.3 1.6 x 0.25	119 7	1	S ... nnp of 2		14.6		*1
09349	14 29.9 + 28 46 042.47 + 67.74	05 − 34 − 065 IC 4450	0070 326	1.2 x 0.5 1.1 x 0.5	48 6	1	S ...		15.4		
09350	14 29.9 + 36 31b 062.49 + 66.83	06 − 32 − 060	0127	1.0 x 0.2 1.0 x 0.2	31 6 − 7	1	S ... f of 2, interaction:		(15.3)		*1 (V)
09351	14 30.0 + 05 54 355.74 + 57.78	01 − 37 − 000	1418	1.1 x 0.1 1.0 x 0.1	66 7	2	... d − m		17.		*
09352	14 30.0 + 08 18 359.09 + 59.43	01 − 37 − 024 NGC 5665	1418 330	2.3 x 1.4 2.2 x 1.4	145 4	2	S[c] Peculiar		12.6	+ 2249 + 2256	*12 3468A
09353	14 30.3 + 10 06 001.92 + 60.55	02 − 37 − 021 NGC 5669	0065 330	4.5 x 3.3 4.3: x 3.3:	50 3	5	Sc Sc +	SAB(rs)c	13.2		*12 3468
09354	14 30.4 + 31 53 050.67 + 67.57	05 − 34 − 068 NGC 5672	0070	0.9 x 0.6 0.9 x 0.7	50 ...	1	... Peculiar?		14.5	+ 3701 + 3796	*12 3
09355	14 30.4 + 79 28 117.39 + 36.81	13 − 10 − 020	1374	1.2 x 0.25 1.2 x 0.25	40 7	1	S ...		14.9		
09356	14 30.5 + 11 49 004.77 + 61.57	02 − 37 − 022	0065	1.6 x 0.8 1.7 x 0.8	105 5	1	S ...		14.3		
09357	14 30.6 + 36 31 062.39 + 66.69	06 − 32 − 062 NGC 5675	0127	2.6 x 1.0 2.9 x 1.1	137 6	1	S ...		14.0		3
09358	14 30.6 + 58 09 100.06 + 54.50	10 − 21 − 005 NGC 5678	1425	3.5: x 1.7 3.3: x 1.7	5 5	3	Sb Disturbed?		12.1	+ 2300 + 2472	*12 3468
09359	14 30.7 − 00 56 347.95 + 52.58	00 − 37 − 000	1440	1.5 x 0.15 1.1: x 0.2	175 7	2	... d − m		15.6		
09360	14 30.7 + 10 43 003.03 + 60.86	02 − 37 − 023 NGC 5666	0065	0.9 x 0.7 0.9 x 0.7	(155) ...	1	... Very compact		13.5		3
09361	14 30.7 + 50 07 089.43 + 60.15	08 − 26 − 042 IC 1029	0120	2.8 x 0.5 2.8 x 0.5	152 6	3	Sb P w 9347		13.7		*12
09362	14 30.8 + 04 07 353.71 + 56.37	01 − 37 − 027	1418 327	1.1 x 1.0 1.1 x 1.1	−− 1	2	Sb − c sf of 2		14.9		*1
09363	14 30.9 + 04 40 354.43 + 56.75	01 − 37 − 028 NGC 5668	1418	3.5 x 3.0 3.3 x 3.0	−− 1	4	Sc	SA d	12.7	+ 1737 + 1732	*12 3468
09364	14 30.9 + 07 05 357.65 + 58.44	01 − 37 − 029	1418	1.8 x 0.8 1.8 x 0.8	20 5 − 6	3	...	S B dm	15.0		
09365	14 31.0 + 03 54 353.50 + 56.18	01 − 37 − 030	1418 327	1.1 x 0.9 1.0 x 0.7	(175) 2:	2	Sb − c		15.0		*
09366	14 31.0 + 49 40 088.66 + 60.40	08 − 26 − 043 NGC 5676	0120	4.0 x 1.7 3.8: x 1.5	47 6	4	Sc P w 9325, Sc −		11.7	+ 2244 + 2395	*12 3468
09367	14 31.2 + 34 57 058.46 + 66.95	06 − 32 − 065	0127	1.0 x 0.6 1.0 x 0.6	40 4:	2	S0 − a		15.6		*1
09368	14 31.3 + 41 52 074.40 + 64.62	07 − 30 − 025 IC 1028	0145	1.2 x 0.6 1.1 x 0.4	15 5	1	S ...		14.8		
09369	14 31.4 + 05 40 355.89 + 57.37	01 − 37 − 031 NGC 5674	1418	1.2 x 1.1 1.2 x 1.1	−− 1	3	Sb/S B c Sc −		13.7		13
09370	14 31.4 + 53 14 093.80 + 57.98	09 − 24 − 011	0715 328	1.0 x 0.5 1.0 x 0.5	12 5	3	Sb np of 2		16.0:		*
09371	14 31.5 + 04 00 353.79 + 56.17	01 − 37 − 032	1418 327	[1.0 x 0.9] ...	... ...	2	Double system Contact		15.0		*1
09372	14 31.5 + 36 10 061.42 + 66.61	06 − 32 − 067	0127	1.0 x 0.9 0.8 x 0.7	−− 1 − 2	3	S B b		15.5		
09373	14 31.6 + 41 01 072.54 + 64.94	07 − 30 − 026	0145	1.0 x 0.8 1.0 x 0.6	140: ...	1	S ...		16.0:		1
09374	14 31.7 + 10 26 002.89 + 60.49	02 − 37 − 000	0065 330	1.1 x 0.7 1.1 x 0.5	132 ...	1	S ...		15.4		
09375	14 31.7 + 28 38 042.17 + 67.34	05 − 34 − 000	0070 326	1.0 x 0.2 1.0 x 0.2	169 7	2	Sb		16.0:		*
09376	14 31.8 + 40 17 070.92 + 65.21	07 − 30 − 027	0145	1.6: x 0.8 1.7 x 0.9	140: 5	2	S ... ssp of 2, disturbed		15.0		*1
09377	14 31.8 + 44 18 079.17 + 63.37	07 − 30 − 000	0145	1.0 x 0.1 0.8 x 0.1	158 7	1	Sc		16.5:		
09378	14 32.0 + 25 41 034.64 + 66.94	04 − 34 − 046 NGC 5677	0061 324	1.0 x 0.7 1.0 x 0.8	135: 2:	1	S ... Singular?		14.8		*3
09379	14 32.0 + 40 28 071.28 + 65.10	07 − 30 − 029	0145	1.7 x 0.9 1.6 x 0.9	145 5	3	Sc		15.6		*
09380	14 32.1 + 04 29 354.59 + 56.41	01 − 37 − 000	1418	2.1 x 1.1 1.7: x 1.0:	... ...	4	Dwarf irregular Ir IV − V		15.3		

N G C 5661, 5665, 5666, 5668, 5669, 5672 − 78   I C 1028, 1029, 4550

1	2	3	4	5	6	7	8	9	10	11	12
09381	14 32.5 + 36 30 062.10 + 66.33	06 - 32 - 000	0127	1.5:: x 1.3:: 1. x 1.	--- ...	4	Dwarf spiral S V		17.		
09382	14 32.6 + 03 45 353.84 + 55.79	01 - 37 - 000	1418 327	1.3 x 0.2 1.3 x 0.20	166 7	1	Sa - b		15.2		
09383	14 32.6 + 05 34 356.14 + 57.09	01 - 37 - 035 NGC 5679	1418	1.1 x 0.6 1.2 x 0.6	127 5	3	Sb 2nd of 3, contact		(14.2)		*13 A
09384	14 32.8 + 26 46 037.45 + 66.93	05 - 34 - 77 +78 IC 4461 + 4462	70 324	[1.0 x 0.9] ...	... ...	3	Double system Bridge:, in group of 3		(15.3)		*1V A
09385	14 32.9 + 05 30 356.15 + 56.99	01 - 37 - 037	1418	1.6:: x 1.5:: 0.9: x 0.9:	---- ...	4	Dwarf V		17.		
09386	14 32.9 + 40 58 072.20 + 64.74	07 - 30 - 030	0145	1.2 x 0.7 1.2 x 0.7	140 4:	2	S B a - b		14.9		
09387	14 33.0 + 23 11 028.64 + 66.16	04 - 34 - 000	0061 335	1.0 x 0.5 1.0 x 0.5	170 5	2	S0		15.5		*
09388	14 33.0 + 48 53 087.01 + 60.63	08 - 27 - 002 NGC 5682	1368	2.0 x 0.4 1.8 x 0.4	127 6	2	S B b disturbed, NGC 5689 group	P w NGC 5683,	15.1 + 2288 + 2438		*12 3
09389	14 33.1 + 13 07 007.78 + 61.81	02 - 37 - 024	0065	2.2 x 0.6 2.2 x 0.6	48 6	2	S B b P w 9394		14.6		*
09390	14 33.2 + 22 37 027.33 + 65.96	04 - 34 - 000	0061 335	1.0 x 0.05 0.7 x 0.05	30 7	1	...	d - m	17. - 18.		
09391	14 33.2 + 59 34 101.20 + 53.21	10 - 21 - 011	1425 329	1.8 x 1.1 1.5: x 1.0	20: ...	3	...	S B dm	15.5		*
09392	14 33.3 + 03 15 353.45 + 55.31	01 - 37 - 000	1418	1.4 x 0.2 1. x 0.2	145 7	3	...	m	17.		
09393	14 33.3 + 08 31 000.46 + 58.96	02 - 37 - 025 NGC 5681	0065 330	0.9 x 0.6 0.9 x 0.6	5 3	1	S ...		14.3		3
09394	14 33.3 + 13 23 008.30 + 61.92	02 - 37 - 026	0065	2.3 x 0.6 2.3 x 0.6	46 6	2	Sc: + superimposed companions, p w 9389		14.9		*1V
09395	14 33.3 + 54 42 095.39 + 56.76	09 - 24 - 020 NGC 5687	0715	2.3 x 1.7 (2.7 x 1.9)	105 3	4	S0		13.3 + 2119 + 2284		*123 4678
09396	14 33.5 + 24 56 032.94 + 66.47	04 - 34 - 047	0061 335	1.0 x 0.9 1.0 x 0.9	-- 1	3	S B a		15.0		*1
09397	14 33.5 + 37 01 063.20 + 66.00	06 - 32 - 071	0127	1.3 x 0.4 1.6 x 0.5	28 6	2	S0		15.3		*
09398	14 33.7 - 01 00 348.64 + 51.92	00 - 37 - 018	1440	1.1: x 0.8 0.6 x 0.6	... ...	2	S B b - c		15.7		1
09399	14 33.7 + 48 57 086.97 + 60.49	08 - 27 - 004 NGC 5689	1368	4.0 x 1.1 4.2 x 1.1	85 6	3	S B a Brightest in group		12.7 + 2205 + 2355		*12 3468
09400	14 33.8 + 05 33 356.50 + 56.86	01 - 37 - 038	1418	(1.1 x 0.8) (1.1 x 0.9)	-- ...	2	S0		14.9		
09401	14 33.8 + 22 01 026.04 + 65.65	04 - 34 - 048	0061 335	1.7 x 0.7 1.5 x 0.6	85 6	1	S ...		14.6		*1
09402	14 33.8 + 36 45 062.52 + 66.01	06 - 32 - 073 NGC 5684	0127	(1.9 x 1.6) (2.1 x 1.7)	(105) 1 - 2	4	S0		14.2		*3
09403	14 34.0 + 30 07 046.02 + 66.88	05 - 34 - 081 NGC 5685	0070 326	1.1 x 1.0 1.0 x 1.0	-- ...	2	[E] Very compact		14.9		
09404	14 34.0 + 56 02 096.96 + 55.75	09 - 24 - 000	0715	1.0 x 0.8 1.0 x 0.8	(10) ...	2	S B c		16.5:		*
09405	14 34.0 + 57 28 098.69 + 54.71	10 - 21 - 013	1425	2.5:: x 1.5:: 1.7:: x 1.0::	[145] ...	4	Dwarf V		17.		*
09406	14 34.4 + 48 47 086.56 + 60.50	08 - 27 - 006 NGC 5693	1368	1.7 x 1.7 1.7 x 1.5	-- 1	3	S B c NGC 5689 group	S B d	14.5		*12 36
09407	14 34.6 + 41 54 073.84 + 64.06	07 - 30 - 031 NGC 5697 = 4471*	0145	1.1 x 0.7 1.1 x 0.7	25 ...	1	... Peculiar		14.6		*3
09408	14 34.7 + 41 18 072.58 + 64.29	07 - 30 - 032	0145	1.1 x 0.3 0.7 x 0.15	31 6 - 7	1	S? ...		16.0:		
09409	14 34.7 + 41 22 072.72 + 64.27	07 - 30 - 033	0145	1.0 x 0.25 0.9 x 0.2	87 7	1	S ...		15.6		
09410	14 34.8 + 08 51 001.42 + 58.89	02 - 37 - 000	0065 330	1.4 x 0.4 1.1 x 0.4	135 6	2	Sb - c		15.3		*
09411	14 34.8 + 10 13 003.50 + 59.76	02 - 37 - 027	0065 330	1.0 x 0.7 1.0 x 0.7	65 3:	2	Sa - b np of 2		14.9		*
09412	14 34.9 + 59 01 100.32 + 53.48	10 - 21 - 000	1425	0.7 x 0.7 0.7 x 0.6	-- ...	2	... Very compact		14.3		
09413	14 34.9 + 78 06 116.31 + 37.89	13 - 11 - 001	0768	1.0 x 0.7 1.0 x 0.7	120 3:	2	Sb - c ssp of 2		14.8		*1
09414	14 35.0 + 18 28 018.50 + 64.08	03 - 37 - 026 IC 4469	1417	1.6 x 0.2 1.6 x 0.15	110 7	3	Sc		15.6		
09415	14 35.0 + 42 02 074.04 + 63.93	07 - 30 - 036 NGC 5696	0145	2.0 x 1.5 1.8 x 1.5	(45) 2	3	Sb - c		14.1		3

N G C   5679, 5681, 5682, 5684, 5685, 5687   I C   4461, 4462, 4469, 4471
          5689, 5693, 5696, 5697

1	2	3	4	5	6	7	8	9	10	11	12
09416	14 35.2 + 02 28 353.12 + 54.42	00 – 37 – 019 NGC 5690	1440 333	3.6 x 1.1 3.6 x 1.1	143 6	3	Sc		13.1		*123 4678
09417	14 35.2 + 22 11 026.66 + 65.39	04 – 34 – 000	0061 335	1.2 x 0.15 0.7 x 0.15	115 7	1	S ...		16.5:		*
09418	14 35.2 + 25 59 035.69 + 66.28	04 – 34 – 049	0061	1.0 x 1.0 1.0 x 1.0	— 1	3	S B a	(R)S B a	15.2		1
09419	14 35.2 + 38 40 066.76 + 65.18	07 – 30 – 038 NGC 5698	0145	1.9 x 1.0 1.9 x 1.0	70 5	4	S B b	(R')S B(s)b	14.0		3
09420	14 35.3 – 00 12 350.07 + 52.33	00 – 37 – 020 NGC 5691	1440	(1.9 x 1.6) (2.1 x 1.7)	(110) ...	3	S ... Peculiar, NGC 5713 group		12.9		*12 3478
09421	14 35.3 + 36 46 062.37 + 65.72	06 – 32 – 077 NGC 5695	0127	1.6 x 1.1 1.5 x 1.1	150 3	1	S ...		13.9		3
09422	14 35.3 + 43 55 077.74 + 63.02	07 – 30 – 039	0145	2.0 x 0.2 1.7 x 0.2	161 7	2	Sc		15.4		*
09423	14 35.3 + 48 45 086.33 + 60.40	08 – 27 – 007 NGC 5700	1368	1.0 x 0.4 1.0 x 0.4	38 6	1	S B ... NGC 5689 group		15.2		1
09424	14 35.6 + 21 35 025.37 + 65.11	04 – 34 – 050	0061	1.1 x 0.6 0.8 x 0.3	0: 5:	1	Sc		15.6		*
09425	14 35.6 + 30 42 047.48 + 66.53	05 – 34 – 083	0070 326	[1.0 x 0.4] ...	... ...	2	Double system Connected, plumes		15.0		*1V A
09426	14 35.7 + 48 49 086.36 + 60.31	08 – 27 – 008	1368	2. x 1. 1. x 1.	... ...	3	Dwarf V, NGC 5689 group		17.		
09427	14 35.8 + 03 37 354.69 + 55.14	01 – 37 – 039 NGC 5692	1418 332	1.0 x 0.6 1.0 x 0.6	35 ...	1	... Peculiar		13.3		
09428	14 35.9 + 51 48 090.91 + 58.43	09 – 24 – 023 NGC 5707	0715 328	2.6 x 0.4 2.7 x 0.4	35 7	2	Sa – b		13.3		*2
09429	14 36.0 + 40 19 070.26 + 64.46	07 – 30 – 042	0145	1.1 x 0.9 1.2 x 0.9	90: ...	2	S ... Singular, distorted?		14.8		*1
09430	14 36.3 + 40 40 070.96 + 64.27	07 – 30 – 044 NGC 5708	0145	1.7 x 0.6 1.6 x 0.5	177 6	2	S – Irr		13.9		*3
09431	14 36.4 + 46 51 082.91 + 61.34	08 – 27 – 011 NGC 5714	1368	3.0 x 0.35 3.0 x 0.35	82 7	3	Sc		14.2		*13
09432	14 36.5 + 03 09 354.34 + 54.69	01 – 37 – 040	1418 332	1.6:: x 1.5:: 1. x 1.	— ...	4	Dwarf V		15.4		*
09433	14 36.5 + 28 43 042.53 + 66.30	05 – 35 – 001 IC 4479	1390 326	1.5 x 1.5 1.2 x 1.1	— 1	3	Sc Disturbed:, nf of 2		14.8		*1
09434	14 36.6 + 20 43 023.63 + 64.60	04 – 35 – 002 NGC 5702	0102	1.1 x 0.8 1.1 x 0.8	150 3	3	S0		14.5		*3
09435	14 36.6 + 30 39 047.34 + 66.31	05 – 35 – 003 NGC 5709	1390	1.6 x 0.45 1.5 x 0.45	105 6	2	S B a sf of 2		14.5		*1
09436	14 36.7 + 05 35 357.46 + 56.37	01 – 37 – 042 NGC 5701	1418	4.5 x 4.5 4.6 x 4.6	— 1	4	S B 0	(R)S B 0/a	12.9		*12 3468
09437	14 36.8 + 18 55 019.84 + 63.88	03 – 37 – 030	1417 334	1.0 x 0.5 0.7 x 0.3	3 5	1	S ...		16.0:		
09438	14 36.8 + 31 10 048.62 + 66.25	05 – 35 – 005	1390 326	1.1 x 0.4 0.9 x 0.4	175 6	2	S0		15.3		
09439	14 36.8 + 51 04 089.65 + 58.79	09 – 24 – 025 NGC 5720	0715	2.2 x 1.5 1.3 x 1.0	140: 3	3	Sb		14.7		
09440	14 36.9 + 20 15 022.68 + 64.37	03 – 37 – 032 NGC 5710	1417	(1.0 x 0.9) (1.4 x 1.1)	— ...	3	E P w 9445		14.5		*3
09441	14 36.9 + 41 14 072.05 + 63.94	07 – 30 – 045 Mark 476	0145	1.4 x 0.25 1.1 x 0.25	33 7	2	Sa		14.6		
09442	14 36.9 + 52 50 092.20 + 57.64	09 – 24 – 000	0715	1.0 x 0.8 1.0 x 0.6	(70) ...	2	S B b		16.0:		
09443	14 37.0 + 09 27 002.99 + 58.86	02 – 37 – 000	0065 330	1.5: x 0.5 0.9 x 0.3	80 6	1	S ...		15.4		
09444	14 37.0 + 17 14 016.48 + 63.11	03 – 37 – 031	1417	(1.3 x 1.1) (1.3 x 1.1)	— (1)	1	S0?		14.8		
09445	14 37.0 + 20 12 022.59 + 64.33	03 – 37 – 033 NGC 5711	1417	1.1 x 0.6 1.2 x 0.5	73 5 – 6	1	S ... P w 9440		15.1		*3
09446	14 37.1 + 38 28 066.03 + 64.89	06 – 32 – 081	0127	1.2 x 0.5 1.0 x 0.4	28 6	3	Sc	Sc +	17.		
09447	14 37.2 – 00 32 350.40 + 51.85	00 – 37 – 021 NGC 5705	1440	2.9 x 1.8 2.9 x 1.9:	75 4	4	S B c S B d NGC 5713 group		14.5		*2
09448	14 37.2 + 51 22 090.03 + 58.56	09 – 22 – 028	0715 328	1.7 x 0.2 1.7 x 0.2	88 7	3	Sb		15.3		*
09449	14 37.4 + 24 45 032.95 + 65.56	04 – 35 – 000	0102 335	1.0 x 0.15 1.1 x 0.2	75 6 – 7	1	Sb – c		15.4		
09450	14 37.5 + 23 37 030.31 + 65.28	04 – 35 – 000	0102 335	1.0 x 0.12 1.0 x 0.1	50 7	1	Sc – Irr Disturbed?		17.		*

N G C   5690 – 92, 5695, 5698, 5700 – 02
5705, 5707 – 11, 5714, 5720

I C   4479

1	2	3	4	5	6	7	8	9	10	11	12
09451	14 37.6 - 00 05 351.01 + 52.12	00 - 37 - 022 NGC 5713	1440	3.3 x 2.5 3.0: x 2.6:	(10) 2 - 3	3	Sc Singular, brightest in group		11.7	+ 1948 + 1930	*123 46̲7̲8
09452	14 37.6 + 54 08 093.86 + 56.70	09 - 24 - 029	0715	1.4 x 0.5 1.2: x 0.5:	145 (6)	3	Dwarf (spiral), S: m S: IV - V		16.5:		
09453	14 37.6 + 79 07 116.81 + 36.95	13 - 11 - 000	0768	1.3 x 0.35 1.3 x 0.35	87 6	2	Sb		16.0:		*
09454	14 37.9 + 06 31 359.07 + 56.78	01 - 37 - 044	1418	1.4: x 1.1: 1.2 x 1.0	--- 2	2	Sc		14.9		
09455	14 38.0 + 16 54 016.07 + 62.75	03 - 37 - 035	1417	1.6 x 0.5 1.5 x 0.4	26 6	3	Sb		15.0		
09456	14 38.0 + 42 58 075.38 + 63.03	07 - 30 - 046 NGC 5730:	0145	1.9 x 0.4 1.9 x 0.4	88 (7)	3	Irr P w 9460		14.7		*13
09457	14 38.1 + 03 40 355.46 + 54.78	01 - 37 - 046 IC 1042	1418 332	(1.1 x 1.1) (1.2 x 1.2)	--- ...	1	SO? P w 9459, contact, disturbed		14.9		*1
09458	14 38.1 + 24 56 033.46 + 65.45	04 - 35 - 000	0102 335	1.0 x 0.4 1.0 x 0.3	20 6	2	SO - a		15.5		
09459	14 38.2 + 03 40 355.49 + 54.76	01 - 37 - 047 NGC 5718	1418 332	(1.4 x 1.1) (1.6 x 1.3)	... ...	1	E - SO P w 9457, contact, disturbed		14.6		*1
09460	14 38.3 + 43 00 075.39 + 62.96	07 - 30 - 047 NGC 5731:	0145	1.7 x 0.4 1.7 x 0.5	116 6	1	S ... P w 9456		14.0		*13
09461	14 38.3 + 62 13 103.22 + 50.79	10 - 21 - 021 IC 1049	1425	1.2 x 0.8 0.9 x 0.8	(65) ...	3	S B a		14.8		
09462	14 38.4 - 00 07 351.21 + 51.96	00 - 37 - 024 NGC 5719	1440	3.2: x 1.2 3.4 x 1.3	107 6	4	Sb S ab Sb -		13.8		*12̲ 3
09463	14 38.4 + 03 21 355.17 + 54.50	01 - 37 - 000	1418 332	1.1 x 0.25 0.9 x 0.2	42 7	2	SO - a		15.2		
09464	14 38.4 + 31 35 049.61 + 65.89	05 - 35 - 006	1390 326	1.0 x 0.4 0.9 x 0.2	62 7	2	SO - a		14.9		
09465	14 38.4 + 34 12 055.96 + 65.63	06 - 32 - 083 NGC 5727	0127	2.3 x 1.2 2.3 x 1.2	135 5	4	... SAB dm		14.6		
09466	14 38.5 + 02 22 354.03 + 53.78	00 - 37 - 025 NGC 5725	1440 333	1.1 x 0.8 1.0 x 0.8	40: 2 - 3	1	S B c?		14.5		3
09467	14 38.6 + 38 51 066.67 + 64.49	07 - 30 - 048 NGC 5732	0145	1.5 x 0.8 1.5 x 0.8	40 5	2	Sb - c		14.4		3
09468	14 38.8 + 10 15 004.74 + 59.00	02 - 37 - 000	0065 330	1.0 x 0.1 0.7 x 0.1	176 7	1	S ... nnf of 2		16.0:		*
09469	14 39.1 - 01 37 349.85 + 50.72	00 - 37 - 026	1440	1.6:: x 1.6:: 1.5: x 1.2:	--- ...	4	Dwarf V		15.7		
09470	14 39.2 + 00 52 352.53 + 52.56	00 - 37 - 027	1440 333	1.2 x 0.7 1.2: x 0.7	52 4:	3	... S B(s)m		14.9		
09471	14 39.2 + 06 10 358.99 + 56.31	01 - 37 - 048	1418	1.2 x 0.5 1.1 x 0.5	165 6	4	S B b		14.9		
09472	14 39.4 - 01 13 350.36 + 50.97	00 - 37 - 000	1440	1.1: x 0.5: 0.8: x 0.3:	45 ...	2	... S dm - m		16.5:		
09473	14 39.5 + 39 04 067.01 + 64.26	07 - 30 - 049	0145	1.4 x 1.2 1.4 x 1.2	(130) 1 - 2	1	SO:		13.8		1
09474	14 39.6 + 08 41 002.62 + 57.87	02 - 37 - 030	0065 330	1.1 x 0.9 0.8 x 0.8	--- 1 - 2	2	Sc		15.0		
09475	14 39.6 + 12 17 008.21 + 60.03	02 - 37 - 000	0065	1.1 x 0.15 1.1 x 0.15	65 7	3	Sc		15.7		
09476	14 39.7 + 44 44 078.46 + 61.94	08 - 27 - 019	1368	1.6 x 1.0 1.6 x 1.0	137 4	3	Sc SAB(rs)c nf of 2		13.9		*1
09477	14 39.9 + 59 32 100.11 + 52.67	10 - 21 - 023 IC	1425	1.2:: x 1.2:: 1.2:: x 1.2::	--- ...	3	Dwarf V		17.		
09478	14 40 + 85 31 120.56 + 31.35	14 - 07 - 008	1363	1.1 x 0.10 0.9 x 0.10	172 7	1	Sc - Irr		17.		
09479	14 40.0 + 04 38 357.24 + 55.12	01 - 37 - 049	1418 332	1.5 x 0.5 1.5 x 0.5	27 6	3	Sb		15.2		
09480	14 40.1 + 22 33 028.27 + 64.42	04 - 35 - 003	0102 335	1.0 x 0.5 1.0 x 0.5	80 5	2	S B O - a		14.8		*
09481	14 40.2 + 28 56 043.17 + 65.50	05 - 35 - 007 NGC 5735	1390 326	2.8 x 2.0 2.2: x 1.8:	(40) 3	4	S B b S B(rs)bc Sb +		13.8		*13
09482	14 40.3 + 00 53 352.88 + 52.39	00 - 38 - 000	1613 333	1.2 x 0.15 1.2 x 0.1	66 7	3	Sc Sc +		15.4		
09483	14 40.4 + 05 06 357.95 + 55.36	01 - 37 - 051 IC 1048	1418 332	2.5 x 0.8 2.6 x 0.6	163 7	1	S ...		14.0		1
09484	14 40.5 + 42 42 074.42 + 62.73	07 - 30 - 000	0145	1.0 x 0.12 1.0 x 0.12	50 7	2	Sc		17.		*
09485	14 40.6 + 04 59 357.86 + 55.25	01 - 38 - 000	1421 332	1.6: x 0.4 1.5 x 0.3	148 6	2	...		15.4		*

N G C   5713, 5718, 5719, 5725, 5727, 5730 - 32   I C   1042, 1048, 1049
        5735

1	2	3	4	5	6	7	8 9	10	11	12
09486	14 40.6 + 42 03 073.10 + 62.98	07 – 30 – 052 NGC 5739	0145	(2.0 x 2.0) (3. x 3.)	-- 1	3	S0/Sa Singular	13.7		*12 34$\overline{8}$
09487	14 40.7 + 21 38 026.36 + 64.01	04 – 35 – 000	0102 335	1.4 x 0.4 1.3 x 0.4	5 6	1	Sa – b	15.7		
09488	14 40.8 + 19 05 021.03 + 63.08	03 – 37 – 039 NGC 5737	1417 334	1.4 x 0.9 1.1 x 0.8	170 3 – 4	3	S B b	14.6		3
09489	14 40.8 + 25 23 034.80 + 64.93	04 – 35 – 000	0102 335	[1.1 x 0.25] ...	[48] ...	2	Double system Connected	16.0:		*
09490	14 41.2 + 11 22 007.17 + 59.19	02 – 38 – 000	1087	1.4 x 0.2 1.2 x 0.15	81 7	3	Sc 2nd of 3	15.6		*
09491	14 41.7 + 04 26 357.50 + 54.67	01 – 38 – 001	1421 332	1.3: x 1.1: 1.2: x 1.0:	(50) 1 – 2	1	S ...	15.3		
09492	14 41.7 + 08 10 002.49 + 57.15	01 – 38 – 002	1421	1.2 x 0.5 1.8 x 0.9	50 5	3	Sa	15.0		
09493	14 41.9 + 01 54 354.51 + 52.85	00 – 38 – 003 NGC 5740	1613 (333)	3.2 x 1.8 3.2: x 1.6:	160 4	4	Sb P w 9499 Sb +	13.2		*12– –57$\overline{8}$9
09494	14 41.9 + 20 50 024.86 + 63.48	04 – 35 – 007 IC 1052	0102	1.1 x 0.25 1.1 x 0.25	30 6	1	S B: c:	15.3		*1
09495	14 41.9 + 41 27 071.68 + 62.99	07 – 30 – 053	0145	1.4 x 1.0 1.4 x 1.0	5 3	3	Sc	16.0:		
09496	14 42.0 + 12 21 008.97 + 59.59	02 – 38 – 002 IC 4493	1087	1.0 x 0.9 0.9 x 0.9	-- 1	1	S ... Companion superimposed	14.4		*
09497	14 42.2 + 42 50 074.39 + 62.39	07 – 30 – 055	0145	1.0 x 0.2 1.1 x 0.2	[64] ...	2	... Peculiar	16.0:		*
09498	14 42.3 + 53 37 092.32 + 56.52	09 – 24 – 033 NGC 5751	0715	1.5 x 0.7 1.3 x 0.7	55 5	2	Sc	13.9		3
09499	14 42.5 + 02 10 355.00 + 52.94	00 – 38 – 005 NGC 5746	1613 (333)	7.4 x 1.1 7.0 x 1.2	170 7	4	Sb P w 9493	12.3	+ 1831 + 1826	*12– –9
09500	14 42.8 + 08 05 002.69 + 56.88	01 – 38 – 000	1421	3. x 3. 2. x 2.	-- ...	4	Dwarf spiral S B V	18.		*
09501	14 42.8 + 62 27 102.83 + 50.27	10 – 21 – 029	1425	1.0 x 0.12 1.0 x 0.20	163 7	2	Sa – b	16.5:		*
09502	14 43.0 + 41 06 070.80 + 62.93	07 – 30 – 059	0145	1.0 x 0.12 0.9 x 0.12	75 7	2	Sc In chain of 4	16.5:		*1
09503	14 43.1 + 19 41 022.70 + 62.81	03 – 38 – 007	0054 334	1.5 x 0.4 1.4 x 0.4	88 6	2	Sb + 2 small companions Connected	15.1		*1A
09504	14 43.3 + 31 39 049.66 + 64.85	05 – 35 – 12 + 13	1390	[1.3 x 0.7] ...	... ...	2	Double system Common envelope	15.3		*1
09505	14 43.3 + 38 56b 066.21 + 63.60	07 – 30 – 061 NGC 5754	0145	2.0 x 1.8 2.0 x 1.7:	-- 1	4	S B b S B(rs)b f of 2, contact, 2nd of 4	(14.1)		*13 A
09506	14 43.4 + 31 38 049.62 + 64.83	05 – 35 – 000	1390	1.0 x 0.4 0.7: x 0.3:	[63] ...	4	Dwarf irregular Ir V, 3rd of 3	18.		*
09507	14 43.4 + 38 59b 066.30 + 63.56	07 – 30 – 063 NGC 5755	0145	1.7 x 1.3 0.9 x 0.8	... ...	3	S B ... Distorted, sf of 2, 4th of 4	(15.1)		*13 A
09508	14 43.4 + 49 18 085.69 + 59.00	08 – 27 – 022	1368	1.3 x 0.9: 1.0:: x 0.6::	[165] ...	4	Irr I A m sp of 2	16.0:		*
09509	14 43.5 + 08 43 003.79 + 57.14	02 – 38 – 3 + 4	1087 330	[1.2 x 0.3] ...	[100] ...	2	Double system Contact, in group of 3	15.4	+ 10470 + 10491	*12– V
09510	14 43.5 + 32 50 052.42 + 64.73	06 – 32 – 094	0127	1.1 x 0.3 1.1 x 0.3	62 6	2	Sb – c	16.0:		
09511	14 43.5 + 51 35 089.20 + 57.65	09 – 24 – 000	0715	1.3 x 0.4 1.3 x 0.4	... ...	3	Strongly peculiar Long jet	16.5:		*C
09512	14 43.7 – 00 01 352.90 + 51.15	00 – 38 – 006 NGC 5750	1613 333	(3.0 x 1.7) (3.2 x 1.8)	65 4	4	S B a S B(r)0/a	13.1		*12– 34$\overline{7}$8
09513	14 43.7 + 13 14 010.89 + 59.72	02 – 38 – 000	1087	1.1: x 0.9: 0.9 x 0.6	... ...	1	... P w 9515	15.4		*
09514	14 44.0 + 01 29 354.65 + 52.19	00 – 38 – 000 IC 1054	1613 333	1.0 x 0.6 (1.2 x 0.8)	(10) ...	3	S0	15.2		
09515	14 44.0 + 13 14 010.96 + 59.66	02 – 38 – 000	1087	1.6: x 1.1: 0.7 x 0.5	(155) ...	1	S ... P w 9513	15.4		*
09516	14 44.1 + 50 38 087.66 + 58.14	08 – 27 – 023 IC 1056 = 1057	1368	2.0: x 1.7: 1.6: x 1.0:	(20) ...	2	Sb	14.4		
09517	14 44.3 + 12 49 010.35 + 59.37	02 – 38 – 006	1087	1.0 x 0.6 1.0 x 0.6	50: ...	1	Sb:	15.0		1
09518	14 44.3 + 32 59 052.74 + 64.55	06 – 32 – 096	0127	(1.0 x 0.8) (1.4 x 1.0)	(0) ...	3	E	14.5		
09519	14 44.3 + 34 35 056.43 + 64.37	06 – 32 – 097	0127	0.8 x 0.7 0.7 x 0.6	75 ...	1	S0:	14.4		
09520	14 44.5 + 33 37 054.19 + 64.45	06 – 32 – 099 IC 4505	0127	1.1 x 0.9 1.1 x 0.9	-- ...	2	E – S0 1st of 3	15.1		*

N G C   5737,   5739,   5740,   5746,   5750,   5751   I C   1052,   1054,   1056,   1057,   4493,   4505
      5754,   5755

1	2	3	4	5	6	7	8 9	10	11	12
09521	14 44.6 + 11 48 008.78 + 58.75	02 - 38 - 009	1087 337	1.3 x 0.7 1.1 x 0.7	0: 5	2	S B a: P w 9523, in group	14.7		*1
09522	14 44.6 + 54 53 093.65 + 55.44	09 - 24 - 036	0715 344	1.1 x 0.18 1.0 x 0.20	45 7	2	Sa - b	16.0:		*
09523	14 44.7 + 11 48 008.80 + 58.73	02 - 38 - 010	1087 337	1.0 x 0.9 (1.0 x 1.0)	— 1	2	S0 - a P w 9521, in group	14.7		*1
09524	14 44.7 + 13 52 012.21 + 59.85	02 - 38 - 011 NGC 5758	1087 336	(1.0 x 0.9) (1.0 x 0.9)	— ...	2	E - S0 npp of 2	15.0		*
09525	14 44.9 + 13 40 011.92 + 59.70	02 - 38 - 012 NGC 5759	1087 336	[1.5 x 0.7] ...	... ...	1	Double system Bridge, distorted	14.9		*1
09526	14 45.0 - 00 18 352.98 + 50.72	00 - 38 - 000	1613	1.1 x 0.1 1.0 x 0.1	61 7	3	Sc	17.		
09527	14 45.0 + 25 04 034.51 + 63.94	04 - 35 - 000	0102	1.0 x 0.2 0.9 x 0.2	124 7	2	Sc - Irr	16.0:		
09528	14 45.2 + 30 07 046.07 + 64.46	05 - 35 - 000	1390	1.0 x 0.15 0.9 x 0.15	110 7	1	S ...	15.7		
09529	14 45.2 + 69 10 108.88 + 44.92	12 - 14 - 012	1442	1.6 x 0.3 1.6 x 0.3	162 7	2	Sb	15.2		
09530	14 45.3 + 09 52 005.98 + 57.49	02 - 38 - 013	1087	1.0 x 0.7 0.9 x 0.6	... ...	1	S: ...	15.0		*
09531	14 45.3 + 18 43 021.21 + 61.95	03 - 38 - 015 NGC 5760	0054 334	1.6 x 0.6 1.7 x 0.7	96 6	3	Sa	14.3		*3
09532	14 45.6 + 19 16 022.35 + 62.10	03 - 38 - 18-22 334	0054	[1.7 x 0.8] ...	... ...	3	Multiple system Connected	15.1		*1
09533	14 46.0 + 14 10 013.04 + 59.73	02 - 38 - 000	1087	1.0 x 0.25 0.9 x 0.22	102 7	1	S ... ssf of 2	15.2		*
09534	14 46.2 + 11 32 008.78 + 58.28	02 - 38 - 000	1087 337	(1.0 x 0.7) (1.0 x 0.6)	(140) ...	1	E? 	15.3		*
09535	14 46.3 + 12 40 010.62 + 58.88	02 - 38 - 014 NGC 5762	1087	(2.0 x 1.5) (1.6 x 1.2)	140: ...	1	S ... spp of 2	14.3		*
09536	14 46.3 + 32 00 050.41 + 64.20	05 - 35 - 016 IC 4509	1390	1.0 x 0.22 0.9 x 0.2	46 6	1	S ...	14.7		*
09537	14 46.4 + 35 12 057.67 + 63.85	06 - 33 - 001	1610	2.5 x 0.5 2.5 x 0.5	140 6	2	Sb	14.9		
09538	14 46.8 + 17 14 018.70 + 61.01	03 - 38 - 029 IC 1058	0054 339	1.0 x 0.25 1.6 x 0.5	115 6	1	S ...	14.8		*
09539	14 46.8 + 22 24 028.95 + 62.89	04 - 35 - 009	0102	1.2 x 0.2 1.3 x 0.2	67 7	2	Sa nf of 2	15.0		*
09540	14 46.8 + 34 55 057.01 + 63.81	06 - 33 - 000	1610	1.0 x 0.3 0.8: x 0.3	... ...	1	Triple system?	17.		*
09541	14 46.9 + 29 57 045.70 + 64.09	05 - 35 - 000	1390	1.4 x 0.12 1.1 x 0.15	152 7	2	Sb - c	15.7		
09542	14 47.1 + 42 40 073.30 + 61.63	07 - 30 - 066	0145	2.0 x 0.6 1.6 x 0.4	32 6	3	Sc Sc +	15.3		
09543	14 47.2 + 38 37 065.05 + 62.96	07 - 30 - 067	0145	1.3 x 0.2 1.5 x 0.2	103 7	2	Sa - b	15.4		
09544	14 47.3 + 25 35 035.88 + 63.53	04 - 35 - 010	0102 335	1.3 x 1.3 1.6 x 1.6	— 1	3	S ... Companion superimposed, interaction?	14.8		*
09545	14 47.3 + 50 09 086.35 + 58.00	08 - 27 - 025	1368	1.5 x 1.0 0.8: x 0.5:	80: 3	1	Sc nf of 2	16.0:		*
09546	14 47.4 + 23 47 031.97 + 63.12	04 - 35 - 000	0102	1.0 x 0.10 1.0 x 0.10	44 7	3	Sc Sc +	17.		
09547	14 47.9 + 10 33 007.71 + 57.38	02 - 38 - 000	1087	1.4 x 0.2 0.9 x 0.15	163 7	1	Sb: f of 2	15.6		*
09548	14 47.9 + 16 55 018.35 + 60.64	03 - 38 - 033	0054 339	1.0 x 0.4 0.9 x 0.3	7 6	2	Sa - b P w 9550	15.3		*
09549	14 47.9 + 47 35 082.08 + 59.29	08 - 27 - 000	1368 340	1.0 x 0.7 1.0 x 0.7	150: 3	2	S B: a - b	15.1		*
09550	14 48.0 + 16 57 018.43 + 60.63	03 - 38 - 034	0054 339	1.3 x 0.2 1.1 x 0.2	22 6 - 7	1	S ... P w 9548	15.2		*
09551	14 48.1 - 01 35 352.51 + 49.25	00 - 38 - 008	1613	1.0 x 0.8 0.6 x 0.6	(150) ...	2	... S d - dm 1st of 3	15.6		*
09552	14 48.2 + 26 17 037.50 + 63.45	04 - 35 - 000	0102 335	1.1 x 0.3 1.1 x 0.3	167 6 - 7	2	Sa - b n of 2	15.6		*
09553	14 48.2 + 63 29 103.20 + 49.09	11 - 18 - 008 IC 1065	1575	1.0 x 0.9 0.9 x 0.9	— 1	3	S B 0	15.0		*
09554	14 48.4 + 05 19 000.55 + 54.03	01 - 38 - 4 + 5 NGC 5765	1421	[1.1 x 0.9] ...	... ...	2	Double system Contact	14.6		*13
09555	14 48.4 + 10 19 007.49 + 57.14	02 - 38 - 16-18	1087	[1.7 x 0.6] ...	... ...	1	Triple system Envelopes in contact	15.7		*1

N G C   5758 - 60,   5762,   5765   I C   1058,   1065,   4509

1	2	3	4	5	6	7	8	9	10	11	12
09556	14 48.5 + 60 35 100.04 + 51.17	10 - 21 - 033	1425	2.0: x 0.15 1.8:: x 0.15	136 7	1	S B? c:		15.3		
09557	14 48.7 + 27 46 040.83 + 63.54	05 - 35 - 019 IC 4514	1390	1.0 x 0.7 0.8 x 0.7	(125) ...	2	S B a - b	S B(s) ...	15.3		
09558	14 48.8 + 17 24 019.43 + 60.65	03 - 38 - 037	0054 339	1.1 x 0.6 1.1 x 0.7	170 4 - 5	4	Sc		15.0		
09559	14 48.8 + 42 56 073.56 + 61.24	07 - 30 - 068	0145	1.2 x 0.4 1.1 x 0.3	150 6	1	S ...		14.6		1
09560	14 48.9 + 35 47 058.78 + 63.25	06 - 33 - 002	1610	0.8 x 0.25 0.7 x 0.25	[58] ...	2	... Peculiar, p w 9562		14.5	+ 1110 + 1232	*1C
09561	14 49.0 + 09 32 006.49 + 56.56	02 - 38 - 019	1087	1.5 x	... ...	3	... ssp of 2, disturbed, very long tail	(14.9)			*1V A
09562	14 49.2 + 35 45 058.68 + 63.20	06 - 33 - 004	1610	1.1 x 1.1 0.9 x 0.9	-- ...	2	Strongly peculiar, jets Eruptive? P w 9560		14.2	+ 1200 + 1322	*1V C
09563	14 49.2 + 54 37 092.55 + 55.10	09 - 24 - 044 IC 1069	0715 344	(1.6 x 0.8) (1.6 x 0.7)	45 5	3	S0		15.1		
09564	14 49.6 - 02 20 352.16 + 48.45	00 - 38 - 009 NGC 5768	1613	2.0 x 1.7 2.0 x 1.6	-- 1 - 2	3	Sc		14.2		*12 34
09565	14 49.7 + 04 53 000.36 + 53.50	01 - 38 - 007 IC 1063	1421	1.1 x 0.7 1.1 x 0.8	162 3 - 4	3	S B b		14.8		*
09566	14 49.7 + 40 49 069.28 + 61.84	07 - 31 - 001 NGC 5772	1371	2.3 x 1.4 2.3 x 1.4	35 4	3	Sb	Sb -	13.9		*12 3
09567	14 49.9 + 43 51 075.12 + 60.70	07 - 31 - 002	1371	1.2 x 0.6 0.9 x 0.5	[137] ...	2	Irr P w 9569		14.9		*
09568	14 50.0 + 59 10 098.19 + 52.02	10 - 21 - 034 NGC 5777	1425	3.5 x 0.4 3.1 x 0.4	144 6	4	Sb P w 9570		14.2		*3
09569	14 50.1 + 43 56 075.25 + 60.63	07 - 31 - 003	1371	1.7 x 1.4 1.3 x 1.1	-- ...	4	S B c P w 9567	S B(rs)d	14.4		*
09570	14 50.2 + 59 09 098.14 + 52.01	10 - 21 - 035	1425	1.5:: x 1.3:: 1.3:: x 1.3::	-- ...	4	Dwarf spiral S IV - V, p w 9568		16.5:		*1
09571	14 50.3 + 30 00 045.87 + 63.35	05 - 35 - 022 NGC 5773	1390	(1.1 x 1.1) (1.0 x 1.0)	-- 1	1	S ... P w NGC 5771		14.5		*3
09572	14 50.4 + 51 29 087.88 + 56.84	09 - 24 - 047 IC 1074	0715	1.0 x 0.35 0.9 x 0.35	118 6	1	S ...		15.2		
09573	14 50.5 + 03 30 358.87 + 52.43	01 - 38 - 009 IC 1066	1421	1.4 x 0.8 1.6 x 0.8	70 4	1	S ... P w 9574		14.2		*1
09574	14 50.6 + 03 32 358.94 + 52.43	01 - 38 - 010 IC 1067	1421	2.2 x 1.7 2.2 x 1.7	(110) 2	4	S B b P w 9573	S B(r)b	13.6		*12
09575	14 50.8 + 04 09 359.75 + 52.81	01 - 38 - 011 NGC 5770	1421	(1.4 x 1.2) (1.6 x 1.3)	-- 2:	4	S B 0		13.3		*3
09576	14 51.1 + 03 46 359.38 + 52.51	01 - 38 - 013 NGC 5774	1421	3.4 x 2.8 3.0: x 2.6:	(145) 2	4	Sc P w 9579	SAB(rs)d	13.9	+ 1525 + 1534	*12 3569
09577	14 51.2 + 18 18 021.58 + 60.51	03 - 38 - 046	0054 341	1.0 x 0.5 1.0 x 0.5	170 5:	3	S B c np of 2		15.7		*
09578	14 51.4 + 20 19 025.46 + 61.21	03 - 38 - 047	0054 345	1.1 x 1.0 1.2 x 1.0	-- 1 - 2	3	S B b		15.4		
09579	14 51.5 + 03 45 359.45 + 52.41	01 - 38 - 014 NGC 5775	1421	4.2 x 0.9 4.0 x 1.1	146 7	3	Sc P w 9576		13.0	+ 1565 + 1574	*12- -9
09580	14 51.6 + 10 18 008.29 + 56.49	02 - 38 - 000	1087	0.7 x 0.15 0.7 x 0.15	40 (7)	1			15.3		
09581	14 51.7 + 10 17 008.29 + 56.46	02 - 38 - 000	1087	0.9: x 0.2 0.8 x 0.2	... ...	1	Double system, distorted, bridge:, in group of 3		15.3		*
09582	14 51.7 + 04 57 000.99 + 53.17	01 - 38 - 015 IC 1071	1421	1.0 x 0.8 1.2 x 1.0	(150) 2:	3	S0 1st of 3		14.4		*
09583	14 51.7 + 42 38 072.57 + 60.86	07 - 31 - 000	1371	1.4 x 0.2 0.8 x 0.2	105 7	1	S ...		15.7		
09584	14 51.8 + 04 43 000.72 + 53.00	01 - 38 - 000	1421	1.1 x 0.4 1.1 x 0.4	142 6	2	Sb		15.3		
09585	14 51.9 + 42 45 072.76 + 60.79	07 - 31 - 005	1371	1.1 x 0.5 1.0 x 0.4	135 6	1	S B b - c		16.0:		
09586	14 51.9 + 52 17 088.82 + 56.19	09 - 24 - 050 NGC 5783 = 5785	0715	2.8 x 1.6 2.8: x 1.6:	0 4	4	S B c	SAB(s)c	14.0		*12
09587	14 52.0 + 16 34 018.59 + 59.60	03 - 38 - 049 IC 4516	0054 339	(1.3 x 1.1) (1.6 x 1.3)	-- ...	2	E		14.9		*
09588	14 52.0 + 30 25 046.81 + 63.00	05 - 35 - 023	1390	[1.3: x 0.9] ...	... ...	2	Double system, contact Strongly disrupted		15.3		*1
09589	14 52.0 + 54 24 091.81 + 54.92	09 - 24 - 052	0715 344	1.1 x 0.3 1.2 x 0.3	62 ...	2	S B ... Singular		17.		*
09590	14 52.1 + 18 50 022.74 + 60.52	03 - 38 - 050	0054 341	1.1 x 0.7 1.3 x 0.9	10 ...	1	E?		15.6		*1

N G C   5768, 5770, 5772 - 75, 5777, 5783, 5785   I C   1063, 1066, 1067, 1069, 1071, 1074
                                                                       4514, 4516

1	2	3	4	5	6	7	8	9	10	11	12
09591	14 52.3 + 64 04b 103.29 + 48.35	11 - 18 - 013	1575	1.3 x 0.5 1.0 x 0.4	166 6	2	Sb - c n of 2		(15.6)		*
09592	14 52.4 + 42 45 072.69 + 60.70	07 - 31 - 006 NGC 5784	1371	(1.9 x 1.8) (1.8 x 1.8)	— 1	3	S0		13.7		*13
09593	14 52.5 + 18 18 021.83 + 60.22	03 - 38 - 053 IC 1075	0054	1.2 x 0.6 1.3 x 0.5	156 5	3	S B b P w 9595		14.9		*1
09594	14 52.5 + 24 17 033.62 + 62.11	04 - 35 - 014	0102 335	1.1 x 0.9 1.0 x 0.8	... ...	3	S B c Singular		15.2		*1
09595	14 52.6 + 18 14 021.72 + 60.17	03 - 38 - 055 IC 1076, Mark 479	0054	1.2 x 0.7 0.8 x 0.35	(5) ...	1	... P w 9593		13.9		*1
09596	14 52.6 + 26 00 037.25 + 62.43	04 - 35 - 015	0102 335	1.0 x 0.7 1.0 x 0.7	0 3	2	Sb - c nnf of 2		15.3		*
09597	14 52.8 + 31 02 048.17 + 62.83	05 - 35 - 025	1390	1.1 x 0.7 1.1 x 0.6	(20) ...	3	Dwarf spiral S V		17.		
09598	14 53.3 + 44 00 074.90 + 60.07	07 - 31 - 009	1371	1.6 x 0.6 1.6 x 0.6	124 6	3	Sc		15.0		
09599	14 53.4 + 42 42 072.46 + 60.55	07 - 31 - 008 NGC 5787	1371	1.1 x 1.0 1.0 x 0.9	— ...	1	...		14.1	+ 5460 + 5602	*3C
09600	14 53.4 + 51 15 087.05 + 56.59	09 - 24 - 000	0715	1.0 x 0.12 1.0 x 0.15	50 7	3	Sb		17.		
09601	14 53.5 - 01 12 354.41 + 48.61	00 - 38 - 011	1613	1.5 x 1.4 1.5: x 1.3:	— 1 - 2	2	...	S B dm:	14.6		
09602	14 53.5 + 12 04 011.45 + 57.08	02 - 38 - 022	1087	(1.0 x 0.9) (1.0 x 0.9)	— (1)	2	S0 ssp of 2		14.9		*1
09603	14 53.5 + 33 01 052.49 + 62.63	06 - 33 - 10 + 11	1610	[1.2: x 0.5] ...	... ...	2	Double system Common envelope, in group of 3		16.0:		*1V
09604	14 53.5 + 37 38 062.34 + 61.99	06 - 33 - 012	1610	1.1 x 0.10 1.1 x 0.10	167 7	3	Sc		16.5:		*
09605	14 53.5 + 48 33 082.78 + 58.00	08 - 27 - 030	1368 343	1.0 x 0.12 0.8 x 0.12	52 7	3	Sc nf of 2		16.0:		*
09606	14 53.6 + 24 55 035.06 + 62.00	04 - 35 - 000	0102 335	1.3: x 0.2 1.3 x 0.25	143 7	2	Sb		16.0:		
09607	14 53.8 + 62 29 101.43 + 49.38	10 - 21 - 038	1425	(1.2 x 1.1) (0.9 x 0.8)	— ...	1	...		16.0:		*
09608	14 54.0 + 09 33 007.80 + 55.57	02 - 38 - 025 IC 1078	1087 342	1.2 x 1.0 1.1 x 1.0	— 1 - 2	3	Sa P w 9611		14.8		*1
09609	14 54.0 + 73 20 111.60 + 41.19	12 - 14 - 000 NGC 5819	1442	1.0 x 1.0 1.1 x 1.1	— 1	3	S B b	SAB(rs)bc Sb +	14.3		*
09610	14 54.1 + 49 56 084.91 + 57.21	08 - 27 - 032 NGC 5794	1368 343	(1.1 x 1.1) (1.1 x 1.1)	— ...	1	...		14.5		*3
09611	14 54.2 + 09 34 007.88 + 55.54	02 - 38 - 026 IC 1079	1087 342	(1.6 x 1.1) (1.5 x 0.9)	80 ...	4	E P w 9608		14.8		*1
09612	14 54.2 + 45 37 077.69 + 59.26	08 - 27 - 034	1368	1.5 x 1.3 1.2 x 1.1	— 1	4	Sc/S B c	SAB(s)c	15.2		
09613	14 54.3 + 05 47b 002.75 + 53.22	01 - 38 - 000	1421	1.1 x 0.5 0.9 x 0.4	47 5 - 6	1	S ... sf of 2		(15.2)		*
09614	14 54.3 + 09 41 008.07 + 55.59	02 - 38 - 000	1087 342	1.6 x 1.4 1.5: x 1.3:	(5) ...	3	Dwarf IV - V or V		16.5:		
09615	14 54.4 + 30 25 046.84 + 62.48	05 - 35 - 026 NGC 5789	1390	1.0 x 0.9 0.9 x 0.8	— 1	3	... P w 9628	S dm	13.9		*13
09616	14 54.5 + 09 28 007.81 + 55.42	02 - 38 - 028	1087 342	1.1 x 0.9 1.0 x 0.9	— ...	2	Sa - b [+ companion?]		14.8		*1V
09617	14 54.6 + 49 36 084.30 + 57.31	08 - 27 - 035	1368 343	1.6 x 0.25 1.6 x 0.25	64 7	1	S ...		14.7		
09618	14 54.7 + 24 48 034.93 + 61.73	04 - 35 - 18 + 19	102 335	[1.6 x 0.6] ...	... ...	3	Double system Contact		14.3		*1V A
09619	14 54.7 + 49 54 084.76 + 57.14	08 - 27 - 036 NGC 5797	1368 343	1.1 x 0.6 (1.5 x 1.0)	110 4	2	S0 - a		13.6		3
09620	14 54.8 + 19 54 025.23 + 60.32	03 - 38 - 059	0054 345	1.3 x 0.8 1.3 x 0.5	... ...	1	S ... P w 9622, distorted		14.7		*1
09621	14 55 + 82 45 118.46 + 33.51	14 - 07 - 000	1363	1.2 x 0.7 1.2 x 0.7	85 4	3	Sc	SA(s)c	15.4		*
09622	14 55.0 + 19 52 025.20 + 60.26	03 - 38 - 060	0054 345	1.3 x 0.8 1.3 x 0.8	178 4	1	S ... P w 9620		14.0		*1
09623	14 55.0 + 38 56 064.87 + 61.41	07 - 31 - 011	1371	1.3 x 0.9 1.1 x 0.8	155: 3	2	S0		14.4		*1
09624	14 55.1 + 08 29 006.57 + 54.73	01 - 38 - 022 NGC 5790	1421 342	1.1 x 0.8 1.2 x 0.9	77 3	3	S0/Sa	(R)SA 0/a	15.1		
09625	14 55.3 + 06 50 004.38 + 53.68	01 - 38 - 023	1421	1.7 x 0.5 1.7 x 0.5	178 6	2	S B: a - b		14.7		

N G C 5784, 5787, 5789, 5790, 5794, 5797   I C 1075, 1076, 1078, 1079
5819

1	2	3	4	5			6	7	8	9	10	11	12
09626	14 55.4 + 48 50 082.95 + 57.59	08 – 27 – 037	1368 343	1.1 x 0.35 1.1 x 0.35			95 6	2	S B b		15.6		
09627	14 55.4 + 49 52 084.60 + 57.06	08 – 27 – 038 NGC 5804	1368 343	1.3 x 1.1 1.3: x 1.1			(0) 1	4	S B b	S B(s)b	14.0		*3
09628	14 55.5 + 30 10 046.32 + 62.23	05 – 35 – 028 NGC 5798	1390	1.4 x 0.9 1.3 x 0.8			42 ...	2	Irr P w 9615		13.5		*3
09629	14 55.6 + 52 32 088.61 + 55.59	09 – 24 – 058	0715 344	1.4 x 0.4 1.4 x 0.4			152 6	3	Sa		14.6		
09630	14 55.7 + 53 04 089.37 + 55.27	09 – 24 – 000	0715	1.1: x 1.1: ...			--- ...	0	Dwarf V		17.		
09631	14 55.8 – 00 55 355.33 + 48.41	00 – 38 – 012 NGC 5792	1613	8.0 x 2.0 8.2 x 2.0			84 6	4	S B b	Sb +	13.5	+ 1985 + 1980	*12 347$\overline{8}$
09632	14 56.4 + 53 59 090.56 + 54.65	09 – 24 – 059	0715 344	1.5 x 1.1 1.5 x 1.1			135 3	4	Sc NGC 5820 group	SA d	15.3		
09633	14 56.6 + 08 53 007.51 + 54.66	02 – 38 – 000	1087 342	(1.5 x 1.0) (1.4 x 0.8)			... ...	2	Double system Common envelope		16.0:		*
09634	14 56.6 + 20 15 026.18 + 60.05	03 – 38 – 062	0054 345	1.0 x 0.5 1.0 x 0.5			102 5	4	S B b	S B(s)b	15.5		
09635	14 56.7 + 19 47 025.33 + 59.86	03 – 38 – 063	0054	1.1 x 0.9 1.0 x 0.8			(20) 2	2	Sa – b		15.7		*
09636	14 56.8 + 13 37b 014.66 + 57.19	02 – 38 – 000	1087	1.1 x 0.3 1.1 x 0.2			20 7	1	S ... f of 2, disturbed, in group		(15.5)		*
09637	14 56.8 + 41 58 070.63 + 60.21	07 – 31 – 000	1371	1.4: x 0.3 1.4: x 0.3:			44 ...	3	Irr		16.0		
09638	14 56.8 + 59 04 097.12 + 51.44	10 – 21 – 040	1425	2.0: x 1.3: 1.6:: x 1.2::			(40) ...	3	Dwarf irregular Ir IV – V or V		15.5		
09639	14 56.9 + 45 09 076.48 + 59.03	08 – 27 – 042	1368	1.0 x 0.9 0.8 x 0.5			--- 1 – 2	2	Sa – b		14.6		
09640	14 57.1 + 16 50 020.09 + 58.62	03 – 38 – 064	0054	(1.3 x 1.1) (1.8 x 1.6)			(110) ...	4	E		14.5		*
09641	14 57.1 + 41 38 069.96 + 60.25	07 – 31 – 014	1371	1.5 x 1.1 0.8 x 0.7			100 ...	1	S ... Peculiar?		14.9		
09642	14 57.1 + 54 05 090.60 + 54.51	09 – 25 – 001 NGC 5820	1096 345	(2.2 x 2.0) (2.0 x 1.6)			... ...	2	E – S0 P w 9648, distorted, brightest of 3		13.0	+ 3269 + 3445	*12 347$\overline{8}$A
09643	14 57.2 + 50 01 084.56 + 56.74	08 – 27 – 046 NGC 5818	1368	(1.2 x 0.9) (1.2 x 0.9)			170: 2 – 3	3	S0		15.0		
09644	14 57.3 + 27 19 040.36 + 61.59	05 – 35 – 030	1390	1.4 x 1.4 1.1: x 1.1:			--- 1	2	S B a	S B(r)a	14.9		
09645	14 57.4 + 02 05 359.07 + 50.20	00 – 38 – 014 NGC 5806	1613 352	3.0 x 1.5: 3.5: x 2.0:			170 5:	3	Sb NGC 5846 group		12.9	+ 1301 + 1309	*123 467$\overline{8}$
09646	14 57.5 + 27 31 040.79 + 61.58	05 – 35 – 000	1390	1.1 x 0.10 0.8 x 0.1			165 7	3	Sc spp of 2		16.0:		*
09647	14 57.5 + 33 02 052.43 + 61.79	06 – 33 – 014	1610	1.7 x 1.1 1.4: x 1.0:			95: ...	3	Dwarf spiral S IV – V		15.7		*1
09648	14 57.5 + 54 07 090.59 + 54.45	09 – 25 – 002 NGC 5821	1096 345	1.7 x 0.9 1.7 x 0.8			148 5	1	S ... P w 9642, NGC 5820 group		14.9		*13
09649	14 57.6 + 71 53 110.10 + 42.15	12 – 14 – 015 NGC 5832	1442	3.7: x 2.5: 3.7: x 2.5:			45 3	1	S B c? Singular		13.3		*12 3$\overline{}$
09650	14 58 + 83 48 119.08 + 32.59	14 – 07 – 012	1363 388	1.6 x 0.25 1.6 x 0.30			6 7	3	Sa P w		14.1		*1
09651	14 58.0 + 49 43 083.97 + 56.78	08 – 27 – 000	1368	1.1: x 0.3: 1.1: x 0.3:			118 6	1	S B c:		16.0:		
09652	14 58.2 + 16 22 019.50 + 58.18	03 – 38 – 065	0054	1.1 x 0.6 1.2 x 0.7			50 4	3	S B a Companion superimposed		15.1		*1
09653	14 58.2 + 61 14 099.49 + 49.87	10 – 21 – 042	1425	1.2 x 0.35 1.1: x 0.35			153 6	2	S B b		15.7		
09654	14 58.5 + 11 43 012.09 + 55.86	02 – 38 – 000	1087	1.1 x 0.8 1.1 x 0.7			(75) 3:	2	Sa – b		15.2		*
09655	14 58.6 + 01 53 359.15 + 49.85	00 – 38 – 016 NGC 5813	1613 352	(3.4 x 2.7) (4.0 x 3.0)			(145) ...	4	E NGC 5846 group		12.5	+ 1882 + 1891	*123 467$\overline{8}$
09656	14 58.9 + 09 21 008.73 + 54.47	02 – 38 – 000	1087	1.1 x 0.3 1.1 x 0.3			88 6	1	S ...		15.2		
09657	14 59.0 + 48 33 081.96 + 57.20	08 – 27 – 050	1368 343	1.5: x 0.2 0.7 x 0.2			43 7	2	S B? c – Irr P w 9665		15.5		*1
09658	14 59.0 + 50 11b 084.56 + 56.40	08 – 27 – 051 NGC 5828	1368	0.60 x 0.45 0.60 x 0.45			50 ...	1	S ... n of 2		(14.3)		*
09659	14 59.4 + 05 08 003.25 + 51.82	01 – 38 – 000	1421	1.0 x 0.35 1.1 x 0.35			84 6	3	Sc		15.3		
09660	14 59.4 + 44 54 075.69 + 58.72	08 – 27 – 053	1368	0.8 x 0.35 0.7 x 0.35			[83] ...	1	... Peculiar?		14.3		

N G C   5792, 5798, 5804, 5806, 5813, 5818
5820, 5821, 5828, 5832

1	2	3	4	5	6	7	8	9	10	11	12
09661	14 59.5 + 02 01 359.54 + 49.77	00 - 38 - 018 352	1613 352	1.4 x 1.3 1.2 x 1.2	— 1	3	...	S B dm	15.0		*
09662	14 59.6 + 26 10 038.16 + 60.91	04 - 35 - 024 NGC 5827	0102 346	1.2 x 0.9 1.1 x 0.8	135 ...	2	S ... Disturbed:		13.7		*1
09663	14 59.6 + 52 48 088.40 + 54.94	09 - 25 - 005	1096 344	1.5: x 1.3: 1.4: x 0.9:	— ...	3	Dwarf IV - V		15.2		*
09664	14 59.7 + 74 05 111.80 + 40.35	12 - 14 - 016 NGC 5836	1442	1.3 x 1.2 1.3 x 1.2	— 1	4	S B b	S B(rs)b	14.9		*3C
09665	14 59.8 + 48 31 081.79 + 57.10	08 - 27 - 054 343	1368 343	1.6 x 0.25 1.6 x 0.25	142 7	2	Sb - c P w 9657		14.9		*1
09666	14 59.8 + 60 27 098.38 + 50.25	10 - 21 - 043	1425	1.2 x 0.15 0.6 x 0.15	48 7	1	S ...		16.0:		
09667	14 59.9 + 05 51 004.28 + 52.18	01 - 38 - 026 352	1421 352	2.0 x 1.0 (2.1 x 1.2)	87 5	2	S0 - a		14.6		
09668	15 00 + 83 44 118.98 + 32.61	14 - 07 - 014 388	1363 388	1.5 x 0.8 1.6 x 0.8	85: ...	1	... P w 9650		13.8		*
09669	15 00 + 85 13 119.97 + 31.39	14 - 07 - 013	1363	(1.1 x 0.8) (1.3 x 1.0)	— ...	1	S? ... P w 9720		15.7		*
09670	15 00.1 + 48 04 081.01 + 57.27	08 - 27 - 056 NGC 5830	1368 343	1.1 x 0.7 0.9 x 0.7	170: 4	2	Sb		15.2		
09671	15 00.2 + 42 23 071.01 + 59.48	07 - 31 - 019	1371	1.5 x 1.2 0.7 x 0.7	(160) 2:	4	Sc	SA cd	15.6		
09672	15 00.4 + 20 00 026.33 + 59.12	03 - 38 - 000	0054	1.6: x 0.9: 1.4: x 0.7:	35 5:	3	...	S m	17.		
09673	15 00.4 + 23 31b 032.97 + 60.17	04 - 35 - 027 NGC 5829 = 4526*	0102	2.0 x 1.6 1.6 x 1.4:	— 1 - 2	4	Sc sf of 2, connected		(14.6)		*12 VA$^-$
09674	15 00.7 + 49 04 082.55 + 56.71	08 - 27 - 057 NGC 5835	1368 343	1.3 x 0.9 1.2 x 0.7	160: 4	2	Sa		15.7		
09675	15 01.0 + 10 51 011.39 + 54.88	02 - 38 - 000	1087	1.2: x 1.0: 1.0 x 0.8	— ...	1	S B ... nnp of 2		15.3		*
09676	15 01.3 + 28 00 042.00 + 60.80	03 - 35 - 000	1390	1.1 x 0.8 1.0 x 0.6	17 3	3	S B c		15.7		
09677	15 01.4 + 22 53 031.87 + 59.79	04 - 36 - 000	0087	1.4 x 0.2 1.2 x 0.2	5 7	2	Sb		16.0:		
09678	15 01.6 + 01 24 359.38 + 48.97	00 - 38 - 020 NGC 5831	1613 352	(2.0 x 1.8) (2.4 x 2.2)	— ...	4	E		13.1	+ 1648 + 1693	*123 4678
09679	15 01.6 + 26 17b 038.55 + 60.49	04 - 36 - 001 IC 4530	0087	1.0 x 0.15 1.0 x 0.2	19 6 - 7	1	S ... ssf of 2, disturbed		(15.3)		*1
09680	15 01.7 + 18 51 024.47 + 58.42	03 - 38 - 000	0054	1.3 x 0.2 1.2 x 0.15	25 7	2	...	d - m	16.5:		
09681	15 01.7 + 42 19 070.71 + 59.24	07 - 31 - 023 351	1371 351	1.5 x 0.1 1.5 x 0.1	177 7	3	Sc P w 9684		16.5:		*1
09682	15 01.9 - 00 40 357.19 + 47.50	00 - 38 - 021	1613	1.9 x 0.6 2.0: x 0.5	165 6	4	...	S m	15.3		
09683	15 02 + 81 37 117.45 + 34.30	14 - 07 - 015	1363	1.0 x 1.0 1.1 x 1.0	— 1	3	S B a	S B(r)a	14.8		*
09684	15 02.0 + 42 18 070.65 + 59.19	07 - 31 - 024 351	1371 351	1.7 x 1.1: 1.2: x 0.4	102 ...	2	S B a - b P w 9681		14.8		*
09685	15 02.3 + 08 22 008.19 + 53.22	01 - 38 - 000 349	1421 349	1.0 x 0.8 1.0 x 0.8	(100) 2	3	Sc		15.1		
09686	15 02.3 + 12 50 014.64 + 55.64	02 - 38 - 036 NGC 5837	1087	1.0 x 0.6 1.0 x 0.55	25 4	1	S ... Brightest in group		14.5		*
09687	15 02.3 + 28 00 042.05 + 60.58	05 - 36 - 004 IC 4533	1092	1.0 x 0.8 1.0 x 0.8	160 2	2	Sa		14.9		
09688	15 02.5 + 53 08 088.47 + 54.39	09 - 25 - 010 344	1096 344	1.3 x 0.4 1.3 x 0.4	8 6	2	S0 - a		14.8		*
09689	15 02.6 + 08 02 007.81 + 52.96	01 - 38 - 000 349	1421 349	1.0: x 0.8: 0.9: x 0.8	— 1 - 2	2	Sb - c		15.3		
09690	15 02.7 + 38 13 062.76 + 60.10	06 - 33 - 000	1610	1.1 x 0.2 1.1 x 0.2	0 7	2	Sb - c		16.5:		
09691	15 02.7 + 40 34 067.31 + 59.56	07 - 31 - 026 351	1371 351	1.2 x 0.9 1.2 x 0.9	85: ...	3	S B b	Sb +	15.1		
09692	15 02.8 + 02 16 000.67 + 49.32	00 - 38 - 022 NGC 5838	1613 352	3.5 x 1.5 3.8 x 1.6	43 6	4	S0		12.1	+ 1427 + 1441	*123 4678
09693	15 02.9 + 01 48 000.16 + 48.99	00 - 38 - 023 NGC 5839	1613 352	(1.2 x 1.2) (1.5 x 1.5)	— 1	2	S0 NGC 5846 group		13.9		*23
09694	15 02.9 + 04 41 003.57 + 50.86	01 - 38 - 000 352	1421 352	1.1 x 0.15 1.0 x 0.20	45 7	1	Sb		15.3		
09695	15 03.0 + 14 37 017.58 + 56.36	03 - 38 - 085	0054	1.0 x 0.5 0.8 x 0.4	32 5	2	S B b		15.1		

N G C 5827, 5829 - 31, 5835 - 39   I C 4526, 4530, 4533

1	2	3	4	5	6	7	8	9	10	11	12
09696	15 03.1 + 08 42 008.84 + 53.25	02 − 38 − 037	1087 349	0.9 x 0.8 0.8 x 0.8	--- 1	2	Sa		14.5		
09697	15 03.2 − 00 32 357.67 + 47.36	00 − 38 − 000	1613	1.1 x 0.4 0.6 x 0.3	107 6	2	...	dm − m	16.0:		
09698	15 03.2 + 23 53 033.99 + 59.64	04 − 36 − 007	0087	1.1 x 0.5 1.0 x 0.5	[50] ...	2	... Peculiar, jets		15.7		1
09699	15 03.4 + 09 43 010.32 + 53.76	02 − 38 − 000	1087 349	1.0 x 0.4 0.9 x 0.3	132 6	2	Sc		15.4		
09700	15 03.5 + 01 48 000.32 + 48.88	00 − 38 − 024 NGC 5845	1613 352	0.6 x 0.35 0.7 x 0.4	150 ...	1	E NGC 5846 group		13.8		*23
09701	15 03.5 + 31 21 048.89 + 60.55	05 − 36 − 007	1092	1.0 x 0.5 1.0 x 0.5	70 5	3	Sc		15.3		
09702	15 03.6 + 51 22 085.71 + 55.18	09 − 25 − 014	1096	1.2 x 0.15 1.0 x 0.15	163 7	2	Sb − c		15.7		
09703	15 03.7 + 46 46 078.35 + 57.29	08 − 27 − 064	1368	1.2 x 0.25 1.2 x 0.25	167 6	1	S ...		15.0		1
09704	15 03.8 + 57 30 094.24 + 51.72	10 − 21 − 045	1425	1.0 x 0.12 0.9 x 0.1	50 7	2	Sc		16.5:		
09705	15 03.9 + 25 58 038.11 + 59.92	04 − 36 − 011	0087	1.0 x 0.5 1.0 x 0.5	70 5:	1	S0 − a		15.0		1
09706	15 04.0 + 01 46b 000.40 + 48.77	00 − 38 − 025 NGC 5846	1613 352	(3.0 x 3.0) (4.5 x 4.5)	--- ...	4	E n of 2 Brightest in group		(11.9) + 1771 + 1784		*12− −9
09707	15 04.0 + 39 43 065.57 + 59.53	07 − 31 − 030 NGC 5853	1371 351	1.5 x 1.0 1.5 x 1.0:	(150) ...	1	S B ...		14.8		1
09708	15 04.1 + 09 38 010.36 + 53.57	02 − 38 − 040	1087 349	1.1 x 1.0 0.9 x 0.7	--- 1	3	Sb P w 9711		15.2		*1
09709	15 04.1 + 42 25 070.63 + 58.79	07 − 31 − 032	1371	1.2 x 1.0 1.2: x 1.0:	(135) ...	3	Dwarf spiral S IV − V		17.		1
09710	15 04.2 + 03 58 003.02 + 50.16	01 − 38 − 032 IC 1088	1421 352	1.0 x 0.2 0.9 x 0.2	95 7	1	Sb P w IC 1087		(15.1)		*1
09711	15 04.2 + 09 37 010.36 + 53.54	02 − 38 − 041	1087 349	1.3 x 0.22 1.3 x 0.22	108 7	3	Sa P w 9708		15.3		*1
09712	15 04.4 + 12 46 015.00 + 55.16	02 − 38 − 000	1087	1.5 x 0.3 1.4 x 0.3	82 6	3	Sc NGC 5837 group		15.2		*
09713	15 04.4 + 23 50 034.03 + 59.36	04 − 36 − 013	0087	1.4 x 0.9 1.5 x 0.8	160 4	3	S B 0		14.7		
09714	15 04.5 + 13 04 015.47 + 55.29	02 − 38 − 044 NGC 5851	1087	1.1 x 0.25 1.1 x 0.25	43 ...	1	... 2nd of 3, disturbed NGC 5837 group		14.9		*13
09715	15 04.7 + 01 44 000.54 + 48.61	00 − 39 − 002 NGC 5850	1402 352	5.0 x 4.5 4.0: x 3.5:	(140) 1	5	S B b S B(r)b NGC 5846 group		13.6 + 2372 + 2385		*12− −789
09716	15 04.7 + 37 29 061.18 + 59.84	06 − 33 − 021	1610	1.0 x 0.7 0.7 x 0.6	(135) ...	2	Sb − c		16.0:		
09717	15 04.7 + 42 50 071.33 + 58.55	07 − 31 − 033 NGC 5860, Mark 480	1371	[1.0 x 1.0] ...	--- ...	2	Double system Contact, common envelope		14.2		*13 C
09718	15 04.8 + 21 25 029.57 + 58.61	04 − 36 − 015	0087	1.4 x 0.5 1.0 x 0.5	96 6	3	Sb		15.2		
09719	15 04.8 + 40 46 067.49 + 59.12	07 − 31 − 000	1371 351	(1.0 x 0.9) (1.0 x 0.8)	--- ...	2	... Compact		15.7		*
09720	15 05 + 85 12 119.86 + 31.34	14 − 07 − 000	1363	1.2: x 0.7: 1.0: x 0.6:	(165) ...	1	...		16.5:		*
09721	15 05.0 + 12 46 015.12 + 55.04	02 − 39 − 000	1422	1.0 x 0.8 0.8 x 0.7	(95) ...	1	... NGC 5837 group		15.3		
09722	15 05.0 + 52 06 086.62 + 54.61	09 − 25 − 018	1096	1.1 x 1.0 1.0 x 0.7	--- 1 − 2	4	Sc	Sc +	16.0:		
09723	15 05.1 + 55 57 092.03 + 52.49	09 − 25 − 017 NGC 5866 = M 102	1096 344	(6.5 x 3.0) (6.5 x 3.0)	128 5	4	S0		11.1 + 788 + 972		*123 4678
09724	15 05.2 + 19 47 026.69 + 57.99	03 − 39 − 004 NGC 5857	0091 345	1.3 x 0.6 1.3 x 0.5	137 5 − 6	3	Sa P w 9728		13.6 + 4705 + 4785		*12 3
09725	15 05.2 + 55 40 091.64 + 52.65	09 − 25 − 016 NGC 5870	1096 344	(1.2 x 0.9) (1.1 x 0.7)	(25) ...	1	S0?		15.3		
09726	15 05.3 + 02 45 001.86 + 49.17	01 − 39 − 001 NGC 5854	1429 352	2.2 x 0.6 2.3 x 0.7	55 6	4	S0		13.1 + 1626 + 1644		*12 3468
09727	15 05.3 + 14 44 018.23 + 55.92	03 − 39 − 002 IC 1093	0091	1.0 x 0.8 0.7 x 0.5	(90) ...	3	Sb/S B c		14.9		*
09728	15 05.3 + 19 46 026.67 + 57.96	03 − 39 − 005 NGC 5859	0091 345	2.9 x 0.7 2.8 x 0.7	136 6	4	S B b S B(s)bc P w 9724 Sb +		13.1 + 4664 + 4745		*12 3
09729	15 05.3 + 63 12 100.85 + 47.93	11 − 18 − 025 IC 1100	1575	1.0 x 0.7 0.9 x 0.7	[60] ...	1	S ...		14.1		
09730	15 05.3 + 77 50 114.47 + 37.22	13 − 11 − 006	0768	1.8: x 1.8: 1.4: x 1.4:	--- 1	4	S B c S B(s)d		15.7		*1

N G C   5845,   5846,   5850,   5851,   5853,   5854
      5857,   5859,   5860,   5866,   5870

I C   1088,   1093,   1100

1	2	3	4	5	6	7	8	9	10	11	12
09731	15 05.6 + 56 44 093.00 + 51.98	09 - 25 - 021 IC 1099	1096	1.2 x 1.0 1.1 x 0.9	--- 1 - 2	3	Sc		15.0		*12
09732	15 05.7 + 01 26 000.45 + 48.23	00 - 39 - 005	1402 352	1.3 x 0.7 1.2 x 0.7	62 5:	3	S B b nff of 2		14.7		*
09733	15 06 + 81 23 117.13 + 34.40	14 - 07 - 000	1363	1.0: x 0.8: 1.2:: x 0.9::	--- ...	2	Dwarf IV - V:		17.		
09734	15 06.1 + 66 21 104.12 + 45.71	11 - 18 - 026	1575	1.5 x 1.2 1.5 x 1.2	60: 2	3	...		15.5		1
09735	15 06.2 + 19 22 026.11 + 57.62	03 - 39 - 010 IC 1097	0091 345	1.1 x 0.45 0.9 x 0.45	58 6	1	S ... Brightest in group		14.7		*1
09736	15 06.4 + 54 56 090.48 + 52.92	09 - 25 - 024 NGC 5874	1096 344	2.6 x 1.7 2.5 x 1.7	53 3	4	Sc	SA(s)c Sc -	14.1		*2
09737	15 06.4 + 55 22 091.07 + 52.68	09 - 25 - 023	1096 344	1.3 x 0.8 (1.1 x 0.7)	90: ...	1	S ...		14.7		
09738	15 06.7 + 19 28 026.36 + 57.55	03 - 39 - 012	0091	1.0 x 0.8 0.6 x 0.3	--- ...	2	Sc IC 1097 group		15.7		
09739	15 06.7 + 25 55 038.24 + 59.30	04 - 36 - 000	0087	1.0 x 0.25 1.0 x 0.3	136 (6)	3	...	S:B d - m	15.6		
09740	15 07.0 + 03 15 002.86 + 49.16	01 - 39 - 002 NGC 5864	1429 352	2.4 x 0.8 2.5 x 0.8	68 6	3	S B 0		12.9	+ 1618 + 1639	*12 3468
09741	15 07.0 + 52 29 086.91 + 54.15	09 - 25 - 026	1096 344	0.45 x 0.40 0.45 x 0.4	--- ...	1	... Peculiar		14.1		*
09742	15 07.3 + 00 39 359.98 + 47.41	00 - 39 - 006 NGC 5865 = 5869	1402 352	(2.4 x 1.7) (2.4 x 1.7)	125 3	2	S0 P w 9743		13.5		*13
09743	15 07.3 + 00 43 000.05 + 47.45	00 - 39 - 007 NGC 5868	1402 352	1.0 x 0.8 1.1 x 0.9	--- ...	2	S0 P w 9742		15.2		*3
09744	15 07.4 + 01 58 001.47 + 48.26	00 - 39 - 000	1402 352	1.1 x 1.0 0.8 x 0.8	--- 1	2	S B a - b	S B(r) ...	15.3		
09745	15 07.7 + 52 44 087.18 + 53.94	09 - 25 - 027 NGC 5875	1096 344	2.6 x 1.3 2.5 x 1.3	145 5	4	Sb		13.4		*23
09746	15 07.8 + 02 07 001.74 + 48.28	00 - 39 - 008	1402 352	1.4 x 0.3 1.4 x 0.3	140 6	1	Sb - c		15.0		
09747	15 08.0 + 54 41 089.92 + 52.86	09 - 25 - 028 NGC 5876	1069 344	2.8 x 1.3 2.8 x 1.3	50 6	4	S B a		13.9		*12
09748	15 08.1 + 76 15 113.05 + 38.34	13 - 11 - 007	0768	1.3 x 1.1 1.3 x 1.1	(5) 1	4	S B 0 P w 9750	(R)S B(r)0/a	15.1		*1
09749	15 08.2 + 67 23 104.97 + 44.84	11 - 18 - 030 Ursa Minor	1575	40. x 25. 35. x 20.	(70) ...	5	Dwarf elliptical Local group		13.6		*1
09750	15 08.3 + 76 21 113.12 + 38.26	13 - 11 - 008	0768	1.5 x 0.4 1.6 x 0.4	2 6	3	Sb P w 9748		15.3		*
09751	15 08.4 + 01 37 001.32 + 47.84	00 - 39 - 000	1402 352	1.1 x 0.2 0.9 x 0.2	121 7	3	Sc		15.6		
09752	15 08.4 + 05 56 006.46 + 50.56	01 - 39 - 000	1429 352	1.2 x 0.6 (1.4 x 0.7)	23 (5)	2	E - S0		15.4		*
09753	15 08.4 + 57 12 093.25 + 51.40	10 - 22 - 001 NGC 5879	0755	4.8 x 1.7 4.0 x 1.6	0 6	3	Sb	Sb +	11.9	+ 876 + 1064	12 3468
09754	15 08.6 + 04 29 004.72 + 49.63	01 - 39 - 000 IC 1102	1429 352	1.2 x 0.6 1.1 x 0.6	[25] ...	1	S[b] Singular, disturbed?		15.3		*
09755	15 08.8 + 10 38 012.82 + 53.14	02 - 39 - 004	1422	1.1 x 0.9 1.0 x 0.8	35: 2	3	Sc s of 2, contact		14.8		*1
09756	15 08.9 + 06 32 007.34 + 50.82	01 - 39 - 000	1429 352	[1.2 x 0.5] ...	[62] ...	3	Double system Contact		15.5		*
09757	15 09.1 + 05 35 006.19 + 50.21	01 - 39 - 004	1429 352	1.3 x 0.6 1.3 x 0.6	60 5	3	Sb	Sb +	15.0		
09758	15 09.1 + 13 40 017.33 + 54.61	02 - 39 - 007	1422	1.3 x 1.3 1.2 x 1.2	--- 1	3	S0		15.0		
09759	15 09.3 + 55 32 090.92 + 52.24	09 - 25 - 030	1096 344	2.0 x 0.35 1.5 x 0.35	51 7	1	S ...		15.4		
09760	15 09.5 + 01 52 001.87 + 47.80	00 - 39 - 011	1402 352	2.8 x 0.2 3. x 0.2	57 7	3	Sc	S d	14.9		
09761	15 09.5 + 46 20 076.90 + 56.54	08 - 28 - 005	0706	1.3: x 1.3: 1.3 x 1.2	--- 	4	S B a	S B(r)0/a	14.1		*
09762	15 09.7 + 32 51 051.89 + 59.24	06 - 33 - 000	1610	1.1: x 1.0: 0.8:: x 0.8::	--- ...	3	Dwarf spiral S V		17.		
09763	15 09.8 + 21 29 030.36 + 57.52	04 - 36 - 025	0087	2.0: x 0.4 1.3: x 0.3	... ...	3	S? ... Peculiar, long streamers		15.6		*
09764	15 09.8 + 65 05 102.42 + 46.30	11 - 18 - 029	1575	2.6: x 1.6: 1.6:: x 1.3::	80: 4:	3	...	S B(s)dm	15.7		*
09765	15 10.2 + 20 50 029.26 + 57.23	04 - 36 - 026	0087	1.2 x 0.6 1.1 x 0.6	144 5	2	Sa - b npp of 2, contact		(15.3)		*

N G C   5864, 5865, 5868, 5869, 5874 - 76, 5879   I C   1097, 1099, 1102

1	2	3	4	5	6	7	8	9	10	11	12
09766	15 10.2 + 61 19 098.14 + 48.72	10 − 22 − 003	0755	1.6 x 0.5 1.5: x 0.5	167 6	3	S B b Sb +				
09767	15 10.3 + 07 37 009.06 + 51.17	01 − 39 − 000	1429	1.0 x 0.9 (1.1 x 1.0)	−− ...	1	... In group		16.0:		*
09768	15 10.6 + 60 00 096.51 + 49.50	10 − 22 − 004 NGC 5894	0755	3.3 x 0.5 3.5 x 0.45	13 6	1	S B? c − Irr		13.2		3
09769	15 10.7 + 55 59 091.35 + 51.83	09 − 25 − 034	1096	3.0: x 2.3: 1.0 x 0.6	20: 2 − 3:	4	...	SA(s)dm	15.7		
09770	15 11.2 + 25 23 037.60 + 58.20	04 − 36 − 027	0087	1.1 x 0.9 1.3 x 0.7	(90) ...	2	Sb f of 2, disturbed:	SAB b	15.6		*
09771	15 11.2 + 41 27 068.17 + 57.78	07 − 31 − 038 NGC 5888	1371 351	1.5 x 0.9 1.4 x 0.8	158 4	2	S B b		14.3		*3
09772	15 11.3 + 34 21 054.78 + 58.85	06 − 33 − 028	1610	1.0: x 0.7: 0.7: x 0.7:	... ...	2	S B c Disturbed:, Sc +		16.5:		*1
09773	15 11.5 + 67 33 104.83 + 44.50	11 − 19 − 001 IC 1110	1426	1.3 x 0.4 1.0 x 0.35	77 7	2	Sa		14.9		
09774	15 11.7 + 42 09 069.39 + 57.51	07 − 31 − 042 NGC 5893	1371 351	1.4 x 1.3 1.3 x 1.3	−− 1	3	S B b		14.1		*12 3
09775	15 11.8 + 20 10b 028.33 + 56.66	03 − 39 − 016	0091	1.1 x 0.5 0.7 x 0.3	148 5 − 6	1	Sa nff of 2		(15.5)		*
09776	15 11.8 + 57 10 092.79 + 51.04	10 − 22 − 005	0755	1.0: x 0.5: 0.8: x 0.4:	... ...	2	Dwarf irregular Ir IV − V or V		17.		
09777	15 12.0 + 20 40 029.22 + 56.78	04 − 36 − 028	0087	1.6 x 0.9 1.7 x 0.9	150 4	1	S ... Singular		15.0		*1
09778	15 12.1 + 75 20 112.04 + 38.87	13 − 11 − 010 NGC 5909	0768	1.1 x 0.6 1.1 x 0.4	52 ...	1	S ... P w NGC 5912, disturbed		14.7		*13
09779	15 12.2 + 01 20 001.92 + 46.94	00 − 39 − 012 NGC 5887	1402 352	1.2 x 1.0 1.3 x 1.1	(160) 2	2	S0 − a		15.2		*
09780	15 12.2 + 44 47 073.97 + 56.64	08 − 28 − 009	0706	2.1 x 0.2 2.1 x 0.2	148 7	3	Sc		15.5		
09781	15 12.3 + 05 43 007.10 + 49.65	01 − 39 − 009	1429 352	1.4 x 1.4 0.9 x 0.8	−− 1	1	Sb − c		15.7		1
09782	15 12.4 + 01 23 002.02 + 46.93	00 − 39 − 000	1402 352	1.0: x 0.1 0.8 x 0.1	18 7	1	Sc		16.5:		*
09783	15 12.5 + 46 32 076.89 + 55.99	08 − 28 − 010	0706	1.0 x 0.4 0.6 x 0.2	58 (6)	3	...	S d − dm	16.5:		
09784	15 12.6 + 62 52 099.67 + 47.52	11 − 19 − 002	1426	1.4: x 0.9: 1.0: x 0.7:	(165) ...	3	Dwarf IV − V or V		16.5:		
09785	15 12.8 + 05 23 006.80 + 49.35	01 − 39 − 000	1429 352	1.0 x 0.5 1.0 x 0.5	167 5	2	Sa sf of 2, disturbed		15.4		*
09786	15 13.1 + 71 25 108.46 + 41.68	12 − 14 − 020	1442	1.5 x 0.45 1.3 x 0.4	158 6	2	Sb − c		15.6		
09787	15 13.2 + 01 37 002.47 + 46.93	00 − 39 − 015	1402 352	1.0 x 0.8 0.9 x 0.7	45 2:	3	...	S m	15.4		
09788	15 13.2 + 08 29 010.84 + 51.06	01 − 39 − 000	1429	1.0 x 0.4 0.9 x 0.4	125 ...	1	Irr? sff of 2		15.4		*
09789	15 13.2 + 42 14 069.40 + 57.22	07 − 31 − 045 NGC 5899	1371 351	2.8 x 1.2 2.5 x 1.0	18 6	2	S B c		12.6	+ 2549 + 2706	*12 3468
09790	15 13.2 + 42 24 069.70 + 57.18	07 − 31 − 046 NGC 5900	1371 351	1.5 x 0.4 1.5 x 0.4	131 7	2	Sb		15.0		*12 3
09791	15 13.3 + 59 02 095.00 + 49.80	10 − 22 − 008	0755	1.1 x 0.6 1.1 x 0.5	20 4	3	S B b	S B(r)b	15.5		
09792	15 13.5 + 00 03 000.82 + 45.85	00 − 39 − 016 Serpens	1402 (352)	15. : x 13. : 12. : x 12. :	−− ...	5	Dwarf		15.1		*1
09793	15 13.5 + 18 34 025.89 + 55.72	03 − 39 − 000	0091	1.2 x 0.20 1.2 x 0.20	51 7	3	Sb		15.7		*
09794	15 13.8 + 10 42 013.98 + 52.13	02 − 39 − 009	1422	3.0 x 0.8 2.6 x 0.8	[42] 6	3	S B c − Irr Disrupted?		14.3		*
09795	15 14.0 + 07 56 010.30 + 50.59	01 − 39 − 000	1429	1.2 x 0.35 1.1 x 0.35	115 6	2	Sa − b		15.3		
09796	15 14.0 + 43 22 071.33 + 56.77	07 − 31 − 048	1371	1.6 x 0.4 1.6 x 0.3	[17] ...	1	... Peculiar, in group		16.0:		*1
09797	15 14.0 + 55 42 090.56 + 51.60	09 − 25 − 038 NGC 5905	1096	4.7 x 3.6 4.7 x 3.6	(135) ...	5	S B b P w 9805	S B(r)b Sb +	13.6		*12 3468
09798	15 14.2 + 07 35 009.89 + 50.35	01 − 39 − 000	1429	1.1 x 0.5 1.0 x 0.4	168 6	2	S0 − a		15.4		
09799	15 14.3 + 07 12 009.42 + 50.11	01 − 39 − 012	1429	(1.8 x 0.9) (1.9 x 1.1)	30 ...	4	E Brightest in cluster		14.8		*1
09800	15 14.6 + 54 40 089.07 + 52.06	09 − 25 − 039 IC 1111	1096	1.3 x 0.6 1.1 x 0.5	55 5	3	Sb		15.5		*

N G C   5887, 5888, 5893, 5894, 5899, 5900
        5905, 5909   I C   1110, 1111

1	2	3	4	5	6	7	8	9	10	11	12
09801	15 14.6 + 56 30 091.57 + 51.09	09 - 25 - 040 NGC 5906 = 5907	1096	12.8 x 1.8 12.6 x 1.8	155 7	4	Sc	Sc -	11.4	+ 535 + 725	*12- -789
09802	15 14.7 + 43 09 070.88 + 56.71	07 - 31 - 051	1371	1.3 x 0.9 1.1: x 0.9:	50 ...	3	Sc		16.0:		1
09803	15 15.2 + 29 36 045.74 + 57.94	05 - 36 - 000	1092	1. x 1. ...	--- ...	3	Dwarf V		18.		
09804	15 15.3 + 04 20 006.10 + 48.22	01 - 39 - 015	1429 352	[1.1 x 0.6] ...	[165] ...	2	Double system Symbiotic		14.9		1
09805	15 15.3 + 55 35 090.25 + 51.50	09 - 25 - 041 NGC 5908	1096	3.0 x 1.3 3.0 x 1.6	154 7	3	Sb Singular		13.5		*123 4678
09806	15 15.5 + 38 50 063.07 + 57.52	07 - 31 - 000	1371	1.0 x 0.3 0.8 x 0.2	104 6	3	Sc		16.5:		
09807	15 15.6 + 11 05 014.88 + 51.95	02 - 39 - 000	1422	1.4 x 0.5 1.1 x 0.4	105 6	2	Sa - b		15.1		
09808	15 15.8 + 14 00 019.13 + 53.32	02 - 39 - 011	1422	1.4 x 1.0 1.4 x 1.0	(110) 2 - 3	3	Sb	SAB b	15.2		*1
09809	15 15.9 + 30 52 048.14 + 57.88	05 - 36 - 012	1092	1.3 x 1.1 1.2 x 1.0	--- 1	4	S B c s of 2	S B(r)c	(14.7)		*1
09810	15 16.2 + 08 01 010.88 + 50.19	01 - 39 - 000	1429	1.1 x 0.3 0.9: x 0.2	71 (6)	1	S: ...		16.0:		*
09811	15 16.2 + 65 22 102.09 + 45.61	11 - 19 - 004	1426	1.0 x 0.4 0.9 x 0.4	143 6	1	S ...		16.0:		
09812	15 16.7 + 09 59 013.60 + 51.15	02 - 39 - 000	1422	1.0 x 0.15 1.0 x 0.12	142 7	1	Sc: n of 2		15.5		*
09813	15 16.7 + 21 00 030.42 + 55.84	04 - 36 - 034	0087	1.1 x 0.7 1.1 x 0.7	155 4	3	S B b P w NGC 5910		15.6		*
09814	15 17.0 + 11 14 015.37 + 51.73	02 - 39 - 014	1422	1.1 x 0.4 0.9 x 0.4	45 6	2	Sc - Irr		14.7		
09815	15 17.5 + 45 06 073.96 + 55.65	08 - 28 - 015	0706	1.6: x 0.6 1.6: x 0.5:	5 6	3	...	S dm - m	15.7		
09816	15 17.5 + 59 40 095.32 + 49.00	10 - 22 - 012	0755	1.5 x 0.3 1.5 x 0.3	62 7	2	Irr		16.0:		*
09817	15 17.7 + 46 04 075.56 + 55.31	08 - 28 - 017 NGC 5918	0706	1.9 x 0.8 1.9 x 0.8	85 6	4	Sc		14.0		*3
09818	15 18.3 - 02 25 359.36 + 43.31	00 - 39 - 021 NGC 5913	1402	2.0 x 0.8 2.5 x 1.2	168 5:	3	S B ... Singular	S B(r) ...	14.6		*3
09819	15 18.5 + 73 07 109.66 + 40.19	12 - 14 - 023	1442	1.4? x 1.1? 1.3 x 0.8	--- ...	1	... Peculiar, s of 2		16.0:		*
09820	15 19.2 + 28 44 044.31 + 56.98	05 - 36 - 014	1092 359	1.2 x 1.0 0.8 x 0.6	(110) 2:	1	S0 - a		15.3		*
09821	15 19.3 + 08 36 012.30 + 49.87	02 - 36 - 016	1422 358	1.5 x 1.5 1.4 x 1.4	--- 1	4	S B b p of 2	S B(r)b	15.1		*
09822	15 19.4 + 07 53 011.39 + 49.45	01 - 39 - 000	1429	(1.1 x 0.9) (1.1 x 0.8)	--- ...	1	S0:		15.5		*
09823	15 19.4 + 41 54 068.30 + 56.19	07 - 32 - 001 NGC 5923	1376 351	2.2 x 2.2 2.2: x 2.2:	--- 1	4	Sb/S B c	SAB(s)bc Sc -	14.7		13
09824	15 19.5 + 05 17 008.14 + 47.93	01 - 39 - 021 NGC 5921	1429	5.3 x 4.6 4.5 x 4.0	(130) 1	5	S B b	S B(r)bc Sb +	12.7	+ 1389 + 1429	*123 4678
09825	15 19.5 + 23 25 034.88 + 55.91	04 - 36 - 040	0087	1.1 x 0.9 1.0 x 0.9	--- 1 - 2	4	Sc	SA(rs)c	15.4		
09826	15 19.7 + 39 22 063.79 + 56.63	07 - 32 - 002	1376	1.1 x 0.7 1.0 x 0.7	163 ...	3	S B c	Sc -	15.3		1
09827	15 20.1 + 41 18 067.19 + 56.19	07 - 32 - 003	1376 350	1.0 x 0.20 1.0 x 0.15	49 (7)	1	...		15.2		
09828	15 20.2 + 19 26 028.30 + 54.55	03 - 39 - 022	0091	1.7 x 0.12 1.7 x 0.15	178 7	3	Sc		15.7		*
09829	15 20.5 - 01 10 001.15 + 43.73	00 - 39 - 023	1402	2.3 x 1.0 2.2 x 0.8	[167] ...	4	S ... 3rd of 3, distorted		15.3		*1
09830	15 20.5 + 04 42 007.70 + 47.40	01 - 39 - 000	1429 352	1.5 x 0.2 1.3 x 0.2	30 6	2	Sc		15.5		
09831	15 20.7 + 29 57 046.59 + 56.78	05 - 36 - 025	1092	1.0 x 0.7 0.3 x 0.2	100: 3	2	Sa		14.9		*1
09832	15 21.5 + 14 03 020.24 + 52.11	02 - 39 - 000	1422	1.1: x 0.7 1.1 x 0.5	... ...	1	S ... ssf of 2, distorted		16.5:		*
09833	15 21.6 + 23 44 035.64 + 55.52	04 - 36 - 042	0087	1.2 x 1.0 1.1 x 0.9	(0) 2	4	...	SA m	15.7		
09834	15 22.1 + 38 18 061.75 + 56.32	06 - 34 - 008	0107	1.0 x 0.9 1.0 x 0.8	--- 1:	1	S0 - a		15.3		
09835	15 22.2 + 46 47 076.28 + 54.34	08 - 28 - 024	0706	1.1 x 0.9 1.1 x 0.9	(90) 2	2	Sb np of 2		14.8		*

N G C   5906 - 08,   5913,   5918,   5921,   5923

1	2	3	4	5	6	7	8	9	10	11	12
09836	15 22.3 + 30 27 047.54 + 56.48	05 – 36 – 000	1092	1.0 x 0.2 0.9 x 0.2	92 7	1	Sa – b		15.4		
09837	15 22.6 + 58 14 092.94 + 49.25	10 – 22 – 013	0755	2.0 x 1.8 2.0: x 1.8	(155) 1	4	Sc	SAB(s)c	14.6		
09838	15 22.8 + 07 20 011.40 + 48.45	01 – 39 – 000	1429	1.4 x 0.4 1.2 x 0.35	13 6	1	Sa – b		15.3		
09839	15 22.7 – 01 39a 001.15 + 43.00	00 – 39 – 000	1402	1.0 x 0.2 1.1 x 0.2	17 7	1	S ... npp of 2		(15.4)		*
09840	15 23.0 + 42 22 068.84 + 55.43	07 – 32 – 005	1376 362	1.1 x 0.9 0.9 x 0.8	[55] ...	2	Sc – Irr		16.0:		
09841	15 23.2 + 18 22 027.03 + 53.51	03 – 39 – 025	0091	2.5 x 0.4 2.3 x 0.35	55 7	2	Sb – c P w 9847		14.6		*
09842	15 23.3 + 38 08 061.40 + 56.11	06 – 34 – 009	0107	1.6 x 0.5 1.7 x 0.4	67 6	4	S B b		15.4		
09843	15 23.4 + 20 58 031.21 + 54.35	04 – 36 – 045	0087	1.1 x 0.7 1.1 x 0.5	(155) ...	2	S ... nf of 2, distorted		15.5		*1
09844	15 23.5 + 08 00 012.39 + 48.67	01 – 39 – 022	1429	1.6 x 0.55 1.5 x 0.55	67 6	4	S B b	S B(s)b Sb +	14.8		
09845	15 23.7 + 09 23 014.21 + 49.37	02 – 39 – 000	1422 358	1.6 x 0.15 1.1 x 0.2	114 7	3	Sc		15.3		
09846	15 23.7 + 16 30 024.24 + 52.68	03 – 39 – 026	0091	1.1 x 0.9 1.0 x 0.8	(160) 2	3	Sb/Sc	SAB c Sc –	14.7		
09847	15 23.7 + 18 15 026.92 + 53.36	03 – 39 – 027 NGC 5928	0091	(2.2 x 1.6) (2.0 x 1.5)	105 3	4	S0 P w 9841		13.8		*3
09848	15 23.7 + 72 00 108.25 + 40.68	12 – 15 – 004	0752	1.0 x 0.18 0.9 x 0.18	146 6	1	Sb – c		16.0:		
09849	15 23.9 + 53 35 086.47 + 51.40	09 – 25 – 049	1096	1.6 x 0.4 1.7: x 0.4	82 6	3	Sc		15.7		
09850	15 24.1 + 48 40 079.09 + 53.39	08 – 28 – 030	0706 357	1.3 x 0.5 1.3 x 0.4	93 6	3	S B b	S B(s)b	15.7		
09851	15 24.3 + 41 51a 067.86 + 55.31	07 – 32 – 006 NGC 5929	1376 362	1.0 x 0.9 1.1 x 1.0	— ...	2	E – S0 P w 9852, contact		13.0	+ 2520 + 2683	*12 3A$\overline{C}$
09852	15 24.3 + 41 51b 067.86 + 55.31	07 – 32 – 007 NGC 5930	1376 362	2.0 x 0.8 2.2 x 0.9	163 6	3	Sa P w 9851, contact			+ 2705 + 2868	*12 3A$\overline{C}$
09853	15 24.4 + 52 37 085.02 + 51.76	09 – 25 – 050	1096	1.8 x 0.3 1.7 x 0.3	93 7	3	Sb		15.1		
09854	15 24.4 + 68 55 105.12 + 42.70	12 – 15 – 007 NGC 5939	0752	0.9 x 0.45 0.9 x 0.45	35 5	1	S ...		13.7		*
09855	15 24.5 + 66 25 102.46 + 44.29	11 – 19 – 007	1426	1.6 x 0.6 1.4 x 0.6	[37] ...	3	Irr		15.0		*
09856	15 24.6 + 41 28 067.18 + 55.33	07 – 32 – 008	1376 362	2.5 x 0.3 2.1: x 0.2	152 7	3	Sc	Sc +	15.5		1
09857	15 24.7 + 41 55 067.94 + 55.22	07 – 32 – 009	1376 362	1.9 x 0.6 1.8: x 0.5:	[145] ...	4	Irr	I B m Ir I	16.0:		
09858	15 24.8 + 40 44 065.90 + 55.44	07 – 32 – 010	1376	4.5 x 0.8 4.2: x 0.8	83 6	4	S B b/Sc	SAB bc Sb +	14.0		1
09859	15 24.9 + 26 53' 041.35 + 55.46	05 – 36 – 000	1092 355	(1.0 x 0.8) (0.6 x 0.6)	— ...	1	... Compact		15.2		
09860	15 25.2 + 31 08 048.85 + 55.90	05 – 36 – 000	1092	1.1 x 0.30 1.1 x 0.3	70 6	3	Sc np of 2		15.6		*
09861	15 25.7 + 29 05bc 045.24 + 55.61	05 – 36 – 033	1092 359	[1.5 x 0.6] ...	... ...	2	Double system, common envelope, in chain of 3		(15.4)		*1
09862	15 26.4 + 43 05 069.81 + 54.66	07 – 32 – 011 NGC 5934	1376 362	0.7 x 0.3 1.3 x 0.6	[2] ...	1	... P w NGC 5935, distorted		14.5		*1
09863	15 26.9 + 03 46 007.98 + 45.57	01 – 39 – 000	1429	1.0 x 0.15 1.0 x 0.20	160 7	2	Sa In group		15.6		*
09864	15 27.0 + 05 16 009.75 + 46.42	01 – 39 – 000	1429	1.7 x 0.6 1.5 x 0.5	25 6	1	S0? spp of 2		15.3		*
09865	15 27.2 + 75 16 111.10 + 38.26	13 – 11 – 014	0768	1.1 x 0.3 1.1: x 0.35	116 6	2	Sb – c		15.7		*
09866	15 27.4 + 64 55 100.54 + 44.98	11 – 19 – 008 NGC 5949	1426	2.3 x 1.1 2.3 x 1.1	147 5	2	Sc		12.7	+ 380 + 589	*12 3468
09867	15 27.7 + 13 10 020.07 + 50.37	02 – 39 – 030 NGC 5936	1422	1.3 x 1.2 1.3 x 1.2	— 1	3	S B b	S B(rs)b	13.0		*123 46$\overline{7}$8
09868	15 27.8 + 53 03 085.29 + 51.11	09 – 25 – 055	1096	(1.0 x 0.6) (1.0 x 0.6)	(30) ...	2	...		16.5:		*
09869	15 27.9 + 23 48 036.37 + 54.14	04 – 37 – 001 IC 1124	1119	0.9 x 0.35 0.9 x 0.35	80 6	1	S ...		14.5		
09870	15 28.0 + 42 57 069.47 + 54.41	07 – 32 – 016 NGC 5943	1376 362	1.1 x 1.1 1.2 x 1.2	— (1)	1	S0?		14.6		

N G C   5928 – 30,   5934,   5936,   5939,   5943,   5949   I C   1124

1	2	3	4	5	6	7	8	9	10	11	12
09871	15 28.0 + 43 05 069.69 + 54.38	07 - 32 - 017 NGC 5945	1376 362	3.3: x 2.5: 3.2: x 2.2:	(105) 2:	4	S B a	S B(rs)ab	14.1		1
09872	15 28.0 + 68 54 104.82 + 42.46	12 - 15 - 009	0752	1.1 x 0.15 1.1 x 0.15	22 7	3	Sc		16.0:		
09873	15 28.1 + 42 48 069.21 + 54.42	07 - 32 - 018	1376 362	1.6 x 0.25 1.6 x 0.25	127 6	3	Sc	Sc -	15.6		*
09874	15 28.5 + 77 20 112.89 + 36.76	13 - 11 - 000	0768	(1.3 x 0.7) (1.4 x 0.8)	0 ...	3	E		14.6		
09875	15 28.6 + 23 14 035.51 + 53.84	04 - 37 002	1119	1.7: x 1.7: 1.4:: x 1.2::	-- ...	4	Dwarf spiral S V		17.		1
09876	15 28.8 + 07 38 012.99 + 47.37	01 - 39 - 025 NGC 5940	1429	0.8 x 0.8 0.8 x 0.8	-- 1	2	S B a - b		14.3		1
09877	15 28.9 + 42 53 069.30 + 54.26	07 - 32 - 019 NGC 5947	1376 362	1.3 x 1.3 1.3 x 1.3	-- 1	3	S B b	Sb +	14.8		*1
09878	15 29 + 83 15 117.86 + 32.48	14 - 07 - 018	1363 388	1.0 x 0.4 1.0 x 0.4	90 6	1	S ...		15.2		
09879	15 29.0 + 36 58 059.16 + 55.10	06 - 34 - 012	0107	1.1 x 0.35 1.0 x 0.3	5 6	1	Sa - b		15.6		
09880	15 29.0 + 47 30 076.80 + 53.01	08 - 28 - 000	0706	1.2: x 0.9: 1. x 0.5::	... ...	2	Dwarf IV - V or V		17.		
09881	15 29.0 + 79 58 115.16 + 34.87	13 - 11 - 000	0768	1.2 x 0.4 1.1 x 0.4	125 6	2	Sb		16.0:		
09882	15 29.5 + 40 27 065.15 + 54.60	07 - 32 - 020	1376	1.2 x 0.2 1.2 x 0.2	146 6 - 7	2	Sc		16.0:		
09883	15 29.6 + 47 13 076.30 + 53.00	08 - 28 - 000	0706	1.0 x 0.12 1.0 x 0.15	97 7	2	Sb		16.0:		*
09884	15 29.7 + 40 36 065.39 + 54.54	07 - 32 - 021 NGC 5950	1376 362	1.6 x 0.8 1.6 x 0.8	37 5	3	Sb		14.8		1
09885	15 29.8 + 70 43 106.55 + 41.17	12 - 15 - 010	0752	1.0 x 0.7 0.9 x 0.5	60 3:	3	Sc		16.0:		
09886	15 30.0 + 04 51 009.88 + 45.57	01 - 40 - 000	1082 360	(1.1 x 1.0) (1.1 x 1.0)	-- ...	2	E - S0 f of 2, in group		15.3		*
09887	15 30.2 + 04 55 010.00 + 45.57	01 - 40 - 005	1082 360	1.3 x 0.4 1.1 x 0.4	[153] ...	1	S: ... In group, disrupted		15.1		*1
09888	15 30.5 - 01 28 003.05 + 41.63	00 - 40 - 003 IC 1125	0151	1.7 x 1.0 1.7 x 1.0	[147] ...	1	S - Irr Interacting systems?		14.5		1
09889	15 30.6 + 00 44 005.38 + 43.00	00 - 40 - 000	0151	1.0 x 0.4 1.1 x 0.3	67 6	1	S ... Brightest in group		15.4		*
09890	15 30.7 + 42 10 067.98 + 54.08	07 - 32 - 023	1376 362	1.3 x 0.5 1.3 x 0.5:	45 6	3	Sc		16.0:		1
09891	15 30.9 + 40 24 064.99 + 54.35	07 - 32 - 024	1376 362	1.1 x 0.15 0.6 x 0.15	158 7	1	S ...		15.6		
09892	15 31.0 + 41 22 066.62 + 54.17	07 - 32 - 025	1376 362	1.7 x 0.3 1.6 x 0.3	104 6	2	Sb		15.3		
09893	15 31.3 + 46 37 075.20 + 52.91	08 - 28 - 038	0706	1.2 x 0.4 1.0: x 0.4	42 ...	2	Peculiar Double system?		15.2		*1C
09894	15 31.4 + 05 27a 010.87 + 45.62	01 - 40 - 000	1082 360	1.0 x 0.7 1.0 x 0.6	(10) 3	3	S B a 2nd of 3	S B(r)a	(15.3)		*
09895	15 31.4 + 15 10 023.51 + 50.45	03 - 40 - 003 NGC 5951	0082	3.5 x 0.7 3.4 x 0.7	5 6	3	Sc		13.8		*12 3<u> </u>
09896	15 31.4 + 67 44 103.31 + 42.95	11 - 19 - 009	1426	1.6 x 1.2 1.5 x 1.2	(10) 2:	4	Sc		14.4		
09897	15 31.5 + 11 11 018.06 + 48.62	02 - 40 - 002	0136	1.1 x 0.8 0.9 x 0.8	(0) ...	3	S B a	S B(r)a	15.0		
09898	15 31.7 + 16 05 024.88 + 50.76	03 - 40 - 000	0082	1.0: x 0.10 0.9 x 0.12	42 7	1	S ...		16.0:		
09899	15 31.7 + 68 25 104.02 + 42.50	11 - 19 - 010 IC 1129	1426	1.2 x 1.0 1.1 x 0.9	(170) 1 - 2	3	Sc Disturbed		13.7		*1
09900	15 31.8 + 14 34 022.73 + 50.10	03 - 40 - 004	0082	1.1 x 1.0 0.9 x 0.8	-- 1	4	...	SA: dm	15.2		
09901	15 32.1 + 12 26 019.84 + 49.08	02 - 40 - 000	0136	1.7 x 0.25 1.2 x 0.2	178 7	2	Sb - c		15.2		*
09902	15 32.2 + 15 17 023.81 + 50.32	03 - 40 - 000	0082	1.0: x 0.3 0.5 x 0.2	105 ...	2	Dwarf	S?B dm - m	17.		
09903	15 32.2 + 15 21a 023.90 + 50.35	03 - 40 - 005 NGC 5953	0082	(1.7 x 1.3) (1.8 x 1.6)	-- ...	3	S0 P w 9904, contact, distorted		13.3	+ 2100 + 2188	*12 3V$\overline{\text{A}}$
09904	15 32.2 + 15 21b 023.90 + 50.35	03 - 40 - 006 NGC 5954	0082	1.1 x 0.5 1.1 x 0.5	[5] ...	1	S[c] P w 9903, contact, distorted		13.7	+ 2140 + 2228	*12 3V$\overline{\text{A}}$
09905	15 32.3 + 08 30 014.75 + 47.09	01 - 40 - 000	1082	1. x 0.1 0.6 x 0.1	120 7	1	Sc sff of 2		15.6		*

N G C   5940, 5945, 5947, 5950, 5951, 5953, 5954   I C   1125, 1129

1	2	3	4	5	6	7	8 9	10	11	12
09906	15 32.3 + 56 45 089.98 + 48.87	09 - 25 - 058 NGC 5963	1096	4. x 3. 2.5:: x 2.0::	(55) ...	3	S ... Peculiar, p w 9914	13.0		*3
09907	15 32.5 + 09 45 016.37 + 47.69	02 - 40 - 000	0136	1.1 x 0.15 1.1 x 0.15	18 7	1	S ...	15.3		
09908	15 32.6 + 11 55 019.23 + 48.73	02 - 40 - 003 NGC 5956	0136	1.7 x 1.7 1.6 x 1.6	-- 1	1	Sc?	13.3		3
09909	15 32.7 + 28 50 045.15 + 54.05	05 - 37 - 003 NGC 5958	0121	1.1 x 1.0 1.1 x 1.0	-- 1	1	S ... Peculiar?	13.2		3
09910	15 32.7 + 31 14 049.23 + 54.31	05 - 37 - 000	0121 364	1.8: x 1.5: 0.5 x 0.4	-- ...	4	... S m	15.5		
09911	15 32.7 + 41 19 066.45 + 53.87	07 - 32 - 028	1376 362	1.3 x 0.8 1.3 x 0.7	170 4	4	Sc Sc +	16.0:		*
09912	15 32.8 + 16 43 025.96 + 50.77	03 - 40 - 007	0082	1.7: x 1.5: 1.3 x 1.1	-- 1:	4	... S B dm	15.0		*1V
09913	15 32.8 + 23 40 036.63 + 53.02	04 - 37 - 005 IC 4553, 4554	1119	2.0 x 1.8 1.7 x 1.5	-- ...	4	Strongly peculiar, double 14.4 nucleus; symbiotic pair?			*A
09914	15 32.8 + 56 52 090.09 + 48.75	10 - 22 - 020 NGC 5965	0755	6.0 x 0.9 5.2 x 0.8	53 6	4	Sb P w 9906	13.4		*
09915	15 33.0 + 12 13 019.70 + 48.79	02 - 40 - 004 NGC 5957	0136	2.8 x 2.8 2.8: x 2.8:	-- 1	4	S B b S B(r)b	13.3		*12 3̄
09916	15 33.0 + 24 15 037.59 + 53.12	04 - 37 - 000	1119	1.0 x 0.2 0.8 x 0.15	117 7	1	S ... sf of 2	15.7		*
09917	15 33.1 + 21 00 032.44 + 52.21	04 - 37 - 006	1119 365	1.1 x 0.6 1.0 x 0.5	55 5	1	S B: ...	15.2		
09918	15 33.2 + 31 01 048.87 + 54.19	05 - 37 - 005 NGC 5961	0121 364	0.9 x 0.35 0.8 x 0.30	100 ...	1	... P w 9920	14.0		*1
09919	15 33.3 + 12 46 020.49 + 48.98	02 - 40 - 000	0136	1.6 x 0.15 1.7 x 0.2	171 7	3	Sc	15.1		*
09920	15 33.3 + 30 58 048.79 + 54.16	05 - 37 - 006	0121 364	1.4 x 0.15 1.4 x 0.20	44 7	2	Sb P w 9918	15.1		*1
09921	15 33.5 + 26 30b 041.30 + 53.50	05 - 37 - 000	0121 363	1.1 x 0.3 0.8: x 0.2	155 6	1	S ... Brightest of 3	(15.6)		*
09922	15 34.0 + 38 50 062.21 + 53.95	07 - 32 - 30+31	1376	[0.9 x 0.3] ...	... ...	2	Double system Contact	14.4	+ 5500 + 5666	*1C
09923	15 34.0 + 39 57 064.09 + 53.82	07 - 32 - 032 NGC 5966	1376	(1.8 x 1.1) (1.4 x 0.9)	90 ...	2	E	13.9		*13
09924	15 34.0 + 45 00 072.41 + 52.88	08 - 28 - 000	0706	0.8 x 0.7 0.7 x 0.7	-- ...	1	S[B?] ...	14.4		
09925	15 34.2 + 16 36 026.00 + 50.41	03 - 40 - 010	0082	1.4: x 0.6: 1.2 x 0.5	10 6	2	Sc	15.4		
09926	15 34.2 + 16 46 026.24 + 50.48	03 - 40 - 011 NGC 5962	0082	2.8 x 2.0 2.8 x 2.0:	110 3	4	Sc Sc -	12.2	+ 1993 + 2088	*12 3468̄
09927	15 34.2 + 22 40 035.17 + 52.45	04 - 37 - 010	1119 365	1.0 x 0.9 (1.0 x 0.9)	-- 1	3	S B 0	14.9		*
09928	15 34.3 + 43 39 070.21 + 53.14	07 - 32 - 034 IC 4562	1376 362	1.1 x 1.1 1.1 x 1.1	-- ...	1	E? sp of 2, IC 4567 group	13.8	+ 5860 + 6036	*C
09929	15 34.4 + 56 39 089.64 + 48.66	09 - 26 - 002 NGC 5971	1413	1.7 x 0.6 1.7 x 0.6	136 6	2	Sa	14.9		*
09930	15 34.8 + 43 41 070.23 + 53.04	07 - 32 - 036 IC 4564	1376 362	1.5 x 0.5 1.5 x 0.5	70 6	1	S ... IC 4567 group	14.4		*
09931	15 34.9 + 43 35 070.06 + 53.05	07 - 32 - 037 IC 4565	1376 362	1.0 x 0.5 0.9 x 0.5	8 (5)	1	... IC 4567 group	14.8		
09932	15 35 + 82 38 117.20 + 32.80	14 - 07 - 022 IC 1143	1363 388	(1.5 x 1.4) (1.5 x 1.4)	-- ...	3	E	14.7		*
09933	15 35.0 + 43 43 070.27 + 53.00	07 - 32 - 038 IC 4566	1376 362	1.9 x 1.3 1.8 x 1.1:	165 3	3	Sb SAB ab IC 4567 group, Sb -	14.3		*
09934	15 35.0 + 59 44 093.61 + 47.11	10 - 22 - 023	0755	1.2 x 1.2 1.0 x 1.0	-- 1	2	S B a - b	15.4		
09935	15 35.1 + 06 09 012.42 + 45.25	01 - 40 - 008 NGC 5964	1082	4.4 x 3.4 4.5: x 3.5:	(145) 1 - 2	5	S B c S B(rs)d	14.2		*23̄
09936	15 35.1 + 44 24 071.37 + 52.83	07 - 32 - 039	1376	1.7 x 0.9 1.7 x 0.9	165 5:	4	Dwarf spiral, S m S IV - V	15.5		*1
09937	15 35.2 + 20 43 032.25 + 51.65	04 - 37 - 013	1119 365	1.1 x 0.22 1.2 x 0.22	15 7	2	S0 - a	14.7		
09938	15 35.2 + 30 15 047.65 + 53.69	05 - 37 - 000	0121 364	1.5:: x 1.5:: 1. x 1.	-- ...	3	Dwarf (spiral) S: V	17.		
09939	15 35.3 - 01 35 003.94 + 40.63	00 - 40 - 004 IC 1128	0151	(1.1 x 0.9) (1.4 x 1.1)	(0) ...	2	[S0] Compact, nff of 2	15.5		*
09940	15 35.5 + 43 28 069.83 + 52.96	07 - 32 - 040 IC 4567	1376 362	1.6 x 1.2 1.6 x 0.9	125 2 - 3	2	S[c?] Peculiar? Brightest in group	13.5		*

N G C   5956 - 58,   5961 - 66,   5971

I C   1128, 1143, 4553, 4554, 4562
      4564 - 67

1	2	3	4	5	6	7	8	9	10	11	12
09941	15 36.0 + 13 07 021.42 + 48.55	02 - 40 - 005	0136	1.5: x 1.3: 1.3: x 1.0	--- ...	3	Dwarf IV - V		15.7		
09942	15 36.0 + 71 35 107.00 + 40.23	12 - 15 - 000	0752	1.0 x 0.15 1.0 x 0.18	67 7	2	Sa		16.0:		*
09943	15 36.1 + 12 21 020.41 + 48.18	02 - 40 - 006 NGC 5970	0136	2.9 x 2.0 2.9 x 1.9	88 3	4	S B c	S B(r)c	12.2	+ 2055 + 2136	*123 46$\overline{78}$
09944	15 36.3 + 73 37 109.01 + 38.91	12 - 15 - 000	0752	1.4 x 0.4 1.5 x 0.4	174 6	1	S ...		14.9		
09945	15 36.4 + 04 45 011.04 + 44.20	01 - 40 - 009	1082	1.3 x 1.2 1.2 x 1.1	--- 1	3	Sc	Sc -	14.4		
09946	15 36.6 + 17 12 027.21 + 50.12	03 - 40 - 016 NGC 5972	0082	1.1 x 0.7 1.1 x 0.7	(5) 4:	2	S0 - a		14.8		
09947	15 36.8 + 41 12 066.05 + 53.12	07 - 32 - 041	1376 362	1.5: x 1.3: 0.9: x 0.7:	(110) ...	3	...	S B(r)dm	16.5:		1
09948	15 36.8 + 59 33 093.22 + 47.00	10 - 22 - 027 NGC 5981	0755	2.8 x 0.35 2.7 x 0.35	140 7	2	Sb - c NGC 5985 group		14.2		*12 36$\overline{}$
09949	15 36.9 + 14 19 023.20 + 48.88	02 - 40 - 000	0136	1.0 x 0.10 0.9 x 0.10	109 7	3	Sc In group		15.5		*
09950	15 37 + 82 25 116.96 + 32.90	14 - 07 - 023	1363 388	1.2 x 0.7 1.1 x 0.6	88 4	2	Sb - c		15.2		*
09951	15 37.0 + 15 33 024.92 + 49.38	03 - 40 - 017	0082	1.4 x 0.8 1.3 x 0.7	130 4	4	Sc	Sc +	15.7		1
09952	15 37.0 + 31 55 050.50 + 53.45	05 - 37 - 010 NGC 5974	0121 364	0.65 x 0.35 0.65 x 0.35	[110] ...	1	... s of 2		14.3		*3
09953	15 37.1 + 03 22 009.60 + 43.27	01 - 40 - 011	1082	1.4 x 0.9 1.1 x 0.7	150: 3 - 4	4	S B c		15.0		*
09954	15 37.1 + 24 37 038.54 + 52.30	04 - 37 - 015	1119	1.1 x 0.5 0.9 x 0.3	13 6	2	Sa - b		15.1		*
09955	15 37.1 + 79 48 114.71 + 34.73	13 - 11 - 000	0768	1.2 x 0.35 1.0 x 0.2	88 6	1	S ...		16.5:		
09956	15 37.3 + 10 23 018.05 + 46.98	02 - 40 - 000	0136	1.0 x 0.2 1.0 x 0.25	117 6	1	S B b		15.4		*
09957	15 37.4 + 40 19 064.57 + 53.13	07 - 32 - 000	1376	1.0 x 0.10 1.0 x 0.10	18 7	3	Sc		17.		
09958	15 37.5 + 21 57 034.41 + 51.51	04 - 37 - 018	1119 365	1.1 x 0.9 (1.2 x 1.1)	--- ...	2	E - S0		15.0		*1
09959	15 37.5 + 44 02 070.63 + 52.49	07 - 32 - 042	1376 362	2.0 x 1.2 2.0: x 1.1	130: ...	4	S B b	S B(s)b	14.6		
09960	15 37.6 + 02 05 008.28 + 42.42	00 - 40 - 000	0151	1.6 x 0.5 1.5 x 0.45	42 6	2	Sa		15.5		
09961	15 37.6 + 59 31 093.10 + 46.93	10 - 22 - 029 NGC 5982	0755	(3.0 x 2.1) (3.3 x 2.2)	110 ...	4	E NGC 5985 group		12.4	+ 2866 + 3072	*12 34$\overline{68}$
09962	15 37.7 + 14 07 023.05 + 48.62	02 - 40 - 000	0136	1.0 x 0.15 0.9 x 0.2	173 ...	3	Irr In group		15.3		*
09963	15 37.8 + 21 38 033.95 + 51.36	04 - 37 - 019 NGC 5975	1119 365	1.0 x 0.25 1.1 x 0.3	171 6	1	S ... sp of 2		14.7		*1
09964	15 37.9 + 07 27 014.51 + 45.36	01 - 40 - 000	1082	1.3 x 1.1 (1. x 1.)	--- 1	4	S B 0/S B a	(R)S B(r)0/a	15.2		*
09965	15 37.9 + 20 51 032.77 + 51.09	04 - 37 - 020 IC 1132	1119 365	1.3 x 1.1 1.2 x 1.1	--- 1 - 2	4	Sc	SA(rs)c Sc -	14.4		
09966	15 38 + 83 15 117.64 + 32.29	14 - 07 - 024	1363 388	1.1 x 0.35 1.1 x 0.3	153 6	2	S B a p of 2		15.6		*
09967	15 38.2 + 17 17 027.56 + 49.79	03 - 40 - 023 NGC 5977	0082	1.1 x 0.9 1.1 x 0.9	(155) 1 - 2	2	S B 0		15.1		*
09968	15 38.6 - 01 33 004.65 + 40.01	00 - 40 - 006	0151	1.3 x 1.0 1.4 x 1.0	(150) 2	3	Sb	SAB(rs)b	15.2		*
09969	15 38.6 + 59 30 092.99 + 46.83	10 - 22 - 030 NGC 5985	0755	5.8 x 3.1 5.6 x 3.0	13 5	4	Sb Brightest of 3, Sb +	SAB(rs)b	12.0	+ 2467 + 2673	*12 34$\overline{68}$
09970	15 38.8 + 51 15 081.64 + 50.26	09 - 26 - 005	1413	1.2 x 0.15 0.8 x 0.12	0 7	2	Sc		16.5:		
09971	15 38.8 + 58 15 091.35 + 47.40	10 - 22 - 032 NGC 5987	0755	5.1 x 1.8 4.7 x 1.7	165 6	4	Sb	Sb -	13.3		3
09972	15 38.8 + 59 59 093.59 + 46.57	10 - 22 - 031	0755	1.0 x 0.6: 0.9 x 0.5	135 4:	2	S B a		15.6		*
09973	15 38.9 + 15 44 025.47 + 49.03	03 - 40 - 025 IC 1133	0082	1.3 x 0.45 1.3 x 0.45	130 6	2	Sc P w 9974		14.8		*
09974	15 39.1 + 15 57 025.80 + 49.08	03 - 40 - 026 NGC 5980	0082	1.9 x 0.6 1.7 x 0.6	13 6	1	S ... P w 9973		13.3		*13
09975	15 39.2 + 28 24 044.78 + 52.58	05 - 37 - 014 IC 4570	0121 366	1.0 x 0.9 0.9 x 0.8	--- 1	3	Sc	SAB: ...	15.1		

N G C 5970, 5972, 5974, 5975, 5977, 5980 - 82   I C 1132, 1133, 4570
      5985, 5987

1	2	3	4	5	6	7	8	9	10	11	12
09976	15 39.4 + 06 00 013.07 + 44.27	01 - 40 - 000	1082	1.1: x 0.8: 0.5 x 0.3	-- 3:	3	S B b	S B(r)b	15.5		
09977	15 39.5 + 00 52 007.35 + 41.31	00 - 40 - 007	0151	3.9 x 0.45 3.9 x 0.5	77 7	4	Sc		14.7		
09978	15 39.7 + 06 49 014.09 + 44.65	01 - 40 - 000	1082	1.3 x 0.5 1.0: x 0.4	35: 6	3	Sc		15.2		*
09979	15 39.8 + 00 37 007.14 + 41.10	00 - 40 - 008	0151	1.5: x 0.5: 1.5: x 0.5:	[140] ...	4	Dwarf irregular Ir V		15.7		*
09980	15 40.0 + 02 10 008.85 + 41.98	00 - 40 - 000	0151	1.0 x 0.6 1.1 x 0.6	175: 4:	3	S B O		15.2		
09981	15 40.2 + 45 43 073.12 + 51.65	08 - 29 - 003	1404	1.0 x 0.3 1.0 x 0.3	105 7	1	SO: ssf of 2		15.5		*
09982	15 40.2 + 70 55 106.04 + 40.38	12 - 15 - 011	0752	1.2 x 1.2 1.2: x 1.2:	-- 1	4	Sc	S cd	15.3		*
09983	15 40.3 + 08 24 016.11 + 45.35	01 - 40 - 012 NGC 5983	1082	1.0 x 1.0 (0.9 x 0.8)	-- ...	3	E		15.1		*3
09984	15 40.5 + 23 02 036.38 + 51.15	04 - 37 - 023	1119 365	1.0 x 0.8 0.9: x 0.7	-- 2:	3	S B b	S B(r)b	15.3		
09985	15 40.5 + 59 55 093.35 + 46.41	10 - 22 - 034 NGC 5989	0755	1.0 x 0.9 1.0 x 0.9	-- 1	1	Sc: Singular		13.6		
09986	15 40.5 + 68 58 103.96 + 41.53	12 - 15 - 012	0752	1.0 x 0.6 0.8 x 0.6	140 4	2	Sa - b		15.6		
09987	15 40.6 + 14 23 023.87 + 48.10	02 - 40 - 011 NGC 5984	0136	2.9 x 0.6 2.9 x 0.7	144 6	3	S B c P w 9991	Sc +	13.5		*12 3478
09988	15 40.6 + 41 48 066.86 + 52.33	07 - 32 - 000	1376 362	1.0 x 0.1 0.9 x 0.1	67 7	2	Sc 2nd of 3		17.		*
09989	15 40.9 + 09 53 018.05 + 45.97	02 - 40 - 000	0136	1.0: x 0.25 1.0 x 0.3	... ...	1	S? ... Peculiar, streamer		15.3		*
09990	15 41.0 + 04 57 012.16 + 43.36	01 - 40 - 000	1082	1.4 x 0.35 1.1 x 0.3	164 6	3	Sc In group		15.3		*
09991	15 41.0 + 14 36 024.22 + 48.10	03 - 40 - 030	0082	1.7 x 0.5 1.6 x 0.4	70 6	3	Sc P w 9987		15.1		*
09992	15 41.4 + 67 25 102.20 + 42.37	11 - 19 - 011	1426	1.8: x 1.2: 1.6:: x 1.0::	... ...	3	Dwarf irregular Ir IV - V		15.5˙		1
09993	15 41.7 + 08 00 015.87 + 44.85	01 - 40 - 000	1082	1.1 x 0.1 0.6: x 0.1	... ...	1	... ssf of 2, distorted, acicular?		16.5:		*
09994	15 41.7 + 33 27 053.14 + 52.55	06 - 34 - 000	0107	1.1 x 0.1 0.9 x 0.1	155 7	2	Sc		16.5:		
09995	15 41.7 + 47 54 076.40 + 50.85	08 - 29 - 006	1404 367	1.0 x 0.5 0.7 x 0.4	132 5	2	Sa - b		15.4		
09996	15 42.1 + 11 42 020.56 + 46.58	02 - 40 - 000	0136	1.7 x 0.15 1.7 x 0.20	111 7	4	Sb np of 2		15.3		*
09997	15 42.1 + 43 57 070.23 + 51.70	07 - 32 - 047	1376 362	1.0 x 0.2 1.0 x 0.2	172 6	2	Sb - c n of 2, in chain		15.6		*
09998	15 42.2 + 10 27 018.99 + 45.96	02 - 40 - 012 NGC 5988	0136	1.3 x 1.0 1.2 x 0.8	115: 2 - 3	4	Sc		15.3		
09999	15 42.2 + 25 29 040.33 + 51.38	04 - 37 - 026	1119 368	(1.4 x 0.8) (1.4 x 0.8)	(35) 4	3	SO		14.7		*
10000	15 42.3 + 04 06 011.44 + 42.62	01 - 40 - 000	1082	1.2 x 0.12 0.7 x 0.1	124 7	3	Sc		15.7		
10001	15 42.3 + 44 28 071.03 + 51.56	07 - 32 - 048	1376	1.5 x 1.1 1.5: x 1.1:	(110) ...	2	S ... Singular		15.4		*1
10002	15 42.5 + 57 25 089.93 + 47.33	10 - 22 - 035	0755	1.2 x 0.15 0.9: x 0.15	134 7	1	S B? ... sf of 2, contact, disturbed		16.0:		*1
10003	15 42.6 + 41 15 065.88 + 52.04	07 - 32 - 049 NGC 5992, Mark 489	1376 362	1.0 x 0.8 1.0 x 0.8	130: ...	1	S ... P w 10007		14.2		*13
10004	15 42.6 + 47 26 075.62 + 50.83	08 - 29 - 007	1404	1.0 x 0.1 1.1: x 0.1	3 7	3	Sc		17.		
10005	15 42.7 + 00 55 008.04 + 40.70	00 - 40 - 009	0151	1.6: x 1.5: 1.4: x 1.4:	-- (1)	1	Sc?		15.7		
10006	15 42.7 + 32 36 051.77 + 52.29	06 - 35 - 000	0071	1.1 x 0.25 0.9 x 0.2	118 7	1	Sb - c		16.0:		
10007	15 42.7 + 41 17 065.93 + 52.01	07 - 32 - 050 NGC 5993	1376 362	1.2 x 0.9 1.1 x 0.9	140: ...	2	S B: b P w 10003	S (r)b	13.9		*13
10008	15 43 + 82 30 116.87 + 32.71	14 - 07 - 027	1363 388	1.1 x 1.1 1.1 x 1.1	-- 1	4	Sc sff of 2	SA(s)c Sc +	15.7		*
10009	15 43.1 + 04 19 011.83 + 42.57	01 - 40 - 000	1082	1.1: x 0.4: 1. x 0.4::	3 ...	3	Dwarf V		17.		
10010	15 43.1 + 46 14 073.74 + 51.04	08 - 29 - 009	1404	1.6 x 0.8 1.4 x 0.5	5 ...	4	Irr		15.4		

N G C   5983, 5984, 5988, 5989, 5992, 5993

1	2	3	4	5	6	7	8	9	10	11	12
10011	15 43.2 + 25 36 040.59 + 51.18	04 – 37 – 029	1119 368	1.1 x 0.35 1.1 x 0.35	145 6	2	Sb – c		15.4		
10012	15 43.2 + 36 15 057.74 + 52.28	06 – 35 – 003	0071	(1.2 x 0.8) (1.4 x 1.0)	(145) ...	3	E Brightest in cluster		16.0:		*1
10013	15 43.2 + 55 49 087.72 + 47.92	09 – 26 – 013	1413	1.1 x 0.55 1.0 x 0.6	12 5	2	S B: a – b		14.9		
10014	15 43.4 + 12 39 022.00 + 46.73	02 – 40 – 013	0136	1.3:: x 1.1:: 1. x 1.	— ...	4	Dwarf V		15.7		
10015	15 43.4 + 21 11 033.90 + 49.98	04 – 37 – 000	1119	1.1: x 0.2 0.7: x 0.2	34 (6)	2	Sc – Irr		16.5:		
10016	15 43.4 + 26 46 042.43 + 51.38	05 – 37 – 000	0121	1.0 x 0.15 1.1 x 0.20	0 7	1	Sa – b		15.5		
10017	15 43.4 + 21 35 034.49 + 50.10	04 – 37 – 000	1119 371	1.2 x 0.6 0.8: x 0.6:	... ...	3	Dwarf IV – V		17.		
10018	15 43.4 + 67 55 102.61 + 41.92	11 – 19 – 012	1426 370	1.1 x 1.0 1.1 x 1.0	— 1	4	S B b	(R')S B(s)bc Sb +	15.0		
10019	15 43.5 + 30 18b 048.06 + 51.91	05 – 37 – 021	0121	1.4 x 0.4 1.0 x 0.2	168 6	2	Sb ssf of 2, 2nd of 3		(15.4)		*1
10020	15 43.6 + 20 44 033.26 + 49.79	04 – 37 – 031	1119	2.3 x 2.2 1.8: x 1.8:	— 1	4	Sc	SA d	14.5		*
10021	15 43.6 + 28 15 044.79 + 51.60	05 – 37 – 020 IC 4582	0121	1.4 x 0.35 1.3 x 0.3	171 6 – 7	1	S ...		14.9		*1
10022	15 43.6 + 57 32 089.99 + 47.15	10 – 22 – 036	0755	1.5 x 0.4 1.3: x 0.2	88 7	3	Sc sp of 2		16.5:		*
10023	15 43.7 + 07 03 015.10 + 43.93	01 – 40 – 000	1082	1.1 x 0.7 1.0 x 0.6	[93] ...	4	Irr Ir IV	I m Ir I	15.2		
10024	15 43.8 + 02 34 010.02 + 41.44	01 – 40 – 014 NGC 5990	1082	1.6 x 0.9 1.7 x 1.0	115 4 – 5	1	Sa?		13.1		*3
10025	15 43.9 + 02 59 010.50 + 41.66	01 – 40 – 000	1082	1.0 x 0.1 0.8 x 0.1	83 7	2	...	d – m	17.		
10026	15 43.9 + 10 55 019.86 + 45.82	02 – 40 – 000	0136	1.0 x 0.4 0.9 x 0.35	162 6	2	Sc		15.3		
10027	15 44.0 + 04 35 012.30 + 42.54	01 – 40 – 000	1082	1.0 x 0.10 0.9 x 0.10	125 7	2	Sb – c		16.0:		*
10028	15 44.1 + 44 24 070.82 + 51.26	07 – 32 – 052	1376 362	1.0 x 0.3 0.9: x 0.2	163 ...	1	...		15.7		*1
10029	15 44.4 + 06 03 014.06 + 43.25	01 – 40 – 000	1082	(1.3 x 0.9) (1.2 x 0.8)	100 3	2	S0		15.1		*
10030	15 44.5 – 00 50 006.56 + 39.29	00 – 40 – 010	0151	1.4: x 0.8 1.4 x 0.7	175 ...	2	S B a – b		15.4		*
10031	15 44.5 + 61 45 095.31 + 45.10	10 – 22 – 000	0755	1.7: x 1.5: 1.5:: x 1.0::	— ...	4	Dwarf spiral S B V		17.		
10032	15 44.6 + 72 35 107.49 + 39.09	12 – 15 – 015 IC 1145	0752	1.3 x 0.45 1.4 x 0.4	168 6	2	Sb – c		15.2		
10033	15 44.7 + 18 02 029.50 + 48.63	03 – 40 – 38 + 39 NGC 5994 + 5996	082	[3.8 x 2.8] ...	... ...	3	S B ... + companion Distorted		13.2		*1V A
10034	15 44.7 + 31 10 049.50 + 51.75	05 – 37 – 024	0121	1.2 x 1.1 1.2 x 1.1	— 1	3	S B b/Sb	SAB(r)b Sb +	15.3		
10035	15 45.5 + 26 13 041.72 + 50.81	04 – 37 – 033	1119 368	1.1 x 0.7 0.8 x 0.6	— ...	1	... Peculiar?		14.9		
10036	15 45.7 + 28 48 045.78 + 51.23	05 – 37 – 027 NGC 6001 = 6002	0121	1.1 x 1.1 1.1 x 1.1	— 1	4	Sc	Sc –	14.4		*13
10037	15 45.8 + 11 26 020.83 + 45.65	02 – 40 – 000	0136	1.0 x 1.0 1.2 x 1.0	— 1	4	S B b	(R')S B(r:) ...	15.2		*
10038	15 46.1 + 26 21 041.97 + 50.70	04 – 37 – 000 IC 1138	1119 368	1.0 x 0.18 0.9 x 0.18	41 7	2	S0 – a		15.3		*
10039	15 46.2 + 11 48 021.36 + 45.73	02 – 40 – 000	0136	1.1 x 0.25 0.9 x 0.25	76 6	1	S ...		15.4		
10040	15 46.4 + 18 01 029.70 + 48.25	03 – 40 – 043	0082 382	1.0 x 0.3 1.0 x 0.3	160 6	1	S ... Disturbed?		15.3		*
10041	15 46.5 + 05 20 013.62 + 42.42	01 – 40 – 015	1082	3.0 x 1.7 3.3: x 1.8:	[165] ...	4	...	SAB dm	15.0		*1
10042	15 46.5 + 07 22 015.97 + 43.50	01 – 40 – 016	1082	1.6 x 0.25 1.4 x 0.20	169 7	2	Sc		14.8		
10043	15 46.5 + 22 02 035.49 + 49.54	04 – 37 – 034	1119 371	2.4 x 0.30 2.2 x 0.30	151 7	4	Sb/Sc		15.4		*1
10044	15 46.6 + 18 15 030.05 + 48.29	03 – 40 – 044	0082 382	1.3 x 0.12 1.2 x 0.12	115 7	3	Sc	Sc +	16.0:		
10045	15 46.8 + 00 19 008.20 + 39.52	00 – 40 – 000	0151	1.2 x 0.15 1.0 x 0.25	90 7	1	S ... 10046 group		15.5		

N G C   5990, 5994, 5996, 6001, 6002   I C   1138, 1145, 4582

1	2	3	4	5	6	7	8	9	10	11	12
10046	15 46.9 + 00 22 008.27 + 39.53	00 − 40 − 011	0151	1.4 x 0.9 1.7 x 1.0	[178] ...	3	Double system, E + E Common envelope		15.1		*1
10047	15 47.0 + 72 19 107.07 + 39.11	12 − 15 − 016 NGC 6011	0752	1.8 x 0.7 1.9 x 0.7	110 6	4	Sb		14.5		3
10048	15 47.1 + 19 11 031.43 + 48.51	03 − 40 − 048 NGC 6003	0082 382	1.1 x 1.1 1.1 x 1.1	--- (1)	1	SO?		14.4		
10049	15 47.1 + 21 59 035.47 + 49.39	04 − 37 − 037	1119 371	[1.3 x 0.7] ...	... ...	3	Chain of 4 galaxies		15.1		*1
10050	15 47.4 + 18 41 030.76 + 48.27	03 − 40 − 049	0082 382	(1.6 x 1.0) (1.7 x 1.1)	154 ...	2	SO		14.6		*1
10051	15 47.5 + 12 33 022.52 + 45.79	02 − 40 − 014 IC 1141	0136	0.50 x 0.45 0.50 x 0.50	--- ...	1	...		14.5		
10052	15 47.8 + 20 58 034.07 + 48.93	04 − 37 − 40+41	1119 371	[1.1 x 0.8] ...	... ...	3	Double system, S + S Contact		14.4		*1V AC
10053	15 47.9 + 69 38 104.18 + 40.61	12 − 15 − 017	0752	1.1 x 0.8 1.1 x 0.6	... ...	3	... npp of 2, distorted, in group		15.2		*1
10054	15 48 + 81 59 116.28 + 32.93	14 − 07 − 029	1363 388	3.2: x 1.4 2.5: x 1.3	150 5 − 6	4	... S B dm S B IV, sf of 2		14.9		*
10055	15 48.1 + 18 17 030.29 + 47.97	03 − 40 − 050 IC 1142	0082 382	1.5 x 1.3 1.5: x 1.3:	--- 1	3	Sc		15.1		*1
10056	15 48.1 + 19 05 031.41 + 48.25	03 − 40 − 051 NGC 6004	0082	1.8 x 1.6 1.8 x 1.6	(105) 1	4	S B c SAB(rs)bc Sc −		13.4		*1
10057	15 48.2 + 68 22 102.77 + 41.30	11 − 19 − 017	1426 370	1.6 x 0.7 1.6 x 0.7	92 ...	4	S B b (R')S B(s)bc Sb +		15.3		*
10058	15 48.3 + 26 05 041.73 + 50.17	04 − 37 − 000	1119 368	1.2: x 0.9: 0.6: x 0.5:	--- ...	4	Dwarf spiral, S B(s)m S B IV − V		17.		
10059	15 48.6 + 22 24 036.24 + 49.18	04 − 37 − 045	1119 371	1.2 x 0.8 0.9 x 0.4	(115 3:	3	Sc		15.7		
10060	15 48.8 + 20 20 033.27 + 48.51	08 − 40 − 052	0082 382	1.0 x 0.7 1.0 x 0.7	4 3:	4	S B b S B(s)b		15.7		*
10061	15 48.8 + 16 27 027.86 + 47.13	03 − 40 − 000	0082	2.3: x 2.0: 2. x 2.	--- ...	4	Dwarf irregular Ir V		16.5:		*
10062	15 49.0 + 22 06 035.84 + 49.01	04 − 37 − 047	1119 371	1.5 x 0.7 1.4 x 0.7	167 5	3	Sc		15.2		*
10063	15 49.0 + 25 53 041.48 + 49.97	04 − 37 − 000	1119 368	1.2: x 0.3 0.7: x 0.3:	130 ...	3	... S B dm − m P w 10064, distorted		16.5:		*
10064	15 49.1 + 25 51 041.43 + 49.94	04 − 37 − 048	1119 368	1.0 x 0.2 0.8 x 0.2	10 6 − 7	2	SO P w 10063, interaction		15.0		*1
10065	15 49.3 + 37 10 059.18 + 51.04	06 − 35 − 000	0071 391	1.0 x 0.7 0.9 x 0.7	(160) 2:	2	Sb		15.1		
10066	15 49.4 + 23 57 038.60 + 49.42	04 − 37 − 000	1119 373	1.2 x 0.12 0.7 x 0.12	3 7	1	S ...		15.4		*
10067	15 49.5 − 01 06 007.25 + 38.13	00 − 40 − 000	0151	1.2 x 0.2 1.2 x 0.2	89 7	2	Sc		15.6		*
10068	15 49.7 + 13 03 023.50 + 45.53	02 − 40 − 016	0136 374	1.0 x 0.8 1.0 x 0.8	(100) 2	3	SO		15.0		
10069	15 49.7 + 43 33 069.22 + 50.41	07 − 33 − 001 IC 1144, Mark 491	1369	0.7 x 0.5 0.7 x 0.5	100 ...	2	E − SO		14.4		
10070	15 49.7 + 47 23 075.08 + 49.69	08 − 29 − 014	1404	1.4 x 0.7 1.1 x 0.6	120 5	1	S ...		13.6		
10071	15 49.7 + 51 57 081.79 + 48.43	09 − 26 − 019	1413	1.1 x 0.6 0.7: x 0.6	95: ...	1	S ...		15.4		
10072	15 49.8 + 71 25 105.97 + 39.46	12 − 15 − 026	0752	1.9 x 1.2 1.6 x 1.1	(140) ...	3	S B b/Sc 3rd of 3, disturbed		15.4		*
10073	15 50.1 + 24 46 039.88 + 49.47	04 − 37 − 050	1119 373	1.2 x 0.6 1.2 x 0.6	147 5	3	S B b/Sc		14.7		1
10074	15 50.6 + 24 32 039.58 + 49.30	04 − 37 − 053	1119 373	1.1 x 0.7 1.0 x 0.5	100: 4:	3	Sc		15.2		
10075	15 50.6 + 62 27 095.69 + 44.13	10 − 23 − 003 NGC 6015	1100	7.0 x 2.4 5.0 x 2.0	28 6	4	Sc SA(s)cd	11.7	+ 674 + 889	*12− −689
10076	15 50.7 + 21 16 034.81 + 48.38	04 − 37 − 052 NGC 6008	1119 371	1.5 x 1.5 1.3: x 1.3:	--- 1	4	S B b S B(r)b Disturbed?, npp of 2		14.2		*1
10077	15 50.8 + 13 23 024.10 + 45.43	02 − 40 − 017	0136 374	1.1 x 0.5 1.1 x 0.5	158 5	2	Sb − c		15.0		
10078	15 50.8 + 67 19a 101.41 + 41.67	11 − 19 − 022	1426 370	(1.4 x 1.3) (1.4 x 1.3)	--- ...	2	E Brightest in group		(14.6)		*1
10079	15 51.0 + 12 06 022.50 + 44.82	02 − 40 − 018 NGC 6007	0136	1.7 x 1.2 1.5 x 1.2	65 3	4	Sc SAB: bc Sc −		14.1		*3
10080	15 51.1 + 40 47 064.86 + 50.50	07 − 33 − 004 NGC 6013	1369 378	1.5 x 0.8 1.5 x 0.8	174 5	3	S B b		14.6		

N G C   6003, 6004, 6007, 6008, 6011, 6013, 6015   I C   1141, 1142, 1144

1	2	3	4	5	6	7	8	9	10	11	12
10081	15 51.8 + 00 40	00 – 40 – 013	0151	2.1 x 0.45	105	2	Sa		13.3		3
	009.51 + 38.71	NGC 6010		2.1 x 0.45	7						
10082	15 51.9 + 12 16	02 – 41 – 000	0136	1.1 x 0.45	[70]	2	S B ... + companion		15.5		*
	022.85 + 44.70		374	1.1: x 0.5	...		Bridge, distorted				
10083	15 51.9 + 14 44	03 – 40 – 059	0082	2.1 x 1.6	168	3	S B a	S B(rs)ab:	13.1		3
	026.01 + 45.76	NGC 6012		2.0 x 1.3	2 – 3						
10084	15 52.1 + 18 47	03 – 40 – 061	0082	1.2 x 0.7	...	1	S0?		14.8		*1
	031.48 + 47.26		382	(1.2 x 0.7)	...		Streamer:				
10085	15 52.2 + 18 40	03 – 40 – 062	0082	1.0 x 0.7	65	3	Sc		14.7		*
	031.33 + 47.20		382	0.9 x 0.6	3		Brightest in group				
10086	15 52.3 + 16 45	03 – 40 – 063	0082	0.7 x 0.20	[177]	1	...		14.5		*
	028.73 + 46.47			0.6 x 0.20	...		nf of 2, disturbed?				
10087	15 52.4 + 41 45b	07 – 33 – 007	1369	1.1 x 1.0	—	4	Sc		(15.1)		*1
	066.33 + 50.16		378	1.0 x 0.8:	1		f of 2				
10088	15 52.6 + 70 32	12 – 15 – 035	0752	(1.8 x 1.2)	(145)	3	E		14.8		
	104.86 + 39.78			(1.8 x 1.2)	...						
10089	15 52.7 + 21 17	04 – 37 – 000	1119	1.1 x 0.2	178	2	Sc		15.4		*
	035.05 + 47.94		371	1.1 x 0.15	7						
10090	15 53.1 + 61 13	10 – 23 – 000	1100	1.0:: x 1.0::	—	3	Dwarf spiral		17.		*
	093.97 + 44.44			0.6: x 0.6:	...		S V				
10091	15 53.4 + 06 05	01 – 41 – 002	1067	(2.0 x 1.6)	—	2	S0		13.8		*3
	015.70 + 41.37	NGC 6014		(1.6 x 1.4)	...						
10092	15 53.4 + 18 25	03 – 41 – 001	0083	1.1 x 0.30	105	3	Irr		14.7		*
	031.13 + 46.84		382	1.1 x 0.30	...						
10093	15 53.6 + 17 18	03 – 41 – 002	0083	1.6 x 1.3	—	3	S B c/Sc	SAB(s)bc	15.0		1
	029.64 + 46.39		382	1.6: x 1.3:	1 – 2			Sc –			
10094	15 53.6 + 24 38	04 – 38 – 000	0113	1.0 x 0.12	75	1	Sb – c		15.5		*
	039.98 + 48.67			1.0 x 0.15	7		np of 2				
10095	15 53.8 + 45 32	08 – 29 – 022	1404	1.4: x 0.6	154	4	Dwarf irregular		16.5:		*
	072.07 + 49.37			1.0: x 0.5:	...		Ir V				
10096	15 53.9 + 27 07	05 – 38 – 001	0134	1.1 x 0.45	26	2	Sc		15.1		3
	043.69 + 49.15	NGC 6016		1.0 x 0.4	6						
10097	15 54.2 + 48 00	08 – 29 – 023	1404	(1.3 x 1.1)	(130)	3	S0		14.1		1
	075.74 + 48.80		389	(1.4 x 1.2)	1 – 2						
10098	15 54.8 + 06 08	01 – 41 – 003	1067	0.8 x 0.8	(140)	1	...		13.8		13
	015.99 + 41.10	NGC 6017		0.7 x 0.6	...						
10099	15 54.9 + 42 01	07 – 33 – 000	1369	0.4 x 0.25	—	1	...		14.3	+ 10330	*C
	066.67 + 49.67		378	0.3 x 0.3	...		Extremely compact, sf of 2			+ 10516	
10100	15 55.0 + 22 33	04 – 38 – 002	0113	(1.4 x 1.0)	(140)	3	E		14.5		3
	037.09 + 47.81	NGC 6020 = 1148*		(1.7 x 1.3)	...						
10101	15 55.2 + 16 00	03 – 41 – 006	0083	1.4 x 0.7	75:	1	S0		14.6		*13
	028.13 + 45.54	NGC 6018	382	1.6 x 0.7	5:		P w 10102				
10102	15 55.2 + 16 05	03 – 41 – 005	0083	(1.6 x 0.8)	160:	3	E		14.1		*13
	028.24 + 45.57	NGC 6021	382	(1.5 x 0.8)	...		P w 10101				
10103	15 55.3 + 48 14	08 – 29 – 024	1404	(1.1 x 1.1)	—	3	E		14.4		*1
	076.02 + 48.57	IC 1152	389	(0.9 x 0.9)	...						
10104	15 55.4 + 30 12	05 – 38 – 004	0134	2.8 x 2.6	—	4	Sc	SA(rs)bc	14.9		1
	048.43 + 49.35			2.2: x 2.0:	1			Sc –			
10105	15 55.4 + 58 51	10 – 23 – 008	1100	1.3: x 1.0	—	3	Dwarf		17.		
	090.77 + 45.19			0.7:: x 0.7::	...		V				
10106	15 55.5 + 16 27	03 – 41 – 010	0083	1.1 x 0.8	(70)	3	E		14.7		*1
	028.76 + 45.65	NGC 6023	382	(1.2 x 0.9)	...		Brightest in group				
10107	15 55.6 + 48 18	08 – 29 – 026	1404	1.2 x 1.1	—	3	S0		13.6		*1
	076.10 + 48.51	IC 1153	389	1.1 x 1.0	1						
10108	15 55.8 + 12 13	02 – 41 – 001	0168	1.3 x 1.1	(160)	2	Sb – c		14.1		*1
	023.38 + 43.83	IC 1149		1.0 x 0.9	2						
10109	15 55.8 + 47 18	08 – 29 – 028	1404	1.4 x 0.6	145	3	Sb/S B b	SAB(r)b	14.9		
	074.62 + 48.69		389	1.2 x 0.6	5						
10110	15 55.9 + 28 57	05 – 38 – 005	0134	(1.2 x 0.8)	...	1	S0?		16.0:		*1
	046.56 + 49.05		379	(1.0 x 0.7)	...						
10111	15 56.0 + 13 27	02 – 41 – 000	0168	1.8: x 0.1:	40	1	...		16.0:		*
	024.95 + 44.32			1.5: x 0.1:	7		Peculiar?				
10112	15 56.0 + 20 06	03 – 41 – 014	0083	1.0 x 0.4	52	2	S0 – a		15.2		
	033.75 + 46.84		382	1.0 x 0.35	6						
10113	15 56.2 + 17 35	03 – 41 – 015	0083	2.5 x 0.7	28	4	S B c	S B(r)c	13.4		
	030.35 + 45.92	IC 1151		2.6: x 0.7	6						
10114	15 56.5 + 37 01	06 – 35 – 016	0071	1.3 x 0.3	147	3	Sc		16.0:		
	058.93 + 49.61			1.0 x 0.25	6						
10115	15 56.5 + 64 04	11 – 19 – 031	1426	(1.3 x 0.8)	140:	2	E		13.9		*
	097.22 + 42.79			(1.4 x 0.9)	...						

N G C   6010, 6012, 6014, 6016 – 18, 6020   I C   1148, 1149, 1151 – 53
        6021, 6023

1	2	3	4	5	6	7	8	9	10	11	12
10116	15 57.0 + 20 55 034.99 + 46.88	04 – 38 – 5-10 NGC 6027 + companions	0113	[1.9 x 1.0] ...	... ...	3	Compact group of 5 or 6 galaxies, Seyfert's sextet		13.4	See Notes	*12 VC
10117	15 57.2 + 21 45 036.18 + 47.09	04 – 38 – 011	0113	1.0 x 0.3 1.2 x 0.25	133 6	1	Sa – b		15.4		
10118	15 57.2 + 48 49 076.77 + 48.13	08 – 29 – 029	1404 389	1.1 x 0.6 0.9 x 0.5	100 .	2	E – S0		14.3		*
10119	15 57.2 + 79 18 113.52 + 34.37	13 – 11 – 018	0768	1.2 x 0.5 1.2: x 0.3	160 ...	1	S? ...		15.5		
10120	15 57.3 + 35 10 056.08 + 49.41	06 – 35 – 017 Mark 493	0071	1.3 x 1.3 1.3 x 1.3	-- 1	4	S B b	S B(r)b Sb –	14.9		
10121	15 57.5 + 18 56 032.33 + 46.11	03 – 41 – 020	0083 382	1.3 x 0.6 1.1 x 0.5	175 5 – 6	2	S B b – c		14.6		
10122	15 57.7 + 39 58 063.46 + 49.30	07 – 33 – 022	1369	1.3 x 1.3 1.3 x 1.3	-- 1	2	Sa: Singular?		15.5		
10123	15 57.7 + 51 28 080.56 + 47.39	09 – 26 – 035	1413	1.3 x 0.25 1.4 x 0.3	57 6	1	Sa – b		15.0		
10124	15 57.8 + 70 50 104.88 + 39.25	12 – 15 – 038 NGC 6048	0752	(2.2 x 1.7) (2.2 x 1.7)	(140) ...	3	E np of 2		13.6		*3
10125	15 58 + 84 00 117.84 + 31.38	14 – 07 – 000	1363	1.4: x 0.4: 0.9:: x 0.3:	[57] ...	2	Dwarf:		17.		
10126	15 58.0 + 79 08 113.34 + 34.44	13 – 11 – 019 NGC 6068	0768	1.1 x 0.8 1.1 x 0.8	155: ...	2	S B ... nff of 2, distorted		13.3	+ 3929 + 4153	*12 3
10127	15 58.2 + 21 00 035.24 + 46.64	04 – 38 – 012	0113	1.4 x 0.8 1.6 x 0.9	70 4	3	Sb		14.2		
10128	15 58.4 + 30 31 049.04 + 48.75	05 – 38 – 8 + 9 Mark 494	0134	[1.2 x 0.6] ...	... ...	2	Double system Connected		15.2		*1
10129	15 58.8 + 51 58 081.19 + 47.09	09 – 26 – 039	1413	1.1: x 0.9 1.2 x 0.9	(30) .	2	... Peculiar		15.7		*C
10130	15 58.9 + 08 58 019.94 + 41.64	02 – 41 – 000	0168	1.4 x 0.8 1.3 x 0.8	170 4	2	Sa		15.1		
10131	15 58.9 + 14 14 026.36 + 44.01	02 – 41 – 000	0168	1.1 x 0.10 0.9 x 0.12	111 7	1	Sb		16.0:		
10132	15 58.9 + 78 37 112.81 + 34.72	13 – 11 – 020	0768	1.2 x 0.8 1.1 x 0.8	(120) ...	3	S B b/Sc SAB(rs)bc nf of 2, disturbed		16.0:		*1
10133	15 59.0 + 01 51 012.05 + 37.90	00 – 41 – 002 IC 1158	0761	2.8 x 1.7 2.8 x 1.7	137 4	4	Sc/S B c	SAB(s)c	14.4		
10134	15 59.0 + 16 27 029.22 + 44.88	03 – 41 – 037	0083 382	1.0 x 0.8 1.0 x 0.8	-- 2	3	S B b	S B(rs)b	15.2		*
10135	15 59.2 + 19 29 033.27 + 45.92	03 – 41 – 043 NGC 6028	0083 382	1.3 x 1.1 1.3 x 1.1	(30) 2:	4	S0	(R)SAB: 0 +	14.8	+ 4480 + 4604	*1C
10136	15 59.2 + 36 44 058.50 + 49.07	06 – 35 – 022	0071 390	1.0 x 0.35 0.9 x 0.3	44 6	1	S ...		15.6		
10137	15 59.4 + 08 17 019.22 + 41.20	01 – 41 – 005	1067	1.3 x 0.7 0.9 x 0.4	40 5	4	Sc	Sc –	15.1		
10138	15 59.5 + 21 30 036.06 + 46.50	04 – 38 – 000	0113 381	1.4 x 0.6 1.6 x 0.8	178 5 – 6	2	Sa		15.2		
10139	15 59.6 + 18 06 031.46 + 45.35	03 – 41 – 044 NGC 6030	0083 382	1.1 x 0.7 1.4 x 0.7	43 4 – 5	1	S0 – a		14.5		*
10140	15 59.6 + 18 53 032.51 + 45.63	03 – 41 – 000	0083 382	1. x 1. 0.9: x 0.7:	-- ...	4	Dwarf V		18.		
10141	15 59.6 + 35 57 057.31 + 48.97	06 – 35 – 023	0071	1.0 x 0.15 0.9 x 0.15	79 7	1	S ...		16.0:		
10142	15 59.7 + 70 29 104.38 + 39.31	12 – 15 – 041	0752	1.6: x 0.5: 1.7 x 0.6:	10 6	1	S B a		15.4		*
10143	16 00.0 + 16 06 028.89 + 44.52	03 – 41 – 054	0083 382	(2.0 x 1.1) (2.3 x 1.2)	10 ...	3	E Brightest in chain of 4		14.9		*1 V A
10144	16 00.0 + 16 29 029.39 + 44.67	03 – 41 – 055	0083 382	(1.7 x 1.0) (1.7 x 1.1)	... ...	3	E ssf of 2		14.6		*
10145	16 00.5 + 47 14 074.28 + 47.92	08 – 29 – 035	1404 389	(1.0 x 0.5) (1.0 x 0.5)	0 ...	2	E Companion superimposed		(15.2)		*
10146	16 00.6 + 05 15 015.99 + 39.41	01 – 41 – 007	1067	1.0: x 0.9: 0.8: x 0.8:	-- 1	3	...	S d – dm	15.6		
10147	16 00.8 + 05 47 016.62 + 39.64	01 – 41 – 008	1067	[1.2 x 0.8] ...	... ...	3	Double system Connected		14.7		*1
10148	16 00.8 + 21 06 035.65 + 46.09	04 – 38 – 016 NGC 6032	0113 382	1.7 x 0.8 1.7 x 0.9	0 ...	1	S B ... P w 10154		15.0		*
10149	16 00.8 + 37 29 059.65 + 48.76	06 – 35 – 026 NGC 6038	0071 391	1.2 x 1.1 1.2 x 1.1	-- 1	3	Sc	Sc –	14.4		3
10150	16 00.8 + 49 20 077.32 + 47.44	08 – 29 – 037	1404 399	1.0 x 0.5 1.0 x 0.5	102 5	3	S B c		15.6		*

N G C   6027, 6028, 6030, 6032, 6038, 6048, 6068   I C   1158

1	2	3	4	5	6	7	8	9	10	11	12
10151	16 00.9 + 27 09 044.21 + 47.63	05 – 38 – 011	0134	1.0 x 0.8 1.0 x 0.8	(170) 2	3	Sc		14.8		
10152	16 00.9 + 41 20 065.48 + 48.61	07 – 33 – 29 + 30	1369	[1.0 x 0.5] ...	... ...	1	Double system Contact		14.5		*1
10153	16 01.1 + 20 47 035.25 + 45.93	04 – 38 – 017	0113 382	1.1 x 0.8 1.1 x 0.9	125: 3:	3	S B b	S B(r)b	15.2		
10154	16 01.2 + 21 02 035.60 + 45.98	04 – 38 – 018 NGC 6035	0113 382	1.1 x 0.9 1.0 x 1.0	— 1	3	Sc P w 10148	Sc –	14.7		*
10155	16 01.2 + 39 46 063.10 + 48.64	07 – 33 – 031	1369	1.0 x 0.9 1.0: x 0.9	— 1	2	Sc		14.5		*
10156	16 01.3 + 47 21 074.42 + 47.76	08 – 29 – 38 + 39	1404 389	[1.6 x 0.6] ...	... ...	2	Double system Bridge:		14.6		*1
10157	16 01.3 + 70 45 104.59 + 39.05	12 – 15 – 043 NGC 6071	0752	1.0 x 0.9 1.0 x 0.9	— 1	4	S B b	S B(s)b	15.0		
10158	16 01.6 + 11 52 023.81 + 42.40	02 – 41 – 000	0168	1.1 x 0.7 1.1 x 0.8	20 3	1	S0		15.2		
10159	16 01.8 – 02 00 008.63 + 35.12	00 – 41 – 003 NGC 6033	0761 384	(1.0 x 0.9) (1.1 x 1.1)	— ...	2	... brightest in cluster	Compact,	15.3		3
10160	16 01.8 + 25 10 041.42 + 46.99	04 – 38 – 019	0113	1.5 x 0.6 1.4 x 0.5	95 6	2	S B a		15.2		*
10161	16 01.9 + 13 53 026.33 + 43.21	02 – 41 – 004 IC 1169	0168	1.0 x 0.7 1.0 x 0.7	10 3	1	S ...		14.1		
10162	16 01.9 + 71 32 105.41 + 38.60	12 – 15 – 046	0752	1.2 x 0.4 1.3 x 0.5	62 6	3	S B b		15.4		*
10163	16 02.0 + 04 00 014.86 + 38.45	01 – 41 – 010 NGC 6036	1067	1.1 x 0.4 1.1 x 0.4	146 6	2	S0 – a P w NGC 6037		13.9		*3
10164	16 02.1 + 14 55 027.66 + 43.59	03 – 41 – 069	0083 382	(1.0 x 0.7) (1.1 x 0.7)	... ...	1	E? P w 10169, disturbed		15.5		*1V A
10165	16 02.1 + 17 53 031.48 + 44.72	03 – 41 – 074 NGC 6040	0083 382	[1.8 x 0.8] ...	... ...	3	Double system Contact		14.6		*13 VA
10166	16 02.1 + 40 07b 063.62 + 48.45	07 – 33 – 034	1369	1.1 x 0.6 1.1 x 0.5	155 ...	2	S B ... sf of 2, disturbed		(14.8)		*1
10167	16 02.1 + 41 39 065.93 + 48.36	07 – 33 – 033	1369	1.5 x 0.7 1.3: x 0.6	135 5	2	S0 – a		15.4		1
10168	16 02.1 + 49 28 077.44 + 47.20	08 – 29 – 041	1404 389	1.6 x 1.3 1.6 x 1.3	160 2	4	S B 0	(R)SAB 0/a	14.3		*1
10169	16 02.2 + 14 57 027.71 + 43.58	03 – 41 – 070	0083 382	2.5 x 0.4 2.3 x 0.4	... ...	3	... peculiar, long streamer	P w 10164,	15.2		*1
10170	16 02.3 + 17 51 031.46 + 44.66	03 – 41 – 078 NGC 6041 + 1170*	0083 382	[1.2 x 1.1] ...	— ...	2	E Companion superimposed		14.9	+ 10457 + 10578	*12 3V̄
10171	16 02.3 + 43 02 067.99 + 48.21	07 – 33 – 000	1369	1.2 x 0.1 1.1 x 0.1	60 7	3	Sc		17.		
10172	16 02.5 + 14 25 027.08 + 43.29	02 – 41 – 000	0168	1.2 x 0.2 1.2 x 0.2	8 6	2	Sb – c		15.2		
10173	16 02.5 + 68 20 101.82 + 40.20	11 – 20 – 002	1410	1.0 x 0.2 1.0 x 0.2	58 6 – 7	1	S ...		15.6		*
10174	16 02.6 + 23 28 039.10 + 46.38	04 – 38 – 000	0113 382	1.0 x 0.15 1.1 x 0.20	92 7	2	Sa – b		16.5		*
10175	16 02.6 + 30 51 049.73 + 47.90	05 – 38 – 014	0134 386	1.1 x 0.5 1.0 x 0.4	[30] ...	2	Irr		15.7		
10176	16 02.7 + 13 50 026.38 + 43.01	02 – 41 – 005	0168	1.5 x 0.15 1.5 x 0.18	66 7	3	Sc		15.2		*
10177	16 02.8 + 17 54 031.59 + 44.57	03 – 41 – 088 NGC 6045	0083 382	1.1 x 0.22 1.1 x 0.22	82 6	1	Sb:		14.8	+ 9912 + 10034	*12
10178	16 02.8 + 24 04 039.95 + 46.50	04 – 38 – 021 NGC 6051	0113 382	(1.6 x 1.0) (2.1 x 1.5)	165 ...	2	E		14.9		*1
10179	16 02.8 + 31 07 050.13 + 47.90	05 – 38 – 000	0134	[1.1 x 0.5] ...	... ...	1	... Compact, streamers		16.5:		*
10180	16 02.9 + 17 33 031.14 + 44.42	03 – 41 – 089 IC 1173	0083 382	1.1 x 0.7 1.0 x 0.5	55 5:	1	S ...		15.6	+ 10840 + 10962	2
10181	16 03 + 81 50 115.74 + 32.62	14 – 07 – 034	1363 388	1.3 x 0.2 1.0 x 0.2	70 7	2	Sa		15.3		
10182	16 03.0 + 20 41 035.31 + 45.47	04 – 38 – 022 NGC 6052 = 6064	0113 382	0.8 x 0.55 0.9 x 0.55	... ...	1	Double system Mark 297 Contact, distorted		14.1	+ 4671 + 4803	*12 3VĀ
10183	16 03.0 + 80 00 114.00 + 33.74	13 – 11 – 022	0768	1.2 x 0.5 1.1 x 0.6	170 5 – 6	2	S B 0 – a		15.3		
10184	16 03.1 + 03 11 014.18 + 37.78	01 – 41 – 000	1067	1.0: x 0.2 0.9: x 0.2	130 6	1	Sa		15.6		
10185	16 03.1 + 15 10 028.11 + 43.47	03 – 41 – 091 IC 1174	0083	1.0 x 0.8 1.0 x 0.8	(50) 2	1	Sa:		14.5		

N G C   6033, 6035, 6036, 6040, 6041, 6045
6051, 6052, 6064, 6071

I C   1169, 1170, 1173, 1174

1	2	3	4	5	6	7	8	9	10	11	12
10186	16 03.1 + 17 54 031.62 + 44.51	03 − 41 − 92 + 93 NGC 6050 = 1179*	0083 382	[1.0 x 0.9] ...	... ...	3	Double system Contact		14.9		*1V
10187	16 03.2 + 16 35 029.93 + 44.00	03 − 41 − 94 + 95	0083 382	[1.2 x 0.6] ...	[124] ...	2	Double system Envelopes in contact		15.4		*1
10188	16 03.2 + 17 44a 031.42 + 44.42	03 − 41 − 097 IC 1178	0083 382	(1.2 x 0.8) (1.2 x 1.1)	−− ...	2		Double system, bridge: Large plumes	15.0	+ 10324 + 10446	*12 VA
10189	16 03.2 + 17 44b 031.42 + 44.42	03 − 41 − 098 IC 1181	0083 382	(1.1 x 0.8) (1.0 x 0.8)	... ...	2				+ 10690 + 10812	*12 VA
10190	16 03.2 + 17 51 031.57 + 44.47	03 − 41 − 000	0083 382	1.0 x 0.1 0.8 x 0.12	158 7	2	Sc		16.0:		
10191	16 03.2 + 18 17 032.14 + 44.62	03 − 41 − 101 NGC 6055	0083 382	(1.0 x 0.5) (1.4 x 0.7)	40 5	3	S0		15.4	+ 11363 + 11487	*12
10192	16 03.3 + 17 56 031.69 + 44.47	03 − 41 − 104 IC 1182 = 1184?	0083 382	1.6? x 0.5 1.1 x 0.7	... ...	1	... Mark 298 Peculiar, jet; Seyfert?		15.2	+ 10245 + 10308	*12
10193	16 03.5 + 16 20 029.64 + 43.83	03 − 41 − 109	0083 382	[1.1 x 0.2] ...	[133] (7)	2	S ... + companion Contact		15.5		*1
10194	16 03.5 + 63 51 096.48 + 42.20	11 − 20 − 004	1410 387	1.8 x 0.2 1.7 x 0.2	50 7	3	... S d − dm		16.0:		
10195	16 03.6 + 18 21 032.27 + 44.56	03 − 41 − 113	0083 382	1.7 x 0.4 1.6 x 0.4	120 6	1	Sb:		15.7		*
10196	16 03.7 + 21 38 036.67 + 45.61	04 − 38 − 025 NGC 6060	0113 382	2.1 x 1.1 2.2 x 1.1	105 5	4	Sc Sc −		14.3		3
10197	16 03.9 + 20 55a 035.72 + 45.35	04 − 38 − 027	0113 382	1.3 x 0.6 1.4 x 0.6	172 5	2	... S dm P w 10198		15.2		*1V
10198	16 03.9 + 20 55b 035.72 + 45.35	04 − 38 − 026	0113 382	1.2 x 0.6 1.4 x 0.6	42 5	3	... SA d P w 10197				*1V
10199	16 04.0 + 18 23 032.36 + 44.48	03 − 41 − 118 NGC 6061	0083 382	(1.0 x 0.9) (1.2 x 1.1)	−− 1	3	S0		15.0	+ 11169 + 11294	*1
10200	16 04.0 + 41 29 065.64 + 48.02	07 − 33 − 039	1369	0.8 x 0.6 0.7 x 0.6	−− ...	1	... n of 2		13.6		*1
10201	16 04.1 + 15 49 029.06 + 43.50	03 − 41 − 123 + 124	83 382	[1.1 x 0.7] ...	... ...	2	Double system Contact		14.3		*1V
10202	16 04.1 + 19 55 034.40 + 44.98	03 − 41 − 125 NGC 6062	0083 382	1.1 x 0.9 1.1 x 0.7	(10) 2	4	S B b S B(rs)bc nf of 2 Sb +		14.4		*1
10203	16 04.3 + 03 55 015.16 + 37.92	01 − 41 − 011	1067	1.0 x 0.25 0.8 x 0.22	115 6 − 7	2	S0		15.1		
10204	16 04.4 + 16 27 029.91 + 43.68	03 − 41 − 127	0083 382	1.1 x 0.7 1.8 x 1.2	... ...	2	S0? Singular, disturbed?		14.3		*1
10205	16 04.6 + 30 14 048.92 + 47.39	05 − 38 − 017	0134 386	1.7 x 0.9 1.7 x 1.2	132 7	3	Sa		14.4		*1
10206	16 04.6 + 69 50 103.39 + 39.29	12 − 15 − 050 NGC 6079	0752	(1.4 x 1.0) (1.4 x 1.0)	150 ...	3	E P w 10221		13.9		*3
10207	16 04.7 + 32 01a 051.54 + 47.61	05 − 38 − 015	0134 386	1.9 x 0.8 1.7 x 0.8	157 5 − 6	2	S B b S B(s)bc: P w 10208		15.3		
10208	16 04.7 + 32 01b 051.54 + 47.61	05 − 38 − 016	0134 386	1.2 x 0.3 0.8 x 0.2	149 ...	2	... P w 10207				*1
10209	16 04.7 + 36 48 058.64 + 47.97	06 − 35 − 033	0071 391	1.1 x 0.15 1.1 x 0.15	165 7	2	Sb		16.0:		
10210	16 04.8 + 08 06 019.87 + 39.95	01 − 41 − 012 NGC 6063	1067	1.9 x 0.9 1.5 x 0.8	159 5	3	Sc		14.1		
10211	16 04.9 + 22 12 037.57 + 45.52	04 − 38 − 028	0113 382	1.2 x 0.4 1.2 x 0.4	13 6	2	Sa − b		15.0		
10212	16 04.9 + 77 45 111.73 + 34.99	13 − 11 − 023	0768	1.5 x 0.2 1.3 x 0.2	138 7	2	Sb − c		16.0:		
10213	16 05.0 + 10 33 022.74 + 41.06	02 − 41 − 007	0168	1.1 x 0.9 1.0 x 0.8	−− 2	1	S ... n of 2		15.0		*
10214	16 05.0 + 55 33 085.74 + 45.15	09 − 26 − 056	1413 392	[4.1 x 0.8] ...	4 ...	4	S B ... Very long streamer		15.0		*1V A
10215	16 05.2 + 19 41 034.21 + 44.66	03 − 41 − 133	0083 382	1.0: x 0.7: 0.6 x 0.4	(40) ...	1	S ...		15.7		
10216	16 05.3 + 13 51 026.76 + 42.44	02 − 41 − 000	0168	1.0 x 0.3 0.9 x 0.2	28 6	1	S ...		15.2		
10217	16 05.5 + 22 29 038.01 + 45.47	04 − 38 − 029	0113 382	1.0 x 0.9 1.0 x 0.9	−− 1 − 2	3	S B b S B(r)b		16.0:		
10218	16 05.6 + 10 55 023.26 + 41.10	02 − 41 − 009 IC 1196	0168	1.1 x 0.5 1.1 x 0.5	3 5	1	Sa		14.8		
10219	16 05.9 + 07 40 019.55 + 39.50	01 − 41 − 013	1067	2.8 x 0.5 2.8 x 0.4	56 7	3	Sc		14.7		
10220	16 05.9 + 20 20 035.15 + 44.72	03 − 41 − 135	0083 382	1.0 x 0.5 1.0 x 0.5	175 5	3	S B c		16.0:		

N G C   6050, 6055, 6060 − 63, 6079

I C   1178, 1179, 1181, 1182, 1184?
1196

1	2	3	4	5	6	7	8	9	10	11	12
10221	16 05.9 + 69 45	12 – 15 – 051	0752	1.1 x 0.3	118	2	Sb		15.6		*
	103.22 + 39.23	IC 1201		1.1 x 0.25	6		P w 10206				
10222	16 06 + 82 01	14 – 07 – 000	1363	[0.45 x 0.45]	...	1	Double system		14.0		*C
	115.84 + 32.43			...	...		Compacts, plume		+ 14.0		
10223	16 06.0 + 76 58	13 – 12 – 000	1433	1.0 x 0.4	75	3	S B b	S B(s)b	16.0:		
	110.90 + 35.38			0.9 x 0.4	6						
10224	16 06.6 + 22 11	04 – 38 – 033	0113	1.0 x 0.6	20	2	Sb – c		15.5		
	037.71 + 45.13		382	1.1 x 0.5	5						
10225	16 06.9 + 08 53	02 – 41 – 012	0168	1.8: x 1.7:	––	2	...	S B dm	14.8		
	021.09 + 39.87			1.5: x 1.5:	1						
10226	16 07.2 + 35 55	06 – 35 – 038	0071	1.1 x 0.5	142	2	Sb		15.4		
	057.36 + 47.43			1.1 x 0.4	6						
10227	16 07.2 + 36 45	06 – 35 – 037	0071	2.2 x 0.2	169	3	S B? c		15.6		
	058.59 + 47.47		391	2.2 x 0.2	7						
10228	16 07.2 + 72 38	12 – 15 – 052	0752	(1.8 x 1.4)	(120)	3	S0		14.6		13
	106.32 + 37.67	NGC 6094		(2.0 x 1.5)	2						
10229	16 07.3 00 00	00 – 41 – 000	0761	1.1: x 0.8:	...	3	Dwarf irregular		17.		
	011.59 + 35.15		385	0.6: x 0.3:	...		Ir IV – V:				
10230	16 07.4 + 00 50	00 – 41 – 004	0761	3.9 x 2.0	62	4	Sc	SA(s)c	13.0	+ 2110	*12
	012.46 + 35.59	NGC 6070	385	4.0: x 2.0	5					+ 2175	3468
10231	16 07.4 + 64 28	11 – 20 – 000	1410	1.6 x 0.9	125:	1	S ...		16.0:		
	096.98 + 41.55		387	0.7 x 0.4	...						
10232	16 07.6 + 20 18	03 – 41 – 138	0083	1.5 x 0.12	130	3	Sc		15.7		
	035.29 + 44.33		382	1.6 x 0.12	7						
10233	16 07.7 + 22 45	04 – 38 – 035	0113	1.3 x 0.15	61	3	Sc		15.7		*
	038.58 + 45.06		382	1.5 x 0.15	7		P w 10236				
10234	16 07.7 + 30 35	05 – 38 – 000	0134	1.3 x 1.0	(15)	3	S0	S 0 +	15.4		
	049.57 + 46.78		386	1.2 x 1.0	2						
10235	16 07.9 + 16 50	03 – 41 – 139	0083	1.2 x 0.6	130	2	Sc		14.5		13
	030.83 + 43.05	NGC 6073		1.2 x 0.6	5						
10236	16 07.9 + 22 47	04 – 38 – 036	0113	1.3 x 0.15	16	2	Sb		15.6		*
	038.64 + 45.02		382	1.5 x 0.20	7		P w 10233				
10237	16 08.0 + 79 45	13 – 12 – 001	1433	1.3 x 1.1	––	3	Sc	SAB cd	15.2		1
	113.60 + 33.71			1.2 x 1.0	1						
10238	16 08.1 + 12 26	02 – 41 – 000	0168	1.6 x 0.8	70	4	S B b	(R')S B(s)b	15.1		
	025.43 + 41.22			1.6 x 0.8	5:						
10239	16 08.1 + 12 57	02 – 41 – 000	0168	1.0 x 0.9	––	1	S B a		15.1		
	026.05 + 41.44			0.9 x 0.3	1						
10240	16 08.1 + 32 15	05 – 38 – 000	0134	1.0 x 0.15	121	2	Sa – b		16.0:		*
	052.01 + 46.93		386	1.0 x 0.18	7						
10241	16 08.1 + 42 27	07 – 33 – 047	1369	1.1 x 0.8	140	2	Sb – c		15.3		
	067.00 + 47.19			0.9 x 0.7	2:						
10242	16 08.2 + 10 10	02 – 41 – 013	0168	1.3 x 0.4	157	1	S ...		14.6		
	022.77 + 40.19	IC 1199		1.3 x 0.45	6						
10243	16 08.2 + 20 05	03 – 41 – 140	0083	1.0 x 0.9	––	2	S B c		15.4		
	035.07 + 44.13		382	0.9 x 0.7	1:						
10244	16 08.3 + 43 15	07 – 33 – 048	1369	1.2 x 0.15	128	1	S ...		15.4		
	068.16 + 47.10			1.0 x 0.2	7						
10245	16 08.4 + 49 17	08 – 29 – 049	1404	1.0 x 0.7	74	2	Sb – c		16.0:		*
	076.87 + 46.24		389	0.8 x 0.15	7						
10246	16 08.5 + 27 38	05 – 38 – 022	0134	1.2: x 0.3	120	1	Sa – b		16.0:		*
	045.41 + 46.08		386	1.1 x 0.35	7		f of 2				
10247	16 08.5 + 60 13	10 – 23 – 027	1100	1.0 x 0.7	50	3	...	S B: m	15.6		
	091.64 + 43.13		403	1.0 x 0.7	...						
10248	16 08.6 + 29 39	05 – 38 – 000	0134	1.0 x 0.15	66	1	Sb		16.0:		
	048.28 + 46.44		386	1.0 x 0.15	7						
10249	16 08.9 + 13 59	02 – 41 – 000	0168	1.1 x 0.5	79	2	S B a – b		15.2		*
	027.41 + 41.70			1.2 x 0.5	6						
10250	16 08.9 + 26 30	04 – 38 – 000	0113	1.0: x 0.9:	––	1	S ...		17.		
	043.84 + 45.75		386	0.9 x 0.8	1 – 2						
10251	16 08.9 + 56 44	09 – 26 – 059	1413	1.2 x 0.3	112	1	Sb		16.0:		
	087.09 + 44.27		392	1.1 x 0.3	6 – 7						
10252	16 09.0 + 41 16	07 – 33 – 050	1369	1.2 x 0.35	173	2	Sa – b		15.3		
	065.24 + 47.09			1.1 x 0.35	6						
10253	16 09.1 + 27 00	05 – 38 – 023	0134	[1.2 x 0.4]	[63]	2	Double system		15.4		*13
	044.56 + 45.82	NGC 6076	386	...	...		Symbiotic, p w 10254				
10254	16 09.1 + 27 04	05 – 38 – 024	0134	(1.4 x 1.1)	––	3	E		14.9		*3
	044.65 + 45.83	NGC 6077	386	(1.4 x 1.1)	...		P w 10253				
10255	16 09.1 + 36 07	06 – 35 – 042	0071	1.1 x 0.7	25:	3	Sb	(R:)SA:(r)b	15.4		*
	057.68 + 47.06		391	0.9 x 0.6	...		np of 2	Sb –			

N G C   6070, 6073, 6076, 6077, 6094   I C   1199, 1201

1	2	3	4	5	6	7	8	9	10	11	12
10256	16 09.3 + 23 56 040.33 + 45.03	04 − 38 − 039	0113 382	1.3 x 0.25 1.3 x 0.25	143 6	2	Sb		15.5		*
10257	16 09.3 + 38 23 061.01 + 47.09	06 − 35 − 043	0071 391	1.8: x 0.35 1.5 x 0.3	165 6	2	Sb − c		15.0		
10258	16 09.4 + 29 29 048.08 + 46.24	05 − 38 − 025	0134 386	1.2 x 0.5 1.1 x 0.4	30 6	1	Sa − b		15.7		
10259	16 09.7 + 29 53 048.67 + 46.24	05 − 38 − 000	0134 386	[1.8 x 0.6] ...	... ...	3	Chain of 5 galaxies Connected, plumes		16.0:		*
10260	16 09.8 + 21 03 036.43 + 44.26	04 − 38 − 000	0113 382	1.1 x 0.7 0.9 x 0.5	150: 3 − 4	1	S ...		15.4		
10261	16 09.8 + 52 35 081.42 + 45.30	09 − 26 − 061	1413 389	(1.1 x 0.8) (1.2 x 0.9)	(170) ...	2	E − S0 1st in chain of 4		15.7		*
10262	16 09.9 + 29 58 048.80 + 46.21	05 − 38 − 029	0134 386	(1.1 x 0.9) (1.3 x 1.2)	— 1 − 2	3	S0 nf of 2		(14.8)		*1
10263	16 10 + 86 20 119.75 + 29.71	14 − 08 − 001 14 − 07 − 035	0775 1363	1.6:: x 1.4:: 1.5:: x 1.5::	— 1:	3	Dwarf spiral, SA m S IV − V		16.5		*
10264	16 10.3 + 00 01 012.12 + 34.54	00 − 41 − 006	0761	1.3 x 0.7 1.3 x 0.7	35: 5	1	S ...		14.9		
10265	16 10.3 + 61 24 093.03 + 42.48	10 − 23 − 033 NGC 6095	1100 403	(1.8 x 1.6) (2.0 x 1.8)	— ...	2	E − S0		14.5		*
10266	16 10.4 + 49 02 076.43 + 45.96	08 − 29 − 052	1404 389	1.3: x 0.7: 0.3:: x 0.3::	... ...	3	Dwarf (spiral?) S? IV − V or V		17.		
10267	16 10.4 + 52 35 081.39 + 45.21	09 − 26 − 064 NGC 6090, Mark 496	1430 389	[2.8 x 1.5] ...	... ...	3	Double system, contact, long curved plumes; in chain	14.0	+ 8730 + 8945	*1C
10268	16 10.5 + 02 18a 014.49 + 35.75	00 − 41 − 007 NGC 6080	0761	(1.2 x 0.9) (1.3 x 1.1)	— ...	1	Sa: sp of 2		(14.1)		*1
10269	16 10.5 + 29 30 048.17 + 46.01	05 − 38 − 034 NGC 6085	0134 386	1.6 x 1.2 1.7 x 1.2	165: 2 − 3	3	Sa P w 10270		14.5		*13
10270	16 10.5 + 29 38 048.36 + 46.03	05 − 38 − 035 NGC 6086	0134 386	(1.8 x 1.5) (2.2 x 1.6)	0 ...	3	E P w 10269		14.8		*3
10271	16 10.5 + 58 55 089.85 + 43.35	10 − 23 − 034	1100 392	1.3 x 0.5 1.3 x 0.5	172 ...	1	S B ... Peculiar?		15.0		
10272	16 10.6 + 10 00 022.93 + 39.59	02 − 41 − 019 NGC 6081 = 1202*	0168	1.7 x 0.5 1.5 x 0.6	131 6	3	S0		14.4		*3
10273	16 10.7 + 28 25b 046.65 + 45.76	05 − 38 − 036	0134 386	1.5 x 0.20 1.4 x 0.20	[159] ...	2	Strongly peculiar Brightest of 3		(15.2)		*1
10274	16 10.7 + 38 23 061.01 + 46.81	06 − 36 − 000	1093	1.4 x 0.2 1.3: x 0.1	52 7	3	Sc		16.0:		
10275	16 10.8 + 28 07 046.23 + 45.68	05 − 38 − 037	0134 386	(1.4 x 1.2) (1.5 x 1.3)	(10) ...	2	E		15.0		
10276	16 10.8 + 32 07 051.92 + 46.34	05 − 38 − 033	0134 386	1.4: x 0.2 1.4 x 0.2	135 7	2	Sb − c		15.6		
10277	16 10.8 + 57 15 087.66 + 43.86	10 − 23 − 037	1100 392	1.0 x 0.5 0.9 x 0.4	105: ...	1	S ...		16.0:		*
10278	16 10.9 + 49 31 077.09 + 45.80	08 − 29 − 053	1404 389	1.0 x 0.4 1.0 x 0.4	170 6	1	Sa − b		14.8		
10279	16 10.9 + 60 43 092.13 + 42.67	10 − 23 − 038	1100 403	[1.0 x 0.9] ...	— ...	2	Double system Distorted, faint bridge		14.5		*1
10280	16 11 + 81 26 115.16 + 32.63	14 − 07 − 036	1363 388	1.6 x 0.8 1.6 x 0.8	170 5	3	Sb/S B b SAB(s)b		15.0		*1
10281	16 11.0 + 17 20 031.84 + 42.55	03 − 41 − 000	0083	2. x 1. 1. x 1.	— ...	3	Dwarf V		18. − 19.		
10282	16 11.0 + 32 38 052.67 + 46.36	06 − 36 − 002	1093 386	1.0 x 0.6 0.8 x 0.6	92 4	3	Sc SA(s)c		15.3		1
10283	16 11.1 + 30 57a 050.26 + 46.12	05 − 38 − 038	0134 386	(1.0 x 0.5) (1.8 x 0.9)	175 ...	2	E np of 2, bridge		(15.5)		*
10284	16 11.1 + 61 18a 092.85 + 42.43	10 − 23 − 040	1100 403	1.0 x 0.2 1.0 x 0.25	51 6 − 7	2	S0 − a		(15.4)		*
10285	16 11.2 + 31 01 050.36 + 46.10	05 − 38 −40 +41	0134 386	[1.5 x 0.7] ...	... ...	2	E: + companion Envelopes in contact		15.7 +16.5:		*1
10286	16 11.4 + 32 25 052.37 + 46.25	05 − 38 − 000	0134 386	1.2 x 0.2 1.0 x 0.2	177 6 − 7	2	Sc		15.7		
10287	16 11.7 + 14 24 028.29 + 41.25	02 − 41 − 021	0168	1.3 x 0.9 1.3 x 1.0	(15) 3	4	S B b		14.7		
10288	16 11.8 − 00 05 012.28 + 34.17	00 − 41 − 009	0761	5.3 x 0.6 4.8: x 0.6	91 7	2	Sc		14.7		
10289	16 11.8 + 48 20 075.38 + 45.86	08 − 29 − 000	1404 389	1.2 x 0.15 0.9 x 0.15	42 7	3	Sc		16.5:		*
10290	16 12.0 + 00 56 013.34 + 34.69	00 − 41 − 010	0761	2.0 x 1.9: 2.0: x 1.8:	— 1:	4	Dwarf spiral, S m S IV − V		15.0		1

N G C 6080, 6081, 6085, 6086, 6090, 6095

I C 1202

1	2	3	4	5	6	7	8	9	10	11	12
10291	16 12.0 + 17 53 032.65 + 42.52	03 – 41 – 143 NGC 6084	0083 382	1.0 x 0.5 1.1 x 0.6	30 5	3	Sa		15.4		
10292	16 12.6 + 56 25 086.45 + 43.88	09 – 27 – 001	1101 392	1.1 x 0.12 0.9 x 0.12	172 7	3	Sb		16.0:		
10293	16 12.9 + 11 25 024.91 + 39.72	02 – 41 – 023 IC 1206	0168	1.1 x 0.6 1.1 x 0.7	2 4	2	Sa – b nnf of 2		14.8		*1
10294	16 12.9 + 65 33 097.96 + 40.56	11 – 20 – 000	1410	2.0 x 1.3 1.8: x 1.0:	5: ...	3	Dwarf (spiral?) S? IV – V		16.5:		
10295	16 13.0 + 31 12 050.70 + 45.75	05 – 38 – 000	0134 386	1.2 x 0.5 0.9 x 0.3	34 6	1	S ...		15.5		
10296	16 13.1 + 39 38 062.84 + 46.35	07 – 33 – 052	1369	1.4 x 0.7 1.0 x 0.6	40 5	3	Sc		15.7		
10297	16 13.2 + 19 01 034.23 + 42.66	03 – 41 – 144	0083	2.1 x 0.25 1.8 x 0.25	1 7	1	S ...		15.1		
10298	16 13.2 + 63 50 095.87 + 41.24	11 – 20 – 005	1410 387	1.2 x 0.35 1.2 x 0.35	103 6	1	S ...		14.5		*
10299	16 13.4 + 19 35 034.98 + 42.81	03 – 41 –145+146 NGC 6098 + 6099	083	[1.6 x 0.9] ...	... ...	3	Double system, E + E Envelopes in contact		14.6		*1V
10300	16 13.6 + 28 17 046.64 + 45.11	05 – 38 – 047 NGC 6102	0134 386	1.5 x 0.9 1.1 x 0.8	70: 4:	1	S ...		15.4		3
10301	16 13.6 + 31 26 051.06 + 45.66	05 – 38 – 000	0134 386	1.4 x 0.15 1.2 x 0.15	8 7	1	S B: b: P w 10303		15.7		*
10302	16 13.7 + 32 05 051.99 + 45.73	05 – 38 – 049 NGC 6103	0134 386	0.8 x 0.45 0.8 x 0.45	80 ...	1	...		14.4		13
10303	16 13.9 + 31 31 051.19 + 45.61	05 – 38 – 048	0134 386	1.0 x 0.35 1.0 x 0.35	87 6	1	S ... P w 10301		15.2		*
10304	16 13.9 + 62 40 094.39 + 41.61	10 – 23 – 048 IC 1210	1100	1.6 x 0.4 1.4 x 0.35	168 6	2	Sa – b		13.8		
10305	16 14.0 + 31 50 051.64 + 45.63	05 – 38 – 000	0134 386	1.0 x 0.35 0.9 x 0.3	0 6	1	S ...		16.0:		
10306	16 14.2 + 00 21 013.11 + 33.91	00 – 41 – 011	0761	2.4 x 0.8 2.1: x 0.8	35 6	2	Sb		14.8		
10307	16 14.4 + 00 56 013.73 + 34.19	00 – 41 – 012 NGC 6100	0761	1.6 x 0.9 1.6 x 0.9	120 4	3	SO	(R)SAB 0/a	14.7		
10308	16 14.4 + 52 22 080.89 + 44.67	09 – 27 – 005	1101 389	1.0 x 0.2 0.8: x 0.1	130 6	3	Sc		16.5:		
10309	16 14.7 + 35 50 057.37 + 45.91	06 – 36 – 011 NGC 6104	1093 393	0.9 x 0.7 0.8 x 0.6	... ...	2	... Peculiar, ring – shaped, p of 2		14.1		*3
10310	16 14.8 + 47 10 073.62 + 45.53	08 – 30 – 002	1375	3.2: x 2.4 2.5: x 2.2	[165] ...	4	Dwarf irregular, I B(s)m Ir IV – V		14.9		*A
10311	16 15.4 + 35 01 056.22 + 45.70	06 – 36 – 014 NGC 6107	1093 393	(1.5 x 1.1) (1.5 x 1.1)	40: ...	3	[E] Compact		14.7		*
10312	16 15.5 + 31 19 050.98 + 45.24	05 – 38 – 051	0134 386	1.7: x 1.1: 1.7 x 1.1	110: 3 – 4	1	S ...		15.1		
10313	16 15.6 + 31 43 051.55 + 45.28	05 – 38 – 000	0134 386	1.1: x 0.8: 1.1: x 0.9:	(10) ...	3	... S B(s)dm		17.		
10314	16 15.6 + 53 08 081.88 + 44.32	09 – 27 – 009 IC 1211	1101 389	(1.2 x 1.2) 0.9 x 0.9	-- ...	2	[E] Compact		13.8		*C
10315	16 15.6 + 68 32 101.31 + 39.04	11 – 20 – 09a IC 1215	1410	1.3 x 0.7 1.0 x 0.7	175: ...	3	S B ... Peculiar, p w 10326		14.0		*
10316	16 15.7 + 35 07 056.37 + 45.65	06 – 36 – 016 NGC 6109	1093 393	(1.0 x 1.0) (1.1 x 1.1)	-- ...	2	... Compact		14.9		1
10317	16 15.8 + 39 03 062.00 + 45.83	07 – 33 – 055	1369	1.5: x 0.2 1.3 x 0.2	40 7	1	S ... Disturbed?		16.5:		
10318	16 15.8 + 64 13 096.19 + 40.82	11 – 20 – 000	1410 387	1.0 x 0.25 1.0 x 0.25	97 6	1	S ...		15.5		*
10319	16 15.9 + 12 55 027.08 + 39.71	02 – 41 – 000	0168	[1.2 x 0.4] ...	... ...	2	Compact group of 4 or 5 galaxies		16.0:		*
10320	16 15.9 + 21 12 037.32 + 42.78	04 – 38 – 049	0113	1.4 x 0.35 1.2 x 0.3	57 6	2	Sa – b		15.4		*
10321	16 15.9 + 21 41 037.95 + 42.93	04 – 38 –46+47	0113	[1.7 x 1.6] ...	-- ...	3	Quintiple group		15.5		*1V
10322	16 15.9 + 22 20 038.79 + 43.12	04 – 38 – 048	0113	1.1 x 0.35 1.2 x 0.35	97 6	1	S ... P w 10327		14.9		*
10323	16 15.9 + 66 05 098.43 + 40.06	11 – 20 – 009 IC 1214	1410	1.1 x 0.5 1.1 x 0.5	18 6	1	SO – a		15.0		
10324	16 15.9 + 66 58 099.47 + 39.69	11 – 20 – 000	1410	1.5 x 0.6 0.8 x 0.3	87 6	2	... S d – dm		16.0:		
10325	16 16.0 + 46 13 072.24 + 45.44	08 – 30 – 000	1375	... ...	... ...	1	Double system		14.5		*

N G C   6084, 6098 – 6100, 6102 – 04, 6107, 6109   I C   1206, 1210, 1211, 1214, 1215

1	2	3	4	5	6	7	8	9	10	11	12
10326	16 16.0 + 68 29 101.24 + 39.03	11 – 20 – 010 IC 1216	1410	1.1 x 1.0 1.0 x 1.0	-- 1	3	Sc P w 10315		14.9		*
10327	16 16.3 + 22 17 038.77 + 43.02	04 – 37 – 050	0113	1.0 x 0.15 1.1 x 0.15	132 7	2	Sb – c P w 10322		15.7		*
10328	16 16.4 + 07 32 021.02 + 37.16	01 – 41 – 016 NGC 6106	1067	2.5 x 1.3 2.5: x 1.3:	140 5	4	Sc		13.4	+ 1470 + 1568	*123 46$\overline{7}$8
10329	16 16.4 + 15 40 030.43 + 40.72	03 – 41 – 149 IC 1209	0083	(1.6 x 1.0) (1.6 x 1.0)	(5) 3 – 4	3	SO		15.1		*
10330	16 16.4 + 40 12 063.65 + 45.72	07 – 33 – 056	1369	1.1 x 0.9 0.9 x 0.9	-- ...	3	S B b Disturbed?	SAB(s)b	15.4		*1
10331	16 16.4 + 59 27 090.19 + 42.46	10 – 23 – 056	1100	1.6 x 0.25 1.6 x 0.25	142 6	1	S ... npp of 2, disturbed		14.8		*1
10332	16 16.6 + 29 57 049.13 + 44.78	05 – 38 – 000	0134 386	[1.3: x 0.6:] ...	... ...	2	Double system Distorted, streamers		16.5:		*
10333	16 16.6 + 62 04 093.49 + 41.54	10 – 23 – 060 NGC 6123	1100 403	0.8 x 0.2 0.9 x 0.2	4 7	1	SO – a		14.4		
10334	16 16.7 + 63 58 095.83 + 40.83	11 – 20 – 000	1410 387	1.6 x 0.8 1.6 x 0.8	[150] ...	4	Irr	I B m	15.4		*
10335	16 16.9 + 16 36 031.63 + 40.96	03 – 41 – 150	0083	1.1 x 0.4 1.2 x 0.4	92 (6 – 7)	2	E – SO		15.5		
10336	16 17.0 + 35 17 056.64 + 45.40	06 – 36 – 021 NGC 6116	1093	1.0 x 0.6 0.9 x 0.55	12 4	2	Sb		15.3		1
10337	16 17.1 + 07 24 020.98 + 36.94	01 – 42 – 001	0088	1.2 x 0.35 1.1 x 0.35	66 6	2	Sb P w 10328		14.9		*
10338	16 17.5 + 37 13 059.40 + 45.43	06 – 36 – 022 NGC 6117	1003 391	1.3 x 1.3 1.2 x 1.0	-- 1	3	Sc nnp of 2	SA(s)bc Sc –	14.7		*
10339	16 17.6 + 02 08 015.48 + 34.16	01 – 42 – 001	0143	0.7 x 0.45 0.6 x 0.5	(70) ...	1	...		14.3		
10340	16 17.6 + 36 24 058.24 + 45.36	06 – 36 – 025	1093 391	1.1 x 0.15 1.1 x 0.15	127 6 – 7	1	S ...		15.6		
10341	16 17.7 + 60 55 091.99 + 41.82	10 – 23 – 063	1100 403	1.2 x 0.7 0.7 x 0.7	70: ...	2	Sb – c		16.0:		
10342	16 17.8 + 36 12 057.96 + 45.31	06 – 36 – 027	1093	1.0 x 0.8 1.0: x 0.8	(130) 2:	3	S B b	(R)SAB b Sb –	15.3		
10343	16 18.0 + 37 54 060.37 + 45.36	06 – 36 – 029 NGC 6120	1093 391	0.5 x 0.4 0.5 x 0.4	... ...	1	... Peculiar, p w NGC 6119		14.3	+ 4998	*13 c
10344	16 18.0 + 38 21 061.01 + 45.38	06 – 36 – 028	1093 391	1.4 x 1.0 0.8 x 0.5	25: 3	1	S ...		15.4		
10345	16 18.2 + 58 06 088.35 + 42.66	10 – 23 – 065 NGC 6127 = 6128	1100 392	(1.4 x 1.4) (1.4 x 1.4)	-- ...	4	E		13.0		*1C
10346	16 18.3 + 37 42 060.09 + 45.30	06 – 36 – 000	1093	1.2 x 0.1 1.2 x 0.10	48 7	2	Sc	Sc +	17.		
10347	16 18.5 + 57 45 087.88 + 42.72	10 – 23 – 066 NGC 6130	1100 392	1.1 x 0.7 1.1 x 0.7	25 4	2	S B: b – c		14.2		*
10348	16 18.7 + 56 20 086.01 + 43.09	09 – 27 – 000	1101	1.0 x 0.15 1.0 x 0.15	135 7	3	Sb ssp of 2		16.0:		*
10349	16 19.2 + 40 14 063.70 + 45.18	07 – 34 – 001	0743 395	1.5 x 0.5 1.4 x 0.5	65 6	3	S B a – b		15.2		*1
10350	16 19.3 – 02 10 011.47 + 31.45	00 – 42 – 002 NGC 6118	0143	5.0 x 2.0 5.0 x 2.0	58 6	4	Sc	SA(s)cd	13.2		12 347$\overline{8}$
10351	16 19.4 + 28 46 047.67 + 43.96	05 – 39 – 001	1097	1.0 x 0.5 0.8 x 0.5	40 ...	1	S – Irr		14.9		
10352	16 19.6 – 01 24 012.27 + 31.82	00 – 42 – 003 IC 1213	0143	(0.9 x 0.9) (1.0 x 1.0)	-- ...	2	E Compact		13.9		
10353	16 19.6 + 36 30 058.42 + 44.97	06 – 36 – 035 NGC 6126	1093	(0.9 x 0.9) (1.0 x 1.0)	-- ...	2	... Compact		14.5		*C
10354	16 19.7 + 40 56 064.69 + 45.09	07 – 34 – 003	0743 395	1.3: x 1.1: 0.8 x 0.7	-- 1 – 2	3	Sc/S B c 	SAB(s)c Sc –	15.7		*1
10355	16 20.0 + 13 57 028.83 + 39.23	02 – 42 – 000	1372	1.1 x 0.12 1.0 x 0.12	50 7	2	Sc		15.7		*
10356	16 20.2 + 39 03 062.03 + 44.97	07 – 34 – 004 NGC 6131	0743 395	1.1 x 1.1 1.1 x 1.0	-- 1:	4	Sc/S B c Singular, np of 2, distorted?		14.2		*1
10357	16 20.3 + 40 34 064.17 + 44.98	07 – 34 – 006	0743 395	1.3: x 1.1: 0.7 x 0.7	-- ...	1	...		15.1		
10358	16 20.5 + 55 12 084.42 + 43.13	09 – 27 – 024 NGC 6143	1101	1.0 x 1.0 0.9 x 0.8	-- 1	1	S ...		13.9		*2$\overline{3}$
10359	16 20.6 + 65 30 097.48 + 39.86	11 – 20 – 012 NGC 6140	1410	8.0 x 5.8 5.5: x 4.0:	95: ...	4	S B c	S B(s)cd	12.6		*12 3$\overline{}$
10360	16 20.9 + 17 03 032.65 + 40.24	03 – 42 – 003	0055	1.0 x 1.0 1.1 x 1.0	-- 1	3	SO		15.1		

N G C   6106, 6116 – 18, 6120, 6123, 6126 – 28
6130, 6131, 6140, 6143

I C   1209, 1213, 1216

1	2	3	4	5	6	7	8	9	10	11	12
10361	16 21.1 + 57 24 087.29 + 42.48	10 – 23 – 069	1100 392	1.4 x 0.7 1.4 x 0.7	60: 5	2	S0	SAB 0	15.2		*C
10362	16 21.2 + 39 55 063.25 + 44.80	07 – 34 – 008	0743 395	1.6 x 0.8 1.5 x 0.7	65: 5	3	S B b	S B(r)b	15.4		*1
10363	16 21.3 + 11 54 026.62 + 38.08	02 – 42 – 002 NGC 6132	1372	1.6 x 0.5 1.2 x 0.4	127 6	2	Sa – b		14.8		3
10364	16 21.3 + 38 02 060.60 + 44.72	06 – 36 – 039 NGC 6137	1093 391	(1.9 x 1.2) (2.1 x 1.3)	175 ...	3	E ssf of 2		14.1		*13
10365	16 21.6 + 04 49 018.89 + 34.69	01 – 42 – 000	0088	1.0 x 0.3 1.0 x 0.25	109 6	1	S ...		16.0:		
10366	16 21.6 + 37 22 059.67 + 44.63	06 – 36 – 041	1093 391	1.9 x 0.45 1.9 x 0.5	165 6	3	Sb nnf of 2		14.8		*
10367	16 21.6 + 40 02 063.42 + 44.72	07 – 34 – 011	0743 395	1.3 x 1.3 1.3 x 1.1	-- 1	3	S B b	S B(r)b	14.6		
10368	16 21.6 + 73 58 107.12 + 36.11	12 – 15 – 059	0752	1.1 x 0.35 1.3 x 0.5	150 (6)	2	E – S0		14.8		*
10369	16 21.7 + 67 25 099.70 + 39.00	11 – 20 – 013	1410	1.2 x 0.3: 0.9 x 0.3	55 6	2	Sc		16.5:		
10370	16 22.1 + 45 51 071.58 + 44.43	08 – 30 – 000	1375	1.1 x 0.25 1.0 x 0.25	83 7	2	Sa		15.3		
10371	16 22.2 + 19 35 035.91 + 40.85	03 – 42 – 005 IC 1219	0055	1.0 x 0.4 (1.4 x 0.7)	125 5	1	S0 sp of 2		15.4		*
10372	16 22.2 + 30 17 049.89 + 43.65	05 – 39 – 002	1097 386	1.1 x 0.7 1.1 x 0.7	170: 4	2	Sb	SAB(rs:)b – bc	15.4		
10373	16 22.3 + 07 17 021.42 + 35.76	01 – 42 – 000	0088	1.0 x 0.15 1.1 x 0.15	9 7	1	Sc		15.3		
10374	16 22.3 + 51 06 078.82 + 43.72	09 – 27 – 032	1101 389	1.3 x 0.9 1.2: x 0.7	135: ...	4	S B a	(R)S B(r)a	14.6		*
10375	16 22.4 + 09 43 024.31 + 36.87	02 – 42 – 000	1372	1.4 x 0.2 1.2 x 0.2	136 7	2	Sb n of 2		15.5		*
10376	16 22.4 + 65 33 097.45 + 39.66	11 – 20 – 000	1410	1.5: x 1.4: 1.2: x 1.0:	-- 1:	4	Dwarf spiral S IV – V		16.5:		*
10377	16 22.9 + 23 11 040.54 + 41.83	04 – 39 – 000	1438	1.4? x 0.5? 0.6 x 0.4	... ...	1	Peculiar?, streamers?		17.		
10378	16 23 + 84 08 117.51 + 30.78	14 – 08 – 000	0775	1.0 x 0.5: 0.9 x 0.5:	163 5:	2	Sc	Sc +	17.		
10379	16 23.5 + 41 01 064.80 + 44.37	07 – 34 – 024 NGC 6146	0743 395	(1.3 x 1.0) (1.5 x 1.3)	(75) ...	3	E		13.8		*13
10380	16 23.6 + 16 41 032.52 + 39.51	03 – 42 – 009	0055	1.7 x 0.25 1.7 x 0.25	111 7	3	Sb		15.5		*
10381	16 23.8 + 39 59 063.36 + 44.30	07 – 34 – 025	0743 395	1.3 x 0.7 1.2 x 0.7	95 5	2	S0 – a		14.9		
10382	16 24.2 + 49 57 077.19 + 43.60	08 – 30 – 012 NGC 6154	1357	(2.2 x 2.2) (1.8 x 1.8)	-- 1	4	S B a	S B(r)a	14.0		3
10383	16 24.3 + 64 37 096.22 + 39.82	11 – 20 – 014	1410	1.6 x 0.8 1.4 x 0.6	115 ...	3	...	S dm – m	15.3		
10384	16 24.4 + 11 41 026.79 + 37.31	02 – 42 – 005	1372	1.4 x 0.25 1.2 x 0.25	91 6 – 7	1	S ...		15.0		
10385	16 24.7 + 48 28 075.14 + 43.72	08 – 30 – 013 NGC 6155	1375 389	1.4 x 0.9 1.4 x 0.9	145: ...	1	S ...		13.0		3
10386	16 24.8 + 02 01 016.50 + 32.57	00 – 42 – 000	0143	1.3 x 0.2 1.4 x 0.2	41 7	1	Sc		15.2		
10387	16 24.8 + 13 05 028.44 + 37.81	02 – 42 – 000	1372	1.0 x 0.5 0.7 x 0.5	23 ...	3	S B c		15.3		
10388	16 24.8 + 16 30 032.44 + 39.17	03 – 42 – 010	0055	1.7 x 0.6 1.7 x 0.6	130 6	3	Sa		15.1		
10389	16 25.0 + 39 14 062.32 + 44.05	07 – 34 – 000	0743 395	1.1 x 0.4 0.6 x 0.3	45 ...	2	S B ...		16.0:		
10390	16 25.0 + 44 06 069.10 + 44.02	07 – 34 – 000	0743	1.1 x 0.4 0.7 x 0.4	166 6:	1	S ...		16.0:		
10391	16 25.2 + 19 42 036.37 + 40.23	03 – 42 – 011	0055 396	1.1 x 0.8 1.3 x 0.9	22 3	3	S0		14.8		
10392	16 25.2 + 45 42 071.31 + 43.90	08 – 30 – 014	1375	1.0 x 0.4 0.9 x 0.3	140 6	2	Sb		16.0:		*
10393	16 25.2 + 66 24 098.32 + 39.08	11 – 20 – 000	1410	1.0 x 0.9 0.8 x 0.8	-- 1:	2	S B b – c	S B(r) ...	16.5:		
10394	16 25.3 – 01 40 012.93 + 30.48	00 – 42 – 000	0143	1.6 x 0.6 1.3 x 0.6	170 6	4	Sc	Sc +	15.3		
10395	16 25.3 + 70 19 102.89 + 37.50	12 – 15 – 064	0752	1.3 x 1.0 1.0 x 1.0	(140) 1 – 2	2	Sa/S B a		15.5		

N G C   6132, 6137, 6146, 6154, 6155   I C   1219

1	2	3	4	5	6	7	8	9	10	11	12
10396	16 25.6 + 51 41 079.49 + 43.11	09 - 27 - 040	1101	1.2 x 0.7 1.2 x 0.7	156 4	3	Sb	Sb +	14.7		*A
10397	16 25.8 + 42 47 067.26 + 43.93	07 - 34 - 038 NGC 6159	0743	1.1 x 1.0 1.1 x 1.0	-- ...	2	E - S0 np of 2		15.2		*
10398	16 25.9 + 17 45 034.07 + 39.39	03 - 42 - 000	0055	1.0: x 0.8: 0.9: x 0.7:	-- ...	3	...	S dm - m	16.0:		
10399	16 25.9 + 48 14 074.79 + 43.55	08 - 13 - 017	1375 389	[1.2 x 0.8] ...	... ...	1	Strongly peculiar Triple system of compacts?		16.0:		*1
10400	16 26.0 + 41 02 064.83 + 43.90	07 - 34 - 042 NGC 6160	0743 395	(1.8 x 1.5) (2.3 x 1.8)	(65) ...	4	E		14.8		*13
10401	16 26.0 + 57 00 086.54 + 41.95	10 - 23 - 000	1100	1.4: x 0.8: 1.2: x 0.8:	(65) ...	3	Dwarf spiral S V		17.		
10402	16 26.3 + 12 52 028.38 + 37.39	02 - 42 - 006	1372	[1.3 x 0.6] ...	... ...	3	Double system, contact Strongly distorted		15.2		1
10403	16 26.5 + 32 58 053.74 + 43.18	06 - 36 - 047 NGC 6162	1093 397	1.0 x 0.8 1.1 x 0.9	60: 2	3	S0 Brightest of 3		15.2		*13
10404	16 26.5 + 39 56 063.31 + 43.78	07 - 34 - 046	0743 395	1.5 x 1.1 1.2 x 0.8	160: 3	1	S B ...		15.5		
10405	16 26.6 + 18 00 034.45 + 39.33	03 - 42 - 012	0055	1.7 x 1.3 1.6 x 1.5	-- 1 - 2	4	Sc		15.2		
10406	16 26.8 + 03 13 018.02 + 32.77	01 - 41 - 000	0088	1.7 x 0.4 1.6 x 0.3	120 7	2	Sb - c		15.7		
10407	16 26.8 + 41 20 065.25 + 43.75	07 - 34 - 053	0743	[0.55 x 0.45] ...	... ...	1	Compact triple group		14.3		1
10408	16 26.8 + 56 08 085.37 + 42.05	09 - 27 - 043	1101	1.4 x 0.2 1.3 x 0.2	75 7	3	Sb		16.0:		
10409	16 26.9 + 39 40 062.94 + 43.70	07 - 34 - 060 NGC 6166	0743 395	(2.2 x 1.7) (2.3 x 1.8)	35: ...	4	E Brightest in cluster (13.9) Companion superimposed		+ 8882 + 9082		*12 3
10410	16 27.0 + 25 49 044.27 + 41.65	04 - 39 - 006	1438 401	(1.1 x 0.8) (1.1 x 0.8)	(90) ...	1	E:		15.3		
10411	16 27.2 + 62 47 093.83 + 40.15	10 - 23 - 077	1100 403	1.4 x 0.35 1.1 x 0.35	60 6	2	Sa - b nff of 2		15.3		*
10412	16 27.3 + 15 46 031.87 + 38.34	03 - 42 - 013	0055	1.0 x 0.5 0.9 x 0.5	81 5	1	S0		15.0		
10413	16 27.3 + 21 27 038.75 + 40.34	04 - 39 - 007	1438 396	2.5 x 0.7 2.0: x 0.6	173 6	3	Sc		15.2		
10414	16 27.4 + 08 45 023.94 + 35.33	01 - 42 - 000	0088	1.0: x 0.8: (1.0 x 0.8)	-- 1:	1	S B: 0 - a P w 10416		15.4		*
10415	16 27.4 + 41 24 065.34 + 43.64	07 - 34 - 080	0743 395	1.0 x 0.8 1.0 x 0.8	(135) ...	2	Sb/S B b		14.8		
10416	16 27.9 + 08 44 023.99 + 35.21	01 - 42 - 006	0088	1.1 x 0.8 1.1 x 0.8	... ...	3	Sb/S B c P w 10414	SAB(s)bc	15.2		*
10417	16 28.0 + 40 49 064.54 + 43.52	07 - 34 - 000	0743 395	1.1 x 0.2 1.0 x 0.15	124 7	2	Sb - c		16.5:		
10418	16 28.0 + 75 00 108.00 + 35.25	13 - 12 - 000	1433	0.25 x 0.22 0.25 x 0.22	... ...	1	... Compact		14.2		*C
10419	16 28.1 + 27 50 046.98 + 41.89	05 - 39 - 000	1097	1.5:: x 1.3:: 1. x 1.	-- ...	4	Dwarf V		17.		
10420	16 28.1 + 39 53 063.25 + 43.48	07 - 34 - 086	0743 395	1.8 x 1.2 1.6 x 1.1	(155) 3:	4	S B b	S B(r)b	15.6		*1
10421	16 28.1 + 40 55 064.68 + 43.50	07 - 34 - 083 NGC 6173	0743 395	(1.9 x 1.4) (2.2 x 1.6)	(140) ...	4	E		14.0		*13
10422	16 28.3 + 40 45 064.45 + 43.46	07 - 34 - 087 NGC 6175	0743 395	[1.2 x 0.7] ...	... ...	2	Double system		15.0		*13
10423	16 28.3 + 41 43 065.78 + 43.47	07 - 34 - 088	0743 395	1.0: x 0.4 1.1 x 0.35	55 ...	1	... nf of 2, contact		16.0:		*1
10424	16 28.5 + 55 37 084.62 + 41.93	09 - 27 - 048 NGC 6182	1101	1.9 x 0.6 1.7 x 0.5	146 6	2	Sa		14.6		*13
10425	16 28.5 + 64 19 095.65 + 39.50	11 - 20 - 015	1410	1.8 x 0.3 1.7 x 0.25	155 6	2	Sb		15.4		
10426	16 28.6 + 16 21 032.71 + 38.27	03 - 42 - 015	0055	1.1 x 0.4 0.9 x 0.3	131 6	1	S ...		15.5		
10427	16 28.7 + 41 13 065.09 + 43.39	07 - 34 - 091	0743 395	1.5: x 1.3: 1.2: x 1.0:	-- 1	3	S B c		15.4		
10428	16 28.8 + 35 10 056.81 + 42.99	06 - 36 - 049 NGC 6177	1093	1.7 x 1.1 1.7 x 1.1	10: 4	3	S B b	(R')S B(s)b	14.8		*3
10429	16 28.9 + 39 57 063.35 + 43.32	07 - 34 - 094	0743 395	1.1 x 0.9 (1.0 x 0.7)	(135) ...	1	...		15.3		1
10430	16 28.9 + 41 36 065.62 + 43.36	07 - 34 - 093	0743 395	1.1 x 1.0 1.1 x 0.9	-- 1	3	S B b	S B(rs)bc Sb +	15.4		*

N G C 6159, 6160, 6162, 6166, 6173, 6175
6177, 6182

1	2	3	4	5	6	7	8	9	10	11	12
10431	16 29 + 81 55 115.21 + 31.82	14 – 08 – 000	0775 388	1.0 x 0.2 0.9 x 0.15	138 7	1	Sc		16.5:		
10432	16 29.0 + 41 19 065.23 + 43.34	07 – 34 – 098	0743 395	1.5 x 0.2 1.5 x 0.2	1 7	3	Sb		16.0:		
10433	16 29.1 + 39 20 062.51 + 43.26	07 – 34 – 100	0743 395	1.5 x 0.3 1.5 x 0.35	113 6	1	S ...		15.7		*1
10434	16 29.2 + 20 17 037.50 + 39.54	03 – 42 – 016	0055 396	1.6 x 0.4 1.6 x 0.5	111 (6)	3	Sc – Irr		14.7		*
10435	16 29.2 + 22 48 040.61 + 40.33	04 – 39 – 012	1438	1.1 x 0.7 0.9 x 0.7	70 3 – 4	3	S B b	S B(r)b – bc	15.3		
10436	16 29.4 + 41 16 065.16 + 43.26	07 – 34 – 103	0743 395	1.4 x 1.4 1.3 x 1.3	--- 1	3	Sc	Sc –	14.8		
10437	16 29.6 + 43 27 068.16 + 43.22	07 – 34 – 106	0743	2.2 x 1.9 2.0: x 1.7:	--- 1	3	S ... Singular, unilateral?		15.4		
10438	16 30 + 81 32 114.80 + 31.98	14 – 08 – 000	0775 388	1.0 x 0.4 1.0 x 0.4	16 6	2	S0 – a		16.0:		
10439	16 30.1 + 19 56 037.16 + 39.22	03 – 42 – 020 NGC 6181	0055	2.5 x 1.0 2.8 x 1.2	175 6	4	Sc	Sc –	12.7	+ 2158 + 2309	*12 3468
10440	16 30.5 + 19 32 036.72 + 39.00	03 – 42 – 021	0055	1.2 x 1.1 1.2 x 1.2	--- 1	4	Sc	Sc +	16.0:		
10441	16 30.6 – 00 07 015.27 + 30.21	00 – 42 – 000	0143	1.0 x 0.6 0.9 x 0.5	105 4	3	Sc		15.3		*
10442	16 30.8 + 59 45 089.86 + 40.63	10 – 23 – 081 NGC 6189	1100 403	1.8 x 0.8 1.8 x 0.8	20 6	3	Sc or Sc/S B c		13.3		
10443	16 31.2 + 58 33 088.31 + 40.89	10 – 23 – 082 NGC 6190	1100 403	1.6 x 1.4 1.7 x 1.5	60: 1	3	Sc		13.2		*1
10444	16 31.4 + 35 26 057.25 + 42.49	06 – 36 – 052 NGC 6185	1093	1.2 x 0.8 1.1 x 0.7	0 ...	1	...	I 0?	14.5		13
10445	16 31.8 + 29 06 048.87 + 41.37	05 – 39 – 005	1097	2.9 x 1.9 2.3: x 2.0:	145: ...	2	Sc: Singular		14.2		*
10446	16 31.8 + 79 36 112.77 + 32.88	13 – 12 – 006	1433	1.8 x 0.2 1.5 x 0.15	150 7	2	Sb – c		16.0:		
10447	16 32.0 + 80 28 113.66 + 32.45	13 – 12 – 005	1433	2.6: x 0.5 2.0: x 0.4	137 6	3	Sb		15.5		1
10448	16 32.2 + 21 39 039.47 + 39.32	04 – 39 – 015 NGC 6186	1438	1.7 x 1.7 1.8 x 1.6	--- 1	4	S B a	(R')S B(s)a	14.2		3
10449	16 32.2 + 62 46 093.58 + 39.61	10 – 23 – 083	1100 403	1.5: x 1.5: 0.8:: x 0.8::	--- 1	4	...	S dm	15.5		
10450	16 32.3 + 36 17 058.42 + 42.40	06 – 36 – 000	1093	1.1 x 0.12 1.1 x 0.12	146 7	2	Sc		16.5:		
10451	16 32.4 + 30 04 050.16 + 41.44	05 – 39 – 000	1097	1.0 x 0.3 1.0 x 0.3	98 ...	1	...		17.		
10452	16 32.4 + 67 52 099.73 + 37.88	11 – 20 – 000	1410	1.2 x 0.12 0.8 x 0.12	28 7	2	Sb – c		16.5:		
10453	16 32.8 + 20 41 038.34 + 38.87	03 – 42 – 000	0055 396	1.5 x 0.15 1.5 x 0.15	50 7	3	Sc	S d	15.6		*
10454	16 32.8 + 56 13 085.23 + 41.22	09 – 27 – 051	1101	1.3 x 0.35 0.9 x 0.25	54 6	2	Sa – b		15.0		
10455	16 32.9 + 25 46 044.66 + 40.35	04 – 39 – 000	1438 401	1.1 x 1.1 0.9 x 0.8	--- 1	4	S B b	S B(r)b Sb +	15.6		
10456	16 33.0 + 53 03 081.07 + 41.77	09 – 27 – 052	1101	1.6 x 0.9 1.2 x 0.4	140 5	3	Sb		15.1		*
10457	16 33.2 + 45 25 070.82 + 42.52	08 – 30 – 031	1375	1.0 x 1.0 0.9 x 0.9	--- 1	3	Sc		15.3		
10458	16 33.2 + 46 30 072.28 + 42.46	08 – 30 – 030 IC 1221	1375	1.3 x 1.1 1.1 x 1.0	--- 1 – 2	4	Sc	Sc +	14.7		*
10459	16 33.5 + 41 06 064.96 + 42.49	07 – 34 – 000	0743	1.6 x 0.15 1.6 x 0.15	120 7	3	Sc		16.0:		
10460	16 33.5 + 57 26 086.78 + 40.86	10 – 24 – 002	1414	1.2 x 0.12 1.0 x 0.12	123 7	3	Sc		16.0:		
10461	16 33.6 + 46 19 072.03 + 42.40	08 – 30 – 032 IC 1222	1375	1.7 x 1.2 1.6 x 1.2	50: ...	4	S B b	SAB(s)bc Sb +	14.6		*A
10462	16 33.8 + 05 29 021.39 + 32.38	01 – 42 – 000	0088	1.0: x 0.2 0.6 x 0.15	31 6 – 7	1	S ...		16.5:		
10463	16 33.9 + 10 28 026.70 + 34.68	02 – 42 – 000	1372	1.0 x 0.5 0.9 x 0.6	75 4 – 5	2	S B b		15.3		*
10464	16 34 + 83 00 116.20 + 31.12	14 – 08 – 6+7	0775 388	[1.3 x 0.5] ...	... ...	2	Double system Connected, plume?		15.5		*1
10465	16 34.2 + 01 45 017.66 + 30.43	00 – 42 – 006	0143	1.1 x 1.0 1.1 x 1.1	--- 1	1	Sb – c		14.7		

N G C   6181, 6185, 6186, 6189, 6190   I C   1221, 1222

1	2	3	4	5	6	7	8	9	10	11	12
10466	16 34.2 + 76 51 109.79 + 34.07	13 – 12 – 007	1433	1.2 x 0.8 1.2 x 0.8	44 ...	3	Sb/S B b		14.9		
10467	16 34.5 + 57 37 086.98 + 40.69	10 – 24 – 003 NGC 6198	1414	(1.0 x 0.7) 0.8 x 0.5	90: ...	2	E		14.8		
10468	16 34.7 + 44 43 069.86 + 42.28	07 – 34 – 115	0743	1.1 x 0.15 1.1 x 0.15	176 7	2	Sc		16.0:		
10469	16 34.8 + 39 08 062.32 + 42.15	07 – 34 – 118 NGC 6195	0743 395	1.6 x 1.1 1.5 x 1.0	45: 3	3	Sb		14.7		
10470	16 34.9 + 78 18 111.32 + 33.37	13 – 12 – 008 NGC 6217	1433 411	3.6 x 3.6 3.3: x 3.2:	-- 1	4	Sb	SAB(s)bc	12.1	+ 1385 + 1616	*12 3468A
10471	16 35 + 81 40 114.83 + 31.76	14 – 08 – 009	0775 388	2.0: x 1.3: 1.6: x 1.2:	155: ...	1	...		15.2		*1
10472	16 35 + 84 39 117.84 + 30.26	14 – 08 – 000	0775	1.0 x 0.2 0.9 x 0.15	76 7	2	Sc		16.5:		
10473	16 35.1 + 36 31 058.82 + 41.87	06 – 36 – 055	1093	1.6 x 0.4 1.5 x 0.4	170 6	2	S B a		15.0		*
10474	16 35.3 – 01 16 014.88 + 28.60	00 – 42 – 007	0143	1.1 x 0.7 1.1 x 0.8	(10) 3	1	S ...		15.1		
10475	16 35.5 + 07 40 023.91 + 33.05	01 – 42 – 000	0088	1.1 x 0.15 1.1 x 0.2	22 6	1	Sb – c ssp of 2		15.7		*
10476	16 35.6 + 68 37 100.48 + 37.34	12 – 16 – 002	0756	1.4 x 0.2 1.5 x 0.3	37 7	1	Sa P w 10478		15.7		*
10477	16 35.8 + 37 22 059.97 + 41.81	06 – 36 – 056	1093	1.8 x 0.3 1.2 x 0.2	26 7	1	S ...		15.4		
10478	16 35.8 + 68 31 100.35 + 37.35	11 – 20 – 021	1410	1.7 x 0.25 1.7 x 0.2	59 7	2	Sb – c P w 10476		16.0:		
10479	16 36.0 + 40 59 064.82 + 42.01	07 – 34 – 121	0743	1.1 x 0.25 1.1 x 0.2	14 (7)	1	... Brightest in group, distorted		16.0:		*1
10480	16 36.0 + 44 14 069.20 + 42.05	07 – 34 – 125	0743 400	1.2 x 0.6 1.2 x 0.6	45 ...	1	E?		14.9		
10481	16 36.0 + 55 00 083.53 + 41.01	09 – 27 – 057	1101	1.0: x 0.8: 0.7: x 0.6:	(100) ...	1	...		16.5:		
10482	16 36.1 + 36 10 058.38 + 41.63	06 – 36 – 058	1093	1.4 x 1.0 1.3 x 0.9	140 3	3	S0 Brightest in group		14.2		*1
10483	16 36.1 + 76 09 108.97 + 34.28	13 – 12 – 000	1433	1.2 x 0.5 1.3 x 0.6	126 6	1	S ...		16.0:		
10484	16 36.2 + 26 22 045.67 + 39.79	04 – 39 – 000	1438 401	1.2 x 0.5 1.2 x 0.5	103 6	2	Sb		15.6		*
10485	16 36.2 + 39 21 062.63 + 41.89	07 – 34 – 000	0743	1.5 x 0.15 1.3 x 0.15	62 7	2	Sb		16.0:		
10486	16 36.2 + 50 26 077.51 + 41.64	08 – 30 – 040	1375	(1.0 x 0.8) (1.1 x 0.9)	90: ...	2	E – S0		13.9		
10487	16 36.3 + 49 32 076.31 + 41.71	08 – 30 – 041	1375	1.2: x 0.5: 1.4 x 0.5	125 6 – 7	2	Sa – b		15.7		*
10488	16 36.5 + 40 01 063.53 + 41.87	07 – 34 – 124	0743	1.5 x 0.5 1.4 x 0.4	174 6	2	Sa – b		15.4		
10489	16 36.5 + 62 53 093.53 + 39.11	10 – 24 – 008	1414 403	1.5 x 0.10 1.3 x 0.10	73 7	2	...		16.5		
10490	16 36.6 + 17 27 034.90 + 36.91	03 – 42 – 025	0055	1.1 x 0.9 0.9: x 0.9:	(140) 2:	4	Sc Sc +		15.3		
10491	16 36.7 + 42 02 066.24 + 41.91	07 – 34 – 127	0743	[1.1 x 0.5] ...	... ...	2	Double system		15.5		*1A C
10492	16 36.8 – 02 20 014.09 + 27.71	00 – 42 – 008	0143	1.0 x 0.9 (1.7 x 1.6)	-- 1	1	Sa		15.5		
10493	16 36.9 + 55 31 084.17 + 40.79	09 – 27 – 059	1101	(1.1 x 0.8) (1.2 x 0.9)	(135) ...	3	[E] Compact		14.9		
10494	16 36.9 + 67 44 099.38 + 37.53	11 – 20 – 022 IC 1225	1410	1.5 x 0.45 1.4 x 0.4	73 6	2	S B? b		15.5		
10495	16 37.0 + 07 23 023.82 + 32.59	01 – 42 – 000	0088	1.0 x 0.3 1.0 x 0.25	120 6	1	S ...		15.5		
10496	16 37.1 + 72 06 104.44 + 35.90	12 – 16 – 003	0756	1.5 x 0.7 1.2 x 0.6	100 5:	3	... S B(s)dm		16.0:		
10497	16 37.3 + 72 30 104.88 + 35.73	12 – 16 – 004	0756	1.2 x 0.45 1.1 x 0.4	96 6	1	S ... Singular, p w 10502		14.4		*
10498	16 37.4 + 29 28 049.68 + 40.26	05 – 39 – 000	1097 402	1.0 x 0.1 0.8 x 0.1	179 7	1	Sb – c		17.		
10499	16 38.0 + 49 08 075.74 + 41.47	08 – 30 – 000	1375	1.0: x 0.2 1.0 x 0.1	27 7	1	S ...		16.5:		
10500	16 38.0 + 57 50 087.12 + 40.18	10 – 24 – 013	1414 403	1.6 x 1.5 1.6 x 1.5	-- 1	3	S0/Sa		14.1		

N G C   6195, 6198, 6217   I C   1225

1	2	3	4	5	6	7	8	9	10	11	12
10501	16 38 + 82 39 115.77 + 31.19	14 – 08 – 010 NGC 6251	0775 388	(1.8 x 1.5) (2.8 x 2.3)	––– ...	4	E s of 2		14.0		*3
10502	16 38.2 + 72 28 104.81 + 35.68	12 – 16 – 005	0756	2.6 x 2.2 2.6 x 2.2	(130) 1:	4	Sc P w 10497	SA(rs)c Sc –	13.1		*1
10503	16 38.3 + 71 37 103.84 + 36.01	12 – 16 – 006	0756	1.8 x 0.4 1.3: x 0.3	[30] ...	4	Irr	Ir I	16.0:		*1
10504	16 38.5 + 33 46 055.30 + 40.82	06 – 37 – 003	1069	1.1 x 0.5 1.0 x 0.4	169 6	1	Sa – b		14.9		
10505	16 39.0 – 00 02 016.63 + 28.47	00 – 42 – 000	0143	1.0 x 0.5 1.1 x 0.6	100 5:	1	S ...		15.7		
10506	16 39.3 + 58 44 088.22 + 39.82	10 – 24 – 019 NGC 6206	1414 403	0.7 x 0.7 0.7 x 0.7	––– ...	2	[S0] Compact		14.5		*
10507	16 39.3 + 66 07 097.36 + 37.85	11 – 20 – 024 NGC 6214	1410	1.1 x 0.9 1.1 x 1.0	(145) 1 – 2	1	S ...		14.3		*
10508	16 39.4 + 62 05 092.42 + 39.00	10 – 24 – 020	1414 403	1.2 x 0.2 1.1 x 0.2	57 7	2	Sa – b		15.7		*
10509	16 39.4 + 78 14a 111.13 + 33.20	13 – 12 – 009	1433 411	1.0 x 0.9 1.0 x 0.7	(125) ...	3	S B b of 3, distorted, Sb +	Brightest	(15.1)		*1
10510	16 39.6 + 58 13 087.55 + 39.90	10 – 24 – 021	1414 403	1.6 x 1.5: 1.1 x 0.9	––– 1	4	Sc	Sc –	14.9		1
10511	16 40.0 + 51 18 078.57 + 40.95	09 – 27 – 063	1101	(1.0 x 0.8) (0.9 x 0.7)	(20) ...	2	S0		15.2		*
10512	16 40.1 + 40 05a 063.67 + 41.19	07 – 34 – 137	0743	(1.1 x 1.1) (1.3 x 1.3)	––– ...	3	E sp of 2		(14.8)		*1
10513	16 40.3 + 00 44 017.57 + 28.60	00 – 43 – 000	0143	1.2: x 0.6 1.0 x 0.5	168 5:	3	Sc		16.0:		
10514	16 40.3 + 25 10 044.49 + 38.58	04 – 39 – 020	1438 401	1.7 x 0.5 1.8 x 0.5	[163] ...	2	S – Irr Peculiar?		15.4		
10515	16 40.3 + 59 26 089.07 + 39.54	10 – 24 – 024	1414 403	1.0 x 0.9 0.8 x 0.45	(175) 1	2	Sb		15.1		
10516	16 40.5 + 57 54 087.12 + 39.85	10 – 24 – 027 NGC 6211	1414 403	1.7 x 1.2 1.5 x 1.1	105 2	4	S0 Brightest in chain of 4		13.8		*12 C
10517	16 40.6 + 61 25 091.55 + 39.04	10 – 24 – 028	1414 403	1.6 x 0.25 1.5 x 0.3	148 7	3	Sa		13.7		*
10518	16 40.6 + 62 29 092.87 + 38.76	10 – 24 – 029	1414 403	1.1 x 0.9 1.1 x 0.8	(145) 2	3	Sb/Sc		14.7		
10519	16 40.7 + 65 52 097.00 + 37.79	11 – 20 – 025	1410	1.1 x 0.25 1.1 x 0.25	33 6 – 7	1	Sa – b		15.7		*
10520	16 41.2 + 48 01 074.21 + 41.02	08 – 30 – 047	1375	1.1 x 0.9 0.8 x 0.6	(40) 2:	3	Sc		16.0:		
10521	16 41.3 + 36 55 059.53 + 40.68	06 – 37 – 007 NGC 6207	1069	3.3 x 1.2 2.4 x 1.2	15: ...	1	S ...		11.9	+ 869 + 1072	*12 3478
10522	16 41.5 + 61 55 092.13 + 38.81	10 – 24 – 036	1414 403	1.0 x 0.35 1.0 x 0.35	145 6	2	Sc		15.5		
10523	16 41.6 + 42 16 066.59 + 41.01	07 – 34 – 141	0743	1.3 x 0.35 1.2 x 0.3.	174 6	1	S ...		15.3		
10524	16 41.8 + 65 40 096.71 + 37.74	11 – 20 – 026 IC 1228	1410	1.6 x 1.5 1.6 x 1.4	––– 1	4	S B b	(R')S B(s)ab Sb –	14.5		1
10525	16 41.9 + 23 29 042.57 + 37.75	04 – 39 – 000	1438	1.0 x 0.9 0.7: x 0.6:	––– 1	2	Sb		15.7		
10526	16 42.3 + 25 13 044.71 + 38.16	04 – 39 – 000	1438 401	1.0 x 0.6 0.8 x 0.5	0 4	2	Sb – c		15.5		
10527	16 42.5 + 61 40 091.79 + 38.76	10 – 24 – 040 NGC 6223	1414 403	(3.5 x 2.5) (3.0 x 2.5)	––– ...	3	... Peculiar		13.1		*3C
10528	16 42.6 + 22 37 041.59 + 37.33	04 – 39 – 021	1438	2.4 x 1.3 2.2 x 1.3	65 5	4	S0	SA 0 +	13.5		
10529	16 42.6 + 32 18 053.60 + 39.73	05 – 39 – 007	1097	1.2 x 1.1 0.9: x 0.9:	––– 1	2	Sb – c		15.4		*1
10530	16 42.6 + 49 36 076.28 + 40.69	08 – 30 – 000	1375	1.0:: x 0.4: 1.2 x 0.3	35 6	1	S ...		17.		
10531	16 42.9 + 43 50 068.67 + 40.81	07 – 34 – 144	0743	1.0 x 0.7 0.7 x 0.6	110: ...	2	Sb – c		16.5:		1
10532	16 42.9 + 62 04 092.27 + 38.61	10 – 24 – 043 NGC 6226	1414 403	0.8 x 0.40 0.8 x 0.40	[68] ...	1	... Peculiar		13.8		*3
10533	16 43.2 + 18 18b 036.59 + 35.75	03 – 43 – 002	0122	1.0 x 0.35 0.9 x 0.3	112 6	1	S ... n of 2, disturbed		(15.3)		*
10534	16 43.4 + 58 58 088.37 + 39.26	10 – 24 – 044	1414 403	1.0 x 0.08 1.0 x 0.1	67 7	2	Sc		17.		
10535	16 43.6 + 06 34 023.87 + 30.76	01 – 43 – 000	1056	1.2 x 0.2 1.1 x 0.2	24 7	2	Sa – b		15.2		

N G C   6206, 6207, 6211, 6214, 6223, 6226, 6251   I C   1228

1	2	3	4	5	6	7	8	9	10	11	12
10536	16 43.6 + 62 10 092.37 + 38.51	10 - 24 - 046	1414 403	2.5: x 1.1: 1.8: x 0.8:	123 6	3	Sb		15.0		
10537	16 43.7 + 70 44 102.63 + 35.93	12 - 16 - 007 NGC 6232	0756 412	1.8 x 1.8 1.8 x 1.8	--- 1	4	S B a	(R´)S B(s)a	13.5		
10538	16 43.8 + 51 22 078.57 + 40.35	09 - 27 - 073 IC 1230	1101	1.0: x 1.0: 1.0: x 1.0:	--- ...	1	...		15.5		*
10539	16 44.0 + 58 10 087.34 + 39.34	10 - 24 - 048	1414 403	1.0 x 0.2 0.8 x 0.12	172 7	2	Sb - c		16.5:		
10540	16 44.2 + 30 04 050.86 + 38.95	05 - 39 - 000	1097	1.1 x 0.9 0.5 x 0.4	--- 1 - 2	1	Sb		16.5:		
10541	16 44.6 - 00 11 017.31 + 27.20	00 - 43 - 000 NGC 6220	0155	(1.7 x 0.8) (2.0 x 1.2)	135: 5:	1	Sa?		15.5		*
10542	16 44.6 + 61 55 092.02 + 38.45	10 - 24 - 049	1414 403	1.6 x 0.2 1.6 x 0.2	88 7	3	Sc		15.7		
10543	16 44.7 + 27 02 047.12 + 38.12	05 - 39 - 008	1097	1.0 x 1.0 0.9 x 0.9	--- 1	2	S B a	S B(r)a	15.5		
10544	16 44.9 + 36 10 058.67 + 39.87	06 - 37 - 008	1069	1.0 x 0.45 1.0 x 0.5	177 5	2	Sa		15.4		
10545	16 45.0 + 34 30 056.53 + 39.61	06 - 37 - 009	1069	1.0 x 0.4 1.0 x 0.4	156 6	2	Sb - c		15.1		
10546	16 45.0 + 70 53 102.76 + 35.77	12 - 16 - 008 NGC 6236	0756 412	3.0 x 1.8 2.8 x 1.7	15 4	4	Sc/S B c	SAB(s)cd	12.7		
10547	16 45.1 + 34 15 056.21 + 39.55	06 - 37 - 010	1069	1.5 x 0.7 1.1 x 0.6	10 5	2	S B b	S B(s)b	14.7		
10548	16 45.1 + 59 44 089.28 + 38.89	10 - 24 - 051	1414 403	1.6 x 0.8 1.5 x 0.8	143 5	3	S B b Sb -		14.7		*
10549	16 45.3 + 21 13 040.19 + 36.29	04 - 40 - 000	1377	1.1 x 0.9 1.0 x 0.8	... ...	4	Irr Ir I		15.7		*
10550	16 45.4 + 40 13 063.94 + 40.19	07 - 34 - 148	0743	1.4 x 0.2 1.1 x 0.15	34 7	2	Sc		16.0:		
10551	16 45.5 + 63 35 094.04 + 37.94	11 - 20 - 000	1410	1.0 x 0.15 1.0 x 0.15	157 7	2	Sc np of 2		16.0:		*
10552	16 45.5 + 78 29 111.24 + 32.81	13 - 12 - 000	1433 411	(1.1 x 0.8) (1.1 x 0.8)	(165) ...	1	...		15.4		*
10553	16 45.6 + 40 20 064.10 + 40.16	07 - 34 - 149	0743	1.1 x 0.5 1.1 x 0.35	153 6	2	S B: a - b		15.0		
10554	16 45.8 - 01 32 016.20 + 26.24	00 - 43 - 000	0155	1.5 x 0.5 1.4 x 0.4	155 6	3	Sc Sc +		15.7		
10555	16 45.8 + 06 24 024.00 + 30.20	01 - 43 - 002 NGC 6224	1056 405	(1.1 x 1.1) (1.6 x 1.6)	--- ...	1	E - S0: P w 10556		15.0		*
10556	16 45.9 + 06 19 023.93 + 30.14	01 - 43 - 003 NGC 6225	1056 405	(1.1 x 0.8) (1.2 x 0.9)	--- ...	1	E - S0: P w 10555		15.0		*
10557	16 46.0 + 13 58 032.03 + 33.48	02 - 43 - 000	0251	1.1 x 0.9 0.9 x 0.9	--- 1 - 2	2	S B b - c		16.5:		
10558	16 46.0 + 26 18 046.32 + 37.65	04 - 40 - 001 NGC 6228	1377	1.1 x 0.55 1.0 x 0.55	[130] ...	3	S ... Peculiar, nf of 2, distorted		15.3		*13
10559	16 46.0 + 59 12 088.58 + 38.88	10 - 24 - 054	1414 403	1.1 x 0.2 1.1 x 0.18	161 (7)	1	...		15.4		
10560	16 46.1 + 58 32 087.73 + 39.00	10 - 24 - 056 IC 1231	1414 403	2.4 x 1.0 2.3 x 1.0	155 6	3	Sc		13.7		
10561	16 46.1 + 62 54 093.18 + 38.05	10 - 24 - 055	1414 403	2.5 x 0.25 2.5 x 0.25	150 7	2	Sb		15.2		
10562	16 46.1 + 75 47 108.27 + 33.88	13 - 12 - 012	1433	1.4 x 0.2 1.2 x 0.2	177 (7)	1	... nf of 2, disturbed?		15.0		*
10563	16 46.6 + 62 14 092.34 + 38.15	10 - 24 - 057 NGC 6238	1414 403	0.50 x 0.30 0.45 x 0.25	[17] ...	1	... Peculiar		14.4		
10564	16 46.8 + 70 28 102.21 + 35.77	12 - 16 - 009	0756 412	3.3 x 1.2 2.6: x 1.2:	150 6	3	S B c Sc +		14.1		1
10565	16 46.9 + 48 49 075.20 + 40.04	08 - 31 - 001	1370	1.0 x 0.9 0.9 x 0.8	(70) ...	2	Dwarf		17.		*
10566	16 47.1 + 36 16 058.87 + 39.44	06 - 37 - 012	1069	2.0 x 0.35 1.9 x 0.3	74 7:	2	S ... P w 10567, disturbed		16.0:		*1
10567	16 47.3 + 36 18 058.92 + 39.40	06 - 37 - 013	1069	1.1 x 0.6: 1.1: x 0.6:	[55] ...	2	... P w 10566, disturbed		15.2		*1
10568	16 47.5 + 62 17 092.37 + 38.04	10 - 24 - 059 NGC 6244	1414 403	1.6 x 0.35 1.7 x 0.35	140 6	3	S B: a		14.3		*1
10569	16 47.6 + 09 31 027.46 + 31.23	02 - 43 - 003	0251	1.7 x 0.20 1.9 x 0.25	137 7	2	Sb		15.4		1
10570	16 47.7 + 59 38 089.07 + 38.58	10 - 24 - 061	1414 403	1.3 x 0.5 0.6 x 0.3	133 6	2	Sc		16.0:		

N G C   6220, 6224, 6225, 6228, 6232, 6236   I C   1230, 1231
      6238, 6244

1	2	3	4	5	6	7	8	9	10	11	12
10571	16 47.8 + 48 43	08 – 31 – 002	1370	1.3 x 0.35	12	2	S B: c		15.4		
	075.06 + 39.89			1.1 x 0.4	6						
10572	16 47.8 + 63 04	11 – 20 – 000	1410	1.1 x 0.30	58	1	...		13.5		3
	093.33 + 37.82	NGC 6247		0.9 x 0.30	...		Peculiar				
10573	16 48.1 + 23 40	04 – 40 – 002	1377	(1.4 x 1.0)	33	4	S0		14.9		*1
	043.33 + 36.45	NGC 6233	409	(1.6 x 1.1)	3:						
10574	16 48.1 + 27 07	05 – 40 – 000	0129	1.1 x 0.5	150:	1	Sa – b		16.0:		
	047.47 + 37.41			1.0 x 0.5	5						
10575	16 48.4 + 04 41b	01 – 43 – 005	1056	(1.0 x 0.9)	—	1	E?		(15.5)		*
	022.62 + 28.82	NGC 6230b		0.9 x 0.7	...		f of 2				
10576	16 48.4 + 08 53	01 – 43 – 004	1056	1.1 x 0.9	(155)	2	Sa – b		15.0		
	026.90 + 30.77	IC 4621		1.1 x 0.9	2:						
10577	16 48.5 + 42 50	07 – 35 – 001	1135	2.5 x 1.1	118	3	S B ...		12.9 + 964		*12
	067.39 + 39.76	NGC 6239		2.3 x 1.1	...		Peculiar		+ 1183		3468
10578	16 48.8 – 02 23	00 – 43 – 002	0155	1.2 x 0.5	[48]	1	...		15.7		
	015.83 + 25.15		408	0.9 x 0.4	...						
10579	16 48.8 + 54 02	09 – 27 – 097	1101	(1.2 x 0.7)	12	2	E		14.4		
	081.92 + 39.31		406	(1.0 x 0.5)	...						
10580	16 48.9 + 55 38	09 – 27 – 098	1101	1.5 x 0.7	43	2	S B? b		14.2		*2
	083.97 + 39.10	NGC 6246		0.8 x 0.5	5						
10581	16 49 + 82 43	14 – 08 – 000	0775	1.0: x 0.1	52	1	...	d – m	17. – 18.		*
	115.65 + 30.85		388	0.9: x 0.1	7						
10582	16 49.1 + 24 02	04 – 40 – 000	1377	1.1 x 0.25	28	1	Sb – c		16.5:		*
	043.85 + 36.34		409	0.9 x 0.2	6						
10583	16 49.1 + 53 39	09 – 27 – 100	1101	1.2 x 0.7	70:	1	S B a		15.5		*
	081.42 + 39.31		406	0.9 x 0.7	...						
10584	16 49.2 + 55 28	09 – 27 – 101	1101	2.5 x 2.5	—	4	Sc/S B c	SAB(rs)c	14.7		
	083.75 + 39.08			2.3: x 2.1:	1			Sc –			
10585	16 49.3 + 19 30	03 – 43 – 007	0122	1.1 x 0.2	133	2	S0 – a		14.8		*
	038.59 + 34.83			1.2 x 0.22	7		p of 2				
10586	16 49.3 + 45 29	08 – 31 – 008	1370	1.5: x 1.5:	—	1	Sb		15.1		*
	070.84 + 39.69			1.1: x 1.1:	1		spp of 2, in group				
10587	16 49.3 + 68 50	12 – 16 – 10a	0756	1.4 x 0.5	92	2	Sa – b		15.4		
	100.22 + 36.09		412	1.2 x 0.45	6						
10588	16 49.4 + 21 58	04 – 40 – 000	1377	1.5 x 0.35	63	2	S B: b – c		16.5:		
	041.44 + 35.63			1.4 x 0.3	6						
10589	16 49.4 + 55 55	09 – 27 – 102	1101	1.8 x 0.7	62	1	...		15.4		
	084.32 + 38.99			1.5: x 0.4	(6)		Peculiar				
10590	16 50.2 + 59 49	10 – 24 – 069	1414	0.9 x 0.9	—	2	S B [c]		14.1		1
	089.22 + 38.24		403	0.9 x 0.9	1		Singular				
10591	16 50.3 + 23 25	04 – 40 – 004	1377	1.3 x 0.5	154	2	Sa		15.1		*1
	043.22 + 35.89	NGC 6243	409	1.5 x 0.7	6						
10592	16 50.5 + 02 29	00 – 43 – 004	0155	2.2 x 0.9	[20]	4	Strongly peculiar		14.7		*12
	020.73 + 27.28	NGC 6240		2.0 x 1.0	...		Jet + loop, eruptive?				3
10593	16 51.3 + 55 59	09 – 28 – 001	0765	1.5 x 0.5	87	2	Sa – b		15.2		
	084.35 + 38.72			1.2 x 0.3	6						
10594	16 51.5 + 51 10	09 – 28 – 003	0765	1.1 x 0.2	102	1	S – Irr		15.3		*
	078.19 + 39.17			0.9 x 0.2	...		p of 2, disturbed?				
10595	16 51.7 + 60 37	10 – 24 – 073	1414	(0.9 x 0.7)	(70)	2	E		14.5		*
	090.17 + 37.90	NGC 6258	403	(0.9 x 0.7)	...						
10596	16 52.0 + 10 30	02 – 43 – 000	0251	1.1 x 0.9	—	1	S B: ...		16.0:		
	029.05 + 30.69			1.0 x 0.8	1 – 2						
10597	16 52.0 + 23 30	04 – 40 – 000	1377	1.1 x 0.7	155	4	Sb		15.7		*
	043.47 + 35.54		409	1.0 x 0.5	4:						
10598	16 52.2 + 27 15	05 – 40 – 005	0129	1.1 x 0.5	145	1	S ...		15.7		
	047.93 + 36.57			1.0 x 0.4	6						
10599	16 52.2 + 39 50	07 – 35 – 002	1135	(1.2 x 1.0)	(160)	2	...		13.7		*
	063.59 + 38.86	Mark 501	420	(1.2 x 1.0)	...		Extremely compact				
10600	16 52.5 + 36 05	06 – 37 – 000	1069	1.1 x 0.15	50	1	S ...		15.7		
	058.84 + 38.33			1.0 x 0.15	7						
10601	16 52.5 + 54 41	09 – 28 – 006	0765	1.6: x 0.9:	70:	3	Sb		16.0:		
	082.67 + 38.71			0.7 x 0.3	...						
10602	16 52.7 + 41 25	07 – 35 – 003	1135	1.0 x 0.9	(40)	2	S B b – c		14.9		*1
	065.63 + 38.90			1.0 x 0.9	1						
10603	16 52.8 + 22 14	04 – 40 – 000	1377	1.2 x 0.2	52	2	Sb – c		16.0:		
	042.06 + 34.97			1.1 x 0.2	7						
10604	16 53 + 81 56	14 – 08 – 014	0775	1.8 x 0.35	152	1	S ...		15.7		*
	114.76 + 31.07		388	1.7 x 0.35	6						
10605	16 53 + 83 23	14 – 08 – 000	0775	1.5:: x 1.3::	—	3	Dwarf		17.		
	116.29 + 30.45			1.0:: x 0.9::	...						

N G C 6230, 6233, 6239, 6240, 6243, 6246
6247. 6258

I C   4621

1	2	3	4	5		6	7	8	9	10	11	12
10606	16 53.0 + 36 35 059.49 + 38.31	06 - 37 - 014 NGC 6255	1069	3.5 x 1.4 3.5 x 1.4		85 6	4	S B c		13.8		*3C
10607	16 53.1 + 26 44 047.38 + 36.24	04 - 40 - 007 IC 4630	1377	1.3 x 0.9 1.2 x 0.9		... ...	3	... Peculiar, two jets		14.8		1
10608	16 53.3 + 53 13 080.79 + 38.73	09 - 28 - 007	0765	1.3 x 0.9 1.2: x 0.9:		100: ...	3	...	S B dm	16.5:		*
10609	16 53.3 + 69 58 101.41 + 35.41	12 - 16 - 000	0756	1.2: x 1.0: 0.8:x 0.8::		(35) ...	3	Dwarf spiral		16.5:		
10610	16 53.4 + 43 08 067.84 + 38.88	07 - 35 -4 + 5	1135	[2.1 x 1.8] ...		... ...	3	Double system Disrupted, streamers		15.0		*1
10611	16 53.5 + 36 58 059.99 + 38.26	06 - 37 - 015	1069	1.1 x 0.3 1.1 x 0.3		83 6	1	S ...		16.0:		
10612	16 54.0 + 80 26 113.14 + 31.65	13 - 12 - 000	1433	1.1 x 0.3 0.8 x 0.2		[144] ...	1	...		16.0:		
10613	16 54.1 + 53 28 081.09 + 38.59	09 - 28 - 008	0765	1.0 x 0.4 0.9 x 0.3		165 6	2	Sa - b		16.0:		
10614	16 54.1 + 65 23 095.94 + 36.59	11 - 21 - 001	0745	1.4 x 1.1 1.5: x 1.0:		95: ...	3	...	S dm	16.0:		1
10615	16 54.3 + 34 55 057.45 + 37.79	06 - 37 - 016	1069	1.1 x 0.5 0.9 x 0.2		70 6	2	S B c		15.6		
10616	16 54.3 + 58 47 087.81 + 37.91	10 - 24 - 000	1414 403	1.1 x 0.10 0.9 x 0.08		16 7	2	...	d - m	17.		*
10617	16 54.5 + 28 04 049.07 + 36.29	05 - 40 - 006 NGC 6261	0129 415	(1.4 x 0.5) (1.5 x 0.7)		88 5	2	S0 - a		15.2		
10618	16 54.7 + 27 54 048.89 + 36.20	05 - 40 - 008 NGC 6263	0129 415	(1.0 x 1.0) (1.1 x 1.1)		--- ...	3	E		15.1		*13
10619	16 54.8 + 54 30 082.40 + 38.39	09 - 28 - 009	0765	1.5 x 0.3 1.4: x 0.1		16 7	3	Sc	Sc +	16.5:		
10620	16 55.2 + 31 18 053.03 + 36.90	05 - 40 - 000	0129	1.0 x 0.4 0.9 x 0.4		98 6	1	S ...		16.0:		
10621	16 55.2 + 55 06 083.15 + 38.27	09 - 28 - 010 IC 1237	0765	2.6: x 0.7 1.5: x 0.7		20 ...	3	Sb Disturbed:, Sb +		14.7		*12
10622	16 55.2 + 70 31 102.00 + 35.09	12 - 16 - 015	0756	1.4 x 0.2 1.4 x 0.2		79 7	3	Sc		15.3		*
10623	16 55.4 + 02 33 021.48 + 26.25	00 - 43 - 005	0155	1.0 x 0.8 1.0: x 0.8:		(135) 1 - 2:	1	Sc	SAB(rs) ...	15.7		
10624	16 55.5 + 27 55 048.96 + 36.04	05 - 40 - 011 NGC 6265	0129 415	(1.0 x 0.7) (1.0 x 0.7)		(35) ...	1	S0		15.4		*3
10625	16 55.7 + 38 45 062.31 + 38.07	06 - 37 - 000	1093	1.4 x 0.1 1.0: x 0.1		132 7	2	Sc	Sc +	16.5:		
10626	16 55.7 + 58 08 086.96 + 37.82	10 - 24 - 077	1414	1.4 x 0.7 0.8 x 0.4		147 ...	1	S ...		15.4		1
10627	16 55.8 + 46 56 072.73 + 38.58	08 - 31 - 015	1370	1.7 x 0.25 1.6 x 0.25		142 7	3	Sc		15.6		
10628	16 56.0 + 23 03 043.29 + 34.53	04 - 40 - 009 NGC 6267	1377	1.4 x 1.1 1.3 x 1.0.		(35) 2	3	S B c	S B(r)bc Sc -	14.0		13
10629	16 56.0 + 27 56 049.02 + 35.93	05 - 40 - 012 NGC 6269	0129 415	(2.0 x 1.6) (2.2 x 1.7)		(80) ...	3	E		14.4		*3
10630	16 56.0 + 32 03 053.99 + 36.90	05 - 40 - 000	0129	1.0 x 0.5 0.9 x 0.5:		140: ...	1	Sa - b		16.0:		
10631	16 56.0 + 36 30 059.50 + 37.70	06 - 37 - 017	1069	1.0 x 0.3 0.8 x 0.3		147 6	1	S ...		15.4		
10632	16 56.0 + 79 09 111.71 + 32.08	13 - 12 - 013	1433 411	1.7 x 0.25 1.4 x 0.2		29 7	3	Sc	Sc +	15.5		
10633	16 56.3 + 20 07 039.98 + 33.50	03 - 43 - 010 IC 1236	0122	1.2 x 1.1 1.0 x 1.0		(35) 1	4	S B c	SAB(s)c	14.6		1
10634	16 56.3 + 15 17 034.64 + 31.71	03 - 43 - 000	0122	1.0 x 0.1 0.7 x 0.12		132 7	1	S ...		17.		
10635	16 56.3 + 38 17 061.74 + 37.89	06 - 37 - 000	1069	0.7 x 0.7 0.7 x 0.7		--- ...	1	... Compact		13.5	+ 10075? *C + 10291?	
10636	16 56.7 + 70 31 101.95 + 34.97	12 - 16 - 016	0756	1.1 x 1.0 1.1: x 1.0:		--- ...	3	Dwarf irregular		16.5:		
10637	16 56.8 + 32 02 054.02 + 36.73	05 - 40 - 000	0129	1.1 x 0.7 1.0 x 0.6		(20) 4:	1	S ...		16.5:		*
10638	16 57 + 81 40 114.41 + 31.05	14 - 08 - 13...	0775	8. x		... ...		Chain of 11 galaxies		...		
10639	16 57.1 + 29 05 050.47 + 35.99	05 - 40 - 018	0129 415	1.0 x 0.5 1.0 x 0.5		108 5	1	S B b		15.3		*
10640	16 57.2 + 02 34 021.74 + 25.87	00 - 43 - 000	0155	1.0 x 1.0 1.0 x 0.9		--- 1	2	S B ... attached, disturbed	Companion	15.3		*

N G C 6255, 6261, 6263, 6265, 6267, 6269   I C 1236, 1237, 4630

1	2	3	4	5	6	7	8	9	10	11	12
10641	16 57.2 + 58 59 087.99 + 37.50	10 - 24 - 079	1414 403	1.5 x 0.1 1.3 x 0.10	88 7	3	Sc		16.5:		
10642	16 57.3 - 00 28 018.85 + 24.33	00 - 43 - 006	0155	20.? x 10.? ...	... ...	5	Resolved dwarf system		15.7		
10643	16 57.4 + 30 01 051.61 + 36.15	05 - 40 - 19 + 20	0129 415	[1.0 x 0.5] ...	... ...	2	Double system Contact		14.5		*1
10644	16 57.5 + 54 56 082.90 + 37.96	09 - 28 - 012	0765	1.0 x 0.6 ...	50 ...	1	...		14.8		
10645	16 57.6 + 47 18 073.20 + 38.28	08 - 31 - 017 NGC 6279	1370	1.0 x 0.8 0.9 x 0.8	(10) 2:	3	S B 0		14.9		
10646	16 57.7 + 59 12 088.24 + 37.41	10 - 24 - 082	1414 403	(1.4 x 0.6) (1.4 x 0.6)	25 ...	3	E		15.0		
10647	16 57.8 + 59 02 088.05 + 37.42	10 - 24 - 084 NGC 6286	1414 403	1.5 x 1.3 1.3 x 1.2	-- ...	3	... p w NGC 6285, disrupted?	Peculiar,	14.2		*1A
10648	16 57.8 + 59 50 089.03 + 37.30	10 - 24 - 083	1414 403	1.4 x 1.0 1.0: x 0.9	(80) ...	3	S B b	S B(s)b Sb -	15.3		
10649	16 58.0 + 23 00 043.41 + 34.08	04 - 40 - 000	1377	1.0: x 0.1 1.0 x 0.15	53 7	1	Sc		17.		
10650	16 58.1 + 23 11 043.63 + 34.11	04 - 40 - 000	1377	1.7 x 0.4 1.6 x 0.4	[23] ...	2	Irr or Peculiar		15.3		*
10651	16 58.2 + 42 37 067.25 + 37.97	07 - 35 - 014	1135	1.3: x 1.1: 1.1: x 0.8:	-- ...	1	S ...		15.6		*
10652	16 58.2 + 50 00 076.63 + 38.16	08 - 31 - 018 NGC 6283	1370	1.2 x 1.2 1.2 x 1.0	-- ...	2	S ... Peculiar		13.7		13
10653	16 58.3 + 27 40 048.87 + 35.37	05 - 40 - 022	0129 415	1.5 x 0.8 1.3 x 0.8	... ...	2	S B b		15.4		
10654	16 58.6 + 29 48 051.43 + 35.85	05 - 40 - 000	0129 415	1.0 x 0.4 1.1 x 0.45	170 6	2	S0 - a		15.5		
10655	16 58.6 + 56 08 084.39 + 37.69	09 - 28 - 013	0765	1.0 x 0.35 0.8 x 0.25	108 6	1	Sb - c		15.4		*
10656	16 58.7 + 23 05 043.57 + 33.95	04 - 40 - 011 NGC 6278	1377	2.1 x 1.2 (2.0 x 1.1)	130 4	4	S0 P w NGC 6276 = 6277		13.8		*13
10657	16 58.7 + 39 13 062.99 + 37.54	07 - 35 - 000	1135 420	1.0 x 0.4 0.4 x 0.15	55 6 - 7	2	Sa - b		16.5:		
10658	16 58.9 + 28 05 049.40 + 35.35	05 - 40 - 000	0129 415	1.2 x 0.7 1.3 x 0.8	... 4	1	S0 - a f of 2		15. - 16.		*
10659	16 59.0 + 07 00 026.35 + 27.58	01 - 43 - 000	1056	1.3 x 0.6 1.1 x 0.6	35 5	1	S ...		15.2		
10660	16 59.1 + 25 56 049.89 + 34.72	04 - 40 - 000	1377	1.0 x 0.4 0.8 x 0.3	137: ...	1	S ... p of 2		16.0:		*
10661	16 59.6 + 28 00 049.35 + 35.18	05 - 40 - 000	0129 415	1.0 x 0.15 1.0 x 0.2	14 7	3	Sb		16.5:		*
10662	16 59.6 + 30 15 052.03 + 35.75	05 - 40 - 029	0129 419	(1.5 x 0.9) (1.8 x 1.1)	75 ...	2	E		15.2		*1
10663	16 59.6 + 30 24 052.21 + 35.78	05 - 40 - 028	0129 419	1.3 x 1.1 1.4 x 1.3	-- 1 - 2	3	S B b	SAB(r)b	15.3		
10664	17 00.1 + 41 18 065.63 + 37.51	07 - 35 - 016	1135 420	1.5 x 0.2 1.5 x 0.2	148 7	3	Sc		15.4		
10665	17 00.1 + 59 04 088.02 + 37.12	10 - 24 - 088 NGC 6290	1414 403	1.2 x 1.0 1.1 x 1.0	30: 1 - 2	4	S B a Brightest of 3		14.3		*1
10666	17 00.3 + 30 49 052.75 + 35.73	05 - 40 - 032	0129 419	1.1 x 1.0 1.0 x 0.7	-- 1:	1	S ... 10668 group		16.0:		*
10667	17 00.4 + 21 38 042.07 + 33.11	04 - 40 - 000	1377	1.1 x 0.5 0.8 x 0.35	79 6	1	S ...		16.0:		
10668	17 00.4 + 30 47 052.72 + 35.70	05 - 40 - 033	0129 419	(1.0 x 0.9) (1.1 x 1.0)	-- 1	1	S0 Brightest in group		15.3		*
10669	17 00.7 + 70 22 101.66 + 34.70	12 - 16 - 000	0756	1.4:: x 1.4:: 1.2:: x 1.2::	-- ...	4	Dwarf		17. - 18.		
10670	17 01.0 + 63 46 093.78 + 36.23	11 - 21 - 000 IC 1241	0745	1.7: x 1.2 1.3 x 1.1	(165) ...	2	S ... Peculiar		14.2		*
10671	17 01.1 + 36 44 059.99 + 36.73	06 - 37 - 022	1069	1.5 x 0.2 1.3 x 0.2	68 7	3	Sb s of 2		16.0:		*
10672	17 01.2 + 25 00 045.98 + 34.00	04 - 40 - 012	1377 414	2.0 x 0.35 2.1 x 0.35	82 7	2	S0 - a f of 2		15.1		*1
10673	17 01.2 + 29 56 051.75 + 35.33	05 - 40 - 000	0129	(1.6 x 0.7) (1.7 x 0.8)	65 ...	1	...		16.0:		*
10674	17 01.3 + 09 22 029.03 + 28.14	02 - 43 - 000	0251	1.7 x 0.3 1.7 x 0.3	163 6	1	S ...		15.5		
10675	17 01.4 + 31 31 053.66 + 35.66	05 - 40 - 034	0129 419	[2.5 x ...] ...	... ...	2	Double system Plume, very long bridge		15.4		*1

N G C   6278, 6279, 6283, 6286, 6290   I C   1241

1	2	3	4	5			6	7	8	9	10	11	12
10676	17 01.8 + 26 35 047.86 + 34.33	05 – 40 – 000	1377	2. ...	x	1.	... ...	0	Dwarf		18.?		
10677	17 01.8 + 36 29 059.71 + 36.55	06 – 37 – 024	1069 416	(0.8 (0.9.	x x	0.8) 0.9)	-- ...	1	... Peculiar		14.5		
10678	17 01.9 + 24 15a 045.17 + 33.62	04 – 40 – 013	1377 414	1.0 (1.3	x x	1.0 1.3)	-- 1	3	S0 npp of 2		(15.1)		*1
10679	17 02.0 + 31 34 053.75 + 35.55	05 – 40 – 035	0129 419	1.6 1.2	x x	0.3 0.25	52 7	2	Sb		15.5		*
10680	17 02.0 + 51 28 078.47 + 37.51	09 – 28 – 018	0765	1.2 1.0	x x	0.5 0.5	99 ...	1	S B ...		15.3		
10681	17 02.1 + 45 32 070.99 + 37.45	08 – 31 – 022	1370	1.3 1.1	x x	0.2 0.2	40 7	3	Sc		15.7		
10682	17 02.3 + 60 26 089.66 + 36.65	10 – 24 – 092	1414 403	1.1 1.1	x x	0.5 0.5	80 6	1	S ...		16.0:		
10683	17 02.4 – 01 29 018.60 + 22.72	00 – 43 – 000	0155	[1.6 ...	x	0.7:]	... ...	2	Double system Bridge		16.0:		*
10684	17 02.4 + 61 07 090.50 + 36.53	10 – 24 – 093 NGC 6292	1414 403	1.8 1.7:	x x	0.8 0.7	105 6	2	Sb – c		14.4		
10685	17 02.6 + 13 00 032.93 + 29.40	02 – 43 – 000	0251	1.3 1.0	x x	0.5 0.4	119 6	1	S ...		15.2		
10686	17 02.7 + 42 00 066.57 + 37.09	07 – 35 – 022	1135 420	1.2 1.2	x x	0.3 0.3	[160] ...	1	...		15.5		
10687	17 02.7 + 59 48 088.87 + 36.69	10 – 24 – 094	1414 403	1.0 1.0	x x	0.7 0.7	35: 3:	3	S B c	Sc +	15.4		*1
10688	17 02.9 + 34 38 057.50 + 36.00	06 – 37 – 025	1069 416	1.3 1.2:	x x	1.2 1.1:	-- 1	4	Sc		15.5		
10689	17 03.0 + 56 05 084.25 + 37.08	09 – 28 – 020	0765	1.0 0.9	x x	0.3 0.2	88 6	2	Sa – b		15.7		*
10690	17 03.0 + 62 06 091.69 + 36.30	10 – 24 – 000 NGC 6297	1414	0.7 0.6	x x	0.5 0.4	(90) ...	1	S0		14.4		*
10691	17 03.2 + 30 28 052.51 + 35.04	05 – 40 – 037	0129 419	1.0 0.9	x x	0.7 0.7	... ...	2	S ... + companion Connected		15.5		*1
10692	17 03.3 + 23 14 044.14 + 33.00	04 – 40 – 017	1377	2.2 2.0	x x	0.4 0.4	179 6	2	Sb		15.4		*1
10693	17 03.3 + 41 56 066.50 + 36.97	07 – 35 – 023	1135 420	(1.8 (1.8	x x	1.2) 1.2)	105 ...	3	E		14.2		*1
10694	17 03.3 + 43 18 068.20 + 37.10	07 – 35 – 000	1135 420	1.0 1.0	x x	0.05 0.1	138 7	3	Sc	Sc +	17.		
10695	17 03.5 + 43 06b 067.96 + 37.04	07 – 35 – 024	1135 420	(1.4 (1.1	x x	1.0) 0.9)	110 ...	3	E sf of 2, contact		(14.9)		*1
10696	17 03.6 + 25 36 046.86 + 33.66	04 – 40 – 000	1377	1.1 1.0	x x	0.15 0.15	54 7	1	S ... Brightest in group		15.6		*
10697	17 03.7 + 72 57 104.58 + 33.74	12 – 16 – 000	0756	1.2 1.2	x x	0.5 0.5	[137] ...	2	S ... Disturbed?		15.7		*
10698	17 03.8 + 32 33 055.03 + 35.39	05 – 40 – 000	0129	1.4: 1.0	x x	0.2 0.15	116 7	1	S ...		17.		
10699	17 03.9 + 10 28 030.46 + 28.04	02 – 43 – 007	0251	0.55 0.6	x x	0.45 0.45	(140) ...	2	S B ...	S B(s) ...	14.5		
10700	17 04.0 + 25 58 047.31 + 33.68	04 – 40 – 000	1377	1.1 1.1:	x x	0.2 0.2	107 7	1	S ...		16.0:		
10701	17 04.0 + 52 48 080.13 + 37.15	09 – 28 – 022	0765	1.1 0.9	x x	0.12 0.15	48 7	2	Sb		16.0:		
10702	17 04.5 + 24 50 046.06 + 33.23	04 – 40 – 018	1377 414	1.0 1.3:	x x	0.9 0.9:	(0) ...	1	... Disturbed or eruptive?		15.4		*1
10703	17 04.6 + 38 27 062.23 + 36.30	06 – 37 – 026	1069 420	1.1: 0.9:	x x	0.9: 0.7:	-- ...	2	Sc P w 10706		15.5	1	
10704	17 04.6 + 79 06 111.48 + 31.72	13 – 12 – 014	1433 411	1.0 1.1	x x	0.7 0.7	166 ...	1	S ...		15.6		*
10705	17 04.8 + 11 29 031.61 + 28.27	02 – 43 – 000	505	1.0 0.9	x x	0.9 0.8	-- 1	2	S B a		15.2		*
10706	17 04.8 + 38 25 062.19 + 36.26	06 – 37 – 027	1069 420	1.1 1.2	x x	0.4 0.5	32 6	2	Sb – c P w 10703		15.1		*1
10707	17 04.8 + 41 07 065.52 + 36.61	07 – 35 – 028	1135 420	1.1 1.1	x x	0.9 0.9	(20) 2	4	Sc	SA(rs)cd	15.6		*
10708	17 05.0 + 35 43 058.91 + 35.78	06 – 37 – 030	1069 416	1.1 1.1	x x	0.3 0.3	125 6	1	S ...		14.8		*
10709	17 05.2 + 31 29 053.84 + 34.87	05 – 40 – 039	0129 419	1.0 0.9	x x	0.9 0.9	-- ...	2	S B ... + companion Connected, p w 10712		15.6		*1
10710	17 05.3 + 43 11 068.10 + 36.72	07 – 35 – 031	1135 420	1.6 1.6	x x	0.35 0.35	150 7	2	Sb		15.2		*1

N G C   6292, 6297

1	2	3	4	5	6	7	8	9	10	11	12
10711	17 05.3 + 68 53 099.79 + 34.69	12 – 16 – 017 NGC 6303	0756 412	(1.3 x 0.8) (1.4 x 0.9)	60 ...	3	E		15.1		*
10712	17 05.4 + 31 30 053.87 + 34.83	05 – 40 – 040	0129 419	1.0 x 0.8 0.9 x 0.8	(60) 1 – 2	3	Sa	(R)SAB a	15.5		*
10713	17 05.4 + 72 31 104.03 + 33.74	12 – 16 – 018	0756 430	2.0 x 0.35 2.0 x 0.35	8 6	1	Sb		13.8		
10714	17 05.5 + 30 17 052.44 + 34.52	05 – 40 – 041	0129 419	1.7 x 0.2 1.7 x 0.2	159 7	2	Sb – c 2nd of 3		15.5		*1
10715	17 05.5 + 31 18 053.64 + 34.76	05 – 40 – 042	0129 419	1.1 x 0.8 1.1 x 0.8	(55) ...	4	S B a		15.0		*1
10716	17 05.8 + 30 23 052.58 + 34.48	05 – 40 – 000	0129 419	1.3 x 0.15 1.3 x 0.15	14 7	3	Sc 3rd of 3		16.0:		*
10717	17 06.0 + 26 26 048.01 + 33.38	04 – 40 – 000	1377	1.0 x 0.8 0.9 x 0.7	100 2	3	Sc	Sc +	15.5		
10718	17 06.2 + 04 06 024.42 + 24.64	01 – 44 – 001 IC 1242	0169	1.0 x 0.7 0.9 x 0.6	130: ...	1	S ... P w 10719		14.9		*
10719	17 06.3 + 03 57 024.29 + 24.55	01 – 44 – 002 NGC 6296	0169	1.0 x 0.7 0.8 x 0.7	130 ...	2	Sb – c P w 10718		14.2		*12 3
10720	17 06.3 + 09 40 029.94 + 27.16	02 – 44 – 000	0505	1.1 x 0.8 0.6 x 0.3	90: ...	3	Sc		15.5		
10721	17 06.3 + 25 35 047.06 + 33.07	04 – 40 – 019	1377	1.1 x 0.6 1.2 x 0.6	110 5	1	Sc?		14.2		*
10722	17 06.8 + 43 01 067.92 + 36.44	07 – 35 – 032	1135 420	(1.2 x 0.8) (1.0 x 0.8)	30 ...	2	[E] Compact		15.4		
10723	17 07.0 + 42 24 067.16 + 36.34	07 – 35 – 034 NGC 6301 =4643*	1135 420	2.5 x 1.6 1.8 x 1.4	115 4	3	Sc		14.6		*3
10724	17 07.0 + 60 47 089.99 + 36.03	10 – 24 – 098 NGC 6306	1414 403	1.0 x 0.3 1.0 x 0.3	166 7	1	S ... P w 10727		14.3	+ 3064 + 3308	*12
10725	17 07.0 + 75 29 107.39 + 32.77	13 – 12 – 016 NGC 6324	1433	1.0 x 0.55 1.0 x 0.6	78 4	3	S ... Peculiar		13.5		*3
10726	17 07.0 + 78 42 111.00 + 31.75	13 – 12 – 018	1433 421	(1.1 x 1.0) (1.4 x 1.2)	— ...	2	E		15.6		
10727	17 07.1 + 60 48 090.01 + 36.02	10 – 24 – 099 NGC 6307	1414 403	1.4 x 1.1 1.4 x 1.1	145: 2	2	S0 – a		14.0	+ 3283 + 3527	*23
10728	17 07.2 + 22 17 043.42 + 31.84	04 – 40 – 000	1377	1.6 x 0.8 1.6 x 0.7	55 5	1	Sb		15.6		*
10729	17 07.4 + 31 40 054.19 + 34.45	05 – 40 – 044	0129 419	1.4: x 0.3 0.9 x 0.2	128 7	1	S – Irr		15.4		
10730	17 07.4 + 61 03 090.31 + 35.95	10 – 24 – 100 NGC 6310	1414 403	1.9 x 0.35 1.7 x 0.35	69 6	1	S ...		13.8		23
10731	17 07.7 + 63 43a 093.55 + 35.51	11 – 21 – 000	0745	1.1 x 0.7 0.9 x 0.6	35: ...	1	S ... superimposed, disturbed	Companion	(13.1)		*C
10732	17 08.0 + 36 28 059.95 + 35.32	06 – 38 – 002	1134 416	1.0 x 0.7 0.9: x 0.7:	30: 3	2	Sc		15.4		
10733	17 08.2 + 32 33 055.28 + 34.49	05 – 40 – 045	0129	(1.1 x 0.7) (1.4 x 0.8)	(125) ...	3	E		15.1		
10734	17 08.2 + 77 11 109.28 + 32.18	13 – 12 – 000	1433	(1.1 x 1.1) (1.3 x 1.3)	— ...	1	... Compact		15.6		
10735	17 08.3 + 32 20 055.03 + 34.42	05 – 40 – 000	0129	1.3 x 0.2 1.1 x 0.2	5 7	1	Sb – c		16.5:		*
10736	17 08.4 + 69 33 100.50 + 34.27	12 – 16 – 020	0756 412	3.5 x 1.2 3.5: x 1.2::	155 6	4	...	SAB dm	15.0		*
10737	17 08.5 + 49 22 075.83 + 36.49	08 – 31 – 024	1370	1.1 x 0.2 1.0 x 0.2	28 6	2	S B: a – b		15.4		
10738	17 08.6 + 05 55 026.50 + 24.96	01 – 44 – 000	0169 422	1.9 x 0.3 2.0 x 0.25	49 7	1	Sb – c		15.5		*
10739	17 08.8 + 36 21 059.84 + 35.14	06 – 38 – 003	1132 416	(1.1 x 1.1) (1.2 x 1.2)	— ...	1	...		14.7		
10740	17 09 + 86 40 119.55 + 28.77	14 – 08 – 022	0775 420	(1.4 x 1.1) 1.4 x 1.1	(140) ...	2	S B 0		15.2		
10741	17 09.0 + 41 43 066.37 + 35.90	07 – 35 – 039 NGC 6311	1135 420	(1.1 x 1.1) (1.2 x 1.2)	— ...	3	E		14.9		*13
10742	17 09.0 + 48 23 074.61 + 36.39	08 – 31 – 025 NGC 6313	1370	1.3 x 0.4 1.3 x 0.4	156 6	2	Sa – b		14.8		
10743	17 09.1 + 08 03 028.66 + 25.82	01 – 44 – 003	0169	1.2 x 0.5 1.3 x 0.5	113 6	1	Sa?		14.8		
10744	17 09.2 + 63 03 092.71 + 35.45	11 – 21 – 010 NGC 6319	0745	0.4 x 0.4 1.1 x 1.1	— ...	1	...		14.4		*
10745	17 09.4 + 61 28 090.78 + 35.65	10 – 24 – 000	1414 403	1.0 x 0.9: 0.9: x 0.7:	— ...	3	...	S dm	16.5:		

N G C   6296, 6301, 6303, 6306, 6307, 6310
      6311, 6313, 6319, 6324

I C   1242, 4643

1	2	3	4	5	6	7	8	9	10	11	12
10746	17 09.4 + 45 56 071.58 + 36.20	08 – 31 – 026	1370	1.1 x 0.10 1.0 x 0.10	0 7	3	Sc		16.5:		
10747	17 09.8 + 23 26 044.93 + 31.65	04 – 40 – 021 NGC 6308	1377	1.3 x 1.0 1.4: x 1.0	(150) ...	3	Sc/S B c Disturbed? NGC 6314 group		14.4		*12 3
10748	17 10.2 + 28 29 050.67 + 33.06	05 – 40 – 000	0129	[1.3 x 0.5] ...	... ...	1	Triple system Connecting streamers		17.		
10749	17 10.3 + 30 14 052.70 + 33.50	05 – 40 – 048	0129 419	1.1 x 0.7 1.1 x 0.7	10: ...	3	S B b		15.2		*
10750	17 10.3 + 49 37 076.14 + 36.21	08 – 31 – 028	1370	1.3 x 0.5 0.9 x 0.3	135 6	2	Sb – c		15.5		
10751	17 10.3 + 54 26 082.11 + 36.15	09 – 28 – 000	0765	1.0 x 0.3 1.0 x 0.2	80 6	3	Sc		16.5:		
10752	17 10.5 + 23 20 044.88 + 31.46	04 – 40 – 022 NGC 6314	1377	1.8 x 0.8 1.8 x 0.9	175 5	3	Sa Brightest of 3		14.3	+ 6748 + 6939	*12 3
10753	17 10.8 + 34 45 058.03 + 34.43	06 – 38 – 006	1132 416	1.0 x 0.2 0.9 x 0.2	100 6	2	Sc		15.5		
10754	17 10.8 + 46 36 072.42 + 36.00	08 – 31 – 031	1370	1.2 x 0.5 0.9 x 0.4	15 6	1	S ...		16.0:		*
10755	17 10.9 + 38 05 062.02 + 35.02	06 – 38 – 007 IC 1245	1132 420	(1.7 x 0.8) (1.7 x 0.8)	125 5:	2	S0		15.0		
10756	17 11.0 + 60 03 089.02 + 35.63	10 – 24 – 106 IC 1248	1414 403	1.4 x 1.4 0.8 x 0.5	-- 1	3	S B c	S B (r)c Sc –	14.6		2
10757	17 11.0 + 72 28 103.84 + 33.35	12 – 16 – 021 IC 1251	0756 430	1.7 x 1.1 1.7 x 1.1	70 4	2	Sc NGC 6340 group		14.4		*
10758	17 11.1 + 53 11 080.56 + 36.07	09 – 28 – 029	0765 425	1.1:: x 1.1:: 0.6 x 0.6	-- 1	1	S ...		15.3		
10759	17 11.2 + 42 52 067.84 + 35.62	07 – 35 – 045	1135 420	1.0 x 0.9 1.0 x 1.0	-- ...	2	... Compact		14.5		
10760	17 11.2 + 44 13 069.49 + 35.75	07 – 35 – 046	1135 420	1.4 x 1.2 1.4 x 1.2	-- 1	1	S ...		15.3		
10761	17 11.3 + 40 20 064.75 + 35.29	07 – 35 – 044 NGC 6320	1135 420	1.3 x 0.9 1.3 x 0.9	85 3	2	Sb – c		14.9		3
10762	17 11.3 + 72 21 103.70 + 33.36	12 – 16 – 023 NGC 6340	0756 430	(3.3 x 3.0) (3.7 x 3.5)	(120) 1	4	Sa Brightest of 3	SA 0/a	11.9	+ 2109 + 2351	*12 3468
10763	17 11.4 + 40 32 065.00 + 35.30	07 – 35 – 000	1135 420	1.5: x 0.4: 0.9 x 0.2	59 ...	1	... Disturbed?		16.5:		*
10764	17 11.7 + 43 50 069.03 + 35.63	07 – 35 – 048 NGC 6323	1135 420	1.1 x 0.35 1.1 x 0.35	172 6	2	Sa – b		14.8		3
10765	17 11.8 + 42 57 067.95 + 35.52	07 – 35 – 049	1135 420	1.1 x 1.0 1.0 x 0.9	-- 1	3	Sc		15.0		*1
10766	17 11.8 + 74 29 106.15 + 32.76	12 – 16 – 000	0756	1.0 x 0.10 0.9 x 0.10	96 7	2	Sc		17.		
10767	17 12.0 + 73 40 105.20 + 32.97	12 – 16 – 000	0756	1.1 x 0.5: 1.0 x 0.4	55 6	1	Sb – c		16.0:		
10768	17 12.3 + 20 22 041.79 + 30.07	03 – 44 – 002 NGC 6321	1127	1.1 x 1.0 1.1 x 1.1	-- 1	3	S B b	S B (rs)bc Sb +	14.5		*13
10769	17 12.4 + 72 27 103.79 + 33.25	12 – 16 – 024 IC 1254	0756 430	1.8: x 0.9 1.9 x 1.1	32 ...	1	S? ... NGC 6340 group		14.7		*
10770	17 12.5 + 59 23 088.18 + 35.51	10 – 24 – 107	1414	[1.7 x 0.7] ...	... ...	2	Double system Contact, disrupted		14.3		*1V A
10771	17 12.7 + 43 45 068.95 + 35.44	07 – 35 – 051 NGC 6329	1135 420	(1.1 x 1.1) (1.1 x 1.1)	-- ...	2	E		14.3		*13
10772	17 13.0 + 29 13 051.71 + 32.67	05 – 41 – 002	0152	1.1 x 0.8 0.9 x 0.7	... ...	3	S ... Unilateral		15.4		*1
10773	17 13.4 + 43 43 068.93 + 35.31	07 – 35 – 054 NGC 6332	1135 420	1.3 x 0.8 1.3 x 0.8	42 4	3	Sa		14.6		*3
10774	17 13.4 + 58 52 087.53 + 35.44	10 – 24 – 108	1414	1.0 x 0.45 0.9 x 0.4	165 6	1	S ...		15.3		
10775	17 13.6 + 06 54 028.09 + 24.31	01 – 44 – 000	0169 422	1.3 x 1.1 1.2 x 1.2	-- 1	2	S B a		15.3		*
10776	17 13.8 + 29 27 052.03 + 32.56	05 – 41 – 005 NGC 6330	0152	1.7 x 0.5 1.5: x 0.5	160 6	2	S B b		15.3		
10777	17 13.8 + 37 52 061.88 + 34.42	06 – 38 – 008	1132	1.1: x 0.7: 0.8: x 0.4:	(170) ...	3	S B b	S B (r)b	16.0:		
10778	17 13.9 + 06 37 027.85 + 24.11	01 – 44 – 004	0169 422	1.0 x 0.8 (1.0 x 0.8)	-- ...	1	E – S0?		14.7		
10779	17 14.0 + 06 29 027.73 + 24.03	01 – 44 – 005	0169 422	2.0 x 0.9 2.0 x 0.8	70 6	3	Sb		14.8		*
10780	17 14.2 + 07 22 028.62 + 24.39	01 – 44 – 000	0169 422	1.0 x 0.1 1.2 x 0.15	51 7	2	Sc		16.5:		

N G C   6308, 6314, 6320, 6321, 6323, 6329, 6330, 6332, 6340   I C   1245, 1248, 1251, 1254

1	2	3	4	5	6	7	8	9	10	11	12
10781	17 14.3 + 31 51 054.82 + 33.07	05 – 41 – 006	0152 423	1.0 x 0.35 1.0 x 0.3	65 6	1	S: ...		15.1		
10782	17 14.3 + 36 26 060.19 + 34.06	06 – 38 – 010	1132	1.2:: x 0.3 0.8: x 0.35	71 6	1	Sb – c		16.0:		*1
10783	17 14.5 + 29 30 052.13 + 32.43	05 – 41 – 000	0152	1.3 x 0.5 0.9 x 0.4	130 6	1	S ...		15.6		*
10784	17 14.5 + 57 29 085.83 + 35.40	10 – 24 – 116 NGC 6338	1414	(1.5 x 1.0) 1.5 x 1.0	(15) ...	2	SO Brightest in group		14.2		*3 (C)
10785	17 14.7 + 37 23 061.34 + 34.16	06 – 38 – 011	1132	1.5 x 0.5 1.3 x 0.5	111 6	1	S ...		14.9		*
10786	17 14.7 + 43 52 069.14 + 35.09	07 – 35 – 057 NGC 6336	1135 420	1.1 x 0.7 1.0 x 0.7	170: 3	3	S B a	SAB(s)a	14.5		*3
10787	17 14.8 + 21 40 043.44 + 29.97	04 – 41 – 002	0263	1.1 x 0.30 1.0 x 0.25	37 6	2	S B ... sf of 2, distorted		14.6		*1
10788	17 14.9 + 57 26 085.76 + 35.35	10 – 24 – 120 IC 1252 = 4649*	1414 429	1.0 x 0.2 1.1 x 0.2	142 7	2	Sa – b NGC 6338 group		15.7		*1
10789	17 15.3 + 07 39 029.03 + 24.27	01 – 44 – 000	0169 422	[1.7 x 0.4] ...	... ...	1	Double system Connected, in chain		15.7		*
10790	17 15.5 + 40 54 065.57 + 34.58	07 – 35 – 059 NGC 6339	1135 420	3.3 x 1.8 2.7 x 1.7	10 4	4	S B c	Sc +	13.7		*1
10791	17 15.5 + 72 27 103.72 + 33.02	12 – 16 – 026	0756	2.3:: x 2.3:: 2. x 2.	— ...	3	Dwarf (spiral) S? V		15.5		
10792	17 15.6 + 75 17 106.98 + 32.31	13 – 12 – 000	1433	2. x 2. 1. x 1.	— ...	4	Dwarf irregular Ir V		17.		*
10793	17 16.0 + 11 00 032.43 + 25.58	02 – 44 – 000	0505	1.0 x 0.5 0.8 x 0.3	67 5:	1	Sb – c		15.7		
10794	17 16.0 + 40 30 065.11 + 34.43	07 – 35 – 000	1135 420	1.1 x 0.25 0.8 x 0.2	58 ...	2	Sc – Irr		16.5:		
10795	17 16.2 + 30 58 053.93 + 32.46	05 – 41 – 000	0152	1.6:: x 0.9:: 1. x 1.	... ...	3	Dwarf V		17.		
10796	17 16.3 + 61 58 091.26 + 34.78	10 – 24 – 124	1414	2.0 x 1.2 2.0 x 1.2	(20) ...	4	S B b P w 10804	S B(s)b Sb –	14.8		*
10797	17 16.5 + 08 29 029.99 + 24.38	01 – 44 – 006	0169	1.1 x 0.35 1.1 x 0.35	16 6	3	S B: c		15.0		*1
10798	17 16.7 + 19 34 041.36 + 28.82	03 – 44 – 000	1127	1.0 x 0.2 1.0 x 0.2	105 6	2	Sb – c		15.4		
10799	17 16.9 + 29 55 052.77 + 32.04	05 – 41 – 000	0152	1.2 x 0.12 1.2 x 0.12	84 7	1	Sc		15.7		
10800	17 17.1 + 41 45 066.65 + 34.40	07 – 35 – 064 NGC 6350	1135 420	(1.0 x 1.0) (1.0 x 1.0)	— (1)	2	SO		14.3		*1
10801	17 17.2 + 32 19 055.54 + 32.59	05 – 41 – 008	0152	1.6 x 0.6 1.3 x 0.5	7 6	4	Sc	S d	16.0:		1
10802	17 17.3 + 11 16 032.85 + 25.41	02 – 44 – 000	0505	1.0 x 1.0 1.0 x 1.0	— 1	1	SO s of 2		15.5		*
10803	17 17.3 + 73 29 104.88 + 32.65	12 – 16 – 028	0756 430	1.3 x 0.8 1.1 x 0.7	15 (4)	1	...		12.9		
10804	17 17.4 + 61 50 091.08 + 34.66	10 – 25 – 001 NGC 6359	1148	(1.0 x 0.7) (1.0 x 0.7)	145 3:	3	SO		13.6	+ 2948 + 3196	*12 3
10805	17 17.6 + 14 27 036.11 + 26.66	02 – 44 – 002	0505	1.8 x 1.6 1.6: x 1.6:	— 1	4	Dwarf spiral, S B m S B V		15.3		*
10806	17 17.6 + 49 55 076.56 + 35.04	08 – 31 – 040	1370	2.4 x 1.0 2.2: x 1.0	77 6	4	...	S B(s)dm	13.9		*
10807	17 17.7 + 16 43 038.46 + 27.53	03 – 44 – 004 NGC 6347	1127	1.2 x 0.6 1.2 x 0.6	100 5	3	S B b		14.6		
10808	17 17.7 + 28 22 051.08 + 31.45	05 – 41 – 000	0152	1.0: x 0.6: 1.1: x 0.8:	... ...	2	Dwarf (spiral)		17.		*
10809	17 17.8 + 24 58 047.30 + 30.41	04 – 41 – 000	0263	[1.4 x 0.5] ...	... ...	1	Double system		15.7		*
10810	17 17.8 + 52 40 079.92 + 35.07	09 – 28 – 033 NGC 6358	0765 425	1.0 x 0.45 0.9 x 0.4	110 (6)	1	...		15.1		1
10811	17 17.8 + 58 12 086.67 + 34.92	10 – 25 – 002	1148 429	1.8 x 0.5 1.6: x 0.5	93 6	2	S B? a – b		15.1		
10812	17 17.9 + 40 59 065.75 + 34.14	07 – 35 – 065	1135 420	1.2 x 0.6 0.9: x 0.4	125 ...	2	S B b		15.0		*
10813	17 18.0 + 23 53 046.13 + 30.02	04 – 41 – 004	0263	(0.9 x 0.9) (0.9 x 0.9)	— ...	1	...		14.5		
10814	17 18.0 + 49 05 075.54 + 34.93	08 – 31 – 043	1370	[3.8: x 0.9] ...	... ...	3	S ... nnf of 2, very long streamer		14.9	+ 7182 + 7425	*12 VA
10815	17 18.0 + 60 40 089.66 + 34.71	10 – 25 – 004 NGC 6361	1148 429	2.3 x 0.7 2.2 x 0.7	54 6	3	Sb nf of 2		13.9		*A

N G C   6336, 6338, 6339, 6347, 6350, 6358
6359, 6361

I C   1252, 4649

1	2	3	4	5	6	7	8	9	10	11	12
10816	17 18.1 + 49 59 076.64 + 34.96	08 - 31 - 000	1370	1.1 x 0.12 0.8 x 0.1	62 7	2	Sc		16.5:		*
10817	17 18.1 + 61 17 090.41 + 34.64	10 - 25 - 000	1148 429	1.1 x 0.10 1.0 x 0.08	8 7	3	Sc	Sc +	16.5:		
10818	17 19.2 + 61 22 090.49 + 34.50	10 - 25 - 006	1148	1.3 x 0.5 0.8 x 0.4	25 ...	1	...		15.7		
10819	17 19.3 + 60 12 089.08 + 34.59	10 - 25 - 007	1148 429	1.1 x 0.5 0.9 x 0.4	56 6	1	Sa - b		15.3		
10820	17 19.4 + 42 13 067.28 + 34.04	07 - 36 - 003	0753 420	1.6 x 0.2 1.6: x 0.20	139 7	2	Sc		16.5:		
10821	17 19.4 + 52 42 079.96 + 34.83	09 - 28 - 035	0765 425	1.2: x 1.0: 0.9 x 0.8	(120) ...	3	S B b	S B(r)b	15.5		
10822	17 19.4 + 57 58 086.37 + 34.72	10 - 25 - 008 Draco system	1148 429	50. : x 30. : ...	... ...	5	Dwarf elliptical Local group		12.4		*12
10823	17 19.5 + 22 53 045.18 + 29.36	04 - 41 - 000	0263	1.1 x 0.5 1.0 x 0.5	74 5 - 6	1	S B ... sp of 2		15.7		*
10824	17 19.8 + 02 03 024.22 + 20.68	00 - 44 - 003	1154	1.1 x 0.4 1.1 x 0.4	119 6	1	S ... np of 2		15.1		*1
10825	17 20.0 + 01 00 023.27 + 20.13	00 - 44 - 000	1154	1.1 x 0.6 1.2 x 0.7	95 4 - 5	1	S ...		15.4		
10826	17 20.8 + 12 44 034.72 + 25.25	02 - 44 - 003 IC 1255	0505	1.1 x 0.5 1.1 x 0.5	12 ...	1	S ... Disturbed?		14.2		
10827	17 21.1 + 41 10 066.08 + 33.57	07 - 36 - 005 NGC 6363	0753 420	(1.1 x 0.9) 1.0 x 0.9	(140) ...	2	E		14.5		*1
10828	17 21.3 + 23 41 046.20 + 29.24	04 - 41 - 006	0263	1.4 x 1.2 1.3 x 1.1	--- ...	1	S0? + companion Contact, interaction		14.9		*1
10829	17 21.8 + 26 31 049.33 + 30.03	04 - 41 - 007 IC 1256	0263	1.7 x 1.3 1.7 x 1.2	97 2	3	Sb		14.5		
10830	17 22.2 + 60 30 089.41 + 34.21	10 - 25 - 015	1148 429	1.2 x 0.5 1.0 x 0.5	[97] ...	1	... Unilateral spiral?		14.4		
10831	17 22.3 + 25 00 047.71 + 29.45	04 - 41 - 008	0263	1.1 x 0.9 1.4: x 1.0:	--- 2:	3	Sc P w 10837, Sc +		15.2		*1
10832	17 22.3 + 62 13a 091.48 + 34.06	10 - 25 - 019 NGC 6365a	1148	1.3 x 1.1 0.9 x 0.9	--- 1	4	S B c P w 10833, connnected		14.6		*1V A
10833	17 22.3 + 62 13b 091.48 + 34.06	10 - 25 - 018 NGC 6365b	1148	1.1 x 0.2 1.1: x 0.2	31 7	2	Sc - Irr P w 10832, connected				*1V A
10834	17 22.4 + 20 26 042.83 + 27.88	03 - 44 - 007	1127	1.0 x 0.6 1.0 x 0.6	95 4:	3	Sc		15.3		
10835	17 22.5 + 29 26 052.62 + 30.74	05 - 41 - 013 NGC 6364	0152	1.5 x 1.2 (1.6 x 1.4)	5 2	3	S0		14.4		*
10836	17 22.5 + 57 02 085.22 + 34.34	10 - 25 - 020 NGC 6370	1148 429	(1.4 x 1.4) (1.2 x 1.2)	--- ...	3	E		14.2		*
10837	17 22.6 + 25 00 047.74 + 29.39	04 - 41 - 009	0263	2.0 x 0.9 1.5 x 0.7	145 6	3	S B b P w 10831, Sb +		15.0		*1
10838	17 22.6 + 48 31 074.91 + 34.14	08 - 32 - 002	0769	1.2 x 0.2 1.0 x 0.2	88 7	2	Sb		15.4		
10839	17 22.6 + 52 04 079.20 + 34.33	09 - 28 - 041	0765 425	1.0 x 0.9 0.7: x 0.5:	--- 1 - 2	1	S ... P w 10843		14.8		*
10840	17 22.7 + 23 47 046.43 + 28.97	04 - 41 - 010	0263	(1.2 x 0.9) (1.0 x 1.0)	--- ...	2	E		14.9		
10841	17 22.8 + 10 42 032.92 + 23.95	02 - 44 - 000	0505	1.0 x 0.25 1.1 x 0.25	37 7	2	Sc - Irr		15.6		
10842	17 22.8 + 28 07 051.17 + 30.30	05 - 41 - 014	0152	1.3 x 0.6 1.0 x 0.5	40 5	2	S B b		15.7		
10843	17 22.9 + 52 00 079.12 + 34.28	09 - 28 - 042	0765 425	1.1 x 0.15 0.9 x 0.2	69 6 - 7	1	Sa - b P w 10839		15.5		*
10844	17 23.0 + 10 20 032.58 + 23.74	02 - 44 - 000	0505	1.0 x 0.5 (0.8 x 0.4)	67 5:	1	S0		15.4		
10845	17 23.0 + 45 00 070.70 + 33.74	08 - 32 - 003	0769	1.1 x 1.0 1.0 x 0.9	--- 1	3	Sc		14.4		
10846	17 23.1 + 49 59 076.69 + 34.16	08 - 32 - 004	0769	1.0 x 0.18 0.9 x 0.18	170 7	2	Sb		15.7		
10847	17 23.2 + 23 21 046.01 + 28.72	04 - 41 - 011	0263	1.1 x 0.45 1.0 x 0.45	150 ...	1	Sc - Irr		15.3		1
10848	17 23.4 + 75 54 107.54 + 31.69	13 - 12 - 023 IC 4660	1433	1.4 x 0.35 1.2 x 0.2	170 6	1	S ...		14.3		
10849	17 23.5 + 20 49 043.33 + 27.78	03 - 44 - 000	1127	1.0 x 0.5 0.9 x 0.5	10 5	1	S B ...		15.1		
10850	17 23.5 + 59 03 087.65 + 34.13	10 - 25 - 023 NGC 6373	1148 429	1.6 x 1.0 1.6 x 1.0	90 ...	3	Sc/S B c	SAB(s)c	14.5		*1

N G C   6363 - 65, 6370, 6373   I C   1255, 1256, 4660

1	2	3	4	5	6	7	8	9	10	11	12
10851	17 23.7 + 49 23 075.97 + 34.02	08 – 32 – 000	0769	1.0 x 0.12 0.9 x 0.1	5 7	2	Sc		17.		
10852	17 23.9 + 11 21 033.69 + 23.98	02 – 44 – 000	0505	1.1 x 0.10 1.0 x 0.1	164 7	2	...	d – m	15.7		
10853	17 23.9 + 74 25 105.83 + 31.99	12 – 16 – 033	0756 430	(1.2 x 0.9) (1.6 x 1.4)	(0) ...	2	E + two companions Brightest in group		15.3		*1
10854	17 24.0 + 58 15 086.68 + 34.10	10 – 25 – 024	1148 429	1.2 x 0.7 1.1 x 0.5	93 5	2	Sc Disturbed?		16.0:		*
10855	17 24.6 + 58 52 087.42 + 34.00	10 – 25 – 25 + 26 NGC 6376 + 6377	1148 429	[1.0 x 0.6] ...	... ...	1	Double or triple system Distorted		14.1		*1C
10856	17 24.8 + 11 35 034.02 + 23.88	02 – 44 – 004 NGC 6368	0505	4.0 x 0.9 3.8 x 0.8	42 6	4	Sb		13.7		13
10857	17 25.0 + 53 37 081.09 + 34.00	09 – 28 – 000	0765	1.3 x 0.2 0.4 x 0.2	... ...	1	... f of 2, distorted, long jet		16.5:		*
10858	17 25.0 + 59 30 088.18 + 33.92	10 – 25 – 029	1148 429	1.0 x 0.12 0.9 x 0.12	93 7	2	Sc		16.5:		*
10859	17 25.1 + 59 40 088.38 + 33.90	10 – 25 – 030	1148 429	1.2 x 0.6 0.6: x 0.5	77 5:	2	S B b – c		15.6		*1
10860	17 25.5 + 06 31 029.17 + 21.50	01 – 44 – 000	0169	1.1 x 0.9 0.9 x 0.25	102 7	2	S0 – a		15.5		
10861	17 25.5 + 26 30 049.61 + 29.24	04 – 41 – 013 NGC 6372	0263	1.8 x 1.2 1.7 x 1.3	[90] ...	3	S ... sf of 2, disturbed		14.1		*13
10862	17 25.8 + 07 28 030.12 + 21.86	01 – 44 – 007	0169	3.3: x 3.0: 3. x 3.	–– 1	4	S B c	S B(rs)c	15.0		
10863	17 25.9 + 62 11 091.39 + 33.64	10 – 25 – 033	1148	1.5 x 0.2 1.4 x 0.2	105 7	1	S ...		15.5		
10864	17 26.1 + 14 13 036.80 + 24.68	02 – 44 – 005	0505	1.1 x 0.5 0.6 x 0.3	175 5:	1	S0		14.3		
10865	17 26.1 + 72 56 104.08 + 32.15	12 – 16 – 038	0756 430	1.3 x 1.0 0.9 x 0.9	(110) 1 – 2	3	S B b/Sb		15.4		
10866	17 26.5 + 26 35 049.78 + 29.05	04 – 41 – 016	0263	1.1 x 0.9 1.0 x 0.8	(160) 2:	3	Sc		15.4		
10867	17 26.5 + 58 33 087.03 + 33.76	10 – 25 – 035 IC 1258	1148 429	1.0 x 0.8 0.9 x 0.7	65: 2:	2	S0 – a P w 10869		14.4		*1
10868	17 26.6 + 06 22 029.16 + 21.19	01 – 44 – 008	0169	[2.1 x 0.5] ...	... ...	2	Double system Contact		15.2		*1
10869	17 26.6 + 58 35 087.07 + 33.75	10 – 25 – 37 + 37a IC 1259	1148 429	[1.2 x 1.2] ...	–– ...	1	Double system Common envelope		14.0		*1V A
10870	17 26.6 + 60 04 088.85 + 33.69	10 – 25 – 039	1148 429	1.1 x 0.5 0.9: x 0.4	85 6	1	S ... P w 10871		15.6		*
10871	17 26.7 + 60 03 088.83 + 33.68	10 – 25 – 038 NGC 6381	1148 429	1.4 x 1.0 1.4 x 1.0	25 3	4	Sc P w 10872		13.6		*2
10872	17 26.7 + 61 14 090.25 + 33.61	10 – 25 – 041	1148	1.2 x 0.10 0.9 x 0.1	58 7	1	...	d – m	16.5:		
10873	17 27.0 + 16 12 038.90 + 25.28	03 – 44 – 008	1127	1.5 x 0.2 1.6 x 0.2	4 7	1	S B? ... P w 10875, disturbed		15.2		*
10874	17 27.0 + 29 21 052.85 + 29.77	05 – 41 – 000	0152	1.3 x 0.10 1.4 x 0.1	49 7	3	Sc		16.0:		
10875	17 27.1 + 16 14 038.94 + 25.27	03 – 44 – 009 NGC 6375	1127	(1.6 x 1.6) (1.7 x 1.7)	–– ...	2	E P w 10873 Brightest of 3		14.5		*13
10876	17 27.1 + 71 08 101.97 + 32.40	12 – 16 – 039 NGC 6395	0756	2.7 x 0.8 2.7 x 0.8	15 6	2	Sc Singular		12.8		
10877	17 27.2 + 57 35 085.86 + 33.70	10 – 25 – 044 NGC 6385	1148 429	1.4 x 1.4 1.5 x 1.4	–– 1	4	S B a	S B(s)a	14.2		*
10878	17 27.7 + 23 22 046.42 + 27.75	04 – 41 – 019	0263	1.1 x 0.55 1.1 x 0.7	... ...	3	... Peculiar, ring		15.4		1
10879	17 27.8 + 24 55 048.08 + 28.24	04 – 41 – 020	0263	1.0 x 0.8 0.9 x 0.8	(100) 2	2	Sc f of 2		14.7		*
10880	17 27.8 + 60 59 089.94 + 33.49	10 – 25 – 048	1148	1.0 x 0.8 1.0 x 0.8	(75) ...	2	Sc – Irr		15.5		
10881	17 27.9 + 60 08 088.92 + 33.52	10 – 25 – 047 NGC 6390	1148 429	1.7 x 0.35 1.5 x 0.35	8 6	2	Sb – c		14.5		*
10882	17 28.0 + 58 35 087.06 + 33.57	10 – 25 – 050	1148 429	1.1 x 0.6 0.9 x 0.4	27 5	3	Sc		16.0:		
10883	17 28.1 + 52 15 079.46 + 33.49	09 – 29 – 005	1102 425	1.0 x 0.2 1.0 x 0.2	98 7	2	S0 – a		15.2		
10884	17 28.2 + 06 19 029.30 + 20.81	01 – 44 – 009 NGC 6378	0169	1.4 x 1.1 1.3 x 1.0	5 2	1	S ...		15.1		*3
10885	17 28.3 + 35 24 059.71 + 31.08	06 – 38 – 023	1132	1.0 x 0.25 1.0 x 0.25	141 6	2	Sb		15.3		*

N G C   6368, 6372, 6375 – 78, 6381, 6385
6390, 6395

I C   1258, 1259

1	2	3	4	5	6	7	8	9	10	11	12
10886	17 28.4 + 16 19 039.16 + 25.01	03 − 44 − 010 NGC 6379	1127	1.2 x 1.1 1.1 x 1.0	-- 1	3	Sc NGC 6375 group		14.6		3
10887	17 28.4 + 77 45 109.57 + 30.97	13 − 12 − 024	1433 421	1.2 x 0.25 1.1 x 0.2	124 7	3	Sc nnp of 2		15.4		*
10888	17 29.4 + 60 24 089.23 + 33.33	10 − 25 − 053	1148 429	1.1 x 0.7 1.1 x 0.7	150 4:	3	Sb	(R')S B(r)b	14.2		*1
10889	17 29.7 + 59 41 088.37 + 33.32	10 − 25 − 055 NGC 6393	1148 429	1.5 x 0.4 1.4 x 0.4	42 6	3	S B b		15.4		
10890	17 29.8 + 32 16 056.27 + 30.00	05 − 41 − 000	0152	1.9 x 0.10 1.7 x 0.10	88 7	3	Sc	Sc +	15.6		
10891	17 30.0 + 07 06 030.27 + 20.77	01 − 45 − 001 NGC 6384	0780	7.0 x 4.6 6.5: x 4.5:	30 3	5	Sb	SAB(r)bc Sb +	13.2	+ 1754 + 1907	*12− −689
10892	17 30.3 + 74 18 105.59 + 31.60	12 − 16 − 040	0756 430	1.8 x 0.12 1.6 x 0.12	135 7	3	...	d − m	16.0:		
10893	17 30.5 + 16 26 039.50 + 24.59	03 − 45 − 001 NGC 6389	0274	3.2 x 2.0 3.0 x 1.6:	130 4	3	Sb	Sb +	13.6		*1
10894	17 31.0 + 27 37 051.25 + 28.42	05 − 41 − 000	0152	1.5: x 0.7: 1.6: x 0.5	167 ...	1	S ...		15.7		*
10895	17 31.2 + 59 30 088.15 + 33.13	10 − 25 − 060	1148 429	1.1 x 0.15 1.1 x 0.15	145 7	2	S B: b − c		16.0:		
10896	17 31.2 + 59 40 088.34 + 33.13	10 − 25 − 059 NGC 6399	1148 429	1.2 x 0.6 1.2 x 0.6	5 5	2	S0 − a		14.8		
10897	17 31.2 + 75 45 107.25 + 31.25	13 − 12 − 026 NGC 6412	1433	2.3 x 2.3 2.4 x 2.3	-- 1	4	Sc	SA(s)c	12.4	+ 1508 + 1751	*12− −689A
10898	17 31.3 + 51 56 079.12 + 32.98	09 − 29 − 008	1102	1.1 x 0.15 0.8: x 0.15	164 7	1	S ...		16.5		
10899	17 31.5 + 20 48 044.07 + 26.02	03 − 45 − 002	0274	1.8 x 1.1 1.3 x 0.9	0: 3 − 4:	1	S: ...		15.0		*
10900	17 31.5 + 43 47 069.51 + 32.08	07 − 36 − 020 IC 1262	0753 427	(1.5 x 0.8) (1.4 x 0.8)	85 ...	3	E		14.9		*
10901	17 31.6 + 05 30 028.94 + 19.69	01 − 45 − 000	0780	1.2: x 0.8: 1. x 1.	... ...	1	...	d − m	17.		
10902	17 31.6 + 43 51 069.59 + 32.07	07 − 36 − 021 IC 1263	0753 427	1.7 x 0.7 1.7 x 0.6	[168] ...	3	S B ... S B(r) ... Peculiar, disturbed?		14.8		*1
10903	17 31.7 + 23 54 047.33 + 27.07	04 − 41 − 000	0263	1.0 x 0.6 1. x 0.6::	... ...	4	Dwarf irregular Ir V		17.		*
10904	17 31.8 + 43 40 069.38 + 32.01	07 − 36 − 022 IC 1264	0753 427	1.4: x 1.2: 1.2 x 1.2	-- ...	1	S B 0?		15.6		*1
10905	17 32.0 + 25 22 048.92 + 27.49	04 − 41 − 021	0263	(1.5 x 0.8) (1.5 x 0.8)	170 5	2	S0 − a		14.6		
10906	17 32.0 + 74 25 105.70 + 31.46	12 − 16 − 041 NGC 6414	0756 430	1.2 x 0.7 (1.2 x 0.7)	145: ...	1	... Brightest in group		15.6		*
10907	17 32.2 + 77 26 109.15 + 30.84	13 − 12 − 027	1433	1.2 x 0.6 1.2 x 0.6	46 5	2	Sb − c		15.3		
10908	17 32.4 + 50 25 077.33 + 32.70	08 − 32 − 010	0769	1.9 x 0.8 1.5:: x 0.8:	160 ...	2	S B c/Irr? or distorted pair?		16.0:		*1V
10909	17 32.5 + 25 36 049.21 + 27.46	04 − 41 − 000	0263	1.1 x 0.7 1.1 x 0.6	155 4	3	Sc		15.5		
10910	17 32.9 + 33 57 058.34 + 29.81	06 − 38 − 000	1132	1.3:: x 1.3:: ...	-- ...	2	Dwarf V		17. − 18.		
10911	17 32.9 + 69 23 099.84 + 32.18	12 − 16 − 043	0756	1.0 x 0.35 1.0 x 0.30	96 6	2	Sc		16.0:		*
10912	17 34.2 + 70 52 101.56 + 31.88	12 − 17 − 000	0801	2.3? x 0.6: 0.7 x 0.5	... ...	1	S ... + streamer or bridge?, disturbed?		15.7		*
10913	17 34.3 + 15 20 038.73 + 23.32	03 − 45 − 004	0274	1.1 x 1.0 0.9: x 0.8:	-- 1	4	S B c	Sc +	15.6		
10914	17 34.7 + 40 10 065.45 + 30.86	07 − 36 − 000	0753	1.0 x 0.8 0.9 x 0.8	(55) 1 − 2	3	Sc		15.4		
10915	17 34.8 + 24 57 048.71 + 26.76	04 − 41 − 000	0263	1.2:: x 0.8:: 0.9: x 0.8:	... ...	2	...	B dm − m	17. − 18.		
10916	17 35.0 + 60 50 089.72 + 32.63	10 − 25 − 068 NGC 6411	1148	(2.3 x 1.8) (2.0 x 1.6)	70: ...	4	E		13.2		*3
10917	17 35.1 + 42 07 067.70 + 31.15	07 − 36 − 027 IC 1265	0753	2.1 x 0.9 2.0: x 0.9:	80 6	2	Sa − b		14.3		
10918	17 35.2 + 11 10 034.79 + 21.39	02 − 45 − 001	1580	1.1 x 0.9 (1.3 x 1.1)	170: ...	2	E		14.8		
10919	17 35.3 + 17 34 041.13 + 23.98	03 − 45 − 006	0274	1.4 x 0.8 1.3 x 0.8	7 4	1	S ...		14.5		
10920	17 35.5 + 49 55 076.79 + 32.16	08 − 32 − 000	0769	1.0: x 0.9: 0.5: x 0.3:	-- ...	2	...	S B dm	17.		*

N G C   6379,   6384,   6389,   6393,   6399,   6411   6412,   6414   I C   1262 − 65

1	2	3	4	5	6	7	8	9	10	11	12
10921	17 35.7 + 17 24 041.01 + 23.83	03 – 45 – 000	0274	1.0 x 0.8 0.5 x 0.4	— ...	3	Dwarf		17.		
10922	17 35.8 + 60 19 089.10 + 32.54	10 – 25 – 069	1148 429	1.4: x 1.1: 0.7 x 0.6	(0) 1	4	Sc	Sc +	16.0:		
10923	17 36 + 86 47 119.53 + 28.36	14 – 08 – 24 + 25	775	[1.4 x 1.1] ...	— ...	3	Double system Strongly distorted		14.3		*1C
10924	17 36.1 + 68 10 098.38 + 32.04	11 – 21 – 012 NGC 6419	0745 435	1.1 x 0.25 1.1 x 0.25	134 7	2	Sa		15.5		*
10925	17 36.3 + 34 58 059.68 + 29.39	06 – 39 – 002	0277	1.0 x 0.4 0.7: x 0.3	[85] ...	2	S ... P w 10927, disturbed		16.0:		*1
10926	17 36.3 + 24 59 048.87 + 26.45	04 – 41 – 023	0263	1.2 x 0.5 1.1 x 0.3	150: 6	1	S B ...		15.1		
10927	17 36.4 + 34 57 059.66 + 29.36	06 – 39 – 003	0277	1.1: x 0.7: 0.6: x 0.6:	(90) (1)	2	S0 P w 10925, disturbed		16.0:		*1
10928	17 36.5 + 21 43 045.49 + 25.27	04 – 41 – 024	0263	1.6 x 0.7 1.2: x 0.4	114 6	3	Sb		15.2		
10929	17 36.5 + 35 33 060.34 + 29.49	06 – 39 – 004	0277 431	1.2 x 0.9 1.1 x 0.9	175 3	2	Sa – b		14.9		*
10930	17 36.6 + 18 54 042.61 + 24.21	03 – 45 – 007 NGC 6408	0274	1.6 x 1.5 1.6 x 1.5	— 1	3	S B a	S B(rs) ...	14.0		3
10931	17 36.6 + 60 28 089.28 + 32.44	10 – 25 – 071	1148 429	1.2 x 0.12 0.8 x 0.10	6 7	3	Sc		16.5:		
10932	17 36.6 + 70 01 100.53 + 31.79	12 – 17 – 001 NGC 6424	0801	0.8 x 0.8 0.9 x 0.8	— ...	2	... Compact		14.5		*
10933	17 36.9 + 39 15 064.51 + 30.26	07 – 36 – 000	0753	1.6 x 0.4 1.6 x 0.4	[12] ...	3	... Peculiar, streamer		15.4		
10934	17 37.6 + 72 07 102.97 + 31.44	12 – 17 – 002 NGC 6434	0801 430	2.4 x 1.0: 2.4 x 0.8	100 6	2	S B: b – c		13.2		3
10935	17 37.8 + 57 17 085.51 + 32.27	10 – 25 – 075	1148 429	1.2 x 0.4 0.9 x 0.25	116 7:	1	S0		16.0:		
10936	17 37.8 + 61 04 089.99 + 32.28	10 – 25 – 076	1148	1.7 x 0.8 1.5 x 0.7	170: 5	1	S ...		15.1		1
10937	17 38.0 + 59 25 088.03 + 32.27	10 – 25 – 077 IC 1267	1167 429	1.5 x 1.0 1.5 x 1.1	50: ...	4	S B b		14.5		*
10938	17 38.3 + 49 38 076.52 + 31.68	08 – 32 – 012	0769	1.4 x 1.0 1.4 x 1.0	0: 3:	3	S0	(R)SAB 0 +	15.0		*1
10939	17 38.5 + 19 32 043.44 + 24.03	03 – 45 – 000	0274	1.0 x 0.9 0.8: x 0.7:	— 1 – 2	2	S B b – c		16.0:		
10940	17 38.9 + 63 59 093.43 + 32.06	11 – 21 – 000	0745	1.0 x 0.7 0.7 x 0.4	100: ...	1	...		15.3		
10941	17 39.0 + 17 23 041.32 + 23.10	03 – 45 – 011	0274	1.2: x 0.7: 1.2 x 0.5	10 ...	1	S ...		15.4		*
10942	17 39.0 + 51 04 078.21 + 31.72	09 – 29 – 015	1102	1.1 x 0.9 1.1 x 0.9	[116] ...	2	S B ... s of 2, distorted		15.7		*1
10943	17 39.7 + 03 12 027.77 + 16.84	01 – 45 – 000	0780	1.6 x 0.6 1.6 x 0.5	118 6	1	S ... f of 2		15.5		*
10944	17 39.7 + 18 19 042.33 + 23.30	03 – 45 – 012	0274	(1.0 x 0.8) (1.0 x 0.9)	(170) ...	3	E		15.2		*
10945	17 39.7 + 23 42 047.83 + 25.28	04 – 42 – 001 NGC 6417	0260 432	1.6 x 1.4 1.5 x 1.5	20 1	2	S B b: Singular	S B(r) ...	14.4		*3
10946	17 39.7 + 45 10 071.37 + 30.85	08 – 32 – 014	0769	1.5 x 0.8 1.5 x 0.7	25 5	3	S B b	Sb +	14.9		
10947	17 39.7 + 62 39 091.85 + 32.02	10 – 25 – 080 NGC 6435	1148	(1.1 x 0.6) (1.1 x 0.6)	(5) ...	2	... Compact		14.9		*
10948	17 40.0 + 00 14 025.08 + 15.38	00 – 45 – 001	1144	1.2: x 0.2 2. x 0.5:	111 7:	1	S ...		15.5		
10949	17 40.3 + 74 52 106.11 + 30.84	12 – 17 – 000	0801 430	(1.0 x 0.8) (0.9 x 0.7)	(60) ...	2	E		14.9		*
10950	17 40.6 + 16 14 040.34 + 22.29	03 – 45 – 15 + 16	274	[1.6 x 1.0] ...	... ...	2	Double system		15.1		*1
10951	17 40.6 + 60 29 089.29 + 31.94	10 – 25 – 082 NGC 6436	1148 429	1.5 x 0.9 1.4: x 0.8	177: 4	4	Sc		14.9		
10952	17 41.2 + 56 10 084.22 + 31.76	09 – 29 – 018	1102 433	1.3: x 1.2: 0.7 x 0.7	— 1	1	S B ...	S B(r) ...	15.6		
10953	17 41.2 + 66 42 096.60 + 31.67	11 – 21 – 018	0745 435	1.1 x 0.7 1.1 x 0.7	145 4	2	S B a – b		15.3		
10954	17 41.2 + 74 03 105.16 + 30.90	12 – 17 – 004	0801 430	1.0 x 0.5 1.0 x 0.5	20: 5	2	Sa – b		15.6		
10955	17 41.3 + 56 00 084.02 + 31.74	09 – 29 – 019	1102 433	1.0 x 0.6 0.9 x 0.6	165 4:	3	Sc		15.6		*

N G C   6408, 6417, 6419, 6424, 6434, 6435
6436

I C   1267

1	2	3	4	5	6	7	8 9	10	11	12
10956	17 41.4 + 04 34 029.24 + 17.09	01 - 45 - 003	0780	1.5 x 1.4 1.5: x 1.5:	-- 1	2	Sb - c SAB(rs) ...	15.7		
10957	17 41.6 + 25 31 049.88 + 25.50	04 - 42 - 003 NGC 6427	0260	1.5 x 0.6 1.5 x 0.5	36 7	2	E - S0 P w 10960	14.6		*3
10958	17 41.8 + 66 53 096.81 + 31.60	11 - 21 - 000	0745 435	1.5 x 1.0 1.3: x 0.8:	(10) ...	4	Dwarf spiral S IV - V	16.5:		*
10959	17 42 + 84 40 117.17 + 28.82	14 - 08 - 000	0775	1.2: x 1.0: 1.0: x 0.8:	-- ...	1	...	16.5:		*
10960	17 42.0 + 25 23 049.77 + 25.37	04 - 42 - 004 NGC 6429	0260	2.3 x 0.8 1.7 x 0.7	23 6	2	S B a P w 10957	14.3		*3
10961	17 42.1 + 58 50 087.35 + 31.74	10 - 25 - 083	1148 429	1.1 x 0.5 1.1 x 0.5	165 6	3	Sc	15.7		
10962	17 42.2 + 36 48 062.05 + 28.68	06 - 39 - 015 NGC 6433	0277 434	2.1 x 0.5 2.3 x 0.5	163 6	2	Sb	14.1		*3
10963	17 42.4 + 68 22 098.55 + 31.44	11 - 21 - 019	0745 435	1.5 x 0.8 1.8 x 0.8	60: 5 - 6	3	S B b S B(s)b Compact core	15.3		*C
10964	17 42.8 + 66 30 096.36 + 31.53	11 - 21 - 021 NGC 6457	0745 435	1.2 x 0.9 1.3 x 1.0	140: ...	2	E - S0	14.7		*
10965	17 42.9 + 56 49 084.99 + 31.56	09 - 29 - 020 NGC 6449	1102	1.1 x 0.8 1.0 x 0.7	135 3	3	Sc	14.6		
10966	17 43.0 + 18 09 042.49 + 22.52	03 - 45 - 019	0274	1.8 x 0.6 1.8 x 0.6	97 6	2	Sa - b	14.8		
10967	17 43.2 + 48 07 074.87 + 30.69	08 - 32 - 018 NGC 6443	0769	1.3 x 0.5 1.2 x 0.5	128 6	2	Sa - b	14.8		
10968	17 43.3 + 58 01 086.40 + 31.55	10 - 25 - 084	1148 429	1.0 x 0.4 0.9 x 0.3	152 6	2	Sb - c	16.0:		
10969	17 43.6 + 38 55 064.47 + 28.92	06 - 39 - 017	0277 434	1.5 x 0.2 1.2 x 0.2	108 7	1	S ...	15.6		
10970	17 43.7 + 24 23 048.88 + 24.66	04 - 42 - 000	0260 432	1.3 x 0.6 1.0 x 0.5	150 5	1	Sa - b	15.7		*
10971	17 43.9 + 55 21 083.28 + 31.34	09 - 29 - 025	1102 433	1.1 x 0.35 1.1 x 0.35	173 6	3	Sa	15.0		
10972	17 44.3 + 26 34 051.20 + 25.28	04 - 42 - 005	0260	2.6 x 0.5 1.9 x 0.5	57 6	1	Sc	14.5		
10973	17 44.3 + 67 36a 097.63 + 31.31	11 - 21 - 023 NGC 6470	0745 435	1.4 x 0.2 1.2 x 0.15	173 7	2	Sc P w NGC 6471	(15.4)		*1
10974	17 44.4 + 67 38 097.67 + 31.30	11 - 21 - 025 NGC 6472	0745 435	1.2 x 0.7 1.2 x 0.7	150: 4:	3	S B b	15.2		*1
10975	17 44.5 + 35 35 060.84 + 27.93	06 - 39 - 019 NGC 6447	0277	1.7: x 0.9 1.3 x 0.9	[145] ...	2	S B? ... Disturbed?	13.8		*13
10976	17 44.6 + 30 43a 055.60 + 26.53	05 - 42 - 002	0532	0.9 x 0.5 0.9 x 0.5	[47] ...	1	S B ... P w 10977, distorted	14.4		*1
10977	17 44.6 + 30 43b 055.60 + 26.53	05 - 42 - 003	0532	1.1 x 0.25 1.1 x 0.2	97 (7)	1	S: ... P w 10976, disturbed			*1
10978	17 44.7 + 20 47 045.30 + 23.15	03 - 45 - 021 NGC 6442	0274	(1.3 x 1.3) (1.8 x 1.4)	-- ...	3	E P w 10979	14.5		*3
10979	17 45.0 + 20 53 045.43 + 23.12	03 - 45 - 022	0274	1.0 x 1.0 1.0 x 1.0	-- 1	3	Sc P w 10978, Sc -	15.1		*
10980	17 45.1 + 57 18 085.57 + 31.28	10 - 25 - 086	1148	(1.0 x 0.8) (0.9 x 0.5)	40: ...	2	... Compact	15.3		
10981	17 45.2 + 09 34 034.38 + 18.49	02 - 45 - 000	1580	1.0 x 0.5 1.0 x 0.5	57 5	2	Sb - c	15.3		
10982	17 45.2 + 59 22 087.99 + 31.35	10 - 25 - 088	1148 429	1.2 x 0.15 1.1 x 0.15	35 7	3	Sb	16.0:		
10983	17 45.5 + 45 42 072.18 + 29.93	08 - 32 - 019	0769	1.2 x 0.4 1.2 x 0.4	154 6	2	S B: a - b	15.6		*
10984	17 45.5 + 51 20 078.64 + 30.73	09 - 29 - 030	1102	1.4 x 0.6 1.3 x 0.5	162 6	1	S B: a - b	15.1		
10985	17 45.8 + 14 45 039.43 + 20.54	02 - 45 - 003	1580	1.4 x 1.0 1.5 x 1.0	110 3:	4	... S B(s)dm nf of 2	13.9		*
10986	17 46.0 + 57 07 085.37 + 31.15	10 - 25 - 000	1148	1.2 x 0.10 1.0: x 0.08	177 7	3	Sc Sc +	17.		
10987	17 46.1 + 17 58 042.62 + 21.77	03 - 45 - 028	0274	[1.3 x 1.0] ...	... ...	1	Triple system Contact	15.1		*1
10988	17 46.1 + 59 16 087.88 + 31.23	10 - 25 - 091	1148 429	(1.3 x 0.9) (1.3 x 0.9)	(15) ...	3	E	14.4		*
10989	17 46.2 + 57 20 085.62 + 31.14	10 - 25 - 092	1148	1.0 x 0.6 0.8 x 0.5	75 4:	1	S ...	15.0		
10990	17 46.4 + 34 06 059.35 + 27.15	06 - 39 - 020	1132	1.8 x 0.45 1.7: x 0.4	177 6	2	Sa - b	15.5		*

N G C   6427, 6429, 6433, 6442, 6443, 6447
        6449, 6457, 6470, 6472

1	2	3	4	5	6	7	8	9	10	11	12
10991	17 46.5 + 67 21 097.33 + 31.12	11 – 22 – 001	0550 435	1.6: x 1.2: 1.0: x 1.0:	— ...	4	Dwarf irregular Ir V		17.		
10992	17 46.6 + 61 27 090.43 + 31.22	10 – 25 – 000	1148	1.4 x 0.8 0.7 x 0.3	115 ...	4	S B b	S B(s)b	15.1		
10993	17 46.7 + 55 12 083.14 + 30.93	09 – 29 – 000	1102	1.0 x 0.4 1.0 x 0.4	102 6	2	S B? ...		16.0:		
10994	17 47.0 + 20 49 045.55 + 22.66	03 – 45 – 029 NGC 6458	0274	1.1 x 0.8 1.1 x 0.7	155 ...	2	S0 P w 10997		14.9		*3
10995	17 47.1 + 64 02 093.45 + 31.16	11 – 22 – 002	0550	1.1 x 0.4 0.9 x 0.35	7 6	1	S ...		14.8		*
10996	17 47.3 + 54 10 081.95 + 30.75	09 – 29 – 000 NGC 6479	1102	1.1 x 1.0: 0.8 x 0.7	— 1	3	Sc	Sc –	14.5		
10997	17 47.4 + 20 46 045.54 + 22.56	03 – 45 – 031 NGC 6460	0274	1.9 x 1.0 1.8 x 1.0	157 5	3	Sc P w 10994		14.4		*3
10998	17 47.5 + 51 12 078.53 + 30.41	09 – 29 – 032 NGC 6478	1102	1.8 x 0.5 1.6 x 0.5	37 6	1	Sc		14.1	+ 6857 + 7114	*12<u>_</u>
10999	17 47.7 + 25 48 050.69 + 24.30	04 – 42 – 006	0260	1.0: x 0.5 0.8 x 0.45	160 ...	1	...		14.8		
11000	17 47.7 + 36 09 061.65 + 27.45	06 – 39 – 022	0277	0.8 x 0.35 0.8 x 0.30	156 ...	1	...		14.0		
11001	17 47.9 + 14 18 039.22 + 19.90	02 – 45 – 004	1580	1.4 x 0.5 1.5 x 0.6	133 (6)	3	Sc – Irr		14.4		1
11002	17 47.9 + 58 26 086.92 + 30.97	10 – 25 – 097	1148 429	1.2 x 0.8 1.0 x 0.7	98 4	3	Sb		15.4		*
11003	17 48.4 + 14 50 039.79 + 20.01	02 – 45 – 006	1580	1.1 x 0.2 1.1 x 0.2	165 7	2	Sc		15.1		1
11004	17 48.5 + 17 33 042.45 + 21.08	03 – 45 – 035 NGC 6467 = 6468	0274	2.3: x 1.7: 1.6: x 1.0:	77 3:	1	S ...		14.5		13
11005	17 48.9 + 63 46 093.14 + 30.97	11 – 22 – 004	0550	1.2 x 1.0 0.9 x 0.8	— 1	3	S B b	S B(r)b	15.5		1
11006	17 49.4 + 51 04 078.42 + 30.10	09 – 29 – 038	1102	1.2 x 0.7 0.6: x 0.5	[110] ...	3	S ... Disturbed		15.6		*
11007	17 49.4 + 61 37 090.63 + 30.89	10 – 25 – 102	1148	1.0 x 0.6 0.9: x 0.6	— ...	3	S B ... Distorted?	S B(s) ...	16.5:		*
11008	17 49.5 + 61 32 090.54 + 30.88	10 – 25 – 103 NGC 6491	1148	1.3 x 0.5 1.3 x 0.5	39 6	2	Sa – b		14.5		*
11009	17 49.7 + 23 05 048.09 + 22.92	04 – 42 – 008 NGC 6482	0260	(2.1 x 1.8) (2.2 x 1.8)	70 ...	3	E		12.8	+ 3922 + 4137	*123 46<u>7</u>8
11010	17 49.7 + 24 30 049.53 + 23.43	04 – 42 – 007 NGC 6484	0260	2.0 x 1.9 1.9 x 1.8	— 1	2	Sb: Brightest of 3	Singular	13.5		*13
11011	17 49.8 + 61 34 090.58 + 30.84	10 – 25 – 105 NGC 6493	1148	1.2 x 1.2 1.2: x 1.1:	1 1	4	S B: c	Sc +	15.5		*1
11012	17 49.9 + 70 10 100.58 + 30.65	12 – 17 – 009 NGC 6503	0801	8.0: x 2.6: 6.2: x 2.2:	123 6	3	Sc	Sc –	10.9	+ 26 + 279	*12– –7<u>8</u>9
11013	17 50.0 + 21 35 046.61 + 22.30	04 – 42 – 009 IC 1269	0260	1.7 x 1.3 1.7 x 1.4	[125] ...	2	Sb/Sc (+ companions?) Disturbed?		14.5		*
11014	17 50.0 + 31 28 056.78 + 25.66	05 – 42 – 004 NGC 6485	0532	1.4 x 1.3 1.2 x 1.1	— 1	3	Sb	Sb +	14.2		3
11015	17 50.0 + 37 22 063.11 + 27.32	06 – 39 – 024	0277	1.0 x 0.10 0.7 x 0.1	3 7	3	Sc s of 2		16.5:		*1
11016	17 50.1 + 28 55 054.11 + 24.84	05 – 42 – 000	0532 438	1.4 x 0.3 1.2 x 0.3	162 6	2	Sc		15.7		
11017	17 50.2 + 29 52 055.11 + 25.12	05 – 42 – 005	0532 438	1.8 x 1.0 1.7: x 0.9:	165: ...	3	S B c – Irr		15.5		
11018	17 50.3 + 22 53 047.94 + 22.72	04 – 42 – 000	0260	1.0: x 0.6: 0.9:: x 0.6::	... ...	1	...	S B dm – m	17.		
11019	17 50.3 + 62 53 092.11 + 30.81	10 – 25 – 108	1148 437	1.0 x 0.2 1.0 x 0.2	63 7	2	Sa – b		15.5		
11020	17 50.6 + 59 30 088.18 + 30.67	10 – 25 – 109 NGC 6497	1148 429	1.5 x 0.7 1.6 x 0.8	113 5	4	S B b	S B(r)b	14.3		
11021	17 50.7 + 29 05 054.33 + 24.77	05 – 42 – 000	0532 438	1.2 x 0.2 1.1 x 0.2	35 7	2	Sb		15.7		
11022	17 50.7 + 29 51 055.13 + 25.01	05 – 42 – 008 NGC 6487	0532 438	(1.9 x 1.7) (1.8 x 1.6)	— ...	3	E P w NGC 6486		14.0		*1
11023	17 50.7 + 72 02 102.74 + 30.44	12 – 17 – 010 NGC 6508	0801 430	(1.3 x 1.3) (1.4 x 1.4)	— ...	3	[E] Compact		14.0		
11024	17 50.8 + 23 14 048.34 + 22.74	04 – 42 – 012	0260	1.7 x 0.35 1.5 x 0.3	143 (6)	1	S? ...		15.0		
11025	17 50.9 + 27 41 052.90 + 24.27	05 – 42 – 009	0532 438	(1.1 x 0.9) (1.0 x 0.8)	(0) 2:	3	S0		14.6		

N G C   6458, 6460, 6467, 6468, 6478, 6479, 6482,
6484, 6485, 6487, 6491, 6493, 6497, 6503, 6508

I C   1269

1	2	3	4	5	6	7	8	9	10	11	12
11026	17 51.1 + 65 33 095.21 + 30.72	11 – 22 – 007 NGC 6505	0550 435	(1.1 x 1.0) (1.1 x 0.9)	-- ...	1	... .		15.4		*
11027	17 51.2 + 24 35 049.74 + 23.14	04 – 42 – 013	0260	1.7 x 1.1 1.6: x 1.1	... ...	3	Irr NGC 6484 group		14.9		*1
11028	17 51.5 + 60 37 089.48 + 30.61	10 – 25 – 111	1148	1.8 x 0.6 1.2 x 0.4	53 6	2	Sb		15.6		
11029	17 51.8 + 24 28 049.68 + 22.97	04 – 42 – 014	0260	1.4 x 1.4 1.2 x 1.1	-- 1	4	S B c NGC 6484 group, Sc +		15.0		*
11030	17 52.1 + 02 53 028.97 + 13.95	00 – 46 – 001	1144	1.8 x 0.5 1.5 x 0.5	61 6	1	Sc		14.9		
11031	17 52.1 + 30 42 056.13 + 25.00	05 – 42 – 011	0532 433	1.0: x 1.0: 0.7 x 0.6	-- ...	3	Irr Disrupted		15.6		*1
11032	17 52.2 + 51 29 078.96 + 29.72	09 – 29 – 000	1102	1.1: x 0.8 0.7: x 0.7:	(30) ...	3	... S m 3rd of 3		17.		*
11033	17 52.3 + 18 23 043.65 + 20.58	03 – 45 – 038 NGC 6490	0274 436	1.0 x 0.8 0.9 x 0.7	115: (2)	2	E – S ... P w 11034		14.7		*3
11034	17 52.6 + 18 20 043.63 + 20.49	03 – 45 – 039 NGC 6495	0274 436	(1.6 x 1.2) (1.4 x 1.1)	-- ...	3	E P w 11033		13.8		*3
11035	17 52.6 + 32 53 058.46 + 25.56	05 – 42 – 013	0532	1.7 x 1.1 1.7 x 1.2:	... ...	4	Strongly peculiar		14.3		*1C
11036	17 52.7 + 58 23 086.90 + 30.34	10 – 25 – 113	1148	1.8 x 0.10 1.4 x 0.10	38 7	2	... d – m		16.5:		
11037	17 52.8 + 18 17 043.60 + 20.43	03 – 45 – 000	0274 436	1.3 x 0.2 0.9 x 0.2	62 (6 – 7)	1	S – Irr		16.0:		*
11038	17 52.8 + 72 09 102.86 + 30.27	12 – 17 – 000	0801 430	1.0 x 0.7 1.0 x 0.3	30: 3:	2	Sc		15.1		
11039	17 52.9 + 18 33 043.88 + 20.51	03 – 46 – 001	0529 436	1.0 x 1.0 0.9: x 0.5:	-- ...	3	Dwarf irregular Ir IV – V		15.7		1
11040	17 53.0 + 28 31 053.93 + 24.11	05 – 42 – 000	0532 438	1.5 x 0.20 1. x 0.2	163 7	1	S ...		15.4		
11041	17 53.0 + 34 47 060.51 + 26.04	06 – 39 – 025	0277	1.3 x 0.8 0.8 x 0.4	65 4	2	Sa – b		13.9		
11042	17 53.2 + 26 22 051.73 + 23.34	04 – 42 – 015	0260	1.2 x 1.2 1.2 x 1.1	-- 1	1	Sa? + two companions		14.9		*1
11043	17 53.3 + 25 26 050.79 + 22.99	04 – 42 – 000	0260	[1.8: x 1.3:] ...	... ...	1	Triple system Extensive halo, in group		15.5		*
11044	17 53.4 + 18 56 044.30 + 20.55	03 – 46 – 002	0529	[1.4 x 1.0] ...	... ...	3	Triple system Disrupted		15.0		*1
11045	17 53.4 + 31 51 057.43 + 25.09	05 – 42 – 015	0532	1.1 x 0.8 1.1 x 0.8	120 3	4	Sa		15.1		1
11046	17 53.5 + 28 50 054.29 + 24.11	05 – 42 – 016	0532 438	1.2 x 0.8 1.3 x 0.9	135 3	4	S B a (R´)S B(s)a		14.6		*1
11047	17 53.7 + 30 01 055.54 + 24.45	05 – 42 – 000	0532 438	1.1 x 0.3 0.9 x 0.3	159 6	1	S ...		15.7		
11048	17 53.8 + 18 21 043.77 + 20.24	03 – 46 – 003 NGC 6500	0529 436	2.5: x 2.0: 1.7 x 1.5	50: ...	1	Sa: P w 11049, disturbed		13.4		*12 3
11049	17 53.9 + 18 23 043.81 + 20.23	03 – 46 – 004 NGC 6501	0529 436	(1.7 x 1.5) (1.8 x 1.6)	-- 1	3	S0/Sa P w 11048		13.4		*12 3
11050	17 53.9 + 34 35 060.36 + 25.80	06 – 39 – 026	0277	1.6 x 0.2 1.2 x 0.2	56 6 – 7	2	Sb – c		15.7		*
11051	17 54.0 + 60 50 089.75 + 30.31	10 – 25 – 114 NGC 6510=6511	1148	1.0 x 0.6 1.0 x 0.5	30 ...	3	S B ...		14.3		*
11052	17 54.1 + 18 49 044.26 + 20.35	03 – 46 – 005	0529 436	1.1 x 0.30 1.1 x 0.30	128 6	1	S ...		15.1		1
11053	17 54.2 + 33 12 058.89 + 25.33	06 – 39 – 027 NGC 6504	0277	2.2 x 0.4 2.0 x 0.4	94 7	1	S ...		13.4		3
11054	17 54.2 + 71 34 102.18 + 30.20	12 – 17 – 011	0801	1.1: x 0.6: 0.7 x 0.2	170 ...	1	... P w 11066		16.5:		*
11055	17 54.7 + 12 15 037.98 + 17.53	02 – 46 – 001	0123	1.0 x 0.8 1.0 x 0.7	105 2 – 3	2	S B ... P w 11057		14.4		*
11056	17 54.8 + 28 05 053.63 + 23.59	05 – 42 – 000	0532 438	1.1 x 0.25 0.9 x 0.2	131 7	1	S0 – a		15.6		
11057	17 54.9 + 12 11 037.94 + 17.46	02 – 46 – 002	0123	2.0 x 0.8 2.0 x 0.8	90 6	3	Sc P w 11055		13.8		*
11058	17 55.0 + 32 38 058.37 + 25.01	05 – 42 – 018	0532	1.6 x 1.2 1.6 x 1.1	45 2 – 3	4	S B b S B(s)b Sb +		14.4		
11059	17 55.2 + 11 45 037.56 + 17.21	02 – 46 – 000	0123	1.0: x 0.15 1.1: x 0.15	90 7	1	S ...		15.7		
11060	17 55.2 + 27 58 053.54 + 23.47	05 – 42 – 019	0532 438	1.5 x 0.4: 1.4 x 0.4	130 7:	2	Sa P w 11064		14.9		*1

N G C   6490, 6495, 6500, 6501, 6504, 6505
6510, 6511

1	2	3	4	5	6	7	8	9	10	11	12
11061	17 55.4 + 62 37 091.82 + 30.22	10 – 25 – 119 NGC 6521	1148 437	(1.6 x 1.3) (1.6 x 1.3)	(160) ...	4	E		14.3		*3
11062	17 55.4 + 73 25 104.30 + 29.99	12 – 17 – 012 NGC 6538	0801 430	1.1 x 0.6 1.0 x 0.5	48 4 – 5	1	S ...		14.1		*
11063	17 55.5 + 23 02 048.57 + 21.66	04 – 42 – 016	0260	1.1 x 1.1 1.0 x 1.0	–– 1	4	Sc	SAB(s)c	15.5		
11064	17 55.7 + 27 51 053.46 + 23.32	05 – 42 – 020	0532 438	2.0: x 2.0: 2.0: x 2.0:	–– 1	4	Sc P w 11060		14.5		*
11065	17 55.7 + 47 45 074.84 + 28.57	08 – 33 – 002	0130	1.0 x 0.9 0.9 x 0.8	–– 1	1	S ...		14.9		*
11066	17 55.7 + 71 33 102.15 + 30.09	12 – 17 – 013	0801	1.6: x 0.8: 1.3: x 0.7:	10 ...	3	... P w 11054	SAB(s)dm	15.5		*
11067	17 56.0 + 09 42 035.72 + 16.15	02 – 46 – 003	0123	1.0 x 1.0 1.0 x 1.0	–– 1	1	S ...	(R)S ...	15.0		
11068	17 56.0 + 28 15 053.89 + 23.40	05 – 42 – 022	0532 438	1.6: x 1.5: 1.0 x 0.9	–– 1	1	S ...		15.5		
11069	17 56.0 + 64 39 094.17 + 30.20	11 – 22 – 000	0550	1.3 x 0.2 0.5 x 0.12	98 7	1	S ...		16.5:		*
11070	17 56.1 + 27 16 052.90 + 23.04	05 – 42 – 023	0532 438	1.1 x 0.8: 1.1 x 0.6	... ...	2	S B b	S B(r) ...	15.4		*1
11071	17 56.2 + 50 44 078.21 + 28.99	08 – 33 – 003 NGC 6515	0130	1.6 x 1.0 (1.7 x 1.1)	10 ...	3	E		14.3		*
11072	17 56.4 + 43 23 070.02 + 27.57	07 – 37 – 003	0423 439	1.1 x 0.15 1.0 x 0.15	115 7	2	Sa – b		16.0:		*
11073	17 56.6 + 10 33 036.59 + 16.38	02 – 46 – 005	0123	1.5: x 0.7: 1.2 x 0.6	157 5	2	Sb – c		14.8		
11074	17 56.7 + 07 09 033.43 + 14.87	01 – 46 – 001	0164	2.7 x 1.1 2.5 x 1.0	142 6	3	...	S d – dm	15.2		
11075	17 57.0 + 06 17 032.66 + 14.41	01 – 46 – 002 NGC 6509	0164	1.6 x 1.3 1.7 x 1.4	105: 2	3	Sc	Sc +	13.4		
11076	17 57.0 + 34 00 059.95 + 25.02	06 – 39 – 028	0277	1.3 x 0.6 1.1 x 0.45	63 6	2	Sb – c		14.8		
11077	17 57.1 + 64 56 094.50 + 30.09	11 – 22 – 016 NGC 6536	0550 435	1.3 x 1.2 1.4 x 1.2	–– 1	4	S B b	S B(r)b Sb –	14.3		
11078	17 57.5 + 24 54 050.62 + 21.91	04 – 42 – 018 NGC 6513	0260	1.2 x 0.9 1.4 x 0.9	(40) 4:	2	S B: 0		14.7		13
11079	17 57.8 + 45 55 072.88 + 27.87	08 – 33 – 005 NGC 6524	0130	(1.7 x 1.3) 1.4 x 0.9	155 ...	1	E – S0		14.0		
11080	17 57.9 + 26 20 052.10 + 22.34	04 – 42 – 019	0260	1.0 x 0.5 0.9: x 0.4	15: ...	1	S ... P w 11082		15.7		*1
11081	17 58 + 81 08 113.10 + 29.01	14 – 08 – 000	0775	1.0: x 0.8: 1.0: x 0.7:	–– ...	2	Dwarf V, nf of 2		17.		*
11082	17 58.0 + 26 21 052.12 + 22.32	04 – 42 – 020	0260	0.8 x 0.8 0.7 x 0.7	–– (1)	1	S0 P w 11080		14.4		*1
11083	17 58.0 + 57 15 085.67 + 29.54	10 – 25 – 000	1148	1.0 x 0.5 0.3 x 0.2	[165] ...	2	S – Irr		17.		*
11084	17 58.2 + 57 08 085.53 + 29.50	10 – 25 – 000	1148	1.1 x 0.12 1.0: x 0.10	158 7	2	Sc		16.5:		
11085	17 58.3 + 56 13 084.49 + 29.39	09 – 29 – 045 NGC 6532	1102	2.1 x 0.9 2.0: x 0.8	123 6	4	Sc	SA c	15.0		
11086	17 58.4 + 28 47 054.63 + 23.08	05 – 42 – 026	0532 438	1.4 x 0.2 1.3 x 0.2	35 7	1	S ... P w 11090		15.0		*1
11087	17 58.5 + 34 38 060.71 + 24.91	06 – 39 – 031	0277	0.65 x 0.55 0.6 x 0.5	(135) ...	1	S ... Peculiar?		14.4		*1
11088	17 58.5 + 44 51 071.73 + 27.52	07 – 37 – 012	0324	1.1 x 0.5 1.1 x 0.5	16 6	2	S B b:	S B(s) ...	15.3		1
11089	17 58.6 + 58 33 087.16 + 29.58	10 – 25 – 125	1148	1.0 x 0.6 1.0 x 0.5	25 5	1	S ...		15.2		1
11090	17 58.7 + 28 43 054.59 + 22.99	05 – 42 – 027	0532 438	1.5: x 1.0 1.2 x 1.0	20 ...	3	Sc P w 11086		14.8		*
11091	17 58.9 + 35 00 061.13 + 24.94	06 – 39 – 033	0277	1.1 x 0.6 0.8 x 0.5	143 5	3	Sc		15.7		
11092	17 59.1 + 61 21 090.38 + 29.72	10 – 25 – 126 NGC 6542	1148	1.4 x 0.4 1.5 x 0.4	98 6	1	S0 – a		14.0		
11093	17 59.5 + 06 58 033.58 + 14.16	01 – 46 – 003	0164	3.4 x 0.6 3.4 x 0.6	4 6	3	Sc		14.6		
11094	17 59.6 + 19 44 045.69 + 19.52	03 – 46 – 009 NGC 6527	0529	1.6 x 1.1 (2.0 x 1.4)	150 3	1	Sa:		14.9		
11095	17 59.6 + 22 35b 048.50 + 20.61	04 – 42 – 000	0260	1.2: x 0.5 1.2: x 0.5	30 6	1	S0 nf of 2, contact, in group		(15.7)		*

N G C   6509, 6513, 6515, 6521, 6524, 6527
        6532, 6536, 6538, 6542

1	2	3	4	5	6	7	8	9	10	11	12
11096	18 00.1 + 66 36 096.43 + 29.80	11 – 22 – 018 NGC 6552	0550 435	1.0 x 0.7 0.9 x 0.7	105 3?	2	S B ... Singular		14.6		*3
11097	18 00.3 + 26 02 052.00 + 21.73	04 – 42 – 023	0260	1.4 x 0.4 1.1 x 0.35	157 ...	1	...		14.6		
11098	18 00.8 + 29 18 055.35 + 22.76	05 – 42 – 028	0532 438	1.3 x 0.7 1.4: x 0.9:	... ...	1	S ... 1st of 3, disturbed		15.6		*1
11099	18 01.3 + 67 25 097.37 + 29.69	11 – 22 – 026	0550	1.3 x 0.45 1.3 x 0.4	82 6	2	S B? c		15.0		*
11100	18 01.4 + 29 05 055.18 + 22.56	05 – 42 – 030	0532 438	1.0 x 0.5: 1.0 x 0.3	62 ...	1	S ...		15.3		*
11101	18 01.6 + 45 15 072.30 + 27.30	08 – 33 – 012	0130	1.3 x 0.3 1.3 x 0.3	42 6	3	Sc		16.0:		
11102	18 01.8 + 52 08 079.95 + 28.34	09 – 29 – 000	1102	1.3 x 0.1 1.1 x 0.1	34 7	1	Sb – c		17.		
11103	18 02.0 + 29 30 055.65 + 22.58	10 – 42 – 000	0532 438	1.0 x 0.2 1.0 x 0.2	154 6	1	Sb – c		15.7		
11104	18 02.4 + 20 05 046.31 + 19.05	03 – 46 – 010	0529	1.1 x 0.5 1.4 x 0.6	95 6	1	S B: ... P w 11108		15.3		*
11105	18 02.4 + 21 38 047.82 + 19.65	04 – 42 – 024	0260	2.5: x 1.8: 2. x 2.	-- ...	4	...	S dm	15.7		
11106	18 02.4 + 69 43 100.02 + 29.59	12 – 17 – 014	0801 443	1.2 x 1.2 1.1 x 1.1	-- 1	3	S B b	S B(r)bc Sb +	15.2		1
11107	18 02.8 + 17 16 043.63 + 17.84	03 – 46 – 011	0529	1.1 x 1.1 1.0 x 1.0	-- 1	3	S B b	S B(r)b – bc Sb(+)	15.6		1
11108	18 02.8 + 20 02 046.30 + 18.94	03 – 46 – 000	0529	1.0 x 0.10 0.9 x 0.10	11 7	2	Sc P w 11104		16.5:		*
11109	18 02.8 + 46 45 074.00 + 27.19	08 – 33 – 000	0130	1.1: x 1.0: 0.5: x 0.3:	-- ...	3	Dwarf V		17.		
11110	18 03.1 + 25 13 051.43 + 20.85	04 – 43 – 001 NGC 6547	1089	1.5 x 0.4 1.5 x 0.4	136 6	1	S0 – a		14.3		3
11111	18 03.3 + 23 07 049.36 + 20.02	04 – 43 – 002	1089	1.2:: x 1.0:: 0.8:: x 0.7::	-- ...	3	Dwarf (spiral) S: B IV – V, ssp of 2		17.		*
11112	18 03.3 + 46 16 073.49 + 27.01	08 – 33 – 016	0130	1.2 x 0.8 1.2 x 0.7	105: 3 – 4	3	Sb/S B b nnf of 2		15.3		*
11113	18 03.4 + 23 16 049.52 + 20.06	04 – 43 – 003	1089 444	1.7 x 1.7 1.7: x 1.5:	-- 1	4	Sc/S B c	Sc +	15.7		*
11114	18 03.6 + 18 32 044.93 + 18.18	03 – 46 – 012 NGC 6549	0529	1.4 x 0.5 1.4 x 0.4	53 6	1	S ... Jet?, p w 11115		14.8		*13
11115	18 03.8 + 18 35 044.99 + 18.15	03 – 46 – 013 NGC 6548 = 6550	0529	(2.8 x 2.5) (2.8 x 2.5)	-- 1	4	S B 0 P w 11114		13.1		*13
11116	18 03.8 + 67 02 096.93 + 29.45	11 – 22 – 000	0550 435	1.2 x 0.6 0.8 x 0.4	94 5	3	Sb		15.3		
11117	18 03.9 + 46 52 074.17 + 27.03	08 – 33 – 019 NGC 6560	0130	1.2 x 0.9 1.1 x 0.6	55: ...	1	S ... Peculiar?		14.2		*
11118	18 04.0 + 34 00 060.43 + 23.64	06 – 40 – 004	0197	1.1 x 0.6 1.0 x 0.6	7 5	3	Sb		15.3		
11119	18 04.0 + 60 13 089.13 + 29.04	10 – 26 – 007	0512	1.3 x 0.15 1.2 x 0.15	18 7	3	Sc	Sc +	16.5:		*
11120	18 04.9 + 20 29 046.94 + 18.67	03 – 46 – 014	0529	1.4 x 0.3 1.1: x 0.3	112 ...	3	Irr		15.6		
11121	18 05.6 + 17 35 044.21 + 17.36	03 – 46 – 015 NGC 6555	0529	2.1 x 1.8 2.0 x 1.7	110 1 – 2	4	Sc	S (rs)c	12.7		*12 3
11122	18 05.7 + 09 42 036.81 + 14.00	02 – 46 – 000	0123	1.0 x 0.9 0.7 x 0.5	-- ...	2	Sc		15.7		
11123	18 05.7 + 28 28 054.91 + 21.47	05 – 43 – 001	0282 438	1.1 x 0.9 1.0 x 0.8	75: ...	3	... Companion superimposed?	S dm	15.4		*1
11124	18 05.7 + 35 33 062.16 + 23.78	06 – 40 – 005	0197	2.6 x 2.3 2.3 x 2.3	-- 1	4	S B c	S B(s)cd	13.9		
11125	18 05.8 + 11 41 038.66 + 14.84	02 – 46 – 006 IC 4688	0123	1.6 x 1.1 1.5: x 1.1	170: 3	2	Sc		14.7		
11126	18 06.4 + 25 43 052.22 + 20.34	04 – 43 – 000	1089	1.4: x 0.08 1.2: x 0.1	46 7	2	...	S: d – m	17.		
11127	18 06.8 + 28 02 054.56 + 21.09	05 – 43 – 003	0282 438	[1.1 x 0.4] ...	... ...	1	Double system Contact, distorted?		14.8		*1
11128	18 07.1 + 18 17 045.04 + 17.32	03 – 46 – 016	0529	1.1 x 0.2 1.0 x 0.2	40 7	2	Sa – b		15.2		
11129	18 07.2 + 30 18 056.88 + 21.79	05 – 43 – 004	0282	1.3 x 0.8 1.2 x 0.8	83 3	1	Sa		14.9		1
11130	18 07.2 + 69 49 100.13 + 29.18	12 – 17 – 000	0801 443	0.15 x 0.15 0.18 x 0.18	-- ...	1	... Extremely compact		14.4 + 15300 + 15557		*C

N G C   6547 – 50, 6552, 6555, 6560

I C   4688

1	2	3	4	5	6	7	8	9	10	11	12
11131	18 07.8 + 01 34 029.63 + 09.86	00 – 46 – 003	0773	1.4: x 0.6: 1.2 x 0.6	... ...	2	...	d – m	15.7		
11132	18 07.8 + 38 46 065.66 + 24.33	06 – 40 – 008	0197	2.2 x 0.2 1.9 x 0.2	145 7	3	Sb		15.6		
11133	18 07.8 + 78 33 110.12 + 28.85	13 – 13 – 000	0776	1.2 x 0.6 0.7 x 0.4	40: 5:	1	S ...		16.0:		
11134	18 08.1 + 05 43 033.42 + 11.69	01 – 46 – 000	0164	1.4 x 0.3 1.4 x 0.3	41 6	1	S ...		17.		
11135	18 08.6 + 30 59 057.68 + 21.74	05 – 43 – 005 IC 1277	0282	1.7 x 1.4 1.2: x 1.0:	(25) 1 – 2	3	Sc P w 11138		15.0		*
11136	18 08.7 + 25 11 051.89 + 19.66	04 – 43 – 000	1089 441	1.2 x 0.2 1.2 x 0.2	89 7	2	Sb		15.7		*
11137	18 08.8 + 14 05 041.23 + 15.20	02 – 46 – 008 NGC 6570	0123	1.8 x 1.0 1.7 x 1.0	30 ...	3	Sc – Irr		13.2		*12 3
11138	18 09.0 + 31 05 057.81 + 21.69	05 – 43 – 006 NGC 6575	0282	(1.8 x 1.3) (1.6 x 1.1)	65 ...	3	E P w 11135		14.4		*
11139	18 09.2 + 69 04 099.28 + 28.99	12 – 17 – 018 NGC 6598	0801 443	2.0 x 1.4 2.2: x 1.6:	40: ...	3	... Peculiar		14.2		*1
11140	18 09.4 + 39 10 066.18 + 24.14	07 – 37 – 022	0324	2.5 x 0.4 2.3: x 0.4	31 6	3	Sc		15.4		
11141	18 09.5 + 12 04 039.42 + 14.19	02 – 46 – 009	0123	2.0: x 2.0: 2. x 1.6:	–– ...	1	...	S d – m	15.4		*1
11142	18 09.5 + 25 38 052.41 + 19.66	04 – 43 – 007	1089 441	1.7 x 0.15 1.4 x 0.15	14 7	3	Sc		15.4		*
11143	18 09.5 + 35 59 062.87 + 23.18	06 – 40 – 009 IC 1279	0197	2.5 x 0.5 2.5 x 0.5	159 6	3	Sb		14.5		*
11144	18 09.6 + 14 57 042.13 + 15.39	02 – 46 – 010 NGC 6574	0123	1.3 x 0.9 1.4 x 1.0	160 3	1	S ...		12.5	+ 2368 + 2572	12 368
11145	18 09.7 + 49 30 077.28 + 26.64	08 – 33 – 028	0130 442	(1.2 x 0.6) (1.1 x 0.5)	(55) ...	1	...		15.5		
11146	18 09.8 + 49 53 077.71 + 26.70	08 – 33 – 29 + 30 NGC 6582	130 442	[1.8 x 1.2] ...	... ...	2	Double system Envelopes in contact		14.8		*1
11147	18 09.8 + 53 20 081.51 + 27.35	09 – 30 – 004	0789	1.1 x 0.45 1.0 x 0.5	2 6	3	Sb		15.6		
11148	18 09.9 + 21 28 048.38 + 17.98	04 – 43 – 009 NGC 6577	1089 444	(1.5 x 1.3) (1.3 x 1.1)	–– ...	2	E		14.7		*3
11149	18 09.9 + 49 51 077.67 + 26.68	08 – 33 – 000	0130 442	(1.3 x 0.7) (1.3 x 0.8)	(160) ...	1	...		16.0:		*
11150	18 10.0 + 25 35 052.40 + 19.53	04 – 43 – 000	1089 441	1.7 x 0.25 1.1 x 0.22	7 7?	2	S? ... Peculiar? or disrupted spiral?		15.5		*
11151	18 10.2 + 29 08 055.94 + 20.78	05 – 43 – 007	0282	1.7 x 0.3 1.7 x 0.3	0 6 – 7	1	S ...		14.9		
11152	18 10.3 + 18 35 045.64 + 16.74	03 – 46 – 018	0529	2.1 x 1.1 2.2 x 1.1	140 5:	3	... S B dm sp of 2		15.0		*1
11153	18 10.4 + 21 25 048.38 + 17.85	04 – 43 – 11 + 12 NGC 6579 + 6580	1089 444	[1.5 x 1.3] ...	... ...	2	Triple system Two galaxies connected		14.5		*13
11154	18 10.4 + 42 38 069.72 + 24.87	07 – 37 – 023	0324	1.1 x 0.5 1.0 x 0.4	1 6	2	Sa – b		16.0:		
11155	18 10.5 + 25 25 052.28 + 19.37	04 – 43 – 014	1089 441	1.6: x 1.3: 1.3: x 1.1:	–– 1 – 2	2	Sc? nff of 2, disturbed		15.5		*1
11156	18 10.5 + 25 31 052.38 + 19.40	04 – 43 – 015	1089 441	(1.1 x 1.0) (1.3 x 1.1)	–– ...	1	E?		14.7		*
11157	18 10.5 + 30 11 057.02 + 21.08	05 – 43 – 008	0282	1.5 x 0.5 1.4: x 0.4	156 6	2	...	S d – dm	15.6		
11158	18 10.7 + 20 40 047.68 + 17.49	03 – 46 – 019	0529	1.0 x 0.25 0.9 x 0.22	65 6 – 7	1	S ...		15.1		*1
11159	18 10.7 + 39 37 066.73 + 24.03	07 – 37 – 024 NGC 6585	0324	2.0 x 0.40 2.0 x 0.40	50 6	1	S ...		13.6		
11160	18 10.7 + 56 47 085.38 + 27.77	09 – 30 – 005	0789 440	1.1 x 0.4 0.7 x 0.20	158 6	3	Sb		15.7		*
11161	18 11.1 + 28 25 055.30 + 20.34	05 – 43 – 000	0282	1.0 x 0.5 0.9 x 0.4	110 5:	1	S ...		15.7		
11162	18 11.1 + 29 41 056.57 + 20.79	05 – 43 – 010	0282	1.5 x 0.8 1.6 x 0.8	115 5:	3	S ... Singular, disturbed?		15.3		*1
11163	18 11.4 + 33 48 060.76 + 22.12	06 – 40 – 010	0197	1.2 x 1.0 1.1 x 1.0	–– ...	1	...B ... P w 11167		14.7		*1
11164	18 11.5 + 21 05 048.16 + 17.49	04 – 43 – 016 NGC 6586	1089 444	1.0 x 0.5 0.9 x 0.5	105 5	1	S ...		14.6		3
11165	18 11.5 + 68 42 098.87 + 28.77	11 – 22 – 000	0550	1.6 x 0.5 1.6 x 0.5	84 6	3	Sc		16.0:		

N G C   6570, 6574, 6575, 6577, 6579, 6580   I C   1277, 1279
6582, 6585, 6586, 6598

1	2	3	4	5	6	7	8	9	10	11	12
11166	18 11.6 + 18 49 046.00 + 16.56	03 – 46 – 020 NGC 6587	0529	(2. x 2.) (1.8 x 1.8)	--- 1	3	S0		14.3		*23
11167	18 11.6 + 33 43 060.68 + 22.05	06 – 40 – 000	0197	1.0 x 0.6 1.0 x 0.5	148 4 – 5	1	S ... P w 11163		15.7		*
11168	18 11.7 + 13 15 040.76 + 14.21	02 – 46 – 011	0123	1.4 x 1.1 1.2 x 1.0	(90) ...	1	Sc – Irr		14.2		
11169	18 11.7 + 64 59 094.64 + 28.55	11 – 22 – 000	0550	1.1 x 0.2 1.1 x 0.2	19 7	2	Sa – b		16.0:		
11170	18 11.9 + 21 30 048.60 + 17.57	04 – 43 – 017	1089 444	(1.4 x 1.2) (1.4 x 1.1)	--- ...	1	... Irregular plume or condensations in envelope		15.6		*
11171	18 12.2 + 30 40 057.64 + 20.90	05 – 43 – 011	0282	1.5 x 0.7 1.1 x 0.45	150 5	1	S ...		14.8		
11172	18 12.2 + 69 55 100.26 + 28.75	12 – 17 – 000	0801 443	0.20 x 0.20 0.15 x 0.15	--- ...	1	... Very blue very compact		14.0		*C
11173	18 12.8 + 49 02 076.89 + 26.04	08 – 33 – 031	0130 442	1.2 x 0.25 1.2 x 0.25	87 6	1	S ...		15.4		
11174	18 13.1 + 43 15 070.70 + 24.60	07 – 37 – 025 NGC 6606	0324	1.0 x 0.8 0.9 x 0.8	(105) ...	1	S ...		14.4		1
11175	18 13.2 + 68 20 098.46 + 28.60	11 – 22 – 30 + 31 NGC 6621 + 6622	550	[2.0 x 0.9] ...	[145] ...	3	Double system Disrupted, bridge + loop		13.6	+ 6230 + 6490	*12 VAC
11176	18 13.5 + 61 18 090.50 + 28.00	10 – 26 – 029 NGC 6617	0512	1.4 x 1.1 1.3: x 0.8:	80: 3	4	Sc	SA(s)d	15.6		
11177	18 13.6 + 06 44 034.97 + 10.93	01 – 46 – 000	0164	1.5: x 1.4: 1.5: x 1.3:	--- 1	3	S B b	S B(s)b	15.5		
11178	18 13.6 + 24 54 052.06 + 18.52	04 – 43 – 019 NGC 6599	1089 441	1.3 x 1.2 (1.3 x 1.2)	--- (1)	2	S0		13.7		
11179	18 13.7 + 25 20 052.49 + 18.66	04 – 43 – 020	1089 441	1.1 x 0.8 1.1: x 0.8:	10: 2 – 3	3	Sc		15.6		
11180	18 13.9 + 47 31 075.30 + 25.52	08 – 33 – 032	0130 453	1.3 x 0.2 1.2 x 0.2	90 6 – 7	1	S ...		15.6		*
11181	18 14.0 + 43 18 070.80 + 24.45	07 – 37 – 27 + 28 + 000	324	[1.6: x 0.8:] ...	... ...	1	Multiple system Bridge + plumes?		16.0:		*1
11182	18 14.0 + 56 28 085.10 + 27.27	09 – 30 – 008	0789	1.0 x 0.12 0.9 x 0.12	105 7	2	Sb – c		16.5:		
11183	18 14.4 + 68 00 098.08 + 28.47	11 – 22 – 035	0550	1.4 x 0.8 1.0: x 0.7:	135 ...	3	... nnp of 2	S B: dm – m	16.5:		*
11184	18 14.5 + 25 01 052.25 + 18.38	04 – 43 – 021	1089 441	1.2 x 0.9 0.7 x 0.5	0 3:	1	S ...		14.6		
11185	18 14.5 + 42 37 070.10 + 24.18	07 – 37 – 30 + 31 324		[1.7 x 1.1] ...	... ...	2	Double system Bridge:		15.9 + + 16.1		*1C
11186	18 15.0 + 47 49 075.67 + 25.41	08 – 33 – 033	0130 453	1.0 x 0.7 0.8 x 0.6	(60) 3	1	S ...		14.6		
11187	18 15.2 + 62 15 091.59 + 27.91	10 – 26 – 033	0512	1.1 x 0.2 1.0 x 0.2	149 7	1	S ...		16.0:		
11188	18 15.3 + 18 53 046.43 + 15.78	03 – 46 – 000	0529	1.1 x 0.15 1.1 x 0.2	119 7	3	Sc	Sc +	15.6		
11189	18 15.3 + 38 46 066.13 + 22.93	06 – 40 – 000	0197	1.2 x 0.3 1.1 x 0.2	45 7	1	Sb – c		16.0:		
11190	18 15.3 + 40 47 068.23 + 23.52	07 – 37 – 032	0324	1.1 x 1.1 0.8 x 0.8	--- 1	3	Sc		16.0:		
11191	18 15.3 + 55 33 084.12 + 26.94	09 – 30 – 010 IC 1286	0789	1.5 x 0.5 1.2 x 0.3	87 6	2	Sa – b		14.8		*
11192	18 15.6 + 22 13 049.64 + 17.06	04 – 43 – 022 NGC 6616	1089 444	1.5 x 0.6 1.5 x 0.6	59 6	2	Sa – b		15.2		*
11193	18 15.7 + 71 00 101.50 + 28.49	12 – 17 – 020 NGC 6651	0801	1.6 x 1.0 1.6: x 1.0:	... ...	3	Dwarf (irregular) Ir IV or IV – V		15.1		*(2)
11194	18 15.9 + 26 44 054.05 + 18.73	04 – 43 – 023	1089	1.7 x 1.0 1.6 x 1.0	112 4:	2	S B b – c		14.4		
11195	18 16.0 + 30 37 057.90 + 20.12	05 – 43 – 014	0282	1.2 x 0.5 1.0 x 0.5	[114] ...	3	S ... + companion Disrupted		14.6		*1
11196	18 16.2 + 13 15 041.24 + 13.22	02 – 46 – 013 NGC 6615	0123	(1.1 x 0.9) 1.2 x 0.9	165 2:	3	S B 0		14.8		*23
11197	18 16.2 + 21 16 048.79 + 16.55	04 – 43 – 024	1089	(1.1 x 1.1) (1.1 x 1.1)	--- (1)	1	S0		14.8		
11198	18 16.7 + 16 14 044.08 + 14.38	03 – 46 – 024	0808	1.3 x 0.25 1.1 x 0.2	50 7	1	Sa – b		15.1		
11199	18 16.7 + 36 17 063.67 + 21.89	06 – 40 – 012	0197	1.4 x 0.7 1.4 x 0.7	163 5	3	Sa		14.6		
11200	18 16.8 + 23 38 051.12 + 17.36	04 – 43 – 025 NGC 6619	1089 444	(1.2 x 1.1) (1.5 x 1.2)	--- ...	1	E?		14.3		3

N G C   6587, 6599, 6006, 6615 – 17, 6619
6621, 6622, 6651

I C   1286

1	2	3	4	5	6	7	8	9	10	11	12
11201	18 17.1 + 54 28 082.98 + 26.49	09 - 30 - 011	0789	1.4 x 0.35 1.1 x 0.3	3 6	2	Sa - b		16.0:		*
11202	18 17.4 + 50 15 078.39 + 25.58	08 - 33 - 035	0130	1.1 x 0.9 1.1 x 0.9	(175) 2:	2	S0		14.4		*
11203	18 17.5 + 23 41a 051.24 + 17.23	04 - 43 - 026 NGC 6623	1089 444	(1.4 x 1.1) (1.5 x 1.1)	(155) ...	2	E Brightest in triple system		(14.4)		*13
11204	18 17.6 + 47 44 075.70 + 24.97	08 - 33 - 036	0130 453	1.1: x 1.0: 0.9: x 0.9:	-- 1	2	Sc		15.6		
11205	18 17.7 + 42 46 070.44 + 23.65	07 - 37 - 037	0324	1.2 x 0.7 1.0 x 0.6	125 4	3	Sc	Sc +	15.5		*1
11206	18 18.3 + 48 32 076.58 + 25.05	08 - 33 - 037	0130	0.9 x 0.7 0.7 x 0.5	160: ...	1	S ...		14.5		
11207	18 18.8 + 47 54 075.93 + 24.81	08 - 33 - 038	0130 453	1.0 x 0.9 1.0 x 0.9	-- 1	1	S ...		16.5:		*
11208	18 18.9 + 68 19 098.47 + 28.07	11 - 22 - 040	0550	1.1 x 1.0 1.1 x 1.0	-- 1	4	Sc		15.1		*1
11209	18 19.5 + 26 56 054.57 + 18.06	04 - 43 - 000	1089	1.0 x 0.25 0.9 x 0.2	118 6 - 7	1	S ...		16.5:		
11210	18 19.8 + 21 08 049.01 + 15.73	04 - 43 - 000	1089	1.2: x 0.08 0.8: x 0.08	131 7	2	...	d - m	17.		
11211	18 20.2 + 23 27 051.26 + 16.57	04 - 43 - 029 NGC 6628	1089 444	(1.7 x 1.1) (1.9 x 1.2)	90 5:	1	S0?		14.8		*3
11212	18 20.4 + 15 40 043.94 + 13.34	03 - 47 - 001 NGC 6627	0808	1.5 x 1.3 1.4 x 1.2	(70) 1	4	S B b	S B(s)b	14.5	+ 5206 + 5419	*12 3
11213	18 20.5 + 52 20 080.77 + 25.57	09 - 30 - 012	0789	1.2 x 0.35 1.3 x 0.5	18 6	2	S B a - b		15.4		
11214	18 20.6 + 12 24 040.94 + 11.89	02 - 47 - 002	0184	1.5 x 1.1 1.3: x 1.2:	175: ...	3	Sc		15.		
11215	18 20.8 + 60 51 090.12 + 27.06	10 - 26 - 034	0512	1.3: x 1.3: 1.0 x 0.7	-- 1	3	Sc		16.0:		
11216	18 20.9 + 21 13 049.20 + 15.53	04 - 43 - 000	1089	1.0: x 0.5: 0.7 x 0.4	160: ...	1	S B: ...		16.5:		
11217	18 20.9 + 48 05 076.22 + 24.52	08 - 33 - 040	0130 453	1.0 x 0.8 0.9 x 0.8	(15) 2	2	Sb		14.6		*
11218	18 21.2 + 74 32 105.52 + 28.18	12 - 17 - 021 NGC 6643	0801	4.0 x 1.9 3.6 x 1.7	38 5	4	Sc		11.8	+ 1538 + 1790	*12- -5789
11219	18 21.3 + 30 14 057.95 + 18.92	05 - 43 - 017	0282 445	[1.6 x 0.4] ...	... ...	2	Double system, distorted Connecting spiral arm		15.3		*1
11220	18 21.8 + 40 55 068.76 + 22.39	07 - 38 - 000	1445	1.4:: x 1.4:: 0.7:: x 0.7::	-- ...	3	Dwarf irregular Ir V		17.		
11221	18 22.0 + 66 36 096.57 + 27.63	11 - 22 - 46+47 NGC 6636	550	[2.1 x 0.5] ...	[177] ...	3	Double system Contact, disrupted		14.2		*1C
11222	18 22.1 + 68 05 098.24 + 27.76	11 - 22 - 048	0550	1.5 x 0.9 0.9 x 0.7	80: 4:	3	S B c		14.9		
11223	18 22.3 + 36 26 064.21 + 20.88	06 - 40 - 000	0197	1.0 x 0.4 1.0 x 0.4	65 6	2	Sb - c		16.0:		
11224	18 22.4 + 36 46 064.56 + 20.97	06 - 40 - 000	0197	1.1: x 1.1: 0.5:: x 0.5::	-- ...	3	Dwarf V		17.		
11225	18 22.7 + 60 43 090.01 + 26.82	10 - 26 - 036	0512	1.3 x 0.20 1.3 x 0.20	79 7	3	Sc - Irr		15.4		
11226	18 23.0 + 27 30 055.43 + 17.56	05 - 43 - 018 NGC 6632	0282	3.0 x 1.3 3.0 x 1.3	155 6	4	Sb	SA(rs)bc Sb +	13.2		3
11227	18 23.1 + 62 14 091.70 + 26.99	10 - 26 - 037	0512	1.0 x 0.3 1.0 x 0.25	137 6 - 7	1	S0 - a		16.0:		
11228	18 23.2 + 41 28 069.41 + 22.30	07 - 38 - 003	1445	1.0 x 0.6 1.0 x 0.6	0: ...	2	S B 0		14.5		
11229	18 23.5 + 22 18 050.48 + 15.42	04 - 43 - 032	1089	1.0 x 0.5 0.9 x 0.5	74 5	2	S0 - a		14.8		
11230	18 23.5 + 65 18 095.13 + 27.34	11 - 22 - 049	0550	2.6 x 0.2 2.0 x 0.2	14 7	3	Sc	Sc +	16.0:		
11231	18 23.6 + 29 22 057.30 + 18.13	05 - 43 - 000	0282	1.1: x 0.10 1.0 x 0.10	149 7	2	Sc		17.		
11232	18 23.8 + 30 43 058.64 + 18.59	05 - 43 - 000	0282	1.1 x 0.2 1.0 x 0.2	159 7	3	Sc		16.5:		
11233	18 23.8 + 32 08 060.03 + 19.10	05 - 43 - 021	0282	1.2 x 0.8 1.1 x 0.8	(75) 3	2	S0		15.1		*
11234	18 24.4 + 22 23 050.65 + 15.26	04 - 43 - 000	1089	1.1 x 0.5 0.7 x 0.25	132 (6)	1	...		15.5		
11235	18 24.4 + 46 08 074.33 + 23.43	08 - 33 - 042	0130	1.5 x 0.3 1.2 x 0.2	101 ...	2	S B c - Irr		16.0:		

N G C   6623, 6627, 6628, 6632, 6636, 6643

1	2	3	4	5	6	7	8	9	10	11	12
11236	18 25.1 + 71 35 102.20 + 27.76	12 – 17 – 022 (NGC 6651)	0801	1.8 x 0.9 1.5 x 0.8	30 5	2	Sb – c		13.7		*2
11237	18 25.2 + 24 50 053.06 + 16.07	04 – 43 – 000	1089	1.4 x 0.9 1.4 x 0.7	125: 4	4	Sc		15.4		*
11238	18 25.2 + 73 09 103.97 + 27.85	12 – 17 – 023 NGC 6654	0801	2.8 x 2.3 2.9 x 2.3	0 2	4	S B 0 Compact core		12.7	+ 1924 + 2180	*2C
11239	18 25.3 + 14 44 043.59 + 11.87	02 – 47 – 003 NGC 6635	0184	(1.0 x 0.9) (1.5 x 1.3)	-- 1	3	S0		14.5:	+ 5071 + 5284	*12 3
11240	18 25.5 + 33 54 061.92 + 19.40	06 – 40 – 000	0197 446	1.0: x 0.1 0.9 x 0.1	2 7	2	Sc		17.		
11241	18 25.6 + 51 07 079.66 + 24.52	09 – 30 – 014	0789	1.2: x 1.1: 1.1 x 0.9	-- 1	3	Sb		15.2		
11242	18 25.7 + 16 06 044.89 + 12.37	03 – 47 – 003	0808	1.2 x 0.6 1.1 x 0.5	88 5	3	Sc	Sc +	15.7		
11243	18 25.8 + 22 41 051.07 + 15.08	04 – 43 – 000	1089	1.1 x 0.10 1.0 x 0.10	163 7	3	Sc		17.		
11244	18 25.9 + 46 53 075.19 + 23.39	08 – 33 – 043	0130	1.7 x 0.8 1.3 x 0.7	150 5:	3	Sc – Irr S dm:		15.3		
11245	18 26.1 + 56 02 084.97 + 25.54	09 – 30 – 015	0789	1.1 x 0.5 0.9 x 0.3	15 6	2	Sa – b		15.4		
11246	18 26.2 + 22 42 051.12 + 15.01	04 – 43 – 034	1089	1.3 x 0.4 1.3 x 0.4	152 6	1	Sa – b		15.1		
11247	18 26.3 + 34 16 062.34 + 19.37	06 – 40 – 018 NGC 6640	0197 446	1.2 x 0.8 1.2 x 0.8	153 3	2	Sc		14.2		1
11248	18 26.6 + 48 10 076.57 + 23.62	08 – 33 – 000	0130 453	1.2 x 0.2 1.2 x 0.7	79 7	2	Sc s of 2, disturbed		16.0:		*
11249	18 26.7 + 34 07 062.22 + 19.24	06 – 40 – 000	0197 445	1.0 x 0.2 0.9 x 0.2	38 7	1	S ... p of 2, disturbed?		16.5:		*
11250	18 26.8 + 22 52 051.34 + 14.95	04 – 43 – 035 NGC 6641	1089	1.1 x 1.0 1.0 x 0.9	-- 1?	1	... Peculiar?		14.3		
11251	18 27.0 + 38 00 066.13 + 20.51	06 – 40 – 019	0197	1.5 x 0.9 1.0 x 0.6	[65] ...	3	... Peculiar, blue arc		15.7		
11252	18 27.4 + 48 13 076.67 + 23.50	08 – 33 – 045	0130 453	1.3 x 0.45 1.0 x 0.45	164 6	1	Sc Companion attached, disturbed		14.9		*
11253	18 27.5 + 23 01 051.55 + 14.86	04 – 43 – 000	1089	1.0 x 0.15 1.0 x 0.15	111 7	2	Sc nnp of 2		15.7		*
11254	18 27.7 + 30 25 058.67 + 17.70	05 – 43 – 000	0282	1.3 x 0.1 1.7: x 0.1	105 7	3	Sc		15.7		
11255	18 27.7 + 51 37 080.28 + 24.32	09 – 30 – 016	0789	1.5 x 0.15 1.6 x 0.15	177 7	3	Sc		15.6		
11256	18 27.8 + 39 41 067.88 + 20.91	07 – 38 – 007 IC 1288	1445	1.3 x 0.7 1.3 x 0.7	2 5	2	S B a		14.3		
11257	18 27.9 + 53 53 082.71 + 24.83	09 – 30 – 017	0789	1.2 x 0.6 1.1 x 0.6	142 5	2	Sb		16.0:		
11258	18 28.0 + 39 50 068.05 + 20.93	07 – 38 – 008 NGC 6646	1445	1.3 x 1.3 1.5 x 1.5	-- 1	3	Sa		13.7		*13
11259	18 28.5 + 33 52 062.12 + 18.80	06 – 40 – 000	0197 446	1.1 x 1.1 0.8: x 0.8:	-- 1	1	Sc		16.5:		
11260	18 28.6 + 34 03 062.30 + 18.85	06 – 40 – 021	0197 446	1.3 x 1.0 1.3 x 1.0	-- 2	2	S B a		15.5		*
11261	18 29.0 + 22 23 051.10 + 14.29	04 – 44 – 001	0284	1.0 x 0.6 0.8 x 0.4	150: ...	1	... S d – m		15.5		
11262	18 29.0 + 42 40 071.00 + 21.64	07 – 38 – 010	1445	1.7 x 0.7 1.4: x 0.5	48 6	4	Sc	Sc +	15.5		
11263	18 29.3 + 30 57 059.32 + 17.58	05 – 43 – 000	0282 447	1.0 x 0.7 1.0 x 0.7:	(70) 3	3	Sc 1st of 3		16.0:		
11264	18 29.8 + 31 02 059.45 + 17.51	05 – 44 – 001	0267 447	1.2 x 0.45 1.2 x 0.4	55 6	2	Sb – c 2nd of 3		15.2		*
11265	18 29.8 + 33 53 062.23 + 18.56	06 – 41 – 001	0148	0.8 x 0.8 0.9 x 0.9	-- ...	1	... Compact		13.9		*
11266	18 29.9 + 67 51 098.06 + 27.01	11 – 22 – 000	0550	1.0 x 0.12 1.0 x 0.12	143 7	2	Sc		16.5:		
11267	18 30 + 87 49 120.55 + 27.56	15 – 01 – 021	0570	1.8: x 0.3 1.3 x 0.15	130 7	3	Sc		16.0:		*
11268	18 30.4 + 37 34 065.93 + 19.73	06 – 41 – 002	0148	1.4 x 0.5 1.1 x 0.5	169 6	2	Sb – c		14.7		1
11269	18 30.8 + 67 56 098.16 + 26.93	11 – 22 – 053 NGC 6667	0550	3.2:: x 2.3:: 2.5: x 1.3:	105: ...	4	... Peculiar		13.7		*12
11270	18 31.0 + 78 44 110.29 + 27.71	13 – 13 – 001	0776 450	1.6 x 1.1 1.5: x 1.1	5: 3	2	Sa – b		15.2		

N G C   6635,   6640,   6641,   6646,   (6651),   6654   I C   1288
       6667

1	2	3	4	5	6	7	8	9	10	11	12
11271	18 31.2 + 34 00 062.46 + 18.33	06 - 41 - 003 NGC 6657	0148 446	0.8 x 0.4 1.0 x 0.4	138 6	1	S B ...		14.2		
11272	18 31.5 + 54 57 083.98 + 24.56	09 - 30 - 019	0789	1.0 x 0.3 0.8 x 0.15	32 6	2	Sb - c		16.5:		
11273	18 31.7 + 19 03 048.25 + 12.33	03 - 47 - 000	0808	1.0 x 0.2 1.0 x 0.2	64 6	1	S ...		15.7		
11274	18 31.8 + 22 51 051.81 + 13.89	04 - 44 - 002 NGC 6658	0284	1.5 x 0.3 1.4 x 0.3	5 7	3	S0 - a P w 11282		15.0	+ 4270 + 4508	*12 3
11275	18 31.8 + 32 06 060.65 + 17.51	05 - 44 - 000	0267	1.7 x 0.10 1.7 x 0.12	25 7	2	Sc 11280 group		16.5:		
11276	18 31.9 + 40 01 068.49 + 20.28	07 - 38 - 011 NGC 6663	1445	1.1 x 0.9 1.0 x 0.9	(140) 1 - 2	3	Sc	SAB(r) ...	14.8		
11277	18 32.0 + 52 49 081.73 + 23.98	09 - 30 - 020	0789	1.1 x 0.10 1.1 x 0.12	147 7	3	Sb		16.0:		1
11278	18 32.1 + 38 34 067.05 + 19.76	06 - 41 - 000	0148 449	1.0 x 0.4 0.9 x 0.4	57 6	2	S0 P w 11281		15.1		*
11279	18 32.2 + 37 09 065.65 + 19.25	06 - 41 - 000	0148	1.0: x 0.9: 0.7: x 0.7:	-- ...	2	Dwarf IV - V or V		17.		
11280	18 32.3 + 32 01 060.61 + 17.38	05 - 44 - 003	0267 447	1.7 x 0.6 1.8 x 0.6	20 6	2	S B? a - b Brightest in group		14.9		*
11281	18 32.4 + 38 33 067.06 + 19.70	06 - 41 - 004	0148 449	1.6 x 0.3 1.6 x 0.3	6 6 - 7	1	Sb P w 11278		14.9		*
11282	18 32.5 + 22 52 051.89 + 13.75	04 - 44 - 003 NGC 6661	0284	1.9 x 1.2 2.1 x 1.3	145 4	4	S0 P w 11274		14.1	+ 4316 + 4554	*12 3
11283	18 32.6 + 49 14 077.98 + 22.17	08 - 34 - 004 IC 1291	0770 453	2.3 x 2.0 1.8: x 1.5:	[30] ...	4	... S B(s)dm? Singular, brightest of 3		14.2		*
11284	18 32.9 + 59 51 089.31 + 25.42	10 - 26 - 044 NGC 6670	0512	[1.0 x 0.4] ...	... ...	2	Triple system Contact, disrupted		15.3		*1C
11285	18 33.1 + 22 28 051.57 + 13.46	04 - 44 - 004	0284	1.6 x 0.4 1.5: x 0.35	42 6	2	Sc - Irr P w 11289		15.3		*
11286	18 33.2 + 66 55 097.07 + 26.57	11 - 22 - 054 NGC 6676	0550	1.6 x 0.3 1.6 x 0.35	142 6	2	Sb - c		15.3		*
11287	18 33.4 + 47 00 075.70 + 22.19	08 - 34 - 009	0770 453	1.8 x 1.2 1.6: x 1.0	(15) ...	2	S ... Disturbed?		15.5		*1
11288	18 33.6 + 67 06 097.28 + 26.56	11 - 22 - 55 + 56 NGC 6677 + 6679	550	[0.6 x 0.2] ...	... ...	1	Double system Bridge, in group of 3		13.6		*1C
11289	18 33.7 + 22 25 051.59 + 13.31	04 - 44 - 005	0284	1.7 x 1.0 1.6: x 0.8	0: 4	1	S ... P w 11285		15.1		*
11290	18 33.7 + 67 04 097.24 + 26.55	11 - 22 - 057 IC 4763	0550	0.9 x 0.35 0.9 x 0.35	103 6	2	S ... Peculiar, 3rd of 3		13.9		*
11291	18 34.1 + 30 47 059.57 + 16.56	05 - 44 - 006	0267	2.1: x 0.5 1.7: x 0.5	116	3	... S d - dm		15.4		
11292	18 34.2 + 52 41 081.68 + 23.62	09 - 30 - 021	0789	(1.0 x 0.8) (1.0 x 0.7)	(135) ...	1	... Compact, jet, disturbed?		14.3		*
11293	18 34.3 + 10 23 040.60 + 07.99	02 - 47 - 004	0184	1.1: x 0.6: 1.1 x 0.6	80: 6	1	... sff of 2		15.		*1
11294	18 34.5 + 19 53 049.31 + 12.08	03 - 47 - 007	0808	(1.4 x 1.0) (1.4 x 1.1)	40: 5 - 6	1	S? ... Halo, in group		15.0		*
11295	18 34.5 + 75 20 106.47 + 27.35	13 - 13 - 000	0776	1.7 x 0.3 1.4 x 0.3	138 7:	2	... d - m		16.5:		
11296	18 35.0 + 30 34 059.44 + 16.30	05 - 44 - 007	0267	1.0 x 0.7 1.0 x 0.7	40: 3:	1	S ... p of 2		15.1		*
11297	18 35.1 + 19 56 049.42 + 11.97	03 - 47 - 000	0808	1.4 x 0.2 1.4: x 0.2	93 6	2	Sb - c In group		15.6		*
11298	18 35.2 + 51 26 080.41 + 23.15	09 - 30 - 025	0789 453	(1.3 x 0.7) (1.4 x 0.8)	(105) ...	3	E		14.9		*
11299	18 35.4 + 26 22 055.47 + 14.57	04 - 44 - 006 NGC 6671	0284	1.9 x 1.9 1.5: x 1.5:	-- 1	2	S ... SA(r) ... Singular		13.8		3
11300	18 35.4 + 70 29 101.05 + 26.83	12 - 17 - 026 NGC 6689 = 6690	0801	4.2 x 1.5 4.6 x 1.4	171 6	3	Sc SA d		12.6		*12 3
11301	18 35.7 + 17 29 047.22 + 10.80	03 - 47 - 009	0808	2.4 x 0.2 2.4 x 0.25	110 7	1	S ...		15.5		1
11302	18 35.7 + 22 03 051.44 + 12.74	04 - 44 - 000	0284	1.0: x 1.0: 0.9: x 0.9:	-- (1)	2	S B c - Irr		16.0:		
11303	18 35.7 + 33 14 062.07 + 17.17	06 - 41 - 000	0148 448	1.3 x 0.7 1.0 x 0.4	118 5	2	Sb		14.8		*
11304	18 35.8 + 36 48 065.57 + 18.46	06 - 41 - 008	0148 449	1.5 x 0.6 1.6 x 0.6	128 6	3	Sa		14.6		
11305	18 35.8 + 40 02 068.77 + 19.58	07 - 38 - 013 NGC 6675	1445	2.0 x 1.4 1.4 x 0.9	130 3	2	Sb - c		13.3		13

N G C   6657, 6658, 6661, 6663, 6670, 6671
6675 - 77, 6679, 6689, 6690

I C   1291, 4763

1	2	3	4	5	6	7	8	9	10	11	12
11306	18 35.8 + 53 43 082.84 + 23.65	09 – 30 – 028	0789	1.1 x 0.15 0.7 x 0.12	24 7	1	Sa – b np of 2		16.0:		*
11307	18 35.9 + 27 45 056.82 + 15.02	05 – 44 – 008	0267	1.5: x 0.4 1.2 x 0.4	50 6	3	S B c		15.2		
11308	18 36.5 + 25 20 054.59 + 13.93	04 – 44 – 007 NGC 6674	0284	4.5 x 2.1 2.4 x 1.3	143 5	4	S B b	S B(r)b Sb +	13.7	+ 3502 + 3748	*12 3 –
11309	18 36.6 + 59 35 089.12 + 24.91	10 – 26 – 046 NGC 6687	0512	1.8: x 1.6: 1.8: x 1.6:	-- 1	4	Sc	SA d	14.9		1
11310	18 37.0 + 22 10 051.68 + 12.51	04 – 44 – 000	0284	1.0 x 0.35 0.9 x 0.3	153 6	1	S ...		16.0:		
11311	18 37.0 + 37 00 065.86 + 18.30	06 – 41 – 009	0148 449	1.6: x 0.5 1.8: x 0.5	131 (6)	1	S: ...		15.4		
11312	18 37.1 + 37 54 066.75 + 18.61	06 – 41 – 010	0148 449	1.2 x 0.7 1.2 x 0.8	5:	1	S0		15.1		*
11313	18 37.7 + 38 52 067.75 + 18.84	06 – 41 – 011	0148	2.3: x 0.35 2.3 x 0.35	23 7	3	Sb		15.5		*
11314	18 38.0 + 21 27 051.11 + 12.00	04 – 44 – 008	0284	1.0 x 0.4 0.9: x 0.4	97 ...	2	Irr		15.7		
11315	18 38.2 + 24 09 053.65 + 13.09	04 – 44 – 009	0284	2.3: x 1.1 1.9: x 1.0	25 5	3	Sb		14.2		*
11316	18 38.3 + 24 37 054.09 + 13.26	04 – 44 – 000	0284	1.0 x 0.4 1.0: x 0.3:	94 6	1	S ...		16.5:		
11317	18 38.3 + 39 57 068.87 + 19.10	07 – 38 – 015 IC 4772	1445	(1.1 x 0.9) 1.0 x 0.8	-- ...	2	E – S0 P w NGC 6685		14.7		*
11318	18 38.3 + 55 35 084.91 + 23.77	09 – 30 – 031 NGC 6691	0789	1.6 x 1.6 1.5 x 1.5	-- 1	4	S B b	S B(rs)bc Sb +	14.1		*
11319	18 38.4 + 40 05 069.00 + 19.13	07 – 38 – 016	1445	1.7: x 0.25 1.3: x 0.25	... ...	1	S ... P w NGC 6686, distorted		16.0:		*
11320	18 38.7 + 23 38 053.21 + 12.77	04 – 44 – 010	0284	1.9 x 0.25 1.8 x 0.25	89 7	2	Sb – c		15.3		*1
11321	18 38.7 + 44 13 073.17 + 20.46	07 – 38 – 000	1445	1.4: x 1.2: 1. x 1.	-- ...	2	Dwarf IV – V		17. – 18.		
11322	18 38.8 + 36 04 065.08 + 17.63	06 – 41 – 013	0148 449	1.1 x 0.35 1.0 x 0.35	61 6	2	Sb – c P w 11325, NGC 6688 group		15.1		*
11323	18 38.9 + 23 02 052.67 + 12.48	04 – 44 – 011	0284	1.6 x 0.6 1.3: x 0.6	85 6	2	S B b – c	S B(s) ...	15.5		*
11324	18 38.9 + 36 14 065.25 + 17.67	06 – 41 – 015 NGC 6688	0148 449	1.6 x 1.6 (1.8 x 1.8)	-- 1	3	S0 Brightest in group	SA 0 +	13.9		*13
11325	18 39.0 + 36 06 065.13 + 17.60	06 – 41 – 014	0148 449	2.3 x 0.5 2.3 x 0.5	136 6 – 7	2	Sb P w 11322, NGC 6688 group		14.6		*1
11326	18 39.2 + 08 05 039.08 + 05.87	01 – 47 – 001	0166	2.0 x 0.9 2.0: x 0.8:	120: ...	0	...		15.		*
11327	18 39.2 + 36 19 065.35 + 17.65	06 – 41 – 016	0148 449	1.0 x 0.3 0.9 x 0.3	57 6	1	S ... NGC 6688 group		15.1		*1
11328	18 39.2 + 37 57 066.95 + 18.24	06 – 41 – 000	0148 449	1.3 x 0.12 1.3 x 0.12	154 7	3	Sc		16.0:		
11329	18 39.5 + 45 13 074.23 + 20.64	08 – 34 – 013	0770	1.9 x 0.7 1.4: x 0.7	17 6	2	Sb – c		15.5		
11330	18 39.9 + 34 47 063.92 + 16.95	06 – 41 – 018 NGC 6692	0148 449	1.1 x 0.7 1.1 x 0.7	110: ...	1	...		14.3		
11331	18 40.0 + 73 34 104.53 + 26.82	12 – 17 – 028	0801	1.5: x 1.1: 1.4:x 1.1:	[30] ...	3	Dwarf (spiral) S: IV – V, 11332 group		15.4		*
11332	18 40.4 + 73 32 104.50 + 26.79	12 – 17 – 029	0801	2.5 x 0.9 2.5 x 0.9	63 ...	2	... [s dm]? Brightest of 3		13.6		*
11333	18 40.5 + 32 19 061.59 + 15.89	05 44 – 000	0267	1.4 x 0.5 0.9 x 0.3	151 6	1	S ...		15.4		
11334	18 40.5 + 73 48 104.80 + 26.82	12 – 18 – 000	1149	2.1: x 0.4 1.1: x 0.3	105 7?	1	... Peculiar?, 11332 group:		16.0:		
11335	18 40.7 + 50 11 079.37 + 21.97	08 – 34 – 015	0770 453	1.1 x 0.15 1.0 x 0.15	144 7	3	Sb		16.0:		
11336	18 40.8 + 34 25 063.63 + 16.64	06 – 41 – 019	0148	1.0 x 0.7 1.1 x 0.6	50 ...	3	Sc		15.5		
11337	18 40.9 + 18 40 048.85 + 10.20	03 – 48 – 001	0287	1.6 x 1.5 (1.6 x 1.6)	-- 1	3	S B a	(R)S B a	15.		
11338	18 40.9 + 35 34 064.76 + 17.05	06 – 41 – 000	0148 449	0.50 x 0.45 0.50 x 0.45	-- ...	1	... Peculiar		14.5		
11339	18 41 + 86 52 119.47 + 27.46	14 – 08 – 000	0775	1.0 x 0.8 (1.2 x 1.1)	-- ...	1	E – S0		15.4		
11340	18 41.1 + 40 20 069.44 + 18.73	07 – 38 – 018 NGC 6695	1445	1.1 x 0.7 1.1 x 0.7	12 4:	4	S B b	S B(s)b	14.3		

N G C   6674, 6687, 6688, 6691, 6692, 6695   I C   4772

1	2	3	4	5	6	7	8	9	10	11	12
11341	18 41.2 + 39 30 068.63 + 18.42	07 – 38 – 019	1445	1.5 x 0.4 1.2 x 0.2	39 6	2	Sc		16.5:		
11342	18 41.6 + 70 30 101.15 + 26.32	12 – 17 – 030	0801	1.8 x 0.9 0.5 x 0.4	132 5:	1	...		15.5		
11343	18 41.8 + 52 59 082.32 + 22.60	09 – 30 – 034	0789	1.2 x 0.9 0.8 x 0.4	(75) 2 – 3	1	S ...		15.5		
11344	18 42.1 + 24 05 053.96 + 12.25	04 – 44 – 012	0284	2.2 x 1.1 1.3 x 0.6	160 5	2	Sa – b		14.8		
11345	18 42.1 + 58 51 088.51 + 24.06	10 – 26 – 049	0512	1.1 x 0.15 1.2 x 0.15	71 7	2	Sb – c		16.0:		
11346	18 42.2 + 25 12 055.01 + 12.70	04 – 44 – 013	0284	1.1 x 0.8 1.1 x 0.8	77 3:	2	S B b		15.6		
11347	18 42.2 + 56 08 085.64 + 23.38	09 – 30 – 035	0789	1.0 x 0.2 0.9 x 0.1	161 7	3	Sc		16.5:		
11348	18 42.6 + 60 37 090.41 + 24.40	10 – 26 – 050 NGC 6701	0512	1.6 x 1.2 1.6 x 1.2	(25) 2:	4	S B a	(R´)S B(s)a	12.9		*
11349	18 43.2 + 25 27 055.34 + 12.59	04 – 44 – 014 NGC 6697	0284	(1.2 x 1.0) (1.2 x 1.0)	— ...	2	E		14.5		3
11350	18 44.1 + 22 34 052.76 + 11.19	04 – 44 – 015	0284	1.5: x 1.3: 1.3 x 0.9	165: ...	4	Sc		15.7		*
11351	18 44.2 + 32 13 061.81 + 15.13	05 – 44 – 010 NGC 6700	0267	1.5 x 1.0 1.6 x 1.0	115: 3	3	S B c	S B(rs)c	14.2		
11352	18 44.8 + 48 47 078.15 + 20.91	08 – 34 – 000	0770 453	1.1 x 0.4 1.0 x 0.35	152 6	2	S B b – c		15.4		
11353	18 45.6 + 23 17 053.57 + 11.19	04 – 44 – 016	0284	1.2 x 0.25 1.3 x 0.25	153 7	2	S0 – a		15.0		
11354	18 45.6 + 45 40 075.05 + 19.78	08 – 34 – 019 NGC 6702	0770 453	(1.9 x 1.5) (2.1 x 1.7)	(65) ...	3	E P w 11356		13.8	+ 4726 + 5003	*12 3
11355	18 45.8 + 22 53 053.22 + 10.97	04 – 44 – 017	0284	1.7: x 0.7 1.5: x 0.7	126 6	1	S ...		15.4		
11356	18 45.9 + 45 30 074.90 + 19.67	08 – 34 – 020 NGC 6703	0770 453	(2.3 x 2.3) (2.5 x 2.5)	— 1	4	S0 P w 11354		12.4	+ 2333 + 2609	*12 3
11357	18 46.5 + 45 22 074.81 + 19.53	08 – 34 – 024	0770 453	1.6:: x 0.3 1.1: x 0.2	126 7	2	...	d – m	16.5:		
11358	18 46.8 + 31 00 060.88 + 14.14	05 – 44 – 000	0267	1.1 x 0.4 1.1 x 0.4	148 (6)	1	S? ...		15.6		
11359	18 47 + 81 21 113.27 + 27.14	14 – 09 – 000	0802	1.4: x 1.2: 1.4: x 1.2:	— ...	1	...		16.0:		
11360	18 47.3 + 18 39 049.50 + 08.82	03 – 48 – 000	0287	1.1: x 1.0: 1.0: x 1.0:	— ...	2	Irr		17.		
11361	18 47.6 + 47 36 077.11 + 20.09	08 – 34 – 025 NGC 6711	0770 453	1.1 x 1.0 1.1 x 1.0	— 1	2	S B: b – c Singular?		14.1		
11362	18 47.8 + 23 12 053.71 + 10.69	04 – 44 – 018	0284	1.0 x 0.6 0.7: x 0.6:	(20) ...	2	... S d – m n of 2, disturbed?		16.5:		*1
11363	18 48.0 + 70 41b 101.44 + 25.82	12 – 18 – 000	1149	1.2 x 0.9 (1.2 x 0.9)	(115) (2)	2	S0 Companion superimposed		(14.9)		*
11364	18 48.6 + 26 46 057.09 + 12.04	04 – 44 – 019 NGC 6710	0284	1.6 x 0.9 1.8 x 0.9	40: 4:	1	Sa?		14.6	+ 4556 + 4810	*12 3
11365	18 48.9 + 33 53 063.80 + 14.89	06 – 41 – 000 NGC 6713	0148	0.4 x 0.35 0.5 x 0.35	— ...	1	... Peculiar		14.2		3
11366	18 49 + 81 40 113.63 + 27.09	14 – 09 – 000	0802	1.2 x 0.1 1.0 x 0.12	158 7	1	Sb – c In group		16.5:		*
11367	18 49.0 + 35 18 065.15 + 15.42	06 – 41 – 000	0148	[1.7 x 0.6] ...	... ...	3	S B ... + two companions Connected, distorted		16.0:		*
11368	18 49.4 + 26 25 056.84 + 11.73	04 – 44 – 020	0284	1.6 x 0.6 1.6 x 0.6	142 6	2	S0 – a Brightest in group		15.1		*1
11369	18 49.5 + 23 34 054.22 + 10.50	04 – 44 – 021	0284	1.1 x 0.6 1.0 x 0.6	158 5:	2	S B a:		15.2		
11370	18 49.7 + 26 29 056.93 + 11.70	04 – 44 – 022	0284	1.5 x 0.6 1.4 x 0.6	48 6	2	... S d – m 11367 group		15.7		*1
11371	18 49.9 + 26 25 056.89 + 11.63	04 – 44 – 023	0284	1.9 x 0.7 1.8 x 0.5	44 6	1	S ... 11367 group		15.5		*1
11372	18 50.7 + 33 46 063.84 + 14.50	06 – 41 – 000	0148	1.3 x 0.2 1.7 x 0.3	154 7	2	S ...		15.0		*
11373	18 50.7 + 58 02 087.96 + 22.76	10 – 27 – 003	0830	1.5:: x 0.45 1.0 x 0.45	146 6	2	Sb		14.8		
11374	18 51.4 + 33 00 063.17 + 14.05	06 – 41 – 022 IC 1296	0148	1.2 x 0.9 1.3 x 1.0	80: 2 – 3	3	S B b	Sb +	15.4		
11375	18 52.2 + 24 35 055.42 + 10.38	04 – 44 – 024	0284	1.3 x 0.7 1.3 x 0.7	150 5	3	Sc		15.6		

N G C   6697, 6700 – 03, 6710, 6711, 6713   I C   1296

1	2	3	4	5	6	7	8	9	10	11	12
11376	18 52.9 + 48 51 078.68 + 19.67	08 – 34 – 027	0770 453	0.7 x 0.45 0.7 x 0.45	85 ...	1	S ...		14.3		
11377	18 53.4 + 73 07 104.18 + 25.81	12 – 18 – 000	1149 451	1.6 x 0.15 1.6 x 0.15	10 7	3	Sc		15.6		*
11378	18 54.5 + 52 05 082.02 + 20.49	09 – 31 – 000	0542	1.1 x 0.15 0.7 x 0.15	[167] ...	1	... Peculiar		15.5		
11379	18 54.6 + 25 10 056.20 + 10.14	04 – 44 – 025	0284	1.0 x 0.45 0.9 x 0.4	93 6	2	Sc		15.3		
11380	18 55.1 + 36 33 066.85 + 14.78	06 – 41 – 023	0148	1.6 x 0.6 1.6 x 0.6	172 6	2	Sa – b		13.8		*1
11381	18 55.2 + 52 18 082.27 + 20.45	09 – 31 – 011 NGC 6732	0542	(0.9 x 0.6) (0.8 x 0.6)	100: ...	1	E? Companion superimposed		14.4		*
11382	18 55.2 + 74 36 105.83 + 25.93	12 – 18 – 000	1149	1.2 x 0.7 1.2: x 0.7:	100: 4	3	Irr		16.5:		
11383	18 55.4 + 59 03 089.20 + 22.45	10 – 27 – 000	0830	1.1 x 0.7 0.2 x 0.2	(20) 4:	3	Sc 2nd of 3 Sc +		16.5:		*
11384	18 57.12 + 45 42 075.83 + 17.89	08 – 34 – 000	0770 453	1.0 x 0.12 1.0 x 0.12	70 7	2	Sb – c In group of 5		16.5:		
11385	18 57.4 + 19 22 051.22 + 07.00	03 – 48 – 002	0287	1.2 x 0.8 1.1 x 0.9	25: ...	2	Sc – Irr		16.0:		
11386	18 58.2 + 73 13 104.36 + 25.49	12 – 18 – 000	1149	1.3: x 1.2: 0.8 x 0.5	-- 1:	3	...	S d – dm	16.0:		
11387	18 58.3 + 42 13 072.52 + 16.41	07 – 39 – 000	0340	1.0 x 0.08 1.0 x 0.1	130 7	2	Sc		17.		
11388	18 58.8 + 28 42 059.85 + 10.82	05 – 45 – 001	0153	1.0 x 0.8 1.0 x 0.8	30: 2:	1	S ...		15.5:		
11389	18 59.9 + 59 06 089.43 + 21.91	10 – 27 – 006 NGC 6750	0830	1.0 x 0.6 1.0 x 0.6	5: 4	2	S ... Singular		13.7		*
11390	19 00 + 84 30 116.83 + 27.00	14 – 09 – 001	0802	1.4 x 0.2 1.4 x 0.2	95 7	3	Sc		16.5:		
11391	19 00.0 + 40 41 071.16 + 15.52	07 – 39 – 000 NGC 6745	0340	[1.6 x 0.7] ...	... ...	2	Triple system Bridges, disrupted		13.3		*
11392	19 00.5 + 34 43 065.58 + 13.03	06 – 42 – 000	1434	1.0: x 0.2 1.0: x 0.2	83 (7)	1	Irr		17.		
11393	19 00.7 + 27 14 058.69 + 09.81	05 – 45 – 002	0153	1.3: x 1.3: 1.3 x 1.3	-- 1	3	Sc		15.5:		
11394	19 01.5 + 27 32 059.04 + 09.78	05 – 45 – 003	0153	2.0 x 0.2 1.8 x 0.2	37 7	3	Sc		16.0:		
11395	19 01.5 + 71 42 102.77 + 24.94	12 – 18 – 000	1149	1.0 x 0.10 1.0 x 0.10	159 7	2	Sc		16.5:		
11396	19 02.0 + 24 17 056.14 + 08.24	04 – 45 – 000	0149	1.0:: x 0.5:: 0.8:: x 0.3::	... ...	0	...		17.		
11397	19 02.0 + 33 45 064.81 + 12.35	06 – 42 – 000	1434	1.2 x 0.7 1.2 x 0.8	95 3	3	S B a		14.0		
11398	19 02.0 + 54 00 084.34 + 20.04	09 – 31 – 015	0542	1.0: x 0.9: 0.9 x 0.9	-- 1 – 2	3	Sc		16.5:		
11399	19 03.1 + 34 22 065.48 + 12.40	06 – 42 – 000	1434	1.6 x 0.25 1.5 x 0.25	124 6	1	S ...		14.7		*
11400	19 03.2 + 78 58 110.71 + 26.18	13 – 13 – 000	0776	1.1 x 0.6 0.8 x 0.4	... ...	3	Irr		16.5:		
11401	19 04.1 + 55 39 086.11 + 20.30	09 – 31 – 019 NGC 6757	0542	1.6 x 1.1 1.8 x 1.2	105: 3	4	S B 0/S B a	S B(r)0/a	14.0		1
11402	19 04.2 + 72 58 104.18 + 25.01	12 – 18 – 000	1149	1.1 x 1.1 1.0: x 1.0:	-- 1	1	S ...		15.2		*
11403	19 04.5 + 40 49 071.65 + 14.80	07 – 39 – 009	0340 453	1.4 x 1.1 1.4 x 1.0	(55) 2:	1	Sb? Disturbed?		15.7		*1
11404	19 05.0 + 28 56 060.66 + 09.70	05 – 45 – 004	0153	2.1 x 1.4 2.2 x 1.4	60 3	3	Sb	Sb +	14.0:		1
11405	19 05.2 + 63 68 094.56 + 22.68	11 – 23 – 000 NGC 6762 = 6763	1090	1.6 x 0.4 1.6 x 0.3	119 6 – 7	2	S0 – a		14.2		
11406	19 06.8 + 43 00 073.90 + 15.28	07 – 39 – 013	0340 453	1.6 x 1.1 2.0 x 1.2	(15) ...	3	S B ... Singular	S B(s) ...	14.7		*1
11407	19 07.0 + 50 50 081.48 + 18.23	08 – 35 – 003 NGC 6764	0814 453	2.3 x 1.4 2.3 x 1.4	62 	3	S B b	S B(s)bc	13.2		*
11408	19 08.2 + 41 30 072.59 + 14.44	07 – 39 – 016	0340 453	1.4 x 0.5 1.3 x 0.5	106 6	2	Sa – b		14.9		
11409	19 08.9 + 37 33 068.95 + 12.68	06 – 42 – 000	1434	1.2: x 0.2 1.2 x 0.2	30 6	1	S ...		16.5:		
11410	19 08.9 + 73 00 104.50 + 23.97	12 – 18 – 000	1149	1.2 x 0.30 1.2 x 0.30	135 6	2	S B: a		15.7		

N G C   6732, 6745, 6750, 6757, 6762 – 64

1	2	3	4	5	6	7	8	9	10	11	12
11411	19 09.0 + 70 12 101.32 + 24.00	12 - 18 - 000	1149	0.9 x 0.7 0.7 x 0.6	[150] ...	1	Peculiar Symbiotic pair?		14.4		*
11412	19 09.6 + 52 05 082.86 + 18.31	09 - 31 - 025	0542 453	1.0 x 0.5 1.2 x 0.7	165 5	2	Sb - c		14.6		
11413	19 11.3 + 53 06 083.96 + 18.43	09 - 31 - 026	0542 453	1.3 x 0.9 0.8 x 0.7	95: 2 - 3	3	Sc		15.6		
11414	19 11.9 + 73 18 104.68 + 24.54	12 - 18 - 000 NGC 6786	1149	1.2 x 1.0 1.2 x 1.0	40: ...	2	S B ... P w 11415, distorted		13.7		*C
11415	19 12.0 + 73 19 104.70 + 24.54	12 - 18 - 000	1149	1.3 x 0.9 1.3: x 0.8	15 ...	1	S ... P w 11414, distorted		15.1		*C
11416	19 12.2 + 50 46 081.73 + 17.44	08 - 35 - 005	0814 453	1.1: x 0.7 1.0: x 0.7:	130: 4	2	Sc		16.0:		
11417	19 12.4 + 29 54 062.26 + 08.68	05 - 45 - 000	0153	1.1: x 0.7 1.0 x 0.8	25: 2:	3	Sc		16.0:		
11418	19 12.5 + 40 05 071.61 + 13.11	07 - 39 - 018	0340 453	1.0 x 0.8 1.1: x 0.8:	(70) ...	2	Sb - c		15.7		1
11419	19 13.2 + 53 40 084.62 + 18.38	09 - 31 - 028	0542 453	1.0: x 0.3 0.8 x 0.25	101 6	1	Sa - b		15.6		*
11420	19 13.9 + 42 49 074.27 + 14.01	07 - 39 - 019	0340 453	1.5 x 0.35 1.5 x 0.3	5 6	2	Sb		15.1		
11421	19 14.0 + 44 01 075.41 + 14.49	07 - 39 - 020	0340 453	1.1 x 1.1 1.1 x 1.1	-- ...	1	[S B 0]? Peculiar?		14.6		
11422	19 14.2 + 43 21 074.80 + 14.19	07 - 39 - 021	0340 453	1.2 x 0.12 1.2 x 0.12	116 7	2	Sc		15.7		
11423	19 14.7 + 77 50 109.59 + 25.40	13 - 14 - 002	0832 450	1.1 x 0.7 1.0 x 0.5	[65] ...	2	... Peculiar, loop		15.4		*
11424	19 15.5 + 60 19 091.34 + 20.43	10 - 27 - 009 NGC 6787	0830	1.2 x 1.1 1.1 x 1.0	-- 1:	3	Sb/S B b Sb +	SAB(s)b	14.9		
11425	19 16.3 + 63 54 094.99 + 21.52	11 - 23 - 001 NGC 6789	1090	1.8: x 1.8:: 1.4: x 1.4::	---- ...	4	Irr	I m Ir I	13.7		
11426	19 16.5 + 34 45 067.05 + 10.08	06 - 42 - 000	1434	0.9 x 0.7 0.9 x 0.7	55: 2	1	S ...		14.4		
11427	19 16.8 + 72 40 104.11 + 24.04	12 - 18 - 001	1149	1.1 x 0.2 1.1 x 0.2	69 (6)	1	S: ...		14.7		
11428	19 18.5 + 30 44 063.62 + 07.89	05 - 45 - 006	0153	1.2 x 1.2 1.2 x 1.1	-- 1	3	Sc		14.5:		
11429	19 19.4 + 43 02 074.91 + 13.19	07 - 40 - 002 NGC 6792	0281 453	2.4 x 1.3 2.4 x 1.4	25	3	S B b Disturbed?, p w 11430		13.4		*
11430	19 19.6 + 43 14 075.11 + 13.25	07 - 40 - 003	0281 453	1.1 x 1.1 1.1 x 1.1	-- 1	4	Sc P w 11429	SA(s)c	14.4		*
11431	19 19.8 + 73 42 105.26 + 24.10	12 - 18 - 000	1149	1.2 x 0.3 1.3 x 0.3	44 6 - 7	1	Sa - b sp of 2		16.0:		*
11432	19 20.8 + 61 03 092.31 + 20.08	10 - 27 - 010 NGC 6796	0830	2.1 x 0.45 1.8 x 0.35	179 6	1	S ...		13.5		*2
11433	19 22.0 + 34 42 067.52 + 09.05	06 - 42 - 000	1434	1.0 x 0.1 1.0 x 0.1	25 7	1	S ...		15.6		
11434	19 22.8 + 53 32 085.06 + 17.01	09 - 32 - 002 NGC 6798	0774 453	1.6 x 0.9 1.6 x 0.9	150: 4	3	S0		14.5		
11435	19 22.8 + 55 53 087.33 + 17.93	09 - 32 - 001	0774 453	2.0 x 0.35 2.0 x 0.35	82 6	3	Sc		15.1		
11436	19 23.6 + 70 16 101.77 + 22.84	12 - 18 - 000	1149	(1.2 x 1.1) 1.5 x 1.1	-- ...	1	S B ...		15.5		
11437	19 25.2 + 50 01 081.88 + 15.24	08 - 35 - 010 IC 4867	0814 453	1.0 x 0.4 1.0 x 0.35	17 6	2	S0 - a sp of 2		14.3		*
11438	19 25.3 + 49 39 081.54 + 15.07	08 - 35 - 009 IC 1301	0814 453	1.3: x 0.9 1.1 x 0.8	155: 3:	4	Sc		14.7		
11439	19 25.6 + 43 46 076.09 + 12.50	07 - 40 - 007	0281 453	1.0: x 0.5 1.0 x 0.6:	159 ...	1	S ... sp of 2		14.9		*
11440	19 25.8 + 65 57 094.43 + 21.25	11 - 23 - 000	1090	1.0 x 0.25 1.0 x 0.25	166 6	3	Sc		17.		
11441	19 26.0 + 80 55 113.05 + 25.59	13 - 14 - 000	0832	1.1 x 0.3 1.2 x 0.2	28 7	2	Sa - b		16.0:		
11442	19 26.3 + 54 43 086.41 + 17.01	09 - 32 - 004	0774 453	1.0 x 0.5 1.0 x 0.5	49 5	2	Sa - b		16.0:		
11443	19 26.5 + 54 16 085.99 + 16.80	09 - 32 - 005 NGC 6801	0774 453	1.4 x 0.7 1.5 x 0.8	44 5	4	Sc	SA cd	14.8		
11444	19 26.9 + 52 47 084.61 + 16.15	09 - 32 - 000	0774 453	1.0 x 0.8 0.9 x 0.8	(170) 1	4	Sc	SAB(s)c	15.3		*
11445	19 27.5 + 56 35 088.28 + 17.60	09 - 32 - 006	0774 453	1.1 x 0.55 1.0 x 0.5	177 ...	1	S ...		16.0:		

N G C 6786, 6787, 6789, 6792, 6796, 6798
6801

I C 1301, 4867

1	2	3	4	5	6	7	8	9	10	11	12
11446	19 27.9 + 41 13 073.95 + 10.98	07 – 40 – 008	0281 453	1.4 x 1.4 1.4 x 1.1	-- 1	3	S B ... (R´)SAB(s)...		14.2		
11447	19 27.9 + 50 18 082.33 + 14.97	08 – 35 – 000	0814 453	1.0: x 0.1 0.9 x 0.1	10 7	2	Sc		17.		
11448	19 28.8 + 35 40 069.04 + 08.26	06 – 43 – 001	0530	1.2 x 0.9 0.9 x 0.8	-- ...	2	Sc P w IC 1302, IC 1302 group		16.0:		*
11449	19 28.9 + 64 50 096.43 + 20.56	11 – 23 – 000	1090	1.0 x 0.2 0.9 x 0.12	126 7	2	Sc		16.5:		
11450	19 29.5 + 46 36 079.01 + 13.13	08 – 35 – 013	0814 453	1.0 x 0.6 1.0 x 0.6	20: 4	2	S B: a – b		15.1		
11451	19 29.6 + 42 06 074.90 + 11.10	07 – 40 – 000	0281	1.5 x 0.15 1.4: x 0.1	92 7	2	Sc		16.5:		*
11452	19 29.7 + 35 46 069.22 + 08.15	06 – 43 – 004 IC 1303	0530	1.4 x 0.8 1.3 x 0.7	(115) 4	3	Sc IC 1302 group		15.0:		*12
11453	19 29.9 + 54 00a 085.95 + 16.24	09 – 32 – 007	0774 453	1.7 x 1.3 1.7 x 1.3:	80: 2	3	Sb Companion attached, disturbed		(14.4)		*1
11454	19 30.2 + 49 45 081.98 + 14.40	08 – 35 – 014	0814 453	1.0 x 0.5 0.8 x 0.4	115 5:	2	... S dm – m		17.		
11455	19 30.5 + 72 00 103.74 + 22.85	12 – 18 – 003	1149	2.8 x 0.5 2.7 x 0.5	61 6 – 7	2	Sc sff of 2		15.3		*
11456	19 31.0 + 37 46 071.13 + 08.86	06 – 43 – 005	0530	1.4: x 0.5 1.3 x 0.5	37 6	1	S ...		15.5:		
11457	19 33.3 + 52 59 085.21 + 15.35	09 – 32 – 000	0774 453	1.3: x 0.5: 0.7 x 0.5	[113] ...	1	... d – m		17.		
11458	19 35.1 + 69 54 101.74 + 21.79	12 – 18 – 000	1149	1.0: x 0.9: 1. x 1.	-- ...	2	Dwarf irregular Ir V		17.		
11459	19 35.7 + 40 36 074.09 + 09.39	07 – 40 – 011	0281	2.4: x 1.5 2.3 x 1.0	12 4	4	Sc Sc +		14.4		
11460	19 36.2 + 40 53 074.39 + 09.44	07 – 40 – 012	0281	1.4 x 0.8 1.2 x 0.8	120: ...	2	E – S0		14.0		
11461	19 36.7 + 08 42 046.28 – 06.42	01 – 50 – 000	0171	1.3 x 0.2 1.3 x 0.2	160 7	2	Sc		15.5		
11462	19 36.8 + 71 14 103.14 + 22.12	12 – 14 – 000	1149	1.1 x 0.6 0.9 x 0.5	173 5:	1	S ...		16.0:		
11463	19 37.6 + 61 29a 093.54 + 18.37	10 – 28 – 007	0545	1.0 x 1.0 0.5 x 0.5	-- 1	3	Sb np of 2		(15.5)		*1
11464	19 40.0 + 51 43 084.51 + 13.88	09 – 32 – 010	0774 453	1.5: x 0.5 0.9 x 0.5	... ...	2	S ... [+ companion?] Distorted, streamer, f of 2		15.2		*
11465	19 40.4 + 50 32 083.45 + 13.28	08 – 36 – 002	0283 453	(1.2 x 1.2) (1.5 x 1.5)	-- ...	2	...		14.4		*
11466	19 41.4 + 45 11 078.69 + 10.64	08 – 36 – 000	0283	2.2 x 1.3 2.1 x 1.3	[35] ...	3	... Peculiar		13.5		
11467	19 41.6 + 41 55 075.80 + 09.04	07 – 40 – 014	0281	1.3: x 0.9: 0.9: x 0.7:	[100] ...	2	Irr P w 11468		15.5		*
11468	19 41.7 + 41 49 075.72 + 08.98	07 – 40 – 016	0281	1.0 x 1.0 1.0 x 0.9	-- 1	1	S ... P w 11467		15.3		*
11469	19 42.5 + 47 53 081.21 + 11.75	08 – 36 – 000	0283 453	1.2 x 1.1 1.2 x 1.1	-- 1	3	S B a P w 11471		16.0:		*
11470	19 42.6 + 55 59 088.63 + 15.45	09 – 32 – 012 NGC 6824	0774 453	(2.6 x 1.7) (2.0 x 1.5)	60: 2	1	Sa – b		13.1	+ 3386 + 3676	*12 3
11471	19 43.2 + 47 53 081.27 + 11.65	08 – 36 – 005	0283 453	1.3 x 0.2 1.2 x 0.25	121 6 – 7	1	Sa – b P w 11469		16.0:		*
11472	19 43.3 + 72 56 105.05 + 22.26	12 – 18 – 000	1149	1.3: x 1.2: 1.2: x 1.2:	-- 1	2	S B b		16.0:		
11473	19 43.9 + 43 01 076.98 + 09.20	07 – 40 – 018	0281	2.2: x 1.3: 1.5: x 1.3:	125: 4:	3	Sc Sc +		14.6		
11474	19 44.9 + 48 55 082.33 + 11.89	08 – 36 – 000	0283 453	1.0 x 0.15 1.0 x 0.15	28 7	1	Sa – b		16.5:		
11475	19 45.8 + 59 35 092.18 + 16.65	10 – 28 – 009	0545	1.6: x 0.8: 0.8 x 0.5	85: ...	2	S B c NGC 6831 group, Sc +		15.4		
11476	19 45.8 + 67 51 100.09 + 20.13	11 – 24 – 001	0777	(1.6 x 0.7) 1.6 x 0.7	140: 6	1	S ...		15.6		
11477	19 46.0 + 56 47 089.59 + 15.38	09 – 32 – 016	0774 453	1.4 x 0.2 1.3 x 0.2	87 7	3	Sc Sc +		15.4		
11478	19 46.3 + 59 47 092.40 + 16.69	10 – 28 – 010 NGC 6829	0545	1.6 x 0.35 1.7 x 0.35	31 7	3	Sb NGC 6831 group		15.0		*C
11479	19 46.6 + 78 14 110.52 + 23.92	13 – 14 – 000	0832 454	1.1 x 0.8 0.9 x 0.7	18: 3	3	Sc		16.0:		*
11480	19 46.8 + 46 18 080.15 + 10.35	08 – 36 – 000	0283	1.2 x 0.9 1.2 x 0.9	165: ...	3	Sc Disturbed?		15.5		

N G C 6824, 6829

1	2	3	4	5	6	7	8	9	10	11	12
11481	19 46.9 + 04 02 043.40 – 10.89	01 – 50 – 000	0171	1.1 x 0.35 1.1 x 0.35	127 6	1	S ...		15.1		
11482	19 46.9 + 07 02 046.06 – 09.43	01 – 50 – 001	0171	2.0 x 0.8 1.5: x 0.8:	[103] ...	3	Irr		15.7		
11483	19 47.1 + 59 46 092.43 + 16.59	10 – 28 – 011 NGC 6831	0545	1.6 x 1.6 (1.5 x 1.5)	— 1	4	S0 Brightest in group		14.7		*1
11484	19 47.2 + 54 05 087.19 + 14.00	09 – 32 – 017	0774 453	1.1 x 0.4 1.0 x 0.4	172 6	2	Sc – Irr		16.5:		
11485	19 47.6 + 50 11 083.68 + 12.10	08 – 36 – 000	0283 453	1.4 x 0.7 1.4 x 0.7	25 5	3	S B a	(R)S B a	14.5		
11486	19 48.5 + 50 35 084.11 + 12.17	08 – 36 – 000	0283 453	1.0 x 0.5 1.0 x 0.5	75 5	1	S ...		15.5		
11487	19 48.8 + 63 23 095.93 + 17.99	11 – 24 – 002	0777	1.4: x 1.1 1.2: x 0.9	(45) ...	2	S B b	Sb +	15.4		
11488	19 49.0 + 03 26 043.12 – 11.64	01 – 50 – 000	0171	1.3 x 1.0 0.7: x 0.6:	25: 2	2	Sb – c		15.6		
11489	19 49.6 + 04 40 044.29 – 11.18	01 – 50 – 002	0171	1.5: x 1.5: 1.1 x 1.0	— 1	2	Sc		15.5		
11490	19 50.8 + 57 52 090.89 + 15.30	10 – 28 – 000	0545	1.2: x 0.6: 0.7: x 0.5:	(170) ...	2	Dwarf: d – m		17.		
11491	19 51.4 + 58 51 091.84 + 15.68	10 – 28 – 013	0545	1.2 x 0.3 1.1 x 0.2	162 6	3	Sc		16.0:		*
11492	19 51.6 + 57 20 090.45 + 14.96	10 – 28 – 015	0545 453	2.1 x 1.1 1.8 x 1.0	70 5:	4	S B b/Sc	SAB(s)bc Sc –	14.2		
11493	19 52.5 + 05 45 045.62 – 11.28	01 – 50 – 003	0171	2.3: x 1.6: 1.4: x 0.9:	170 3	3	Sc		15.2		
11494	19 52.6 + 49 48 083.74 + 11.22	08 – 36 – 007	0283 453	[1.9: x 0.9:] ...	... ...	2	Double system Contact, distorted		15.3		*1
11495	19 53 + 87 09 119.94 + 26.58	15 – 01 – 024	0570	1.3: x 1.3: 1.3: x 1.3:	— 1	3	Sc		16.0:		
11496	19 53.0 + 67 33 100.10 + 19.38	11 – 24 – 000	0777	2.3: x 2.3: 2.3: x 2.3:	— 1:	3	Dwarf spiral S V		17.		
11497	19 53.1 + 02 02 042.37 – 13.21	00 – 51 – 000	0297	1.2 x 0.9 0.5 x 0.3	(80) ...	3	S B b		14.9		
11498	19 54.8 + 05 45 045.90 – 11.77	01 – 51 – 001	0805	3.4 x 1.1 3.0 x 1.0	75 6	3	S B: b		14.6		
11499	19 55.4 + 49 45 083.92 + 10.80	08 – 36 – 008	0283 453	1.2 x 0.8: 1.1: x 0.5:	[150] ...	1	Sc – Irr		15.7		
11500	19 55.6 + 50 03 084.20 + 10.92	08 – 36 – 011	0283 453	1.0 x 0.9 1.0 x 0.9	— 1	3	Sc Brightest of 4		15.5		*
11501	19 56.1 + 02 28 043.13 – 13.66	00 – 51 – 003	0297	1.5 x 1.4 1.3: x 1.3:	— 1	4	Sb Disturbed?, streamer?	SAB(rs)bc	14.5		*
11502	19 56.5 + 52 44 086.65 + 12.13	09 – 32 – 018	0774	(1.0 x 0.9) (1.0 x 0.8)	— ...	1	... Compact, halo		15.3		*
11503	19 57.2 + 49 54 084.20 + 10.62	08 – 36 – 012	0283 453	0.6 x 0.6 0.7 x 0.7	— ...	1	...		14.5		*
11504	19 59.7 + 07 39 048.21 – 11.88	01 – 51 – 000	0805	1.1: x 0.8: 0.8:: x 0.6::	... ...	1	Dwarf:		17.		
11505	19 59.9 + 01 30 042.73 – 14.96	00 – 51 – 005	0297	1.8 x 0.6 1.6 x 0.6	163 6	1	S ...		15.2		
11506	20 00.3 + 66 05 099.05 + 18.09	11 – 24 – 004 NGC 6869	0777	(1.3 x 1.3) (1.6 x 1.6)	— (1)	1	S0		12.8		1
11507	20 00.4 + 49 00 083.69 + 09.72	08 – 36 – 015	0283 453	1.0 x 0.8 1.0 x 0.8	75: 2:	3	Sc		15.6		
11508	20 00.7 + 53 39 087.79 + 12.05	09 – 33 – 002	0543	1.1 x 0.9 0.9 x 0.8	105: 1 – 2	3	Sc P w 11510		16.0:		*1
11509	20 00.7 + 56 52 090.64 + 13.65	09 – 33 – 001	0543	1.0 x 0.3 0.9 x 0.3	175 6	2	Sb – c		15.2		
11510	20 01.6 + 53 44 087.93 + 11.97	09 – 33 – 003	0543	2.1 x 0.7 2.0: x 0.7	170 6	4	Sc P w 11508, Sc +		15.6		*
11511	20 01.8 + 07 16 048.13 – 12.52	01 – 51 – 000	0805	1.8: x 1.6: 1. x 1.	— 1	2	Sc		17.		
11512	20 02.2 + 12 35 052.85 – 09.89	02 – 51 – 000	0325	1.1 x 0.5 1.1 x 0.4	166 6	1	S B? ...		14.4		
11513	20 02.4 + 13 58 054.07 – 09.22	02 – 51 – 000	0325	1.3 x 1.2 1.2 x 1.2	— 1	4	S B c	S B(s)c: Sc +	15.7		
11514	20 03.5 + 55 32 089.66 + 12.65	09 – 33 – 000	0543	1.4: x 0.8 1.2 x 0.8	125 ...	1	...		16.5:		
11515	20 04.2 + 62 38 096.07 + 16.09	10 – 28 – 016	0545	1.6 x 1.6 1.6 x 1.5	— 1	4	Sc	Sc +	14.8		1

N G C   6831, 6869

1	2	3	4	5	6	7	8	9	10	11	12
11516	20 05.3 - 00 23 041.69 - 17.05	00 - 51 - 008	0297	1.0 x 0.3 0.9 x 0.3	115 ...	1	...		15.5		
11517	20 06.1 + 05 49 047.40 - 14.18	01 - 51 - 004	0805	1.1 x 0.35 1.1 x 0.3	104 (6 - 7)	1	S - Irr		15.6		
11518	20 06.1 + 14 52 055.32 - 09.51	02 - 51 - 000	0325	1.0 x 0.6 0.9 x 0.5	115 4:	1	S ...		15.5		
11519	20 06.5 + 07 38 049.06 - 13.34	01 - 51 - 000	0805	1.0: x 0.5: 0.9: x 0.4:	105: ...	1	Sc:		15.6		
11520	20 06.8 + 73 31 106.35 + 20.94	12 - 19 - 000	0809	1.1 x 0.2 0.9 x 0.2	165 6	3	Sc		16.5:		*
11521	20 08.3 + 01 59 044.23 - 16.56	00 - 51 - 000	0297	(1.0 x 0.7) 1.1 x 1.0	(135) ...	3	S0		14.9		
11522	20 09.4 + 05 37 047.65 - 14.99	01 - 51 - 005	0805	1.1 x 0.6 1.0 x 0.6	3 4 - 5	2	Sb - c P w 11524, Brightest in group		14.8		*
11523	20 09.5 + 05 22 047.44 - 15.13	01 - 51 - 006	0805	1.1 x 1.0 0.9: x 0.9:	--- 1	2	Sc 11522 group		15.4		
11524	20 09.6 + 05 37 047.67 - 15.03	01 - 51 - 007	0805	1.1 x 1.0 1.1 x 1.0	--- 1	3	Sc Disturbed: P w 11522, Sc -		14.9		*
11525	20 09.8 + 01 48 044.26 - 16.98	00 - 51 - 009	0297	1.6 x 0.3 1.7 x 0.25	53 6 - 7	1	Sa - b		15.4		
11526	20 11.1 + 07 07 049.21 - 14.59	01 - 51 - 008	0805	1.6 x 0.7 1.8 x 0.8	20 6	3	Sa		15.3		
11527	20 11.3 - 01 18 041.60 - 18.81	00 - 51 - 010	0297	1.8 x 0.35 1.5 x 0.4	21 6	1	S ...		14.4		
11528	20 11.3 + 55 25 090.14 + 11.64	09 - 23 - 000	0543	1.1 x 0.7 1.1 x 0.7	30: 4:	1	S ...		16.5:		*
11529	20 12.9 + 66 08 099.73 + 16.99	11 - 24 - 005	0777	1.8 x 0.5 1.8 x 0.5	17 6	2	Sb - c		16.5:		*
11530	20 13.1 + 73 32 106.59 + 20.55	12 - 19 - 000	0809	1.1 x 0.5 0.9: x 0.5:	150 ...	2	Dwarf (spiral) S: V		17.		*
11531	20 14.0 + 58 18 092.85 + 12.85	10 - 29 - 000	0593	1.0: x 0.1 1.0: x 0.1	172 7	1	...		17. - 18.		
11532	20 14.2 + 07 14 049.72 - 15.19	01 - 51 - 000	0805	1.0 x 0.9 (1.0 x 0.9)	--- ...	3	[S0] Compact		15.1		
11533	20 14.4 + 00 27 043.61 - 18.64	00 - 51 - 011	0297	1.8: x 0.9: 1.6: x 0.8:	[80] ...	4	Dwarf irregular Ir V		17.		
11534	20 14.8 + 05 08 047.92 - 16.39	01 - 51 - 000	0805	1.2 x 0.3 1.1 x 0.25	178 ...	2	Sc - Irr P w 11535		16.0:		*
11535	20 14.9 + 05 05 047.89 - 16.44	01 - 51 - 010	0805	1.1 x 1.0 1.1 x 1.0	--- 1	2	S B: a - b S (r)... P w 11534		15.5		*
11536	20 15.0 + 79 15 112.13 + 23.04	13 - 14 - 000	0832 454	1.1 x 1.0 1.1 x 0.9	--- 1	3	S B a (R´)S B(s)a		15.3		*C
11537	20 16.1 - 00 18 043.14 - 19.38	00 - 51 - 013	0297	2.3 x 0.7 2.2 x 0.7	82 6	2	Sc		14.8		*
11538	20 17.4 + 54 15 089.62 + 10.27	09 - 33 - 000	0543	1.0 x 0.25 0.9 x 0.3	170 6:	2	... S dm s of 2		17.		*
11539	20 17.7 + 62 31 096.78 + 14.69	10 - 29 - 001	0593	(1.2 x 1.2) (1.0 x 1.0)	--- ...	3	[E] Compact		15.2		*C
11540	20 19.2 + 66 35 100.47 + 16.68	11 - 24 - 006 NGC 6911	0777	2.2 x 1.2 2.2 x 1.2	115: ...	1	S B b:		15.7		
11541	20 19.8 + 00 08 044.02 - 19.97	00 - 52 - 003	0794	1.9: x 0.9: 1.8: x 0.9:	140 5	4	S B c S B d s of 2		16.5:		*
11542	20 19.9 + 06 16 049.62 - 16.90	01 - 52 - 002 IC 1316 = 5000*	0315	1.2 x 0.4 1.1 x 0.4	63 6	3	S B b		14.9		*2
11543	20 20.2 + 09 25 052.45 - 15.31	02 - 52 - 002	0812	1.1 x 0.9 1.1 x 0.8	70: ...	2	Sb npp of 2		15.4		*
11544	20 20.4 + 07 58 051.20 - 16.12	01 - 52 - 000	0315	1.0 x 0.3 0.9 x 0.25	8 6	2	Sc		16.5:		
11545	20 20.5 + 52 15 088.20 + 08.78	09 - 33 - 000	0543	1.3 x 0.12 1.2 x 0.12	61 7	1	S ...		16.5:		
11546	20 20.7 + 00 30 044.48 - 19.99	00 - 52 - 004 IC 1317	0794	0.7 x 0.6 0.8 x 0.6	--- ...	1	... Compact		14.5	+ 3975 + 4148	*12C
11547	20 21.0 + 60 35 095.31 + 13.33	10 - 29 - 000	0593	1.0 x 1.0 0.7 x 0.7	--- 1	2	S B ...		16.0:		
11548	20 21.1 + 06 17 049.79 - 17.15	01 - 52 - 003 NGC 6906	0315	1.6 x 0.8 1.6 x 0.8	36 5	3	Sb		13.7		*12 3
11549	20 21.9 + 07 00 050.54 - 16.94	01 - 52 - 004	0315	1.1 x 0.5 1.0 x 0.4	110 6	2	Sb - c		15.3		
11550	20 22.1 + 12 16 055.20 - 14.18	02 - 52 - 000	0812	1.4 x 0.2 1.0 x 0.2	58 7	2	Sc P w 11552		16.5:		*

N G C   6906,   6911   I C   1316,   1317,   5000

1	2	3	4	5	6	7	8	9	10	11	12
11551	20 22.2 + 06 45 050.36 - 17.14	01 - 52 - 005	0315	1.7 x 0.4 1.6 x 0.35	109 6	2	Sb - c		15.1		
11552	20 22.2 + 12 17 055.23 - 14.20	02 - 52 - 004	0812	2.1 x 0.5 2.0 x 0.6	16 7	2	Sa - b P w 11550 Brightest in group		15.4		*
11553	20 22.3 + 01 16 045.40 - 19.95	00 - 52 - 005	0794	1.5 x 0.3 1.5 x 0.3	141 6	1	Sc	Sc +	16.0:		
11554	20 22.4 + 58 12 093.37 + 11.87	10 - 28 - 004 NGC 6916	0545	2.0 x 1.4 1.8 x 1.3	(90) 3	2	S B b - c		15.3		
11555	20 22.8 + 05 06 048.96 - 18.12	01 - 52 - 006	0315	1.9: x 0.9 1.7: x 0.9	... ...	4	Sb/Sc spp of 2, disturbed	SA(s)bc	14.7		*1
11556	20 22.8 + 12 14 055.27 - 14.35	02 - 52 - 005	0812	1.0: x 1.0: 0.8: x 0.8:	— 1:	2	... 11552 group	S d - dm	17.		
11557	20 23.0 + 60 02 094.98 + 12.82	10 - 29 - 005	0545	2.4 x 2.2 2.2 x 2.0	— 1	4	...	SAB(s)dm	14.5		
11558	20 23.5 - 01 31 042.96 - 21.58	00 - 52 - 000	0794	1.0 x 0.9 0.7: x 0.7:	— ...	0	...		16.0:		
11559	20 23.6 + 01 00 045.32 - 20.37	00 - 52 - 007	0794	(1.3 x 1.0) (1.3 x 1.0)	— ...	2	S0 nnp of 2		14.5		*
11560	20 23.9 + 02 45 046.97 - 19.56	00 - 52 - 009 IC 1320	0794	1.0 x 0.6 0.9 x 0.5	87 ...	1	S? ... Brightest in group		14.5		*
11561	20 24.1 + 02 29 046.76 - 19.73	00 - 52 - 010	0794	1.1 x 1.0 1.3 x 1.1	— 1?	1	S0? P w 11562 IC 1320 group		14.9		*
11562	20 24.2 + 02 32 046.82 - 19.73	00 - 52 - 011	0794	1.4 x 1.3 1.3 x 1.3	— 1	2	Sb P w 11561, IC 1320 group		14.7		*
11563	20 25.0 + 07 56 051.79 - 17.11	01 - 52 - 007 NGC 6917	0315	1.6 x 1.1 1.6 x 1.1	40: 3	3	S ... Singular		14.3		*3
11564	20 25.2 + 10 35 054.16 - 15.73	02 - 52 - 008	0812	1.6 x 0.2 1.6 x 0.25	[102] ...	1	Double system? Bridge?,3rd in chain		15.1		*1 C
11565	20 25.6 + 04 48 049.07 - 18.87	01 - 52 - 008	0315	1.2 x 0.15 1.3 x 0.15	40 (7)	2	Irr		15.0		
11566	20 25.7 + 00 07 044.78 - 21.26	00 - 52 - 015	0794	0.7 x 0.5 0.8 x 0.5	[55] ...	1	...		14.2		
11567	20 25.7 + 00 18 044.95 - 21.17	00 - 52 - 014	0794	1.0 x 0.9 0.9 x 0.9	— ...	1	... s of 2		15.5		*
11568	20 25.9 + 10 35 054.25 - 15.88	02 - 52 - 009	0812	2.3 x 0.6 2.3 x 0.7	37 6	1	Sc: 4th in chain		15.0		
11569	20 26.4 + 10 28 054.22 - 16.04	02 - 52 - 000	0812	1.1 x 0.7 0.8 x 0.6	165: ...	2	Sc 5th in chain		16.0:		
11570	20 26.4 + 25 33 066.99 - 07.57	04 - 48 - 001 NGC 6921	1608	1.1 x 0.2 1.7 x 0.5	141 6	1	S ... spp of 2		15.0:	+ 4317 + 4595	*12 3
11571	20 26.7 + 10 31 054.31 - 16.08	02 - 52 - 011	0812	2.1 x 0.5 1.9 x 0.5	18 6	2	S B c 6th in chain		15.4		*
11572	20 26.8 + 10 35 054.38 - 16.06	02 - 52 - 012	0812	(1.0 x 0.9) (1.3 x 1.2)	— ...	2	E 7th in chain		14.9		*
11573	20 27.1 + 14 10 057.53 - 14.16	02 - 52 - 000	0812	1.1 x 0.4 0.8 x 0.3	69 6	2	Sc		16.0:		
11574	20 27.3 - 02 21 042.67 - 22.82	00 - 52 - 018 NGC 6922	0794 458	1.2 x 0.9 1.2 x 0.9	(150) ...	3	Sc Disturbed?		14.4		*3
11575	20 27.6 + 02 54 047.61 - 20.28	00 - 52 - 020	0794	1.6 x 0.5 1.5 x 0.5	122 (6)	2	Sc - Irr Disturbed?		14.8		*1
11576	20 27.8 - 01 10 043.86 - 22.35	00 - 52 - 021	0794 458	(1.2 x 1.1) (1.6 x 1.5)	— ...	1	...		15.2		*
11577	20 28.1 + 01 13 046.12 - 21.23	00 - 52 - 025	0794	1.7 x 1.1 1.7 x 1.3:	5 2:	1	S ...		14.6		
11578	20 28.3 + 09 00 053.19 - 17.23	01 - 52 - 000	0315	1.7: x 1.6: 0.7 x 0.5	— 1	4	...	S dm	16.5:		
11579	20 28.4 - 00 49 044.26 - 22.31	00 - 52 - 026	0794	1.4 x 0.2 1.6 x 0.2	30 7	3	Sc		16.0:		
11580	20 28.4 + 07 55 052.25 - 17.83	01 - 52 - 000	0315	1.1 x 0.2 1.1 x 0.15	29 7	1	Sc		16.0:		
11581	20 28.5 - 02 05 043.08 - 22.95	00 - 52 - 027	0794 458	1.3 x 0.8 1.3 x 0.8	40 4	4	S B 0/S B a		15.4		*
11582	20 28.6 + 20 07 062.78 - 11.11	03 - 52 - 001	0276	2.1 x 1.2 1.5: x 1.1:	30: ...	2	Sc - Irr nf of 2		15.6		*
11583	20 29.2 + 60 17 095.63 + 12.32	10 - 29 - 000	0593	2.5: x 0.6: 2.5:: x 0.6::	93 ...	0	...		17.		
11584	20 29.3 + 01 22 046.43 - 21.42	00 - 52 - 031	0794	1.8 x 0.2 1.8 x 0.2	53 7	2	Sc		15.7		
11585	20 29.7 - 02 24 042.94 - 23.37	00 - 52 - 032	0794 458	1.9: x 1.8: 2. x 2.	— 1	3	Sb	SAB(s)b - bc Sb +	14.8		

N G C   6916, 6917, 6921, 6922   I C   1320

1	2	3	4	5	6	7	8	9	10	11	12
11586	20 29.7 + 04 41 049.52 – 19.81	01 – 52 – 012	0315	1.1 x 0.5 1.1 x 0.5	18 5	3	Sa		15.6		*
11587	20 29.9 + 11 12 055.35 – 16.37	02 – 52 – 014	0812	1.8 x 0.5 1.8 x 0.5	166 7:	3	S0 s of 2		14.7		*
11588	20 30.5 – 02 11 043.25 – 23.44	00 – 52 – 033 NGC 6926	0794 458	2.1 x 1.4 2.0 x 1.4	[0] ...	3	Irr p of 2, distorted?		13.6		*13
11589	20 30.5 + 09 45 054.16 – 17.28	02 – 52 – 017 NGC 6928	0812	2.2 x 0.6 2.2 x 0.7	106 6	2	S B a – b Brightest in group		13.7	+ 4754 + 4993	*12 3
11590	20 30.6 + 09 41 054.12 – 17.34	02 – 52 – 018 NGC 6930	0812	1.2 x 0.5 1.3 x 0.4	8 ...	2	S ... + companion		14.3	+ 4182 + 4421	*12 3
11591	20 30.7 + 01 35 046.82 – 21.61	00 – 52 – 034	0794	1.2 x 0.6 1.1 x 0.6	143 ...	2	S ... Companion superimposed, disturbed		15.3		*
11592	20 31.5 + 54 27 090.97 + 08.70	09 – 33 – 000	0543	1.1 x 0.5 1.1 x 0.5	40: 5	1	S ...		16.0:		
11593	20 32.2 + 07 48 052.67 – 18.69	01 – 52 – 015	0315	[1.2 x 0.4] ...	... ...	2	Double system Contact		14.7		*1
11594	20 32.2 + 12 46 057.03 – 15.97	02 – 52 – 019	0812	1.1 x 0.20 1.2 x 0.20	41 7	1	S ...		15.6		
11595	20 32.5 + 01 46 047.23 – 21.91	00 – 52 – 036	0794	2.1 x 0.25 1.8 x 0.3	12 7	2	Sb – c		15.6		
11596	20 33.3 + 01 13 046.83 – 22.36	00 – 52 – 039	0794	1.0 x 1.0 1.0 x 1.0	--- 1	2	S ... S (r) ...		16.0:		*
11597	20 33.8 + 59 59 095.72 + 11.68	10 – 29 – 006 NGC 6946	0593	14. x 14. 14. x 14.	--- 1	4	Sc SA(rs)cd		10.5	+ 80 + 371	*12 –68A
11598	20 34.1 + 13 15 057.72 – 16.09	02 – 52 – 020	0812	1.0 x 0.7 1.0 x 0.8	150: 3	3	Sc		15.5		
11599	20 34.2 + 11 20 056.07 – 17.17	02 – 52 – 022	0812	2.5 x 1.9 2.3 x 1.8	95: 2	4	Sb Sb –		14.2		
11600	20 34.3 + 64 38 099.65 + 14.31	11 – 25 – 001 NGC 6949	0546	1.7 x 1.6 1.7 x 1.6	--- 1?	1	S ...		15.0		
11601	20 35.4 – 02 12 043.89 – 24.51	00 – 52 – 041	0794 458	1.0 x 0.2 1.0 x 0.2	41 6 – 7	1	S ...		15.7		
11602	20 35.5 + 10 27 055.48 – 17.93	02 – 52 – 023	0812	1.3 x 1.0 1.5 x 1.1	0 ...	3	S0 Brightest in group		15.1		
11603	20 35.5 + 63 31 098.79 + 13.56	11 – 25 – 000	0546	1.4: x 0.1 1.2 x 0.1	76 7	1	S ...		17.		
11604	20 36.6 + 65 55 100.88 + 14.85	11 – 25 – 002 NGC 6951	0546	4.0 x 3.6 4.0 x 3.6	(170) 1	4	Sb/S B c SAB(rs)bc Sb +		12.3	+ 1364 + 1647	*124 5689
11605	20 36.9 + 10 37 055.83 – 18.12	02 – 52 – 000	0812	(1.2 x 1.1) (1.2 x 1.2)	--- ...	3	E 11602 group		15.2		*
11606	20 37.2 + 00 54 047.07 – 23.36	00 – 52 – 045	0794	1.0 x 0.5 1.0 x 0.5	100 5	3	Sc		15.3		
11607	20 37.2 + 01 52 047.97 – 22.86	00 – 52 – 044	0794	1.0 x 0.3 0.8 x 0.25	28 (6)	1	S? ... Brightest of 3		14.8		*
11608	20 37.2 + 27 04a 069.66 – 08.63	05 – 48 – 000	0332				Pair of exceedingly		10.5		
11609	20 37.2 + 27 04b 069.66 – 08.63	05 – 48 – 000	0332	[0.55 x 0.35] ...	... ...	0	compact galaxies, or stars?		10.5		*C
11610	20 37.6 + 07 05 052.80 – 20.20	01 – 52 – 000	0315	1.2 x 0.7 1.3 x 0.8	138 4	2	Sb – c		14.9		
11611	20 38.1 + 14 05 059.01 – 16.42	02 – 52 – 026	0812	1.3 x 0.7 1.2 x 0.5	55 5	3	S B c		15.4		
11612	20 38.3 + 00 29 046.83 – 23.81	00 – 52 – 046	0794	1.1 x 0.12 1.1 x 0.15	45 7	2	Sb – c		15.7		
11613	20 38.9 + 63 27 098.98 + 13.22	11 – 25 – 000	0546	1.3 x 0.2 1.2: x 0.1	140 7	2	Sc		17.		
11614	20 39.1 + 12 53 058.12 – 17.30	02 – 52 – 028	0812	1.1 x 0.25 1.1 x 0.25	69 7	1	Sb		15.6		
11615	20 39.3 + 19 01 063.36 – 13.80	03 – 52 – 002	0276	1.4 x 0.5 1.3 x 0.5	153 ...	3	Irr		15.7		*
11616	20 40.3 + 63 19 098.95 + 13.01	11 – 25 – 000	0546	1.1: x 0.9: 1.0: x 0.7:	(90) ...	2	Sc		17.		
11617	20 41.4 + 14 08 059.53 – 17.05	02 – 53 – 000	0558	1.0 x 0.6 0.8: x 0.5:	122 4:	3	Sc		16.0:		
11618	20 41.6 + 03 02 049.68 – 23.20	00 – 53 – 001 NGC 6954	0305	0.9 x 0.5 1.0 x 0.6	68 4:	1	S: ...		14.2	+ 4011 + 4228	*12 3
11619	20 41.6 + 12 20 058.00 – 18.11	02 – 53 – 001 NGC 6956	0558	1.7 x 1.6 1.7 x 1.5	--- 1	4	S B b Brightest of 3		13.5		*13
11620	20 41.8 + 12 14 057.95 – 18.21	02 – 53 – 002	0558	0.7 x 0.4 0.7 x 0.4	25 ...	1	... NGC 6956 group		14.5		*1

N G C   6926, 6928, 6930, 6946, 6949, 6951
      6954, 6956

1	2	3	4	5	6	7	8	9	10	11	12
11621	20 41.8 + 02 25 049.13 – 23.56	00 – 52 – 000 NGC 6955	0305	1.5 x 1.5 1.5: x 1.2:	— 1	3	Sb		14.9		3
11622	20 42.1 – 01 54 045.08 – 25.82	00 – 53 – 000	0305 458	1.1 x 0.4 0.8 x 0.2	48 6:	1	S B ...		15.2		
11623	20 42.1 + 12 19 058.06 – 18.22	02 – 53 – 003	0558	1.1 x 0.7 1.0 x 0.6	(40) 4:	4	S B a S B(r) ... NGC 6956 group		14.5		*1
11624	20 43.8 + 06 30 053.17 – 21.81	01 – 53 – 000	0836	1.1: x 0.6: 0.9: x 0.4:	... ...	2	Dwarf m		17.		
11625	20 43.9 + 28 22 071.62 – 09.02	05 – 49 – 000	0757	1.2 x 1.1: 1.2 x 0.7	— 1:	1	S ...		16.5:		*
11626	20 44.0 + 00 03 047.22 – 25.25	00 – 53 – 000	0305	1.0: x 1.0: 0.9: x 0.8:	— 1:	2	Sc – Irr NGC 6962 group		16.0:		
11627	20 44.4 + 05 27 052.30 – 22.50	01 – 53 – 000	0836	1.3: x 1.2: 1. x 1.	— 1:	3	... SAB(s)m		17.		
11628	20 44.8 + 00 08 047.41 – 25.38	00 – 53 – 003 NGC 6962	0305	3.0 x 2.3 3.0 x 2.3	75 2	4	Sa SAB ab Brightest in group		13.5	+ 4183 + 4390	*12 3
11629	20 44.9 + 00 07 047.41 – 25.41	00 – 53 – 005 NGC 6964	0305	(1.6 x 1.1) 1.6 x 1.1	168 ...	3	E P w 11628 NGC 6962 group		14.2	+ 3832 + 4039	*12 3
11630	20 45.0 + 00 13 047.52 – 25.38	00 – 53 – 006 NGC 6965	0305	1.0 x 0.7 1.0 x 0.7	100 6	1	S? ... NGC 6962 group		14.3		*3
11631	20 45.5 – 00 22 047.03 – 25.79	00 – 53 – 007	0305	1.1 x 0.25 1.1 x 0.2	63 (7)	2	Irr or Peculiar		15.0		*
11632	20 45.8 + 69 06 104.10 + 15.99	12 – 19 – 003	0546	1.1 x 0.4 1.1 x 0.4	19 6	2	S B a – b		16.0:		
11633	20 46.0 + 07 33 054.43 – 21.69	01 – 53 – 001 NGC 6969	0836	1.1 x 0.3 1.2 x 0.35	15 6	2	Sa		15.0		*3
11634	20 46.0 + 16 32 062.24 – 16.55	03 – 53 – 000	0831	[1.5 x 0.8] ...	... ...	1	Double system Contact		14.7		*
11635	20 46.0 + 79 58 113.55 + 22.18	13 – 15 – 001	1232	3.5 x 1.3 3.5 x 1.3	35 6	3	Sb Sb +		14.2		
11636	20 46.5 + 64 17 100.16 + 13.04	11 – 25 – 000	0546	1.4: x 1.3: 1.5: x 1.4:	— 1:	3	S B c		16.0:		
11637	20 47.0 + 05 48 053.00 – 22.86	01 – 53 – 002 NGC 6971	0836	1.2 x 1.0 1.2 x 1.0	(60) 2	3	Sb		14.8		3
11638	20 47.3 + 16 40 062.54 – 16.73	03 – 53 – 003	0831	2.1 x 1.2 1.7 x 1.2	135 3 – 4:	2	S B b:		14.3		
11639	20 47.4 + 06 02 053.27 – 22.81	01 – 53 – 003	0836	(1.2 x 1.0) 1.2 x 1.2	— ...	2	... Compact		14.6		*
11640	20 47.6 + 09 43 056.60 – 20.81	02 – 43 – 004 NGC 6972	0558	1.2 x 0.5 1.2 x 0.5	143 6	2	S0 – a Brightest of 3		14.3	+ 4442 + 4682	*3
11641	20 48.6 + 21 41 066.89 – 13.96	04 – 49 – 000	1103	1.0 x 0.4 0.8 x 0.4	163 6	1	S B ...		15.7		
11642	20 48.6 + 70 03 105.05 + 16.35	12 – 19 – 005	0809	1.3: x 0.8: 1.3 x 0.8	55 4	1	S ...		16.0:		
11643	20 49.2 + 18 46 064.58 – 15.84	03 – 53 – 005	0831	1.1 x 0.4 1.1 x 0.4	65 6	3	S B b sf of 2, disturbed		14.8		*1
11644	20 50.6 + 06 57 054.58 – 22.97	01 – 53 – 005	0836	1.2 x 0.35 1.2 x 0.35	174 6	3	S B b		15.1		*
11645	20 50.9 + 00 27 048.58 – 26.53	00 – 53 – 000	0305	1.0 x 0.2 0.9 x 0.15	110 7	2	Sb – c		15.5		*
11646	20 51.6 + 00 35 048.81 – 26.61	00 – 53 – 000	0305	1.0 x 0.3 1.0 x 0.35	167 6	1	S ...		15.2		*
11647	20 51.9 + 17 35 064.00 – 17.06	03 – 53 – 000	0831	1.5 x 0.12 1.4 x 0.15	141 7	3	Sc		15.6		
11648	20 52.3 + 66 58 102.72 + 14.21	11 – 25 – 000	0546	1.6: x 0.3 1.7 x 0.3	[153] ...	0	Irr		17.		
11649	20 52.9 – 01 25 047.07 – 27.92	00 – 53 – 009	0305	1.5 x 1.2 1.4 x 1.2	85 2	3	S B a S B(r)a		14.5		*
11650	20 54.1 + 11 51 059.45 – 20.89	02 – 53 – 000	0558	1.4: x 0.3 1.3 x 0.15	143 6	2	Sc		15.7		*
11651	20 55.1 + 25 46 071.13 – 12.59	04 – 49 – 005	1103	3.5: x 0.9 4.0: x 1.1:	165 6	4	... S dm		15.4		
11652	20 55.2 + 14 02 061.50 – 19.82	02 – 53 – 000	0558	1.0 x 0.5 0.9 x 0.4	38 5:	1	S B ... Disturbed?		15.3		*
11653	20 55.2 + 18 37 065.35 – 17.06	03 – 53 – 006	0831	1.5: x 1.2: 1.0:: x 0.8::	0: 1	2	... S d:		15.7		
11654	20 55.7 + 04 17 052.90 – 25.50	01 – 53 – 000	0836	1.0 x 0.5 0.9 x 0.4	... ...	1	... Peculiar, disrupted pair?		16.5:		
11655	20 55.7 + 25 27 070.97 – 12.90	04 – 49 – 000	1103	1.3: x 0.3 0.9 x 0.3	95 6	2	... S d – dm		16.5:		

N G C   6955, 6962, 6964, 6965, 6969, 6971
6972

1	2	3	4	5	6	7	8	9	10	11	12
11656	20 56.7 + 11 06 059.19 – 21.85	02 – 53 – 000	0558	1.0 x 0.15 0.9 x 0.1	20 7	3	Sc 2nd of 3		15.7		*
11657	20 57.2 – 02 04a 047.05 – 29.17	00 – 53 – 014	0305 458	[2.6 x 1.5] ...	... ...	4	Double system Disrupted, large plume		14.4		*1C
11658	20 57.2 – 02 04b 047.05 – 29.17	00 – 53 – 015	0305 458								
11659	20 57.6 + 09 23 057.82 – 23.02	01 – 53 – 009	0836	1.6 x 0.6 1.0: x 0.6	[125] ...	2	S ... 1st of 3, disturbed?		15.2		*
11660	20 58.1 + 13 19 061.33 – 20.81	02 – 53 – 000	0558	1.4 x 0.2 1.2 x 0.1	[25] ...	2	S ... Integral		16.0:		*
11661	20 58.1 + 16 39 064.16 – 18.80	03 – 53 – 007	0831	1.2 x 1.0 0.6 x 0.5	115: 2	2	Sb – c		14.8		1
11662	20 58.4 + 17 36 065.00 – 18.28	03 – 53 – 008 NGC 7003	0831	1.1 x 0.7 1.0 x 0.7	120 4	2	Sb – c		13.8		3
11663	20 58.6 – 00 23 048.91 – 28.61	00 – 53 – 016 NGC 7001	0305	1.3 x 1.0 1.3 x 0.9	162 2	2	Sb		14.0		3
11664	20 59 + 83 53 117.35 + 24.00	14 – 09 – 000	0802	[1.0 x 0.3] ...	... ...	2	Double system Distorted, bridge		16.0:		*
11665	20 59.8 + 07 15 056.25 – 24.69	01 – 53 – 000	0836	1.1 x 0.4 1.0 x 0.4	100 6	1	S ...		16.0:		
11666	20 59.9 + 14 57a 063.01 – 20.18	02 – 53 – 000	0558	1.1 x 0.1 0.4 x 0.1	4 7	3	Sc np of 2, Sc +		(15.1)		*
11667	21 00.4 + 17 51 065.52 – 18.50	03 – 53 – 000	0831	1.1: x 0.1 0.6 x 0.1	41 7	1	Sc:		17.		
11668	21 00.4 + 36 30 080.18 – 06.51	06 – 46 – 000	0279	[0.45 x 0.30] ...	... ...	1	Pair of compacts Contact		15.5 + + 14.0		*C
11669	21 00.7 + 12 58 061.44 – 21.53	02 – 53 – 000	0558	1.1: x 1.0: 0.9: x 0.8:	— 1:	1	S ... In group		16.0:		*
11670	21 01.4 + 29 42 075.13 – 11.14	05 – 49 – 001 NGC 7013	0757	5.2 x 1.6 5.0 x 1.6	157 6	4	Sa	SA 0/a	12.9 + 570 + 857		*12 3
11671	21 02.1 + 09 26 058.57 – 23.90	01 – 53 – 000	0836	[1.2 x 0.9] ...	... ...	3	Double system Contact, disrupted		14.4		*
11672	21 02.1 + 15 53 064.14 – 20.03	03 – 53 – 013	0831 460	[1.5 x 1.2] ...	... ...	3	Double system Symbiotic		15.1		*1
11673	21 02.2 – 00 25 049.41 – 29.39	00 – 53 – 000	0305	[1.0 x 0.7] ...	... ...	2	Double system Bridge, plumes		15.6		*C
11674	21 03.3 + 11 13 060.33 – 23.08	02 – 53 – 012 NGC 7015	0558	1.8 x 1.7 1.7 x 1.6	— 1	4	Sb Sb +		13.2		*
11675	21 03.4 + 07 27 057.00 – 25.31	01 – 53 – 012	0836	1.0 x 0.3 0.9 x 0.3	46 6	1	Sa sf of 2		15.0		*
11676	21 03.4 + 18 16 066.33 – 18.80	03 – 53 – 015	0831	1.0 x 0.4 0.6 x 0.25	2 (6)	1	S? ... Interacting pair?		14.8		1
11677	21 03.7 + 15 45 064.28 – 20.42	03 – 53 – 000	0831 460	1.4 x 0.6 0.9 x 0.4	105 ...	1	...		16.0:		
11678	21 04.4 + 66 35 103.20 + 13.05	11 – 25 – 004	0546	(1.3 x 1.3) (1.6 x 1.6)	— ...	1			15.5		*
11679	21 05 + 86 43 119.98 + 25.53	14 – 09 – 000	0802	1.2 x 0.1 0.9 x 0.1	23 7	2	Sc Sc +		17.		
11680	21 05.2 + 03 40 053.79 – 27.82	01 – 54 – 000	0552	[2.1 x 0.8] ...	[70] ...	3	Sc + compact companion Connected		14.5 (+ 7858)		*C
11681	21 05.5 + 16 08 064.89 – 20.52	03 – 54 – 001 NGC 7025	1141 460	2.3 x 1.5 (1.8 x 1.5)	33 4	4	Sa		14.1		*13
11682	21 06.1 + 17 36 066.20 – 19.72	03 – 54 – 003	1141	1.7: x 0.5 1.6 x 0.6	170 ...	1	S? ...		14.6		
11683	21 06.2 + 17 59 066.54 – 19.50	03 – 54 – 004	1141	1.3 x 0.3 1.2 x 0.25	168 6	2	Sc sff of 2		15.3		*
11684	21 06.3 + 12 15 061.71 – 23.05	02 – 54 – 001 IC 1359	0298 461	1.1 x 0.3 0.9 x 0.25	172 6	1	S ... Peculiar?		14.8		*C
11685	21 06.7 – 02 00 048.53 – 31.18	00 – 54 – 001	0575	1.1 x 0.15 1.0 x 0.15	82 7	1	S ...		15.0		*
11686	21 07.0 + 65 32 102.56 + 12.16	11 – 25 – 000	0546	1.3: x 1.1: 0.5 x 0.4	— ...	2	S B c		17.		
11687	21 07.3 + 14 57 064.18 – 21.60	02 – 54 – 003 NGC 7034	0298 461	(1.0 x 0.6) 1.0 x 0.6	125 ...	3	E P w NGC 7033		15.3		*13
11688	21 08.7 + 04 12 054.84 – 28.24	01 – 54 – 001	0552	1.3 x 0.5 1.1 x 0.4	40 6	1	Sb		15.7		*
11689	21 08.8 + 65 57 103.01 + 12.30	11 – 25 – 005	0546	1.4 x 1.3 1.4 x 1.4	— 1	3	S B b	S B(r)b	15.1		
11690	21 08.9 + 22 43 070.80 – 16.96	04 – 50 – 000	0290	1.3 x 0.2 1.3 x 0.2	20 7	1	S0 – a		16.0:		

N G C   7001, 7003, 7013, 7015, 7025, 7034   I C   1359

1	2	3	4	5	6	7	8	9	10	11	12
11691	21 09.0 - 02 14 048.64 - 31.79	00 - 54 - 003 IC 5090	0575	1.1 x 0.4 0.9 x 0.35	26 6	1	S ... Peculiar, disturbed?		14.5		*
11692	21 09.0 + 04 51 055.50 - 27.94	01 - 54 - 002 IC 1361	0552	1.1 x 1.1 0.8 x 0.7	--- ...	1	S ... Peculiar?		15.4		
11693	21 09.2 + 37 42 082.26 - 07.01	06 - 46 - 000	0279	1.5 x 0.3 1.5 x 0.3	149 6	2	Sa - b		15.5:		
11694	21 09.4 + 11 03 061.16 - 24.38	02 - 54 - 004	0298 461	(1.8 x 1.1) (1.5 x 1.2)	... ...	2	... Peculiar? Companion attached?		15.2		*1
11695	21 09.6 - 01 40 049.30 - 31.62	00 - 54 - 004	0575	1.4 x 0.8 1.1 x 0.7	115: ...	3	Sb n of 2, disturbed, bridge:		14.7		*1
11696	21 09.6 + 11 09 061.28 - 24.36	02 - 54 - 000	0298 461	1.0 x 0.15 0.9 x 0.15	45 7	2	Sc - Irr		16.0:		*
11697	21 09.7 + 11 26 061.55 - 24.21	02 - 54 - 007	0298 461	1.0 x 0.55 1.0 x 0.5	152 5	1	S B ...		15.4		
11698	21 09.8 + 12 48 062.75 - 23.40	02 - 54 - 8 + 9	0298 461	[1.0 x 0.7] ...	... ...	2	S ... + companion Contact		14.6		*1
11699	21 09.9 + 12 24 062.42 - 23.66	02 - 54 - 010	0298 461	1.0 x 0.9 0.7: x 0.6	--- 1:	3	S B: c Sc +		15.6		
11700	21 10.0 + 11 11 061.38 - 24.42	02 - 54 - 011	0298 461	1.0 x 0.8 1.1 x 0.7	13 4:	4	S B b		14.9		
11701	21 10.8 + 08 39 059.26 - 26.09	01 - 54 - 004 NGC 7040	0552	1.1 x 0.8 1.0 x 0.7	150 3	1	S ... Companion superimposed		14.9		*
11702	21 11.4 + 13 21 063.48 - 23.37	02 - 54 - 013 NGC 7042	0298 461	2.1 x 1.8 1.8 x 1.6	(140) 1 - 2	3	Sb P w 11704		13.0		*3
11703	21 11.7 + 01 58 053.20 - 30.11	00 - 54 - 008 IC 1368	0575 462	1.3 x 0.5 1.1 x 0.4	48 6:	1	S: ...		14.3		
11704	21 11.7 + 13 24 063.57 - 23.39	02 - 54 - 014 NGC 7043	0298 461	1.1 x 0.8 0.8 x 0.6	135 3	1	S B: a P w 11702		14.9		*3
11705	21 12.0 + 12 27 062.80 - 24.04	02 - 54 - 000	0298 461	1.0: x 0.8: 0.4: x 0.4:	--- ...	2	... S dm:		16.5:		
11706	21 12.0 + 15 02 065.01 - 22.43	02 - 54 - 015	0298	1.7 x 0.2 1.5: x 0.2	55 7	3	Sb		15.7		
11707	21 12.3 + 26 32 074.31 - 15.03	04 - 50 - 001	0290	3.6: x 2.0: 3.5: x 2.0:	55 4 - 5	4	... SA dm		15.6	1	
11708	21 12.4 + 02 37 053.94 - 29.89	00 - 54 - 009 NGC 7046	0575 462	1.8 x 1.5 2.0 x 1.3:	115: ...	2	S B ...		14.2	23	
11709	21 12.8 + 25 40 073.72 - 15.69	04 - 50 - 002	0290	1.6:: x 1.4:: 1.5:: x 1.3::	--- ...	3	Dwarf IV - V:		16.5:	1	
11710	21 13.2 + 24 09 072.60 - 16.77	04 - 50 - 003	0290	1.5: x 1.4: 1.3:: x 1.2::	--- ...	2	... SAB dm:		16.5:		
11711	21 13.4 + 17 10 067.02 - 21.34	03 - 54 - 000	1141	1.1 x 0.10 0.9 x 0.10	64 7	2	Sc		16.0:		
11712	21 13.9 - 01 02 050.60 - 32.20	00 - 54 - 010 NGC 7047	0575	1.2 x 0.6 1.3 x 0.6	107 5	1	S ...		14.3	3	
11713	21 15.0 + 15 20 065.76 - 22.81	02 - 54 - 000	0298	1.6 x 0.5 0.8: x 0.4	16 6	2	S B c Brightest of 3		15.3		*
11714	21 15.2 + 02 09 053.94 - 30.74	00 - 54 - 011	0575 462	1.0 x 0.6 0.8 x 0.5	40 4:	2	Sc		15.7		
11715	21 15.4 + 13 22 064.16 - 24.12	02 - 54 - 000	0298 461	1.0 x 0.15 0.6 x 0.1	112 7	1	Sc:		15.6		
11716	21 15.5 + 25 45 074.21 - 16.09	04 - 50 - 005	0290	1.5: x 1.2: 1.3: x 1.3:	--- ...	2	... S dm:		16.5:		*1
11717	21 16.2 + 19 30 069.39 - 20.35	03 - 54 - 006	1141	1.4 x 0.5 1.4 x 0.5	42 6	1	S ... Brightest in group		15.1		*1
11718	21 16.3 + 26 14 074.71 - 15.90	04 - 50 - 006 NGC 7052	0290	(2.5 x 1.4) (3.0 x 1.6)	64 ...	3	E		14.0	3	
11719	21 16.5 + 15 27 066.10 - 23.01	02 - 54 - 019	0298	1.5 x 0.15 1.4 x 0.15	163 7	2	Sa - b		15.7		
11720	21 16.8 + 05 48 057.66 - 28.98	01 - 54 - 006	0552	1.0 x 0.8 1.1 x 0.9	130 2	4	S B c S B(rs)cd		14.8		
11721	21 17.0 + 13 51 064.83 - 24.12	02 - 54 - 020	0298	1.5 x 0.15 1.3 x 0.15	163 (7)	2	Irr f of 2, distorted?		15.3		*1
11722	21 17.4 + 21 45 071.38 - 19.08	04 - 50 - 008	0290	1.7 x 0.5 1.7 x 0.5	110 6	1	S ...		15.3		
11723	21 17.7 - 01 53 050.35 - 33.46	00 - 54 - 012	0575	1.8 x 0.25 1.7 x 0.25	33 7	3	Sb		14.9		
11724	21 18.1 + 00 50 053.13 - 32.07	00 - 54 - 014	0575	1.7 x 0.35 1.5 x 0.3	145 6	3	Sc P w IC 1373, Sc +		15.3		*
11725	21 18.5 + 08 58 060.83 - 27.43	01 - 54 - 007	0552	1.1 x 0.35 1.0 x 0.3	148 6	2	Sc - Irr		15.4	1	

N G C   7040, 7042, 7043, 7046, 7047, 7052   I C   1361, 1368, 5090

1	2	3	4	5	6	7	8	9	10	11	12
11726	21 18.5 + 37 22 083.30 – 08.57	06 – 47 – 001	0269	1.4 x 0.4 1.1 x 0.3	160 6	2	Sb – c		15.3		
11727	21 18.9 + 22 52 072.51 – 18.60	04 – 50 – 009 NGC 7053	0290	1.6 x 1.5 1.6 x 1.5	--- ...	1	... Compact, halo		14.3		*13 C
11728	21 18.9 + 26 03 074.98 – 16.45	04 – 50 – 010	0290	1.5: x 1.5: 1.2: x 1.2:	--- 1	4	Sc	Sc +	15.6		1
11729	21 19.0 + 24 49 074.04 – 17.30	04 – 50 – 000	0290	1.2 x 0.3 0.8 x 0.2	108 6	2	S B: c		16.5:		
11730	21 19.1 + 29 15 077.43 – 14.30	05 – 50 – 000	0803	1. x 1. 1. x 1.	--- ...	2	Dwarf P w 11732, contact?		17. – 18.		*
11731	21 19.2 + 21 01 071.10 – 19.88	03 – 54 – 007 IC 5104	1141	1.8 x 0.45 1.6 x 0.4	173 6	2	S B? a – b		14.4		
11732	21 19.2 + 29 15 077.45 – 14.31	05 – 50 – 000	0803	1.2: x 0.3 1.2: x 0.2	128 6 – 7	1	S: ... P w 11730, contact?		16.5:		*
11733	21 19.5 + 06 56 059.15 – 28.85	01 – 54 – 000	0552	1.2 x 0.25 1.1 x 0.2	172 6:	1	S B? b: npp of 2		16.0:		*
11734	21 19.8 + 18 27 069.13 – 21.68	03 – 54 – 008 NGC 7056	1141	1.0 x 1.0 1.0 x 1.0	--- 1	2	S B b		13.8		3
11735	21 20.1 + 15 04 066.39 – 23.92	02 – 54 – 022	0298	(1.0 x 0.8) 0.9 x 0.8	--- ...	1	E?		14.9		*
11736	21 20.4 + 03 18 055.90 – 31.16	00 – 54 – (15)	0575	[3.8 x ...] ...	[170] ...	2	Chain of 4 or 5 galaxies		...		*
11737	21 20.4 + 29 12 077.60 – 14.54	05 – 50 – 000	0803	1.5: x 0.6 1.4: x 0.6	163 6	1	S ... nnp of 2, contact		16.5:		*
11738	21 21 + 83 37 117.52 + 23.36	14 – 10 – 001	1216	1.1 x 0.6 1.0 x 0.5	65 5	2	Sb – c		15.2		
11739	21 22.3 + 20 28 071.17 – 20.79	03 – 54 – 000	1141	1.0 x 0.7 0.8 x 0.7	(170) ...	2	...	S d – dm	16.5:		
11740	21 23.8 + 09 35 062.28 – 28.08	02 – 54 – 026	0298	1.6 x 0.5 1.0 x 0.3	122 6	1	S ...		14.7		
11741	21 23.8 + 13 57 066.07 – 25.33	02 – 25 – 025 NGC 7066	0298	1.1 x 1.1 0.8 x 0.7	--- 1:	1	S ... Companion superimposed		15.2		*1C
11742	21 23.9 + 01 53 055.12 – 32.68	00 – 54 – 000	0575	1.2: x 1.2: 0.8: x 0.7:	--- 1	1	S ...		16.5:		
11743	21 23.9 + 30 16 078.93 – 14.34	05 – 50 – 001	0803	1.2 x 1.0 1.2 x 1.0	--- ...	--	... Disturbed?		15.5		*1
11744	21 24.9 + 37 28 084.29 – 09.40	06 – 47 – 000	0269	1.0 x 0.9 0.6: x 0.5:	--- 1:	3	Sc		16.5:		
11745	21 24.9 + 37 43 084.46 – 09.22	06 – 47 – 000	0269	1.4 x 0.6 1.5 x 0.6	0 6	2	Sb – c		16.0:		*
11746	21 25 + 82 06 116.37 + 22.27	14 – 10 – 000	1216	1.1 x 0.2 1.1 x 0.2	43 7	3	Sb		16.0:		
11747	21 25.5 – 01 52 051.63 – 35.09	00 – 54 – 019 NGC 7069	0575	(1.3 x 0.9) (1.3 x 0.9)	(20) ...	2	E – S0		14.8		3
11748	21 25.8 + 45 16 089.86 – 03.90	07 – 44 – 000	0806	1.4:: x 0.5: 1.4: x 0.5;	103 6	1	S ...		17.		
11749	21 26.4 + 20 17 071.72 – 21.62	03 – 54 – 012	1141	1.5 x 1.4 1.3 x 1.1	--- 1 – 2	2	S B b		14.6		
11750	21 26.6 + 02 35 056.26 – 32.83	00 – 54 – 024	0575	1.1 x 0.3 1.0 x 0.2	10 6	1	S ...		16.0:		
11751	21 26.6 + 11 10b 064.17 – 27.63	02 – 54 – 000	0298	1.0 x 0.6 0.9 x 0.5	170: ...	2	S B ... nnf of 2, disturbed		(14.7)		*
11752	21 26.6 + 24 48 075.27 – 18.57	04 – 50 – 000	0290	1.1 x 0.2 1.0 x 0.2	58 7	2	Sb – c		16.5:		*
11753	21 26.7 + 31 36 080.35 – 13.83	05 – 50 – 002	0803	1.4 x 1.4 1.2 x 1.1	--- 1	3	S B b	S B(r)b Sb –	15.6		
11754	21 27.3 + 27 06 077.12 – 17.09	04 – 50 – 011	0290	2.1: x 1.8: 1.7: x 1.6:	--- 1	4	Sc	SAB cd	15.6		1
11755	21 27.4 + 02 12 056.02 – 33.22	00 – 54 – 028 NGC 7077	0575	0.6 x 0.6 (0.8 x 0.7)	--- ...	1	E?		14.3		*3
11756	21 27.8 + 26 30 076.75 – 17.59	04 – 50 – 012 NGC 7080	0290	1.8 x 1.8 1.8 x 1.7	--- 1	4	S B b	S B(r)b Sb +	14.1		3
11757	21 28.4 + 41 38 087.69 – 06.86	07 – 44 – 000	0806 463	1.6 x 1.1 1.6 x 1.1	85: 3	2	S B a – b		15.7		
11758	21 28.5 + 13 46 066.74 – 26.32	02 – 54 – 028	0298	1.6 x 0.2 1.2 x 0.2	60 7	1	S ...		15.5		*
11759	21 28.8 + 02 16 056.33 – 33.47	00 – 54 – 030 NGC 7081	0575	1.1 x 1.1 0.8 x 0.7	--- 1:	1	S? ... P w 11760		13.7		*3
11760	21 29.2 + 02 15 056.38 – 33.56	00 – 55 – 002	1130	1.1 x 0.9 1.3: x 1.0	(145) 2:	1	S ... P w 11759		14.6		*1

N G C 7053, 7056, 7066, 7069, 7077, 7080
7081

I C 5104

1	2	3	4	5	6	7	8	9	10	11	12
11761	21 29.6 + 34 18 082.73 - 12.33	06 - 47 - 000	0269	0.18 x 0.18 0.25 x 0.25	— ...	1	... Compact		13.8		*C
11762	21 29.8 + 29 55 079.61 - 15.50	05 - 50 - 000	0803	0.15 x 0.15 0.15 x 0.15	— ...	1	... Extremely compact		13.0		*C
11763	21 30.0 + 09 56 063.68 - 29.06	02 - 55 - 000	0799	0.5 x 0.20 0.5 x 0.20	65 ...	1	... Seyfert, very compact		14.3	+ 18315 + 18547	*C
11764	21 30.7 + 07 47 061.87 - 30.54	01 - 55 - 000	0860	1. x 1. 1. x 1.	— ...	3	Dwarf spiral S V		18.		
11765	21 31.1 + 08 33 062.64 - 30.14	01 - 55 - 002	0860	1.2 x 0.7 0.9 x 0.5	80 4	2	Sb 1st of 3		15.3		*1
11766	21 31.7 + 21 36 073.66 - 21.63	04 - 51 - 000 IC 5119	0286	1.0 x 0.25 0.9 x 0.25	150 6	1	Sa		15.3		*
11767	21 31.8 + 41 36 088.14 - 07.32	07 - 44 - 000	0806 463	1.2 x 0.5 1.1 x 0.5	[155] ...	2	... sp of 2	S d - m	17.		*
11768	21 32.2 + 18 44 071.48 - 23.68	03 - 55 - 000	0375 473	1.1 x 0.12 1.1 x 0.12	51 7	2	Sb - c		16.0:		
11769	21 32.2 + 26 08 077.21 - 18.55	04 - 51 - 001	0286	1.6: x 0.6 1.5 x 0.6	142 6	3	Sb		15.5		
11770	21 32.5 + 01 14 055.96 - 34.82	00 - 55 - 006	1130	1.0 x 0.15 1.1 x 0.2	177 6	1	S ...		15.6		
11771	21 33.2 + 23 15 075.19 - 20.73	04 - 51 - 000	0286	1.5 x 0.3 1.3: x 0.2	143 7:	2	Sc		16.0:		
11772	21 33.5 + 35 11 083.95 - 12.24	06 - 47 - 003 IC 1392	0269	(1.6 x 1.3) (1.7 x 1.4)	75: ...	2	E - S0 P w 11775		13.0		*
11773	21 33.7 + 09 26 063.90 - 30.08	01 - 55 - 000	0799	1.0 x 0.3 0.9 x 0.25	23 6	1	Sa		15.3		
11774	21 33.8 + 16 50 070.23 - 25.24	03 - 55 - 001	0375	1.0: x 0.7: 0.9 x 0.6	(150) 3:	3	...	S d - dm	16.0:		
11775	21 33.8 + 35 08 083.96 - 12.32	06 - 47 - 004	0269	1.0 x 0.4 1.2 x 0.5	50 6	2	S0 P w 11772		14.5		*
11776	21 33.9 + 12 01 066.20 - 28.45	02 - 55 - 003	0799	1.1 x 0.15 1.1 x 0.15	172 7	2	Sc		15.6		
11777	21 33.9 + 27 53 078.78 - 17.57	05 - 51 - 000	1212	1.4:: x 1.1:: 1.3:: x 1.0::	... ...	3	Dwarf (spiral) S: V		17.		
11778	21 34.0 + 05 31 060.37 - 32.58	01 - 55 - 000	0860	1.1 x 0.10 0.9 x 0.10	146 7	1	S ...		16.5:		
11779	21 34.1 + 70 34 108.08 + 13.83	12 - 20 - 000	0772	1.2 x 0.3 0.9 x 0.2	164 6	2	Sc - Irr		16.5:		
11780	21 34.5 + 00 13 055.29 - 35.81	00 - 55 - 008	1130	1.0 x 0.6 0.9 x 0.7	65 ...	1	S ... sp of 2		15.6		*
11781	21 34.6 + 35 28 084.32 - 12.19	06 - 47 - 005	0269	1.4 x 1.0 1.4 x 1.0	(75) 3	4	S0	SAB 0	13.7		
11782	21 35.7 + 08 45 063.65 - 30.89	01 - 55 - 006	0860	2.5 x 1.4 2.3: x 1.4:	35: 5?	4	...	S B(s)m	15.3		
11783	21 35.8 + 30 58 081.33 - 15.64	05 - 51 - 000	1212	1.3 x 0.6 1.1 x 0.5	173 5 - 6	2	Sa - b		16.0:		
11784	21 36.8 + 38 17 086.59 - 10.41	06 - 47 - 000	0269	1.0: x 0.5: 1.0: x 0.5:	55: ...	2	...	S d - m	17.		
11785	21 36.9 + 02 36 058.09 - 34.91	00 - 55 - 000	1130	1.4 x 0.15 1.4 x 0.15	61 7	3	Sc		15.4		
11786	21 37.2 + 06 04b 061.46 - 32.87	01 - 55 - 008 NGC 7102	0860	1.7 x 0.9 1.6: x 1.0	(153) ...	4	S B b nnf of 2, disturbed	S B(r)b	(14.4)		*12 3
11787	21 37.8 + 24 49 077.17 - 20.37	04 - 51 - 002	0286	1.0 x 0.7 0.9 x 0.5	155: 3	1	Sa		15.7		
11788	21 38.1 + 24 57 077.32 - 20.33	04 - 51 - 003	0286	[1.8 x 1.7] ...	— ...	4	S ... + companion Connected, distorted		15.7		*1
11789	21 38.5 + 01 06 056.89 - 36.12	00 - 55 - 010	1130	1.0: x 0.45 1.0 x 0.5	108 5	2	Sb		14.9		*
11790	21 39.0 + 00 40 056.54 - 36.47	00 - 55 - 011	1130	1.6 x 1.1 1.6: x 1.1	160: 3:	4	Sc	Sc +	15.3		
11791	21 39.5 + 25 11 077.74 - 20.38	04 - 51 - 000	0286	1.0 x 0.3 1.0 x 0.3	122 6	2	Sc		16.5:		*
11792	21 39.7 + 05 23 061.28 - 33.78	01 - 55 - 000	0860	1.6 x 0.15 1.2 x 0.2	160 7	3	Sc		15.6		
11793	21 39.7 + 22 29 075.74 - 22.34	04 - 51 - 004	0286	1.4 x 1.3 1.3 x 1.2	— 1	4	S B b	S B(s)b Sb +	15.0		*
11794	21 39.9 + 12 16 067.52 - 29.40	02 - 55 - 010 NGC 7112	0799	1.2 x 0.4 1.1 x 0.3	81 6	2	Sa - b P w NGC 7113		15.5		*
11795	21 40.2 + 23 24 076.52 - 21.77	04 - 51 - 000	0286	1.2: x 0.9: 0.6: x 0.4:	... ...	3	...	I m	16.0:		

N G C   7102, 7112

I C   1392, 5119

1	2	3	4	5	6	7	8 9	10	11	12
11796	21 40.4 + 28 43 080.47 – 17.96	05 – 51 – 001	1212	1.2 x 0.45 1.1 x 0.5	105 6	1	S ...	14.2		
11797	21 41.5 + 43 19a 090.64 – 07.23	07 – 44 – 001	0806 463	1.4: x 0.7: 1.6: x 0.9:	[98] ...	1	... 1st of 3, distorted, halo			*1
11798	21 41.5 + 43 19b 090.64 – 07.23	07 – 44 – 002	0806 463	1.9 x 0.3 2. x 0.25	137 6 – 7	1	S ... 2nd of 3, disturbed	15.2		
11799	21 41.5 + 43 27 090.72 – 07.13	07 – 44 – 004	0806 463	1.6: x 1.1: 1.7: x 1.0:	(40) ...	1	S B ...	16.0:		
11800	21 41.6 + 46 01 092.42 – 05.19	08 – 39 – 000	0589	1.5 x 0.7 1.3 x 0.6	20 5	1	S ... 1st of 3	15.7		*
11801	21 41.6 + 43 18 090.64 – 07.25	07 – 44 – 003	0806 463	1.0: x 0.7 0.8 x 0.6	(55) ...	1	S ... 3rd of 3	15.3		*1
11802	21 42.0 + 46 24 092.72 – 04.95	08 – 39 – 001	0589	1.0 x 0.7 0.9 x 0.6	(120) ...	1	S B: c: P w 11806, disturbed	15.1		*1
11803	21 42.1 + 14 40 069.96 – 28.18	02 – 55 – 011	0799	1.0 x 0.9 0.9 x 0.8	— 1	2	Sb – c	15.4		
11804	21 42.3 + 46 02a 092.52 – 05.26	08 – 39 – 02a	0589	1.5: x 0.6: 1.3 x 0.5	[85] ...	1	... P w 11805, disrupted, 2nd of 3			*1
11805	21 42.3 + 46 02b 092.52 – 05.26	08 – 39 – 02b	0589	1.6: x 0.4: 1.6 x 0.5	[136] ...	1	... P w 11804, disrupted, 3rd of 3	15.1		*1
11806	21 42.3 + 46 24 092.76 – 04.98	08 – 39 – 003	0589	1.7 x 0.6 1.7 x 0.5	[58] ...	3	S B b P w 11802, distorted	15.0		*1
11807	21 43.8 + 06 52 063.44 – 33.64	01 – 55 – 000	0860	1.0: x 0.5: ...	... ...	2	Dwarf irregular Ir V	18.		
11808	21 43.8 + 41 02 089.46 – 09.24	07 – 44 – 000	0806	2.3:: x 1.7:: 1.8: x 1.5:	(160) ...	1	S ...	16.5:		*
11809	21 44.1 + 08 04 064.60 – 32.93	01 – 55 – 012	0860	1.0 x 0.45 1.0 x 0.45	24 6	1	Sb	15.4		
11810	21 44.4 + 01 28 058.34 – 37.09	00 – 55 – 015 IC 1401	1130	2.1 x 0.7 2.0 x 0.7	175 6	3	Sb	14.7		
11811	21 45.0 + 18 31 073.62 – 26.02	03 – 55 – 002	0375	1.0 x 0.3 0.9 x 0.2	74 (6 – 7)	2	Sc – Irr	16.0:		
11812	21 45.0 + 31 53 083.48 – 16.30	05 – 51 – 000	1212	1. x 1. 0.7? x 0.7:	— ...	3	Dwarf V	17. – 18.		
11813	21 45.2 + 21 56 076.31 – 23.62	04 – 51 – 000	0286	1.0 x 0.4 0.8: x 0.3:	[98] ...	2	Irr	17.		
11814	21 45.8 – 01 54 055.12 – 39.34	00 – 55 – 016	1130	0.9 x 0.5 0.6 x 0.35	160 ...	1	...	14.3		
11815	21 45.9 + 21 56 076.43 – 23.74	04 – 51 – 005 NGC 7137	0286	1.6 x 1.3 1.4 x 1.3	— 1:	2	S ... Singular	13.3	+ 1505 + 1774	*12 3468
11816	21 46.5 + 00 13 057.46 – 38.25	00 – 55 – 018	1130	1.6: x 1.5: 1.4: x 1.3:	— 1	2	S B b – c S B(rs)bc:	14.8		
11817	21 46.6 + 12 16 068.79 – 30.61	02 – 55 – 014 NGC 7138	0799	1.1 x 0.5 1.2 x 0.5	177 6	2	S B a	15.4		3
11818	21 46.8 + 72 15 110.02 + 14.40	12 – 20 – 000	0772	1.5:: x 1.3:: 2.0: x 1.3:	... ...	1	S ...	15.7		
11819	21 47.0 + 41 41 090.37 – 09.11	07 – 44 – 000	0806	1.6: x 0.6: 1.4: x 0.5:	168 ...	3	Dwarf IV – V or V	17.		
11820	21 47.1 + 13 59 070.33 – 29.53	02 – 55 – 015	0799	1.5: x 1.0: 1.1: x 1.0:	... ...	4	Dwarf spiral, S m S V	17.		
11821	21 47.1 + 15 33 071.63 – 28.45	03 – 55 – 003	0375	1.8 x 0.35 1.9 x 0.35	79 7	3	Sa	15.3		*
11822	21 47.6 + 40 26 089.62 – 10.16	07 – 44 – 000	0806	(1.0 x 0.6) (1.1 x 0.7)	20: ...	2	S B 0	16.5:		
11823	21 47.8 + 34 43 085.87 – 14.55	06 – 48 – 000	0815	0.20 x 0.15 0.3 x 0.2	— ...	1	... Extremely compact	14.0		*C
11824	21 48.0 + 30 37 083.10 – 17.67	05 – 51 – 000	1212	1.0 x 0.3 1.0 x 0.20	52 7	2	S B? c	16.0:		*
11825	21 48.1 + 00 41 058.24 – 38.30	00 – 55 – 019	1130	1.1 x 0.3 1.1 x 0.25	18 7	1	Sa? sf of 2	15.4		*1
11826	21 48.3 + 01 47 059.40 – 37.68	00 – 55 – 020 IC 1405	1130	1.0 x 0.8 1.0 x 0.8	(115) ...	2	S B a – b P w IC 1406	15.0		*
11827	21 48.3 + 22 37 077.39 – 23.63	04 – 51 – 000	0286	1.0: x 0.4 0.5 x 0.4	... ...	1	... Peculiar? Double or multiple?	15.2		*
11828	21 48.5 + 02 32 060.19 – 37.26	00 – 55 – 000	1130 464	1.0 x 0.9 0.9 x 0.7	— 1:	1	Sa – b	15.4		*
11829	21 48.6 + 44 28 092.37 – 07.17	07 – 44 – 000	0806	1.5? x 1.5? 1. x 1.	— ...	3	Dwarf V	18.		
11830	21 48.8 + 25 37 079.69 – 21.51	04 – 51 – 006	0286	1.4 x 1.4 1.1 x 1.0	— 1	3	Sb	15.0		

N G C   7137,   7138   I C   1401,   1405

1	2	3	4	5	6	7	8	9	10	11	12
11831	21 48.9 + 44 19 092.32 - 07.32	07 - 45 - 001	0580	(1.0 x 0.5) 1.1 x 0.6	125: ...	2	E		16.0:		*
11832	21 49.3 + 11 21 068.52 - 31.72	02 - 55 - 017	0799	1.2 x 0.12 1.1 x 0.12	64 7	2	Sb - c		16.0:		
11833	21 49.3 + 33 53 085.55 - 15.38	06 - 48 - 000	0815	1.0 x 0.1 1.0 x 0.1	6 7	1	Sc		17.		
11834	21 49.4 + 25 01 079.36 - 22.04	04 - 51 - 009	0286	1.1 x 1.0 1.0 x 1.0	-- 1?	1	...		14.5		
11835	21 49.6 + 03 04 060.93 - 37.15	00 - 55 - 026 NGC 7149	1130 464	(1.2 x 0.9) (1.5 x 1.0)	(25) ...	3	E		14.6		*3
11836	21 49.9 + 46 25 093.79 - 05.80	08 - 39 - 000	0589	1.6 x 0.3 1.4 x 0.3	29 ...	1	S - Irr		17.		
11837	21 50.1 + 35 52 087.01 - 13.98	06 - 48 - 001	0815	1.4 x 0.9 1.3 x 0.8	153 4	4	Sc	Sc +	16.0:		
11838	21 50.3 + 28 04 081.71 - 19.91	05 - 51 - 002	1212	2.1 x 0.20 2.0 x 0.20	23 7	3	Sc	Sc +	15.4		
11839	21 50.6 + 36 13 087.32 - 13.77	06 - 48 - 000	0815	1.0 x 0.15 0.9 x 0.15	53 7	2	Sb - c		15.7		
11840	21 50.7 + 04 00 062.05 - 36.78	01 - 55 - 000	0860 464	1.0 x 0.9 0.2 x 0.2	-- ...	1	S ... sf of 2		17.		*
11841	21 50.8 + 38 42 088.98 - 11.89	06 - 48 - 000	0815	3.0: x 0.2 2.8: x 0.2	66 7	1	...	d - m	16.5:		
11842	21 51.0 + 25 16 079.83 - 22.10	04 - 51 - 011	0286	1.0 x 0.2 1.0 x 0.2	39 6	1	Sa - b		15.5		
11843	21 52.0 + 02 42 061.03 - 37.85	00 - 55 - 029 NGC 7156	1130 464	1.6 x 1.4 1.7 x 1.4	(105) 2	3	Sc	Sc +	13.5	+ 3966 + 4177	*12 3
11844	21 52.0 + 14 54 072.04 - 29.75	02 - 55 - 028 IC 5145	0799	1.6 x 0.8 1.8 x 0.9	170 5	2	Sa - b		14.7		
11845	21 52.1 + 14 47 071.96 - 29.85	02 - 55 - 029	0799	1.1 x 0.1 0.9: x 0.12	150 7	2	Sb - c In group		16.0:		*
11846	21 52.2 + 06 10 064.42 - 35.69	01 - 55 - 016	0795	1.0 x 0.5 1.0 x 0.5	42 5	2	Sb - c		15.2		*
11847	21 53.0 + 30 16 083.71 - 18.64	05 - 51 - 000	1212	1.0 x 0.6 0.5:: x 0.3::	(110) ...	2	Dwarf		17.		
11848	21 53.2 + 10 14 068.31 - 33.18	02 - 56 - 000	1137	1.1 x 0.5 0.8 x 0.4	124 5:	3	...	S dm	16.0:		
11849	21 53.3 + 24 39 079.81 - 22.91	04 - 51 - 013	0286	1.7 x 1.0 1.7 x 1.0	73 4	2	S B c		14.8		
11850	21 53.5 - 01 45 056.73 - 40.82	00 - 56 - 001 IC 1411	1146	0.9 x 0.5 0.9 x 0.7	(40) ...	1	E		14.5		*
11851	21 53.5 + 05 40a 064.20 - 36.26	01 - 56 - 002	0795	1.0 x 0.7 1.0 x 0.7	10 3	4	Sa sp of 2, contact		(14.7)		*1
11852	21 53.7 + 27 40 082.03 - 20.71	05 - 51 - 000	1212	1.6 x 0.7 1.4 x 0.7	15: ...	1	S B a?		15.0		
11853	21 53.8 - 01 25 057.14 - 40.69	00 - 56 - 003	1146	1.1 x 0.8 1.1 x 0.8	75: 3	3	Sc	S d	15.4		*1
11854	21 54.1 + 15 43 073.12 - 29.53	03 - 56 - 000	0187	1.1 x 0.4 0.9 x 0.3	12 6	1	S ... Companion superimposed?		16.0:		
11855	21 54.2 + 30 33 084.12 - 18.59	05 - 51 - 003	1212	1.5 x 0.4 1.3 x 0.3	116 7	2	Sb - c		15.7		
11856	21 54.3 + 06 41 065.32 - 35.75	01 - 56 - 000	0795	1.0 x 0.4 0.9 x 0.4	155 6	1	S ...		16.0:		
11857	21 54.5 + 12 06 070.19 - 32.14	02 - 56 - 001	1137	1.1 x 0.5 0.9 x 0.5	105 5:	1	S ...		15.6		
11858	21 55.4 + 41 00 091.15 - 10.66	07 - 45 - 002	0580	1.8 x 0.8 1.7 x 0.7	47 6	2	S0 - a		15.1		
11859	21 55.6 + 00 46 059.78 - 39.75	00 - 56 - 004	1146	2.5: x 0.2 1.6 x 0.25	63 7	1	Sb		15.6		
11860	21 55.7 + 24 01 079.79 - 23.74	04 - 51 - 014	0286	1.6 x 0.7 1. x 0.5:	133 ...	3	...	S dm	16.0:		
11861	21 55.7 + 73 01 111.05 + 14.57	12 - 20 - 000	0772	3.8: x 2.8: 3.5: x 2.6:	(20) ...	4	...	SAB dm	15.2		
11862	21 55.8 + 38 42 089.75 - 12.50	06 - 48 - 000	0815	1.8 x 0.2 1.7: x 0.2	132 7	2	...	S d - m	16.5:		
11863	21 55.9 - 00 58 058.03 - 40.85	00 - 56 - 000	1146	1.0 x 0.8 1.0 x 0.8	(35) ...	3	Dwarf spiral, S dm - m S IV - V		15.7		
11864	21 55.9 + 42 03 091.89 - 09.89	07 - 45 - 003	0580	2.5: x 1.6: 2.0: x 1.0:	0: ...	4	...	S B(s)dm	15.7		
11865	21 56.2 + 11 48 070.27 - 32.64	02 - 56 - 000	1137	0.5 x 0.5 0.35 x 0.30	-- ...	1	...		14.3		

1	2	3	4	5	6	7	8	9	10	11	12
11866	21 56.2 + 13 53 072.03 − 31.19	02 − 56 − 002	1137	1.5 x 0.8 1.3: x 0.7	178 ...	2	... Peculiar, jet		15.6		*1
11867	21 56.3 − 02 14 056.74 − 41.67	00 − 56 − 005	1146	1.1 x 0.5 1.1 x 0.45	140 6	2	Sb − c		15.4		*
11868	21 56.7 + 17 56 075.41 − 28.37	03 − 56 − 001	0187	2.6: x 2.3: 2.2:: x 2.2::	-- 1:	4	...	S B(s)m	15.1		*C
11869	21 57.7 + 44 08 093.45 − 08.45	07 − 45 − 000	0580	1.0 x 0.2 0.8 x 0.25	49 6	1	S ... 3rd of 3		16.5:		*
11870	21 58.0 + 06 12 065.62 − 36.76	01 − 56 − 007	0795	1.0 x 0.15 1.1 x 0.15	172 7	1	S ... np of 2		15.6		*
11871	21 58.2 + 10 19 069.39 − 34.02	02 − 56 − 005	1137	[1.0 x 0.6] ...	[160] ...	2	Double system Symbiotic		14.7		*1
11872	21 58.3 + 17 30 075.37 − 28.96	03 − 56 − 003 NGC 7177	0187	3.3 x 2.0 2.7 x 1.9	90 4	4	Sb		12.2	+ 1105 + 1360	*12 3468
11873	21 58.4 + 19 26 076.89 − 27.56	03 − 56 − 002	0187	1.1 x 0.2 1.0 x 0.2	103 6 − 7	1	S ...		16.0:		
11874	21 59.0 + 51 20 098.00 − 02.85	08 − 40 − 000	0599	1.0 x 0.3 1.3 x 0.5	45 6	1	S ...		15.6		
11875	21 59.4 + 09 58 069.33 − 34.48	02 − 56 − 000	1137	1.0 x 0.25 0.9 x 0.2	71 ...	3	Irr		16.0:		
11876	21 59.9 + 02 36 062.52 − 39.45	00 − 56 − 000	1146	1.0 x 0.10 0.9 x 0.10	141 7	1	Sc		16.0:		
11877	21 59.9 + 11 37 070.87 − 33.42	02 − 56 − 007	1137	1.0: x 0.4: 0.6 x 0.3	15 6	3	Sc	Sc +	15.7		*
11878	22 00.0 + 18 00 076.10 − 28.87	03 − 56 − 004	0187	1.4 x 1.1 1.1 x 1.1	-- ...	1	S B? + superimposed companion		14.8		*1
11879	22 00.2 + 15 01b 073.76 − 31.06	02 − 56 − 000	1137	1.0 x 0.2 0.9 x 0.15	50 7	1	S ... f of 2, bridge		(15.3)		*
11880	22 00.2 + 19 30 077.29 − 27.80	03 − 56 − 005 IC 1420	0187	1.5 x 1.5 1.4 x 1.3	-- 1:	1	S B ... Peculiar		14.5		*
11881	22 00.5 + 17 40 075.94 − 29.19	03 − 56 − 000	0187	1.1 x 0.25 1.0 x 0.2	64 (7)	1	...		16.0:		
11882	22 00.7 + 00 20 060.36 − 41.02	00 − 56 − 007 NGC 7189	1146	0.9 x 0.6 0.9 x 0.6	115 3	3	S B b		14.4		3
11883	22 00.7 + 04 03 064.11 − 38.67	01 − 56 − 010 IC 1423	0795	1.1 x 0.5 1.0 x 0.5	35 5	3	S B b	Sb −	14.8		
11884	22 00.7 + 10 41 070.22 − 34.21	02 − 56 − 000 NGC 7190	1137	1.1 x 0.8 1.1 x 0.6	22 3	3	S B 0		15.0		3
11885	22 00.7 + 10 58 070.47 − 34.02	02 − 56 − 000 IC 5160	1137	(1.0 x 0.5) 1.0 x 0.6	66 4 − 5	2	S0 − a		15.0		*
11886	22 00.8 + 44 28 094.10 − 08.52	07 − 45 − 004	0580	1.0 x 0.5 1.1 x 0.5	60 ...	3	Sb/S B b	SAB(s)... Sb +	16.0:		*1
11887	22 00.9 + 40 49 091.87 − 11.44	07 − 45 − 005 NGC 7197	0580 467	2.1 x 1.0 1.8 x 0.9	112 5	1	Sa		14.5		3
11888	22 01.1 + 12 23 071.76 − 33.09	02 − 56 − 008 NGC 7194	1137	(1.3 x 0.9) (1.3 x 0.9)	(20) ...	3	E Brightest in group		14.5		*1
11889	22 01.2 + 14 52 073.85 − 31.34	02 − 56 − 010 IC 1427	1137	('.2 x 0.9) (1.3 x 1.0)	(105) ...	3	E		15.2		1
11890	22 01.4 + 38 18 090.39 − 13.50	06 − 48 − 003	0815	1.0 x 0.4 1.0 x 0.4	40 6	2	Sa − b		14.8		
11891	22 01.5 + 43 30 093.61 − 09.37	07 − 45 − 006	0580	4. x 3. 2.5:: x 2.0::	-- ...	4	Dwarf irregular Ir V		15.5		*1C
11892	22 01.7 + 35 45 088.82 − 15.55	06 − 48 − 004	0815	(1.1 x 0.8) (1.1 x 0.8)	(120) ...	2	E − S0 P w 11893		15.1		*1
11893	22 02.0 + 35 42 088.84 − 15.63	06 − 48 − 005	0815	2.3 x 0.3 2.3 x 0.3	118 7	3	Sc P w 11892, Sc +		16.0:		*1
11894	22 02.0 + 41 35 092.51 − 10.95	07 − 45 − 007	0580 467	1.4 x 0.6 1.3 x 0.6	47 5	3	Sa		15.7		*1
11895	22 02.2 + 39 30 091.26 − 12.64	06 − 48 − 008	0580	1.4 x 0.9 0.9 x 0.5	20: 4	2	Sb		14.8		
11896	22 02.4 + 15 33 074.64 − 31.05	03 − 56 − 006	0187 466	1.0 x 0.7 0.9 x 0.6	118 3	2	Sb − c		15.5		
11897	22 02.4 + 41 10 092.32 − 11.33	07 − 45 − 009	0580 467	1.5 x 1.5 1.4 x 1.1	-- 1	2	S B: b − c Disturbed?		14.3		*1
11898	22 02.5 + 42 05 092.90 − 10.61	07 − 45 − 011	0580 467	1.0 x 0.5 1.0 x 0.5	175 5	1	S B a − b		15.7		
11899	22 02.6 + 44 44 094.52 − 08.50	07 − 45 − 012	0580	1.0 x 0.3 1.0 x 0.35	61 6	1	S ... sf of 2		16.0:		*1
11900	22 02.8 + 43 50 094.00 − 09.24	07 − 45 − 000	0580	1.0: x 0.4 1.0 x 0.4	66 6	2	...	S d − dm	17. − 18.		

N G C 7177, 7189, 7190, 7194, 7197

I C 1420, 1423, 1427, 5160

1	2	3	4	5	6	7	8	9	10	11	12
11901	22 02.9 + 04 56 065.43 – 38.51	01 – 56 – 012	0795	1.0 x 0.15 1.0 x 0.15	144 7	1	S ...		15.4		
11902	22 03.1 + 16 29 075.53 – 30.48	03 – 56 – 000	0187 466	1.1: x 0.15 1.0 x 0.2	37 7	1	S ... NGC 7206 group		16.0:		
11903	22 03.1 + 34 45 088.42 – 16.51	06 – 48 – 006	0815	1.0 x 1.0 0.8 x 0.8	-- 1	1	S B ...	S B(r) ...	16.0:		
11904	22 03.2 + 16 33 075.60 – 30.45	03 – 56 – 007 NGC 7206	0187 466	(1.2 x 1.2) (1.8 x 1.7)	-- ...	2	[SO] Compact, brightest of 3		14.8		*13
11905	22 03.5 + 20 24 078.62 – 27.65	03 – 56 – 008	0187	1.9 x 0.8 2.0 x 0.7	... ...	3	... Peculiar, long streamer		15.1		*C
11906	22 03.5 + 44 38 094.59 – 08.67	07 – 45 – 000	0580	1.0 x 0.3 1.0 x 0.3	28 6	1	Sb – c		17.		
11907	22 03.8 + 02 06 062.82 – 40.52	00 – 56 – 000	1146	1.1 x 0.7 1.1 x 0.7	110: 4	2	SO – a		15.0		
11908	22 04.2 + 15 49 075.22 – 31.15	03 – 56 – 010	0187 466	1.2: x 1.2: 1.0 x 1.0	-- ...	1	...		16.0:		1
11909	22 04.3 + 47 00 096.11 – 06.85	08 – 40 – 001	0599	3.2 x 0.7 2.9 x 0.5	3 7	1	S ... Singular		13.3		
11910	22 04.6 + 10 00 070.44 – 35.38	02 – 56 – 011 NGC 7212	1137	1.6: x 0.7 1.3 x 0.5	... ...	3	S ... Companion superimposed, distorted		15.1		*1
11911	22 05.1 + 44 03 094.47 – 09.31	07 – 45 – 013	0580	(1.7 x 1.6) (1.5 x 1.5)	-- ...	2	... Compact		15.7		*1
11912	22 05.3 + 38 30 091.14 – 13.80	06 – 48 – 007	0815 468	1.8: x 0.7: 1.7 x 0.6	10: (6)	1	S: ...		16.0:		
11913	22 05.3 + 40 48 092.54 – 11.95	07 – 45 – 000	0580 467	1.0: x 0.8: 0.9: x 0.7:	-- ...	2	...	S B d – dm	17.		
11914	22 05.6 + 31 07 086.50 – 19.70	05 – 52 – 001 NGC 7217	0383	3.8 x 3.3 4.0 x 3.5	95: 2	4	Sb	Sb –	11.0	+ 911 + 1192	*12– –689
11915	22 05.7 + 04 27a 065.55 – 39.35	01 – 56 – 015	0795	1.4 x 1.0 1.1: x 0.9:	-- ...	2	S ... nnp of 2, distorted		(14.7)		*
11916	22 05.9 + 18 13 077.45 – 29.65	03 – 56 – 012	0187	1.0 x 0.5 0.7 x 0.25	50: ...	1	...		15.6		*C
11917	22 06.0 + 45 21 095.38 – 08.35	08 – 40 – 000	0599	1.0: x 0.5: 1.0 x 0.4	135 5 – 6	1	S ...		17.		
11918	22 06.0 + 74 51 112.80 + 15.57	12 – 21 – 000	0559	1.1 x 0.2 1.1 x 0.2	41 6 – 7	1	S ...		17.		
11919	22 06.1 + 40 56 092.75 – 11.93	07 – 45 – 014	0580 467	1.8 x 1.3 1.6 x 1.0	35 ...	3	Sb/S B c NGC 7223 group		14.8		1
11920	22 06.5 + 48 12 097.12 – 06.09	08 – 40 – 002	0599	2.5 x 1.6 2.8 x 1.7	45 4	3	SO/S B a		13.0		*
11921	22 06.8 + 14 07 074.39 – 32.82	02 – 56 – 012	1137	1.9 x 0.8 1.8 x 0.8	127 ...	3	Irr	I B m	14.7		
11922	22 06.8 + 40 07 092.36 – 12.67	07 – 45 – 015	0580	1.1: x 1.0: 1.0: x 1.0:	-- 1	3	Sc		16.0:		
11923	22 06.8 + 44 18 094.87 – 09.29	07 – 45 – 016	0580	1.3: x 0.8: 1.2: x 0.8:	(140) 3 – 4:	1	S ...		16.0:		
11924	22 06.9 + 21 16 079.94 – 27.51	03 – 56 – 014	0187	1.7 x 1.0 1.4 x 1.0	120 4	4	Sc	Sc +	14.8		1
11925	22 07.0 + 41 10 093.03 – 11.85	07 – 45 – 000	0580 467	1.0 x 0.35 1.0 x 0.4	39 6	1	S B c		16.5:		
11926	22 07.2 + 18 27 077.89 – 29.69	03 – 56 – 015	0187	1.4 x 1.0 0.3 x 0.3	120: ...	2	...	d – m	16.0:		*
11927	22 07.3 + 40 45 092.83 – 12.22	07 – 45 – 017	0580 467	1.8 x 1.2 1.3 x 1.0	130: 3:	3	Sb NGC 7223 group		15.4		*
11928	22 07.4 + 16 38 076.53 – 31.07	03 – 56 – 016	0187	1.0: x 0.8: 0.9 x 0.7	-- ...	1	S B ...		15.7		
11929	22 07.4 + 39 02 091.80 – 13.61	06 – 48 – 009	0815 468	1.0 x 0.7 1.1 x 0.8	35: 3:	2	SO		14.4		*
11930	22 07.7 + 20 43 079.70 – 28.05	03 – 56 – 000	0187	1.0 x 0.2 0.8 x 0.2	65 7	2	Sa – b		16.0:		
11931	22 08.0 + 40 46 092.95 – 12.28	07 – 45 – 018 NGC 7223	0580 467	2.0: x 1.3 1.7: x 1.3	... ...	4	S B c S B(rs)bc Brightest of 3, Sc –		13.5		*13
11932	22 08.0 + 44 19 095.06 – 09.40	07 – 45 – 000	0580	1.0: x 0.5:: 0.9: x 0.5	(80) ...	2	Dwarf		18.		
11933	22 08.1 + 41 20 093.30 – 11.83	07 – 45 – 019	0580 467	1.1 x 0.8 1.1 x 0.7	175: 3:	1	S: ...		16.0:		
11934	22 08.4 + 01 51 063.54 – 41.56	00 – 56 – 012 NGC 7222	1146	1.1 x 1.1 1.1 x 1.1	-- 1	2	S B a – b p of 2		15.1		*3
11935	22 08.5 + 41 47 093.63 – 11.51	07 – 45 – 020	0580 467	(1.0 x 0.9) (1.0 x 0.9)	-- ...	2	[SO] Compact		15.0		

N G C   7206, 7212, 7217, 7222, 7223

1	2	3	4	5	6	7	8	9	10	11	12
11936	22 08.7 + 41 01 093.21 – 12.16	07 – 45 – 000	0580 467	1.0 x 0.2 0.9 x 0.2	34 6	2	Sc P w 11937		16.5:		*
11937	22 08.9 + 41 03 093.26 – 12.15	07 – 45 – 000	0580 467	1.0 x 0.20 0.9 x 0.25	136 7	2	Sa P w 11936		15.3		*
11938	22 09.0 + 38 40 091.84 – 14.09	06 – 48 – 012 IC 5180	0815 468	(1.0 x 0.8) (1.1 x 0.9)	--- ...	2	[E] Compact		14.8		*
11939	22 09.1 + 11 33 072.74 – 35.06	02 – 56 – 015 IC 5177	1137	1.6 x 0.8 1.5 x 0.6	25 5?	4	S B b		15.2		*1
11940	22 09.2 + 25 36 083.47 – 24.52	04 – 52 – 004 NGC 7224	0204	(1.6 x 1.0) (1.6 x 1.0)	110 ...	3	E		14.7		*13
11941	22 09.3 + 29 37 086.19 – 21.38	05 – 52 – 000	0383	1.1 x 0.12 0.9 x 0.12	169 7	2	Sb – c		16.5:		
11942	22 09.3 + 38 28 091.77 – 14.29	06 – 48 – 015 NGC 7227	0815 468	1.3 x 0.6 1.2 x 0.5	8 5 – 6	3	S0		15.0		*3
11943	22 09.5 + 25 53 083.72 – 24.34	04 – 52 – 005	0204	1.0 x 0.35 1.2 x 0.3	156 6	2	Sa		16.0:		
11944	22 09.6 + 17 40 077.79 – 30.65	03 – 56 – 018	0187	2.3 x 0.8 1.8: x 0.5	28 ...	3	Dwarf (irregular) Irr: IV – V		16.0:		
11945	22 09.6 + 38 27 091.80 – 14.34	06 – 48 – 016 NGC 7228	0815 468	1.6: x 1.2 1.7 x 1.2	(150) 2 – 3	4	S B a	(R´)S B(s)a	15.0		*13
11946	22 09.6 + 46 04 096.31 – 08.13	08 – 40 – 003	0599	1.1 x 0.8 1.1 x 0.8	80: 3	4	Sc	SAB(s)c	14.7		
11947	22 09.8 + 11 15 072.63 – 35.39	02 – 56 – 016	1137	1.2 x 1.0 1.1 x 1.0	--- 1	4	S B b	SAB(s)b	15.1		
11948	22 09.9 + 13 47 074.77 – 33.58	02 – 56 – 017	1137 469	1.2 x 0.8 1.0 x 0.7	85 3:	3	S B b		15.2		*
11949	22 10.2 + 39 02 092.25 – 13.93	06 – 48 – 017	0815 468	1.1 x 0.2 1.2 x 0.2	0 7	2	Sa – b		14.8		
11950	22 10.3 + 38 25 091.90 – 14.44	06 – 48 – 018	0815	(1.3 x 0.9) (1.7 x 1.2)	40: ...	2	[E] Compact		14.3		*
11951	22 10.4 + 45 05 095.86 – 09.02	07 – 45 – 000 NGC 7231	0580	2.1 x 0.7 1.8 x 0.6	88 6	2	S B a		14.0		
11952	22 11.2 + 13 58 075.20 – 33.66	02 – 56 – 000	1137 469	1.7 x 0.2 1.0 x 0.15	143 7	2	S B? c		15.7		
11953	22 11.5 + 13 30 074.88 – 34.05	02 – 56 – 000	1137 469	1.0 x 0.25 1.0 x 0.25	169 6	2	Sa – b sf of 2		15.4		*
11954	22 11.5 + 40 32 093.36 – 12.86	07 – 45 – 000	0580 467	1.0 x 0.5 0.6: x 0.3:	... ...	2	Dwarf V		18.		
11955	22 11.6 + 38 59 092.45 – 14.13	06 – 48 – 019	0815 468	1.2 x 0.6 1.2 x 0.6	[35] ...	1	Sc Disturbed?		14.7		*
11956	22 11.7 + 28 40 086.01 – 22.45	05 – 52 – 000	0383	1.2 x 0.9 1.0: x 0.9:	(0) 2:	3	...	S B dm	16.5:		
11957	22 12.0 + 04 47 067.24 – 40.30	01 – 56 – 018	0795	1.1 x 0.6 1.1 x 0.5	130 5	2	Sb – c Disturbed?		15.5		*
11958	22 12.3 + 13 35 075.12 – 34.12	02 – 56 – 23 + 24 NGC 7236 + 7237	1137 469	[2.3 x 1.3] ...	[131] ...	2	Triple system Plumes		14.3	+ 7854 + 8096	*13 AC
11959	22 12.3 + 48 11 097.91 – 06.66	08 – 40 – 000	0599	1.0 x 0.8 0.9: x 0.8:	(20) ...	3	...	S d – dm	17.		
11960	22 12.5 + 41 12 093.91 – 12.42	07 – 45 – 000	0580 467	1.0 x 0.6 0.8: x 0.5:	65: 	3	S B: c		16.5:		
11961	22 12.6 + 41 56 094.36 – 11.83	07 – 45 – 021	0580 467	1.6 x 0.2 1.4: x 0.2	48 7	3	Sb		15.7		
11962	22 12.7 + 41 45 094.27 – 11.99	07 – 45 – 000	0580 467	1.0 x 0.5 0.9 x 0.4	7 5:	1	S ...		16.5:		
11963	22 12.8 + 37 03 091.49 – 15.84	06 – 48 – 021	0815 468	1.0 x 0.2 1.1 x 0.2	65 7	2	S0 – a NGC 7242 group		15.0		*1
11964	22 13.1 + 18 58 079.50 – 30.21	03 – 56 – 019	0187	2.1 x 0.2 2.0 x 0.2	45 7	3	Sc P w 11968		16.0:		*
11965	22 13.2 + 01 48 064.54 – 42.50	00 – 56 – 016 IC 1437	1146	1.1 x 1.0 1.1 x 1.1	--- 1	2	S0 – a		14.9		
11966	22 13.3 + 08 24 070.90 – 38.03	01 – 56 – 019	0795	1.0 x 0.5 0.9 x 0.4	177 ...	3	Irr	Ir I	15.4		
11967	22 13.3 + 33 23 089.33 – 18.88	05 – 42 – 005	0383	1.2 x 0.2 1.2 x 0.2	23 7	2	Sc – Irr sf of 2		15.5		*1
11968	22 13.4 + 18 59 079.58 – 30.24	03 – 56 – 020 NGC 7241	0187	3.5 x 0.8 2.9 x 0.8	20 ...	2	S – Irr		13.8		*13 C
11969	22 13.4 + 37 03a 091.59 – 15.91	06 – 48 – 025 NGC 7242	0815 468	(2.3 x 1.7) (2.7 x 1.8)	30: 3	4	E Brightest in group Companion superimposed		(14.6)	+ 5684 + 5971	*12 3
11970	22 14.0 + 03 09 066.08 – 41.76	00 – 56 – 017	1146	1.0 x 0.5 1.0 x 0.5	83 5:	1	S B b:		15.3		

N G C   7224, 7227, 7228, 7231, 7236, 7237
7241, 7242

I C   1437, 5177, 5180

1	2	3	4	5	6	7	8	9	10	11	12
11971	22 14.0 + 18 34 079.39 - 30.65	03 - 56 - 000	0187	1.1 x 0.5 0.8 x 0.3	57 6	1	Sb - c		16.0:		
11972	22 14.7 + 40 15 093.71 - 13.44	07 - 45 - 022 NGC 7248	0580 467	1.8 x 0.9 1.8 x 0.9	133 5	3	S0		13.6		*12 3
11973	22 14.7 + 41 15 094.29 - 12.62	07 - 45 - 023	0580 467	3.3 x 0.9 3.2 x 0.9	42 6	4	Sb/S B c	SAB(s)bc	13.5		
11974	22 15.0 + 33 16 089.56 - 19.18	05 - 52 - 007	0383	1.2 x 1.0 1.2 x 1.0	(45) 2	3	S B b		14.7		
11975	22 15.1 + 35 20 090.84 - 17.51	06 - 49 - 002	0778	1.6 x 1.2 1.6 x 1.2	(150) 2 - 3	3	S0/Sa		15.2		*1
11976	22 15.3 + 28 43 086.73 - 22.88	05 - 52 - 000	0383	1.1 x 0.2 0.9 x 0.2	85 6 - 7	1	S ...		16.5:		
11977	22 15.3 + 33 01 089.46 - 19.42	05 - 52 - 000	0383	1.2 x 0.9 1.1 x 0.8	(105) ...	3	...	SAB dm	16.5:		
11978	22 15.4 + 28 02 086.30 - 23.44	05 - 52 - 008	0383	1.3 x 1.3 1.3 x 1.3	--- 1	2	Sc	SAB(r) ...	15.7		
11979	22 16.0 + 45 28 096.89 - 09.27	08 - 40 - 004	0599	2. x 2. 1. x 1.	--- ...	3	...	S B: m	17.		
11980	22 16.1 + 40 19 093.97 - 13.54	07 - 45 - 024 NGC 7250	0580 467	1.3 x 0.6 1.2 x 0.5	157 5?	1	S - Irr Disturbed?, Ir II?		13.1		*13
11981	22 16.2 + 29 00 087.08 - 22.77	05 - 52 - 009	0383	1.6 x 1.3 1.6 x 1.4:	(170) 2	4	Sc	SA(s)c	15.4		*
11982	22 16.3 - 01 18 061.93 - 45.09	00 - 56 - 000	1146	1.3 x 0.15 1.1 x 0.1	170 7	2	Sc		15.7		
11983	22 16.8 + 33 05 089.77 - 19.55	05 - 52 - 000	0383	1.0 x 0.2 1.1 x 0.2	70 7	1	Sb - c		15.6		*
11984	22 17.2 + 29 08a 087.36 - 22.80	05 - 52 - 010 NGC 7253a	0383	1.8 x 0.8: 1.8 x 0.8:	... ...	4					*13
11985	22 17.2 + 29 08b 087.36 - 22.80	05 - 52 - 011 NGC 7253b	0383	1.7 x 0.7: 1.7 x 0.6:	... ...	4	Double system Contact, disrupted		14.4		VA
11986	22 17.3 + 41 00 094.56 - 13.10	07 - 45 - 000	0580	1.3 x 0.2 0.9 x 0.1	9 7	1	...	d - m	16.5:		
11987	22 17.6 + 08 45 072.18 - 38.53	01 - 57 - 001	1157	1.6 x 0.3 0.7 x 0.25	39 6	1	S B? ...		15.4		
11988	22 17.7 + 32 55 089.83 - 19.79	05 - 52 - 000	0383	1.0: x 0.7: 1.0: x 0.6:	160: ...	2	... np pf 2	S d - dm	17.		*
11989	22 18.2 + 32 42 089.79 - 20.03	05 - 52 - 000	0383	1. x 1. 0.4: x 0.4:	--- ...	3	Dwarf spiral S V		17. - 18.		
11990	22 18.3 + 35 44 091.64 - 17.56	06 - 49 - 000	0778	(1.0 x 0.8) (1.1 x 0.9)	(110) ...	2	[E] Compact		15.6		
11991	22 18.3 + 47 27 098.34 - 07.83	08 - 40 - 006	0599	2.1 x 1.0 1.8 x 1.0	88 5	4	Sc		14.4		
11992	22 18.4 + 13 59 076.78 - 34.81	02 - 57 - 000	0875	1.1: x 1.0: 1.0: x 0.9:	--- 1:	3	...	S d - dm	16.0:		
11993	22 18.5 + 34 58 091.22 - 18.21	06 - 49 - 000	0778	1.2 x 0.1 1.1 x 0.12	113 7	2	Sc		16.0:		
11994	22 18.6 + 33 03 090.08 - 19.79	05 - 52 - 012	0383	2.5 x 0.3 2.3 x 0.25	122 7	2	Sb - c		15.0		
11995	22 18.9 + 36 20 092.10 - 17.13	06 - 49 - 000	0778 471	1.6 x 0.2 1.5 x 0.2	141 7	3	Sa		14.8		*
11996	22 19.1 + 12 14 075.51 - 36.23	02 - 57 - 000	0875	1.0: x 0.12 0.9 x 0.15	154 7	1	S ... Brightest of 3		16.5:		*
11997	22 19.3 + 45 15 097.27 - 09.76	08 - 40 - 007	0537	(1.0 x 0.7) 1.0 x 0.7	70 (3)	2	[S0] Compact		15.5		
11998	22 19.8 + 04 53 069.10 - 41.64	01 - 57 - 000	1157	1.1 x 0.45 1.1 x 0.5	... ...	2	S B c: 1st of 3, disturbed		16.0:		*
11999	22 19.8 + 42 43 095.93 - 11.93	07 - 46 - 000	0537	1.0 x 0.15 0.9 x 0.10	129 7	2	Sc		17.		
12000	22 19.9 + 02 36 066.87 - 43.22	00 - 57 - 000	0364	1.0 x 0.5 0.8 x 0.4	160 5	1	S ... In group		15.7		*
12001	22 20.0 + 36 08 092.18 - 17.42	06 - 49 - 005 NGC 7264	0778 471	2.2 x 0.30 2.2 x 0.30	57 7	4	Sb P w NGC 7263		14.7		*3
12002	22 20.0 + 41 05 095.04 - 13.31	07 - 46 - 001	0537	(1.0 x 0.9) (1.0 x 0.9)	--- ...	2	E		15.7		
12003	22 20.1 + 73 24 112.75 + 13.83	12 - 21 - 000	0559	1. x 1. 1. x 1.	--- ...	1	Dwarf?		17. - 18.		
12004	22 20.2 + 35 58 092.11 - 17.58	06 - 49 - 006 NGC 7265	0778 471	(2.4 x 1.9) (2.6 x 2.2)	170: 2	4	S0 P w 12007	S 0 -	13.7		*13
12005	22 20.3 + 35 46 092.01 - 17.76	06 - 49 - 000	0778 471	1.0 x 1.0 1.0 x 1.0	--- ...	1	...		16.5:		*

N G C   7248, 7250, 7253, 7264, 7265

1	2	3	4	5	6	7	8	9	10	11	12
12006	22 20.3 + 36 51 092.65 − 16.86	06 − 49 − 007	0778 471	(1.5 x 0.8) (1.5 x 0.8)	150 5	2	S B: 0 − a		14.5		
12007	22 20.5 + 35 57 092.16 − 17.63	06 − 49 − 009	0778 471	(1.4 x 1.1) (1.3 x 1.1)	... ...	1	... P w 12004		15.3		*1C
12008	22 20.5 + 37 27 093.04 − 16.39	06 − 49 − 000	0779 471	1.0 x 0.9 0.9: x 0.9:	--- 1	4	Sc	Sc +	16.0:		
12009	22 20.5 + 37 44 093.20 − 16.15	06 − 49 − 008	0778 471	(1.5 x 0.8) (1.2 x 0.6)	176 ...	1	...		13.8		*
12010	22 20.8 + 05 17 069.72 − 41.54	01 − 57 − 002	1157	1.1 x 0.2 1.1 x 0.2	101 7	1	Sa − b		15.4		
12011	22 20.8 + 30 40 089.01 − 22.01	05 − 52 − 000	0383	[0.6 x 0.6] ...	--- ...	1	Pair of compacts Disrupted		14.0	See Notes	*C
12012	22 20.8 + 35 35 091.99 − 17.97	06 − 49 − 010	0778 471	1.0 x 0.6 1.0 x 0.6	133 4	3	Sb nf of 2		15.3		*
12013	22 20.8 + 35 48 092.12 − 17.79	06 − 49 − 000	0778 471	1.1 x 0.2 0.8 x 0.2	118 6 − 7	1	S ...		15.3		
12014	22 20.9 + 48 23 099.22 − 07.28	08 − 40 − 008	0599	1.0: x 0.8: 1.1: x 0.9:	--- ...	1	... npp of 2	d − m	17.		*
12015	22 21.0 + 19 35 081.63 − 30.92	03 − 57 − 001	0842	1.3 x 1.1 1.2: x 0.9:	(90) 1 − 2	3	S B c	SAB d	15.7		
12016	22 21.1 + 39 48 094.49 − 14.50	07 − 46 − 002	0537	(1.2 x 0.6) (1.2 x 0.6)	126 ...	1	... Compact		14.0		
12017	22 21.2 + 11 36 075.46 − 37.05	02 − 57 − 001	0875	1.2 x 0.9 1.2 x 0.9	140: 3:	4	S B b	S B(s)b	15.2		*1
12018	22 21.3 + 30 36 089.07 − 22.13	05 − 52 − 014	0383	1.8 x 0.35 1.8 x 0.4	34 6	2	S B? b		14.9		
12019	22 21.4 + 32 08 090.03 − 20.88	05 − 52 − 015 NGC 7270	0383	1.1 x 0.6 1.1 x 0.6	(90) (5)	1	S? ... In group		15.0		*3
12020	22 21.6 + 35 09 091.88 − 18.42	06 − 49 − 011	0778 471	1.4 x 1.1 1.4 x 1.1	160: 2	4	S B a		15.6		
12021	22 21.7 + 05 44 070.36 − 41.39	01 − 57 − 003	1157	1.6: x 0.7 0.9 x 0.4	100 6	3	Sb		15.3		
12022	22 21.7 + 33 11 090.72 − 20.05	05 − 52 − 017	0383	1.6 x 0.5 1.5 x 0.4	96 6	1	S B? ...		15.1		
12023	22 21.8 + 05 06 069.78 − 41.85	01 − 57 − 004	1157	1.1 x 1.1 1.0 x 0.9	--- 1	3	S B b	S B(r)b:	15.6		
12024	22 21.8 + 41 14 095.42 − 13.37	07 − 46 − 004	0537	1.0 x 0.9 1.0 x 0.8	--- 1 − 2	3	S0 s of 2		16.0:		*
12025	22 21.9 + 32 11 090.15 − 20.90	05 − 52 − 019 NGC 7275	0383	1.0 x 0.2 1.0 x 0.2	37 7	1	Sa In group		15.0		*
12026	22 22.0 + 35 53 092.38 − 17.86	06 − 49 − 013 NGC 7274	0778 471	(1.5 x 1.5) (1.7 x 1.7)	--- ...	3	E Brightest of 3		14.2		*13
12027	22 22.0 + 41 00 095.32 − 13.59	07 − 46 − 000	0537	1. x 1. 0.5: x 0.5:	--- ...	1	...	S: m	17.		
12028	22 22.1 + 16 20b 079.45 − 33.60	03 − 57 − 003 NGC 7272	0842	1.0 x 0.9 1.1 x 1.0,	--- 1	2	S B a		(15.0)		*13
12029	22 22.3 + 22 42 084.11 − 28.63	04 − 52 − 000	0204	1.6: x 0.5 0.6: x 0.3:	[50] ...	2	Dwarf V		17.		
12030	22 22.5 + 44 39 097.42 − 10.57	07 − 46 − 005	0537	1.0 x 1.0 1.0 x 1.0	--- 1	4	Sc		16.5:		
12031	22 23.1 + 39 15 094.51 − 15.17	06 − 49 − 000	0778	1.1 x 0.15 1.2 x 0.15	81 7	2	Sb − c		16.0:		*
12032	22 23.2 + 42 27 096.32 − 12.49	07 − 46 − 006	0537	2.2: x 0.3 2.1 x 0.3	108 6 − 7	2	S ... f of 2, distorted		15.3		*1
12033	22 23.6 + 46 49 098.76 − 08.85	08 − 42 − 000	1162	1.0: x 0.7: 0.8: x 0.8:	(120) ...	1	...	S d − m	17.		
12034	22 23.8 + 40 03 095.08 − 14.57	07 − 46 − 007 NGC 7282	0537 471	2.6 x 1.0 2.3 x 1.0:	0: 6	3	S B b	S B(r) ...	15.3		
12035	22 24.1 + 15 54 079.56 − 34.24	03 − 57 − 005 NGC 7280	0842	1.9 x 1.3 2.0 x 1.4	78 3	4	S0/Sa spp of 2		13.6	+ 1812 + 2056	*23
12036	22 24.2 + 24 52 085.97 − 27.16	04 − 53 − 001	0817	1.1 x 0.9 1.1 x 0.9	--- ...	2	E − S0		15.0		*
12037	22 24.3 + 36 29 093.14 − 17.61	06 − 49 − 016	0778 471	1.5: x 1.3: 0.8 x 0.6	(50) 1	4	S B b	S B(r)b	15.5		
12038	22 24.5 + 19 17 082.17 − 31.66	03 − 57 − 007	0842	1.2 x 0.3 1.3 x 0.3	55 6	2	Sb		15.1		
12039	22 24.5 + 35 17 092.48 − 18.64	06 − 49 − 018	0778 471	1.7 x 1.2 1.6 x 1.1	(0) 3:	4	S B b	S B(r)b Sb −	14.1		*
12040	22 24.8 + 36 06 093.00 − 17.99	06 − 49 − 019	0778 471	1.4 x 1.1 1.2 x 1.1	(175) 2:	4	S B 0	(R)S B 0	14.6		*

N G C   7270, 7272, 7274, 7275, 7280, 7282

1	2	3	4	5	6	7	8	9	10	11	12
12041	22 24.8 + 39 32 094.96 - 15.11	07 - 46 - 010	0537 471	1.2 x 0.5 1.2 x 0.4	62 6	1	S ...		16.0:		*
12042	22 25.5 + 37 12 093.76 - 17.14	06 - 49 - 000	0778 471	1.1: x 1.1: 0.9 x 0.8	--- ...	1	...		16.5:		*
12043	22 25.6 + 28 51 088.81 - 24.10	05 - 53 - 002 NGC 7286	1174	1.9 x 0.7 1.6 x 0.6	98 6	2	S0 - a		13.4		3
12044	22 25.6 + 38 20 094.42 - 16.20	06 - 49 - 020	0778 471	1.8 x 0.7 1.7: x 0.7	81 6	2	Sa - b		15.5		*
12045	22 26.0 + 16 53 080.74 - 33.77	03 - 57 - 009 NGC 7290	0842	1.7 x 1.0 1.6 x 0.9	161 4	4	Sb		13.8		*12 3
12046	22 26.0 + 29 37 089.36 - 23.51	05 - 53 - 000	1174	1.2 x 1.2 0.7 x 0.7	--- 1	2	Sb		16.0:		
12047	22 26.1 + 16 31 080.48 - 34.07	03 - 57 - 008 NGC 7291	0842	(1.3 x 1.2) 1.2 x 1.2	--- (1)	1	S0		14.8		*3
12048	22 26.1 + 30 03 089.65 - 23.17	05 - 53 - 003 NGC 7292	1174	2.3 x 1.8 2.0: x 2.0:	--- ...	4	Irr	I B m Ir I	13.1		*3
12049	22 26.4 + 16 45 080.73 - 33.93	03 - 57 - 000	0842	1.1 x 0.5 1.0 x 0.4	[100] ...	3	Dwarf irregular Ir V		17.		*
12050	22 26.6 + 18 51 082.32 - 32.31	03 - 57 - 013	0842	1.0 x 0.5 0.9 x 0.4	82 5 - 6	1	S ...		15.0		
12051	22 27.0 + 37 30 094.19 - 17.05	06 - 49 - 000	0778 471	1.0 x 0.9 0.8 x 0.7	--- 1	2	Sb/S B c:	(R′)SAB(s)bc:	15.1		
12052	22 27.0 + 38 16a 094.62 - 16.41	06 - 49 - 022	0778 471	1.0 x 0.4 1.0 x 0.4	161 6	3	Sc P w 12053		15.1		*1
12053	22 27.0 + 38 16b 094.62 - 16.41	06 - 49 - 023	0778 471	1.2 x 0.15 1.2 x 0.15	153 7	3	Sb P w 12052				*1
12054	22 27.1 + 07 27 073.23 - 41.09	01 - 57 - 005	1157	1.7 x 0.2 1.2 x 0.2	46 7	1	S ...		15.0		
12055	22 27.3 + 07 43 073.51 - 40.93	01 - 57 - 000 IC 5223	1157	[1.1 x 0.3] ...	... ...	1	Triple system		15.6		*
12056	22 27.7 + 36 28 093.73 - 18.00	06 - 49 - 024	0778 471	1.6 x 1.0 1.3: x 1.0	160: 3:	2	Sa		15.5		
12057	22 28.1 + 42 32 097.15 - 12.90	07 - 46 - 011	0537	1.2 x 0.9 (1.1 x 0.8)	55: ...	1	[S0] Compact		15.2		
12058	22 28.2 + 50 50 101.53 - 05.82	08 - 41 - 000	0590	1.0 x 0.8 1.0 x 0.8	(35) 1 - 2	1	S ...		16.5:		
12059	22 28.3 + 22 17 085.10 - 29.80	04 - 53 - 002	0817	1.2 x 0.8 1.1 x 0.7	10: 3	3	S B b		15.0		*
12060	22 28.3 + 33 34 092.17 - 20.50	06 - 49 - 025	0778 471	2. x 2. 1.6: x 0.7:	--- 1?	4	Irr	I B m Ir I	15.5		
12061	22 28.3 + 37 38 094.49 - 17.08	06 - 49 - 026	0778 471	1.1 x 0.2 1.2 x 0.2	171 7	2	Sa - b		15.4		
12062	22 28.7 + 39 13 095.44 - 15.78	06 - 49 - 000	0778 471	1.1 x 0.5 1.0 x 0.5	162 5 - 6	2	S0 - a		16.0:		*
12063	22 28.9 + 35 08 093.19 - 19.25	06 - 49 - 028	0778 471	1.5 x 0.5 1.3 x 0.4	146 6	2	Sa - b		15.6		
12064	22 29.1 + 39 06a 095.45 - 15.92	06 - 49 - 029	0778 471	(1.1 x 1.1) (1.2 x 1.2)	--- ...	2	E - S0 Brightest of 3		(14.6)		*1
12065	22 29.2 + 30 42 090.65 - 23.00	05 - 53 - 004 NGC 7303 = 7304	1174	1.6 x 1.2 1.7 x 1.4	[125] ...	3	S ... Peculiar, plume		13.7		*13
12066	22 29.4 + 19 25 083.35 - 32.26	03 - 57 - 015 Mark 305 + 306	0842 473	[1.0 x 0.8] ...	... ...	3	Double system Contact, disrupted		14.6 (+ 5700)		*
12067	22 29.4 + 20 22 084.02 - 31.50	03 - 57 - 016	0842 473	1.0 x 0.6 0.7 x 0.6	50: 4:	1	Sa s of 2		15.3		*
12068	22 29.6 - 00 13 066.23 - 46.91	00 - 57 - 003	0364 472	1.1 x 0.6 1.1 x 0.6	175 4 - 5	2	Sb		15.6		
12069	22 30.0 + 76 15 114.90 + 15.88	13 - 16 - 000	1210	2.0 x 1.3 1.6 x 1.2	(100) ...	4	...	SAB(s)dm	15.7		
12070	22 30.0 + 77 52 118.76 + 15.92	13 - 16 - 000	1210	1.1 x 0.4 0.9: x 0.3	[150] ...	3	Irr		16.5:		
12071	22 30.1 + 30 35 090.75 - 23.21	05 - 53 - 005	1174	1.3 x 0.4 1.1 x 0.3	[35] ...	2	S B b Singular?		15.1		
12072	22 30.3 + 23 40 086.45 - 28.95	04 - 53 - 000	0817	1.0 x 0.2 0.7 x 0.2	45 6 - 7	1	S ... In group		16.5:		*
12073	22 30.3 + 38 58 095.58 - 16.16	06 - 49 - 032	0778 471	2.2 x 0.8 1.7 x 0.8	100 6	4	S B b Brightest of 4	S B(s)b	14.8		*
12074	22 30.6 + 07 50 074.42 - 41.40	01 - 57 - 008	1157	0.60 x 0.40 0.55 x 0.40	145: ...	1	... Compact, loops		14.2		*C
12075	22 30.6 + 38 58 095.63 - 16.19	06 - 49 - 033	0778 471	1.6 x 1.2 1.6 x 1.2	... ...	4	S B c 12073 group, distorted		15.1		*1

N G C   7286, 7290 - 92, 7303, 7304

I C   5223

1	2	3	4	5	6	7	8 9	10	11	12
12076	22 30.9 + 43 17 098.00 - 12.53	07 - 46 - 012	0537	1.2 x 0.5 0.7 x 0.3	42 6	2	Sa - b	16.0:		
12077	22 31.1 + 38 42 095.57 - 16.47	06 - 49 - 000	0778 471	1.0: x 0.7 1.0: x 0.7:	... ...	2	S ... Brightest of 3, distorted?	16.0:		*
12078	22 31.2 + 30 35 090.97 - 23.34	05 - 53 - 000	1174	1.5 x 0.3 1.3: x 0.3	168 6	1	S ...	16.0:		
12079	22 31.3 + 38 43 095.61 - 16.47	06 - 49 - 000	0778 471	1.0: x 0.8: 0.9: x 0.7:	... ...	2	Dwarf IV - V:, 12077 group	16.5:		*
12080	22 31.6 + 05 19 072.34 - 43.41	01 - 57 - 009 NGC 7311	1157	1.8 x 0.9 1.8 x 0.9	10 5	2	Sa - b P w 12083	13.4		*3
12081	22 31.6 + 09 54 076.47 - 40.01	02 - 57 - 000	0875	1.3 x 0.25 1.3 x 0.25	132 7	2	Sa - b	15.5		
12082	22 31.9 + 32 37 092.30 - 21.71	05 - 53 - 006	1174	3.5:x 3.0:: 3. x 2.5::	--- ...	4	Dwarf spiral, S m S V	15.6		*
12083	22 32.1 + 05 33 072.69 - 43.32	01 - 57 - 010 NGC 7312	1157	1.7 x 1.0 1.3 x 0.8	83 4	4	S B b S B(s)b P w 12080, Sb(-)	14.5		*3
12084	22 32.1 + 24 46 087.55 - 28.28	04 - 53 - 005	0817	1.6 x 1.2 1.4 x 1.2	120: 2 - 3	2	Sb/S B b	15.3		
12085	22 32.1 + 34 53 093.63 - 19.82	06 - 49 - 000	0778 471	1.0 x 0.35 0.9 x 0.3	0 6	2	Sb	15.6		
12086	22 32.1 + 41 03 097.01 - 14.56	07 - 46 - 013	0573 471	1.0 x 0.8 1.0 x 0.8	85: 2:	2	S0 - a	14.8		
12087	22 32.2 + 41 17 097.15 - 14.27	07 - 46 - 000	0537 471	1.0 x 0.2 1.0 x 0.2	117 7	1	S ...	16.5:		
12088	22 32.2 + 43 32 098.34 - 12.44	07 - 46 - 014	0537	1.1 x 0.9 1.1 x 0.9	(20) 2	3	Sa	15.2		
12089	22 32.2 + 50 09 101.74 - 06.73	08 - 41 - 001	0590	1.1 x 0.4 1.1 x 0.4	165 6	2	Sb - c P w 12091, 12091 group	15.7		*
12090	22 32.3 + 15 42 081.29 - 35.63	03 - 57 - 017	0842	1.0 x 0.4 0.7 x 0.4	100 ...	3	Irr	15.7		
12091	22 32.3 + 50 11 101.77 - 06.71	08 - 41 - 002	0590	(1.2 x 0.7) (2.0 x 1.5)	178 ...	2	[E] Compact, p w 12089 Brightest in group	15.7		*
12092	22 32.4 + 06 03 073.23 - 43.01	01 - 57 - 011	1157	1.0: x 1.0: 0.7: x 0.5:	--- 1:	2	... S d - dm	15.7		
12093	22 32.4 + 18 23 083.29 - 33.51	03 - 57 - 018	0842 473	1.5 x 0.7 1.4 x 0.7	110 5	2	Sc	15.4		*
12094	22 32.4 + 22 17 085.99 - 30.36	04 - 53 - 000	0817	1.0: x 0.8: 0.7: x 0.7:	(0) 1 - 2	3	... S dm	16.5:		
12095	22 32.6 + 49 55a 101.67 - 06.97	08 - 41 - 003	0590	1.0: x 0.6: 0.9: x 0.5:	... ...	3	Irr P w 12096, contact	15.6		*1
12096	22 32.6 + 49 55b 101.67 - 06.97	08 - 41 - 004	0590	1.1 x 0.15 1.1 x 0.2	117 (7)	1	S? ... P w 12095, contact			
12097	22 33.2 + 34 34 093.66 - 20.20	06 - 49 - 037 NGC 7315	0778 471	(1.6 x 1.6) (1.7 x 1.7)	--- 1	4	S0	13.8		3
12098	22 33.5 + 20 04 084.72 - 32.31	03 - 57 - 020 NGC 7316, Mark 307	0842 473	1.1 x 0.9 1.0 x 0.7	60: 2 - 3	1	S ...	13.7	+ 7200	3
12099	22 33.7 + 33 43a 093.27 - 20.98	06 - 49 - 040 NGC 7318A	0778 471	[1.7 x 1.2]	...	3	E + S B(s) ...	14.9	+ 6689 + 6967	*12 3VĀ
12100	22 33.7 + 33 43b 093.27 - 20.98	06 - 49 - 040 NGC 7318B	0778 471	...	...		Loops, Stephan's Quintet	14.4	+ 5727 + 6005	*12 3VĀ
12101	22 33.8 + 33 41 093.27 - 21.02	06 - 49 - 042 NGC 7320	0778 471	1.9 x 1.0 1.8 x 0.9	132 5	4	Sc Brightest in Stephan's Quintet, Sc -	13.8	+ 795 + 1073	*12 3VĀ
12102	22 33.8 + 33 43 093.29 - 20.99	06 - 49 - 041 NGC 7319	0778 471	1.6 x 1.2 1.4 x 1.2	... ...	4	S B ... Stephan's Quintet, distorted	14.8	+ 6657 + 6935	*12 3VĀ
12103	22 34.0 + 21 22 085.73 - 31.32	03 - 57 - 021 NGC 7321	0842	1.7 x 1.0 1.6 x 0.9	12 4	3	S B b S B(r)b	14.0		*13
12104	22 34.1 + 02 09 069.85 - 46.11	00 - 57 - 000	0364 472	1.2: x 0.15 1.1 x 0.15	93 7	1	S ...	16.0:		
12105	22 34.1 + 75 23 118.22 + 13.48	13 - 16 - 000 IC 1502	1210	1.1 x 0.30 1.2 x 0.45	52 7:	2	S0	14.7		
12106	22 34 2 + 25 30 088.47 - 27.94	04 - 53 - 007 IC 5233	0817 474	1.1 x 0.9 1.1 x 0.8	175: 2 - 3	1	S ...	14.8		
12107	22 34.3 + 19 07 084.24 - 33.19	03 - 57 - 024	0842 473	(1.3 x 0.7) 1.4 x 0.7	135 5:	1	S0	15.3		
12108	22 34.4 + 18 53 084.10 - 33.39	03 - 57 - 025 NGC 7323	0842 473	1.3 x 1.1 1.2 x 1.0	170 2	2	Sb p of 2, disturbed:	14.0		*13
12109	22 34.4 + 38 35 096.08 - 16.90	06 - 49 - 000	0778 471	1.0 x 1.0 0.8: x 0.8:	--- ...	1	...	16.0:		*
12110	22 34.5 + 14 09 080.63 - 37.18	02 - 57 - 005	0875	1.2 x 0.2 1.1 x 0.2	137 6 - 7	1	S ... Brightest of 3	15.4		*1

N G C   7311, 7312, 7315, 7316, 7318 - 21   I C   1502, 5233
        7323

1	2	3	4	5	6	7	8	9	10	11	12
12111	22 34.7 + 38 17 095.97 – 17.19	06 – 49 – 046 NGC 7330	0778 471	(1.8 x 1.7) (1.8 x 1.7)	— ...	2	E		13.6		*3
12112	22 34.8 + 11 42 078.75 – 39.14	02 – 57 – 000	0875	1.0 x 0.10 1.0 x 0.10	137 7	2	Sc		16.0:		
12113	22 34.8 + 34 10 093.73 – 20.72	06 – 49 – 045 NGC 7331	0778 (471)	11.4 x 4.0 10.3 x 4.0	171 6	5	Sb	SA bc Sb +	10.4	+ 794 + 1072	*12– –689
12114	22 34.9 – 01 17 066.37 – 48.61	00 – 57 – 005	0364 475	1.0: x 0.9: 0.8: x 0.7:	— 1:	2	...	S B d – dm	16.5:		*
12115	22 35.0 + 23 32 087.38 – 29.67	04 – 53 – 008 NGC 7332	0817	3.6 x 0.9 3.6 x 0.9	155 7	3	S0 P w 12122		12.0	+ 1204 + 1463	*12 3468
12116	22 35.0 + 34 12 093.79 – 20.71	06 – 49 – 047 NGC 7335	0778 471	1.4 x 0.6 1.1 x 0.6	151 5 – 6	2	S0 – a Brightest of 4		14.7	+ 6298 + 6576	*12 3
12117	22 35.0 + 49 56 102.02 – 07.14	08 – 41 – 000	0590	1.0: x 0.9: 0.9: x 0.9:	— ...	1	... p of 2, disturbed?		16.0:		*
12118	22 35.1 + 10 16 077.64 – 40.29	02 – 57 – 007 NGC 7328	0875	2.1 x 0.7 2.0 x 0.7	88 6	2	Sa – b		14.3		3
12119	22 35.1 + 18 32 084.01 – 33.77	03 – 57 – 027	0852 473	1.0 x 0.5 0.7 x 0.5	150 5	1	S ...		16.0:		
12120	22 35.1 + 34 07 093.76 – 20.79	06 – 49 – 050 NGC 7337	0778 471	1.2 x 1.0 1.2 x 1.0	— 1 – 2	3	S B a NGC 7335 group		15.7		*12 3
12121	22 35.3 + 34 36 094.07 – 20.40	06 – 49 – 051	0778 471	1.1 x 0.5 1.1 x 0.5	70 5 – 6	2	S0		14.9		
12122	22 35.4 + 23 32 087.46 – 29.74	04 – 53 – 009 NGC 7339	0817	2.8 x 0.8 2.9 x 0.8	93 6	1	Sb – c P w 12115		13.1	+ 1271 + 1530	*12 36
12123	22 35.5 + 24 56 088.39 – 28.57	04 – 53 – 000	0817 474	1.1 x 0.1 0.8 x 0.1	36 7	2	S B? c		15.6		
12124	22 35.5 + 25 05 088.48 – 28.45	04 – 53 – 000	0817 474	1.0 x 0.5 1.0 x 0.5	113 (5)	1	S0? Companion superimposed?		15.0		*
12125	22 35.5 + 40 04 097.06 – 15.73	07 – 46 – 015	0537 471	1.5 x 0.3 1.2 x 0.25	5 6	1	Sa		16.0:		
12126	22 35.9 + 35 15 094.54 – 19.91	06 – 49 – 054 NGC 7342	0778 471	1.5 x 1.4 1.5 x 1.4	— 1	4	S B a 12127 group		15.3		*3
12127	22 36.2 + 35 05 094.50 – 20.08	06 – 49 – 058 NGC 7343	0778 471	(1.5 x 1.5) (1.5 x 1.5)	— ...	3	E Brightest in group		15.0		*1
12128	22 36.3 + 28 23 090.68 – 25.79	05 – 53 – 000	1174	1.1: x 0.6: 1.0: x 0.5:	137 (5)	1	S? ... Singular?		17.		*
12129	22 36.3 + 33 48 093.81 – 21.19	06 – 49 – 059 NGC 7343	0778 471	1.0 x 0.8 1.0 x 0.8	(160) 2	4	S B b 	(R')S B(s)b Sb –	14.3	+ 1206 + 1493	*12 3
12130	22 36.4 + 35 17 094.65 – 19.93	06 – 49 – 064 NGC 7345	0778 471	1.2 x 0.2 1.2 x 0.2	39 7	2	Sa 12127 group		15.1		*3
12131	22 36.4 + 37 20 095.77 – 18.19	06 – 49 – 065	0778 471	1.1 x 0.8 1.0 x 0.8	(45) 2 – 3	2	Sb		15.2		
12132	22 36.5 + 34 03 093.99 – 21.00	06 – 49 – 000	0778 471	1.0: x 0.5 0.9 x 0.4	167 5	2	S B? b		15.4		
12133	22 37.0 + 08 22 076.48 – 42.04	01 – 57 – 014	1157	1.8 x 0.2 1.8 x 0.2	44 7	2	Sc 12139 group		15.7		*
12134	22 37.1 + 11 31 079.17 – 39.64	02 – 57 – 008	0875	1.8 x 0.7 1.7 x 0.7	145 6	2	Sb – c		14.9		1
12135	22 37.1 + 72 15 113.23 + 12.19	12 – 21 – 000	0559	1.4 x 0.6 1.4 x 0.6	89 6	3	Sc		16.0:		
12136	22 37.5 + 10 46 078.65 – 40.28	02 – 57 – 009 NGC 7347	0875	1.7 x 0.35 1.7 x 0.3	133 7	1	S ...		14.7		*3
12137	22 37.6 + 37 57 096.31 – 17.77	06 – 49 – 069	0778 471	1.9 x 1.6 1.9 x 1.7	(25) 1 – 2	4	S B b/Sc	SAB(s)bc Sb +	14.0		1
12138	22 37.7 + 07 47 076.15 – 42.60	01 – 57 – 016	1157	0.9 x 0.8 0.8 x 0.6	— 1	3	S B a		14.3		
12139	22 37.7 + 08 18 076.60 – 42.21	01 – 57 – 015	1157	1.1 x 0.2 1.0 x 0.2	175 6	1	S B ... Brightest in group		15.5		*
12140	22 38.0 + 15 41 082.63 – 36.48	03 – 57 – 000	0842	1.1 x 0.4 1.1 x 0.3	31 6	1	S ...		15.3		
12141	22 38.0 + 80 25 117.53 + 19.24	13 – 16 – 000	1210	1.4 x 0.3 1.3 x 0.3	175 6	2	Sc		15.6		
12142	22 38.1 + 11 39 079.53 – 39.69	02 – 57 – 010 NGC 7348	0875	1.3 x 0.7 1.0 x 0.6	12 5:	2	Sc		14.8		*23
12143	22 38.1 + 31 35 092.91 – 23.28	05 – 53 – 008	1174	1.6 x 0.6 1.3 x 0.5	64 6	1	Sa – b		15.2		
12144	22 38.2 + 33 12 093.84 – 21.91	05 – 53 – 000	1174	1.2 x 0.9 1.0 x 0.9	(120) 2:	4	S B b	S B(s)b	15.4		*
12145	22 38.3 + 33 19 093.92 – 21.82	05 – 53 – 000	1174	1.1 x 0.3 0.5 x 0.15	74 7?	1	S ... Disturbed?		16.0:		*

N G C   7328, 7330 – 32, 7335, 7337, 7339
7342, 7343, 7345, 7347, 7348

1	2	3	4	5	6	7	8	9	10	11	12
12146	22 38.5 + 22 13 087.30 - 31.21	04 - 53 - 000	0817	1.1 x 0.3 1.1 x 0.3	107 6	2	Sb - c		15.7		
12147	22 38.7 + 18 32 084.85 - 34.27	03 - 57 - 030	0842 473	1.1 x 0.5 1.1 x 0.5	30 ...	1	S ... Disturbed?		15.4		*1
12148	22 38.8 + 23 08 087.96 - 30.49	04 - 53 - 010 IC 5242	0817	1.0 x 0.9 1.0 x 0.9	--- ...	1	... Compact, p w 12153		14.7		*1C
12149	22 38.8 + 31 55 093.24 - 23.07	05 - 53 - 009	1174	1.1 x 1.1 1.0 x 1.0	--- 1	4	S B a	(R´)S B(s)a	14.5		
12150	22 38.9 + 34 00 094.42 - 21.30	06 - 49 - 071	0778 471	1.1 x 0.6 1.0 x 0.5	25 5	2	S B 0 - a		15.0		
12151	22 39.0 + 00 08 068.99 - 48.39	00 - 57 - 007	0364 475	3. x 2. 3. x 2.	(0) ...	4	Dwarf IV - V		16.0:		
12152	22 39.0 + 02 22 071.33 - 46.81	00 - 57 - 008 IC 5241	0364	1.0 x 0.9 0.9 x 0.8	--- 1:	1	S? ... Peculiar?		14.9		*1
12153	22 39.0 + 23 07 087.99 - 30.53	04 - 53 - 011 IC 5243	0817	0.7 x 0.6 0.7 x 0.6	--- ...	1	Pair of compacts Contact, p w 12148		14.3	+ 7105 + 7362	*1C
12154	22 39.0 + 71 34 113.01 + 11.53	12 - 21 - 000	0559	1.6 x 0.2 1.6 x 0.2	24 7	3	Sb		16.0:		
12155	22 39.2 + 35 01 095.03 - 20.46	06 - 49 - 073	0778 471	1.1 x 0.4 1.1 x 0.4	59 6	3	S B b		15.4		
12156	22 39.2 + 39 02 097.16 - 16.99	06 - 49 - 074	0778 471	1.1 x 1.1 1.2 x 1.0	--- 1	4	S B c	S B(s)c	14.8		
12157	22 39.6 + 34 40a 094.91 - 20.80	06 - 49 - 075	0778 471	1.6 x 0.5 1.5 x 0.5	102 6	1	S ... p of 2, disturbed:		(15.5)		*C
12158	22 39.7 + 19 44 085.91 - 33.42	03 - 57 - 032	0842	1.3 x 1.1 1.3 x 1.1	--- 2	2	Sb		15.3		
12159	22 39.7 + 30 27 092.58 - 24.43	05 - 53 - 010 NGC 7356	1174	1.1 x 0.5 1.1 x 0.55	76 5	2	Sb - c		15.0		
12160	22 39.7 + 74 53 114.72 + 14.40	12 - 21 - 000	0599	2.3: x 1.8: 1.9: x 1.5:	17: ...	4	Sc		15.5:		
12161	22 39.9 + 32 57 094.03 - 22.31	05 - 53 - 000	1174	1.0 x 1.0 0.7 x 0.5	--- 1	1	S ...		15.3		
12162	22 40.0 + 29 55 092.34 - 24.92	05 - 53 - 011 NGC 7357	1174	1.8 x 0.7 1.3 x 0.7	120 5 - 6	2	Sb		15.1		1
12163	22 40.3 + 29 28 092.14 - 25.33	05 - 53 - 012	1174	0.8 x 0.45 0.7 x 0.45	118 ...	1	S B ...		14.4		
12164	22 40.5 + 30 15 092.63 - 24.69	05 - 53 - 013	1174	1.6 x 0.8 1.5: x 0.8	132 5	2	Sb		15.6		
12165	22 40.5 + 32 44 094.03 - 22.56	05 - 53 - 000	1174	1.1 x 0.15 1.0 x 0.15	124 7	1	Sb - c		15.7		
12166	22 40.6 + 04 47 074.13 - 45.32	01 - 57 - 000	1157	1.0 x 0.2 0.6 x 0.2	30 ...	1	Dwarf irregular		17.		
12167	22 41.0 + 03 53 073.37 - 46.06	01 - 58 - 001 NGC 7360	0821	0.7 x 0.3 0.7 x 0.3	153 (6)	1	...		14.5		3
12168	22 41.1 + 05 52 075.29 - 44.60	01 - 57 - 000	1157	1.0: x 0.12 0.7 x 0.12	32 7	1	S ...		16.0:		
12169	22 41.2 + 23 41 088.84 - 30.33	04 - 53 - 013	0817	1.1 x 0.4 1.1 x 0.4	26 6	2	Sb nff of 2		15.7		*
12170	22 41.2 + 78 42 116.76 + 17.68	13 - 16 - 000	1210	1.1 x 0.2 0.9 x 0.2	145 6	2	Sc		16.0:		
12171	22 41.3 + 08 27 077.66 - 42.67	01 - 58 - 002 NGC 7362	0821	(1.2 x 0.8) (1.5 x 1.0)	175 ...	3	[E] Compact		14.9		*
12172	22 41.5 + 07 07 076.53 - 43.72	01 - 58 - 003	0821	1.2 x 0.6 1.1 x 0.5	98 5	2	Sc		15.5		
12173	22 41.6 + 38 06 097.11 - 18.03	06 - 49 - 079	0778	2.1 x 1.3 2.2 x 1.2	80 5	4	Sc	SAB(rs)c	13.7		
12174	22 41.9 - 00 23 069.19 - 49.27	00 - 58 - 001 NGC 7364	0905	(1.7 x 1.2) (1.8 x 1.3)	65: 3	3	Sa		13.8		*3
12175	22 42.1 + 03 25 073.20 - 46.58	00 - 58 - 002 NGC 7367	0905	1.6 x 0.4 1.5 x 0.4	128 6	2	Sa - b		14.9		3
12176	22 42.2 + 48 30 102.36 - 08.97	08 - 41 - 005	0590	1.2 x 1.1 1.1 x 1.0	--- 1	4	Sc	Sc +	16.0:		
12177	22 42.4 + 33 12 094.66 - 22.36	05 - 53 - 014	1174	1.0 x 1.0 1.1 x 1.1	--- 1	1	S ... sp of 2		13.5		*1
12178	22 42.6 + 06 10 075.96 - 44.62	01 - 58 - 004	0821	3.3 x 1.7 3.1 x 1.6	10 5	4	...	SAB(s)dm	14.2		
12179	22 42.8 + 33 44 095.02 - 21.94	06 - 50 - 002	0838	(1.3 x 1.3) 1.1 x 1.1	--- 1	3	S0		14.5		*
12180	22 43.0 + 21 33 087.88 - 32.35	04 - 53 - 000 IC 5253	0817	1.0 x 0.3 0.9 x 0.25	163 6 - 7	1	S0 - a		15.5		*^

N G C   7356, 7357, 7360, 7362, 7364, 7367   I C   5241 - 43, 5253

1	2	3	4	5	6	7	8	9	10	11	12
12181	22 43.9 + 37 48 097.37 - 18.51	06 - 50 - 003	0838	1.1 x 0.9 1.1 x 0.9	(155) 1 - 2	3	Sc/S B c npp of 2, disturbed		14.5		*
12182	22 44.0 + 72 53 114.00 + 12.50	12 - 21 - 001	0559	1.1: x 0.9: 1.0: x 0.9:	(15) ...	1	S ...		15.5:		
12183	22 44.5 - 00 18 069.98 - 49.68	00 - 58 - 003	0905	1.0 x 0.3 1.0 x 0.3	29 6 - 7	1	S ...		15.5		
12184	22 44.6 + 11 20 080.92 - 40.92	02 - 58 - 010	0800 476	1.0 x 1.0 1.0: x 0.9:	--- 1	4	Sc	Sc +	15.7		
12185	22 45.0 + 31 07 094.03 - 24.44	05 - 53 - 015	1174	1.8 x 0.9 1.6 x 0.9	150 4 - 5	3	S B b	S B(s)ab Sb -	14.8		
12186	22 45.1 + 49 52 103.43 - 07.98	08 - 41 - 000	0590	1.0 x 0.8: 0.9: x 0.8:	--- ...	2	...	S m	16.0:		
12187	22 45.2 + 39 58 098.69 - 16.74	07 - 46 - 018 NGC 7379	0537 471	1.2 x 0.9 1.4 x 0.9	(90) 3:	3	S B a		14.4		3
12188	22 45.3 + 39 37 098.54 - 17.05	07 - 46 - 019	0537	[0.9 x 0.7] ...	... ...	2	S ... + companion Contact		14.4		*1
12189	22 45.6 + 03 39 074.38 - 47.00	01 - 58 - 007	0821	1.0 x 0.6 0.9 x 0.6	(130) ...	1	S B ...		16.0:		
12190	22 45.7 + 28 01 092.43 - 27.19	05 - 53 - 000	1174	2.2 x 0.2 2.0 x 0.2	172 7	3	Sc	Sc -	15.6		
12191	22 46.3 + 27 21 092.18 - 27.83	04 - 53 - 015	0817	1.5 x 0.8 1.1 x 0.6	2 5	1	S ... 12193 group		14.9		*1
12192	22 46.3 + 36 57 097.37 - 19.49	06 - 50 - 000	0838	1.0: x 0.9: 1. x 1.	--- 1:	2	...	S dm - m	17.		
12193	22 46.4 + 27 19 092.18 - 27.87	04 - 53 - 016	0817	1.3 x 1.2 1.1 x 1.0	--- 1	4	S B c Brightest of 3, Sc -	S B(s)bc	14.1		*1
12194	22 46.4 + 27 42 092.40 - 27.54	05 - 53 - 000	1174	1.1 x 0.5 1.0 x 0.4	85: 6	1	S ...		15.7		
12195	22 46.4 + 32 58 095.32 - 22.98	05 - 53 - 000	1174	1.1 x 0.8 1.1 x 0.9:	--- 1	2	S B a - b		15.7		
12196	22 46.5 + 06 57b 077.71 - 44.64	01 - 58 - 009	0821	1.1 x 0.3 1.1 x 0.35	156 6	1	Sa s of 2		(15.2)		*1
12197	22 46.5 + 24 35 090.58 - 30.22	04 - 53 - 000	0817	1.2: x 0.2 1.1 x 0.2	119 7	1	Sb - c		17.		
12198	22 46.8 + 27 36 092.43 - 27.67	05 - 53 - 000	1174	1.0: x 0.25 0.9 x 0.25	176 6	1	S ...		15.7		
12199	22 46.8 + 39 44 098.86 - 17.09	07 - 46 - 021	0537	1.5 x 0.9 1.3 x 0.9	90: 4:	4	S B b	S B(rs)b	14.5		
12200	22 46.9 + 19 01 087.13 - 34.97	03 - 58 - 006	1161	1.0 x 0.55 0.9 x 0.5	25 4 - 5	2	Sb		14.8		
12201	22 46.9 + 34 44 096.35 - 21.49	06 - 50 - 004	0838	1.6 x 0.3 1.4 x 0.3	48 7	2	Sa - b		15.0		*
12202	22 47.0 + 11 03 081.31 - 41.50	02 - 58 - 013	0800 476	1.1 x 0.8 1.2 x 0.8	135 3	2	Sa		14.8		*1
12203	22 47.1 + 32 33 095.23 - 23.41	05 - 53 - 000	1174	1.0 x 0.12 1.0 x 0.12	173 7	1	S ...		16.5:		
12204	22 47.1 + 39 58 099.03 - 16.91	07 - 46 - 000	0537	1.6 x 0.12 1.4 x 0.12	28 7	2	Sb - c		16.0:		
12205	22 47.2 + 14 49 084.26 - 38.48	02 - 58 - 015	0800	0.9 x 0.3 0.9 x 0.3	165 ...	3	Irr		15.4		
12206	22 47.2 + 33 06 095.55 - 22.94	05 - 53 - 016	1174	3.0 x 1.7 (2.3 x 1.3)	125: 4	3	S B a Streamer?	S B(rs)a	15.5		*
12207	22 47.4 + 11 20 081.64 - 41.33	02 - 58 - 017 NGC 7385	0800 476	(1.3 x 1.1) (1.9 x 1.7)	--- ...	3	E Brightest in cluster		14.4	+ 7829 + 8051	*12 <u>3</u>
12208	22 47.6 - 01 17 069.75 - 50.92	00 - 58 - 005	0905	1.1 x 0.8 1.1 x 0.7	70 3	3	S B c	Sc +	15.2		
12209	22 47.6 + 11 25 081.76 - 41.29	02 - 58 - 018 NGC 7386	0800 476	(1.8 x 1.5) (2.1 x 1.6)	150: ...	2	E - S0		14.6	+ 7198 + 7420	*12 <u>3</u>
12210	22 47.8 + 31 07 094.60 - 24.73	05 - 53 - 018	1174	1.3 x 1.2 1.2 x 1.1	--- 1	4	S B a	S B(r)a	15.4		*
12211	22 48.1 - 01 47 069.32 - 51.36	00 - 58 - 006 NGC 7391	0905	(1.8 x 1.7) (2.0 x 1.8)	--- ...	4	E		13.7		3
12212	22 48.1 + 28 52 093.43 - 26.72	05 - 53 - 019	1174	2.5:: x 1.8:: 2. x 1.	... ...	4	Dwarf spiral S V		16.0:		
12213	22 48.5 + 07 02 078.32 - 44.89	01 - 58 - 010	0821	2.0: x 1.9: 2.1:: x 2.0::	--- 1	4	...	SAB: dm	16.0:		
12214	22 48.6 + 31 06 094.76 - 24.83	05 - 53 - 020	1174 477	(1.4 x 0.9) (1.2 x 0.8)	160: ...	2	... Compact, p of 2		14.1		*1
12215	22 48.6 + 34 35 096.60 - 21.79	06 - 50 - 005	0838	(1.5 x 1.0) (1.6 x 0.9)	[65] ...	1	... Compact, halo		15.5		*C

N G C   7379, 7385, 7386, 7391

1	2	3	4	5	6	7	8	9	10	11	12
12216	22 48.8 + 36 49 097.77 – 19.84	06 – 50 – 006 NGC 7395	0838	1.2 x 1.1 (1.2 x 1.1)	-- (1)	1	SO?		15.3		*13
12217	22 49.0 + 22 49 090.07 – 32.02	04 – 53 – 017 IC 5258	0817	(1.6 x 1.0) (1.5 x 1.0)	105: ...	2	E – SO		14.1		
12218	22 49.0 + 32 05 095.37 – 24.02	05 – 53 – 023	1174	1.0 x 0.45 1.0 x 0.4	[22] ...	1	Irr? or S B ...?		14.9		
12219	22 49.3 + 17 00 086.34 – 36.96	03 – 58 – 000	1161	[1.0 x 0.7] ...	... ...	2	Double system In group of 3		15.6		*
12220	22 49.9 + 00 50 072.71 – 49.80	00 – 58 – 007 NGC 7396	0905	1.8 x 1.0 (2.0 x 1.3)	103 7	2	Sa Brightest in group		14.1		*13
12221	22 50 + 82 38 119.09 + 20.95	14 – 01 – 001	0568	2.5 x 0.8 2.2 x 0.8	95 6	4	Sc	SA(s)d Sc +	14.9		
12222	22 50.0 + 11 23 082.37 – 41.67	02 – 58 – 031	0800 476	1.2 x 0.8 1.2 x 0.8	175 3	3	Sb		15.2		*
12223	22 50.0 + 49 11 103.83 – 08.95	08 – 41 – 000	0590	1.1: x 0.7: 1.2 x 0.8	(175) 3	1	S ...		16.0:		
12224	22 50.1 + 05 49 077.67 – 46.09	01 – 58 – 011	0821	2.1: x 2.1: 1.8 x 1.8	-- 1	4	Sc		15.3		*
12225	22 50.3 + 00 57 072.94 – 49.79	00 – 58 – 009 NGC 7398	0905	1.2 x 0.8 1.1 x 0.8	75 3	3	Sa NGC 7396 group		14.9		*13
12226	22 50.4 + 15 33 085.60 – 38.31	03 – 58 – 009	1161	1.1 x 0.5 1.1 x 0.5:	130 6	2	S B c	Sc +	15.3		
12227	22 50.5 + 34 13 096.78 – 22.30	06 – 50 – 000	0838	1.1: x 0.9: 1.1: x 1.0:	-- 1	4	Sc f of 2	SAB(s)d	16.5:		*
12228	22 50.6 + 50 17 104.41 – 08.01	08 – 41 – 000	0590	1.0 x 0.4 1.0 x 0.4	148 6	2	...	S dm:	17.		
12229	22 50.8 + 32 30 095.95 – 23.84	05 – 54 – 001	1184 477	1.1 x 0.4 1.1 x 0.4	178 6	1	S ...		15.6		
12230	22 50.9 + 31 52 095.64 – 24.40	05 – 54 – 002 NGC 7407	1184 477	2.1 x 1.0 2.0 x 0.9	152 5	3	Sc Singular,	S bc Sc –	14.2		*3
12231	22 51.2 + 31 21 095.43 – 24.89	05 – 54 – 004	1184 477	1.5 x 0.7 1.3 x 0.7	13 5?	3	S B c ssf of 2, distorted		15.5		*1
12232	22 51.3 + 01 07 073.28 – 49.83	00 – 58 – 011 IC 1455	0905	1.0 x 0.6 1.0 x 0.6	42 4 – 5	3	S B a		14.9		
12233	22 51.3 + 25 35 092.25 – 29.92	04 – 54 – 001	1175	1.1: x 1.0 0.8 x 0.7	-- 1	3	S B b	SAB(s)b	15.4		
12234	22 51.3 + 33 26 096.54 – 23.06	05 – 54 – 005	1184 477	1.3 x 0.7 1.3 x 0.7	130 5	2	Sb – c		13.3		*
12235	22 51.7 + 32 06 095.93 – 24.28	05 – 54 – 006	1184 477	1.1 x 0.7 1.2 x 0.7	45: 4	2	SO		15.1		
12236	22 51.7 + 35 58 097.90 – 20.87	06 – 50 – 008	0838	1.6: x 1.6: 1.5: x 1.5:	-- 1	4	...	SA m	16.5:		*
12237	22 51.8 + 11 30 082.94 – 41.83	02 – 58 – 032	0800 476	1.3 x 0.4 1.5: x 0.2	92 6 – 7	2	Sb – c P w 12243		15.7		*
12238	22 51.9 + 31 59 095.91 – 24.40	05 – 54 – 007	1184 477	1.2 x 1.1 1.2 x 1.1,	(70) 1	3	SO	SAB 0	14.9		*
12239	22 51.9 + 48 53 103.98 – 09.36	08 – 41 – 000	0590	1.0: x 1.0: 0.5: x 0.5:	-- 1:	2	...	S d – dm	16.5:		
12240	22 52.0 + 20 55 089.59 – 34.01	03 – 58 – 000	1161 476	1.2 x 0.2 1.1 x 0.2	68 6	2	S B? a – b		16.0:		
12241	22 52.1 + 19 58 089.01 – 34.83	03 – 58 – 010 NGC 7411	1161 479	(1.0 x 0.9) (1.1 x 1.0)	-- ...	3	[E] Compact 4th in chain of 6		14.8		*3
12242	22 52.1 + 32 10 096.04 – 24.26	05 – 54 – 009	1184 477	(1.3 x 1.3) (1.4 x 1.4)	-- ...	2	[E] Compact		14.5		*
12243	22 52.2 + 11 27 083.01 – 41.93	02 – 58 – 034	0800 476	1.0 x 0.25 1.0 x 0.25	150 7	2	Sa – b P w 12237		15.6		*
12244	22 52.4 + 19 59b 089.10 – 34.85	03 – 58 – 012 NGC 7415	1161 479	1.1 x 0.15 0.9 x 0.12	128 7	2	Sa – b 6th in chain of 6		(15.0)		*13
12245	22 52.4 + 21 33 090.09 – 33.51	04 – 54 – 002	1175	1.0 x 0.45 1.0 x 0.45	128 6	2	Sc Singular?		15.6		*
12246	22 52.4 + 37 58 099.01 – 19.16	06 – 50 – 009	0838	1.0 x 0.7 0.8 x 0.6	(55) ...	2	...	S dm:	16.5:		
12247	22 52.4 + 74 57 115.51 + 14.07	12 – 21 – 000	0559	1.2 x 0.25 1.1 x 0.2	20 6	2	Sc		16.0:		
12248	22 52.5 + 36 16 098.20 – 20.68	06 – 50 – 000	0838	1.0 x 0.9 0.8 x 0.7	-- ...	3	... Peculiar		16.5:		*
12249	22 53.0 + 28 04 094.04 – 27.95	05 – 54 – 000	1184	1.0 x 0.12 0.9 x 0.12	55 7	3	Sc		16.0:		
12250	22 53.1 + 12 32 084.08 – 41.17	02 – 58 – 036	0800 478	1.6 x 0.9 1.6 x 0.9	13 4	4	S B b		14.4		*1

N G C   7395, 7396, 7398, 7407, 7411, 7415   I C   1455, 5258

1	2	3	4	5	6	7	8	9	10	11	12
12251	22 53.4 + 06 07 078.86 – 46.37	01 – 58 – 012	0821	1.0 x 0.4: 1.1 x 0.35	172: ...	1	S0?		15.4		
12252	22 53.4 + 31 30 095.96 – 24.98	05 – 54 – 000	1184 477	1.6 x 0.12 1.3: x 0.12	173 7	3	Sc	Sc +	16.0:		*
12253	22 53.5 + 12 31 084.18 – 41.24	02 – 58 – 037	0800 478	1.7 x 0.15 1.7 x 0.15	145 7	3	Sb Disturbed?		16.0:		*1
12254	22 53.6 + 03 40 076.63 – 48.30	01 – 58 – 013 NGC 7422	0821	0.9 x 0.7 0.9 x 0.7	140: ...	1	S ... Complex, disturbed?		14.3		3
12255	22 53.7 + 05 07 078.03 – 47.19	01 – 58 – 014	0821	1.2 x 1.0 0.7 x 0.7	— ...	1	S ...		15.5		
12256	22 53.7 + 36 05 098.34 – 20.95	06 – 50 – 012 NGC 7426	0838	(1.7 x 1.4) (1.8 x 1.4)	72 ...	3	E		13.6		*3
12257	22 53.8 + 19 06 088.87 – 35.77	03 – 58 – 014	1161 479	1.1 x 0.9: 0.9 x 0.8	— 1 – 2	2	Sc		15.7		*
12258	22 54.0 + 17 31 087.87 – 37.14	03 – 58 – 015	1161	1.6 x 0.3 1.5 x 0.25	85 7	2	Sa		15.1		*
12259	22 54.0 + 18 27 088.49 – 36.35	03 – 58 – 000	1161	1.6 x 0.3 1.3: x 0.3	7 6	1	S B ...		16.0:		
12260	22 54.2 + 37 28 099.11 – 19.76	06 – 50 – 013	9838	1.7 x 0.3 1.6 x 0.3	21 6	3	Sc	Sc +	15.7		
12261	22 54.6 + 71 11 113.96 + 10.61	12 – 21 – 000	0599	1.0 x 0.9 1.0 x 0.9	— ...	0	Irr		16.0:		
12262	22 54.7 – 01 18 071.76 – 52.18	00 – 58 – 014 NGC 7428	0905	2.5 x 1.3 2.5 x 1.2	160 5	4	S B a	SAB(r)a	13.8		*3
12263	22 55.0 + 72 25 114.53 + 11.71	12 – 21 – 000	0559	3. ? x 2. ? 3. x 2.	... ...	1	Dwarf?		17.		
12264	22 55.1 + 12 39 084.70 – 41.35	02 – 58 – 000	0800 478	1.0 x 0.10 0.9 x 0.10	105 7	2	Sc		17.		
12265	22 55.1 + 19 31 089.46 – 35.58	03 – 58 – 000	1161 479	[0.8 x 0.5] ...	... ...	1	Double system		14.5		*
12266	22 55.4 + 05 48 079.14 – 46.93	01 – 58 – 018	0821 481	1.2 x 1.0 1.4 x 1.1	173: 2	2	S B a – b		14.9		*
12267	22 55.5 + 25 52 093.36 – 30.14	04 – 54 – 004 NGC 7435	1175 480	1.5: x 0.7 1.3 x 0.7	122 5	3	S B a NGC 7436 group	S B(s)a	15.5		*13
12268	22 55.6 + 12 52 085.00 – 41.24	02 – 58 – 040 NGC 7432	0800 478	(1.5 x 1.0) (1.6 x 1.2)	(40) ...	2	E Companion superimposed?		15.1		*13
12269	22 55.6 + 25 53a 093.40 – 30.14	04 – 54 – 006 NGC 7436	1175 480	(2. x 2.) (2.0 x 2.0)	— ...	3	E Brightest in group NGC 7433 superimposed		(14.0)		*13 V
12270	22 55.7 + 14 02 085.88 – 40.29	02 – 58 – 041 NGC 7437	0800 478	1.8 x 1.8 1.7 x 1.7	— 1	4	Sc	SAB(rs)d	14.4		1
12271	22 55.8 + 02 02 075.65 – 49.90	00 – 58 – 015	0905	1.6 x 1.5 1.4: x 1.4:	— 1	4	Sc	SAB(s)c Sc –	14.9		
12272	22 55.8 + 25 30 093.23 – 30.49	04 – 54 – 008	1175 480	1.0 x 0.8 1.2 x 0.9	75: 1	3	Sa		15.0		
12273	22 55.8 + 28 58 095.13 – 27.45	05 – 54 – 021 NGC 7439	1184	(1.3 x 0.7) (1.0 x 0.6)	150: 5:	2	S B 0		15.2		3
12274	22 55.9 + 25 48 093.42 – 30.24	04 – 54 – 010	1175 480	1.3 x 0.5 1.1 x 0.4	143 6	1	S ...		15.4		*
12275	22 56.0 + 18 39 089.12 – 36.43	03 – 58 – 000 IC 5274	1161 479	1.1 x 1.0 (1.0 x 0.9)	— ...	2	[S0] Very compact		15.3		*
12276	22 56.2 + 35 32 098.56 – 21.67	06 – 50 – 014 NGC 7440	0838	1.7 x 1.7 1.7 x 1.6	— 1	4	S B a	S B(r)a	14.6		*3
12277	22 56.2 + 50 33 105.34 – 08.15	08 – 41 – 000	0590	1.0: x 0.8: 0.5 x 0.3	— ...	2	Sc Brightest in group		16.5:		*
12278	22 56.3 + 20 02 090.08 – 35.28	03 – 58 – 016	1161 479	1.0 x 0.7 0.8 x 0.6	175 3	2	Sc		15.3		*
12279	22 56.3 + 21 27 090.96 – 34.06	03 – 58 – 000	1161	1.0 x 0.12 1.1 x 0.12	109 7	2	Sc		16.0:		
12280	22 56.3 + 24 22 092.69 – 31.54	04 – 54 – 011	1175 477	[1.0 x 0.5] ...	[70] ...	3	Double system Common envelope		15.5		*1
12281	22 56.7 + 13 19 085.63 – 41.02	02 – 58 – 043	0800 478	3.5 x 0.2 3.2 x 0.2	30 7	2	...	S d – m	15.2		
12282	22 56.7 + 40 40 101.07 – 17.11	07 – 47 – 002	0876	1.9 x 0.5 2.1 x 0.5	1 6	2	Sa		14.7		
12283	22 56.8 + 23 51 092.51 – 32.04	04 – 54 – 012	1175 477	1.2 x 1.2 1.1 x 1.0	— 1	2	S B a – b P w 12289		15.7		*
12284	22 56.8 + 28 41 095.20 – 27.81	05 – 54 – 022	1184	(1.3 x 0.8) (1.1 x 0.7)	... ...	1	...		15.6		
12285	22 56.9 + 12 26 085.03 – 41.78	02 – 58 – 044	0800 478	1.0 x 0.5 0.9 x 0.4	10: 5:	1	S ...		15.6		

N G C 7422, 7426, 7428, 7432, 7435 – 37
7439, 7440

I C 5274

1	2	3	4	5	6	7	8	9	10	11	12
12286	22 56.9 + 15 17 087.09 – 39.40	02 – 58 – 045 NGC 7442	0800 478	1.0 x 1.0 1.0 x 1.0	-- 1	4	Sc		14.2		23
12287	22 57.1 + 53 26 106.69 – 05.60	09 – 37 – 001	0872	1.6: x 0.8: 1.6 x 0.8	5 5	3	Sc		15.5:		
12288	22 57.2 + 19 53 090.21 – 35.52	03 – 58 – 017	1161 479	1.1 x 0.8 0.7 x 0.5	... ...	1	S ... + companion Distorted		15.7		*1
12289	22 57.2 + 23 50 092.59 – 32.10	04 – 54 – 013	1175 480	1.4 x 1.3 1.2: x 1.2:	-- 1	4	Sc P w 12283, Sc +		16.0:		*
12290	22 57.2 + 24 35 093.03 – 31.45	04 – 54 – 000	1175 480	1.5 x 0.2 1.6 x 0.2	153 7	3	Sb		15.7		*
12291	22 57.3 + 26 02 093.87 – 30.19	04 – 54 – 014	1175 480	1.6 x 1.1 1.3: x 1.0:	75: 3	3	Sc		15.6		*
12292	22 57.3 + 38 52 100.35 – 18.78	06 – 50 – 016 NGC 7449	0838	(1.0 x 0.8) (1.1 x 0.9)	(130) ...	2	[E] Compact, 12298 group		15.4		*
12293	22 57.4 + 25 45 093.73 – 30.45	04 – 54 – 015	1175 480	1.1 x 1.0 1.1 x 1.0	-- 1	2	Sa		15.5		*
12294	22 57.6 + 15 42 087.56 – 39.14	03 – 58 – 018 NGC 7448	1161	2.7 x 1.1 2.3 x 1.0	170 6	3	Sc Brightest in group, Sc –		12.0	+ 2419 + 2650	*12 3468A
12295	22 57.8 + 01 22 075.56 – 50.73	00 – 58 – 019	0905	1.3 x 0.9 1.1 x 0.8	95 3	4	Sc	SAB(s)cd	15.5		
12296	22 57.8 + 18 51 089.71 – 36.48	03 – 58 – 019	1161 479	1.2 x 0.45 1.1 x 0.4	72 6	3	Sc		15.5		
12297	22 57.8 + 31 08 096.69 – 25.74	05 – 54 – 024	1184 477	(1.3 x 0.6) 1.2 x 0.5	138 5 – 6	1	S0? Brightest of 3		15.0		*1
12298	22 57.9 + 38 58 100.50 – 18.74	06 – 50 – 017	0838	1.5 x 0.4 1.4 x 0.4	79 6	2	S B a – b Brightest in group		15.0		*
12299	22 58.1 + 08 12 082.01 – 45.42	01 – 58 – 020 NGC 7451	0821 481	1.1 x 0.5 1.0 x 0.5	67 6	2	S B b – c		15.0		3
12300	22 58.1 + 12 31 085.42 – 41.87	02 – 58 – 049	0800 478	1.3 x 0.4 1.3 x 0.4	75 6	3	S B : a P w 12307		15.3		*
12301	22 58.2 + 02 14 076.56 – 50.13	00 – 58 – 000	0905	[1.3 x 0.4] ...	... ...	2	Double system Bridge		15.7		*
12302	22 58.3 + 06 34 080.66 – 46.76	01 – 58 – 000	0821 481	(2. x 1.) (2.3 x 1.2)	... ...	1	Double system Common halo		16.0:		*
12303	22 58.4 + 26 28 094.36 – 29.93	04 – 54 – 016	1175 480	1.7 x 1.3 1.5: x 1.5:	-- 1	3	Sb		14.8		*
12304	22 58.6 + 05 23 079.69 – 47.74	01 – 58 – 000	0821	1.4 x 0.2 1.4 x 0.2	125 7	1	S ...		15.4		
12305	22 58.6 + 16 07 088.11 – 38.91	03 – 58 – 020 NGC 7454	1161	(1.8 x 1.5) (1.9 x 1.6)	150 ...	3	E NGC 7448 group		13.6		*12 3
12306	22 58.6 + 29 53 096.22 – 26.93	05 – 54 – 026 NGC 7457	1184	(4.0 x 2.4) (4.2 x 2.5)	130 4	3	S0		12.3	+ 525 + 788	*123 4678
12307	22 58.7 + 12 28 085.54 – 41.99	02 – 58 – 053	0800 478	1.6: x 0.3 1.6: x 0.4:	156 ...	2	Dwarf irregular Ir IV – V:, p w 12300		16.0:		*1
12308	22 58.8 + 14 04 086.74 – 40.67	02 – 58 – 052	0800 478	2.5 x 0.5 2.5: x 0.5	122 6	3	Sc		15.3		
12309	22 58.9 + 01 29 076.01 – 50.82	00 – 58 – 020 NGC 7458	0905	(1.1 x 0.9) (1.5 x 1.1)	15: ...	4	E p of 2		13.9		*3
12310	22 59.0 + 28 08 095.39 – 28.52	05 – 54 – 027	1184	1.0 x 0.9 1.0 x 0.9	-- ...	1	... Compact		15.0		
12311	22 59.0 + 29 58 096.35 – 26.90	05 – 54 – 028	1184	1.6 x 0.4 1.2 x 0.4	143 ...	1	S? ...		15.4		*
12312	22 59.2 + 02 00 076.63 – 50.47	00 – 58 – 021 NGC 7460	0905	1.1 x 1.0 1.5 x 1.1	-- 3	3	Sb Companion superimposed, disturbed		14.2		*3
12313	22 59.2 + 15 48 088.05 – 39.26	03 – 58 – 021	1161	1.8: x 0.5 0.9: x 0.3:	117 ...	3	Dwarf IV – V, NGC 7448 group?		16.0:		
12314	22 59.3 + 15 19 087.75 – 39.68	02 – 58 – 056 NGC 7461	0800 478	0.8 x 0.8 0.9 x 0.8	-- 1	3	S B 0		14.5		*3
12315	22 59.3 + 15 42 088.01 – 39.35	03 – 58 – 023 NGC 7464	1161	0.5 x 0.5 0.5 x 0.4	-- ...	1	... NGC 7448 group		14.5	+ 1872 + 2102	*12 3
12316	22 59.3 + 15 43 088.02 – 39.34	03 – 58 – 022 NGC 7463	1161	3.2 x 0.6 2.4: x 0.6	90 6	1	S ... Distorted, NGC 7448 group		13.5	+ 2440 + 2670	*12 3
12317	22 59.5 + 15 42 088.06 – 39.38	03 – 58 – 024 NGC 7465, Mark 313	1161	1.2 x 0.7 1.1 x 0.7	72 ...	2	S B 0 NGC 7448 group		13.3	+ 1994 + 2224	*12 3
12318	22 59.6 + 25 24 094.04 – 30.99	04 – 54 – 000	1175 480	1.2 x 0.4 1.2 x 0.4	167 6	2	S B b s of 2		16.0:		*
12319	22 59.6 + 26 47 094.80 – 29.78	04 – 54 – 017 NGC 7466	1175 480	1.6 x 0.5 1.6 x 0.5	26 6	3	Sb		14.4		3
12320	22 59.7 + 30 30 096.77 – 26.49	05 – 54 – 000	1184	1.1 x 0.12 1.1 x 0.12	91 7	3	Sc		16.0:		

N G C 7442, 7448, 7449, 7451, 7454, 7457
7458, 7460, 7461, 7463 – 66

1	2	3	4	5	6	7	8	9	10	11	12
12321	22 59.8 + 15 45 088.18 - 39.37	03 - 58 - 000	1161	1.0 x 0.12 0.7 x 0.10	71 7	1	Sb - c		15.7		
12322	23 00.2 + 20 04 091.07 - 35.72	03 - 58 - 000	1161 479	1.0 x 0.3 1.0 x 0.15	110 7:	1	S ... spp of 2, disturbed		16.0:		*
12323	23 00.2 + 32 20 097.80 - 24.90	05 - 54 - 029	1184	1.1 x 1.0 1.2 x 1.1	--- 1	3	Sc	Sc -	15.0		*1
12324	23 00.3 + 08 19 082.73 - 45.64	01 - 58 - 000	0821 481	1.3 x 0.1 1.2 x 0.15	48 7	2	Sc		16.0:		
12325	23 00.3 + 21 37 092.04 - 34.38	04 - 54 - 019 IC 5282	1175	1.5 x 0.7 1.3 x 0.7	173 5	4	Sc	Sc +	15.2		
12326	23 00.3 + 21 50 092.17 - 34.19	04 - 54 - 018	1175	1.2 x 1.0 1.2 x 0.9	[55] ...	3	Sc Distorted		15.6		*1
12327	23 00.3 + 25 46 094.41 - 30.75	04 - 54 - 000	1175 480	1.5 x 0.10 1.3: x 0.10	176 7	1	S ...		16.5:		
12328	23 00.4 + 26 34 094.87 - 30.05	04 - 54 - 000	1175 480	1.1 x 0.2 1.0 x 0.2	3 7	1	S0 - a		15.3		
12329	23 00.5 + 16 20 088.76 - 38.97	03 - 58 - 000 NGC 7468, Mark 314	1161	0.9 x 0.6 0.7 x 0.4	... ...	1	... Peculiar, two jets		14.0	+ 1800	*3
12330	23 00.6 + 09 05 083.45 - 45.05	01 - 58 - 000	0821 481	1.0 x 0.9 1.0 x 0.9	--- (1)	1	S? ...		15.6		*
12331	23 00.6 + 34 28 098.91 - 23.03	06 - 50 - 019	0838	1.0 x 0.6 1.1 x 0.7	155: 4	1	S ...		15.5		
12332	23 00.7 + 08 35 083.07 - 45.48	01 - 58 - 025 NGC 7469	0821 481	1.6 x 1.1 1.8 x 1.2	125 3	3	S B a P w IC 5283; Seyfert		13.0	+ 4807 + 5015	*12 3468A
12333	23 00.8 + 19 42b 091.00 - 36.11	03 - 58 - 025	1161 479	1.4 x 0.4 1.4 x 0.45	160 6	3	S B b nf of 2		(15.5)		*1
12334	23 01.5 + 27 02 095.37 - 29.75	04 - 54 - 021	1175	(1.3 x 1.1) 0.6 x 0.4	--- 1:	2	S B 0		15.5		*
12335	23 01.5 + 29 54 096.86 - 27.20	05 - 54 - 030 NGC 7473	1184	1.1 x 0.5 1.1 x 0.5	45 6	2	S B 0		14.8		*3
12336	23 01.6 + 01 38 076.98 - 51.14	00 - 58 - 025	0905	(1.0 x 0.9) (1.1 x 1.0)	--- ...	2	E - S0		15.2		*
12337	23 01.7 + 19 47 091.28 - 36.14	03 - 58 - 27 +28 NGC 7475	1161 479	[1.5 x 0.9] ...	... ...	2	Double system, symbiotic In group of 3		15.1		*13
12338	23 01.8 + 17 38 089.95 - 38.01	03 - 58 - 000	1161	1.1 x 0.8 1.0 x 0.7	(40) 3:	4	Sc	Sc +	16.0:		*
12339	23 02.1 - 01 45 073.49 - 53.77	00 - 58 - 000	0905	0.25 x 0.22 0.22 x 0.22	--- ...	0	... Compact		14.0		*C
12340	23 02.1 + 26 53 095.43 - 29.95	04 - 54 - 024	1175	1.1 x 0.4 1.0 x 0.4	74 6	1	S ...		15.6		
12341	23 02.2 + 43 57 103.47 - 14.57	07 - 47 - 004	0876	1.7 x 1.0 1.6 x 0.8	75: 4 - 5	4	S B c		15.7		*
12342	23 02.3 + 16 24 089.28 - 39.13	03 - 58 - 030	1161	1.5 x 0.5 1.5 x 0.5	... ...	2	Strongly peculiar Disrupted pair?		15.0		1
12343	23 02.4 + 12 03 086.26 - 42.84	02 - 58 - 060 NGC 7479	0800	4.4 x 3.4 4.0 x 3.3	25: 2	4	S B b	S B(s)bc Sb +	11.7	+ 2441 + 2659	*12 3468
12344	23 02.5 + 18 27 090.65 - 37.39	03 - 58 - 029	1161 479	2.3 x 0.9 2.1: x 0.6:	177 6	2	...	S B d - dm	16.0:		
12345	23 02.5 + 44 17 103.67 - 14.29	07 - 47 - 000	0876	1.0: x 1.0: 1.4: x 1.3:	--- 1?	1	S B: ...		17.		
12346	23 02.6 + 00 33 076.17 - 52.13	00 - 58 - 026	0905	1.4 x 1.4 0.9 x 0.7	--- 1	1	Sc		15.7		
12347	23 02.6 + 18 35 090.76 - 37.29	03 - 58 - 031	1161 479	1.0 x 0.8 0.9 x 0.6	... ...	3	Irr		15.5		*
12348	23 02.7 - 00 05 075.52 - 52.63	00 - 58 - 028	0905	1.2: x 0.3 1.1: x 0.3	138 7	3	Sa		15.5		
12349	23 02.7 + 02 16 077.96 - 50.82	00 - 58 - 027 NGC 7480	0905	1.5 x 0.3 1.5 x 0.3	105 7	3	Sa Brightest of 3		15.1		*13
12350	23 02.8 + 16 35 089.53 - 39.04	03 - 58 - 032	1161	3.0 x 1.0 2.8: x 0.9	95 6	4	...	S m	15.1		*
12351	23 03.2 + 18 43 091.00 - 37.24	03 - 58 - 033	1161 479	1.6 x 0.7 1.2: x 0.5:	38 ...	2	Irr		15.6		
12352	23 03.2 + 27 23 095.94 - 29.61	04 - 54 - 025	1175	1.8 x 0.9 1.8 x 0.9	10: 5	1	S ...		14.7		
12353	23 03.3 + 03 16 079.12 - 50.13	00 - 58 - 030 NGC 7483	0905	1.6 x 0.9 1.5 x 1.0	110: ...	4	S B a		14.3		3
12354	23 03.4 + 14 05 088.00 - 41.25	02 - 58 - 062	0800	1.1 x 0.4 1.1 x 0.5	159 6	3	Sc		15.3		
12355	23 03.4 + 24 20 094.35 - 32.34	04 - 54 - 000	1175	1.2 x 0.3 1.1 x 0.3	139 6	3	Sc		16.5:		

N G C   7469, 7473, 7475, 7479, 7480, 7483   I C   5282

1	2	3	4	5	6	7	8	9	10	11	12
12356	23 03.4 + 30 50 097.74 – 26.55	05 – 54 – 032	1184	1.0 x 1.0 0.9 x 0.8	— ...	2	... Compact, p w 12357		14.6		*1
12357	23 03.5 + 30 48 097.74 – 26.59	05 – 54 – 033	1184	1.0 x 0.25 1.0 x 0.20	[22] 7:	1	... P w 12356, disturbed		15.7		*1
12358	23 03.5 + 33 53 099.23 – 23.81	06 – 50 – 000	0838	1.0 x 0.2 0.9 x 0.2	97 6 – 7	2	Sb – c P w 12360		16.0:		*
12359	23 03.7 + 14 36 088.41 – 40.83	02 – 58 – 063	0800	(1.4 x 1.2) (1.2 x 1.2)	— 1:	1	S ...		16.5:		*1
12360	23 03.7 + 33 50 099.24 – 23.88	06 – 50 – 022 NGC 7485	0838	1.2 x 0.6 1.2 x 0.6	146 5:	2	S0 P w 12358		14.2		*3
12361	23 03.8 + 11 00 085.87 – 43.91	02 – 58 – 000	0800	1.0 x 0.4 1.0 x 0.3	78 ...	3	Irr		16.0:		
12362	23 03.8 + 31 57 098.20 – 25.88	05 – 54 – 034	1184	1.7 x 0.6 1.6 x 0.6	118 6	3	S B b	S B(s)b	15.3		*1
12363	23 03.9 – 00 07 075.86 – 52.85	00 – 58 – 032	0905	1.1 x 0.55 1.1 x 0.5	30 5	2	S ... Disturbed		14.6		*
12364	23 04.2 + 18 50 091.33 – 37.26	03 – 58 – 034 IC 5284	1161 479	1.1 x 0.25 1.0 x 0.25	141 7	1	Sa		14.9		12
12365	23 04.4 + 22 40 093.65 – 33.98	04 – 54 – 026 IC 5285	1175	1.7 x 1.3' 1.7 x 1.3	(100) ...	2	... Compact, ring, p w 12378		14.4	+ 6427	*1C
12366	23 04.5 + 05 41 081.72 – 48.37	01 – 58 – 000	0821	1.2 x 0.3 0.9 x 0.25	116 6	1	S ... In group		16.0:		*
12367	23 04.5 + 16 52 090.17 – 39.00	03 – 58 – 000	1161	1.0 x 0.2 0.9 x 0.1	45 7	3	Sc		16.5:		
12368	23 04.5 + 27 55 096.51 – 29.27	05 – 54 – 035 NGC 7487	1184	(2.0 x 1.8) (2.3 x 2.0)	— ...	2	... Compact		14.8		*
12369	23 04.6 + 05 43 081.78 – 48.37	01 – 58 – 000	0821	1.0: x 0.6: 1.0: x 0.6:	... ...	3	Dwarf V, in group		17.		
12370	23 04.6 + 09 40 085.07 – 45.13	02 – 58 – 065	0800	1.5 x 0.3 0.9 x 0.25	56 6	1	Sc:		15.5		
12371	23 04.6 + 16 51 090.18 – 39.02	03 – 59 – 000	0320	1.0: x 0.15 0.9 x 0.1	44 7	2	Sc		16.5:		
12372	23 04.6 + 35 30 100.20 – 22.45	06 – 50 – 023	0838	0.8 x 0.7 0.8 x 0.7	— ...	1	S ... Peculiar		14.5		*
12373	23 04.8 + 13 33 088.01 – 41.88	02 – 58 – 000	0800	1.0 x 0.6 1.0 x 0.6	30 ...	2	...	S dm – m	16.0:		
12374	23 04.8 + 15 05 089.07 – 40.57	02 – 58 – 066	0800	1.2 x 0.5 1.1 x 0.4	97 6	1	S B b:		15.7		1
12375	23 04.8 + 34 59 100.00 – 22.93	06 – 50 – 024	0838	1.2 x 0.5 1.2 x 0.5	123 5	1	S ...		16.0:		*
12376	23 04.9 + 15 36 089.44 – 40.14	00 – 58 – 000	1161	0.2 x 0.2 0.18 x 0.18	— ...	1	... Compact		14.0		*C
12377	23 05 + 86 30 121.32 + 24.24	14 – 01 – 002	0568	1.0: x 0.8: 1.1: x 0.9:	(145) ...	1	... P w 12387		16.0:		*
12378	23 05.0 + 22 44 093.84 – 33.93	04 – 54 – 028 NGC 7489	1175	2.2 x 1.1 1.9 x 1.0	170 5	4	Sc P w 12365, Sc +		14.3		*13
12379	23 05.0 + 32 06 098.69 – 25.56	05 – 54 – 036 NGC 7490	1184	3.0 x 2.8 2.3 x 2.3	— 1	4	Sb	Sb +	13.5		
12380	23 05.1 + 11 16 086.44 – 43.86	02 – 58 – 000	0800	1.1 x 0.6 1.0 x 0.5	5 5	1	Sc		15.7		
12381	23 05.1 + 43 20 103.71 – 15.35	07 – 47 – 008	0876	1.7 x 1.4 1.8 x 1.5	— 1 – 2	3	S B c	S B(s)c Sc –	15.5		*1C
12382	23 05.3 + 04 53 081.24 – 49.15	01 – 59 – 000	0316	1.2 x 0.1 0.5 x 0.1	49 7	2	Sc		16.0:		
12383	23 05.3 + 22 26 093.74 – 34.22	04 – 54 – 000	1175	1.5: x 1.0: 1. x 1.	175: ...	3	S B b/Sb		15.7		*
12384	23 05.4 + 24 41 095.01 – 32.24	04 – 54 – 000	1175	1.0: x 0.8: 0.7 x 0.6	— 1 – 2	2	S B c P w 12386		16.0:		*
12385	23 05.4 + 30 03 097.78 – 27.44	05 – 54 – 037	1184	1.0 x 0.9 1.0 x 1.0	— 1	2	S B c		15.5		
12386	23 05.6 + 24 42 095.07 – 32.25	04 – 54 – 000	1175	1.2: x 0.6: 0.8 x 0.4	20 5:	3	Sc P w 12384		16.0:		*
12387	23 06 + 86 29 121.33 + 24.22	14 – 01 – 003	0568	1.2 x 1.2 1.2: x 1.1	— 1	3	S B a P w 12377		15.6		*
12388	23 06.0 + 12 33 087.64 – 42.89	02 – 59 – 001	0313	1.8 x 0.7 1.8 x 0.8	45 6	2	...	S d – dm	16.0:		*
12389	23 06.1 + 46 38 105.23 – 12.40	08 – 42 – 000	1162	0.7 x 0.5 0.6 x 0.4	... ...	1	Pair of compacts Contact, halo		13.8		*C
12390	23 06.3 + 23 07 094.37 – 33.73	04 – 54 – 029	1175	1.1 x 1.0 1.1 x 0.9	— (1)	2	S0		16.0:		

N G C   7485, 7487, 7489, 7490   I C   5284, 5285

1	2	3	4	5	6	7	8	9	10	11	12
12391	23 06.5 + 11 48 087.23 – 43.59	02 – 59 – 003 NGC 7495	0313	2.0 x 1.8 2.0 x 1.7	(5) 1	4	Sc	SAB(s)c	14.7		*1
12392	23 06.7 + 17 54 091.41 – 38.36	03 – 59 – 002 NGC 7497	0320	4.4 x 1.5: 4.5: x 1.5:	48 6	3	Sc	S d Sc +	13.3		*23
12393	23 06.8 + 00 29 077.42 – 52.85	00 – 59 – 003 IC 5287	0834	1.1 x 1.0 1.1 x 1.0	— 1	4	S B b	S B(r)b	15.2		
12394	23 07.1 + 32 24 099.28 – 25.48	05 – 54 – 040	1184	1.0 x 0.8 1.0 x 0.8	(45) 2	2	S B a – b		14.8		
12395	23 07.2 + 15 24 089.94 – 40.59	02 – 59 – 000	0313	1.0 x 0.15 1.0 x 0.15	138 7	2	Sc – Irr		15.7		
12396	23 07.7 + 42 18 103.73 – 16.49	07 – 47 – 009	0876	(1.1 x 0.9) (1.2 x 1.0)	(0) ...	2	... Compact		15.5		
12397	23 07.8 + 07 18 084.11 – 47.53	01 – 59 – 005 NGC 7499	0316 482	1.1 x 0.6 1.6 x 1.0	10 ...	2	S0		15.0	+ 11916 + 12116	*12 3
12398	23 07.9 + 02 22 079.68 – 51.55	00 – 59 – 004	0834	1.1 x 0.12 0.9 x 0.15	109 7	1	S ...		15.7		
12399	23 08.0 + 10 45 086.89 – 44.67	02 – 59 – 004 NGC 7500	0313	(2.1 x 1.1) (2.2 x 1.3)	(125) ...	3	S0		14.9		
12400	23 08.1 + 21 26 093.87 – 35.41	03 – 59 – 006	0320	1.0: x 1.0: 1.1 x 1.0	— 1	3	S B a		15.5		
12401	23 08.1 + 34 08 100.29 – 23.99	06 – 50 – 000	0838	1.0: x 0.8: 1. x 1.	— ...	3	Dwarf irregular Ir V		17.		
12402	23 08.2 + 48 21 106.25 – 10.95	08 – 42 – 000	1162	1.0: x 0.6: 0.9: x 0.5:	(90) ...	2	...	d – m	17.		
12403	23 08.6 + 12 30 088.34 – 43.26	02 – 59 – 000	0313	1.1 x 0.4 0.9 x 0.3	128 6	2	S B: c		16.0:		
12404	23 08.7 + 29 22 098.18 – 28.37	05 – 54 – 043 IC 1473	1184 484	2.1 x 1.0 1.7 x 0.9	176 5	3	S0		14.2		*
12405	23 08.8 + 05 58 083.28 – 48.77	01 – 59 – 000	0316 487	1.0 x 0.5 0.9 x 0.4	154 5	1	S ...		16.0:		
12406	23 09.2 – 02 26 074.98 – 55.46	00 – 59 – 005 NGC 7506	0834 483	1.7 x 1.1 (2.1 x 1.3)	103 ...	3	S0		14.3		3
12407	23 09.3 + 09 14 086.12 – 46.12	01 – 59 – 010	0316 487	1.2 x 0.6 0.6 x 0.35	125 5	1	S ...		14.2		
12408	23 09.3 + 12 42 088.68 – 43.18	01 – 59 – 005 NGC 7508	0313	1.1 x 0.3 0.7 x 0.25	160 ...	1	...		15.7		13
12409	23 09.4 + 23 47 095.50 – 33.46	04 – 54 – 000	1175	1.1 x 0.4 1.0 x 0.3	63 6	1	Sa		16.0:		
12410	23 09.5 + 30 45 099.02 – 27.19	05 – 54 – 044	1184	1.8 x 0.2 1.8 x 0.2	48 7	3	Sc		16.0:		
12411	23 09.8 + 48 32 106.57 – 10.88	08 – 42 – 000	1162	2.3 x 0.12 2.3 x 0.12	175 7	2	...	S m:	16.5:		
12412	23 09.9 + 13 27 089.38 – 42.61	02 – 59 – 007 NGC 7511	0313	1.1: x 0.5 1.1 x 0.4	133 ...	1	S? ...		15.1		
12413	23 09.9 + 28 27 098.01 – 29.31	05 – 54 – 045	1184 484	1.0 x 1.0 1.1 x 1.0	— (1)	1	S B 0? Singular		15.7		
12414	23 09.9 + 30 52 099.16 – 27.12	05 – 54 – 046 NGC 7512	1184	(1.8 x 1.5) (2.1 x 1.7)	— ...	3	E spp of 2		14.1		*
12415	23 10.0 + 34 37 100.89 – 23.71	06 – 50 – 026 NGC 7514	0838	1.6 x 1.0 1.6 x 1.0	132 4	1	S ...		13.5		
12416	23 10.2 + 10 28 087.33 – 45.20	02 – 59 – 000	0313 487	1.1 x 0.25 0.9 x 0.25	... ...	3	... Peculiar, streamer		15.4		*
12417	23 10.3 + 05 31 083.36 – 49.36	01 – 59 – 011 IC 1474	0316 487	1.1 x 0.5 1.0 x 0.4	150 5 – 6	2	Sc		14.9		
12418	23 10.3 + 12 25 088.77 – 43.55	02 – 59 – 008 NGC 7515	0313	1.7 x 1.4: 1.7 x 1.6	(15) 1 – 2	1	S ...		14.0		3
12419	23 10.3 + 15 38 090.94 – 40.76	03 – 59 – 000	0320	0.2 x 0.2 0.2 x 0.2	— ...	1	... Compact		14.0		*C
12420	23 10.4 + 19 58 093.62 – 36.96	03 – 59 – 010 NGC 7516	0320	1.1 x 0.9 1.1 x 0.9	(110) ...	2	S0		14.6		*3
12421	23 10.4 + 77 19 117.54 + 15.77	13 – 16 – 001	1210	1.4 x 0.5 1.4 x 0.6	62 6	3	Sc		15.4		
12422	23 10.6 + 06 03 083.92 – 48.96	01 – 59 – 012 NGC 7518	0316 487	1.5 x 1.4 1.5 x 1.5	— 1	2	Sa	(R)SAB(r)a	14.5		*3
12423	23 10.6 + 06 08 083.99 – 48.89	01 – 59 – 013	0316 487	3.6 x 0.4 3.6 x 0.4	145 7	2	Sc	Sc –?	14.8		*
12424	23 10.7 + 10 31 087.51 – 45.22	02 – 59 – 009 NGC 7519	0313 487	1.4 x 1.0 1.4 x 1.0	165 3	1	Sb:		15.2		13
12425	23 10.7 + 23 58 095.91 – 33.43	04 – 54 – 030	1175	1.6 x 0.25 1.7 x 0.25	172 7	2	Sa – b		15.0		

N G C   7495, 7497, 7499, 7500, 7506, 7508
7511, 7512, 7514 – 16, 7518, 7519

I C   1473, 1474, 5287

1	2	3	4	5	6	7	8	9	10	11	12
12426	23 11.0 + 06 17 084.24 – 48.82	01 – 59 – 000	0316 487	1.3 x 0.2 1.1 x 0.2	171 7	2	Sc		16.0:		
12427	23 11.0 + 28 41 098.37 – 29.20	05 – 54 – 048	1184 484	1.1 x 0.3 1.0 x 0.3	146 6	2	Sb – c		15.4		
12428	23 11.2 + 24 38 096.39 – 32.88	04 – 54 – 031 NGC 7527	1175	(1.3 x 0.9) (1.2 x 0.8)	165: ...	2	E P w 12432		14.7		*
12429	23 11.3 + 22 25 095.23 – 34.87	04 – 54 – 032	1175	1.5 x 1.0 1.4 x 0.9:	110: 3	4	Sc	Sc +	15.6		*
12430	23 11.3 + 28 44 098.46 – 29.18	05 – 54 – 049	1184 484	2.5 x 0.2 2.4 x 0.2	164 7	3	Sc	Sc +	14.9		
12431	23 11.5 + 08 42 086.37 – 46.87	01 – 59 – 014 NGC 7529	0316 487	1.1 x 1.0 1.0 x 0.9	-- 1	1	S ...		14.6		
12432	23 11.5 + 24 38 096.46 – 32.91	04 – 54 – 000	1175	1.0 x 0.15 1.0 x 0.15	109 7	3	Sc P w 12428, Sc +		16.0:		*
12433	23 11.5 + 49 24 107.16 – 10.19	08 – 42 – 000	1162	(2. x 2.) (1.5 x 1.5)	-- 1	2	S B 0 – a		16.0:		*
12434	23 11.6 + 12 54 089.49 – 43.29	02 – 59 – 000	0313	1.1 x 0.25 0.6 x 0.25	156 6	1	Sa?		15.4		
12435	23 11.6 + 31 47 099.96 – 26.43	05 – 54 – 000	1184	1.0 x 0.7 0.9 x 0.6	45: 3	2	S B a		16.0:		
12436	23 11.7 + 05 05 083.42 – 49.91	01 – 59 – 000	0316 487	[1.0 x 0.2] ...	[27] ...	2	Double system Contact		15.6		*
12437	23 11.7 + 13 10 089.70 – 43.07	02 – 59 – 011 NGC 7536	0313	2.2 x 0.8 2.1 x 0.7	56 6	2	S B b P w 12438, Sb +		14.8		*1
12438	23 11.7 + 13 19 089.80 – 42.94	02 – 59 – 010 NGC 7535	0313	1.7 x 1.7 1.7: x 1.6:	-- 1	4	Sc P w 12437, Sc +		15.6		*1
12439	23 11.7 + 48 03 106.68 – 11.45	08 – 42 – 000	1162	1.3 x 1.1 1.2 x 1.1	(0) 1	3	S B b		16.0:		*
12440	23 11.8 + 16 11 091.71 – 40.45	03 – 59 – 012	0320	1.0 x 0.4 1.0 x 0.3	150 6	1	Sa – b		15.6		
12441	23 11.9 + 30 52 099.60 – 27.30	05 – 54 – 000	1184	1.0 x 0.15 1.0 x 0.15	14 7	2	S B? b – c		16.0:		*
12442	23 12.0 + 04 13 082.74 – 50.66	01 – 59 – 016 NGC 7537	0316 487	2.1 x 0.5 2.1 x 0.5	79 6	3	Sb P w 12447		13.8	+ 2682 + 2860	*12 36⁻
12443	23 12.0 + 23 24 095.93 – 34.07	04 – 54 – 035 NGC 7539	1175	1.5 x 1.2 1.5 x 1.3	(165) 2:	2	S0		13.7		*3
12444	23 12.0 + 31 17 099.82 – 26.92	05 – 54 – 051	1184	1.0 x 0.9 1.1 x 1.1	-- ...	1	...		14.5		*
12445	23 12.0 + 49 32 107.29 – 10.09	08 – 42 – 000	1162	(1.0 x 0.9) (1.2 x 1.1)	-- (1)	1	S0:		15.7		*
12446	23 12.1 – 00 02 078.61 – 54.09	00 – 59 – 011	0834	0.6 x 0.6 0.7 x 0.5	-- ...	1	...		14.4		
12447	23 12.2 + 04 15 082.84 – 50.67	01 – 59 – 017 NGC 7541	0316 487	3.4 x 1.1 3.3 x 1.1	102 6	3	Sc P w 12442, Sc –		12.7	+ 2672 + 2859	*12 3468
12448	23 12.2 + 14 38 090.82 – 41.86	02 – 59 – 000	0313	1.0 x 0.2 1.0 x 0.2	134 7	2	Sc		16.0:		*
12449	23 12.2 + 23 03 095.80 – 34.40	04 – 54 – 000	1175	1.0 x 0.15 1.0 x 0.15	142 7	1	S ...		16.0:		
12450	23 12.2 + 28 03 098.34 – 29.89	05 – 54 – 052 NGC 7543	1184 484	1.2 x 1.0 1.5 x 1.3	(140) ...	1	S ...		14.1		*
12451	23 12.3 + 05 09 083.67 – 49.94	01 – 59 – 018	0316 487	1.6 x 0.4 1.6: x 0.3	[155] ...	4	Dwarf irregular, I m		16.0:		1
12452	23 12.5 + 01 10 079.99 – 53.20	00 – 59 – 012	0834 485	2.0 x 0.15 1.9 x 0.2	137 7	3	Sc		15.7		
12453	23 12.5 + 18 42 093.43 – 38.31	03 – 59 – 013 NGC 7547	0320 486	1.1 x 0.5 1.1 x 0.4	106 6	2	S B a NGC 7550 group		14.9		*13 A
12454	23 12.6 + 09 24a 087.24 – 46.42	01 – 59 – 020	0316 487	1.1 x 0.4 1.4 x 0.5	152 6 – 7	2	S0 nnp of 2		(15.0)		*1C
12455	23 12.7 + 25 00 096.94 – 32.69	04 – 54 – 036 NGC 7548	1175	1.0 x 0.8 1.0 x 0.8	15: 2	3	S0	(R)SAB:(r)0	14.5		3
12456	23 12.8 + 18 41 093.51 – 38.36	03 – 59 – 015 NGC 7550	0320 486	(1.4 x 1.4) (1.6 x 1.6)	-- ...	2	E – S0 Brightest in group		13.9		*13 A
12457	23 12.8 + 18 46 093.55 – 38.29	03 – 59 – 014 NGC 7549	0320 486	2.8 x 0.7 2.8 x 0.7	[8] ...	3	S B ... NGC 7550 group, distorted		14.1		*13 A
12458	23 12.8 + 30 41 099.72 – 27.54	05 – 54 – 000	1184	1.0 x 0.12 1.0 x 0.15	136 7	2	Sb – c		16.5:		
12459	23 13.2 + 24 37 096.87 – 33.09	04 – 54 – 000	1175	1.1 x 0.4 1.0 x 0.4	136 6	2	Sb – c		15.6		
12460	23 13.2 + 24 50 096.98 – 32.89	04 – 54 – 037 IC 5296	1175	1.0 x 0.6 1.0 x 0.6	25: 4:	3	S B b		15.5		*

N G C   7527, 7529, 7535 – 37, 7539, 7541
       7543, 7547, 7548 – 50

I C   5296

1	2	3	4	5	6	7	8	9	10	11	12
12461	23 13.3 + 04 51 083.73 - 50.33	01 - 59 - 022	0316 487	(1.0 x 0.8) (0.8 x 0.7)	-- ...	1	SO?		15.6		*
12462	23 13.3 + 09 14 087.33 - 46.65	01 - 59 - 023	0316 487	1.2 x 0.3 1.2 x 0.2	31 6	1	Sb		15.5		
12463	23 13.3 + 13 01b 090.06 - 43.40	02 - 59 - 014 NGC 7559b	0313	1.3 x 1.2 1.2 x 1.1	-- ...	2	E - SO P w 12465		(14.7)		*13
12464	23 13.4 + 06 24 085.09 - 49.05	01 - 59 - 024 NGC 7562	0316 487	(2.1 x 1.6) (2.3 x 1.7)	83 ...	4	E		13.0	+ 3806 + 4000	*12 3
12465	23 13.4 + 12 55 090.02 - 43.49	02 - 59 - 015 NGC 7563	0313	2.1 x 0.9 1.9 x 0.9	155 6	4	S B a P w 12463		14.5		*13
12466	23 13.5 - 02 07 076.79 - 55.92	00 - 59 - 015	0834 483	1.0 x 0.6 1.0 x 0.6	148 4	3	S B b		14.9		
12467	23 13.5 + 06 22 085.10 - 49.10	01 - 59 - 000	0316 487	1.5 x 0.4 1.3: x 0.4	0 6	2	...	S dm	16.0:		*
12468	23 13.7 + 15 34 091.85 - 41.21	03 - 59 - 019 NGC 7567	0320	1.0 x 0.25 0.9 x 0.25	76 ...	1	...		15.4		*3
12469	23 13.9 + 24 13 096.83 - 33.52	04 - 54 - 000 NGC 7568	1175	1.0 x 0.6 0.8 x 0.7	120: ...	1	S ...		14.5		*3
12470	23 13.9 + 28 13 098.81 - 29.89	05 - 54 - 055	1184 484	1.1 x 1.0 1.0 x 0.9	-- 1 - 2	2	Sa		15.4		
12471	23 14.0 + 18 17 093.59 - 38.85	03 - 59 - 021	0320 486	1.0 x 0.2 1.0 x 0.2	70 6	1	S ...		16.0:		
12472	23 14.2 + 08 37 087.13 - 47.29	01 - 59 - 026	0316 487	1.0 x 0.7 1.0 x 0.6	(88) ...	1	SO? Jet		14.8		*C
12473	23 14.2 + 13 13 090.46 - 43.33	02 - 59 - 018 NGC 7570	0313	1.6 x 0.8 1.7 x 0.9	30 ...	3	S B a		14.3		*13
12474	23 14.3 + 33 43 101.40 - 24.89	06 - 51 - 002	0873	1.2 x 0.4 1.1 x 0.4	88 6	2	Sa		14.1		
12475	23 14.5 + 03 26 082.84 - 51.66	00 - 59 - 016	0834	1.6 x 0.2 0.9 x 0.2	81 7	2	Sb		15.7		
12476	23 14.6 + 30 04 099.84 - 28.26	05 - 54 - 057	1184	1.4 x 0.8 1.3 x 0.7	92 4	2	SO - a		14.1		*
12477	23 14.7 + 18 25 093.86 - 38.80	03 - 59 - 024 NGC 7578a	0320 486	[1.8 x 1.2]	...	2	Double system Common envelope		15.3 15.0		*13 VA
12478	23 14.8 + 18 26 093.90 - 38.80	03 - 59 - 025 NGC 7578b	0320 486	...	...						
12479	23 14.9 - 01 52 077.56 - 55.95	00 - 59 - 017	0834 483	1.0 x 0.35 1.0 x 0.3	33 6	2	Sa		15.3		
12480	23 14.9 + 07 22 086.36 - 48.44	01 - 59 - 000	0316 487	1.0 x 1.0 1. x 1.	-- ...	3	Dwarf irregular Ir V		17.		
12481	23 15.1 + 13 44 091.06 - 42.98	02 - 59 - 019 NGC 7580, Mark 318	0313	0.8 x 0.6 0.8 x 0.6	45: ...	1	S? ...		14.8	+ 4800	*1
12482	23 15.1 + 28 45 099.34 - 29.51	05 - 54 - 059	1184	(1.4 x 1.2) (1.6 x 1.4)	-- ...	2	[E] Compact, brightest of 3		14.4		*
12483	23 15.4 + 11 05 089.33 - 45.33	02 - 59 - 020 NGC 7593	0313	1.0 x 0.5 1.0 x 0.5	104 ...	1	S ... Peculiar		14.6		*3
12484	23 15.5 + 09 24 088.13 - 46.79	01 - 59 - 037 NGC 7587	0316 487	1.2 x 0.25 1.2 x 0.35	123 7	3	Sa n of 2		14.9		*12 3
12485	23 15.7 + 10 02 088.66 - 46.27	02 - 59 - 023 IC 1478	0313 487	1.7 x 1.1 1.6 x 0.9	30 4	4	Sb		15.3		*1
12486	23 15.8 + 06 18 085.78 - 49.46	01 - 59 - 038 NGC 7591	0316 487	1.9 x 0.8 1.8 x 0.9	145 5 - 6	2	S B b	Sb +	13.8		*3
12487	23 16.2 + 08 57 088.01 - 47.26	01 - 59 - 039 NGC 7601	0316 487	1.3 x 1.0 1.0 x 0.9	(90) 2:	2	Sb - c		14.7		*2
12488	23 16.2 + 20 42 095.55 - 36.92	03 - 59 - 000	0320 486	1.1 x 0.3 1.1 x 0.3	60 (6)	2	...	m	16.5:		
12489	23 16.2 + 22 37 096.58 - 35.19	04 - 55 - 000	0843	1.3 x 0.15 1.3 x 0.15	176 7	2	Sc		16.0:		
12490	23 16.2 + 24 57 097.77 - 33.08	04 - 55 - 001 Mark 319	0843	1.0 x 0.8 0.9 x 0.7	(75) 2:	3	S B a s of 2		14.0	+ 8400	*
12491	23 16.3 + 42 41 105.42 - 16.74	07 - 47 - 011	0876	(1.0 x 0.8) (1.2 x 1.0)	(40) ...	3	[E] Compact		15.3		*
12492	23 16.4 - 01 20 078.67 - 55.77	00 - 59 - 020	0834	(1.1 x 0.8) 1.1 x 0.9	98 ...	3	SO		14.9		
12493	23 16.4 - 00 01 080.09 - 54.73	00 - 59 - 021 NGC 7603	0834 485	1.4 x 0.9 1.6: x 0.9	165 ...	2	S ... Seyfert, np of 2, bridge		14.4	+ 8700	*3A
12494	23 16.4 + 06 35b 086.20 - 49.30	01 - 59 - 040	0316 487	1.5 x 0.5 1.4 x 0.5	35 6	3	Sc/S B c s of 2	SAB d	(15.0)		*1
12495	23 16.4 + 16 21 093.10 - 40.83	03 - 59 - 035	0320	1.4 x 0.8 1.0 x 0.7	100: ...	2	Sc - Irr		15.2		

N G C   7559, 7562, 7563, 7567, 7568, 7570
7578, 7580, 7587, 7591, 7593, 7601, 7603

I C   1478

1	2	3	4	5	6	7	8	9	10	11	12
12496	23 16.5 + 41 48 105.11 − 17.57	07 − 47 − 012	0876	1.0 x 0.25 1.1 x 0.25	18 7	3	Sa		14.8		
12497	23 16.6 + 07 25 086.94 − 48.62	01 − 59 − 045	0316 487	1.1 x 0.25 0.9 x 0.3	101 ...	2	Irr		15.6		
12498	23 16.6 + 07 50 087.27 − 48.27	01 − 59 − 042 IC 5309	0316	1.5 x 0.6 1.3 x 0.6	23 ⟋ 6	2	Sb		15.0		
12499	23 16.6 + 25 53 098.33 − 32.26	04 − 55 − 000	0843	1.1 x 0.7 1.1: x 0.7:	0: ...	3	...	S dm	17.		
12500	23 16.7 + 08 04 087.49 − 48.08	01 − 59 − 044 NGC 7608	0316 487	1.5 x 0.4 1.6 x 0.4	20 6	1	S ...		15.2		3
12501	23 16.7 + 10 32 089.34 − 45.96	02 − 59 − 024	0313 487	1.3: x 0.15 0.9 x 0.20	87 7	1	S ...		15.7		1
12502	23 16.7 + 22 09 096.46 − 35.66	04 − 55 − 000	0843	1.0 x 0.35 1.1 x 0.35	121 6	2	Sa		15.5		*
12503	23 16.8 + 00 40 080.93 − 54.25	00 − 59 − 023	0843 485	1.3 x 0.15 1.2 x 0.2	139 7	1	S ...		15.7		*
12504	23 16.8 + 77 03 117.78 + 15.39	13 − 01 − 001	1213	1.6: x 1.0 1.5 x 1.2	30: ...	3	Sc		15.6		
12505	23 16.9 + 05 38 085.57 − 50.17	01 − 59 − 000	0316 487	0.9 x 0.8 0.8 x 0.7	— ...	1	... Peculiar?		14.5		
12506	23 17.0 + 15 48 092.93 − 41.38	03 − 59 − 036	0320	2.8 x 0.2 2.8 x 0.2	81 7	3	Sc P w 12519		15.6		*1
12507	23 17.0 + 43 42 105.92 − 15.84	07 − 47 − 000	0876 488	1.2 x 0.2 1.2 x 0.2	152 7	3	Sa P w 12517		14.2		*
12508	23 17.1 + 01 12b 081.57 − 53.86	00 − 59 − 024	0834 485	0.9 x 0.4 0.9 x 0.4	141 5 − 6	1	S ... ssf of 2		(14.5)		*
12509	23 17.1 + 07 46 087.38 − 48.39	01 − 59 − 049 NGC 7611	0316 487	1.2 x 0.6 1.2 x 0.6	139 5	2	S0		14.0	+ 3383 + 3580	*12 3
12510	23 17.1 + 07 59 087.55 − 48.21	01 − 59 − 000	0316 487	(1.3 x 0.9) 1.3 x 0.9	65 ...	2	Dwarf? elliptical		16.0:		
12511	23 17.1 + 09 54 089.00 − 46.56	02 − 59 − 025 NGC 7610	0313 487	2.7 x 2.2 2.5 x 2.1	... 1 − 2	4	Sc Disturbed?		14.9		*1
12512	23 17.2 + 08 17 087.81 − 47.96	01 − 59 − 050 NGC 7612	0316 487	1.5 x 0.7 1.6 x 0.8	2 5	3	S0		14.3		*3
12513	23 17.3 + 00 56 081.37 − 54.10	00 − 59 − 000	0834 485	1.0 x 0.1 0.5 x 0.1	152 7	1	S ...		16.0:		
12514	23 17.4 + 25 44 098.45 − 32.47	04 − 55 − 000	0843	1.1 x 0.5 1.1 x 0.6	2 ...	2	Sc − Irr		15.6		*
12515	23 17.4 + 26 00 098.58 − 32.23	04 − 55 − 003	0843	1.3 x 1.0 (1.4 x 1.1)	55: 2	2	S0		14.1		*
12516	23 17.4 + 42 35 105.58 − 16.90	07 − 47 − 013 NGC 7618	0876 488	(1.2 x 1.0) (1.5 x 1.2)	(5) ...	2	E		14.3		*
12517	23 17.5 + 43 41 106.01 − 15.88	07 − 47 − 014	0876 488	(1.4 x 1.3) (1.4 x 1.4)	— ...	1	E? P w 12507		14.5		*
12518	23 17.6 + 07 39 087.45 − 48.56	01 − 59 − 053	0316 487	1.3 x 0.20 1.3 x 0.20	24 7	3	Sb		15.7		
12519	23 17.6 + 15 41 093.03 − 41.55	03 − 59 − 037	0320	1.4 x 0.4 1.5 x 0.4	158 6	1	S ... P w 12506		14.3		*
12520	23 17.6 + 23 56 097.61 − 34.13	04 − 55 − 000 NGC 7620, Mark 321	0843	1.3: x 1.2: 1.0: x 0.9:	— 1	2	Sc		13.5	+ 9600	3
12521	23 17.7 − 02 07 078.25 − 56.59	00 − 59 − 026	0834	1.6 x 0.9 1.6 x 0.9	167 4	3	Sc		14.4		
12522	23 17.7 + 07 43 087.53 − 48.51	01 − 59 − 000	0316 487	1.7 x 1.6 1.2: x 1.2:	— 1:	3	Dwarf spiral S IV − V		16.5:		
12523	23 17.7 + 07 55 087.69 − 48.34	01 − 59 − 052 NGC 7619	0316 487	(2.8 x 2.5) (3.0 x 2.6)	(30) ...	4	E P w 12531		12.7	+ 3757 + 3954	*123 4678
12524	23 17.8 + 42 41 105.68 − 16.84	07 − 47 − 015	0876 488	1.4: x 0.6: 1.4: x 0.6:	20: 6	2	S B b: S B(r) ...		15.7		
12525	23 17.9 + 01 17 081.93 − 53.91	00 − 59 − 027	0834	[1.0 x 0.5] ...	... ...	1	Triple system		15.1		*1
12526	23 17.9 + 08 06 087.89 − 48.21	01 − 59 − 056 NGC 7623	0316 487	1.6 x 1.0 1.7 x 1.1	175: ...	2	S0		13.9	+ 3463 + 3661	*12 36
12527	23 17.9 + 27 03 099.20 − 31.32	04 − 55 − 004 NGC 7624, Mark 323	0843	1.0 x 0.7 1.1 x 0.7	30: 3	2	Sc		13.7	+ 4500	*
12528	23 17.9 + 35 14 102.79 − 23.77	06 − 51 − 003	0873	1.2 x 0.4 1.2 x 0.6	109 6	2	Sb − c		15.3		
12529	23 18.0 + 16 57 093.91 − 40.47	03 − 59 − 038 NGC 7625	0320 486	1.5 x 1.5 1.5 x 1.5	— ...	1	...		12.8	+ 1784 + 2009	*12V 3468AC
12530	23 18.1 + 29 02 100.16 − 29.51	05 − 55 − 002	0914 484	1.5 x 1.4 1.4 x 1.4	— 1	4	S B c		14.9		

N G C 7608, 7610 − 12, 7618 − 20, 7623 − 25

1	2	3	4	5	6	7	8	9	10	11	12
12531	23 18.2 + 07 56 087.86 - 48.39	01 - 59 - 057 NGC 7626	0316 487	(2.4 x 1.8) (2.9 x 2.5)	— ...	4	E P w 12523		12.8	+ 3357 + 3554	*12 348
12532	23 18.3 + 02 14 082.99 - 53.19	00 - 59 - 000	0834	1.1: x 0.8: 0.8: x 0.3:	... ...	2	... m		17.		
12533	23 18.4 + 23 32 097.61 - 34.57	04 - 55 - 000	0843	1.5 x 0.2 1.6 x 0.2	21 7	3	Sb		15.6		
12534	23 18.4 + 25 37 098.64 - 32.67	04 - 55 - 005 NGC 7628	0843	(1.1 x 0.9) (1.3 x 1.1)	— ...	3	E		13.8		*
12535	23 18.5 + 07 54 087.93 - 48.46	01 - 59 - 000	0316 487	1.1 x 0.15 1.0 x 0.15	156 7	2	Sb - c npp of 2		16.0:		*
12536	23 18.5 + 43 22 106.06 - 16.25	07 - 47 - 000	0876 488	1.0 x 0.1 1.0 x 0.1	15 7	3	... S d - m		17.		
12537	23 18.7 + 29 17 100.41 - 29.34	05 - 55 - 003	0914 484	1.3 x 0.7 1.2 x 0.7	140 ...	2	... Compact nucleus		15.4		*1C
12538	23 18.7 + 33 08 102.08 - 25.78	05 - 55 - 004	0914	1.0 x 0.3 1.0 x 0.3	20 ...	1	... Peculiar		14.7		
12539	23 18.8 + 07 56 088.05 - 48.47	01 - 59 - 060 NGC 7631	0316 487	1.8 x 0.8 1.8 x 0.8	79 6	3	Sb		13.8		*23
12540	23 18.8 + 11 08 090.40 - 45.70	02 - 59 - 027 NGC 7630	0313 487	1.1 x 0.4 0.9 x 0.35	162 6	1	S ...		15.5		
12541	23 18.8 + 25 06 098.48 - 33.18	04 - 55 - 006 IC 5315	0843	(1. x 1.) (1. x 1.)	— ...	1	... Compact		14.6		*C
12542	23 19.1 + 08 36 088.66 - 47.93	01 - 59 - 062 NGC 7634	0316 487	1.2 x 0.9 1.3 x 1.0	95 2	3	S B 0 Brightest of 3		13.7		3
12543	23 19.1 + 26 51 099.39 - 31.61	04 - 55 - 007	0843	1.0 x 0.3 1.0 x 0.3	75 6	3	Sc 1st of 2		15.4		*
12544	23 19.2 + 08 47 088.83 - 47.79	01 - 59 - 064	0316 487	1.2: x 1.1: 0.9 x 0.9	— 1:	3	... I B(s)m NGC 7634 group		15.5		*
12545	23 19.2 + 26 48 099.39 - 31.66	04 - 55 - 009	0843	1.3 x 0.6 1.4 x 0.7	85: ...	3	S B c P w 12546, 2nd of 3		15.3		*1
12546	23 19.2 + 26 49 099.40 - 31.65	04 - 55 - 008	0843	1.4 x 0.4 1.4 x 0.4	19 6	2	Sb - c P w 12545, 3rd of 3		15.3		*1
12547	23 19.3 + 04 43 085.58 - 51.27	01 - 59 - 066	0316	1.2 x 0.6 1.2 x 0.5	(150) ...	2	S ... P w 12548, disturbed		15.2		*1
12548	23 19.4 + 04 45 085.65 - 51.25	01 - 59 - 067	0316	1.0 x 0.20 1.1 x 0.22	118 7	1	Sa? P w 12547		14.9		*1
12549	23 19.4 + 17 10 094.43 - 40.42	03 - 59 - 000	0320 486	1.1: x 0.5: 1.1: x 0.4:	138: ...	2	Irr nnp of 2		16.0:		*
12550	23 19.4 + 43 44 106.36 - 15.96	07 - 48 - 001	0554 488	1.1: x 1.0: 1.2 x 1.2	— (1)	1	S B 0?		16.0:		
12551	23 19.6 + 09 00 089.12 - 47.65	01 - 59 - 068	0316 487	1.1 x 0.35 1.2 x 0.3	2 6	2	S B b		15.7		*
12552	23 19.6 + 12 46 091.76 - 44.36	02 - 59 - 028	0313	1.8 x 0.25 1.0 x 0.25	169 7	2	Sa - b		15.7		*
12553	23 19.7 + 09 07 089.23 - 47.56	01 - 59 - 069	0316 487	1.4: x 1.1: 1.3:: x 1.1::	... ...	4	Dwarf V		17.		
12554	23 19.7 + 40 35 105.24 - 18.93	07 - 48 - 002 NGC 7640	0554	11. x 2.3 11. x 2.1	167 6	4	S B c S B(s)c Sc +		11.6	+ 423 + 692	*12- -689
12555	23 20.0 + 04 51 085.93 - 51.25	01 - 59 - 000	0316	1.1 x 0.3 0.9: x 0.2	144 7	2	Sc		16.0:		
12556	23 20.0 + 11 37 091.11 - 45.42	02 - 59 - 029 NGC 7641	0313 487	1.7 x 0.5 1.7 x 0.6	144 7	1	Sa NGC 7643 group		15.2		3
12557	23 20.0 + 28 55 100.55 - 29.78	05 - 55 - 005	0914 484	1.5: x 0.4 1.4: x 0.2	11 7	1	S ... P w 12566		15.3		*
12558	23 20.0 + 49 54 108.45 - 10.21	08 - 42 - 000	1162	1.1: x 0.7 1.1 x 0.7	0: 4	3	Sc		14.9		*
12559	23 20.1 + 43 44 106.48 - 16.00	07 - 48 - 000	0554 488	1.2 x 0.6 1.1 x 0.6	170 5:	2	... S d - m		17.		*
12560	23 20.3 + 01 10 082.65 - 54.35	00 - 59 - 035 NGC 7642	0834	0.45 x 0.45 0.45 x 0.35	— 1:	1	S ... np of 2		14.5		*3
12561	23 20.3 + 08 41 089.10 - 48.01	01 - 59 - 070	0316 487	1.7 x 0.4 1.1: x 0.4	175 6	2	... S d - dm		16.0:		
12562	23 20.3 + 11 29 091.11 - 45.57	02 - 59 - 031	0313 487	1.3: x 0.3 1.0 x 0.3	84 (6)	2	Sc - Irr NGC 7643 group		16.0:		
12563	23 20.3 + 11 43 091.27 - 45.36	02 - 59 - 033 NGC 7643	0313 487	1.4 x 0.8 1.5 x 0.8	45 5	1	S ... Brightest of 3		14.8		*3
12564	23 20.3 + 22 39 097.65 - 35.56	04 - 55 - 000	0843	1.0 x 0.35 1.0 x 0.35	80 6	3	Sc npp of 2, Sc +		15.3		*
12565	23 20.3 + 22 56 097.79 - 35.30	04 - 55 - 010	0843	1.1 x 0.7 1.0 x 0.7	[52] ...	1	S ...		15.4		*1

N G C   7626, 7628, 7630, 7631, 7634, 7640 - 43   I C   5315

1	2	3	4	5	6	7	8	9	10	11	12
12566	23 20.3 + 28 52 100.60 − 29.86	05 − 55 − 006	0914 484	1.8 x 1.4 0.7 x 0.6	(140) ...	2	Sa − b		15.0		*
12567	23 20.3 + 43 42 106.50 − 16.05	07 − 48 − 003	0554 488	1.3: x 1.1: 1.3: x 1.0:	-- ...	1	S B ...	S B(s) ...	16.0:		*
12568	23 20.6 + 42 25 106.09 − 17.27	07 − 48 − 000	0554 488	1. x 1. 0.9: x 0.8:	-- 1:	2	...	S dm − m	17.		
12569	23 20.7 + 14 38 093.28 − 42.83	02 − 59 − 000	0313 489	1.0 x 0.9 0.8 x 0.7	-- 1 − 2	1	S ...		16.0:		
12570	23 20.7 + 32 15 102.15 − 26.76	05 − 55 − 008	0914 490	0.6 x 0.4 0.50 x 0.35	... ...	1	... Peculiar		14.5		*
12571	23 20.8 + 13 03 092.30 − 44.24	01 − 59 − 034	0313 487	2.0 x 1.1 1.0 x 0.5	95: 5	1	S B ...	(R:)S B(r) ...	15.5		1
12572	23 20.8 + 24 38 098.76 − 33.79	04 − 55 − 000	0843	1.0 x 0.2 0.9 x 0.25	6 7	1	Sa − b		15.6		*
12573	23 20.8 + 43 41 106.59 − 16.10	07 − 48 − 004	0554 488	(1.1 x 0.6) (1.5 x 0.9)	100: ...	2	[E] Compact		15.0		*
12574	23 21.1 + 19 17 096.09 − 38.69	03 − 59 − 052	0320 486	1.2: x 0.6 1.1 x 0.6	35: ...	3	S B b In group, distorted		15.5		*1 VA
12575	23 21.3 + 09 23 089.94 − 47.52	01 − 59 − 072 NGC 7648 = 1486*	0316 487	1.6 x 1.0 1.5 x 1.1	85 3 − 4	2	S0		13.5		*3
12576	23 21.5 + 16 31 094.64 − 41.23	03 − 59 − 055 NGC 7647	0320	1.1 x 0.8 (1.6 x 1.1)	(170) ...	2	E		15.2		*13
12577	23 21.5 + 20 26 096.82 − 37.69	03 − 59 − 054	0320 486	1.3 x 0.2 1.3 x 0.20	58 7	3	Sb		16.0:		*
12578	23 21.8 − 00 23 081.61 − 55.83	00 − 59 − 038	0834	1.6 x 1.1 1.5: x 1.0:	... ...	4	S ... Disrupted, plume, p w 12589		15.1		*1
12579	23 21.8 + 14 22 093.44 − 43.19	02 − 59 − 034 NGC 7649	0313 489	(1.2 x 0.8) (1.6 x 1.1)	(80) ...	3	E		15.7		*1
12580	23 22.0 + 08 20 089.39 − 48.52	01 − 59 − 073	0316 487	1.3: x 0.3 1.2 x 0.25	177 7	1	Sa?		15.6		
12581	23 22.0 + 08 59 089.87 − 47.96	01 − 59 − 074	0316 487	1.3 x 1.0 1.5: x 1.1:	(135) ...	2	S B a − b		15.6		
12582	23 22.0 + 16 36 094.84 − 41.21	03 − 59 − 056	0320	1.1: x 1.0: 0.7 x 0.5	-- 1:	1	S B ...		15.6		
12583	23 22.0 + 23 42 098.61 − 34.76	04 − 55 − 000	0843	1.3 x 0.3 1.1 x 0.2	53 6 − 7	3	Sc		15.7		
12584	23 22.1 − 02 16 079.67 − 57.38	00 − 59 − 040	0834	[1.5 x 1.1] ...	... ...	3	Double system Bridge		15.6		*1
12585	23 22.1 + 08 08 089.27 − 48.71	01 − 59 − 075	0316 487	1.6: x 1.5: 1.4: x 1.3:	-- ...	2	...	S dm:	15.5		
12586	23 22.3 + 15 00 093.97 − 42.67	02 − 59 − 038 NGC 7653 = 1488*?	0313 489	1.7 x 1.5 1.7 x 1.5	-- 1	4	Sb		13.8		*3
12587	23 22.3 + 26 22 099.94 − 32.33	04 − 55 − 011	0843	1.1 x 0.8 1.2 x 1.0	(155) 2	2	Sb		15.6		1
12588	23 22.3 + 41 04 105.91 − 18.65	07 − 48 − 005	0554	1.8 x 1.8 1.6 x 1.6	-- 1	4	...	S dm	14.0		1
12589	23 22.5 − 00 16 081.98 − 55.84	00 − 59 − 042	0834	1.4 x 0.5 1.3 x 0.3	[57] ...	2	S − Irr ssp of 2, disrupted, p w 12578		15.6		*1
12590	23 22.7 + 14 55 094.04 − 42.79	02 − 59 − 000	0313 489	1.0: x 0.18 0.8 x 0.2	155 7	2	S0 − a		15.4		
12591	23 22.9 + 28 13 100.92 − 30.67	05 − 55 − 015	0914 484	1.6 x 0.7 1.6 x 0.7	58 6	2	S0 − a		14.0		
12592	23 23.0 + 05 11 087.23 − 51.36	01 − 59 − 000	0316	1. x 1. 0.7: x 0.7:	-- ...	3	Dwarf V		17.		
12593	23 23.0 + 32 24 102.72 − 26.80	05 − 55 − 016	0914 490	1.3 x 1.0 1.0: x 0.9:	(120) 2:	4	Sc	SA d	15.6		
12594	23 23.3 + 26 45 100.36 − 32.06	04 − 55 − 012 NGC 7660	0843	(1.1 x 0.8) (1.5 x 1.0)	35: ...	3	E		13.9		*13
12595	23 23.4 + 13 55 093.63 − 43.76	02 − 59 − 040 NGC 7659	0313 489	1.0 x 0.4 1.1 x 0.5	110 6	1	S0 − a		15.1		3
12596	23 23.5 + 12 55 093.03 − 44.66	02 − 59 − 041	0313 487	1.1 x 0.5 1.0 x 0.4	166 6	1	S ...		15.4		
12597	23 23.7 + 21 29 097.95 − 36.94	03 − 59 − 000	320	1.0: x 0.3: 0.7 x 0.3	15 ...	1	...		16.5 :		
12598	23 24.2 + 24 48 099.69 − 33.94	04 − 55 − 013 NGC 7664	0843 493	3.3 x 1.8 2.6: x 1.5:	90 5	2	Sc	Sc −?	13.3		3
12599	23 24.3 + 25 24 099.99 − 33.40	04 − 55 − 000	0843 493	1.1 x 0.2 1.1 x 0.25	152 6 − 7	2	S0 − a nf of 2		15.4		*
12600	23 24.4 + 48 55 109.00 − 11.38	08 − 42 − 000	1162	1.3: x 0.3 1.1 x 0.2	178 ...	1	...		17.		

N G C   7647 − 49, 7653, 7659, 7660, 7664

I C   1486, 1488

1	2	3	4	5	6	7	8	9	10	11	12
12601	23 24.6 + 14 46 094.51 - 43.13	02 - 59 - 043	0313	1.6: x 1.0: 1.2: x 0.9:	(60) ...	2	... S B(s)dm:		16.0:		
12602	23 24.8 + 12 12 092.97 - 45.45	02 - 59 - 044 NGC 7671	0313 487	1.4 x 0.8 1.6 x 0.9	138 4	4	S0 P w NGC 7672		14.3	+ 4129 + 4336	*12 3
12603	23 25.0 + 29 35 102.00 - 29.58	05 - 55 - 000	0914	1.1 x 0.3 1.0 x 0.3	28 6	1	S ...		16.5:		
12604	23 25.1 + 22 29 098.82 - 36.16	04 - 55 - 000	0843 493	1.0 x 0.7 1.0 x 0.7	50: 3:	3	S B a		15.6		
12605	23 25.1 + 36 07 104.62 - 23.48	06 - 51 - 000	0873	1.0 x 0.5 1.0 x 0.5	25: ...	3	S B: c		16.5:		
12606	23 25.1 + 43 52 107.41 - 16.19	07 - 48 - 000	0554	1.0 x 0.2 1.1 x 0.2	130 6 - 7	1	S ...		16.5:		
12607	23 25.2 + 23 19 099.25 - 35.40	04 - 55 - 014 NGC 7673, Mark 325	0843 493	(1.7 x 1.6) (1.3 x 1.1)	-- ...	1	... Compact, p w 12610		12.7	+ 3255	*13 C
12608	23 25.4 + 08 30a 090.62 - 48.79	01 - 59 - 080 NGC 7674	0316 487	1.0 x 0.9 1.0 x 0.9	-- 1 - 2	3	S B b P w NGC 7675, disturbed		(13.6)		*13
12609	23 25.5 + 30 44 102.59 - 28.55	05 - 55 - 020	0914 490	1.3 x 0.6 1.3 x 0.6	120: ...	3	... S dm		15.7		
12610	23 25.6 + 23 15 099.32 - 35.50	04 - 55 - 015 NGC 7677, Mark 326	0843 493	1.7 x 1.1 1.6 x 1.1	35 ...	3	S B? b		13.9	+ 3900	*13
12611	23 25.7 + 27 24 101.23 - 31.66	04 - 55 - 016	0843	1.1 x 0.2 1.2 x 0.2	177 7	2	Sa		15.5		
12612	23 25.8 + 35 02 104.36 - 24.54	06 - 51 - 000	0873	1.0 x 0.1 0.8 x 0.1	175 7	3	Sc Sc +		17.		
12613	23 26.0 + 14 27 094.74 - 43.56	02 - 59 - 046	0313	5. x 3. 5. x 3.	(120) ...	5	Dwarf irregular Ir V		15.5		*12
12614	23 26.0 + 22 08 098.88 - 36.56	04 - 55 - 017 NGC 7678	0843 493	2.5 x 1.8 2.8: x 1.9	5: 3	4	Sc/S B c SAB(rs)c		12.7	+ 3446 + 3680	*12 3468
12615	23 26.0 + 28 30 101.78 - 30.67	05 - 55 - 000	0914	1.0: x 0.7: 0.8: x 0.7:	... ...	2	... S dm		17.		
12616	23 26.1 + 32 08 103.29 - 27.28	05 - 55 - 023 NGC 7680	0914 490	(1.6 x 1.6) (2.0 x 2.0)	-- ...	2	E - S0		13.5		*13
12617	23 26.1 + 35 10 104.47 - 24.44	06 - 51 - 000	0873	1.0 x 0.3 0.9 x 0.3	164 6 - 7	1	S ...		16.5:		
12618	23 26.2 + 03 14 086.67 - 53.44	00 - 59 - 046 NGC 7679	0834	1.7 x 1.1 1.6 x 1.0	93 4:	2	S0 P w 12622, disturbed		13.2	+ 5154 + 5330	*12 348VA
12619	23 26.2 + 21 57 098.85 - 36.75	04 - 55 - 018	0843 493	1.6 x 1.0 1.6 x 1.0:	173 ...	4	... SAB(s)dm		15.4		
12620	23 26.4 + 17 03 096.36 - 41.25	03 - 59 - 063	0320	(1.4 x 1.2) (1.8 x 1.6)	-- ...	3	S0 Companion superimposed, disturbed		14.2		*3
12621	23 26.4 + 49 43 109.58 - 10.72	08 - 42 - 000	1162	1.5: x 0.9: 1.0: x 0.8:	(115) ...	1	S B ...		16.0:		*
12622	23 26.5 + 03 15 086.79 - 53.46	00 - 59 - 047 NGC 7682	0834	1.1 x 0.9 (1.3 x 1.0)	-- 2:	3	S B a P w 12618, disturbed		14.3		*12 3VA
12623	23 26.5 + 11 10 092.83 - 46.56	02 - 59 - 048 NGC 7683	0313 487	(2.0 x 0.9) (2.1 x 1.1)	140 5 - 6	4	S0		14.6		13
12624	23 26.6 + 21 17 098.63 - 37.40	03 - 59 - 062	0320	1.2 x 0.3 1.0 x 0.3	61 (6)	3	Irr		16.0:		
12625	23 26.6 + 29 30 102.34 - 29.78	05 - 55 - 025	0914	1.6 x 0.2 1.5 x 0.2	176 7	3	Sb		15.0		*
12626	23 26.7 + 26 06 100.91 - 32.95	04 - 55 - 000	0843 493	1.2 x 0.4 1.1: x 0.4	0 6	1	S ...		15.7		
12627	23 26.7 + 34 47 104.45 - 24.84	06 - 51 - 006	0873	1.0 x 0.35 1.0 x 0.4	5 6	2	S0 - a		16.0:		
12628	23 26.8 + 03 07 086.78 - 53.62	00 - 59 - 048	0834	1.5 x 1.3 1.5 x 1.1	(80) 1 - 2	4	S B c		15.6		
12629	23 27.3 + 32 25 103.67 - 27.11	05 - 55 - 029	0914 490	1.1 x 0.2 1.2 x 0.2	22 7	2	Sa s of 2		15.6		*
12630	23 27.4 + 08 40 091.40 - 48.88	01 - 59 - 086	0316 487	1.0 x 0.4 1.0 x 0.3	75 6	1	Sa - b		15.4		
12631	23 27.5 + 26 49 101.42 - 32.35	04 - 55 - 019	0843	1.1 x 0.25 1.0 x 0.25	145 6	3	Sb		14.8		*
12632	23 27.6 + 40 43 106.78 - 19.31	07 - 48 - 007	0554	5. x 4. 4. x 2.	-- ...	4	Dwarf spiral S V		15.7		*
12633	23 27.7 + 15 29 095.86 - 42.80	02 - 59 - 050	0313	1.0 x 0.6 0.9 x 0.5	... ...	1	S B ...		15.4		
12634	23 27.7 + 39 57 106.53 - 20.04	07 - 48 - 008	0554	1.6:: x 0.3 1.2: x 0.3	175 6	3	Sc		15.7		
12635	23 27.8 - 00 07 084.09 - 56.47	00 - 59 - 049	0834	1.1 x 1.0 1.0 x 1.0	-- 1	3	Sc P w 12637		15.6		*

N G C   7671, 7673, 7674, 7677 - 80, 7682, 7683

1	2	3	4	5	6	7	8	9	10	11	12
12636	23 27.8 + 26 51 101.51 – 32.35	04 – 55 – 020	0843	(1.0 x 1.0) (1.1 x 1.1)	— (1)	2	S0		15.7		*1
12637	23 27.9 – 00 12 084.04 – 56.55	00 – 59 – 050 NGC 7684	0834	1.5 x 0.4 1.7 x 0.4	21 6	2	Sa P w 12635		14.8		*3
12638	23 27.9 + 03 37 087.61 – 53.33	01 – 59 – 087 NGC 7685	0316	1.9 x 1.6 2.0 x 1.7	170: 2	4	Sc		15.0		23
12639	23 27.9 + 29 57 102.83 – 29.46	05 – 55 – 031	0914	1.4: x 1.1 1.2 x 1.0	(20) 2:	4	S B b	S B(s)b Sb +	14.9		
12640	23 28.4 + 15 13 095.91 – 43.12	02 – 59 – 000	0313	1.0 x 0.12 0.8 x 0.10	115 7	2	...	d – m	15.6		
12641	23 28.4 + 19 58 098.46 – 38.77	03 – 59 – 065	0320	1.1 x 0.2 1.3 x 0.25	144 6	2	Sb – c		15.5		
12642	23 28.7 + 08 57 092.04 – 48.78	01 – 59 – 000	0316 487	1.2 x 0.1 1.1 x 0.10	153 7	2	...	d – m	16.5:		
12643	23 29.0 + 24 56 100.97 – 34.23	04 – 55 – 000	0843	1.4 x 0.6: 1.0 x 0.4:	[100] ...	3	...	S B dm	16.5:		
12644	23 29.0 + 28 55 102.67 – 30.51	05 – 55 – 032	0914	1.3 x 0.7 1.1 x 0.7	110: ...	4	S B c	Sc +	15.7		
12645	23 29.0 + 32 12 103.97 – 27.43	05 – 55 – 033	0914 490	1.0 x 0.5 1.1 x 0.6:	97 ...	1	S ...		15.1		
12646	23 29.1 + 25 40 101.32 – 33.56	04 – 55 – 022	0843 493	1.9 x 1.7 1.7 x 1.6	— 1	4	S B b	S B(r)b	14.4		
12647	23 29.3 + 47 10 109.23 – 13.30	08 – 42 – 000	1162	1.0 x 0.15 0.8 x 0.15	154 6 – 7	1	Sa – b		16.5:		
12648	23 29.4 – 02 26 082.25 – 58.60	00 – 60 – 000	0431 491	1.0 x 0.9 0.8 x 0.8	— ...	3	Irr		15.1		
12649	23 29.4 + 44 33 108.40 – 15.79	07 – 48 – 009	0554	(1.1 x 1.0) (1.1 x 1.0)	— ...	2	... Compact		15.7		
12650	23 29.5 + 32 08 104.06 – 27.53	05 – 55 – 034	0914 490	1.4 x 0.2 1.4 x 0.2	160 7	2	Sc		15.5		
12651	23 29.7 + 35 07 105.22 – 24.74	06 – 51 – 007	0873	1.0: x 0.8: 1.1: x 0.7:	(120) ...	2	...	S dm – m	16.5:		
12652	23 29.9 + 44 42 108.54 – 15.67	07 – 48 – 010	0554	1. x 1. 1.3 x 0.9	— ...	1	... Compact		15.5		*C
12653	23 30.0 + 14 34 096.02 – 43.87	02 – 60 – 001	0318	1.6 x 0.5 1.6 x 0.5	103 6	1	S ...		15.3		
12654	23 30.0 + 15 34 096.59 – 42.96	03 – 60 – 001 NGC 7691	1182	2.4 x 1.7 2.0 x 1.6	175: 3	4	S B b/Sc	SAB(rs)bc Sc –	14.2		13
12655	23 30.0 + 23 39 100.65 – 35.51	04 – 55 – 025	0843 493	1.5 x 0.9 1.3 x 0.8	135 4	3	S0		14.0		*
12656	23 30.1 – 02 01 082.98 – 58.36	00 – 60 – 001	0431	1.0 x 0.2 0.9 x 0.2	126 6	1	S ... P w 12659		15.5		*
12657	23 30.2 + 29 11 103.06 – 30.36	05 – 55 – 035	0914	1.9: x 0.6 2.1 x 0.8	140 6	3	S0		14.5		
12658	23 30.2 + 30 52 103.72 – 28.77	05 – 55 – 000	0914 490	1.3: x 0.5 1.3: x 0.5	147 ...	1	S B b?		16.0:		
12659	23 30.3 – 02 03 083.02 – 58.42	00 – 60 – 002	0431 491	1.3 x 0.15 1.3 x 0.15	105 7	3	Sc P w 12656		15.6		*
12660	23 30.8 + 20 57 099.61 – 38.08	03 – 60 – 002 IC 5329	1182	1.8 x 0.25 1.6 x 0.2	111 7	3	Sc P w 12662		15.7		*
12661	23 31.0 – 02 15b 083.08 – 58.68	00 – 60 – 007	0431 491	1.1 x 0.2 1.1 x 0.2	147 7	2	Sa – b sf of 2		15.0		*1
12662	23 31.0 + 20 51 099.62 – 38.19	03 – 60 – 003 IC 5331	1182	1.1 x 0.3 1.1 x 0.3	18 6	2	Sa – b P w 12660		(15.1)		*
12663	23 31.0 + 23 44 100.95 – 35.51	04 – 55 – 027	0843 493	1.1 x 1.0 0.8 x 0.8	— ...	1	...		16.0:		
12664	23 31.1 + 15 17 096.76 – 43.32	02 – 60 – 000	0318	1.2 x 0.6 1.0 x 0.5	157 5	3	Sc		15.7		
12665	23 31.2 + 29 46 103.53 – 29.88	05 – 55 – 036	0914	1.3 x 0.9 1.3 x 0.8	... ...	2	S ... Companion attached, p w 12667		15.1		*1VA
12666	23 31.2 + 32 06 104.42 – 27.69	05 – 55 – 037	0914 490	1.5 x 0.7 1.5 x 0.7	128 5	4	Sc		14.5		
12667	23 31.3 + 29 48 103.56 – 29.86	05 – 55 – 038	0914	1.6 x 0.9 1.6 x 0.9	142 4	4	Sc P w 12665		13.5		*1
12668	23 31.5 + 24 40 101.49 – 34.68	04 – 55 – 029 NGC 7698	0843 493	1.0 x 0.8 1.1 x 0.9	170: 2	3	S0		14.5		
12669	23 31.5 + 44 25 108.73 – 16.03	07 – 48 – 011	0554	1.2 x 1.1 1.1 x 0.9	— 1 – 2	2	Sb – c	SAB ...	15.3		
12670	23 31.8 + 12 17 095.23 – 46.12	02 – 60 – 002	0318	1.0 x 0.5 0.9 x 0.5	67 5	1	Sa – b		15.5		

N G C   7684, 7685, 7691, 7698   I C   5329, 5331

1	2	3	4	5	6	7	8	9	10	11	12
12671	23 31.8 + 43 07 108.36 – 17.28	07 – 48 – 000	0554	1.1 x 0.7 1.0 x 0.5	105 ...	1	...		16.0:		
12672	23 31.9 + 34 21 105.41 – 25.61	06 – 51 – 008	0873	1.1 x 0.8 1.1 x 0.6	... ...	3	S ... Peculiar, unilateral?		14.6		*1
12673	23 32.0 + 14 54 096.82 – 43.76	02 – 60 – 000	0318	1.0 x 0.1 0.9 x 0.1	111 7	2	Sc nnf of 2		16.0:		*
12674	23 32.0 + 28 27 103.20 – 31.18	05 – 55 – 000	0914	1.1 x 0.4 1.1 x 0.4	[5] 6:	2	...	S d – m	16.5:		
12675	23 32.1 + 22 57 100.89 – 36.33	04 – 55 – 000	0843 493	1.0: x 0.2 1.1 x 0.2	38 (7)	1	...		15.7		
12676	23 32.3 + 15 48 097.41 – 42.97	03 – 60 – 004 NGC 7703	1182	2.3 x 0.4 2.3 x 0.4	147 7	2	S0		14.8		3
12677	23 32.3 + 16 35 097.83 – 42.25	03 – 60 – 005	1182	1.0 x 0.7 0.8 x 0.4	0 3:	1	Sb – c		15.4		*
12678	23 32.3 + 26 02 102.28 – 33.47	04 – 55 – 000	0843 493	1.2 x 0.2 1.2 x 0.2	84 7	3	Sb		15.4		
12679	23 32.3 + 47 13 109.74 – 13.40	08 – 42 – 000	1162	(1.0 x 0.8) 1.1 x 0.9	— ...	0	S0?		16.0:		*
12680	23 32.4 + 17 01 098.09 – 41.86	03 – 60 – 006	1182	1.3 x 0.15 1.3 x 0.15	121 7	2	S B? b – c		15.7		1
12681	23 32.4 + 17 48 098.50 – 41.14	03 – 60 – 008	1182	1.3 x 0.7 1.1 x 0.6	[15] ...	1	S B ...		15.2		
12682	23 32.4 + 17 57 098.58 – 41.00	03 – 60 – 007	1182	1.6: x 1.3: 1.5: x 1.1:	[40] ...	4	Irr	I m Ir I	14.7		*1
12683	23 32.4 + 44 02 108.76 – 16.44	07 – 48 – 012 NGC 7707	0554	(1.2 x 0.9) (1.3 x 1.1)	— ...	2	E – S0		14.8		3
12684	23 32.6 + 04 38 090.15 – 53.04	01 – 60 – 005 NGC 7704	0796	(1.1 x 0.9) (1.2 x 0.9)	67 ...	2	E – S0 P w 12686		14.8		*3
12685	23 32.7 – 00 13 085.87 – 57.22	00 – 60 – 012	0431	1.1 x 1.1 0.9: x 0.8:	— 1	3	Sc		16.0:		
12686	23 32.7 + 04 43 090.26 – 52.98	01 – 60 – 006 NGC 7706	0796	1.1 x 0.9 1.1 x 0.9	120 2	3	S0 P w 12684		14.6		*
12687	23 32.8 + 12 40 095.78 – 45.88	02 – 60 – 003	0318	2.0: x 1.4: 1.2 x 0.9	125: 3:	4	S B b	Sb +	14.9		
12688	23 32.9 + 07 03 092.11 – 50.94	01 – 60 – 007	0796	1.6 x 0.4 1.5 x 0.4	[88] ...	3	... Peculiar, sickle–shaped		14.4		*1
12689	23 33.0 + 04 57 090.55 – 52.81	01 – 60 – 008	0796	1.7 x 0.3 1.9 x 0.3	149 7	2	Sa – b		14.5		*
12690	23 33.1 + 00 57 087.14 – 56.29	00 – 60 – 015	0431	2. x 2. 1.3:: x 1.0::	— ...	3	Dwarf V		17.		
12691	23 33.1 + 15 02 097.23 – 43.75	02 – 60 – 004 NGC 7711	0318	(4.0 x 1.6) (4. x 1.6)	95 5 – 6	2	S0 Singular		14.0		*3
12692	23 33.2 + 20 01 099.83 – 39.16	03 – 60 – 010	1182 492	1.2 x 0.3 0.9 x 0.25	52 (7)	1	S? ...		15.6		
12693	23 33.2 + 32 06 104.88 – 27.83	05 – 55 – 039	0914	2.1 x 0.15 2.0: x 0.1	2 7	3	S B? c		15.6		
12694	23 33.3 + 23 20 101.38 – 36.08	04 – 55 – 030 NGC 7712	0843 493	0.9 x 0.8 0.8 x 0.7	— ...	1	... Peculiar		13.7		3
12695	23 33.5 + 12 35 095.95 – 46.02	02 – 60 – 005	0318	1.2: x 1.2: 1. x 1.	— ...	2	Dwarf spiral		17.		
12696	23 33.5 + 22 46 101.18 – 36.62	04 – 55 – 031	0843 493	1.0 x 0.9 1.0 x 0.9	— 1	3	S B b/S B c		15.1		
12697	23 33.5 + 35 40 106.22 – 24.47	06 – 51 – 000	0873	1.0: x 0.6: 1. x 0.5::	... ...	2	Dwarf V		17.		
12698	23 33.5 + 48 36 110.36 – 12.14	08 – 42 – 000	1162	1.0 x 0.3 1.0 x 0.3	15 6	3	Sc	Sc +	16.5:		
12699	23 33.7 + 01 53 088.23 – 55.56	00 – 60 – 017 NGC 7714	0431	1.8 x 1.3 2.3 x 1.8	... ...	3	S ... P w 12700, bridge, disturbed		13.1	+ 2833 + 3000	*12 3VĀ
12700	23 33.8 + 01 53 088.27 – 55.58	00 – 60 – 018 NGC 7715	0431	3.2: x 0.4 3.3: x 0.9:	73 ...	1	S ... P w 12699, bridge		14.9	+ 2795 + 2962	*12 3VĀ
12701	23 33.8 + 27 41 103.33 – 32.04	05 – 55 – 040	0914	1.0 x 0.1 0.9: x 0.1	176 7	3	Sc		16.5:		1
12702	23 34.0 + 00 02 086.62 – 57.19	00 – 60 – 019 NGC 7716	0431	2.3 x 1.8 2.2 x 1.8	35: 2	4	S B b		12.9	+ 2546 + 2706	*12 3468
12703	23 34.0 + 20 52 100.45 – 38.44	03 – 60 – 013 IC 5338	1182 492	(1.1 x 0.6) (1.4 x 1.1)	(30) ...	1	E? P w IC 5337		15.5		*1
12704	23 34.0 + 50 00 110.86 – 10.83	08 – 43 – 001	0368	(1.1 x 0.8) (1.0 x 0.7)	— ...	1	E?		16.0:		*
12705	23 34.1 + 13 53 096.90 – 44.90	02 – 60 – 006	0318	1.5: x 1.0: 1.0: x 0.7:	147 3:	3	S B c		15.7		

N G C   7703, 7704, 7706, 7707, 7711, 7712
7714 – 16   I C   5338

1	2	3	4	5	6	7	8	9	10	11	12
12706	23 34.1 + 75 23 118.22 + 13.48	13 – 01 – 002 IC 1502	1213	1.3 x 0.5 1.2 x 0.5	53 6 – 7	2	S0 – a		14.7		
12707	23 34.2 + 17 15 098.75 – 41.81	03 – 60 – 014	1182	1.7 x 1.1 1.6: x 1.0:	40: ...	1	S B ...		15.0		*1
12708	23 34.3 + 26 13 102.86 – 33.45	04 – 55 – 000	0843 493	1.1 x 0.2 1.1 x 0.3	67 ...	1	...		15.5		
12709	23 34.9 + 00 08 087.07 – 57.22	00 – 60 – 022	0431	3.0 x 1.8 3. x 2.	145 ...	4	Dwarf spiral, S m S IV – V		15.7		*
12710	23 35.0 + 17 44 099.23 – 41.43	03 – 60 – 015	1182	1.5 x 0.9 1.2 x 0.9	[140] ...	3	Irr or Peculiar		14.7		*1
12711	23 35.4 + 31 43 105.24 – 28.34			(1.2 x 1.1)? (0.9 x 0.7)	-- ...	1	...		14.1		
12712	23 35.5 + 25 25 102.84 – 34.30	04 – 55 – 034 NGC 7718	0843 493	1.2: x 0.8: 1.4 x 0.9	(160) ...	1	S: ... sf of 2		15.2		*3
12713	23 35.7 + 30 25 104.83 – 29.59	05 – 55 – 044	0914	1.2 x 0.6 1.2 x 0.6	... ...	1	...		15.1		
12714	23 35.7 + 32 03 105.43 – 28.04	05 – 55 – 043	0914	1.3: x 0.2 1.1 x 0.1	165 7	2	Sc sf of 2		15.6		*
12715	23 35.9 + 04 32 091.28 – 53.52	01 – 60 – 016 IC 1503	0796	1.0 x 0.45 0.9 x 0.45	[65] ...	1	S – Irr		14.3		
12716	23 36.0 + 26 45 103.50 – 33.08	04 – 55 – 036 NGC 7720	0843 493	(1.6 x 1.3) (2.8 x 1.7)	... ...	2	E + companion Symbiotic		13.9		*13
12717	23 36.1 + 05 10 091.85 – 52.98	01 – 60 – 017	0796	1.1 x 1.0 1.1 x 1.1	-- 1	1	Sc P w 12720		15.3		*
12718	23 36.2 + 15 41 098.53 – 43.44	03 – 60 – 017 NGC 7722	1182	(2.2 x 1.9) (2.2 x 1.8)	(150) 4:	2	S0 – a		13.7		*13
12719	23 36.2 + 26 30 103.46 – 33.33	04 – 55 – 000	0843 493	1.1 x 0.5 1.1 x 0.4	70 6	2	S0 – a		15.7		*
12720	23 36.7 + 05 13 092.11 – 53.00	01 – 60 – 019	0796	1.0 x 0.20 1.1 x 0.20	156 6	1	S ... P w 12717		15.3		*
12721	23 36.7 + 26 50 103.71 – 33.05	04 – 55 – 040	0843 493	1.5 x 0.6 1.5 x 0.6	60 6	4	S B b	S B(s)b	15.0		*
12722	23 36.8 + 08 57 094.78 – 49.66	01 – 60 – 015?	0796 487	1.0 x 0.4 1.0 x 0.4	20 6	1	S ...		15.3		*
12723	23 36.8 + 10 35 095.82 – 48.18	02 – 60 – 000	0318	1.0 x 0.12 0.9 x 0.12	77 7	2	...	d – m	15.7		
12724	23 37.0 + 28 33 104.45 – 31.45	05 – 55 – 000	0914	1.0: x 0.6: 0.8 x 0.4	[130] ...	1	Dwarf:		17.		
12725	23 37.1 + 21 39 101.67 – 37.96	04 – 55 – 000	0843 493	1.0: x 0.4 1.1 x 0.4	16 6 – 7	3	S0		15.1		
12726	23 37.3 + 31 07 105.46 – 29.04	05 – 55 – 000	0914	1.3 x 0.2 1.2 x 0.2	161 7	2	Sa – b		16.0:		*
12727	23 37.4 + 26 52 103.90 – 33.07	04 – 55 – 041 NGC 7728	0843 493	(1.0 x 0.8) (1.5 x 1.2)	75: ...	3	E		14.3		13
12728	23 37.7 + 47 42 110.78 – 13.21	08 – 43 – 003	0368	(1.0 x 0.9) 0.8 x 0.8	-- ...	1	... Compact		15.4		
12729	23 37.9 + 00 58 089.04 – 56.88	00 – 60 – 029	0431	1.3 x 0.5 1.2 x 0.5	74 6	1	S0 – a		15.2		
12730	23 38.0 + 28 55 104.83 – 31.17	05 – 55 – 046 NGC 7729	0914	1.9 x 0.6 1.9 x 0.6	7 6	2	Sa		14.6		*
12731	23 38.1 + 20 10 101.28 – 39.43	03 – 60 – 018	1182	1.5 x 0.22 1.5 x 0.22	107 7	3	Sb		15.7		*1
12732	23 38.1 + 25 57 103.72 – 33.99	04 – 55 – 042	0843 493	3. x 3. 3. x 3.	-- ...	4	Dwarf spiral S IV – V		14.5		*1
12733	23 38.2 + 26 33 103.98 – 33.43	04 – 55 – 043	0843 493	(1.0 x 0.8) (1.1 x 0.9)	(0) ...	2	E npp of 2		15.1		*
12734	23 38.7 + 03 45 091.71 – 54.53	01 – 60 – 020 IC 1504	0796	1.9 x 0.5 1.8 x 0.5	94 6	2	Sb		14.5		
12735	23 38.8 + 15 52 099.43 – 43.51	03 – 60 – 019	1182	1.0 x 0.5 1.0 x 0.5	19 5	2	S B a		15.3		
12736	23 38.8 + 47 33 110.92 – 13.40	08 – 43 – 000	0368	1.1 x 0.15 1.1 x 0.15	51 6 – 7	1	S B: b		16.0:		*
12737	23 39.0 + 03 28 091.60 – 54.82	00 – 60 – 034 NGC 7731	0431	1.5 x 1.0 1.4 x 1.0	95: 3:	3	S B a P w 12738		14.3	+ 2850 + 3016	*13 C
12738	23 39.1 + 03 27 091.63 – 54.85	00 – 60 – 035 NGC 7732	0431	1.9 x 0.7 1.8 x 0.5	96 (6)	2	Sc – Irr P w 12737, disturbed:		14.5		*13 C
12739	23 39.3 – 01 37 087.12 – 59.29	00 – 60 – 000	0431	0.9 x 0.8 0.9 x 0.7	-- ...	1	...		14.2		*
12740	23 39.3 + 23 33 103.07 – 36.35	04 – 55 – 000	0843 493	1.1: x 1.1: 1.0: x 1.0:	-- 1	1	Sc? np of 2		16.5:		*

N G C   7718, 7720, 7722, 7728, 7729, 7731   I C   1502 – 04
7732

1	2	3	4	5	6	7	8	9	10	11	12
12741	23 39.4 + 30 19 105.67 - 29.94	05 - 55 - 047	0914	0.9 x 0.25 1.0 x 0.25	102 6	1	Sa		14.4		1
12742	23 39.4 + 44 42 110.21 - 16.17	07 - 48 - 017	0554	2.3: x 1.3: 1.8: x 1.2:	80: ...	3	S B c	S B(s) ...	15.5		
12743	23 39.6 + 35 52 107.58 - 24.65	06 - 52 - 002	0405	1.0 x 0.4 0.4 x 0.3	[125] ...	3	S B ... f of 2, distorted, in group		16.5:		*
12744	23 39.7 + 25 57 104.13 - 34.10	04 - 55 - 046	0843 493	(1.2 x 1.0) (1.3 x 1.1)	90: ...	3	E		15.0		3
12745	23 40.2 + 26 46 104.57 - 33.37	04 - 55 - 048 NGC 7737	0843 493	1.1 x 0.5 1.1 x 0.5	147 6	2	S0 - a sf of 2		14.8		*
12746	23 40.3 + 27 01 104.69 - 33.14	04 - 55 - 049	0843 493	1.4 x 0.2 1.4 x 0.2	50 7	3	Sc		15.0		*
12747	23 40.5 + 19 09 101.52 - 40.58	03 - 60 - 020 Mark 330	1182	1.1 x 0.7 1.0 x 0.5	110 ...	2	Irr		14.6	+ 4200	*
12748	23 40.7 + 10 07 096.89 - 48.99	02 - 60 - 000	0318	1.2 x 0.4 0.9 x 0.3	69 6	1	S ...		15.5		
12749	23 40.8 + 08 26 095.81 - 50.53	01 - 60 - 021	0796 494	1.8: x 0.6 1.8: x 0.9:	34 5 - 6	1	S ...		16.0:		
12750	23 40.8 + 49 42 111.84 - 11.42	08 - 43 - 004	0368	1.8 x 0.8 1.6 x 0.8	7 6	3	Sa		15.0		
12751	23 41.0 + 04 49 093.41 - 53.84	01 - 60 - 000	0796	1.1 x 1.1 1.1 x 1.0	-- 1	3	...	SAB(s)dm	16.5:		
12752	23 41.0 + 11 23 097.71 - 47.86	02 - 60 - 008	0318	(1.0 x 0.6) 0.6 x 0.5	135 ...	1	...		15.5		
12753	23 41.0 + 12 55 098.57 - 46.44	02 - 60 - 007	0318	1.5 x 1.3 1.7 x 1.4	(10) 1 - 2	3	Sb	SAB(r)bc Sb +	15.1		*1
12754	23 41.3 + 25 48 104.49 - 34.36	04 - 55 - 050 NGC 7741	0843 (493)	4.4 x 2.9 4.3 x 3.0	(170) 3	4	S B c	S B(s)cd	11.8	+ 729 + 964	*12- -689
12755	23 41.3 + 28 04 105.33 - 32.21	05 - 55 - 049	0914 493	1.6 x 1.0 1.1 x 0.6	105: ...	1	S B ...		15.0		
12756	23 41.4 + 11 15 097.77 - 48.02	02 - 60 - 009	0318	1.2 x 0.9 0.9 x 0.6	150: ...	3	S B b	S B(r) ...	15.7		
12757	23 41.5 + 00 14 089.82 - 57.96	00 - 60 - 038 NGC 7738 = 7739	0431	2.1 x 1.5 1.7 x 1.5	(80) 3:	4	S B b	S B(s)b Sb +	14.4		3
12758	23 41.7 + 02 00 091.45 - 56.43	00 - 60 - 039	0431	1.3 x 1.0 1.1 x 0.9	155 2	2	Sb		15.2		
12759	23 41.8 + 09 39 096.96 - 49.53	02 - 60 - 011 NGC 7743	0318 494	2.5 x 2.1 2.7 x 2.2	80: 2	4	S B 0/S B a P w 12760		12.9	+ 1802 + 1992	*12 3468
12760	23 41.8 + 10 30 097.47 - 48.75	02 - 60 - 010 NGC 7742	0318	(2.0 x 1.8) (1.8 x 1.8)	-- 1:	1	S0? P w 12759		12.5	+ 1678 + 1870	*12 3468
12761	23 42.0 + 05 44 094.46 - 53.12	01 - 60 - 027	0796 495	1.0 x 0.15 1.0 x 0.15	25 7	3	Sc p of 2		16.5:		*
12762	23 42.0 + 21 40 103.04 - 38.33	04 - 56 - 002	0843	1.3 x 0.7 1.3: x 0.7:	125: ...	1	S ...		15.6		*1
12763	23 42.2 + 12 36 098.79 - 46.84	02 - 60 - 014	0318	1.0: x 0.7: 0.6 x 0.3	... ...	2	...	S dm:	16.5:		
12764	23 42.2 + 21 44 103.12 - 38.28	04 - 55 - 003	0843	1.0 x 1.0 0.9 x 0.9	-- 1	3	Sc		15.6		*
12765	23 42.3 + 44 20 110.63 - 16.66	07 - 48 - 018	0554	1.3 x 0.4 1.3 x 0.4	74 6	2	S B? b		15.2		
12766	23 42.5 + 25 15 104.59 - 34.97	04 - 56 - 000	0779 493	1.1 x 0.2 0.7 x 0.2	101 6 - 7	1	S ...		16.0:		
12767	23 42.6 + 06 46 095.40 - 52.25	01 - 60 - 031	0796 495	2.0 x 2.0 2.1 x 2.0	-- 1	4	S B b	S B(rs)b	14.8		
12768	23 42.8 - 01 57 088.27 - 60.02	00 - 60 - 043 NGC 7746	0431	(1.4 x 1.1) (1.2 x 1.0)	(160) 2	2	S0		14.5		
12769	23 42.9 - 01 32 088.73 - 59.67	00 - 60 - 042	0431	1.5 x 0.6 1.5: x 0.6:	23 ...	2	...	dm - m	16.0:		*
12770	23 43.0 + 01 25 091.49 - 57.10	00 - 60 - 044 IC 1507	0431	1.4 x 0.5 1.4 x 0.5	134 6	2	S0 - a		14.8		*
12771	23 43.0 + 16 58 101.26 - 42.84	03 - 60 - 000	1182	1.1: x 0.7: 1.0: x 0.5:	[124] ...	4	Irr	I m Ir I	16.5:		
12772	23 43.0 + 27 06 105.40 - 33.24	04 - 56 - 005 NGC 7747	0779 493	1.7 x 0.6 1.6 x 0.6	36 6	3	S B b	(R')S B(s) ... Sb -	14.5		*3
12773	23 43.4 + 11 47 098.75 - 47.71	02 - 60 - 016 IC 1508	0318	1.8 x 0.4 1.8 x 0.45	168 ...	1	S - Irr		14.6		
12774	23 43.5 - 02 10 088.35 - 60.30	00 - 60 - 045	0431	1.0 x 0.9 1.0 x 0.9	-- 1	3	S B a		14.7		
12775	23 43.6 + 05 36 094.96 - 53.41	01 - 60 - 033	0796 495	1.0 x 0.4 1.0 x 0.4	100 6	1	S ...		15.5		

N G C   7737 - 39, 7741 - 43, 7746, 7747   I C   1507, 1508

1	2	3	4	5	6	7	8	9	10	11	12
12776	23 43.7 + 33 06 107.60 – 27.54	05 – 56 – 002	0606	2.8 x 2.2 2.7 x 2.2	(90) ...	3	S B b Singular	S B(rs)b	14.2		*1
12777	23 44.1 + 03 30 093.61 – 55.36	01 – 60 – 034 NGC 7750	0796	1.7 x 0.8 1.7 x 0.8	171 5	3	Sc	S bc:	13.8		3
12778	23 44.4 + 06 35 095.94 – 52.60	01 – 60 – 035 NGC 7751	0796 495	(1.6 x 1.5) (1.6 x 1.6)	— ...	1	...		13.9		*3
12779	23 44.5 + 29 10 106.50 – 31.36	05 – 56 – 004 NGC 7752	0606	0.45 x 0.2 0.45 x 0.2	[113] ...	1	... Connected with 12780		14.3	+ 4845 + 5085	*12 3VA̅
12780	23 44.6 + 29 12 106.54 – 31.34	05 – 56 – 005 NGC 7753	0606	3.5: x 1.8 3.5: x 1.8	[50] ...	4	Sb Connected with 12779	SAB(rs)b	13.2	+ 4868 + 5108	*12 3VA̅
12781	23 44.7 + 32 31 107.64 – 28.16	05 – 56 – 006 IC 5355	0606	1.1 x 0.6 1.0 x 0.6	10 4 – 5	2	S B c		14.4		
12782	23 45.0 + 50 43 112.77 – 10.61	08 – 43 – 000	0368	1.1: x 0.8: 1.0 x 0.6	... ...	1	S B? ...		16.0:		*
12783	23 45.1 + 18 19 102.51 – 41.73	03 – 60 – 026	1182	1. x 1. 0.8: x 0.8:	— ...	1	Dwarf:		16.5:		
12784	23 45.5 + 17 12 102.13 – 42.82	03 – 60 – 027	1182	1.6: x 1.6: 1.2 x 1.2	— 1	3	S B b	SAB(s)bc	15.2		
12785	23 45.5 + 27 06 106.04 – 33.41	04 – 56 – 008	0779 493	1.2 x 0.4 1.2 x 0.3	100 6 – 7	3	S0		15.0		*
12786	23 45.5 + 49 58 112.66 – 11.36	08 – 43 – 000	0368	1.1: x 0.3 0.8 x 0.3	42 6	1	S ...		16.5:		
12787	23 45.7 + 20 58 103.80 – 39.27	03 – 60 – 000	1182	1.0: x 0.2 1.0 x 0.2	178 6	2	Sb – c		16.0:		
12788	23 46.2 + 03 55 094.76 – 55.21	01 – 60 – 037 NGC 7757	0796	2.5 x 2.2 2.2 x 1.8	(115) 1 – 2	4	Sc		13.9		*12 3—
12789	23 46.2 + 26 08 105.88 – 34.38	04 – 56 – 000	0779 493	1.1 x 0.2 1.0 x 0.2	45 7	2	Sb – c		16.5:		
12790	23 46.2 + 50 10 112.82 – 11.19	08 – 43 – 000	0368	1.0 x 0.5 0.9 x 0.4	168 5	1	S ...		16.5:		
12791	23 46.3 + 25 57 105.84 – 34.56	04 – 56 – 011	0779 493	1.7 x 0.6 1.6: x 0.5:	[83] ...	3	Dwarf (irregular) Ir: IV – V or V		15.2		*
12792	23 46.5 + 26 31 106.09 – 34.03	04 – 56 – 012	0779 493	1.5 x 0.7 1.3 x 0.7	49 5	3	S B b		14.8		
12793	23 46.7 + 11 32 099.73 – 48.24	02 – 60 – 019	0318	1.1: x 0.9: 1.0: x 0.9:	— 1 – 2	3	Sc		15.6		
12794	23 46.7 + 30 42 107.53 – 30.03	05 – 56 – 008 NGC 7760	0606	(1.1 x 1.1) (1.5 x 1.5)	— ...	1	... Very compact		14.8		
12795	23 46.7 + 46 17 111.93 – 14.98	08 – 43 – 006	0368 497	1.0 x 0.5 0.9 x 0.4	111 5	1	S ...		14.9		
12796	23 46.8 + 47 39 112.29 – 13.66	08 – 43 – 007	0368 497	1.4 x 1.0 1.0 x 0.9	... ...	1	S B ...		15.2		
12797	23 47.1 + 45 51 111.88 – 15.41	08 – 43 – 000	0368	1.0 x 1.0 0.9 x 0.8	— 1	3	S B c Disturbed?		16.5:		
12798	23 47.2 + 29 40 107.32 – 31.05	05 – 56 – 009	0606	1.2 x 0.3 1.2 x 0.3	69 7	1	S ...		15.5		1
12799	23 47.4 + 02 25 094.10 – 56.70	00 – 60 – 052	0431	1.5 x 0.3 0.9 x 0.3	32 7	3	Sc		15.7		
12800	23 47.8 + 10 29 099.54 – 49.32	02 – 60 – 020	0318 496	1.2 x 0.5 1.1 x 0.45	16 6	3	S B a		15.4		
12801	23 47.9 + 25 53 106.24 – 34.73	04 – 56 – 000	0779 493	(1.0 x 0.7) (1.0 x 0.7)	(10) ...	2	E – S0 Companion attached		16.0:		*
12802	23 48.0 + 14 13 101.53 – 45.83	02 – 60 – 000	0318	1.1 x 0.10 1.0 x 0.10	121 7	3	Sc		16.5:		
12803	23 48.0 + 28 43 107.21 – 32.01	05 – 56 – 013	0606	1.4: x 1.1: 1.5: x 1.1:	— ...	1	...		15.2		
12804	23 48.1 + 24 16 105.72 – 36.29	04 – 56 – 013	0779	(1.0 x 0.8) (1.1 x 0.9)	— ...	3	E		15.6		*1
12805	23 48.3 + 26 49 106.66 – 33.86	04 – 56 – 016 NGC 7767 = 1511*	0779 493	1.1 x 0.2 1.1 x 0.2	142 7	2	S0 – a		14.2		*1
12806	23 48.4 + 26 53 106.71 – 33.80	04 – 56 – 018 NGC 7768	0779 493	(1.6 x 1.3) (1.9 x 1.5)	(60) ...	3	E		14.0		*13
12807	23 48.4 + 35 30 109.37 – 25.49	06 – 52 – 000	0405	1.0 x 0.10 0.8 x 0.1	84 7	2	...	d – m	17.		
12808	23 48.5 + 19 52 104.17 – 40.52	03 – 60 – 030 NGC 7769	1182	1.8 x 1.8 2.5 x 2.1	— 1:	2	Sa – b		12.9	+ 4349 + 4565	*12 3468̅
12809	23 48.5 + 46 27 112.28 – 14.89	08 – 01 – 001	0839 497	1.8: x 1.7: 1.5: x 1.4:	— 1	2	Sb		15.3		
12810	23 48.6 + 00 46 093.27 – 58.31	00 – 60 – 055	0431	1.8 x 0.7 1.8 x 0.8	52 6	3	Sb	Sb +	14.4		

N G C   7750 – 53, 7757, 7760, 7767 – 69   I C   1511, 5355

1	2	3	4	5	6	7	8	9	10	11	12
12811	23 48.6 + 08 27 098.67 – 51.28	01 – 60 – 041	0796 495	1.2 x 0.3 1.3 x 0.4	125 7	2	S0		15.4		*
12812	23 48.7 + 20 17a 104.39 – 40.13	03 – 60 – 032	1182	1.0 x 0.2 0.9 x 0.15	155 6 – 7	1	S ... npp of 2, bridge, 1st of 3		(15.5)		*1
12813	23 48.8 + 19 49 104.24 – 40.59	03 – 60 – 034 NGC 7770	1182	1.0 x 0.9 (1.0 x 0.9)	— ...	1	... P w 12815, disturbed, NGC 7769 group		14.5	+ 4338 + 4554	*12 3
12814	23 48.8 + 48 48 112.91 – 12.62	08 – 01 – 002	0839	1.2 x 0.2 1.1 x 0.25	80 6	3	Sc		15.6		
12815	23 48.9 + 19 50 104.27 – 40.58	03 – 60 – 035 NGC 7771	1182	2.5 x 1.2: 2.8 x 1.4	68 5:	2	S B a P w 12813, disturbed, NGC 7769 group		13.1	+ 4282 + 4498	*12 36
12816	23 49.3 + 02 48 095.18 – 56.55	00 – 60 – 056	0431	1.9 x 0.9 1.8 x 0.9	147 5	2	Sc		15.0		*
12817	23 49.3 + 29 00 107.63 – 31.82	05 – 56 – 014	0606	1.2 x 0.3 1.2 x 0.3	136 6	2	Sb		15.7		
12818	23 49.6 + 08 07 098.84 – 51.68	01 – 60 – 042	0796 495	1.3 x 0.7 1.3 x 0.7	20 5	4	Sc		15.1		
12819	23 49.6 + 11 12 100.56 – 48.80	02 – 60 – 022 NGC 7774	0318 496	(1.3 x 1.2) (1.6 x 1.3)	— ...	2	E + companion		14.6		*1
12820	23 49.7 + 31 00 108.34 – 29.91	05 – 56 – 015 NGC 7773	0606	1.2 x 1.2 1.1 x 1.1	— 1	4	S B b	Sb +	14.5		3
12821	23 49.9 + 28 30 107.62 – 32.33	05 – 56 – 016 NGC 7775	0606	1.1 x 0.9 1.0 x 0.9	(20) 1	3	Sc Disturbed?		13.9		*1
12822	23 50.0 + 14 17 102.22 – 45.93	02 – 60 – 023	0318	(1.0 x 0.7) (1.1 x 0.8)	(145) ...	2	E – S0		15.0		
12823	23 50.2 + 26 52 107.17 – 33.92	04 – 56 – 020	0779 493	1.0 x 0.8 0.9 x 0.8	(0) 1 – 2	3	S B 0		15.4		*
12824	23 50.5 + 17 35 103.84 – 42.84	03 – 60 – 000	1182	1.1 x 0.2 1.0 x 0.2	... ...	1	S ... nff of 2, distorted		16.0:		*
12825	23 50.6 + 10 38 100.62 – 49.42	02 – 60 – 000	0318 496	1.0 x 0.10 1.1: x 0.10	169 7	2	Sb – c		16.5:		*
12826	23 50.6 + 29 18 108.04 – 31.60	05 – 56 – 000	0606	1.1 x 0.6: 1.2: x 0.3	93 ...	1	... sf of 2, disturbed?		15.7		*
12827	23 50.7 + 07 36 098.95 – 52.26	01 – 60 – 043 NGC 7778	0796 495	1.1 x 1.0 1.1 x 1.1	— ...	3	E P w 12831, NGC 7782 group		13.8		*13
12828	23 50.7 + 19 07 104.52 – 41.39	03 – 60 – 000	1182	1.2 x 0.12 1.1 x 0.12	73 7	2	Sc		16.0:		
12829	23 50.7 + 28 01 107.67 – 32.85	05 – 56 – 018 NGC 7777	0606	1.1 x 0.8 1.2 x 0.9	48 3	3	S0		14.5		*3
12830	23 50.8 + 10 41 100.71 – 49.39	02 – 60 – 000	0318 496	1.0 x 0.15 0.9 x 0.2	41 7	2	Sa – b		16.5:		
12831	23 50.9 + 07 36 099.02 – 52.28	01 – 60 – 045 NGC 7779	0796 495	1.4 x 1.0 1.6 x 1.1	10 3	4	S0/Sa P w 12827, NGC 7782 group		13.6		*12 3
12832	23 50.9 + 11 03 100.94 – 49.05	02 – 60 – 024 IC 1513	0318 496	1.0 x 0.35 1.0 x 0.4	107 6	1	S ...		15.1		*
12833	23 51.0 + 07 50 099.20 – 52.07	01 – 60 – 046 NGC 7780	0796 495	1.1 x 0.6 1.3 x 0.5	3 5 – 6	1	Sa – b sf of 2		14.8		*
12834	23 51.3 + 07 42 099.23 – 52.22	01 – 60 – 048 NGC 7782	0796	2.1 x 1.3 2.1 x 1.2	175 4	4	Sb Brightest of 3		13.2		*12 348
12835	23 51.4 + 28 13 107.91 – 32.69	05 – 56 – 021	0606	(1.0 x 0.8) (1.2 x 0.9)	(0) ...	2	[E] Compact		14.4		*
12836	23 51.4 + 51 53 114.04 – 09.71	09 – 01 – 001	0636	1.2 x 0.4 1.0 x 0.4	170 6	2	Sc		16.5:		
12837	23 51.6 + 00 06 094.01 – 59.24	00 – 60 – 58 – 60 NGC 7783	431	[1.6 x 0.5] ...	... ...	2	Double system Contact, in group		14.1		*13 VA
12838	23 51.9 – 02 13 092.08 – 61.35	00 – 60 – 061	0431	1.0 x 0.9 0.9 x 0.8	— 1	2	Sc		14.9		
12839	23 52.0 + 28 02 108.00 – 32.91	05 – 01 – 001	1257	(1.5 x 0.4) (1.4 x 1.4)	— 1	2	S0		14.7		
12840	23 52.0 + 28 36 108.18 – 32.36	01 – 01 – 002	1257	1.2 x 1.0 1.2 x 1.0	(15) 2:	3	S B 0	(R)SAB(s)0	14.3		1
12841	23 52.8 + 05 38 098.55 – 54.28	01 – 60 – 049 NGC 7785	0796	(1.8 x 1.2) (2.3 x 1.3)	143 ...	4	E		13.0	+ 3846 + 4015	12 348
12842	23 52.8 + 21 19 105.98 – 39.42	03 – 60 – 038 NGC 7786	1182	0.6 x 0.4 0.55 x 0.35	5 ...	1	...		13.9		
12843	23 53.0 + 17 38 104.64 – 42.97	03 – 60 – 039	1182	3.0 x 1.3 2.8 x 1.2	27 5	3	...	SAB dm	14.4		*1
12844	23 53.1 + 19 15 105.30 – 41.43	03 – 01 – 001	1195	1.6 x 0.5 0.9: x 0.4	138 6	2	Sc		15.4		
12845	23 53.2 + 31 38 109.36 – 29.49	05 – 01 – 006	1257	2.4 x 1.8 2.4: x 1.8:	70: 3	4	Sc	Sc +	15.4		*

N G C   7770, 7771, 7773 – 75, 7777 – 80, 7782
7783, 7785, 7786   I C   1513

1	2	3	4	5	6	7	8	9	10	11	12
12846	23 53.3 + 18 10 104.94 – 42.48	03 – 01 – 002	1195	2.0: x 1.8: 1.5:: x 1.5::	— ...	3	Dwarf (spiral) S: V		17.		*
12847	23 53.4 + 00 17 094.96 – 59.26	00 – 01 – 001	0319	(1. x 1.) (1.0 x 1.0)	— ...	1	... Very compact, p w 12849		15.7		*
12848	23 53.6 – 01 16 093.74 – 60.69	00 – 01 – 004 IC 1515	0319	1.1 x 0.9 1.2 x 1.0	(160) ...	3	S B b P w 12852, disturbed		14.8		*
12849	23 53.6 + 00 17 095.05 – 59.28	00 – 01 – 005 NGC 7787	0319	2.0 x 0.5 2. x 0.6	104 6	1	S ... P w 12847		15.7		*
12850	23 53.6 + 29 07 108.73 – 31.95	05 – 01 – 007	1257	1.1 x 0.4 0.9 x 0.4	86 6	1	S ... 1st of 3		15.6		*1
12851	23 53.6 + 49 13 113.79 – 12.39	08 – 01 – 008	0839 497	1.2 x 1.1 1.1 x 1.0	— 1:	4	S B b Compact nucleus		15.5		*C
12852	23 53.7 – 01 12 093.84 – 60.64	00 – 01 – 006 IC 1516	0319	1.3 x 1.2 1.2 x 1.1	— 1	2	Sb – c P w 12848, disturbed		14.3		*
12853	23 54.0 + 45 08 112.93 – 16.39	07 – 01 – 000	1243	1.2: x 0.6: 1.1 x 0.5	90: ...	1	S ...		16.0:		
12854	23 54.1 + 13 30 103.23 – 46.99	02 – 01 – 001	0010	0.9 x 0.5 0.9 x 0.5	163 4 – 5	1	S ...		14.3		*
12855	23 54.1 + 26 50 108.17 – 34.18	04 – 01 – 005	0205	1.4 x 0.7 1.2 x 0.7	170 5	3	S0/Sa		14.7		*
12856	23 54.2 + 16 33 104.58 – 44.09	03 – 01 – 004	1195	2.6 x 0.9: 2.0 x 0.9	[12] ...	3	Irr I B (s)m Disturbed:, Ir I		14.5		*1 VA
12857	23 54.3 + 01 05 096.00 – 58.63	00 – 01 – 009	0319	1.9 x 0.4 1.9 x 0.4	34 6	1	S ...		14.7		
12858	23 54.3 + 44 21 112.81 – 17.17	07 – 01 – 001	1243	1.1 x 0.7 1.1 x 0.7	(95) 4:	3	Sc		15.6		*1
12859	23 54.4 + 05 15 098.95 – 54.79	01 – 01 – 001	1465	1.4 x 0.5 1.1 x 0.4	98 6	2	Sb – c		15.5		*
12860	23 54.4 + 10 33 101.93 – 49.81	02 – 01 – 002	0010	1.5 x 0.35 1.6 x 0.4	99 6	1	Sb		15.2		
12861	23 54.5 + 29 35 109.09 – 31.54	05 – 01 – 010	1257	1.0 x 0.6 0.9 x 0.5	140 4	2	Sa – b		15.2		1
12862	23 54.5 + 47 15 113.50 – 14.34	08 – 01 – 010	0839 497	1.0 x 0.7 1.1 x 0.7	70: 3	1	S B ...		15.7		*
12863	23 54.8 + 08 14 100.86 – 52.03	01 – 01 – 002	1465	1.0 x 0.3 1.1 x 0.3	13 6	2	Sb – c		16.0:		*
12864	23 54.8 + 30 43 109.49 – 30.46	05 – 01 – 011	1257	1.8 x 1.0 1.8 x 1.0	110 4 – 5	2	S B b		14.7		
12865	23 55.2 + 00 22 095.83 – 59.37	00 – 01 – 010	0319	1.1 x 0.15 1.0 x 0.18	136 7	1	Sa – b s of 2		15.7		*
12866	23 55.3 + 22 04 106.96 – 38.86	04 – 01 – 000	0205	1.0 x 0.3 1.0 x 0.2	139 6	2	Sc		16.0:		
12867	23 55.7 + 29 06 109.25 – 32.08	05 – 01 – 000	1257	(1.0 x 0.4) (1.2 x 0.5)	0 6	2	S0 – a		15.4		*
12868	23 55.8 + 43 47 112.96 – 17.78	07 – 01 – 000	1243	1.1 x 0.3 1.0 x 0.25	2 6	1	S ...		16.5:		*
12869	23 55.9 + 31 58 110.09 – 29.30	05 – 01 – 016	1257	1.0 x 1.0 1.0 x 1.0	— 1	3	S B c		15.3		
12870	23 55.9 + 44 17 113.08 – 17.29	07 – 01 – 002	1243	1.0 x 0.5 0.8 x 0.5	150 4 – 5	1	S ...		15.5		
12871	23 56.0 + 09 41 102.08 – 50.76	02 – 01 – 003	0010	1.1: x 0.8: 1.1: x 0.8	— 2	3	Sc		15.3		
12872	23 56.0 + 10 26 102.45 – 50.05	02 – 01 – 004 NGC 7794	0010	1.4 x 1.3 1.4 x 1.3	— 1	2	S ... Peculiar		13.8		*13
12873	23 56.0 + 25 56 108.40 – 35.16	04 – 01 – 006	0205	1.5 x 1.3 1.3: x 1.2:	— ...	3	... S m		15.6		
12874	23 56.1 + 26 52 108.70 – 34.26	04 – 01 – 000	0205	1.1: x 0.8: 0.8 x 0.7	— 3:	3	Sc Disturbed?		15.7		*
12875	23 56.2 + 08 07 101.33 – 52.25	01 – 01 – 004	1465 2	1.0 x 0.4 1.1 x 0.5	130 6	1	S ...		15.6		
12876	23 56.3 + 12 19 103.45 – 48.28	02 – 01 – 005	0010	1.1: x 0.8: 1. x 1.	— ...	3	Dwarf irregular Ir V		17. – 18.		*
12877	23 56.4 + 03 22 098.55 – 56.72	00 – 01 – 011 NGC 7797	0319	1.2 x 1.0 1.0 x 0.9	(10) 2	1	S ...		14.8		*3
12878	23 56.4 + 41 17 112.50 – 20.24	07 – 01 – 003	1243	1.1 x 0.7 1.1 x 0.7	155: 4	2	Sb/S B b		16.5:		*
12879	23 56.5 + 18 34 106.08 – 42.31	03 – 01 – 000	1195	1.0: x 0.7: 0.6 x 0.4	... ...	1	S ... Singular		15.0		*
12880	23 56.6 + 51 19 114.72 – 10.44	09 – 01 – 001	0636	1.0 x 0.3 0.9 x 0.3	171 6	1	Sc		16.5:		*

N G C 7787, 7794, 7797   I C 1515, 1516

1	2	3	4	5	6	7	8	9	10	11	12
12881	23 56.8 + 04 29 099.44 – 55.72	01 – 01 – 005	1465	1.8 x 0.3 1.4 x 0.3	160 (7)	1	S – Irr		15.5		
12882	23 56.8 + 31 00 110.05 – 30.29	05 – 01 – 000	1257	1.0 x 0.12 1.0 x 0.12	16 7	2	Sc		16.0:		
12883	23 56.8 + 46 37 113.76 – 15.05	08 – 01 – 016 IC 1525	0839 497	1.9 x 1.3 1.9 x 1.4	20 3:	4	S B b		13.3		*
12884	23 56.9 + 20 29 106.88 – 40.48	03 – 01 – 010 NGC 7798, Mark 332	1195	1.4 x 1.3 1.6 x 1.4	--- 1	1	S ...		12.7	+ 2700	3
12885	23 57.0 + 14 32 104.66 – 46.21	02 – 01 – 007 NGC 7800	0010	2.3 x 1.6 2.1 x 1.7	[42] ...	3	Irr	Ir II?	13.4		*13
12886	23 57.0 + 17 56 106.00 – 42.95	03 – 01 – 009	1195	1.6 x 1.1 0.8 x 0.6	110: ...	2	Sb – c		15.1		*
12887	23 57.2 + 48 15 114.17 – 13.46	08 – 01 – 000	0839 497	1.5: x 1.0: 1.5: x 1.1:	[35] ...	2	Irr		17.		
12888	23 57.3 + 46 37 113.85 – 15.06	08 – 01 – 017	0839 497	1.3 x 0.9 1.4: x 0.8:	165 3:	2	Sc		15.1		*
12889	23 57.5 + 47 00 113.96 – 14.70	08 – 01 – 018	0839 497	2.3 x 1.8 2.3 x 1.8	165: 2	4	S B b	S B(rs)b	14.0		1
12890	23 57.6 + 08 00 101.80 – 52.48	01 – 01 – 006	1465 2	(1.0 x 0.9) (1.1 x 1.0)	--- ...	1	E:		16.0:		*
12891	23 57.8 + 22 43 107.89 – 38.38	04 – 01 – 007	0205	[1.1 x 0.7] ...	... ...	2	Double system Contact, jet		15.1		*1V AC
12892	23 57.9 + 07 34 101.69 – 52.91	01 – 01 – 007	1465 2	1.0 x 0.6 1.0 x 0.9	... 1:	2	S B a – b		15.7		
12893	23 57.9 + 16 57 105.91 – 43.95	03 – 01 – 011	1195	2.0: x 1.8: 2.0: x 2.0:	--- 1	3	...	SA dm	15.3		
12894	23 57.9 + 39 14 112.33 – 22.31	06 – 01 – 000	1247	1. x 1. 1. x 1.	--- ...	4	Dwarf irregular Ir V		17.		
12895	23 58.0 + 19 47 106.98 – 41.21	03 – 01 – 000	1195	1.2: x 0.9: 0.5 x 0.5	(100) ...	2	...	S d – dm	16.5:		
12896	23 58.0 + 26 03 108.96 – 35.16	04 – 01 – 008	0205	1.0: x 0.9: 1.0: x 0.9:	--- ...	1	S ... n of 2, disturbed		14.7		*1
12897	23 58.0 + 28 07 109.56 – 33.15	05 – 01 – 020	1257	1.2 x 0.35 1.1 x 0.35	11 6 – 7	2	Sa – b 12899 group		14.9		*1
12898	23 58.1 + 33 20 110.96 – 28.08	05 – 01 – 000	1257	1.0 x 0.4: 0.9 x 0.3	10 6	2	Sc		16.5:		
12899	23 58.2 + 28 08 109.61 – 33.15	05 – 01 – 022	1257	0.7 x 0.7 0.8 x 0.8	--- ...	2	... Compact, brightest of 3		14.4		*1
12900	23 58.4 + 20 04 107.19 – 40.98	03 – 01 – 012	1195	2.0 x 0.15 1.8 x 0.15	111 7	3	Sc		15.7		
12901	23 58.4 + 28 39 109.81 – 32.65	05 – 01 – 023	1257	1.8 x 0.7 1.5 x 0.5	48 6	3	S B b		14.8		*
12902	23 58.5 + 05 57 101.02 – 54.49	01 – 01 – 008 NGC 7802	1465 1	1.0 x 0.5 1.1 x 0.5	52 5	2	S0 P w 12903		14.7		*3
12903	23 58.7 + 06 03 101.16 – 54.41	01 – 01 – 011	1465 1	1.1 x 0.2 1.1 x 0.2	155 6 – 7	2	Sb – c P w 12902		15.7		
12904	23 58.7 + 34 24 111.36 – 27.06	06 – 01 – 005	1247	1.0 x 0.7 1.1 x 0.7	(40) 3:	3	S B a – b Disturbed		15.4		*1
12905	23 58.7 + 80 23 120.81 + 18.00	13 – 01 – 000	1213	1.2 x 0.12 1.1 x 0.12	55 7	3	Sc		16.5:		
12906	23 58.8 + 12 50 104.55 – 47.97	02 – 01 – 011 NGC 7803	0010	1.0 x 0.7 1.0 x 0.6	85 4	1	S0 – a Brightest in chain of 6		13.8		*1
12907	23 58.8 + 17 55 106.56 – 43.08	03 – 01 – 000	1195	[1.1 x 0.5] ...	[132] ...	1	Double system Contact, halo?		17.		*
12908	23 58.8 + 31 09 110.57 – 30.24	05 – 01 – 024 NGC 7805, Mark 333	1257	1.0 x 0.7 1.1 x 1.0	--- ...	1	... P w 12911, contact, disturbed		14.3		*13 VA
12909	23 58.8 + 34 15 111.35 – 27.21	06 – 52 – 007 IC 5376	1247	2.1 x 0.4 1.8 x 0.5	4 6	2	Sa – b		14.7		
12910	23 58.9 + 05 05 100.68 – 55.34	01 – 01 – 000	1465 1	1.1: x 1.1: 1. x 1.	--- 1:	3	...	S B(r)m	17.		*
12911	23 58.9 + 31 10 110.60 – 30.23	05 – 01 – 025 NGC 7806	1257	1.1 x 0.8 1.3 x 0.8	[20] ...	3	S ... P w 12908, contact, disturbed		14.4		*13 VA
12912	23 59.0 + 08 44 102.72 – 51.89	01 – 01 – 000	1465 2	1.0: x 0.8: 0.4 x 0.4	--- ...	1	S ...		15.6		
12913	23 59.1 + 03 14 099.61 – 57.09	00 – 01 – 018	0319	1.4 x 0.2 1.2 x 0.2	5 7	3	Sc		16.0:		
12914	23 59.1 + 23 13 108.41 – 37.97	04 – 01 – 010	0205	2.7 x 1.3 2.7: x 1.3	[160] ...	3	S ... P w 12915, disrupted		13.2		*1V
12915	23 59.2 + 23 14 108.45 – 37.96	04 – 01 – 011	0205	1.6 x 0.5 1.6 x 0.4	[137] ...	1	... P w 12914, disrupted		13.9		*1V C

N G C 7798, 7800, 7802, 7803, 7805, 7806   I C 1525, 5376

1	2	3	4	5	6	7	8	9	10	11	12
12916	23 59.4 + 17 17 106.52 - 43.73	03 - 01 - 000	1195	1.1 x 0.7 0.8 x 0.6	(170) ...	2	...	S: d - dm	16.5:		*
12917	23 59.4 + 40 03 112.82 - 21.57	07 - 01 - 000	1243	1.2 x 0.9 1.1 x 0.8	95: 2 - 3	3	S B b	S B(r)b	16.0:		
12918	23 59.5 + 16 19 106.19 - 44.67	03 - 01 - 013 IC 5377	1195	1.3 x 0.7 0.9: x 0.6:	... ...	3	Irr		15.6		
12919	23 59.8 + 12 41 104.84 - 48.18	02 - 01 - 015 NGC 7810	0010	1.0 x 0.7 1.0 x 0.7	80 3	1	S0		14.3		*3
12920	23 59.8 + 26 56 109.69 - 34.40	04 - 01 - 012	0205	1.3 x 0.2 1.2 x 0.2	47 7	2	Sb - c		15.5		
12921	23 59.9 + 76 59 120.16 + 14.65	13 - 01 - 003	1213	1.6: x 1.4: 1.7: x 1.4:	-- ...	3	...	SA dm - m	15.6		

N G C   7810   I C   5377

NOTES

00001   U 1) VV 263, Arp 130; a) 1.1 x 0.9, E?, b) 1.2 x 0.9, S B c, sep 0.5; comp 2.7, 165 from a), 0.7 x 0.35, double
system -- U 2) P w comp at 1.8, 171, 0.35 x 0.25 -- U 3) P w 00 00.5 + 18 35 at 4.0, 102, 0.7 x 0.5, spir, m =
= 15.5; comp 3.1, 147, 0.5 x 0.3; 5.0, 5, 0.6 x 0.35 -- U 5) comp 3.7, 148, 0.30 x 0.15 -- U 6) IV Zw 1, loop
+ fragment of spir arm; "poss post-eruptive, B neutral disc + disr bluish core, d = 13"; halo, d = 40"" (Zw) --
U 7) B v red centr reg 0.45 x 0.15, no indic of spir str; comp 1.0, 310, 0.4 x 0.2, dif -- U 8) SA(s)ab: (de V) --

00010   U 10) comp attached to tip of vF arm at 0.75, 230, 0.6 x 0.2; comp 4.4, 76, 0.7 x 0.3 -- U 11) comp 3.2, 112,
0.5 x 0.35 -- U 13) comp 3.3, 343, 0.3 x 0.3, vF -- U 15) comp 3.8, 197, 0.7: x 0.6:, dif; 3.9, 268, 0.3 x 0.2
-- U 16) thin alm circ arms inside 1.3 x 1.3, dif outer disc; P w U 21) at 7.8, 141; comp 3.5, 138, 0.45 x 0.15
-- U 17) DDO 222, S V (vdB) -- U 19) SA bc (de V) --

00020   U 21) P w U 16 at 7.8, 321 -- U 22) comp 4.6, 168, 0.5 x 0.4 -- U 23) comp 4.6, 35, 0.5 x 0.1, F -- U 24) arm
fragment to cond at 0.5, 42, 0.05 x 0.05 -- U 26) comp 3.0, 42, 0.4 x 0.2 -- U 27) comp 3.9, 319, 0.4 x 0.2 --
U 28) brightest in group, comp 3.2, 125, 0.35 x 0.25; 3.7, 55, 0.9 x 0.5; 5.0, 51, 0.8: x 0.6:, dif --

00030   U 30) 1st and brightest of 3, comp 1.8, 131, 0.3 x 0.2; 3.7, 77, 0.12 x 0.12 + F halo -- U 31) comp 4.4, 56,
0.7 x 0.4 -- U 34) P w U 36 at 9.2, 168 -- U 35) P w comp at 1.6, 52, 0.8 x 0.6 -- U 36) (r) 0.7 x 0.3, vF disc,
P w U 34 at 9.2, 348 -- U 37) MCG ident N 7827 is wrong; P w U 38 at 2.8, 3; spir pattern appears disturbed --
U 38) MCG ident N 7825 is wrong; P w U 37 at 2.8, 183 --

00040   U 41) P w comp at 2.7, 342, 0.7 x 0.2 -- U 42) F cond at 0.4 p center, comp? -- U 46) a) 0.55 x 0.18, b) 0.40 x
x 0.20, sep 0.4; broad dif bridge; a) is a spir -- U 47) dif spindle; comp ·3.0, 165, 0.35 x 0.20 -- U 49) 00 04.4
+ 08 04 = N 7838 is a pair at 3.5, 106, 0.35 x 0.25 and 4.1, 104, 0.9: x 0.6:, sep 0.6, m = 15.3; comp 4.2, 31,
0.7 x 0.3 --

00050   U 53) blue cond s center, comp? -- U 57) P w U 59 at 1.9, 172 -- U 58) comp 4.8, 24, 0.6 x 0.2 -- U 59) P w U 57
at 1.9, 352 --

00060   U 60) comp 2.1, 175, 0.3 x 0.1 -- U 61) brightest of 3; 00 05.0 + 46 46 at 1.0, 96, 0.30 x 0.22; 00 04.9 + 46 43
at 3.2, 178, 0.8 x 0.4; m = 15.6 and 15.3 -- U 62) IV Zw 7; "d = 13", halo 45" x 25" (Zw) -- U 64) curved chain of
blue cond in dif env; disr pair or triple system? -- U 65) comp 2.5, 287, 0.4 x 0.3, v dif -- U 66) comp 4.5,
264, 0.9 x 0.2 -- U 68) one spir arm s center, dif ext of the bar northw; probl disturbed by comp at 0.9, 110,
0.45 x 0.12, high surf brightn -- U 69) comp 5.0, 257, 0.5 x 0.2, F --

00070   U 70) P w comp at 3.4, 245, 0.6 x 0.6, S B b - c; interaction? -- U 72) comp 1.0, 14, 0.4 x 0.3; 1.3, 294, 0.3 x
x 0.1, F -- U 74) comp 3.7, 275, 0.5 x 0.3 -- U 75) VV 80, Arp 235; amorph main body, irr on blue map, sev conds
in centr reg -- U 77) SO (de V) -- U 78) B main body 0.45 x 0.25, bridge to cond at 0.40, 155, 0.25 x 0.20; eF
env w tend to (R); N8 at 2.4, 302 appears to be a * --

00080   U 80) SAB 0- (de V); comp 3.3, 347, 0.4 x 0.3, vF, interacting pair -- U 83) comp 1.4, 16, 0.4 x 0.1, vF --
U 84) E (de V); comp 3.8, 122, 0.5 x 0.2 -- U 86) MCG descr "E? + *" is incorrect; comp 3.3, 342, 0.4 x 0.25
-- U 87) sev obj in vic, brightest at 3.5, 155, 0.6 x 0.12 -- U 88) b) at 0.45, 166, 0.15 x 0.12 -- U 89) S B(s)a
(de V); P w U 94 at 9.2 --

00090   U 91) MCG δ = 27 59, but no galaxy exists there -- U 92) disturbed? by comp at 1.4, 27, 0.5 x 0.5, dif -- U 94)
SA(rs)ab (de V); P w U 89 at 9.2 -- U 95) P w U 96 at 1.5, 68; probl not disturbed -- U 96) see U 95 --

00100   U 100) SAB b: (de V); asym spir pattern, no disturbing obj vis -- U 101) poss later than S B a; comp 0.8, 282,
0.7: x 0.7:, dwarf, e dif; 1.3, 38, 0.4: x 0.2:, vF -- U 102) comp 1.5, 329, 0.45 x 0.12 -- U 106) bridge? to
comp at 1.0, 83, 0.8 x 0.3; comp 4.5, 54, 0.6 x 0.1 --

00110   U 113) P w comp at 1.7, 288, 0.5 x 0.5; MCG nrs 37 and 38 shall be reversed -- U 115) comp 3.8, 295, 0.3 x 0.2
-- U 118) P w comp at 1.0, 77, 0.9: x 0.8:, dwarf, e dif; 00 10.2 + 21 46 = N 41 at 5.0, 203, 0.8 x 0.6, m =
= 14.6 --

00120   U 122) comp superimp at 0.9 sf center, 0.7: x 0.5:, dif, irr: -- U 123) pronouncedly S-shaped; massive irr arms,
one double; late -- U 125) V Zw 6; "Sb, compact core, F halo 60" x 25" " (Zw); in group of 6 w U 138 brightest
at 10.4; U 131 at 2.0 -- U 126) comp 1.2, 274, 0.4 x 0.2 --

00130   U 130) comp 3.2, 344, 0.5 x 0.1, spir -- U 131) in group of 6 w U 138 brightest at 8.6; U 125 at 2.0 -- U 132)
P w U 134 at 4.1, 171 -- U 133) asym (r) or (rs) 0.55 x 0.35; disturbed by U 136?; 2nd brightest in group of 6,
U 138 brightest at 5.6; U 136 at 3.4 -- U 134) P w U 132 at 4.1,351 -- U 136) in group of 6 w U 138 brightest at
2.1; U 133 at 3.4, interaction? -- U 137) 00 11.0 + 28 25 at 7.7, 0.45 x 0.30, spir:, m = 15.1; comp 4.0, 67,
0.4 x 0.3 -- U 138) brightest in group of 6; U 136 at 2.1, U 133 at 5.6; 4 fainter galaxies belong to the
group; 00 11.7 + 47 52 = I 1536 at 7.3, 203, (0.8 x 0.7), E:, m = 15.7; comp 2.6, 306, 0.40 x 0.35 -- U 139)
comp 4.1, 241, 0.3 x 0.2 --

00140) U 140) comp 2.4, 340, 0.3 x 0.1 -- U 141) comp 1.9, 155, 0.5 x 0.2, F; 4.1, 337, 0.4 x 0.4 -- U 142) P w comp
at 3.4, 110, 0.8 x 0.6, Sb - c -- U 147) comp 0.50, 131, 0.35 x 0.12; 4.7, 266, 0.8 x 0.10 -- U 148) comp 4.5,
148, 0.4 x 0.2 --

00150   U 150) dif chaotic spir str, two stronger arm fragments s center, comp superimp at 0.40, 105, 0.1 x 0.1,
disturbing?; comp 1.9, 242, 0.3 x 0.2 -- U 153) P w comp at 1.5, 28, 0.7 x 0.18, m appr 16.; interaction? --
U 155) comp 2.6, 329, 0.5 x 0.2 -- U 157) P w 00 14.3 + 27 37 at 2.6 n, 0.7 x 0.4, m = 14.7, compact -- U 159)
P w U 162 at 4.6, 172 --

00160   U 161) P w U 163 at 2.4, 83 -- U 162) early; P w U 159 at 4.6, 352 -- U 163) P w U 161 at 2.4, 263 -- U 164)
knotty uneven bar, spir str v ill def -- U 165) comp 1.2, 163, 0.5 x 0.4, F -- U 166) 00 15.2 + 29 53 at 3.9,
144, 0.7 x 0.15, m = 15.7, early; 00 14.7 + 29 56 at 4.0, 272, 0.6 x 0.5, spir, m = 15.3 -- -- U 168) comp
1.6, 19, 0.5 x 0.2 -- U 169) knotty; GC 364 at 0.7 n, m = 8.7 --

00170   U 170) VV 166, Arp 113; see U 173 and U 174; 1st in intercon triplet, members of compact group; in contact w U 173;
6 S alm stell objs near;N 72A at 00 16.0 + 29 45, compact (CGCG), m = 15.1, v = 6807, v₀ = 7031 -- U 171)
utterly chaotic -- U 173) see U 170; E (de V) -- U 174) see U 170; in contact w U 170 and U 173 -- U 175) M 51 case?,
app a vS comp at tip of spir arm at 0.55 sp, the arm appears disturbed; comp 2.3, 87, 0.4: x 0.3:; MCG α =
= 00 16.4 -- U 176) in a compact group, incl U 170-173-174; S B ab (de V); see U 170 -- U 177) a) 0.9 x 0.45, b)
0.2 x 0.2, sep 0.65, intercon by bridge or distorted arm --

00180   U 182) comp 4.6, 327, 0.35 x 0.10, F -- U 185) complex nucl reg?; comp 1.1, 86, 0.5 x 0.3 -- U 186) S B a?; comp
1.5, 67, 0.3 x 0.1, eF; 2.2, 179, 0.3 x 0.2, eF -- U 188) disturbed by 00 17.3 + 19 42 at 1.5, 201, 0.4 x 0.2,
m = 15.5, pec, asym --

00190   U 191) DDO 2; blue cond 0.1 x 0.1 at 0.3 p center; fainter cond at 0.55 n center; no nucl cond -- U 192) I B m
(de V); strongly obscured, v uncert diam; not in CGCG; integr magn B(0) = 13.33 in BG -- U 193) in contact w

U 194 at 0.6, 45 -- U 194) see U 193 -- U 195) brightest in triple group w U 201 at 11.0 and comp at 8.1, 87, 0.6 x 0.15 --

00200   U 201) see U 195; 1st group member at 5.4, 343 -- U 202) eF smooth (R) -- U 203) S0 (de V); brightest in a dense part of the cl, v ext env or disc; 6 S comp w high surf brightn in vic; U 206 at 5.4 -- U 205) late S B w comp at 0.7, 5, 0.4: x 0.4:, dif, probl a v dif bridge -- U 206) E0 (de V), ident in MCG as N 82-83; see U 203 -- U 207) last in group of 3 or 4; 00 18.8 + 29 21 at 2.2, 208, 0.7 x 0.7, spir, m = 15.1; 00 18.6 + 29 22 at 4.2, 255, 0.6 x 0.4, m = 15.3, compact; comp 6.3, 298, 0.4 x 0.15, m approx 17. -- U 208) Arp 65; in a dense part of the cl, may be interacting w U 209 at 2.7; vF comp appears attached at tip of arm at 2.4 n -- U 209) in a dense part of the cl, see U 208 --

00210   U 210) sev obj in vic belong to cl 00 20.0 + 23 24, MC, VD -- U 212) VV 257, Arp 257; blue asym bar, 2 dif arms; listed in MCG as 2 syst in contact, "N, 1 Sb + N, 1 Sb", which appears to be a mistake; disturbed by comp at 2.6, 187, 0.4 x 0.15 -- U 213) comp 3.6, 70, 0.30 x 0.25 -- U 214) blue patches along edge, asym uneven arms, loop on sp side -- U 219) 2 vF cond close to center; comp 4.0, 46, 0.9:: x 0.7:, dwarf --

00220   U 220) P w comp at 4.9, 298, 0.7 x 0.6, S B ?; 4.7, 288, 0.2 x 0.2, sep 0.8 betw these obj; comp 4.5, 211, 0.4 x x 0.3 -- U 221) blue complex obj, spir str?; comp 3.3, 57, 0.3 x 0.1, F -- U 224) VV 38, Arp 201; sep 0.4 -- U 226) probl strongly pec; comp 2.7, 110, 0.4 x 0.3, dif -- U 227) ext transp env --

00230   U 230) v blue cond at 0.45 sf center; complex asym spir pattern w crossing arms; disturbed? -- U 233) complex centr str, F spir arc sp center?, e dif outer parts; comp 2.7, 122, 0.4 x 0.3, dif -- U 235) F v red comp at 0.45, 115, 0.1 x 0.1, bridge? --

00240   U 240) main body 1.0 x 0.5 w 2 long asym arms; e dif irr extensions; disturbed by comp at 1.1, 68, 0.45 x 0.15; comp 3.6, 207, 0.5 x 0.4, dif; 4.7, 280, 0.9 x 0.4, Sc -- U 241) in contact w comp at 0.6, 352, 0.6 x 0.2; SA a (de V) -- U 242) complex irr spir str, brightest northw; disturbed by comp at 2.3, 2, 0.8: x 0.7:; red, v asym on red PA map -- U 244) sev patches on blue PA map, superimp ** ? -- U 246) smooth eF (R) -- U 248) spir disrupted by comp 0.30 x 0.25 superimp sp center; 00 23.5 + 25 23 at 3.9, 0.35 x 0.30, m = 15.7 -- U 249) disrupted spir, irr arm fragments, dif bridge to comp at 0.7, 257, 0.22 x 0.18, high surf brightn; comp 4.4, 177, 0.45 x 0.35, SA(r) ..., only inner region vis --

00250   U 250) comp 3.6, 91, 0.5 x 0.4, dif -- U 251) P w 00 23.4 + 21 31 at 2.8, 255, 0.7 x 0.5, Sc, m = 15.7 -- U 253) P w comp at 3.2, 10, 0.6 x 0.5 -- U 255) comp 4.7, 182, 0.5 x 0.4 -- U 256) comp 2.5, 199, 0.3 x 0.3, F -- U 257) smooth F disc or env, S0?; comp 3.4, 187, 0.6 x 0.1; 00 24.6 + 01 03 = N 117 at 8.0, 163, 0.8 x 0.4, S0-a, m = 15.5 -- U 258) P w comp at 2.5, 290, 0.8 x 0.7, SAB(s)b -- U 259) U 264 at 4.5 --

00260   U 260) P w 00 24.3 + 11 18 at 2.5, 257, 0.8 x 0.25, m = 15.7, interaction; comp 3.5, 324, 0.3 x 0.2 -- U 261) B ecc nucl region, hints of spir pattern in s part, v dif "disc"; comp 1.4, 60, 0.30 x 0.15, high surf brightn; 1.8, 328, 0.4: x 0.4: -- U 262) IV Zw 20; "streamer" (CGCG); "post-eruptive fuzzy blue spherical disc, jet 35" to ENE" (Zw); "Somewhat irr B core 16" = 11.1 kpc diam w a F plume 30" = 21 kpc ext to the E" (Sargent); sev obj in vic, brightest at 3.0, 22, 0.9 x 0.4, early; 00 24.8 + 39 37 at 6.7, 0.20 x 0.18, m = 15.7, very compact -- U 263) asym, dif, disturbed?; a vF blue cond at n edge;   comp 1.0, 135 (from nucl), 0.35 x 0.1; 2.5, 132, 0.5 x 0.3, vF, patchy -- U 264) III Zw 9; "blue disc spherical compact, vF F halo 45"" (Zw); vF asym env; U 259 at 4.5 -- U 267) comp 5.0, 307, 0.8 x 0.5 -- U 268) S-shaped; centr region 0.5 x 0.5, 2 vF open slightly asym arms, no disturbing obj vis --

00270   U 271) SA(s)c (de V); Pettit classif E1, which is wrong; comp superimp at 0.30, 200, 0.12 x 0.08 -- U 274) comp 3.8, 330, 0.7 x 0.7, dif, IV - V: -- U 275) 2 v broad neb arms, earlier than Sc; P w U 281 at 6.0, 79; ident as IC 17 in MCG -- U 276) P w U 279 at 4.1, 55; double system at 1.8, 119, 0.20 x 0.12 and 2.2, 105, 0.20 x 0.18 -- U 279) see U 276; asym, curved, convex southw; disturbed by U 276? --

00280   U 280) comp 1.7, 190, 0.6 x 0.4; 3.9, 281, 0.3 x 0.2 -- U 281) see U 275 -- U 282) in contact w U 283 at 0.9, 64 ; asym; probl interaction -- U 283) see U 282) -- U 284) 00 25.6 + 33 00 (double system) at 4.1, 267, 0.45 x 0.25, spir, and 4.9, 269, 0.45 x 0.12, total m = 15.6 -- U 286) in a group; U 292 (w N 127 and N 130) at 6.4, 75, probl interaction; not ident as N 125 in CGCG; NGC descr "D * sp" for N 125 is not correct; An 26 in BG at 7.6, 337, 0.8 x 0.2, spir, m approx 16.; comp 2.5, 339, 0.4: x 0.3:, v dif; 4.6, 205, 0.7 x 0.2 -- U 287) S red cond at 0.25 np center, disturbing obj?; slightly asym disc; comp 3.0, 302, 0.4 x 0.4, dif -- U 288) intragalactic?; stellar nucl cond; v dif blue env; F chain of conds in pos angle 118 from center -- U 289) P w U 290 at 4.8, 155 --

00290   U 290) see U 289); v dif spindle -- U 292) see U 286; brightest in group; box-shaped nucl bulge; comp N 127 at 0.8, 304, 0.9: x 0.6, disturbed, S0 (de V); and N 130 at 1.0, 70, 0.8 x 0.5, disturbed; CGCG magn incl these 2; comp 3.6, 208, (at 4.7, 111 from U 286), 0.8 x 0.7; S0 (de V) -- U 293) S stellar nucl cond in F ring; v rare obj; chain of 3 E or S0 sf -- U 295) comp 4.8, 257, 0.5 x 0.4, dif -- U 296) spir? 1.0: x 0.7: w comp 0.20 x 0.12 p at 0.65 and 0.15 x 0.12 sp at 0.45; sev vF obj in vic --

00300   U 301) 3 blue conds in contact s center; 2 asym arms, one irr and much broader outw -- U 303) P w U 306 at 6.8, 43 -- U 305) slightly asym; vF red obj at 0.9 sp, at tip of spir arm? -- U 306) see U 303; blue cond superimp at 0.30, 85, 0.05 x 0.05, invis on red PA map, disturbing comp? -- U 308) brightest in subgroup; 00 28.8 + 04 55 = = N 141 at 4.7, 75, 0.9 x 0.6, m = 15.4, double?; 00 28.4 + 04 49 = N 139 at 5.2, 160, 0.9 x 0.5, S B a - b, m = 15.5, α (CGCG and UGC) should be 00 28.5; comp 0.65, 273, 0.25 x 0.15; 3.2, 135, 0.55 x 0.12; 4.3, 89, 0.4 x 0.4

00310   U 312) brightest in group of 4; B asym bar, p arm has chain of B blue patches; disturbed by 00 28.7 + 08 12 at 1.4, 290, 0.9 x 0.9, early, m = 15.0; U 314 at 4.2, 162 -- U 313) B centr reg 0.45 x 0.25; comp 3.1, 355, 0.4 x x 0.4 -- U 314) 3rd brightest in group of 4; v asym SAB(s)...; ecc nucl reg; one v broad arm on p side; disturbed by U 312?; P w U 315 at 2.2, 105 -- U 315) faintest in group of 4; see U 314, U 312 -- U 318) (r) 0.55 x 0.50; eF outside (r); thin slightly curved bar -- U 319) slight internal asym, one well def arm, fragments of 2 almost dissolved arms; SN 1954d; Sb (Kowal) --

00320   U 320) comp 0.5, 78, 0.3 x 0.2 F; 0.7, 320, 0.2 x 0.2, F -- U 322) spindle w dif ext eastw and southw; probably interacting w comp at 0.30, 47, 0.10 x 0.08 -- U 323) P w comp at 2.2, 244, 0.45 x 0.15 -- U 324) P w comp at 4.0, 121, 0.7 x 0.7, S B b: -- U 326) DDO 3; E5 pec (de V), E pec (Ho); P w U 396 = N 185 at 58'; "eSB stellar nucl, e low surf brightn dwarf" (de V) -- U 327) P w 00 30.8 + 07 35 at 3.8, 140, 0.9 x 0.6, m = 15.6, Sc --

00330   U 330) 1st in group of 3 w U 338 at 8.4 and U 344 at 13.2; prob late S0 -- U 331) 1st of 3; U 335ab at 2.7, 62 -- U 333) v red centr reg; fragments of broad spir arms -- U 335) in group of 3 w U 331 at 2.7, 244; a) 0.9 x 0.5, b) (1.6 x 1.2), E, sep 0.8, contact; sev F obj in vic -- U 338) see U 330 -- U 339) nucl conds sep 0.15, connected main bodies, strongly distorted --

00340   U 340) comp 3.8, 308, 0.5 x 0.3, dif -- U 341) smooth F disc or env, probl v early spir; P w comp at 0.9, 175, 0.7 x 0.15; comp 4.3, 287, 0.4 x 0.4 -- U 343) not cert extragalactic -- U 344) see U 330 -- U 349) brightest in group; P w comp at 2.1, 7, 0.9 x 0.5, S B a - b, M 51 case?, F cond at 0.6 n, prob at tip of spir arm; one long thin arm towards 136, disturbed by the comp?; comp 1.5, 330, 0.3 x 0.1; 2.7, 281, 0.35 x 0.15 --

00350   U 350) comp 2.6, 230, 0.4 x 0.4, F; 3.7, 329, 0.4 x 0.3, dif; 4.1, 252, 0.5 x 0.4 -- U 351) (r) 1.1 x 0.45; P w U 353 at 7.7, 175 -- U 352) S B III - IV? -- U 353) see U 351 -- U 354) see U 356 -- U 355) comp 3.9, 220, 0.7 x 0.2 -- U 356) (R)SA 0+pec (de V), "vS nucl in center of F double (R) ..."; U 365 = N 169 + I 1559 at 11.0; comp

2.7, 109, 0.5 x 0.15; U 354 at 5.0, 353 -- U 357) comp 1.5, 342, 0.35 x 0.35, spir; 2.6, 206, 0.5 x 0.3 -- U 358)
MCG α = 01 36 is wrong; comp 1.8, 278, 0.3: x 0.2 --

00360   U 360) main body 0.7 x 0.5; prob an eF plume northw, (may be illusory); P w comp at 2.0 p, 0.9 x 0.4, disturbing?
-- U 364) P w 00 34.0 + 21 20 at 2.6, 349, 0.5 x 0.3, double system in contact, m = 15.6 -- U 365) Arp 282; P w
I 1559 at 0.35 s center, 0.7 x 0.35; "comp appears to rain into the nucleus of spir" (Arp); Sb (de V); I 1559 E
(de V) -- U 366) eF env; comp 2.0, 77, 0.30 x 0.15; 00 34.4 + 28 34 at 4.6 sf, 0.5 x 0.25, pec, m = 15.0 -- U 367)
P w comp at 0.7 sf, (0.8 x 0.8), E:, MCG 4-2-37, env in contact; sev obj in vic -- U 369) prob late Sb; 00 34.1 +
+ 01 35 = N 170 at 7.5, 243, 0.4 x 0.3, m = 15.4 --

00370   U 373) spindle?; hints of a vF env -- U 378) blue "spindle" w cond at f side, double system? --

00380   U 380) (rs) 1.0 x 0.6; 2 thin inner arms, multiple arms in the outer disc -- U 382) Sa (de V); brightest in group
w comp at 1.5, 251, 0.6 x 0.5; 3.3, 5, 0.3 x 0.2, double, dif; 4.2, 20, 0.5: x 0.5:, dif; sev S obj in vic --
U 385) B centr reg 0.4 x 0.2; e dif outer parts -- U 387) in a dense region in the center of the cl; brightest
comps are 00 35.7 + 29 12 = N 181 at 2.7, 0.8 x 0.2, m = 15.4; 00 35.9 + 29 10 = N 184 at 4.1, 0.8 x 0.2, m =
= 15.5; 00 36.2 + 29 15 at 4.7, 0.2 x 0.2, compact, m = 15.7 --

00390   U 390) triple comp around 2.6, 223, 0.5 x 0.3, 0.3: x 0.2:, 0.2 x 0.2, sep less than 0.4 -- U 392) asym spindle
w ext southw -- U 396) E3 pec (de V), E pec (Ho); P w U 326 = N 147 at 58'; "low surf brightn dwarf, dark markings
in B middle, no def nucl" (de V) -- U 397) III Zw 10; Sb: 1.1 x 0.8 w comp 0.15 x 0.15 superimposed 0.35 s center
at tip of spir arm; comp 0.8, 320, 0.2 x 0.2 and 0.9, 178, 0.4 x 0.15, both incl in CGCG magn -- U 398) comp
2.6, 161, 0.35 x 0.30 -- U 399) (r) or (rs) 0.7 x 0.4; v dif outer disc --

00400   U 401) U 406 at 2.1, 240; U 405 at 3.1, 19; U 419 at 5.4, 94 -- U 402) P w U 416 at 3.7, 108; comp 4.1, 26, 0.5 x
x 0.3; 4.1, 286, 0.5 x 0.5 -- U 403) P w U 404 at 1.2, 44 -- U 404) see U 403 -- U 405) U 406 at 1.3, 169,
interaction?; U 401 at 3.1, 199; U 419 at 5.3, 128 -- U 406) prob early spir; see U 405 and 401; U 419 at 4.4,
117 -- U 407) tend to spir str in the env on red PA map?; E1 (de V); U 415 at 7.1, 31; U 420 at 9.9, 156; comp
3.0, 177, 0.9: x 0.8:, e dif, dwarf -- U 408) P w U 423 at 6.6, 106; comp 1.4, 152, 0.3: x 0.3:, e dif, double;
3.8, 240, 0.8 x 0.4; 4.8, 144, 0.6 x 0.3 -- U 409) cond 0.2 x 0.1 superimp sf center, comp?; comp 3.2, 113, 0.35 x
x 0.30 --

00410   U 410) vF ext env; in a dense region of the cl, 8 obj inside 5.0 -- U 414) SA(r)c (de V); sev thin tightly wound
arms; U 420 at 6.1, 29; comp 3.6, 16, 0.8 x 0.4 -- U 415) see U 407; comp 4.7, 251, 0.4 x 0.3; 4.4, 105, 0.4 x
x 0.1, F; 5.0, 45, 0.7 x 0.3 -- U 416) see U 402 -- U 418) P w U 422 at 2.7, 28 -- U 419) see U 401, 405, 406 --

00420   U 420) S B(s)bc (de V); 2 B v knotty arms; see U 407, 414; comp 2.8, 225, 0.8 x 0.4 (common w U 414); 3.6, 271,
0.5 x 0.1 -- U 421) 00 37.1 + 03 10 = N 203 at 5.6, 180, 0.9 x 0.5, S0-a, m = 15.0; comp 1.9, 298, 0.7 x 0.3,
dif; 3.1, 262, 0.6 x 0.6 -- U 422) see U 418 -- U 423) see U 408; comp 4.0, 14, 0.4 x 0.3; 4.1, 241, 0.6 x 0.3
(common w U 408) -- U 424) brightest galaxy in group, 3 eF red obj in vic -- U 425) B massive arms; vB blue cond
superimp nf center, comp or * ? -- U 426) E5 pec (de V), E pec (Ho); comp to U 454 = M 31 at 37, 134; ident
M 110 suggested by Kenneth Glynn Jones, see Sky and Tel vol XXXIII p 156 -- U 427) asym; no disturbing obj vis --
U 428) brightest in a region w a v large nr of F galaxies, cl 00 38.0 + 29 14, MC, MD -- U 429) comp 3.8, 240,
0.35 x 0.15; 4.8, 235, 0.5 x 0.2; 5.0, 206, 0.35 x 0.35, distorted, streamer, comp at 1.0 --

00430   U 432) 2 long well def arms; cond superimp np center, comp?; comp 2.4, 342, 0.6 x 0.6 -- U 434) measurements and
classif disturbed by a nearby *; comp 3.3, 62, 0.45 x 0.05 -- U 435) early spir?; P w U 439 at 7.4, 132 -- U 436)
comp 1.8, 15, 0.3 x 0.1; 4.7, 135, 0.3 x 0.2 -- U 438) SAB(r)c (de V); a large nr of vF galaxies near, in CGCG cl
00 38.4 + 25 15, compact, VD -- U 439) see U 435 --

00440   U 440) dif; hints of asym spir pattern, thin bar on blue PA map -- U 441) comp 4.6, 0.9 x 0.1, dif -- U 442) comp
3.0, 103, 0.3 x 0.1 --   U 443) P w comp at 1.7, 138, 0.5:: x 0.3::, e dif, dwarf: -- U 444) overexp on PA
maps; one spir arm in n part, or absorption marking? -- U 448) v red elong nucl region, one spir arm on s side,
one blue spir arc on n side; P w U 449 at 3.5, 2; comp 2.4, 315, 0.3 x 0.3 -- U 449) slightly curved spindle,
brighter westw, B nucl region; see U 448 --

00450   U 450) 00 39.7 + 00 38 = N 219 at 3.7, 341, 0.25 x 0.15 + env, compact, m = 15.6 -- U 451) double nucl?, vF ext
env -- U 452) Arp 168; E2 (de V), E (Ho); comp to U 454 = M 31 at 24, 1; "F dif plume curved away from M 31 disc"
(Arp); main body 4 x 3; sharp transition to a transparent env superimp M 31 disc, v uncert diam -- U 453)
00 40.1 + 29 39 at 3.4, 32, 0.6 x 0.6, m = 14.7; sev obj in vic -- U 454) The Andromeda Nebula, brightest in
Local group; SA(s)b (de V), Sb- (Ho); comp U 426 = N 205 at 37, 314 and U 452 = N 221 at 24, 181; SN 1885;
possible SN 1664b (?), 1898 (prob nova), 1919 (prob nova) MCG 0-2-132 is a superimp * -- U 456) E4
(de V) -- U 457) comp 0.9, 189, 0.12 x 0.12, bridge from the s end of the spindle? (may be illusory); "w jet" (CGCG),
but   no jet vis on PA maps   -- U 458) 00 40.1 + 23 13 at 1.5, 237, 0.8 x 0.2, pec?, m = 15.7; 00 40.4 + 23 14
= N 229 at 2.4, 80, 0.9 x 0.2, S0, m = 14.7 -- U 459) vB cent region; dif, asym outer parts, no spir str vis;
blue?; comp 3.0, 222, 0.4 x 0.25 --

00460   U 460) 3-armed spir; 2 long thin arms in n part; 1 shorter, massive s center; comp 4.0, 78, 0.8 x 0.2 -- U 461)
SAB(rs)cd (de V); appears earlier on PA maps   -- U 462) 2 asym arms w sev B blue patches; comp 3.5, 321, 0.8 x 0.5,
E:; 4.2, 328, 0.7 x 0.5, S0: -- U 463) MCG diam 1.4 x 0.4 are not correct -- U 464) asym env, or dwarf comp
attached at 0.75, 330? -- U 468) disturbed?, irr loop f center; * or compact comp 0.10 x 0.10 superimposed 0.5
f center -- U 469) comp 3.4, 250, 0.7 x 0.5; 3.8, 200, 0.7 x 0.6, v dif --

00470   U 470) main body overexp on PA maps, irr border; comp 4.9, 311, 0.9 x 0.12, Sc -- U 473) (r) 0.7 x 0.6, eF
outer disc; comp 2.0, 182, 0.3 x 0.1 -- U 476) v chaotic asym spir; blue; vB and massive spir arcs and patches;
appears strongly distorted, but no disturbing obj vis; comp 3.0, 337, 0.4 x 0.3; sev F obj in vic -- U 477) vF
slightly ecc cent cond; spir str?; sev F red comp n, prob background --

00480   U 480) interacting w 00 43.9 + 36 04 at 1.4, 77, 0.8 x 0.5, distorted spir, m = 15.6 -- U 481) comp 0.9, 179,
0.4 x 0.1; 2.5, 194, 0.7 x 0.3, spir -- U 484) P w 00 44.4 + 32 15 at 3.1, 69, 0.7 x 0.4, spir, m = 15.1 --
U 485) 00 44.2 + 30 07 at 5.1, 221, 0.6 x 0.2, m = 15.7 -- U 488) comp 3.6, 68, 0.7: x 0.4:, dif --

00490   U 490) amorph disc, no clear spir pattern; comp 3.5, 249, 0.3 x 0.2; SN 1916 -- U 491) P w U 497 at 8.4; comp
3.2, 51, 0.5 x 0.4 -- U 492) comp 2.8, 349, 0.7 x 0.2, E-S0 -- U 495) comp 0.9, 152, 0.5 x 0.2, dif; 2.2, 278,
0.3 x 0.1 -- U 496) sep 0.35, env in contact; a large nr of S obj in vic, bel to cl 00 47.5 + 00 51, MC, MD --
U 497) see U 491 -- U 498) comp 3.6, 170, 0.7 x 0.5, dif -- U 499) (R)?; comp 1.1, 326, 0.3 x 0.1; 1.2, 86,
0.5 x 0.25 --

00500   U 500) comp 2.7, 218, 0.6 x 0.2 -- U 501) comp 4.0, 277, 0.35 x 0.30, pec -- U 503) P w comp at 1.8, 307, 0.6 x
x 0.2 -- U 504) P w 00 46.6 + 11 52 at 3.3, 237, 0.8 x 0.25, compact, m = 15.7 -- U 506) v ext transp env;
absorption markings in nucl region?; 3 F comps superimp (or vis through env) -- U 508) comp 3.6, 219, 0.4 x 0.1:

00510   U 510) comp 2.7, 336, 0.3 x 0.2 -- U 512) ecc nucl region; broad spir arcs in sf part; double system? -- U 516)
comp 4.2, 214, 0.3 x 0.2; 4.7, 12, 0.35 x 0.25 -- U 517) vF disc or env; sev F obj in vic -- U 518) earlier than
S B c? --

00520   U 521) v blue, chaotic; comp 4.0, 247, 0.7: x 0.12 -- U 522) asym; 4 blue cond superimp, comp?; comp 3.0, 220,

0.5 x 0.20; 4.7, 62, 0.8 x 0.20; sev eF obj in vic -- U 526) P w U 527 at 4.3, 158 -- U 527) fragments of spir arms in s part; see U 526 -- U 528) SAB(rs)b (de V); "large B nucl, sev knotty massive arms. B part has sharp edge" (de V) --

00530   U 530) comp 1.9, 352, 0.4 x 0.2, F -- U 531) M 51 case, comp 0.9, 150, 0.15 x 0.15, F, at tip of spir arm; comp 1.0, 23, 0.5 x 0.15 -- U 533) comp 0.8, 65, 0.3 x 0.2 and 1.0, 54, 0.20 x 0.12, sep 0.30; 4.7, 258, 0.8: x 0.1 (or 1.2? x 0.5?) -- U 534) nucl region; 1 long thin smooth arm in n part; 1 short dif arm s; comp superimp at 0.40, 247 from nucl, 0.12 x 0.12; sev objs near, in CGCG cl 00 50.2 + 24 06, MC, MD -- U 537) U 538 at 2.9, 53; comp 1.4, 148, 0.6 x 0.4, early -- U 538) brightest of 4; U 537 at 2.9, 233; comp 3.2, 207, 0.6 x 0.4, early (common w U 537); 2.8, 122, 0.5 x 0.5 -- U 539) patchy ill def arms; slightly asym, hints of dif ext southw --

00540   U 540) compl overexp on PA maps; hints of short v massive arms; blue? -- U 543) P w comp at 1.4, 265, 0.9 x 0.5 -- U 545) I Zw 1; compact, not in CGCG; 2 curved plumes or spir arms w F cond at the tips; "Variable, blue, v compact, patchy halo, $m_p \leqslant 14.3$. Spect: broad em lines, Balmer, FeII, no forbidden lines -- poss sharp abs at λ 3500" (Zw); "The nucl is stellar and radiates a remarkable spect composed of broad em lines of H and FeII on a blue continuum. The two plumes which emerge from the nucl extend out to 12" = 14 kpc. They appear to terminate at two F stars. A spectrogram of the brighter star (the one to the north of the nucl), badly contaminated w night-sky em, revealed an absorption line at λ 5172 on a red continuum. This is either the Mg I b-lines at zero redshift or Hβ at a redshift of 19300 km sec^{-1}. Lynds has discovered Hα and λ 3525 em from the plumes" (Sargent) --

00550   U 550) main body 1.3 x 0.4; e dif outer disc or env; hints of spir str in main body -- U 553) comp 2.5, 168, 0.3 x x 0.2, F; 3.4, 352, 0.5: x 0.3, double system, sep 0.15 -- U 554) 00 52.0 + 28 30 at 2.9, 338, 0.35 x 0.22, compl overexposed on PA maps, sharp edge, m = 15.3; sev vF comps   -- U 556) asym distr of light, comp superimp?; comp 1.7, 8, 0.4 x 0.10 -- U 557) asym; ecc nucl region; disturbed by a S comp superimp n center -- U 559) comp 2.9, 246, 0.35 x 0.25; 3.0, 47, 0.5 x 0.3 --

00560   U 563) comp 3.4, 146, 0.3 x 0.1; 4.6, 183, 0.6: x 0.6:, dif; 4.8, 111, 0.4 x 0.2 -- U 564) spir arc or ring fragment on f side; elong nucl region -- U 566) P w U 567 at 1.1n; U 565 at 3.2 -- U 567) see U 566 -- U 568) ext eF env; P w 00 52.8 - 01 20 at 2.1, 147, 0.7 x 0.2, m = 15.7 -- U 569) comp 4.9, 294, 0.3 x 0.2, F --

00570   U 570) comp superimp at 0.15 sp center, 0.10 x 0.10; comp 1.2, 347, 0.4 x 0.2; 4.6, 306, 0.30 x 0.25 -- U 573) prob disc-shaped obj w vB nucl region; transparent e dif env; v thin jet from nucl reg vis on red PA map; P w 00 53.1 + 23 53 at 4.2, 288, 0.7 x 0.25, m = 15.4 -- U 578) sharp transition from main body to F smooth env; comp 2.5, 180, 0.3? x 0.1; 4.7, 85, 0.5 x 0.2 -- U 579) 2nd brightest in the v dense central region of the cl; U 583 at 2.4, 100; 00 53.8 - 01 28 at 3.2, 32, 0.7 x 0.4, m = 15.3 --

00580   U 580) early; in contact w U 582 at 1.0, 32 -- U 581) classif difficult, may be earlier -- U 582) see U 580 -- U 583) brightest in the v dense central region of the cl; "N_n or (N)" (MCG); 00 53.9 - 01 32 at 0.75, 173, 0.22 x x 0.18, compact, m = 15.6; U 579 at 2.4, 280; 00 53.8 - 01 28 at 3.2, 240, 0.7 x 0.4, m = 15.3; 00 54.1 - 01 33 at 3.9 sf, m = 15.5, is prob MCG 0-3-36, but MCG magn is incorrect -- U 584) comp 1.5, 338, 0.30 x 0.18; 3.0, 215, 0.40 x 0.20 -- U 585) P w comp at 1.4, 323, 0.5 x 0.15, MCG 4-3-21 -- U 588) 2 comp superimp at 0.3 s center, 0.12 x 0.10, and 0.3 nf center, 0.10 x 0.10; sev S obj in vic -- U 589) P w comp at 3.0, 202, 0.8 x 0.6, spir; comp 3.5, 357, 0.3 x 0.2 --

00590   U 591) B main body 0.35 x 0.15; irr amorph asym env; no indication of spir str -- U 592) P w U 597 at 5.5, 39 -- U 593) V Zw 42; contact w U 594 at 0.55, 155 -- U 594) see U 593; these 2 form triple group w comp at 5.4, 177, 0.8 x x 0.5, Sb, m approx 16., 2 long well def arms -- U 596) brightest in cl 00 55.0 + 07 45, open, MD; comp 2.7, 276, 0.5 x 0.2; 3.5, 322, 0.7 x 0.6, dif; 3.6, 135, 0.30 x 0.15; sev fainter comps   -- U 597) see U 592; 00 54.8 + + 30 09 = N 318 at 5.6, 40, 0.5 x 0.3, m = 15.2; N 313 and N 316 are ** -- U 598) prob (R)SA 0+ --

00600   U 600) 2 main arms split in 4 after about 90° -- U 601) IV Zw 35; a large nr of F objs near ; MCG ident galaxy at 4.4, 294 as N 326, 0.7 x 0.4, m approx 17. -- U 603) b) at 0.40, 38, 0.22 x 0.18, bridge; 00 55.9 + 11 20 at 2.0, 27, 0.5 x 0.5, spir, m = 15.5 -- U 605) pronouncedly S-shaped, thin smooth arms; P w comp at 1.3, 286, 0.55 x x 0.45, ring, thin bar?; interaction -- U 606) comp 4.0, 93, 0.5 x 0.4 -- U 607) S F nucl cond, hints of broad spir arms; rare; comp 1.3, 192, 0.3 x 0.2; 2.9, 83, 0.5 x 0.5; 3.5, 194, 0.7 x 0.5 -- U 609) comp 3.0, 32, 0.35 x 0.22, F --

00610   U 610) a large nr of obj in vic, about 15 comps inside 5.0; 00 56.5 + 12 43 at 3.8, 90, 0.7 x 0.4, compact, m = = 15.7 -- U 613) e dif env; no E; comp 4.7, 258, (0.9 x 0.9) -- U 614) irr broad arms; blue cond attached n center -- U 615) comp 3.9, 325, 0.5: x 0.5:, dif -- U 616) (r) 0.6 x 0.4; sev F comps -- U 617) U 620 at 3.8, 108; comp 4.2, 35, 0.9 x 0.15 --

00620   U 620) see U 617; 2nd group member at 4.8, 345 -- U 622) comp 4.1, 140, 0.4 x 0.1 -- U 626) P w 00 58.0 - 01 57 at 1.3, 133, 0.9 x 0.7, spir, m = 15.6, dif bridge -- U 628) comp 2.8, 153, 0.5: x 0.3:, dif --

00630   U 630) eFS nucl region, dif disc; group of background galaxies sf; interacting background pair at 2.1 sp -- U 631) comp 2.8, 281, 0.40 x 0.35; 3.0, 87, 0.7: x 0.5:, dif -- U 632) P w 00 58.8 + 29 53 at 1.4, 15, 0.8 x 0.6, very compact, m = 14.9 -- U 634) DDO 7 -- U 636) comp 4.8, 337, 0.4 x 0.2 -- U 638) dif transp env; surrounded by sev S objs w high surf brightn; P w comp at 3.2, 27, 0.9 x 0.3 -- U 639) P w U 641 at 6.8, 101 --

00640   U 640) see U 639 -- U 642) P w comp at 0.45, 189, 0.10 x 0.10 -- U 644) b) at 0.70, 124, 0.7 x 0.20, spir, hints of vF bridge -- U 647) in group w 01 00.4 - 01 43 brightest at 8.8 np, spir, m = 15.0; comp 2.6, 217, 0.9 x 0.7 -- U 648) prob one-armed, long arm curved 400° -- U 649) comp 0.9, 317, 0.35 x 0.15, F; 3.5, 143, 0.40 x 0.25 --

00650   U 650) in triple group w comp at 2.3, 347, (0.9 x 0.7), and 2.8, 7, 0.12 x 0.10 -- U 651) chain of 4 galaxies at 4. p, length 1.9 -- U 656) dif irr streamer from f side in direction 155; no disturbing obj vis; 2 blue patches p center -- U 659) blue "spindle"; 2 conds w sep 0.20 in center --

00660   U 660) P w 01 01.8 - 01 22 at 2.4, 191, 0.5 x 0.3, m = 15.7 -- U 662) P w U 666 at 6.3, 111; comp 4.8, 232, (0.7 x x 0.5), E; 4.8, 248, 0.8 x 0.15 -- U 664) comp 2.1, 203, 0.35 x 0.25 -- U 666) see U 662 -- U 667) dif; measurement disturbed by GC 1301 at 1.4 p, m = 6.17 -- U 668) DDO 8; I m (de V), Ir I (Ho), Ir V (vdB) -- U 669) earlier than Sc? --

00670   U 671) P w comp at 1.9, 218, 0.7 x 0.3, S0 - a, m approx 16.; 01 02.8 + 32 10 at 4.7 p, (0.8 x 0.7), S0, m = 15.7 -- U 673) SN 1960p; comp 1.8, 275, 0.5 x 0.2 and 0.2 x 0.1, interacting pair, sep 0.25, vF -- U 677) comp 1.0, 217, 0.3 x 0.3, v dif -- U 678) (rs) 0.5 x 0.4; v dif outer disc, no spir pattern vis -- U 679) ident at N 375 in MCG, which is prob a mistake; N 375 is presumably an E (compact?) galaxy at 01 04.4 + 32 04, 2.6 sf, m approx 16.; de V classif N 375 as E5 --

00680   U 680) prob (R)SA 0+ -- U 681) S B(r)?; a) at 1.0, 264, 0.6 x 0.20, prob vF bridge; comp 2.7, 273, 0.35 x 0.20 -- U 682) E2 (de V); Pisces cloud; nr 7 from s in a chain of 8 galaxies, see U 689 -- U 683) S0 (de V); Pisces cloud;   ; nr 8 from s in a chain of 8 galaxies, see U 689 -- U 685) blue; no nucl cond; fragments of spir str vis on red PA map; S cond superimp sf center -- U 686) S0 (de V); Pisces cloud; 1st from s in a chain of 8 galaxies, see U 689 -- U 687) E3 (de V); Pisces cloud; 2nd from s in a chain of 8 galaxies, see U 689 -- U 688) IV Zw 38; E0 (de V); Pisces cloud; nr 5 from s in a chain of 8 galaxies, P w U 689 at 0.55, 23; see U 689 -- U 689) VV 193 (w U 688); brightest in a chain of 8 galaxies, this chain is Arp 331; S0 (de V), Pisces cloud; P w U 688 at 0.55,

2.3; 01 04.8 + 32 05 = N 386 at 2.1, 127, 0.15 x 0.12, E3 (de V), m = 15.4, v = 5555, v_0 = 5753; at least 3 fainter group members; see U 682, 683, 686, 687, 688 ---

00690   U 692) comp 2.6, 134, 0.3 x 0.1, vF -- U 693) comp 1.1, 295, 0.4: x 0.3:, v dif; 1.5, 107, 0.5: x 0.4:, v dif; 2.0, 178, 0.45 x 0.12; 4.7, 5, 0.8: x 0.6:, v dif, dwarf -- U 694) b) at 0.65, 22, 0.20 x 0.12; in a group, 8 obj inside 5.0 -- U 695) v blue, app complex centr region, no evid of spir str; comp 4.4, 75, 0.4 x 0.3 -- U 696) U 669 at 2.6, 120; comp 2.0, 68, 0.5: x 0.3:, e dif, dwarf -- U 697) bar in F ring, no arms vis -- U 698) comp 2.0, 173, 0.3 x x 0.2 -- U 699) U 696 at 2.6, 300 --

00700   U 700) 01 05.7 + 32 53 = N 394 at 1.0, 32, 0.45 x 0.22, m = 14.8; 01 05.8 + 32 50 = N 397 at 2.1, 132, 0.7: x 0.5:, nucl region 0.15 x 0.15, very compact, m = 15.7 -- U 701) nucl cond 0.15 x 0.12 and 0.12 x 0.12, sep 0.50; sev obj in vic -- U 702) 2 vF red nucl cond, sep 0.10; loop, long F tail -- U 703) P w U707 at 3.3, 156 -- U 704) comp 2.0, 71, 0.4 x 0.2; 3.3, 40, 0.35 x 0.20 -- U 705) (r) 0.55 x 0.35; v broad dif arms; comp 0.9, 39, 0.25 x 0.15 -- U 706) (r) 0.50 x 0.22; comp 2.0, 139, 0.55 x 0.10; 4.2, 93, 0.5 x 0.2 -- U 707) V Zw 52; "large E in group, compact v red core" (Zw); see U 703 -- U 709) P w 01 06.5 + 01 07 at 5.5, 77, 0.8 x 0.2, S0-a, m = 15.1 --

00710   U 711) comp 2.5, 221, 0.35 x 0.15 -- U 712) S B O? -- U 714) 2 or 3 vS vF patches n center; 01 06.8 + 31 54 at 4.2, 72, 0.7 x 0.2, m = 15.6 -- U 715) "N 400, 401, 402 are **" (de V); comp 2.1, 143, 0.6 x 0.4, spir; 2.8, 65, 0.5 x 0.2 -- U 716) "N; H" (MCG) -- U 717) VV 348, Arp 11; "Outer arms do not start at termination of bar" (Arp); P w U 719 at 2.0, 54; fragmentary (r) 0.50 x 0.45; 2 strongly distorted arms; comp 1.1, 127, 0.4 x 0.2, brightest of 8 comps ; 4.1, 351, 0.5 x 0.4, S B ... -- U 718) SA(s)0- (de V), S0 (Ho); measurements disturbed by GC 1400, m = 2.37, the galaxy is not vis on the red PA map -- U 719) see U 717; comp 0.5, 162, 0.4 x 0.2, vF --

00720   U 720) nucl cond 0.15 x 0.12 and 0.15 x 0.15, sep 0.45, bridge?, dif streamer from b) away from a) -- U 724) SN 1961m; classif E0 (Kearns) is not correct -- U 725) P w U728 at 11.3, 17 -- U 726) S blue cond at np edge, irr ridge from nucl region; double nucl, or F cond in the ridge close to center -- U 727) v ext env w B centr cond, 3 conds inside 0.3 and sev obj in vic -- U 728) see U 725; comp 3.8, 260, 0.6: x 0.4:, v dif --

00730   U 730) P w U735 at 4.9, 70; comp 3.0, 103, 0.3 x 0.2 -- U 731) DD0 9 -- U 732) comp 1.5, 35, 0.7 x 0.1; 2.1, 80, 0.3 x 0.05, dif -- U 734) comp 1.5, 303, 0.5 x 0.2, dif -- U 735) E2 (de V); see U 730; 01 08.3 + 32 58 at 4.9, 21, 0.9 x 0.4, E - S0, m = 15.7; U 744 at 4.5, 122 -- U 739) 2 E-type obj w sep 0.7 in center of cl 01 08.2 + + 17 26, compact, MD; a large nr of obj in vic --

00740   U 743) comp 3.3, 269, 0.4 x 0.2; 3.8, 154, 0.4 x 0.2, F -- U 744) IV Zw 39; pair of compacts, blue plumes -- U 747) P w U749 at 3.2, 34, bridge? -- U 749) see U 747; integral-shaped, patchy, spir arc southw (bridge to U 747?); blue cond 0.15 x 0.15 attached at n side, comp? --

00750   U 750) vB nucl region in dif env; MCG ident I 1640 is not correct; P w 01 09.3 - 00 53 = I 1640 at 2.4, 30, 0.22 x x 0.22 + dif env, m = 15.1 -- U 751) "Nn; H" (MCG); in cl 01 08.2 + 17 26, compact, MD; sev obj in vic, brightest at 2.4, 190, 0.9 x 0.4, S0-a -- U 753) B nucl region in F env; comp 1.2, 203, 0.3 x 0.3 -- U 755) comp 2.9, 323, 0.4: x 0.3:, v dif -- U 757) 2 v broad arms; comp 3.4, 77, 0.3 x 0.1 -- U 758) asym spir; v massive B short arms -- U 759) 1st of 3; U 766 at 3.8, 139; 3rd member at 7.0, see U 766 --

00760   U 760) 1st of 3; U 765 at 3.6, 51; U 762 at 4.0, 147 -- U 761) comp 2.7, 328, 0.6 x 0.25, Sc -- U 762) 2nd of 3; see U 760; U 765 at 4.0, 327 -- U 763) SAB(s)m (de V), Sc+ (Ho); patchy centr region, alm dissolved spir str, patchy loop s and sf center -- U 765) 3rd of 3; see U 760, 762 -- U 766) 2nd of 3; see U 759; 3rd member at 3.4, 110, 0.5 x 0.25 -- U 768) v dif asym env; comp 0.6, 115, 0.12 x 0.10, interaction? --

00770   U 770) comp 0.7, 268, 0.5 x 0.4; 1.1, 265, 0.2 x 0.2, double? -- U 771) "similar to M 81" (MCG) -- U 772) blue, chaotic; 2 vF cond w sep 0.7; poss 2 dwarf in contact -- U 774) B nucl cond, 1 long F arm s, 1 short broken arm n w comp at the tip at 0.25, 30, 0.12 x 0.12 -- U 778) comp 3.4, 90, 0.3: x 0.2:, F, double? --

00780   U 781) comp 2.1, 65, 0.6: x 0.4:, dif -- U 784) 01 11.6 - 02 07 at 6.9 n, 0.8 x 0.7, early spir? or pec?, m = 15.1 -- U 785) fragments of (R) -- U 786) comp 2.3, 200, 0.40 x 0.25 -- U 787) comp 4.5, 4, 0.8 x 0.4 -- U 789) S 0/a (de V); B main body w short irr ext, jets?, v red --

00790   U 790) P w 01 12.2 + 00 51 at 4.4, 146, 0.7 x 0.2, m = 15.1 -- U 791) 01 12.4 + 01 32 at 5.0, 112, 0.7 x 0.6, m = 15.2; 01 12.3 + 01 40 = N 445 at 6.3, 34, (0.9 x 0.9), m = 15.5 -- U 793) a) 0.5 x 0.4 connected w b) 0.35 f, 0.25 x 0.15; comp 3.6, 260 from a), 0.40 x 0.35 -- U 796) IV Zw 42; compact core, spir str in halo -- U 797) poss member of CGCG cl 01 13.2 - 00 04, medium compact, medium distant, sev galaxies near, brightest at 2.1, 136, 0.8 x 0.7, S B a, m approx 16. -- U 798) (r) 0.8 x 0.5; comp 3.1, 37, 0.6 x 0.4 -- U 799) asym late spir, no disturbing obj vis; U 805 at 6.9, 77 --

00800   U 800) P w comp at 2.3, 357, 0.7 x 0.12 -- U 801) comp 2.6, 359, 0.6 x 0.4 -- U 804) identif as N 447 in MCG; 01 13.4 + 32 50, m = 15.2, is N 449 in MCG but I 1661 in CGCG -- U 805) F blue cond np center; see U 799 -- U 806) sev vB patches in f and nf part; 2 v broad main arms branching outwards; U 807 visible through the nf edge at 1.4, 57 from center; SAB(s)cd: (de V) -- U 807) see U 806 --

00810   U 810) P w U820; comp 2.9, 121, 0.35 x 0.30? 4.3, 290, 0.30 x 0.15 -- U 811) no spir str vis; comp 2.1, 156, 0.6 x 0.3, F, Sc: -- U 812) comp 2.7, 157, 0.4 x 0.3 -- U 813) 2nd of 3; interacting w U816 at 0.8 nf; 01 13.1 + 46 28 at 2.1, 253, 0.9 x 0.3. Sa:, m = 15.1 -- U 815) Arp 164; F smooth streamer nortw from np side, eF ext southw, loop?, no disturbing obj vis; comp 3.8, 101, 0.7 x 0.6, late spir; 4.6, 211, 0.7 x 0.1, dif -- U 816) 3rd of 3, 2 v open distorted arms; see U 813 -- U 817) comp 4.0, 157, 0.9: x 0.9:, v dif, dwarf --

00820   U 820) see U 810 -- U 821) comp 3.6, 333, 0.6 x 0.6 -- U 823) comp 0.7, 205, 0.3 x 0.1, vF; 3.8, 132, 0.5 x 0.2, F -- U 826) V Zw 61; v blue; ecc nucl,1 short blue spir arc, 1 dif arm prob cont in a long dif arm on p side; P w 01 14.0 + 43 21 at 7.8, 258, 0.9 x 0.4, Sc, m = 15.7 -- U 827) VV 205, Arp 128; a) prob early spir, b) E-type obj, sep 0.35 -- U 829) double system?; S B nucl cond 0.12 x 0.10 w ext towards 25 and 145; eF disc w hints of a broad arc or ring fragment n center --

00830   U 830) dif (r) 0.8 x 0.6; 2 long dif arms; vF nucl cond; comp 4.2, 147, 0.5 x 0.3 -- U 831) S-shaped; 1 v long dif arm n; comp 1.4, 180, 0.5 x 0.3, F, interaction? -- U 834) comp 1.2, 268, 0.3 x 0.2, F -- U 836) 2nd of 3; comp 1.8, 84, 0.35 x 0.30; 2.8, 189, 0.8 x 0.10 -- U 838) high surf brightn, overexp on PA maps -- U 839) late spir, 2 F nucl conds sep 0.10 on blue PA map, dif asym disc; comp 4.9, 312, 0.9 x 0.7, early --

00840   U 840) S0: w cond 0.20 x 0.12 superimp 0.15 n center -- U 842) "N; H" (MCG); 01 16.2 - 01 10 at 6.2, 339, 0.8 x x 0.6, S0:, m = 15.4 -- U 843) P w comp at 2.8, 325, 0.9: x 0.5, e dif, dwarf -- U 844) comp 3.1, 188, 0.8 x 0.1, dif; 3.8, 234, F -- U 847) comp 0.9, 310, 0.4 x 0.3, v dif, dwarf::; 3.5, 65, 0.50 x 0.18, early -- U 848) SA(s)0 pec? (de V); 1st of 3; U 858 at 10.6, 52; see U 864; comp 1.6, 115, 0.8 x 0.3, dif -- U 849) VV 347, Arp 119; "Some material seems attracted, some repelled" (Arp); late strongly distorted spir, complex nucl region in n part, short filaments northw; P w 01 16.7 + 12 12 at 1.0, 5 from center, 0.8 x 0.6, S0:, m = 15.2, interaction; comp 2.6, 189, 0.45 x 0.25 --

00850   U 850) (r) 0.22 x 0.22; comp 4.3, 17, 0.45 x 0.10; 4.5, 177, 0.4 x 0.3; sev fainter obj in vic -- U 851) F cond in bar sf center; 2 arms form angular ring -- U 853) comp 0.7, 233, 0.1 x 0.1; one arm appears disturbed, cont in a bridge to comp? -- U 858) SA(rs)b (de V); main body 2.1 x 1.2 w knotty ill def arms; 2nd of 3, P w U864 at 5.4, 87, prob interaction; see U 848 -- U 859) SAB(r)0:(de V) --

00860 U 860) comp 1.6, 104, 0.7 x 0.10 -- U 861) dominating in quadruple group, comp incl in CGCG magn; comp 0.20, 155, 0.12 x 0.10; 0.50, 340, 0.12 x 0.10; 0.75, 60, 0.12 x 0.12 -- U 864) Arp 227; E-type main body w extensive eF corona, hints of sev concentric rings or arcs, strongest straight opposite U 858, interaction?; "vF rings extend t diameter of 7.4" (Arp); SA(s)0 (de V); see U 848, 858 -- U 865) nucl cond 0.15 x 0.15 in F v smooth env; "Fp?; H" (MCG) -- U 866) comp 3.5, 247, 0.35 x 0.15 -- U 867) comp 3.5, 99, 0.35 x 0.20 -- U 868) comp 2.7, 36, 0.4 x 0.12 --

00870 U 872) main body 0.6 x 0.35; P w 01 17.9 + 29 22 at 5.0, 193, 0.8 x 0.35, spir:, m = 14.7 -- U 874) U 890 at 6.8, 52 -- U 875) misprint in MCG, listed as 0-4-98; U 890 at 7.0, 129 --

00880 U 880) compact comp or * superimp at 0.25 sp center; comp 2.1, 52, 0.45 x 0.10 -- U 881) comp 2.7, 240, 0.4 x 0.2 -- U 883) thin F bar, no nucl cond; 01 18.6 + 16 46 at 4.2, 131, 0.6 x 0.3, "elliptical compact" (CGCG), m = 15.6, III Zw 25, "E, neutral elliptical compact disc, blue halo" (Zw), v = 21710, v_0 = 21843, not in MCG, Zw ident MCG 4-4-5 is not correct -- U 884) vB nucl region 0.12 x 0.12; wedge-shaped, v dif, broader northw, disturbed by comp at 1.3, 264, 0.8 x 0.3:; comp 0.8, 165, 0.3 x 0.1, F -- U 886) 2 main arms split in sev dif secondary arms; 01 18.2 + 40 12 at 4.6, 244, 0.8 x 0.45, spir, m = 14.8; comp 4.4, 235, 0.45 x 0.12, m approx 16. -- U 888) U 907 at 13.3 sf --

00890 U 890) see U 874, 875 -- U 891) DDO 10; S(B) IV-V (vdB); tend to centr cond and asym bar; comp 3.3, 349, 0.6 x 0.3 -- U 892) Arp 67; "comps lie on inner and outer spir arms" (Arp); asym (r) 1.0 x 0.7, vF slightly asym disc w ill def spir pattern, 2 F cond in the disc sf, one prob background, one disturbing?; comp 2.6, 287, 0.6 x 0.6, Sb-c; 3.2, 176, 0.7 x 0.6, SO: -- U 894) patchy; S cond np center is prob a superimp comp; comp 2.4, 357, 0.35 x 0.30 -- U 895) patchy main body 0.8 x 0.25; comp 3.2, 8, 0.9 x 0.2, dif; 3.3, 22, 0.5 x 0.1; 3.9, 287, 0.3 x 0.2 -- U 896) U 908 at 6.7 --

00900 U 900) P w U905 at 3.1, 65; interaction?, app an eF spir arc from s side westw-northw -- U 901) spindle completely overexp on PA maps, v large nucl bulge -- U 902) comp 1.3, 113, 0.4 x 0.3; 3.5, 101, (0.6 x 0.6); sev F obj p -- U 904) comp 2.1, 307, 0.3 x 0.2 -- U 905) see U 900 -- U 906) vB nucl region 0.25 x 0.25, hints of spir str in the disc?; in a v dense part of the cl; comp 1.2, 124, 0.5 x 0.4; 3.0, 236, 0.50 x 0.45; 3.5, 177, 0.5 x 0.2 -- U 907) SA(r)b (de V), Sb- (Ho) -- U 908) U 896 at 6.7 -- U 909) slightly asym; complex central str, tend to ecc (rs), 2 blue cond in nf part --

00910 U 910) 2 main arms, one broader, split in 2 secondary -- U 912) in a v dense part of the cl; 01 19.0 + 33 02 = = I 1680 at 4.8 p, 0.8 x 0.7, SO:, m = 14.9 -- U 914) v patchy, chaotic spir pattern, prob (r) or (rs); SN 1971 Nov 15 (Pigatto) -- U 915) Arp 8; split arms; comp 2.0, 3, 0.35 x 0.15; 2.6, 94, 0.9: x 0.8:, v dif, dwarf -- U 917) comp 3.5, 331, 0.8 x 0.15 --

00920 U 920) SO (de V); in a triple subgroup in a v dense part of the cl; U 926 at 3.3 p; U 927 at 4.7 np -- U 922) comp 4.3, 280, 0.35 x 0.35, E:; 4.4, 7, 0.8 x 0.6, E-SO -- U 924) U 932 at 6.9 -- U 926) SO (de V); see U 920; U 927 at 4.2 n -- U 927) see U 920, 926 -- U 928) P w U931 at 3.6, 172 -- U 929) comp 3.2, 4, 0.35 x 0.12; 3.2, 220, 0.3 x 0.2; 4.6, 323, 0.5 x 0.3 --

00930 U 931) see U 928 -- U 932) main body 1.4 x 0.6, F corona; U 924 at 6.9 -- U 933) comp 5.0, 268, 0.6 x 0.5, dif -- U 934) VV 341, Arp 70; distorted spir, disturbing comp 0.80, 205 from center, 0.35 x 0.15, high surf brightn, one arm cont in dif bridge -- U 935) in a v dense part of the cl; U 938 at 4.0 -- U 937) bar 0.3 x 0.1, no evid of spir str; 01 21.0 + 32 25 at 3.5, 60, (0.9 x 0.6), m = 15.7 -- U 938) VV 207, Arp 229; E3 (de V); "circular or near circular rings of small density difference" (Arp); P w U939 at 1.5, 6 (inside env) -- U 939) see U 938 --

00940 U 940) P w 01 20.7 + 34 19 at 1.8, 270, 0.25 x 0.25, extremely compact, m = 14.7, I Zw 4 + II Zw 2, "blue, two stellar knots, ns 4", in spherical matrix, d = 7"" (Zw), v = 6870, v_0 = 7019; comp 2.5, 328, 0.3 x 0.3 -- U 941) comp 1.1, 265, 0.5 x 0.2; 4.6, 160, 0.5 x 0.4 -- U 944) P w 01 21.2 + 33 36 at 2.5, 174, 0.7 x 0.6, spir, m = 14.9 -- U 947) SAB(rs)c (de V) -- U 948) P w comp at 3.0, 88, 0.7 x 0.4, vF --

00950 U 950) early-type obj, prob non-regular; S v red comp attached at p side; MCG δ = - 00 47, but no galaxy exists there -- U 952) comp 1.5, 347, 0.3 x 0.1; 3.1, 128, 0.4 x 0.1, F -- U 953) comp 1.7, 137, 0.35 x 0.30, dif; 1.9, 195, 0.4 x 0.4; 2.5, 143, 0.35 x 0.10 -- U 954) U 949 at 5.1; U 968 at 8.0 -- U 956) P w U960 at 2.9, 157; sev F obj in vic -- U 957) 6.0 from geom center of U 966 -- U 959) asym, prob interacting w comp at 0.85, 256, 0.15 x 0.15; 01 22.0 + 31 59 at 4.2, 14, 0.7 x 0.4, spir, m = 15.1 --

00960 U 960) VV 36; see U 956; sev F comps, one attached at 0.7, 358, 0.2 x 0.2 -- U 962) S B(r)bc (de V); SN 1966g; P w U 992 at 14.4, 83; comp 2.9, 292, 0.9 x 0.7, 0.8 x 0.7, E? -- U 963) 01 22.1 + 01 49 = I 105 at 3.1, 53, 0.8 x 0.6, m = 15.6; comp 2.7, 200, 0.9 x 0.7, early spir; sev F comps   -- U 964) comp 2.4, 293, 0.4 x 0.3, dif; 3.2, 70, 0.7 x 0.5; 3.8, 140, 0.7 x 0.35 -- U 965) P w U973 at 3.2, 125 -- U 966) VV 231, Arp 157; Pec (de V), Ir II (Ho); chaotic, dif streamer np; "inspection of the original plate suggests that N 520 is not a collision of two galaxies but rather a system of the M 82 type" (Sandage); see U 957 -- U 968) SA(rs)0+ (de V) -- U 969) P w U 972 at 3.8, 153 --

00970 U 970) comp 3.1, 52, 0.5 x 0.2; 3.8, 70, 0.6 x 0.5, dif -- U 972) see U 969 -- U 973) ident I 1696 = N 530 in MCG, but IC descr states "530 np"; see U 965; this is N 530 = I 106 -- U 976) early-type obj, disturbed?, vF v red cond at 0.45 f center; asym env; comp 1.5, 312, 0.3 x 0.3 -- U 977) central obj in a v rich field of galaxies; cl 01 22.4 + 08 13, MC, D -- U 978) 1st of 3; U 983 at 2.9, 134; U 986 at 2.7, 100 -- U 979) Arp 158, IV Zw 45; in Arp's class "disturbed w interior absorption"; "post-eruptive, blue, 3 compact knots connected by B bar, fan-shaped jets and matrix" (Zw) --

00980 U 980) comp 1.7, 182, 0.5 x 0.1 -- U 982) main body 2.1 x 0.9, no spir str vis, red cond in disc sp and nf center -- U 983) 2nd of 3, see U 978; U 986 at 1.7, 22 -- U 984) in a dense part of the cl; U 991 at 2.8, 135; U 996 at 3.6, 77 -- U 986) 3rd of 3; see U 978, 983; asym env, interaction w U983 -- U 988) comp 3.4, 188, 0.40 x 0.20, high surf brightn --

00990 U 991) in a dense part of the cl; see U 984; U 996 at 3.2, 27 -- U 992) E3 (de V); see U 962; many S galaxies near, 10 inside 5.0 -- U 993) a) 0.35 x 0.30, b) 0.9 x 0.5, disrupted, streamer eastw; sev objs near , brightest comp at 3.6, 37, 0.8 x 0.12 -- U 995) P w U1013 at 8.5, 95; comp 1.5, 331, 0.5 x 0.15 -- U 996) in a dense part of the cl; see U 984, 991; U 1003 at 4.1, 47 -- U 997) in a dense part of the cl; U 1004 at 3.7, 63 -- U 999) double system?, F cond in disc at 0.2 f center --

01000 U 1002) nucl cond 0.20 x 0.12 in smooth env -- U 1003) in a dense part of the cl; see U 996, 997 -- U 1004) see U 997; 2 F obj in the env at 1.0 f and 0.9 nf center; Arp 133 -- U 1005) S cond 0.15 p center, * or comp? -- U 1006) comp 2.9, 137, 0.3 x 0.2; 3.5, 131, pec, v blue -- U 1007 + U 1009) Arp 308; SA 0- and E1 (de V); sep 0.50, pos angle 128, common env; U 1009 is brightest in cl; in triple group w U1004 at 4.4, 241 from geom center; ident in MCG as N 545ab, which is incorrect --

01010 U 1010) U 1007 + 1009 at 7.1 -- U 1012) U 1013 at 3.1, 171 -- U 1013) (r) 1.5 x 0.55; 2 v broad F arms; see U 995, 1012; 01 23.7 + 34 25 = N 542 at 2.6, 133, 0.8 x 0.2, m = 15.4; SN 1963n -- U 1016) centr region 0.8 x 0.4, v dif asym disc, tend to spir pattern close to center; SN 1963e, Sb (Rosino); comp 4.2, 23, (0.9 x 0.7) --

01020 U 1020) comp 3.8, 236, 0.7 x 0.4, Irr or pec, v blue; 5.0, 55, 0.9: x 0.7:, dif -- U 1021) Sb: (de V); SN 1961q, Sb (Humason); comp 1.9, 47, 0.35 x 0.30, dif -- U 1022) comp 3.5, 57, 0.35 x 0.12 -- U 1023) 1st of 3; U 1029 at

8.4, 110; 2nd group member at 7.3, 51, 0.7 x 0.4; comp 2.1, 238, 0.35 x 0.12 -- U 1029) 3rd of 3, see U 1023; 2nd group member at 4.8, 324; comp 0.8, 308, 0.6 x 0.4, Sc - Irr --

01030   U 1030) comp 2.4, 56, 0.35 x 0.25, double; 4.5, 79, 0.55 x 0.55, (R)SAB 0 -- U 1031) measurements disturbed by nearby
         * GC 1756, m = 6.36; comp 4.8, 98, 0.35 x 0.15 -- U 1032) main body 0.55 x 0.35, vB elong centr region; sev galaxies
         near-- U 1033) comp 2.7, 189, 0.4: x 0.3, v dif -- U 1036) S0 (de V); U 1044 at 6.0, 71; 01 24.7 - 02 14 = N 558 at
         4.2, 214, 0.50 x 0.22, m = 15.0, v = 5018 -- U 1037) P w U 1038 at 5.9, 2; comp 4.9, 23, 0.35 x 0.10 -- U 1038) see
         U 1037, same comp at 2.2, 129 --

01040   U 1040) U 1043 at 2.7, 140 -- U 1041) MCG 2-4-55 is the brightest galaxy, CGCG magn prob for nrs 1 - 4 from s; sev
         galaxies near   -- U 1043) see U 1040 -- U 1044) E3 (de V); see U 1036 -- U 1047) measurements disturbed by nearby
         * GC 1772 at 1.6 f, m = 6.87 -- U 1049) poss SAB(rs)c; comp 4.4, 144, 0.4 x 0.2 --

01050   U 1052) comp 4.5, 278, 0.5 x 0.10; 4.7, 229, 0.5 x 0.3, double -- U 1056) a) blue, brightest in sp part; b) superimp
         outer s part of a), 0.1 x 0.1 -- 1059) v open, 2 v thin arms, one pointing towards 01 26.2 + 39 09 at 2.3, 237,
         0.55 x 0.25, m = 15.4, IV Zw 47, "blue intercon pair of B and F compact knots, sep 8" n-s" (Zw) --

01060   U 1061) dwarf 4.0, 37, ..., e dif, dwarf -- U 1063) chaotic, prob disrupted spir; P w U 1065 at 1.1, 60; MCG α =
         01 24.5 is not correct -- U 1065) see U 1063 -- U 1066) blue nucl region, v dif disc; comp 1.0, 167, 0.5 x 0.2, F --
         U 1068) P w 01 27.2 + 45 22 at 4.3, 76, 0.7 x 0.6, S B c, m = 15.1; comp 3.3, 63, 0.45 x 0.10 --

01070   U 1070) 2 comp superimp at 0.40, 21, 0.20 x 0.15 and 0.45, 247, 0.20 x 0.15, blue bridges from centr region; 2 smaller
         conds in nf part; comp 2.8, 231, 0.35 x 0.20 -- U 1072) comp 5.0, 300, 0.9 x 0.2, spir -- U 1073) DDO 11, S(B) V: (vdB)
         -- U 1077) comp 3.8, 93, 0.4 x 0.2; 4.2, 209, 0.5: x 0.4:; 4.3, 2, 0.4 x 0.2 -- U 1078) overexp on PA maps; sharp
         edges,   short jets northw and westw; comp 1.4, 76, 0.6 x 0.3, v dif --

01080   U 1081) (rs) 0.7 x 0.6, bar 0.40 -- U 1086) asym, no disturbing obj vis -- U 1089) main body 1.1 x 1.0, e dif outside;
         P w U 1094 at 8.7, 163, interaction? --

01090   U 1090) comp 2.8, 232, 0.3 x 0.3 -- U 1092) nucl bulge 0.50 x 0.35; comp 3.5, 269, 0.3 x 0.2 -- U 1093) S0?, contact
         w a) at 0.8, 320, 0.9: x 0.8:, interaction; c) at 1.5, 114, 0.8 x 0.8 -- U 1094) see U 1089; comp 1.9, 42, 0.3 x 0.2;
         4.4, 49, 0.1 x 0.1 + eF env -- U 1095) VV 301, Arp 98, IV Zw 51; "Pair of intercon compacts. No 1, red spherical w
         halo... v = 12567 ... No 2, 60" east of No 1, blue core, thin spir arms... v = 12532" (Zw); "high surf brightn s
         inside spir, similar to Arp 96" (Arp) -- U 1097) smooth spir arc northw on blue PA map; abs marking in centr region?;
         comp 3.0, 110, 0.7 x 0.3; 4.5, 33, 0.7 x 0.2 -- U 1098) (r) 0.35 x 0.35, only 1 arm vis; comp 4.5, 342, 0.8 x 0.15
         -- U 1099) comp 3.8, 117, 0.35 x 0.12 --

01100   U 1101) V Zw 77; compact disc, blue halo-- U 1102) VV 173, Arp 306a; S B m w blue irr cond 0.35 x 0.20 in s part of
         disc; P w U 1105 at 3.5, 47; "Resolution; dif, hooked countertail" (Arp) -- U 1104) blue, irr, sev B blue conds --
         U 1105) VV 174, Arp 306b; see U 1102 -- U 1106) comp 4.3, 64, 0.7 x 0.6, double system, contact; 4.5, 46, 0.7 x 0.6
         -- U 1108) comp 4.5, 117, 0.5 x 0.1 -- U 1109) P w 01 30.4 + 44 41 at 2.3, 291, 0.7 x 0.35, m = 15.4 --

01110   U 1111) B bar 0.7 x 0.35 -- U 1112) blue, asym, hints of spir pattern, prob v late spir -- U 1114) comp 3.5, 356,
         0.5: x 0.2 -- U 1115) v blue, brightest in p part -- U 1116) 1st of 3; U 1120 at 7.1, 83; U 1123 at 9.1, 67; comp
         1.5, 218, 0.4 x 0.3 -- U 1117) SA(s)cd (de V), Sc + (Ho); M 31 sub-group; SN? 1928 (Baade), prob nova -- U 1119) comp
         4.3, 163, 0.9 x 0.4, double system, bridge --

01120   U 1120) 2nd of 3; see U 1116; U 1123 at 3.0, 29 -- U 1121) dominating in triple group, comp 0.6, 180, 0.15 x 0.10;
         0.9, 138, 0.6 x 0.4 -- U 1122) vF disc, tend to bar on red PA map; comp 1.6, 157, 0.4 x 0.15, F -- U 1123) 3rd of 3;
         see U 1116, 1120 -- U 1124) P w comp at 2.0, 296, 0.8 x 0.7, Sc -- U 1125) massive v tightly wound arms, blue? --
         U 1126) angular (r) 0.35 x 0.35; asym branching arms attached to the ring at 3 points -- U 1127) sym vF disc, F red
         nucl cond; comp 3.1, 212, 0.35 x 0.12 --

01130   U 1132) S m? -- U 1133) DDO 12; S V: (vdB) -- U 1135) P w U 1140 at 5.2, 72 -- U 1136) no nucl cond; chaotic; sev vF
         obj in vic -- U 1137) comp 3.9, 222, 0.5: x 0.5:, Irr -- U 1138) chaotic, sev blue patches; comp 4.4, 260, 0.40 x
         x 0.25 --

01140   U 1140) v ext transp env; see U 1135; comp 1.6, 265, 0.3 x 0.2 -- U 1143) (r) 0.7 x 0.5 -- U 1145) complex bar, 2
         eF arms; singular? -- U 1147) IV Zw 54; "E, neutral knotted elliptical disc, blue halo 80"" (Zw) -- U 1149) SA(s)c
         (de V), Sc - (Ho); e dif outskirts appear vis to 14.5 x 14. on blue PA map --

01150   U 1150) V Zw 81; "blue post-eruptive Sc, B core, jets" (Zw) -- U 1151) P w comp at 0.55, 249, 0.4 x 0.15 -- U 1152)
         comp 1.6, 282, 0.35 x 0.12, double?; 4.6, 163, 0.45 x 0.15 -- U 1153) P w U 1157 at 7.9, 71 -- U 1154) vB main body,
         irr outskirts, fragments of spir arms? -- U 1155) main body 0.6 x 0.35, F irr fringes -- U 1156) (rs) 0.30 x 0.25,
         v dif disc -- U 1157) see U 1153; comp 1.5, 10, 0.35 x 0.15 -- U 1158) IV Zw 4; "neutral spherical compact, extended
         uniform halo, m_p = 17.3" (Zw) --

01160   U 1163) comp 2.6, 344, 0.8 x 0.12, early --

01170   U 1170) overexp inner region 0.55 x 0.40 on PA maps -- U 1171) P w U 1176 at 6. f; U 1149 = N 628 at 44. -- U 1172)
         late spir?, P w U 1177 at 8.3, 115 -- U 1174) S cond nf center, *? -- U 1176) DDO 13, Ir V (vdB); see U 1171; U 1149
         = N 628 at 50. -- U 1177) 2 massive smooth arms, rare; see U 1172 -- U 1178) patchy, poss v late -- U 1179) V Zw 86;
         "E, v red disc in halo, m_p = 15.6" (Zw); not in CGCG --

01180   U 1180) stell nucl cond in vF ext env -- U 1181) P w comp at 2.0, 146, 0.5 x 0.3 -- U 1182) comp 2.6, 311, 0.4 x 0.3
         -- U 1184) comp 1.4, 110, 0.4 x 0.15; 3.5, 4, 0.3 x 0.2, F -- U 1185) sharp transition to eF env -- U 1186) vB nucl
         region 0.12 x 0.12, vF slightly asym irr disc or env, prob no S0; MCG δ = 48 38 is not correct; P w comp at 1.5, 302,
         0.7 x 0.4, interaction? -- U 1187) S0 or early Sa?; comp 3.6, 260, 0.5 x 0.12; 6.6, 312, 0.9 x 0.9, S B(s)dm, m approx
         16. -- U 1188) P w comp at 2.1, 34, 0.5 x 0.3, spir -- U 1189) comp 2.3, 259, 0.4 x 0.4 --

01190   U 1191) VV 177; P w VV 177 at 1.0, 202, E-type obj 0.6 x 0.6 w comp attached at sf side, 0.18 x 0.15; sev obj in vic
         - U 1192) disc appears disturbed in nf part, no disturbing obj vis -- U 1195) no spir str, slightly arc-shaped, ecc
         blue nucl region; P w U 1201 at 22., 157, interaction? -- U 1196) comp 2.0, 229, 0.7: x 0.5:; 2.1, 192, 0.8: x 0.7: --
         U 1197) blue, strongly asym, brightest eastw -- U 1198) VII Zw 3; "eruptive galaxy" (CGCG); "blue post-eruptive,
         similar to M 82, v compact bar (jet) (northwest - southeast) crossing main axis (northeast - southwest), halo
         and jets" (Zw) --

01200   U 1200) B patchy main body 0.8 x 0.5, dif outskirts; comp 1.6, 351, 0.5 x 0.4, red, background -- U 1201) S B(s)a pec
         (de V); bar 2.8 x 0.9 in pos angle 217, thin abs lane, complex abs regions in np part of bar; 2 dif v broad arms,
         eF outskirts to 11.8 x 5.2 -- U 1202) 2 v blue cond sf center; 1st of 3; U 1204 at 6.0, 37; U 1210 at 12.1, 83 --
         U 1204) 2nd of 3; see U 1202; U 1210 at 9.1, 111 -- U 1205) comp 3.7, 65, 0.4 x 0.1 -- U 1209) cond 0.1 x 0.1 in
         sp part, disturbing comp?; comp 4.0, 251, 0.7: x 0.4 --

01210   U 1210) 3rd of 3; see U 1202, 1204 -- U 1214) comp 4.9, 15, 0.5 x 0.3, dif -- U 1215) comp 3.8, 261, 0.6 x 0.5, v
         dif -- U 1217) comp 3.9, 206, 0.3 x 0.2; 4.4, 260, 0.3 x 0.2; 4.5, 188, 0.35 x 0.22, F; 4.8, 275, 0.3 x 0.1, F --
         U 1219) bar 0.45 x 0.20, massive blue arms, appears unilateral; comp 3.2, 274, 0.7: x 0.5:, dif; 3.7, 130, 0.6 x 0.6;
         4.1, 10, 0.35 x 0.35, F --

01220 U 1220) V Zw 98; "blue post-eruptive compound containing many compact knots and sharp knots" (Zw) -- U 1221) comp 3.7, 169, 0.3 x 0.08; 4.3, 344, 0.5 x 0.25; 4.5, 328, 0.7 x 0.12 -- U 1223) diam of main body, eF (R) or (R') approx 3.0 x 2.5; brightest in triple group w U 1229 at 14.4 and U 1231 at 11.2; U 1226 at 3.7, 158; comp 2.0, 53, 0.50 x 0.35; 5.0, 173, 0.7 x 0.25; 01 42.4 + 10 06 at 6.5, 134, 0.8 x 0.3, m=15.5 -- U 1224) U 1227 at 4.0 s -- U 1226) see U 1223 -- U 1227) see U 1224 -- U 1228) strongly asym, dif; v dif bar and arms in direction 205; disrupted pair? -- U 1229) see U 1223; U 1231 at 6.1 --

01230 U 1231) see U 1223, 1229 -- U 1233) comp 3.1, 272, 0.4 x 0.4, halo; 3.8, 124, 0.3 x 0.3, F -- U 1234) strongly asym open spir; 2 dif arms, one stronger; no disturbing obj vis -- U 1236) VI Zw 26; "E w compact neutral disc" (Zw) -- U 1237) brightest of 4; comp 0.35, 15, 0.12 x 0.12; 2.1, 23, 0.8 x 0.12; 2.7, 333, 0.4 x 0.1, F --

01240 U 1240) prob distorted spir; blue irr cond attached to f side; cond 0.7 p is prob a *; comp 5.0, 112, 0.9: x 0.1, dif -- U 1242) B blue cond in s part -- U 1243) hints of eF corona outside 1.5 x 0.8 on blue PA map; comp 2.0, 105, 0.55 x 0.25, dif spir; 4.1, 321, 0.4 x 0.1 -- U 1245) B nucl region 0.35 x 0.25, "compact nucleus" (CGCG); comp 2.6, 30, 0.6 x 0.5 -- U 1246) VV 93, VV α = 01 43.3; sev blue conds -- U 1248) a large nr of F obj in vic -- U 1249) VV 338; S B(s)m (de V), Ir I (Ho), S B IV (vdB); no nucl cond, patchy bar, hints of spir pattern northw; P w U 1256 at 8.1, 42; comp 5.3, 120, 0.9 x 0.12 --

01250 U 1250) E6 (de V) -- U 1251) drop-shaped, B nucl region in n part; comp 0.50, 355, 0.35 x 0.08; 1.1, 100, 0.5: x 0.1, asym; 2.5, 16, 0.6 x 0.2 -- U 1252) no nucl cond, (rs) 0.55 x 0.45, 2 vF arms -- U 1253) prob late; comp 2.0, 156, 0.3 x 0.1 -- U 1256) S B(s)cd (de V), Sc + (Ho), S B III (vdB); see U 1249 -- U 1258) a) 0.7 x 0.2, b) 0.2 x 0.1 at 0.8 n a), bridge -- U 1259) 2 B main arms; asym, one arm crosses the disc southw and terminates at right angle to the edge ; comp 2.7, 123, 0.7 x 0.2 --

01260 U 1260) comp 0.9, 110, 0.5 x 0.4, v dif; 4.5, 258, 0.5 x 0.1 -- U 1261) comp 2.7, 207, 0.5 x 0.3; 4.9, 95, 0.3 x 0.2 -- U 1262) P w U 1264 at 4.2, 12; chain of 3 smaller obj f; sev F obj in vic -- U 1263) U 1266 at 5.1, 146; U 1267 at 5.7, 121; U 1268 at 4.8, 76 -- U 1264) see U 1262 -- U 1265) a) 0.9: x 0.6, b) 0.5: x 0.4 superimp at 0.30 s center of a) -- U 1266) VV 54; vB centr region; F comp 2.0 x 0.2 attached to sf edge ; I 161 does not exist; see U 1263; U 1267 at 5.1, 326; U 1268 at 5.4, 23; comp 3.2, 85, 0.9 x 0.25 -- U 1267) VV 53, Arp 228 (ident by Arp as I 162); in Arp's class "concentric rings"; see U 1263, 1266; U 1268 at 4.1, 357; comp 1.0, 123, 0.9 x 0.25 (same as for U 1266) -- U 1268) see U 1263, 1266, 1267 -- U 1269) 1st of 4; U 1272 at 5.8; this group includes 01 46.3 + 34 43, 0.25 x 0.15, very compact, m = 15.4, and 01 46.5 + 34 44, 0.50 x 0.45, m = 15.5 --

01270 U 1270) brightest in triple chain N 676-693-706; U 1304 = N 693 at 26.; U 1334 = N 706 at 49. -- U 1272) main body 0.65 x 0.35, sharp edge, eF outer disc or corona, (R)?; see U 1269 -- U 1273) P w U 1275 at 1.4, 101; comp 2.1, 351, 0.7 x 0.5 -- U 1274) asym on red PA map; comp 2.3, 301, 0.4 x 0.3; 4.9, 334, 0.9 x 0.8, early -- U 1275) see U 1273; comp 2.9, 324, 0.7 x 0.5 -- U 1276) B irr ecc bar, knotty ill-def arms -- U 1278) comp 3.5, 222, 0.2 x 0.1 and 3.5, 218, 0.3 x 0.2, sep 0.3 --

01280 U 1280) P w U 1286 at 5.5, 107 -- U 1281) P w 01 46.8 + 32 20 at 0.8, 135, 0.25 x 0.25, compact, m=15.6; comp 3.3, 45, 0.5 x 0.5; 4.4, 178, 0.3 x 0.1 -- U 1282) comp 1.5, 98, 0.6: x 0.5:, e dif, dwarf; 2.6, 25, 0.5: x 0.2:, e dif, dwarf -- U 1283) V Zw 114; "neutral E0, flares in halo" (Zw) -- U 1286) E2: (de V); see U 1280; slightly asym env, disturbed?; comp 3.5, 46, 0.7 x 0.5, E - S0 -- U 1288) P w comp at 0.8, 67, 0.7 x 0.3, dif; comp 2.0, 237, 0.35 x 0.15

01290 U 1292) S0: (de V) -- U 1295) comp 3.4, 257, 0.5: x 0.4:, e dif, dwarf -- U 1299) U 1302 at 5.1, 154 --

01300 U 1300) a) 0.8 x 0.4, b) 0.5 x 0.3 at 0.8 n a), bridge -- U 1301) sev objs near ; brightest comp at 1.9, 73, 0.9 x x 0.4; 01 47.7 + 32 48 at 3.1, 272, 0.4 x 0.3, compact, m=15.4 -- U 1302) bar 0.9 x 0.35 (elongated ring); see U 1299 -- U 1304) see U 1270; U 1334 = N 706 at 21.6 -- U 1305) Sc (de V); rare, smoother than a normal Sc; comp 2.1, 83, 0.5: x 0.5:, e dif -- U 1306) a large nr of S comps --

01310 U 1310) V Zw 122; Sc (de V); "post-eruptive spir, compact blue core and various knots" (Zw); N 697 group -- U 1313) Arp 31, in Arp's class "integral sign"; v open bar spir, short bar (0.5), slightly irr arms; "high surface brightn irregularity is 5'n" (Arp); may be disturbed by U 1310 at 5.5, 336 -- U 1314) P w 01 48.2 + 12 18 at 3.2, 217, 0.7 x 0.5, m=15.2 -- U 1315) V Zw 123; E: (de V); "tight blue spir, compact core and knots" (Zw); U 1317 = N 697 at 13.5; comp 2.6, 74, 0.3 x 0.1 -- U 1316) e dif outside 0.6 x 0.2; P w 01 48.4 + 34 37 at 1.5, 10, 0.9 x 0.6, S0-a:, m = =15.6 -- U 1317) S bc (de V) -- U 1319) complex; B centr region, asym spir arcs or arm fragments; comp 2.1, 219, 0.5 x 0.5 --

01320 U 1320) P w comp at 5.4, 183, 0.8 x 0.5, S d - dm, v dif; comp 2.5, 178, 0.3 x 0.1, vF -- U 1321) P w 01 49.7 - 01 19 at 2.0, 161, 0.5 x 0.5, compact, m=15.5 -- U 1322) comp 2.6, 234, 0.4 x 0.2, F -- U 1324) (rs): 0.35 x 0.30, ecc bar -- U 1325) P w U 1326 at 2.8, 16 -- U 1326) see U 1325 -- U 1327) P w comp at 2.0, 105, 0.6 x 0.3 -- U 1329) DDO 16; prob open S B; dif loop southw from sf side; disturbed?; chain of 4 galaxies s, but no certainly disturbing obj vis --

01330 U 1332) P w U 1340 at 3.7, 85 -- U 1333) B v red nucl cond 0.12 x 0.12; appears unilateral, one broad arm vis -- U 1334) v patchy main body 1.0 x 0.7; 3rd in chain of 3, see U 1270, 1304 -- U 1335) P w U 1342 at 6.4, 102 -- U 1336) V Zw 133; "neutral spherical compact, plumes n and s" (Zw); in a dense part of the cluster, 01 49.1 + 35 53 at 2.9 np, 0.8 x 0.5, S Bc, m=15.6, this is MCG 6-5-24, ident as N 700 in MCG; 01 49.3 + 35 47 at 3.7 sf, 0.9: x x 0.7:, S0:, m=15.5 -- U 1338) P w U 1339 at 3.6, 10 -- U 1339) see U 1338 --

01340 U 1340) see U 1332 -- U 1341) comp 2.9, 287, 0.4 x 0.25, vF; 3.1, 1, 0.3 x 0.2, vF -- U 1342) see U 1335; comp 3.6, 81, 0.5 x 0.2 -- U 1343) VI Zw 134; "southern neutral spherical component of pair [N 704], mp= 15.6" (Zw); U 1345 at 1.4; U 1348 at 2.3; U 1346 at 2.7; 01 49.6 + 35 52 at 1.2 p, 0.25 x 0.25 (env 0.9? x 0.8?), compact, m=15.7; sev fainter obj in the group -- U 1345) IV Zw 90; "neutral spir edge on" (Zw); U 1348 at 1.0 (inside the halo of U 1345); U 1343 at 1.4; U 1346 at 1.7 -- U 1346) see U 1343, 1345; U 1348 at 1.8 -- U 1347) double; 2 thin B main arms, sev patchy arm fragments in the outer regions; "pair of intertwined Sc w red compact core and blue patchy compact, 10" ssp" (Zw) -- U 1348) see 1343, 1345, 1346; the group appears to be surrounded by an e dif halo -- U 1349) patchy; comp 3.4, 20, 0.5 x 0.12 --

01350 U 1350) (r) 0.8 x 0.5, dif outside -- U 1352) comp 4.4, 67, 0.4: x 0.4:; 4.5, 225, 0.5 x 0.12; 4.7, 295, 0.7 x 0.1, vF -- U 1353) VI Zw 93; "E, elliptical red compact disc" (Zw); comp 3.2, 15, 0.3 x 0.15, F -- U 1354) comp 2.7, 120, 0.35 x 0.20; 4.2, 192, 0.8 x 0.15 -- U 1356) SAB(s)a (de V); 2 smooth arms, one slightly stronger; (R') 1.7 x x 1.7, smooth eF corona; early S B a --

01360 U 1361) P w 01 50.9 + 36 18 at 1.5, 204, 0.9 x 0.7, Sc:; m=15.3; 01 51.0 + 36 23 at 3.3, 13, 0.6 x 0.4, spir, m = =15.4 -- U 1365) P w U 1367 at 2.2, 77 -- U 1366) 01 51.0 + 36 23 at 4.6, 279, see U 1361 -- U 1367) see U 1365 -- U 1368) comp 3.7, 248, 0.8 x 0.2; 4.0, 268, 0.4 x 0.3 -- U 1369) prob late SAB; patchy; irr abs markings; 1st of 3, U 1384 at 8.2, 72; 01 52.0 + 17 48 at 7.9, 114, 0.6 x 0.5, m=15.0 --

01370 U 1373) 1st of 3; comp 4.6, 87, 0.7 x 0.4; 11.0, 105, 0.9 x 0.6 -- U 1375) (r) or (rs)? 0.40 x 0.35 -- U 1376) S B b (de V) -- U 1379) measurements disturbed by GC 2309, m=2.72 --

01380 U 1380) v compact comp at 5.0, 117, 0.15 x 0.15 -- U 1381) P w comp at 4.5, 114, 0.7: x 0.3: -- U 1382) appears double on red PA map; 01 51.8 - 00 28 at 5.5, 222, 0.7 x 0.20, early, m=15.7 -- U 1383) comp 4.5, 153, 0.3 x 0.2, m approx 16.5 -- U 1385) 1st of 3; UCG α is incorrect, shall be 01 52.0; 01 52.1 + 36 41 at 1.6, 86, 0.9 x 0.45,

SAB(s)b, m=15.2; 01 52.1 + 36 38 at 3.1, 162, 0.7 x 0.6, m=15.2; comp 3.1, 318, 0.4 x 0.15 -- U 1386) P w a) at 0.9, 180, 0.9 x 0.6, S B c, MCG 2-5-57, diam for 56 and 57 are interchanged in MCG -- U 1387) chaotic, fragments of spir arms; may be disturbed by comp at 1.8, 220, 0.6 x 0.4, E:, m approx 16.; U 1390 at 3.5, 38 -- U 1388) complex center, slightly asym env --

01390   U 1390) see U 1387 -- U 1392) comp 4.0, 289, 0.3 x 0.2 -- U 1393) P w U 1396 at 3.3, 55 -- U 1394) comp 4.8, 171, 0.5 x 0.4, dif -- U 1395) ecc (rs) 0.6 x 0.5 -- U 1396) see U 1393 -- U 1397) comp 2.6, 34, 0.4 x 0.1 -- U 1399) P w 01 53.0 + 17 45 at 3.5, 219, 0.8 x 0.5, m=15.2 --

01400   U 1400) comp 1.5, 22, 0.5 x 0.3 -- U 1401) comp 1.9, 169, 0.9: x 0.7:, v dif, dwarf; 4.3, 36, 0.4 x 0.2 -- U 1402) comp 2.5, 254, 0.5: x 0.3:, v dif; 3.3, 312, 0.5 x 0.4 -- U 1404) S B c?; comp 3.1, 165, 0.45 x 0.08 -- U 1405) comp 3.5, 359, 0.5 x 0.4 -- U 1409) B nucl region 0.25 x 0.22, extensive vF corona; comp 0.8, 5, 0.5: x 0.5:, e dif, dwarf --

01410   U 1411) brightest in triple system V Zw 146; "No 1, red elliptical compact, 85" nf N 735, m_p=16.4. No 2, N 735, blue S B b, split compact core... No 3, red spherical compact, 85" np N 735, m_p=16.2" (Zw); comp 3.6, 260, 0.5 x x 0.2 -- U 1412) comp 5.0, 296, 0.3 x 0.2, dif -- U 1413) VV 175, III Zw 38; E0 (de V); "red elliptical, stellar nucleus" (Zw); ext eF corona outside 3. x 3.; central obj in a group; 01 53.6 + 05 21 at 3.3, 223, 0.2 x 0.2, compact, m=15.0; 01 53.7 + 05 25 = I 1751 at 1.5, 342, 0.8 x 0.3, nucl cond 0.2 x 0.2, compact, m=15.1; 01 53.9 + + 05 23 = N 742 superimp p part of U 1413 at 0.8, 100, 0.2 x 0.2 + env, prob interaction, m=14.8; N 741 is not I 1751 (as stated in BG); * at 2.4, 312 appears to be superimp a galaxy -- U 1414) VI Zw 111; "neutral spherical compact, extended halo" (Zw); P w U 1421 at 3.4, 120; comp 1.4, 47, 0.35 x 0.25; 3.6, 290, 0.5 x 0.15; E1 (de V) -- U 1416) * superimp?; comp 2.0, 301, 0.5 x 0.3, vF --

01420   U 1420) asym; dif ext southw; F cond 0.08 x 0.08 superimp sp center, * or disturbing comp? -- U 1421) see U 1412; spindle?, v red; F pair at 1.5, 130, 0.1 x 0.1 and 0.1 x 0.1, sep 0.15 -- U 1422) centr region appears double; comp 1.7, 286, 0.35 x 0.25, v dif -- U 1425) main body 0.9 x 0.6, E or S0, w eF tail northw; P w comp at 1.1, 70, 0.8 x x 0.25, interaction; 01 54.1 + 05 34 at 4.2, 295, (0.9 x 0.8), nucl region 0.18 x 0.18, compact, m=15.4 -- U 1426) comp 1.5, 151, 0.3 x 0.2, F; 3.6, 125, 0.4 x 0.2 -- U 1428) comp 1.7, 217, 0.35 x 0.12; 2.7, 199, 0.4: x 0.1 -- U 1429) P w I 1757 at 1.5, 100, 0.6 x 0.25, pec, B nucl region in p part, F loop f; MCG ident 0-6-5 as I 1757 is incorrect --

01430   U 1430-31) VV 189; Arp 166; VI Zw 123; Epec + E pec (de V); sep 0.45; "interconnected red pair... compact cores... broad intergalactic bridge, l=150", connecting to red E, m_p=18.6" (Zw); "S spir at end of plume" (Arp); comp at tip of plume appears peculiar -- U 1432) VV 12; M 51 case, comp 0.45, 240, 0.1 x 0.1, v blue, at tip of distorted arm -- U 1433) main body 0.4 x 0.4, eF outer disc, spir pattern disturbed; 1st of 3, U 1436 at 2.9, 151; comp 2.4, 140, 0.9: x 0.7:, S d - dm -- U 1434) U 1440 at 5.8, 24 -- U 1436) see U 1433; 3rd group member at 0.75, 8 -- U 1437) SAB(rs)bc (de V); SN 1954 e; U 1440 at 23.5; comp 3.2, 109, 0.7 x 0.2 -- U 1439) ident as N 760 - 761 in MCG, but N 760 is a double *; comp 1.9, 0, 0.3 x 0.1 --

01440   U 1440) see U 1434, 1437; E0: (de V) -- U 1442) U 1426 at 11.3; comp 3.6, 358, 0.4 x 0.3 -- U 1444) spindle w strong conc of light to the center, asym on blue PA map; early S B on edge? -- U 1445) U 1466 at 16.3 -- U 1446) I m or S m 1.0: x 0.4 w comp 0.6 x 0.5 attached nf, blue, Irr; comp 7.4, 355, 0.9 x 0.4, S B a, * superimp, m approx 16. -- U 1447) comp 2.0, 301, 0.25 x 0.20, F -- U 1449) VV 122, Arp 126; a) 0.45 x 0.35, high surf brightn; b) 2.2: x 0.9:, chaotic, strongly disrupted, complex nucl region at 0.45 from center of a) --

01450   U 1450) B nucl cond 0.20 x 0.15 in eF env; "N" (MCG) -- U 1451) P w U 1462 at 9.0 nf -- U 1452) F main body 0.7 x 0.2, no nucl cond; v late bar spir? -- U 1455) e low surf brightn outside 2.3 x 2.3 -- U 1456) comp 2.2, 0, 0.9 x 0.2 -- U 1457) P w I 1761 = 01 56.3 + 00 19 at 3.7, 50, 0.9 x 0.5, S0:, m=15.5, interaction?; comp 2.7, 195, 0.35 x 0.20 -- U 1458) comp 0.9, 240, 0.15 x 0.15; 2.9, 85, 0.6 x 0.3:; 3.5, 172, 0.9 x 0.6 --

01460   U 1462) see U 1451; S fairly B nucl region, v dif disc -- U 1463) S B a (de V); B main body 0.35 x 0.25, smooth disc or env; P w U 1466 at 3.6, 25, prob interaction; comp 1.5, 283, 0.10 x 0.10 and 3.5, 277, 0.5: x 0.3:; these obj appear to be connected w U 1466, e dif plumes -- U 1466) sev tightly coiled arms, one strongly asym crossing the disc westw; e dif ext not incl in diam; Arp 78, in Arp's class "S, high surf brightn comp" ; see U 1463, broad vF arc opposite U 1463 suggests interaction; comp at 4.1 and 5.7, prob e dif plumes, "F material tow each of the two dwarf comp" (Arp). These obj have anomalous redshifts (Astrophys. Letters 5, pp 257 - 260), v = 20174 and 19680 -- U 1467) asym; v massive arms; P w comp at 1.1, 195, 0.6 x 0.10; comp 3.1, 180, 0.7: x 0.4:, arc-shaped, v dif -- U 1468) fairly B red nucl region, dif disc, alm dissolved spir pattern; P w U 1469 at 4.0, 14 -- U 1469) see U 1468 --

01470   U 1470) 01 57.3 + 31 48 at 7.2 sf, 0.9 x 0.4, S0-a:, m=15.4 -- U 1471) 01 57.2 + 23 22 = I 180 at 2.0, 63, 0.5 x 0.4, E:, m=15.3; 01 57.2 + 23 25 at 2.8, 150, 0.9 x 0.4:, S0:, m=15.7 -- U 1473) (rs) 0.8 x 0.6, e dif disc -- U 1476) E1 (de V); P w U 1480 at 7.1, 172 -- U 1478) slightly asym; P w comp at 5.8, 308, 0.8 x 0.15 -- U 1479) U 1468 at 6.5 --

01480   U 1480) see U 1476 -- U 1481) P w 01 57.2 - 00 31 at 4.5, 267, 0.8 x 0.35, m=15.6 -- U 1485) chaotic, brightest southw, no disturbing obj vis; symbiotic pair? -- U 1486) (r) 0.7 x 0.45, vF arms -- U 1488) V Zw 164; "neutral post-eruptive split Sa, 2 sharp jets n, plumes np, sf" (Zw) --

01490   U 1490) P w 01 57.5 + 21 04 at 3.0, 287, 0.6 x 0.25 -- U 1493) comp 1.2, 20, 0.45 x 0.15; 2.5, 214, 0.4 x 0.3, dif; 3.8, 80, 0.3 x 0.05 -- U 1497) centr region 0.35 x 0.35, 2 B knotty arms; U 1499 at 4.7, 8 -- U 1498) comp 2.4, 248, 0.5 x 0.12 -- U 1499) bar?; see U 1497 --

01500   U 1500) 2nd of 3, comp 4.2, 299, 0.9 x 0.1, Sc; 4.2, 106, 0.7 x 0.2, early spir -- U 1501) comp 6.6, 160, 0.40 x 0.35, v high surf brightn -- U 1504) comp 2.8, 40, 0.6 x 0.2; U 1508 at 7.4, 156 -- U 1507) comp 0.7 x 0.25 superimp f center; P w U 1510 at 5.3, 41 -- U 1508) dif on blue PA map, v red, spir?; 1st of 3, comp 2.2, 151, 0.5 x 0.5; 2.3, 19, 0.9 x 0.5, S dm:; see U 1504 --

01510   U 1507) see U 1511 -- U 1511) 2nd of 3, 01 59.0 + 08 14 at 2.4, 238, 0.7 x 0.2, m=15.7; U 1513 at 1.5, 154 -- U 1513) U 1511 -- U 1514) B main body 0.35 x 0.30, eF disc or env; comp 3.5, 6, 0.7 x 0.3 -- U 1516) comp 2.7, 324, 0.3 x 0.2, vF; U 1508 at 8.6; U 1532 at 5.9 -- U 1517) comp 3.2, 209, 0.5 x 0.3, U 1506 at 12.7 -- U 1518) comp 1.4, 198, 0.3 x 0.2, sev much fainter objs near ; U 1519 at 6.8, 356 -- U 1519) see U 1518 --

01520   U 1522) P w 01 59.6 + 09 43 = I 1771 at 0.8, 153, 0.7 x 0.7, m=15.4; "N" (MCG) -- U 1523) comp 2.5, 247, 0.35 x 0.10 -- U 1524) distorted E?, app asym env; comp superimp at 0.18 sp center, 0.08 x 0.08 -- U 1525) P w comp at 1.8, 36, 0.8 x 0.6, E? -- U 1526) P w U 1527 at 1.8, 4 -- U 1527) see U 1526 -- U 1528) comp (double system) at 4.0, 153, 0.35 x 0.15 and 4.3, 161, 0.3 x 0.12, sep 0.6 -- U 1529) irr (rs) 0.9 x 0.8 w broad inner spir segments; amorph outer disc; classif uncertain --

01530   U 1530) see U 1516 -- U 1537) in a chain of 7 w U 1552 brightest --

01540   U 1541) V Zw 170; "S B c w elliptical compact comp at tip of spir arm sp, m_p=16.9" (Zw), comp at 0.70, 246, 0.15 x x 0.15; P w U 1550 at 9.2, 20 -- U 1543) comp 0.9, 127, 0.10 x 0.10, compact; 2.0, 221, 0.3 x 0.2, early -- U 1544) vF nucl cond, thin F broken arms; in chain of 7 w U 1552 brightest at 8.5 -- U 1547) DDO 17; Ir V or Pec (vdB); chain of blue patches 0.55 x 0.10 -- U 1549) 02 00.8 + 26 02 at 2.6, 133, 0.6 x 0.35, spir, m=14.9 --

01550   U 1550) see U 1541 -- U 1551) comp 3.1, 155, 0.8 x 0.6 -- U 1552) V Zw 172 ; brightest in chain of 7; U 1563 at 4.8, 118 -- U 1554) comp 3.5, 216, 0.3 x 0.1, F -- U 1555) VV 309b, Arp 290a; in Arp's class "double galaxies -

- wind effects"; VV δ = - 13 37 is incorrect; P w U 1556 at 2.1, 35 -- U 1556) VV 309a, Arp 290b, see U 1555; bar
in pos angle 188 w ext towards U 1555 (or vF cond at edge), smooth asym disc; comp 1.3, 17, 0.2 x 0.2, compact:,
interaction?; note error in VV pos -- U 1557) comp 1.8, 140, 0.4 x 0.3, F -- U 1558) vF S nucl cond; S red cond at
0.15 f center, * or comp?; asym spir pattern -- U 1559) S cond at 0.2 p center; sev blue cond in the outskirts --

01560   U 1560) P w 02 01.1+19 25 at 1.3, 233, 0.8 x 0.2, S B? b-c, m=15.5, interaction?; comp 4.7, 257, 0.5 x 0.4,
S B 0-a -- U 1561) V Zw 173; "irr blue post-eruptive, mottled, blue knots" (Zw) -- U 1562) VV 37, V Zw 175; 1st
and 2nd in group of 4; 02 01.4+45 32 at 1.5, 97 from geom center, 0.9 x 0.7, m=15.4, comp 0.2 x 0.2 attached
0.50 np; "4 intercon galaxies w compact neutral cores in row [lenght = 95" west-east]. No 1, S B c, m_p=16.2
(U 1562a=VV 37b). No 2, Sb, m_p=15.0 (U 1562b=VV 37a). No 3, spherical compact, m_p=17.1. No 4, Sc, m_p=15.4" (Zw)
U 1563) in chain of 7 w U 1552 brightest at 4.8, 298 -- U 1564) P w comp at 0.6 n, 0.8 x 0.1, prob interaction --
U 1567) comp 3.2, 358, 0.3 x 0.1, vF -- U 1569) P w comp at 1.2, 179, 0.7 x 0.6, S B a:; MCG δ for 6-5-81 and
6-5-82 shall be changed round --

01570   U 1573) P w comp at 2.3, 257, 0.8 x 0.4, Sc -- U 1574) 02 02.5+37 23 at 3.1, 101, 0.22 x 0.20, compact, m=15.7 --
U 1577) (r) 0.8 x 0.8; one double arm eastw from n side, hints of one e dif arm on s side, poss also double --
U 1579) comp 2.2, 168, 0.3 x 0.3, F, dif --

01580   U 1581) irr abs line along edge, late spir?, prob not regular; 02 02.5+34 42 at 2.4, 348, 0.5 x 0.25, S B ...; m=
= 15.5; comp 1.2, 222, 0.35 x 0.10 -- U 1583) E or S0 w comp superimp 0.25 f center, 0.15 x 0.12; "E? +*" (MCG);
brightest in group of 5 or 6, U 1584 at 4.6, 355, U 1593 at 9.3, 77; comp 2.1, 283, 0.15 x 0.12 + env? -- U 1584)
see U 1593 -- U 1585) S-shaped; broad massive arms; P w 02 02.6+44 58 at 2.8, 275, 0.5 x 0.4, m=15.6 -- U 1586)
measurements disturbed by sev superimp ** -- U 1587) v high surf brightn; comp 1.7, 283, 0.1 x 0.1, v blue --
U 1589) comp 4.8, 170, 0.5 x 0.4; sev eF obj in vic, members of cl 02 04.0+14 52, open, ED --

01590   U 1590) comp 2.8, 308, 0.5 x 0.4 -- U 1591) Prob disturbed by V Zw 182 at 0.7, 322, 0.4: x 0.3:, "red pear-shaped
compact, m_p=17.6" (Zw); U 1596 at 4.2, 68 -- U 1592) P w U 1594 at 5.8, 137 -- U 1593) U 1583 = N 810 at 9.3, 257 --
U 1594) see U 1592; comp 3.2, 57, 0.7 x 0.5; 3.9, 73, 0.35 x 0.30 -- U 1596) see U 1591 -- U 1598) strongly asym
late spir, no disturbing obj vis --

01600   U 1603) double nucl? or F * superimp sp center? -- U 1604) VI Zw 169; "pair of blue post-eruptive Sa and spherical
compact, sep = 10"" (Zw) -- U 1605) vB nucl region 0.30 x 0.18, F disc, no spir str -- U 1606) comp 4.6, 124, 0.6 x
x 0.4, dif --

01610   U 1610) comp 2.8, 175, 0.6 x 0.5; 4.6, 225, 0.6 x 0.6 -- U 1611) v high surf brightn, blue? -- U 1616) P w 01 59+
+ 85 08 at 3.8, 333, 0.9 x 0.5, S0?, m=15.7 -- U 1617) 02 04.9+01 52 at 2.2, 0.9 x 0.9, E-S0, m=15.5; U 1619
at 4.0; U 1620 at 2.5; U 1624 at 5.6; "(N)" (MCG) -- U 1618) b) at 0.6, 5, 0.55 x 0.30, spir; see U 1617, 02 04.9
+ 0152 at 4.9; U 1620 at 2.0; U 1624 at 4.0; "(N)" (MCG) prob V Zw 188; "pair of neutral spherical
compacts, sep = 25" east-west, imbedded in post-eruptive cloud formation, each m_p=17.8" (Zw); spir at 4.4, 194,
0.7 x 0.5 is prob V Zw 189, "Sc, neutral compact core, m_p=17.3" (Zw), note error in Zw coord --

01620   U 1620) see U 1617, 1618; 02 04.9+01 52 at 4.1; U 1624 at 3.5; "N" (MCG) -- U 1622) comp 4.2, 298, 0.30 x 0.15 --
U 1623) 02 04.7+1857 at 6.9, 282, 0.65 x 0.25, asym, B blue nucl cond in sp part -- U 1624) see U 1617, 1618,
1620; "(N)_n" (MCG) -- U 1625) dif spindle; comp 0.8, 295, 0.4 x 0.3; 1.2, 58, 0.3 x 0.3 -- U 1626) (rs) 0.7 x 0.5,
2 dif asym arms; comp 0.75, 169, 0.20 x 0.15, interaction; the arms appear to join close to the comp -- U 1627)
comp 1.1, 17, 0.25 x 0.20 and 1.1, 352, 0.45 x 0.2, Ir I, sep 0.45 --

01630   U 1630) B main body 0.7 x 0.35, bar?, vF ext -- U 1631) E+6 (de V); "vF arms?" (Morgan) -- U 1632) 02 05.2 +
29 01 = N 816 at 5.7, 284, 0.35 x 0.35, v high surf brightn, m=15.3; comp 2.4, 227, 0.4 x 0.1, F; 2.8, 295, 0.35 x
x 0.08, F; 4.9, 39, 0.8 x 0.5, v dif spir -- U 1633) SAB c; (de V) -- U 1634) (r) 0.7 x 0.7; fragments of spir str
in the outskirts -- U 1635) P w U 1636 at 4.4, 346 -- U 1636) see U 1635 --

01640   U 1642) 2nd of 3; U 1650 at 4.2, 45; comp 3.8, 261, 0.30 x 0.20 -- U 1643) comp 3.1, 219, 0.5 x 0.3, late spir; 3.2,
21, 0.5 x 0.3, E: -- U 1645) bar ecc in (r), v dif disc; sev galaxies near --

01650   U 1650) main body 0.8 x 0.10, thin vF disc; 3rd of 3, disturbed by U 1642? -- U 1651) V Zw 191.1; V Zw 191.2 at
0.70, 107, 0.15 x 0.15; "pair of compacts. No 1, post-eruptive twisted Sa... No 2, spherical neutral compact 40" s
of Sa, m_p=16.6" (Zw); 02 07.0+35 29 at 6.2 sf, 0.8 x 0.7, m=14.9; comp 2.7, 205, 0.7 x 0.6; 3.4, 139, 0.9 x 0.8,
S B ... -- U 1652) asym disc, one arm is straight, poss disturbed; vF S blue cond nf center, disturbing comp? --
U 1653) chaotic; curved tail sf; comp 0.8, 350 from center, 0.3 x 0.15, bridge?; 1.0, 90, 0.4 x 0.4, F; sev S obj
in vic -- U 1655) VI Zw 177; chaotic; arc-shaped abs markings close to center; "post-eruptive distorted Sa, B disc,
red and blue flare formations" (Zw) -- U 1656) 1st of 3, U 1661 at 2.5, 167; 02 07.6+41 17 at 5.7, 117, 0.9 x 0.25,
S0-a, m=15.4 -- U 1657) spindle, * superimp -- U 1659) broad irr approx 0.6 x 0.6 --

01660   U 1661) 2nd of 3, see U 1656, 02 07.6+41 17 at 4.5, 91 -- U 1664) (r) 0.7 x 0.4; comp 2.6, 182, 0.5 x 0.3, B nucl
region; 3.3, 243, 0.6 x 0.2, F; U 1663 at 12.2 -- U 1666) nucl region 0.12 x 0.12, vF corona; 3rd of 3, disturbed
by comp at 0.85, 185, 0.8 x 0.3, Irr w F nucl cond, or distorted spir; comp 2.5, 324, 0.6 x 0.5 -- U 1667) U 1668
at 3.7 n -- U 1668) V Zw 193; complex obj in smooth env; "cluster of sev compacts in luminous matrix" (Zw); see
U 1667; sev F galaxies near --

01670   U 1670) DDO 18; Ir V (vdB); tend to spir str -- U 1671) P w 02 08.0+32 30 at 1.9, 10, 0.9 x 0.15 -- U 1672) high
surf brightn, massive arms; see U 1673; U 1676 at 10.6 -- U 1673) ident as N 834 in MCG -- U 1675) slightly curved
elong main body, dif disc, sev B patches close to center, no regular bar spir; sev objs near , interacting pair at
2.0, 75 and 2.3, 62, 0.12 x 0.12 and 0.10 x 0.10 + envelopes; comp 3.8, 40, 0.5 x 0.4 -- U 1676) V Zw 194; B centr
region 0.60 x 0.30, 2 thin tightly coiled arms; "Sa pec, B blue elliptical disc, distinct blue spiral arms" (Zw);
see U 1672 -- U 1678) P w U 1680 at 4.5, 167 --

01680   U 1680) see U 1678 -- U 1681) P w comp at 3.3, 82, 0.7 x 0.5, late spir; comp 2.9, 140, 0.7: x 0.6:, v dif, dwarf --
U 1682) measurements disturbed by nearby * GC 2613, m=6.20 --   U 1683) U 1684 at 3.4, 356 (foreground?); 02 08.6
+ 15 40 at 1.7, 237, 0.5 x 0.3, E-S0, m=15.4; comp 2.5, 307, 0.6 x 0.15 -- U 1684) see U 1683; 02 08.6+15 40 at
4.4, 195; comp 2.5, 223 (same for U 1683) -- U 1687) "E" (MCG); P w 02 08.8+14 04 at 2.0, 274, 0.8 x 0.5, m=15.6,
this is VV 44a, VV 44b is an eF stellar obj at 0.3 sp a, connection by spir arm or bridge -- U 1688) comp 1.5,
134, 0.35 x 0.15 -- U 1689) P w U 1693 at 5.3, 7;

01690   U 1691) VI Zw 182; "pair of large Sa w blue disc and halo, plus blue spherical compact comp 13" p, m_p=18.0" (Zw) —
U 1692) comp 2.1, 327, 0.3 x 0.2 -- U 1693) see U 1689 -- U 1694) P w comp at 1.6, 199, 0.40 x 0.25, m approx 17. --
U 1695) comp 4.0, 40, 0.9 x 0.8, v dif, dwarf:; U 1676 at 12.4 -- U 1697) comp 0.7, 283, 0.12 x 0.10; 4.6, 276,
0.4 x 0.4 -- U 1698) (r) 0.6 x 0.5; eF (R) or spir arcs -- U 1699) "compact nucleus" (CGCG); α = 2 10.0, note
error in UGC α --

01700   U 1700) P w b) at 0.8, 130, 0.12 x 0.12; comp 2.0, 176, 0.3 x 0.2, early; 3.9, 220, 0.3 x 0.15, asym -- U 1701) "blu
nucleus in elliptical disc, surrounded by two large distinct circular halos" (Zw) -- U 1702) S v red nucl cond;
U 1689 at 10.6 -- U 1705) P w 02 10.4+ 09 15 at 1.3, 200, 0.4 x 0.3, compact, bridge?, vF jet southw, m=15.2;
one arm points tow comp, one dif arm northw-eastw -- U 1709) P w comp 1.5, 143, 0.8 x 0.5 --

01710   U 1713) P w U 1727 at 14., 102 -- U 1716) III Zw 43; "patchy blue v compact" (Zw); "elliptical core 3" x 5" = 0.8 x
x 1.4 kpc in irr F halo 9" = 2.5 kpc diam" (Sargent); v = 3430 (Zw), 4320 (Sargent) -- U 1717) comp 4.1, 280,

0.7 x 0.2, early; 6.6, 319, 0.6 x 0.1, late spir; 6.7, 310, 0.3 x 0.2, vB nucl region -- U 1718) comp 2.1, 168, 0.3 x 0.1, F --

01720　U 1720) B arc-shaped main body, streamers northw; P w comp at 2.5, 260, 0.9 x 0.3 -- U 1725) dif curved spindle, interacting w comp at 0.5 nf, 0.12 x 0.12, prob vF bridge -- U 1726) comp 2.5, 202, 0.3 x 0.1 -- U 1727) see U 1713 -- U 1729) U 1730 at 3.5, 120 --

01730　U 1730) vB bar 0.2 x 0.1; see U 1729 -- U 1736) SAB(rs)c (de V) -- U 1737) comp 4.9, 218, 0.4 x 0.14 -- U 1738) P w U 1743 at 3.3, 91 -- U 1739) wedge-shaped, broader northw --

01740　U 1741) P w U 1746 at 8.1, 29 -- U 1742) 1st of 3, comp 2.8, 107, 0.7 x 0.4, Irr; 3.8, 174, 0.6 x 0.5, Sc: -- U 1743) see U 1738 -- U 1744) Sb+ (Ho); e dif outside 1.1 x 0.7; P w I 1785 = 2 13.4 + 32 26 at 2.0, 59, 0.7 x 0.5, E or early spir, m = 15.6 -- U 1746) see U 1741 -- U 1747) asym, patchy, prob late; comp 3.0, 55 from center, 0.7 x x 0.4, S B c, blue --

01750　U 1750) P w comp at 1.3, 20, 0.8 x 0.10 -- U 1754) curved spindle or v open S B; interacting w comp at 1.5, 248, F bridge -- U 1759) S B(s)c: (de V); 2nd brightest in N 877 group; U 1768 = N 877 at 11.8; N 870 is a comp at 1.5, 192, 0.3 x 0.3, m approx 16.5 --

01760　U 1760) P w I 218 = 02 14.5 + 01 02 at 2.4, 14, 0.9 x 0.3, Sb:, m = 15.7, comp 0.3 x 0.1 f at 0.8 -- U 1761) U 1768 = = N 877 at 8.3; U 1759 at 4.2 -- U 1762) P w U 1764 at 2.5, 161 -- U 1764) see U 1762; comp 2.2, 104, 0.3 x 0.2, v dif; 4.8, 258, 0.3 x 0.15, double? -- U 1765) P w 02 14.8 + 35 33 at 3.4, 321, 0.55 x 0.45, spir, m = 15.2 -- U 1766) SA c: (de V); v red spindle, strong internal absorption, late Sb?; P w U 1768 at 2.1, 48 -- U 1767) chaotic; nucl cond sp center w smaller cond n, disturbing comp?; disrupted spir?; P w U 1772 at 5.3, 235 from geom center -- U 1768) SAB(rs)bc (de V); stellar cond 0.1 x 0.1 at tip of spir arm 0.7 n center; see U 1766; comp 1.8, 108, 0.4:: x x 0.4::, e dif, dwarf; 2.8, 316, 0.5 x 0.4, v dif, dwarf -- U 1769) comp 0.9, 40, 0.9: x 0.2, v dif, Irr; 3.0, 309, 0.4 x 0.2; sev fainter obj in vic --

01770　U 1772) blue vB main body 0.50 x 0.45, e dif outskirts, no evid of spir str; see U 1767 -- U 1773) brightest nf center; prob late -- U 1774) U 1775 at 6.0 n -- U 1775) Arp 10; elliptical (r) 0.80 x 0.55, brightest on np side, B exc nucl region, smooth disc w tend to asym spir arcs; "nucl off center in ring" (Arp) -- U 1779) P w comp at 0.9, 180, 0.30 x 0.15 --

01780　U 1781) sev comps, brightest at 2.5, 3.3, 4.2 and 4.5 -- U 1784) S-shaped, blue, v dif, alm invis on red PA map; sev obj in vic, U 1788 at 6.8 sf -- U 1786) P w comp at 2.2, 126, 0.8 x 0.25, * superimp? -- U 1787) asym main body 0.8 x 0.22, dif ext eastw, disturbed by F comp at 0.5, 122, 0.20 x 0.15; U 1793 at 5.5 --

01790　U 1790) S B 0?; comp 2.2, 133, 0.4 x 0.2, F; 3.9, 277, 0.4 x 0.1, vF -- U 1792) (r) 0.45 x 0.35 -- U 1793) see U 1787; 02 17.3 + 37 39 at 4.6 nf, 0.50 x 0.35, m = 15.1 -- U 1796) comp 3.5, 56, 0.7 x 0.4; 3.5, 181, 0.6 x 0.3 -- U 1797) P w comp at 0.9, 328, 0.4 x 0.4; sev vF obj in vic -- U 1798) P w comp at 1.2, 101, 0.45 x 0.40 -- U 1799) S-shaped, smooth arms, much broader and fainter outw; disturbed by comp at 1.3, 286, 0.45 x 0.12 --

01800　U 1801) 2 arms, one stronger; cond 0.12 x 0.12 at 0.20 sf center, prob * -- U 1804) blue amorph disc; nucl cond 0.15 x 0.12, double? -- U 1806) P w comp at 1.6, 139, 0.9:: x 0.5::, e dif, dwarf -- U 1809) b) superimp at 0.30, 90, 0.15 x 0.12 --

01810　U 1810) VV 323, Arp 273, V Zw 223; An 0218A in BG, SA(s)b pec (de V); smooth outer spir loop forms a closed ring, ecc inner spir loop, poss also closed, hints of fainter arms inside; "B long well def arms, but smooth, not patchy" (Arp); interacting w U 1813 sep 1.4 betw nucl regions; An 0218B in BG, S B(s)a pec (de V), v open bar spir, tend to F ring on s side; "pair of barred spirals in tidal interaction" (Zw) -- U 1814) comp appears attached to spir arm on sf side at 1.1, 160, 0.6 x 0.2, red, background?; major diam 2.2, comp excluded; cond 0.25 sp center, superimp comp?; comp 3.4, 2, 0.5 x 0.2 -- U 1817) chain of 5 galaxies p, dominating obj at 5.7, 236, 0.8 x 0.7, Sb: -- U 1818) comp at tip of spir arm at 0.18 p, 0.12 x 0.12; one straight arm sf -- U 1819) comp 6.0, 24, 0.9:: x 0.9::, dwarf; U 1814 at 11.7, 346 --

01820　U 1823) SAB(r)0-: (de V); sev F obj in vic, some inside the env -- U 1826) may be disturbed by comp at 0.35, 290, 0.1 x 0.1, F -- U 1828) P w U 1833 at 2.6, 116 --

01830　U 1830) V Zw 227; "v compact core, double ring halo" -- U 1831) SA(s)b? (de V), Sb+ (Ho) -- U 1833) see U 1828 -- U 1834) B knotty blue bar; P w b) at 1.1, 51, 0.8 x 0.6, blue S B, distorted, spir arcs or loop eastw -- U 1835) 2 broad patchy asym main arcs, hints of 2 fainter arms; comp 1.7, 117, 0.3 x 0.1 -- U 1836) e dif, blue diam may be overestimated (not smaller than 0.7 x 0.5); comp 2.2, 288, 0.3: x 0.3:, F -- U 1837) P w U 1841 at 2.7, 116 --

01840　U 1840) Arp 145, V Zw 229; main body 0.9 x 0.9 w smooth env; loop westw, 2 F cond, 2 ** superimp; similar to VV 117 = N 2444 + 2445; "post-eruptive distorted neutral Sc w compact core..., interconnected w spherical compact (m$_p$ = 17.8), 40" west of Sc core" (Zw) -- U 1841) V Zw 230; ext env; surrounded by many S galaxies; comp in the env at 0.40, 133, 0.15 x 0.15, incl in CGCG magn; see U 1837 -- U 1842) comp 1.9, 205, 0.30 x 0.20, early; 3.9, 142, 0.3 x 0.2, F -- U 1843) N 901 in MCG; sev obj in vic, brightest at 2.8, 8, 0.9 x 0.6, S0? -- U 1847) a) 0.8: x 0.2, prob spir; b) at 0.45 sf a), 0.3 x 0.2, bridge -- U 1849) early; P w U 1854 at 2.5, 137; comp 1.8, 185, 0.5 x 0.2, F--

01850　U 1852) P w comp at 1.4, 307, 0.7 x 0.4 -- U 1854) see U 1849; comp 1.6, 273, same as for U 1849 -- U 1859) comp 3.4, 218, 0.8 x 0.2 --

01860　U 1860) P w 02 21.9 + 25 18 at 1.8, 170, 0.5 x 0.5, spir, m = 15.7 -- U 1862) main body 1.1 x 0.8; 2 broad dif arms, alm dissolved in the disc; no nucl cond vis on blue PA map; v dif ext outside main body -- U 1865) DDO 19, S V: (vdB); MCG diam are incorrect -- U 1866) in a dense part of the cl, U 1875 at 4.0, 114; comp 1.5, 231, 0.40 x 0.20 -- U 1867) comp 3.7, 280, 0.4 x 0.1, F -- U 1868) 2 smooth arms w thin ridges; in a dense part of the cl, U 1872 at 3.5 --

01870　U 1870) v red; P w comp at 1.1, 7, 0.3 x 0.1, F v red -- U 1871) red, early, slightly asym; comp 3.6, 278, 0.5: x 0.4: -- U 1872) in a dense part of the cl, see U 1868 -- U 1873) B nucl region 0.15 x 0.15; ext vF env; "N, Da" (MCG) -- U 1875) in a dense part of the cl; surrounded by a ring of 6 or 7 galaxies; see U 1866; 02 22.6 + 41 33 at 4.1, 134, 0.9 x 0.8, nucl cond 0.20 x 0.20, very compact, m = 15.0 -- U 1877) tend to spir str? comp 2.2, 212, 0.5 x 0.12 -- U 1878) in a dense part of the cl -- U 1879) vS blue cond in outer arm sp center, the arm appears slightly disturbed; P w comp at 2.9, 350, 0.8 x 0.6, early S B? --

01880　U 1880) early; comp 3.0, 167, 0.6 x 0.2 -- U 1881) M 51 case, comp 0.70, 205, 0.12 x 0.12 at tip of dif arm -- U 1883) comp 2.5, 307, 0.4 x 0.1, dif, red -- U 1884) comp 5.0, 206, 0.8 x 0.12 -- U 1887) in a dense part of the cl; comp 2.1, 75, 0.35 x 0.10 -- U 1888) SAB(s)c (de V); 2 main arms; 2 longer vF secondary arms --

01890　U 1890) comp 3.6, 63, 0.35 x 0.10; 4.2, 237, 0.35 x 0.10 -- U 1891) P w 02 23.3 + 25 53 at 3.2, 16, 0.8 x 0.3, m = 15.1 -- U 1895) comp 2.7, 205, 0.3 x 0.3; 4.2, 265, 0.7 x 0.12 -- U 1898) comp 1.9, 183, 0.35 x 0.18 -- U 1899) B nucl region, S stell comp attached nf; v dif prob asym env, no indication of spir str --

01900　U 1900) vF outside 0.55 x 0.12; comp 2.5, 183, 0.6 x 0.2; 4.4, 290, 0.4 x 0.3 -- U 1902) comp 3.1, 72, 0.5: x 0.5:, v dif -- U 1903) SN 1964n -- U 1908) 3 B arms; P w comp at 1.7, 258, 0.4 x 0.2, high surf brightn; comp 0.9, 350, 0.15 x 0.12, appears compact -- U 1909) comp 4.5, 206, 0.45 x 0.15 --

01910　U 1910) comp 1.9, 295, 0.7: x 0.1, vF; 2.4, 96, 0.5: x 0.2:, vF; 4.0, 153, 0.3: x 0.2: -- U 1912) P w 02 24.1 + 20 18 at 3.1, 46, 0.9 x 0.7, late S B c, asym, m = 15.7; in a chain of 5 -- U 1913) SAB(s)d (de V), Sc + (Ho); note position

error in BG — U 1915) 2 thin asym main arms, one abruptly cut off, the other about 3 times longer; 2nd in a chain
of 3, 02 24.3+ 41 42 at 2.9, 193, 0.9 x 0.35, S B..., m=15.7; 02 24.4+41 47 at 1.9, 11, 0.45 x 0.40, E – S0?, m=
=15.0 —

01920   U 1920) 1st of 3, U 1943 at 9.5, 80; U 1956 at 13.9, 97 — U 1921) V Zw 245; S-shaped, strongly asym, one short
arm on f side, hints of one v dif arm pointing westw on p side; "blue post-eruptive galaxy w pair of compact cores,
sep = 5" northwest–southeast" (Zw); comp 3.1, 263, 0.3 x 0.2, F — U 1922) vF disc, spir arc on nf side; U 1932 at
4.4, 116 — U 1925) V Zw 246; high surf brightn regions 0.15 x 0.12 and 0.18 x 0.12, sep 0.20, plumes; "multiple
patchy compact (possibly two pairs of blue compact cores in luminous matrix), total m_p=15.5" (Zw); not in CGCG —
U 1929) F (r) 1.3 x 1.0, v dif outside, no evid of spir str; comp 3.1, 74, 0.4 x 0.2 —

01930   U 1930) S B(rs)0+ (de V), Sa (Ho); no spir str vis on PA maps; U 1954 at 12.4; U 1945 at 14.3 — U 1931) smooth alm
circ spir pattern; Reinmuth ident this galaxy as N 932 and a S red cond 0.15 x 0.15 at nf edge   (0.6, 55 from center
as N 930; NGC states that N 930 is 1′ sf N 932, which is impossible; 3rd in chain of 5 — U 1932) see U 1922 —
U 1933) asym spir pattern vis on red PA map; MCG diam for nucl region; sev obj in vic, nearest comp at 1.3, 126,
0.45 x 0.45, disturbing? — U 1934) comp 3.4, 274, 0.35 x 0.30 — U 1935) comp superimp at 0.35, 353, 0.2 x 0.1 —
U 1936) VV 238b, Arp 276; in contact w U 1937 at 1.0, 340; "both intersecting edges seem dimmed" (Arp) — U 1937)
VV 238a, see U 1936 — U 1939) v massive bar, v dif disc; comp 3.3, 249, 0.8: x 0.5, dif —

01940   U 1941) B nucl region 0.12 x 0.12, sym vF disc or env — U 1943) 2nd of 3, U 1956 at 6.6, 260; see U 1920 — U 1945)
see U 1830; U 1954 at 12.3 — U 1947) nr 4 in a chain of 5 — U 1948) VV 107; asym, ext eastw; stell comp 0.12 x 0.12
attached to f side; U 1972 at 4.7 sf —

01950   U 1951) E-type obj w comp in the halo at 0.30, 110, 0.12 x 0.12; comp (double system) at 4.1, 343, 0.9: x 0.7: and
4.2, 333, 0.7: x 0.5:, double nucleus?, sep 0.7 — U 1952) P w comp at 5.5, 123, 0.8 x 0.2, late spir, m approx 17. —
U 1953) F nucl region; strongly asym, dif loop or spir arc, disrupted?; P w comp at 5.6, 256, 0.5: x 0.5:, v dif,
dwarf — U 1954) SAB(rs)c (de V), Sc + (Ho); see U 1930, 1945 — U 1955) asym, vF outer arms; comp 3.4, 110, 0.5 x
x 0.20; 4.1, 127, 0.35 x 0.30 — U 1956) 3rd of 3, see U 1920, 1943 — U 1958) dif blue slightly curved spindle;
P w comp at 0.9, 299, 0.5 x 0.3, interaction — U 1959) P w comp at 0.9, 37, 0.3 x 0.2, v dif, dwarf: —

01960   U 1960) P w 02 27.3+38 09 at 1.7, 132, 0.8: x 0.5:, m=15.5 — U 1963) SN 1965k, Sa (Hoffmeister); 02 26.8+31 13
at 3.8 sf, 0.7 x 0.30, m=15.3 — U 1965) red spindle, prob strong internal absorption; nr 5 in chain of 5 —
U 1966) B slightly elong nucl region, F smooth arms, earlier than Sb?; brightest in chain of 4, 02 27.1+09 59 at
5.7, 76, 0.8 x 0.5, spir, m=15.5; comp 3.9, 83, 0.8 x 0.5, S B c; 6.5, 280, nucl region 0.15 x 0.15 + env approx
0.8 x 0.7 —

01970   U 1971) main body 0.45 x 0.35, sharp transition to F halo; 02 26.8 + 28 31 at 5.1 np, 0.6 x 0.6, m=15.4 — U 1972)
comp superimp at 0.20, 90, 0.10 x 0.10; see U 1948 — U 1977) comp 1.0, 98, 0.3 x 0.1 — U 1979) smooth disc, F arms?,
abs markings?; comp 1.8, 75, 0.4 x 0.2, F —

01980   U 1983) SAB(rs)b:? (de V); main body 1.1 x 0.5, smooth F outer disc, no evid of spir str; comp 3.5, 352, 0.35 x 0.30;
3.9, 6, 0.50 x 0.45 — U 1984) U 1987 at 2.6, 102; comp 3.6, 35, 0.9 x 0.2; 3.6, 81; 0.6 x 0.3 — U 1985) in group,
comp 2.8, 167, 0.8 x 0.5, spir; 3.3, 88, 0.5: x 0.3: — U 1986) Sb: (de V); comp 5.0, 216, 0.6 x 0.1, F — U 1987)
brightest in subgroup of 4 in 1997 group, see U 1984, comp 1.2, 355, 0.6 x 0.3 and 3.9, 336, 0.9 x 0.2, same as for
U 1984 — U 1989) U 1992 at 1.3, 28; comp 1.4, 300, 0.7 x 0.3, spir —

01990   U 1990) spindle 0.8 x 0.15 w streamer in pos angle 10; no disturbing obj vis — U 1992) invis on red PA map, non-
existing?; see U 1989 — U 1994) P w comp at 1.3, 160, 0.8 x 0.3; U 1995 at 6.0 — U 1995) see U 1994; U 2005 at
9.8 — U 1996) III Zw 48; "red S B b w sev cores on bar, v ext (d = 2′) elliptical blue halo" (Zw); comp 4.2, 189,
0.3 x 0.1 — U 1998) comp 1.5, 42, 0.4 x 0.3, v dif —

02000   U 2000) P w comp at 1.9, 87, 0.6 x 0.3, asym, B center, m approx 16.5 — U 2002) v patchy, irr, blue; comp 1.8,
115, 0.4 x 0.3, e dif, dwarf — U 2004) 02 29.1+00 41 at 4.1, 266, 0.9: x 0.9:; comp 0.45 f, 0.4 x 0.1, F — U 2005)
v ext F env; "N; L?" (MCG); 02 29.2+01 02 at 2.3, 292, (0.9 x 0.9), nucl region 0.18 x 0.18, very compact, m=15.4;
sev other obj in vic, brightest at 2.5, 205, 0.9 x 0.7, m approx 16. — U 2008) P w U 2011 at 3.0, 155 — U 2009)
P w U 2010 at 1.5, 340 —

02010   U 2010) e dif ext to approx 2.0 x 1.3; see U 2009 — U 2011) see U 2008; comp 2.4, 63, 0.7 x 0.08 — U 2012) comp
2.6, 332, 0.5 x 0.4 — U 2013) comp 4.2, 49, 0.50 x 0.15; 4.2, 258, 0.5 x 0.4 — U 2014) DDO 22; V (vdB) —

02020   U 2023) DDO 25; Ir V (vdB) — U 2028) comp 2.1, 238, 0.3 x 0.2, vF; 2.8, 244, 0.8: x 0.1, vF —

02030   U 2030) 02 30.5+09 28 at 6.1, 336, 0.7 x 0.5, Sa – b, m=15.5; U 2041 at 7.9; sev S obj in vic — U 2031) comp 3.0,
29, 0.5 x 0.5, early; 4.2, 74, 0.6 x 0.3, early — U 2032) comp 4.9, 290, 0.4 x 0.4, dif — U 2033) 3 arms, 2 start
from the eastern end of the bar — U 2034) DDO 24; Ir IV– V or Ir V (vdB); U 2038 at 4.9, 104; comp 4.3, 204, 0.35 x
x 0.20 — U 2036) comp 4.4, 66, 0.35 x 0.15 — U 2037) P w 02 30.9 + 06 19 at 3.7, 330, 0.6 x 0.4, m=15.2 — U 2038)
see U 2034; comp 3.1, 68, 0.5 x 0.4; 3.4, 343, 0.3 x 0.3 — U 2039) P w U 2049 at 3.9 f; 02 31.2+32 46=N 970 at
2.0 nf, double system, m=15.7 —

02040   U 2040) possible SN 1856 (Argelander, visual obs; interpreted as SN by Lundmark); P w 02 31.2 + 34 15 at 1.8, 113,
0.5: x 0.35, * superimp, m=15.6 — U 2041) see U 2030; 2nd brightest in N 975 group — U 2042) SA(rs)c: (de V)
comp 1.6, 25, 0.22 x 0.15 and 0.12 x 0.10, double system, sep 0.15; 4.5, 36, 0.9: x 0.2: — U 2043) P w U 2050 at
6.0, 155 — U 2045) I O (de V) — U 2047) P w U 2048 at 4.6 n; see U 2048; comp 1.5, 103, 0.9 x 0.10; 1.9, 128, 0.7: x
x 0.12 — U 2048) strongly asym spindle; brightest in subgroup, prob 6 members; see U 2047; comp 3.7, 25, 0.6 x 0.22
— U 2049) (rs) 1.6 x 0.8, eF disc w hints of 2 arms; see U 2039; 02 31.2 + 32 46=N 970 at 3.3, 0.6 x 0.2, double
system, m=15.7 —

02050   U 2050) see U 2043 — U 2053) DDO 26; Ir V (vdB) — U 2056) comp 1.8, 80, 0.5 x 0.4, F — U 2057) comp attached at
0.35, 135, 0.5 x 0.20 (incl in CGCG magn) — U 2058) 2 dif asym arms; P w 02 31.3+40 56 at 5.5, 290, 0.55 x 0.45,
pec, m=15.2 — U 2059) V Zw 257; "elliptical, compact core, extended halo, broad jet east" (Zw) —

02060   U 2060) comp 4.9, 264, 0.4 x 0.3 — U 2063) P w U 2066 at 3.6, 197; see U 2066 — U 2064) eF ext to approx 3. x 2. —
U 2065) VV 96; main body 0.8 x 0.7 w high surf brightn; P w U 2067 at 2.5, 35; U 2069 at 9.6; comp 2.2, 143, 0.3 x
x 0.2, e dif, dwarf: — U 2066) see U 2063; ident as N 982 in MCG; NGC and earlier catalogues list the southern
obj in the pair as p, but it is f; N 980–982 is GC 565–566 = Herschel III 572–573; UGC ident from NP distances in
NGC; U 2068 at 1.6, 17, satellite in pair 2063–2066; comp 2.1, 84, 0.5 x 0.4 — U 2067) see U 2065; U 2069 at 7.4 —
U 2068) v dif spindle, one of 2 satellites in pair U 2063–2066; see U 2063, 2066 — U 2069) SN 1961p, Sc (Wild);
see U 2065, 2067 —

02070   U 2070) comp 1.9, 99, 0.3 x 0.2, F — U 2073) asym; vB nucl region, * or comp superimp? — U 2074) comp 1.5, 288,
0.4 x 0.1; 4.8, 285, 0.3 x 0.3 — U 2075) comp 3.6, 130, 0.4 x 0.3 — U 2076) comp 2.0, 292, 0.5: x 0.2; 3.5, 42,
0.45 x 0.1 — U 2077) P w U 2090 at 5.5, 82 — U 2078) "(Nc)?; or Np; H" (MCG); a large nr of S and vS obj in vic —
U 2079) open, S-shaped, thin arms; comp 4.3, 196, 0.7 x 0.15 —

02080   U 2080) SAB(rs)cd (de V), Sc – (Ho) — U 2084) P w U 2094 at 3.9, 110 — U 2085) comp 2.8, 312, 0.5 x 0.4, dif —
U 2086) comp 2.6, 194, 0.3 x 0.3 — U 2088) P w comp at 1.4, 67, 0.4 x 0.1 —

02090   U 2090) see U 2077 — U 2091) comp 3.6, 205, 0.6 x 0.15 — U 2092) U 2102 = N 997 at 10.6; comp 1.9, 234, 0.4 x 0.2 — U 2094) see U 2084 —

02100   U 2101) slightly integral-shaped, prob disturbed by comp at 2.0, 19, 0.40 x 0.15; U 2108 at 5.9 — U 2102) b) superimp at 0.30, 350, 0.12 x 0.10, prob interaction; P w comp at 1.9, 15, 0.9 x 0.7, Sa:; m approx 16.; 02 34.6 + 07 13 = N 998 at 8.3 n, 0.6 x 0.5, m = 15.6 — U 2103) 02 34.7 + 20 56 at 2.6, 14, 0.20 x 0.20, extremely compact, m = 15.0, completely stellar on PA maps — U 2105) An 0235 in BG; S B a (de V); SN 1938a — U 2106) U 2119 at 8.6; comp 2.6, 75, 0.3 x 0.3 — U 2108) see U 2101 —

02110   U 2110) vF red cond at tip of spir arm 0.50 nf center — U 2112) in a dense part of the cl; comp 3.9, 83, 0.5 x 0.12; 4.9, 67, 0.4 x 0.3 — U 2114) in a dense part of the cl; 02 35.2 + 01 56 = N 1007 at 4.6, 350, 0.7 x 0.2, S0 – a; m = = 15.7 — U 2115) "N?" (MCG); U 2129 at 6.2, 100 — U 2117) 02 35.2 + 32 56 at 3.2, 0.9 x 0.4, spir:, m = 15.7 — U 2118) smooth; v strong abs along the edge, no spir pattern vis; in a loose cond of galaxies around the pair U 2123- 2127 = N 996-999, in vic; sev obj w m approx 17. - 19. in vic; 02 35.7 + 41 15 = N 1000 at 5.3, 141, (0.8 x 0.7), compact, halo, m = 15.6; chain of 4 galaxies betw U 2123 and 2127 — U 2119) see U 2106 —

02120   U 2121) in a dense part of the cl; comp 2.6, 179, 0.3 x 0.2; 3.3, 329, 0.3 x 0.1 — U 2123) P w U 2127 at 2.1, 47; see U 2118 — U 2125) F complex outer arm pattern, 4 arms start from the ring close to the ends of the bar — U 2126) comp 3.0, 224, 0.9 x 0.08, vF spindle; 3.5, 110, 0.3 x 0.3, vF — U 2127) see U 2118, 2123; 02 36.0 + 41 28 = N 1001 at 4.7, 89, 0.9 x 0.35, S0:, m = 14.7 — U 2128) in a dense part of the cl; comp 2.9, 310, 0.7 x 0.3; 4.9, 234, 0.5 x x 0.3 — U 2129) see U 2115; comp 1.7, 53, 0.3 x 0.1, F —

02130   U 2130) not error in MCG pos — U 2132) in a dense part of the cl; 02 35.8 + 01 45 at 3.7, 351, 0.6 x 0.45, m = 15.6, distorted, prob interacting w 2 S comp — U 2133) comp 2.2, 247, 0.3 x 0.3; 2.5, 279, 0.8: x 0.5: — U 2135) comp 1.5, 228, 0.4 x 0.05, F; 1.6, 104, 0.3 x 0.08, F — U 2137) SA(s)cd (de V), Sc+ (Ho); SN 1937d; main body 4.0 x 1.4; patchy ill def arms; comp 2.4, 205, 0.9 x 0.4, S B c; 2.7, 120, 0.6 x 0.4; 3.1, 80, 0.3 x 0.3; 4.1, 129, 0.5 x 0.4; 4.2, 332, 0.4: x 0.3:; v dif; 02 36.4 + 40 35 at 6.6, 150, 0.45 x 0.45, m = 15.7 — U 2138) comp 2.0, 233, 0.3 x 0.1 —

02140   U 2140) An 0236 in BG; VV 143, listed as Arp 254 in Arp's table II, but nr 254 is 15 19.9 – 07 15 (see table I); I: (de V); P w b) at 1.5, 142 from geometrical center, 0.9 x 0.2, Sa:; in group incl U 2153 and U 2170 — U 2141) high surf brightn, fragments of spir arms in the outskirts, large nucl bulge — U 2142) Arp 333; "thin circular arms, * in southeast superimp on wisp" (Arp); U 2149 at 6.9, 118; 02 36.9 + 10 38 = N 1028 at 6.1, 92, 0.9 x 0.5; S B:; m = 15.3 — U 2143) V Zw 266; "patchy blue disc and compact cores" (Zw); comp 4.9, 252, 0.5 x 0.08 — U 2145) comp in the env at 0.45, 160, 0.15 x 0.15 — U 2147) late Sa; comp 4.3, 313, 0.6 x 0.3 — U 2148) blue, vF blue cond s center; prob interacting w 02 36.7 + 12 28 = III Zw 51 at 1.2, 267, 0.12 x 0.12 + halo 0.6 x 0.5, "neutral spherical compact, F halo and jet east, m_p = 16.5. 70" west of S B c" (Zw), v_o = 13305 (Sargent) — U 2149) see U 2142, 02 36.9 + + 10 38 at 3.1, 0 —

02150   U 2152) P w U 2158 at 5.6, 121 — U 2153) wedge-shaped, poss alm edge-on spir, abs line slightly inclined to the major axis; comp 3.5, 296, 0.9 x 0.4, spir; 4.4, 63, 0.5 x 0.3, disrupted spir?; in group incl U 2140 and U 2170 — U 2154) Arp 135; S B(rs)0 – (de V), S0 (Ho); F dif comp attached f; "similar nebulosity about one diameter further east" (Arp); 8 dwarf members known in N 1023 group; 02 37.8 + 14 05 at 8.1, 125, (0.9 x 0.9), vB center 0.15 x 0.15; sev vS obj in vic — U 2158) see U 2152 —

02160   U 2160) S0-type obj but asym; B prob complex centr region — U 2162) DDO 27; Ir V (vdB) — U 2169) VI Zw 218; "compact elliptical disc, extended halo" (Zw) —

02170   U 2170) see U 2140, 2153 — U 2172) B complex centr region, asym, no evidence of spir str — U 2173) Sb (de V), Sb+ (Ho); U 2188 at 30. — U 2175) 02 39.1 + 42 08 at 3.4, 174, (0.9 x 0.6), nucl region 0.18 x 0.15, compact, m = 15.7 — U 2179) comp 2.4, 194, 0.3 x 0.2 —

02180   U 2180) prob earlier than Sc; ident by Jonckheere as a planetary nebula, see Perek-Kohoutek "Catalogue of Galactic Planetary Nebulae", Prague 1967, p 10 ; 02 39.8 + 32 28 at 3.5 np, 0.9 x 0.20, Sa:, spindle w large nucl bulge, m = = 15.7 — U 2183) v large nucl bulge, strong abs along the edge; eF ext env — U 2184) in a dense part of the cl; U 2191 at 4.8; 02 40.2 + 32 15 = N 1061 at 3.1 sf, 0.8 x 0.5, pec?, m = 15.2 — U 2185) P w comp at 3.3, 326, 0.8 x 0.5 — U 2186) comp 4.1, 59, 0.4 x 0.1; 4.6, 206, 0.3 x 0.15 — U 2187) comp 0.8, 354, 0.7 x 0.4; U 2194 at 6.2, 155 — U 2188) Arp 37; (R)SA(rs)b (de V), Sb – (Ho); brightest Seyfert galaxy; in Arp's class "low surf brightn comp", "S knot in arm" (Arp); see U 2173 — U 2189) main body 0.7 x 0.7; poss SAB(rs)b; comp 2.4, 321, 0.3 x 0.2 —

02190   U 2190) P w comp at 0.9, 262, 0.7: x 0.6:, E: — U 2191) in a dense part of the cl; 02 40.2 + 32 15 = N 1061 at 2.5 n, see U 2184; comp 3.0, 158, 0.4 x 0.4 — U 2193) SA(rs)c (de V), Sc – (Ho); SN 1961v, 19691; patchy main body, multiple arm pattern; hints of eF ext to approx 5. x 5.; "irr, thready, B features" (Morgan) — U 2194) 2nd brightest in group; see U 2187; comp 5.7, 66, 0.8 x 0.7, asym spir — U 2195) comp 3.6, 190, 0.6 x 0.3, E:, m approx 16. — U 2196) comp 2.0, 335, 0.55 x 0.08; 2.1, 69, 0.30 x 0.25 —

02200   U 2200) eF (R), approx 4.0 x 3.5 — U 2201) dif spindle; in a dense part of the cl; in subgroup of 3 w U 2203, 2204; comp 1.1, 268, 0.3 x 0.2 — U 2203) see U 2201; U 2201 at 2.2, U 2204 at 1.8 — U 2204) S-shaped; well def arms; see U 2201, 2203; U 2201 at 1.6; comp 2.3, 64, 0.5 x 0.5 — U 2207) P w comp at 0.9, 354, 0.5 x 0.15 —

02210   U 2210) S B(rs)c (de V), Sc+ (Ho); SN 19621; (rs) 2.5 x 2.0, inner diam 1.7 x 1.4; U 2188 = N 1068 at 1° 24 ' — U 2211) comp 1.0, 58, 0.4 x 0.4, E:; 2.1, 70, 0.8 x 0.7, S0 — U 2212) blue? massive elong main body, prob pec — U 2216) 02 41.5 + 00 34 at 7.4 np, 0.9 x 0.7, S B(s)b, m approx 16.5, SN 1961a — U 2217) curved spindle, no disturbing obj vis; comp 3.7, 69, 0.35 x 0.20, prob background —

02220   U 2222) asym, S0-type obj, P w U 2225 at 1.0, 130, interaction — U 2225) see U 2222 — U 2226) P w 02 42.3 + 41 57 at 0.8, 269, 0.35 x 0.20, spir, m = 15.6, disturbed; comp 2.5, 343, 0.6 x 0.1; U 2231 at 6.3 — U 2227) (r) 0.5 x 0.4, amorph outside; comp 4.0, 182, 0.8 x 0.8, S B c — U 2229) a) in the env at 0.35, 225, 0.22 x 0.20 —

02230   U 2230) b) attached at 0.50, 68, 0.7: x 0.5:, Sc, listed as f in MCG, which is incorrect; U 2230 is N 1077b in MCG — U 2231) blue; asym bar, brighter westw; see U 2226 — U 2232) ext transp env; a large nr of obj in vic — U 2233) comp 1.0, 69, 0.3 x 0.2, F — U 2234) P w U 2235 at 3.9, 27 — U 2235) v dif; see U 2234 — U 2237) comp 4.7, 267, 0.40 x 0.25; a vF obj 0.35 s center may be a disturbing comp — U 2239) comp 1.8, 292, 0.9: x 0.2:, e dif, dwarf —

02240   U 2241) asym outskirts, eF ext southw; stell obj 1.0 sf center, disturbing comp? — U 2243) comp 1.6, 265, 0.3 x 0.15 — U 2244) main body 0.7 x 0.6, e dif outside; a) at 0.45, 354, 0.9: x 0.20, interaction — U 2245) SAB(rs)c (de V), Sc – (Ho); P w U 2247 at 15.2, 8; comp 3.5, 45, 0.6 x 0.6 — U 2247) S B(rs)bc (de V), Sb + (Ho); see U 2245 —

02250   U 2251) P w comp at 3.4, 206, 0.9: x 0.8:, e dif, dwarf? — U 2252) 02 44.5 + 15 11 at 4.4, 118, 0.8 x 0.5, m = 15.6; comp 3.4, 303, 0.5 x 0.1, spir — U 2253) b) superimp at 0.25, 65, 0.10 x 0.10, m approx 16., a) appears disturbed — U 2258) B prob complex centr region 0.35 x 0.35 w dif asym ext; 2 thin knotty arms; comp 2.5, 211, 0.6 x 0.3; 4.6, 4, 0.3 x 0.2 — U 2259) S-shaped; broad irr arms, dissolved in the outskirts; P w 02 44.8 + 37 22 at 1.9, 10, 0.9 x 0.7, very compact, m = 15.6; SN 19631 —

02260   U 2261) P w U 2270 at 3.5, 143 — U 2262) SAB(s)ab (de V); P w 02 44.9 – 00 28 at 1.0, 10, 0.8 x 0.22, spir, m = 15.6; U 2245 at 20.; U 2247 at 13.4 — U 2265) comp 1.9, 175, 0.4 x 0.4, spir; 2.3, 110, 0.3 x 0.2; sev fainter obj in vic —

02270   U 2270) see U 2261 — U 2273) P w comp at 1.8, 203, 0.8 x 0.5, S0:, interaction? — U 2279) strongly curved spindle or integral-shaped system; vS blue cond in s part; comp 0.4, 115, 0.3 x 0.1, vF, prob interaction --

02280   U 2281) U 2294 at 6.0; U 2308 at 6.6; comp 1.5, 82, 0.5: x 0.5:, v dif, dwarf:; 2.1, 0, 0.30 x 0.20 -- U 2282) P w
02 45.8 + 14 03 at 3.3, 205, 0.9 x 0.6, early spir, m=15.5 -- U 2283) SAB(s)bc?; asym spir pattern, uneven distr
of light; no disturbing obj vis -- U 2285) F red cond superimp 0.40 f center, *? -- U 2288) sev vF comp; group of
3 S high surf brightn obj at 4.7 nf -- U 2289) comp 4.6, 61, 0.8 x 0.4 --

02290   U 2294) early; P w comp at 0.65, 193, 0.15 x 0.15, bridge; see U 2281 -- U 2295) comp 2.3, 30, 0.4 x 0.3, dif,
interacting pair? -- U 2296) F halo, plumes? -- U 2297) P w U 2300 at 2.6, 48; comp 1.5, 166, 0.4 x 0.10 -- U 2298)
V Zw 280 is a group of 3 spherical compacts around 02 46.4 + 46 47, m=18.1, 17.9 and 17.7 --

02300   U 2300) see U 2297 -- U 2302) DDO 29; V (vdB) -- U 2303) earlier than Sb? -- U 2306) P w 02 46.5 + 40 51 = I 258 at
0.9, 283, 0.7 x 0.6, E:, m=15.5 -- U 2307) comp (double system) at 3.7, 137, 0.3 x 0.2 and 4.7, 147, 0.6 x 0.3,
late spir -- U 2308) double nucl cond in dif env, S comp superimp at 0.40f, 0.55f and 0.50 sf geom center, 10 comp
inside 5.0; see U 2281 --

02310   U 2313) P w 02 46.6 + 41 16 at 5.4, 324, (0.8 x 0.6), E:; m=14.7, comp 0.50, 0.5 x 0.1 -- U 2314) comp 2.5, 87,
0.5 x 0.4 -- U 2316) comp 3.8, 259, (0.5 x 0.4), * superimp -- U 2317) comp 2.5, 216, 0.4 x 0.3; 3.3, 172, 0.5 x
x 0.3 --

02320   U 2320) VV 221, Arp 190; a) 0.7 x 0.6, b) 0.6 x 0.5, sep 0.40; stell obj at 0.30 p a), plume tow northwest 1.6,
"filament seems to originate from stell obj" (Arp); in group of 3 w 02 47.4 + 12 38 at 2.5, 193 from geom center,
(0.7 x 0.6), nucl region 0.18 x 0.18, compact, m=15.4 -- U 2322) v late S0 (or early Sa); comp 1.7, 349, 0.4 x 0.10;
3.7, 29, (0.8 x 0.8) -- U 2323) (r) 0.5 x 0.4; P w comp at 1.3, 243, 0.5 x 0.4; comp 3.8, 12, 0.7 x 0.2, double? --
U 2324) a) at 0.55, 283, 0.7 x 0.1, m approx 16.5 -- U 2325) comp 1.8, 285, 0.18 x 0.12 + halo; 3.4, 359, 0.5 x 0.1
-- U 2326) 02 47.7 + 13 03 = N 1115 at 4.7, 212, 0.6 x 0.4, early spir, m=15.6 -- U 2327) asym; blue spir arc north
and east; one strongly curved arm extends from nucl region; no disturbing obj vis -- U 2329) central obj in a
constellation of S F galaxies; P w comp at 1.7, 239, 0.8 x 0.12, m approx 17.; comp 4.0, 134, 0.5: x 0.4:, late spir
or Irr; 4.4, 312, 0.45 x 0.20, early; U 2339 at 8.5, 76 --

02330   U 2334) 2nd of 3; comp 0.8, 50, 0.8 x 0.2, distorted, tail eastw; 1.1, 298, 0.5 x 0.2 -- U 2336) comp 3.8, 295,
0.35 x 0.20 -- U 2337) 3 obj incl in CGCG magn, c) is a stell obj attached to nf side of a); sev obj in vic, comp
1.0, 255 (from geom center), 0.3 x 0.12; 1.4, 76, 0.3 x 0.2, F -- U 2339) see U 2329 --

02340   U 2341) comp 4.1, 351, 0.4 x 0.3, F -- U 2343) 02 49.4 - 01 02 at 8.1, 140, 0.8 x 0.8, S0 - a, m=14.9 -- U 2344)
U 2348 at 6.0, 26 -- U 2345) U 2359 at 15.9 -- U 2348) see U 2344 -- U 2349) P w U 2350 at 4.3, 355 --

02350   U 2350) see U 2349 -- U 2351) comp 4.9, 252, 0.6 x 0.3, E: -- U 2353) William Herschel's original observation
(Herschel II 601) listed as nr 1123 in NGC; see U 2354 -- U 2354) comp to U 2353 at 2.5, 202 -- U 2358) 02 50.9 +
+74 53 at 4.5, 148, 0.7 x 0.3, E or early spir, m=15.7 -- U 2359) double system 02 49.8 - 01 30 = N 1126 and
02 49.9 - 01 29 at 8.3 p and 6.9 p, 0.9 x 0.2, early spir and 0.9 x 0.3, spir, sep 1.5, m=15.4 and 15.7; comp 3.4,
345, 0.3 x 0.3, F; see U 2345 --

02360   U 2362) U 2365 at 7.1, 98; U 2368 at 14.4, 140; comp 4.7, 297, 0.8 x 0.4:, comp 0.25 nf -- U 2364) P w comp at
4.6, 150, 0.8 x 0.7, Sb - c; comp 1.4 f, 0.4 x 0.3, e dif, blue, dwarf -- U 2365) Arp 200, in Arp's class "material
ejected from nuclei", "splash appearance on western side of galaxy points to low surf brightn comp 7's" (Arp);
see U 2362; U 2368 at 10.3, 168 -- U 2366) sev obj in vic, 7 inside 5.0 -- U 2367) P w U 2375 at 5.6, 95 -- U 2368)
bar 1.7 x 0.6 in F ring; 2nd brightest in N 1134 group, see U 2365 -- U 2369) a) 0.45 x 0.40, b) 0.7 x 0.6, sep
0.35 --

02370   U 2370) P w 02 51.0 + 42 32 at 4.9, 340, 0.8 x 0.45, spir, m=14.7 -- U 2372) comp 2.6, 344, 0.3 x 0.3, dif --
U 2373) VV 85; v ext env; a large nr of obj in vic, about 10 inside 5.0; 02 51.2 + 41 25 = N 1130 at 1.7, 341, 0.40 x
x 0.35, compact, m=15.6; 02 51.4 + 41 22 = N 1131 at 1.8, 136, 0.35 x 0.30, very compact, m=15.6; 02 51.5 + 41 20 at
4.5, 134, (0.9 x 0.8), nucl cond 0.18 x 0.18, compact, m=14.8; 02 51.5 + 41 28 = I 265 at 5.5, 34, 0.40 x 0.35, E:,
m=15.7; MCG ident as N 1130, which is prob incorrect -- U 2375) see U 2367 -- U 2379) a) 0.6 x 0.3, b) 1.1 x 0.5,
spir, sep 0.50, bridge or connecting spir arm --

02380   U 2384) MCG ident as N 1145, which is incorrect -- U 2388) + U 2389) VV 331, Arp 118; a) 0.9 x 0.9, E, b) 0.65 x 0.50
at 0.65 sf a), spir w patchy loop; "arms and loops seem attracted to E galaxy" (Arp); U 2388, S0 (de V) --

02390   U 2391) eF outer disc, max diam 2.0? x 1.5?; comp 3.8, 310, 0.7 x 0.2, F; 4.7, 162, 0.4 x 0.3, F -- U 2394) v dif
spindle; in group of 5, brightest member at 4.3, 131, 0.7 x 0.3, S Bb: -- U 2395) sev e dif S obj in vic, dwarfs?;
comp 3.7, 218, 0.4 x 0.4, E: -- U 2396) U 2397 at 5.0 n; U 2401 at 4.0, 138 -- U 2397) 3 vF blue cond nf center,
dif streamers southw; see U 2396; U 2401 at 7.9 -- U 2398) P w comp at 4.6, 280, 0.7 x 0.3 --

02400   U 2400) v red B nucl region; prob S0 or early Sa; MCG ident as I 1868, but I 1868 is prob a double * at 2.3 nf;
"Nn; H" (MCG); comp 1.5, 160, 0.7: x 0.5:, dif; 2.6, 286, 0.5 x 0.3 -- U 2401) ext env; see U 2396, 2397 -- U 2403)
comp 1.2, 157, 0.4: x 0.2:, prob interaction, hints of bridge -- U 2409) P w comp at 4.9, 136, 0.9 x 0.7, S0: --

02410   U 2410) comp 3.5, 77, 0.5: x 0.3:, F -- U 2413) P w 02 53.7 + 15 43 at 1.0, 290, 0.35 x 0.35, E:, m=15.7; comp 3.5,
126, 0.5 x 0.3, E; 3.8, 37, 0.5 x 0.15, early -- U 2414) 02 53.8 + 04 25 at 5.6, 28, 0.6 x 0.5, S0 - a, m=15.2;
comp 0.8, 20, 0.3 x 0.2, vF -- U 2415) * superimp; comp 3.0, 33, 0.4 x 0.3, E or early spir; 3.5, 224, 0.4 x 0.4,
S B ... -- U 2419) comp 4.7, 323, 0.4 x 0.3, late spir --

02420   U 2420) in a dense part of the cl, about 10 obj inside 5.0; U 2413 at 6.6 -- U 2421) comp 2.0, 170, 0.5 x 0.5,
S B0?; 3.1, 154, 0.8: x 0.8:, dif -- U 2425) comp 4.4, 271, 0.6 x 0.5, spir, F; 4.5, 172, 0.9 x 0.15, F -- U 2428)
02 54.8 - 00 30 = N 1149 at 5.8, 150, 0.9 x 0.9, early spir, m=15.1 --

02430   U 2430) P w U 2433 at 1.6, 146 -- U 2433) 2 vF v red cond sf center, ** or disturbing comp?; see U 2430 -- U 2437)
02 55.5 + 10 55 at 5.8, 121, 0.6 x 0.4, vB nucl region, m=15.4, 1st of 2, comp 1.0, 148, 0.6 x 0.3, spir -- U 2438)
F red nucl region in F v ext corona; surrounded by a large nr of obj in cl 02 54.4 + 12 40, compact, distant --
U 2439) prob v early spir; * superimp at 0.25 sp center; U 2446 at 8.8 --

02440   U 2440) 02 55.3 - 02 32 = I 1870 at 6.9, 227, 2.9 x 1.8, I Bm, m approx 13.5 -- U 2441) P w comp at 2.5, 297, 0.7 x
x 0.5, S0:, m approx 16. -- U 2444) main body 0.7 x 0.4; sev obj in vic, nearest comp at 2.5, 27, 0.7 x 0.6, S0: --
U 2447) comp 1.9, 126, 0.5 x 0.1 -- U 2448) dif; blue nucl region; poss late distorted spir; F v blue cond np
center; P w a) at 0.9, 256, 0.45 x 0.18; comp 3.9, 6, 0.4 x 0.3 -- U 2449) B ecc main body 0.25 x 0.18; v dif env,
ext tow northwest --

02450   U 2450) F red nucl region in v ext F corona, similar to U 2438; in the center of cl 02 56.2 + 13 27, compact, distant;
a large nr of S galaxies in vic -- U 2454) comp 2.7, 197, 0.5 x 0.15; 2.8, 233, 0.5 x 0.3; 5.1, 112, 0.9: x 0.8:,
S dm: -- U 2455) I B(s)m (de V), Ir I (Ho); main body 1.2 x 0.6 --

02460   U 2460) comp 0.9, 10, 0.3 x 0.2, interaction? -- U 2461) P w U 2462 at 2.6, 142 -- U 2462) see U 2461 -- U 2465)
V Zw 308; "edge-on neutral Sc, compact core" (Zw) -- U 2466) comp 3.2, 164, 0.4 x 0.4 -- U 2468) v red; a large nr
of vF obj in vic -- U 2469) B prob complex nucl region, hints of spir arms or filaments; comp 2.7, 269, 0.9 x 0.6,
Sc; 02 57.5 + 05 36 at 6.2, 0.7 x 0.6, S0, m=15.3 --

02470   U 2470) P w comp at 0.35, 230, 0.12 x 0.10 + halo -- U 2471) P w U 2476 at 5.2 sf -- U 2473) 2 dif asym arms; e dif streamer in direction 330 to 2.0, prob an eF red obj at termination of streamer -- U 2474) P w U 2475 at 3.5, 357 -- U 2475) sev short patchy arms; see U 2474 -- U 2476) see U 2471 -- U 2477) comp 3.4, 258, 0.4: x 0.3: -- U 2478) comp 2.8, 279, 0.6: x 0.6: --

02480   U 2480) comp attached at 0.5, 295, 0.9 x 0.4: -- U 2484) comp 3.7, 26, 0.3 x 0.08; 4.8, 25, 0.55 x 0.12 -- U 2486) (r) 0.50 x 0.45; comp 3.1, 215, 0.35 x 0.30, early; 3.5, 156, 0.4: x 0.2:, v dif, Irr: -- U 2487) comp 2.9, 208, 0.6 x 0.15; 4.0, 130, 0.9 x 0.2 -- U 2489) V Zw 311, in cl 02 57.8 + 35 42, compact, MC; "group of 6 or more neutral compacts in luminous matrix plus several isolated compacts around... individual m_p's from 17.6 to 19.4" (Zw), see photo in CSCG page 102 --

02490   U 2490) comp 3.8, 332, 0.4 x 0.3, pec? -- U 2491) (r) 1.1 x 0.9, eF outside; P w U 2493 at 2.2, 44 -- U 2493) sickle-shaped; nucl cond in eastern part, comp in the west at 0.50, symbiotic pair; see U 2491 -- U 2494) comp 3.3, 78, 0.55 x 0.15; sev fainter obj in vic -- U 2496) dif spindle; comp 1.6, 38, 0.8 x 0.2; 2.3, 313, 0.9: x 0.7:, v dif, dwarf; 2.5, 58, 0.2 x 0.2; 3.1, 35, 0.5 x 0.1 -- U 2499) e dif; 2 vF cond n center; interacting pair w loop?

02500   U 2501) complex centr region, B blue cond sf center; multiple?; comp 1.5, 27, 0.4 x 0.4; 2.0, 5, 0.4 x 0.4 -- U 2503) SAB(r)b (de V); sev F thin arms; comp 3.2, 52, 0.7 x 0.3 -- U 2504) spir w curved plane; 2 spir arcs vis, dif ext northwestw; no disturbing obj vis -- U 2509) comp 3.9, 50, 0.4 x 0.3; 03 00.4 + 04 12 at 7.9, 0.7 x 0.4, double system, connecting loop, m = 15.3 --

02510   U 2510) comp 3.8, 21, 0.4: x 0.4:; 3.8, 344, 0.5: x 0.3: -- U 2514) strong abs on p side; dif streamer (vis in red) to comp at 2.7, 312, 0.4 x 0.4, pec?, m approx 16.; P w U 2517 at 7.9, 132 -- U 2515) prob later than Sa; P w 03 01.4 + 42 10 = N 1177 = I 281 at 1.7, 33, (0.7 x 0.7), E, m = 15.7; comp 5.0, 132, 0.5: x 0.2:; U 2531 at 18.1; SA(r)0+ (de V) -- U 2517) see U 2514; comp 4.6, 274, 0.5 x 0.5, late spir, dif -- U 2518) comp 4.7, 179, 0.8: x x 0.8:; dwarf irregular, m approx 18. --

02520   U 2527) asym; comp 0.8, 70, 0.1 x 0.1, interaction?; 1.4, 259, 0.4 x 0.2 -- U 2528) comp 3.7, 249, 0.4 x 0.1, F -- U 2529) v open blue spir; P w 03 02.9 - 00 36 at 1.5, 168, (0.9 x 0.6), E, m = 14.9, interaction --

02530   U 2531) F stell nucl cond; a large nr of v thin arms; V Zw 319 superimp at 0.7, 244, 0.15 x 0.15 in blue, 0.7 x 0.6 in red; m = 17.5 ; see U 2515; U 2544 at 15.3 -- U 2534) U 2536 at 4.4, 72 -- U 2536) see U 2534 --

02540   U 2540) smooth spindle w vB centr region; comp 4.1, 333, 0.5 x 0.2 -- U 2544) early; comp 1.5, 205, 0.4 x 0.2, F; 2.5, 246, double system, 0.1 x 0.1 and 0.08 x 0.08, sep 0.15; see U 2531 -- U 2545) v smooth transp disc, no spir pattern, (r) 1.0 x 0.8; comp 2.1, 120, 0.3 x 0.2 -- U 2548) 03 04.5 + 38 10 at 5.7, 263, 0.8 x 0.5, S B ..., m = 14.9 --

02550   U 2550) comp 4.3, 190, 0.5 x 0.3 -- U 2552) may be interacting w comp at 0.65, 323, 0.3 x 0.3; comp 2.3, 282, 0.9 x x 0.7, S0 - a -- U 2553) asym; 2 arms, one vF vS red cond at the tip of each arm -- U 2554) comp 1.5, 80, 0.3 x 0.2, vF; 4.3, 11, 0.5: x 0.5:, e dif, dwarf -- U 2557) comp 2.5, 59, 0.5 x 0.4, dif --

02560   U 2560) in CGCG is U 2560 = I 1884 ident as I 290; IC coord support this ident, but IC descr "eeF, S, R" suggests that U 2560 is I 290; U 2561 at 4.9, 181; measurements disturbed by nearby * GC 3733 = β Per -- U 2561) see U 2560 -- U 2562) U 2564 at 4.0, U 2568 at 7.0 -- U 2563) comp 1.6, 197, 0.4 x 0.3 -- U 2564) see U 2562; U 2568 at 3.0 -- U 2565) P w comp at 1.2, 90, 0.7 x 0.6 -- U 2568) see U 2562, 2564; comp 4.6, 296, (0.5 x 0.4), E: --

02570   U 2571) comp 3.5, 209, 0.4 x 0.2, vF -- U 2572) inside the "ghost" from * GC 3733 = β Per -- U 2574) ident as I 294 in MCG and as I 296 in CGCG, but these IC obj are identical; U 2567 at 12.9 -- U 2575) brightest in a subgroup, U 2577 at 3.7, 172, U 2579 at 6.1; comp 3.5, 76, 0.4 x 0.1, vF, * superimp; 4.1, 50, 0.5 x 0.3 -- U 2577) see U 2575 -- U 2578) comp 1.5, 0, 0.4: x 0.3:, F -- U 2579) tend to disc, early S0?; see U 2575; comp 2.0, 214, 0.5 x x 0.3 --

02580   U 2580) P w comp at 1.0 sp, 0.18 x 0.12, early -- U 2583) comp 5.2, 94, 0.5 x 0.3 -- U 2586) B centr region 0.35 x x 0.25; main body 1.3 x 0.5, e dif outside --

02590   U 2590) B core 0.18 x 0.15, eF corona; comp 1.5, 145, 0.5 x 0.1, interacting pair; 3.0, 235, 0.5 x 0.2 -- U 2592) comp 1.9, 140, 0.4 x 0.1, F -- U 2596) comp 2.2, 14, 0.4: x 0.3:, e dif; 3.7, 90, 0.3 x 0.2, F; 4.3, 12, 0.5 x 0.5 -- U 2598) "neutral spherical compact, extended plumes south by east and north by west" (Zw) -- U 2599) comp 0.8, 27, 0.3 x 0.2; 3.5, 137, 0.35 x 0.25 --

02600   U 2600) B red arc across nucl region, prob no bar; (r)?; comp 0.7 f, 0.4 x 0.1, background -- U 2602) P w comp at 3.1, 34, 0.8 x 0.6 -- U 2603) comp 4.6, 176, 0.4 x 0.3 -- U 2608) bar 0.45 x 0.20, 2 asym arms form loops, one dif; P w U 2612 at 4.1, 145; comp 3.3, 221, 0.5 x 0.3 -- U 2609) P w 03 11.9 + 37 41 = I 305 at 1.4, 161, (0.9 x 0.6), nucl region 0.20 x 0.18, compact, m = 15.7, comp at 0.40, 0.3 x 0.12 --

02610   U 2612) see U 2608 -- U 2613) comp 4.5, 296, 0.6 x 0.5, v dif; sev eF comp -- U 2617) P w 03 12.8 + 40 38 = I 309 at 5.1, 168, 0.9 x 0.9, S0, m = 14.9 -- U 2618) comp 1.3, 43, 0.3 x 0.3 -- U 2619) comp 2.0, 92, 0.4: x 0.3:; 4.0, 302, 0.4: x 0.4: --

02620   U 2624) Perseus cl; tend to spir str; U 2626 at 3.7, 59 -- U 2625) prob edge-on; broad abs line in pos angle 113; F globular halo; may be similar to N 4594 = M 104; comp 3.8, 346, 0.8 x 0.4, S B b - c -- U 2626) Perseus cl; see U 2624; N 1259 at 3.7, 62, (0.9 x 0.9), E?, m approx 16. -- U 2627) P w U 2629 at 1.9, 60, interaction -- U 2629) see U 2627 --

02630   U 2634) Perseus cl; ident as N 1259 in MCG, but N 1259 is prob a galaxy at 2.2 sp, see U 2626; 5 comp inside 5.0; may be interacting w comp at 1.0, 68, 0.7 x 0.3 -- U 2637) comp 1.6, 311, 0.4 x 0.2, vF; 3.8, 324, 0.9: x 0.4:, F -- U 2639) Perseus cl; comp 2.0, 202, 0.3: x 0.3:; v dif; 4.3, 161, 0.5 x 0.3 --

02640   U 2640) P w U 2655 at 10.4, 112 -- U 2641) in chain of 5; 03 14.7 + 03 26 at 1.4, 30, 0.45 x 0.25, E - S0, m = 15.2 -- U 2642) comp 1.8, 193, 0.5: x 0.4:, F -- U 2643) Perseus cl; comp 1.4, 196, 0.4 x 0.3; sev other comp inside 5.0 -- U 2644) Perseus cl; P w comp at 1.9, 128, 0.6 x 0.22, E - S0 -- U 2645) interacting w comp at 0.50, 60, 0.25 x 0.12, bridge?; comp 1.8, 304, 0.7 x 0.12, pec? -- U 2646) SN 1964b --

02650   U 2650) 03 15.1 - 01 54 at 1.6, 167, 0.6 x 0.25, m = 15.6; comp 2.5, 86, 0.55 x 0.15, m approx 16. -- U 2651) Perseus cl; comp 2.7, 246, 0.7 x 0.4, dif; 4.2, 308, (0.5 x 0.4) -- U 2652) asym; thin vB arms, alm circular -- U 2654) 3 cond in common env, disrupted multiple system?; 2 comp p at 0.5 from p edge , 0.3 x 0.3 and 0.25 x 0.25 -- U 2655) 2 F uneven arms; see U 2640 -- U 2656) Perseus cl; 4 obj inside 5.0 -- U 2657) Perseus cl; * superimp close to center; U 2658 at 1.3, 2; U 2660 at 2.5, 87 -- U 2658) Perseus cl, in a cond around U 2669 = N 1275; see U 2657; 03 15.3 + 41 17 at 1.9, 247, 0.9 x 0.3, early, m = 15.7; U 2660 at 2.6, 115 --

02660   U 2660) Perseus cl, see U 2657, 2658; U 2669 = N 1275 at 9.6; E3 (de V) -- U 2662) Perseus cl; v ext env; U 2669 = = N 1275 at 5.1; S0 (de V) -- U 2663) comp 2.2, 278, 0.3 x 0.15; 2.5, 40, 0.3 x 0.2 -- U 2665) Perseus cl; U 2669 = = N 1275 at 8.4 -- U 2666) comp 2.2, 138, 0.6: x 0.4: -- U 2667) disturbed by comp at 0.55, 78, 0.10 x 0.10 -- U 2668) measurements disturbed by nearby *; dif disc, no str; F nucl cond; in group of 4 in Perseus cl s U 2669 = = N 1275; comp 2.7, 128, 0.8: x 0.5:, dif; 3.8, 273, 0.9 x 0.6; 4th member at 5.9 -- U 2669) brightest in Perseus cl, centr obj in a dense cond of galaxies; chaotic, appears early; Pec (de V); "F irr B filaments. Brightest in nest of around 12 k systems" (Morgan); radio source Perseus A; Seyfert galaxy, only non-spiral obj among classical Seyferts --

02670   U 2670) Perseus cl; E1 (de V); U 2669 = N 1275 at 3.3; P w 03 16.6 + 41 23 = N 1277 at 0.8, 321, 0.6 x 0.22, early
spir?, distorted, m = 14.9; sev fainter obj in vic; V Zw 339 at 03 16.7 + 41 23 (1'n), "red spherical compact, m_p =
= 16.3. Prob member of cl 03 03.0 + 41 25 (Perseus cl)" (Zw) -- U 2673) Perseus cl; U 2669 = N 1275 at 16.0 --
U 2675) Perseus cl; P w U 2676 at 2.0, 18, 2 smaller galaxies are comp to this pair; E3 (de V) -- U 2676) Perseus
cl; see U 2675 --

02680   U 2682) Perseus cl; * superimp close to center; P w U 2688 at 4.7, 62 -- U 2683) U 2687 at 2.7 -- U 2687) ident as
N 1298 in MCG; comp 0.6, 165, 0.25 x 0.20; see U 2683 -- U 2688) B core 0.20 x 0.18 in smooth F disc, tend to (R);
b) superimp at 0.20 s a), 0.5 x 0.1, this is MCG 7-7-74; see U 2682 -- U 2689) vB core, F disc or env; U 2686 at
5.0 --

02690   U 2690) brightest in group of 7; patchy disc, ill def spir pattern, but prob earlier than Sc; 2 S comp superim s
center, interaction? -- U 2691) comp 1.8, 290, 0.25 x 0.20; 2.0, 124, 0.20 x 0.18; 2.4, 343, 0.25 x 0.22 and 2.6,
347, 0.5 x 0.18, double system, sep 0.30 -- U 2692) asym spir pattern, one arm B and well def; alm dissolved
spir pattern on p side; ** superimp -- U 2693) B core, smooth (R), S0?; comp 2.5, 250, 0.3 x 0.2, F -- U 2694)
Perseus cl; B core in vF env; P w 03 18.4 + 41 13 = N 1293 at 2.1, 342, (0.9 x 0.9), m = 15.0, E3 (de V); E (de V) --
U 2696) spindle w high surf brightn, no nucl bulge -- U 2698) comp 3.3, 198, 0.3 x 0.1 --

02700   U 2700) MCG δ = 41 21 is incorrect -- U 2702) S B core 0.22 x 0.22; smooth vF v ext corona, no clear edge; hints of
ring around the core, illusory?; U 2706 at 11.8; U 2707 at 9.2 -- U 2703) P w 03 19.8 + 14 47 at 2.8, 7, 0.35 x
x 0.35, compact, m = 15.7; 5 or 6 galaxies in the group -- U 2706) ecc F nucl region, hints of spir str, may be a
late spir; see U 2702, 2706 -- U 2707) dif, S-shaped; see U 2702, 2706 -- U 2708) comp 1.5, 12, 0.5 x 0.2;
3.0, 305, 0.3: x 0.3:, v dif --

02710   U 2710) comp 3.4, 31, 0.3 x 0.1, v dif -- U 2715) U 2717 at 6.4; comp 3.5, 65, 0.4: x 0.1, * superimp -- U 2716)
asym, blue; spir arc or streamer eastw; comp 1.7, 280, 0.4 x 0.3, eF red comp ssf at 0.7 -- U 2717) see U 2715;
comp 4.2, 270, 0.3 x 0.3; 4.6, 165, 0.6: x 0.4:, dif -- U 2719) comp 3.9, 177, 0.9: x 0.1, vF; 4.5, 304, 0.5: x
x 0.4:, vF --

02720   U 2723) comp 2.6, 27, (0.7 x 0.3), core 0.12 x 0.12; 4.7, 315, 0.5: x 0.5:, v dif -- U 2724) measurements disturbed
by nearby *; U 2730 at 5.3, 107 -- U 2725) P w U 2733 at 6.3 f; comp 2.5, 85, 0.6 x 0.2, spir -- U 2726) F spir w
comp at 0.8, 103, 0.3 x 0.2, vF, connected by spir arm (distorted M 51 case) -- U 2728) P w U 2731 at 3.8, 126 --
U 2729) strongly obscured by galactic abs, alm invis in blue --

02730   U 2730) see U 2724; U 2732 at 2.8, 210 -- U 2731) see U 2728; comp 3.6, 157, 0.3 x 0.3; 4.0, 104, 0.25 x 0.20 --
U 2732) (rs) 0.6 x 0.5, smooth disc, may be earlier than S B b; see U 2730 -- U 2733) see U 2725; comp 3.8, 263,
0.6 x 0.2, spir -- U 2735) bridge to comp at 0.9, 79, 0.3 x 0.1; similar to Arp 98 or Arp 99 -- U 2736) comp 1.8,
15, 0.5 x 0.2, dif -- U 2737) 2nd brightest of 4; comp 1.7, 147, 0.8 x 0.10; 3.3, 223, 0.5 x 0.12; 03 24.3 + 46 47
at 7.0, 105, 0.7 x 0.7, Sc:, m = 15.5 -- U 2738) v dif; comp 4.6, 350, 0.5 x 0.20, S0: --

02740   U 2743) comp 2.7, 148, 0.55 x 0.18; U 2749 at 8.4 -- U 2744) P w U 2748 at 9.6 sf -- U 2746) comp 2.7, 309, 0.35 x
x 0.1, vF; 03 25.3 + 36 31 at 5.5 s, 0.7 x 0.2 and 0.25 x 0.15, sep 0.20, total m = 15.5 -- U 2747) * or S comp
superimp nucl region; comp 3.0, 94, 0.7 x 0.2, v dif; 3.4, 264, 0.7 x 0.3, early -- U 2748) see U 2744; sev S F
obj in vic -- U 2749) asym, blue, prob v late spir; see U 2743; U 2751 at 4.7 --

02750   U 2751) smooth vF disc, stell core 0.12 x 0.12; see U 2749 -- U 2752) P w U 2756 at 6.7 nf -- U 2755) stell nucl region
0.2 x 0.2, vF ext corona; b) at 0.40, 25, 0.15 x 0.12; comp 3.1, 47, 0.30 x 0.25; 3.9, 247, 0.7 x 0.7, S B c?, dif;
6.4, 65, 0.8 x 0.8, S B c -- U 2756) V Zw 348; "interconnected pair of Sc's, w blue (s) and red (n) elliptical cores"
(Zw); sep 0.45 -- U 2759) red asym main body, spir arc on f side; curved chain of 6 ** superimp; no disturbing obj
vis --

02760   U 2760) comp 2.9, 8, 0.5 x 0.5, v red -- U 2765) strongly obscured by galactic abs -- MCG 7 - 8 - 22 at 03 27.9 + 43 44
is intragalactic, presumably a planetary nebula; diam 1.1 x 1.0 -- U 2769) P w comp at 5.1, 119, 0.9 x 0.7, d - m --

02770   U 2771) comp 2.4, 184, 0.8: x 0.6:, v dif, dwarf; 03 27.9 + 39 28 at 7.3, 0.9 x 0.7, S0, m = 15.4; comp 8.7, 130,
0.9 x 0.9, S0 -- U 2773) core 0.15 x 0.15 w comp 0.15 x 0.15 attached to p side; irr disc, no evid of spir pattern --
U 2775) classif uncertain; comp 4.8, 330, 0.4: x 0.4:, v dif -- U 2778) comp 2.4, 28, (0.7 x 0.7); U 2780 at 5.1 --
U 2779) 03 29.5 + 00 06 = I 329 at 4.8, 200, 0.6 x 0.5, S0:, m = 15.1; 03 29.8 + 00 07 = I 331 at 5.2, 147, (0.9 x 0.9),
E:, m = 14.9 --

02780   U 2780) vF nucl region in ext vF "disc", see U 2778 -- U 2782) P w comp at 1.6, 250, 0.9 x 0.9, (R')SAB 0/a, m
approx 16.5 -- U 2783) comp 2.9, 336, 0.5 x 0.05, eF; 3.9, 45, 0.6: x 0.2:, eF -- U 2784) 03 31.0 + 39 26 at 2.9,
359, 0.5 x 0.3, pec:, m = 15.0 -- U 2785) P w comp at 1.5, 237, 0.18 x 0.18, appears compact -- U 2786) S-shaped,
no nucl cond; 2 well def but vF arms -- U 2787) U 2790 at 5.4; U 2791 at 9.6 -- U 2788) v ext transp env, comp
0.18 x 0.15 superimposed close to center --

02790   U 2790) see U 2787; U 2791 at 5.4 -- U 2791) see U 2787, 2790; comp 4.2, 18, 0.4 x 0.2 -- U 2792) VII Zw 8; "heavy
ringshaped nucleus" (CGCG); "B blue neutral ring-shaped core, large elliptical halo w plume northeast" (Zw) --
U 2798) in group of 3 w U 2805 at 7.9 and 03 36.3 + 40 55 at 17.5, S B(r)0, m = 15.1 --

02800   U 2801) comp 2.2, 12, 0.3 x 0.1, F; 3.5, 147, 0.4 x 0.4, dif -- U 2802) P w 03 35.1 + 04 48 at 2.4, 277, 0.9: x 0.9:,
spir, m = 15.7 -- U 2805) 03 36.3 + 40 55 at 10.2 -- U 2806) P w comp at 5.2, 70, 0.4 x 0.2 -- U 2808)
comp 4.2, 62, 0.3: x 0.3:, e dif --

02810   U 2812) P w U 2814 at 3.8, 47 -- U 2814) see U 2812 -- U 2817) strongly obscured by galactic abs; comp 1.8, 103,
0.3 x 0.2, F; 2.6, 231, 0.3 x 0.1, vF; 4.5, 99, 0.3: x 0.3:, vF -- U 2818) S-shaped, asym, ecc nucl region; U 2828
at 9.5 -- U 2819) comp 4.5, 114, 0.5 x 0.1 --

02820   U 2821) III Zw 55; "Post-eruptive pair of galaxies. N 1409, red, patchy, rectangular extended halo, m_p = 15.4 .
N 1410, 10" north by east of N 1409, blue elliptical, m_p = 15.4". (Zw); N 1409 Seyfert galaxy? (Sargent) -- U 2824)
B complex nucl region, asym dif env, no clear edge , abs lane from center eastw -- U 2825) amorph disc, no spir
pattern vis; * or comp superimp at 0.18 sp center -- U 2828) sharp (rs) 0.35 x 0.25 cont in 2 open arms, much broader
outw; comp 1.4, 222, 0.3 x 0.1; see U 2818 --

02830   U 2831) P w comp at 1.0, 309, 0.4 x 0.3, early -- U 2835) P w comp at 3.3, 149, 0.9: x 0.1, Sc:, m approx 18. --
U 2836) vB core; in group w comp at 3.0, 84, 0.7 x 0.35 and U 2841 at 4.8, 65 -- U 2837) 2 thin ill def arms, (rs)
1.1 x 0.7; comp 3.5, 259, 0.6: x 0.3:, F -- U 2839) P w comp at 1.4, 348, 0.7 x 0.2, early --

02840   U 2841) see U 2836 -- U 2845) comp 3.2, 338, 0.5 x 0.3 -- U 2847) SAB(rs)cd (de V) --

02850   U 2851) P w comp at 2.1, 6, 0.6 x 0.5, Sc: -- U 2855) MCG δ = 69 49 is incorrect; P w U 2866 at 9.7 sf -- U 2859)
comp 3.8, 167, 0.6 x 0.3, F --

02860   U 2861) (r) 0.6 x 0.4, e dif outside; comp 1.5, 336, 0.12 x 0.12 + halo -- U 2862) P w U 2871 at 18., 113; comp 2.5,
350, 0.3 x 0.2, early -- U 2863) P w U 2867 at 5.8, 134 -- U 2866) early spir?; see U 2855 -- U 2867) see U 2863;
comp 4.3, 147, 0.9:: x 0.8::, v dif, dwarf -- U 2868) comp 1.5, 354, 0.5: x 0.3:, vF; 4.5, 282, 0.5: x 0.4:, vF --
U 2869) in group of 3 w U 2870 at 2.0, 134 and comp at 3.3, 143, 0.4 x 0.2 --

02870   U 2870) see U 2869, 3rd group member at 1.3, 155, 0.4 x 0.2 -- U 2871) prob SA(rs)b, (rs) or (r) 0.35 x 0.35; see
U 2862; comp 1.0, 54, 0.4 x 0.2, F -- U 2872) P w 03 46.5+01 00 at 2.3, 103, 0.8 x 0.4, pec, double?, m = 14.8 --
U 2873) P w U 2874 at 9.3, 167 -- U 2874) see U 2875) comp 4.3, 172, 0.7: x 0.08, eF --

02880   U 2881) comp 4.6, 343, 0.7 x 0.3, F -- U 2883) S red core in smooth disc or env -- U 2885) comp 2.4, 249, 0.55 x
x 0.10; 4.4, 335, 0.3 x 0.2, v dif -- U 2886) P w U 2887 at 1.0, 130 -- U 2887) see U 2886 -- U 2888) P w 03 50.1+
+ 32 09 at 2.0, 198, 0.45 x 0.30, m = 15.7; in group of 3 w U 2891 at 10.2 -- U 2889) strongly obscured by galactic
abs, in np part of the nebulosity N 1499; in group of 3 w comp at 2.0, 150, 0.7 x 0.3 and 2.8, 8, 0.8:: x 0.7::, eF,
spir --

02890   U 2891) see U 2888 -- U 2892) asym, v dif ext southw; no disturbing obj vis -- U 2893) comp superim at 0.35, 320,
0.12 x 0.08, prob at tip of spir arm -- U 2894) P w 03 51.4+15 47 at 4.3, 152, 0.7 x 0.6, spir, m = 15.3 -- U 2895)
vS vF red comp at tip of spir arm at 0.45, 215, M 51 case; comp 1.9, 160, 0.3 x 0.3 --

02900   U 2901) F disc w ring or alm circ arms; comp 3.3, 2, 0.6 x 0.1, F; 3.9, 23, 0.4 x 0.1, vF -- U 2903) comp 1.5, 245,
0.3 x 0.2 -- U 2904) P w U 2905 at 2.3, 15 -- U 2905) see U 2904 -- U 2907) VII Zw 10; "blue post-eruptive spir
w sev dense nuclei, jets, and disrupted halo" (Zw); ecc nucl cond 0.45 x 0.30 sp geom center -- U 2908) in group of
3 w U 2911 at 5.4, 113 and comp at 2.6, 233, 0.8 x 0.7, dif --

02910   U 2911) see U 2908, 1st group member at 7.1, 275 -- U 2913) 2nd brightest in group, about 7 members -- U 2914)
03 56.8+06 32 at 2.0, 72, 0.6 x 0.6, early; 03 56.8+06 33 at 1.8, 106, 0.8 x 0.8, S0; comp 3.2, 88, 0.5 x 0.2,
early -- U 2916) P w 03 57.9+71 34 at 3.4, 75, 0.9 x 0.20, late spir, m = 15.7 -- U 2917) comp 4.7, 108, 0.4 x 0.4

02920   U 2923) P w comp at 1.0, 72, 0.4 x 0.2, disturbed -- U 2927) U 2928 at 5.7; U 2929 at 6.1 -- U 2928) see U 2927;
U 2929 at 7.5 -- U 2929) see U 2927, 2928 --

02930   U 2930) sep 0.30; comp 1.0, 199 from a), 0.3 x 0.2 -- U 2935) comp 2.7, 251, 0.4 x 0.1, F -- U 2938) comp 4.1, 92,
0.5: x 0.3, v dif, * superimp --

02940   U 2940) comp 3.7, 263, 0.45 x 0.2, early -- U 2942) + U 2943) sep 0.40; comp 0.9, 358 from U 2942, 0.7 x 0.5, v dif,
dwarf; U 2943 is a spindle-shaped obj, dif in blue -- U 2944) chain of vF galaxies n -- U 2947) S B(s)m pec: (de V)

02950   U 2953) Arp 213; "F straight abs lanes lead tow nucl, become triple" (Arp) -- U 2954) P w 04 02.9+04 17 at 2.0,
180, 0.8 x 0.8, Sa:, m = 15.4 -- U 2958) comp 4.7, 127, 0.4 x 0.2 -- U 2959) dif blue disc, late spir?; P w comp
at 3.8, 169, 0.9 x 0.15, dif --

02960   U 2963) n part appears disturbed; stell obj at 0.50, 220, 0.10 x 0.10, * or disturbing comp? -- U 2967) prob earlier
than S B c; U 2971 at 3.4, 104 --

02970   U 2971) see U 2967; comp 2.9, 103, 0.9 x 0.5 -- U 2978) dif, asym, prob late; vis behind the s outskirts of the
nebulosity N 1499 --

02980   U 2983) comp 3.9, 255, 0.5 x 0.12 -- U 2984) P w comp at 4.2, 174, 0.22 x 0.22 (1.1 x 0.9 in red), S0-a -- U 2985)
P w U 2986 at 2.6, 19 -- U 2986) see U 2985; δ's for U 2985 and 2986 appear confused in MCG -- U 2987) comp 2.2,
358, 0.6 x 0.15, F -- U 2989) V Zw 372; "v red spherical compact in ext uniform red elliptical halo and sharp jets
east and west, m_p = 15.8" (Zw); "B nucl 10" = 3.4 kpc in diam in a F elliptical halo 90" x 30" = 31 x 10 kpc"
(Sargent) --

02990   U 2990) comp 4.2, 149, 0.3 x 0.3, F -- U 2992) sep 0.30 -- U 2996) P w comp at 3.6, 279, 0.5: x 0.5:, v red, early --

03000   U 3004) U 3005 at 1.0, 13; P w U 3006 at 4.0, 158 -- U 3005) see U 3004 -- U 3006) see U 3004 --

03010   U 3010) comp 4.6, 106, 0.4 x 0.4 -- U 3012) P w 04 17.1+02 14 = I 366 at 3.2, 162, 0.35 x 0.22, E:, m = 15.6 -- U 3013)
VII Zw 12; S B(rs)b (de V); "split compact core, many blue cond in v long spir arm" (Zw) -- U 3014) blue, asym spir
pattern, no disturbing obj vis -- U 3015) comp 5.4, 129, 0.9 x 0.2, spir -- U 3016) v dif spindle; P w U 3019 at 4.4,
113 -- U 3019) see U 3016 --

03020   U 3021) e dif in blue, strongly obscured; comp superim at 0.35, 218, 0.08 x 0.08, interaction? -- U 3023) smooth
eF disc, v uncertain diam; MCG 0-12-16 is a superim * --

03030   U 3031) P w U 3032 at 1.3, 111; in group of 3 w U 3029 at 7.7, 225 -- U 3032) see U 3031; U 3029 at 8.2, 233; pair
U 3031 + 3032 is II Zw 10, "pair of neutral intercon Sb's, sep 75" northwest-southeast... filamentary blue twisted
jet (d = 80") to northeast" (Zw) -- U 3035) P w U 3038 at 6.0, 119 -- U 3036) P w U 3043 at 5.6 nf -- U 3038) see
U 3035 --

03040   U 3042) S-shaped; P w U 3048 at 5.9, 38 -- U 3043) see U 3036 -- U 3045) P w comp at 5.9, 63, 0.8 x 0.4, Sa-b --
U 3046) P w 04 23.4+69 29 at 4.2, 316, 0.4 x 0.3, "compact, twisted jet" (CGCG), m = 15.6, may be interacting w
distorted S B at 0.50, 0.9: x 0.9: -- U 3048) see U 3042 --

03050   U 3051) dif, late; P w comp at 10.2, 252, 0.9: x 0.8:, S dm:, m approx 16.5 -- U 3054) P w comp at 1.0, 152, 0.4 x
x 0.3 -- U 3056) Arp 210, VII Zw 16; I B m (de V), Ir I (Ho); "eruptive system" (CGCG); "much abs, resolution into
**" (Arp) -- U 3057) 2 v open arms; comp 3.7, 358, 0.7 x 0.2, F -- U 3058) comp superim at 0.45, 245, 0.12 x 0.12,
bridge --

03060   U 3060) SA(s)d (de V), Sc + (Ho); core 0.30 x 0.10, dif halo, sev comp superim -- U 3062) comp 1.2, 20,
0.8 x 0.25, spir -- U 3063) see U 3064; E1 (de V) -- U 3064) P w U 3063 at 1.0, 257; U 3065 at 12.0; U 3063 +
+ U 3064 are II Zw 12; "N 1588... Dwarf galaxy 14" = 2.7 kpc diam and 54" = 10.6 kpc east of brighter elliptical,
N 1587. A plume extends to the south of N 1588" (Sargent) -- U 3065) pair U 3063 + 3064 at 12.2; U 3072 at 6.0,
103 -- U 3067) P w U 3075 at 7.6, 54 -- U 3068) a) 2 vS vF intercon obj w long curved tail p dif spindle --
U 3069) U 3077 at 4.3 --

03070   U 3070) asym, ill def spir pattern; comp superim nf center?; nucl region appears blue -- U 3071) II Zw 13; "twisted
blue post-eruptive galaxy w sev knots and jets" (Zw); "spir (Sc) 30" x 24" = 7.3 x 6.1 kpc w v thick inner arm"
(Sargent); comp 3.5, 70, 0.5 x 0.1, F; 4.0, 272, 0.5 x 0.2, F -- U 3072) see U 3065 -- U 3075) see U 3067 --
U 3077) VII Zw 18; "intercon pair of red E's w compact discs... [U 3069] 270" sp N 1573" (Zw), connection U 3069 -
- U 3077?; see U 3069; 04 28.2 + 72 11 at 4.5 np, 0.8 x 0.35, spir, m = 15.5 --

03080   U 3085) comp 4.7, 308, 0.4 x 0.12 -- U 3087) II Zw 14; "3C 120, blue elliptical compact and ext halo... Identical
w PKS 0430 + 05. Noted as a variable * (BW Tau) by Hanley and Shapley" (Zw); Seyfert spectrum; not in CGCG --
U 3088) comp 4.8, 232, double system, 0.40 x 0.12 and 0.30 x 0.15, sep 0.20 --

03090   U 3091) comp 3.8, 274, 0.7 x 0.5, late Sc -- U 3096) S0 (de V) -- U 3097) comp 1.5, 267, 0.4 x 0.2; 2.3, 198, 0.6 x
x 0.3, dif -- U 3098) P w comp at 1.2, 148, 0.9 x 0.4, Sb-c, both obj appear disturbed --

03100   U 3101) P w U 3112 at 5.5 nf -- U 3104) comp? superim at 0.40, 315, 0.15 x 0.12; P w comp at 1.4, 235, 0.4 x 0.4,
E:, m approx 16.0; U 3105 at 6.9, 98 -- U 3105) see U 3104 -- U 3108) strongly obscured, alm invis in blue, magn
approx 15.5 in red -- U 3109) P w 04 34.9-00 23 at 2.0, 310, 0.8 x 0.5, E or early spir, m = 14.8 --

03110   U 3111) comp 2.4, 49, 0.4 x 0.1 -- U 3112) see U 3101 -- U 3116) comp 3.2, 86, 0.3 x 0.12, blue, double --

03120   U 3122) (rs) 0.8 x 0.6, S-shaped pseudo-bar, long F arms -- U 3125) P w 04 37.4+07 14 = N 1634 at 0.7, 163, 0.5 x x 0.4, m=15.0 -- U 3127) strongly asym; 2 vF vS cond in s part -- U 3128) An 04 38 in BG, Sa (de V); no spir pattern vis on PA maps --

03130   U 3130) N 1530 A in BG -- U 3133) SA 0-: (de V); comp 2.6, 50, 0.4 x 0.2 -- U 3134) P w 04 38.9-04 26 at 4.9, 240, 0.9 x 0.6, late Sc, m=15.6 -- U 3136) P w 04 39.3+00 57 at 0.9, 195, 0.55 x 0.30, m=15.5, prob connecting spir arm -- U 3138) P w comp at 2.8, 308, 0.8 x 0.4 --

03140   U 3140) SA c (de V); U 3141 at 8.4, 24 -- U 3141) II Zw 21; see U 3140; comp superimp at 0.30, 15, 0.1 x 0.1, curved bridge from center; "intercon pair of red Sc and E0, both w compact nuclei" (Zw) -- U 3144) DDO 33; S B IV - V or S B V (vdB); "plume + streamer" (CGCG); no spir pattern; comp 3.2, 295, 0.4 x 0.4, F --

03150   U 3152) comp 3.8, 97, 0.4 x 0.2, F; 3.8, 353, 0.35 x 0.15, double system, contact -- U 3154) U 3156 at 4.6, 85 -- U 3156) see U 3154 -- U 3157) S-shaped, one blue cond in each arm, the arms are much brighter inside --

03160   U 3162) a large nr of vF obj in vic -- U 3167) comp 2.2, 271, 0.3 x 0.1 --

03170   U 3170) P w 04 45.0-01 18 at 3.5, 207, 0.9 x 0.6, S B 0:, m=15.7 -- U 3171) asym, appears disturbed, no disturbing obj vis -- U 3174) DDO 34; V (vdB); P w 04 45.8+00 08 at 3.1, 250, 0.6 x 0.5, E:, m=15.7 -- U 3179) II Zw 23; "blue fuzzy elliptical compact w long blue jet n and short B jet s" (Zw); "the 4 straight jets or plumes emerging from the nucl may be traced out to 45" = 25 kpc. The sharp em lines in the spectrogram come from the B nucl regions only" (Sargent); P w 04 47.0+03 14 at 1.7, 252, 0.7 x 0.3, early, "pec" (Sargent), m=15.7 --

03180   U 3184) measurements disturbed by * GC 5911, m=3.78 -- U 3188) high surf brightn; jets? --

03190   U 3190) SA(s)c (de V); vB main body 0.9 x 0.8, e dif outside, blue -- U 3191) U 3195 at 6.2, 98; comp 4.6, 237, 0.4 x 0.3, early -- U 3192) U 3194 at 5.0, 28; comp 2.0, 95, 0.7 x 0.6, spir; 2.1, 250, 0.5 x 0.2, early -- U 3193) comp 0.8, 77, 0.30 x 0.15; 4.3, 27, 0.5 x 0.4 -- U 3194) see U 3192 -- U 3195) see U 3191 -- U 3196) P w comp at 2.1, 218, 0.5: x 0.5:, m approx 16. -- U 3197) complex core?; ext corona, no regular E; sev vF comps (at 1.4, 1.5, 2.6, 3.1 and 3.6) -- U 3198) P w U 3199 at 1.8, 338 -- U 3199) see U 3198; ident as N 1690 in MCG, presumably incorrect --

03200   U 3206) U 3209 at 9.1, 50 -- U 3207) P w 04 53.3+02 09 at 5.3, 329, 0.55 x 0.35, m=14.7 -- U 3209) see U 3206 --

03210   U 3210) core 0.12 x 0.10; smooth disc, no spir pattern vis -- U 3214) comp 3.2, 286, 0.3 x 0.2, early -- U 3215) P w 04 55.4+00 43 at 6.4, 301, 0.9 x 0.4, E-S0, m=15.0 -- U 3219) P w U 3220 at 8.3, 358 --

03220   U 3220) see U 3219 -- U 3222) P w 04 56.2-00 33 = N 1709 at 2.7, 284, 0.9 x 0.7, compact, m=15.6 -- U 3226) ident as N 1717 = 1719 in MCG, but N 1717 is presumably a * 1.7 np; N 1717 "np h340 (? F *)" (NGC), h340 = N 1719 --

03230   U 3230) late S0?, no spir str vis; comp 3.8, 331, 0.8 x 0.18 -- U 3231) P w U 3233 at 7.5, 85 -- U 3233) see U 3231; comp 2.4, 28, 0.3 x 0.2 -- U 3234) DDO 35; V: (vdB); dif; comp 3.5, 223, 0.4 x 0.1 --

03240   U 3249) P w comp at 1.2, 246, 0.12 x 0.12, appears compact --

03250   U 3252) S-shaped; P w 05 06.0+67 30 at 6.4 np, 0.9 x 0.3, spir, m=15.7 -- U 3256) P w 05 07.5+00 52 at 2.1, 280, 0.35 x 0.35, m=15.4; comp 3.7, 20, 0.5 x 0.12 -- U 3258) vB asym arms form ring 0.60 x 0.55 --

03260   U 3262) P w U 3264 at 1.5, 118 -- U 3264) see U 3262 -- U 3267) comp 1.5, 276, 0.4 x 0.2, dif; 4.4, 33, 0.4 x 0.4 -- -- MCG 5-13-1, 5-13-2, 5-13-3 and 5-13-4 are presumably ** surrounded by nebulosities --

03270   U 3274) VV 161; in a dense part of the cl; this chain includes CGCG obj 05 14.0+06 23, double system, m=15.1 -- U 3275) a nest of early-type galaxies s around 05 14+06 25 -- U 3276) comp 4.0, 210, 0.55 x 0.12 -- U 3278) in group w comp at 1.3, 283, 0.3 x 0.15 and 1.7, 256, 0.5 x 0.3; sev other galaxies in vic, 05 14.6+06 05 at 3.9, 123, 0.7 x 0.5, S0-a, m=15.7 --

03280   U 3280) dif, 2 ** superimp; comp 2.0, 313, 0.5 x 0.3 -- U 3283) P w comp at 3.1, 94, 0.5 x 0.3 -- MCG 2-14-1 is intra-galactic -- U 3287) a) at 0.7 sp, 0.55 x 0.4, distorted -- U 3289) U 3291 at 5.1, 98; 05 17.3+06 37 at 6.1, 346, 0.5 x 0.4, S0-a:, m=15.5 --

03290   U 3291) see U 3289 -- U 3293) comp 4.4, 40, 0.7: x 0.3, spir; 4.8, 292, (0.4 x 0.3); 5.0, 112, 0.5 x 0.4, vF, late -- U 3298) + U 3299) VV 225, VV α = 15 19.25 is incorrect --

03300   --

03310   U 3317) DDO 38; Ir V (vdB) -- U 3319) comp 3.7, 185, 0.45 x 0.15 --

03320   U 3322) P w comp at 5.7, 202, 0.8: x 0.2, vF -- U 3324) comp 4.5, 288, 0.4 x 0.4, spir --

03330   U 3334) Arp 184; SAB(s)c (de V), Sb+ (Ho); "2 long straight arms or filaments tangent to nf side of galaxy" (Arp) -- U 3337) red spindle; comp 0.9, 102, 0.2 x 0.1; 1.7, 190, 0.5: x 0.4:, v dif; 4.1, 326, 0.3 x 0.3, v dif --

03340   U 3342) U 3334 at 13.6 -- U 3344) U 3334 at 20.0 -- U 3346) B core, dif disc, p end curved southw; no disturbing obj vis; comp 3.6, 197, 0.3 x 0.3:, dif -- U 3349) U 3334 at 30.6; 05 40.2+69 02 = VII Zw 45 at 4.8 p, 0.9 x 0.9, m=15.1, "blue irr post-eruptive galaxy w ecc patchy compact disc" (Zw) --

03350   U 3350) U 3351 at 6.3 n -- U 3351) see U 3350 -- U 3352) in group w comp at 2.0, 82, 0.5 x 0.12 and 2.7, 335, 0.8 x 0.8, E-S0 -- U 3355) comp 3.5, 250, 0.4 x 0.2 -- U 3356) P w comp at 1.4, 66, 0.7 x 0.4, pec -- MCG 5-14-2 is presumably a * surrounded by nebulosity -- U 3357) comp 3.4, 83, 0.5 x 0.3, dif --

03360   U 3360) U 3362 at 12.0, U 3363 at 9.7 -- U 3362) P w U 3363 at 2.2, 1; see U 3360 -- U 3363) vB core, dif env; see U 3360, 3362 --

03370   U 3370) U 3373 at 7.4 f -- U 3371) DDO 39; V (vdB) -- U 3374) B stell nucl region in dif bar -- U 3378) B core in in smooth ext corona; sev comps, brightest at 2.0, 160, 0.12 x 0.12 + env --

03380   U 3381) asym disc?, fragments of spir arms; comp 2.1, 164, 0.4 x 0.3 -- U 3389) P w comp at 3.8, 293, 0.55 x 0.15 --

03390   U 3393) II Zw 42; "red spherical compact, m_p=16.0" (Zw); "structureless image w diam 4" = 1.4 kpc" (Sargent); m = = 14.5 acc to Sargent; not in CGCG -- U 3396) P w U 3397 at 3.9 sf -- U 3397) see U 3396 --

03400   U 3401) P w U 3413 at 8.4, 62 -- U 3404) comp 2.0, 188, 0.4 x 0.25; 2.9, 121, 0.7 x 0.5, Sc -- U 3405) P w U 3410 at 2.0 f -- U 3408) elong main body in pos angle 12; prob 2 comp attached to s end; P w U 3412 at 3.9 f --

03410   U 3410) see U 3405; comp 0.8, 40, 0.22 x 0.20 -- MCG 2-16-1 is presumably intragalactic -- U 3412) see U 3408 -- U 3413) see U 3401 -- U 3414) VII Zw 61; "blue post-eruptive Sc, sev compact cores" (Zw); comp 1.9, 312, 0.7 x 0.5, Sc -- U 3416) comp 1.7, 302, 0.4 x 0.2; 2.6, 200, 0.3 x 0.3 -- U 3418) comp 2.8, 46, 0.45 x 0.15, dif --

03420   U 3420) comp 6.1, 120, 0.8: x 0.7:, dwarf -- U 3422) P w U 3426 at 6.5 sf -- U 3426) abs lane close to center?; see U 3422 -- U 3428) comp 4.4, 139, 0.5 x 0.3 -- U 3429) S B(s)ab pec (de V), S pec (Ho); e broad spir arms, prob in different planes (about 45°?); P w U 3439 at 18.9, interaction? --

03430   U 3439) N 2146 A in BG; Sc: (de V); see U 3429 --

03440   U 3440) S v red obj superimp s center -- U 3445) + U 3446) VII Zw 68; sep 0.7; "pair of intercon neutral Sa's" (Zw)
-- U 3448) comp 2.8, 111, 0.5 x 0.15 --

03450   U 3451) (r) 0.6 x 0.5), v dif outside -- U 3454) comp 1.0, 212, 0.4 x 0.10; 2.4, 160, 0.3 x 0.10; 2.8, 92, 0.35 x
0.35 -- U 3455) in group w comp at 3.5, 185, 0.5 x 0.1 and 6.0, 0.5 x 0.2 -- U 3458) comp 1.4, 242, 0.8 x 0.12;
3.3, 290, 0.7 x 0.5; 4.6, 113, 0.35 x 0.15, high surf brightn --

03460   U 3460) 3 blue conds in arm np center; comp 1.4, 67, 0.25 x 0.12 --

03470   U 3470) "double nucl in halo" (CGCG); no spir pattern vis, slightly asym; comp 3.5, 250, 0.35 x 0.18 -- U 3471)
06 23.9 + 74 28 at 4.5, 276, 0.7: x 0.6:, core 0.15 x 0.15, E - S0, m = 15.6 -- U 3473) comp 4.2, 107, 0.4 x 0.4, (R)SB...
-- U 3478) comp 2.2, 244, 0.3 x 0.1 --

03480   U 3484) P w comp at 5.5, 268, 0.8 x 0.12, Sc -- U 3486) 06 30.4 + 74 18 at 4.6, 139, 0.35 x 0.22, high surf brightn,
m = 15.3; comp 4.5, 165, 0.35 x 0.35 -- U 3487) P w comp at 1.3, 177, 0.7 x 0.15, late Sc -- U 3488) no spir pattern
vis; comp 2.4, 332, 0.4 x 0.8 --

03490   U 3492) (r) 0.50 x 0.45 -- U 3497) B core 0.55 x 0.45, vF disc or halo; 06 32.4 + 67 53 at 2.4, 117, 0.3 x 0.2, m =
= 15.7 -- U 3499) disturbed?; comp 2.9, 362, 0.5 x 0.3, spir --

03500   U 3500) irr "spindle", brightest in s part, disrupted multiple system? -- U 3507) F disc, hints of F arc at the
edge; comp 3.3, 326, 0.4 x 0.2 --

03510   U 3510) a) at 1.1, 310, 0.7 x 0.1, interaction, F obj 0.15 x 0.15 betw a) and b) -- U 3511) v blue comp at tip of
spir arm at 0.5, 120, 0.25 x 0.15 -- U 3514) comp 1.7, 56, 0.55 x 0.12; 4.7, 255, 0.4 x 0.08 -- U 3519) early S0?;
comp 3.9, 205, 0.4 x 0.3; 4.7, 252, 0.4 x 0.4 --

03520   U 3521) P w U 3528 at 6.2, 38 -- U 3522) VII Zw 92; "F irr network halo" (Zw); "compact nucl in spir" (CGCG); comp 2.5,
280, 0.3 x 0.1, vF -- U 3523) 06 40.3 + 74 34 at 4.3, 310, 0.6 x 0.25, E - S0:, m = 15.4 -- U 3526) comp 1.6, 233,
0.3 x 0.2, F -- U 3527) P w comp at 6.4, 343, 0.7 x 0.7, Sc -- U 3528) see U 3521 --

03530   U 3530) N 2273 B in BG; S B...: (de V); asym, v broad knotty arms s, dif broad vF arm n -- U 3531) vS red obj at
0.85 ssf, poss at tip of spir arm -- U 3532) in group of 3 w U 3535 at 2.9 nf and 06 42.4 + 43 53 at 3.6 n, 0.30 x
x 0.20, m = 15.7 -- U 3533) (r) 0.7 x 0.6, v dif outside -- U 3535) "compact w plume" (CGCG); see U 3532; 06 42.4 +
+ 43 53 at 3.9 p -- U 3536) 06 24.3 + 29 16 at 9.8, 219, 0.7 x 0.6, SAB(s)a - b, m = 15.4 -- U 3537) v thin bar, broad
dif arms forming a ring; U 3541 at 7.3, 115 --

03540   U 3541) P w U 3542 at 1.9, 2; note ident error in MCG, N 2274 and N 2275 are confused -- U 3542) see U 3541 --
U 3543) v uncertain if this obj is extragalactic -- U 3544) U 3541 at 10.6, 267 -- U 3546) (r) 0.9 x 0.7 w high
surf brightn; see note to N 2273 B in BG; Sa (de V) -- U 3547) P w U 3550 at 4.2 -- U 3548) comp 1.4, 199,
0.45 x 0.12 -- U 3549) P w 06 46 + 81 07 at 1.4, 225, 0.9 x 0.6, S B(s)b: --

03550   U 3550) see U 3547 -- U 3551) P w 06 46.7 + 29 35 at 2.0, 95, 0.9 x 0.8, S B O - a, m = 15.5 -- U 3553) U 3558 at 8.3 --
U 3545) prob (R')S B(r)b -- U 3555) brightest of 6; P w comp at 0.6, 299, 0.55 x 0.35, spir, interaction; 06 47.0 +
+ 25 43 nf, 0.4 x 0.25, m = 15.0 -- U 3556) in triple system w comp at 0.6, 220, 0.15 x 0.10 and 0.6, 330, 0.3 x 0.08
-- U 3558) in group w 4 or more members, see U 3553 --

03560   U 3560) 06 47.5 + 33 31 = N 2288 at 1.0, 197, 0.3 x 0.2, m = 15.5; U 3562 at 2.6, 164; 06 47.6 + 33 30 = N 2291 at 3.0,
20, (0.9 x 0.9), compact, m = 15.0; 06 47.8 + 33 35 = N 2294 at 4.8, 52, 0.8 x 0.25, m = 15.0; U 3560 and 3562 are
brightest in the group -- U 3562) see U 3560; 06 47.5 + 33 31 at 1.9, 325; 06 47.8 + 33 30 at 5.3, 3; 06 47.8 + 33 35
at 6.2, 29 -- U 3563) VII Zw 105; "post-eruptive triple system, face-on red Sc, edge-on red Sc and blue spherical
compact" (Zw) -- U 3567) comp 0.55, 246, 0.18 x 0.10; 3.5, 0, 0.6 x 0.1 --

03570   U 3573) P w 06 48.9 + 27 32 at 3.6, 111, compact, m = 14.9 -- U 3574) U 3569 at 6.0 p -- U 3578) P w 06 49.4 + 15 18
at 2.0, 248, 0.9 x 0.4, E - S0, m = 15.2 --

03580   U 3582) prob a S comp attached to tip of spir arm sp -- U 3583) star cluster? -- U 3585) comp 3.6, 238, 0.4 x 0.2 --
U 3586) in group of 3 w comp at 5.5, 322, 0.35 x 0.20 and 8.6, 297, 0.8 x 0.35, E - S0, m approx 16. (brightest) --
U 3589) comp attached at 0.55, 145, 0.3 x 0.1, vF --

03590   U 3593) P w 06 51.6 + 41 04 at 2.7, 343, (0.8 x 0.8), core 0.22 x 0.20, compact, m = 14.6 -- U 3597) P w comp at 1.6,
118, 0.15 x 0.12, e dif bridge?; comp 4.8, 135, 0.3 x 0.2 -- U 3599) P w 06 52.2 + 24 19 at 2.1, 17, 0.5 x 0.3, m =
= 15.5 --

03600   U 3603) S comp superimp close to center -- U 3604) massive arm fragments, prob no regular spir pattern -- U 3608)
vB main body 0.7 x 0.45; comp 4.9, 274, 0.4 x 0.2 --

03610   U 3613) P w U 3617 at 7.6, 72 -- U 3615) comp 2.8, 49, 0.45 x 0.3; 06 55.1 + 35 55 at 8.2 nf, 0.8 x 0.6, E:, m = 14.8
-- U 3616) e dif; comp 2.0, 63, 0.3 x 0.1, vF; 2.5, 62, 0.4 x 0.1 -- U 3617) see U 3613 --

03620   U 3620) comp 4.9, 35, 0.6 x 0.3 -- U 3625) prob a v late bar spir w S comp attached to s end --

03630   U 3630) Sb or later -- U 3631) short bar, 2 straight smooth arms; P w comp at 3.9, 40, 0.6 x 0.3, Sc: -- U 3632)
comp 3.1, 4, 0.7 x 0.3; a large nr of vF obj in vic -- U 3633) double system 06 59.0 + 50 40 at 3.0 and 3.5 f,
cores 0.18 x 0.18 and 0.12 x 0.12, sep 0.50, m = 14.9 --

03640   U 3640) comp 3.9, 183, 0.6 x 0.1, spir -- U 3642) eF outer disc may be vis out to approx 3. x 3.; 06 59.0 + 64 07 at
3.8 p, 0.7 x 0.22, m = 15.6; 06 59.4 + 64 04 at 2.9 sp, 0.6 x 0.15, m = 15.7 -- U 3647) 06 59.8 + 56 33 at 7.5 sp, 0.9 x
x 0.6, S B..., m = 15.3 -- U 3648) 07 00.6 + 64 08 at 3.1, 189, 0.55 x 0.45, core 0.15 x 0.15, compact, m = 15.4 -- U 3649)
P w 07 00.9 + 29 20 at 1.3, 98, 0.5 x 0.4, early, m = 15.4; comp 2.1, 65, 0.3 x 0.2; 4.5, 314, 0.4 x 0.4 --

03650   U 3653) SAB(r)bc, Sc - (Ho) -- U 3656) in group w 07 01.5 + 18 40 brightest at 3.0, 22, (0.9 x 0.9), E - S0, m = 15.2;
comp 6.8, 13, 0.6 x 0.5, m approx 16. -- U 3659) U 3662 at 5.3 --

03660   U 3662) see U 3659 -- U 3666) P w U 3677 at 5.8 f --

03670   U 3673) 07 03.5 + 44 53 at 2.9 sf, 0.9 x 0.3, m = 14.7; U 3679 at 6.1 -- U 3676) P w comp at 2.2, 120, 0.5 x 0.12 --
U 3677) E3 (de V); see U 3666 -- U 3679) 07 03.5 + 44 53 at 3.2 nf, see U 3673 --

03680   U 3681) U 3687 at 4.8 nf -- U 3687) see U 3681; N 2326 A in BG, E3 (de V) -- U 3688) in group of 10 or more galaxies;
07 05.2 + 13 14 is a subgroup of 4 at 5.0 np, m = 15.4 --

03690   U 3690) DDO 41; Ir V: (vdB) -- U 3691) sev blue patches along edge -- U 3693) SAB(rs)bc (de V); 07 04 + 18 50 at
8.1 p, 0.5 x 0.5, spir, m = 15.3 -- U 3694) comp 4.8, 278, 0.6 x 0.5 -- U 3695) P w U 3696 at 2.8 nf; sev obj in vic
U 3696) see U 3695 -- U 3697) smooth integral, presumably interacting w U 3714 at 7.6 -- U 3699) 07 05.6 + 50 13 =
= I 457 at 2.0 sp, 0.4 x 0.3, E, m = 15.7 --

03700   U 3700) late Sb or Sc; comp 2.4, 240, 0.30 x 0.25, brightest in pair, comp 0.20 x 0.10 at 0.45 -- U 3701) comp 4.8,
354, 0.7 x 0.25 -- U 3702) P w U 3708 at 5.3, 112 -- U 3703) 07 06.3 + 36 27 at 6.6 nf, 0.8 x 0.22, m = 15.5 --
U 3706) a) 0.5 x 0.12 superimp dif streamer from b), sep 0.4 betw nucl regions; 07 05.9 + 48 00 at 2.7 p, 0.6 x 0.2,
m = 15.2 -- U 3708) P w U 3709 at 2.5, 36 -- U 3709) see U 3708 --

03710   U 3710) see U 3702; double nucl in e dif env — U 3713) 07 06.8 + 50 14 = I 459 at 3.6 n, 0.15 x 0.15, extremely
     compact, m=15.5; 07 06.9 + 50 09 = I 461 at 2.9 sf, 0.6 x 0.3, m=15.7 — U 3714) see U 3697; comp 3.5, 244, 0.5 x
     x 0.4, asym, pec? or interacting pair?; 07 07.3 + 71 45 at 5.4, 151, 0.40 x 0.18, high surf brightn, m=15.7 —
     U 3715) comp 3.8, 70, 0.3 x 0.3 — U 3716) P w U 3718 at 1.5 sf — U 3718) see U 3716; v red * superimp; comp 2.8,
     247, 0.6 x 0.08, F — U 3719) comp 1.2, 355, 0.35 x 0.15 —

03720   U 3720) 07 07.1 + 50 13 = I 464 at 2.5 sp, 0.8 x 0.4, E, m=14.8 — U 3728) P w comp at 1.5, 32, 0.7 x 0.15 —

03730   U 3730) VV 123, Arp 141, An 0708 in BG; in Arp's group "Material emanating from E galaxies" ; "extensive jets +
     + plume" (CGCG); E-type obj in n part, chaotic plumes southw, disrupted triple system?; U 3705 at 9.4 — U 3734)
     S B c (de V) —

03740   U 3740) Arp 25 (and 114, see U 3798), VII Zw 134; SAB(rs)c (de V), Sc + (Ho); "Tubular arm, straight at first, then
     bent. Secondary arm from straight portion" (Arp); "Spir somewhat pec, may be perturbed" (Arp, comment to nr 114);
     "blue, badly disrupted post-eruptive Sc w many compact knots" (Zw); P w U 3798 at 7.5, 115, interaction; SN 1962q,
     1968v, 1968w — U 3743) comp 3.2, 292, 0.45 x 0.1 —

03750   U 3750) E+ 1 (de V); tend to spir pattern in halo?; no normal E; P w U 3759 at 13.3 sf — U 3754) comp 3.4, 227,
     0.6 x 0.4 — U 3758) comp 2.4, 53, 0.7 x 0.5, Sc — U 3759) see U 3750; Pec SA(r)b: (de V) —

03760   U 3760) E? w stell comp superimp f center — U 3764) sharp (r), amorph outside, no normal S B c — U 3765) 07 12.0 +
     + 56 54 at 1.2 f, 0.5 x 0.15, m=15.7 —

03770   U 3771) comp 3.4, 197, 0.3 x 0.12 — U 3772) comp 0.8, 225, 0.4: x 0.4:, vF, interaction? — U 3774) comp 5.0, 77,
     0.9 x 0.1 — U 3776) in group of 3 w U 3779 at 11.0 and U 3780 at 11.7 — U 3779) see U 3776; U 3780 at 6.3 —

03780   U 3780) see U 3776, 3779 — U 3781) a) 0.15 x 0.15, b) 0.12 x 0.10, sep 0.50, broad dif bridge, long dif plume eastw
     from b) — U 3782) comp 3.6, 166, 0.3 x 0.2 + smaller comp — U 3785) P w comp at 0.55, 105, 0.8 x 0.3, SO: —
     U 3787) P w comp at 0.8, 87, 0.4 x 0.2 — U 3789) VII Zw 140; "blue post-eruptive multiple (elliptical) ring spir,
     w compact elliptical core" (Zw); (r) 0.85 x 0.45; U 3797 at 4.3 —

03790   U 3790) 2 spirals 0.9 x 0.3 and 0.7 x 0.4, sep 0.45 — U 3793) P w U 3795 at 3.1,129 — U 3794) MCG α = 07 17.4 is
     incorrect — U 3795) see U 3793 — U 3796) P w comp at 1.9, 224, 0.55 x 0.55, S B c; double system 07 15.3 + 60 24
     at 3.9, 303, 0.22 x 0.15 and 0.30 x 0.10, contact, m=15.7 — U 3797) see U 3789 — U 3798) Arp 114, see U 3740;
     E+ 2 (de V), E (Ho) — U 3799) thin knotty arms; comp 2.8, 342, 0.5 x 0.2 —

03800   U 3800) stell nucl in smooth corona ; comp 0.50, 234, 0.15 x 0.10, interaction; comp 4.3, 198, 0.3 x 0.3, F — U 3801)
     P w comp at 1.8, 344, 0.7 x 0.15, interaction? — U 3805) comp 2.5, 25, 0.3 x 0.3 — U 3807) comp 0.40, 80, 0.1 x
     x 0.1, interaction? — U 3808) comp 4.9, 66, 0.3: x 0.3:, v dif — U 3809) SAB(r)bc, Sc - (Ho); P w U 3834 at 20.6;
     comp 3.4, 10, 0.45 x 0.40 —

03810   U 3810) "system w jet" (CGCG); comp superimp at 0.45, 198, 0.35 x 0.08 — U 3812) U 3814 at 1.3 sf — U 3814) see
     U 3812 — U 3816) 07 18.8 + 58 13 at 2.2 np, (0.9 x 0.5), E:, m=15.2 —

03820   U 3820) comp 4.3, 56, 0.9: x 0.6:, v dif — U 3821) P w 07 19.4 + 22 08 at 2.6, 207, 0.7 x 0.4, m=15.4 — U 3829)
     a) 1.1 x 0.7, b) 0.7: x 0.12; a) strongly asym —

03830   U 3831) strongly asym; curved bar in p part; comp 3.7, 358, 0.4 x 0.2 — U 3832) comp 2.8, 155, 0.6 x 0.1; 3.2,
     106, 0.4 x 0.2 — U 3834) see U 3809; SAB(s)c: (de V) — U 3838) VII Zw 153; "blue post-eruptive, horseshoe-shaped
     galaxy" (Zw) —

03840   U 3840) P w U 3842 at 4.9, 127 — U 3841) a) 0.35 x 0.25, E?, b) 0.45 x 0.05, spir:, sep 0.55; 1st member in the group
     at 1.5, 273 from geom center, 0.45 x 0.12, spir — U 3842) see U 3840 — U 3845) comp 0.9, 306, 0.3 x 0.1, dif —
     U 3847) "a B em patch at the sp end of N 2366 (U 3851)" (CGCG); P w U 3851 at 2.2 from geom center — U 3849) a)
     0.9: x 0.35, b) 0.8 x 0.3, sep 0.7 —

03850   U 3850) alm closed (R'); similar to N 2217 — U 3851) see U 3847; DDO 42; S IV-V or Ir IV-V (vdB); I B(s)m (de V),
     Ir I (Ho) — U 3852) VII Zw 156; "triple system in halo" (CGCG); "blue post-eruptive quadruple of 2 bar-shaped and
     2 spherical compacts" (Zw) — U 3854) P w U 3857 at 3.7, 110 — U 3857) SO: (de V); see U 3854 — U 3858) P w
     U 3859 at 4.7 f; comp 2.1, 30, 0.4 x 0.3, common w U 3859 — U 3859) see U 3858, comp 3.0, 163 —

03860   U 3860) DDO 43; Ir IV-V or Ir V — U 3861) comp 3.3 nf, 0.5 x 0.5 — U 3864) VV 141, listed as quadrupel system
     by VV — U 3866) asym; comp 0.7, 235, 0.35 x 0.1 — U 3867) VII Zw 169; "pair of red post-eruptive patchy compacts,
     imbedded in irr luminous cloud" (Zw) —

03870   U 3870) U 3872 at 3.5 nf — U 3872) SAB(rs)c (de V); patchy, ill-def arms; see U 3870 — U 3874) B main body 0.5 x
     x 0.35, dif plumes in pos angles 38 and 160 — U 3877) P w comp at 2.4, 320, 0.5 x 0.2, m approx 16. — U 3879)
     U 3872 at 13.1 —

03880   U 3884) U 3872 at 16.1 — U 3885) prob late spir; bar?; comp 2.1, 283, 0.3 x 0.1 —

03890   U 3890) "long eF jets" (CGCG); hints of one eF jet in direction 117 to d = 0.6 — U 3891) a) E:, (0.8 x 0.8), sp
     b) at 0.50, 0.8 x 0.5, early spir?, a) brightest — U 3892) 07 28.0 + 13 10 at 4.3, 266, 0.9 x 0.7, m=15.4; 07 28.1 +
     + 13 12 at 3.3, 316, double system, total diam 0.5 x 0.4, m=15.5; 07 28.7 + 13 16 at 7.7, 50, 0.7 x 0.6, m=15.3 —
     U 3893) (r) 0.35 x 0.35, no spir arms vis in the disc outside ring; comp 4.9, 180, 0.6 x 0.12; U 3894 at 5.3 —
     U 3894) comp 2.3, 245, 0.5 x 0.12; see U 3893 — U 3896) 07 29.2 + 18 26 at 2.4, 102, 0.6 x 0.6, Sb:, m=15.4;
     07 28.9 + 18 24 = N 2406 at 3.4, 217, (0.7 x 0.6), m=15.0; comp 1.7, 152, 0.35 x 0.25 — U 3897) ring 1.1 x 0.9, eF
     outer disc not incl in diam; comp 4.9, 201, 0.45 x 0.35 —

03900   U 3900) comp 1.6, 150, 0.3 x 0.3 — U 3904) P w 07 30.2 + 30 37 at 4.5, 196, 0.9 x 0.12, m=15.7 — U 3906) disrupted
     obj 0.6: x 0.3 attached to large Irr w B main body 0.55 x 0.12 — U 3907) comp 1.5, 93, 0.3 x 0.12 — U 3909)
     07 32.0 + 73 51 at 5.9, 68, 0.9 x 0.5, spir, m=15.5, this galaxy and U 3909 are confused in MCG; comp 5.0, 125,
     0.4 x 0.4, dif —

03910   U 3910) U 3915 at 12.6; comp 3.5, 74, 0.7 x 0.2 — U 3914) b) at 1.0, 37, 0.55 x 0.30 — U 3915) comp 3.6, 58, 0.35 x
     x 0.12 — U 3918) SAB(s)cd (de V), Sc + (Ho); possible SN 1910 (prob illusory, emulsion defect) —

03920   U 3922) dif spindle, distorted in s part?, no disturbing obj vis — U 3924) U 3925 at 5.5, 9 — U 3925) asym late
     spir; see U 3924; 07 32.9 + 11 49 at 6.9, 5, 0.9 x 0.5, asym spir, m=15.0; comp 3.5, 281, 0.4 x 0.15; 4.8, 281,
     0.3 x 0.2 — U 3926) 07 33.5 + 33 14 at 6.1 f, 0.7 x 0.20, m=15.5 — U 3928) "strong jets" (CGCG); B core in smooth
     ext halo, sev comps — U 3929) VII Zw 170; "neutral elliptical compact, F halo" (Zw) —

03930   U 3930) Haro 1; high surf brightn, irr fringes; comp 1.0, 270, 0.9: x 0.3:, Irr, dif — U 3931) slightly asym env,
     sev vF comps , brightest at 2.0, 10, 0.2 x 0.2 + eF ext northw — U 3934) U 3937 at 3.3 s — U 3937) see U 3934;
     07 33.9 + 35 46 at 5.8 nf, 0.7 x 0.25, early spir, m=15.3 — U 3938) 07 34.5 + 10 01 at 1.9, 85, triple system, total
     diam 0.9 x 0.5, m=15.7; 07 34.0 + 10 01 at 5.0, 275, 0.6 x 0.6, spir, m=15.3 — U 3939) F outer arc in n part;
     comp 5.0, 245, 0.6 x 0.4, Irr —

03940   U 3943) comp 2.7, 313, 0.9 x 0.7, v dif spir; 4.8, 262, 0.3 x 0.2 — U 3946) comp 0.8 np, 0.3 x 0.2, interaction? —

03950 U 3953) comp 3.5, 232, 0.3 x 0.2 -- U 3956) comp 2.7, 193, 0.3 x 0.2, dif -- U 3957) comp 0.45, 150, 0.12 x 0.12; 0.8, 15, 0.10 x 0.10; 1.3, 322, 0.12 x 0.12; 2.2, 150, 0.25 x 0.25 -- U 3958) P w U 3965 at 9.3 f -- U 3959) S B(r)b: (de V) --

03960 U 3965) see U 3958 -- U 3966) DDO 46; V (vdB) -- U 3967) bar-shaped main body; one patchy vF arm, hints of another long e dif arm, no normal S B c -- U 3969) comp 3.5, 196, 0.3 x 0.2 --

03970 U 3971) 07 38.4 + 60 19 at 3.0, 324, 0.35 x 0.35, m = 15.5; comp 4.5, 219, 0.4 x 0.10 -- U 3972) VV 349; late bar spir w comp superimp at 0.30, 356 from center, 0.3 x 0.12, interaction, the comp is MCG 12-8-8b -- U 3974) DDO 47; Ir V? (vdB); MCG diam 2.0 x 0.2: are incorrect -- U 3977) U 3983 at 5.0 nf -- U 3978) VII Zw 180; "neutral S B c w compact core and sym knots on bar" (Zw) --

03980 U 3982) P w U 3985 at 3.9 sf -- U 3983) distorted early spir interacting w comp at 0.8, 133, 0.4 x 0.2, smooth bridge; see U 3977 -- U 3985) see U 3982 -- U 3986) P w 07 40.2 + 31 41 at 2.2, 336, 0.4 x 0.12, m = 15.7 -- U 3988) comp 1.9, 110, 0.3 x 0.1 -- U 3989) 2 broad eF arms; comp 3.2, 122, 0.4 x 0.1; 4.1, 65, 0.3 x 0.1 --

03990 U 3990) P w U 3997 at 3.2 f -- U 3992) dif neb "disc", B alm stell core, * superimp; intragalactic? -- U 3993) eF ring? -- U 3995) prob a companion superimp at 0.5, 285, interaction?; 07 41.2 + 29 23 at 4.0, 70, 0.6: x 0.25, double system, jets, m = 15.5 -- U 3997) see U 3990 --

04000 U 4003) in group of 3 w U 4012 at 7.6 and U 4020 at 12.1 -- U 4005) S B w twisted plane?; P w 07 42.8 + 08 03 at 4.9, 91, 0.7 x 0.35, spir, m = 14.8 -- U 4007) comp 2.2, 192, 0.30 x 0.18 -- U 4008) a) 0.8 x 0.15, b) 0.8 x 0.35, sep 0.9 -- U 4009) arc-shaped, may be interacting w comp at 0.5, 45 from center, 0.08 x 0.08 --

04010 U 4011) comp 3.2, 121, 0.3 x 0.2 -- U 4012) see U 4003; U 4020 at 4.8 -- U 4013) comp 2.1, 91, 0.5 x 0.25, S B:... -- U 4014) 07 44.2 + 74 26 at 3.4 f, double system, 0.9 x 0.4 and 0.5 x 0.4, sep 0.45, m = 15.6; U 4028 at 5.9 f -- U 4016) + U 4017) VV 117, Arp 143; in Arp's group "Material emanating from E galaxies"; E w distorted env, ext knotty plume, poss a disrupted comp; "dif counter filament" (Arp) -- U 4018) a) at 0.65, 203, 0.4 x 0.18 --

04020 U 4020) see U 4003, 4012; comp 1.5, 255, 0.3 x 0.15 -- U 4023) MCG major diam 0.85 is incorrect -- U 4026) P w 07 44.2 + 27 04 = I 476 at 1.5, 322, 0.5 x 0.4, S B..., m = 15.5 -- U 4027) comp 3.8, 35, 0.35 x 0.25 -- U 4028) see U 4014, 07 44.2 + 74 26 at 3.3, 227 --

04030 U 4030) a) 0.30 x 0.15, b) 0.18 x 0.18, sep 0.55 -- U 4033) comp 4.1, 206, 0.7 x 0.20 -- U 4035) P w 07 46.2 + 55 32 at 1.9, 37, 0.8 x 0.18, spir, m = 14.9 -- U 4036) SAB(r)b: (de V); comp 4.7, 230, 0.5 x 0.15 -- U 4039) P w U 4042 at 6.5, 40 --

04040 U 4041) early spir; U 4037 at 5.7 -- U 4042) see U 4039 -- U 4044) short patchy F arms; comp 1.9, 213, 0.3 x 0.1, F -- U 4045) "very compact, long jet" (CGCG) -- U 4047) 07 46.8 + 30 52 at 2.1, 300, 0.7 x 0.2, m = 15.4; U 4046 at 2.9, 176 --

04050 U 4050) P w comp at 5.0, 261, 0.7 x 0.4, spir -- U 4051) 07 47.5 + 50 18 at 1.0 np, 0.5 x 0.18, m = 15.4; U 4052 at 3.5 n -- U 4052) see U 4051; 07 47.5 + 5018 at 2.8 -- U 4053) comp 3.3, 62, 0.4 x 0.2 -- U 4056) comp 3.9, 112, 0.4 x 0.12 --

04060 U 4066) dif arms, chain of B patches in one; comp 3.8, 109, 0.5 x 0.10 --

04070 U 4070) comp 2.5, 96, 0.8: x 0.7:, dif; 2.9, 204, 0.6 x 0.6 -- U 4072) comp 1.0, 198, 0.4 x 0.05; 2.2, 352, 0.5 x x 0.25; 3.4, 310, 0.3 x 0.1, vF -- U 4077) comp 3.6, 353, 0.4 x 0.1, F -- U 4079) vB core, short spir arc from center --

04080 U 4082) env appears asym; comp 3.6, 272, 0.4 x 0.12 -- U 4083) dif spindle; P w U 4088 at 1.7 f -- U 4085) asym, vB nucl region, one B arm -- U 4086) comp 3.8, 172, 0.4 x 0.1, vF -- U 4088) dif spindle; see U 4083 --

04090 U 4093) B center, chaotic disc; P w U 4097 at 5.4 nf; S B(rs)b: (de V) -- U 4095) main body 0.60 x 0.35 w v high surf brightn, smooth F disc or env -- U 4097) SA(s)a (de V); see U 4093; comp 3.5, 20, 0.3 x 0.2, F; 4.5, 184, 0.7 x 0.15 -- U 4098) blue ecc spir, high surf brightn --

04100 U 4101) 07 52.6 + 45 37 at 5.8 sp, 0.6 x 0.5, m = 15.7 -- U 4109) comp 4.8, 347, 0.35 x 0.30 --

04110 U 4110) S0: 0.9 x 0.5 w comp at 0.5, 200, 0.20 x 0.15 -- U 4112) P w comp at 0.8, 275, 0.35 x 0.30, double -- U 4114) a) = N 2474 0.45 x 0.45, b) = N 2475 0.6 x 0.6, E3 (de V), sep 0.40 -- U 4116) P w U 4118 at 1.0, 135 -- U 4118) see U 4116; high surf brightn, irr fringes -- U 4119) comp 1.0, 0, 0.6 x 0.15 --

04120 U 4121) DDO 48; S (B) IV-V: (vdB); MCG α = 07 55.8 is incorrect -- U 4122) P w U 4124 at 1.3 f -- U 4123) P w U 4126 at 5.4, 97 -- U 4124) see U 4122 -- U 4126) see U 4123 -- U 4127) 07 55.8 + 08 07 = N 2491 at 3.7, 222, 0.30 x 0.20, m = 15.6 -- U 4129) 2 S cores w sep 0.15; e dif strongly asym env --

04130 U 4132) comp 4.3, 42, 0.6 x 0.3; 4.7, 310, 0.4: x 0.2; 07 55.5 + 33 00 at 6.9 sp, 0.7 x 0.15, S0-a, m = 15.6 -- U 4133) P w U 4134 at 1.0 sf -- U 4134) see U 4133 -- U 4138) 07 56.2 + 27 13 = N 2490 at 4.1, 319, 0.45 x 0.25, compact, m = 15.3; comp 4.3, 353, 0.3 x 0.2 --

04140 U 4142) vB bar 0.50 x 0.22, loop westw -- U 4145) comp 1.5, 103, 0.4 x 0.1, vF -- U 4147) comp 0.30, 306, 0.15 x 0.12; 2.6, 353, 0.35 x 0.25 --

04150 U 4150) 07 57.1 + 39 59 = N 2495 at 1.9 nf, 0.30 x 0.18, m = 15.6; 07 57.2 + 40 00 at 4.2 nf, 0.25 x 0.20, m = 15.6 -- U 4153) U 4159 at 2.9 -- U 4154) comp 1.8, 85, 0.4 x 0.2 -- U 4155) 07 56.9 + 26 48 = I 484 at 5.4, 305, 0.7 x 0.35, Sa:, m = 15.7 -- U 4156) P w 07 57.1 + 26 50 at 1.3, 260, 0.3 x 0.2, m = 15.7; 07 56.9 + 26 48 at 4.7, 242, see U 4155; U 4155 at 5.3, 176 -- U 4159) P w U 4169 at 5.2 f --

04160 U 4161) comp 4.6, 290, 0.5 x 0.08, w comp 0.1 x 0.1, sep 0.20; U 4164 at 5.9 -- U 4164) see U 4161 -- U 4165) S B(rs)d (de V), Sc + (Ho) -- U 4166) N 2523A in BG -- U 4167) a) at 0.9, 345, 0.9 x 0.3, late spir or Irr -- U 4168) comp 1.5, 143, 0.50 x 0.10; 1.8, 207, 0.40 x 0.15 -- U 4169) see U 4159

04170 U 4172) comp 3.5, 116, 0.4 x 0.15; 4.0, 169, 0.35 x 0.30 -- U 4173) chaotic, dif, comp superimp? -- U 4178) U 4184 = = N 2513 at 5.6, 142 -- U 4179) spir w 3 stell comps superimp --

04180 U 4182) comp 0.55, 350, 0.2 x 0.1, v long dif plume northeast; P w 07 59.6 + 61 25 at 3.5, 182, 0.8 x 0.5, spir, m = = 15.1, disturbed?, comp 1.6 np, 0.4 x 0.15, e dif bridge? -- U 4183) comp 1.5, 359, 0.45 x 0.1, spir -- U 4184) see U 4178; 07 59.5 + 09 32 = N 2511 at 2.9, 244, 0.8 x 0.5, S0:, m = 15.0; 07 59.4 + 09 25 at 4.6, 227, 0.5 x 0.4, S B 0:, m = 15.6 -- U 4185) U 4186 at 0.7 f; 2nd group member at 1.2, 168, 0.8 x 0.15 -- U 4186) see U 4185; 2nd group member at 1.1, 202; comp 4.5, 132, 0.15 x 0.15 + env -- U 4187) P w U 4195 at 3.3 f --

04190 U 4190) P w 08 00.2 + 16 29 at 2.2, 28, 0.50 x 0.15, m = 15.6 -- U 4195) see U 4187 -- U 4196) comp 2.9, 285, 0.3 x 0.12; 3.6, 291, 0.6 x 0.3 -- U 4197) P w U 4198 at 8.1, 142; 08 01.0 + 10 12 at 4.6, 90, 0.6: x 0.3:, spir:, m = 15.5 -- U 4198) see U 4197; 08 01.3 + 08 09 at 5.1, 53, 0.6 x 0.5, spir, m = 15.2 -- U 4199) MCG α = 08 05.0 is incorrect; P w 08 00.3 + 73 28 at 2.2, 248, 0.7 x 0.3, m = 15.6 --

04200 U 4200) 08 00.9 + 40 20 at 2.5 sp, 0.30 x 0.25, compact, m = 15.0 -- U 4209) comp 1.2, 297, 0.35 x 0.35 --

04210 U 4211) 08 02.2 + 10 59 at 5.0, 28, 0.45 x 0.35, m = 15.1 -- U 4212) S B(r) ? (de V) -- U 4216) P w U 4220 at 6.3, 52 --

U 4218) 1st in chain of 5 galaxies; U 4232 at 16.5, 80 --

04220   U 4220) see U 4216 -- U 4221) comp 1.8, 322, 0.35 x 0.10, spir -- U 4223) in group w comps at 1.1, 225, 0.8 x 0.4, SO?, and 1.6, 354, 0.25 x 0.20 --

04230   U 4230) P w 08 04.4+08 11 = I 2228 at 3.3, 36, 0.7 x 0.6, S B..., m=15.3 -- U 4232) 2nd in chain of 5, see U4218; U 4237 at 8.9, 63 -- U 4235) VII Zw 212; "intercon w blue disc-like compact" (Zw) at 0.30, 340, 0.20 x 0.10; U4241 at 4.6 -- U 4237) sev S blue conds in np part; 3rd in chain of 5 w U 4218 brightest at 25.3; see U 4232 -- U4238) ecc bar; v blue; comp 5.1, 136, 0.5 x 0.15, double -- U 4239) 08 05.6+05 21 at 4.6, 83, 0.5 x 0.4, early, m=15.7; comp 4.3, 178, 0.35 x 0.20 --

04240   U 4241) sev v dif obj n; 08 05.2+57 59 at 4.3 n, 0.8 x 0.6, m=15.3; comp 5.1 s, 0.9 x 0.45 -- U 4243) MCG δ = 67 34 is incorrect -- U 4245) knotty arm n; comp 3.3, 282, 0.30 x 0.20; nr 5 in chain of 5 w U 4218 brightest at 46.; 08 05.8+18 14 at 6.6, 190, 0.8 x 0.6, SO-a, m=15.3 -- U 4247) comp 4.4, 163, 0.5 x 0.3 -- U 4248) broad alm closed (rs) 0.9 x 0.8; P w U4251 at 4.9, 103; comp 2.7, 302, 0.4 x 0.4, F --

04250   U 4251) see U 4248 -- U 4255) strongly deformed spir pattern, B arc in northwest, exc nucl region; no disturbing obj vis; comp 1.7, 32, 0.3 x 0.3 -- U 4256) SAB(rs)c (de V) -- U 4257) in group w 08 07.0+25 03 at 2.1, 325, 0.8 x 0.4, SO-a, m=15.2 (I 497) and 08 07.1+25 01 at 1.0, 187, 0.4 x 0.4, Sb-c, m=15.6 -- U 4259) N 2523 B in BG; SA(s)b: (de V); U 4271 at 8.8 f --

04260   U 4260) DDO 49; Ir IV-V: (vdB) -- U 4261) "system w plume + jet" (CGCG) -- U 4262) comp superimp at 0.55, 190, 0.5 x 0.1, no evidence of interaction -- U 4263) VII Zw 216; "post-eruptive, blue, split oval compact" (Zw) -- U 4264) VV 9, Arp 82; SA(r)c pec (de V); 2 long v open arms; N 2536 is a comp at the tip of one arm at 1.8, 162, 0.8 x 0.5, S... pec (de V), v = 4072, v₀ = 3983; "arm opposite comp e long" (Arp); SN 1901a -- U 4265) P w comp at 0.9, 153, 0.20 x 0.20, appears compact, prob incl in CGCG magn; comp 2.9, 146, 0.6 x 0.4 -- U 4267) early spir? -- U 4268) vB complex centr region, multiple?; E1 (de V) --

04270   U 4270) comp 3.6, 302, 0.35 x 0.30 -- U 4271) Arp 9; in Arp's class "spir galaxies w split arms"; S B(r)bc (de V); "bifurcated arm does not start at end of bar" (Arp); comp 4.5, 190, 0.7 :: x 0.6 ::, e dif; see U 4259 -- U 4273) S B(r)b (de V); comp 3.4, 178, 0.35 x 0.15 -- U 4274) VV 138, Arp 6; in Arp's class "spir galaxies w low surf brightn"; I B s)m pec (de V); comp 4.6, 90, 0.6 x 0.6, N 2537 A in BG, S B(rs)c (de V); U 4278 at 17.0 -- U 4278) see U 4274; S d: (de V), S... (Ho) --

04280   U 4280) comp 4.3, 245, 0.3: x 0.25 --- 08 10.6+58 07) VII Zw 220; "red spherical compact (poss plus *), m_p = 15.6. 15" northeast of E galaxy. Total brightn of the pair, m_p = 14.4" (Zw); in CGCG, omitted in UGC -- U 4281) in a dense part of the cl; comp superimp at 0.35, 240, prob F bridge; "double system, contact + jet" (CGCG) -- U 4282) P w comp at 2.0, 188, 0.6 x 0.22, Sc -- U 4284) SA(s)cd (de V), Sc + (Ho); "in group w N 2500, 2552" (de V) -- U 4286) P w comp at 1.4, 264, 0.35 x 0.12, prob interaction -- U 4287) (R)S B(r)ab (de V); "one-armed spir starting from inner loop" (Morgan) -- U 4289) in a dense part of the cl; 08 11.6+58 27 at 1.7, 208, 0.8 x 0.7, S B 0, m=15.3; comp 2.3, 71, 0.8 x 0.10 --

04290   U 4290) N 2523 C in BG; E4: (de V); SO?; comp 2.7, 314, 0.5 x 0.25 -- U 4295) P w U 4302 at 2.4, 53 -- U 4299) P w 08 13.1+23 16 = I 2248 at 4.2, 162, 0.9: x 0.7:, S O-a, m=15.1; comp 2.6, 305, 0.5: x 0.3 --

04300   U 4302) see U 4295 -- U 4303) early; P w comp at 3.3, 347, 0.5 x 0.4 -- U 4305) Arp 268, VII Zw 223, DDO 50; in Arp's class "galaxies w irr clumps"; An 0814 in BG; I m (de V), Ir I (Ho), Ir IV-V (vdB); "resolution of **. Note linear loop of em regions" (Arp); "large post-eruptive blue Irr w compact core and many knots" (Zw) -- U 4308) SN 1962f; comp 2.9, 228, 0.4 x 0.4, v dif; 3.3, 68, 0.4 x 0.15 --

04310   U 4311) comp 1.7, 131, 0.35 x 0.1 -- U 4312) vF slightly asym disc w spir arcs; comp 1.5, 279, 0.5: x 0.5: -- U 4313) comp in env at 0.8, 215, 0.4 x 0.1, F; SA(r)0/a (de V) -- U 4314) VII Zw 224; blue obj w a v blue comp superimp p center, sep 0.20; "blue post-eruptive galaxy w red nucl and blue peripheral knots" (Zw) -- U 4315) spindle w comp attached to s end, 0.2: x 0.2:; comp 1.6, 19, 0.3 x 0.2; sev obj in vic, 08 15.1+24 57 = I 2268 at 3.8, 18, 0.25 x 0.25, m=15.3, B disc, sharp border -- U 4316) comp 3.9, 91, 0.7 x 0.3:, v dif -- U 4318) P w comp at 0.8, 43, 0.20 x 0.15, prob interaction -- U 4319) comp 3.1, 296, 0.3 x 0.3 --

04320   U 4320) comp 3.6, 298, (0.8 x 0.7), v red -- U 4321) comp 1.5, 79, 0.3 x 0.1 -- U 4323) "eF jets" (CGCG) -- U 4324) prob early spir; P w 08 15.6+20 56 at 1.8, 332, 0.45 x 0.35, m=15.6 -- U 4325) SA(s)m? (de V) -- U 4327) S B 0?; MCG minor diam 0.2 is incorrect; P w 08 16.2+74 09 at 1.3, 62, 0.9 x 0.18, m=15.5; SO: (de V) -- U 4329) 08 16.0+ + 21 22 at 3.6, 304, 0.7 x 0.22, S B:..., m=15.5 --

04330   U 4330) 08 19.6+21 33 = I 2293 at 5.5, 117, 0.9 x 0.7, S B(rs)a-b, m=15.2 -- U 4333) comp 1.8, 149, 0.3 x 0.1, spir; 4.7, 264, 0.4 x 0.3, dif -- U 4334) SN 1960m; S B b (de V); 08 16.8+22 22 at 1.8, 290, 0.45 x 0.22, m=15.5 -- U 4335) P w 08 16.5+54 17 at 3.1, 286, 0.7 x 0.18, m=15.4 -- U 4336) short massive arms, thick bar; early, prob non-regular -- U 4337) in a dense part of the cl -- U 4338) 08 17.1+55 20 at 4.6 s, 0.4 x 0.3, core 0.12 x 0.12, very compact, m=15.3; comp 4.2, 98, 0.3 x 0.12 -- U 4339) Sc w abnormal colour? (much too red); comp 4.3, 185, 0.45 x 0.10 --

04340   U 4340) comp 3.3, 126, 0.45 x 0.15; U 4346 at 9.6, 138 -- U 4342) comp 1.3, 126, double system, 0.15 x 0.15 and 0.15 x x 0.05, sep 0.20; 3.9, 79, 0.8 x 0.7, Sc -- U 4344) SN 1960d -- U 4345) in a dense part of the cl; comp 1.7, 206, 0.5 x 0.4, Sc; U 4347 at 4.7, 144; Sa (de V) -- U 4346) comp 4.4 see U 4340 -- U 4347) SO (de V); in a dense part of the cl; see U 4345 -- U 4348) v low surf brightn, tend to bar and spir pattern, no spir arms; P w 08 17 + 86 09 at 2.3, 349, 0.9 x 0.7, Sc:, m=15.2 -- U 4349) P w triple system U 4352 at 2.9, 140 --

04350   U 4350) P w comp at 1.0, 69, 0.45 x 0.25, prob no interaction -- U 4351) brightest obj in a cl not incl in CGCG -- U 4352) spindle 0.8 x 0.12 in contact w double system 0.8 x 0.8; see U 4349 -- U 4354) P w 08 18.5+21 02 = N 2569 at 2.6, 188, 0.6 x 0.5, early spir, m=15.3 -- U 4357) comp 5.0, 221, 0.8 x 0.7, S B a -- U 4358) comp 6.0 f, 0.1 x x 0.1, bridge, incl in diam; in group w comps at 1.6, 138, 0.40 x 0.25; 4.3, 102, 0.9 x 0.5, dif, chaotic; sev vF S obj in the group --

04360   U 4361) bar? -- U 4362) SA(s)0 (de V); see U 4363; comp 2.0, 285, 0.4 x 0.2, v dif -- U 4363) DDO 51; S B IV-V: (vdB); P w 08 19.4 + 74 40 at 3.8, 7, 0.8 x 0.7, m=14.7, ident as N 2551 in MCG -- U 4365) comp 3.6, 247, 0.3 x 0.2 -- U 4367) P w U 4375 at 9.3, 42 -- U 4368) comp 2.0, 89, 0.35 x 0.1 --

04370   U 4373) P w comp at 2.2, 282, 0.6 x 0.15 -- U 4375) see U 4367; comp 4.3, 102, 0.4 x 0.1 --

04380   U 4380) comp 2.0, 110, 0.35 x 0.12; 4.7, 153, 0.3 x 0.1; 4.8, 184, 0.35 x 0.18, * superimp?; 5.0, 139, 0.3 x 0.3 -- U 4383) Arp 247; a) 0.6 x 0.6, b) 1.0 x 0.6 nf, sep 0.7; "spirals have common arm, n spir arm continues northeast" (Arp); U 4384 at 5.7, 18 -- U 4385) v blue, irr; high surf brightn, filaments -- U 4386) P w 08 21.0 + 21 08 at 3.5, 205, 0.7 x 0.7, S B O-a, m=15.6 -- U 4388) comp 3.7, 7, 0.3 x 0.1 -- U 4389) VV 157; spir w (double?) cond in the disc at 0.2 f center; listed as triple system by VV --

04390   U 4391) S-shaped, alm circular arms; comp 4.2, 21, 0.45 x 0.2 -- U 4394) early spir?, high surf brightn; P w U 4402 at 7.8, 105 -- U 4395) U 4402 at 17. -- U 4396) MCG δ = 83 28 is incorrect -- U 4397) slightly asym patchy arms; U 4313 at 7.0; U 4413 at 8.1 -- U 4399) SN 1960 n --

04400   U 4402) see U 4395 -- U 4405) comp 3.5, 175, 0.4 x 0.1 -- U 4406) 08 23.4+23 04 at 4.6, 228, 0.9 x 0.15, m=15.6, vF ring -- U 4408) comp 1.6, 272, 0.3 x 0.1; 2.8, 328, 0.3 x 0.2 --

04410   U 4410) irregularity in the outskirts; comp 0.8, 335, 0.3 x 0.2, v dif; 08 24.2 + 55 15 at 4.1 sf, 0.55 x 0.40, m =
= 14.8 -- U 4411) comp 1.3, 99, 0.6 x 0.1; in triple group w 08 24.3 + 26 02 = N 2594 at 5.3, 160, (0.8 x 0.7),
E - S0, m = 15.0 and U 4118 at 15.5, 162 -- U 4413) N 2550 A in BG; comp 1.0, 353, 0.3 x 0.3; see U 4397 -- U 4414)
08 24.1 + 21 53 at 3.9, 343, 0.55 x 0.18, lens w high surf brightn, m = 14.8 -- U 4416) irregularity in n part, S
comp superimp?; comp 1.6, 314, 0.5: x 0.2:, v dif, dwarf; 08 24.2 + 23 00 at 3.0, 233, 0.8 x 0.25, early spir,
m = 15.4 -- U 4417) I Zw 14; "blue patchy spherical compact" (Zw); "B core less than 1" = 0.6 kpc in diam in disc
of uniform brightn 7" = 4.1 kpc diam. The em comes from the whole disc" (Sargent); alm stell on PA maps -- U 4418)
see U 4411 --

04420   U 4420) comp 2.7, 197, 0.9 x 0.12; 4.8, 283, 0.4 x 0.2, F -- U 4422) III Zw 59; slightly asym spir pattern, no
disturbing obj vis; S B...: (de V); "S Bc w compact blue core and blue spir arms" (Zw); "barred spir w prominent
nucl and distorted outer regions which extend to 70" = 18 kpc radius. There are no other galaxies nearby" (Sargent)
U 4423) asym disc; cond 0.30 from center may be a disturbing comp -- U 4424) comp 3.8, 13, 0.3 x 0.2 -- U 4425)
integral-shaped spindle, or v open S B? -- U 4426) DDO 52; V (vdB) -- U 4427) P w 08 25.4 + 55 40 at 0.8, 107,
0.6 x 0.3, m = 14.9 -- U 4429) 08 25.1 + 40 48 at 4.6 sp, 0.5 x 0.3, E:, m = 15.0 --

04430   U 4430) brightest in group of 5 or more galaxies; comp 1.5, 223, 0.35 x 0.1 -- U 4433) vB core, asym disc, tend to
spir pattern but no arms -- U 4438) patchy; B (r) or (rs) 0.6 x 0.25, broad arms --

04440   U 4440) P w comp at 1.5, 30, 0.35 x 0.15; comp 1.4, 184, 0.5: x 0.4:, v dif, dwarf -- U 4441) P w comp 1.5, 46,
0.3 x 0.3 -- U 4447) 08 27.6 + 19 54 at 6.9, 222, 0.9 x 0.10, (late) Sc, m = 15.6 --

04450   U 4451) b) = Mark 15 at 1.0, 145, 0.5 x 0.20 -- U 4452) P w 08 28.4 + 09 38 at 6.7, 186, (0.5 x 0.5), S0?, m = 15.3,
eF v ext disc or corona; comp 3.3, 87, 0.5 x 0.2, v dif -- U 4454) asym; B nucl cond s center -- U 4457) Arp 58; 2 long
v open arms, one broad and broken, "comp on end of broken arm nearly star-like" (Arp) -- U 4458) SN 1965 p -- U 4459)
VII Zw 238, DDO 53; Ir V (vdB); "resolved dwarf Irr w formation of blue knots n" (Zw) --

04460   U 4460) S B(s)b: w comp in spir arm at 0.25 p center, 0.15 x 0.12 -- U 4461) slightly asym, may be interacting w
stell comp at 0.8, 218, 0.08 x 0.08 (at tip of spir arm); 2 conds n center may be superimp comps -- U 4463) comp
2.0, 85, 0.4 x 0.15, F -- U 4464) P w comp 1.5, 280, 0.4 x 0.3, E, appears disturbed -- U 4465) comp 3.9, 310,
0.8 x 0.4, disrupted, surrounded by sev S obj -- U 4469) 08 30.5 + 29 40 at 3.6, 130, 0.9 x 0.7, Irr, m = 15.6; blue
cond in the disc at 0.7, 216, 0.15 x 0.15; sev vF objs near --

04470   U 4470) Sb, or poss Sc; sev S comps -- U 4471) halo?; 08 30.7 + 41 35 at 3.0 sf, 0.45 x 0.22, spir, m = 15.7 --
U 4472) patchy; prob late Sc; U 4466 at 11.8, 187 -- U 4473) B core 0.15 x 0.15 in circular dif disc 0.7 x 0.7,
e dif outer disc -- U 4477) P w comp at 4.0, 248, 0.8: x 0.3:, core 0.15 x 0.15 -- U 4478) comp 3.3, 201, 0.3 x 0.1;
sev smaller galaxies near --

04480   U 4480) lens 0.3 x 0.3 attached to arm of S B(s)b nf, 0.9 x 0.8 -- U 4483) dif dwarf w F blue nucl cond, comp 0.2 x 0.1
superimp n center -- U 4484) Arp 12; S B(s)b: (de V); in Arp's class "spir galaxies w split arms"; 2 asym arms v
high surf brightn; 2 ** superimp bar; no disturbing obj vis; "nucl may be double or superposed *" (Arp); SN 1920 a --
U 4486) P w comp at 2.4, 318, 0.35 x 0.25 -- U 4487) P w comp at 0.6, 165, 0.5 x 0.25 -- U 4488) U 4489 at 3.2, 13 --
U 4489) see U 4488; 08 33.1 - 01 35 at 6.0, 18, 0.9 x 0.8, Sb:, m = 15.3 --

04490   U 4490) comp 1.5, 190, 0.4 x 0.1, vF -- U 4491) vB nucl region; prob similar to N 210 -- U 4492) v dif outside 2.0 x
x 1.7; U 4493 at 5.4, 162 -- U 4492 -- U 4494) comp 2.4, 308, 0.3 x 0.2, dif -- U 4496) comp 2.5, 122,
0.4 x 0.4, dif -- U 4497) blue, dif, strongly asym; hints of spir pattern?; comp 3.7, 127, 0.5 x 0.05, vF -- U 4499)
asym, w alm dissolved spir pattern; blue pretty B nucl cond; Mark 94 superimp 0.7 sf center, 0.12 x 0.12, v = 750 --

04500   U 4500) comp superimp at 0.15 from center, 0.4 x 0.1 -- U 4501) comp 3.5, 237, 0.6 x 0.3, v dif, dwarf; 4.2, 190,
0.7 x 0.15; 08 34.7 + 25 10 = N 2621 at 3.8 nf, 0.9 x 0.8 -- U 4503) many v thin arms, may be later than
S bc; P w U 4484 at 33., 243 -- U 4506) P w 08 35.5 + 19 53 = N 2625 at 3.2, 100, 0.35 x 0.30, m = 15.1, * superimp? --
U 4507) comp 3.5, 156, 0.35 x 0.2 -- U 4509) VV 79, Arp 243; S B b (de V); in Arp's class "galaxies w appearance of
fission"; triple centr region 0.55 x 0.45, 2 long curved streamers (or arms); "some vS B knots resolved in interior"
(Arp) --

04510   U 4511) comp 4.7, 238, 0.3 x 0.1 -- U 4516) blue; hints of spir pattern; comp 4.1, 275, 0.3 x 0.1, F -- U 4518)
asym, pec? --

04520   U 4520) comp 2.0, 265, 0.3 x 0.2; sev F galaxies near --

04530   U 4533) strongly asym; spir arcs n, filaments s -- U 4538) 2 well def v thin arms; 08 39.6 + 65 10 at 4.5 s, (0.6 x
x 0.6), core 0.22 x 0.22, very compact, m = 15.4; 08 39.7 + 65 10 at 4.5 sf, 0.45 x 0.12, m = 15.5 -- U 4539) ecc nucl
region, bar?; v smooth outer disc, well def border --

04540   U 4540) v blue cond superimp at 0.25 nf center, *?; comp 3.0, 334, 0.3 x 0.2 -- U 4541) P w 08 40.0 + 14 27 at 2.5,
122, 0.7 x 0.20, spir, m = 15.0; this pair is Arp 89; "Position of larger spir. Abs lanes in comp. Dif arms extend
beyond comp" (Arp) -- U 4544) (R)SA(r)a:? (de V), "poss SA(r)0 + " (de V); "vF thin spir arm" (Morgan) -- U 4545)
one v broad dif arm curved northw; centr region double? -- U 4547) ident as N 2629 in MCG; U 4523 at 12.3; U 4569
at 7.4 -- U 4548) comp 4.1, 165, 0.6 x 0.1, Sc, SN 1950 d, m approx 17. --

04550   U 4553) comp 2.1, 247, 0.30 x 0.20; 2.4, 192, 0.7: x 0.6:, v dif, dwarf: -- U 4555) SA ab (de V) -- U 4556) high surf
brightn, jet in pos angle 20; comp 2.9, 121, 0.4 x 0.2 -- U 4557) comp 2.0, 230, 0.5 x 0.4, dif --

04560   U 4563) P w U 4566 at 1.0 sf -- U 4564) early spir, interacting w comp at 0.6, 217, 0.3 x 0.3 -- U 4566) see U 4563
-- U 4567) P w comp at 4.4, 109, 0.4 x 0.3 -- U 4568) "spindle" w 3 sym conds, one in center -- U 4569) early S0?;
N 2630 - 1 in MCG; see U 4547; U 4577 at 6.3; comp 2.6, 78, 0.4 x 0.2 --

04570   U 4570) in triple group w comp at 1.0, 22, 0.7: x 0.4:, dif, and 1.7, 93, 0.3 x 0.2 -- U 4574) Arp 80; S B(s)b
(de V); in Arp's class "galaxies w large high surf brightn comps on arms"; brightest in a group incl U 4581 at
8.2; broad blue arms, centr region not compl developed to bar; "end of one arm heavy; abs break in same arm near
nucl" (Arp); (Arp's class "galaxies w one heavy arm" appears more appropiate) -- U 4576) S B(s)...? (de V); in a
group incl U 4583 at 8.; comp (triple or double system) at 0.6, 307, 0.25 x 0.18; 0.7, 147, 0.20 x 0.15 (or **?) --
U 4577) comp 2.1, 270, 0.45 x 0.20; see U 4569 -- U 4579) comp 1.0, 329, 0.3 x 0.1; 3.2, 344, 0.4 x 0.12 --

04580   U 4580) comp 3.5, 164, 0.3 x 0.1; sev smaller comps -- U 4581) E 1: (de V); P w U 4585 at 1.9 sf; see U 4574; comp
4.3, 357, 0.5 x 0.3 (common w U 4493) -- U 4582) P w 08 43.3 + 12 59 at 1.5, 83, 0.9 x 0.20, Sb:, m = 15.0 --
U 4583) E 0 (de V); disc w v high surf brightn -- U 4584) patchy; asym broken arms, no disturbing obj vis -- U 4585)
N 2634 A in BG; S B...: (de V); see U 4581 -- U 4588) comp 4.1, 79, 0.7 x 0.1, spir -- U 4589) in triple group w
U 4591 at 4.5, 26 and U 4595 at 5.6, 45 --

04590   U 4590) comp 3.5, 150, 0.7 x 0.15, spir -- U 4591) see U 4589; U 4596 at 4.5, 206; sev S galaxies near -- U 4592)
S-shaped, v open; comp 3.9, 232, 0.5 x 0.08 -- U 4593) v massive, v tightly wound arms; 08 44.1 + 70 21 at 3.9, 32,
0.35 x 0.20, compact, m = 15.5 -- U 4595) see U 4589, 4591 -- U 4597) asym; P w U 4602 at 5.2, 136; comp 0.45, 127,
0.4: x 0.4:; 2.6, 204, 0.3 x 0.3 --

04600   U 4602) see U 4597 -- U 4603) S B(r)b: (de V) -- U 4604) S B(r)0/a: (de V); note that "N 2646" in S - A is I 520 =
= U 4630 -- U 4605) S a? (de V) -- U 4606) P w U 4607 at 5.6, 170 -- U 4607) see U 4606 -- U 4609) stell core, e dif
irr env; comp 2.6, 125, 0.35 x 0.15 --

04610   U 4610) P w U 4613 at 3.8, 116 -- U 4612) comp to U 4646 at 4.5, 112 -- U 4613) see U 4610 -- U 4615) comp 1.4, 72,
0.55 x 0.05 -- U 4619) + U 4620) Arp 167; E 1 -2(U 4619; de V) and E + 0: (U 4620; de V); U 4620 is an E-type galaxy
in the halo of U 4619 at 0.6, 95 from center, (1.2 x 1.2); in Arp's class "galaxies w dif counter-tails"; "comp
galaxy v condensed, has curved plume" (Arp); comp 3.3, 277 from U 4619, 0.6 x 0.3, spir --

04620   U 4620) see U 4619 -- U 4627) vB core, e dif disc or corona; comp 3.0, 137, 0.35 x 0.12 --

04630   U 4630) SAB(rs)ab? (de V); ident as N 2646 in S - A; P w U 4634 at 2.6 -- U 4632) prob early S B 0; comp superimp at
0.2 f center, 0.20 x 0.15 -- U 4633) 08 48.4 + 53 07 at 2.9 s, 0.4 x 0.4, w comp 0.12 x 0.10 at 0.3, m = 15.3 --
U 4634) see U 4630 -- U 4635) B core 0.30 x 0.18 in asym halo, long curved plume northw - eastw -- U 4637) Arp 225;
SAB(s)0/a (de V); in Arp's class "galaxies w amorphous arms"; "vF dif outer arms, abs one side of nucl" (Arp);
"vB v large nucl w asym dark matter in lens (?) 1.9 x 1.4, vF smooth outer whorls" (de V) -- U 4638) VV 41, Arp
257; disrupted spir (?) w chaotic outer disc, main body 0.8 x 0.5; comp at 0.8 n, 0.35 x 0.30; "galaxies joined by
segment of thin arc" (Arp) --

04640   U 4641) SA(rs)b (de V), Sb - (Ho); "S vB nucl in peanut-shaped bulge, many filamentary arms w dark lanes on one
side, pseudo (R) 7.3 x 0.85" (de V); comp 3.2, 184, 0.7 x 0.1 -- U 4643) comp 2.4, 129, 0.4 x 0.05; 2.4, 283, 0.4 x
x 0.3, v dif -- U 4645) Pec SAB(rs)0/a (de V), Sa (Ho) -- U 4646) amorphous disc, v broad arms in the outskirts;
see U 4612 --

04650   U 4650) comp 5.0, 298, 0.4 x 0.3 -- U 4653) VV 243, Arp 195; S Bb w complex double (?) system superimp n center,
jet 1.0 in direction 280 (or dif spindle?) -- U 4656) bar?; P w comp at 2.4, 232, 0.6 x 0.5 -- U 2658) 08 51.3 +
+ 32 49 at 3.2 s, 0.6 x 0.15, m = 15.3 --

04660   U 4660) comp 1.5, 2, 0.3 x 0.1 -- U 4661) comp 0.4, 127, 0.3 x 0.1; 08 51.3 + 36 41 at 3.5 n, 0.5 x 0.5, S B...,
m = 15.7 -- U 4662) group of S galaxies p; vB centr region, asym disc -- U 4664) comp 3.1, 75, 0.35 x 0.35 -- U 4665)
comp 1.2, 83, 0.3 x 0.2; 08 52.8 + 68 27 at 7.2, 103, 0.35 x 0.18, m = 15.7 -- U 4666) Arp 336; (R)S B 0 (de V),
S0 pec (Ho); appears cigar-shaped, w equatorial rings -- U 4667) 08 51.5 + 20 25 = I 2422 at 5.3, 273, 0.22 x 0.20,
extremely compact, m = 15.6, eF halo? --

04670   U 4670) U 4677 at 7.6, 112; U 4685 at 16.2, 97 -- U 4671) vB core in dif slightly irr disc; P w U 4675 at 3.4 sf,
interaction? -- U 4672) comp 3.1, 291, 0.3 x 0.3, dif -- U 4674) E 3:(de V); CGCG δ = 51 03 is incorrect; P w
N 2694 at 0.9 s, 0.22 x 0.20, very compact, E (de V), m = 15.2, CGCG δ = 51 02 is incorrect -- U 4675) E: (de V); see
4671 -- U 4677) see U 4670 -- U 4685 at 9.1, 83 -- U 4678) prob = N 2704; (r) 0.55 x 0.35, v broad dif arms outside;
comp 3.8, 139, 0.5 x 0.2 -- U 4679) dif spindle; comp 5.0, 284, 0.7 x 0.6, S B a: --

04680   U 4681) comp 0.80, 115, 0.20 x 0.12, bridge; 1.0, 250, 0.25 x 0.20, dif; 1.1, 304, 0.45 x 0.05, dif; sev vF obj in vic
-- U 4683) sev comps , group of 3 dif obj at 3.0 sf -- U 4685) see U 4670, 4677 -- U 4686) comp 2.1, 92, 0.35 x 0.35,
spir -- U 4687) red obj w high surf brightn, well-def edge;   comp 2.1, 166, 0.7 x 0.15; 3.8, 225, 0.4 x 0.2 --

04690   U 4691) S B(rs)ab (de V); SN 1968 e; U 4692 at 10.9, 20 -- U 4692) (R)S B(r)0 + (de V); see U 4691 -- U 4694) P w
comp at 3.6, 270, 0.2 x 0.2, early -- U 4695) SAB(rs)c: (de V) -- U 4696) U 4695 at 8.3 n; comp 3.1, 123, 0.40 x
0.35, core + ring -- U 4699) vB complex centr region, dif env; "double nucl, F jet" (CGCG) --

04700   U 4703) brightest component 0.4 x 0.3 w thin straight slightly irr bridge to comp at 1.5, 304, 0.3 x 0.2; "post-
eruptive pair of compacts intercon w thin straight bridge. No 1, blue patchy, w jets and plumes, m_p = 15.5 ...
v = 3556 ... No 2, blue patchy, w various jets, 85" northwest of No 1, m_p = 18.5" (Zw) -- U 4704) sev comps , nearest
2.1, 153, 0.5 x 0.5 -- U 4706) elong ring 0.55 x 0.35 in bar; comp 3.2, 180, 0.45 x 0.15 -- U 4708) S B(r)b: (de V) --
U 4709) massive v tightly wound spir arms; sev cl members in vic --

04710   U 4710) P w U 4712 at 3.9, 103 -- U 4712) see U 4710 -- U 4715) core 0.12 x 0.12 w S stell obj attached n, long
curved plume northeast; comp 3.8, 68, 0.5 x 0.15 -- U 4718) Arp 202; high surf brightn, core + loop, e dif bridge
to comp at 0.5, 175, 0.45 x 0.35, spindle? w high surf brightn; the comp is N 2719 A in BG; in Arp's class "galaxies
w material ejected from nuclei"; "F tail from smaller galaxy" (Arp); in triple group w U 4726 at 9.6, 76 -- U 4719)
comp 1.7, 105, 0.35 x 0.30; 1.9, 307, 0.3 x 0.1, dif; sev smaller comps --

04720   U 4722) e dif plume from f side out to 2. from center -- U 4723) S0 (de V) -- U 4724) P w 08 57.4 + 17 36 at 4.3,
295, 0.6 x 0.5, m = 15.2; 2 smaller obj in the group -- U 4726) see U 4718 -- U 4727) P w U 4730 f -- U 4728) P w
08 57.6 + 17 22 at 5.3, 267, 0.9 x 0.2, Sb - c, m = 15.7 --

04730   U 4730) B main body 0.4 x 0.2, smooth outer disc, no spir pattern; see U 4727 -- U 4737) comp 4.6, 293, 0.3 x 0.3 --

04740   U 4742) in triple group w 08 59.4 + 14 43 at 2.2, 244, 0.5 x 0.15, m = 15.6, and 08 59.6 + 14 40 at 4.3, 170, 0.5 x 0.4,
m = 15.7 -- U 4743) chaotic patchy spir pattern; S blue cond in p part, disturbing comp?; comp 3.1, 146, 0.4 x 0.3,
Irr (dwarf?) -- U 4744) "red S B a w B core... v = 2670" (Zw); disturbed by comp at 0.8, 75, 0.4 x 0.3, bridge, "bluish
compact... m_p = 16.8... v = 2830" (Zw); this pair is VV 40, Arp 287 -- U 4748) spir 1.1 x 0.1 w comp 0.45 x 0.1 attached
0.30 ssf; in group w 09 00.5 + 10 21 brightest, at 8.0, 350, 0.9 x 0.5, S B b, m = 15.3 --

04750   U 4751) in a S group, incl U 4752 at 3.8, 2; comp 4.2, 229, 0.45 x 0.1 -- U 4752) see U 4751; 09 01.2 + 22 15 at 5.7,
10, 0.7 x 0.35, spir, m = 15.4 -- U 4757) SA(s)...pec (de V); U 4763 at 13.6, 132 -- U 4758) U 4752 at 11.4, 251 --
U 4759) SAB(rs)c (de V), Sc - (Ho); comp 4.3, 167, 0.5 x 0.3 --

04760   U 4761) sev comps , 7 obj inside 5.0 -- U 4763) E 3 (de V); U 4772 at 5.4, 72; see U 4757; 09 02.7 + 18 27 = N 2751
at 4.0, 320, 0.9 x 0.5, Sa:, m = 15.1 -- U 4764) B asym main body, e dif outside, outer ring?; U 4769 at 8.4; comp
3.5, 100, 0.3 x 0.2, double -- U 4765) main body 0.45 x 0.35, strongly asym dif env; a) at 1.0 ssp, 0.45 x 0.15 --
U 4769) slightly asym; distorted arm at p side brighter outw w chain of patches; comp 3.3, 154, 0.3 x 0.3; 4.8,
142, 0.7 x 0.2 --

04770   U 4770) sev F obj in vic, members in cl 09 03.2 + 35 38, medium compact, v distant -- U 4772) in triple group w
U 4763 at 5.4, 252, and U 4757 at 17.1 -- U 4773) early; 2 F vS conds attached at p side -- U 4778) comp 2.6, 273,
0.35 x 0.08 -- U 4779) SA(s)c: (de V) --

04780   U 4780) comp attached at 0.8, 188, 0.5: x 0.4:, dif, dwarf; comp 2.0, 216, 0.5: x 0.4:, dif -- U 4784) in chain w
09 03.9 + 46 25 at 1.8, 247, (0.9 x 0.9), prob bridge, compact, m = 15.6; and comp 4.2, 257, 0.35 x 0.35 -- U 4785)
B curved bar, asym outskirts; prob interacting w comp at 1.4, 318, 0.15 x 0.10 -- U 4788) smooth disc, appears v
early but the colour is abnormal (blue); comp 0.7, 163, 0.35 x 0.12 -- U 4789) comp 3.6, 25, 0.3 x 0.2; 3.8, 15,
0.35 x 0.15 --

04790   U 4790) comp 1.2, 93, 0.6: x 0.05; 1.9, 50, 0.9 x 0.08 -- U 4794) prob early spir -- U 4796) 09 05.4 + 54 06 at 2.8
n, 0.35 x 0.25, m = 15.5 -- U 4797) DDO 54; S V (vdB) -- U 4799) strongly asym, one long thin straight arm in direction
218; comp 3.1, 102, 0.6: x 0.4 --

04800   U 4801) (r) or (rs) 0.7 x 0.18 -- U 4802) U 4804 at 9.3, 157; 09 05.2 - 01 27 at 11.9, 194, 0.8 x 0.3, m = 15.6 --
U 4803) N 2742 A in BG; asym, early; amorphous disc; 2 comps superimp f and sff center -- U 4804) see U 4802;
09 05.2 - 01 27 at 7.1, 245 -- U 4806) SA(s)c (de V); comp 3.2, 273, 0.8 x 0.3, 2 nuclei in common env --

04810 U 4810) smooth F disc, no spir pattern; no S0 — U 4811) comp 0.9, 316, 0.3 x 0.2; 1.0, 177, 0.6 x 0.1; 1.1, 58, 0.5 x 0.1 — U 4812) a) 0.8 x 0.6, Sc; b) 0.35 x 0.30 at 0.9, 150 from a) — U 4813) S0?; MCG δ = 44 49 is incorrect; U 4816 at 3.7; U 4817 at 4.7; 09 06.4 + 50 38 = N 2762 at 2.9 p, 0.35 x 0.25, m=15.7 — U 4814) MCG 8-17-46 at this position, but MCG data do not refer to U 4814 — U 4815) comp 3.1, 132, 0.5 x 0.3 — U 4816) P w U4817 at 3.5, 161; see U 4813 — U 4817) see U 4813, 4816 — U 4818) S0 (de V); P w U4832 at 4.1 f —

04820 U 4820) SA(r)ab (de V) — U 4821) E+ 6 (de V) — U 4822) 09 07.6 + 19 40 at 3.7 np, 0.5 x 0.20, m=15.6 — U4823) comp 3.5, 251, 0.5 x 0.2, double; U 4820 at 11.4 — U 4825) SA bc (de V); comp 3.0, 340, 0.7: x 0.1, vF —

04830 U 4830) P w 09 08.4 + 19 36 at 6.3, 350, 0.8 x 0.5, spir, m=15.4; comp 3.0, 332, 0.6: x 0.4:, dwarf — U 4832) see U 4818; e dif halo — U 4833) P w comp at 3.0, 292, 0.20 x 0.15 — U 4836) VII Zw 266; "neutral oval ring barred spir, compact spherical core" (Zw); comp 2.1, 64, 0.6 x 0.2 — U 4837) DDO 55; IV-V or V (vdB) — U 4838) SAB(rs)c (de V) —

04840 U 4840) P w 09 09.3 + 35 15 = N 2779 at 1.8 n, 0.7 x 0.7, acicular?, m=15.5 — U 4841) SAB(s)d (de V), Sc + (Ho) — U 4843) massive v tightly wound arms — U 4847) U 4832 at 6.2 —

04850 U 4851) comp 5.0, 182, 0.5 x 0.15 — U 4852) comp 4.5, 140, 0.9 x 0.10, Sb - c — U 4856) P w U4859 at 1.5, 105; sev comps to the pair, see U 4859 — U 4857) comp 0.50, 240, 0.25 x 0.12, at tip of spir arm? — U 4858) DDO 58; Ir V (vdB) — U 4859) see U 4856; comp 0.60, 312 (inside env), 0.3 x 0.25; 1.8, 317, 0.25 x 0.22; 3.2, 337, 0.25 x 0.2

04860 U 4860) P w comp at 1.2, 75, 0.4 x 0.2, Irr? — U 4862) Arp 215, in Arp's class "galaxies w adjacent loops"; "dif outer arms" (Arp); SAB(rs)a pec (de V); v broad asym arm nf — U 4864) (r) 0.8 x 0.8, dif arms outside; comp 1.0, 339, 0.4 x 0.3 — U 4865) P w comp at 5.4, 349, 0.9: x 0.8:, dwarf — U 4866) P w 09 11.3 + 36 18 at 2.3, 285, 0.9 x 0.8, (R')S B(s)a, m=15.0 — U 4867) S-shaped; P w U4876 at 6.6, 68 — U 4868) P w U4874 at 5.9, 42 — U 4869) P w 09 11.7 +30 23 at 1.7, 38, 0.8 x 0.3, m=15.7 —

04870 U 4870) SN 1966a; data for MCG 8-7-62 and 8-7-63 are confused in MCG — U 4871) DDO 59; S(B) V (vdB) — U 4874) VV 131 (note errors in VV coord); VV 131 b is prob a superimp *; see U 4868 — U 4876) see U 4867 — U 4878) P w comp at 2.9, 197, 0.6 x 0.2, double; comp (interacting pair) at 2.9, 200 and 3.0, 195, 0.12 x 0.10 and 0.10 x 0.10, bridge — U 4879) VV 124; sev F v blue conds superimp; comp 4.2, 107, 0.35 x 0.12 —

04880 U 4880) comp 1.4, 207, 0.4 x 0.3, early — U 4881) VV 155, Arp 55; VV gives coord for a compact galaxy at 09 12.4 + + 44 26, m=15.4; "arm has four separate conds in line" (Arp) — U 4882) b) at 0.8 n, 0.40 x 0.12 — U 4884) comp 2.3, 29, 0.4 x 0.3; 2.5, 292, 0.5 x 0.4 — U 4885) main body 0.7 x 0.7, amorphous w dif outside; bridge to comp?; a large nr of S obj in vic, brightest in a subgroup of about 10 galaxies — U 4886) early, asym — U 4887) a large nr of S galaxies in vic; 09 13.4 + 17 52 at 2.4, 38, 0.5 x 0.25, m=15.5 — U 4888) Sc - (Ho) — U 4889) vB red nucl region, amorphous disc, v broad arms; conflicting criteria —

04890 U 4890) P w comp at 1.2, 54, 0.8: x 0.3: — U 4891) 09 13.3 + 17 57 at 3.4, 277, 0.3 x 0.3, compact, m=15.7 — U 4892) B nucl region; chaotic, jet?; "eruptive system" (CGCG) — U 4893) a) at 1.0, 234, 0.4 x 0.25; c) at 1.0, 108, 0.4 x 0.3; both incl in CGCG magn — U 4894) S B(s)m pec (de V); blue patchy spir arc on n edge; short B bar ecc in sf part; comp 1.5, 313, 0.3 x 0.2, interaction? — U 4896) U 4918 at 4.7, 96; comp 4.6, 322, 0.7 x 0.1; 6.0, 333, 0.5 x 0.15; 09 11.8 + 70 05 at 11.7 np, 0.8 x 0.7, S B..., m=15.7 — U 4897 + U 4898) sep 0.8, broad bridge; a) has curved filament opposite bridge — U 4899) sev comps ; U 4910 at 9.7 —

04900 U 4904) pair U 4905 + U 4909 at 5.4 n — U 4905) VV 50a, Arp 283a (see U 4909); S B(s)a pec (de V); P w U 4909 at 1.5 f, bridge; "arc of barely resolved knots curves into nucl of larger galaxy" (Arp) — U 4909) VV 50b, Arp 283b; see U 4905; S B(s)m? (de V) —

04910 U 4910) 6 galaxies inside 3.7; 09 14.2 + 20 14 = N 2807 at 2.8, 220, double system, 0.35 x 0.35 and 0.7 x 0.25, sep 0.8, m=15.1 — U 4913) cond in the disc 0.45 p center, comp?; comp 2.7, 312, 0.3: x 0.3:, dif — U 4914) S B(r)0 + (de V) — U 4916) P w 09 14.8 + 20 07 = N 2812 at 1.3, 205, 0.55 x 0.12, m=15.7; 09 14.7 + 20 04 at 4.3, 226, 0.35 x x 0.30, m=15.7 — U 4917) U 4922 at 5.8 sf — U 4918) see U 4896 —

04920 U 4926) comp 0.7, 212, 0.3 x 0.2 — U 4927) ext dif corona; comp 2.6, 260, 0.5: x 0.4:, spir — U 4928) P w 09 15.4 + + 51 35 at 2.8, 355, 0.9 x 0.2, spindle w high surf brightn, m=15.2 —

04930 U 4931) comp 3.6, 99, 0.6: x 0.6:, dwarf — U 4934) P w 09 16.2 + 51 04 at 1.6, 152, 0.35 x 0.30, early spir, m=15.7 — U 4936) SAB(rs)d (de V), Sc + (Ho); slightly asym, row of B patches opposite comps ; U 4952 at 10.7; U 4961 at 13.2; 09 17.5 + 64 27 = I 2458 at 11.2 nf, 0.5 x 0.2, m=15.1 —

04940 U 4941) N 2830 and N 2831 are confused in MCG; U 4942 (=N 2832) + N 2831 at 1.4 nff — U 4942) Arp 315; E 2 (de V); N 2831 at 0.4, 220, (0.6 x 0.6), E 1 (de V), v = 5155, v₀ = 5106, m=14.7; "companion E is quite compact" (Arp); 09 16.3 + 35 57 = N 2825 at 5.1, 265, 0.9 x 0.4, spir: (de V), m=15.3 — U 4949) P w comp at 1.6, 281, 0.45 x 0.30; in a group in s part of cl (w U 4947) —

04950 U 4952) I 0? (de V); see U 4936; U 4961 at 3.7, 83; 09 17.5 + 64 27 = I 2458 = Mark 108 at 2.2, 123, see U 4961 — U 4956) ident by Jonckheere as a planetary nebula, see Perek and Kohoutek, Catalogue of Planetary Nebulae, Prague 1967, p 10; triple group at 5.5 s, 09 17.3 + 01 09 brightest, 0.7 x 0.4, m=15.4 —

04960 U 4961) S B(s)c pec (de V), Sc+ (Ho); in contact w 09 17.5 + 64 27 = I 2458 = Mark 108 at 2.0 p, 0.5 x 0.2, complex, m=15.1, v = 1200, interaction?; see U 4936, 4952 — U 4962) P w comp at 2.0, 57, 0.35 x 0.22, high surf brightn — U 4963) sev galaxies near, 09 18.5 + 18 26 brightest at 9.1, 0.9 x 0.5, S B a, m=14.9 — U 4964) 2 v blue conds in disc; P w comp at 3.8, 40, 0.8 x 0.1, S comp nnp, m=15.6 — U 4966) SA(r)b: (de V), Sb - (Ho); SN 1912, SN 1957 a —

04970 U 4971) SA(r)a: (de V) — U 4974) comp 1.8, 301, 0.6 x 0.2; 3.2, 356, 0.3 x 0.2 —

04980 U 4981) comp 4.3, 314, 0.6 x 0.2; 4.6, 150, 0.4: x 0.3: — U 4985) comp 1.3, 325, 0.3: x 0.1, interaction?, bridge?; 2.7, 253, 0.6 x 0.1 — U 4986) S B 0: (de V), notice that data for N 2852 (U 4986) and N 2853 (U 4987) are confused in BG; P w U4987 at 2.2 — U 4987) SAB(r)a? (de V); see U 4986 —

04990 U 4991) VV 171; a) (0.9 x 0.8), E; b) 0.20 x 0.18 w eF corona; sep 0.45; in triple group w 09 20.6 + 22 33 at 0.6, 204 from b), 0.4 x 0.2, m=15.1; total diam of chain 1.7 x 0.8; in a dense part of the cl — U 4993) P w 09 20.5 + + 02 42 at 7.2, 191, 0.9 x 0.8, Sc, m=15.2 — U 4994) 09 20.9 + 25 01 at 4.5, 42, 0.30 x 0.12, m=15.6 — U 4995) Arp 285a (see U 4997); P w U4997 at 3.5 nf; in Arp's class "double galaxies, infall and attraction" — U 4996) comp 1.2, 58, 0.4 x 0.2, dif — U 4997) Arp 285b, see U 4995; "narrow tail leads away from n nucl" (Arp) — U 4999) 09 21.2 + 02 19 at 4.3, 117, 0.9 x 0.9, S B b, m=15.4, interaction?; 09 21.5 + 02 20 at 7.1, 98, 0.6 x 0.3, spir:, m=15.5; group member? at 8.2, 114, 0.35 x 0.15 —

05000 U 5000) Arp 1, in Arp's class "spir galaxies w low surf brightn"; 2 v long thin arms; 09 21.6 + 49 35 at 3.9 f, (0.6 x 0.2), core 0.18 x 0.18, very compact, m=15.6; comp 3.1, 37, 0.6 x 0.2 — U 5001) (R)S B(r)0 + (de V) —

05010 U 5016) in group of 10 or more galaxies; 09 22.4 + 49 32 at 3.3 np, 0.9: x 0.3, w comp 0.12 x 0.12 at 0.7, total m= = 15.6; 09 22.5 + 49 38 at 7.6, 343, double system, 0.35 x 0.25 and 0.8 x 0.10, sep 0.40, total m=15.5; comp 2.6, 168, 0.3: x 0.2: — U 5018) Arp 307a (see U 5021); E? (de V); P w U 5021 at 1.4, 110; comp 2.0, 47, 0.7 x 0.15 —

05020 U 5020) comp 1.7, 209, 0.7 x 0.5, Sc — U 5021) Arp 307b; see U 5018; "poss not interacting" (Arp); comp 1.9, 8, common w U 5018 — U 5024) comp 2.1, 176, 0.6 x 0.3 — U 5025) vB nucl region in ring — U 5028) VII Zw 280; "pair of blue post-eruptive spirals, sep = 70" east-west... p component has 2 compact cores and close pair of red compacts,

1' s, each m$_p$= 18.5" (Zw); VII Zw 280 is VV 106b, Arp 300a -- U 5029) VV 106a, Arp 300b; P w U 5028 at 1.3, 246; "note elongated feature pointing tow nucl of larger spir" (Arp) --

05030   U 5037) brightest in group of 6; 09 24.2 + 23 12 at 3.2, 230, 0.45 x 0.25, m=15.5; 09 24.5 + 23 15 at 1.8, 69, 0.4 x 0.4, m=15.6, this is prob I 538 (U 5037 is ident as I 538 in MCG) --

05040   U 5040) vF nucl cond, v dif disc, no spir pattern -- U 5042) asym knotty bar, dif disc w broad alm dissolved arms; cond 0.12 x 0.08 at 0.4 s center, disturbing comp? -- U 5043) central obj in subgroup of 7; 09 24.5 + 30 12 at 6.5, 0.6 x 0.25, m=15.6; 09 24.5 + 30 15 at 5.9, 0.5 x 0.2, m=15.7; 09 25.0 + 30 16 = I 2478 at 3.5, 29, 0.5 x 0.3, compact, m=15.6; 09 25.1 + 30 14 = I 2479 at 2.4, 82, 0.45 x 0.25, m=15.5 -- U 5044) Arp 237; 2 B nuclei sep 0.25, spir arc s; "knots in arm as large, not quite as B as nucl" (Arp); in group w comp at 2.6, 244, 0.8 x 0.15, 09 24.9 + 12 29, spir, m=15.6, and 2.6, 290, 0.4 x 0.2 -- U 5046) comp 2.1, 37, 0.3 x 0.1; 2.9, 178, 0.25 x 0.25 -- U 5048) irr spindle; comp 1.7, 17, 0.45 x 0.2 --

05050   U 5050) VV 58, Arp 207; asym;   short jet-shaped feature in pos angle 50; 09 26.3 + 76 46 at 5.6, 25, 0.55 x 0.50, m=15.6 -- U 5051) S B O - (de V); 09 25.7 + 62 46 at 3.2, 10, 0.35 x 0.35, very compact, m = 15.1 -- U 5055) rare; bar 0.55 x 0.22 continues in 2 long slightly irr arms, these arms form a closed ring -- U 5056) comp 2.7, 357, 0.3 x x 0.2, dif -- U 5057) slightly curved chain with galaxies separated 0.40, 0.35 and C.30, hints of bridge betw obj's in center; poss group member at 1.8 sf center of chain -- U 5058) comp 2.6, 189, 0.30 x 0.12; U 5061 at 9.6; U 5073 at 13.2 --

05060   U 5061) see U 5058; U 5073 at 7.5 -- U 5065) filaments from the bar; appears too red to be a normal S B c -- U 5067) comp 3.4, 130, 0.35 x 0.18 -- U 5069) not in CGCG, but is prob brighter than 15.7 --

05070   U 5071) appears larger than 2' x 2'; not vis in red; not certainly real -- U 5073) comp? (*?) superimp at 0.30, 50, 0.12 x 0.12; see U 5058, U 5061 -- U 5077) sev comps, largest at 1.0, 240, 0.35 x 0.30; 1.2, 17, 0.35 x 0.35, vF -- U 5078) comp 3.7, 165, 0.4 x 0.1, vF -- U 5079) SAB(rs)bc (de V), Sc - (Ho); U 5086 at 9.6, 105 --

05080   U 5082) P w 09 29.7 + 10 59 at 3.0, 160, 0.4 x 0.4, E:, m=15.5 -- U 5084) P w 09 30.3 + 27 50 at 6.6, 16, 0.8 x x 0.5, (R')S B b, m=15.5 -- U 5086) see U 5079 --

05090   U 5091) MCG α = 09 29.5 is incorrect; asym; comp 4.1, 260, 0.6 x 0.3, distorted S B ... w comp -- U 5092) SA(s)... pec (de V); E-type galaxy w ext corona; P w U 5102 at 17.0, 63; group members U 5093 at 8.4, 156 and U 5096 at 4.8, 134; U 5092 is Arp 232, "abs lane reaching away from galaxy" (Arp) -- U 5093) see U 5092 -- U 5094) P w 09 31.6 + 24 26 at 2.9, 90, 0.45 x 0.30, m=15.5 -- U 5096) Arp 137; S B(s)ab (de V), in Arp's class "material emanating from E galaxies" -- U 5097) v high surf brightn; P w comp at 1.0, 355, 0.45 x 0.25, incl in CGCG magn -- U 5099) stell obj superimp p center, *?; comp 2.8, 166, 0.5 x 0.15 --

05100   U 5101) "eruptive galaxy, filaments" (CGCG) -- U 5102) SAB(r)b: (de V); see U 5092 -- U 5103) asym; radial filaments n, terminating on a patchy arc; spir pattern s center; comp 2.3, 140, 0.4 x 0.2; 09 32.3 + 21 52 at 5.0, 157, 0.8 x x 0.18, m=15.6 -- U 5107) comp 4.7, 62, 0.5: x 0.4:, dif -- U 5108) B main body 0.45 x 0.35, 2 smooth arms; comp 0.8, 347, 0.3: x 0.2: -- U 5109) asym, spindle? --

05110   U 5112) comp 3.3, 297, 0.4 x 0.1 -- U 5116) comp 0.8, 158, 0.1 x 0.1, vF, prob interaction -- U 5117) a) 0.3 x 0.3; b) 0.5 x 0.2, vF comp 0.35 np; in triple group w comp at 2.4, 275, 0.3 x 0.1 --

05120   U 5122) S blue cond in spir arm sp center -- U 5126) in group w 09 34.7 + 23 26 = N 2930 at 2.8, 15, 0.8 x 0.40, m = = 14.7; 09 34.8 + 23 28 = N 2931 at 5.2, 20, 0.7 x 0.6, m=14.9; 09 35.1 + 23 22 at 5.6, 97, 0.6 x 0.35, double, m= = 15.5 -- U 5127) P w comp at 1.3, 346, 0.6: x 0.4, dif; 3rd group member at 4.0, 127, 0.30 x 0.25 --

05130   U 5130) + U 5131) VV 316, Arp 142; E 0.8 x 0.5, P w multiple system 1.6 x 0.8; comp 2.8 np, 0.7 x 0.2; VV δ = = 03 48 is incorrect -- U 5132) comp 0.8, 15, 0.5 x 0.05, blue spindle -- U 5134) 09 35.4 + + 09 50 = N 2940 at 5.6, 354, 0.9 x 0.7, E - SO, m=14.8; 09 35.6 + 09 40 = I 548 at 5.4, 150, 0.7 x 0.2, early, m= = 15.3 -- U 5136) sev comps, 09 35.6 + 17 17 = N 2941 brightest at 2.2, 292, 0.55 x 0.45, SO - a, m=15.1; possible SN 1855 -- U 5139) DDO 63; Ir I (Ho), Ir V (vdB); An 0936 in BG --

05140   U 5140) SA(s)c (de V); comp 2.8, 250, 0.3 x 0.2 -- U 5143) U 5136 at 6.9, 283 -- U 5144) VV 82, Arp 63; "blue S B c w 2 compact knots" (Zw); double system U 5146 at 3.8, 25 -- U 5145) asym, alm on edge; no disturbing obj vis -- U 5146) VV 83, Arp 129, III Zw 60; a) disrupted spir 0.9 x 0.8, b) early spir? 0.6 x 0.5, sep 0.45; "post-eruptive pair of blue and red spherical compacts, intercon by blue spir arms, total m$_p$=13.3 [m$_p$=14.3 in CGCG]" (Zw) -- U 5147) patchy slightly curved spindle; sev F obj in vic --

05150   U 5153) 2 long arms w B knots; comp 4.0, 218, 0.3 x 0.1 -- U 5156) P w 09 38.1 + 25 41 at 6.0, 102, 0.9 x 0.2, m= = 15.4 --

05160   U 5161) P w U 5162 at 0.85, 123; comp 0.35 sp, 0.1 x 0.1, bridge?; stell obj 0.2 nnf -- U 5162) see U 5161 -- U 5163) P w comp 09 37.5 + 68 37 at 3.5, 250, 0.5 x 0.3, m=15.2; in group? w comp at 5.4 f, 0.9: x 0.5:, dwarf; vF spindle 9.9 f -- U 5164) P w comp at 0.45, 180, 0.15 x 0.12; 09 38.4 + 11 44 at 5.7, 124, 0.8 x 0.1, Sc, m=15.6 -- U 5165) comp 0.9, 249, 0.15 x 0.15; 1.8, 342, 0.3 x 0.2; 4 group members brighter than 15.8; 09 38.0 + 21 28 = Mark 403 at 4.3, 305, 0.45 x 0.40, m=15.4; 09 38.3 + 21 27 at 2.0, 33, 0.6 x 0.2, m=15.2, brightest in the group -- U 5166) Pec SA(r)b (de V) -- U 5167) (R)SAB(rs)0+ (de V) --

05170   U 5171) P w 09 38.9 + 10 52 at 3.7, 95, 0.7 x 0.5, SO - a, m=15.4 -- U 5172) v low surf brightn; sev galaxies near; brightest at 1.4, 197, 0.5 x 0.1 and 2.5, 270, 0.55 x 0.55 -- U 5173) 09 38.9 + 11 41 at 3.6, 15, 0.6 x 0.25, m=15.5; comp 4.2, 24, 0.3 x 0.1; 09 39.1 + 11 42 at 5.5, 41, 0.8 x 0.6, SO, m=15.0 -- U 5176) (R)S B(r)0/a (de V) --

05180   U 5180) SA(s)c (de V) -- U 5181) 2 v broad ill def arms; 3 or 4 stell obj superimp -- U 5182) stell comp superimp 0.30 sp center; comp 2.9, 258, 0.55 x 0.15; 4.4, 349, 0.5 x 0.3 -- U 5183) SAB(r)bc: (de V), Sc - (Ho); P w U 5190 at 6.3, 37; comp 4.2, 160, 0.8: x 0.7:, dwarf; Mark 404 is a cond in disc 0.6 f center -- U 5185) a) at 1.5, 327, 0.22 x 0.15; 09 39.5 + 29 11 at 6.3, 254, 0.6 x 0.4, SO - a, m=15.6; b) 0.3 x 0.3 -- U 5186) comp 3.7, 14, 0.4 x 0.3; 4.9, 23, 0.3 x 0.3 -- U 5189) chaotic arc-shaped obj, brightest in np part; prob chain of disrupted galaxies --

05190   U 5190) I 0 (de V), Ir II (Ho); comp 0.7 sf, 0.3 x 0.1, dif; 09 40.6 + 32 13 = N 2970 = Mark 405 at 5.1, 54, (0.9 x 0.9), E:, m=14.7 -- U 5191) U 5197 at 6.8, 126; 09 39.8 + 36 26 at 5.4, 247, 0.3 x 0.3, compact, m=15.7 -- U 5193) cores 0.18 x 0.15 and 0.12 x 0.10, sep 1.1, in v dif asym env -- U 5196) late-type obj in blue, listed in CGCG and MCG, non-existing on red PA map! -- U 5197) see U 5191 -- U 5198) comp 3.5, 356, 0.4 x 0.1, double -- U 5199) a) at 1.4, 215, 0.8 x 0.2, early spir; comp 2.5, 170, 0.4 x 0.15 --

05200   U 5200) prob in cl 151; comp 1.9, 120, 0.4 x 0.15 -- U 5202) SO? (de V); P w 09 41.2 + 68 50 = N 2961 at 1.5, 55, 0.7 x 0.2, m=15.6 -- U 5203) prob brighter than CGCG limit 15.7; U 5152 at 11.2 -- U 5205) v blue centr region; P w 09 41.5 + 00 25 at 1.7, 307, 0.5 x 0.3:, pec, m=15.5 -- U 5208) comp 1.7, 220, 0.35 x 0.05 --

05210   U 5210) comp 5.0, 166, 0.5 x 0.2 -- U 5211) asym v ext corona -- U 5213) comp 5.0, 122, 0.8 x 0.15, dif --

05220   U 5221) SA c pec (de V), S pec (Ho); B main body 3.4 x 1.4, F outer disc; comp 4.0, 98, 0.3 x 0.2, vF -- U 5222) P w 09 42.7 + 73 13 = N 2957 at 2.8, 297, double system, 0.8 x 0.3 and 0.4 x 0.2, sep 0.25, total m=15.3; comp 2.7, 102, 0.5 x 0.3 -- U 5225) I Zw 21; "neutral spherical compact, uniform circular disc halo... v = 4870" (Zw) -- -- U 5229) Sc (de V); v high surf brightn, sharp edge; comp (double system) at 1.3, 142, 0.4 x 0.15 and 1.5, 148, 0.4 x 0.2, sep 0.30 --

05230 U 5230) Arp 303 (w I 563); P w 09 43.8 + 03 16 = I 563 at 1.6, 185, 0.9 x 0.4, S B..., m = 14.7, v = 6030, v_0 = 5841,
An 0944 B in BG; U 5230 is An 0944 A in BG; in triple group? w U 5224 at 9.2, 231 -- U 5231) comp 2.3, 190, 0.3 x
x 0.3, S B... -- U 5232) early, in contact w comp at 0.6, 3, 0.3: x 0.1, vF; comp 3.3, 119, 0.3 x 0.2 -- U 5233)
a) = N 2988 at 0.50, 255, 0.7 x 0.2; U 5239 at 7.6, 53; sev alm stell obj s in vic -- U 5234) comp 1.9, 39, 0.5 x
x 0.3, 3.2, 161, 0.4: x 0.3:, e dif, dwarf -- U 5235) comp 3.4, 86, 0.30 x 0.20; 4.8, 26, 0.3 x 0.15; 09 44.5 + 23 15
at 5.8, 102, 0.4 x 0.15, m = 15.7 -- U 5239) see U 5233 --

05240 U 5240) in triple group w comp at 6.3, 231, 0.9: x 0.7:, Sc:, and 09 43.6 + 25 46 at 17.0, 223, 0.9: x 0.8:, m = 15.6
-- U 5241) 09 44.9 + 54 14 at 3.7, 95, 0.7 x 0.7, spir, m = 15.4 -- U 5243) massive bar, vF arms, disturbed? by
comp at 1.0, 187, 0.4 x 0.2, F; sev obj in vic, prob members of cl 09 43.7 + 58 35, medium compact, distant --
U 5244) 09 45.4 + 64 27 at 3.2, 56, 0.7 x 0.18, m = 15.7 -- U 5248) S stell comp superim close to center --

05250 U 5250) SAB(rs)c (de V); 09 46.1 + 44 15 = N 3006 at 6.9, 120, 0.7 x 0.15, m = 15.6; 09 46.2 + 44 08 at 11.7, 147,
0.9 x 0.3, m = 15.2; 09 46.4 + 44 19 = N 3008 at 14.2, 84, 0.5 x 0.3, m = 15.4; N 3000 is a double * nf -- U 5251) S bc?
(de V), Sc (Ho); SN 1961 f -- U 5253) Pec SA(rs)ab (de V), Sb+ (Ho); P w U 5316 at 24.8, 102; eF outskirts to approx
7. x 6. -- U 5258) comp 1.1, 197, 0.3 x 0.1, dif --

05260 U 5261) comp 0.8, 310, 0.6 x 0.5, dif spir, Sc:; 3.1, 310, 0.5 x 0.25, early -- U 5263) asym; MCG δ = 46 22 is
incorrect -- U 5264) U 5264 at 4.6 nff; U 5250 at 20. sp -- U 5265) P w U 5269 at 2.7, 97 -- U 5266) U 5271 at 8.2,
28; U 5275 at 9.8, 64; 09 47.4 + 12 58 = N 3019 at 5.1, 51, 0.9 x 0.6, spir, m = 15.0; 09 47.0 + 12 54 at 3.1, 225,
0.6 x 0.3, m = 15.6 -- U 5267) U 5270 at 4.2, 6; 09 47.0 + 09 14 at 6.7, 216, 0.9 x 0.4, S0:, m = 14.7; 09 47.3 + 09 23
at 4.1, 359, 0.4 x 0.2, double, m = 14.6 -- U 5269) see U 5265 --

05270 U 5270) P w comp at 0.45, 250, 0.5 x 0.2; see U 5267 -- U 5271) patchy (r) 1.1 x 0.5, B asym bar, sev ill def filam
arms; see U 5266; U 5275 at 5.7, 120; 09 47.4 + 12 58 = N 3019 at 4.1, 178, 0.9 x 0.6, spir, m = 15.0 -- U 5272)
DDO 64; Ir IV-V: (vdB); comp 2.0, 192, 0.3 x 0.2 -- U 5273) a) 0.7 x 0.4, m = 15.3; b) 0.4 x 0.2 at 0.55 nf a), m =
= 15.4; c) 0.4 x 0.2 at 1.0 f b), m = 15.8 -- U 5275) see U 5266, 5272; 09 47.4 + 12 58 = N 3019 at 5.0, 256, 0.9 x 0.6,
spir, m = 15.0 -- U 5276) core 0.35 x 0.25; eF env; U 5281 at 7., 35 --

05280 U 5280) SA(rs)bc: (de V); U 5251 at 30. spp -- U 5281) a) 0.9: x 0.15; b) 0.6 x 0.2, sep 1.3; see U 5276 -- U 5283)
blue spir; comp 1.9, 135, 0.6 x 0.4, dwarf -- U 5285) (rs) or (r)? 0.6 x 0.6; smooth eF arms, outskirts vis to approx
1.8 x 1.4 in blue; comp 2.9, 66, 0.35 x 0.30; 3.7, 135, 0.5 x 0.1 -- U 5287) asym; no disturbing obj vis; vF blue cond
in s part -- U 5289) late spindle 0.9 x 0.1 r.p dif spir 0.7: x 0.6: --

05290 U 5291) comp 0.75, 102, 0.2 x 0.2; 3.2, 320, 0.7 x 0.3 -- U 5292) SAB(r)0 (de V) -- U 5294) main body 0.6 x 0.3;
prob a vF jet in pos angle 137, l = 0.9 -- U 5295) comp 2.9, 309, 0.7 x 0.05; 5.2, 250, 0.7 x 0.7, dif; U 5301 at
4.9, 152 -- U 5296) U 5306 at 11. sf -- U 5297) comp 2.0, 323, 0.3 x 0.15 --

05300 U 5300) a) 0.7 x 0.2, early spir; b) 0.8 x 0.4, E, brightest in chain; sep 0.8; 3rd member in chain at 1.1 s b),
0.4 x 0.4; 09 50.2 + 19 36 at 4.3 s b), 0.8 x 0.7, S B c, m = 15.4 -- U 5301) see U 5295 -- U 5302) U 5318 = M 81 at
45. n -- U 5303) SAB(rs)c -- U 5304) VV 342, Arp 255; a) 0.9 x 0.8, late spir; b) 0.5 x 0.2,
connected by spir arm from a); U 5308 at 4.3, 91; comp 2.1, 226, 0.5 x 0.35; "F arm extends beyond high surf brightn
comp" (Arp) -- U 5306) P w comp at 3.7, 348, 0.7 x 0.35, spir; see U 5296 -- U 5307) comp 1.4, 2, 0.3 x 0.2 -- U 5308)
asym; see U 5304; comp 2.0, 100, 0.7 x 0.15 --

05310 U 5311) S B(s)c? (de V) -- U 5313) P w U 5320 at 7.5, 140 -- U 5315) CGCG δ = 37 46 is incorrect -- U 5316) see U 5253;
S B(rs)d: (de V), Sc+ (Ho); comp 2.7, 296, 0.35 x 0.1 -- U 5318) SA(s)ab (de V), Sb- (Ho); U 5336 (= Ho IX = DDO
66) at 10.8, 95; U 5322 (= M 82) at 37.; U 5398 at 46.5 --

05320 U 5320) see U 5313 -- U 5322) Arp 337; I 0 (de V), Ir II (Ho); see U 5318; U 5398 at 70. -- U 5323) a) 0.35 x 0.35;
b) 0.45 x 0.45; high surf brightn; sep 0.65; b) is much brighter than a) -- U 5326) blue, chaotic, approx circular;
comp 2.8, 56, 0.3 x 0.3 -- U 5328) SAB(s)c (de V) -- U 5329) U 5332 at 3.8, 108 --

05330 U 5330) main body 0.7 x 0.6; v ext e dif disc or env, prob strongly asym -- U 5332) see U 5329 -- U 5333) may be
interacting w comp at 0.9, 90, 0.15 x 0.15 -- U 5334) P w U 5334 at 3.1, 164 -- U 5335) high surf brightn, sharp
edge;   comp 4.3, 64, 0.7 x 0.3, early -- U 5336) DDO 66; Ir I (Ho); Ir V (vdB); see U 5318 -- U 5337) see U 5334 --
U 5338) P w U 5343 at 6.4, 101 -- U 5339) brightest in group of 5; 09 53.5 + 20 44 at 2.8, 342, 0.4 x 0.2, m = 15.5;
09 53.6 + 20 45 at 1.9, 296, 0.8 x 0.35, m = 15.7 --

05340 U 5340) DDO 68; pec (vdB); blue bar-shaped main body 0.6 x 0.3; no nucl cond; chaotic outer parts, spir arc or curved
plume northw -- U 5342) P w U 5344 at 6.2, 125 -- U 5343) see U 5338 -- U 5344) see U 5342 -- U 5345) data for MCG
8-18-46 and 8-18-47 are confused in MCG; U 5346 at 1.9, 29 -- U 5346) see U 5345 -- U 5348) blue spindle; comp 5.0,
116, 0.4 x 0.4, dif --

05350 U 5350) comp 4.2, 88, 0.6 x 0.3; 09 55.3 + 10 41 = I 580 at 5.1, 330, 0.7 x 0.18, S0:, m = 15.0 -- U 5351) SAB(s)ab?
(de V) -- U 5322) surrounded by a ring of 5 comps inside 2.0, diam from 0.45 to 0.50; 2 comps incl in CGCG magn --
U 5353) a) at 0.60, 228, 0.20 x 0.15; dif bridge, continues in a n e dif plume to 3. or 4. in southwest -- U 5355)
comp superimp at 0.45, 275, 0.10 x 0.10 and 0.30, 115, 0.7 x 0.1; dif asym outer disc -- U 5355) 4 F obj's inside 4.0
-- U 5336) comp 1.7, 145, 0.3 x 0.3 -- U 5358) S-shaped, dif; ecc nucl; comp 2.6, 318, 0.5 x 0.4, double --

05360 U 5360) comp 3.5, 242, 0.3 x 0.1, v dif -- U 5361) B core in F corona; sev F objs in vic -- U 5362) P w U 5363 at
1.1, 78; 3rd group member at 3.1, 98, 0.6 x 0.1, Irr -- U 5363) see U 5362; 3rd group member at 2.0, 110 -- U 5364)
DDO 69, An 0956 in BG; Ir I (Ho); Ir V (vdB) -- U 5365) P w U 5376 at 18.2, 79 -- U 5366) SN 1965 n; comp 4.7, 122,
0.5 x 0.2 -- U 6367) VV 321; a) 1.1 x 0.45 nnp b) at 0.6, 0.9 x 0.4 --

05370 U 5370) P w comp at 2.7, 359, 0.45 x 0.45, triple -- U 5371) P w U 5372 at 4.4, 139 -- U 5372) see U 5371 -- U 5373)
DDO 70, An 0957 in BG; Ir IV-V (vdB) -- U 5374) E pec ? (de V); U 5387 at 9.9 -- U 5375) VII Zw 303; SA(r)0/a
(de V); "red spherical disc compact, huge uniform halo" (Zw); P w U 5379 at 2.9, 155 -- U 5376) 10 at 10.4, 174;
U 5399 at 25.5, 110 -- U 5377) see U 5376 -- U 5379) asym main body 0.8 x 0.6; dif arm fragments; "blue post-eruptive,
compact cores and streamers" (Zw); Pec SAB(s)bc pec (de V); see U 5375 --

05380 U 5381) P w U 5384 at 6.9, 37 -- U 5382) 09 58.4 + 36 54 at 5.7 nf, 0.7 x 0.6, Sa-b, m = 15.3 -- U 5383) comp 2.1,
30, 0.45 x 0.35 -- U 5384) a) 0.9 x 0.8, S0; b) 0.7 x 0.2, spir; see U 5381 -- U 5387) S B(s)m (de V), Sc- (Ho);
see U 5374; 09 57.8 + 55 57 at 6.5 nf, 0.9 x 0.4, m = 14.6 -- U 5388) vB core; comp 2.6, 255, 0.35 x 0.15 --

05390 U 5391) 09 58.5 + 37 26 = I 2530 at 3.3 np, 0.5 x 0.5, compact, m = 15.5 -- U 5394) comp 2.2, 250, 0.30 x 0.05 --
U 5395) v dif objs at 0.7 and 1.1 nf, prob dwarfs; S stell obj at 0.65 nf, *? -- U 5396) comp 1.7, 54, 0.35 x 0.25,
spir -- U 5397) S0 (de V) -- U 5398) I 0 pec (de V), Ir II (Ho); "dust streamers tending to radial direction" (Morgan);
see U 5318, 5322 -- U 5399) see U 5376; comp 1.3, 103, 0.6: x 0.1 --

05400 U 5403) 09 59.5 + 19 27 at 5.1 np, 0.4 x 0.3, m = 15.7 -- U 5404) DDO 67; S IV (vdB) -- U 5405) comp 4.9, 274, 0.7 x
x 0.5 -- U 5408) VII Zw 308; "blue fuzzy spherical disc compact" (Zw) --

05410 U 5410) F comp at s edge, interaction? -- U 5413) comp 3.5, 221, 0.3 x 0.2 -- U 5414) VV 119, Arp 264; "F dif outer
material" (Arp) -- U 5417) open, S-shaped; arms w low surf brightn; 10 01.2 + 53 41 at 3.3 nf, 0.45 x 0.35, m = 15.3 --
U 5419) 10 00.7 + 30 24 at 5.6 f, 0.2 x 0.2, compact, m = 15.7 --

05420 U 5420) a) at 0.75, 185, 0.7 x 0.5 -- U 5423) P w 10 01.2 + 70 35 at 2.5, 210, 0.6 x 0.3, m = 15.7; S cond nf center --

U 5426) 10 02.0 + 15 01 at 4.4, 82, 0.5 x 0.3, core 0.1 x 0.1, very compact, m=15.6; comp 3.5, 337, 0.25 x 0.15 --
U 5428) DDO 71; V (vdB) --

05430   U 5430) comp 2.1, 199, 0.7 x 0.1 -- U 5431) P w U 5434 at 6.0, 147 -- U 5433) P w comp at 2.7, 262, 0.9 x 0.35,
dwarf -- U 5434) see U 5431 -- U 5436) 10 02.6 + 19 30 at 0.9, 159, 0.5 x 0.4, m=15.2, interaction; U 5450 at 5.9,
133 --

05440   U 5441) comp 2.1, 20, 0.4 x 0.2, dif -- U 5443) early, sep 0.20; P w 10 02.8 + 00 47 at 8.5, 230, 0.9 x 0.3, spir,
m=15.0 -- U 5447) a) E? or S0?, 0.6 x 0.4; b) 0.5 x 0.4, spir:; sep 0.40; sev vF objs in vic -- U 5448) VV 240;
a) at 0.50, 215, 0.15 x 0.10, high surf brightn, bridge; P w U 5450 at 5.9, 227 -- U 5449) comp 0.7, 138, 0.3 x 0.2 --

05450   U 5450) S stell comp superimp nnf center; comp 0.9, 175, 0.2 x 0.2; 1.4, 358, 0.30 x 0.30; see U 5448 -- U 5456) vB
main body 1.1 x 0.25, 3 B knots -- U 5458) comp 2.0, 207, 0.4 x 0.1, spir --

05460   U 5460) comp 5.0, 106, 0.35 x 0.20 -- U 5461) P w U 5464 at 5.7, 30 -- U 5464) DDO 72; S V: (vdB); see U 5461 --
U 5465) P w U 5469 at 5.1, 110; 10 05.0 - 02 17 at 5.9, 252, 0.25 x 0.20, high surf brightn, m=15.5 -- U 5468)
measurements disturbed by nearby * GC 13911 at 4.6, m=4.58 -- U 5469) see U 5465 --

05470   U 5470) DDO 74, An 1006 in BG; dE (vdB), E4 pec (de V); listed as Leo II in MCG -- U 5471) comp 1.5, 152, 0.25 x
0.05 -- U 5473) P w comp at 2.0, 61, 0.5 x 0.2, Sc: -- U 5474) comp 3.5, 326, 0.9 x 0.4, dif -- U 5475) U 5480 at
6.1, 70 -- U 5476) earlier than Sb?; comp 2.1, 287, 0.35 x 0.20, interaction? -- U 5477) P w comp at 2.0, 115, 0.4 x
x 0.3, early spir (S B?) -- U 5478) DDO 73; Ir V (vdB); P w U 5481 at 11.2, 23 -- U 5479) interacting w quadrupel
system 10 06.9 + 54 44 at 0.9 f, 0.6 x 0.6, m=15.7; sev galaxies inside 5.0 --

05480   U 5480) P w 10 07.0 + 58 43 at 1.7, 122, 0.45 x 0.30, m=15.7; see U 5475 -- U 5481) see U 5478 -- U 5483) sev obj
in vic, prob members of cl 170 (but outside cl contour in CGCG) -- U 5485) 2 F conds in streamer sf; main body
0.45 x 0.15; 2 short jets from main body; comp 3.2, 125 (from center of main body), 0.35 x 0.35; 4.1, 138, 0.3 x 0.12
-- U 5487) slightly asym; comp 1.7, 269, 0.35 x 0.30, dif; 3.3, 211, 0.3 x 0.3, early S B? --

05490   U 5491) P w 10 08.5 + 59 08 at 1.8, 330, 0.35 x 0.20, asym, pec?, m=15.7; comp 3.7, 257, 0.4 x 0.12 -- U 5492) comp
5.3, 134, 0.9 x 0.10 -- U 5493) P w comp at 0.9, 285, 0.3: x 0.3: -- U 5498) spindle 1.5 x 0.25 w comp 0.3 x 0.2
superimp f center --

05500   U 5500) comp 2.0, 235, 0.35 x 0.15; 3.4, 351, 0.6 x 0.25 -- U 5503) S0: (de V) -- U 5505) P w 10 10.5 + 12 54 at 5.3,
96, 0.9 x 0.5, Sc:, m=15.5 -- U 5506) smooth disc, hints of spir pattern, comp 0.2 x 0.05 superimp at 0.2 sf center;
P w 10 10.1 + 05 04 at 3.0, 305, 0.9 x 0.8, (R')S B(s)a - b, m=14.9; comp 3.2, 175, 0.6: x 0.4:, dif -- U 5507) a)
at 0.7, 252, 0.55 x 0.10; c) at 1.2, 63, 0.25 x 0.12; dif pair at 1.6, 334, total diam 0.6 x 0.3 -- U 5508) MCG δ =
= 60 02 is incorrect --

05510   U 5510) SAB(rs)bc (de V) -- U 5511) E 3: (de V); see U 5513; sev galaxies in vic -- U 5512) U 5516 at 4.6, 50 --
U 5513) slightly curved spindle, may be disturbed by U 5511 at 4.8, 190; 10 10.6 + 39 06=N 3152 at 4.1, 269, 0.8 x
x 0.7, S B 0:, m=15.5 -- U 5514) P w 10 11.1 + 18 22 at 1.4, 86, 0.8 x 0.5, spir, m=15.3 -- U 5515) a v large nr of
galaxies in vic; 10 22.1 - 00 40 at 2.2, 73, 0.7 x 0.12, S0?, m=15.5 -- U 5516) SAB(rs)0/a (de V), Sa (Ho); P w
U 5525 at 7.7, 250; U 5512 at 4.6, 230 -- U 5517) E1 (de V); sev objs in vic; 10 11.0 + 38 55=N 3161 at 1.5, 260,
(0.8 x 0.5), E, v = 6204, v_o = 6188; m = 15.3; 10 10.9 + 38 55=N 3159 at 2.8, 252, (0.8 x 0.7), E, v = 6950, v_o =
6934, m=14.9; 10 10.5 + 38 53=N 3151 at 7.7, 255, 0.7 x 0.4, S0:, v = 7140, v_o = 7124, m=15.1; 10 10.4 + 38 55=
≃ N 3150 at 6.3, 272, 0.9: x 0.6:, Sa:, m=15.4 -- U 5518) P w 10 11.5 + 39 45 at 4.7, 50, 0.7 x 0.6, S0, m=15.4 --

05520   U 5522) main body 1.2 x 0.6; eF outer disc -- U 5525) SA(s)a pec (de V), Sa (Ho); eF corona vis to approx 8. x 6.,
disturbed on f side; see U 5516 -- U 5527) comp 4.3, 176, 0.6 x 0.3 -- U 5528) prob v early spir; sev comps , stell
obj superimp sf center --

05530   U 5532) SA(rs)bc (de V) -- U 5534) P w U 5541 at 6.2, 110; comp 3.2, 330, 0.35 x 0.25 -- U 5536) in triple group w
U 5542 at 4.8, 52, and 10 12.9 + 60 34 at 4.7, 300, 0.55 x 0.25, m=15.7; sev F objs in vic -- U 5539) comp 1.1 np,
0.4 x 0.2, v dif, incl in CGCG magn --

05540   U 5541) chaotic; broad dif plume or disrupted arm northw; 6 B blue knots; see U 5534 -- U 5542) U 5536; 10 12.9 +
+ 60 34 at 4.7 w -- U 5544) SA(rs)b (de V); SN 1947 a -- U 5545) asym; dif np part, sharp sf edge ; comp 1.9, 275,
0.6 x 0.1 -- U 5548) P w 10 13.9 + 12 54 at 4.6, 349, (0.7 x 0.3), spir, m=15.5 --

05550   U 5551) P w comp at 1.1, 325, 0.6: x 0.5:, dwarf, m approx 18. - 19. -- U 5554) (R)S B(r)a (de V), Sa (Ho) -- U 5556)
VV 307 b, Arp 316 a; S B(s)c pec (de V), S... (Ho); see U 5559; P w U 5559 at 4.9, 120 -- U 5557) SAB(rs)cd (de V),
Sc - (Ho); SN 1921 b, 1921 c, 1937 f; listed as N 3180=3181=3184 in MCG -- U 5559) VV 307 a, Arp 316 b; SA(s)a pec
(de V), Sa (Ho); see U 5556; "edge-on spir shows sign of interaction" (Arp) --

05560   U 5561) P w 10 15.6 + 07 17=I 601 at 1.3, 237, 0.7 x 0.2, pec, m=15.0 -- U 5562) E2 (de V), E (Ho); U 5559 at
5.9, 230 -- U 5564) comp 0.50, 38, 0.15 x 0.12; 1.9, 215, 0.5 x 0.3; 3.0, 196, 0.4 x 0.1, F -- U 5565) massive, v
tightly wound arms; P w comp at 1.3, 268, 0.2 x 0.2, alm stell -- U 5566) B asym main body, spir?; comp 1.7, 147,
0.7 x 0.25 -- U 5567) P w comp at 1.3, 203, 0.7 x 0.1, curved spindle -- U 5568) SA(r)...? (de V) -- U 5569) comp
0.7, 246, 0.45 x 0.30, no evid of interaction --

05570   U 5570) P w comp at 4.8, 88, 0.5 x 0.3 -- U 5572) S B(rs)c (de V), Sc - (Ho); SN 1966 j -- U 5574) P w U 5575 at 8.6,
5 -- U 5575) see U 5574 -- U 5577) ecc bar-shaped main body, disrupted spir pattern? --

05580   U 5580) comp 2.8, 122, 0.3 x 0.2; 4.2, 161, 0.35 x 0.15 -- U 5581) in triple group w U 5585 at 4.4, 132, and U 5587
at 5.7, 112 -- U 5582) S B b (de V), Sb+ (Ho); main body 1.1 x 0.45; 2 smooth arms, one stronger; comp 1.0, 18,
0.12 x 0.12, alm stell; 2.3, 28, 0.5: x 0.3 -- U 5583) U 5584 at 7.4; U 5588 at 5.1, 100 -- U 5584) 10 18.2 + 25 36
at 4.7, 77, 0.6 x 0.4, early, m=15.2; comp 4.8, 40, 0.5 x 0.2 -- U 5585) see U 5581; U 5587 at 2.0, 67 -- U 5586)
stell cores sep 0.80; ext vF coronas; sev F objs in vic -- U 5587) main body 1.1 x 0.45; 2 stell, 5585 -- U 5588) high surf brightn,
sharp edge ; see U 5583; U 5584 at 9.4 -- U 5589) S B(s)cd (de V); one main arm northw - eastw, another arm is
strongly disturbed by 2 blue superimp objs sp center --

05590   U 5593) U 5597 at 4.6, 145; comp 2.0, 325, 0.7 x 0.12 -- U 5596) comp 3.6, 28, 0.7 x 0.15; sev fainter comps --
U 5597) see U 5593 --

05600   U 5600) VV 330 b; main body 0.45 x 0.30, amorphous disc; P w U 5609, interaction; "F halo + streamers" (CGCG) --
U 5601) SN 1961 l; comp 2.0, 45, 0.7 x 0.2; 2.1, 142, 0.4 x 0.3, B core; 3.2, 177, 0.3 x 0.2 -- U 5603) b) at 0.85,
155, 0.25 x 0.20; P w 10 19.4 + 36 50=I 2566 at 2.4, 245, 0.8 x 0.5, S0, m=14.9 -- U 5607) comp 2.4, 203, 0.4 x
x 0.10 -- U 5608) comp 2.1, 0, 0.4 x 0.05; 4.3, 331, 0.4 x 0.1 -- U 5609) VV 330 a; see U 5600; "ecc nucl" (CGCG) --

05610   U 5610) S B 0 (de V); quadrupel group (indiv m approx 16.5) around 5. s -- U 5611) comp 2.6, 245, 0.4 x 0.3 --
U 5612) DDO 77; S IV-V (vdB); comp 2.4, 295, 0.7: x 0.4: -- U 5614) comp 2.7, 42, 0.6 x 0.12; 10 19.8 + 51 18 =
= N 3214 at 5.0 p, 0.8 x 0.4, early spir, m=15.2 -- U 5615) VV 312; a) 0.8 x 0.7; b) 0.7 x 0.5, sep 0.55 -- U 5617)
VV 209 b, Arp 94 (see U 5620); E+ 2 pec: (de V); P w U 5620 at 2.3 sf; "comp on edge of large vF loop extending
opposite galaxy" (Arp) -- U 5619) comp 4.3, 286, 0.8 x 0.3, prob double --

05620   U 5620) VV 209 a, Arp 94; see U 5617; SAB(s)a pec (de V); blue diam incl a vF env surrounding both systems --
U 5622) P w U 5623 at 2.5, 35 -- U 5623) see U 5622 -- U 5624) U 5625 at 3.9, 17 -- U 5625) see U 5624 -- U 5626)

v blue, chaotic; disrupted pair? -- U 5627) P w 10 21.5 + 13 51 at 2.7, 352, 0.8 x 0.4, early spir, m = 15.2 --
U 5628) Arp 43; late S B b; "one side of ring obscured or disrupted; other side has low surf brightn comp" (Arp) --
U 5629) comp 3.1, 310, 0.5 x 0.4, blue --

05630   U 5630) a) 0.25 x 0.20; b) 1.4 x 0.7, sep 0.3, streamer from b) southw -- U 5633) DDO 79; S B IV-V (vdB) -- U 5634)
comp 1.3, 139, 0.3: x 0.2: -- U 5635) U 5636 at 4.7, 23; 10 21.5 + 28 17 = N 3232 at 7.8, 272, 0.7 x 0.6, m = 15.4;
sev S obj near   -- U 5636) see U 5635 -- U 5637) VV 95, Arp 263; S B m (de V); ecc v patchy main body 1.6 x 0.6,
2 dif asym spir arcs southw; "dif outer filaments, B knots inside" (Arp); sev galaxies in cl 173 in vic; 10 22.6 +
+ 17 24 at 3. sff, 0.9 x 0.5, S B b - c, m = 15.2; Sc + (Ho) -- U 5639) in group of 8 or more galaxies in vic of U
5637; 10 22.6 + 17 30 at 3.4, 264, (0.7 x 0.7), compact, m = 15.1 --

05640   U 5640) comp 2.2, 316, 0.35 x 0.25 -- U 5641) VV 354a, Arp 44; asym, s part F; comp (VV 354 b) at 1.4, 170, 0.3: x
x 0.3:, F, interaction? -- U 5643) VV 319 b, Arp 181 (see U 5659); main body 0.50 x 0.25, asym F arms; P w U 5659 at
1.2 sf, hints of an eF plume opposite U 5659 -- U 5644) b) at 0.85, 40, (0.6 x 0.5); in quadrupel group w comp at
2.0, 10, 0.5 x 0.4, late spir:, and 2.3, 17, 0.3 x 0.2, sep 0.4 -- U 5649) 10 23.0 + 57 33 at 5.4 np, 0.9 x 0.5,
compact, m = 15.1 --

05650   U 5654) comp 2.5, 47, 0.4: x 0.1 -- U 5655) in group of 6 or 7 galaxies; 10 24.2 + 16 23 at 4.4, 102, (0.9 x 0.8),
E, m = 15.1 -- U 5657) comp 2.9, 81, 0.3 x 0.2 -- U 5659) VV 319 a, Arp 181; see U 5643 --

05660   U 5660) 10 24.6 + 01 13 at 6.6, 111, 0.7 x 0.6, S B ., m = 15.4 -- U 5662) N 3245 A in BG; S B(s)b (de V); P w
U 5663 at 9.0, 155 -- U 5663) SA(r)0/a:? (de V); see U 5662 -- U 5665) in a dense part of the cl; 10 24.5 + 11 16 =
= I 613 at 4.2, (0.8 x 0.8), E:, m = 15.1; 10 24.5 + 11 18 = I 612 at 5.4, 0.6 x 0.4, m = 15.3 -- U 5666) Coddington
nebula, DDO 81, VII Zw 330; SAB(s)m (de V), S IV-V (vdB), Ir I (Ho); "v large blue resolved Irr, many knots on
nf periphery" (Zw) -- U 5667) a) 0.9 x 0.4, b) 0.8 x 0.7, brightest, sep 0.45; a) distorted w plume s; 10 24.9 +
+ 01 30 at 1.1, 255 from a), 0.30 x 0.20, m = 15.4 --

05670   U 5672) comp 3.0, 221, 0.3 x 0.1; 4.0, 196, 0.4: x 0.3:; 4.7, 284, 0.5 x 0.2 -- U 5674) comp 2.1, 229, 0.6 x 0.1 --
U 5676) blue, strongly asym, tend to spir pattern; comp 2.6, 262, 0.35 x 0.10 -- U 5677) "* 14^m.3 superimp" (CGCG);
v blue; P w U 5678 at 7.2, 7 -- U 5678) see U 5677 -- U 5679) v open early spir, no disturbing obj vis; U 5684 at
15.4, 160 --

05680   U 5681) P w 10 26.4 + 20 00 at 2.5, 104, 0.7 x 0.6, m = 15.6 -- U 5682) P w 10 26.1 + 79 10 at 2.8, 0.7 x 0.7, outer
disc 1.5? x 1.5?, m = 15.6 -- U 5683) M 51 case; distorted S B(s) or SAB(s) w straight arm to comp at 0.8, 141,
0.2 x 0.1; comp 1.0, 0 from a), 0.40 x 0.05 -- U 5684) prob v late; comp 2.1, 140, 0.55 x 0.15; see U 5685)
SA(s)bc (de V), Sb + (Ho); SN 1941 b -- U 5688) VII Zw 331, VV 294; "disrupted chain of galaxies" (CGCG); multiple
system, centr obj prob v late S B; 2 S conds nf are prob interacting galaxies, comp 2.0, 143, 0.5 x 0.12; "blue
Irr, w spir chain of 7 compact knots and blue compact bar at sf end" (Zw) -- U 5689) comp 4.0, 266, 0.20 x 0.15;
5.0, 285, 0.6 x 0.4 --

05690   U 5692) DDO 82; S V: (vdB) -- U 5694) a) 0.25 x 0.20 nnp b) 0.7 x 0.2, sep 0.70, vF bridge -- U 5696) double spir
pattern, arms in centr region wound opposite arms in outer disc; comp 4.1, 198, 0.8 x 0.7, spir, v dif --

05700   U 5703) sev S objs inside 5.0 -- U 5704) comp attached at 0.7, 79, 0.2 x 0.2, incl in blue major diam --

05710   U 5711) long arms w low surf brightn; comp 2.6, 121, 0.25 x 0.12; 3.9, 101, 0.4: x 0.4: -- U 5714) no nucl cond, one
S cond at n end of bar; comp 3.1, 309, 0.5 x 0.5, dif -- U 5717) SAB(rs)bc: (de V); comp 1.2, 198, 0.3 x 0.15 --
U 5719) late chaotic S B, v blue; S stell cond in n part, comp 2.9 n, 0.4 x 0.2 --

05720   U 5720) Arp 233, Haro 2; "narrow F abs lane in sf direction" (Arp); comp 4.7, 63, 0.4 x 0.2 -- U 5721) spir-type
obj w B main body, asym disc w a large nr of blue patches; comp 1.5, 305, 0.3 x 0.2 -- U 5723) main body 0.9 x
x 0.45, spir pattern, long streamer nf vis in red; a) at 1.0, 242, 0.20 x 0.15; 10 29.7 + 01 18 at 3.8, 164, 0.8 x
x 0.25, spir, m = 15.4 -- U 5724) cores 0.22 x 0.18 and 0.15 x 0.15, sep 0.85 in common env; 3rd group member (MCG
11-13-31) at 1.1 f a), 0.35 x 0.25 -- U 5725) SAB0/a? (de V); SN 1950 c -- U 5729) comp 3.5, 290, 0.35 x 0.1 --

05730   U 5731) SA(r)ab (de V) -- U 5734) sev S objs in vic; comp 0.6, 215, 0.15 x 0.12, streamer towards U 5734 -- U 5737)
P w U 5741 at 12.1, 93 --

05740   U 5740) comp 1.8, 320, 0.5 x 0.2 -- U 5741) see U 5734 -- U 5742) S B(s)d (de V); strongly asym, chaotic; blue
patches; comp 5.2, 122, 0.4 x 0.12; U 5767 at 33. -- U 5744) alm stell in blue, not listed as compact in CGCG --
U 5747) thin spindle w B nucl region -- U 5749) not listed as compact in CGCG --

05750   U 5750) P w U 5751 at 3.5, 4 -- U 5751) see U 5750; 10 33.1 + 21 21 at 2.1, 30, 0.6 x 0.4, compact, m = 15.5; sev
objs in vic -- U 5752) P w 10 33.1 + 58 53 = N 3286 at 3.9, 352, (0.8 x 0.5), E, m = 14.6; comp 3.9, 86, 0.3 x 0.15
-- U 5753) SA(s)c (de V) -- U 5756) P w comp at 1.5, 248, 0.7 x 0.2 -- U 5757) "F jets" (CGCG) -- U 5758) P w
10 33.7 + 13 42 at 1.7, 42, 0.8 x 0.7, S B: b:, m = 15.2 --

05760   U 5760) comp 2.5, 68, 0.3 x 0.3, dif -- U 5761) SAB(s)dm (de V); U 5774 at 11.8 -- U 5762) P w comp at 3.1, 253,
0.7 x 0.5 -- U 5764) Arp 267, DDO 83; Ir V (vdB); "semi-stell nucl, F oval ring outside" (Arp) -- U 5766) SAB(r)
0/a:? (de V) -- U 5767) Pec S B(rs)0/a (de V), S0 (Ho); U 5742 at 33., 245 -- U 5768) P w 10 33.7 + 60 16 at 3.8,
282, 0.40 x 0.20, m = 15.7 -- U 5769) b) = U 5770 at 0.8, 40 --

05770   U 5770) see U 5769 -- U 5771) P w 10 34.5 + 43 55 at 5.2, 29, 0.9 x 0.2, m = 15.7 -- U 5773) VV 71, Arp 192; main
body 0.5 x 0.4 w stell comp superimp, loop + sharp jet, enormous irr plumes; "dif F arms off both sides, spike
comes from stell comp" (Arp) -- U 5774) S B(s)m ? (de V); see U 5761; 10 34.4 + 13 02 at 7.2, 351, 0.9 x 0.9, Sc,
m = 15.2 -- U 5776) VII Zw 339; "neutral fuzzy spherical compact" (Zw) -- U 5777) S B(s)a ? (de V) -- U 5778) P w
comp at 0.9, 280, 0.7 x 0.6 -- U 5779) P w U 5780 at 1.7, 153 --

05780   U 5780) see U 5779 -- U 5782) comp 3.7, 144, 0.4 x 0.1 -- U 5786) Arp 217; SAB(r)bc pec (de V), Sb + (Ho); B main
body 1.8 x 1.7; vF outer disc; see Perek and Kohoutek, Catalogue of Galactic Planetary Nebulae, Prague 1967, p 10;
"much Hα em incl half arc outside galaxy" (Arp) -- U 5787) bar alm along line of sight; P w 10 35.6 - 02 24 at
6.2, 225, 0.9 x 0.5, E - S0 m = 15.3 -- U 5788) P w comp at 2.0, 145, 0.35 x 0.20 -- U 5789) S B(rs)cd (de V), Sc +
(Ho) --

05790   U 5791) P w U 5798 at 3.5, 105 -- U 5792) P w U 5796 at 3.3, 67; U 5795 at 12.7 -- U 5794) S cd: (de V); comp 2.0,
196, 0.5 x 0.1 -- U 5795) see U 5792; U 5796 at 11.3 -- U 5796) see U 5792, 5795 -- U 5797) sev B blue patches --
U 5798) see U 5791 --

05800   U 5800) v blue, strongly asym -- U 5801) S vB nucl region in dif disc or corona -- U 5806) U 5817 at 10.5, 96 --

05810   U 5814) Arp 156; B nucl region w broad abs lane; "vF oval loop in nf - sp direction" (Arp) -- U 5816) U 5818 at 4.6,
18; sev fainter objs in vic -- U 5817) see U 5806 -- U 5818) P w 10 38.6 + 06 38 at 0.8, 308, (0.8 x 0.7), E?,
m = 15.2, brightest in group -- U 5819) in triple group w comps at 0.60, 278, contact, 0.40 x 0.10, and 1.2, 80,
0.4 x 0.12 --

05820   U 5822) in triple group w 10 38.9 + 21 28 at 4.6, 207, 0.8 x 0.3, m = 15.6, and 10 39.0 + 21 35 at 3.6, 350, 0.9 x
x 0.8, spir, m = 15.1 -- U 5823) VV 113 (note error in VV α); sev S v blue conds in centr region, hints of spir
pattern in the disc -- U 5826) SA(s)c (de V), Sc - (Ho) -- U 5828) a) at 0.30, 350, 0.1 x 0.1; P w 10 39.9 + 16 00

at 1.5, 115, 0.7 x 0.7, Sb, m=15.0 -- U 5829) DDO 84; Ir V (vdB) --

05830 U 5832) VV 112, Arp 291; in Arp's class "double galaxies w wind effects"; strongly asym, B patchy bar, broad arm on s side, hints of smooth eF arm on n side; "main body has cylindrical appearance" (Arp) -- U 5834) P w U 5835 at 3.5, 356 -- U 5835) comp 0.50, 283, 0.12 x 0.12 incl in CGCG magn; see U 5834 -- U 5837) (R)SA(r)b: (de V); brightest in a subgroup of 12 or more galaxies; U 5841 at 7.5 -- U 5838) comp 2.4, 352, 0.3 x 0.3 --

05840 U 5840) (R)SAB(r)bc (de V), Sc - (Ho) -- U 5841) see U 5837 -- U 5842) S B(rs)cd (de V) -- U 5846) DDO 86; Ir V (vdB); sev objs in cl 10 40.3 + 60 58, medium compact, very distant, in vic -- U 5847) P w comp at 5.1, 102, 0.6 x x 0.5, dif -- U 5848) P w 10 41.6 + 56 38 at 3.7, 326, 0.9 x 0.5, S B a - b, m=15.1 -- U 5849) B main body 0.55 x x 0.35; eF outer parts, hints of spir pattern?; stell obj 0.1 x 0.1 at 0.45 nnp --

05850 U 5850) S B(r)b (de V), Sb + (Ho); "hot spots in nucl" (Morgan); (r) 2.3 x 2.1; U 5882 at 41.5, 80 -- U 5852) P w U 5857 at 13.8, 135; comp 1.7, 60, 0.25 x 0.15, disturbing?; 3.0, 197, 0.3 x 0.2, F; triple system 10 41.2 + 07 01 = N 3349 at 5.3, 270, total diam 1.0 x 0.7, a) 0.45 x 0.35, b) 0.7 x 0.4, double, disrupted, total m = 15.2 -- U 5853) P w U 5862 at 10.0, 35 -- U 5856) MCG descr is not valid for U 5856 -- U 5857) see U 5852 --

05860 U 5860) Haro 3, "muy violeta" (Haro); asym B main body, sharp edge; F smooth env -- U 5862) slightly asym spir, blue; row of blue patches along edge; see U 5853 -- U 5866) comp 2.4, 230, 0.8 x 0.7, v dif -- U 5867) 3-armed system, vS red comp at np edge, probl at tip of spir arm; brightest in chain of 4; comp 2.1, 148, 0.6: x 0.5:, v dif; 6.2 sf, 0.8 x 0.1, v dif; 8.2 sf, 0.5:: x 0.5::, e dif, dwarf -- U 5868) P w U 5871 at 2.1, 139 --

05870 U 5870) comp superimp at 0.25 nf center -- U 5871) see U 5868 -- U 5873) S B(rs)c (de V) -- U 5874) 10 43.1 + 26 13 at 4.4, 309, 0.9 x 0.7, compact, m=15.2; U 5881 at 8.1, 79 -- U 5875) E 0 (de V); comp 2.6, 115, 0.6 x 0.4, S B ... -- U 5877) P w 10 44.4 + 77 05 at 7.7, 138, 0.9 x 0.4, Sc, m=15.0 -- U 5879) comp 2.5, 343, 0.35 x 0.20 --

05880 U 5880) S B(rs)c (de V); complex spir str, strongly asym; 3 main arms; (rs) 0.9 x 0.6; appears disturbed by U 5899 at 21.5 nf, but redshift for U 5899 is 4 times larger than for U 5880 -- U 5881) comp 1.0, 149, 0.3 x 0.2; in triple group w U 5874 at 8.1, 259 and 10 43.1 + 26 13 at 11.4 -- U 5882) SAB(rs)ab (de V), Sa (Ho); "one-armed spir" (Morgan); comp attached at 3.8, 205, 0.4 x 0.4, E:; U 5850 at 41.5, 260 -- U 5884) comp 0.9, 5, 0.2 x 0.1, double?; U 5894 at 17.2, 151; U 5912 at 19.2, 83 -- U 5885) comp 4.5, 24, 0.7 x 0.3, late spir -- U 5886) U 5896 at 10.4 sf -- U 5897) SA(s)c (de V); comp 3.9, 269, 0.9: x 0.7:, e dif -- U 5889) N 3377 A in BG; SAB(s)m (de V); comp to U 5899 at 7.0, 137 --

05890 U 5890) comp 3.5, 268, 0.7 x 0.3; 4.4, 214, 0.4 x 0.12; Sc - (Ho) -- U 5892) in triple group w 10 44.5 + 07 30 at 5.9 and 10 44.5 + 07 32 at 5.3, 0.4 x 0.3, m = 15.5 and 0.55 x 0.12, m = 15.1, sep 1.5 -- U 5893) a) at 0.6, 220, 0.5 x 0.35, contact; 10 44.5 + 39 12 at 2.7, 277, 0.6 x 0.5, distorted S B ..., m=15.1 -- U 5894) U 5884 at 17.2, 331; 10 45.1 + 26 31 at 4.0, 0.40 x 0.20, m=15.4; U 5912 at 20.5, 32 -- U 5896) P w comp at 2.9, 150, 0.6 x 0.15; see U 5886 -- U 5898) P w 10 45.3 + 34 00 at 4.4, 78, 0.4 x 0.2, m=15.6 -- U 5899) E 5 - 6 (de V); see U 5889 --

05900 U 5901) 10 45.0 + 43 24 at 2.3, 197, 0.35 x 0.18, m=15.6 -- U 5902) E + 1 (de V); P w U 5911 at 7.2; U 5914 at 10.0, 108 -- U 5904) VII Zw 346; a) at 0.85, 328, 0.7 x 0.35; "pair of interacting neutral E and 'split' edge-on spir" (Zw); comp 2.2, 68, 0.3 x 0.10 -- U 5906) main body 0.8 x 0.45, massive spir arms; smooth vF outer disc -- U 5907) P w comp at 1.1, 258, 0.15 x 0.15 -- U 5908) P w 10 45.6 + 05 16 = N 3386 at 4.3, 0, (0.8 x 0.7), E - S0, m=14.8; sev vF objs in vic -- U 5909) asym v broad arms, one stronger --

05910 U 5911) In triple group w U 5902 brightest at 7.2, 247; U 5914 at 6.5, 156; S B(s)0 - : (de V); S0 (Ho) -- U 5912) slightly asym, prob disturbed by comp at 3.6, 290, 0.5 x 0.4, m approx 16.; see U 5884, 5994 -- U 5914) SA(s)c (de V); SN 1967 c; in triple group w U 5902 brightest, see U 5902, 5911 -- U 5918) DDO 87, V (vdB) --

05920 U 5922) P w 10 46.8 - 00 24 at 5.1, 110, 0.6 x 0.6, m=14.9; comp 1.9, 156, 0.3 x 0.2; 4.6, 178, 0.6 x 0.1, dif -- U 5925) P w comp at 3.3, 315, 0.6 x 0.1 -- U 5926) P w U 5939 at 2.9, 112, prob interaction -- U 5927) Irr: (de V) -- U 5928) 10 47.0 + 52 09 at 2.2, 134, 0.2 x 0.1, m=15.6 --

05930 U 5930) spir approx 0.8 x 0.45 w F curved bridge to comp at 0.9, 352, 0.1 x 0.1; U 5924 at 5.8 -- U 5931) VV 246 a, Arp 270 a; "note arc form of emission knots" (Arp); SAB(rs)cd pec : (de V); P w U 5935 at 1.5 nf -- U 5933) a) 0.7 x 0.7, S0?; b) 0.20 x 0.18 at 0.35 nf; P w 10 46.8 + 16 29 = N 3399 at 4.2, 252, (0.9 x 0.9), E:, m=14.7 -- U 5935) VV 246 b, Arp 270 b; see U 5931; I B m pec (de V) -- U 5937) P w 10 47.6 + 66 02 = N 3392 at 4.0, 35, 0.8 x x 0.6, E:, m=14.8 -- U 5938) comet-shaped, interacting w U 5942, sep 1.3 -- U 5939) see U 5926 --

05940 U 5941) a) 0.7 x 0.45, b) 0.2 x 0.2, sep 0.45; comp 2.7 sp a), 0.7 x 0.1 -- U 5942) comet-shaped obj, see U 5938 -- U 5943) M 51 case, comp 0.1 x 0.1 at tip of spir arm 0.40 nf center; F thin arm to comp, brighter broad arm northw -- U 5944) U 5952 = N 3412 in Leo group at 11.8 nf -- U 5946) a) (0.6 x 0.6), b) (0.7 x 0.6), sep 0.50; listed as triple system in CGCG -- U 5947) DDO 89; pec (vdB); elongated obj w 4 blue patches s center -- U 5949) S B 0? (de V) --

05950 U 5952) S B(s)0 (de V) -- U 5954) ident as I 644 by sev authors, which is prob incorrect; N 3398? (see U 5976); 10 48.5 + 55 44 = I 646 at 4.6, 7, 0.45 x 0.35, m=15.7 -- U 5955) comp 1.2, 50, 0.4 x 0.08; 5.0, 222, 0.3 x 0.2 -- U 5958) in triple subgroup w U 5959 at 7.6, 3, and U 5963; comp 5.0, 110, 0.3 x 0.3, v dif, dwarf -- U 5959) S B(s)0? (de V); Arp 162; in Arp's class "galaxies w dif counter-tails, dif filaments"; see U 5958; U 5963 at 8.5, 12; comp 2.0, 338, 0.4 x 0.4 --

05960 U 5960) S0 (de V) -- U 5961) comp 4.3, 287, 0.4 x 0.1 -- U 5962) SA(s)cd (de V), Sc + (Ho) -- U 5963) S B(s)0 + (de V); see U 5958, 5959 -- U 5964) (R)SAB(rs)0 + (de V); U 5965 at 4.7, 6 -- U 5965) N 3419 A in BG, S B(s)b: (de V); see U 5964; comp 3.3, 103, 0.5 x 0.05 -- U 5967) P w comp at 0.65, 202, 0.25 x 0.15 -- U 5969) 10 48.9 + 44 02 = = N 3416 at 3.3, 20, 0.40 x 0.20, pec, m=15.2 --

05970 U 5970) a) strongly disturbed early obj w b) superimp nf center, sep 0.20; 10 48.9 + 51 17 = N 3410 at 1.8, 123, 0.7 x 0.5, spir, m=14.8; comp 2.8, 253, 0.6 x 0.1 -- U 5972) S B(s)b:? (de V); U 5982 at 6.0, 60 -- U 5973) comp 2.4, 277, 0.30 x 0.25, vF disc or env -- U 5976) uncertain identification --

05980 U 5980) B (r) 0.55 x 0.45; one arm vis through approx 540°, prob disturbed; in triple group w 10 49.2 + 04 30 at 2.3, 250, 0.5 x 0.3, m=15.6, and 10 49.4 + 04 05 at 2.7, 12, 0.6 x 0.4:, m=15.6 -- U 5981) SA(s)c (de V); chain of 5 background galaxies s and sp, brightest 2.0, 105, 0.5 x 0.12; 2.7, 183, 0.20 x 0.15; 3.9, 220, 0.6 x 0.15; 5.1, 236, 0.8 x 0.3 -- U 5982) SAB(rs)c (de V); see U 5972 -- U 5983) comp to U 5986, sep 3.2 -- U 5984) VV 233, Arp 107; a) 1.8 x 1.2, unilateral w S B center; b) 1.1: x 0.5:, ecc nucl reg 0.20 x 0.15, w streamer; "double arm leads to E galaxy, dif material out other side of E galaxy" (Arp) -- U 5986) VV 11, Arp 206; S B(s)m (de V), Sc + (Ho); see U 5983 -- U 5988) P w 10 49.8 + 10 46 at 3.6, 169, 0.6 x 0.5, S0 - a:, m=15.5; comp 3.9, 27, 0.9 x 0.1, Sc --

05990 U 5991) U 6013 at 12.0, 77 -- U 5994) comp 1.6, 259, 0.35 x 0.25, dif -- U 5995) SAB(rs)c: (de V) -- U 5997) SAbc: (de V) --

06000 U 6004) comp 1.5, 298, 0.4 x 0.2 -- U 6005) U 6015 at 6.1, 86; comp 4.0, 225, 0.9 x 0.7, dwarf spir, m approx 17 - - 18 -- U 6006) VV 252 a; P w U 6007 at 1.7, 65; comp 1.7, 150, 0.3 x 0.3, Sc - Irr -- U 6007) VV 252 b; see U 6006 -- U 6009) S B(s) ? (de V) --

06010 U 6010) comp (double system) at 2.5, 180, 0.15 x 0.15 and 0.3 x 0.15, sep 0.25 -- U 6011) in triple group w comp at 4.9, 179, 0.5 x 0.3, spir, m approx 16, and comp 8.0 sp, m approx 16. -- U 6012) in group w comps at 4.0, 91,

0.7 x 0.5, B core w env, and 4.1, 67, 0.3 x 0.1; 10 50.4 + 27 09 brightest at 6.7, 254, 0.6 x 0.25, m=15.1 --
U 6013) P w 10 51.0 + 49 52 at 3.3, 178, 0.8 x 0.5, spir, m=14.8 -- U 6016) S B(s)d pec (de V, note to N 3448);
U 6024 = N 3448 at 4.1; see U 6024 --

06020 U 6020) P w 10 51.8 + 21 26 at 4.0, 62, 0.8 x 0.7, Sa, m=15.0 -- U 6021) VV 14, Arp 24; in Arp's class "one-armed
(unilateral) spiral galaxies"; SAB(s)m (de V); interacting w comp at 1.3, 105, 0.4 x 0.15, bridge -- U 6024) Arp
205; in Arp's class "galaxies w material ejected from nuclei"; B main body 2.7 x 1.0, vF ext along major axis;
U 6016 may be a part of this galaxy -- U 6025) B main body 0.7 x 0.3; F disc, alm dissolved arms -- U 6026) S B(s)c?
(de V); B spindle w plume, disturbed? by U 6028 at 3.7, 175 -- U 6028) Pec SAB(rs)b (de V); see U 6026 --

06030 U 6030) spherical obj w smooth edge, neither E nor S0 -- U 6031) v dif disc, invis on red map; comp 2.9, 267,
0.3 x 0.3, dif; 3.1, 144, 0.4 x 0.2 -- U 6033) P w 10 53.3 + 42 36 at 8.2, 75, 0.8 x 0.2, E:, m=15.4 -- U 6034)
comp 2.8, 255, 0.25 x 0.20; broad plume vis in red straight opposite comp; 2 other comps at 7.4 and 8.5 in dir
257, vF, v red -- U 6037) SAB 0: (de V) --

06040 U 6042) Comp superimp at 0.35, 85, 0.15 x 0.10; comp 2.0, 83, 0.4 x 0.3, Sc; 10 53.2 + 10 00 at 2.9, 253, 0.8 x 0.1,
Sc, m=15.7; U 6045 = N 3467 at 7.0, 87; see U 6045 -- U 6045) U 6042 at 7.0, 267; 10 53.9 + 10 12 at 10.8 n,
0.8 x 0.6, SAB(s)a, m=14.7 -- U 6046) VV 149; B cond sp center, disturbing comp?; comp 2.9, 73, 0.35 x 0.25 --
U 6048) comp 1.9, 243, 0.4 x 0.2 --

06050 U 6050) comp 4.1, 25, 0.3 x 0.3; 4.9, 2, 0.4 x 0.3 -- U 6052) P w 10 55.5 + 17 21 = N 3474 at 1.9, 154, 0.7 x 0.6,
S0:, m=14.9; sev fainter objs in vic, comp 1.1, 348, 0.5 x 0.1, v red spindle -- U 6054) * superimp near center;
comp 0.8, 84, 0.4 x 0.3, dif, dwarf: (Irr?) -- U 6056) U 6090 at 9.0 f -- U 6057) a) (0.7 x 0.7), E:; b) 0.25 x 0.20;
c) 0.15 x 0.15; in a dense part of the cl; 4th obj at 1.3 nf a, 0.2 x 0.2 -- U 6058) 10 55.8 + 24 28 at 1.9, 167,
0.20 x 0.20, very compact, m=15.7; comp 2.5, 206, 0.8 x 0.2, early -- U 6059) 10 55.3 + 55 53 at 3.1, 293, 0.7 x
x 0.4, S B b:, m=15.7; 10 55.8 + 55 54 at 2.6, 33, 0.15 x 0.15 + env, m=15.7, interaction w U 6059?; sev F obj
in vic --

06060 U 6060) 10 55.8 + 59 45 at 1.5, 168, 0.35 x 0.12, m=15.6, high surf brightn, pec -- U 6064) prob early Sa; comp
2.4, 55, 0.8 x 0.2 -- U 6065) "F annex" (CGCG); 3 vF objs at 0.9 np, one blue, dif, two red -- U 6069) S-shaped,
2 main arms split outwards; comp 2.0, 195, 0.7 x 0.7, dif, Dwarf --

06070 U 6071) P w 10 56.9 + 50 17 at 2.6, 202, 0.7 x 0.6, hints of outskirts to 2.2? x 1.4?, m=14.8; comp 4.2, 354,
0.3 x 0.2; 10 56.6 + 50 15 at 3.8, 224, 0.6 x 0.5, compact, m=15.5 -- U 6072) comp 2.7, 63, 0.45 x 0.15; sev F comps
-- U 6073) Arp 198, VV 267; a) 0.9: x 0.15; b) 0.9 x 0.7; U 6092 = N 3487 at 15.0, 105 -- U 6074) bar? or B ridge,
plumes?; comp 3.5, 157, 0.9 x 0.5, Irr, chaotic w F bar -- U 6077) S B(r)b: (de V); broad, split, uneven arms,
asym (r) 0.9 x 0.7 -- U 6078) U 6088 at 8.8 sf -- U 6079) SAB(r)c (de V), Sc - (Ho) --

06080 U 6081) this pair is brightest in quadrupel group; U 6086 at 6.6, 145 -- U 6082) SAB(rs)0+ (de V) -- U 6086)
comp 1.9, 264, 0.4 x 0.3, spir; U 6081 at 6.6, 325; U 6091 at 5.0, 165 -- U 6087) MCG α = 10 58.8 is incorrect --
U 6088) see U 6078 --

06090 U 6090) see U 6056; U 6105 at 7.2 sf -- U 6091) see U 6086 -- U 6092) prob late; b) at 0.9, 64, 0.25 x 0.20; in
triple group w U 6073 and U 6074 + 17 42 at 14.0, 225, 0.9 x 0.35, Sa - b, m=14.8; see U 6073 -- U 6093) U 6094
at 13.6 s -- U 6094) S0 or Sa?; early-type obj w comp superimp sp center; 10 58.0 + 10 43 = I 663 at 6.5, 314,
(0.9 x 0.6), pec, filaments?, v ext env on red map, m=15.6; 10 58.1 + 10 49 = I 664 at 4.2, 314, 0.4 x 0.4, m=14.8;
10 58.7 + 10 45 = I 666 at 4.7, 109, 0.5 x 0.3, m=15.3 -- U 6096) S c:(de V) -- U 6097) comp 4.0, 332, 0.5 x 0.2 --
U 6098) S d: (de V), Sc (Ho); 10 59.4 + 03 51 at 3.9, 178, 0.7 x 0.2, spir, m=15.1; comp 4.6, 191, 0.4 x 0.3 --
U 6099) comp 1.7, 170, 0.4 x 0.1 --

06100 U 6103) complex; v high surf brightn, sharp edge,   osculating pair?; comp 3.0, 157, 0.4 x 0.3 -- U 6104) P w
U 6112 at 13.1, 54 -- U 6105) see U 6090 -- U 6106) comp 4.3, 259, 0.8: x 0.8:, v dif, Dwarf -- U 6107) vB core
0.20 x 0.20; in group w 10 59.3 + 38 56 = I 2617 at 10.2, 343, 0.9 x 0.5, early, m=15.0; and 10 59.2 + 39 03 =
= I 2616 at 17.5, 350, 0.35 x 0.25, m=15.6; sev fainter objs in vic -- U 6109) bar in S (r) 0.2 x 0.2, F core --

06110 U 6112) see U 6104 -- U 6114) a) (0.9 x 0.8), E; b) 0.5 x 0.5, at 0.7, 15 from a); sev S objs in vic -- U 6116)
no nucl bulge; P w U 6123 at 12.7, 45; comp 1.8, 10, 0.35 x 0.25, v dif; 4.5, 247, 0.8: x 0.2, v dif, Dwarf --
U 6118) (R)SAB(s)ab (de V); P w U 6128 at 12.0, 72; comp 4.0, 55, 0.4 x 0.1 --

06120 U 6120) Leo group?; S c (de V); acc to de V a possible member of M 66 subgroup in Leo group; asym, patchy, high
surf brightn; P w U 6122 at 5.3, 62 -- U 6122) dif; thin spir arms in the inner region; see U 6120 -- U 6123)
see U 6116 -- U 6125) comp 0.9, 236, 0.4 x 0.2, prob phys comp -- U 6126) S B(s)m (de V); Sc (Ho); Haro 26,
"violeta" (Haro) -- U 6127) knotty B arms; "vF comp" (CGCG); comp 0.9, 223, 0.5 x 0.3 -- U 6128) SAB(rs)c (de V);
see U 6118 --

06130 U 6131) asym env?; comp 3.6, 100, 0.5 x 0.2 -- U 6132) P w U 6140 at 3.2, 78 -- U 6134) VV 75, Arp 335; S...pec
(de V); "large luminous system" (Arp) -- U 6135) 11 01.6 + 45 24 at 1.7 p, 0.5 x 0.2, E - S0, m=15.3 -- U 6136)
plume northw, distorted spir on edge?; no disturbing obj vis -- U 6138) 2 eF conds in center; may be non-regular;
comp 2.6, 282, 0.5: x 0.5: --

06140 U 6140) see U 6132; comp 3.5, 12, 0.35 x 0.15 -- U 6142) in triple group w 11 02.2 + 04 33 at 0.9, 85, 0.5 x 0.3,
m=15.2, and 11 02.4 + 04 33 at 4.3, 96, 0.6 x 0.6, S B..., m=15.1 -- U 6142) comp 4.8, 171, 0.4 x 0.2 --
U 6144) in contact w b) at 0.6, 5, 0.7 x 0.2; b) is much fainter than a); sev S comps --

06150 U 6150) SAB(rs)bc (de V), Sb - (Ho); texture indicates Sb+ but colour Sb- -- U 6151) DDO 91; Ir V (vdB); hints
of spir pattern; MCG magn applies to 11 03.0 + 20 04 at 2.9, 252, 0.4 x 0.2, E, m=15.5; de V lists U 6151 as a
possible member of M 66 subgroup in Leo group -- U 6153) (R)S B(s)0/a: (de V); 11 02.5 + 72 46 at 4.6, 236, 0.35 x
x 0.30, m=15.4 -- U 6154) VII Zw 367; "neutral S B w compact elliptical bar, F disc halo" (Zw) -- U 6155) P w
11 03.6 + 04 36 at 6.0, 160, 0.9 x 0.7, S0?, m=14.7 -- U 6156) distorted?; Mark 181 at 1.2, 220, 0.3 x 0.1 -- U 6158)
P w 11 03.8 + 11 41 at 2.0, 324, 0.55 x 0.30, spir, m=15.7 -- U 6159) comp 2.1, 255, 0.4 x 0.1 --

06160 U 6160) P w a) at 1.2, 329, 0.9 x 0.4, S0: -- U 6161) comp 3.3, 66, 0.35 x 0.15 -- U 6163) P w U 6164 at 1.9, 108
U 6164) see U 6163; comp 3.5, 78, 0.4 x 0.2; pair U 6172 at 5.5, 130 -- U 6165) comp 3.8, 84, 0.35 x 0.35, spir --
U 6166) P w U 6170 at 4.4, 153 --

06170 U 6170) (r) 0.8 x 0.6, vF arms; see U 6166 -- U 6172) a) 0.4 x 0.4, dif; b) 0.7 x 0.6, compact; sep 0.50 -- U 6173)
main body 0.35 x 0.35, thin broken arms, loop northw; S blue cond in n part -- U 6175) VV 239, Arp 191; a) main
body 0.20 x 0.15; b) main body 0.40 x 0.20; sep 0.40; "acute bend in link between galaxies; plumes from stellar-like
images" (Arp); comp 4.5, 73, 0.8 x 0.5, S B..., m approx 16; U 6171 at 8.6, 343 -- U 6177) comp 1.6, 137, 0.4 x 0.2 --

06180 U 6182) P w U 6186 at 12.6, 10 -- U 6183) comp 4.2, 205, 0.8 x 0.2 -- U 6186) see U 6182; comp 3.3, 260, 0.3 x 0.2 --

06190 U 6190) P w U 6193 at 0.9, 160; 11 06.4 + 26 54 at 2.9 f, 0.7 x 0.3, m=15.6 -- U 6191) (r) 0.6 x 0.45, vF disc,
no spir pattern vis -- U 6193) see U 6190, 11 06.4 + 26 54 at 3.1, 52 -- U 6194) complex centr str -- U 6197)
a) 0.12 x 0.12, compact; b) 0.9: x 0.45, SAB(s)...; dif bridge, sep 0.80 -- U 6198) comp 2.6, 250, 0.35 x 0.15 --
U 6199) comp 2.0, 17, 0.3: x 0.3:; v dif; 11 06.3 + 62 31 at 4.3, 228, 0.9 x 0.15, late spir, m=15.7 --

06200 U 6200) diameters too small?, 2.6? x 1.1? in blue; comp 1.6, 230, 0.4 x 0.3, dif -- U 6203) double system 11 06.9 +

+ 22 01 at 1.4, 213 and 1.5, 193, 0.20 x 0.20 and 0.20 x 0.15, total m = 15.7; 11 07.1 + 22 05 at 3.4, 23, 0.35 x 0.20, m = 15.7 -- U 6204) VV 229a, Arp 301a; in contact w U6207 at 0.8, 116; MCG δ = 24 42.5 is incorrect -- U 6206) asym spir, listed as "double system" in CGCG; comp 1.7, 348, 0.45 x 0.12 -- U 6207) VV 229b, 301b; see U 6204 -- U 6208) 11 07.0 + 07 31 at 3.8, 280, 0.9: x 0.8:, S B..., m = 15.6; 11 07.5 + 07 36 at 4.3, 42, 0.6: x 0.5:, dif, m = 15.4 -- U 6209) Sb: (de V); vB main body 0.9 x 0.45, dif asym outer disc; listed by de V as a member of M 66 subgroup in Leo group --

06210   U 6210) comp 3.2, 220, 0.30 x 0.15 -- U 6212) P w a) at 0.6, 202, 0.20 x 0.15; group of 5 S B galaxies at 3' sp, a) belongs prob to this group; U 6216 at 5.1, 80 -- U 6213) vB main body, asym outskirts, dif ext northw -- U 6214) brightest in a dense region in cl 206; 11 07.9 + 29 00 at 2.0, 186, 0.6 x 0.4, very compact, m = 15.5 -- U 6215) SA(s)c: (de V) -- U 6216) see U 6212; 11 08.4 + 05 08 at 4.8, 73, 0.8 x 0.5, spir, m = 15.1 -- U 6217) spir-type obj w complex str -- U 6218) a large nr of brighter objs in cl 206 in vic; 11 07.9 + 28 35 at 4.2, 312, 0.8 x 0.8, E?, m = 15.1; 11 08.0 + 28 33 at 1.4, 291, 0.7 x 0.4, asym, early, m = 15.7 -- U 6219) P w 11 08.4 + 19 28 at 1.6, 70, 0.55 x 0.30, spir, m = 15.6 --

06220   U 6220) a) strongly distorted, spir?, 0.9: x 0.4:, b) 0.7 x 0.20, sep 0.9 -- U 6222) P w comp at 5.7, 67, 0.6 x 0.5, v late spir -- U 6223) 11 08.2 + 62 25 at 3.6, 207, 0.9 x 0.20, m = 15.6 -- U 6224) VV 237, Arp 105; a) 0.9 x 0.7, E:, b) 0.7 x 0.6, spir, sep 0.9, SN 1953a in the disc of b), Sb pec (Zw); "Ambarzumian's knot" is a S cond at 0.55 s a), 0.1 x 0.1; broad irr plume to 3.6 from b); in a dense region of the cl; "group of 3 intercon v blue, v compact galaxies emerging in straight row from N 3561 to the South. No 1, Ambarzumian knot, 32" s N 3561, m_p = 18.8, M_p = - 15.9, v = 8839. No 2. Zw knot A, 8".4 n No 1, m_p = 21.0, M_p = - 13.7, v = 8780. No 3. Zw knot B, 24" n No 1, m_p = 21.5, M_p = - 13.2, v = 8650. Knots A and B are prob the first galaxies discovered to appear completely stellar when observed with the 200" telescope." (Zw, abbr quotation) -- U 6225) S B(s)cd (de V), Sc + (Ho); SN 1969b -- U 6228) broken knotty ring w bar; comp 0.7, 155, 0.4 x 0.1, contact; 1.7, 224, 0.7? x 0.6? --

06230   U 6230) S vB nucl reg in smooth vF env; P w 11 08.8 + 06 06 at 0.65, 125, smooth asym anv 0.9? x 0.7?, m = 15.1 -- U 6233) comp 3.4, 313, 0.4 x 0.1 -- U 6234) P w 11 08.8 + 27 14a at 0.35, 255, 0.35 x 0.20, incl in CGCG magn -- U 6237) blue cond nff center -- U 6238) comp 1.9, 248, 0.25 x 0.15, B lens -- U 6239) P w 11 09.5 + 03 26 at 2.0, 77, 0.4 x 0.15, early, m = 15.3; 11 09.4 + 03 19 at 5.6, 174, 0.15 x 0.15, compact, m = 15.4 --

06240   U 6240) in a dense region of the cl; 11 09.5 + 27 53 = N 3574 at 2.9, 44, 0.4 x 0.3, m = 15.7 -- U 6242) 11 10.6 + + 73 11 at 4.9, 67, 0.7 x 0.15 -- U 6243) comp 0.9, 76, 0.4 x 0.05 and 1.3, 75, 0.15 x 0.10, sep 0.3 -- U 6244) a) 0.5 x 0.15 w irr ext, b) 0.45 x 0.20, sep 0.8, dif bridge -- U 6246) in group w 11 10.0 + 23 35 at 3.7 nf, 0.6 x x 0.6, spir, m = 15.7, and 11 09.8 + 23 33 at 4.8 f, 0.6 x 0.6 and 0.12 x 0.12, sep 0.40, E + compact, m = 15.0 -- U 6249) comp 1.0, 1, 0.25 x 0.20; 1.9, 353, 0.7 x 0.3, E - S0; 1.2, 96, 0.3 x 0.3, early; 2.8, 25, 0.3 x 0.2; one vS blue cond at tip of spir arm sp center, prob comp --

06250   U 6251) DDO 92 -- U 6253) DDO 93; An 1111 in BG; E 0 pec (de V), E pec (Ho); Sculptor-type dwarf; Ho m_{pg} = 12.85 -- U 6255) comp 2.7, 263, 0.5 x 0.15 -- U 6257) S B(r)a (de V); P w U6263 at 5.1, 57; comp 2.8, 322, 0.6 x 0.5, S B ... -- U 6258) v blue dif integral-shaped obj, blue cond n center --

06260   U 6262) slightly asym; cond nf center, superimp comp?; U 6272 in M 66 group at 32', U 6262 is prob a background obj -- U 6263) S B(s)b (de V); S comp attached at 0.9 nf, 0.25 x 0.25 -- U 6264) main bodies in contact, crossing plumes, similar to N 4038 + 4039 = VV 245 = Arp 244 -- U 6265) early; comp 1.2, 128, 0.15 x 0.12; 3.4, 265, 0.20 x 0.20 -- U 6266) curved bar-like cond in centr reg -- U 6267) P w U6278 at 10.4, 90 --

06270   U 6272) SA(s)0/a: (de V) -- U 6273) comp 0.8, 159, 0.5 x 0.4, early, incl in CGCG magn; U 6279 at 6.2, 91 -- U 6274) chaotic spir; stell cond np center, disturbing obj? -- U 6277) SAB(rs)c (de V); B inner region 1.9 x 1.8; a large nr of S blue patches; prob member of M 66 subgroup in Leo group -- U 6278) see U 6267 -- U 6279) see U 6273 --

06280   U 6281) SA 0: (de V); smooth v ext vF env, hints of spir str in centr reg?; U 6297 at 20.7, 85 -- U 6282) comp 3.8, 107, 0.5 x 0.4, dif; 4.4, 85, 0.45 x 0.1 -- U 6283) disturbed?; comp 1.8, 305, 0.3 x 0.15, F; 2.7, 288, 0.4 x 0.2, B core -- U 6286) comp 2.2, 219, 0.45 x 0.40, S B(r)... --

06290   U 6291) 11 12.1 + 69 35 at 7.7, 276, 0.5 x 0.4, m = 15.5 -- U 6292) in a dense subgroup, prob disturbed by comp at 1.4, 128, 0.5 x 0.2 -- U 6293) comp at tip of spir arm at 0.8, 254, 0.30 x 0.20; comp superimp near center? -- U 6295) E 4 -5(de V); U 6297 at 3.0, 60 -- U 6297) SA(s)0/a (de V); P w U6299 at 6.2, 10; see U 6295 -- U 6298) P w U6303 at 7.6, 128 -- U 6299) E 2 (de V); see U 6297; comp 5.9, 251, 0.8 x 0.3 --

06300   U 6301) eF comp at 0.6, 254, 0.1 x 0.1 -- U 6303) see U 6298 -- U 6305) P w U6306 at 3.1, 348; smooth vF disc w spir arc or ring fragment at 1.7 in direction towards U 6306, interaction?; SA(s)a pec (de V) -- U 6306) S B(s)m (de V, see note to N 3611); see U 6305 -- U 6307) P w U6326 at 11.6, 122 -- U 6309) (r) 0.8 x 0.7, F arc attached in pos angle 110; 11 14.6 + 51 41 at 3.5 s, 0.2 x 0.2, compact, m = 15.6 --

06310   U 6310) P w U6321 at 5.4, 274; ecc (r) or (rs)-- U 6313) 11 15.6 + 23 42 at 3.2, 63, 0.3 x 0.15, compact, m = 15.6; U 3627 at 7.4, 54 -- U 6314) early spir w comp 0.18 x 0.15 superimp at 0.35 nf center, prob at tip of spir arm; P w 11 15.5 + 30 40 at 0.8, 153, 0.9 x 0.9, SAB b, m = 15.2 -- U 6315) a) 0.6 x 0.35, b) 0.7 x 0.35, sep 0.55; sev obj's in vic -- U 6316) a) 0.6 x 0.2, b) 0.8 x 0.35, sep 0.9 -- U 6317) S B(s)m (de V, see note to N 3611); U 6305 at 10.2 -- U 6318) SAB(r)c (de V); P w comp at 2.6, 220, 0.9 x 0.7, dwarf, dif, patchy, N 3614A in BG, S B(s)m (de V) -- U 6319) E + 5:(de V); comp 3.7, 232, 0.25 x 0.25 --

06320   U 6320) B disc w sharp edge, uneven surf brightn; P w U6324 at 6.7, 171 -- U 6321) see U 6310 -- U 6322) smooth disc, F outer arcs; SN 1966k; P w 11 15.6 + 28 30 at 2.1, 174, 0.6 x 0.5, S B(r)b, m = 14.8; comp 3.3, 162, 0.7 x 0.1; 4.4, 131, 0.3 x 0.1 -- U 6323) E + 6 (de V); P w U6330 at 15.7, 158 -- U 6324) see U 6320 -- U 6326) see U 6307 -- U 6327) see U 6313 -- U 6328) SAB(rs)a (de V), Sa (Ho); M 66 subgroup in Leo group; U 6346 = M 66 at 20.5, 108; U 6350 = N 3628 at 35.5, 33 --

06330   U 6330) (R)SA(s)0 + : (de V); see U 6323; U 6348 at 9.5 -- U 6331) asym spindle -- U 6333) v late; slightly ecc nucl reg; in group w 11 16.3 + 23 10 at 4.3, 285, 0.7 x 0.5, late spir, m = 14.7, and comp at 2.8, 160, 0.6 x 0.5, S B(s) ... -- U 6334) ecc nucl reg -- U 6336) P w 11 16.8 + 25 15 at 4.2, 330, 0.4 x 0.3, very compact, m = 15.3 -- U 6337) nucl reg 0.25 x 9.22 ecc in red env -- U 6338) comp 4.0, 281, 0.6 x 0.4, early --

06340   U 6340) wisps south-eastw vis on PA 1400-0 -- U 6341) U 6343 at 5.8, ·8 -- U 6343) (R)SA(r)0+ (de V);comp 5.8, 187, 0.7 x 0.15; see U 6341 -- U 6344) no nucl cond, the bar is much brighter than the arms; U 6348 at 2.6, 25 -- U 6345) DDO 94; Ir V (vdB) -- U 6346) VV 308a, Arp 16 and 317; SAB(s)b (de V), Sb+ (Ho); eF corona; brightest in a subgroup in Leo group; "large concentration at end of s arm" (Arp); VV picture is incorrect, shows U 6350 = = N 3628; see U 6328; U 6350 at 37., 0 -- U 6347) comp 3.5, 304, 0.3 x 0.2 -- U 6348) SAB(s)b: (de V); sev comps, see U 6330, 6344 -- U 6349) S0 (de V); comp 2.6, 135, 0.30 x 0.15 --

06350   U 6350) VV 308 b, Arp 317; Sb pec (de V), Sb+ (Ho); square-shaped abs lane slightly inclined to the symmetry plane; VV picture shows only this obj; see U 6346 -- U 6351) S0: (de V) -- U 6352) SA(s)cd: (de V) -- U 6353) comp 0.15 x x 0.12 superimp at 0.25, 90, no hints of distortion -- U 6355) P w U6367 at 5.7, 72; comp 1.0, 154, 0.2 x 0.1; 2.8, 151, 0.4 x 0.2, high surf brightn -- U 6358) v open spir w comp 0.20 x 0.15 at tip of spir arm 0.7 sf center; P w 11 18.3 + 73 06 at 1.4, 50, 0.9: x 0.7:, plus comp 0.15 x 0.12, total m = 15.3 -- U 6359) P w b) at 0.8, 40, 0.35 x 0.30 --

06360   U 6360) Arp 27; SA(s)c (de V), Sc- (Ho); in Arp's class "spir galaxies w one heavy arm"; "note straight arm, abs tube crossing from inside to outside of s arm" (Arp); SN 1964a, 1965I -- U 6361) a) 0.5 x 0.20, high surf brightn; b) 0.8 x 0.15, early spindle; c) 0.18 x 0.18, compact: -- U 6362) sev S objs in vic, prob background; comp (double syst) at 1.3, 237, 0.5 x 0.15, and 1.6, 222, 0.1 x 0.1, sep 0.45; 3.1, 156, 0.5 x 0.1 -- U 6363) P w U6366 at 1.6, 51 -- U 6364) in a dense part of the cl; nearest comp at 1.4, 73, 0.5 x 0.4, Sc: -- U 6366) see U 6363 -- U 6367) see U 6355 -- U 6368) E3 (de V); P w U6370 at 2.5, 170 -- U 6369) vF nucl cond, hints of spir pattern; comp 4.7, 175, 0.9: x 0.8:, dwarf spiral, m approx 17.; U 6348 at 7.5, 277 --

06370   U 6370) E 1 ? (de V); not in CGCG; see U 6368 -- U 6371) comp (double syst) at 3.1, 350, 0.15 x 0.10 and 0.15 x 0.10, sep 0.25 -- U 6373) MCG magn 15.5 is a misprint; many objs in vis; 11 18.8 + 03 01 = I 683 at 4.2, 202, 0.4 x 0.3, m = 15.5; 11 18.9 + 03 01 at 3.6, 184, 0.8 x 0.6, m = 15.0, S B nucl reg in F env; 11 18.9 + 03 09 at 4.6, 359, 0.4 x x 0.3, vB nucl reg, m = 15.5; 11 19.0 + 03 09 at 5.0, 7, 0.3 x 0.3, m = 15.4; 11 19.0 + 03 10 = N 3647 at 5.2, 26, 0.3 x 0.3, m = 15.6 -- U 6376) Pec SA(r)bc pec? (de V), Sc- (Ho); amorphous disc w patchy broken arms, nucl ecc in (R) 2.5 x 1.3; U 6386 at 7.8, 72 --

06380   U 6383) comp (double syst) at 1.5, 144, 0.4 x 0.25, and 1.9, 156, 0.20 x 0.12, sep 0.50 -- U 6385) SA(r)bc: (de V), Sc- (Ho) -- U 6386) SB(s)a (de V); see U 6376 -- U 6387) b) = I 2767 at 1.4, 58, 0.5 x 0.15, m approx 17. -- U 6388) comp 0.4: x 0.2 superimp s center; 11 19.5 + 24 35 = I 2759 at 2.7, 271, 0.9 x 0.4, spir, m = 15.5; 11 19.5 + 24 36 at 3.1, 290, 0.4 x 0.3, m = 15.6; 11 19.9 + 24 32 = N 3653 at 1.5, 326, (0.9 x 0.6), E - SO, m = = 15.0; comp 1.0, 87, 0.5 x 0.5 --

06390   U 6391) SA b (de V); comp 1.9, 137, 0.4 x 0.2; 3.8, 52, 0.9: x 0.1 -- U 6392) Z-shaped late spir, arms w broad dif ext; comp 1.5, 63, 0.6: x 0.3:, v dif; 3.9, 343, 0.35 x 0.10 and 0.10 x 0.10, sep 0.30 -- U 6393) in a dense part of the cl, sev S galaxies near -- U 6394) surrounded by a large nr of vS objs; comp or *? superimp 0.2 sp center -- U 6396) SA(s)c: (de V) -- U 6397) P w 11 20.4 + 34 46 at 1.9, 105, 0.9 x 0.2, spir, m = 15.6 --

06400   U 6400) comp 4.2, 286, 0.4 x 0.2 -- U 6403) VV 22, Arp 155; E: (de V); VV α = 10 20.8 is incorrect; B main body 0.6 x 0.5, irr loop in pos angle 170, dif halo; in Arp's class "disturbed galaxies w interior abs"; Arp's picture shows hints of outer ring -- U 6405) SB(s)m (de V) -- U 6406) main body 0.55 x 0.45 -- U 6409) P w U6426 at 14.4, 35; comp 2.6, 295, 0.5 x 0.15 --

06410   U 6413) P w comp (prob background) at 2.7, 283, 0.50 x 0.45, (R) ... -- U 6418) N 3664 A in BG; S B m pec?(de V) -- U 6419) VV 251, Arp 5, DDO 95; SB(s)m pec (de V); strongly asym, thin patchy bar, sev loops; in Arp's class "spir galaxies w low surf brightn"; U 6419 at 6.4 s --

06420   U 6420) SA(rs)c: (de V); prob M 66 subgroup in Leo group -- U 6426) SA(s)0/a (de V); see U 6409 -- U 6427) 11 22.0 + + 24 16 at 3.7, 335, 0.25 x 0.15, m = 15.7 -- U 6429) P w U6430 at 16.8, 176; comp 4.0, 8, 0.4 x 0.1, F --

06430   U 6430) long well def arms; see U 6429; comp superimp at 0.6, 315, 0.12 x 0.12; comp 1.1, 36, 0.35 x 0.2 -- U 6431) v high surf brightn, patchy, brightest in the s part, no hints of spir str -- U 6434) in group of 6 -- U 6435) thin jet in pos angle 310 vis on PA 1400-0; comp 4.6, 43, 0.7 x 0.1, Sc:, v dif -- U 6436) P w comp at 1.2, 114, 0.8: x x 0.2 -- U 6439) SA(s)b (de V) --

06440   U 6441) vB core in F halo; 11 23.8 + 47 17 at 2.7, 56, 0.7 x 0.3, m = 15.5 -- U 6442) P w comp at 1.8, 261, 0.2 x 0.2, E:, m approx 16. -- U 6444) SO (de V); 11 22.6 + 57 18 at 7.8, 258, 0.9 x 0.7, SB(s)ab:, m = 15.6 -- U 6445) SAB(r)bc (de V); B inner reg 1.2 x 1.2; U 6453 at 14., 44 -- U 6447) "compact, halo + jet" (CGCG); F short jet northw -- U 6448) e blue main body 0.55 x 0.25 ecc in dif env --

06450   U 6450) S ecc nucl reg, F cond 0.30 nf; in contact w U 6451 at 0.7, 84 -- U 6451) see U 6450; U 6481 at 5.7, 114 -- U 6452) patchy; a cond in n part may be a disturbing comp -- U 6453) SA(rs)bc (de V); see U 6445; U 6460 at 14., 33 -- U 6455) comp 2.8, 21, 0.4 x 0.05; 3.6, 17, 0.4 x 0.3, core + env -- U 6456) VII Zw 403; "post-eruptive group of 5 or more blue compact galaxies" (Zw); comp 4.4, 10, 0.6 x 0.2 -- U 6458) SB(s)c ? (de V) -- U 6459) SA(s)0/a: ? (de V) --

06460   U 6460) SB(s)bc (de V); see U 6453 -- U 6462) in group w 11 25.4 + 08 15 at 3.5, 93, 0.15 x 0.15, compact, m = 15.6, and 11 25.6 + 08 17 at 6.0, 77, 0.8 x 0.15, Sb:, m = 15.3 -- U 6463) comp superimp on the disc at 0.8, 93, 0.1 x 0.1, prob at tip of spir arm -- U 6464) SB ... (de V); U 6445 at 19.2, 352 -- U 6465) late spir, chaotic spir pattern, 2 or 3 conds may be disturbing objs -- U 6466) P w 11 25.7 + 04 37 at 1.0, 50, 0.5 x 0.3, early, m = 15.3 -- U 6467) SAB(rs)c (de V); high surf brightn; comp 4.7, 347, 0.4 x 0.1 --

06470   U 6470) U 6475 at 4.7, 122 -- U 6471+6472) VV 118, Arp 299; a) approx 1.1 x 0.9:, b) 2.9: x ...; Arp 296 (ident as N 3690 + I 694 by Arp) are F objs np U 6471+6472, m approx 17 and 21; ident of I 694 uncertain, may be a S obj np and inside the outer parts of the double system, 0.30 x 0.25, MCG 10 - 17 - 2; "B internal knots" (Arp); "long straight filament alm to attachment w arm of spir" (Arp, note to 296) -- U 6473) 11 25.1 + 72 13 at 6.0, 207, 0.35 x 0.18, high surf brightn, m = 15.3 -- U 6474) poss member of M 66 subgroup in Leo group?; U 6498 at 26., 106 -- U 6475) P w U6477 at 2.2, 99; see U 6470; 11 25.6 + 09 20 = I 2850 at 5.1, 241, 0.5 x 0.2, pec, m = 14.8 -- U 6477) see U 6475 -- U 6478) P w 11 26.5 + 20 02 at 7.9, 72, 0.8 x 0.4, E - SO, m = 14.9 -- U 6479) 11 26.2 + 21 00 at 3.3, 147, 0.7 x 0.5, spir, distorted, m = 15.3, and 11 26.2 + 21 01 at 4.0, 146, 0.8 x 0.5, SO:, m = 15.3, sep 0.7; comp 3.5, 169, 0.5: x 0.4:, dwarf; U 6487 at 13.9, 155 --

06480   U 6480) comp 4.8, 56, 0.3 x 0.1 -- U 6481) see U 6451 -- U 6482) double syst 11 26.4 + 09 21 = I 2867 at 1.6, 210, 0.45 x 0.25, and 2.0, 229, 0.2 x 0.1, total m = 15.6 -- U 6483) U 6460 at 18.5, 89 -- U 6484) N 3683 A in BG; S B(rs)c (de V) -- U 6487) see U 6479 -- U 6488) P w comp at 0.70, 97, 0.5: x 0.2; VV 60 = MCG 7 - 24 - 14 at 11 27.9 + 42 10 in position 6.0, 21, [0.8 x 0.4], triple syst, E + E + vF vS red obj p --

06490   U 6490) P w U 6494 at 5.7, 132; comp 3.0, 288, 0.3 x 0.2 -- U 6492) in a loose group of 9, incl 11 27.2 + 52 28 at 5.3 sf, 0.6 x 0.2, m = 15.7, and comp at 4.5, 92, 0.4 x 0.15; 2 vF objs near U 6492 -- U 6494) see U 6490; comp 1.5, 70, 0.35 x 0.12 -- U 6496) measurements disturbed by a nearby * -- U 6498) SAB(r)ab (de V); poss member of M 66 subgroup in Leo group; see U 6474 -- U 6499) prob irr; comp 1.9, 30, 0.7 x 0.2 --

06500   U 6500) P w 11 27.6 + 44 26 at 3.3, 257, 0.8 x 0.5, spir disturbed by S comp, total m = 15.4 -- U 6502) comp 3.1, 246, (0.5 x 0.3), core 0.12 x 0.12; 4.3, 306, 0.3 x 0.15 -- U 6503) VV 3, Arp 197; main body 0.7 x 0.5, streamer to 0.9 in pos angle 47; P w 11 28.2 + 20 43 at 4.7, 195, 0.9 x 0.15, spir, m = 15.7; in Arp's class "galaxies w material ejected from nuclei"; "straight filament off one end of bar, kink at end of filament" (Arp) -- U 6506) Arp 203; thin patchy bar w dif broad plumes, hints of spir str on one side; in Arp's class "galaxies w material ejected from nuclei" -- U 6508) in triple group w 11 28.5 + 26 30 at 5.0, 213, 0.8 x 0.15, spir, m = 15.5, and comp at 6.5, 286, 0.9 x 0.15 -- U 6509) comp 4.6, 222, 0.3 x 0.2; 4.8, 199, 0.6 x 0.35, spir --

06510   U 6511) 11 28.6 + 28 28 at 5.6, 302, 0.45 x 0.35, m = 15.2 -- U 6514) VV 172, Arp 329, An 1129 in BG; famous chain w one discordant redshift; v = (from n) + 16070, + 36880, + 15820, + 15690, + 15480 (Sargent, ApJ 153, L 135) -- U 6516) vB centr reg 0.35 x 0.35, eF smooth env -- U 6517) comp 2.4, 340, 0.35 x 0.2; 2.4, 83, 0.3 x 0.2; 3.9, 50, 0.3 x 0.2 --

06520   U 6520) prob early bar spir; B centr reg 0.55 x 0.35, hints of spir pattern in the disc; P w comp at 3.2, 155, 0.15 x x 0.12; 11 29.8 + 62 43 at 4.9, 159, 0.9 x 0.2, m = 15.2 -- U 6521) P w U6523 at 2.2, 115 -- U 6522) asym; B elong nucl reg; may be disturbed by comp at 1.1, 60, 0.4 x 0.3; comp 3.0, 278, 0.6 x 0.1; 11 29.8 + 29 23 at 4.4 n, 0.9 x x 0.3, m = 15.7 -- U 6523) see U 6521 -- U 6524) VV 150, Arp 214; S B(s)a pec (de V), SO pec (Ho); in Arp's class "galaxies w irregularities, absorption and resolution"; "barred spir, sharp nucl, narrow abs lanes through center" (Arp); P w U6547 at 11.8, 72 -- U 6525) comp 2.4, 237, 0.30x0.15; 2.9, 121, 0.4 x 0.12 -- U 6527) VV 150, Arp 322, I Zw 27; "quintiple system" (CGCG); diameters for VV 150b, c and d; CGCG magn incl VV 150a at 0.9: np, (0.4 x 0.3), E:, and VV 150e at 1.5: sf, 0.9 x 0.15, Sc:; "The obj VV 150 is a nonlinear chain of 4 galaxies... These four are connected and cover a length of 90" = 48 kpc. The fifth galaxy is about 65" from the nearest member of the chain and is possibly at the same distance." (Sargent); "Quintiple system of neutral galaxies w sev compact cores, knots and jets" (Zw); see U 6524 -- U 6528) P w U6542 at 7.6, 62 --

06530   U 6530) comp 3.5, 143, 0.30x0.10 -- U 6532) comp 3.7, 200, 0.4 x 0.2, spir; 4.7, 77, 0.5: x 0.5:, dif; 11 30.2 + 70 34 at 6.1, 179, 0.6 x 0.5, pec?, m=15.6 -- U 6535) 11 30.2 + 50 33=I 705 at 4.6, 216, 0.35x0.20, compact, m=15.1 -- U 6537) SAB(r)c (de V), Sc- (Ho); slightly asym, one arm off at right angle from the ring -- U 6538) U 6541 at 3.5, 145 -- U 6539) P w U6545 at 4.6, 67 --

06540   U 6540) comp 2.0, 228, 0.4 x 0.12; 11 30.6 + 34 32=I 2925 at 4.7, 229, 0.5 x 0.20, m=15.6, w comp 0.15x0.12 at 0.55 -- U 6541) prob double syst, cond 0.3 x 0.2 superimp s center; see U 6538 -- U 6542) see U 6528 -- U 6545) see U 6539 -- U 6546) comp 4.2, 291, 0.7 x 0.3, Sc: -- U 6547) S B(r)a pec (de V), Sc - (Ho); see U 6524 -- U 6548) a) at 0.50, 180, 0.12x0.12, compact: -- U 6549) sev S galaxies near --

06550   U 6550) U 6540 at 8.7, 271 -- U 6553) comp 0.55, 173, 0.18x0.15 -- U 6554) Sc: (de V); in triple chain w U6565 at 20.9 and U 6579 at 16.4; comp 4.4, 304, 0.7 x 0.12 -- U 6555) 11 32.5 + 51 27 at 4.0, 147, 0.22x0.20, compact, m=15.3; 6 more galaxies inside 5.0 -- U 6556) P w U6559 at 10.0, 161 -- U 6557) elong dif ring, nucl cond on p side, plume southw? -- U 6559) see U 6556 --

06560   U 6560) comp 3.5, 21, 0.4 x 0.1; 3.8, 71, 0.8 x 0.12, Sc: -- U 6563) brightest in group of 9, incl 11 32.7 + 55 12 at 1.3, 213, 0.8 x 0.3, m=15.4; many S systems in vic -- U 6565) Arp 234; Im (de V), Ir I (Ho); in Arp's class "galaxies w appearance of fission"; "considerable resolution into ** and abs tubes" (Arp) -- U 6567) SA c: (de V)

06570   U 6572) v blue, brightest in s part; double? -- U 6575) U 6566 at 5.6, 271 -- U 6576) comp 1.5, 48, 0.8: x 0.7:, v dif -- U 6577) SAB(rs)c pec (de V); asym, no disturbing obj vis -- U 6578) listed as double nebula in CGCG -- U 6579) SAB(rs)bc (de V), Sc- (Ho); see U 6554, 6565 --

06580   U 6580) 11 33.5 + 70 27 at 5.8, 337, core 0.25x0.18, extremely compact, m=15.5 -- U 6581) comp in the env at 0.35, 225, 0.12x0.12; P w 11 33.9 + 55 08=I 2943 at 2.1, 310, 0.35x0.35, m=15.0 -- U 6582) N 3759A in BG; asym spir, may be disturbed by comp at 1.7, 195, 0.55x0.15; comp 4.6, 158, 0.45x0.30; sev S galaxies near -- U 6583) v high surf brightn; in triple group w 11 34.3 + 20 16 at 1.6, 358, 0.55x0.35, distorted spir, m=14.6, and 11 34.2 + 20 17 at 2.1, 341, 0.6 x 0.5, B core in halo, m=15.3 ; Mark 182 at 2.7, 180, 0.15x0.12 -- U 6585) stell core 0.20x x 0.20 w S comp attached to sf side; asym ext disc or halo -- U 6586) comp 2.8, 207, 0.3 x 0.15, high surf brightn, v blue -- U 6587) P w 11 34.5 + 03 07 at 0.8, 303, 0.45x0.20, pec, m=15.3; comp 2.4, 250, 0.4: x 0.2:, dif -- U 6588) U 6586 at 8.6, 342 -- U 6589) 11 34.3 + 18 10a at 5.8, 302, 0.7 x 0.7, late spir; b) at 5.6, 302, 0.25x0.20, b) is superimp a), this pair is N 3764, m=14.9 --

06590   U 6590) hints of smooth spir arc or (R); comp 1.3, 178, 0.5 x 0.1, late, blue -- U 6593) vB main body 0.55x0.25; asym outer reg, eF streamer southw-eastw -- or chain of background galaxies? -- U 6594) patchy spindle, prob late; comp 1.9, 200, 0.7: x 0.2:, e dif, dwarf -- U 6595) Arp 280a; S B(r)b: (de V); P w 11 35.1 + 48 09=N 3769A (in BG)=Arp 280 b, 0.9 x 0.3, v = 712, v_0 = 768, S B m pec (de V), m = 14.7; in Arp's class "interacting double galaxies" U 6596) v chaotic; knotty ecc bar -- U 6597)(VV 282), Arp 320 a (Arp 320 incl 8 galaxies); U 6602 at 2.9, 126 -- U 6599) 11 34.5 + 24 23=N 3765 at 8.8, 156, 0.9 x 0.6, S B a - b, m=15.1 --

06600   U 6601) U 6602 at 2.7, 0 -- U 6602) VV 282a, Arp 320 (see U 6597); b) = VV 282 b at 0.35, 35, 0.35x0.30, pec; 11 35.3 + 22 15 = VV 282c at 0.7, 230, (0.8 x 0.6), E, m=15.2; 11 35.2 + 22 18 at 2.9, 337, 0.6 x 0.20, m=15.5; see U 6597, 6601 -- U 6605) SA 0: (de V) -- U 6606) large chaotic obj interacting w comp at 0.9, 270, 0.25x0.15, stell; sev objs in vic; pair 11 35.3 + 47 45 at 4.5, 293, total diam 1.5 x 0.7, total m = 15.1 -- U 6607) U 6609 at 12.6, 175 -- U 6609) S comp superimp 0.25 sf center; 11 35.8 + 20 49 at 1.5, 263, 0.35x0.25, m=15.5, listed as following 4-28-13 in MCG --

06610   U 6610) P w 11 35.9 + 34 08 at 4.1, 337, 0.8 x 0.6, E-SO, m=15.1; comp 2.0, 48, 0.6 x 0.12; 2.3, 68, 0.5 x 0.1, F -- U 6611) 11 36.4 + 43 31 at 5.7, 27, 0.6 x 0.5, pec?, m=15.3 -- U 6613) curved bar; comp 1.0, 58, 0.3 x 0.2, dif -- U 6615) SA(s)c: (de V); P w U6640 at 13.2, 108 -- U 6617) P w b) at 0.7, 165, 0.2 x 0.2 -- U 6618) SAB(s)cd: (de V); asym, v knotty, hints of e dif outer reg to 2.? x 1.5? --

06620   U 6620) eF smooth ring segment or spir arc at 1.2 f and sf; P w 11 36.8 + 26 36=N 3784 at 0.8, 302, 0.9 x 0.6, m=15.2; comp 1.5, 14, 0.4 x 0.42; 2.8, 277, 0.5 x 0.15 -- U 6621) VV 228 b, Arp 294 a; SAB(rs)a pec (de V); asym, P w U6623 at 1.5, 20; in Arp's class "double galaxies w long filaments"; "pec filaments" (Arp); 11 36.8 + 32 09 at 4.7 sp, 0.8 x 0.4, Irr or Pec?, m=15.7 -- U 6623) VV 228 a, Arp 294 b; v high surf brightn; massive vB arms; see U 6621; SAB(rs)ab pec (de V) -- U 6625) blue spir w high surf brightn; near the border of cl Abell 1367; comp 2.1, 196, 0.30x0.25; 3.6, 117, 0.25x0.20, v high surf brightn -- U 6626) v patchy, asym; P w U6633 at 10.3, 38 --

06630   U 6630) VV 350 b, Arp 83 a; S B a: (de V); P w U6634 at 1.4 nf; in Arp's class "spir galaxies w large high surf brightn comps on arms"; see U 6634 -- U 6632) in triple group w U 6648 at 14.2, 58, and U 6654 at 20.3, 72 -- U 6632) see U 6626 -- U 6634) VV 350a, Arp 83 b; Sb (de V); see U 6630; "some hazy material at juncture of two arms; high surf brightn, S shape inside comp" (Arp) -- U 6635) early, w abs lanes in sf part, no spir str vis; P w U 6636 at 2.3, 191; see U 6624 at 7.0, 262 -- U 6636) see U 6635 -- U 6637) Haro 27; "violeta" (Haro) --

06640   U 6640) see U 6615 -- U 6641) U 6635 at 8.1, 240 -- U 6643) VV 300, Arp 87; a) 1.6: x 0.7, distorted spir, V = 7050; b) 0.7: x 0.7:, bridge partly included, v = 7200, acicular?; in Arp's class "spirs w large high surf brightn comps on arms"; "arm appears wrapped around cylindrical comp" (Arp) -- U 6644) SA(rs)c (de V), Sc- (Ho); approx 5°.5 p M 66 subgroup in Leo group, background? -- U 6648) see U 6632; U 6654 at 6.7, 100 --

06650   U 6650) strongly asym; blue cond in spir arm at 0.45 p nucl; SN 1969c -- U 6652) broad asym arms w v low surf brightn; comp 1.9, 119, 0.1 x 0.1 + env; 11 38.2 + 33 57 at 6.2, 267, 0.9 x 0.6, Sb, m=15.2 -- U 6653) P w U6655 at 2.6, 80; comp 1.7, 271, 0.5 x 0.1 -- U 6654) see U 6632, 6648; U 6674 at 13.6, 85 -- U 6655) see U 6653 -- U 6656) slightly asym; 11 39.0 + 20 24 at 7.4, 83, 0.9 x 0.7, spir, disturbed? -- U 6657) Comp superimp at 0.35, 252, 0.1 x 0.1 and 0.40, 81, 0.15x0.1, both incl in CGCG magn; 11 39.5 + 10 38=N 3819 at 4.2, 48, (0.8 x 0.7), E, m=14.8; 11 39.5 + 10 40=N 3820 at 5.5, 31, 0.7 x 0.5, spir?, asym, disturbed by N 3819?, m=14.9 --

06660   U 6660) asym spindle; sev vF comps -- U 6661) U 6668 at 3.2, 105 -- U 6663) comp 3.1, 275, 0.8 x 0.7, SAB ... -- U 6665) vB main body 0.40x0.25; short B jet southw; hints of a v ext e dif plume s -- U 6666) in group w U 6653 and 6655; U 6655 at 7.4, 253; comp 5.0, 4, 0.8 x 0.7, Sc -- U 6668) see U 6661 --

06670   U 6670) B elong conc in np part; 11 39.6 + 18 41 at 5.8, 317, (0.9 x 0.9), very compact, core 0.20x0.20 in smooth env, m=15.5 -- U 6671) U 6677 at 5.4 nf; comp 3.2, 353, 0.3 x 0.3, dif -- U 6673) sev S objs in vic; comp 4.0, 50, 0.30x0.20, member of a group -- U 6674) see U 6654 -- U 6676) prob pec; spindle-shaped main body 0.7 x 0.2 w v dif asym ext -- U 6677) 11 40.6 + 26 51 at 4.6 f, double syst, 0.5 x 0.3 and 0.25x0.15, sep 1.0, m=15.4 and 15.5, total m=14.7, this pair is N 3830 -- U 6678) VV 353a, Arp 115; P w 11 40.3 + 26 34 at 1.0, 13, [0.5x0.5], double syst, a) 0.15x0.15, stell, b) 0.5 x 0.5, contact, sep 0.2, this pair is incl in Arp 115, "... just perceptibly non-stellar" (Arp, remark to a)), total m=15.1 --

06680   U 6681) in group w galaxies at 3.9, 22, 0.8 x 0.5, late spir, and 4.8, 122, 0.8 x 0.5 -- U 6682) DDO 96; S V (vdB) -- U 6683) P w U6688 at 2.0, 88 -- U 6684) P w 11 40.9 + 31 45 at 2.5, 89, 0.4 x 0.3, m=15.3 -- U 6686) 11 40.4 + + 16 46 at 5.9, 271, 0.7 x 0.5, S0 - a, m=15.6; P w U6712 at 16. nf -- U 6688) in a dense part of the cl, see U 6683 -- U 6689) many S galaxies in vic --

06690   U 6690) SB... (de V); U 6676 at 7.6, 304 -- U 6693) the ring is open on f side -- U 6695) 11 40.7 + 09 15 at 3.8, 290, 0.5: x 0.4, m=15.7 -- U 6696) P w comp at 1.9, 143, (0.7 x 0.6) -- U 6697) irr "spindle" w high surf brightn, in a dense part of the cl (Abell 1367); U 6704 at 3.2, 112 --

06700   U 6701) in a dense part of the cl (Abell 1367); U 6704 at 3.6, 21 -- U 6702) in a dense part of the cl (Abell 1367) -- U 6704) in a dense part of the cl (Abell 1367); see U 6697, 6701; 11 41.3 + 20 13 at 1.0, 281, 0.22 x 0.18 + + 0.10 x 0.10, sep 0.10; U 6705 at 4.7, 356; 11 41.5 + 20 16 = N 3841 at 2.9, 15, 0.7 x 0.3, early, m=15.1 -- U 6705) in a dense part of the cl (Abell 1367); 11 41.4 + 58 11.5 = VV 241) no galaxy at this position -- U 6706) VV 320, N 3846 A in BG; B patchy reg 0.8 x 0.4 ecc in dif disc, no spir str, red; 11 41.0 + 55 19 at 4.2, 280, 0.7 x 0.5, S B a:, m=15.3 -- U 6707) 11 42.1 + 58 10 at 6.7, 314, 0.8 x 0.5, m=15.3 -- U 6708) 11 41.4 + + 33 48 at 2.6, 307, 0.35 x 0.15 and 0.35 x 0.15, sep 0.15, interaction; comp 3.5, 270, 0.25 x 0.15 -- U 6709) 11 41.6 + 33 37 = I 2952 at 1.8, 264, 0.40 x 0.18, m=15.6; 11 42.1 + 33 35 = N 3855 = N 3856 at 4.6, 118, 0.9 x 0.7, m=15.6 --

06710   U 6710) SA c: (de V); prob in a group incl comp at 1.4, 195, 0.6 x 0.12; comp 4.5, 182, 0.4 x 0.08 -- U 6712) P w U 6686 at 16. sp; comp 1.3, 252, 0.4 x 0.3 -- U 6715) P w 11 42.0 + 11 04 at 2.0, 84, 0.8 x 0.2, m=15.6, prob interaction -- U 6717) tend to bar; U 6730 at 10.4, 107 -- U 6718) in a dense part of the cl (Abell 1367), sev S galaxies in vic -- U 6719) in a dense part of the cl (Abell 1367) --

06720   U 6721) in a dense part of the cl (Abell 1367); 11 42.3 + 19 46 at 2.5, 19, 0.4 x 0.15, early, m=15.7 -- U 6723) in a dense part of the cl (Abell 1367); 11 42.5 + 19 54 = I 2955 at 0.9, 343, 0.25 x 0.20, red env 0.6 x 0.6, m=15.2 -- U 6724) in a dense part of the cl (Abell 1367); in contact w comp at 0.8, 117, 0.7 x 0.25, spir -- U 6729) I 2954 may be a galaxy at 11 42.4 + 27 07 (in pos 4.8, 321 from U 6729), 0.4 x 0.3, very compact, m=15.3 --

06730   U 6730) B patchy spir arc in n part; P w U6734 at 3.5, 138 -- U 6731) in a dense part of the cl (Abell 1367); 11 42.8 + 19 43 = N 3868 at 2.7, 2, 0.6 x 0.3, E - S0, m=14.8; 11 42.6 + 19 40 = N 3864 at 3.3, 262, 0.6 x 0.4, spir, m=15.5 -- U 6732) VII Zw 421; "neutral oval disc compact w ext halo" (Zw) -- U 6733) SB(s)c: (de V) -- U 6734) see U 6730 -- U 6735) in a dense part of the cl (Abell 1367), sev galaxies in vic; P w U6739 at 0.9, 120; comp 1.2, 259, 0.2 x 0.2 w vS comp attached to p side -- U 6738) E 5 (de V); v ext asym env; P w U6758 at 19.0, 101; comp 2.7, 58, 0.4 x 0.3, spir:; 4.6, 22, 0.5 x 0.1 -- U 6739) in a dense part of the cl (Abell 1367); see U 6735; MCG magn for 3 - 30 - 105 and 3 - 30 - 106 are confused --

06740   U 6741) a) 0.30 x 0.20; b) 0.1 x 0.1, sep 1.1, thin irr bridge continues beyond b) -- U 6742) S blue cond in nf part of the env -- U 6745) SA(s)c: (de V) -- U 6746) double syst 11 43.5 + 20 44 = I 732 at 4.0, 316 and 4.5, 318, 0.8 x x 0.5 and 0.5 x 0.2, contact, total m=15.1; comp 3.4, 348, 0.5 x 0.2 -- U 6747) v thin "spindle" w dif plume eastw; U 6738 at 9.0, 246 -- U 6748) 2nd in group of 5; comp 0.9, 245, 0.4 x 0.2, prob disturbing: 1.5, 89, 0.45 x 0.22 --

06750   U 6751) P w comp at 1.5, 350, 0.8 x 0.6, S d - dm, dif -- U 6753) 11 44.2 + 14 45 at 4.1, 169, [0.8 x 0.3], chain of 3, total m=15.5 -- U 6754) B slightly elong nucl reg, F disc, thin filam arms -- U 6758) 2 short arms in amorph disc; see U 6738; comp 1.5, 286, 0.4: x 0.2 --

06760   U 6761) comp 1.8, 77, 0.5 x 0.2, v dif; 3.0, 1, 0.3 x 0.15 -- U 6765) SAB(rs)c (de V); asym outer arms; comp 3.7, 30, 0.5 x 0.4, early -- U 6766) P w comp at 1.0, 164, 0.45 x 0.25, MCG 9 - 19 - 186 -- U 6767) "double nebula, halo" (CGCG); 2 galaxies in contact, a curved vB jet eastw from the s galaxy -- U 6768) in group incl 11 44.7 + 44 02 at 4.2, 279, 0.5 x 0.4, spir, m=15.4 --

06770   U 6770) a) 0.9 x 0.25, Sc; b) 0.7 x 0.18, Sc:, sep 0.55 -- U 6774) comp 1.9, 343, 0.6 x 0.4 -- U 6778) SAB(rs)c: (de V); P w U6781 at 3.9, 125 -- U 6779) E4-5 (de V); P w U6785 at 2.0, 60; comp 3.5, 284, 0.8 x 0.3 --

06780   U 6780) pretty B nucl reg 0.20 x 0.15; dif disc, only hints of spir pattern -- U 6781) see U 6778; interaction? -- U 6782) DDO 97; Ir V (vdB) -- U 6784) comp 1.8, 34, 0.20 x 0.15, high surf brightn, double *?; 2.5, 1, 0.4 x 0.1 -- U 6785) S B(rs)a: (de V); see U 6779 -- U 6786) SA(r)0+ (de V); B smooth (r) -- U 6787) SA(s)ab (de V), Sa (Ho); comp 5.0, 60, 0.7 x 0.12, early --

06790   U 6793) P w U6796 at 1.7, 95 -- U 6795) P w U6803 at 9.9, 97 -- U 6796) see U 6793; 11 46.5 - 00 50 at 3.8, 263, 0.8 x 0.4, m=15.3 -- U 6797) ecc bar; a large nr of thin arms --

06800   U 6800) comp 1.8, 181, 0.35 x 0.20; 5.1, 96, 0.5 x 0.5, pec spir -- U 6802) U 6815 at 6.2, 107; comp 4.6, 242, 0.30 x 0.10 -- U 6803) main body (ring?) 0.45 x 0.45, vF "disc" w F cond on f side; see U 6795 -- U 6806) P w U6807 at 2.0, 36 -- U 6807) see U 6806 --

06810   U 6810) 11 48.2 + 20 19 at 3.0, 48, 0.5 x 0.2, early, m=15.7 -- U 6811) P w comp at 1.0, 25, 0.7 x 0.15; U 6812 at 3.1, 4; sev vF systems in vic -- U 6812) see U 6811 -- U 6813) SN 1963 j -- U 6814) P w comp at 3.0, 118, 0.7 x x 0.5 -- U 6815) SA cd: (de V); see U 6802; comp 4.9, 204, 0.3 x 0.3, dif -- U 6816) VV 273, DDO 98; S (B) IV - V: (vdB) -- U 6817) DDO 99; "dif + disrupted" (CGCG); Ir IV - V or Ir V (vdB) -- U 6818) comp 0.7, 295, 0.45 x 0.15 -- U 6819) P w U6823 at 4.5, 115 --

06820   U 6820) P w U6821 at 6.4, 159; comp 2.5, 16, 0.5 x 0.4 -- U 6821) see U 6820 -- U 6823) VV 31, Arp 224, I Zw 28; vB core 0.45 x 0.35, broad dif loop southw; in Arp's class "galaxies w adjacent loops"; "straight filament leads to B offset nucl" (Arp); "blue disc-like post-eruptive compact w large external loop and ext jets" (Zw) -- U 6824) comp 4.7, 60, 0.4 x 0.15 -- U 6825) SA 0- (de V); comp 3.8, 140, 0.45 x 0.35 -- U 6829) VV 218; 11 49.0 + + 22 20 at 2.6, 42, 0.8 x 0.4, early, m=15.7 --

06830   U 6833) dif patchy S-shaped spir; comp 3.3, 75, 0.6 x 0.4 -- U 6834) E (de V) -- U 6835) U 6839 at 2.7, 143 -- U 6836) P w 11 49.0 + 36 52 at 1.6, 262, 0.9 x 0.7, Sa - b, m=15.7 -- U 6837) in triple group w 11 49.6 + 18 56 at 8.1, 39, 0.50 x 0.30, m=15.7, and 11 50.1 + 18 54 at 12.7, 70, 0.8 x 0.4, m=15.7 --

06840   U 6840) DDO 100; S B IV - V (vdB) -- U 6841) Type N 2685?; U 6839 at 3.5, 224; comp 1.7, 356, 0.4: x 0.4:, irr -- U 6846) sev S galaxies in vic; 11 49.8 + 23 54 at 3.8, 312, 0.30 x 0.20, very compact, m=15.6 -- U 6848) P w 11 48.8 + 35 39 at 1.9, 278, 0.9 x 0.3, spir, m approx 16.5 --

06850   U 6851) I 2968?; 11 49.9 + 20 55 = I 2958 at 2.8, 262, 0.9 x 0.45, S0 - a, m=15.5; comp 5.0, 284, 0.5 x 0.25; sev F obj in vic -- U 6853) slightly asym, may be interacting w comp at 0.8, 205, 0.12 x 0.08 -- U 6854) VV 105; strongly asym; comp 0.1 x 0.1 superimp spir arm at 0.30 f center -- U 6856) SA(s)c (de V), Sc - (Ho); SN 1961u, 1964 l -- U 6857) S B(s)0/a (de V) --

06860   U 6860) SB(rs)0+ (de V) -- U 6863) comp 1.6, 232, 0.35 x 0.15; 3.0, 250, 0.7: x 0.4:, dwarf:; 3.9, 152, 0.4 x 0.2 -- U 6864) comp 2.1, 302, 0.5 x 0.2, S B: b -- U 6865) VV 286, Arp 62, II Zw 55; a) 0.15 x 0.12, b) 1.1: x 0.4; "high surf brightn comp" (Arp); "pair of neutral post-eruptive S B and elliptical compact, sep 40" northeast to southwest" (Zw) -- U 6869) SA(s)bc: (de V); comp 4.4, 273, 0.3 x 0.1 --

06870   U 6870) S B(r)bc (de V), Sb + (Ho); comp 3.9, 190, 0.3 x 0.1; 4.6, 156, 0.4 x 0.4, dif -- U 6873) asym spindle -- U 6875) comp 1.8, 245, 0.3 x 0.2, dif, incl in CGCG magn? -- U 6876) comp 1.6, 216, 0.4 x 0.2; 2.9, 41, 0.4: x x 0.4:, dif -- U 6877) comp 0.50, 280, 0.20 x 0.12, double (or * superimp) --

06880   U 6880) S B(s)a (de V); U 6884 at 8.3, 23; comp 0.40, 295, 0.1 x 0.1 -- U 6881) comp 4.0, 88, (0.4 x 0.2), S B core -- U 6882) comp 0.55, 295, 0.12 x 0.10, contact -- U 6883) listed also w coord 11 52.4 + 26 29 and magn 15.6 in CGCG -- U 6884) SAB(rs)bc (de V); a beautiful obj w 2 sym arms vis through 360°; see U 6880 -- U 6887) * superimp --

06890   U 6891) P w comp at 3.4, 330, 0.4 x 0.3, early -- U 6892) P w 11 52.7 + 33 25 at 1.6, 317, 0.55 x 0.12, m = 15.6, high surf brightn -- U 6893) P w 11 53.2 P 39 30 at 4.4, 85, 0.9 x 0.8, Sc, m = 15.6 -- U 6895) 11 52.9 + 12 13 at 2.8, 48, 0.4 x 0.2, m = 15.7; 11 53.0 + 12 17 = N 3973 at 2.6, 48, 0.4 x 0.2, early, m = 15.4 -- U 6896) VII Zw 429; "Quartet of intercon blue post-eruptive galaxies... No 1, S B c, m_p = 17.7. No 2, spherical compact, 25" nnf No 1, m_p = 16.8, v = 13070. No 3, 20" n No 2, patchy compact, m_p = 16.3, v = 12560. No 4, spherical compact, 11" n No 3, m_p = 18.0" (Zw) -- U 6899) U 6905 at 4.0, 147 --

06900   U 6900) DDO 101; V (vdB) -- U 6902) a) 0.5: x 0.3, b) 0.5 x 0.2, sep 0.35, bridge -- U 6904) SA(s)bc (de V); U 6909 at 5.3, 37 -- U 6905) see U 6899 -- U 6906) SAB(s)b (de V); major diam incl a F cond, which may be a dwarf comp; 11 53.4 + 06 58 at 4.6, 243, 0.5 x 0.1, Sb - c, m = 15.5 -- U 6907) early-type obj w 2 S comps superimp -- U 6908) comp 1.2, 20, 0.3 x 0.2 -- U 6909) (R)SA(rs)ab: (de V); SN 1946 a; see U 6904 --

06910   U 6910) comp 2.0, 292, 0.6 x 0.25, spir, N 3975?; 3.8, 122, 0.5 x 0.12 -- U 6912) VII Zw 430; "distorted blue post-eruptive spir, exploded core, blue knots on spir arm" (Zw) VV 57 (note error in VV α, listed w α = 10 40.75) -- U 6914) comp 4.7, 122, 0.8 x 0.15, dif -- U 6915) v blue; P w U 6920 at 4.6, 104; comp 1.0, 250, 0.4: x 0.1, dif -- U 6916) comp 4.9, 261, 0.7 x 0.3, late; 11 54.8 + 40 02 at 11.0 f, 0.7 x 0.6, no spir str, late?, m = 14.9 -- U 6919) comp 4.7, 127, 0.7: x 0.4:, dif

06920   U 6920) see U 6915; SO (de V) -- U 6921) S B(s)m: (de V) -- U 6923) U 6937 at 14.8 nf -- U 6925) comp 5.0, 250, 0.9: x 0.2, chaotic, double or triple system? -- U 6926) 11 55.0 + 57 50 at 7.1 sf, 0.5 x 0.5, core 0.20 x 0.20, m = 14.6 -- U 6928) U 6935 at 4.6, 55; 11 54.8 + 25 30 = N 3989 at 2.6, 30, 0.7 x 0.3, m = 15.7 --

06930   U 6931) comp 4.9, 135, 0.3 x 0.2 -- U 6933) Im pec (de V); Haro 5; "muy violeta" (Haro); see Arp's picture 313; "double system, bridge" (CGCG); v high surf brightn; in triple group w U 6936 at 3.8, 160, and U 6944 at 3.9, 135, this pair is VV 249, Arp 313 -- U 6934) comp 4.2, 178, 0.8 x 0.1 -- U 6935) see U 6928; U 6942 at 2.9, 53 -- U 6936) VV 249b, Arp 313a; see U 6933; U 6944 at 2.0, 242; SA(r)c pec? (de V); see U 6944; SN 1956a -- U 6938) SO-: (de V); P w U 6946 at 2.9, 96 --

06940   U 6942) early asym S B..., hints of a smooth eF arm or ring fragment n center; S blue cond 0.4 s center; see U 6935 -- U 6944) VV 249a, Arp 313 b; SA m pec (de V); see U 6933, 6938; "linear strings of knots like deformed spir arms. Strong [O II] em" (Arp) -- U 6945) VV 126, Arp 194; in Arp's class "galaxies w material ejected from nuclei"; a) 0.8 x 0.6, spir; b) 0.35 x 0.35, sep 0.7; "Outer material connected by thin filament to v hard nucl" (Arp); comp 2.0, 203, 0.7: x 0.2 -- U 6946) SA(r)0/a? (de V); see U 6938; comp 4.7, 145, 0.3: x 0.3:, dif -- U 6947) 11 55.2+ + 22 33 at 4.3, 344, 0.6 x 0.4, compact, m = 15.5 -- U 6948) P w 11 55.4 + 23 29 = N 4002 at 4.6, 1, 0.9 x 0.5, SO, m = 14.7 -- U 6949) U 6952 at 3.2, 116 --

06950   U 6950) VV 230; P w 11 55.2 + 28 08 = I 2982 at 3.2, 258, 0.5 x 0.3, E - SO, m = 15.2 -- U 6951) P w U 6958 at 10.5, 155; 11 56.0 - 01 50 at 7.0, 98, 0.8 x 0.7, S B a, m = 14.7; comp 3.7, 198, 0.6 x 0.6, v dif -- U 6952) see U 6949 -- U 6953) E 5 (de V); U 6968 at 8.5 nf -- U 6954) Arp 305 a; P w U 6967 at 5.9, 141; distorted, patchy, hints of spir str in the outskirts -- U 6955) DDO 105; Ir V (vdB) -- U 6956) DDO 102; S (B) V (vdB); listed in MCG at position 11 56.6 + 51 04; U 6985 at 9.8 f -- U 6957) red obj w thin wedge-shaped bar; 2 F ill-def arms, prob not in same plane; no disturbing obj vis -- U 6958) see U 6951 -- U 6959) a) 0.5 x 0.3, double?; b) 0.3 x 0.1; c) 0.4: x 0.3; sep 0.45 and 0.45 --

06960   U 6961) comp 2.6, 311, 0.35 x 0.3, spir -- U 6962) P w U 6973 at 3.3, 100 -- U 6963) Sb (de V) -- U 6964) 11 55.0+ + 47 37 = N 4001 at 6.7, 312, 0.6 x 0.3, m = 15.6; compact obj at 7.7, 147, 0.30 x 0.30, alm stellar, m approx 15.5 -- U 6965) VV 216, Arp 138; in Arp's class "material emanating from E galaxies"; a) (1.1 x 0.9), E; b) 0.8 x 0.2, Irr:; "abs leads directly into E galaxy" (Arp); U 6952 at 8.9, 306; 11 56.4 + 25 22 = N 4021 at 5.3, 68, 0.7 x 0.5, compact, m = 15.3; U 6977 at 5.8, 120 -- U 6966) U 6952 at 13.4, 210; U 6975 at 7.3, 142 -- U 6967) Arp 305 b; see U 6954; Z - shaped centr region, long B arms; "Segment breaking from arm of S galaxy, weak filaments reach to n galaxy, which has figure 8 loops" (Arp); ident as N 4016 in CGCG, MCG prefers N 4017 and ident tentatively U 6954 as N 4016; NGC lists 4016 and 4017 with the same α -- U 6968) S B core in vF disc, 2 long v thin arms emanate from the edge of the disc; no regular spir -- U 6969) tend to spir str?, blue; U 6937 at 11.2 p --

06970   U 6970) tend to spir str?; b) at 0.4, 160, 0.20 x 0.12, contact -- U 6972) bar?; 11 56.7 + 42 50 = I 752 at 4.1, 92, 0.55 x 0.50, pec?, m = 15.2 -- U 6973) v red obj; smooth halo or outer disc; see U 6962 -- U 6975) see U 6952, 6966 -- U 6977) see U 6965 --

06980   U 6981) B bar-shaped main body w plumes; P w 11 56.5 + 30 58 = I 2984 at 2.4, 207, 0.9: x 0.6:, spir, m = 15.6, interaction? -- U 6982) S B(s)cd (de V); DDO 107; S (B) IV: (vdB); anon I B m listed in BG at position 11 57.2 + + 37 48 (An 1157), but no galaxy exists there -- U 6983) SN 1964 e -- U 6984) comp 0.9, 291, 0.25 x 0.10; 4.4, 224, 0.6 x 0.4, Irr, v blue -- U 6989) S B ecc nucl region --

06990   U 6990) complex spir str, arm on f side appears broken; comp 0.9, 22, 0.3 x 0.1, F, interaction? -- U 6993) SA(s)bc (de V) -- U 6995) N 4042 observed by Marth (see introduction to NGC) is prob identical w N 4032 --

07000   U 7001) asym spindle w B drop-shaped nucl reg; 11 58.9 + 14 18 at 5.9, 135, 0.45 x 0.35, m = 15.2; 11 58.2 + 14 26 at 6.9, 299, 0.4 x 0.15, m = 15.7 -- U 7002) S B(rs)b:? (de V); amorph; nucl ecc in bar, ill-def spir str -- U 7003) 11 59.1 + 14 45 at 4.5, 63, 0.9 x 0.5, spir, m = 15.7 -- U 7005) SO- (de V), SO (Ho); U 7014 at 15.6, 19 -- U 7006) incorrectly listed as N 4043 in CGCG -- U 7009) U 7014 at 11.9; U 7019 at 7.7 --

07010   U 7011) comp 0.6, 185, 0.2 x 0.1 -- U 7012) patchy centr reg 0.6 x 0.3, amorph outer disc; P w U 7017 at 4.3, 79 -- U 7013) 11 59.2 + 18 10 at 6.6, 313, 0.9 x 0.7, spir, m = 15.1; 11 59.5 + 18 11 at 5.3, 359, 0.4 x 0.3, m = 15.6 -- U 7014) SA(rs)bc: (de V); Sc - (Ho); B centr reg 0.8 x 0.7, multiple arm pattern, uneven arms; see U 7005 -- U 7017) see U 7012 -- U 7019) VV 259; dif, chaotic, no spir pattern, sev F conds; see U 7009; U 7056 at 13.0 --

07020   U 7020) 2 thin arms vis through 360°; comp 3.5, 299, 0.3 x 0.15 -- U 7021) SAB(r)a (de V); P w 12 00.2 + 02 14 at 1.5, 179, 0.6 x 0.2, m = 15.2 -- U 7023) prob spir; 2 vS comps at 0.55, 300 and 0.70, 130; "quadrupel system, F bridges" (CGCG) -- U 7025) (R)SA(rs)b: (de V); comp 1.6, 8, 0.3 x 0.1, F -- U 7028) U 7032 at 4.4, 68; comp 2.5, 306, 0.7 x 0.15, spir --

07030   U 7030) SAB(rs)bc (de V), Sc - (Ho) -- U 7032) brightest of 5; P w 12 00.8 + 16 47 at 2.2, 313, 0.50 x 0.45, m = 14.6, SN 1961 k (Humason, classif Sc); see U 7028; vS F comp attached to sp side -- U 7034) MCG α = 12 00.1 is incorrect; sev S objs in vic -- U 7039) U 7032 at 6.9, 234 --

07040   U 7040) VV 276; chain of blue patches on p side -- U 7041) comp 3.6, 253, 0.6 x 0.2 -- U 7042) in a dense part of the cl; 12 01.5 + 02 08 = N 4063 at 3.9, 132, 0.9 x 0.3, SO, m = 15.0 -- U 7044) VV 179 b; P w U 7050 at 1.0, 82 -- U 7045) SA(s)c (de V) -- U 7048) E ? (de V); this galaxy is certainly a spiral! -- U 7049) P w 12 01.7 + 20 29 = = N 4072 at 1.8, 34, 0.6 x 0.2, m = 15.6; U 7050 at 3.1, 345 --

07050   U 7050) VV 179 a; see U 7044; comp 0.8, 338, 0.15 x 0.15 -- U 7051) 12 01.5 + 20 37 = N 4060 at 2.0, 252, 0.35 x 0.25, m = 15.6; U 7050 at 6.9, 187; U 7052 at 3.8, 7 -- U 7052) see U 7051 -- U 7053) DDO 108; Ir V (vdB) -- U 7054)

S B(s)a pec: (de V), Sa (Ho) -- U 7055) VV 46; "System w jet" (CGCG); S0 or E w F comp sp -- U 7056) blue bar, slightly
asym; U 7019 at 13.0 -- U 7057) comp 1.5, 214, 0.35 x 0.12 -- U 7058) P w U 7059 at 1.8, 20, no indication of interaction
-- U 7059) see U 7058 --

07060   U 7060) S comp in env at 0.50 f center -- U 7061) comp 2.1, 51, 0.35 x 0.15, SF comp nf; U 7050 at 6.5, 286 -- U 7062)
Sa? (de V)   -- U 7063) P w 12 02.0 + 02 05 = I 2989 at 1.3, 310, 0.9 x 0.35, m = 14.8; 12 02.0 + 02 01 at 3.1, 198,
0.35 x 0.25, m = 15.7; in a dense part of the cl -- U 7064) vB main body 0.40 x 0.25; smooth outer disc, no spir str
clearly   vis; 12 02.2 + 31 26 at 1.1, 160, 0.45 x 0.15, pec, m = 15.2; 12 02.2 + 31 28 at 0.9, 25, double system,
0.25 x 0.15 and 0.12 x 0.10, sep 0.15, m = 15.7 -- U 7066)sev galaxies near; 12 02.7 + 10 52 = N 4078 at 6.6, 81, (0.9 x
x 0.7), E?, m = 15.1; 12 02.7 + 10 55 = I 2991 at 6.7, 67, 0.7 x 0.25, spir, m = 15.5; 12 02.7 + 10 57 = N 4082 at 7.4,
52, 0.7 x 0.20, early spir, m = 15.1 -- U 7069) in a subgroup w U 7103 brightest; 12 02.7 + 43 26 at 3.4, 70, (0.9 x
x 0.8), E- S0, m = 15.7 --

07070   U 7070) VV 270 (listed at α = 11 45.1); chaotic spir w e dif bridge to dwarf comp 0.4:: x 0.4:: at 1.0 ssp center --
U 7073) P w U 7074 at 2.0, 359 -- U 7074) see U 7073 -- U 7075) U 7081 at 11.4, 10; SAB(s)c:? (de V), Sc- (Ho) --
U 7076) U 7077 at 3.8, 356 -- U 7077) see U 7076; comp 3.0, 271, 0.4 x 0.2; 4.7, 77, 0.8 x 0.3, spir -- U 7079)
12 02.4 + 63 25 at 3.7, 247, 0.6 x 0.6, spir, m = 15.1; comp 1.8, 341, 0.45 x 0.10; 2.0, 20, 0.45 x 0.35; 3.6, 134,
0.8 x 0.2 --

07080   U 7081) Arp 18; SAB(rs)bc (de V), Sc- (Ho); asym late spir; massive knotty arms; in Arp's class "spir galaxies w
detached segments"; "end of one spir arm partially disconnected" (Arp); comp 5.0, 122, 0.6 x 0.4; see U 7075 --
U 7082) a large nr of F comps, brightest at 1.6, 203, 0.12 x 0.12 -- U 7083) P w 12 03.0 + 20 50 = N 4089 at 0.6,
272, (0.8 x 0.8), E:, m = 14.9; sev objs in vic -- U 7085) P w 12 03.1 + 09 16 at 1.5, 265, 0.7 x 0.3, m = 15.0,
bridge -- U 7087) 12 03.3 + 20 49 = N 4093 at 2.6, 6, (0.8 x 0.7), compact, m = 15.3; sev objs in vic -- U 7088)
N 4108 A in BG; in a triple subgroup w U 7101 at 7.7 and U 7106 at 8.0 -- U 7089) in a subgroup w U 7103 brightest
at 12.6, 112 --

07090   U 7090) SAB(rs)c (de V), Sc + (Ho); SN 1960 h -- U 7091) 12 03.4 + 20 51 = N 4095 at 3.1, 226, (0.9 x 0.9), very
compact, m = 14.6; 12 03.3 + 20 49 = N 4093 at 5.9, 209, (0.8 x 0.7), compact, m = 15.3; VV 61 -- U 7095) Pec SA(rs)bc
(de V) -- U 7096) SAB(s)b? (de V) -- U 7098) asym main body 0.7 x 0.35, dif plumes in pos angles 90 and 315; comp
1.1 nf, 0.5 x 0.12, prob background -- U 7099) 12 04.0 + 28 25 at 2.8, 219, 0.4 x 0.4, compact, m = 15.3; sev obj
in vic --

07100   U 7101) in a triple subgroup w U 7088 at 7.7, 318, and U 7106 at 5.0, 32 -- U 7103) SA(r)0+ : (de V), S0 (Ho);
brightest in a subgroup; 12 04.5 + 43 16 = N 4109 at 4.6, 206, 0.6 x 0.5, m = 15.1 -- U 7104) measurements disturbed
by superimp *; in group w 12 04.6 + 17 16 brightest at 4.5, 1, 0.6 x 0.6, spir, m = 14.9, An 1205 in BG, Sb: (de V),
v = 6740, v₀ = 6670, SN 1960 c (Humason, classif Sa); 12 04.7 + 17 15 at 3.0, 15, 0.6 x 0.25, double system, m = 15.6
-- U 7106) VII Zw 439; see U 7088, 7101; N 4108 B in BG, S... pec (de V); "blue post-eruptive disrupted spir w
compact knots and bars" (Zw) -- U 7109) In group w 12 04.6 + 17 16 brightest, see U 7104 --

07110   U 7111) S B(rs)dm (de V); Sc+ (Ho); P w U 7116 at 14., 37 -- U 7112) 12 05.3 + 43 24 = N 4118 at 1.5, 137, 0.6 x 0.3,
m = 15.7; in a subgroup w U 7103 brightest at 8.5, 245 -- U 7114) 2nd and 3rd in group of 4; 1st group member at 1.2
np, 0.2 x 0.2; c) at 0.9 nf, 0.2 x 0.1, incl in CGCG magn -- U 7115) many S F galaxies in vic -- U 7116) S B(r)c
(de V), Sc + (Ho); see U 7111 -- U 7117) SA(r)0+ (de V) -- U 7118) E+6 (de V); 12 05.5 + 65 23 = N 4121 at 3.9,
195, 0.40 x 0.35, E 2 (de V), m = 14.6 --

07120   U 7120) SA 0: (de V); no NGC ident in CGCG; comp 2.0, 302, core 0.12 x 0.12, F smooth env 0.8 x 0.8; 3.1, 70, 0.30 x
x 0.25 -- U 7122) v dif disc outside 1.7 x 0.9; comp 4.0, 234, 0.5 x 0.4 -- U 7124) 12 05.6 + 25 14 at 7.5, 269,
0.9 x 0.25, m = 15.1; 12 06.7 + 25 15 at 7.3, 80, 0.6 x 0.45, m = 15.0; comp 3.6, 82, 0.4 x 0.3, E: -- U 7126)
12 06.5 + 29 31 at 4.5, 138, see U 7130; U 7130 at 9.1, 148 -- U 7129) 12 06.7 + 41 53 at 8.7, 155, 0.8 x 0.7,
compact, m = 15.2 --

07130   U 7130) 12 06.5 + 29 31 = N 4132 at 4.8, 338, 0.9 x 0.35, spir, m = 14.6, P w comp at 2.2, 55, 0.8: x 0.2, spir; see
U 7126 -- U 7134) SAB(r)c (de V); SN 1941 c -- U 7135) M 51 case, comp 0.55, 272, 0.1 x 0.1, at tip of spir arm;
P w comp at 1.9, 124, 0.35 x 0.25; 12 06.6 + 44 16 = N 4135 at 5.5, 196, spir, distorted?, m = 14.8 -- U 7137) a)
0.4 x 0.18; b) 1.0 x 0.25, sep 1.4, vF bridge -- U 7138) comp 2.0, 248, 0.45 x 0.25; sev S v dif obj in vic --
U 7139) SA(r)0+ (de V) --

07140   U 7142) SAB(s)0/a (de V) -- U 7143) comp 2.9, 100, 0.5 x 0.1, double system -- U 7147) B bar in amorph centr reg,
2 thin knotty arms; comp 1.1, 155, 0.4 x 0.1, F, asym --

07150   U 7151) SAB(s)cd? (de V) -- U 7152) asym; early bar spir on edge?; comp 1.9, 7, 0.6 x 0.12; 2.2, 140, 0.6 x 0.15 --
U 7154) SAB(rs)d (de V); chains of H II regions in the arms; P w U 7166 at 29., 170 -- U 7155) P w comp at 2.1, 224,
0.6 x 0.5, MCG 8 - 22 - 78 -- U 7158) comp 2.3, 97, 0.3 x 0.2; 4.6, 56, 0.3 x 0.2 --

07160   U 7160) U 7168 at 3.0, 136 -- U 7162) comp 4.8, 166, 0.45 x 0.12 -- U 7163) SN 1963 d; comp 4.3, 155, 0.30 x 0.18 --
U 7165) SA(r)0/a ? (de V) -- U 7166) SAB(rs)ab: (de V), Sa (Ho); see U 7154 -- U 7167) spindle w high surf brightn;
comp 2.4, 233, 0.8: x 0.8:, irr dwarf, v dif -- U 7168) see U 7160 -- U 7169) SAB(rs)c (de V) --

07170   U 7170) v slightly integral-shaped -- U 7171) interacting? w comp at 1.1, 172, 0.20 x 0.20, v blue -- U 7172) sev
S comps, brightest at 2.6, 235, 0.3 x 0.3 -- U 7173) S B(rs)b (de V); U 7166 at 5.3 sp -- U 7175) P w comp at 0.6,
200, 0.4 x 0.3, dif, contact -- U 7176) DDO 111; Ir V (vdB); comp to U 7183 at 12.0 n -- U 7178) MCG α = 12 07.75
is incorrect -- U 7179) spindle w high surf brightn; MCG major diam 0.65 is incorrect --

07180   U 7182) SA(r)b: (de V) -- U 7183) SAB(s)b? (de V); SN 1937 a, 1955 a; comp 3.5, 210, 0.3 x 0.1 -- U 7185) DDO 110;
V (vdB) -- U 7186) DDO 112; Ir V (vdB) -- U 7187) MCG 6 - 27 - 24 is a B cond in the disc at 0.30, 110 --

07190   U 7192) P w U 7209 at 10.8, 95 -- U 7193) (R)SA(rs)bc (de V); SN 1965 g -- U 7197) comp 4.6, 348, 0.8 x 0.35, spir--
07200   U 7201) SAB(r)a:? (de V), Sc- (Ho); b) at 0.9, 353, 0.3 x 0.1; P w U 7203 at 2.8, 152; SN 1971 g -- VV 147) α =
12 09.7; listed at 16 09.68 by VV; MCG 3 - 31 - 69, 0.8 x 0.1 -- U 7202) U 7204 at 1.7, 15; U 7206 at 2.5, 137;
U 7211 at 2.9, 105 -- U 7203) comp 12 09.5 + 13 28 at 2.8, 270, 0.2 x 0.2, E:, m = 15.7; E 2 (de V), E (Ho) --
U 7204) see U 7202; U 7206 at 3.5, 164; U 7211 at 3.2, 138 -- U 7205) comp 0.30, 177, 0.1 x 0.1; 3.0, 77, 0.4 x 0.1
-- U 7206) comp 0.7, 232, 0.1 x 0.1; see U 7202, 7204; U 7211 at 1.6, 37 -- U 7208) I Zw 32; "blue post-eruptive
disrupted spir, compact nucl, broken jet to the southeast" (Zw); dif jet in pos angle 157 towards comp at 1.9,
160, 0.35 x 0.12 -- U 7209) see U 7192 --

07210   U 7211) see U 7202, 7204, 7206 -- U 7212) VV 345 b (listed at δ = 35 59); a) = VV 345 a at 0.9, 352, 0.7 x 0.3 --
U 7214) S0 (de V) -- U 7215) S B(rs)dm (de V), Sc+ (Ho); SN 1963 i; comp 4.4, 112, 0.5 x 0.4, late spir -- U 7216)
chaotic spir w B blue nucl reg; comp 2.8, 274, 0.7 x 0.1, Sb - c -- U 7218) chaotic; 3 B blue patches on f side --

07220   U 7220) P w comp at 0.50, 340, 0.15 x 0.12 -- U 7222) SA(s)cd? (de V); SN 1968 u; data for N 4160 in Publ. David
Dunlap Obs. Vol. II No. 6 refer to U 7222 = N 4183 -- U 7223) comp to U 7231 at 9.4, 32 -- U 7224) sev S galaxies
near-- U 7226) comp 0.8, 112, 0.2 x 0.2; 2.2, 340, 0.4 x 0.3, dif, F -- U 7229) sev comps near the galaxy, brightest
at 3.3, 290, 0.45 x 0.15 --

07230   U 7230) VV 128, Arp 260; (VV α = 16 11.0 is incorrect); large S B(s) or SAB(s) w distorted arms, smaller pec spir
nf; "poss lines of F conds extending s" (Arp) -- U 7231) SAB(s)ab (de V), Sb + (Ho); see U 7223; U 7240 at 11.4,

156 -- U 7232) VV 104 -- U 7234) SAB(s)c:? (de V), Sb+ (Ho) -- U 7235) SAB(rs)cd? (de V), Sc- (Ho); SN 1966 e --

07240 U 7240) comp to U 7231 at 11.4, 336 -- U 7241) VV 261, Arp 160, I Zw 33; S B 0 pec (de V); "broken ring, large plume" (CGCG); chaotic main body w short jet and broad plume northw; in Arp's class "disturbed galaxies w interior abs"; "blue post-eruptive Sa w elliptical patchy compact disc, jets, plumes and knots" (Zw) -- U 7242) * superimp; U7264 at 9.6; U 7288 at 14.0 -- U 7243) in triple group w 12 11.6 + 32 43 at 2.0, 282, 0.8 x 0.5, S0, m=15.1, and comp at 2.2, 348, 0.7: x 0.5: -- U 7244) v thin bar, 2 F main arms split in multiple arm pattern outwards; comp 4.8, 68, 0.40x0.25, S B a - b -- U 7247) listed as An 1213 in BG, w α = 12 12.9 -- U 7249) DDO 114; Ir IV-V: (vdB) --

07250 U 7250) P w 12 12.0 + 01 06 at 6.5, 322, 0.7 x 0.25, m=15.0 -- U 7253) VV 183 -- U 7255) U 7275 at 11.4, 132 -- U 7256) SAB 0-: (de V) -- U 7257) P w comp 0.6, 335, 0.12x0.12, prob interaction -- U 7258) B main body 1.1 x 0.25 w sharp edge, S smooth outer disc; comp 3.5, 258, 0.5 x 0.4 --

07260 U 7260) SA(s)bc: (de V), Sc- (Ho); dwarf comps at 3.1, 277, 0.9:: x 0.9::, e dif, and 8.0, 297, 0.5: x 0.5:; P w U 7284 at 11.6, 48 -- U 7262) U 7254 at 5.2, 261 -- U 7264) SB(r)b (de V); see U 7242; U 7288 at 15.6 -- U 7266) U 7276 at 7.0, 164 -- U 7268) dwarf comp at 7.7, 259, 0.8:: x 0.8::, e dif --

07270 U 7270) B nucl bulge w broad abs lane; P w 12 13.1 + 22 11 at 4.7, 7, 0.7 x 0.15, m=15.1 -- U 7274) DDO 115; Ir IV-V or Ir V (vdB); diameters for MCG 2-31-65 = I 3073 refers to this galaxy -- U 7275) SA(rs)bc? (de V), Sc- (Ho); see U 7255 -- U 7276) sev galaxies near, see U 7266 -- U 7277) VV 199, Arp 106, An 1214 A and 1214 B in BG; S0 pec and Sb pec (de V); in Arp's class "E and E-like galaxies connected to spirals"; sep 0.55 -- U 7278) IAB(s)m (de V), Ir I (Ho); SN 1954 a --

07280 U 7281) SA(r)0+ (de V) -- U 7282) Sb (de V); comp 3.8, 70, 0.3 x 0.2, F; U 7297 at 7.3 -- U 7283) S0? (de V): Haro 28; "violeta" (Haro) -- U 7284) SAB(s)b: (de V), Sb- (Ho); see U 7260; U 7291 at 11.8, 36 -- U 7288) (R)SB(r)0+ (de V); comp 1.8, 100, 0.5 x 0.15; see U 7242, 7264 --

07290 U 7290) SA(r)0+ (de V) -- U 7291) Sd: (de V); see U 7284 -- U 7292) SA(s)a: (de V), Sa (Ho) -- U 7293) comp 0.55, 307, 0.15x0.12 -- U 7295) U 7296 at 4.4 n; comp 1.3, 249, 0.3 x 0.1; 4.7, 265, 0.5 x 0.25, dif -- U 7296) P w U 7299 at 2.6, 25 -- U 7297) B ecc main body; 2 S v red galaxies p at 0.4 and 0.5 -- U 7299) see U 7296 --

07300 U 7300) DDO 117; Ir V (vdB); comp 1.4, 8, 0.5 x 0.05; U 7289 at 6.7, 286 -- U 7303) P w U7304 at 1.1, 0; comp 4.8, 187, 0.35x0.30 -- U 7304) see U 7303 -- U 7306) SB(s)dm (de V), Sc+ (Ho); VII Zw 446 is a "blue patchy compact" (Zw) w m$_p$ = 15.7 at the edge of U 7306 ssf center -- U 7308) comp 4.6, 254, 0.5 x 0.3, v dif, dwarf -- U 7309) I...? (de V); Haro 7; "violeta" (Haro) --

07310 U 7310) SA(s)a (de V), Sa (Ho) -- U 7311) S0 (Ho) -- U 7312) U 7313 at 9.8, 5; U 7326 at 9.0, 49 -- U 7313) see U 7312; U 7326 at 7.1, 123 -- U 7315) SAB(rs)bc (de V) -- U 7316) comp 0.6, 345, 0.3: x 0.1, interaction? -- U 7317) comp 1.6, 67, 0.5 x 0.1 -- U 7318) comp 2.5, 297, 0.3 x 0.2 -- U 7319) Sa (Ho); P w U7333 at 8.6, 105 --

07320 U 7322) SA(s)cd: (de V), Sc+ (Ho) -- U 7323) SAB(s)dm (de V), Sc+ (Ho) -- U 7324) interacting pair?; MCG δ = = 46 41 -- U 7326) v blue; 2 S blue comps superimp?; see U 7312, 7313 -- U 7328) SB(r)0/a: (de V) -- U 7329) VII Zw 447; SAB(r)0+ (de V); "blue sym S B c w dense compact elliptical core" (Zw); Mark 205 at 0.7 s center, v =21000

07330 U 7331) comp 2.3, 332, 0.45x0.25; 2.9, 234, 0.4 x 0.2 -- U 7333) Sc- (Ho); see U 7319 -- U 7334) SA(s)c (de V), Sc- (Ho); 12 15.4 + 07 33=N 4247 at 5.3, 359, 0.6 x 0.5, m=14.7, (R)SAB(s)abpec? (de V) -- U 7338) S B 0 ? (de V); comp 2.7, 75, 0.3 x 0.2; 2.9, 250, 0.25x0.15 --

07340 U 7340) a) 0.35x0.3; b) 0.6 x 0.5, spir, massive arms; F bridge -- U 7342) distorted?; P w comp at 1.6, 266, 0.3x x 0.2 -- U 7345) SA(s)c (de V), Sc- (Ho); SN 1967 h; slightly asym, no disturbing obj vis -- U 7347) comp 1.5, 165, 0.2 x 0.2, F comp 0.2 s -- U 7349) Sc: (de V) --

07350 U 7350) a) at 1.2, 220, 0.45x0.25; comp 4.3, 290, 0.6 x 0.2, early; 4.6, 110, 0.9: x 0.4:, dwarf -- U 7351) SA(s)b: (de V); comp 2.0, 188, 0.4 x 0.1 -- U 7352) F comp superimp?; P w 12 16.4 + 09 15=I 3134 at 6.6, 346, 0.8 x 0.3, S0-a, m=15.4 -- U 7353) SAB(s)bc (de V), Sb+ (Ho) -- U 7354) An 1217 in BG; Haro 8; "muy violeta" (Haro); "osculating blue post-eruptive pair of large irr galaxy and oval patchy compact" (Zw) -- U 7355) a) 1.0: x 1.0:, dwarf; b) 1.0: x 0.5:; both dif; sep 0.7 -- U 7359) S0 (de V), S0 (Ho) --

07360 U 7360) E 2 (de V); U 7364 at 3.3, 67; 12 16.5 + 06 00=N 5257 at 7.1, 214, 0.9: x 0.25, spir, m=15.0 -- U 7361) S B(s)a (de V) -- U 7364) S B(rs)0+ (de V); see U 7360 -- U 7365) S B(s)0-? (de V) -- U 7369) comp 3.1, 15, 0.35x x 0.25 --

07370 U 7370) 12 17.1 + 02 19 at 4.2, 187, 0.35x0.15, m=15.6 -- U 7371) S0: (de V), S0 (Ho) -- U 7372) P w 12 17.1 + + 06 17=I 3135 at 1.1, 240, 0.8 x 0.4, early spir, S0: (de V), m=15.0 -- U 7373) S B(s)0-? (de V) -- U 7374) 12 17.2 + 29 00 at 8.3, 183, 0.5 x 0.5, compact, m=15.5; 12 17.2 + 29 04 at 4.7, 187, 0.3 x 0.2, compact, m=15.7; 12 17.2 + 29 07 at 2.3, 194, 0.5 x 0.5, m=15.3 -- U 7375) comp 3.0, 250, 0.55x0.25 -- U 7376) S0 (de V), S0 (Ho) -- U 7377) (R)SB(r)ab (de V), Sa (Ho); comp 2.5, 214, 0.4 x 0.3; 3.2, 331, 0.5: x 0.4:, dif, dwarf -- U 7378) sev S galaxies near --

07380 U 7380) S B(s)c (de V), Sc- (Ho); 12 17.5 + 05 37=N 4277 at 2.0, 96, 0.7 x 0.6, S B 0, m=15.0; SN 1936 a -- U 7383) 12 17.6 + 08 55 at 3.1, 45, 0.9 x 0.45, m=15.1; 12 17.7 + 08 49 at 6.2, 137, 0.45x0.45, m=15.6 -- U7384) 12 17.8 + 28 12 at 4.5, 138, 0.7 x 0.4, m=15.7 -- U 7386) E 1- 2 (de V), E (Ho); in a triple chain w U7390 at 3.6, 58, and U 7398 at 8.8, 62 -- U 7389) S0+: (de V), S0 (Ho) --

07390 U 7390) E 0 (de V), E (Ho); see U 7386; U 7398 at 5.0, 65 -- U 7393) P w U7402 at 4.6, 90 ; Sc? (de V, note to N 4290 -- U 7397) E 2-3 (de V); P w U7429 at 6.2, 119 -- U 7398) SA(r)0/a: (de V); see U 7386, 7390 -- U 7399) DDO 119; S (B) IV? (vdB); comp 2.2, 169, 0.4 x 0.4 --

07400 U 7402) S B(rs)ab: (de V); see U 7393; comp 3.8, 97, 0.9 x 0.2 -- U 7404) comp 2.2, 2, 0.35x 0.35 -- U 7405) (R)S B(s)0/a (de V), Sa (Ho) -- U 7406) comp 3.1, 210, 0.55x0.35; 4.9, 278, 0.12x0.12 and 0.20 x 0.20, sep 0.40, prob interacting pair in triple group -- U 7407) S B(s)cd (de V), Sc- (Ho); P w U7414 at 5.6, 97 -- U 7408) DDO 120; Ir IV-V (vdB) -- U 7409) P w N4297 at 1.1, 346, 0.5 x 0.2, m approx 16. --

07410 U 7412) SA(rs)c (de V), Sc- (Ho); P w U7418 at 2.4, 102 -- U 7414) S B d pec? (de V), Sc+ (Ho); see U 7407 -- U 7415) S0: (de V) -- U 7418) Sc: (de V); see U 7412; comp 2.1, 58, 0.4 x 0.2, Sc: --

07420 U 7420) SAB(rs)bc (de V), Sc- (Ho); SN 1926 a, 1961 i, 1964 f -- U 7426) U 7438 at 5.0, 99 -- U 7429) S B(rs)ab: (de V); (r) 1.4 x 0.7; see U 7397, disturbed? --

07430 U 7430) P w U7431 at 3.2, 355 -- U 7431) see U 7430 -- U 7432) P w U7433 at 2.8, 3 -- U 7433) see U 7432 -- U 7435) SAB(r)0+ (de V) -- U 7436) P w comp at 1.5, 126, 0.5 x 0.3, v dif, blue -- U 7437) comp 1.6, 223, 0.2 x 0.2, E:; 2.1, 130, 0.1 x 0.1 and 2.4, 134, 0.3 x 0.2, pair, sep 0.25 -- U 7438) see U 7426 -- U 7439) S B cd: (de V), Sc+ (Ho) --

07440 U 7441) P w U7448 at 3.3, 86 -- U 7442) S 0/a (de V) -- U 7443) S B(rs)a (de V), Sa (Ho) -- U 7445) SA ab: (de V) -- U 7448) see U 7441 -- U 7449) VII Zw 451; "blue post-eruptive oval patchy compact w comet-like plume, d = 2' n - s immersed double knot" (Zw) --

07450 U 7450) SAB(s)bc (de V), Sc- (Ho); SN 1901 b, 1914 a, 1959 e; 12 20.5 + 16 11=N 4322 =N 4323 at 5.2, 19, 0.9: x x 0.7:, late spir?, S0: (de V), m=15.7; 12 20.8 + 16 05=N 4328 at 6.1, 92, (0.9 x 0.8), S0:, m=15.2 -- U 7451)

SA(r)0+ (de V) -- U 7452) B nucl reg w 2 spir arcs southw; P w 12 20.6 + 10 54 = N 4325 at 4.9, 27, (0.9 x 0.7),
E, m = 15.0; comp 4.3, 178, [0.7 x 0.3], pec, or double system; B cond w dif annex, disturbing comp? -- U 7453)
S B(s)a (de V) -- U 7454) SAB(r)ab? (de V); U 7461 at 5.7, 84 -- U 7455) SN 1955 e; comp 3.8, 294, 0.3 x 0.15;
4.1, 87, 0.5 x 0.20 -- U 7458) S B 0: (de V) --

07460   U 7461) E 0 (de V); 12 20.8 + 06 19 = N 4333 at 4.0, 232, 0.9 x 0.8, S B a, m = 14.8, S B(s)ab (de V); see U 7454 --
U 7463) S0 (de V) -- U 7465) SA(rs)b: (de V), Sa (Ho, ident as N 4341); see BG pages 18 - 19 and note to N 4343
for discussion of identification -- U 7466) S0- (de V), E (Ho, see U 7465) -- U 7467) S B(r)0 + (de V); P w U 7473
at 5.7, 110; comp 2.3, 265, 0.4 x 0.1, dif -- U 7469) S B 0: (de V), Sc- (Ho); smooth disc w low surf brightn,
prob non-regular --

07470   U 7470) S B m: (de V) -- U 7472) S0 (Ho, listed as N 4343, see U 7465) -- U 7473) SA 0 (de V); see U 7467 -- U 7474)
Pec S B 0/a: (de V), Sc- (Ho); smooth disc, but may be late -- U 7476) SA(s)ab: (de V) -- U 7479) N 4358 is an obj
at 0.7, 245, 0.20 x 0.15, incl in CGCG magn; P w 12 21.8 + 58 38 = N 4362 at 1.9, 140, 0.8 x 0.25, m = 15.2 --

07480   U 7480) comp 2.2, 80, 0.3 x 0.2, F -- U 7484) 12 21.7 + 09 29 at 4.4, 212, 0.35 x 0.25, m = 15.7; 12 21.7 + 09 33 =
= I 3274 at 2.3, 228, 0.4 x 0.3, m = 15.2; 12 21.7 + 09 37 at 2.7, 338, 0.4 x 0.3, m = 15.3 -- U 7488) E 3 (de V) --
U 7489) (R)SA(rs)a (de V) --

07490   U 7490) DDO 122; S V (vdB) -- U 7491) E 6-7 (de V) -- U 7492) SA a (de V) -- U 7493) S B(r)0 + (de V), S0 (Ho) --
U 7494) E + 1 (de V), S0 (Ho); SN 1957 b; see A. J. 66, 555 -- U 7496) Sa: (de V); SN 1960 j -- U 7497) (R)SA(s)a
(de V) --

07500   U 7501) III Zw 65; SA 0- (de V); "Group of 3. No. 1, NGC 4377, red E w compact elliptical disc... v = 1330. No. 2,
25" p No. 1, neutral elliptical compact, m_p = 16.8. No. 3, 15" nf No. 1, neutral elliptical compact, m_p = 16.9"
(Zw); "...Elliptical galaxy 50" x 40" = 4.6 x 3.4 kpc w a fainter comp 15" = 1.3 kpc from its center. The comp has
a diameter of 7" = 0.6 kpc. The redshift given is for the brighter component." (Sargent) -- U 7503) SA(rs)b:?
(de V) -- U 7504) prob late spir; P w U 7507 at 2.6, 20 -- U 7505) prob interacting w comp at 1.0, 244, 0.4: x 0.2
-- U 7507) early spir?; comp 2.6, 200, 0.8: x 0.3:, dif; S0: (de V) -- U 7508) SA(s)0 + pec (de V), S0 (Ho); SN
1960 r; P w U 7523 at 7.8, 80; comp 3.1, 180, 0.4 x 0.4 --

07510   U 7510) P w comp at 3.5, 26, 0.9: x 0.8:, dwarf -- U 7511) VII Zw 454; S B ... (de V); "large blue spherical disc
compact w extended uniform spherical halo" (Zw) -- U 7513) I 3322 A in BG; S d: (de V); comp 1.8, 217, 0.35 x 0.25;
2.9, 93, 0.35 x 0.30; 2.9, 184, 0.45 x 0.15 -- U 7514) S B(rs)... (de V) -- U 7515) S B(rs)0 +: (de V) -- U 7517)
E + 3: (de V) -- U 7518) Sc: (de V) --

07520   U 7520) S B c (de V) -- U 7523) (R)S B(r)b (de V), Sb- (Ho); see U 7508 -- U 7524) SA(s)m: (de V), Sc + (Ho); low
surf brightn; N 4399, 4400 and 4401 are conds in the disc -- U 7525) P w 12 23.3 + 45 34 at 1.5, 290, 0.7 x 0.6,
B core 0.18 x 0.18, S0?, m = 15.4 --

07530   U 7530) P w comp at 2,8, 292, 0.8 x 0.40, Sc -- U 7531) comp 4.0, 42, 0.9 x 0.4, double system? -- U 7532) E + 3
(de V), E (Ho); comp superimp at 1.4, 35, 0.8: x 0.6:; see U 7494 -- U 7534) DDO 123; IV- V (vdB) -- U 7535) prob
2 spirs in close contact; SN 1965 a; sev galaxies near; 12 24.0 + 09 18 = I 790 at 1.9, 60, 0.9 x 0.5, E- S0, m =
= 15.2; 12 24.2 + 09 19 at 4.2, 65, 0.55 x 0.45, S B ..., m = 15.3 -- U 7536) S B(r) ...pec (de V) -- U 7537) Sc + (Ho);
P w U 7546 at 4.3, 81; many smaller galaxies in vic -- U 7538) S B ab: (de V) -- U 7539) SA(rs)c? (de V), Sc- (Ho) --

07540   U 7542) S B(s)0 : (de V), S0 (Ho); comp 2.8, 150, 0.4 x 0.2 -- U 7544) P w 12 23.6 + 62 43 at 6.0, 300, 0.9 x 0.5,
spir, m = 15.0 -- U 7545) P w 12 24.5 - 00 37 at 3.0, 126, 0.8 x 0.5, comp superimp, m = 15.0 -- U 7546) Sc + (Ho);
see U 7537 -- U 7547) e dif; IC descr "B middle" refers to superimp *; U 7550 at 6.5, 7 -- U 7549) S B(r)bc: (de V)

07550   U 7550) see U 7549 -- U 7551) S B(s)a (de V) -- U 7554) S B a: (de V) -- U 7558) comp 1.7, 291, 0.4 x 0.1; 2.1,
293, 0.3 x 0.1; 2.3, 318, 0.3 x 0.1; parallel spindles --

07560   U 7560) vB core; 2 open arms, one stronger; SN 1960 i (Humason, classif S B b) -- U 7561) S B(s)a: (de V); chaotic
early spir; SN 1895 a; comp 0.6, 175, 0.30 x 0.15 -- U 7562) S0 : (de V) -- 12 24.7 + 14 24) III Zw 66; core 0.20 x
x 0.20, halo 0.6? x 0.6?; "neutral e compact, F halo (but more probably, * superposed on Irr galaxy)"(Zw) --
U 7564) P w 12 24.9 + 03 32 at 2.7, 261, 0.5 x 0.5, m = 15.4 -- U 7566) ecc (rs) 0.9 x 0.6; 2 broad smooth arms;
P w U 7570 at 2.4, 138, interaction -- U 7567) comp 3.2, 299, 0.5 x 0.5, dwarf; 4.0, 266, 0.35 x 0.1, v dif --
U 7568) SA(r)0 + (de V), S0 (Ho) -- U 7569) S0? (de V); in group w U 7573 at 3.8, 65, and U 7581 at 6.4, 189 --

07570   U 7570) see U 7566 -- U 7572) B core in ext chaotic env; comp 1.5, 126, 0.3 x 0.3; 3.5, 258, 0.3 x 0.2; 3.6, 200,
0.4 x 0.3; 4.3, 264, 0.5 x 0.4 -- U 7573) S0 (de V); see U 7569; U 7581 at 3.3, 115 -- U 7574) VV 188 (a), Arp 120 a;
P w U 7575 at 4.4, 343; see U 7494; SA(s)0/a pec: (de V), Sa pec (Ho); in Arp's class "E and E-like galaxies close
to and pertubing spirals"; "E galaxy breaking up a spiral" (Arp) -- U 7575) VV 188 (b), Arp 120 b; S B(s)0/a (de V),
S0 (Ho); see U 7574, 7494 -- U 7577) DDO 125; Ir IV- V (vdB); P w U 7608 at 20., 142 --

07580   U 7581) S B(rs)a (de V); see U 7569, 7573 -- U 7582) comp 2.0, 196, 0.3: x 0.3:, dif; 4.6, 287, 0.3 x 0.1; sev
smaller galaxies near -- U 7583) S B(s)0/a (de V), S0 (Ho); 12 25.6 + 10 03 at 3.2, 145, 0.8 x 0.4, m = 15.6 --
U 7585) sev S galaxies near -- U 7586) P w 12 25.7 + 14 10 = N 4447 at 1.6, 119, 0.9 x 0.7, S0:, m = 15.2; one arm of
U 7586 cont in a bridge to N 4447 -- U 7587) S B 0/a: (de V) -- U 7588) comp 2.1, 282, 0.4 x 0.3 -- U 7589) P w
U 7601 at 7.3, 107 --

07590   U 7590) P w U 7596 at 6.6, 146 -- U 7591) S B(r)ab (de V), Sb- (Ho); sev galaxies in vic -- U 7592) I B m (de V),
Ir I (Ho); box-shaped chaotic centr reg, dif outskirts -- U 7593) I Zw 37; "Group of 3. Nos. 1,2, pair of intertwined neutral pinwheel Sc's w equal compact cores, sep 12" east- west... No. 3, 170" np No. 1, neutral spherical,
m_p = 16.0" (Zw) -- U 7594) SA(s)ab (de V), Sb- (Ho); comp 3.8, 342, 0.3 x 0.2, spir -- U 7596) see U 7590 --
U 7597) U 7591 at 4.2, 218 -- U 7599) DDO 127; V (vdB) --

07600   U 7601) see U 7589 -- U 7606) (R)S B(r)0/a (de V); comp 2.0, 45, 0.3 x 0.2 -- U 7608) DDO 129; Ir V (vdB); see U 7577
-- U 7609) (R)SAB(rs)0/a (de V) --

07610   U 7610) E 0-1 (de V); P w U 7613 at 3.7, 160; see U 7494 -- U 7611) S B(s)0 +? (de V); comp 0.9, 328, 0.1 x 0.1 --
U 7612) DDO 128; V pec (vdB) -- U 7613) S B(s)0 +: (de V); see U 7610, 7494 -- U 7614) SA(r)0 + (de V); comp 2.2,
44, 0.5: x 0.5: -- U 7617) 12 26.0 + 44 54 at 7.4 p, 0.9 x 0.7, spir, m = 15.1 -- U 7618) listed at δ = 58 01 in MCG
-- U 7619) E 2 (de V) --

07620   U 7620) E (de V; data listed for I 3418 in BG prob refers to this galaxy; position for I 3418 in BG agrees w position
for U 7630); P w U 7630 at 5.6, 110 -- U 7622) similar to N 4594; 12 26.7 + 09 09 at 4.7, 321, double system, 0.4 x 0.3
and 0.1 x 0.1, total [0.8 x 0.4], m = 15.6; S B(s)0/a? (de V) -- U 7625) comp 5.0, 255, 0.6 x 0.5 -- U 7628) E 4 :
(de V); double system?, S comp or superimp * at 0.25, 75; U 7634 at 5.6, 78 -- U 7629) Arp 134; E 2 (de V), E (Ho);
v uncertain diameter values, corona appears vis to 12.? x 10.?, see U 7636 ; in Arp's class "E and E-like galaxies
w nearby fragments"; 12 26.8 + 08 18 = N 4465 at 5.8, 285, 0.25 x 0.20, m = 15.4; 12 26.9 + 08 16 = N 4467 at 4.0, 264,
0.45 x 0.35, E 2 (de V), v = 1474, v_0 = 1381, m = 15.2; U 7636 at 5.4, 142 --

07630   U 7630) see U 7620 -- U 7631) E 5 (de V); see U 7494 -- U 7634) S0 (de V); see U 7628 -- U 7635) vB core 0.35 x 0.25;
12 28.0 + 57 04 at 5.4 f, 0.6 x 0.4, spir, m = 15.6 -- U 7636) comp to U 7629 at 5.4, 322, prob superimp the corona
of this galaxy -- U 7637) P w U 7645 at 4.6, 105; SA(r)0-: (de V) -- U 7638) S B(s)0/a:? (de V); U 7646 at 5.3, 133;
see U 7494 --

07640   U 7640) in a subgroup w U 7633 at 10.5, 199, and 12 27.1 + 11 04 = I 3416 at 8.6, 276, 0.9 x 0.6, Irr, m=15.3 --
U 7643) U 7649 at 7.1, 126 -- U 7645) E 2 (de V); see U 7637; U 7654 at 8.6, 65 -- U 7646) S B(s)0/a:? (de V); see
U 7638 -- U 7647) Sc- (Ho) -- 12 28.0 + 12 46) I Zw 38; 0.35 x 0.35; "red oval compact, m_p = 14.5 ... v = 1650..."
(Zw) -- U 7648) VV 30 b, Arp 269 a; I B(s)m pec (de V), Sc+ (Ho); in Arp's class "double galaxies w connected arms";
"resolution of knots, em regions and dust lanes only" (Arp); P w U 7651 at 3.6, 165 -- U 7649) sev F galaxies near --

07650   U 7651) VV 30 a, Arp 269 b; S B(s)d pec (de V), Sc+ (Ho); late chaotic spir; see U 7648 -- U 7652) vis through the
corona of U 7654 = M 87; a) 0.3 x 0.15, b) 0.3 x 0.15, sep 0.65 -- U 7654) Arp 152; E 0-1 pec (de V), E (Ho);
12 28.0 + 12 46 = I Zw 38 = N 4486 B (in BG) at 7.4, 325, 0.35 x 0.35, compact, m=14.5, E 0 (de V); see U 7645;
U 7658 at 7.4, 164; in Arp's class "galaxies w jets"; Virgo A radio source; Arp's short exp shows jet; SN 1919 a,
poss SN 1922 (Wolf) -- U 7655) U 7669 at 12.6, 64 -- U 7657) S B 0?; 12 28.8 + 11 46 = I 3446 at 6.4, 85, 0.7 x 0.3,
m=15.3 -- U 7659) P w 12 28.3 + 57 35 at 1.5, 310, 0.50 x 0.25, m=15.3, interaction? --

07660   U 7660) comp 3.8, 136, 0.5 x 0.1 -- U 7661) P w 12 29.5 + 52 41 at 7.0, 98, 0.7 x 0.6, m=15.4 -- U 7662) E 1-2
(de V); comp 4.7, 264, 0.4: x 0.2: -- U 7663) comp 3.0, 11, 0.5 x 0.4, S B... -- U 7667) S B(s)a (de V) -- U 7668)
VV 76; S B(rs)m (de V, I ? for comp), Sc+ (Ho, Sc+ for comp); N 4496 a is a v late S B... in contact w N 4496 b at
0.9, 160, 0.9 x 0.7, Irr or v late spir, v = 1773, v_0 = 1665; SN 1960 f -- U 7669) see U 7655; comp 4.5, 296, 0.6 x
x 0.1 --

07670   U 7672) comp 3.3, 134, 0.5:: x 0.3:, e dif, dwarf -- U 7673) DDO 131; Ir V (vdB) -- U 7675) SA(rs)b (de V), Sb+ (Ho);
SN? 1961 (Rosino, prob var *); comp 6.9, 190, 0.9: x 0.6:, dwarf -- U 7679) comp 1.7, 280, 0.10 x 0.10, compact,
stellar; 3.5, 76, 0.45 x 0.22, spir --

07680   U 7680) S B 0-: (de V) -- U 7683) MCG classif "E" is incorrect; broad e dif (R) vis in blue; sev comp , 12 29.8 +
+ 66 40 at 4.3, 5, 0.45 x 0.30, ring-shaped, similar to Arp 146 and 147, m=15.6; 12 30.0 + 66 40 at 4.3, 15, 0.20 x
x 0.18, plus 2 fainter comp , total m=15.7; these 4 galaxies form a quadrupel group -- U 7685) S B(rs)dm: (de V),
Sc+ (Ho); N 4517 a in BG; S V (vdB) -- U 7687) Sc: (de V) -- U 7688) comp 3.0, 17, 0.6 x 0.3 -- U 7689) comp 0.6,
274, 0.2 x 0.1, dif, F bridge to U 7689? --

07690   U 7691) sharp edge, no spir pattern vis; comp 2.2, 327, 0.35 x 0.25; a large nr of F galaxies in the background, in
CGCG cl 12 29.8 + 56 59, open, v distant -- U 7692) DDO 132; "spheroidal?" (vdB); E (de V) -- U 7694) SA(s)cd:
(de V), Sc- (Ho); P w U 7685 at 17. -- U 7695) Sm: (de V) ; SN 1970 a -- U 7698) DDO 133; Ir V (vdB) --

07700   U 7700) P w U 7706 at 4.1 nf -- U 7704) P w comp at 3.7, 324, 0.8 x 0.35, S B b: -- U 7705) comp 3.4, 298, double
syst, 0.12 x 0.12 and 0.25 x 0.1, sep 0.25 -- U 7706) see U 7700 -- U 7707) P w 12 30.6 + 17 40 = I 3484 at 4.6, 224,
0.7 x 0.6, spir, m=15.4 -- U 7708) comp 0.7, 3, 0.1 x 0.1 -- U 7709) S B(rs)d (de V), Sc- (Ho); 12 30.9 + 08 57 at
2.6, 327, 0.3 x 0.2, m=15.3 --

07710   U 7710) DDO 134; Ir V (vdB) -- U 7713) SAB(s)cd? (de V); DDO 135; S: V: (vdB) -- U 7716) S B(s)cd: (de V) -- U 7718)
SAB(s)0/a: (de V); SN 1969 e --

07720   U 7721) SAB(s)bc (de V), Sb- (Ho); SN 1915 a; P w U 7732 at 29., 171 -- U 7722) S0 (de V) -- U 7726) I B m (de V),
Ir I (Ho) -- U 7727) SAB(s)c (de V), Sc- (Ho) --

07730   U 7731) alm compl stell on PA prints, prob e compact galaxy -- U 7732) SAB(rs)bc (de V), Sc- (Ho); see U 7721 --
U 7735) Sa (Ho) -- U 7737) DDO 136, Ir V (vdB) -- U 7739) Ir I (Ho); An 1232 in BG; DDO 137; V (vdB) --

07740   U 7740) comp 0.4, 328, 0.1 x 0.1, F; 0.4, 175, 0.1 x 0.1; 4.8, 267, 0.4 x 0.15 -- U 7741) comp 2.1, 117, 0.30 x 0.15;
3.0, 3, 0.18 x 0.18 + halo; sev other galaxies in vic, prob in CGCG cl 12 32.7 + 70 05, medium compact, distant --
U 7742) SAB(rs)cd (de V); P w 12 32.4 + 15 51 = I 3528 at 1.6, 55, 0.5 x 0.4, SAB(r)b (de V), m=15.2; comp 4.3,
317, 0.4 x 0.3, dif, S d-dm -- U 7745) U 7767 at 3.8, 82 -- U 7746) prob early, Sa?; comp 1.0, 22, 0.3 x 0.12; 3.2,
131, 0.7: x 0.2; sev fainter comps -- U 7747) SN 1940 d --

07750   U 7753) S B(rs)b (de V), Sb+ (Ho); 12 33.2 + 14 41 at 6.6, 141, 0.8 x 0.3, spir, m=15.7 -- U 7755) in triple group
w double system 12 33.1 + 00 03 at 1.0, 119, 0.35 x 0.22 and 1.6, 130, 0.5: x 0.3:, sep 0.6, interaction -- U 7757)
P w U 7759 at 3.2, 35; S0 (de V) -- U 7759) E 3 (de V); see U 7757 --

07760   U 7760) E 0 (de V) -- U 7761) massive bar, patchy blue arms; appears late, but no normal SBc -- U 7762) E : (de V) --
U 7764) P w 12 33.2 + 26 30 = I 3546 at 4.0, 180, 0.9 x 0.5, spir, m=15.3; I 3543 is listed as I 3546 in BG, see
correction in P.A.S.P. 78, 262; "I 3546 is a compact E galaxy 24: s" (de V); spir str vis in I 3546 on PA prints --
U 7765) brightest in a subgroup; comp 0.8, 269, 0.12 x 0.12; 12 33.6 + 27 11 = I 3561 at 4.3, 97, 0.5 x 0.2, m=15.7;
12 33.5 + 27 15 = N 4536 at 4.4 nf, 0.8 x 0.4, m=15.6; sev other galaxies in CGCG cl 12 17.5 + 29 15, medium compact,
near, in vic -- U 7766) SAB(rs)cd (de V), Sc+ (Ho); SN 1941 a; I 3550, 3551, 3552, 3554, 3555, 3563, 3564 are
cond in this galaxy -- U 7767) see U 7745 -- U 7768) S B(rs)dm (de V); disrupted filamentary arms; vB cond in the
disc at 0.30, 125, 0.22 x 0.15, disturbing comp? --

07770   U 7772) SA(s)b? (de V), Sb+ (Ho); comp 6.1, 359, 0.6: x 0.5:, dwarf; U 7758 at 13.4 -- U 7773) E 6 (de V), E (Ho);
SN 1961 h -- U 7775) integral-shaped main body, e dif disc; P w U 7797 at 7.5 nf -- U 7776) VV 219 a; SA(rs)bc (de V),
Sc- (Ho); P w U 7777 at 1.4, 210, contact -- U 7777) VV 219 b; SA(rs)bc (de V), Sc- (Ho); see U 7776 -- U 7779)
P w 12 34.3 + 32 23 at 4.1, 58, 0.7 x 0.25, pec, m=15.4 --

07780   U 7781) DDO 138; S IV-V or S V (vdB) -- U 7784) I B m (de V); P w U 7786 at 6.0, 166 -- U 7785) S0 (de V) -- U 7786)
Arp 76; SAB(rs)ab (de V); Sb+ (Ho); see U 7784; in Arp's class "spir galaxies w S high surf brightn comps on arms";
"apparent gap between arm and comp" (Arp) -- U 7787) comp 1.4, 333, 0.25 x 0.08 -- U 7788) SA(r)cd: (de V); Sc+ (Ho);
comp 4.6, 314, 0.25 x 0.25 --

07790   U 7791) S0: (de V); comp 3.7, 184, 0.3 x 0.2; 4.8, 355, 0.3 x 0.3 -- U 7793) SA 0: (de V) -- U 7794) SAB(rs)a pec
(de V) -- U 7795) DDO 139; Ir V (vdB) -- U 7796) SAB(rs)b (de V), Sb- (Ho) -- U 7797) E 2 (de V); see U 7775 --
U 7799) comp 0.8, 272, 0.15 x 0.12; 1.2, 83, 0.7 x 0.3 --

07800   U 7802) MCG 1-32-121 at this position, but MCG data are incorrect -- U 7803) S B(s)ab: (de V) -- U 7804) SA(s)0/a:
(de V), Sa (Ho) -- U 7806) U 7807 at 4.3, 9 --

07810   U 7811) F plume in pos angle 115; comp 2.2, 334, 0.4 x 0.2 -- U 7813) 12 36.2 + 00 35 at 4.4, 234, 0.6 x 0.2, m=15.7
-- U 7819) SA(s)dm: (de V) --

07820   U 7822) DDO 140; Ir IV-V: (vdB) -- U 7828) S B(r)0+ (de V), Sa (Ho); P w U 7842 at 18.9, 94 --

07830   U 7831) S B(s)c pec (de V); B main body 3.2 x 0.9, smooth outer disc -- U 7835) P w 12 38.7 + 27 01 at 4.8, 85, 0.7 x
x 0.7, spir, m=15.5 -- U 7836) B patches in the arms; 12 38.6 + 28 49 at 5.2, 34, [0.35 x 0.35], double system,
contact, m=15.7 -- U 7839) S B(s)a (de V), Sa (Ho); P w U 7843 at 4.0, 115 --

07840   U 7842) S B(r)0/a (de V), S0 (Ho); see U 7828 -- U 7843) S B(s)...? (de V); see U 7839 -- U 7849) P w U 7857 at 7.5,
69 --

07850   U 7850) (R)SAB 0 (de V); hints of (R) approx 3. x 3. -- U 7851) in triple group w U 7852 at 2.4, 37, at 12 39.0 +
+ 26 21 = N 4613 at 2.8, 348, 0.5 x 0.4, m=15.5 -- U 7852) see U 7851; 12 39.0 + 26 21 = N 4613 at 2.1, 296 --
U 7853) VV 73 ab, Arp 23; S B(rs)m (de V), Sc+ (Ho); asym disc w patchy spir arc, alm dissolved arms; in Arp's class
"one-armed spir galaxies"; P w U 7861 at 8.3, 26 -- U 7857) see U 7849 -- U 7858) E 5 (de V); SN 1939 b --

07860   U 7860) E + pec (de V); S... (Ho); P w U7865 at 2.2 s -- U 7861) SAB(rs)m pec (de V), Sc + (Ho); asym; one short, thin arm and one long, massive; see U 7853 -- U 7865) Arp 281; S B(s)d (de V), Sc + (Ho); in Arp's class "double galaxies, infall and attraction"; see U 7860; "knots resolved w 48-inch. Dif counter tail on comp" (Arp) -- U 7866) DDO 141; Ir IV- V (vdB); B cond 0.15 x 0.10 approx 0.6 s center, superimp comp? -- U 7868) E + 3 (de V) --

07870   U 7870) SA c (de V); SN 1946 b -- U 7873) comp 2.8, 339, 0.3 x 0.2 -- U 7874) S B(s)d: (de V), Sc + (Ho); P w U7875 at 3.8, 163 -- U 7875) S d: (de V); see U 7874 -- U 7878) E + 0 - 1 (de V); SN 1939 a --

07880   U 7880) S O- (de V); b) = U7881 at 1.7, 99 -- U 7881) m approx 16., incl in CGCG magn for U 7880 + 7881; see U 7880 -- U 7883) 12 40.5 - 01 00 at 3.5, 156, 0.9 x 0.8, Sc, m = 15.5 -- U 7884) Sb+ (Ho); comp 2.9, 280, 0.7 x 0.3; P w U 7902 at 17.5 -- U 7887) in a chain of galaxies w 7946 brightest; P w b) at 1.5, 90, 0.9 x 0.12, interaction? -- U 7888) P w b) at 0.9, 96, 0.15 x 0.12 --

07890   U 7891) VV 151 (listed at α = 10 40.7 by VV); a) 0.6 x 0.15; b) 0.8 x 0.6 -- U 7892) high surf brightn, sharp border; in U 7946 chain; comp 2.8, 28, 0.4 x 0.1 -- U 7893) P w U7900 at 9.6, 59 -- U 7895) S B(rs)0/a (de V) -- U 7896) VV 206 b, Arp 116 a; SAB(rs)c (de V), Sc- (Ho); P w U7898 at 2.6, 136, contact; see U 7898 -- U 7898) VV 206 a, Arp 116 b; E 2 (de V), E (Ho); see U 7896; in Arp's class "E and E-like galaxies close to and perturbing spirals"; "abs heavier on spir side away from E galaxy" (arp); 12 41.0 + 11 52 at 5.7, 153, 0.9 x 0.4, spir, m = 15.7 -- U 7899) S blue comp attached sp?; sev F comps --

07900   U 7900) SAB(rs)cd (de V), Sc- (Ho); see U 7893; U 7926 at 20., 74 -- U 7901) VV 56, Arp 189; SA(rs)c (de V), Sc- (Ho); "F spur and blob attached" (de V); in Arp's class "galaxies w narrow filaments"; "radio source near tail apparently not associated" (Arp) -- U 7902) SAB(rs)cd (de V), Sc- (Ho); asym -- U 7905) I Zw 41; "intercon blue post-eruptive pair, w compact cores and extended plumes" (Zw); "the n component has m_p = 14.7 and has an elliptical main body 7" x 12" = 2.2 x 3.8 kpc w a wide plume 15" long to the north. The s component is 35" = 11 kpc away and has a main body 12" x 25" = 3.8 x 8 kpc and a F wide plume extending 15" to the west" (Sargent); in a chain of galaxies w U7946 brightest -- U 7907) S B(s)m pec + I m (de V); Ir I (Ho); N 4657 is a comp 1.1 x 0.7 superimp N 4656; "prob interacting pair of Magell systems" (de V); U 7865 at 32.5 -- U 7909) VV 64; a) 0.45 x 0.25; b) 0.5 x 0.4; sep 0.50 --

07910   U 7910) at least 3 interacting systems, bridges; bar spir w comps ? -- U 7911) DDO 144; V (vdB) -- U 7912) a) 1.4: x x 0.7; b) 0.15 x 0.15 attached sf a), sep 0.50 -- U 7914) E6 (de V); comp 5.5, 262, 0.9: x 0.8:, v late spir -- U 7916) VV 127, DDO 143; Ir V (vdB); listed as group of 6 interacting systems by VV -- U 7918) blue, strongly chaotic; sev comps -- U 7919) DDO 145; E (de V), dE? (vdB) --

07920   U 7921) comp 3.4, 288, 0.3 x 0.1; 12 42.1 + 41 01 = I 3723 at 4.2 sp, 0.3 x 0.2, m = 15.7 -- U 7924) S B(s)0/a (de V), SO (Ho) -- U 7925) in a chain of galaxies w U 7946 brightest -- U 7926) SAB c: (de V), Sc- (Ho); SN 1965 h; P w U 7931 at 7.3, 127; see U 7900 -- U 7928) P w comp at 0.6, 286, 0.3 x 0.2, contact --

07930   U 7930) Arp 163, Haro 9; in Arp's classes "galaxies w dif filaments, dif counter-tails"; "muy violeta" (Haro); P w U 7933 at 5.6, 136 -- U 7931) S B(s)d: (de V), Ir I (Ho); see U 7926 -- U 7932) P w comp at 1.5, 97, 0.8 x 0.4 -- U 7933) E1-2 (de V); see U 7930 -- U 7934) comp 0.6, 252, 0.2 x 0.1, F; 2.9, 333, 0.45 x 0.15 -- U 7935) B bar 0.8 x 0.20, arms w low surf brightn; in a chain w U7946 brightest -- U 7936) B arc-shaped obj interacting w stellar obj 0.15 x 0.15, sep 0.8; "disrupted spir" (CGCG) -- U 7938 + 7939) VV 224, Arp 242; a) 2.2 x 0.35; b) 1.9 x 0.8:; SO pec + S B(s)0/a pec (de V); in Arp's classes "galaxies w adjacent loops, appearance of fission"; "v thin B tail from n nucl which has strong abs" (Arp); total diam 4.3 x 1.1 --

07940   U 7944) 12 44.4 + 10 08 = I 817 = I 3764 at 2.6, 81, double system, E? 0.6 x 0.5 w S comp superimp, total m = 15.4 -- U 7947) comp 3.8, 46, 0.4 x 0.3 -- U 7949) DDO 14.7; V (vdB); sev S galaxies in vic; 12 44.6 + 36 47 = I 3772 at 3.2, 345, 0.6 x 0.4, SB a:, m = 15.7 --

07950   U 7950) Haro 36; "violeta" (Haro); see A.J. 75, 1143 -- U 7952) comp 5.0, 255, 0.9 x 0.4, late spir, m approx 16. -- U 7953) P w comp at 0.50, 100, 0.4 x 0.1 -- U 7955) comp 1.5, 259, 0.3 x 0.2 -- U 7956) asym, one arm stronger; comp 4.4, 352, 0.6 x 0.12; measurements on blue PA print disturbed by ** GC 17440, 17444 -- U 7959) An 1245 in BG --

07960   U 7961) S B(s)bc (de V); SN 1966 b -- U 7962) comp 6.1, 282, 0.8 x 0.2 -- U 7965) SA(rs)bc (de V), Sc- (Ho) -- U 7966) in a chain w U7946 brightest; comp 0.7, 143, 0.25 x 0.08; 4.7, 40, 0.5 x 0.1 -- U 7967) comp 3.3, 125, 0.4 x x 0.1, F; 3.8, 26, 0.40 x 0.10 -- U 7968) comp 4.4, 208, 0.4 x 0.4, dif -- U 7969) S B 0 pec (de V) --

07970   U 7970) SA(s)ab (de V), Sa (Ho) -- U 7971) I Zw 43, DDO 150; "v blue Irr... w many blue knots (poss neutral stellar nucl, or * superposed, m_p = 15.0)" (Zw); S IV- V or S V (vdB) -- U 7975) SA(s)cd (de V) -- U 7977) SA(s)bc (de V); U 7989 at 12.0 --

07980   U 7980) SA(r)0+? (de V), SO (Ho) -- U 7985) SAB(rs)d (de V) -- U 7989) SAB(r)ab pec (de V), Sb+ (Ho); SN 1940b; see U 7977 --

07990   U 7992) P w 12 48.0 + 27 43 = N 4728 at 2.1, 287, (0.8 x 0.6), F halo, v = 6522, v_0 = 6526, m = 15.6; comp 3.6, 358, 0.3 x 0.2 -- U 7994) (R)SA(rs)ab (de V); comp 3.2, 312, 0.3 x 0.12; 12 47.6 + 73 16 at 7.2, 341, 0.9 x 0.4, Sc, m = = 15.6 -- U 7996) (R)SA(r)ab (de V), Sb- (Ho) -- U 7997) E1 (de V) -- U 7998) S c: (de V) --

08000   U 8001) P w comp at 3.0, 259, 0.25 x 0.25, compact -- U 8002) earlier than Sb?; comp 3.1, 45, 0.3 x 0.10; 3.9, 197, 0.45 x 0.18 -- U 8003) in a chain w U7946 brightest -- U 8005) Arp 159; S B(s)c? (de V), Sc- (Ho); e dif ext to diam 6.? x 2.5?; in Arp's class "disturbed galaxies w interior abs"; "vF plume extending northeast" (Arp) -- U 8009) I 0 (de V); SN 1965 i; broad abs lanes --

08010   U 8010) S B(r)0-: (de V), SO (Ho); U 8016 at 10.6, 119 -- U 8012) MCG δ = 52 07 -- U 8016) S B(r)0/a? (de V), SO (Ho); see U 8010 --

08020   U 8020) SA d? (de V) -- U 8021) SA(s)a (de V) -- U 8023) VV 266, Arp 265; in Arp's class "galaxies w adjacent loops"; "resolution of stars or knots" (Arp) -- U 8024) DDO 154, N 4789 A in BG; I m (de V), Ir IV- V (vdB); U 8026 at 4.8, 182; U 8028 at 5.8, 120 -- U 8025) P w 12 51.4 + 29 51 at 3.3, 255, 0.8 x 0.7, S B..., m = 15.3 -- U 8026) see U 8024 -- U 8028) SO: (de V) see U 8024 --

08030   U 8033) SAB(rs)c (de V); comp 1.4, 128, 0.6 x 0.3 -- U 8034) VV 313a, Arp 277; P w 12 52.3 + 02 55 = N 4809 = VV 313 a at 0.8, 180, 0.7 x 0.45, I m pec (de V), m = 14.8; in Arp's class "interacting double galaxies" -- U 8035) SA(rs)b (de V) -- U 8037) N 4796 superimp 0.45 f center, 0.22 x 0.22; 12 52.5 + 08 19 = N 4791 at 4.8, 261, 0.5 x 0.35, m = = 15.1 -- U 8038) E 2 (de V) --

08040   U 8040) in quadrupel group incl U 8046 at 2.9 f; comp 1.0, 305, 0.7: x 0.5:, e dif, dwarf -- U 8041) An 1253 in BG; S B(s)d (de V) -- U 8042) P w U8045 at 3.3, 99 -- U 8044) P w 12 52.9 + 39 28 = I 3895 at 1.3, 150, 0.9 x 0.9, Sb- c, m = 15.7 -- U 8045) see U 8052 -- U 8046) see U 8040; the group incl 12 52.2 + 59 10 at 8.4 np, 0.9 x 0.9, m = 15.0, and 12 52.6 + 59 09 at 5.8 np, 0.8 x 0.6, m = 15.1 -- U 8047) S B nucl reg, late-type disc; comp 4.0, 325, 0.4 x 0.3, dif; 5.0, 224, 0.5 x 0.3 -- U 8048) comp 3.5, 354, 0.3 x 0.1 -- U 8049) comp 1.4, 15, 0.35 x 0.1 --

08050   U 8051) SA(s)b (de V) -- U 8054) SA(s)cd: (de V) -- U 8056) comp 4.1, 255, 0.7 x 0.25 -- U 8057) 12 54.0 + 28 01 at 1.9, 77, 0.30 x 0.25, m = 15.5 -- U 8058) VII Zw 490; "blue post-eruptive Sc, compact elliptical core, fine streamers" (Zw) -- U 8059) MCG major diam 0.65 is incorrect; comp 2.2, 212, 0.4 x 0.2 --

08060   U 8060) 12 54.1 + 27 13 = N 4821 at 1.9, 170, 0.5 x 0.3, v = 6974, v_0 = 6980, m = 15.0 -- U 8062) (R)SA(rs)ab (de V), Sb+ (Ho) -- U 8063) comp 5.0, 111, 0.8 x 0.2 -- U 8064) P w 12 54.3 + 11 26 at 5.1, 4, 0.6 x 0.3, S0- a, m = 15.7 --

U 8065) E (de V) -- U 8068) I Zw 46; "blue post-eruptive intercon pair of galaxies, w extended plumes and jets" (Zw);
"irr obj 50" x 25" = 28 x 14 kpc... The centers are 26" = 14.7 kpc apart. The individual redshifts are 9158 km sec^{-1}
for MCG 8-24-11, 8813 km sec^{-1} for the center of the 'bridge', and 8510 km sec^{-1} for MCG 8-24-12" (Sargent); comp
2.0, 331, 0.3 x 0.3, dif; 4.7, 126, 0.3 x 0.2 -- U 8069) 12 54.6 + 29 19 at 2.4, 296, 0.45 x 0.20, m = 15.4 --

08070   U 8070) sev Coma I cl galaxies near; comp superimp at 0.40, 225, 0.20 x 0.12; double syst 12 55.2 + 27 45 = N 4842
at 2.6 f, 0.4 x 0.3 and 0.3 x 0.2, sep 0.45, v = 7512, v_0 = 7521, total m = 14.9 -- U 8071) P w comp at 0.8, 206,
0.5 x 0.25; Coma I cl -- U 8072 + 8073) a) (1.4 x 1.4); b) (1.0 x 1.0), sep 0.50; E 1 and E 0 (de V) -- U 8074) DDO
156; Ir V (vdB) at 11., 57 -- U 8075) P w comp at 2.4, 312, 0.45 x 0.20 -- U 8076) v blue -- U 8077) B
core 0.25 x 0.22, 2 thin open arms; comp 4.6, 32, 0.7 x 0.45, m approx 16. -- U 8078) SA(s)ab (de V) --

08080   U 8080) III Zw 68; "blue compact elliptical disc, surrounded by uniform blue halo" (Zw); Coma I cl -- U 8081) DDO
157; S IV: (vdB) -- U 8084) DDO 158; S B IV-V (vdB); see U 8074 -- U 8086) listed as I 838 in CGCG, which is
incorrect; I 838 is a comp at 1.9, 5, 0.9 x 0.7, Sc; Coma I cl -- U 8089) S 8093 at 8.3, 38 --

08090   U 8091) DDO 155; strongly chaotic, sev B knots; Ir V (vdB) -- U 8092) II Zw 67; E 1 (de V); "neutral elliptical
compact, w F ring halo... Possible member of the Coma cl" (Zw) -- U 8093) 12 55.9 + 09 54 at 6.3, 257, 0.8 x 0.15,
S B: a - b, m = 15.6; U 8089 at 8.3, 218; eF dwarf comp at 1.1, 223? -- U 8096) α = 12 56 32, δ = 28 06.4 (CGCG);
12 56 24 + 28 05.0 = I 3946 at 2.0, 230, 0.7 x 0.3, S0 (de V), v = 6101, v_0 = 6112, m = 15.3 -- U 8098) Arp 266,
I Zw 49; S B(s)m: (de V); I 3961 is a large dif syst w m = 14.1; N 4861 is a B knot 0.35 x 0.30, m = 13.2, superimp the
s part of I 3961; "blue Irr, plus blue patchy compact sp tip... (Blue spherical comp 4'5 nnf.)" (Zw) -- U 8099) P w
U 8125 at 18.6, 90 --

08100   U 8100) α = 12 56 55, δ = 28 21.2 (CGCG); Coma I cl; a large nr of galaxies near -- U 8102) SA(r)0+ : (de V); U 8091
at 11.8, 284 -- U 8103) S0 (de V); p of 2 B galaxies in center of Coma I cl, a large nr of galaxies near; U 8110 at
7.2, 80; SN 1968 b -- U 8105) a) 0.1 x 0.1; b) 1.0: x 0.6:; a) superimp b) -- U 8106) E 1 (de V); α = 12 57 33, δ =
= 28 31.0 (CGCG); 18' n center of Coma I cl -- U 8109) S0 (de V) --

08110   U 8110) E 4 (de V); α = 12 57 42, δ = 28 14.7 (CGCG); f of 2 B galaxies in the center of Coma I cl, see U 8103; a
large nr of cl galaxies near -- U 8111) VV 222; a) 0.3 x 0.2; b) 0.6 x 0.5, sep 0.35; I 4073 is a * 0.50 np b) --
U 8113) S0 (de V); α = 12 57 53, δ = 28 28.1 (CGCG) -- U 8114) blue, chaotic; loop westw -- U 8116) S B(rs)c (de V)
-- U 8117) S0 (de V); Coma cl -- U 8119) comp 2.4, 211, 0.30 x 0.20; 2.7, 264, 0.5 x 0.15 --

08120   U 8121) SA(s)cd (de V) -- U 8123) comp 3.3, 238, double syst, 0.30 x 0.30 and 0.6 x 0.15, sep 0.6 -- U 8125) see
U 8099; comp 4.7, 352, 0.3 x 0.12 -- U 8126) comp 1.9, 347, 0.6 x 0.1; U 8131 at 6.8 nf -- U 8127) a) at 1.6, 205,
0.7: x 0.6:, dwarf irr -- U 8128) S b (de V); 18' sf center of Coma I cl, a large nr of galaxies near -- U 8129)
α = 12 58 29, δ = 28 16.5 (CGCG); Coma I cl, a large nr of galaxies near; 12 58.5 + 28 19 = N 4908 at 2.2, 341,
(0.9 x 0.8), E 4 (de V), v = 8838, v_0 = 8851, S comp superimp, m = 14.9 --

08130   U 8131) CGCG ident I 4062 is incorrect; I 4062 is a galaxy at 1.7, 305, 0.45 x 0.35; comp 4.3, 38, 0.4 x 0.3 --
U 8132) comp 1.9, 332, 0.5 x 0.5, MCG 9-21-82 -- U 8133) 21' sf center of Coma I cl, a large nr of galaxies near --
U 8134) S a (de V); SN 1959 b; Coma I cl; 12 59.1 + 28 06 = N 4923 at 2.6, 150, (0.9 x 0.8), E, m = 14.7 -- U 8135)
"double syst w halo" (CGCG) -- U 8136) comp 3.2, 288, 0.4 x 0.3 -- U 8138) 12 59.2 + 05 18 at 2.9, 327, 0.6 x 0.3,
E:, m = 15.6; 12 59.5 + 05 20 at 5.3, 20, 0.4 x 0.4, compact, m = 15.5 -- U 8139) comp 4.8, 155, 0.8: x 0.5, Sc --

08140   U 8140) Coma I cl; comp 2.1, 22, 0.4 x 0.3, F pair, contact; 2.7, 178, 0.55 x 0.12 -- U 8142) Coma I cl; 12 59.7 +
+ 27 55 at 3.5, 65, 0.6 x 0.5, m = 15.1; E (de V) -- U 8145) comp 2.0, 38, double syst, [0.4 x 0.2], sep 0.20 --
U 8147) CGCG magn 15.4 may refer to U 8148 (at 3.2, 5); 13 01.0 + 80 17 at 3.3, 135, 0.4 x 0.3, pec, m = 15.3 --
U 8148) see U 8147; not in CGCG but appears brighter than U 8147 --

08150   U 8150) comp 2.3, 265, 0.6 x 0.3, prob interacting pair -- U 8151) a) at 1.2, 254, 0.6 x 0.3, S0 - a -- U 8154) E 4
(de V); Coma I cl -- U 8155) late-type spir w vB nucl reg; 13 00.5 + 08 07 at 5.1, 312, 0.8 x 0.4, spir, m = 15.7 --
U 8156) a) at 1.5, 260, 0.6 x 0.5, Sc: -- U 8157) a) at 1.2, 290, 0.35 x 0.12; c) at 0.8, 154, 0.35 x 0.15; d) at 2.0,
152, 0.7 x 0.35 -- U 8159) similar to N 210 --

08160   U 8164) comp 4.0, 303, 0.3 x 0.2; 4.8, 266, 0.25 x 0.10 -- U 8167) P w comp at 1.1, 84, 0.4 x 0.3 --

08170   U 8173) comp 4.5, 90, 0.4 x 0.2, pec, multiple?; 13 02.2 + 04 10 at 5.2, 340, 0.9 x 0.15, spir, m = 15.6 -- U 8174)
comp 2.6, 310, 0.30 x 0.12 -- U 8175) E 5 (de V); Coma I cl -- U 8178) E 3 (de V); Coma I cl -- U 8179) P w comp at
1.7, 188, 0.4 x 0.3 --

08180   U 8185) S B(s)cd (de V); B nucl reg; chaotic disc w filaments and B patches -- U 8187) U 8191 at 3.8 sf; comp 2.8,
78, 0.3 x 0.2; 3.6, 339, 0.35 x 0.35 -- U 8188) S m (de V); SN 1937 c --

08190   U 8190) comp 3.3, 344, 0.6 x 0.20 -- U 8192) P w 13 03.9 + 10 42 at 3.7, 20, 0.8 x 0.5, Irr, m = 15.7 -- U 8194)
a) 0.50, 280, 0.22 x 0.20; Coma I cl -- U 8197) P w comp at 1.1, 45, 0.8:: x 0.6::, e dif -- U 8198) comp 2.8, 215,
0.3 x 0.2 -- U 8199) VV 292 a; P w U 8200 at 2.9, 50 --

08200   U 8200) VV 292 b; see U 8199 -- U 8201) VII Zw 499, DDO 165; "blue resolved dwarf Irr, w compact condensations" (Zw);
Ir IV-V (vdB) -- U 8202) P w 13 05.2 - 00 35 = I 850 at 5.4, 32, 0.7 x 0.15, spir:, m = 14.8 -- U 8203) P w U 8210
at 6.0, 153 -- U 8205) comp 0.8, 233, 0.1 x 0.1, F; 1.7, 172, 0.3 x 0.3 -- U 8206) P w comp at 1.5, 297, 0.35 x 0.15
-- U 8209) P w U 8220 at 13., 122 --

08210   U 8210) see U 8209 -- U 8212) comp 3.8, 81, 0.35 x 0.10; 4.2, 117, 0.3 x 0.15 -- U 8214) in quadrupel group w U 8240
at 11., U 8234 at 8.7 and U 8237 at 8.3 --

08220   U 8220) see U 8209 -- U 8221) (r) 0.55 x 0.55; 2 long F arms; comp 2.9, 157, 0.25 x 0.18 -- U 8222) 13 06.5 + 52 17
at 3.8 f, 0.7 x 0.4, m = 14.7 -- U 8228) comp 1.7, 333, 0.4 x 0.18 --

08230   U 8231) asym bar spir; 2 arms point eastw -- U 8234) P w U 8237 at 2.3, 25; U 8240 at 6.7 sp -- U 8236) S B(r)b (de V)
-- U 8237) outer arc or ring w large blue pathes on p side; see U 8234 -- U 8239) P w comp at 1.0, 295, 0.2 x 0.2;
4.5, 343, 0.4 x 0.12 --

08240   U 8240) P w 13 07.6 + 62 25 at 2.1, 113, 0.25 x 0.20, m = 15.6; in group w U 8214 at 11.6, U 8234 at 13.8 and U 8237
at 8.3 -- U 8241) M 51 case?, vF cond at tip of spir arm at 0.9, 165, comp or *?; comp 1.0, 248, 0.4 x 0.3, interaction? -- U 8242) comp 0.5, 284, 0.1 x 0.1, v red, interaction? -- U 8243) asym spir w 2 open massive arms --
U 8248) comp 2.1, 244, 0.3 x 0.1, pec? or irr w superimp *? --

08250   U 8251) comp 3.0, 45, 0.4 x 0.15 -- U 8252) 13 08.5 + 44 17 at 3.7, 57, 0.45 x 0.25, pec?, m = 15.3 -- U 8253) comp
3.0, 64, 0.35 x 0.15, pec? -- U 8254) comp 2.0, 212, 0.15 x 0.15, compact -- U 8256) SAB(rs)bc (de V), Sb+ (Ho); P w
U 8307 at 41., 133 -- U 8257) comp 2.2, 152, 0.4 x 0.3; 2.9, 258, 0.35 x 0.22; 3.8, 292, 0.35 x 0.25 -- U 8259) P w
U 8260 at 2.4, 0 --

08260   U 8260) see U 8259; 13 08.4 + 28 59 = I 4210 at 5.3, 326, 0.8 x 0.5, late spir, m = 15.3 -- U 8261) P w comp at 2.0,
176, 0.8: x 0.6: -- U 8263) An 1309 in BG; S B c (de V); SN 1959 c -- U 8264) VII Zw 501; B main body 0.55 x 0.50 w
dif broad plume in position angle 90; "blue post-eruptive patchy compact w attached F comp (m_p = 18.0) sp" (Zw) --
U 8267) comp 4.7, 331, 0.50 x 0.35 -- U 8269) in triple group incl comps at 3.1, 353, 0.8 x 0.25 and 4.5, 210,
0.9 x 0.20 --

08270   U 8270) SAB(rs)c (de V); comp 2.1, 43, 0.3:: x 0.3::, e dif, dwarf; 2.7, 20, 0.35 x 0.10; 4.7, 0, 0.30 x 0.10 -- U 8272)
comp 1.7, 263, 0.30 x 0.22; 1.9, 147, 0.7 x 0.15 -- U 8278) sev S galaxies near; comp 1.6, 355, 0.4 x 0.1 -- U 8279)
SAB(rs)c (de V) --

08280   U 8281) comp 0.8, 242, 0.5 x 0.5, contact; 0.9, 272, 0.35 x 0.20; 13 10.0 + 45 07 at 3.0 nf, (0.9 x 0.9), core 0.20 x
x 0.20, compact, m=15.5 -- U 8283) comp 4.8, 310, 0.55 x 0.55, Sc -- U 8288) many blue patches along the arms --

08290   U 8290) Pec or S m (de V, note to N 5012); U 8270 at 16. np -- U 8294) S B ecc nucl, or superimp *? -- U 8296) b)
at 0.8, 160, 0.3 x 0.1, m approx 17. -- U 8299) comp 3.4, 145, 0.35 x 0.10 --

08300   U 8300) P w 13 11.0 + 28 01 at 2.4, 187, 0.9 x 0.4, m=15.4 -- U 8301) comp 3.2, 322, 0.35 x 0.20 -- U 8303) DDO 166,
An 1311 in BG; Ir I (Ho), Ir IV-V (vdB); comp to U 8307 = N 5033 at 23. n -- U 8305) comp 3.4, 149, 0.6 x 0.4, e dif
-- U 8307) SA(s)c (de V), Sc- (Ho); SN? 1957 (Carpenter, unconfirmed discovery); see U 8256 -- U 8308) DDO 167;
Ir V -- U 8309) comp 4.7, 145, 0.35 x 0.2 --

08310   U 8315) "double syst, bridge + jet" (CGCG) -- U 8316) comp 1.2, 198, 0.25 x 0.1 -- U 8319) comp 3.7, 25, 0.4 x 0.2;
3.7, 39, 0.4 x 0.08 --

08320   U 8321) P w 13 12.5 + 17 29 = I 859 at 1.3 f, 0.35 x 0.30, prob v ext eF halo, m=15.2; 2 short blue spikes towards
I 859 -- U 8324) 13 12.5 + 03 17 at 2.0, 210, 0.7 x 0.5, S0- a, m=15.3 -- U 8327) MCG 8-24-96 is Mark 248 --

08330   U 8330) comp 1.3, 150, 0.4 x 0.3; 4.1, 167, 0.5 x 0.15 -- U 8331) DDO 169; Ir IV-V or Ir V (vdB) -- U 8333) DDO
170; V (vdB); P w U8336 at 5.5, 110 -- U 8334) SA(rs)bc (de V), Sb + (Ho); SN 1971i -- U 8335) VV 250, Arp 238,
VII Zw 506; in Arp's class "galaxies w appearance of fission"; "double nuclei, n nucl has third arm" (Arp); "post-
eruptive pair of intercon stretched blue Sc's w compact patchy cores" (Zw) -- U 8336) see U 8333 -- U 8337) U 8342
at 6.0, 35 --

08340   U 8340) P w U8348 at 12.5, 146 -- U 8342) see U 8337 -- U 8345) comp 2.4, 93, 0.4 x 0.4, E- S0 -- U 8347) ext F
env; comp 1.7, 138, 0.4 x 0.12 -- U 8348) see U 8340 -- U 8349) comp 4.4, 52, 0.25 x 0.22, compact --

08350   U 8353) U 8354 at 3.2, 348; 13 14.9 + 20 53 = I 868 at 2.6, 128, 0.7 x 0.5, m=15.4; 13 15.0 + 20 52 = I 870 at 3.5,
133, 0.8 x 0.45, m=15.4 -- U 8354) see U 8353 -- U 8356) P w 13 15.0 + 31 18 at 2.9, 222, 0.45 x 0.25, m=15.4 --
U 8357) a) 1.2: x 0.5, disrupted; b) 0.4: x 0.4:, S B ...; c) 0.12 x 0.10, eF --

08360   U 8365) DDO 172; S B IV-V (vdB) -- U 8366) comp superimp at 0.9, 267, 0.15 x 0.15 -- U 8368) * or compact comp
superimp at 0.30, 6 -- U 8369) comp 1.7, 59, 0.4 x 0.1, dif --

08370   U 8371) dif irr outer disc, prob twisted plane -- U 8375) P w 13 17.7 + 16 10 = I 882 at 3.8, 42, 0.8 x 0.6, E, m =
= 15.1, interaction? -- U 8377) vF ext env, no E; not listed as compact in CGCG --

08380   U 8380) P w comp at 4.0, 310, 0.8 x 0.2, Sa? -- U 8385) DDO 173; S (B) IV-V: (vdB) -- U 8387) Arp 193, I Zw 56; in
Arp's class "galaxies w narrow filaments"; "F straight outer spikes, hard knots in main body" (Arp); "blue post-
eruptive patchy compact w sharp jets southwest and southeast" (Zw); "B irr core 30" x 15" = 14 x 7 kpc, w 2 jets
each about 40" = 18 kpc long" (Sargent); Burbidge (Ap.J. 140, 1617) gives v_0 = 6918, Sargent (Ap.J. 160, 405) v_0 =
= 6986; UGC v_0 is a mean value -- U 8388) similar to U 4666 = N 2685; acicular? main body w arc-shaped filaments,
poss rings -- U 8389) a) 0.8: x 0.5, spir; b) 0.8 x 0.4; both distorted --

08390   U 8393) 13 19.4 + 57 58 = N 5113 at 5.3 nf, 0.9 x 0.3, pec?, m=15.2 -- U 8394) VII Zw 509; a) 0.30 x 0.20; b) 0.7 x
x 0.15, sep 0.60; "pair of intercon galaxies. No. 1, patchy blue compact, m_p = 15.7... No. 2, neutral Sb, 40" nf
No. 1, m_p = 17.0" (Zw); in group, incl U 8454 at 6.8 -- U 8395) 13 18.8 + 12 29 at 3.7, 295, 0.5 x 0.3, m=15.6 --
U 8396) P S B ... (de V); U 8403 at 13.4 -- U 8397) 13 19.0 + 31 48 at 5.2, 296, 0.7 x 0.45, m=15.4 --

08400   U 8401) asym dif spir pattern; F v red nucl cond, or *? -- U 8403) S B(rs)cd (de V) -- U 8406) chain of 5 or 6 S
galaxies around 2.0, 40 -- U 8409) comp 1.4, 13, 0.3 x 0.12 --

08410   U 8410) S B(s)c: (de V); comp 3.1, 10, 0.3 x 0.2 -- U 8413) 13 21,1 + 06 40 at 2.5, 70, 0.5 x 0.4, S B ..., m=15.5
U 8416) VV 235, I Zw 57; a) 0.6 x 0.4; b) 1.0 x 0.35; "twisted streamer" (CGCG); "neutral post-eruptive intercon
pair of Sc's w one long straight jet east northeast" (Zw) -- U 8419) E 2 (de V); S cond in halo at 0.50, 78; comp
3.0, 190, 0.5 x 0.3; 13 21.3 + 31 54 at 4.7, 352, 0.8 x 0.6, spir, m=15.5 --

08420   U 8420) VII Zw 511; S d pec ? (de V); "blue post-eruptive badly distorted spir w numerous conds and jets" (Zw);
cond 0.30 s center is prob a superimp comp; U 8434 at 6.7 f -- U 8421) P w 13 21.7 + 09 55 at 3.8, 143, 0.9 x 0.3,
Sb- c, m=15.4 -- U 8422) comp 2.5, 340, 0.25 x 0.22; 3.5, 170, 0.7 x 0.25 -- U 8423) comp 0.8, 236, 0.2 x 0.1;
13 21.8 + 14 12 at 2.8, 159, double syst, [0.9 x 0.15], total m=15.6; 13 21.9 + 14 12 at 4.2, 65, 0.8 x 0.2, m=
= 15.4 -- U 8424) "* 13^m superimp" (CGCG) -- U 8425) P w 13 21.8 + 15 50 at 1.5, 173, 0.7 x 0.2, S0- a, m=15.7 --

08430   U 8433) P w U8435 at 2.3, 57 -- U 8435) see U 8433; 13 22.8 + 36 42 = N 5143 at 2.3 n, 0.45 x 0.30, pec?, m=15.5 --
U 8437) comp 3.2, 181, 0.4 x 0.2 -- U 8439) main body 1.1 x 0.9 in blue; extensive F outer disc --

08440   U 8441) DDO 175; V (vdB) -- U 8443) S B(s)dm (de V) -- U 8444) P w U8447 at 6.1, 40 -- U 8445) comp 2.6, 290,
0.6 x 0.5, S0: -- U 8447) see U 8444 -- U 8448) P w comp at 2.8, 34, 0.7: x 0.7:, dwarf irregular -- U 8449) comp
4.9, 79, 0.7 x 0.3 --

08450   U 8451) comp 3.7, 270, 0.3: x 0.2:, dif, dwarf? -- U 8452) P w U8456 at 8.4, 46 -- U 8454) VV 39, Arp 204, VII Zw
514; a) 0.8 x 0.4; b) 0.6 x 0.2, sep 1.4, dif bridge; p component blue, double; f component cloud-like, v red; in
Arp's class "galaxies w material ejected from nuclei"; "blue post-eruptive triple galaxy, intercon by pronounced
thin bridge, length 2' south-southwest to north-northeast" (Zw) -- U 8455) hints of eF outer disc to approx 4. x
x 3. -- U 8456) see U 8452 -- U 8458) comp 3.9, 281, 0.5 x 0.3 -- U 8459) "N 5188" in CGCG is incorrect --

08460   U 8461) S stell core, dif asym halo; P w comp at 1.2, 174, 0.9 x 0.7, spir, m approx 16., MCG 6-30-13 -- U 8463)
13 26.2 + 31 20 at 4.5, 60, 0.7 x 0.5, m=15.5; 13 25.7 + 32 13 at 5.2, 210, 0.8 x 0.2, m=15.7 -- U 8465) S B(rs)b:
(de V); P w U8468 at 5.5, 152 -- U 8466) comp 4.3, 263, 0.4: x 0.4:, dwarf -- U 8468) E 0: (de V); see U 8465 --

08470   U 8471) b) at 0.8, 106, 0.4 x 0.2; a) brightest -- U 8475) listed in CGCG as "double nebula", which is incorrect;
* superimp s center -- U 8476) in a dense part of the cl; 13 26.9 + 12 03 = N 5176 at 2.9, 15, 0.4 x 0.4, E:, m =
= 15.4; 13 27.0 + 12 01 = N 5179 at 2.4, 74, (0.9 x 0.5), E- S0, m=14.9 -- U 8477) SAB(rs)bc: (de V); 13 27.3 +
+ 17 16 at 6.8, 115, 0.6 x 0.2, spir, m=15.7 -- U 8479) vB core in F ext halo; comp 4.3, 233, 0.2 x 0.2 and 4.4,
226, 0.25 x 0.12, double system --

08480   U 8480) complex spir pattern in centr reg, 2 smooth asym outer arms; no disturbing obj vis; comp 1.2, 275, 0.3 x
x 0.1, prob background -- U 8482) P w comp at 0.9, 38, 0.5 x 0.2 -- U 8485) P w U8487 at 3.7, 20 -- U 8487) prob
SAB bc; see U 8485 -- U 8488) P w 13 27.2 + 13 34 = N 5181 at 8.4, 216, 0.9 x 0.8, S0:, m=14.7 -- U 8489) DDO 176;
S (B) IV-V (vdB) --

08490   U 8490) SA(s)m (de V), Sc + (Ho); v late spir, only fragments of spir pattern; comp 5.6, 40, 0.5 x 0.1; 13 28.6 +
58 35 at 8.7 sf, 0.30 x 0.25, m=15.7 -- U 8491) asym blue spindle; MCG diam 1.5 x 0.95 are incorrect; comp 2.3, 292
(from center), 0.35 x 0.25 -- U 8492) U 8496 at 3.4, 145 -- U 8493) VV 1, Arp 85; SA(s)bc pec (de V), Sc- (Ho); P w
U 8494 at 4.6, 16; total diam for U 8493 + 8484 approx 15. x 7.5; in Arp's class "galaxies w large, high surf
brightn comps on arms"; "F plumes and extensions from comp" (Arp) -- U 8494) VV 1 b, Arp 85; I 0 pec (de V),
Ir II (Ho); see U 8493; SN 1945a -- U 8496) VV 69; see U 8492 -- U 8497) comp approx 0.4 x 0.2 superimp p center --

U 8498) Arp 334; compact obj superimp 0.9 s center, 0.10 x 0.10, stell on PA prints; "second 'star' s not quite stellar" (Arp) -- U 8499) I Zw 59; E+1-2: (de V); "red E w compact core" (Zw) --

08500   U 8501) prob SAB(rs)b: or SB(rs)b: -- U 8502) VV 326; a) 0.6 x 0.5; b) 0.60 x 0.40,   Mark 455, strongly asym distr of light, curved vB main body; in group? w comp 2.5 p, 0.35 x 0.30 -- U 8507) VV 88; v blue; "nucl ecc" (CGCG); cigar-shaped? main body in irr env; eruptive? -- U 8508) I Zw 60; "blue resolved Irr. dwarf galaxy w compact knots" (Zw) --

08510   U 8511) b) at 1.0, 10, 0.35 x 0.35 -- U 8512) U 8512) P w U8513 at 6.1, 4; sev cl galaxies near -- U 8513) see U8512 -- U 8515) F quadrupel syst around 1.6, 0 -- U 8519) 13 30.0 + 07 34 at 1.7, 142, 0.7 x 0.2, early spir, m=15.5; U 8522 at 3.7, 81 --

08520   U 8522) see U 8519; 13 30.3 + 07 35 at 4.1, 126, 0.6 x 0.4, compact, m=15.6; U 8523 at 9.7, 170 -- U 8523) see U 8522; 13 31.0 + 07 24 at 4.3, 135, 0.22 x 0.18, sharp edge, m=15.5 -- U 8525) asym env, hints of short dif jet northw -- U 8526) 13 30.4 - 01 00 at 6.0, 185, 0.25 x 0.25, compact (not listed as compact in CGCG), m=15.4; U 8530 at 7.9, 118 -- U 8528) VV 33b, Arp 104; B core 0.35 x 0.35, vF asym corona, curved plume towards southwest, thin bridge to U 8529 at 4.0, 5; in Arp's class "E and E-like galaxies connected to spirals"; "known as Keenans system" (Arp) -- U 8529) VV 33a, Arp 104; see U 8528; complex asym spir pattern --

08530   U 8530) similar to N 210; see U 8526 -- U 8531) P w comp at 0.5, 215, 0.45 x 0.12 -- U 8532) 3 vS vF conds attached in sf quadrant, poss background objs -- U 8533) P w 13 31.0 + 05 45 at 1.6, 85, 0.55 x 0.25, spir, m=15.3 -- U 8536) early spir?; N 5223 subgroup in CGCG cl 13 31.5 + 34 32, medium compact, medium distant; comp 2.8, 288, 0.5 x 0.1 -- U 8537) comp 1.9, 190, 0.4 x 0.1 -- U 8539) in triple subgroup in CGCG cl 13 31.5 + 34 32 w 13 31.0 + 33 22 at 5.2, 320, 0.9 x 0.3, pec?, m=15.6, and 13 30.9 + 33 25 at 8.1, 325, 0.7 x 0.5, spir, m=15.5 --

08540   U 8543) P w U8545 at 7.1, 14 -- U 8545) see U 8543 -- U 8546) brightest in group w at least 6 galaxies; 13 31.9 + + 1806 at 3.2, 100, 0.6 x 0.15, m=15.7; 13 31.5 + 18 10 at 3.7, 317, 0.8 x 0.2, m=15.7 -- U 8548) VV 4, Arp 36; in Arp's class "integral-sign spir galaxies"; "knots in arms approach appearance of S comp" (Arp); P w U8560 at 8.9, 102 --

08550   U 8552) VV 18a; streamer or disturbed arm southeastw from n part, VV 18b is an eF cond at the tip -- U 8553) brightest in subgroup in CGCG cl 13 31.5 + 34 32, medium compact, medium distant; U 8556 at 5.7, 22 -- U 8556) see U 8553 -- U 8557) P w comp at 3.4, 166, 0.7 x 0.4, w dif comp 0.4: x 0.4: at 0.6 -- U 8558) VV 315a,c, Arp 288; a) (1.3 x 1.0); c) 0.6 x 0.3; prob interacting w U8559 at 5.4, 0; c) is ident by VV as N 5221, which is incorrect; in Arp's class "double galaxies w wind effects"; "streamer in both directions from edge of spir" (Arp) -- U 8559) VV 315b, Arp 288; main body 1.2 x 0.45; broad streamers from s side southeastw and from n side northwestw; prob disturbed by U 8558 at 5.4, 180; see U 8558 --

08560   U 8560) Arp 183; in Arp's class "galaxies w narrow filaments"; "3 F patches constitute third arm or filament" (Arp); see U 8548 -- U 8561) P w comp at 1.5, 299, 0.9 x 0.2, spir -- U 8562 + 8563) VV 211; a) 0.35 x 0.20; b) (0.8 x 0.8), sep 0.40 -- U 8564) comp 2.3, 115, 0.3 x 0.2 -- U 8565) I Zw 64; "pair of blue post-eruptive Sc's w compact nuclei" (Zw)

08570   U 8573) slightly asym, similar to U 8981=M 101, may be disturbed by pair U 8558 at 9.5, 295; SA(s)c (de V); sev S v blue conds at edge, apparently at tips of spir arms; comp 3.6, 135, 0.5 x 0.15 -- U 8574) comp 1.3, 277, 0.20 x x 0.15, alm stell -- U 8576) prob disturbed by 13 33.3 + 13 35 = I 901 at 1.3, 78, 0.7 x 0.4, SB?b:, m=15.2; 13 13.5 + 13 36 at 4.8, 78, 0.7 x 0.12, spir, m=15.6; 13 33.4 + 13 41 at 6.4, 29, 0.7 x 0.3, spir, m=15.4 -- U 8578) Haro 38; "muy violeta" (Haro), DuPuy (A.J. 75, 1143) dimen 34" x 7" = 1.5 x 0.3 kpc, B - V = 0.59, U - B = = - 0.27 -- U 8579) I 4304 subgroup in CGCG cl 13 31.5 + 34 32, medium compact, medium distant; U 8586 = I 4394 at 5.7, 54; comp 0.8, 4, 0.4: x 0.1 --

08580   U 8580) I 4304 subgroup, see U 8579 (at 5.4, 303) -- U 8582) comp 0.50, 90, 0.2 x 0.1; 13 33.5 + 06 47 at 3.6, 185, 0.5 x 0.3, m=15.7 -- U 8583) N 5223 subgroup, see U 8584) a) 0.5 x 0.12; b) 0.12 x 0.12; c) 0.5 x 0.20; b) and c) in contact; long streamer from a) vis in red -- U 8585) (r) 0.35 x 0.30; 2 long thin arms -- U 8586) brightest in subgroup, see U 8579; 13 33.7 + 33 44 = I 4305 at 2.7, 3, 0.7 x 0.6, SO:, m=15.1; 13 34.0 + 33 40 = I 4306 at 4.6, 95, 0.9: x 0.2, m=15.7; see U 8579; U 8580 at 5.4, 123 -- U 8587) comp 4.1, 298, 0.3 x 0.2 -- U 8588) DDO 178; S V (vdB) -- U 8589) dif irr annex n, disrupted spir arm? --

08590   U 8591) sev vF galaxies in CGCG cl 13 34.1 + 16 08, medium compact, extremely distant, in vic -- U 8595) P w U8606 at 1.9, 70 --

08600   U 8600) N 5223 subgroup, see U 8553 -- U 8601) DDO 177; SB V (vdB) -- U 8602) P w U8605 at 1.9, 85 -- U 8603) P w comp at 2.0, 90, 0.7: x 0.3:, dif, pec -- U 8605) see U 8602 -- U 8606) see U 8595 --

08610   U 8610) 3rd from s in chain of 5 = VV 6 = Arp 326; 13 40.3 at 3.2, 160; comp 1.9, 38, 0.55 x 0.35, spir, No 4 in this group, No 5 at 3.6, 7, 0.35 x 0.30 -- U 8613) VV 6a, Arp 33, 326; MCG diam 1.2 x 1.2 are incorrect; 2nd from s, VV ident 3 S near conds as companion galaxies; comp 1.9, 150, 0.55 x 0.35, E?, m approx 16., 1st from s in the group, vis on Arp plate 33, double syst?; see U 8610 -- U 8614) DDO 179; IV- V (vdB) -- U 8616) SAB(rs)bc (de V), Sc- (Ho) -- U 8617) P w comp at 2.0, 77, 0.6 x 0.4 -- U 8618) strong abs n center -- U 8619) a) at 0.9, 248, 0.40 x 0.15; b) brightest --

08620   U 8622) 13 36.2 + 04 42 at 7.7, 126, 0.8 x 0.7, late spir, m=15.4 -- U 8623) P w comp at 2.4, 38, 0.5 x 0.4, dif, late spir -- U 8627) comp 2.0, 163, 0.4 x 0.05 -- U 8628) 13 35.9 + 00 45 at 2.0, 207, 0.9 x 0.10, spir, m=15.7; U 8631 at 1.9, 80 --

08630   U 8630) strongly asym "spindle"; comp 5.0, 170 (from center), 0.35 x 0.30, double syst -- U 8631) see U 8628 -- U 8632) I Zw 67; "post-eruptive pair of patchy compacts, w ext plumes, jets and halo... v = 8340 and 8130" (Zw) -- U 8635) VV 108; vF comp 0.3 x 0.1 attached p, thin disturbed arm on opposite side -- U 8636) comp 3.7, 132, 0.4 x x 0.2 -- U 8637) P w comp at 1.0, 343, 0.5 x 0.2 -- U 8638) VV 133; chaotic, w sev blue patches --

08640   U 8640) in group w 13 36.6 + 31 38 brightest, 0.6 x 0.2, m=15.0; comp 2.7, 344, [0.4 x 0.3], double syst, sep 0.20 -- U 8641 + 8645) VV 55, Arp 240; U 8645 in pos 1.3, 110 from U 8641; in Arp's classes "galaxies w appearance of fission, adjacent loops"; SAB(s)bpec and SA(s)b pec: (de V) -- U 8643) comp 2.3, 0, 0.7 x 0.25 -- U 8645) see U 8641 -- U 8646) in triple group incl comp at 1.5, 115, 0.5 x 0.1 and 13 37.7 + 30 22 at 2.5, 112, 0.8 x 0.5, E or SO, m=15.4 -- U 8647) v blue; in group w 13 36.6 + 31 38 brightest, 0.6 x 0.2, m=15.0; 13 37.2 + 31 34 at 4.0, 97, 0.6 x 0.25, m=15.5 -- U 8648) comp 2.5, 120, 0.4 x 0.10 --

08650   U 8650) P w comp at 2.4, 153, 0.8 x 0.12, spir, prob background giant -- U 8651) DDO 181; Ir IV- V or Ir V (vdB) -- U 8652) comp 3.9, 132, 0.6 x 0.2, SB... + S comp -- U 8653) in group w 13 36.6 + 31 38 brightest, m=15.0 -- U 8656) comp 0.3, 70, 0.25 x 0.12 -- U 8657) comp 3.6, 151, 0.3 x 0.2 -- U 8658) An 1339 in BG; Sc- (Ho) --

08660   U 8660) P w 13 38.8 + 14 01 at 5.2, 201, 0.5 x 0.4, spir:, m=15.5; 13 38.5 + 24 44 = I 909 at 6.3, 258, 0.6 x 0.6, spir?, m=14.9, diam prob for main body only, hints of eF outer disc -- U 8663) 13 38.8 + + 05 17 at 4.2, 276, 0.8 x 0.15, spir, m=15.4 -- U 8664) 13 39.3 + 23 26 = I 916 at 2.9, 61, 0.60 x 0.35, spir, m = = 15.2; sev galaxies near; pair U 8665 at 4.8 n -- U 8665) 13 39.3 + 23 26 at 4.6 sf, sev galaxies near; P w U 8664; 13 38.8 + 23 32 = = I 910 at 4.9 np, double syst, 0.45 x 0.35, SB... and 0.35 x 0.15, sep 0.40, total m=15.1 -- U 8666) VV 170; a) 0.6 x 0.5; b) 0.4 x 0.3, sep 0.7; c) 0.15 x 0.15, sep b) - c) 0.35; in CGCG cl 13 39.5 + 02 33, compact, medium distant -- U 8669) VV 195; VV α = 13 37.8 is incorrect; 2 conds approx 0.2 x 0.2 w sep 0.35; a large nr of vS F galaxies near --

08670   U 8671) brighter component in pair U 8677 at 2.8; comp 2.2, 345, 0.6 x 0.1 -- U 8673) 13 39.1 + 04 30 at 7.1, 275, 0.9 x 0.4, pec, double system?, m=15.5 -- U 8674) comp 1.2, 200, 0.3 x 0.2 -- U 8675) SA(s)0/a (de V); P w U 8680 at 3.4, 124 -- U 8677+8678) VV 19, Arp 239, I Zw 69; SA(s)...pec and S B(s)a pec(de V); "pair of blue doubly inter- con Sc's w cross-jets and compact cores" (Zw); "smaller galaxy is fairly sym spir" (Arp) --

08680   U 8680) SAB(s)b (de V); see U 8675 -- U 8682) in a subgroup in the center of cl; 13 40.1 + 30 06 = N 5274 at 2.9, 261, (0.6 x 0.6), compact, m=15.7; 13 40.1 + 30 05 = N 5275 at 3.4, 239, 0.6 x 0.6 w comp 0.5 x 0.1 attached p, m= = 15.4; 13 40.6 + 30 07 = N 5280 at 4.1, 77, 0.8 x 0.8, compact, m=15.1; comp 1.2, 308, 0.3 x 0.3; 3.6, 86, 0.9 x x 0.12, spir -- U 8683) DDO 182; Ir V (vdB) -- U 8684) comp 4.0, 64, 0.3 x 0.2 -- U 8686) P w 13 41.2 + 04 08 at 1.4, 110, 0.45 x 0.35, spir, m=15.5; 13 41.6 + 04 09 at 6.1, 85, 0.9 x 0.4, S0, m=15.2 -- U 8687) B core 0.22 x x 0.22, dif asym env ext eastw --

08690   U 8691) VV 163 a - c; edge-on spir w 2 comps , both 0.10 x 0.10, attached nf; F curved bridge to outer comp; P w 13 42.0 + 20 38 at 2.2, 177, 0.40 x 0.25, E? or early spir, m=15.2 -- U 8692) in chain w comps at 2.0, 327, 0.5 x x 0.3; 1.1, 322, 0.2 x 0.2, vF halo?; and 13 42.1 + 30 08 at 1.4, 187, 0.7 x 0.7 -- U 8693) P w U 8698 at 8.7, 80 -- U 8695) P w 13 42.4 + 37 25 at 2.4, 283, 0.25 x 0.18, m=15.6 -- U 8696) I Zw 71; "post-eruptive blue patchy compact w long blue hairline jet" (Zw); comp 3.4, 235, 0.7 x 0.12; 4.8, 324, 0.7 x 0.23 -- U 8697) comp 2.2, 336, 0.55 x x 0.15, S0-a; 2.5, 173, 0.4 x 0.25 -- U 8698) see U 8693 -- U 8699) B centr reg surrounded by a B ring --

08700   U 8700) S B...? pec (de V); strongly asym abs features -- U 8701) asym; in group w 13 43.2 + 22 18 at 2.5, 205, 0.6 x x 0.6, compact, m=15.7, comp at 3.0, 223, 0.5 x 0.1, and 13 43.7 + 22 22 at 4.6, 66, 0.8 x 0.6, compact, m=15.3 -- U 8704) comp 1.8, 56, 0.3 x 0.1; 2.9, 55, 0.35 x 0.12 -- U 8706) slightly curved spindle, sev galaxies near; P w 13 44.2 + 23 20 at 1.6, 189, 0.6 x 0.4, E, m=15.4, interaction? -- U 8707) comp 3.2, 54, 0.7 x 0.1 -- U 8708) B nucl reg; chaotic, w tend to spir pattern; comp 2.9, 305, 0.3: x 0.2:; dif; 4.4, 112, 0.45 x 0.10 -- U 8709) SAB(s)b: (de V); P w 13 44.2 + 44 05 = N 5296 at 1.5, 215, 0.9 x 0.5, S0 (de V), m=15.0 --

08710   U 8711) SA(s)b: (de V) -- U 8713) VV 317 b, An 1345 B in BG; P w U 8715 at 1.3, 126; diameters for U 8713, 8715 in BG Notes are incorrect -- U 8714) 3 alm stell comps ; U 8716 at 2.6, 167 -- U 8715) VV 317 a, An 1345 A in BG; see U 8713 -- U 8716) asym; F nucl reg; disturbed by stell obj at 0.45, 216, 0.12 x 0.10 -- U 8718) P w comp at 3.9, 202, 0.9 x 0.2, m approx 16.; comp 2.2, 85, 0.6 x 0.3 -- U 8719) comp 0.8, 213, 0.45 x 0.25; sev F galaxies near, prob members in CGCG cl 13 44.2 + 77 03, compact, e distant --

08720   U 8720) slightly asym disc; chain of blue patches northw from nucl reg; 2 vF comps p at 0.30 and 0.50 -- U 8722) S0- (de V); slightly asym, bar?; comp 2.0, 293, 0.4 x 0.2 -- U 8723) comp 3.7, 132, 0.3 x 0.3 -- U 8724) streamers? northeastw and southwestw; U 8729 at 4.7, 14 -- U 8725) hints of 2 short jets or streamers on sf side; P w 13 45.6 + + 38 30 at 2.8, 180, 0.6 x 0.2, m=15.3 -- U 8728) VV 306 a; P w 13 45.6 + 07 38 = VV 306 b at 1.9, 266, 0.40 x 0.35 w comp = VV 306 c sf at 0.35, 0.25 x 0.15, total m=14.7 -- U 8729) see U 8724; comp 3.5, 340, 0.40 x 0.15 --

08730   U 8732) P w 13 46.8 + 72 17 = I 945 at 3.3, 113, 0.7 x 0.35, S B(s)b-c, "eF jets" (CGCG), prob interaction -- U 8733) complex spir pattern w arms in opposite directions; F cond in sp part a disturbing obj? -- U 8737) comp 3.1, 78, 0.4 x 0.05, vF -- U 8738) a large nr of vS galaxies near, F double syst at 0.8 sf -- U 8739) spindle w patches in centr reg; comp 1.7, 346, 0.4 x 0.2; 4.5, 0.4 x 0.4; 4.1, 316, 0.4 x 0.4, F; 4.2, 94, 0.50 x 0.45 --

08740   U 8740) comp 4.1, 288, 0.3: x 0.2: -- U 8745) E 3 - 4 (de V) -- U 8746) comp 3.5, 153, 0.8 x 0.3, spir -- U 8747) comp 3.1, 291, 0.30 x 0.15 -- U 8748) comp 2.3, 244, 0.45 x 0.08 --

08750   U 8750) a) 0.3 x 0.2; b) 1.0 x 0.35, distorted spir; in group w 13 48.2 + 02 33 brightest, at 2.0, 240, 0.35 x 0.25, m=15.1; comp 1.3, 15, 0.25 x 0.10; 2.6, 72, 0.6 x 0.1 -- U 8751) hints of spir pattern on red PA map; comp 1.9, 347, 0.7 x 0.15, spir; 3.5, 17, 0.6 x 0.25; 13 48.5 + 33 53 = N 5321 at 4.7 sf, 0.7 x 0.5, S B 0-a, m=15.3; sev other galaxies near -- U 8752) P w U 8754 at 5.8, 347 -- U 8753) comp 2.2, 177, 0.45 x 0.15; 3.7, 191, 0.35 x 0.30 -- U 8754) see U 8752 -- U 8758) P w 13 48.9 + 21 47 at 4.2, 90, 0.45 x 0.45, m=15.2 -- U 8759) comp 2.8, 100, 0.4 x 0.2 --

08760   U 8760) DDO 183; Ir IV-V -- U 8762) P w 13 48.8 + 24 15 at 4.8, 163, 0.7 x 0.35, m=15.3 -- U 8763) a) 0.18 x 0.18, stell; b) 0.9 x 0.8, S0?, streamer or bridge towards a; comp 2.5, 15 (from a), 0.35 x 0.15 -- U 8765) VII Zw 527; "double nucl" (CGCG); "blue post-eruptive patchy compact, w streamers" (Zw) -- U 8766) P w 13 49.1 + 14 21 at 1.2, 24, 0.6 x 0.4, SAB(s)b:, m=15.4; comp 3.3, 287, 0.35 x 0.1 -- U 8768) 13 49.5 - 01 53 at 4.8, 359, 0.8 x 0.3, Irr, m=15.3; 13 49.3 - 01 52 at 6.7, 337, 0.5 x 0.4, spir:, m=15.7 --

08770   U 8771) comp 0.7, 150, 0.2 x 0.2; 2.4, 29, 0.4 x 0.2; 3.4, 69, 0.5 x 0.25, vB core; 4.8, 4, 0.9: x 0.4: -- U 8773) comp 3.0, 97, 0.3 x 0.1; 13 49.4 + 17 11 at 3.8, 252, 0.5 x 0.2, m=15.7 -- U 8774) VV 253; spirals; a) 0.55 x 0.25; b) 0.45 x 0.25; long v dif streamer on p side; in triple group w 13 49.8 + 02 20 at 1.3 spp, 0.20 x 0.18, m=15.7 -- U 8775) comp 4.7, 350, 0.8 x 0.12 -- U 8779) U 8772 at 4.7, 288 --

08780   U 8780) main body 0.45 x 0.35 w vF asym outskirts; P w a) at 0.3 sp, 0.20 x 0.12 -- U 8781) brightest in subgroup; 13 50.0 + 21 50 at 2.8, 8, 0.8 x 0.2, compact, m=15.5; comp 1.5, 355, 0.45 x 0.15; 2.8, 45, 0.3 x 0.3; 3.4, 54, 0.6 x x 0.2 -- U 8782) vS F comp attached sp; comp 2.3, 12, 0.35 x 0.30, dif -- U 8784) measurements disturbed by superimp *; comp 2.2, 115, 0.45 x 0.35 -- U 8786) comp 2.7, 166, 0.6 x 0.6; 3.8, 208, 0.35 x 0.15 -- U 8787) major diam 2.5?; 13 50.5 + 02 30 at 2.7, 94, 0.9 x 0.2, Irr or Pec?, m=15.4 --

08790   U 8790) S B(rs)c: (de V) -- U 8793) comp 4.4, 17, 0.4 x 0.2, blue double syst, contact -- U 8795) group of compact? galaxies at 13 50.3 + 37 48, at least 7 members -- U 8797) sev S galaxies near --

08800   U 8800) a v large nr of S F galaxies near, members of CGCG cl 13 50.1 + 05 24, medium compact, medium distant -- U 8803) P w U 8809 at 3.5, 57; sev objs near -- U 8807) e blue systems; a) 0.2 x 0.2; b) 1.0 x 0.4 -- U 8809) SA(r)b: (de V); see U 8803 --

08810   U 8810) S B(r)b (de V); U 8813 at 5.0, 179 -- U 8811) vF arms, one alm invis; comp 4.1, 49, 0.3 x 0.15, F; 4.5, 326, 0.4 x 0.2; 4.7, 110, 0.4 x 0.2 -- U 8813) S0 (de V); P w U 8814 at 1.3 n; see U 8810 -- U 8814) S0 (de V); see U 8813 -- U 8817) VV 202 a + b; a) 0.6 x 0.3; b) = VV 202 a + b 1.6: x 1.0:, B nuclei in common env; sep 0.55; VV 203 a + b are 2 S objs in contact in the env f c) -- U 8818) comp 0.15 x 0.15 attached nf --

08820   U 8820) main body 1.0 x 0.9 w (r); eF smooth disc, no spir str -- U 8823) B core, sharp edge; in triple group incl comp at 0.75, 75, 0.5: x 0.4: and 13 52.2 + 69 33 at 2.5, 110, 0.7 x 0.4, m=15.7 -- U 8825) I Zw 75; "contiguous pair of post-eruptive blue compacts w plumes" (Zw) -- U 8828) strong wedge-shaped abs feature; comp 3.2, 139, 0.6 x x 0.2; 13 52.4 + 22 05 at 4.4, 91, 0.6 x 0.6, core 0.15 x 0.15, extremely compact, m=15.5 -- U 8829) comp 2.5, 131, 0.4 x 0.1, v dif; 2.6, 242, 0.4 x 0.2, dif; 4.6, 357, 0.3 x 0.1; 4.7, 200, 0.4 x 0.4, v dif --

08830   U 8835) short streamer s -- or superimp comp? -- U 8837) An 1353 in BG; IB m (de V); DDO 185; Ir IV- V (vdB) -- U 8838) I 0 (de V) -- U 8839) DDO 184; V (vdB) --

08840   U 8842) VV 281; a) 0.15 x 0.10; b) 1.4 x 0.4, distorted; in triple group w c) at 0.7 f, 0.25 x 0.20; in a dense part of the cl; 13 52.9 + 25 15 = I 4344 at 4.9, 232, 0.9 x 0.8, S0?, m=15.5; 13 52.9 + 25 17 = I 4345 at 3.8, 251, (0.9 x x 0.9), E, m=14.7; comp 1.8, 145, 0.6 x 0.4; 3.2, 10, 0.6 x 0.4; sev other galaxies near -- U 8846) SAB(rs)bc (de V) -- U 8847) I 0 ? (de V), Ir II (Ho); P w U 8853 at 14.5, 176 -- U 9949) VV 335; a) 0.9 x 0.6:, disrupted spir; b) 0.7 x 0.20, sep 0.7 --

08850   U 8850) main body 0.45 x 0.25 w dif streamer southwestw -- U 8852) SAB(r)b ? (de V); comp 2.6, 83, 0.7 x 0.10 -- U 8853) SA(rs)bc pec (de V), Sc - (Ho); see U 8847; U 8838 at 8.7, 258 -- U 8854) comp 2.7, 74, 0.4 x 0.2; 4.1, 356,

0.4 x 0.05, spir; 4.9, 43, 0.35 x 0.1 -- U 8856) a) 1.5: x 0.3:; b) 0.8: x 0.7:, sep 1.1, broad dif bridge; long
e dif streamer from b) northeastw --

08860   U 8860) SAB(r)b ? (de V); P w U 8866 at 4.1   f -- U 8862) P w 13 54.5 + 20 22 at 3.3, 120, 0.35 x 0.20, Pec?, m = 15.4;
comp 4.3, 109, 0.25 x 0.10 -- U 8863) (R)SB(s)a (de V) -- U 8866) SAB(r)0/a ? (de V) -- U 8868) 13 54.7 + 12 15 at
1.4, 175, 0.8 x 0.2, spir:, m = 14.9; 13 54.8 + 12 14 at 3.1, 153, 0.55 x 0.55, m = 15.2 --

08870   U 8870) S 0/a (de V) -- U 8872) hidden behind the reflection of Arcturus on the red PA map -- U 8873) comp 3.2,
269, 0.3: x 0.2:, dif -- U 8875) SB(rs)... pec (de V); U 8877) at 3.1 sf -- U 8877) S-shaped spir w v low surf
brightn; see U 8875 --

08880   U 8880) in group of 8 w 13 55.3 + 07 39 brightest, 0.9 x 0.8, SB..., m = 14.8, p in subgroup of 3 -- U 8883) 2 S
blue conds in the disc, one connected w centr reg by thin filament; in the center of the reflection of Arcturus
on PA maps -- U 8885) P w U 8890 at 5.2, 122; sev B galaxies near -- U 8888) a) 0.3 x 0.3; b) 0.1 x 0.1; c) 0.15 x
x 0.15; d) 0.25 x 0.20; prob members of CGCG cl 13 55.4 + 20 59, medium compact, medium distant, sev cl galaxies
near -- U 8889) comp 3.8, 278, 0.35 x 0.12 --

08890   U 8890) see U 8885 -- U 8892) dif, irr, interacting w stell obj at 0.55 from center, 2 curved jets -- U 8896) in
group of 8, see U 8880; 13 55.8 + 07 29 at 4.6, 276, 0.9 x 0.7, SB..., disturbed?, m = 15.1; 13 56.5 + 07 30 at
5.3, 65, 0.8 x 0.35, spir, m = 15.4 -- U 8898) VV 48 b, Arp 84; SB(s)b pec (de V); "open Sc w compact blue core"(Zw);
center of U 8900 at 1.9, 156; 2 smooth v open arms, one appears to continue behind U 8900; see U 8900 --

08900   U 8900) VV 48 a, Arp 84; I Zw 77; SA(s)b pec (de V); see U 8898; in Arp's class "spir galaxies w large high surf brightn
comps on arms"; "arcs of high surf brightn around nucl of comp [U 8898]" (Arp); twisted plane; sev S galaxies
near -- U 8902) slightly asym; P w 13 56.5 + 15 52 at 4.7, 337, 0.8 x 0.4, SB..., m = 15.0; comp 1.5, 107, 0.55 x
x 0.10, spir, prob background; 1.0, 58, 0.3 x 0.2, double syst, prob background -- U 8904) doubtful diam in blue,
larger than 1.1 x 0.9 -- U 8905) comp 2.2, 265, 0.6: x 0.5: dif; 3.7, 342, 0.25 x 0.22 -- U 8907) P w 13 57.3 + 12 59
at 5.1, 128, 0.7 x 0.5, E?, m = 15.2 -- U 8909) only one arm clearly vis; comp 4.4, 239, 0.6 x 0.5 --

08910   U 8911) prob disturbed by comp at 2.0, 343, 0.3 x 0.2 -- U 8917) comp attached at 0.9, 288, 0.7 x 0.7, Irr --
U 8919) VV 310 a; strong slightly asym abs feature; P w 13 57.9 + 38 26 = VV 310 b at 1.6, 50, 0.55 x 0.20, m = 15.2 --

08920   U 8920) VV 339 b; late-type asym spir; vS blue cond at 0.12 sp nucl; P w 13 57.8 + 13 12 = VV 339 a at 1.4, 273,
0.8 x 0.35, m = 15.2 -- U 8921) asym "spindle", brightest in s part -- U 8923) 13 58.3 + 38 45 at 2.1, 60, 0.5 x
x 0.3, w S comp, m = 15.4 -- U 8926) 13 58.9 + 32 05 at 5.3, 112, 0.5 x 0.2, m = 15.6 -- U 8929) VV 277; a) 0.15 x
x 0.15, stell; b) 0.6 x 0.4, distorted, curved tail; eF bridge --

08930   U 8931) VV 256 a; asym; ecc bar; blue patches along edge; P w U 8932 at 1.2 nf -- U 8932) VV 256 b; see U 8931 --
U 8934) P w a) at 1.8, 288, 0.5 x 0.10, Sb-c; comp 4.1, 36, 0.3 x 0.2 -- U 8935) S0 (de V) -- U 8936) comp 2.3,
67, 0.30 x 0.15 -- U 8937) SB(s)b (de V); v broad, slightly curved bar; 3 arms, one off center; 13 58.4 + 59 30 at
6.7 sp, 0.8 x 0.8, core 0.12 x 0.12, very compact, m = 15.5 -- U 8938) U 8944 at 6.9, 117 -- U 8939) slightly ecc
nucl reg; 13 59.5 - 01 07 at 2.3, 171, (0.6 x 0.6), very compact, m = 15.3; 13 59.6 - 01 08 at 2.6, 127, 0.7 x 0.15,
m = 15.6 --

08940   U 8940) P w comp at 0.8, 300, 0.30 x 0.15, asym, m approx 17. -- U 8941) VV 120 a-c; Arp 111, I Zw 78; a) = VV 120
a+b (supposed by VV to be double) 1.1 x 1.1, SB?...; b) = VV 120 c 0.20 x 0.20, stell, superimp ssf part of a);
VV 120 d+e at 1.0 s, 0.5 x 0.3, double syst, prob background; "N 5421, blue post-eruptive pair of open Sc and
spherical compact in matrix" (Zw); in Arp's class "E and E-like galaxies repelling spir arms"; "E galaxy [b)]
apparently bending arm at root" (Arp) -- U 8942) vB main body 0.45 x 0.30 w asym env, tend to streamer southeastw;
S v blue cond at s edge; a) at 1.8, 323, 0.3 x 0.1; b) at 2.1, 356, 0.6 x 0.3, S0? -- U 8944) chaotic spir pattern;
see U 8938 -- U 8946) only bar vis on PA prints -- U 8948) brightest in subgroup; 14 00.1 + 09 22 at 2.4, 15, 0.9 x
x 0.15, slightly curved spindle, m = 15.6; U 8950 at 5.6, 125 -- U 8949) P w 14 00.1 + 14 46 at 1.1, 167, 0.9 x 0.5,
spir, m = 15.6 --

08950   U 8950) see U 8948; 14 00.1 + 09 22 at 3.2, 212 -- U 8952) 14 00.2 + 09 36 at 1.6, 282, 0.40 x 0.22, m = 15.5; 14 00.4
+ 09 36 at 1.7, 73, 0.7 x 0.3, m = 15.5; 14 00.6 + 09 37 = N 5431 at 4.8, 75, 0.55 x 0.45, SB 0-a, m = 14.8 -- U 8953)
14 00.1 + 18 48 at 3.3, 326, 0.35 x 0.20, m = 15.7; comp 2.7, 247, 0.3: x 0.3:, dif, late spir -- U 8954) 13 59.6 +
+ 32 42 at 9.1, 251, 0.4 x 0.2, m = 15.7; 14 00.0 + 32 41 at 5.0, 222, 0.9 x 0.2, m = 15.6 -- U 8955) U 8963 at 7.8,
140 --

08960   U 8960) P w U 8962 at 2.0, 167 -- U 8962) see U 8960; comp 0.8, 134, 0.15 x 0.12 -- U 8963) comp 5.0, 154, 0.50 x
x 0.45, spir; see U 8955 -- U 8964) comp 3.8, 158, 0.4: x 0.2, dif -- U 8965) P w U 8967 at 1.5, 42 -- U 8967) see
U 8965 -- U 8968) integral-shaped, w 2 centr conds, one blue, one red; P w comp at 1.3, 155, 0.5 x 0.4, B nucl
cond, interaction w U 8969) (R)SAB(r)a (de V) --

08970   U 8971) 14 01.3 + 09 45 = N 5437 at 3.4, 154, 0.8 x 0.25, m = 15.1 -- U 8972) P w U 8973 at 0.7, 350, contact --
U 8973) see U 8972 -- U 8974) U 8976 at 6.6, 168 -- U 8976) see U 8974 -- U 8978) in triple group w 14 01.4 + 15 58
2.7, 351, 0.45 x 0.20, m = 15.5, and 14 01.3 + 16 02 at 2.5, 342, 0.7 x 0.5, spir, m = 15.6 -- U 8979) double system
14 01.2 + 06 42 at 5.1, 262, [0.7 x 0.35], a) 0.3: x 0.2:, SB..., b) 0.45 x 0.12, b) superimp a), m = 15.6 --

08980   U 8980) comp 3.2, 10, 0.3 x 0.25 -- U 8981) VV 344 a, Arp 26; SAB(rs)cd (de V), Sc- (Ho); N 5447, 5449, 5450, 5451,
5453, 5455, 5458, 5461, 5462 and 5471 are conds in this galaxy; SN 1909; in Arp's class "spir galaxies w one heavy
arm"; "note straight arm, B knot on east appears almost stellar" (Arp) -- U 8983) comp 3.1, 97, 0.5 x 0.2; 4.3,
81, 0.6 x 0.25 -- U 8984) 14 02.1 + 36 02 at 6.8, 59, 0.7 x 0.5, E-S0, m = 15.4, surrounded by sev F galaxies --
U 8987) P w 14 02.0 + 16 32 at 2.5, 125, 0.45 x 0.35, m = 14.6, broad dif bridge --

08990   U 8990) b) at 0.50, 358, 0.4 x 0.4; comp 1.3, 248, 0.45 x 0.30, disturbed; 1.3, 222, 0.45 x 0.35; 2.4, 115, 0.8 x 0.2,
spir w comp 0.4 x 0.1 attached nf; 14 01.9 + 33 33 = I 4369 at 1.8, 174, 0.9 x 0.6, S0:, m = 15.3 -- U 8991) P w b) at
0.55, 67, 0.45 x 0.30, dif -- U 8993) P w U 8994 at 1.8, 6 -- U 8994) see U 8993 -- U 8996) P w U 8997 at 6.3, 6;
comp 4.6, 110, 0.3 x 0.1 -- U 8997) see U 8996 -- U 8999) P w comp at 1.5, 194, 0.8 x 0.15 --

09000   U 9000) P w U 9001 at 0.8, 125 -- U 9001) see U 9000 -- U 9002) VV 328 a; P w 14 02.4 + 12 57 = VV 328 b at 1.0,
165, 0.35 x 0.30, m = 15.2 -- U 9003) comp 4.6, 330, 0.8 x 0.6, spir; 14 02.8 + 35 51 at 5.8, 48, 0.4 x 0.2, m = 15.7 --

09010   U 9010) U 9012 at 5.5, 117 -- U 9011) SAB(s)0- : (de V); comp 3.7, 315, 0.15 x 0.12, plume southw? -- U 9012) see
U 9010; comp 1.7, 213, 0.5 x 0.3; 14 02.5 + 30 59 at 8.8, 260, 0.8 x 0.4, S0-a:, m = 15.5 -- U 9013) SA(s)cd pec
(de V), Sc+ (Ho); VV 344 b; amorphous; ecc nucl, prob interacting w U 8981 = M 101 at 45.; -- U 9016) comp 2.6, 196,
0.3 x 0.1; 4.7, 215, 0.5: x 0.4:, dif -- U 9017) b) at 0.7, 47, 0.25 x 0.25 -- U 9018) DDO 186; S d: (de V); Ir IV-V
(vdB); *? superimp near nucl; U 8981 = M 101 at 22., 250 --

09020   U 9023) blue nucl reg -- U 9024) S vB nucl cond, ext eF disc w spir pattern in outer part -- U 9025) P w 14 04.7 +
+ 12 50 at 2.8, 54, 0.8 x 0.2, S0:, m = 14.8 -- U 9026) SA(s)c: (de V); P w U 9029 at 3.2, 94; patchy, asym -- U 9029)
see U 9026 --

09030   U 9031) 14 05.0 + 15 09 at 4.8, 315, 0.8: x 0.4, m = 15.6 -- U 9032) comp 1.9, 213, 0.4 x 0.3, F -- U 9033) SA 0 pec
(de V); 14 05.1 + 55 16 = N 5484 at 3.7, core 0.22 x 0.15, E0: (de V), m = 15.6; comp 4.7, 267, 0.9: x 0.8:, SB d-dm;
U 9036 at 6.5 -- U 9036) S m? (de V); see U 9033 -- U 9039) P w 14 06.3 + 72 25 at 2.9, 14, 0.8 x 0.4, spir, m =
= 15.2 --

09040   U 9042) complex spir str, F arm pattern; comp 1.1, 304, 0.8: x 0.8:, S m or I m; 2.3, 145, 0.3 x 0.1 -- U 9049) comp
        0.8, 31, 0.15 x 0.15, alm stell; *? or comp? 0.12 x 0.12 superimp 0.20 p center --

09050   U 9053) P w U 9055 at 5.3, 62 -- U 9055) see U 9053 -- U 9056) comp 4.2, 85, 0.4: x 0.4:, v dif, dwarf -- U 9058)
        E 2 - 3 (de V); 14 07.7 + 17 47 at 1.5, 78, 0.8 x 0.4, spir, m=15.7; 14 07.5 + 17 49 at 2.9, 330, 0.5 x 0.4, m=15.7;
        comp 4.3, 248, 0.4 x 0.12; 14 07.8 + 17 51 at 4.8, 30, 0.8 x 0.6, S B b:, Arp 79, in Arp's class "spir galaxies w
        large high surf brightn comps on arms"; "small sep between two knots in arm" (Arp), m=15.2; U 9061 at 11.4,
        8 -- U 9059) Arp 117; SA 0 (de V); P w U 9061 at 2.6, 28; in Arp's class "E and E-like galaxies close to and dis-
        turbing spirals"; see U 9061 --

09060   U 9060) comp 2.0, 123, 0.3 x 0.10 -- U 9061) Arp 117; S B (r)ab (de V); see U 9059; "flattening of spirals nucleus
        appears to be in different plane than arms" (Arp) -- U 9064) comp 4.8, 52, 0.9: x 0.4:, S d - dm; 4.8, 140, 0.5 x
        x 0.3, S B c -- U 9065) asym spindle; comp 0.6, 108, 0.2 x 0.2; 2.7, 253, 0.3 x 0.3, spir -- U 9069) comp attached
        at 0.3, 350, 0.10 x 0.10 (*?); P w comp 2.0, 260, 0.8 x 0.5; 3.0, 357, 0.6 x 0.1 --

09070   U 9071) comp 3.3, 290, 0.4 x 0.10 -- U 9072) P w comp at 0.45, 353, 0.3 x 0.2 -- U 9073) asym; sev smaller galaxies
        near; 14 08.8 + 25 45 = I 4382 at 1.9, 42, 0.75 x 0.18, m=15.4; comp 2.0, 116, 0.6 x 0.5, spir; 3.4, 133, 0.15 x 0.15,
        compact?; 3.5, 349, 0.3 x 0.2, pec, streamers -- U 9076) comp 1.6, 325, 0.3 x 0.12; 2.5, 260, 0.4 x 0.2 -- U 9079)
        S d: (de V) --

09080   U 9080) P w comp at 1.1, 332, 0.22 x 0.22, compact:, double? -- U 9083) VV 125 -- U 9084) 14 10.0 + 08 53 at 4.4,
        102, 0.9 x 0.8, S0, m=15.1 -- U 9085) 14 09.8 + 16 06 at 1.8, 335, 0.5 x 0.5, spir, m=15.4; U 9086 at 2.2, 0;
        comp 2.2, 67, 0.6 x 0.5, v dif -- U 9086) P w 14 09.8 + 16 06 at 1.1, 232, see U 9085 -- U 9087) P w U 9090 at 7.0,
        15; comp 2.8, 332, 0.6 x 0.35, late spir --

09090   U 9090) see U 9087 -- U 9094) comp 3.5, 299, 0.4 x 0.4 -- U 9095) a) 0.50 x 0.35, spir?; b) 0.20 x 0.20, compact,
        sep 0.7 -- U 9096) comp 3.3, 89, 0.7 x 0.2; 4.0, 205, 0.3 x 0.2; 4.2, 72, 0.4 x 0.12 -- U 9098) a) 0.45 x 0.30; b)
        1.0: x 0.4; sep 0.5 -- U 9099) comp 1.3, 72, 0.3 x 0.2; 1.4, 225, 0.8 x 0.2, E - S0, m approx 16.; sev S F galaxies
        near; 14 10.3 + 20 38 at 7.0, 256, 0.9 x 0.6, compact, m=15.2 --

09100   U 9102) VV 70; dif curved plume southeast; P w U 9111 at 13.6, 130 -- U 9103) VV 223 c; VV 223 a + b = 14 11.3 +
        + 08 27 at 1.0 f, a) 0.40 x 0.20; b) 0.45 x 0.20; contact, m=14.7; hints of broad e dif bridge from VV 223 a + b --
        U 9105) P w 14 11.4 + 44 05 at 3.1, 244, 0.45 x 0.25, m=15.6; comp 3.7, 322, 0.4 x 0.15 -- U 9107) v open; 2 long
        thin arms; comp 2.4, 124, 0.35 x 0.25; 3.3, 87, 0.30 x 0.20 -- U 9108) a) + b) [0.30 x 0.20], contact; c) 0.12 x 0.12,
        F bridge southw; sev galaxies near -- U 9109) a) 1.0 x 0.2, spir; b) 0.2 x 0.1, attached n part of a) --

09110   U 9110) chain of blue patches in n part; 14 12.0 + 15 50 at 3.4, 101, 0.55 x 0.20, m=15.3 -- U 9111) see U 9102 --
        U 9113) comp 2.3, 226, 0.4 x 0.2 -- U 9114) MCG diam refer to U 9118; comp 2.4, 239, 0.4 x 0.2; 14 12.0 + 03 25 =
        = I 988 at 5.8, 307, 0.55 x 0.35, m=14.8; U 9118 at 4.3, 95 -- U 9115) P w comp at 0.50, 263, 0.8 x 0.7, S b - c --
        U 9118) MCG diam listed for 1 - 36 - 7 = U 9114 refer to this galaxy -- U 9119) SA(s)cd: (de V) --

09120   U 9120) brightest of 4 spirals, group members at 2.9, 359, 0.9 x 0.9, Sc; 6.2, 354, 0.7 x 0.5; and 6.6, 318, 0.5 x
        x 0.4 -- U 9124) strong abs along edge -- U 9126) DDO 188; V (vdB) -- U 9127) S c: (de V); 14 13.6 + 36 24 at 3.8,
        146, 0.6 x 0.35, m=15.7; comp 3.5, 244, 0.9: x 0.6:, dwarf irregular -- U 9128) DDO 187; Pec (vdB); U 9130 at
        4.0, 132 --

09130   U 9130) spir 0.9 x 0.9 w cond 0.1 x 0.1 n; see U 9128 -- U 9131) dif spindle; comp 4.6, 336, 0.4 x 0.15; 4.9, 276,
        0.3 x 0.2, F -- U 9133) SA(rs)ab (de V); BG v and v_o refer to N 5529 = U 9127; comp 2.6, 48, 0.3 x 0.1 -- U 9136)
        one-armed spir w massive bar 0.7 x 0.3; P w U 9139 at 5.5, 18 -- U 9137) P w comp at 0.55, 168, 0.3 x 0.2; 14 14.3
        + 11 07 = N 5531 at 5.2, 333, 0.9 x 0.9, S0 w S comp superimp ssf center -- U 9139) see U 9136; sev galaxies near;
        comp 3.9, 345, 0.8 x 0.4, Sc --

09140   U 9140) in triple group w comp at 1.7, 331, 0.5 x 0.2, and 2.5, 355, 0.3 x 0.3 -- U 9141) P w comp at 0.50, 282,
        0.35 x 0.15 -- U 9142) VV 210 a, Arp 199; (R)S B (rs)0/a (de V); P w U 9143 at 0.6, 70; in Arp's class "galaxies w
        material ejected from nuclei"; "spirals appear disturbed" (Arp) -- U 9143) VV 210 b, Arp 199; SA(s)bc: (de V);
        see U 9142 -- U 9145) comp 2.7, 339, 0.8 x 0.2, F -- U 9147) comp 2.3, 266, 0.3 x 0.2, F; 2.8, 290, 0.3 x 0.3, F
        -- U 9148) 14 15.6 + 07 53 = N 5543 at 3.5, 265, 0.55 x 0.25, m=15.3 -- U 9149) Pec SA(s)0/a (de V); slightly asym

09150   U 9150) v blue chaotic B arms; U 9157 at 5.9, 107 -- U 9151) MCG identification N 5561 is incorrect; N 5561 is a
        galaxy at 14 15.8 + 58 58 (position 2.6, 17), 0.35 x 0.35, m=15.5 -- U 9153) P w 14 15.8 + 11 27 = I 993 at 1.6,
        321, 0.4 x 0.4, m=15.4 -- U 9156) comp 1.6, 207, 0.35 x 0.15 -- U 9157) see U 9150; U 9160 at 5.9, 287 --

09160   U 9160) see U 9157 -- U 9161) E 1 (de V) -- U 9162) 2 long thin arms, unequal bending; may be disturbed by comp
        at 1.7, 138, 0.3 x 0.1 -- U 9164) chaotic, v blue; disrupted multiple system or eruptive; P w 14 16.1 + 22 01 at
        3.9, 263, 0.7 x 0.3, S B b: m=15.7 -- U 9165) comp 1.6, 228, 0.25 x 0.20; 14 16.7 + 25 10 at 3.1, 86, 0.8 x 0.25,
        S b - c, m=15.7 -- U 9168) in triple chain w U 9170 at 2.2, 124, and U 9171 at 5.0, 118 --

09170   U 9170) see U 9168; U 9171 at 2.9, 112 -- U 9171) see U 9168, 9170 -- U 9172) Arp 286 (w U 9175, U 9176); S B(s)b pec
        (de V); see U 9175 -- U 9175) Arp 286, see U 9172; S B (r)ab (de V); U 9172 at 5.3, 313; U 9176 at 4.1, 46; in Arp's
        class "double galaxies, infall and attraction"; "connection not vis" (Arp) -- U 9176) SAB(rs)b: (de V); Arp 286,
        see U 9172, 9175 -- U 9178) VV 2, Arp 45; a) 1.2: x 0.7, S B...; S blue cond in p arm, 2 blue conds in n part;
        b) 0.4 x 0.3; dif bridge; a) is MCG 9 - 23 - 64; in Arp's class "spir galaxies w low surf brightn comps on arms";
        "One arm leads towards large comp, other towards small comp" (Arp); comp 4.1, 258 from a), 0.7 x 0.4, dif spir --
        U 9179) SAB(s)d (de V), Sc + (Ho) --

09180   U 9180) VV 142, Arp 69; asym 3-armed spir, one arm broken; no disturbing obj vis; VV 142 b is a part of a spir arm;
        comp 1.7, 171, 0.4 x 0.1 -- U 9181) S B: 0 - ? (de V) -- U 9183) E + 3 (de V); in triple group w U 9181 at 2.9, 225,
        and U 9187 at 10.3, 14 -- U 9187) SA bc (de V) -- U 9189) VII Zw 547; "post-eruptive blue spir w compact spherical
        disc and F conds" (Zw) --

09190   U 9190) comp 2.9, 273, 0.35 x 0.25, spir -- U 9191) P w comp at 0.9, 178, 0.6 x 0.2, distorted spir -- U 9195) comp
        2.4, 140, 0.3 x 0.3, dif, dwarf; 3.8, 243, 0.3 x 0.3, dif -- U 9196) vB main body 0.45 x 0.35 -- U 9197) P w U 9200
        at 4.8, 146; comp 3.8, 293, 0.8 x 0.12 --

09200   U 9200) see U 9197 -- U 9201) SAB(rs)cd (de V) -- U 9204) P w comp at 5.3, 258, 0.9: x 0.2:, S nucl cond in v dif
        disc -- U 9206) B main body 0.55 x 0.25, smooth vF outer region -- U 9207) arc-shaped disrupted obj w compact comp
        0.12 x 0.12 superimp -- U 9208) comp 3.8, 118, 0.55 x 0.15 -- U 9209) 14 20.9 + 40 32 = N 5601 at 4.8 f, 0.9 x 0.3,
        E - S0, m=15.6 --

09210   U 9211) DDO 189; Ir V (vdB) -- U 9213) P w comp at 1.0, 338, 0.4 x 0.15; comp 4.3, 46, 0.4: x 0.2; sev S galaxies
        near -- U 9215) 14 20.6 + 01 54 at 5.4, 235, 0.5 x 0.5, compact, m=15.2 -- U 9216) MCG ident N 5603 is incorrect,
        N 5603 is U 9217 at 2.6, 157 -- U 9217) I Zw 86; MCG ident N 5601 is incorrect, see U 9216; "neutral large spheri-
        cal disc compact, extended halo" (Zw) -- U 9219) hints of spir pattern?, ecc nucl; comp 3.1, 198, 0.3 x 0.12 --

09220   U 9220) S c pec (de V) -- U 9221) P w U 9222 at 0.9, 60 -- U 9222) see U 9221 -- U 9223) comp 3.7, 116, 0.35 x 0.2;
        3.9, 343, 0.4 x 0.2 -- U 9226) VV 77 a, Arp 178; N 5615 = VV 77 b is a cond 0.20 x 0.15 at 0.45 np center; a broad
        curved plume leaves the galaxy near this point, connection w N 5615?; in Arp's class "galaxies w narrow counter-
        tail"; "ring off center, broad ejected plume from cond in ring"; P w U 9228 at 2.0, 351 -- U 9227) comp 2.3,

10, 0.35 x 0.15; U 9232 at 7.5, 141 -- U 9228) VV 77 c, Arp 178; (R)SAB(r)0+ (de V); see U 9226 -- U 9229) P w b) at 1.1, 112, 0.8 x 0.5: --

09230   U 9230) comp 1.3, 104, 0.15 x 0.10 -- U 9231) 2 or 3 conds in n part; asym; comp 4.9, 223, 0.8 x 0.2 -- U 9233) in triple group w U 9235 at 1.7, 120, and U 9238 at 3.1, 85 -- U 9234) S vB nucl reg, 2 vF open arms; P w comp at 1.2, 353, 0.30 x 0.25 -- U 9235) see U 9233; U 9235 at 1.9, 57; sev galaxies near -- U 9236) comp 2.4, 337, 0.5 x 0.2, Sb - c -- U 9237) in open group incl 14 21.9 + 27 56 at 12.7, 240, 0.9 x 0.45, S B b, m = 15.4; 14 22.6 + 28 07 at 4.2, 343, 0.9 x 0.7, Sa - b, m = 15.6; 14 22.8 + 27 59 at 3.7, 194, 0.8 x 0.7, spir, m = 15.6; 14 23.0 + 28 05 at 4.4, 104, 0.8 x 0.25, S0 - a, m = 15.7 -- U 9238) see U 9233, 9235 --

09240   U 9240) I Zw 87, DDO 190; "resolved blue dwarf galaxy w compact conds at south edge" (Zw); Ir IV or Ir IV- V (vdB) -- U 9241) comp 2.0, 199, 0.4 x 0.3, double system -- U 9242) I Zw 88; "extremely flattened blue galaxy, d_1 = = 250", d_2 = 15", w many F stellar knots" (Zw) -- U 9243) comp 0.8, 8, 0.3 x 0.2, dif -- U 9244) P w comp at 1.3, 27, 0.5 x 0.3, spir -- U 9245) DDO 191; S (B) V (vdB) -- U 9247) comp 1.4, 314, 0.35 x 0.12 --

09250   U 9251) VV 152 -- U 9254) comp 2.9, 52, 0.45 x 0.15 -- U 9255) U 9258 at 3.1, 117; 14 25.0 + 05 02 = I 4424 at 3.7, 73, 0.9 x 0.4, S0 - a, m = 14.8 --

09260   U 9261) SA(s)0/a (de V) -- U 9262) P w comp at 1.7, 224, 0.35 x 0.30 -- U 9266) 2 S blue conds in disc sp and nf center --

09270   U 9271) I Zw 89; (R)SA(rs)b (de V); "neutral Sb w elliptical blue compact core and F blue spir arms" (Zw) -- U 9274) strongly asym S B(s), brightest in sp part; S v red obj superimp nf edge, 0.05 x 0.05; P w U 9282 at 9.3, 77 -- U 9276) U 9281 at 2.4, 120; sev galaxies near, see U 9281 --

09280   U 9280) brightest in subgroup; 14 25.9 + 11 35 at 3.8, 310, 0.8 x 0.5, S0, m = 14.8, comp at 0.55 ssp, 0.3 x 0.1, incl in CGCG magn; 14 25.9 + 11 35 at 3.6, 252, 0.7 x 0.3, spir, m = 15.4; 14 26.2 + 11 36 at 1.7, 112, 0.60 x 0.55, S B(r) ..., m = 15.3 -- U 9281) 14 26.0 + 26 03 = I 1018 at 1.4, 215, extremely compact, m = 15.6; see U 9276; comp 2.8, 246, 0.4 x 0.4; 3.1, 78, 0.35 x 0.35; 4.0, 285, 0.5 x 0.2; 14 25.6 + 26 04 at 5.8, 264, 0.7 x 0.4, Sa:, m = = 15.4; 14 26.0 + 26 10 = I 1019 at 6.0, 353, 0.8 x 0.5, S0:, m = 15.3 -- U 9282) see U 9274 -- U 9283) Haro 41; "vio- leta" (Haro); v, v_0 from A.J. 75, 1143 -- U 9287) VV 15; listed by VV as triple system; 2 v red conds in disc np and sf center -- U 9289) sev vS galaxies near --

09290   U 9293) slightly asym; high surf brightn -- U 9295) VII Zw 551; "blue elliptical disc compact" (Zw); P w 14 26.4 + + 70 08 at 2.8, 219, 0.65 x 0.45, spir:, m = 14.8 -- U 9297) 2 unequal arms, bent in opposite directions? -- U 9299) comp 1.5, 186, 0.6 x 0.3, v dif --

09300   U 9300) comp 3.9, 88, 0.3 x 0.1 -- U 9301) P w 14 26.9 + 30 18 at 3.0, 352, 0.50 x 0.45, m = 15.5 -- U 9304) SAB(r)0+ (de V); U 9308 at 2.0, 170 -- U 9308) E 1 (de V); see U 9304; U 9310 at 5.2, 98 --

09310   U 9310) see U 9308 -- U 9314) comp 1.6, 349, 0.35 x 0.25 -- U 9315) b) at 0.6, 115, 0.6 x 0.2 -- U 9318) Pec SA(rs)b (de V) --

09320   U 9321) P w U 9329 at 4.1, 141 -- U 9322) comp 0.9, 273, 0.15 x 0.15, compact; 1.5, 354, 0.25 x 0.15; 14 27.9 + 23 15 at 3.0, 225, (0.8 x 0.6), compact, m = 15.3 -- U 9324) DDO 192; arc-shaped, strongly asym; no disturbing obj vis; S B V (vdB) -- U 9325) SAB(rs)c (de V); comp 2.6, 290, 0.8 x 0.4, Irr; P w U 9366 at 30.5, 110 -- U 9326) P w U 9327 at 0.8, 116 -- U 9327) see U 9326 -- U 9328) S B(s)d (de V) -- U 9329) b) at 0.7, 100, 0.2 x 0.2; see U 9321 --

09330   U 9330) P w U 9333 at 5.5, 127 -- U 9331) B core in vF disc; sev F comps -- U 9332) comp 2.3, 155, 0.4 x 0.10; 2.5, 177, 0.4 x 0.15 -- U 9333) see U 9330 -- U 9334) comp 4.1, 327, 0.5 x 0.1 -- U 9336) comp 1.3, 249, 0.6 x 0.3 -- U 9338) P w comp at 3.1, 310, 0.7 x 0.2, spir, m approx 16. --

09340   U 9340) 14 28.3 + 25 44 at 6.3, 283, 0.7 x 0.45, Sa, m = 15.3 -- U 9342) comp 2.5, 346, 0.3 x 0.2 -- U 9346) P w comp at 0.5, 240, 0.15 x 0.15, compact; U 9345 at 6.5, 211; comp 4.5, 145, 0.8 x 0.25 -- U 9347) P w U 9361 at 9.7, 290 -- U 9348) P w comp at 1.4, 155, 0.6 x 0.2 --

09350   U 9350) P w a) = VV 262 a + b at 1.0, 270, 0.8 x 0.4, main body 0.20 x 0.15 w broad irr plume, and 0.65, 260, 0.15 x x 0.15, compact; VV 262 b brightest -- U 9351) prob in CGCG cl 14 29.4 + 05 36, open, distant; sev cl members near -- U 9352) Arp 49; S B(s)m pec: (de V); in Arp's class "spir galaxies w S high surf brightn comp on arms"; "appearance of wake from stellar obj in eastern arm" (Arp) -- U 9353) SAB(rs)cd (de V); 14 29.9 + 10 09 at 6.2, 294, 0.7 x 0.6, S0 - a, m = 15.3 -- U 9354) S b (de V); appears acicular; comp 1.4, 187, 0.4: x 0.3:, dif in blue, v red -- U 9358) SAB(rs)b (de V); slightly asym; complex inner str; may be disturbed by comp at 1.9, 0. 0.25 x 0.2, E: --

09360   U 9361) see U 9347 -- U 9362) P w 14 30.6 + 04 10 at 3.4, 323, 0.9 x 0.3, S B ..., cond in s arm, m = 14.9; comp 4.6, 69, 0.4 x 0.4; 4.7, 264, 0.3 x 0.2 -- U 9363) SA(s)d (de V); SN 1954 b -- U 9365) comp 2.0, 74, 0.4 x 0.3; sev galaxies near -- U 9366) SA(rs)bc (de V); see U 9325 -- U 9367) comp 1.8, 104, 0.4: x 0.1; 2.1, 36, 0.3 x 0.1; 3.2, 262, 0.9 x 0.3, spir; many vF galaxies in vic --

09370   U 9370) P w comp at 1.3, 141, (0.4 x 0.3) -- U 9371) disrupted pair; a large nr of S galaxies near -- U 9375) comp 2.4, 70, 0.4: x 0.1; 4.1, 92, 0.3 x 0.1; 4.8, 250, 0.30 x 0.15 -- U 9376) contact w 14 31.8 + 40 18 at 0.85, 20, 0.8 x 0.6, spir, m = 14.8, eF (R)?; comp 6.9, 320, 0.9 x 0.7, Irr, disrupted pair?; U 9379 at 10.1 -- U 9378) slight- ly asym; B knotty arms, Sc? -- U 9379) comp 3.8, 117, 0.35 x 0.25, E: --

09380   U 9383) Arp 274; triple system, a) at 0.7, 258, 0.7 x 0.5, Sb; c) at 0.7, 117, 0.30 x 0.15; in Arp's class "double galaxies w connected arms"; "perturbation of arm by S galaxy to east" (Arp) -- U 9384) VV 303, Arp 95; a) = VV 303 a 0.8 x 0.45, S B ...; b) = VV 303 b 0.7 x 0.2, vB core; sep 0.50; in triple group w VV 303 c at 0.8, 210 from a), 0.4 x 0.2, incl in CGCG magn -- U 9387) vB core in F disc or halo; sev S galaxies near; comp 4.4, 250, 0.5 x 0.3 -- U 9388) S B(s)b (de V); N 5682 and 5683 are confused in MCG; P w 14 33.1 + 48 52 = N 5683 at 1.4, 114, 0.45 x 0.40, S B(s)0/a ? (de V), m = 15.5; comp 1.6, 216, 0.3 x 0.1, F; U 9399 at 8.5, 60 -- U 9389) P w U 9394 at 15.8, 6 --

09390   U 9391) DDO 193; S (B) V (vdB); DDO position incorrect -- U 9394) VV 146 a; VV 146 b and c are S blue comps super- imp sp part, total diam 0.22 x 0.12; see U 9389 -- U 9395) S 0 - ? (de V) -- U 9396) comp 1.7, 10, 0.7 x 0.4, early spir; 3.7, 49, 0.5 x 0.3; distant quadrupel group around 3. f -- U 9397) 14 33.9 + 36 56 at 7.4, 144, 0.5 x 0.5, E - S, m = 15.3 -- U 9399) S B(s)0/a: (de V); see U 9388; U 9406 at 15., 145 --

09400   U 9401) comp 0.8, 300, contact, 0.6: x 0.3, dif, v red; 1.9, 323, 0.5 x 0.2; 3.4, 338, 0.5 x 0.4, E - S0: -- U 9402) comp 1.2, 10, 0.3 x 0.2; 1.9, 103, 0.5 x 0.4, dif; 14 34.0 + 36 43 = N 5684 at 3.5, 134, 0.45 x 0.45, m = 15.2 -- U 9404) comp 2.8, 30, 0.4 x 0.2 -- U 9405) DDO 194; V (vdB) -- U 9406) S B(rs)d (de V) -- U 9407) B wedge-shaped main body in asym disc --

09410   U 9410) comp 1.1, 325, 0.3 x 0.1 -- U 9411) P w 14 34.9 + 10 12 at 1.4, 127, 0.5 x 0.25, m = 15.1 -- U 9413) P w 14 35.3 + 78 11 at 4.3, 20, 0.9 x 0.4, S0, m = 14.9 -- U 9416) S c? (de V); δ = 02 30 in CGCG I, p 175 -- U 9417) comp 0.55, 352, 0.10 x 0.08; 14 35.6 + 22 11 at 4.9, 93, 0.5 x 0.3, m = 15.2 --

09420   U 9420) SAB(s)...pec (de V); U 9451 at 35., 79 -- U 9422) comp 2.2, 52, 0.6:x 0.6:, e dif, dwarf -- U 9424) comp 2.6, 344, 0.3 x 0.2, pec? -- U 9425) VV 264, Arp 241; 2 B connected cores; curved plumes; in Arp's class "galaxies w appearance of fission" -- U 9428) comp 1.5, 26, 0.25 x 0.15 -- U 9429) patchy; short disrupted arms; comp (double syst) attached at 0.5, 254, 0.1 x 0.1 and 0.1 x 0.05, sep 0.15; comp 0.8, 183, 0.3 x 0.1; prob disturbed --

09430   U 9430) B patches; prob v late spir -- U 9431) 14 36.8 + 46 53 = N 5717 at 4.8 nf, 0.9 x 0.5, m=15.4 -- U 9432) CGCG magn appears too bright -- U 9433) may be interacting w comp at 1.0, 244, 0.3 x 0.1; 14 36.3 + 28 40 = I 4477 at 3.6, 220, 0.6 x 0.3, spir, m=15.5 -- U 9434) comp 5.4, 254, 0.6 x 0.3 -- U 9435) 14 36.4 + 30 41 = N 5706 at 2.1, 311, 0.4 x 0.3, E?, m=15.7, * superimp -- U 9436) (R)SB(rs)0/a (de V); comp 0.9, 221 (superimp), 0.35 x x 0.30; 4.1, 263, 0.5 x 0.12, spir, m approx 16. --

09440   U 9440) comp 2.1, 10, 0.6 x 0.5; P w U 9445 at 3.5, 155 -- U 9445) see U 9440 -- U 9447) S B d: (de V); comp 2.4, 134, 0.9 x 0.2, dif; 4.0, 94, 0.35 x 0.10; U 9451 at 26.5, 12 -- U 9448) comp 2.7, 183, 0.7 x 0.2; 3.0, 244, 0.3 x x 0.12; 4.2, 5, 0.35 x 0.15 --

09450   U 9450) asym spindle; comp 0.7 nf, 0.15 x 0.10, F, disturbing? -- U 9451) SAB(rs)bc pec (de V); asym patchy centr reg, neb outer disc; P w U 9462 at 11.5, 100, interaction? -- U 9453) comp 4.5, 226, 0.5 x 0.4, Sc -- U 9456) chaotic "spindle", may be interacting w S obj attached to p side; listed as N 5731 in CGCG, ident uncertain, N 5730 appears most probable; P w U 9460 at 4.0, 55 -- U 9457) Arp 171; P w U 9459 at 1.0, 105; 14 38.0 + 03 38 = = I 1039 at 3.3, 227, 0.8 x 0.4, S0, m=15.6; 14 38.1 + 03 35 = I 1041 at 5.6, 182, 0.9 x 0.5, Sa, m=15.1 -- U 9459) Arp 171; see U 9457 --

09460   U 9460) listed as N 5730 in CGCG, see U 9456 -- U 9462) SAB(s)ab pec (de V); strong abs along edge; see U 9451 -- U 9468) P w comp at 1.2, 204, 0.30 x 0.25 --

09470   U 9476) 14 39.4 + 44 42 at 3.3, 233, 0.8 x 0.7, core 0.20 x 0.20, S0:, m=14.8 --

09480   U 9480) comp 1.9, 202, 0.4: x 0.3:, dif, v red; 3.7, 334, 0.4 x 0.4, spir; pair 14 40.2 + 22 34 at 2.1, 60, 0.50 x x 0.20 and 0.5: x 0.15, contact, total m=15.4 -- U 9481) 14 40.3 + 28 52 at 4.3, 174, 0.6 x 0.4, core 0.15 x 0.12, m=15.7 -- U 9484) comp 4.2, 116, 0.3 x 0.10 -- U 9485) U 9483 at 7.7, 352 -- U 9486) S0? (de V); slightly asym; main body 1.5 x 1.4 -- U 9489) a) 0.5: x 0.15; b) 0.5: x 0.25, sep 0.50, connected, bridge; --

09490   U 9490) in triple group w 14 41.1 + 11 25 at 4.4, 332, 0.9 x 0.7, S0, m=14.9, and comp at 4.9, 5, 0.8 x 0.5, spir, m=15.2 -- U 9493) SAB(rs)b (de V), Sb+ (Ho); P w U 9499 at 18.4, 26 -- U 9494) comp 0.4 p, 0.20 x 0.15, dif, interaction? -- U 9496) B main body 0.45 x 0.45, ring?; comp 0.25 x 0.20 superimp at 0.3 s center -- U 9497) v blue; 2 conds near center -- U 9499) SAB(rs)b? (de V), Sb- (Ho); see U 9493 --

09500   U 9500) DDO 196; Ir V (vdB) -- U 9501) comp 2.2, 79, 0.4 x 0.15 -- U 9502) in chain of 4, incl comp at 3.9, 213, 0.7 x 0.2, spir -- U 9503) Arp 64; S blue obj at 0.9 p, 0.1 x 0.1, at tip of spir arm; comp 0.6, 117, 0.3 x 0.1, at tip of spir arm; 1.4, 160, 0.45 x 0.25; 14 42.8 + 19 39 at 4.7, 245, 0.7 x 0.6, compact, m=15.3; in Arp's class "spir galaxies w S high surf brightn comps on arms"; "both arms lead toward comps" (Arp) -- U 9504) a) 0.35 x x 0.15; b) 0.45 x 0.15, sep 0.65, both v blue; irr env; U 9506 at 1.5 sf -- U 9505) Arp 297; v long arms w low surf brightn; a) = N 5752 at 1.0, 262, 0.6 x 0.25; in Arp's class "double galaxies w long filaments"; "comp on arm has long tail extending westw"; P w U 9507 at 3.1, 19; see U 9507 -- U 9506) see U 9504 -- U 9507) broad asym arms w low surf brightn, bar?; Arp 297, see U 9505; a) = N 5753 at 1.9, 325, 0.5 x 0.3 -- U 9508) P w comp at 2.5, 48, 0.7 x 0.5, spir -- U 9509) VV 109; An 1444 in BG; Pec: (de V); VV 109 a is an integral-shaped obj 1.2 x 0.1 w comp 0.4 x 0.25 attached np; VV 109 b at 1.5 nnf, 0.35 x 0.15 --

09510   U 9511) I Zw 96; main body 0.5 x 0.4 w thin F jet in pos angle 202; S knot approx 19^m at 0.55 np nucl; sev F galaxies near; brightest at 2.2, 48, 0.20 x 0.15; "red post-eruptive globular compact w extended halo, pencil jets and associated F stell knots" (Zw); "unresolved core 10" = 12.5 kpc diam, surrounded by a F halo 40 kpc diam. A remarkable F, thin jet 48" = 84 kpc long extends to the southwest, w a prominent knot of 19^m about 30" from nucl" (Sargent) -- U 9512) S B(r)0/a (de V) -- U 9513) vB main body 0.18 x 0.15, F asym env; P w U 9515 at 2.8, 101 -- U 9515) see U 9513 --

09520   U 9520) P w 14 44.6 + 33 36 = I 4506 at 1.4, 109, 0.4 x 0.4, m=15.1; 2.9, 106, 0.45 x 0.35 -- U 9521) in group of 5 or 6 galaxies; P w U 9523 at 1.5, 88; 14 44.4 + 11 47 at 3.5, 248, (0.9 x 0.7), E:, m=15.0; 14 44.5 + 11 50 at 2.6, 323, (0.6 x 0.4), E:, m=15.0; comp 3.7, 272, 0.35 x 0.15; sev S galaxies near -- U 9522) comp 4.0, 300, 0.35 x x 0.35 -- U 9523) see U 9521; 14 44.8 + 11 48 at 1.3, 111, 0.4 x 0.2, E-S, m=15.5 -- U 9524) P w comp at 1.8, 119, 0.9 x 0.3, S0, m approx 15. (appears brighter than CGCG limit 15.7); 14 44.3 + 13 53 at 5.6, 275, 0.25 x 0.25, compact, m=15.2; sev S B objs near -- U 9525) a) 0.20 x 0.15; b) 0.6 x 0.45, distorted spir; sep 0.7, broad curved bridge; 14 44.8 + 13 37 at 3.5, 185, 0.5 x 0.2, m=15.5 --

09530   U 9530) comp 2.0, 260, 0.30 x 0.20; 3.5, 181, 0.5 x 0.1 -- U 9531) 14 45.3 + 18 40 = I 4507 at 2.8, 183, 0.8 x 0.4, m=15.7 -- U 9532) VV 165 b - g, Arp 328; compact group of 5, bridges + streamers; the group incl MCG 3 - 38 - 17 = = VV 165 a at 1. np 3 - 38 - 18; "6 galaxies more or less in line; center one has semi-stellar component" (Arp) -- U 9533) P w 14 46.0 + 14 12 at 1.2, 332, 0.6 x 0.3, E- S0, m=15.3 -- U 9534) 14 46.2 + 11 29 at 3.9, 177, 0.45 x x 0.25, spir:, m=15.7, comp 0.15 x 0.12 0.7 p -- U 9535) B main body 0.45 x 0.35; P w 14 46.0 + 12 41 = N 5763 at 4.4, 64, (0.5 x 0.5), E?, m=15.3; comp 3.2, 249, 0.4 x 0.3 -- U 9536) comp 2.8, 167, 0.35 x 0.30, dif, P w comp at 0.5 s, 0.2: x 0.2: -- U 9538) comp 3.1, 117, 0.40 x 0.10; 3.5, 318, 0.5 x 0.4 -- U 9539) P w 14 46.7 + 22 20 at 4.2, 214, 0.7 x 0.6, compact, m=15.7 --

09540   U 9540) 3 vF conds , prob connected -- U 9544) early spir w comp 0.10 x 0.10 superimp at 0.55 f center, prob at tip of spir feature; comp 4.8, 255, 0.8 x 0.1, spir -- U 9545) P w comp at 5.1, 225, 0.6 x 0.3 -- U 9547) P w comp at 1.1, 277, 0.3 x 0.2, E?; comp 2.6, 199, 0.55 x 0.10; 2.3, 65, 0.4 x 0.3, dif -- U 9548) P w U 9550 at 3.0, 57; comp 4.3, 287, 0.5 x 0.2 -- U 9550) comp 3.4, 318, 0.30 x 0.10 --

09550   U 9550) see U 9548 -- U 9551) pair 14 48.1 - 01 31 at 3.0, 41 0.8 x 0.2, dif, and 3.7, 6, 0.7 x 0.25, sep 0.7, total m=15.3 -- U 9552) 14 48.2 + 26 13 at 3.6, 179, (0.8 x 0.7), core 0.18 x 0.18, compact, m=15.7 -- U 9553) comp 2.7, 52, 0.35 x 0.30, spir -- U 9554) a) 0.9 x 0.4, Sb:; b) 0.7 x 0.6, Sa - b, sep 0.35 -- U 9555) a) (0.7 x 0.7), B core, F halo; b) + c) 0.7 x 0.6, 2 B cores in common halo; sep 0.7 between a) and b) + c) --

09560   U 9560) II Zw 70, VV 324 b; "blue post-eruptive oval compact w jets southwest and northeast" (Zw); P w U 9562 at 4.1, 117 -- U 9561) VV 296 b, Arp 173; main body 0.3 x 0.2, thin tail bent southwestw; P w b) = MCG 2 - 38 - 20 at 0.8, 26, 0.6 x 0.4, S0:, VV 296 a; in Arp's class "galaxies w narrow counter-tails" -- U 9562) II Zw 71, VV 324 a; "blue post-eruptive oval compact w 2 extended obliquely crossed knotty jets" (Zw); see U 9560 -- U 9564) SA(rs)c: (de V) -- U 9565) comp 2.2, 105, 0.4 x 0.1, dif; 3.6, 32, 0.45 x 0.1, dif -- U 9566) SA(r)b: (de V) -- U 9567) P w U 9569 at 5.3, 28 -- U 9568) P w U 9570 at 2.8, 125 -- U 9569) see U 9567 --

09570   U 9570) see U 9568 -- U 9571) S0?; main body 0.4 x 0.4 w spir str, smooth outer disc; P w 14 50.0 + 30 03 = N 5771 at 4.2, 303, (0.8 x 0.7), E, m=14.6 -- U 9574) P w U 9573 at 2.2, 16 -- U 9573) S B(s)b (de V); see U 9573; de V classification is incorrect -- U 9575) comp 2.0, 178, 0.9: x 0.5:, v dif, dwarf -- U 9576) SAB(rs)d (de V), Sc - (Ho); P w U 9579 at 4.5, 123 -- U 9577) P w comp at 2.2, 139, 0.8 x 0.4, E?; sev F galaxies near -- U 9579) S B c? (de V), Sb+ (Ho); see U 9576; 14 51.4 + 03 41 = I 1070 at 3.8, 204, 0.7 x 0.25, m=15.4 --

09580   U 9580 + 9581) v blue spindle and v blue distorted spir, prob bridge; in group incl 14 51.4 + 10 20 at 3.6, 312, 0.30 x 0.20, m=15.7 -- U 9582) 14 51.7 + 05 00 = I 1073 at 2.9, 9, 0.4 x 0.3, early, m=15.3; 14 51.7 + 05 03 = = I 1072 at 5.4, 2, 0.6 x 0.4, early, m=15.1 -- U 9586) 14 51.7 + 52 15 at 2.6 sp, 0.4 x 0.25, m=15.6; Sc: (de V) -- U 9587) comp 4.3, 101, 0.5 x 0.3, double system, contact -- U 9588) 2 B cores in contact, strongly asym outer parts, long v dif plume towards northeast --

09590   U 9590) 14 52.4 + 18 46 at 6.1, 141, 0.8 x 0.4, m=15.7; many S galaxies near -- U 9591) b) = MCG 11 - 18 - 13 at

1.2x 186, 0.55 x 0.15 -- U 9592) U 9585 at 4.4, 274 -- U 9593) P w U 9595 at 4.8, 210; 14 52.8 + 18 19 at 5.4, 82,
0.50 x 0.45, m = 15.6 -- U 9594) 3-armed system, one arm off center of bar; asym bar; 14 52.3 + 24 13 at 5.1, 216,
0.4 x 0.4, compact, m = 15.4 -- U 9595) see U 9593 -- U 9596) P w 14 52.4 + 25 55 at 5.2, 200, 0.9 x 0.15, m = 15.6
-- U 9599) I Zw 98; "sharply delineated neutral elliptical disc compact, w uniform elliptical halo" (Zw); "a B
nucl 20" = 7 kpc in diam is surrounded by a much fainter halo 35" = 12 kpc in diam" (Sargent) --

09600   U 9602) P w 14 53.6 + 12 04 at 0.8, 20, 0.55 x 0.45, S0:, m = 15.3 -- U 9603) VV 275; 2 S cores in dif env; VV 274
0.7 np, 0.6: x 0.5:, S B..., listed by VV as double system -- U 9604) comp 4.9, 238, 0.4 x 0.1; 14 52.8 + 37 37 =
= I 4519 at 7.9, 267, 0.9 x 0.5 -- U 9605) P w comp at 2.5, 214, 0.35 x 0.18 -- U 9608) P w U 9611 at 2.0, 62;
VV 26 is ident as I 1078-79, which is incorrect, see U 9616 -- U 9609) comp 3.2, 130, 0.6 x 0.3 --

09610   U 9610) B core in vF halo, no S0; comp 4.8, 185, 0.6 x 0.4, Sb-c; U 9619 at 5.3 -- U 9611) see U 9608 -- U 9613)
P w a) at 0.9, 312, 0.3 x 0.3; sev S F galaxies in vic -- U 9615) v blue, v patchy spir, S B cond on sf edge;
P w U 9628 at 21., 139 -- U 9616) VV 26 a; VV 26 b is a S F obj attached s, * ?-- U 9618) VV 340, Arp 302; a) =
= VV 340 b 0.6 x 0.6, spir, disturbed; b) = VV 340 a 0.9 x 0.3, slightly curved spindle, sep 0.7 --

09620   U 9620) P w U 9622 at 2.8, 123 -- U 9621) comp 2.6, 348, 0.4: x 0.2:, core 0.15 x 0.12, jet? -- U 9622) see U 9620;
comp 4.7, 168, (0.4 x 0.4), E-S0 -- U 9623) comp 4.5, 103, 0.7 x 0.5 -- U 9627) comp 2.5, 164, 0.4 x 0.2; 4.4,
100, 0.5 x 0.2; spir at 7.1 -- U 9628) see U 9615 --

09630   U 9631) S B(rs)b (de V) -- U 9633) 2 B cores in F asym halo, sep 0.40 -- U 9635) comp 1.0, 336, 0.4: x 0.4: --
U 9636) P w a) at 0.45, 270, 0.2 x 0.2; comp 1.9, 138, 0.45 x 0.12; sev vF galaxies in vic --

09640   U 9640) comp 1.9, 43, 0.5: x 0.18, S0:; 5.0, 69, 0.3 x 0.2 -- U 9642) Arp 136; S0 (de V); main body 0.7 x 0.35,
strongly asym env; in Arp's class "E and E-like galaxies w nearby fragments"; "F streamer off one end of E galaxy"
(Arp); comp 1.7, 175, 0.45 x 0.20; 1.7, 318, 0.3 x 0.2, v dif; sev S galaxies near -- U 9645) SAB(s)b (de V) --
U 9646) P w 14 57.7 + 27 32 at 3.3, 64, 0.9 x 0.7, spir, m = 15.7 -- U 9647) patchy dwarf; comp 1.4, 73, 0.3 x 0.1;
4.0, 126, 0.7 x 0.4, F -- U 9648) P w U 9648 at 3.7, 233 -- U 9649) unusual obj; bar and spir pattern vis on blue
PA map only; nucl reg appears too B for class Sc; S B(rs)b? (de V) --

09650   U 9650) P w U 9668 at 5.6, 139 -- U 9652) comp 0.15 x 0.12 superimp 0.25 s center -- U 9654) 14 58.2 + 11 44 at 6.1,
278, 0.45 x 0.35, asym pec?, m = 15.3 -- U 9655) E 1-2 (de V); 14 58.8 + 01 49 = N 5814 at 4.7, 147, 0.9 x 0.5, spir,
m = 14.7 -- U 9657) U 9665 at 7.8, 109 -- U 9658) P w comp 0.7, 190, 0.35 x 0.22 --

09660   U 9661) MCG minor diam 0.3 is incorrect -- U 9662) asym Sb or late Sa; S F cond at np edge,   disturbing comp? --
U 9663) DDO 198; Ir IV-V -- U 9664) VII Zw 576; "blue S B c w compact spherical core" (Zw); comp 5.7, 75, 0.9 x 0.6,
dif, dwarf: -- U 9665) see U 9657 -- U 9668) see U 9650 -- U 9669) S0?; P w U 9720 at 5.6, 6 --

09670   U 9673) VV 7, Arp 42; SA(s)c (de V); np arm split in two; interacting w comp at 1.4, 320, 0.5 x 0.4; in Arp's class
"galaxies w low surf brightn comps on arms"; "F bifurcated arm to comp, one F arm on comp coiled in same direction
as parent" (Arp) -- U 9675) P w 15 01.1 + 10 48 at 3.7, 157, 0.45 x 0.35, m = 15.3 -- U 9678) E 3 (de V) -- U 9679)
a) at 0.9, 335, 0.25 x 0.12 --

09680   U 9681) P w U 9684 at 2.1 f -- U 9683) comp 3.2, 356, 0.4 x 0.15 -- U 9684) see U 9681 -- U 9686) early, patchy,
pec?; comp 1.5, 39, 0.30 x 0.20 -- U 9688) comp 4.4, 97, 0.45 x 0.10 --

09690   U 9692) SA 0- (de V) -- U 9693) S0 (de V) --

09700   U 9700) E 3 (de V) -- U 9706) E + 0-1 (de V), E (Ho); P w a) at 0.7, 180, N 5846 A in BG, 0.35 x 0.35, E 2-3 (de V),
v = 2291, v_o = 2304, incl in CGCG magn; U 9715 at 10.3, 112 -- U 9708) P w U 9711 at 1.2, 124 --

09710   U 9710) P w I 1087 at 0.55, 301, 0.5 x 0.25, S0-a:; sev F galaxies near -- U 9711) see U 9708 -- U 9712) 15 04.4 +
+ 12 53 at 7.1, 353, 0.8 x 0.8, Sb, m = 14.9 -- U 9714) in triple group w 15 04.6 + 13 03 = N 5852 brightest at 1.0,
135, 0.9 x 0.7, S0, m = 14.7, and 15 04.4 + 13 03 at 1.8, 253, 0.55 x 0.25, spir, m = 15.3 -- U 9715) S B(r)b (de V),
Sb + (Ho); see U 9706 -- U 9717) I Zw 102; contiguous compacts -- U 9719) comp 3.8, 170, 0.4 x 0.3, Irr --

09720   U 9720) see U 9669 -- U 9723) SA 0+ (de V), S0 (Ho) -- U 9724) S B(s)b (de V); P w U 9728 at 2.0, 120 -- U 9726)
S B(s)0 + (de V) -- U 9727) 15 05.4 + 14 49 = I 1094 at 4.9, 20, triple system, total diam 0.55 x 0.30, total m = 15.2
-- U 9728) see U 9724 --

09730   U 9730) comp 5.0, 175, 0.3: x 0.15 -- U 9731) SN 1940 c -- U 9732) P w 15 05.5 + 01 25 at 2.8, 249, 0.45 x 0.40,
m = 14.9 -- U 9735) 15 06.0 + 19 23 at 2.4, 282, 0.5 x 0.3, S0-a, m = 15.6; 15 06.1 + 19 24 = I 1096 at 2.2, 309,
0.9 x 0.9, S0, m = 15.1; comp 4.5, 37, 0.4 x 0.3 --

09740   U 9740) S B(s)0/a ? (de V) -- U 9741) v blue chaotic spir; 15 06.6 + 52 31 at 4.4, 293, 0.45 x 0.20, m = 15.3 -- U 9742)
P w U 9743 at 3.6, 0 -- U 9743) see U 9742 -- U 9745) Sb (de V); 15 07.4 + 52 38 at 5.6 sp, 0.7 x 0.3, S B 0?, m =
= 15.7; comp 3.1, 190, 0.3 x 0.2, F, dif -- U 9748) P w U 9750 at 6.0 n -- U 9749) DDO 199; resolved dwarf; "spheroidal" (vdB) --

09750   U 9750) see U 9748 -- U 9752) surrounded by a v large nr of vF galaxies -- U 9753) SA(rs)bc:? (de V), Sb + (Ho);
SN 1954 c -- U 9754) asym disc; no disturbing obj vis -- U 9755) P w 15 08.8 + 10 39 at 1.4, 359, 0.7 x 0.15, m =
= 15.3 -- U 9756) a) 0.7 x 0.4, S B: a; 0.6 x 0.4, E-S, compl 0.60 --

09760   U 9761) (r) 0.6 x 0.6 ecc in smooth disc; comp 1.5, 331, 0.7 x 0.2, dif; 1.9, 285, 0.3 x 0.1, vF -- U 9763) main
body 0.9 x 0.2 in pos angle 193; dif streamers toward northeast and southwest; appearently one or two stell obj
at tip of streamer in northeast -- U 9764) MCG α = 15 07.9 is incorrect -- U 9765) P w comp at 0.40, 110, 0.12 x 0.12,
compact -- U 9766) comp 3.6, 295, 0.35 x 0.35, w comp 0.15 x 0.12 at 0.55 -- U 9767) B core 0.20 x 0.18, ext asym disc
or halo; brightest in a dense group, about 10 galaxies inside 3: --

09770   U 9770) prob interacting w comp at 1.3, 273, 0.15 x 0.12, compact -- U 9771) 15 10.9 + 41 25 = N 5886 at 4.6 sp, 0.9 x
x 0.4, E-S0, m = 15.1; comp 4.2, 23, 0.8 x 0.2 -- U 9772) comp 0.9, 227, 0.2 x 0.2, F, interaction? -- U 9774) pair
15 12.0 + 42 11 = N 5895 + 5896 at 4.2 and 4.9 nf, 0.8 x 0.2 and 0.20 x 0.15, sep 1.0, vF bridge, total m = 15.5 --
U 9775) P w a) at 0.9, 254, 0.6 x 0.5; comp 5.0, 3, 0.6 x 0.25 -- U 9777) abnormally B prob complex center --
U 9778) P w 15 12.3 + 75 20 = N 5912 at 0.8, 89, (0.9 x 0.8), core 0.20 x 0.20, E, m = 14.6, * superimp -- U 9779)
U 9782 at 3.8, 58; comp 5.0, 239, 0.5 x 0.1 --

09780   U 9782) see U 9779 -- U 9785) P w comp at 1.5, 327, 0.3 x 0.2 -- U 9788) P w comp at 1.3, 298, 0.45 x 0.10 -- U 9789)
SAB(rs)c (de V); U 9790 at 9.8 n --

09790   U 9790) see U 9789 -- U 9792) resolved dwarf; v uncertain diam values -- U 9793) comp 3.5, 50, 0.3 x 0.2; 5.0, 37,
0.3 x 0.12 -- U 9794) strongly asym bar spir; a S red cond n center may be a disturbing comp -- U 9796) dif "spindle"
(dwarf?) w elong "core", prob superimp comp; P w 15 14.2 + 43 20 at 1.5, 107, 0.7 x 0.6, patchy, Irr or multiple
syst; in group of 5 w 15 13.3 + 43 20 brightest, 0.9 x 0.6, I 0?, m = 15.2; sev S galaxies in vic may belong to this
group -- U 9797) S B(r)b (de V); (r) 1.1 x 1.1; 2 thin arms superimp dif arcs or concentric rings; SN 1963 o; P w
U 9805 at 13.2 -- U 9799) brightest in cl or v rich group not incl in CGCG; comp 0.9, 304, 0.25 x 0.18; 15 14.2 +
+ 07 11 at 1.1, 164, 0.50 x 0.45, spir, m = 15.6; 15 14.3 + 07 14 at 2.0, 4, 0.20 x 0.18, m = 15.6 --

09800   U 9800) comp 2.0, 148, 0.4 x 0.08 -- U 9801) SA(s)c: (de V), Sb + (Ho); no nucl bulge; SN 1940 a -- U 9805) SA(s)b:
(de V); B nucl bulge w F v ext halo, similar to N 4594; see U 9797; comp 4.8, 62, 0.45 x 0.4 -- U 9808) comp 2.0,

318, 0.4 x 0.2, dif -- U 9809) P w comp at 1.4, 0, 0.5 x 0.2 --

09810   U 9810) comp 3.2, 297, 0.9 x 0.1, Sc, m approx 16.; sev S galaxies near, members of CGCG cl 15 18.8 + 07 47, medium compact, medium distant -- U 9812) P w 15 16.7 + 09 57 at 1.5, 180, 0.6 x 0.4, S B ..., m=15.5 -- U 9813) P w 15 17.1 + 21 05 = N 5910 = VV 139 at 8.0, 63, triple system, S0? 0.9 x 0.8 w comp 0.18 x 0.15 0.40 s and 0.12 x 0.12 0.35 nf, total m=14.9 -- U 9816) "spindle" w row of B blue patches -- U 9817) not marked on CGCG map; comp 3.0, 153, 0.25 x 0.1; 3 patches in disc may be superimp ** -- U 9818) broad complex bar; thin (r) 1.0 x 0.4, smooth disc outside, no spir arms -- U 9819) P w comp at 3.8, 14, 0.6 x 0.6, S B: c --

09820   U 9820) comp 1.6, 120, 0.7 x 0.4, Irr; sev vF galaxies near -- U 9821) P w 15 19.5 + 08 37 = I 1116 at 3.7, 83, (0.9 x 0.7), E, m=14.9 -- U 9822) brightest in group of about 10 galaxies in CGCG cl 15 18.8 + 07 47, medium compact, medium distant; 15 19.2 + 07 54 at 2.7, 289, 0.15 x 0.15, very compact, m=15.7 -- U 9824) S B(r)bc (de V) -- U 9828) comp 3.5, 268, triple system, total diam 0.4 x 0.2; 4.3, 221, double system, total diam 0.35 x 0.30 -- U 9829) v open spir; thin broken arm w streamer on n side; in triple group w 15 20.3 - 01 11 at 1.6, 249, 0.20 x 0.18, compact, m=15.7, and 15 20.4 - 01 13 at 2.3, 209, 0.4 x 0.4, compact, m=15.6 --

09830   U 9831) SN 1962b; comp 2.4, 115, 0.45 x 0.15, F -- U 9832) P w comp 0.7, 332, 0.25 x 0.20 -- U 9835) P w comp at 2.0, 140, 0.9 x 0.3 -- U 9839) P w b) at 1.1, 115, 0.6 x 0.5, spir --

09840   U 9841) P w U 9847 at 14.0, 151 -- U 9843) VV 227; P w comp at 1.0, 234, 0.5 x 0.05, prob interaction -- U 9847) see U 9841 --

09850   U 9851) Arp 90, I Zw 112; P w U 9852 at 0.5, 230; E2 (de V); in Arp's class "spir galaxies w large high surf brightn comps on arms"; "abs lanes around comp" (Arp); "post-eruptive pair of intercon neutral E and Sb in common matrix, w jets" (Zw) -- U 9852) Arp 90, I Zw 112, see U 9851; E0 (de V, this classification is incorrect) -- U 9854) 15 23.4 + 68 50 at 8.2, 225, 0.5 x 0.3, E:, m=15.5 -- U 9855) blue elong cond superimp nf center; comp 2.6, 302, 0.3 x 0.2; 4.5, 125, 0.9 x 0.1 --

09860   U 9860) P w comp at 4.6, 140, 0.7 x 0.6, Sc -- U 9861) b) 0.12 x 0.12; c) 0.15 x 0.15, sep 0.60, large smooth halo; a) at 0.70, 222 (from b)), 0.12 x 0.12, no halo; a large nr of cl members near -- U 9862) P w 15 26.5 + 43 06 = = N 5934 at 1.1, 328, 0.6 x 0.2, E-S0, m=15.1, interaction; 15 26.4 + 43 11 at 4.7, 354, 0.8 x 0.35, m=15.7 -- U 9863) in group of 4 or 5 incl 15 27.3 + 03 41 at 9.3, 125, (0.9 x 0.8), E-S0, m=14.9 -- U 9864) P w 15 26.8 + + 05 15 at 3.1, 240, 0.5 x 0.4, S0-a, m=15.2 -- U 9865) comp 2.4, 239, 0.55 x 0.12 -- U 9866) SA(r)bc? (de V) -- U 9867) S B(rs)b (de V) -- U 9868) a large nr of S galaxies near --

09870   U 9873) 15 27.9 + 42 50 at 2.4 np, 0.5 x 0.3, B core 0.30 x 0.25, m=15.2 -- U 9877) comp 2.0, 84, 0.4 x 0.12, prob member of CGCG cl 15 28.2 + 42 59, medium compact, distant --

09880   U 9883) * superimp; comp 2.6, 309, 0.6 x 0.4, S B b:; 4.0, 208, 0.25 x 0.22 -- U 9886) comp 0.9, 291, 0.4 x 0.3, E- - S0; many S galaxies near; U 9887 at 5.0, 138 -- U 9887) prob disrupted spir; a large nr of galaxies in vic; pair 15 30.0 + 04 57 at 2.9 np, 0.30 x 0.15 and 0.35 x 0.15, sep 0.50; comp 2.0, 301, 0.35 x 0.25, E:, compact?; see U 9886 -- U 9889) in group incl 15 30.3 + 00 38, 0.4 x 0.3, E-S0, m=15.6, and 15 30.4 + 00 36, 0.6 x 0.6, E, m=15.5; sev smaller galaxies near --

09890   U 9893) I Zw 115; "blue post-eruptive galaxy w sev compact knots on central disc" (Zw); "irr, elongated, F outer env 65" x 20" = 3.3 x 1.0 kpc w brighter inner core 20" = 1 kpc diam" (Sargent) -- U 9894) b) at 0.7, 129, 0.4 x x 0.12; P w 15 31.3 + 05 28 at 1.3, 326, (0.8 x 0.6), S0, m=15.4 -- U 9895) U 9903 + 9904 at 16.4, 47 -- U 9899) complex spir str, bar?; disturbed by comp at 0.6, 172, 0.20 x 0.10, contact; comp 2.1, 182, 0.5 x 0.1, F --

09900   U 9901) comp 3.6, 194, 0.7 x 0.12, spir -- U 9902) U 9903 + 9904 at 3.7 n -- U 9903) VV 244 a, Arp 91; E (de V); P w U 9904 at 0.75 nf; total m=12.7; in Arp's class "spir galaxies w large high surf brightn comps on arms"; "broad pec arm to comp, then abs; F extension from comp" (Arp) -- U 9904) VV 244 b, Arp 91; Sc (de V); see U 9903 -- U 9905) P w comp at 2.5, 295, 0.2 x 0.2, E:, compact? -- U 9906) vB main body 0.7 x 0.45; comp 3.1, 352, 0.5 x 0.2; P w U 9914 at 9.1, 33 --

09910   U 9911) comp 2.7, 248, 0.3 x 0.1 -- U 9912) VV 132 -- U 9913) Arp 220; main bodies 0.25 x 0.18 and 0.25 x 0.15, sep 0.25; blue chaotic env, loop? or curved plume?; in Arp's class "galaxies w adjacent loops"; comp 2.2, 133, 0.45 x x 0.25, E: -- U 9914) see U 9906 -- U 9915) S B(r)b: (de V) -- U 9916) P w 15 32.8 + 24 17 at 3.5, 309, (0.6 x 0.4), core 0.18 x 0.12, compact, m=15.6 -- U 9918) P w U 9920 at 3.7, 176 -- U 9919) comp 2.9, 215, 0.5 x 0.1, w comp 0.15 x 0.12 0.30 nf --

09920   U 9920) see U 9918 -- U 9921) P w a) at 0.6, 332, 0.20 x 0.15; comp 4.8, 51, 0.7 x 0.10 -- U 9922) I Zw 117; "v blue intercon pair of patchy compacts" (Zw); "an elong, elliptical body 33" x 14" = 12 x 5 kpc in size is nearly in contact w a fainter satellite 13" x 10" = 4.7 x 3.6 kpc in size, at n end" (Sargent) -- U 9923) comp 2.8, 173, 0.4 x 0.3, pec?; 15 34.2 + 40 00 = I 4563 at 4.4 nf, 0.8 x 0.4, spir, m=15.1 -- U 9926) SA(r)c (de V) -- U 9927) comp 1.7, 243, 0.5 x 0.1 -- U 9928) I Zw 118 (w 15 34.4 + 43 40); "Pair of red compacts. No 1, I 4562, fuzzy spherical disc... v = 5860 ... No 2, 70" nf No 1, very compact, m=15.3 ... v = 5795" (Zw); U 9930 at 5.6, 75 -- U 9929) 15 33.7 + 56 38 = N 5969 at 6.4 p, 0.5 x 0.3, core 0.25 x 0.20, compact, m=15.4 --

09930   U 9930) see U 9928; U 9933 at 3.0, 66 -- U 9932) comp 1.9, 240, 0.4 x 0.1; 2.7, 225, 0.7 x 0.2; 3.3, 234, core 0.20 x 0.18 + halo; these galaxies form a triple group -- U 9933) see U 9930 -- U 9935) S B(rs)d (de V) -- U 9936) DDO 200; S V (vdB) -- U 9939) P w 15 35.1 - 01 36 at 2.6, 248, 0.6 x 0.2, E-S0, m=15.3 --

09940   U 9940) B knotty central part, short blue vB arm segment n center; brightest in a subgroup in cl 302 -- U 9942) comp 1.0, 212, 0.3 x 0.1 -- U 9943) 15 36.5 + 12 14 = I 1131 at 8.3, 141, 0.7 x 0.45, S0: (de V), m=14.8; comp 2.2, 304, 0.6 x 0.3, spir -- U 9948) Sc? (de V); in triple group w U 9961 at 6.2, 110, and U 9969 at 13.6, 64; 15 35.6 + + 59 33 = N 5976 at 8.4, 172, 0.7 x 0.25, S0:, m=15.7 -- U 9949) in group w 15 37.5 + 14 20 brightest, 0.9 x 0.6, spir, m=14.9 --

09950   U 9950) comp 2.7, 297, 0.35 x 0.12 -- U 9952) P w 15 37.1 + 31 59 at 3.8, 13, 0.35 x 0.35, m=15.6 -- U 9953) comp 2.7, 325, 0.7 x 0.1, Sc, m approx 16.5 -- U 9956) comp 4.9, 286, 0.50 x 0.45, spir -- U 9958) surrounded by a large nr of cl members; 15 37.2 + 21 55 at 4.6, 135, (0.6 x 0.3), E-S0, m=15.6; 15 37.8 + 21 59 at 4.9, 60, double syst, 0.15 x 0.10 and 0.2 x 0.2, sep 0.30, m=15.4 --

09960   U 9961) see U 9948; U 9969 at 6.2, 290; E 3 (de V) -- U 9962) in group w 15 37.5 + 14 20 brightest, 0.9 x 0.6, spir, m=14.9; comp 4.6, 198, 0.45 x 0.35 -- U 9963) P w 15 38.0 + 21 41 at 3.7, 44, 0.8 x 0.8, (R)S B(r)a, m=15.2 -- U 9964) comp 2.9, 163, 0.5 x 0.3; 4.0, 193, 0.30 x 0.15; 4.1, 123, 0.3 x 0.05 -- U 9966) P w comp at 5.3, 97, 0.7 x x 0.4, Sc-Irr -- U 9967) comp 4.0, 82, 0.6 x 0.35, early spir -- U 9968) comp 1.5, 165, 0.3 x 0.2, dif; 2.9, 117, 0.8 x 0.1, dif; 3.7, 233, v dif -- U 9969) SA B(r)b (de V); see U 9948, 9961 --

09970   U 9972) comp 2.0, 203, 0.35 x 0.05 -- U 9973) P w U 9974 at 13.6, 19 -- U 9974) see U 9973 -- U 9978) comp 4.5, 245, 0.45 x 0.10 -- U 9979) DDO 201; Ir IV-V (vdB) --

09980   U 9981) P w comp at 1.6, 340, 0.8 x 0.12 -- U 9982) comp 2.8, 55, 0.3 x 0.2 -- U 9983) comp 3.4, 217, 0.45 x 0.10 -- 09987) S B(rs)d: (de V); P w U 9991 at 14.4, 31 -- U 9988) comp 4.4, 113, 0.9 x 0.7 -- U 9989) main body 0.55 x 0.20 w e dif streamer toward northwest --

09990   U 9990) in group w 15 41.1 + 05 01 brightest at 4.8, 13, (0.9 x 0.6), compact, m=15.1; 15 41.6 + 04 56 at 8.5,

101, (0.9 x 0.8), S B 0/SB a:, m=15.4, may be interacting w comp at 1.7 sf, 0.3 x 0.2 -- U 9991) see U 9987 --
U 9993) acicular, or v open spir; interacting w 15 41.5 + 08 00 at 1.8, 285, 0.45 x 0.35, m=15.5 -- U 9996) P w
15 42.4 + 11 39 at 4.6, 137, 0.45 x 0.40, Sa?, m=15.2 -- U 9997) 2nd brightest in chain of galaxies in cl 362; P
w 15 42.1 + 43 56 at 1.5, 186, 0.8 x 0.6, m=15.3, comp 0.20 x 0.12 at 0.8 s; comp 2.7, 323, 0.45 x 0.05; 3.6, 328,
0.35 x 0.10 -- U 9999) comp 1.9, 208, 0.30 x 0.25 --

10000   U 10001) only one arm vis, long, thin, w low surf brightn -- U 10002) P w comp at 0.7, 330, 0.35 x 0.12, contact --
U 10003) P w U 10007 at 2.5, 29 -- U 10007) see U 10003 -- U 10008) P w 15 41 + 82 32 at 4.5, 296, 0.8 x 0.8, Sc,
m=15.7 --

10010   U 10012) D galaxy, in CGCG cl 15 42.1 + 36 20, compact, very distant -- U 10019) P w comp at 0.9, 332, 0.4 x 0.2;
15 43.8 + 30 21 at 4.6, 43, 0.6 x 0.5, spir, m=15.6 --

10020   U 10020) * superimp near center -- U 10021) comp 3.6, 236, 0.3 x 0.2 -- U 10022) P w comp at 4.4, 50, 0.9 x 0.15
U 10024) P w 15 43.7 + 02 36 at 2.3, 342, 0.9 x 0.6, v late spir, m=15.7; 15 43.2 + 02 34 at 7.7, 268, 0.8 x 0.2,
spir, m=15.0 -- U 10027) comp 2.9, 35, 0.35 x 0.15, spir -- U 10028) early spir?; comp 1.6, 52, 0.5 x 0.3 --
U 10029) comp 0.9, 121, 0.4 x 0.05, vF; 15 44.8 + 06 04 at 5.5, 72, (0.5 x 0.3), E:, m=15.2 --

10030   U 10030) comp 3.9, 36, 0.7 x 0.1 -- U 10033) VV 16, Arp 72; N 5996 = VV 16 a distorted late S B, main body 1.7 x
x 0.8, F ring segment f; N 5994 = VV 16 b 0.35 x 0.15, sep 1.5; in Arp's class "spir galaxies w S high surf brightn
comps on arms"; "F material from arm to and around comp. Opposite arm F, sweeps around East of galaxy" (Arp) --
U 10036) comp 1.1, 233, 0.3 x 0.1 -- U 10037) comp 1.1, 132, 0.3 x 0.2; 4.3, 48, 0.35 x 0.30, spir -- U 10038) comp
4.4, 359, double system, total diam [0.35 x 0.30] --

10040   U 10040) comp 0.3 f, 0.2 x 0.1, disturbing comp? -- U 10041) comp 1.5, 321, 0.5 x 0.1, spir, m approx 16.5 --
U 10043) comp 2.7, 110, 0.9: x 0.9:, dwarf; 4.0, 94, 0.55 x 0.10 -- U 10046) brightest in group; 15 47.5 + 00 20 at
8.1, 103, 0.9 x 0.4, S0, m=15.4 -- U 10049) + b) total diam 0.5 x 0.5, close double system; c) + d) 0.8 x 0.4,
c) 0.12 x 0.12 superimp d); sep 0.6 between the pairs --

10050   U 10050) 15 47.0 + 18 43 at 5.6, 290, 0.6 x 0.4, S0:, m=15.5 -- U 10052) a) 0.7 x 0.6, spir; b) 0.5 x 0.5, spir,
0.45 -- U 10053) VV 291a, Arp 109, VII Zw 623; hard core 0.15 x 0.15 surrounded by thin B spir arc, 2 dif spir
arcs w unequal bending in northwest; "irr blue spir, compact elliptical core, two spir arms to the west" (Zw); in
group w 15 48.4 + 69 33 = I 1146 brightest at 5.7, 153, 0.9 x 0.8, S0:, m=14.7; prob disturbed by 15 48.1 + 69 37 =
= VV 291b at 1.6, 115, (0.6 x 0.4), E:, m=15.1; in Arp's class "E and E-like galaxies repelling spir arms" --
U 10054) DDO 203; "spir + F outer loop" (CGCG); S B IV-V (vdB); may be disturbed by comp at 4.1, 308, 0.7 x 0.6,
S0?; F loop on comp side -- U 10055) comp 1.0, 257, 0.3 x 0.05 -- U 10056) comp 3.1, 235, 0.35 x 0.25, m approx 16.,
compact? or *?; 4.0, 290, 0.3 x 0.2; sev F galaxies near, members of CGCG cl 15 48.0 + 19 14, compact, distant --
U 10057) 15 48.0 + 68 21 at 1.1, 225, 0.6 x 0.15, m=15.2; comp 1.7, 59, 0.5 x 0.4; 2.8, 126, 0.6 x 0.1, dif --

10060   U 10060) comp 1.5, 259, 0.3 x 0.2; 1.9, 87, 0.4 x 0.3 -- U 10061) Ir V (vdB) -- U 10062) comp 3.9, 224,
0.8 :: x 0.5 ::, e dif, dwarf -- U 10063) P w U 10064 at 1.7, 140; sev galaxies near -- U 10064) see U 10063 --
U 10066) comp 0.45, 265, 0.3: x 0.3:, v dif, dwarf -- U 10067) comp 1.0, 90, 0.3 x 0.1 --

10070   U 10072) in group incl 15 49.3 + 71 24 at 2.4, 259, 0.9 x 0.8, m=15.4, w comp 0.15 x 0.12 at 0.45, and 15 49.7 +
+ 71 20 at 4.5, 188, 0.7 x 0.45, Sb-c, m=15.3 -- U 10075) SA(s)cd (de V), Sc + (Ho) -- U 10076) e blue stell obj
superimp 0.35 np center; one arm appears disturbed; P w 15 50.9 + 21 15 at 3.3, 119, 0.6 x 0.3, E, m=15.4 --
U 10078) surrounded by many S galaxies; comp 1.1, 115, 0.3 x 0.2, incl in CGCG magn; 15 50.3 + 67 20 at 3.1, 279,
0.4 x 0.2, m=15.3 -- U 10079) comp 3.2, 152, 0.40 x 0.25; 15 50.6 + 12 10 = N 6006 at 5.8, 259, 0.7 x 0.4, m=15.3;
15 51.1 + 12 13 = N 6009 at 6.0, 2, 0.6 x 0.25, spir, m=15.4 --

10080   U 10082) S B(s) w B bar; vF s comp at tip of bridge or spir arm at 0.7 nf; comp 3.0, 162, 0.6 x 0.2 -- U 10084)
15 51.8 + 18 47 at 4.3, 267, 0.9 x 0.7, spir, m=15.1; the "streamer" may be an attached S dwarf -- U 10085) comp
2.0, 137, 0.4 x 0.3, late spir or Irr -- U 10086) may be interacting w comp at 0.8, 237, 0.18 x 0.12, dif bridge? --
U 10087) a) at 0.9, 262, 0.4 x 0.15 -- U 10089) v blue spindle; comp 3.0, 1, 0.5 x 0.2, v dif, dwarf:; 3.9, 180,
0.4 x 0.1 --

10090   U 10090) comp 1.2, 262, 0.3 x 0.2 -- U 10091) prob late S0 -- U 10092) comp 4.3, 290, 0.6 x 0.4, S B(s)b, m approx
16. -- U 10094) P w 15 54.0 + 24 35 at 5.1, 124, (0.6 x 0.4), E, m=15.4; 3 F galaxies inside 3.0 -- U 10095)
15 53.5 + 45 38 at 4.5 np, (0.9 x 0.9), E, m=15.1 -- U 10099) I Zw 129; "red elliptical disc-like galaxy, F halo"
(Zw); "structureless image 10" = 6.8 kpc diam" (Sargent); P w 15 54.8 + 42 02 at 1.3, 330, 0.7 x 0.4, m=15.0 --

10100   U 10101) prob late S0; comp 2.6, 89, 0.3 x 0.3, compact; P w U 10102 at 5.0, 3 -- U 10102) see U 10101 -- U 10103)
U 10107 at 5.4, 36; 15 54.6 + 48 14 at 6.0, 270, 0.4 x 0.25, m=14.9 -- U 10106) 15 55.5 + 16 25 = N 6022 at 1.7,
197, 0.7 x 0.45, (R')S B(s)b, m=15.2; 15 55.7 + 16 29 at 3.9, 55, 0.8 x 0.8, Sb, m=15.3; sev smaller galaxies
near -- U 10107) 15 55.7 + 48 19 at 1.5, 47, 0.55 x 0.55, S B a, m=15.1; see U 10103 -- U 10108) comp 1.5, 168,
0.4 x 0.1, spir --

10110   U 10110) comp 1.1, 304, 0.4 x 0.1 -- U 10111) B core 0.15 x 0.15, dif disc -- U 10115) comp 4.5, 68, 0.3 x 0.2;
15 56.4 + 63 59 at 4.7, 188, double system, total diam 0.35 x 0.25, m=15.3; 15 57.2 + 64 09 at 7.1, 43, 0.5 x 0.5,
spir, m=15.5 -- U 10116) VV 115, VII Zw 631; N 6027 0.35 x 0.18; a) 0.6 x 0.42; b) 0.3 x 0.15; c) 0.8 x 0.15;
d) 0.18 x 0.18; N 6027 + e) 1.0 x 0.3; (designations from Seyfert, see P.A.S.P. 63, 72); e) is not certainly a separate galaxy; Zw magn for N 6027 14.7, "bar-like compact"; a) 15.1, "fuzzy E"; b) 15.3, "patchy compact"; c) 15.6,
"spherical compact"; d) 16.0, "S B c"; e) 16.5, "Irr"; v = 4468, 4141, 4430, 4581 and 19930 (see O'Connel (Ed.),
"Nuclei of Galaxies", Pontificiae Academiae Scientiarum 1971), note discordant redshift for d) -- U 10118) comp
3.5, 220, 0.5 x 0.2, dif --

10120   U 10124) P w comp at 2.5, 130, 0.7 x 0.2 -- U 10126) S bc: (de V); P w 15 57.3 + 79 07 at 2.0, 250, 0.9 x 0.15, m=
= 14.7 -- U 10128) a) 0.5 x 0.4, Mark 494; b) 0.7 x 0.6; sep 0.45 -- U 10129) I Zw 132; "neutral to blue post-eruptive
galaxy, compact pear-shaped core, spiral jet, extended halo" (Zw) --

10130   U 10132) P w comp at 0.7, 235, 0.15 x 0.10, interaction; MCG δ = 78 47 is incorrect -- U 10134) comp 4.3, 88) 0.7 x
x 0.15, spir; pair 15 58.6 + 16 28 at 6.9, 290, a) 0.45 x 0.35, b) 0.7 x 0.45, S B(s)b, sep 0.7, total m=15.0 --
U 10135) I Zw 133; "red spherical compact w huge ring-like halo" (Zw); 15 58.8 + 19 35 at 7.6, 317, 0.9 x 0.8, E
or S0, m=15.0 -- U 10139) comp 2.3, 260, 0.30 x 0.15; 2.3, 358, 0.8 x 0.12, early spir --

10140   U 10142) comp 2.5, 73, 0.25 x 0.25, core 0.15 x 0.15; 15 58.7 + 70 33 at 6.7, 309, 0.9 x 0.8, spir, m=15.6 --
U 10143) VV 159 a, Arp 324; brightest in chain incl 15 59.9 + 16 04 = VV 159 b at 2.7, 202, 0.6 x 0.5, m=15.3;
15 59.9 + 16 02 = VV 159 c at 4.3, 195, (0.8 x 0.6), m=15.5, E-S0; VV 159 f at 5.6, 200, 0.50 x 0.12; "dif elonga-
tion of E's along line joining them" (Arp) -- U 10144) P w 16 00.0 + 16 30 at 1.4, 340, 0.6 x 0.5, E-S0, m=15.7;
sev galaxies in vic -- U 10145) comp 0.10 x 0.10 superimp at 0.15, 330 -- U 10147) M 51 case; a) 0.9 x 0.7, Sc; b)
0.4 x 0.3, v blue, at tip of vF distorted arm from a), sep 0.6 -- U 10148) P w U 10154 at 6.6, 127; comp 1.7, 357,
0.4 x 0.2, dif --

10150   U 10150) comp 1.5, 126, 0.3 x 0.1; 3.6, 280, 0.3 x 0.2 -- U 10152) a) 0.55 x 0.45, distorted spir; b) 0.35 x 0.20,
high surf brightn, sep 0.45 -- U 10154) see U 10148 -- U 10155) comp 1.0, 179, 0.4 x 0.15, dif -- U 10156) a) 0.7 x
x 0.3, late spir; b) 0.9: x 0.6:, S B: ..., sep 0.50, thin bridge or distorted arm from a) to b) --

10160 U 10160) CGCG δ = 25 13 is incorrect; comp 4.9, 220, 0.3 x 0.2 -- U 10162) 16 01.5 + 71 38 at 5.9, 345, 0.6 x 0.30,
m = 15.1 -- U 10163) P w 16 02.0 + 03 56 = N 6037 at 3.2, 183, 0.7 x 0.6, S B a, m = 15.2 -- U 10164) VV 318 a, Arp 101;
P w U 10169 at 2.4, 22 -- U 10165) VV 212, Arp 122; a) 0.8 x 0.7, S0; b) 1.6: x 0.6, S B ..., contact, sep 0.50
between centers; pair U 10170 at 2.7 sf; MCG and Arp ident N 6039 is prob incorrect; in Arp's class "E and E-like
galaxies close to and perturbing spirals" -- U 10166) comp 0.40, 313, 0.2 x 0.1; 4.3, 20, 0.9 x 0.1 -- U 10168)
comp 3.9, 212, 0.3 x 0.15; 4.1, 237, 0.5 x 0.10 -- U 10169) VV 318 b, Arp 101; main body 0.55 x 0.30, long F tail
toward northeast; see U 10164 --

10170 U 10170) VV 213; b) (1.2 x 1.1); a) 0.20 x 0.15 superimposed 0.30 sp center of b); see U 10165; in a dense part of
Hercules cl -- U 10173) 16 01.1 + 68 24 at 8.4, 301, 0.15 x 0.12, extremely compact, m = 15.5 -- U 10174) comp 2.5,
216, 0.3 x 0.2; 3.2, 325, 0.3 x 0.2 -- U 10176) comp 4.0, 279, 0.3 x 0.2; 5.1, 20, 0.8 x 0.2 -- U 10177) Sb (de V);
16 02.8 + 17 52 = N 6047 at 1.7, 173, (0.8 x 0.8), E0 (de V), v = 9470, v₀ = 9594, m = 15.4; in a dense part of
Hercules cl -- U 10178) IC note to I 4588 shall be "6051 p", not "5051 p"; I 4588 is a comp at 2.0, 120, 0.4 x 0.3;
about 15 galaxies inside 3.0; brightest comps at 2.5, 17, 0.6 x 0.15, and 2.9, 332, 0.2 x 0.2, E? -- U 10179)
vB core 0.15 x 0.15, streamer in northwest w 2 vS conds at 0.6; F streamer in southwest tp 0.3; triple system?; comp
2.2, 254, 0.4 x 0.3, S B(s)b --

10180 U 10182) VV 86, Arp 209; see P.A.S.P. 81, 637; "chaotic w loops" (Arp) -- U 10186) VV 220; a) = VV 220 b 0.5 x 0.4,
S B c; b) = VV 220 a 0.8 x 0.5, Sb - c; in a dense part of Hercules cl -- U 10187) a) 0.7 x 0.6; b) (0.7 x 0.6), both
E or early spirs , sep 0.8; a v large nr of vS vF objs near, members of CGCG cl 16 03.0 + 16 39, compact, distant
-- U 10188 + 10189) VV 194, Arp 172; I 1181 A + B in BG; U 10188 1.2: x 0.8:, U 10189 1.1: x 0.8:, sep 0.50, prob a
broad vF bridge; in Arp's class "galaxies w dif counter-tails" --

10190 U 10191) S B 0 (de V); P w 16 03.4 + 18 18 = N 6057 at 1.7, 82, (0.5 x 0.5), B core 0.18 x 0.18, compact, m = 15.7 --
U 10192) Pec: (de V); long jet eastw; 2 or 3 S v blue conds attached f; st£ll obj 0.55 f center is prob a superim
*; hints of short counter-jet; in a dense part of Hercules cl; 16 03.4 + 17 55 = I 1183 at 2.0, 171, 0.7 x 0.3, E1
(de V), v = 10038, v₀ = 10161, m = 15.6, CGCG ident I 1184 is prob incorrect; 16 03.5 + 17 51 = I 1185 at 5.4, 160,
0.7 x 0.5, Sa (de V), v = 10452, v₀ = 10575, m = 15.1 -- U 10193) Sa? 1.0: x 0.2 w comp attached sf; sev galaxies in
vic, brightest at 1.9, 97, 0.7 x 0.15, Sa? -- 10195) 16 03.8 + 18 20 at 2.4, 130, 0.9 x 0.4, Sc, m = 15.7; 16 03.3 +
+ 18 25 at 4.9, 311, 0.7 x 0.5, spir, m = 15.6 -- U 10197) VV 327 a; U 10198 at 1.1, 340; VV 327 c at 0.8, 214, 0.2 x
x 0.2, dif -- U 10198) VV 327 b; see U 10197; comp 2.8, 343, 0.4 x 0.2 -- U 10199) 2 vS red conds in env on opposite
sides of center; sev B galaxies near --

10200 U 10200) P w 16 04.0 + 41 27 at 1.8, 172, 0.8 x 0.4, pec, "F plume" (CGCG), m = 15.0, interaction?; comp 1.1, 138,
0.15 x 0.08, eF disrupted env?; 2.6, 343, 0.4: x 0.2:, v dif -- U 10201) VV 215; a) = VV 215 a (0.8 x 0.7), E:; b) =
= VV 215 b (0.6 x 0.4), E:, sep 0.50; sev galaxies near; VV 214 at 1.9, 248 from a), double system, total diam
0.25 x 0.10, m approx 19. -- U 10202) P w comp at 1.3, 227, 0.45 x 0.40, Sc: -- U 10204) asym; v broad dif streamer
toward northeast -- U 10205) similar to N 4594; v large nucl bulge, F halo -- U 10206) P w U 10221 at 7.7, 126 --
U 10207) P w U 10208 at 1.9, 152 -- U 10208) see U 10207 --

10210 U 10213) P w 16 04.9 + 10 29 at 3.7, 193, 0.8 x 0.2, Sb, m = 15.6 -- U 10214) VV 29, Arp 188; S B(s)bc: w v long dif
streamer toward northeast; no disturbing obj vic; "disturbance inside western arm, filament may originate there"
(Arp); 16 04.6 + 55 30 at 4.3, 212, (0.6 x 0.6), core 0.12 x 0.12, compact, m = 15.5 --

10220 U 10221) see U 10206; comp 4.9, 338, 0.30 x 0.12, common w U 10206 -- U 10222) VII Zw 636; a) 0.20 x 0.18; b) 0.15 x
x 0.15, sep 0.20; "pair of equal blue elliptical compacts, sep = 12" east - west, curved plume north to west, each
m_p = 14.0 (poss **)" (Zw) --

10230 U 10230) 16 07.6 + 00 53 at 4.3, 36, distorted pair, total diam 0.7 x 0.4, m = 15.6, in background group of 6 galaxies;
SA(s)cd (de V) -- U 10233) P w U 10236 at 3.8, 56; comp 2.3, 160, 0.3 x 0.15 -- U 10236) see U 10233 --

10240 U 10240) comp 3.0, 75, 0.35 x 0.30 -- U 10245) comp 2.5, 106, 0.35 x 0.10; sev F galaxies near -- U 10246) P w comp
at 1.6, 259, 0.35 x 0.10 -- U 10249) 16 08.6 + 13 53 at 8.6, 216, 0.9 x 0.6, S0, m = 14.8; comp 4.7, 346, 0.8: x 0.4,
S0? --

10250 U 10253) P w U 10254 at 3.1, 4 -- U 10254) see U 10253 -- U 10255) (r) 0.35 x 0.35; 16 09.4 + 36 05 at 4.2, 142,
0.7 x 0.4, "compact, jet" (CGCG), m = 15.3, comp 0.12 x 0.08 at 0.40 f -- U 10256) comp 1.5, 328, 0.3 x 0.2, dif;
3.2, 306, 0.5 x 0.4 -- U 10259) a) 0.5 x 0.2, early; b) 0.15 x 0.15, stell, 0.22 nf a); c) + d) [0.8 x 0.6], 2 B
cores sep 0.20, plumes, 0.7 nnp a) + b); e) 0.15 x 0.15, alm stell, 1.3 nnp a) + b) --

10260 U 10261) U 10267 at 5.6, 85; 16 10.1 + 52 35 at 2.9, 85, 0.35 x 0.25, m = 15.5; a large nr of F galaxies near, prob
members of CGCG cl 16 09.6 + 52 19, medium compact, distant -- U 10262) P w comp at 0.80, 230, 0.2 x 0.2, incl in
CGCG magn; comp 2.1, 220, 0.4 x 0.3 -- U 10263) MCG 14 - 07 - 35 and 14 - 08 - 01 are identical -- U 10265) comp 3.0,
153, 0.4 x 0.2 -- U 10267) I Zw 135; "post-eruptive galaxy w one B red and two fainter blue compact cores in contact, enormous curved wings" (Zw); "B nucl 15" = 8.6 kpc diam composed of three B conds in contact. Two F, curved
plumes emerge from the nucl; they are detectable out to 80" = 46 kpc" (Sargent); see U 10261; 16 10.1 + 52 35 at
2.6, 275, 0.35 x 0.25, m = 15.5; U 10267 is similar to Arp 244 = N 4038 - 39 -- U 10268) P w b) at 0.40, 48, 0.20 x 0.20,
compact -- U 10269) P w U 10270 at 6.7, 0 --

10270 U 10270) see U 10269 -- U 10272) comp 4.1, 122, 0.5 x 0.2 -- U 10273) strongly asym "spindle"; a) at 0.3 from s
border, 0.2 x 0.1; c) 0.3 x 0.15 attached nf; 16 10.5 + 28 27 at 3. np, 0.3 x 0.3, E: + stell comp 0.10 x 0.10 0.2
np, total m = 15.6 -- U 10277) comp 4.7, 125, 0.5 x 0.3 -- U 10279) a) 0.3 x 0.15; b) 0.8 x 0.4, sep 0.3 --

10280 U 10280) broad curved bar, B core 0.15 x 0.15; comp 2.2, 203, 0.45 x 0.25; sev vF galaxies inside 5.0 -- U 10283)
brightest in subgroup in cl 386; comp 1.0, 135, 0.20 x 0.15, dif thin bridge; sev galaxies in vic, U 10285 at 4.7,
15 -- U 10284) P w comp at 0.45, 144, 0.18 x 0.12 -- U 10287) subgroup in cl 386; E or S0 0.9 x 0.7 w comp
(MCG 5 - 38 - 40) at 0.7 p (not incl in CGCG magn); triple system 16 11.0 + 31 01 (MCG 05 - 38 - 39) at 1.8 p, total
diam 0.9 x 0.6, m = 15.6; MCG magn for 39 and 40 should be interchanged -- U 10289) comp 4.2, 136, 0.6 x 0.6, S B c —

10290 U 10293) P w 16 12.8 + 11 24 at 1.4, 208, 0.6 x 0.4, m = 15.7 -- U 10298) 2 S blue comps superimp, or **?; comp
3.1, 112, 0.3 x 0.3, dif -- U 10299) VV 192; both (0.8 x 0.5), sep 0.60; VV 192 a is sf --

10300 U 10301) P w U 10303 at 6.5, 36 -- U 10303) see U 10301; comp 3.2, 1, 0.3 x 0.3 -- U 10309) massive B ring w bar;
16 15.0 + 35 50 at 3.9, 96, 0.7 x 0.6, asym, m = 15.4 --

10310 U 10310) Arp 2, DDO 204; "Low surf brightn dwarf. Large B knot in arm appears almost stellar" (Arp); IV- V (vdB) --
U 10311) 16 15.2 + 35 00 = N 6105 at 2.6, 238, 0.55 x 0.5, m = 15.3; sev galaxies near, U 10316 at 7.5 nf -- U 10314)
I Zw 139; "neutral E galaxy, compact core" (Zw) -- U 10315) bar w high surf brightn, 2 thin arms, blue knot at
tip of s arm; P w U 10326 at 3.4, 150 -- U 10318) 16 15.0 + 64 21 = I 1212 at 9.0, 327, (0.8 x 0.7), core 0.22 x 0.22,
m = 15.2 -- U 10319) 16 15.9 + 12 50 at 4.7 s, 0.4 x 0.4, E?, m = 15.4 --

10320 U 10320) comp 4.6, 39, 0.4 x 0.2 -- U 10321) VV 129; a) = VV 129 a + b 0.8 x 0.5, distorted S B: c w comp 0.2 x 0.1
superimp nf center; b) = VV 129 c 0.2 x 0.1 at 0.6 n a); c) = VV 129 d 0.4 x 0.1 at 1.0 n a); d) = VV 129 e 0.6 x
x 0.5 at 1.1 nf a), Sc: -- U 10322) P w U 10327 at 6.4, 127; comp 4.4, 242, 0.4 x 0.2 -- U 10325) a) 0.55 x 0.25;
b) 0.22 x 0.20 at 1.2, 120 from a) -- U 10326) see U 10315; 16 16.6 + 68 20 = I 1218 at 9.7 sf, 0.9 x 0.2, m = 14.6 --
U 10327) see U 10322 -- U 10328) SA(s)c (de V); P w U 10337 at 13.0, 128 -- U 10329) U 10329) comp 4.9, 5, 0.5 x 0.1—

10330   U 10330) comp 0.9, 160, 0.25 x 0.1, interaction? -- U 10331) 16 16.5 + 59 27 at 1.0, 112, 0.55 x 0.15, m=15.7, interaction; comp 4.0, 359, 0.6 x 0.1 -- U 10332) a) 0.5: x 0.5:, S B core; b) 0.3 x 0.3, B core, long curved streamer
southwards, sep 0.55; comp 3.4 nf, 0.6 x 0.4, S B ... -- U 10334) DDO 205; IV- V: (vdB) -- U 10337) see U 10328 --
U 10338) P w comp at 1.7, 157, 0.5 x 0.3; 2.5, 18, 0.3: x 0.3:, dif --

10340   U 10343) I Zw 141; "post-eruptive pear-shaped patchy blue compact, extended F halo" (Zw); "hard image 26" x 20" =
= 15.4 x 11.9 kpc" (Sargent); 16 17.9 + 37 56 = N 6119 at 2.3, 328, 0.9 x 0.7, Sb, m=15.4; comp 3.3, 67, 0.3 x 0.3,
double, F; 4.5, 73, 0.9: x 0.2 -- U 10345) I Zw 142; "Pair. No 1, N 6127 = 6128, neutral spherical galaxy, compact
core, extended halo ... No 2, red spherical compact, 100" p No 1, m_p = 17.6" (Zw) -- U 10347) comp 4.7, 155, 0.7 x
x 0.7 -- U 10348) P w comp at 0.9, 27, 0.4 x 0.2, SB(s)b -- U 10349) SN 1953 b; comp 0.8, 282, 0.3 x 0.2 --

10350   10353) I Zw 144; "red elliptical disc compact" (Zw) -- U 10354) comp 1.0, 185, 0.3 x 0.2; 4.2, 158, 0.4 x 0.4,
pec, streamers?; 4.2, 197, 0.4 x 0.3, Irr; sev galaxies in vic -- U 10355) comp 3.3, 10, 0.4 x 0.4, dif -- U 10356)
one long blue arm in southeast, e blue stell obj 0.10 x 0.10 at tip, invis in red (non-existing?); chaotic arm fragment in northwest, 2 vS F knots in outer part; comp 1.9, 126, 0.45 x 0.25, disturbing; sev S galaxies near --
U 10358) SAB(rs)bc: (de V); comp 2.5, 150, 0.25 x 0.20, high surf brightn; 4.8, 121, 0.6 x 0.12 -- U 10359) SA(s)c
(de V) --

10360   U 10361) I Zw 146; "Pair. No 1, neutral elliptical compact core in uniform elliptical halo ... No 2, neutral spherical compact, 160" np No 1, m_p = 16.9" (Zw) -- U 10362) comp 2.6, 40, 0.4 x 0.25 -- U 10364) P w 16 21.2 + 38 04
at 1.7, 338, 0.7 x 0.6, E- S0, m=15.5; comp 3.9, 340, 0.30 x 0.15; sev galaxies in vic -- U 10366) P w 16 21.5 +
+ 37 19 at 2.7, 196, 0.55 x 0.50, spir, m=15.6 -- U 10368) comp 3.3, 196, 0.8 x 0.15; 4.4, 260, 0.35 x 0.12 --

10370   U 10371) P w 16 22.4 + 19 36 at 2.6, 58, 0.8 x 0.35, spir:, m=15.4 -- U 10374) (r) 0.45 x 0.40; comp 1.9, 10, 0.3 x
x 0.1; 3.6, 240, 0.5 x 0.15; 4.8, 317, 0.4 x 0.1 -- U 10375) P w 16 22.4 + 09 41 at 1.9, 187, 0.6 x 0.5, E?, m =
= 15.3 -- U 10376) U 10359 at 12.1 -- U 10379) comp 2.3, 217, 0.20 x 0.20; 2.3, 338, 0.45 x 0.40; 16 23.4 + 41 04 =
= N 6145 at 3.6, 335, 0.9 x 0.35, m=15.1; I Zw 148 at 9.7, 323 (16 23.0 + 41 08), 0.35 x 0.35, short jet in northwest, "blue post-eruptive galaxy w compact knots, m_p = 16.0 ... v = 8080" (Zw), Sargent's ident MCG 7 - 34 - 20 is
incorrect --

10380   U 10380) comp 3.9, 306, 0.6 x 0.3; 16 23.4 + 16 35 at 7.9, 203, 0.7 x 0.7, S0:, m=14.9 --

10390   U 10392) comp 4.5, 90, 0.8 x 0.25, Irr -- U 10396) Arp 66; blue cond 0.12 x 0.10 at 0.35 n center, prob at tip of
spir arm; in Arp's class "spir galaxies w S high surf brightn comps on arms"; comp 3.3, 314, 0.9: x 0.8:, dif;
4.1, 28, 0.8 x 0.08 -- U 10397) 16 26.0 + 42 45 at 3.4, 138, 0.3 x 0.2, m=15.5; comp 1.7, 307, 0.3 x 0.25 --
U 10399) 3 stell objs superimp a blue chaotic material, prob group of compacts w outflow of matter --

10400   U 10400) comp 1.8, 70, 0.2 x 0.2, v red, compact?; 3.2, 55, 0.45 x 0.20; 3.2, 185, 0.3 x 0.3 -- U 10403) in triple
group w 16 26.4 + 32 55 = N 6161 at 2.3, 190, 0.8 x 0.25, spir, m=15.6, comp 0.4 x 0.15 0.8 np; and 16 26.6 + 32 58
= N 6163 at 1.1, 100, 0.8 x 0.45, S B 0, m=15.4 -- U 10409) E pec (de V); comp superimp nf center; a large nr of
S galaxies near --

10410   U 10411) P w 16 26.7 + 62 46 at 3.8, 241, 0.9 x 0.5, SB(s)b, m=15.5 -- U 10414) P w U 10416 at 8.2, 96 -- U 10416)
see U 10414 -- U 10418) VII Zw 649; "blue elliptical patchy compact" (Zw) --

10420   U 10420) comp 2.2, 240, 0.30 x 0.15; 3.0, 191, 0.35 x 0.35; sev galaxies in vic -- U 10421) comp 3.7, 8, double syst,
total diam 0.7 x 0.4; 4.0, 104, 0.6 x 0.3, spir; sev galaxies in vic -- U 10422) a) 1.2 x 0.5, S0?; b) (0.6 x 0.6),
superimp sf part of a) -- U 10423) in contact w comp at 0.7, 240, 0.3 x 0.1, F -- U 10424) comp 0.7, 200, 0.30 x
x 0.15; 16 28.4 + 55 35 at 2.1, 214, 0.45 x 0.15, m=15.7 -- U 10428) 16 28.9 + 35 13 = N 6179 at 3.3, 31, 0.15 x 0.12,
extremely compact, m=15.7 --

10430   U 10430) comp 2.9, 157, 0.3 x 0.1 -- U 10433) * superimp; prob early spir; comp 1.5, 207, (0.8 x 0.6), E- S0; 3.6,
103, 0.7 x 0.5, Sc: -- U 10434) comp 1.8, 333, 0.4: x 0.2:, dif, vF -- U 10439) SAB(rs)c (de V); SN 1926 b; thin
knotty arms w high surf brightn; strong abs along arm on f side; comp 3.9, 108, 0.5: x 0.3:, dwarf --

10440   U 10441) comp 2.0, 320, 0.4 x 0.4, Sc - Irr -- U 10443) (r) 0.40 x 0.40; comp 4.0, 298, 0.4 x 0.3 -- U 10445) blue
disc w amorph arm fragments, sev blue patches; B blue prob complex nucl reg --

10450   U 10453) dif comp 2.2, 308, 0.4: x 0.3:, dwarf -- U 10456) hints of e dif outer ring; comp 3.9, 153, 0.4 x 0.08 --
U 10458) slightly asym, spir pattern in inner part only; a large nr of vF galaxies near; U 10461 at 11.9 --

10460   U 10461) Arp 73; S-shaped, 2 arms continue a curved bar; comp 2.9, 118, 0.2 x 0.2; "arm leads toward, but not up
to comp" (Arp); U 10458 at 11.9 -- U 10463) comp 3.9, 308, 0.3 x 0.15, dif -- U 10464) a) 0.6 x 0.4; b) 0.6 x 0.4,
one spir arm vis, unilateral?, prob disturbed; sep 0.55; hints of eF plume northw from a) -- U 10466) complex spir
pattern, asym arms, one arm appears distorted; no disturbing obj vis --

10470   U 10470) Arp 185; (R)SB(rs)bc (de V); in Arp's class "galaxies w narrow filaments"; "condensed nucl, F outer arms
less curved than inner arms" (Arp); inner reg 2.2 x 1.7 w high surf brightn; e low surf brightn in outer arms --
U 10471) dif disc, no spir str clearly vis; * or S comp superim near center; may be interacting w comp at 1.9,
342, 0.8 x 0.2; "double system, F bridge" (CGCG; the meaning of this note is not obvious, no bridge vis on PA maps)
-- U 10473) 16 35.2 + 36 33 at 2.1 nf, 0.4 x 0.3, eF outer arms?, m=15.5 -- U 10475) P w 16 35.6 + 07 43 at 3.6,
26, 0.7 x 0.4, m=15.6 -- U 10476) P w U 10478 at 6.5, 169; comp 3.7, 231, 0.4 x 0.1, v dif; sev smaller comps --
U 10478) see U 10476 -- U 10479) asym spindle, brightest in sp part; P w comp at 1.2, 305 (from center of brightest
part), 0.9 x 0.15, Sa - b, prob interaction; 2 smaller, F galaxies near --

10480   U 10482) NGC and IC ident in this area are uncertain; this galaxy appears to be identified as N 6196 = I 4613 in
BG; E 5 : (de V, N 6196); comp 1.3, 235, 0.20 x 0.20, v high surf brightn; 16 36.0 + 36 13 = I 4614 at 2.9, 332, 0.35 x
x 0.35, hints of disc 0.9: x 0.8:, SA(r)a - b, m=15.3; 16 36.3 + 36 05 = I 4616 at 4.8, 167, 0.7 x 0.2, m=15.4 --
U 10484) 16 36.1 + 26 25 at 3.5, 348, 0.4 x 0.3, m=15.7; comp 4.7, 29, 0.5 x 0.4, dif; 4.9, 256, (0.7 x 0.4) --
U 10487) broad abs lane along edge; comp 2.6, 306, 0.4 x 0.35 --

10490   U 10491) Arp 125; comet-shaped obj 1.1 x 0.45 interacting w S triple system s; in Arp's class "E and E-like galaxies
close to and perturbing spirals"; I Zw 162; "post-eruptive pair of neutral comet-shaped galaxy w many stellar knots
... and patchy red companion 25" s, m_p = 17.4" (Zw) -- U 10497) P w U 10502 at 4.3, 115 --

10500   U 10501) P w 16 38 + 82 42 = N 6252 at 2.4, 10, 0.8 x 0.5, early spir, m=15.1 -- U 10502) see U 10497; comp 4.0,
53, 0.30 x 0.10 -- U 10503) curved asym obj, * superimp; disturbing comp? in sf part -- U 10506) comp 3.0, 0, 0.8 x
x 0.2 -- U 10507) comp 4.6, 256, 0.8 x 0.7, Sc -- U 10508) bar?; comp 2.6, 40, 0.30 x 0.20 -- U 10509) interacting
w comp at 0.9, 0, 0.22 x 0.15, bridge; comp 1.1, 28, 0.25 x 0.10; both incl in CGCG magn --

10510   U 10511) comp 2.6, 162, 0.5 x 0.5 -- U 10512) comp superimp at 0.45, 40, 0.15 x 0.12 -- U 10516) VII Zw 655; "neutral
patchy disc compact in extended B red halo" (Zw); E 1 (de V); brightest in chain of 4, incl 16 40.7 + 57 56 = N 6213
at 2.2 nf, 0.7 x 0.35, early spir, m=15.2; 16 40.9 + 57 28 at 4.6 nf, 0.55 x 0.18, m=15.7; and 16 41.1 + 58 01 at
8.4 nf, 0.9 x 0.15, m=15.6 -- U 10517) 16 40.3 + 61 23 at 2.7, 217, 0.8 x 0.2, m=15.6 -- U 10519) sev S galaxies
near; comp 4.1, 92, 0.35 x 0.15; U 10524 at 14.3 --

10520   U 10521) SA(s)c (de V) -- U 10527) VII Zw 657; "post-eruptive neutral patchy disc compact w jet to s in tremendous
red halo" (Zw); comp 3.8, 103, 0.35 x 0.25 -- U 10529) comp 1.3, 106, 0.25 x 0.1; 2.0, 87, 0.3 x 0.1 --

10530 U 10532) high surf brightn; comp 1.5, 160, 0.7 x 0.10; 3.8, 293, 0.35 x 0.35 -- U 10533) slightly asym; P w comp at 0.45, 188, 0.35 x 0.25, early spir -- U 10538) brightest in quadrupel system; B stell core, eF corona; 2 compact comps 0.10 x 0.10 and one spir? 0.35 x 0.25; comp? or *? 0.08 x 0.08 superimp near center; I 1225 at 2.9, 354, 0.4 x 0.3 --

10540 U 10541) comp 3.0, 19, 0.5 x 0.4 -- U 10548) 2 long smooth arms; prob SB ab or SAB ab; comp 3.1, 170, 0.3 x 0.1; 3.2, 180, 0.3 x 0.2, F -- U 10549) v patchy, blue; comp 4.2, 242, 0.8 x 0.5, strongly disrupted pair --

10550 U 10551) P w comp at 4.9, 148, 0.50 x 0.50 -- U 10552) stell core in dif halo; comp 1.6, 234, 0.25 x 0.15; 3.2, 177, 0.3 x 0.1 -- U 10555) P w U 10556 at 5.5, 173; these galaxies are brightest in cl 405 -- U 10556) see U 10555; comp 2.3, 302, 0.25 x 0.25, E:; 3.2, 346, 0.25 x 0.25, E:, 2.4 from U 10555 -- U 10558) disturbed? by comp at 1.2, 216, 0.20 x 0.15 --

10560 U 10562) B main body 0.55 x 0.15; v dif plume northw; P w comp at 2.5, 234, 0.9: x 0.3, SB b?, interaction?; comp 0.9, 274, 0.12 x 0.10, interaction? -- U 10565) 16 47.0 + 48 47 = I Zw 166 = Mark 499 at 1.7 ssf, 0.22 x 0.22 + vF halo, "post-eruptive v blue patchy v compact, 2 thin curved jets east to north and west... v = 7630" (Zw), m = 14.6, listed as "extremely compact" in CGCG; Mark 500 at 2.6, 115, 0.35 x 0.12 -- U 10566) P w U 10567 at 2.4, 41, interaction, in group -- U 10567) see U 10566 -- U 10568) comp 1.4, 103, 0.5 x 0.3; U 10563 at 6.4 --

10570 U 10573) comp 3.8, 260, 0.45 x 0.15; sev S comps near -- U 10575) P w a) at 1.0, 274, 0.5 x 0.4, E?, no evidence of interaction -- U 10577) SB(s)b pec ? (de V); asym; vB bar, massive inner arms --

10580 U 10580) comp 2.6, 165, 0.5: x 0.4:, late SB...; U 10584 at 9.8 -- U 10581) v dif spindle; comp 2.8, 6, 0.35 x 0.12 -- U 10582) 16 49.5 + 24 03 at 6.7, 78, 0.6 x 0.5, spir, m = 15.4 -- U 10583) MCG 9 - 27 - 100, note misprint in MCG -- U 10585) P w comp at 3.2, 78, 0.20 x 0.20, compact -- U 10586) comp 1.2, 60, 0.25 x 0.25; 16 48.7 + 45 30 = N 6241 at 6.7, 291, 0.9 x 0.8, spir, m = 14.8; 16 48.0 + 45 33 = An 16 48 A in BG at 13.3, 285, 0.8 x 0.7, w comp 0.15 x 0.15 attached sf, this pair is connected w comp at 2.4, 23, 0.4 x 0.3, thin vF bridge, this triplet is Arp 103, "incomplete connection, blue knots in s member" (Arp) --

10590 U 10591) comp 2.9, 228, 0.3 x 0.2; 16 50.1 + 23 23 at 3.4, 245, 0.7 x 0.25, SB a - b, m = 15.7 -- U 10592) Pec (de V); irr loop southw, large dif plume northeast; comp 2.9, 170, 0.3 x 0.1; 4.3, 30, 0.55 x 0.10 -- U 10594) comp 0.8, 75, 0.1 x 0.1, interaction? -- U 10595) comp 4.3, 64, 0.45 x 0.20 -- U 10597) comp 0.8, 58, 0.3 x 0.1, dif -- U 10599) sev S galaxies near --

10600 U 10602) asym, S-shaped; cond attached sf may be a disturbing comp; comp 4.8, 270, 0.6 x 0.15 -- U 10604) comp 5.0, 184, 0.45 x 0.10 -- U 10606) "post-eruptive blue barred spir w blue compact patchy knot or comp 75" f center" (Zw) -- U 10608) DDO 206; S (B) V (vdB) --

10610 U 10610) a) 2.1 x 1.2:, disrupted, twisted streamer s, thin blue streamer n; b) 0.6 x 0.35; sep 0.7 -- U 10616) blue spindle; comp 1.1, 140, 0.25: x 0.20:; 5.0, 165, 0.4 x 0.08 -- U 10618) comp 0.7, 343, 0.20 x 0.15 --

10620 U 10621) SB c: (de V); 2 v long arms w B inner sections; may be interacting w comp at 3.0, 154, 0.7 x 0.2; comp 2.8, 170, 0.3 x 0.1 -- U 10622) comp 1.6, 171, 0.35 x 0.12; U 10636 at 7.4 -- U 10624) 16 55.3 + 27 56 = N 6264 at 2.9, 276, 0.8 x 0.5, spir, m = 15.5 -- U 10629) a large nr of vS vF galaxies in vic; U 10624 at 6.5, 265 --

10630 U 10635) II Zw 75; "large sharply defined neutral spherical disc compact (or * and galaxy)... v = 10075?" (Zw) -- U 10637) comp 1.1, 288, 0.4 x 0.1; 2.1, 235, 0.7 x 0.1; sev S galaxies in vic -- U 10639) comp 2.8, 145, 0.6 x 0.15 --

10640 U 10640) (r) w low surf brightn, hints of eF arms outside; comp 0.12 x 0.10 attached p -- U 10643) a) 0.55 x 0.40; b) 0.7 x 0.15; sev galaxies near -- U 10647) Arp 293; N 6285 0.9 x 0.5, distorted bar? spir; N 6286 spindle 1.1 x x 0.25 w large asym halo approx 1.2 x 1.0; "dif arc sf brighter galaxy [N 6286]" (Arp); sep 1.5, N 6285 in pos angle 320 from N 6286 --

10650 U 10650) blue asym bar-shaped galaxy; vS blue cond in n part may be a disturbing comp; comp 2.3, 312, 0.6: x 0.3, long streamer -- U 10651) comp 0.9, 8, 0.2: x 0.2:, v dif, broad bridge -- U 10655) comp 1.1, 82, 0.3 x 0.1 -- U 10656) 16 58.6 + 23 07 = N 6276 = N 6277 at 2.3, 329, 0.6 x 0.6, S0:, m = 15.2 -- U 10658) appears brighter than CGCG magn limit; comp 1.7, 277, 0.6 x 0.5, E: --

10660 U 10660) P w comp at 3.0, 93, 0.40 x 0.35 -- U 10661) comp 4.6, 1, 0.4 x 0.3 -- U 10662) comp superimp at 0.50, 120, 0.12 x 0.10 -- U 10665) ident as N 6291 in MCG; in triple group incl 17 00.1 + 59 02 = N 6291 at 2.0 s, 0.5 x 0.4, m = 14.8; and comp at 3.7, 216, 0.7 x 0.45, SB a - b -- U 10666) B stell core, 2 broad v dif arms; U 10668 at 2.2, 163 -- U 10668) see U 10666; 2 S objs w high surf brightn in the group --

10670 U 10670) core 0.1 x 0.1 surrounded by massive patchy spir arcs 0.7 x 0.5; outer disc w low surf brightn -- U 10671) P w comp at 2.4, 353, 0.45 x 0.18 -- U 10672) P w comp at 1.5, 278, 0.25 x 0.20; comp 3.7, 55, 0.45 x 0.10 -- U 10673) comp 2.6, 352, 0.4 x 0.1; 2.7, 223, 0.3 x 0.12 -- U 10675) a) 0.6: x 0.4:, compact, plume southwest; b) at 1.9, 17 from a), 0.30 x 0.20, F, thin vF bridge -- U 10678) P w b) at 0.9, 116, 0.30 x 0.20; comp 3.8, 308, 0.6 x 0.1 -- U 10679) 17 02.3 + 31 33 at 3.5, 107, (0.5 x 0.5), compact, m = 15.7; 17 01.7 + 31 26 at 5.0, 215, 0.6 x 0.25, m = 15.4

10680 U 10683) a) 0.4 x 0.2; b) 0.8 x 0.5, sep 1.1, hints of broad curved bridge -- U 10687) comp 4.7, 327, 0.55 x 0.45 -- U 10689) comp 3.9, 66, 0.5 x 0.5, Sc --

10690 U 10690) triple group around 4.8, 74, total m approx 16.5 -- U 10691) M 51 case; distorted spir w comp 0.20 x 0.12 at 0.50 ssf, at tip of spir arm -- U 10692) P w double system 17 02.8 + 23 15 at 6.0, 272, 0.50 x 0.40, spir, and 6.5, 279, 0.50 x 0.35, spir, sep 0.85, total m = 15.2; comp 1.1, 235, 0.35 x 0.12; 3.6, 262, 0.4: x 0.3:, dwarf -- U 10693) sev comps , nearest at 1.4, 340, 0.35 x 0.15 -- U 10695) P w comp at 0.7, 310, 0.25 x 0.25 -- U 10696) comp 1.3, 193, 0.45 x 0.08; 2.5, 312, 0.8 x 0.15 -- U 10697) asym spir; comp 1.3, 345, 0.12 x 0.12, compact, interaction? --

10700 U 10702) comp 0.45, 335, 0.12 x 0.10, prob thin bridge, or jet toward comp; sev S galaxies near -- U 10704) comp 3.6, 323, 0.35 x 0.10; 3.9, 333, 0.25 x 0.10 -- U 10705) 17 05.2 + 11 31 at 6.1, 80, 0.8 x 0.6, spir, m = 15.3 -- U 10706) P w U 10703 at 3.0 np -- U 10707) comp 1.6, 277, 0.45 x 0.30 -- U 10708) comp 3.6, 239, 0.8 x 0.7, dif late spir (S cd - dm) -- U 10709) distorted SB... w comp attached at 0.45, 260, 0.5 x 0.4, Ir II?; P w U 10712 at 3.2, 70 --

10710 U 10710) 17 05.1 + 43 09 at 2.9 sp, 0.8: x 0.5, m = 15.6 -- U 10711) comp 4.1, 245, 0.5: x 0.4: -- U 10712) see U 10712 -- U 10714) comp 0.35, 130, 0.2 x 0.05; in triple group w U 10716 at 7.4, 34; and 17 05.0 + 30 20 at 7.1, 293, 0.9 x 0.8, SB a, m = 14.7 -- U 10715) comp 4.3, 62, 0.3 x 0.2, background -- U 10716) see U 10714; 17 05.5 + + 30 20 at 11.2, 253; comp 2.5, 100, 0.3 x 0.3, spir -- U 10718) P w U 10719 at 9.4, 178 -- U 10719) see U 10718; SB(s)... : (de V); comp 3.6, 248, 0.6 x 0.5, dif --

10720 U 10721) comp 4.8, 122, 0.3 x 0.2 -- U 10723) comp 4.1, 148, 0.4 x 0.3 -- U 10724) P w U 10727 at 1.4 nf -- U 10725) vB main body 0.45 x 0.35; one dif spir arm on p side; comp 1.6, 289, 0.3: x 0.2:; 3.0, 126, 0.5 x 0.08 -- U 10527) E4 : (de V); see U 10724 -- U 10728) comp 4.2, 271, 0.6: x 0.2 --

10730 U 10731) VII Zw 681; b) superimp at 0.30, 167, 0.20 x 0.15; "double system, loop" (CGCG); "post-eruptive neutral spir w compact disc, distorted arms and double compact knot s" (Zw) -- U 10735) 17 08.4 + 32 17 at 3.8 s, 0.2 x x 0.2, F halo?, compact, m = 15.6 -- U 10736) comp 2.1, 176, 0.50 x 0.18 -- U 10738) comp 2.0, 283, 0.5: x 0.3: --

10740 U 10741) 17 08.8 + 41 40 at 4.8 sp, (0.4 x 0.4), core 0.15 x 0.15, very compact, m = 15.7 -- U 10744) P w comp at

6.7, 228, 0.9 x 0.5, m approx 16. -- U 10747) SAB(r)c (de V); S stell comp (or *?) at 0.60 np, 0.1 x 0.1, poss at tip of thin disturbed spir arm; comp 1.4 sf, 0.1 x 0.1; U 10752 at 11.1, 150 -- U 10749) comp 4.8, 335, (0.6 x 0.5), S0?, vF halo --

10750   U 10752) SA a: (de V); strong abs along eastern side of nucl bulge; P w 17 10.6 + 23 17 = N 6315 at 3.3, 150, 0.9 x x 0.7, S B bc: (de V), m = 15.4; see U 10747 -- U 10754) comp 2.8, 300, 0.5 x 0.3 -- U 10757) prob v late; in triple subgroup w U 10762 at 6.4 and U 10769 at 6.1 --

10760   U 10762) see U 10757; U 10769 at 6.8; SA(s)0/a (de V) -- U 10763) comp 1.5, 42, 0.4 x 0.2, interaction?; 0.7, 187, 0.3 x 0.1, dif -- U 10765) 17 11.5 + 42 55 at 2.6 sp, 0.3 x 0.3, core 0.18 x 0.18, very compact, m = 15.6; comp 2.9, 148, 0.3 x 0.1 -- U 10768) comp 2.5, 50, 0.3 x 0.3, spir -- U 10769) nucl reg 0.45 x 0.15, dif disc, no spir pattern vis; see U 10757, 10762; sev comps --

10770   U 10770) VV 89, Arp 32; a) 1.1 x 0.5, irr, blue; b) 0.8 x 0.2, spir?; in Arp's class "integral sign spir galaxies" -- U 10771) comp 0.1 x 0.1 superimp at 0.4, 74; sev galaxies near; 17 12.4 + 43 42 = N 6327 at 3.3 sp, 0.18 x 0.18, compact, * superimp, m = 15.7 -- U 10772) one v long asym arm; prob interacting w comp at 0.70, 0, 0.4: x 0.3:; comp 1.3, 183, 0.4: x 0.3: -- U 10773) comp 3.5, 102, 0.35 x 0.12 -- U 10775) compact comp, or *?, superimp nf center -- U 10779) U 10778 at 7.9, 348 --

10780   U 10782) one B spir arm in centr part, (r)?, e dif disc; comp 1.5, 310, 0.55 x 0.15; 3.3, 53, 0.4 x 0.1 -- U 10783) comp 5.0, 69, 0.5 x 0.1 -- U 10784) brightest in subgroup, 8 galaxies brighter than 15.8, a large nr of F members; 17 14.5 + 57 30 = VII Zw 700 at 1.3 n, 0.5? x 0.3?, "pair of red spherical compacts, sep 7" n-s, imbedded in blue luminous cloud extending north from N 6338... m_p's = 16.0 and 16.5" (Zw) -- U 10785) comp 1.4, 114, 0.6 x 0.1 -- U 10786) comp 3.7, 148, 0.45 x 0.1 -- U 10787) P w comp at 0.9 np, 0.8 x 0.30, early S B ...?, distorted, interaction, bridge? -- U 10788) U 10784 at 4.6 -- U 10789) a) 0.6: x 0.4:; b) 1.0: x 0.3:; nr 4 and 5 in chain of 5; 17 15.1 + + 07 43 at 4.0, 316 from b), 0.7 x 0.5, m = 15.2 --

10790   U 10790) short B bar 0.40 x 0.18, v broad irr arms; comp 3.0, 57, 0.9 x 0.1 -- U 10792) comp 2.8, 158, 0.7: x 0.7:, dwarf (spir?) -- U 10796) comp 4.3, 187, 0.4: x 0.3:, v dif -- U 10804 at 11.5 -- U 10797) comp 1.0, 43, 0.25 x 0.20 --

10800   U 10800) comp 2.4, 230, 0.8 x 0.1; 17 16.8 + 41 43 = N 6348 at 4.8 sp, 0.8 x 0.6, m = 15.6 -- U 10802) P w 17 17.2 + + 11 20 at 4.4, 352, 0.6 x 0.6, spir, m = 15.6 -- U 10804) SA 0 - (de V); see U 10796 -- U 10805) DDO 207; S (B) V: (vdB); comp 3.3, 124, 0.3 x 0.2, dif; 3.3, 294, 0.5: x 0.4, v dif -- U 10808) comp 1.5, 17, 0.4 x 0.15 -- U 10809) a) B bar 0.4 x 0.15, v dif arms or filaments; b) 0.5 x 0.15, sep 0.7, bridge? --

10810   U 10812) comp 2.5, 217, 0.5 x 0.12 -- U 10814) VV 10, Arp 102; An 1718 B in BG; spir 1.0 x 0.6 w long broad dif streamer n; 17 17.9 + 49 01 = An 1718 A in BG at 3.8 sp, (0.9 x 0.7), E0 pec (de V), v = 7250, v_o = 7493; "VV pos wrong. Note loop eastern side of spir; dif vF connection to E galaxy" (Arp); S b pec (de V) -- U 10815) Arp 124; in Arp's class "E and E-like galaxies close to and perturbing spir's"; P w 17 17.9 + 60 38 at 1.8, 220, 0.25 x x 0.20, extremely compact, m = 15.5 -- U 10816) 17 18.0 + 50 00 at 1.2 np, 0.3 x 0.3, core 0.18 x 0.18, compact, m = = 15.6; U 10806 at 6.5 --

10820   U 10822) DDO 208, An 1719 in BG; "spheroidal" (vdB) -- U 10823) P w comp at 3.8, 40, 0.7 x 0.6, Sc -- U 10824) P w comp at 1.2, 27, 0.55 x 0.15 -- U 10827) comp 1.5, 147, 0.25 x 0.20 -- U 10828) comp 0.15 x 0.12 attached at 0.60, 163, incl in CGCG magn; 17 21.4 + 23 43 at 2.4, 32, 0.35 x 0.25, compact, m = 15.5; 17 21.1 + 23 43 at 3.3, 306, 0.55 x 0.45, m = 15.3 --

10830   U 10831) P w U 10837 at 3.9, 90 -- U 10832) VV 232 b, Arp 30; slightly asym spir; P w U 10833, sep 0.45; in Arp's class "spir galaxies w one heavy arm"; "comp appears physically connected to flat-on spir system" (Arp) -- U 10833) VV 232 a, Arp 30; see U 10832; appears superimp U 10832 -- U 10835) comp 3.0, 274, 0.4: x 0.2 -- U 10836) comp 3.3, 50, 0.55 x 0.45, S B a - b; 4.2, 292, 0.3 x 0.2 -- U 10837) see U 10831 -- U 10839) P w U 10843 at 4.2, 139

10840   U 10843) see U 10839 --

10850   U 10850) asym, 2 unequal arms, low surf brightn in p part; comp 3.8, 350, 0.5 x 0.4 -- U 10853) large dif halo; comps at 0.70, 0.15 x 0.12, and 0.8, 0.35 x 0.20 incl in CGCG magn; at least 8 members in the group, all appear v red -- U 10854) blue; a cond in the disc may be a superimp comp, app slightly distorted; comp 4.1, 142, 0.6 x 0.3 -- U 10855) VII Zw 712; a) 0.55 x 0.25; b) 0.55 x 0.20, sep 0.6; "blue post-eruptive interacting group of 3 galaxies. No 1, disc-shaped. Nos 2, 3, form pair of equal spherical compacts, sep = 5" n-s, w disrupted plumes" (Zw) -- U 10857) a) 0.4 x 0.15; b) 1.3 x 0.2, thin jet to 0.8 from main body; sep 1.2 -- U 10858) comp 3.6, 358, 0.4 x 0.1 -- U 10859) centr body 0.35 x 0.20, 2 long arms w v low surf brightn; sev galaxies near --

10860   U 10861) Sb?; 3-armed asym galaxy, sev blue patches in arms; P w 17 25.2 + 26 32 = N 6371 at 3.1, 306, 0.7 x 0.35, m = 15.2 -- U 10867) P w U 10869 at 2.3 nf; a large nr of E - S0 galaxies w magn approx 16.5 in vic -- U 10868) a) 0.8 x 0.3; b) 1.6 x 0.3, distorted, Irr or late spir -- U 10869) VV 101, Arp 310 - 311; see U 10867; " v close E galaxies" (Arp) --

10870   U 10870) P w U 10871 at 1.3, 135 -- U 10871) see U 10870; sev comps -- U 10873) P w U 10875 at 3.0, 29; slightly integralshaped, prob interaction; in triple group w U 10886 at 20.5 -- U 10875) see U 10873; U 10886 at 18.3; comp 0.9, 296, 0.3 x 0.15 -- U 10877) hints of eF outer disc, diam 1.7? x 1.7?; S-shaped, v broad arms; pair 17 27.6 + + 57 36 = N 6387 at 3.3 f, 0.30 x 0.22 and 0.25 x 0.18, contact, "compact, jet" (CGCG), brighter component compact, total m = 15.0; comp 1.9, 270, 0.9 x 0.4, F -- U 10879) P w 17 27.2 + 24 55 at 7.1, 269, (0.9 x 0.8), S0, m = 14.8 --

10880   U 10881) 17 27.3 + 60 09 at 4.4 p, (0.8 x 0.6), core 0.20 x 0.20, extremely compact, m = 15.5; 17 28.4 + 60 07 at 4.2 f, 0.9 x 0.7, Sb:, m = 15.6; sev F galaxies near -- U 10884) comp 3.0, 57, 0.5 x 0.25 -- U 10885) comp 2.6, 138, 0.35 x 0.20 -- U 10886) see U 10873, 10875 -- U 10887) blue spindle; comp 5.4, 160, 0.8 x 0.4, Sc -- U 10888) comp 4.9, 212, 0.3 x 0.3 --

10890   U 10891) SAB(r)bc (de V), Sb + (Ho) -- U 10893) comp 2.7, 100, 0.3 x 0.2 -- U 10894) comp 1.0, 354, 0.3 x 0.3 -- U 10897) Arp 38; SA(s)c (de V), Sc - (Ho); "S ring in arm on n side, part of large ring on f side shows in H α only" (Arp) -- U 10899) comp 4.5, 10, 0.4 x 0.10, background spir --

10900   U 10900) a large nr of galaxies near -- U 10902) asym (r), F alm straight arms; may be interacting w U 10900 at 4.0 s; many galaxies in vic -- U 10903) comp 0.5 sf, 0.4 x 0.1, background -- U 10904) comp 0.9, 68, 0.20 x 0.20 -- U 10906) stell core in smooth F halo; brightest in subgroup, comps at 2.3, 2.4, 3.4, 3.6, 4.2 -- U 10908) VV 268; interpreted as triple system by VV; appears S-shaped, w thin irr arms, disrupted pair?; comp 4.2, 273, 0.7 x 0.10 --

10910   U 10911) comp 3.1, 60, 0.5 x 0.4, v dif -- U 10912) v long dif curved streamer, may be connected w compact obj at 2.8, 34, 0.18 x 0.18; short counter-streamer; 17 33.8 + 70 48 at 4.6, 207, core 0.18 x 0.18, compact, m = 15.6 -- U 10916) MCG δ = 60 20 is incorrect --

10920   U 10920) tend to bar in irr (r); comp 2.5, 167, 0.3 x 0.2, core 0.12 x 0.12 -- U 10923) VII Zw 729; "post-eruptive blue pair of disrupted spirs , w distorted arms, jets and sev red and blue F compact knots in field" (Zw); sep 0.65 between nucl regions -- U 10924) 17 36.9 + 68 11 = N 6423 at 4.5, 80, (0.8 x 0.7), core 0.18 x 0.15, compact, m = 15.6 -- U 10925) v open S-shaped spir, or integral; P w U 10927 at 1.3 sp, interaction -- U 10927) see U 10925 -- U 10929) comp 4.5, 277, 0.55 x 0.55, S B c --

10930   U 10932) vB stell core; comp 3.8, 222, 0.3 x 0.2, F -- U 10937) comp 2.0, 138, 0.35 x 0.30 -- U 10938) v smooth disc, no spir pattern, (R) not completely closed; comp 4.8, 153, 0.6 x 0.2 --

10940 U 10941) comp 5.0, 111, 0.6: x 0.6:; sev smaller comps -- U 10942) interacting w 17 39.0 + 51 06 at 1.7, 2, 0.9 x
       x 0.30, Sa - b, m=15.4 -- U 10943) prob late spir; P w comp at 2.9, 270, 0.9: x 0.9:, dwarf -- U 10944) comp 3.2,
       224, 0.4: x 0.3: -- U 10945) uneven (r), distorted arm in sf part -- U 10947) comp 3.5, 28, 0.5 x 0.15; 17 40.4 +
       + 62 37 at 5.2 sf, 0.8 x 0.6, Sb:, m=15.2 -- U 10949) comp 2.0, 204, 0.55 x 0.10 --

10950 U 10950) a) 0.9: x 0.9:, spir:, disturbed; b) 0.8 x 0.2, sep 0.8 x 0.2; sev S vF objs in vic -- U 10955) comp
       4.5, 224, 0.30 x 0.12 -- U 10957) P w U 10960 at 10.5, 325 -- U 10958) comp 3.4, 58, 0.6 x 0.15 -- U 10959) distorted
       pair?; P w comp at 1.1, 188, 0.7: x 0.6:, v dif, dwarf --

10960 U 10960) see U 10957; comp 2.6, 47, 0.3 x 0.2, dif -- U 10962) 17 42.2 + 36 52 at 4.2 f, 0.45 x 0.30, m=15.2 --
       U 10963) VII Zw 738; "red long thin S B c" (Zw) -- U 10964) comp 0.50, 265, 0.12 x 0.10; 4.5, 190, 0.5 x 0.2 --

10970 U 10970) comp 4.2, 146, 0.4: x 0.3:, dif -- U 10973) in a dense part of the cl; b) at 0.50, 105, 0.35 x 0.30;
       U 10974 at 1.6 -- U 10974) see U 10973 -- U 10975) centr reg 0.50 x 0.35 w high surf brightn, 2 long thin arms;
       17 44.3 + 35 35 = N 6446 at 2.0 p, 0.6 x 0.3, m=15.5, interaction? -- U 10976) P w U 10977 at 0.9, 113; one thin
       B arm -- U 10977) v blue; see U 10976 -- U 10978) P w U 10979 at 7.2, 33; prob in group w U 10997 brightest --
       U 10979) see U 10978 --

10980 U 10983) 17 45.9 + 45 43 at 4.2 f, (0.9 x 0.8), E - S0, m=15.3 -- U 10985) P w 17 45.5 + 14 42 at 5.2, 231, 0.7 x
       x 0.45, S B(r) ..., m=15.2 -- U 10987) a) 0.4 x 0.4; b) + c) 0.9 x 0.9, b) superimp c), early systems, E? --
       U 10988) comp 2.0, 10, 0.45 x 0.10; 2.5, 168, 0.45 x 0.25 -- U 10982 at 7.4 --

10990 U 10990) 17 46.6 + 34 05 at 2.5 sf, 0.7 x 0.3, pec?, B main body 0.25 x 0.25 w dif extensions, m=15.5; 17 46.5 +
       + 34 01 at 5.5 s, 0.55 x 0.40, spir:, pec?, m=15.6 -- U 10994) P w U 10997 at 5.2, 118 -- U 10995) early:; comp 1.9,
       124, 0.8 x 0.1 -- U 10997) see U 10994 -- U 10998) S c (de V); comp 0.7, 90, 0.2 x 0.1 --

11000 U 11002) 17 47.6 + 58 31 at 6.0 np, 0.25 x 0.22, compact, m=15.2 -- U 11006) 2 v open arms; prob interacting w
       17 49.8 + 51 03 at 3.7 f, core 0.20 x 0.18 + halo, compact, m=15.5; comp 1.4, 122, 0.3 x 0.1 -- U 11007) one thin
       alm straight arm, hints of vF second arm; appears distorted, but no disturbing obj vis; U 11008 at 4.8; U 11011 at
       5.0 -- U 11008) P w U 11011 at 3.1 nf; see U 11007; sev F comps -- U 11009) E 2 ? (de V); comp 2.9, 275, 0.3 x 0.2

11010 U 11010) main body 0.7 x 0.45 w high surf brightn; 2 long uneven arms w B knots; prob in group w U 11027 and U 11029;
       pair 17 50.1 + 24 32 at 5.9, 68, 0.55 x 0.12 and 0.45 x 0.12, sep 0.40, total m=15.1 -- U 11011) see U 11007, 11008;
       sev F comps -- U 11012) SA(s)cd (de V), Sc - (Ho) -- U 11013) strongly asym open spir; s arm w sev ramifications;
       vS vF blue cond at tip of straight arm p center; stell cond attached nf, prob at tip of arm -- U 11015) P w comp at
       1.0, 0, 0.65 x 0.55, spir --

11020 U 11022) P w 17 50.6 + 29 50 = N 6486 at 1.9, 229, (0.8 x 0.8), core 0.20 x 0.20, compact, m=15.0; U 11017 at 7.4,
       281 -- U 11026) S B core in large halo; sev S galaxies near -- U 11027) asym; B red cond 0.45 x 0.25 in sf part,
       superimp comp?; P w U 11029 at 10.4, 130; see U 11010 -- U 11029) see U 11027 --

11030 U 11031) chaotic bar system; cond 0.1 x 0.1 in n part, disturbing comp? -- U 11032) in triple group incl comps at
       0.8, 354, 0.4 x 0.3, and 1.5, 218, 0.6 x 0.12 -- U 11033) P w U 11034 at 5.7, 121; 17 52.3 + 18 27 at 4.3 n, 0.35 x
       x 0.22, S B ? 0 - a:, m=15.7 -- U 11034) see U 10033; U 11037 at 5.4, 140 -- U 11035) "blue post-eruptive chain of
       compacts, extended plumes" (Zw) -- U 11037) see U 11034 --

11040 U 11042) stell comps superimp at 0.35, 248, 0.15 x 0.15; and 0.6, 55, 0.18 x 0.10, prob at tip of spir arm --
       U 11043) a) 0.22 x 0.20; b) 0.15 x 0.10; c) 0.22 x 0.20; sep a) - c) 0.6, F ext halo emanates from c); CGCG magn incl
       a) and c); comp 0.25 x 0.15 at 1.5 n c); sev galaxies near, in group or S cluster -- U 11044) chaotic, v blue; sev
       dif galaxies in vic; comp 2.8, 76, 0.8: x 0.3:, Irr; 2.8, 240, 0.6 x 0.2, late spir -- U 11046) comp 3.5, 214,
       0.45 x 0.15 -- U 11048) S 0 ? (de V); P w U 11049 at 2.3, 26; asym, one F distorted arm points toward U 11049 --
       U 11049) S 0 ? (de V); see U 11048 --

11050 U 11050) comp 3.6, 329, 0.3 x 0.1 -- U 11051) S-shaped; massive bar and arms; comp 2.0, 265, 0.3 x 0.2 -- U 11054)
       P w U 11066 at 6.1, 94 -- U 11055) P w U 11057 at 4.4, 147 -- U 11057) see U 11055 --

11060 U 11060) P w U 11064 at 10.2, 139; comp 1.3, 126, 0.6: x 0.3:, v dif -- U 11061) brightest in a dense part of the
       cl; a large nr of galaxies near, 4 brighter than 15.8 inside 8.0 -- U 11062) comp 2.2, 228, 0.35 x 0.10; 3.7, 62,
       0.3 x 0.10 -- U 11064) see U 11060 -- U 11065) Sa?; comp 3.4, 165, 0.3 x 0.2; 4.3, 164, 0.35 x 0.1 -- U 11066) see
       U 11054 -- U 11069) comp 2.4, 163, 0.4 x 0.3, v dif --

11070 U 11070) blue obj 0.25 x 0.1 attached at 0.40 ssp -- U 11071) comp 2.9, 182, 0.3 x 0.2 -- U 11072) 17 56.6 + 43 25
       at 2.5 nf, core 0.18 x 0.18, compact, m=15.6 --

11080 U 11080) P w U 11082 at 1.9, 72 -- U 11081) P w comp at 4.7, 240, 0.8 x 0.1 -- U 11082) see U 11080 -- U 11083)
       asym, w sev v blue patches; F blue nucl cond; sev vF comps ; U 11084 at 7.5 -- U 11086) P w U 11090 at 6.3, 139 --
       U 11087) comp superimp?; comp 0.7, 55, 0.25 x 0.10 --

11090 U 11090) see U 11086; comp 3.1, 174, double syst, total diam 0.6 x 0.4 -- U 11095) comp 0.7, 214, 0.4 x 0.10; sev
       galaxies near -- U 11096) bar 0.7 x 0.20 in elongated smooth ring -- U 11098) P w 18 00.9 + 29 18 at 1.7, 94,
       (0.9 x 0.7), S0 - a, m=15.2, prob interaction; 18 01.0 + 29 22 at 4.9, 36, 0.7 x 0.5, spir, m=15.5 -- U 11099)
       comp 5.9, 291, 0.8 x 0.35, S B(s)b - c --

11100 U 11100) comp 4.2, 14, 0.5 x 0.15 -- U 11104) P w U 11108 at 6.4, 123 -- U 11108) see U 11104 --

11110 U 11111) P w 18 03.3 + 23 08 at 1.6, 24, 0.35 x 0.25, m=15.7 -- U 11112) P w comp at 1.7, 200, 0.9 x 0.8, dif --
       U 11113) S-shaped, dif; comp 4.3, 67, 0.3: x 0.2:, dif -- U 11114) short jet in southeast?; P w U 11115 at 3.7, 38
       -- U 11115) see U 11114 -- U 11117) massive blue arms; comp 4.6, 260, 0.8 x 0.3, Irr, dif -- U 11119) small triple
       group around 4.6, 248 --

11120 U 11121) SAB(rs)c (de V); knotty (rs) 0.9 x 0.8; comp 3.8, 255, 0.5 x 0.10 -- U 11123) blue cond superimp 0.2 n
       center; comp 2.5, 77, 0.3 x 0.1 --   U 11127) a) 0.7 x 0.4; b) 0.5 x 0.1; a) Pec or distorted --

11130 U 11130) VII Zw 768; 3C 371; "large blue spherical compact... variable, intercon w sev elliptical galaxies of about
       the same redshift" (Zw) -- U 11135) P w U 11138 at 9.5 nf -- U 11136) 18 08.6 + 25 06 at 5.1 s, 0.7: x 0.3, Pec,
       m=15.2 -- U 11137) S B(rs)m: (de V); asym patchy arm fragments; v blue cond in nf part -- U 11138) see U 11135 --
       U 11139) B core 0.22 x 0.20, hints of abs features outside; large dif halo --

11140 U 11141) short B bar; e dif, patchy, prob S-shaped -- U 11142) in group w U 11156 brightest; comp 3.9, 67, 0.6 x
       x 0.3, spir, distorted? -- U 11143) 18 09.8 + 35 58 = I 1281 at 4.7 f, double system, 0.6 x 0.3 and 0.4 x 0.15, sep
       0.6, total m=15.5 -- U 11146) a) 0.7 x 0.7, S0; b) 0.9 x 0.9, S0; a large nr of galaxies near; U 11149 at 2.8,
       160 -- U 11148) in subgroup in cl 444, sev galaxies in vic; 18 09.7 + 21 30 at 3.0, 318, 0.7 x 0.25, m=15.7;
       18 09.6 + 21 25 = N 6576 at 3.3, 236, 0.7 x 0.5, m=15.5 -- U 11149) see U 11146; large halo; a large nr of galaxies
       in vic --

11150 U 11150) in group w U 11156 brightest; comp 1.8, 228, 0.8: x 0.1; 2.2, 323, 0.2? x 0.15, * superimp; 18 10.2 +
       + 25 39 = I 1280 at 4.9, 35, 0.5 x 0.5, core 0.15 x 0.15, compact, m=15.4 -- U 11152) P w 18 10.5 + 18 38 at 3.7,
       46, 0.35 x 0.30, m=15.4 -- U 11153) a) = N 6579, 0.5? x 0.4?, B core; b) 0.35 x 0.15, background?, 0.65 s a); c) =

= N 6580, 1.5: x 1.0?; sep a)-c) 0.50; galaxy at 18 10.2 + 21 35 (0.5 x 0.3, S B a - b, m=15.7) is identified in
CGCG as N 6579+6580, which is prob incorrect -- U 11155) in group w U 11156 brightest; interacting w 18 10.4 +
+ 25 25=I 4697 at 1.3, 147, 0.8 x 0.2, Pec, B main body w streamer; 18 10.7 + 25 24 at 3.6, 112, 0.45 x 0.25, m=
= 15.5 -- U 11156) brightest in central part of cl 441; prob eF ext env outside 1.1 x 1.0; 18 10.8 + 25 29 at 3.5,
116, 0.4 x 0.25, early, m=15.7 -- U 11158) comp 1.4, 261, 0.35 x 0.10; 4.0, 253, 0.5 x 0.25 --

11160   U 11160) comp 1.8, 57, 0.4: x 0.2, interacting w comp 0.1 x 0.1 at 0.4, bridge; sev smaller galaxies near --
U 11162) asym late spir; 3 S conds near center, one v blue, one v red, comps ?; comp 3.6, 119, 0.3 x 0.3 --
U 11163) P w U11167 at 5.5, 164 -- U 11166) SA 0-: (de V); main body 0.6 x 0.3; vF ext corona; P w comp at 2.6,
235, 0.25 x 0.20, high surf brightn -- U 11167) early; see U 11163 --

11170   U 11170) core 0.25 x 0.20 in large dif asym halo; 2 F irr conds f -- U 11172) VII Zw 776; "very blue very compact"
(Zw) -- U 11175) VV 247, Arp 81, VII Zw 778; interacting pair w sep 0.65, bridge, broad loop; "post-eruptive pair
of distorted spir w v blue compact core intercon w neutral Irr comp" (Zw); "comp resembles M 51 comp" (Arp) --

11180   U 11180) comp 1.5, 83, 0.3 x 0.2 -- U 11181) 2 S B cores w sep 0.70 in ext halo, prob F bridge + plumes; sev vS
objs near; 3rd component at 2. f, 0.5 x 0.5, B core in disc, spir? -- U 11183) P w comp at 3.4, 157, 0.9 x 0.4,
dif -- U 11185) I Zw 205; a) 0.3 x 0.2 (main body), vF ext halo vis in red; b) 1.1 x 1.1, B core, ext smooth env;
sep 0.50, hints of bridge; "pair of neutral compacts in huge mutual halo. No 1, patchy elliptical, m_p = 15.9. No
2, 25" nf No 1, spherical, m_p = 16.1" (Zw) --

11190   U 11191) comp 1.5, 2, 0.4: x 0.2: e dif -- U 11192) strong abs lane across nucl region; comp 2.2, 287, 0.4 x 0.2 --
U 11193) S b: (de V); see U 11236; comp 4.4, 188 (from center), 0.5 x 0.4; 4.5, 147 (from center), 0.4 x 0.05 --
U 11195) disrupted bar spir; comp superimp at 0.3, 160, 0.2 x 0.2, high surf brightn -- U 11196) SB(s)a: (de V) --

11200   U 11201) appears brighter than CGCG magn limit; comp 3.4, 38, 0.4: x 0.2, v dif -- U 10202) comp 2.0, 228, 0.3 x 0.1,
F -- U 11203) comps at 0.30, 184, 0.15 x 0.15, and 0.8, 65, 0.7: x 0.3:, dif, incl in CGCG magn -- U 11205) comp
4.2, 115, 0.4 x 0.25 -- U 11207) core ecc in F disc, tend to (R); comp 1.0, 50, 0.3 x 0.1 -- U 11208) comp 3.1, 159,
0.25 x 0.10; 4.9, 154, 0.4 x 0.3; 2 stell objs in vic may be galaxies --

11210   U 11211) comp 3.3, 200, 0.5 x 0.12 -- U 11212) SB(s)b (de V) -- U 11217) comp 3.4, 183, 0.3: x 0.3:, F -- U 11218)
SA(rs)c (de V), S c- (Ho) -- U 11219) a) 0.4: x 0.2; b) 0.9: x 0.4, distorted spir; sep 0.8 --

11220   U 11221) VII Zw 790; a) 2.1 x 0.25, in pos angle 3; b) 0.35 x 0.30, attached to nf side of a); "pair of contiguous
galaxies... No 1, edge-on post-eruptive S B c. No 2, spherical compact, 25" nnf No 1" (Zw) --

11230   U 11233) comp 3.2, 312, 0.9: x 0.5, v dif, dwarf: -- U 11236) ident as N 6651 in MCG; comp 2.7, 276, 0.5 x 0.12 --
U 11237) comp 4.7, 113, 0.5 x 0.10 -- U 11238) VII Zw 793; Pec S B(s)0/a (de V); "red compact elliptical core, bar
and enormous uniform halo" (Zw); Zw lists v = 1294, which is a misprint, correct v = 1924 -- U 11239) S0 (de V) --

11240   U 11248) curved spindle; P w comp at 0.7, 356, 0.35 x 0.12, interaction -- U 11249) P w comp at 0.3, 105, 0.30 x 0.2;
comp 3.4, 305, 0.7 x 0.2 --

11250   U 11252) may be interacting w comp at 0.3, 5, 0.2 x 0.1; SN 1953 c -- U 11253) P w 18 27.6 + 22 58 at 4.2, 150,
0.7 x 0.6, spir, m=15.2 -- U 11258) comp? 1.0, 320, 0.12 x 0.12, *?; 3.8, 10, 0.3 x 0.1 --

11260   U 11260) comp 3.3, 185, 0.4 x 0.15 -- U 11264) 18 30.2 + 31 00 at 5.2, 124, 0.5 x 0.35, m=14.9; U 11263 at 10.0,
222 -- U 11265) "compact, halo" (CGCG); comp 1.6, 254, 0.5 x 0.1 -- U 11267) comp 4.8, 210, 0.6 x 0.08 -- U 11269)
red complex centr reg ecc in main body 1.6 x 0.7 w low surf brightn; v dif outer parts, uncertain diameter values;
no spir str; U 11266 at 6.4, 232 --

11270   U 11274) S0 (de V); P w U11282 at 9.6, 84 -- U 11278) P w U11281 at 4.3, 106 --

11280   U 11280) comp 2.5, 136, 0.25 x 0.1; 3.6, 168, 0.3 x 0.2 -- U 11281) see U 11278 -- U 11282) SA(s)0/a (de V); see U
11274; comp 7.5, 215, 0.9 x 0.4, S B ..., m approx 16.0 -- U 11283) S-shaped, v patchy; in triple group w comps at
3.8, 50, 0.4 x 0.3; and 5.9, 12, 0.8: x 0.3, Irr -- U 11284) VII Zw 812; a) 0.45 x 0.12; b) 0.5 x 0.2; c) 0.12 x 0.12;
"post-eruptive quadruple system of three brush-like and one spherical compact, d = 65"" (Zw); 18 33.5 + 59 49 at
5.0, 110, 0.35 x 0.25, m=15.2 -- U 11285) P w U11289 at 8.0, 103 -- U 11286) comp 5.1, 338, 0.9 x 0.7; 5.9, 229,
0.9 x 0.1; 6.2, 204, 0.8 x 0.1 -- U 11287) double nucl? or superimp *?; asym spir pattern -- U 11288) VII Zw 814;
a) 0.35 x 0.25, double?, or * superimp?; b) 0.22 x 0.15; sep 0.55; "post-eruptive blue patchy compacts intercon by
broad uniform bridge, sep = 30" northnortheast - southsouthwest" (Zw); ident as I 4763 in MCG; in triple group w
U 11290 at 1.6 from a) -- U 11289) see U 11285 --

11290   U 11290) see U 11288 -- U 11292) comp 0.7, 312, 0.10 x 0.10, interaction?; "compact, F jet" (CGCG) -- U 11293) in
a rich *-field; comp 1.0, 285, 0.6 x 0.2, Irr -- U 11294) second brightest in group w 18 34.5 + 19 41 brightest,
(0.9 x 0.9), E?, m=14.8, 7 galaxies brighter than 15.8 -- U 11296) 18 35.2 + 30 35 at 2.8, 77, 0.5 x 0.3, spir,
m=15.7 -- U 11297) in group w 18 34.5 + 19 41 brightest, see U 11294 -- U 11298) 18 34.8 + 51 25 at 3.5 p, 0.9 x
x 0.7, m=15.7; comp 4.0, 252, 0.7 x 0.35, spir, 1.0 from 18 34.8 + 51 25; a large nr of vF galaxies near --

11300   U 11300) S d? (de V) -- U 11303) comp 2.3, 248, 0.35 x 0.15 -- U 11306) P w 18 35.3 + 53 47 at 0.6, 136, 0.9 x 0.25,
spir, m=15.3 -- U 11308) SB(r)b (de V) --

11310   U 11312) comp 2.6, 23, 0.35 x 0.15 -- U 11313) comp 3.2, 335, 0.5 x 0.05; 4.5, 10, 0.55 x 0.3 -- U 11315) comp 2.1,
33, 0.3: x 0.1, F -- U 11317) 18 38.2 + 40 00=N 6685 at 2.8, 352, 0.25 x 0.15, compact, m=15.0, comp 0.15 x 0.15
at 0.4 -- U 11318) comp 2.5, 189, 0.4: x 0.2 -- U 11319) B core 0.20 x 0.15, v long thin alm straight arms, stell
cond at tip of n arm, *?; P w 18 38.5 + 40 06 at 1.8, 45, (0.8 x 0.7), core 0.15 x 0.15, compact, m=14.9, prob
interaction --

11320   U 11320) comp superimp at 0.9, 270, 0.5: x ... -- U 11322) P w U11325 at 3.7, 50 -- U 11323) comp 2.3, 80, 0.3 x
x 0.2 -- U 11324) comp 2.4, 171, 0.6 x 0.5; U 11325 at 7.7 s -- U 11325) see U 11322; comp 2.6, 157, 0.4 x 0.12 --
U 11326) appears to be a late spir, or Irr; in center of a hole in a rich *-field, not certainly extra-galactic --
U 11327) U 11324 at 7.2 nf --

11330   U 11331) P w U11332 at 2.6, 133; tend to bar; F blue cond sf center; comp 3.4, 330, 0.9 x 0.12 -- U 11332) blue,
sev B patches; late spir?; see U 11331 --

11340   U 11348) comp 4.5, 265, 0.3 x 0.2 --

11350   U 11350) 3-armed galaxy -- U 11354) E4 (de V); P w U11356 at 10.1, 159 -- U 11356) S0 (de V); see U 11354 --

11360   U 11362) dif spindle; comp 0.8, 190, 0.2 x 0.2, prob background -- U 11363) P w a) at 0.45, 195, 0.15 x 0.12 --
U 11364) SA(s)0+; 11367 group?; comp 4.4, 333, 0.4 x 0.3 -- U 11366) at least 6 members, incl comp at 1.5,
334, 0.8: x 0.4:, distorted? -- U 11367) open S B(s), one alm stell comp at tip of spir arm 0.7 np center; one comp
attached to straight arm at 0.8, 70, 0.2 x 0.1 -- U 11368) in triple group w U 11370 at 5.4, 36, and U 11371 at
5.9, 88; U 11364 may belong to this group; comp 2.1, 292, 0.4 x 0.15 --

11370   U 11370) see U 11368; 11371 at 4.9, 146 -- U 11371) see U 11268, 11370 -- U 11372) inside the reflection of α Lyr
on PA maps -- U 11377) comp 3.6, 299, 0.3 x 0.15; 4.1, 126, 0.3 x 0.3 --

11380   U 11380) comp 1.9, 295, 0.45 x 0.15 -- U 11381) stell obj at 0.35, 100, 0.10 x 0.10 incl in CGCG magn; comp 1.6, 235,

0.7: x 0.4, v dif -- U 11383) in triple group w comps at 1.6, 215, 0.3 x 0.3, core 0.12 x 0.12; and 2.0, 168, core 0.15 x 0.15 -- U 11389) v massive arms, prob late; comp 4.6, 262, 0.25 x 0.25 --

11390   U 11391) a) + b) 1.3 x 0.5, strongly distorted pair; c) 0.4 x 0.15 -- U 11399) comp 2.6, 259, 0.45 x 0.1 --

11400   U 11402) Sc?; tend to bar; sev B patches; asymmetry suggests distortion -- U 11403) comp 1.3, 187, 0.2 x 0.1, interaction? -- U 11406) S-shaped, w thin uneven arms, B core; late S B b? -- U 11407) broad uneven slightly curved bar; 2 amorphous arms; comp 2.6, 110, 0.4 x 0.3, E-S0; 4.5, 32, 0.8: x 0.2 --

11410   U 11411) 2 knots in blue irr halo -- U 11414) P w U 11415 at 1.2, 42; "distorted blue spir, one pronounced and one washed-out arm" (Zw); thin B arm in direction of U 11415; comp 2.6, 170, 0.5 x 0.5 -- U 11415) see U 11414; dif disrupted spir pattern -- U 11419) comp 3.3, 40, 0.35 x 0.30 --

11420   U 11423) asym main body 0.45 x 0.25, loop on eastern side; distorted?, sev F stell objs near may be disturbing comps ; comp 3.9, 235, 0.35 x 0.35, w comp 0.20 x 0.15 at 0.30 -- U 11429) complex str, 3 fairly B asym arms superimp dif background w hints of spir pattern; P w U 11430 at 11.8, 11 --

11430   U 11430) see U 11429 -- U 11431) P w comp at 4.6, 57, 0.8: x 0.8:, Sc -- U 11432) S b: (de V); ident as N 6797 in CGCG -- U 11437) P w comp at 1.0, 43, 0.5: x 0.3, asym env, prob interaction -- U 11439) comp 2.4, 37, 0.6 x 0.4 --

11440   U 11444) comp 2.6, 237, 0.3 x 0.2 -- U 11448) P w I 1302 at 4.0, 83, 0.8 x 0.7, S b (de V), m = 15. :, brightest in group of 4, v = 4575, v_0 = 4857 --

11450   U 11452) I 1302 at 9.5, 234, see U 11448 -- U 11453) ecc nucl reg, hints of F spir pattern, spir arc at n and nf edge; P w comp at 0.7, 150, 0.25 x 0.20 -- U 11455) P w 19 29.8 + 72 01 at 3.3, 298, 0.8 x 0.2, m = 15.7 --

11460   U 11463) P w b) at 0.55, 120, 0.35 x 0.30; fainter comp at 0.7, 95, 0.1 x 0.1 -- U 11464) main body 0.8 x 0.4, dif streamer in southwest, disturbing comp is probably superimp s part of main body; comp 1.4, 264, 0.4 x 0.1 -- U 11465) vB core 0.25 x 0.20 in large halo; 19 40.5 + 50 30 at 2.4 sf, (0.9 x 0.5), core 0.20 x 0.20, compact, m = = 14.9; comp 0.8, 150, 0.18 x 0.15; sev fainter galaxies near; in a rich *-field -- U 11467) P w U 11468 at 6.9, 168 -- U 11468) see U 11467 -- U 11469) P w U 11471 at 7.0, 87 --

11470   U 11470) SA(s)b: (de V); comp 1.3, 26, 0.12 x 0.10; 3.0, 133, 0.4 x 0.2, dif; 4.2, 15, 0.6: x 0.1 -- U 11471) see U 11469 -- U 11478) VII Zw 915; "red 'sandwich' galaxy, heavy abs lane, twisted tips, internal compact knots" (Zw) -- U 11479) comp 1.5, 78, 0.4 x 0.25; 2.8, 252, 0.8 x 0.35 --

11480   U 11483) 4 group members brighter than 15.8; slightly asym; U 11478 at 6.3 --

11490   U 11491) comp 1.0, 146, 0.3: x 0.1; 4.2, 238, 0.3: x 0.3:, v dif -- U 11494) a) 1.5: x 0.9:, amorph disc w 2 S conds , B center; b) 0.6: x 0.4:, Irr, sep 0.7; * or compact galaxy between these systems; "double system, jets" (CGCG) --

11500   U 11500) P w 19 55.5 + 50 02 at 1.5, 332, 0.7 x 0.4, compact, m = 15.2; comp 2.1, 192, 0.9 x 0.2, spir?; 2.8, 193, 0.6 x 0.15; these 4 galaxies form a quadrupel group -- U 11501) comp 1.8, 308, 0.1 x 0.1; 2.7, 296, 0.3 x 0.2; e dif streamer from U 11501 toward these comps? -- U 11502) "compact in extended halo" (CGCG); comp 0.7, 353, 0.3 x 0.3, dif -- U 11503) comp 2.4, 42, 0.35 x 0.22 -- U 11508) P w U 11510 at 9.5, 57 --

11510   U 11510) dif spir; see U 11508 --

11520   U 11520) comp 4.0, 57, 0.7 x 0.25, Sc? -- U 11522) P w U 11524 at 3.6, 90; comp 2.2, 235, 0.35 x 0.15, spir; 20 09.5 + 05 41 at 4.1, 28, 0.3 x 0.2, m = 15.7 -- U 11524) 2 thin asym arms; B cond n center, *?; fainter blue cond in arm nf center; see U 11522 -- U 11528) comp 3.2, 129, 0.4 x 0.1 -- U 11529) MCG diam values 1.3 x 1.3 are incorrect, also α = 20 11.9 --

11530   U 11530) P w comp at 2.0, 60, 0.4: x 0.4:, v dif, dwarf -- U 11534) P w U 11535 at 3.3, 137 -- U 11535) see U 11534 -- U 11536) VII Zw 929; "neutral spherical very compact uniform disc plus very extended very faint blue spir arms" (Zw) -- U 11537) 2 ** superimp, one near center -- U 11538) P w brighter comp at 1.6, 352, 0.9 x 0.3 -- U 11539) VII Zw 931; "pair of large red fuzzy spherical compacts, m_p = 15.2, and neutral stell comp 10" sf, m_p = 18.6" (Zw)

11540   U 11541) P w comp at 3.5, 357, 0.4 x 0.4, dif -- U 11542) S B(r)ab (de V) -- U 11543) P w 20 20.6 + 09 22 at 6.7, 113, 0.5 x 0.3, m = 15.7 -- U 11546) II Zw 82; S B 0 (de V); "neutral large disc-like elliptical compact, F halo... v = 3955" (Zw) -- U 11548) S (r)b: (de V); comp 3.7, 20, 0.35 x 0.35, spir --

11550   U 11550) P w U 11552 at 2.3, 63 -- U 11552) see U 11550; at least 4 members in the group -- U 11555) interacting w comp at 1.9, 61, 0.22 x 0.12, one arm points toward this obj -- U 11559) P w 20 23.7 + 00 57 at 3.4, 158, 0.8 x x 0.5, m = 15.7 --

11560   U 11560) U 11561 at 17.5; U 11562 at 13.6 -- U 11561) P w U 11562 at 3.4, 17; see U 11560 -- U 11562) see U 11560, 11561; comp 3.7, 111, 0.8 x 0.10 -- U 11563) asym (r) or (rs), dif disc, no spir str -- U 11564) II Zw 92; "blue post-eruptive pair of Sb on edge and wedge-like compact connected by spotty bridge" (Zw); a) 0.25 x 0.10; b) 0.8: x x 0.2, sep 0.7; 20 23.1 + 10 25 1st in chain, 0.9 x 0.8, S B b, m = 14.9 -- U 11567) P w comp at 1.1, 6, 0.5 x 0.15 --

11570   U 11570) SA(r)0/a: (de V); P w comp at 1.5, 68, 0.9 x 0.3 -- U 11571) U 11572 at 4.4, 27; comp 5.1, 233, 0.9 x 0.7, SAB d -- U 11572) see U 11571; comp 1.5, 252, 0.4 x 0.1 -- U 11574) S comp superimp? -- U 11575) cond in sf part, disturbing comp? -- U 11576) prob E or early S0; 20 28.0 - 01 10 at 2.4, 99, 0.25 x 0.25, m = 15.5, w comp 0.7 n, 0.5: x 0.2, v dif; sev S galaxies near --

11580   U 11581) 20 28.9 - 02 07 at 6.8, 111, 0.7 x 0.5, E, m = 15.5 -- U 11582) P w 20 28.3 + 20 02 at 7.5, 215, 0.7 x 0.6, Sc, m = 15.6 -- U 11586) comp 3.1, 228, 0.5 x 0.1, Sc -- U 11587) P w comp at 2.2, 348, 0.9: x 0.7: -- U 11588) B v patchy arcs, tend to disrupted spir pattern; P w 20 30.8 - 02 12 = N 6929 at 4.0, 98, 0.7 x 0.6, m = 14.9 -- U 11589) S B(s)ab (de V); P w U 11590 at 3.9, 148; 20 30.2 + 09 45 = N 6927 at 3.0, 257, 0.4 x 0.15, S0 (de V), v = = 4277, v_0 = 4516, m = 15.6 --

11590   U 11590) S B(s)...? (de V); see U 11589; comp 1.6, 123, 0.4 x 0.3 -- U 11591) comp 0.1 x 0.1 in sf part, interaction -- U 11593) a) 0.4 x 0.4, E?, superimp p tip of b); b) 1.0: x 0.2, prob late spir -- U 11596) 20 33.6 + 01 10 at 6.4, 111, 0.35 x 0.35, m = 15.7 -- U 11597) Arp 29; SAB(rs)cd (de V), Sc - (Ho); SN 1917 a, 1939 c, 1948 b; "supernova once observed in tip of thick arm" (Arp) --

11600   U 11604) SAB(rs)bc (de V), Sb + (Ho) -- U 11605) comp 3.2, 105, 0.5 x 0.3, early spir; 3.8, 132, 0.4 x 0.3 -- U 11607) 20 36.9 + 01 51 at 4.7, 261, 0.7 x 0.3, early spir?, m = 15.2 -- U 11608 + 11609) IV Zw 64; "pair of exceedingly compact blue spherical galaxies (or stars), sep = 15" southsouthwest - northnortheast" (Zw) --

11610   U 11615) comp 2.3, 71, 0.4 x 0.05 -- U 11618) S b (de V) -- U 11619) in triple group w U 11619 at 7.0, 146, and U 11623 at 8.2 --

11620   U 11620) see U 11619; U 11623 at 6.4, 41 -- U 11623) see U 11619, 11623 -- U 11625) comp 1.8, 175, 0.5 x 0.1 -- U 11628) SAB(r)ab (de V); P w U 11629 at 1.9, 135; 20 44.5 + 00 15 = N 6959 at 7.3, 337, 0.50 x 0.25, m = 14.6, comp 0.4 x 0.3 1.0 npp; 20 44.6 + 00 11 = N 6961 at 3.3, 321, 0.4 x 0.4, m = 14.8 -- U 11629) E4 (de V); see U 11628 --

11630   U 11630) 20 44.8 + 00 18 = N 6963 at 5.5, 323, 0.5 x 0.5, E0 (de V), v = 4351, v_0 = 4558, m = 15.2 -- U 11631) blue asym spindle -- U 11633) CGCG note "extremely compact" is prob incorrect, no non-stell obj near -- U 11634) a)

0.6 x 0.4; b) 1.5 x 0.4; sep 0.35 -- U 11639) B centr reg in smooth asym env; comp 5.7, 270, 0.8 x 0.2 --

11640 U 11640) 20 47.5 + 09 46 at 3.9, 338, 0.8 x 0.7, m=15.5; 20 48.1 + 09 38 at 9.0, 125, 0.9 x 0.3, SB: a - b, m= = 15.4 -- U 11643) comp 0.3 x 0.1 attached s, pec?; comp superimp nf center?; P w 20 49.1 + 18 47 at 1.1, 302, 0.9: x 0.3, plume? or v dif comp attached p, m=15.7 -- U 11644) comp 3.3, 193, 0.35 x 0.30 -- U 11645) P w U 11646 at 12.6, 63 -- U 11646) see U 11645 -- U 11649) comp 1.9, 308, 0.5 x 0.2, Irr, v dif --

11650 U 11650) comp 1.6, 111, 0.4 x 0.2 -- U 11652) stell cond in nf part, disturbing comp? -- U 11656) in triple group w 20 55.7 + 10 52 at 15.4, 228, 0.9 x 0.9, m=15.1, and 20 56.7 + 10 57 at 7.0, 178, 0.8 x 0.8, m=15.7 -- U 11657 + + 11658) a) 1.0 x 0.7, S core w filaments; b) 1.6 x 1.3:, large plume; sep 0.9; II Zw 97; "post-eruptive violently disrupted group of 3 blue galaxies w compact cores, sharp jets and one extended spir arm" (Zw) -- U 11659) vB nucl reg, or superimp *?; pair 20 57.8 + 09 22 at 2.5, 125, 0.20 x 0.15, and 2.8, 110, 0.25 x 0.20, sep 0.7, total magn 15.0 --

11660 U 11660) comp 4.2, 60, 0.5 x 0.2 -- U 11664) distorted spir 0.5 x 0.3 w comp 0.2 x 0.2 sp -- U 11666) P w brighter spir at 0.50, 122, 0.9 x 0.4, SAB: b -- U 11668) a) 0.15 x 0.15; b) 0.30 x 0.30 nnf a); IV Zw 67; "pair of blue fuzzy oval compacts. No 1, m_p = 15.5. No 2, 12" nnf No 1, m_p = 14.0" (Zw) -- U 11669) comp 2.5, 307, 0.8 x 0.1; 4.5, 237, (0.4 x 0.4); at least 2 more members in the group --

11670 U 11670) SA(r)0/a (de V) -- U 11671) a) 0.8 x 0.3, arc-shaped, v blue; b) 0.9: x 0.3, patchy -- U 11672) VV 102; VV δ = - 15 53 is incorrect; 2 B conds in center, sep 0.25, disc w spir str; comp 4.8, 70, 0.3 x 0.2 -- U 11673) II Zw 98; a) 0.5 x 0.4, distorted spir w plume; b) 0.3 x 0.2, F plume northw; sep 0.45 -- U 11674) Sc-?; comp 3.6, 85, 0.6 x 0.15; 5.2, 270, 0.7 x 0.15 -- U 11675) P w 21 03.2 + 07 30 at 5.1, 306, 0.6 x 0.3, early spir, m=14.9 -- U 11678) "vF ejecta" (CGCG) --

11680 U 11680) II Zw 101, II Zw 102; M 51 case; "pair of Sb and blue spherical compact 70" nff at tip of spir arm... spectrum of compact: ... v = 7680" (Zw); "compact obj 10" = 5 kpc diam on the end of a drawn-out spir arm of an Sc or SB c galaxy" (Sargent); see Astr. and Ap. 3, 418 -- U 11681) comp? superimp at 0.3, 260, 0.2: x 0.05 -- U 11683) P w 21 06.1 + 18 00 at 1.5, 294, (0.8 x 0.8), E:, m=14.9 -- U 11684) II Zw 103; "elliptical disc galaxy" (Zw); comp 2.0, 194, 0.4: x 0.4:, v dif -- U 11685) 21 06.9 - 02 03 at 3.8, 130, 0.5 x 0.3, m=15.4 -- U 11687) P w 21 07.2 + 14 55 = N 7033 at 1.6, 197, 0.7 x 0.5, E, m=15.2 -- U 11688) ecc nucl cond, or superimp *? --

11690 U 11691) massive arms, one appears disturbed; comp 2.0, 47, 0.3 x 0.2, v dif; 21 08.6 - 02 15 at 5.7, 266, 0.3 x x 0.3, E, m=15.2 -- U 11694) early; pec E or S0?; ext asym env; cond np center may be a comp galaxy -- U 11695) asym open spir, prob interacting w comp at 1.3, 189, 0.6 x 0.5, SB(s)c:; comp 3.9, 63, 0.7 x 0.4, late spir -- U 11696) comp 3.5, 128, 0.5 x 0.2 -- U 11698) a) 0.7 x 0.15; b) 0.8 x 0.45, SB? ...; 21 09.7 + 12 47 at 2.4, 67, 0.9 x 0.8, S0:, m=15.4; sev S objs in vic --

11700 U 11701) comp 0.25 x 0.12 superimp at 0.35 s center; 21 10.3 + 08 34 at 8.7, 235, 0.8 x 0.6, SB a, m=15.5 -- U 11702) P w U 11704 at 5.4, 56 -- U 11704) see U 11702 --

11710 U 11713) in triple group w 21 15.1 + 15 15 at 6.4, 172, 0.6 x 0.6, compact, m=15.3, and comp at 3.9, 354, 0.5 x x 0.3, late spir -- U 11716) comp 3.5, 340, 0.8: x 0.4:, dif -- U 11717) comp 0.65, 200, 0.5: x 0.3:, open spir, disturbed?; 3.3, 54, (0.4 x 0.4), E; cond (comp?) in disc at 0.30, 30, 0.15: x 0.1 --

11720 U 11721) P w comp at 1.6, 278, 0.55 x 0.12 -- U 11724) P w 21 18.1 + 00 52 at 2.7, 355, (0.6 x 0.5), E, m=15.4, S comp 0.55 sp -- U 11727) II Zw 124; "neutral large patchy compact, extended halo" (Zw); main body 0.55 x 0.45 w high surf brightn; comp 0.7, 89, 0.35 x 0.15; 2.6, 18, 0.50 x 0.45, E:; sev fainter galaxies near, prob members of CGCG cl 21 20.1 + 22 56, medium compact, distant --

11730 U 11730) P w U 11732 at 1.5 sf, contact? -- U 11732) see U 11730; spindle, prob early; e dif streamer np? -- U 11733) P w 21 19.7 + 06 53 at 4.7, 118, compact, m=15.6 -- U 11735) comp 0.9, 273, 0.15 x 0.15, alm stell -- U 11736) a) 0.5 x 0.3, pec, comet-shaped (1st from s); b) = 00 - 54 - 15 0.4 x 0.4, E or early spir, 0.7 from a); c) 0.4 x 0.1, 1.5 from b); c) 0.2: x 0.1:, 0.6 from c), e dif; e) 0.2 x 0.2, 0.8 from d) -- U 11737) interacting w comp at 1.0, 160, 0.7 x 0.3, early, contact --

11740 U 11741) II Zw 130; S comp at 0.2, 146; comp 2.4, 312, 0.18 x 0.15 and 3.9, 290, 0.18 x 0.15, both have high surf brightn -- U 11743) cond in disc n center may be a disturbing comp, diam approx 0.6 x 0.2; comp 0.75, 37, core 0.12 x 0.10, interaction?; comp 0.9, 358, 0.12 x 0.10, F -- U 11745) comp 3.8, 183, 0.3 x 0.3, vF --

11750 U 11751) B v blue conds in arms on both sides of center; one arm connected w a) at 1.0, 197, 0.7 x 0.7, SB a:; S stell obj superimp a) at nf side of disc -- U 11752) comp 1.7, 89, 0.3: x 0.2: -- U 11755) poss member of a group w U 11759 brightest -- U 11758) comp 3.1, 283, 0.25 x 0.20, background double system? -- U 11759) P w U 11760 at 4.6, 122; poss brightest obj in a group incl U 11755 --

11760 U 11760) see U 11759; comp 4.7, 37, 0.6 x 0.4, spir -- U 11761) IV Zw 71; "patchy red compact, but absolutely stell on blue plate (prob *)" (Zw) -- U 11762) IV Zw 72; "blue elliptical extremely compact" (Zw) -- U 11763) II Zw 136; "variable very blue spherical, very compact... v = 18315. Intrinsically B Seyfert galaxy, M_p = - 22.0" (Zw); "on the 200-inch plate, which is underexposed, there is only a stellar core less than 3" = 3.5 kpc in diam" (Sargent) -- U 11765) in triple group w 21 31.7 + 08 33 at 7.6, 83, 0.9 x 0.4, m=15.2, spir, disturbed?, and 21 31.8 + 08 26 at 9.0, 88, 0.5 x 0.3, compact, m=15.6, these galaxies are prob interacting -- U 11766) comp 2.4, 266, 0.5: x 0.4: -- U 11767) P w comp at 1.9, 38, 0.9 x 0.3, spir --

11770 U 11772) P w U 11775 at 4.2, 132 -- U 11775) see U 11772 --

11780 U 11780) P w 21 34.6 + 00 14 at 1.9, 54, (0.5 x 0.4), m=15.2, S comp 0.45 np incl in CGCG magn; S comp 0.25 sf incl in CGCG magn -- U 11786) S B(rs)b ? (de V); P w comp at 1.2, 198, 0.45 x 0.10, spindle, hints of F thin bridge; sev vS galaxies near -- U 11788) v open asym 2-armed galaxy; S stell obj in arm f center is prob a disturbing comp -- U 11789) comp 4.0, 218, 0.4 x 0.2 --

11790 U 11791) comp 4.9, 240, double system, a) 0.4 x 0.25, b) 0.5 x 0.2, sep 0.15, total diam 0.7 x 0.3 -- U 11793) comp 3.7, 44, 0.40 x 0.10 -- U 11794) P w 21 40.0 + 12 21 = N 7113 at 4.4, 12, (0.9 x 0.9), E, m=15.2 -- U 11797) v red spindle in e dif asym halo; interacting w b) at 1.9, 137; streamer from the plane of b) appears to continue in the halo around a); in triple group w U 11801 at 2.9, 124 -- U 11798) see U 11797 --

11800 U 11800) U 11804 at 7.2, 86; U 11801) see U 11797; comp 1.0, 78, 0.25 x 0.20 -- U 11802) P w U 11806 at 3.3, 86, interaction; in a rich *- field -- U 11804) interacting w b) at 0.55, 30; F arc curved westw; see U 11800; in a rich *- field -- U 11805) see U 11804, 11800; dif plume sf -- U 11806) see U 11802 -- U 11808) (r) or (rs) 0.8 x 0.6, slightly elong nucl reg; strong galactic reddening --

11810 U 11815) SAB(rs)c (de V) --

11820 U 11821) 21 47.8 + 15 38 at 11.2 nf, double system, a) 0.8 x 0.6, Sc; b) 0.45 x 0.25, Sb - c, sep 0.9, total m=15.7; comp 4.9, 278, 0.3 x 0.1 -- U 11823) IV Zw 78; "neutral patchy spherical compact, F halo" (Zw) -- U 11824) comp 1.0, 330, 0.30 x 0.15 -- U 11825) strong abs along edge; P w comp at 0.7, 317, 0.35 x 0.12 -- U 11826) ident as I 1406 in MCG; P w 21 48.5 + 01 45 = I 1406 at 4.2, 119, 0.3 x 0.25, m=15.5 -- U 11827) complex, prob double or triple; * superimp?; dif extension southw -- U 11828) 21 48.6 + 02 28 at 4.4, 174, 0.25 x 0.20, m=15.3; comp 5.2, 302, 0.7 x 0.7, S0, P w spir 0.5 x 0.2 1.1 sp --

11830   U 11831) comp 2.3, 105, 0.3 x 0.1; 3.1, 40, 0.3 x 0.1 -- U 11835) 21 49.4 + 02 50 = N 7147 at 14.2 s, 0.9 x 0.8,
        S B a, m = 14.8, brightest in group of 6; 21 49.2 + 02 47 = N 7146 at 4.3 sp, 0.9 x 0.5, early spir, m = 15.5 --

11840   U 11840) nucl reg 0.3 x 0.2, surrounded by eF ring; P w 21 50.5 + 04 03 at 4.5, 316, 0.9 x 0.3, Sc, m = 15.6 --
        U 11843) SAB(rs)cd: (de V) -- U 11845) in group of 6; 21 51.7 + 14 47 = I 5144 at 4.6, 286, 0.3 x 0.2, m = 15.4,
        distorted triple group (prob background) p; 21 51.9 + 14 47 at 2.4, 314, 0.5 x 0.4, dif, m = 15.7 -- U 11846)
        comp 2.0, 57, 0.3 x 0.1; 4.8, 146, 0.5 x 0.2 --

11850   U 11850) 21 53.7 - 01 53 at 6.6, 163, 0.6 x 0.6, spir, m = 15.5; comp 4.7, 17, 0.4 x 0.1, vF -- U 11851) P w b)
        at 0.55, 42, 0.4 x 0.15; 21 53.3 + 05 34 at 6.9, 214, 0.6 x 0.6, Sa?, m = 15.5 -- U 11853) comp 1.3, 301, 0.55 x
        x 0.10 --

11860   U 11866) dif blue disc; F blue centr reg, prob complex; short F jet -- U 11867) comp 3.1, 48, 0.30 x 0.12 -- U 11868)
        II Zw 158 -- U 11869) in triple group w comps at 4.4, 250, 0.1 x 0.1, hints of halo; and 5.7, 247, 0.9: x 0.9: --

11870   U 11870) P w comp at 5.2, 133, (0.3 x 0.3), E:, m approx 16. -- U 11871) 2 stell cores in common halo, loop s,
        sep 0.40 -- U 11872) SAB(r)b (de V) -- U 11877) 21 59.3 + 11 37 at 9.4, 270, 0.8 x 0.5, Sa:, m = 15.6; comp 2.7,
        107, 0.35 x 0.10 -- U 11878) prob S B 0 or S B a w B comp 0.2 x 0.2 superimp near center -- U 11879) P w a) at 0.55
        p, 0.6: x 0.1, main body 0.45 x 0.10, F bridge to s part of b) --

11880   U 11880) B massive bar w short arm embryos at the tips, no spir str in env -- U 11885) comp 1.6, 116, 0.4 x 0.3,
        dif -- U 11886) v open S-shaped galaxy; comp 0.50, 288, 0.2 x 0.1, blue; measurements disturbed by * GC 30848 --
        U 11888) in triple group w 22 00.9 + 12 24 at 2.1, 290, 0.4 x 0.2, m = 15.7; and 22 01.0 + 12 25 = N 7195 at 1.5,
        354, 0.45 x 0.40, E - S0, m = 15.6; comp 0.7, 170, 0.12 x 0.12, alm stell; 0.9, 39, 0.3 x 0.1 --

11890   U 11891) V Zw 380; "blue large irr dwarf galaxy w compact conds" (Zw) -- U 11892) P w U 11893 at 4.3, 137 --
        U 11893) see U 11892 -- U 11894) comp 1.8, 145, 0.3 x 0.2, vF -- U 11897) ·cond 0.1 x 0.1 at 0.3 sf center, distur-
        bing comp? -- U 11899) P w comp at 1.2, 306, 0.8 x 0.8, E - Sa, prob interaction --

11900   U 11904) vF v ext disc or halo; 22 03.3 + 16 32 at 1.6, 134, 0.6 x 0.3, S0?, m = 15.6 -- U 11905) II Zw 163 --

11910   U 11910) disturbed by comp superimp 0.3 n center, 0.4 x 0.2; comp 3.2, 19, 0.3 x 0.2 -- U 11911) core 0.15 x 0.15,
        smooth asym halo; v smooth F arc in northeast -- U 11914) (R)SA(r)ab (de V), Sb - (Ho) -- U 11915) S-shaped Irr;
        P w b) at 1.0, 164, 0.15 x 0.15 -- U 11916) II Zw 166; "badly disrupted neutral compact w sharp spikes and halo" (Zw) --

11920   U 11920) comp 4.5, 140, 0.5: x 0.3:, F -- U 11926) comp 4.2, 60, 0.7 x 0.15 -- U 11927) low surf brightn; comp
        2.5, 166, 0.6: x 0.6: v dif; U 11931 at 8.0, 78 -- U 11929) hints of outer disc to 1.7? x 1.7?; late S0 or early
        Sa --

11930   U 11931) asym spir w 3 main arms, 2 v blue conds close to each other at tip of secondary arm at 0.65, 268; comp
        1.0, 316, 0.8 x 0.12, S0?, background?; see U 11927 -- U 11934) P w 22 08.6 + 01 52 at 4.5, 90, 0.7 x 0.5, Sa?,
        m = 15.6 -- U 11936) P w U 11937 at 3.0, 240 -- U 11937) see U 11936 -- U 11938) comp 2.4, 175, 0.8 x 0.55, (R)SB a
        or S B(r)a - b; 2.6, 89, 0.35 x 0.12; 3.2, 143, 0.6 x 0.12 -- U 11939) open S-shaped galaxy; comp 3.2, 255, 0.4 x
        x 0.1; 5.5, 273, 0.4 x 0.3, E?, m approx 16. --

11940   U 11940) comp 3.8, 239, 0.55 x 0.30 -- U 11942) U 11945 at 3.6, 113 -- U 11945) see U 11942; early S B b?; comp 1.7,
        85, 0.5 x 0.1 -- U 11948) comp 3.4, 97, 0.25 x 0.20; 4.3, 213, 0.3 x 0.2; many S galaxies near --

11950   U 11950) measurements disturbed by nearby *; comp 1.0, 108, 0.45 x 0.15; 5.0, 86, 0.6 x 0.5, core 0.12 x 0.12 --
        U 11953) P w comp at 3.1, 314, 0.9 x 0.15, Sa: -- U 11955) S stell obj at tip of arm at 0.55, 216, 0.05 x 0.05;
        blue cond 0.3 nf center -- U 11957) disturbed?; S stell obj superimp 0.35 np center; comp 4.2, 44, 0.45 x 0.25 --
        U 11958) Arp 169, II Zw 172; chain of 3 objs , plumes from a) and b) form dif halo; a) 0.25 x 0.25; b) 0.20 x 0.20
        at 0.5 sf a); c) 0.15 x 0.12 at 0.6 sf b); "F dif plumes coming away from two galaxies. 3C 442" (Arp); "3 blue com-
        pacts, one spherical, 2 patchy in straight row, d = 100" northwest - southeast... v = 7854" (Zw); N 7236, S 0: (de
        V); N 7237, S 0: (de V); comp 2.2, 330 from a), 0.20 x 0.15; 22 12.1 + 13 32 at 5.5, 215, (0.9 x 0.9), 0.12 x 0.12
        (core), compact, m = 15.7 --

11960   U 11963) ident as I 1441 in MCG; CGCG lists 22 13.1 + 37 03 as I 1441; U 11969 at 7.3 -- U 11964) P w U 11968 at
        5.0, 82 -- U 11967) P w comp at 1.5, 308, 0.9: x 0.5, late spir, hints of outer disc to 1.5? x ... -- U 11968)
        II Zw 174; "edge-on spir, tremendous abs lane and various large blue compact patches" (Zw); see U 11964 --
        U 11969) b) superimp at 0.50, 57, 0.15 x 0.15; 22 13.1 + 37 03 = I 1441 at 4.0, 0.8 x 0.25, m = 15.3; 22 13.2 + 37 02
        = N 7240 at 3.5, 0.5 x 0.5, core 0.20 x 0.20, compact, m = 15.6, S0 (de V); U 11963 at 7.3 --

11970   U 11972) comp 0.7, 336, 0.12 x 0.12; 2.9, 169, 0.4 x 0.1, F; sev smaller galaxies near -- U 11975) comp 4.0, 231,
        0.35 x 0.15 --

11980   U 11980) prob a comp superimp at 0.4, 335, 0.25 x 0.18 -- U 11981) comp 1.2, 232, 0.5 x 0.1, vF; 4.4, 162, 0.4 x
        x 0.2 -- U 11983) comp 3.8, 66, 0.3 x 0.3 -- U 11984 + 11985) VV 242, Arp 278; strongly disrupted pair; "dif mate-
        rial between galaxies, many internal abs lanes" (Arp) -- U 11988) comp 3.8, 142, 0.7: x 0.4:, dwarf: --

11990   U 11995) comp 3.3, 190, 0.6 x 0.1 -- U 11996) in triple group w comps at 0.5 p, 0.1 x 0.1, and 1.5, 60, 0.6 x 0.1
        -- U 11998) in triple group w 22 19.9 + 04 52 = I 1444 at 2.5, 118, 0.6 x 0.4, m = 15.4; and comp at 4.5, 25, 0.7 x
        x 0.25 --

12000   U 12000) in group w 22 20.0 + 02 32 at 5.4, 204, 0.7 x 0.5, Sb, m = 15.4, brightest in the group; and comps at 5.6,
        320, 0.4 x 0.3, spir; 7.6, 180, 0.45 x 0.25 -- U 12001) 22 19.6 + 36 06 = N 7263 at 6.1, 250, 0.8 x 0.7, core 0.15 x
        x 0.15, compact, m = 15.7 -- U 12004) P w U 12007 at 2.8, 112; comp 2.5, 303, 0.5 x 0.20; 3.0, 16, 0.6 x 0.15 --
        U 12005) SN 1970 h -- U 12007) IV Zw 98; "post-eruptive pear-shaped compact" (Zw); see U 12004 -- U 12009) vB main
        body 0.55 x 0.55, e dif outer parts; comp 2.6, 208, 0.8 x 0.2 --

12010   U 12011) "intercon pair of disrupted patchy blue compacts, sep = 15" east - west... [a] v = 6692... [b] v = 6707"
        (Zw); a) 0.25 x 0.20; b) 0.20 x 0.15 + plume -- U 12012) P w comp at 3.9, 211, 0.7 x 0.3, S B: a -- U 12014) P w comp
        at 1.1, 115, 0.8 x 0.4, Irr; comp 4.1, 332, 0.6: x 0.4:, v dif -- U 12017) comp 2.8, 124, 0.5 x 0.4:, S B ... --
        U 12019) one of the two brightest galaxies in group w at least 6 members --

12020   U 12024) P w comp at 4.5, 345, 0.8 x 0.7, S0; comp 3.9, 138, 0.4 x 0.1, F -- U 12025) one of the two brightest
        galaxies in group, see U 12019; comp 2.8, 232, 0.8 x 0.1; sev other galaxies near -- U 12026) in triple group w
        22 22.0 + 35 53 = N 7276 at 2.4, 196, 0.9 x 0.9, core 0.18 x 0.18, very compact, m = 15.0; and 22 21.9 + 35 57 =
        = N 7273 at 4.5, 356, 0.8 x 0.5, m = 14.8 -- U 12028) P w a) at 0.9, 203, 0.20 x 0.15; comp 1.9, 54, 0.30 x 0.15 --

12030   U 12031) comp 3.1, 98, 0.35 x 0.15 -- U 12032) interacting w comp at 1.5, 273, 0.6: x 0.5:, core 0.22 x 0.15, dif
        bridge -- U 12035) S 0? (de V); P w 22 24.4 + 15 55 at 4.9, 64, 0.9 x 0.6, dwarf irregular, m = 15.7 -- U 12036)
        P w comp at 1.6, 199, 0.4 x 0.4, core 0.12 x 0.12 -- U 12039) comp 3.8, 280, 0.8 x 0.5, double system; 4.2, 20,
        0.7 x 0.4, spir --

12040   U 12040) comp 4.5, 121, 0.7 x 0.3 - U 12041) comp 0.9, 22, 0.3 x 0.3, vF; 3.6, 86, 0.3 x 0.2; 4.2, 95, 0.3 x 0.2,
        vF -- U 12042) F stell core in v large dif halo, hints of extensions to diam 2.? x 1.9?; comp 4.0, 297, 0.3 x 0.15,
        double?, or * superimp -- U 12044) comp 0.7, 127, 0.4 x 0.3, e dif, dwarf -- U 12045) SA(r)bc (de V) -- U 12047)
        6 S comps w high surf brightn inside 5.0 -- U 12048) SN 1964 h -- U 12049) comp 2.9, 353, 0.7 x 0.5, S0? --

12050   U 12052) P w U 12053 at 0.75, 107 -- U 12053) see U 12052 -- U 12055) a) 0.5: x 0.1, Sc; b) 0.1 x 0.1; c) 0.45x
         x 0.15, S0?, 0.2 nf b); sep a) and b)+c) 0.7 -- U 12059) P w comp at 4.8, 281, 0.45x0.35, core 0.15x0.15, E -
         - S0; comp 3.8, 55, 0.25x0.20 --

12060   U 12062) comp 2.6, 114, 0.4 x 0.1; 4.0, 50, 0.4 x 0.12 -- U 12064) b) at 0.60, 11, 0.20x0.18; comp 1.6, 290,
         0.8 x 0.12 -- U 12065) strongly asym bar spir, broad loop in northwest; Reinmuth ident N 7304 w a * 3′ nf this
         galaxy -- U 12066) a) 0.15x0.15, vB core; b) 1.0 x 0.6:, disrupted spir; bridge or connecting spir arm --
         U 12067) P w 22 29.3 + 20 27 at 6.9, 348, 0.8 x 0.25, late spir, m=15.4 --

12070   U 12072) in group of 6, w members at 0.55 (F), 2.2, 3.1, 5.6 and 7.8 (F) -- U 12073) brightest in quadrupel group,
         see U 12075; P w U 12075 at 3.0, 96 -- U 12074) II Zw 181; "neutral post-eruptive large patchy oval compact, w
         blue loops and red conds " (Zw) -- U 12075) see U 12073; one arm distorted, prob interaction w comp at 1.2, 5,
         0.12x0.12, alm stell; comp 1.1, 98, 0.8: x 0.35, dif, 2 S cores, symbiotic pair?; these galaxies form quadrupel
         group incl 22 30.9 + 38 54 at 5.1, 130, (0.8 x 0.8), core 0.18x0.18, very compact, m=15.5 -- U 12077) P w U
         12079 at 2.2, 50; comp 1.7, 17, 0.3: x 0.3:, F, dif; these galaxies form triple group -- U 12079) see U 12077 --

12080   U 12080) P w U 12083 at 16.5, 25 -- U 12082) DDO 213; S V (vdB) -- U 12083) see U 12080 -- U 12089) second brigh-
         test in group of 9, see U 12091; P w U 12091 at 2.0, 29 --

12090   U 12091) brightest in group of 9, in center; see U 12089; group members at 2.5, 17, 0.5 x 0.3; 2.6, 121, 0.10 x
         x 0.10, and 2.8, 128, 0.40x0.35, sep 0.45; 3.8, 235, 0.15x0.10; 4.8, 181, 0.9: x 0.2; 4.8, 247, 0.3 x 0.1 --
         U 12095 + 12096) total diam 1.6 x 1.0; comp 1.4 np geom center, 0.5 x 0.1; MCG α = 22 33.6 is incorrect --
         U 12099) see U 12100 --

12100   U 12100 + 12099) VV 288, Arp 319; N 7318 A, E 2 pec (de V); N 7318 B, S B(s)c pec (de V); in Stephan's Quintet; B
         cores sep 0.35, connecting loops; 22 33.6 + 33 41 = N 7317 at 1.8, 219 from U 12099, 0.7 x 0.5, E4: (de V), v =
         = 6736, v_0 = 7013, m=15.3, S stell comp 0.45 p -- U 12101) SA(s)bc pec (de V); brightest in Stephan's Quintet;
         foreground obj?, redshift indicates the same distance as for N 7331 = U 12113; VV 288, Arp 319, see U 12100 --
         U 12102) S B(s)bc pec (de V); Stephan's Quintet; double arm east; s arm continues in dif streamer east; VV 288,
         Arp 319, see U 12100 -- U 12103) comp 1.7, 177, 0.3 x 0.2 -- U 12108) P w 22 34.6 + 18 53 = N 7324 at 1.8, 87,
         0.8 x 0.4, S0:, m=15.1 -- U 12109) core 0.12x0.10, dif prob asym halo; comp 0.8, 113, 0.3 x 0.2, F --

12110   U 12110) in triple group w 22 34.4 + 14 07 at 2.0, 217, 0.25x0.15, m=15.7; and comp at 1.7, 109, 0.2 x 0.2 --
         U 12111) comp 4.9, 101, 0.7 x 0.2 -- U 12113) SA(s)bc (de V), Sb+ (Ho); SN? 1899 (prob non-existing; emulsion
         defect); SN 1959 d; see U 12101; quadrupel group p, incl U 12116 = N 7335, U 12120 = N 7337, and N 7340 (see
         U 12116) -- U 12114) comp 2.7, 319, 0.3 x 0.1, F -- U 12115) S0 pec (de V); P w U 12122 at 5.2, 98 -- U 12116)
         S a (de V); in quadrupel group f U 12113 = N 7331, 3.6 from center of that galaxy; U 12120 at 4.7; 22 35.4 +
         + 34 10 = N 7340 at 5.5 sf, (0.9 x 0.6), core 0.22x0.20, compact, m=14.9; comp 2.1, 13, 0.8 x 0.3, last in group
         -- U 12117) core 0.12x0.12, dif disc or halo; P w comp at 1.0, 77, 0.20x0.10, hints of halo w diam 1.0? x 0.5?;
         rich *- field --

12120   U 12120) in quadrupel group f U 12113 = N 7331, 5.0 from center of that galaxy; see U 12115; 22 35.4 + 34 10 at
         4.2 nf -- U 12122) SAB(s)bc:? (de V); see U 12115 -- U 12124) "double nebula" (CGCG); comp 4.0, 201, 0.8 x 0.15
         -- U 12126) third brightest in subgroup, U 12127 brightest, at 10.8; U 12130 at 6.8 -- U 12127) brightest in sub-
         group, prob 11 members, 6 inside 5.0; 22 36.1 + 35 07 at 2.4, 338, 0.8 x 0.2, S0-a, m=15.4 -- U 12128) S stellar
         B nucl reg; eF disc -- U 12129) S B(s)b (de V); comp 4.3, 107, 0.7: x 0.4:, v dif, dwarf: --

12130   U 12130) second brightest in 12127 subgroup; U 12127 at 13.0 -- U 12133) comp 5.4, 48, 0.5 x 0.4, spir -- U 12136)
         22 37.1 + 10 49 = N 7346 at 6.1, 303, 0.4 x 0.3, E:, m=15.6 -- U 12139) 22 35.8 + 08 22 possible group member,
         0.6 x 0.5, S0-a, m=15.7 --

12140   U 12142) Scd ? (de V); comp 4.6, 328, 0.3 x 0.2, F -- U 12144) ring-shaped bar w ridge along major axis; U 12145
         at 6.6, 157 -- U 12145) prob v open bar spir; see U 12144 -- U 12147) broad dif plume in pos angle 37, short
         spir arm sp; comp 0.9, 128, 0.2 x 0.1, vF, prob not disturbing -- U 12148) "patchy red compact, disrupted halo"
         (Zw); P w U 12153 at 2.9, 131 --

12150   U 12152) main body 0.55x0.45; asym arcs, sev cond   -- U 12153) "very blue post-eruptive osculating pair of com-
         pacts, curved tail and disrupted halo... v = 7105" (Zw); II Zw 185; "core about 15" = 7 kpc diam w tail about 12
         kpc long" (Sargent), listed by Sargent as I 5442, which is incorrect; see U 12148 -- U 12157) P w b) at 0.85,
         102, 0.12x0.10; * superimp; comp 2.9, 350, 0.6 x 0.10; 4.0, 61, 0.6: x 0.2; "distorted Sc... compact core" (Zw) --

12160   U 12169) P w comp at 2.0, 243, 0.40x0.22; comp 1.5, 207, 0.3 x 0.2 --

12170   U 12171) sev galaxies near; 22 40.9 + 08 25 at 5.8, 158, 0.25x0.15, high surf brightn, m=15.5; comp 1.7, 35,
         0.5 x 0.2; 1.7, 174, 0.3 x 0.1; 4.2, 174, 0.6 x 0.5; 4.6, 340, 0.3 x 0.1 -- U 12174) comp 3.0, 353, 0.4 x 0.1 --
         U 12177) P w comp at 1.2, 45, 0.7 x 0.5, Sc -- U 12179) 22 43.0 + 33 47 at 5.0, 41, 0.35x0.22, m=15.6 --

12180   U 12180) comp 2.9, 286, 0.3 x 0.3, dif -- U 12181) asym B arms; may be interacting w comp at 0.9, 116, 0.18x0.10
         -- U 12188) a) 0.9 x 0.45, spir, complex center, 2 outer arms, one stronger; b) 0.5: x 0.5:, v dif --

12190   U 12191) in triple group w U 12193 at 1.8, 155; and 22 46.5 + 27 18 at 2.3, 107, 0.5 x 0.3, very compact, m=15.7
         -- U 12193) see U 12193 -- U 12196) a) at 1.2, 347, 0.5 x 0.25, spir --

12200   U 12201) group of 5 f, nearest galaxy at 4.3, 76, 0.30x0.30 -- U 12202) comp 3.2, 118, 0.7: x 0.2, dif -- U 12206)
         * or stell comp superimp arm nf center, eF streamer northw -- U 12207) E0 (de V); P w U 12209 at 5.8, 19; 22 47.1 +
         + 11 20 = N 7383 at 5.6, 236, 0.8 x 0.7, S B0, m=15.1; 22 47.8 + 11 19 = N 7389 at 5.9, 117, 0.9 x 0.5, S B0, m=
         = 15.2; 22 47.8 + 11 22 = N 7387 at 5.9, 115, 0.5 x 0.4, E-S0, m=15.3; 22 47.9 + 11 16 = N 7390 at 7.7, 128, 0.8 x
         x 0.5, S0-a, m=15.7 -- U 12209) S0 (de V); see U 12207 --

12210   U 12210) (r) 0.7 x 0.7; comp 1.6, 171, 0.6 x 0.35 -- U 12214) P w 22 48.7 + 31 06 at 0.50, 95, 0.35x0.25, compact,
         m=15.7; comp 3.0, 76, 0.5 x 0.3 -- U 12215) IV Zw 118; "post-eruptive patchy red oval compact w large F halo and
         'ejected' knots" (Zw); core 0.25x0.25 in smooth halo; comp 1, 283, 0.4 x 0.2, v dif, Irr -- U 12216) double nucl?
         or abs? -- U 12219) a) 0.7 x 0.5, S0-a:; b) 0.5 x 0.3, S B..., sep 0.5; comp 3.3 nf a), 0.4 x 0.3 --

12220   U 12220) strong abs in plane, large nucl bulge, halo; comp 3.1, 234, 0.30x0.25, E?; 5 or 6 members in the group
         brighter than 15.8; 22 50.3 + 00 52 = N 7397 at 6.5, 68, 0.8 x 0.4, S B a, m=15.3; U 12232 at 9.3, 46; 22 50.5 +
         + 00 53 at 9.5, 72, 0.7 x 0.6, early S B .., disturbed by comp at 1.5 f, m=15.7 -- U 12222) comp 3.5, 232, 0.5 x
         x 0.2, dif -- U 12224) trace of minor planet on PA maps at 3., 79, 0.18 on blue map, 0.45 on red -- U 12225)
         U 12220 at 9.3, 226 -- U 12227) P w comp at 3.8, 267, 0.9 x 0.10, Sc --

12230   U 12230) inner S str; ecc ring 0.8 x 0.4, slightly asym outer disc; comp 2.5, 60, 0.50x0.25 -- U 12231) P w
         22 51.1 + 31 23 at 1.7, 335, 0.55x0.5, m=14.7 -- U 12234) sev comps, largest at 2.6, 190, 0.9 x 0.7, v dif --
         U 12236) knotty disc; prob a large nr of F irr arms; comp 2.5, 144, 0.7: x 0.5:, v dif -- U 12237) P w U 12243 at
         7.2, 126 -- U 12238) in a dense part of the cl --

12240   U 12241) U 12244 at 4.7, 70; comp 2.7, 292, 0.6 x 0.15, early spir; 22 51.7 + 19 56 at 6.4, 259, 0.5 x 0.4, spir,
         m=15.7; 22 51.3 + 19 56 = N 7409 at 11.1, 262, 0.5 x 0.4, S B 0, m=15.6 -- U 12242) in a dense part of the cl,
         7 galaxies inside 5.0 -- U 12243) see U 12237 -- U 12244) a) at 0.45, 277, 0.45x0.25; see U 12241 -- U 12245) sev

blue patches, S red core -- U 12248) chaotic; irr bar-shaped main body, dif spir arc n and f; blue --

12250 U 12250) P w U 12253 at 6.5, 103 -- U 12252) sev F galaxies near; comp attached at 0.40, 155, 0.3 x 0.1 -- U 12253) see U 12250; slightly integral-shaped; sev S galaxies near; comp 1.0, 78, pair, total diam 0.2 x 0.2; 1.7, 310, 0.35 x 0.05; 2.2, 49, 0.6 x 0.2, spir -- U 12256) measurements disturbed by nearby * GC 31976 at 3.8; 22 54.1 + + 35 58 at 8.5 sp, 0.9 x 0.35, spir?, m=15.6 -- U 12257) comp 4.0, 72, 0.7 x 0.6, SBa: -- U 12258) comp 3.9, 85, 0.4 x 0.2 --

12260 U 12262) comp 4.5, 287, 0.3 x 0.2; sev S F galaxies in vic -- U 12265) a) 0.20 x 0.15, high surf brightn; b) 0.5 x x 0.2; sep 0.5 -- U 12266) comp 2.0, 287, 0.25 x 0.20 -- U 12267) VV 84, see U 12269; Pair U 12269 at 0.9 nf; sev galaxies near -- U 12268) S cond 0.2 f center; comp 2.1, 350, 0.3 x 0.1 -- U 12269) VV 84a; VV 84 b superimp, 0.9 x x 0.20; U 12267 at 0.9 sp; 22 55.4 + 25 54 = N 7431 at 1.5 np, 0.8 x 0.2, m=15.6, listed as N 7433 in CGCG (U 12269 as N 7436 a + b) --

12270 U 12274) sev galaxies near (at 1.4, 1.9, 2.2, 3.5 and 4.7) -- U 12275) 22 56.2 + 18 33 = I 5276 at 6.6, 155, 0.8 x x 0.35, Sa?, m=15.2; comp 2.5, 290, 0.5 x 0.4, spir -- U 12276) comp 3.3, 62, 0.4 x 0.2, dif; a large nr of eF galaxies p, prob members of CGCG cl 22 55.0 + 35 31, medium compact, very distant -- U 12277) brightest in group of 5, incl comps at 3.9, 270, 0.7: x 0.7:; 4.1, 258, core 0.15 x 0.15, halo 0.5? x 0.5?; 5.8, 235, 0.6 x 0.4; 6.0, 253, 0.25 x 0.20, F -- U 12278) 22 56.5 + 20 07 at 5.9, 17, 0.8 x 0.2, spir, m=15.4 --

12280 U 12280) 2 cores, sep 0.18, in common env; P w comp at 1.5, 305, 0.5 x 0.3, E-SO; comp 4.7, 148, 0.4 x 0.4 -- U 12283) P w U 12289 at 5.3, 112 -- U 12288) late spir w S comp superimp p center, long streamer northw from comp -- U 12289) see U 12283 --

12290 U 12290) comp 3.9, 237, 0.4 x 0.2; 4.5, 299, 0.4 x 0.1 -- U 12291) comp 1.3, 50, 0.3 x 0.2 -- U 12292) 2nd brightest in 12298 group, which incl 22 57.1 + 38 50 = N 7445 at 3.7 sp, 0.7 x 0.2, m=15.6; and 22 57.2 + 38 48 = N 7446 at 4.2 s, (0.9 x 0.9), core 0.18 x 0.18, m=15.7 -- U 12293) comp 0.9, 199, 0.3: x 0.1, eF; 3.5, 323, 0.4 x 0.4, Sc: -- U 12294) Arp 13; SA(rs)bc (de V); in Arp's class "spir galaxies w detached segments" -- U 12297) in triple group w 22 57.8 + 31 06 at 2.1, 179, 0.9 x 0.30, m=15.5; and comp at 0.9, 54, 0.6 x 0.2; fainter comp at 2.2, 141, 0.22 x x 0.10 -- U 12298) brightest of 6; see U 12292; U 12292 at 8.9 sp --

12300 U 12300) P w U 12307 at 9.3, 114 -- U 12301) a) 0.2 x 0.1; b) 0.4 x 0.3; sep 1.0 -- U 12302) cores 0.20 x 0.20 and 0.15 x 0.15, sep 0.20, in vF ext halo; in center of a concentration of galaxies, about 15 inside 5.0 -- U 12303) comp 2.8, 162, 0.35 x 0.25, high surf brightn; 2.9, 6, 0.4 x 0.3, dif -- U 12305) E + 4 (de V); comp 1.7, 203, 0.4 x x 0.4 -- U 12306) SA(rs)0- ? (de V) -- U 12307) see U 12300 -- U 12309) P w comp at 4.9, 92, 0.9 x 0.5, Sc --

12310 U 12311) asym spindle, early spir?; U 12306 at 7.8 -- U 12312) 3-armed galaxy; cond at tip of spir arm 0.35 n center -- U 12314) 22 59.9 + 15 17 = N 7467 at 9.6, 100, 0.4 x 0.3, very compact, m=15.6 -- U 12315) SO? (de V); U 12316 at 0.8 np; in triple subgroup w U 12317 at 1.9, 106 -- U 12316) see U 12315; U 12317 at 2.6, 115 -- U 12317) E ? (de V); SN? 1950 (photografic effect?); brightest in subgroup of 3, see U 12315, 12316 -- U 12318) P w 22 59.6 + 25 29 at 5.5, 355, 0.7 x 0.2, m=15.5 --

12320 U 12322) P w comp at 1.0, 73, 0.5 x 0.2 -- U 12323) comp 4.2, 197, 0.45 x 0.45, S B 0-a -- U 12326) M 51 case; comp at tip of F dif arm at 0.55 nf, vF; comp 1.1, 1, 0.3 x 0.2 -- U 12329) vB centr reg 0.45 x 0.25; 2 thin jets --

12330 U 12330) 23 01.0 + 09 01 at 8.5, 128, 0.8 x 0.4, S B ..., m=14.9 -- U 12332) Arp 298; Pec SAB(rs)a (de V); P w 23 00.8 + 08 36 = I 5283 at 1.4, 25, 0.7 x 0.25, distorted spir, v = 4875, v_0 = 5083, m=15.2; in Arp's class "double galaxies w wind effects"; "abs, knots. Note apparent re-entrant spir arm on s galaxy [U 12332]" (Arp) -- U 12333) P w a) at 0.45, 224, 0.3 x 0.1; sev galaxies near; 23 00.9 + 19 40 at 2.4, 125, 0.8 x 0.15, m=15.6; 23 00.7 + + 19 45 at 4.3, 331, 0.7 x 0.3, m=15.7 -- U 12334) comp 3.6, 54, 0.30 x 0.20, high surf brightn -- U 12335) comp 1.9, 126, 0.4 x 0.08 -- U 12336) 23 01.1 + 01 35 at 7.7, 243, 0.7 x 0.6, compact, m=15.4 -- U 12337) both E or SO; 23 01.6 + 19 46 = N 7474 at 1.6, 240, 0.3 x 0.3, SO:, m=15.2 -- U 12338) 23 01.8 + 17 41 at 2.9, 37, (0.6 x 0.5), m=15.5 -- U 12339) III Zw 92; "oval spherical compact (or * and F galaxy)" (Zw) --

12340 U 12341) comp 3.9, 210, 0.9 x 0.8, core 0.15 x 0.15 -- U 12343) S B(s)c (de V) -- U 12347) comp 3.7, 160, 0.3 x 0.3 -- U 12349) comp 1.8, 260, 0.6 x 0.25; 4.5, 292, 0.4 x 0.4, S B 0-a --

12350 U 12350) comp 4.8, 104, pair in contact, total diam 0.3 x 0.2 -- U 12356) P w U 12357 at 1.5, 160, interaction; δ's for this pair are interchanged in MCG; v blue comp at 1.7, see U 12357 -- U 12357) see U 12356; comp 1.3, 267, 0.4 x 0.1, Irr, v blue -- U 12358) P w U 12360 at 4.6, 130 -- U 12359) comp 2.2, 313, 0.7 x 0.4, dif --

12360 U 12360) see U 12358; comp 2.1, 155, 0.3 x 0.2, F; 2.1, 171, 0.25 x 0.12 -- U 12361) comp 1.5, 90, 0.6 x 0.3 -- U 12363) blue cond at tip of spir arm; asym spir str, prob twisted plane; comp 1.9, 251, 0.3 x 0.2 -- U 12365) II Zw 188; (r) 0.55 x 0.35; "neutral triple elliptical ring galaxy, compact core... v = 6220" (Zw); "An elliptical obj of high surf brightn 30" x 20" = 12 x 8 kpc is surrounded by a distorted but uniform ring about 90" = 36 kpc diam and 15" = 6 kpc thick. The obj resembles I Zw 155" (Sargent); v_0 = 6427 (Sargent); P w U 12378 at 8.6, 64 -- U 12366) in group of 4 or 5 galaxies, incl comps at 0.5 s, 0.6 x 0.1, Irr; 2.6, 180, 0.7 x 0.5, Sa-b; and U 12369 at 2.4, 28 -- U 12368) vF ext halo; comp 2.9, 101, 0.4 x 0.1 --

12370 U 12372) v high surf brightn; asym, one B massive arm p; symbiotic pair? -- U 12375) comp 2.6, 172, 0.3 x 0.2 -- U 12376) III Zw 93; "neutral fuzzy spherical compact" (Zw) -- U 12377) P w U 12387 at 1.2, 144; 2 smaller galaxies near, see U 12387 -- U 12378) see U 12365 --

12380 U 12381) V Zw 397; "neutral large S B c spir w compact spherical core" (Zw) -- U 12383) ring-shaped bar 0.8 x 0.35; comp 4.5, 245, 0.35 x 0.20 -- U 12384) P w U 12386 at 3.4, 78 -- U 12386) see U 12384 -- U 12387) see U 12377; comp 1.6, 15, 0.9: x 0.7:, bar, dwarf spir?; 1.9, 354, 0.4 x 0.1, dif spindle; quadrupel group?; comp 4.5, 82, 0.5 x 0.3, dif -- U 12388) comp 3.1, 237, 0.3 x 0.2, F -- U 12389) V Zw 398; "osculating pair of blue spherical very compacts (or **) in large, uniform halo" (Zw); cores w total diam 0.22 x 0.18 --

12390 U 12391) comp 3.6, 203, 0.6 x 0.2, early spir -- U 12392) S B(s)c (de V) -- U 12397) S O (de V); 23 08.0 + 07 18 = = N 7501 at 2.0, 77, (0.8 x 0.6), E3 (de V), v = 12714, v_0 = 12914, m=15.3; 23 08.2 + 07 17 = N 7503 at 5.0, 99, (0.8 x 0.7), E1 (de V), m=14.9, brightest in cluster, v = 13229, v_0 = 13429; a large nr of S galaxies w high surf brightn near --

12400 U 12404) comp 1.8, 324, 0.4 x 0.12 --

12410 U 12414) P w 23 10.0 + 30 52 at 1.7, 74, 0.22 x 0.20, very compact, m=15.7 -- U 12416) B main body 0.55 x 0.20, dif streamer n -- U 12419) III Zw 95; "neutral spherical compact" (Zw) --

12420 U 12420) comp 3.2, 160, 0.5: x 0.4:, v dif -- U 12422) P w U 12423 at 6.6, 1 -- U 12423) Sb?; see U 12422 -- U 12428) P w U 12432 at 3.7, 96; comp 4.7, 306, 0.5 x 0.10 -- U 12429) amorphous disc w knots and patches; comp 4.6, 299, 0.7 x 0.6, dif --

12430 U 12432) see U 12428 -- U 12433) (r) 0.7 x 0.6; P w U 12445 at 9.1, 37 -- U 12436) a) 0.8 x 0.1; b) 0.2 x 0.2; sep 0.5 -- U 12437) P w U 12438 at 9.4, 0, in group -- U 12438) see U 12437 -- U 12439) comp 4.5, 72, 0.35 x 0.15 --

12440 U 12441) 23 12.1 + 30 54 at 4.6, 51, 0.7 x 0.2, m=15.7 -- U 12442) SA bc ? (de V); P w U 12447 at 3.1, 47 -- U 12443) slightly asym disc or env; comp 2.0, 95, 0.5 x 0.08 -- U 12444) 23 11.8 + 31 13 at 4.3, 227, 0.5 x 0.3, m=15.6 -- U 12445) see U 12442 -- U 12447) S B bc:? (de V) -- U 12448) slightly brighter comp at 6.5, 27, 0.9 x x 0.8, Sc —

12450   U 12450) B asym centr reg, F asym disc, hints of spir pattern; comp 4.6, 287, 0.35 x 0.25 — U 12453) Arp 99; in group w U 12456 at 3.1, 104, and U 12457 at 5.3, 39; see U 12457 — U 12454) III Zw 96; P w b) at 0.9, 152, 0.15 x 0.12   — U 12456) Arp 99; see U 12453, U 12457 at 4.8, 4; see U 12457 — U 12457) Arp 99; in Arp's class "spir galaxies w E galaxy comps on arms"; "connection not seen, but note difference in arms toward and away from E galaxy. Note also material between p spir and E galaxy" (Arp); 23 13.1 + 18 47 at 3.7, 84, (0.7 x 0.6), E, m=15.7 —

12460   U 12460) comp 0.85, 175, 0.20 x 0.20, alm stell; P w comp at 3.7, 296, 0.8 x 0.7, SB 0 — U 12461) comp 3.3, 114, 0.3: x 0.1; 4.9, 313, 0.35 x 0.15 — U 12463) P w U 12465 at 6.2, 159; in group; comp attached at 0.45, 343, 0.25 x x 0.15 — U 12464) U 12467 at 2.3, 157; comp 3.6, 173, 0.3 x 0.3; 23 13.1 + 06 25 = N 7557 at 4.7, 286, 0.8 x 0.6, S0:, m=15.0; E2-3 (de V) — U 12465) see U 12463; comp 3.7, 130, 0.5 x 0.4, v dif — U 12467) see U 12464 — U 12468) CGCG δ = 15 36 is incorrect; 23 13.9 + 15 35 at 3.2, 81, 0.7 x 0.4, m=15.7; 23 14.2 + 15 37 at 7.3, 67, 0.8 x 0.25, m=15.4 — U 12469) comp 1.1, 219, 0.35 x 0.35, late bar spir; 2.4, 317, 0.5 x 0.4 —

12470   U 12472) III Zw 100; "compact patches and jets" (Zw) — U 12473) hints of ext outer disc; F blue knots at tips of bar; in group — U 12476) comp 2.2, 142, 0.4 x 0.15; 4.5, 255, 0.6 x 0.25; P w 12477 + 12478) VV 181, Arp 170; a) approx 1.2 x 1.2; b) 0.8 x 0.7, sep 0.55; sev S galaxies near; Arp's class "galaxies w dif counter-tails"

12480   U 12481) comp 3.1, 150, 0.9 x 0.15, S0-a — U 12482) in triple group w 23 15.3 + 28 45 at 2.8, 80, 0.30 x 0.22, compact, m=15.6; and 23 15.0 + 28 47 at 3.0, 326, 0.22 x 0.18, compact, hints of halo, m=15.7 — U 12483) strongly asym; main body 0.60 x 0.55; comp 4.8, 312, 0.5 x 0.3, dif; 5.0, 338, 0.5 x 0.3, dif — U 12484) P w 23 15.4 + + 09 23 at 0.9, 166, 0.8 x 0.2, pair in contact, m=15.6 — U 12485) sev galaxies in vic; 23 15.6 + 10 02 = I 5305 at 1.9, 273, 0.50 x 0.35, early spir, m=15.4; 23 15.6 + 09 58 = I 5306 at 3.3, 182, double system, 0.7 x 0.1 + + 0.6 x 0.25, total m=15.6; comp 4.3, 152, 0.30 x 0.25 — U 12486) comp 1.8, 201, 0.35 x 0.12 — U 12487) comp 4.0, 49, 0.5 x 0.1, vF —

12490   U 12490) massive slightly curved bar; smooth arms w low surf brightn; P w comp at 2.0, 11, 0.9 x 0.9, SAB(rs)c — U 12491) comp 3.2, 323, 0.6 x 0.3; 4.4, 190, 0.6 x 0.4, spir? — U 12493) Arp 92; P w comp at 1.0, 120, 0.2 x 0.2, bridge; strongly discordant redshifts, v = 8700 for U 12493, v = 16800 for comp; see Evans, D.E. (ed.), "External Galaxies and Quasi-stellar Objects", I A U Symp. 44, 380; in Arp's class "spir galaxies w E galaxy comp's on arms"; "vF connection shows better in red" (Arp) — U 12494) a) at 1.1, 347, 0.5 x 0.4 —

12500   U 12502) comp 1.1, 221, 0.4 x 0.15 — U 12503) comp 3.8, 345, 0.35 x 0.15; 5.0, 32, 0.45 x 0.15 — U 12506) comp 0.1 x 0.1 attached 0.15 s center; P w U 12519 at 10.7, 133 — U 12507) P w U 12517 at 6.2, 99 — U 12508) P w a) at 0.9, 342, 0.20 x 0.15 — U 12509) S0 (de V); comp 1.5, 4, 0.6 x 0.1 —

12510   U 12511) slightly asym; thin arms w chains of S blue knots; comp 3.6, 335, 0.3 x 0.3; 4.0, 242, 0.40 x 0.35 — U 12512) 23 17.3 + 08 17 at 2.8, 110, 0.6 x 0.5, S B 0?, m=15.7 — U 12514) comp 3.4, 80, 0.45 x 0.22, high surf brightn — U 12515) 23 17.6 + 25 57 at 3.9, 137, 0.30 x 0.25, compact, m=15.0 — U 12516) comp 4.0, 321, 0.4 x 0.2; U 12524 at 8.0, 30 — U 12517) F v ext halo; hints of spir pattern?; see U 12507 — U 12519) see U 12506 —

12520   U 12523) E 3 (de V); P w U 12531 at 6.9, 84; 23 17.6 + 07 53 = N 7617 at 2.8, 209, 0.7 x 0.5, S0 (de V), v = 4072, v_0 = 4269, m=15.1 — U 12525) a) 0.1 x 0.1; b) 0.6 x 0.4; c) 0.3 x 0.15 — U 12526) E4 (de V); comp 2.0, 94, 0.5 x 0.4; 23 17.8 + 08 04 = N 7621 at 2.3, 217, 0.6 x 0.15, m=15.6 — U 12527) S-shaped; arms w v high surf brightn; comp 4.8, 209, 0.4 x 0.4 — U 12529) VV 280, Arp 212, III Zw 102;   SA(rs)a pec (de V); in Arp's class "galaxies w irregularities, absorption and resolution"; "narrow chaotic abs tubes across one end" (Arp) —

12530   U 12531) E 1 (de V); see U 12523 — U 12534) comp 3.5, 133, 0.4 x 0.15; 4.7, 75, 0.3 x 0.2 — U 12535) P w 23 18.6 + + 07 54 at 1.2, 107, 0.5 x 0.2, compact, m=15.6 — U 12537) IV Zw 145; core 0.2 x 0.2; F halo; "e dif, compact nucl" (CGCG); P w comp at 1.9, 240, 0.9 x 0.7, v dif — U 12539) comp 2.6, 305, 0.4 x 0.3; sev smaller comps —

12540   U 12543) in triple group w U 12545 at 2.5, 147, and U 12546 at 3.5 — U 12544) 23 19.1 + 08 42 at 5.5, 191, 0.8 x x 0.15, Irr?, m=15.5 — U 12545) see U 12543; P w U 12546 at 1.0, 355, interaction? — U 12546) see U 12543, 12545 — U 12547) P w U 12548 at 2.4, 38 — U 12548) see U 12547 — U 12549) P w comp at 4.5, 163, 0.8 x 0.6, Sa? —

12550   U 12551) 23 19.4 + 09 01 at 2.5, 309, 0.5 x 0.4, S0, m=15.7 — U 12552) III Zw 104 at 4.7, 259 (23 20.6 + 12 46); "pair of osculating blue fuzzy compacts w blue B pencil jet to northwest, m_p = 16.3 ... v = 12780" (Zw); total diam 0.3 x 0.2; "distorted spir obj 8" = 6.6 kpc diam. The 'jet' seen on the Sky Survey is prob a background edge-on spir" (Sargent) — U 12554) S B(s)c (de V); Sc + (Ho) — U 12557) P w U 12566 at 4.6, 126 — U 12558) sev thin arms; patchy; comp 3.0, 155, 0.4: x 0.1; 5.4, 269, 0.8: x 0.5:, dif — U 12559) U 12550 at 7.4, 265; U 12567 at 3.1, 138 —

12560   U 12560) P w 23 20.4 + 01 07 at 2.8, 143, 0.7 x 0.3, m=15.2 — U 12563) comp 0.9, 24, 0.3 x 0.3, v dif, dwarf; 2.3, 318, 0.4 x 0.2 — U 12564) P w comp at 2.5, 105, 0.35 x 0.25 — U 12565) comp 0.60, 106, 0.15 x 0.10, F — U 12566) (r) 0.7 x 0.6; see U 12557 — U 12567) see U 12559; U 12573 at 5.8, 94 —

12570   U 12570) one short massive arm southw from f side — U 12572) comp 2.9, 201, 0.3 x 0.3, dif — U 12573) see U 12567; comp 3.7, 167, 0.35 x 0.22 — U 12574) VV 305; VV δ = 18 18.5 is incorrect; 23 21.0 + 19 19 at 2.7, 312, 0.45 x 0.40, long streamer southeast, S v red cend at tip, m=15.4; comp 3.4, 29, 0.35 x 0.12 — U 12575) comp 4.9, 345, 0.3 x x 0.2 — U 12576) a large nr of S comps — U 12577) comp 3.0, 136, 0.4 x 0.3 — U 12578) P w U 12589 at 11.8, 56 — U 12579) a large nr of S galaxies near; 23 22.2 + 14 21 = I 1487 at 5.1, 100, 0.5 x 0.4, compact, m=15.7 —

12580   U 12584) a) 0.35 x 0.30, Irr?; b) 1.2 x 1.1, S d-dm; sep 0.6 — U 12586) comp 3.2, 24, 0.35 x 0.30; 4.8, 72, 0.6 x x 0.4, Sc, dif — U 12589) see U 12578; comp 1.3, 25, 0.35 x 0.25 —

12590   U 12594) comp 1.7, 146, 0.6 x 0.15, high surf brightn; 2.1, 73, 0.9 x 0.8, dif; in triple group? — U 12599) P w 23 24.1 + 25 22 at 4.1, 230, 0.9 x 0.35, Sc, m=15.7 —

12600   U 12602) P w 23 25.0 + 12 07 = N 7672 at 5.8 sf, 0.8 x 0.7, Sb, m=14.8 — U 12607) IV Zw 149; "blue post-eruptive large patchy compact w v large blue ring halo... v = 3460" (Zw); P w U 12610 at 6.7, 123 — U 12608) VV 343; listed as N 7675 in MCG, which is incorrect; P w b) at 0.6, 64, 0.20 x 0.15; 23 25.5 + 08 29 = N 7675 at 2.5, 105, 0.5 x 0.3, E, m=14.8; comp 3.7, 250, 0.4 x 0.2 —

12610   U 12610) (r) 0.50 x 0.45; see U 12607 — U 12613) DDO 216; Ir V (vdB); An 2326 in BG; I m (de V); comp 1.6, 215, 0.3: x 0.1, early spir — U 12614) Arp 28; SAB(rs)c (de V); slightly asym; one massive arm s center; in Arp's class "spir galaxies w one heavy arm" — U 12616) comp 4.1, 218, 0.7 x 0.2; sev smaller comps — U 12618) VV 329, Arp 216; S B 0 pec: (de V); in Arp's class "galaxies w adjacent loops"; P w U 12622 at 4.6, 74; "patches n disturbed spir, em strong" (Arp) —

12620   U 12620) asym disc; stell comp 0.12 x 0.12 superimp 0.4 nf center; 23 26.1 + 17 00 at 5.0, 240, 0.7 x 0.2, m=15.7; 23 26.0 + 17 00 = N 7681 at 6.2, 251, 0.8 x 0.4, m=15.0 — U 12621) S B c?; comp 3.1, 193, 0.35 x 0.35 — U 12622) VV 329, Arp 216; S B(r)b: (de V); see U 12618 — U 12625) 23 26.3 + 29 27 at 4.5, 239, 0.8 x 0.5, E:, m=15.0 — U 12629) P w comp at 3.3, 345, 0.8 x 0.35 —

12630   U 12631) comp 2.2, 50, 0.4: x 0.3:, and 2.6, 43, 0.3 x 0.2, sep 0.40; 4.4, 82, 0.6 x 0.6, v late spir; U 12636 at 4.9, 54 — U 12632) DDO 217; S V: (vdB) — U 12635) P w U 12637 at 4.8, 161 — U 12636) see U 12631; sev comps —

U 12637) see U 12635 --

12640 ---

12650 U 12652) V Zw 410; "pair of osculating elliptical compacts in luminous matrix" (Zw) -- U 12655) sev S galaxies near;
double system at 3.0, 267, 0.35 x 0.2 and 0.4 x 0.3, sep 0.40; 23 29.5 + 23 38 at 6.1, 259, 0.8 x 0.5, m=15.3,
3 comp   incl in CGCG magn -- U 12656) P w U12659 at 3.9, 133 -- U 12659) see U 12656 --

12660 U 12660) P w U12662 at 7.4, 151 -- U 12661) a) at 1.3, 327, 0.45 x 0.20 -- U 12662) see U 12660 -- U 12665) VV
314, Arp 46; interacting w S comp = VV 314 b at 0.55, 32; "comp connected to main spir" (Arp); P w U12667 at
2.6, 61 -- U 12667) irr patchy arms w high surf brightn; see U 12665 --

12670 U 12672) one long thin smooth arm; no disturbing obj vic; hints of vF short second arm on p side -- U 12673) P w
comp at 2.2, 268, 0.30 x 0.25, spir -- U 12677) comp 4.8, 240, 0.4 x 0.2, high surf brightn -- U 12679) in a reg
w a v large nr of vF and eF galaxies, prob a background cl not incl in CGCG --

12680 U 12682) DDO 218; pec (vdB) -- U 12684) P w U12686 at 4.7, 30; comp 1.9, 253, 0.3 x 0.2; 23 32.6 + 04 32 =
= N 7705 at 5.7, 183, 0.60 x 0.45, S B: 0, m=15.4 -- U 12686) see U 12684 -- U 12688) double or triple system? --
U 12689) comp 1.9, 323, 0.3 x 0.2, v dif; 2.0, 20, 0.4 x 0.2, v dif --

12690 U 12691) ecc nucl reg, vF disc   -- U 12699) VV 51 a, Arp 284; S B(s)b pec: (de V); P w U12700 at 1.9 f; "some vS
knots in connecting streamer" (Arp) --

12700 U 12700) VV 51 b, Arp 284; see U 12699; I m pec (de V) -- U 12702) SAB(r)b: (de V) -- U 12703) * superimp? or ecc
nucl?; P w 23 33.9 + 20 52 = I 5337 at 1.3, 285, 0.8 x 0.2, m=15.4; 23 34.2 + 20 50 at 3.5, 141, 0.35 x 0.35, m =
= 15.6 -- U 12704) slightly asym halo; comp 1.6, 173, 0.5 x 0.12; 5.0, 144, 0.7 x 0.3 -- U 12707) comp 4.0, 174,
0.5 x 0.2, Irr:, 6.2, 11, 0.9 x 0.4, v dif -- U 12709) DDO 219; S V: (vdB) --

12710 U 12710) v blue, v chaotic, hints of spir arcs; blue cond at 0.3 n center -- U 12712) P w comp at 3.2, 312, 0.9 x
x 0.8, SAB(s)c -- U 12714) comp 1.8, 315, 0.8 x 0.1 -- U 12716) F large halo; brightest in subgroup; comp 0.15 x
x 0.15 superimp n center; 23 35.9 + 26 43 = I 5341 at 2.9, 192, 0.25 x 0.25, compact, m=15.5; 23 36.0 + 26 42 at
3.3, 183, 0.25 x 0.25, compact, m=15.3; 23 36.1 + 26 42 at 3.3, 162, 0.20 x 0.18, very compact, m=15.6; 23 36.2 +
+ 26 44 = I 5342 at 2.4, 122, 0.25 x 0.25, compact, m=15.4; comp 1.5, 97, 0.6 x 0.2, S0? -- U 12717) P w U12720
at 9.1, 71 -- U 12718) comp 3.8, 258, 0.6 x 0.10, spir -- U 12719) in a dense part of the cl, sev S galaxies near --

12720 U 12720) see U 12717 -- U 12721) comp 2.7, 0, 0.45 x 0.15 -- U 12722) blue cond attached nf -- U 12726) comp 1.9,
62, 0.30 x 0.08; 3.9, 183, 0.4 x 0.1 --

12730 U 12730) comp 1.5, 35, 0.3: x 0.3:, v dif; 4.6, 100, 0.6 x 0.4 -- U 12731) comp 2.1, 46, 0.5 x 0.1; 2.2, 229, 0.3 x
x 0.1 -- U 12732) hints of ring approx 0.5 x 0.5; one long thin arm w v low surf brightn -- U 12733) P w comp at
0.90, 108, 0.25 x 0.22, high surf brightn; comp 3.3, 292, 0.4 x 0.1 -- U 12736) slightly integral-shaped; comp 3.7,
237, 0.3 x 0.3; 4.8, 197, 0.5 x 0.1 -- U 12737) "neutral S B 0 w compact core and large halo" (Zw); P w U12738 at
1.5, 128 -- U 12738) "disrupted and badly 'dented' Sc" (Zw); see U 12737 -- U 12739) comp 1.9, 126, 0.25 x 0.22 --

12740 U 12740) comp 4.1, 136, 0.5: x 0.5: -- U 12743) prob interacting w comp at 1.7, 269, 0.5 x 0.4; in group of 5 --
U 12745) P w comp at 2.9, 323, 0.6 x 0.25; sev S galaxies in vic -- U 12746) comp 4.8, 269, 0.5 x 0.1, F --
U 12747) comps at 2.0, 337, 0.3 x 0.3, E?, and 2.2, 348, 0.6 x 0.3, long streamer nf, sep 0.40, interaction --

12750 U 12753) comp superimp at 0.45, 73, 0.20 x 0.10 -- U 12754) S B(s)cd (de V), Sc + (Ho); comp 3.1, 344, 0.5: x 0.4:,
dif, dwarf: -- U 12759) (R)S B(s)0 + (de V); P w U12760 at 49., 359 --

12760 U 12760) SA(r)b (de V); see U 12759 -- U 12761) P w comp at 0.7, 98, 0.6 x 0.3, v dif -- U 12762) P w U12764 at
4.3, 22 -- U 12764) see U 12762 -- U 12769) comp 0.9, 157, 0.4 x 0.3 --

12770 U 12770) comp 3.6, 29, 0.35 x 0.35 -- U 12772) S B(s)b?; 23 43.3 + 27 04 at 3.6, 111, 0.6 x 0.4, core 0.22 x 0.20,
compact, m=15.5 -- U 12776) bar 0.8 x 0.35, vF disc; 2 F arc-shaped arms not connected to bar -- U 12778) core
0.45 x 0.40, smooth vF halo; comp 1.9, 265, 0.9: x 0.8:, dwarf -- U 12779) VV 5 b, Arp 86; Pec (de V); connected
w U 12780; see U 12780; IV Zw 165; "compact ragged blue disc..." (Zw) --

12780 U 12780) VV 5 a, Arp 86; SAB(rs)b (de V); M 51 case, U 12779 at tip of spir arm at 2.0, 225; in Arp's class "spir
galaxies w large high surf brightn comps on arms"; "double arm leading to comp" (Arp) -- U 12782) comp 2.8,
152, 0.5 x 0.1 -- U 12785) comp 1.8, 130, 0.4 x 0.1, vF; 23 45.2 + 27 09 at 5.0, 308, 0.8 x 0.2, spir, m=15.3 --
U 12788) Arp 68; SA(r)c: (de V); listed as N 7756 by Arp and in CGCG, but N 7756 is a *; "many star-like knots
lined up along straight arm" (Arp); comp 1.0, 288, 0.3 x 0.1 --

12790 U 12791) DDO 220; IV- V or V (vdB) --

12800 U 12801) v ext halo; comp 0.45, 203, 0.4 x 0.3; 3.0, 40, 0.3 x 0.2, F -- U 12804) sev comps , brightest at 1.3,
30, 0.25 x 0.15 -- U 12805) in a dense part of the cl, U 12806 at 3.7 n -- U 12806) see U 12805 -- U 12808)
(R)SA(rs)b (de V); 23 48.7 + 19 57 at 5.3, 25, 0.45 x 0.45, S B a, m=15.7; comp 3.0, 70, 0.3 x 0.2; U 12815 at 5.4,
115 --

12810 U 12811) 23 48.4 + 08 24 at 3.6, 215, 0.9: x 0.5:, S B..., m=15.5; 23 49.0 + 08 27 at 5.0, 85, 0.5 x 0.3, m=15.5
-- U 12812) b) = MCG 3 - 60 - 33 at 0.7, 105, 0.5 x 0.4, Irr, bridge; 23 48.9 + 20 18 at 2.0, 78, 0.8 x 0.5, Sa:,
m=14.9 -- U 12813) S b : (de V); asym halo; P w U12815 at 1.1, 25 -- U 12815) S B(s)a (de V); see U 12808, 12813;
comp 2.7, 257, 0.8 x 0.2 -- U 12816) comp 4.8, 157, 0.4 x 0.3 -- U 12819) comp 0.5 x 0.3 superimp f center --

12820 U 12821) chaotic spir; S cond attached sp, disturbing comp?; short disrupted arm in northeast -- U 12823) comp
3.7, 281, 0.8 x 0.3, 2 long v open arms, interacting w S comp at 0.40 -- U 12824) integral-shaped, interacting w
comp at 0.8, 252, 0.1 x 0.1; comp 2.6, 145, 0.5 x 0.1 -- U 12825) sev S galaxies near -- U 12826) dif asym outer
disc, may be interacting w comp at 0.9, 313, 0.35 x 0.20 -- U 12827) P w U12831 at 1.8, 80 -- U 12829) comp 1.7,
33, 0.3 x 0.12; 3.8, 32, 0.25 x 0.25 --

12830 U 12831) see U 12827; 23 51.2 + 07 35 = N 7781 at 4.9, 101, 0.8 x 0.2, S0 - a, m=15.2 -- U 12832) comp 4.1, 191,
0.4 x 0.4, spir -- U 12833) P w comp at 3.1, 307, 0.9 x 0.5, S B... -- U 12834) SA(s)b (de V) -- U 12835) large
dif halo; 6 comp inside 5.0, brightest at 2.9, 183, 0.8 x 0.2, early spir -- U 12837) VV 208, Arp 323; a) 1.0 x
x 0.5, * superimp nf part; b) 0.4 x 0.3; in Arp's class "chains of galaxies"; comp 0.7 n, 0.20 x 0.15; 1.9 sf,
0.20 x 0.20; 1.4 sf, 0.12 x 0.12, * or compact? --

12840 U 12843) amorphous disc w blue patches, spir str in centr part; * superimp 0.4 nf center -- U 12845) comp 3.1, 266,
0.3 x 0.3; 3.7, 122, 0.6 x 0.2 -- U 12846) F blue cond 0.3 nf center -- U 12847) P w U12849 at 4.0, 97 -- U 12848)
P w U12852 at 4.4, 11 -- U 12849) see U 12847 --

12850 U 12850) in triple group w 23 53.7 + 29 09 at 2.9, 50, 0.6 x 0.6, spir, m=15.4; and comp at 2.4, 75, 0.7 x 0.5
U 12851) V Zw 400; "blue Sc" (Zw) -- U 12852) see U 12848 -- U 12854) comp 1.5, 117, 0.3 x 0.2, F -- U 12855) comp 2.2,
192, 0.4 x 0.05, F; 3.2, 179, 0.3 x 0.3, F -- U 12856) VV 255, Arp 262; v blue; cond (prob comp galaxy) 0.45 x 0.20
superimp s part at 0.5, 193; "some resolution into knots" (Arp) -- U 12858) comp 0.40, 145, 0.2 x 0.1, F, prob
background -- U 12859) comp 4.1, 130, 0.25 x 0.25 --

12860 U 12862) comp 1.7, 170, 0.8: x 0.8:, e dif, dwarf irr; 23 55.0 + 47 10 at 6.8, 133, (0.9 x 0.9), core 0.15 x 0.15, compact, m=15.5 -- U 12863) comp 2.5, 38, 0.4 x 0.1 -- U 12865) P w comp at 2.4, 345, 0.5 x 0.4, spir; comp 2.7, 156, 0.3 x 0.15 -- U 12867) comp 1.7, 123, 0.3 x 0.2 -- U 12868) comp 1.1, 273, 0.3 x 0.2, eF; 3.5, 280, 0.3 x 0.3, F --

12870 U 12872) ecc (r) or (rs) 0.65 x 0.50; chaotic outer disc w irr patches, no spir arms; cond 0.1 x 0.1 at 0.35 nf center, disturbing comp, or *? -- U 12874) (r) 0.4 x 0.4; one eF outer arm; comp 2.0, 341, 0.5 x 0.3; 2.2, 15, 0.5 x 0.4; interaction? -- U 12876) comp 3.2, 111, 0.5 x 0.2, dif -- U 12877) sev F comps , brightest at 3.7, 20, 0.4 x 0.2 -- U 12878) comp 3.2, 68, 0.7: x 0.5:, e dif -- U 12879) loop northeast; comp 3.8, 6, 0.7 x 0.3 --

12880 U 12880) P w comp at 10.9, 75, 0.8 x 0.25, Sc - Irr -- U 12883) (rs) 0.55 x 0.50; 2 main arms; slightly asym; U 12888 at 6.5, 92 -- U 12885) centr reg 0.8 x 0.25; chaotic outer parts w blue filaments; no spir str -- U 12886) comp 1.6, 340, 0.9: x 0.4:, v dif, dwarf -- U 12888) comp 2.9, 235, 0.3: x 0.2:, F; 2.5, 188, 0.3 x 0.1; 3.3, 219, 0.30 x 0.18; 23 57.7 + 46 41 at 5.7, 34, 0.9 x 0.8, S (r)a:, m=14.8; see U 12883 --

12890 U 12890) sev S galaxies in vic; comp 3.0, 244, 0.45 x 0.15 -- U 12891) VV 186, Arp 249; 2 cores w bridge, jet and halo, sep 0.20; IV Zw 177, "pair of neutral equal spherical compacts, sep = 12" southwest - northeast" (Zw); "straight connection from F material on east to middle galaxy" (Arp) -- U 12896) prob interacting w comp at 1.2, 177, core 0.25 x 0.15 -- U 12897) in triple group w 23 58.1 + 28 08 at 1.6, 54, 0.6 x 0.5, m=15.5; and U 12899 at 2.2, 63 -- U 12899) see U 12897; 23 58.1 + 28 08 at 0.7, 267 --

12900 U 12901) comp 2.7, 300, 0.3 x 0.1 -- U 12902) P w U 12903 at 5.9, 210 -- U 12903) see U 12901 -- U 12904) comp 0.2 x 0.1 attached at 0.45, 355; P w 23 58.5 + 34 22 at 2.5, 239, 0.8 x 0.7, SAB a:, m=15.7; sev vF galaxies around this pair, prob phys comps -- U 12906) brightest in chain of 6; 23 58.9 + 12 50 at 1.5, 87, 0.9 x 0.6, distorted, m=15.3; comp 1.2, 276, 0.5 x 0.15; 2.6, 321, 0.8 x 0.3; 6.2, 97, (0.4 x 0.4), core 0.18 x 0.18, compact:; sev much fainter objs in the chain, brightest at 2.5, 110, 0.5 x 0.4, v dif, dwarf: -- U 12907) 2 cores 0.08 x 0.08 in contact; dif irr halo; a large nr of vF galaxies near -- U 12908) VV 226, Arp 112; see U 12911 --

12910 U 12910) F blue (r) 0.8 x 0.8; chaotic -- U 12911) VV 226 a, Arp 112; P w U 12908 = VV 226 b at 0.80 sp; in Arp's class "E and E-like galaxies repelling spir arms"; in triple group w comp at 1.0, 83, 0.45 x 0.1, curved -- U 12914) VV 254 a; III Zw 125, see U 12915; (r) 0.8 x 0.35, smooth dif (R); curved B ridge in n part, superimp comp?; P w U 12915 at 1.1, 50; VV α = 13 59.05 is incorrect -- U 12915) VV 254 b, III Zw 125; "interacting pair of [post-eruptive] galaxies, w B red conds . No 1, twisted Sc... No 2, Irr, 70" nf No 1" (Zw) -- U 12916) comp 2.0, 114, 0.5 x 0.3, dif -- U 12919) comp 2.1, 223, 0.4 x 0.4, dif; 3.3, 192, 0.3 x 0.2 --

12920 ----

F I N I S

ADDENDA

1	2	3	4	5			6	7	8	9	10	11	12
01493A	01 57.9 + 50 16 134.11 − 10.84	08 − 04 − 000	0861	0.8 0.8	x x	0.6 0.6	90 ...	1	...		14.4		
02886A	03 50 + 88 37 124.10 + 26.42	15 − 01 − 000	0570	1.2 0.9	x x	0.2 0.15	112 7	2	Sc		17.		
03211A	04 55 + 89 17 123.71 + 27.06	15 − 01 − 002	0570	1.4: 1.0:	x x	0.9: 0.6:	85 4	2	Sc		16.0:		
03457A	06 20 + 87 33 125.73 + 27.06	15 − 01 − 004	0570	1.3: 1.3	x x	0.4: 0.35	52 6	2	...	S d − m	16.5:		
03460A	06 21.9 + 06 14 204.34 − 03.10	01 − 19 − 000	0999	1.0 0.8	x x	0.1 0.1	107 7	2	...	d − m	17.		
03528A	06 42 + 86 40 126.75 + 27.25	14 − 04 − 010	1328								15.6		*1V
03536A	06 43 + 86 38 126.79 + 27.26	14 − 04 − 011	1328	[2.3 ...	x	0.9]	...	3	Double system Connected?		14.8		*1V
03617A	06 55 + 87 49 125.46 + 27.44	15 − 01 − 006	0570	1.5: 1.4:	x x	1.0: 0.9:	(125) ...	3	Dwarf (spiral) S: IV− V or V		16.5:		
04280A	08 10.6 + 58 07 159.50 + 33.66	10 − 12 − 000	0960 116	[0.8 ...	x	0.8]	... ...	2	Double system		14.4		*C
05040A	09 24.8 + 04 09 229.13 + 36.15	01 − 24 − 022 IC 2481	0028	0.9 0.9	x x	0.5 0.6	160 ...	2	S[a − b] Peculiar?		14.5		
05083A	09 30 + 88 23 124.40 + 28.44	15 − 01 − 009	0570	2.4 2.2:	x x	2.1 2.2:	-- 1	4	...	SA dm	16.0:		
05854A	10 41.7 + 14 21 230.26 + 57.66	02 − 28 − 002 NGC 3357	0976	(0.9 (1.0	x x	0.8) 0.8)	-- ...	2	S0		14.3		*
06063A	10 50.4 + 25 57 209.29 + 63.64	04 − 26 − 000	1366	1.1 0.9	x x	0.12 0.15	158 7	1	S ...		16.5:		
07020A	12 00.1 + 64 39 131.46 + 51.94	11 − 15 − 012	0674 255	1.2 1.0	x x	0.6 0.5	100 (5)	1	S0?		14.3		
07064A	12 02.1 + 60 57 133.04 + 55.53	10 − 17 − 000	0723	[2. ...	x	0.6:]	... ...	1	Multiple system Interaction		17.		*
07085A	12 03.2 + 31 20 188.74 + 79.28	05 − 29 − 10 +11	1398 263	[2.3 ...	x	0.6]	... ...	3	Double system Bridge, plume		15.1		*12 VAC̄
07399A	12 18.3 + 17 46 265.11 + 78.05	03 − 32 − 004	1576	(1.4 (1.4	x x	0.7) 0.7)	85: ...	1	Dwarf elliptical?		15.1		
07956A	12 45 + 88 05 123.04 + 29.32	15 − 01 − 013	0570	1.3: 0.8	x x	1.1: 0.5	150: 1 − 2	2	Sc		16.0:		
11411A	19 09.1 + 60 03 090.79 + 21.10	10 − 27 − 008	0830	2.2 1.7	x x	0.3 0.2	1 7	3	Sc		16.0:		

N G C   3357

I C   2481

PRECESSION IN RIGHT ASCENSION 1950.0 - 2000.0

α	00h00m 12 00	00 10 11 50	00 20 11 40	00 30 11 30	00 40 11 20	00 50 11 10	01 00 11 00	01 10 10 50	01 20 10 40	01 30 10 30	01 40 10 20	01 50 10 10	02 00 10 00
δ													
+ 00°	+2ᵐ6	2.6	2.6	2.6	2.6	2.6	2.6	2.6	2.6	2.6	2.6	2.6	2.6
+ 05	2.6	2.6	2.6	2.6	2.6	2.6	2.6	2.6	2.6	2.6	2.6	2.6	2.6
+ 10	2.6	2.6	2.6	2.6	2.6	2.6	2.6	2.6	2.7	2.7	2.7	2.7	2.7
+ 15	2.6	2.6	2.6	2.6	2.6	2.7	2.7	2.7	2.7	2.7	2.7	2.7	2.7
+ 20	2.6	2.6	2.6	2.6	2.7	2.7	2.7	2.7	2.7	2.7	2.7	2.7	2.7
+ 25	2.6	2.6	2.6	2.7	2.7	2.7	2.7	2.7	2.7	2.7	2.7	2.8	2.8
+ 30	2.6	2.6	2.6	2.7	2.7	2.7	2.7	2.7	2.7	2.8	2.8	2.8	2.9
+ 32	2.6	2.6	2.7	2.7	2.7	2.7	2.7	2.7	2.8	2.8	2.8	2.9	2.9
+ 34	2.6	2.6	2.7	2.7	2.7	2.7	2.7	2.7	2.8	2.8	2.9	2.9	2.9
+ 36	2.6	2.6	2.7	2.7	2.7	2.7	2.7	2.8	2.8	2.8	2.9	2.9	3.0
+ 38	2.6	2.6	2.7	2.7	2.7	2.7	2.7	2.8	2.8	2.9	2.9	3.0	3.0
+ 40	2.6	2.6	2.7	2.7	2.7	2.7	2.8	2.8	2.9	2.9	3.0	3.0	3.0
+ 42	2.6	2.6	2.7	2.7	2.7	2.7	2.8	2.8	2.9	2.9	3.0	3.0	3.1
+ 44	2.6	2.6	2.7	2.7	2.7	2.8	2.8	2.9	2.9	3.0	3.0	3.1	3.1
+ 46	2.6	2.6	2.7	2.7	2.7	2.8	2.8	2.9	3.0	3.0	3.1	3.1	3.2
+ 48	2.6	2.6	2.7	2.7	2.7	2.8	2.9	2.9	3.0	3.0	3.1	3.2	3.2
+ 50	2.6	2.6	2.7	2.7	2.8	2.8	2.9	3.0	3.0	3.1	3.2	3.2	3.2
+ 51	2.6	2.7	2.7	2.7	2.8	2.8	2.9	3.0	3.0	3.1	3.2	3.2	3.2
+ 52	2.6	2.7	2.7	2.7	2.8	2.8	2.9	3.0	3.1	3.1	3.2	3.2	3.2
+ 53	2.6	2.7	2.7	2.7	2.8	2.9	2.9	3.0	3.1	3.2	3.2	3.2	3.3
+ 54	2.6	2.7	2.7	2.7	2.8	2.9	3.0	3.1	3.1	3.2	3.2	3.2	3.3
+ 55	2.6	2.7	2.7	2.7	2.8	2.9	3.0	3.1	3.1	3.2	3.2	3.3	3.3
+ 56	2.6	2.7	2.7	2.7	2.8	2.9	3.0	3.1	3.2	3.2	3.2	3.3	3.4
+ 57	2.6	2.7	2.7	2.7	2.8	2.9	3.0	3.1	3.2	3.2	3.2	3.3	3.4
+ 58	2.6	2.7	2.7	2.8	2.8	2.9	3.0	3.1	3.2	3.2	3.3	3.4	3.4
+ 59	2.6	2.7	2.7	2.8	2.9	3.0	3.1	3.1	3.2	3.2	3.3	3.4	3.5
+ 60	2.6	2.7	2.7	2.8	2.9	3.0	3.1	3.2	3.2	3.3	3.4	3.4	3.5
+ 61	2.6	2.7	2.7	2.8	2.9	3.0	3.1	3.2	3.3	3.3	3.4	3.5	3.6
+ 62	2.6	2.7	2.7	2.8	2.9	3.0	3.1	3.2	3.2	3.3	3.4	3.5	3.6
+ 63	2.6	2.7	2.7	2.8	2.9	3.0	3.2	3.2	3.3	3.4	3.5	3.6	3.7
+ 64	2.6	2.7	2.7	2.8	2.9	3.1	3.2	3.2	3.3	3.4	3.5	3.6	3.7
+ 65	2.6	2.7	2.7	2.8	3.0	3.1	3.2	3.2	3.4	3.5	3.6	3.7	3.8
+ 66	2.6	2.7	2.7	2.9	3.0	3.1	3.2	3.3	3.4	3.5	3.6	3.8	3.8
+ 67	2.6	2.7	2.7	2.9	3.0	3.2	3.2	3.3	3.5	3.6	3.7	3.8	3.8
+ 68	2.6	2.7	2.8	2.9	3.1	3.2	3.2	3.4	3.5	3.6	3.8	3.8	3.9
+ 69	2.6	2.7	2.8	2.9	3.1	3.2	3.3	3.4	3.6	3.7	3.8	3.9	4.0
+ 70	2.6	2.7	2.8	3.0	3.1	3.2	3.3	3.5	3.6	3.8	3.8	4.0	4.1
+ 71	2.6	2.7	2.8	3.0	3.1	3.2	3.4	3.5	3.7	3.8	3.9	4.1	4.2
+ 72	2.6	2.7	2.8	3.0	3.2	3.3	3.4	3.6	3.8	3.8	4.0	4.2	4.2
+ 73	2.6	2.7	2.9	3.0	3.2	3.3	3.5	3.7	3.9	4.0	4.1	4.2	4.4
+ 74	2.6	2.7	2.9	3.1	3.2	3.4	3.6	3.8	3.9	4.1	4.2	4.3	4.5
+ 75	2.6	2.7	2.9	3.1	3.2	3.5	3.7	3.8	4.0	4.2	4.3	4.5	4.7
+ 76	2.6	2.7	2.9	3.2	3.3	3.5	3.8	3.9	4.1	4.2	4.4	4.6	4.8
+ 77	2.6	2.7	3.0	3.2	3.4	3.6	3.8	4.0	4.2	4.4	4.6	4.7	5.0
+ 78	2.6	2.7	3.0	3.2	3.5	3.7	3.9	4.2	4.3	4.6	4.7	5.0	5.2
+ 79	2.6	2.8	3.1	3.3	3.6	3.8	4.1	4.2	4.5	4.7	5.0	5.2	5.4
+ 80	2.6	2.8	3.1	3.4	3.7	3.9	4.2	4.5	4.7	5.0	5.2	5.5	+ 5.7

PRECESSION IN DECLINATION 1950.0 - 2000.0

	± 17'	17	17	17	17	16	16	16	16	15	15	15	15

PRECESSION IN RIGHT ASCENSION 1950.0 - 2000.0

α	02h00m 10 00	02 10 09 50	02 20 09 40	02 30 09 30	02 40 09 20	02 50 09 10	03 00 09 00	03 10 08 50	03 20 08 40	03 30 08 30	03 40 08 20	03 50 08 10	04 00 08 00
δ													
+ 00°	+2ᵐ.6	2.6	2.6	2.6	2.6	2.6	2.6	2.6	2.6	2.6	2.6	2.6	2.6
+ 05	2.6	2.6	2.6	2.6	2.6	2.7	2.7	2.7	2.7	2.7	2.7	2.7	2.7
+ 10	2.7	2.7	2.7	2.7	2.7	2.7	2.7	2.7	2.7	2.7	2.7	2.7	2.7
+ 15	2.7	2.7	2.7	2.7	2.7	2.7	2.7	2.7	2.7	2.8	2.8	2.8	2.8
+ 20	2.7	2.7	2.8	2.8	2.8	2.8	2.8	2.8	2.8	2.9	2.9	2.9	2.9
+ 25	2.8	2.8	2.8	2.9	2.9	2.9	2.9	2.9	3.0	3.0	3.0	3.0	3.0
+ 30	2.9	2.9	2.9	2.9	3.0	3.0	3.0	3.0	3.1	3.1	3.1	3.1	3.1
+ 32	2.9	2.9	3.0	3.0	3.0	3.0	3.0	3.0	3.1	3.1	3.1	3.1	3.1
+ 34	2.9	2.9	3.0	3.0	3.1	3.1	3.1	3.1	3.2	3.2	3.2	3.2	3.2
+ 36	3.0	3.0	3.0	3.1	3.1	3.1	3.2	3.2	3.2	3.2	3.2	3.2	3.2
+ 38	3.0	3.0	3.1	3.1	3.1	3.2	3.2	3.2	3.2	3.2	3.2	3.3	3.3
+ 40	3.0	3.1	3.1	3.2	3.2	3.2	3.2	3.2	3.2	3.3	3.3	3.3	3.3
+ 42	3.1	3.1	3.2	3.2	3.2	3.2	3.2	3.3	3.3	3.3	3.3	3.3	3.4
+ 44	3.1	3.2	3.2	3.2	3.2	3.2	3.3	3.3	3.4	3.4	3.4	3.5	3.5
+ 46	3.2	3.2	3.2	3.2	3.3	3.3	3.4	3.4	3.4	3.5	3.5	3.5	3.6
+ 48	3.2	3.2	3.2	3.3	3.3	3.4	3.4	3.5	3.5	3.6	3.6	3.6	3.7
+ 50	3.2	3.2	3.3	3.3	3.4	3.4	3.5	3.5	3.6	3.6	3.7	3.7	3.8
+ 51	3.2	3.3	3.3	3.4	3.4	3.5	3.5	3.6	3.6	3.7	3.7	3.8	3.8
+ 52	3.2	3.3	3.4	3.4	3.5	3.5	3.6	3.6	3.7	3.7	3.8	3.8	3.8
+ 53	3.3	3.3	3.4	3.5	3.5	3.6	3.6	3.7	3.7	3.8	3.8	3.8	3.8
+ 54	3.3	3.4	3.4	3.5	3.6	3.6	3.7	3.7	3.8	3.8	3.8	3.8	3.9
+ 55	3.3	3.4	3.5	3.5	3.6	3.7	3.7	3.8	3.8	3.8	3.8	3.9	3.9
+ 56	3.4	3.4	3.5	3.6	3.6	3.7	3.8	3.8	3.8	3.9	3.9	3.9	4.0
+ 57	3.4	3.5	3.6	3.6	3.7	3.8	3.8	3.8	3.8	3.9	4.0	4.0	4.1
+ 58	3.4	3.5	3.6	3.7	3.7	3.8	3.8	3.9	4.0	4.0	4.1	4.2	4.2
+ 59	3.5	3.5	3.6	3.7	3.8	3.8	3.8	3.9	4.0	4.0	4.1	4.2	4.2
+ 60	3.5	3.6	3.7	3.8	3.8	3.9	3.9	4.0	4.0	4.1	4.2	4.2	4.2
+ 61	3.6	3.7	3.8	3.8	3.8	3.9	4.0	4.1	4.1	4.2	4.2	4.2	4.3
+ 62	3.6	3.7	3.8	3.8	3.9	4.0	4.1	4.2	4.2	4.2	4.2	4.3	4.3
+ 63	3.7	3.8	3.8	3.9	4.0	4.0	4.1	4.2	4.2	4.3	4.3	4.4	4.4
+ 64	3.7	3.8	3.8	3.9	4.0	4.1	4.2	4.2	4.3	4.3	4.4	4.5	4.5
+ 65	3.8	3.8	3.9	4.0	4.1	4.2	4.2	4.3	4.4	4.4	4.5	4.6	4.7
+ 66	3.8	3.9	4.0	4.1	4.2	4.2	4.3	4.4	4.5	4.6	4.6	4.7	4.7
+ 67	3.8	4.0	4.1	4.2	4.2	4.3	4.4	4.5	4.6	4.7	4.7	4.7	4.8
+ 68	3.9	4.1	4.2	4.2	4.3	4.4	4.5	4.6	4.7	4.7	4.8	4.9	4.9
+ 69	4.0	4.1	4.2	4.3	4.4	4.5	4.6	4.7	4.7	4.8	4.9	5.0	5.1
+ 70	4.1	4.2	4.3	4.4	4.5	4.6	4.7	4.8	4.9	5.0	5.1	5.2	5.2
+ 71	4.2	4.3	4.4	4.5	4.7	4.7	4.8	4.9	5.0	5.1	5.2	5.3	5.4
+ 72	4.2	4.4	4.5	4.7	4.7	4.9	5.0	5.1	5.2	5.2	5.3	5.4	5.5
+ 73	4.4	4.5	4.6	4.7	4.9	5.0	5.2	5.2	5.3	5.4	5.6	5.7	5.7
+ 74	4.5	4.7	4.7	4.9	5.1	5.2	5.3	5.4	5.5	5.7	5.7	5.8	5.9
+ 75	4.7	4.8	4.9	5.1	5.2	5.3	5.5	5.7	5.7	5.8	6.0	6.1	6.2
+ 76	4.8	4.8	5.1	5.2	5.4	5.6	5.7	5.8	6.0	6.1	6.2	6.3	6.4
+ 77	5.0	5.2	5.3	5.5	5.7	5.8	6.0	6.2	6.2	6.4	6.5	6.7	6.7
+ 78	5.2	5.4	5.6	5.7	5.9	6.1	6.2	6.4	6.6	6.7	6.8	7.0	7.1
+ 79	5.4	5.7	5.8	6.1	6.2	6.4	6.6	6.7	6.9	7.1	7.2	7.4	7.5
+ 80	5.7	5.9	6.2	6.4	6.6	6.8	7.0	7.2	7.4	7.6	7.7	7.9	+ 8.0

PRECESSION IN DECLINATION 1950.0 - 2000.0

	02h00m 10 00	02 10 09 50	02 20 09 40	02 30 09 30	02 40 09 20	02 50 09 10	03 00 09 00	03 10 08 50	03 20 08 40	03 30 08 30	03 40 08 20	03 50 08 10	04 00 08 00
± 15'	14	14	13	13	12	12	11	11	10	10	9	8	

PRECESSION IN RIGHT ASCENSION 1950.0 - 2000.0

α	04^{h}00^m 08 00	04 10 07 50	04 20 07 40	04 30 07 30	04 40 07 20	04 50 07 10	05 00 07 00	05 10 06 50	05 20 06 40	05 30 06 30	05 40 06 20	05 50 06 10	06 00 06 00
δ													
+ 00	+2^m.6	2.6	2.6	2.6	2.6	2.6	2.6	2.6	2.6	2.6	2.6	2.6	2.6
+ 05	2.7	2.7	2.7	2.7	2.7	2.7	2.7	2.7	2.7	2.7	2.7	2.7	2.7
+ 10	2.7	2.7	2.7	2.7	2.7	2.7	2.7	2.7	2.7	2.7	2.7	2.7	2.7
+ 15	2.8	2.8	2.8	2.8	2.8	2.8	2.8	2.8	2.8	2.8	2.8	2.8	2.8
+ 20	2.9	2.9	2.9	2.9	2.9	2.9	2.9	2.9	2.9	2.9	2.9	2.9	2.9
+ 25	3.0	3.0	3.0	3.0	3.1	3.1	3.1	3.1	3.1	3.1	3.1	3.1	3.1
+ 30	3.1	3.2	3.2	3.2	3.2	3.2	3.2	3.2	3.2	3.2	3.2	3.2	3.2
+ 32	3.1	3.2	3.2	3.2	3.2	3.2	3.2	3.2	3.2	3.2	3.2	3.2	3.2
+ 34	3.2	3.2	3.2	3.2	3.2	3.2	3.2	3.3	3.3	3.3	3.3	3.3	3.3
+ 36	3.2	3.2	3.3	3.3	3.3	3.3	3.3	3.3	3.3	3.3	3.3	3.3	3.3
+ 38	3.3	3.3	3.3	3.3	3.4	3.4	3.4	3.4	3.4	3.4	3.4	3.4	3.4
+ 40	3.3	3.4	3.4	3.4	3.4	3.4	3.5	3.5	3.5	3.5	3.5	3.5	3.5
+ 42	3.4	3.4	3.5	3.5	3.5	3.5	3.5	3.5	3.6	3.6	3.6	3.6	3.6
+ 44	3.5	3.5	3.5	3.6	3.6	3.6	3.6	3.6	3.6	3.7	3.7	3.7	3.7
+ 46	3.6	3.6	3.6	3.7	3.7	3.7	3.7	3.7	3.7	3.7	3.8	3.8	3.8
+ 48	3.7	3.7	3.7	3.7	3.8	3.8	3.8	3.8	3.8	3.8	3.8	3.8	3.8
+ 50	3.8	3.8	3.8	3.8	3.8	3.8	3.8	3.8	3.8	3.9	3.9	3.9	3.9
+ 51	3.8	3.8	3.8	3.8	3.8	3.8	3.9	3.9	3.9	3.9	3.9	3.9	3.9
+ 52	3.8	3.8	3.8	3.9	3.9	3.9	3.9	3.9	4.0	4.0	4.0	4.0	4.0
+ 53	3.8	3.8	3.9	3.9	3.9	4.0	4.0	4.0	4.0	4.0	4.0	4.0	4.0
+ 54	3.9	3.9	3.9	4.0	4.0	4.0	4.0	4.1	4.1	4.1	4.1	4.1	4.1
+ 55	3.9	4.0	4.0	4.0	4.1	4.1	4.1	4.1	4.2	4.2	4.2	4.2	4.2
+ 56	4.0	4.0	4.1	4.1	4.1	4.2	4.2	4.2	4.2	4.2	4.2	4.2	4.2
+ 57	4.1	4.1	4.1	4.2	4.2	4.2	4.2	4.2	4.2	4.2	4.2	4.2	4.2
+ 58	4.1	4.2	4.2	4.2	4.2	4.2	4.2	4.3	4.3	4.3	4.3	4.3	4.3
+ 59	4.2	4.2	4.2	4.2	4.3	4.3	4.3	4.3	4.4	4.4	4.4	4.4	4.4
+ 60	4.2	4.2	4.3	4.3	4.3	4.4	4.4	4.4	4.5	4.5	4.5	4.5	4.5
+ 61	4.3	4.3	4.4	4.4	4.4	4.5	4.5	4.5	4.5	4.6	4.6	4.6	4.6
+ 62	4.3	4.4	4.5	4.5	4.5	4.6	4.6	4.6	4.6	4.7	4.7	4.7	4.7
+ 63	4.4	4.5	4.6	4.6	4.6	4.7	4.7	4.7	4.7	4.7	4.7	4.7	4.7
+ 64	4.5	4.6	4.7	4.7	4.7	4.7	4.7	4.7	4.8	4.8	4.8	4.8	4.8
+ 65	4.7	4.7	4.7	4.8	4.8	4.8	4.8	4.9	4.9	4.9	4.9	4.9	4.9
+ 66	4.7	4.7	4.8	4.8	4.9	4.9	5.0	5.0	5.0	5.0	5.1	5.1	5.1
+ 67	4.8	4.9	4.9	5.0	5.0	5.1	5.1	5.1	5.2	5.2	5.2	5.2	5.2
+ 68	4.9	5.0	5.1	5.1	5.1	5.2	5.2	5.2	5.2	5.3	5.3	5.3	5.3
+ 69	5.1	5.2	5.2	5.2	5.2	5.3	5.4	5.4	5.4	5.4	5.4	5.5	5.5
+ 70	5.2	5.2	5.3	5.3	5.4	5.5	5.5	5.6	5.6	5.6	5.6	5.6	5.6
+ 71	5.4	5.4	5.5	5.6	5.6	5.7	5.7	5.7	5.7	5.7	5.7	5.8	5.8
+ 72	5.5	5.6	5.7	5.7	5.7	5.8	5.8	5.8	5.9	6.0	6.0	6.0	6.0
+ 73	5.7	5.8	5.8	5.9	6.0	6.0	6.1	6.1	6.2	6.2	6.2	6.2	6.2
+ 74	5.9	6.0	6.0	6.2	6.2	6.2	6.3	6.3	6.4	6.4	6.4	6.4	6.4
+ 75	6.2	6.3	6.3	6.4	6.4	6.5	6.6	6.6	6.7	6.7	6.7	6.7	6.7
+ 76	6.4	6.5	6.6	6.7	6.7	6.8	6.8	6.9	6.9	7.0	7.0	7.0	7.0
+ 77	6.7	6.8	6.9	7.0	7.1	7.2	7.2	7.2	7.3	7.3	7.3	7.4	7.4
+ 78	7.1	7.2	7.3	7.4	7.5	7.6	7.6	7.7	7.7	7.7	7.7	7.8	7.8
+ 79	7.5	7.6	7.7	7.8	7.8	8.0	8.1	8.2	8.2	8.2	8.2	8.2	8.3
+ 80	8.0	8.2	8.2	8.4	8.6	8.7	8.8	8.8	8.8	8.8	8.8	8.8	+8.9

PRECESSION IN DECLINATION 1950.0 - 2000.0

± 8'	8	7	6	6	5	4	4	3	2	1	1	0

PRECESSION IN RIGHT ASCENSION 1950.0 - 2000.0

α	12^h00^m 24 00	12 10 23 50	12 20 23 40	12 30 23 30	12 40 23 20	12 50 23 10	13 00 23 00	13 10 22 50	13 20 22 40	13 30 22 30	13 40 22 20	13 50 22 10	14 00 22 00
δ													
+ 00°	$+2^m.6$	2.6	2.6	2.6	2.6	2.6	2.6	2.6	2.6	2.6	2.6	2.6	2.6
+ 05	2.6	2.6	2.6	2.6	2.6	2.5	2.5	2.5	2.5	2.5	2.5	2.5	2.5
+ 10	2.6	2.6	2.6	2.5	2.5	2.5	2.5	2.5	2.5	2.5	2.5	2.5	2.5
+ 15	2.6	2.6	2.5	2.5	2.5	2.5	2.5	2.5	2.5	2.4	2.4	2.4	2.4
+ 20	2.6	2.6	2.5	2.5	2.5	2.5	2.4	2.4	2.4	2.4	2.4	2.3	2.3
+ 25	2.6	2.5	2.5	2.5	2.5	2.4	2.4	2.4	2.4	2.3	2.3	2.3	2.3
+ 30	2.6	2.5	2.5	2.5	2.4	2.4	2.4	2.3	2.3	2.3	2.3	2.3	2.3
+ 32	2.6	2.5	2.5	2.5	2.4	2.4	2.4	2.3	2.3	2.3	2.3	2.3	2.3
+ 34	2.6	2.5	2.5	2.5	2.4	2.4	2.3	2.3	2.3	2.3	2.3	2.3	2.2
+ 36	2.6	2.5	2.5	2.4	2.4	2.4	2.3	2.3	2.3	2.3	2.3	2.2	2.2
+ 38	2.6	2.5	2.5	2.4	2.4	2.3	2.3	2.3	2.3	2.3	2.2	2.2	2.2
+ 40	2.6	2.5	2.5	2.4	2.4	2.3	2.3	2.3	2.3	2.2	2.2	2.2	2.1
+ 42	2.6	2.5	2.5	2.4	2.4	2.3	2.3	2.3	2.3	2.2	2.2	2.1	2.1
+ 44	2.6	2.5	2.5	2.4	2.4	2.3	2.3	2.3	2.2	2.2	2.1	2.1	2.0
+ 46	2.6	2.5	2.5	2.4	2.3	2.3	2.3	2.3	2.2	2.1	2.1	2.0	2.0
+ 48	2.6	2.5	2.4	2.4	2.3	2.3	2.3	2.2	2.2	2.1	2.0	2.0	1.9
+ 50	2.6	2.5	2.4	2.4	2.3	2.3	2.3	2.2	2.1	2.1	2.0	1.9	1.9
+ 51	2.6	2.5	2.4	2.4	2.3	2.3	2.2	2.2	2.1	2.0	2.0	1.9	1.8
+ 52	2.6	2.5	2.4	2.4	2.3	2.3	2.2	2.2	2.1	2.0	2.0	1.9	1.8
+ 53	2.6	2.5	2.4	2.3	2.3	2.3	2.2	2.1	2.1	2.0	1.9	1.9	1.8
+ 54	2.6	2.5	2.4	2.3	2.3	2.3	2.2	2.1	2.0	2.0	1.9	1.8	1.8
+ 55	2.6	2.5	2.4	2.3	2.3	2.3	2.2	2.1	2.0	2.0	1.9	1.8	1.7
+ 56	2.6	2.5	2.4	2.3	2.3	2.2	2.2	2.1	2.0	1.9	1.8	1.8	1.7
+ 57	2.6	2.5	2.4	2.3	2.3	2.2	2.1	2.1	2.0	1.9	1.8	1.7	1.7
+ 58	2.6	2.5	2.4	2.3	2.3	2.2	2.1	2.0	1.9	1.9	1.8	1.7	1.7
+ 59	2.6	2.5	2.4	2.3	2.3	2.2	2.1	2.0	1.9	1.8	1.7	1.7	1.7
+ 60	2.6	2.5	2.4	2.3	2.3	2.2	2.1	2.0	1.9	1.8	1.7	1.7	1.6
+ 61	2.6	2.5	2.4	2.3	2.3	2.2	2.1	1.9	1.9	1.8	1.7	1.7	1.6
+ 62	2.6	2.5	2.4	2.3	2.2	2.1	2.0	1.9	1.8	1.7	1.7	1.6	1.5
+ 63	2.6	2.5	2.3	2.3	2.2	2.1	2.0	1.9	1.8	1.7	1.7	1.6	1.5
+ 64	2.6	2.5	2.3	2.3	2.2	2.1	2.0	1.9	1.8	1.7	1.6	1.5	1.4
+ 65	2.6	2.5	2.3	2.3	2.2	2.1	1.9	1.8	1.7	1.7	1.6	1.5	1.3
+ 66	2.6	2.4	2.3	2.3	2.2	2.0	1.9	1.8	1.7	1.6	1.5	1.4	1.3
+ 67	2.6	2.4	2.3	2.3	2.2	2.0	1.9	1.7	1.7	1.6	1.5	1.3	1.3
+ 68	2.6	2.4	2.3	2.2	2.1	2.0	1.9	1.7	1.7	1.5	1.4	1.3	1.2
+ 69	2.6	2.4	2.3	2.2	2.1	1.9	1.8	1.7	1.6	1.4	1.3	1.3	1.1
+ 70	2.6	2.4	2.3	2.2	2.0	1.8	1.7	1.7	1.5	1.4	1.3	1.2	1.0
+ 71	2.6	2.4	2.3	2.2	2.0	1.8	1.7	1.6	1.5	1.3	1.2	1.1	0.9
+ 72	2.6	2.4	2.3	2.1	2.0	1.8	1.7	1.6	1.4	1.3	1.1	1.0	0.8
+ 73	2.6	2.4	2.3	2.1	1.9	1.7	1.6	1.5	1.3	1.2	1.0	0.9	0.7
+ 74	2.6	2.4	2.3	2.1	1.9	1.7	1.6	1.4	1.3	1.1	0.9	0.7	0.6
+ 75	2.6	2.4	2.2	2.0	1.8	1.7	1.5	1.3	1.2	1.0	0.8	0.7	0.5
+ 76	2.6	2.4	2.2	2.0	1.7	1.6	1.4	1.3	1.0	0.8	0.7	0.5	0.3
+ 77	2.6	2.3	2.2	1.9	1.7	1.5	1.3	1.1	0.9	0.7	0.5	0.3	+ 0.2
+ 78	2.6	2.3	2.1	1.9	1.7	1.4	1.3	1.0	0.7	0.6	0.3	+ 0.2	- 0.1
+ 79	2.6	2.3	2.1	1.8	1.6	1.3	1.1	0.8	0.6	0.3	+ 0.2	- 0.1	- 0.3
+ 80	2.6	2.3	2.0	1.7	1.5	1.2	0.9	0.7	0.4	+ 0.2	- 0.1	- 0.3	- 0.6

PRECESSION IN DECLINATION 1950.0 - 2000.0

	$\mp 17'$	17	17	17	17	16	16	16	16	15	15	15	15

PRECESSION IN RIGHT ASCENSION 1950.0 - 2000.0

α	14^h00^m 22 00	14 10 21 50	14 20 21 40	14 30 21 30	14 40 21 20	14 50 21 10	15 00 21 00	15 10 20 50	15 20 20 40	15 30 20 30	15 40 20 20	15 50 20 10	16 00 20 00
δ													
+ 00°	$+2^m_.6$	2.6	2.6	2.6	2.6	2.6	2.6	2.6	2.6	2.6	2.6	2.6	2.6
+ 05	2.5	2.5	2.5	2.5	2.5	2.5	2.5	2.5	2.5	2.5	2.5	2.5	2.5
+ 10	2.5	2.5	2.4	2.4	2.4	2.4	2.4	2.4	2.4	2.4	2.4	2.4	2.4
+ 15	2.4	2.4	2.4	2.4	2.3	2.3	2.3	2.3	2.3	2.3	2.3	2.3	2.3
+ 20	2.3	2.3	2.3	2.3	2.3	2.3	2.3	2.3	2.3	2.3	2.3	2.3	2.3
+ 25	2.3	2.3	2.3	2.3	2.3	2.3	2.2	2.2	2.2	2.2	2.2	2.2	2.1
+ 30	2.3	2.3	2.2	2.2	2.2	2.2	2.1	2.1	2.1	2.1	2.0	2.0	2.0
+ 32	2.3	2.2	2.2	2.2	2.1	2.1	2.1	2.1	2.0	2.0	2.0	2.0	2.0
+ 34	2.2	2.2	2.2	2.1	2.1	2.1	2.0	2.0	2.0	2.0	1.9	1.9	1.9
+ 36	2.2	2.2	2.1	2.1	2.1	2.0	2.0	2.0	1.9	1.9	1.9	1.9	1.8
+ 38	2.2	2.1	2.1	2.0	2.0	2.0	1.9	1.9	1.9	1.8	1.8	1.8	1.8
+ 40	2.1	2.1	2.0	2.0	2.0	1.9	1.9	1.9	1.8	1.8	1.8	1.7	1.7
+ 42	2.1	2.1	2.0	1.9	1.9	1.9	1.8	1.8	1.8	1.7	1.7	1.7	1.7
+ 44	2.0	2.0	1.9	1.9	1.9	1.8	1.8	1.8	1.7	1.7	1.7	1.7	1.7
+ 46	2.0	1.9	1.9	1.8	1.8	1.7	1.7	1.7	1.7	1.7	1.6	1.6	1.6
+ 48	1.9	1.9	1.8	1.8	1.7	1.7	1.7	1.7	1.6	1.6	1.6	1.5	1.5
+ 50	1.9	1.8	1.8	1.7	1.7	1.7	1.7	1.6	1.6	1.5	1.5	1.4	1.4
+ 51	1.8	1.8	1.7	1.7	1.7	1.7	1.6	1.6	1.5	1.5	1.4	1.4	1.3
+ 52	1.8	1.8	1.7	1.7	1.7	1.6	1.6	1.5	1.5	1.4	1.4	1.3	1.3
+ 53	1.8	1.7	1.7	1.7	1.6	1.6	1.5	1.5	1.4	1.4	1.3	1.3	1.3
+ 54	1.8	1.7	1.7	1.7	1.6	1.5	1.5	1.4	1.4	1.3	1.3	1.3	1.3
+ 55	1.7	1.7	1.7	1.6	1.6	1.5	1.4	1.4	1.3	1.3	1.3	1.3	1.2
+ 56	1.7	1.7	1.6	1.6	1.5	1.4	1.4	1.3	1.3	1.3	1.3	1.2	1.2
+ 57	1.7	1.7	1.6	1.5	1.5	1.4	1.3	1.3	1.3	1.2	1.2	1.1	1.1
+ 58	1.7	1.6	1.6	1.5	1.4	1.3	1.3	1.3	1.2	1.2	1.1	1.1	1.0
+ 59	1.7	1.6	1.5	1.4	1.3	1.3	1.3	1.2	1.2	1.1	1.1	1.0	0.9
+ 60	1.6	1.5	1.5	1.4	1.3	1.3	1.2	1.2	1.1	1.0	1.0	0.9	0.9
+ 61	1.6	1.5	1.4	1.3	1.3	1.2	1.2	1.2	1.1	1.0	0.9	0.8	0.8
+ 62	1.5	1.4	1.3	1.3	1.3	1.2	1.1	1.0	1.0	0.9	0.8	0.8	0.7
+ 63	1.5	1.4	1.3	1.3	1.2	1.1	1.0	0.9	0.9	0.8	0.7	0.7	0.7
+ 64	1.4	1.3	1.3	1.2	1.1	1.0	0.9	0.9	0.8	0.7	0.7	0.7	0.6
+ 65	1.3	1.3	1.2	1.1	1.0	0.9	0.8	0.8	0.7	0.7	0.6	0.6	0.5
+ 66	1.3	1.3	1.2	1.1	0.9	0.9	0.8	0.7	0.7	0.6	0.5	0.4	0.4
+ 67	1.3	1.2	1.1	1.0	0.9	0.8	0.7	0.7	0.6	0.5	0.4	0.3	0.3
+ 68	1.2	1.1	1.0	0.9	0.8	0.7	0.6	0.5	0.5	0.4	0.3	0.2	0.2
+ 69	1.1	1.0	0.9	0.8	0.7	0.6	0.5	0.4	0.3	0.2	0.2	+ 0.1	+ 0.1
+ 70	1.0	0.9	0.8	0.7	0.6	0.5	0.4	0.3	0.2	+ 0.2	+ 0.1	0.0	- 0.1
+ 71	0.9	0.8	0.7	0.6	0.5	0.4	0.2	+ 0.2	+ 0.1	0.0	- 0.1	- 0.2	- 0.2
+ 72	0.8	0.7	0.6	0.5	0.3	0.2	+ 0.2	0.0	- 0.1	- 0.2	- 0.2	- 0.3	- 0.4
+ 73	0.7	0.6	0.5	0.3	0.2	+ 0.1	0.0	- 0.2	- 0.2	- 0.3	- 0.4	- 0.5	- 0.6
+ 74	0.6	0.5	0.3	+ 0.2	+ 0.1	- 0.1	- 0.2	- 0.3	- 0.4	- 0.5	- 0.6	- 0.7	- 0.8
+ 75	0.5	0.3	+ 0.2	0.0	- 0.1	- 0.2	- 0.4	- 0.5	- 0.7	- 0.7	- 0.8	- 0.9	- 1.0
+ 76	0.3	+ 0.2	0.0	- 0.2	- 0.3	- 0.6	- 0.6	- 0.7	- 0.8	- 1.0	- 1.1	- 1.3	- 1.3
+ 77	+ 0.2	0.0	- 0.2	- 0.4	- 0.6	- 0.7	- 0.8	- 1.0	- 1.2	- 1.3	- 1.4	- 1.5	- 1.6
+ 78	- 0.1	- 0.2	- 0.4	- 0.7	- 0.8	- 1.0	- 1.2	- 1.3	- 1.3	- 1.4	- 1.7	- 1.8	- 2.0
+ 79	- 0.3	- 0.5	- 0.7	- 0.9	- 1.1	- 1.3	- 1.5	- 1.7	- 1.8	- 2.0	- 2.2	- 2.3	- 2.4
+ 80	- 0.6	- 0.8	- 1.1	- 1.3	- 1.5	- 1.7	- 1.9	- 2.1	- 2.3	- 2.4	- 2.6	- 2.7	- 2.9

PRECESSION IN DECLINATION 1950.0 - 2000.0

	$\mp 15'$	14	14	13	13	12	12	11	11	10	10	9	8

PRECESSION IN RIGHT ASCENSION 1950.0 - 2000.0

α	16h00m / 20 00	16 10 / 19 50	16 20 / 19 40	16 30 / 19 30	16 40 / 19 20	16 50 / 19 10	17 00 / 19 00	17 10 / 18 50	17 20 / 18 40	17 30 / 18 30	17 40 / 18 20	17 50 / 18 10	18 00 / 18 00
δ													
+ 00°	+ 2.6	2.6	2.6	2.6	2.6	2.6	2.6	2.6	2.6	2.6	2.6	2.6	2.6
+ 05	2.5	2.5	2.5	2.5	2.5	2.5	2.5	2.5	2.5	2.5	2.5	2.5	2.5
+ 10	2.4	2.4	2.4	2.4	2.4	2.4	2.4	2.3	2.3	2.3	2.3	2.3	2.3
+ 15	2.3	2.4	2.4	2.4	2.4	2.4	2.3	2.3	2.3	2.3	2.3	2.3	2.3
+ 20	2.3	2.2	2.2	2.2	2.2	2.2	2.2	2.2	2.2	2.2	2.2	2.2	2.2
+ 25	2.1	2.1	2.1	2.1	2.1	2.1	2.1	2.1	2.1	2.1	2.1	2.1	2.1
+ 30	2.0	2.0	2.0	2.0	2.0	1.9	1.9	1.9	1.9	1.9	1.9	1.9	1.9
+ 32	2.0	1.9	1.9	1.9	1.9	1.9	1.9	1.9	1.9	1.9	1.8	1.8	1.8
+ 34	1.9	1.9	1.9	1.8	1.8	1.8	1.8	1.8	1.8	1.8	1.8	1.8	1.8
+ 36	1.8	1.8	1.8	1.8	1.8	1.8	1.7	1.7	1.7	1.7	1.7	1.7	1.7
+ 38	1.8	1.8	1.7	1.7	1.7	1.7	1.7	1.7	1.7	1.7	1.7	1.7	1.7
+ 40	1.7	1.7	1.7	1.7	1.7	1.7	1.7	1.7	1.7	1.7	1.7	1.7	1.7
+ 42	1.7	1.7	1.7	1.7	1.6	1.6	1.6	1.6	1.6	1.6	1.6	1.6	1.6
+ 44	1.7	1.6	1.6	1.6	1.6	1.5	1.5	1.5	1.5	1.5	1.5	1.5	1.5
+ 46	1.6	1.6	1.5	1.5	1.5	1.5	1.4	1.4	1.4	1.4	1.4	1.4	1.4
+ 48	1.5	1.5	1.4	1.4	1.4	1.4	1.3	1.3	1.3	1.3	1.3	1.3	1.3
+ 50	1.4	1.4	1.4	1.3	1.3	1.3	1.3	1.3	1.3	1.3	1.3	1.3	1.3
+ 51	1.3	1.3	1.3	1.3	1.3	1.3	1.3	1.3	1.3	1.2	1.2	1.2	1.2
+ 52	1.3	1.3	1.3	1.3	1.3	1.2	1.2	1.2	1.2	1.2	1.2	1.2	1.2
+ 53	1.3	1.3	1.3	1.2	1.2	1.2	1.2	1.1	1.1	1.1	1.1	1.1	1.1
+ 54	1.3	1.2	1.2	1.2	1.1	1.1	1.1	1.1	1.1	1.1	1.0	1.0	1.0
+ 55	1.2	1.2	1.1	1.1	1.1	1.1	1.0	1.0	1.0	1.0	1.0	1.0	1.0
+ 56	1.2	1.1	1.1	1.0	1.0	1.0	1.0	0.9	0.9	0.9	0.9	0.9	0.9
+ 57	1.1	1.1	1.0	1.0	0.9	0.9	0.9	0.9	0.9	0.8	0.8	0.8	0.8
+ 58	1.0	1.0	0.9	0.9	0.9	0.8	0.8	0.8	0.8	0.8	0.7	0.7	0.7
+ 59	0.9	0.9	0.9	0.8	0.8	0.8	0.7	0.7	0.7	0.7	0.7	0.7	0.7
+ 60	0.9	0.8	0.8	0.7	0.7	0.7	0.7	0.7	0.7	0.7	0.7	0.7	0.7
+ 61	0.8	0.7	0.7	0.7	0.7	0.7	0.6	0.6	0.6	0.6	0.6	0.6	0.6
+ 62	0.7	0.7	0.7	0.7	0.6	0.6	0.6	0.5	0.5	0.5	0.5	0.5	0.5
+ 63	0.7	0.7	0.6	0.6	0.6	0.5	0.4	0.4	0.4	0.4	0.4	0.4	0.4
+ 64	0.6	0.6	0.5	0.4	0.4	0.4	0.3	0.3	0.3	0.3	0.2	0.2	0.2
+ 65	0.5	0.4	0.4	0.3	0.3	0.2	0.2	0.2	0.2	0.2	0.2	0.2	0.2
+ 66	0.4	0.3	0.3	0.2	0.2	0.2	+ 0.2	+ 0.1	+ 0.1	+ 0.1	+ 0.1	+ 0.1	+ 0.1
+ 67	0.3	0.2	0.2	+ 0.2	+ 0.1	+ 0.1	0.0	0.0	0.0	- 0.1	- 0.1	- 0.1	- 0.1
+ 68	0.2	+ 0.1	+ 0.1	0.0	0.0	- 0.1	- 0.1	- 0.2	- 0.2	- 0.2	- 0.2	- 0.2	- 0.2
+ 69	+ 0.1	0.0	- 0.1	- 0.1	- 0.2	- 0.2	- 0.2	- 0.2	- 0.3	- 0.3	- 0.3	- 0.3	- 0.3
+ 70	- 0.1	- 0.2	- 0.2	- 0.3	- 0.3	- 0.3	- 0.4	- 0.4	- 0.4	- 0.5	- 0.5	- 0.5	- 0.5
+ 71	- 0.2	- 0.3	- 0.4	- 0.4	- 0.5	- 0.5	- 0.6	- 0.6	- 0.7	- 0.7	- 0.7	- 0.7	- 0.7
+ 72	- 0.4	- 0.5	- 0.6	- 0.6	- 0.7	- 0.7	- 0.7	- 0.7	- 0.8	- 0.8	- 0.8	- 0.8	- 0.8
+ 73	- 0.6	- 0.7	- 0.7	- 0.8	- 0.8	- 0.9	- 1.0	- 1.0	- 1.0	- 1.1	- 1.1	- 1.1	- 1.1
+ 74	- 0.8	- 0.9	- 1.0	- 1.0	- 1.1	- 1.2	- 1.2	- 1.3	- 1.3	- 1.3	- 1.3	- 1.3	- 1.3
+ 75	- 1.0	- 1.2	- 1.3	- 1.3	- 1.3	- 1.4	- 1.4	- 1.4	- 1.5	- 1.6	- 1.6	- 1.6	- 1.6
+ 76	- 1.3	- 1.4	- 1.5	- 1.6	- 1.7	- 1.7	- 1.7	- 1.8	- 1.8	- 1.8	- 1.9	- 1.9	- 1.9
+ 77	- 1.6	- 1.7	- 1.8	- 1.9	- 2.0	- 2.1	- 2.1	- 2.2	- 2.2	- 2.3	- 2.3	- 2.3	- 2.3
+ 78	- 2.0	- 2.1	- 2.2	- 2.3	- 2.3	- 2.4	- 2.5	- 2.6	- 2.6	- 2.7	- 2.7	- 2.7	- 2.7
+ 79	- 2.4	- 2.5	- 2.7	- 2.7	- 2.8	- 2.9	- 3.0	- 3.1	- 3.1	- 3.1	- 3.2	- 3.2	- 3.2
+ 80	- 2.9	- 3.1	- 3.1	- 3.2	- 3.3	- 3.5	- 3.6	- 3.7	- 3.7	- 3.7	- 3.8	- 3.8	- 3.8

PRECESSION IN DECLINATION 1950.0 - 2000.0

∓ 8'	8	7	6	6	5	4	4	3	2	2	1	0

UGC COORDINATES WITH $\delta_{1950.0} > 80°00'$ PRECESSED TO 2000.0

U G C	α, δ (2000.0)	U G C	α, δ (2000.0)	U G C	α, δ (2000.0)	U G C	α, δ (2000.0)
00020	00^h04^m+ 80°18'	03521	06 54 + 84 03	05309	09 57 + 80 45	10181	15 59 + 81 42
00115	00 15 + 88 23	03522	06 56 + 84 55	05404	10 06 + 80 17	10183	16 00 + 79 52
00392	00 40 + 83 13	03528	06 55 + 84 05	05438	10 12 + 84 49	10222	16 02 + 81 53
00481	00 49 + 83 45	03540	06 55 + 82 29	05643	10 28 + 79 48	10263	15 58 + 86 12
00642	01 04 + 80 40	03549	06 55 + 80 58	05659	10 29 + 79 47	10280	16 07 + 81 18
01039	01 32 + 85 01	03581	06 59 + 80 00	05820	10 44 + 80 54	10378	16 16 + 84 01
01141	01 38 + 80 29	03604	07 02 + 80 57	06694	11 44 + 83 28	10431	16 24 + 81 48
01148	01 41 + 83 58	03653	07 15 + 84 23	06896	11 56 + 80 13	10438	16 26 + 81 25
01198	01 48 + 85 16	03654	07 18 + 85 43	07097	12 06 + 80 55	10447	16 28 + 80 22
01207	01 47 + 82 13	03661	07 19 + 85 46	07410	12 21 + 82 41	10464	16 28 + 82 54
01285	01 58 + 86 42	03668	07 12 + 80 59	07624	12 29 + 81 53	10471	16 31 + 81 34
01616	02 15 + 85 19	03669	07 19 + 85 28	07664	12 30 + 84 47	10472	16 27 + 84 33
01985	02 36 + 82 12	03670	07 20 + 85 36	07731	12 33 + 82 35	10501	16 33 + 82 33
02583	03 17 + 80 48	03715	07 26 + 86 14	07956	12 46 + 83 26	10581	16 43 + 82 38
02603	03 19 + 81 21	03740	07 27 + 85 45	08002	12 50 + 80 37	10604	16 48 + 81 51
02620	03 20 + 80 15	03798	07 33 + 85 43	08029	12 53 + 82 24	10605	16 46 + 83 18
02753	03 35 + 82 25	03809	07 27 + 80 10	08101	12 56 + 84 07	10612	16 50 + 80 21
02935	04 11 + 83 28	03815	07 35 + 85 33	08147	13 01 + 80 03	10638	16 52 + 81 35
03132	04 50 + 82 39	03890	07 40 + 83 47	08148	13 01 + 80 07	10740	16 53 + 86 36
03160	05 03 + 86 14	03992	07 54 + 84 37	08158	13 02 + 79 57	10923	17 19 + 86 45
03197	05 00 + 80 10	03993	07 55 + 84 56	08264	13 08 + 84 37	10959	17 32 + 84 38
03210	05 04 + 82 05	03994	07 56 + 85 10	08369	13 17 + 81 23	11081	17 53 + 81 08
03253	05 20 + 84 03	04063	08 05 + 85 49	08380	13 17 + 84 41	11267	18 03 + 87 50
03257	05 22 + 84 29	04078	08 04 + 84 38	08394	13 17 + 84 30	11339	18 23 + 86 54
03336	05 55 + 85 55	04100	08 06 + 84 47	08414	13 20 + 84 26	11359	18 42 + 81 24
03345	05 58 + 85 50	04129	08 13 + 86 09	08454	13 23 + 84 31	11366	18 44 + 81 43
03378	06 06 + 83 50	04173	08 07 + 80 06	08520	13 30 + 80 30	11390	18 51 + 84 34
03381	06 08 + 83 43	04262	08 19 + 83 16	08615	13 31 + 86 01	11441	19 22 + 81 01
03385	06 05 + 80 08	04297	08 28 + 85 27	09204	14 18 + 82 38	11495	19 35 + 87 16
03400	06 12 + 80 38	04348	08 34 + 85 57	09205	14 14 + 85 49	11664	20 54 + 84 05
03401	06 14 + 81 04	04396	08 35 + 84 19	09298	14 23 + 84 29	11679	20 53 + 86 55
03404	06 14 + 80 00	04454	08 37 + 80 34	09478	14 34 + 85 18	11738	21 17 + 83 50
03405	06 14 + 80 28	04463	08 44 + 85 44	09621	14 52 + 82 33	11746	21 22 + 82 19
03410	06 15 + 80 27	04474	08 43 + 84 38	09650	14 54 + 83 36	12141	22 38 + 80 41
03413	06 17 + 81 09	04557	08 52 + 84 18	09668	14 56 + 83 32	12221	22 50 + 82 54
03427	06 19 + 80 04	04601	08 57 + 84 49	09669	14 53 + 85 01	12377	23 03 + 86 46
03435	06 25 + 82 20	04612	08 59 + 85 33	09683	14 59 + 81 25	12387	23 04 + 86 45
03441	06 28 + 83 18	04646	09 03 + 85 30	09720	14 58 + 85 00	12905	00 01 + 80 40
03442	06 31 + 84 55	04682	09 08 + 86 08	09733	15 03 + 81 11	02886A	04 36 + 88 45
03455	06 35 + 85 21	04852	09 17 + 80 03	09878	15 24 + 83 05	03211A	06 28 + 89 18
03461	06 34 + 85 53	04948	09 28 + 85 19	09932	15 31 + 82 28	03457A	06 48 + 87 31
03466	06 35 + 83 32	05068	09 35 + 81 09	09950	15 33 + 82 15	03528A	07 03 + 86 36
03470	06 37 + 82 58	05114	09 40 + 82 07	09966	15 33 + 83 05	03536A	07 04 + 86 34
03496	06 50 + 85 50	05128	09 44 + 83 48	10008	15 38 + 82 20	03617A	07 25 + 87 44
03500	06 47 + 84 10	05203	09 47 + 79 48	10054	15 44 + 81 50	05083A	09 53 + 88 09
03509	06 55 + 85 39	05298	09 57 + 82 18	10125	15 52 + 83 51	07956A	12 42 + 87 49

"N E A R"   C L U S T E R S   O F   G A L A X I E S   I N   C G C G

1	00^{h}00^m.8 + 04°52'	51	03 21.0 + 06 48	99	07 31.9 + 31 25	149	09 27.2 + 34 46
2	00 02.4 + 07 44	52	03 37.8 + 15 36	100	07 33.4 + 61 02	150	09 35.3 + 17 01
3	00 13.4 + 18 05	53	03 54.0 + 79 00	101	07 33.8 + 23 56	151	09 38.2 + 11 30
4	00 13.6 + 29 27	54	04 00.5 + 13 35	102	07 35.0 + 85 45	152	09 38.7 + 39 57
5	00 14.5 + 23 15	55	04 03.1 + 30 40	103	07 39.8 + 49 49	153	09 41.6 + 15 40
6	00 24.4 + 30 14	56	04 14.5 + 02 16	104	07 43.5 + 31 10	154	09 41.7 + 24 30
7	00 32.6 + 02 07	57	04 17.7 + 35 57	105	07 44.3 + 18 39	155	09 43.7 + 54 54
8	00 33.8 + 05 38	58	04 25.7 + 41 20	106	07 45.5 + 40 20	156	09 56.4 + 37 30
9	00 34.4 + 25 32	59	04 26.4 – 00 02	107	07 46.5 + 23 15	157	09 57.9 + 11 11
10	00 36.3 + 29 14	59a	04 30.8 – 04 24	108	07 52.9 + 28 33	158	09 58.9 + 00 38
11	00 36.4 + 20 55	60	04 34.5 + 21 12	109	07 53.2 + 41 01	159	10 03.6 + 14 43
12	00 46.5 + 23 00	61	04 35.9 + 30 53	110	07 54.0 + 46 19	160	10 06.0 + 00 14
13	00 48.2 + 43 05	62	04 44.7 + 08 28	111	07 55.9 + 08 05	161	10 10.5 + 39 22
14	00 54.6 – 01 27	62a	04 49.3 – 04 37	112	07 56.1 + 56 16	162	10 12.0 – 00 47
15	00 55.0 + 12 12	63	04 51.3 + 01 59	113	08 00.0 + 09 46	163	10 12.8 + 53 37
16	00 56.9 + 26 36	64	04 52.2 + 73 05	114	08 01.3 + 39 54	164	10 14.1 + 22 15
17	01 03.7 + 39 42	65	04 56.1 – 01 03	115	08 06.8 + 05 14	165	10 19.3 + 15 27
18	01 05.5 + 36 50	66	05 06.6 + 16 49	116	08 10.1 + 58 13	166	10 19.4 + 25 01
19	01 06.9 + 00 28	67	05 07.9 – 01 10	117	08 14.8 + 56 48	166a	10 19.5 + 00 41
20	01 07.5 + 32 12	68	05 10.0 + 04 58	118	08 19.6 + 22 09	167	10 19.8 + 36 53
21	01 10.5 + 15 15	69	05 17.8 + 51 08	119	08 20.1 – 00 29	168	10 20.1 + 13 06
22	01 19.6 + 50 35	70	05 21.2 + 64 18	120	08 20.6 + 04 36	169	10 20.1 + 20 46
23	01 21.5 + 01 13	71	05 29.2 + 05 16	121	08 22.4 + 54 53	170	10 20.4 – 03 16
24	01 23.0 + 09 53	72	05 31.3 + 01 27	122	08 24.8 + 17 31	171	10 20.4 + 42 10
25	01 23.6 – 01 33	73	05 41.3 + 01 01	123	08 26.2 + 30 39	172	10 21.0 + 77 28
26	01 27.6 + 18 28	74	05 44.4 + 50 36	124	08 29.6 + 52 45	173	10 21.8 + 17 25
27	01 40.1 + 31 44	75	05 58.5 + 59 51	125	08 32.2 + 38 45	174	10 23.8 + 10 56
28	01 43.8 + 23 23	76	06 03.0 + 79 22	126	08 32.6 – 02 35	175	10 26.1 + 04 12
29	01 44.0 + 12 30	77	06 11.6 + 46 01	127	08 36.3 + 41 47	176	10 26.2 + 22 15
30	01 45.8 + 47 40	78	06 13.9 + 47 53	128	08 37.0 + 25 06	177	10 26.9 + 40 23
31	01 50.8 + 36 15	79	06 20.4 + 46 20	129	08 37.1 + 64 45	178	10 27.9 + 61 50
32	01 50.9 + 30 50	80	06 28.9 + 52 32	130	08 42.3 + 28 20	179	10 28.2 + 43 57
33	02 02.6 + 18 52	81	06 31.6 + 26 09	131	08 46.3 + 19 10	180	10 29.3 + 57 36
34	02 05.5 + 01 10	82	06 32.9 + 22 44	132	08 51.1 + 49 25	181	10 29.8 + 20 23
35	02 08.0 + 15 15	83	06 32.7 + 63 63	133	08 52.7 + 00 54	182	10 33.3 + 38 04
36	02 10.9 + 50 38	84	06 40.4 + 25 59	134	08 53.5 – 03 12	183	10 34.9 + 29 12
37	02 16.0 + 36 25	85	06 42.0 + 73 34	135	08 55.0 + 52 48	184	10 35.1 + 72 26
38	02 26.0 + 26 00	86	06 42.2 + 41 30	136	08 56.3 + 45 54	184a	10 35.3 + 50 13
39	02 33.0 + 01 24	87	06 47.4 + 33 23	137	09 00.0 + 13 43	185	10 37.4 + 21 56
40	02 36.2 + 32 49	88	06 54.8 + 27 53	138	09 01.2 + 66 40	186	10 39.3 + 11 09
41	02 40.6 + 07 40	89	07 00.4 + 48 01	139	09 03.9 + 37 16	187	10 39.4 + 16 49
42	02 41.2 + 35 58	90	07 01.0 + 18 58	140	09 09.7 + 18 14	188	10 42.4 + 39 10
43	02 46.1 – 00 45	91	07 05.4 + 36 42	141	09 11.0 + 30 25	189	10 43.9 + 56 35
44	02 48.0 + 13 07	92	07 06.6 + 32 21	142	09 15.6 + 34 09	190	10 45.8 + 05 10
45	02 54.7 + 05 55	93	07 09.5 + 06 42	143	09 16.7 + 49 52	191	10 46.3 + 00 38
46	02 54.7 + 16 06	94	07 10.5 + 42 22	144	09 17.8 + 07 04	192	10 46.8 + 27 47
47	03 03.0 + 41 25	95	07 18.6 + 32 49	145	09 20.4 + 40 37	193	10 47.6 + 16 23
48	03 07.7 + 19 07	96	07 18.9 + 54 12	146	09 26.6 + 23 54	194	10 48.6 + 23 58
49	03 10.0 – 01 30	97	07 20.6 + 22 59	147	09 23.9 + 73 53	195	10 49.2 + 09 02
50	03 12.2 + 15 51	98	07 30.1 + 18 58	148	09 26.5 + 30 26	196	10 55.0 + 17 25

197	10 55.4 + 01 42	251	12 02.0 + 20 28
198	10 56.1 + 06 44	252	12 02.9 + 10 51
199	10 56.5 + 12 40	253	12 04.7 + 22 46
200	10 56.9 + 09 22	254	12 04.7 + 33 19
201	10 57.2 + 38 04	255	12 04.8 + 65 20
202	10 58.6 + 10 49	256	12 05.4 + 25 15
203	10 58.6 + 46 11	257	12 07.6 + 57 35
204	11 02.1 + 31 36	258	12 08.1 + 36 03
205	11 05.2 + 13 42	259	12 11.4 + 60 13
206	11 05.3 + 28 35	260	12 12.0 + 24 09
207	11 05.6 + 23 23	261	12 16.5 + 61 48
208	11 06.2 + 05 16	262	12 16.7 + 55 42
209	11 07.6 + 10 41	263	12 17.5 + 29 15
210	11 07.7 + 36 10	264	12 21.4 + 24 11
211	11 09.1 + 38 06	265	12 22.6 + 45 48
212	11 12.7 + 72 59	266	12 24.1 + 09 14
213	11 12.9 + 26 00	267	12 24.6 + 31 31
214	11 14.3 + 54 57	268	12 26.0 + 08 59
215	11 15.2 + 30 13	269	12 30.3 + 74 50
215a	11 16.0 - 04 10	270	12 31.4 + 50 44
216	11 17.0 + 46 53	271	12 31.8 + 48 10
217	11 17.6 + 33 52	272	12 39.5 + 09 49
218	11 19.7 + 03 05	273	12 43.2 + 41 43
219	11 22.3 + 63 17	274	12 45.0 - 00 02
220	11 23.5 + 22 56	275	12 55.6 + 78 58
221	11 23.9 + 35 41	276	12 57.1 + 28 06
222	11 25.5 + 27 59	277	12 57.3 + 39 25
223	11 26.3 + 09 13	278	13 01.9 + 50 01
224	11 26.4 + 36 56	279	13 02.2 + 62 43
225	11 30.9 + 34 35	280	13 07.0 + 39 44
226	11 31.2 + 49 23	281	13 08.2 + 35 31
227	11 31.6 + 44 21	282	13 08.3 + 44 56
228	11 38.3 + 10 24	283	13 09.3 + 22 55
229	11 38.7 + 56 50	284	13 12.3 + 41 15
230	11 40.0 + 27 15	285	13 13.0 + 54 10
231	11 41.7 - 01 58	286	13 17.9 + 33 50
232	11 41.7 + 31 13	287	13 19.6 + 31 35
233	11 41.9 + 30 04	288	13 21.4 + 13 58
234	11 42.1 + 21 26	289	13 26.6 + 37 50
235	11 42.2 + 34 56	290	13 27.3 + 11 45
236	11 44.0 + 55 55	291	13 30.7 + 13 47
237	11 44.6 + 54 52	292	13 38.4 + 24 20
238	11 46.8 + 12 37	293	13 39.9 + 30 30
239	11 48.6 + 06 42	294	13 41.0 + 59 30
240	11 50.0 + 03 39	295	13 41.6 + 26 14
241	11 51.8 + 09 44	296	13 42.2 + 11 49
242	11 53.0 + 25 22	297	13 42.5 + 78 41
243	11 54.9 + 28 06	298	13 43.3 + 22 45
244	11 55.0 + 31 27	299	13 46.2 + 28 14
245	11 56.2 + 22 01	300	13 47.5 + 18 15
246	11 56.8 + 10 54	301	13 48.7 + 02 49
247	11 56.8 + 56 27	302	13 49.2 + 09 40
248	11 57.2 + 43 32	303	13 50.9 + 21 42
249	12 01.3 + 01 51	304	13 51.3 + 33 33
250	12 01.5 + 39 16	305	13 52.0 + 31 07

306	13 52.9 + 38 56	361	15 33.5 + 27 24
307	13 53.2 + 25 08	362	15 34.0 + 42 22
308	13 53.7 + 05 53	363	15 34.4 + 25 53
309	13 55.1 + 22 37	364	15 34.4 + 31 00
310	13 57.1 + 28 36	365	15 46.5 + 21 47
311	13 57.6 + 32 44	366	15 39.1 + 28 20
312	13 58.7 + 15 21	367	15 40.9 + 47 58
313	14 00.4 + 09 49	368	15 44.4 + 25 43
314	14 04.8 + 12 54	369	15 46.0 + 08 53
315	14 06.4 - 03 34	370	15 46.0 + 67 22
316	14 06.4 + 55 13	371	15 48.6 + 21 36
317	14 10.0 + 25 09	372	15 48.6 + 28 55
318	14 15.3 + 00 38	373	15 49.6 + 24 17
319	14 16.0 + 07 52	374	15 50.4 + 12 43
320	14 16.2 + 36 06	375	15 52.5 + 34 55
321	14 20.2 + 48 27	376	15 52.7 + 11 12
322	14 20.7 + 40 25	377	15 54.5 + 39 55
323	14 22.0 + 17 32	378	15 55.1 + 41 46
324	14 24.0 + 26 13	379	15 56.1 + 28 29
325	14 26.4 + 11 32	380	15 56.5 + 02 45
326	14 26.8 + 29 47	381	15 59.0 + 53 53
327	14 29.9 + 03 36	382	16 00.4 + 19 25
328	14 29.9 + 52 56	383	16 00.9 + 25 28
329	14 31.9 + 60 20	384	16 01.5 - 02 01
330	14 36.0 + 09 26	385	16 03.7 + 00 06
331	14 37.1 + 12 35	386	16 08.5 + 30 44
332	14 38.4 + 04 05	387	16 09.0 + 64 11
333	14 40.3 + 01 28	388	16 09.0 + 82 12
334	14 41.2 + 19 09	389	16 10.3 + 49 55
335	14 42.0 + 24 14	390	16 10.7 + 11 10
336	14 45.2 + 13 56	391	16 11.6 + 37 17
337	14 45.4 + 11 25	392	16 13.8 + 56 32
338	14 46.6 + 18 09	393	16 15.8 + 35 05
339	14 48.7 + 16 51	394	16 19.0 + 24 20
340	14 48.8 + 46 54	395	16 25.5 + 40 06
341	14 51.6 + 18 55	396	16 26.2 + 20 45
342	14 54.3 + 09 15	397	16 26.6 + 33 26
343	14 56.2 + 49 01	398	16 28.0 + 24 38
344	14 57.5 + 54 15	399	16 29.7 + 50 27
345	14 59.8 + 20 43	400	16 34.4 + 44 12
346	15 00.6 + 25 59	401	16 35.5 + 26 08
347	15 01.5 + 35 50	402	16 35.9 + 29 39
348	15 02.0 + 28 41	403	16 38.4 + 60 38
349	15 03.8 + 08 53	404	16 42.4 + 14 28
350	15 04.6 + 16 39	405	16 45.1 + 06 20
351	15 08.8 + 40 54	406	16 47.6 + 53 37
352	15 10.0 + 03 15	407	16 48.2 + 12 03
353	15 11.1 + 21 02	408	16 48.6 + 03 07
354	15 18.2 + 30 40	409	16 49.9 + 23 43
355	15 19.4 + 26 10	410	16 49.9 + 33 20
356	15 19.7 + 21 07	411	16 53.9 + 78 56
357	15 21.0 + 48 44	412	16 55.8 + 68 44
358	15 21.2 + 08 51	413	16 56.0 - 01 45
359	15 24.8 + 28 50	414	16 58.5 + 24 59
360	15 30.9 + 04 54	415	17 01.4 + 28 30

416	17 02.9 + 35 10	437	17 54.9 + 62 30	458	20 44.9 − 03 15	479	22 56.5 + 19 33
417	17 03.4 + 18 16	438	17 56.5 + 29 04	459	20 46.1 + 22 42	480	22 56.8 + 24 45
418	17 03.5 + 22 00	439	17 59.9 + 42 38	460	21 00.0 + 16 23	481	22 59.6 + 07 46
419	17 04.9 + 30 56	440	18 07.2 + 56 33	461	21 10.2 + 12 38	482	23 07.6 + 07 13
420	17 07.6 + 40 45	441	18 08.9 + 25 31	462	21 10.9 + 02 03	483	23 12.5 − 02 29
421	17 12.1 + 77 40	442	18 10.2 + 49 49	463	21 34.6 + 42 53	484	23 15.5 + 28 54
422	17 12.6 + 06 16	443	18 11.4 + 69 41	464	21 49.6 + 03 19	485	23 16.5 + 00 46
423	17 15.9 + 32 18	444	18 12.3 + 22 37	465	22 01.1 + 08 08	486	23 18.0 + 19 10
424	17 18.1 − 01 08	445	18 22.2 + 30 09	466	22 02.7 + 16 28	487	23 20.0 + 08 45
425	17 18.6 + 52 29	446	18 26.4 + 34 10	467	22 07.8 + 41 14	488	23 20.2 + 43 09
426	17 22.8 + 31 20	447	18 31.2 + 31 54	468	22 10.0 + 37 45	489	23 22.4 + 14 27
427	17 28.5 + 43 53	448	18 36.8 + 33 06	469	22 12.0 + 13 26	490	23 24.7 + 31 45
428	17 29.3 + 40 39	449	18 37.1 + 36 33	470	22 23.0 − 02 32	491	23 29.0 − 02 24
429	17 30.4 + 58 29	450	18 47.2 + 77 11	471	22 31.2 + 37 32	492	23 32.8 + 20 27
430	17 30.9 + 73 04	451	18 53.0 + 72 26	472	22 31.5 + 00 52	493	23 35.5 + 24 49
431	17 38.0 + 35 16	452	18 53.5 + 37 44	473	22 33.3 + 19 11	494	23 43.4 + 08 45
432	17 42.2 + 23 45	453	19 16.8 + 48 55	474	22 36.7 + 24 42	495	23 47.5 + 07 07
433	17 43.8 + 55 28	454	19 54.8 + 78 24	475	22 37.5 − 01 00	496	23 49.0 + 10 45
434	17 44.5 + 38 46	455	20 06.0 + 03 04	476	22 47.3 + 11 07	497	23 52.1 + 47 18
435	17 45.6 + 67 03	456	20 26.5 + 74 25	477	22 52.6 + 31 35	498	23 54.9 + 52 12
436	17 52.6 + 18 42	457	20 42.9 + 05 31	478	22 55.8 + 13 50		

CATALOGUES OF GALAXIES
A SHORT SURVEY OF THEIR HISTORY

Star catalogues in earlier times usually included some "nebulous stars" or "nebulae". *Almagest* contains seven of them. Tycho Brahe, in his *Astronomiae Instaurate Progymnasmata,* listed six, but only one (Praesepe) was also listed in the catalogue of *Almagest.* In *Prodromus Astronomiae* Johannes Hevelius listed sixteen entries as nebulous. This ancient term "nebula" or "nebulosa" covers — as was gradually recognized up to 1925 — a variety of phenomena: gaseous nebulae in the Milky Way, clusters of stars, galaxies, small groups of stars, and sometimes separate stars, which the observer for some reason had perceived as "nebulous". Only three extragalactic objects had been observed before the invention of the telescope: the Andromeda nebula and the Magellanic Clouds. Thus, when galaxies (as we now recognize them) were recorded by and by in the astronomical annals, it was as a part of the whole process in which non-stellar, "nebulous" objects of all kinds were discovered and described.

The Andromeda nebula was mentioned by Al-Sûfi in the 10th century (and its discovery is usually ascribed to him). It was re-discovered by Simon Marius in 1612. Al-Sûfi probably also knew the Large Magellanic Cloud, apparently mentioned by him as "El-Bakar", "the (White) Ox" (cf. Humboldt 1850).

The first separate list of nebulae was compiled by Edmond Halley. In 1715, he published in the *Philosophical Transactions* of the Royal Society a memoir called "Of Nebulae or lucid Spots among the Fix't Stars". Only six objects were included, and the Andromeda nebula has turned out to be the sole extragalactic object among them. The remaining ones are the Orion nebula (M 42, discovered by Nicholas Pieresc in 1610), the globular cluster M 22 (discovered by Abraham Ihle in 1665 or possibly somewhat earlier by Hevelius), the globular clusters ω Centauri and M 13 (discovered by Halley in 1677 and 1714), and "Kirch's nebula", the open cluster M 11 (discovered by Gottfried Kirch in 1681).

The second early list of nebulae alone was prepared by the English divine William Derham in 1733; it included Halley's six nebulae and sixteen other objects listed in Hevelius's *Prodromus Astronomiae.* As was pointed out by Herman Schultz in 1866, only thirteen of these "Hevelian" nebulae are really to be found in the work of the Polish astronomer, and only two of them (the Andromeda nebula and Praesepe) are in the classical sense "nebulous". Derham states that he had made "some good Observations" of five of Halley's objects, but he never observed the remaining ones from Hevelius's catalogue. This list was a poor work (from an observer's point of view), but it appears to have inspired others to make their own observations of nebulae. It also attained a somewhat undeserved credit, as it was reprinted by Maupertuis in his *Discours sur les différents Figures des Astres* (2nd edition, 1742) and was commended by Le Gentil, who also claimed that most of the objects had been observed by himself and by the Swiss astronomer Philippe Luis de Chésaux. Derham's list reached Immanuel Kant through Maupertuis, and this information supported the philosopher in his speculations concerning the structure of the Universe.

De Chésaux — also well known in the early history of Olber's paradox — is a name of interest in this connection, as he took the first step towards a classification of the objects. He divided them roughly into two groups: the star clusters, clearly resolvable into faint stars through the telescope, and the true nebulae, those which never looked like anything else than "white clouds". De Chésaux observed at Lausanne, and he reported on his observations of 20 nebulae in a letter written about 1746 to his grandfather in Paris (Hogg 1947). According to Kenneth Glyn Jones, eight or possibly nine of these objects were discoveries by de Chésaux himself.

Up to the middle of the 18th century, a total of about 60 nebulous objects had been discovered. Among the newcomers since Halley's and Derham's lists, only two later on turned out to be extragalactic: M 32, discovered by Le Gentil in 1749, and the bright southern spiral M 83, discovered by Lacaille about 1752. Lacaille, a remarkably assiduous observer, worked at the Cape of Good Hope from 1751 to 1753, and his observations resulted in a catalogue of some 10,000 stars, as well as in the first catalogue of southern nebulous objects, published in the *Mémoires de l'Academie Royale des Sciences* in 1755. It was reprinted in *Connoissance des Temps* for 1783, 1784 and 1787, along with Messier's lists of nebulae. Lacaille divided his objects into three groups: "Nebulae of the First Class" (i.e. nebulae without stars), "Nebulous Stars in Clusters", and "Stars accompanied by Nebulosity". This division happens to agree approximately with the modern division into globular clusters, open clusters and gaseous nebulae. He did not intentionally look for nebulae; like Messier, he was occupied with other observations when the nebulous objects happened to enter his field of view. He made the illustrative statement that his "Nebulae of the First Class" might all be faint comets, as time did not allow him to check whether the objects remained in the same position.

The famous Messier catalogue originated from a list of 45 nebulae, which Charles Messier in 1771 published in the *Mémoires de l'Académie Royale des Sciences.* It is probably well known that Messier's interest in nebulae (or dislike of them, as he mistook them for comets) originated in 1758, when he accidentally found the Crab nebula during his observations of a comet. (The nebula had been discovered by John Bevis at Greenwich in 1731.) Three of these early Messier objects have proved to be galaxies: to M 31 and M 32 was added the Triangulum spiral M 33, which Messier had discovered in 1764. A similar list was published six years later by J.E. Bode in *Astronomisches Jahrbuch* for 1779: "Vollständiges Verzeichnis der bisher bemerkten Nebelsterne und Sternhäuflein bis zur 38sten Grad südlicher Abweichung". It contains 75 entries, but several of the objects are not nebulous, and, moreover, there are many obvious errors in the list and some objects are impossible to identify. However, Bode had in 1774 discovered the Ursa Major galaxies M 81 and M 82 ("Bode's nebula"), and later on, in 1779, he also discovered M 64 in Coma Berenices. He claimed to have discovered M 51, but this galaxy was observed by Messier for the first time in 1773.

During the following years, several new nebulae were subsequently found, and some of them later on turned out to be extragalactic: M 59 and M 60 in Virgo were discovered by Koehler in Dresden (cf. Bode 1779), and M 61 (also in Virgo) by Oriano in Milan. Messier included these discoveries, and several others, in his second list, which appeared in *Connoissance des Temps* for 1783, published in 1780. The southern nebulae observed by Lacaille were not listed among Messier's objects, but Lacaille's list was reprinted along with Messier's own. Among the Messier objects, three nebulae discovered by Pierre Méchain proved to be galaxies: the spiral M 63 in Canes Venatici and the pair M 65 and M 66 in Leo.

The final Messier catalogue was published in 1781 (*Connoissance des Temps* for 1784). It included 103 objects, of which we now recognize 32 to be galaxies. However, one object is duplicated (M 102 is identical with M 101), one is a double star (M 40), one is merely an "asterism" (M 73; four stars close together), and one is non-existent (M 91). The same list appeared again in the same calendar three years later. Most of the new nebulae were, in fact, discoveries by Pierre Méchain, and the last three entries were listed under the note "Par M. Méchain, que M. Messier n'a pas encore vue" ([Objects reported] "By Mr. Méchain, which have not yet been seen by Mr. Messier"). Méchain, in a letter published in *Astronomisches Jahrbuch* for 1786, first pointed out that nebula No. 102 is the same objects as No. 101. Moreover, he listed in this letter six more nebulae, among them the "Sombrero" spiral, now recognized as M 104. These additional objects have been provided with Messier numbers by Camille Flammarion, Helen Sawyer Hogg and Owen Gingerich, so the number of Messier objects, including this "third supplement", is 109. Finally, in 1967, Kenneth Glyn Jones argued that the M 31 companion NGC 205 should bear the Messier number 110. This identification has been accepted in the present Catalogue.

Thus 106 separate Messier objects ("nebulae and star clusters") exist, of which 39 are galaxies.

The publication of Messier's final list in 1781 marked the end of an epoch in the search for nebulae. In March of that year, William Herschel discovered the planet Uranus. When he got *Connoissance des Temps* in his hands, the list by Messier and Méchain inspired him to start his famous "sweeps" for nebulae and clusters of stars, which were to make him the most successful nebula discoverer in the pre-photographic period. Messier himself had found 39 of the nebulae he listed. Méchain discovered 28, Lacaille 24, de Chéseaux 8 or 9, Bode 5 and Le Gentil 4. Including Méchain's "third supplement" and the southern objects found by Lacaille, a total of 137 nebulae and clusters of stars were known before the beginning of Herschel's work. It may be of some interest to sum up the total number of extra-galactic objects—as we now recognize them—known at different times:

Year	Event	Number of extra-galactic objects
1609	Galileo's telescope	3
1715	Halley's list	3
1749	Le Gentil's discovery of M 32	4
1755	Lacaille's list	5
1771	Messier's first list	6
1780	Messier's second list	19
1781	Messier's final list, including Méchain's discoveries but before Herschel's sweeps	39

As concerns nebula discoveries in general, the reader is referred to the more extensive table and diagram by Kenneth Glyn Jones (1969e, p. 453).

William Herschel's sweeps for nebulae started on 28 October 1783, and, after some experiments, his long series of nebula observations was commenced in December that year and continued until 30 December 1802. They ended with sweep No. 1112. During those years, Herschel discovered about 2500 nebulae or clusters of stars, i.e. the number of such objects grew by 1800 per cent during this period of Herschel's work.

His first catalogue appeared in the *Philosophical Transactions*, vol. LXXVI, in 1786. It contained 1000 objects divided into eight classes: "bright nebulae", "faint nebulae", "very faint nebulae", "planetary nebulae", "very large nebulae", "very compressed and rich clusters of stars", "pretty much compressed and rich clusters of stars", and "coarsely scattered clusters of stars". The second and third classes contained the main part (778 objects). In this catalogue, Herschel introduced a notation system that has been used (in slightly revised versions) by most nebula observers up to the present time: "B" for "bright", "S" for "small", and so on. Herschel's next catalogue was published in the *Philosophical Transactions* three years later and also included 1000 entries, and his third, with 500 entries, in 1802. The total numbers of objects in the eight Herschelian classes were 288, 910, 985, 78, 52, 42, 67 and 88. The objects in Herschel's catalogues are identified by the number of the class (in Roman numerals) and the number inside the class, for example III 868. (Later on, William Herschel's discoveries were usually identified by an "H", to distinguish them from those made by his son, which were denoted by an "h".) The number of separate objects seen by Herschel is, of course – owing to duplications and other errors – not quite as high as 2500.

William Herschel's opinion concerning the nature of the nebulae changed quite radically towards the end of his life. He originally regarded them as separate stellar systems outside the Milky Way and thus as evidence supporting the theory of "island universes", but the appearance of some planetary nebulae made him subsequently change his mind and he come to believe that they were all "true nebulae" inside our own stellar system. His publications after 1802 contain no new discoveries, although he discussed the nature of the objects in papers dealing with "the Construction of the Heavens". The next large surveys for nebulae were carried out by John Herschel, who observed at Slough from 1825 to 1833. With a 20-ft. reflector, he found about 500 objects which had not been recognized by his father. His observations in those years covered in all 2307 objects (some duplications); they are published in a catalogue which appeared in the *Philosophical Transactions* for 1833. His descriptions of the nebulae are in general more detailed than his father's, and he included 91 drawings of interesting or typical objects. William Herschel had also published drawings of nebulae and star clusters; the oldest such pictures in the astronomical literature are probably those published by Le Gentil in 1755. However, the drawings by John Herschel are those which appear to have been most frequently reproduced in astronomical textbooks up to the beginning of this century.

William Herschel had measured the positions of his nebulae in relation to nearby stars. John Herschel arranged the objects in right ascension and used the epoch 1830.0. As in most early catalogues of nebulae and star clusters, the positions are given with an unrealistic precision (to tenth of seconds in right ascension and to seconds of arc in declination). Of course, many of these positions were very erroneous due to misidentifications, writing errors and instrumental shortcomings. The positions were later on critically studied, especially by Henrik Ludvig D'Arrest in Leipzig (later in Copenhagen). In 1855, he began a survey of objects in this first catalogue by John Herschel, and these observations were published with detailed descriptions in 1856. About 320 objects were included, but some of them were discoveries by D'Arrest himself. In Copenhagen he continued these observations in a very careful manner, of which more hereafter.

From 1834 to 1838 John Herschel made his famous journey to the Cape of Good Hope, which resulted, among other things, in the discovery of about 1700 new southern nebulae (besides the clusters and nebulosities in the Magellanic Clouds). Back in England, Herschel some years later started to compile a general catalogue of all nebulae and star clusters discovered up to that time. The only "general catalogue" prepared so far was the old list by Bode dating from 1777. As several observers were now in possession of telescopes which allowed them to perceive fairly faint nebulae, a general catalogue was highly desirable, in order to avoid already known objects being reported as new discoveries; it also happened now and then that nebulae were mistaken for comets. When John Herschel started this work, he had at his disposal a manuscript catalogue prepared by Caroline Herschel during the first years of her retirement at Hanover after the death of her brother. This catalogue contained all the nebulae observed by William Herschel, with positions reduced to the common epoch 1800.0 and arranged in declination zones of 1° in breadth. The first step was to reduce these positions to 1830.0 and combine this catalogue and John Herschel's own from 1833 into one. While this work on the new catalogue was in progress, Arthur Auwers at Königsberg published (in 1862) a sort of "general catalogue" which had been compiled quite independently of John Herschel's project: it was a list, including all of William Herschel's discoveries, arranged in right ascension and precessed to 1830.0, as well as similar lists for Messier's, Méchain's and Lacaille's objects and a list of 50 nebulae discovered by other observers. John Herschel was able to add some new nebulae to his catalogue from Auwer's lists; however, the data from Caroline Herschel's manuscript (in which *all* separate observations were listed) permitted a greater accuracy in the positions than Auwers had been able to achieve using William Herschel's printed catalogues. Moreover, Auwers did not list John Herschel's own observations. Thus, his ambitious work was bound to be overshadowed by John Herschel's, as *A General Catalogue of Nebulae and Clusters of Stars* was published by the latter in the *Philosophical Transactions* two years later, in 1864. It contained 5079 entries, arranged in right ascension and reduced to the epoch 1860.0. This catalogue, known among nebula observers as "GC", was an indispensable standard work for more than two decades, and it did not entirely lose this position even when more extensive catalogues were published. GC has been considered a more homogeneous work than its successors NGC and IC.

Three years after the publication of John Herschel's *General Catalogue*, D'Arrest published his *Siderum Nebulosorum Observationes Havnienses*, containing observations made with the 16-ft. Merz refractor at the Observatory in Copenhagen from 1861 to 1867. This work, written wholly in Latin, contains 4800 observations of 1942 objects, with a separate determination of the position on each occasion and careful descriptions of the appearances of the objects. (On the recommendation of Knut Lundmark, D'Arrest's Latin text was translated into English by Per Collinder in the 1930's; this translation — arranged as a card catalogue — is now preserved at the Uppsala Observatory.) D'Arrest's careful work revealed several corrections to the positions listed in GC. An analysis of errors in the catalogues by Lacaille, Messier and William Herschel is also to be found in D'Arrest's list of 1856 (*Resultate aus Beobachtungen der Nebelflecken und Sternhaufen*).

In these years, the determinations of accurate positions of nebulae attracted much attention, as the nature of the objects was unknown and the possibility of proper motion could not be excluded. Thus, the positions were usually determined with great care, and the natural way was to measure the positions of the nebulae in relation to nearby stars. The most extensive observation programs of this kind were undertaken by Schönfeld at Mannheim and by Schultz at Uppsala, who studied about 500 objects each. Dreyer, in his introduction to NGC, states that, next to D'Arrest's work, the publications by Schönfeld and Schultz furnished most corrections to John Herschel's *General Catalogue* when Dreyer's *New General Catalogue* was compiled. A discussion of several other lists of nebulae published up to the late 1880's is to be found in this introduction by Dreyer.

During the 1880's it became increasingly clear that a revised general catalogue of nebulae was needed, partly because more accurate positions were available for many objects in GC, and partly because a large number of new nebulae and clusters of stars had been discovered since 1864. The Danish-born John Louis Emil Dreyer had worked as an assistant to the Earl of Rosse at Parsontown from 1874 and made observations with the 72-in. reflector, the largest telescope in the world at that time. The power of this instrument caused him to focus his attention especially on the nebulae and star clusters, and he became very much aware of the errors in GC. During his time at Parsontown, he compiled for his own use a supplement to GC, published in 1878 in the *Transactions of the Royal Irish Academy*. Dreyer was able to add 1172 objects to those listed by John Herschel, besides corrections of errors. In 1879, Lord Rosse himself published a list of observations made from 1848 to 1878, together with observations made at several other places. Dreyer arranged a new supplementary list, which he submitted to the Council of the Royal Astronomical Society. The Council, however, proposed that Dreyer should compile an entirely new general catalogue, combining all existing catalogues into one. This was the origin of Dreyer's famous work *A New General Catalogue of Nebulae and Clusters of Stars* (NGC), published in the *Memoirs* of the Royal Astronomical Society in 1888.

NGC contains positions to the epoch 1860.0 (with precessions for 1880.0), "summary descriptions", according to William Herschel's notation system, and references to GC, to John Herschel's catalogue from 1833, to William Herschel's catalogues and to lists by various other observers. It contains 7840 entries. Dreyer did a skilful job in comparing and weighing the observations, whose quality varied very much as between the different sources. In the introduction he states that "with regard to the very numerous new nebulae recorded in late years, it was frequently a matter of some difficulty to decide about the identity of objects announced by several observers, and differing little as regards place, but often much as to description". Later investigations have shown, of course, that several duplications exist in NGC, and that some position errors from the late 18th century have still survived in this catalogue. Among the objects in the northern sky, about 85 per cent have been identified as galaxies in the *Catalogue of Galaxies and of Clusters of Galaxies* by Zwicky *et al.* The identification of the faintest objects is, however, often difficult and sometimes impossible. Even for the brighter objects a few difficulties exist; the Virgo area, including NGC 4341, 4342 and 4343, is a well-known case.

Seven years later, Dreyer was ready to publish his *Index Catalogue of Nebulae found in the Years 1888 to 1894* (IC I). He listed 1529 entries, together with notes and corrections to NGC. During the next few years, a rapidly growing number of nebulae were recorded photographically. About 1890, Max Wolf had already started a photographic "Durchmusterung" for nebulae at his private observatory at Heidelberg, and he continued this work on a larger scale when he became director of the Königstuhl-Sternwarte. The "Königstuhlnebellisten" (lists of nebulae from the Königstuhl) finally included about 6000 nebulae in selected fields, and only a small percentage of them are listed in NGC. Among the striking results of these surveys was the discovery of the Coma cluster of galaxies ("Wolfs Nebelneste") in 1901. Wolf also discovered the Perseus cluster (in 1905). The well-known classification system for nebulae, the Wolf code *a—w* with 23 standard objects, was presented for the first time in 1909.

Although Heidelberg may be regarded as the centre of the early photographic studies of the nebulae, similar observations were subsequently made at several other stations, especially at Harvard and its station at Arequipa in Peru (from about 1900). Moreover, several hundreds of nebulae were discovered by visual observers (Burnham, Finlay, Howe, Swift and others). In 1908, Dreyer published the *Second Index Catalogue of Nebulae and Clusters of Stars,* containing objects found in the years 1895 to 1907 (IC II), with 3857 entries. About 1300 objects were taken from the "Königstuhlnebellisten" already published; however, the Coma cluster was not included, as Dreyer believed the nebulae there to be merely "conspicuous points of condensations or 'knots' in one great mass of nebulosity". The notation "spir" was introduced in the description column to mark objects with spiral structure. Because of the selected use of photographic investigations, IC II is a very inhomogeneous catalogue, which for some areas lists objects down to a limiting magnitude of about 17.

Dreyer, an indefatigable compiler for more than three decades, then turned his attention to the history of astronomy and prepared his splendid editions of William Herschel's and Tycho Brahe's works, *The Scientific Papers of Sir William Herschel* and *Tychonis Brahe Dani Opera Omnia;* the later project occupied him for most of the remainder of his life. For the observers of "nebulae", NGC and IC have remained standard reference works up to the present time.

The growing importance of photographic methods during these years makes it easy to forget the enormous amount of work which was done visually at the Paris Observatory by M.G. Bigourdan and his assistants. The project consisted in the determination of very accurate positions in relation to nearby stars for all nebulae visible through the 12-in. equatorial telescope in the western tower of the Observatory. These observations commenced in 1884 and continued up to 1909; they included about 6600 separate objects, and the results – covering some 2600 pages – appeared gradually in the *Annales de l'Observatoire de Paris* from 1886 to 1911. The nebulae themselves are carefully described, sometimes also with diameters and magnitudes estimated. The extensive introduction to *Observations de Nébuleuses et d'Amas Stellaires* contains, among many other things, a history of early nebula research and an elaborate bibliography of publications concerning nebulae and clusters of stars up to 1912. However, the work by Bigourdan appeared too late to influence the development of nebula research in any important way. On the other hand, Bigourdan's catalogue contained a vast amount of information, and it was frequently used as a standard reference work; it appears also to have been the most homogeneous observational work concerning nebulae which was done during the "visual" epoch.

In 1912, Max Wolf advised his assistant Adam Massinger to start a photographic survey of all the objects included in John Herschel's *General Catalogue*. Massinger began this work; however, at the outbreak of World War I he was called up for military service, and on 21 October 1914 he was killed at Ypern. The project was in abeyance during the war, but in 1919 the work was taken up by Karl Reinmuth, and it was decided that the study should cover all of Herschel's objects which could be reached from Heidelberg. In all, 4445 such objects had positions to the north of δ = −20°, and of those only 247 had not been recorded on plates taken at the Königstuhl-Sternwarte when the project was brought to an end in 1924. Plates taken as early as 1900 were used in this survey. In 1926, Reinmuth published the results in a catalogue called *Die Herschel-Nebel nach*

Aufnahmen der Königstuhl-Sternwarte. This catalogue contains positions for the epoch 1875.0, galactic coordinates, classifications in the Wolf system, position angles, diameters, and descriptions (in English). Reinmuth's catalogue was the only one before the present Catalogue which contained position angles for a large number of galaxies in the northern skies. Magnitudes were estimated in the Herschel code only, but several attempts were later made to evaluate this code numerically.

Several other projects of a similar kind were undertaken during the first few decades of the century, although Reinmuth's catalogue was the largest and most important one. Among the many publications on nebulae from the Harvard College Observatory, the catalogue by Solon Baily may be mentioned as an example: it contained all NGC objects which could be seen on plates taken with a 1-in. lens with an exposure time of one hour. Baily used a fairly elaborate classification system with four main groups and several subgroups. Another similar project was started in England at approximately the same time as Massinger commenced his work on the Herschel nebulae: J.A. Hardcastle, on the advice of A.R. Hinks, surveyed the sky, using the Franklin-Adams plates. These forerunners of the Palomar Sky Survey, 206 in number, were exposed through an 10-in. telescope and the plate scale was about 180″/mm. Hardcastle's catalogue contained 785 NGC and IC objects classified as "spirals", "elongated", "diffused" or "small"; the number of galaxies among them was later estimated by Shapley and Ames at 682. This catalogue contains, besides positions and classifications, only NGC or IC numbers and position angles.

In fact, the number of catalogues or lists published during these years is astonishingly large. Several appeared from the Harvard Observatory under the directorships of Edward Pickering, Solon Baily and Harlow Shapley. Seven of these catalogues were published in the famous volume 88 of *Annals of the Astronomical Observatory of Harvard College,* in which the most important item is that by Harlow Shapley and Adelaide Ames: *A Survey of the External Galaxies brighter than the thirteenth Magnitude* (1932). This was the first catalogue which was fairly complete to a certain photographic magnitude limit, and, moreover, it was the first survey which intentionally listed galaxies only. It contained 1249 objects, of which 1025 were considered to brighter than the 13th magnitude. The magnitudes had been estimated on small-scale plates (usually 600″/mm), in such a way that the galaxies were compared with stars in standard magnitude sequences. Besides the magnitudes and the positions for the epoch 1950.0, this "Shapley-Ames Catalogue" contains diameters and classifications in the Hubble system.

Early in this century, it became clear that the number of nebulae recorded on photographic plates was extremely large. J.E. Keeler used the Crossley reflector at Mount

Wilson for nebula photography from 1898 to 1900, and he estimated the total number of photographable nebulae at 120,000; he also supposed that most of them would show spiral form. It was no longer meaningful to compile "general" catalogues of the objects, if the term "general" meant that all known objects were recorded. However, Knut Lundmark hoped to compile a "general" catalogue, including all published data — apparently in the form of a "data bank" at the Lund Observatory — but this project presented unsurmountable difficulties and was never finished. A separate account of the *Lund General Catalogue of Nebulae and Clusters of Stars* (LGC) will be given elsewhere.

The first extensive counts of galaxies — to reveal the distribution of galaxies over the sky—were made in the early 1930's. (Earlier "counts" had included mainly objects listed in different catalogues or were made on a fairly small scale; cf. Fath 1914.) Hubble's counts in selected areas included 44,000 objects. On Harvard plates, about 10^6 galaxies had been recorded up to 1957, and magnitudes had been estimated for some 170,000 of them (Shapley 1957). The Lick counts (Shane and Wirtanen 1967) included about 800,000 separate galaxies. Finally, Zwicky states (1971 *a*) that he plotted the positions of approximately $15 \cdot 10^6$ galaxies in defining the clusters of galaxies in CGCG (see below).

The modern catalogues of galaxies will be treated briefly, as they are assumed to be well known to all workers in this field. Four important catalogues of galaxies have been published since 1960. The first which should be mentioned is the *Reference Catalogue of Bright Galaxies* (usually called BG or RCBG) by Gérard and Antoinette de Vaucouleurs, published in 1964. Its aim is explained by the continuation of the title: "... being the Harvard Survey of Galaxies brighter than the 13th magnitude of H. Shapley and A. Ames, revised, corrected and enlarged, with Notes, Bibliography and Appendices". The project of compiling this catalogue started as early as 1949. A description of the compilation procedure, which included the comparison, weighing and checking of a vast amount of data, is described in the Introduction to BG. The catalogue contains 2599 entries (i.e. about twice as many as the Shapley-Ames catalogue). The authors expect a 50 per cent completeness level to be reached near $B = -13.0$. A second edition is announced but had not yet appeared when this survey was written in February, 1973.

The Palomar Sky Survey was completed in 1956. One of the first surveys of this material was Abell's search for clusters of galaxies, made on the original plates. His catalogue appeared in 1958 and contained 2512 entries. This was the first extensive catalogue of clusters only. Between 1964 and 1968 appeared the *Morfologičeskij Katalog Galaktik* (Morphological Catalogue of Galaxies, MGC) in four volumes, compiled in Moscow by B.E.

Voroncov-Vel'jaminov and his co-workers. This catalogue contains in all about 29,000 galaxies (with additional information concerning some 5000 objects) for the sky to the north of $\delta = -33°$. The limiting magnitude is approximately 15.0 (partly estimated magnitudes, partly from CGCG; see below), with the exception of the Virgo area, where the limiting magnitude is 15.2. Total diameters and diameters for "the bright inner part" are listed, also intensities for the outer and inner regions on a scale from 1 to 6, and estimated inclinations for spiral galaxies. No classifications are made (classifications by other observers are listed in the notes), but the objects are described by a detailed notation system from the centre outwards. This system is well suited for the appearance of the galaxies on the Sky Survey prints, and it is naturally adjustable to the varying information contents of their images. However, like other pure description systems, it is impossible to use for statistical investigations. Similar galaxies may be listed with different descriptions (owing to different orientation or apparent magnitude), and different types of galaxies may be listed with the same descriptions.

The *Catalogue of Galaxies and of Clusters of Galaxies* (CGCG; sometimes called CAT) by Fritz Zwicky and his co-workers was published in six volumes from 1961 to 1968. It includes 31,350 separate galaxies and 9700 clusters of galaxies. The catalogue is designed to be complete to the photographic magnitude 15.5, and the limiting magnitude is 15.7 (measured by the "jiggle-camera" technique), which allows for a possible error of ± 0.2 magnitudes. Besides positions, only magnitudes and sometimes short descriptions ("diffuse", "compact", "double system") are listed. No classifications are made; like Voroncov-Vel'jaminov, Zwicky has no confidence in the existing classification systems, which he regards as too simple to cover the large variety of extragalactic objects. The positions of the galaxies and the cluster's contours are recorded on maps covering the same fields as the Palomar Sky Survey.

Although some galaxies are listed as "compact" (or "very compact", "extremely compact") in CGCG, these remarks are used rather selectively and only, as Zwicky puts it, "if they [the galaxies] might easily be mistaken for stars". In 1964, Zwicky presented to the IAU Assembly at Hamburg his first list of 210 "compact" galaxies, including the Humason-Zwicky "star" HZ 46, which had been found in 1938 and was the first stellar object recognized as extragalactic. In the years 1963 to 1969, Zwicky circulated in all seven lists of "compact galaxies, compact parts of galaxies, eruptive and post-eruptive galaxies", including some 2300 objects. These galaxies were random discoveries made during the compilation of CGCG. A complete survey of twelve fields covering about 450 square degrees led Zwicky to draw the conclusion that some 200,000 — 300,000 such objects could be found with the Hale 48-in. Schmidt telescope. In 1971, Zwicky summed up his observations in the *Catalogue of Selected Compact Galaxies and of Post-*

Eruptive Galaxies, including about 3700 entries, with descriptions and magnitudes, as well as radial velocities for 250 objects.

Finally, Robert S. Dixon, at the Ohio State University Radio Observatory, has announced the preparation of a "Master List" of all known non-stellar objects, similar to the list of radio sources published by the same Observatory in 1970. Although the details of this project are not known at present, it may be expected that this "Master List" will be the first literally "general" catalogue of such objects prepared since Dreyer's time.

BIBLIOGRAPHY

Abell, G.O. 1958, *Astrophys. J. Suppl.,* 3, 211.

D'Arrest, H.L. 1856, *Abhandlungen der Königlichen Sachsischen Gesellschaft der Wissenschaften,* 5, 293.

D'Arrest, H.L. 1867, *Siderum Nebulosorum Observationes Havnienses,* Societatis Regiae Scientiarum Daniae, Copenhagen.

Auwers, M. 1862, *Astr. Beob. Königsberg,* 34, 155.

Bailey, S.J. 1912, *Ann. Harv. Coll. Obs.,* 60, No. 8.

Bigourdan, M.G. 1906, 1907a, 1907b, 1908, 1911, *Observations des Nébuleuses et d'Amas Stellaires,* I–V, Gauthier-Villars, Paris.

Bode, J.E. 1777, *Astr. Jahrb. für 1779,* Zweyter Theil, p. 65, G.J. Decker, Berlin.

Bode, J.E. 1779, *Astr. Jahrb. für 1782,* Zweyter Theil, p. 151, G.J. Decker, Berlin.

Brahe, T. 1602, *Astronomiae Instaurate Progymnasmata,* Uranienburg and Prague.

Derham, W. 1733, *Phil. Trans. R. Soc. London,* 38, 70.

Dixon, R.S. 1970, *Astrophys. J. Suppl. 20,* 1.

Dreyer, J.L.E. 1878, *R. Irish Acad.. Trans.,* 26, 381.

Dreyer, J.L.E. 1888, *Mem. R. astr. Soc.,* 49, 1 (reprinted in 1953, RAS, London).

Dreyer, J.L.E. 1895, *Mem. R. astr. Soc.,* 51, 185 (reprinted in 1953, RAS, London).

Dreyer, J.L.E. 1908, *Mem. R. astr. Soc.,* 59, 105 (reprinted in 1953, RAS, London).

Dreyer, J.L.E. 1912, *The Scientific Papers of Sir William Herschel,* I–II, The Royal Society and the Royal Astronomical Society, London.

Dreyer, J.L.E. 1913–1926, *Tychonis Brahe Dani Opera Omnia,* I–XV, Libraria Gyldendaliana, Copenhagen.

Evans, D.E. *et al.* 1969, *Herschel at the Cape. Diaries and Correspondence of Sir John Herschel, 1834–1838,* University of Texas Press, Austin.

Fath, E.A. 1914, *Astr. J.,* 28, 75.

Freiesleben, H.C. 1962, *Max Wolf. Der Bahnbrecher der Himmelsphotographie.* No. 26 in the series: Grosse Naturforscher. Wissenschaftliche Verlagsgesellschaft m.b. H., Stuttgart.

Gingerich, O. 1960, *Sky Telesc.* 20, 196.

Halley, E. 1715, *Phil. Trans. R. Soc. London,* 29, 390.

Hardcastle, J.A. 1914, *Mon. Not. R. astr. Soc.,* 74, 699.

Herschel, J.F.W. 1833, *Phil. Trans. R. Soc. London,* 120, 359.

Herschel, J.F.W. 1864, *Phil. Trans. R. Soc. London,* 154, 1.

Hevelius, J. 1690, Catalogus stellarum fixarum, in *Prodromus Astronomiae,* p. 143, Gdansk.

Hogg, H.S. 1947, *Commun. David Dunlap Obs.,* 1, No. 14.

Hoskin, M.A. 1963, *William Herschel and the Construction of the Heavens.* Oldbourne, London.

Hubble, E. 1934, *Astrophys. J.,* 79, 8.

Humboldt, A. von 1850, *Kosmos. Entwurf einer physischen Weltbeschreibung,* Dritter Band, p. 311. J.G. Cotta'scher Verlag, Stuttgart.

Jaki, S.L. 1973, *The Milky Way. An elusive Road for Science.* Science History Publications, New York.

Jones, B.Z., Boud, L.G. 1971, *The Harvard College Observatory. The First Four Directorships, 1839–1919.* The Belknap Press of Harvard University Press, Cambridge, Massachusetts.

Jones, K.G. 1967, *Sky Telesc.,* 33, 156.

Jones, K.G. 1968a, *J. Br. astr. Ass.,* 78, 256.

Jones, K.G. 1968b, *ibid.,* 78, 360.

Jones, K.G. 1968c, *ibid.,* 78, 446.

Jones, K.G. 1968d, *ibid.,* 79, 19.

Jones, K.G. 1969a, *ibid.,* 79, 105.

Jones, K.G. 1969b, *ibid.,* 79, 213.

Jones, K.G. 1969c, *ibid.,* 79, 268.

Jones, K.G. 1969d, *ibid.,* 79, 357.

Jones, K.G. 1969e, *ibid.,* 79, 450.

Lindsay, E.M. 1965, *Astr. Soc. Pacific Leafl.,* No. 436.

Lundmark, K. 1930, *Publ. astr. Soc. Pacific,* 42, 31.

Maupertuis, P.L.M. de 1742, *Discours sur les Differentes Figures des Astres,* 2nd ed., Paris.

Messier, C. 1771, *Mem. Acad. R. Sci.,* 435.

Messier, C. 1780, Catalogue des Nébuleuses et des Amas d'Etoiles, in *Connoissance des Temps... pour l'Année commune 1783,* l'Imprimerie Royale, Paris, p. 225.

Messier, C. 1781, Catalogue des Nébuleuses et des Amas d'Etoiles, in *Connoissance des Temps... pour l'Année bissextile 1784,* l'Imprimerie Royale, Paris, p. 227.

Reinmuth, K. 1926, *Veröff. Sternw. Heidelberg,* 9, 1.

Shane, C.D., Wirtanen, C.A. 1967, *Publ. Lick Obs.,* 22.

Schultz, H. 1866, *Astr. Nachr.,* 67, 1.

Shapley, H., Ames, A. 1932, *Ann. Harv. Coll. Obs.,* 88, No. 2.

Shapley, H. 1957, *The Inner Metagalaxy,* Yale University Press, New Haven.

Vaucouleurs, G. de, Vaucouleurs, A. de 1964, *Reference Catalogue of Bright Galaxies,* No. 1 in the series: The University of Texas Monographs in Astronomy, The University of Texas Press, Austin.

Voroncov-Vel'jaminov, B.A. *et al.* 1962, 1963, 1964, 1968, *Morfologičeskij Katalog Galaktik* (Morphological Catalogue of Galaxies), I–IV, Moscow State University, Moscow.

Wolf, M. 1909, *Publ. astrophys. Inst. Königstuhl-Heidelberg, 3,* 109.

Zwicky, F. *et al.* 1961, 1963, 1965, 1966, 1968a, 1968b, *Catalogue of Galaxies and of Clusters of Galaxies,* I–VI, California Institute of Technology, Pasadena.

Zwicky, F. 1964, 1965a, 1965b, 1966, 1967, 1968a, 1968b, Seven lists of compact galaxies, compact parts of galaxies, eruptive and post-eruptive galaxies. California Institute of Technology.

Zwicky, F., 1971a, *Jeder ein Genie,* Herbert Lang, Berne.

Zwicky, F. 1971b, *Catalogue of Selected Compact Galaxies and of Post-Eruptive Galaxies.* L. Speich, Zürich.

Notes, Addenda

U 3528A + 3536A) VV 248; a) 0.9 x 0.8, distorted S B(s)...; b) 0.8 x 0.6, sep 1.5 — U 4280A) VII Zw 220; a) (0.8 x 0.5), E; b) 0.22 x 0.20, sep 0.35; "red spherical compact (possibly plus star), m=15.6 northeast of E galaxy. Total brightn of the pair, m=14.4" (Zw) — U 5854A) comp 2.0, 310, 0.3 x 0.2; 2.2, 259, 0.3: x 0.1; 2.9, 230, 0.3 x 0.1 — U 7064A) three compact conds, one v red, alm invis on blue print; dif jets and bridges; sev F comps — U 7085A) VV 13, Arp 97; An 1203 A, B in BG; a) v = 7010, v_0 = 7002; b) v = 6894, v_0 = 6886; see ApJ *132,* 627 —

Errata

U 1371) l^{II} = − 54.65, not + 54.65 —

Page 182) add "3640" to NGC numbers at the bottom of the page —

U 8511) position angle 171, not 191 —

U 12493) radial velocity + 8700 in column 11, line *a,* not line *b* —